Fundamentals of Physics

HALLIDAY & RESNICK

Volume 2
Tenth Edition

JEARL WALKER

General Physics SP 211-212

Wiley Custom Learning Solutions

ISBN 978-1-118-95951-0
Printed and bound by Quad/Graphics.

V10010399_052619

Custom Brief Contents

MATHEMATICAL FORMULAS*

Quadratic Formula

If $ax^2 + bx + c = 0$, then $x = \dfrac{-b \pm \sqrt{b^2 - 4ac}}{2a}$

Binomial Theorem

$(1 + x)^n = 1 + \dfrac{nx}{1!} + \dfrac{n(n-1)x^2}{2!} + \cdots \qquad (x^2 < 1)$

Products of Vectors

Let θ be the smaller of the two angles between $\vec{a}$ and $\vec{b}$. Then

$$\vec{a} \cdot \vec{b} = \vec{b} \cdot \vec{a} = a_x b_x + a_y b_y + a_z b_z = ab \cos\theta$$

$$\vec{a} \times \vec{b} = -\vec{b} \times \vec{a} = \begin{vmatrix} \hat{i} & \hat{j} & \hat{k} \\ a_x & a_y & a_z \\ b_x & b_y & b_z \end{vmatrix}$$

$$= \hat{i} \begin{vmatrix} a_y & a_z \\ b_y & b_z \end{vmatrix} - \hat{j} \begin{vmatrix} a_x & a_z \\ b_x & b_z \end{vmatrix} + \hat{k} \begin{vmatrix} a_x & a_y \\ b_x & b_y \end{vmatrix}$$

$$= (a_y b_z - b_y a_z)\hat{i} + (a_z b_x - b_z a_x)\hat{j} + (a_x b_y - b_x a_y)\hat{k}$$

$$|\vec{a} \times \vec{b}| = ab \sin\theta$$

Trigonometric Identities

$$\sin\alpha \pm \sin\beta = 2 \sin\tfrac{1}{2}(\alpha \pm \beta) \cos\tfrac{1}{2}(\alpha \mp \beta)$$

$$\cos\alpha + \cos\beta = 2 \cos\tfrac{1}{2}(\alpha + \beta) \cos\tfrac{1}{2}(\alpha - \beta)$$

*See Appendix E for a more complete list.

Derivatives and Integrals

$$\dfrac{d}{dx} \sin x = \cos x \qquad \int \sin x \, dx = -\cos x$$

$$\dfrac{d}{dx} \cos x = -\sin x \qquad \int \cos x \, dx = \sin x$$

$$\dfrac{d}{dx} e^x = e^x \qquad \int e^x \, dx = e^x$$

$$\int \dfrac{dx}{\sqrt{x^2 + a^2}} = \ln(x + \sqrt{x^2 + a^2})$$

$$\int \dfrac{x \, dx}{(x^2 + a^2)^{3/2}} = -\dfrac{1}{(x^2 + a^2)^{1/2}}$$

$$\int \dfrac{dx}{(x^2 + a^2)^{3/2}} = \dfrac{x}{a^2(x^2 + a^2)^{1/2}}$$

Cramer's Rule

Two simultaneous equations in unknowns x and y,

$$a_1 x + b_1 y = c_1 \quad \text{and} \quad a_2 x + b_2 y = c_2,$$

have the solutions

$$x = \dfrac{\begin{vmatrix} c_1 & b_1 \\ c_2 & b_2 \end{vmatrix}}{\begin{vmatrix} a_1 & b_1 \\ a_2 & b_2 \end{vmatrix}} = \dfrac{c_1 b_2 - c_2 b_1}{a_1 b_2 - a_2 b_1}$$

and

$$y = \dfrac{\begin{vmatrix} a_1 & c_1 \\ a_2 & c_2 \end{vmatrix}}{\begin{vmatrix} a_1 & b_1 \\ a_2 & b_2 \end{vmatrix}} = \dfrac{a_1 c_2 - a_2 c_1}{a_1 b_2 - a_2 b_1}.$$

SI PREFIXES*

Factor	Prefix	Symbol	Factor	Prefix	Symbol
10^{24}	yotta	Y	10^{-1}	deci	d
10^{21}	zetta	Z	10^{-2}	centi	c
10^{18}	exa	E	10^{-3}	milli	m
10^{15}	peta	P	10^{-6}	micro	μ
10^{12}	tera	T	10^{-9}	nano	n
10^{9}	giga	G	10^{-12}	pico	p
10^{6}	mega	M	10^{-15}	femto	f
10^{3}	kilo	k	10^{-18}	atto	a
10^{2}	hecto	h	10^{-21}	zepto	z
10^{1}	deka	da	10^{-24}	yocto	y

*In all cases, the first syllable is accented, as in ná-no-mé-ter.

VOLUME TWO

Halliday & Resnick

FUNDAMENTALS OF PHYSICS

TENTH EDITION

JEARL WALKER
CLEVELAND STATE UNIVERSITY

WILEY

EXECUTIVE EDITOR Stuart Johnson
SENIOR PRODUCT DESIGNER Geraldine Osnato
CONTENT EDITOR Alyson Rentrop
ASSOCIATE MARKETING DIRECTOR Christine Kushner
TEXT and COVER DESIGNER Madelyn Lesure
PAGE MAKE-UP Lee Goldstein
PHOTO EDITOR Jennifer Atkins
COPYEDITOR Helen Walden
PROOFREADER Lilian Brady
SENIOR PRODUCTION EDITOR Elizabeth Swain

COVER IMAGE © 2007 CERN

This book was set in 10/12 Times Ten by cMPreparé, CSR Francesca Monaco, and was printed and bound by Quad Graphics. The cover was printed by Quad Graphics.

This book is printed on acid free paper.

Library of Congress Cataloging-in-Publication Data

Walker, Jearl
    Fundamentals of physics / Jearl Walker, David Halliday, Robert Resnick—10th edition.
       volumes cm
    Includes index.
ISBN 978-1-118-23073-2 (Volume 2)
Also catalogued as
Extended edition ISBN 978-1-118-23072-5
    1. Physics—Textbooks. I. Resnick, Robert. II. Halliday, David. III. Title.
       QC21.3.H35 2014
       530—dc23

                                                    2012035307

Printed in the United States of America

BRIEF CONTENTS

CONTENTS

viii CONTENTS

WHY I WROTE THIS BOOK

Fun with a big challenge. That is how I have regarded physics since the day when Sharon, one of the students in a class I taught as a graduate student, suddenly demanded of me, "What has any of this got to do with my life?" Of course I immediately responded, "Sharon, this has everything to do with your life—this is physics."

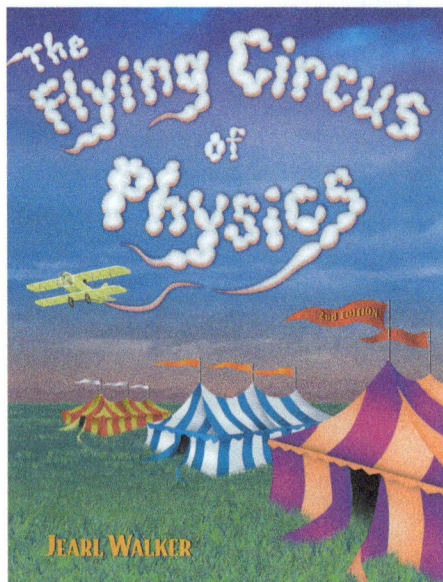

She asked me for an example. I thought and thought but could not come up with a single one. That night I began writing the book *The Flying Circus of Physics* (John Wiley & Sons Inc., 1975) for Sharon but also for me because I realized her complaint was mine. I had spent six years slugging my way through many dozens of physics textbooks that were carefully written with the best of pedagogical plans, but there was something missing. Physics is the most interesting subject in the world because it is about how the world works, and yet the textbooks had been thoroughly wrung of any connection with the real world. The fun was missing.

I have packed a lot of real-world physics into *Fundamentals of Physics*, connecting it with the new edition of *The Flying Circus of Physics*. Much of the material comes from the introductory physics classes I teach, where I can judge from the faces and blunt comments what material and presentations work and what do not. The notes I make on my successes and failures there help form the basis of this book. My message here is the same as I had with every student I've met since Sharon so long ago: "Yes, you *can* reason from basic physics concepts all the way to valid conclusions about the real world, and that understanding of the real world is where the fun is."

I have many goals in writing this book but the overriding one is to provide instructors with tools by which they can teach students how to effectively read scientific material, identify fundamental concepts, reason through scientific questions, and solve quantitative problems. This process is not easy for either students or instructors. Indeed, the course associated with this book may be one of the most challenging of all the courses taken by a student. However, it can also be one of the most rewarding because it reveals the world's fundamental clockwork from which all scientific and engineering applications spring.

Many users of the ninth edition (both instructors and students) sent in comments and suggestions to improve the book. These improvements are now incorporated into the narrative and problems throughout the book. The publisher John Wiley & Sons and I regard the book as an ongoing project and encourage more input from users. You can send suggestions, corrections, and positive or negative comments to John Wiley & Sons or Jearl Walker (mail address: Physics Department, Cleveland State University, Cleveland, OH 44115 USA; or the blog site at www.flyingcircusofphysics.com). We may not be able to respond to all suggestions, but we keep and study each of them.

WHAT'S NEW?

Modules and Learning Objectives "What was I supposed to learn from this section?" Students have asked me this question for decades, from the weakest student to the strongest. The problem is that even a thoughtful student may not feel confident that the important points were captured while reading a section. I felt the same way back when I was using the first edition of Halliday and Resnick while taking first-year physics.

To ease the problem in this edition, I restructured the chapters into concept modules based on a primary theme and begin each module with a list of the module's learning objectives. The list is an explicit statement of the skills and learning points that should be gathered in reading the module. Each list is following by a brief summary of the key ideas that should also be gathered. For example, check out the first module in Chapter 16, where a student faces a truck load of concepts and terms. Rather than depending on the student's ability to gather and sort those ideas, I now provide an explicit checklist that functions somewhat like the checklist a pilot works through before taxiing out to the runway for takeoff.

Links Between Homework Problems and Learning Objectives   In *WileyPLUS*, every question and problem at the end of the chapter is linked to a learning objective, to answer the (usually unspoken) questions, "Why am I working this problem? What am I supposed to learn from it?" By being explicit about a problem's purpose, I believe that a student might better transfer the learning objective to other problems with a different wording but the same key idea. Such transference would help defeat the common trouble that a student learns to work a particular problem but cannot then apply its key idea to a problem in a different setting.

Rewritten Chapters   My students have continued to be challenged by several key chapters and by spots in several other chapters and so, in this edition, I rewrote a lot of the material. For example, I redesigned the chapters on Gauss' law and electric potential, which have proved to be tough-going for my students. The presentations are now smoother and more direct to the key points. In the quantum chapters, I expanded the coverage of the Schrödinger equation, including reflection of matter waves from a step potential. At the request of several instructors, I decoupled the discussion of the Bohr atom from the Schrödinger solution for the hydrogen atom so that the historical account of Bohr's work can be bypassed. Also, there is now a module on Planck's blackbody radiation.

New Sample Problems and Homework Questions and Problems   Sixteen new sample problems have been added to the chapters, written so as to spotlight some of the difficult areas for my students. Also, about 250 problems and 50 questions have been added to the homework sections of the chapters. Some of these problems come from earlier editions of the book, as requested by several instructors.

Video Illustrations   In the eVersion of the text available in *WileyPLUS*, David Maiullo of Rutgers University has created video versions of approximately 30 of the photographs and figures from the text. Much of physics is the study of things that move and video can often provide a better representation than a static photo or figure.

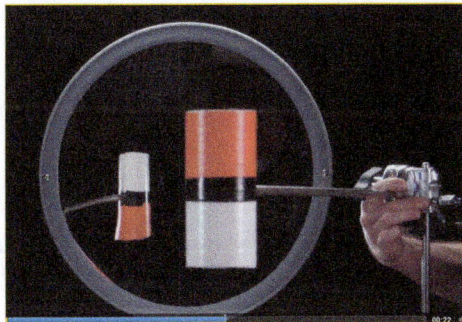

Online Aid   *WileyPLUS* is not just an online grading program. Rather, it is a dynamic learning center stocked with many different learning aids, including just-in-time problem-solving tutorials, embedded reading quizzes to encourage reading, animated figures, hundreds of sample problems, loads of simulations and demonstrations, and over 1500 videos ranging from math reviews to mini-lectures to examples. More of these learning aids are added every semester. For this 10th edition of HRW, some of the photos involving motion have been converted into videos so that the motion can be slowed and analyzed.

These thousands of learning aids are available 24/7 and can be repeated as many times as desired. Thus, if a student gets stuck on a homework problem at, say, 2:00 AM (which appears to be a popular time for doing physics homework), friendly and helpful resources are available at the click of a mouse.

LEARNING TOOLS

When I learned first-year physics in the first edition of Halliday and Resnick, I caught on by repeatedly rereading a chapter. These days we better understand that students have a wide range of learning styles. So, I have produced a wide range of learning tools, both in this new edition and online in *WileyPLUS*:

Animations of one of the key figures in each chapter. Here in the book, those figures are flagged with the swirling icon. In the online chapter in *WileyPLUS*, a mouse click begins the animation. I have chosen the figures that are rich in information so that a student can see the physics in action and played out over a minute or two instead of just being flat on a printed page. Not only does this give life to the physics, but the animation can be repeated as many times as a student wants.

Videos I have made well over 1500 instructional videos, with more coming each semester. Students can watch me draw or type on the screen as they hear me talk about a solution, tutorial, sample problem, or review, very much as they would experience were they sitting next to me in my office while I worked out something on a notepad. An instructor's lectures and tutoring will always be the most valuable learning tools, but my videos are available 24 hours a day, 7 days a week, and can be repeated indefinitely.

Starts from rest.

In a certain time interval, it rotates $\pi/4$ rad at constant angular acceleration 4.0 rad/s², reaching angular speed 4.5 rad/s.

How much time (from rest) to reach that time interval?

Interval 2: This time inteval with given data
Interval 1: From rest to the start of that time interval

01:54 / 07:32

- **Video tutorials on subjects in the chapters.** I chose the subjects that challenge the students the most, the ones that my students scratch their heads about.

- **Video reviews of high school math**, such as basic algebraic manipulations, trig functions, and simultaneous equations.

- **Video introductions to math**, such as vector multiplication, that will be new to the students.

- **Video presentations of every Sample Problem** in the textbook chapters . My intent is to work out the physics, starting with the Key Ideas instead of just grabbing a formula. However, I also want to demonstrate how to read a sample problem, that is, how to read technical material to learn problem-solving procedures that can be transferred to other types of problems.

- **Video solutions to 20% of the end-of chapter problems.** The availability and timing of these solutions are controlled by the instructor. For example, they might be available after a homework deadline or a quiz. Each solution is not simply a plug-and-chug recipe. Rather I build a solution from the Key Ideas to the first step of reasoning and to a final solution. The student learns not just how to solve a particular problem but how to tackle any problem, even those that require *physics courage*.

- **Video examples of how to read data from graphs** (more than simply reading off a number with no comprehension of the physics).

Problem-Solving Help I have written a large number of resources for *WileyPLUS* designed to help build the students' problem-solving skills.

- **Every sample problem in the textbook** is available online in both reading and video formats.

- **Hundreds of additional sample problems.** These are available as standalone resources but (at the discretion of the instructor) they are also linked out of the homework problems. So, if a homework problem deals with, say, forces on a block on a ramp, a link to a related sample problem is provided. However, the sample problem is not just a replica of the homework problem and thus does not provide a solution that can be merely duplicated without comprehension.

- **GO Tutorials** for 15% of the end-of-chapter homework problems. In multiple steps, I lead a student through a homework problem, starting with the Key Ideas and giving hints when wrong answers are submitted. However, I purposely leave the last step (for the final answer) to the student so that they are responsible at the end. Some online tutorial systems trap a student when wrong answers are given, which can generate a lot of frustration. My GO Tutorials are not traps, because at any step along the way, a student can return to the main problem.

GO Tutorial   Close

This GO Tutorial will provide you with a step-by-step guide on how to approach this problem. When you are finished, go back and try the problem again on your own. To view the original question while you work, you can just drag this screen to the side. (This GO Tutorial consists of 4 steps).

Step 1 : Solution Step 1 of GO Tutorial 10-30

KEY IDEAS:
(1) When an object rotates at constant angular acceleration, we can use the constant-acceleration equations of Table 10-1 modified for angular motion:

$(1)\ \omega = \omega_0 + \alpha t$

$(2)\ \theta - \theta_0 = \omega_0 t + \frac{1}{2}\alpha t^2$

$(3)\ \omega^2 = \omega_0^2 + 2\alpha(\theta - \theta_0)$

$(4)\ \theta - \theta_0 = \frac{1}{2}(\omega_0 + \omega)t$

$(5)\ \theta - \theta_0 = \omega t - \frac{1}{2}\alpha t^2$

Counterclockwise is the positive direction of rotation, and clockwise is the negative direction.
(2) If a particle moves around a rotation axis at radius r, the magnitude of its radial (centripetal) acceleration ar at any moment is related to its tangential speed v (the speed along the circular path) and its angular speed at that moment by

$a_r = \dfrac{v^2}{r} = \omega^2 r$

(3) If a particle moves around a rotation axis at radius r, the magnitude of its tangential acceleration at (the acceleration along the circular path) at any moment is related to angular acceleration a at that moment by
$a_t = r\alpha$

(4) If a particle moves around a rotation axis at radius r, the angular displacement through which it rotates is related to the distance s it moves along its circular path by
$s = r\Delta\theta$

GETTING STARTED: What is the radius of rotation (in meters) of a point on the rim of the flywheel?

Number   Unit   ▼

exact number, no tolerance

Check Your Input

Step 2 : Solution Step 2 of GO Tutorial 10-30

What is the final angular speed in radians per second?

Number   Unit   ▼

the tolerance is +/-2%

Check Your Input

Step 3 : Solution Step 3 of GO Tutorial 10-30

What was the initial angular speed?

Number   Unit   ▼

exact number, no tolerance

Check Your Input

Step 4 : Solution Step 4 of GO Tutorial 10-30

Through what angular distance does the flywheel rotate to reach the final angular speed?

Number   Unit   ▼

the tolerance is +/-2%

Check Your Input

Now that you know how to solve the problem, go back and try again on your own.   Close

- **Hints on every end-of-chapter homework problem** are available (at the discretion of the instructor). I wrote these as true hints about the main ideas and the general procedure for a solution, not as recipes that provide an answer without any comprehension.

Evaluation Materials

- **Reading questions are available within each online section.** I wrote these so that they do not require analysis or any deep understanding; rather they simply test whether a student has read the section. When a student opens up a section, a randomly chosen reading question (from a bank of questions) appears at the end. The instructor can decide whether the question is part of the grading for that section or whether it is just for the benefit of the student.

- **Checkpoints are available within most sections.** I wrote these so that they require analysis and decisions about the physics in the section. *Answers to all checkpoints are in the back of the book.*

> ☑ **Checkpoint 1**
>
> Here are three pairs of initial and final positions, respectively, along an x axis. Which pairs give a negative displacement: (a) -3 m, $+5$ m; (b) -3 m, -7 m; (c) 7 m, -3 m?

- **All end-of-chapter homework Problems** in the book (and many more problems) are available in *WileyPLUS*. The instructor can construct a homework assignment and control how it is graded when the answers are submitted online. For example, the instructor controls the deadline for submission and how many attempts a student is allowed on an answer. The instructor also controls which, if any, learning aids are available with each homework problem. Such links can include hints, sample problems, in-chapter reading materials, video tutorials, video math reviews, and even video solutions (which can be made available to the students after, say, a homework deadline).

- **Symbolic notation problems** that require algebraic answers are available in every chapter.

- **All end-of-chapter homework Questions** in the book are available for assignment in *WileyPLUS*. These Questions (in a multiple choice format) are designed to evaluate the students' conceptual understanding.

Icons for Additional Help When worked-out solutions are provided either in print or electronically for certain of the odd-numbered problems, the statements for those problems include an icon to alert both student and instructor as to where the solutions are located. There are also icons indicating which problems have GO Tutorial, an Interactive LearningWare, or a link to the *The Flying Circus of Physics*. An icon guide is provided here and at the beginning of each set of problems.

GO	Tutoring problem available (at instructor's discretion) in *WileyPLUS* and WebAssign	
SSM	Worked-out solution available in Student Solutions Manual	**WWW** Worked-out solution is at
• – •••	Number of dots indicates level of problem difficulty	**ILW** Interactive solution is at
	Additional information available in *The Flying Circus of Physics* and at flyingcircusofphysics.com	

http://www.wiley.com/college/halliday

VERSIONS OF THE TEXT

To accommodate the individual needs of instructors and students, the ninth edition of *Fundamentals of Physics* is available in a number of different versions.

The **Regular Edition** consists of Chapters 1 through 37 (ISBN 9781118230718).

The **Extended Edition** contains seven additional chapters on quantum physics and cosmology, Chapters 1–44 (ISBN 9781118230725).

Volume 1 — Chapters 1–20 (Mechanics and Thermodynamics), hardcover, ISBN 9781118233764

Volume 2 — Chapters 21–44 (E&M, Optics, and Quantum Physics), hardcover, ISBN 9781118230732

INSTRUCTOR SUPPLEMENTS

Instructor's Solutions Manual by Sen-Ben Liao, Lawrence Livermore National Laboratory. This manual provides worked-out solutions for all problems found at the end of each chapter. It is available in both MSWord and PDF.

Instructor Companion Site http://www.wiley.com/college/halliday

• **Instructor's Manual** This resource contains lecture notes outlining the most important topics of each chapter; demonstration experiments; laboratory and computer projects; film and video sources; answers to all Questions, Exercises, Problems, and Checkpoints; and a correlation guide to the Questions, Exercises, and Problems in the previous edition. It also contains a complete list of all problems for which solutions are available to students (SSM,WWW, and ILW).

• **Lecture PowerPoint Slides** These PowerPoint slides serve as a helpful starter pack for instructors, outlining key concepts and incorporating figures and equations from the text.

• **Classroom Response Systems ("Clicker") Questions** by David Marx, Illinois State University. There are two sets of questions available: Reading Quiz questions and Interactive Lecture questions. The Reading Quiz questions are intended to be relatively straightforward for any student who reads the assigned material. The Interactive Lecture questions are intended for use in an interactive lecture setting.

• **Wiley Physics Simulations** by Andrew Duffy, Boston University and John Gastineau, Vernier Software. This is a collection of 50 interactive simulations (Java applets) that can be used for classroom demonstrations.

• **Wiley Physics Demonstrations** by David Maiullo, Rutgers University. This is a collection of digital videos of 80 standard physics demonstrations. They can be shown in class or accessed from *WileyPLUS*. There is an accompanying Instructor's Guide that includes "clicker" questions.

• **Test Bank** For the 10th edition, the Test Bank has been completely over-hauled by Suzanne Willis, Northern Illinois University. The Test Bank includes more than 2200 multiple-choice questions. These items are also available in the Computerized Test Bank which provides full editing features to help you customize tests (available in both IBM and Macintosh versions).

• **All text illustrations** suitable for both classroom projection and printing.

Online Homework and Quizzing. In addition to *WileyPLUS*, *Fundamentals of Physics*, tenth edition, also supports WebAssignPLUS and LON-CAPA, which are other programs that give instructors the ability to deliver and grade homework and quizzes online. WebAssign PLUS also offers students an online version of the text.

STUDENT SUPPLEMENTS

Student Companion Site. The web site http://www.wiley.com/college/halliday was developed specifically for *Fundamentals of Physics*, tenth edition, and is designed to further assist students in the study of physics. It includes solutions to selected end-of-chapter problems (which are identified with a www icon in the text); simulation exercises; tips on how to make best use of a programmable calculator; and the Interactive LearningWare tutorials that are described below.

Student Study Guide (ISBN 9781118230787) by Thomas Barrett of Ohio State University. The Student Study Guide consists of an overview of the chapter's important concepts, problem solving techniques and detailed examples.

Student Solutions Manual (ISBN 9781118230664) by Sen-Ben Liao, Lawrence Livermore National Laboratory. This manual provides students with complete worked-out solutions to 15 percent of the problems found at the end of each chapter within the text. The Student Solutions Manual for the 10th edition is written using an innovative approach called TEAL which stands for Think, Express, Analyze, and Learn. This learning strategy was originally developed at the Massachusetts Institute of Technology and has proven to be an effective learning tool for students. These problems with TEAL solutions are indicated with an SSM icon in the text.

Interactive Learningware. This software guides students through solutions to 200 of the end-of-chapter problems. These problems are indicated with an ILW icon in the text. The solutions process is developed interactively, with appropriate feedback and access to error-specific help for the most common mistakes.

Introductory Physics with Calculus as a Second Language: (ISBN 9780471739104) *Mastering Problem Solving* by Thomas Barrett of Ohio State University. This brief paperback teaches the student how to approach problems more efficiently and effectively. The student will learn how to recognize common patterns in physics problems, break problems down into manageable steps, and apply appropriate techniques. The book takes the student step by step through the solutions to numerous examples.

ACKNOWLEDGMENTS

A great many people have contributed to this book. Sen-Ben Liao of Lawrence Livermore National Laboratory, James Whitenton of Southern Polytechnic State University, and Jerry Shi, of Pasadena City College, performed the Herculean task of working out solutions for every one of the homework problems in the book. At John Wiley publishers, the book received support from Stuart Johnson, Geraldine Osnato and Aly Rentrop, the editors who oversaw the entire project from start to finish. We thank Elizabeth Swain, the production editor, for pulling all the pieces together during the complex production process. We also thank Maddy Lesure for her design of the text and the cover; Lee Goldstein for her page make-up; Helen Walden for her copyediting; and Lilian Brady for her proofreading. Jennifer Atkins was inspired in the search for unusual and interesting photographs. Both the publisher John Wiley & Sons, Inc. and Jearl Walker would like to thank the following for comments and ideas about the recent editions:

Jonathan Abramson, *Portland State University*; Omar Adawi, *Parkland College*; Edward Adelson, *The Ohio State University*; Steven R. Baker, *Naval Postgraduate School*; George Caplan, *Wellesley College*; Richard Kass, *The Ohio State University*; M. R. Khoshbin-e-Khoshnazar, *Research Institution for Curriculum Development & Educational Innovations (Tehran)*; Craig Kletzing, *University of Iowa*, Stuart Loucks, *American River College*; Laurence Lurio, *Northern Illinois University*; Ponn Maheswaranathan, *Winthrop University;* Joe McCullough, *Cabrillo College*; Carl E. Mungan, *U. S. Naval Academy*, Don N. Page, *University of Alberta*; Elie Riachi, *Fort Scott Community College*; Andrew G. Rinzler, *University of Florida*; Dubravka Rupnik, *Louisiana State University*; Robert Schabinger, *Rutgers University*; Ruth Schwartz, *Milwaukee School of Engineering*; Carol Strong, *University of Alabama at Huntsville*, Nora Thornber, *Raritan Valley Community College*; Frank Wang, *LaGuardia Community College*; Graham W. Wilson, *University of Kansas*; Roland Winkler, *Northern Illinois University*; William Zacharias, *Cleveland State University*; Ulrich Zurcher, *Cleveland State University*.

Finally, our external reviewers have been outstanding and we acknowledge here our debt to each member of that team.

Maris A. Abolins, *Michigan State University*
Edward Adelson, *Ohio State University*
Nural Akchurin, *Texas Tech*
Yildirim Aktas, *University of North Carolina-Charlotte*
Barbara Andereck, *Ohio Wesleyan University*
Tetyana Antimirova, *Ryerson University*
Mark Arnett, *Kirkwood Community College*
Arun Bansil, *Northeastern University*
Richard Barber, *Santa Clara University*
Neil Basecu, *Westchester Community College*
Anand Batra, *Howard University*
Kenneth Bolland, *The Ohio State University*
Richard Bone, *Florida International University*
Michael E. Browne, *University of Idaho*
Timothy J. Burns, *Leeward Community College*
Joseph Buschi, *Manhattan College*
Philip A. Casabella, *Rensselaer Polytechnic Institute*
Randall Caton, *Christopher Newport College*
Roger Clapp, *University of South Florida*
W. R. Conkie, *Queen's University*
Renate Crawford, *University of Massachusetts-Dartmouth*
Mike Crivello, *San Diego State University*
Robert N. Davie, Jr., *St. Petersburg Junior College*
Cheryl K. Dellai, *Glendale Community College*
Eric R. Dietz, *California State University at Chico*
N. John DiNardo, *Drexel University*
Eugene Dunnam, *University of Florida*
Robert Endorf, *University of Cincinnati*

F. Paul Esposito, *University of Cincinnati*
Jerry Finkelstein, *San Jose State University*
Robert H. Good, *California State University-Hayward*
Michael Gorman, *University of Houston*
Benjamin Grinstein, *University of California, San Diego*
John B. Gruber, *San Jose State University*
Ann Hanks, *American River College*
Randy Harris, *University of California-Davis*
Samuel Harris, *Purdue University*
Harold B. Hart, *Western Illinois University*
Rebecca Hartzler, *Seattle Central Community College*
John Hubisz, *North Carolina State University*
Joey Huston, *Michigan State University*
David Ingram, *Ohio University*
Shawn Jackson, *University of Tulsa*
Hector Jimenez, *University of Puerto Rico*
Sudhakar B. Joshi, *York University*
Leonard M. Kahn, *University of Rhode Island*
Sudipa Kirtley, *Rose-Hulman Institute*
Leonard Kleinman, *University of Texas at Austin*
Craig Kletzing, *University of Iowa*
Peter F. Koehler, *University of Pittsburgh*
Arthur Z. Kovacs, *Rochester Institute of Technology*
Kenneth Krane, *Oregon State University*
Hadley Lawler, *Vanderbilt University*
Priscilla Laws, *Dickinson College*
Edbertho Leal, *Polytechnic University of Puerto Rico*

Vern Lindberg, *Rochester Institute of Technology*
Peter Loly, *University of Manitoba*
James MacLaren, *Tulane University*
Andreas Mandelis, *University of Toronto*
Robert R. Marchini, *Memphis State University*
Andrea Markelz, *University at Buffalo, SUNY*
Paul Marquard, *Caspar College*
David Marx, *Illinois State University*
Dan Mazilu, *Washington and Lee University*
James H. McGuire, *Tulane University*
David M. McKinstry, *Eastern Washington University*
Jordon Morelli, *Queen's University*
Eugene Mosca, *United States Naval Academy*
Eric R. Murray, *Georgia Institute of Technology, School of Physics*
James Napolitano, *Rensselaer Polytechnic Institute*
Blaine Norum, *University of Virginia*
Michael O'Shea, *Kansas State University*
Patrick Papin, *San Diego State University*
Kiumars Parvin, *San Jose State University*
Robert Pelcovits, *Brown University*
Oren P. Quist, *South Dakota State University*
Joe Redish, *University of Maryland*
Timothy M. Ritter, *University of North Carolina at Pembroke*
Dan Styer, *Oberlin College*
Frank Wang, *LaGuardia Community College*
Robert Webb, *Texas A&M University*
Suzanne Willis, *Northern Illinois University*
Shannon Willoughby, *Montana State University*

Coulomb's Law

21-1 COULOMB'S LAW

Learning Objectives

After reading this module, you should be able to . . .

21.01 Distinguish between being electrically neutral, negatively charged, and positively charged and identify excess charge.

21.02 Distinguish between conductors, nonconductors (insulators), semiconductors, and superconductors.

21.03 Describe the electrical properties of the particles inside an atom.

21.04 Identify conduction electrons and explain their role in making a conducting object negatively or positively charged.

21.05 Identify what is meant by "electrically isolated" and by "grounding."

21.06 Explain how a charged object can set up induced charge in a second object.

21.07 Identify that charges with the same electrical sign repel each other and those with opposite electrical signs attract each other.

21.08 For either of the particles in a pair of charged particles, draw a free-body diagram, showing the electrostatic force (Coulomb force) on it and anchoring the tail of the force vector on that particle.

21.09 For either of the particles in a pair of charged particles, apply Coulomb's law to relate the magnitude of the electrostatic force, the charge magnitudes of the particles, and the separation between the particles.

21.10 Identify that Coulomb's law applies only to (point-like) particles and objects that can be treated as particles.

21.11 If more than one force acts on a particle, find the net force by adding all the forces as vectors, not scalars.

21.12 Identify that a shell of uniform charge attracts or repels a charged particle that is outside the shell as if all the shell's charge were concentrated as a particle at the shell's center.

21.13 Identify that if a charged particle is located inside a shell of uniform charge, there is no net electrostatic force on the particle from the shell.

21.14 Identify that if excess charge is put on a spherical conductor, it spreads out uniformly over the external surface area.

21.15 Identify that if two identical spherical conductors touch or are connected by conducting wire, any excess charge will be shared equally.

21.16 Identify that a nonconducting object can have any given distribution of charge, including charge at interior points.

21.17 Identify current as the rate at which charge moves through a point.

21.18 For current through a point, apply the relationship between the current, a time interval, and the amount of charge that moves through the point in that time interval.

Key Ideas

● The strength of a particle's electrical interaction with objects around it depends on its electric charge (usually represented as q), which can be either positive or negative. Particles with the same sign of charge repel each other, and particles with opposite signs of charge attract each other.

● An object with equal amounts of the two kinds of charge is electrically neutral, whereas one with an imbalance is electrically charged and has an excess charge.

● Conductors are materials in which a significant number of electrons are free to move. The charged particles in nonconductors (insulators) are not free to move.

● Electric current i is the rate dq/dt at which charge passes a point:

$$i = \frac{dq}{dt}.$$

● Coulomb's law describes the electrostatic force (or electric force) between two charged particles. If the particles have charges q_1 and q_2, are separated by distance r, and are at rest (or moving only slowly) relative to each other, then the magnitude of the force acting on each due to the other is given by

$$F = \frac{1}{4\pi\varepsilon_0} \frac{|q_1|\,|q_2|}{r^2} \quad \text{(Coulomb's law),}$$

where $\varepsilon_0 = 8.85 \times 10^{-12}\ \mathrm{C^2/N \cdot m^2}$ is the permittivity constant. The ratio $1/4\pi\varepsilon_0$ is often replaced with the electrostatic constant (or Coulomb constant) $k = 8.99 \times 10^9\ \mathrm{N \cdot m^2/C^2}$.

● The electrostatic force vector acting on a charged particle due to a second charged particle is either directly toward the second particle (opposite signs of charge) or directly away from it (same sign of charge).

● If multiple electrostatic forces act on a particle, the net force is the vector sum (not scalar sum) of the individual forces.

● Shell theorem 1: A charged particle outside a shell with charge uniformly distributed on its surface is attracted or repelled as if the shell's charge were concentrated as a particle at its center.

● Shell theorem 2: A charged particle inside a shell with charge uniformly distributed on its surface has no net force acting on it due to the shell.

● Charge on a conducting spherical shell spreads uniformly over the (external) surface.

What Is Physics?

You are surrounded by devices that depend on the physics of electromagnetism, which is the combination of electric and magnetic phenomena. This physics is at the root of computers, television, radio, telecommunications, household lighting, and even the ability of food wrap to cling to a container. This physics is also the basis of the natural world. Not only does it hold together all the atoms and molecules in the world, it also produces lightning, auroras, and rainbows.

The physics of electromagnetism was first studied by the early Greek philosophers, who discovered that if a piece of amber is rubbed and then brought near bits of straw, the straw will jump to the amber. We now know that the attraction between amber and straw is due to an electric force. The Greek philosophers also discovered that if a certain type of stone (a naturally occurring magnet) is brought near bits of iron, the iron will jump to the stone. We now know that the attraction between magnet and iron is due to a magnetic force.

From these modest origins with the Greek philosophers, the sciences of electricity and magnetism developed separately for centuries—until 1820, in fact, when Hans Christian Oersted found a connection between them: an electric current in a wire can deflect a magnetic compass needle. Interestingly enough, Oersted made this discovery, a big surprise, while preparing a lecture demonstration for his physics students.

The new science of electromagnetism was developed further by workers in many countries. One of the best was Michael Faraday, a truly gifted experimenter with a talent for physical intuition and visualization. That talent is attested to by the fact that his collected laboratory notebooks do not contain a single equation. In the mid-nineteenth century, James Clerk Maxwell put Faraday's ideas into mathematical form, introduced many new ideas of his own, and put electromagnetism on a sound theoretical basis.

Our discussion of electromagnetism is spread through the next 16 chapters. We begin with electrical phenomena, and our first step is to discuss the nature of electric charge and electric force.

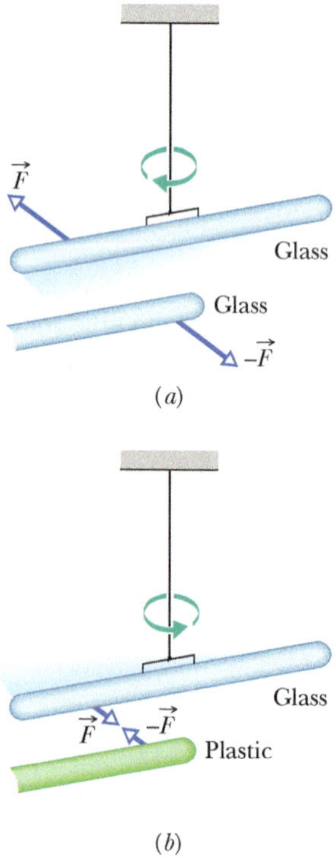

Electric Charge

Here are two demonstrations that seem to be magic, but our job here is to make sense of them. After rubbing a glass rod with a silk cloth (on a day when the humidity is low), we hang the rod by means of a thread tied around its center (Fig. 21-1a). Then we rub a second glass rod with the silk cloth and bring it near the hanging rod. The hanging rod magically moves away. We can see that a force repels it from the second rod, but how? There is no contact with that rod, no breeze to push on it, and no sound wave to disturb it.

In the second demonstration we replace the second rod with a plastic rod that has been rubbed with fur. This time, the hanging rod moves toward the nearby rod (Fig. 21-1b). Like the repulsion, this attraction occurs without any contact or obvious communication between the rods.

In the next chapter we shall discuss how the hanging rod knows of the presence of the other rods, but in this chapter let's focus on just the forces that are involved. In the first demonstration, the force on the hanging rod was *repulsive*, and

Figure 21-1 (a) The two glass rods were each rubbed with a silk cloth and one was suspended by thread. When they are close to each other, they repel each other. (b) The plastic rod was rubbed with fur. When brought close to the glass rod, the rods attract each other.

in the second, *attractive*. After a great many investigations, scientists figured out that the forces in these types of demonstrations are due to the *electric charge* that we set up on the rods when they are in contact with silk or fur. Electric charge is an intrinsic property of the fundamental particles that make up objects such as the rods, silk, and fur. That is, charge is a property that comes automatically with those particles wherever they exist.

Two Types. There are two types of electric charge, named by the American scientist and statesman Benjamin Franklin as positive charge and negative charge. He could have called them anything (such as cherry and walnut), but using algebraic signs as names comes in handy when we add up charges to find the net charge. In most everyday objects, such as a mug, there are about equal numbers of negatively charged particles and positively charged particles, and so the net charge is zero, the charge is said to be *balanced*, and the object is said to be *electrically neutral* (or just *neutral* for short).

Excess Charge. Normally you are approximately neutral. However, if you live in regions where the humidity is low, you know that the charge on your body can become slightly unbalanced when you walk across certain carpets. Either you gain negative charge from the carpet (at the points of contact between your shoes with the carpet) and become negatively charged, or you lose negative charge and become positively charged. Either way, the extra charge is said to be an *excess charge*. You probably don't notice it until you reach for a door handle or another person. Then, if your excess charge is enough, a spark leaps between you and the other object, eliminating your excess charge. Such sparking can be annoying and even somewhat painful. Such *charging* and *discharging* does not happen in humid conditions because the water in the air *neutralizes* your excess charge about as fast as you acquire it.

Two of the grand mysteries in physics are (1) *why* does the universe have particles with electric charge (what is it, really?) and (2) *why* does electric charge come in two types (and not, say, one type or three types). We just do not know. Nevertheless, with lots of experiments similar to our two demonstrations scientists discovered that

> Particles with the same sign of electrical charge repel each other, and particles with opposite signs attract each other.

In a moment we shall put this rule into quantitative form as Coulomb's law of *electrostatic force* (or *electric force*) between charged particles. The term *electrostatic* is used to emphasize that, relative to each other, the charges are either stationary or moving only very slowly.

Demos. Now let's get back to the demonstrations to understand the motions of the rod as being something other than just magic. When we rub the glass rod with a silk cloth, a small amount of negative charge moves from the rod to the silk (a transfer like that between you and a carpet), leaving the rod with a small amount of excess positive charge. (Which way the negative charge moves is not obvious and requires a lot of experimentation.) We *rub* the silk over the rod to increase the number of contact points and thus the amount, still tiny, of transferred charge. We hang the rod from the thread so as to *electrically isolate* it from its surroundings (so that the surroundings cannot neutralize the rod by giving it enough negative charge to rebalance its charge). When we rub the second rod with the silk cloth, it too becomes positively charged. So when we bring it near the first rod, the two rods repel each other (Fig. 21-2a).

Next, when we rub the plastic rod with fur, it gains excess negative charge from the fur. (Again, the transfer direction is learned through many experiments.) When we bring the plastic rod (with negative charge) near the hanging glass rod (with positive charge), the rods are attracted to each other (Fig. 21-2b). All this is subtle. You cannot see the charge or its transfer, only the results.

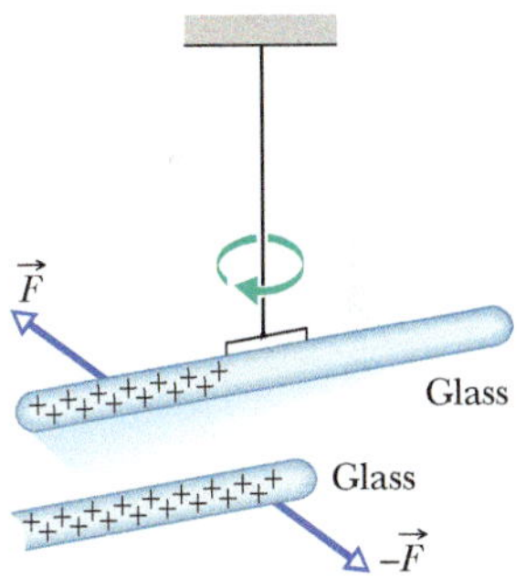

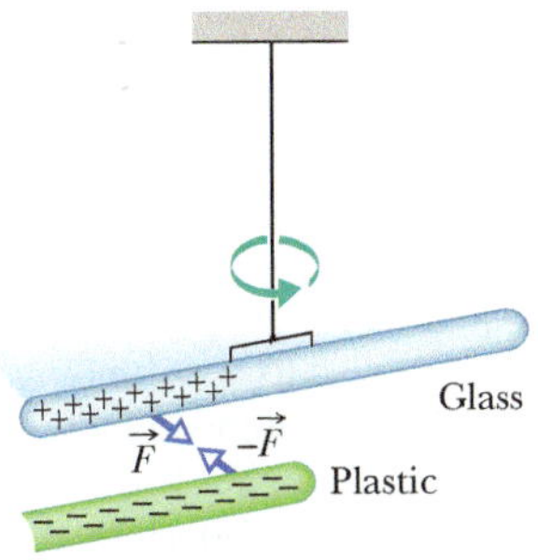

(a)

(b)

Figure 21-2 (a) Two charged rods of the same sign repel each other. (b) Two charged rods of opposite signs attract each other. Plus signs indicate a positive net charge, and minus signs indicate a negative net charge.

Conductors and Insulators

We can classify materials generally according to the ability of charge to move through them. **Conductors** are materials through which charge can move rather freely; examples include metals (such as copper in common lamp wire), the human body, and tap water. **Nonconductors**—also called **insulators**—are materials through which charge cannot move freely; examples include rubber (such as the insulation on common lamp wire), plastic, glass, and chemically pure water. **Semiconductors** are materials that are intermediate between conductors and insulators; examples include silicon and germanium in computer chips. **Superconductors** are materials that are *perfect* conductors, allowing charge to move without *any* hindrance. In these chapters we discuss only conductors and insulators.

Conducting Path. Here is an example of how conduction can eliminate excess charge on an object. If you rub a copper rod with wool, charge is transferred from the wool to the rod. However, if you are holding the rod while also touching a faucet, you cannot charge the rod in spite of the transfer. The reason is that you, the rod, and the faucet are all conductors connected, via the plumbing, to Earth's surface, which is a huge conductor. Because the excess charges put on the rod by the wool repel one another, they move away from one another by moving first through the rod, then through you, and then through the faucet and plumbing to reach Earth's surface, where they can spread out. The process leaves the rod electrically neutral.

In thus setting up a pathway of conductors between an object and Earth's surface, we are said to *ground* the object, and in neutralizing the object (by eliminating an unbalanced positive or negative charge), we are said to *discharge* the object. If instead of holding the copper rod in your hand, you hold it by an insulating handle, you eliminate the conducting path to Earth, and the rod can then be charged by rubbing (the charge remains on the rod), as long as you do not touch it directly with your hand.

Charged Particles. The properties of conductors and insulators are due to the structure and electrical nature of atoms. Atoms consist of positively charged *protons*, negatively charged *electrons*, and electrically neutral *neutrons*. The protons and neutrons are packed tightly together in a central *nucleus*.

The charge of a single electron and that of a single proton have the same magnitude but are opposite in sign. Hence, an electrically neutral atom contains equal numbers of electrons and protons. Electrons are held near the nucleus because they have the electrical sign opposite that of the protons in the nucleus and thus are attracted to the nucleus. Were this not true, there would be no atoms and thus no you.

When atoms of a conductor like copper come together to form the solid, some of their outermost (and so most loosely held) electrons become free to wander about within the solid, leaving behind positively charged atoms (*positive ions*). We call the mobile electrons *conduction electrons*. There are few (if any) free electrons in a nonconductor.

Induced Charge. The experiment of Fig. 21-3 demonstrates the mobility of charge in a conductor. A negatively charged plastic rod will attract either end of an isolated neutral copper rod. What happens is that many of the conduction electrons in the closer end of the copper rod are repelled by the negative charge on the plastic rod. Some of the conduction electrons move to the far end of the copper rod, leaving the near end depleted in electrons and thus with an unbalanced positive charge. This positive charge is attracted to the negative charge in the plastic rod. Although the copper rod is still neutral, it is said to have an *induced charge*, which means that some of its positive and negative charges have been separated due to the presence of a nearby charge.

Similarly, if a positively charged glass rod is brought near one end of a neutral copper rod, induced charge is again set up in the neutral copper rod but now the near end gains conduction electrons, becomes negatively charged, and is attracted to the glass rod, while the far end is positively charged.

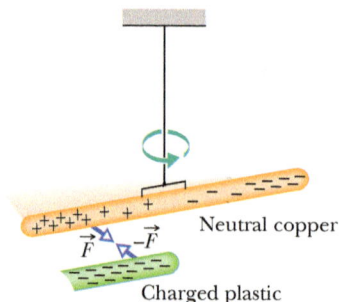

Figure 21-3 A neutral copper rod is electrically isolated from its surroundings by being suspended on a nonconducting thread. Either end of the copper rod will be attracted by a charged rod. Here, conduction electrons in the copper rod are repelled to the far end of that rod by the negative charge on the plastic rod. Then that negative charge attracts the remaining positive charge on the near end of the copper rod, rotating the copper rod to bring that near end closer to the plastic rod.

Note that only conduction electrons, with their negative charges, can move; positive ions are fixed in place. Thus, an object becomes positively charged only through the *removal of negative charges*.

Blue Flashes from a Wintergreen LifeSaver

Indirect evidence for the attraction of charges with opposite signs can be seen with a wintergreen LifeSaver (the candy shaped in the form of a marine lifesaver). If you adapt your eyes to darkness for about 15 minutes and then have a friend chomp on a piece of the candy in the darkness, you will see a faint blue flash from your friend's mouth with each chomp. Whenever a chomp breaks a sugar crystal into pieces, each piece will probably end up with a different number of electrons. Suppose a crystal breaks into pieces A and B, with A ending up with more electrons on its surface than B (Fig. 21-4). This means that B has positive ions (atoms that lost electrons to A) on its surface. Because the electrons on A are strongly attracted to the positive ions on B, some of those electrons jump across the gap between the pieces.

As A and B move away from each other, air (primarily nitrogen, N_2) flows into the gap, and many of the jumping electrons collide with nitrogen molecules in the air, causing the molecules to emit ultraviolet light. You cannot see this type of light. However, the wintergreen molecules on the surfaces of the candy pieces absorb the ultraviolet light and then emit blue light, which you *can* see—it is the blue light coming from your friend's mouth.

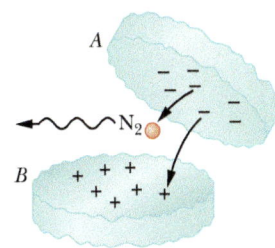

Figure 21-4 Two pieces of a wintergreen LifeSaver candy as they fall away from each other. Electrons jumping from the negative surface of piece A to the positive surface of piece B collide with nitrogen (N_2) molecules in the air.

✓ Checkpoint 1

The figure shows five pairs of plates: A, B, and D are charged plastic plates and C is an electrically neutral copper plate. The electrostatic forces between the pairs of plates are shown for three of the pairs. For the remaining two pairs, do the plates repel or attract each other?

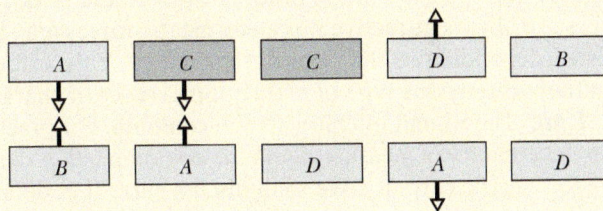

Coulomb's Law

Now we come to the equation for Coulomb's law, but first a caution. This equation works for only charged particles (and a few other things that can be treated as particles). For extended objects, with charge located in many different places, we need more powerful techniques. So, here we consider just charged particles and not, say, two charged cats.

If two charged particles are brought near each other, they each exert an **electrostatic force** on the other. The direction of the force vectors depends on the signs of the charges. If the particles have the same sign of charge, they repel each other. That means that the force vector on each is directly away from the other particle (Figs. 21-5a and b). If we release the particles, they accelerate away from each other. If, instead, the particles have opposite signs of charge, they attract each other. That means that the force vector on each is directly toward the other particle (Fig. 21-5c). If we release the particles, they accelerate toward each other.

The equation for the electrostatic forces acting on the particles is called **Coulomb's law** after Charles-Augustin de Coulomb, whose experiments in 1785 led him to it. Let's write the equation in vector form and in terms of the particles shown in Fig. 21-6, where particle 1 has charge q_1 and particle 2 has charge q_2. (These symbols can represent either positive or negative charge.) Let's also focus on particle 1 and write the force acting on it in terms of a unit vector $\hat{r}$ that points along a radial

Figure 21-5 Two charged particles repel each other if they have the same sign of charge, either (a) both positive or (b) both negative. (c) They attract each other if they have opposite signs of charge.

Figure 21-6 The electrostatic force on particle 1 can be described in terms of a unit vector $\hat{r}$ along an axis through the two particles, radially away from particle 2.

axis extending through the two particles, radially away from particle 2. (As with other unit vectors, $\hat{r}$ has a magnitude of exactly 1 and no unit; its purpose is to point, like a direction arrow on a street sign.) With these decisions, we write the electrostatic force as

$$\vec{F} = k\,\frac{q_1 q_2}{r^2}\,\hat{r} \quad \text{(Coulomb's law)}, \tag{21-1}$$

where r is the separation between the particles and k is a positive constant called the *electrostatic constant* or the *Coulomb constant*. (We'll discuss k below.)

Let's first check the direction of the force on particle 1 as given by Eq. 21-1. If q_1 and q_2 have the same sign, then the product $q_1 q_2$ gives us a positive result. So, Eq. 21-1 tells us that the force on particle 1 is in the direction of $\hat{r}$. That checks, because particle 1 is being repelled from particle 2. Next, if q_1 and q_2 have opposite signs, the product $q_1 q_2$ gives us a negative result. So, now Eq. 21-1 tells us that the force on particle 1 is in the direction opposite $\hat{r}$. That checks because particle 1 is being attracted toward particle 2.

An Aside. Here is something that is very curious. The form of Eq. 21-1 is the same as that of Newton's equation (Eq. 13-3) for the gravitational force between two particles with masses m_1 and m_2 and separation r:

$$\vec{F} = G\,\frac{m_1 m_2}{r^2}\,\hat{r} \quad \text{(Newton's law)}, \tag{21-2}$$

where G is the gravitational constant. Although the two types of forces are wildly different, both equations describe inverse square laws (the $1/r^2$ dependences) that involve a product of a property of the interacting particles—the charge in one case and the mass in the other. However, the laws differ in that gravitational forces are always attractive but electrostatic forces may be either attractive or repulsive, depending on the signs of the charges. This difference arises from the fact that there is only one type of mass but two types of charge.

Unit. The SI unit of charge is the **coulomb**. For practical reasons having to do with the accuracy of measurements, the coulomb unit is derived from the SI unit *ampere* for electric current i. We shall discuss current in detail in Chapter 26, but here let's just note that current i is the rate dq/dt at which charge moves past a point or through a region:

$$i = \frac{dq}{dt} \quad \text{(electric current)}. \tag{21-3}$$

Rearranging Eq. 21-3 and replacing the symbols with their units (coulombs C, amperes A, and seconds s) we see that

$$1\,\text{C} = (1\,\text{A})(1\,\text{s}).$$

Force Magnitude. For historical reasons (and because doing so simplifies many other formulas), the electrostatic constant k in Eq. 21-1 is often written as $1/4\pi\varepsilon_0$. Then the magnitude of the electrostatic force in Coulomb's law becomes

$$F = \frac{1}{4\pi\varepsilon_0}\,\frac{|q_1||q_2|}{r^2} \quad \text{(Coulomb's law)}. \tag{21-4}$$

The constants in Eqs. 21-1 and 21-4 have the value

$$k = \frac{1}{4\pi\varepsilon_0} = 8.99 \times 10^9\,\text{N}\cdot\text{m}^2/\text{C}^2. \tag{21-5}$$

The quantity ε_0, called the **permittivity constant,** sometimes appears separately in equations and is

$$\varepsilon_0 = 8.85 \times 10^{-12}\,\text{C}^2/\text{N}\cdot\text{m}^2. \tag{21-6}$$

Working a Problem. Note that the charge magnitudes appear in Eq. 21-4, which gives us the force magnitude. So, in working problems in this chapter, we use Eq. 21-4 to find the magnitude of a force on a chosen particle due to a second particle and we separately determine the direction of the force by considering the charge signs of the two particles.

   Multiple Forces. As with all forces in this book, the electrostatic force obeys the principle of superposition. Suppose we have *n* charged particles near a chosen particle called particle 1; then the net force on particle 1 is given by the vector sum

$$\vec{F}_{1,net} = \vec{F}_{12} + \vec{F}_{13} + \vec{F}_{14} + \vec{F}_{15} + \cdots + \vec{F}_{1n}, \qquad (21\text{-}7)$$

in which, for example, $\vec{F}_{14}$ is the force on particle 1 due to the presence of particle 4.

   This equation is the key to many of the homework problems, so let's state it in words. If you want to know the net force acting on a chosen charged particle that is surrounded by other charged particles, first clearly identify that chosen particle and then find the force on it due to each of the other particles. Draw those force vectors in a free-body diagram of the chosen particle, with the tails anchored on the particle. (That may sound trivial, but failing to do so easily leads to errors.) Then add all those forces *as vectors* according to the rules of Chapter 3, not as scalars. (You cannot just willy-nilly add up their magnitudes.) The result is the net force (or resultant force) acting on the particle.

   Although the vector nature of the forces makes the homework problems harder than if we simply had scalars, be thankful that Eq. 21-7 works. If two force vectors did not simply add but for some reason amplified each other, the world would be very difficult to understand and manage.

   Shell Theories. Analogous to the shell theories for the gravitational force (Module 13-1), we have two shell theories for the electrostatic force:

> Shell theory 1. A charged particle outside a shell with charge uniformly distributed on its surface is attracted or repelled as if the shell's charge were concentrated as a particle at its center.

> Shell theory 2. A charged particle inside a shell with charge uniformly distributed on its surface has no net force acting on it due to the shell.

(In the first theory, we assume that the charge on the shell is much greater than the particle's charge. Thus the presence of the particle has negligible effect on the distribution of charge on the shell.)

Spherical Conductors

If excess charge is placed on a spherical shell that is made of conducting material, the excess charge spreads uniformly over the (external) surface. For example, if we place excess electrons on a spherical metal shell, those electrons repel one another and tend to move apart, spreading over the available surface until they are uniformly distributed. That arrangement maximizes the distances between all pairs of the excess electrons. According to the first shell theorem, the shell then will attract or repel an external charge as if all the excess charge on the shell were concentrated at its center.

   If we remove negative charge from a spherical metal shell, the resulting positive charge of the shell is also spread uniformly over the surface of the shell. For example, if we remove *n* electrons, there are then *n* sites of positive charge (sites missing an electron) that are spread uniformly over the shell. According to the first shell theorem, the shell will again attract or repel an external charge as if all the shell's excess charge were concentrated at its center.

✓ Checkpoint 2

The figure shows two protons (symbol p) and one electron (symbol e) on an axis. On the central proton, what is the direction of (a) the force due to the electron, (b) the force due to the other proton, and (c) the net force?

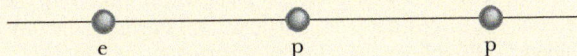

Sample Problem 21.01 Finding the net force due to two other particles

This sample problem actually contains three examples, to build from basic stuff to harder stuff. In each we have the same charged particle 1. First there is a single force acting on it (easy stuff). Then there are two forces, but they are just in opposite directions (not too bad). Then there are again two forces but they are in very different directions (ah, now we have to get serious about the fact that they are vectors). The key to all three examples is to draw the forces correctly *before* you reach for a calculator, otherwise you may be calculating nonsense on the calculator. (Figure 21-7 is available in *WileyPLUS* as an animation with voiceover.)

(a) Figure 21-7a shows two positively charged particles fixed in place on an x axis. The charges are $q_1 = 1.60 \times 10^{-19}$ C and $q_2 = 3.20 \times 10^{-19}$ C, and the particle separation is $R = 0.0200$ m. What are the magnitude and direction of the electrostatic force $\vec{F}_{12}$ on particle 1 from particle 2?

KEY IDEAS

Because both particles are positively charged, particle 1 is repelled by particle 2, with a force magnitude given by Eq. 21-4. Thus, the direction of force $\vec{F}_{12}$ on particle 1 is *away from* particle 2, in the negative direction of the x axis, as indicated in the free-body diagram of Fig. 21-7b.

Two particles: Using Eq. 21-4 with separation R substituted for r, we can write the magnitude F_{12} of this force as

$$F_{12} = \frac{1}{4\pi\varepsilon_0} \frac{|q_1||q_2|}{R^2}$$

$$= (8.99 \times 10^9 \, \text{N} \cdot \text{m}^2/\text{C}^2)$$

$$\times \frac{(1.60 \times 10^{-19} \, \text{C})(3.20 \times 10^{-19} \, \text{C})}{(0.0200 \, \text{m})^2}$$

$$= 1.15 \times 10^{-24} \, \text{N}.$$

Thus, force $\vec{F}_{12}$ has the following magnitude and direction (relative to the positive direction of the x axis):

$$1.15 \times 10^{-24} \, \text{N} \quad \text{and} \quad 180°. \qquad \text{(Answer)}$$

We can also write $\vec{F}_{12}$ in unit-vector notation as

$$\vec{F}_{12} = -(1.15 \times 10^{-24} \, \text{N})\hat{i}. \qquad \text{(Answer)}$$

(b) Figure 21-7c is identical to Fig. 21-7a except that particle 3 now lies on the x axis between particles 1 and 2. Particle 3 has charge $q_3 = -3.20 \times 10^{-19}$ C and is at a distance $\frac{3}{4}R$ from particle 1. What is the net electrostatic force $\vec{F}_{1,\text{net}}$ on particle 1 due to particles 2 and 3?

KEY IDEA

The presence of particle 3 does not alter the electrostatic force on particle 1 from particle 2. Thus, force $\vec{F}_{12}$ still acts on particle 1. Similarly, the force $\vec{F}_{13}$ that acts on particle 1 due to particle 3 is not affected by the presence of particle 2. Because particles 1

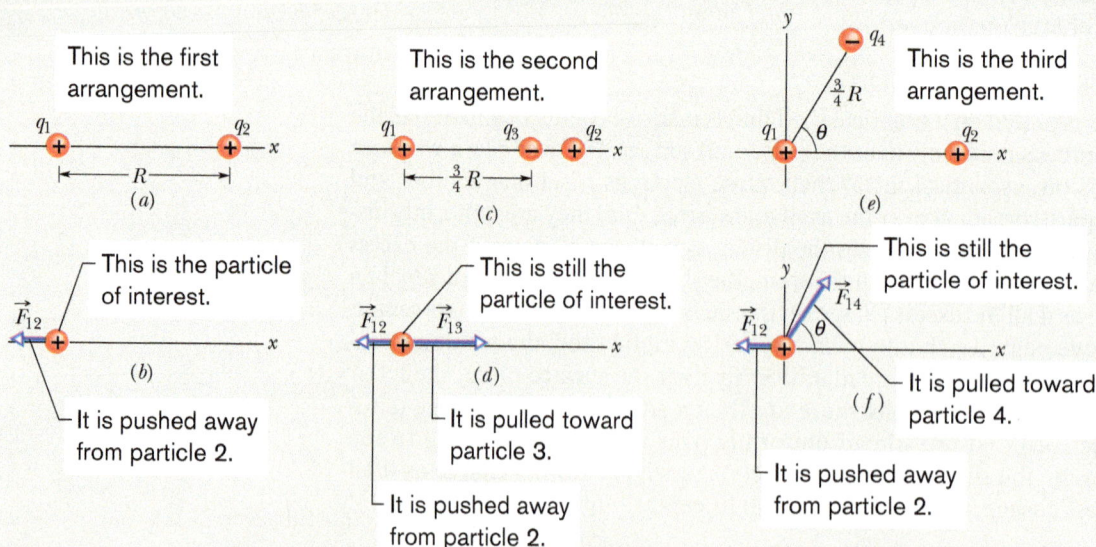

This is the first arrangement.

This is the second arrangement.

This is the third arrangement.

This is the particle of interest.

It is pushed away from particle 2.

This is still the particle of interest.

It is pulled toward particle 3.

It is pushed away from particle 2.

This is still the particle of interest.

It is pulled toward particle 4.

It is pushed away from particle 2.

Figure 21-7 (a) Two charged particles of charges q_1 and q_2 are fixed in place on an x axis. (b) The free-body diagram for particle 1, showing the electrostatic force on it from particle 2. (c) Particle 3 included. (d) Free-body diagram for particle 1. (e) Particle 4 included. (f) Free-body diagram for particle 1.

and 3 have charge of opposite signs, particle 1 is attracted to particle 3. Thus, force $\vec{F}_{13}$ is directed *toward* particle 3, as indicated in the free-body diagram of Fig. 21-7d.

Three particles: To find the magnitude of $\vec{F}_{13}$, we can rewrite Eq. 21-4 as

$$F_{13} = \frac{1}{4\pi\varepsilon_0} \frac{|q_1||q_3|}{(\frac{3}{4}R)^2}$$

$$= (8.99 \times 10^9 \, \text{N·m}^2/\text{C}^2)$$

$$\times \frac{(1.60 \times 10^{-19} \, \text{C})(3.20 \times 10^{-19} \, \text{C})}{(\frac{3}{4})^2(0.0200 \, \text{m})^2}$$

$$= 2.05 \times 10^{-24} \, \text{N}.$$

We can also write $\vec{F}_{13}$ in unit-vector notation:

$$\vec{F}_{13} = (2.05 \times 10^{-24} \, \text{N})\hat{\text{i}}.$$

The net force $\vec{F}_{1,\text{net}}$ on particle 1 is the vector sum of $\vec{F}_{12}$ and $\vec{F}_{13}$; that is, from Eq. 21-7, we can write the net force $\vec{F}_{1,\text{net}}$ on particle 1 in unit-vector notation as

$$\vec{F}_{1,\text{net}} = \vec{F}_{12} + \vec{F}_{13}$$

$$= -(1.15 \times 10^{-24} \, \text{N})\hat{\text{i}} + (2.05 \times 10^{-24} \, \text{N})\hat{\text{i}}$$

$$= (9.00 \times 10^{-25} \, \text{N})\hat{\text{i}}. \qquad \text{(Answer)}$$

Thus, $\vec{F}_{1,\text{net}}$ has the following magnitude and direction (relative to the positive direction of the *x* axis):

$$9.00 \times 10^{-25} \, \text{N} \quad \text{and} \quad 0°. \qquad \text{(Answer)}$$

(c) Figure 21-7e is identical to Fig. 21-7a except that particle 4 is now included. It has charge $q_4 = -3.20 \times 10^{-19}$ C, is at a distance $\frac{3}{4}R$ from particle 1, and lies on a line that makes an angle $\theta = 60°$ with the *x* axis. What is the net electrostatic force $\vec{F}_{1,\text{net}}$ on particle 1 due to particles 2 and 4?

KEY IDEA

The net force $\vec{F}_{1,\text{net}}$ is the vector sum of $\vec{F}_{12}$ and a new force $\vec{F}_{14}$ acting on particle 1 due to particle 4. Because particles 1 and 4 have charge of opposite signs, particle 1 is attracted to particle 4. Thus, force $\vec{F}_{14}$ on particle 1 is directed *toward* particle 4, at angle $\theta = 60°$, as indicated in the free-body diagram of Fig. 21-7f.

Four particles: We can rewrite Eq. 21-4 as

$$F_{14} = \frac{1}{4\pi\varepsilon_0} \frac{|q_1||q_4|}{(\frac{3}{4}R)^2}$$

$$= (8.99 \times 10^9 \, \text{N·m}^2/\text{C}^2)$$

$$\times \frac{(1.60 \times 10^{-19} \, \text{C})(3.20 \times 10^{-19} \, \text{C})}{(\frac{3}{4})^2(0.0200 \, \text{m})^2}$$

$$= 2.05 \times 10^{-24} \, \text{N}.$$

Then from Eq. 21-7, we can write the net force $\vec{F}_{1,\text{net}}$ on particle 1 as

$$\vec{F}_{1,\text{net}} = \vec{F}_{12} + \vec{F}_{14}.$$

Because the forces $\vec{F}_{12}$ and $\vec{F}_{14}$ are not directed along the same axis, we *cannot* sum simply by combining their magnitudes. Instead, we must add them as vectors, using one of the following methods.

Method 1. *Summing directly on a vector-capable calculator.* For $\vec{F}_{12}$, we enter the magnitude 1.15×10^{-24} and the angle 180°. For $\vec{F}_{14}$, we enter the magnitude 2.05×10^{-24} and the angle 60°. Then we add the vectors.

Method 2. *Summing in unit-vector notation.* First we rewrite $\vec{F}_{14}$ as

$$\vec{F}_{14} = (F_{14} \cos \theta)\hat{\text{i}} + (F_{14} \sin \theta)\hat{\text{j}}.$$

Substituting 2.05×10^{-24} N for F_{14} and 60° for θ, this becomes

$$\vec{F}_{14} = (1.025 \times 10^{-24} \, \text{N})\hat{\text{i}} + (1.775 \times 10^{-24} \, \text{N})\hat{\text{j}}.$$

Then we sum:

$$\vec{F}_{1,\text{net}} = \vec{F}_{12} + \vec{F}_{14}$$

$$= -(1.15 \times 10^{-24} \, \text{N})\hat{\text{i}}$$

$$+ (1.025 \times 10^{-24} \, \text{N})\hat{\text{i}} + (1.775 \times 10^{-24} \, \text{N})\hat{\text{j}}$$

$$\approx (-1.25 \times 10^{-25} \, \text{N})\hat{\text{i}} + (1.78 \times 10^{-24} \, \text{N})\hat{\text{j}}.$$

$$\text{(Answer)}$$

Method 3. *Summing components axis by axis.* The sum of the *x* components gives us

$$F_{1,\text{net},x} = F_{12,x} + F_{14,x} = F_{12} + F_{14} \cos 60°$$

$$= -1.15 \times 10^{-24} \, \text{N} + (2.05 \times 10^{-24} \, \text{N})(\cos 60°)$$

$$= -1.25 \times 10^{-25} \, \text{N}.$$

The sum of the *y* components gives us

$$F_{1,\text{net},y} = F_{12,y} + F_{14,y} = 0 + F_{14} \sin 60°$$

$$= (2.05 \times 10^{-24} \, \text{N})(\sin 60°)$$

$$= 1.78 \times 10^{-24} \, \text{N}.$$

The net force $\vec{F}_{1,\text{net}}$ has the magnitude

$$F_{1,\text{net}} = \sqrt{F_{1,\text{net},x}^2 + F_{1,\text{net},y}^2} = 1.78 \times 10^{-24} \, \text{N}. \quad \text{(Answer)}$$

To find the direction of $\vec{F}_{1,\text{net}}$, we take

$$\theta = \tan^{-1} \frac{F_{1,\text{net},y}}{F_{1,\text{net},x}} = -86.0°.$$

However, this is an unreasonable result because $\vec{F}_{1,\text{net}}$ must have a direction between the directions of $\vec{F}_{12}$ and $\vec{F}_{14}$. To correct θ, we add 180°, obtaining

$$-86.0° + 180° = 94.0°. \qquad \text{(Answer)}$$

PLUS Additional examples, video, and practice available at *WileyPLUS*

Checkpoint 3

The figure here shows three arrangements of an electron e and two protons p. (a) Rank the arrangements according to the magnitude of the net electrostatic force on the electron due to the protons, largest first. (b) In situation c, is the angle between the net force on the electron and the line labeled d less than or more than 45°?

(a)　　　(b)　　　(c)

Sample Problem 21.02 — Equilibrium of two forces on a particle

Figure 21-8a shows two particles fixed in place: a particle of charge $q_1 = +8q$ at the origin and a particle of charge $q_2 = -2q$ at $x = L$. At what point (other than infinitely far away) can a proton be placed so that it is in *equilibrium* (the net force on it is zero)? Is that equilibrium *stable* or *unstable*? (That is, if the proton is displaced, do the forces drive it back to the point of equilibrium or drive it farther away?)

KEY IDEA

If $\vec{F}_1$ is the force on the proton due to charge q_1 and $\vec{F}_2$ is the force on the proton due to charge q_2, then the point we seek is where $\vec{F}_1 + \vec{F}_2 = 0$. Thus,

$$\vec{F}_1 = -\vec{F}_2. \tag{21-8}$$

This tells us that at the point we seek, the forces acting on the proton due to the other two particles must be of equal magnitudes,

$$F_1 = F_2, \tag{21-9}$$

and that the forces must have opposite directions.

Reasoning: Because a proton has a positive charge, the proton and the particle of charge q_1 are of the same sign, and force $\vec{F}_1$ on the proton must point away from q_1. Also, the proton and the particle of charge q_2 are of opposite signs, so force $\vec{F}_2$ on the proton must point toward q_2. "Away from q_1" and "toward q_2" can be in opposite directions only if the proton is located on the x axis.

If the proton is on the x axis at any point between q_1 and q_2, such as point P in Fig. 21-8b, then $\vec{F}_1$ and $\vec{F}_2$ are in the same direction and not in opposite directions as required. If the proton is at any point on the x axis to the left of q_1, such as point S in Fig. 21-8c, then $\vec{F}_1$ and $\vec{F}_2$ are in opposite directions. However, Eq. 21-4 tells us that $\vec{F}_1$ and $\vec{F}_2$ cannot have equal magnitudes there: F_1 must be greater than F_2, because F_1 is produced by a closer charge (with lesser r) of greater magnitude ($8q$ versus $2q$).

Finally, if the proton is at any point on the x axis to the right of q_2, such as point R in Fig. 21-8d, then $\vec{F}_1$ and $\vec{F}_2$ are again in opposite directions. However, because now the charge of greater magnitude (q_1) is *farther* away from the proton than the charge of lesser magnitude, there is a point at which F_1 is equal to F_2. Let x be the coordinate of this point, and let q_p be the charge of the proton.

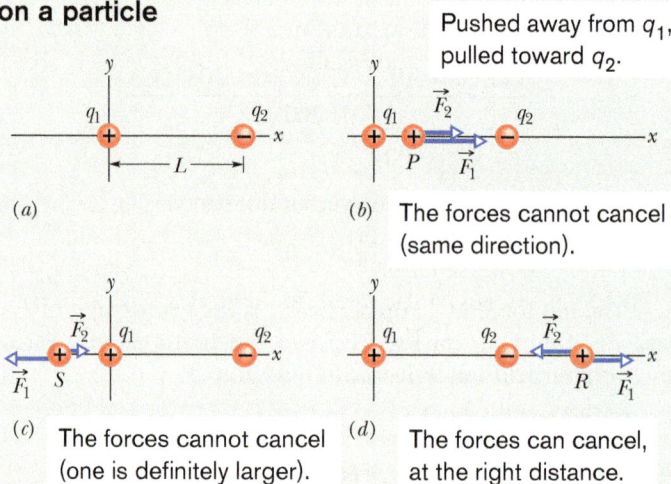

Pushed away from q_1, pulled toward q_2.

(a)　(b) The forces cannot cancel (same direction).

(c) The forces cannot cancel (one is definitely larger).　(d) The forces can cancel, at the right distance.

Figure 21-8 (a) Two particles of charges q_1 and q_2 are fixed in place on an x axis, with separation L. (b)–(d) Three possible locations P, S, and R for a proton. At each location, $\vec{F}_1$ is the force on the proton from particle 1 and $\vec{F}_2$ is the force on the proton from particle 2.

Calculations: With Eq. 21-4, we can now rewrite Eq. 21-9:

$$\frac{1}{4\pi\varepsilon_0}\frac{8qq_p}{x^2} = \frac{1}{4\pi\varepsilon_0}\frac{2qq_p}{(x-L)^2}. \tag{21-10}$$

(Note that only the charge magnitudes appear in Eq. 21-10. We already decided about the directions of the forces in drawing Fig. 21-8d and do not want to include any positive or negative signs here.) Rearranging Eq. 21-10 gives us

$$\left(\frac{x-L}{x}\right)^2 = \frac{1}{4}.$$

After taking the square roots of both sides, we find

$$\frac{x-L}{x} = \frac{1}{2}.$$

and

$$x = 2L. \tag{Answer}$$

The equilibrium at $x = 2L$ is unstable; that is, if the proton is displaced leftward from point R, then F_1 and F_2 both increase but F_2 increases more (because q_2 is closer than q_1), and a net force will drive the proton farther leftward. If the proton is displaced rightward, both F_1 and F_2 decrease but F_2 decreases more, and a net force will then drive the proton farther rightward. In a stable equilibrium, if the proton is displaced slightly, it returns to the equilibrium position.

Sample Problem 21.03 Charge sharing by two identical conducting spheres

In Fig. 21-9a, two identical, electrically isolated conducting spheres A and B are separated by a (center-to-center) distance a that is large compared to the spheres. Sphere A has a positive charge of +Q, and sphere B is electrically neutral. Initially, there is no electrostatic force between the spheres. (The large separation means there is no induced charge.)

(a) Suppose the spheres are connected for a moment by a conducting wire. The wire is thin enough so that any net charge on it is negligible. What is the electrostatic force between the spheres after the wire is removed?

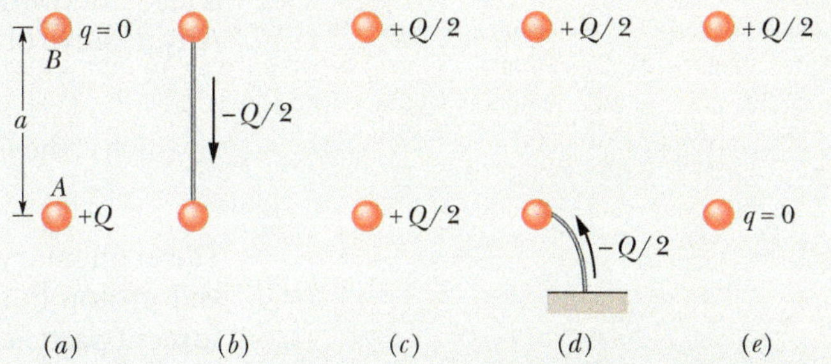

KEY IDEAS

(1) Because the spheres are identical, connecting them means that they end up with identical charges (same sign and same amount). (2) The initial sum of the charges (including the signs of the charges) must equal the final sum of the charges.

Reasoning: When the spheres are wired together, the (negative) conduction electrons on B, which repel one another, have a way to move away from one another (along the wire to positively charged A, which attracts them—Fig. 21-9b). As B loses negative charge, it becomes positively charged, and as A gains negative charge, it becomes *less* positively charged. The transfer of charge stops when the charge on B has increased to $+Q/2$ and the charge on A has decreased to $+Q/2$, which occurs when $-Q/2$ has shifted from B to A.

After the wire has been removed (Fig. 21-9c), we can assume that the charge on either sphere does not disturb the uniformity of the charge distribution on the other sphere, because the spheres are small relative to their separation. Thus, we can apply the first shell theorem to each sphere. By Eq. 21-4 with $q_1 = q_2 = Q/2$ and $r = a$,

Figure 21-9 Two small conducting spheres A and B. (a) To start, sphere A is charged positively. (b) Negative charge is transferred from B to A through a connecting wire. (c) Both spheres are then charged positively. (d) Negative charge is transferred through a grounding wire to sphere A. (e) Sphere A is then neutral.

$$F = \frac{1}{4\pi\varepsilon_0}\frac{(Q/2)(Q/2)}{a^2} = \frac{1}{16\pi\varepsilon_0}\left(\frac{Q}{a}\right)^2. \qquad \text{(Answer)}$$

The spheres, now positively charged, repel each other.

(b) Next, suppose sphere A is grounded momentarily, and then the ground connection is removed. What now is the electrostatic force between the spheres?

Reasoning: When we provide a conducting path between a charged object and the ground (which is a huge conductor), we neutralize the object. Were sphere A negatively charged, the mutual repulsion between the excess electrons would cause them to move from the sphere to the ground. However, because sphere A is positively charged, electrons with a total charge of $-Q/2$ move *from* the ground up onto the sphere (Fig. 21-9d), leaving the sphere with a charge of 0 (Fig. 21-9e). Thus, the electrostatic force is again zero.

PLUS Additional examples, video, and practice available at *WileyPLUS*

21-2 CHARGE IS QUANTIZED

Learning Objectives

After reading this module, you should be able to . . .

21.19 Identify the elementary charge.

21.20 Identify that the charge of a particle or object must be a positive or negative integer times the elementary charge.

Key Ideas

● Electric charge is quantized (restricted to certain values).

● The charge of a particle can be written as ne, where n is a positive or negative integer and e is the elementary charge,

which is the magnitude of the charge of the electron and proton ($\approx 1.602 \times 10^{-19}$ C).

Charge Is Quantized

In Benjamin Franklin's day, electric charge was thought to be a continuous fluid—an idea that was useful for many purposes. However, we now know that fluids themselves, such as air and water, are not continuous but are made up of atoms and molecules; matter is discrete. Experiment shows that "electrical fluid" is also not continuous but is made up of multiples of a certain elementary charge. Any positive or negative charge q that can be detected can be written as

$$q = ne, \qquad n = \pm1, \pm2, \pm3, \dots, \tag{21-11}$$

in which e, the **elementary charge,** has the approximate value

$$e = 1.602 \times 10^{-19} \text{ C}. \tag{21-12}$$

The elementary charge e is one of the important constants of nature. The electron and proton both have a charge of magnitude e (Table 21-1). (Quarks, the constituent particles of protons and neutrons, have charges of $\pm e/3$ or $\pm 2e/3$, but they apparently cannot be detected individually. For this and for historical reasons, we do not take their charges to be the elementary charge.)

You often see phrases—such as "the charge on a sphere," "the amount of charge transferred," and "the charge carried by the electron"—that suggest that charge is a substance. (Indeed, such statements have already appeared in this chapter.) You should, however, keep in mind what is intended: *Particles are the substance and charge happens to be one of their properties, just as mass is.*

When a physical quantity such as charge can have only discrete values rather than any value, we say that the quantity is **quantized.** It is possible, for example, to find a particle that has no charge at all or a charge of $+10e$ or $-6e$, but not a particle with a charge of, say, $3.57e$.

The quantum of charge is small. In an ordinary 100 W lightbulb, for example, about 10^{19} elementary charges enter the bulb every second and just as many leave. However, the graininess of electricity does not show up in such large-scale phenomena (the bulb does not flicker with each electron).

Table 21-1 The Charges of Three Particles

Particle	Symbol	Charge
Electron	e or e$^-$	$-e$
Proton	p	$+e$
Neutron	n	0

✓ Checkpoint 4

Initially, sphere A has a charge of $-50e$ and sphere B has a charge of $+20e$. The spheres are made of conducting material and are identical in size. If the spheres then touch, what is the resulting charge on sphere A?

Sample Problem 21.04 Mutual electric repulsion in a nucleus

The nucleus in an iron atom has a radius of about 4.0×10^{-15} m and contains 26 protons.

(a) What is the magnitude of the repulsive electrostatic force between two of the protons that are separated by 4.0×10^{-15} m?

KEY IDEA

The protons can be treated as charged particles, so the magnitude of the electrostatic force on one from the other is given by Coulomb's law.

Calculation: Table 21-1 tells us that the charge of a proton is $+e$. Thus, Eq. 21-4 gives us

$$F = \frac{1}{4\pi\varepsilon_0} \frac{e^2}{r^2}$$

$$= \frac{(8.99 \times 10^9 \text{ N} \cdot \text{m}^2/\text{C}^2)(1.602 \times 10^{-19} \text{ C})^2}{(4.0 \times 10^{-15} \text{ m})^2}$$

$$= 14 \text{ N}. \qquad \text{(Answer)}$$

No explosion: This is a small force to be acting on a macroscopic object like a cantaloupe, but an enormous force to be acting on a proton. Such forces should explode the nucleus of any element but hydrogen (which has only one proton in its nucleus). However, they don't, not even in nuclei with a great many protons. Therefore, there must be some enormous attractive force to counter this enormous repulsive electrostatic force.

(b) What is the magnitude of the gravitational force between those same two protons?

KEY IDEA

Because the protons are particles, the magnitude of the gravitational force on one from the other is given by Newton's equation for the gravitational force (Eq. 21-2).

Calculation: With m_p ($= 1.67 \times 10^{-27}$ kg) representing the mass of a proton, Eq. 21-2 gives us

$$F = G \frac{m_p^2}{r^2}$$

$$= \frac{(6.67 \times 10^{-11}\,\text{N} \cdot \text{m}^2/\text{kg}^2)(1.67 \times 10^{-27}\,\text{kg})^2}{(4.0 \times 10^{-15}\,\text{m})^2}$$

$$= 1.2 \times 10^{-35}\,\text{N.} \qquad \text{(Answer)}$$

Weak versus strong: This result tells us that the (attractive) gravitational force is far too weak to counter the repulsive electrostatic forces between protons in a nucleus. Instead, the protons are bound together by an enormous force called (aptly) the *strong nuclear force*—a force that acts between protons (and neutrons) when they are close together, as in a nucleus.

Although the gravitational force is many times weaker than the electrostatic force, it is more important in large-scale situations because it is always attractive. This means that it can collect many small bodies into huge bodies with huge masses, such as planets and stars, that then exert large gravitational forces. The electrostatic force, on the other hand, is repulsive for charges of the same sign, so it is unable to collect either positive charge or negative charge into large concentrations that would then exert large electrostatic forces.

PLUS Additional examples, video, and practice available at *WileyPLUS*

21-3 CHARGE IS CONSERVED

Learning Objectives

After reading this module, you should be able to . . .

21.21 Identify that in any isolated physical process, the net charge cannot change (the net charge is always conserved).

21.22 Identify an annihilation process of particles and a pair production of particles.

21.23 Identify mass number and atomic number in terms of the number of protons, neutrons, and electrons.

Key Ideas

● The net electric charge of any isolated system is always conserved.

● If two charged particles undergo an annihilation process, they have opposite signs of charge.

● If two charged particles appear as a result of a pair production process, they have opposite signs of charge.

Charge Is Conserved

If you rub a glass rod with silk, a positive charge appears on the rod. Measurement shows that a negative charge of equal magnitude appears on the silk. This suggests that rubbing does not create charge but only transfers it from one body to another, upsetting the electrical neutrality of each body during the process. This hypothesis of **conservation of charge,** first put forward by Benjamin Franklin, has stood up under close examination, both for large-scale charged bodies and for atoms, nuclei, and elementary particles. No exceptions have ever been found. Thus, we add electric charge to our list of quantities—including energy and both linear momentum and angular momentum—that obey a conservation law.

Important examples of the conservation of charge occur in the *radioactive decay* of nuclei, in which a nucleus transforms into (becomes) a different type of nucleus. For example, a uranium-238 nucleus (^{238}U) transforms into a thorium-234 nucleus (^{234}Th) by emitting an *alpha particle*. Because that particle has the same makeup as a helium-4 nucleus, it has the symbol ^{4}He. The number used in the name of a nucleus and as a superscript in the symbol for the nucleus is called the *mass number* and is the total number of the protons and neutrons in the nucleus. For example, the total number in ^{238}U is 238. The number of protons in a nucleus is the *atomic number Z*, which is listed for all the elements in Appendix F. From that list we find that in the decay

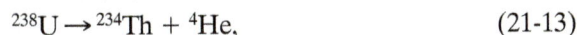

$$^{238}\text{U} \rightarrow {}^{234}\text{Th} + {}^4\text{He,} \qquad (21\text{-}13)$$

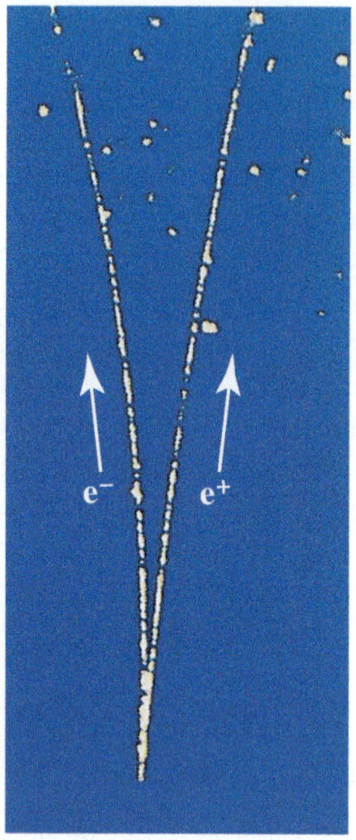

the *parent* nucleus ^{238}U contains 92 protons (a charge of $+92e$), the *daughter* nucleus ^{234}Th contains 90 protons (a charge of $+90e$), and the emitted alpha particle ^{4}He contains 2 protons (a charge of $+2e$). We see that the total charge is $+92e$ before and after the decay; thus, charge is conserved. (The total number of protons and neutrons is also conserved: 238 before the decay and $234 + 4 = 238$ after the decay.)

Another example of charge conservation occurs when an electron e^{-} (charge $-e$) and its antiparticle, the *positron* e^{+} (charge $+e$), undergo an *annihilation process,* transforming into two *gamma rays* (high-energy light):

$$e^{-} + e^{+} \rightarrow \gamma + \gamma \quad \text{(annihilation).} \quad (21\text{-}14)$$

In applying the conservation-of-charge principle, we must add the charges algebraically, with due regard for their signs. In the annihilation process of Eq. 21-14 then, the net charge of the system is zero both before and after the event. Charge is conserved.

In *pair production,* the converse of annihilation, charge is also conserved. In this process a gamma ray transforms into an electron and a positron:

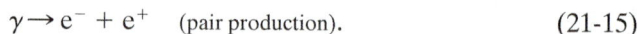

$$\gamma \rightarrow e^{-} + e^{+} \quad \text{(pair production).} \quad (21\text{-}15)$$

Figure 21-10 shows such a pair-production event that occurred in a bubble chamber. (This is a device in which a liquid is suddenly made hotter than its boiling point. If a charged particle passes through it, tiny vapor bubbles form along the particle's trail.) A gamma ray entered the chamber from the bottom and at one point transformed into an electron and a positron. Because those new particles were charged and moving, each left a trail of bubbles. (The trails were curved because a magnetic field had been set up in the chamber.) The gamma ray, being electrically neutral, left no trail. Still, you can tell exactly where it underwent pair production—at the tip of the curved V, which is where the trails of the electron and positron begin.

Courtesy Lawrence Berkeley Laboratory

Figure 21-10 A photograph of trails of bubbles left in a bubble chamber by an electron and a positron. The pair of particles was produced by a gamma ray that entered the chamber directly from the bottom. Being electrically neutral, the gamma ray did not generate a telltale trail of bubbles along its path, as the electron and positron did.

Review & Summary

Electric Charge The strength of a particle's electrical interaction with objects around it depends on its **electric charge** (usually represented as q), which can be either positive or negative. Particles with the same sign of charge repel each other, and particles with opposite signs of charge attract each other. An object with equal amounts of the two kinds of charge is electrically neutral, whereas one with an imbalance is electrically charged and has an excess charge.

Conductors are materials in which a significant number of electrons are free to move. The charged particles in **nonconductors** (**insulators**) are not free to move.

Electric current i is the rate dq/dt at which charge passes a point:

$$i = \frac{dq}{dt} \quad \text{(electric current).} \quad (21\text{-}3)$$

Coulomb's Law Coulomb's law describes the electrostatic force (or electric force) between two charged particles. If the particles have charges q_1 and q_2, are separated by distance r, and are at rest (or moving only slowly) relative to each other, then the magnitude of the force acting on each due to the other is given by

$$F = \frac{1}{4\pi\varepsilon_0} \frac{|q_1|\,|q_2|}{r^2} \quad \text{(Coulomb's law),} \quad (21\text{-}4)$$

where $\varepsilon_0 = 8.85 \times 10^{-12}$ C^2/N·m^2 is the **permittivity constant.** The ratio $1/4\pi\varepsilon_0$ is often replaced with the **electrostatic constant** (or **Coulomb constant**) $k = 8.99 \times 10^9$ N·m^2/C^2.

The electrostatic force vector acting on a charged particle due to a second charged particle is either directly toward the second particle (opposite signs of charge) or directly away from it (same sign of charge). As with other types of forces, if multiple electrostatic forces act on a particle, the net force is the vector sum (not scalar sum) of the individual forces.

The two shell theories for electrostatics are

Shell theorem 1: A charged particle outside a shell with charge uniformly distributed on its surface is attracted or repelled as if the shell's charge were concentrated as a particle at its center.

Shell theorem 2: A charged particle inside a shell with charge uniformly distributed on its surface has no net force acting on it due to the shell.

Charge on a conducting spherical shell spreads uniformly over the (external) surface.

The Elementary Charge Electric charge is quantized (restricted to certain values). The charge of a particle can be written as ne, where n is a positive or negative integer and e is the elementary charge, which is the magnitude of the charge of the electron and proton ($\approx 1.602 \times 10^{-19}$ C).

Conservation of Charge The net electric charge of any isolated system is always conserved.

Questions

1 Figure 21-11 shows four situations in which five charged particles are evenly spaced along an axis. The charge values are indicated except for the central particle, which has the same charge in all four situations. Rank the situations according to the magnitude of the net electrostatic force on the central particle, greatest first.

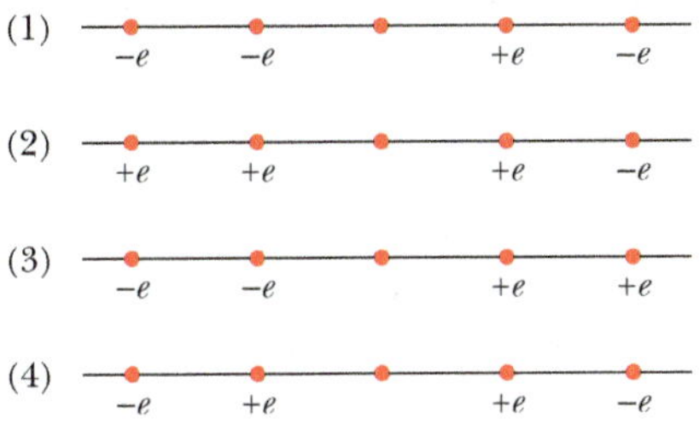

Figure 21-11 Question 1.

2 Figure 21-12 shows three pairs of identical spheres that are to be touched together and then separated. The initial charges on them are indicated. Rank the pairs according to (a) the magnitude of the charge transferred during touching and (b) the charge left on the positively charged sphere, greatest first.

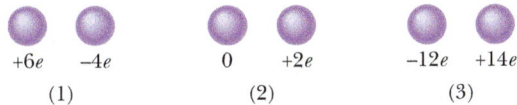

Figure 21-12 Question 2.

3 Figure 21-13 shows four situations in which charged particles are fixed in place on an axis. In which situations is there a point to the left of the particles where an electron will be in equilibrium?

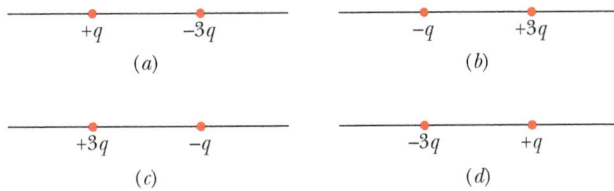

Figure 21-13 Question 3.

4 Figure 21-14 shows two charged particles on an axis. The charges are free to move. However, a third charged particle can be placed at a certain point such that all three particles are then in equilibrium. (a) Is that point to the left of the first two particles, to their right, or between them? (b) Should the third particle be positively or negatively charged? (c) Is the equilibrium stable or unstable?

Figure 21-14 Question 4.

5 In Fig. 21-15, a central particle of charge $-q$ is surrounded by two circular rings of charged particles. What are the magnitude and direction of the net electrostatic force on the central particle due to the other particles? (*Hint:* Consider symmetry.)

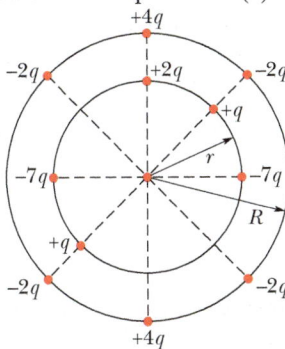

Figure 21-15 Question 5.

6 A positively charged ball is brought close to an electrically neutral isolated conductor. The conductor is then grounded while the ball is kept close. Is the conductor charged positively, charged negatively, or neutral if (a) the ball is first taken away and then the ground connection is removed and (b) the ground connection is first removed and then the ball is taken away?

7 Figure 21-16 shows three situations involving a charged particle and a uniformly charged spherical shell. The charges are given, and the radii of the shells are indicated. Rank the situations according to the magnitude of the force on the particle due to the presence of the shell, greatest first.

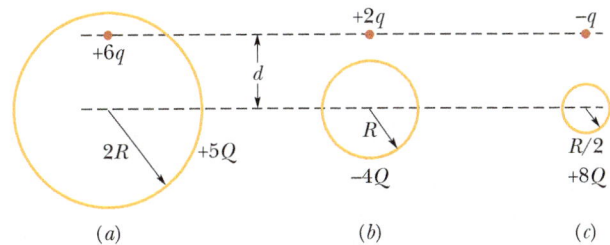

Figure 21-16 Question 7.

8 Figure 21-17 shows four arrangements of charged particles. Rank the arrangements according to the magnitude of the net electrostatic force on the particle with charge $+Q$, greatest first.

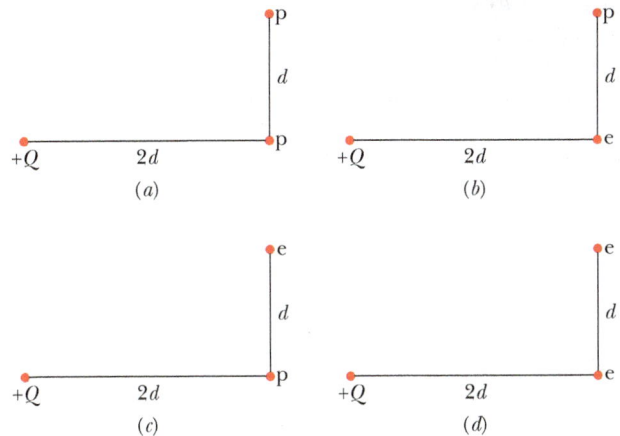

Figure 21-17 Question 8.

9 Figure 21-18 shows four situations in which particles of charge $+q$ or $-q$ are fixed in place. In each situation, the parti-

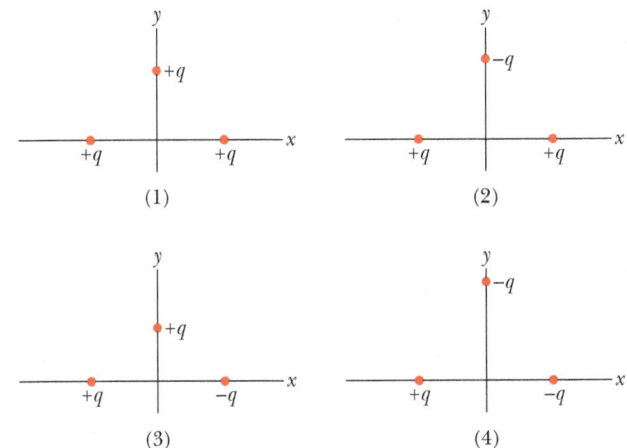

Figure 21-18 Question 9.

cles on the x axis are equidistant from the y axis. First, consider the middle particle in situation 1; the middle particle experiences an electrostatic force from each of the other two particles. (a) Are the magnitudes F of those forces the same or different? (b) Is the magnitude of the net force on the middle particle equal to, greater than, or less than $2F$? (c) Do the x components of the two forces add or cancel? (d) Do their y components add or cancel? (e) Is the direction of the net force on the middle particle that of the canceling components or the adding components? (f) What is the direction of that net force? Now consider the remaining situations: What is the direction of the net force on the middle particle in (g) situation 2, (h) situation 3, and (i) situation 4? (In each situation, consider the symmetry of the charge distribution and determine the canceling components and the adding components.)

10 In Fig. 21-19, a central particle of charge $-2q$ is surrounded by a square array of charged particles, separated by either distance d or $d/2$ along the perimeter of the square. What are the magnitude and direction of the net electrostatic force on the central particle due to the other particles? (*Hint:* Consideration of symmetry can greatly reduce the amount of work required here.)

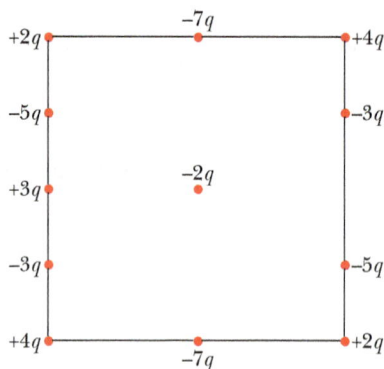

Figure 21-19 Question 10.

11 Figure 21-20 shows three identical conducting bubbles A, B, and C floating in a con-

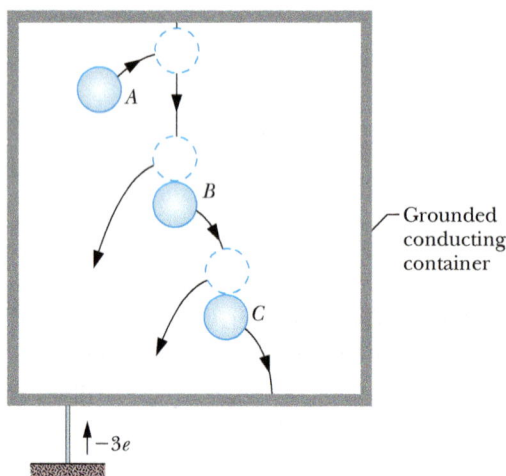

Figure 21-20 Question 11.

ducting container that is grounded by a wire. The bubbles initially have the same charge. Bubble A bumps into the container's ceiling and then into bubble B. Then bubble B bumps into bubble C, which then drifts to the container's floor. When bubble C reaches the floor, a charge of $-3e$ is transferred upward through the wire, from the ground to the container, as indicated. (a) What was the initial charge of each bubble? When (b) bubble A and (c) bubble B reach the floor, what is the charge transfer through the wire? (d) During this whole process, what is the total charge transfer through the wire?

12 Figure 21-21 shows four situations in which a central proton is partially surrounded by protons or electrons fixed in place along a half-circle. The angles θ are identical; the angles ϕ are also. (a) In each situation, what is the direction of the net force on the central proton due to the other particles? (b) Rank the four situations according to the magnitude of that net force on the central proton, greatest first.

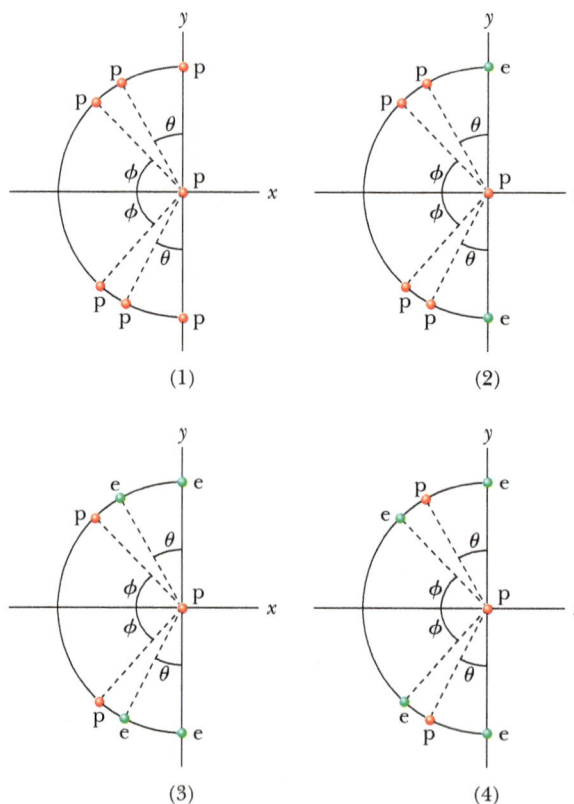

Figure 21-21 Question 12.

Problems

Module 21-1 Coulomb's Law

•**1** **SSM** **ILW** Of the charge Q initially on a tiny sphere, a portion q is to be transferred to a second, nearby sphere. Both spheres can be treated as particles and are fixed with a certain separation. For what value of q/Q will the electrostatic force between the two spheres be maximized?

•2 Identical isolated conducting spheres 1 and 2 have equal charges and are separated by a distance that is large compared with their diameters (Fig. 21-22a). The electrostatic force acting on sphere 2 due to sphere 1 is $\vec{F}$. Suppose now that a third identical sphere 3, having an insulating handle and initially neutral, is touched first to sphere 1 (Fig. 21-22b), then to sphere 2 (Fig. 21-22c), and finally removed (Fig. 21-22d). The electrostatic force that now acts on sphere 2 has magnitude F'. What is the ratio F'/F?

Figure 21-22 Problem 2.

•3 SSM What must be the distance between point charge $q_1 = 26.0\ \mu C$ and point charge $q_2 = -47.0\ \mu C$ for the electrostatic force between them to have a magnitude of 5.70 N?

•4 In the return stroke of a typical lightning bolt, a current of 2.5×10^4 A exists for 20 μs. How much charge is transferred in this event?

•5 A particle of charge $+3.00 \times 10^{-6}$ C is 12.0 cm distant from a second particle of charge -1.50×10^{-6} C. Calculate the magnitude of the electrostatic force between the particles.

•6 ILW Two equally charged particles are held 3.2×10^{-3} m apart and then released from rest. The initial acceleration of the first particle is observed to be 7.0 m/s² and that of the second to be 9.0 m/s². If the mass of the first particle is 6.3×10^{-7} kg, what are (a) the mass of the second particle and (b) the magnitude of the charge of each particle?

••7 In Fig. 21-23, three charged particles lie on an x axis. Particles 1 and 2 are fixed in place. Particle 3 is free to move, but the net electrostatic force on it from particles 1 and 2 happens to be zero. If $L_{23} = L_{12}$, what is the ratio q_1/q_2?

Figure 21-23 Problems 7 and 40.

••8 In Fig. 21-24, three identical conducting spheres initially have the following charges: sphere A, $4Q$; sphere B, $-6Q$; and sphere C, 0. Spheres A and B are fixed in place, with a center-to-center separation that is much larger than the spheres. Two experiments are conducted. In experiment 1, sphere C is touched to sphere A and then (separately) to sphere B, and then it is removed. In experiment 2, starting with the same initial states, the procedure is reversed: Sphere C is touched to sphere B and then (separately) to sphere A, and then it is removed. What is the ratio of the electro-

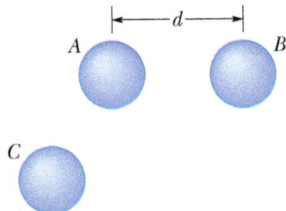

Figure 21-24 Problems 8 and 65.

static force between A and B at the end of experiment 2 to that at the end of experiment 1?

••9 SSM WWW Two identical conducting spheres, fixed in place, attract each other with an electrostatic force of 0.108 N when their center-to-center separation is 50.0 cm. The spheres are then connected by a thin conducting wire. When the wire is removed, the spheres repel each other with an electrostatic force of 0.0360 N. Of the initial charges on the spheres, with a positive net charge, what was (a) the negative charge on one of them and (b) the positive charge on the other?

••10 GO In Fig. 21-25, four particles form a square. The charges are $q_1 = q_4 = Q$ and $q_2 = q_3 = q$. (a) What is Q/q if the net electrostatic force on particles 1 and 4 is zero? (b) Is there any value of q that makes the net electrostatic force on each of the four particles zero? Explain.

••11 ILW In Fig. 21-25, the particles have charges $q_1 = -q_2 = 100$ nC and $q_3 = -q_4 = 200$ nC, and distance $a = 5.0$ cm. What are the (a) x and (b) y components of the net electrostatic force on particle 3?

Figure 21-25
Problems 10, 11, and 70.

••12 Two particles are fixed on an x axis. Particle 1 of charge 40 μC is located at $x = -2.0$ cm; particle 2 of charge Q is located at $x = 3.0$ cm. Particle 3 of charge magnitude 20 μC is released from rest on the y axis at $y = 2.0$ cm. What is the value of Q if the initial acceleration of particle 3 is in the positive direction of (a) the x axis and (b) the y axis?

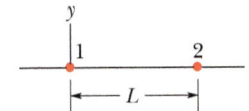

••13 GO In Fig. 21-26, particle 1 of charge $+1.0\ \mu$C and particle 2 of charge $-3.0\ \mu$C are held at separation $L = 10.0$ cm on an x axis. If particle 3 of unknown charge q_3 is to be located such that the net electrostatic force on it from particles 1 and 2 is zero, what must be the (a) x and (b) y coordinates of particle 3?

Figure 21-26 Problems 13, 19, 30, 58, and 67.

••14 Three particles are fixed on an x axis. Particle 1 of charge q_1 is at $x = -a$, and particle 2 of charge q_2 is at $x = +a$. If their net electrostatic force on particle 3 of charge $+Q$ is to be zero, what must be the ratio q_1/q_2 when particle 3 is at (a) $x = +0.500a$ and (b) $x = +1.50a$?

••15 GO The charges and coordinates of two charged particles held fixed in an xy plane are $q_1 = +3.0\ \mu$C, $x_1 = 3.5$ cm, $y_1 = 0.50$ cm, and $q_2 = -4.0\ \mu$C, $x_2 = -2.0$ cm, $y_2 = 1.5$ cm. Find the (a) magnitude and (b) direction of the electrostatic force on particle 2 due to particle 1. At what (c) x and (d) y coordinates should a third particle of charge $q_3 = +4.0\ \mu$C be placed such that the net electrostatic force on particle 2 due to particles 1 and 3 is zero?

••16 GO In Fig. 21-27a, particle 1 (of charge q_1) and particle 2 (of charge q_2) are fixed in place on an x axis, 8.00 cm apart. Particle 3 (of

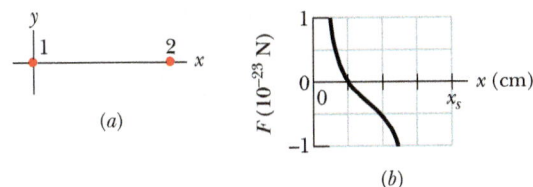

Figure 21-27 Problem 16.

charge $q_3 = +8.00 \times 10^{-19}$ C) is to be placed on the line between particles 1 and 2 so that they produce a net electrostatic force $\vec{F}_{3,net}$ on it. Figure 21-27b gives the x component of that force versus the coordinate x at which particle 3 is placed. The scale of the x axis is set by $x_s = 8.0$ cm. What are (a) the sign of charge q_1 and (b) the ratio q_2/q_1?

••17 In Fig. 21-28a, particles 1 and 2 have charge 20.0 μC each and are held at separation distance $d = 1.50$ m. (a) What is the magnitude of the electrostatic force on particle 1 due to particle 2? In Fig. 21-28b, particle 3 of charge 20.0 μC is positioned so as to complete an equilateral triangle. (b) What is the magnitude of the net electrostatic force on particle 1 due to particles 2 and 3?

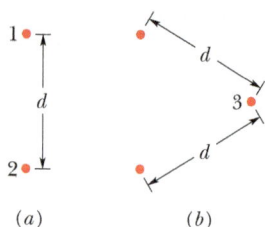

Figure 21-28 Problem 17.

••18 In Fig. 21-29a, three positively charged particles are fixed on an x axis. Particles B and C are so close to each other that they can be considered to be at the same distance from particle A. The net force on particle A due to particles B and C is 2.014×10^{-23} N in the negative direction of the x axis. In Fig. 21-29b, particle B has been moved to the opposite side of A but is still at the same distance from it. The net force on A is now 2.877×10^{-24} N in the negative direction of the x axis. What is the ratio q_C/q_B?

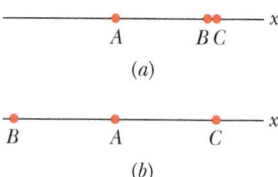

Figure 21-29 Problem 18.

••19 SSM WWW In Fig. 21-26, particle 1 of charge $+q$ and particle 2 of charge $+4.00q$ are held at separation $L = 9.00$ cm on an x axis. If particle 3 of charge q_3 is to be located such that the three particles remain in place when released, what must be the (a) x and (b) y coordinates of particle 3, and (c) the ratio q_3/q?

•••20 GO Figure 21-30a shows an arrangement of three charged particles separated by distance d. Particles A and C are fixed on the x axis, but particle B can be moved along a circle centered on particle A. During the movement, a radial line between A and B makes an angle θ relative to the positive direction of the x axis (Fig. 21-30b). The curves in Fig. 21-30c give, for two situations, the magnitude F_{net} of the net electrostatic force on particle A due to the other particles. That net force is given as a function of angle θ and as a multiple of a basic amount F_0. For example on curve 1, at $\theta = 180°$, we see that $F_{net} = 2F_0$. (a) For the situation corresponding to curve 1, what is the ratio of the charge of particle C to that of particle B (including sign)? (b) For the situation corresponding to curve 2, what is that ratio?

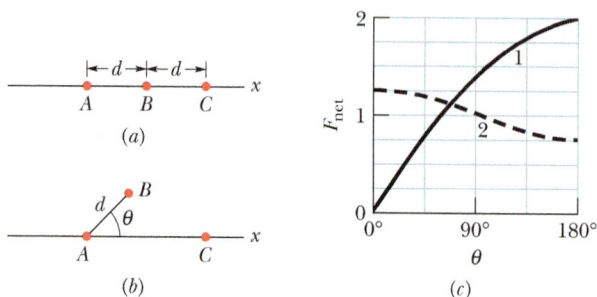

Figure 21-30 Problem 20.

•••21 GO A nonconducting spherical shell, with an inner radius of 4.0 cm and an outer radius of 6.0 cm, has charge spread nonuniformly through its volume between its inner and outer surfaces. The *volume charge density* ρ is the charge per unit volume, with the unit coulomb per cubic meter. For this shell $\rho = b/r$, where r is the distance in meters from the center of the shell and $b = 3.0$ μC/m^2. What is the net charge in the shell?

•••22 GO Figure 21-31 shows an arrangement of four charged particles, with angle $\theta = 30.0°$ and distance $d = 2.00$ cm. Particle 2 has charge $q_2 = +8.00 \times 10^{-19}$ C; particles 3 and 4 have charges $q_3 = q_4 = -1.60 \times 10^{-19}$ C. (a) What is distance D between the origin and particle 2 if the net electrostatic force on particle 1 due to the other particles is zero? (b) If particles 3 and 4 were moved closer to the x axis but maintained their symmetry about that axis, would the required value of D be greater than, less than, or the same as in part (a)?

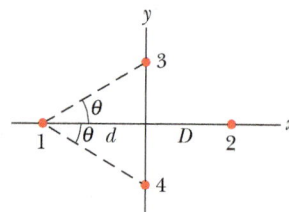

Figure 21-31 Problem 22.

•••23 GO In Fig. 21-32, particles 1 and 2 of charge $q_1 = q_2 = +3.20 \times 10^{-19}$ C are on a y axis at distance $d = 17.0$ cm from the origin. Particle 3 of charge $q_3 = +6.40 \times 10^{-19}$ C is moved gradually along the x axis from $x = 0$ to $x = +5.0$ m. At what values of x will the magnitude of the electrostatic force on the third particle from the other two particles be (a) minimum and (b) maximum? What are the (c) minimum and (d) maximum magnitudes?

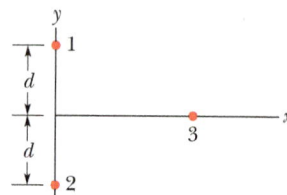

Figure 21-32 Problem 23.

Module 21-2 Charge Is Quantized

•24 Two tiny, spherical water drops, with identical charges of -1.00×10^{-16} C, have a center-to-center separation of 1.00 cm. (a) What is the magnitude of the electrostatic force acting between them? (b) How many excess electrons are on each drop, giving it its charge imbalance?

•25 ILW How many electrons would have to be removed from a coin to leave it with a charge of $+1.0 \times 10^{-7}$ C?

•26 What is the magnitude of the electrostatic force between a singly charged sodium ion (Na$^+$, of charge $+e$) and an adjacent singly charged chlorine ion (Cl$^-$, of charge $-e$) in a salt crystal if their separation is 2.82×10^{-10} m?

•27 SSM The magnitude of the electrostatic force between two identical ions that are separated by a distance of 5.0×10^{-10} m is 3.7×10^{-9} N. (a) What is the charge of each ion? (b) How many electrons are "missing" from each ion (thus giving the ion its charge imbalance)?

•28 A current of 0.300 A through your chest can send your heart into fibrillation, ruining the normal rhythm of heartbeat and disrupting the flow of blood (and thus oxygen) to your brain. If that current persists for 2.00 min, how many conduction electrons pass through your chest?

••29 GO In Fig. 21-33, particles 2 and 4, of charge $-e$, are fixed in place on a y axis, at $y_2 = -10.0$ cm

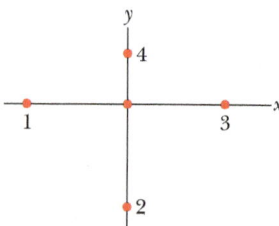

Figure 21-33 Problem 29.

and $y_4 = 5.00$ cm. Particles 1 and 3, of charge $-e$, can be moved along the x axis. Particle 5, of charge $+e$, is fixed at the origin. Initially particle 1 is at $x_1 = -10.0$ cm and particle 3 is at $x_3 = 10.0$ cm. (a) To what x value must particle 1 be moved to rotate the direction of the net electric force $\vec{F}_{net}$ on particle 5 by 30° counterclockwise? (b) With particle 1 fixed at its new position, to what x value must you move particle 3 to rotate $\vec{F}_{net}$ back to its original direction?

••30 In Fig. 21-26, particles 1 and 2 are fixed in place on an x axis, at a separation of $L = 8.00$ cm. Their charges are $q_1 = +e$ and $q_2 = -27e$. Particle 3 with charge $q_3 = +4e$ is to be placed on the line between particles 1 and 2, so that they produce a net electrostatic force $\vec{F}_{3,net}$ on it. (a) At what coordinate should particle 3 be placed to minimize the magnitude of that force? (b) What is that minimum magnitude?

••31 ILW Earth's atmosphere is constantly bombarded by *cosmic ray protons* that originate somewhere in space. If the protons all passed through the atmosphere, each square meter of Earth's surface would intercept protons at the average rate of 1500 protons per second. What would be the electric current intercepted by the total surface area of the planet?

••32 GO Figure 21-34a shows charged particles 1 and 2 that are fixed in place on an x axis. Particle 1 has a charge with a magnitude of $|q_1| = 8.00e$. Particle 3 of charge $q_3 = +8.00e$ is initially on the x axis near particle 2. Then particle 3 is gradually moved in the positive direction of the x axis. As a result, the magnitude of the net electrostatic force $\vec{F}_{2,net}$ on particle 2 due to particles 1 and 3 changes. Figure 21-34b gives the x component of that net force as a function of the position x of particle 3. The scale of the x axis is set by $x_s = 0.80$ m. The plot has an asymptote of $F_{2,net} = 1.5 \times 10^{-25}$ N as $x \to \infty$. As a multiple of e and including the sign, what is the charge q_2 of particle 2?

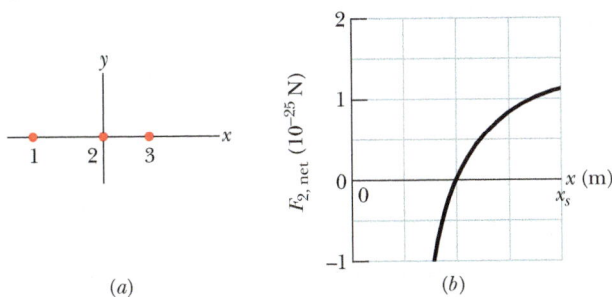

Figure 21-34 Problem 32.

••33 Calculate the number of coulombs of positive charge in 250 cm³ of (neutral) water. (*Hint:* A hydrogen atom contains one proton; an oxygen atom contains eight protons.)

•••34 GO Figure 21-35 shows electrons 1 and 2 on an x axis and charged ions 3 and 4 of identical charge $-q$ and at identical angles θ. Electron 2 is free to move; the other three particles are fixed in place at horizontal distances R from electron 2 and are intended to hold electron 2 in place. For physically possible values of $q \leq 5e$, what are the (a) smallest, (b) second smallest, and (c) third smallest values of θ for which electron 2 is held in place?

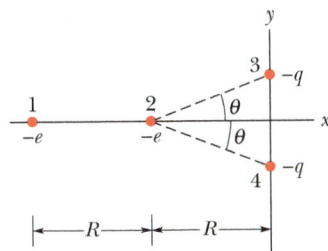

Figure 21-35 Problem 34.

•••35 SSM In crystals of the salt cesium chloride, cesium ions Cs^+ form the eight corners of a cube and a chlorine ion Cl^- is at the cube's center (Fig. 21-36). The edge length of the cube is 0.40 nm. The Cs^+ ions are each deficient by one electron (and thus each has a charge of $+e$), and the Cl^- ion has one excess electron (and thus has a charge of $-e$). (a) What is the magnitude of the net electrostatic force exerted on the Cl^- ion by the eight Cs^+ ions at the corners of the cube? (b) If one of the Cs^+ ions is missing, the crystal is said to have a *defect*; what is the magnitude of the net electrostatic force exerted on the Cl^- ion by the seven remaining Cs^+ ions?

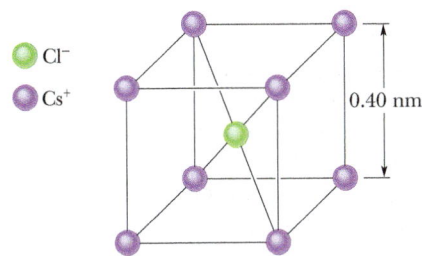

Figure 21-36 Problem 35.

Module 21-3 Charge Is Conserved

•36 Electrons and positrons are produced by the nuclear transformations of protons and neutrons known as *beta decay*. (a) If a proton transforms into a neutron, is an electron or a positron produced? (b) If a neutron transforms into a proton, is an electron or a positron produced?

•37 SSM Identify X in the following nuclear reactions: (a) ^{1}H + ^{9}Be $\rightarrow$ X + n; (b) ^{12}C + ^{1}H $\rightarrow$ X; (c) ^{15}N + ^{1}H $\rightarrow$ ^{4}He + X. Appendix F will help.

Additional Problems

38 GO Figure 21-37 shows four identical conducting spheres that are actually well separated from one another. Sphere W (with an initial charge of zero) is touched to sphere A and then they are separated. Next, sphere W is touched to sphere B (with an initial charge of $-32e$) and then they are separated. Finally, sphere W is touched to sphere C (with an initial charge of $+48e$), and then they are separated. The final charge on sphere W is $+18e$. What was the initial charge on sphere A?

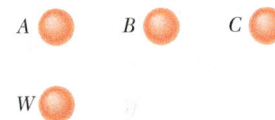

Figure 21-37 Problem 38.

39 SSM In Fig. 21-38, particle 1 of charge $+4e$ is above a floor by distance $d_1 = 2.00$ mm and particle 2 of charge $+6e$ is on the floor, at distance $d_2 = 6.00$ mm horizontally from particle 1. What is the x component of the electrostatic force on particle 2 due to particle 1?

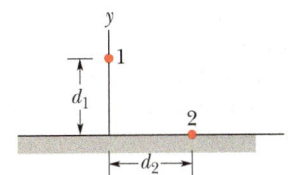

Figure 21-38 Problem 39.

40 In Fig. 21-23, particles 1 and 2 are fixed in place, but particle 3 is free to move. If the net electrostatic force on particle 3 due to particles 1 and 2 is zero and $L_{23} = 2.00L_{12}$, what is the ratio q_1/q_2?

41 (a) What equal positive charges would have to be placed on Earth and on the Moon to neutralize their gravitational attraction? (b) Why don't you need to know the lunar distance to solve this problem? (c) How many kilograms of hydrogen ions (that is, protons) would be needed to provide the positive charge calculated in (a)?

42 In Fig. 21-39, two tiny conducting balls of identical mass m and identical charge q hang from nonconducting threads of length L. Assume that θ is so small that $\tan \theta$ can be replaced by its approximate equal, $\sin \theta$. (a) Show that

$$x = \left(\frac{q^2 L}{2\pi\varepsilon_0 mg} \right)^{1/3}$$

gives the equilibrium separation x of the balls. (b) If $L = 120$ cm, $m = 10$ g, and $x = 5.0$ cm, what is $|q|$?

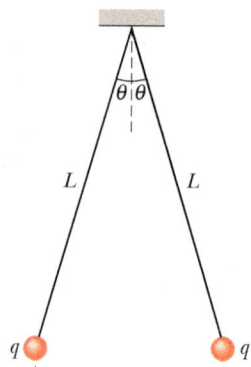

Figure 21-39
Problems 42 and 43.

43 (a) Explain what happens to the balls of Problem 42 if one of them is discharged (loses its charge q to, say, the ground). (b) Find the new equilibrium separation x, using the given values of L and m and the computed value of $|q|$.

44 SSM How far apart must two protons be if the magnitude of the electrostatic force acting on either one due to the other is equal to the magnitude of the gravitational force on a proton at Earth's surface?

45 How many megacoulombs of positive charge are in 1.00 mol of neutral molecular-hydrogen gas (H_2)?

46 In Fig. 21-40, four particles are fixed along an x axis, separated by distances $d = 2.00$ cm. The charges are $q_1 = +2e$, $q_2 = -e$, $q_3 = +e$, and $q_4 = +4e$, with $e = 1.60 \times 10^{-19}$ C. In unit-vector notation, what is the net electrostatic force on (a) particle 1 and (b) particle 2 due to the other particles?

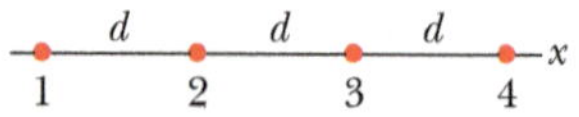

Figure 21-40 Problem 46.

47 GO Point charges of $+6.0\ \mu$C and $-4.0\ \mu$C are placed on an x axis, at $x = 8.0$ m and $x = 16$ m, respectively. What charge must be placed at $x = 24$ m so that any charge placed at the origin would experience no electrostatic force?

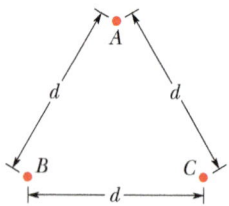

48 In Fig. 21-41, three identical conducting spheres form an equilateral triangle of side length $d = 20.0$ cm. The sphere radii are much smaller than d, and the sphere charges are $q_A = -2.00$ nC, $q_B = -4.00$ nC, and $q_C = +8.00$ nC. (a) What is the magnitude of the electrostatic force between spheres A and C? The following steps are then taken: A and B are connected by a thin wire and then disconnected; B is grounded by the wire, and the wire is then removed; B and C are connected by the wire and then disconnected. What now are the magnitudes of the electrostatic force (b) between spheres A and C and (c) between spheres B and C?

Figure 21-41
Problem 48.

49 A neutron consists of one "up" quark of charge $+2e/3$ and two "down" quarks each having charge $-e/3$. If we assume that the down quarks are 2.6×10^{-15} m apart inside the neutron, what is the magnitude of the electrostatic force between them?

50 Figure 21-42 shows a long, nonconducting, massless rod of length L, pivoted at its center and balanced with a block of weight W at a distance x from the left end. At the left and right ends of the rod are attached small conducting spheres with positive charges q and $2q$, respectively. A distance h directly beneath each of these spheres is a fixed sphere with positive charge Q. (a) Find the distance x when the rod is horizontal and balanced. (b)

What value should h have so that the rod exerts no vertical force on the bearing when the rod is horizontal and balanced?

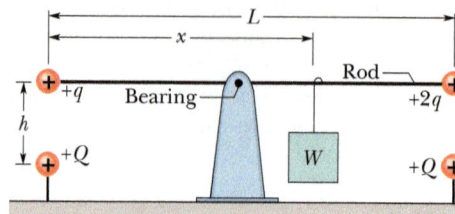

Figure 21-42 Problem 50.

51 A charged nonconducting rod, with a length of 2.00 m and a cross-sectional area of 4.00 cm^2, lies along the positive side of an x axis with one end at the origin. The *volume charge density* ρ is charge per unit volume in coulombs per cubic meter. How many excess electrons are on the rod if ρ is (a) uniform, with a value of $-4.00\ \mu$C/m^3, and (b) nonuniform, with a value given by $\rho = bx^2$, where $b = -2.00\ \mu$C/m^5?

52 A particle of charge Q is fixed at the origin of an xy coordinate system. At $t = 0$ a particle ($m = 0.800$ g, $q = 4.00\ \mu$C) is located on the x axis at $x = 20.0$ cm, moving with a speed of 50.0 m/s in the positive y direction. For what value of Q will the moving particle execute circular motion? (Neglect the gravitational force on the particle.)

53 What would be the magnitude of the electrostatic force between two 1.00 C point charges separated by a distance of (a) 1.00 m and (b) 1.00 km if such point charges existed (they do not) and this configuration could be set up?

54 A charge of 6.0 μC is to be split into two parts that are then separated by 3.0 mm. What is the maximum possible magnitude of the electrostatic force between those two parts?

55 Of the charge Q on a tiny sphere, a fraction α is to be transferred to a second, nearby sphere. The spheres can be treated as particles. (a) What value of α maximizes the magnitude F of the electrostatic force between the two spheres? What are the (b) smaller and (c) larger values of α that put F at half the maximum magnitude?

56 If a cat repeatedly rubs against your cotton slacks on a dry day, the charge transfer between the cat hair and the cotton can leave you with an excess charge of $-2.00\ \mu$C. (a) How many electrons are transferred between you and the cat?

You will gradually discharge via the floor, but if instead of waiting, you immediately reach toward a faucet, a painful spark can suddenly appear as your fingers near the faucet. (b) In that spark, do electrons flow from you to the faucet or vice versa? (c) Just before the spark appears, do you induce positive or negative charge in the faucet? (d) If, instead, the cat reaches a paw toward the faucet, which way do electrons flow in the resulting spark? (e) If you stroke a cat with a bare hand on a dry day, you should take care not to bring your fingers near the cat's nose or you will hurt it with a spark. Considering that cat hair is an insulator, explain how the spark can appear.

57 We know that the negative charge on the electron and the positive charge on the proton are equal. Suppose, however, that these magnitudes differ from each other by 0.00010%. With what force would two copper coins, placed 1.0 m apart, repel each other? Assume that each coin contains 3×10^{22} copper atoms. (*Hint:* A neutral copper atom contains 29 protons and 29 electrons.) What do you conclude?

58 In Fig. 21-26, particle 1 of charge $-80.0\ \mu C$ and particle 2 of charge $+40.0\ \mu C$ are held at separation $L = 20.0$ cm on an x axis. In unit-vector notation, what is the net electrostatic force on particle 3, of charge $q_3 = 20.0\ \mu C$, if particle 3 is placed at (a) $x = 40.0$ cm and (b) $x = 80.0$ cm? What should be the (c) x and (d) y coordinates of particle 3 if the net electrostatic force on it due to particles 1 and 2 is zero?

59 What is the total charge in coulombs of 75.0 kg of electrons?

60 GO In Fig. 21-43, six charged particles surround particle 7 at radial distances of either $d = 1.0$ cm or $2d$, as drawn. The charges are $q_1 = +2e, q_2 = +4e, q_3 = +e, q_4 = +4e, q_5 = +2e, q_6 = +8e, q_7 = +6e$, with $e = 1.60 \times 10^{-19}$ C. What is the magnitude of the net electrostatic force on particle 7?

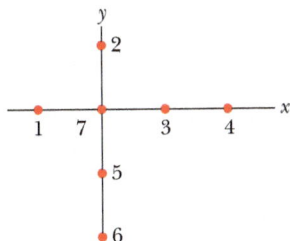

Figure 21-43 Problem 60.

61 Three charged particles form a triangle: particle 1 with charge $Q_1 = 80.0$ nC is at xy coordinates (0, 3.00 mm), particle 2 with charge Q_2 is at (0, -3.00 mm), and particle 3 with charge $q = 18.0$ nC is at (4.00 mm, 0). In unit-vector notation, what is the electrostatic force on particle 3 due to the other two particles if Q_2 is equal to (a) 80.0 nC and (b) -80.0 nC?

62 SSM In Fig. 21-44, what are the (a) magnitude and (b) direction of the net electrostatic force on particle 4 due to the other three particles? All four particles are fixed in the xy plane, and $q_1 = -3.20 \times 10^{-19}$ C, $q_2 = +3.20 \times 10^{-19}$ C, $q_3 = +6.40 \times 10^{-19}$ C, $q_4 = +3.20 \times 10^{-19}$ C, $\theta_1 = 35.0°, d_1 = 3.00$ cm, and $d_2 = d_3 = 2.00$ cm.

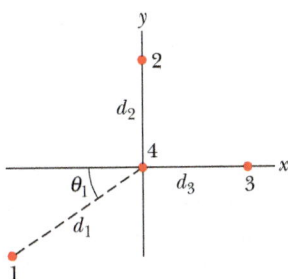

Figure 21-44 Problem 62.

63 Two point charges of 30 nC and -40 nC are held fixed on an x axis, at the origin and at $x = 72$ cm, respectively. A particle with a charge of 42 μC is released from rest at $x = 28$ cm. If the initial acceleration of the particle has a magnitude of 100 km/s², what is the particle's mass?

64 Two small, positively charged spheres have a combined charge of 5.0×10^{-5} C. If each sphere is repelled from the other by an electrostatic force of 1.0 N when the spheres are 2.0 m apart, what is the charge on the sphere with the smaller charge?

65 The initial charges on the three identical metal spheres in Fig. 21-24 are the following: sphere A, Q; sphere B, $-Q/4$; and sphere C, $Q/2$, where $Q = 2.00 \times 10^{-14}$ C. Spheres A and B are fixed in place, with a center-to-center separation of $d = 1.20$ m, which is much larger than the spheres. Sphere C is touched first to sphere A and then to sphere B and is then removed. What then is the magnitude of the electrostatic force between spheres A and B?

66 An electron is in a vacuum near Earth's surface and located at $y = 0$ on a vertical y axis. At what value of y should a second electron be placed such that its electrostatic force on the first electron balances the gravitational force on the first electron?

67 SSM In Fig. 21-26, particle 1 of charge $-5.00q$ and particle 2 of charge $+2.00q$ are held at separation L on an x axis. If particle 3 of unknown charge q_3 is to be located such that the net electrostatic force on it from particles 1 and 2 is zero, what must be the (a) x and (b) y coordinates of particle 3?

68 Two engineering students, John with a mass of 90 kg and Mary with a mass of 45 kg, are 30 m apart. Suppose each has a 0.01% imbalance in the amount of positive and negative charge, one student being positive and the other negative. Find the order of magnitude of the electrostatic force of attraction between them by replacing each student with a sphere of water having the same mass as the student.

69 In the radioactive decay of Eq. 21-13, a ^{238}U nucleus transforms to ^{234}Th and an ejected ^{4}He. (These are nuclei, not atoms, and thus electrons are not involved.) When the separation between ^{234}Th and ^{4}He is 9.0×10^{-15} m, what are the magnitudes of (a) the electrostatic force between them and (b) the acceleration of the ^{4}He particle?

70 In Fig. 21-25, four particles form a square. The charges are $q_1 = +Q$, $q_2 = q_3 = q$, and $q_4 = -2.00Q$. What is q/Q if the net electrostatic force on particle 1 is zero?

71 In a spherical metal shell of radius R, an electron is shot from the center directly toward a tiny hole in the shell, through which it escapes. The shell is negatively charged with a *surface charge density* (charge per unit area) of 6.90×10^{-13} C/m². What is the magnitude of the electron's acceleration when it reaches radial distances (a) $r = 0.500R$ and (b) $2.00R$?

72 An electron is projected with an initial speed $v_i = 3.2 \times 10^5$ m/s directly toward a very distant proton that is at rest. Because the proton mass is large relative to the electron mass, assume that the proton remains at rest. By calculating the work done on the electron by the electrostatic force, determine the distance between the two particles when the electron instantaneously has speed $2v_i$.

73 In an early model of the hydrogen atom (the *Bohr model*), the electron orbits the proton in uniformly circular motion. The radius of the circle is restricted (*quantized*) to certain values given by

$$r = n^2 a_0, \quad \text{for } n = 1, 2, 3, \ldots,$$

where $a_0 = 52.92$ pm. What is the speed of the electron if it orbits in (a) the smallest allowed orbit and (b) the second smallest orbit? (c) If the electron moves to larger orbits, does its speed increase, decrease, or stay the same?

74 A 100 W lamp has a steady current of 0.83 A in its filament. How long is required for 1 mol of electrons to pass through the lamp?

75 The charges of an electron and a positron are $-e$ and $+e$. The mass of each is 9.11×10^{-31} kg. What is the ratio of the electrical force to the gravitational force between an electron and a positron?

CHAPTER 22

Electric Fields

22-1 THE ELECTRIC FIELD

Learning Objectives

After reading this module, you should be able to . . .

22.01 Identify that at every point in the space surrounding a charged particle, the particle sets up an electric field $\vec{E}$, which is a vector quantity and thus has both magnitude and direction.

22.02 Identify how an electric field $\vec{E}$ can be used to explain how a charged particle can exert an electrostatic force $\vec{F}$

on a second charged particle even though there is no contact between the particles.

22.03 Explain how a small positive test charge is used (in principle) to measure the electric field at any given point.

22.04 Explain electric field lines, including where they originate and terminate and what their spacing represents.

Key Ideas

● A charged particle sets up an electric field (a vector quantity) in the surrounding space. If a second charged particle is located in that space, an electrostatic force acts on it due to the magnitude and direction of the field at its location.

● The electric field $\vec{E}$ at any point is defined in terms of the electrostatic force $\vec{F}$ that would be exerted on a positive test charge q_0 placed there:

$$\vec{E} = \frac{\vec{F}}{q_0}.$$

● Electric field lines help us visualize the direction and magnitude of electric fields. The electric field vector at any point is tangent to the field line through that point. The density of field lines in that region is proportional to the magnitude of the electric field there. Thus, closer field lines represent a stronger field.

● Electric field lines originate on positive charges and terminate on negative charges. So, a field line extending from a positive charge must end on a negative charge.

What Is Physics?

Figure 22-1 shows two positively charged particles. From the preceding chapter we know that an electrostatic force acts on particle 1 due to the presence of particle 2. We also know the force direction and, given some data, we can calculate the force magnitude. However, here is a leftover nagging question. How does particle 1 "know" of the presence of particle 2? That is, since the particles do not touch, how can particle 2 push on particle 1—how can there be such an *action at a distance*?

One purpose of physics is to record observations about our world, such as the magnitude and direction of the push on particle 1. Another purpose is to provide an explanation of what is recorded. Our purpose in this chapter is to provide such an explanation to this nagging question about electric force at a distance.

The explanation that we shall examine here is this: Particle 2 sets up an **electric field** at all points in the surrounding space, even if the space is a vacuum. If we place particle 1 at any point in that space, particle 1 knows of the presence of particle 2 because it is affected by the electric field particle 2 has already set up at that point. Thus, particle 2 pushes on particle 1 not by touching it as you would push on a coffee mug by making contact. Instead, particle 2 pushes by means of the electric field it has set up.

q_1 q_2

Figure 22-1 How does charged particle 2 push on charged particle 1 when they have no contact?

Our goals in this chapter are to (1) define electric field, (2) discuss how to calculate it for various arrangements of charged particles and objects, and (3) discuss how an electric field can affect a charged particle (as in making it move).

The Electric Field

A lot of different fields are used in science and engineering. For example, a *temperature field* for an auditorium is the distribution of temperatures we would find by measuring the temperature at many points within the auditorium. Similarly, we could define a *pressure field* in a swimming pool. Such fields are examples of *scalar fields* because temperature and pressure are scalar quantities, having only magnitudes and not directions.

In contrast, an electric field is a *vector field* because it is responsible for conveying the information for a force, which involves both magnitude and direction. This field consists of a distribution of electric field vectors $\vec{E}$, one for each point in the space around a charged object. In principle, we can define $\vec{E}$ at some point near the charged object, such as point P in Fig. 22-2a, with this procedure: At P, we place a particle with a small positive charge q_0, called a *test charge* because we use it to test the field. (We want the charge to be small so that it does not disturb the object's charge distribution.) We then measure the electrostatic force $\vec{F}$ that acts on the test charge. The electric field at that point is then

$$\vec{E} = \frac{\vec{F}}{q_0} \quad \text{(electric field).} \tag{22-1}$$

Because the test charge is positive, the two vectors in Eq. 22-1 are in the same direction, so the direction of $\vec{E}$ is the direction we measure for $\vec{F}$. The magnitude of $\vec{E}$ at point P is F/q_0. As shown in Fig. 22-2b, we always represent an electric field with an arrow with its tail anchored on the point where the measurement is made. (This may sound trivial, but drawing the vectors any other way usually results in errors. Also, another common error is to mix up the terms *force* and *field* because they both start with the letter f. Electric force is a push or pull. Electric field is an abstract property set up by a charged object.) From Eq. 22-1, we see that the SI unit for the electric field is the newton per coulomb (N/C).

We can shift the test charge around to various other points, to measure the electric fields there, so that we can figure out the distribution of the electric field set up by the charged object. That field exists independent of the test charge. It is something that a charged object sets up in the surrounding space (even vacuum), independent of whether we happen to come along to measure it.

For the next several modules, we determine the field around charged particles and various charged objects. First, however, let's examine a way of visualizing electric fields.

Electric Field Lines

Look at the space in the room around you. Can you visualize a field of vectors throughout that space—vectors with different magnitudes and directions? As impossible as that seems, Michael Faraday, who introduced the idea of electric fields in the 19th century, found a way. He envisioned lines, now called **electric field lines**, in the space around any given charged particle or object.

Figure 22-3 gives an example in which a sphere is uniformly covered with negative charge. If we place a positive test charge at any point near the sphere (Fig. 22-3a), we find that an electrostatic force pulls on it toward the center of the sphere. Thus at every point around the sphere, an electric field vector points radially inward toward the sphere. We can represent this electric field with

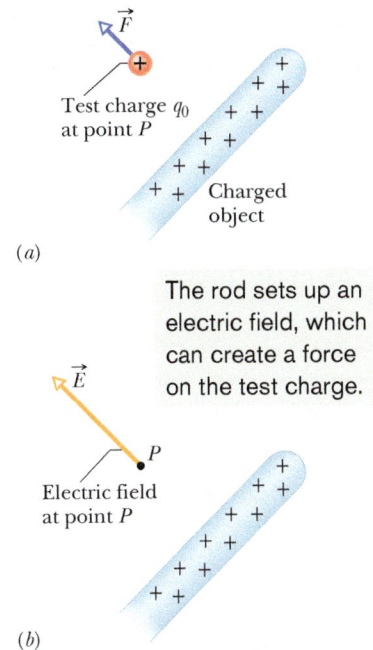

(a)

The rod sets up an electric field, which can create a force on the test charge.

(b)

Figure 22-2 (a) A positive test charge q_0 placed at point P near a charged object. An electrostatic force $\vec{F}$ acts on the test charge. (b) The electric field $\vec{E}$ at point P produced by the charged object.

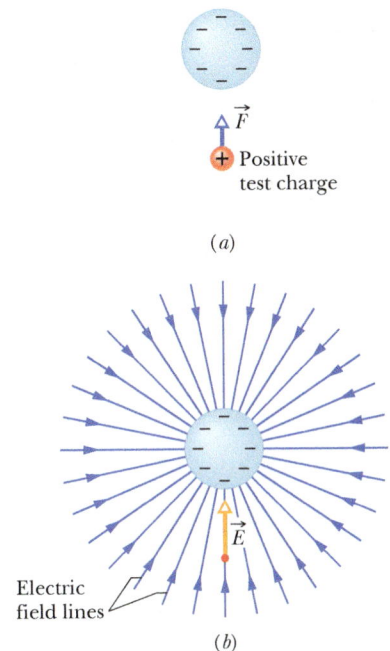

(a)

(b)

Figure 22-3 (a) The electrostatic force $\vec{F}$ acting on a positive test charge near a sphere of uniform negative charge. (b) The electric field vector $\vec{E}$ at the location of the test charge, and the electric field lines in the space near the sphere. The field lines extend *toward* the negatively charged sphere. (They originate on distant positive charges.)

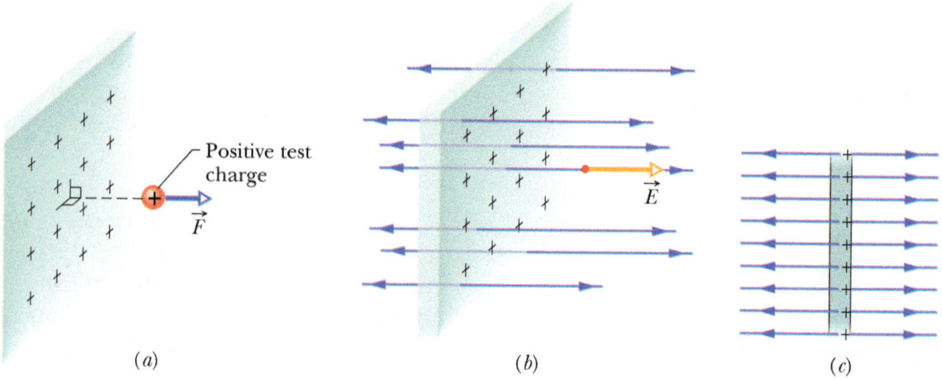

Figure 22-4 (*a*) The force on a positive test charge near a very large, nonconducting sheet with uniform positive charge on one side. (*b*) The electric field vector $\vec{E}$ at the test charge's location, and the nearby electric field lines, extending away from the sheet. (*c*) Side view.

electric field lines as in Fig. 22-3*b*. At any point, such as the one shown, the direction of the field line through the point matches the direction of the electric vector at that point.

The rules for drawing electric fields lines are these: (1) At any point, the electric field vector must be tangent to the electric field line through that point and in the same direction. (This is easy to see in Fig. 22-3 where the lines are straight, but we'll see some curved lines soon.) (2) In a plane perpendicular to the field lines, the relative density of the lines represents the relative magnitude of the field there, with greater density for greater magnitude.

If the sphere in Fig. 22-3 were uniformly covered with positive charge, the electric field vectors at all points around it would be radially outward and thus so would the electric field lines. So, we have the following rule:

Electric field lines extend away from positive charge (where they originate) and toward negative charge (where they terminate).

In Fig. 22-3*b*, they originate on distant positive charges that are not shown.

For another example, Fig. 22-4*a* shows part of an infinitely large, nonconducting *sheet* (or plane) with a uniform distribution of positive charge on one side. If we place a positive test charge at any point near the sheet (on either side), we find that the electrostatic force on the particle is outward and perpendicular to the sheet. The perpendicular orientation is reasonable because any force component that is, say, upward is balanced out by an equal component that is downward. That leaves only outward, and thus the electric field vectors and the electric field lines must also be outward and perpendicular to the sheet, as shown in Figs. 22-4*b* and *c*.

Because the charge on the sheet is uniform, the field vectors and the field lines are also. Such a field is a *uniform electric field*, meaning that the electric field has the same magnitude and direction at every point within the field. (This is a lot easier to work with than a *nonuniform field*, where there is variation from point to point.) Of course, there is no such thing as an infinitely large sheet. That is just a way of saying that we are measuring the field at points close to the sheet relative to the size of the sheet and that we are not near an edge.

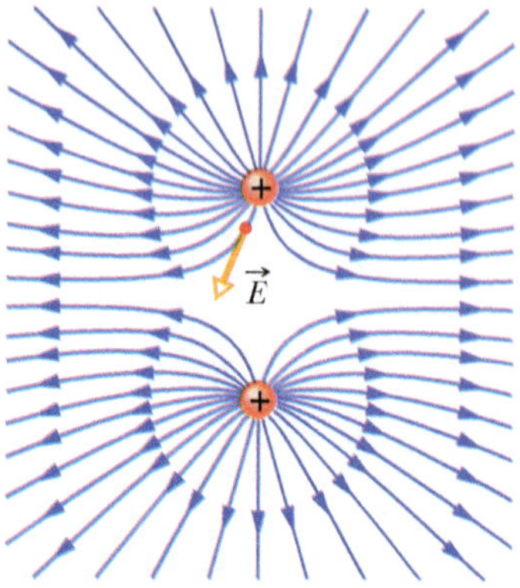

Figure 22-5 shows the field lines for two particles with equal positive charges. Now the field lines are curved, but the rules still hold: (1) the electric field vector at any given point must be tangent to the field line at that point and in the same direction, as shown for one vector, and (2) a closer spacing means a larger field magnitude. To imagine the full three-dimensional pattern of field lines around the particles, mentally rotate the pattern in Fig. 22-5 around the *axis of symmetry*, which is a vertical line through both particles.

Figure 22-5 Field lines for two particles with equal positive charge. Doesn't the pattern itself suggest that the particles repel each other?

22-2 THE ELECTRIC FIELD DUE TO A CHARGED PARTICLE

Learning Objectives

After reading this module, you should be able to . . .

22.05 In a sketch, draw a charged particle, indicate its sign, pick a nearby point, and then draw the electric field vector $\vec{E}$ at that point, with its tail anchored on the point.

22.06 For a given point in the electric field of a charged particle, identify the direction of the field vector $\vec{E}$ when the particle is positively charged and when it is negatively charged.

22.07 For a given point in the electric field of a charged particle, apply the relationship between the field

magnitude E, the charge magnitude $|q|$, and the distance r between the point and the particle.

22.08 Identify that the equation given here for the magnitude of an electric field applies only to a particle, not an extended object.

22.09 If more than one electric field is set up at a point, draw each electric field vector and then find the net electric field by adding the individual electric fields as vectors (not as scalars).

Key Ideas

● The magnitude of the electric field $\vec{E}$ set up by a particle with charge q at distance r from the particle is

$$E = \frac{1}{4\pi\varepsilon_0}\frac{|q|}{r^2}.$$

● The electric field vectors set up by a positively charged particle all point directly away from the particle. Those set up

by a negatively charged particle all point directly toward the particle.

● If more than one charged particle sets up an electric field at a point, the net electric field is the *vector* sum of the individual electric fields—electric fields obey the superposition principle.

The Electric Field Due to a Point Charge

To find the electric field due to a charged particle (often called a *point charge*), we place a positive test charge at any point near the particle, at distance r. From Coulomb's law (Eq. 21-4), the force on the test charge due to the particle with charge q is

$$\vec{F} = \frac{1}{4\pi\varepsilon_0}\frac{qq_0}{r^2}\hat{r}.$$

As previously, the direction of $\vec{F}$ is directly away from the particle if q is positive (because q_0 is positive) and directly toward it if q is negative. From Eq. 22-1, we can now write the electric field set up by the particle (at the location of the test charge) as

$$\vec{E} = \frac{\vec{F}}{q_0} = \frac{1}{4\pi\varepsilon_0}\frac{q}{r^2}\hat{r} \quad \text{(charged particle).} \tag{22-2}$$

Let's think through the directions again. The direction of $\vec{E}$ matches that of the force on the positive test charge: directly away from the point charge if q is positive and directly toward it if q is negative.

So, if given another charged particle, we can immediately determine the directions of the electric field vectors near it by just looking at the sign of the charge q. We can find the magnitude at any given distance r by converting Eq. 22-2 to a magnitude form:

$$E = \frac{1}{4\pi\varepsilon_0}\frac{|q|}{r^2} \quad \text{(charged particle).} \tag{22-3}$$

We write $|q|$ to avoid the danger of getting a negative E when q is negative, and then thinking the negative sign has something to do with direction. Equation 22-3 gives magnitude E only. We must think about the direction separately.

Figure 22-6 gives a number of electric field vectors at points around a positively charged particle, but be careful. Each vector represents the vector

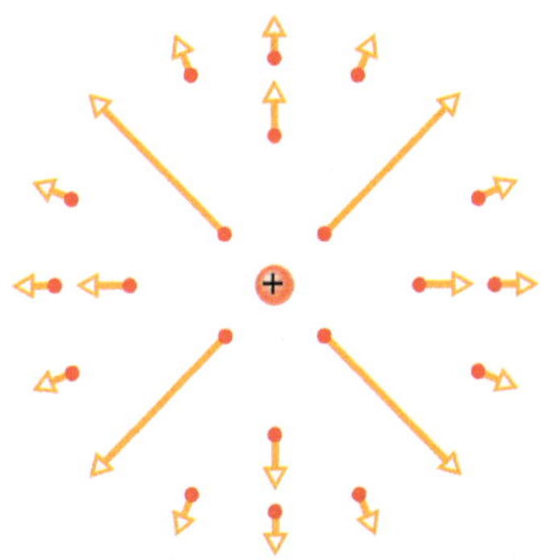

Figure 22-6 The electric field vectors at various points around a positive point charge.

quantity at the point where the tail of the arrow is anchored. The vector is not something that stretches from a "here" to a "there" as with a displacement vector.

In general, if several electric fields are set up at a given point by several charged particles, we can find the net field by placing a positive test particle at the point and then writing out the force acting on it due to each particle, such as $\vec{F}_{01}$ due to particle 1. Forces obey the principle of superposition, so we just add the forces as vectors:

$$\vec{F}_0 = \vec{F}_{01} + \vec{F}_{02} + \cdots + \vec{F}_{0n}.$$

To change over to electric field, we repeatedly use Eq. 22-1 for each of the individual forces:

$$\vec{E} = \frac{\vec{F}_0}{q_0} = \frac{\vec{F}_{01}}{q_0} + \frac{\vec{F}_{02}}{q_0} + \cdots + \frac{\vec{F}_{0n}}{q_0}$$

$$= \vec{E}_1 + \vec{E}_2 + \cdots + \vec{E}_n. \tag{22-4}$$

This tells us that electric fields also obey the principle of superposition. If you want the net electric field at a given point due to several particles, find the electric field due to each particle (such as $\vec{E}_1$ due to particle 1) and then sum the fields as vectors. (As with electrostatic forces, you cannot just willy-nilly add up the magnitudes.) This addition of fields is the subject of many of the homework problems.

☑ **Checkpoint 1**

The figure here shows a proton p and an electron e on an x axis. What is the direction of the electric field due to the electron at (a) point S and (b) point R? What is the direction of the net electric field at (c) point R and (d) point S?

Sample Problem 22.01   Net electric field due to three charged particles

Figure 22-7a shows three particles with charges $q_1 = +2Q$, $q_2 = -2Q$, and $q_3 = -4Q$, each a distance d from the origin. What net electric field $\vec{E}$ is produced at the origin?

KEY IDEA

Charges q_1, q_2, and q_3 produce electric field vectors $\vec{E}_1$, $\vec{E}_2$, and $\vec{E}_3$, respectively, at the origin, and the net electric field is the vector sum $\vec{E} = \vec{E}_1 + \vec{E}_2 + \vec{E}_3$. To find this sum, we first must find the magnitudes and orientations of the three field vectors.

Magnitudes and directions: To find the magnitude of $\vec{E}_1$, which is due to q_1, we use Eq. 22-3, substituting d for r and $2Q$ for q and obtaining

$$E_1 = \frac{1}{4\pi\varepsilon_0} \frac{2Q}{d^2}.$$

Similarly, we find the magnitudes of $\vec{E}_2$ and $\vec{E}_3$ to be

$$E_2 = \frac{1}{4\pi\varepsilon_0} \frac{2Q}{d^2} \quad \text{and} \quad E_3 = \frac{1}{4\pi\varepsilon_0} \frac{4Q}{d^2}.$$

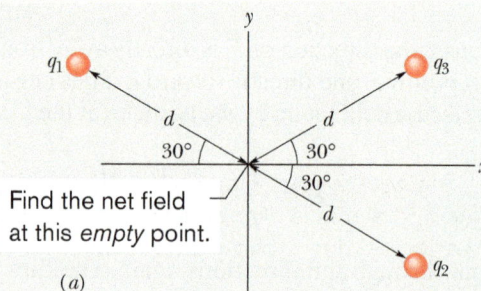

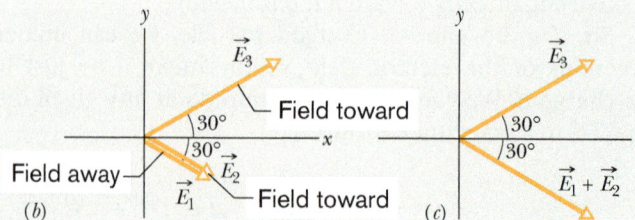

Figure 22-7 (a) Three particles with charges q_1, q_2, and q_3 are at the same distance d from the origin. (b) The electric field vectors $\vec{E}_1$, $\vec{E}_2$, and $\vec{E}_3$, at the origin due to the three particles. (c) The electric field vector $\vec{E}_3$ and the vector sum $\vec{E}_1 + \vec{E}_2$ at the origin.

We next must find the orientations of the three electric field vectors at the origin. Because q_1 is a positive charge, the field vector it produces points directly *away* from it, and because q_2 and q_3 are both negative, the field vectors they produce point directly *toward* each of them. Thus, the three electric fields produced at the origin by the three charged particles are oriented as in Fig. 22-7*b*. (*Caution:* Note that we have placed the tails of the vectors at the point where the fields are to be evaluated; doing so decreases the chance of error. Error becomes very probable if the tails of the field vectors are placed on the particles creating the fields.)

Adding the fields: We can now add the fields vectorially just as we added force vectors in Chapter 21. However, here we can use symmetry to simplify the procedure. From Fig. 22-7*b*, we see that electric fields $\vec{E}_1$ and $\vec{E}_2$ have the same direction. Hence, their vector sum has that direction and has the magnitude

$$E_1 + E_2 = \frac{1}{4\pi\varepsilon_0}\frac{2Q}{d^2} + \frac{1}{4\pi\varepsilon_0}\frac{2Q}{d^2}$$
$$= \frac{1}{4\pi\varepsilon_0}\frac{4Q}{d^2},$$

which happens to equal the magnitude of field $\vec{E}_3$.

We must now combine two vectors, $\vec{E}_3$ and the vector sum $\vec{E}_1 + \vec{E}_2$, that have the same magnitude and that are oriented symmetrically about the *x* axis, as shown in Fig. 22-7*c*. From the symmetry of Fig. 22-7*c*, we realize that the equal *y* components of our two vectors cancel (one is upward and the other is downward) and the equal *x* components add (both are rightward). Thus, the net electric field $\vec{E}$ at the origin is in the positive direction of the *x* axis and has the magnitude

$$E = 2E_{3x} = 2E_3 \cos 30°$$
$$= (2)\frac{1}{4\pi\varepsilon_0}\frac{4Q}{d^2}(0.866) = \frac{6.93Q}{4\pi\varepsilon_0 d^2}. \qquad \text{(Answer)}$$

PLUS Additional examples, video, and practice available at *WileyPLUS*

22-3 THE ELECTRIC FIELD DUE TO A DIPOLE

Learning Objectives

After reading this module, you should be able to . . .

22.10 Draw an electric dipole, identifying the charges (sizes and signs), dipole axis, and direction of the electric dipole moment.

22.11 Identify the direction of the electric field at any given point along the dipole axis, including between the charges.

22.12 Outline how the equation for the electric field due to an electric dipole is derived from the equations for the electric field due to the individual charged particles that form the dipole.

22.13 For a single charged particle and an electric dipole, compare the rate at which the electric field magnitude

decreases with increase in distance. That is, identify which drops off faster.

22.14 For an electric dipole, apply the relationship between the magnitude p of the dipole moment, the separation d between the charges, and the magnitude q of either of the charges.

22.15 For any distant point along a dipole axis, apply the relationship between the electric field magnitude E, the distance z from the center of the dipole, and either the dipole moment magnitude p or the product of charge magnitude q and charge separation d.

Key Ideas

● An electric dipole consists of two particles with charges of equal magnitude q but opposite signs, separated by a small distance d.

● The electric dipole moment $\vec{p}$ has magnitude qd and points from the negative charge to the positive charge.

● The magnitude of the electric field set up by an electric dipole at a distant point on the dipole axis (which runs through both particles) can be written in terms of either the product qd or the magnitude p of the dipole moment:

$$E = \frac{1}{2\pi\varepsilon_0}\frac{qd}{z^3} = \frac{1}{2\pi\varepsilon_0}\frac{p}{z^3},$$

where z is the distance between the point and the center of the dipole.

● Because of the $1/z^3$ dependence, the field magnitude of an electric dipole decreases more rapidly with distance than the field magnitude of either of the individual charges forming the dipole, which depends on $1/r^2$.

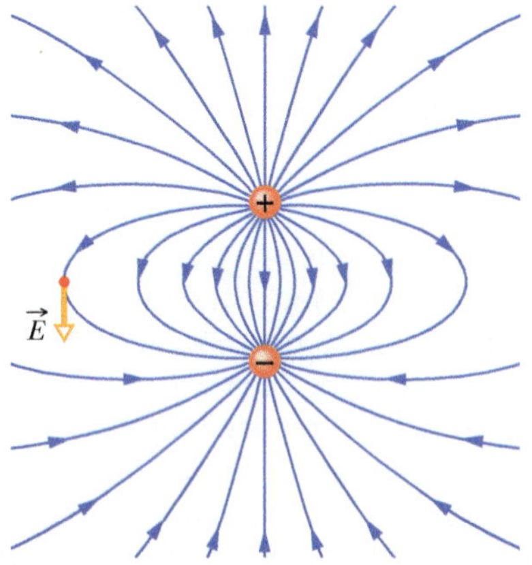

Figure 22-8 The pattern of electric field lines around an electric dipole, with an electric field vector $\vec{E}$ shown at one point (tangent to the field line through that point).

The Electric Field Due to an Electric Dipole

Figure 22-8 shows the pattern of electric field lines for two particles that have the same charge magnitude q but opposite signs, a very common and important arrangement known as an **electric dipole**. The particles are separated by distance d and lie along the *dipole axis*, an axis of symmetry around which you can imagine rotating the pattern in Fig. 22-8. Let's label that axis as a z axis. Here we restrict our interest to the magnitude and direction of the electric field $\vec{E}$ at an arbitrary point P along the dipole axis, at distance z from the dipole's midpoint.

Figure 22-9a shows the electric fields set up at P by each particle. The nearer particle with charge $+q$ sets up field $E_{(+)}$ in the positive direction of the z axis (directly away from the particle). The farther particle with charge $-q$ sets up a smaller field $E_{(-)}$ in the negative direction (directly toward the particle). We want the net field at P, as given by Eq. 22-4. However, because the field vectors are along the same axis, let's simply indicate the vector directions with plus and minus signs, as we commonly do with forces along a single axis. Then we can write the magnitude of the net field at P as

$$
\begin{aligned}
E &= E_{(+)} - E_{(-)} \\
&= \frac{1}{4\pi\varepsilon_0}\frac{q}{r_{(+)}^2} - \frac{1}{4\pi\varepsilon_0}\frac{q}{r_{(-)}^2} \\
&= \frac{q}{4\pi\varepsilon_0(z - \frac{1}{2}d)^2} - \frac{q}{4\pi\varepsilon_0(z + \frac{1}{2}d)^2}.
\end{aligned}
\tag{22-5}
$$

After a little algebra, we can rewrite this equation as

$$
E = \frac{q}{4\pi\varepsilon_0 z^2}\left(\frac{1}{\left(1 - \dfrac{d}{2z}\right)^2} - \frac{1}{\left(1 + \dfrac{d}{2z}\right)^2}\right).
\tag{22-6}
$$

After forming a common denominator and multiplying its terms, we come to

$$
E = \frac{q}{4\pi\varepsilon_0 z^2}\frac{2d/z}{\left(1 - \left(\dfrac{d}{2z}\right)^2\right)^2} = \frac{q}{2\pi\varepsilon_0 z^3}\frac{d}{\left(1 - \left(\dfrac{d}{2z}\right)^2\right)^2}.
\tag{22-7}
$$

We are usually interested in the electrical effect of a dipole only at distances that are large compared with the dimensions of the dipole—that is, at distances such that $z \gg d$. At such large distances, we have $d/2z \ll 1$ in Eq. 22-7. Thus, in our approximation, we can neglect the $d/2z$ term in the denominator, which leaves us with

$$
E = \frac{1}{2\pi\varepsilon_0}\frac{qd}{z^3}.
\tag{22-8}
$$

The product qd, which involves the two intrinsic properties q and d of the dipole, is the magnitude p of a vector quantity known as the **electric dipole moment** $\vec{p}$ of the dipole. (The unit of $\vec{p}$ is the coulomb-meter.) Thus, we can write Eq. 22-8 as

$$
E = \frac{1}{2\pi\varepsilon_0}\frac{p}{z^3} \quad \text{(electric dipole).}
\tag{22-9}
$$

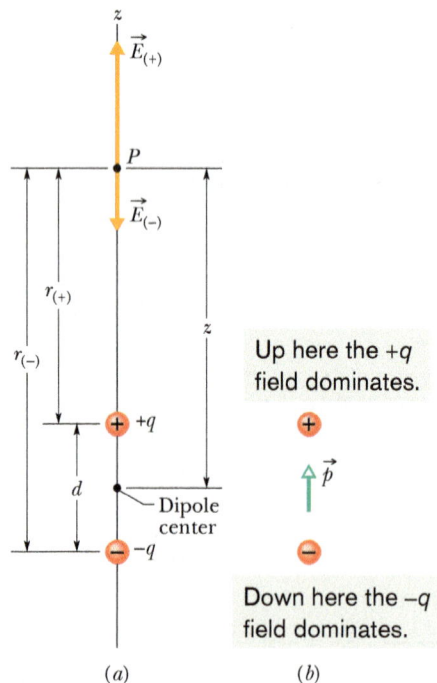

Up here the +q field dominates.

$+$

$\vec{p}$

Down here the $-q$ field dominates.

(a) (b)

Figure 22-9 (a) An electric dipole. The electric field vectors $\vec{E}_{(+)}$ and $\vec{E}_{(-)}$ at point P on the dipole axis result from the dipole's two charges. Point P is at distances $r_{(+)}$ and $r_{(-)}$ from the individual charges that make up the dipole. (b) The dipole moment $\vec{p}$ of the dipole points from the negative charge to the positive charge.

The direction of $\vec{p}$ is taken to be from the negative to the positive end of the dipole, as indicated in Fig. 22-9b. We can use the direction of $\vec{p}$ to specify the orientation of a dipole.

Equation 22-9 shows that, if we measure the electric field of a dipole only at distant points, we can never find q and d separately; instead, we can find only their product. The field at distant points would be unchanged if, for example, q

were doubled and d simultaneously halved. Although Eq. 22-9 holds only for distant points along the dipole axis, it turns out that E for a dipole varies as $1/r^3$ for *all* distant points, regardless of whether they lie on the dipole axis; here r is the distance between the point in question and the dipole center.

Inspection of Fig. 22-9 and of the field lines in Fig. 22-8 shows that the direction of $\vec{E}$ for distant points on the dipole axis is always the direction of the dipole moment vector $\vec{p}$. This is true whether point P in Fig. 22-9a is on the upper or the lower part of the dipole axis.

Inspection of Eq. 22-9 shows that if you double the distance of a point from a dipole, the electric field at the point drops by a factor of 8. If you double the distance from a single point charge, however (see Eq. 22-3), the electric field drops only by a factor of 4. Thus the electric field of a dipole decreases more rapidly with distance than does the electric field of a single charge. The physical reason for this rapid decrease in electric field for a dipole is that from distant points a dipole looks like two particles that almost—but not quite—coincide. Thus, because they have charges of equal magnitude but opposite signs, their electric fields at distant points almost—but not quite—cancel each other.

Sample Problem 22.02 Electric dipole and atmospheric sprites

Sprites (Fig. 22-10a) are huge flashes that occur far above a large thunderstorm. They were seen for decades by pilots flying at night, but they were so brief and dim that most pilots figured they were just illusions. Then in the 1990s sprites were captured on video. They are still not well understood but are believed to be produced when especially powerful lightning occurs between the ground and storm clouds, particularly when the lightning transfers a huge amount of negative charge $-q$ from the ground to the base of the clouds (Fig. 22-10b).

Just after such a transfer, the ground has a complicated distribution of positive charge. However, we can model the electric field due to the charges in the clouds and the ground by assuming a vertical electric dipole that has charge $-q$ at cloud height h and charge $+q$ at below-ground depth h (Fig. 22-10c). If $q = 200$ C and $h = 6.0$ km, what is the magnitude of the dipole's electric field at altitude $z_1 = 30$ km somewhat above the clouds and altitude $z_2 = 60$ km somewhat above the stratosphere?

KEY IDEA

We can approximate the magnitude E of an electric dipole's electric field on the dipole axis with Eq. 22-8.

Calculations: We write that equation as

$$E = \frac{1}{2\pi\varepsilon_0}\frac{q(2h)}{z^3},$$

where $2h$ is the separation between $-q$ and $+q$ in Fig. 22-10c. For the electric field at altitude $z_1 = 30$ km, we find

$$E = \frac{1}{2\pi\varepsilon_0}\frac{(200\ \text{C})(2)(6.0 \times 10^3\ \text{m})}{(30 \times 10^3\ \text{m})^3}$$

$$= 1.6 \times 10^3\ \text{N/C}. \qquad \text{(Answer)}$$

Similarly, for altitude $z_2 = 60$ km, we find

$$E = 2.0 \times 10^2\ \text{N/C}. \qquad \text{(Answer)}$$

As we discuss in Module 22-6, when the magnitude of

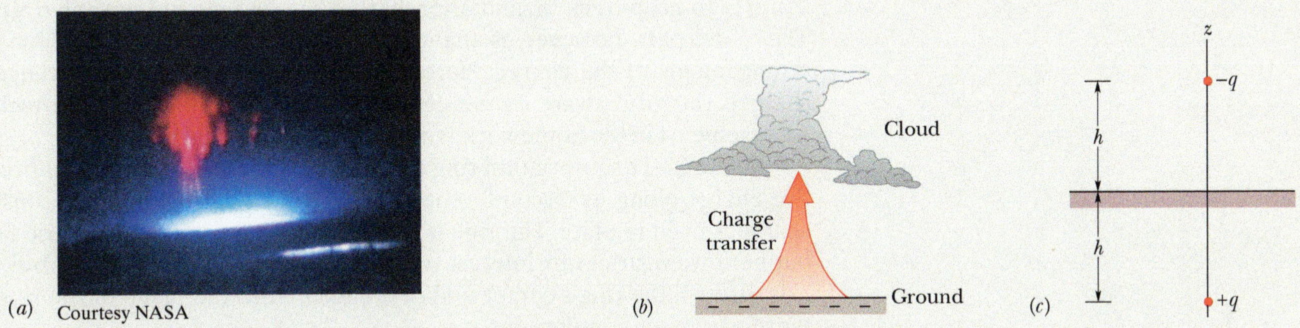

(a) Courtesy NASA (b) (c)

Figure 22-10 (a) Photograph of a sprite. (b) Lightning in which a large amount of negative charge is transferred from ground to cloud base. (c) The cloud–ground system modeled as a vertical electric dipole.

an electric field exceeds a certain critical value E_c, the field can pull electrons out of atoms (ionize the atoms), and then the freed electrons can run into other atoms, causing those atoms to emit light. The value of E_c depends on the density of the air in which the electric field exists. At altitude $z_2 = 60$ km the density of the air is so low that $E = 2.0 \times 10^2$ N/C exceeds E_c, and thus light is emitted by the atoms in the air. That light forms sprites. Lower down, just above the clouds at $z_1 = 30$ km, the density of the air is much higher, $E = 1.6 \times 10^3$ N/C does not exceed E_c, and no light is emitted. Hence, sprites occur only far above storm clouds.

WILEY PLUS Additional examples, video, and practice available at *WileyPLUS*

22-4 THE ELECTRIC FIELD DUE TO A LINE OF CHARGE

Learning Objectives

After reading this module, you should be able to . . .

22.16 For a uniform distribution of charge, find the linear charge density λ for charge along a line, the surface charge density σ for charge on a surface, and the volume charge density ρ for charge in a volume.

22.17 For charge that is distributed uniformly along a line, find the net electric field at a given point near the line by splitting the distribution up into charge elements dq and then summing (by integration) the electric field vectors $d\vec{E}$ set up at the point by each element.

22.18 Explain how symmetry can be used to simplify the calculation of the electric field at a point near a line of uniformly distributed charge.

Key Ideas

● The equation for the electric field set up by a particle does not apply to an extended object with charge (said to have a continuous charge distribution).

● To find the electric field of an extended object at a point, we first consider the electric field set up by a charge element dq in the object, where the element is small enough for us to apply the equation for a particle. Then we sum, via integration, components of the electric fields $d\vec{E}$ from all the charge elements.

● Because the individual electric fields $d\vec{E}$ have different magnitudes and point in different directions, we first see if symmetry allows us to cancel out any of the components of the fields, to simplify the integration.

The Electric Field Due to a Line of Charge

So far we have dealt with only charged particles, a single particle or a simple collection of them. We now turn to a much more challenging situation in which a thin (approximately one-dimensional) object such as a rod or ring is charged with a huge number of particles, more than we could ever even count. In the next module, we consider two-dimensional objects, such as a disk with charge spread over a surface. In the next chapter we tackle three-dimensional objects, such as a sphere with charge spread through a volume.

Heads Up. Many students consider this module to be the most difficult in the book for a variety of reasons. There are lots of steps to take, a lot of vector features to keep track of, and after all that, we set up and then solve an integral. The worst part, however, is that the procedure can be different for different arrangements of the charge. Here, as we focus on a particular arrangement (a charged ring), be aware of the general approach, so that you can tackle other arrangements in the homework (such as rods and partial circles).

Figure 22-11 shows a thin ring of radius R with a uniform distribution of positive charge along its circumference. It is made of plastic, which means that the charge is fixed in place. The ring is surrounded by a pattern of electric field lines, but here we restrict our interest to an arbitrary point P on the central axis (the axis through the ring's center and perpendicular to the plane of the ring), at distance z from the center point.

The charge of an extended object is often conveyed in terms of a charge density rather than the total charge. For a line of charge, we use the *linear charge* density λ (the charge per unit length), with the SI unit of coulomb per meter. Table 22-1 shows the other charge densities that we shall be using for charged surfaces and volumes.

First Big Problem. So far, we have an equation for the electric field of a particle. (We can combine the field of several particles as we did for the electric dipole to generate a special equation, but we are still basically using Eq. 22-3). Now take a look at the ring in Fig. 22-11. That clearly is not a particle and so Eq. 22-3 does not apply. So what do we do?

The answer is to mentally divide the ring into differential elements of charge that are so small that we can treat them as though they *are* particles. Then we *can* apply Eq. 22-3.

Second Big Problem. We now know to apply Eq. 22-3 to each charge element dq (the front d emphasizes that the charge is very small) and can write an expression for its contribution of electric field $d\vec{E}$ (the front d emphasizes that the contribution is very small). However, each such contributed field vector at P is in its own direction. How can we add them to get the net field at P?

The answer is to split the vectors into components and then separately sum one set of components and then the other set. However, first we check to see if one set simply all cancels out. (Canceling out components saves lots of work.)

Third Big Problem. There is a huge number of dq elements in the ring and thus a huge number of $d\vec{E}$ components to add up, even if we can cancel out one set of components. How can we add up more components than we could even count? The answer is to add them by means of integration.

Do It. Let's do all this (but again, be aware of the general procedure, not just the fine details). We arbitrarily pick the charge element shown in Fig. 22-11. Let ds be the arc length of that (or any other) dq element. Then in terms of the linear density λ (the charge per unit length), we have

$$dq = \lambda \, ds. \qquad (22\text{-}10)$$

An Element's Field. This charge element sets up the differential electric field $d\vec{E}$ at P, at distance r from the element, as shown in Fig. 22-11. (Yes, we are introducing a new symbol that is not given in the problem statement, but soon we shall replace it with "legal symbols.") Next we rewrite the field equation for a particle (Eq. 22-3) in terms of our new symbols dE and dq, but then we replace dq using Eq. 22-10. The field magnitude due to the charge element is

$$dE = \frac{1}{4\pi\varepsilon_0}\frac{dq}{r^2} = \frac{1}{4\pi\varepsilon_0}\frac{\lambda\,ds}{r^2}. \qquad (22\text{-}11)$$

Notice that the illegal symbol r is the hypotenuse of the right triangle displayed in Fig. 22-11. Thus, we can replace r by rewriting Eq. 22-11 as

$$dE = \frac{1}{4\pi\varepsilon_0}\frac{\lambda\,ds}{(z^2+R^2)}. \qquad (22\text{-}12)$$

Because every charge element has the same charge and the same distance from point P, Eq. 22-12 gives the field magnitude contributed by each of them. Figure 22-11 also tells us that each contributed $d\vec{E}$ leans at angle θ to the central axis (the z axis) and thus has components perpendicular and parallel to that axis.

Canceling Components. Now comes the neat part, where we eliminate one set of those components. In Fig. 22-11, consider the charge element on the opposite side of the ring. It too contributes the field magnitude dE but the field vector leans at angle θ in the opposite direction from the vector from our first charge

Table 22-1 Some Measures of Electric Charge

Name	Symbol	SI Unit
Charge	q	C
Linear charge density	λ	C/m
Surface charge density	σ	C/m^2
Volume charge density	ρ	C/m^3

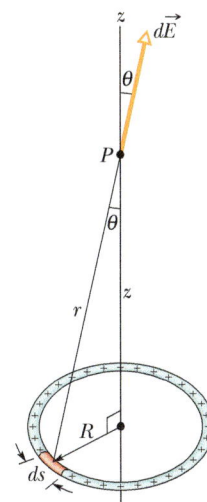

Figure 22-11 A ring of uniform positive charge. A differential element of charge occupies a length ds (greatly exaggerated for clarity). This element sets up an electric field $d\vec{E}$ at point P.

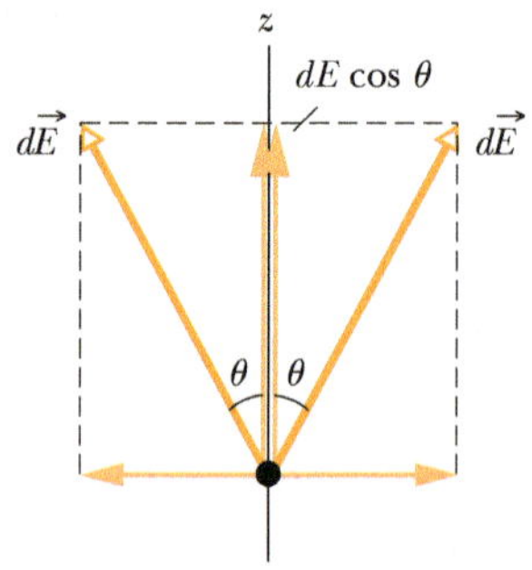

Figure 22-12 The electric fields set up at P by a charge element and its symmetric partner (on the opposite side of the ring). The components perpendicular to the z axis cancel; the parallel components add.

element, as indicated in the side view of Fig. 22-12. Thus the two perpendicular components cancel. All around the ring, this cancelation occurs for every charge element and its *symmetric partner* on the opposite side of the ring. So we can neglect all the perpendicular components.

Adding Components. We have another big win here. All the remaining components are in the positive direction of the z axis, so we can just add them up as scalars. Thus we can already tell the direction of the net electric field at P: directly away from the ring. From Fig. 22-12, we see that the parallel components each have magnitude $dE \cos \theta$, but θ is another illegal symbol. We can replace $\cos \theta$ with legal symbols by again using the right triangle in Fig. 22-11 to write

$$\cos \theta = \frac{z}{r} = \frac{z}{(z^2 + R^2)^{1/2}}. \qquad (22\text{-}13)$$

Multiplying Eq. 22-12 by Eq. 22-13 gives us the parallel field component from each charge element:

$$dE \cos \theta = \frac{1}{4\pi\varepsilon_0} \frac{z\lambda}{(z^2 + R^2)^{3/2}} \, ds. \qquad (22\text{-}14)$$

Integrating. Because we must sum a huge number of these components, each small, we set up an integral that moves along the ring, from element to element, from a starting point (call it $s = 0$) through the full circumference ($s = 2\pi R$). Only the quantity s varies as we go through the elements; the other symbols in Eq. 22-14 remain the same, so we move them outside the integral. We find

$$E = \int dE \cos \theta = \frac{z\lambda}{4\pi\varepsilon_0(z^2 + R^2)^{3/2}} \int_0^{2\pi R} ds$$

$$= \frac{z\lambda(2\pi R)}{4\pi\varepsilon_0(z^2 + R^2)^{3/2}}. \qquad (22\text{-}15)$$

This is a fine answer, but we can also switch to the total charge by using $\lambda = q/(2\pi R)$:

$$E = \frac{qz}{4\pi\varepsilon_0(z^2 + R^2)^{3/2}} \quad \text{(charged ring).} \qquad (22\text{-}16)$$

If the charge on the ring is negative, instead of positive as we have assumed, the magnitude of the field at P is still given by Eq. 22-16. However, the electric field vector then points toward the ring instead of away from it.

Let us check Eq. 22-16 for a point on the central axis that is so far away that $z \gg R$. For such a point, the expression $z^2 + R^2$ in Eq. 22-16 can be approximated as z^2, and Eq. 22-16 becomes

$$E = \frac{1}{4\pi\varepsilon_0} \frac{q}{z^2} \quad \text{(charged ring at large distance).} \qquad (22\text{-}17)$$

This is a reasonable result because from a large distance, the ring "looks like" a point charge. If we replace z with r in Eq. 22-17, we indeed do have the magnitude of the electric field due to a point charge, as given by Eq. 22-3.

Let us next check Eq. 22-16 for a point at the center of the ring—that is, for $z = 0$. At that point, Eq. 22-16 tells us that $E = 0$. This is a reasonable result because if we were to place a test charge at the center of the ring, there would be no net electrostatic force acting on it; the force due to any element of the ring would be canceled by the force due to the element on the opposite side of the ring. By Eq. 22-1, if the force at the center of the ring were zero, the electric field there would also have to be zero.

Sample Problem 22.03 Electric field of a charged circular rod

Figure 22-13a shows a plastic rod with a uniform charge $-Q$. It is bent in a 120° circular arc of radius r and symmetrically paced across an x axis with the origin at the center of curvature P of the rod. In terms of Q and r, what is the electric field $\vec{E}$ due to the rod at point P?

KEY IDEA

Because the rod has a continuous charge distribution, we must find an expression for the electric fields due to differential elements of the rod and then sum those fields via calculus.

An element: Consider a differential element having arc length ds and located at an angle θ above the x axis (Figs. 22-13b and c). If we let λ represent the linear charge density of the rod, our element ds has a differential charge of magnitude

$$dq = \lambda \, ds. \qquad (22\text{-}18)$$

The element's field: Our element produces a differential electric field $d\vec{E}$ at point P, which is a distance r from the element. Treating the element as a point charge, we can

rewrite Eq. 22-3 to express the magnitude of $d\vec{E}$ as

$$dE = \frac{1}{4\pi\varepsilon_0} \frac{dq}{r^2} = \frac{1}{4\pi\varepsilon_0} \frac{\lambda \, ds}{r^2}. \qquad (22\text{-}19)$$

The direction of $d\vec{E}$ is toward ds because charge dq is negative.

Symmetric partner: Our element has a symmetrically located (mirror image) element ds' in the bottom half of the rod. The electric field $d\vec{E}'$ set up at P by ds' also has the magnitude given by Eq. 22-19, but the field vector points toward ds' as shown in Fig. 22-13d. If we resolve the electric field vectors of ds and ds' into x and y components as shown in Figs. 22-13e and f, we see that their y components cancel (because they have equal magnitudes and are in opposite directions). We also see that their x components have equal magnitudes and are in the same direction.

Summing: Thus, to find the electric field set up by the rod, we need sum (via integration) only the x components of the differential electric fields set up by all the differential elements of the rod. From Fig. 22-13f and Eq. 22-19, we can write

Figure 22-13 Available in *WileyPLUS* as an animation with voiceover. (a) A plastic rod of charge $-Q$ is a circular section of radius r and central angle 120°; point P is the center of curvature of the rod. (b)–(c) A differential element in the top half of the rod, at an angle θ to the x axis and of arc length ds, sets up a differential electric field $d\vec{E}$ at P. (d) An element ds', symmetric to ds about the x axis, sets up a field $d\vec{E}'$ at P with the same magnitude. (e)–(f) The field components. (g) Arc length ds makes an angle $d\theta$ about point P.

Labels within figures:

(a) This negatively charged rod is obviously not a particle. Plastic rod of charge $-Q$. 60° 60° r

(b) But we can treat this element as a particle. ds

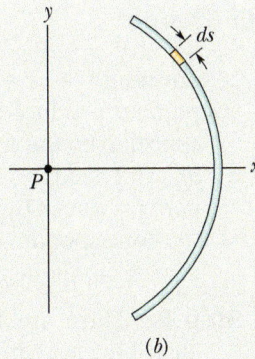

(c) Here is the field the element creates. ds $d\vec{E}$ θ

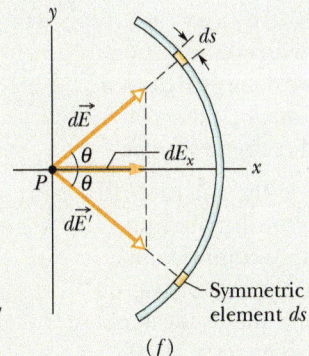

(d) Here is the field created by the symmetric element, same size and angle. ds $d\vec{E}$ θ θ $d\vec{E}'$ Symmetric element ds'

(e) These y components just cancel, so neglect them. ds dE_y $d\vec{E}$ θ θ $d\vec{E}'$ Symmetric element ds'

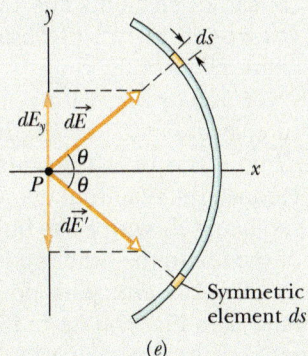

(f) These x components add. Our job is to add all such components. ds $d\vec{E}$ θ dE_x θ $d\vec{E}'$ Symmetric element ds'

(g) We use this to relate the element's arc length to the angle that it subtends. ds $d\theta$ r

the component dE_x set up by ds as

$$dE_x = dE \cos \theta = \frac{1}{4\pi\varepsilon_0} \frac{\lambda}{r^2} \cos \theta \, ds. \qquad (22\text{-}20)$$

Equation 22-20 has two variables, θ and s. Before we can integrate it, we must eliminate one variable. We do so by replacing ds, using the relation

$$ds = r \, d\theta,$$

in which $d\theta$ is the angle at P that includes arc length ds (Fig. 22-13g). With this replacement, we can integrate Eq. 22-20 over the angle made by the rod at P, from $\theta = -60°$ to $\theta = 60°$; that will give us the field magnitude at P:

$$E = \int dE_x = \int_{-60°}^{60°} \frac{1}{4\pi\varepsilon_0} \frac{\lambda}{r^2} \cos \theta \, r \, d\theta$$

$$= \frac{\lambda}{4\pi\varepsilon_0 r} \int_{-60°}^{60°} \cos \theta \, d\theta = \frac{\lambda}{4\pi\varepsilon_0 r} \left[\sin \theta \right]_{-60°}^{60°}$$

$$= \frac{\lambda}{4\pi\varepsilon_0 r} [\sin 60° - \sin(-60°)]$$

$$= \frac{1.73\lambda}{4\pi\varepsilon_0 r}. \qquad (22\text{-}21)$$

(If we had reversed the limits on the integration, we would have gotten the same result but with a minus sign. Since the integration gives only the magnitude of $\vec{E}$, we would then have discarded the minus sign.)

Charge density: To evaluate λ, we note that the full rod subtends an angle of 120° and so is one-third of a full circle. Its arc length is then $2\pi r/3$, and its linear charge density must be

$$\lambda = \frac{\text{charge}}{\text{length}} = \frac{Q}{2\pi r/3} = \frac{0.477Q}{r}.$$

Substituting this into Eq. 22-21 and simplifying give us

$$E = \frac{(1.73)(0.477Q)}{4\pi\varepsilon_0 r^2}$$

$$= \frac{0.83Q}{4\pi\varepsilon_0 r^2}. \qquad \text{(Answer)}$$

The direction of $\vec{E}$ is toward the rod, along the axis of symmetry of the charge distribution. We can write $\vec{E}$ in unit-vector notation as

$$\vec{E} = \frac{0.83Q}{4\pi\varepsilon_0 r^2} \hat{\mathbf{i}}.$$

Problem-Solving Tactics   A Field Guide for Lines of Charge

Here is a generic guide for finding the electric field $\vec{E}$ produced at a point P by a line of uniform charge, either circular or straight. The general strategy is to pick out an element dq of the charge, find $d\vec{E}$ due to that element, and integrate $d\vec{E}$ over the entire line of charge.

Step 1. If the line of charge is circular, let ds be the arc length of an element of the distribution. If the line is straight, run an x axis along it and let dx be the length of an element. Mark the element on a sketch.

Step 2. Relate the charge dq of the element to the length of the element with either $dq = \lambda \, ds$ or $dq = \lambda \, dx$. Consider dq and λ to be positive, even if the charge is actually negative. (The sign of the charge is used in the next step.)

Step 3. Express the field $d\vec{E}$ produced at P by dq with Eq. 22-3, replacing q in that equation with either $\lambda \, ds$ or $\lambda \, dx$. If the charge on the line is positive, then at P draw a vector $d\vec{E}$ that points directly away from dq. If the charge is negative, draw the vector pointing directly toward dq.

Step 4. Always look for any symmetry in the situation. If P is on an axis of symmetry of the charge distribution, resolve the field $d\vec{E}$ produced by dq into components that are perpendicular and parallel to the axis of symmetry. Then consider a second element dq' that is located symmetrically to dq about the line of symmetry. At P draw the vector $d\vec{E}'$ that this symmetrical element pro-

duces and resolve it into components. One of the components produced by dq is a *canceling component;* it is canceled by the corresponding component produced by dq' and needs no further attention. The other component produced by dq is an *adding component;* it adds to the corresponding component produced by dq'. Add the adding components of all the elements via integration.

Step 5. Here are four general types of uniform charge distributions, with strategies for the integral of step 4.

Ring, with point P on (central) axis of symmetry, as in Fig. 22-11. In the expression for dE, replace r^2 with $z^2 + R^2$, as in Eq. 22-12. Express the adding component of $d\vec{E}$ in terms of θ. That introduces $\cos \theta$, but θ is identical for all elements and thus is not a variable. Replace $\cos \theta$ as in Eq. 22-13. Integrate over s, around the circumference of the ring.

Circular arc, with point P at the center of curvature, as in Fig. 22-13. Express the adding component of $d\vec{E}$ in terms of θ. That introduces either $\sin \theta$ or $\cos \theta$. Reduce the resulting two variables s and θ to one, θ, by replacing ds with $r \, d\theta$. Integrate over θ from one end of the arc to the other end.

Straight line, with point P on an extension of the line, as in Fig. 22-14a. In the expression for dE, replace r with x. Integrate over x, from end to end of the line of charge.

Straight line, with point P at perpendicular distance y from the line of charge, as in Fig. 22-14b. In the expression for dE, replace r with an expression involving x and y. If P is on the perpendicular bisector of the line of charge, find an expression for the adding component of $d\vec{E}$. That will introduce either sin θ or cos θ. Reduce the resulting two variables x and θ to one, x, by replacing the trigonometric function with an expression (its definition) involving x and y. Integrate over x from end to end of the line of charge. If P is not on a line of symmetry, as in Fig. 22-14c, set up an integral to sum the components dE_x, and integrate over x to find E_x. Also set up an integral to sum the components dE_y, and integrate over x again to find E_y. Use the components E_x and E_y in the usual way to find the magnitude E and the orientation of $\vec{E}$.

Step 6. One arrangement of the integration limits gives a positive result. The reverse gives the same result with a mi-

nus sign; discard the minus sign. If the result is to be stated in terms of the total charge Q of the distribution, replace λ with Q/L, in which L is the length of the distribution.

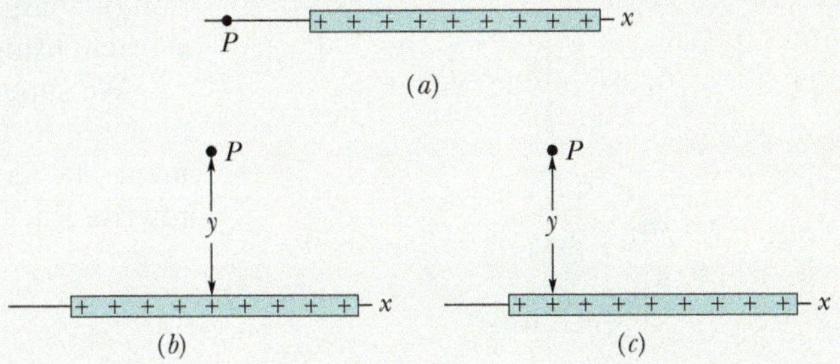

Figure 22-14 (*a*) Point P is on an extension of the line of charge. (*b*) P is on a line of symmetry of the line of charge, at perpendicular distance y from that line. (*c*) Same as (*b*) except that P is not on a line of symmetry.

WILEY PLUS Additional examples, video, and practice available at *WileyPLUS*

✓ Checkpoint 2

The figure here shows three nonconducting rods, one circular and two straight. Each has a uniform charge of magnitude Q along its top half and another along its bottom half. For each rod, what is the direction of the net electric field at point P?

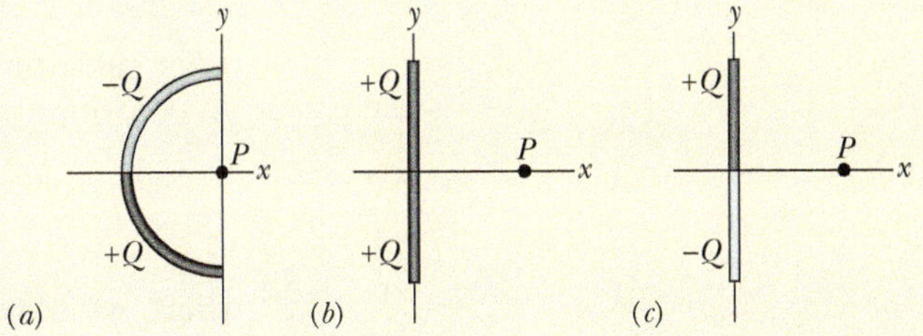

22-5 THE ELECTRIC FIELD DUE TO A CHARGED DISK

Learning Objectives

After reading this module, you should be able to . . .

22.19 Sketch a disk with uniform charge and indicate the direction of the electric field at a point on the central axis if the charge is positive and if it is negative.

22.20 Explain how the equation for the electric field on the central axis of a uniformly charged ring can be used to find

the equation for the electric field on the central axis of a uniformly charged disk.

22.21 For a point on the central axis of a uniformly charged disk, apply the relationship between the surface charge density σ, the disk radius R, and the distance z to that point.

Key Idea

● On the central axis through a uniformly charged disk,

$$E = \frac{\sigma}{2\varepsilon_0}\left(1 - \frac{z}{\sqrt{z^2 + R^2}}\right)$$

gives the electric field magnitude. Here z is the distance along the axis from the center of the disk, R is the radius of the disk, and σ is the surface charge density.

The Electric Field Due to a Charged Disk

Now we switch from a line of charge to a surface of charge by examining the electric field of a circular plastic disk, with a radius R and a uniform surface charge density σ (charge per unit area, Table 22-1) on its top surface. The disk sets up a

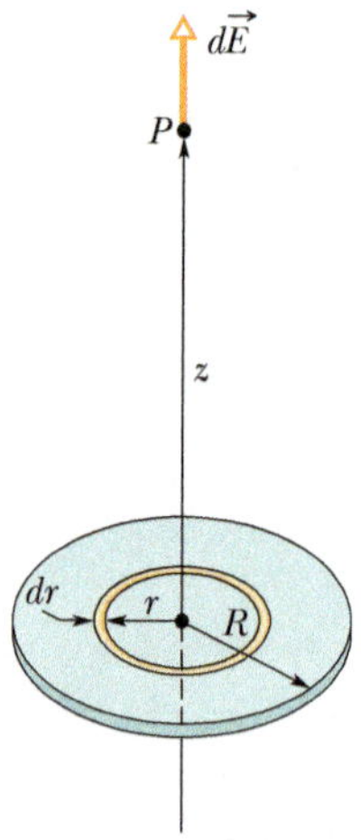

Figure 22-15 A disk of radius R and uniform positive charge. The ring shown has radius r and radial width dr. It sets up a differential electric field $d\vec{E}$ at point P on its central axis.

pattern of electric field lines around it, but here we restrict our attention to the electric field at an arbitrary point P on the central axis, at distance z from the center of the disk, as indicated in Fig. 22-15.

We could proceed as in the preceding module but set up a two-dimensional integral to include all of the field contributions from the two-dimensional distribution of charge on the top surface. However, we can save a lot of work with a neat shortcut using our earlier work with the field on the central axis of a thin ring.

We superimpose a ring on the disk as shown in Fig. 22-15, at an arbitrary radius $r \le R$. The ring is so thin that we can treat the charge on it as a charge element dq. To find its small contribution dE to the electric field at point P, we rewrite Eq. 22-16 in terms of the ring's charge dq and radius r:

$$dE = \frac{dq\, z}{4\pi\varepsilon_0(z^2 + r^2)^{3/2}}. \tag{22-22}$$

The ring's field points in the positive direction of the z axis.

To find the total field at P, we are going to integrate Eq. 22-22 from the center of the disk at $r = 0$ out to the rim at $r = R$ so that we sum all the dE contributions (by sweeping our arbitrary ring over the entire disk surface). However, that means we want to integrate with respect to a variable radius r of the ring.

We get dr into the expression by substituting for dq in Eq. 22-22. Because the ring is so thin, call its thickness dr. Then its surface area dA is the product of its circumference $2\pi r$ and thickness dr. So, in terms of the surface charge density σ, we have

$$dq = \sigma\, dA = \sigma\, (2\pi r\, dr). \tag{22-23}$$

After substituting this into Eq. 22-22 and simplifying slightly, we can sum all the dE contributions with

$$E = \int dE = \frac{\sigma z}{4\varepsilon_0} \int_0^R (z^2 + r^2)^{-3/2}(2r)\, dr, \tag{22-24}$$

where we have pulled the constants (including z) out of the integral. To solve this integral, we cast it in the form $\int X^m\, dX$ by setting $X = (z^2 + r^2)$, $m = -\frac{3}{2}$, and $dX = (2r)\, dr$. For the recast integral we have

$$\int X^m\, dX = \frac{X^{m+1}}{m+1},$$

and so Eq. 22-24 becomes

$$E = \frac{\sigma z}{4\varepsilon_0} \left[\frac{(z^2 + r^2)^{-1/2}}{-\frac{1}{2}} \right]_0^R. \tag{22-25}$$

Taking the limits in Eq. 22-25 and rearranging, we find

$$E = \frac{\sigma}{2\varepsilon_0}\left(1 - \frac{z}{\sqrt{z^2 + R^2}}\right) \quad \text{(charged disk)} \tag{22-26}$$

as the magnitude of the electric field produced by a flat, circular, charged disk at points on its central axis. (In carrying out the integration, we assumed that $z \ge 0$.)

If we let $R \to \infty$ while keeping z finite, the second term in the parentheses in Eq. 22-26 approaches zero, and this equation reduces to

$$E = \frac{\sigma}{2\varepsilon_0} \quad \text{(infinite sheet).} \tag{22-27}$$

This is the electric field produced by an infinite sheet of uniform charge located on one side of a nonconductor such as plastic. The electric field lines for such a situation are shown in Fig. 22-4.

We also get Eq. 22-27 if we let $z \to 0$ in Eq. 22-26 while keeping R finite. This shows that at points very close to the disk, the electric field set up by the disk is the same as if the disk were infinite in extent.

22-6 A POINT CHARGE IN AN ELECTRIC FIELD

Learning Objectives

After reading this module, you should be able to . . .

22.22 For a charged particle placed in an external electric field (a field due to other charged objects), apply the relationship between the electric field $\vec{E}$ at that point, the particle's charge q, and the electrostatic force $\vec{F}$ that acts on the particle, and identify the relative directions of the force

and the field when the particle is positively charged and negatively charged.

22.23 Explain Millikan's procedure of measuring the elementary charge.

22.24 Explain the general mechanism of ink-jet printing.

Key Ideas

● If a particle with charge q is placed in an external electric field $\vec{E}$, an electrostatic force $\vec{F}$ acts on the particle:

$$\vec{F} = q\vec{E}.$$

● If charge q is positive, the force vector is in the same direction as the field vector. If charge q is negative, the force vector is in the opposite direction (the minus sign in the equation reverses the force vector from the field vector).

A Point Charge in an Electric Field

In the preceding four modules we worked at the first of our two tasks: given a charge distribution, to find the electric field it produces in the surrounding space. Here we begin the second task: to determine what happens to a charged particle when it is in an electric field set up by other stationary or slowly moving charges.

What happens is that an electrostatic force acts on the particle, as given by

$$\vec{F} = q\vec{E}, \tag{22-28}$$

in which q is the charge of the particle (including its sign) and $\vec{E}$ is the electric field that other charges have produced at the location of the particle. (The field is *not* the field set up by the particle itself; to distinguish the two fields, the field acting on the particle in Eq. 22-28 is often called the *external field*. A charged particle or object is not affected by its own electric field.) Equation 22-28 tells us

⭐ The electrostatic force $\vec{F}$ acting on a charged particle located in an external electric field $\vec{E}$ has the direction of $\vec{E}$ if the charge q of the particle is positive and has the opposite direction if q is negative.

Measuring the Elementary Charge

Equation 22-28 played a role in the measurement of the elementary charge e by American physicist Robert A. Millikan in 1910–1913. Figure 22-16 is a representation of his apparatus. When tiny oil drops are sprayed into chamber A, some of them become charged, either positively or negatively, in the process. Consider a drop that drifts downward through the small hole in plate P_1 and into chamber C. Let us assume that this drop has a negative charge q.

If switch S in Fig. 22-16 is open as shown, battery B has no electrical effect on chamber C. If the switch is closed (the connection between chamber C and the positive terminal of the battery is then complete), the battery causes an excess positive charge on conducting plate P_1 and an excess negative charge on conducting plate P_2. The charged plates set up a downward-directed electric field $\vec{E}$ in chamber C. According to Eq. 22-28, this field exerts an electrostatic force on any charged drop that happens to be in the chamber and affects its motion. In particular, our negatively charged drop will tend to drift upward.

By timing the motion of oil drops with the switch opened and with it closed and thus determining the effect of the charge q, Millikan discovered that the

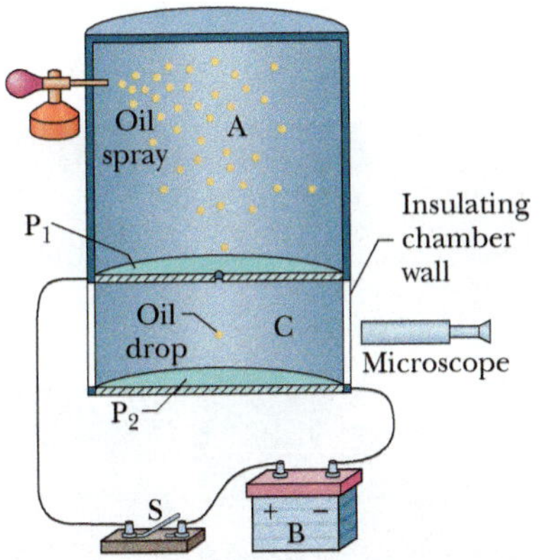

Figure 22-16 The Millikan oil-drop apparatus for measuring the elementary charge e. When a charged oil drop drifted into chamber C through the hole in plate P_1, its motion could be controlled by closing and opening switch S and thereby setting up or eliminating an electric field in chamber C. The microscope was used to view the drop, to permit timing of its motion.

Figure 22-17 Ink-jet printer. Drops shot from generator G receive a charge in charging unit C. An input signal from a computer controls the charge and thus the effect of field $\vec{E}$ on where the drop lands on the paper.

values of q were always given by

$$q = ne, \quad \text{for } n = 0, \pm1, \pm2, \pm3, \ldots, \quad (22\text{-}29)$$

in which e turned out to be the fundamental constant we call the *elementary charge*, 1.60×10^{-19} C. Millikan's experiment is convincing proof that charge is quantized, and he earned the 1923 Nobel Prize in physics in part for this work. Modern measurements of the elementary charge rely on a variety of interlocking experiments, all more precise than the pioneering experiment of Millikan.

Ink-Jet Printing

The need for high-quality, high-speed printing has caused a search for an alternative to impact printing, such as occurs in a standard typewriter. Building up letters by squirting tiny drops of ink at the paper is one such alternative.

Figure 22-17 shows a negatively charged drop moving between two conducting deflecting plates, between which a uniform, downward-directed electric field $\vec{E}$ has been set up. The drop is deflected upward according to Eq. 22-28 and then strikes the paper at a position that is determined by the magnitudes of $\vec{E}$ and the charge q of the drop.

In practice, E is held constant and the position of the drop is determined by the charge q delivered to the drop in the charging unit, through which the drop must pass before entering the deflecting system. The charging unit, in turn, is activated by electronic signals that encode the material to be printed.

Electrical Breakdown and Sparking

If the magnitude of an electric field in air exceeds a certain critical value E_c, the air undergoes *electrical breakdown*, a process whereby the field removes electrons from the atoms in the air. The air then begins to conduct electric current because the freed electrons are propelled into motion by the field. As they move, they collide with any atoms in their path, causing those atoms to emit light. We can see the paths, commonly called sparks, taken by the freed electrons because of that emitted light. Figure 22-18 shows sparks above charged metal wires where the electric fields due to the wires cause electrical breakdown of the air.

Adam Hart-Davis/Photo Researchers, Inc.

Figure 22-18 The metal wires are so charged that the electric fields they produce in the surrounding space cause the air there to undergo electrical breakdown.

Checkpoint 3

(a) In the figure, what is the direction of the electrostatic force on the electron due to the external electric field shown? (b) In which direction will the electron accelerate if it is moving parallel to the y axis before it encounters the external field? (c) If, instead, the electron is initially moving rightward, will its speed increase, decrease, or remain constant?

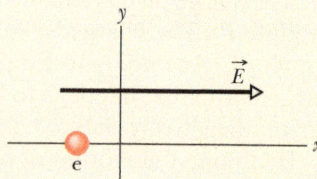

Sample Problem 22.04 Motion of a charged particle in an electric field

Figure 22-19 shows the deflecting plates of an ink-jet printer, with superimposed coordinate axes. An ink drop with a mass m of 1.3×10^{-10} kg and a negative charge of magnitude $Q = 1.5 \times 10^{-13}$ C enters the region between the plates, initially moving along the x axis with speed $v_x = 18$ m/s. The length L of each plate is 1.6 cm. The plates are charged and thus produce an electric field at all points between them. Assume that field $\vec{E}$ is downward directed, is uniform, and has a magnitude of 1.4×10^6 N/C. What is the vertical deflection of the drop at the far edge of the plates? (The gravitational force on the drop is small relative to the electrostatic force acting on the drop and can be neglected.)

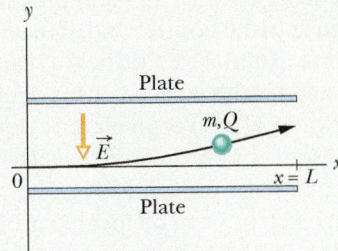

Figure 22-19 An ink drop of mass m and charge magnitude Q is deflected in the electric field of an ink-jet printer.

KEY IDEA

The drop is negatively charged and the electric field is directed *downward*. From Eq. 22-28, a constant electrostatic force of magnitude QE acts *upward* on the charged drop. Thus, as the drop travels parallel to the x axis at constant speed v_x, it accelerates upward with some constant acceleration a_y.

Calculations: Applying Newton's second law ($F = ma$) for components along the y axis, we find that

$$a_y = \frac{F}{m} = \frac{QE}{m}. \qquad (22\text{-}30)$$

Let t represent the time required for the drop to pass through the region between the plates. During t the vertical and horizontal displacements of the drop are

$$y = \tfrac{1}{2}a_y t^2 \quad \text{and} \quad L = v_x t, \qquad (22\text{-}31)$$

respectively. Eliminating t between these two equations and substituting Eq. 22-30 for a_y, we find

$$y = \frac{QEL^2}{2mv_x^2}$$

$$= \frac{(1.5 \times 10^{-13}\ \text{C})(1.4 \times 10^6\ \text{N/C})(1.6 \times 10^{-2}\ \text{m})^2}{(2)(1.3 \times 10^{-10}\ \text{kg})(18\ \text{m/s})^2}$$

$$= 6.4 \times 10^{-4}\ \text{m}$$

$$= 0.64\ \text{mm.} \qquad \text{(Answer)}$$

WILEY PLUS Additional examples, video, and practice available at *WileyPLUS*

22-7 A DIPOLE IN AN ELECTRIC FIELD

Learning Objectives

After reading this module, you should be able to . . .

22.25 On a sketch of an electric dipole in an external electric field, indicate the direction of the field, the direction of the dipole moment, the direction of the electrostatic forces on the two ends of the dipole, and the direction in which those forces tend to rotate the dipole, and identify the value of the net force on the dipole.

22.26 Calculate the torque on an electric dipole in an external electric field by evaluating a cross product of the dipole moment vector and the electric field vector, in magnitude-angle notation and unit-vector notation.

22.27 For an electric dipole in an external electric field, relate the potential energy of the dipole to the work done by a torque as the dipole rotates in the electric field.

22.28 For an electric dipole in an external electric field, calculate the potential energy by taking a dot product of the dipole moment vector and the electric field vector, in magnitude-angle notation and unit-vector notation.

22.29 For an electric dipole in an external electric field, identify the angles for the minimum and maximum potential energies and the angles for the minimum and maximum torque magnitudes.

Key Ideas

● The torque on an electric dipole of dipole moment $\vec{p}$ when placed in an external electric field $\vec{E}$ is given by a cross product:

$$\vec{\tau} = \vec{p} \times \vec{E}.$$

● A potential energy U is associated with the orientation of the dipole moment in the field, as given by a dot product:

$$U = -\vec{p} \cdot \vec{E}.$$

● If the dipole orientation changes, the work done by the electric field is

$$W = -\Delta U.$$

If the change in orientation is due to an external agent, the work done by the agent is $W_a = -W$.

A Dipole in an Electric Field

We have defined the electric dipole moment $\vec{p}$ of an electric dipole to be a vector that points from the negative to the positive end of the dipole. As you will see, the behavior of a dipole in a uniform external electric field $\vec{E}$ can be described completely in terms of the two vectors $\vec{E}$ and $\vec{p}$, with no need of any details about the dipole's structure.

A molecule of water (H_2O) is an electric dipole; Fig. 22-20 shows why. There the black dots represent the oxygen nucleus (having eight protons) and the two hydrogen nuclei (having one proton each). The colored enclosed areas represent the regions in which electrons can be located around the nuclei.

In a water molecule, the two hydrogen atoms and the oxygen atom do not lie on a straight line but form an angle of about 105°, as shown in Fig. 22-20. As a result, the molecule has a definite "oxygen side" and "hydrogen side." Moreover, the 10 electrons of the molecule tend to remain closer to the oxygen nucleus than to the hydrogen nuclei. This makes the oxygen side of the molecule slightly more negative than the hydrogen side and creates an electric dipole moment $\vec{p}$ that points along the symmetry axis of the molecule as shown. If the water molecule is placed in an external electric field, it behaves as would be expected of the more abstract electric dipole of Fig. 22-9.

To examine this behavior, we now consider such an abstract dipole in a uniform external electric field $\vec{E}$, as shown in Fig. 22-21a. We assume that the dipole is a rigid structure that consists of two centers of opposite charge, each of magnitude q, separated by a distance d. The dipole moment $\vec{p}$ makes an angle θ with field $\vec{E}$.

Electrostatic forces act on the charged ends of the dipole. Because the electric field is uniform, those forces act in opposite directions (as shown in Fig. 22-21a) and with the same magnitude $F = qE$. Thus, *because the field is uniform,* the net force on the dipole from the field is zero and the center of mass of the dipole does not move. However, the forces on the charged ends do produce a net torque $\vec{\tau}$ on the dipole about its center of mass. The center of mass lies on the line connecting the charged ends, at some distance x from one end and thus a distance $d - x$ from the other end. From Eq. 10-39 ($\tau = rF \sin \phi$), we can write the magnitude of the net torque $\vec{\tau}$ as

$$\tau = Fx \sin \theta + F(d - x) \sin \theta = Fd \sin \theta. \qquad (22\text{-}32)$$

We can also write the magnitude of $\vec{\tau}$ in terms of the magnitudes of the electric field E and the dipole moment $p = qd$. To do so, we substitute qE for F and p/q for d in Eq. 22-32, finding that the magnitude of $\vec{\tau}$ is

$$\tau = pE \sin \theta. \qquad (22\text{-}33)$$

We can generalize this equation to vector form as

$$\vec{\tau} = \vec{p} \times \vec{E} \qquad \text{(torque on a dipole).} \qquad (22\text{-}34)$$

Vectors $\vec{p}$ and $\vec{E}$ are shown in Fig. 22-21b. The torque acting on a dipole tends to rotate $\vec{p}$ (hence the dipole) into the direction of field $\vec{E}$, thereby reducing θ. In Fig. 22-21, such rotation is clockwise. As we discussed in Chapter 10, we can represent a torque that gives rise to a clockwise rotation by including a minus sign with the magnitude of the torque. With that notation, the torque of Fig. 22-21 is

$$\tau = -pE \sin \theta. \qquad (22\text{-}35)$$

Potential Energy of an Electric Dipole

Potential energy can be associated with the orientation of an electric dipole in an electric field. The dipole has its least potential energy when it is in its equilibrium orientation, which is when its moment $\vec{p}$ is lined up with the field $\vec{E}$ (then $\vec{\tau} = \vec{p} \times \vec{E} = 0$). It has greater potential energy in all other orientations. Thus the dipole is like a pendulum, which has *its* least gravitational potential

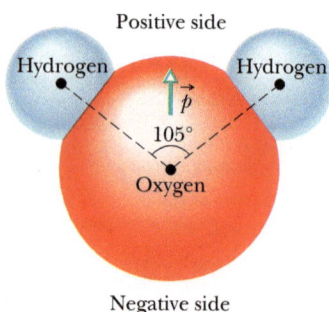

Figure 22-20 A molecule of H_2O, showing the three nuclei (represented by dots) and the regions in which the electrons can be located. The electric dipole moment $\vec{p}$ points from the (negative) oxygen side to the (positive) hydrogen side of the molecule.

energy in *its* equilibrium orientation — at its lowest point. To rotate the dipole or the pendulum to any other orientation requires work by some external agent.

In any situation involving potential energy, we are free to define the zero-potential-energy configuration in an arbitrary way because only differences in potential energy have physical meaning. The expression for the potential energy of an electric dipole in an external electric field is simplest if we choose the potential energy to be zero when the angle θ in Fig. 22-21 is 90°. We then can find the potential energy U of the dipole at any other value of θ with Eq. 8-1 ($\Delta U = -W$) by calculating the work W done by the field on the dipole when the dipole is rotated to that value of θ from 90°. With the aid of Eq. 10-53 ($W = \int \tau \, d\theta$) and Eq. 22-35, we find that the potential energy U at any angle θ is

$$U = -W = -\int_{90°}^{\theta} \tau \, d\theta = \int_{90°}^{\theta} pE \sin \theta \, d\theta. \qquad (22\text{-}36)$$

Evaluating the integral leads to

$$U = -pE \cos \theta. \qquad (22\text{-}37)$$

We can generalize this equation to vector form as

$$U = -\vec{p} \cdot \vec{E} \quad \text{(potential energy of a dipole).} \qquad (22\text{-}38)$$

Equations 22-37 and 22-38 show us that the potential energy of the dipole is least ($U = -pE$) when $\theta = 0$ ($\vec{p}$ and $\vec{E}$ are in the same direction); the potential energy is greatest ($U = pE$) when $\theta = 180°$ ($\vec{p}$ and $\vec{E}$ are in opposite directions).

When a dipole rotates from an initial orientation θ_i to another orientation θ_f, the work W done on the dipole by the electric field is

$$W = -\Delta U = -(U_f - U_i), \qquad (22\text{-}39)$$

where U_f and U_i are calculated with Eq. 22-38. If the change in orientation is caused by an applied torque (commonly said to be due to an external agent), then the work W_a done on the dipole by the applied torque is the negative of the work done on the dipole by the field; that is,

$$W_a = -W = (U_f - U_i). \qquad (22\text{-}40)$$

Microwave Cooking

Food can be warmed and cooked in a microwave oven if the food contains water because water molecules are electric dipoles. When you turn on the oven, the microwave source sets up a rapidly oscillating electric field $\vec{E}$ within the oven and thus also within the food. From Eq. 22-34, we see that any electric field $\vec{E}$ produces a torque on an electric dipole moment $\vec{p}$ to align $\vec{p}$ with $\vec{E}$. Because the oven's $\vec{E}$ oscillates, the water molecules continuously flip-flop in a frustrated attempt to align with $\vec{E}$.

Energy is transferred from the electric field to the thermal energy of the water (and thus of the food) where three water molecules happened to have bonded together to form a group. The flip-flop breaks some of the bonds. When the molecules reform the bonds, energy is transferred to the random motion of the group and then to the surrounding molecules. Soon, the thermal energy of the water is enough to cook the food.

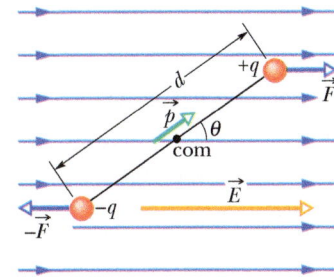

(a)

The dipole is being torqued into alignment.

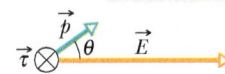

(b)

Figure 22-21 (a) An electric dipole in a uniform external electric field $\vec{E}$. Two centers of equal but opposite charge are separated by distance d. The line between them represents their rigid connection. (b) Field $\vec{E}$ causes a torque $\vec{\tau}$ on the dipole. The direction of $\vec{\tau}$ is into the page, as represented by the symbol $\otimes$.

✓ Checkpoint 4

The figure shows four orientations of an electric dipole in an external electric field. Rank the orientations according to (a) the magnitude of the torque on the dipole and (b) the potential energy of the dipole, greatest first.

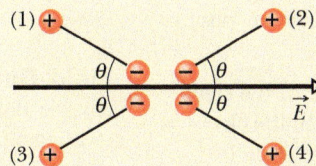

Sample Problem 22.05 Torque and energy of an electric dipole in an electric field

A neutral water molecule (H_2O) in its vapor state has an electric dipole moment of magnitude 6.2×10^{-30} C·m.

(a) How far apart are the molecule's centers of positive and negative charge?

KEY IDEA

A molecule's dipole moment depends on the magnitude q of the molecule's positive or negative charge and the charge separation d.

Calculations: There are 10 electrons and 10 protons in a neutral water molecule; so the magnitude of its dipole moment is

$$p = qd = (10e)(d),$$

in which d is the separation we are seeking and e is the elementary charge. Thus,

$$d = \frac{p}{10e} = \frac{6.2 \times 10^{-30} \text{ C·m}}{(10)(1.60 \times 10^{-19} \text{ C})}$$
$$= 3.9 \times 10^{-12} \text{ m} = 3.9 \text{ pm}. \qquad \text{(Answer)}$$

This distance is not only small, but it is also actually smaller than the radius of a hydrogen atom.

(b) If the molecule is placed in an electric field of 1.5×10^4 N/C, what maximum torque can the field exert on it? (Such a field can easily be set up in the laboratory.)

KEY IDEA

The torque on a dipole is maximum when the angle θ between $\vec{p}$ and $\vec{E}$ is 90°.

Calculation: Substituting $\theta = 90°$ in Eq. 22-33 yields

$$\tau = pE \sin \theta$$
$$= (6.2 \times 10^{-30} \text{ C·m})(1.5 \times 10^4 \text{ N/C})(\sin 90°)$$
$$= 9.3 \times 10^{-26} \text{ N·m}. \qquad \text{(Answer)}$$

(c) How much work must an *external agent* do to rotate this molecule by 180° in this field, starting from its fully aligned position, for which $\theta = 0$?

KEY IDEA

The work done by an external agent (by means of a torque applied to the molecule) is equal to the change in the molecule's potential energy due to the change in orientation.

Calculation: From Eq. 22-40, we find

$$W_a = U_{180°} - U_0$$
$$= (-pE \cos 180°) - (-pE \cos 0)$$
$$= 2pE = (2)(6.2 \times 10^{-30} \text{ C·m})(1.5 \times 10^4 \text{ N/C})$$
$$= 1.9 \times 10^{-25} \text{ J}. \qquad \text{(Answer)}$$

PLUS Additional examples, video, and practice available at *WileyPLUS*

Review & Summary

Electric Field To explain the electrostatic force between two charges, we assume that each charge sets up an electric field in the space around it. The force acting on each charge is then due to the electric field set up at its location by the other charge.

Definition of Electric Field The *electric field* $\vec{E}$ at any point is defined in terms of the electrostatic force $\vec{F}$ that would be exerted on a positive test charge q_0 placed there:

$$\vec{E} = \frac{\vec{F}}{q_0}. \qquad (22\text{-}1)$$

Electric Field Lines *Electric field lines* provide a means for visualizing the direction and magnitude of electric fields. The electric field vector at any point is tangent to a field line through that point. The density of field lines in any region is proportional to the magnitude of the electric field in that region. Field lines originate on positive charges and terminate on negative charges.

Field Due to a Point Charge The magnitude of the electric field $\vec{E}$ set up by a point charge q at a distance r from the charge is

$$E = \frac{1}{4\pi\varepsilon_0} \frac{|q|}{r^2}. \qquad (22\text{-}3)$$

The direction of $\vec{E}$ is away from the point charge if the charge is positive and toward it if the charge is negative.

Field Due to an Electric Dipole An *electric dipole* consists of two particles with charges of equal magnitude q but opposite sign, separated by a small distance d. Their **electric dipole moment** $\vec{p}$ has magnitude qd and points from the negative charge to the positive charge. The magnitude of the electric field set up by the dipole at a distant point on the dipole axis (which runs through both charges) is

$$E = \frac{1}{2\pi\varepsilon_0} \frac{p}{z^3}, \qquad (22\text{-}9)$$

where z is the distance between the point and the center of the dipole.

Field Due to a Continuous Charge Distribution The electric field due to a *continuous charge distribution* is found by treating charge elements as point charges and then summing, via integration, the electric field vectors produced by all the charge elements to find the net vector.

Field Due to a Charged Disk   The electric field magnitude at a point on the central axis through a uniformly charged disk is given by

$$E = \frac{\sigma}{2\varepsilon_0}\left(1 - \frac{z}{\sqrt{z^2 + R^2}}\right), \tag{22-26}$$

where z is the distance along the axis from the center of the disk, R is the radius of the disk, and σ is the surface charge density.

Force on a Point Charge in an Electric Field   When a point charge q is placed in an external electric field $\vec{E}$, the electrostatic force $\vec{F}$ that acts on the point charge is

$$\vec{F} = q\vec{E}. \tag{22-28}$$

Force $\vec{F}$ has the same direction as $\vec{E}$ if q is positive and the opposite direction if q is negative.

Dipole in an Electric Field   When an electric dipole of dipole moment $\vec{p}$ is placed in an electric field $\vec{E}$, the field exerts a torque $\vec{\tau}$ on the dipole:

$$\vec{\tau} = \vec{p} \times \vec{E}. \tag{22-34}$$

The dipole has a potential energy U associated with its orientation in the field:

$$U = -\vec{p} \cdot \vec{E}. \tag{22-38}$$

This potential energy is defined to be zero when $\vec{p}$ is perpendicular to $\vec{E}$; it is least ($U = -pE$) when $\vec{p}$ is aligned with $\vec{E}$ and greatest ($U = pE$) when $\vec{p}$ is directed opposite $\vec{E}$.

Questions

1   Figure 22-22 shows three arrangements of electric field lines. In each arrangement, a proton is released from rest at point A and is then accelerated through point B by the electric field. Points A and B have equal separations in the three arrangements. Rank the arrangements according to the linear momentum of the proton at point B, greatest first.

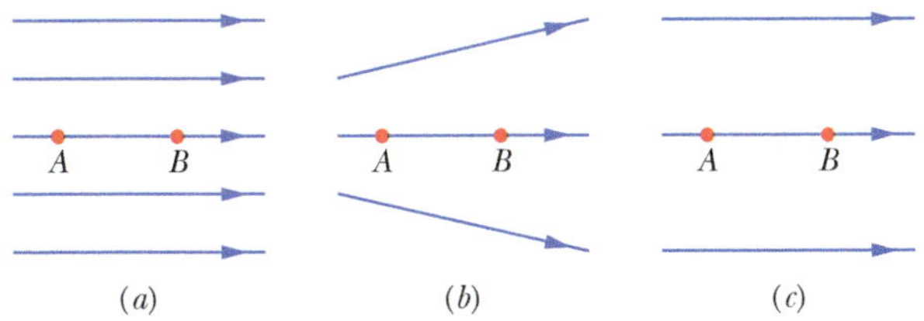

Figure 22-22   Question 1.

2   Figure 22-23 shows two square arrays of charged particles. The squares, which are centered on point P, are misaligned. The particles are separated by either d or $d/2$ along the perimeters of the squares. What are the magnitude and direction of the net electric field at P?

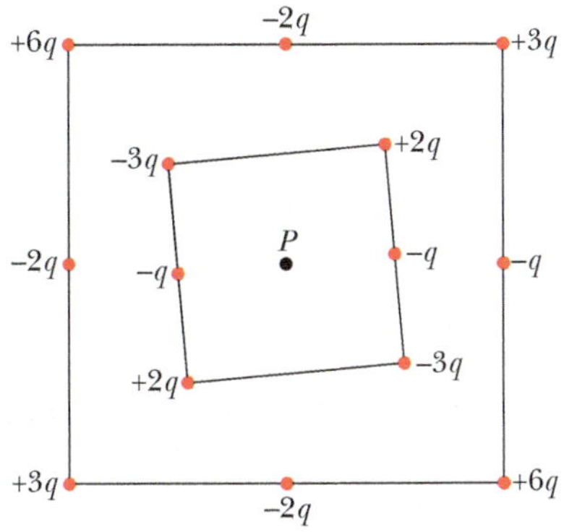

Figure 22-23   Question 2.

3   In Fig. 22-24, two particles of charge $-q$ are arranged symmetrically about the y axis; each produces an electric field at point P on that axis. (a) Are the magnitudes of the fields at P equal? (b) Is each electric field directed toward or away from the charge producing it? (c) Is the magnitude of the net electric field at P equal to the sum of the magnitudes E of the two field vectors (is it equal to $2E$)? (d) Do the x components of those two field vectors add or cancel? (e) Do their y components add or cancel? (f) Is the direction of the net field at P that of the canceling components or the adding components? (g) What is the direction of the net field?

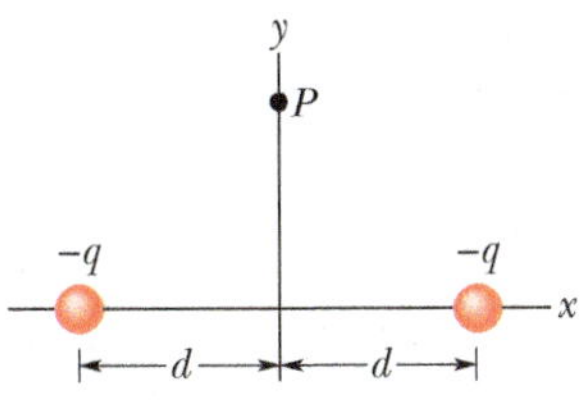

Figure 22-24   Question 3.

4   Figure 22-25 shows four situations in which four charged particles are evenly spaced to the left and right of a central point. The charge values are indicated. Rank the situations according to the magnitude of the net electric field at the central point, greatest first.

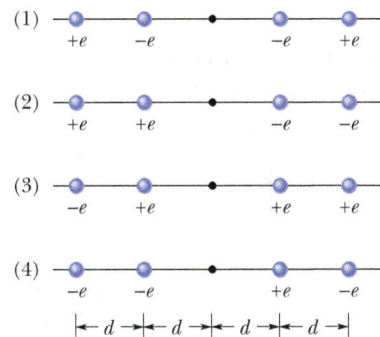

Figure 22-25   Question 4.

5   Figure 22-26 shows two charged particles fixed in place on an axis. (a) Where on the axis (other than at an infinite distance) is there a point at which their net electric field is zero: between the charges, to their left, or to their right? (b) Is there a point of zero net electric field anywhere *off* the axis (other than at an infinite distance)?

Figure 22-26   Question 5.

6   In Fig. 22-27, two identical circular nonconducting rings are centered on the same line with their planes perpendicular to the line. Each ring has charge that is uniformly distributed along its circumference. The rings each produce electric fields at points along the line. For three situations, the charges on rings A and B are, respectively, (1) q_0 and q_0, (2) $-q_0$ and $-q_0$, and (3) $-q_0$ and q_0. Rank the situations according to the magnitude of the net electric field at (a) point P_1 midway between the rings, (b) point P_2 at the center of ring B, and (c) point P_3 to the right of ring B, greatest first.

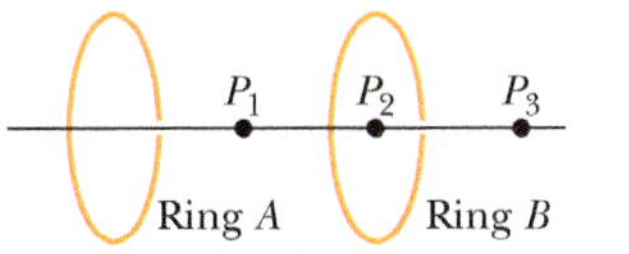

Figure 22-27   Question 6.

7   The potential energies associated with four orientations of an electric dipole in an electric field are (1) $-5U_0$, (2) $-7U_0$, (3) $3U_0$, and (4) $5U_0$, where U_0 is positive. Rank the orientations according to (a) the angle between the electric dipole moment $\vec{p}$ and the electric field $\vec{E}$ and (b) the magnitude of the torque on the electric dipole, greatest first.

8   (a) In Checkpoint 4, if the dipole rotates from orientation 1 to orientation 2, is the work done on the dipole by the field positive, negative, or zero? (b) If, instead, the dipole rotates from orientation 1 to orientation 4, is the work done by the field more than, less than, or the same as in (a)?

9 Figure 22-28 shows two disks and a flat ring, each with the same uniform charge Q. Rank the objects according to the magnitude of the electric field they create at points P (which are at the same vertical heights), greatest first.

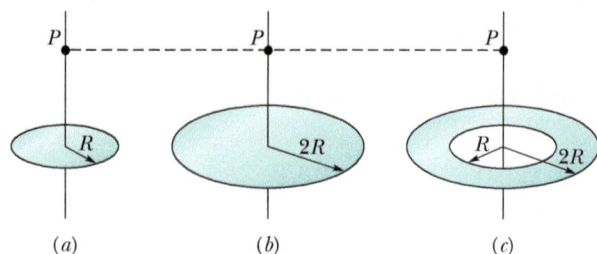

Figure 22-28 Question 9.

10 In Fig. 22-29, an electron e travels through a small hole in plate A and then toward plate B. A uniform electric field in the region between the plates then slows the electron without deflecting it. (a) What is the direction of the field? (b) Four other particles similarly travel through small holes in either plate A or plate B and then into the region between the plates. Three have charges $+q_1$, $+q_2$, and $-q_3$. The fourth (labeled n) is a neutron, which is electrically neutral. Does the speed of each of those four other particles increase, decrease, or remain the same in the region between the plates?

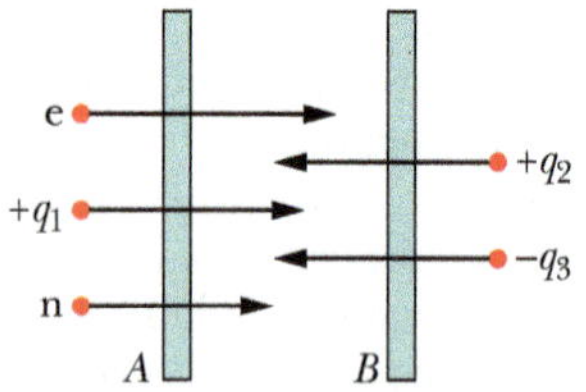

Figure 22-29 Question 10.

11 In Fig. 22-30a, a circular plastic rod with uniform charge $+Q$ produces an electric field of magnitude E at the center of curvature (at the origin). In Figs. 22-30b, c, and d, more circular rods, each with identical uniform charges $+Q$, are added until the circle is complete. A fifth arrangement (which would be labeled e) is like that in d except the rod in the fourth quadrant has charge $-Q$. Rank the five arrangements according to the magnitude of the electric field at the center of curvature, greatest first.

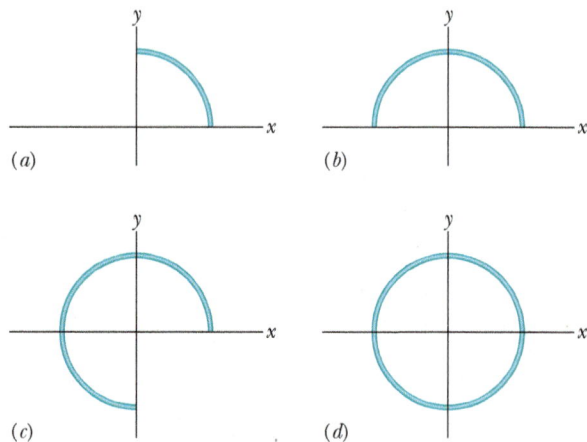

Figure 22-30 Question 11.

12 When three electric dipoles are near each other, they each experience the electric field of the other two, and the three-dipole system has a certain potential energy. Figure 22-31 shows two arrangements in which three electric dipoles are side by side. Each dipole has the same magnitude of electric dipole moment, and the spacings between adjacent dipoles are identical. In which arrangement is the potential energy of the three-dipole system greater?

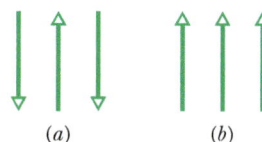

Figure 22-31 Question 12.

13 Figure 22-32 shows three rods, each with the same charge Q spread uniformly along its length. Rods a (of length L) and b (of length $L/2$) are straight, and points P are aligned with their midpoints. Rod c (of length $L/2$) forms a complete circle about point P. Rank the rods according to the magnitude of the electric field they create at points P, greatest first.

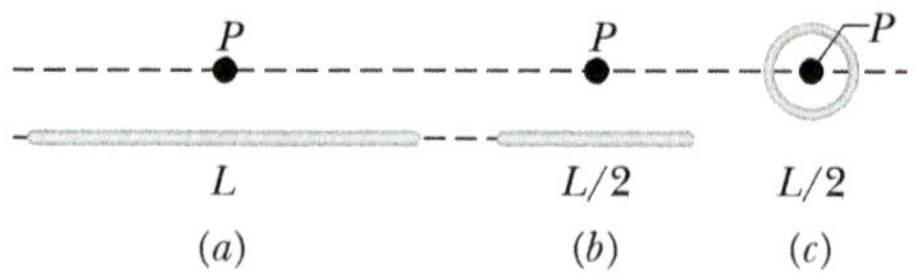

Figure 22-32 Question 13.

14 Figure 22-33 shows five protons that are launched in a uniform electric field $\vec{E}$; the magnitude and direction of the launch velocities are indicated. Rank the protons according to the magnitude of their accelerations due to the field, greatest first.

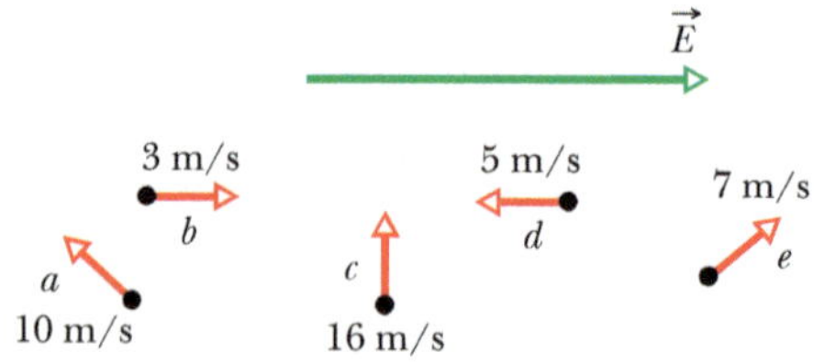

Figure 22-33 Question 14.

Problems

GO Tutoring problem available (at instructor's discretion) in *WileyPLUS* and WebAssign
SSM Worked-out solution available in Student Solutions Manual
• – ••• Number of dots indicates level of problem difficulty
Additional information available in *The Flying Circus of Physics* and at flyingcircusofphysics.com
WWW Worked-out solution is at
ILW Interactive solution is at
http://www.wiley.com/college/halliday

Module 22-1 The Electric Field

•**1** Sketch qualitatively the electric field lines both between and outside two concentric conducting spherical shells when a uniform positive charge q_1 is on the inner shell and a uniform negative charge $-q_2$ is on the outer. Consider the cases $q_1 > q_2$, $q_1 = q_2$, and $q_1 < q_2$.

·2 In Fig. 22-34 the electric field lines on the left have twice the separation of those on the right. (a) If the magnitude of the field at A is 40 N/C, what is the magnitude of the force on a proton at A? (b) What is the magnitude of the field at B?

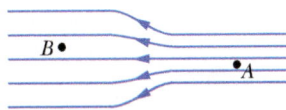

Figure 22-34 Problem 2.

Module 22-2   The Electric Field Due to a Charged Particle

·3 SSM The nucleus of a plutonium-239 atom contains 94 protons. Assume that the nucleus is a sphere with radius 6.64 fm and with the charge of the protons uniformly spread through the sphere. At the surface of the nucleus, what are the (a) magnitude and (b) direction (radially inward or outward) of the electric field produced by the protons?

·4 Two charged particles are attached to an x axis: Particle 1 of charge -2.00×10^{-7} C is at position $x = 6.00$ cm and particle 2 of charge $+2.00 \times 10^{-7}$ C is at position $x = 21.0$ cm. Midway between the particles, what is their net electric field in unit-vector notation?

·5 SSM A charged particle produces an electric field with a magnitude of 2.0 N/C at a point that is 50 cm away from the particle. What is the magnitude of the particle's charge?

·6 What is the magnitude of a point charge that would create an electric field of 1.00 N/C at points 1.00 m away?

··7 SSM ILW WWW In Fig. 22-35, the four particles form a square of edge length $a = 5.00$ cm and have charges $q_1 = +10.0$ nC, $q_2 = -20.0$ nC, $q_3 = +20.0$ nC, and $q_4 = -10.0$ nC. In unit-vector notation, what net electric field do the particles produce at the square's center?

Figure 22-35 Problem 7.

··8 GO In Fig. 22-36, the four particles are fixed in place and have charges $q_1 = q_2 = +5e$, $q_3 = +3e$, and $q_4 = -12e$. Distance $d = 5.0$ μm. What is the magnitude of the net electric field at point P due to the particles?

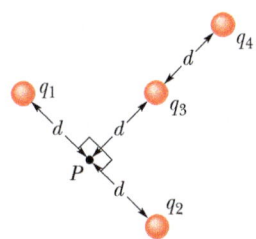

Figure 22-36 Problem 8.

··9 GO Figure 22-37 shows two charged particles on an x axis: $-q = -3.20 \times 10^{-19}$ C at $x = -3.00$ m and $q = 3.20 \times 10^{-19}$ C at $x = +3.00$ m. What are the (a) magnitude and (b) direction (relative to the positive direction of the x axis) of the net electric field produced at point P at $y = 4.00$ m?

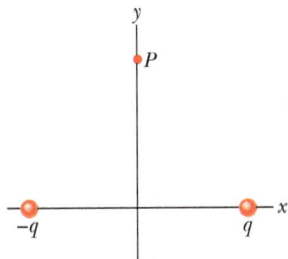

Figure 22-37 Problem 9.

··10 GO Figure 22-38a shows two charged particles fixed in place on an x axis with separation L. The ratio q_1/q_2 of their charge magnitudes is 4.00. Figure 22-38b shows the x component $E_{net,x}$ of their net electric field along the x axis just to the right of particle 2. The x axis scale is set by $x_s = 30.0$ cm. (a) At what value of $x > 0$ is $E_{net,x}$ maximum? (b) If particle 2 has charge $-q_2 = -3e$, what is the value of that maximum?

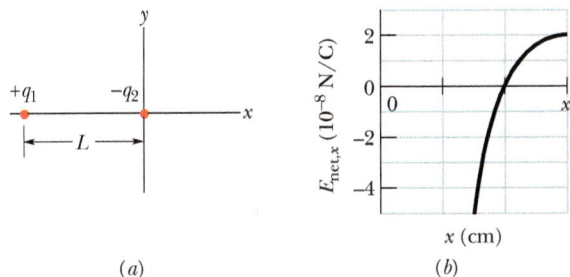

Figure 22-38 Problem 10.

··11 SSM Two charged particles are fixed to an x axis: Particle 1 of charge $q_1 = 2.1 \times 10^{-8}$ C is at position $x = 20$ cm and particle 2 of charge $q_2 = -4.00q_1$ is at position $x = 70$ cm. At what coordinate on the axis (other than at infinity) is the net electric field produced by the two particles equal to zero?

··12 GO Figure 22-39 shows an uneven arrangement of electrons (e) and protons (p) on a circular arc of radius $r = 2.00$ cm, with angles $\theta_1 = 30.0°$, $\theta_2 = 50.0°$, $\theta_3 = 30.0°$, and $\theta_4 = 20.0°$. What are the (a) magnitude and (b) direction (relative to the positive direction of the x axis) of the net electric field produced at the center of the arc?

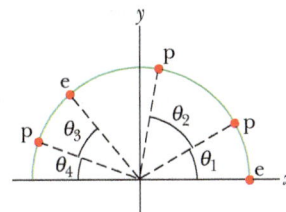

Figure 22-39 Problem 12.

··13 GO Figure 22-40 shows a proton (p) on the central axis through a disk with a uniform charge density due to excess electrons. The disk is seen from an edge-on view. Three of those electrons are shown: electron e_c at the disk center and electrons e_s at opposite sides of the disk, at radius R from the center. The proton is initially at distance $z = R = 2.00$ cm from the disk. At that location, what are the magnitudes of (a) the electric field $\vec{E}_c$ due to electron e_c and (b) the net electric field $\vec{E}_{s,net}$ due to electrons e_s? The proton is then moved to $z = R/10.0$. What then are the magnitudes of (c) $\vec{E}_c$ and (d) $\vec{E}_{s,net}$ at the proton's location? (e) From (a) and (c) we see that as the proton gets nearer to the disk, the magnitude of $\vec{E}_c$ increases, as expected. Why does the magnitude of $\vec{E}_{s,net}$ from the two side electrons decrease, as we see from (b) and (d)?

Figure 22-40 Problem 13.

··14 In Fig. 22-41, particle 1 of charge $q_1 = -5.00q$ and particle 2 of charge $q_2 = +2.00q$ are fixed to an x axis. (a) As a multiple of distance L, at what coordinate on the axis is the net electric field of the particles zero? (b) Sketch the net electric field lines between and around the particles.

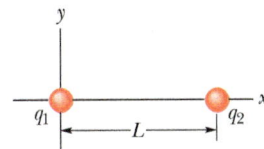

Figure 22-41 Problem 14.

••15 In Fig. 22-42, the three particles are fixed in place and have charges $q_1 = q_2 = +e$ and $q_3 = +2e$. Distance $a = 6.00\ \mu m$. What are the (a) magnitude and (b) direction of the net electric field at point P due to the particles?

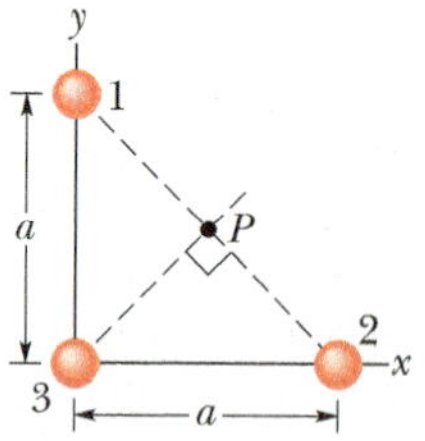

Figure 22-42
Problem 15.

•••16 Figure 22-43 shows a plastic ring of radius $R = 50.0$ cm. Two small charged beads are on the ring: Bead 1 of charge $+2.00\ \mu C$ is fixed in place at the left side; bead 2 of charge $+6.00\ \mu C$ can be moved along the ring. The two beads produce a net electric field of magnitude E at the center of the ring. At what (a) positive and (b) negative value of angle θ should bead 2 be positioned such that $E = 2.00 \times 10^5$ N/C?

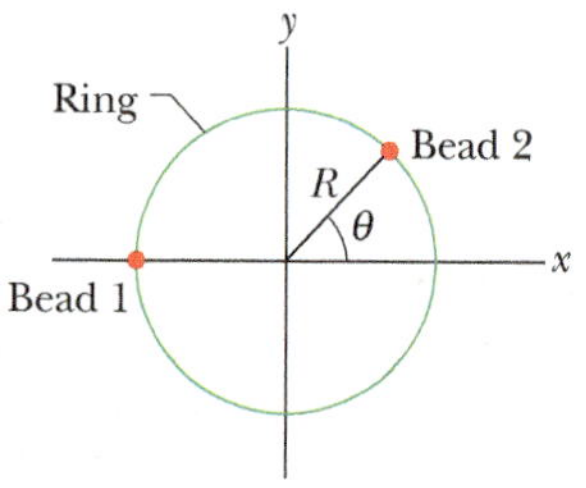

Figure 22-43 Problem 16.

•••17 Two charged beads are on the plastic ring in Fig. 22-44a. Bead 2, which is not shown, is fixed in place on the ring, which has radius $R = 60.0$ cm. Bead 1, which is not fixed in place, is initially on the x axis at angle $\theta = 0°$. It is then moved to the opposite side, at angle $\theta = 180°$, through the first and second quadrants of the xy coordinate system. Figure 22-44b gives the x component of the net electric field produced at the origin by the two beads as a function of θ, and Fig. 22-44c gives the y component of that net electric field. The vertical axis scales are set by $E_{xs} = 5.0 \times 10^4$ N/C and $E_{ys} = -9.0 \times 10^4$ N/C. (a) At what angle θ is bead 2 located? What are the charges of (b) bead 1 and (c) bead 2?

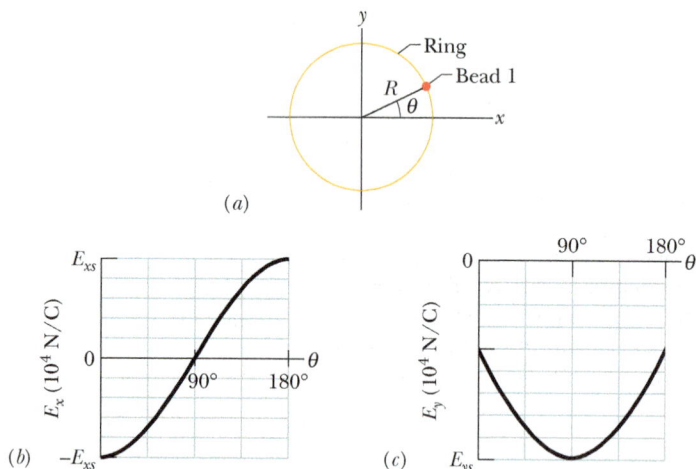

Figure 22-44 Problem 17.

Module 22-3 The Electric Field Due to a Dipole

••18 The electric field of an electric dipole along the dipole axis is approximated by Eqs. 22-8 and 22-9. If a binomial expansion is made of Eq. 22-7, what is the next term in the expression for the dipole's electric field along the dipole axis? That is, what is E_{next} in the expression

$$E = \frac{1}{2\pi\varepsilon_0}\frac{qd}{z^3} + E_{next}?$$

••19 Figure 22-45 shows an electric dipole. What are the (a) magnitude and (b) direction (relative to the positive direction of the x axis) of the dipole's electric field at point P, located at distance $r \gg d$?

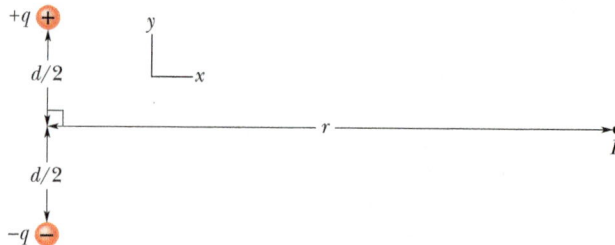

Figure 22-45 Problem 19.

••20 Equations 22-8 and 22-9 are approximations of the magnitude of the electric field of an electric dipole, at points along the dipole axis. Consider a point P on that axis at distance $z = 5.00d$ from the dipole center (d is the separation distance between the particles of the dipole). Let E_{appr} be the magnitude of the field at point P as approximated by Eqs. 22-8 and 22-9. Let E_{act} be the actual magnitude. What is the ratio E_{appr}/E_{act}?

•••21 SSM *Electric quadrupole.* Figure 22-46 shows a generic electric quadrupole. It consists of two dipoles with dipole moments that are equal in magnitude but opposite in direction. Show that the value of E on the axis of the quadrupole for a point P a distance z from its center (assume $z \gg d$) is given by

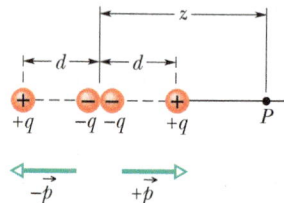

Figure 22-46 Problem 21.

$$E = \frac{3Q}{4\pi\varepsilon_0 z^4},$$

in which $Q\ (= 2qd^2)$ is known as the *quadrupole moment* of the charge distribution.

Module 22-4 The Electric Field Due to a Line of Charge

•22 *Density, density, density.* (a) A charge $-300e$ is uniformly distributed along a circular arc of radius 4.00 cm, which subtends an angle of 40°. What is the linear charge density along the arc? (b) A charge $-300e$ is uniformly distributed over one face of a circular disk of radius 2.00 cm. What is the surface charge density over that face? (c) A charge $-300e$ is uniformly distributed over the surface of a sphere of radius 2.00 cm. What is the surface charge density over that surface? (d) A charge $-300e$ is uniformly spread through the volume of a sphere of radius 2.00 cm. What is the volume charge density in that sphere?

•23 Figure 22-47 shows two parallel nonconducting rings with their central axes along a common line. Ring 1 has uniform charge q_1 and radius R; ring 2 has uniform charge q_2 and the same radius R. The rings are separated by distance $d = 3.00R$. The net electric field at point P on the common line, at distance R from ring 1, is zero. What is the ratio q_1/q_2?

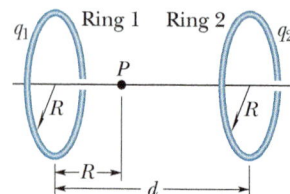

Figure 22-47 Problem 23.

••24 A thin nonconducting rod with a uniform distribution of positive charge Q is bent into a complete circle of radius R

(Fig. 22-48). The central perpendicular axis through the ring is a z axis, with the origin at the center of the ring. What is the magnitude of the electric field due to the rod at (a) $z = 0$ and (b) $z = \infty$? (c) In terms of R, at what positive value of z is that magnitude maximum? (d) If $R = 2.00$ cm and $Q = 4.00\ \mu C$, what is the maximum magnitude?

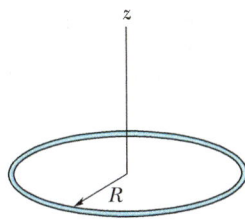

Figure 22-48 Problem 24.

••25 Figure 22-49 shows three circular arcs centered on the origin of a coordinate system. On each arc, the uniformly distributed charge is given in terms of $Q = 2.00\ \mu C$. The radii are given in terms of $R = 10.0$ cm. What are the (a) magnitude and (b) direction (relative to the positive x direction) of the net electric field at the origin due to the arcs?

Figure 22-49 Problem 25.

••26 GO ILW In Fig. 22-50, a thin glass rod forms a semicircle of radius $r = 5.00$ cm. Charge is uniformly distributed along the rod, with $+q = 4.50$ pC in the upper half and $-q = -4.50$ pC in the lower half. What are the (a) magnitude and (b) direction (relative to the positive direction of the x axis) of the electric field $\vec{E}$ at P, the center of the semicircle?

Figure 22-50
Problem 26.

••27 GO In Fig. 22-51, two curved plastic rods, one of charge $+q$ and the other of charge $-q$, form a circle of radius $R = 8.50$ cm in an xy plane. The x axis passes through both of the connecting points, and the charge is distributed uniformly on both rods. If $q = 15.0$ pC, what are the (a) magnitude and (b) direction (relative to the positive direction of the x axis) of the electric field $\vec{E}$ produced at P, the center of the circle?

Figure 22-51
Problem 27.

••28 Charge is uniformly distributed around a ring of radius $R = 2.40$ cm, and the resulting electric field magnitude E is measured along the ring's central axis (perpendicular to the plane of the ring). At what distance from the ring's center is E maximum?

••29 GO Figure 22-52a shows a nonconducting rod with a uniformly distributed charge $+Q$. The rod forms a half-circle with radius R and produces an electric field of magnitude E_{arc} at its center of curvature P. If the arc is collapsed to a point at distance R from P (Fig. 22-52b), by what factor is the magnitude of the electric field at P multiplied?

Figure 22-52 Problem 29.

••30 GO Figure 22-53 shows two concentric rings, of radii R and $R' = 3.00R$, that lie on the same plane. Point P lies on the central z axis, at distance $D = 2.00R$ from the center of the rings. The smaller ring has uniformly distributed charge $+Q$. In terms of Q, what is the uniformly distributed charge on the larger ring if the net electric field at P is zero?

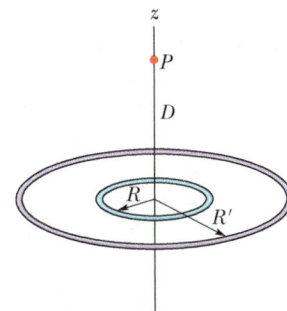

Figure 22-53 Problem 30.

••31 SSM ILW WWW In Fig. 22-54, a nonconducting rod of length $L = 8.15$ cm has a charge $-q = -4.23$ fC uniformly distributed along its length. (a) What is the linear charge density of the rod? What are the (b) magnitude and (c) direction (relative to the positive direction of the x axis) of the electric field produced at point P, at distance $a = 12.0$ cm from the rod? What is the electric field magnitude produced at distance $a = 50$ m by (d) the rod and (e) a particle of charge $-q = -4.23$ fC that we use to replace the rod? (At that distance, the rod "looks" like a particle.)

Figure 22-54 Problem 31.

•••32 GO In Fig. 22-55, positive charge $q = 7.81$ pC is spread uniformly along a thin nonconducting rod of length $L = 14.5$ cm. What are the (a) magnitude and (b) direction (relative to the positive direction of the x axis) of the electric field produced at point P, at distance $R = 6.00$ cm from the rod along its perpendicular bisector?

Figure 22-55 Problem 32.

•••33 GO In Fig. 22-56, a "semi-infinite" nonconducting rod (that is, infinite in one direction only) has uniform linear charge density λ. Show that the electric field $\vec{E}_p$ at point P makes an angle of 45° with the rod and that this result is independent of the distance R. (*Hint:* Separately find the component of $\vec{E}_p$ parallel to the rod and the component perpendicular to the rod.)

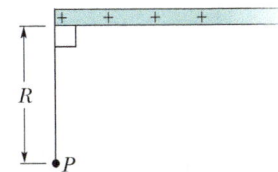

Figure 22-56 Problem 33.

Module 22-5 The Electric Field Due to a Charged Disk

•34 A disk of radius 2.5 cm has a surface charge density of $5.3\ \mu C/m^2$ on its upper face. What is the magnitude of the electric field produced by the disk at a point on its central axis at distance $z = 12$ cm from the disk?

•35 SSM WWW At what distance along the central perpendicular axis of a uniformly charged plastic disk of radius 0.600 m is the magnitude of the electric field equal to one-half the magnitude of the field at the center of the surface of the disk?

••36 A circular plastic disk with radius $R = 2.00$ cm has a uniformly distributed charge $Q = +(2.00 \times 10^6)e$ on one face. A circular ring of width 30 μm is centered on that face, with the center of that width at radius $r = 0.50$ cm. In coulombs, what charge is contained within the width of the ring?

••37 Suppose you design an apparatus in which a uniformly charged disk of radius R is to produce an electric field. The field magnitude is most important along the central perpendicular axis of the disk, at a point P at distance $2.00R$ from the disk (Fig. 22-57a). Cost analysis suggests that you switch to a ring of the same outer radius R but with inner radius $R/2.00$ (Fig. 22-57b). Assume that the ring will have the same surface charge density as the original disk. If you switch to the ring, by what percentage will you decrease the electric field magnitude at P?

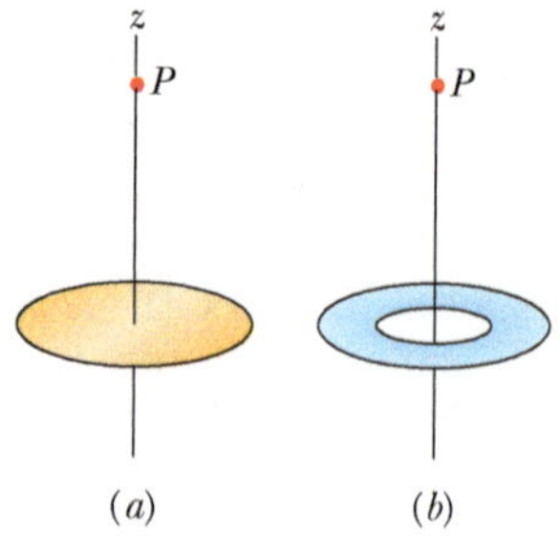

Figure 22-57 Problem 37.

••38 Figure 22-58a shows a circular disk that is uniformly charged. The central z axis is perpendicular to the disk face, with the origin at the disk. Figure 22-58b gives the magnitude of the electric field along that axis in terms of the maximum magnitude E_m at the disk surface. The z axis scale is set by $z_s = 8.0$ cm. What is the radius of the disk?

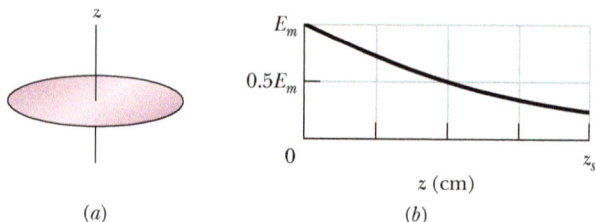

Figure 22-58 Problem 38.

Module 22-6 A Point Charge in an Electric Field

•39 In Millikan's experiment, an oil drop of radius 1.64 μm and density 0.851 g/cm³ is suspended in chamber C (Fig. 22-16) when a downward electric field of 1.92×10^5 N/C is applied. Find the charge on the drop, in terms of e.

•40 [GO] An electron with a speed of 5.00×10^8 cm/s enters an electric field of magnitude 1.00×10^3 N/C, traveling along a field line in the direction that retards its motion. (a) How far will the electron travel in the field before stopping momentarily, and (b) how much time will have elapsed? (c) If the region containing the electric field is 8.00 mm long (too short for the electron to stop within it), what fraction of the electron's initial kinetic energy will be lost in that region?

•41 SSM A charged cloud system produces an electric field in the air near Earth's surface. A particle of charge -2.0×10^{-9} C is acted on by a downward electrostatic force of 3.0×10^{-6} N when placed in this field. (a) What is the magnitude of the electric field? What are the (b) magnitude and (c) direction of the electrostatic force $\vec{F}_{el}$ on the proton placed in this field? (d) What is the magnitude of the gravitational force $\vec{F}_g$ on the proton? (e) What is the ratio F_{el}/F_g in this case?

•42 Humid air breaks down (its molecules become ionized) in an electric field of 3.0×10^6 N/C. In that field, what is the magnitude of the electrostatic force on (a) an electron and (b) an ion with a single electron missing?

•43 SSM An electron is released from rest in a uniform electric field of magnitude 2.00×10^4 N/C. Calculate the acceleration of the electron. (Ignore gravitation.)

•44 An alpha particle (the nucleus of a helium atom) has a mass of 6.64×10^{-27} kg and a charge of $+2e$. What are the (a) magnitude and (b) direction of the electric field that will balance the gravitational force on the particle?

•45 ILW An electron on the axis of an electric dipole is 25 nm from the center of the dipole. What is the magnitude of the electrostatic force on the electron if the dipole moment is 3.6×10^{-29} C·m? Assume that 25 nm is much larger than the separation of the charged particles that form the dipole.

•46 An electron is accelerated eastward at 1.80×10^9 m/s² by an electric field. Determine the field (a) magnitude and (b) direction.

•47 SSM Beams of high-speed protons can be produced in "guns" using electric fields to accelerate the protons. (a) What acceleration would a proton experience if the gun's electric field were 2.00×10^4 N/C? (b) What speed would the proton attain if the field accelerated the proton through a distance of 1.00 cm?

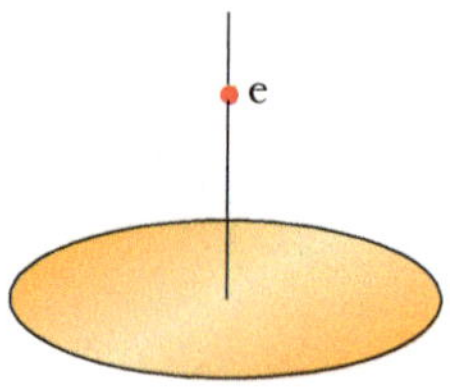

••48 In Fig. 22-59, an electron (e) is to be released from rest on the central axis of a uniformly charged disk of radius R. The surface charge density on the disk is $+4.00$ μC/m². What is the magnitude of the electron's initial acceleration if it is released at a distance (a) R, (b) $R/100$, and (c) $R/1000$ from the center of the disk? (d) Why does the acceleration magnitude increase only slightly as the release point is moved closer to the disk?

Figure 22-59 Problem 48.

••49 A 10.0 g block with a charge of $+8.00 \times 10^{-5}$ C is placed in an electric field $\vec{E} = (3000\hat{i} - 600\hat{j})$ N/C. What are the (a) magnitude and (b) direction (relative to the positive direction of the x axis) of the electrostatic force on the block? If the block is released from rest at the origin at time $t = 0$, what are its (c) x and (d) y coordinates at $t = 3.00$ s?

••50 At some instant the velocity components of an electron moving between two charged parallel plates are $v_x = 1.5 \times 10^5$ m/s and $v_y = 3.0 \times 10^3$ m/s. Suppose the electric field between the plates is uniform and given by $\vec{E} = (120$ N/C$)\hat{j}$. In unit-vector notation, what are (a) the electron's acceleration in that field and (b) the electron's velocity when its x coordinate has changed by 2.0 cm?

••51 Assume that a honeybee is a sphere of diameter 1.000 cm with a charge of $+45.0$ pC uniformly spread over its surface. Assume also that a spherical pollen grain of diameter 40.0 μm is electrically held on the surface of the bee because the bee's charge induces a charge of -1.00 pC on the near side of the grain and a charge of $+1.00$ pC on the far side. (a) What is the magnitude of the net electrostatic force on the grain due to the bee? Next, assume that the bee brings the grain to a distance of 1.000 mm from the tip of a flower's stigma and that the tip is a particle of charge -45.0 pC. (b) What is the magnitude of the net electrostatic force on the grain due to the stigma? (c) Does the grain remain on the bee or does it move to the stigma?

••52 An electron enters a region of uniform electric field with an initial velocity of 40 km/s in the same direction as the electric field, which has magnitude $E = 50$ N/C. (a) What is the speed of the electron 1.5 ns after entering this region? (b) How far does the electron travel during the 1.5 ns interval?

••53 GO Two large parallel copper plates are 5.0 cm apart and have a uniform electric field between them as depicted in Fig. 22-60. An electron is released from the negative plate at the same time that a proton is released from the positive plate. Neglect the force of the particles on each other and find their distance from the positive plate when they pass each other. (Does it surprise you that you need not know the electric field to solve this problem?)

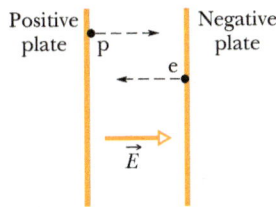

Figure 22-60 Problem 53.

••54 GO In Fig. 22-61, an electron is shot at an initial speed of $v_0 = 2.00 \times 10^6$ m/s, at angle $\theta_0 = 40.0°$ from an x axis. It moves through a uniform electric field $\vec{E} = (5.00 \text{ N/C})\hat{j}$. A screen for detecting electrons is positioned parallel to the y axis, at distance $x = 3.00$ m. In unit-vector notation, what is the velocity of the electron when it hits the screen?

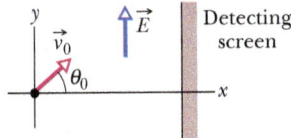

Figure 22-61 Problem 54.

••55 ILW A uniform electric field exists in a region between two oppositely charged plates. An electron is released from rest at the surface of the negatively charged plate and strikes the surface of the opposite plate, 2.0 cm away, in a time 1.5×10^{-8} s. (a) What is the speed of the electron as it strikes the second plate? (b) What is the magnitude of the electric field $\vec{E}$?

Module 22-7   A Dipole in an Electric Field

•56 An electric dipole consists of charges $+2e$ and $-2e$ separated by 0.78 nm. It is in an electric field of strength 3.4×10^6 N/C. Calculate the magnitude of the torque on the dipole when the dipole moment is (a) parallel to, (b) perpendicular to, and (c) antiparallel to the electric field.

•57 SSM An electric dipole consisting of charges of magnitude 1.50 nC separated by 6.20 μm is in an electric field of strength 1100 N/C. What are (a) the magnitude of the electric dipole moment and (b) the difference between the potential energies for dipole orientations parallel and antiparallel to $\vec{E}$?

••58 A certain electric dipole is placed in a uniform electric field $\vec{E}$ of magnitude 20 N/C. Figure 22-62 gives the potential energy U of the dipole versus the angle θ between $\vec{E}$ and the dipole moment $\vec{p}$. The vertical axis scale is set by $U_s = 100 \times 10^{-28}$ J. What is the magnitude of $\vec{p}$?

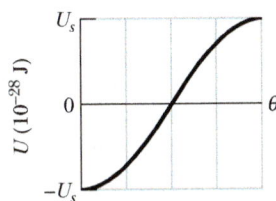

Figure 22-62 Problem 58.

••59 How much work is required to turn an electric dipole 180° in a uniform electric field of magnitude $E = 46.0$ N/C if the dipole moment has a magnitude of $p = 3.02 \times 10^{-25}$ C·m and the initial angle is 64°?

••60 A certain electric dipole is placed in a uniform electric field $\vec{E}$ of magnitude 40 N/C. Figure 22-63 gives the magnitude τ of the torque on the dipole versus the angle θ between field $\vec{E}$ and the dipole moment $\vec{p}$. The vertical axis scale is set by $\tau_s = 100 \times 10^{-28}$ N·m. What is the magnitude of $\vec{p}$?

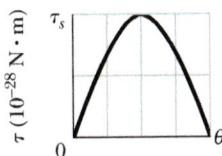

Figure 22-63 Problem 60.

••61 Find an expression for the oscillation frequency of an electric dipole of dipole moment $\vec{p}$ and rotational inertia I for small amplitudes of oscillation about its equilibrium position in a uniform electric field of magnitude E.

Additional Problems

62 (a) What is the magnitude of an electron's acceleration in a uniform electric field of magnitude 1.40×10^6 N/C? (b) How long would the electron take, starting from rest, to attain one-tenth the speed of light? (c) How far would it travel in that time?

63 A spherical water drop 1.20 μm in diameter is suspended in calm air due to a downward-directed atmospheric electric field of magnitude $E = 462$ N/C. (a) What is the magnitude of the gravitational force on the drop? (b) How many excess electrons does it have?

64 Three particles, each with positive charge Q, form an equilateral triangle, with each side of length d. What is the magnitude of the electric field produced by the particles at the midpoint of any side?

65 In Fig. 22-64a, a particle of charge $+Q$ produces an electric field of magnitude E_{part} at point P, at distance R from the particle. In Fig. 22-64b, that same amount of charge is spread uniformly along a circular arc that has radius R and subtends an angle θ. The charge on the arc produces an electric field of magnitude E_{arc} at its center of curvature P. For what value of θ does $E_{arc} = 0.500 E_{part}$? (*Hint:* You will probably resort to a graphical solution.)

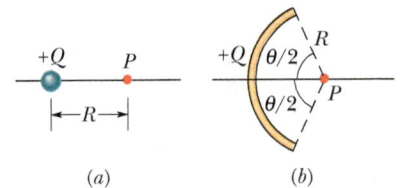

Figure 22-64 Problem 65.

66 A proton and an electron form two corners of an equilateral triangle of side length 2.0×10^{-6} m. What is the magnitude of the net electric field these two particles produce at the third corner?

67 A charge (uniform linear density = 9.0 nC/m) lies on a string that is stretched along an x axis from $x = 0$ to $x = 3.0$ m. Determine the magnitude of the electric field at $x = 4.0$ m on the x axis.

68 In Fig. 22-65, eight particles form a square in which distance $d = 2.0$ cm. The charges are $q_1 = +3e$, $q_2 = +e$, $q_3 = -5e$, $q_4 = -2e$, $q_5 = +3e$, $q_6 = +e$, $q_7 = -5e$, and $q_8 = +e$. In unit-vector notation, what is the net electric field at the square's center?

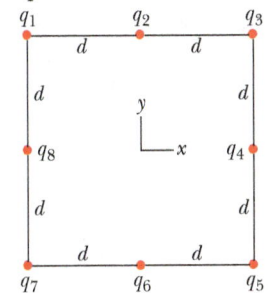

69 Two particles, each with a charge of magnitude 12 nC, are at two of the vertices of an equilateral triangle with edge length 2.0 m. What is the magnitude of the electric field at the third vertex if (a) both charges are positive and (b) one charge is positive and the other is negative?

Figure 22-65 Problem 68.

70 The following table gives the charge seen by Millikan at different times on a single drop in his experiment. From the data, calculate the elementary charge e.

6.563×10^{-19} C	13.13×10^{-19} C	19.71×10^{-19} C
8.204×10^{-19} C	16.48×10^{-19} C	22.89×10^{-19} C
11.50×10^{-19} C	18.08×10^{-19} C	26.13×10^{-19} C

71 A charge of 20 nC is uniformly distributed along a straight rod of length 4.0 m that is bent into a circular arc with a radius of 2.0 m. What is the magnitude of the electric field at the center of curvature of the arc?

72 An electron is constrained to the central axis of the ring of charge of radius R in Fig. 22-11, with $z \ll R$. Show that the electrostatic force on the electron can cause it to oscillate through the ring center with an angular frequency

$$\omega = \sqrt{\frac{eq}{4\pi\varepsilon_0 mR^3}},$$

where q is the ring's charge and m is the electron's mass.

73 SSM The electric field in an xy plane produced by a positively charged particle is $7.2(4.0\hat{i} + 3.0\hat{j})$ N/C at the point $(3.0, 3.0)$ cm and $100\hat{i}$ N/C at the point $(2.0, 0)$ cm. What are the (a) x and (b) y coordinates of the particle? (c) What is the charge of the particle?

74 (a) What total (excess) charge q must the disk in Fig. 22-15 have for the electric field on the surface of the disk at its center to have magnitude 3.0×10^6 N/C, the E value at which air breaks down electrically, producing sparks? Take the disk radius as 2.5 cm. (b) Suppose each surface atom has an effective cross-sectional area of 0.015 nm². How many atoms are needed to make up the disk surface? (c) The charge calculated in (a) results from some of the surface atoms having one excess electron. What fraction of these atoms must be so charged?

75 In Fig. 22-66, particle 1 (of charge $+1.00$ μC), particle 2 (of charge $+1.00$ μC), and particle 3 (of charge Q) form an equilateral triangle of edge length a. For what value of Q (both sign and magnitude) does the net electric field produced by the particles at the center of the triangle vanish?

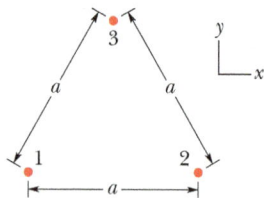

Figure 22-66 Problems 75 and 86.

76 In Fig. 22-67, an electric dipole swings from an initial orientation i ($\theta_i = 20.0°$) to a final orientation f ($\theta_f = 20.0°$) in a uniform external electric field $\vec{E}$. The electric dipole moment is 1.60×10^{-27} C·m; the field magnitude is 3.00×10^6 N/C. What is the change in the dipole's potential energy?

Figure 22-67 Problem 76.

77 A particle of charge $-q_1$ is at the origin of an x axis. (a) At what location on the axis should a particle of charge $-4q_1$ be placed so that the net electric field is zero at $x = 2.0$ mm on the axis? (b) If, instead, a particle of charge $+4q_1$ is placed at that location, what is the direction (relative to the positive direction of the x axis) of the net electric field at $x = 2.0$ mm?

78 Two particles, each of positive charge q, are fixed in place on a y axis, one at $y = d$ and the other at $y = -d$. (a) Write an expression that gives the magnitude E of the net electric field at points on the x axis given by $x = \alpha d$. (b) Graph E versus α for the range $0 < \alpha < 4$. From the graph, determine the values of α that give (c) the maximum value of E and (d) half the maximum value of E.

79 A clock face has negative point charges $-q$, $-2q$, $-3q$, . . . , $-12q$ fixed at the positions of the corresponding numerals. The clock hands do not perturb the net field due to the point charges. At what time does the hour hand point in the same direction as the electric field vector at the center of the dial? (*Hint:* Use symmetry.)

80 Calculate the electric dipole moment of an electron and a proton 4.30 nm apart.

81 An electric field $\vec{E}$ with an average magnitude of about 150 N/C points downward in the atmosphere near Earth's surface. We wish to "float" a sulfur sphere weighing 4.4 N in this field by charging the sphere. (a) What charge (both sign and magnitude) must be used? (b) Why is the experiment impractical?

82 A circular rod has a radius of curvature $R = 9.00$ cm and a uniformly distributed positive charge $Q = 6.25$ pC and subtends an angle $\theta = 2.40$ rad. What is the magnitude of the electric field that Q produces at the center of curvature?

83 SSM An electric dipole with dipole moment

$$\vec{p} = (3.00\hat{i} + 4.00\hat{j})(1.24 \times 10^{-30} \text{ C·m})$$

is in an electric field $\vec{E} = (4000$ N/C$)\hat{i}$. (a) What is the potential energy of the electric dipole? (b) What is the torque acting on it? (c) If an external agent turns the dipole until its electric dipole moment is

$$\vec{p} = (-4.00\hat{i} + 3.00\hat{j})(1.24 \times 10^{-30} \text{ C·m}),$$

how much work is done by the agent?

84 In Fig. 22-68, a uniform, upward electric field $\vec{E}$ of magnitude 2.00×10^3 N/C has been set up between two horizontal plates by charging the lower plate positively and the upper plate negatively. The plates have length $L = 10.0$ cm and separation $d = 2.00$ cm. An electron is then shot between the plates from the left edge of the lower plate. The initial velocity $\vec{v}_0$ of the electron makes an angle $\theta = 45.0°$ with the lower plate and has a magnitude of 6.00×10^6 m/s. (a) Will the electron strike one of the plates? (b) If so, which plate and how far horizontally from the left edge will the electron strike?

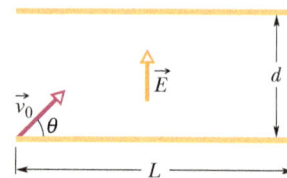

Figure 22-68 Problem 84.

85 For the data of Problem 70, assume that the charge q on the drop is given by $q = ne$, where n is an integer and e is the elementary charge. (a) Find n for each given value of q. (b) Do a linear regression fit of the values of q versus the values of n and then use that fit to find e.

86 In Fig. 22-66, particle 1 (of charge $+2.00$ pC), particle 2 (of charge -2.00 pC), and particle 3 (of charge $+5.00$ pC) form an equilateral triangle of edge length $a = 9.50$ cm. (a) Relative to the positive direction of the x axis, determine the direction of the force $\vec{F}_3$ on particle 3 due to the other particles by sketching electric field lines of the other particles. (b) Calculate the magnitude of $\vec{F}_3$.

87 In Fig. 22-69, particle 1 of charge $q_1 = 1.00$ pC and particle 2 of charge $q_2 = -2.00$ pC are fixed at a distance $d = 5.00$ cm apart. In unit-vector notation, what is the net electric field at points (a) A, (b) B, and (c) C? (d) Sketch the electric field lines.

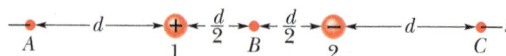

Figure 22-69 Problem 87.

Gauss' Law

23-1 ELECTRIC FLUX

Learning Objectives

After reading this module, you should be able to . . .

23.01 Identify that Gauss' law relates the electric field at points on a closed surface (real or imaginary, said to be a Gaussian surface) to the net charge enclosed by that surface.

23.02 Identify that the amount of electric field piercing a surface (not skimming along the surface) is the electric flux Φ through the surface.

23.03 Identify that an area vector for a flat surface is a vector that is perpendicular to the surface and that has a magnitude equal to the area of the surface.

23.04 Identify that any surface can be divided into area elements (patch elements) that are each small enough and flat enough for an area vector $d\vec{A}$ to be assigned to it, with the vector perpendicular to the element and having a magnitude equal to the area of the element.

23.05 Calculate the flux Φ through a surface by integrating the dot product of the electric field vector $\vec{E}$ and the area vector $d\vec{A}$ (for patch elements) over the surface, in magnitude-angle notation and unit-vector notation.

23.06 For a closed surface, explain the algebraic signs associated with inward flux and outward flux.

23.07 Calculate the *net* flux Φ through a *closed* surface, algebraic sign included, by integrating the dot product of the electric field vector $\vec{E}$ and the area vector $d\vec{A}$ (for patch elements) over the full surface.

23.08 Determine whether a closed surface can be broken up into parts (such as the sides of a cube) to simplify the integration that yields the net flux through the surface.

Key Ideas

● The electric flux Φ through a surface is the amount of electric field that pierces the surface.

● The area vector $d\vec{A}$ for an area element (patch element) on a surface is a vector that is perpendicular to the element and has a magnitude equal to the area dA of the element.

● The electric flux $d\Phi$ through a patch element with area vector $d\vec{A}$ is given by a dot product:

$$d\Phi = \vec{E} \cdot d\vec{A}.$$

● The total flux through a surface is given by

$$\Phi = \int \vec{E} \cdot d\vec{A} \quad \text{(total flux)},$$

where the integration is carried out over the surface.

● The net flux through a closed surface (which is used in Gauss' law) is given by

$$\Phi = \oint \vec{E} \cdot d\vec{A} \quad \text{(net flux)},$$

where the integration is carried out over the entire surface.

What Is Physics?

In the preceding chapter we found the electric field at points near extended charged objects, such as rods. Our technique was labor-intensive: We split the charge distribution up into charge elements dq, found the field $d\vec{E}$ due to an element, and resolved the vector into components. Then we determined whether the components from all the elements would end up canceling or adding. Finally we summed the adding components by integrating over all the elements, with several changes in notation along the way.

One of the primary goals of physics is to find simple ways of solving such labor-intensive problems. One of the main tools in reaching this goal is the use of symmetry. In this chapter we discuss a beautiful relationship between charge and

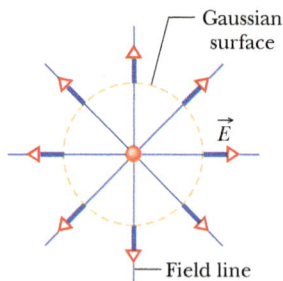

Figure 23-1 Electric field vectors and field lines pierce an imaginary, spherical Gaussian surface that encloses a particle with charge $+Q$.

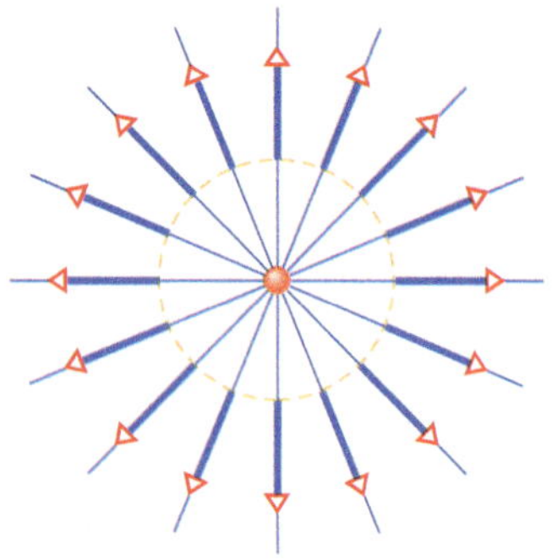

Figure 23-2 Now the enclosed particle has charge $+2Q$.

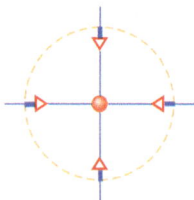

Figure 23-3 Can you tell what the enclosed charge is now?

electric field that allows us, in certain symmetric situations, to find the electric field of an extended charged object with a few lines of algebra. The relationship is called **Gauss' law,** which was developed by German mathematician and physicist Carl Friedrich Gauss (1777–1855).

Let's first take a quick look at some simple examples that give the spirit of Gauss' law. Figure 23-1 shows a particle with charge $+Q$ that is surrounded by an imaginary concentric sphere. At points on the sphere (said to be a *Gaussian surface*), the electric field vectors have a moderate magnitude (given by $E = kQ/r^2$) and point radially away from the particle (because it is positively charged). The electric field lines are also outward and have a moderate density (which, recall, is related to the field magnitude). We say that the field vectors and the field lines *pierce* the surface.

Figure 23-2 is similar except that the enclosed particle has charge $+2Q$. Because the enclosed charge is now twice as much, the magnitude of the field vectors piercing outward through the (same) Gaussian surface is twice as much as in Fig. 23-1, and the density of the field lines is also twice as much. That sentence, in a nutshell, is Gauss' law.

> Guass' law relates the electric field at points on a (closed) Gaussian surface to the net charge enclosed by that surface.

Let's check this with a third example with a particle that is also enclosed by the same spherical Gaussian surface (a *Gaussian sphere*, if you like, or even the catchy *G-sphere*) as shown in Fig. 23-3. What is the amount and sign of the enclosed charge? Well, from the inward piercing we see immediately that the charge must be negative. From the fact that the density of field lines is half that of Fig. 23-1, we also see that the charge must be $0.5Q$. (Using Gauss' law is like being able to tell what is inside a gift box by looking at the wrapping paper on the box.)

The problems in this chapter are of two types. Sometimes we know the charge and we use Gauss' law to find the field at some point. Sometimes we know the field on a Gaussian surface and we use Gauss' law to find the charge enclosed by the surface. However, we cannot do all this by simply comparing the density of field lines in a drawing as we just did. We need a quantitative way of determining how much electric field pierces a surface. That measure is called the electric flux.

Electric Flux

Flat Surface, Uniform Field. We begin with a flat surface with area A in a uniform electric field $\vec{E}$. Figure 23-4a shows one of the electric field vectors $\vec{E}$ piercing a small square patch with area ΔA (where Δ indicates "small"). Actually, only the x component (with magnitude $E_x = E \cos \theta$ in Fig. 23-4b) pierces the patch. The y component merely skims along the surface (no piercing in that) and does not come into play in Gauss' law. The *amount* of electric field piercing the patch is defined to be the **electric flux $\Delta\Phi$** through it:

$$\Delta\Phi = (E \cos \theta) \, \Delta A.$$

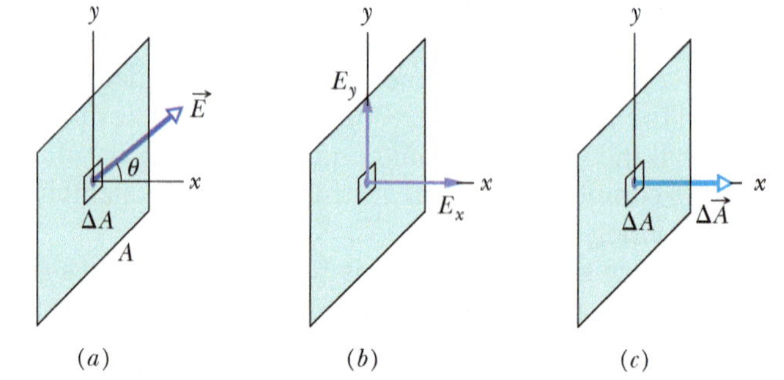

Figure 23-4 (a) An electric field vector pierces a small square patch on a flat surface. (b) Only the x component actually pierces the patch; the y component skims across it. (c) The area vector of the patch is perpendicular to the patch, with a magnitude equal to the patch's area.

(a)   (b)   (c)

There is another way to write the right side of this statement so that we have only the piercing component of $\vec{E}$. We define an area vector $\Delta\vec{A}$ that is perpendicular to the patch and that has a magnitude equal to the area ΔA of the patch (Fig. 23-4c). Then we can write

$$\Delta\Phi = \vec{E} \cdot \Delta\vec{A},$$

and the dot product automatically gives us the component of $\vec{E}$ that is parallel to $\Delta\vec{A}$ and thus piercing the patch.

To find the total flux Φ through the surface in Fig. 23-4, we sum the flux through every patch on the surface:

$$\Phi = \sum \vec{E} \cdot \Delta\vec{A}. \tag{23-1}$$

However, because we do not want to sum hundreds (or more) flux values, we transform the summation into an integral by shrinking the patches from small squares with area ΔA to *patch elements* (or *area elements*) with area dA. The total flux is then

$$\Phi = \int \vec{E} \cdot d\vec{A} \quad \text{(total flux).} \tag{23-2}$$

Now we can find the total flux by integrating the dot product over the full surface.

Dot Product. We can evaluate the dot product inside the integral by writing the two vectors in unit-vector notation. For example, in Fig. 23-4, $d\vec{A} = dA\hat{i}$ and $\vec{E}$ might be, say, $(4\hat{i} + 4\hat{j})$ N/C. Instead, we can evaluate the dot product in magnitude-angle notation: $E \cos\theta\, dA$. When the electric field is uniform and the surface is flat, the product $E \cos\theta$ is a constant and comes outside the integral. The remaining $\int dA$ is just an instruction to sum the areas of all the patch elements to get the total area, but we already know that the total area is A. So the total flux in this simple situation is

$$\Phi = (E \cos\theta)A \quad \text{(uniform field, flat surface).} \tag{23-3}$$

Closed Surface. To use Gauss' law to relate flux and charge, we need a closed surface. Let's use the closed surface in Fig. 23-5 that sits in a nonuniform electric field. (Don't worry. The homework problems involve less complex surfaces.) As before, we first consider the flux through small square patches. However, now we are interested in not only the piercing components of the field but also on whether the piercing is inward or outward (just as we did with Figs. 23-1 through 23-3).

Directions. To keep track of the piercing direction, we again use an area vector $\Delta\vec{A}$ that is perpendicular to a patch, but now we always draw it pointing outward from the surface (*away from the interior*). Then if a field vector pierces outward, it and the area vector are in the same direction, the angle is $\theta = 0$, and $\cos\theta = 1$. Thus, the dot product $\vec{E} \cdot \Delta\vec{A}$ is positive and so is the flux. Conversely, if a field vector pierces inward, the angle is $\theta = 180°$ and $\cos\theta = -1$. Thus, the dot product is negative and so is the flux. If a field vector skims the surface (no piercing), the dot product is zero (because $\cos 90° = 0$) and so is the flux. Figure 23-5 gives some general examples and here is a summary:

⭐ An inward piercing field is negative flux. An outward piercing field is positive flux. A skimming field is zero flux.

Net Flux. In principle, to find the **net flux** through the surface in Fig. 23-5, we find the flux at every patch and then sum the results (with the algebraic signs included). However, we are not about to do that much work. Instead, we shrink the squares to patch elements with area vectors $d\vec{A}$ and then integrate:

$$\Phi = \oint \vec{E} \cdot d\vec{A} \quad \text{(net flux).} \tag{23-4}$$

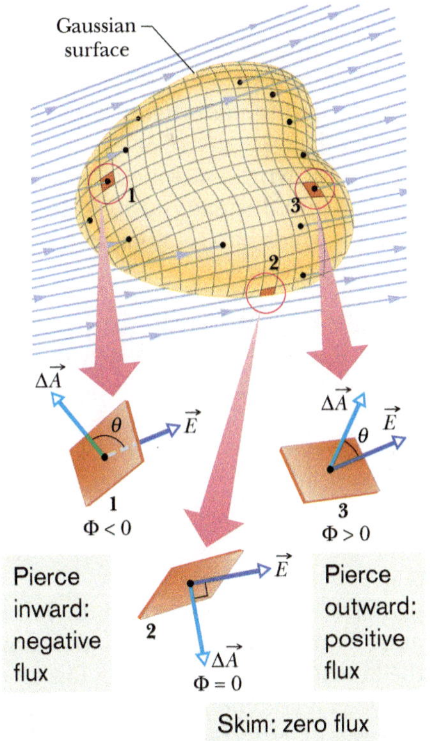

Figure 23-5 A Gaussian surface of arbitrary shape immersed in an electric field. The surface is divided into small squares of area ΔA. The electric field vectors $\vec{E}$ and the area vectors $\Delta\vec{A}$ for three representative squares, marked 1, 2, and 3, are shown.

The loop on the integral sign indicates that we must integrate over the entire closed surface, to get the *net* flux through the surface (as in Fig. 23-5, flux might enter on one side and leave on another side). Keep in mind that we want to determine the net flux through a surface because that is what Gauss' law relates to the charge enclosed by the surface. (The law is coming up next.) Note that flux is a scalar (yes, we talk about field vectors but flux is the *amount* of piercing field, not a vector itself). The SI unit of flux is the newton–square-meter per coulomb $(N \cdot m^2/C)$.

✓ Checkpoint 1

The figure here shows a Gaussian cube of face area A immersed in a uniform electric field $\vec{E}$ that has the positive direction of the z axis. In terms of E and A, what is the flux through (a) the front face (which is in the xy plane), (b) the rear face, (c) the top face, and (d) the whole cube?

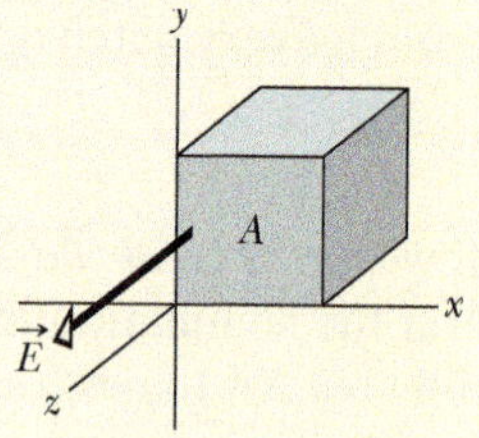

Sample Problem 23.01 Flux through a closed cylinder, uniform field

Figure 23-6 shows a Gaussian surface in the form of a closed cylinder (a Gaussian cylinder or G-cylinder) of radius R. It lies in a uniform electric field $\vec{E}$ with the cylinder's central axis (along the length of the cylinder) parallel to the field. What is the net flux Φ of the electric field through the cylinder?

KEY IDEAS

We can find the net flux Φ with Eq. 23-4 by integrating the dot product $\vec{E} \cdot d\vec{A}$ over the cylinder's surface. However, we cannot write out functions so that we can do that with one integral. Instead, we need to be a bit clever: We break up the surface into sections with which we can actually evaluate an integral.

Calculations: We break the integral of Eq. 23-4 into three terms: integrals over the left cylinder cap a, the curved cylindrical surface b, and the right cap c:

$$\Phi = \oint \vec{E} \cdot d\vec{A}$$

$$= \int_a \vec{E} \cdot d\vec{A} + \int_b \vec{E} \cdot d\vec{A} + \int_c \vec{E} \cdot d\vec{A}. \qquad (23\text{-}5)$$

Pick a patch element on the left cap. Its area vector $d\vec{A}$ must be perpendicular to the patch and pointing away from the interior of the cylinder. In Fig. 23-6, that means the angle between it and the field piercing the patch is 180°. Also, note that the electric field through the end cap is uniform and thus E can be pulled out of the integration. So, we can write the flux through the left cap as

$$\int_a \vec{E} \cdot d\vec{A} = \int E(\cos 180°)\, dA = -E \int dA = -EA,$$

where $\int dA$ gives the cap's area A $(= \pi R^2)$. Similarly, for the right cap, where $\theta = 0$ for all points,

$$\int_c \vec{E} \cdot d\vec{A} = \int E(\cos 0)\, dA = EA.$$

Finally, for the cylindrical surface, where the angle θ is 90° at all points,

$$\int_b \vec{E} \cdot d\vec{A} = \int E(\cos 90°)\, dA = 0.$$

Substituting these results into Eq. 23-5 leads us to

$$\Phi = -EA + 0 + EA = 0. \qquad \text{(Answer)}$$

The net flux is zero because the field lines that represent the electric field all pass entirely through the Gaussian surface, from the left to the right.

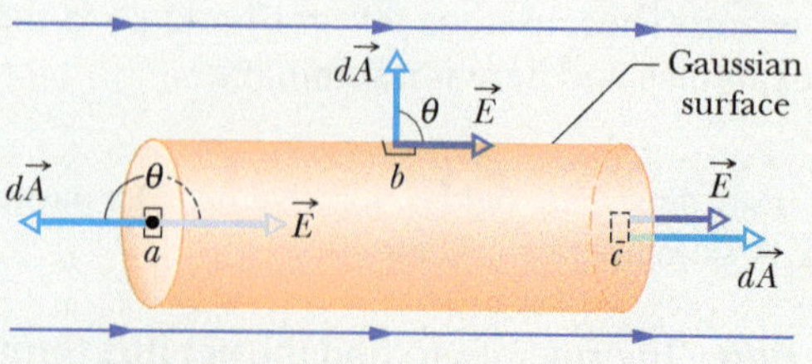

Figure 23-6 A cylindrical Gaussian surface, closed by end caps, is immersed in a uniform electric field. The cylinder axis is parallel to the field direction.

Sample Problem 23.02 Flux through a closed cube, nonuniform field

A *nonuniform* electric field given by $\vec{E} = 3.0x\hat{i} + 4.0\hat{j}$ pierces the Gaussian cube shown in Fig. 23-7a. (*E* is in newtons per coulomb and *x* is in meters.) What is the electric flux through the right face, the left face, and the top face? (We consider the other faces in another sample problem.)

KEY IDEA

We can find the flux Φ through the surface by integrating the scalar product $\vec{E} \cdot d\vec{A}$ over each face.

Right face: An area vector $\vec{A}$ is always perpendicular to its surface and always points away from the interior of a Gaussian surface. Thus, the vector $d\vec{A}$ for any patch element (small section) on the right face of the cube must point in the positive direction of the *x* axis. An example of such an element is shown in Figs. 23-7b and c, but we would have an identical vector for any other choice of a patch element on that face. The most convenient way to express the vector is in unit-vector notation,

$$d\vec{A} = dA\hat{i}.$$

From Eq. 23-4, the flux Φ_r through the right face is then

$$\Phi_r = \int \vec{E} \cdot d\vec{A} = \int (3.0x\hat{i} + 4.0\hat{j}) \cdot (dA\hat{i})$$

$$= \int [(3.0x)(dA)\hat{i} \cdot \hat{i} + (4.0)(dA)\hat{j} \cdot \hat{i}]$$

$$= \int (3.0x\, dA + 0) = 3.0 \int x\, dA.$$

We are about to integrate over the right face, but we note that *x* has the same value everywhere on that face—namely, *x* = 3.0 m. This means we can substitute that constant value for *x*. This can be a confusing argument. Although *x* is certainly a variable as we move left to right across the figure, because the right face is perpendicular to the *x* axis, every point on the face has the same *x* coordinate. (The *y* and *z* coordinates do not matter in our integral.) Thus, we have

$$\Phi_r = 3.0 \int (3.0)\, dA = 9.0 \int dA.$$

The integral $\int dA$ merely gives us the area $A = 4.0$ m² of the right face, so

$$\Phi_r = (9.0 \text{ N/C})(4.0 \text{ m}^2) = 36 \text{ N} \cdot \text{m}^2/\text{C}. \quad \text{(Answer)}$$

Left face: We repeat this procedure for the left face. However,

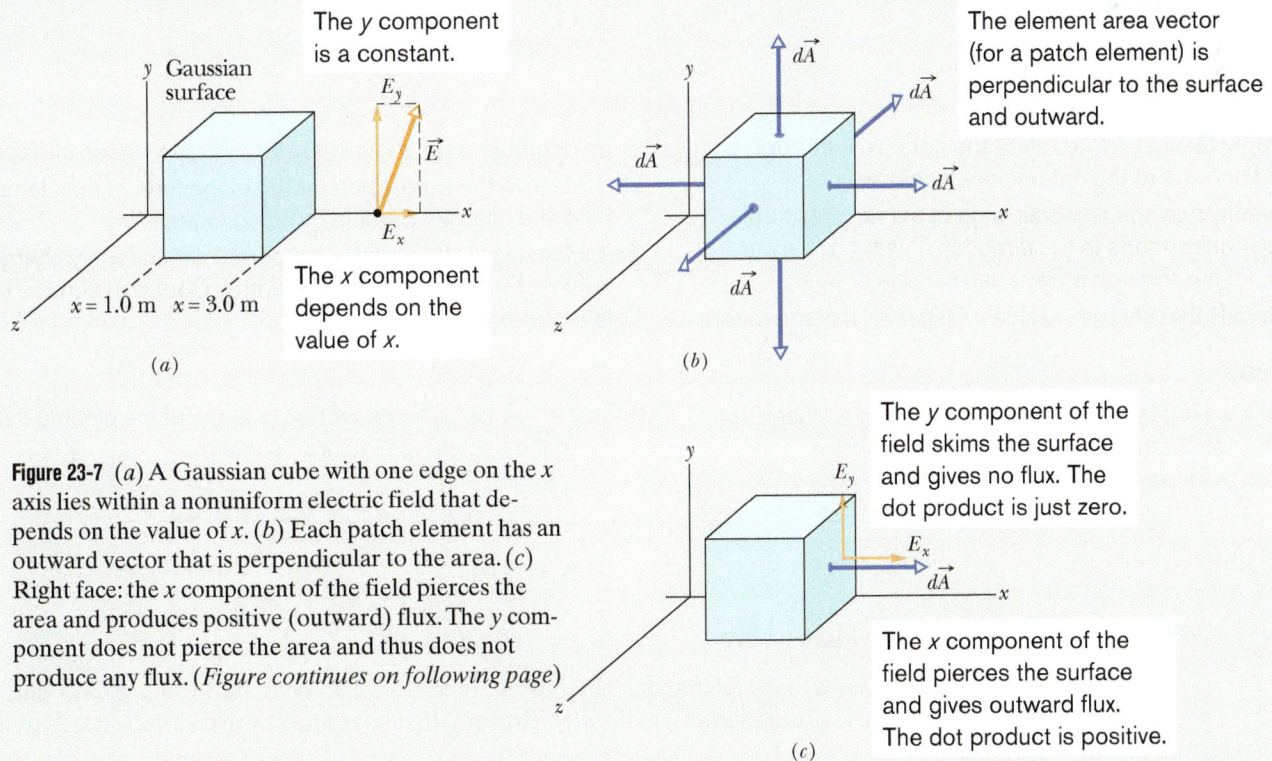

The *y* component is a constant. The *x* component depends on the value of *x*.

The element area vector (for a patch element) is perpendicular to the surface and outward.

The *y* component of the field skims the surface and gives no flux. The dot product is just zero. The *x* component of the field pierces the surface and gives outward flux. The dot product is positive.

Figure 23-7 (*a*) A Gaussian cube with one edge on the *x* axis lies within a nonuniform electric field that depends on the value of *x*. (*b*) Each patch element has an outward vector that is perpendicular to the area. (*c*) Right face: the *x* component of the field pierces the area and produces positive (outward) flux. The *y* component does not pierce the area and thus does not produce any flux. (*Figure continues on following page*)

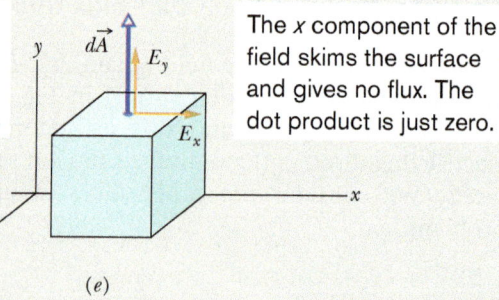

The y component of the field skims the surface and gives no flux. The dot product is just zero.

The y component of the field pierces the surface and gives outward flux. The dot product is positive.

The x component of the field skims the surface and gives no flux. The dot product is just zero.

The x component of the field pierces the surface and gives inward flux. The dot product is negative.

(d)

(e)

Figure 23-7 (*Continued from previous page*) (*d*) Left face: the x component of the field produces negative (inward) flux. (*e*) Top face: the y component of the field produces positive (outward) flux.

two factors change. (1) The element area vector $d\vec{A}$ points in the negative direction of the x axis, and thus $d\vec{A} = -dA\hat{i}$ (Fig. 23-7d). (2) On the left face, $x = 1.0$ m. With these changes, we find that the flux Φ_l through the left face is

$$\Phi_l = -12\ \text{N} \cdot \text{m}^2/\text{C}. \qquad \text{(Answer)}$$

Top face: Now $d\vec{A}$ points in the positive direction of the y axis, and thus $d\vec{A} = dA\hat{j}$ (Fig. 23-7e). The flux Φ_t is

$$\Phi_t = \int (3.0x\hat{i} + 4.0\hat{j}) \cdot (dA\hat{j})$$

$$= \int [(3.0x)(dA)\hat{i} \cdot \hat{j} + (4.0)(dA)\hat{j} \cdot \hat{j}]$$

$$= \int (0 + 4.0\ dA) = 4.0 \int dA$$

$$= 16\ \text{N} \cdot \text{m}^2/\text{C}. \qquad \text{(Answer)}$$

WILEY PLUS Additional examples, video, and practice available at *WileyPLUS*

23-2 GAUSS' LAW

Learning Objectives

After reading this module, you should be able to . . .

23.09 Apply Gauss' law to relate the net flux Φ through a closed surface to the net enclosed charge q_{enc}.

23.10 Identify how the algebraic sign of the net enclosed charge corresponds to the direction (inward or outward) of the net flux through a Gaussian surface.

23.11 Identify that charge outside a Gaussian surface makes

no contribution to the *net* flux through the closed surface.

23.12 Derive the expression for the magnitude of the electric field of a charged particle by using Gauss' law.

23.13 Identify that for a charged particle or uniformly charged sphere, Gauss' law is applied with a Gaussian surface that is a concentric sphere.

Key Ideas

● Gauss' law relates the net flux Φ penetrating a closed surface to the net charge q_{enc} enclosed by the surface:

$$\varepsilon_0 \Phi = q_{\text{enc}} \quad \text{(Gauss' law)}.$$

● Gauss' law can also be written in terms of the electric field piercing the enclosing Gaussian surface:

$$\varepsilon_0 \oint \vec{E} \cdot d\vec{A} = q_{\text{enc}} \quad \text{(Gauss' law)}.$$

Gauss' Law

Gauss' law relates the net flux Φ of an electric field through a closed surface (a Gaussian surface) to the *net* charge q_{enc} that is *enclosed* by that surface. It tells us that

$$\varepsilon_0 \Phi = q_{\text{enc}} \quad \text{(Gauss' law)}. \qquad (23\text{-}6)$$

By substituting Eq. 23-4, the definition of flux, we can also write Gauss' law as

$$\varepsilon_0 \oint \vec{E} \cdot d\vec{A} = q_{enc} \quad \text{(Gauss' law)}. \qquad (23\text{-}7)$$

Equations 23-6 and 23-7 hold only when the net charge is located in a vacuum or (what is the same for most practical purposes) in air. In Chapter 25, we modify Gauss' law to include situations in which a material such as mica, oil, or glass is present.

In Eqs. 23-6 and 23-7, the net charge q_{enc} is the algebraic sum of all the *enclosed* positive and negative charges, and it can be positive, negative, or zero. We include the sign, rather than just use the magnitude of the enclosed charge, because the sign tells us something about the net flux through the Gaussian surface: If q_{enc} is positive, the net flux is *outward*; if q_{enc} is negative, the net flux is *inward*.

Charge outside the surface, no matter how large or how close it may be, is not included in the term q_{enc} in Gauss' law. The exact form and location of the charges inside the Gaussian surface are also of no concern; the only things that matter on the right side of Eqs. 23-6 and 23-7 are the magnitude and sign of the net enclosed charge. The quantity $\vec{E}$ on the left side of Eq. 23-7, however, is the electric field resulting from *all* charges, both those inside and those outside the Gaussian surface. This statement may seem to be inconsistent, but keep this in mind: The electric field due to a charge outside the Gaussian surface contributes zero net flux *through* the surface, because as many field lines due to that charge enter the surface as leave it.

Let us apply these ideas to Fig. 23-8, which shows two particles, with charges equal in magnitude but opposite in sign, and the field lines describing the electric fields the particles set up in the surrounding space. Four Gaussian surfaces are also shown, in cross section. Let us consider each in turn.

Surface S_1. The electric field is outward for all points on this surface. Thus, the flux of the electric field through this surface is positive, and so is the net charge within the surface, as Gauss' law requires. (That is, in Eq. 23-6, if Φ is positive, q_{enc} must be also.)

Surface S_2. The electric field is inward for all points on this surface. Thus, the flux of the electric field through this surface is negative and so is the enclosed charge, as Gauss' law requires.

Surface S_3. This surface encloses no charge, and thus $q_{enc} = 0$. Gauss' law (Eq. 23-6) requires that the net flux of the electric field through this surface be zero. That is reasonable because all the field lines pass entirely through the surface, entering it at the top and leaving at the bottom.

Surface S_4. This surface encloses no *net* charge, because the enclosed positive and negative charges have equal magnitudes. Gauss' law requires that the net flux of the electric field through this surface be zero. That is reasonable because there are as many field lines leaving surface S_4 as entering it.

What would happen if we were to bring an enormous charge Q up close to surface S_4 in Fig. 23-8? The pattern of the field lines would certainly change, but the net flux for each of the four Gaussian surfaces would not change. Thus, the value of Q would not enter Gauss' law in any way, because Q lies outside all four of the Gaussian surfaces that we are considering.

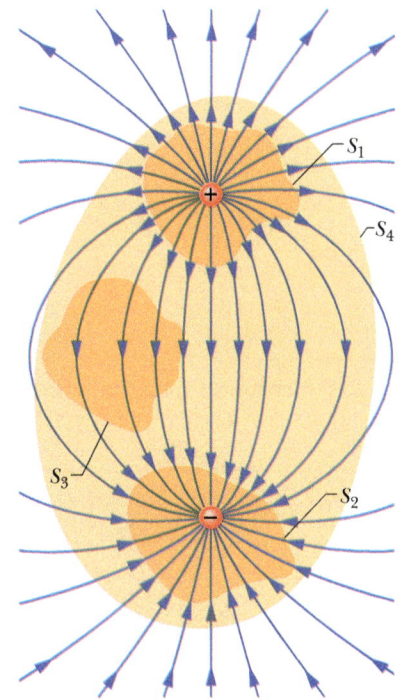

Figure 23-8 Two charges, equal in magnitude but opposite in sign, and the field lines that represent their net electric field. Four Gaussian surfaces are shown in cross section. Surface S_1 encloses the positive charge. Surface S_2 encloses the negative charge. Surface S_3 encloses no charge. Surface S_4 encloses both charges and thus no net charge.

✓ **Checkpoint 2**

The figure shows three situations in which a Gaussian cube sits in an electric field. The arrows and the values indicate the directions of the field lines and the magnitudes (in $N \cdot m^2/C$) of the flux through the six sides of each cube. (The lighter arrows are for the hidden faces.) In which situation does the cube enclose (a) a positive net charge, (b) a negative net charge, and (c) zero net charge?

Gauss' Law and Coulomb's Law

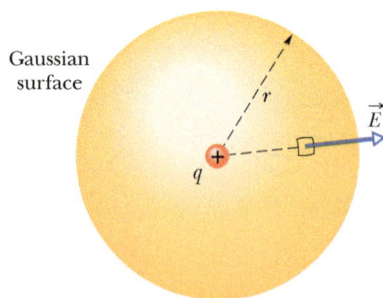

Gaussian surface

Figure 23-9 A spherical Gaussian surface centered on a particle with charge q.

One of the situations in which we can apply Gauss' law is in finding the electric field of a charged particle. That field has spherical symmetry (the field depends on the distance r from the particle but not the direction). So, to make use of that symmetry, we enclose the particle in a Gaussian sphere that is centered on the particle, as shown in Fig. 23-9 for a particle with positive charge q. Then the electric field has the same magnitude E at any point on the sphere (all points are at the same distance r). That feature will simplify the integration.

The drill here is the same as previously. Pick a patch element on the surface and draw its area vector $d\vec{A}$ perpendicular to the patch and directed outward. From the symmetry of the situation, we know that the electric field $\vec{E}$ at the patch is also radially outward and thus at angle $\theta = 0$ with $d\vec{A}$. So, we rewrite Gauss' law as

$$\varepsilon_0 \oint \vec{E} \cdot d\vec{A} = \varepsilon_0 \oint E\, dA = q_{enc}. \qquad (23\text{-}8)$$

Here $q_{enc} = q$. Because the field magnitude E is the same at every patch element, E can be pulled outside the integral:

$$\varepsilon_0 E \oint dA = q. \qquad (23\text{-}9)$$

The remaining integral is just an instruction to sum all the areas of the patch elements on the sphere, but we already know that the total area is $4\pi r^2$. Substituting this, we have

$$\varepsilon_0 E(4\pi r^2) = q$$

or

$$E = \frac{1}{4\pi\varepsilon_0} \frac{q}{r^2}. \qquad (23\text{-}10)$$

This is exactly Eq. 22-3, which we found using Coulomb's law.

> ✓ **Checkpoint 3**
>
> There is a certain net flux Φ_i through a Gaussian sphere of radius r enclosing an isolated charged particle. Suppose the enclosing Gaussian surface is changed to (a) a larger Gaussian sphere, (b) a Gaussian cube with edge length equal to r, and (c) a Gaussian cube with edge length equal to $2r$. In each case, is the net flux through the new Gaussian surface greater than, less than, or equal to Φ_i?

Sample Problem 23.03 Using Gauss' law to find the electric field

Figure 23-10a shows, in cross section, a plastic, spherical shell with uniform charge $Q = -16e$ and radius $R = 10$ cm. A particle with charge $q = +5e$ is at the center. What is the electric field (magnitude and direction) at (a) point P_1 at radial distance $r_1 = 6.00$ cm and (b) point P_2 at radial distance $r_2 = 12.0$ cm?

KEY IDEAS

(1) Because the situation in Fig. 23-10a has spherical symmetry, we can apply Gauss' law (Eq. 23-7) to find the electric field at a point if we use a Gaussian surface in the form of a sphere concentric with the particle and shell. (2) To find the electric field at a point, we put that point on a Gaussian surface (so that the $\vec{E}$ we want is the $\vec{E}$ in the dot product inside the integral in Gauss' law). (3) Gauss' law relates the net electric flux through a closed surface to the net enclosed charge. Any external charge is not included.

Calculations: To find the field at point P_1, we construct a Gaussian sphere with P_1 on its surface and thus with a radius of r_1. Because the charge enclosed by the Gaussian sphere is positive, the electric flux through the surface must be positive and thus outward. So, the electric field $\vec{E}$ pierces the surface outward and, because of the spherical symmetry, must be *radially* outward, as drawn in Fig. 23-10b. That figure does not include the plastic shell because the shell is not enclosed by the Gaussian sphere.

Consider a patch element on the sphere at P_1. Its area vector $d\vec{A}$ is radially outward (it must always be outward from a Gaussian surface). Thus the angle θ between $\vec{E}$ and $d\vec{A}$ is zero. We can now rewrite the left side of Eq. 23-7 (Gauss' law) as

$$\varepsilon_0 \oint \vec{E} \cdot d\vec{A} = \varepsilon_0 \oint E \cos 0\, dA = \varepsilon_0 \oint E\, dA = \varepsilon_0 E \oint dA,$$

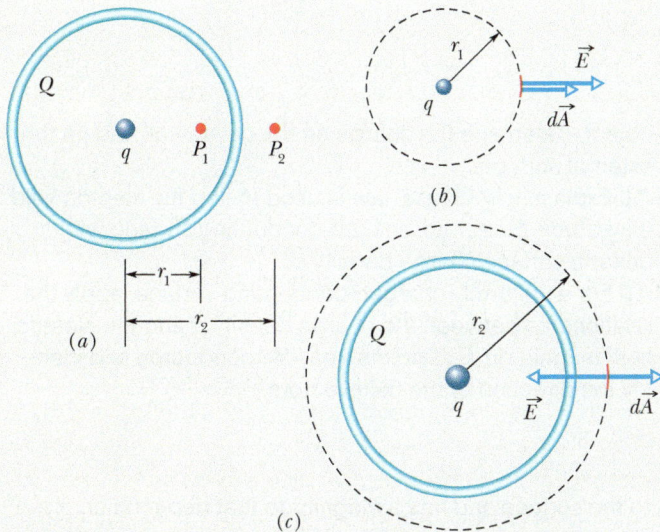

Figure 23-10 (*a*) A charged plastic spherical shell encloses a charged particle. (*b*) To find the electric field at P_1, arrange for the point to be on a Gaussian sphere. The electric field pierces outward. The area vector for the patch element is outward. (*c*) P_2 is on a Gaussian sphere, $\vec{E}$ is inward, and $d\vec{A}$ is still outward.

where in the last step we pull the field magnitude E out of the integral because it is the same at all points on the Gaussian sphere and thus is a constant. The remaining integral is simply an instruction for us to sum the areas of all the patch elements on the sphere, but we already know that the surface area of a sphere is $4\pi r^2$. Substituting these results, Eq. 23-7 for Gauss' law gives us

$$\varepsilon_0 E 4\pi r^2 = q_{enc}.$$

The only charge enclosed by the Gaussian surface through P_1 is that of the particle. Solving for E and substituting $q_{enc} = 5e$ and $r = r_1 = 6.00 \times 10^{-2}$ m, we find that the magnitude of the electric field at P_1 is

$$E = \frac{q_{enc}}{4\pi\varepsilon_0 r^2}$$
$$= \frac{5(1.60 \times 10^{-19}\,\text{C})}{4\pi(8.85 \times 10^{-12}\,\text{C}^2/\text{N}\cdot\text{m}^2)(0.0600\,\text{m})^2}$$
$$= 2.00 \times 10^{-6}\,\text{N/C}. \qquad \text{(Answer)}$$

To find the electric field at P_2, we follow the same procedure by constructing a Gaussian sphere with P_2 on its surface. This time, however, the net charge enclosed by the sphere is $q_{enc} = q + Q = 5e + (-16e) = -11e$. Because the net charge is negative, the electric field vectors on the sphere's surface pierce inward (Fig. 23-10*c*), the angle θ between $\vec{E}$ and $d\vec{A}$ is 180°, and the dot product is $E(\cos 180°)\,dA = -E\,dA$. Now solving Gauss' law for E and substituting $r = r_2 = 12.00 \times 10^{-2}$ m and the new q_{enc}, we find

$$E = \frac{-q_{enc}}{4\pi\varepsilon_0 r^2}$$
$$= \frac{-[-11(1.60 \times 10^{-19}\,\text{C})]}{4\pi(8.85 \times 10^{-12}\,\text{C}^2/\text{N}\cdot\text{m}^2)(0.120\,\text{m})^2}$$
$$= 1.10 \times 10^{-6}\,\text{N/C}. \qquad \text{(Answer)}$$

Note how different the calculations would have been if we had put P_1 or P_2 on the surface of a Gaussian cube instead of mimicking the spherical symmetry with a Gaussian sphere. Then angle θ and magnitude E would have varied considerably over the surface of the cube and evaluation of the integral in Gauss' law would have been difficult.

Sample Problem 23.04 Using Gauss' law to find the enclosed charge

What is the net charge enclosed by the Gaussian cube of Sample Problem 23.02?

KEY IDEA

The net charge enclosed by a (real or mathematical) closed surface is related to the total electric flux through the surface by Gauss' law as given by Eq. 23-6 ($\varepsilon_0\Phi = q_{enc}$).

Flux: To use Eq. 23-6, we need to know the flux through all six faces of the cube. We already know the flux through the right face ($\Phi_r = 36$ N·m²/C), the left face ($\Phi_l = -12$ N·m²/C), and the top face ($\Phi_t = 16$ N·m²/C).

For the bottom face, our calculation is just like that for the top face *except* that the element area vector $d\vec{A}$ is now directed downward along the y axis (recall, it must be *outward* from the Gaussian enclosure). Thus, we have

$d\vec{A} = -dA\hat{j}$, and we find
$$\Phi_b = -16\,\text{N}\cdot\text{m}^2/\text{C}.$$

For the front face we have $d\vec{A} = dA\hat{k}$, and for the back face, $d\vec{A} = -dA\hat{k}$. When we take the dot product of the given electric field $\vec{E} = 3.0x\hat{i} + 4.0\hat{j}$ with either of these expressions for $d\vec{A}$, we get 0 and thus there is no flux through those faces. We can now find the total flux through the six sides of the cube:

$$\Phi = (36 - 12 + 16 - 16 + 0 + 0)\,\text{N}\cdot\text{m}^2/\text{C}$$
$$= 24\,\text{N}\cdot\text{m}^2/\text{C}.$$

Enclosed charge: Next, we use Gauss' law to find the charge q_{enc} enclosed by the cube:

$$q_{enc} = \varepsilon_0\Phi = (8.85 \times 10^{-12}\,\text{C}^2/\text{N}\cdot\text{m}^2)(24\,\text{N}\cdot\text{m}^2/\text{C})$$
$$= 2.1 \times 10^{-10}\,\text{C}. \qquad \text{(Answer)}$$

Thus, the cube encloses a *net* positive charge.

23-3 A CHARGED ISOLATED CONDUCTOR

Learning Objectives

After reading this module, you should be able to . . .

23.14 Apply the relationship between surface charge density σ and the area over which the charge is uniformly spread.

23.15 Identify that if excess charge (positive or negative) is placed on an isolated conductor, that charge moves to the surface and none is in the interior.

23.16 Identify the value of the electric field inside an isolated conductor.

23.17 For a conductor with a cavity that contains a charged object, determine the charge on the cavity wall and on the external surface.

23.18 Explain how Gauss' law is used to find the electric field magnitude E near an isolated conducting surface with a uniform surface charge density σ.

23.19 For a uniformly charged conducting surface, apply the relationship between the charge density σ and the electric field magnitude E at points near the conductor, and identify the direction of the field vectors.

Key Ideas

● An excess charge on an isolated conductor is located entirely on the outer surface of the conductor.

● The internal electric field of a charged, isolated conductor is zero, and the external field (at nearby points) is perpendicular to the surface and has a magnitude that depends on the surface charge density σ:

$$E = \frac{\sigma}{\varepsilon_0}.$$

A Charged Isolated Conductor

Gauss' law permits us to prove an important theorem about conductors:

⭐ If an excess charge is placed on an isolated conductor, that amount of charge will move entirely to the surface of the conductor. None of the excess charge will be found within the body of the conductor.

This might seem reasonable, considering that charges with the same sign repel one another. You might imagine that, by moving to the surface, the added charges are getting as far away from one another as they can. We turn to Gauss' law for verification of this speculation.

Figure 23-11a shows, in cross section, an isolated lump of copper hanging from an insulating thread and having an excess charge q. We place a Gaussian surface just inside the actual surface of the conductor.

The electric field inside this conductor must be zero. If this were not so, the field would exert forces on the conduction (free) electrons, which are always present in a conductor, and thus current would always exist within a conductor. (That is, charge would flow from place to place within the conductor.) Of course, there is no such perpetual current in an isolated conductor, and so the internal electric field is zero.

(An internal electric field *does* appear as a conductor is being charged. However, the added charge quickly distributes itself in such a way that the net internal electric field—the vector sum of the electric fields due to all the charges, both inside and outside—is zero. The movement of charge then ceases, because the net force on each charge is zero; the charges are then in *electrostatic equilibrium*.)

If $\vec{E}$ is zero everywhere inside our copper conductor, it must be zero for all points on the Gaussian surface because that surface, though close to the surface of the conductor, is definitely inside the conductor. This means that the flux through the Gaussian surface must be zero. Gauss' law then tells us that the net charge inside the Gaussian surface must also be zero. Then because the excess charge is not inside the Gaussian surface, it must be outside that surface, which means it must lie on the actual surface of the conductor.

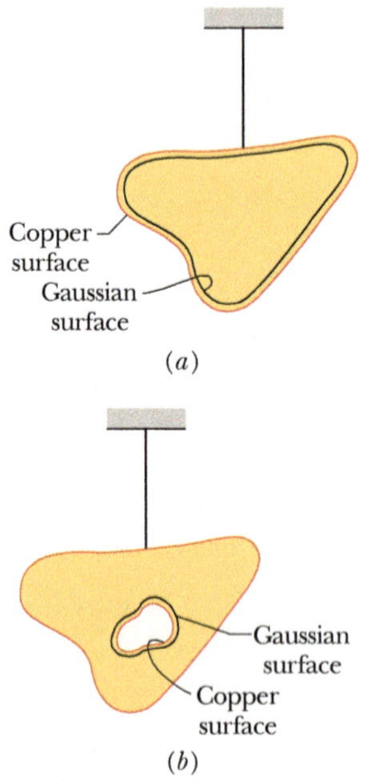

Figure 23-11 (a) A lump of copper with a charge q hangs from an insulating thread. A Gaussian surface is placed within the metal, just inside the actual surface. (b) The lump of copper now has a cavity within it. A Gaussian surface lies within the metal, close to the cavity surface.

Copper surface
Gaussian surface
(a)

Gaussian surface
Copper surface
(b)

An Isolated Conductor with a Cavity

Figure 23-11b shows the same hanging conductor, but now with a cavity that is totally within the conductor. It is perhaps reasonable to suppose that when we scoop out the electrically neutral material to form the cavity, we do not change the distribution of charge or the pattern of the electric field that exists in Fig. 23-11a. Again, we must turn to Gauss' law for a quantitative proof.

We draw a Gaussian surface surrounding the cavity, close to its surface but inside the conducting body. Because $\vec{E} = 0$ inside the conductor, there can be no flux through this new Gaussian surface. Therefore, from Gauss' law, that surface can enclose no net charge. We conclude that there is no net charge on the cavity walls; all the excess charge remains on the outer surface of the conductor, as in Fig. 23-11a.

The Conductor Removed

Suppose that, by some magic, the excess charges could be "frozen" into position on the conductor's surface, perhaps by embedding them in a thin plastic coating, and suppose that then the conductor could be removed completely. This is equivalent to enlarging the cavity of Fig. 23-11b until it consumes the entire conductor, leaving only the charges. The electric field would not change at all; it would remain zero inside the thin shell of charge and would remain unchanged for all external points. This shows us that the electric field is set up by the charges and not by the conductor. The conductor simply provides an initial pathway for the charges to take up their positions.

The External Electric Field

You have seen that the excess charge on an isolated conductor moves entirely to the conductor's surface. However, unless the conductor is spherical, the charge does not distribute itself uniformly. Put another way, the surface charge density σ (charge per unit area) varies over the surface of any nonspherical conductor. Generally, this variation makes the determination of the electric field set up by the surface charges very difficult.

However, the electric field just outside the surface of a conductor is easy to determine using Gauss' law. To do this, we consider a section of the surface that is small enough to permit us to neglect any curvature and thus to take the section to be flat. We then imagine a tiny cylindrical Gaussian surface to be partially embedded in the section as shown in Fig. 23-12: One end cap is fully inside the conductor, the other is fully outside, and the cylinder is perpendicular to the conductor's surface.

The electric field $\vec{E}$ at and just outside the conductor's surface must also be perpendicular to that surface. If it were not, then it would have a component along the conductor's surface that would exert forces on the surface charges, causing them to move. However, such motion would violate our implicit assumption that we are dealing with electrostatic equilibrium. Therefore, $\vec{E}$ is perpendicular to the conductor's surface.

We now sum the flux through the Gaussian surface. There is no flux through the internal end cap, because the electric field within the conductor is zero. There is no flux through the curved surface of the cylinder, because internally (in the conductor) there is no electric field and externally the electric field is parallel to the curved portion of the Gaussian surface. The only flux through the Gaussian surface is that through the external end cap, where $\vec{E}$ is perpendicular to the plane of the cap. We assume that the cap area A is small enough that the field magnitude E is constant over the cap. Then the flux through the cap is EA, and that is the net flux Φ through the Gaussian surface.

The charge q_{enc} enclosed by the Gaussian surface lies on the conductor's surface in an area A. (Think of the cylinder as a cookie cutter.) If σ is the charge per unit area, then q_{enc} is equal to σA. When we substitute σA for q_{enc} and EA for Φ,

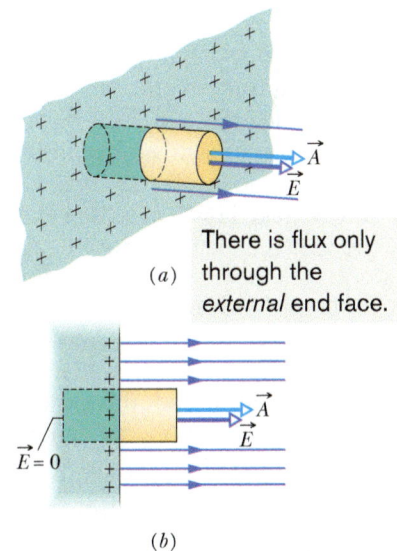

(a) There is flux only through the *external* end face.

(b)

Figure 23-12 (a) Perspective view and (b) side view of a tiny portion of a large, isolated conductor with excess positive charge on its surface. A (closed) cylindrical Gaussian surface, embedded perpendicularly in the conductor, encloses some of the charge. Electric field lines pierce the external end cap of the cylinder, but not the internal end cap. The external end cap has area A and area vector $\vec{A}$.

Gauss' law (Eq. 23-6) becomes

$$\varepsilon_0 EA = \sigma A,$$

from which we find

$$E = \frac{\sigma}{\varepsilon_0} \quad \text{(conducting surface).} \qquad (23\text{-}11)$$

Thus, the magnitude of the electric field just outside a conductor is proportional to the surface charge density on the conductor. The sign of the charge gives us the direction of the field. If the charge on the conductor is positive, the electric field is directed away from the conductor as in Fig. 23-12. It is directed toward the conductor if the charge is negative.

The field lines in Fig. 23-12 must terminate on negative charges somewhere in the environment. If we bring those charges near the conductor, the charge density at any given location on the conductor's surface changes, and so does the magnitude of the electric field. However, the relation between σ and E is still given by Eq. 23-11.

Sample Problem 23.05 Spherical metal shell, electric field and enclosed charge

Figure 23-13a shows a cross section of a spherical metal shell of inner radius R. A particle with a charge of $-5.0\ \mu C$ is located at a distance $R/2$ from the center of the shell. If the shell is electrically neutral, what are the (induced) charges on its inner and outer surfaces? Are those charges uniformly distributed? What is the field pattern inside and outside the shell?

KEY IDEAS

Figure 23-13b shows a cross section of a spherical Gaussian surface within the metal, just outside the inner wall of the shell. The electric field must be zero inside the metal (and thus on the Gaussian surface inside the metal). This means that the electric flux through the Gaussian surface must also be zero. Gauss' law then tells us that the *net* charge enclosed by the Gaussian surface must be zero.

Reasoning: With a particle of charge $-5.0\ \mu C$ within the shell, a charge of $+5.0\ \mu C$ must lie on the inner wall of the shell in order that the net enclosed charge be zero. If the particle were centered, this positive charge would be uniformly distributed along the inner wall. However, since the particle is off-center, the distribution of positive charge is skewed, as suggested by Fig. 23-13b, because the positive charge tends to collect on the section of the inner wall nearest the (negative) particle.

Because the shell is electrically neutral, its inner wall can have a charge of $+5.0\ \mu C$ only if electrons, with a total charge of $-5.0\ \mu C$, leave the inner wall and move to the outer wall. There they spread out uniformly, as is also suggested by Fig. 23-13b. This distribution of negative charge is

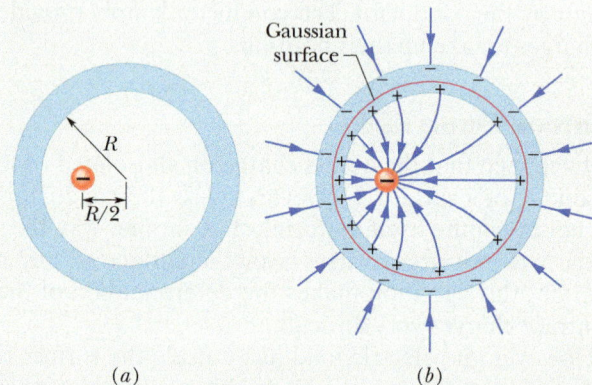

Figure 23-13 (a) A negatively charged particle is located within a spherical metal shell that is electrically neutral. (b) As a result, positive charge is nonuniformly distributed on the inner wall of the shell, and an equal amount of negative charge is uniformly distributed on the outer wall.

uniform because the shell is spherical and because the skewed distribution of positive charge on the inner wall cannot produce an electric field in the shell to affect the distribution of charge on the outer wall. Furthermore, these negative charges repel one another.

The field lines inside and outside the shell are shown approximately in Fig. 23-13b. All the field lines intersect the shell and the particle perpendicularly. Inside the shell the pattern of field lines is skewed because of the skew of the positive charge distribution. Outside the shell the pattern is the same as if the particle were centered and the shell were missing. In fact, this would be true no matter where inside the shell the particle happened to be located.

Additional examples, video, and practice available at *WileyPLUS*

23-4 APPLYING GAUSS' LAW: CYLINDRICAL SYMMETRY

Learning Objectives

After reading this module, you should be able to . . .

23.20 Explain how Gauss' law is used to derive the electric field magnitude outside a line of charge or a cylindrical surface (such as a plastic rod) with a uniform linear charge density λ.

23.21 Apply the relationship between linear charge density λ

on a cylindrical surface and the electric field magnitude E at radial distance r from the central axis.

23.22 Explain how Gauss' law can be used to find the electric field magnitude *inside* a cylindrical nonconducting surface (such as a plastic rod) with a uniform volume charge density ρ.

Key Idea

● The electric field at a point near an infinite line of charge (or charged rod) with uniform linear charge density λ is perpendicular to the line and has magnitude

$$E = \frac{\lambda}{2\pi\varepsilon_0 r} \quad \text{(line of charge),}$$

where r is the perpendicular distance from the line to the point.

Applying Gauss' Law: Cylindrical Symmetry

Figure 23-14 shows a section of an infinitely long cylindrical plastic rod with a uniform charge density λ. We want to find an expression for the electric field magnitude E at radius r from the central axis of the rod, outside the rod. We could do that using the approach of Chapter 22 (charge element dq, field vector $d\vec{E}$, etc.). However, Gauss' law gives a much faster and easier (and prettier) approach.

The charge distribution and the field have cylindrical symmetry. To find the field at radius r, we enclose a section of the rod with a concentric Gaussian cylinder of radius r and height h. (If you want the field at a certain point, put a Gaussian surface through that point.) We can now apply Gauss' law to relate the charge enclosed by the cylinder and the net flux through the cylinder's surface.

First note that because of the symmetry, the electric field at any point must be radially outward (the charge is positive). That means that at any point on the end caps, the field only skims the surface and does not pierce it. So, the flux through each end cap is zero.

To find the flux through the cylinder's curved surface, first note that for any patch element on the surface, the area vector $d\vec{A}$ is radially outward (away from the interior of the Gaussian surface) and thus in the same direction as the field piercing the patch. The dot product in Gauss' law is then simply $E\,dA\cos 0 = E\,dA$, and we can pull E out of the integral. The remaining integral is just the instruction to sum the areas of all patch elements on the cylinder's curved surface, but we already know that the total area is the product of the cylinder's height h and circumference $2\pi r$. The net flux through the cylinder is then

$$\Phi = EA\cos\theta = E(2\pi rh)\cos 0 = E(2\pi rh).$$

On the other side of Gauss's law we have the charge q_{enc} enclosed by the cylinder. Because the linear charge density (charge per unit length, remember) is uniform, the enclosed charge is λh. Thus, Gauss' law,

$$\varepsilon_0\Phi = q_{enc},$$

reduces to

$$\varepsilon_0 E(2\pi rh) = \lambda h,$$

yielding

$$E = \frac{\lambda}{2\pi\varepsilon_0 r} \quad \text{(line of charge).} \tag{23-12}$$

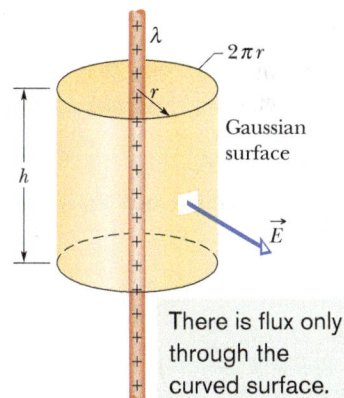

Figure 23-14 A Gaussian surface in the form of a closed cylinder surrounds a section of a very long, uniformly charged, cylindrical plastic rod.

This is the electric field due to an infinitely long, straight line of charge, at a point that is a radial distance r from the line. The direction of $\vec{E}$ is radially outward from the line of charge if the charge is positive, and radially inward if it is negative. Equation 23-12 also approximates the field of a *finite* line of charge at points that are not too near the ends (compared with the distance from the line).

If the rod has a uniform volume charge density ρ, we could use a similar procedure to find the electric field magnitude *inside* the rod. We would just shrink the Gaussian cylinder shown in Fig. 23-14 until it is inside the rod. The charge q_{enc} enclosed by the cylinder would then be proportional to the volume of the rod enclosed by the cylinder because the charge density is uniform.

Sample Problem 23.06 Gauss' law and an upward streamer in a lightning storm

Upward streamer in a lightning storm. The woman in Fig. 23-15 was standing on a lookout platform high in the Sequoia National Park when a large storm cloud moved overhead. Some of the conduction electrons in her body were driven into the ground by the cloud's negatively charged base (Fig. 23-16a), leaving her positively charged. You can tell she was highly charged because her hair strands repelled one another and extended away from her along the electric field lines produced by the charge on her.

Lightning did not strike the woman, but she was in extreme danger because that electric field was on the verge of causing electrical breakdown in the surrounding air. Such a breakdown would have occurred along a path extending away from her in what is called an *upward streamer*. An upward streamer is dangerous because the resulting ionization of molecules in the air suddenly frees a tremendous number of electrons from those molecules. Had the woman in Fig. 23-15 developed an upward streamer, the free electrons in the air would have moved to neutralize her (Fig. 23-16b), producing a large, perhaps fatal, charge flow through her body. That charge flow is dangerous because it could have interfered with or even stopped her breathing (which is obviously necessary for oxygen) and the steady beat of her heart (which is obviously necessary for the blood flow that carries the oxygen). The charge flow could also have caused burns.

Let's model her body as a narrow vertical cylinder of height $L = 1.8$ m and radius $R = 0.10$ m (Fig. 23-16c). Assume that charge Q was uniformly distributed along the cylinder and that electrical breakdown would have occurred if the electric

Courtesy NOAA

Figure 23-15 This woman has become positively charged by an overhead storm cloud.

field magnitude along her body had exceeded the critical value $E_c = 2.4$ MN/C. What value of Q would have put the air along her body on the verge of breakdown?

KEY IDEA

Because $R \ll L$, we can approximate the charge distribution as a long line of charge. Further, because we assume that the charge is uniformly distributed along this line, we can approximate the magnitude of the electric field along the side of her body with Eq. 23-12 ($E = \lambda/2\pi\varepsilon_0 r$).

Calculations: Substituting the critical value E_c for E, the cylinder radius R for radial distance r, and the ratio Q/L for linear charge density λ, we have

$$E_c = \frac{Q/L}{2\pi\varepsilon_0 R},$$

or

$$Q = 2\pi\varepsilon_0 RLE_c.$$

Substituting given data then gives us

$$Q = (2\pi)(8.85 \times 10^{-12} \text{ C}^2/\text{N}\cdot\text{m}^2)(0.10 \text{ m})$$
$$\times (1.8 \text{ m})(2.4 \times 10^6 \text{ N/C})$$
$$= 2.402 \times 10^{-5} \text{ C} \approx 24 \ \mu\text{C}. \qquad \text{(Answer)}$$

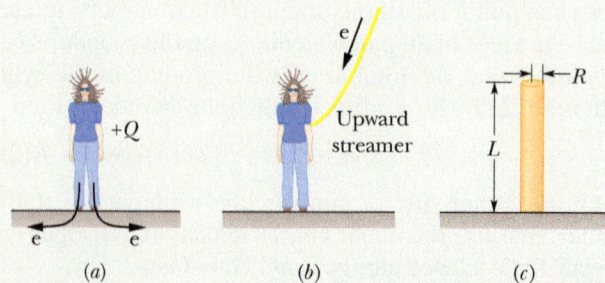

Figure 23-16 (a) Some of the conduction electrons in the woman's body are driven into the ground, leaving her positively charged. (b) An upward streamer develops if the air undergoes electrical breakdown, which provides a path for electrons freed from molecules in the air to move to the woman. (c) A cylinder represents the woman.

23-5 APPLYING GAUSS' LAW: PLANAR SYMMETRY

Learning Objectives

After reading this module, you should be able to . . .

23.23 Apply Gauss' law to derive the electric field magnitude E near a large, flat, nonconducting surface with a uniform surface charge density σ.

23.24 For points near a large, flat *nonconducting* surface with a uniform charge density σ, apply the relationship be-tween the charge density and the electric field magnitude E and also specify the direction of the field.

23.25 For points near two large, flat, parallel, *conducting* sur-faces with a uniform charge density σ, apply the relation-ship between the charge density and the electric field magnitude E and also specify the direction of the field.

Key Ideas

● The electric field due to an infinite nonconducting sheet with uniform surface charge density σ is perpendicular to the plane of the sheet and has magnitude

$$E = \frac{\sigma}{2\varepsilon_0} \quad \text{(nonconducting sheet of charge)}.$$

● The external electric field just outside the surface of an iso-lated charged conductor with surface charge density σ is per-pendicular to the surface and has magnitude

$$E = \frac{\sigma}{\varepsilon_0} \quad \text{(external, charged conductor)}.$$

Inside the conductor, the electric field is zero.

Applying Gauss' Law: Planar Symmetry

Nonconducting Sheet

Figure 23-17 shows a portion of a thin, infinite, nonconducting sheet with a uni-form (positive) surface charge density σ. A sheet of thin plastic wrap, uniformly charged on one side, can serve as a simple model. Let us find the electric field $\vec{E}$ a distance r in front of the sheet.

A useful Gaussian surface is a closed cylinder with end caps of area A, arranged to pierce the sheet perpendicularly as shown. From symmetry, $\vec{E}$ must be perpendicular to the sheet and hence to the end caps. Furthermore, since the charge is positive, $\vec{E}$ is directed *away* from the sheet, and thus the electric field lines pierce the two Gaussian end caps in an outward direction. Because the field lines do not pierce the curved surface, there is no flux through this portion of the Gaussian surface. Thus $\vec{E} \cdot d\vec{A}$ is simply $E\,dA$; then Gauss' law,

$$\varepsilon_0 \oint \vec{E} \cdot d\vec{A} = q_{\text{enc}},$$

becomes

$$\varepsilon_0 (EA + EA) = \sigma A,$$

where σA is the charge enclosed by the Gaussian surface. This gives

$$E = \frac{\sigma}{2\varepsilon_0} \quad \text{(sheet of charge)}. \tag{23-13}$$

Since we are considering an infinite sheet with uniform charge density, this result holds for any point at a finite distance from the sheet. Equation 23-13 agrees with Eq. 22-27, which we found by integration of electric field components.

Figure 23-17 (*a*) Perspective view and (*b*) side view of a portion of a very large, thin plastic sheet, uni-formly charged on one side to sur-face charge density σ. A closed cylindrical Gaussian surface passes through the sheet and is perpendi-cular to it.

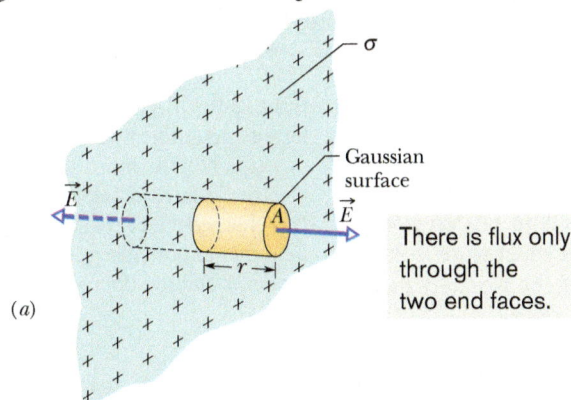

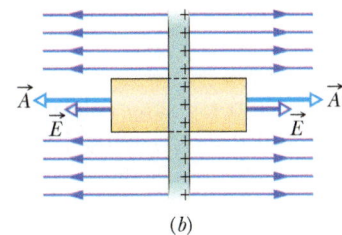

Gaussian surface

There is flux only through the two end faces.

(a)

(b)

(a)   (b)

(c)

Figure 23-18 (a) A thin, very large conducting plate with excess positive charge. (b) An identical plate with excess negative charge. (c) The two plates arranged so they are parallel and close.

Two Conducting Plates

Figure 23-18a shows a cross section of a thin, infinite conducting plate with excess positive charge. From Module 23-3 we know that this excess charge lies on the surface of the plate. Since the plate is thin and very large, we can assume that essentially all the excess charge is on the two large faces of the plate.

If there is no external electric field to force the positive charge into some particular distribution, it will spread out on the two faces with a uniform surface charge density of magnitude σ_1. From Eq. 23-11 we know that just outside the plate this charge sets up an electric field of magnitude $E = \sigma_1/\varepsilon_0$. Because the excess charge is positive, the field is directed away from the plate.

Figure 23-18b shows an identical plate with excess negative charge having the same magnitude of surface charge density σ_1. The only difference is that now the electric field is directed toward the plate.

Suppose we arrange for the plates of Figs. 23-18a and b to be close to each other and parallel (Fig. 23-18c). Since the plates are conductors, when we bring them into this arrangement, the excess charge on one plate attracts the excess charge on the other plate, and all the excess charge moves onto the inner faces of the plates as in Fig. 23-18c. With twice as much charge now on each inner face, the new surface charge density (call it σ) on each inner face is twice σ_1. Thus, the electric field at any point between the plates has the magnitude

$$E = \frac{2\sigma_1}{\varepsilon_0} = \frac{\sigma}{\varepsilon_0}. \qquad (23\text{-}14)$$

This field is directed away from the positively charged plate and toward the negatively charged plate. Since no excess charge is left on the outer faces, the electric field to the left and right of the plates is zero.

Because the charges moved when we brought the plates close to each other, the charge distribution of the two-plate system is not merely the sum of the charge distributions of the individual plates.

One reason why we discuss seemingly unrealistic situations, such as the field set up by an infinite sheet of charge, is that analyses for "infinite" situations yield good approximations to many real-world problems. Thus, Eq. 23-13 holds well for a finite nonconducting sheet as long as we are dealing with points close to the sheet and not too near its edges. Equation 23-14 holds well for a pair of finite conducting plates as long as we consider points that are not too close to their edges. The trouble with the edges is that near an edge we can no longer use planar symmetry to find expressions for the fields. In fact, the field lines there are curved (said to be an *edge effect* or *fringing*), and the fields can be very difficult to express algebraically.

Sample Problem 23.07 Electric field near two parallel nonconducting sheets with charge

Figure 23-19a shows portions of two large, parallel, nonconducting sheets, each with a fixed uniform charge on one side. The magnitudes of the surface charge densities are $\sigma_{(+)} = 6.8\ \mu C/m^2$ for the positively charged sheet and $\sigma_{(-)} = 4.3\ \mu C/m^2$ for the negatively charged sheet.

Find the electric field $\vec{E}$ (a) to the left of the sheets, (b) between the sheets, and (c) to the right of the sheets.

KEY IDEA

With the charges fixed in place (they are on nonconductors), we can find the electric field of the sheets in Fig. 23-19a by (1) finding the field of each sheet as if that sheet were isolated and (2) algebraically adding the fields of the isolated sheets

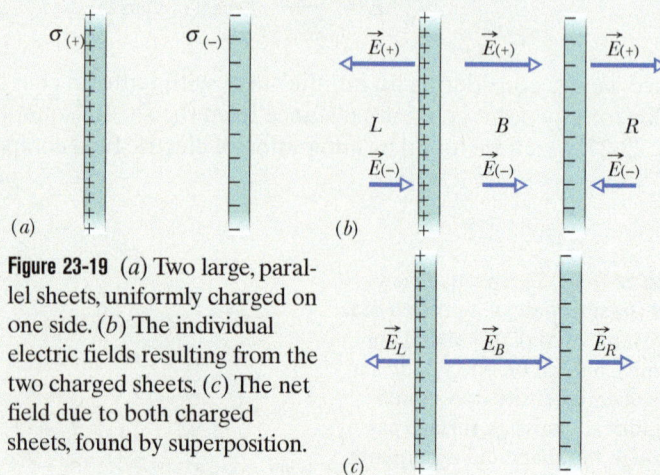

Figure 23-19 (a) Two large, parallel sheets, uniformly charged on one side. (b) The individual electric fields resulting from the two charged sheets. (c) The net field due to both charged sheets, found by superposition.

via the superposition principle. (We can add the fields algebraically because they are parallel to each other.)

Calculations: At any point, the electric field $\vec{E}_{(+)}$ due to the positive sheet is directed *away* from the sheet and, from Eq. 23-13, has the magnitude

$$E_{(+)} = \frac{\sigma_{(+)}}{2\varepsilon_0} = \frac{6.8 \times 10^{-6}\ \text{C/m}^2}{(2)(8.85 \times 10^{-12}\ \text{C}^2/\text{N} \cdot \text{m}^2)}$$
$$= 3.84 \times 10^5\ \text{N/C}.$$

Similarly, at any point, the electric field $\vec{E}_{(-)}$ due to the negative sheet is directed *toward* that sheet and has the magnitude

$$E_{(-)} = \frac{\sigma_{(-)}}{2\varepsilon_0} = \frac{4.3 \times 10^{-6}\ \text{C/m}^2}{(2)(8.85 \times 10^{-12}\ \text{C}^2/\text{N} \cdot \text{m}^2)}$$
$$= 2.43 \times 10^5\ \text{N/C}.$$

Figure 23-19b shows the fields set up by the sheets to the left of the sheets (L), between them (B), and to their right (R).

The resultant fields in these three regions follow from the superposition principle. To the left, the field magnitude is

$$E_L = E_{(+)} - E_{(-)}$$
$$= 3.84 \times 10^5\ \text{N/C} - 2.43 \times 10^5\ \text{N/C}$$
$$= 1.4 \times 10^5\ \text{N/C}. \qquad \text{(Answer)}$$

Because $E_{(+)}$ is larger than $E_{(-)}$, the net electric field $\vec{E}_L$ in this region is directed to the left, as Fig. 23-19c shows. To the right of the sheets, the net electric field has the same magnitude but is directed to the right, as Fig. 23-19c shows.

Between the sheets, the two fields add and we have

$$E_B = E_{(+)} + E_{(-)}$$
$$= 3.84 \times 10^5\ \text{N/C} + 2.43 \times 10^5\ \text{N/C}$$
$$= 6.3 \times 10^5\ \text{N/C}. \qquad \text{(Answer)}$$

The electric field $\vec{E}_B$ is directed to the right.

WILEY PLUS Additional examples, video, and practice available at *WileyPLUS*

23-6 APPLYING GAUSS' LAW: SPHERICAL SYMMETRY

Learning Objectives

After reading this module, you should be able to . . .

23.26 Identify that a shell of uniform charge attracts or repels a charged particle that is outside the shell as if all the shell's charge is concentrated at the center of the shell.

23.27 Identify that if a charged particle is enclosed by a shell of uniform charge, there is no electrostatic force on the particle from the shell.

23.28 For a point outside a spherical shell with uniform charge, apply the relationship between the electric field magnitude E, the charge q on the shell, and the distance r from the shell's center.

23.29 Identify the magnitude of the electric field for points enclosed by a spherical shell with uniform charge.

23.30 For a uniform spherical charge distribution (a uniform ball of charge), determine the magnitude and direction of the electric field at interior and exterior points.

Key Ideas

● Outside a spherical shell of uniform charge q, the electric field due to the shell is radial (inward or outward, depending on the sign of the charge) and has the magnitude

$$E = \frac{1}{4\pi\varepsilon_0} \frac{q}{r^2} \quad \text{(outside spherical shell)},$$

where r is the distance to the point of measurement from the center of the shell. The field is the same as though all of the charge is concentrated as a particle at the center of the shell.

● Inside the shell, the field due to the shell is zero.

● Inside a sphere with a uniform volume charge density, the field is radial and has the magnitude

$$E = \frac{1}{4\pi\varepsilon_0} \frac{q}{R^3} r \quad \text{(inside sphere of charge)},$$

where q is the total charge, R is the sphere's radius, and r is the radial distance from the center of the sphere to the point of measurement.

Applying Gauss' Law: Spherical Symmetry

Here we use Gauss' law to prove the two shell theorems presented without proof in Module 21-1:

> A shell of uniform charge attracts or repels a charged particle that is outside the shell as if all the shell's charge were concentrated at the center of the shell.

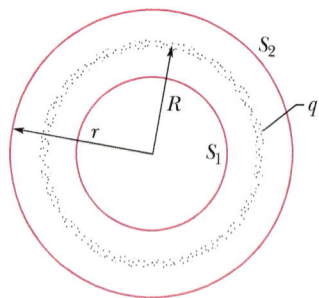

Figure 23-20 A thin, uniformly charged, spherical shell with total charge q, in cross section. Two Gaussian surfaces S_1 and S_2 are also shown in cross section. Surface S_2 encloses the shell, and S_1 encloses only the empty interior of the shell.

Figure 23-20 shows a charged spherical shell of total charge q and radius R and two concentric spherical Gaussian surfaces, S_1 and S_2. If we followed the procedure of Module 23-2 as we applied Gauss' law to surface S_2, for which $r \geq R$, we would find that

$$E = \frac{1}{4\pi\varepsilon_0} \frac{q}{r^2} \quad \text{(spherical shell, field at } r \geq R\text{).} \tag{23-15}$$

This field is the same as one set up by a particle with charge q at the center of the shell of charge. Thus, the force produced by a shell of charge q on a charged particle placed outside the shell is the same as if all the shell's charge is concentrated as a particle at the shell's center. This proves the first shell theorem.

Applying Gauss' law to surface S_1, for which $r < R$, leads directly to

$$E = 0 \quad \text{(spherical shell, field at } r < R\text{),} \tag{23-16}$$

because this Gaussian surface encloses no charge. Thus, if a charged particle were enclosed by the shell, the shell would exert no net electrostatic force on the particle. This proves the second shell theorem.

⭐ If a charged particle is located inside a shell of uniform charge, there is no electrostatic force on the particle from the shell.

Any spherically symmetric charge distribution, such as that of Fig. 23-21, can be constructed with a nest of concentric spherical shells. For purposes of applying the two shell theorems, the volume charge density ρ should have a single value for each shell but need not be the same from shell to shell. Thus, for the charge distribution as a whole, ρ can vary, but only with r, the radial distance from the center. We can then examine the effect of the charge distribution "shell by shell."

In Fig. 23-21a, the entire charge lies within a Gaussian surface with $r > R$. The charge produces an electric field on the Gaussian surface as if the charge were that of a particle located at the center, and Eq. 23-15 holds.

Figure 23-21b shows a Gaussian surface with $r < R$. To find the electric field at points on this Gaussian surface, we separately consider the charge inside it and the charge outside it. From Eq. 23-16, the outside charge does not set up a field on the Gaussian surface. From Eq. 23-15, the inside charge sets up a field as though it is concentrated at the center. Letting q' represent that enclosed charge, we can then rewrite Eq. 23-15 as

$$E = \frac{1}{4\pi\varepsilon_0} \frac{q'}{r^2} \quad \text{(spherical distribution, field at } r \leq R\text{).} \tag{23-17}$$

If the full charge q enclosed within radius R is uniform, then q' enclosed within radius r in Fig. 23-21b is proportional to q:

$$\frac{\left(\begin{array}{c}\text{charge enclosed by}\\ \text{sphere of radius } r\end{array}\right)}{\left(\begin{array}{c}\text{volume enclosed by}\\ \text{sphere of radius } r\end{array}\right)} = \frac{\text{full charge}}{\text{full volume}}$$

or

$$\frac{q'}{\frac{4}{3}\pi r^3} = \frac{q}{\frac{4}{3}\pi R^3}. \tag{23-18}$$

This gives us

$$q' = q \frac{r^3}{R^3}. \tag{23-19}$$

Substituting this into Eq. 23-17 yields

$$E = \left(\frac{q}{4\pi\varepsilon_0 R^3}\right) r \quad \text{(uniform charge, field at } r \leq R\text{).} \tag{23-20}$$

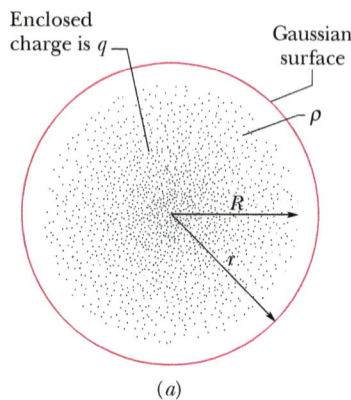

(a)

(b)

The flux through the surface depends on only the *enclosed* charge.

Figure 23-21 The dots represent a spherically symmetric distribution of charge of radius R, whose volume charge density ρ is a function only of distance from the center. The charged object is not a conductor, and therefore the charge is assumed to be fixed in position. A concentric spherical Gaussian surface with $r > R$ is shown in (a). A similar Gaussian surface with $r < R$ is shown in (b).

Checkpoint 4

The figure shows two large, parallel, nonconducting sheets with identical (positive) uniform surface charge densities, and a sphere with a uniform (positive) volume charge density. Rank the four numbered points according to the magnitude of the net electric field there, greatest first.

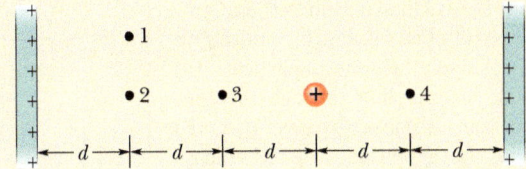

Review & Summary

Gauss' Law *Gauss' law* and Coulomb's law are different ways of describing the relation between charge and electric field in static situations. Gauss' law is

$$\varepsilon_0 \Phi = q_{enc} \quad \text{(Gauss' law)}, \quad (23\text{-}6)$$

in which q_{enc} is the net charge inside an imaginary closed surface (a *Gaussian surface*) and Φ is the net *flux* of the electric field through the surface:

$$\Phi = \oint \vec{E} \cdot d\vec{A} \quad \begin{array}{l}\text{(electric flux through a}\\ \text{Gaussian surface).}\end{array} \quad (23\text{-}4)$$

Coulomb's law can be derived from Gauss' law.

Applications of Gauss' Law Using Gauss' law and, in some cases, symmetry arguments, we can derive several important results in electrostatic situations. Among these are:

1. An excess charge on an isolated *conductor* is located entirely on the outer surface of the conductor.

2. The external electric field near the *surface of a charged conductor* is perpendicular to the surface and has a magnitude that depends on the surface charge density σ:

$$E = \frac{\sigma}{\varepsilon_0} \quad \text{(conducting surface).} \quad (23\text{-}11)$$

Within the conductor, $E = 0$.

3. The electric field at any point due to an infinite *line of charge*

with uniform linear charge density λ is perpendicular to the line of charge and has magnitude

$$E = \frac{\lambda}{2\pi\varepsilon_0 r} \quad \text{(line of charge)}, \quad (23\text{-}12)$$

where r is the perpendicular distance from the line of charge to the point.

4. The electric field due to an *infinite nonconducting sheet* with uniform surface charge density σ is perpendicular to the plane of the sheet and has magnitude

$$E = \frac{\sigma}{2\varepsilon_0} \quad \text{(sheet of charge).} \quad (23\text{-}13)$$

5. The electric field *outside a spherical shell of charge* with radius R and total charge q is directed radially and has magnitude

$$E = \frac{1}{4\pi\varepsilon_0}\frac{q}{r^2} \quad \text{(spherical shell, for } r \geq R). \quad (23\text{-}15)$$

Here r is the distance from the center of the shell to the point at which E is measured. (The charge behaves, for external points, as if it were all located at the center of the sphere.) The field *inside* a uniform spherical shell of charge is exactly zero:

$$E = 0 \quad \text{(spherical shell, for } r < R). \quad (23\text{-}16)$$

6. The electric field *inside a uniform sphere of charge* is directed radially and has magnitude

$$E = \left(\frac{q}{4\pi\varepsilon_0 R^3}\right)r. \quad (23\text{-}20)$$

Questions

1 A surface has the area vector $\vec{A} = (2\hat{i} + 3\hat{j})$ m². What is the flux of a uniform electric field through the area if the field is (a) $\vec{E} = 4\hat{i}$ N/C and (b) $\vec{E} = 4\hat{k}$ N/C?

2 Figure 23-22 shows, in cross section, three solid cylinders, each of length L and uniform charge Q. Concentric with each cylinder is a cylindrical Gaussian surface, with all three surfaces having the same radius. Rank the Gaussian surfaces according to the electric field at any point on the surface, greatest first.

3 Figure 23-23 shows, in cross section, a central metal ball, two spherical metal shells, and three spherical Gaussian surfaces of radii $R, 2R$, and $3R$, all with the same center. The uniform charges on the three objects are: ball, Q; smaller shell, $3Q$; larger shell, $5Q$. Rank the Gaussian surfaces according to the magnitude of the electric field at any point on the surface, greatest first.

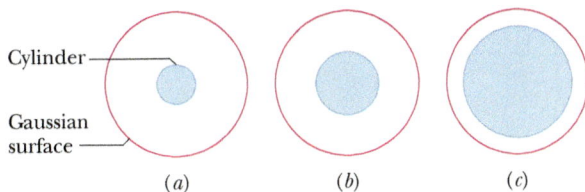

Figure 23-22 Question 2.

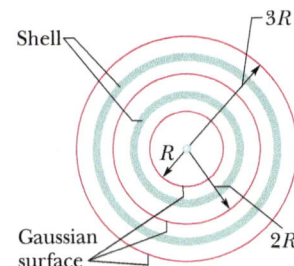

Figure 23-23 Question 3.

4 Figure 23-24 shows, in cross section, two Gaussian spheres and two Gaussian cubes that are centered on a positively charged particle. (a) Rank the net flux through the four Gaussian surfaces, greatest first. (b) Rank the magnitudes of the electric fields on the surfaces, greatest first, and indicate whether the magnitudes are uniform or variable along each surface.

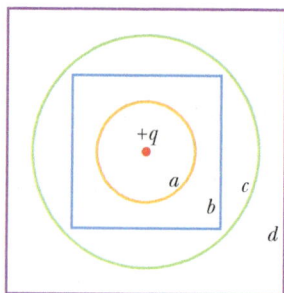

Figure 23-24 Question 4.

5 In Fig. 23-25, an electron is released between two infinite nonconducting sheets that are horizontal and have uniform surface charge densities $\sigma_{(+)}$ and $\sigma_{(-)}$, as indicated. The electron is subjected to the following three situations involving surface charge densities and sheet separations. Rank the magnitudes of the electron's acceleration, greatest first.

Situation	$\sigma_{(+)}$	$\sigma_{(-)}$	Separation
1	$+4\sigma$	-4σ	d
2	$+7\sigma$	$-\sigma$	$4d$
3	$+3\sigma$	-5σ	$9d$

Figure 23-25 Question 5.

6 Three infinite nonconducting sheets, with uniform positive surface charge densities σ, 2σ, and 3σ, are arranged to be parallel like the two sheets in Fig. 23-19a. What is their order, from left to right, if the electric field $\vec{E}$ produced by the arrangement has magnitude $E = 0$ in one region and $E = 2\sigma/\varepsilon_0$ in another region?

7 Figure 23-26 shows four situations in which four very long rods extend into and out of the page (we see only their cross sections). The value below each cross section gives that particular rod's uniform charge density in microcoulombs per meter. The rods are separated by either d or $2d$ as drawn, and a central point is shown midway between the inner rods. Rank the situations according to the magnitude of the net electric field at that central point, greatest first.

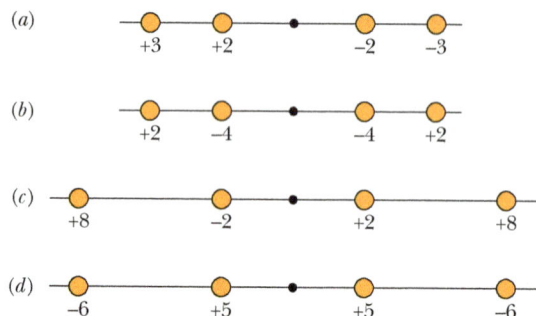

Figure 23-26 Question 7.

8 Figure 23-27 shows four solid spheres, each with charge Q uniformly distributed through its volume. (a) Rank the spheres according to their volume charge density, greatest first. The figure also shows a point P for each sphere, all at the same distance from the center of the sphere. (b) Rank the spheres according to the magnitude of the electric field they produce at point P, greatest first.

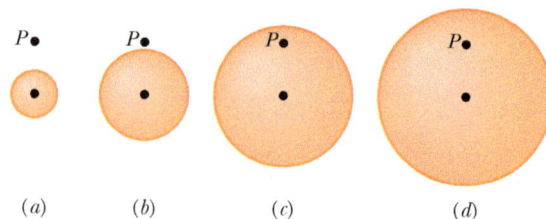

Figure 23-27 Question 8.

9 A small charged ball lies within the hollow of a metallic spherical shell of radius R. For three situations, the net charges on the ball and shell, respectively, are (1) $+4q$, 0; (2) $-6q$, $+10q$; (3) $+16q$, $-12q$. Rank the situations according to the charge on (a) the inner surface of the shell and (b) the outer surface, most positive first.

10 Rank the situations of Question 9 according to the magnitude of the electric field (a) halfway through the shell and (b) at a point $2R$ from the center of the shell, greatest first.

11 Figure 23-28 shows a section of three long charged cylinders centered on the same axis. Central cylinder A has a uniform charge $q_A = +3q_0$. What uniform charges q_B and q_C should be on cylinders B and C so that (if possible) the net electric field is zero at (a) point 1, (b) point 2, and (c) point 3?

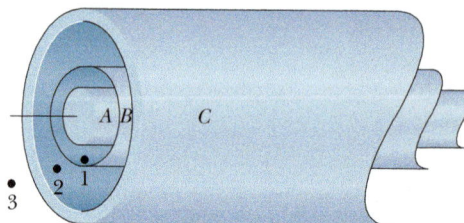

Figure 23-28 Question 11.

12 Figure 23-29 shows four Gaussian surfaces consisting of identical cylindrical midsections but different end caps. The surfaces are in a uniform electric field $\vec{E}$ that is directed parallel to the central axis of each cylindrical midsection. The end caps have these shapes: S_1, convex hemispheres; S_2, concave hemispheres; S_3, cones; S_4, flat disks. Rank the surfaces according to (a) the net electric flux through them and (b) the electric flux through the top end caps, greatest first.

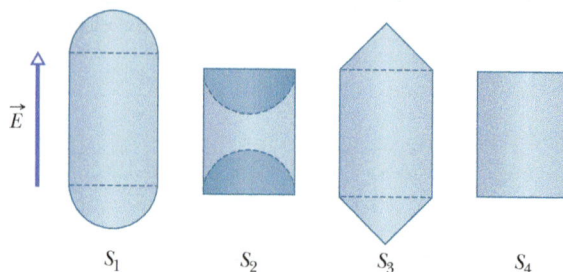

Figure 23-29 Question 12.

Problems

Module 23-1   Electric Flux

•1 **SSM** The square surface shown in Fig. 23-30 measures 3.2 mm on each side. It is immersed in a uniform electric field with magnitude $E = 1800$ N/C and with field lines at an angle of $\theta = 35°$ with a normal to the surface, as shown. Take that normal to be directed "outward," as though the surface were one face of a box. Calculate the electric flux through the surface.

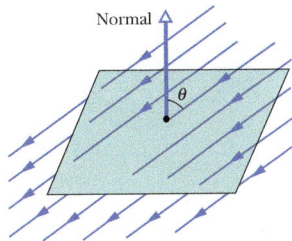
Figure 23-30 Problem 1.

•2 An electric field given by $\vec{E} = 4.0\hat{i} - 3.0(y^2 + 2.0)\hat{j}$ pierces a Gaussian cube of edge length 2.0 m and positioned as shown in Fig. 23-7. (The magnitude E is in newtons per coulomb and the position x is in meters.) What is the electric flux through the (a) top face, (b) bottom face, (c) left face, and (d) back face? (e) What is the net electric flux through the cube?

•3 The cube in Fig. 23-31 has edge length 1.40 m and is oriented as shown in a region of uniform electric field. Find the electric flux through the right face if the electric field, in newtons per coulomb, is given by (a) $6.00\hat{i}$, (b) $-2.00\hat{j}$, and (c) $-3.00\hat{i} + 4.00\hat{k}$. (d) What is the total flux through the cube for each field?

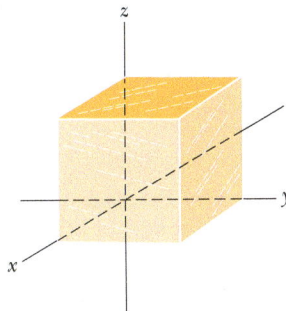
Figure 23-31 Problems 3, 6, and 9.

Module 23-2   Gauss' Law

•4 In Fig. 23-32, a butterfly net is in a uniform electric field of magnitude $E = 3.0$ mN/C. The rim, a circle of radius $a = 11$ cm, is aligned perpendicular to the field. The net contains no net charge. Find the electric flux through the netting.

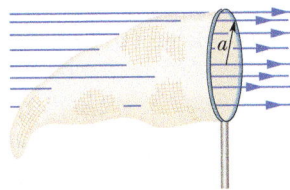
Figure 23-32 Problem 4.

•5 In Fig. 23-33, a proton is a distance $d/2$ directly above the center of a square of side d. What is the magnitude of the electric flux through the square? (*Hint:* Think of the square as one face of a cube with edge d.)

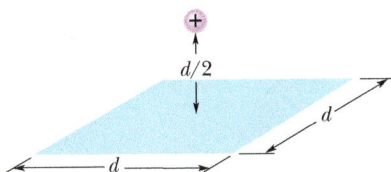
Figure 23-33 Problem 5.

•6 At each point on the surface of the cube shown in Fig. 23-31, the electric field is parallel to the z axis. The length of each edge of the cube is 3.0 m. On the top face of the cube the field is

$\vec{E} = -34\hat{k}$ N/C, and on the bottom face it is $\vec{E} = +20\hat{k}$ N/C. Determine the net charge contained within the cube.

•7 A particle of charge 1.8 μC is at the center of a Gaussian cube 55 cm on edge. What is the net electric flux through the surface?

••8 When a shower is turned on in a closed bathroom, the splashing of the water on the bare tub can fill the room's air with negatively charged ions and produce an electric field in the air as great as 1000 N/C. Consider a bathroom with dimensions 2.5 m × 3.0 m × 2.0 m. Along the ceiling, floor, and four walls, approximate the electric field in the air as being directed perpendicular to the surface and as having a uniform magnitude of 600 N/C. Also, treat those surfaces as forming a closed Gaussian surface around the room's air. What are (a) the volume charge density ρ and (b) the number of excess elementary charges e per cubic meter in the room's air?

••9 **ILW** Fig. 23-31 shows a Gaussian surface in the shape of a cube with edge length 1.40 m. What are (a) the net flux Φ through the surface and (b) the net charge q_{enc} enclosed by the surface if $\vec{E} = (3.00y\hat{j})$ N/C, with y in meters? What are (c) Φ and (d) q_{enc} if $\vec{E} = [-4.00\hat{i} + (6.00 + 3.00y)\hat{j}]$ N/C?

••10 Figure 23-34 shows a closed Gaussian surface in the shape of a cube of edge length 2.00 m. It lies in a region where the nonuniform electric field is given by $\vec{E} = (3.00x + 4.00)\hat{i} + 6.00\hat{j} + 7.00\hat{k}$ N/C, with x in meters. What is the net charge contained by the cube?

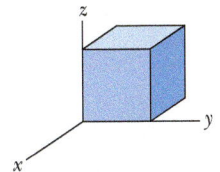
Figure 23-34 Problem 10.

••11 **GO** Figure 23-35 shows a closed Gaussian surface in the shape of a cube of edge length 2.00 m, with one corner at $x_1 = 5.00$ m, $y_1 = 4.00$ m. The cube lies in a region where the electric field vector is given by $\vec{E} = -3.00\hat{i} - 4.00y^2\hat{j} + 3.00\hat{k}$ N/C, with y in meters. What is the net charge contained by the cube?

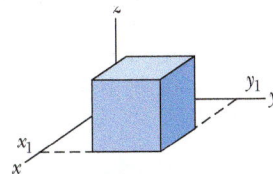
Figure 23-35 Problem 11.

••12 Figure 23-36 shows two nonconducting spherical shells fixed in place. Shell 1 has uniform surface charge density $+6.0$ μC/m² on its outer surface and radius 3.0 cm; shell 2 has uniform surface charge density $+4.0$ μC/m² on its outer surface and radius 2.0 cm; the shell centers are separated by $L = 10$ cm. In unit-vector notation, what is the net electric field at $x = 2.0$ cm?

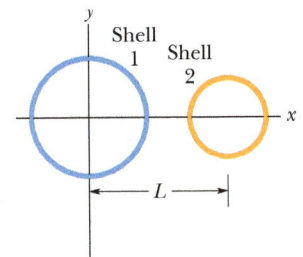
Figure 23-36 Problem 12.

••**13** SSM The electric field in a certain region of Earth's atmosphere is directed vertically down. At an altitude of 300 m the field has magnitude 60.0 N/C; at an altitude of 200 m, the magnitude is 100 N/C. Find the net amount of charge contained in a cube 100 m on edge, with horizontal faces at altitudes of 200 and 300 m.

••**14** GO *Flux and nonconducting shells.* A charged particle is suspended at the center of two concentric spherical shells that are very thin and made of nonconducting material. Figure 23-37a shows a cross section. Figure 23-37b gives the net flux Φ through a Gaussian sphere centered on the particle, as a function of the radius r of the sphere. The scale of the vertical axis is set by $\Phi_s = 5.0 \times 10^5$ N·m²/C. (a) What is the charge of the central particle? What are the net charges of (b) shell A and (c) shell B?

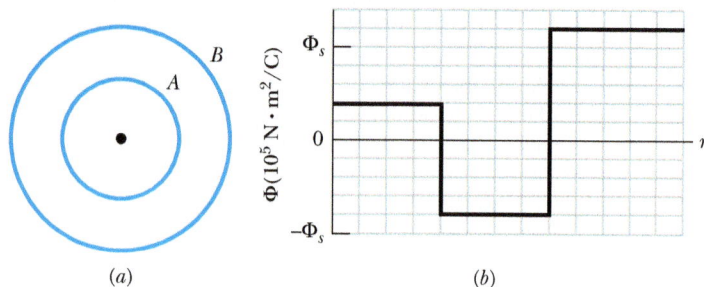

Figure 23-37 Problem 14.

••**15** A particle of charge $+q$ is placed at one corner of a Gaussian cube. What multiple of q/ε_0 gives the flux through (a) each cube face forming that corner and (b) each of the other cube faces?

•••**16** GO The box-like Gaussian surface shown in Fig. 23-38 encloses a net charge of $+24.0\varepsilon_0$ C and lies in an electric field given by $\vec{E} = [(10.0 + 2.00x)\hat{i} - 3.00\hat{j} + bz\hat{k}]$ N/C, with x and z in meters and b a constant. The bottom face is in the xz plane; the top face is in the horizontal plane passing through $y_2 = 1.00$ m. For $x_1 = 1.00$ m, $x_2 = 4.00$ m, $z_1 = 1.00$ m, and $z_2 = 3.00$ m, what is b?

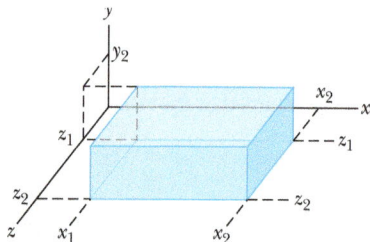

Figure 23-38 Problem 16.

Module 23-3 A Charged Isolated Conductor

•**17** SSM A uniformly charged conducting sphere of 1.2 m diameter has surface charge density 8.1 μC/m². Find (a) the net charge on the sphere and (b) the total electric flux leaving the surface.

•**18** The electric field just above the surface of the charged conducting drum of a photocopying machine has a magnitude E of 2.3×10^5 N/C. What is the surface charge density on the drum?

•**19** Space vehicles traveling through Earth's radiation belts can intercept a significant number of electrons. The resulting charge buildup can damage electronic components and disrupt operations. Suppose a spherical metal satellite 1.3 m in diameter accumulates 2.4 μC of charge in one orbital revolution. (a) Find the resulting surface charge density. (b) Calculate the magnitude of the electric field just outside the surface of the satellite, due to the surface charge.

•**20** GO *Flux and conducting shells.* A charged particle is held at the center of two concentric conducting spherical shells. Figure 23-39a shows a cross section. Figure 23-39b gives the net flux Φ through a Gaussian sphere centered on the particle, as a function of the radius r of the sphere. The scale of the vertical axis is set by $\Phi_s = 5.0 \times 10^5$ N·m²/C. What are (a) the charge of the central particle and the net charges of (b) shell A and (c) shell B?

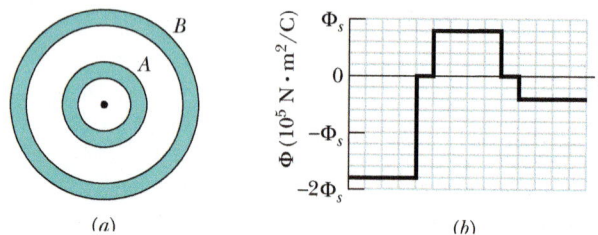

Figure 23-39 Problem 20.

••**21** An isolated conductor has net charge $+10 \times 10^{-6}$ C and a cavity with a particle of charge $q = +3.0 \times 10^{-6}$ C. What is the charge on (a) the cavity wall and (b) the outer surface?

Module 23-4 Applying Gauss' Law: Cylindrical Symmetry

•**22** An electron is released 9.0 cm from a very long nonconducting rod with a uniform 6.0 μC/m. What is the magnitude of the electron's initial acceleration?

•**23** (a) The drum of a photocopying machine has a length of 42 cm and a diameter of 12 cm. The electric field just above the drum's surface is 2.3×10^5 N/C. What is the total charge on the drum? (b) The manufacturer wishes to produce a desktop version of the machine. This requires reducing the drum length to 28 cm and the diameter to 8.0 cm. The electric field at the drum surface must not change. What must be the charge on this new drum?

•**24** Figure 23-40 shows a section of a long, thin-walled metal tube of radius $R = 3.00$ cm, with a charge per unit length of $\lambda = 2.00 \times 10^{-8}$ C/m. What is the magnitude E of the electric field at radial distance (a) $r = R/2.00$ and (b) $r = 2.00R$? (c) Graph E versus r for the range $r = 0$ to $2.00R$.

•**25** SSM An infinite line of charge produces a field of magnitude 4.5 × 10^4 N/C at distance 2.0 m. Find the linear charge density.

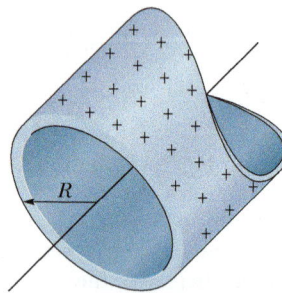

Figure 23-40 Problem 24.

••**26** Figure 23-41a shows a narrow charged solid cylinder that is coaxial with a larger charged cylindrical shell. Both are noncon-

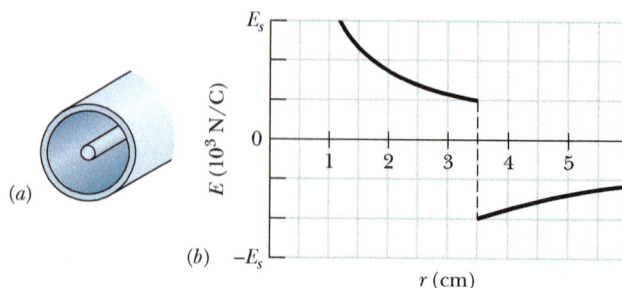

Figure 23-41 Problem 26.

ducting and thin and have uniform surface charge densities on their outer surfaces. Figure 23-41*b* gives the radial component E of the electric field versus radial distance r from the common axis, and $E_s = 3.0 \times 10^3$ N/C. What is the shell's linear charge density?

••27 GO A long, straight wire has fixed negative charge with a linear charge density of magnitude 3.6 nC/m. The wire is to be enclosed by a coaxial, thin-walled nonconducting cylindrical shell of radius 1.5 cm. The shell is to have positive charge on its outside surface with a surface charge density σ that makes the net external electric field zero. Calculate σ.

••28 GO A charge of uniform linear density 2.0 nC/m is distributed along a long, thin, nonconducting rod. The rod is coaxial with a long conducting cylindrical shell (inner radius = 5.0 cm, outer radius = 10 cm). The net charge on the shell is zero. (a) What is the magnitude of the electric field 15 cm from the axis of the shell? What is the surface charge density on the (b) inner and (c) outer surface of the shell?

••29 SSM WWW Figure 23-42 is a section of a conducting rod of radius $R_1 = 1.30$ mm and length $L = 11.00$ m inside a thin-walled coaxial conducting cylindrical shell of radius $R_2 = 10.0R_1$ and the (same) length L. The net charge on the rod is $Q_1 = +3.40 \times 10^{-12}$ C; that on the shell is $Q_2 = -2.00Q_1$. What are the (a) magnitude E and (b) direction (radially inward or outward) of the electric field at radial

Figure 23-42 Problem 29.

distance $r = 2.00R_2$? What are (c) E and (d) the direction at $r = 5.00R_1$? What is the charge on the (e) interior and (f) exterior surface of the shell?

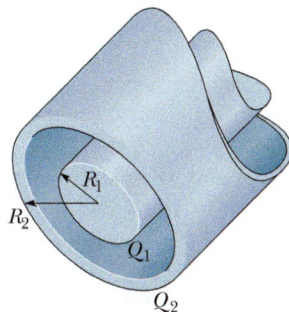

••30 In Fig. 23-43, short sections of two very long parallel lines of charge are shown, fixed in place, separated by $L = 8.0$ cm. The uniform linear charge densities are $+6.0$ μC/m for line 1 and -2.0 μC/m for line 2. Where along the x axis shown is the net electric field from the two lines zero?

Figure 23-43 Problem 30.

••31 ILW Two long, charged, thin-walled, concentric cylindrical shells have radii of 3.0 and 6.0 cm. The charge per unit length is 5.0×10^{-6} C/m on the inner shell and -7.0×10^{-6} C/m on the outer shell. What are the (a) magnitude E and (b) direction (radially inward or outward) of the electric field at radial distance $r = 4.0$ cm? What are (c) E and (d) the direction at $r = 8.0$ cm?

•••32 GO A long, nonconducting, solid cylinder of radius 4.0 cm has a nonuniform volume charge density ρ that is a function of radial distance r from the cylinder axis: $\rho = Ar^2$. For $A = 2.5$ μC/m^5, what is the magnitude of the electric field at (a) $r = 3.0$ cm and (b) $r = 5.0$ cm?

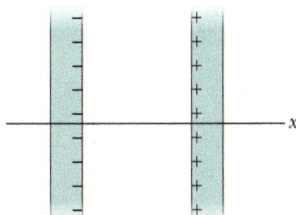

Module 23-5 Applying Gauss' Law: Planar Symmetry

•33 In Fig. 23-44, two large, thin metal plates are parallel and close to each other. On their inner faces,

Figure 23-44 Problem 33.

the plates have excess surface charge densities of opposite signs and magnitude 7.00×10^{-22} C/m^2. In unit-vector notation, what is the electric field at points (a) to the left of the plates, (b) to the right of them, and (c) between them?

•34 In Fig. 23-45, a small circular hole of radius $R = 1.80$ cm has been cut in the middle of an infinite, flat, nonconducting surface that has uniform charge density $\sigma = 4.50$ pC/m^2. A z axis, with its origin at the hole's center, is perpendicular to the surface. In unit-vector notation, what is the electric field at point P at $z = 2.56$ cm? (*Hint:* See Eq. 22-26 and use superposition.)

Figure 23-45 Problem 34.

•35 GO Figure 23-46*a* shows three plastic sheets that are large, parallel, and uniformly charged. Figure 23-46*b* gives the component of the net electric field along an x axis through the sheets. The scale of the vertical axis is set by $E_s = 6.0 \times 10^5$ N/C. What is the ratio of the charge density on sheet 3 to that on sheet 2?

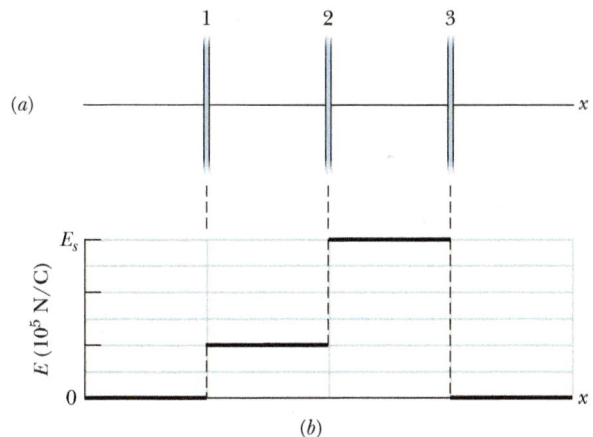

Figure 23-46 Problem 35.

•36 Figure 23-47 shows cross sections through two large, parallel, nonconducting sheets with identical distributions of positive charge with surface charge density $\sigma = 1.77 \times 10^{-22}$ C/m^2. In unit-vector notation, what is $\vec{E}$ at points (a) above the sheets, (b) between them, and (c) below them?

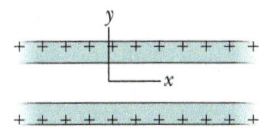

Figure 23-47 Problem 36.

•37 SSM WWW A square metal plate of edge length 8.0 cm and negligible thickness has a total charge of 6.0×10^{-6} C. (a) Estimate the magnitude E of the electric field just off the center of the plate (at, say, a distance of 0.50 mm from the center) by assuming that the charge is spread uniformly over the two faces of the plate. (b) Estimate E at a distance of 30 m (large relative to the plate size) by assuming that the plate is a charged particle.

••38 [GO] In Fig. 23-48a, an electron is shot directly away from a uniformly charged plastic sheet, at speed $v_s = 2.0 \times 10^5$ m/s. The sheet is nonconducting, flat, and very large. Figure 23-48b gives the electron's vertical velocity component v versus time t until the return to the launch point. What is the sheet's surface charge density?

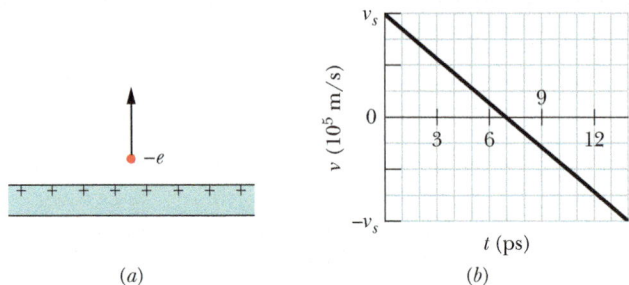

(a)   (b)

Figure 23-48 Problem 38.

••39 SSM In Fig. 23-49, a small, nonconducting ball of mass $m = 1.0$ mg and charge $q = 2.0 \times 10^{-8}$ C (distributed uniformly through its volume) hangs from an insulating thread that makes an angle $\theta = 30°$ with a vertical, uniformly charged nonconducting sheet (shown in cross section). Considering the gravitational force on the ball and assuming the sheet extends far vertically and into and out of the page, calculate the surface charge density σ of the sheet.

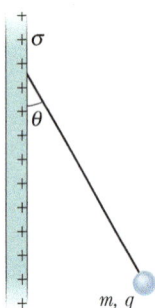

Figure 23-49
Problem 39.

••40 Figure 23-50 shows a very large nonconducting sheet that has a uniform surface charge density of $\sigma = -2.00$ μC/m^2; it also shows a particle of charge $Q = 6.00$ μC, at distance d from the sheet. Both are fixed in place. If $d = 0.200$ m, at what (a) positive and (b) negative coordinate on the x axis (other than infinity) is the net electric field $\vec{E}_{net}$ of the sheet and particle zero? (c) If $d = 0.800$ m, at what coordinate on the x axis is $\vec{E}_{net} = 0$?

Figure 23-50 Problem 40.

••41 [GO] An electron is shot directly toward the center of a large metal plate that has surface charge density -2.0×10^{-6} C/m^2. If the initial kinetic energy of the electron is 1.60×10^{-17} J and if the electron is to stop (due to electrostatic repulsion from the plate) just as it reaches the plate, how far from the plate must the launch point be?

••42 Two large metal plates of area 1.0 m^2 face each other, 5.0 cm apart, with equal charge magnitudes $|q|$ but opposite signs. The field magnitude E between them (neglect fringing) is 55 N/C. Find $|q|$.

•••43 [GO] Figure 23-51 shows a cross section through a very large nonconducting slab of thickness $d = 9.40$ mm and uniform volume charge density $\rho = 5.80$ fC/m^3. The origin of an x axis is at the slab's center. What is the magnitude of the slab's electric field at an x coordinate of (a) 0, (b) 2.00 mm, (c) 4.70 mm, and (d) 26.0 mm?

Figure 23-51
Problem 43.

Module 23-6   Applying Gauss' Law: Spherical Symmetry

•44 Figure 23-52 gives the magnitude of the electric field inside and outside a sphere with a positive charge distributed uniformly throughout its volume. The scale of the vertical axis is set by $E_s = 5.0 \times 10^7$ N/C. What is the charge on the sphere?

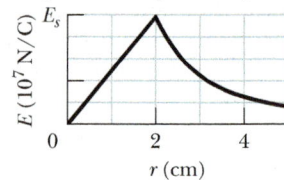

Figure 23-52 Problem 44.

•45 Two charged concentric spherical shells have radii 10.0 cm and 15.0 cm. The charge on the inner shell is 4.00×10^{-8} C, and that on the outer shell is 2.00×10^{-8} C. Find the electric field (a) at $r = 12.0$ cm and (b) at $r = 20.0$ cm.

•46 Assume that a ball of charged particles has a uniformly distributed negative charge density except for a narrow radial tunnel through its center, from the surface on one side to the surface on the opposite side. Also assume that we can position a proton anywhere along the tunnel or outside the ball. Let F_R be the magnitude of the electrostatic force on the proton when it is located at the ball's surface, at radius R. As a multiple of R, how far from the surface is there a point where the force magnitude is $0.50F_R$ if we move the proton (a) away from the ball and (b) into the tunnel?

•47 SSM An unknown charge sits on a conducting solid sphere of radius 10 cm. If the electric field 15 cm from the center of the sphere has the magnitude 3.0×10^3 N/C and is directed radially inward, what is the net charge on the sphere?

•••48 [GO] A positively charged particle is held at the center of a spherical shell. Figure 23-53 gives the magnitude E of the electric field versus radial distance r. The scale of the vertical axis is set by $E_s = 10.0 \times 10^7$ N/C. Approximately, what is the net charge on the shell?

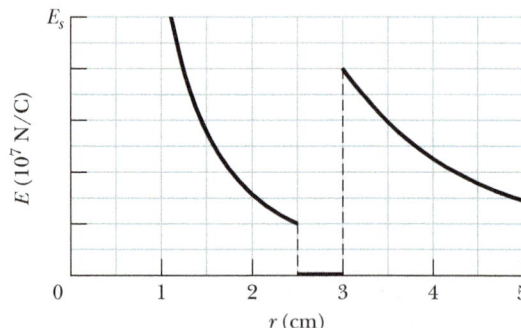

Figure 23-53 Problem 48.

••49 In Fig. 23-54, a solid sphere of radius $a = 2.00$ cm is concentric with a spherical conducting shell of inner radius $b = 2.00a$ and outer radius $c = 2.40a$. The sphere has a net uniform charge $q_1 = +5.00$ fC; the shell has a net charge $q_2 = -q_1$. What is the magnitude of the electric field at radial distances (a) $r = 0$, (b) $r = a/2.00$, (c) $r = a$, (d) $r = 1.50a$, (e) $r = 2.30a$, and (f) $r = 3.50a$? What is the net charge on the (g) inner and (h) outer surface of the shell?

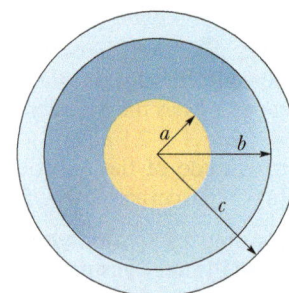

Figure 23-54 Problem 49.

••**50** **GO** Figure 23-55 shows two non-conducting spherical shells fixed in place on an x axis. Shell 1 has uniform surface charge density $+4.0\ \mu C/m^2$ on its outer surface and radius 0.50 cm, and shell 2 has uniform surface charge density $-2.0\ \mu C/m^2$ on its outer surface and radius 2.0 cm; the centers are separated by $L = 6.0$ cm. Other than at $x = \infty$, where on the x axis is the net electric field equal to zero?

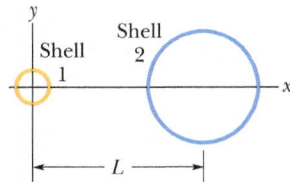

Figure 23-55 Problem 50.

••**51** **SSM** **WWW** In Fig. 23-56, a nonconducting spherical shell of inner radius $a = 2.00$ cm and outer radius $b = 2.40$ cm has (within its thickness) a positive volume charge density $\rho = A/r$, where A is a constant and r is the distance from the center of the shell. In addition, a small ball of charge $q = 45.0$ fC is located at that center. What value should A have if the electric field in the shell ($a \le r \le b$) is to be uniform?

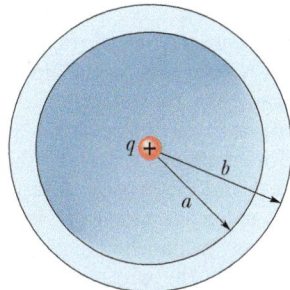

Figure 23-56 Problem 51.

••**52** **GO** Figure 23-57 shows a spherical shell with uniform volume charge density $\rho = 1.84$ nC/m^3, inner radius $a = 10.0$ cm, and outer radius $b = 2.00a$. What is the magnitude of the electric field at radial distances (a) $r = 0$; (b) $r = a/2.00$, (c) $r = a$, (d) $r = 1.50a$, (e) $r = b$, and (f) $r = 3.00b$?

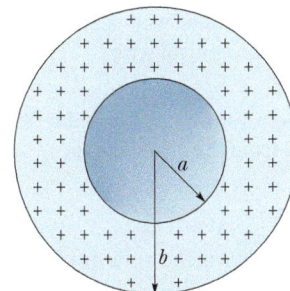

Figure 23-57 Problem 52.

•••**53** **ILW** The volume charge density of a solid nonconducting sphere of radius $R = 5.60$ cm varies with radial distance r as given by $\rho = (14.1\ pC/m^3)r/R$. (a) What is the sphere's total charge? What is the field magnitude E at (b) $r = 0$, (c) $r = R/2.00$, and (d) $r = R$? (e) Graph E versus r.

•••**54** Figure 23-58 shows, in cross section, two solid spheres with uniformly distributed charge throughout their volumes. Each has radius R. Point P lies on a line connecting the centers of the spheres, at radial distance $R/2.00$ from the center of sphere 1. If the net electric field at point P is zero, what is the ratio q_2/q_1 of the total charges?

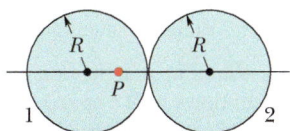

Figure 23-58 Problem 54.

•••**55** A charge distribution that is spherically symmetric but not uniform radially produces an electric field of magnitude $E = Kr^4$, directed radially outward from the center of the sphere. Here r is the radial distance from that center, and K is a constant. What is the volume density ρ of the charge distribution?

Additional Problems

56 The electric field in a particular space is $\vec{E} = (x + 2)\hat{i}$ N/C, with x in meters. Consider a cylindrical Gaussian surface of radius 20 cm that is coaxial with the x axis. One end of the cylinder is at $x = 0$. (a) What is the magnitude of the electric flux through the other end of the cylinder at $x = 2.0$ m? (b) What net charge is enclosed within the cylinder?

57 A thin-walled metal spherical shell has radius 25.0 cm and charge 2.00×10^{-7} C. Find E for a point (a) inside the shell, (b) just outside it, and (c) 3.00 m from the center.

58 A uniform surface charge of density 8.0 nC/m^2 is distributed over the entire xy plane. What is the electric flux through a spherical Gaussian surface centered on the origin and having a radius of 5.0 cm?

59 Charge of uniform volume density $\rho = 1.2$ nC/m^3 fills an infinite slab between $x = -5.0$ cm and $x = +5.0$ cm. What is the magnitude of the electric field at any point with the coordinate (a) $x = 4.0$ cm and (b) $x = 6.0$ cm?

60 ✈ *The chocolate crumb mystery.* Explosions ignited by electrostatic discharges (sparks) constitute a serious danger in facilities handling grain or powder. Such an explosion occurred in chocolate crumb powder at a biscuit factory in the 1970s. Workers usually emptied newly delivered sacks of the powder into a loading bin, from which it was blown through electrically grounded plastic pipes to a silo for storage. Somewhere along this route, two conditions for an explosion were met: (1) The magnitude of an electric field became 3.0×10^6 N/C or greater, so that electrical breakdown and thus sparking could occur. (2) The energy of a spark was 150 mJ or greater so that it could ignite the powder explosively. Let us check for the first condition in the powder flow through the plastic pipes.

Suppose a stream of *negatively* charged powder was blown through a cylindrical pipe of radius $R = 5.0$ cm. Assume that the powder and its charge were spread uniformly through the pipe with a volume charge density ρ. (a) Using Gauss' law, find an expression for the magnitude of the electric field $\vec{E}$ in the pipe as a function of radial distance r from the pipe center. (b) Does E increase or decrease with increasing r? (c) Is $\vec{E}$ directed radially inward or outward? (d) For $\rho = 1.1 \times 10^{-3}$ C/m^3 (a typical value at the factory), find the maximum E and determine where that maximum field occurs. (e) Could sparking occur, and if so, where? (The story continues with Problem 70 in Chapter 24.)

61 **SSM** A thin-walled metal spherical shell of radius a has a charge q_a. Concentric with it is a thin-walled metal spherical shell of radius $b > a$ and charge q_b. Find the electric field at points a distance r from the common center, where (a) $r < a$, (b) $a < r < b$, and (c) $r > b$. (d) Discuss the criterion you would use to determine how the charges are distributed on the inner and outer surfaces of the shells.

62 A particle of charge $q = 1.0 \times 10^{-7}$ C is at the center of a spherical cavity of radius 3.0 cm in a chunk of metal. Find the electric field (a) 1.5 cm from the cavity center and (b) anyplace in the metal.

63 A proton at speed $v = 3.00 \times 10^5$ m/s orbits at radius $r = 1.00$ cm outside a charged sphere. Find the sphere's charge.

64 Equation 23-11 ($E = \sigma/\varepsilon_0$) gives the electric field at points near a charged conducting surface. Apply this equation to a conducting sphere of radius r and charge q, and show that the electric field outside the sphere is the same as the field of a charged particle located at the center of the sphere.

65 Charge Q is uniformly distributed in a sphere of radius R. (a) What fraction of the charge is contained within the radius $r = R/2.00$? (b) What is the ratio of the electric field magnitude at $r = R/2.00$ to that on the surface of the sphere?

66 A charged particle causes an electric flux of -750 N·m^2/C to pass through a spherical Gaussian surface of 10.0 cm radius centered on the charge. (a) If the radius of the Gaussian surface were doubled, how much flux would pass through the surface? (b) What is the charge of the particle?

67 SSM The electric field at point P just outside the outer surface of a hollow spherical conductor of inner radius 10 cm and outer radius 20 cm has magnitude 450 N/C and is directed outward. When a particle of unknown charge Q is introduced into the center of the sphere, the electric field at P is still directed outward but is now 180 N/C. (a) What was the net charge enclosed by the outer surface before Q was introduced? (b) What is charge Q? After Q is introduced, what is the charge on the (c) inner and (d) outer surface of the conductor?

68 The net electric flux through each face of a die (singular of dice) has a magnitude in units of 10^3 N·m^2/C that is exactly equal to the number of spots N on the face (1 through 6). The flux is inward for N odd and outward for N even. What is the net charge inside the die?

69 Figure 23-59 shows, in cross section, three infinitely large nonconducting sheets on which charge is uniformly spread. The surface charge densities are $\sigma_1 = +2.00$ μC/m^2, $\sigma_2 = +4.00$ μC/m^2, and $\sigma_3 = -5.00$ μC/m^2, and distance $L = 1.50$ cm. In unit-vector notation, what is the net electric field at point P?

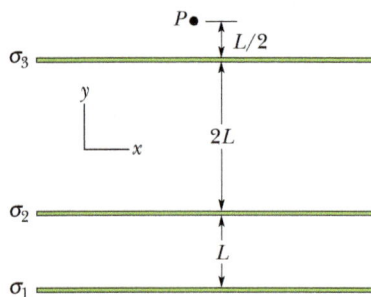

Figure 23-59 Problem 69.

70 Charge of uniform volume density $\rho = 3.2$ μC/m^3 fills a nonconducting solid sphere of radius 5.0 cm. What is the magnitude of the electric field (a) 3.5 cm and (b) 8.0 cm from the sphere's center?

71 A Gaussian surface in the form of a hemisphere of radius $R = 5.68$ cm lies in a uniform electric field of magnitude $E = 2.50$ N/C. The surface encloses no net charge. At the (flat) base of the surface, the field is perpendicular to the surface and directed into the surface. What is the flux through (a) the base and (b) the curved portion of the surface?

72 What net charge is enclosed by the Gaussian cube of Problem 2?

73 A nonconducting solid sphere has a uniform volume charge density ρ. Let $\vec{r}$ be the vector from the center of the sphere to a general point P within the sphere. (a) Show that the electric field at P is given by $\vec{E} = \rho \vec{r}/3\varepsilon_0$. (Note that the result is independent of the radius of the sphere.) (b) A spherical cavity is hollowed out of the sphere, as shown in Fig. 23-60. Using superposition concepts, show that the electric field at all points within the cavity is uniform and equal to $\vec{E} = \rho \vec{a}/3\varepsilon_0$, where $\vec{a}$ is the position vector from the center of the sphere to the center of the cavity.

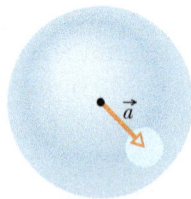

Figure 23-60 Problem 73.

74 A uniform charge density of 500 nC/m^3 is distributed throughout a spherical volume of radius 6.00 cm. Consider a cubical Gaussian surface with its center at the center of the sphere. What is the electric flux through this cubical surface if its edge length is (a) 4.00 cm and (b) 14.0 cm?

75 Figure 23-61 shows a Geiger counter, a device used to detect ionizing radiation, which causes ionization of atoms. A thin, posi-tively charged central wire is surrounded by a concentric, circular, conducting cylindrical shell with an equal negative charge, creating a strong radial electric field. The shell contains a low-pressure inert gas. A particle of radiation entering the device through the shell wall ionizes a few of the gas atoms. The resulting free electrons (e) are drawn to the positive wire. However, the electric field is so intense that, between collisions with gas atoms, the free electrons gain energy sufficient to ionize these atoms also. More free electrons are thereby created, and the process is repeated until the electrons reach the wire. The re-sulting "avalanche" of electrons is collected by the wire, generating a signal that is used to record the passage of the original particle of radiation. Suppose that the radius of the central wire is 25 μm, the inner radius of the shell 1.4 cm, and the length of the shell 16 cm. If the electric field at the shell's inner wall is 2.9×10^4 N/C, what is the total positive charge on the central wire?

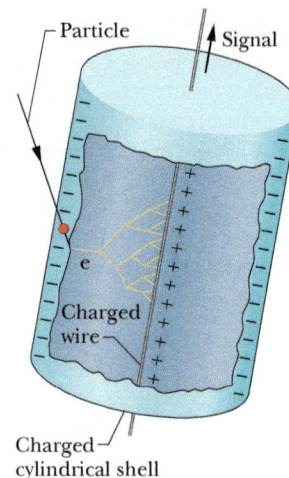

Figure 23-61 Problem 75.

76 Charge is distributed uniformly throughout the volume of an infinitely long solid cylinder of radius R. (a) Show that, at a distance $r < R$ from the cylinder axis,

$$E = \frac{\rho r}{2\varepsilon_0},$$

where ρ is the volume charge density. (b) Write an expression for E when $r > R$.

77 SSM A spherical conducting shell has a charge of -14 μC on its outer surface and a charged particle in its hollow. If the net charge on the shell is -10 μC, what is the charge (a) on the inner surface of the shell and (b) of the particle?

78 A charge of 6.00 pC is spread uniformly throughout the volume of a sphere of radius $r = 4.00$ cm. What is the magnitude of the electric field at a radial distance of (a) 6.00 cm and (b) 3.00 cm?

79 Water in an irrigation ditch of width $w = 3.22$ m and depth $d = 1.04$ m flows with a speed of 0.207 m/s. The *mass flux* of the flowing water through an imaginary surface is the product of the water's density (1000 kg/m^3) and its volume flux through that surface. Find the mass flux through the following imaginary surfaces: (a) a surface of area wd, entirely in the water, perpendicular to the flow; (b) a surface with area $3wd/2$, of which wd is in the water, perpendicular to the flow; (c) a surface of area $wd/2$, entirely in the water, perpendicular to the flow; (d) a surface of area wd, half in the water and half out, perpendicular to the flow; (e) a surface of area wd, entirely in the water, with its normal 34.0° from the direction of flow.

80 Charge of uniform surface density 8.00 nC/m^2 is distributed over an entire xy plane; charge of uniform surface density 3.00 nC/m^2 is distributed over the parallel plane defined by $z = 2.00$ m. Determine the magnitude of the electric field at any point having a z coordinate of (a) 1.00 m and (b) 3.00 m.

81 A spherical ball of charged particles has a uniform charge density. In terms of the ball's radius R, at what radial distances (a) inside and (b) outside the ball is the magnitude of the ball's electric field equal to $\frac{1}{4}$ of the maximum magnitude of that field?

Electric Potential

24-1 ELECTRIC POTENTIAL

Learning Objectives

After reading this module, you should be able to . . .

24.01 Identify that the electric force is conservative and thus has an associated potential energy.

24.02 Identify that at every point in a charged object's electric field, the object sets up an electric potential V, which is a scalar quantity that can be positive or negative depending on the sign of the object's charge.

24.03 For a charged particle placed at a point in an object's electric field, apply the relationship between the object's electric potential V at that point, the particle's charge q, and the potential energy U of the particle–object system.

24.04 Convert energies between units of joules and electron-volts.

24.05 If a charged particle moves from an initial point to a final point in an electric field, apply the relationships

between the change ΔV in the potential, the particle's charge q, the change ΔU in the potential energy, and the work W done by the electric force.

24.06 If a charged particle moves between two given points in the electric field of a charged object, identify that the amount of work done by the electric force is path independent.

24.07 If a charged particle moves through a change ΔV in electric potential without an applied force acting on it, relate ΔV and the change ΔK in the particle's kinetic energy.

24.08 If a charged particle moves through a change ΔV in electric potential while an applied force acts on it, relate ΔV, the change ΔK in the particle's kinetic energy, and the work W_{app} done by the applied force.

Key Ideas

● The electric potential V at a point P in the electric field of a charged object is

$$V = \frac{-W_\infty}{q_0} = \frac{U}{q_0},$$

where W_∞ is the work that would be done by the electric force on a positive test charge q_0 were it brought from an infinite distance to P, and U is the electric potential energy that would then be stored in the test charge–object system.

● If a particle with charge q is placed at a point where the electric potential of a charged object is V, the electric potential energy U of the particle–object system is

$$U = qV.$$

● If the particle moves through a potential difference ΔV, the change in the electric potential energy is

$$\Delta U = q \, \Delta V = q(V_f - V_i).$$

● If a particle moves through a change ΔV in electric potential without an applied force acting on it, applying the conservation of mechanical energy gives the change in kinetic energy as

$$\Delta K = -q \, \Delta V.$$

● If, instead, an applied force acts on the particle, doing work W_{app}, the change in kinetic energy is

$$\Delta K = -q \, \Delta V + W_{\text{app}}.$$

● In the special case when $\Delta K = 0$, the work of an applied force involves only the motion of the particle through a potential difference:

$$W_{\text{app}} = q \, \Delta V.$$

What Is Physics?

One goal of physics is to identify basic forces in our world, such as the electric force we discussed in Chapter 21. A related goal is to determine whether a force is conservative—that is, whether a potential energy can be associated with it. The motivation for associating a potential energy with a force is that we can then

Figure 24-1 Particle 1 is located at point P in the electric field of particle 2.

apply the principle of the conservation of mechanical energy to closed systems involving the force. This extremely powerful principle allows us to calculate the results of experiments for which force calculations alone would be very difficult. Experimentally, physicists and engineers discovered that the electric force is conservative and thus has an associated electric potential energy. In this chapter we first define this type of potential energy and then put it to use.

For a quick taste, let's return to the situation we considered in Chapter 22: In Figure 24-1, particle 1 with positive charge q_1 is located at point P near particle 2 with positive charge q_2. In Chapter 22 we explained how particle 2 is able to push on particle 1 without any contact. To account for the force $\vec{F}$ (which is a vector quantity), we defined an electric field $\vec{E}$ (also a vector quantity) that is set up at P by particle 2. That field exists regardless of whether particle 1 is at P. If we choose to place particle 1 there, the push on it is due to charge q_1 and that pre-existing field $\vec{E}$.

Here is a related problem. If we release particle 1 at P, it begins to move and thus has kinetic energy. Energy cannot appear by magic, so from where does it come? It comes from the electric potential energy U associated with the force between the two particles in the arrangement of Fig. 24-1. To account for the potential energy U (which is a scalar quantity), we define an **electric potential** V (also a scalar quantity) that is set up at P by particle 2. The electric potential exists regardless of whether particle 1 is at P. If we choose to place particle 1 there, the potential energy of the two-particle system is then due to charge q_1 and that pre-existing electric potential V.

Our goals in this chapter are to (1) define electric potential, (2) discuss how to calculate it for various arrangements of charged particles and objects, and (3) discuss how electric potential V is related to electric potential energy U.

Electric Potential and Electric Potential Energy

We are going to define the electric potential (or *potential* for short) in terms of electric potential energy, so our first job is to figure out how to measure that potential energy. Back in Chapter 8, we measured gravitational potential energy U of an object by (1) assigning $U = 0$ for a reference configuration (such as the object at table level) and (2) then calculating the work W the gravitational force does if the object is moved up or down from that level. We then defined the potential energy as being

$$U = -W \quad \text{(potential energy).} \qquad (24\text{-}1)$$

(a)

The rod sets up an electric potential, which determines the potential energy.

(b)

Figure 24-2 (a) A test charge has been brought in from infinity to point P in the electric field of the rod. (b) We define an electric potential V at P based on the potential energy of the configuration in (a).

Let's follow the same procedure with our new conservative force, the electric force. In Fig. 24-2a, we want to find the potential energy U associated with a positive test charge q_0 located at point P in the electric field of a charged rod. First, we need a reference configuration for which $U = 0$. A reasonable choice is for the test charge to be infinitely far from the rod, because then there is no interaction with the rod. Next, we bring the test charge in from infinity to point P to form the configuration of Fig. 24-2a. Along the way, we calculate the work done by the electric force on the test charge. The potential energy of the final configuration is then given by Eq. 24-1, where W is now the work done by the electric force. Let's use the notation W_∞ to emphasize that the test charge is brought in from infinity. The work and thus the potential energy can be positive or negative depending on the sign of the rod's charge.

Next, we define the electric potential V at P in terms of the work done by the electric force and the resulting potential energy:

$$V = \frac{-W_\infty}{q_0} = \frac{U}{q_0} \quad \text{(electric potential).} \qquad (24\text{-}2)$$

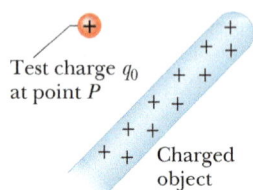

That is, the electric potential is the amount of electric potential energy per unit charge when a positive test charge is brought in from infinity. The rod sets up this potential V at P regardless of whether the test charge (or anything else) happens to be there (Fig. 24-2b). From Eq. 24-2 we see that V is a scalar quantity (because there is no direction associated with potential energy or charge) and can be positive or negative (because potential energy and charge have signs).

Repeating this procedure we find that an electric potential is set up at every point in the rod's electric field. In fact, every charged object sets up electric potential V at points throughout its electric field. If we happen to place a particle with, say, charge q at a point where we know the pre-existing V, we can immediately find the potential energy of the configuration:

$$(\text{electric potential energy}) = (\text{particle's charge})\left(\frac{\text{electric potential energy}}{\text{unit charge}}\right),$$

or

$$U = qV, \qquad (24\text{-}3)$$

where q can be positive or negative.

Two Cautions. (1) The (now very old) decision to call V a *potential* was unfortunate because the term is easily confused with *potential energy*. Yes, the two quantities are related (that is the point here) but they are very different and not interchangeable. (2) Electric potential is a scalar, not a vector. (When you come to the homework problems, you will rejoice on this point.)

Language. A potential energy is a property of a system (or configuration) of objects, but sometimes we can get away with assigning it to a single object. For example, the gravitational potential energy of a baseball hit to outfield is actually a potential energy of the baseball–Earth system (because it is associated with the force between the baseball and Earth). However, because only the baseball noticeably moves (its motion does not noticeably affect Earth), we might assign the gravitational potential energy to it alone. In a similar way, if a charged particle is placed in an electric field and has no noticeable effect on the field (or the charged object that sets up the field), we usually assign the electric potential energy to the particle alone.

Units. The SI unit for potential that follows from Eq. 24-2 is the joule per coulomb. This combination occurs so often that a special unit, the *volt* (abbreviated V), is used to represent it. Thus,

$$1 \text{ volt} = 1 \text{ joule per coulomb}.$$

With two unit conversions, we can now switch the unit for electric field from newtons per coulomb to a more conventional unit:

$$1 \text{ N/C} = \left(1\frac{\text{N}}{\text{C}}\right)\left(\frac{1 \text{ V}}{1 \text{ J/C}}\right)\left(\frac{1 \text{ J}}{1 \text{ N}\cdot\text{m}}\right)$$
$$= 1 \text{ V/m}.$$

The conversion factor in the second set of parentheses comes from our definition of volt given above; that in the third set of parentheses is derived from the definition of the joule. From now on, we shall express values of the electric field in volts per meter rather than in newtons per coulomb.

Motion Through an Electric Field

Change in Electric Potential. If we move from an initial point i to a second point f in the electric field of a charged object, the electric potential changes by

$$\Delta V = V_f - V_i.$$

If we move a particle with charge q from i to f, then, from Eq. 24-3, the potential energy of the system changes by

$$\Delta U = q\,\Delta V = q(V_f - V_i). \tag{24-4}$$

The change can be positive or negative, depending on the signs of q and ΔV. It can also be zero, if there is no change in potential from i to f (the points have the same value of potential). Because the electric force is conservative, the change in potential energy ΔU between i and f is the same for all paths between those points (it is *path independent*).

Work by the Field. We can relate the potential energy change ΔU to the work W done by the electric force as the particle moves from i to f by applying the general relation for a conservative force (Eq. 8-1):

$$W = -\Delta U \quad \text{(work, conservative force).} \tag{24-5}$$

Next, we can relate that work to the change in the potential by substituting from Eq. 24-4:

$$W = -\Delta U = -q\,\Delta V = -q(V_f - V_i). \tag{24-6}$$

Up until now, we have always attributed work to a force but here can also say that W is the work done on the particle by the electric field (because it, of course, produces the force). The work can be positive, negative, or zero. Because ΔU between any two points is path independent, so is the work W done by the field. (If you need to calculate work for a difficult path, switch to an easier path—you get the same result.)

Conservation of Energy. If a charged particle moves through an electric field with no force acting on it other than the electric force due to the field, then the mechanical energy is conserved. Let's assume that we can assign the electric potential energy to the particle alone. Then we can write the conservation of mechanical energy of the particle that moves from point i to point f as

$$U_i + K_i = U_f + K_f, \tag{24-7}$$

or

$$\Delta K = -\Delta U. \tag{24-8}$$

Substituting Eq. 24-4, we find a very useful equation for the change in the particle's kinetic energy as a result of the particle moving through a potential difference:

$$\Delta K = -q\,\Delta V = -q(V_f - V_i). \tag{24-9}$$

Work by an Applied Force. If some force in addition to the electric force acts on the particle, we say that the additional force is an *applied force* or *external force*, which is often attributed to an *external agent*. Such an applied force can do work on the particle, but the force may not be conservative and thus, in general, we cannot associate a potential energy with it. We account for that work W_{app} by modifying Eq. 24-7:

(initial energy) + (work by applied force) = (final energy)

or

$$U_i + K_i + W_{app} = U_f + K_f. \tag{24-10}$$

Rearranging and substituting from Eq. 24-4, we can also write this as

$$\Delta K = -\Delta U + W_{app} = -q\,\Delta V + W_{app}. \tag{24-11}$$

The work by the applied force can be positive, negative, or zero, and thus the energy of the system can increase, decrease, or remain the same.

In the special case where the particle is stationary before and after the move, the kinetic energy terms in Eqs. 24-10 and 24-11 are zero and we have

$$W_{app} = q\,\Delta V \quad \text{(for } K_i = K_f\text{).} \tag{24-12}$$

In this special case, the work W_{app} involves the motion of the particle through the potential difference ΔV and not a change in the particle's kinetic energy.

By comparing Eqs. 24-6 and 24-12, we see that in this special case, the work by the applied force is the negative of the work by the field:

$$W_{app} = -W \quad \text{(for } K_i = K_f \text{)}. \tag{24-13}$$

Electron-volts. In atomic and subatomic physics, energy measures in the SI unit of joules often require awkward powers of ten. A more convenient (but non-SI unit) is the *electron-volt* (eV), which is defined to be equal to the work required to move a single elementary charge e (such as that of an electron or proton) through a potential difference ΔV of exactly one volt. From Eq. 24-6, we see that the magnitude of this work is $q \, \Delta V$. Thus,

$$1 \text{ eV} = e(1 \text{ V})$$
$$= (1.602 \times 10^{-19} \text{ C})(1 \text{ J/C}) = 1.602 \times 10^{-19} \text{ J}. \tag{24-14}$$

✓ Checkpoint 1

In the figure, we move a proton from point i to point f in a uniform electric field. Is positive or negative work done by (a) the electric field and (b) our force? (c) Does the electric potential energy increase or decrease? (d) Does the proton move to a point of higher or lower electric potential?

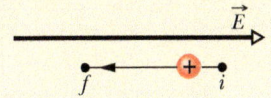

Sample Problem 24.01 Work and potential energy in an electric field

Electrons are continually being knocked out of air molecules in the atmosphere by cosmic-ray particles coming in from space. Once released, each electron experiences an electric force $\vec{F}$ due to the electric field $\vec{E}$ that is produced in the atmosphere by charged particles already on Earth. Near Earth's surface the electric field has the magnitude $E = 150$ N/C and is directed downward. What is the change ΔU in the electric potential energy of a released electron when the electric force causes it to move vertically upward through a distance $d = 520$ m (Fig. 24-3)? Through what potential change does the electron move?

KEY IDEAS

(1) The change ΔU in the electric potential energy of the electron is related to the work W done on the electron by the electric field. Equation 24-5 ($W = -\Delta U$) gives the relation. (2) The work done by a constant force $\vec{F}$ on a particle undergoing a displacement $\vec{d}$ is

$$W = \vec{F} \cdot \vec{d}.$$

(3) The electric force and the electric field are related by the force equation $\vec{F} = q\vec{E}$, where here q is the charge of an electron ($= -1.6 \times 10^{-19}$ C).

Calculations: Substituting the force equation into the work equation and taking the dot product yield

$$W = q\vec{E} \cdot \vec{d} = qEd \cos \theta,$$

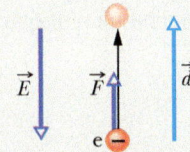

Figure 24-3 An electron in the atmosphere is moved upward through displacement $\vec{d}$ by an electric force $\vec{F}$ due to an electric field $\vec{E}$.

where θ is the angle between the directions of $\vec{E}$ and $\vec{d}$. The field $\vec{E}$ is directed downward and the displacement $\vec{d}$ is directed upward; so $\theta = 180°$. We can now evaluate the work as

$$W = (-1.6 \times 10^{-19} \text{ C})(150 \text{ N/C})(520 \text{ m}) \cos 180°$$
$$= 1.2 \times 10^{-14} \text{ J}.$$

Equation 24-5 then yields

$$\Delta U = -W = -1.2 \times 10^{-14} \text{ J}. \quad \text{(Answer)}$$

This result tells us that during the 520 m ascent, the electric potential energy of the electron *decreases* by 1.2×10^{-14} J. To find the change in electric potential, we apply Eq. 24-4:

$$\Delta V = \frac{\Delta U}{-q} = \frac{-1.2 \times 10^{-14} \text{ J}}{-1.6 \times 10^{-19} \text{ C}}$$
$$= 4.5 \times 10^4 \text{ V} = 45 \text{ kV}. \quad \text{(Answer)}$$

This tells us that the electric force does work to move the electron to a *higher* potential.

WILEY PLUS Additional examples, video, and practice available at *WileyPLUS*

24-2 EQUIPOTENTIAL SURFACES AND THE ELECTRIC FIELD

Learning Objectives

After reading this module, you should be able to . . .

24.09 Identify an equipotential surface and describe how it is related to the direction of the associated electric field.

24.10 Given an electric field as a function of position, calculate the change in potential ΔV from an initial point to a final point by choosing a path between the points and integrating the dot product of the field $\vec{E}$ and a length element $d\vec{s}$ along the path.

24.11 For a uniform electric field, relate the field magnitude E and the separation Δx and potential difference ΔV between adjacent equipotential lines.

24.12 Given a graph of electric field E versus position along an axis, calculate the change in potential ΔV from an initial point to a final point by graphical integration.

24.13 Explain the use of a zero-potential location.

Key Ideas

● The points on an equipotential surface all have the same electric potential. The work done on a test charge in moving it from one such surface to another is independent of the locations of the initial and final points on these surfaces and of the path that joins the points. The electric field $\vec{E}$ is always directed perpendicularly to corresponding equipotential surfaces.

● The electric potential difference between two points i and f is

$$V_f - V_i = -\int_i^f \vec{E} \cdot d\vec{s},$$

where the integral is taken over any path connecting the points. If the integration is difficult along any particular path,

we can choose a different path along which the integration might be easier.

● If we choose $V_i = 0$, we have, for the potential at a particular point,

$$V = -\int_i^f \vec{E} \cdot d\vec{s}.$$

● In a uniform field of magnitude E, the change in potential from a higher equipotential surface to a lower one, separated by distance Δx, is

$$\Delta V = -E\,\Delta x.$$

Equipotential Surfaces

Adjacent points that have the same electric potential form an **equipotential surface,** which can be either an imaginary surface or a real, physical surface. No net work W is done on a charged particle by an electric field when the particle moves between two points i and f on the same equipotential surface. This follows from Eq. 24-6, which tells us that W must be zero if $V_f = V_i$. Because of the path independence of work (and thus of potential energy and potential), $W = 0$ for *any* path connecting points i and f on a given equipotential surface regardless of whether that path lies entirely on that surface.

Figure 24-4 shows a *family* of equipotential surfaces associated with the electric field due to some distribution of charges. The work done by the electric field on a charged particle as the particle moves from one end to the other of paths

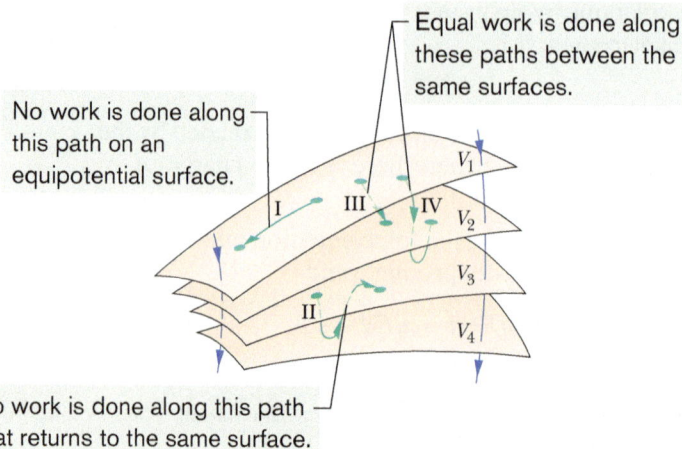

Figure 24-4 Portions of four equipotential surfaces at electric potentials $V_1 = 100$ V, $V_2 = 80$ V, $V_3 = 60$ V, and $V_4 = 40$ V. Four paths along which a test charge may move are shown. Two electric field lines are also indicated.

I and II is zero because each of these paths begins and ends on the same equipotential surface and thus there is no net change in potential. The work done as the charged particle moves from one end to the other of paths III and IV is not zero but has the same value for both these paths because the initial and final potentials are identical for the two paths; that is, paths III and IV connect the same pair of equipotential surfaces.

From symmetry, the equipotential surfaces produced by a charged particle or a spherically symmetrical charge distribution are a family of concentric spheres. For a uniform electric field, the surfaces are a family of planes perpendicular to the field lines. In fact, equipotential surfaces are always perpendicular to electric field lines and thus to $\vec{E}$, which is always tangent to these lines. If $\vec{E}$ were *not* perpendicular to an equipotential surface, it would have a component lying along that surface. This component would then do work on a charged particle as it moved along the surface. However, by Eq. 24-6 work cannot be done if the surface is truly an equipotential surface; the only possible conclusion is that $\vec{E}$ must be everywhere perpendicular to the surface. Figure 24-5 shows electric field lines and cross sections of the equipotential surfaces for a uniform electric field and for the field associated with a charged particle and with an electric dipole.

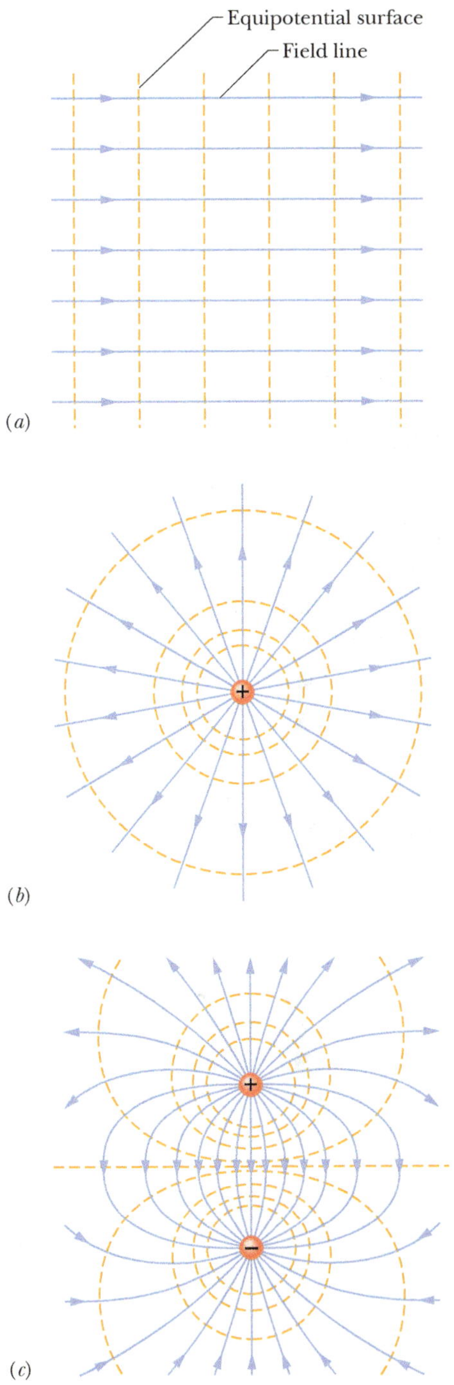

Calculating the Potential from the Field

We can calculate the potential difference between any two points i and f in an electric field if we know the electric field vector $\vec{E}$ all along any path connecting those points. To make the calculation, we find the work done on a positive test charge by the field as the charge moves from i to f, and then use Eq. 24-6.

Consider an arbitrary electric field, represented by the field lines in Fig. 24-6, and a positive test charge q_0 that moves along the path shown from point i to point f. At any point on the path, an electric force $q_0\vec{E}$ acts on the charge as it moves through a differential displacement $d\vec{s}$. From Chapter 7, we know that the differential work dW done on a particle by a force $\vec{F}$ during a displacement $d\vec{s}$ is given by the dot product of the force and the displacement:

$$dW = \vec{F}\cdot d\vec{s}. \quad (24\text{-}15)$$

For the situation of Fig. 24-6, $\vec{F} = q_0\vec{E}$ and Eq. 24-15 becomes

$$dW = q_0\vec{E}\cdot d\vec{s}. \quad (24\text{-}16)$$

To find the total work W done on the particle by the field as the particle moves from point i to point f, we sum—via integration—the differential works done on the charge as it moves through all the displacements $d\vec{s}$ along the path:

$$W = q_0\int_i^f \vec{E}\cdot d\vec{s}. \quad (24\text{-}17)$$

If we substitute the total work W from Eq. 24-17 into Eq. 24-6, we find

$$V_f - V_i = -\int_i^f \vec{E}\cdot d\vec{s}. \quad (24\text{-}18)$$

(a)
(b)
(c)

Figure 24-5 Electric field lines (purple) and cross sections of equipotential surfaces (gold) for (a) a uniform electric field, (b) the field due to a charged particle, and (c) the field due to an electric dipole.

Figure 24-6 A test charge q_0 moves from point i to point f along the path shown in a nonuniform electric field. During a displacement $d\vec{s}$, an electric force $q_0\vec{E}$ acts on the test charge. This force points in the direction of the field line at the location of the test charge.

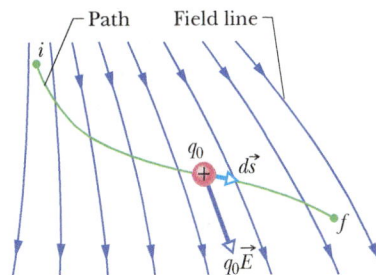

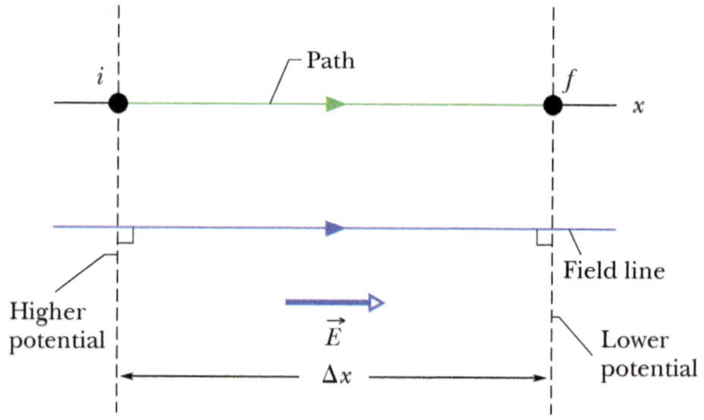

Figure 24-7 We move between points i and f, between adjacent equipotential lines in a uniform electric field $\vec{E}$, parallel to a field line.

Thus, the potential difference $V_f - V_i$ between any two points i and f in an electric field is equal to the negative of the *line integral* (meaning the integral along a particular path) of $\vec{E} \cdot d\vec{s}$ from i to f. However, because the electric force is conservative, all paths (whether easy or difficult to use) yield the same result.

Equation 24-18 allows us to calculate the difference in potential between any two points in the field. If we set potential $V_i = 0$, then Eq. 24-18 becomes

$$V = -\int_i^f \vec{E} \cdot d\vec{s}, \tag{24-19}$$

in which we have dropped the subscript f on V_f. Equation 24-19 gives us the potential V at any point f in the electric field *relative to the zero potential* at point i. If we let point i be at infinity, then Eq. 24-19 gives us the potential V at any point f relative to the zero potential at infinity.

Uniform Field. Let's apply Eq. 24-18 for a uniform field as shown in Fig. 24-7. We start at point i on an equipotential line with potential V_i and move to point f on an equipotential line with a lower potential V_f. The separation between the two equipotential lines is Δx. Let's also move along a path that is parallel to the electric field $\vec{E}$ (and thus perpendicular to the equipotential lines). The angle between $\vec{E}$ and $d\vec{s}$ in Eq. 24-18 is zero, and the dot product gives us

$$\vec{E} \cdot d\vec{s} = E\, ds \cos 0 = E\, ds.$$

Because E is constant for a uniform field, Eq. 24-18 becomes

$$V_f - V_i = -E \int_i^f ds. \tag{24-20}$$

The integral is simply an instruction for us to add all the displacement elements ds from i to f, but we already know that the sum is length Δx. Thus we can write the change in potential $V_f - V_i$ in this uniform field as

$$\Delta V = -E\, \Delta x \quad \text{(uniform field)}. \tag{24-21}$$

This is the change in voltage ΔV between two equipotential lines in a uniform field of magnitude E, separated by distance Δx. If we move in the direction of the field by distance Δx, the potential decreases. In the opposite direction, it increases.

The electric field vector points from higher potential toward lower potential.

✓ Checkpoint 2

The figure here shows a family of parallel equipotential surfaces (in cross section) and five paths along which we shall move an electron from one surface to another. (a) What is the direction of the electric field associated with the surfaces? (b) For each path, is the work we do positive, negative, or zero? (c) Rank the paths according to the work we do, greatest first.

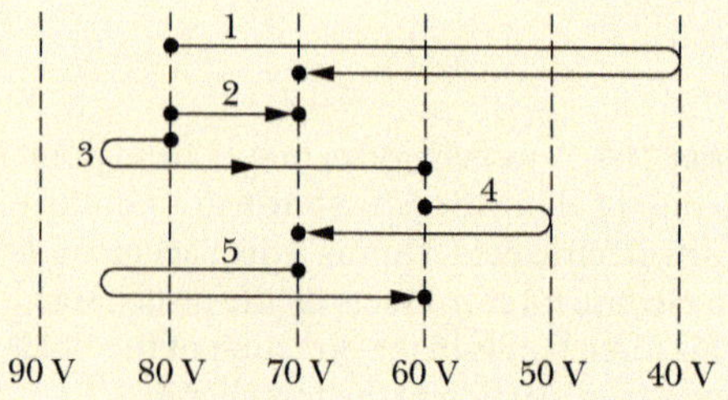

Sample Problem 24.02 Finding the potential change from the electric field

(a) Figure 24-8a shows two points i and f in a uniform electric field $\vec{E}$. The points lie on the same electric field line (not shown) and are separated by a distance d. Find the potential difference $V_f - V_i$ by moving a positive test charge q_0 from i to f along the path shown, which is parallel to the field direction.

KEY IDEA

We can find the potential difference between any two points in an electric field by integrating $\vec{E} \cdot d\vec{s}$ along a path connecting those two points according to Eq. 24-18.

Calculations: We have actually already done the calculation for such a path in the direction of an electric field line in a uniform field when we derived Eq. 24-21. With slight changes in notation, Eq. 24-21 gives us

$$V_f - V_i = -Ed. \qquad \text{(Answer)}$$

(b) Now find the potential difference $V_f - V_i$ by moving the positive test charge q_0 from i to f along the path icf shown in Fig. 24-8b.

Calculations: The Key Idea of (a) applies here too, except now we move the test charge along a path that consists of two lines: ic and cf. At all points along line ic, the displace-

ment $d\vec{s}$ of the test charge is perpendicular to $\vec{E}$. Thus, the angle θ between $\vec{E}$ and $d\vec{s}$ is 90°, and the dot product $\vec{E} \cdot d\vec{s}$ is 0. Equation 24-18 then tells us that points i and c are at the same potential: $V_c - V_i = 0$. Ah, we should have seen this coming. The points are on the same equipotential surface, which is perpendicular to the electric field lines.

For line cf we have $\theta = 45°$ and, from Eq. 24-18,

$$V_f - V_i = -\int_c^f \vec{E} \cdot d\vec{s} = -\int_c^f E(\cos 45°)\, ds$$

$$= -E(\cos 45°)\int_c^f ds.$$

The integral in this equation is just the length of line cf; from Fig. 24-8b, that length is $d/\cos 45°$. Thus,

$$V_f - V_i = -E(\cos 45°)\frac{d}{\cos 45°} = -Ed. \quad \text{(Answer)}$$

This is the same result we obtained in (a), as it must be; the potential difference between two points does not depend on the path connecting them. Moral: When you want to find the potential difference between two points by moving a test charge between them, you can save time and work by choosing a path that simplifies the use of Eq. 24-18.

The electric field points *from* higher potential *to* lower potential.

The field is perpendicular to this *ic* path, so there is no change in the potential.

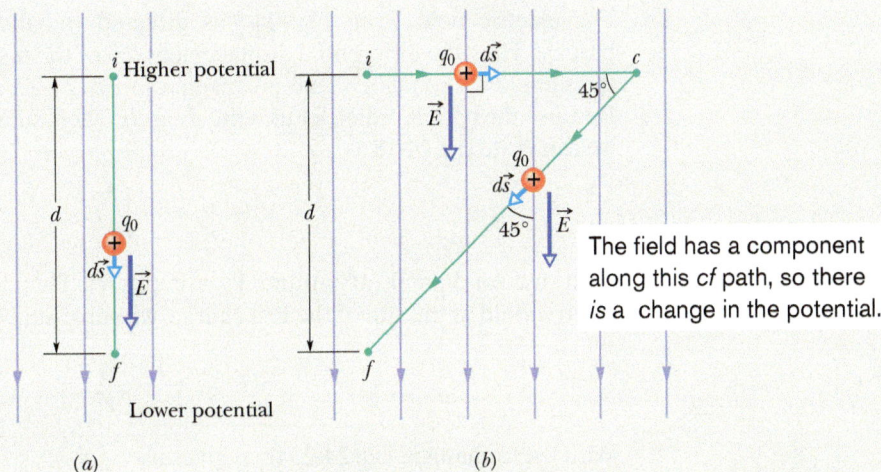

The field has a component along this *cf* path, so there is a change in the potential.

Figure 24-8 (a) A test charge q_0 moves in a straight line from point i to point f, along the direction of a uniform external electric field. (b) Charge q_0 moves along path icf in the same electric field.

24-3 POTENTIAL DUE TO A CHARGED PARTICLE

Learning Objectives

After reading this module, you should be able to . . .

24.14 For a given point in the electric field of a charged particle, apply the relationship between the electric potential V, the charge of the particle q, and the distance r from the particle.

24.15 Identify the correlation between the algebraic signs of the potential set up by a particle and the charge of the particle.

24.16 For points outside or on the surface of a spherically

symmetric charge distribution, calculate the electric potential as if all the charge is concentrated as a particle at the center of the sphere.

24.17 Calculate the net potential at any given point due to several charged particles, identifying that algebraic addition is used, not vector addition.

24.18 Draw equipotential lines for a charged particle.

Key Ideas

● The electric potential due to a single charged particle at a distance r from that charged particle is

$$V = \frac{1}{4\pi\varepsilon_0}\frac{q}{r},$$

where V has the same sign as q.

● The potential due to a collection of charged particles is

$$V = \sum_{i=1}^{n} V_i = \frac{1}{4\pi\varepsilon_0}\sum_{i=1}^{n}\frac{q_i}{r_i}.$$

Thus, the potential is the algebraic sum of the individual potentials, with no consideration of directions.

To find the potential of the charged particle, we move this test charge out to infinity.

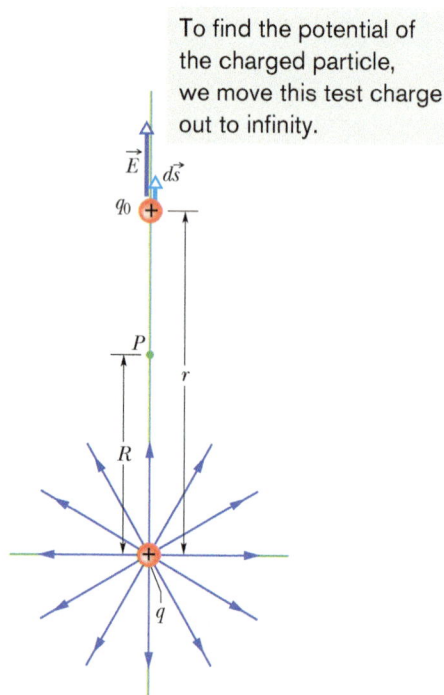

Figure 24-9 The particle with positive charge q produces an electric field $\vec{E}$ and an electric potential V at point P. We find the potential by moving a test charge q_0 from P to infinity. The test charge is shown at distance r from the particle, during differential displacement $d\vec{s}$.

Potential Due to a Charged Particle

We now use Eq. 24-18 to derive, for the space around a charged particle, an expression for the electric potential V relative to the zero potential at infinity. Consider a point P at distance R from a fixed particle of positive charge q (Fig. 24-9). To use Eq. 24-18, we imagine that we move a positive test charge q_0 from point P to infinity. Because the path we take does not matter, let us choose the simplest one— a line that extends radially from the fixed particle through P to infinity.

To use Eq. 24-18, we must evaluate the dot product

$$\vec{E}\cdot d\vec{s} = E\cos\theta\,ds. \tag{24-22}$$

The electric field $\vec{E}$ in Fig. 24-9 is directed radially outward from the fixed particle. Thus, the differential displacement $d\vec{s}$ of the test particle along its path has the same direction as $\vec{E}$. That means that in Eq. 24-22, angle $\theta = 0$ and $\cos\theta = 1$. Because the path is radial, let us write ds as dr. Then, substituting the limits R and ∞, we can write Eq. 24-18 as

$$V_f - V_i = -\int_R^\infty E\,dr. \tag{24-23}$$

Next, we set $V_f = 0$ (at ∞) and $V_i = V$ (at R). Then, for the magnitude of the electric field at the site of the test charge, we substitute from Eq. 22-3:

$$E = \frac{1}{4\pi\varepsilon_0}\frac{q}{r^2}. \tag{24-24}$$

With these changes, Eq. 24-23 then gives us

$$0 - V = -\frac{q}{4\pi\varepsilon_0}\int_R^\infty \frac{1}{r^2}\,dr = \frac{q}{4\pi\varepsilon_0}\left[\frac{1}{r}\right]_R^\infty$$

$$= -\frac{1}{4\pi\varepsilon_0}\frac{q}{R}. \tag{24-25}$$

Solving for V and switching R to r, we then have

$$V = \frac{1}{4\pi\varepsilon_0} \frac{q}{r} \qquad (24\text{-}26)$$

as the electric potential V due to a particle of charge q at any radial distance r from the particle.

Although we have derived Eq. 24-26 for a positively charged particle, the derivation holds also for a negatively charged particle, in which case, q is a negative quantity. Note that the sign of V is the same as the sign of q:

> A positively charged particle produces a positive electric potential. A negatively charged particle produces a negative electric potential.

Figure 24-10 shows a computer-generated plot of Eq. 24-26 for a positively charged particle; the magnitude of V is plotted vertically. Note that the magnitude increases as $r \rightarrow 0$. In fact, according to Eq. 24-26, V is infinite at $r = 0$, although Fig. 24-10 shows a finite, smoothed-off value there.

Equation 24-26 also gives the electric potential either *outside or on the external surface of* a spherically symmetric charge distribution. We can prove this by using one of the shell theorems of Modules 21-1 and 23-6 to replace the actual spherical charge distribution with an equal charge concentrated at its center. Then the derivation leading to Eq. 24-26 follows, provided we do not consider a point within the actual distribution.

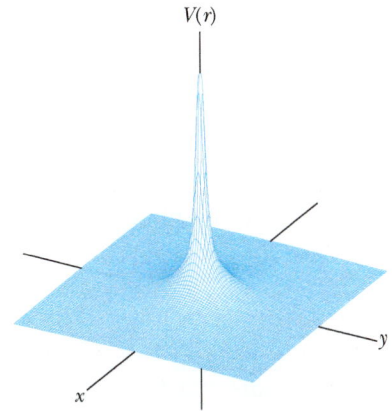

Figure 24-10 A computer-generated plot of the electric potential $V(r)$ due to a positively charged particle located at the origin of an xy plane. The potentials at points in the xy plane are plotted vertically. (Curved lines have been added to help you visualize the plot.) The infinite value of V predicted by Eq. 24-26 for $r = 0$ is not plotted.

Potential Due to a Group of Charged Particles

We can find the net electric potential at a point due to a group of charged particles with the help of the superposition principle. Using Eq. 24-26 with the plus or minus sign of the charge included, we calculate separately the potential resulting from each charge at the given point. Then we sum the potentials. Thus, for n charges, the net potential is

$$V = \sum_{i=1}^{n} V_i = \frac{1}{4\pi\varepsilon_0} \sum_{i=1}^{n} \frac{q_i}{r_i} \quad (n \text{ charged particles}). \qquad (24\text{-}27)$$

Here q_i is the value of the ith charge and r_i is the radial distance of the given point from the ith charge. The sum in Eq. 24-27 is an *algebraic sum,* not a vector sum like the sum that would be used to calculate the electric field resulting from a group of charged particles. Herein lies an important computational advantage of potential over electric field: It is a lot easier to sum several scalar quantities than to sum several vector quantities whose directions and components must be considered.

✓ Checkpoint 3

The figure here shows three arrangements of two protons. Rank the arrangements according to the net electric potential produced at point P by the protons, greatest first.

Sample Problem 24.03 Net potential of several charged particles

What is the electric potential at point P, located at the center of the square of charged particles shown in Fig. 24-11a? The distance d is 1.3 m, and the charges are

$$q_1 = +12 \text{ nC}, \qquad q_3 = +31 \text{ nC},$$
$$q_2 = -24 \text{ nC}, \qquad q_4 = +17 \text{ nC}.$$

KEY IDEA

The electric potential V at point P is the algebraic sum of the electric potentials contributed by the four particles.

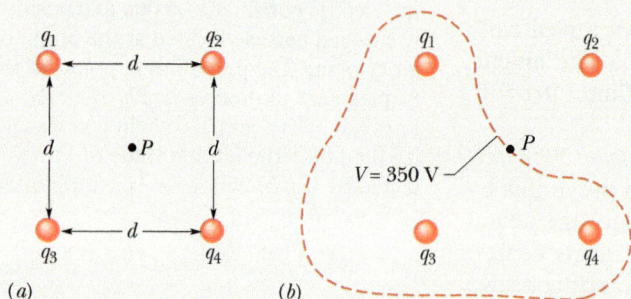

Figure 24-11 (a) Four charged particles. (b) The closed curve is a (roughly drawn) cross section of the equipotential surface that contains point P.

(Because electric potential is a scalar, the orientations of the particles do not matter.)

Calculations: From Eq. 24-27, we have

$$V = \sum_{i=1}^{4} V_i = \frac{1}{4\pi\varepsilon_0} \left(\frac{q_1}{r} + \frac{q_2}{r} + \frac{q_3}{r} + \frac{q_4}{r} \right).$$

The distance r is $d/\sqrt{2}$, which is 0.919 m, and the sum of the charges is

$$q_1 + q_2 + q_3 + q_4 = (12 - 24 + 31 + 17) \times 10^{-9} \text{ C}$$
$$= 36 \times 10^{-9} \text{ C}.$$

Thus, $$V = \frac{(8.99 \times 10^9 \text{ N}\cdot\text{m}^2/\text{C}^2)(36 \times 10^{-9} \text{ C})}{0.919 \text{ m}}$$

$$\approx 350 \text{ V}. \qquad \text{(Answer)}$$

Close to any of the three positively charged particles in Fig. 24-11a, the potential has very large positive values. Close to the single negative charge, the potential has very large negative values. Therefore, there must be points within the square that have the same intermediate potential as that at point P. The curve in Fig. 24-11b shows the intersection of the plane of the figure with the equipotential surface that contains point P.

Sample Problem 24.04 Potential is not a vector, orientation is irrelevant

(a) In Fig. 24-12a, 12 electrons (of charge $-e$) are equally spaced and fixed around a circle of radius R. Relative to $V = 0$ at infinity, what are the electric potential and electric field at the center C of the circle due to these electrons?

KEY IDEAS

(1) The electric potential V at C is the algebraic sum of the electric potentials contributed by all the electrons. Because

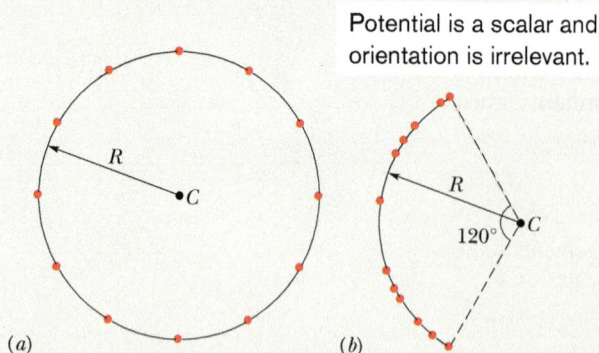

Potential is a scalar and orientation is irrelevant.

Figure 24-12 (a) Twelve electrons uniformly spaced around a circle. (b) The electrons nonuniformly spaced along an arc of the original circle.

electric potential is a scalar, the orientations of the electrons do not matter. (2) The electric field at C is a vector quantity and thus the orientation of the electrons *is* important.

Calculations: Because the electrons all have the same negative charge $-e$ and are all the same distance R from C, Eq. 24-27 gives us

$$V = -12 \frac{1}{4\pi\varepsilon_0} \frac{e}{R}. \qquad \text{(Answer)} \quad (24\text{-}28)$$

Because of the symmetry of the arrangement in Fig. 24-12a, the electric field vector at C due to any given electron is canceled by the field vector due to the electron that is diametrically opposite it. Thus, at C,

$$\vec{E} = 0. \qquad \text{(Answer)}$$

(b) The electrons are moved along the circle until they are nonuniformly spaced over a 120° arc (Fig. 24-12b). At C, find the electric potential and describe the electric field.

Reasoning: The potential is still given by Eq. 24-28, because the distance between C and each electron is unchanged and orientation is irrelevant. The electric field is no longer zero, however, because the arrangement is no longer symmetric. A net field is now directed toward the charge distribution.

24-4 POTENTIAL DUE TO AN ELECTRIC DIPOLE

Learning Objectives

After reading this module, you should be able to . . .

24.19 Calculate the potential V at any given point due to an electric dipole, in terms of the magnitude p of the dipole moment or the product of the charge separation d and the magnitude q of either charge.

24.20 For an electric dipole, identify the locations of positive potential, negative potential, and zero potential.

24.21 Compare the decrease in potential with increasing distance for a single charged particle and an electric dipole.

Key Idea

● At a distance r from an electric dipole with dipole moment magnitude $p = qd$, the electric potential of the dipole is

$$V = \frac{1}{4\pi\varepsilon_0}\frac{p\cos\theta}{r^2}$$

for $r \gg d$; the angle θ lies between the dipole moment vector and a line extending from the dipole midpoint to the point of measurement.

Potential Due to an Electric Dipole

Now let us apply Eq. 24-27 to an electric dipole to find the potential at an arbitrary point P in Fig. 24-13a. At P, the positively charged particle (at distance $r_{(+)}$) sets up potential $V_{(+)}$ and the negatively charged particle (at distance $r_{(-)}$) sets up potential $V_{(-)}$. Then the net potential at P is given by Eq. 24-27 as

$$V = \sum_{i=1}^{2} V_i = V_{(+)} + V_{(-)} = \frac{1}{4\pi\varepsilon_0}\left(\frac{q}{r_{(+)}} + \frac{-q}{r_{(-)}}\right)$$

$$= \frac{q}{4\pi\varepsilon_0}\frac{r_{(-)} - r_{(+)}}{r_{(-)}r_{(+)}}. \qquad (24\text{-}29)$$

Naturally occurring dipoles—such as those possessed by many molecules—are quite small; so we are usually interested only in points that are relatively far from the dipole, such that $r \gg d$, where d is the distance between the charges and r is the distance from the dipole's midpoint to P. In that case, we can approximate the two lines to P as being parallel and their length difference as being the leg of a right triangle with hypotenuse d (Fig. 24-13b). Also, that difference is so small that the product of the lengths is approximately r^2. Thus,

$$r_{(-)} - r_{(+)} \approx d\cos\theta \quad \text{and} \quad r_{(-)}r_{(+)} \approx r^2.$$

If we substitute these quantities into Eq. 24-29, we can approximate V to be

$$V = \frac{q}{4\pi\varepsilon_0}\frac{d\cos\theta}{r^2},$$

where θ is measured from the dipole axis as shown in Fig. 24-13a. We can now write V as

$$V = \frac{1}{4\pi\varepsilon_0}\frac{p\cos\theta}{r^2} \quad \text{(electric dipole)}, \qquad (24\text{-}30)$$

in which $p \ (= qd)$ is the magnitude of the electric dipole moment $\vec{p}$ defined in Module 22-3. The vector $\vec{p}$ is directed along the dipole axis, from the negative to the positive charge. (Thus, θ is measured from the direction of $\vec{p}$.) We use this vector to report the orientation of an electric dipole.

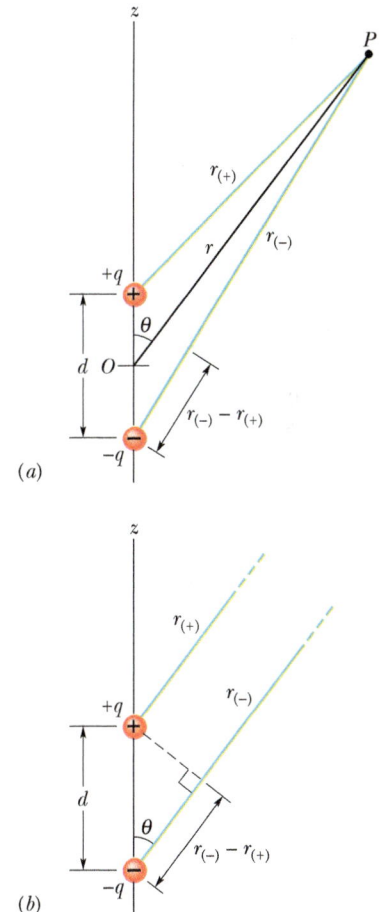

Figure 24-13 (a) Point P is a distance r from the midpoint O of a dipole. The line OP makes an angle θ with the dipole axis. (b) If P is far from the dipole, the lines of lengths $r_{(+)}$ and $r_{(-)}$ are approximately parallel to the line of length r, and the dashed black line is approximately perpendicular to the line of length $r_{(-)}$.

The electric field shifts the positive and negative charges, creating a dipole.

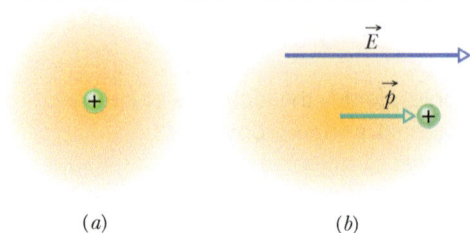

(a)   *(b)*

Figure 24-14 (*a*) An atom, showing the positively charged nucleus (green) and the negatively charged electrons (gold shading). The centers of positive and negative charge coincide. (*b*) If the atom is placed in an external electric field $\vec{E}$, the electron orbits are distorted so that the centers of positive and negative charge no longer coincide. An induced dipole moment $\vec{p}$ appears. The distortion is greatly exaggerated here.

☑ **Checkpoint 4**

Suppose that three points are set at equal (large) distances r from the center of the dipole in Fig. 24-13: Point a is on the dipole axis above the positive charge, point b is on the axis below the negative charge, and point c is on a perpendicular bisector through the line connecting the two charges. Rank the points according to the electric potential of the dipole there, greatest (most positive) first.

Induced Dipole Moment

Many molecules, such as water, have *permanent* electric dipole moments. In other molecules (called *nonpolar molecules*) and in every isolated atom, the centers of the positive and negative charges coincide (Fig. 24-14*a*) and thus no dipole moment is set up. However, if we place an atom or a nonpolar molecule in an external electric field, the field distorts the electron orbits and separates the centers of positive and negative charge (Fig. 24-14*b*). Because the electrons are negatively charged, they tend to be shifted in a direction opposite the field. This shift sets up a dipole moment $\vec{p}$ that points in the direction of the field. This dipole moment is said to be *induced* by the field, and the atom or molecule is then said to be *polarized* by the field (that is, it has a positive side and a negative side). When the field is removed, the induced dipole moment and the polarization disappear.

24-5 POTENTIAL DUE TO A CONTINUOUS CHARGE DISTRIBUTION

Learning Objective

After reading this module, you should be able to . . .

24.22 For charge that is distributed uniformly along a line or over a surface, find the net potential at a given point by splitting the distribution up into charge elements and summing (by integration) the potential due to each one.

Key Ideas

● For a continuous distribution of charge (over an extended object), the potential is found by (1) dividing the distribution into charge elements dq that can be treated as particles and then (2) summing the potential due to each element by integrating over the full distribution:

$$V = \frac{1}{4\pi\varepsilon_0} \int \frac{dq}{r}.$$

● In order to carry out the integration, dq is replaced with the product of either a linear charge density λ and a length element (such as dx), or a surface charge density σ and area element (such as $dx\,dy$).

● In some cases where the charge is symmetrically distributed, a two-dimensional integration can be reduced to a one-dimensional integration.

Potential Due to a Continuous Charge Distribution

When a charge distribution q is continuous (as on a uniformly charged thin rod or disk), we cannot use the summation of Eq. 24-27 to find the potential V at a point P. Instead, we must choose a differential element of charge dq, determine the potential dV at P due to dq, and then integrate over the entire charge distribution.

Let us again take the zero of potential to be at infinity. If we treat the element of charge dq as a particle, then we can use Eq. 24-26 to express the potential dV at point P due to dq:

$$dV = \frac{1}{4\pi\varepsilon_0} \frac{dq}{r} \quad \text{(positive or negative } dq\text{)}. \tag{24-31}$$

Here r is the distance between P and dq. To find the total potential V at P, we integrate to sum the potentials due to all the charge elements:

$$V = \int dV = \frac{1}{4\pi\varepsilon_0} \int \frac{dq}{r}.$$ (24-32)

The integral must be taken over the entire charge distribution. Note that because the electric potential is a scalar, there are *no vector components* to consider in Eq. 24-32.

We now examine two continuous charge distributions, a line and a disk.

Line of Charge

In Fig. 24-15a, a thin nonconducting rod of length L has a positive charge of uniform linear density λ. Let us determine the electric potential V due to the rod at point P, a perpendicular distance d from the left end of the rod.

We consider a differential element dx of the rod as shown in Fig. 24-15b. This (or any other) element of the rod has a differential charge of

$$dq = \lambda \, dx.$$ (24-33)

This element produces an electric potential dV at point P, which is a distance $r = (x^2 + d^2)^{1/2}$ from the element (Fig. 24-15c). Treating the element as a point charge, we can use Eq. 24-31 to write the potential dV as

$$dV = \frac{1}{4\pi\varepsilon_0} \frac{dq}{r} = \frac{1}{4\pi\varepsilon_0} \frac{\lambda \, dx}{(x^2 + d^2)^{1/2}}.$$ (24-34)

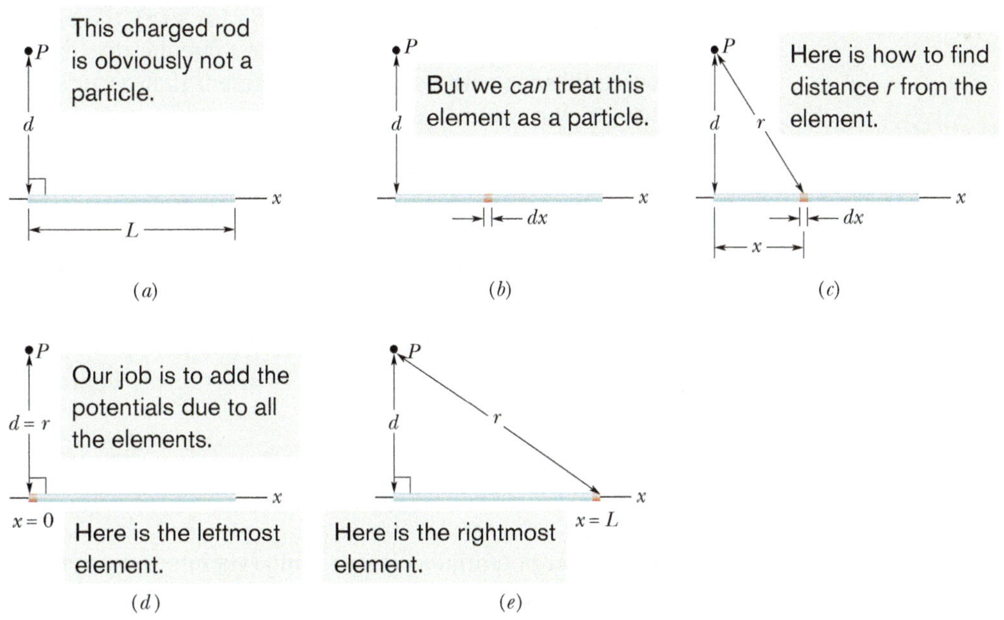

Figure 24-15 (a) A thin, uniformly charged rod produces an electric potential V at point P. (b) An element can be treated as a particle. (c) The potential at P due to the element depends on the distance r. We need to sum the potentials due to all the elements, from the left side (d) to the right side (e).

Since the charge on the rod is positive and we have taken $V = 0$ at infinity, we know from Module 24-3 that dV in Eq. 24-34 must be positive.

We now find the total potential V produced by the rod at point P by integrating Eq. 24-34 along the length of the rod, from $x = 0$ to $x = L$ (Figs. 24-15d and e), using integral 17 in Appendix E. We find

$$V = \int dV = \int_0^L \frac{1}{4\pi\varepsilon_0} \frac{\lambda}{(x^2 + d^2)^{1/2}} \, dx$$

$$= \frac{\lambda}{4\pi\varepsilon_0} \int_0^L \frac{dx}{(x^2 + d^2)^{1/2}}$$

$$= \frac{\lambda}{4\pi\varepsilon_0} \left[\ln\left(x + (x^2 + d^2)^{1/2} \right) \right]_0^L$$

$$= \frac{\lambda}{4\pi\varepsilon_0} \left[\ln\left(L + (L^2 + d^2)^{1/2} \right) - \ln d \right].$$

We can simplify this result by using the general relation $\ln A - \ln B = \ln(A/B)$. We then find

$$V = \frac{\lambda}{4\pi\varepsilon_0} \ln\left[\frac{L + (L^2 + d^2)^{1/2}}{d} \right]. \tag{24-35}$$

Because V is the sum of positive values of dV, it too is positive, consistent with the logarithm being positive for an argument greater than 1.

Charged Disk

In Module 22-5, we calculated the magnitude of the electric field at points on the central axis of a plastic disk of radius R that has a uniform charge density σ on one surface. Here we derive an expression for $V(z)$, the electric potential at any point on the central axis. Because we have a circular distribution of charge on the disk, we could start with a differential element that occupies angle $d\theta$ and radial distance dr. We would then need to set up a two-dimensional integration. However, let's do something easier.

In Fig. 24-16, consider a differential element consisting of a flat ring of radius R' and radial width dR'. Its charge has magnitude

$$dq = \sigma(2\pi R')(dR'),$$

in which $(2\pi R')(dR')$ is the upper surface area of the ring. All parts of this charged element are the same distance r from point P on the disk's axis. With the aid of Fig. 24-16, we can use Eq. 24-31 to write the contribution of this ring to the electric potential at P as

$$dV = \frac{1}{4\pi\varepsilon_0} \frac{dq}{r} = \frac{1}{4\pi\varepsilon_0} \frac{\sigma(2\pi R')(dR')}{\sqrt{z^2 + R'^2}}. \tag{24-36}$$

We find the net potential at P by adding (via integration) the contributions of all the rings from $R' = 0$ to $R' = R$:

$$V = \int dV = \frac{\sigma}{2\varepsilon_0} \int_0^R \frac{R' \, dR'}{\sqrt{z^2 + R'^2}} = \frac{\sigma}{2\varepsilon_0} (\sqrt{z^2 + R^2} - z). \tag{24-37}$$

Note that the variable in the second integral of Eq. 24-37 is R' and not z, which remains constant while the integration over the surface of the disk is carried out. (Note also that, in evaluating the integral, we have assumed that $z \geq 0$.)

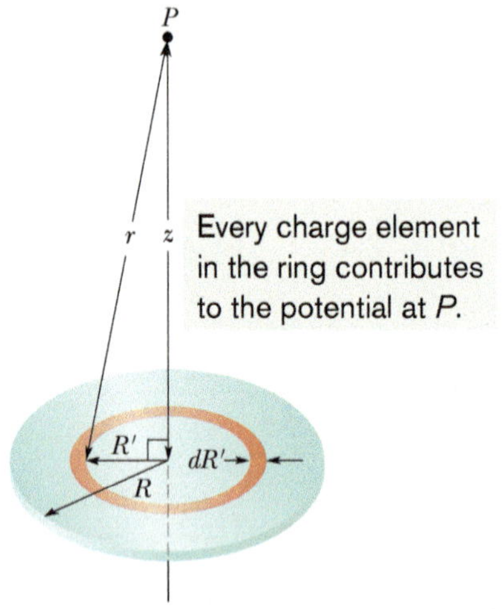

Every charge element in the ring contributes to the potential at P.

Figure 24-16 A plastic disk of radius R, charged on its top surface to a uniform surface charge density σ. We wish to find the potential V at point P on the central axis of the disk.

24-6 CALCULATING THE FIELD FROM THE POTENTIAL

Learning Objectives

After reading this module, you should be able to . . .

24.23 Given an electric potential as a function of position along an axis, find the electric field along that axis.

24.24 Given a graph of electric potential versus position along an axis, determine the electric field along the axis.

24.25 For a uniform electric field, relate the field magnitude E

and the separation Δx and potential difference ΔV between adjacent equipotential lines.

24.26 Relate the direction of the electric field and the directions in which the potential decreases and increases.

Key Ideas

● The component of $\vec{E}$ in any direction is the negative of the rate at which the potential changes with distance in that direction:

$$E_s = -\frac{\partial V}{\partial s}.$$

● The x, y, and z components of $\vec{E}$ may be found from

$$E_x = -\frac{\partial V}{\partial x}; \qquad E_y = -\frac{\partial V}{\partial y}; \qquad E_z = -\frac{\partial V}{\partial z}.$$

When $\vec{E}$ is uniform, all this reduces to

$$E = -\frac{\Delta V}{\Delta s},$$

where s is perpendicular to the equipotential surfaces.

● The electric field is zero parallel to an equipotential surface.

Calculating the Field from the Potential

In Module 24-2, you saw how to find the potential at a point f if you know the electric field along a path from a reference point to point f. In this module, we propose to go the other way—that is, to find the electric field when we know the potential. As Fig. 24-5 shows, solving this problem graphically is easy: If we know the potential V at all points near an assembly of charges, we can draw in a family of equipotential surfaces. The electric field lines, sketched perpendicular to those surfaces, reveal the variation of $\vec{E}$. What we are seeking here is the mathematical equivalent of this graphical procedure.

Figure 24-17 shows cross sections of a family of closely spaced equipotential surfaces, the potential difference between each pair of adjacent surfaces being dV. As the figure suggests, the field $\vec{E}$ at any point P is perpendicular to the equipotential surface through P.

Suppose that a positive test charge q_0 moves through a displacement $d\vec{s}$ from one equipotential surface to the adjacent surface. From Eq. 24-6, we see that the work the electric field does on the test charge during the move is $-q_0\, dV$. From Eq. 24-16 and Fig. 24-17, we see that the work done by the electric field may also be written as the scalar product $(q_0\vec{E}) \cdot d\vec{s}$, or $q_0 E(\cos\theta)\, ds$. Equating these two expressions for the work yields

$$-q_0\, dV = q_0 E(\cos\theta)\, ds, \qquad (24\text{-}38)$$

or

$$E\cos\theta = -\frac{dV}{ds}. \qquad (24\text{-}39)$$

Since $E\cos\theta$ is the component of $\vec{E}$ in the direction of $d\vec{s}$, Eq. 24-39 becomes

$$E_s = -\frac{\partial V}{\partial s}. \qquad (24\text{-}40)$$

We have added a subscript to E and switched to the partial derivative symbols to emphasize that Eq. 24-40 involves only the variation of V along a specified axis (here called the s axis) and only the component of $\vec{E}$ along that axis. In words, Eq. 24-40 (which is essentially the reverse operation of Eq. 24-18) states:

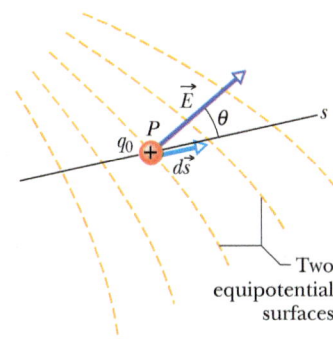

Figure 24-17 A test charge q_0 moves a distance $d\vec{s}$ from one equipotential surface to another. (The separation between the surfaces has been exaggerated for clarity.) The displacement $d\vec{s}$ makes an angle θ with the direction of the electric field $\vec{E}$.

The component of $\vec{E}$ in any direction is the negative of the rate at which the electric potential changes with distance in that direction.

If we take the s axis to be, in turn, the x, y, and z axes, we find that the x, y, and z components of $\vec{E}$ at any point are

$$E_x = -\frac{\partial V}{\partial x}; \qquad E_y = -\frac{\partial V}{\partial y}; \qquad E_z = -\frac{\partial V}{\partial z}. \qquad (24\text{-}41)$$

Thus, if we know V for all points in the region around a charge distribution — that is, if we know the function $V(x, y, z)$ — we can find the components of $\vec{E}$, and thus $\vec{E}$ itself, at any point by taking partial derivatives.

For the simple situation in which the electric field $\vec{E}$ is uniform, Eq. 24-40 becomes

$$E = -\frac{\Delta V}{\Delta s}, \qquad (24\text{-}42)$$

where s is perpendicular to the equipotential surfaces. The component of the electric field is zero in any direction parallel to the equipotential surfaces because there is no change in potential along the surfaces.

Checkpoint 5

The figure shows three pairs of parallel plates with the same separation, and the electric potential of each plate. The electric field between the plates is uniform and

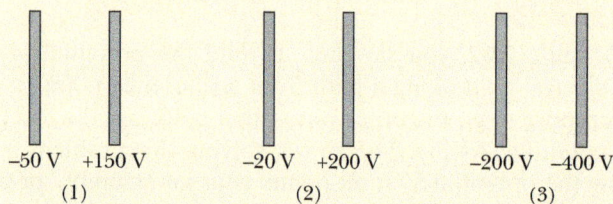

−50 V +150 V   −20 V +200 V   −200 V −400 V
    (1)   (2)   (3)

perpendicular to the plates. (a) Rank the pairs according to the magnitude of the electric field between the plates, greatest first. (b) For which pair is the electric field pointing rightward? (c) If an electron is released midway between the third pair of plates, does it remain there, move rightward at constant speed, move leftward at constant speed, accelerate rightward, or accelerate leftward?

Sample Problem 24.05 Finding the field from the potential

The electric potential at any point on the central axis of a uniformly charged disk is given by Eq. 24-37,

$$V = \frac{\sigma}{2\varepsilon_0}(\sqrt{z^2 + R^2} - z).$$

Starting with this expression, derive an expression for the electric field at any point on the axis of the disk.

KEY IDEAS

We want the electric field $\vec{E}$ as a function of distance z along the axis of the disk. For any value of z, the direction of $\vec{E}$ must be along that axis because the disk has circular symmetry about that axis. Thus, we want the component E_z of $\vec{E}$ in the direction of z. This component is the negative of the rate at which the electric potential changes with distance z.

Calculation: Thus, from the last of Eqs. 24-41, we can write

$$E_z = -\frac{\partial V}{\partial z} = -\frac{\sigma}{2\varepsilon_0}\frac{d}{dz}(\sqrt{z^2 + R^2} - z)$$

$$= \frac{\sigma}{2\varepsilon_0}\left(1 - \frac{z}{\sqrt{z^2 + R^2}}\right). \qquad \text{(Answer)}$$

This is the same expression that we derived in Module 22-5 by integration, using Coulomb's law.

24-7 ELECTRIC POTENTIAL ENERGY OF A SYSTEM OF CHARGED PARTICLES

Learning Objectives

After reading this module, you should be able to . . .

24.27 Identify that the total potential energy of a system of charged particles is equal to the work an applied force must do to assemble the system, starting with the particles infinitely far apart.

24.28 Calculate the potential energy of a pair of charged particles.

24.29 Identify that if a system has more than two charged parti-

cles, then the system's total potential energy is equal to the sum of the potential energies of every pair of the particles.

24.30 Apply the principle of the conservation of mechanical energy to a system of charged particles.

24.31 Calculate the escape speed of a charged particle from a system of charged particles (the minimum initial speed required to move infinitely far from the system).

Key Idea

● The electric potential energy of a system of charged particles is equal to the work needed to assemble the system with the particles initially at rest and infinitely distant from each other. For two particles at separation r,

$$U = W = \frac{1}{4\pi\varepsilon_0} \frac{q_1 q_2}{r}.$$

Electric Potential Energy of a System of Charged Particles

In this module we are going to calculate the potential energy of a system of two charged particles and then briefly discuss how to expand the result to a system of more than two particles. Our starting point is to examine the work we must do (as an external agent) to bring together two charged particles that are initially infinitely far apart and that end up near each other and stationary. If the two particles have the same sign of charge, we must fight against their mutual repulsion. Our work is then positive and results in a positive potential energy for the final two-particle system. If, instead, the two particles have opposite signs of charge, our job is easy because of the mutual attraction of the particles. Our work is then negative and results in a negative potential energy for the system.

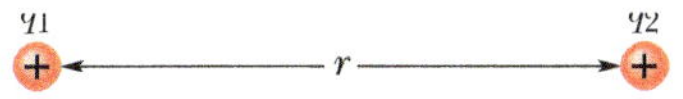

Let's follow this procedure to build the two-particle system in Fig. 24-18, where particle 1 (with positive charge q_1) and particle 2 (with positive charge q_2) have separation r. Although both particles are positively charged, our result will apply also to situations where they are both negatively charged or have different signs.

We start with particle 2 fixed in place and particle 1 infinitely far away, with an initial potential energy U_i for the two-particle system. Next we bring particle 1 to its final position, and then the system's potential energy is U_f. Our work changes the system's potential energy by $\Delta U = U_f - U_i$.

With Eq. 24-4 ($\Delta U = q(V_f - V_i)$), we can relate ΔU to the change in potential through which we move particle 1:

$$U_f - U_i = q_1(V_f - V_i). \qquad (24\text{-}43)$$

Let's evaluate these terms. The initial potential energy is $U_i = 0$ because the particles are in the reference configuration (as discussed in Module 24-1). The two potentials in Eq. 24-43 are due to particle 2 and are given by Eq. 24-26:

$$V = \frac{1}{4\pi\varepsilon_0} \frac{q_2}{r}. \qquad (24\text{-}44)$$

This tells us that when particle 1 is initially at distance $r = \infty$, the potential at its location is $V_i = 0$. When we move it to the final position at distance r, the potential at its location is

$$V_f = \frac{1}{4\pi\varepsilon_0} \frac{q_2}{r}. \qquad (24\text{-}45)$$

q_1 r q_2

Figure 24-18 Two charges held a fixed distance r apart.

Substituting these results into Eq. 24-43 and dropping the subscript f, we find that the final configuration has a potential energy of

$$U = \frac{1}{4\pi\varepsilon_0}\frac{q_1 q_2}{r} \quad \text{(two-particle system)}. \quad (24\text{-}46)$$

Equation 24-46 includes the signs of the two charges. If the two charges have the same sign, U is positive. If they have opposite signs, U is negative.

If we next bring in a third particle, with charge q_3, we repeat our calculation, starting with particle 3 at an infinite distance and then bringing it to a final position at distance r_{31} from particle 1 and distance r_{32} from particle 2. At the final position, the potential V_f at the location of particle 3 is the algebraic sum of the potential V_1 due to particle 1 and the potential V_2 of particle 2. When we work out the algebra, we find that

> The total potential energy of a system of particles is the sum of the potential energies for every pair of particles in the system.

This result applies to a system for any given number of particles.

Now that we have an expression for the potential energy of a system of particles, we can apply the principle of the conservation of energy to the system as expressed in Eq. 24-10. For example, if the system consists of many particles, we might consider the kinetic energy (and the associated *escape speed*) required of one of the particles to escape from the rest of the particles.

Sample Problem 24.06 Potential energy of a system of three charged particles

Figure 24-19 shows three charged particles held in fixed positions by forces that are not shown. What is the electric potential energy U of this system of charges? Assume that $d = 12$ cm and that

$$q_1 = +q, \quad q_2 = -4q, \quad \text{and} \quad q_3 = +2q,$$

in which $q = 150$ nC.

KEY IDEA

The potential energy U of the system is equal to the work we must do to assemble the system, bringing in each charge from an infinite distance.

Calculations: Let's mentally build the system of Fig. 24-19, starting with one of the charges, say q_1, in place and the others at infinity. Then we bring another one, say q_2, in from infinity and put it in place. From Eq. 24-46 with d substituted for r, the potential energy U_{12} associated with the pair of charges q_1 and q_2 is

$$U_{12} = \frac{1}{4\pi\varepsilon_0}\frac{q_1 q_2}{d}.$$

We then bring the last charge q_3 in from infinity and put it in

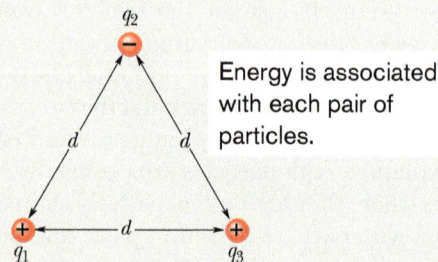

Figure 24-19 Three charges are fixed at the vertices of an equilateral triangle. What is the electric potential energy of the system?

place. The work that we must do in this last step is equal to the sum of the work we must do to bring q_3 near q_1 and the work we must do to bring it near q_2. From Eq. 24-46, with d substituted for r, that sum is

$$W_{13} + W_{23} = U_{13} + U_{23} = \frac{1}{4\pi\varepsilon_0}\frac{q_1 q_3}{d} + \frac{1}{4\pi\varepsilon_0}\frac{q_2 q_3}{d}.$$

The total potential energy U of the three-charge system is the sum of the potential energies associated with the three pairs of charges. This sum (which is actually independent of the order in which the charges are brought together) is

$$U = U_{12} + U_{13} + U_{23}$$

$$= \frac{1}{4\pi\varepsilon_0}\left(\frac{(+q)(-4q)}{d} + \frac{(+q)(+2q)}{d} + \frac{(-4q)(+2q)}{d}\right)$$

$$= -\frac{10q^2}{4\pi\varepsilon_0 d}$$

$$= -\frac{(8.99 \times 10^9 \text{ N·m}^2/\text{C}^2)(10)(150 \times 10^{-9} \text{ C})^2}{0.12 \text{ m}}$$

$$= -1.7 \times 10^{-2} \text{ J} = -17 \text{ mJ}. \qquad \text{(Answer)}$$

The negative potential energy means that negative work would have to be done to assemble this structure, starting with the three charges infinitely separated and at rest. Put another way, an external agent would have to do 17 mJ of positive work to disassemble the structure completely, ending with the three charges infinitely far apart.

The lesson here is this: If you are given an assembly of charged particles, you can find the potential energy of the assembly by finding the potential energy of every possible pair of the particles and then summing the results.

Sample Problem 24.07 Conservation of mechanical energy with electric potential energy

An alpha particle (two protons, two neutrons) moves into a stationary gold atom (79 protons, 118 neutrons), passing through the electron region that surrounds the gold nucleus like a shell and headed directly toward the nucleus (Fig. 24-20). The alpha particle slows until it momentarily stops when its center is at radial distance $r = 9.23$ fm from the nuclear center. Then it moves back along its incoming path. (Because the gold nucleus is much more massive than the alpha particle, we can assume the gold nucleus does not move.) What was the kinetic energy K_i of the alpha particle when it was initially far away (hence external to the gold atom)? Assume that the only force acting between the alpha particle and the gold nucleus is the (electrostatic) Coulomb force and treat each as a single charged particle.

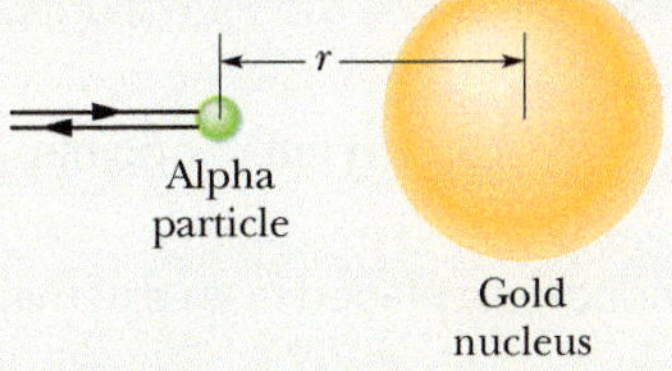

KEY IDEA

During the entire process, the mechanical energy of the *alpha particle + gold atom* system is conserved.

Reasoning: When the alpha particle is outside the atom, the system's initial electric potential energy U_i is zero because the atom has an equal number of electrons and protons, which produce a *net* electric field of zero. However, once the alpha particle passes through the electron region surrounding the nucleus on its way to the nucleus, the electric field due to the electrons goes to zero. The reason is that the electrons act like a closed spherical shell of uniform negative charge and, as discussed in Module 23-6, such a shell produces zero electric field in the space it encloses. The alpha particle still experiences the electric field of the protons in the nucleus, which produces a repulsive force on the protons within the alpha particle.

Figure 24-20 An alpha particle, traveling head-on toward the center of a gold nucleus, comes to a momentary stop (at which time all its kinetic energy has been transferred to electric potential energy) and then reverses its path.

As the incoming alpha particle is slowed by this repulsive force, its kinetic energy is transferred to electric potential energy of the system. The transfer is complete when the alpha particle momentarily stops and the kinetic energy is $K_f = 0$.

Calculations: The principle of conservation of mechanical energy tells us that

$$K_i + U_i = K_f + U_f. \qquad (24\text{-}47)$$

We know two values: $U_i = 0$ and $K_f = 0$. We also know that the potential energy U_f at the stopping point is given by the right side of Eq. 24-46, with $q_1 = 2e$, $q_2 = 79e$ (in which e is the elementary charge, 1.60×10^{-19} C), and $r = 9.23$ fm. Thus, we can rewrite Eq. 24-47 as

$$K_i = \frac{1}{4\pi\varepsilon_0}\frac{(2e)(79e)}{9.23 \text{ fm}}$$

$$= \frac{(8.99 \times 10^9 \text{ N·m}^2/\text{C}^2)(158)(1.60 \times 10^{-19} \text{ C})^2}{9.23 \times 10^{-15} \text{ m}}$$

$$= 3.94 \times 10^{-12} \text{ J} = 24.6 \text{ MeV}. \qquad \text{(Answer)}$$

PLUS Additional examples, video, and practice available at *WileyPLUS*

24-8 POTENTIAL OF A CHARGED ISOLATED CONDUCTOR

Learning Objectives

After reading this module, you should be able to . . .

24.32 Identify that an excess charge placed on an isolated conductor (or connected isolated conductors) will distribute itself on the surface of the conductor so that all points of the conductor come to the same potential.

24.33 For an isolated spherical conducting shell, sketch graphs of the potential and the electric field magnitude versus distance from the center, both inside and outside the shell.

24.34 For an isolated spherical conducting shell, identify that internally the electric field is zero and the electric potential

has the same value as the surface and that externally the electric field and the electric potential have values as though all of the shell's charge is concentrated as a particle at its center.

24.35 For an isolated cylindrical conducting shell, identify that internally the electric field is zero and the electric potential has the same value as the surface and that externally the electric field and the electric potential have values as though all of the cylinder's charge is concentrated as a line of charge on the central axis.

Key Ideas

● An excess charge placed on a conductor will, in the equilibrium state, be located entirely on the outer surface of the conductor.

● The entire conductor, including interior points, is at a uniform potential.

● If an isolated conductor is placed in an external electric

field, then at every internal point, the electric field due to the conduction electrons cancels the external electric field that otherwise would have been there.

● Also, the net electric field at every point on the surface is perpendicular to the surface.

Potential of a Charged Isolated Conductor

In Module 23-3, we concluded that $\vec{E} = 0$ for all points inside an isolated conductor. We then used Gauss' law to prove that an excess charge placed on an isolated conductor lies entirely on its surface. (This is true even if the conductor has an empty internal cavity.) Here we use the first of these facts to prove an extension of the second:

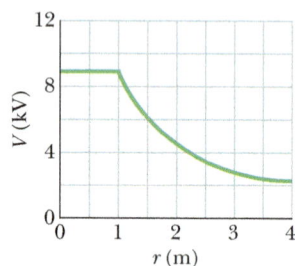

An excess charge placed on an isolated conductor will distribute itself on the surface of that conductor so that all points of the conductor—whether on the surface or inside—come to the same potential. This is true even if the conductor has an internal cavity and even if that cavity contains a net charge.

Our proof follows directly from Eq. 24-18, which is

$$V_f - V_i = -\int_i^f \vec{E} \cdot d\vec{s}.$$

Since $\vec{E} = 0$ for all points within a conductor, it follows directly that $V_f = V_i$ for all possible pairs of points i and f in the conductor.

Figure 24-21a is a plot of potential against radial distance r from the center for an isolated spherical conducting shell of 1.0 m radius, having a charge of 1.0 μC. For points outside the shell, we can calculate $V(r)$ from Eq. 24-26 because the charge q behaves for such external points as if it were concentrated at the center of the shell. That equation holds right up to the surface of the shell. Now let us push a small test charge through the shell—assuming a small hole exists—to its center. No extra work is needed to do this because no net electric force acts on the test charge once it is inside the shell. Thus, the potential at all points inside the shell has the same value as that on the surface, as Fig. 24-21a shows.

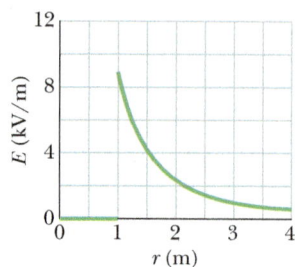

Figure 24-21 (*a*) A plot of $V(r)$ both inside and outside a charged spherical shell of radius 1.0 m. (*b*) A plot of $E(r)$ for the same shell.

Figure 24-21*b* shows the variation of electric field with radial distance for the same shell. Note that $E = 0$ everywhere inside the shell. The curves of Fig. 24-21*b* can be derived from the curve of Fig. 24-21*a* by differentiating with respect to r, using Eq. 24-40 (recall that the derivative of any constant is zero). The curve of Fig. 24-21*a* can be derived from the curves of Fig. 24-21*b* by integrating with respect to r, using Eq. 24-19.

Spark Discharge from a Charged Conductor

On nonspherical conductors, a surface charge does not distribute itself uniformly over the surface of the conductor. At sharp points or sharp edges, the surface charge density—and thus the external electric field, which is proportional to it— may reach very high values. The air around such sharp points or edges may become ionized, producing the corona discharge that golfers and mountaineers see on the tips of bushes, golf clubs, and rock hammers when thunderstorms threaten. Such corona discharges, like hair that stands on end, are often the precursors of lightning strikes. In such circumstances, it is wise to enclose yourself in a cavity inside a conducting shell, where the electric field is guaranteed to be zero. A car (unless it is a convertible or made with a plastic body) is almost ideal (Fig. 24-22).

Isolated Conductor in an External Electric Field

If an isolated conductor is placed in an *external electric field,* as in Fig. 24-23, all points of the conductor still come to a single potential regardless of whether the conductor has an excess charge. The free conduction electrons distribute themselves on the surface in such a way that the electric field they produce at interior points cancels the external electric field that would otherwise be there. Furthermore, the electron distribution causes the net electric field at all points on the surface to be perpendicular to the surface. If the conductor in Fig. 24-23 could be somehow removed, leaving the surface charges frozen in place, the internal and external electric field would remain absolutely unchanged.

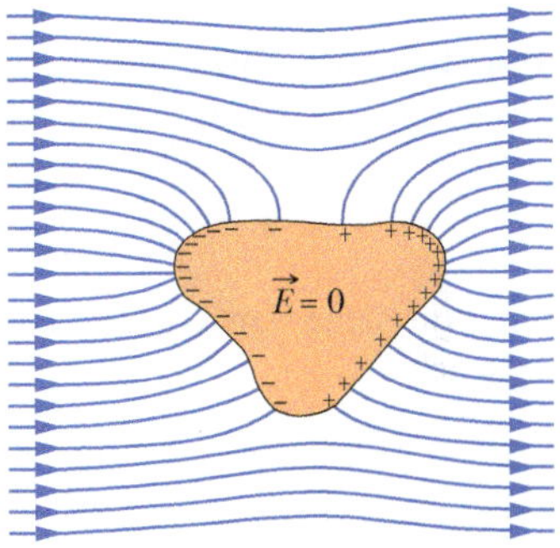

Courtesy Westinghouse Electric Corporation

Figure 24-22 A large spark jumps to a car's body and then exits by moving across the insulating left front tire (note the flash there), leaving the person inside unharmed.

Figure 24-23 An uncharged conductor is suspended in an external electric field. The free electrons in the conductor distribute themselves on the surface as shown, so as to reduce the net electric field inside the conductor to zero and make the net field at the surface perpendicular to the surface.

Review & Summary

Electric Potential The electric potential V at a point P in the electric field of a charged object is

$$V = \frac{-W_\infty}{q_0} = \frac{U}{q_0}, \tag{24-2}$$

where W_∞ is the work that would be done by the electric force on a positive test charge were it brought from an infinite distance to P, and U is the potential energy that would then be stored in the test charge–object system.

Electric Potential Energy If a particle with charge q is placed at a point where the electric potential of a charged object is V, the electric potential energy U of the particle–object system is

$$U = qV. \tag{24-3}$$

If the particle moves through a potential difference ΔV, the change in the electric potential energy is

$$\Delta U = q\,\Delta V = q(V_f - V_i). \tag{24-4}$$

Mechanical Energy If a particle moves through a change ΔV in electric potential without an applied force acting on it, applying the conservation of mechanical energy gives the change in kinetic energy as

$$\Delta K = -q\,\Delta V. \tag{24-9}$$

If, instead, an applied force acts on the particle, doing work W_{app}, the change in kinetic energy is

$$\Delta K = -q\,\Delta V + W_{app}. \tag{24-11}$$

In the special case when $\Delta K = 0$, the work of an applied force involves only the motion of the particle through a potential difference:

$$W_{app} = q \, \Delta V \quad \text{(for } K_i = K_f\text{)}. \tag{24-12}$$

Equipotential Surfaces The points on an **equipotential surface** all have the same electric potential. The work done on a test charge in moving it from one such surface to another is independent of the locations of the initial and final points on these surfaces and of the path that joins the points. The electric field $\vec{E}$ is always directed perpendicularly to corresponding equipotential surfaces.

Finding V from $\vec{E}$ The electric potential difference between two points i and f is

$$V_f - V_i = -\int_i^f \vec{E} \cdot d\vec{s}, \tag{24-18}$$

where the integral is taken over any path connecting the points. If the integration is difficult along any particular path, we can choose a different path along which the integration might be easier. If we choose $V_i = 0$, we have, for the potential at a particular point,

$$V = -\int_i^f \vec{E} \cdot d\vec{s}. \tag{24-19}$$

In the special case of a uniform field of magnitude E, the potential change between two adjacent (parallel) equipotential lines separated by distance Δx is

$$\Delta V = -E \, \Delta x. \tag{24-21}$$

Potential Due to a Charged Particle The electric potential due to a single charged particle at a distance r from that particle is

$$V = \frac{1}{4\pi\varepsilon_0} \frac{q}{r}, \tag{24-26}$$

where V has the same sign as q. The potential due to a collection of charged particles is

$$V = \sum_{i=1}^n V_i = \frac{1}{4\pi\varepsilon_0} \sum_{i=1}^n \frac{q_i}{r_i}. \tag{24-27}$$

Potential Due to an Electric Dipole At a distance r from an electric dipole with dipole moment magnitude $p = qd$, the electric potential of the dipole is

$$V = \frac{1}{4\pi\varepsilon_0} \frac{p \cos\theta}{r^2} \tag{24-30}$$

for $r \gg d$; the angle θ is defined in Fig. 24-13.

Potential Due to a Continuous Charge Distribution For a continuous distribution of charge, Eq. 24-27 becomes

$$V = \frac{1}{4\pi\varepsilon_0} \int \frac{dq}{r}, \tag{24-32}$$

in which the integral is taken over the entire distribution.

Calculating $\vec{E}$ from V The component of $\vec{E}$ in any direction is the negative of the rate at which the potential changes with distance in that direction:

$$E_s = -\frac{\partial V}{\partial s}. \tag{24-40}$$

The x, y, and z components of $\vec{E}$ may be found from

$$E_x = -\frac{\partial V}{\partial x}; \quad E_y = -\frac{\partial V}{\partial y}; \quad E_z = -\frac{\partial V}{\partial z}. \tag{24-41}$$

When $\vec{E}$ is uniform, Eq. 24-40 reduces to

$$E = -\frac{\Delta V}{\Delta s}, \tag{24-42}$$

where s is perpendicular to the equipotential surfaces.

Electric Potential Energy of a System of Charged Particles The electric potential energy of a system of charged particles is equal to the work needed to assemble the system with the particles initially at rest and infinitely distant from each other. For two particles at separation r,

$$U = W = \frac{1}{4\pi\varepsilon_0} \frac{q_1 q_2}{r}. \tag{24-46}$$

Potential of a Charged Conductor An excess charge placed on a conductor will, in the equilibrium state, be located entirely on the outer surface of the conductor. The charge will distribute itself so that the following occur: (1) The entire conductor, including interior points, is at a uniform potential. (2) At every internal point, the electric field due to the charge cancels the external electric field that otherwise would have been there. (3) The net electric field at every point on the surface is perpendicular to the surface.

Questions

1 Figure 24-24 shows eight particles that form a square, with distance d between adjacent particles. What is the net electric potential at point P at the center of the square if we take the electric potential to be zero at infinity?

2 Figure 24-25 shows three sets of cross sections of equipotential surfaces in uniform electric fields; all three cover the same size region of space. The electric potential is indi-

cated for each equipotential surface. (a) Rank the arrangements according to the magnitude of the electric field present in the region, greatest first. (b) In which is the electric field directed down the page?

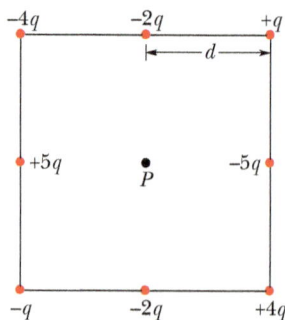

Figure 24-24 Question 1.

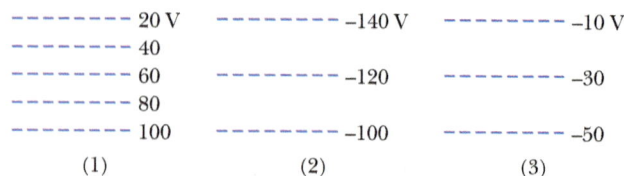

Figure 24-25 Question 2.

3 Figure 24-26 shows four pairs of charged particles. For each pair, let $V = 0$ at infinity and consider V_{net} at points on the x axis. For which pairs is there a point at which $V_{net} = 0$ (a) between the particles and (b) to the right of the particles? (c) At such a point is $\vec{E}_{net}$ due to the particles equal to zero? (d) For each pair, are there off-axis points (other than at infinity) where $V_{net} = 0$?

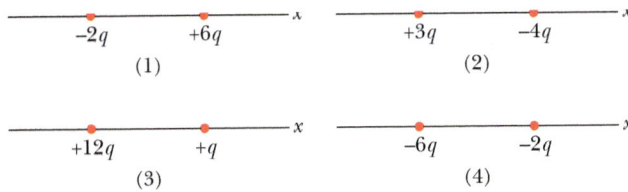

Figure 24-26 Questions 3 and 9.

4 Figure 24-27 gives the electric potential V as a function of x. (a) Rank the five regions according to the magnitude of the x component of the electric field within them, greatest first. What is the direction of the field along the x axis in (b) region 2 and (c) region 4?

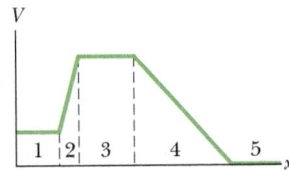

Figure 24-27 Question 4.

5 Figure 24-28 shows three paths along which we can move the positively charged sphere A closer to positively charged sphere B, which is held fixed in place. (a) Would sphere A be moved to a higher or lower electric potential? Is the work done (b) by our force and (c) by the electric field due to B positive, negative, or zero? (d) Rank the paths according to the work our force does, greatest first.

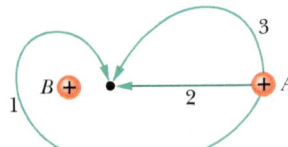

Figure 24-28 Question 5.

6 Figure 24-29 shows four arrangements of charged particles, all the same distance from the origin. Rank the situations according to the net electric potential at the origin, most positive first. Take the potential to be zero at infinity.

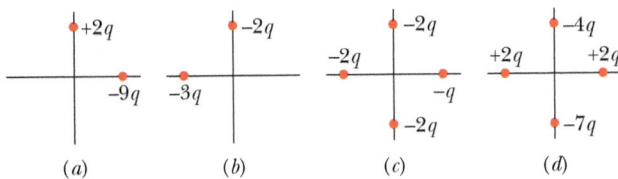

Figure 24-29 Question 6.

7 Figure 24-30 shows a system of three charged particles. If you move the particle of charge $+q$ from point A to point D, are the following quantities positive, negative, or zero: (a) the change in the electric potential energy of the three-particle system, (b) the work done by the net electric force on the particle you moved (that is, the net force due to the other two particles), and (c) the work done by your force? (d) What are the answers to (a) through (c) if, instead, the particle is moved from B to C?

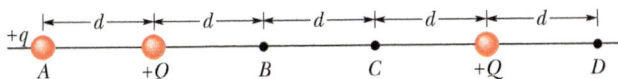

Figure 24-30 Questions 7 and 8.

8 In the situation of Question 7, is the work done by your force positive, negative, or zero if the particle is moved (a) from A to B, (b) from A to C, and (c) from B to D? (d) Rank those moves according to the magnitude of the work done by your force, greatest first.

9 Figure 24-26 shows four pairs of charged particles with identical separations. (a) Rank the pairs according to their electric potential energy (that is, the energy of the two-particle system), greatest (most positive) first. (b) For each pair, if the separation between the particles is increased, does the potential energy of the pair increase or decrease?

10 (a) In Fig. 24-31a, what is the potential at point P due to charge Q at distance R from P? Set $V = 0$ at infinity. (b) In Fig. 24-31b, the same charge Q has been spread uniformly over a circular arc of radius R and central angle 40°. What is the potential at point P, the center of curvature of the arc? (c) In Fig. 24-31c, the same charge Q has been spread uniformly over a circle of radius R. What is the potential at point P, the center of the circle? (d) Rank the three situations according to the magnitude of the electric field that is set up at P, greatest first.

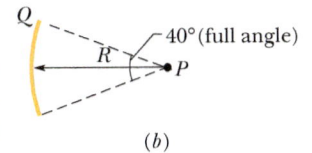

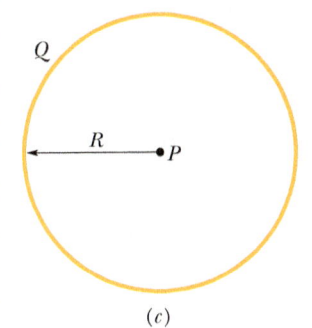

Figure 24-31 Question 10.

11 Figure 24-32 shows a thin, uniformly charged rod and three points at the same distance d from the rod. Rank the magnitude of the electric potential the rod produces at those three points, greatest first.

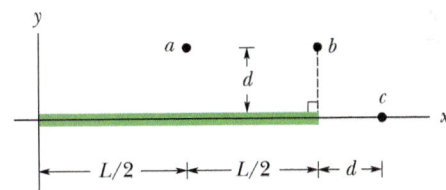

Figure 24-32 Question 11.

12 In Fig. 24-33, a particle is to be released at rest at point A and then is to be accelerated directly through point B by an electric field. The potential difference between points A and B is 100 V. Which point should be at higher electric potential if the particle is (a) an electron, (b) a proton, and (c) an alpha particle (a nucleus of two protons and two neutrons)? (d) Rank the kinetic energies of the particles at point B, greatest first.

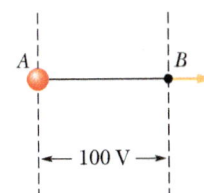

Figure 24-33
Question 12.

Problems

Module 24-1 Electric Potential

•1 SSM A particular 12 V car battery can send a total charge of 84 A·h (ampere-hours) through a circuit, from one terminal to the other. (a) How many coulombs of charge does this represent? (*Hint:* See Eq. 21-3.) (b) If this entire charge undergoes a change in electric potential of 12 V, how much energy is involved?

•2 The electric potential difference between the ground and a cloud in a particular thunderstorm is 1.2×10^9 V. In the unit electron-volts, what is the magnitude of the change in the electric potential energy of an electron that moves between the ground and the cloud?

•3 Suppose that in a lightning flash the potential difference between a cloud and the ground is 1.0×10^9 V and the quantity of charge transferred is 30 C. (a) What is the change in energy of that transferred charge? (b) If all the energy released could be used to accelerate a 1000 kg car from rest, what would be its final speed?

Module 24-2 Equipotential Surfaces and the Electric Field

•4 Two large, parallel, conducting plates are 12 cm apart and have charges of equal magnitude and opposite sign on their facing surfaces. An electric force of 3.9×10^{-15} N acts on an electron placed anywhere between the two plates. (Neglect fringing.) (a) Find the electric field at the position of the electron. (b) What is the potential difference between the plates?

•5 SSM An infinite nonconducting sheet has a surface charge density $\sigma = 0.10$ $\mu C/m^2$ on one side. How far apart are equipotential surfaces whose potentials differ by 50 V?

•6 When an electron moves from A to B along an electric field line in Fig. 24-34, the electric field does 3.94×10^{-19} J of work on it. What are the electric potential differences (a) $V_B - V_A$, (b) $V_C - V_A$, and (c) $V_C - V_B$?

••7 The electric field in a region of space has the components $E_y = E_z = 0$ and $E_x = (4.00$ N/C$)x$. Point A is on the y axis at $y = 3.00$ m, and point B is on the x axis at $x = 4.00$ m. What is the potential difference $V_B - V_A$?

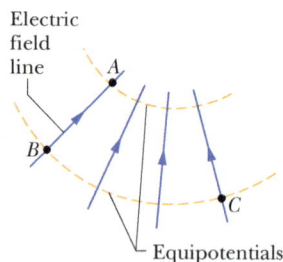

Figure 24-34 Problem 6.

••8 A graph of the x component of the electric field as a function of x in a region of space is shown in Fig. 24-35. The scale of the vertical axis is set by $E_{xs} = 20.0$ N/C. The y and z components of the electric field are zero in this region. If the electric potential at the origin is 10 V, (a) what is the electric potential at $x = 2.0$ m, (b) what is the greatest positive value of the electric potential for points on the x axis for which $0 \leq x \leq 6.0$ m, and (c) for what value of x is the electric potential zero?

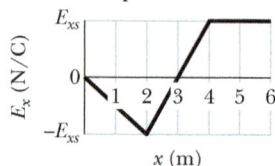

Figure 24-35 Problem 8.

••9 An infinite nonconducting sheet has a surface charge density $\sigma = +5.80$ pC/m². (a) How much work is done by the electric field due to the sheet if a particle of charge $q = +1.60 \times 10^{-19}$ C is moved from the sheet to a point P at distance $d = 3.56$ cm from the sheet? (b) If the electric potential V is defined to be zero on the sheet, what is V at P?

•••10 GO Two uniformly charged, infinite, nonconducting planes are parallel to a yz plane and positioned at $x = -50$ cm and $x = +50$ cm. The charge densities on the planes are -50 nC/m² and $+25$ nC/m², respectively. What is the magnitude of the potential difference between the origin and the point on the x axis at $x = +80$ cm? (*Hint:* Use Gauss' law.)

•••11 A nonconducting sphere has radius $R = 2.31$ cm and uniformly distributed charge $q = +3.50$ fC. Take the electric potential at the sphere's center to be $V_0 = 0$. What is V at radial distance (a) $r = 1.45$ cm and (b) $r = R$. (*Hint:* See Module 23-6.)

Module 24-3 Potential Due to a Charged Particle

•12 As a space shuttle moves through the dilute ionized gas of Earth's ionosphere, the shuttle's potential is typically changed by -1.0 V during one revolution. Assuming the shuttle is a sphere of radius 10 m, estimate the amount of charge it collects.

•13 What are (a) the charge and (b) the charge density on the surface of a conducting sphere of radius 0.15 m whose potential is 200 V (with $V = 0$ at infinity)?

•14 Consider a particle with charge $q = 1.0$ μC, point A at distance $d_1 = 2.0$ m from q, and point B at distance $d_2 = 1.0$ m. (a) If A and B are diametrically opposite each other, as in Fig. 24-36a, what is the electric potential difference $V_A - V_B$? (b) What is that electric potential difference if A and B are located as in Fig. 24-36b?

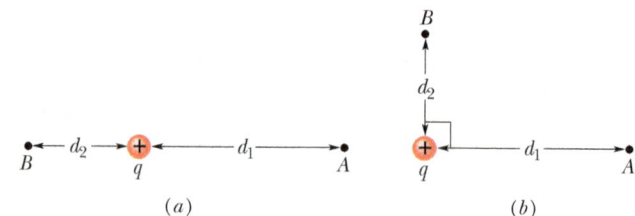

Figure 24-36 Problem 14.

••15 SSM ILW A spherical drop of water carrying a charge of 30 pC has a potential of 500 V at its surface (with $V = 0$ at infinity). (a) What is the radius of the drop? (b) If two such drops of the same charge and radius combine to form a single spherical drop, what is the potential at the surface of the new drop?

••16 GO Figure 24-37 shows a rectangular array of charged particles fixed in place, with distance $a = 39.0$ cm and the charges shown as integer multiples of $q_1 = 3.40$ pC and $q_2 = 6.00$ pC. With $V = 0$ at infinity, what

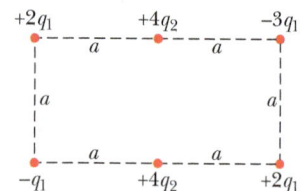

Figure 24-37 Problem 16.

is the net electric potential at the rectangle's center? (*Hint:* Thoughtful examination of the arrangement can reduce the calculation.)

••17 **GO** In Fig. 24-38, what is the net electric potential at point P due to the four particles if $V = 0$ at infinity, $q = 5.00$ fC, and $d = 4.00$ cm?

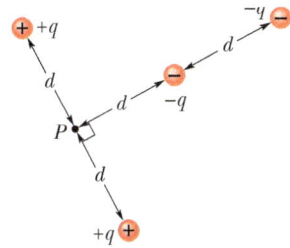

••18 **GO** Two charged particles are shown in Fig. 24-39a. Particle 1, with charge q_1, is fixed in place at distance d. Particle 2, with charge q_2, can be moved along the x axis. Figure 24-39b gives the net electric potential V at the origin due to the two particles as a function of the x coordinate of particle 2. The scale of the x axis is set by $x_s = 16.0$ cm. The plot has an asymptote of $V = 5.76 \times 10^{-7}$ V as $x \to \infty$. What is q_2 in terms of e?

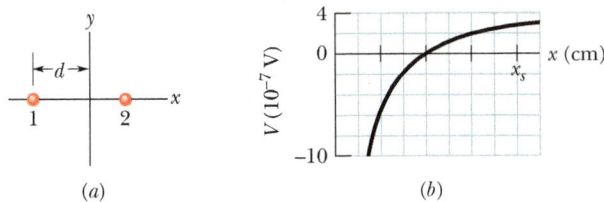

Figure 24-38 Problem 17.

Figure 24-39 Problem 18.

••19 In Fig. 24-40, particles with the charges $q_1 = +5e$ and $q_2 = -15e$ are fixed in place with a separation of $d = 24.0$ cm. With electric potential defined to be $V = 0$ at infinity, what are the finite (a) positive and (b) negative values of x at which the net electric potential on the x axis is zero?

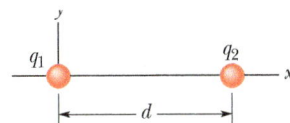

Figure 24-40 Problems 19 and 20.

••20 Two particles, of charges q_1 and q_2, are separated by distance d in Fig. 24-40. The net electric field due to the particles is zero at $x = d/4$. With $V = 0$ at infinity, locate (in terms of d) any point on the x axis (other than at infinity) at which the electric potential due to the two particles is zero.

Module 24-4   Potential Due to an Electric Dipole

•21 **ILW** The ammonia molecule NH_3 has a permanent electric dipole moment equal to 1.47 D, where 1 D = 1 debye unit = 3.34×10^{-30} C·m. Calculate the electric potential due to an ammonia molecule at a point 52.0 nm away along the axis of the dipole. (Set $V = 0$ at infinity.)

Figure 24-41 Problem 22.

••22 In Fig. 24-41a, a particle of elementary charge $+e$ is initially at coordinate $z = 20$ nm on the dipole axis (here a z axis) through

an electric dipole, on the positive side of the dipole. (The origin of z is at the center of the dipole.) The particle is then moved along a circular path around the dipole center until it is at coordinate $z = -20$ nm, on the negative side of the dipole axis. Figure 24-41b gives the work W_a done by the force moving the particle versus the angle θ that locates the particle relative to the positive direction of the z axis. The scale of the vertical axis is set by $W_{as} = 4.0 \times 10^{-30}$ J. What is the magnitude of the dipole moment?

Module 24-5   Potential Due to a Continuous Charge Distribution

•23 (a) Figure 24-42a shows a nonconducting rod of length $L = 6.00$ cm and uniform linear charge density $\lambda = +3.68$ pC/m. Assume that the electric potential is defined to be $V = 0$ at infinity. What is V at point P at distance $d = 8.00$ cm along the rod's perpendicular bisector? (b) Figure 24-42b shows an identical rod except that one half is now negatively charged. Both halves have a linear charge density of magnitude 3.68 pC/m. With $V = 0$ at infinity, what is V at P?

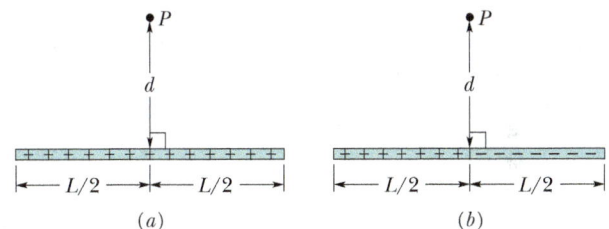

Figure 24-42 Problem 23.

•24 In Fig. 24-43, a plastic rod having a uniformly distributed charge $Q = -25.6$ pC has been bent into a circular arc of radius $R = 3.71$ cm and central angle $\phi = 120°$. With $V = 0$ at infinity, what is the electric potential at P, the center of curvature of the rod?

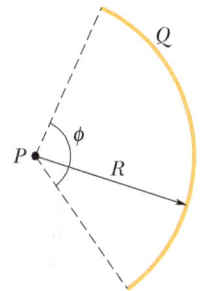

•25 A plastic rod has been bent into a circle of radius $R = 8.20$ cm. It has a charge $Q_1 = +4.20$ pC uniformly distributed along one-quarter of its circumference and a charge $Q_2 = -6Q_1$ uniformly distributed along the rest of the circumference (Fig. 24-44). With $V = 0$ at infinity, what is the electric potential at (a) the center C of the circle and (b) point P, on the central axis of the circle at distance $D = 6.71$ cm from the center?

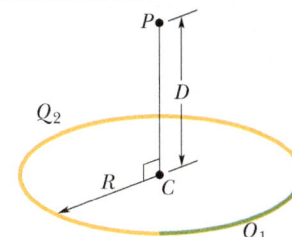

Figure 24-43
Problem 24.

Figure 24-44 Problem 25.

••26 **GO** Figure 24-45 shows a thin rod with a uniform charge density of 2.00 μC/m. Evaluate the electric potential at point P if $d = D = L/4.00$. Assume that the potential is zero at infinity.

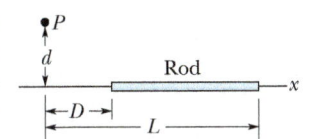

Figure 24-45 Problem 26.

•27 In Fig. 24-46, three thin plastic rods form quarter-circles with a common center of curvature at the origin. The uniform charges on the three rods are $Q_1 = +30$ nC, $Q_2 = +3.0Q_1$, and $Q_3 = -8.0Q_1$. What is the net electric potential at the origin due to the rods?

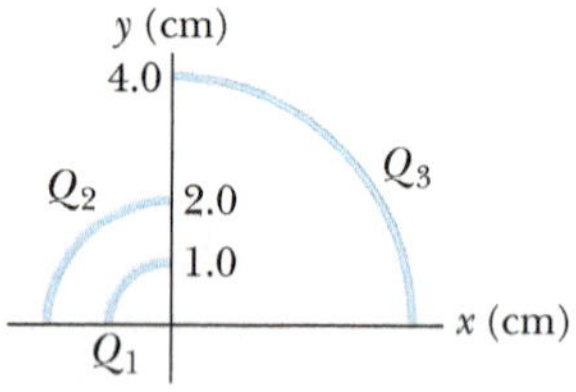

Figure 24-46 Problem 27.

••28 GO Figure 24-47 shows a thin plastic rod of length $L = 12.0$ cm and uniform positive charge $Q = 56.1$ fC lying on an x axis. With $V = 0$ at infinity, find the electric potential at point P_1 on the axis, at distance $d = 2.50$ cm from the rod.

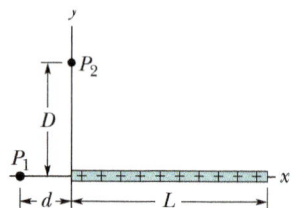

••29 In Fig. 24-48, what is the net electric potential at the origin due to the circular arc of charge $Q_1 = +7.21$ pC and the two particles of charges $Q_2 = 4.00Q_1$ and $Q_3 = -2.00Q_1$? The arc's center of curvature is at the origin and its radius is $R = 2.00$ m; the angle indicated is $\theta = 20.0°$.

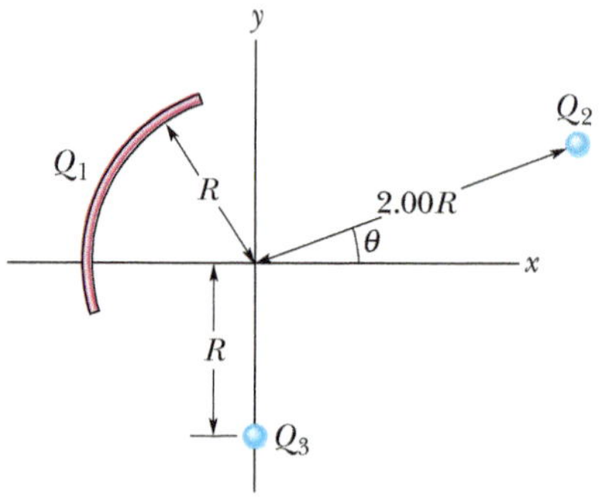

Figure 24-47 Problems 28, 33, 38, and 40.

Figure 24-48 Problem 29.

••30 GO The smiling face of Fig. 24-49 consists of three items:

1. a thin rod of charge -3.0 μC that forms a full circle of radius 6.0 cm;

2. a second thin rod of charge 2.0 μC that forms a circular arc of radius 4.0 cm, subtending an angle of 90° about the center of the full circle;

3. an electric dipole with a dipole moment that is perpendicular to a radial line and has a magnitude of 1.28×10^{-21} C·m.

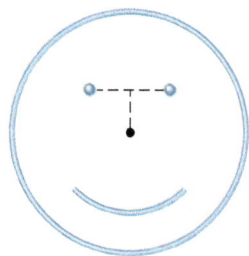

What is the net electric potential at the center?

Figure 24-49 Problem 30.

••31 SSM WWW A plastic disk of radius $R = 64.0$ cm is charged on one side with a uniform surface charge density $\sigma = 7.73$ fC/m^2, and then three quadrants of the disk are removed. The remaining quadrant is shown in Fig. 24-50. With $V = 0$ at infinity, what is the potential due to the remaining quadrant at point P, which is on the central axis of the original disk at distance $D = 25.9$ cm from the original center?

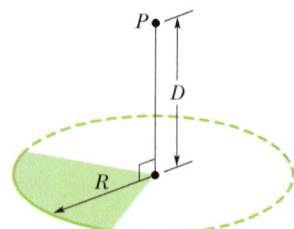

Figure 24-50 Problem 31.

•••32 GO A nonuniform linear charge distribution given by $\lambda = bx$, where b is a constant, is located along an x axis from $x = 0$ to $x = 0.20$ m. If $b = 20$ nC/m^2 and $V = 0$ at infinity, what is the electric potential at (a) the origin and (b) the point $y = 0.15$ m on the y axis?

•••33 GO The thin plastic rod shown in Fig. 24-47 has length $L = 12.0$ cm and a nonuniform linear charge density $\lambda = cx$, where $c = 28.9$ pC/m^2. With $V = 0$ at infinity, find the electric potential at point P_1 on the axis, at distance $d = 3.00$ cm from one end.

Module 24-6 Calculating the Field from the Potential

•34 Two large parallel metal plates are 1.5 cm apart and have charges of equal magnitudes but opposite signs on their facing surfaces. Take the potential of the negative plate to be zero. If the potential halfway between the plates is then $+5.0$ V, what is the electric field in the region between the plates?

•35 The electric potential at points in an xy plane is given by $V = (2.0 \text{ V/m}^2)x^2 - (3.0 \text{ V/m}^2)y^2$. In unit-vector notation, what is the electric field at the point (3.0 m, 2.0 m)?

•36 The electric potential V in the space between two flat parallel plates 1 and 2 is given (in volts) by $V = 1500x^2$, where x (in meters) is the perpendicular distance from plate 1. At $x = 1.3$ cm, (a) what is the magnitude of the electric field and (b) is the field directed toward or away from plate 1?

••37 SSM What is the magnitude of the electric field at the point $(3.00\hat{i} - 2.00\hat{j} + 4.00\hat{k})$ m if the electric potential in the region is given by $V = 2.00xyz^2$, where V is in volts and coordinates x, y, and z are in meters?

••38 Figure 24-47 shows a thin plastic rod of length $L = 13.5$ cm and uniform charge 43.6 fC. (a) In terms of distance d, find an expression for the electric potential at point P_1. (b) Next, substitute variable x for d and find an expression for the magnitude of the component E_x of the electric field at P_1. (c) What is the direction of E_x relative to the positive direction of the x axis? (d) What is the value of E_x at P_1 for $x = d = 6.20$ cm? (e) From the symmetry in Fig. 24-47, determine E_y at P_1.

••39 An electron is placed in an xy plane where the electric potential depends on x and y as shown, for the coordinate axes, in Fig. 24-51 (the potential does not depend on z). The scale of the vertical axis is set by $V_s = 500$ V. In unit-vector notation, what is the electric force on the electron?

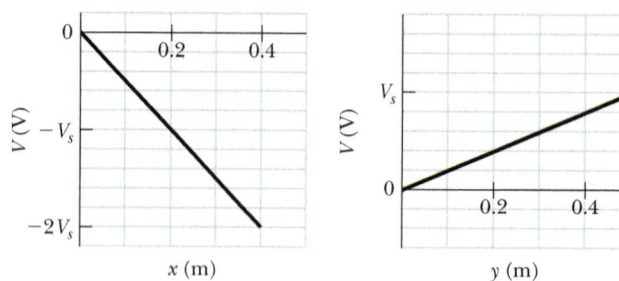

Figure 24-51 Problem 39.

•••40 GO The thin plastic rod of length $L = 10.0$ cm in Fig. 24-47 has a nonuniform linear charge density $\lambda = cx$, where $c = 49.9$ pC/m^2. (a) With $V = 0$ at infinity, find the electric potential at point P_2 on the y axis at $y = D = 3.56$ cm. (b) Find the electric field component E_y at P_2. (c) Why cannot the field component E_x at P_2 be found using the result of (a)?

Module 24-7   Electric Potential Energy of a System of Charged Particles

•**41**   A particle of charge $+7.5\ \mu C$ is released from rest at the point $x = 60$ cm on an x axis. The particle begins to move due to the presence of a charge Q that remains fixed at the origin. What is the kinetic energy of the particle at the instant it has moved 40 cm if (a) $Q = +20\ \mu C$ and (b) $Q = -20\ \mu C$?

•**42**   (a) What is the electric potential energy of two electrons separated by 2.00 nm? (b) If the separation increases, does the potential energy increase or decrease?

•**43 SSM ILW WWW**   How much work is required to set up the arrangement of Fig. 24-52 if $q = 2.30$ pC, $a = 64.0$ cm, and the particles are initially infinitely far apart and at rest?

•**44**   In Fig. 24-53, seven charged particles are fixed in place to form a square with an edge length of 4.0 cm. How much work must we do to bring a particle of charge $+6e$ initially at rest from an infinite distance to the center of the square?

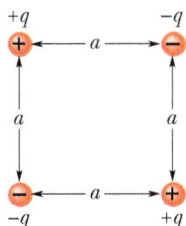

Figure 24-52
Problem 43.

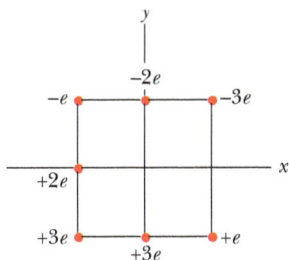

Figure 24-53 Problem 44.

••**45 ILW**   A particle of charge q is fixed at point P, and a second particle of mass m and the same charge q is initially held a distance r_1 from P. The second particle is then released. Determine its speed when it is a distance r_2 from P. Let $q = 3.1\ \mu C$, $m = 20$ mg, $r_1 = 0.90$ mm, and $r_2 = 2.5$ mm.

••**46**   A charge of -9.0 nC is uniformly distributed around a thin plastic ring lying in a yz plane with the ring center at the origin. A -6.0 pC particle is located on the x axis at $x = 3.0$ m. For a ring radius of 1.5 m, how much work must an external force do on the particle to move it to the origin?

••**47 GO**   What is the *escape speed* for an electron initially at rest on the surface of a sphere with a radius of 1.0 cm and a uniformly distributed charge of 1.6×10^{-15} C? That is, what initial speed must the electron have in order to reach an infinite distance from the sphere and have zero kinetic energy when it gets there?

••**48**   A thin, spherical, conducting shell of radius R is mounted on an isolating support and charged to a potential of -125 V. An electron is then fired directly toward the center of the shell, from point P at distance r from the center of the shell ($r \gg R$). What initial speed v_0 is needed for the electron to just reach the shell before reversing direction?

••**49 GO**   Two electrons are fixed 2.0 cm apart. Another electron is shot from infinity and stops midway between the two. What is its initial speed?

••**50**   In Fig. 24-54, how much work must we do to bring a particle, of charge $Q = +16e$ and initially at rest, along the dashed line from infinity to the indicated point near two fixed particles of charges $q_1 = +4e$ and $q_2 = -q_1/2$? Distance $d = 1.40$ cm, $\theta_1 = 43°$, and $\theta_2 = 60°$.

••**51 GO**   In the rectangle of Fig. 24-55, the sides have lengths 5.0 cm and 15 cm, $q_1 = -5.0\ \mu C$, and $q_2 = +2.0\ \mu C$. With $V = 0$ at infinity, what is the electric potential at (a) corner A and (b) corner B? (c) How much work is required to move a charge $q_3 = +3.0\ \mu C$ from B to A along a diagonal of the rectangle? (d) Does this work increase or decrease the electric potential energy of the three-charge system? Is more, less, or the same work required if q_3 is moved along a path that is (e) inside the rectangle but not on a diagonal and (f) outside the rectangle?

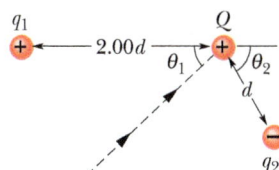

Figure 24-54 Problem 50.

Figure 24-55 Problem 51.

••**52**   Figure 24-56a shows an electron moving along an electric dipole axis toward the negative side of the dipole. The dipole is fixed in place. The electron was initially very far from the dipole, with kinetic energy 100 eV. Figure 24-56b gives the kinetic energy K of the electron versus its distance r from the dipole center. The scale of the horizontal axis is set by $r_s = 0.10$ m. What is the magnitude of the dipole moment?

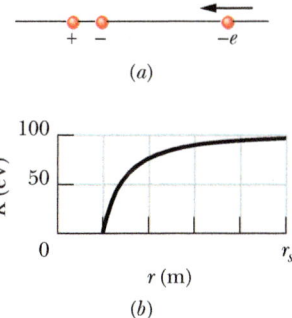

Figure 24-56 Problem 52.

••**53**   Two tiny metal spheres A and B, mass $m_A = 5.00$ g and $m_B = 10.0$ g, have equal positive charge $q = 5.00\ \mu C$. The spheres are connected by a massless nonconducting string of length $d = 1.00$ m, which is much greater than the radii of the spheres. (a) What is the electric potential energy of the system? (b) Suppose you cut the string. At that instant, what is the acceleration of each sphere? (c) A long time after you cut the string, what is the speed of each sphere?

••**54 GO**   A positron (charge $+e$, mass equal to the electron mass) is moving at 1.0×10^7 m/s in the positive direction of an x axis when, at $x = 0$, it encounters an electric field directed along the x axis. The electric potential V associated with the field is given in Fig. 24-57. The scale of the vertical axis is set by $V_s = 500.0$ V. (a) Does the positron emerge from the field at $x = 0$ (which means its motion is reversed) or at $x = 0.50$ m (which means its motion is not reversed)? (b) What is its speed when it emerges?

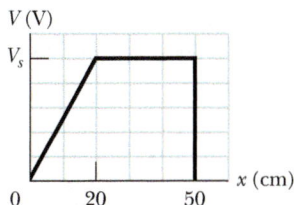

Figure 24-57 Problem 54.

••**55**   An electron is projected with an initial speed of 3.2×10^5 m/s directly toward a proton that is fixed in place. If the electron is initially a great distance from the proton, at what distance from the proton is the speed of the electron instantaneously equal to twice the initial value?

••**56**   Particle 1 (with a charge of $+5.0\ \mu C$) and particle 2 (with a charge of $+3.0\ \mu C$) are fixed in place with separation $d = 4.0$ cm on the *x* axis shown in Fig. 24-58*a*. Particle 3 can be moved along the *x* axis to the right of particle 2. Figure 24-58*b* gives the electric potential energy *U* of the three-particle system as a function of the *x* coordinate of particle 3. The scale of the vertical axis is set by $U_s = 5.0$ J. What is the charge of particle 3?

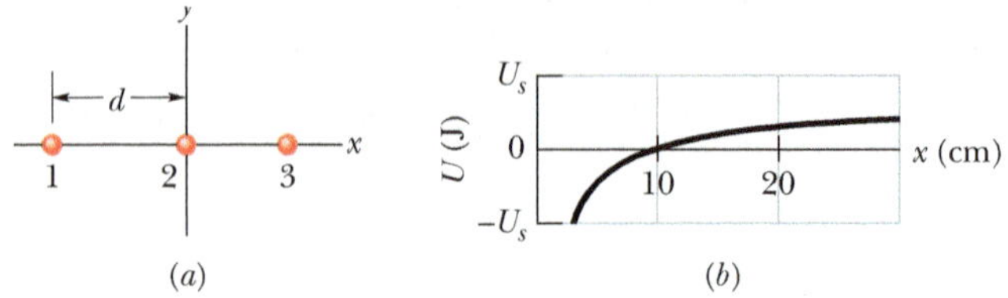

(a)   (b)

Figure 24-58 Problem 56.

··57 SSM Identical 50 µC charges are fixed on an *x* axis at $x = \pm 3.0$ m. A particle of charge $q = -15$ µC is then released from rest at a point on the positive part of the *y* axis. Due to the symmetry of the situation, the particle moves along the *y* axis and has kinetic energy 1.2 J as it passes through the point $x = 0$, $y = 4.0$ m. (a) What is the kinetic energy of the particle as it passes through the origin? (b) At what negative value of *y* will the particle momentarily stop?

··58 GO *Proton in a well.* Figure 24-59 shows electric potential *V* along an *x* axis. The scale of the vertical axis is set by $V_s = 10.0$ V. A proton is to be released at $x = 3.5$ cm with initial kinetic energy 4.00 eV. (a) If it is initially moving in the negative direction of the axis, does it reach a turning point (if so, what is the *x* coordinate of that point) or does it escape from the plotted region (if so, what is its speed at $x = 0$)? (b) If it is initially moving in the positive direction of the axis, does it reach a turning point (if so, what is the *x* coordinate of that point) or does it escape from the plotted region (if so, what is its speed at $x = 6.0$ cm)? What are the (c) magnitude *F* and (d) direction (positive or negative direction of the *x* axis) of the electric force on the proton if the proton moves just to the left of $x = 3.0$ cm? What are (e) *F* and (f) the direction if the proton moves just to the right of $x = 5.0$ cm?

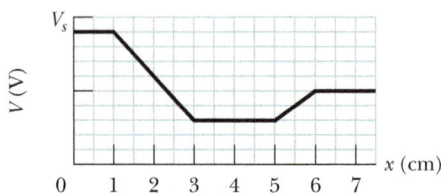

Figure 24-59 Problem 58.

··59 In Fig. 24-60, a charged particle (either an electron or a proton) is moving rightward between two parallel charged plates separated by distance $d = 2.00$ mm. The plate potentials are $V_1 = -70.0$ V and $V_2 = -50.0$ V. The particle is slowing from an initial speed of 90.0 km/s at the left plate. (a) Is the particle an electron or a proton? (b) What is its speed just as it reaches plate 2?

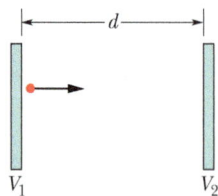

Figure 24-60 Problem 59.

··60 In Fig. 24-61*a*, we move an electron from an infinite distance to a point at distance $R = 8.00$ cm from a tiny charged ball. The move requires work $W = 2.16 \times 10^{-13}$ J by us. (a) What is the charge *Q* on the ball? In Fig. 24-61*b*, the ball has been sliced up and the slices spread out so that an equal amount of charge is at the hour positions on a circular clock face of radius $R = 8.00$ cm. Now the electron is brought from an infinite distance to the center of

the circle. (b) With that addition of the electron to the system of 12 charged particles, what is the change in the electric potential energy of the system?

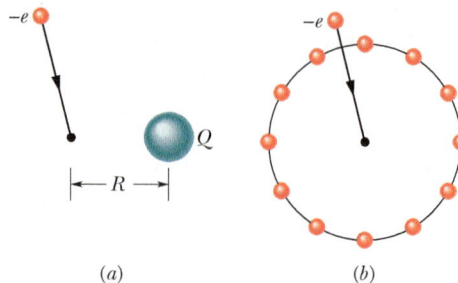

(a)   (b)

Figure 24-61 Problem 60.

···61 Suppose *N* electrons can be placed in either of two configurations. In configuration 1, they are all placed on the circumference of a narrow ring of radius *R* and are uniformly distributed so that the distance between adjacent electrons is the same everywhere. In configuration 2, $N - 1$ electrons are uniformly distributed on the ring and one electron is placed in the center of the ring. (a) What is the smallest value of *N* for which the second configuration is less energetic than the first? (b) For that value of *N*, consider any one circumference electron—call it e_0. How many other circumference electrons are closer to e_0 than the central electron is?

Module 24-8 Potential of a Charged Isolated Conductor

·62 Sphere 1 with radius R_1 has positive charge *q*. Sphere 2 with radius $2.00R_1$ is far from sphere 1 and initially uncharged. After the separated spheres are connected with a wire thin enough to retain only negligible charge, (a) is potential V_1 of sphere 1 greater than, less than, or equal to potential V_2 of sphere 2? What fraction of *q* ends up on (b) sphere 1 and (c) sphere 2? (d) What is the ratio σ_1/σ_2 of the surface charge densities of the spheres?

·63 SSM WWW Two metal spheres, each of radius 3.0 cm, have a center-to-center separation of 2.0 m. Sphere 1 has charge $+1.0 \times 10^{-8}$ C; sphere 2 has charge -3.0×10^{-8} C. Assume that the separation is large enough for us to say that the charge on each sphere is uniformly distributed (the spheres do not affect each other). With $V = 0$ at infinity, calculate (a) the potential at the point halfway between the centers and the potential on the surface of (b) sphere 1 and (c) sphere 2.

·64 A hollow metal sphere has a potential of $+400$ V with respect to ground (defined to be at $V = 0$) and a charge of 5.0×10^{-9} C. Find the electric potential at the center of the sphere.

·65 SSM What is the excess charge on a conducting sphere of radius $r = 0.15$ m if the potential of the sphere is 1500 V and $V = 0$ at infinity?

··66 Two isolated, concentric, conducting spherical shells have radii $R_1 = 0.500$ m and $R_2 = 1.00$ m, uniform charges $q_1 = +2.00$ µC and $q_2 = +1.00$ µC, and negligible thicknesses. What is the magnitude of the electric field *E* at radial distance (a) $r = 4.00$ m, (b) $r = 0.700$ m, and (c) $r = 0.200$ m? With $V = 0$ at infinity, what is *V* at (d) $r = 4.00$ m, (e) $r = 1.00$ m, (f) $r = 0.700$ m, (g) $r = 0.500$ m, (h) $r = 0.200$ m, and (i) $r = 0$? (j) Sketch $E(r)$ and $V(r)$.

··67 A metal sphere of radius 15 cm has a net charge of 3.0×10^{-8} C. (a) What is the electric field at the sphere's surface? (b) If $V = 0$ at infinity, what is the electric potential at the sphere's surface? (c) At what distance from the sphere's surface has the electric potential decreased by 500 V?

Additional Problems

68 Here are the charges and coordinates of two charged particles located in an xy plane: $q_1 = +3.00 \times 10^{-6}$ C, $x = +3.50$ cm, $y = +0.500$ cm and $q_2 = -4.00 \times 10^{-6}$ C, $x = -2.00$ cm, $y = +1.50$ cm. How much work must be done to locate these charges at their given positions, starting from infinite separation?

69 SSM A long, solid, conducting cylinder has a radius of 2.0 cm. The electric field at the surface of the cylinder is 160 N/C, directed radially outward. Let A, B, and C be points that are 1.0 cm, 2.0 cm, and 5.0 cm, respectively, from the central axis of the cylinder. What are (a) the magnitude of the electric field at C and the electric potential differences (b) $V_B - V_C$ and (c) $V_A - V_B$?

70 _The chocolate crumb mystery._ This story begins with Problem 60 in Chapter 23. (a) From the answer to part (a) of that problem, find an expression for the electric potential as a function of the radial distance r from the center of the pipe. (The electric potential is zero on the grounded pipe wall.) (b) For the typical volume charge density $\rho = -1.1 \times 10^{-3}$ C/m^3, what is the difference in the electric potential between the pipe's center and its inside wall? (The story continues with Problem 60 in Chapter 25.)

71 SSM Starting from Eq. 24-30, derive an expression for the electric field due to a dipole at a point on the dipole axis.

72 The magnitude E of an electric field depends on the radial distance r according to $E = A/r^4$, where A is a constant with the unit volt–cubic meter. As a multiple of A, what is the magnitude of the electric potential difference between $r = 2.00$ m and $r = 3.00$ m?

73 (a) If an isolated conducting sphere 10 cm in radius has a net charge of 4.0 μC and if $V = 0$ at infinity, what is the potential on the surface of the sphere? (b) Can this situation actually occur, given that the air around the sphere undergoes electrical breakdown when the field exceeds 3.0 MV/m?

74 Three particles, charge $q_1 = +10$ μC, $q_2 = -20$ μC, and $q_3 = +30$ μC, are positioned at the vertices of an isosceles triangle as shown in Fig. 24-62. If $a = 10$ cm and $b = 6.0$ cm, how much work must an external agent do to exchange the positions of (a) q_1 and q_3 and, instead, (b) q_1 and q_2?

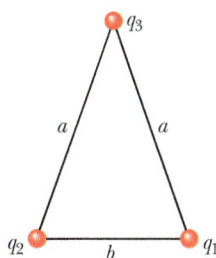

Figure 24-62 Problem 74.

75 An electric field of approximately 100 V/m is often observed near the surface of Earth. If this were the field over the entire surface, what would be the electric potential of a point on the surface? (Set $V = 0$ at infinity.)

76 A Gaussian sphere of radius 4.00 cm is centered on a ball that has a radius of 1.00 cm and a uniform charge distribution. The total (net) electric flux through the surface of the Gaussian sphere is $+5.60 \times 10^4$ N·m^2/C. What is the electric potential 12.0 cm from the center of the ball?

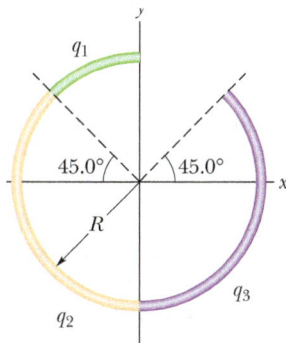

77 In a Millikan oil-drop experiment (Module 22-6), a uniform electric field of 1.92×10^5 N/C is maintained in the region between two plates separated by 1.50 cm. Find the potential difference between the plates.

78 Figure 24-63 shows three circular, nonconducting arcs of radius $R = 8.50$ cm. The charges on the arcs are $q_1 = 4.52$

Figure 24-63 Problem 78.

pC, $q_2 = -2.00q_1$, $q_3 = +3.00q_1$. With $V = 0$ at infinity, what is the net electric potential of the arcs at the common center of curvature?

79 An electron is released from rest on the axis of an electric dipole that has charge e and charge separation $d = 20$ pm and that is fixed in place. The release point is on the positive side of the dipole, at distance $7.0d$ from the dipole center. What is the electron's speed when it reaches a point $5.0d$ from the dipole center?

80 Figure 24-64 shows a ring of outer radius $R = 13.0$ cm, inner radius $r = 0.200R$, and uniform surface charge density $\sigma = 6.20$ pC/m^2. With $V = 0$ at infinity, find the electric potential at point P on the central axis of the ring, at distance $z = 2.00R$ from the center of the ring.

Figure 24-64 Problem 80.

81 GO _Electron in a well._ Figure 24-65 shows electric potential V along an x axis. The scale of the vertical axis is set by $V_s = 8.0$ V. An electron is to be released at $x = 4.5$ cm with initial kinetic energy 3.00 eV. (a) If it is initially moving in the negative direction of the axis, does it reach a turning point (if so, what is the x coordinate of that point) or does it escape from the plotted region (if so, what is its speed at $x = 0$)? (b) If it is initially moving in the positive direction of the axis, does it reach a turning point (if so, what is the x coordinate of that point) or does it escape from the plotted region (if so, what is its speed at $x = 7.0$ cm)? What are the (c) magnitude F and (d) direction (positive or negative direction of the x axis) of the electric force on the electron if the electron moves just to the left of $x = 4.0$ cm? What are (e) F and (f) the direction if it moves just to the right of $x = 5.0$ cm?

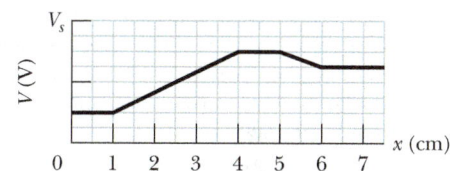

Figure 24-65 Problem 81.

82 (a) If Earth had a uniform surface charge density of 1.0 electron/m^2 (a very artificial assumption), what would its potential be? (Set $V = 0$ at infinity.) What would be the (b) magnitude and (c) direction (radially inward or outward) of the electric field due to Earth just outside its surface?

83 In Fig. 24-66, point P is at distance $d_1 = 4.00$ m from particle 1 ($q_1 = -2e$) and distance $d_2 = 2.00$ m from particle 2 ($q_2 = +2e$), with both particles fixed in place. (a) With $V = 0$ at infinity, what is V at P? If we bring a particle of charge $q_3 = +2e$ from infinity to P, (b) how much work do we do and (c) what is the potential energy of the three-particle system?

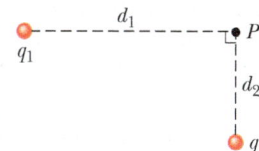

Figure 24-66 Problem 83.

84 A solid conducting sphere of radius 3.0 cm has a charge of 30 nC distributed uniformly over its surface. Let A be a point 1.0 cm from the center of the sphere, S be a point on the surface of the sphere, and B be a point 5.0 cm from the center of the sphere. What are the electric potential differences (a) $V_S - V_B$ and (b) $V_A - V_B$?

85 In Fig. 24-67, we move a particle of charge $+2e$ in from infinity to the x axis. How much work do we do? Distance D is 4.00 m.

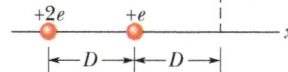

Figure 24-67 Problem 85.

86 Figure 24-68 shows a hemi-sphere with a charge of 4.00 μC distributed uniformly through its volume. The hemisphere lies on an xy plane the way half a grapefruit might lie face down on a kitchen table. Point P is located on the plane, along a radial line from the hemisphere's center of curvature, at radial distance 15 cm. What is the electric potential at point P due to the hemisphere?

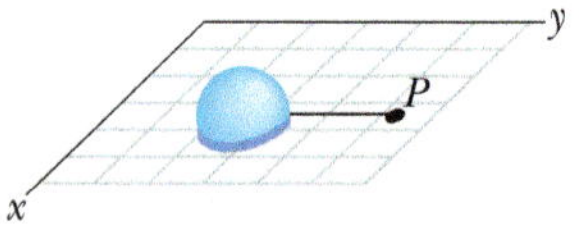

Figure 24-68 Problem 86.

87 SSM Three +0.12 C charges form an equilateral triangle 1.7 m on a side. Using energy supplied at the rate of 0.83 kW, how many days would be required to move one of the charges to the midpoint of the line joining the other two charges?

88 Two charges $q = +2.0$ μC are fixed a distance $d = 2.0$ cm apart (Fig. 24-69). (a) With $V = 0$ at infinity, what is the electric potential at point C? (b) You bring a third charge $q = +2.0$ μC from infinity to C. How much work must you do? (c) What is the potential energy U of the three-charge configuration when the third charge is in place?

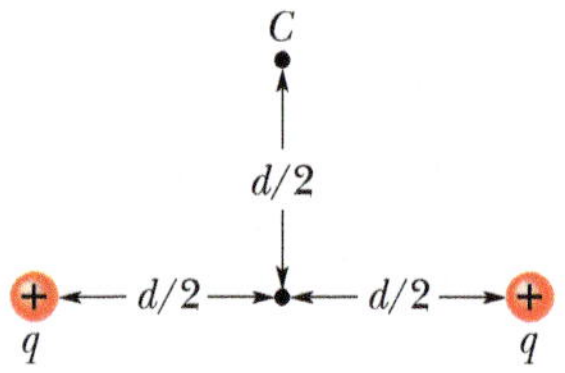

Figure 24-69 Problem 88.

89 Initially two electrons are fixed in place with a separation of 2.00 μm. How much work must we do to bring a third electron in from infinity to complete an equilateral triangle?

90 A particle of positive charge Q is fixed at point P. A second particle of mass m and negative charge $-q$ moves at constant speed in a circle of radius r_1, centered at P. Derive an expression for the work W that must be done by an external agent on the second particle to increase the radius of the circle of motion to r_2.

91 Two charged, parallel, flat conducting surfaces are spaced $d = 1.00$ cm apart and produce a potential difference $\Delta V = 625$ V between them. An electron is projected from one surface directly toward the second. What is the initial speed of the electron if it stops just at the second surface?

92 In Fig. 24-70, point P is at the center of the rectangle. With $V = 0$ at infinity, $q_1 = 5.00$ fC, $q_2 = 2.00$ fC, $q_3 = 3.00$ fC, and $d = 2.54$ cm, what is the net electric potential at P due to the six charged particles?

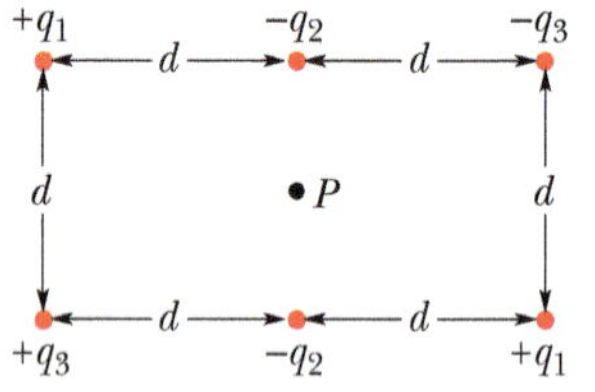

Figure 24-70 Problem 92.

93 SSM A uniform charge of +16.0 μC is on a thin circular ring lying in an xy plane and centered on the origin. The ring's radius is 3.00 cm. If point A is at the origin and point B is on the z axis at $z = 4.00$ cm, what is $V_B - V_A$?

94 Consider a particle with charge $q = 1.50 \times 10^{-8}$ C, and take $V = 0$ at infinity. (a) What are the shape and dimensions of an equipotential surface having a potential of 30.0 V due to q alone? (b) Are surfaces whose potentials differ by a constant amount (1.0 V, say) evenly spaced?

95 SSM A thick spherical shell of charge Q and uniform volume charge density ρ is bounded by radii r_1 and $r_2 > r_1$. With $V = 0$ at infinity, find the electric potential V as a function of distance r from the center of the distribution, considering regions (a) $r > r_2$, (b) $r_2 > r > r_1$, and (c) $r < r_1$. (d) Do these solutions agree with each other at $r = r_2$ and $r = r_1$? (*Hint:* See Module 23-6.)

96 A charge q is distributed uniformly throughout a spherical volume of radius R. Let $V = 0$ at infinity. What are (a) V at radial distance $r < R$ and (b) the potential difference between points at $r = R$ and the point at $r = 0$?

97 SSM A solid copper sphere whose radius is 1.0 cm has a very thin surface coating of nickel. Some of the nickel atoms are radioactive, each atom emitting an electron as it decays. Half of these electrons enter the copper sphere, each depositing 100 keV of energy there. The other half of the electrons escape, each carrying away a charge $-e$. The nickel coating has an activity of 3.70×10^8 radioactive decays per second. The sphere is hung from a long, non-conducting string and isolated from its surroundings. (a) How long will it take for the potential of the sphere to increase by 1000 V? (b) How long will it take for the temperature of the sphere to increase by 5.0 K due to the energy deposited by the electrons? The heat capacity of the sphere is 14 J/K.

98 In Fig. 24-71, a metal sphere with charge $q = 5.00$ μC and radius $r = 3.00$ cm is concentric with a larger metal sphere with charge $Q = 15.0$ μC and radius $R = 6.00$ cm. (a) What is the potential difference between the spheres? If we connect the spheres with a wire, what then is the charge on (b) the smaller sphere and (c) the larger sphere?

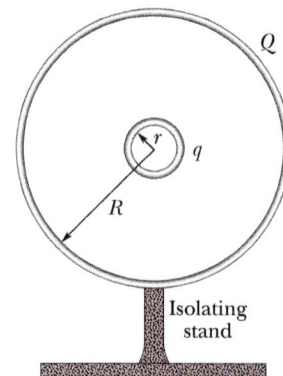

99 (a) Using Eq. 24-32, show that the electric potential at a point on the central axis of a thin ring (of charge q and radius R) and at distance z from the ring is

$$V = \frac{1}{4\pi\varepsilon_0} \frac{q}{\sqrt{z^2 + R^2}}.$$

Figure 24-71 Problem 98.

(b) From this result, derive an expression for the electric field magnitude E at points on the ring's axis; compare your result with the calculation of E in Module 22-4.

100 An alpha particle (which has two protons) is sent directly toward a target nucleus containing 92 protons. The alpha particle has an initial kinetic energy of 0.48 pJ. What is the least center-to-center distance the alpha particle will be from the target nucleus, assuming the nucleus does not move?

101 In the quark model of fundamental particles, a proton is composed of three quarks: two "up" quarks, each having charge $+2e/3$, and one "down" quark, having charge $-e/3$. Suppose that the three quarks are equidistant from one another. Take that separation distance to be 1.32×10^{-15} m and calculate the electric potential energy of the system of (a) only the two up quarks and (b) all three quarks.

102 A charge of 1.50×10^{-8} C lies on an isolated metal sphere of radius 16.0 cm. With $V = 0$ at infinity, what is the electric potential at points on the sphere's surface?

103 In Fig. 24-72, two particles of charges q_1 and q_2 are fixed to an x axis. If a third particle, of charge $+6.0$ μC, is brought from an infinite distance to point P, the three-particle system has the same electric potential energy as the original two-particle system. What is the charge ratio q_1/q_2?

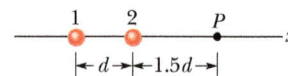

Figure 24-72 Problem 103.

Capacitance

25-1 CAPACITANCE

Learning Objectives

After reading this module, you should be able to . . .

25.01 Sketch a schematic diagram of a circuit with a parallel-plate capacitor, a battery, and an open or closed switch.

25.02 In a circuit with a battery, an open switch, and an uncharged capacitor, explain what happens to the conduction electrons when the switch is closed.

25.03 For a capacitor, apply the relationship between the magnitude of charge q on either plate ("the charge on the capacitor"), the potential difference V between the plates ("the potential across the capacitor"), and the capacitance C of the capacitor.

Key Ideas

● A capacitor consists of two isolated conductors (the plates) with charges $+q$ and $-q$. Its capacitance C is defined from

$$q = CV,$$

where V is the potential difference between the plates.

● When a circuit with a battery, an open switch, and an uncharged capacitor is completed by closing the switch, conduction electrons shift, leaving the capacitor plates with opposite charges.

What Is Physics?

One goal of physics is to provide the basic science for practical devices designed by engineers. The focus of this chapter is on one extremely common example—the capacitor, a device in which electrical energy can be stored. For example, the batteries in a camera store energy in the photoflash unit by charging a capacitor. The batteries can supply energy at only a modest rate, too slowly for the photoflash unit to emit a flash of light. However, once the capacitor is charged, it can supply energy at a much greater rate when the photoflash unit is triggered—enough energy to allow the unit to emit a burst of bright light.

The physics of capacitors can be generalized to other devices and to any situation involving electric fields. For example, Earth's atmospheric electric field is modeled by meteorologists as being produced by a huge spherical capacitor that partially discharges via lightning. The charge that skis collect as they slide along snow can be modeled as being stored in a capacitor that frequently discharges as sparks (which can be seen by nighttime skiers on dry snow).

The first step in our discussion of capacitors is to determine how much charge can be stored. This "how much" is called capacitance.

Paul Silvermann/Fundamental Photographs
Figure 25-1 An assortment of capacitors.

Capacitance

Figure 25-1 shows some of the many sizes and shapes of capacitors. Figure 25-2 shows the basic elements of *any* capacitor—two isolated conductors of any

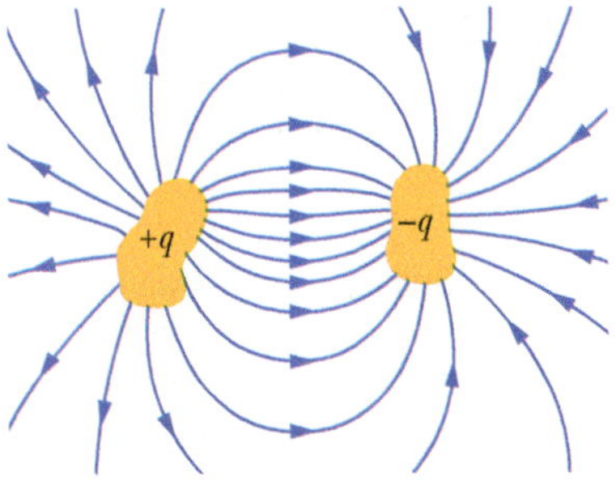

Figure 25-2 Two conductors, isolated electrically from each other and from their surroundings, form a *capacitor*. When the capacitor is charged, the charges on the conductors, or *plates* as they are called, have the same magnitude q but opposite signs.

Figure 25-3 (*a*) A parallel-plate capacitor, made up of two plates of area *A* separated by a distance *d*. The charges on the facing plate surfaces have the same magnitude *q* but opposite signs. (*b*) As the field lines show, the electric field due to the charged plates is uniform in the central region between the plates. The field is not uniform at the edges of the plates, as indicated by the "fringing" of the field lines there.

shape. No matter what their geometry, flat or not, we call these conductors *plates*.

Figure 25-3*a* shows a less general but more conventional arrangement, called a *parallel-plate capacitor*, consisting of two parallel conducting plates of area *A* separated by a distance *d*. The symbol we use to represent a capacitor (⊣⊢) is based on the structure of a parallel-plate capacitor but is used for capacitors of all geometries. We assume for the time being that no material medium (such as glass or plastic) is present in the region between the plates. In Module 25-5, we shall remove this restriction.

When a capacitor is *charged*, its plates have charges of equal magnitudes but opposite signs: $+q$ and $-q$. However, we refer to the *charge of a capacitor* as being *q*, the absolute value of these charges on the plates. (Note that *q* is not the net charge on the capacitor, which is zero.)

Because the plates are conductors, they are equipotential surfaces; all points on a plate are at the same electric potential. Moreover, there is a potential difference between the two plates. For historical reasons, we represent the absolute value of this potential difference with *V* rather than with the ΔV we used in previous notation.

The charge *q* and the potential difference *V* for a capacitor are proportional to each other; that is,

$$q = CV. \tag{25-1}$$

The proportionality constant *C* is called the **capacitance** of the capacitor. Its value depends only on the geometry of the plates and *not* on their charge or potential difference. The capacitance is a measure of how much charge must be put on the plates to produce a certain potential difference between them: The *greater the capacitance, the more charge is required.*

The SI unit of capacitance that follows from Eq. 25-1 is the coulomb per volt. This unit occurs so often that it is given a special name, the *farad* (F):

$$1 \text{ farad} = 1 \text{ F} = 1 \text{ coulomb per volt} = 1 \text{ C/V}. \tag{25-2}$$

As you will see, the farad is a very large unit. Submultiples of the farad, such as the microfarad ($1 \ \mu\text{F} = 10^{-6} \text{ F}$) and the picofarad ($1 \text{ pF} = 10^{-12} \text{ F}$), are more convenient units in practice.

Charging a Capacitor

One way to charge a capacitor is to place it in an electric circuit with a battery. An *electric circuit* is a path through which charge can flow. A *battery* is a device that maintains a certain potential difference between its *terminals* (points at which charge can enter or leave the battery) by means of internal electrochemical reactions in which electric forces can move internal charge.

In Fig. 25-4*a*, a battery B, a switch S, an uncharged capacitor C, and interconnecting wires form a circuit. The same circuit is shown in the *schematic diagram* of Fig. 25-4*b*, in which the symbols for a battery, a switch, and a capacitor represent those devices. The battery maintains potential difference *V* between its terminals. The terminal of higher potential is labeled + and is often called the *positive* terminal; the terminal of lower potential is labeled − and is often called the *negative* terminal.

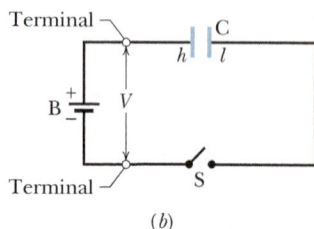

Figure 25-4 (*a*) Battery B, switch S, and plates *h* and *l* of capacitor C, connected in a circuit. (*b*) A schematic diagram with the *circuit elements* represented by their symbols.

The circuit shown in Figs. 25-4a and b is said to be *incomplete* because switch S is *open;* that is, the switch does not electrically connect the wires attached to it. When the switch is *closed,* electrically connecting those wires, the circuit is complete and charge can then flow through the switch and the wires. As we discussed in Chapter 21, the charge that can flow through a conductor, such as a wire, is that of electrons. When the circuit of Fig. 25-4 is completed, electrons are driven through the wires by an electric field that the battery sets up in the wires. The field drives electrons from capacitor plate *h* to the positive terminal of the battery; thus, plate *h*, losing electrons, becomes positively charged. The field drives just as many electrons from the negative terminal of the battery to capacitor plate *l*; thus, plate *l*, gaining electrons, becomes negatively charged *just as much* as plate *h*, losing electrons, becomes positively charged.

Initially, when the plates are uncharged, the potential difference between them is zero. As the plates become oppositely charged, that potential difference increases until it equals the potential difference *V* between the terminals of the battery. Then plate *h* and the positive terminal of the battery are at the same potential, and there is no longer an electric field in the wire between them. Similarly, plate *l* and the negative terminal reach the same potential, and there is then no electric field in the wire between them. Thus, with the field zero, there is no further drive of electrons. The capacitor is then said to be *fully charged,* with a potential difference *V* and charge *q* that are related by Eq. 25-1.

In this book we assume that during the charging of a capacitor and afterward, charge cannot pass from one plate to the other across the gap separating them. Also, we assume that a capacitor can retain (or *store*) charge indefinitely, until it is put into a circuit where it can be *discharged.*

✓ Checkpoint 1

Does the capacitance C of a capacitor increase, decrease, or remain the same (a) when the charge q on it is doubled and (b) when the potential difference V across it is tripled?

25-2 CALCULATING THE CAPACITANCE

Learning Objectives

After reading this module, you should be able to . . .

25.04 Explain how Gauss' law is used to find the capacitance of a parallel-plate capacitor.

25.05 For a parallel-plate capacitor, a cylindrical capacitor, a spherical capacitor, and an isolated sphere, calculate the capacitance.

Key Ideas

● We generally determine the capacitance of a particular capacitor configuration by (1) assuming a charge q to have been placed on the plates, (2) finding the electric field $\vec{E}$ due to this charge, (3) evaluating the potential difference V between the plates, and (4) calculating C from $q = CV$. Some results are the following:

● A parallel-plate capacitor with flat parallel plates of area A and spacing d has capacitance

$$C = \frac{\varepsilon_0 A}{d}.$$

● A cylindrical capacitor (two long coaxial cylinders) of length

L and radii a and b has capacitance

$$C = 2\pi\varepsilon_0 \frac{L}{\ln(\frac{b}{a})}.$$

● A spherical capacitor with concentric spherical plates of radii a and b has capacitance

$$C = 4\pi\varepsilon_0 \frac{ab}{b - a}.$$

● An isolated sphere of radius R has capacitance

$$C = 4\pi\varepsilon_0 R.$$

Calculating the Capacitance

Our goal here is to calculate the capacitance of a capacitor once we know its geometry. Because we shall consider a number of different geometries, it seems wise to develop a general plan to simplify the work. In brief our plan is as follows: (1) Assume a charge q on the plates; (2) calculate the electric field $\vec{E}$ between the plates in terms of this charge, using Gauss' law; (3) knowing $\vec{E}$, calculate the potential difference V between the plates from Eq. 24-18; (4) calculate C from Eq. 25-1.

Before we start, we can simplify the calculation of both the electric field and the potential difference by making certain assumptions. We discuss each in turn.

Calculating the Electric Field

To relate the electric field $\vec{E}$ between the plates of a capacitor to the charge q on either plate, we shall use Gauss' law:

$$\varepsilon_0 \oint \vec{E} \cdot d\vec{A} = q. \tag{25-3}$$

Here q is the charge enclosed by a Gaussian surface and $\oint \vec{E} \cdot d\vec{A}$ is the net electric flux through that surface. In all cases that we shall consider, the Gaussian surface will be such that whenever there is an electric flux through it, $\vec{E}$ will have a uniform magnitude E and the vectors $\vec{E}$ and $d\vec{A}$ will be parallel. Equation 25-3 then reduces to

$$q = \varepsilon_0 EA \quad \text{(special case of Eq. 25-3)}, \tag{25-4}$$

in which A is the area of that part of the Gaussian surface through which there is a flux. For convenience, we shall always draw the Gaussian surface in such a way that it completely encloses the charge on the positive plate; see Fig. 25-5 for an example.

Calculating the Potential Difference

In the notation of Chapter 24 (Eq. 24-18), the potential difference between the plates of a capacitor is related to the field $\vec{E}$ by

$$V_f - V_i = -\int_i^f \vec{E} \cdot d\vec{s}, \tag{25-5}$$

in which the integral is to be evaluated along any path that starts on one plate and ends on the other. We shall always choose a path that follows an electric field line, from the negative plate to the positive plate. For this path, the vectors $\vec{E}$ and $d\vec{s}$ will have opposite directions; so the dot product $\vec{E} \cdot d\vec{s}$ will be equal to $-E\,ds$. Thus, the right side of Eq. 25-5 will then be positive. Letting V represent the difference $V_f - V_i$, we can then recast Eq. 25-5 as

$$V = \int_-^+ E\,ds \quad \text{(special case of Eq. 25-5)}, \tag{25-6}$$

in which the $-$ and $+$ remind us that our path of integration starts on the negative plate and ends on the positive plate.

We are now ready to apply Eqs. 25-4 and 25-6 to some particular cases.

A Parallel-Plate Capacitor

We assume, as Fig. 25-5 suggests, that the plates of our parallel-plate capacitor are so large and so close together that we can neglect the fringing of the electric field

We use Gauss' law to relate q and E. Then we integrate the E to get the potential difference.

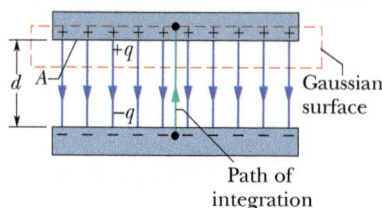

Figure 25-5 A charged parallel-plate capacitor. A Gaussian surface encloses the charge on the positive plate. The integration of Eq. 25-6 is taken along a path extending directly from the negative plate to the positive plate.

at the edges of the plates, taking $\vec{E}$ to be constant throughout the region between the plates.

We draw a Gaussian surface that encloses just the charge q on the positive plate, as in Fig. 25-5. From Eq. 25-4 we can then write

$$q = \varepsilon_0 EA, \tag{25-7}$$

where A is the area of the plate.

Equation 25-6 yields

$$V = \int_-^+ E\, ds = E \int_0^d ds = Ed. \tag{25-8}$$

In Eq. 25-8, E can be placed outside the integral because it is a constant; the second integral then is simply the plate separation d.

If we now substitute q from Eq. 25-7 and V from Eq. 25-8 into the relation $q = CV$ (Eq. 25-1), we find

$$C = \frac{\varepsilon_0 A}{d} \quad \text{(parallel-plate capacitor).} \tag{25-9}$$

Thus, the capacitance does indeed depend only on geometrical factors—namely, the plate area A and the plate separation d. Note that C increases as we increase area A or decrease separation d.

As an aside, we point out that Eq. 25-9 suggests one of our reasons for writing the electrostatic constant in Coulomb's law in the form $1/4\pi\varepsilon_0$. If we had not done so, Eq. 25-9—which is used more often in engineering practice than Coulomb's law—would have been less simple in form. We note further that Eq. 25-9 permits us to express the permittivity constant ε_0 in a unit more appropriate for use in problems involving capacitors; namely,

$$\varepsilon_0 = 8.85 \times 10^{-12}\ \text{F/m} = 8.85\ \text{pF/m}. \tag{25-10}$$

We have previously expressed this constant as

$$\varepsilon_0 = 8.85 \times 10^{-12}\ \text{C}^2/\text{N} \cdot \text{m}^2. \tag{25-11}$$

A Cylindrical Capacitor

Figure 25-6 shows, in cross section, a cylindrical capacitor of length L formed by two coaxial cylinders of radii a and b. We assume that $L \gg b$ so that we can neglect the fringing of the electric field that occurs at the ends of the cylinders. Each plate contains a charge of magnitude q.

As a Gaussian surface, we choose a cylinder of length L and radius r, closed by end caps and placed as is shown in Fig. 25-6. It is coaxial with the cylinders and encloses the central cylinder and thus also the charge q on that cylinder. Equation 25-4 then relates that charge and the field magnitude E as

$$q = \varepsilon_0 EA = \varepsilon_0 E(2\pi rL),$$

in which $2\pi rL$ is the area of the curved part of the Gaussian surface. There is no flux through the end caps. Solving for E yields

$$E = \frac{q}{2\pi\varepsilon_0 Lr}. \tag{25-12}$$

Substitution of this result into Eq. 25-6 yields

$$V = \int_-^+ E\, ds = -\frac{q}{2\pi\varepsilon_0 L}\int_b^a \frac{dr}{r} = \frac{q}{2\pi\varepsilon_0 L}\ln\left(\frac{b}{a}\right), \tag{25-13}$$

where we have used the fact that here $ds = -dr$ (we integrated radially inward).

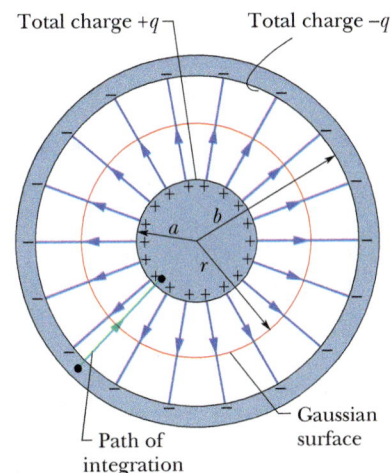

Figure 25-6 A cross section of a long cylindrical capacitor, showing a cylindrical Gaussian surface of radius r (that encloses the positive plate) and the radial path of integration along which Eq. 25-6 is to be applied. This figure also serves to illustrate a spherical capacitor in a cross section through its center.

From the relation $C = q/V$, we then have

$$C = 2\pi\varepsilon_0 \frac{L}{\ln(b/a)} \qquad \text{(cylindrical capacitor)}. \qquad (25\text{-}14)$$

We see that the capacitance of a cylindrical capacitor, like that of a parallel-plate capacitor, depends only on geometrical factors, in this case the length L and the two radii b and a.

A Spherical Capacitor

Figure 25-6 can also serve as a central cross section of a capacitor that consists of two concentric spherical shells, of radii a and b. As a Gaussian surface we draw a sphere of radius r concentric with the two shells; then Eq. 25-4 yields

$$q = \varepsilon_0 EA = \varepsilon_0 E(4\pi r^2),$$

in which $4\pi r^2$ is the area of the spherical Gaussian surface. We solve this equation for E, obtaining

$$E = \frac{1}{4\pi\varepsilon_0} \frac{q}{r^2}, \qquad (25\text{-}15)$$

which we recognize as the expression for the electric field due to a uniform spherical charge distribution (Eq. 23-15).

If we substitute this expression into Eq. 25-6, we find

$$V = \int_{-}^{+} E\,ds = -\frac{q}{4\pi\varepsilon_0} \int_{b}^{a} \frac{dr}{r^2} = \frac{q}{4\pi\varepsilon_0}\left(\frac{1}{a} - \frac{1}{b}\right) = \frac{q}{4\pi\varepsilon_0}\frac{b-a}{ab}, \quad (25\text{-}16)$$

where again we have substituted $-dr$ for ds. If we now substitute Eq. 25-16 into Eq. 25-1 and solve for C, we find

$$C = 4\pi\varepsilon_0 \frac{ab}{b-a} \qquad \text{(spherical capacitor)}. \qquad (25\text{-}17)$$

An Isolated Sphere

We can assign a capacitance to a *single* isolated spherical conductor of radius R by assuming that the "missing plate" is a conducting sphere of infinite radius. After all, the field lines that leave the surface of a positively charged isolated conductor must end somewhere; the walls of the room in which the conductor is housed can serve effectively as our sphere of infinite radius.

To find the capacitance of the conductor, we first rewrite Eq. 25-17 as

$$C = 4\pi\varepsilon_0 \frac{a}{1 - a/b}.$$

If we then let $b \to \infty$ and substitute R for a, we find

$$C = 4\pi\varepsilon_0 R \qquad \text{(isolated sphere)}. \qquad (25\text{-}18)$$

Note that this formula and the others we have derived for capacitance (Eqs. 25-9, 25-14, and 25-17) involve the constant ε_0 multiplied by a quantity that has the dimensions of a length.

✓ Checkpoint 2

For capacitors charged by the same battery, does the charge stored by the capacitor increase, decrease, or remain the same in each of the following situations? (a) The plate separation of a parallel-plate capacitor is increased. (b) The radius of the inner cylinder of a cylindrical capacitor is increased. (c) The radius of the outer spherical shell of a spherical capacitor is increased.

Sample Problem 25.01 Charging the plates in a parallel-plate capacitor

In Fig. 25-7a, switch S is closed to connect the uncharged capacitor of capacitance $C = 0.25\ \mu F$ to the battery of potential difference $V = 12$ V. The lower capacitor plate has thickness $L = 0.50$ cm and face area $A = 2.0 \times 10^{-4}\ m^2$, and it consists of copper, in which the density of conduction electrons is $n = 8.49 \times 10^{28}$ electrons/m³. From what depth d within the plate (Fig. 25-7b) must electrons move to the plate face as the capacitor becomes charged?

KEY IDEA

The charge collected on the plate is related to the capacitance and the potential difference across the capacitor by Eq. 25-1 ($q = CV$).

Calculations: Because the lower plate is connected to the negative terminal of the battery, conduction electrons move up to the face of the plate. From Eq. 25-1, the total charge

magnitude that collects there is

$$q = CV = (0.25 \times 10^{-6}\ F)(12\ V)$$
$$= 3.0 \times 10^{-6}\ C.$$

Dividing this result by e gives us the number N of conduction electrons that come up to the face:

$$N = \frac{q}{e} = \frac{3.0 \times 10^{-6}\ C}{1.602 \times 10^{-19}\ C}$$
$$= 1.873 \times 10^{13}\ \text{electrons}.$$

These electrons come from a volume that is the product of the face area A and the depth d we seek. Thus, from the density of conduction electrons (number per volume), we can write

$$n = \frac{N}{Ad},$$

or

$$d = \frac{N}{An} = \frac{1.873 \times 10^{13}\ \text{electrons}}{(2.0 \times 10^{-4}\ m^2)(8.49 \times 10^{28}\ \text{electrons/m}^3)}$$

$$= 1.1 \times 10^{-12}\ m = 1.1\ \text{pm}. \qquad \text{(Answer)}$$

We commonly say that electrons move from the battery to the negative face but, actually, the battery sets up an electric field in the wires and plate such that electrons very close to the plate face move up to the negative face.

Figure 25-7 (a) A battery and capacitor circuit. (b) The lower capacitor plate.

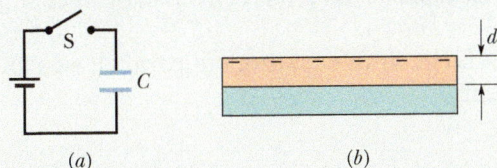

(a)   *(b)*

PLUS Additional examples, video, and practice available at *WileyPLUS*

25-3 CAPACITORS IN PARALLEL AND IN SERIES

Learning Objectives

After reading this module, you should be able to . . .

25.06 Sketch schematic diagrams for a battery and (a) three capacitors in parallel and (b) three capacitors in series.

25.07 Identify that capacitors in parallel have the same potential difference, which is the same value that their equivalent capacitor has.

25.08 Calculate the equivalent of parallel capacitors.

25.09 Identify that the total charge stored on parallel capacitors is the sum of the charges stored on the individual capacitors.

25.10 Identify that capacitors in series have the same charge, which is the same value that their equivalent capacitor has.

25.11 Calculate the equivalent of series capacitors.

25.12 Identify that the potential applied to capacitors in series is equal to the sum of the potentials across the individual capacitors.

25.13 For a circuit with a battery and some capacitors in parallel and some in series, simplify the circuit in steps by finding equivalent capacitors, until the charge and potential on the final equivalent capacitor can be determined, and then reverse the steps to find the charge and potential on the individual capacitors.

25.14 For a circuit with a battery, an open switch, and one or more uncharged capacitors, determine the amount of charge that moves through a point in the circuit when the switch is closed.

25.15 When a charged capacitor is connected in parallel to one or more uncharged capacitors, determine the charge and potential difference on each capacitor when equilibrium is reached.

Key Idea

● The equivalent capacitances C_{eq} of combinations of individual capacitors connected in parallel and in series can be found from

$$C_{eq} = \sum_{j=1}^{n} C_j \qquad (n\ \text{capacitors in parallel})$$

and

$$\frac{1}{C_{eq}} = \sum_{j=1}^{n} \frac{1}{C_j} \qquad (n\ \text{capacitors in series}).$$

Equivalent capacitances can be used to calculate the capacitances of more complicated series–parallel combinations.

(a)

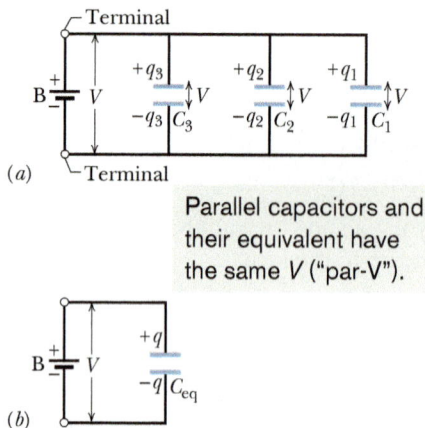

Parallel capacitors and their equivalent have the same V ("par-V").

(b)

Figure 25-8 (a) Three capacitors connected in parallel to battery B. The battery maintains potential difference V across its terminals and thus across *each* capacitor. (b) The equivalent capacitor, with capacitance C_{eq}, replaces the parallel combination.

Capacitors in Parallel and in Series

When there is a combination of capacitors in a circuit, we can sometimes replace that combination with an **equivalent capacitor**—that is, a single capacitor that has the same capacitance as the actual combination of capacitors. With such a replacement, we can simplify the circuit, affording easier solutions for unknown quantities of the circuit. Here we discuss two basic combinations of capacitors that allow such a replacement.

Capacitors in Parallel

Figure 25-8a shows an electric circuit in which three capacitors are connected *in parallel* to battery B. This description has little to do with how the capacitor plates are drawn. Rather, "in parallel" means that the capacitors are directly wired together at one plate and directly wired together at the other plate, and that the same potential difference V is applied across the two groups of wired-together plates. Thus, each capacitor has the same potential difference V, which produces charge on the capacitor. (In Fig. 25-8a, the applied potential V is maintained by the battery.) In general:

> When a potential difference V is applied across several capacitors connected in parallel, that potential difference V is applied across each capacitor. The total charge q stored on the capacitors is the sum of the charges stored on all the capacitors.

When we analyze a circuit of capacitors in parallel, we can simplify it with this mental replacement:

> Capacitors connected in parallel can be replaced with an equivalent capacitor that has the same *total* charge q and the same potential difference V as the actual capacitors.

(You might remember this result with the nonsense word "par-V," which is close to "party," to mean "capacitors in parallel have the same V.") Figure 25-8b shows the equivalent capacitor (with equivalent capacitance C_{eq}) that has replaced the three capacitors (with actual capacitances C_1, C_2, and C_3) of Fig. 25-8a.

To derive an expression for C_{eq} in Fig. 25-8b, we first use Eq. 25-1 to find the charge on each actual capacitor:

$$q_1 = C_1 V, \quad q_2 = C_2 V, \quad \text{and} \quad q_3 = C_3 V.$$

The total charge on the parallel combination of Fig. 25-8a is then

$$q = q_1 + q_2 + q_3 = (C_1 + C_2 + C_3)V.$$

The equivalent capacitance, with the same total charge q and applied potential difference V as the combination, is then

$$C_{eq} = \frac{q}{V} = C_1 + C_2 + C_3,$$

a result that we can easily extend to any number n of capacitors, as

$$C_{eq} = \sum_{j=1}^{n} C_j \quad (n \text{ capacitors in parallel}). \tag{25-19}$$

Thus, to find the equivalent capacitance of a parallel combination, we simply add the individual capacitances.

Capacitors in Series

Figure 25-9a shows three capacitors connected *in series* to battery B. This description has little to do with how the capacitors are drawn. Rather, "in series" means that the capacitors are wired serially, one after the other, and that a potential difference V is applied across the two ends of the series. (In Fig. 25-9a, this potential difference V is maintained by battery B.) The potential differences that then exist across the capacitors in the series produce identical charges q on them.

> When a potential difference V is applied across several capacitors connected in series, the capacitors have identical charge q. The sum of the potential differences across all the capacitors is equal to the applied potential difference V.

We can explain how the capacitors end up with identical charge by following a *chain reaction* of events, in which the charging of each capacitor causes the charging of the next capacitor. We start with capacitor 3 and work upward to capacitor 1. When the battery is first connected to the series of capacitors, it produces charge $-q$ on the bottom plate of capacitor 3. That charge then repels negative charge from the top plate of capacitor 3 (leaving it with charge $+q$). The repelled negative charge moves to the bottom plate of capacitor 2 (giving it charge $-q$). That charge on the bottom plate of capacitor 2 then repels negative charge from the top plate of capacitor 2 (leaving it with charge $+q$) to the bottom plate of capacitor 1 (giving it charge $-q$). Finally, the charge on the bottom plate of capacitor 1 helps move negative charge from the top plate of capacitor 1 to the battery, leaving that top plate with charge $+q$.

Here are two important points about capacitors in series:

1. When charge is shifted from one capacitor to another in a series of capacitors, it can move along only one route, such as from capacitor 3 to capacitor 2 in Fig. 25-9a. If there are additional routes, the capacitors are not in series.

2. The battery directly produces charges on only the two plates to which it is connected (the bottom plate of capacitor 3 and the top plate of capacitor 1 in Fig. 25-9a). Charges that are produced on the other plates are due merely to the shifting of charge already there. For example, in Fig. 25-9a, the part of the circuit enclosed by dashed lines is electrically isolated from the rest of the circuit. Thus, its charge can only be redistributed.

When we analyze a circuit of capacitors in series, we can simplify it with this mental replacement:

> Capacitors that are connected in series can be replaced with an equivalent capacitor that has the same charge q and the same *total* potential difference V as the actual series capacitors.

(a)

Series capacitors and their equivalent have the same q ("seri-q").

(b)

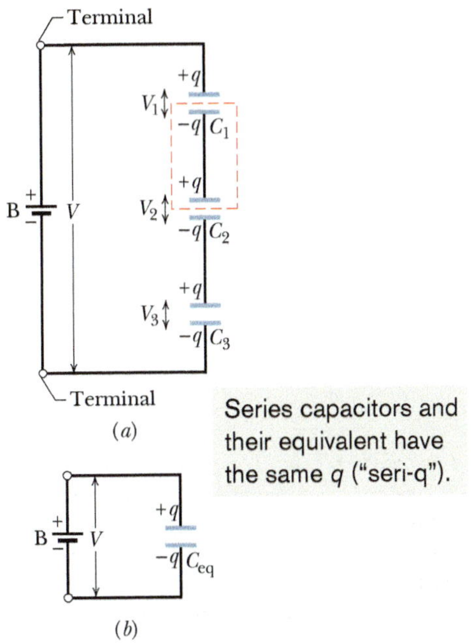

Figure 25-9 (a) Three capacitors connected in series to battery B. The battery maintains potential difference V between the top and bottom plates of the series combination. (b) The equivalent capacitor, with capacitance C_{eq}, replaces the series combination.

(You might remember this with the nonsense word "seri-q" to mean "capacitors in series have the same q.") Figure 25-9b shows the equivalent capacitor (with equivalent capacitance C_{eq}) that has replaced the three actual capacitors (with actual capacitances C_1, C_2, and C_3) of Fig. 25-9a.

To derive an expression for C_{eq} in Fig. 25-9b, we first use Eq. 25-1 to find the potential difference of each actual capacitor:

$$V_1 = \frac{q}{C_1}, \quad V_2 = \frac{q}{C_2}, \quad \text{and} \quad V_3 = \frac{q}{C_3}.$$

The total potential difference V due to the battery is the sum

$$V = V_1 + V_2 + V_3 = q\left(\frac{1}{C_1} + \frac{1}{C_2} + \frac{1}{C_3}\right).$$

The equivalent capacitance is then

$$C_{eq} = \frac{q}{V} = \frac{1}{1/C_1 + 1/C_2 + 1/C_3},$$

or

$$\frac{1}{C_{eq}} = \frac{1}{C_1} + \frac{1}{C_2} + \frac{1}{C_3}.$$

We can easily extend this to any number n of capacitors as

$$\frac{1}{C_{eq}} = \sum_{j=1}^{n} \frac{1}{C_j} \qquad (n \text{ capacitors in series}). \qquad (25\text{-}20)$$

Using Eq. 25-20 you can show that the equivalent capacitance of a series of capacitances is always *less* than the least capacitance in the series.

☑ Checkpoint 3

A battery of potential V stores charge q on a combination of two identical capacitors. What are the potential difference across and the charge on either capacitor if the capacitors are (a) in parallel and (b) in series?

Sample Problem 25.02 Capacitors in parallel and in series

(a) Find the equivalent capacitance for the combination of capacitances shown in Fig. 25-10a, across which potential difference V is applied. Assume

$$C_1 = 12.0 \ \mu F, \quad C_2 = 5.30 \ \mu F, \quad \text{and} \quad C_3 = 4.50 \ \mu F.$$

KEY IDEA

Any capacitors connected in series can be replaced with their equivalent capacitor, and any capacitors connected in

parallel can be replaced with their equivalent capacitor. Therefore, we should first check whether any of the capacitors in Fig. 25-10a are in parallel or series.

Finding equivalent capacitance: Capacitors 1 and 3 are connected one after the other, but are they in series? No. The potential V that is applied to the capacitors produces charge on the bottom plate of capacitor 3. That charge causes charge to shift from the top plate of capacitor 3. However, note that the shifting charge can move to the bot-

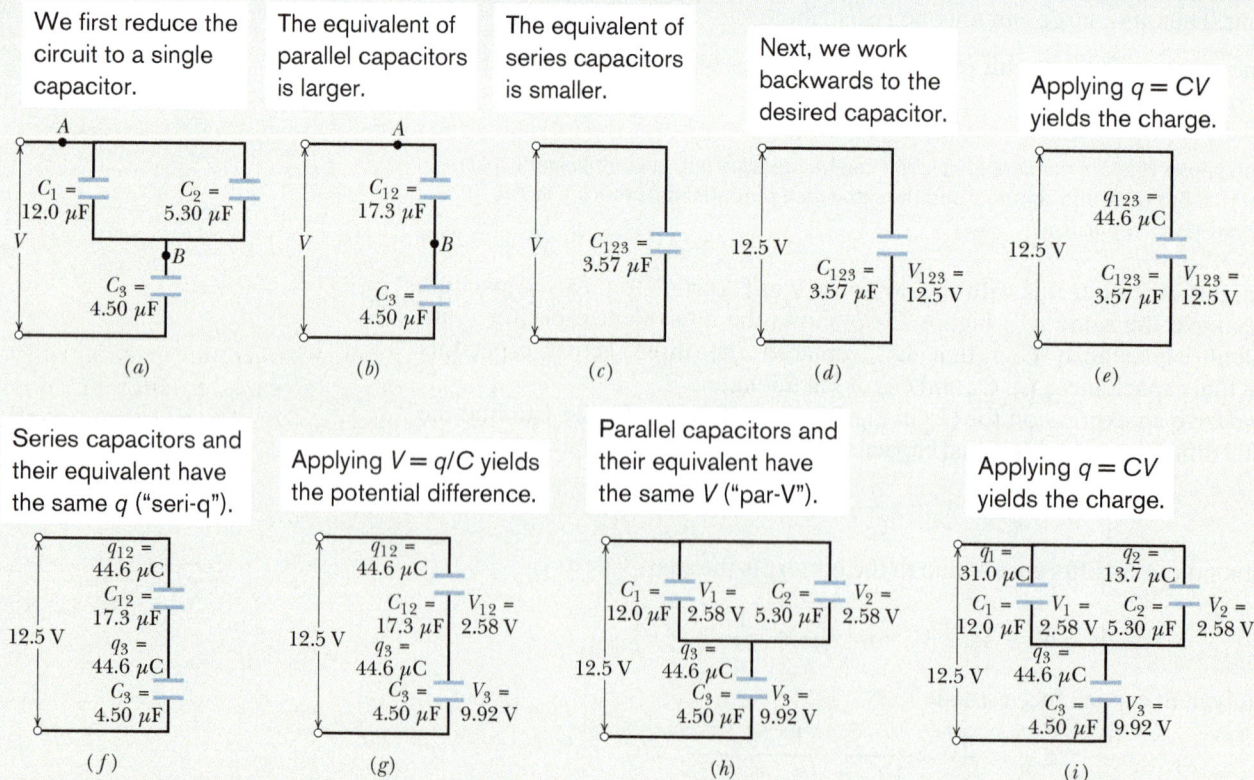

Figure 25-10 (a)–(d) Three capacitors are reduced to one equivalent capacitor. (e)–(i) Working backwards to get the charges.

tom plates of both capacitor 1 and capacitor 2. Because there is more than one route for the shifting charge, capacitor 3 is not in series with capacitor 1 (or capacitor 2). Any time you think you might have two capacitors in series, apply this check about the shifting charge.

Are capacitor 1 and capacitor 2 in parallel? Yes. Their top plates are directly wired together and their bottom plates are directly wired together, and electric potential is applied between the top-plate pair and the bottom-plate pair. Thus, capacitor 1 and capacitor 2 are in parallel, and Eq. 25-19 tells us that their equivalent capacitance C_{12} is

$$C_{12} = C_1 + C_2 = 12.0 \ \mu\text{F} + 5.30 \ \mu\text{F} = 17.3 \ \mu\text{F}.$$

In Fig. 25-10b, we have replaced capacitors 1 and 2 with their equivalent capacitor, called capacitor 12 (say "one two" and not "twelve"). (The connections at points A and B are exactly the same in Figs. 25-10a and b.)

Is capacitor 12 in series with capacitor 3? Again applying the test for series capacitances, we see that the charge that shifts from the top plate of capacitor 3 must entirely go to the bottom plate of capacitor 12. Thus, capacitor 12 and capacitor 3 are in series, and we can replace them with their equivalent C_{123} ("one two three"), as shown in Fig. 25-10c. From Eq. 25-20, we have

$$\frac{1}{C_{123}} = \frac{1}{C_{12}} + \frac{1}{C_3}$$

$$= \frac{1}{17.3 \ \mu\text{F}} + \frac{1}{4.50 \ \mu\text{F}} = 0.280 \ \mu\text{F}^{-1},$$

from which

$$C_{123} = \frac{1}{0.280 \ \mu\text{F}^{-1}} = 3.57 \ \mu\text{F}. \qquad \text{(Answer)}$$

(b) The potential difference applied to the input terminals in Fig. 25-10a is $V = 12.5$ V. What is the charge on C_1?

KEY IDEAS

We now need to work backwards from the equivalent capacitance to get the charge on a particular capacitor. We have two techniques for such "backwards work": (1) Seri-q: Series capacitors have the same charge as their equivalent capacitor. (2) Par-V: Parallel capacitors have the same potential difference as their equivalent capacitor.

Working backwards: To get the charge q_1 on capacitor 1, we work backwards to that capacitor, starting with the equivalent capacitor 123. Because the given potential difference V (= 12.5 V) is applied across the actual combination of three capacitors in Fig. 25-10a, it is also applied across C_{123} in Figs. 25-10d and e. Thus, Eq. 25-1 ($q = CV$) gives us

$$q_{123} = C_{123}V = (3.57 \ \mu\text{F})(12.5 \ \text{V}) = 44.6 \ \mu\text{C}.$$

The series capacitors 12 and 3 in Fig. 25-10b each have the same charge as their equivalent capacitor 123 (Fig. 25-10f). Thus, capacitor 12 has charge $q_{12} = q_{123} = 44.6 \ \mu\text{C}$. From Eq. 25-1 and Fig. 25-10g, the potential difference across capacitor 12 must be

$$V_{12} = \frac{q_{12}}{C_{12}} = \frac{44.6 \ \mu\text{C}}{17.3 \ \mu\text{F}} = 2.58 \ \text{V}.$$

The parallel capacitors 1 and 2 each have the same potential difference as their equivalent capacitor 12 (Fig. 25-10h). Thus, capacitor 1 has potential difference $V_1 = V_{12} = 2.58$ V, and, from Eq. 25-1 and Fig. 25-10i, the charge on capacitor 1 must be

$$q_1 = C_1V_1 = (12.0 \ \mu\text{F})(2.58 \ \text{V})$$

$$= 31.0 \ \mu\text{C}. \qquad \text{(Answer)}$$

Sample Problem 25.03   One capacitor charging up another capacitor

Capacitor 1, with $C_1 = 3.55 \ \mu\text{F}$, is charged to a potential difference $V_0 = 6.30$ V, using a 6.30 V battery. The battery is then removed, and the capacitor is connected as in Fig. 25-11 to an uncharged capacitor 2, with $C_2 = 8.95 \ \mu\text{F}$. When switch S is closed, charge flows between the capacitors. Find the charge on each capacitor when equilibrium is reached.

KEY IDEAS

The situation here differs from the previous example because here an applied electric potential is *not* maintained across a combination of capacitors by a battery or some other source. Here, just after switch S is closed, the only applied electric potential is that of capacitor 1 on capacitor 2, and that potential is decreasing. Thus, the capacitors in Fig. 25-11 are not connected *in series;* and although they are drawn parallel, in this situation they are not *in parallel.*

As the electric potential across capacitor 1 decreases, that across capacitor 2 increases. Equilibrium is reached when the two potentials are equal because, with no potential difference between connected plates of the capacitors, there

After the switch is closed, charge is transferred until the potential differences match.

Figure 25-11 A potential difference V_0 is applied to capacitor 1 and the charging battery is removed. Switch S is then closed so that the charge on capacitor 1 is shared with capacitor 2.

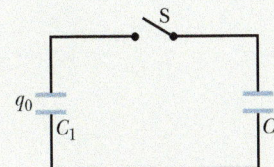

is no electric field within the connecting wires to move conduction electrons. The initial charge on capacitor 1 is then shared between the two capacitors.

Calculations: Initially, when capacitor 1 is connected to the battery, the charge it acquires is, from Eq. 25-1,

$$q_0 = C_1 V_0 = (3.55 \times 10^{-6}\,\text{F})(6.30\,\text{V})$$
$$= 22.365 \times 10^{-6}\,\text{C}.$$

When switch S in Fig. 25-11 is closed and capacitor 1 begins to charge capacitor 2, the electric potential and charge on capacitor 1 decrease and those on capacitor 2 increase until

$$V_1 = V_2 \quad \text{(equilibrium).}$$

From Eq. 25-1, we can rewrite this as

$$\frac{q_1}{C_1} = \frac{q_2}{C_2} \quad \text{(equilibrium).}$$

Because the total charge cannot magically change, the total after the transfer must be

$$q_1 + q_2 = q_0 \quad \text{(charge conservation);}$$

thus

$$q_2 = q_0 - q_1.$$

We can now rewrite the second equilibrium equation as

$$\frac{q_1}{C_1} = \frac{q_0 - q_1}{C_2}.$$

Solving this for q_1 and substituting given data, we find

$$q_1 = 6.35\ \mu\text{C}. \qquad \text{(Answer)}$$

The rest of the initial charge ($q_0 = 22.365\ \mu\text{C}$) must be on capacitor 2:

$$q_2 = 16.0\ \mu\text{C}. \qquad \text{(Answer)}$$

WILEY PLUS Additional examples, video, and practice available at *WileyPLUS*

25-4 ENERGY STORED IN AN ELECTRIC FIELD

Learning Objectives

After reading this module, you should be able to . . .

25.16 Explain how the work required to charge a capacitor results in the potential energy of the capacitor.

25.17 For a capacitor, apply the relationship between the potential energy U, the capacitance C, and the potential difference V.

25.18 For a capacitor, apply the relationship between the

potential energy, the internal volume, and the internal energy density.

25.19 For any electric field, apply the relationship between the potential energy density u in the field and the field's magnitude E.

25.20 Explain the danger of sparks in airborne dust.

Key Ideas

● The electric potential energy U of a charged capacitor,

$$U = \frac{q^2}{2c} = \tfrac{1}{2}CV^2,$$

is equal to the work required to charge the capacitor. This energy can be associated with the capacitor's electric field $\vec{E}$.

● Every electric field, in a capacitor or from any other source, has an associated stored energy. In vacuum, the energy density u (potential energy per unit volume) in a field of magnitude E is

$$u = \tfrac{1}{2}\varepsilon_0 E^2.$$

Energy Stored in an Electric Field

Work must be done by an external agent to charge a capacitor. We can imagine doing the work ourselves by transferring electrons from one plate to the other, one by one. As the charges build, so does the electric field between the plates, which opposes the continued transfer. So, greater amounts of work are required. Actually, a battery does all this for us, at the expense of its stored chemical energy. We visualize the work as being stored as electric potential energy in the electric field between the plates.

Suppose that, at a given instant, a charge q' has been transferred from one plate of a capacitor to the other. The potential difference V' between the plates at that instant will be q'/C. If an extra increment of charge dq' is then transferred, the increment of work required will be, from Eq. 24-6,

$$dW = V' \, dq' = \frac{q'}{C} \, dq'.$$

The work required to bring the total capacitor charge up to a final value q is

$$W = \int dW = \frac{1}{C} \int_0^q q' \, dq' = \frac{q^2}{2C}.$$

This work is stored as potential energy U in the capacitor, so that

$$U = \frac{q^2}{2C} \quad \text{(potential energy).} \tag{25-21}$$

From Eq. 25-1, we can also write this as

$$U = \tfrac{1}{2}CV^2 \quad \text{(potential energy).} \tag{25-22}$$

Equations 25-21 and 25-22 hold no matter what the geometry of the capacitor is.

To gain some physical insight into energy storage, consider two parallel-plate capacitors that are identical except that capacitor 1 has twice the plate separation of capacitor 2. Then capacitor 1 has twice the volume between its plates and also, from Eq. 25-9, half the capacitance of capacitor 2. Equation 25-4 tells us that if both capacitors have the same charge q, the electric fields between their plates are identical. And Eq. 25-21 tells us that capacitor 1 has twice the stored potential energy of capacitor 2. Thus, of two otherwise identical capacitors with the same charge and same electric field, the one with twice the volume between its plates has twice the stored potential energy. Arguments like this tend to verify our earlier assumption:

> The potential energy of a charged capacitor may be viewed as being stored in the electric field between its plates.

Explosions in Airborne Dust

As we discussed in Module 24-8, making contact with certain materials, such as clothing, carpets, and even playground slides, can leave you with a significant electrical potential. You might become painfully aware of that potential if a spark leaps between you and a grounded object, such as a faucet. In many industries involving the production and transport of powder, such as in the cosmetic and food industries, such a spark can be disastrous. Although the powder in bulk may not burn at all, when individual powder grains are airborne and thus surrounded by oxygen, they can burn so fiercely that a cloud of the grains burns as an explosion. Safety engineers cannot eliminate all possible sources of sparks in the powder industries. Instead, they attempt to keep the amount of energy available in the sparks below the threshold value U_t ($\approx$ 150 mJ) typically required to ignite airborne grains.

Suppose a person becomes charged by contact with various surfaces as he walks through an airborne powder. We can roughly model the person as a spherical capacitor of radius $R = 1.8$ m. From Eq. 25-18 ($C = 4\pi\varepsilon_0 R$) and Eq. 25-22 ($U = \tfrac{1}{2}CV^2$), we see that the energy of the capacitor is

$$U = \tfrac{1}{2}(4\pi\varepsilon_0 R)V^2.$$

From this we see that the threshold energy corresponds to a potential of

$$V = \sqrt{\frac{2U_t}{4\pi\varepsilon_0 R}} = \sqrt{\frac{2(150 \times 10^{-3}\,\text{J})}{4\pi(8.85 \times 10^{-12}\,\text{C}^2/\text{N}\cdot\text{m}^2)(1.8\,\text{m})}}$$
$$= 3.9 \times 10^4\,\text{V}.$$

Safety engineers attempt to keep the potential of the personnel below this level by "bleeding" off the charge through, say, a conducting floor.

Energy Density

In a parallel-plate capacitor, neglecting fringing, the electric field has the same value at all points between the plates. Thus, the **energy density** u—that is, the potential energy per unit volume between the plates—should also be uniform. We can find u by dividing the total potential energy by the volume Ad of the space between the plates. Using Eq. 25-22, we obtain

$$u = \frac{U}{Ad} = \frac{CV^2}{2Ad}. \tag{25-23}$$

With Eq. 25-9 ($C = \varepsilon_0 A/d$), this result becomes

$$u = \tfrac{1}{2}\varepsilon_0 \left(\frac{V}{d}\right)^2. \tag{25-24}$$

However, from Eq. 24-42 ($E = -\Delta V/\Delta s$), V/d equals the electric field magnitude E; so

$$u = \tfrac{1}{2}\varepsilon_0 E^2 \quad \text{(energy density)}. \tag{25-25}$$

Although we derived this result for the special case of an electric field of a parallel-plate capacitor, it holds for any electric field. If an electric field $\vec{E}$ exists at any point in space, that site has an electric potential energy with a density (amount per unit volume) given by Eq. 25-25.

Sample Problem 25.04 Potential energy and energy density of an electric field

An isolated conducting sphere whose radius R is 6.85 cm has a charge $q = 1.25$ nC.

(a) How much potential energy is stored in the electric field of this charged conductor?

KEY IDEAS

(1) An isolated sphere has capacitance given by Eq. 25-18 ($C = 4\pi\varepsilon_0 R$). (2) The energy U stored in a capacitor depends on the capacitor's charge q and capacitance C according to Eq. 25-21 ($U = q^2/2C$).

Calculation: Substituting $C = 4\pi\varepsilon_0 R$ into Eq. 25-21 gives us

$$U = \frac{q^2}{2C} = \frac{q^2}{8\pi\varepsilon_0 R}$$

$$= \frac{(1.25 \times 10^{-9}\,\text{C})^2}{(8\pi)(8.85 \times 10^{-12}\,\text{F/m})(0.0685\,\text{m})}$$

$$= 1.03 \times 10^{-7}\,\text{J} = 103\,\text{nJ}. \qquad \text{(Answer)}$$

(b) What is the energy density at the surface of the sphere?

KEY IDEA

The density u of the energy stored in an electric field depends on the magnitude E of the field, according to Eq. 25-25 ($u = \tfrac{1}{2}\varepsilon_0 E^2$).

Calculations: Here we must first find E at the surface of the sphere, as given by Eq. 23-15:

$$E = \frac{1}{4\pi\varepsilon_0}\frac{q}{R^2}.$$

The energy density is then

$$u = \tfrac{1}{2}\varepsilon_0 E^2 = \frac{q^2}{32\pi^2\varepsilon_0 R^4}$$

$$= \frac{(1.25 \times 10^{-9}\,\text{C})^2}{(32\pi^2)(8.85 \times 10^{-12}\,\text{C}^2/\text{N}\cdot\text{m}^2)(0.0685\,\text{m})^4}$$

$$= 2.54 \times 10^{-5}\,\text{J/m}^3 = 25.4\,\mu\text{J/m}^3. \qquad \text{(Answer)}$$

PLUS Additional examples, video, and practice available at *WileyPLUS*

25-5 CAPACITOR WITH A DIELECTRIC

Learning Objectives

After reading this module, you should be able to . . .

25.21 Identify that capacitance is increased if the space between the plates is filled with a dielectric material.

25.22 For a capacitor, calculate the capacitance with and without a dielectric.

25.23 For a region filled with a dielectric material with a given dielectric constant κ, identify that all electrostatic equations containing the permittivity constant ε_0 are modified by multiplying that constant by the dielectric constant to get $\kappa\varepsilon_0$.

25.24 Name some of the common dielectrics.

25.25 In adding a dielectric to a charged capacitor, distinguish the results for a capacitor (a) connected to a battery and (b) not connected to a battery.

25.26 Distinguish polar dielectrics from nonpolar dielectrics.

25.27 In adding a dielectric to a charged capacitor, explain what happens to the electric field between the plates in terms of what happens to the atoms in the dielectric.

Key Ideas

● If the space between the plates of a capacitor is completely filled with a dielectric material, the capacitance C in vacuum (or, effectively, in air) is multiplied by the material's dielectric constant κ, which is a number greater than 1.

● In a region that is completely filled by a dielectric, all electrostatic equations containing the permittivity constant ε_0 must be modified by replacing ε_0 with $\kappa\varepsilon_0$.

● When a dielectric material is placed in an external electric field, it develops an internal electric field that is oriented opposite the external field, thus reducing the magnitude of the electric field inside the material.

● When a dielectric material is placed in a capacitor with a fixed amount of charge on the surface, the net electric field between the plates is decreased.

Capacitor with a Dielectric

If you fill the space between the plates of a capacitor with a *dielectric,* which is an insulating material such as mineral oil or plastic, what happens to the capacitance? Michael Faraday—to whom the whole concept of capacitance is largely due and for whom the SI unit of capacitance is named—first looked into this matter in 1837. Using simple equipment much like that shown in Fig. 25-12, he found that the capacitance *increased* by a numerical factor κ, which he called

The Royal Institute, England/Bridgeman Art Library/NY

Figure 25-12 The simple electrostatic apparatus used by Faraday. An assembled apparatus (second from left) forms a spherical capacitor consisting of a central brass ball and a concentric brass shell. Faraday placed dielectric materials in the space between the ball and the shell.

Table 25-1 Some Properties of Dielectrics[a]

Material	Dielectric Constant κ	Dielectric Strength (kV/mm)
Air (1 atm)	1.00054	3
Polystyrene	2.6	24
Paper	3.5	16
Transformer oil	4.5	
Pyrex	4.7	14
Ruby mica	5.4	
Porcelain	6.5	
Silicon	12	
Germanium	16	
Ethanol	25	
Water (20°C)	80.4	
Water (25°C)	78.5	
Titania ceramic	130	
Strontium titanate	310	8

For a vacuum, κ = unity.

[a]Measured at room temperature, except for the water.

the **dielectric constant** of the insulating material. Table 25-1 shows some dielectric materials and their dielectric constants. The dielectric constant of a vacuum is unity by definition. Because air is mostly empty space, its measured dielectric constant is only slightly greater than unity. Even common paper can significantly increase the capacitance of a capacitor, and some materials, such as strontium titanate, can increase the capacitance by more than two orders of magnitude.

Another effect of the introduction of a dielectric is to limit the potential difference that can be applied between the plates to a certain value V_{max}, called the *breakdown potential.* If this value is substantially exceeded, the dielectric material will break down and form a conducting path between the plates. Every dielectric material has a characteristic *dielectric strength,* which is the maximum value of the electric field that it can tolerate without breakdown. A few such values are listed in Table 25-1.

As we discussed just after Eq. 25-18, the capacitance of any capacitor can be written in the form

$$C = \varepsilon_0 \mathcal{L}, \tag{25-26}$$

in which $\mathcal{L}$ has the dimension of length. For example, $\mathcal{L} = A/d$ for a parallel-plate capacitor. Faraday's discovery was that, with a dielectric *completely* filling the space between the plates, Eq. 25-26 becomes

$$C = \kappa \varepsilon_0 \mathcal{L} = \kappa C_{air}, \tag{25-27}$$

where C_{air} is the value of the capacitance with only air between the plates. For example, if we fill a capacitor with strontium titanate, with a dielectric constant of 310, we multiply the capacitance by 310.

Figure 25-13 provides some insight into Faraday's experiments. In Fig. 25-13a the battery ensures that the potential difference V between the plates will remain constant. When a dielectric slab is inserted between the plates, the charge q on the plates increases by a factor of κ; the additional charge is delivered to the capacitor plates by the battery. In Fig. 25-13b there is no battery, and therefore the charge q must remain constant when the dielectric slab is inserted; then the potential difference V between the plates decreases by a factor of κ. Both these observations are consistent (through the relation $q = CV$) with the increase in capacitance caused by the dielectric.

Comparison of Eqs. 25-26 and 25-27 suggests that the effect of a dielectric can be summed up in more general terms:

> In a region completely filled by a dielectric material of dielectric constant κ, all electrostatic equations containing the permittivity constant ε_0 are to be modified by replacing ε_0 with $\kappa\varepsilon_0$.

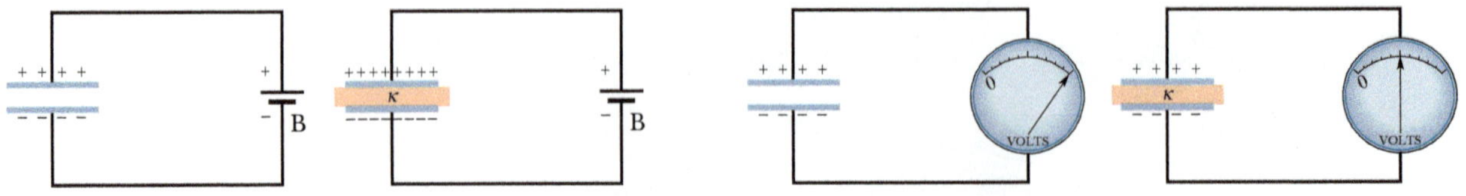

V = a constant

(a)

q = a constant

(b)

Figure 25-13 (a) If the potential difference between the plates of a capacitor is maintained, as by battery B, the effect of a dielectric is to increase the charge on the plates. (b) If the charge on the capacitor plates is maintained, as in this case, the effect of a dielectric is to reduce the potential difference between the plates. The scale shown is that of a *potentiometer,* a device used to measure potential difference (here, between the plates). A capacitor cannot discharge through a potentiometer.

Thus, the magnitude of the electric field produced by a point charge inside a dielectric is given by this modified form of Eq. 23-15:

$$E = \frac{1}{4\pi\kappa\varepsilon_0}\frac{q}{r^2}. \qquad (25\text{-}28)$$

Also, the expression for the electric field just outside an isolated conductor immersed in a dielectric (see Eq. 23-11) becomes

$$E = \frac{\sigma}{\kappa\varepsilon_0}. \qquad (25\text{-}29)$$

Because κ is always greater than unity, both these equations show that *for a fixed distribution of charges, the effect of a dielectric is to weaken the electric field* that would otherwise be present.

Sample Problem 25.05 Work and energy when a dielectric is inserted into a capacitor

A parallel-plate capacitor whose capacitance C is 13.5 pF is charged by a battery to a potential difference $V = 12.5$ V between its plates. The charging battery is now disconnected, and a porcelain slab ($\kappa = 6.50$) is slipped between the plates.

(a) What is the potential energy of the capacitor before the slab is inserted?

KEY IDEA

We can relate the potential energy U_i of the capacitor to the capacitance C and either the potential V (with Eq. 25-22) or the charge q (with Eq. 25-21):

$$U_i = \tfrac{1}{2}CV^2 = \frac{q^2}{2C}.$$

Calculation: Because we are given the initial potential V (= 12.5 V), we use Eq. 25-22 to find the initial stored energy:

$$U_i = \tfrac{1}{2}CV^2 = \tfrac{1}{2}(13.5 \times 10^{-12}\text{ F})(12.5\text{ V})^2$$
$$= 1.055 \times 10^{-9}\text{ J} = 1055\text{ pJ} \approx 1100\text{ pJ}. \quad \text{(Answer)}$$

(b) What is the potential energy of the capacitor–slab device after the slab is inserted?

KEY IDEA

Because the battery has been disconnected, the charge on the capacitor cannot change when the dielectric is inserted. However, the potential *does* change.

Calculations: Thus, we must now use Eq. 25-21 to write the final potential energy U_f, but now that the slab is within the capacitor, the capacitance is κC. We then have

$$U_f = \frac{q^2}{2\kappa C} = \frac{U_i}{\kappa} = \frac{1055\text{ pJ}}{6.50}$$
$$= 162\text{ pJ} \approx 160\text{ pJ}. \qquad \text{(Answer)}$$

When the slab is introduced, the potential energy decreases by a factor of κ.

The "missing" energy, in principle, would be apparent to the person who introduced the slab. The capacitor would exert a tiny tug on the slab and would do work on it, in amount

$$W = U_i - U_f = (1055 - 162)\text{ pJ} = 893\text{ pJ}.$$

If the slab were allowed to slide between the plates with no restraint and if there were no friction, the slab would oscillate back and forth between the plates with a (constant) mechanical energy of 893 pJ, and this system energy would transfer back and forth between kinetic energy of the moving slab and potential energy stored in the electric field.

PLUS Additional examples, video, and practice available at *WileyPLUS*

Dielectrics: An Atomic View

What happens, in atomic and molecular terms, when we put a dielectric in an electric field? There are two possibilities, depending on the type of molecule:

1. *Polar dielectrics.* The molecules of some dielectrics, like water, have permanent electric dipole moments. In such materials (called *polar dielectrics*), the

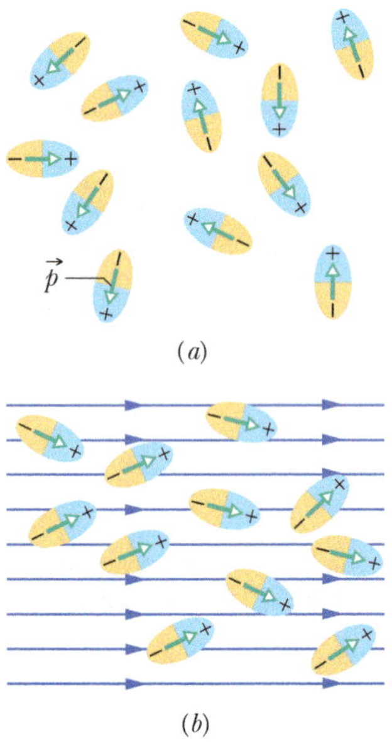

(a)

(b)

Figure 25-14 (a) Molecules with a permanent electric dipole moment, showing their random orientation in the absence of an external electric field. (b) An electric field is applied, producing partial alignment of the dipoles. Thermal agitation prevents complete alignment.

electric dipoles tend to line up with an external electric field as in Fig. 25-14. Because the molecules are continuously jostling each other as a result of their random thermal motion, this alignment is not complete, but it becomes more complete as the magnitude of the applied field is increased (or as the temperature, and thus the jostling, are decreased). The alignment of the electric dipoles produces an electric field that is directed opposite the applied field and is smaller in magnitude.

2. *Nonpolar dielectrics.* Regardless of whether they have permanent electric dipole moments, molecules acquire dipole moments by induction when placed in an external electric field. In Module 24-4 (see Fig. 24-14), we saw that this occurs because the external field tends to "stretch" the molecules, slightly separating the centers of negative and positive charge.

Figure 25-15a shows a nonpolar dielectric slab with no external electric field applied. In Fig. 25-15b, an electric field $\vec{E}_0$ is applied via a capacitor, whose plates are charged as shown. The result is a slight separation of the centers of the positive and negative charge distributions within the slab, producing positive charge on one face of the slab (due to the positive ends of dipoles there) and negative charge on the opposite face (due to the negative ends of dipoles there). The slab as a whole remains electrically neutral and—within the slab—there is no excess charge in any volume element.

Figure 25-15c shows that the induced surface charges on the faces produce an electric field $\vec{E}'$ in the direction opposite that of the applied electric field $\vec{E}_0$. The resultant field $\vec{E}$ inside the dielectric (the vector sum of fields $\vec{E}_0$ and $\vec{E}'$) has the direction of $\vec{E}_0$ but is smaller in magnitude.

Both the field $\vec{E}'$ produced by the surface charges in Fig. 25-15c and the electric field produced by the permanent electric dipoles in Fig. 25-14 act in the same way—they oppose the applied field $\vec{E}$. Thus, the effect of both polar and nonpolar dielectrics is to weaken any applied field within them, as between the plates of a capacitor.

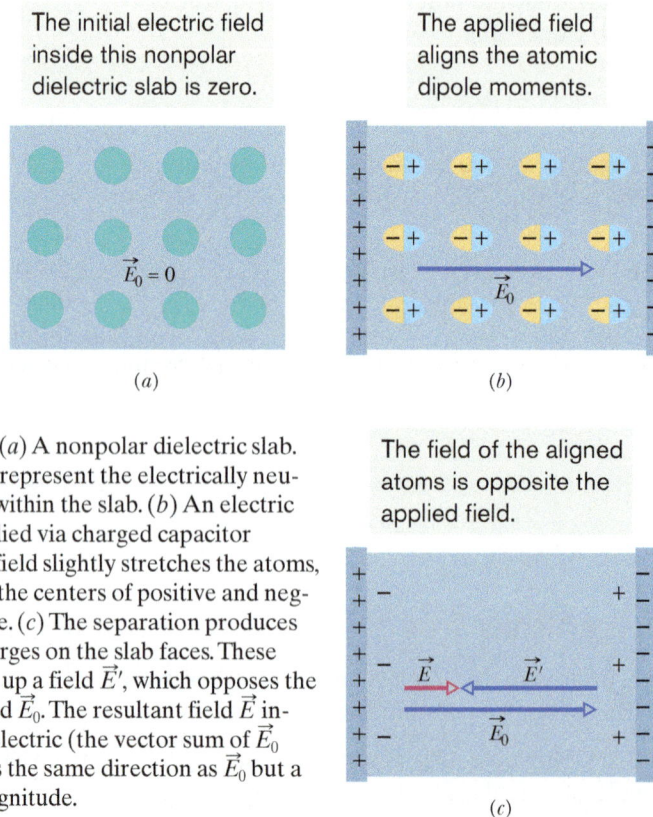

The initial electric field inside this nonpolar dielectric slab is zero.

The applied field aligns the atomic dipole moments.

(a)

(b)

Figure 25-15 (a) A nonpolar dielectric slab. The circles represent the electrically neutral atoms within the slab. (b) An electric field is applied via charged capacitor plates; the field slightly stretches the atoms, separating the centers of positive and negative charge. (c) The separation produces surface charges on the slab faces. These charges set up a field $\vec{E}'$, which opposes the applied field $\vec{E}_0$. The resultant field $\vec{E}$ inside the dielectric (the vector sum of $\vec{E}_0$ and $\vec{E}'$) has the same direction as $\vec{E}_0$ but a smaller magnitude.

The field of the aligned atoms is opposite the applied field.

(c)

25-6 DIELECTRICS AND GAUSS' LAW

Learning Objectives

After reading this module, you should be able to . . .

25.28 In a capacitor with a dielectric, distinguish free charge from induced charge.

25.29 When a dielectric partially or fully fills the space in a capacitor, find the free charge, the induced charge, the electric field between the plates (if there is a gap, there is more than one field value), and the potential between the plates.

Key Ideas

● Inserting a dielectric into a capacitor causes induced charge to appear on the faces of the dielectric and weakens the electric field between the plates.

● The induced charge is less than the free charge on the plates.

● When a dielectric is present, Gauss' law may be generalized to

$$\varepsilon_0 \oint \kappa \vec{E} \cdot d\vec{A} = q,$$

where q is the free charge. Any induced surface charge is accounted for by including the dielectric constant κ inside the integral.

Dielectrics and Gauss' Law

In our discussion of Gauss' law in Chapter 23, we assumed that the charges existed in a vacuum. Here we shall see how to modify and generalize that law if dielectric materials, such as those listed in Table 25-1, are present. Figure 25-16 shows a parallel-plate capacitor of plate area A, both with and without a dielectric. We assume that the charge q on the plates is the same in both situations. Note that the field between the plates induces charges on the faces of the dielectric by one of the methods described in Module 25-5.

For the situation of Fig. 25-16a, without a dielectric, we can find the electric field $\vec{E}_0$ between the plates as we did in Fig. 25-5: We enclose the charge $+q$ on the top plate with a Gaussian surface and then apply Gauss' law. Letting E_0 represent the magnitude of the field, we find

$$\varepsilon_0 \oint \vec{E} \cdot d\vec{A} = \varepsilon_0 EA = q, \tag{25-30}$$

or

$$E_0 = \frac{q}{\varepsilon_0 A}. \tag{25-31}$$

In Fig. 25-16b, with the dielectric in place, we can find the electric field between the plates (and within the dielectric) by using the same Gaussian surface. However, now the surface encloses two types of charge: It still encloses charge $+q$ on the top plate, but it now also encloses the induced charge $-q'$ on the top face of the dielectric. The charge on the conducting plate is said to be *free charge* because it can move if we change the electric potential of the plate; the induced charge on the surface of the dielectric is not free charge because it cannot move from that surface.

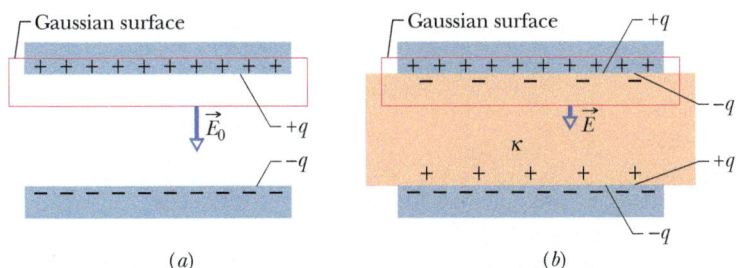

Figure 25-16 A parallel-plate capacitor (a) without and (b) with a dielectric slab inserted. The charge q on the plates is assumed to be the same in both cases.

The net charge enclosed by the Gaussian surface in Fig. 25-16b is $q - q'$, so Gauss' law now gives

$$\varepsilon_0 \oint \vec{E} \cdot d\vec{A} = \varepsilon_0 EA = q - q', \qquad (25\text{-}32)$$

or

$$E = \frac{q - q'}{\varepsilon_0 A}. \qquad (25\text{-}33)$$

The effect of the dielectric is to weaken the original field E_0 by a factor of κ; so we may write

$$E = \frac{E_0}{\kappa} = \frac{q}{\kappa \varepsilon_0 A}. \qquad (25\text{-}34)$$

Comparison of Eqs. 25-33 and 25-34 shows that

$$q - q' = \frac{q}{\kappa}. \qquad (25\text{-}35)$$

Equation 25-35 shows correctly that the magnitude q' of the induced surface charge is less than that of the free charge q and is zero if no dielectric is present (because then $\kappa = 1$ in Eq. 25-35).

By substituting for $q - q'$ from Eq. 25-35 in Eq. 25-32, we can write Gauss' law in the form

$$\varepsilon_0 \oint \kappa \vec{E} \cdot d\vec{A} = q \quad \text{(Gauss' law with dielectric)}. \qquad (25\text{-}36)$$

This equation, although derived for a parallel-plate capacitor, is true generally and is the most general form in which Gauss' law can be written. Note:

1. The flux integral now involves $\kappa \vec{E}$, not just $\vec{E}$. (The vector $\varepsilon_0 \kappa \vec{E}$ is sometimes called the *electric displacement* $\vec{D}$, so that Eq. 25-36 can be written in the form $\oint \vec{D} \cdot d\vec{A} = q$.)

2. The charge q enclosed by the Gaussian surface is now taken to be the *free charge only*. The induced surface charge is deliberately ignored on the right side of Eq. 25-36, having been taken fully into account by introducing the dielectric constant κ on the left side.

3. Equation 25-36 differs from Eq. 23-7, our original statement of Gauss' law, only in that ε_0 in the latter equation has been replaced by $\kappa \varepsilon_0$. We keep κ inside the integral of Eq. 25-36 to allow for cases in which κ is not constant over the entire Gaussian surface.

Sample Problem 25.06 Dielectric partially filling the gap in a capacitor

Figure 25-17 shows a parallel-plate capacitor of plate area A and plate separation d. A potential difference V_0 is applied between the plates by connecting a battery between them. The battery is then disconnected, and a dielectric slab of thickness b and dielectric constant κ is placed between the plates as shown. Assume $A = 115$ cm^2, $d = 1.24$ cm, $V_0 = 85.5$ V, $b = 0.780$ cm, and $\kappa = 2.61$.

(a) What is the capacitance C_0 before the dielectric slab is inserted?

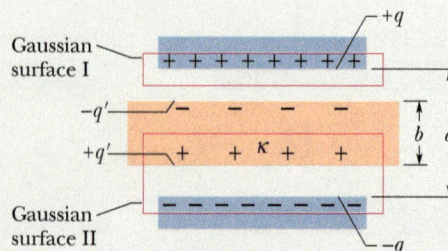

Figure 25-17 A parallel-plate capacitor containing a dielectric slab that only partially fills the space between the plates.

Calculation: From Eq. 25-9 we have

$$C_0 = \frac{\varepsilon_0 A}{d} = \frac{(8.85 \times 10^{-12}\,\text{F/m})(115 \times 10^{-4}\,\text{m}^2)}{1.24 \times 10^{-2}\,\text{m}}$$

$$= 8.21 \times 10^{-12}\,\text{F} = 8.21\,\text{pF}. \qquad \text{(Answer)}$$

(b) What free charge appears on the plates?

Calculation: From Eq. 25-1,

$$q = C_0 V_0 = (8.21 \times 10^{-12}\,\text{F})(85.5\,\text{V})$$

$$= 7.02 \times 10^{-10}\,\text{C} = 702\,\text{pC}. \qquad \text{(Answer)}$$

Because the battery was disconnected before the slab was inserted, the free charge is unchanged.

(c) What is the electric field E_0 in the gaps between the plates and the dielectric slab?

KEY IDEA

We need to apply Gauss' law, in the form of Eq. 25-36, to Gaussian surface I in Fig. 25-17.

Calculations: That surface passes through the gap, and so it encloses *only* the free charge on the upper capacitor plate. Electric field pierces only the bottom of the Gaussian surface. Because there the area vector $d\vec{A}$ and the field vector $\vec{E}_0$ are both directed downward, the dot product in Eq. 25-36 becomes

$$\vec{E}_0 \cdot d\vec{A} = E_0\,dA\cos 0° = E_0\,dA.$$

Equation 25-36 then becomes

$$\varepsilon_0 \kappa E_0 \oint dA = q.$$

The integration now simply gives the surface area A of the plate. Thus, we obtain

$$\varepsilon_0 \kappa E_0 A = q,$$

or $\qquad\qquad E_0 = \dfrac{q}{\varepsilon_0 \kappa A}.$

We must put $\kappa = 1$ here because Gaussian surface I does not pass through the dielectric. Thus, we have

$$E_0 = \frac{q}{\varepsilon_0 \kappa A} = \frac{7.02 \times 10^{-10}\,\text{C}}{(8.85 \times 10^{-12}\,\text{F/m})(1)(115 \times 10^{-4}\,\text{m}^2)}$$

$$= 6900\,\text{V/m} = 6.90\,\text{kV/m}. \qquad \text{(Answer)}$$

Note that the value of E_0 does not change when the slab is introduced because the amount of charge enclosed by Gaussian surface I in Fig. 25-17 does not change.

(d) What is the electric field E_1 in the dielectric slab?

KEY IDEA

Now we apply Gauss' law in the form of Eq. 25-36 to Gaussian surface II in Fig. 25-17.

Calculations: Only the free charge $-q$ is in Eq. 25-36, so

$$\varepsilon_0 \oint \kappa \vec{E}_1 \cdot d\vec{A} = -\varepsilon_0 \kappa E_1 A = -q. \qquad (25\text{-}37)$$

The first minus sign in this equation comes from the dot product $\vec{E}_1 \cdot d\vec{A}$ along the top of the Gaussian surface because now the field vector $\vec{E}_1$ is directed downward and the area vector $d\vec{A}$ (which, as always, points outward from the interior of a closed Gaussian surface) is directed upward. With 180° between the vectors, the dot product is negative. Now $\kappa = 2.61$. Thus, Eq. 25-37 gives us

$$E_1 = \frac{q}{\varepsilon_0 \kappa A} = \frac{E_0}{\kappa} = \frac{6.90\,\text{kV/m}}{2.61}$$

$$= 2.64\,\text{kV/m}. \qquad \text{(Answer)}$$

(e) What is the potential difference V between the plates after the slab has been introduced?

KEY IDEA

We find V by integrating along a straight line directly from the bottom plate to the top plate.

Calculation: Within the dielectric, the path length is b and the electric field is E_1. Within the two gaps above and below the dielectric, the total path length is $d - b$ and the electric field is E_0. Equation 25-6 then yields

$$V = \int_-^+ E\,ds = E_0(d - b) + E_1 b$$

$$= (6900\,\text{V/m})(0.0124\,\text{m} - 0.00780\,\text{m})$$

$$+ (2640\,\text{V/m})(0.00780\,\text{m})$$

$$= 52.3\,\text{V}. \qquad \text{(Answer)}$$

This is less than the original potential difference of 85.5 V.

(f) What is the capacitance with the slab in place?

KEY IDEA

The capacitance C is related to q and V via Eq. 25-1.

Calculation: Taking q from (b) and V from (e), we have

$$C = \frac{q}{V} = \frac{7.02 \times 10^{-10}\,\text{C}}{52.3\,\text{V}}$$

$$= 1.34 \times 10^{-11}\,\text{F} = 13.4\,\text{pF}. \qquad \text{(Answer)}$$

This is greater than the original capacitance of 8.21 pF.

PLUS Additional examples, video, and practice available at *WileyPLUS*

Review & Summary

Capacitor; Capacitance   A **capacitor** consists of two isolated conductors (the *plates*) with charges $+q$ and $-q$. Its **capacitance** C is defined from

$$q = CV, \qquad (25\text{-}1)$$

where V is the potential difference between the plates.

Determining Capacitance   We generally determine the capacitance of a particular capacitor configuration by (1) assuming a charge q to have been placed on the plates, (2) finding the electric field $\vec{E}$ due to this charge, (3) evaluating the potential difference V, and (4) calculating C from Eq. 25-1. Some specific results are the following:

A *parallel-plate capacitor* with flat parallel plates of area A and spacing d has capacitance

$$C = \frac{\varepsilon_0 A}{d}. \qquad (25\text{-}9)$$

A *cylindrical capacitor* (two long coaxial cylinders) of length L and radii a and b has capacitance

$$C = 2\pi\varepsilon_0 \frac{L}{\ln(b/a)}. \qquad (25\text{-}14)$$

A *spherical capacitor* with concentric spherical plates of radii a and b has capacitance

$$C = 4\pi\varepsilon_0 \frac{ab}{b-a}. \qquad (25\text{-}17)$$

An *isolated sphere* of radius R has capacitance

$$C = 4\pi\varepsilon_0 R. \qquad (25\text{-}18)$$

Capacitors in Parallel and in Series   The **equivalent capacitances** C_{eq} of combinations of individual capacitors connected in **parallel** and in **series** can be found from

$$C_{\text{eq}} = \sum_{j=1}^{n} C_j \quad (n \text{ capacitors in parallel}) \qquad (25\text{-}19)$$

and

$$\frac{1}{C_{\text{eq}}} = \sum_{j=1}^{n} \frac{1}{C_j} \quad (n \text{ capacitors in series}). \qquad (25\text{-}20)$$

Equivalent capacitances can be used to calculate the capacitances of more complicated series–parallel combinations.

Potential Energy and Energy Density   The **electric potential energy** U of a charged capacitor,

$$U = \frac{q^2}{2C} = \tfrac{1}{2}CV^2, \qquad (25\text{-}21, 25\text{-}22)$$

is equal to the work required to charge the capacitor. This energy can be associated with the capacitor's electric field $\vec{E}$. By extension we can associate stored energy with any electric field. In vacuum, the **energy density** u, or potential energy per unit volume, within an electric field of magnitude E is given by

$$u = \tfrac{1}{2}\varepsilon_0 E^2. \qquad (25\text{-}25)$$

Capacitance with a Dielectric   If the space between the plates of a capacitor is completely filled with a dielectric material, the capacitance C is increased by a factor κ, called the **dielectric constant,** which is characteristic of the material. In a region that is completely filled by a dielectric, all electrostatic equations containing ε_0 must be modified by replacing ε_0 with $\kappa\varepsilon_0$.

The effects of adding a dielectric can be understood physically in terms of the action of an electric field on the permanent or induced electric dipoles in the dielectric slab. The result is the formation of induced charges on the surfaces of the dielectric, which results in a weakening of the field within the dielectric for a given amount of free charge on the plates.

Gauss' Law with a Dielectric   When a dielectric is present, Gauss' law may be generalized to

$$\varepsilon_0 \oint \kappa \vec{E} \cdot d\vec{A} = q. \qquad (25\text{-}36)$$

Here q is the free charge; any induced surface charge is accounted for by including the dielectric constant κ inside the integral.

Questions

1 Figure 25-18 shows plots of charge versus potential difference for three parallel-plate capacitors that have the plate areas and separations given in the table. Which plot goes with which capacitor?

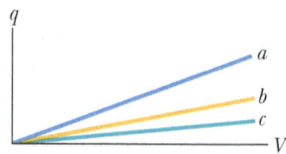

Figure 25-18 Question 1.

Capacitor	Area	Separation
1	A	d
2	$2A$	d
3	A	$2d$

2 What is C_{eq} of three capacitors, each of capacitance C, if they are connected to a battery (a) in series with one another and (b) in parallel? (c) In which arrangement is there more charge on the equivalent capacitance?

3 (a) In Fig. 25-19a, are capacitors 1 and 3 in series? (b) In the same

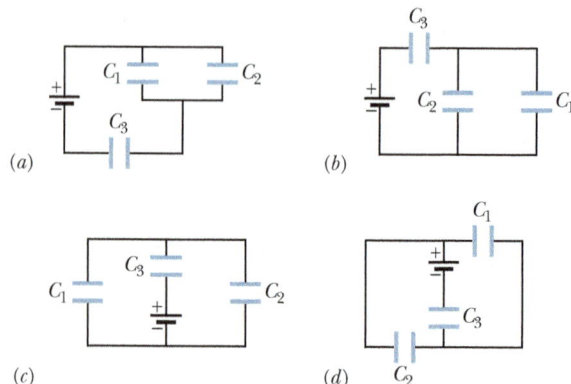

Figure 25-19 Question 3.

figure, are capacitors 1 and 2 in parallel? (c) Rank the equivalent capacitances of the four circuits shown in Fig. 25-19, greatest first.

4 Figure 25-20 shows three circuits, each consisting of a switch and two capacitors, initially charged as indicated (top plate positive). After the switches have been closed, in which circuit (if any) will the charge on the left-hand capacitor (a) increase, (b) decrease, and (c) remain the same?

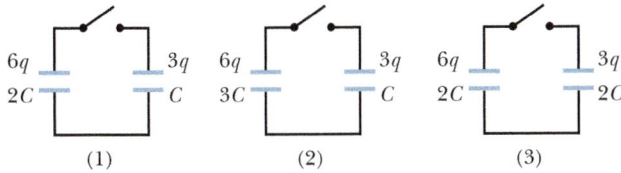

Figure 25-20 Question 4.

5 Initially, a single capacitance C_1 is wired to a battery. Then capacitance C_2 is added in parallel. Are (a) the potential difference across C_1 and (b) the charge q_1 on C_1 now more than, less than, or the same as previously? (c) Is the equivalent capacitance C_{12} of C_1 and C_2 more than, less than, or equal to C_1? (d) Is the charge stored on C_1 and C_2 together more than, less than, or equal to the charge stored previously on C_1?

6 Repeat Question 5 for C_2 added in series rather than in parallel.

7 For each circuit in Fig. 25-21, are the capacitors connected in series, in parallel, or in neither mode?

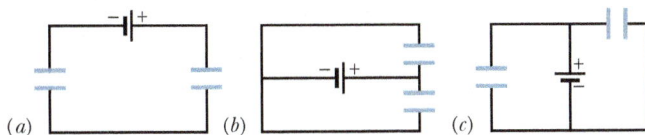

Figure 25-21 Question 7.

8 Figure 25-22 shows an open switch, a battery of potential difference V, a current-measuring meter A, and three identical uncharged capacitors of capacitance C. When the switch is closed and the circuit reaches equilibrium, what are (a) the potential difference across each capacitor and (b) the charge on the left plate of each capacitor? (c) During charging, what net charge passes through the meter?

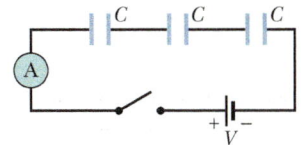

Figure 25-22 Question 8.

9 A parallel-plate capacitor is connected to a battery of electric potential difference V. If the plate separation is decreased, do the following quantities increase, decrease, or remain the same: (a) the capacitor's capacitance, (b) the potential difference across the capacitor, (c) the charge on the capacitor, (d) the energy stored by the capacitor, (e) the magnitude of the electric field between the plates, and (f) the energy density of that electric field?

10 When a dielectric slab is inserted between the plates of one of the two identical capacitors in Fig. 25-23, do the following properties of that capacitor increase, decrease, or remain the same: (a) capacitance, (b) charge, (c) potential difference, and (d) potential energy? (e) How about the same properties of the other capacitor?

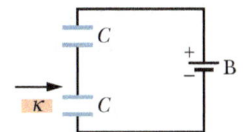

Figure 25-23 Question 10.

11 You are to connect capacitances C_1 and C_2, with $C_1 > C_2$, to a battery, first individually, then in series, and then in parallel. Rank those arrangements according to the amount of charge stored, greatest first.

Problems

Module 25-1   Capacitance

•1 The two metal objects in Fig. 25-24 have net charges of +70 pC and −70 pC, which result in a 20 V potential difference between them. (a) What is the capacitance of the system? (b) If the charges are changed to +200 pC and −200 pC, what does the capacitance become? (c) What does the potential difference become?

Figure 25-24 Problem 1.

•2 The capacitor in Fig. 25-25 has a capacitance of 25 μF and is initially uncharged. The battery provides a potential difference of 120 V. After switch S is closed, how much charge will pass through it?

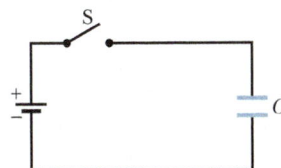

Figure 25-25 Problem 2.

Module 25-2   Calculating the Capacitance

•3 SSM A parallel-plate capacitor has circular plates of 8.20 cm radius and 1.30 mm separation. (a) Calculate the capacitance. (b) Find the charge for a potential difference of 120 V.

•4 The plates of a spherical capacitor have radii 38.0 mm and 40.0 mm. (a) Calculate the capacitance. (b) What must be the plate area of a parallel-plate capacitor with the same plate separation and capacitance?

•5 What is the capacitance of a drop that results when two mercury spheres, each of radius $R = 2.00$ mm, merge?

•6 You have two flat metal plates, each of area 1.00 m², with which to construct a parallel-plate capacitor. (a) If the capacitance of the device is to be 1.00 F, what must be the separation between the plates? (b) Could this capacitor actually be constructed?

•7 If an uncharged parallel-plate capacitor (capacitance C) is connected to a battery, one plate becomes negatively charged as electrons move to the plate face (area A). In Fig. 25-26, the depth d from which the electrons come in the plate in a particular capacitor is plotted against a range of values for the potential difference V of the battery. The density of conduction electrons in the copper plates is 8.49×10^{28} electrons/m³. The vertical scale is set by $d_s = 1.00$ pm, and the horizontal scale is set by $V_s = 20.0$ V. What is the ratio C/A?

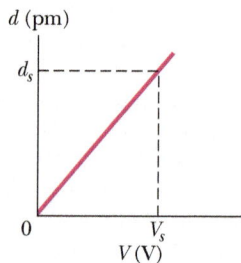

Figure 25-26 Problem 7.

Module 25-3 Capacitors in Parallel and in Series

•8 How many 1.00 μF capacitors must be connected in parallel to store a charge of 1.00 C with a potential of 110 V across the capacitors?

•9 Each of the uncharged capacitors in Fig. 25-27 has a capacitance of 25.0 μF. A potential difference of $V = 4200$ V is established when the switch is closed. How many coulombs of charge then pass through meter A?

Figure 25-27 Problem 9.

•10 In Fig. 25-28, find the equivalent capacitance of the combination. Assume that C_1 is 10.0 μF, C_2 is 5.00 μF, and C_3 is 4.00 μF.

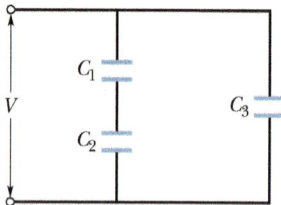

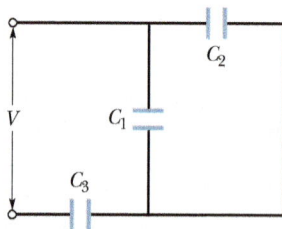

Figure 25-28 Problems 10 and 34.

•11 **ILW** In Fig. 25-29, find the equivalent capacitance of the combination. Assume that $C_1 = 10.0$ μF, $C_2 = 5.00$ μF, and $C_3 = 4.00$ μF.

Figure 25-29 Problems 11, 17, and 38.

••12 Two parallel-plate capacitors, 6.0 μF each, are connected in parallel to a 10 V battery. One of the capacitors is then squeezed so that its plate separation is 50.0% of its initial value. Because of the squeezing, (a) how much additional charge is transferred to the capacitors by the battery and (b) what is the increase in the total charge stored on the capacitors?

••13 **SSM** **ILW** A 100 pF capacitor is charged to a potential difference of 50 V, and the charging battery is disconnected. The capacitor is then connected in parallel with a second (initially uncharged) capacitor. If the potential difference across the first

capacitor drops to 35 V, what is the capacitance of this second capacitor?

••14 **GO** In Fig. 25-30, the battery has a potential difference of $V = 10.0$ V and the five capacitors each have a capacitance of 10.0 μF. What is the charge on (a) capacitor 1 and (b) capacitor 2?

Figure 25-30 Problem 14.

••15 **GO** In Fig. 25-31, a 20.0 V battery is connected across capacitors of capacitances $C_1 = C_6 = 3.00$ μF and $C_3 = C_5 = 2.00C_2 = 2.00C_4 = 4.00$ μF. What are (a) the equivalent capacitance C_{eq} of the capacitors and (b) the charge stored by C_{eq}? What are (c) V_1 and (d) q_1 of capacitor 1, (e) V_2 and (f) q_2 of capacitor 2, and (g) V_3 and (h) q_3 of capacitor 3?

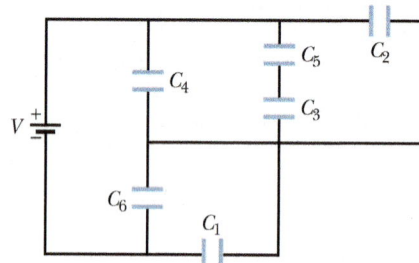

Figure 25-31 Problem 15.

••16 Plot 1 in Fig. 25-32a gives the charge q that can be stored on capacitor 1 versus the electric potential V set up across it. The vertical scale is set by $q_s = 16.0$ μC, and the horizontal scale is set by $V_s = 2.0$ V. Plots 2 and 3 are similar plots for capacitors 2 and 3, respectively. Figure 25-32b shows a circuit with those three capacitors and a 6.0 V battery. What is the charge stored on capacitor 2 in that circuit?

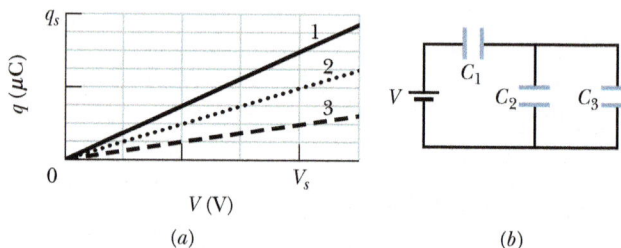

Figure 25-32 Problem 16.

••17 **GO** In Fig. 25-29, a potential difference of $V = 100.0$ V is applied across a capacitor arrangement with capacitances $C_1 = 10.0$ μF, $C_2 = 5.00$ μF, and $C_3 = 4.00$ μF. If capacitor 3 undergoes electrical breakdown so that it becomes equivalent to conducting wire, what is the increase in (a) the charge on capacitor 1 and (b) the potential difference across capacitor 1?

••18 Figure 25-33 shows a circuit section of four air-filled capacitors that is connected to a larger circuit. The graph below the section shows the electric potential $V(x)$ as a function of position x along the lower part of the section, through capacitor 4. Similarly, the graph above the section shows the electric potential $V(x)$ as a function of position x along the upper part of the section, through capacitors 1, 2, and 3.

Capacitor 3 has a capacitance of $0.80 \, \mu F$. What are the capacitances of (a) capacitor 1 and (b) capacitor 2?

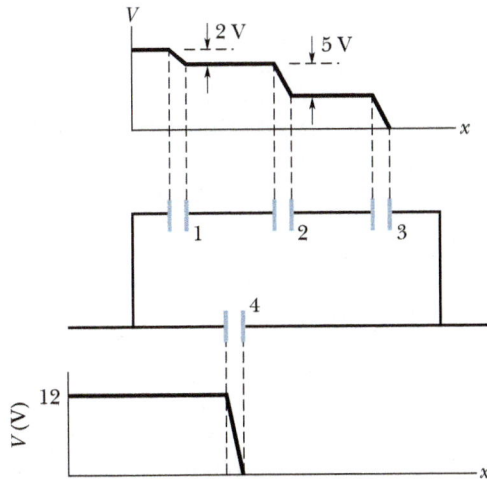

Figure 25-33
Problem 18.

••19 GO In Fig. 25-34, the battery has potential difference $V = 9.0$ V, $C_2 = 3.0 \, \mu F$, $C_4 = 4.0 \, \mu F$, and all the capacitors are initially uncharged. When switch S is closed, a total charge of $12 \, \mu C$ passes through point a and a total charge of $8.0 \, \mu C$ passes through point b. What are (a) C_1 and (b) C_3?

Figure 25-34 Problem 19.

••20 Figure 25-35 shows a variable "air gap" capacitor for manual tuning. Alternate plates are connected together; one group of plates is fixed in position, and the other group is capable of rotation. Consider a capacitor of $n = 8$ plates of alternating polarity, each plate having area $A = 1.25 \, cm^2$ and separated from adjacent plates by distance $d = 3.40$ mm. What is the maximum capacitance of the device?

Figure 25-35 Problem 20.

••21 SSM WWW In Fig. 25-36, the capacitances are $C_1 = 1.0 \, \mu F$ and $C_2 = 3.0 \, \mu F$, and both capacitors are charged to a potential difference of $V = 100$ V but with opposite polarity as shown. Switches S_1 and S_2 are now closed. (a) What is now the potential difference between points a and b? What now is the charge on capacitor (b) 1 and (c) 2?

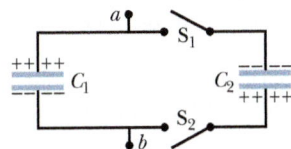

Figure 25-36 Problem 21.

••22 In Fig. 25-37, $V = 10$ V, $C_1 = 10$ μF, and $C_2 = C_3 = 20 \, \mu F$. Switch S is first thrown to the left side until capacitor 1 reaches equilibrium. Then the switch is thrown to the right. When equilibrium is again reached, how much charge is on capacitor 1?

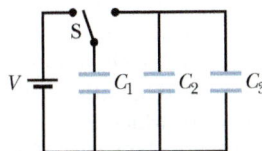

Figure 25-37 Problem 22.

••23 The capacitors in Fig. 25-38 are initially uncharged. The capacitances are $C_1 = 4.0 \, \mu F$, $C_2 = 8.0 \, \mu F$, and $C_3 = 12 \, \mu F$, and the battery's potential difference is $V = 12$ V. When switch S is closed, how many electrons travel through (a) point a, (b) point b, (c) point c, and (d) point d? In the figure, do the electrons travel up or down through (e) point b and (f) point c?

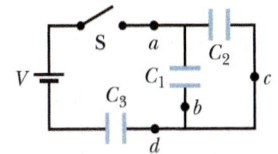

Figure 25-38 Problem 23.

••24 GO Figure 25-39 represents two air-filled cylindrical capacitors connected in series across a battery with potential $V = 10$ V. Capacitor 1 has an inner plate radius of 5.0 mm, an outer plate radius of 1.5 cm, and a length of 5.0 cm. Capacitor 2 has an inner plate radius of 2.5 mm, an outer plate radius of 1.0 cm, and a length of 9.0 cm. The outer plate of capacitor 2 is a conducting organic membrane that can be stretched, and the capacitor can be inflated to increase the plate separation. If the outer plate radius is increased to 2.5 cm by inflation, (a) how many electrons move through point P and (b) do they move toward or away from the battery?

Figure 25-39 Problem 24.

••25 GO In Fig. 25-40, two parallel-plate capacitors (with air between the plates) are connected to a battery. Capacitor 1 has a plate area of $1.5 \, cm^2$ and an electric field (between its plates) of magnitude 2000 V/m. Capacitor 2 has a plate area of $0.70 \, cm^2$ and an electric field of magnitude 1500 V/m. What is the total charge on the two capacitors?

Figure 25-40
Problem 25.

•••26 GO Capacitor 3 in Fig. 25-41a is a *variable capacitor* (its capacitance C_3 can be varied). Figure 25-41b gives the electric potential V_1 across capacitor 1 versus C_3. The horizontal scale is set by $C_{3s} = 12.0 \, \mu F$. Electric potential V_1 approaches an asymptote of 10 V as $C_3 \rightarrow \infty$. What are (a) the electric potential V across the battery, (b) C_1, and (c) C_2?

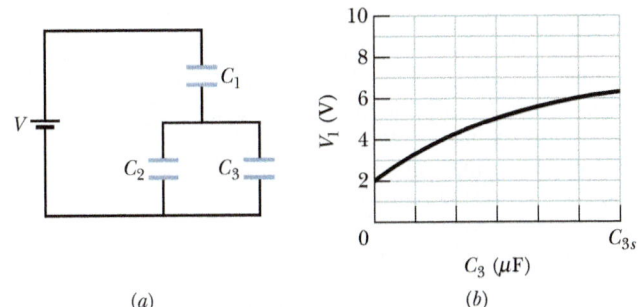

(a)   (b)

Figure 25-41 Problem 26.

•••27 GO Figure 25-42 shows a 12.0 V battery and four uncharged capacitors of capacitances $C_1 = 1.00 \, \mu F$, $C_2 = 2.00 \, \mu F$, $C_3 = 3.00 \, \mu F$, and $C_4 = 4.00 \, \mu F$. If only switch S_1 is closed, what is the charge on (a) capacitor 1, (b) capacitor 2, (c) capacitor 3, and (d) capacitor 4? If both switches are closed, what is the charge on (e) capacitor 1, (f) capacitor 2, (g) capacitor 3, and (h) capacitor 4?

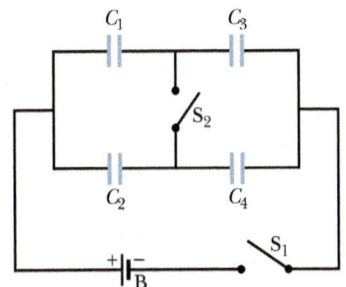

Figure 25-42 Problem 27.

•••**28** GO Figure 25-43 displays a 12.0 V battery and 3 uncharged capacitors of capacitances $C_1 = 4.00$ μF, $C_2 = 6.00$ μF, and $C_3 = 3.00$ μF. The switch is thrown to the left side until capacitor 1 is fully charged. Then the switch is thrown to the right. What is the final charge on (a) capacitor 1, (b) capacitor 2, and (c) capacitor 3?

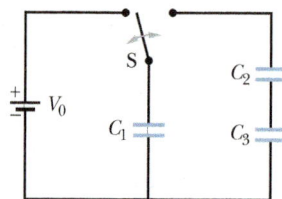

Figure 25-43 Problem 28.

Module 25-4 Energy Stored in an Electric Field

•**29** What capacitance is required to store an energy of 10 kW·h at a potential difference of 1000 V?

•**30** How much energy is stored in 1.00 m^3 of air due to the "fair weather" electric field of magnitude 150 V/m?

•**31** SSM A 2.0 μF capacitor and a 4.0 μF capacitor are connected in parallel across a 300 V potential difference. Calculate the total energy stored in the capacitors.

•**32** A parallel-plate air-filled capacitor having area 40 cm^2 and plate spacing 1.0 mm is charged to a potential difference of 600 V. Find (a) the capacitance, (b) the magnitude of the charge on each plate, (c) the stored energy, (d) the electric field between the plates, and (e) the energy density between the plates.

••**33** A charged isolated metal sphere of diameter 10 cm has a potential of 8000 V relative to $V = 0$ at infinity. Calculate the energy density in the electric field near the surface of the sphere.

••**34** In Fig. 25-28, a potential difference $V = 100$ V is applied across a capacitor arrangement with capacitances $C_1 = 10.0$ μF, $C_2 = 5.00$ μF, and $C_3 = 4.00$ μF. What are (a) charge q_3, (b) potential difference V_3, and (c) stored energy U_3 for capacitor 3, (d) q_1, (e) V_1, and (f) U_1 for capacitor 1, and (g) q_2, (h) V_2, and (i) U_2 for capacitor 2?

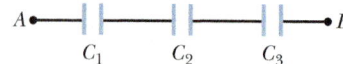

••**35** Assume that a stationary electron is a point of charge. What is the energy density u of its electric field at radial distances (a) $r = 1.00$ mm, (b) $r = 1.00$ μm, (c) $r = 1.00$ nm, and (d) $r = 1.00$ pm? (e) What is u in the limit as $r \rightarrow 0$?

••**36** ✈ As a safety engineer, you must evaluate the practice of storing flammable conducting liquids in nonconducting containers. The company supplying a certain liquid has been using a squat, cylindrical plastic container of radius $r = 0.20$ m and filling it to height $h = 10$ cm, which is not the container's full interior height (Fig. 25-44). Your investigation reveals that during handling at the company, the exterior surface of the container commonly acquires a negative charge density of magnitude 2.0 μC/m^2 (approximately uniform). Because the liquid is a conducting material, the charge on the container induces charge separation within the liquid. (a) How much negative charge is induced in the center of the liquid's bulk? (b) Assume the capacitance of the central portion of the liquid relative to ground is 35 pF. What is the potential energy associated with the negative charge in that effective capacitor? (c) If a spark occurs between the ground and the central portion of the liquid (through the venting port), the potential energy can be fed into the spark. The minimum spark energy needed to ignite the liquid is 10 mJ. In this situation, can a spark ignite the liquid?

Figure 25-44 Problem 36.

••**37** SSM ILW WWW The parallel plates in a capacitor, with a plate area of 8.50 cm^2 and an air-filled separation of 3.00 mm, are charged by a 6.00 V battery. They are then disconnected from the battery and pulled apart (without discharge) to a separation of 8.00 mm. Neglecting fringing, find (a) the potential difference between the plates, (b) the initial stored energy, (c) the final stored energy, and (d) the work required to separate the plates.

••**38** In Fig. 25-29, a potential difference $V = 100$ V is applied across a capacitor arrangement with capacitances $C_1 = 10.0$ μF, $C_2 = 5.00$ μF, and $C_3 = 15.0$ μF. What are (a) charge q_3, (b) potential difference V_3, and (c) stored energy U_3 for capacitor 3, (d) q_1, (e) V_1, and (f) U_1 for capacitor 1, and (g) q_2, (h) V_2, and (i) U_2 for capacitor 2?

••**39** GO In Fig. 25-45, $C_1 = 10.0$ μF, $C_2 = 20.0$ μF, and $C_3 = 25.0$ μF. If no capacitor can withstand a potential difference of more than 100 V without failure, what are (a) the magnitude of the maximum potential difference that can exist between points A and B and (b) the maximum energy that can be stored in the three-capacitor arrangement?

Figure 25-45 Problem 39.

Module 25-5 Capacitor with a Dielectric

•**40** An air-filled parallel-plate capacitor has a capacitance of 1.3 pF. The separation of the plates is doubled, and wax is inserted between them. The new capacitance is 2.6 pF. Find the dielectric constant of the wax.

•**41** SSM A coaxial cable used in a transmission line has an inner radius of 0.10 mm and an outer radius of 0.60 mm. Calculate the capacitance per meter for the cable. Assume that the space between the conductors is filled with polystyrene.

•**42** A parallel-plate air-filled capacitor has a capacitance of 50 pF. (a) If each of its plates has an area of 0.35 m^2, what is the separation? (b) If the region between the plates is now filled with material having $\kappa = 5.6$, what is the capacitance?

•**43** Given a 7.4 pF air-filled capacitor, you are asked to convert it to a capacitor that can store up to 7.4 μJ with a maximum potential difference of 652 V. Which dielectric in Table 25-1 should you use to fill the gap in the capacitor if you do not allow for a margin of error?

•••**44** You are asked to construct a capacitor having a capacitance near 1 nF and a breakdown potential in excess of 10 000 V. You think of using the sides of a tall Pyrex drinking glass as a dielectric, lining the inside and outside curved surfaces with aluminum foil to act as the plates. The glass is 15 cm tall with an inner radius of 3.6 cm and an outer radius of 3.8 cm. What are the (a) capacitance and (b) breakdown potential of this capacitor?

•••**45** A certain parallel-plate capacitor is filled with a dielectric for which $\kappa = 5.5$. The area of each plate is 0.034 m^2, and the plates are separated by 2.0 mm. The capacitor will fail (short out and burn up) if the electric field between the plates exceeds 200 kN/C. What is the maximum energy that can be stored in the capacitor?

•••**46** In Fig. 25-46, how much charge is stored on the parallel-plate capacitors by the 12.0 V battery? One is filled with air, and the other is filled with a dielectric for which $\kappa = 3.00$; both capacitors have a plate area of 5.00×10^{-3} m^2 and a plate separation of 2.00 mm.

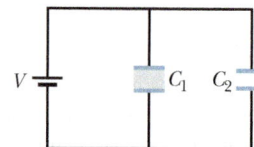

Figure 25-46 Problem 46.

••47 SSM ILW A certain substance has a dielectric constant of 2.8 and a dielectric strength of 18 MV/m. If it is used as the dielectric material in a parallel-plate capacitor, what minimum area should the plates of the capacitor have to obtain a capacitance of $7.0 \times 10^{-2} \ \mu F$ and to ensure that the capacitor will be able to withstand a potential difference of 4.0 kV?

••48 Figure 25-47 shows a parallel-plate capacitor with a plate area $A = 5.56 \ cm^2$ and separation $d = 5.56$ mm. The left half of the gap is filled with material of dielectric constant $\kappa_1 = 7.00$; the right half is filled with material of dielectric constant $\kappa_2 = 12.0$. What is the capacitance?

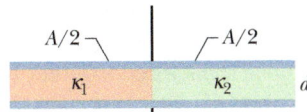

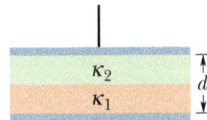

Figure 25-47 Problem 48.

••49 Figure 25-48 shows a parallel-plate capacitor with a plate area $A = 7.89 \ cm^2$ and plate separation $d = 4.62$ mm. The top half of the gap is filled with material of dielectric constant $\kappa_1 = 11.0$; the bottom half is filled with material of dielectric constant $\kappa_2 = 12.0$. What is the capacitance?

Figure 25-48
Problem 49.

••50 GO Figure 25-49 shows a parallel-plate capacitor of plate area $A = 10.5$ cm^2 and plate separation $2d = 7.12$ mm. The left half of the gap is filled with material of dielectric constant $\kappa_1 = 21.0$; the top of the right half is filled with material of dielectric constant $\kappa_2 = 42.0$; the bottom of the right half is filled with material of dielectric constant $\kappa_3 = 58.0$. What is the capacitance?

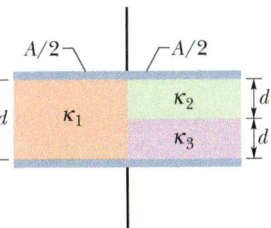

Figure 25-49 Problem 50.

Module 25-6 Dielectrics and Gauss' Law

•51 SSM WWW A parallel-plate capacitor has a capacitance of 100 pF, a plate area of 100 cm², and a mica dielectric ($\kappa = 5.4$) completely filling the space between the plates. At 50 V potential difference, calculate (a) the electric field magnitude E in the mica, (b) the magnitude of the free charge on the plates, and (c) the magnitude of the induced surface charge on the mica.

•52 For the arrangement of Fig. 25-17, suppose that the battery remains connected while the dielectric slab is being introduced. Calculate (a) the capacitance, (b) the charge on the capacitor plates, (c) the electric field in the gap, and (d) the electric field in the slab, after the slab is in place.

••53 A parallel-plate capacitor has plates of area 0.12 m² and a separation of 1.2 cm. A battery charges the plates to a potential difference of 120 V and is then disconnected. A dielectric slab of thickness 4.0 mm and dielectric constant 4.8 is then placed symmetrically between the plates. (a) What is the capacitance before the slab is inserted? (b) What is the capacitance with the slab in place? What is the free charge q (c) before and (d) after the slab is inserted? What is the magnitude of the electric field (e) in the space between the plates and dielectric and (f) in the dielectric itself? (g) With the slab in place, what is the potential difference across the plates? (h) How much external work is involved in inserting the slab?

••54 Two parallel plates of area 100 cm² are given charges of equal magnitudes 8.9×10^{-7} C but opposite signs. The electric field within the dielectric material filling the space between the plates is 1.4×10^6 V/m. (a) Calculate the dielectric constant of the material. (b) Determine the magnitude of the charge induced on each dielectric surface.

••55 The space between two concentric conducting spherical shells of radii $b = 1.70$ cm and $a = 1.20$ cm is filled with a substance of dielectric constant $\kappa = 23.5$. A potential difference $V = 73.0$ V is applied across the inner and outer shells. Determine (a) the capacitance of the device, (b) the free charge q on the inner shell, and (c) the charge q' induced along the surface of the inner shell.

Additional Problems

56 In Fig. 25-50, the battery potential difference V is 10.0 V and each of the seven capacitors has capacitance 10.0 μF. What is the charge on (a) capacitor 1 and (b) capacitor 2?

Figure 25-50
Problem 56.

57 SSM In Fig. 25-51, $V = 9.0$ V, $C_1 = C_2 = 30$ μF, and $C_3 = C_4 = 15 \ \mu F$. What is the charge on capacitor 4?

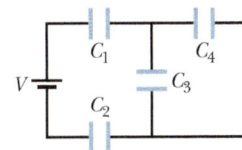

Figure 25-51 Problem 57.

58 (a) If $C = 50 \ \mu F$ in Fig. 25-52, what is the equivalent capacitance between points A and B? (*Hint:* First imagine that a battery is connected between those two points.) (b) Repeat for points A and D.

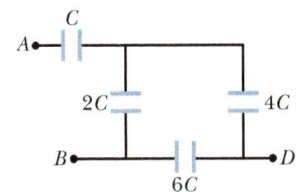

Figure 25-52 Problem 58.

59 In Fig. 25-53, $V = 12$ V, $C_1 = C_4 = 2.0 \ \mu F$, $C_2 = 4.0 \ \mu F$, and $C_3 = 1.0 \ \mu F$. What is the charge on capacitor 4?

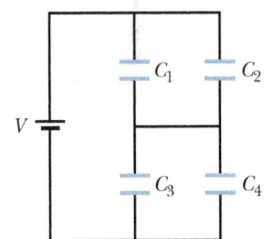

60 The chocolate crumb mystery. This story begins with Problem 60 in Chapter 23. As part of the investigation of the biscuit factory explosion, the electric potentials of the workers were measured as they emptied sacks of chocolate crumb powder into the loading bin, stirring up a cloud of the powder around themselves. Each worker had an

Figure 25-53 Problem 59.

electric potential of about 7.0 kV relative to the ground, which was taken as zero potential. (a) Assuming that each worker was effectively a capacitor with a typical capacitance of 200 pF, find the energy stored in that effective capacitor. If a single spark between the worker and any conducting object connected to the ground neutralized the worker, that energy would be transferred to the spark. According to measurements, a spark that could ignite a cloud of chocolate crumb powder, and thus set off an explosion, had to have an energy of at least 150 mJ. (b) Could a spark from a worker have set off an explosion in the cloud of powder in the loading bin? (The story continues with Problem 60 in Chapter 26.)

61 Figure 25-54 shows capacitor 1 ($C_1 = 8.00 \ \mu F$), capacitor 2 ($C_2 = 6.00 \ \mu F$), and capacitor 3 ($C_3 = $

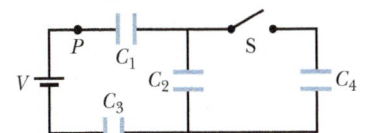

Figure 25-54 Problem 61.

8.00 μF) connected to a 12.0 V battery. When switch S is closed so as to connect uncharged capacitor 4 ($C_4 = 6.00$ μF), (a) how much charge passes through point P from the battery and (b) how much charge shows up on capacitor 4? (c) Explain the discrepancy in those two results.

62 Two air-filled, parallel-plate capacitors are to be connected to a 10 V battery, first individually, then in series, and then in parallel. In those arrangements, the energy stored in the capacitors turns out to be, listed least to greatest: 75 μJ, 100 μJ, 300 μJ, and 400 μJ. Of the two capacitors, what is the (a) smaller and (b) greater capacitance?

63 Two parallel-plate capacitors, 6.0 μF each, are connected in series to a 10 V battery. One of the capacitors is then squeezed so that its plate separation is halved. Because of the squeezing, (a) how much additional charge is transferred to the capacitors by the battery and (b) what is the increase in the *total* charge stored on the capacitors (the charge on the positive plate of one capacitor plus the charge on the positive plate of the other capacitor)?

64 **GO** In Fig. 25-55, $V = 12$ V, $C_1 = C_5 = C_6 = 6.0$ μF, and $C_2 = C_3 = C_4 = 4.0$ μF. What are (a) the net charge stored on the capacitors and (b) the charge on capacitor 4?

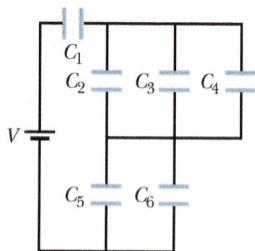

Figure 25-55 Problem 64.

65 **SSM** In Fig. 25-56, the parallel-plate capacitor of plate area 2.00×10^{-2} m^2 is filled with two dielectric slabs, each with thickness 2.00 mm. One slab has dielectric constant 3.00, and the other, 4.00. How much charge does the 7.00 V battery store on the capacitor?

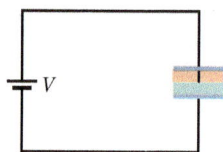

Figure 25-56
Problem 65.

66 A cylindrical capacitor has radii a and b as in Fig. 25-6. Show that half the stored electric potential energy lies within a cylinder whose radius is $r = \sqrt{ab}$.

67 A capacitor of capacitance $C_1 = 6.00$ μF is connected in series with a capacitor of capacitance $C_2 = 4.00$ μF, and a potential difference of 200 V is applied across the pair. (a) Calculate the equivalent capacitance. What are (b) charge q_1 and (c) potential difference V_1 on capacitor 1 and (d) q_2 and (e) V_2 on capacitor 2?

68 Repeat Problem 67 for the same two capacitors but with them now connected in parallel.

69 A certain capacitor is charged to a potential difference V. If you wish to increase its stored energy by 10%, by what percentage should you increase V?

70 A slab of copper of thickness $b = 2.00$ mm is thrust into a parallel-plate capacitor of plate area $A = 2.40$ cm^2 and plate separation $d = 5.00$ mm, as shown in Fig. 25-57; the slab is exactly halfway between the plates. (a) What is the capacitance after the slab is introduced? (b) If a charge $q = 3.40$ μC is maintained on the plates, what is the ratio of the stored energy before to that after the slab is inserted? (c) How much work is done on the slab as it is inserted? (d) Is the slab sucked in or must it be pushed in?

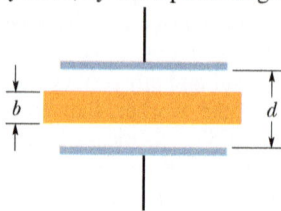

Figure 25-57
Problems 70 and 71.

71 Repeat Problem 70, assuming that a potential difference $V = 85.0$ V, rather than the charge, is held constant.

72 A potential difference of 300 V is applied to a series connection of two capacitors of capacitances $C_1 = 2.00$ μF and $C_2 = 8.00$ μF. What are (a) charge q_1 and (b) potential difference V_1 on capacitor 1 and (c) q_2 and (d) V_2 on capacitor 2? The *charged* capacitors are then disconnected from each other and from the battery. Then the capacitors are reconnected with plates of the *same* signs wired together (the battery is not used). What now are (e) q_1, (f) V_1, (g) q_2, and (h) V_2? Suppose, *instead,* the capacitors charged in part (a) are reconnected with plates of *opposite* signs wired together. What now are (i) q_1, (j) V_1, (k) q_2, and (l) V_2?

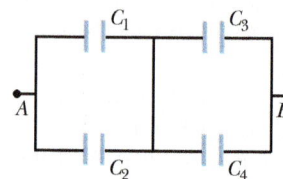

73 Figure 25-58 shows a four-capacitor arrangement that is connected to a larger circuit at points A and B. The capacitances are $C_1 = 10$ μF and $C_2 = C_3 = C_4 = 20$ μF. The charge on capacitor 1 is 30 μC. What is the magnitude of the potential difference $V_A - V_B$?

Figure 25-58 Problem 73.

74 You have two plates of copper, a sheet of mica (thickness = 0.10 mm, $\kappa = 5.4$), a sheet of glass (thickness = 2.0 mm, $\kappa = 7.0$), and a slab of paraffin (thickness = 1.0 cm, $\kappa = 2.0$). To make a parallel-plate capacitor with the largest C, which sheet should you place between the copper plates?

75 A capacitor of unknown capacitance C is charged to 100 V and connected across an initially uncharged 60 μF capacitor. If the final potential difference across the 60 μF capacitor is 40 V, what is C?

76 A 10 V battery is connected to a series of n capacitors, each of capacitance 2.0 μF. If the total stored energy is 25 μJ, what is n?

77 **SSM** In Fig. 25-59, two parallel-plate capacitors A and B are connected in parallel across a 600 V battery. Each plate has area 80.0 cm^2; the plate separations are 3.00 mm. Capacitor A is filled with air; capacitor B is filled with a dielectric of dielectric constant $\kappa = 2.60$. Find the magnitude of the electric field within (a) the dielectric of capacitor B and (b) the air of capacitor A. What are the free charge densities σ on the higher-potential plate of (c) capacitor A and (d) capacitor B? (e) What is the induced charge density σ' on the top surface of the dielectric?

Figure 25-59 Problem 77.

78 You have many 2.0 μF capacitors, each capable of withstanding 200 V without undergoing electrical breakdown (in which they conduct charge instead of storing it). How would you assemble a combination having an equivalent capacitance of (a) 0.40 μF and (b) 1.2 μF, each combination capable of withstanding 1000 V?

79 A parallel-plate capacitor has charge q and plate area A. (a) By finding the work needed to increase the plate separation from x to $x + dx$, determine the force between the plates. (*Hint:* See Eq. 8-22.) (b) Then show that the force per unit area (the *electrostatic stress*) acting on either plate is equal to the energy density $\varepsilon_0 E^2/2$ between the plates.

80 A capacitor is charged until its stored energy is 4.00 J. A second capacitor is then connected to it in parallel. (a) If the charge distributes equally, what is the total energy stored in the electric fields? (b) Where did the missing energy go?

C H A P T E R   2 6

Current and Resistance

26-1 ELECTRIC CURRENT

Learning Objectives

After reading this module, you should be able to . . .

26.01 Apply the definition of current as the rate at which charge moves through a point, including solving for the amount of charge that passes the point in a given time interval.

26.02 Identify that current is normally due to the motion of conduction electrons that are driven by electric fields (such as those set up in a wire by a battery).

26.03 Identify a junction in a circuit and apply the fact that (due to conservation of charge) the total current into a junction must equal the total current out of the junction.

26.04 Explain how current arrows are drawn in a schematic diagram of a circuit, and identify that the arrows are not vectors.

Key Ideas

● An electric current i in a conductor is defined by

$$i = \frac{dq}{dt},$$

where dq is the amount of positive charge that passes in time dt.

● By convention, the direction of electric current is taken as the direction in which positive charge carriers would move even though (normally) only conduction electrons can move.

What Is Physics?

In the last five chapters we discussed electrostatics—the physics of stationary charges. In this and the next chapter, we discuss the physics of **electric currents**—that is, charges in motion.

Examples of electric currents abound and involve many professions. Meteorologists are concerned with lightning and with the less dramatic slow flow of charge through the atmosphere. Biologists, physiologists, and engineers working in medical technology are concerned with the nerve currents that control muscles and especially with how those currents can be reestablished after spinal cord injuries. Electrical engineers are concerned with countless electrical systems, such as power systems, lightning protection systems, information storage systems, and music systems. Space engineers monitor and study the flow of charged particles from our Sun because that flow can wipe out telecommunication systems in orbit and even power transmission systems on the ground. In addition to such scholarly work, almost every aspect of daily life now depends on information carried by electric currents, from stock trades to ATM transfers and from video entertainment to social networking.

In this chapter we discuss the basic physics of electric currents and why they can be established in some materials but not in others. We begin with the meaning of electric current.

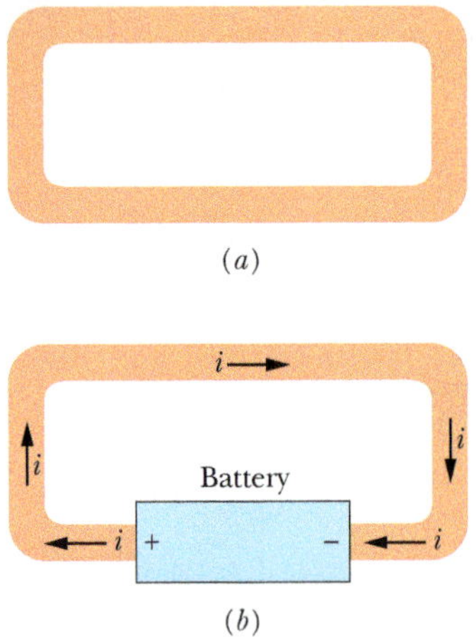

(a)

Battery

(b)

Figure 26-1 (a) A loop of copper in electrostatic equilibrium. The entire loop is at a single potential, and the electric field is zero at all points inside the copper. (b) Adding a battery imposes an electric potential difference between the ends of the loop that are connected to the terminals of the battery. The battery thus produces an electric field within the loop, from terminal to terminal, and the field causes charges to move around the loop. This movement of charges is a current i.

Electric Current

Although an electric current is a stream of moving charges, not all moving charges constitute an electric current. If there is to be an electric current through a given surface, there must be a net flow of charge through that surface. Two examples clarify our meaning.

1. The free electrons (conduction electrons) in an isolated length of copper wire are in random motion at speeds of the order of 10^6 m/s. If you pass a hypothetical plane through such a wire, conduction electrons pass through it *in both directions* at the rate of many billions per second—but there is *no net transport* of charge and thus *no current* through the wire. However, if you connect the ends of the wire to a battery, you slightly bias the flow in one direction, with the result that there now is a net transport of charge and thus an electric current through the wire.

2. The flow of water through a garden hose represents the directed flow of positive charge (the protons in the water molecules) at a rate of perhaps several million coulombs per second. There is no net transport of charge, however, because there is a parallel flow of negative charge (the electrons in the water molecules) of exactly the same amount moving in exactly the same direction.

In this chapter we restrict ourselves largely to the study—within the framework of classical physics—of *steady* currents of *conduction electrons* moving through *metallic conductors* such as copper wires.

As Fig. 26-1a reminds us, any isolated conducting loop—regardless of whether it has an excess charge—is all at the same potential. No electric field can exist within it or along its surface. Although conduction electrons are available, no net electric force acts on them and thus there is no current.

If, as in Fig. 26-1b, we insert a battery in the loop, the conducting loop is no longer at a single potential. Electric fields act inside the material making up the loop, exerting forces on the conduction electrons, causing them to move and thus establishing a current. After a very short time, the electron flow reaches a constant value and the current is in its *steady state* (it does not vary with time).

Figure 26-2 shows a section of a conductor, part of a conducting loop in which current has been established. If charge dq passes through a hypothetical plane (such as aa') in time dt, then the current i through that plane is defined as

$$i = \frac{dq}{dt} \quad \text{(definition of current).} \tag{26-1}$$

We can find the charge that passes through the plane in a time interval extending from 0 to t by integration:

$$q = \int dq = \int_0^t i\, dt, \tag{26-2}$$

in which the current i may vary with time.

Under steady-state conditions, the current is the same for planes aa', bb', and cc' and indeed for all planes that pass completely through the conductor, no matter what their location or orientation. This follows from the fact that charge is conserved. Under the steady-state conditions assumed here, an electron must pass through plane aa' for every electron that passes through plane cc'. In the same way, if we have a steady flow of water through a garden hose, a drop of water must leave the nozzle for every drop that enters the hose at the other end. The amount of water in the hose is a conserved quantity.

The SI unit for current is the coulomb per second, or the ampere (A), which is an SI base unit:

$$1 \text{ ampere} = 1 \text{ A} = 1 \text{ coulomb per second} = 1 \text{ C/s.}$$

The formal definition of the ampere is discussed in Chapter 29.

The current is the same in any cross section.

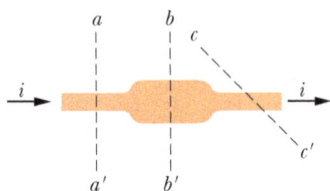

Figure 26-2 The current i through the conductor has the same value at planes aa', bb', and cc'.

Current, as defined by Eq. 26-1, is a scalar because both charge and time in that equation are scalars. Yet, as in Fig. 26-1b, we often represent a current with an arrow to indicate that charge is moving. Such arrows are not vectors, however, and they do not require vector addition. Figure 26-3a shows a conductor with current i_0 splitting at a junction into two branches. Because charge is conserved, the magnitudes of the currents in the branches must add to yield the magnitude of the current in the original conductor, so that

$$i_0 = i_1 + i_2. \qquad (26\text{-}3)$$

As Fig. 26-3b suggests, bending or reorienting the wires in space does not change the validity of Eq. 26-3. Current arrows show only a direction (or sense) of flow along a conductor, not a direction in space.

The Directions of Currents

In Fig. 26-1b we drew the current arrows in the direction in which positively charged particles would be forced to move through the loop by the electric field. Such positive *charge carriers,* as they are often called, would move away from the positive battery terminal and toward the negative terminal. Actually, the charge carriers in the copper loop of Fig. 26-1b are electrons and thus are negatively charged. The electric field forces them to move in the direction opposite the current arrows, from the negative terminal to the positive terminal. For historical reasons, however, we use the following convention:

A current arrow is drawn in the direction in which positive charge carriers would move, even if the actual charge carriers are negative and move in the opposite direction.

We can use this convention because in *most* situations, the assumed motion of positive charge carriers in one direction has the same effect as the actual motion of negative charge carriers in the opposite direction. (When the effect is not the same, we shall drop the convention and describe the actual motion.)

The current into the junction must equal the current out (charge is conserved).

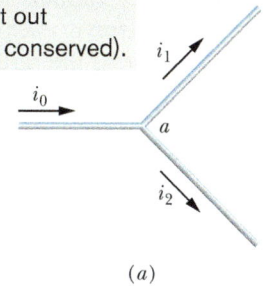

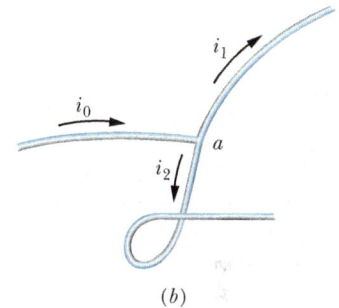

(a)

(b)

Figure 26-3 The relation $i_0 = i_1 + i_2$ is true at junction a no matter what the orientation in space of the three wires. Currents are scalars, not vectors.

✓ **Checkpoint 1**

The figure here shows a portion of a circuit. What are the magnitude and direction of the current i in the lower right-hand wire?

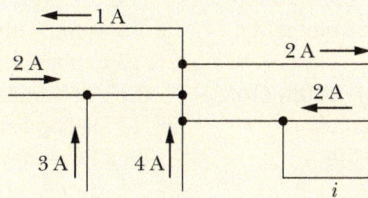

Sample Problem 26.01   Current is the rate at which charge passes a point

Water flows through a garden hose at a volume flow rate dV/dt of 450 cm³/s. What is the current of negative charge?

KEY IDEAS

The current i of negative charge is due to the electrons in the water molecules moving through the hose. The current is the rate at which that negative charge passes through any plane that cuts completely across the hose.

Calculations: We can write the current in terms of the number of molecules that pass through such a plane per second as

$$i = \left(\begin{array}{c}\text{charge}\\\text{per}\\\text{electron}\end{array}\right)\left(\begin{array}{c}\text{electrons}\\\text{per}\\\text{molecule}\end{array}\right)\left(\begin{array}{c}\text{molecules}\\\text{per}\\\text{second}\end{array}\right)$$

or

$$i = (e)(10)\frac{dN}{dt}.$$

We substitute 10 electrons per molecule because a water (H_2O) molecule contains 8 electrons in the single oxygen atom and 1 electron in each of the two hydrogen atoms.

We can express the rate dN/dt in terms of the given volume flow rate dV/dt by first writing

$$\begin{pmatrix} \text{molecules} \\ \text{per} \\ \text{second} \end{pmatrix} = \begin{pmatrix} \text{molecules} \\ \text{per} \\ \text{mole} \end{pmatrix} \begin{pmatrix} \text{moles} \\ \text{per unit} \\ \text{mass} \end{pmatrix}$$
$$\times \begin{pmatrix} \text{mass} \\ \text{per unit} \\ \text{volume} \end{pmatrix} \begin{pmatrix} \text{volume} \\ \text{per} \\ \text{second} \end{pmatrix}.$$

"Molecules per mole" is Avogadro's number N_A. "Moles per unit mass" is the inverse of the mass per mole, which is the molar mass M of water. "Mass per unit volume" is the (mass) density ρ_{mass} of water. The volume per second is the volume flow rate dV/dt. Thus, we have

$$\frac{dN}{dt} = N_A \left(\frac{1}{M} \right) \rho_{mass} \left(\frac{dV}{dt} \right) = \frac{N_A \rho_{mass}}{M} \frac{dV}{dt}.$$

Substituting this into the equation for i, we find

$$i = 10 e N_A M^{-1} \rho_{mass} \frac{dV}{dt}.$$

We know that Avogadro's number N_A is 6.02×10^{23} molecules/mol, or 6.02×10^{23} mol^{-1}, and from Table 14-1 we know that the density of water ρ_{mass} under normal conditions is 1000 kg/m^3. We can get the molar mass of water from the molar masses listed in Appendix F (in grams per mole): We add the molar mass of oxygen (16 g/mol) to twice the molar mass of hydrogen (1 g/mol), obtaining 18 g/mol = 0.018 kg/mol. So, the current of negative charge due to the electrons in the water is

$$i = (10)(1.6 \times 10^{-19} \text{ C})(6.02 \times 10^{23} \text{ mol}^{-1})$$
$$\times (0.018 \text{ kg/mol})^{-1}(1000 \text{ kg/m}^3)(450 \times 10^{-6} \text{ m}^3/\text{s})$$
$$= 2.41 \times 10^7 \text{ C/s} = 2.41 \times 10^7 \text{ A}$$
$$= 24.1 \text{ MA.} \qquad \text{(Answer)}$$

This current of negative charge is exactly compensated by a current of positive charge associated with the nuclei of the three atoms that make up the water molecule. Thus, there is no net flow of charge through the hose.

WILEY **PLUS** Additional examples, video, and practice available at *WileyPLUS*

26-2 CURRENT DENSITY

Learning Objectives

After reading this module, you should be able to . . .

26.05 Identify a current density and a current density vector.

26.06 For current through an area element on a cross section through a conductor (such as a wire), identify the element's area vector $d\vec{A}$.

26.07 Find the current through a cross section of a conductor by integrating the dot product of the current density vector $\vec{J}$ and the element area vector $d\vec{A}$ over the full cross section.

26.08 For the case where current is uniformly spread over a cross section in a conductor, apply the relationship

between the current i, the current density magnitude J, and the area A.

26.09 Identify streamlines.

26.10 Explain the motion of conduction electrons in terms of their drift speed.

26.11 Distinguish the drift speeds of conduction electrons from their random-motion speeds, including relative magnitudes.

26.12 Identify charge carrier density n.

26.13 Apply the relationship between current density J, charge carrier density n, and charge carrier drift speed v_d.

Key Ideas

● Current i (a scalar quantity) is related to current density $\vec{J}$ (a vector quantity) by

$$i = \int \vec{J} \cdot d\vec{A},$$

where $d\vec{A}$ is a vector perpendicular to a surface element of area dA and the integral is taken over any surface cutting across the conductor. The current density $\vec{J}$ has the same direction as the velocity of the moving charges if

they are positive and the opposite direction if they are negative.

● When an electric field $\vec{E}$ is established in a conductor, the charge carriers (assumed positive) acquire a drift speed v_d in the direction of $\vec{E}$.

● The drift velocity $\vec{v}_d$ is related to the current density by

$$\vec{J} = (ne)\vec{v}_d,$$

where ne is the carrier charge density.

Current Density

Sometimes we are interested in the current i in a particular conductor. At other times we take a localized view and study the flow of charge through a cross section of the conductor at a particular point. To describe this flow, we can use the **current density** $\vec{J}$, which has the same direction as the velocity of the moving charges if they are positive and the opposite direction if they are negative. For each element of the cross section, the magnitude J is equal to the current per unit area through that element. We can write the amount of current through the element as $\vec{J} \cdot d\vec{A}$, where $d\vec{A}$ is the area vector of the element, perpendicular to the element. The total current through the surface is then

$$i = \int \vec{J} \cdot d\vec{A}. \tag{26-4}$$

Figure 26-4 Streamlines representing current density in the flow of charge through a constricted conductor.

If the current is uniform across the surface and parallel to $d\vec{A}$, then $\vec{J}$ is also uniform and parallel to $d\vec{A}$. Then Eq. 26-4 becomes

$$i = \int J \, dA = J \int dA = JA,$$

so

$$J = \frac{i}{A}, \tag{26-5}$$

where A is the total area of the surface. From Eq. 26-4 or 26-5 we see that the SI unit for current density is the ampere per square meter (A/m^2).

In Chapter 22 we saw that we can represent an electric field with electric field lines. Figure 26-4 shows how current density can be represented with a similar set of lines, which we can call *streamlines*. The current, which is toward the right in Fig. 26-4, makes a transition from the wider conductor at the left to the narrower conductor at the right. Because charge is conserved during the transition, the amount of charge and thus the amount of current cannot change. However, the current density does change — it is greater in the narrower conductor. The spacing of the streamlines suggests this increase in current density; streamlines that are closer together imply greater current density.

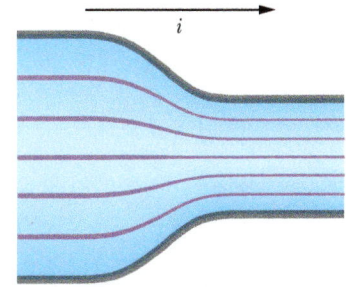

Drift Speed

When a conductor does not have a current through it, its conduction electrons move randomly, with no net motion in any direction. When the conductor does have a current through it, these electrons actually still move randomly, but now they tend to *drift* with a **drift speed** v_d in the direction opposite that of the applied electric field that causes the current. The drift speed is tiny compared with the speeds in the random motion. For example, in the copper conductors of household wiring, electron drift speeds are perhaps 10^{-5} or 10^{-4} m/s, whereas the random-motion speeds are around 10^6 m/s.

We can use Fig. 26-5 to relate the drift speed v_d of the conduction electrons in a current through a wire to the magnitude J of the current density in the wire. For

Figure 26-5 Positive charge carriers drift at speed v_d in the direction of the applied electric field $\vec{E}$. By convention, the direction of the current density $\vec{J}$ and the sense of the current arrow are drawn in that same direction.

Current is said to be due to positive charges that are propelled by the electric field.

convenience, Fig. 26-5 shows the equivalent drift of *positive* charge carriers in the direction of the applied electric field $\vec{E}$. Let us assume that these charge carriers all move with the same drift speed v_d and that the current density J is uniform across the wire's cross-sectional area A. The number of charge carriers in a length L of the wire is nAL, where n is the number of carriers per unit volume. The total charge of the carriers in the length L, each with charge e, is then

$$q = (nAL)e.$$

Because the carriers all move along the wire with speed v_d, this total charge moves through any cross section of the wire in the time interval

$$t = \frac{L}{v_d}.$$

Equation 26-1 tells us that the current i is the time rate of transfer of charge across a cross section, so here we have

$$i = \frac{q}{t} = \frac{nALe}{L/v_d} = nAev_d. \tag{26-6}$$

Solving for v_d and recalling Eq. 26-5 ($J = i/A$), we obtain

$$v_d = \frac{i}{nAe} = \frac{J}{ne}$$

or, extended to vector form,

$$\vec{J} = (ne)\vec{v}_d. \tag{26-7}$$

Here the product ne, whose SI unit is the coulomb per cubic meter (C/m^3), is the *carrier charge density*. For positive carriers, ne is positive and Eq. 26-7 predicts that $\vec{J}$ and $\vec{v}_d$ have the same direction. For negative carriers, ne is negative and $\vec{J}$ and $\vec{v}_d$ have opposite directions.

✓ Checkpoint 2

The figure shows conduction electrons moving leftward in a wire. Are the following leftward or rightward: (a) the current i, (b) the current density $\vec{J}$, (c) the electric field $\vec{E}$ in the wire?

Sample Problem 26.02 Current density, uniform and nonuniform

(a) The current density in a cylindrical wire of radius $R = 2.0$ mm is uniform across a cross section of the wire and is $J = 2.0 \times 10^5$ A/m². What is the current through the outer portion of the wire between radial distances $R/2$ and R (Fig. 26-6a)?

KEY IDEA

Because the current density is uniform across the cross section, the current density J, the current i, and the cross-sectional area A are related by Eq. 26-5 ($J = i/A$).

Calculations: We want only the current through a reduced cross-sectional area A' of the wire (rather than the entire

area), where

$$A' = \pi R^2 - \pi \left(\frac{R}{2}\right)^2 = \pi \left(\frac{3R^2}{4}\right)$$

$$= \frac{3\pi}{4} (0.0020 \text{ m})^2 = 9.424 \times 10^{-6} \text{ m}^2.$$

So, we rewrite Eq. 26-5 as

$$i = JA'$$

and then substitute the data to find

$$i = (2.0 \times 10^5 \text{ A/m}^2)(9.424 \times 10^{-6} \text{ m}^2)$$

$$= 1.9 \text{ A.} \tag{Answer}$$

(b) Suppose, instead, that the current density through a cross section varies with radial distance r as $J = ar^2$, in which $a = 3.0 \times 10^{11}$ A/m^4 and r is in meters. What now is the current through the same outer portion of the wire?

KEY IDEA

Because the current density is not uniform across a cross section of the wire, we must resort to Eq. 26-4 ($i = \int \vec{J} \cdot d\vec{A}$) and integrate the current density over the portion of the wire from $r = R/2$ to $r = R$.

Calculations: The current density vector $\vec{J}$ (along the wire's length) and the differential area vector $d\vec{A}$ (perpendicular to a cross section of the wire) have the same direction. Thus,

$$\vec{J} \cdot d\vec{A} = J\, dA \cos 0 = J\, dA.$$

We need to replace the differential area dA with something we can actually integrate between the limits $r = R/2$ and $r = R$. The simplest replacement (because J is given as a function of r) is the area $2\pi r\, dr$ of a thin ring of circumference $2\pi r$ and width dr (Fig. 26-6b). We can then integrate with r as the variable of integration. Equation 26-4 then gives us

$$i = \int \vec{J} \cdot d\vec{A} = \int J\, dA$$

$$= \int_{R/2}^{R} ar^2\, 2\pi r\, dr = 2\pi a \int_{R/2}^{R} r^3\, dr$$

$$= 2\pi a \left[\frac{r^4}{4} \right]_{R/2}^{R} = \frac{\pi a}{2} \left[R^4 - \frac{R^4}{16} \right] = \frac{15}{32} \pi a R^4$$

$$= \frac{15}{32} \pi (3.0 \times 10^{11} \text{ A/m}^4)(0.0020 \text{ m})^4 = 7.1 \text{ A}.$$

(Answer)

We want the current in the area between these two radii.

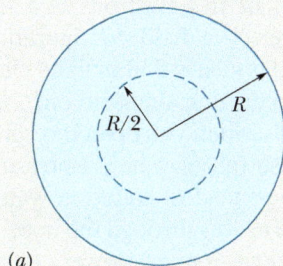

(a)

If the current is nonuniform, we start with a ring that is so thin that we can approximate the current density as being uniform within it.

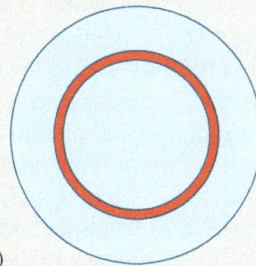

(b)

Its area is the product of the circumference and the width.

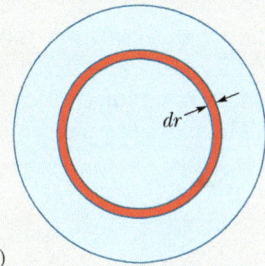

(c)

The current within the ring is the product of the current density and the ring's area.

Our job is to sum the current in all rings from this smallest one ...

(d)

... to this largest one.

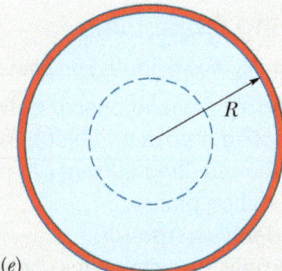

(e)

Figure 26-6 (a) Cross section of a wire of radius R. If the current density is uniform, the current is just the product of the current density and the area. (b)–(e) If the current is nonuniform, we must first find the current through a thin ring and then sum (via integration) the currents in all such rings in the given area.

Sample Problem 26.03   In a current, the conduction electrons move very slowly

What is the drift speed of the conduction electrons in a copper wire with radius $r = 900$ μm when it has a uniform current $i = 17$ mA? Assume that each copper atom contributes one conduction electron to the current and that the current density is uniform across the wire's cross section.

KEY IDEAS

1. The drift speed v_d is related to the current density $\vec{J}$ and the number n of conduction electrons per unit volume according to Eq. 26-7, which we can write as $J = nev_d$.

2. Because the current density is uniform, its magnitude J is related to the given current i and wire size by Eq. 26-5 ($J = i/A$, where A is the cross-sectional area of the wire).

3. Because we assume one conduction electron per atom, the number n of conduction electrons per unit volume is the same as the number of atoms per unit volume.

Calculations: Let us start with the third idea by writing

$$n = \begin{pmatrix} \text{atoms} \\ \text{per unit} \\ \text{volume} \end{pmatrix} = \begin{pmatrix} \text{atoms} \\ \text{per} \\ \text{mole} \end{pmatrix}\begin{pmatrix} \text{moles} \\ \text{per unit} \\ \text{mass} \end{pmatrix}\begin{pmatrix} \text{mass} \\ \text{per unit} \\ \text{volume} \end{pmatrix}.$$

The number of atoms per mole is just Avogadro's number N_A ($= 6.02 \times 10^{23}$ mol^{-1}). Moles per unit mass is the inverse of the mass per mole, which here is the molar mass M of copper. The mass per unit volume is the (mass) density ρ_{mass} of copper. Thus,

$$n = N_A\left(\frac{1}{M}\right)\rho_{mass} = \frac{N_A\rho_{mass}}{M}.$$

Taking copper's molar mass M and density ρ_{mass} from Appendix F, we then have (with some conversions of units)

$$n = \frac{(6.02 \times 10^{23}\text{ mol}^{-1})(8.96 \times 10^3\text{ kg/m}^3)}{63.54 \times 10^{-3}\text{ kg/mol}}$$

$$= 8.49 \times 10^{28}\text{ electrons/m}^3$$

or

$$n = 8.49 \times 10^{28}\text{ m}^{-3}.$$

Next let us combine the first two key ideas by writing

$$\frac{i}{A} = nev_d.$$

Substituting for A with πr^2 ($= 2.54 \times 10^{-6}$ m^2) and solving for v_d, we then find

$$v_d = \frac{i}{ne(\pi r^2)}$$

$$= \frac{17 \times 10^{-3}\text{ A}}{(8.49 \times 10^{28}\text{ m}^{-3})(1.6 \times 10^{-19}\text{ C})(2.54 \times 10^{-6}\text{ m}^2)}$$

$$= 4.9 \times 10^{-7}\text{ m/s,} \qquad \text{(Answer)}$$

which is only 1.8 mm/h, slower than a sluggish snail.

Lights are fast: You may well ask: "If the electrons drift so slowly, why do the room lights turn on so quickly when I throw the switch?" Confusion on this point results from not distinguishing between the drift speed of the electrons and the speed at which *changes* in the electric field configuration travel along wires. This latter speed is nearly that of light; electrons everywhere in the wire begin drifting almost at once, including into the lightbulbs. Similarly, when you open the valve on your garden hose with the hose full of water, a pressure wave travels along the hose at the speed of sound in water. The speed at which the water itself moves through the hose—measured perhaps with a dye marker—is much slower.

WILEY PLUS   Additional examples, video, and practice available at *WileyPLUS*

26-3 RESISTANCE AND RESISTIVITY

Learning Objectives

After reading this module, you should be able to . . .

26.14 Apply the relationship between the potential difference V applied across an object, the object's resistance R, and the resulting current i through the object, between the application points.

26.15 Identify a resistor.

26.16 Apply the relationship between the electric field magnitude E set up at a point in a given material, the material's resistivity ρ, and the resulting current density magnitude J at that point.

26.17 For a uniform electric field set up in a wire, apply the relationship between the electric field magnitude E,

the potential difference V between the two ends, and the wire's length L.

26.18 Apply the relationship between resistivity ρ and conductivity σ.

26.19 Apply the relationship between an object's resistance R, the resistivity of its material ρ, its length L, and its cross-sectional area A.

26.20 Apply the equation that approximately gives a conductor's resistivity ρ as a function of temperature T.

26.21 Sketch a graph of resistivity ρ versus temperature T for a metal.

Key Ideas

● The resistance R of a conductor is defined as

$$R = \frac{V}{i},$$

where V is the potential difference across the conductor and i is the current.

● The resistivity ρ and conductivity σ of a material are related by

$$\rho = \frac{1}{\sigma} = \frac{E}{J},$$

where E is the magnitude of the applied electric field and J is the magnitude of the current density.

● The electric field and current density are related to the resistivity by

$$\vec{E} = \rho\vec{J}.$$

● The resistance R of a conducting wire of length L and uniform cross section is

$$R = \rho\frac{L}{A},$$

where A is the cross-sectional area.

● The resistivity ρ for most materials changes with temperature. For many materials, including metals, the relation between ρ and temperature T is approximated by the equation

$$\rho - \rho_0 = \rho_0\alpha(T - T_0).$$

Here T_0 is a reference temperature, ρ_0 is the resistivity at T_0, and α is the temperature coefficient of resistivity for the material.

Resistance and Resistivity

If we apply the same potential difference between the ends of geometrically similar rods of copper and of glass, very different currents result. The characteristic of the conductor that enters here is its electrical **resistance.** We determine the resistance between any two points of a conductor by applying a potential difference V between those points and measuring the current i that results. The resistance R is then

$$R = \frac{V}{i} \quad \text{(definition of } R\text{)}. \tag{26-8}$$

The SI unit for resistance that follows from Eq. 26-8 is the volt per ampere. This combination occurs so often that we give it a special name, the **ohm** (symbol Ω); that is,

$$1 \text{ ohm} = 1\ \Omega = 1 \text{ volt per ampere}$$
$$= 1 \text{ V/A}. \tag{26-9}$$

A conductor whose function in a circuit is to provide a specified resistance is called a **resistor** (see Fig. 26-7). In a circuit diagram, we represent a resistor and a resistance with the symbol -⋀⋀⋀- . If we write Eq. 26-8 as

$$i = \frac{V}{R},$$

we see that, for a given V, the greater the resistance, the smaller the current.

The resistance of a conductor depends on the manner in which the potential difference is applied to it. Figure 26-8, for example, shows a given potential difference applied in two different ways to the same conductor. As the current density streamlines suggest, the currents in the two cases—hence the measured resistances—will be different. Unless otherwise stated, we shall assume that any given potential difference is applied as in Fig. 26-8b.

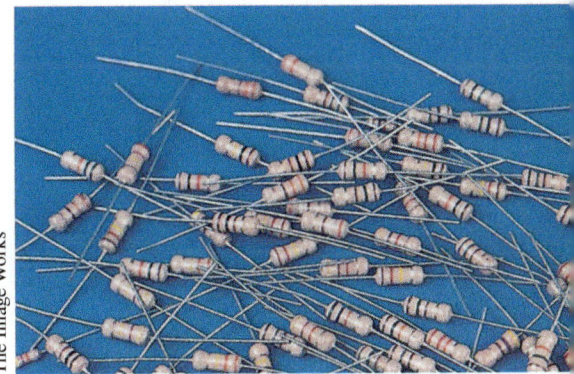

The Image Works

Figure 26-7 An assortment of resistors. The circular bands are color-coding marks that identify the value of the resistance.

(a)   (b)

Figure 26-8 Two ways of applying a potential difference to a conducting rod. The gray connectors are assumed to have negligible resistance. When they are arranged as in (a) in a small region at each rod end, the measured resistance is larger than when they are arranged as in (b) to cover the entire rod end.

Table 26-1 Resistivities of Some Materials at Room Temperature (20°C)

Material	Resistivity, ρ ($\Omega \cdot$m)	Temperature Coefficient of Resistivity, α (K^{-1})
Typical Metals		
Silver	1.62×10^{-8}	4.1×10^{-3}
Copper	1.69×10^{-8}	4.3×10^{-3}
Gold	2.35×10^{-8}	4.0×10^{-3}
Aluminum	2.75×10^{-8}	4.4×10^{-3}
Manganin^a	4.82×10^{-8}	0.002×10^{-3}
Tungsten	5.25×10^{-8}	4.5×10^{-3}
Iron	9.68×10^{-8}	6.5×10^{-3}
Platinum	10.6×10^{-8}	3.9×10^{-3}
Typical Semiconductors		
Silicon, pure	2.5×10^3	-70×10^{-3}
Silicon, n-type^b	8.7×10^{-4}	
Silicon, p-type^c	2.8×10^{-3}	
Typical Insulators		
Glass	$10^{10} - 10^{14}$	
Fused quartz	$\sim 10^{16}$	

^{a}An alloy specifically designed to have a small value of α.
^{b}Pure silicon doped with phosphorus impurities to a charge carrier density of 10^{23} m^{-3}.
^{c}Pure silicon doped with aluminum impurities to a charge carrier density of 10^{23} m^{-3}.

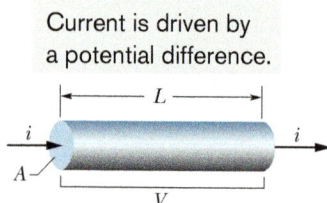

As we have done several times in other connections, we often wish to take a general view and deal not with particular objects but with materials. Here we do so by focusing not on the potential difference V across a particular resistor but on the electric field $\vec{E}$ at a point in a resistive material. Instead of dealing with the current i through the resistor, we deal with the current density $\vec{J}$ at the point in question. Instead of the resistance R of an object, we deal with the **resistivity** ρ of the *material:*

$$\rho = \frac{E}{J} \quad \text{(definition of } \rho). \tag{26-10}$$

(Compare this equation with Eq. 26-8.)

If we combine the SI units of E and J according to Eq. 26-10, we get, for the unit of ρ, the ohm-meter ($\Omega \cdot$m):

$$\frac{\text{unit } (E)}{\text{unit } (J)} = \frac{\text{V/m}}{\text{A/m}^2} = \frac{\text{V}}{\text{A}} \text{ m} = \Omega \cdot \text{m}.$$

(Do not confuse the *ohm-meter*, the unit of resistivity, with the *ohmmeter*, which is an instrument that measures resistance.) Table 26-1 lists the resistivities of some materials.

We can write Eq. 26-10 in vector form as

$$\vec{E} = \rho \vec{J}. \tag{26-11}$$

Equations 26-10 and 26-11 hold only for *isotropic* materials—materials whose electrical properties are the same in all directions.

We often speak of the **conductivity** σ of a material. This is simply the reciprocal of its resistivity, so

$$\sigma = \frac{1}{\rho} \quad \text{(definition of } \sigma). \tag{26-12}$$

The SI unit of conductivity is the reciprocal ohm-meter, $(\Omega \cdot \text{m})^{-1}$. The unit name mhos per meter is sometimes used (mho is ohm backwards). The definition of σ allows us to write Eq. 26-11 in the alternative form

$$\vec{J} = \sigma \vec{E}. \tag{26-13}$$

Calculating Resistance from Resistivity

We have just made an important distinction:

Resistance is a property of an object. Resistivity is a property of a material.

If we know the resistivity of a substance such as copper, we can calculate the resistance of a length of wire made of that substance. Let A be the cross-sectional area of the wire, let L be its length, and let a potential difference V exist between its ends (Fig. 26-9). If the streamlines representing the current density are uniform throughout the wire, the electric field and the current density will be constant for all points within the wire and, from Eqs. 24-42 and 26-5, will have the values

$$E = V/L \quad \text{and} \quad J = i/A. \tag{26-14}$$

We can then combine Eqs. 26-10 and 26-14 to write

$$\rho = \frac{E}{J} = \frac{V/L}{i/A}. \tag{26-15}$$

Current is driven by a potential difference.

Figure 26-9 A potential difference V is applied between the ends of a wire of length L and cross section A, establishing a current i.

However, V/i is the resistance R, which allows us to recast Eq. 26-15 as

$$R = \rho \frac{L}{A}. \qquad (26\text{-}16)$$

Equation 26-16 can be applied only to a homogeneous isotropic conductor of uniform cross section, with the potential difference applied as in Fig. 26-8b.

The macroscopic quantities V, i, and R are of greatest interest when we are making electrical measurements on specific conductors. They are the quantities that we read directly on meters. We turn to the microscopic quantities E, J, and ρ when we are interested in the fundamental electrical properties of materials.

✓ Checkpoint 3

The figure here shows three cylindrical copper conductors along with their face areas and lengths. Rank them according to the current through them, greatest first, when the same potential difference V is placed across their lengths.

Variation with Temperature

The values of most physical properties vary with temperature, and resistivity is no exception. Figure 26-10, for example, shows the variation of this property for copper over a wide temperature range. The relation between temperature and resistivity for copper—and for metals in general—is fairly linear over a rather broad temperature range. For such linear relations we can write an empirical approximation that is good enough for most engineering purposes:

$$\rho - \rho_0 = \rho_0 \alpha (T - T_0). \qquad (26\text{-}17)$$

Here T_0 is a selected reference temperature and ρ_0 is the resistivity at that temperature. Usually $T_0 = 293$ K (room temperature), for which $\rho_0 = 1.69 \times 10^{-8}$ $\Omega \cdot$m for copper.

Because temperature enters Eq. 26-17 only as a difference, it does not matter whether you use the Celsius or Kelvin scale in that equation because the sizes of degrees on these scales are identical. The quantity α in Eq. 26-17, called the *temperature coefficient of resistivity,* is chosen so that the equation gives good agreement with experiment for temperatures in the chosen range. Some values of α for metals are listed in Table 26-1.

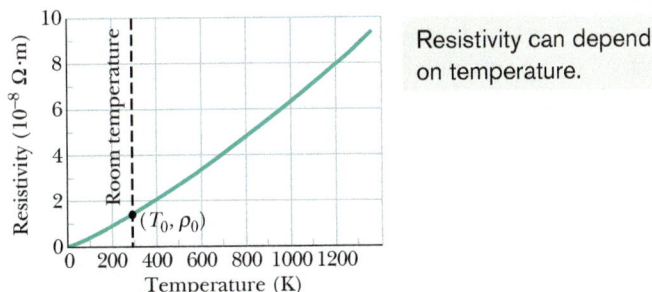

Resistivity can depend on temperature.

Figure 26-10 The resistivity of copper as a function of temperature. The dot on the curve marks a convenient reference point at temperature $T_0 = 293$ K and resistivity $\rho_0 = 1.69 \times 10^{-8}$ $\Omega \cdot$m.

Sample Problem 26.04 **A material has resistivity, a block of the material has resistance**

A rectangular block of iron has dimensions 1.2 cm × 1.2 cm × 15 cm. A potential difference is to be applied to the block between parallel sides and in such a way that those sides are equipotential surfaces (as in Fig. 26-8b). What is the resistance of the block if the two parallel sides are (1) the square ends (with dimensions 1.2 cm × 1.2 cm) and (2) two rectangular sides (with dimensions 1.2 cm × 15 cm)?

KEY IDEA

The resistance R of an object depends on how the electric potential is applied to the object. In particular, it depends on the ratio L/A, according to Eq. 26-16 ($R = \rho L/A$), where A is the area of the surfaces to which the potential difference is applied and L is the distance between those surfaces.

Calculations: For arrangement 1, we have $L = 15$ cm = 0.15 m and

$$A = (1.2 \text{ cm})^2 = 1.44 \times 10^{-4} \text{ m}^2.$$

Substituting into Eq. 26-16 with the resistivity ρ from Table 26-1, we then find that for arrangement 1,

$$R = \frac{\rho L}{A} = \frac{(9.68 \times 10^{-8} \, \Omega \cdot \text{m})(0.15 \text{ m})}{1.44 \times 10^{-4} \text{ m}^2}$$
$$= 1.0 \times 10^{-4} \, \Omega = 100 \, \mu\Omega. \qquad \text{(Answer)}$$

Similarly, for arrangement 2, with distance $L = 1.2$ cm and area $A = (1.2 \text{ cm})(15 \text{ cm})$, we obtain

$$R = \frac{\rho L}{A} = \frac{(9.68 \times 10^{-8} \, \Omega \cdot \text{m})(1.2 \times 10^{-2} \text{ m})}{1.80 \times 10^{-3} \text{ m}^2}$$
$$= 6.5 \times 10^{-7} \, \Omega = 0.65 \, \mu\Omega. \qquad \text{(Answer)}$$

WILEY
PLUS Additional examples, video, and practice available at *WileyPLUS*

26-4 OHM'S LAW

Learning Objectives

After reading this module, you should be able to . . .

26.22 Distinguish between an *object* that obeys Ohm's law and one that does not.

26.23 Distinguish between a *material* that obeys Ohm's law and one that does not.

26.24 Describe the general motion of a conduction electron in a current.

26.25 For the conduction electrons in a conductor, explain the relationship between the mean free time τ, the effective speed, and the thermal (random) motion.

26.26 Apply the relationship between resistivity ρ, number density n of conduction electrons, and the mean free time τ of the electrons.

Key Ideas

● A given device (conductor, resistor, or any other electrical device) obeys Ohm's law if its resistance R ($= V/i$) is independent of the applied potential difference V.

● A given material obeys Ohm's law if its resistivity ρ ($= E/J$) is independent of the magnitude and direction of the applied electric field $\vec{E}$.

● The assumption that the conduction electrons in a metal are free to move like the molecules in a gas leads to an

expression for the resistivity of a metal:

$$\rho = \frac{m}{e^2 n \tau}.$$

Here n is the number of free electrons per unit volume and τ is the mean time between the collisions of an electron with the atoms of the metal.

● Metals obey Ohm's law because the mean free time τ is approximately independent of the magnitude E of any electric field applied to a metal.

Ohm's Law

As we just discussed, a resistor is a conductor with a specified resistance. It has that same resistance no matter what the magnitude and direction (*polarity*) of the applied potential difference are. Other conducting devices, however, might have resistances that change with the applied potential difference.

Figure 26-11*a* shows how to distinguish such devices. A potential difference *V* is applied across the device being tested, and the resulting current *i* through the device is measured as *V* is varied in both magnitude and polarity. The polarity of *V* is arbitrarily taken to be positive when the left terminal of the device is at a higher potential than the right terminal. The direction of the resulting current (from left to right) is arbitrarily assigned a plus sign. The reverse polarity of *V* (with the right terminal at a higher potential) is then negative; the current it causes is assigned a minus sign.

Figure 26-11*b* is a plot of *i* versus *V* for one device. This plot is a straight line passing through the origin, so the ratio *i/V* (which is the slope of the straight line) is the same for all values of *V*. This means that the resistance $R = V/i$ of the device is independent of the magnitude and polarity of the applied potential difference *V*.

Figure 26-11*c* is a plot for another conducting device. Current can exist in this device only when the polarity of *V* is positive and the applied potential difference is more than about 1.5 V. When current does exist, the relation between *i* and *V* is not linear; it depends on the value of the applied potential difference *V*.

We distinguish between the two types of device by saying that one obeys Ohm's law and the other does not.

Ohm's law is an assertion that the current through a device is *always* directly proportional to the potential difference applied to the device.

(This assertion is correct only in certain situations; still, for historical reasons, the term "law" is used.) The device of Fig. 26-11*b*—which turns out to be a 1000 Ω resistor—obeys Ohm's law. The device of Fig. 26-11*c*—which is called a *pn* junction diode—does not.

A conducting device obeys Ohm's law when the resistance of the device is independent of the magnitude and polarity of the applied potential difference.

It is often contended that $V = iR$ is a statement of Ohm's law. That is not true! This equation is the defining equation for resistance, and it applies to all conducting devices, whether they obey Ohm's law or not. If we measure the potential difference *V* across, and the current *i* through, any device, even a *pn* junction diode, we can find its resistance *at that value of V* as $R = V/i$. The essence of Ohm's law, however, is that a plot of *i* versus *V* is linear; that is, *R* is independent of *V*. We can generalize this for conducting materials by using Eq. 26-11 ($\vec{E} = \rho\vec{J}$):

A conducting material obeys Ohm's law when the resistivity of the material is independent of the magnitude and direction of the applied electric field.

All homogeneous materials, whether they are conductors like copper or semiconductors like pure silicon or silicon containing special impurities, obey Ohm's law within some range of values of the electric field. If the field is too strong, however, there are departures from Ohm's law in all cases.

Figure 26-11 (*a*) A potential difference *V* is applied to the terminals of a device, establishing a current *i*. (*b*) A plot of current *i* versus applied potential difference *V* when the device is a 1000 Ω resistor. (*c*) A plot when the device is a semiconducting *pn* junction diode.

✓ **Checkpoint 4**

The following table gives the current *i* (in amperes) through two devices for several values of potential difference *V* (in volts). From these data, determine which device does not obey Ohm's law.

Device 1		Device 2	
V	i	V	i
2.00	4.50	2.00	1.50
3.00	6.75	3.00	2.20
4.00	9.00	4.00	2.80

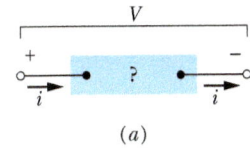

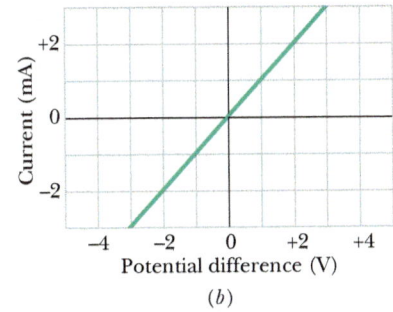

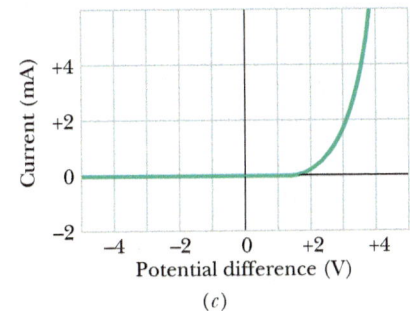

A Microscopic View of Ohm's Law

To find out *why* particular materials obey Ohm's law, we must look into the details of the conduction process at the atomic level. Here we consider only conduction in metals, such as copper. We base our analysis on the *free-electron model*, in which we assume that the conduction electrons in the metal are free to move throughout the volume of a sample, like the molecules of a gas in a closed container. We also assume that the electrons collide not with one another but only with atoms of the metal.

According to classical physics, the electrons should have a Maxwellian speed distribution somewhat like that of the molecules in a gas (Module 19-6), and thus the average electron speed should depend on the temperature. The motions of electrons are, however, governed not by the laws of classical physics but by those of quantum physics. As it turns out, an assumption that is much closer to the quantum reality is that conduction electrons in a metal move with a single effective speed v_{eff}, and this speed is essentially independent of the temperature. For copper, $v_{eff} \approx 1.6 \times 10^6$ m/s.

When we apply an electric field to a metal sample, the electrons modify their random motions slightly and drift very slowly—in a direction opposite that of the field—with an average drift speed v_d. The drift speed in a typical metallic conductor is about 5×10^{-7} m/s, less than the effective speed (1.6×10^6 m/s) by many orders of magnitude. Figure 26-12 suggests the relation between these two speeds. The gray lines show a possible random path for an electron in the absence of an applied field; the electron proceeds from A to B, making six collisions along the way. The green lines show how the same events *might* occur when an electric field $\vec{E}$ is applied. We see that the electron drifts steadily to the right, ending at B' rather than at B. Figure 26-12 was drawn with the assumption that $v_d \approx 0.02 v_{eff}$. However, because the actual value is more like $v_d \approx (10^{-13}) v_{eff}$, the drift displayed in the figure is greatly exaggerated.

The motion of conduction electrons in an electric field $\vec{E}$ is thus a combination of the motion due to random collisions and that due to $\vec{E}$. When we consider all the free electrons, their random motions average to zero and make no contribution to the drift speed. Thus, the drift speed is due only to the effect of the electric field on the electrons.

If an electron of mass m is placed in an electric field of magnitude E, the electron will experience an acceleration given by Newton's second law:

$$a = \frac{F}{m} = \frac{eE}{m}. \tag{26-18}$$

After a typical collision, each electron will—so to speak—completely lose its memory of its previous drift velocity, starting fresh and moving off in a random direction. In the average time τ between collisions, the average electron will acquire a drift speed of $v_d = a\tau$. Moreover, if we measure the drift speeds of all the electrons at any instant, we will find that their average drift speed is also $a\tau$. Thus, at any instant, on average, the electrons will have drift speed $v_d = a\tau$. Then Eq. 26-18 gives us

$$v_d = a\tau = \frac{eE\tau}{m}. \tag{26-19}$$

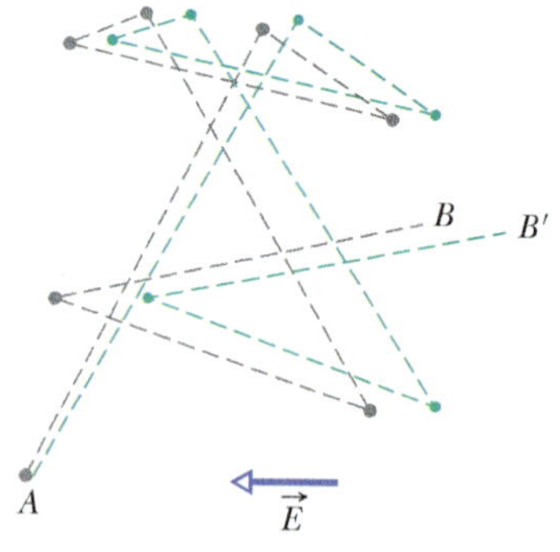

Figure 26-12 The gray lines show an electron moving from A to B, making six collisions en route. The green lines show what the electron's path might be in the presence of an applied electric field $\vec{E}$. Note the steady drift in the direction of $-\vec{E}$. (Actually, the green lines should be slightly curved, to represent the parabolic paths followed by the electrons between collisions, under the influence of an electric field.)

Combining this result with Eq. 26-7 ($\vec{J} = ne\vec{v}_d$), in magnitude form, yields

$$v_d = \frac{J}{ne} = \frac{eE\tau}{m}, \tag{26-20}$$

which we can write as

$$E = \left(\frac{m}{e^2 n\tau}\right) J. \tag{26-21}$$

Comparing this with Eq. 26-11 ($\vec{E} = \rho\vec{J}$), in magnitude form, leads to

$$\rho = \frac{m}{e^2 n\tau}. \tag{26-22}$$

Equation 26-22 may be taken as a statement that metals obey Ohm's law if we can show that, for metals, their resistivity ρ is a constant, independent of the strength of the applied electric field $\vec{E}$. Let's consider the quantities in Eq. 26-22. We can reasonably assume that n, the number of conduction electrons per volume, is independent of the field, and m and e are constants. Thus, we only need to convince ourselves that τ, the average time (or *mean free time*) between collisions, is a constant, independent of the strength of the applied electric field. Indeed, τ can be considered to be a constant because the drift speed v_d caused by the field is so much smaller than the effective speed v_{eff} that the electron speed— and thus τ—is hardly affected by the field. Thus, because the right side of Eq. 26-22 is independent of the field magnitude, metals obey Ohm's law.

Sample Problem 26.05 Mean free time and mean free distance

(a) What is the mean free time τ between collisions for the conduction electrons in copper?

KEY IDEAS

The mean free time τ of copper is approximately constant, and in particular does not depend on any electric field that might be applied to a sample of the copper. Thus, we need not consider any particular value of applied electric field. However, because the resistivity ρ displayed by copper under an electric field depends on τ, we can find the mean free time τ from Eq. 26-22 ($\rho = m/e^2 n\tau$).

Calculations: That equation gives us

$$\tau = \frac{m}{ne^2\rho}. \tag{26-23}$$

The number of conduction electrons per unit volume in copper is 8.49×10^{28} m^{-3}. We take the value of ρ from Table 26-1. The denominator then becomes

$$(8.49 \times 10^{28} \text{ m}^{-3})(1.6 \times 10^{-19} \text{ C})^2(1.69 \times 10^{-8} \, \Omega\cdot\text{m})$$
$$= 3.67 \times 10^{-17} \text{ C}^2\cdot\Omega/\text{m}^2 = 3.67 \times 10^{-17} \text{ kg/s},$$

where we converted units as

$$\frac{\text{C}^2\cdot\Omega}{\text{m}^2} = \frac{\text{C}^2\cdot\text{V}}{\text{m}^2\cdot\text{A}} = \frac{\text{C}^2\cdot\text{J/C}}{\text{m}^2\cdot\text{C/s}} = \frac{\text{kg}\cdot\text{m}^2/\text{s}^2}{\text{m}^2/\text{s}} = \frac{\text{kg}}{\text{s}}.$$

Using these results and substituting for the electron mass m, we then have

$$\tau = \frac{9.1 \times 10^{-31} \text{ kg}}{3.67 \times 10^{-17} \text{ kg/s}} = 2.5 \times 10^{-14} \text{ s}. \quad \text{(Answer)}$$

(b) The mean free path λ of the conduction electrons in a conductor is the average distance traveled by an electron between collisions. (This definition parallels that in Module 19-5 for the mean free path of molecules in a gas.) What is λ for the conduction electrons in copper, assuming that their effective speed v_{eff} is 1.6×10^6 m/s?

KEY IDEA

The distance d any particle travels in a certain time t at a constant speed v is $d = vt$.

Calculation: For the electrons in copper, this gives us

$$\lambda = v_{\text{eff}}\tau \tag{26-24}$$
$$= (1.6 \times 10^6 \text{ m/s})(2.5 \times 10^{-14} \text{ s})$$
$$= 4.0 \times 10^{-8} \text{ m} = 40 \text{ nm}. \quad \text{(Answer)}$$

This is about 150 times the distance between nearest-neighbor atoms in a copper lattice. Thus, on the average, each conduction electron passes many copper atoms before finally hitting one.

WILEY PLUS Additional examples, video, and practice available at *WileyPLUS*

26-5 POWER, SEMICONDUCTORS, SUPERCONDUCTORS

Learning Objectives

After reading this module, you should be able to . . .

26.27 Explain how conduction electrons in a circuit lose energy in a resistive device.

26.28 Identify that power is the rate at which energy is transferred from one type to another.

26.29 For a resistive device, apply the relationships between power P, current i, voltage V, and resistance R.

26.30 For a battery, apply the relationship between power P, current i, and potential difference V.

26.31 Apply the conservation of energy to a circuit with a battery and a resistive device to relate the energy transfers in the circuit.

26.32 Distinguish conductors, semiconductors, and superconductors.

Key Ideas

● The power P, or rate of energy transfer, in an electrical device across which a potential difference V is maintained is

$$P = iV.$$

● If the device is a resistor, the power can also be written as

$$P = i^2 R = \frac{V^2}{R}.$$

● In a resistor, electric potential energy is converted to internal thermal energy via collisions between charge carriers and atoms.

● Semiconductors are materials that have few conduction electrons but can become conductors when they are doped with other atoms that contribute charge carriers.

● Superconductors are materials that lose all electrical resistance. Most such materials require very low temperatures, but some become superconducting at temperatures as high as room temperature.

Power in Electric Circuits

Figure 26-13 shows a circuit consisting of a battery B that is connected by wires, which we assume have negligible resistance, to an unspecified conducting device. The device might be a resistor, a storage battery (a rechargeable battery), a motor, or some other electrical device. The battery maintains a potential difference of magnitude V across its own terminals and thus (because of the wires) across the terminals of the unspecified device, with a greater potential at terminal a of the device than at terminal b.

Because there is an external conducting path between the two terminals of the battery, and because the potential differences set up by the battery are maintained, a steady current i is produced in the circuit, directed from terminal a to terminal b. The amount of charge dq that moves between those terminals in time interval dt is equal to $i\,dt$. This charge dq moves through a decrease in potential of magnitude V, and thus its electric potential energy decreases in magnitude by the amount

$$dU = dq\, V = i\, dt\, V. \qquad (26\text{-}25)$$

The principle of conservation of energy tells us that the decrease in electric potential energy from a to b is accompanied by a transfer of energy to some other form. The power P associated with that transfer is the rate of transfer dU/dt, which is given by Eq. 26-25 as

$$P = iV \quad \text{(rate of electrical energy transfer).} \qquad (26\text{-}26)$$

Moreover, this power P is also the rate at which energy is transferred from the battery to the unspecified device. If that device is a motor connected to a mechanical load, the energy is transferred as work done on the load. If the device is a storage battery that is being charged, the energy is transferred to stored chemical energy in the storage battery. If the device is a resistor, the energy is transferred to internal thermal energy, tending to increase the resistor's temperature.

The battery at the left supplies energy to the conduction electrons that form the current.

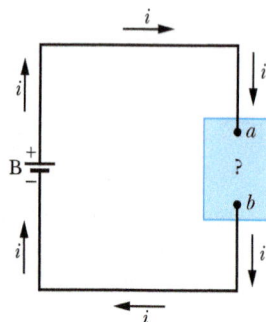

Figure 26-13 A battery B sets up a current i in a circuit containing an unspecified conducting device.

The unit of power that follows from Eq. 26-26 is the volt-ampere (V·A). We can write it as

$$1 \text{ V} \cdot \text{A} = \left(1 \frac{\text{J}}{\text{C}}\right)\left(1 \frac{\text{C}}{\text{s}}\right) = 1 \frac{\text{J}}{\text{s}} = 1 \text{ W}.$$

As an electron moves through a resistor at constant drift speed, its average kinetic energy remains constant and its lost electric potential energy appears as thermal energy in the resistor and the surroundings. On a microscopic scale this energy transfer is due to collisions between the electron and the molecules of the resistor, which leads to an increase in the temperature of the resistor lattice. The mechanical energy thus transferred to thermal energy is *dissipated* (lost) because the transfer cannot be reversed.

For a resistor or some other device with resistance R, we can combine Eqs. 26-8 ($R = V/i$) and 26-26 to obtain, for the rate of electrical energy dissipation due to a resistance, either

$$P = i^2R \quad \text{(resistive dissipation)} \tag{26-27}$$

or

$$P = \frac{V^2}{R} \quad \text{(resistive dissipation).} \tag{26-28}$$

Caution: We must be careful to distinguish these two equations from Eq. 26-26: $P = iV$ applies to electrical energy transfers of all kinds; $P = i^2R$ and $P = V^2/R$ apply only to the transfer of electric potential energy to thermal energy in a device with resistance.

✓ Checkpoint 5

A potential difference V is connected across a device with resistance R, causing current i through the device. Rank the following variations according to the change in the rate at which electrical energy is converted to thermal energy due to the resistance, greatest change first: (a) V is doubled with R unchanged, (b) i is doubled with R unchanged, (c) R is doubled with V unchanged, (d) R is doubled with i unchanged.

Sample Problem 26.06 Rate of energy dissipation in a wire carrying current

You are given a length of uniform heating wire made of a nickel–chromium–iron alloy called Nichrome; it has a resistance R of 72 Ω. At what rate is energy dissipated in each of the following situations? (1) A potential difference of 120 V is applied across the full length of the wire. (2) The wire is cut in half, and a potential difference of 120 V is applied across the length of each half.

KEY IDEA

Current in a resistive material produces a transfer of mechanical energy to thermal energy; the rate of transfer (dissipation) is given by Eqs. 26-26 to 26-28.

Calculations: Because we know the potential V and resistance R, we use Eq. 26-28, which yields, for situation 1,

$$P = \frac{V^2}{R} = \frac{(120 \text{ V})^2}{72 \text{ }\Omega} = 200 \text{ W}. \quad \text{(Answer)}$$

In situation 2, the resistance of each half of the wire is (72 Ω)/2, or 36 Ω. Thus, the dissipation rate for each half is

$$P' = \frac{(120 \text{ V})^2}{36 \text{ }\Omega} = 400 \text{ W},$$

and that for the two halves is

$$P = 2P' = 800 \text{ W}. \quad \text{(Answer)}$$

This is four times the dissipation rate of the full length of wire. Thus, you might conclude that you could buy a heating coil, cut it in half, and reconnect it to obtain four times the heat output. Why is this unwise? (What would happen to the amount of current in the coil?)

Semiconductors

Semiconducting devices are at the heart of the microelectronic revolution that ushered in the information age. Table 26-2 compares the properties of silicon—a typical semiconductor—and copper—a typical metallic conductor. We see that silicon has many fewer charge carriers, a much higher resistivity, and a temperature coefficient of resistivity that is both large and negative. Thus, although the resistivity of copper increases with increasing temperature, that of pure silicon decreases.

Pure silicon has such a high resistivity that it is effectively an insulator and thus not of much direct use in microelectronic circuits. However, its resistivity can be greatly reduced in a controlled way by adding minute amounts of specific "impurity" atoms in a process called *doping*. Table 26-1 gives typical values of resistivity for silicon before and after doping with two different impurities.

We can roughly explain the differences in resistivity (and thus in conductivity) between semiconductors, insulators, and metallic conductors in terms of the energies of their electrons. (We need quantum physics to explain in more detail.) In a metallic conductor such as copper wire, most of the electrons are firmly locked in place within the atoms; much energy would be required to free them so they could move and participate in an electric current. However, there are also some electrons that, roughly speaking, are only loosely held in place and that require only little energy to become free. Thermal energy can supply that energy, as can an electric field applied across the conductor. The field would not only free these loosely held electrons but would also propel them along the wire; thus, the field would drive a current through the conductor.

In an insulator, significantly greater energy is required to free electrons so they can move through the material. Thermal energy cannot supply enough energy, and neither can any reasonable electric field applied to the insulator. Thus, no electrons are available to move through the insulator, and hence no current occurs even with an applied electric field.

A semiconductor is like an insulator *except* that the energy required to free some electrons is not quite so great. More important, doping can supply electrons or positive charge carriers that are very loosely held within the material and thus are easy to get moving. Moreover, by controlling the doping of a semiconductor, we can control the density of charge carriers that can participate in a current and thereby can control some of its electrical properties. Most semiconducting devices, such as transistors and junction diodes, are fabricated by the selective doping of different regions of the silicon with impurity atoms of different kinds.

Let us now look again at Eq. 26-22 for the resistivity of a conductor:

$$\rho = \frac{m}{e^2 n \tau}, \tag{26-29}$$

where n is the number of charge carriers per unit volume and τ is the mean time between collisions of the charge carriers. The equation also applies to semiconductors. Let's consider how n and τ change as the temperature is increased.

In a conductor, n is large but very nearly constant with any change in temperature. The increase of resistivity with temperature for metals (Fig. 26-10) is due to an increase in the collision rate of the charge carriers, which shows up in Eq. 26-29 as a decrease in τ, the mean time between collisions.

Table 26-2 Some Electrical Properties of Copper and Silicon

Property	Copper	Silicon
Type of material	Metal	Semiconductor
Charge carrier density, m^{-3}	8.49×10^{28}	1×10^{16}
Resistivity, $\Omega \cdot m$	1.69×10^{-8}	2.5×10^{3}
Temperature coefficient of resistivity, K^{-1}	$+4.3 \times 10^{-3}$	-70×10^{-3}

In a semiconductor, n is small but increases very rapidly with temperature as the increased thermal agitation makes more charge carriers available. This causes a *decrease* of resistivity with increasing temperature, as indicated by the negative temperature coefficient of resistivity for silicon in Table 26-2. The same increase in collision rate that we noted for metals also occurs for semiconductors, but its effect is swamped by the rapid increase in the number of charge carriers.

Superconductors

In 1911, Dutch physicist Kamerlingh Onnes discovered that the resistivity of mercury absolutely disappears at temperatures below about 4 K (Fig. 26-14). This phenomenon of **superconductivity** is of vast potential importance in technology because it means that charge can flow through a superconducting conductor without losing its energy to thermal energy. Currents created in a superconducting ring, for example, have persisted for several years without loss; the electrons making up the current require a force and a source of energy at start-up time but not thereafter.

Prior to 1986, the technological development of superconductivity was throttled by the cost of producing the extremely low temperatures required to achieve the effect. In 1986, however, new ceramic materials were discovered that become superconducting at considerably higher (and thus cheaper to produce) temperatures. Practical application of superconducting devices at room temperature may eventually become commonplace.

Superconductivity is a phenomenon much different from conductivity. In fact, the best of the normal conductors, such as silver and copper, cannot become superconducting at any temperature, and the new ceramic superconductors are actually good insulators when they are not at low enough temperatures to be in a superconducting state.

One explanation for superconductivity is that the electrons that make up the current move in coordinated pairs. One of the electrons in a pair may electrically distort the molecular structure of the superconducting material as it moves through, creating nearby a short-lived concentration of positive charge. The other electron in the pair may then be attracted toward this positive charge. According to the theory, such coordination between electrons would prevent them from colliding with the molecules of the material and thus would eliminate electrical resistance. The theory worked well to explain the pre-1986, lower temperature superconductors, but new theories appear to be needed for the newer, higher temperature superconductors.

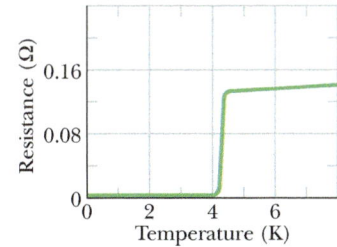

Figure 26-14 The resistance of mercury drops to zero at a temperature of about 4 K.

Courtesy Shoji Tonaka/International Superconductivity Technology Center, Tokyo, Japan

A disk-shaped magnet is levitated above a superconducting material that has been cooled by liquid nitrogen. The goldfish is along for the ride.

Review & Summary

Current An **electric current** i in a conductor is defined by

$$i = \frac{dq}{dt}. \tag{26-1}$$

Here dq is the amount of (positive) charge that passes in time dt through a hypothetical surface that cuts across the conductor. By convention, the direction of electric current is taken as the direction in which positive charge carriers would move. The SI unit of electric current is the **ampere** (A): 1 A = 1 C/s.

Current Density Current (a scalar) is related to **current density** $\vec{J}$ (a vector) by

$$i = \int \vec{J} \cdot d\vec{A}, \tag{26-4}$$

where $d\vec{A}$ is a vector perpendicular to a surface element of area dA and the integral is taken over any surface cutting across the conductor. $\vec{J}$ has the same direction as the velocity of the moving charges if they are positive and the opposite direction if they are negative.

Drift Speed of the Charge Carriers When an electric field $\vec{E}$ is established in a conductor, the charge carriers (assumed positive) acquire a **drift speed** v_d in the direction of $\vec{E}$; the velocity $\vec{v}_d$ is related to the current density by

$$\vec{J} = (ne)\vec{v}_d, \tag{26-7}$$

where ne is the *carrier charge density*.

Resistance of a Conductor The **resistance** R of a conductor is defined as

$$R = \frac{V}{i} \quad \text{(definition of } R\text{),} \tag{26-8}$$

where V is the potential difference across the conductor and i is the current. The SI unit of resistance is the **ohm** (Ω): 1 Ω = 1 V/A. Similar equations define the **resistivity** ρ and **conductivity** σ of a material:

$$\rho = \frac{1}{\sigma} = \frac{E}{J} \quad \text{(definitions of } \rho \text{ and } \sigma\text{),} \tag{26-12, 26-10}$$

where E is the magnitude of the applied electric field. The SI unit of resistivity is the ohm-meter ($\Omega \cdot m$). Equation 26-10 corresponds to the vector equation

$$\vec{E} = \rho \vec{J}. \qquad (26\text{-}11)$$

The resistance R of a conducting wire of length L and uniform cross section is

$$R = \rho \frac{L}{A}, \qquad (26\text{-}16)$$

where A is the cross-sectional area.

Change of ρ with Temperature   The resistivity ρ for most materials changes with temperature. For many materials, including metals, the relation between ρ and temperature T is approximated by the equation

$$\rho - \rho_0 = \rho_0 \alpha (T - T_0). \qquad (26\text{-}17)$$

Here T_0 is a reference temperature, ρ_0 is the resistivity at T_0, and α is the temperature coefficient of resistivity for the material.

Ohm's Law   A given device (conductor, resistor, or any other electrical device) obeys *Ohm's law* if its resistance R, defined by Eq. 26-8 as V/i, is independent of the applied potential difference V. A given *material* obeys Ohm's law if its resistivity, defined by Eq. 26-10, is independent of the magnitude and direction of the applied electric field $\vec{E}$.

Resistivity of a Metal   By assuming that the conduction electrons in a metal are free to move like the molecules of a gas, it is

possible to derive an expression for the resistivity of a metal:

$$\rho = \frac{m}{e^2 n \tau}. \qquad (26\text{-}22)$$

Here n is the number of free electrons per unit volume and τ is the mean time between the collisions of an electron with the atoms of the metal. We can explain why metals obey Ohm's law by pointing out that τ is essentially independent of the magnitude E of any electric field applied to a metal.

Power   The power P, or rate of energy transfer, in an electrical device across which a potential difference V is maintained is

$$P = iV \quad \text{(rate of electrical energy transfer).} \qquad (26\text{-}26)$$

Resistive Dissipation   If the device is a resistor, we can write Eq. 26-26 as

$$P = i^2 R = \frac{V^2}{R} \quad \text{(resistive dissipation).} \qquad (26\text{-}27, 26\text{-}28)$$

In a resistor, electric potential energy is converted to internal thermal energy via collisions between charge carriers and atoms.

Semiconductors   *Semiconductors* are materials that have few conduction electrons but can become conductors when they are *doped* with other atoms that contribute charge carriers.

Superconductors   *Superconductors* are materials that lose all electrical resistance at low temperatures. Some materials are superconducting at surprisingly high temperatures.

Questions

1   Figure 26-15 shows cross sections through three long conductors of the same length and material, with square cross sections of edge lengths as shown. Conductor B fits snugly within conductor A, and conductor C fits snugly within conductor B. Rank the following according to their end-to-end resistances, greatest first: the individual conductors and the combinations of $A + B$ (B inside A), $B + C$ (C inside B), and $A + B + C$ (B inside A inside C).

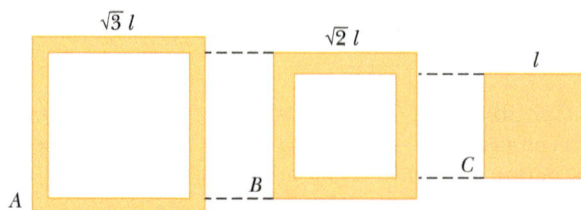

Figure 26-15   Question 1.

2   Figure 26-16 shows cross sections through three wires of identical length and material; the sides are given in millimeters. Rank the wires according to their resistance (measured end to end along each wire's length), greatest first.

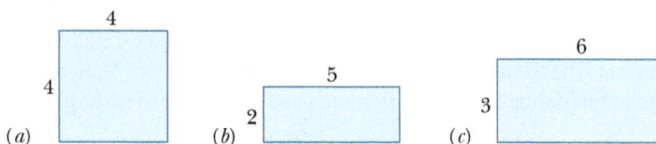

Figure 26-16   Question 2.

3   Figure 26-17 shows a rectangular solid conductor of edge lengths $L, 2L,$ and $3L$. A potential difference V is to be applied uniformly between pairs of opposite faces of the conductor as in Fig. 26-8b. (The potential difference is applied between the entire face on one side and the entire face on the other side.) First V is applied between the left–right faces, then between the top–bottom faces, and then between the front–back faces. Rank those pairs, greatest first, according to the following (within the conductor): (a) the magnitude of the electric field, (b) the current density, (c) the current, and (d) the drift speed of the electrons.

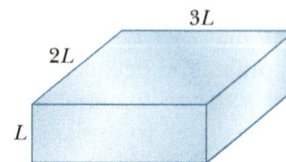

Figure 26-17   Question 3.

4   Figure 26-18 shows plots of the current i through a certain cross section of a wire over four different time periods. Rank the periods according to the net charge that passes through the cross section during the period, greatest first.

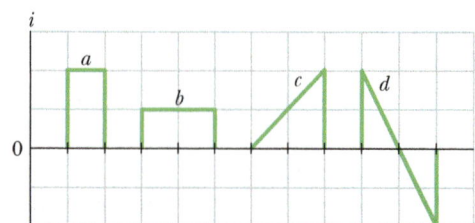

Figure 26-18   Question 4.

5 Figure 26-19 shows four situations in which positive and negative charges move horizontally and gives the rate at which each charge moves. Rank the situations according to the effective current through the regions, greatest first.

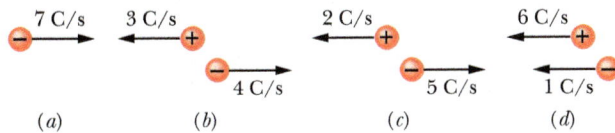

Figure 26-19 Question 5.

6 In Fig. 26-20, a wire that carries a current consists of three sections with different radii. Rank the sections according to the following quantities, greatest first: (a) current, (b) magnitude of current density, and (c) magnitude of electric field.

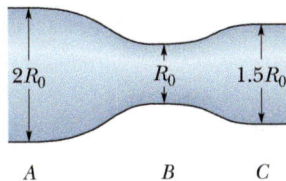

Figure 26-20 Question 6.

7 Figure 26-21 gives the electric potential $V(x)$ versus position x along a copper wire carrying current. The wire consists of three sections that differ in radius. Rank the three sections according to the magnitude of the (a) electric field and (b) current density, greatest first.

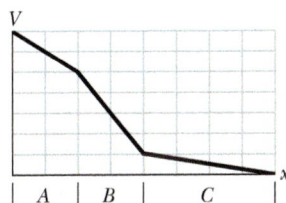

Figure 26-21 Question 7.

8 The following table gives the lengths of three copper rods, their diameters, and the potential differences between their ends. Rank the rods according to (a) the magnitude of the electric field within them, (b) the current density within them, and (c) the drift speed of electrons through them, greatest first.

Rod	Length	Diameter	Potential Difference
1	L	$3d$	V
2	$2L$	d	$2V$
3	$3L$	$2d$	$2V$

9 Figure 26-22 gives the drift speed v_d of conduction electrons in a copper wire versus position x along the wire. The wire consists of three sections that differ in radius. Rank the three sections according to the following quantities, greatest first: (a) radius, (b) number of conduction electrons per cubic meter, (c) magnitude of electric field, (d) conductivity.

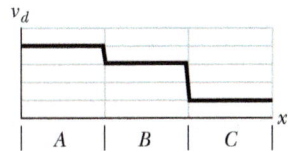

Figure 26-22 Question 9.

10 Three wires, of the same diameter, are connected in turn between two points maintained at a constant potential difference. Their resistivities and lengths are ρ and L (wire A), 1.2ρ and $1.2L$ (wire B), and 0.9ρ and L (wire C). Rank the wires according to the rate at which energy is transferred to thermal energy within them, greatest first.

11 Figure 26-23 gives, for three wires of radius R, the current density $J(r)$ versus radius r, as measured from the center of a circular cross section through the wire. The wires are all made from the same material. Rank the wires according to the magnitude of the electric field (a) at the center, (b) halfway to the surface, and (c) at the surface, greatest first.

Figure 26-23 Question 11.

Problems

Module 26-1 Electric Current

•1 During the 4.0 min a 5.0 A current is set up in a wire, how many (a) coulombs and (b) electrons pass through any cross section across the wire's width?

••2 An isolated conducting sphere has a 10 cm radius. One wire carries a current of 1.000 002 0 A into it. Another wire carries a current of 1.000 000 0 A out of it. How long would it take for the sphere to increase in potential by 1000 V?

••3 A charged belt, 50 cm wide, travels at 30 m/s between a source of charge and a sphere. The belt carries charge into the sphere at a rate corresponding to 100 μA. Compute the surface charge density on the belt.

Module 26-2 Current Density

•4 The (United States) National Electric Code, which sets maximum safe currents for insulated copper wires of various diameters, is given (in part) in the table. Plot the safe current density as a function of diameter. Which wire gauge has the maximum safe current density? ("Gauge" is a way of identifying wire diameters, and 1 mil = 10^{-3} in.)

Gauge	4	6	8	10	12	14	16	18
Diameter, mils	204	162	129	102	81	64	51	40
Safe current, A	70	50	35	25	20	15	6	3

•5 **SSM** **WWW** A beam contains 2.0×10^8 doubly charged positive ions per cubic centimeter, all of which are moving north with a speed of 1.0×10^5 m/s. What are the (a) magnitude and (b) direction of the current density $\vec{J}$? (c) What additional quantity do you need to calculate the total current i in this ion beam?

•6 A certain cylindrical wire carries current. We draw a circle of radius r around its central axis in Fig. 26-24a to determine the current i within the circle. Figure 26-24b shows current i as a function of r^2. The vertical scale is set by $i_s = 4.0$ mA, and the

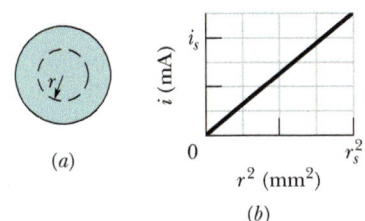

Figure 26-24 Problem 6.

horizontal scale is set by $r_s^2 = 4.0$ mm^2. (a) Is the current density uniform? (b) If so, what is its magnitude?

•7   A fuse in an electric circuit is a wire that is designed to melt, and thereby open the circuit, if the current exceeds a predetermined value. Suppose that the material to be used in a fuse melts when the current density rises to 440 A/cm^2. What diameter of cylindrical wire should be used to make a fuse that will limit the current to 0.50 A?

•8   A small but measurable current of 1.2×10^{-10} A exists in a copper wire whose diameter is 2.5 mm. The number of charge carriers per unit volume is 8.49×10^{28} m^{-3}. Assuming the current is uniform, calculate the (a) current density and (b) electron drift speed.

••9   The magnitude $J(r)$ of the current density in a certain cylindrical wire is given as a function of radial distance from the center of the wire's cross section as $J(r) = Br$, where r is in meters, J is in amperes per square meter, and $B = 2.00 \times 10^5$ A/m^3. This function applies out to the wire's radius of 2.00 mm. How much current is contained within the width of a thin ring concentric with the wire if the ring has a radial width of 10.0 μm and is at a radial distance of 1.20 mm?

••10   The magnitude J of the current density in a certain lab wire with a circular cross section of radius $R = 2.00$ mm is given by $J = (3.00 \times 10^8)r^2$, with J in amperes per square meter and radial distance r in meters. What is the current through the outer section bounded by $r = 0.900R$ and $r = R$?

••11   What is the current in a wire of radius $R = 3.40$ mm if the magnitude of the current density is given by (a) $J_a = J_0 r/R$ and (b) $J_b = J_0(1 - r/R)$, in which r is the radial distance and $J_0 = 5.50 \times 10^4$ A/m^2? (c) Which function maximizes the current density near the wire's surface?

••12   Near Earth, the density of protons in the solar wind (a stream of particles from the Sun) is 8.70 cm^{-3}, and their speed is 470 km/s. (a) Find the current density of these protons. (b) If Earth's magnetic field did not deflect the protons, what total current would Earth receive?

••13   GO ILW How long does it take electrons to get from a car battery to the starting motor? Assume the current is 300 A and the electrons travel through a copper wire with cross-sectional area 0.21 cm^2 and length 0.85 m. The number of charge carriers per unit volume is 8.49×10^{28} m^{-3}.

Module 26-3   Resistance and Resistivity
•14   A human being can be electrocuted if a current as small as 50 mA passes near the heart. An electrician working with sweaty hands makes good contact with the two conductors he is holding, one in each hand. If his resistance is 2000 Ω, what might the fatal voltage be?

•15   SSM A coil is formed by winding 250 turns of insulated 16-gauge copper wire (diameter = 1.3 mm) in a single layer on a cylindrical form of radius 12 cm. What is the resistance of the coil? Neglect the thickness of the insulation. (Use Table 26-1.)

•16   Copper and aluminum are being considered for a high-voltage transmission line that must carry a current of 60.0 A. The resistance per unit length is to be 0.150 Ω/km. The densities of copper and aluminum are 8960 and 2600 kg/m^3, respectively. Compute (a) the magnitude J of the current density and (b) the mass per unit length λ for a copper cable and (c) J and (d) λ for an aluminum cable.

•17   A wire of Nichrome (a nickel–chromium–iron alloy commonly used in heating elements) is 1.0 m long and 1.0 mm^2 in cross-sectional area. It carries a current of 4.0 A when a 2.0 V potential difference is applied between its ends. Calculate the conductivity σ of Nichrome.

•18   A wire 4.00 m long and 6.00 mm in diameter has a resistance of 15.0 mΩ. A potential difference of 23.0 V is applied between the ends. (a) What is the current in the wire? (b) What is the magnitude of the current density? (c) Calculate the resistivity of the wire material. (d) Using Table 26-1, identify the material.

•19   SSM What is the resistivity of a wire of 1.0 mm diameter, 2.0 m length, and 50 mΩ resistance?

•20   A certain wire has a resistance R. What is the resistance of a second wire, made of the same material, that is half as long and has half the diameter?

••21   ILW A common flashlight bulb is rated at 0.30 A and 2.9 V (the values of the current and voltage under operating conditions). If the resistance of the tungsten bulb filament at room temperature (20°C) is 1.1 Ω, what is the temperature of the filament when the bulb is on?

••22   *Kiting during a storm.* The legend that Benjamin Franklin flew a kite as a storm approached is only a legend—he was neither stupid nor suicidal. Suppose a kite string of radius 2.00 mm extends directly upward by 0.800 km and is coated with a 0.500 mm layer of water having resistivity 150 $\Omega \cdot$m. If the potential difference between the two ends of the string is 160 MV, what is the current through the water layer? The danger is not this current but the chance that the string draws a lightning strike, which can have a current as large as 500 000 A (way beyond just being lethal).

••23   When 115 V is applied across a wire that is 10 m long and has a 0.30 mm radius, the magnitude of the current density is 1.4×10^8 A/m^2. Find the resistivity of the wire.

••24   GO Figure 26-25a gives the magnitude $E(x)$ of the electric fields that have been set up by a battery along a resistive rod of length 9.00 mm (Fig. 26-25b). The vertical scale is set by $E_s = 4.00 \times 10^3$ V/m. The rod consists of three sections of the same material but with different radii. (The schematic diagram of Fig. 26-25b does not indicate the different radii.) The radius of section 3 is 2.00 mm. What is the radius of (a) section 1 and (b) section 2?

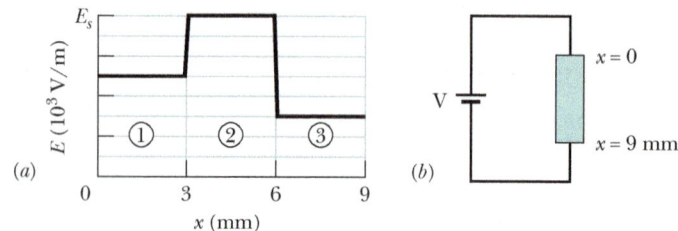

Figure 26-25 Problem 24.

••25   SSM ILW A wire with a resistance of 6.0 Ω is drawn out through a die so that its new length is three times its original length. Find the resistance of the longer wire, assuming that the resistivity and density of the material are unchanged.

••26   In Fig. 26-26a, a 9.00 V battery is connected to a resistive strip that consists of three sections with the same cross-sectional areas but different conductivities. Figure 26-26b gives the electric potential $V(x)$ versus position x along the strip. The horizontal scale is set by $x_s = 8.00$ mm. Section 3 has conductivity $3.00 \times 10^7 \, (\Omega \cdot m)^{-1}$. What is the conductivity of section (a) 1 and (b) 2?

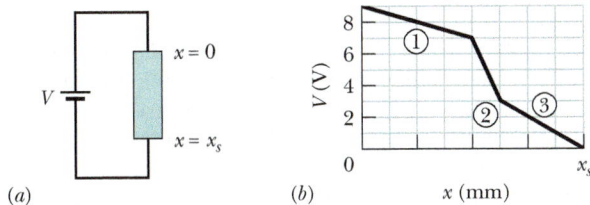

Figure 26-26 Problem 26.

••27 SSM WWW Two conductors are made of the same material and have the same length. Conductor A is a solid wire of diameter 1.0 mm. Conductor B is a hollow tube of outside diameter 2.0 mm and inside diameter 1.0 mm. What is the resistance ratio R_A/R_B, measured between their ends?

••28 GO Figure 26-27 gives the electric potential $V(x)$ along a copper wire carrying uniform current, from a point of higher potential $V_s = 12.0 \, \mu V$ at $x = 0$ to a point of zero potential at $x_s = 3.00$ m. The wire has a radius of 2.00 mm. What is the current in the wire?

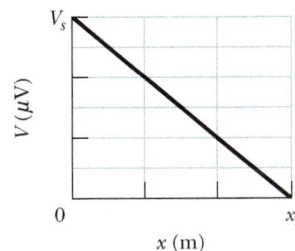

Figure 26-27 Problem 28.

••29 A potential difference of 3.00 nV is set up across a 2.00 cm length of copper wire that has a radius of 2.00 mm. How much charge drifts through a cross section in 3.00 ms?

••30 If the gauge number of a wire is increased by 6, the diameter is halved; if a gauge number is increased by 1, the diameter decreases by the factor $2^{1/6}$ (see the table in Problem 4). Knowing this, and knowing that 1000 ft of 10-gauge copper wire has a resistance of approximately 1.00 Ω, estimate the resistance of 25 ft of 22-gauge copper wire.

••31 An electrical cable consists of 125 strands of fine wire, each having 2.65 $\mu\Omega$ resistance. The same potential difference is applied between the ends of all the strands and results in a total current of 0.750 A. (a) What is the current in each strand? (b) What is the applied potential difference? (c) What is the resistance of the cable?

••32 Earth's lower atmosphere contains negative and positive ions that are produced by radioactive elements in the soil and cosmic rays from space. In a certain region, the atmospheric

Figure 26-28 Problem 32.

electric field strength is 120 V/m and the field is directed vertically down. This field causes singly charged positive ions, at a density of 620 cm^{-3}, to drift downward and singly charged negative ions, at a density of 550 cm^{-3}, to drift upward (Fig. 26-28). The measured conductivity of the air in that region is 2.70×10^{-14} $(\Omega \cdot m)^{-1}$. Calculate (a) the magnitude of the current density and (b) the ion drift speed, assumed to be the same for positive and negative ions.

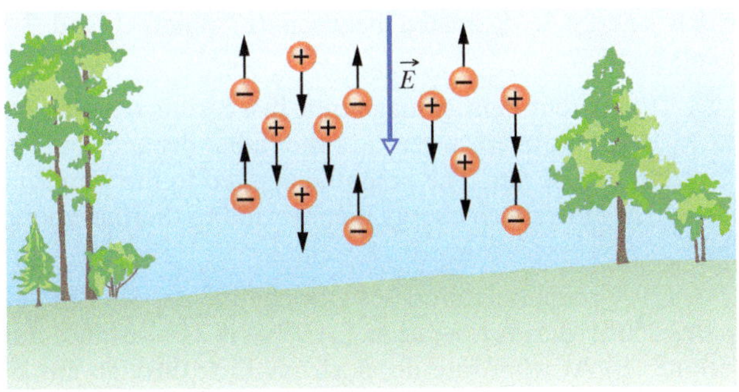

••33 A block in the shape of a rectangular solid has a cross-sectional area of 3.50 cm^2 across its width, a front-to-rear length of 15.8 cm, and a resistance of 935 Ω. The block's material contains 5.33×10^{22} conduction electrons/m^3. A potential difference of 35.8 V is maintained between its front and rear faces. (a) What is the current in the block? (b) If the current density is uniform, what is its magnitude? What are (c) the drift velocity of the conduction electrons and (d) the magnitude of the electric field in the block?

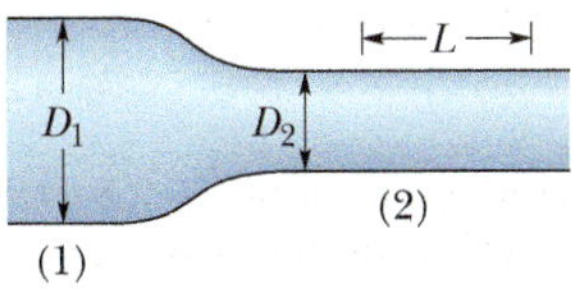

Figure 26-29 Problem 34.

•••34 GO Figure 26-29 shows wire section 1 of diameter $D_1 = 4.00R$ and wire section 2 of diameter $D_2 = 2.00R$, connected by a tapered section. The wire is copper and carries a current. Assume that the current is uniformly distributed across any cross-sectional area through the wire's width. The electric potential change V along the length $L = 2.00$ m shown in section 2 is 10.0 μV. The number of charge carriers per unit volume is 8.49×10^{28} m^{-3}. What is the drift speed of the conduction electrons in section 1?

•••35 GO In Fig. 26-30, current is set up through a truncated right circular cone of resistivity 731 $\Omega \cdot m$, left radius $a = 2.00$ mm, right radius $b = 2.30$ mm, and length $L = 1.94$ cm. Assume that the current density is uniform across any cross section taken perpendicular to the length. What is the resistance of the cone?

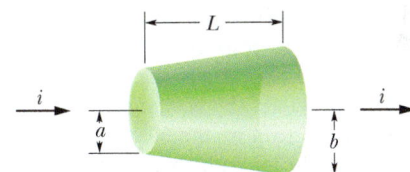

Figure 26-30 Problem 35.

•••36 GO *Swimming during a storm.* Figure 26-31 shows a swimmer at distance $D = 35.0$ m from a lightning strike to the water, with current $I = 78$ kA. The water has resistivity 30 $\Omega \cdot m$, the width of the swimmer along a radial line from the strike is 0.70 m, and his resistance across that width is 4.00 kΩ. Assume that the current spreads through the water over a hemi-

Figure 26-31 Problem 36.

sphere centered on the strike point. What is the current through the swimmer?

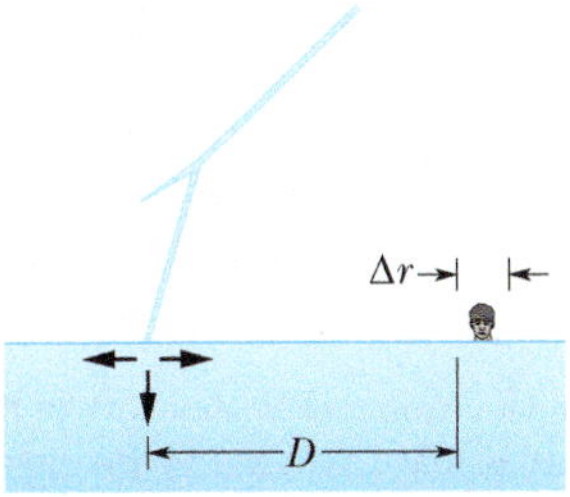

Module 26-4 Ohm's Law
••37 Show that, according to the free-electron model of electrical conduction in metals and classical physics, the resistivity of metals should be proportional to $\sqrt{T}$, where T is the temperature in kelvins. (See Eq. 19-31.)

Module 26-5 Power, Semiconductors, Superconductors

•38 In Fig. 26-32a, a 20 Ω resistor is connected to a battery. Figure 26-32b shows the increase of thermal energy E_{th} in the resistor as a function of time t. The vertical scale is set by $E_{th,s}$ = 2.50 mJ, and the horizontal scale is set by t_s = 4.0 s. What is the electric potential across the battery?

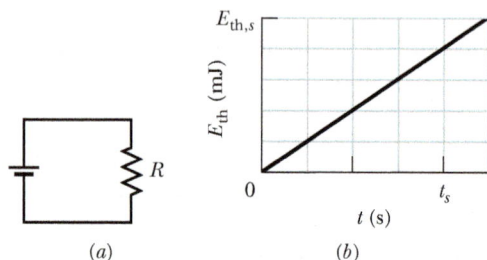

Figure 26-32 Problem 38.

•39 A certain brand of hot-dog cooker works by applying a potential difference of 120 V across opposite ends of a hot dog and allowing it to cook by means of the thermal energy produced. The current is 10.0 A, and the energy required to cook one hot dog is 60.0 kJ. If the rate at which energy is supplied is unchanged, how long will it take to cook three hot dogs simultaneously?

•40 Thermal energy is produced in a resistor at a rate of 100 W when the current is 3.00 A. What is the resistance?

•41 SSM A 120 V potential difference is applied to a space heater whose resistance is 14 Ω when hot. (a) At what rate is electrical energy transferred to thermal energy? (b) What is the cost for 5.0 h at US$0.05/kW·h?

•42 In Fig. 26-33, a battery of potential difference V = 12 V is connected to a resistive strip of resistance R = 6.0 Ω. When an electron moves through the strip from one end to the other, (a) in which direction in the figure does the electron move, (b) how much work is done on the electron by the electric field in the strip, and (c) how much energy is transferred to the thermal energy of the strip by the electron?

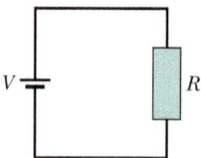

Figure 26-33 Problem 42.

•43 ILW An unknown resistor is connected between the terminals of a 3.00 V battery. Energy is dissipated in the resistor at the rate of 0.540 W. The same resistor is then connected between the terminals of a 1.50 V battery. At what rate is energy now dissipated?

•44 A student kept his 9.0 V, 7.0 W radio turned on at full volume from 9:00 P.M. until 2:00 A.M. How much charge went through it?

•45 SSM ILW A 1250 W radiant heater is constructed to operate at 115 V. (a) What is the current in the heater when the unit is operating? (b) What is the resistance of the heating coil? (c) How much thermal energy is produced in 1.0 h?

••46 GO A copper wire of cross-sectional area 2.00×10^{-6} m² and length 4.00 m has a current of 2.00 A uniformly distributed across that area. (a) What is the magnitude of the electric field along the wire? (b) How much electrical energy is transferred to thermal energy in 30 min?

••47 A heating element is made by maintaining a potential difference of 75.0 V across the length of a Nichrome wire that has a 2.60×10^{-6} m² cross section. Nichrome has a resistivity of 5.00×10^{-7} Ω·m. (a) If the element dissipates 5000 W, what is its length? (b) If 100 V is used to obtain the same dissipation rate, what should the length be?

••48 *Exploding shoes.* The rain-soaked shoes of a person may explode if ground current from nearby lightning vaporizes the water. The sudden conversion of water to water vapor causes a dramatic expansion that can rip apart shoes. Water has density 1000 kg/m³ and requires 2256 kJ/kg to be vaporized. If horizontal current lasts 2.00 ms and encounters water with resistivity 150 Ω·m, length 12.0 cm, and vertical cross-sectional area 15×10^{-5} m², what average current is required to vaporize the water?

••49 A 100 W lightbulb is plugged into a standard 120 V outlet. (a) How much does it cost per 31-day month to leave the light turned on continuously? Assume electrical energy costs US$0.06/kW·h. (b) What is the resistance of the bulb? (c) What is the current in the bulb?

••50 GO The current through the battery and resistors 1 and 2 in Fig. 26-34a is 2.00 A. Energy is transferred from the current to thermal energy E_{th} in both resistors. Curves 1 and 2 in Fig. 26-34b give that thermal energy E_{th} for resistors 1 and 2, respectively, as a function of time t. The vertical scale is set by $E_{th,s}$ = 40.0 mJ, and the horizontal scale is set by t_s = 5.00 s. What is the power of the battery?

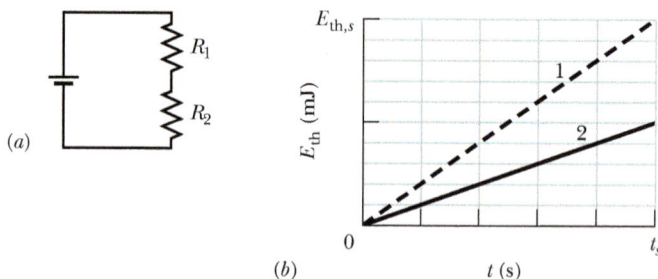

Figure 26-34 Problem 50.

••51 GO SSM WWW Wire C and wire D are made from different materials and have length $L_C = L_D$ = 1.0 m. The resistivity and diameter of wire C are 2.0×10^{-6} Ω·m and 1.00 mm, and those of wire D are 1.0×10^{-6} Ω·m and 0.50 mm. The wires are joined as shown in Fig. 26-35, and a current of 2.0 A is set up in them. What is the electric potential difference between (a) points 1 and 2 and (b) points 2 and 3? What is the rate at which energy is dissipated between (c) points 1 and 2 and (d) points 2 and 3?

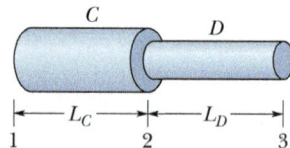

Figure 26-35 Problem 51.

••52 GO The current-density magnitude in a certain circular wire is $J = (2.75 \times 10^{10}$ A/m⁴$)r^2$, where r is the radial distance out to the wire's radius of 3.00 mm. The potential applied to the wire (end to end) is 60.0 V. How much energy is converted to thermal energy in 1.00 h?

••53 A 120 V potential difference is applied to a space heater that dissipates 500 W during operation. (a) What is its resistance during operation? (b) At what rate do electrons flow through any cross section of the heater element?

•••54 GO Figure 26-36a shows a rod of resistive material. The resistance per unit length of the rod increases in the positive direction of the x axis. At any position x along the rod, the resistance dR of a narrow (differential) section of width dx is given by $dR = 5.00x\, dx$, where dR is in ohms and x is in meters. Figure 26-36b shows such a narrow section. You are to slice off a length of the rod between $x = 0$ and some position $x = L$ and then connect that length to a battery with potential difference $V = 5.0$ V (Fig. 26-36c). You want the current in the length to transfer energy to thermal energy at the rate of 200 W. At what position $x = L$ should you cut the rod?

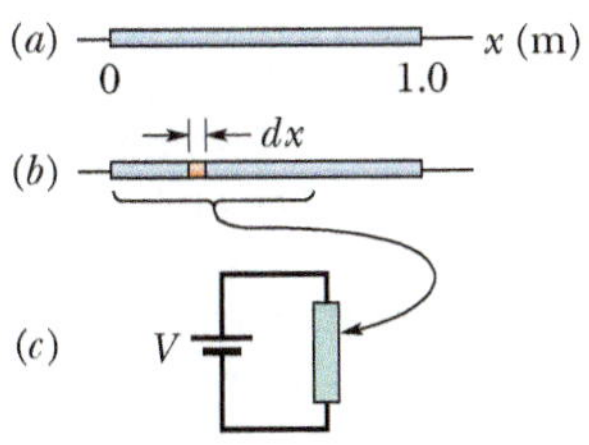

Figure 26-36 Problem 54.

Additional Problems

55 SSM A Nichrome heater dissipates 500 W when the applied potential difference is 110 V and the wire temperature is 800°C. What would be the dissipation rate if the wire temperature were held at 200°C by immersing the wire in a bath of cooling oil? The applied potential difference remains the same, and α for Nichrome at 800°C is 4.0×10^{-4} K^{-1}.

56 A potential difference of 1.20 V will be applied to a 33.0 m length of 18-gauge copper wire (diameter = 0.0400 in.). Calculate (a) the current, (b) the magnitude of the current density, (c) the magnitude of the electric field within the wire, and (d) the rate at which thermal energy will appear in the wire.

57 An 18.0 W device has 9.00 V across it. How much charge goes through the device in 4.00 h?

58 An aluminum rod with a square cross section is 1.3 m long and 5.2 mm on edge. (a) What is the resistance between its ends? (b) What must be the diameter of a cylindrical copper rod of length 1.3 m if its resistance is to be the same as that of the aluminum rod?

59 A cylindrical metal rod is 1.60 m long and 5.50 mm in diameter. The resistance between its two ends (at 20°C) is 1.09×10^{-3} Ω. (a) What is the material? (b) A round disk, 2.00 cm in diameter and 1.00 mm thick, is formed of the same material. What is the resistance between the round faces, assuming that each face is an equipotential surface?

60 ✈ *The chocolate crumb mystery.* This story begins with Problem 60 in Chapter 23 and continues through Chapters 24 and 25. The chocolate crumb powder moved to the silo through a pipe of radius R with uniform speed v and uniform charge density ρ. (a) Find an expression for the current i (the rate at which charge on the powder moved) through a perpendicular cross section of the pipe. (b) Evaluate i for the conditions at the factory: pipe radius $R = 5.0$ cm, speed $v = 2.0$ m/s, and charge density $\rho = 1.1 \times 10^{-3}$ C/m³.

If the powder were to flow through a change V in electric potential, its energy could be transferred to a spark at the rate $P = iV$. (c) Could there be such a transfer within the pipe due to the radial potential difference discussed in Problem 70 of Chapter 24?

As the powder flowed from the pipe into the silo, the electric potential of the powder changed. The magnitude of that change was at least equal to the radial potential difference within the pipe (as evaluated in Problem 70 of Chapter 24). (d) Assuming that value for the potential difference and using the current found in (b) above, find the rate at which energy could have been transferred from the powder to a spark as the powder exited the pipe. (e) If a spark did occur at the exit and lasted for 0.20 s (a reasonable expectation), how much energy would have been transferred to the spark? Recall

from Problem 60 in Chapter 23 that a minimum energy transfer of 150 mJ is needed to cause an explosion. (f) Where did the powder explosion most likely occur: in the powder cloud at the unloading bin (Problem 60 of Chapter 25), within the pipe, or at the exit of the pipe into the silo?

61 SSM A steady beam of alpha particles $(q = +2e)$ traveling with constant kinetic energy 20 MeV carries a current of 0.25 μA. (a) If the beam is directed perpendicular to a flat surface, how many alpha particles strike the surface in 3.0 s? (b) At any instant, how many alpha particles are there in a given 20 cm length of the beam? (c) Through what potential difference is it necessary to accelerate each alpha particle from rest to bring it to an energy of 20 MeV?

62 A resistor with a potential difference of 200 V across it transfers electrical energy to thermal energy at the rate of 3000 W. What is the resistance of the resistor?

63 A 2.0 kW heater element from a dryer has a length of 80 cm. If a 10 cm section is removed, what power is used by the now shortened element at 120 V?

64 A cylindrical resistor of radius 5.0 mm and length 2.0 cm is made of material that has a resistivity of 3.5×10^{-5} Ω·m. What are (a) the magnitude of the current density and (b) the potential difference when the energy dissipation rate in the resistor is 1.0 W?

65 A potential difference V is applied to a wire of cross-sectional area A, length L, and resistivity ρ. You want to change the applied potential difference and stretch the wire so that the energy dissipation rate is multiplied by 30.0 and the current is multiplied by 4.00. Assuming the wire's density does not change, what are (a) the ratio of the new length to L and (b) the ratio of the new cross-sectional area to A?

66 The headlights of a moving car require about 10 A from the 12 V alternator, which is driven by the engine. Assume the alternator is 80% efficient (its output electrical power is 80% of its input mechanical power), and calculate the horsepower the engine must supply to run the lights.

67 A 500 W heating unit is designed to operate with an applied potential difference of 115 V. (a) By what percentage will its heat output drop if the applied potential difference drops to 110 V? Assume no change in resistance. (b) If you took the variation of resistance with temperature into account, would the actual drop in heat output be larger or smaller than that calculated in (a)?

68 The copper windings of a motor have a resistance of 50 Ω at 20°C when the motor is idle. After the motor has run for several hours, the resistance rises to 58 Ω. What is the temperature of the windings now? Ignore changes in the dimensions of the windings. (Use Table 26-1.)

69 How much electrical energy is transferred to thermal energy in 2.00 h by an electrical resistance of 400 Ω when the potential applied across it is 90.0 V?

70 A caterpillar of length 4.0 cm crawls in the direction of electron drift along a 5.2-mm-diameter bare copper wire that carries a uniform current of 12 A. (a) What is the potential difference between the two ends of the caterpillar? (b) Is its tail positive or negative relative to its head? (c) How much time does the caterpillar take to crawl 1.0 cm if it crawls at the drift speed of the electrons in the wire? (The number of charge carriers per unit volume is 8.49×10^{28} m^{-3}.)

71 SSM (a) At what temperature would the resistance of a copper conductor be double its resistance at 20.0°C? (Use 20.0°C as the reference point in Eq. 26-17; compare your answer with

Fig. 26-10.) (b) Does this same "doubling temperature" hold for all copper conductors, regardless of shape or size?

72 A steel trolley-car rail has a cross-sectional area of 56.0 cm². What is the resistance of 10.0 km of rail? The resistivity of the steel is $3.00 \times 10^{-7}\,\Omega\cdot\text{m}$.

73 A coil of current-carrying Nichrome wire is immersed in a liquid. (Nichrome is a nickel–chromium–iron alloy commonly used in heating elements.) When the potential difference across the coil is 12 V and the current through the coil is 5.2 A, the liquid evaporates at the steady rate of 21 mg/s. Calculate the heat of vaporization of the liquid (see Module 18-4).

74 The current density in a wire is uniform and has magnitude 2.0×10^6 A/m², the wire's length is 5.0 m, and the density of conduction electrons is 8.49×10^{28} m^{-3}. How long does an electron take (on the average) to travel the length of the wire?

75 A certain x-ray tube operates at a current of 7.00 mA and a potential difference of 80.0 kV. What is its power in watts?

76 A current is established in a gas discharge tube when a sufficiently high potential difference is applied across the two electrodes in the tube. The gas ionizes; electrons move toward the positive terminal and singly charged positive ions toward the negative terminal. (a) What is the current in a hydrogen discharge tube in which 3.1×10^{18} electrons and 1.1×10^{18} protons move past a cross-sectional area of the tube each second? (b) Is the direction of the current density $\vec{J}$ toward or away from the negative terminal?

77 In Fig. 26-37, a resistance coil, wired to an external battery, is placed inside a thermally insulated cylinder fitted with a frictionless piston and containing an ideal gas. A current $i = 240$ mA flows through the coil, which has a resistance $R = 550\,\Omega$. If the gas temperature remains constant while the 12 kg piston moves upward, what is the limit of the piston's speed?

78 An insulating belt moves at speed 30 m/s and has a width of 50 cm. It carries charge into an experimental device at a rate corresponding to 100 μA. What is the surface charge density on the belt?

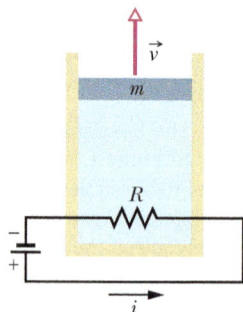

Figure 26-37 Problem 77.

79 In a hypothetical fusion research lab, high temperature helium gas is completely ionized and each helium atom is separated into two free electrons and the remaining positively charged nucleus, which is called an alpha particle. An applied electric field causes the alpha particles to drift to the east at 25.0 m/s while the electrons drift to the west at 88.0 m/s. The alpha particle density is 2.80×10^{15} cm^{-3}. What are (a) the net current density and (b) the current direction?

80 When a metal rod is heated, not only its resistance but also its length and cross-sectional area change. The relation $R = \rho L/A$ suggests that all three factors should be taken into account in measuring ρ at various temperatures. If the temperature changes by 1.0 C°, what percentage changes in (a) L, (b) A, and (c) R occur for a copper conductor? (d) What conclusion do you draw? The coefficient of linear expansion is 1.70×10^{-5} K^{-1}.

81 A beam of 16 MeV deuterons from a cyclotron strikes a copper block. The beam is equivalent to current of 15 μA. (a) At what rate do deuterons strike the block? (b) At what rate is thermal energy produced in the block?

82 A linear accelerator produces a pulsed beam of electrons. The pulse current is 0.50 A, and the pulse duration is 0.10 μs. (a) How many electrons are accelerated per pulse? (b) What is the average current for a machine operating at 500 pulses/s? If the electrons are accelerated to an energy of 50 MeV, what are the (c) average power and (d) peak power of the accelerator?

83 An electric immersion heater normally takes 100 min to bring cold water in a well-insulated container to a certain temperature, after which a thermostat switches the heater off. One day the line voltage is reduced by 6.00% because of a laboratory overload. How long does heating the water now take? Assume that the resistance of the heating element does not change.

84 A 400 W immersion heater is placed in a pot containing 2.00 L of water at 20°C. (a) How long will the water take to rise to the boiling temperature, assuming that 80% of the available energy is absorbed by the water? (b) How much longer is required to evaporate half of the water?

85 A 30 μF capacitor is connected across a programmed power supply. During the interval from $t = 0$ to $t = 3.00$ s the output voltage of the supply is given by $V(t) = 6.00 + 4.00t - 2.00t^2$ volts. At $t = 0.500$ s find (a) the charge on the capacitor, (b) the current into the capacitor, and (c) the power output from the power supply.

Circuits

27-1 SINGLE-LOOP CIRCUITS

Learning Objectives

After reading this module, you should be able to . . .

27.01 Identify the action of an emf source in terms of the work it does.

27.02 For an ideal battery, apply the relationship between the emf, the current, and the power (rate of energy transfer).

27.03 Draw a schematic diagram for a single-loop circuit containing a battery and three resistors.

27.04 Apply the loop rule to write a loop equation that relates the potential differences of the circuit elements around a (complete) loop.

27.05 Apply the resistance rule in crossing through a resistor.

27.06 Apply the emf rule in crossing through an emf.

27.07 Identify that resistors in series have the same current, which is the same value that their equivalent resistor has.

27.08 Calculate the equivalent of series resistors.

27.09 Identify that a potential applied to resistors wired in series is equal to the sum of the potentials across the individual resistors.

27.10 Calculate the potential difference between any two points in a circuit.

27.11 Distinguish a real battery from an ideal battery and, in a circuit diagram, replace a real battery with an ideal battery and an explicitly shown resistance.

27.12 With a real battery in a circuit, calculate the potential difference between its terminals for current in the direction of the emf and in the opposite direction.

27.13 Identify what is meant by grounding a circuit, and draw a schematic diagram for such a connection.

27.14 Identify that grounding a circuit does not affect the current in a circuit.

27.15 Calculate the dissipation rate of energy in a real battery.

27.16 Calculate the net rate of energy transfer in a real battery for current in the direction of the emf and in the opposite direction.

Key Ideas

● An emf device does work on charges to maintain a potential difference between its output terminals. If dW is the work the device does to force positive charge dq from the negative to the positive terminal, then the emf (work per unit charge) of the device is

$$\mathscr{E} = \frac{dW}{dq} \quad \text{(definition of } \mathscr{E}\text{)}.$$

● An ideal emf device is one that lacks any internal resistance. The potential difference between its terminals is equal to the emf.

● A real emf device has internal resistance. The potential difference between its terminals is equal to the emf only if there is no current through the device.

● The change in potential in traversing a resistance R in the direction of the current is $-iR$; in the opposite direction it is $+iR$ (resistance rule).

● The change in potential in traversing an ideal emf device in the direction of the emf arrow is $+\mathscr{E}$; in the opposite direction it is $-\mathscr{E}$ (emf rule).

● Conservation of energy leads to the loop rule:

Loop Rule. The algebraic sum of the changes in potential encountered in a complete traversal of any loop of a circuit must be zero.

Conservation of charge leads to the junction rule (Chapter 26):

Junction Rule. The sum of the currents entering any junction must be equal to the sum of the currents leaving that junction.

● When a real battery of emf $\mathscr{E}$ and internal resistance r does work on the charge carriers in a current i through the battery, the rate P of energy transfer to the charge carriers is

$$P = iV,$$

where V is the potential across the terminals of the battery.

● The rate P_r at which energy is dissipated as thermal energy in the battery is

$$P_r = i^2 r.$$

● The rate P_{emf} at which the chemical energy in the battery changes is

$$P_{\text{emf}} = i\mathscr{E}.$$

● When resistances are in series, they have the same current. The equivalent resistance that can replace a series combination of resistances is

$$R_{\text{eq}} = \sum_{j=1}^{n} R_j \quad (n \text{ resistances in series}).$$

What Is Physics?

You are surrounded by electric circuits. You might take pride in the number of electrical devices you own and might even carry a mental list of the devices you wish you owned. Every one of those devices, as well as the electrical grid that powers your home, depends on modern electrical engineering. We cannot easily estimate the current financial worth of electrical engineering and its products, but we can be certain that the financial worth continues to grow yearly as more and more tasks are handled electrically. Radios are now tuned electronically instead of manually. Messages are now sent by email instead of through the postal system. Research journals are now read on a computer instead of in a library building, and research papers are now copied and filed electronically instead of photocopied and tucked into a filing cabinet. Indeed, you may be reading an electronic version of this book.

The basic science of electrical engineering is physics. In this chapter we cover the physics of electric circuits that are combinations of resistors and batteries (and, in Module 27-4, capacitors). We restrict our discussion to circuits through which charge flows in one direction, which are called either *direct-current circuits* or *DC circuits*. We begin with the question: How can you get charges to flow?

"Pumping" Charges

If you want to make charge carriers flow through a resistor, you must establish a potential difference between the ends of the device. One way to do this is to connect each end of the resistor to one plate of a charged capacitor. The trouble with this scheme is that the flow of charge acts to discharge the capacitor, quickly bringing the plates to the same potential. When that happens, there is no longer an electric field in the resistor, and thus the flow of charge stops.

To produce a steady flow of charge, you need a "charge pump," a device that—by doing work on the charge carriers—maintains a potential difference between a pair of terminals. We call such a device an **emf device,** and the device is said to provide an **emf** $\mathscr{E}$, which means that it does work on charge carriers. An emf device is sometimes called a *seat of emf.* The term *emf* comes from the outdated phrase *electromotive force,* which was adopted before scientists clearly understood the function of an emf device.

In Chapter 26, we discussed the motion of charge carriers through a circuit in terms of the electric field set up in the circuit—the field produces forces that move the charge carriers. In this chapter we take a different approach: We discuss the motion of the charge carriers in terms of the required energy—an emf device supplies the energy for the motion via the work it does.

A common emf device is the *battery,* used to power a wide variety of machines from wristwatches to submarines. The emf device that most influences our daily lives, however, is the *electric generator,* which, by means of electrical connections (wires) from a generating plant, creates a potential difference in our homes and workplaces. The emf devices known as *solar cells,* long familiar as the wing-like panels on spacecraft, also dot the countryside for domestic applications. Less familiar emf devices are the *fuel cells* that powered the space shuttles and the *thermopiles* that provide onboard electrical power for some spacecraft and for remote stations in Antarctica and elsewhere. An emf device does not have to be an instrument—living systems, ranging from electric eels and human beings to plants, have physiological emf devices.

Although the devices we have listed differ widely in their modes of operation, they all perform the same basic function—they do work on charge carriers and thus maintain a potential difference between their terminals.

Courtesy Southern California Edison Company

The world's largest battery energy storage plant (dismantled in 1996) connected over 8000 large lead-acid batteries in 8 strings at 1000 V each with a capability of 10 MW of power for 4 hours. Charged up at night, the batteries were then put to use during peak power demands on the electrical system.

Work, Energy, and Emf

Figure 27-1 shows an emf device (consider it to be a battery) that is part of a simple circuit containing a single resistance R (the symbol for resistance and a resistor is -\/\/\/-). The emf device keeps one of its terminals (called the positive terminal and often labeled +) at a higher electric potential than the other terminal (called the negative terminal and labeled −). We can represent the emf of the device with an arrow that points from the negative terminal toward the positive terminal as in Fig. 27-1. A small circle on the tail of the emf arrow distinguishes it from the arrows that indicate current direction.

When an emf device is not connected to a circuit, the internal chemistry of the device does not cause any net flow of charge carriers within it. However, when it is connected to a circuit as in Fig. 27-1, its internal chemistry causes a net flow of positive charge carriers from the negative terminal to the positive terminal, in the direction of the emf arrow. This flow is part of the current that is set up around the circuit in that same direction (clockwise in Fig. 27-1).

Within the emf device, positive charge carriers move from a region of low electric potential and thus low electric potential energy (at the negative terminal) to a region of higher electric potential and higher electric potential energy (at the positive terminal). This motion is just the opposite of what the electric field between the terminals (which is directed from the positive terminal toward the negative terminal) would cause the charge carriers to do.

Thus, there must be some source of energy within the device, enabling it to do work on the charges by forcing them to move as they do. The energy source may be chemical, as in a battery or a fuel cell. It may involve mechanical forces, as in an electric generator. Temperature differences may supply the energy, as in a thermopile; or the Sun may supply it, as in a solar cell.

Let us now analyze the circuit of Fig. 27-1 from the point of view of work and energy transfers. In any time interval dt, a charge dq passes through any cross section of this circuit, such as aa'. This same amount of charge must enter the emf device at its low-potential end and leave at its high-potential end. The device must do an amount of work dW on the charge dq to force it to move in this way. We define the emf of the emf device in terms of this work:

$$\mathscr{E} = \frac{dW}{dq} \quad \text{(definition of } \mathscr{E}\text{)}. \qquad (27\text{-}1)$$

In words, the emf of an emf device is the work per unit charge that the device does in moving charge from its low-potential terminal to its high-potential terminal. The SI unit for emf is the joule per coulomb; in Chapter 24 we defined that unit as the *volt*.

An **ideal emf device** is one that lacks any internal resistance to the internal movement of charge from terminal to terminal. The potential difference between the terminals of an ideal emf device is equal to the emf of the device. For example, an ideal battery with an emf of 12.0 V always has a potential difference of 12.0 V between its terminals.

A **real emf device,** such as any real battery, has internal resistance to the internal movement of charge. When a real emf device is not connected to a circuit, and thus does not have current through it, the potential difference between its terminals is equal to its emf. However, when that device has current through it, the potential difference between its terminals differs from its emf. We shall discuss such real batteries near the end of this module.

When an emf device is connected to a circuit, the device transfers energy to the charge carriers passing through it. This energy can then be transferred from the charge carriers to other devices in the circuit, for example, to light a bulb. Figure 27-2a shows a circuit containing two ideal rechargeable (*storage*) batteries A and B, a resistance R, and an electric motor M that can lift an object by using

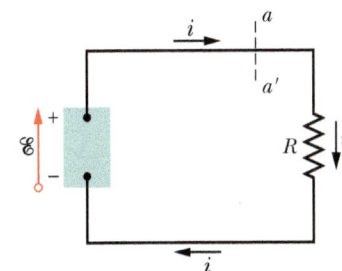

Figure 27-1 A simple electric circuit, in which a device of emf $\mathscr{E}$ does work on the charge carriers and maintains a steady current i in a resistor of resistance R.

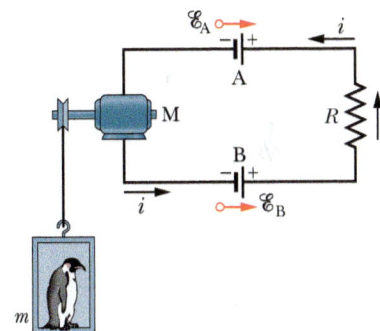

(a)

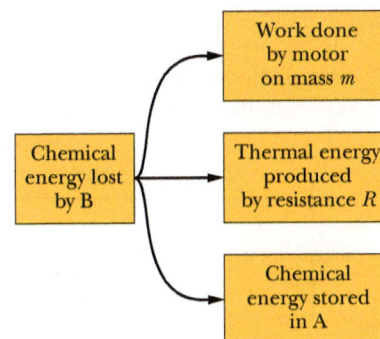

(b)

Figure 27-2 (a) In the circuit, $\mathscr{E}_B > \mathscr{E}_A$; so battery B determines the direction of the current. (b) The energy transfers in the circuit.

energy it obtains from charge carriers in the circuit. Note that the batteries are connected so that they tend to send charges around the circuit in opposite directions. The actual direction of the current in the circuit is determined by the battery with the larger emf, which happens to be battery B, so the chemical energy within battery B is decreasing as energy is transferred to the charge carriers passing through it. However, the chemical energy within battery A is increasing because the current in it is directed from the positive terminal to the negative terminal. Thus, battery B is charging battery A. Battery B is also providing energy to motor M and energy that is being dissipated by resistance R. Figure 27-2b shows all three energy transfers from battery B; each decreases that battery's chemical energy.

Calculating the Current in a Single-Loop Circuit

We discuss here two equivalent ways to calculate the current in the simple *single-loop* circuit of Fig. 27-3; one method is based on energy conservation considerations, and the other on the concept of potential. The circuit consists of an ideal battery B with emf $\mathscr{E}$, a resistor of resistance R, and two connecting wires. (Unless otherwise indicated, we assume that wires in circuits have negligible resistance. Their function, then, is merely to provide pathways along which charge carriers can move.)

Energy Method

Equation 26-27 ($P = i^2 R$) tells us that in a time interval dt an amount of energy given by $i^2 R \, dt$ will appear in the resistor of Fig. 27-3 as thermal energy. As noted in Module 26-5, this energy is said to be *dissipated*. (Because we assume the wires to have negligible resistance, no thermal energy will appear in them.) During the same interval, a charge $dq = i \, dt$ will have moved through battery B, and the work that the battery will have done on this charge, according to Eq. 27-1, is

$$dW = \mathscr{E} \, dq = \mathscr{E} i \, dt.$$

From the principle of conservation of energy, the work done by the (ideal) battery must equal the thermal energy that appears in the resistor:

$$\mathscr{E} i \, dt = i^2 R \, dt.$$

This gives us

$$\mathscr{E} = iR.$$

The emf $\mathscr{E}$ is the energy per unit charge transferred to the moving charges by the battery. The quantity iR is the energy per unit charge transferred *from* the moving charges to thermal energy within the resistor. Therefore, this equation means that the energy per unit charge transferred to the moving charges is equal to the energy per unit charge transferred from them. Solving for i, we find

$$i = \frac{\mathscr{E}}{R}. \tag{27-2}$$

Potential Method

Suppose we start at any point in the circuit of Fig. 27-3 and mentally proceed around the circuit in either direction, adding algebraically the potential differences that we encounter. Then when we return to our starting point, we must also have returned to our starting potential. Before actually doing so, we shall formalize this idea in a statement that holds not only for single-loop circuits such as that of Fig. 27-3 but also for any complete loop in a *multiloop* circuit, as we shall discuss in Module 27-2:

The battery drives current through the resistor, from high potential to low potential.

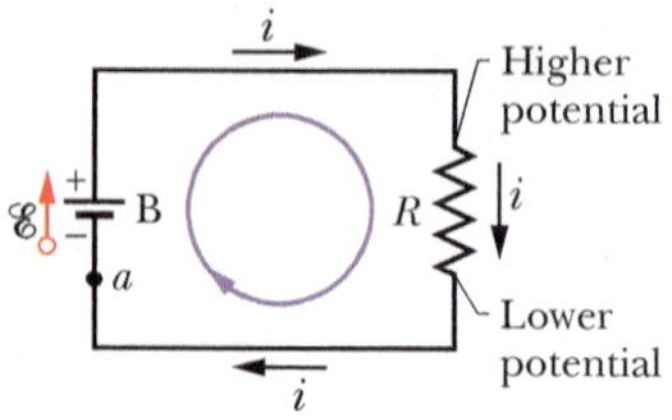

Figure 27-3 A single-loop circuit in which a resistance R is connected across an ideal battery B with emf $\mathscr{E}$. The resulting current i is the same throughout the circuit.

> ⭐ **LOOP RULE:** The algebraic sum of the changes in potential encountered in a complete traversal of any loop of a circuit must be zero.

This is often referred to as *Kirchhoff's loop rule* (or *Kirchhoff's voltage law*), after German physicist Gustav Robert Kirchhoff. This rule is equivalent to saying that each point on a mountain has only one elevation above sea level. If you start from any point and return to it after walking around the mountain, the algebraic sum of the changes in elevation that you encounter must be zero.

In Fig. 27-3, let us start at point a, whose potential is V_a, and mentally walk clockwise around the circuit until we are back at a, keeping track of potential changes as we move. Our starting point is at the low-potential terminal of the battery. Because the battery is ideal, the potential difference between its terminals is equal to $\mathscr{E}$. When we pass through the battery to the high-potential terminal, the change in potential is $+\mathscr{E}$.

As we walk along the top wire to the top end of the resistor, there is no potential change because the wire has negligible resistance; it is at the same potential as the high-potential terminal of the battery. So too is the top end of the resistor. When we pass through the resistor, however, the potential changes according to Eq. 26-8 (which we can rewrite as $V = iR$). Moreover, the potential must decrease because we are moving from the higher potential side of the resistor. Thus, the change in potential is $-iR$.

We return to point a by moving along the bottom wire. Because this wire also has negligible resistance, we again find no potential change. Back at point a, the potential is again V_a. Because we traversed a complete loop, our initial potential, as modified for potential changes along the way, must be equal to our final potential; that is,

$$V_a + \mathscr{E} - iR = V_a.$$

The value of V_a cancels from this equation, which becomes

$$\mathscr{E} - iR = 0.$$

Solving this equation for i gives us the same result, $i = \mathscr{E}/R$, as the energy method (Eq. 27-2).

If we apply the loop rule to a complete *counterclockwise* walk around the circuit, the rule gives us

$$-\mathscr{E} + iR = 0$$

and we again find that $i = \mathscr{E}/R$. Thus, you may mentally circle a loop in either direction to apply the loop rule.

To prepare for circuits more complex than that of Fig. 27-3, let us set down two rules for finding potential differences as we move around a loop:

> ⭐ **RESISTANCE RULE:** For a move through a resistance in the direction of the current, the change in potential is $-iR$; in the opposite direction it is $+iR$.

> ⭐ **EMF RULE:** For a move through an ideal emf device in the direction of the emf arrow, the change in potential is $+\mathscr{E}$; in the opposite direction it is $-\mathscr{E}$.

✓ Checkpoint 1

The figure shows the current i in a single-loop circuit with a battery B and a resistance R (and wires of negligible resistance). (a) Should the emf arrow at B be drawn pointing leftward or rightward? At points a, b, and c, rank (b) the magnitude of the current, (c) the electric potential, and (d) the electric potential energy of the charge carriers, greatest first.

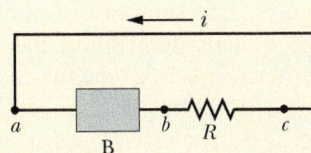

Other Single-Loop Circuits

Next we extend the simple circuit of Fig. 27-3 in two ways.

Internal Resistance

Figure 27-4*a* shows a real battery, with internal resistance *r*, wired to an external resistor of resistance *R*. The internal resistance of the battery is the electrical resistance of the conducting materials of the battery and thus is an unremovable feature of the battery. In Fig. 27-4*a*, however, the battery is drawn as if it could be separated into an ideal battery with emf $\mathscr{E}$ and a resistor of resistance *r*. The order in which the symbols for these separated parts are drawn does not matter.

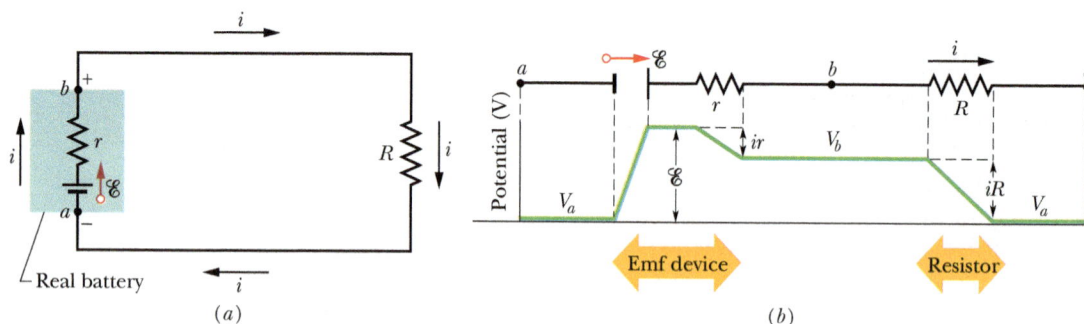

Figure 27-4 (*a*) A single-loop circuit containing a real battery having internal resistance *r* and emf $\mathscr{E}$. (*b*) The same circuit, now spread out in a line. The potentials encountered in traversing the circuit clockwise from *a* are also shown. The potential V_a is arbitrarily assigned a value of zero, and other potentials in the circuit are graphed relative to V_a.

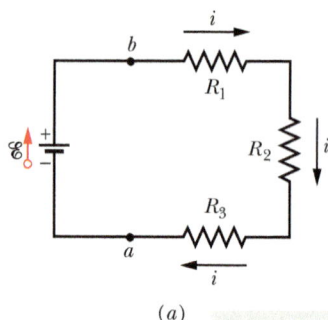

If we apply the loop rule clockwise beginning at point *a*, the *changes* in potential give us

$$\mathscr{E} - ir - iR = 0. \tag{27-3}$$

Solving for the current, we find

$$i = \frac{\mathscr{E}}{R + r}. \tag{27-4}$$

Note that this equation reduces to Eq. 27-2 if the battery is ideal—that is, if $r = 0$.

Figure 27-4*b* shows graphically the changes in electric potential around the circuit. (To better link Fig. 27-4*b* with the *closed circuit* in Fig. 27-4*a*, imagine curling the graph into a cylinder with point *a* at the left overlapping point *a* at the right.) Note how traversing the circuit is like walking around a (potential) mountain back to your starting point—you return to the starting elevation.

In this book, when a battery is not described as real or if no internal resistance is indicated, you can generally assume that it is ideal—but, of course, in the real world batteries are always real and have internal resistance.

Resistances in Series

Figure 27-5*a* shows three resistances connected **in series** to an ideal battery with emf $\mathscr{E}$. This description has little to do with how the resistances are drawn. Rather, "in series" means that the resistances are wired one after another and that a potential difference *V* is applied across the two ends of the series. In Fig. 27-5*a*, the resistances are connected one after another between *a* and *b*, and a potential difference is maintained across *a* and *b* by the battery. The potential differences that then exist across the resistances in the series produce identical currents *i* in them. In general,

Series resistors and their equivalent have the same current ("ser-i").

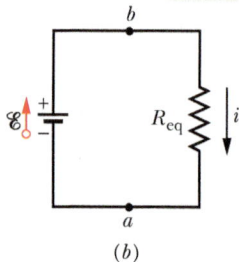

Figure 27-5 (*a*) Three resistors are connected in series between points *a* and *b*. (*b*) An equivalent circuit, with the three resistors replaced with their equivalent resistance R_{eq}.

⭐ When a potential difference V is applied across resistances connected in series, the resistances have identical currents i. The sum of the potential differences across the resistances is equal to the applied potential difference V.

Note that charge moving through the series resistances can move along only a single route. If there are additional routes, so that the currents in different resistances are different, the resistances are not connected in series.

⭐ Resistances connected in series can be replaced with an equivalent resistance R_{eq} that has the same current i and the same *total* potential difference V as the actual resistances.

You might remember that R_{eq} and all the actual series resistances have the same current i with the nonsense word "ser-i." Figure 27-5b shows the equivalent resistance R_{eq} that can replace the three resistances of Fig. 27-5a.

To derive an expression for R_{eq} in Fig. 27-5b, we apply the loop rule to both circuits. For Fig. 27-5a, starting at a and going clockwise around the circuit, we find

$$\mathscr{E} - iR_1 - iR_2 - iR_3 = 0,$$

or

$$i = \frac{\mathscr{E}}{R_1 + R_2 + R_3}. \tag{27-5}$$

For Fig. 27-5b, with the three resistances replaced with a single equivalent resistance R_{eq}, we find

$$\mathscr{E} - iR_{eq} = 0,$$

or

$$i = \frac{\mathscr{E}}{R_{eq}}. \tag{27-6}$$

Comparison of Eqs. 27-5 and 27-6 shows that

$$R_{eq} = R_1 + R_2 + R_3.$$

The extension to n resistances is straightforward and is

$$R_{eq} = \sum_{j=1}^{n} R_j \quad (n \text{ resistances in series}). \tag{27-7}$$

Note that when resistances are in series, their equivalent resistance is greater than any of the individual resistances.

✅ **Checkpoint 2**

In Fig. 27-5a, if $R_1 > R_2 > R_3$, rank the three resistances according to (a) the current through them and (b) the potential difference across them, greatest first.

Potential Difference Between Two Points

We often want to find the potential difference between two points in a circuit. For example, in Fig. 27-6, what is the potential difference $V_b - V_a$ between points a and b? To find out, let's start at point a (at potential V_a) and move through the battery to point b (at potential V_b) while keeping track of the potential changes we encounter. When we pass through the battery's emf, the potential increases by $\mathscr{E}$. When we pass through the battery's internal resistance r, we move in the direction of the current and thus the potential decreases by ir. We are then at the

The internal resistance reduces the potential difference between the terminals.

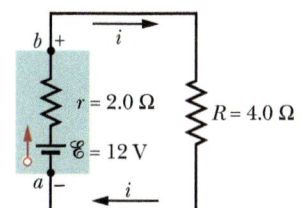

Figure 27-6 Points a and b, which are at the terminals of a real battery, differ in potential.

potential of point b and we have

$$V_a + \mathcal{E} - ir = V_b,$$

or
$$V_b - V_a = \mathcal{E} - ir. \qquad (27\text{-}8)$$

To evaluate this expression, we need the current i. Note that the circuit is the same as in Fig. 27-4a, for which Eq. 27-4 gives the current as

$$i = \frac{\mathcal{E}}{R + r}. \qquad (27\text{-}9)$$

Substituting this equation into Eq. 27-8 gives us

$$V_b - V_a = \mathcal{E} - \frac{\mathcal{E}}{R + r} r$$

$$= \frac{\mathcal{E}}{R + r} R. \qquad (27\text{-}10)$$

Now substituting the data given in Fig. 27-6, we have

$$V_b - V_a = \frac{12 \text{ V}}{4.0 \ \Omega + 2.0 \ \Omega} \, 4.0 \ \Omega = 8.0 \text{ V}. \qquad (27\text{-}11)$$

Suppose, instead, we move from a to b counterclockwise, passing through resistor R rather than through the battery. Because we move opposite the current, the potential increases by iR. Thus,

$$V_a + iR = V_b$$

or
$$V_b - V_a = iR. \qquad (27\text{-}12)$$

Substituting for i from Eq. 27-9, we again find Eq. 27-10. Hence, substitution of the data in Fig. 27-6 yields the same result, $V_b - V_a = 8.0$ V. In general,

> To find the potential between any two points in a circuit, start at one point and traverse the circuit to the other point, following any path, and add algebraically the changes in potential you encounter.

Potential Difference Across a Real Battery

In Fig. 27-6, points a and b are located at the terminals of the battery. Thus, the potential difference $V_b - V_a$ is the terminal-to-terminal potential difference V across the battery. From Eq. 27-8, we see that

$$V = \mathcal{E} - ir. \qquad (27\text{-}13)$$

If the internal resistance r of the battery in Fig. 27-6 were zero, Eq. 27-13 tells us that V would be equal to the emf $\mathcal{E}$ of the battery—namely, 12 V. However, because $r = 2.0 \ \Omega$, Eq. 27-13 tells us that V is less than $\mathcal{E}$. From Eq. 27-11, we know that V is only 8.0 V. Note that the result depends on the value of the current through the battery. If the same battery were in a different circuit and had a different current through it, V would have some other value.

Grounding a Circuit

Figure 27-7a shows the same circuit as Fig. 27-6 except that here point a is directly connected to *ground*, as indicated by the common symbol $\perp$. *Grounding a circuit* usually means connecting the circuit to a conducting path to Earth's surface (actually to the electrically conducting moist dirt and rock below ground). Here, such a connection means only that the potential is defined to be zero at the grounding point in the circuit. Thus in Fig. 27-7a, the potential at a is defined to be $V_a = 0$. Equation 27-11 then tells us that the potential at b is $V_b = 8.0$ V.

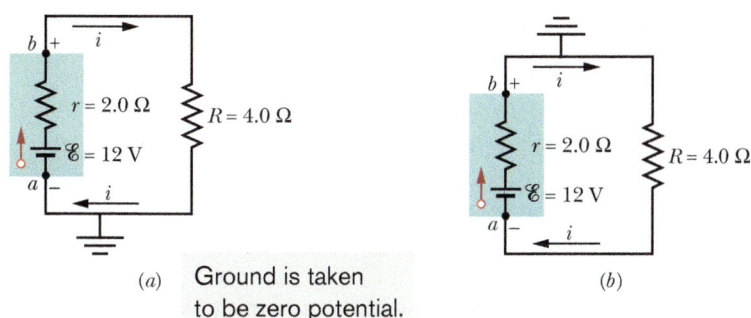

(a) Ground is taken to be zero potential. (b)

Figure 27-7 (a) Point a is directly connected to ground. (b) Point b is directly connected to ground.

Figure 27-7b is the same circuit except that point b is now directly connected to ground. Thus, the potential there is defined to be $V_b = 0$. Equation 27-11 now tells us that the potential at a is $V_a = -8.0$ V.

Power, Potential, and Emf

When a battery or some other type of emf device does work on the charge carriers to establish a current i, the device transfers energy from its source of energy (such as the chemical source in a battery) to the charge carriers. Because a real emf device has an internal resistance r, it also transfers energy to internal thermal energy via resistive dissipation (Module 26-5). Let us relate these transfers.

The net rate P of energy transfer from the emf device to the charge carriers is given by Eq. 26-26:

$$P = iV, \tag{27-14}$$

where V is the potential across the terminals of the emf device. From Eq. 27-13, we can substitute $V = \mathscr{E} - ir$ into Eq. 27-14 to find

$$P = i(\mathscr{E} - ir) = i\mathscr{E} - i^2 r. \tag{27-15}$$

From Eq. 26-27, we recognize the term $i^2 r$ in Eq. 27-15 as the rate P_r of energy transfer to thermal energy within the emf device:

$$P_r = i^2 r \quad \text{(internal dissipation rate).} \tag{27-16}$$

Then the term $i\mathscr{E}$ in Eq. 27-15 must be the rate P_{emf} at which the emf device transfers energy both to the charge carriers and to internal thermal energy. Thus,

$$P_{\text{emf}} = i\mathscr{E} \quad \text{(power of emf device).} \tag{27-17}$$

If a battery is being *recharged*, with a "wrong way" current through it, the energy transfer is then *from* the charge carriers *to* the battery—both to the battery's chemical energy and to the energy dissipated in the internal resistance r. The rate of change of the chemical energy is given by Eq. 27-17, the rate of dissipation is given by Eq. 27-16, and the rate at which the carriers supply energy is given by Eq. 27-14.

✓ Checkpoint 3

A battery has an emf of 12 V and an internal resistance of 2 Ω. Is the terminal-to-terminal potential difference greater than, less than, or equal to 12 V if the current in the battery is (a) from the negative to the positive terminal, (b) from the positive to the negative terminal, and (c) zero?

Sample Problem 27.01 Single-loop circuit with two real batteries

The emfs and resistances in the circuit of Fig. 27-8a have the following values:

$$\mathscr{E}_1 = 4.4 \text{ V}, \quad \mathscr{E}_2 = 2.1 \text{ V},$$

$$r_1 = 2.3 \, \Omega, \quad r_2 = 1.8 \, \Omega, \quad R = 5.5 \, \Omega.$$

(a) What is the current i in the circuit?

KEY IDEA

We can get an expression involving the current i in this single-loop circuit by applying the loop rule, in which we sum the potential changes around the full loop.

Calculations: Although knowing the direction of i is not necessary, we can easily determine it from the emfs of the

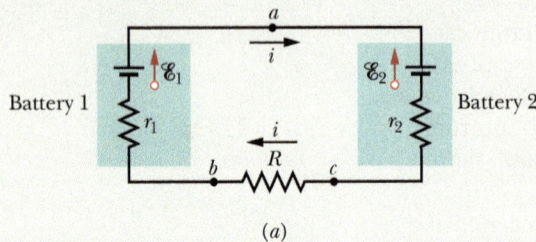

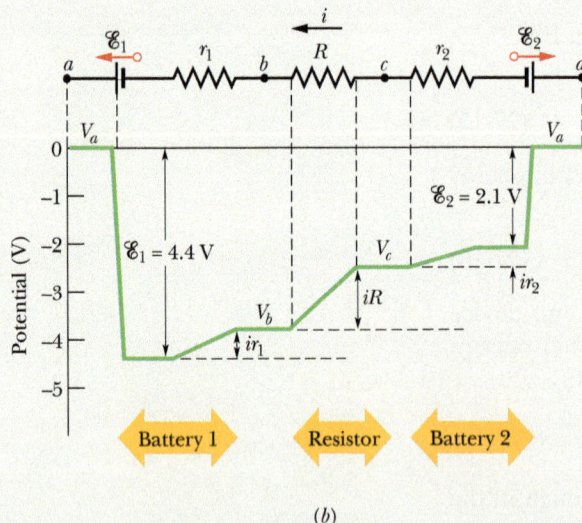

two batteries. Because $\mathscr{E}_1$ is greater than $\mathscr{E}_2$, battery 1 controls the direction of i, so the direction is clockwise. Let us then apply the loop rule by going counterclockwise—against the current—and starting at point a. (These decisions about where to start and which way you go are arbitrary but, once made, you must be consistent with decisions about the plus and minus signs.) We find

$$-\mathscr{E}_1 + ir_1 + iR + ir_2 + \mathscr{E}_2 = 0.$$

Check that this equation also results if we apply the loop rule clockwise or start at some point other than a. Also, take the time to compare this equation term by term with Fig. 27-8b, which shows the potential changes graphically (with the potential at point a arbitrarily taken to be zero).

Solving the above loop equation for the current i, we obtain

$$i = \frac{\mathscr{E}_1 - \mathscr{E}_2}{R + r_1 + r_2} = \frac{4.4 \text{ V} - 2.1 \text{ V}}{5.5 \, \Omega + 2.3 \, \Omega + 1.8 \, \Omega}$$

$$= 0.2396 \text{ A} \approx 240 \text{ mA}. \quad \text{(Answer)}$$

(b) What is the potential difference between the terminals of battery 1 in Fig. 27-8a?

KEY IDEA

We need to sum the potential differences between points a and b.

Calculations: Let us start at point b (effectively the negative terminal of battery 1) and travel clockwise through battery 1 to point a (effectively the positive terminal), keeping track of potential changes. We find that

$$V_b - ir_1 + \mathscr{E}_1 = V_a,$$

which gives us

$$V_a - V_b = -ir_1 + \mathscr{E}_1$$

$$= -(0.2396 \text{ A})(2.3 \, \Omega) + 4.4 \text{ V}$$

$$= +3.84 \text{ V} \approx 3.8 \text{ V}, \quad \text{(Answer)}$$

which is less than the emf of the battery. You can verify this result by starting at point b in Fig. 27-8a and traversing the circuit counterclockwise to point a. We learn two points here. (1) The potential difference between two points in a circuit is independent of the path we choose to go from one to the other. (2) When the current in the battery is in the "proper" direction, the terminal-to-terminal potential difference is low, that is, lower than the stated emf for the battery that you might find printed on the battery.

Figure 27-8 (a) A single-loop circuit containing two real batteries and a resistor. The batteries oppose each other; that is, they tend to send current in opposite directions through the resistor. (b) A graph of the potentials, counterclockwise from point a, with the potential at a arbitrarily taken to be zero. (To better link the circuit with the graph, mentally cut the circuit at a and then unfold the left side of the circuit toward the left and the right side of the circuit toward the right.)

27-2 MULTILOOP CIRCUITS

Learning Objectives

After reading this module, you should be able to . . .

27.17 Apply the junction rule.

27.18 Draw a schematic diagram for a battery and three parallel resistors and distinguish it from a diagram with a battery and three series resistors.

27.19 Identify that resistors in parallel have the same potential difference, which is the same value that their equivalent resistor has.

27.20 Calculate the resistance of the equivalent resistor of several resistors in parallel.

27.21 Identify that the total current through parallel resistors is the sum of the currents through the individual resistors.

27.22 For a circuit with a battery and some resistors in parallel and some in series, simplify the circuit in steps by finding

equivalent resistors, until the current through the battery can be determined, and then reverse the steps to find the currents and potential differences of the individual resistors.

27.23 If a circuit cannot be simplified by using equivalent resistors, identify the several loops in the circuit, choose names and directions for the currents in the branches, set up loop equations for the various loops, and solve these simultaneous equations for the unknown currents.

27.24 In a circuit with identical real batteries in series, replace them with a single ideal battery and a single resistor.

27.25 In a circuit with identical real batteries in parallel, replace them with a single ideal battery and a single resistor.

Key Idea

● When resistances are in parallel, they have the same potential difference. The equivalent resistance that can replace a parallel combination of resistances is given by

$$\frac{1}{R_{eq}} = \sum_{j=1}^{n} \frac{1}{R_j} \quad (n \text{ resistances in parallel}).$$

Multiloop Circuits

Figure 27-9 shows a circuit containing more than one loop. For simplicity, we assume the batteries are ideal. There are two *junctions* in this circuit, at b and d, and there are three *branches* connecting these junctions. The branches are the left branch (*bad*), the right branch (*bcd*), and the central branch (*bd*). What are the currents in the three branches?

We arbitrarily label the currents, using a different subscript for each branch. Current i_1 has the same value everywhere in branch *bad*, i_2 has the same value everywhere in branch *bcd*, and i_3 is the current through branch *bd*. The directions of the currents are assumed arbitrarily.

Consider junction d for a moment: Charge comes into that junction via incoming currents i_1 and i_3, and it leaves via outgoing current i_2. Because there is no variation in the charge at the junction, the total incoming current must equal the total outgoing current:

$$i_1 + i_3 = i_2. \tag{27-18}$$

You can easily check that applying this condition to junction b leads to exactly the same equation. Equation 27-18 thus suggests a general principle:

⭐ **JUNCTION RULE:** The sum of the currents entering any junction must be equal to the sum of the currents leaving that junction.

This rule is often called *Kirchhoff's junction rule* (or *Kirchhoff's current law*). It is simply a statement of the conservation of charge for a steady flow of charge—there is neither a buildup nor a depletion of charge at a junction. Thus, our basic tools for solving complex circuits are the *loop rule* (based on the conservation of energy) and the *junction rule* (based on the conservation of charge).

The current into the junction must equal the current out (charge is conserved).

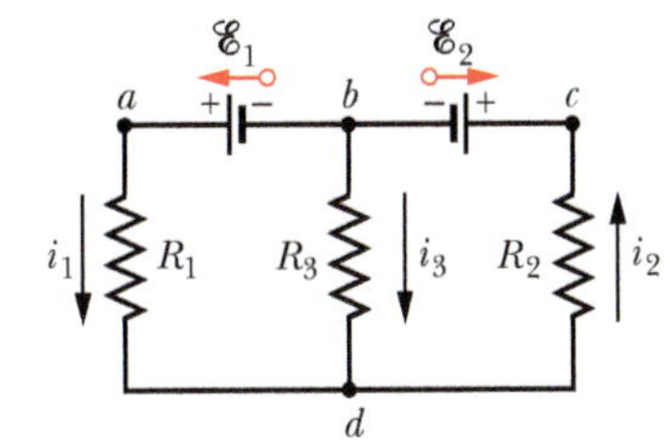

Figure 27-9 A multiloop circuit consisting of three branches: left-hand branch *bad*, right-hand branch *bcd*, and central branch *bd*. The circuit also consists of three loops: left-hand loop *badb*, right-hand loop *bcdb*, and big loop *badcb*.

Equation 27-18 is a single equation involving three unknowns. To solve the circuit completely (that is, to find all three currents), we need two more equations involving those same unknowns. We obtain them by applying the loop rule twice. In the circuit of Fig. 27-9, we have three loops from which to choose: the left-hand loop (*badb*), the right-hand loop (*bcdb*), and the big loop (*badcb*). Which two loops we choose does not matter—let's choose the left-hand loop and the right-hand loop.

If we traverse the left-hand loop in a counterclockwise direction from point *b*, the loop rule gives us

$$\mathscr{E}_1 - i_1 R_1 + i_3 R_3 = 0. \qquad (27\text{-}19)$$

If we traverse the right-hand loop in a counterclockwise direction from point *b*, the loop rule gives us

$$-i_3 R_3 - i_2 R_2 - \mathscr{E}_2 = 0. \qquad (27\text{-}20)$$

We now have three equations (Eqs. 27-18, 27-19, and 27-20) in the three unknown currents, and they can be solved by a variety of techniques.

If we had applied the loop rule to the big loop, we would have obtained (moving counterclockwise from *b*) the equation

$$\mathscr{E}_1 - i_1 R_1 - i_2 R_2 - \mathscr{E}_2 = 0.$$

However, this is merely the sum of Eqs. 27-19 and 27-20.

Resistances in Parallel

Figure 27-10*a* shows three resistances connected *in parallel* to an ideal battery of emf $\mathscr{E}$. The term "in parallel" means that the resistances are directly wired together on one side and directly wired together on the other side, and that a potential difference V is applied across the pair of connected sides. Thus, all three resistances have the same potential difference V across them, producing a current through each. In general,

When a potential difference V is applied across resistances connected in parallel, the resistances all have that same potential difference V.

In Fig. 27-10*a*, the applied potential difference V is maintained by the battery. In Fig. 27-10*b*, the three parallel resistances have been replaced with an equivalent resistance R_{eq}.

Parallel resistors and their equivalent have the same potential difference ("par-V").

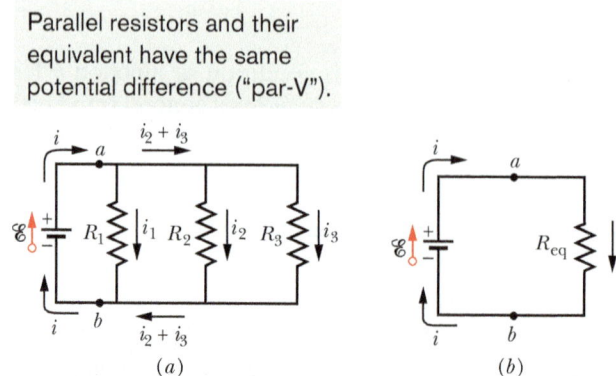

Figure 27-10 (*a*) Three resistors connected in parallel across points *a* and *b*. (*b*) An equivalent circuit, with the three resistors replaced with their equivalent resistance R_{eq}.

Resistances connected in parallel can be replaced with an equivalent resistance R_{eq} that has the same potential difference V and the same *total* current i as the actual resistances.

You might remember that R_{eq} and all the actual parallel resistances have the same potential difference V with the nonsense word "par-V."

To derive an expression for R_{eq} in Fig. 27-10b, we first write the current in each actual resistance in Fig. 27-10a as

$$i_1 = \frac{V}{R_1}, \quad i_2 = \frac{V}{R_2}, \quad \text{and} \quad i_3 = \frac{V}{R_3},$$

where V is the potential difference between a and b. If we apply the junction rule at point a in Fig. 27-10a and then substitute these values, we find

$$i = i_1 + i_2 + i_3 = V\left(\frac{1}{R_1} + \frac{1}{R_2} + \frac{1}{R_3}\right). \qquad (27\text{-}21)$$

If we replaced the parallel combination with the equivalent resistance R_{eq} (Fig. 27-10b), we would have

$$i = \frac{V}{R_{eq}}. \qquad (27\text{-}22)$$

Comparing Eqs. 27-21 and 27-22 leads to

$$\frac{1}{R_{eq}} = \frac{1}{R_1} + \frac{1}{R_2} + \frac{1}{R_3}. \qquad (27\text{-}23)$$

Extending this result to the case of n resistances, we have

$$\frac{1}{R_{eq}} = \sum_{j=1}^{n} \frac{1}{R_j} \quad (n \text{ resistances in parallel}). \qquad (27\text{-}24)$$

For the case of two resistances, the equivalent resistance is their product divided by their sum; that is,

$$R_{eq} = \frac{R_1 R_2}{R_1 + R_2}. \qquad (27\text{-}25)$$

Note that when two or more resistances are connected in parallel, the equivalent resistance is smaller than any of the combining resistances. Table 27-1 summarizes the equivalence relations for resistors and capacitors in series and in parallel.

Table 27-1 Series and Parallel Resistors and Capacitors

Series	Parallel	Series	Parallel
Resistors		Capacitors	
$R_{eq} = \sum_{j=1}^{n} R_j$ Eq. 27-7	$\dfrac{1}{R_{eq}} = \sum_{j=1}^{n} \dfrac{1}{R_j}$ Eq. 27-24	$\dfrac{1}{C_{eq}} = \sum_{j=1}^{n} \dfrac{1}{C_j}$ Eq. 25-20	$C_{eq} = \sum_{j=1}^{n} C_j$ Eq. 25-19
Same current through all resistors	Same potential difference across all resistors	Same charge on all capacitors	Same potential difference across all capacitors

Checkpoint 4

A battery, with potential V across it, is connected to a combination of two identical resistors and then has current i through it. What are the potential difference across and the current through either resistor if the resistors are (a) in series and (b) in parallel?

Sample Problem 27.02 Resistors in parallel and in series

Figure 27-11a shows a multiloop circuit containing one ideal battery and four resistances with the following values:

$$R_1 = 20\ \Omega, \quad R_2 = 20\ \Omega, \quad \mathscr{E} = 12\ \text{V},$$

$$R_3 = 30\ \Omega, \quad R_4 = 8.0\ \Omega.$$

(a) What is the current through the battery?

KEY IDEA

Noting that the current through the battery must also be the current through R_1, we see that we might find the current by applying the loop rule to a loop that includes R_1 because the current would be included in the potential difference across R_1.

Incorrect method: Either the left-hand loop or the big loop should do. Noting that the emf arrow of the battery points upward, so the current the battery supplies is clockwise, we might apply the loop rule to the left-hand loop, clockwise from point a. With i being the current through the battery, we would get

$$+\mathscr{E} - iR_1 - iR_2 - iR_4 = 0 \quad \text{(incorrect)}.$$

However, this equation is incorrect because it assumes that R_1, R_2, and R_4 all have the same current i. Resistances R_1 and R_4 do have the same current, because the current passing through R_4 must pass through the battery and then through R_1 with no change in value. However, that current splits at junction point b—only part passes through R_2, the rest through R_3.

Dead-end method: To distinguish the several currents in the circuit, we must label them individually as in Fig. 27-11b. Then, circling clockwise from a, we can write the loop rule for the left-hand loop as

$$+\mathscr{E} - i_1R_1 - i_2R_2 - i_1R_4 = 0.$$

Unfortunately, this equation contains two unknowns, i_1 and i_2; we would need at least one more equation to find them.

Successful method: A much easier option is to simplify the circuit of Fig. 27-11b by finding equivalent resistances. Note carefully that R_1 and R_2 are *not* in series and thus cannot be replaced with an equivalent resistance. However, R_2 and R_3 are in parallel, so we can use either Eq. 27-24 or Eq. 27-25 to find their equivalent resistance R_{23}. From the latter,

$$R_{23} = \frac{R_2R_3}{R_2 + R_3} = \frac{(20\ \Omega)(30\ \Omega)}{50\ \Omega} = 12\ \Omega.$$

We can now redraw the circuit as in Fig. 27-11c; note that the current through R_{23} must be i_1 because charge that moves through R_1 and R_4 must also move through R_{23}. For this simple one-loop circuit, the loop rule (applied clockwise from point a as in Fig. 27-11d) yields

$$+\mathscr{E} - i_1R_1 - i_1R_{23} - i_1R_4 = 0.$$

Substituting the given data, we find

$$12\ \text{V} - i_1(20\ \Omega) - i_1(12\ \Omega) - i_1(8.0\ \Omega) = 0,$$

which gives us

$$i_1 = \frac{12\ \text{V}}{40\ \Omega} = 0.30\ \text{A}. \qquad \text{(Answer)}$$

(b) What is the current i_2 through R_2?

KEY IDEAS

(1) we must now work backward from the equivalent circuit of Fig. 27-11d, where R_{23} has replaced R_2 and R_3. (2) Because R_2 and R_3 are in parallel, they both have the same potential difference across them as R_{23}.

Working backward: We know that the current through R_{23} is $i_1 = 0.30$ A. Thus, we can use Eq. 26-8 ($R = V/i$) and Fig. 27-11e to find the potential difference V_{23} across R_{23}. Setting $R_{23} = 12\ \Omega$ from (a), we write Eq. 26-8 as

$$V_{23} = i_1R_{23} = (0.30\ \text{A})(12\ \Omega) = 3.6\ \text{V}.$$

The potential difference across R_2 is thus also 3.6 V (Fig. 27-11f), so the current i_2 in R_2 must be, by Eq. 26-8 and Fig. 27-11g,

$$i_2 = \frac{V_2}{R_2} = \frac{3.6\ \text{V}}{20\ \Omega} = 0.18\ \text{A}. \qquad \text{(Answer)}$$

(c) What is the current i_3 through R_3?

KEY IDEAS

We can answer by using either of two techniques: (1) Apply Eq. 26-8 as we just did. (2) Use the junction rule, which tells us that at point b in Fig. 27-11b, the incoming current i_1 and the outgoing currents i_2 and i_3 are related by

$$i_1 = i_2 + i_3.$$

Calculation: Rearranging this junction-rule result yields the result displayed in Fig. 27-11g:

$$i_3 = i_1 - i_2 = 0.30\ \text{A} - 0.18\ \text{A}$$
$$= 0.12\ \text{A}. \qquad \text{(Answer)}$$

PLUS Additional examples, video, and practice available at *WileyPLUS*

The equivalent of parallel resistors is smaller.

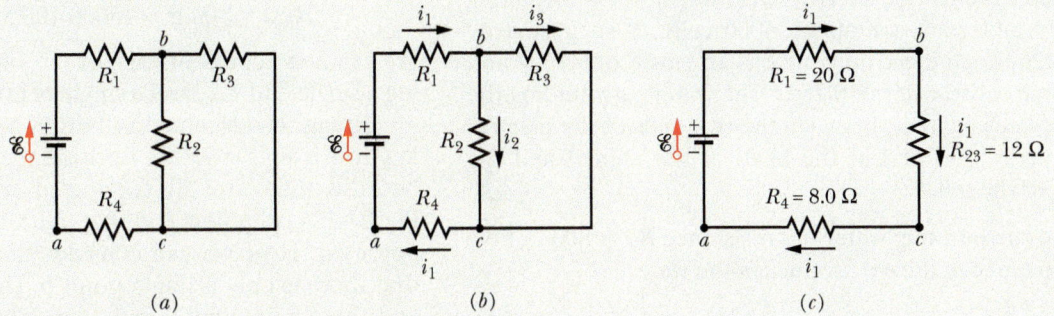

(a)

(b)

(c)

Applying the loop rule yields the current.

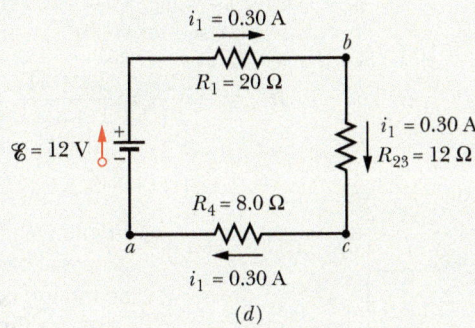

(d)

Applying $V = iR$ yields the potential difference.

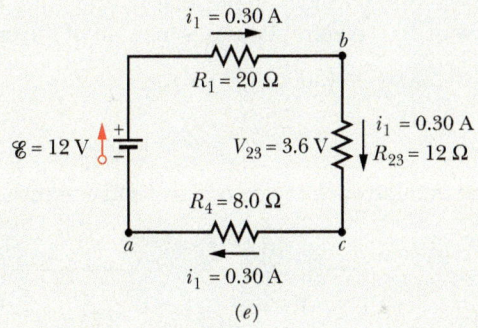

(e)

Parallel resistors and their equivalent have the same V ("par-V").

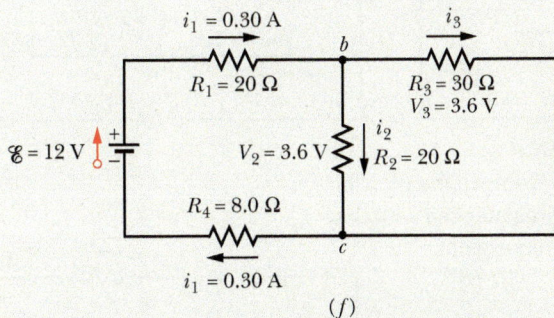

(f)

Applying $i = V/R$ yields the current.

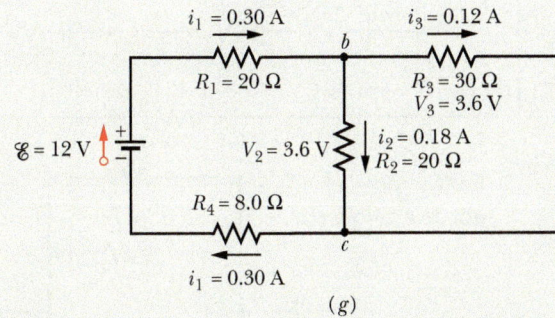

(g)

Figure 27-11 (a) A circuit with an ideal battery. (b) Label the currents. (c) Replacing the parallel resistors with their equivalent. (d)–(g) Working backward to find the currents through the parallel resistors.

Sample Problem 27.03 Many real batteries in series and in parallel in an electric fish

Electric fish can generate current with biological emf cells called *electroplaques*. In the South American eel they are arranged in 140 rows, each row stretching horizontally along the body and each containing 5000 cells, as suggested by Fig. 27-12a. Each electroplaque has an emf $\mathscr{E}$ of 0.15 V and an internal resistance r of 0.25 Ω. The water surrounding the eel completes a circuit between the two ends of the electroplaque array, one end at the head of the animal and the other near the tail.

(a) If the surrounding water has resistance $R_w = 800$ Ω, how much current can the eel produce in the water?

KEY IDEA

We can simplify the circuit of Fig. 27-12a by replacing combinations of emfs and internal resistances with equivalent emfs and resistances.

Calculations: We first consider a single row. The total emf $\mathscr{E}_{row}$ along a row of 5000 electroplaques is the sum of the emfs:

$$\mathscr{E}_{row} = 5000\mathscr{E} = (5000)(0.15 \text{ V}) = 750 \text{ V}.$$

The total resistance R_{row} along a row is the sum of the internal resistances of the 5000 electroplaques:

$$R_{row} = 5000r = (5000)(0.25 \text{ Ω}) = 1250 \text{ Ω}.$$

We can now represent each of the 140 identical rows as having a single emf $\mathscr{E}_{row}$ and a single resistance R_{row} (Fig. 27-12b).

In Fig. 27-12b, the emf between point a and point b on any row is $\mathscr{E}_{row} = 750$ V. Because the rows are identical and because they are all connected together at the left in Fig. 27-12b, all points b in that figure are at the same electric potential. Thus, we can consider them to be connected so that there is only a single point b. The emf between point a and this single point b is $\mathscr{E}_{row} = 750$ V, so we can draw the circuit as shown in Fig. 27-12c.

Between points b and c in Fig. 27-12c are 140 resistances $R_{row} = 1250$ Ω, all in parallel. The equivalent resistance R_{eq} of this combination is given by Eq. 27-24 as

$$\frac{1}{R_{eq}} = \sum_{j=1}^{140} \frac{1}{R_j} = 140 \frac{1}{R_{row}},$$

or

$$R_{eq} = \frac{R_{row}}{140} = \frac{1250 \text{ Ω}}{140} = 8.93 \text{ Ω}.$$

First, reduce each row to one emf and one resistance.

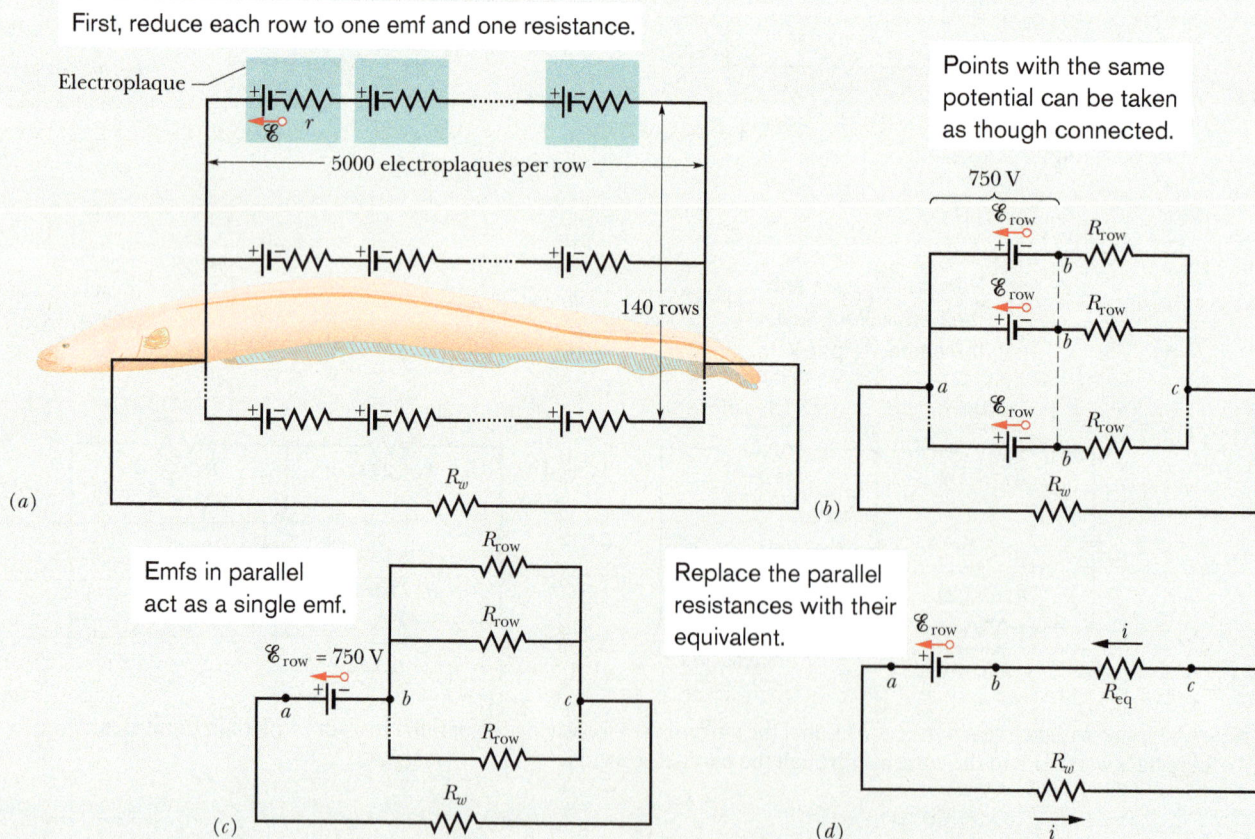

Points with the same potential can be taken as though connected.

Emfs in parallel act as a single emf.

Replace the parallel resistances with their equivalent.

Figure 27-12 (a) A model of the electric circuit of an eel in water. Along each of 140 rows extending from the head to the tail of the eel, there are 5000 electroplaques. The surrounding water has resistance R_w. (b) The emf $\mathscr{E}_{row}$ and resistance R_{row} of each row. (c) The emf between points a and b is $\mathscr{E}_{row}$. Between points b and c are 140 parallel resistances R_{row}. (d) The simplified circuit.

Replacing the parallel combination with R_{eq}, we obtain the simplified circuit of Fig. 27-12d. Applying the loop rule to this circuit counterclockwise from point b, we have

$$\mathscr{E}_{row} - iR_w - iR_{eq} = 0.$$

Solving for i and substituting the known data, we find

$$i = \frac{\mathscr{E}_{row}}{R_w + R_{eq}} = \frac{750\ V}{800\ \Omega + 8.93\ \Omega}$$

$$= 0.927\ A \approx 0.93\ A. \qquad \text{(Answer)}$$

If the head or tail of the eel is near a fish, some of this current could pass along a narrow path through the fish, stunning or killing it.

(b) How much current i_{row} travels through each row of Fig. 27-12a?

KEY IDEA

Because the rows are identical, the current into and out of the eel is evenly divided among them.

Calculation: Thus, we write

$$i_{row} = \frac{i}{140} = \frac{0.927\ A}{140} = 6.6 \times 10^{-3}\ A. \quad \text{(Answer)}$$

Thus, the current through each row is small, so that the eel need not stun or kill itself when it stuns or kills a fish.

Sample Problem 27.04 Multiloop circuit and simultaneous loop equations

Figure 27-13 shows a circuit whose elements have the following values: $\mathscr{E}_1 = 3.0\ V$, $\mathscr{E}_2 = 6.0\ V$, $R_1 = 2.0\ \Omega$, $R_2 = 4.0\ \Omega$. The three batteries are ideal batteries. Find the magnitude and direction of the current in each of the three branches.

KEY IDEAS

It is not worthwhile to try to simplify this circuit, because no two resistors are in parallel, and the resistors that are in series (those in the right branch or those in the left branch) present no problem. So, our plan is to apply the junction and loop rules.

Junction rule: Using arbitrarily chosen directions for the currents as shown in Fig. 27-13, we apply the junction rule at point a by writing

$$i_3 = i_1 + i_2. \qquad (27\text{-}26)$$

An application of the junction rule at junction b gives only the same equation, so we next apply the loop rule to any two of the three loops of the circuit.

Left-hand loop: We first arbitrarily choose the left-hand loop, arbitrarily start at point b, and arbitrarily traverse the loop in the clockwise direction, obtaining

$$-i_1R_1 + \mathscr{E}_1 - i_1R_1 - (i_1 + i_2)R_2 - \mathscr{E}_2 = 0,$$

where we have used $(i_1 + i_2)$ instead of i_3 in the middle branch. Substituting the given data and simplifying yield

$$i_1(8.0\ \Omega) + i_2(4.0\ \Omega) = -3.0\ V. \qquad (27\text{-}27)$$

Right-hand loop: For our second application of the loop rule, we arbitrarily choose to traverse the right-hand loop counterclockwise from point b, finding

$$-i_2R_1 + \mathscr{E}_2 - i_2R_1 - (i_1 + i_2)R_2 - \mathscr{E}_2 = 0.$$

Substituting the given data and simplifying yield

$$i_1(4.0\ \Omega) + i_2(8.0\ \Omega) = 0. \qquad (27\text{-}28)$$

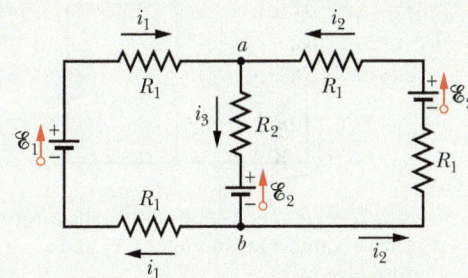

Figure 27-13 A multiloop circuit with three ideal batteries and five resistances.

Combining equations: We now have a system of two equations (Eqs. 27-27 and 27-28) in two unknowns (i_1 and i_2) to solve either "by hand" (which is easy enough here) or with a "math package." (One solution technique is Cramer's rule, given in Appendix E.) We find

$$i_1 = -0.50\ A. \qquad (27\text{-}29)$$

(The minus sign signals that our arbitrary choice of direction for i_1 in Fig. 27-13 is wrong, but we must wait to correct it.) Substituting $i_1 = -0.50\ A$ into Eq. 27-28 and solving for i_2 then give us

$$i_2 = 0.25\ A. \qquad \text{(Answer)}$$

With Eq. 27-26 we then find that

$$i_3 = i_1 + i_2 = -0.50\ A + 0.25\ A$$

$$= -0.25\ A.$$

The positive answer we obtained for i_2 signals that our choice of direction for that current is correct. However, the negative answers for i_1 and i_3 indicate that our choices for those currents are wrong. Thus, as a *last step* here, we correct the answers by reversing the arrows for i_1 and i_3 in Fig. 27-13 and then writing

$$i_1 = 0.50\ A \quad \text{and} \quad i_3 = 0.25\ A. \qquad \text{(Answer)}$$

Caution: Always make any such correction as the last step and not before calculating *all* the currents.

27-3 THE AMMETER AND THE VOLTMETER

Learning Objective

After reading this module, you should be able to . . .

27.26 Explain the use of an ammeter and a voltmeter, including the resistance required of each in order not to affect the measured quantities.

Key Idea

● Here are three measurement instruments used with circuits: An ammeter measures current. A voltmeter measures voltage (potential differences). A multimeter can be used to measure current, voltage, or resistance.

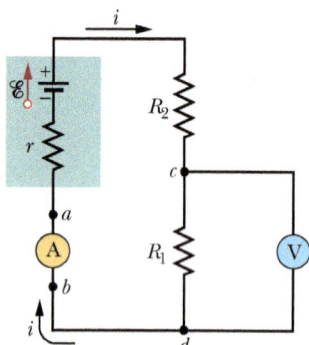

Figure 27-14 A single-loop circuit, showing how to connect an ammeter (A) and a voltmeter (V).

The Ammeter and the Voltmeter

An instrument used to measure currents is called an *ammeter*. To measure the current in a wire, you usually have to break or cut the wire and insert the ammeter so that the current to be measured passes through the meter. (In Fig. 27-14, ammeter A is set up to measure current *i*.) It is essential that the resistance R_A of the ammeter be very much smaller than other resistances in the circuit. Otherwise, the very presence of the meter will change the current to be measured.

A meter used to measure potential differences is called a *voltmeter*. To find the potential difference between any two points in the circuit, the voltmeter terminals are connected between those points without breaking or cutting the wire. (In Fig. 27-14, voltmeter V is set up to measure the voltage across R_1.) It is essential that the resistance R_V of a voltmeter be very much larger than the resistance of any circuit element across which the voltmeter is connected. Otherwise, the meter alters the potential difference that is to be measured.

Often a single meter is packaged so that, by means of a switch, it can be made to serve as either an ammeter or a voltmeter—and usually also as an *ohmmeter*, designed to measure the resistance of any element connected between its terminals. Such a versatile unit is called a *multimeter*.

27-4 RC CIRCUITS

Learning Objectives

After reading this module, you should be able to . . .

27.27 Draw schematic diagrams of charging and discharging RC circuits.

27.28 Write the loop equation (a differential equation) for a charging RC circuit.

27.29 Write the loop equation (a differential equation) for a discharging RC circuit.

27.30 For a capacitor in a charging or discharging RC circuit, apply the relationship giving the charge as a function of time.

27.31 From the function giving the charge as a function of time in a charging or discharging RC circuit, find the capacitor's potential difference as a function of time.

27.32 In a charging or discharging RC circuit, find the resistor's current and potential difference as functions of time.

27.33 Calculate the capacitive time constant τ.

27.34 For a charging RC circuit and a discharging RC circuit, determine the capacitor's charge and potential difference at the start of the process and then a long time later.

Key Ideas

● When an emf $\mathscr{E}$ is applied to a resistance R and capacitance C in series, the charge on the capacitor increases according to

$$q = C\mathscr{E}(1 - e^{-t/RC}) \quad \text{(charging a capacitor)},$$

in which $C\mathscr{E} = q_0$ is the equilibrium (final) charge and $RC = \tau$ is the capacitive time constant of the circuit.

● During the charging, the current is

$$i = \frac{dq}{dt} = \left(\frac{\mathscr{E}}{R}\right)e^{-t/RC} \quad \text{(charging a capacitor)}.$$

● When a capacitor discharges through a resistance R, the charge on the capacitor decays according to

$$q = q_0 e^{-t/RC} \quad \text{(discharging a capacitor)}.$$

● During the discharging, the current is

$$i = \frac{dq}{dt} = -\left(\frac{q_0}{RC}\right)e^{-t/RC} \quad \text{(discharging a capacitor)}.$$

RC Circuits

In preceding modules we dealt only with circuits in which the currents did not vary with time. Here we begin a discussion of time-varying currents.

Charging a Capacitor

The capacitor of capacitance C in Fig. 27-15 is initially uncharged. To charge it, we close switch S on point a. This completes an *RC series circuit* consisting of the capacitor, an ideal battery of emf $\mathscr{E}$, and a resistance R.

From Module 25-1, we already know that as soon as the circuit is complete, charge begins to flow (current exists) between a capacitor plate and a battery terminal on each side of the capacitor. This current increases the charge q on the plates and the potential difference V_C ($= q/C$) across the capacitor. When that potential difference equals the potential difference across the battery (which here is equal to the emf $\mathscr{E}$), the current is zero. From Eq. 25-1 ($q = CV$), the *equilibrium* (final) *charge* on the then fully charged capacitor is equal to $C\mathscr{E}$.

Here we want to examine the charging process. In particular we want to know how the charge $q(t)$ on the capacitor plates, the potential difference $V_C(t)$ across the capacitor, and the current $i(t)$ in the circuit vary with time during the charging process. We begin by applying the loop rule to the circuit, traversing it clockwise from the negative terminal of the battery. We find

$$\mathscr{E} - iR - \frac{q}{C} = 0. \tag{27-30}$$

The last term on the left side represents the potential difference across the capacitor. The term is negative because the capacitor's top plate, which is connected to the battery's positive terminal, is at a higher potential than the lower plate. Thus, there is a drop in potential as we move down through the capacitor.

We cannot immediately solve Eq. 27-30 because it contains two variables, i and q. However, those variables are not independent but are related by

$$i = \frac{dq}{dt}. \tag{27-31}$$

Substituting this for i in Eq. 27-30 and rearranging, we find

$$R\frac{dq}{dt} + \frac{q}{C} = \mathscr{E} \quad \text{(charging equation).} \tag{27-32}$$

This differential equation describes the time variation of the charge q on the capacitor in Fig. 27-15. To solve it, we need to find the function $q(t)$ that satisfies this equation and also satisfies the condition that the capacitor be initially uncharged; that is, $q = 0$ at $t = 0$.

We shall soon show that the solution to Eq. 27-32 is

$$q = C\mathscr{E}(1 - e^{-t/RC}) \quad \text{(charging a capacitor).} \tag{27-33}$$

(Here e is the exponential base, 2.718 . . . , and not the elementary charge.) Note that Eq. 27-33 does indeed satisfy our required initial condition, because at $t = 0$ the term $e^{-t/RC}$ is unity; so the equation gives $q = 0$. Note also that as t goes to infinity (that is, a long time later), the term $e^{-t/RC}$ goes to zero; so the equation gives the proper value for the full (equilibrium) charge on the capacitor—namely, $q = C\mathscr{E}$. A plot of $q(t)$ for the charging process is given in Fig. 27-16a.

The derivative of $q(t)$ is the current $i(t)$ charging the capacitor:

$$i = \frac{dq}{dt} = \left(\frac{\mathscr{E}}{R}\right)e^{-t/RC} \quad \text{(charging a capacitor).} \tag{27-34}$$

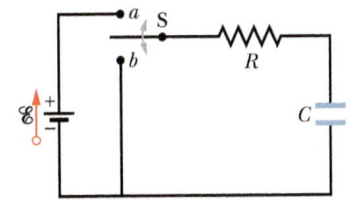

Figure 27-15 When switch S is closed on a, the capacitor is *charged* through the resistor. When the switch is afterward closed on b, the capacitor *discharges* through the resistor.

The capacitor's charge grows as the resistor's current dies out.

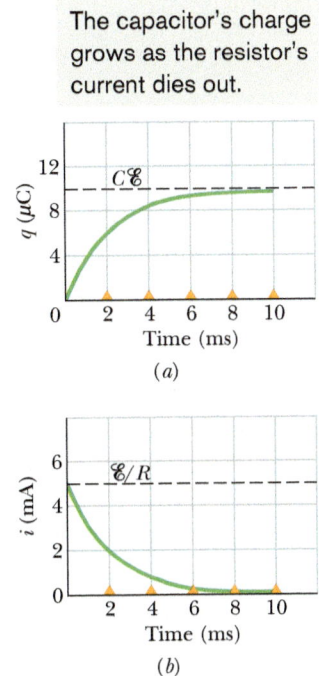

Figure 27-16 (a) A plot of Eq. 27-33, which shows the buildup of charge on the capacitor of Fig. 27-15. (b) A plot of Eq. 27-34, which shows the decline of the charging current in the circuit of Fig. 27-15. The curves are plotted for $R = 2000\,\Omega$, $C = 1\,\mu\text{F}$, and $\mathscr{E} = 10$ V; the small triangles represent successive intervals of one time constant τ.

A plot of $i(t)$ for the charging process is given in Fig. 27-16b. Note that the current has the initial value $\mathscr{E}/R$ and that it decreases to zero as the capacitor becomes fully charged.

A capacitor that is being charged initially acts like ordinary connecting wire relative to the charging current. A long time later, it acts like a broken wire.

By combining Eq. 25-1 ($q = CV$) and Eq. 27-33, we find that the potential difference $V_C(t)$ across the capacitor during the charging process is

$$V_C = \frac{q}{C} = \mathscr{E}(1 - e^{-t/RC}) \quad \text{(charging a capacitor).} \qquad (27\text{-}35)$$

This tells us that $V_C = 0$ at $t = 0$ and that $V_C = \mathscr{E}$ when the capacitor becomes fully charged as $t \rightarrow \infty$.

The Time Constant

The product RC that appears in Eqs. 27-33, 27-34, and 27-35 has the dimensions of time (both because the argument of an exponential must be dimensionless and because, in fact, $1.0 \, \Omega \times 1.0 \, \text{F} = 1.0 \, \text{s}$). The product RC is called the **capacitive time constant** of the circuit and is represented with the symbol τ:

$$\tau = RC \quad \text{(time constant).} \qquad (27\text{-}36)$$

From Eq. 27-33, we can now see that at time $t = \tau \, (= RC)$, the charge on the initially uncharged capacitor of Fig. 27-15 has increased from zero to

$$q = C\mathscr{E}(1 - e^{-1}) = 0.63C\mathscr{E}. \qquad (27\text{-}37)$$

In words, during the first time constant τ the charge has increased from zero to 63% of its final value $C\mathscr{E}$. In Fig. 27-16, the small triangles along the time axes mark successive intervals of one time constant during the charging of the capacitor. The charging times for RC circuits are often stated in terms of τ. For example, a circuit with $\tau = 1 \, \mu\text{s}$ charges quickly while one with $\tau = 100 \, \text{s}$ charges much more slowly,

Discharging a Capacitor

Assume now that the capacitor of Fig. 27-15 is fully charged to a potential V_0 equal to the emf $\mathscr{E}$ of the battery. At a new time $t = 0$, switch S is thrown from a to b so that the capacitor can *discharge* through resistance R. How do the charge $q(t)$ on the capacitor and the current $i(t)$ through the discharge loop of capacitor and resistance now vary with time?

The differential equation describing $q(t)$ is like Eq. 27-32 except that now, with no battery in the discharge loop, $\mathscr{E} = 0$. Thus,

$$R\frac{dq}{dt} + \frac{q}{C} = 0 \quad \text{(discharging equation).} \qquad (27\text{-}38)$$

The solution to this differential equation is

$$q = q_0 e^{-t/RC} \quad \text{(discharging a capacitor),} \qquad (27\text{-}39)$$

where $q_0 \, (= CV_0)$ is the initial charge on the capacitor. You can verify by substitution that Eq. 27-39 is indeed a solution of Eq. 27-38.

Equation 27-39 tells us that q decreases exponentially with time, at a rate that is set by the capacitive time constant $\tau = RC$. At time $t = \tau$, the capacitor's charge has been reduced to $q_0 e^{-1}$, or about 37% of the initial value. Note that a greater τ means a greater discharge time.

Differentiating Eq. 27-39 gives us the current $i(t)$:

$$i = \frac{dq}{dt} = -\left(\frac{q_0}{RC}\right) e^{-t/RC} \quad \text{(discharging a capacitor).} \quad (27\text{-}40)$$

This tells us that the current also decreases exponentially with time, at a rate set by τ. The initial current i_0 is equal to q_0/RC. Note that you can find i_0 by simply applying the loop rule to the circuit at $t = 0$; just then the capacitor's initial potential V_0 is connected across the resistance R, so the current must be $i_0 = V_0/R = (q_0/C)/R = q_0/RC$. The minus sign in Eq. 27-40 can be ignored; it merely means that the capacitor's charge q is decreasing.

Derivation of Eq. 27-33

To solve Eq. 27-32, we first rewrite it as

$$\frac{dq}{dt} + \frac{q}{RC} = \frac{\mathscr{E}}{R}. \quad (27\text{-}41)$$

The general solution to this differential equation is of the form

$$q = q_p + Ke^{-at}, \quad (27\text{-}42)$$

where q_p is a *particular solution* of the differential equation, K is a constant to be evaluated from the initial conditions, and $a = 1/RC$ is the coefficient of q in Eq. 27-41. To find q_p, we set $dq/dt = 0$ in Eq. 27-41 (corresponding to the final condition of no further charging), let $q = q_p$, and solve, obtaining

$$q_p = C\mathscr{E}. \quad (27\text{-}43)$$

To evaluate K, we first substitute this into Eq. 27-42 to get

$$q = C\mathscr{E} + Ke^{-at}.$$

Then substituting the initial conditions $q = 0$ and $t = 0$ yields

$$0 = C\mathscr{E} + K,$$

or $K = -C\mathscr{E}$. Finally, with the values of q_p, a, and K inserted, Eq. 27-42 becomes

$$q = C\mathscr{E} - C\mathscr{E}e^{-t/RC},$$

which, with a slight modification, is Eq. 27-33.

✓ Checkpoint 5

The table gives four sets of values for the circuit elements in Fig. 27-15. Rank the sets according to (a) the initial current (as the switch is closed on *a*) and (b) the time required for the current to decrease to half its initial value, greatest first.

	1	2	3	4
$\mathscr{E}$ (V)	12	12	10	10
R (Ω)	2	3	10	5
C (μF)	3	2	0.5	2

Sample Problem 27.05 Discharging an *RC* circuit to avoid a fire in a race car pit stop

As a car rolls along pavement, electrons move from the pavement first onto the tires and then onto the car body. The car stores this excess charge and the associated electric potential energy as if the car body were one plate of a capacitor and the pavement were the other plate (Fig. 27-17a). When the car stops, it discharges its excess charge and energy through the tires, just as a capacitor can discharge through a resistor. If a conducting object comes within a few centimeters of the car before the car is discharged, the remaining energy can be suddenly transferred to a spark between the car and the object. Suppose the conducting object is a fuel dispenser. The spark will not ignite the fuel and cause a fire if the spark energy is less than the critical value $U_{fire} = 50$ mJ.

When the car of Fig. 27-17a stops at time $t = 0$, the car–ground potential difference is $V_0 = 30$ kV. The car–ground capacitance is $C = 500$ pF, and the resistance of *each* tire is $R_{tire} = 100$ GΩ. How much time does the car take to discharge through the tires to drop below the critical value U_{fire}?

KEY IDEAS

(1) At any time t, a capacitor's stored electric potential energy U is related to its stored charge q according to Eq. 25-21 ($U = q^2/2C$). (2) While a capacitor is discharging, the charge decreases with time according to Eq. 27-39 ($q = q_0 e^{-t/RC}$).

Calculations: We can treat the tires as resistors that are connected to one another at their tops via the car body and at their bottoms via the pavement. Figure 27-17b shows how the four resistors are connected in parallel across the car's capacitance, and Fig. 27-17c shows their equivalent resistance R. From Eq. 27-24, R is given by

$$\frac{1}{R} = \frac{1}{R_{tire}} + \frac{1}{R_{tire}} + \frac{1}{R_{tire}} + \frac{1}{R_{tire}},$$

or $\quad R = \dfrac{R_{tire}}{4} = \dfrac{100 \times 10^9 \, \Omega}{4} = 25 \times 10^9 \, \Omega.$ (27-44)

When the car stops, it discharges its excess charge and energy through R. We now use our two Key Ideas to analyze the discharge. Substituting Eq. 27-39 into Eq. 25-21 gives

$$U = \frac{q^2}{2C} = \frac{(q_0 e^{-t/RC})^2}{2C}$$

$$= \frac{q_0^2}{2C} e^{-2t/RC}. \tag{27-45}$$

From Eq. 25-1 ($q = CV$), we can relate the initial charge q_0 on the car to the given initial potential difference V_0: $q_0 = CV_0$. Substituting this equation into Eq. 27-45 brings us to

$$U = \frac{(CV_0)^2}{2C} e^{-2t/RC} = \frac{CV_0^2}{2} e^{-2t/RC},$$

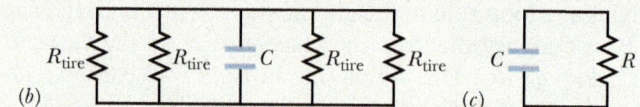

Effective capacitance — Tire resistance

(a)

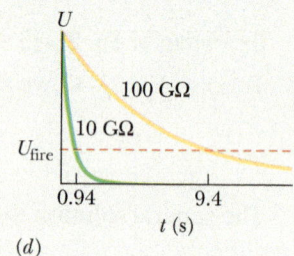

R_{tire} $\quad$ R_{tire} $\quad$ C $\quad$ R_{tire} $\quad$ R_{tire} $\qquad$ C $\quad$ R

(b) $\qquad\qquad\qquad\qquad\qquad$ (c)

Figure 27-17 (a) A charged car and the pavement acts like a capacitor that can discharge through the tires. (b) The effective circuit of the car–pavement capacitor, with four tire resistances R_{tire} connected in parallel. (c) The equivalent resistance R of the tires. (d) The electric potential energy U in the car–pavement capacitor decreases during discharge.

(d)

or $\qquad e^{-2t/RC} = \dfrac{2U}{CV_0^2}.$ (27-46)

Taking the natural logarithms of both sides, we obtain

$$-\frac{2t}{RC} = \ln\!\left(\frac{2U}{CV_0^2}\right),$$

or $\qquad t = -\dfrac{RC}{2} \ln\!\left(\dfrac{2U}{CV_0^2}\right).$ (27-47)

Substituting the given data, we find that the time the car takes to discharge to the energy level $U_{fire} = 50$ mJ is

$$t = -\frac{(25 \times 10^9 \, \Omega)(500 \times 10^{-12} \, F)}{2}$$

$$\times \ln\!\left(\frac{2(50 \times 10^{-3} \, J)}{(500 \times 10^{-12} \, F)(30 \times 10^3 \, V)^2}\right)$$

$$= 9.4 \text{ s.} \tag{Answer}$$

Fire or no fire: This car requires at least 9.4 s before fuel can be brought safely near it. A pit crew cannot wait that long. So the tires include some type of conducting material (such as carbon black) to lower the tire resistance and thus increase the discharge rate. Figure 27-17d shows the stored energy U versus time t for tire resistances of $R = 100$ GΩ (our value) and $R = 10$ GΩ. Note how much more rapidly a car discharges to level U_{fire} with the lower R value.

PLUS Additional examples, video, and practice available at *WileyPLUS*

Review & Summary

Emf An **emf device** does work on charges to maintain a potential difference between its output terminals. If dW is the work the device does to force positive charge dq from the negative to the positive terminal, then the **emf** (work per unit charge) of the device is

$$\mathscr{E} = \frac{dW}{dq} \quad \text{(definition of } \mathscr{E}\text{)}. \tag{27-1}$$

The volt is the SI unit of emf as well as of potential difference. An **ideal emf device** is one that lacks any internal resistance. The potential difference between its terminals is equal to the emf. A **real emf device** has internal resistance. The potential difference between its terminals is equal to the emf only if there is no current through the device.

Analyzing Circuits The change in potential in traversing a resistance R in the direction of the current is $-iR$; in the opposite direction it is $+iR$ (resistance rule). The change in potential in traversing an ideal emf device in the direction of the emf arrow is $+\mathscr{E}$; in the opposite direction it is $-\mathscr{E}$ (emf rule). Conservation of energy leads to the loop rule:

Loop Rule. *The algebraic sum of the changes in potential encountered in a complete traversal of any loop of a circuit must be zero.*

Conservation of charge gives us the junction rule:

Junction Rule. *The sum of the currents entering any junction must be equal to the sum of the currents leaving that junction.*

Single-Loop Circuits The current in a single-loop circuit containing a single resistance R and an emf device with emf $\mathscr{E}$ and internal resistance r is

$$i = \frac{\mathscr{E}}{R + r}, \tag{27-4}$$

which reduces to $i = \mathscr{E}/R$ for an ideal emf device with $r = 0$.

Power When a real battery of emf $\mathscr{E}$ and internal resistance r does work on the charge carriers in a current i through the battery, the rate P of energy transfer to the charge carriers is

$$P = iV, \tag{27-14}$$

where V is the potential across the terminals of the battery. The rate P_r at which energy is dissipated as thermal energy in the battery is

$$P_r = i^2 r. \tag{27-16}$$

The rate P_{emf} at which the chemical energy in the battery changes is

$$P_{emf} = i\mathscr{E}. \tag{27-17}$$

Series Resistances When resistances are in **series,** they have the same current. The equivalent resistance that can replace a series combination of resistances is

$$R_{eq} = \sum_{j=1}^{n} R_j \quad (n \text{ resistances in series}). \tag{27-7}$$

Parallel Resistances When resistances are in **parallel,** they have the same potential difference. The equivalent resistance that can replace a parallel combination of resistances is given by

$$\frac{1}{R_{eq}} = \sum_{j=1}^{n} \frac{1}{R_j} \quad (n \text{ resistances in parallel}). \tag{27-24}$$

RC Circuits When an emf $\mathscr{E}$ is applied to a resistance R and capacitance C in series, as in Fig. 27-15 with the switch at a, the charge on the capacitor increases according to

$$q = C\mathscr{E}(1 - e^{-t/RC}) \quad \text{(charging a capacitor)}, \tag{27-33}$$

in which $C\mathscr{E} = q_0$ is the equilibrium (final) charge and $RC = \tau$ is the **capacitive time constant** of the circuit. During the charging, the current is

$$i = \frac{dq}{dt} = \left(\frac{\mathscr{E}}{R}\right)e^{-t/RC} \quad \text{(charging a capacitor)}. \tag{27-34}$$

When a capacitor discharges through a resistance R, the charge on the capacitor decays according to

$$q = q_0 e^{-t/RC} \quad \text{(discharging a capacitor)}. \tag{27-39}$$

During the discharging, the current is

$$i = \frac{dq}{dt} = -\left(\frac{q_0}{RC}\right)e^{-t/RC} \quad \text{(discharging a capacitor)}. \tag{27-40}$$

Questions

1 (a) In Fig. 27-18a, with $R_1 > R_2$, is the potential difference across R_2 more than, less than, or equal to that across R_1? (b) Is the current through resistor R_2 more than, less than, or equal to that through resistor R_1?

2 (a) In Fig. 27-18a, are resistors R_1 and R_3 in series? (b) Are resistors R_1 and R_2 in parallel? (c) Rank the equivalent resistances of the four circuits shown in Fig. 27-18, greatest first.

3 You are to connect resistors R_1 and R_2, with $R_1 > R_2$, to a battery, first individually, then in series, and then in parallel. Rank those arrangements according to the amount of current through the battery, greatest first.

4 In Fig. 27-19, a circuit consists of a battery and two uniform resistors, and the section lying along an x axis is divided into five segments of equal lengths. (a) Assume that $R_1 = R_2$ and rank the segments according to the magnitude of the average electric

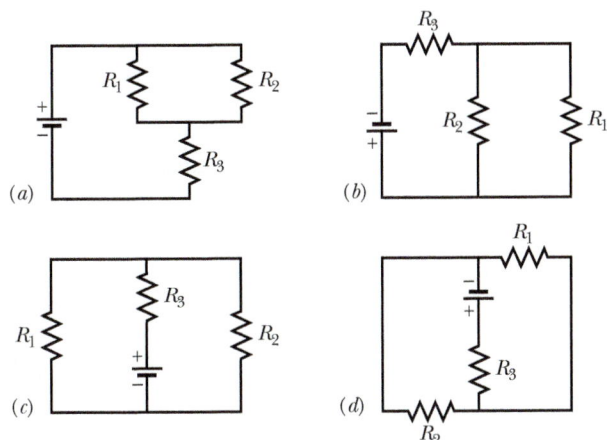

Figure 27-18 Questions 1 and 2.

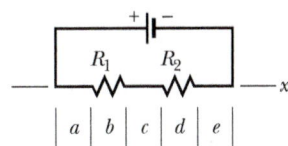

Figure 27-19 Question 4.

field in them, greatest first. (b) Now assume that $R_1 > R_2$ and then again rank the segments. (c) What is the direction of the electric field along the x axis?

5 For each circuit in Fig. 27-20, are the resistors connected in series, in parallel, or neither?

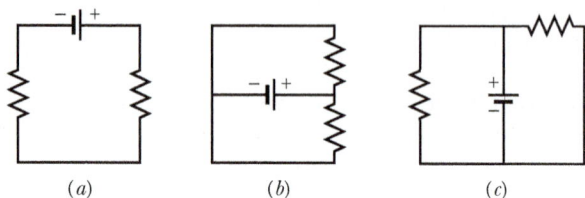

Figure 27-20 Question 5.

6 *Res-monster maze.* In Fig. 27-21, all the resistors have a resistance of 4.0 Ω and all the (ideal) batteries have an emf of 4.0 V. What is the current through resistor R? (If you can find the proper loop through this maze, you can answer the question with a few seconds of mental calculation.)

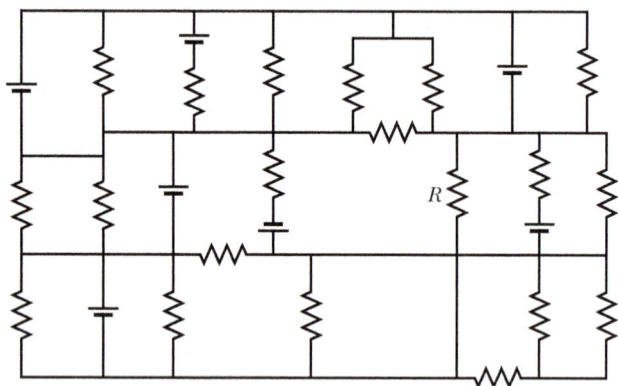

Figure 27-21 Question 6.

7 A resistor R_1 is wired to a battery, then resistor R_2 is added in series. Are (a) the potential difference across R_1 and (b) the current i_1 through R_1 now more than, less than, or the same as previously? (c) Is the equivalent resistance R_{12} of R_1 and R_2 more than, less than, or equal to R_1?

8 What is the equivalent resistance of three resistors, each of resistance R, if they are connected to an ideal battery (a) in series with one another and (b) in parallel with one another? (c) Is the potential difference across the series arrangement greater than, less than, or equal to that across the parallel arrangement?

9 Two resistors are wired to a battery. (a) In which arrangement, parallel or series, are the potential differences across each resistor and across the equivalent resistance all equal? (b) In which arrangement are the currents through each resistor and through the equivalent resistance all equal?

10 *Cap-monster maze.* In Fig. 27-22, all the capacitors have a capacitance of 6.0 μF, and all the batteries have an emf of 10 V. What is the charge on capacitor C? (If you can find the proper loop through this maze, you can answer the question with a few seconds of mental calculation.)

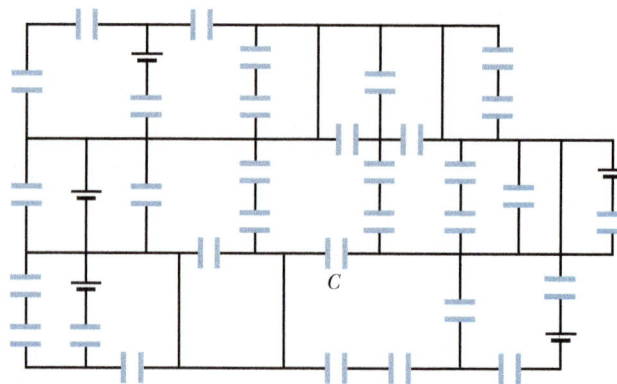

Figure 27-22 Question 10.

11 Initially, a single resistor R_1 is wired to a battery. Then resistor R_2 is added in parallel. Are (a) the potential difference across R_1 and (b) the current i_1 through R_1 now more than, less than, or the same as previously? (c) Is the equivalent resistance R_{12} of R_1 and R_2 more than, less than, or equal to R_1? (d) Is the total current through R_1 and R_2 together more than, less than, or equal to the current through R_1 previously?

12 After the switch in Fig. 27-15 is closed on point a, there is current i through resistance R. Figure 27-23 gives that current for four sets of values of R and capacitance C: (1) R_0 and C_0, (2) $2R_0$ and C_0, (3) R_0 and $2C_0$, (4) $2R_0$ and $2C_0$. Which set goes with which curve?

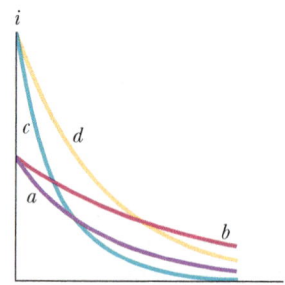

Figure 27-23 Question 12.

13 Figure 27-24 shows three sections of circuit that are to be connected in turn to the same battery via a switch as in Fig. 27-15. The resistors are all identical, as are the capacitors. Rank the sections according to (a) the final (equilibrium) charge on the capacitor and (b) the time required for the capacitor to reach 50% of its final charge, greatest first.

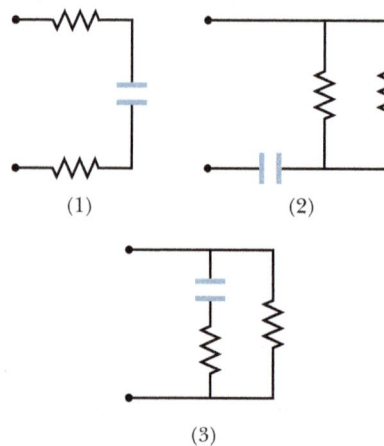

Figure 27-24 Question 13.

Problems

Module 27-1 Single-Loop Circuits

•1 **SSM** **WWW** In Fig. 27-25, the ideal batteries have emfs $\mathscr{E}_1 = 12$ V and $\mathscr{E}_2 = 6.0$ V. What are (a) the current, the dissipation rate in (b) resistor 1 (4.0 Ω) and (c) resistor 2 (8.0 Ω), and the energy transfer rate in (d) battery 1 and (e) battery 2? Is energy being supplied or absorbed by (f) battery 1 and (g) battery 2?

Figure 27-25
Problem 1.

•2 In Fig. 27-26, the ideal batteries have emfs $\mathscr{E}_1 = 150$ V and $\mathscr{E}_2 = 50$ V and the resistances are $R_1 = 3.0$ Ω and $R_2 = 2.0$ Ω. If the potential at P is 100 V, what is it at Q?

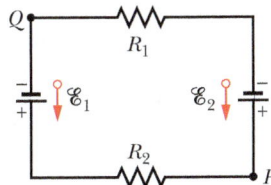

Figure 27-26 Problem 2.

•3 **ILW** A car battery with a 12 V emf and an internal resistance of 0.040 Ω is being charged with a current of 50 A. What are (a) the potential difference V across the terminals, (b) the rate P_r of energy dissipation inside the battery, and (c) the rate P_{emf} of energy conversion to chemical form? When the battery is used to supply 50 A to the starter motor, what are (d) V and (e) P_r?

•4 **GO** Figure 27-27 shows a circuit of four resistors that are connected to a larger circuit. The graph below the circuit shows the electric potential $V(x)$ as a function of position x along the lower branch of the circuit, through resistor 4; the potential V_A is 12.0 V. The graph above the circuit shows the electric potential $V(x)$ versus position x along the upper branch of the circuit, through resistors 1, 2, and 3; the potential differences are $\Delta V_B = 2.00$ V and $\Delta V_C = 5.00$ V. Resistor 3 has a resistance of 200 Ω. What is the resistance of (a) resistor 1 and (b) resistor 2?

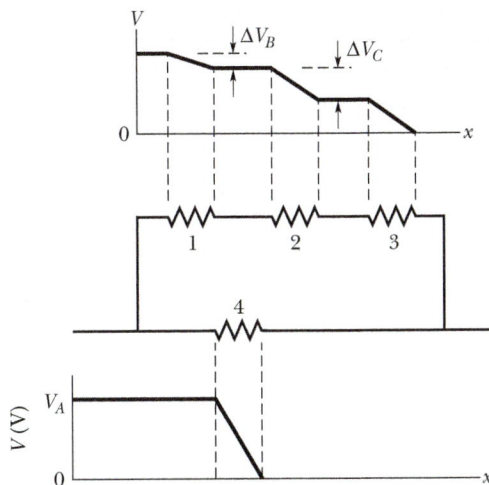

Figure 27-27
Problem 4.

•5 A 5.0 A current is set up in a circuit for 6.0 min by a rechargeable battery with a 6.0 V emf. By how much is the chemical energy of the battery reduced?

•6 A standard flashlight battery can deliver about 2.0 W·h of energy before it runs down. (a) If a battery costs US$0.80, what is the cost of operating a 100 W lamp for 8.0 h using batteries? (b) What is the cost if energy is provided at the rate of US$0.06 per kilowatt-hour?

•7 A wire of resistance 5.0 Ω is connected to a battery whose emf $\mathscr{E}$ is 2.0 V and whose internal resistance is 1.0 Ω. In 2.0 min, how much energy is (a) transferred from chemical form in the battery, (b) dissipated as thermal energy in the wire, and (c) dissipated as thermal energy in the battery?

•8 A certain car battery with a 12.0 V emf has an initial charge of 120 A·h. Assuming that the potential across the terminals stays constant until the battery is completely discharged, for how many hours can it deliver energy at the rate of 100 W?

•9 (a) In electron-volts, how much work does an ideal battery with a 12.0 V emf do on an electron that passes through the battery from the positive to the negative terminal? (b) If 3.40×10^{18} electrons pass through each second, what is the power of the battery in watts?

••10 (a) In Fig. 27-28, what value must R have if the current in the circuit is to be 1.0 mA? Take $\mathscr{E}_1 = 2.0$ V, $\mathscr{E}_2 = 3.0$ V, and $r_1 = r_2 = 3.0$ Ω. (b) What is the rate at which thermal energy appears in R?

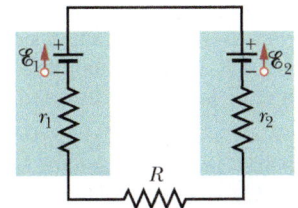

Figure 27-28 Problem 10.

••11 **SSM** In Fig. 27-29, circuit section AB absorbs energy at a rate of 50 W when current $i = 1.0$ A through it is in the indicated direction. Resistance $R = 2.0$ Ω. (a) What is the potential difference between A and B? Emf device X lacks internal resistance. (b) What is its emf? (c) Is point B connected to the positive terminal of X or to the negative terminal?

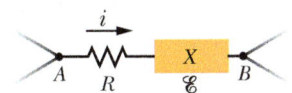

Figure 27-29 Problem 11.

••12 Figure 27-30 shows a resistor of resistance $R = 6.00$ Ω connected to an ideal battery of emf $\mathscr{E} = 12.0$ V by means of two copper wires. Each wire has length 20.0 cm and radius 1.00 mm. In dealing with such circuits in this chapter, we generally neglect the potential differences along the wires and the transfer of energy to thermal energy in them. Check the validity of this neglect for the circuit of Fig. 27-30: What is the potential difference across (a) the resistor and (b) each of the two sections of wire? At what rate is energy lost to thermal energy in (c) the resistor and (d) each section of wire?

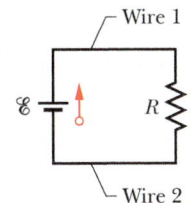

Figure 27-30
Problem 12.

••13 A 10-km-long underground cable extends east to west and consists of two parallel wires, each of which has resistance 13 Ω/km. An electrical short develops at distance x from the west end when a conducting path of resistance R connects the wires (Fig. 27-31). The resistance of the wires and the short is then 100 Ω when measured from the east end and 200 Ω when measured from the west end. What are (a) x and (b) R?

Figure 27-31 Problem 13.

••14 **GO** In Fig. 27-32a, both batteries have emf $\mathcal{E} = 1.20$ V and the external resistance R is a variable resistor. Figure 27-32b gives the electric potentials V between the terminals of each battery as functions of R: Curve 1 corresponds to battery 1, and curve 2 corresponds to battery 2. The horizontal scale is set by $R_s = 0.20$ Ω. What is the internal resistance of (a) battery 1 and (b) battery 2?

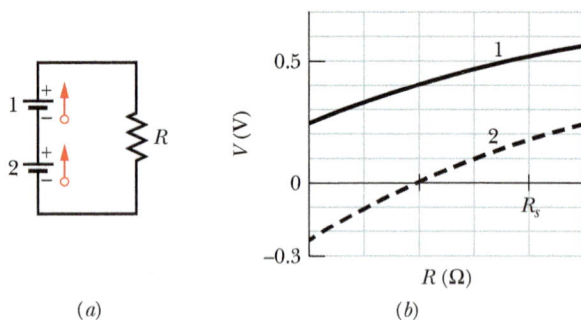

Figure 27-32 Problem 14.

••15 **ILW** The current in a single-loop circuit with one resistance R is 5.0 A. When an additional resistance of 2.0 Ω is inserted in series with R, the current drops to 4.0 A. What is R?

•••16 A solar cell generates a potential difference of 0.10 V when a 500 Ω resistor is connected across it, and a potential difference of 0.15 V when a 1000 Ω resistor is substituted. What are the (a) internal resistance and (b) emf of the solar cell? (c) The area of the cell is 5.0 cm², and the rate per unit area at which it receives energy from light is 2.0 mW/cm². What is the efficiency of the cell for converting light energy to thermal energy in the 1000 Ω external resistor?

•••17 **SSM** In Fig. 27-33, battery 1 has emf $\mathcal{E}_1 = 12.0$ V and internal resistance $r_1 = 0.016$ Ω and battery 2 has emf $\mathcal{E}_2 = 12.0$ V and internal resistance $r_2 = 0.012$ Ω. The batteries are connected in series with an external resistance R. (a) What R value makes the terminal-to-terminal potential difference of one of the batteries zero? (b) Which battery is that?

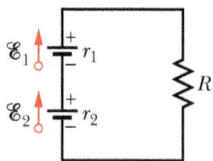

Figure 27-33 Problem 17.

Module 27-2 Multiloop Circuits

•18 In Fig. 27-9, what is the potential difference $V_d - V_c$ between points d and c if $\mathcal{E}_1 = 4.0$ V, $\mathcal{E}_2 = 1.0$ V, $R_1 = R_2 = 10$ Ω, and $R_3 = 5.0$ Ω, and the battery is ideal?

•19 A total resistance of 3.00 Ω is to be produced by connecting an unknown resistance to a 12.0 Ω resistance. (a) What must be the value of the unknown resistance, and (b) should it be connected in series or in parallel?

•20 When resistors 1 and 2 are connected in series, the equivalent resistance is 16.0 Ω. When they are connected in parallel, the equivalent resistance is 3.0 Ω. What are (a) the smaller resistance and (b) the larger resistance of these two resistors?

•21 Four 18.0 Ω resistors are connected in parallel across a 25.0 V ideal battery. What is the current through the battery?

•22 Figure 27-34 shows five 5.00 Ω resistors. Find the equivalent resistance between points (a) F and H and (b) F and G. (*Hint:* For each pair of points, imagine that a battery is connected across the pair.)

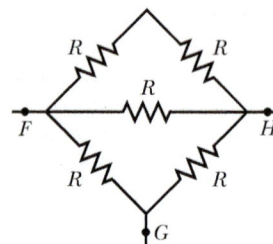

Figure 27-34 Problem 22.

•23 In Fig. 27-35, $R_1 = 100$ Ω, $R_2 = 50$ Ω, and the ideal batteries have emfs $\mathcal{E}_1 = 6.0$ V, $\mathcal{E}_2 = 5.0$ V, and $\mathcal{E}_3 = 4.0$ V. Find (a) the current in resistor 1, (b) the current in resistor 2, and (c) the potential difference between points a and b.

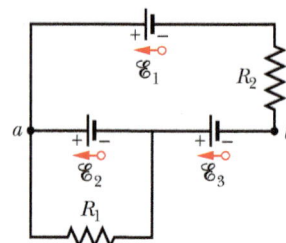

•24 In Fig. 27-36, $R_1 = R_2 = 4.00$ Ω and $R_3 = 2.50$ Ω. Find the equivalent resistance between points D and E. (*Hint:* Imagine that a battery is connected across those points.)

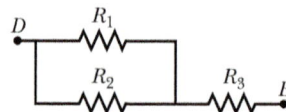

•25 **SSM** Nine copper wires of length l and diameter d are connected in parallel to form a single composite conductor of resistance R. What must be the diameter D of a single copper wire of length l if it is to have the same resistance?

•••26 Figure 27-37 shows a battery connected across a uniform resistor R_0. A sliding contact can move across the resistor from $x = 0$ at the left to $x = 10$ cm at the right. Moving the contact changes how much resistance is to the left of the contact and how much is to the right. Find the rate at which energy is dissipated in resistor R as a function of x. Plot the function for $\mathcal{E} = 50$ V, $R = 2000$ Ω, and $R_0 = 100$ Ω.

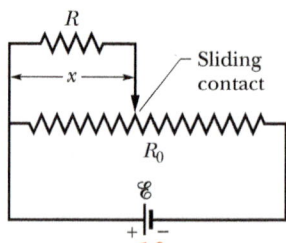

••27 Side flash. Figure 27-38 indicates one reason no one should stand under a tree during a lightning storm. If lightning comes down the side of the tree, a portion can jump over to the person, especially if the current on the tree reaches a dry region on the bark and thereafter must travel through air to reach the ground. In the figure, part of the lightning jumps through distance d in air and then travels through the person (who has negligible resistance relative to that of air because of the highly conducting salty fluids within the body). The rest of the current travels through air alongside the tree, for a distance h. If $d/h = 0.400$ and the total current is $I = 5000$ A, what is the current through the person?

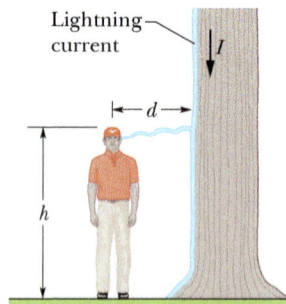

••28 The ideal battery in Fig. 27-39a has emf $\mathcal{E} = 6.0$ V. Plot 1 in Fig. 27-39b gives the electric potential difference V that can appear across resistor 1 versus the current i in that resistor when the resistor

Figure 27-35 Problem 23.

Figure 27-36 Problem 24.

Figure 27-37 Problem 26.

Figure 27-38 Problem 27.

is individually tested by putting a variable potential across it. The scale of the V axis is set by $V_s = 18.0$ V, and the scale of the i axis is set by $i_s = 3.00$ mA. Plots 2 and 3 are similar plots for resistors 2 and 3, respectively, when they are individually tested by putting a variable potential across them. What is the current in resistor 2 in the circuit of Fig. 27-39a?

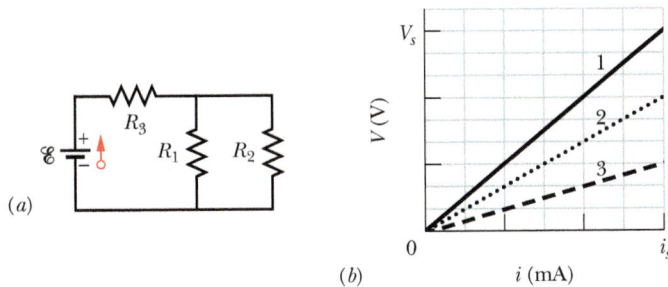

Figure 27-39 Problem 28.

••29 In Fig. 27-40, $R_1 = 6.00$ Ω, $R_2 = 18.0$ Ω, and the ideal battery has emf $\mathscr{E} = 12.0$ V. What are the (a) size and (b) direction (left or right) of current i_1? (c) How much energy is dissipated by all four resistors in 1.00 min?

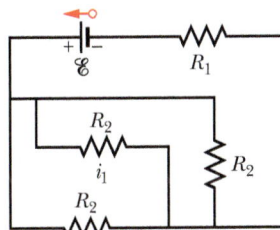

••30 In Fig. 27-41, the ideal batteries have emfs $\mathscr{E}_1 = 10.0$ V and $\mathscr{E}_2 = 0.500\mathscr{E}_1$, and the resistances are each 4.00 Ω. What is the current in (a) resistance 2 and (b) resistance 3?

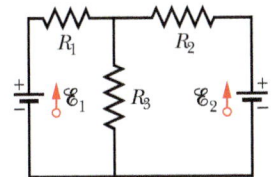

••31 SSM In Fig. 27-42, the ideal batteries have emfs $\mathscr{E}_1 = 5.0$ V and $\mathscr{E}_2 = 12$ V, the resistances are each 2.0 Ω, and the potential is defined to be zero at the grounded point of the circuit. What are potentials (a) V_1 and (b) V_2 at the indicated points?

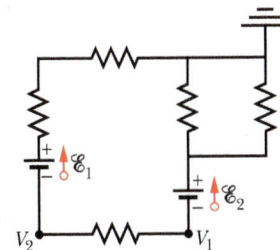

••32 Both batteries in Fig. 27-43a are ideal. Emf $\mathscr{E}_1$ of battery 1 has a fixed value, but emf $\mathscr{E}_2$ of battery 2 can be varied between 1.0 V and 10 V. The plots in Fig. 27-43b give the currents through the two batteries as a function of $\mathscr{E}_2$. The vertical scale is set by $i_s = 0.20$ A. You must decide which plot corresponds to which battery, but for both plots, a negative current occurs when the direction of the current through the

Figure 27-40 Problem 29.

Figure 27-41 Problems 30, 41, and 88.

Figure 27-42 Problem 31.

Figure 27-43 Problem 32.

battery is opposite the direction of that battery's emf. What are (a) emf $\mathscr{E}_1$, (b) resistance R_1, and (c) resistance R_2?

••33 In Fig. 27-44, the current in resistance 6 is $i_6 = 1.40$ A and the resistances are $R_1 = R_2 = R_3 = 2.00$ Ω, $R_4 = 16.0$ Ω, $R_5 = 8.00$ Ω, and $R_6 = 4.00$ Ω. What is the emf of the ideal battery?

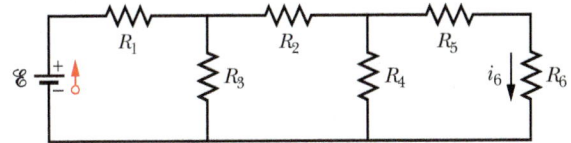

Figure 27-44 Problem 33.

••34 The resistances in Figs. 27-45a and b are all 6.0 Ω, and the batteries are ideal 12 V batteries. (a) When switch S in Fig. 27-45a is closed, what is the change in the electric potential V_1 across resistor 1, or does V_1 remain the same? (b) When switch S in Fig. 27-45b is closed, what is the change in V_1 across resistor 1, or does V_1 remain the same?

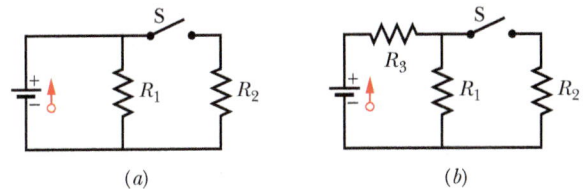

Figure 27-45 Problem 34.

••35 In Fig. 27-46, $\mathscr{E} = 12.0$ V, $R_1 = 2000$ Ω, $R_2 = 3000$ Ω, and $R_3 = 4000$ Ω. What are the potential differences (a) $V_A - V_B$, (b) $V_B - V_C$, (c) $V_C - V_D$, and (d) $V_A - V_C$?

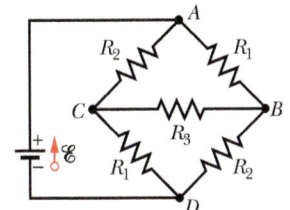

••36 In Fig. 27-47, $\mathscr{E}_1 = 6.00$ V, $\mathscr{E}_2 = 12.0$ V, $R_1 = 100$ Ω, $R_2 = 200$ Ω, and $R_3 = 300$ Ω. One point of the circuit is grounded ($V = 0$). What are the (a) size and (b) direction (up or down) of the current through resistance 1, the (c) size and (d) direction (left or right) of the current through resistance 2, and the (e) size and (f) direction of the current through resistance 3? (g) What is the electric potential at point A?

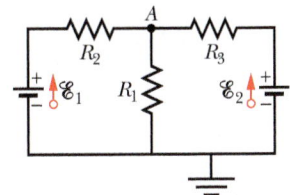

••37 In Fig. 27-48, the resistances are $R_1 = 2.00$ Ω, $R_2 = 5.00$ Ω, and the battery is ideal. What value of R_3 maximizes the dissipation rate in resistance 3?

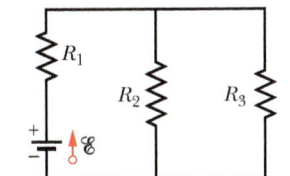

••38 Figure 27-49 shows a section of a circuit. The resistances are $R_1 = 2.0$ Ω, $R_2 = 4.0$ Ω, and $R_3 = 6.0$ Ω, and the indicated current is $i = 6.0$ A. The electric potential difference between points A and B that connect the section to the rest of the circuit is $V_A - V_B = 78$ V. (a) Is the device represented by "Box" absorbing or providing energy to the circuit, and (b) at what rate?

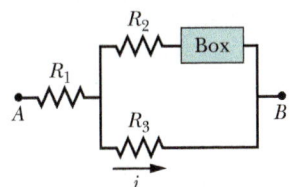

Figure 27-46 Problem 35.

Figure 27-47 Problem 36.

Figure 27-48 Problems 37 and 98.

Figure 27-49 Problem 38.

••39 GO In Fig. 27-50, two batteries with an emf $\mathscr{E} = 12.0$ V and an internal resistance $r = 0.300$ Ω are connected in parallel across a resistance R. (a) For what value of R is the dissipation rate in the resistor a maximum? (b) What is that maximum?

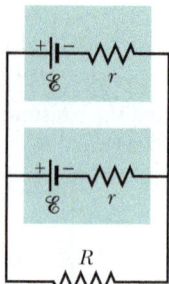

••40 GO Two identical batteries of emf $\mathscr{E} = 12.0$ V and internal resistance $r = 0.200$ Ω are to be connected to an external resistance R, either in parallel (Fig. 27-50) or in series (Fig. 27-51). If $R = 2.00r$, what is the current i in the external resistance in the (a) parallel and (b) series arrangements? (c) For which arrangement is i greater? If $R = r/2.00$, what is i in the external resistance in the (d) parallel arrangement and (e) series arrangement? (f) For which arrangement is i greater now?

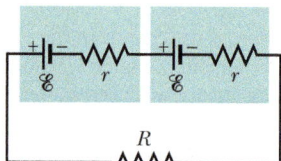

Figure 27-50 Problems 39 and 40.

••41 In Fig. 27-41, $\mathscr{E}_1 = 3.00$ V, $\mathscr{E}_2 = 1.00$ V, $R_1 = 4.00$ Ω, $R_2 = 2.00$ Ω, $R_3 = 5.00$ Ω, and both batteries are ideal. What is the rate at which energy is dissipated in (a) R_1, (b) R_2, and (c) R_3? What is the power of (d) battery 1 and (e) battery 2?

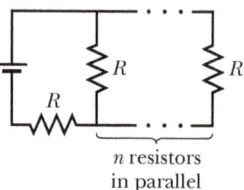

Figure 27-51 Problem 40.

••42 In Fig. 27-52, an array of n parallel resistors is connected in series to a resistor and an ideal battery. All the resistors have the same resistance. If an identical resistor were added in parallel to the parallel array, the current through the battery would change by 1.25%. What is the value of n?

Figure 27-52 Problem 42.

••43 You are given a number of 10 Ω resistors, each capable of dissipating only 1.0 W without being destroyed. What is the minimum number of such resistors that you need to combine in series or in parallel to make a 10 Ω resistance that is capable of dissipating at least 5.0 W?

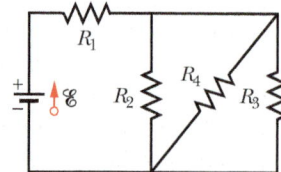

••44 GO In Fig. 27-53, $R_1 = 100$ Ω, $R_2 = R_3 = 50.0$ Ω, $R_4 = 75.0$ Ω, and the ideal battery has emf $\mathscr{E} = 6.00$ V. (a) What is the equivalent resistance? What is i in (b) resistance 1, (c) resistance 2, (d) resistance 3, and (e) resistance 4?

Figure 27-53 Problems 44 and 48.

••45 ILW In Fig. 27-54, the resistances are $R_1 = 1.0$ Ω and $R_2 = 2.0$ Ω, and the ideal batteries have emfs $\mathscr{E}_1 = 2.0$ V and $\mathscr{E}_2 = \mathscr{E}_3 = 4.0$ V. What are the (a) size and (b) direction (up or down) of the current in battery 1, the (c) size and (d) direction of the current in battery 2, and the (e) size and (f) direction of the current in battery 3? (g) What is the potential difference $V_a - V_b$?

Figure 27-54 Problem 45.

••46 In Fig. 27-55a, resistor 3 is a variable resistor and the ideal battery has emf $\mathscr{E} = 12$ V. Figure 27-55b gives the current i through the battery as a function of R_3. The horizontal scale is set by $R_{3s} = 20$ Ω. The curve has an asymptote of 2.0 mA as $R_3 \to \infty$. What are (a) resistance R_1 and (b) resistance R_2?

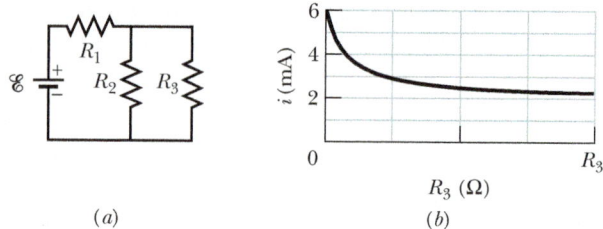

(a)   (b)

Figure 27-55 Problem 46.

•••47 SSM A copper wire of radius $a = 0.250$ mm has an aluminum jacket of outer radius $b = 0.380$ mm. There is a current $i = 2.00$ A in the composite wire. Using Table 26-1, calculate the current in (a) the copper and (b) the aluminum. (c) If a potential difference $V = 12.0$ V between the ends maintains the current, what is the length of the composite wire?

•••48 GO In Fig. 27-53, the resistors have the values $R_1 = 7.00$ Ω, $R_2 = 12.0$ Ω, and $R_3 = 4.00$ Ω, and the ideal battery's emf is $\mathscr{E} = 24.0$ V. For what value of R_4 will the rate at which the battery transfers energy to the resistors equal (a) 60.0 W, (b) the maximum possible rate P_{max}, and (c) the minimum possible rate P_{min}? What are (d) P_{max} and (e) P_{min}?

Module 27-3 The Ammeter and the Voltmeter

••49 ILW (a) In Fig. 27-56, what current does the ammeter read if $\mathscr{E} = 5.0$ V (ideal battery), $R_1 = 2.0$ Ω, $R_2 = 4.0$ Ω, and $R_3 = 6.0$ Ω? (b) The ammeter and battery are now interchanged. Show that the ammeter reading is unchanged.

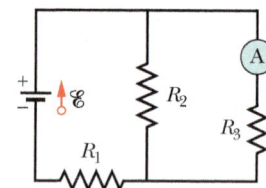

Figure 27-56 Problem 49.

••50 In Fig. 27-57, $R_1 = 2.00R$, the ammeter resistance is zero, and the battery is ideal. What multiple of $\mathscr{E}/R$ gives the current in the ammeter?

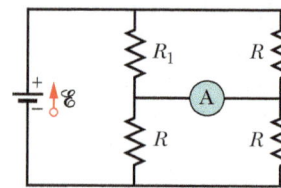

Figure 27-57 Problem 50.

••51 In Fig. 27-58, a voltmeter of resistance $R_V = 300$ Ω and an ammeter of resistance $R_A = 3.00$ Ω are being used to measure a resistance R in a circuit that also contains a resistance $R_0 = 100$ Ω and an ideal battery with an emf of $\mathscr{E} = 12.0$ V. Resistance R is given by $R = V/i$, where V is the potential across R and i is the ammeter reading. The voltmeter reading is V', which is V plus the potential difference across the ammeter. Thus, the ratio of the two meter readings is not R but only an *apparent* resistance $R' = V'/i$. If $R = 85.0$ Ω, what are (a) the ammeter reading, (b) the voltmeter reading, and (c) R'? (d) If R_A is decreased, does the difference between R' and R increase, decrease, or remain the same?

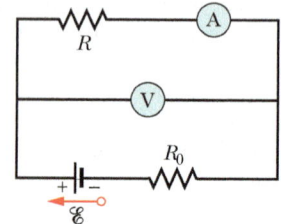

Figure 27-58 Problem 51.

••52 A simple ohmmeter is made by connecting a 1.50 V flashlight battery in series with a resistance R and an ammeter that reads from 0 to 1.00 mA, as shown in Fig. 27-59. Resistance R is adjusted so that when the clip leads are shorted together, the meter deflects to its full-scale value of 1.00 mA. What external resistance across the leads results in a deflection of (a) 10.0%, (b) 50.0%, and (c) 90.0% of full scale? (d) If the ammeter has a resistance of 20.0 Ω and the internal resistance of the battery is negligible, what is the value of R?

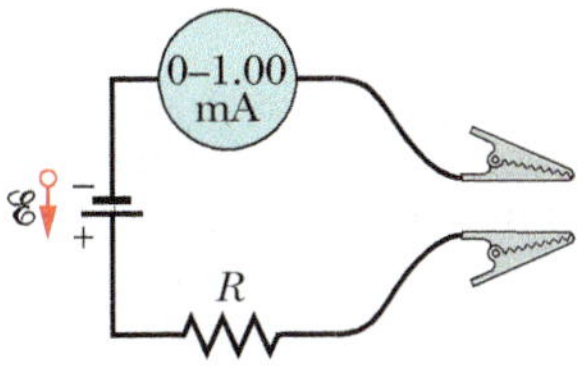

Figure 27-59 Problem 52.

••53 In Fig. 27-14, assume that $\mathscr{E} = 3.0$ V, $r = 100$ Ω, $R_1 = 250$ Ω, and $R_2 = 300$ Ω. If the voltmeter resistance R_V is 5.0 kΩ, what percent error does it introduce into the measurement of the potential difference across R_1? Ignore the presence of the ammeter.

••54 When the lights of a car are switched on, an ammeter in series with them reads 10.0 A and a voltmeter connected across them reads 12.0 V (Fig. 27-60). When the electric starting motor is turned on, the ammeter reading drops to 8.00 A and the lights dim somewhat. If the internal resistance of the battery is 0.0500 Ω and that of the ammeter is negligible, what are (a) the emf of the battery and (b) the current through the starting motor when the lights are on?

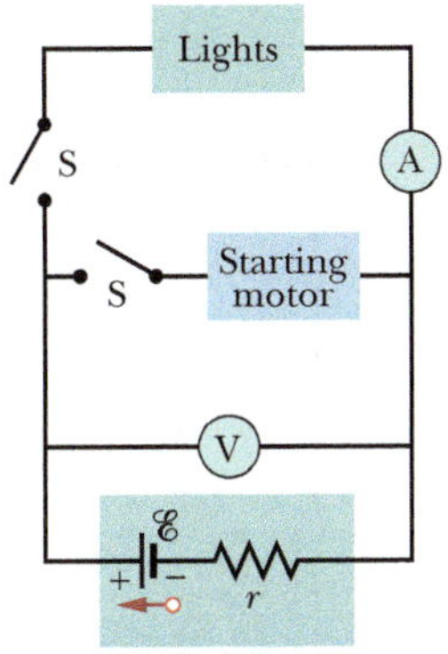

Figure 27-60 Problem 54.

••55 In Fig. 27-61, R_s is to be adjusted in value by moving the sliding contact across it until points a and b are brought to the same potential. (One tests for this condition by momentarily connecting a sensitive ammeter between a and b; if these points are at the same potential, the ammeter will not deflect.) Show that when this adjustment is made, the following relation holds: $R_x = R_s R_2/R_1$. An unknown resistance (R_x) can be measured in terms of a standard (R_s) using this device, which is called a Wheatstone bridge.

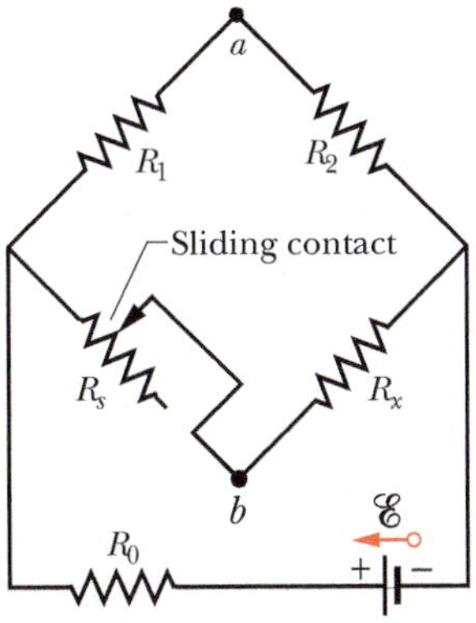

Figure 27-61 Problem 55.

••56 In Fig. 27-62, a voltmeter of resistance $R_V = 300$ Ω and an ammeter of resistance $R_A = 3.00$ Ω are being used to measure a resistance R in a circuit that also contains a resistance $R_0 = 100$ Ω and an ideal battery of emf $\mathscr{E} = 12.0$ V. Resistance R is given by $R = V/i$, where V is the voltmeter reading and i is the current in resistance R. However, the ammeter reading is not i but rather i', which is i plus the current through the voltmeter. Thus, the ratio of the two meter readings is not R but only an *apparent* resistance $R' = V/i'$. If $R = 85.0$ Ω, what are (a) the ammeter reading, (b) the voltmeter reading, and (c) R'? (d) If R_V is increased, does the difference between R' and R increase, decrease, or remain the same?

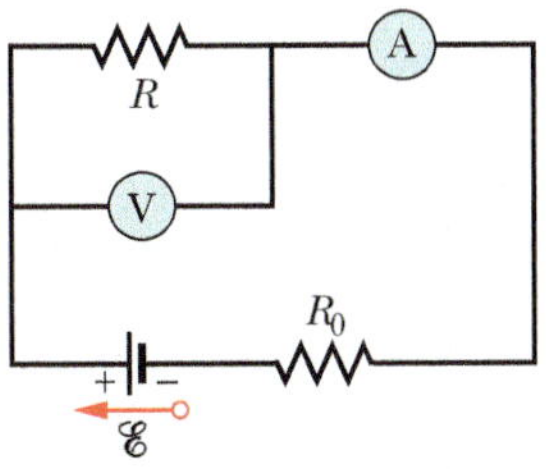

Figure 27-62 Problem 56.

Module 27-4 RC Circuits

•57 Switch S in Fig. 27-63 is closed at time $t = 0$, to begin charging an initially uncharged capacitor of capacitance $C = 15.0$ μF through a resistor of resistance $R = 20.0$ Ω. At what time is the potential across the capacitor equal to that across the resistor?

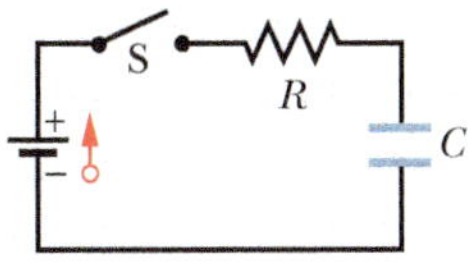

Figure 27-63 Problems 57 and 96.

•58 In an RC series circuit, emf $\mathscr{E} = 12.0$ V, resistance $R = 1.40$ MΩ, and capacitance $C = 1.80$ μF. (a) Calculate the time constant. (b) Find the maximum charge that will appear on the capacitor during charging. (c) How long does it take for the charge to build up to 16.0 μC?

•59 SSM What multiple of the time constant τ gives the time taken by an initially uncharged capacitor in an RC series circuit to be charged to 99.0% of its final charge?

•60 A capacitor with initial charge q_0 is discharged through a resistor. What multiple of the time constant τ gives the time the capacitor takes to lose (a) the first one-third of its charge and (b) two-thirds of its charge?

•61 ILW A 15.0 kΩ resistor and a capacitor are connected in series, and then a 12.0 V potential difference is suddenly applied across them. The potential difference across the capacitor rises to 5.00 V in 1.30 μs. (a) Calculate the time constant of the circuit. (b) Find the capacitance of the capacitor.

••62 Figure 27-64 shows the circuit of a flashing lamp, like those attached to barrels at highway construction sites. The fluorescent lamp L (of negligible capacitance) is connected in parallel across the capacitor C of an RC circuit. There is a current through the lamp only when the potential difference across it reaches the breakdown voltage V_L; then the capacitor discharges completely through the lamp and the lamp flashes briefly. For a lamp with breakdown voltage $V_L = 72.0$ V, wired to a 95.0 V ideal battery and a 0.150 μF capacitor, what resistance R is needed for two flashes per second?

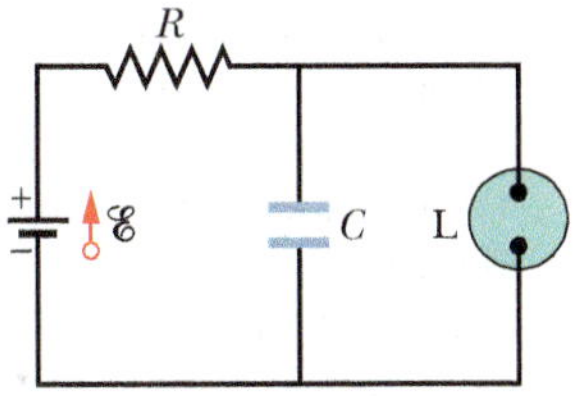

Figure 27-64 Problem 62.

••63 SSM WWW In the circuit of Fig. 27-65, $\mathscr{E} = 1.2$ kV, $C = 6.5$ μF, $R_1 = R_2 = R_3 = 0.73$ MΩ. With C completely uncharged, switch S is suddenly closed (at $t = 0$). At $t = 0$, what are (a) current i_1 in resistor 1, (b) current i_2 in resistor 2, and (c) current i_3 in resistor 3? At $t = \infty$ (that is, after many time constants), what are (d) i_1, (e) i_2, and (f) i_3? What is the potential difference V_2 across resistor 2 at (g) $t = 0$ and (h) $t = \infty$? (i) Sketch V_2 versus t between these two extreme times.

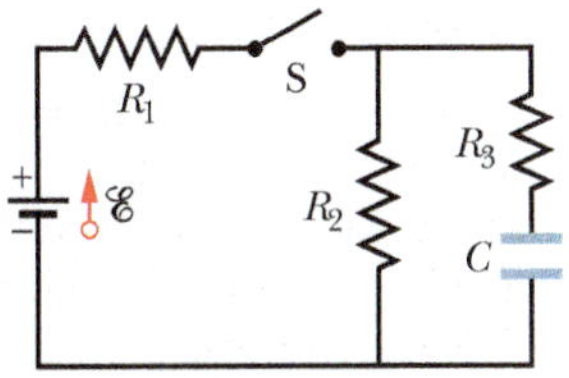

Figure 27-65 Problem 63.

••64 A capacitor with an initial potential difference of 100 V is discharged through a resistor when a switch between them is closed at $t = 0$. At $t = 10.0$ s, the potential difference across the capacitor is 1.00 V. (a) What is the time constant of the circuit? (b) What is the potential difference across the capacitor at $t = 17.0$ s?

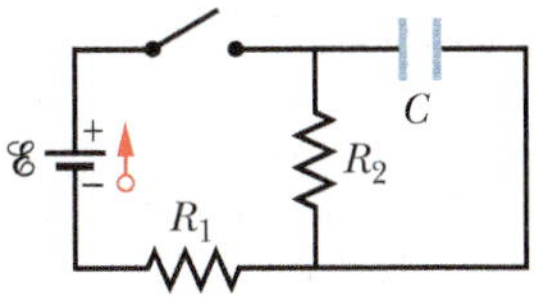

Figure 27-66 Problems 65 and 99.

••65 GO In Fig. 27-66, $R_1 = 10.0$ kΩ, $R_2 = 15.0$ kΩ, $C = 0.400$ μF, and the ideal battery has emf $\mathscr{E} = 20.0$ V. First, the switch is closed a long time so that the steady state is reached. Then the switch is opened at time $t = 0$. What is the current in resistor 2 at $t = 4.00$ ms?

••66 Figure 27-67 displays two circuits with a charged capacitor that is to be discharged through a resistor when a switch is closed. In Fig. 27-67a, $R_1 = 20.0$ Ω and $C_1 = 5.00$ μF. In Fig. 27-67b, $R_2 = 10.0$ Ω and $C_2 = 8.00$ μF. The ratio of the initial charges on the two capacitors is $q_{02}/q_{01} = 1.50$. At time $t = 0$, both switches are closed. At what time t do the two capacitors have the same charge?

Figure 27-67 Problem 66.

••67 The potential difference between the plates of a leaky (meaning that charge leaks from one plate to the other) 2.0 μF capacitor drops to one-fourth its initial value in 2.0 s. What is the equivalent resistance between the capacitor plates?

••68 A 1.0 μF capacitor with an initial stored energy of 0.50 J is discharged through a 1.0 MΩ resistor. (a) What is the initial charge on the capacitor? (b) What is the current through the resistor when the discharge starts? Find an expression that gives, as a function of time t, (c) the potential difference V_C across the capacitor, (d) the potential difference V_R across the resistor, and (e) the rate at which thermal energy is produced in the resistor.

•••69 GO A 3.00 MΩ resistor and a 1.00 μF capacitor are connected in series with an ideal battery of emf $\mathscr{E} = 4.00$ V. At 1.00 s after the connection is made, what is the rate at which (a) the charge of the capacitor is increasing, (b) energy is being stored in the capacitor, (c) thermal energy is appearing in the resistor, and (d) energy is being delivered by the battery?

Additional Problems

70 GO Each of the six real batteries in Fig. 27-68 has an emf of 20 V and a resistance of 4.0 Ω. (a) What is the current through the (external) resistance $R = 4.0$ Ω? (b) What is the potential difference across each battery? (c) What is the power of each battery? (d) At what rate does each battery transfer energy to internal thermal energy?

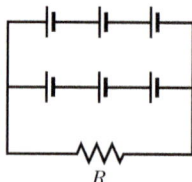

Figure 27-68 Problem 70.

71 In Fig. 27-69, $R_1 = 20.0$ Ω, $R_2 = 10.0$ Ω, and the ideal battery has emf $\mathscr{E} = 120$ V. What is the current at point a if we close (a) only switch S_1, (b) only switches S_1 and S_2, and (c) all three switches?

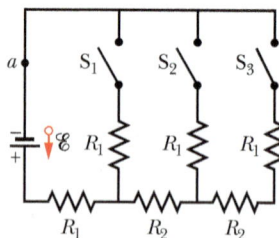

Figure 27-69 Problem 71.

72 In Fig. 27-70, the ideal battery has emf $\mathscr{E} = 30.0$ V, and the resistances are $R_1 = R_2 = 14$ Ω, $R_3 = R_4 = R_5 = 6.0$ Ω, $R_6 = 2.0$ Ω, and $R_7 = 1.5$ Ω. What are currents (a) i_2, (b) i_4, (c) i_1, (d) i_3, and (e) i_5?

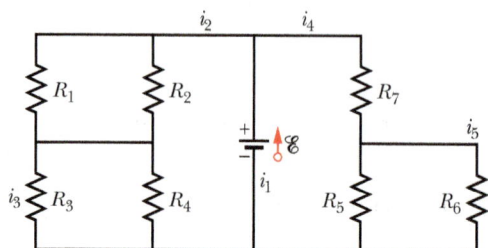

Figure 27-70 Problem 72.

73 SSM Wires A and B, having equal lengths of 40.0 m and equal diameters of 2.60 mm, are connected in series. A potential difference of 60.0 V is applied between the ends of the composite wire. The resistances are $R_A = 0.127$ Ω and $R_B = 0.729$ Ω. For wire A, what are (a) magnitude J of the current density and (b) potential difference V? (c) Of what type material is wire A made (see Table 26-1)? For wire B, what are (d) J and (e) V? (f) Of what type material is B made?

74 What are the (a) size and (b) direction (up or down) of current i in Fig. 27-71, where all resistances are 4.0 Ω and all batteries are ideal and have an emf of 10 V? (*Hint:* This can be answered using only mental calculation.)

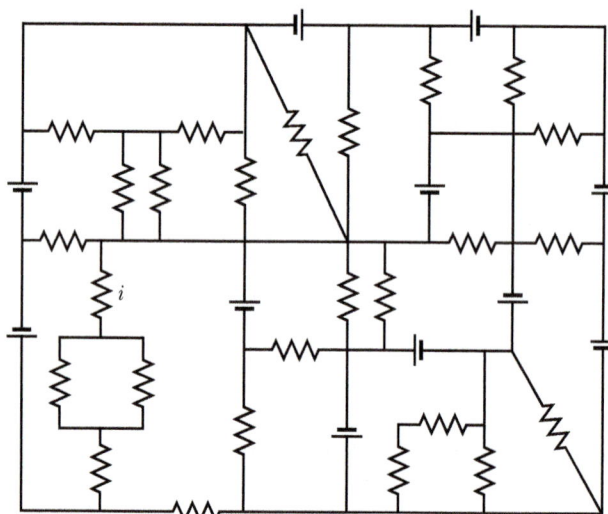

Figure 27-71 Problem 74.

75 Suppose that, while you are sitting in a chair, charge separation between your clothing and the chair puts you at a potential of 200 V, with the capacitance between you and the chair at 150 pF. When you stand up, the increased separation between your body and the chair decreases the capacitance to 10 pF. (a) What then is the potential of your body? That potential is reduced over time, as the charge on you drains through your body and shoes (you are a capacitor discharging through a resistance). Assume that the resistance along that route is 300 GΩ. If you touch an electrical component while your potential is greater than 100 V, you could ruin the component. (b) How long must you wait until your potential reaches the safe level of 100 V?

If you wear a conducting wrist strap that is connected to ground, your potential does not increase as much when you stand up; you also discharge more rapidly because the resistance through the grounding connection is much less than through your body and shoes. (c) Suppose that when you stand up, your potential is 1400 V and the chair-to-you capacitance is 10 pF. What resistance in that wrist-strap grounding connection will allow you to discharge to 100 V in 0.30 s, which is less time than you would need to reach for, say, your computer?

76 GO In Fig. 27-72, the ideal batteries have emfs $\mathscr{E}_1 = 20.0$ V, $\mathscr{E}_2 = 10.0$ V, and $\mathscr{E}_3 = 5.00$ V, and the resistances are each 2.00 Ω. What are the (a) size and (b) direction (left or right) of current i_1? (c) Does battery 1 supply or absorb energy, and (d) what is its power? (e) Does battery 2 supply or absorb energy, and (f) what is its power? (g) Does battery 3 supply or absorb energy, and (h) what is its power?

Figure 27-72 Problem 76.

77 SSM A temperature-stable resistor is made by connecting a resistor made of silicon in series with one made of iron. If the required total resistance is 1000 Ω in a wide temperature range around 20°C, what should be the resistance of the (a) silicon resistor and (b) iron resistor? (See Table 26-1.)

78 In Fig. 27-14, assume that $\mathscr{E} = 5.0$ V, $r = 2.0$ Ω, $R_1 = 5.0$ Ω, and $R_2 = 4.0$ Ω. If the ammeter resistance R_A is 0.10 Ω, what percent error does it introduce into the measurement of the current? Assume that the voltmeter is not present.

79 SSM An initially uncharged capacitor C is fully charged by a device of constant emf $\mathscr{E}$ connected in series with a resistor R. (a) Show that the final energy stored in the capacitor is half the energy supplied by the emf device. (b) By direct integration of i^2R over the charging time, show that the thermal energy dissipated by the resistor is also half the energy supplied by the emf device.

80 In Fig. 27-73, $R_1 = 5.00$ Ω, $R_2 = 10.0$ Ω, $R_3 = 15.0$ Ω, $C_1 = 5.00$ μF, $C_2 = 10.0$ μF, and the ideal battery has emf $\mathscr{E} = 20.0$ V. Assuming that the circuit is in the steady state, what is the total energy stored in the two capacitors?

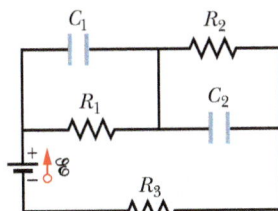

Figure 27-73 Problem 80.

81 In Fig. 27-5a, find the potential difference across R_2 if $\mathscr{E} = 12$ V, $R_1 = 3.0$ Ω, $R_2 = 4.0$ Ω, and $R_3 = 5.0$ Ω.

82 In Fig. 27-8a, calculate the potential difference between a and c by considering a path that contains R, r_1, and $\mathscr{E}_1$.

83 SSM A controller on an electronic arcade game consists of a variable resistor connected across the plates of a 0.220 μF capacitor. The capacitor is charged to 5.00 V, then discharged through the resistor. The time for the potential difference across the plates to decrease to 0.800 V is measured by a clock inside the game. If the range of discharge times that can be handled effectively is from 10.0 μs to 6.00 ms, what should be the (a) lower value and (b) higher value of the resistance range of the resistor?

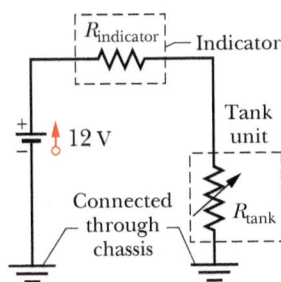

84 An automobile gasoline gauge is shown schematically in Fig. 27-74. The indicator (on the dashboard) has a resistance of 10 Ω. The tank

Figure 27-74 Problem 84.

unit is a float connected to a variable resistor whose resistance varies linearly with the volume of gasoline. The resistance is 140 Ω when the tank is empty and 20 Ω when the tank is full. Find the current in the circuit when the tank is (a) empty, (b) half-full, and (c) full. Treat the battery as ideal.

85 SSM The starting motor of a car is turning too slowly, and the mechanic has to decide whether to replace the motor, the cable, or the battery. The car's manual says that the 12 V battery should have no more than 0.020 Ω internal resistance, the motor no more than 0.200 Ω resistance, and the cable no more than 0.040 Ω resistance. The mechanic turns on the motor and measures 11.4 V across the battery, 3.0 V across the cable, and a current of 50 A. Which part is defective?

86 Two resistors R_1 and R_2 may be connected either in series or in parallel across an ideal battery with emf $\mathscr{E}$. We desire the rate of energy dissipation of the parallel combination to be five times that of the series combination. If $R_1 = 100$ Ω, what are the (a) smaller and (b) larger of the two values of R_2 that result in that dissipation rate?

87 The circuit of Fig. 27-75 shows a capacitor, two ideal batteries, two resistors, and a switch S. Initially S has been open for a long time. If it is then closed for a long time, what is the change in the charge on the capacitor? Assume $C = 10$ μF, $\mathscr{E}_1 = 1.0$ V, $\mathscr{E}_2 = 3.0$ V, $R_1 = 0.20$ Ω, and $R_2 = 0.40$ Ω.

Figure 27-75 Problem 87.

88 In Fig. 27-41, $R_1 = 10.0$ Ω, $R_2 = 20.0$ Ω, and the ideal batteries have emfs $\mathscr{E}_1 = 20.0$ V and $\mathscr{E}_2 = 50.0$ V. What value of R_3 results in no current through battery 1?

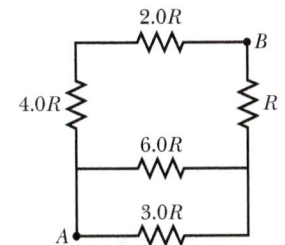

89 In Fig. 27-76, $R = 10$ Ω. What is the equivalent resistance between points A and B? (*Hint:* This circuit section might look simpler if you first assume that points A and B are connected to a battery.)

Figure 27-76 Problem 89.

90 (a) In Fig. 27-4a, show that the rate at which energy is dissipated in R as thermal energy is a maximum when $R = r$. (b) Show that this maximum power is $P = \mathscr{E}^2/4r$.

91 In Fig. 27-77, the ideal batteries have emfs $\mathscr{E}_1 = 12.0$ V and $\mathscr{E}_2 = 4.00$ V, and the resistances are each 4.00 Ω. What are the (a) size and (b) direction (up or down) of i_1 and the (c) size and (d) direction of i_2? (e) Does battery 1 supply or absorb energy, and (f) what is its energy transfer rate? (g) Does battery 2 supply or absorb energy, and (h) what is its energy transfer rate?

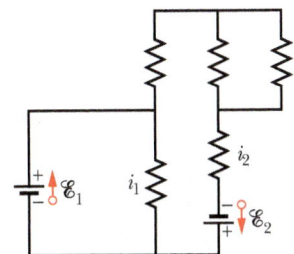

Figure 27-77 Problem 91.

92 Figure 27-78 shows a portion of a circuit through which there is a current $I = 6.00$ A. The resistances are $R_1 = R_2 = 2.00R_3 = 2.00R_4 = 4.00$ Ω. What is the current i_1 through resistor 1?

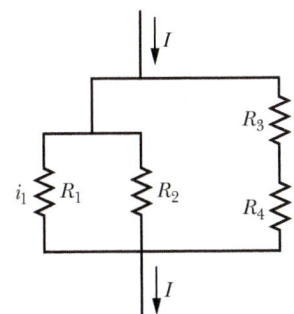

93 Thermal energy is to be generated in a 0.10 Ω resistor at the rate of

Figure 27-78 Problem 92.

10 W by connecting the resistor to a battery whose emf is 1.5 V. (a) What potential difference must exist across the resistor? (b) What must be the internal resistance of the battery?

94 Figure 27-79 shows three 20.0 Ω resistors. Find the equivalent resistance between points (a) A and B, (b) A and C, and (c) B and C. (*Hint:* Imagine that a battery is connected between a given pair of points.)

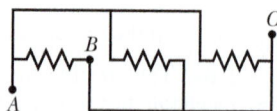

Figure 27-79 Problem 94.

95 A 120 V power line is protected by a 15 A fuse. What is the maximum number of 500 W lamps that can be simultaneously operated in parallel on this line without "blowing" the fuse because of an excess of current?

96 Figure 27-63 shows an ideal battery of emf $\mathscr{E} = 12$ V, a resistor of resistance $R = 4.0$ Ω, and an uncharged capacitor of capacitance $C = 4.0$ μF. After switch S is closed, what is the current through the resistor when the charge on the capacitor is 8.0 μC?

97 SSM A group of N identical batteries of emf $\mathscr{E}$ and internal resistance r may be connected all in series (Fig. 27-80a) or all in parallel (Fig. 27-80b) and then across a resistor R. Show that both arrangements give the same current in R if $R = r$.

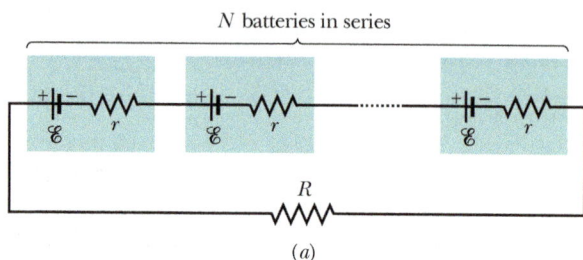

Figure 27-80 Problem 97.

98 SSM In Fig. 27-48, $R_1 = R_2 = 10.0$ Ω, and the ideal battery has emf $\mathscr{E} = 12.0$ V. (a) What value of R_3 maximizes the rate at which the battery supplies energy and (b) what is that maximum rate?

99 SSM In Fig. 27-66, the ideal battery has emf $\mathscr{E} = 30$ V, the resistances are $R_1 = 20$ kΩ and $R_2 = 10$ kΩ, and the capacitor is uncharged. When the switch is closed at time $t = 0$, what is the current in (a) resistance 1 and (b) resistance 2? (c) A long time later, what is the current in resistance 2?

100 In Fig. 27-81, the ideal batteries have emfs $\mathscr{E}_1 = 20.0$ V, $\mathscr{E}_2 = 10.0$ V,

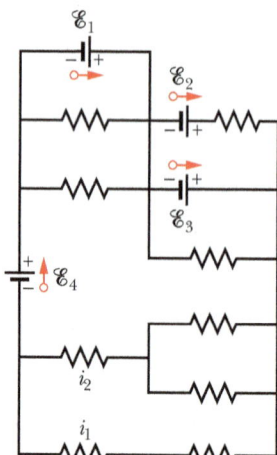

Figure 27-81 Problem 100.

$\mathscr{E}_3 = 5.00$ V, and $\mathscr{E}_4 = 5.00$ V, and the resistances are each 2.00 Ω. What are the (a) size and (b) direction (left or right) of current i_1 and the (c) size and (d) direction of current i_2? (This can be answered with only mental calculation.) (e) At what rate is energy being transferred in battery 4, and (f) is the energy being supplied or absorbed by the battery?

101 In Fig. 27-82, an ideal battery of emf $\mathscr{E} = 12.0$ V is connected to a network of resistances $R_1 = 6.00$ Ω, $R_2 = 12.0$ Ω, $R_3 = 4.00$ Ω, $R_4 = 3.00$ Ω, and $R_5 = 5.00$ Ω. What is the potential difference across resistance 5?

102 The following table gives the electric potential difference V_T across the terminals of a battery as a function of current i being drawn

Figure 27-82 Problem 101.

from the battery. (a) Write an equation that represents the relationship between V_T and i. Enter the data into your graphing calculator and perform a linear regression fit of V_T versus i. From the parameters of the fit, find (b) the battery's emf and (c) its internal resistance.

i(A):	50.0	75.0	100	125	150	175	200
V_T(V):	10.7	9.00	7.70	6.00	4.80	3.00	1.70

103 In Fig. 27-83, $\mathscr{E}_1 = 6.00$ V, $\mathscr{E}_2 = 12.0$ V, $R_1 = 200$ Ω, and $R_2 = 100$ Ω. What are the (a) size and (b) direction (up or down) of the current through resistance 1, the (c) size and (d) direction of the current through resistance 2, and the (e) size and (f) direction of the current through battery 2?

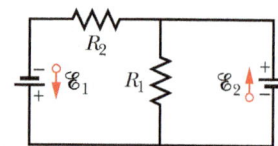

Figure 27-83 Problem 103.

104 A three-way 120 V lamp bulb that contains two filaments is rated for 100-200-300 W. One filament burns out. Afterward, the bulb operates at the same intensity (dissipates energy at the same rate) on its lowest as on its highest switch positions but does not operate at all on the middle position. (a) How are the two filaments wired to the three switch positions? What are the (b) smaller and (c) larger values of the filament resistances?

105 In Fig. 27-84, $R_1 = R_2 = 2.0$ Ω, $R_3 = 4.0$ Ω, $R_4 = 3.0$ Ω, $R_5 = 1.0$ Ω, and $R_6 = R_7 = R_8 = 8.0$ Ω, and the ideal batteries have emfs $\mathscr{E}_1 = 16$ V and $\mathscr{E}_2 = 8.0$ V. What are the (a) size and (b) direction (up or down) of current i_1 and the (c) size and (d) direction of current i_2? What is the energy transfer rate in (e) battery 1 and (f) battery 2? Is energy being supplied or absorbed in (g) battery 1 and (h) battery 2?

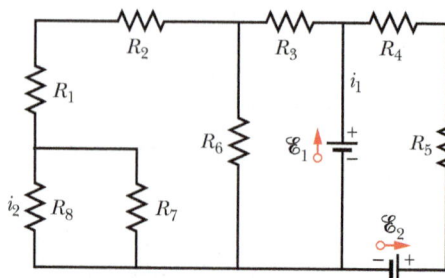

Figure 27-84 Problem 105.

CHAPTER 28

Magnetic Fields

28-1 MAGNETIC FIELDS AND THE DEFINITION OF $\vec{B}$

Learning Objectives

After reading this module, you should be able to . . .

28.01 Distinguish an electromagnet from a permanent magnet.

28.02 Identify that a magnetic field is a vector quantity and thus has both magnitude and direction.

28.03 Explain how a magnetic field can be defined in terms of what happens to a charged particle moving through the field.

28.04 For a charged particle moving through a uniform magnetic field, apply the relationship between force magnitude F_B, charge q, speed v, field magnitude B, and the angle ϕ between the directions of the velocity vector $\vec{v}$ and the magnetic field vector $\vec{B}$.

28.05 For a charged particle sent through a uniform magnetic field, find the direction of the magnetic force $\vec{F}_B$ by (1) applying the right-hand rule to find the direction

of the cross product $\vec{v} \times \vec{B}$ and (2) determining what effect the charge q has on the direction.

28.06 Find the magnetic force $\vec{F}_B$ acting on a moving charged particle by evaluating the cross product $q(\vec{v} \times \vec{B})$ in unit-vector notation and magnitude-angle notation.

28.07 Identify that the magnetic force vector $\vec{F}_B$ must always be perpendicular to both the velocity vector $\vec{v}$ and the magnetic field vector $\vec{B}$.

28.08 Identify the effect of the magnetic force on the particle's speed and kinetic energy.

28.09 Identify a magnet as being a magnetic dipole.

28.10 Identify that opposite magnetic poles attract each other and like magnetic poles repel each other.

28.11 Explain magnetic field lines, including where they originate and terminate and what their spacing represents.

Key Ideas

● When a charged particle moves through a magnetic field $\vec{B}$, a magnetic force acts on the particle as given by

$$\vec{F}_B = q(\vec{v} \times \vec{B}),$$

where q is the particle's charge (sign included) and $\vec{v}$ is the particle's velocity.

● The right-hand rule for cross products gives the direction

of $\vec{v} \times \vec{B}$. The sign of q then determines whether $\vec{F}_B$ is in the same direction as $\vec{v} \times \vec{B}$ or in the opposite direction.

● The magnitude of $\vec{F}_B$ is given by

$$F_B = |q|vB \sin \phi,$$

where ϕ is the angle between $\vec{v}$ and $\vec{B}$.

What Is Physics?

As we have discussed, one major goal of physics is the study of how an *electric field* can produce an *electric force* on a charged object. A closely related goal is the study of how a *magnetic field* can produce a *magnetic force* on a (moving) charged particle or on a magnetic object such as a magnet. You may already have a hint of what a magnetic field is if you have ever attached a note to a refrigerator door with a small magnet or accidentally erased a credit card by moving it near a magnet. The magnet acts on the door or credit card via its magnetic field.

The applications of magnetic fields and magnetic forces are countless and changing rapidly every year. Here are just a few examples. For decades, the entertainment industry depended on the magnetic recording of music and images on audiotape and videotape. Although digital technology has largely replaced

Digital Vision/Getty Images, Inc.

Figure 28-1 Using an electromagnet to collect and transport scrap metal at a steel mill.

magnetic recording, the industry still depends on the magnets that control CD and DVD players and computer hard drives; magnets also drive the speaker cones in headphones, TVs, computers, and telephones. A modern car comes equipped with dozens of magnets because they are required in the motors for engine ignition, automatic window control, sunroof control, and windshield wiper control. Most security alarm systems, doorbells, and automatic door latches employ magnets. In short, you are surrounded by magnets.

The science of magnetic fields is physics; the application of magnetic fields is engineering. Both the science and the application begin with the question "What produces a magnetic field?"

What Produces a Magnetic Field?

Because an electric field $\vec{E}$ is produced by an electric charge, we might reasonably expect that a magnetic field $\vec{B}$ is produced by a magnetic charge. Although individual magnetic charges (called *magnetic monopoles*) are predicted by certain theories, their existence has not been confirmed. How then are magnetic fields produced? There are two ways.

One way is to use moving electrically charged particles, such as a current in a wire, to make an **electromagnet.** The current produces a magnetic field that can be used, for example, to control a computer hard drive or to sort scrap metal (Fig. 28-1). In Chapter 29, we discuss the magnetic field due to a current.

The other way to produce a magnetic field is by means of elementary particles such as electrons because these particles have an *intrinsic* magnetic field around them. That is, the magnetic field is a basic characteristic of each particle just as mass and electric charge (or lack of charge) are basic characteristics. As we discuss in Chapter 32, the magnetic fields of the electrons in certain materials add together to give a net magnetic field around the material. Such addition is the reason why a **permanent magnet**, the type used to hang refrigerator notes, has a permanent magnetic field. In other materials, the magnetic fields of the electrons cancel out, giving no net magnetic field surrounding the material. Such cancellation is the reason you do not have a permanent field around your body, which is good because otherwise you might be slammed up against a refrigerator door every time you passed one.

Our first job in this chapter is to define the magnetic field $\vec{B}$. We do so by using the experimental fact that when a charged particle moves through a magnetic field, a magnetic force $\vec{F}_B$ acts on the particle.

The Definition of $\vec{B}$

We determined the electric field $\vec{E}$ at a point by putting a test particle of charge q at rest at that point and measuring the electric force $\vec{F}_E$ acting on the particle. We then defined $\vec{E}$ as

$$\vec{E} = \frac{\vec{F}_E}{q}.$$ (28-1)

If a magnetic monopole were available, we could define $\vec{B}$ in a similar way. Because such particles have not been found, we must define $\vec{B}$ in another way, in terms of the magnetic force $\vec{F}_B$ exerted on a moving electrically charged test particle.

Moving Charged Particle. In principle, we do this by firing a charged particle through the point at which $\vec{B}$ is to be defined, using various directions and speeds for the particle and determining the force $\vec{F}_B$ that acts on the particle at that point. After many such trials we would find that when the particle's velocity

$\vec{v}$ is along a particular axis through the point, force $\vec{F}_B$ is zero. For all other directions of $\vec{v}$, the magnitude of $\vec{F}_B$ is always proportional to $v \sin \phi$, where ϕ is the angle between the zero-force axis and the direction of $\vec{v}$. Furthermore, the direction of $\vec{F}_B$ is always perpendicular to the direction of $\vec{v}$. (These results suggest that a cross product is involved.)

The Field. We can then define a **magnetic field** $\vec{B}$ to be a vector quantity that is directed along the zero-force axis. We can next measure the magnitude of $\vec{F}_B$ when $\vec{v}$ is directed perpendicular to that axis and then define the magnitude of $\vec{B}$ in terms of that force magnitude:

$$B = \frac{F_B}{|q|v},$$

where q is the charge of the particle.

We can summarize all these results with the following vector equation:

$$\vec{F}_B = q\vec{v} \times \vec{B}; \tag{28-2}$$

that is, the force $\vec{F}_B$ on the particle is equal to the charge q times the cross product of its velocity $\vec{v}$ and the field $\vec{B}$ (all measured in the same reference frame). Using Eq. 3-24 for the cross product, we can write the magnitude of $\vec{F}_B$ as

$$F_B = |q|vB \sin \phi, \tag{28-3}$$

where ϕ is the angle between the directions of velocity $\vec{v}$ and magnetic field $\vec{B}$.

Finding the Magnetic Force on a Particle

Equation 28-3 tells us that the magnitude of the force $\vec{F}_B$ acting on a particle in a magnetic field is proportional to the charge q and speed v of the particle. Thus, the force is equal to zero if the charge is zero or if the particle is stationary. Equation 28-3 also tells us that the magnitude of the force is zero if $\vec{v}$ and $\vec{B}$ are either parallel ($\phi = 0°$) or antiparallel ($\phi = 180°$), and the force is at its maximum when $\vec{v}$ and $\vec{B}$ are perpendicular to each other.

Directions. Equation 28-2 tells us all this plus the direction of $\vec{F}_B$. From Module 3-3, we know that the cross product $\vec{v} \times \vec{B}$ in Eq. 28-2 is a vector that is perpendicular to the two vectors $\vec{v}$ and $\vec{B}$. The right-hand rule (Figs. 28-2a through c) tells us that the thumb of the right hand points in the direction of $\vec{v} \times \vec{B}$ when the fingers sweep $\vec{v}$ into $\vec{B}$. If q is positive, then (by Eq. 28-2) the force $\vec{F}_B$ has the same sign as $\vec{v} \times \vec{B}$ and thus must be in the same direction; that is, for positive q, $\vec{F}_B$ is directed along the thumb (Fig. 28-2d). If q is negative, then

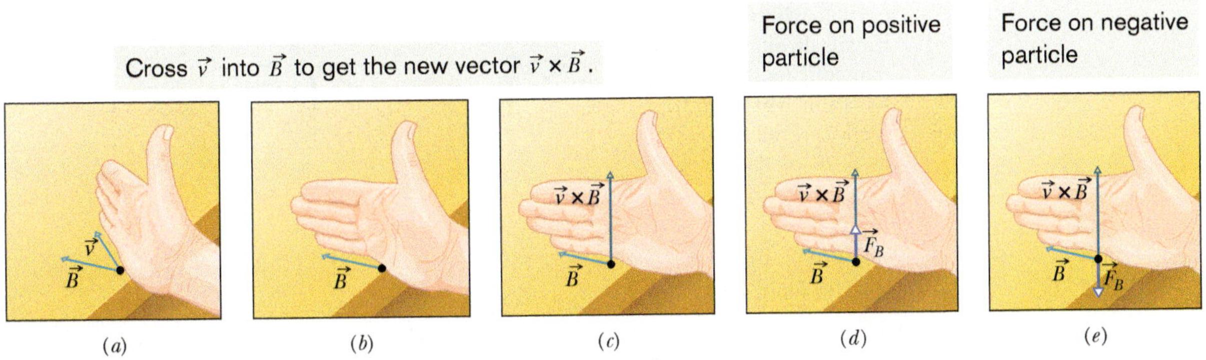

Cross $\vec{v}$ into $\vec{B}$ to get the new vector $\vec{v} \times \vec{B}$.

Force on positive particle

Force on negative particle

(a)   (b)   (c)   (d)   (e)

Figure 28-2 (a)–(c) The right-hand rule (in which $\vec{v}$ is swept into $\vec{B}$ through the smaller angle ϕ between them) gives the direction of $\vec{v} \times \vec{B}$ as the direction of the thumb. (d) If q is positive, then the direction of $\vec{F}_B = q\vec{v} \times \vec{B}$ is in the direction of $\vec{v} \times \vec{B}$. (e) If q is negative, then the direction of $\vec{F}_B$ is opposite that of $\vec{v} \times \vec{B}$.

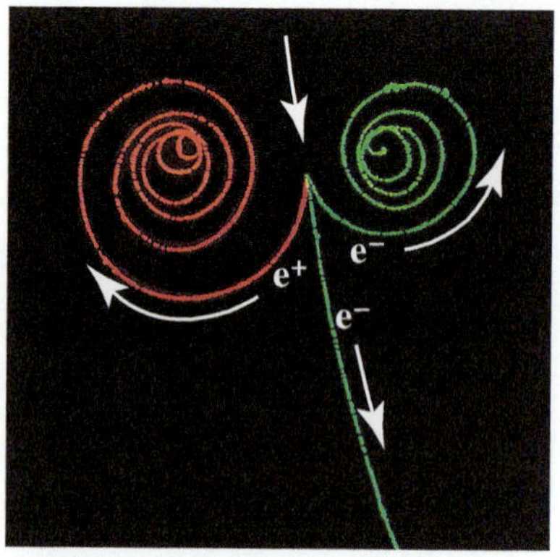

Lawrence Berkeley Laboratory/Photo Researchers, Inc.

Figure 28-3 The tracks of two electrons (e⁻) and a positron (e⁺) in a bubble chamber that is immersed in a uniform magnetic field that is directed out of the plane of the page.

Table 28-1 Some Approximate Magnetic Fields

At surface of neutron star	10^8 T
Near big electromagnet	1.5 T
Near small bar magnet	10^{-2} T
At Earth's surface	10^{-4} T
In interstellar space	10^{-10} T
Smallest value in magnetically shielded room	10^{-14} T

the force $\vec{F}_B$ and cross product $\vec{v} \times \vec{B}$ have opposite signs and thus must be in opposite directions. For negative q, $\vec{F}_B$ is directed opposite the thumb (Fig. 28-2e). *Heads up:* Neglect of this effect of negative q is a very common error on exams.

Regardless of the sign of the charge, however,

The force $\vec{F}_B$ acting on a charged particle moving with velocity $\vec{v}$ through a magnetic field $\vec{B}$ is *always* perpendicular to $\vec{v}$ and $\vec{B}$.

Thus, $\vec{F}_B$ *never* has a component parallel to $\vec{v}$. This means that $\vec{F}_B$ cannot change the particle's speed v (and thus it cannot change the particle's kinetic energy). The force can change only the direction of $\vec{v}$ (and thus the direction of travel); only in this sense can $\vec{F}_B$ accelerate the particle.

To develop a feeling for Eq. 28-2, consider Fig. 28-3, which shows some tracks left by charged particles moving rapidly through a *bubble chamber.* The chamber, which is filled with liquid hydrogen, is immersed in a strong uniform magnetic field that is directed out of the plane of the figure. An incoming gamma ray particle—which leaves no track because it is uncharged—transforms into an electron (spiral track marked e⁻) and a positron (track marked e⁺) while it knocks an electron out of a hydrogen atom (long track marked e⁻). Check with Eq. 28-2 and Fig. 28-2 that the three tracks made by these two negative particles and one positive particle curve in the proper directions.

Unit. The SI unit for $\vec{B}$ that follows from Eqs. 28-2 and 28-3 is the newton per coulomb-meter per second. For convenience, this is called the **tesla** (T):

$$1 \text{ tesla} = 1 \text{ T} = 1 \frac{\text{newton}}{(\text{coulomb})(\text{meter/second})}.$$

Recalling that a coulomb per second is an ampere, we have

$$1 \text{ T} = 1 \frac{\text{newton}}{(\text{coulomb/second})(\text{meter})} = 1 \frac{\text{N}}{\text{A} \cdot \text{m}}. \quad (28\text{-}4)$$

An earlier (non-SI) unit for $\vec{B}$, still in common use, is the *gauss* (G), and

$$1 \text{ tesla} = 10^4 \text{ gauss.} \quad (28\text{-}5)$$

Table 28-1 lists the magnetic fields that occur in a few situations. Note that Earth's magnetic field near the planet's surface is about 10^{-4} T ($= 100 \ \mu$T or 1 G).

✓ **Checkpoint 1**

The figure shows three situations in which a charged particle with velocity $\vec{v}$ travels through a uniform magnetic field $\vec{B}$. In each situation, what is the direction of the magnetic force $\vec{F}_B$ on the particle?

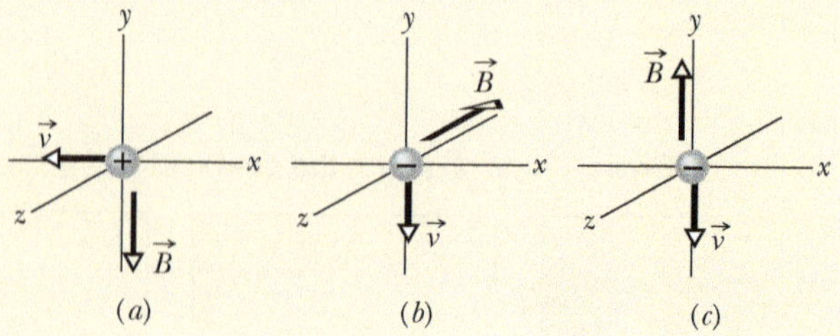

(a)   (b)   (c)

Magnetic Field Lines

We can represent magnetic fields with field lines, as we did for electric fields. Similar rules apply: (1) the direction of the tangent to a magnetic field line at any point gives the direction of $\vec{B}$ at that point, and (2) the spacing of the lines represents the magnitude of $\vec{B}$—the magnetic field is stronger where the lines are closer together, and conversely.

Figure 28-4*a* shows how the magnetic field near a *bar magnet* (a permanent magnet in the shape of a bar) can be represented by magnetic field lines. The lines all pass through the magnet, and they all form closed loops (even those that are not shown closed in the figure). The external magnetic effects of a bar magnet are strongest near its ends, where the field lines are most closely spaced. Thus, the magnetic field of the bar magnet in Fig. 28-4*b* collects the iron filings mainly near the two ends of the magnet.

Two Poles. The (closed) field lines enter one end of a magnet and exit the other end. The end of a magnet from which the field lines emerge is called the *north pole* of the magnet; the other end, where field lines enter the magnet, is called the *south pole*. Because a magnet has two poles, it is said to be a **magnetic dipole.** The magnets we use to fix notes on refrigerators are short bar magnets. Figure 28-5 shows two other common shapes for magnets: a *horseshoe magnet* and a magnet that has been bent around into the shape of a **C** so that the *pole faces* are facing each other. (The magnetic field between the pole faces can then be approximately uniform.) Regardless of the shape of the magnets, if we place two of them near each other we find:

> Opposite magnetic poles attract each other, and like magnetic poles repel each other.

When you hold two magnets near each other with your hands, this attraction or repulsion seems almost magical because there is no contact between the two to visibly justify the pulling or pushing. As we did with the electrostatic force between two charged particles, we explain this noncontact force in terms of a field that you cannot see, here the magnetic field.

Earth has a magnetic field that is produced in its core by still unknown mechanisms. On Earth's surface, we can detect this magnetic field with a compass, which is essentially a slender bar magnet on a low-friction pivot. This bar magnet, or this needle, turns because its north-pole end is attracted toward the Arctic region of Earth. Thus, the *south* pole of Earth's magnetic field must be located toward the Arctic. Logically, we then should call the pole there a south pole. However, because we call that direction north, we are trapped into the statement that Earth has a *geomagnetic north pole* in that direction.

With more careful measurement we would find that in the Northern Hemisphere, the magnetic field lines of Earth generally point down into Earth and toward the Arctic. In the Southern Hemisphere, they generally point up out of Earth and away from the Antarctic—that is, away from Earth's *geomagnetic south pole.*

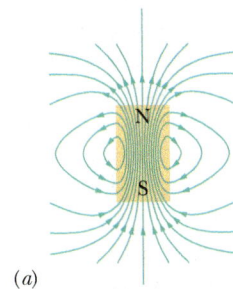

(*a*)

(*b*)

Courtesy Dr. Richard Cannon, Southeast Missouri State University, Cape Girardeau

Figure 28-4 (*a*) The magnetic field lines for a bar magnet. (*b*) A "cow magnet"—a bar magnet that is intended to be slipped down into the rumen of a cow to prevent accidentally ingested bits of scrap iron from reaching the cow's intestines. The iron filings at its ends reveal the magnetic field lines.

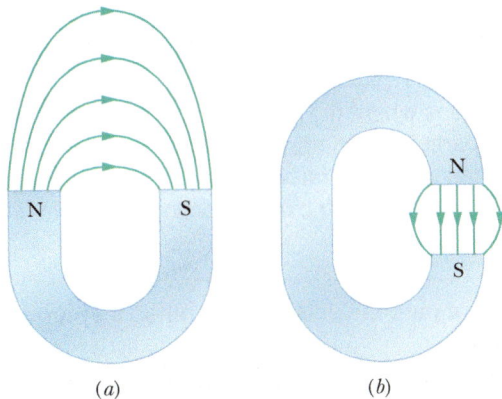

The field lines run from the north pole to the south pole.

(*a*)   (*b*)

Figure 28-5 (*a*) A horseshoe magnet and (*b*) a **C**-shaped magnet. (Only some of the external field lines are shown.)

Sample Problem 28.01 Magnetic force on a moving charged particle

A uniform magnetic field $\vec{B}$, with magnitude 1.2 mT, is directed vertically upward throughout the volume of a laboratory chamber. A proton with kinetic energy 5.3 MeV enters the chamber, moving horizontally from south to north. What magnetic deflecting force acts on the proton as it enters the chamber? The proton mass is 1.67×10^{-27} kg. (Neglect Earth's magnetic field.)

KEY IDEAS

Because the proton is charged and moving through a magnetic field, a magnetic force $\vec{F}_B$ can act on it. Because the initial direction of the proton's velocity is not along a magnetic field line, $\vec{F}_B$ is not simply zero.

Magnitude: To find the magnitude of $\vec{F}_B$, we can use Eq. 28-3 ($F_B = |q|vB \sin \phi$) provided we first find the proton's speed v. We can find v from the given kinetic energy because $K = \frac{1}{2}mv^2$. Solving for v, we obtain

$$v = \sqrt{\frac{2K}{m}} = \sqrt{\frac{(2)(5.3 \text{ MeV})(1.60 \times 10^{-13} \text{ J/MeV})}{1.67 \times 10^{-27} \text{ kg}}}$$

$$= 3.2 \times 10^7 \text{ m/s}.$$

Equation 28-3 then yields

$$\begin{aligned} F_B &= |q|vB \sin \phi \\ &= (1.60 \times 10^{-19} \text{ C})(3.2 \times 10^7 \text{ m/s}) \\ &\quad \times (1.2 \times 10^{-3} \text{ T})(\sin 90°) \\ &= 6.1 \times 10^{-15} \text{ N}. \qquad\qquad \text{(Answer)} \end{aligned}$$

This may seem like a small force, but it acts on a particle of small mass, producing a large acceleration; namely,

$$a = \frac{F_B}{m} = \frac{6.1 \times 10^{-15} \text{ N}}{1.67 \times 10^{-27} \text{ kg}} = 3.7 \times 10^{12} \text{ m/s}^2.$$

Direction: To find the direction of $\vec{F}_B$, we use the fact that $\vec{F}_B$ has the direction of the cross product $q\vec{v} \times \vec{B}$. Because the charge q is positive, $\vec{F}_B$ must have the same direction as $\vec{v} \times \vec{B}$, which can be determined with the right-hand rule for cross products (as in Fig. 28-2d). We know that $\vec{v}$ is directed horizontally from south to north and $\vec{B}$ is directed vertically up. The right-hand rule shows us that the deflecting force $\vec{F}_B$ must be directed horizontally from west to east, as Fig. 28-6 shows. (The array of dots in the figure represents a magnetic field directed out of the plane of the figure. An array of **X**s would have represented a magnetic field directed into that plane.)

If the charge of the particle were negative, the magnetic deflecting force would be directed in the opposite direction— that is, horizontally from east to west. This is predicted automatically by Eq. 28-2 if we substitute a negative value for q.

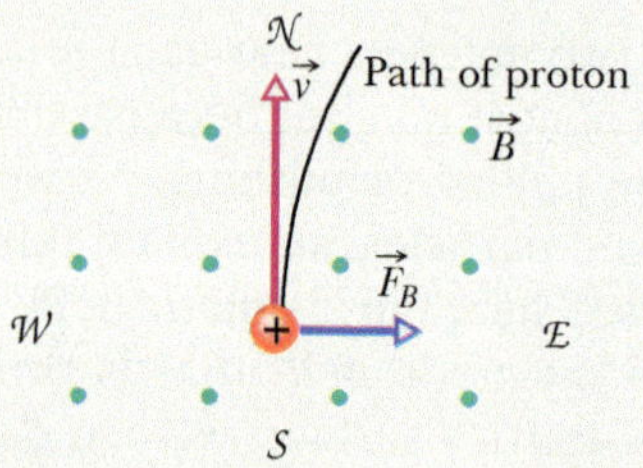

Figure 28-6 An overhead view of a proton moving from south to north with velocity $\vec{v}$ in a chamber. A magnetic field is directed vertically upward in the chamber, as represented by the array of dots (which resemble the tips of arrows). The proton is deflected toward the east.

PLUS Additional examples, video, and practice available at *WileyPLUS*

28-2 CROSSED FIELDS: DISCOVERY OF THE ELECTRON

Learning Objectives

After reading this module, you should be able to . . .

28.12 Describe the experiment of J. J. Thomson.

28.13 For a charged particle moving through a magnetic field and an electric field, determine the net force on the particle in both magnitude-angle notation and unit-vector notation.

28.14 In situations where the magnetic force and electric force on a particle are in opposite directions, determine the speeds at which the forces cancel, the magnetic force dominates, and the electric force dominates.

Key Ideas

● If a charged particle moves through a region containing both an electric field and a magnetic field, it can be affected by both an electric force and a magnetic force.

● If the fields are perpendicular to each other, they are said to be *crossed fields*.

● If the forces are in opposite directions, a particular speed will result in no deflection of the particle.

Crossed Fields: Discovery of the Electron

Both an electric field $\vec{E}$ and a magnetic field $\vec{B}$ can produce a force on a charged particle. When the two fields are perpendicular to each other, they are said to be *crossed fields*. Here we shall examine what happens to charged particles—namely, electrons—as they move through crossed fields. We use as our example the experiment that led to the discovery of the electron in 1897 by J. J. Thomson at Cambridge University.

Two Forces. Figure 28-7 shows a modern, simplified version of Thomson's experimental apparatus—a *cathode ray tube* (which is like the picture tube in an old-type television set). Charged particles (which we now know as electrons) are emitted by a hot filament at the rear of the evacuated tube and are accelerated by an applied potential difference V. After they pass through a slit in screen C, they form a narrow beam. They then pass through a region of crossed $\vec{E}$ and $\vec{B}$ fields, headed toward a fluorescent screen S, where they produce a spot of light (on a television screen the spot is part of the picture). The forces on the charged particles in the crossed-fields region can deflect them from the center of the screen. By controlling the magnitudes and directions of the fields, Thomson could thus control where the spot of light appeared on the screen. Recall that the force on a negatively charged particle due to an electric field is directed opposite the field. Thus, for the arrangement of Fig. 28-7, electrons are forced up the page by electric field $\vec{E}$ and down the page by magnetic field $\vec{B}$; that is, the forces are *in opposition*. Thomson's procedure was equivalent to the following series of steps.

1. Set $E = 0$ and $B = 0$ and note the position of the spot on screen S due to the undeflected beam.

2. Turn on $\vec{E}$ and measure the resulting beam deflection.

3. Maintaining $\vec{E}$, now turn on $\vec{B}$ and adjust its value until the beam returns to the undeflected position. (With the forces in opposition, they can be made to cancel.)

We discussed the deflection of a charged particle moving through an electric field $\vec{E}$ between two plates (step 2 here) in Sample Problem 22.04. We found that the deflection of the particle at the far end of the plates is

$$y = \frac{|q|EL^2}{2mv^2},\qquad(28\text{-}6)$$

where v is the particle's speed, m its mass, and q its charge, and L is the length of the plates. We can apply this same equation to the beam of electrons in Fig. 28-7; if need be, we can calculate the deflection by measuring the deflection of the beam on screen S and then working back to calculate the deflection y at the end of the plates. (Because the direction of the deflection is set by the sign of the particle's charge, Thomson was able to show that the particles that were lighting up his screen were negatively charged.)

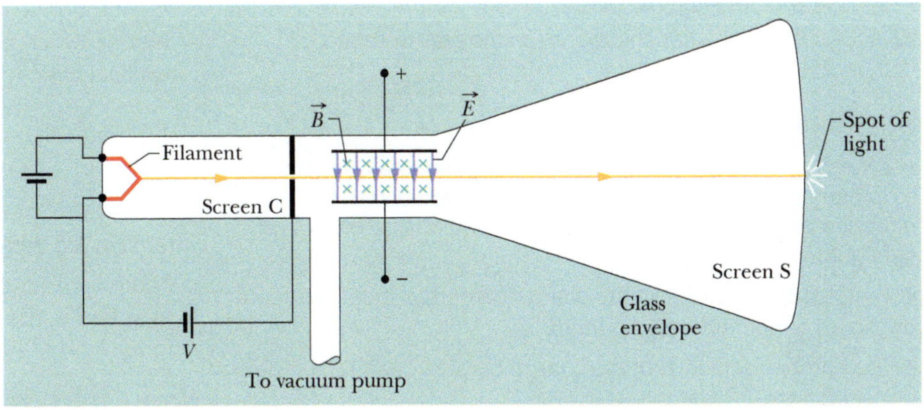

Figure 28-7 A modern version of J. J. Thomson's apparatus for measuring the ratio of mass to charge for the electron. An electric field $\vec{E}$ is established by connecting a battery across the deflecting-plate terminals. The magnetic field $\vec{B}$ is set up by means of a current in a system of coils (not shown). The magnetic field shown is into the plane of the figure, as represented by the array of **X**s (which resemble the feathered ends of arrows).

Canceling Forces. When the two fields in Fig. 28-7 are adjusted so that the two deflecting forces cancel (step 3), we have from Eqs. 28-1 and 28-3

$$|q|E = |q|vB \sin(90°) = |q|vB$$

or
$$v = \frac{E}{B} \quad \text{(opposite forces canceling).}\tag{28-7}$$

Thus, the crossed fields allow us to measure the speed of the charged particles passing through them. Substituting Eq. 28-7 for v in Eq. 28-6 and rearranging yield

$$\frac{m}{|q|} = \frac{B^2L^2}{2yE},\tag{28-8}$$

in which all quantities on the right can be measured. Thus, the crossed fields allow us to measure the ratio $m/|q|$ of the particles moving through Thomson's apparatus. (*Caution:* Equation 28-7 applies only when the electric and magnetic forces are in opposite directions. You might see other situations in the homework problems.)

Thomson claimed that these particles are found in all matter. He also claimed that they are lighter than the lightest known atom (hydrogen) by a factor of more than 1000. (The exact ratio proved later to be 1836.15.) His $m/|q|$ measurement, coupled with the boldness of his two claims, is considered to be the "discovery of the electron."

✓ Checkpoint 2

The figure shows four directions for the velocity vector $\vec{v}$ of a positively charged particle moving through a uniform electric field $\vec{E}$ (directed out of the page and represented with an encircled dot) and a uniform magnetic field $\vec{B}$. (a) Rank directions 1, 2, and 3 according to the magnitude of the net force on the particle, greatest first. (b) Of all four directions, which might result in a net force of zero?

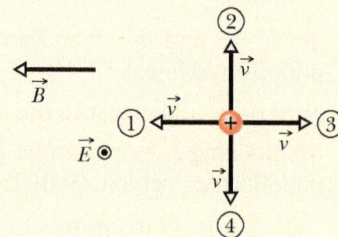

28-3 CROSSED FIELDS: THE HALL EFFECT

Learning Objectives

After reading this module, you should be able to ...

28.15 Describe the Hall effect for a metal strip carrying current, explaining how the electric field is set up and what limits its magnitude.

28.16 For a conducting strip in a Hall-effect situation, draw the vectors for the magnetic field and electric field. For the conduction electrons, draw the vectors for the velocity, magnetic force, and electric force.

28.17 Apply the relationship between the Hall potential

difference V, the electric field magnitude E, and the width of the strip d.

28.18 Apply the relationship between charge-carrier number density n, magnetic field magnitude B, current i, and Hall-effect potential difference V.

28.19 Apply the Hall-effect results to a conducting object moving through a uniform magnetic field, identifying the width across which a Hall-effect potential difference V is set up and calculating V.

Key Ideas

● When a uniform magnetic field $\vec{B}$ is applied to a conducting strip carrying current i, with the field perpendicular to the direction of the current, a Hall-effect potential difference V is set up across the strip.

● The electric force $\vec{F}_E$ on the charge carriers is then balanced by the magnetic force $\vec{F}_B$ on them.

● The number density n of the charge carriers can then be determined from

$$n = \frac{Bi}{Vle},$$

where l is the thickness of the strip (parallel to $\vec{B}$).

● When a conductor moves through a uniform magnetic field $\vec{B}$ at speed v, the Hall-effect potential difference V across it is

$$V = vBd,$$

where d is the width perpendicular to both velocity $\vec{v}$ and field $\vec{B}$.

Crossed Fields: The Hall Effect

As we just discussed, a beam of electrons in a vacuum can be deflected by a magnetic field. Can the drifting conduction electrons in a copper wire also be deflected by a magnetic field? In 1879, Edwin H. Hall, then a 24-year-old graduate student at the Johns Hopkins University, showed that they can. This **Hall effect** allows us to find out whether the charge carriers in a conductor are positively or negatively charged. Beyond that, we can measure the number of such carriers per unit volume of the conductor.

Figure 28-8a shows a copper strip of width d, carrying a current i whose conventional direction is from the top of the figure to the bottom. The charge carriers are electrons and, as we know, they drift (with drift speed v_d) in the opposite direction, from bottom to top. At the instant shown in Fig. 28-8a, an external magnetic field $\vec{B}$, pointing into the plane of the figure, has just been turned on. From Eq. 28-2 we see that a magnetic deflecting force $\vec{F}_B$ will act on each drifting electron, pushing it toward the right edge of the strip.

As time goes on, electrons move to the right, mostly piling up on the right edge of the strip, leaving uncompensated positive charges in fixed positions at the left edge. The separation of positive charges on the left edge and negative charges on the right edge produces an electric field $\vec{E}$ within the strip, pointing from left to right in Fig. 28-8b. This field exerts an electric force $\vec{F}_E$ on each electron, tending to push it to the left. Thus, this electric force on the electrons, which opposes the magnetic force on them, begins to build up.

Equilibrium. An equilibrium quickly develops in which the electric force on each electron has increased enough to match the magnetic force. When this happens, as Fig. 28-8b shows, the force due to $\vec{B}$ and the force due to $\vec{E}$ are in balance. The drifting electrons then move along the strip toward the top of the page at velocity $\vec{v}_d$ with no further collection of electrons on the right edge of the strip and thus no further increase in the electric field $\vec{E}$.

A *Hall potential difference V* is associated with the electric field across strip width d. From Eq. 24-21, the magnitude of that potential difference is

$$V = Ed. \tag{28-9}$$

By connecting a voltmeter across the width, we can measure the potential difference between the two edges of the strip. Moreover, the voltmeter can tell us which edge is at higher potential. For the situation of Fig. 28-8b, we would find that the left edge is at higher potential, which is consistent with our assumption that the charge carriers are negatively charged.

For a moment, let us make the opposite assumption, that the charge carriers in current i are positively charged (Fig. 28-8c). Convince yourself that as these charge carriers move from top to bottom in the strip, they are pushed to the right edge by $\vec{F}_B$ and thus that the *right* edge is at higher potential. Because that last statement is contradicted by our voltmeter reading, the charge carriers must be negatively charged.

Number Density. Now for the quantitative part. When the electric and magnetic forces are in balance (Fig. 28-8b), Eqs. 28-1 and 28-3 give us

$$eE = ev_d B. \tag{28-10}$$

From Eq. 26-7, the drift speed v_d is

$$v_d = \frac{J}{ne} = \frac{i}{neA}, \tag{28-11}$$

in which J ($= i/A$) is the current density in the strip, A is the cross-sectional area of the strip, and n is the *number density* of charge carriers (number per unit volume).

In Eq. 28-10, substituting for E with Eq. 28-9 and substituting for v_d with Eq. 28-11, we obtain

$$n = \frac{Bi}{Vle}, \tag{28-12}$$

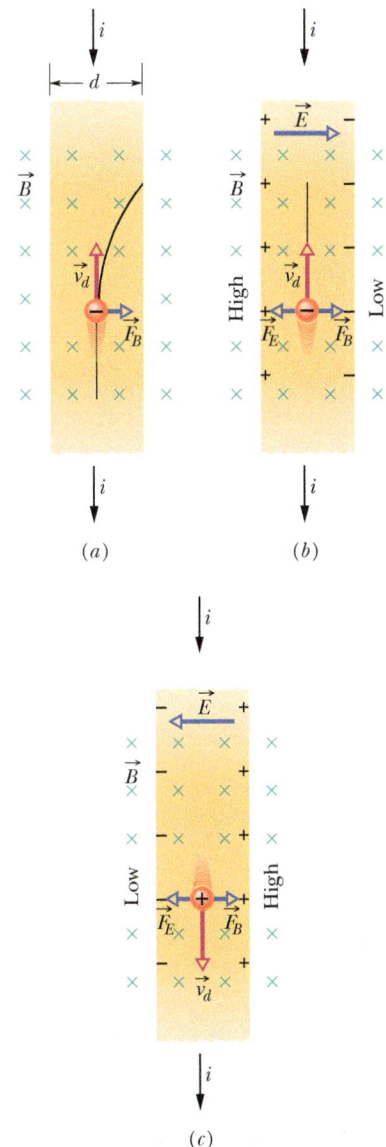

(a)   (b)

(c)

Figure 28-8 A strip of copper carrying a current i is immersed in a magnetic field $\vec{B}$. (a) The situation immediately after the magnetic field is turned on. The curved path that will then be taken by an electron is shown. (b) The situation at equilibrium, which quickly follows. Note that negative charges pile up on the right side of the strip, leaving uncompensated positive charges on the left. Thus, the left side is at a higher potential than the right side. (c) For the same current direction, if the charge carriers were positively charged, *they* would pile up on the right side, and the right side would be at the higher potential.

in which $l (= A/d)$ is the thickness of the strip. With this equation we can find n from measurable quantities.

Drift Speed. It is also possible to use the Hall effect to measure directly the drift speed v_d of the charge carriers, which you may recall is of the order of centimeters per hour. In this clever experiment, the metal strip is moved mechanically through the magnetic field in a direction opposite that of the drift velocity of the charge carriers. The speed of the moving strip is then adjusted until the Hall potential difference vanishes. At this condition, with no Hall effect, the velocity of the charge carriers *with respect to the laboratory frame* must be zero, so the velocity of the strip must be equal in magnitude but opposite the direction of the velocity of the negative charge carriers.

Moving Conductor. When a conductor begins to move at speed v through a magnetic field, its conduction electrons do also. They are then like the moving conduction electrons in the current in Figs. 28-8a and b, and an electric field $\vec{E}$ and potential difference V are quickly set up. As with the current, equilibrium of the electric and magnetic forces is established, but we now write that condition in terms of the conductor's speed v instead of the drift speed v_d in a current as we did in Eq. 28-10:

$$eE = evB.$$

Substituting for E with Eq. 28-9, we find that the potential difference is

$$V = vBd. \tag{28-13}$$

Such a motion-caused circuit potential difference can be of serious concern in some situations, such as when a conductor in an orbiting satellite moves through Earth's magnetic field. However, if a conducting line (said to be an *electrodynamic tether*) dangles from the satellite, the potential produced along the line might be used to maneuver the satellite.

Sample Problem 28.02   Potential difference set up across a moving conductor

Figure 28-9a shows a solid metal cube, of edge length $d = 1.5$ cm, moving in the positive y direction at a constant velocity $\vec{v}$ of magnitude 4.0 m/s. The cube moves through a uniform magnetic field $\vec{B}$ of magnitude 0.050 T in the positive z direction.

(a) Which cube face is at a lower electric potential and which is at a higher electric potential because of the motion through the field?

KEY IDEA

Because the cube is moving through a magnetic field $\vec{B}$, a magnetic force $\vec{F}_B$ acts on its charged particles, including its conduction electrons.

Reasoning: When the cube first begins to move through the magnetic field, its electrons do also. Because each electron has charge q and is moving through a magnetic field with velocity $\vec{v}$, the magnetic force $\vec{F}_B$ acting on the electron is given by Eq. 28-2. Because q is negative, the direction of $\vec{F}_B$ is opposite the cross product $\vec{v} \times \vec{B}$, which is in the posi-

tive direction of the x axis (Fig. 28-9b). Thus, $\vec{F}_B$ acts in the negative direction of the x axis, toward the left face of the cube (Fig. 28-9c).

Most of the electrons are fixed in place in the atoms of the cube. However, because the cube is a metal, it contains conduction electrons that are free to move. Some of those conduction electrons are deflected by $\vec{F}_B$ to the left cube face, making that face negatively charged and leaving the right face positively charged (Fig. 28-9d). This charge separation produces an electric field $\vec{E}$ directed from the positively charged right face to the negatively charged left face (Fig. 28-9e). Thus, the left face is at a lower electric potential, and the right face is at a higher electric potential.

(b) What is the potential difference between the faces of higher and lower electric potential?

KEY IDEAS

1. The electric field $\vec{E}$ created by the charge separation produces an electric force $\vec{F}_E = q\vec{E}$ on each electron

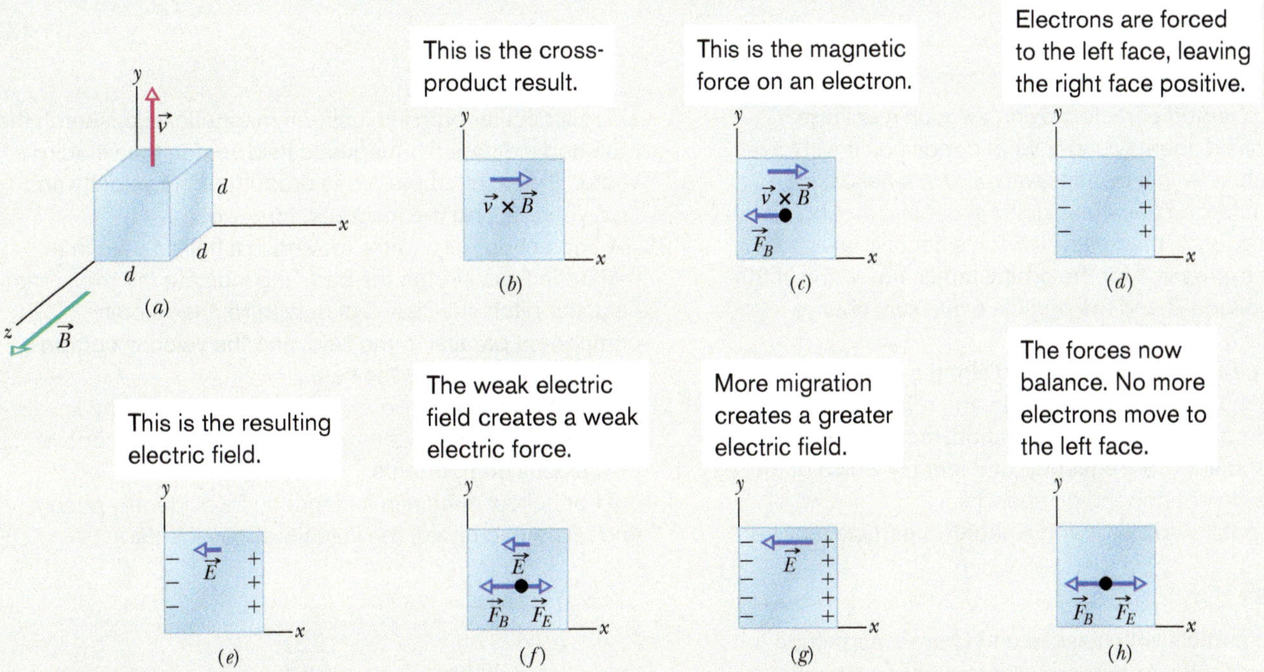

This is the cross-product result.

This is the magnetic force on an electron.

Electrons are forced to the left face, leaving the right face positive.

(a)

(b)

(c)

(d)

This is the resulting electric field.

The weak electric field creates a weak electric force.

More migration creates a greater electric field.

The forces now balance. No more electrons move to the left face.

(e)

(f)

(g)

(h)

Figure 28-9 (a) A solid metal cube moves at constant velocity through a uniform magnetic field. (b)–(d) In these front views, the magnetic force acting on an electron forces the electron to the left face, making that face negative and leaving the opposite face positive. (e)–(f) The resulting weak electric field creates a weak electric force on the next electron, but it too is forced to the left face. Now (g) the electric field is stronger and (h) the electric force matches the magnetic force.

(Fig. 28-9f). Because q is negative, this force is directed opposite the field $\vec{E}$—that is, rightward. Thus on each electron, $\vec{F}_E$ acts toward the right and $\vec{F}_B$ acts toward the left.

2. When the cube had just begun to move through the magnetic field and the charge separation had just begun, the magnitude of $\vec{E}$ began to increase from zero. Thus, the magnitude of $\vec{F}_E$ also began to increase from zero and was initially smaller than the magnitude of $\vec{F}_B$. During this early stage, the net force on any electron was dominated by $\vec{F}_B$, which continuously moved additional electrons to the left cube face, increasing the charge separation between the left and right cube faces (Fig. 28-9g).

3. However, as the charge separation increased, eventually magnitude F_E became equal to magnitude F_B (Fig. 28-9h). Because the forces were in opposite directions, the net force on any electron was then zero, and no additional electrons were moved to the left cube face. Thus, the magnitude of $\vec{F}_E$ could not increase further, and the electrons were then in equilibrium.

Calculations: We seek the potential difference V between the left and right cube faces after equilibrium was reached (which occurred quickly). We can obtain V with Eq. 28-9 ($V = Ed$) provided we first find the magnitude E of the electric field at equilibrium. We can do so with the equation for the balance of forces ($F_E = F_B$).

For F_E, we substitute $|q|E$, and then for F_B, we substitute $|q|vB \sin \phi$ from Eq. 28-3. From Fig. 28-9a, we see that the angle ϕ between velocity vector $\vec{v}$ and magnetic field vector $\vec{B}$ is 90°; thus $\sin \phi = 1$ and $F_E = F_B$ yields

$$|q|E = |q|vB \sin 90° = |q|vB.$$

This gives us $E = vB$; so $V = Ed$ becomes

$$V = vBd.$$

Substituting known values tells us that the potential difference between the left and right cube faces is

$$V = (4.0 \text{ m/s})(0.050 \text{ T})(0.015 \text{ m})$$
$$= 0.0030 \text{ V} = 3.0 \text{ mV}. \qquad \text{(Answer)}$$

WILEY PLUS Additional examples, video, and practice available at *WileyPLUS*

28-4 A CIRCULATING CHARGED PARTICLE

Learning Objectives

After reading this module, you should be able to . . .

28.20 For a charged particle moving through a uniform magnetic field, identify under what conditions it will travel in a straight line, in a circular path, and in a helical path.

28.21 For a charged particle in uniform circular motion due to a magnetic force, start with Newton's second law and derive an expression for the orbital radius r in terms of the field magnitude B and the particle's mass m, charge magnitude q, and speed v.

28.22 For a charged particle moving along a circular path in a uniform magnetic field, calculate and relate speed, centripetal force, centripetal acceleration, radius, period, frequency, and angular frequency, and identify which of the quantities do not depend on speed.

28.23 For a positive particle and a negative particle moving along a circular path in a uniform magnetic field, sketch the path and indicate the magnetic field vector, the velocity vector, the result of the cross product of the velocity and field vectors, and the magnetic force vector.

28.24 For a charged particle moving in a helical path in a magnetic field, sketch the path and indicate the magnetic field, the pitch, the radius of curvature, the velocity component parallel to the field, and the velocity component perpendicular to the field.

28.25 For helical motion in a magnetic field, apply the relationship between the radius of curvature and one of the velocity components.

28.26 For helical motion in a magnetic field, identify pitch p and relate it to one of the velocity components.

Key Ideas

● A charged particle with mass m and charge magnitude $|q|$ moving with velocity $\vec{v}$ perpendicular to a uniform magnetic field $\vec{B}$ will travel in a circle.

● Applying Newton's second law to the circular motion yields

$$|q|vB = \frac{mv^2}{r},$$

from which we find the radius r of the circle to be

$$r = \frac{mv}{|q|B}.$$

● The frequency of revolution f, the angular frequency ω, and the period of the motion T are given by

$$f = \frac{\omega}{2\pi} = \frac{1}{T} = \frac{|q|B}{2\pi m}.$$

● If the velocity of the particle has a component parallel to the magnetic field, the particle moves in a helical path about field vector $\vec{B}$.

A Circulating Charged Particle

If a particle moves in a circle at constant speed, we can be sure that the net force acting on the particle is constant in magnitude and points toward the center of the circle, always perpendicular to the particle's velocity. Think of a stone tied to a string and whirled in a circle on a smooth horizontal surface, or of a satellite moving in a circular orbit around Earth. In the first case, the tension in the string provides the necessary force and centripetal acceleration. In the second case, Earth's gravitational attraction provides the force and acceleration.

Figure 28-10 shows another example: A beam of electrons is projected into a chamber by an *electron gun* G. The electrons enter in the plane of the page with speed v and then move in a region of uniform magnetic field $\vec{B}$ directed out of that plane. As a result, a magnetic force $\vec{F}_B = q\vec{v} \times \vec{B}$ continuously deflects the electrons, and because $\vec{v}$ and $\vec{B}$ are always perpendicular to each other, this deflection causes the electrons to follow a circular path. The path is visible in the photo because atoms of gas in the chamber emit light when some of the circulating electrons collide with them.

We would like to determine the parameters that characterize the circular motion of these electrons, or of any particle of charge magnitude $|q|$ and mass m moving perpendicular to a uniform magnetic field $\vec{B}$ at speed v. From Eq. 28-3, the force acting on the particle has a magnitude of $|q|vB$. From Newton's second law ($\vec{F} = m\vec{a}$) applied to uniform circular motion (Eq. 6-18),

$$F = m\frac{v^2}{r}, \tag{28-14}$$

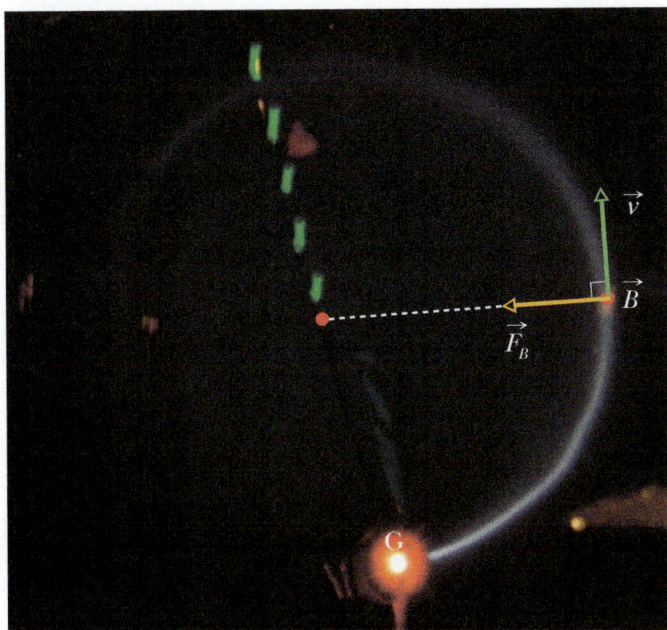

Courtesy Jearl Walker

Figure 28-10 Electrons circulating in a chamber containing gas at low pressure (their path is the glowing circle). A uniform magnetic field $\vec{B}$, pointing directly out of the plane of the page, fills the chamber. Note the radially directed magnetic force $\vec{F}_B$; for circular motion to occur, $\vec{F}_B$ *must* point toward the center of the circle. Use the right-hand rule for cross products to confirm that $\vec{F}_B = q\vec{v} \times \vec{B}$ gives $\vec{F}_B$ the proper direction. (Don't forget the sign of q.)

we have

$$|q|vB = \frac{mv^2}{r}. \qquad (28\text{-}15)$$

Solving for r, we find the radius of the circular path as

$$r = \frac{mv}{|q|B} \quad \text{(radius).} \qquad (28\text{-}16)$$

The period T (the time for one full revolution) is equal to the circumference divided by the speed:

$$T = \frac{2\pi r}{v} = \frac{2\pi}{v}\frac{mv}{|q|B} = \frac{2\pi m}{|q|B} \quad \text{(period).} \qquad (28\text{-}17)$$

The frequency f (the number of revolutions per unit time) is

$$f = \frac{1}{T} = \frac{|q|B}{2\pi m} \quad \text{(frequency).} \qquad (28\text{-}18)$$

The angular frequency ω of the motion is then

$$\omega = 2\pi f = \frac{|q|B}{m} \quad \text{(angular frequency).} \qquad (28\text{-}19)$$

The quantities T, f, and ω do not depend on the speed of the particle (provided the speed is much less than the speed of light). Fast particles move in large circles and slow ones in small circles, but all particles with the same charge-to-mass ratio $|q|/m$ take the same time T (the period) to complete one round trip. Using Eq. 28-2, you can show that if you are looking in the direction of $\vec{B}$, the direction of rotation for a positive particle is always counterclockwise, and the direction for a negative particle is always clockwise.

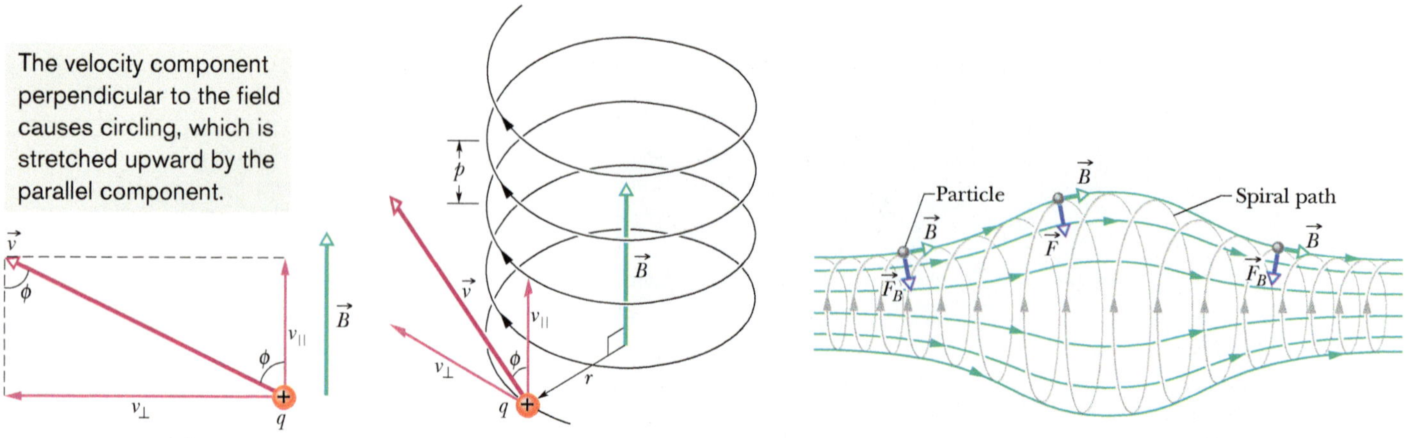

The velocity component perpendicular to the field causes circling, which is stretched upward by the parallel component.

(a)

(b)

(c)

Figure 28-11 (a) A charged particle moves in a uniform magnetic field $\vec{B}$, the particle's velocity $\vec{v}$ making an angle ϕ with the field direction. (b) The particle follows a helical path of radius r and pitch p. (c) A charged particle spiraling in a nonuniform magnetic field. (The particle can become trapped in this *magnetic bottle*, spiraling back and forth between the strong field regions at either end.) Note that the magnetic force vectors at the left and right sides have a component pointing toward the center of the figure.

Helical Paths

If the velocity of a charged particle has a component parallel to the (uniform) magnetic field, the particle will move in a helical path about the direction of the field vector. Figure 28-11a, for example, shows the velocity vector $\vec{v}$ of such a particle resolved into two components, one parallel to $\vec{B}$ and one perpendicular to it:

$$v_\parallel = v \cos \phi \quad \text{and} \quad v_\perp = v \sin \phi. \tag{28-20}$$

The parallel component determines the *pitch p* of the helix—that is, the distance between adjacent turns (Fig. 28-11b). The perpendicular component determines the radius of the helix and is the quantity to be substituted for v in Eq. 28-16.

Figure 28-11c shows a charged particle spiraling in a nonuniform magnetic field. The more closely spaced field lines at the left and right sides indicate that the magnetic field is stronger there. When the field at an end is strong enough, the particle "reflects" from that end.

✓ **Checkpoint 3**

The figure here shows the circular paths of two particles that travel at the same speed in a uniform magnetic field $\vec{B}$, which is directed into the page. One particle is a proton; the other is an electron (which is less massive). (a) Which particle follows the smaller circle, and (b) does that particle travel clockwise or counterclockwise?

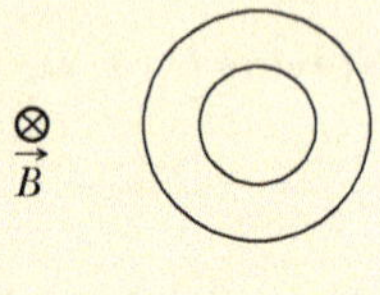

$\otimes$
$\vec{B}$

Sample Problem 28.03   Helical motion of a charged particle in a magnetic field

An electron with a kinetic energy of 22.5 eV moves into a region of uniform magnetic field $\vec{B}$ of magnitude 4.55 × 10^{-4} T. The angle between the directions of $\vec{B}$ and the electron's velocity $\vec{v}$ is 65.5°. What is the pitch of the helical path taken by the electron?

KEY IDEAS

(1) The pitch p is the distance the electron travels parallel to the magnetic field $\vec{B}$ during one period T of circulation. (2) The period T is given by Eq. 28-17 for any nonzero angle between $\vec{v}$ and $\vec{B}$.

Calculations: Using Eqs. 28-20 and 28-17, we find

$$p = v_\parallel T = (v \cos \phi) \frac{2\pi m}{|q|B}. \tag{28-21}$$

Calculating the electron's speed v from its kinetic energy, we find that $v = 2.81 \times 10^6$ m/s, and so Eq. 28-21 gives us

$$p = (2.81 \times 10^6 \text{ m/s})(\cos 65.5°)$$

$$\times \frac{2\pi(9.11 \times 10^{-31} \text{ kg})}{(1.60 \times 10^{-19} \text{ C})(4.55 \times 10^{-4} \text{ T})}$$

$$= 9.16 \text{ cm.} \tag{Answer}$$

Sample Problem 28.04   Uniform circular motion of a charged particle in a magnetic field

Figure 28-12 shows the essentials of a *mass spectrometer*, which can be used to measure the mass of an ion; an ion of mass m (to be measured) and charge q is produced in source S. The initially stationary ion is accelerated by the electric field due to a potential difference V. The ion leaves S and enters a separator chamber in which a uniform magnetic field $\vec{B}$ is perpendicular to the path of the ion. A wide detector lines the bottom wall of the chamber, and the $\vec{B}$ causes the ion to move in a semicircle and thus strike the detector. Suppose that $B = 80.000$ mT, $V = 1000.0$ V, and ions of charge $q = +1.6022 \times 10^{-19}$ C strike the detector at a point that lies at $x = 1.6254$ m. What is the mass m of the individual ions, in atomic mass units (Eq. 1-7: 1 u $= 1.6605 \times 10^{-27}$ kg)?

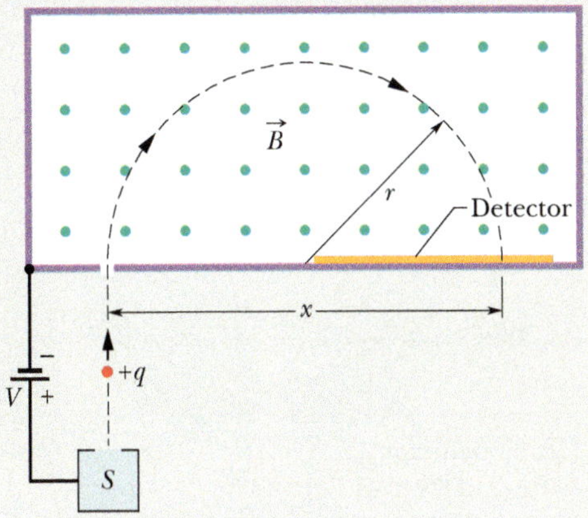

Figure 28-12 A positive ion is accelerated from its source S by a potential difference V, enters a chamber of uniform magnetic field $\vec{B}$, travels through a semicircle of radius r, and strikes a detector at a distance x.

KEY IDEAS

(1) Because the (uniform) magnetic field causes the (charged) ion to follow a circular path, we can relate the ion's mass m to the path's radius r with Eq. 28-16 ($r = mv/|q|B$). From Fig. 28-12 we see that $r = x/2$ (the radius is half the diameter). From the problem statement, we know the magnitude B of the magnetic field. However, we lack the ion's speed v in the magnetic field after the ion has been accelerated due to the potential difference V. (2) To relate v and V, we use the fact that mechanical energy ($E_{mec} = K + U$) is conserved during the acceleration.

Finding speed: When the ion emerges from the source, its kinetic energy is approximately zero. At the end of the acceleration, its kinetic energy is $\frac{1}{2}mv^2$. Also, during the acceleration, the positive ion moves through a change in potential of $-V$. Thus, because the ion has positive charge q, its potential energy changes by $-qV$. If we now write the conservation of mechanical energy as

$$\Delta K + \Delta U = 0,$$

we get

$$\tfrac{1}{2}mv^2 - qV = 0$$

or

$$v = \sqrt{\frac{2qV}{m}}. \qquad (28\text{-}22)$$

Finding mass: Substituting this value for v into Eq. 28-16 gives us

$$r = \frac{mv}{qB} = \frac{m}{qB}\sqrt{\frac{2qV}{m}} = \frac{1}{B}\sqrt{\frac{2mV}{q}}.$$

Thus,

$$x = 2r = \frac{2}{B}\sqrt{\frac{2mV}{q}}.$$

Solving this for m and substituting the given data yield

$$m = \frac{B^2 q x^2}{8V}$$

$$= \frac{(0.080000 \text{ T})^2(1.6022 \times 10^{-19} \text{ C})(1.6254 \text{ m})^2}{8(1000.0 \text{ V})}$$

$$= 3.3863 \times 10^{-25} \text{ kg} = 203.93 \text{ u}. \qquad \text{(Answer)}$$

WILEY PLUS Additional examples, video, and practice available at *WileyPLUS*

28-5 CYCLOTRONS AND SYNCHROTRONS

Learning Objectives

After reading this module, you should be able to . . .

28.27 Describe how a cyclotron works, and in a sketch indicate a particle's path and the regions where the kinetic energy is increased.

28.28 Identify the resonance condition.

28.29 For a cyclotron, apply the relationship between the particle's mass and charge, the magnetic field, and the frequency of circling.

28.30 Distinguish between a cyclotron and a synchrotron.

Key Ideas

● In a cyclotron, charged particles are accelerated by electric forces as they circle in a magnetic field.

● A synchrotron is needed for particles accelerated to nearly the speed of light.

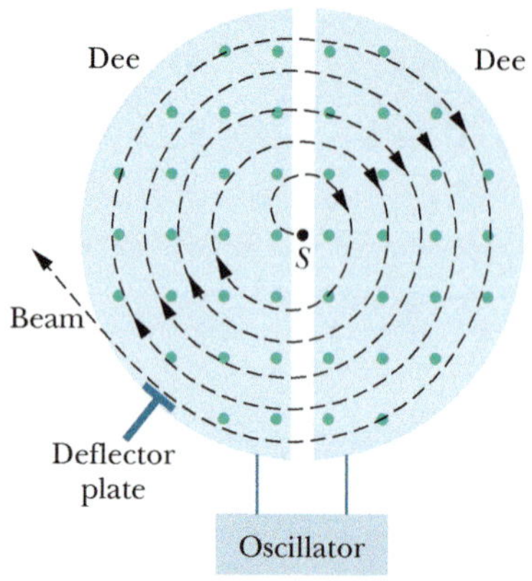

The protons spiral outward in a cyclotron, picking up energy in the gap.

Figure 28-13 The elements of a cyclotron, showing the particle source S and the dees. A uniform magnetic field is directed up from the plane of the page. Circulating protons spiral outward within the hollow dees, gaining energy every time they cross the gap between the dees.

Cyclotrons and Synchrotrons

Beams of high-energy particles, such as high-energy electrons and protons, have been enormously useful in probing atoms and nuclei to reveal the fundamental structure of matter. Such beams were instrumental in the discovery that atomic nuclei consist of protons and neutrons and in the discovery that protons and neutrons consist of quarks and gluons. Because electrons and protons are charged, they can be accelerated to the required high energy if they move through large potential differences. The required acceleration distance is reasonable for electrons (low mass) but unreasonable for protons (greater mass).

A clever solution to this problem is first to let protons and other massive particles move through a modest potential difference (so that they gain a modest amount of energy) and then use a magnetic field to cause them to circle back and move through a modest potential difference again. If this procedure is repeated thousands of times, the particles end up with a very large energy.

Here we discuss two *accelerators* that employ a magnetic field to repeatedly bring particles back to an accelerating region, where they gain more and more energy until they finally emerge as a high-energy beam.

The Cyclotron

Figure 28-13 is a top view of the region of a *cyclotron* in which the particles (protons, say) circulate. The two hollow **D**-shaped objects (each open on its straight edge) are made of sheet copper. These *dees*, as they are called, are part of an electrical oscillator that alternates the electric potential difference across the gap between the dees. The electrical signs of the dees are alternated so that the electric field in the gap alternates in direction, first toward one dee and then toward the other dee, back and forth. The dees are immersed in a large magnetic field directed out of the plane of the page. The magnitude B of this field is set via a control on the electromagnet producing the field.

Suppose that a proton, injected by source S at the center of the cyclotron in Fig. 28-13, initially moves toward a negatively charged dee. It will accelerate toward this dee and enter it. Once inside, it is shielded from electric fields by the copper walls of the dee; that is, the electric field does not enter the dee. The magnetic field, however, is not screened by the (nonmagnetic) copper dee, so the proton moves in a circular path whose radius, which depends on its speed, is given by Eq. 28-16 ($r = mv/|q|B$).

Let us assume that at the instant the proton emerges into the center gap from the first dee, the potential difference between the dees is reversed. Thus, the proton *again* faces a negatively charged dee and is *again* accelerated. This process continues, the circulating proton always being in step with the oscillations of the dee potential, until the proton has spiraled out to the edge of the dee system. There a deflector plate sends it out through a portal.

Frequency. The key to the operation of the cyclotron is that the frequency f at which the proton circulates in the magnetic field (and that does *not* depend on its speed) must be equal to the fixed frequency f_{osc} of the electrical oscillator, or

$$f = f_{osc} \quad \text{(resonance condition)}. \qquad (28\text{-}23)$$

This *resonance condition* says that, if the energy of the circulating proton is to increase, energy must be fed to it at a frequency f_{osc} that is equal to the natural frequency f at which the proton circulates in the magnetic field.

Combining Eqs. 28-18 ($f = |q|B/2\pi m$) and 28-23 allows us to write the resonance condition as

$$|q|B = 2\pi m f_{osc}. \qquad (28\text{-}24)$$

The oscillator (we assume) is designed to work at a single fixed frequency f_{osc}. We then "tune" the cyclotron by varying B until Eq. 28-24 is satisfied, and then many protons circulate through the magnetic field, to emerge as a beam.

The Proton Synchrotron

At proton energies above 50 MeV, the conventional cyclotron begins to fail because one of the assumptions of its design—that the frequency of revolution of a charged particle circulating in a magnetic field is independent of the particle's speed—is true only for speeds that are much less than the speed of light. At greater proton speeds (above about 10% of the speed of light), we must treat the problem relativistically. According to relativity theory, as the speed of a circulating proton approaches that of light, the proton's frequency of revolution decreases steadily. Thus, the proton gets out of step with the cyclotron's oscillator—whose frequency remains fixed at f_{osc}—and eventually the energy of the still circulating proton stops increasing.

There is another problem. For a 500 GeV proton in a magnetic field of 1.5 T, the path radius is 1.1 km. The corresponding magnet for a conventional cyclotron of the proper size would be impossibly expensive, the area of its pole faces being about 4×10^6 m^2.

The *proton synchrotron* is designed to meet these two difficulties. The magnetic field B and the oscillator frequency f_{osc}, instead of having fixed values as in the conventional cyclotron, are made to vary with time during the accelerating cycle. When this is done properly, (1) the frequency of the circulating protons remains in step with the oscillator at all times, and (2) the protons follow a circular—not a spiral—path. Thus, the magnet need extend only along that circular path, not over some 4×10^6 m^2. The circular path, however, still must be large if high energies are to be achieved.

Sample Problem 28.05 Accelerating a charged particle in a cyclotron

Suppose a cyclotron is operated at an oscillator frequency of 12 MHz and has a dee radius $R = 53$ cm.

(a) What is the magnitude of the magnetic field needed for deuterons to be accelerated in the cyclotron? The deuteron mass is $m = 3.34 \times 10^{-27}$ kg (twice the proton mass).

KEY IDEA

For a given oscillator frequency f_{osc}, the magnetic field magnitude B required to accelerate any particle in a cyclotron depends on the ratio $m/|q|$ of mass to charge for the particle, according to Eq. 28-24 ($|q|B = 2\pi m f_{osc}$).

Calculation: For deuterons and the oscillator frequency $f_{osc} = $ 12 MHz, we find

$$B = \frac{2\pi m f_{osc}}{|q|} = \frac{(2\pi)(3.34 \times 10^{-27}\,\text{kg})(12 \times 10^6\,\text{s}^{-1})}{1.60 \times 10^{-19}\,\text{C}}$$

$$= 1.57\,\text{T} \approx 1.6\,\text{T}. \qquad \text{(Answer)}$$

Note that, to accelerate protons, B would have to be reduced by a factor of 2, provided the oscillator frequency remained fixed at 12 MHz.

(b) What is the resulting kinetic energy of the deuterons?

KEY IDEAS

(1) The kinetic energy ($\frac{1}{2}mv^2$) of a deuteron exiting the cyclotron is equal to the kinetic energy it had just before exiting, when it was traveling in a circular path with a radius approximately equal to the radius R of the cyclotron dees. (2) We can find the speed v of the deuteron in that circular path with Eq. 28-16 ($r = mv/|q|B$).

Calculations: Solving that equation for v, substituting R for r, and then substituting known data, we find

$$v = \frac{R|q|B}{m} = \frac{(0.53\,\text{m})(1.60 \times 10^{-19}\,\text{C})(1.57\,\text{T})}{3.34 \times 10^{-27}\,\text{kg}}$$

$$= 3.99 \times 10^7\,\text{m/s}.$$

This speed corresponds to a kinetic energy of

$$K = \tfrac{1}{2}mv^2$$

$$= \tfrac{1}{2}(3.34 \times 10^{-27}\,\text{kg})(3.99 \times 10^7\,\text{m/s})^2$$

$$= 2.7 \times 10^{-12}\,\text{J}, \qquad \text{(Answer)}$$

or about 17 MeV.

PLUS Additional examples, video, and practice available at *WileyPLUS*

28-6 MAGNETIC FORCE ON A CURRENT-CARRYING WIRE

Learning Objectives

After reading this module, you should be able to . . .

28.31 For the situation where a current is perpendicular to a magnetic field, sketch the current, the direction of the magnetic field, and the direction of the magnetic force on the current (or wire carrying the current).

28.32 For a current in a magnetic field, apply the relationship between the magnetic force magnitude F_B, the current i, the length of the wire L, and the angle ϕ between the length vector $\vec{L}$ and the field vector $\vec{B}$.

28.33 Apply the right-hand rule for cross products to find

the direction of the magnetic force on a current in a magnetic field.

28.34 For a current in a magnetic field, calculate the magnetic force $\vec{F}_B$ with a cross product of the length vector $\vec{L}$ and the field vector $\vec{B}$, in magnitude-angle and unit-vector notations.

28.35 Describe the procedure for calculating the force on a current-carrying wire in a magnetic field if the wire is not straight or if the field is not uniform.

Key Ideas

● A straight wire carrying a current i in a uniform magnetic field experiences a sideways force

$$\vec{F}_B = i\vec{L} \times \vec{B}.$$

● The force acting on a current element $i\,d\vec{L}$ in a magnetic

field is

$$d\vec{F}_B = i\,d\vec{L} \times \vec{B}.$$

● The direction of the length vector $\vec{L}$ or $d\vec{L}$ is that of the current i.

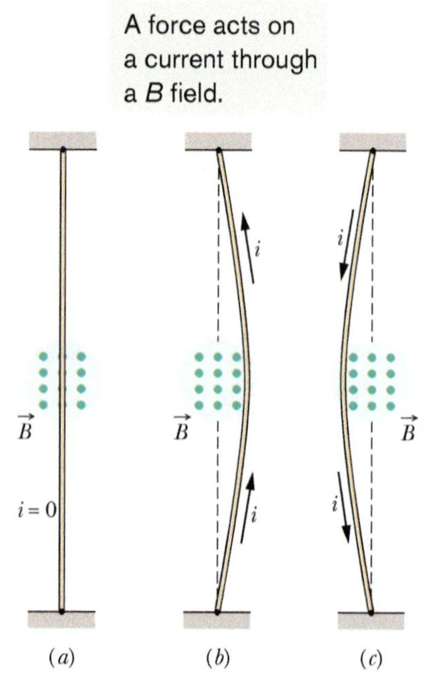

A force acts on a current through a B field.

(a)   (b)   (c)

Figure 28-14 A flexible wire passes between the pole faces of a magnet (only the farther pole face is shown). (*a*) Without current in the wire, the wire is straight. (*b*) With upward current, the wire is deflected rightward. (*c*) With downward current, the deflection is leftward. The connections for getting the current into the wire at one end and out of it at the other end are not shown.

Magnetic Force on a Current-Carrying Wire

We have already seen (in connection with the Hall effect) that a magnetic field exerts a sideways force on electrons moving in a wire. This force must then be transmitted to the wire itself, because the conduction electrons cannot escape sideways out of the wire.

In Fig. 28-14*a*, a vertical wire, carrying no current and fixed in place at both ends, extends through the gap between the vertical pole faces of a magnet. The magnetic field between the faces is directed outward from the page. In Fig. 28-14*b*, a current is sent upward through the wire; the wire deflects to the right. In Fig. 28-14*c*, we reverse the direction of the current and the wire deflects to the left.

Figure 28-15 shows what happens inside the wire of Fig. 28-14*b*. We see one of the conduction electrons, drifting downward with an assumed drift speed v_d. Equation 28-3, in which we must put $\phi = 90°$, tells us that a force $\vec{F}_B$ of magnitude ev_dB must act on each such electron. From Eq. 28-2 we see that this force must be directed to the right. We expect then that the wire as a whole will experience a force to the right, in agreement with Fig. 28-14*b*.

If, in Fig. 28-15, we were to reverse *either* the direction of the magnetic field *or* the direction of the current, the force on the wire would reverse, being directed now to the left. Note too that it does not matter whether we consider negative charges

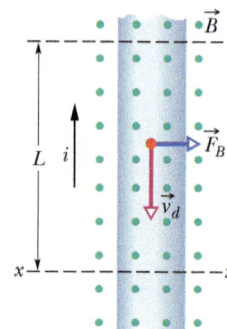

Figure 28-15 A close-up view of a section of the wire of Fig. 28-14*b*. The current direction is upward, which means that electrons drift downward. A magnetic field that emerges from the plane of the page causes the electrons and the wire to be deflected to the right.

drifting downward in the wire (the actual case) or positive charges drifting upward. The direction of the deflecting force on the wire is the same. We are safe then in dealing with a current of positive charge, as we usually do in dealing with circuits.

Find the Force. Consider a length L of the wire in Fig. 28-15. All the conduction electrons in this section of wire will drift past plane xx in Fig. 28-15 in a time $t = L/v_d$. Thus, in that time a charge given by

$$q = it = i\frac{L}{v_d}$$

will pass through that plane. Substituting this into Eq. 28-3 yields

$$F_B = qv_dB \sin \phi = \frac{iL}{v_d} v_dB \sin 90°$$

or
$$F_B = iLB. \tag{28-25}$$

Note that this equation gives the magnetic force that acts on a length L of straight wire carrying a current i and immersed in a uniform magnetic field $\vec{B}$ that is *perpendicular* to the wire.

If the magnetic field is *not* perpendicular to the wire, as in Fig. 28-16, the magnetic force is given by a generalization of Eq. 28-25:

$$\vec{F}_B = i\vec{L} \times \vec{B} \quad \text{(force on a current).} \tag{28-26}$$

Here $\vec{L}$ is a *length vector* that has magnitude L and is directed along the wire segment in the direction of the (conventional) current. The force magnitude F_B is

$$F_B = iLB \sin \phi, \tag{28-27}$$

where ϕ is the angle between the directions of $\vec{L}$ and $\vec{B}$. The direction of $\vec{F}_B$ is that of the cross product $\vec{L} \times \vec{B}$ because we take current i to be a positive quantity. Equation 28-26 tells us that $\vec{F}_B$ is always perpendicular to the plane defined by vectors $\vec{L}$ and $\vec{B}$, as indicated in Fig. 28-16.

Equation 28-26 is equivalent to Eq. 28-2 in that either can be taken as the defining equation for $\vec{B}$. In practice, we define $\vec{B}$ from Eq. 28-26 because it is much easier to measure the magnetic force acting on a wire than that on a single moving charge.

Crooked Wire. If a wire is not straight or the field is not uniform, we can imagine the wire broken up into small straight segments and apply Eq. 28-26 to each segment. The force on the wire as a whole is then the vector sum of all the forces on the segments that make it up. In the differential limit, we can write

$$d\vec{F}_B = i\,d\vec{L} \times \vec{B}, \tag{28-28}$$

and we can find the resultant force on any given arrangement of currents by integrating Eq. 28-28 over that arrangement.

In using Eq. 28-28, bear in mind that there is no such thing as an isolated current-carrying wire segment of length dL. There must always be a way to introduce the current into the segment at one end and take it out at the other end.

The force is perpendicular to both the field and the length.

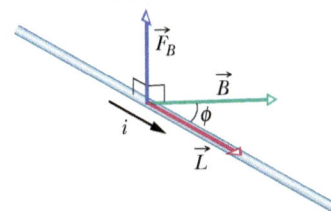

Figure 28-16 A wire carrying current i makes an angle ϕ with magnetic field $\vec{B}$. The wire has length L in the field and length vector $\vec{L}$ (in the direction of the current). A magnetic force $\vec{F}_B = i\vec{L} \times \vec{B}$ acts on the wire.

✓ **Checkpoint 4**

The figure shows a current i through a wire in a uniform magnetic field $\vec{B}$, as well as the magnetic force $\vec{F}_B$ acting on the wire. The field is oriented so that the force is maximum. In what direction is the field?

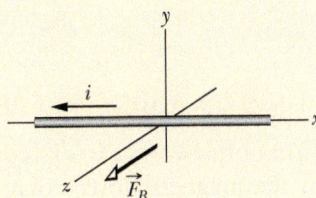

Sample Problem 28.06 Magnetic force on a wire carrying current

A straight, horizontal length of copper wire has a current $i = 28$ A through it. What are the magnitude and direction of the minimum magnetic field $\vec{B}$ needed to suspend the wire—that is, to balance the gravitational force on it? The linear density (mass per unit length) of the wire is 46.6 g/m.

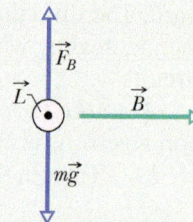

Figure 28-17 A wire (shown in cross section) carrying current out of the page.

KEY IDEAS

(1) Because the wire carries a current, a magnetic force $\vec{F}_B$ can act on the wire if we place it in a magnetic field $\vec{B}$. To balance the downward gravitational force $\vec{F}_g$ on the wire, we want $\vec{F}_B$ to be directed upward (Fig. 28-17). (2) The direction of $\vec{F}_B$ is related to the directions of $\vec{B}$ and the wire's length vector $\vec{L}$ by Eq. 28-26 ($\vec{F}_B = i\vec{L} \times \vec{B}$).

Calculations: Because $\vec{L}$ is directed horizontally (and the current is taken to be positive), Eq. 28-26 and the right-hand rule for cross products tell us that $\vec{B}$ must be horizontal and rightward (in Fig. 28-17) to give the required upward $\vec{F}_B$.

The magnitude of $\vec{F}_B$ is $F_B = iLB \sin \phi$ (Eq. 28-27). Because we want $\vec{F}_B$ to balance $\vec{F}_g$, we want

$$iLB \sin \phi = mg, \qquad (28\text{-}29)$$

where mg is the magnitude of $\vec{F}_g$ and m is the mass of the wire. We also want the minimal field magnitude B for $\vec{F}_B$ to balance $\vec{F}_g$. Thus, we need to maximize $\sin \phi$ in Eq. 28-29. To do so, we set $\phi = 90°$, thereby arranging for $\vec{B}$ to be perpendicular to the wire. We then have $\sin \phi = 1$, so Eq. 28-29 yields

$$B = \frac{mg}{iL \sin \phi} = \frac{(m/L)g}{i}. \qquad (28\text{-}30)$$

We write the result this way because we know m/L, the linear density of the wire. Substituting known data then gives us

$$B = \frac{(46.6 \times 10^{-3} \text{ kg/m})(9.8 \text{ m/s}^2)}{28 \text{ A}}$$

$$= 1.6 \times 10^{-2} \text{ T}. \qquad \text{(Answer)}$$

This is about 160 times the strength of Earth's magnetic field.

WILEY PLUS Additional examples, video, and practice available at *WileyPLUS*

28-7 TORQUE ON A CURRENT LOOP

Learning Objectives

After reading this module, you should be able to . . .

28.36 Sketch a rectangular loop of current in a magnetic field, indicating the magnetic forces on the four sides, the direction of the current, the normal vector $\vec{n}$, and the direction in which a torque from the forces tends to rotate the loop.

28.37 For a current-carrying coil in a magnetic field, apply the relationship between the torque magnitude τ, the number of turns N, the area of each turn A, the current i, the magnetic field magnitude B, and the angle θ between the normal vector $\vec{n}$ and the magnetic field vector $\vec{B}$.

Key Ideas

● Various magnetic forces act on the sections of a current-carrying coil lying in a uniform external magnetic field, but the net force is zero.

● The net torque acting on the coil has a magnitude given by

$$\tau = NiAB \sin \theta,$$

where N is the number of turns in the coil, A is the area of each turn, i is the current, B is the field magnitude, and θ is the angle between the magnetic field $\vec{B}$ and the normal vector to the coil $\vec{n}$.

Torque on a Current Loop

Much of the world's work is done by electric motors. The forces behind this work are the magnetic forces that we studied in the preceding section—that is, the forces that a magnetic field exerts on a wire that carries a current.

Figure 28-18 shows a simple motor, consisting of a single current-carrying loop immersed in a magnetic field $\vec{B}$. The two magnetic forces $\vec{F}$ and $-\vec{F}$ produce a torque on the loop, tending to rotate it about its central axis. Although many essential details have been omitted, the figure does suggest how the action of a magnetic field on a current loop produces rotary motion. Let us analyze that action.

Figure 28-19a shows a rectangular loop of sides a and b, carrying current i through uniform magnetic field $\vec{B}$. We place the loop in the field so that its long sides, labeled 1 and 3, are perpendicular to the field direction (which is into the page), but its short sides, labeled 2 and 4, are not. Wires to lead the current into and out of the loop are needed but, for simplicity, are not shown.

To define the orientation of the loop in the magnetic field, we use a normal vector $\vec{n}$ that is perpendicular to the plane of the loop. Figure 28-19b shows a right-hand rule for finding the direction of $\vec{n}$. Point or curl the fingers of your right hand in the direction of the current at any point on the loop. Your extended thumb then points in the direction of the normal vector $\vec{n}$.

In Fig. 28-19c, the normal vector of the loop is shown at an arbitrary angle θ to the direction of the magnetic field $\vec{B}$. We wish to find the net force and net torque acting on the loop in this orientation.

Net Torque. The net force on the loop is the vector sum of the forces acting on its four sides. For side 2 the vector $\vec{L}$ in Eq. 28-26 points in the direction of the current and has magnitude b. The angle between $\vec{L}$ and $\vec{B}$ for side 2 (see Fig. 28-19c) is $90° - \theta$. Thus, the magnitude of the force acting on this side is

$$F_2 = ibB \sin(90° - \theta) = ibB \cos \theta. \qquad (28\text{-}31)$$

You can show that the force $\vec{F}_4$ acting on side 4 has the same magnitude as $\vec{F}_2$ but the opposite direction. Thus, $\vec{F}_2$ and $\vec{F}_4$ cancel out exactly. Their net force is zero and, because their common line of action is through the center of the loop, their net torque is also zero.

The situation is different for sides 1 and 3. For them, $\vec{L}$ is perpendicular to $\vec{B}$, so the forces $\vec{F}_1$ and $\vec{F}_3$ have the common magnitude iaB. Because these two forces have opposite directions, they do not tend to move the loop up or down. However, as Fig. 28-19c shows, these two forces do *not* share the same line of action; so they *do* produce a net torque. The torque tends to rotate the loop so as to align its normal vector $\vec{n}$ with the direction of the magnetic field $\vec{B}$. That torque has moment arm $(b/2) \sin \theta$ about the central axis of the loop. The magnitude τ' of the torque due to forces $\vec{F}_1$ and $\vec{F}_3$ is then (see Fig. 28-19c)

$$\tau' = \left(iaB \frac{b}{2} \sin \theta\right) + \left(iaB \frac{b}{2} \sin \theta\right) = iabB \sin \theta. \qquad (28\text{-}32)$$

Coil. Suppose we replace the single loop of current with a *coil* of N loops, or *turns*. Further, suppose that the turns are wound tightly enough that they can be

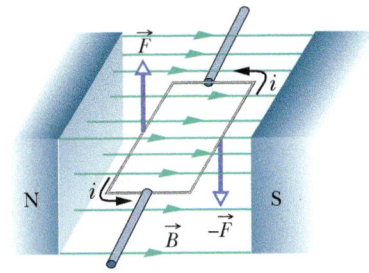

Figure 28-18 The elements of an electric motor. A rectangular loop of wire, carrying a current and free to rotate about a fixed axis, is placed in a magnetic field. Magnetic forces on the wire produce a torque that rotates it. A commutator (not shown) reverses the direction of the current every half-revolution so that the torque always acts in the same direction.

Figure 28-19 A rectangular loop, of length a and width b and carrying a current i, is located in a uniform magnetic field. A torque τ acts to align the normal vector $\vec{n}$ with the direction of the field. (a) The loop as seen by looking in the direction of the magnetic field. (b) A perspective of the loop showing how the right-hand rule gives the direction of $\vec{n}$, which is perpendicular to the plane of the loop. (c) A side view of the loop, from side 2. The loop rotates as indicated.

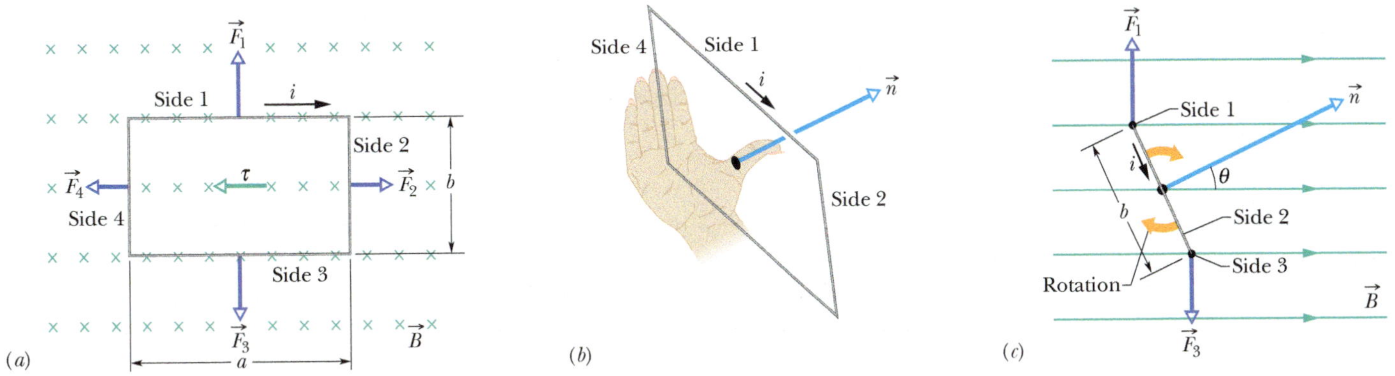

approximated as all having the same dimensions and lying in a plane. Then the turns form a *flat coil,* and a torque τ' with the magnitude given in Eq. 28-32 acts on each of them. The total torque on the coil then has magnitude

$$\tau = N\tau' = NiabB \sin \theta = (NiA)B \sin \theta, \qquad (28\text{-}33)$$

in which $A\ (= ab)$ is the area enclosed by the coil. The quantities in parentheses (NiA) are grouped together because they are all properties of the coil: its number of turns, its area, and the current it carries. Equation 28-33 holds for all flat coils, no matter what their shape, provided the magnetic field is uniform. For example, for the common circular coil, with radius r, we have

$$\tau = (Ni\pi r^2)B \sin \theta. \qquad (28\text{-}34)$$

Normal Vector. Instead of focusing on the motion of the coil, it is simpler to keep track of the vector $\vec{n}$, which is normal to the plane of the coil. Equation 28-33 tells us that a current-carrying flat coil placed in a magnetic field will tend to rotate so that $\vec{n}$ has the same direction as the field. In a motor, the current in the coil is reversed as $\vec{n}$ begins to line up with the field direction, so that a torque continues to rotate the coil. This automatic reversal of the current is done via a commutator that electrically connects the rotating coil with the stationary contacts on the wires that supply the current from some source.

28-8 THE MAGNETIC DIPOLE MOMENT

Learning Objectives

After reading this module, you should be able to . . .

28.38 Identify that a current-carrying coil is a magnetic dipole with a magnetic dipole moment $\vec{\mu}$ that has the direction of the normal vector $\vec{n}$, as given by a right-hand rule.

28.39 For a current-carrying coil, apply the relationship between the magnitude μ of the magnetic dipole moment, the number of turns N, the area A of each turn, and the current i.

28.40 On a sketch of a current-carrying coil, draw the direction of the current, and then use a right-hand rule to determine the direction of the magnetic dipole moment vector $\vec{\mu}$.

28.41 For a magnetic dipole in an external magnetic field, apply the relationship between the torque magnitude τ, the dipole moment magnitude μ, the magnetic field magnitude B, and the angle θ between the dipole moment vector $\vec{\mu}$ and the magnetic field vector $\vec{B}$.

28.42 Identify the convention of assigning a plus or minus sign to a torque according to the direction of rotation.

28.43 Calculate the torque on a magnetic dipole by evaluating a cross product of the dipole moment vector $\vec{\mu}$ and the

external magnetic field vector $\vec{B}$, in magnitude-angle notation and unit-vector notation.

28.44 For a magnetic dipole in an external magnetic field, identify the dipole orientations at which the torque magnitude is minimum and maximum.

28.45 For a magnetic dipole in an external magnetic field, apply the relationship between the orientation energy U, the dipole moment magnitude μ, the external magnetic field magnitude B, and the angle θ between the dipole moment vector $\vec{\mu}$ and the magnetic field vector $\vec{B}$.

28.46 Calculate the orientation energy U by taking a dot product of the dipole moment vector $\vec{\mu}$ and the external magnetic field vector $\vec{B}$, in magnitude-angle and unit-vector notations.

28.47 Identify the orientations of a magnetic dipole in an external magnetic field that give the minimum and maximum orientation energies.

28.48 For a magnetic dipole in a magnetic field, relate the orientation energy U to the work W_a done by an external torque as the dipole rotates in the magnetic field.

Key Ideas

● A coil (of area A and N turns, carrying current i) in a uniform magnetic field $\vec{B}$ will experience a torque $\vec{\tau}$ given by

$$\vec{\tau} = \vec{\mu} \times \vec{B}.$$

Here $\vec{\mu}$ is the magnetic dipole moment of the coil, with magnitude $\mu = NiA$ and direction given by the right-hand rule.

● The orientation energy of a magnetic dipole in a magnetic

field is

$$U(\theta) = -\vec{\mu} \cdot \vec{B}.$$

● If an external agent rotates a magnetic dipole from an initial orientation θ_i to some other orientation θ_f and the dipole is stationary both initially and finally, the work W_a done on the dipole by the agent is

$$W_a = \Delta U = U_f - U_i.$$

The Magnetic Dipole Moment

As we have just discussed, a torque acts to rotate a current-carrying coil placed in a magnetic field. In that sense, the coil behaves like a bar magnet placed in the magnetic field. Thus, like a bar magnet, a current-carrying coil is said to be a *magnetic dipole*. Moreover, to account for the torque on the coil due to the magnetic field, we assign a **magnetic dipole moment** $\vec{\mu}$ to the coil. The direction of $\vec{\mu}$ is that of the normal vector $\vec{n}$ to the plane of the coil and thus is given by the same right-hand rule shown in Fig. 28-19. That is, grasp the coil with the fingers of your right hand in the direction of current i; the outstretched thumb of that hand gives the direction of $\vec{\mu}$. The magnitude of $\vec{\mu}$ is given by

$$\mu = NiA \quad \text{(magnetic moment)}, \quad (28\text{-}35)$$

in which N is the number of turns in the coil, i is the current through the coil, and A is the area enclosed by each turn of the coil. From this equation, with i in amperes and A in square meters, we see that the unit of $\vec{\mu}$ is the ampere–square meter $(\text{A} \cdot \text{m}^2)$.

Torque. Using $\vec{\mu}$, we can rewrite Eq. 28-33 for the torque on the coil due to a magnetic field as

$$\tau = \mu B \sin \theta, \quad (28\text{-}36)$$

in which θ is the angle between the vectors $\vec{\mu}$ and $\vec{B}$.

We can generalize this to the vector relation

$$\vec{\tau} = \vec{\mu} \times \vec{B}, \quad (28\text{-}37)$$

which reminds us very much of the corresponding equation for the torque exerted by an *electric* field on an *electric* dipole—namely, Eq. 22-34:

$$\vec{\tau} = \vec{p} \times \vec{E}.$$

In each case the torque due to the field—either magnetic or electric—is equal to the vector product of the corresponding dipole moment and the field vector.

Energy. A magnetic dipole in an external magnetic field has an energy that depends on the dipole's orientation in the field. For electric dipoles we have shown (Eq. 22-38) that

$$U(\theta) = -\vec{p} \cdot \vec{E}.$$

In strict analogy, we can write for the magnetic case

$$U(\theta) = -\vec{\mu} \cdot \vec{B}. \quad (28\text{-}38)$$

In each case the energy due to the field is equal to the negative of the scalar product of the corresponding dipole moment and the field vector.

A magnetic dipole has its lowest energy $(= -\mu B \cos 0 = -\mu B)$ when its dipole moment $\vec{\mu}$ is lined up with the magnetic field (Fig. 28-20). It has its highest energy $(= -\mu B \cos 180° = +\mu B)$ when $\vec{\mu}$ is directed opposite the field. From Eq. 28-38, with U in joules and $\vec{B}$ in teslas, we see that the unit of $\vec{\mu}$ can be the joule per tesla (J/T) instead of the ampere–square meter as suggested by Eq. 28-35.

Work. If an applied torque (due to "an external agent") rotates a magnetic dipole from an initial orientation θ_i to another orientation θ_f, then work W_a is done on the dipole by the applied torque. *If the dipole is stationary* before and after the change in its orientation, then work W_a is

$$W_a = U_f - U_i, \quad (28\text{-}39)$$

where U_f and U_i are calculated with Eq. 28-38.

The magnetic moment vector attempts to align with the magnetic field.

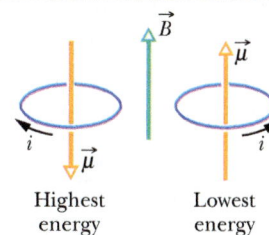

Figure 28-20 The orientations of highest and lowest energy of a magnetic dipole (here a coil carrying current) in an external magnetic field $\vec{B}$. The direction of the current i gives the direction of the magnetic dipole moment $\vec{\mu}$ via the right-hand rule shown for $\vec{n}$ in Fig. 28-19b.

Table 28-2 Some Magnetic Dipole Moments

Small bar magnet	5 J/T
Earth	8.0×10^{22} J/T
Proton	1.4×10^{-26} J/T
Electron	9.3×10^{-24} J/T

So far, we have identified only a current-carrying coil and a permanent magnet as a magnetic dipole. However, a rotating sphere of charge is also a magnetic dipole, as is Earth itself (approximately). Finally, most subatomic particles, including the electron, the proton, and the neutron, have magnetic dipole moments. As you will see in Chapter 32, all these quantities can be viewed as current loops. For comparison, some approximate magnetic dipole moments are shown in Table 28-2.

Language. Some instructors refer to U in Eq. 28-38 as a potential energy and relate it to work done by the magnetic field when the orientation of the dipole changes. Here we shall avoid the debate and say that U is an energy associated with the dipole orientation.

✓ **Checkpoint 5**

The figure shows four orientations, at angle θ, of a magnetic dipole moment $\vec{\mu}$ in a magnetic field. Rank the orientations according to (a) the magnitude of the torque on the dipole and (b) the orientation energy of the dipole, greatest first.

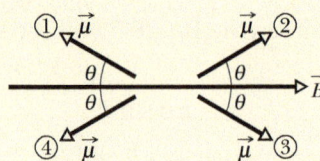

Sample Problem 28.07 **Rotating a magnetic dipole in a magnetic field**

Figure 28-21 shows a circular coil with 250 turns, an area A of 2.52×10^{-4} m², and a current of 100 μA. The coil is at rest in a uniform magnetic field of magnitude $B = 0.85$ T, with its magnetic dipole moment $\vec{\mu}$ initially aligned with $\vec{B}$.

(a) In Fig. 28-21, what is the direction of the current in the coil?

Right-hand rule: Imagine cupping the coil with your right hand so that your right thumb is outstretched in the direction of $\vec{\mu}$. The direction in which your fingers curl around the coil is the direction of the current in the coil. Thus, in the wires on the near side of the coil—those we see in Fig. 28-21—the current is from top to bottom.

(b) How much work would the torque applied by an external agent have to do on the coil to rotate it 90° from its ini-

tial orientation, so that $\vec{\mu}$ is perpendicular to $\vec{B}$ and the coil is again at rest?

KEY IDEA

The work W_a done by the applied torque would be equal to the change in the coil's orientation energy due to its change in orientation.

Calculations: From Eq. 28-39 ($W_a = U_f - U_i$), we find

$$W_a = U(90°) - U(0°)$$
$$= -\mu B \cos 90° - (-\mu B \cos 0°) = 0 + \mu B$$
$$= \mu B.$$

Substituting for μ from Eq. 28-35 ($\mu = NiA$), we find that

$$W_a = (NiA)B$$
$$= (250)(100 \times 10^{-6}\text{ A})(2.52 \times 10^{-4}\text{ m}^2)(0.85\text{ T})$$
$$= 5.355 \times 10^{-6}\text{ J} \approx 5.4\ \mu\text{J}. \qquad \text{(Answer)}$$

Similarly, we can show that to change the orientation by another 90°, so that the dipole moment is opposite the field, another 5.4 μJ is required.

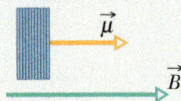

Figure 28-21 A side view of a circular coil carrying a current and oriented so that its magnetic dipole moment is aligned with magnetic field $\vec{B}$.

PLUS Additional examples, video, and practice available at *WileyPLUS*

Review & Summary

Magnetic Field $\vec{B}$ A **magnetic field** $\vec{B}$ is defined in terms of the force $\vec{F}_B$ acting on a test particle with charge q moving through the field with velocity $\vec{v}$:

$$\vec{F}_B = q\vec{v} \times \vec{B}. \tag{28-2}$$

The SI unit for $\vec{B}$ is the **tesla** (T): $1\text{ T} = 1\text{ N/(A·m)} = 10^4$ gauss.

The Hall Effect   When a conducting strip carrying a current i is placed in a uniform magnetic field $\vec{B}$, some charge carriers (with charge e) build up on one side of the conductor, creating a potential difference V across the strip. The polarities of the sides indicate the sign of the charge carriers.

A Charged Particle Circulating in a Magnetic Field   A charged particle with mass m and charge magnitude $|q|$ moving with velocity $\vec{v}$ perpendicular to a uniform magnetic field $\vec{B}$ will travel in a circle. Applying Newton's second law to the circular motion yields

$$|q|vB = \frac{mv^2}{r}, \tag{28-15}$$

from which we find the radius r of the circle to be

$$r = \frac{mv}{|q|B}. \tag{28-16}$$

The frequency of revolution f, the angular frequency ω, and the period of the motion T are given by

$$f = \frac{\omega}{2\pi} = \frac{1}{T} = \frac{|q|B}{2\pi m}. \tag{28-19, 28-18, 28-17}$$

Magnetic Force on a Current-Carrying Wire   A straight wire carrying a current i in a uniform magnetic field experiences a sideways force

$$\vec{F}_B = i\vec{L} \times \vec{B}. \tag{28-26}$$

The force acting on a current element $i\,d\vec{L}$ in a magnetic field is

$$d\vec{F}_B = i\,d\vec{L} \times \vec{B}. \tag{28-28}$$

The direction of the length vector $\vec{L}$ or $d\vec{L}$ is that of the current i.

Torque on a Current-Carrying Coil   A coil (of area A and N turns, carrying current i) in a uniform magnetic field $\vec{B}$ will experience a torque $\vec{\tau}$ given by

$$\vec{\tau} = \vec{\mu} \times \vec{B}. \tag{28-37}$$

Here $\vec{\mu}$ is the **magnetic dipole moment** of the coil, with magnitude $\mu = NiA$ and direction given by the right-hand rule.

Orientation Energy of a Magnetic Dipole   The orientation energy of a magnetic dipole in a magnetic field is

$$U(\theta) = -\vec{\mu} \cdot \vec{B}. \tag{28-38}$$

If an external agent rotates a magnetic dipole from an initial orientation θ_i to some other orientation θ_f and the dipole is stationary both initially and finally, the work W_a done on the dipole by the agent is

$$W_a = \Delta U = U_f - U_i. \tag{28-39}$$

Questions

1   Figure 28-22 shows three situations in which a positively charged particle moves at velocity $\vec{v}$ through a uniform magnetic field $\vec{B}$ and experiences a magnetic force $\vec{F}_B$. In each situation, determine whether the orientations of the vectors are physically reasonable.

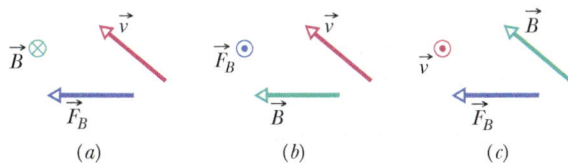

Figure 28-22 Question 1.

2   Figure 28-23 shows a wire that carries current to the right through a uniform magnetic field. It also shows four choices for the direction of that field. (a) Rank the choices according to the magnitude of the electric potential difference that would be set up across the width of the wire, greatest first. (b) For which choice is the top side of the wire at higher potential than the bottom side of the wire?

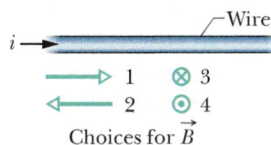

Figure 28-23 Question 2.

3   Figure 28-24 shows a metallic, rectangular solid that is to move at a certain speed v through the uniform magnetic field $\vec{B}$. The dimensions of the solid are multiples of d, as shown. You have six choices for the direction of the velocity: parallel to x, y, or z in ei-

ther the positive or negative direction. (a) Rank the six choices according to the potential difference set up across the solid, greatest first. (b) For which choice is the front face at lower potential?

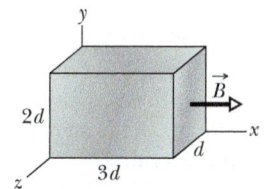

Figure 28-24 Question 3.

4   Figure 28-25 shows the path of a particle through six regions of uniform magnetic field, where the path is either a half-circle or a quarter-circle. Upon leaving the last region, the particle travels between two charged, parallel plates and is deflected toward the plate of higher potential. What is the direction of the magnetic field in each of the six regions?

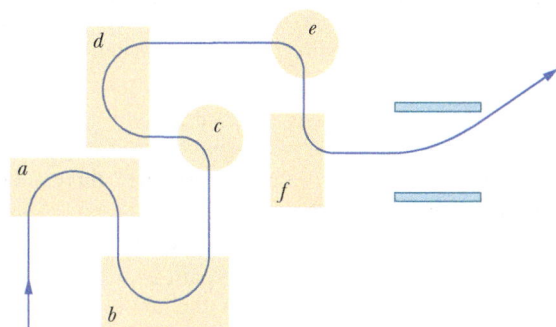

Figure 28-25 Question 4.

5 In Module 28-2, we discussed a charged particle moving through crossed fields with the forces $\vec{F}_E$ and $\vec{F}_B$ in opposition. We found that the particle moves in a straight line (that is, neither force dominates the motion) if its speed is given by Eq. 28-7 ($v = E/B$). Which of the two forces dominates if the speed of the particle is (a) $v < E/B$ and (b) $v > E/B$?

6 Figure 28-26 shows crossed uniform electric and magnetic fields $\vec{E}$ and $\vec{B}$ and, at a certain instant, the velocity vectors of the 10 charged particles listed in Table 28-3. (The vectors are not drawn to scale.) The speeds given in the table are either less than or greater than E/B (see Question 5). Which particles will move out of the page toward you after the instant shown in Fig. 28-26?

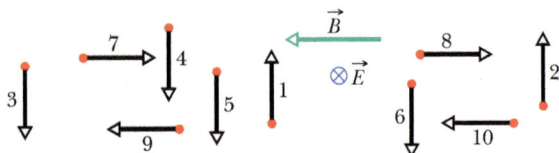

Figure 28-26 Question 6.

Table 28-3 Question 6

Particle	Charge	Speed	Particle	Charge	Speed
1	+	Less	6	−	Greater
2	+	Greater	7	+	Less
3	+	Less	8	+	Greater
4	+	Greater	9	−	Less
5	−	Less	10	−	Greater

7 Figure 28-27 shows the path of an electron that passes through two regions containing uniform magnetic fields of magnitudes B_1 and B_2. Its path in each region is a half-circle. (a) Which field is stronger? (b) What is the direction of each field? (c) Is the time spent by the electron in the $\vec{B}_1$ region greater than, less than, or the same as the time spent in the $\vec{B}_2$ region?

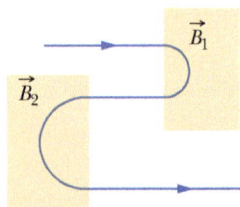

Figure 28-27 Question 7.

8 Figure 28-28 shows the path of an electron in a region of uniform magnetic field. The path consists of two straight sections, each between a pair of uniformly charged plates, and two half-circles. Which plate is at the higher electric potential in (a) the top pair of plates and (b) the bottom pair? (c) What is the direction of the magnetic field?

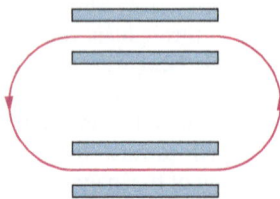

Figure 28-28 Question 8.

9 (a) In Checkpoint 5, if the dipole moment $\vec{\mu}$ is rotated from orientation 2 to orientation 1 by an external agent, is the work done on the dipole by the agent positive, negative, or zero? (b) Rank the work done on the dipole by the agent for these three rotations, greatest first: $2 \rightarrow 1, 2 \rightarrow 4, 2 \rightarrow 3$.

10 *Particle roundabout.* Figure 28-29 shows 11 paths through a region of uniform magnetic field. One path is a straight line; the rest are half-circles. Table 28-4 gives the masses, charges, and speeds of 11 particles that take these paths through the field in the directions shown. Which path in the figure corresponds to which particle in the table? (The direction of the magnetic field can be determined by means of one of the paths, which is unique.)

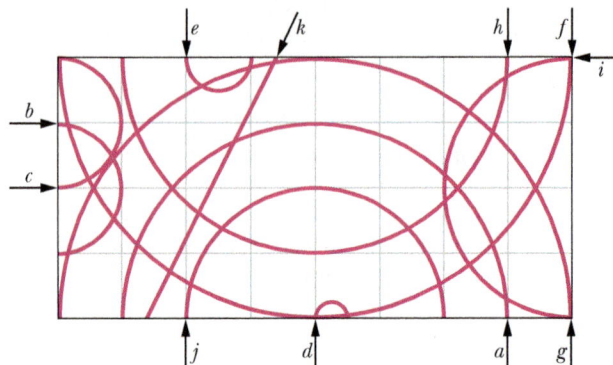

Figure 28-29 Question 10.

Table 28-4 Question 10

Particle	Mass	Charge	Speed
1	$2m$	q	v
2	m	$2q$	v
3	$m/2$	q	$2v$
4	$3m$	$3q$	$3v$
5	$2m$	q	$2v$
6	m	$-q$	$2v$
7	m	$-4q$	v
8	m	$-q$	v
9	$2m$	$-2q$	$3v$
10	m	$-2q$	$8v$
11	$3m$	0	$3v$

11 In Fig. 28-30, a charged particle enters a uniform magnetic field $\vec{B}$ with speed v_0, moves through a half-circle in time T_0, and then leaves the field. (a) Is the charge positive or negative? (b) Is the final speed of the particle greater than, less than, or equal to v_0? (c) If the initial speed had been $0.5v_0$, would the time spent in field $\vec{B}$ have been greater than, less than, or equal to T_0? (d) Would the path have been a half-circle, more than a half-circle, or less than a half-circle?

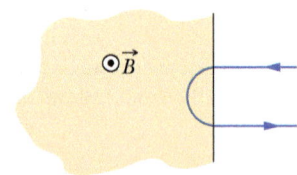

Figure 28-30 Question 11.

12 Figure 28-31 gives snapshots for three situations in which a positively charged particle passes through a uniform magnetic field $\vec{B}$. The velocities $\vec{v}$ of the particle differ in orientation in the three snapshots but not in magnitude. Rank the situations according to (a) the period, (b) the frequency, and (c) the pitch of the particle's motion, greatest first.

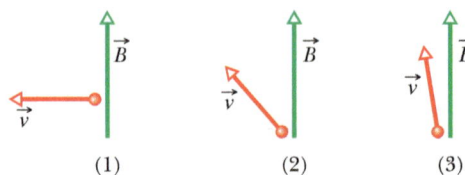

Figure 28-31 Question 12.

Problems

Module 28-1   Magnetic Fields and the Definition of $\vec{B}$

•1   SSM   ILW A proton traveling at 23.0° with respect to the direction of a magnetic field of strength 2.60 mT experiences a magnetic force of 6.50×10^{-17} N. Calculate (a) the proton's speed and (b) its kinetic energy in electron-volts.

•2 A particle of mass 10 g and charge 80 μC moves through a uniform magnetic field, in a region where the free-fall acceleration is $-9.8\hat{j}$ m/s². The velocity of the particle is a constant $20\hat{i}$ km/s, which is perpendicular to the magnetic field. What, then, is the magnetic field?

•3 An electron that has an instantaneous velocity of

$$\vec{v} = (2.0 \times 10^6 \text{ m/s})\hat{i} + (3.0 \times 10^6 \text{ m/s})\hat{j}$$

is moving through the uniform magnetic field $\vec{B} = (0.030 \text{ T})\hat{i} - (0.15 \text{ T})\hat{j}$. (a) Find the force on the electron due to the magnetic field. (b) Repeat your calculation for a proton having the same velocity.

•4 An alpha particle travels at a velocity $\vec{v}$ of magnitude 550 m/s through a uniform magnetic field $\vec{B}$ of magnitude 0.045 T. (An alpha particle has a charge of $+3.2 \times 10^{-19}$ C and a mass of 6.6×10^{-27} kg.) The angle between $\vec{v}$ and $\vec{B}$ is 52°. What is the magnitude of (a) the force $\vec{F}_B$ acting on the particle due to the field and (b) the acceleration of the particle due to $\vec{F}_B$? (c) Does the speed of the particle increase, decrease, or remain the same?

••5   GO An electron moves through a uniform magnetic field given by $\vec{B} = B_x\hat{i} + (3.0B_x)\hat{j}$. At a particular instant, the electron has velocity $\vec{v} = (2.0\hat{i} + 4.0\hat{j})$ m/s and the magnetic force acting on it is $(6.4 \times 10^{-19} \text{ N})\hat{k}$. Find B_x.

••6   GO A proton moves through a uniform magnetic field given by $\vec{B} = (10\hat{i} - 20\hat{j} + 30\hat{k})$ mT. At time t_1, the proton has a velocity given by $\vec{v} = v_x\hat{i} + v_y\hat{j} + (2.0 \text{ km/s})\hat{k}$ and the magnetic force on the proton is $\vec{F}_B = (4.0 \times 10^{-17} \text{ N})\hat{i} + (2.0 \times 10^{-17} \text{ N})\hat{j}$. At that instant, what are (a) v_x and (b) v_y?

Module 28-2   Crossed Fields: Discovery of the Electron

•7 An electron has an initial velocity of $(12.0\hat{j} + 15.0\hat{k})$ km/s and a constant acceleration of $(2.00 \times 10^{12} \text{ m/s}^2)\hat{i}$ in a region in which uniform electric and magnetic fields are present. If $\vec{B} = (400 \ \mu\text{T})\hat{i}$, find the electric field $\vec{E}$.

•8 An electric field of 1.50 kV/m and a perpendicular magnetic field of 0.400 T act on a moving electron to produce no net force. What is the electron's speed?

•9   ILW In Fig. 28-32, an electron accelerated from rest through potential difference $V_1 = 1.00$ kV enters the gap between two parallel plates having separation $d = 20.0$ mm and potential difference

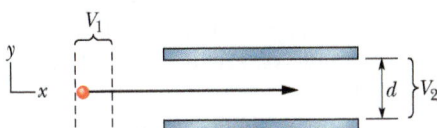

Figure 28-32 Problem 9.

$V_2 = 100$ V. The lower plate is at the lower potential. Neglect fringing and assume that the electron's velocity vector is perpendicular to the electric field vector between the plates. In unit-vector notation, what uniform magnetic field allows the electron to travel in a straight line in the gap?

••10 A proton travels through uniform magnetic and electric fields. The magnetic field is $\vec{B} = -2.50\hat{i}$ mT. At one instant the velocity of the proton is $\vec{v} = 2000\hat{j}$ m/s. At that instant and in unit-vector notation, what is the net force acting on the proton if the electric field is (a) $4.00\hat{k}$ V/m, (b) $-4.00\hat{k}$ V/m, and (c) $4.00\hat{i}$ V/m?

••11   GO An ion source is producing ^{6}Li ions, which have charge $+e$ and mass 9.99×10^{-27} kg. The ions are accelerated by a potential difference of 10 kV and pass horizontally into a region in which there is a uniform vertical magnetic field of magnitude $B = 1.2$ T. Calculate the strength of the electric field, to be set up over the same region, that will allow the ^{6}Li ions to pass through without any deflection.

•••12   GO At time t_1, an electron is sent along the positive direction of an x axis, through both an electric field $\vec{E}$ and a magnetic field $\vec{B}$, with $\vec{E}$ directed parallel to the y axis. Figure 28-33 gives the y component $F_{\text{net},y}$ of the net force on the electron due to the two fields, as a function of the electron's speed v at time t_1. The scale of the velocity axis is set by $v_s = 100.0$ m/s. The x and z components of the net force are zero at t_1. Assuming $B_x = 0$, find (a) the magnitude E and (b) $\vec{B}$ in unit-vector notation.

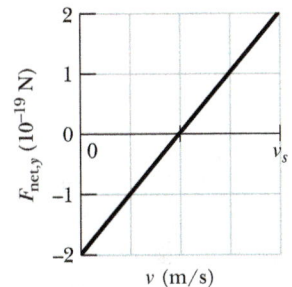

Figure 28-33 Problem 12.

Module 28-3   Crossed Fields: The Hall Effect

•13 A strip of copper 150 μm thick and 4.5 mm wide is placed in a uniform magnetic field $\vec{B}$ of magnitude 0.65 T, with $\vec{B}$ perpendicular to the strip. A current $i = 23$ A is then sent through the strip such that a Hall potential difference V appears across the width of the strip. Calculate V. (The number of charge carriers per unit volume for copper is 8.47×10^{28} electrons/m³.)

•14 A metal strip 6.50 cm long, 0.850 cm wide, and 0.760 mm thick moves with constant velocity $\vec{v}$ through a uniform magnetic field $B = 1.20$ mT directed perpendicular to the strip, as shown in Fig. 28-34. A potential difference of 3.90 μV is measured between points x and y across the strip. Calculate the speed v.

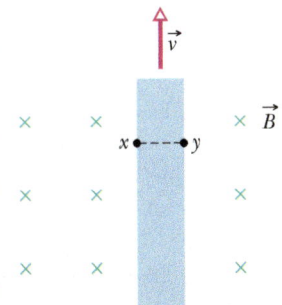

Figure 28-34 Problem 14.

••15   GO A conducting rectangular solid of dimensions $d_x = 5.00$ m, $d_y = 3.00$ m, and $d_z = 2.00$ m moves with a constant velocity $\vec{v} = (20.0 \text{ m/s})\hat{i}$ through a uniform magnetic field

$\vec{B} = (30.0 \text{ mT})\hat{j}$ (Fig. 28-35). What are the resulting (a) electric field within the solid, in unit-vector notation, and (b) potential difference across the solid?

•••**16** GO Figure 28-35 shows a metallic block, with its faces parallel to coordinate axes. The block is in a uniform magnetic field of magnitude 0.020 T. One edge length of the block is 25 cm; the block is *not* drawn to scale. The block is moved at 3.0 m/s parallel to each axis, in turn, and the resulting potential difference V that appears across the block is measured. With the motion parallel to the y axis, V = 12 mV; with the motion parallel to the z axis, V = 18 mV; with the motion parallel to the x axis, V = 0. What are the block lengths (a) d_x, (b) d_y, and (c) d_z?

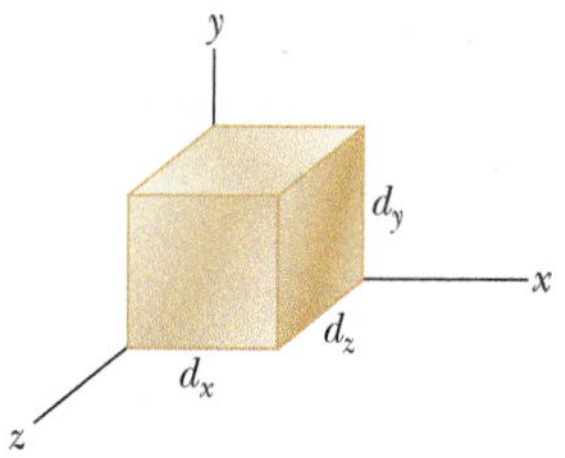

Figure 28-35 Problems 15 and 16.

Module 28-4 A Circulating Charged Particle

•**17** An alpha particle can be produced in certain radioactive decays of nuclei and consists of two protons and two neutrons. The particle has a charge of q = +2e and a mass of 4.00 u, where u is the atomic mass unit, with 1 u = 1.661×10^{-27} kg. Suppose an alpha particle travels in a circular path of radius 4.50 cm in a uniform magnetic field with B = 1.20 T. Calculate (a) its speed, (b) its period of revolution, (c) its kinetic energy, and (d) the potential difference through which it would have to be accelerated to achieve this energy.

•**18** GO In Fig. 28-36, a particle moves along a circle in a region of uniform magnetic field of magnitude B = 4.00 mT. The particle is either a proton or an electron (you must decide which). It experiences a magnetic force of magnitude 3.20×10^{-15} N. What are (a) the particle's speed, (b) the radius of the circle, and (c) the period of the motion?

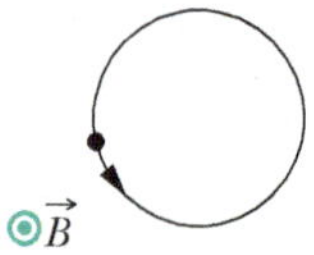

Figure 28-36 Problem 18.

•**19** A certain particle is sent into a uniform magnetic field, with the particle's velocity vector perpendicular to the direction of the field. Figure 28-37 gives the period T of the particle's motion versus the *inverse* of the field magnitude B. The vertical axis scale is set by T_s = 40.0 ns, and the horizontal axis scale is set by B_s^{-1} = 5.0 T^{-1}. What is the ratio m/q of the particle's mass to the magnitude of its charge?

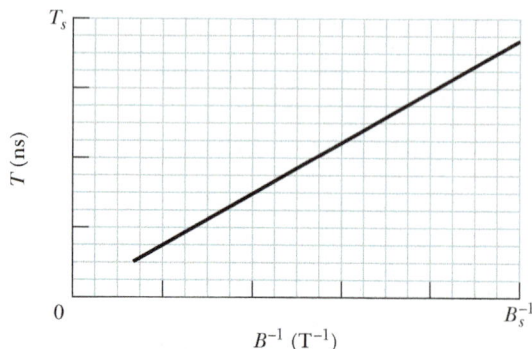

Figure 28-37 Problem 19.

•**20** An electron is accelerated from rest through potential difference V and then enters a region of uniform magnetic field, where it undergoes uniform circular motion. Figure 28-38 gives the radius r of that motion versus $V^{1/2}$. The vertical axis scale is set by r_s = 3.0 mm, and the horizontal axis scale is set by $V_s^{1/2}$ = 40.0 V$^{1/2}$. What is the magnitude of the magnetic field?

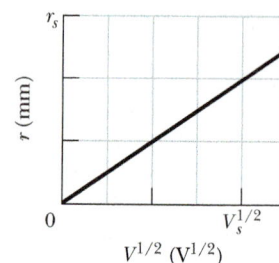

Figure 28-38 Problem 20.

•**21** SSM An electron of kinetic energy 1.20 keV circles in a plane perpendicular to a uniform magnetic field. The orbit radius is 25.0 cm. Find (a) the electron's speed, (b) the magnetic field magnitude, (c) the circling frequency, and (d) the period of the motion.

•**22** In a nuclear experiment a proton with kinetic energy 1.0 MeV moves in a circular path in a uniform magnetic field. What energy must (a) an alpha particle (q = +2e, m = 4.0 u) and (b) a deuteron (q = +e, m = 2.0 u) have if they are to circulate in the same circular path?

•**23** What uniform magnetic field, applied perpendicular to a beam of electrons moving at 1.30×10^6 m/s, is required to make the electrons travel in a circular arc of radius 0.350 m?

•**24** An electron is accelerated from rest by a potential difference of 350 V. It then enters a uniform magnetic field of magnitude 200 mT with its velocity perpendicular to the field. Calculate (a) the speed of the electron and (b) the radius of its path in the magnetic field.

•**25** (a) Find the frequency of revolution of an electron with an energy of 100 eV in a uniform magnetic field of magnitude 35.0 μT. (b) Calculate the radius of the path of this electron if its velocity is perpendicular to the magnetic field.

••**26** In Fig. 28-39, a charged particle moves into a region of uniform magnetic field $\vec{B}$, goes through half a circle, and then exits that region. The particle is either a proton or an electron (you must decide which). It spends 130 ns in the region. (a) What is the magnitude of $\vec{B}$? (b) If the particle is sent back through the magnetic field (along the same initial path) but with 2.00 times its previous kinetic energy, how much time does it spend in the field during this trip?

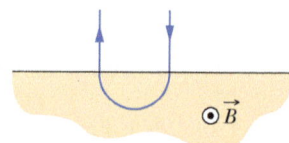

Figure 28-39 Problem 26.

••**27** A mass spectrometer (Fig. 28-12) is used to separate uranium ions of mass 3.92×10^{-25} kg and charge 3.20×10^{-19} C from related species. The ions are accelerated through a potential difference of 100 kV and then pass into a uniform magnetic field, where they are bent in a path of radius 1.00 m. After traveling through 180° and passing through a slit of width 1.00 mm and height 1.00 cm, they are collected in a cup. (a) What is the magnitude of the (perpendicular) magnetic field in the separator? If the machine is used to separate out 100 mg of material per hour, calculate (b) the current of the desired ions in the machine and (c) the thermal energy produced in the cup in 1.00 h.

••**28** A particle undergoes uniform circular motion of radius 26.1 μm in a uniform magnetic field. The magnetic force on the particle has a magnitude of 1.60×10^{-17} N. What is the kinetic energy of the particle?

••**29** An electron follows a helical path in a uniform magnetic field of magnitude 0.300 T. The pitch of the path is 6.00 μm, and the magnitude of the magnetic force on the electron is 2.00×10^{-15} N. What is the electron's speed?

••30 GO In Fig. 28-40, an electron with an initial kinetic energy of 4.0 keV enters region 1 at time $t = 0$. That region contains a uniform magnetic field directed into the page, with magnitude 0.010 T. The electron goes through a half-circle and then exits region 1, headed toward region 2 across a gap of 25.0 cm. There is an electric potential difference $\Delta V = 2000$ V across the gap, with a polarity such that the electron's speed increases uniformly as it traverses the gap. Region 2 contains a uniform magnetic field directed out of the page, with magnitude 0.020 T. The electron goes through a half-circle and then leaves region 2. At what time t does it leave?

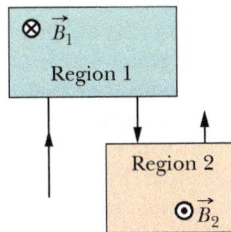

Figure 28-40
Problem 30.

••31 A particular type of fundamental particle decays by transforming into an electron e^- and a positron e^+. Suppose the decaying particle is at rest in a uniform magnetic field $\vec{B}$ of magnitude 3.53 mT and the e^- and e^+ move away from the decay point in paths lying in a plane perpendicular to $\vec{B}$. How long after the decay do the e^- and e^+ collide?

••32 A source injects an electron of speed $v = 1.5 \times 10^7$ m/s into a uniform magnetic field of magnitude $B = 1.0 \times 10^{-3}$ T. The velocity of the electron makes an angle $\theta = 10°$ with the direction of the magnetic field. Find the distance d from the point of injection at which the electron next crosses the field line that passes through the injection point.

••33 SSM WWW A positron with kinetic energy 2.00 keV is projected into a uniform magnetic field $\vec{B}$ of magnitude 0.100 T, with its velocity vector making an angle of 89.0° with $\vec{B}$. Find (a) the period, (b) the pitch p, and (c) the radius r of its helical path.

••34 An electron follows a helical path in a uniform magnetic field given by $\vec{B} = (20\hat{i} - 50\hat{j} - 30\hat{k})$ mT. At time $t = 0$, the electron's velocity is given by $\vec{v} = (20\hat{i} - 30\hat{j} + 50\hat{k})$ m/s. (a) What is the angle ϕ between $\vec{v}$ and $\vec{B}$? The electron's velocity changes with time. Do (b) its speed and (c) the angle ϕ change with time? (d) What is the radius of the helical path?

Module 28-5 Cyclotrons and Synchrotrons

••35 A proton circulates in a cyclotron, beginning approximately at rest at the center. Whenever it passes through the gap between dees, the electric potential difference between the dees is 200 V. (a) By how much does its kinetic energy increase with each passage through the gap? (b) What is its kinetic energy as it completes 100 passes through the gap? Let r_{100} be the radius of the proton's circular path as it completes those 100 passes and enters a dee, and let r_{101} be its next radius, as it enters a dee the next time. (c) By what percentage does the radius increase when it changes from r_{100} to r_{101}? That is, what is

$$\text{percentage increase} = \frac{r_{101} - r_{100}}{r_{100}} 100\%?$$

••36 A cyclotron with dee radius 53.0 cm is operated at an oscillator frequency of 12.0 MHz to accelerate protons. (a) What magnitude B of magnetic field is required to achieve resonance? (b) At that field magnitude, what is the kinetic energy of a proton emerging from the cyclotron? Suppose, instead, that $B = 1.57$ T. (c) What oscillator frequency is required to achieve resonance now? (d) At that frequency, what is the kinetic energy of an emerging proton?

••37 Estimate the total path length traveled by a deuteron in a cyclotron of radius 53 cm and operating frequency 12 MHz during the (entire) acceleration process. Assume that the accelerating potential between the dees is 80 kV.

••38 In a certain cyclotron a proton moves in a circle of radius 0.500 m. The magnitude of the magnetic field is 1.20 T. (a) What is the oscillator frequency? (b) What is the kinetic energy of the proton, in electron-volts?

Module 28-6 Magnetic Force on a Current-Carrying Wire

•39 SSM A horizontal power line carries a current of 5000 A from south to north. Earth's magnetic field (60.0 μT) is directed toward the north and inclined downward at 70.0° to the horizontal. Find the (a) magnitude and (b) direction of the magnetic force on 100 m of the line due to Earth's field.

•40 A wire 1.80 m long carries a current of 13.0 A and makes an angle of 35.0° with a uniform magnetic field of magnitude $B = 1.50$ T. Calculate the magnetic force on the wire.

•41 ILW A 13.0 g wire of length $L = 62.0$ cm is suspended by a pair of flexible leads in a uniform magnetic field of magnitude 0.440 T (Fig. 28-41). What are the (a) magnitude and (b) direction (left or right) of the current required to remove the tension in the supporting leads?

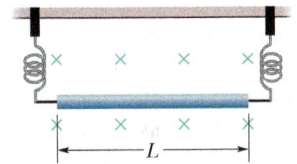

Figure 28-41 Problem 41.

•42 The bent wire shown in Fig. 28-42 lies in a uniform magnetic field. Each straight section is 2.0 m long and makes an angle of $\theta = 60°$ with the x axis, and the wire carries a current of 2.0 A. What is the net magnetic force on the wire in unit-vector notation if the magnetic field is given by (a) $4.0\hat{k}$ T and (b) $4.0\hat{i}$ T?

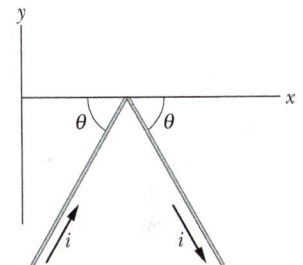

Figure 28-42 Problem 42.

•43 A single-turn current loop, carrying a current of 4.00 A, is in the shape of a right triangle with sides 50.0, 120, and 130 cm. The loop is in a uniform magnetic field of magnitude 75.0 mT whose direction is parallel to the current in the 130 cm side of the loop. What is the magnitude of the magnetic force on (a) the 130 cm side, (b) the 50.0 cm side, and (c) the 120 cm side? (d) What is the magnitude of the net force on the loop?

••44 Figure 28-43 shows a wire ring of radius $a = 1.8$ cm that is perpendicular to the general direction of a radially symmetric, diverging magnetic field. The magnetic field at the ring is everywhere of the same magnitude $B = 3.4$ mT, and its direction at the ring everywhere makes an angle $\theta = 20°$ with a normal to the plane of the ring. The twisted lead wires have no effect on the problem. Find the magnitude of the force the field exerts on the ring if the ring carries a current $i = 4.6$ mA.

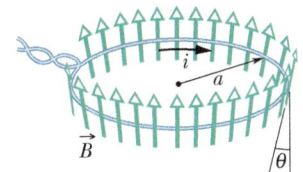

Figure 28-43 Problem 44.

••45 A wire 50.0 cm long carries a 0.500 A current in the positive direction of an x axis through a magnetic field $\vec{B} = (3.00$ mT$)\hat{j} + (10.0$ mT$)\hat{k}$. In unit-vector notation, what is the magnetic force on the wire?

••46 In Fig. 28-44, a metal wire of mass $m = 24.1$ mg can slide with negligible friction on two horizontal parallel rails separated by distance $d = 2.56$ cm. The track lies in a vertical uniform magnetic field of magnitude 56.3 mT. At time $t = 0$, device G is connected to the rails, producing a constant current $i = 9.13$ mA in the wire and rails (even as the wire moves). At $t = 61.1$ ms, what are the wire's (a) speed and (b) direction of motion (left or right)?

Figure 28-44 Problem 46.

•••47 GO A 1.0 kg copper rod rests on two horizontal rails 1.0 m apart and carries a current of 50 A from one rail to the other. The coefficient of static friction between rod and rails is 0.60. What are the (a) magnitude and (b) angle (relative to the vertical) of the smallest magnetic field that puts the rod on the verge of sliding?

•••48 GO A long, rigid conductor, lying along an x axis, carries a current of 5.0 A in the negative x direction. A magnetic field $\vec{B}$ is present, given by $\vec{B} = 3.0\hat{i} + 8.0x^2\hat{j}$, with x in meters and $\vec{B}$ in milliteslas. Find, in unit-vector notation, the force on the 2.0 m segment of the conductor that lies between $x = 1.0$ m and $x = 3.0$ m.

Module 28-7 Torque on a Current Loop

•49 SSM Figure 28-45 shows a rectangular 20-turn coil of wire, of dimensions 10 cm by 5.0 cm. It carries a current of 0.10 A and is hinged along one long side. It is mounted in the xy plane, at angle $\theta = 30°$ to the direction of a uniform magnetic field of magnitude 0.50 T. In unit-vector notation, what is the torque acting on the coil about the hinge line?

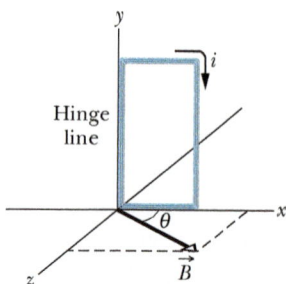

Figure 28-45 Problem 49.

••50 An electron moves in a circle of radius $r = 5.29 \times 10^{-11}$ m with speed 2.19×10^6 m/s. Treat the circular path as a current loop with a constant current equal to the ratio of the electron's charge magnitude to the period of the motion. If the circle lies in a uniform magnetic field of magnitude $B = 7.10$ mT, what is the maximum possible magnitude of the torque produced on the loop by the field?

••51 Figure 28-46 shows a wood cylinder of mass $m = 0.250$ kg and length $L = 0.100$ m, with $N = 10.0$ turns of wire wrapped around it longitudinally, so that the plane of the wire coil contains the long central axis of the cylinder. The cylinder is released on a plane inclined at an angle θ to the horizontal, with the plane of the coil parallel to the incline plane. If there is a vertical uniform magnetic field of magnitude 0.500 T, what is the least current i through the coil that keeps the cylinder from rolling down the plane?

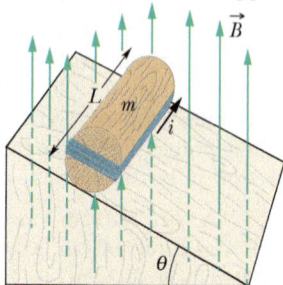

Figure 28-46 Problem 51.

••52 In Fig. 28-47, a rectangular loop carrying current lies in the plane of a uniform magnetic field of magnitude 0.040 T. The loop consists of a single turn of flexible conducting wire that is wrapped around a flexible mount such that the dimensions of the rectangle can be changed. (The total length of the wire is not changed.) As edge length x is varied from approximately zero to its maximum value of approximately 4.0 cm, the magnitude τ of the torque on the loop changes. The maximum value of τ is 4.80×10^{-8} N·m. What is the current in the loop?

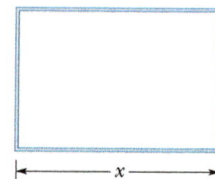

Figure 28-47 Problem 52.

••53 Prove that the relation $\tau = NiAB \sin\theta$ holds not only for the rectangular loop of Fig. 28-19 but also for a closed loop of any shape. (*Hint:* Replace the loop of arbitrary shape with an assembly of adjacent long, thin, approximately rectangular loops that are nearly equivalent to the loop of arbitrary shape as far as the distribution of current is concerned.)

Module 28-8 The Magnetic Dipole Moment

•54 A magnetic dipole with a dipole moment of magnitude 0.020 J/T is released from rest in a uniform magnetic field of magnitude 52 mT. The rotation of the dipole due to the magnetic force on it is unimpeded. When the dipole rotates through the orientation where its dipole moment is aligned with the magnetic field, its kinetic energy is 0.80 mJ. (a) What is the initial angle between the dipole moment and the magnetic field? (b) What is the angle when the dipole is next (momentarily) at rest?

•55 SSM Two concentric, circular wire loops, of radii $r_1 = 20.0$ cm and $r_2 = 30.0$ cm, are located in an xy plane; each carries a clockwise current of 7.00 A (Fig. 28-48). (a) Find the magnitude of the net magnetic dipole moment of the system. (b) Repeat for reversed current in the inner loop.

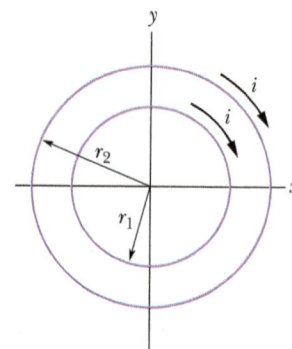

Figure 28-48 Problem 55.

•56 A circular wire loop of radius 15.0 cm carries a current of 2.60 A. It is placed so that the normal to its plane makes an angle of 41.0° with a uniform magnetic field of magnitude 12.0 T. (a) Calculate the magnitude of the magnetic dipole moment of the loop. (b) What is the magnitude of the torque acting on the loop?

•57 SSM A circular coil of 160 turns has a radius of 1.90 cm. (a) Calculate the current that results in a magnetic dipole moment of magnitude 2.30 A·m². (b) Find the maximum magnitude of the torque that the coil, carrying this current, can experience in a uniform 35.0 mT magnetic field.

•58 The magnetic dipole moment of Earth has magnitude 8.00×10^{22} J/T. Assume that this is produced by charges flowing in Earth's molten outer core. If the radius of their circular path is 3500 km, calculate the current they produce.

•59 A current loop, carrying a current of 5.0 A, is in the shape of a right triangle with sides 30, 40, and 50 cm. The loop is in a uniform magnetic field of magnitude 80 mT whose direction is parallel to the current in the 50 cm side of the loop. Find the magnitude of (a) the magnetic dipole moment of the loop and (b) the torque on the loop.

••60 Figure 28-49 shows a current loop *ABCDEFA* carrying a current $i = 5.00$ A. The sides of the loop are parallel to the coordinate axes shown, with $AB = 20.0$ cm, $BC = 30.0$ cm, and $FA = 10.0$ cm. In unit-vector notation, what is the magnetic dipole moment of this loop? (*Hint:* Imagine equal and opposite currents i in the line segment AD; then treat the two rectangular loops *ABCDA* and *ADEFA*.)

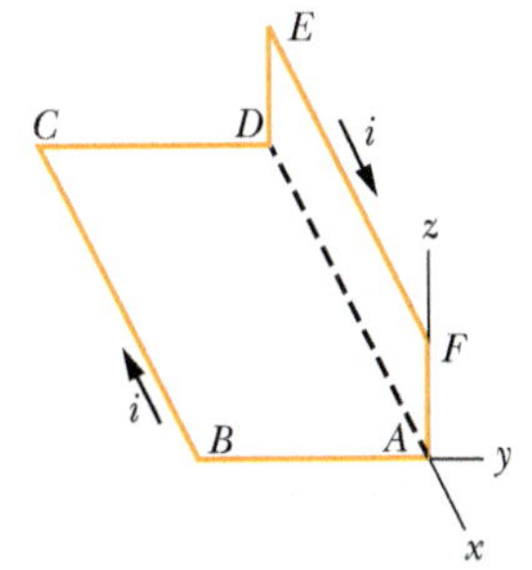

Figure 28-49 Problem 60.

••61 **SSM** The coil in Fig. 28-50 carries current $i = 2.00$ A in the direction indicated, is parallel to an xz plane, has 3.00 turns and an area of 4.00×10^{-3} m², and lies in a uniform magnetic field $\vec{B} = (2.00\hat{i} - 3.00\hat{j} - 4.00\hat{k})$ mT. What are (a) the orientation energy of the coil in the magnetic field and (b) the torque (in unit-vector notation) on the coil due to the magnetic field?

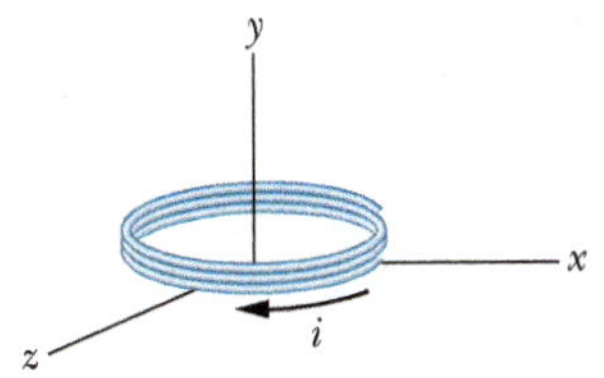

Figure 28-50 Problem 61.

••62 **GO** In Fig. 28-51a, two concentric coils, lying in the same plane, carry currents in opposite directions. The current in the larger coil 1 is fixed. Current i_2 in coil 2 can be varied. Figure 28-51b gives the net magnetic moment of the two-coil system as a function of i_2. The vertical axis scale is set by $\mu_{net,s} = 2.0 \times 10^{-5}$ A·m², and the horizontal axis scale is set by $i_{2s} = 10.0$ mA. If the current in coil 2 is then reversed, what is the magnitude of the net magnetic moment of the two-coil system when $i_2 = 7.0$ mA?

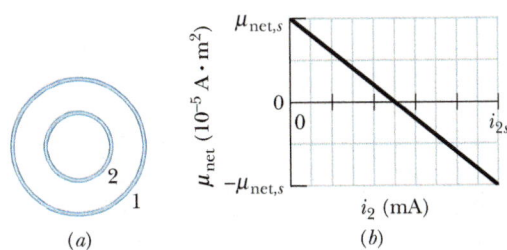

Figure 28-51 Problem 62.

••63 A circular loop of wire having a radius of 8.0 cm carries a current of 0.20 A. A vector of unit length and parallel to the dipole moment $\vec{\mu}$ of the loop is given by $0.60\hat{i} - 0.80\hat{j}$. (This unit vector gives the orientation of the magnetic dipole moment vector.) If the loop is located in a uniform magnetic field given by $\vec{B} = (0.25$ T$)\hat{i} + (0.30$ T$)\hat{k}$, find (a) the torque on the loop (in unit-vector notation) and (b) the orientation energy of the loop.

••64 **GO** Figure 28-52 gives the orientation energy U of a magnetic dipole in an external magnetic field $\vec{B}$, as a function of angle ϕ between the directions of $\vec{B}$ and the dipole moment. The vertical axis scale is set by $U_s = 2.0 \times 10^{-4}$ J. The dipole can be rotated about an axle with negligible friction in order to change ϕ. Counterclockwise rotation from $\phi = 0$ yields positive values of ϕ, and clockwise rotations yield negative values. The dipole is to be released at angle $\phi = 0$ with a rotational kinetic energy of 6.7×10^{-4} J, so that it rotates counterclockwise. To what maximum value of ϕ will it rotate? (In the language of Module 8-3, what value ϕ is the turning point in the potential well of Fig. 28-52?)

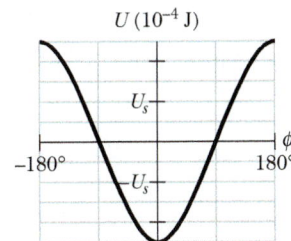

Figure 28-52 Problem 64.

••65 **SSM** **ILW** A wire of length 25.0 cm carrying a current of 4.51 mA is to be formed into a circular coil and placed in a uniform magnetic field $\vec{B}$ of magnitude 5.71 mT. If the torque on the coil from the field is maximized, what are (a) the angle between $\vec{B}$ and the coil's magnetic dipole moment and (b) the number of turns in the coil? (c) What is the magnitude of that maximum torque?

Additional Problems

66 A proton of charge $+e$ and mass m enters a uniform magnetic field $\vec{B} = B\hat{i}$ with an initial velocity $\vec{v} = v_{0x}\hat{i} + v_{0y}\hat{j}$. Find an expression in unit-vector notation for its velocity $\vec{v}$ at any later time t.

67 A stationary circular wall clock has a face with a radius of 15 cm. Six turns of wire are wound around its perimeter; the wire carries a current of 2.0 A in the clockwise direction. The clock is located where there is a constant, uniform external magnetic field of magnitude 70 mT (but the clock still keeps perfect time). At exactly 1:00 P.M., the hour hand of the clock points in the direction of the external magnetic field. (a) After how many minutes will the minute hand point in the direction of the torque on the winding due to the magnetic field? (b) Find the torque magnitude.

68 A wire lying along a y axis from $y = 0$ to $y = 0.250$ m carries a current of 2.00 mA in the negative direction of the axis. The wire fully lies in a nonuniform magnetic field that is given by $\vec{B} = (0.300$ T/m$)y\hat{i} + (0.400$ T/m$)y\hat{j}$. In unit-vector notation, what is the magnetic force on the wire?

69 Atom 1 of mass 35 u and atom 2 of mass 37 u are both singly ionized with a charge of $+e$. After being introduced into a mass spectrometer (Fig. 28-12) and accelerated from rest through a potential difference $V = 7.3$ kV, each ion follows a circular path in a uniform magnetic field of magnitude $B = 0.50$ T. What is the distance Δx between the points where the ions strike the detector?

70 An electron with kinetic energy 2.5 keV moving along the positive direction of an x axis enters a region in which a uniform electric field of magnitude 10 kV/m is in the negative direction of the y axis. A uniform magnetic field $\vec{B}$ is to be set up to keep the electron moving along the x axis, and the direction of $\vec{B}$ is to be chosen to minimize the required magnitude of $\vec{B}$. In unit-vector notation, what $\vec{B}$ should be set up?

71 Physicist S. A. Goudsmit devised a method for measuring the mass of heavy ions by timing their period of revolution in a known magnetic field. A singly charged ion of iodine makes 7.00 rev in a 45.0 mT field in 1.29 ms. Calculate its mass in atomic mass units.

72 A beam of electrons whose kinetic energy is K emerges from a thin-foil "window" at the end of an accelerator tube. A metal plate at distance d from this window is perpendicular to the direction of the emerging beam (Fig. 28-53). (a) Show that we can prevent the beam from hitting the plate if we apply a uniform magnetic field such that

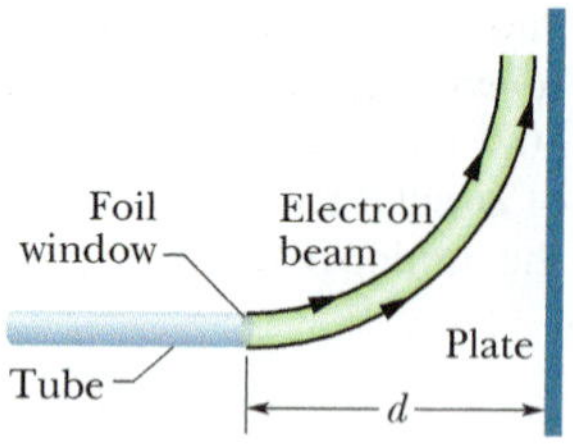

Figure 28-53 Problem 72.

$$B \geq \sqrt{\frac{2mK}{e^2 d^2}},$$

in which m and e are the electron mass and charge. (b) How should $\vec{B}$ be oriented?

73 SSM At time $t = 0$, an electron with kinetic energy 12 keV moves through $x = 0$ in the positive direction of an x axis that is parallel to the horizontal component of Earth's magnetic field $\vec{B}$. The field's vertical component is downward and has magnitude 55.0 μT. (a) What is the magnitude of the electron's acceleration due to $\vec{B}$? (b) What is the electron's distance from the x axis when the electron reaches coordinate $x = 20$ cm?

74 GO A particle with charge 2.0 C moves through a uniform magnetic field. At one instant the velocity of the particle is $(2.0\hat{i} + 4.0\hat{j} + 6.0\hat{k})$ m/s and the magnetic force on the particle is $(4.0\hat{i} - 20\hat{j} + 12\hat{k})$ N. The x and y components of the magnetic field are equal. What is $\vec{B}$?

75 A proton, a deuteron ($q = +e$, $m = 2.0$ u), and an alpha particle ($q = +2e$, $m = 4.0$ u) all having the same kinetic energy enter a region of uniform magnetic field $\vec{B}$, moving perpendicular to $\vec{B}$. What is the ratio of (a) the radius r_d of the deuteron path to the radius r_p of the proton path and (b) the radius r_α of the alpha particle path to r_p?

76 Bainbridge's mass spectrometer, shown in Fig. 28-54, separates ions having the same velocity. The ions, after entering through slits, S_1 and S_2, pass through a velocity selector composed of an electric field produced by the charged plates P and P', and a magnetic field $\vec{B}$ perpendicular to the electric field and the ion path. The ions that then pass undeviated through the crossed $\vec{E}$ and $\vec{B}$ fields enter into a region where a second magnetic field $\vec{B}'$ exists, where they are made to follow circular paths. A photographic plate (or a modern detector) registers their arrival. Show that, for the ions, $q/m = E/rBB'$, where r is the radius of the circular orbit.

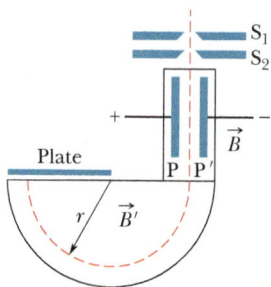

Figure 28-54 Problem 76.

77 SSM In Fig. 28-55, an electron moves at speed $v = 100$ m/s along an x axis through uniform electric and magnetic fields. The magnetic field $\vec{B}$ is directed into the page and has magnitude 5.00 T. In unit-vector notation, what is the electric field?

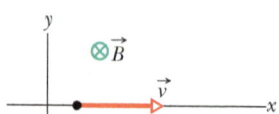

Figure 28-55 Problem 77.

78 (a) In Fig. 28-8, show that the ratio of the Hall electric field magnitude E to the magnitude E_C of the electric field responsible for moving charge (the current) along the length of

the strip is

$$\frac{E}{E_C} = \frac{B}{ne\rho},$$

where ρ is the resistivity of the material and n is the number density of the charge carriers. (b) Compute this ratio numerically for Problem 13. (See Table 26-1.)

79 SSM A proton, a deuteron ($q = +e$, $m = 2.0$ u), and an alpha particle ($q = +2e$, $m = 4.0$ u) are accelerated through the same potential difference and then enter the same region of uniform magnetic field $\vec{B}$, moving perpendicular to $\vec{B}$. What is the ratio of (a) the proton's kinetic energy K_p to the alpha particle's kinetic energy K_α and (b) the deuteron's kinetic energy K_d to K_α? If the radius of the proton's circular path is 10 cm, what is the radius of (c) the deuteron's path and (d) the alpha particle's path?

80 An electron is moving at 7.20×10^6 m/s in a magnetic field of strength 83.0 mT. What is the (a) maximum and (b) minimum magnitude of the force acting on the electron due to the field? (c) At one point the electron has an acceleration of magnitude 4.90×10^{14} m/s². What is the angle between the electron's velocity and the magnetic field?

81 A 5.0 μC particle moves through a region containing the uniform magnetic field $-20\hat{i}$ mT and the uniform electric field $300\hat{j}$ V/m. At a certain instant the velocity of the particle is $(17\hat{i} - 11\hat{j} + 7.0\hat{k})$ km/s. At that instant and in unit-vector notation, what is the net electromagnetic force (the sum of the electric and magnetic forces) on the particle?

82 In a Hall-effect experiment, a current of 3.0 A sent lengthwise through a conductor 1.0 cm wide, 4.0 cm long, and 10 μm thick produces a transverse (across the width) Hall potential difference of 10 μV when a magnetic field of 1.5 T is passed perpendicularly through the thickness of the conductor. From these data, find (a) the drift velocity of the charge carriers and (b) the number density of charge carriers. (c) Show on a diagram the polarity of the Hall potential difference with assumed current and magnetic field directions, assuming also that the charge carriers are electrons.

83 SSM A particle of mass 6.0 g moves at 4.0 km/s in an xy plane, in a region with a uniform magnetic field given by $5.0\hat{i}$ mT. At one instant, when the particle's velocity is directed 37° counterclockwise from the positive direction of the x axis, the magnetic force on the particle is $0.48\hat{k}$ N. What is the particle's charge?

84 A wire lying along an x axis from $x = 0$ to $x = 1.00$ m carries a current of 3.00 A in the positive x direction. The wire is immersed in a nonuniform magnetic field that is given by $\vec{B} = (4.00 \text{ T/m}^2)x^2\hat{i} - (0.600 \text{ T/m}^2)x^2\hat{j}$. In unit-vector notation, what is the magnetic force on the wire?

85 At one instant, $\vec{v} = (-2.00\hat{i} + 4.00\hat{j} - 6.00\hat{k})$ m/s is the velocity of a proton in a uniform magnetic field $\vec{B} = (2.00\hat{i} - 4.00\hat{j} + 8.00\hat{k})$ mT. At that instant, what are (a) the magnetic force $\vec{F}$ acting on the proton, in unit-vector notation, (b) the angle between $\vec{v}$ and $\vec{F}$, and (c) the angle between $\vec{v}$ and $\vec{B}$?

86 An electron has velocity $\vec{v} = (32\hat{i} + 40\hat{j})$ km/s as it enters a uniform magnetic field $\vec{B} = 60\hat{i}$ μT. What are (a) the radius of the helical path taken by the electron and (b) the pitch of that path? (c) To an observer looking into the magnetic field region from the entrance point of the electron, does the electron spiral clockwise or counterclockwise as it moves?

87 Figure 28-56 shows a *homopolar generator*, which has a solid conducting disk as rotor and which is rotated by a motor (not shown). Conducting brushes connect this emf device to a circuit through which the device drives current. The device can produce a greater emf than wire loop rotors because they can spin at a much higher angular speed without rupturing. The disk has radius $R = 0.250$ m and rotation frequency $f = 4000$ Hz, and the device is in a uniform magnetic field of magnitude $B = 60.0$ mT that is perpendicular to the disk. As the disk is rotated, conduction electrons along the conducting path (dashed line) are forced to move through the magnetic field. (a) For the indicated rotation, is the magnetic force on those electrons up or down in the figure? (b) Is the magnitude of that force greater at the rim or near the center of the disk? (c) What is the work per unit charge done by that force in moving charge along the radial line, between the rim and the center? (d) What, then, is the emf of the device? (e) If the current is 50.0 A, what is the power at which electrical energy is being produced?

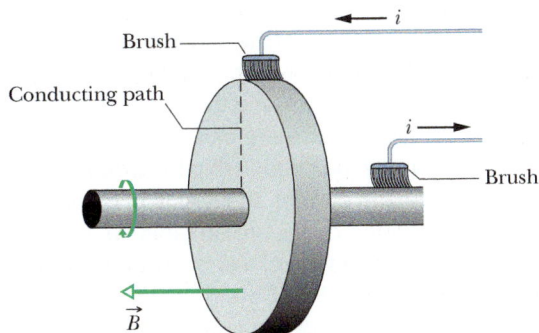

Figure 28-56 Problem 87.

88 In Fig. 28-57, the two ends of a **U**-shaped wire of mass $m = 10.0$ g and length $L = 20.0$ cm are immersed in mercury (which is a conductor). The wire is in a uniform field of magnitude $B = 0.100$ T. A switch (unshown) is rapidly closed and then reopened, sending a pulse of current through the wire, which causes the wire to jump upward. If jump height $h = 3.00$ m, how much charge was in the pulse? Assume that the duration of the pulse is much less than the time of flight. Consider the definition of impulse (Eq. 9-30) and its

relationship with momentum (Eq. 9-31). Also consider the relationship between charge and current (Eq. 26-2).

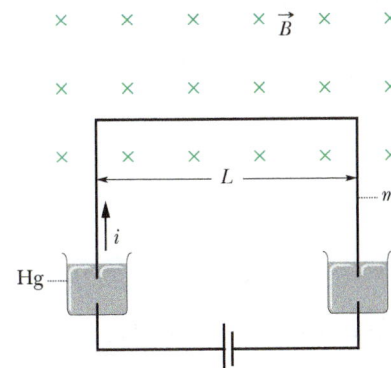

Figure 28-57 Problem 88.

89 In Fig. 28-58, an electron of mass m, charge $-e$, and low (negligible) speed enters the region between two plates of potential difference V and plate separation d, initially headed directly toward the top plate. A uniform magnetic field of magnitude B is normal to the plane of the figure. Find the minimum value of B such that the electron will not strike the top plate.

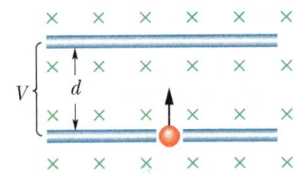

Figure 28-58 Problem 89.

90 A particle of charge q moves in a circle of radius r with speed v. Treating the circular path as a current loop with an average current, find the maximum torque exerted on the loop by a uniform field of magnitude B.

91 In a Hall-effect experiment, express the number density of charge carriers in terms of the Hall-effect electric field magnitude E, the current density magnitude J, and the magnetic field magnitude B.

92 An electron that is moving through a uniform magnetic field has velocity $\vec{v} = (40 \text{ km/s})\hat{i} + (35 \text{ km/s})\hat{j}$ when it experiences a force $\vec{F} = -(4.2 \text{ fN})\hat{i} + (4.8 \text{ fN})\hat{j}$ due to the magnetic field. If $B_x = 0$, calculate the magnetic field $\vec{B}$.

Magnetic Fields Due to Currents

29-1 MAGNETIC FIELD DUE TO A CURRENT

Learning Objectives

After reading this module, you should be able to . . .

29.01 Sketch a current-length element in a wire and indicate the direction of the magnetic field that it sets up at a given point near the wire.

29.02 For a given point near a wire and a given current-length element in the wire, determine the magnitude and direction of the magnetic field due to that element.

29.03 Identify the magnitude of the magnetic field set up by a current-length element at a point in line with the direction of that element.

29.04 For a point to one side of a long straight wire carrying current, apply the relationship between the magnetic field magnitude, the current, and the distance to the point.

29.05 For a point to one side of a long straight wire carrying current, use a right-hand rule to determine the direction of the field vector.

29.06 Identify that around a long straight wire carrying current, the magnetic field lines form circles.

29.07 For a point to one side of the end of a semi-infinite wire carrying current, apply the relationship between the magnetic field magnitude, the current, and the distance to the point.

29.08 For the center of curvature of a circular arc of wire carrying current, apply the relationship between the magnetic field magnitude, the current, the radius of curvature, and the angle subtended by the arc (in radians).

29.09 For a point to one side of a short straight wire carrying current, integrate the Biot–Savart law to find the magnetic field set up at the point by the current.

Key Ideas

● The magnetic field set up by a current-carrying conductor can be found from the Biot–Savart law. This law asserts that the contribution $d\vec{B}$ to the field produced by a current-length element $i\,d\vec{s}$ at a point P located a distance r from the current element is

$$d\vec{B} = \frac{\mu_0}{4\pi} \frac{i\,d\vec{s} \times \hat{r}}{r^2} \quad \text{(Biot–Savart law)}.$$

Here $\hat{r}$ is a unit vector that points from the element toward P. The quantity μ_0, called the permeability constant, has the value

$$4\pi \times 10^{-7}\ \text{T} \cdot \text{m/A} \approx 1.26 \times 10^{-6}\ \text{T} \cdot \text{m/A}.$$

● For a long straight wire carrying a current i, the Biot–Savart law gives, for the magnitude of the magnetic field at a perpendicular distance R from the wire,

$$B = \frac{\mu_0 i}{2\pi R} \quad \text{(long straight wire)}.$$

● The magnitude of the magnetic field at the center of a circular arc, of radius R and central angle ϕ (in radians), carrying current i, is

$$B = \frac{\mu_0 i \phi}{4\pi R} \quad \text{(at center of circular arc)}.$$

What Is Physics?

One basic observation of physics is that a moving charged particle produces a magnetic field around itself. Thus a current of moving charged particles produces a magnetic field around the current. This feature of *electromagnetism*, which is the combined study of electric and magnetic effects, came as a surprise to the people who discovered it. Surprise or not, this feature has become enormously important in everyday life because it is the basis of countless electromagnetic devices. For example, a magnetic field is produced in maglev trains and other devices used to lift heavy loads.

Our first step in this chapter is to find the magnetic field due to the current in a very small section of current-carrying wire. Then we shall find the magnetic field due to the entire wire for several different arrangements of the wire.

Calculating the Magnetic Field Due to a Current

Figure 29-1 shows a wire of arbitrary shape carrying a current i. We want to find the magnetic field $\vec{B}$ at a nearby point P. We first mentally divide the wire into differential elements ds and then define for each element a length vector $d\vec{s}$ that has length ds and whose direction is the direction of the current in ds. We can then define a differential *current-length element* to be $i\,d\vec{s}$; we wish to calculate the field $d\vec{B}$ produced at P by a typical current-length element. From experiment we find that magnetic fields, like electric fields, can be superimposed to find a net field. Thus, we can calculate the net field $\vec{B}$ at P by summing, via integration, the contributions $d\vec{B}$ from all the current-length elements. However, this summation is more challenging than the process associated with electric fields because of a complexity; whereas a charge element dq producing an electric field is a scalar, a current-length element $i\,d\vec{s}$ producing a magnetic field is a vector, being the product of a scalar and a vector.

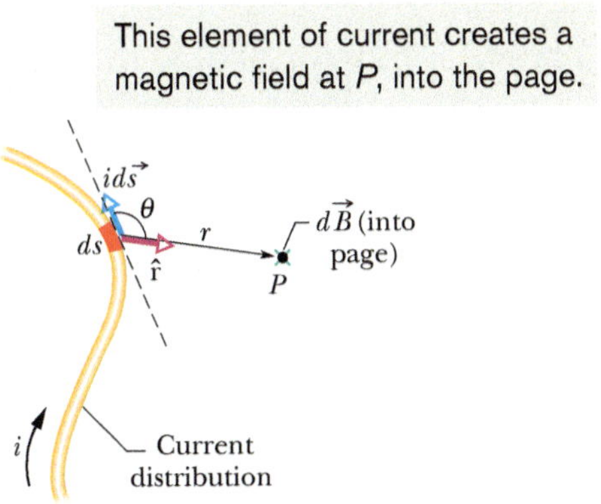

Magnitude. The magnitude of the field $d\vec{B}$ produced at point P at distance r by a current-length element $i\,d\vec{s}$ turns out to be

$$dB = \frac{\mu_0}{4\pi} \frac{i\,ds\,\sin\theta}{r^2}, \tag{29-1}$$

where θ is the angle between the directions of $d\vec{s}$ and $\hat{r}$, a unit vector that points from ds toward P. Symbol μ_0 is a constant, called the *permeability constant*, whose value is defined to be exactly

$$\mu_0 = 4\pi \times 10^{-7}\,\text{T·m/A} \approx 1.26 \times 10^{-6}\,\text{T·m/A}. \tag{29-2}$$

Direction. The direction of $d\vec{B}$, shown as being into the page in Fig. 29-1, is that of the cross product $d\vec{s} \times \hat{r}$. We can therefore write Eq. 29-1 in vector form as

$$d\vec{B} = \frac{\mu_0}{4\pi} \frac{i\,d\vec{s} \times \hat{r}}{r^2} \quad \text{(Biot–Savart law).} \tag{29-3}$$

This vector equation and its scalar form, Eq. 29-1, are known as the **law of Biot and Savart** (rhymes with "Leo and bazaar"). The law, which is experimentally deduced, is an inverse-square law. We shall use this law to calculate the net magnetic field $\vec{B}$ produced at a point by various distributions of current.

Here is one easy distribution: If current in a wire is either directly toward or directly away from a point P of measurement, can you see from Eq. 29-1 that the magnetic field at P from the current is simply zero (the angle θ is either 0° for *toward* or 180° for *away*, and both result in $\sin\theta = 0$)?

Magnetic Field Due to a Current in a Long Straight Wire

Shortly we shall use the law of Biot and Savart to prove that the magnitude of the magnetic field at a perpendicular distance R from a long (infinite) straight wire carrying a current i is given by

$$B = \frac{\mu_0 i}{2\pi R} \quad \text{(long straight wire).} \tag{29-4}$$

The field magnitude B in Eq. 29-4 depends only on the current and the perpendicular distance R of the point from the wire. We shall show in our derivation that the field lines of $\vec{B}$ form concentric circles around the wire, as Fig. 29-2 shows

This element of current creates a magnetic field at P, into the page.

Figure 29-1 A current-length element $i\,d\vec{s}$ produces a differential magnetic field $d\vec{B}$ at point P. The green × (the tail of an arrow) at the dot for point P indicates that $d\vec{B}$ is directed *into* the page there.

The magnetic field vector at any point is tangent to a circle.

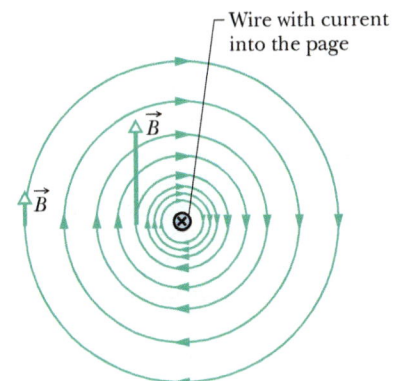

Figure 29-2 The magnetic field lines produced by a current in a long straight wire form concentric circles around the wire. Here the current is into the page, as indicated by the ×.

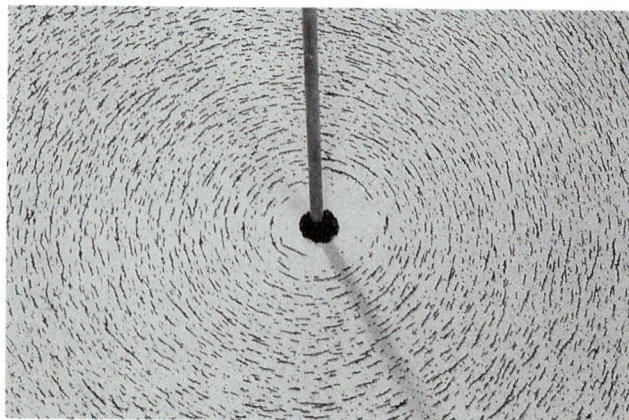

Courtesy Education Development Center

Figure 29-3 Iron filings that have been sprinkled onto cardboard collect in concentric circles when current is sent through the central wire. The alignment, which is along magnetic field lines, is caused by the magnetic field produced by the current.

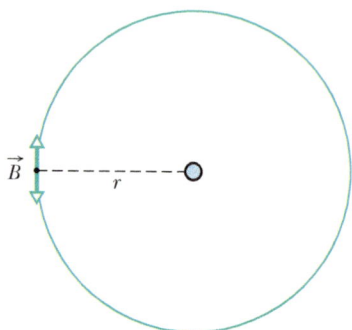

Figure 29-4 The magnetic field vector $\vec{B}$ is perpendicular to the radial line extending from a long straight wire with current, but which of the two perpendicular vectors is it?

and as the iron filings in Fig. 29-3 suggest. The increase in the spacing of the lines in Fig. 29-2 with increasing distance from the wire represents the $1/R$ decrease in the magnitude of $\vec{B}$ predicted by Eq. 29-4. The lengths of the two vectors $\vec{B}$ in the figure also show the $1/R$ decrease.

Directions. Plugging values into Eq. 29-4 to find the field magnitude B at a given radius is easy. What is difficult for many students is finding the direction of a field vector $\vec{B}$ at a given point. The field lines form circles around a long straight wire, and the field vector at any point on a circle must be tangent to the circle. That means it must be perpendicular to a radial line extending to the point from the wire. But there are two possible directions for that perpendicular vector, as shown in Fig. 29-4. One is correct for current into the figure, and the other is correct for current out of the figure. How can you tell which is which? Here is a simple right-hand rule for telling which vector is correct:

⭐ *Curled–straight right-hand rule:* Grasp the element in your right hand with your extended thumb pointing in the direction of the current. Your fingers will then naturally curl around in the direction of the magnetic field lines due to that element.

The result of applying this right-hand rule to the current in the straight wire of Fig. 29-2 is shown in a side view in Fig. 29-5a. To determine the direction of the magnetic field $\vec{B}$ set up at any particular point by this current, mentally wrap your right hand around the wire with your thumb in the direction of the current. Let your fingertips pass through the point; their direction is then the direction of the magnetic field at that point. In the view of Fig. 29-2, $\vec{B}$ at any point is *tangent to a magnetic field line;* in the view of Fig. 29-5, it is *perpendicular to a dashed radial line connecting the point and the current.*

Figure 29-5 A right-hand rule gives the direction of the magnetic field due to a current in a wire. (a) The situation of Fig. 29-2, seen from the side. The magnetic field $\vec{B}$ at any point to the left of the wire is perpendicular to the dashed radial line and directed into the page, in the direction of the fingertips, as indicated by the ×. (b) If the current is reversed, $\vec{B}$ at any point to the left is still perpendicular to the dashed radial line but now is directed out of the page, as indicated by the dot.

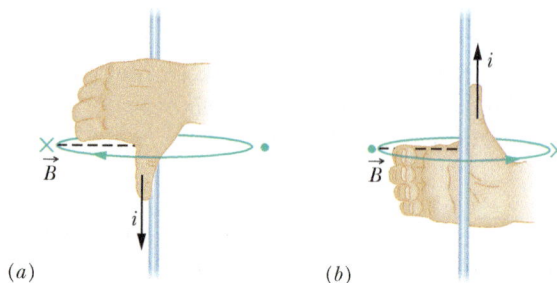

The thumb is in the current's direction. The fingers reveal the field vector's direction, which is tangent to a circle.

(a)   (b)

Proof of Equation 29-4

Figure 29-6, which is just like Fig. 29-1 except that now the wire is straight and of infinite length, illustrates the task at hand. We seek the field $\vec{B}$ at point P, a perpendicular distance R from the wire. The magnitude of the differential magnetic field produced at P by the current-length element $i\,d\vec{s}$ located a distance r from P is given by Eq. 29-1:

$$dB = \frac{\mu_0}{4\pi}\,\frac{i\,ds\,\sin\theta}{r^2}.$$

The direction of $d\vec{B}$ in Fig. 29-6 is that of the vector $d\vec{s} \times \hat{r}$—namely, directly into the page.

Note that $d\vec{B}$ at point P has this same direction for all the current-length elements into which the wire can be divided. Thus, we can find the magnitude of the magnetic field produced at P by the current-length elements in the upper half of the infinitely long wire by integrating dB in Eq. 29-1 from 0 to ∞.

Now consider a current-length element in the lower half of the wire, one that is as far below P as $d\vec{s}$ is above P. By Eq. 29-3, the magnetic field produced at P by this current-length element has the same magnitude and direction as that from element $i\,d\vec{s}$ in Fig. 29-6. Further, the magnetic field produced by the lower half of the wire is exactly the same as that produced by the upper half. To find the magnitude of the *total* magnetic field $\vec{B}$ at P, we need only multiply the result of our integration by 2. We get

$$B = 2\int_0^\infty dB = \frac{\mu_0 i}{2\pi}\int_0^\infty \frac{\sin\theta\,ds}{r^2}. \tag{29-5}$$

The variables $\theta, s,$ and r in this equation are not independent; Fig. 29-6 shows that they are related by

$$r = \sqrt{s^2 + R^2}$$

and

$$\sin\theta = \sin(\pi - \theta) = \frac{R}{\sqrt{s^2 + R^2}}.$$

With these substitutions and integral 19 in Appendix E, Eq. 29-5 becomes

$$B = \frac{\mu_0 i}{2\pi}\int_0^\infty \frac{R\,ds}{(s^2 + R^2)^{3/2}}$$

$$= \frac{\mu_0 i}{2\pi R}\left[\frac{s}{(s^2 + R^2)^{1/2}}\right]_0^\infty = \frac{\mu_0 i}{2\pi R}, \tag{29-6}$$

as we wanted. Note that the magnetic field at P due to either the lower half or the upper half of the infinite wire in Fig. 29-6 is half this value; that is,

$$B = \frac{\mu_0 i}{4\pi R} \quad \text{(semi-infinite straight wire).} \tag{29-7}$$

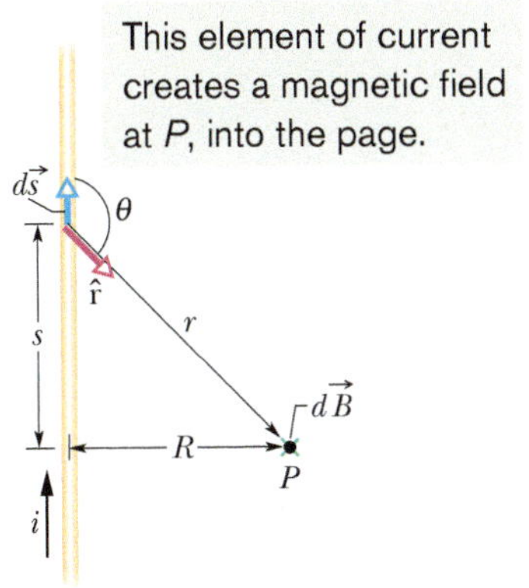

Magnetic Field Due to a Current in a Circular Arc of Wire

To find the magnetic field produced at a point by a current in a curved wire, we would again use Eq. 29-1 to write the magnitude of the field produced by a single current-length element, and we would again integrate to find the net field produced by all the current-length elements. That integration can be difficult, depending on the shape of the wire; it is fairly straightforward, however, when the wire is a circular arc and the point is the center of curvature.

Figure 29-7a shows such an arc-shaped wire with central angle ϕ, radius R, and center C, carrying current i. At C, each current-length element $i\,d\vec{s}$ of the wire produces a magnetic field of magnitude dB given by Eq. 29-1. Moreover, as Fig. 29-7b shows, no matter where the element is located on the wire, the angle θ

This element of current creates a magnetic field at P, into the page.

Figure 29-6 Calculating the magnetic field produced by a current i in a long straight wire. The field $d\vec{B}$ at P associated with the current-length element $i\,d\vec{s}$ is directed into the page, as shown.

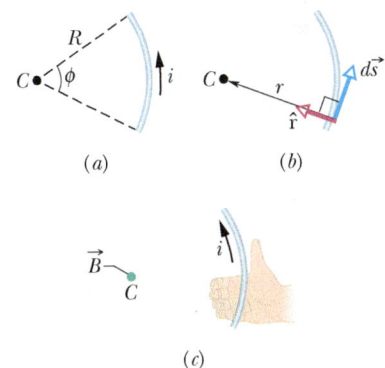

The right-hand rule reveals the field's direction at the center.

Figure 29-7 (a) A wire in the shape of a circular arc with center C carries current i. (b) For any element of wire along the arc, the angle between the directions of $d\vec{s}$ and $\hat{r}$ is 90°. (c) Determining the direction of the magnetic field at the center C due to the current in the wire; the field is out of the page, in the direction of the fingertips, as indicated by the colored dot at C.

between the vectors $d\vec{s}$ and $\hat{r}$ is 90°; also, $r = R$. Thus, by substituting R for r and 90° for θ in Eq. 29-1, we obtain

$$dB = \frac{\mu_0}{4\pi} \frac{i\,ds \sin 90°}{R^2} = \frac{\mu_0}{4\pi} \frac{i\,ds}{R^2}. \tag{29-8}$$

The field at C due to each current-length element in the arc has this magnitude.

Directions. How about the direction of the differential field $d\vec{B}$ set up by an element? From above we know that the vector must be perpendicular to a radial line extending through point C from the element, either into the plane of Fig. 29-7a or out of it. To tell which direction is correct, we use the right-hand rule for any of the elements, as shown in Fig. 29-7c. Grasping the wire with the thumb in the direction of the current and bringing the fingers into the region near C, we see that the vector $d\vec{B}$ due to any of the differential elements is out of the plane of the figure, not into it.

Total Field. To find the total field at C due to all the elements on the arc, we need to add all the differential field vectors $d\vec{B}$. However, because the vectors are all in the same direction, we do not need to find components. We just sum the magnitudes dB as given by Eq. 29-8. Since we have a vast number of those magnitudes, we sum via integration. We want the result to indicate how the total field depends on the angle ϕ of the arc (rather than the arc length). So, in Eq. 29-8 we switch from ds to $d\phi$ by using the identity $ds = R\,d\phi$. The summation by integration then becomes

$$B = \int dB = \int_0^{\phi} \frac{\mu_0}{4\pi} \frac{iR\,d\phi}{R^2} = \frac{\mu_0 i}{4\pi R} \int_0^{\phi} d\phi.$$

Integrating, we find that

$$B = \frac{\mu_0 i\phi}{4\pi R} \quad \text{(at center of circular arc).} \tag{29-9}$$

Heads Up. Note that this equation gives us the magnetic field *only* at the center of curvature of a circular arc of current. When you insert data into the equation, you must be careful to express ϕ in radians rather than degrees. For example, to find the magnitude of the magnetic field at the center of a full circle of current, you would substitute 2π rad for ϕ in Eq. 29-9, finding

$$B = \frac{\mu_0 i (2\pi)}{4\pi R} = \frac{\mu_0 i}{2R} \quad \text{(at center of full circle).} \tag{29-10}$$

Sample Problem 29.01 Magnetic field at the center of a circular arc of current

The wire in Fig. 29-8a carries a current i and consists of a circular arc of radius R and central angle $\pi/2$ rad, and two straight sections whose extensions intersect the center C of the arc. What magnetic field $\vec{B}$ (magnitude and direction) does the current produce at C?

KEY IDEAS

We can find the magnetic field $\vec{B}$ at point C by applying the Biot–Savart law of Eq. 29-3 to the wire, point by point along the full length of the wire. However, the application of Eq. 29-3 can be simplified by evaluating $\vec{B}$ separately for the three distinguishable sections of the wire—namely, (1) the

straight section at the left, (2) the straight section at the right, and (3) the circular arc.

Straight sections: For any current-length element in section 1, the angle θ between $d\vec{s}$ and $\hat{r}$ is zero (Fig. 29-8b); so Eq. 29-1 gives us

$$dB_1 = \frac{\mu_0}{4\pi} \frac{i\,ds \sin \theta}{r^2} = \frac{\mu_0}{4\pi} \frac{i\,ds \sin 0}{r^2} = 0.$$

Thus, the current along the entire length of straight section 1 contributes no magnetic field at C:

$$B_1 = 0.$$

Figure 29-8 (*a*) A wire consists of two straight sections (1 and 2) and a circular arc (3), and carries current *i*. (*b*) For a current-length element in section 1, the angle between $d\vec{s}$ and $\hat{r}$ is zero. (*c*) Determining the direction of magnetic field $\vec{B}_3$ at *C* due to the current in the circular arc; the field is into the page there.

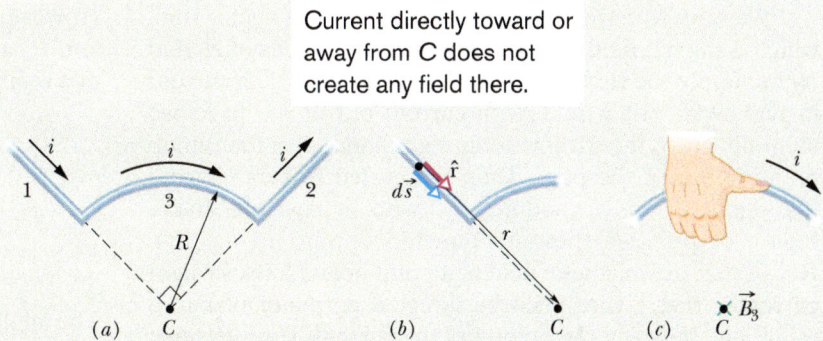

The same situation prevails in straight section 2, where the angle θ between $d\vec{s}$ and $\hat{r}$ for any current-length element is 180°. Thus,

$$B_2 = 0.$$

Circular arc: Application of the Biot–Savart law to evaluate the magnetic field at the center of a circular arc leads to Eq. 29-9 ($B = \mu_0 i\phi/4\pi R$). Here the central angle ϕ of the arc is $\pi/2$ rad. Thus from Eq. 29-9, the magnitude of the magnetic field $\vec{B}_3$ at the arc's center *C* is

$$B_3 = \frac{\mu_0 i(\pi/2)}{4\pi R} = \frac{\mu_0 i}{8R}.$$

To find the direction of $\vec{B}_3$, we apply the right-hand rule displayed in Fig. 29-5. Mentally grasp the circular arc with your right hand as in Fig. 29-8*c*, with your thumb in the direction of the current. The direction in which your fingers curl around the wire indicates the direction of the magnetic field lines around the wire. They form circles around the wire, coming out of the page above the arc and going into the page inside the arc. In the region of point *C* (inside the arc), your fingertips point *into the plane* of the page. Thus, $\vec{B}_3$ is directed into that plane.

Net field: Generally, we combine multiple magnetic fields as vectors. Here, however, only the circular arc produces a magnetic field at point *C*. Thus, we can write the magnitude of the net field $\vec{B}$ as

$$B = B_1 + B_2 + B_3 = 0 + 0 + \frac{\mu_0 i}{8R} = \frac{\mu_0 i}{8R}. \qquad \text{(Answer)}$$

The direction of $\vec{B}$ is the direction of $\vec{B}_3$—namely, into the plane of Fig. 29-8.

Sample Problem 29.02 Magnetic field off to the side of two long straight currents

Figure 29-9*a* shows two long parallel wires carrying currents i_1 and i_2 in opposite directions. What are the magnitude and direction of the net magnetic field at point *P*? Assume the following values: $i_1 = 15$ A, $i_2 = 32$ A, and $d = 5.3$ cm.

KEY IDEAS

(1) The net magnetic field $\vec{B}$ at point *P* is the vector sum of the magnetic fields due to the currents in the two wires. (2) We can find the magnetic field due to any current by applying the Biot–Savart law to the current. For points near the current in a long straight wire, that law leads to Eq. 29-4.

Finding the vectors: In Fig. 29-9*a*, point *P* is distance *R* from both currents i_1 and i_2. Thus, Eq. 29-4 tells us that at point *P* those currents produce magnetic fields $\vec{B}_1$ and $\vec{B}_2$ with magnitudes

$$B_1 = \frac{\mu_0 i_1}{2\pi R} \quad \text{and} \quad B_2 = \frac{\mu_0 i_2}{2\pi R}.$$

In the right triangle of Fig. 29-9*a*, note that the base angles (between sides *R* and *d*) are both 45°. This allows us to write

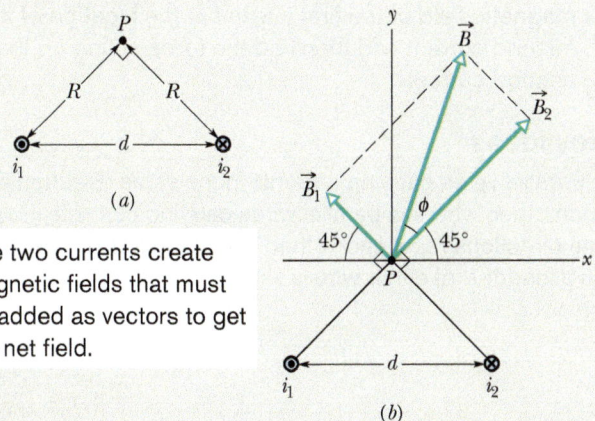

The two currents create magnetic fields that must be added as vectors to get the net field.

Figure 29-9 (*a*) Two wires carry currents i_1 and i_2 in opposite directions (out of and into the page). Note the right angle at *P*. (*b*) The separate fields $\vec{B}_1$ and $\vec{B}_2$ are combined vectorially to yield the net field $\vec{B}$.

$\cos 45° = R/d$ and replace *R* with $d \cos 45°$. Then the field magnitudes B_1 and B_2 become

$$B_1 = \frac{\mu_0 i_1}{2\pi d \cos 45°} \quad \text{and} \quad B_2 = \frac{\mu_0 i_2}{2\pi d \cos 45°}.$$

We want to combine $\vec{B}_1$ and $\vec{B}_2$ to find their vector sum, which is the net field $\vec{B}$ at P. To find the directions of $\vec{B}_1$ and $\vec{B}_2$, we apply the right-hand rule of Fig. 29-5 to each current in Fig. 29-9a. For wire 1, with current out of the page, we mentally grasp the wire with the right hand, with the thumb pointing out of the page. Then the curled fingers indicate that the field lines run counterclockwise. In particular, in the region of point P, they are directed upward to the left. Recall that the magnetic field at a point near a long, straight current-carrying wire must be directed perpendicular to a radial line between the point and the current. Thus, $\vec{B}_1$ must be directed upward to the left as drawn in Fig. 29-9b. (Note carefully the perpendicular symbol between vector $\vec{B}_1$ and the line connecting point P and wire 1.)

Repeating this analysis for the current in wire 2, we find that $\vec{B}_2$ is directed upward to the right as drawn in Fig. 29-9b.

Adding the vectors: We can now vectorially add $\vec{B}_1$ and $\vec{B}_2$ to find the net magnetic field $\vec{B}$ at point P, either by using a vector-capable calculator or by resolving the vectors into components and then combining the components of $\vec{B}$.

However, in Fig. 29-9b, there is a third method: Because $\vec{B}_1$ and $\vec{B}_2$ are perpendicular to each other, they form the legs of a right triangle, with $\vec{B}$ as the hypotenuse. So,

$$B = \sqrt{B_1^2 + B_2^2} = \frac{\mu_0}{2\pi d(\cos 45°)}\sqrt{i_1^2 + i_2^2}$$

$$= \frac{(4\pi \times 10^{-7}\,\text{T·m/A})\sqrt{(15\,\text{A})^2 + (32\,\text{A})^2}}{(2\pi)(5.3 \times 10^{-2}\,\text{m})(\cos 45°)}$$

$$= 1.89 \times 10^{-4}\,\text{T} \approx 190\,\mu\text{T}. \qquad \text{(Answer)}$$

The angle ϕ between the directions of $\vec{B}$ and $\vec{B}_2$ in Fig. 29-9b follows from

$$\phi = \tan^{-1}\frac{B_1}{B_2},$$

which, with B_1 and B_2 as given above, yields

$$\phi = \tan^{-1}\frac{i_1}{i_2} = \tan^{-1}\frac{15\,\text{A}}{32\,\text{A}} = 25°.$$

The angle between $\vec{B}$ and the x axis shown in Fig. 29-9b is then

$$\phi + 45° = 25° + 45° = 70°. \qquad \text{(Answer)}$$

WILEY **PLUS** Additional examples, video, and practice available at *WileyPLUS*

29-2 FORCE BETWEEN TWO PARALLEL CURRENTS

Learning Objectives

After reading this module, you should be able to . . .

29.10 Given two parallel or antiparallel currents, find the magnetic field of the first current at the location of the second current and then find the force acting on that second current.

29.11 Identify that parallel currents attract each other, and antiparallel currents repel each other.

29.12 Describe how a rail gun works.

Key Ideas

● Parallel wires carrying currents in the same direction attract each other, whereas parallel wires carrying currents in opposite directions repel each other. The magnitude of the force on a length L of either wire is

$$F_{ba} = i_b L B_a \sin 90° = \frac{\mu_0 L i_a i_b}{2\pi d},$$

where d is the wire separation, and i_a and i_b are the currents in the wires.

Force Between Two Parallel Currents

Two long parallel wires carrying currents exert forces on each other. Figure 29-10 shows two such wires, separated by a distance d and carrying currents i_a and i_b. Let us analyze the forces on these wires due to each other.

We seek first the force on wire b in Fig. 29-10 due to the current in wire a. That current produces a magnetic field $\vec{B}_a$, and it is this magnetic field that actually causes the force we seek. To find the force, then, we need the magnitude and direction of the field $\vec{B}_a$ *at the site of wire b*. The magnitude of $\vec{B}_a$ at every point of wire b is, from Eq. 29-4,

$$B_a = \frac{\mu_0 i_a}{2\pi d}. \qquad (29\text{-}11)$$

The curled–straight right-hand rule tells us that the direction of $\vec{B}_a$ at wire b is down, as Fig. 29-10 shows. Now that we have the field, we can find the force it produces on wire b. Equation 28-26 tells us that the force $\vec{F}_{ba}$ on a length L of wire b due to the external magnetic field $\vec{B}_a$ is

$$\vec{F}_{ba} = i_b \vec{L} \times \vec{B}_a, \tag{29-12}$$

where $\vec{L}$ is the length vector of the wire. In Fig. 29-10, vectors $\vec{L}$ and $\vec{B}_a$ are perpendicular to each other, and so with Eq. 29-11, we can write

$$F_{ba} = i_b L B_a \sin 90° = \frac{\mu_0 L i_a i_b}{2\pi d}. \tag{29-13}$$

The direction of $\vec{F}_{ba}$ is the direction of the cross product $\vec{L} \times \vec{B}_a$. Applying the right-hand rule for cross products to $\vec{L}$ and $\vec{B}_a$ in Fig. 29-10, we see that $\vec{F}_{ba}$ is directly toward wire a, as shown.

The general procedure for finding the force on a current-carrying wire is this:

To find the force on a current-carrying wire due to a second current-carrying wire, first find the field due to the second wire at the site of the first wire. Then find the force on the first wire due to that field.

We could now use this procedure to compute the force on wire a due to the current in wire b. We would find that the force is directly toward wire b; hence, the two wires with parallel currents attract each other. Similarly, if the two currents were antiparallel, we could show that the two wires repel each other. Thus,

Parallel currents attract each other, and antiparallel currents repel each other.

The force acting between currents in parallel wires is the basis for the definition of the ampere, which is one of the seven SI base units. The definition, adopted in 1946, is this: The ampere is that constant current which, if maintained in two straight, parallel conductors of infinite length, of negligible circular cross section, and placed 1 m apart in vacuum, would produce on each of these conductors a force of magnitude 2×10^{-7} newton per meter of wire length.

Rail Gun

The basics of a rail gun are shown in Fig. 29-11a. A large current is sent out along one of two parallel conducting rails, across a conducting "fuse" (such as a narrow piece of copper) between the rails, and then back to the current source along the second rail. The projectile to be fired lies on the far side of the fuse and fits loosely between the rails. Immediately after the current begins, the fuse element melts and vaporizes, creating a conducting gas between the rails where the fuse had been.

The curled–straight right-hand rule of Fig. 29-5 reveals that the currents in the rails of Fig. 29-11a produce magnetic fields that are directed downward between the rails. The net magnetic field $\vec{B}$ exerts a force $\vec{F}$ on the gas due to the current i through the gas (Fig. 29-11b). With Eq. 29-12 and the right-hand rule for cross products, we find that $\vec{F}$ points outward along the rails. As the gas is forced outward along the rails, it pushes the projectile, accelerating it by as much as $5 \times 10^6 g$, and then launches it with a speed of 10 km/s, all within 1 ms. Someday rail guns may be used to launch materials into space from mining operations on the Moon or an asteroid.

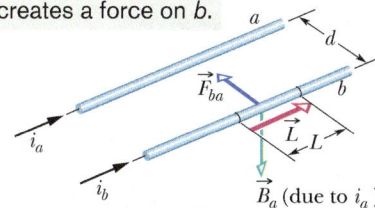

The field due to a at the position of b creates a force on b.

Figure 29-10 Two parallel wires carrying currents in the same direction attract each other. $\vec{B}_a$ is the magnetic field at wire b produced by the current in wire a. $\vec{F}_{ba}$ is the resulting force acting on wire b because it carries current in $\vec{B}_a$.

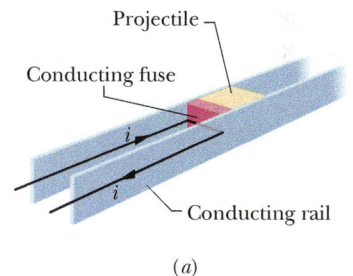

Figure 29-11 (a) A rail gun, as a current i is set up in it. The current rapidly causes the conducting fuse to vaporize. (b) The current produces a magnetic field $\vec{B}$ between the rails, and the field causes a force $\vec{F}$ to act on the conducting gas, which is part of the current path. The gas propels the projectile along the rails, launching it.

Checkpoint 1

The figure here shows three long, straight, parallel, equally spaced wires with identical currents either into or out of the page. Rank the wires according to the magnitude of the force on each due to the currents in the other two wires, greatest first.

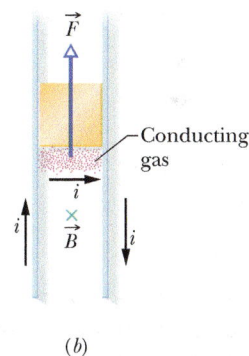

29-3 AMPERE'S LAW

Learning Objectives

After reading this module, you should be able to . . .

29.13 Apply Ampere's law to a loop that encircles current.

29.14 With Ampere's law, use a right-hand rule for determining the algebraic sign of an encircled current.

29.15 For more than one current within an Amperian loop, determine the net current to be used in Ampere's law.

29.16 Apply Ampere's law to a long straight wire with current, to find the magnetic field magnitude inside and outside the wire, identifying that only the current encircled by the Amperian loop matters.

Key Idea

● Ampere's law states that

$$\oint \vec{B} \cdot d\vec{s} = \mu_0 i_{enc} \quad \text{(Ampere's law).}$$

The line integral in this equation is evaluated around a closed loop called an Amperian loop. The current i on the right side is the *net* current encircled by the loop.

Ampere's Law

We can find the net electric field due to *any* distribution of charges by first writing the differential electric field $d\vec{E}$ due to a charge element and then summing the contributions of $d\vec{E}$ from all the elements. However, if the distribution is complicated, we may have to use a computer. Recall, however, that if the distribution has planar, cylindrical, or spherical symmetry, we can apply Gauss' law to find the net electric field with considerably less effort.

Similarly, we can find the net magnetic field due to *any* distribution of currents by first writing the differential magnetic field $d\vec{B}$ (Eq. 29-3) due to a current-length element and then summing the contributions of $d\vec{B}$ from all the elements. Again we may have to use a computer for a complicated distribution. However, if the distribution has some symmetry, we may be able to apply **Ampere's law** to find the magnetic field with considerably less effort. This law, which can be derived from the Biot–Savart law, has traditionally been credited to André-Marie Ampère (1775–1836), for whom the SI unit of current is named. However, the law actually was advanced by English physicist James Clerk Maxwell. Ampere's law is

$$\oint \vec{B} \cdot d\vec{s} = \mu_0 i_{enc} \quad \text{(Ampere's law).} \qquad (29\text{-}14)$$

The loop on the integral sign means that the scalar (dot) product $\vec{B} \cdot d\vec{s}$ is to be integrated around a *closed* loop, called an *Amperian loop*. The current i_{enc} is the *net* current encircled by that closed loop.

To see the meaning of the scalar product $\vec{B} \cdot d\vec{s}$ and its integral, let us first apply Ampere's law to the general situation of Fig. 29-12. The figure shows cross sections of three long straight wires that carry currents i_1, i_2, and i_3 either directly into or directly out of the page. An arbitrary Amperian loop lying in the plane of the page encircles two of the currents but not the third. The counterclockwise direction marked on the loop indicates the arbitrarily chosen direction of integration for Eq. 29-14.

To apply Ampere's law, we mentally divide the loop into differential vector elements $d\vec{s}$ that are everywhere directed along the tangent to the loop in the direction of integration. Assume that at the location of the element $d\vec{s}$ shown in Fig. 29-12, the net magnetic field due to the three currents is $\vec{B}$. Because the wires are perpendicular to the page, we know that the magnetic field at $d\vec{s}$ due to each current is in the plane of Fig. 29-12; thus, their net magnetic field $\vec{B}$ at $d\vec{s}$ must also be in that plane. However, we do not know the orientation of $\vec{B}$ within the plane. In Fig. 29-12, $\vec{B}$ is arbitrarily drawn at an angle θ to the direction of $d\vec{s}$. The scalar product $\vec{B}\cdot d\vec{s}$ on the left side of Eq. 29-14 is equal to $B\cos\theta\, ds$. Thus, Ampere's law can be written as

$$\oint \vec{B}\cdot d\vec{s} = \oint B\cos\theta\, ds = \mu_0 i_{\text{enc}}. \tag{29-15}$$

We can now interpret the scalar product $\vec{B}\cdot d\vec{s}$ as being the product of a length ds of the Amperian loop and the field component $B\cos\theta$ tangent to the loop. Then we can interpret the integration as being the summation of all such products around the entire loop.

Signs. When we can actually perform this integration, we do not need to know the direction of $\vec{B}$ before integrating. Instead, we arbitrarily assume $\vec{B}$ to be generally in the direction of integration (as in Fig. 29-12). Then we use the following curled–straight right-hand rule to assign a plus sign or a minus sign to each of the currents that make up the net encircled current i_{enc}:

Only the currents encircled by the loop are used in Ampere's law.

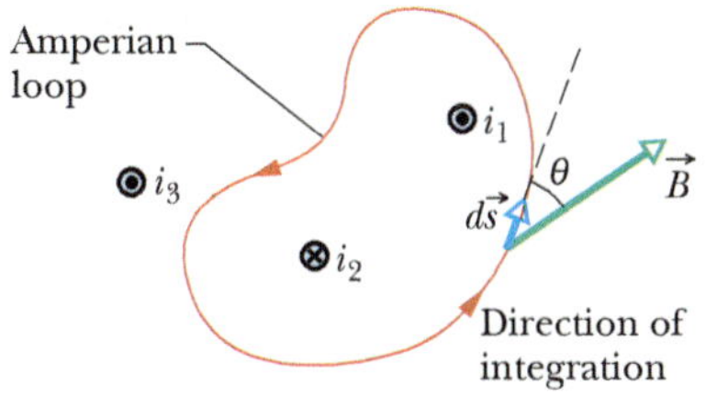

Figure 29-12 Ampere's law applied to an arbitrary Amperian loop that encircles two long straight wires but excludes a third wire. Note the directions of the currents.

> Curl your right hand around the Amperian loop, with the fingers pointing in the direction of integration. A current through the loop in the general direction of your outstretched thumb is assigned a plus sign, and a current generally in the opposite direction is assigned a minus sign.

Finally, we solve Eq. 29-15 for the magnitude of $\vec{B}$. If B turns out positive, then the direction we assumed for $\vec{B}$ is correct. If it turns out negative, we neglect the minus sign and redraw $\vec{B}$ in the opposite direction.

Net Current. In Fig. 29-13 we apply the curled–straight right-hand rule for Ampere's law to the situation of Fig. 29-12. With the indicated counterclockwise direction of integration, the net current encircled by the loop is

$$i_{\text{enc}} = i_1 - i_2.$$

(Current i_3 is not encircled by the loop.) We can then rewrite Eq. 29-15 as

$$\oint B\cos\theta\, ds = \mu_0(i_1 - i_2). \tag{29-16}$$

This is how to assign a sign to a current used in Ampere's law.

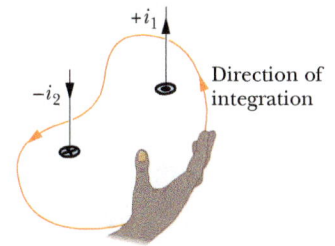

Figure 29-13 A right-hand rule for Ampere's law, to determine the signs for currents encircled by an Amperian loop. The situation is that of Fig. 29-12.

You might wonder why, since current i_3 contributes to the magnetic-field magnitude B on the left side of Eq. 29-16, it is not needed on the right side. The answer is that the contributions of current i_3 to the magnetic field cancel out because the integration in Eq. 29-16 is made around the full loop. In contrast, the contributions of an encircled current to the magnetic field do not cancel out.

We cannot solve Eq. 29-16 for the magnitude B of the magnetic field because for the situation of Fig. 29-12 we do not have enough information to simplify and solve the integral. However, we do know the outcome of the integration; it must be equal to $\mu_0(i_1 - i_2)$, the value of which is set by the net current passing through the loop.

We shall now apply Ampere's law to two situations in which symmetry does allow us to simplify and solve the integral, hence to find the magnetic field.

Magnetic Field Outside a Long Straight Wire with Current

Figure 29-14 shows a long straight wire that carries current i directly out of the page. Equation 29-4 tells us that the magnetic field $\vec{B}$ produced by the current has the same magnitude at all points that are the same distance r from the wire; that is, the field $\vec{B}$ has cylindrical symmetry about the wire. We can take advantage of that symmetry to simplify the integral in Ampere's law (Eqs. 29-14 and 29-15) if we encircle the wire with a concentric circular Amperian loop of radius r, as in Fig. 29-14. The magnetic field then has the same magnitude B at every point on the loop. We shall integrate counterclockwise, so that $d\vec{s}$ has the direction shown in Fig. 29-14.

All of the current is encircled and thus all is used in Ampere's law.

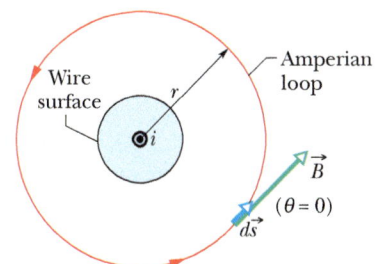

Figure 29-14 Using Ampere's law to find the magnetic field that a current i produces outside a long straight wire of circular cross section. The Amperian loop is a concentric circle that lies outside the wire.

We can further simplify the quantity $B \cos \theta$ in Eq. 29-15 by noting that $\vec{B}$ is tangent to the loop at every point along the loop, as is $d\vec{s}$. Thus, $\vec{B}$ and $d\vec{s}$ are either parallel or antiparallel at each point of the loop, and we shall arbitrarily assume the former. Then at every point the angle θ between $d\vec{s}$ and $\vec{B}$ is $0°$, so $\cos \theta = \cos 0° = 1$. The integral in Eq. 29-15 then becomes

$$\oint \vec{B} \cdot d\vec{s} = \oint B \cos \theta \, ds = B \oint ds = B(2\pi r).$$

Note that $\oint ds$ is the summation of all the line segment lengths ds around the circular loop; that is, it simply gives the circumference $2\pi r$ of the loop.

Our right-hand rule gives us a plus sign for the current of Fig. 29-14. The right side of Ampere's law becomes $+\mu_0 i$, and we then have

$$B(2\pi r) = \mu_0 i$$

or
$$B = \frac{\mu_0 i}{2\pi r} \quad \text{(outside straight wire).} \quad (29\text{-}17)$$

With a slight change in notation, this is Eq. 29-4, which we derived earlier—with considerably more effort—using the law of Biot and Savart. In addition, because the magnitude B turned out positive, we know that the correct direction of $\vec{B}$ must be the one shown in Fig. 29-14.

Magnetic Field Inside a Long Straight Wire with Current

Figure 29-15 shows the cross section of a long straight wire of radius R that carries a uniformly distributed current i directly out of the page. Because the current is uniformly distributed over a cross section of the wire, the magnetic field $\vec{B}$ produced by the current must be cylindrically symmetrical. Thus, to find the magnetic field at points inside the wire, we can again use an Amperian loop of radius r, as shown in Fig. 29-15, where now $r < R$. Symmetry again suggests that $\vec{B}$ is tangent to the loop, as shown; so the left side of Ampere's law again yields

$$\oint \vec{B} \cdot d\vec{s} = B \oint ds = B(2\pi r). \quad (29\text{-}18)$$

Because the current is uniformly distributed, the current i_{enc} encircled by the loop is proportional to the area encircled by the loop; that is,

$$i_{enc} = i \frac{\pi r^2}{\pi R^2}. \quad (29\text{-}19)$$

Our right-hand rule tells us that i_{enc} gets a plus sign. Then Ampere's law gives us

$$B(2\pi r) = \mu_0 i \frac{\pi r^2}{\pi R^2}$$

or
$$B = \left(\frac{\mu_0 i}{2\pi R^2}\right) r \quad \text{(inside straight wire).} \quad (29\text{-}20)$$

Thus, inside the wire, the magnitude B of the magnetic field is proportional to r, is zero at the center, and is maximum at $r = R$ (the surface). Note that Eqs. 29-17 and 29-20 give the same value for B at the surface.

Only the current encircled by the loop is used in Ampere's law.

Figure 29-15 Using Ampere's law to find the magnetic field that a current i produces inside a long straight wire of circular cross section. The current is uniformly distributed over the cross section of the wire and emerges from the page. An Amperian loop is drawn inside the wire.

✓ **Checkpoint 2**

The figure here shows three equal currents i (two parallel and one antiparallel) and four Amperian loops. Rank the loops according to the magnitude of $\oint \vec{B} \cdot d\vec{s}$ along each, greatest first.

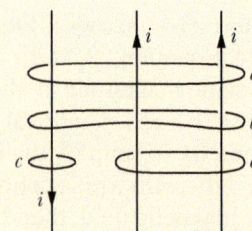

Sample Problem 29.03   Ampere's law to find the field inside a long cylinder of current

Figure 29-16a shows the cross section of a long conducting cylinder with inner radius $a = 2.0$ cm and outer radius $b = 4.0$ cm. The cylinder carries a current out of the page, and the magnitude of the current density in the cross section is given by $J = cr^2$, with $c = 3.0 \times 10^6$ A/m^4 and r in meters. What is the magnetic field $\vec{B}$ at the dot in Fig. 29-16a, which is at radius $r = 3.0$ cm from the central axis of the cylinder?

KEY IDEAS

The point at which we want to evaluate $\vec{B}$ is inside the material of the conducting cylinder, between its inner and outer radii. We note that the current distribution has cylindrical symmetry (it is the same all around the cross section for any given radius). Thus, the symmetry allows us to use Ampere's law to find $\vec{B}$ at the point. We first draw the Amperian loop shown in Fig. 29-16b. The loop is concentric with the cylinder and has radius $r = 3.0$ cm because we want to evaluate $\vec{B}$ at that distance from the cylinder's central axis.

Next, we must compute the current i_{enc} that is encircled by the Amperian loop. However, we *cannot* set up a proportionality as in Eq. 29-19, because here the current is not uniformly distributed. Instead, we must integrate the current density magnitude from the cylinder's inner radius a to the loop radius r, using the steps shown in Figs. 29-16c through h.

Calculations: We write the integral as

$$i_{enc} = \int J \, dA = \int_a^r cr^2 (2\pi r \, dr)$$

$$= 2\pi c \int_a^r r^3 \, dr = 2\pi c \left[\frac{r^4}{4} \right]_a^r$$

$$= \frac{\pi c(r^4 - a^4)}{2}.$$

Note that in these steps we took the differential area dA to be the area of the thin ring in Figs. 29-16d–f and then

We want the magnetic field at the dot at radius r.

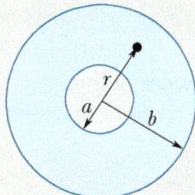

So, we put a concentric Amperian loop through the dot.

We need to find the current in the area encircled by the loop.

We start with a ring that is so thin that we can approximate the current density as being uniform within it.

(a)   (b)   (c)   (d)

Its area dA is the product of the ring's circumference and the width dr.

The current within the ring is the product of the current density J and the ring's area dA.

Our job is to sum the currents in all rings from this smallest one ...

... to this largest one, which has the same radius as the Amperian loop.

(e)   (f)   (g)   (h)

Figure 29-16 (*a*)–(*b*) To find the magnetic field at a point within this conducting cylinder, we use a concentric Amperian loop through the point. We then need the current encircled by the loop. (*c*)–(*h*) Because the current density is nonuniform, we start with a thin ring and then sum (via integration) the currents in all such rings in the encircled area.

replaced it with its equivalent, the product of the ring's circumference $2\pi r$ and its thickness dr.

For the Amperian loop, the direction of integration indicated in Fig. 29-16b is (arbitrarily) clockwise. Applying the right-hand rule for Ampere's law to that loop, we find that we should take i_{enc} as negative because the current is directed out of the page but our thumb is directed into the page.

We next evaluate the left side of Ampere's law as we did in Fig. 29-15, and we again obtain Eq. 29-18. Then Ampere's law,

$$\oint \vec{B} \cdot d\vec{s} = \mu_0 i_{enc},$$

gives us

$$B(2\pi r) = -\frac{\mu_0 \pi c}{2}(r^4 - a^4).$$

Solving for B and substituting known data yield

$$B = -\frac{\mu_0 c}{4r}(r^4 - a^4)$$

$$= -\frac{(4\pi \times 10^{-7}\,\text{T·m/A})(3.0 \times 10^6\,\text{A/m}^4)}{4(0.030\,\text{m})}$$

$$\times [(0.030\,\text{m})^4 - (0.020\,\text{m})^4]$$

$$= -2.0 \times 10^{-5}\,\text{T}.$$

Thus, the magnetic field $\vec{B}$ at a point 3.0 cm from the central axis has magnitude

$$B = 2.0 \times 10^{-5}\,\text{T} \qquad \text{(Answer)}$$

and forms magnetic field lines that are directed opposite our direction of integration, hence counterclockwise in Fig. 29-16b.

WILEY PLUS Additional examples, video, and practice available at *WileyPLUS*

29-4 SOLENOIDS AND TOROIDS

Learning Objectives

After reading this module, you should be able to . . .

29.17 Describe a solenoid and a toroid and sketch their magnetic field lines.

29.18 Explain how Ampere's law is used to find the magnetic field inside a solenoid.

29.19 Apply the relationship between a solenoid's internal magnetic field B, the current i, and the number of turns per

unit length n of the solenoid.

29.20 Explain how Ampere's law is used to find the magnetic field inside a toroid.

29.21 Apply the relationship between a toroid's internal magnetic field B, the current i, the radius r, and the total number of turns N.

Key Ideas

● Inside a long solenoid carrying current i, at points not near its ends, the magnitude B of the magnetic field is

$$B = \mu_0 in \quad \text{(ideal solenoid)},$$

where n is the number of turns per unit length.

● At a point inside a toroid, the magnitude B of the magnetic field is

$$B = \frac{\mu_0 iN}{2\pi}\frac{1}{r} \quad \text{(toroid)},$$

where r is the distance from the center of the toroid to the point.

Solenoids and Toroids

Magnetic Field of a Solenoid

We now turn our attention to another situation in which Ampere's law proves useful. It concerns the magnetic field produced by the current in a long, tightly wound helical coil of wire. Such a coil is called a **solenoid** (Fig. 29-17). We assume that the length of the solenoid is much greater than the diameter.

Figure 29-18 shows a section through a portion of a "stretched-out" solenoid. The solenoid's magnetic field is the vector sum of the fields produced by the individual turns (*windings*) that make up the solenoid. For points very

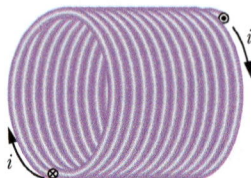

Figure 29-17 A solenoid carrying current i.

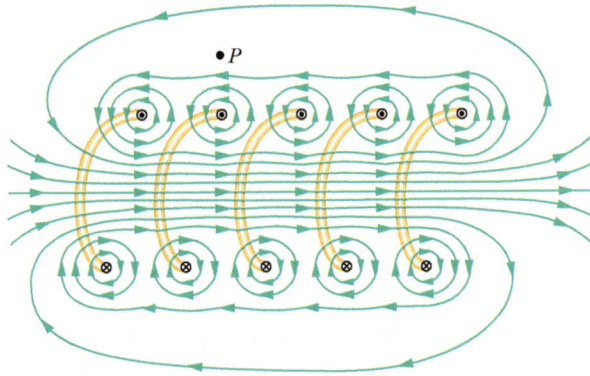

Figure 29-18 A vertical cross section through the central axis of a "stretched-out" solenoid. The back portions of five turns are shown, as are the magnetic field lines due to a current through the solenoid. Each turn produces circular magnetic field lines near itself. Near the solenoid's axis, the field lines combine into a net magnetic field that is directed along the axis. The closely spaced field lines there indicate a strong magnetic field. Outside the solenoid the field lines are widely spaced; the field there is very weak.

close to a turn, the wire behaves magnetically almost like a long straight wire, and the lines of $\vec{B}$ there are almost concentric circles. Figure 29-18 suggests that the field tends to cancel between adjacent turns. It also suggests that, at points inside the solenoid and reasonably far from the wire, $\vec{B}$ is approximately parallel to the (central) solenoid axis. In the limiting case of an *ideal solenoid*, which is infinitely long and consists of tightly packed (*close-packed*) turns of square wire, the field inside the coil is uniform and parallel to the solenoid axis.

At points above the solenoid, such as P in Fig. 29-18, the magnetic field set up by the upper parts of the solenoid turns (these upper turns are marked $\odot$) is directed to the left (as drawn near P) and tends to cancel the field set up at P by the lower parts of the turns (these lower turns are marked $\otimes$), which is directed to the right (not drawn). In the limiting case of an ideal solenoid, the magnetic field outside the solenoid is zero. Taking the external field to be zero is an excellent assumption for a real solenoid if its length is much greater than its diameter and if we consider external points such as point P that are not at either end of the solenoid. The direction of the magnetic field along the solenoid axis is given by a curled–straight right-hand rule: Grasp the solenoid with your right hand so that your fingers follow the direction of the current in the windings; your extended right thumb then points in the direction of the axial magnetic field.

Figure 29-19 shows the lines of $\vec{B}$ for a real solenoid. The spacing of these lines in the central region shows that the field inside the coil is fairly strong and uniform over the cross section of the coil. The external field, however, is relatively weak.

Ampere's Law. Let us now apply Ampere's law,

$$\oint \vec{B} \cdot d\vec{s} = \mu_0 i_{\text{enc}}, \qquad (29\text{-}21)$$

to the ideal solenoid of Fig. 29-20, where $\vec{B}$ is uniform within the solenoid and zero outside it, using the rectangular Amperian loop *abcda*. We write $\oint \vec{B} \cdot d\vec{s}$ as the sum of four integrals, one for each loop segment:

$$\oint \vec{B} \cdot d\vec{s} = \int_a^b \vec{B} \cdot d\vec{s} + \int_b^c \vec{B} \cdot d\vec{s} + \int_c^d \vec{B} \cdot d\vec{s} + \int_d^a \vec{B} \cdot d\vec{s}. \qquad (29\text{-}22)$$

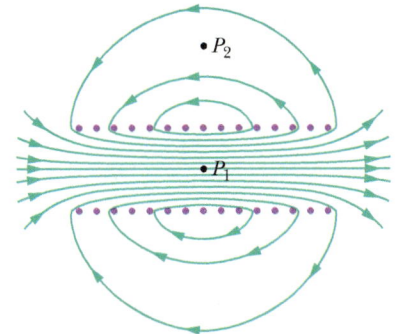

Figure 29-19 Magnetic field lines for a real solenoid of finite length. The field is strong and uniform at interior points such as P_1 but relatively weak at external points such as P_2.

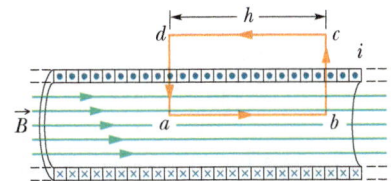

Figure 29-20 Application of Ampere's law to a section of a long ideal solenoid carrying a current i. The Amperian loop is the rectangle *abcda*.

The first integral on the right of Eq. 29-22 is Bh, where B is the magnitude of the uniform field $\vec{B}$ inside the solenoid and h is the (arbitrary) length of the segment from a to b. The second and fourth integrals are zero because for every element ds of these segments, $\vec{B}$ either is perpendicular to ds or is zero, and thus the product $\vec{B} \cdot d\vec{s}$ is zero. The third integral, which is taken along a segment that lies outside the solenoid, is zero because $B = 0$ at all external points. Thus, $\oint \vec{B} \cdot d\vec{s}$ for the entire rectangular loop has the value Bh.

Net Current. The net current i_{enc} encircled by the rectangular Amperian loop in Fig. 29-20 is not the same as the current i in the solenoid windings because the windings pass more than once through this loop. Let n be the number of turns per unit length of the solenoid; then the loop encloses nh turns and

$$i_{enc} = i(nh).$$

Ampere's law then gives us

$$Bh = \mu_0 i n h$$

or
$$B = \mu_0 i n \quad \text{(ideal solenoid)}. \tag{29-23}$$

Although we derived Eq. 29-23 for an infinitely long ideal solenoid, it holds quite well for actual solenoids if we apply it only at interior points and well away from the solenoid ends. Equation 29-23 is consistent with the experimental fact that the magnetic field magnitude B within a solenoid does not depend on the diameter or the length of the solenoid and that B is uniform over the solenoidal cross section. A solenoid thus provides a practical way to set up a known uniform magnetic field for experimentation, just as a parallel-plate capacitor provides a practical way to set up a known uniform electric field.

Magnetic Field of a Toroid

Figure 29-21a shows a **toroid,** which we may describe as a (hollow) solenoid that has been curved until its two ends meet, forming a sort of hollow bracelet. What magnetic field $\vec{B}$ is set up inside the toroid (inside the hollow of the bracelet)? We can find out from Ampere's law and the symmetry of the bracelet.

From the symmetry, we see that the lines of $\vec{B}$ form concentric circles inside the toroid, directed as shown in Fig. 29-21b. Let us choose a concentric circle of radius r as an Amperian loop and traverse it in the clockwise direction. Ampere's law (Eq. 29-14) yields

$$(B)(2\pi r) = \mu_0 i N,$$

where i is the current in the toroid windings (and is positive for those windings enclosed by the Amperian loop) and N is the total number of turns. This gives

$$B = \frac{\mu_0 i N}{2\pi} \frac{1}{r} \quad \text{(toroid)}. \tag{29-24}$$

In contrast to the situation for a solenoid, B is not constant over the cross section of a toroid.

It is easy to show, with Ampere's law, that $B = 0$ for points outside an ideal toroid (as if the toroid were made from an ideal solenoid). The direction of the magnetic field within a toroid follows from our curled–straight right-hand rule: Grasp the toroid with the fingers of your right hand curled in the direction of the current in the windings; your extended right thumb points in the direction of the magnetic field.

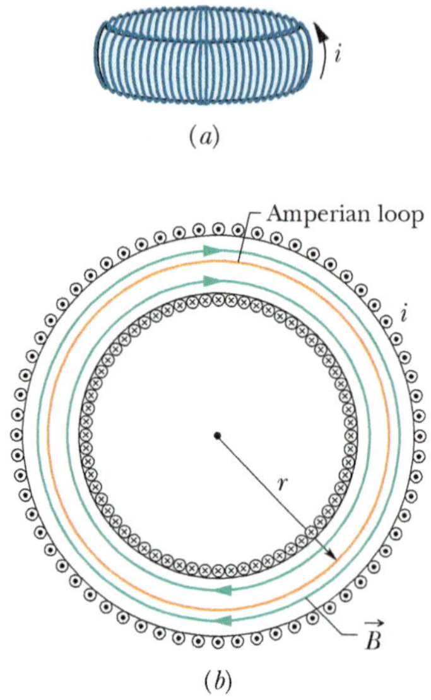

Figure 29-21 (a) A toroid carrying a current i. (b) A horizontal cross section of the toroid. The interior magnetic field (inside the bracelet-shaped tube) can be found by applying Ampere's law with the Amperian loop shown.

Sample Problem 29.04 **The field inside a solenoid (a long coil of current)**

A solenoid has length $L = 1.23$ m and inner diameter $d = 3.55$ cm, and it carries a current $i = 5.57$ A. It consists of five close-packed layers, each with 850 turns along length L. What is B at its center?

KEY IDEA

The magnitude B of the magnetic field along the solenoid's central axis is related to the solenoid's current i and number of turns per unit length n by Eq. 29-23 ($B = \mu_0 in$).

Calculation: Because B does not depend on the diameter of the windings, the value of n for five identical layers is simply five times the value for each layer. Equation 29-23 then tells us

$$B = \mu_0 in = (4\pi \times 10^{-7}\ \text{T} \cdot \text{m/A})(5.57\ \text{A})\frac{5 \times 850\ \text{turns}}{1.23\ \text{m}}$$

$$= 2.42 \times 10^{-2}\ \text{T} = 24.2\ \text{mT}. \qquad \text{(Answer)}$$

To a good approximation, this is the field magnitude throughout most of the solenoid.

PLUS Additional examples, video, and practice available at *WileyPLUS*

29-5 A CURRENT-CARRYING COIL AS A MAGNETIC DIPOLE

Learning Objectives

After reading this module, you should be able to . . .

29.22 Sketch the magnetic field lines of a flat coil that is carrying current.

29.23 For a current-carrying coil, apply the relationship between the dipole moment magnitude μ and the coil's current i, number of turns N, and area per turn A.

29.24 For a point along the central axis, apply the relationship between the magnetic field magnitude B, the magnetic moment μ, and the distance z from the center of the coil.

Key Idea

● The magnetic field produced by a current-carrying coil, which is a magnetic dipole, at a point P located a distance z along the coil's perpendicular central axis is parallel to the axis and is given by

$$\vec{B}(z) = \frac{\mu_0}{2\pi}\frac{\vec{\mu}}{z^3},$$

where $\vec{\mu}$ is the dipole moment of the coil. This equation applies only when z is much greater than the dimensions of the coil.

A Current-Carrying Coil as a Magnetic Dipole

So far we have examined the magnetic fields produced by current in a long straight wire, a solenoid, and a toroid. We turn our attention here to the field produced by a coil carrying a current. You saw in Module 28-8 that such a coil behaves as a magnetic dipole in that, if we place it in an external magnetic field $\vec{B}$, a torque $\vec{\tau}$ given by

$$\vec{\tau} = \vec{\mu} \times \vec{B} \qquad (29\text{-}25)$$

acts on it. Here $\vec{\mu}$ is the magnetic dipole moment of the coil and has the magnitude NiA, where N is the number of turns, i is the current in each turn, and A is the area enclosed by each turn. (*Caution:* Don't confuse the magnetic dipole moment $\vec{\mu}$ with the permeability constant μ_0.)

Recall that the direction of $\vec{\mu}$ is given by a curled–straight right-hand rule: Grasp the coil so that the fingers of your right hand curl around it in the direction of the current; your extended thumb then points in the direction of the dipole moment $\vec{\mu}$.

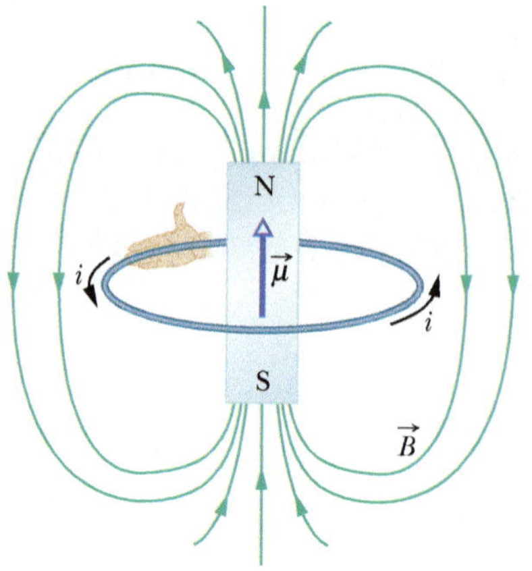

Figure 29-22 A current loop produces a magnetic field like that of a bar magnet and thus has associated north and south poles. The magnetic dipole moment $\vec{\mu}$ of the loop, its direction given by a curled–straight right-hand rule, points from the south pole to the north pole, in the direction of the field $\vec{B}$ within the loop.

Magnetic Field of a Coil

We turn now to the other aspect of a current-carrying coil as a magnetic dipole. What magnetic field does *it* produce at a point in the surrounding space? The problem does not have enough symmetry to make Ampere's law useful; so we must turn to the law of Biot and Savart. For simplicity, we first consider only a coil with a single circular loop and only points on its perpendicular central axis, which we take to be a z axis. We shall show that the magnitude of the magnetic field at such points is

$$B(z) = \frac{\mu_0 i R^2}{2(R^2 + z^2)^{3/2}}, \tag{29-26}$$

in which R is the radius of the circular loop and z is the distance of the point in question from the center of the loop. Furthermore, the direction of the magnetic field $\vec{B}$ is the same as the direction of the magnetic dipole moment $\vec{\mu}$ of the loop.

   Large z. For axial points far from the loop, we have $z \gg R$ in Eq. 29-26. With that approximation, the equation reduces to

$$B(z) \approx \frac{\mu_0 i R^2}{2z^3}.$$

Recalling that πR^2 is the area A of the loop and extending our result to include a coil of N turns, we can write this equation as

$$B(z) = \frac{\mu_0}{2\pi} \frac{NiA}{z^3}.$$

Further, because $\vec{B}$ and $\vec{\mu}$ have the same direction, we can write the equation in vector form, substituting from the identity $\mu = NiA$:

$$\vec{B}(z) = \frac{\mu_0}{2\pi} \frac{\vec{\mu}}{z^3} \quad \text{(current-carrying coil).} \tag{29-27}$$

   Thus, we have two ways in which we can regard a current-carrying coil as a magnetic dipole: (1) it experiences a torque when we place it in an external magnetic field; (2) it generates its own intrinsic magnetic field, given, for distant points along its axis, by Eq. 29-27. Figure 29-22 shows the magnetic field of a current loop; one side of the loop acts as a north pole (in the direction of $\vec{\mu}$)

and the other side as a south pole, as suggested by the lightly drawn magnet in the figure. If we were to place a current-carrying coil in an external magnetic field, it would tend to rotate just like a bar magnet would.

☑ Checkpoint 3

The figure here shows four arrangements of circular loops of radius r or $2r$, centered on vertical axes (perpendicular to the loops) and carrying identical currents in the directions indicated. Rank the arrangements according to the magnitude of the net magnetic field at the dot, midway between the loops on the central axis, greatest first.

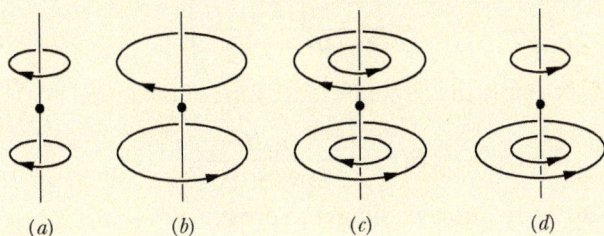

(a) (b) (c) (d)

Proof of Equation 29-26

Figure 29-23 shows the back half of a circular loop of radius R carrying a current i. Consider a point P on the central axis of the loop, a distance z from its plane. Let us apply the law of Biot and Savart to a differential element ds of the loop, located at the left side of the loop. The length vector $d\vec{s}$ for this element points perpendicularly out of the page. The angle θ between $d\vec{s}$ and $\hat{r}$ in Fig. 29-23 is $90°$; the plane formed by these two vectors is perpendicular to the plane of the page and contains both $\hat{r}$ and $d\vec{s}$. From the law of Biot and Savart (and the right-hand rule), the differential field $d\vec{B}$ produced at point P by the current in this element is perpendicular to this plane and thus is directed in the plane of the figure, perpendicular to $\hat{r}$, as indicated in Fig. 29-23.

Let us resolve $d\vec{B}$ into two components: $dB_{\parallel}$ along the axis of the loop and $dB_{\perp}$ perpendicular to this axis. From the symmetry, the vector sum of all the perpendicular components $dB_{\perp}$ due to all the loop elements ds is zero. This leaves only the axial (parallel) components $dB_{\parallel}$ and we have

$$B = \int dB_{\parallel}.$$

For the element $d\vec{s}$ in Fig. 29-23, the law of Biot and Savart (Eq. 29-1) tells us that the magnetic field at distance r is

$$dB = \frac{\mu_0}{4\pi}\frac{i\, ds \sin 90°}{r^2}.$$

We also have

$$dB_{\parallel} = dB \cos \alpha.$$

Combining these two relations, we obtain

$$dB_{\parallel} = \frac{\mu_0 i \cos \alpha\, ds}{4\pi r^2}. \tag{29-28}$$

Figure 29-23 shows that r and α are related to each other. Let us express each in terms of the variable z, the distance between point P and the center of the loop. The relations are

$$r = \sqrt{R^2 + z^2} \tag{29-29}$$

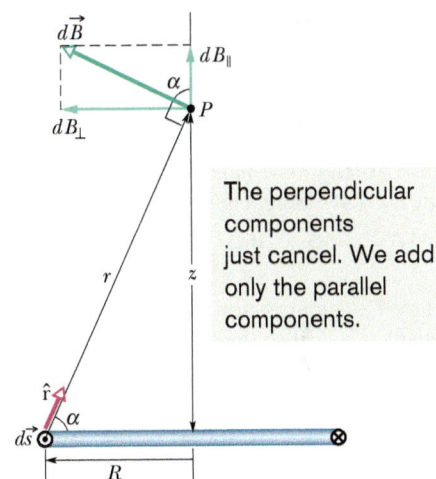

The perpendicular components just cancel. We add only the parallel components.

Figure 29-23 Cross section through a current loop of radius R. The plane of the loop is perpendicular to the page, and only the back half of the loop is shown. We use the law of Biot and Savart to find the magnetic field at point P on the central perpendicular axis of the loop.

and
$$\cos \alpha = \frac{R}{r} = \frac{R}{\sqrt{R^2 + z^2}}. \tag{29-30}$$

Substituting Eqs. 29-29 and 29-30 into Eq. 29-28, we find

$$dB_{\parallel} = \frac{\mu_0 iR}{4\pi(R^2 + z^2)^{3/2}} \, ds.$$

Note that i, R, and z have the same values for all elements ds around the loop; so when we integrate this equation, we find that

$$B = \int dB_{\parallel}$$

$$= \frac{\mu_0 iR}{4\pi(R^2 + z^2)^{3/2}} \int ds$$

or, because $\int ds$ is simply the circumference $2\pi R$ of the loop,

$$B(z) = \frac{\mu_0 iR^2}{2(R^2 + z^2)^{3/2}}.$$

This is Eq. 29-26, the relation we sought to prove.

Review & Summary

The Biot–Savart Law The magnetic field set up by a current-carrying conductor can be found from the *Biot–Savart law*. This law asserts that the contribution $d\vec{B}$ to the field produced by a current-length element $i \, d\vec{s}$ at a point P located a distance r from the current element is

$$d\vec{B} = \frac{\mu_0}{4\pi} \frac{i \, d\vec{s} \times \hat{r}}{r^2} \quad \text{(Biot–Savart law).} \tag{29-3}$$

Here $\hat{r}$ is a unit vector that points from the element toward P. The quantity μ_0, called the *permeability constant*, has the value

$$4\pi \times 10^{-7} \, \text{T·m/A} \approx 1.26 \times 10^{-6} \, \text{T·m/A}.$$

Magnetic Field of a Long Straight Wire For a long straight wire carrying a current i, the Biot–Savart law gives, for the magnitude of the magnetic field at a perpendicular distance R from the wire,

$$B = \frac{\mu_0 i}{2\pi R} \quad \text{(long straight wire).} \tag{29-4}$$

Magnetic Field of a Circular Arc The magnitude of the magnetic field at the center of a circular arc, of radius R and central angle ϕ (in radians), carrying current i, is

$$B = \frac{\mu_0 i\phi}{4\pi R} \quad \text{(at center of circular arc).} \tag{29-9}$$

Force Between Parallel Currents Parallel wires carrying currents in the same direction attract each other, whereas parallel wires carrying currents in opposite directions repel each other. The magnitude of the force on a length L of either wire is

$$F_{ba} = i_b L B_a \sin 90° = \frac{\mu_0 L i_a i_b}{2\pi d}, \tag{29-13}$$

where d is the wire separation, and i_a and i_b are the currents in the wires.

Ampere's Law **Ampere's law** states that

$$\oint \vec{B} \cdot d\vec{s} = \mu_0 i_{\text{enc}} \quad \text{(Ampere's law).} \tag{29-14}$$

The line integral in this equation is evaluated around a closed loop called an *Amperian loop*. The current i on the right side is the *net* current encircled by the loop. For some current distributions, Eq. 29-14 is easier to use than Eq. 29-3 to calculate the magnetic field due to the currents.

Fields of a Solenoid and a Toroid Inside a *long solenoid* carrying current i, at points not near its ends, the magnitude B of the magnetic field is

$$B = \mu_0 in \quad \text{(ideal solenoid),} \tag{29-23}$$

where n is the number of turns per unit length. Thus the internal magnetic field is uniform. Outside the solenoid, the magnetic field is approximately zero.

At a point inside a *toroid*, the magnitude B of the magnetic field is

$$B = \frac{\mu_0 iN}{2\pi} \frac{1}{r} \quad \text{(toroid),} \tag{29-24}$$

where r is the distance from the center of the toroid to the point.

Field of a Magnetic Dipole The magnetic field produced by a current-carrying coil, which is a *magnetic dipole*, at a point P located a distance z along the coil's perpendicular central axis is parallel to the axis and is given by

$$\vec{B}(z) = \frac{\mu_0}{2\pi} \frac{\vec{\mu}}{z^3}, \tag{29-27}$$

where $\vec{\mu}$ is the dipole moment of the coil. This equation applies only when z is much greater than the dimensions of the coil.

Questions

1 Figure 29-24 shows three circuits, each consisting of two radial lengths and two concentric circular arcs, one of radius r and the other of radius $R > r$. The circuits have the same current through them and the same angle between the two radial lengths. Rank the circuits according to the magnitude of the net magnetic field at the center, greatest first.

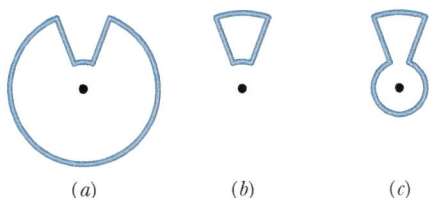

(a)   (b)   (c)

Figure 29-24 Question 1.

2 Figure 29-25 represents a snapshot of the velocity vectors of four electrons near a wire carrying current i. The four velocities have the same magnitude; velocity $\vec{v}_2$ is directed into the page. Electrons 1 and 2 are at the same distance from the wire, as are electrons 3 and 4. Rank the electrons according to the magnitudes of the magnetic forces on them due to current i, greatest first.

Figure 29-25 Question 2.

3 Figure 29-26 shows four arrangements in which long parallel wires carry equal currents directly into or out of the page at the corners of identical squares. Rank the arrangements according to the magnitude of the net magnetic field at the center of the square, greatest first.

(a)   (b)   (c)   (d)

Figure 29-26 Question 3.

4 Figure 29-27 shows cross sections of two long straight wires; the left-hand wire carries current i_1 directly out of the page. If the net magnetic field due to the two currents is to be zero at point P, (a) should the direction of current i_2 in the right-hand wire be directly into or out of the page and (b) should i_2 be greater than, less than, or equal to i_1?

Figure 29-27 Question 4.

5 Figure 29-28 shows three circuits consisting of straight radial lengths and concentric circular arcs (either half- or quarter-circles

(a)   (b)   (c)

Figure 29-28 Question 5.

of radii r, $2r$, and $3r$). The circuits carry the same current. Rank them according to the magnitude of the magnetic field produced at the center of curvature (the dot), greatest first.

6 Figure 29-29 gives, as a function of radial distance r, the magnitude B of the magnetic field inside and outside four wires (a, b, c, and d), each of which carries a current that is uniformly distributed across the wire's cross section. Overlapping portions of the plots (drawn slightly separated) are indicated by double labels. Rank the wires according to (a) radius, (b) the magnitude of the magnetic field on the surface, and (c) the value of the current, greatest first. (d) Is the magnitude of the current density in wire a greater than, less than, or equal to that in wire c?

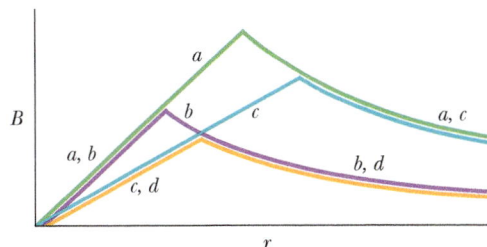

Figure 29-29 Question 6.

7 Figure 29-30 shows four circular Amperian loops (a, b, c, d) concentric with a wire whose current is directed out of the page. The current is uniform across the wire's circular cross section (the shaded region). Rank the loops according to the magnitude of $\oint \vec{B} \cdot d\vec{s}$ around each, greatest first.

Figure 29-30 Question 7.

8 Figure 29-31 shows four arrangements in which long, parallel, equally spaced wires carry equal currents directly into or out of the page. Rank the arrangements according to the magnitude of the net force on the central wire due to the currents in the other wires, greatest first.

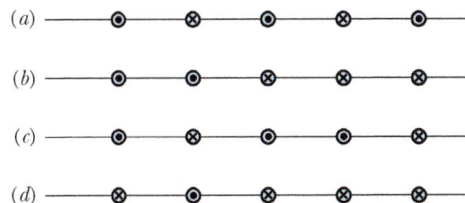

Figure 29-31 Question 8.

9 Figure 29-32 shows four circular Amperian loops (a, b, c, d) and, in cross section, four long circular conductors (the shaded regions), all of which are concentric. Three of the conductors are hollow cylinders; the central conductor is a solid cylinder. The currents in the conductors are, from smallest radius to largest radius, 4 A out of

Figure 29-32 Question 9.

the page, 9 A into the page, 5 A out of the page, and 3 A into the page. Rank the Amperian loops according to the magnitude of $\oint \vec{B} \cdot d\vec{s}$ around each, greatest first.

10 Figure 29-33 shows four identical currents i and five Amperian paths (a through e) encircling them. Rank the paths according to the value of $\oint \vec{B} \cdot d\vec{s}$ taken in the directions shown, most positive first.

Figure 29-33 Question 10.

11 Figure 29-34 shows three arrangements of three long straight wires carrying equal currents directly into or out of the page. (a) Rank the arrangements according to the magnitude of the net force on wire A due to the currents in the other wires, greatest first. (b) In arrangement 3, is the angle between the net force on wire A and the dashed line equal to, less than, or more than 45°?

Figure 29-34 Question 11.

Problems

Module 29-1 Magnetic Field Due to a Current

•1 A surveyor is using a magnetic compass 6.1 m below a power line in which there is a steady current of 100 A. (a) What is the magnetic field at the site of the compass due to the power line? (b) Will this field interfere seriously with the compass reading? The horizontal component of Earth's magnetic field at the site is 20 μT.

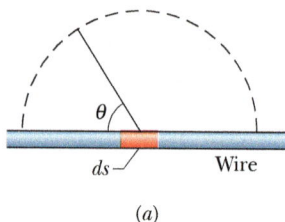

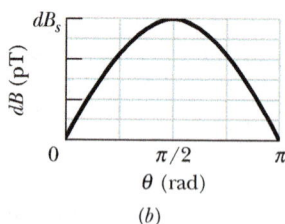

•2 Figure 29-35a shows an element of length $ds = 1.00 \ \mu$m in a very long straight wire carrying current. The current in that element sets up a differential magnetic field $d\vec{B}$ at points in the surrounding space. Figure 29-35b gives the magnitude dB of the field for points 2.5 cm from the element, as a function of angle θ between the wire and a straight line to the point. The vertical scale is set by $dB_s = 60.0$ pT. What is the magnitude of the magnetic field set up by the entire wire at perpendicular distance 2.5 cm from the wire?

Figure 29-35 Problem 2.

•3 **SSM** At a certain location in the Philippines, Earth's magnetic field of 39 μT is horizontal and directed due north. Suppose the net field is zero exactly 8.0 cm above a long, straight, horizontal wire that carries a constant current. What are the (a) magnitude and (b) direction of the current?

•4 A straight conductor carrying current $i = 5.0$ A splits into identical semicircular arcs as shown in Fig. 29-36. What is the magnetic field at the center C of the resulting circular loop?

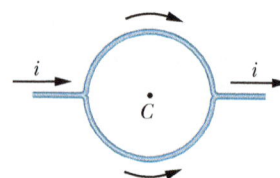

Figure 29-36 Problem 4.

•5 In Fig. 29-37, a current $i = 10$ A is set up in a long hairpin conductor formed by bending a wire into a semicircle of radius $R = 5.0$ mm. Point b is midway between the straight sections and so distant from the semicircle that each straight section can be approximated as being an infinite wire. What are the (a) magnitude and (b) direction (into or out of the page) of $\vec{B}$ at a and the (c) magnitude and (d) direction of $\vec{B}$ at b?

Figure 29-37 Problem 5.

•6 In Fig. 29-38, point P is at perpendicular distance $R = 2.00$ cm from a very long straight wire carrying a current. The magnetic field $\vec{B}$ set up at point P is due to contributions from all the identical current-length elements $i \ d\vec{s}$ along the wire. What is the distance s to the

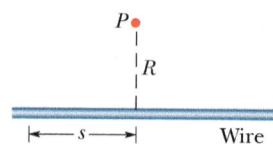

Figure 29-38 Problem 6.

element making (a) the greatest contribution to field $\vec{B}$ and (b) 10.0% of the greatest contribution?

•7 GO In Fig. 29-39, two circular arcs have radii $a = 13.5$ cm and $b = 10.7$ cm, subtend angle $\theta = 74.0°$, carry current $i = 0.411$ A, and share the same center of curvature P. What are the (a) magnitude and (b) direction (into or out of the page) of the net magnetic field at P?

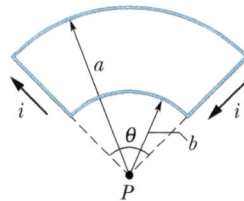

Figure 29-39 Problem 7.

•8 In Fig. 29-40, two semicircular arcs have radii $R_2 = 7.80$ cm and $R_1 = 3.15$ cm, carry current $i = 0.281$ A, and have the same center of curvature C. What are the (a) magnitude and (b) direction (into or out of the page) of the net magnetic field at C?

Figure 29-40 Problem 8.

•9 SSM Two long straight wires are parallel and 8.0 cm apart. They are to carry equal currents such that the magnetic field at a point halfway between them has magnitude 300 μT. (a) Should the currents be in the same or opposite directions? (b) How much current is needed?

•10 In Fig. 29-41, a wire forms a semicircle of radius $R = 9.26$ cm and two (radial) straight segments each of length $L = 13.1$ cm. The wire carries current $i = 34.8$ mA. What are the (a) magnitude and (b) direction (into or out of the page) of the net magnetic field at the semicircle's center of curvature C?

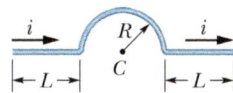

Figure 29-41 Problem 10.

•11 In Fig. 29-42, two long straight wires are perpendicular to the page and separated by distance $d_1 = 0.75$ cm. Wire 1 carries 6.5 A into the page. What are the (a) magnitude and (b) direction (into or out of the page) of the current in wire 2 if the net magnetic field due to the two currents is zero at point P located at distance $d_2 = 1.50$ cm from wire 2?

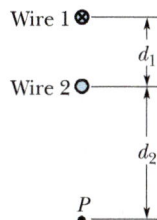

Figure 29-42 Problem 11.

•12 In Fig. 29-43, two long straight wires at separation $d = 16.0$ cm carry currents $i_1 = 3.61$ mA and $i_2 = 3.00i_1$ out of the page. (a) Where on the x axis is the net magnetic field equal to zero? (b) If the two currents are doubled, is the zero-field point shifted toward wire 1, shifted toward wire 2, or unchanged?

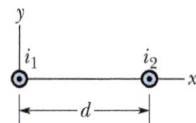

Figure 29-43 Problem 12.

••13 In Fig. 29-44, point P_1 is at distance $R = 13.1$ cm on the perpendicular bisector of a straight wire of length $L = 18.0$ cm carrying

Figure 29-44 Problems 13 and 17.

current $i = 58.2$ mA. (Note that the wire is *not* long.) What is the magnitude of the magnetic field at P_1 due to i?

••14 Equation 29-4 gives the magnitude B of the magnetic field set up by a current in an *infinitely long* straight wire, at a point P at perpendicular distance R from the wire. Suppose that point P is actually at perpendicular distance R from the midpoint of a wire with a *finite* length L. Using Eq. 29-4 to calculate B then results in a certain percentage error. What value must the ratio L/R exceed if the percentage error is to be less than 1.00%? That is, what L/R gives

$$\frac{(B \text{ from Eq. 29-4}) - (B \text{ actual})}{(B \text{ actual})}(100\%) = 1.00\%?$$

••15 Figure 29-45 shows two current segments. The lower segment carries a current of $i_1 = 0.40$ A and includes a semicircular arc with radius 5.0 cm, angle 180°, and center point P. The upper segment carries current $i_2 = 2i_1$ and includes a circular arc with radius 4.0 cm, angle 120°, and the same center point P. What are the (a) magnitude and (b) direction of the net magnetic field $\vec{B}$ at P for the indicated current directions? What are the (c) magnitude and (d) direction of $\vec{B}$ if i_1 is reversed?

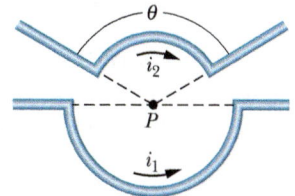

Figure 29-45 Problem 15.

••16 GO In Fig. 29-46, two concentric circular loops of wire carrying current in the same direction lie in the same plane. Loop 1 has radius 1.50 cm and carries 4.00 mA. Loop 2 has radius 2.50 cm and carries 6.00 mA. Loop 2 is to be rotated about a diameter while the net magnetic field $\vec{B}$ set up by the two loops at their common center is measured. Through what angle must loop 2 be rotated so that the magnitude of that net field is 100 nT?

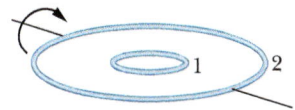

Figure 29-46 Problem 16.

••17 SSM In Fig. 29-44, point P_2 is at perpendicular distance $R = 25.1$ cm from one end of a straight wire of length $L = 13.6$ cm carrying current $i = 0.693$ A. (Note that the wire is *not* long.) What is the magnitude of the magnetic field at P_2?

••18 A current is set up in a wire loop consisting of a semicircle of radius 4.00 cm, a smaller concentric semicircle, and two radial straight lengths, all in the same plane. Figure 29-47a shows the arrangement but is not drawn to scale. The magnitude of the magnetic field produced at the center of curvature is 47.25 μT. The smaller semicircle is then flipped over (rotated) until the loop is again entirely in the same plane (Fig. 29-47b). The magnetic field produced at the (same) center of curvature now has magnitude 15.75 μT, and its direction is reversed from the initial magnetic field. What is the radius of the smaller semicircle?

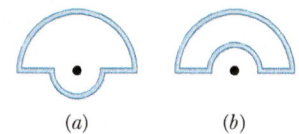

Figure 29-47 Problem 18.

••19 One long wire lies along an x axis and carries a current of 30 A in the positive x direction. A second long wire is perpendicular to the xy plane, passes through the point $(0, 4.0 \text{ m}, 0)$, and carries a current of 40 A in the positive z direction. What is the magnitude of the resulting magnetic field at the point $(0, 2.0 \text{ m}, 0)$?

••20 In Fig. 29-48, part of a long insulated wire carrying current $i = 5.78$ mA is bent into a circular section of radius $R = 1.89$ cm. In unit-vector notation, what is the magnetic field at the center of curvature C if the circular section (a) lies in the plane of the page as shown and (b) is perpendicular to the plane of the page after being rotated 90° counterclockwise as indicated?

Figure 29-48 Problem 20.

••21 GO Figure 29-49 shows two very long straight wires (in cross section) that each carry a current of 4.00 A directly out of the page. Distance $d_1 = 6.00$ m and distance $d_2 = 4.00$ m. What is the magnitude of the net magnetic field at point P, which lies on a perpendicular bisector to the wires?

Figure 29-49 Problem 21.

••22 GO Figure 29-50a shows, in cross section, two long, parallel wires carrying current and separated by distance L. The ratio i_1/i_2 of their currents is 4.00; the directions of the currents are not indicated. Figure 29-50b shows the y component B_y of their net magnetic field along the x axis to the right of wire 2. The vertical scale is set by $B_{ys} = 4.0$ nT, and the horizontal scale is set by $x_s = 20.0$ cm. (a) At what value of $x > 0$ is B_y maximum? (b) If $i_2 = 3$ mA, what is the value of that maximum? What is the direction (into or out of the page) of (c) i_1 and (d) i_2?

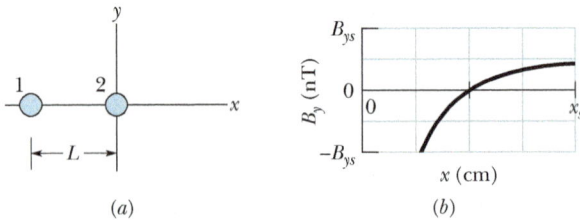

Figure 29-50 Problem 22.

••23 ILW Figure 29-51 shows a snapshot of a proton moving at velocity $\vec{v} = (-200 \text{ m/s})\hat{j}$ toward a long straight wire with current $i = 350$ mA. At the instant shown, the proton's distance from the wire is $d = 2.89$ cm. In unit-vector notation, what is the magnetic force on the proton due to the current?

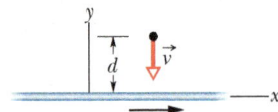

Figure 29-51 Problem 23.

••24 GO Figure 29-52 shows, in cross section, four thin wires that are parallel, straight, and very long. They carry identical currents in the directions indicated. Initially all four wires are at distance $d = 15.0$ cm from the origin of the coordinate system, where they create a net magnetic field $\vec{B}$. (a) To what value of x must you move wire 1 along the x axis in order to rotate $\vec{B}$ counterclockwise by 30°? (b) With wire 1 in that new position, to what value of x

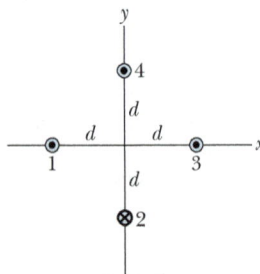

Figure 29-52 Problem 24.

must you move wire 3 along the x axis to rotate $\vec{B}$ by 30° back to its initial orientation?

••25 SSM A wire with current $i = 3.00$ A is shown in Fig. 29-53. Two semi-infinite straight sections, both tangent to the same circle, are connected by a circular arc that has a central angle θ and runs along the circumference of the circle. The arc and the two straight sections all lie in the same plane. If $B = 0$ at the circle's center, what is θ?

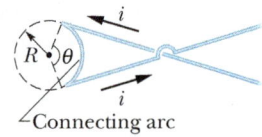

Figure 29-53 Problem 25.

••26 GO In Fig. 29-54a, wire 1 consists of a circular arc and two radial lengths; it carries current $i_1 = 0.50$ A in the direction indicated. Wire 2, shown in cross section, is long, straight, and perpendicular to the plane of the figure. Its distance from the center of the arc is equal to the radius R of the arc, and it carries a current i_2 that can be varied. The two currents set up a net magnetic field $\vec{B}$ at the center of the arc. Figure 29-54b gives the square of the field's magnitude B^2 plotted versus the square of the current i_2^2. The vertical scale is set by $B_s^2 = 10.0 \times 10^{-10}$ T². What angle is subtended by the arc?

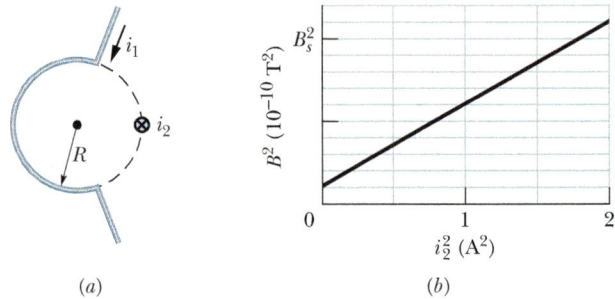

Figure 29-54 Problem 26.

••27 In Fig. 29-55, two long straight wires (shown in cross section) carry the currents $i_1 = 30.0$ mA and $i_2 = 40.0$ mA directly out of the page. They are equal distances from the origin, where they set up a magnetic field $\vec{B}$. To what value must current i_1 be changed in order to rotate $\vec{B}$ 20.0° clockwise?

Figure 29-55 Problem 27.

••28 GO Figure 29-56a shows two wires, each carrying a current. Wire 1 consists of a circular arc of

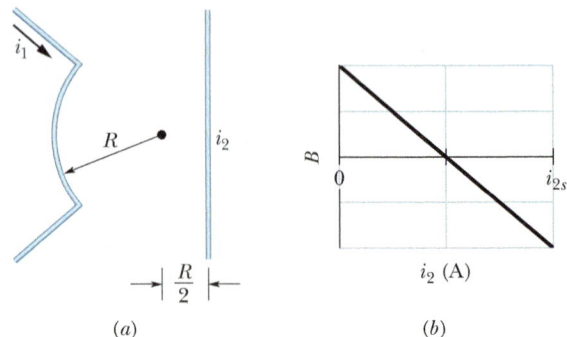

Figure 29-56 Problem 28.

radius R and two radial lengths; it carries current $i_1 = 2.0$ A in the direction indicated. Wire 2 is long and straight; it carries a current i_2 that can be varied; and it is at distance $R/2$ from the center of the arc. The net magnetic field $\vec{B}$ due to the two currents is measured at the center of curvature of the arc. Figure 29-56b is a plot of the component of $\vec{B}$ in the direction perpendicular to the figure as a function of current i_2. The horizontal scale is set by $i_{2s} = 1.00$ A. What is the angle subtended by the arc?

••29 SSM In Fig. 29-57, four long straight wires are perpendicular to the page, and their cross sections form a square of edge length $a = 20$ cm. The currents are out of the page in wires 1 and 4 and into the page in wires 2 and 3, and each wire carries 20 A. In unit-vector notation, what is the net magnetic field at the square's center?

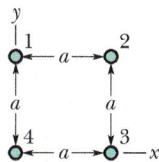

Figure 29-57   Problems 29, 37, and 40.

•••30 GO Two long straight thin wires with current lie against an equally long plastic cylinder, at radius $R = 20.0$ cm from the cylinder's central axis. Figure 29-58a shows, in cross section, the cylinder and wire 1 but not wire 2. With wire 2 fixed in place, wire 1 is moved around the cylinder, from angle $\theta_1 = 0°$ to angle $\theta_1 = 180°$, through the first and second quadrants of the xy coordinate system. The net magnetic field $\vec{B}$ at the center of the cylinder is measured as a function of θ_1. Figure 29-58b gives the x component B_x of that field as a function of θ_1 (the vertical scale is set by $B_{xs} = 6.0$ μT), and Fig. 29-58c gives the y component B_y (the vertical scale is set by $B_{ys} = 4.0$ μT). (a) At what angle θ_2 is wire 2 located? What are the (b) size and (c) direction (into or out of the page) of the current in wire 1 and the (d) size and (e) direction of the current in wire 2?

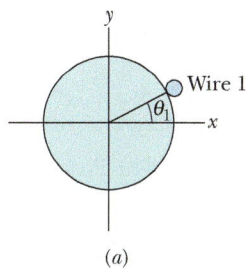

(a)

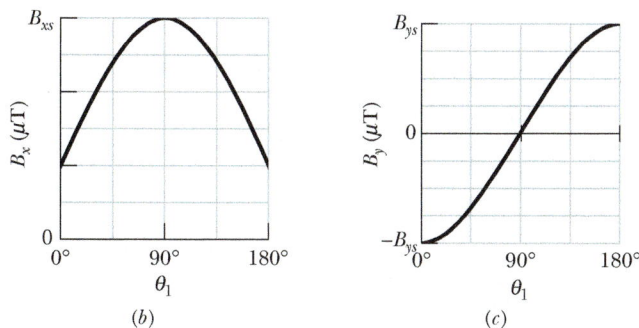

(b)   (c)

Figure 29-58   Problem 30.

•••31 In Fig. 29-59, length a is 4.7 cm (short) and current i is 13 A. What are the (a) magnitude and (b) direction (into or out of the page) of the magnetic field at point P?

•••32 GO The current-carrying wire loop in Fig. 29-60a lies all in one plane and consists of a semicircle of radius 10.0 cm, a smaller semicircle with the same center, and two radial lengths. The smaller semicircle is rotated out of that plane by angle θ, until it is perpendicular to the plane (Fig. 29-60b). Figure 29-60c gives the magnitude of the net magnetic field at the center of curvature versus angle θ. The vertical scale is set by $B_a = 10.0$ μT and $B_b = 12.0$ μT. What is the radius of the smaller semicircle?

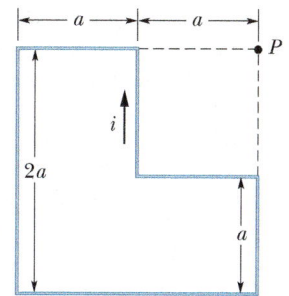

Figure 29-59   Problem 31.

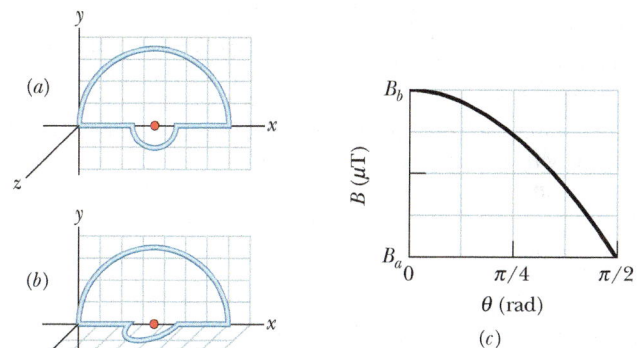

Figure 29-60   Problem 32.

•••33 SSM ILW Figure 29-61 shows a cross section of a long thin ribbon of width $w = 4.91$ cm that is carrying a uniformly distributed total current $i = 4.61$ μA into the page. In unit-vector notation, what is the magnetic field $\vec{B}$ at a point P in the plane of the ribbon at a distance $d = 2.16$ cm from its edge? (*Hint:* Imagine the ribbon as being constructed from many long, thin, parallel wires.)

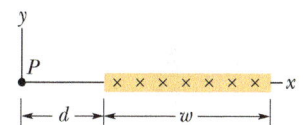

Figure 29-61   Problem 33.

•••34 GO Figure 29-62 shows, in cross section, two long straight wires held against a plastic cylinder of radius 20.0 cm. Wire 1 carries current $i_1 = 60.0$ mA out of the page and is fixed in place at the left side of the cylinder. Wire 2 carries current $i_2 = 40.0$ mA out of the page and can be moved around the cylinder. At what (positive) angle θ_2 should wire 2 be positioned such that, at the origin, the net magnetic field due to the two currents has magnitude 80.0 nT?

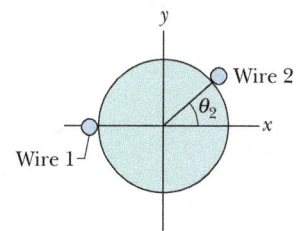

Figure 29-62   Problem 34.

Module 29-2 Force Between Two Parallel Currents

•35 SSM Figure 29-63 shows wire 1 in cross section; the wire is long

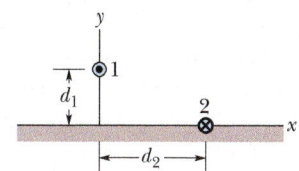

Figure 29-63   Problem 35.

and straight, carries a current of 4.00 mA out of the page, and is at distance $d_1 = 2.40$ cm from a surface. Wire 2, which is parallel to wire 1 and also long, is at horizontal distance $d_2 = 5.00$ cm from wire 1 and carries a current of 6.80 mA into the page. What is the x component of the magnetic force *per unit length* on wire 2 due to wire 1?

••36 In Fig. 29-64, five long parallel wires in an xy plane are separated by distance $d = 8.00$ cm, have lengths of 10.0 m, and carry identical currents of 3.00 A out of the page. Each wire experiences a magnetic force due to the currents in the other wires. In unit-vector notation, what is the net magnetic force on (a) wire 1, (b) wire 2, (c) wire 3, (d) wire 4, and (e) wire 5?

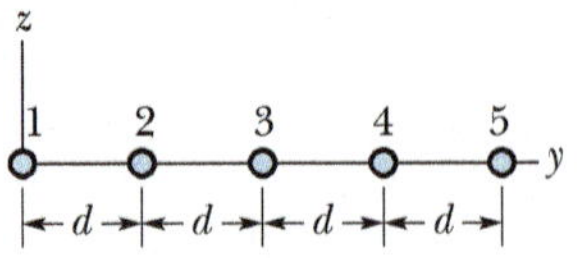

Figure 29-64 Problems 36 and 39.

••37 GO In Fig. 29-57, four long straight wires are perpendicular to the page, and their cross sections form a square of edge length $a = 13.5$ cm. Each wire carries 7.50 A, and the currents are out of the page in wires 1 and 4 and into the page in wires 2 and 3. In unit-vector notation, what is the net magnetic force *per meter of wire length* on wire 4?

••38 GO Figure 29-65a shows, in cross section, three current-carrying wires that are long, straight, and parallel to one another. Wires 1 and 2 are fixed in place on an x axis, with separation d. Wire 1 has a current of 0.750 A, but the direction of the current is not given. Wire 3, with a current of 0.250 A out of the page, can be moved along the x axis to the right of wire 2. As wire 3 is moved, the magnitude of the net magnetic force $\vec{F}_2$ on wire 2 due to the currents in wires 1 and 3 changes. The x component of that force is F_{2x} and the value per unit length of wire 2 is F_{2x}/L_2. Figure 29-65b gives F_{2x}/L_2 versus the position x of wire 3. The plot has an asymptote $F_{2x}/L_2 = -0.627$ μN/m as $x \to \infty$. The horizontal scale is set by $x_s = 12.0$ cm. What are the (a) size and (b) direction (into or out of the page) of the current in wire 2?

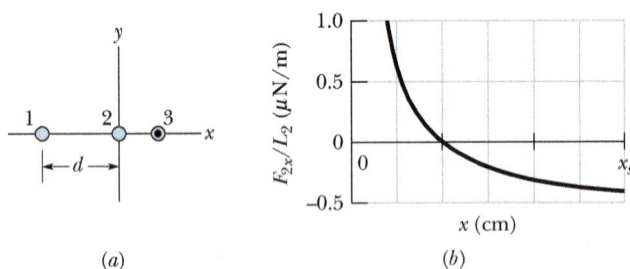

Figure 29-65 Problem 38.

••39 GO In Fig. 29-64, five long parallel wires in an xy plane are separated by distance $d = 50.0$ cm. The currents into the page are $i_1 = 2.00$ A, $i_3 = 0.250$ A, $i_4 = 4.00$ A, and $i_5 = 2.00$ A; the current out of the page is $i_2 = 4.00$ A. What is the magnitude of the net force *per unit length* acting on wire 3 due to the currents in the other wires?

••40 In Fig. 29-57, four long straight wires are perpendicular to the page, and their cross sections form a square of edge length $a = 8.50$ cm. Each wire carries 15.0 A, and all the currents are out of the page. In unit-vector notation, what is the net magnetic force *per meter of wire length* on wire 1?

•••41 ILW In Fig. 29-66, a long straight wire carries a current $i_1 = 30.0$ A and a rectangular loop carries current $i_2 = 20.0$ A. Take the dimensions to be $a = 1.00$ cm, $b = 8.00$ cm, and $L = 30.0$ cm. In unit-vector notation, what is the net force on the loop due to i_1?

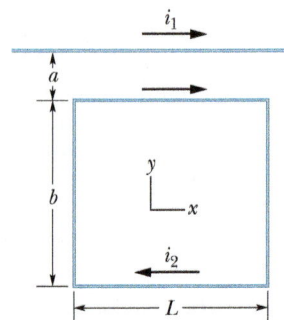

Figure 29-66 Problem 41.

Module 29-3 Ampere's Law

•42 In a particular region there is a uniform current density of 15 A/m² in the positive z direction. What is the value of $\oint \vec{B} \cdot d\vec{s}$ when that line integral is calculated along a closed path consisting of the three straight-line segments from (x, y, z) coordinates $(4d, 0, 0)$ to $(4d, 3d, 0)$ to $(0, 0, 0)$ to $(4d, 0, 0)$, where $d = 20$ cm?

•43 Figure 29-67 shows a cross section across a diameter of a long cylindrical conductor of radius $a = 2.00$ cm carrying uniform current 170 A. What is the magnitude of the current's magnetic field at radial distance (a) 0, (b) 1.00 cm, (c) 2.00 cm (wire's surface), and (d) 4.00 cm?

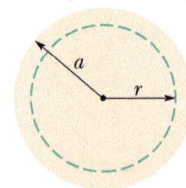

Figure 29-67 Problem 43.

•44 Figure 29-68 shows two closed paths wrapped around two conducting loops carrying currents $i_1 = 5.0$ A and $i_2 = 3.0$ A. What is the value of the integral $\oint \vec{B} \cdot d\vec{s}$ for (a) path 1 and (b) path 2?

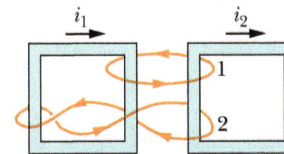

Figure 29-68 Problem 44.

•45 SSM Each of the eight conductors in Fig. 29-69 carries 2.0 A of current into or out of the page. Two paths are indicated for the line integral $\oint \vec{B} \cdot d\vec{s}$. What is the value of the integral for (a) path 1 and (b) path 2?

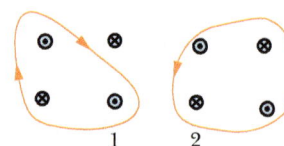

Figure 29-69 Problem 45.

•46 Eight wires cut the page perpendicularly at the points shown in Fig. 29-70. A wire labeled with the integer k ($k = 1, 2, \ldots, 8$) carries the current ki, where $i = 4.50$ mA. For those wires with odd k, the current is out of the page; for those with even k, it is into the page. Evaluate $\oint \vec{B} \cdot d\vec{s}$ along the closed path indicated and in the direction shown.

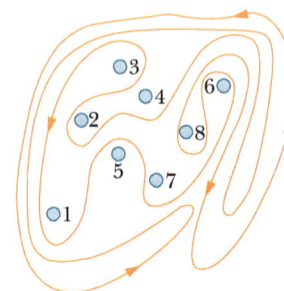

Figure 29-70 Problem 46.

••47 ILW The current density $\vec{J}$ inside a long, solid, cylindrical wire of radius $a = 3.1$ mm is in the direction of the central axis, and its magnitude varies linearly with radial distance r from the axis according to $J = J_0 r/a$, where $J_0 =$

310 A/m². Find the magnitude of the magnetic field at (a) $r = 0$, (b) $r = a/2$, and (c) $r = a$.

••48 In Fig. 29-71, a long circular pipe with outside radius $R = 2.6$ cm carries a (uniformly distributed) current $i = 8.00$ mA into the page. A wire runs parallel to the pipe at a distance of $3.00R$ from center to center. Find the (a) magnitude and (b) direction (into or out of the page) of the current in the wire such that the net magnetic field at point P has the same magnitude as the net magnetic field at the center of the pipe but is in the opposite direction.

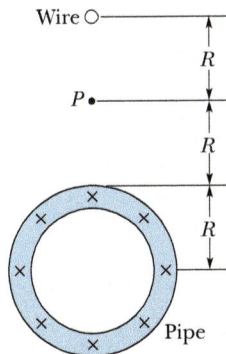

Figure 29-71
Problem 48.

Module 29-4 Solenoids and Toroids

•49 A toroid having a square cross section, 5.00 cm on a side, and an inner radius of 15.0 cm has 500 turns and carries a current of 0.800 A. (It is made up of a square solenoid—instead of a round one as in Fig. 29-17—bent into a doughnut shape.) What is the magnetic field inside the toroid at (a) the inner radius and (b) the outer radius?

•50 A solenoid that is 95.0 cm long has a radius of 2.00 cm and a winding of 1200 turns; it carries a current of 3.60 A. Calculate the magnitude of the magnetic field inside the solenoid.

•51 A 200-turn solenoid having a length of 25 cm and a diameter of 10 cm carries a current of 0.29 A. Calculate the magnitude of the magnetic field $\vec{B}$ inside the solenoid.

•52 A solenoid 1.30 m long and 2.60 cm in diameter carries a current of 18.0 A. The magnetic field inside the solenoid is 23.0 mT. Find the length of the wire forming the solenoid.

••53 A long solenoid has 100 turns/cm and carries current i. An electron moves within the solenoid in a circle of radius 2.30 cm perpendicular to the solenoid axis. The speed of the electron is $0.0460c$ (c = speed of light). Find the current i in the solenoid.

••54 An electron is shot into one end of a solenoid. As it enters the uniform magnetic field within the solenoid, its speed is 800 m/s and its velocity vector makes an angle of 30° with the central axis of the solenoid. The solenoid carries 4.0 A and has 8000 turns along its length. How many revolutions does the electron make along its helical path within the solenoid by the time it emerges from the solenoid's opposite end? (In a real solenoid, where the field is not uniform at the two ends, the number of revolutions would be slightly less than the answer here.)

••55 SSM ILW WWW A long solenoid with 10.0 turns/cm and a radius of 7.00 cm carries a current of 20.0 mA. A current of 6.00 A exists in a straight conductor located along the central axis of the solenoid. (a) At what radial distance from the axis will the direction of the resulting magnetic field be at 45.0° to the axial direction? (b) What is the magnitude of the magnetic field there?

Module 29-5 A Current-Carrying Coil as a Magnetic Dipole

•56 Figure 29-72 shows an arrangement known as a Helmholtz coil. It consists of two circular coaxial coils, each of 200 turns and radius $R = 25.0$ cm, separated by a distance

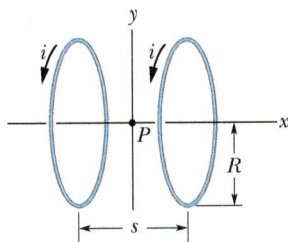

Figure 29-72
Problem 56.

$s = R$. The two coils carry equal currents $i = 12.2$ mA in the same direction. Find the magnitude of the net magnetic field at P, midway between the coils.

•57 SSM A student makes a short electromagnet by winding 300 turns of wire around a wooden cylinder of diameter $d = 5.0$ cm. The coil is connected to a battery producing a current of 4.0 A in the wire. (a) What is the magnitude of the magnetic dipole moment of this device? (b) At what axial distance $z \gg d$ will the magnetic field have the magnitude 5.0 μT (approximately one-tenth that of Earth's magnetic field)?

•58 Figure 29-73a shows a length of wire carrying a current i and bent into a circular coil of one turn. In Fig. 29-73b the same length of wire has been bent to give a coil of two turns, each of half the original radius. (a) If B_a and B_b are the magnitudes of the magnetic fields at the centers of the two coils, what is the ratio B_b/B_a? (b) What is the ratio μ_b/μ_a of the dipole moment magnitudes of the coils?

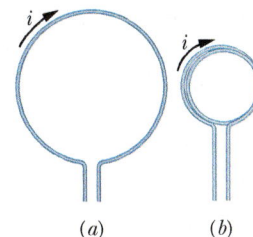

Figure 29-73 Problem 58.

•59 SSM What is the magnitude of the magnetic dipole moment $\vec{\mu}$ of the solenoid described in Problem 51?

••60 GO In Fig. 29-74a, two circular loops, with different currents but the same radius of 4.0 cm, are centered on a y axis. They are initially separated by distance $L = 3.0$ cm, with loop 2 positioned at the origin of the axis. The currents in the two loops produce a net magnetic field at the origin, with y component B_y. That component is to be measured as loop 2 is gradually moved in the positive direction of the y axis. Figure 29-74b gives B_y as a function of the position y of loop 2. The curve approaches an asymptote of $B_y = 7.20$ μT as $y \to \infty$. The horizontal scale is set by $y_s = 10.0$ cm. What are (a) current i_1 in loop 1 and (b) current i_2 in loop 2?

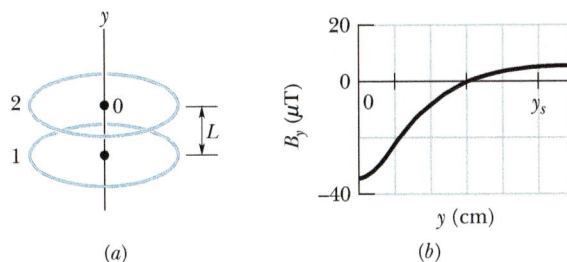

Figure 29-74 Problem 60.

••61 A circular loop of radius 12 cm carries a current of 15 A. A flat coil of radius 0.82 cm, having 50 turns and a current of 1.3 A, is concentric with the loop. The plane of the loop is perpendicular to the plane of the coil. Assume the loop's magnetic field is uniform across the coil. What is the magnitude of (a) the magnetic field produced by the loop at its center and (b) the torque on the coil due to the loop?

••62 In Fig. 29-75, current $i = 56.2$ mA is set up in a loop having two radial lengths and two semicir-

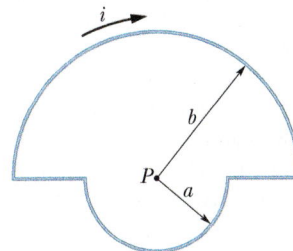

Figure 29-75 Problem 62.

cles of radii $a = 5.72$ cm and $b = 9.36$ cm with a common center P. What are the (a) magnitude and (b) direction (into or out of the page) of the magnetic field at P and the (c) magnitude and (d) direction of the loop's magnetic dipole moment?

••63 In Fig. 29-76, a conductor carries 6.0 A along the closed path *abcdefgha* running along 8 of the 12 edges of a cube of edge length 10 cm. (a) Taking the path to be a combination of three square current loops (*bcfgb*, *abgha*, and *cdefc*), find the net magnetic moment of the path in unit-vector notation. (b) What is the magnitude of the net magnetic field at the *xyz* coordinates of $(0, 5.0 \text{ m}, 0)$?

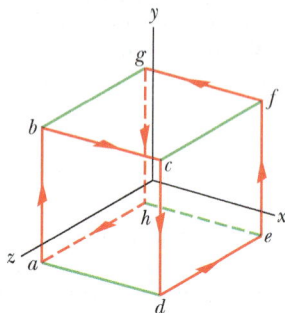

Figure 29-76 Problem 63.

Additional Problems

64 In Fig. 29-77, a closed loop carries current $i = 200$ mA. The loop consists of two radial straight wires and two concentric circular arcs of radii 2.00 m and 4.00 m. The angle θ is $\pi/4$ rad. What are the (a) magnitude and (b) direction (into or out of the page) of the net magnetic field at the center of curvature P?

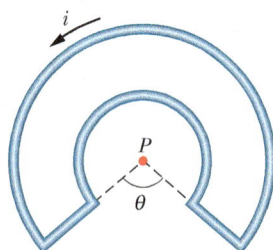

65 A cylindrical cable of radius 8.00 mm carries a current of 25.0 A, uniformly spread over its cross-sectional area. At what distance from the center of the wire is there a point within the wire where the magnetic field magnitude is 0.100 mT?

Figure 29-77 Problem 64.

66 Two long wires lie in an *xy* plane, and each carries a current in the positive direction of the *x* axis. Wire 1 is at $y = 10.0$ cm and carries 6.00 A; wire 2 is at $y = 5.00$ cm and carries 10.0 A. (a) In unit-vector notation, what is the net magnetic field $\vec{B}$ at the origin? (b) At what value of y does $\vec{B} = 0$? (c) If the current in wire 1 is reversed, at what value of y does $\vec{B} = 0$?

67 Two wires, both of length L, are formed into a circle and a square, and each carries current i. Show that the square produces a greater magnetic field at its center than the circle produces at its center.

68 A long straight wire carries a current of 50 A. An electron, traveling at 1.0×10^7 m/s, is 5.0 cm from the wire. What is the magnitude of the magnetic force on the electron if the electron velocity is directed (a) toward the wire, (b) parallel to the wire in the direction of the current, and (c) perpendicular to the two directions defined by (a) and (b)?

69 Three long wires are parallel to a *z* axis, and each carries a current of 10 A in the positive *z* direction. Their points of intersection with the *xy* plane form an equilateral triangle with sides of 50 cm, as shown in Fig. 29-78. A fourth wire (wire *b*) passes through the midpoint of the base of the triangle and is parallel to the other three wires. If the net magnetic force on wire *a* is zero, what are the (a) size and (b) direction ($+z$ or $-z$) of the current in wire *b*?

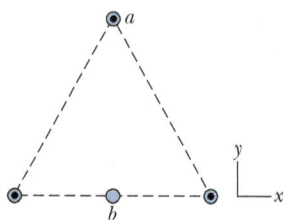

Figure 29-78 Problem 69.

70 Figure 29-79 shows a closed loop with current $i = 2.00$ A. The loop consists of a half-circle of radius 4.00 m, two quarter-circles each of radius 2.00 m, and three radial straight wires. What is the magnitude of the net magnetic field at the common center of the circular sections?

Figure 29-79 Problem 70.

71 A 10-gauge bare copper wire (2.6 mm in diameter) can carry a current of 50 A without overheating. For this current, what is the magnitude of the magnetic field at the surface of the wire?

72 A long vertical wire carries an unknown current. Coaxial with the wire is a long, thin, cylindrical conducting surface that carries a current of 30 mA upward. The cylindrical surface has a radius of 3.0 mm. If the magnitude of the magnetic field at a point 5.0 mm from the wire is 1.0 μT, what are the (a) size and (b) direction of the current in the wire?

73 Figure 29-80 shows a cross section of a long cylindrical conductor of radius $a = 4.00$ cm containing a long cylindrical hole of radius $b = 1.50$ cm. The central axes of the cylinder and hole are parallel and are distance $d = 2.00$ cm apart; current $i = 5.25$ A is uniformly distributed over the tinted area. (a) What is the magnitude of the magnetic field at the center of the hole? (b) Discuss the two special cases $b = 0$ and $d = 0$.

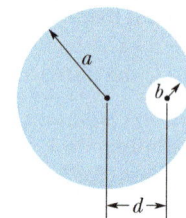

Figure 29-80 Problem 73.

74 The magnitude of the magnetic field at a point 88.0 cm from the central axis of a long straight wire is 7.30 μT. What is the current in the wire?

75 **SSM** Figure 29-81 shows a wire segment of length $\Delta s = 3.0$ cm, centered at the origin, carrying current $i = 2.0$ A in the positive *y* direction (as part of some complete circuit). To calculate the magnitude of the magnetic field $\vec{B}$ produced by the segment at a point several meters from the origin, we can use $B = (\mu_0/4\pi)i\,\Delta s\,(\sin\theta)/r^2$ as the Biot–Savart law. This is because r and θ are essentially constant over the segment. Calculate $\vec{B}$ (in unit-vector notation) at the (x, y, z) coordinates (a) $(0, 0, 5.0 \text{ m})$, (b) $(0, 6.0 \text{ m}, 0)$, (c) $(7.0 \text{ m}, 7.0 \text{ m}, 0)$, and (d) $(-3.0 \text{ m}, -4.0 \text{ m}, 0)$.

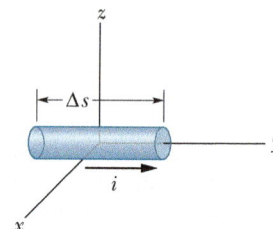

Figure 29-81 Problem 75.

76 **GO** Figure 29-82 shows, in cross section, two long parallel wires spaced by distance $d = 10.0$ cm; each carries 100 A, out of the page in wire 1. Point P is on a perpendicular bisector of the line connecting the wires. In unit-vector notation, what is the net magnetic field at P if the current in wire 2 is (a) out of the page and (b) into the page?

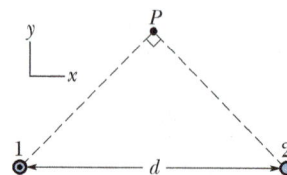

Figure 29-82 Problem 76.

77 In Fig. 29-83, two infinitely long wires carry equal currents i. Each follows a 90° arc on the circumference of the same circle of radius R. Show that the magnetic field $\vec{B}$ at the center of the circle is the same as the field $\vec{B}$ a distance R below an infinite straight wire carrying a current i to the left.

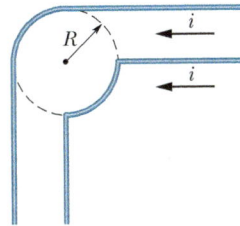

Figure 29-83
Problem 77.

78 A long wire carrying 100 A is perpendicular to the magnetic field lines of a uniform magnetic field of magnitude 5.0 mT. At what distance from the wire is the net magnetic field equal to zero?

79 A long, hollow, cylindrical conductor (with inner radius 2.0 mm and outer radius 4.0 mm) carries a current of 24 A distributed uniformly across its cross section. A long thin wire that is coaxial with the cylinder carries a current of 24 A in the opposite direction. What is the magnitude of the magnetic field (a) 1.0 mm, (b) 3.0 mm, and (c) 5.0 mm from the central axis of the wire and cylinder?

80 A long wire is known to have a radius greater than 4.0 mm and to carry a current that is uniformly distributed over its cross section. The magnitude of the magnetic field due to that current is 0.28 mT at a point 4.0 mm from the axis of the wire, and 0.20 mT at a point 10 mm from the axis of the wire. What is the radius of the wire?

81 SSM Figure 29-84 shows a cross section of an infinite conducting sheet carrying a current per unit x-length of λ; the current emerges perpendicularly out of the page. (a) Use the Biot–Savart law and symmetry to show that for all points P above the sheet and all points P' below it, the magnetic field $\vec{B}$ is parallel to the sheet and directed as shown. (b) Use Ampere's law to prove that $B = \frac{1}{2}\mu_0\lambda$ at all points P and P'.

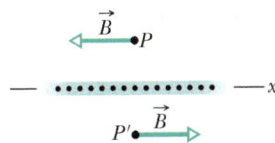

Figure 29-84 Problem 81.

82 Figure 29-85 shows, in cross section, two long parallel wires that are separated by distance $d = 18.6$ cm. Each carries 4.23 A, out of the page in wire 1 and into the page in wire 2. In unit-vector notation, what is the net magnetic field at point P at distance $R = 34.2$ cm, due to the two currents?

Figure 29-85 Problem 82.

83 SSM In unit-vector notation, what is the magnetic field at point P in Fig. 29-86 if $i = 10$ A and $a = 8.0$ cm? (Note that the wires are *not* long.)

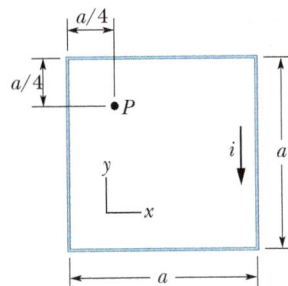

84 Three long wires all lie in an xy plane parallel to the x axis. They are spaced equally, 10 cm apart. The two outer wires each carry a current of 5.0 A in the positive x direction. What is the magnitude of the force on a 3.0 m section of either of the outer wires if the current in the cen-

Figure 29-86 Problem 83.

ter wire is 3.2 A (a) in the positive x direction and (b) in the negative x direction?

85 SSM Figure 29-87 shows a cross section of a hollow cylindrical conductor of radii a and b, carrying a uniformly distributed current i. (a) Show that the magnetic field magnitude $B(r)$ for the radial distance r in the range $b < r < a$ is given by

$$B = \frac{\mu_0 i}{2\pi(a^2 - b^2)} \frac{r^2 - b^2}{r}.$$

Figure 29-87
Problem 85.

(b) Show that when $r = a$, this equation gives the magnetic field magnitude B at the surface of a long straight wire carrying current i; when $r = b$, it gives zero magnetic field; and when $b = 0$, it gives the magnetic field inside a solid conductor of radius a carrying current i. (c) Assume that $a = 2.0$ cm, $b = 1.8$ cm, and $i = 100$ A, and then plot $B(r)$ for the range $0 < r < 6$ cm.

86 Show that the magnitude of the magnetic field produced at the center of a rectangular loop of wire of length L and width W, carrying a current i, is

$$B = \frac{2\mu_0 i}{\pi} \frac{(L^2 + W^2)^{1/2}}{LW}.$$

87 Figure 29-88 shows a cross section of a long conducting coaxial cable and gives its radii (a, b, c). Equal but opposite currents i are uniformly distributed in the two conductors. Derive expressions for $B(r)$ with radial distance r in the ranges (a) $r < c$, (b) $c < r < b$, (c) $b < r < a$, and (d) $r > a$. (e) Test these expressions for all the special cases that occur to you. (f) Assume that $a = 2.0$ cm, $b = 1.8$ cm, $c = 0.40$ cm, and $i = 120$ A and plot the function $B(r)$ over the range $0 < r < 3$ cm.

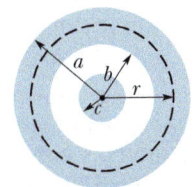

Figure 29-88
Problem 87.

88 Figure 29-89 is an idealized schematic drawing of a rail gun. Projectile P sits between two wide rails of circular cross section; a source of current sends current through the rails and through the (conducting) projectile (a fuse is not used). (a) Let w be the distance between the rails, R the radius of each rail, and i the current. Show that the force on the projectile is directed to the right along the rails and is given approximately by

$$F = \frac{i^2\mu_0}{2\pi}\ln\frac{w + R}{R}.$$

(b) If the projectile starts from the left end of the rails at rest, find the speed v at which it is expelled at the right. Assume that $i = 450$ kA, $w = 12$ mm, $R = 6.7$ cm, $L = 4.0$ m, and the projectile mass is 10 g.

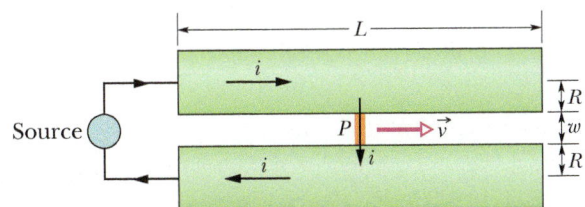

Figure 29-89 Problem 88.

Induction and Inductance

30-1 FARADAY'S LAW AND LENZ'S LAW

Learning Objectives

After reading this module, you should be able to . . .

30.01 Identify that the amount of magnetic field piercing a surface (not skimming along the surface) is the magnetic flux Φ_B through the surface.

30.02 Identify that an area vector for a flat surface is a vector that is perpendicular to the surface and that has a magnitude equal to the area of the surface.

30.03 Identify that any surface can be divided into area elements (patch elements) that are each small enough and flat enough for an area vector $d\vec{A}$ to be assigned to it, with the vector perpendicular to the element and having a magnitude equal to the area of the element.

30.04 Calculate the magnetic flux Φ_B through a surface by integrating the dot product of the magnetic field vector $\vec{B}$ and the area vector $d\vec{A}$ (for patch elements) over the surface, in magnitude-angle notation and unit-vector notation.

30.05 Identify that a current is induced in a conducting loop while the number of magnetic field lines intercepted by the loop is changing.

30.06 Identify that an induced current in a conducting loop is driven by an induced emf.

30.07 Apply Faraday's law, which is the relationship between an induced emf in a conducting loop and the rate at which magnetic flux through the loop changes.

30.08 Extend Faraday's law from a loop to a coil with multiple loops.

30.09 Identify the three general ways in which the magnetic flux through a coil can change.

30.10 Use a right-hand rule for Lenz's law to determine the direction of induced emf and induced current in a conducting loop.

30.11 Identify that when a magnetic flux through a loop changes, the induced current in the loop sets up a magnetic field to oppose that change.

30.12 If an emf is induced in a conducting loop containing a battery, determine the net emf and calculate the corresponding current in the loop.

Key Ideas

● The magnetic flux Φ_B through an area A in a magnetic field $\vec{B}$ is defined as

$$\Phi_B = \int \vec{B} \cdot d\vec{A},$$

where the integral is taken over the area. The SI unit of magnetic flux is the weber, where $1\,\text{Wb} = 1\,\text{T} \cdot \text{m}^2$.

● If $\vec{B}$ is perpendicular to the area and uniform over it, the flux is

$$\Phi_B = BA \quad (\vec{B} \perp A, \vec{B}\,\text{uniform}).$$

● If the magnetic flux Φ_B through an area bounded by a closed conducting loop changes with time, a current and

an emf are produced in the loop; this process is called induction. The induced emf is

$$\mathscr{E} = -\frac{d\Phi_B}{dt} \quad \text{(Faraday's law)}.$$

● If the loop is replaced by a closely packed coil of N turns, the induced emf is

$$\mathscr{E} = -N\frac{d\Phi_B}{dt}.$$

● An induced current has a direction such that the magnetic field *due to the current* opposes the change in the magnetic flux that induces the current. The induced emf has the same direction as the induced current.

What Is Physics?

In Chapter 29 we discussed the fact that a current produces a magnetic field. That fact came as a surprise to the scientists who discovered the effect. Perhaps even more surprising was the discovery of the reverse effect: A magnetic field can produce an electric field that can drive a current. This link between a magnetic field and the electric field it produces (*induces*) is now called *Faraday's law of induction*.

The observations by Michael Faraday and other scientists that led to this law were at first just basic science. Today, however, applications of that basic science are almost everywhere. For example, induction is the basis of the electric guitars that revolutionized early rock and still drive heavy metal and punk today. It is also the basis of the electric generators that power cities and transportation lines and of the huge induction furnaces that are commonplace in foundries where large amounts of metal must be melted rapidly.

Before we get to applications like the electric guitar, we must examine two simple experiments about Faraday's law of induction.

Two Experiments

Let us examine two simple experiments to prepare for our discussion of Faraday's law of induction.

First Experiment. Figure 30-1 shows a conducting loop connected to a sensitive ammeter. Because there is no battery or other source of emf included, there is no current in the circuit. However, if we move a bar magnet toward the loop, a current suddenly appears in the circuit. The current disappears when the magnet stops. If we then move the magnet away, a current again suddenly appears, but now in the opposite direction. If we experimented for a while, we would discover the following:

1. A current appears only if there is relative motion between the loop and the magnet (one must move relative to the other); the current disappears when the relative motion between them ceases.

2. Faster motion produces a greater current.

3. If moving the magnet's north pole toward the loop causes, say, clockwise current, then moving the north pole away causes counterclockwise current. Moving the south pole toward or away from the loop also causes currents, but in the reversed directions.

The current produced in the loop is called an **induced current;** the work done per unit charge to produce that current (to move the conduction electrons that constitute the current) is called an **induced emf;** and the process of producing the current and emf is called **induction.**

Second Experiment. For this experiment we use the apparatus of Fig. 30-2, with the two conducting loops close to each other but not touching. If we close switch S, to turn on a current in the right-hand loop, the meter suddenly and briefly registers a current—an induced current—in the left-hand loop. If we then open the switch, another sudden and brief induced current appears in the left-hand loop, but in the opposite direction. We get an induced current (and thus an induced emf) only when the current in the right-hand loop is changing (either turning on or turning off) and not when it is constant (even if it is large).

The induced emf and induced current in these experiments are apparently caused when something changes—but what is that "something"? Faraday knew.

Faraday's Law of Induction

Faraday realized that an emf and a current can be induced in a loop, as in our two experiments, by changing the *amount of magnetic field* passing through the loop. He further realized that the "amount of magnetic field" can be visualized in terms of the magnetic field lines passing through the loop. **Faraday's law of induction,** stated in terms of our experiments, is this:

> An emf is induced in the loop at the left in Figs. 30-1 and 30-2 when the number of magnetic field lines that pass through the loop is changing.

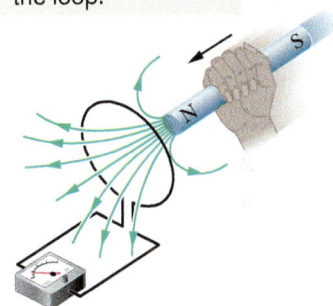

The magnet's motion creates a current in the loop.

Figure 30-1 An ammeter registers a current in the wire loop when the magnet is moving with respect to the loop.

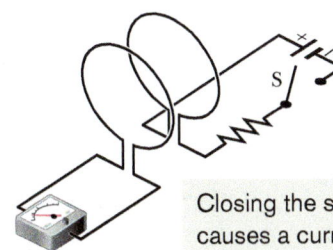

Closing the switch causes a current in the left-hand loop.

Figure 30-2 An ammeter registers a current in the left-hand wire loop just as switch S is closed (to turn on the current in the right-hand wire loop) or opened (to turn off the current in the right-hand loop). No motion of the coils is involved.

The actual number of field lines passing through the loop does not matter; the values of the induced emf and induced current are determined by the *rate* at which that number changes.

In our first experiment (Fig. 30-1), the magnetic field lines spread out from the north pole of the magnet. Thus, as we move the north pole closer to the loop, the number of field lines passing through the loop increases. That increase apparently causes conduction electrons in the loop to move (the induced current) and provides energy (the induced emf) for their motion. When the magnet stops moving, the number of field lines through the loop no longer changes and the induced current and induced emf disappear.

In our second experiment (Fig. 30-2), when the switch is open (no current), there are no field lines. However, when we turn on the current in the right-hand loop, the increasing current builds up a magnetic field around that loop and at the left-hand loop. While the field builds, the number of magnetic field lines through the left-hand loop increases. As in the first experiment, the increase in field lines through that loop apparently induces a current and an emf there. When the current in the right-hand loop reaches a final, steady value, the number of field lines through the left-hand loop no longer changes, and the induced current and induced emf disappear.

A Quantitative Treatment

To put Faraday's law to work, we need a way to calculate the *amount of magnetic field* that passes through a loop. In Chapter 23, in a similar situation, we needed to calculate the amount of electric field that passes through a surface. There we defined an electric flux $\Phi_E = \int \vec{E} \cdot d\vec{A}$. Here we define a *magnetic flux:* Suppose a loop enclosing an area A is placed in a magnetic field $\vec{B}$. Then the **magnetic flux** through the loop is

$$\Phi_B = \int \vec{B} \cdot d\vec{A} \quad \text{(magnetic flux through area } A\text{)}. \tag{30-1}$$

As in Chapter 23, $d\vec{A}$ is a vector of magnitude dA that is perpendicular to a differential area dA. As with electric flux, we want the component of the field that *pierces* the surface (not skims along it). The dot product of the field and the area vector automatically gives us that piercing component.

Special Case. As a special case of Eq. 30-1, suppose that the loop lies in a plane and that the magnetic field is perpendicular to the plane of the loop. Then we can write the dot product in Eq. 30-1 as $B\, dA \cos 0° = B\, dA$. If the magnetic field is also uniform, then B can be brought out in front of the integral sign. The remaining $\int dA$ then gives just the area A of the loop. Thus, Eq. 30-1 reduces to

$$\Phi_B = BA \quad (\vec{B} \perp \text{area } A, \vec{B} \text{ uniform}). \tag{30-2}$$

Unit. From Eqs. 30-1 and 30-2, we see that the SI unit for magnetic flux is the tesla–square meter, which is called the *weber* (abbreviated Wb):

$$1 \text{ weber} = 1 \text{ Wb} = 1 \text{ T} \cdot \text{m}^2. \tag{30-3}$$

Faraday's Law. With the notion of magnetic flux, we can state Faraday's law in a more quantitative and useful way:

The magnitude of the emf $\mathcal{E}$ induced in a conducting loop is equal to the rate at which the magnetic flux Φ_B through that loop changes with time.

As you will see below, the induced emf $\mathcal{E}$ tends to oppose the flux change, so

Faraday's law is formally written as

$$\mathscr{E} = -\frac{d\Phi_B}{dt} \quad \text{(Faraday's law)}, \qquad (30\text{-}4)$$

with the minus sign indicating that opposition. We often neglect the minus sign in Eq. 30-4, seeking only the magnitude of the induced emf.

If we change the magnetic flux through a coil of N turns, an induced emf appears in every turn and the total emf induced in the coil is the sum of these individual induced emfs. If the coil is tightly wound (*closely packed*), so that the same magnetic flux Φ_B passes through all the turns, the total emf induced in the coil is

$$\mathscr{E} = -N\frac{d\Phi_B}{dt} \quad \text{(coil of } N \text{ turns)}. \qquad (30\text{-}5)$$

Here are the general means by which we can change the magnetic flux through a coil:

1. Change the magnitude B of the magnetic field within the coil.

2. Change either the total area of the coil or the portion of that area that lies within the magnetic field (for example, by expanding the coil or sliding it into or out of the field).

3. Change the angle between the direction of the magnetic field $\vec{B}$ and the plane of the coil (for example, by rotating the coil so that field $\vec{B}$ is first perpendicular to the plane of the coil and then is along that plane).

✓ Checkpoint 1

The graph gives the magnitude $B(t)$ of a uniform magnetic field that exists throughout a conducting loop, with the direction of the field perpendicular to the plane of the loop. Rank the five regions of the graph according to the magnitude of the emf induced in the loop, greatest first.

Sample Problem 30.01 Induced emf in coil due to a solenoid

The long solenoid S shown (in cross section) in Fig. 30-3 has 220 turns/cm and carries a current $i = 1.5$ A; its diameter D is 3.2 cm. At its center we place a 130-turn closely packed coil C of diameter $d = 2.1$ cm. The current in the solenoid is reduced to zero at a steady rate in 25 ms. What is the magnitude of the emf that is induced in coil C while the current in the solenoid is changing?

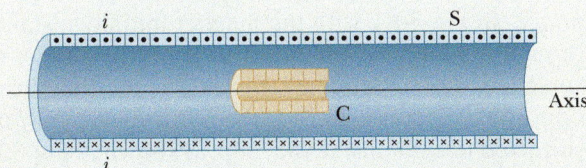

Figure 30-3 A coil C is located inside a solenoid S, which carries current i.

KEY IDEAS

1. Because it is located in the interior of the solenoid, coil C lies within the magnetic field produced by current i in the solenoid; thus, there is a magnetic flux Φ_B through coil C.

2. Because current i decreases, flux Φ_B also decreases.

3. As Φ_B decreases, emf $\mathscr{E}$ is induced in coil C.

4. The flux through each turn of coil C depends on the area A and orientation of that turn in the solenoid's magnetic field $\vec{B}$. Because $\vec{B}$ is uniform and directed perpendicular to area A, the flux is given by Eq. 30-2 ($\Phi_B = BA$).

5. The magnitude B of the magnetic field in the interior of a solenoid depends on the solenoid's current i and its number n of turns per unit length, according to Eq. 29-23 ($B = \mu_0 in$).

Calculations: Because coil C consists of more than one turn, we apply Faraday's law in the form of Eq. 30-5 ($\mathscr{E} = -N\, d\Phi_B/dt$), where the number of turns N is 130 and $d\Phi_B/dt$ is the rate at which the flux changes.

Because the current in the solenoid decreases at a steady rate, flux Φ_B also decreases at a steady rate, and so we can write $d\Phi_B/dt$ as $\Delta\Phi_B/\Delta t$. Then, to evaluate $\Delta\Phi_B$, we need the final and initial flux values. The final flux $\Phi_{B,f}$ is zero because the final current in the solenoid is zero. To find the initial flux $\Phi_{B,i}$, we note that area A is $\frac{1}{4}\pi d^2$ ($= 3.464 \times 10^{-4}\,\text{m}^2$) and the number n is 220 turns/cm, or 22 000 turns/m. Substituting Eq. 29-23 into Eq. 30-2 then leads to

$$\Phi_{B,i} = BA = (\mu_0 in)A$$

$$= (4\pi \times 10^{-7}\,\text{T·m/A})(1.5\,\text{A})(22\,000\,\text{turns/m})$$

$$\times (3.464 \times 10^{-4}\,\text{m}^2)$$

$$= 1.44 \times 10^{-5}\,\text{Wb}.$$

Now we can write

$$\frac{d\Phi_B}{dt} = \frac{\Delta\Phi_B}{\Delta t} = \frac{\Phi_{B,f} - \Phi_{B,i}}{\Delta t}$$

$$= \frac{(0 - 1.44 \times 10^{-5}\,\text{Wb})}{25 \times 10^{-3}\,\text{s}}$$

$$= -5.76 \times 10^{-4}\,\text{Wb/s}$$

$$= -5.76 \times 10^{-4}\,\text{V}.$$

We are interested only in magnitudes; so we ignore the minus signs here and in Eq. 30-5, writing

$$\mathscr{E} = N\frac{d\Phi_B}{dt} = (130\,\text{turns})(5.76 \times 10^{-4}\,\text{V})$$

$$= 7.5 \times 10^{-2}\,\text{V}$$

$$= 75\,\text{mV}. \qquad\text{(Answer)}$$

PLUS Additional examples, video, and practice available at *WileyPLUS*

Lenz's Law

Soon after Faraday proposed his law of induction, Heinrich Friedrich Lenz devised a rule for determining the direction of an induced current in a loop:

> An induced current has a direction such that the magnetic field due to *the current* opposes the change in the magnetic flux that induces the current.

Furthermore, the direction of an induced emf is that of the induced current. The key word in Lenz's law is "opposition." Let's apply the law to the motion of the north pole toward the conducting loop in Fig. 30-4.

1. *Opposition to Pole Movement.* The approach of the magnet's north pole in Fig. 30-4 increases the magnetic flux through the loop and thereby induces a current in the loop. From Fig. 29-22, we know that the loop then acts as a magnetic dipole with a south pole and a north pole, and that its magnetic dipole moment $\vec{\mu}$ is directed from south to north. To *oppose* the magnetic flux increase being caused by the approaching magnet, the loop's north pole (and thus $\vec{\mu}$) must face *toward* the approaching north pole so as to repel it (Fig. 30-4). Then the curled–straight right-hand rule for $\vec{\mu}$ (Fig. 29-22) tells us that the current induced in the loop must be counterclockwise in Fig. 30-4.

   If we next pull the magnet away from the loop, a current will again be induced in the loop. Now, however, the loop will have a south pole facing the retreating north pole of the magnet, so as to oppose the retreat. Thus, the induced current will be clockwise.

2. *Opposition to Flux Change.* In Fig. 30-4, with the magnet initially distant, no magnetic flux passes through the loop. As the north pole of the magnet then nears the loop with its magnetic field $\vec{B}$ directed *downward*, the flux through the loop increases. To oppose this increase in flux, the induced current i must set up its own field $\vec{B}_{\text{ind}}$ directed *upward* inside the loop, as shown in Fig. 30-5a; then the upward flux of field $\vec{B}_{\text{ind}}$ opposes the increasing downward flux of field $\vec{B}$. The curled–straight right-hand rule of Fig. 29-22 then tells us that i must be counterclockwise in Fig. 30-5a.

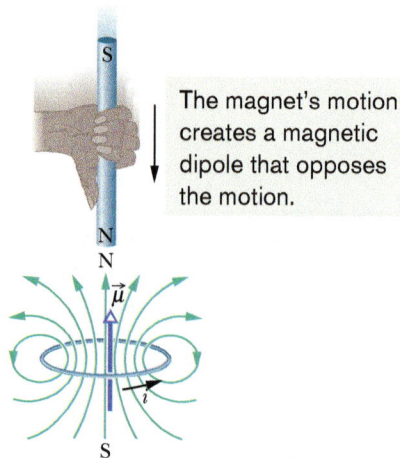

The magnet's motion creates a magnetic dipole that opposes the motion.

Figure 30-4 Lenz's law at work. As the magnet is moved toward the loop, a current is induced in the loop. The current produces its own magnetic field, with magnetic dipole moment $\vec{\mu}$ oriented so as to oppose the motion of the magnet. Thus, the induced current must be counterclockwise as shown.

Heads Up. The flux of $\vec{B}_{ind}$ always opposes the *change* in the flux of $\vec{B}$, but $\vec{B}_{ind}$ is not always opposite $\vec{B}$. For example, if we next pull the magnet away from the loop in Fig. 30-4, the magnet's flux Φ_B is still downward through the loop, but it is now decreasing. The flux of $\vec{B}_{ind}$ must now be downward inside the loop, to oppose that *decrease* (Fig. 30-5b). Thus, $\vec{B}_{ind}$ and $\vec{B}$ are now in the same direction. In Figs. 30-5c and d, the south pole of the magnet approaches and retreats from the loop, again with opposition to change.

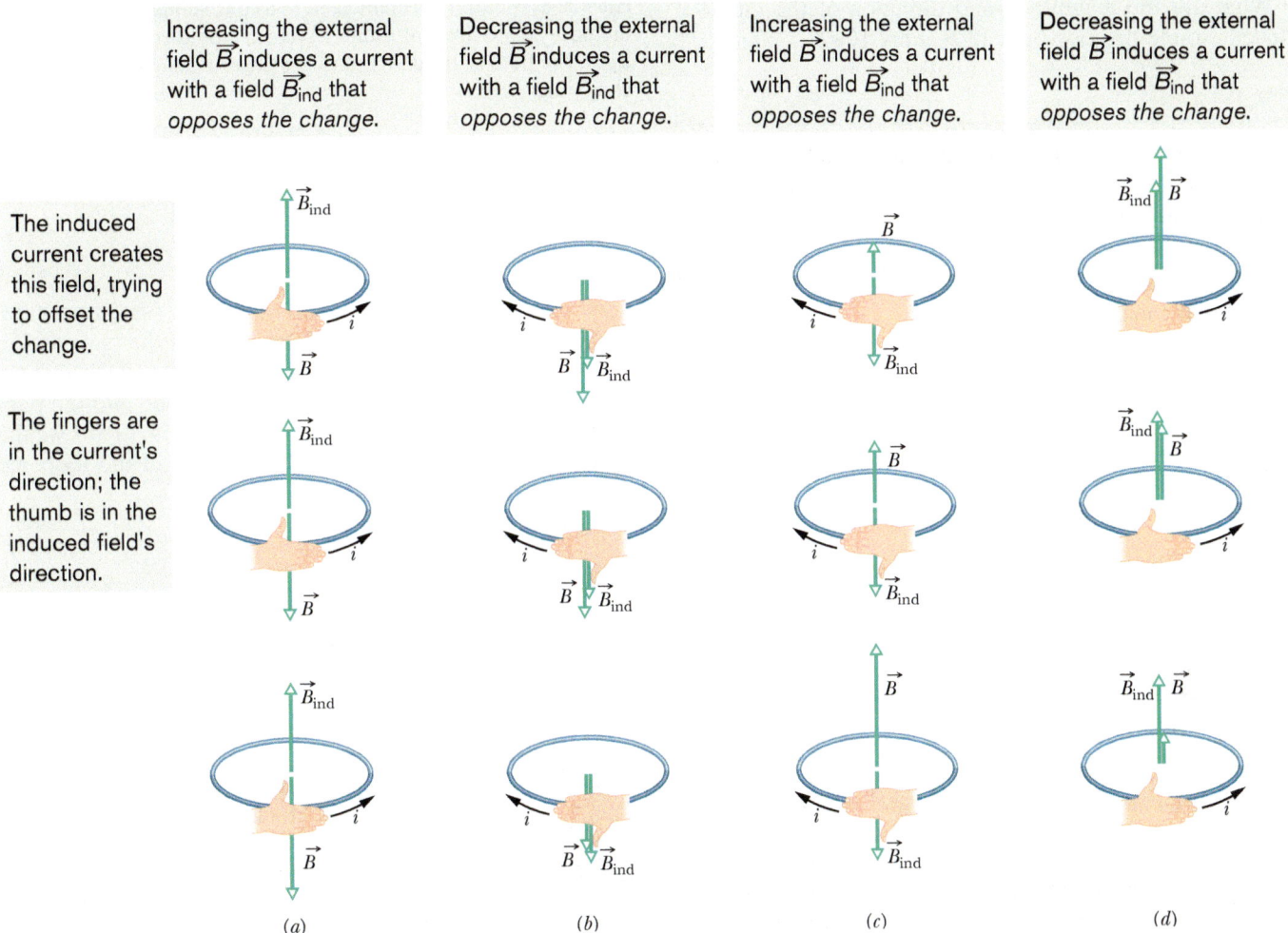

| Increasing the external field $\vec{B}$ induces a current with a field $\vec{B}_{ind}$ that *opposes the change.* | Decreasing the external field $\vec{B}$ induces a current with a field $\vec{B}_{ind}$ that *opposes the change.* | Increasing the external field $\vec{B}$ induces a current with a field $\vec{B}_{ind}$ that *opposes the change.* | Decreasing the external field $\vec{B}$ induces a current with a field $\vec{B}_{ind}$ that *opposes the change.* |

The induced current creates this field, trying to offset the change.

The fingers are in the current's direction; the thumb is in the induced field's direction.

(a)   (b)   (c)   (d)

Figure 30-5 The direction of the current i induced in a loop is such that the current's magnetic field $\vec{B}_{ind}$ opposes the *change* in the magnetic field $\vec{B}$ inducing i. The field $\vec{B}_{ind}$ is always directed opposite an increasing field $\vec{B}$ (a, c) and in the same direction as a decreasing field $\vec{B}$ (b, d). The curled–straight right-hand rule gives the direction of the induced current based on the direction of the induced field.

☑ **Checkpoint 2**

The figure shows three situations in which identical circular conducting loops are in uniform magnetic fields that are either increasing (Inc) or decreasing (Dec) in magnitude at identical rates. In each, the dashed line coincides with a diameter. Rank the situations according to the magnitude of the current induced in the loops, greatest first.

(a)   (b)   (c)

Sample Problem 30.02   Induced emf and current due to a changing uniform B field

Figure 30-6 shows a conducting loop consisting of a half-circle of radius $r = 0.20$ m and three straight sections. The half-circle lies in a uniform magnetic field $\vec{B}$ that is directed out of the page; the field magnitude is given by $B = 4.0t^2 + 2.0t + 3.0$, with B in teslas and t in seconds. An ideal battery with emf $\mathscr{E}_{bat} = 2.0$ V is connected to the loop. The resistance of the loop is $2.0\ \Omega$.

(a) What are the magnitude and direction of the emf $\mathscr{E}_{ind}$ induced around the loop by field $\vec{B}$ at $t = 10$ s?

KEY IDEAS

1. According to Faraday's law, the magnitude of $\mathscr{E}_{ind}$ is equal to the rate $d\Phi_B/dt$ at which the magnetic flux through the loop changes.

2. The flux through the loop depends on how much of the loop's area lies within the flux and how the area is oriented in the magnetic field $\vec{B}$.

3. Because $\vec{B}$ is uniform and is perpendicular to the plane of the loop, the flux is given by Eq. 30-2 ($\Phi_B = BA$). (We don't need to integrate B over the area to get the flux.)

4. The induced field B_{ind} (due to the induced current) must always oppose the *change* in the magnetic flux.

Magnitude: Using Eq. 30-2 and realizing that only the field magnitude B changes in time (not the area A), we rewrite Faraday's law, Eq. 30-4, as

$$\mathscr{E}_{ind} = \frac{d\Phi_B}{dt} = \frac{d(BA)}{dt} = A\frac{dB}{dt}.$$

Because the flux penetrates the loop only within the half-circle, the area A in this equation is $\frac{1}{2}\pi r^2$. Substituting this and the given expression for B yields

$$\mathscr{E}_{ind} = A\frac{dB}{dt} = \frac{\pi r^2}{2}\frac{d}{dt}(4.0t^2 + 2.0t + 3.0)$$

$$= \frac{\pi r^2}{2}(8.0t + 2.0).$$

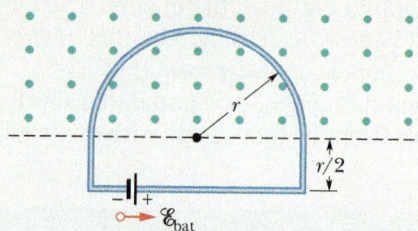

Figure 30-6 A battery is connected to a conducting loop that includes a half-circle of radius r lying in a uniform magnetic field. The field is directed out of the page; its magnitude is changing.

At $t = 10$ s, then,

$$\mathscr{E}_{ind} = \frac{\pi (0.20\text{ m})^2}{2}[8.0(10) + 2.0]$$

$$= 5.152\text{ V} \approx 5.2\text{ V}. \qquad \text{(Answer)}$$

Direction: To find the direction of $\mathscr{E}_{ind}$, we first note that in Fig. 30-6 the flux through the loop is out of the page and increasing. Because the induced field B_{ind} (due to the induced current) must oppose that increase, it must be *into* the page. Using the curled–straight right-hand rule (Fig. 30-5c), we find that the induced current is clockwise around the loop, and thus so is the induced emf $\mathscr{E}_{ind}$.

(b) What is the current in the loop at $t = 10$ s?

KEY IDEA

The point here is that *two* emfs tend to move charges around the loop.

Calculation: The induced emf $\mathscr{E}_{ind}$ tends to drive a current clockwise around the loop; the battery's emf $\mathscr{E}_{bat}$ tends to drive a current counterclockwise. Because $\mathscr{E}_{ind}$ is greater than $\mathscr{E}_{bat}$, the net emf $\mathscr{E}_{net}$ is clockwise, and thus so is the current. To find the current at $t = 10$ s, we use Eq. 27-2 ($i = \mathscr{E}/R$):

$$i = \frac{\mathscr{E}_{net}}{R} = \frac{\mathscr{E}_{ind} - \mathscr{E}_{bat}}{R}$$

$$= \frac{5.152\text{ V} - 2.0\text{ V}}{2.0\ \Omega} = 1.58\text{ A} \approx 1.6\text{ A}. \qquad \text{(Answer)}$$

Sample Problem 30.03   Induced emf due to a changing nonuniform B field

Figure 30-7 shows a rectangular loop of wire immersed in a nonuniform and varying magnetic field $\vec{B}$ that is perpendicular to and directed into the page. The field's magnitude is given by $B = 4t^2x^2$, with B in teslas, t in seconds, and x in meters. (Note that the function depends on *both* time and position.) The loop has width $W = 3.0$ m and height $H = 2.0$ m. What are the magnitude and direction of the induced emf $\mathscr{E}$ around the loop at $t = 0.10$ s?

KEY IDEAS

1. Because the magnitude of the magnetic field $\vec{B}$ is changing with time, the magnetic flux Φ_B through the loop is also changing.

2. The changing flux induces an emf $\mathscr{E}$ in the loop according to Faraday's law, which we can write as $\mathscr{E} = d\Phi_B/dt$.

3. To use that law, we need an expression for the flux Φ_B at

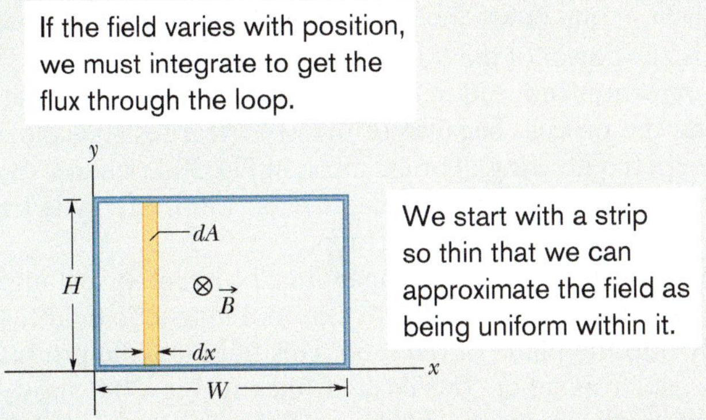

If the field varies with position, we must integrate to get the flux through the loop.

We start with a strip so thin that we can approximate the field as being uniform within it.

Figure 30-7 A closed conducting loop, of width W and height H, lies in a nonuniform, varying magnetic field that points directly into the page. To apply Faraday's law, we use the vertical strip of height H, width dx, and area dA.

any time t. However, because B is *not* uniform over the area enclosed by the loop, we *cannot* use Eq. 30-2 ($\Phi_B = BA$) to find that expression; instead we must use Eq. 30-1 ($\Phi_B = \int \vec{B} \cdot d\vec{A}$).

Calculations: In Fig. 30-7, $\vec{B}$ is perpendicular to the plane of the loop (and hence parallel to the differential area vector $d\vec{A}$); so the dot product in Eq. 30-1 gives $B\,dA$. Because the magnetic field varies with the coordinate x but not with the coordinate y, we can take the differential area

dA to be the area of a vertical strip of height H and width dx (as shown in Fig. 30-7). Then $dA = H\,dx$, and the flux through the loop is

$$\Phi_B = \int \vec{B} \cdot d\vec{A} = \int B\,dA = \int BH\,dx = \int 4t^2 x^2 H\,dx.$$

Treating t as a constant for this integration and inserting the integration limits $x = 0$ and $x = 3.0$ m, we obtain

$$\Phi_B = 4t^2 H \int_0^{3.0} x^2\,dx = 4t^2 H \left[\frac{x^3}{3}\right]_0^{3.0} = 72t^2,$$

where we have substituted $H = 2.0$ m and Φ_B is in webers. Now we can use Faraday's law to find the magnitude of $\mathscr{E}$ at any time t:

$$\mathscr{E} = \frac{d\Phi_B}{dt} = \frac{d(72t^2)}{dt} = 144t,$$

in which $\mathscr{E}$ is in volts. At $t = 0.10$ s,

$$\mathscr{E} = (144 \text{ V/s})(0.10 \text{ s}) \approx 14 \text{ V}. \qquad \text{(Answer)}$$

The flux of $\vec{B}$ through the loop is into the page in Fig. 30-7 and is increasing in magnitude because B is increasing in magnitude with time. By Lenz's law, the field B_{ind} of the induced current opposes this increase and so is directed out of the page. The curled–straight right-hand rule in Fig. 30-5a then tells us that the induced current is counterclockwise around the loop, and thus so is the induced emf $\mathscr{E}$.

Additional examples, video, and practice available at *WileyPLUS*

30-2 INDUCTION AND ENERGY TRANSFERS

Learning Objectives

After reading this module, you should be able to . . .

30.13 For a conducting loop pulled into or out of a magnetic field, calculate the rate at which energy is transferred to thermal energy.

30.14 Apply the relationship between an induced current and the rate at which it produces thermal energy.

30.15 Describe eddy currents.

Key Idea

● The induction of a current by a changing flux means that energy is being transferred to that current. The energy can then be transferred to other forms, such as thermal energy.

Induction and Energy Transfers

By Lenz's law, whether you move the magnet toward or away from the loop in Fig. 30-1, a magnetic force resists the motion, requiring your applied force to do positive work. At the same time, thermal energy is produced in the material of the loop because of the material's electrical resistance to the current that is induced by the motion. The energy you transfer to the closed *loop + magnet* system via your applied force ends up in this thermal energy. (For now, we neglect energy that is radiated away from the loop as electromagnetic waves during the induction.) The faster you move the magnet, the more rapidly your applied force does work and the greater the rate at which your energy is transferred to thermal energy in the loop; that is, the power of the transfer is greater.

Regardless of how current is induced in a loop, energy is always transferred to thermal energy during the process because of the electrical resistance of the loop (unless the loop is superconducting). For example, in Fig. 30-2, when switch S is closed and a current is briefly induced in the left-hand loop, energy is transferred from the battery to thermal energy in that loop.

Figure 30-8 shows another situation involving induced current. A rectangular loop of wire of width L has one end in a uniform external magnetic field that is directed perpendicularly into the plane of the loop. This field may be produced, for example, by a large electromagnet. The dashed lines in Fig. 30-8 show the assumed limits of the magnetic field; the fringing of the field at its edges is neglected. You are to pull this loop to the right at a constant velocity $\vec{v}$.

Flux Change. The situation of Fig. 30-8 does not differ in any essential way from that of Fig. 30-1. In each case a magnetic field and a conducting loop are in relative motion; in each case the flux of the field through the loop is changing with time. It is true that in Fig. 30-1 the flux is changing because $\vec{B}$ is changing and in Fig. 30-8 the flux is changing because the area of the loop still in the magnetic field is changing, but that difference is not important. The important difference between the two arrangements is that the arrangement of Fig. 30-8 makes calculations easier. Let us now calculate the rate at which you do mechanical work as you pull steadily on the loop in Fig. 30-8.

Rate of Work. As you will see, to pull the loop at a constant velocity $\vec{v}$, you must apply a constant force $\vec{F}$ to the loop because a magnetic force of equal magnitude but opposite direction acts on the loop to oppose you. From Eq. 7-48, the rate at which you do work—that is, the power—is then

$$P = Fv, \tag{30-6}$$

where F is the magnitude of your force. We wish to find an expression for P in terms of the magnitude B of the magnetic field and the characteristics of the loop—namely, its resistance R to current and its dimension L.

As you move the loop to the right in Fig. 30-8, the portion of its area within the magnetic field decreases. Thus, the flux through the loop also decreases and, according to Faraday's law, a current is produced in the loop. It is the presence of this current that causes the force that opposes your pull.

Induced emf. To find the current, we first apply Faraday's law. When x is the length of the loop still in the magnetic field, the area of the loop still in the field is Lx. Then from Eq. 30-2, the magnitude of the flux through the loop is

$$\Phi_B = BA = BLx. \tag{30-7}$$

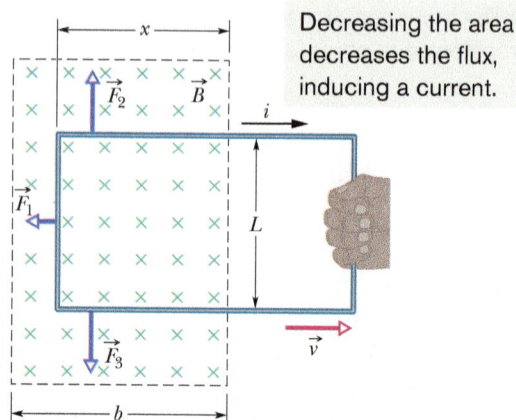

Decreasing the area decreases the flux, inducing a current.

Figure 30-8 You pull a closed conducting loop out of a magnetic field at constant velocity $\vec{v}$. While the loop is moving, a clockwise current i is induced in the loop, and the loop segments still within the magnetic field experience forces $\vec{F}_1$, $\vec{F}_2$, and $\vec{F}_3$.

As x decreases, the flux decreases. Faraday's law tells us that with this flux decrease, an emf is induced in the loop. Dropping the minus sign in Eq. 30-4 and using Eq. 30-7, we can write the magnitude of this emf as

$$\mathcal{E} = \frac{d\Phi_B}{dt} = \frac{d}{dt} BLx = BL \frac{dx}{dt} = BLv, \qquad (30\text{-}8)$$

in which we have replaced dx/dt with v, the speed at which the loop moves.

Figure 30-9 shows the loop as a circuit: induced emf $\mathcal{E}$ is represented on the left, and the collective resistance R of the loop is represented on the right. The direction of the induced current i is obtained with a right-hand rule as in Fig. 30-5b for decreasing flux; applying the rule tells us that the current must be clockwise, and $\mathcal{E}$ must have the same direction.

Induced Current. To find the magnitude of the induced current, we cannot apply the loop rule for potential differences in a circuit because, as you will see in Module 30-3, we cannot define a potential difference for an induced emf. However, we can apply the equation $i = \mathcal{E}/R$. With Eq. 30-8, this becomes

$$i = \frac{BLv}{R}. \qquad (30\text{-}9)$$

Because three segments of the loop in Fig. 30-8 carry this current through the magnetic field, sideways deflecting forces act on those segments. From Eq. 28-26 we know that such a deflecting force is, in general notation,

$$\vec{F}_d = i\vec{L} \times \vec{B}. \qquad (30\text{-}10)$$

In Fig. 30-8, the deflecting forces acting on the three segments of the loop are marked $\vec{F}_1$, $\vec{F}_2$, and $\vec{F}_3$. Note, however, that from the symmetry, forces $\vec{F}_2$ and $\vec{F}_3$ are equal in magnitude and cancel. This leaves only force $\vec{F}_1$, which is directed opposite your force $\vec{F}$ on the loop and thus is the force opposing you. So, $\vec{F} = -\vec{F}_1$.

Using Eq. 30-10 to obtain the magnitude of $\vec{F}_1$ and noting that the angle between $\vec{B}$ and the length vector $\vec{L}$ for the left segment is 90°, we write

$$F = F_1 = iLB \sin 90° = iLB. \qquad (30\text{-}11)$$

Substituting Eq. 30-9 for i in Eq. 30-11 then gives us

$$F = \frac{B^2 L^2 v}{R}. \qquad (30\text{-}12)$$

Because B, L, and R are constants, the speed v at which you move the loop is constant if the magnitude F of the force you apply to the loop is also constant.

Rate of Work. By substituting Eq. 30-12 into Eq. 30-6, we find the rate at which you do work on the loop as you pull it from the magnetic field:

$$P = Fv = \frac{B^2 L^2 v^2}{R} \quad \text{(rate of doing work)}. \qquad (30\text{-}13)$$

Thermal Energy. To complete our analysis, let us find the rate at which thermal energy appears in the loop as you pull it along at constant speed. We calculate it from Eq. 26-27,

$$P = i^2 R. \qquad (30\text{-}14)$$

Substituting for i from Eq. 30-9, we find

$$P = \left(\frac{BLv}{R} \right)^2 R = \frac{B^2 L^2 v^2}{R} \quad \text{(thermal energy rate)}, \qquad (30\text{-}15)$$

which is exactly equal to the rate at which you are doing work on the loop (Eq. 30-13). Thus, the work that you do in pulling the loop through the magnetic field appears as thermal energy in the loop.

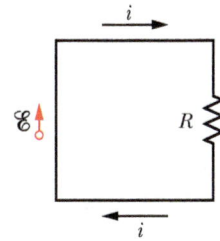

Figure 30-9 A circuit diagram for the loop of Fig. 30-8 while the loop is moving.

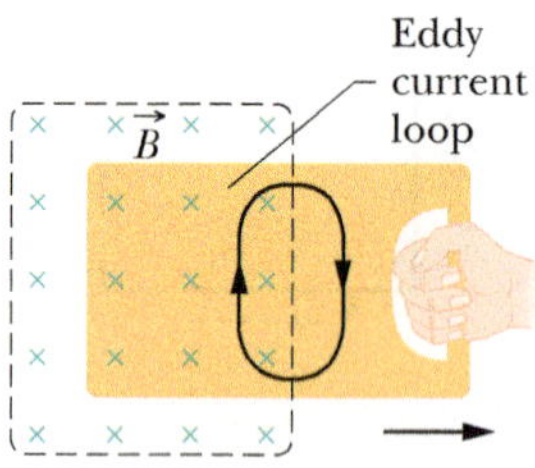

(a)

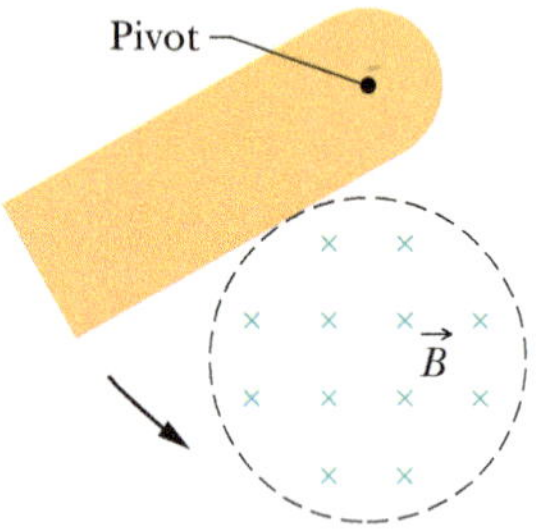

(b)

Figure 30-10 (a) As you pull a solid conducting plate out of a magnetic field, *eddy currents* are induced in the plate. A typical loop of eddy current is shown. (b) A conducting plate is allowed to swing like a pendulum about a pivot and into a region of magnetic field. As it enters and leaves the field, eddy currents are induced in the plate.

Eddy Currents

Suppose we replace the conducting loop of Fig. 30-8 with a solid conducting plate. If we then move the plate out of the magnetic field as we did the loop (Fig. 30-10a), the relative motion of the field and the conductor again induces a current in the conductor. Thus, we again encounter an opposing force and must do work because of the induced current. With the plate, however, the conduction electrons making up the induced current do not follow one path as they do with the loop. Instead, the electrons swirl about within the plate as if they were caught in an eddy (whirlpool) of water. Such a current is called an *eddy current* and can be represented, as it is in Fig. 30-10a, *as if* it followed a single path.

As with the conducting loop of Fig. 30-8, the current induced in the plate results in mechanical energy being dissipated as thermal energy. The dissipation is more apparent in the arrangement of Fig. 30-10b; a conducting plate, free to rotate about a pivot, is allowed to swing down through a magnetic field like a pendulum. Each time the plate enters and leaves the field, a portion of its mechanical energy is transferred to its thermal energy. After several swings, no mechanical energy remains and the warmed-up plate just hangs from its pivot.

✅ **Checkpoint 3**

The figure shows four wire loops, with edge lengths of either L or $2L$. All four loops will move through a region of uniform magnetic field $\vec{B}$ (directed out of the page) at the same constant velocity. Rank the four loops according to the maximum magnitude of the emf induced as they move through the field, greatest first.

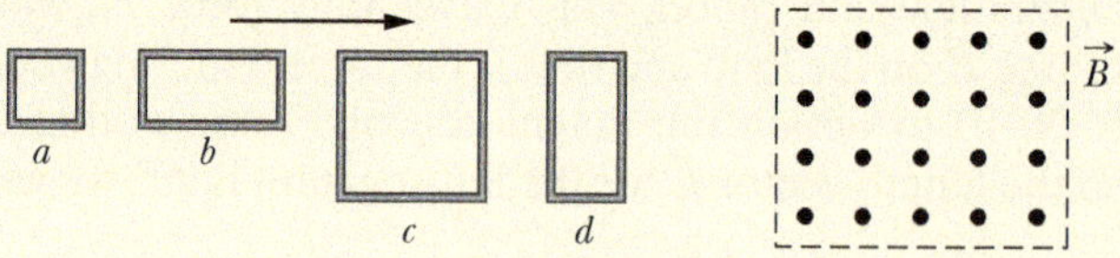

30-3 INDUCED ELECTRIC FIELDS

Learning Objectives

After reading this module, you should be able to . . .

30.16 Identify that a changing magnetic field induces an electric field, regardless of whether there is a conducting loop.

30.17 Apply Faraday's law to relate the electric field $\vec{E}$ induced along a closed path (whether it has conducting material or not) to the rate of change $d\Phi/dt$ of the magnetic flux encircled by the path.

30.18 Identify that an electric potential cannot be associated with an induced electric field.

Key Ideas

● An emf is induced by a changing magnetic flux even if the loop through which the flux is changing is not a physical conductor but an imaginary line. The changing magnetic field induces an electric field $\vec{E}$ at every point of such a loop; the induced emf is related to $\vec{E}$ by

$$\mathscr{E} = \oint \vec{E} \cdot d\vec{s}.$$

● Using the induced electric field, we can write Faraday's law in its most general form as

$$\oint \vec{E} \cdot d\vec{s} = -\frac{d\Phi_B}{dt} \quad \text{(Faraday's law)}.$$

A changing magnetic field induces an electric field $\vec{E}$.

Induced Electric Fields

Let us place a copper ring of radius r in a uniform external magnetic field, as in Fig. 30-11a. The field—neglecting fringing—fills a cylindrical volume of radius R. Suppose that we increase the strength of this field at a steady rate, perhaps by increasing—in an appropriate way—the current in the windings of the electromagnet that produces the field. The magnetic flux through the ring will then change at a steady rate and—by Faraday's law—an induced emf and thus an induced current will appear in the ring. From Lenz's law we can deduce that the direction of the induced current is counterclockwise in Fig. 30-11a.

If there is a current in the copper ring, an electric field must be present along the ring because an electric field is needed to do the work of moving the conduction electrons. Moreover, the electric field must have been produced by the changing magnetic flux. This **induced electric field** $\vec{E}$ is just as real as an electric field produced by static charges; either field will exert a force $q_0\vec{E}$ on a particle of charge q_0.

By this line of reasoning, we are led to a useful and informative restatement of Faraday's law of induction:

> A changing magnetic field produces an electric field.

The striking feature of this statement is that the electric field is induced even if there is no copper ring. Thus, the electric field would appear even if the changing magnetic field were in a vacuum.

To fix these ideas, consider Fig. 30-11b, which is just like Fig. 30-11a except the copper ring has been replaced by a hypothetical circular path of radius r. We assume, as previously, that the magnetic field $\vec{B}$ is increasing in magnitude at a constant rate dB/dt. The electric field induced at various points around the

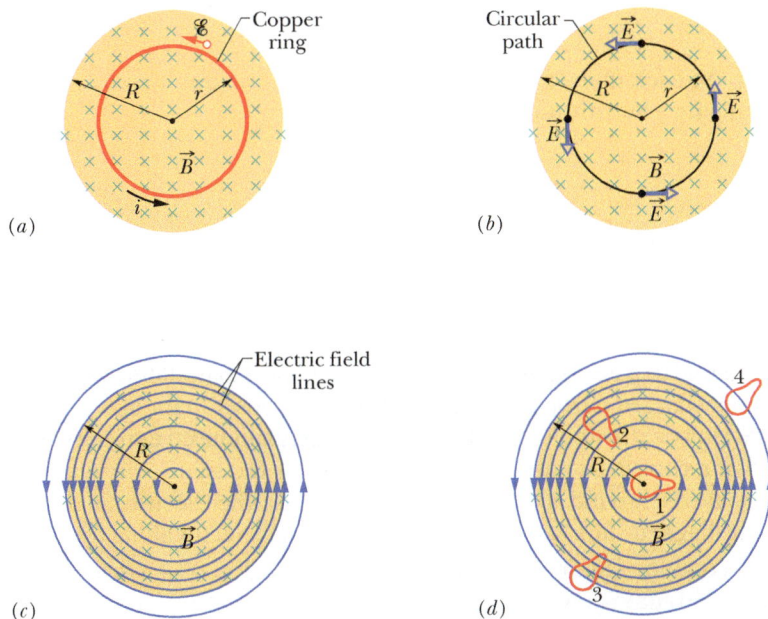

(a)　　(b)

(c)　　(d)

Figure 30-11 (a) If the magnetic field increases at a steady rate, a constant induced current appears, as shown, in the copper ring of radius r. (b) An induced electric field exists even when the ring is removed; the electric field is shown at four points. (c) The complete picture of the induced electric field, displayed as field lines. (d) Four similar closed paths that enclose identical areas. Equal emfs are induced around paths 1 and 2, which lie entirely within the region of changing magnetic field. A smaller emf is induced around path 3, which only partially lies in that region. No net emf is induced around path 4, which lies entirely outside the magnetic field.

circular path must—from the symmetry—be tangent to the circle, as Fig. 30-11b shows.* Hence, the circular path is an electric field line. There is nothing special about the circle of radius r, so the electric field lines produced by the changing magnetic field must be a set of concentric circles, as in Fig. 30-11c.

As long as the magnetic field is *increasing* with time, the electric field represented by the circular field lines in Fig. 30-11c will be present. If the magnetic field remains *constant* with time, there will be no induced electric field and thus no electric field lines. If the magnetic field is *decreasing* with time (at a constant rate), the electric field lines will still be concentric circles as in Fig. 30-11c, but they will now have the opposite direction. All this is what we have in mind when we say "A changing magnetic field produces an electric field."

A Reformulation of Faraday's Law

Consider a particle of charge q_0 moving around the circular path of Fig. 30-11b. The work W done on it in one revolution by the induced electric field is $W = \mathscr{E}q_0$, where $\mathscr{E}$ is the induced emf—that is, the work done per unit charge in moving the test charge around the path. From another point of view, the work is

$$W = \int \vec{F} \cdot d\vec{s} = (q_0E)(2\pi r), \tag{30-16}$$

where q_0E is the magnitude of the force acting on the test charge and $2\pi r$ is the distance over which that force acts. Setting these two expressions for W equal to each other and canceling q_0, we find that

$$\mathscr{E} = 2\pi rE. \tag{30-17}$$

Next we rewrite Eq. 30-16 to give a more general expression for the work done on a particle of charge q_0 moving along any closed path:

$$W = \oint \vec{F} \cdot d\vec{s} = q_0 \oint \vec{E} \cdot d\vec{s}. \tag{30-18}$$

(The loop on each integral sign indicates that the integral is to be taken around the closed path.) Substituting $\mathscr{E}q_0$ for W, we find that

$$\mathscr{E} = \oint \vec{E} \cdot d\vec{s}. \tag{30-19}$$

This integral reduces at once to Eq. 30-17 if we evaluate it for the special case of Fig. 30-11b.

Meaning of emf. With Eq. 30-19, we can expand the meaning of induced emf. Up to this point, induced emf has meant the work per unit charge done in maintaining current due to a changing magnetic flux, or it has meant the work done per unit charge on a charged particle that moves around a closed path in a changing magnetic flux. However, with Fig. 30-11b and Eq. 30-19, an induced emf can exist without the need of a current or particle: An induced emf is the sum—via integration—of quantities $\vec{E} \cdot d\vec{s}$ around a closed path, where $\vec{E}$ is the electric field induced by a changing magnetic flux and $d\vec{s}$ is a differential length vector along the path.

If we combine Eq. 30-19 with Faraday's law in Eq. 30-4 ($\mathscr{E} = -d\Phi_B/dt$), we can rewrite Faraday's law as

$$\oint \vec{E} \cdot d\vec{s} = -\frac{d\Phi_B}{dt} \quad \text{(Faraday's law).} \tag{30-20}$$

---

*Arguments of symmetry would also permit the lines of $\vec{E}$ around the circular path to be *radial*, rather than tangential. However, such radial lines would imply that there are free charges, distributed symmetrically about the axis of symmetry, on which the electric field lines could begin or end; there are no such charges.

This equation says simply that a changing magnetic field induces an electric field. The changing magnetic field appears on the right side of this equation, the electric field on the left.

Faraday's law in the form of Eq. 30-20 can be applied to *any* closed path that can be drawn in a changing magnetic field. Figure 30-11*d*, for example, shows four such paths, all having the same shape and area but located in different positions in the changing field. The induced emfs $\mathscr{E}$ $(= \oint \vec{E} \cdot d\vec{s})$ for paths 1 and 2 are equal because these paths lie entirely in the magnetic field and thus have the same value of $d\Phi_B/dt$. This is true even though the electric field vectors at points along these paths are different, as indicated by the patterns of electric field lines in the figure. For path 3 the induced emf is smaller because the enclosed flux Φ_B (hence $d\Phi_B/dt$) is smaller, and for path 4 the induced emf is zero even though the electric field is not zero at any point on the path.

A New Look at Electric Potential

Induced electric fields are produced not by static charges but by a changing magnetic flux. Although electric fields produced in either way exert forces on charged particles, there is an important difference between them. The simplest evidence of this difference is that the field lines of induced electric fields form closed loops, as in Fig. 30-11*c*. Field lines produced by static charges never do so but must start on positive charges and end on negative charges. Thus, a field line from a charge can never loop around and back onto itself as we see for each of the field lines in Fig. 30-11*c*.

In a more formal sense, we can state the difference between electric fields produced by induction and those produced by static charges in these words:

> Electric potential has meaning only for electric fields that are produced by static charges; it has no meaning for electric fields that are produced by induction.

You can understand this statement qualitatively by considering what happens to a charged particle that makes a single journey around the circular path in Fig. 30-11*b*. It starts at a certain point and, on its return to that same point, has experienced an emf $\mathscr{E}$ of, let us say, 5 V; that is, work of 5 J/C has been done on the particle by the electric field, and thus the particle should then be at a point that is 5 V greater in potential. However, that is impossible because the particle is back at the same point, which cannot have two different values of potential. Thus, potential has no meaning for electric fields that are set up by changing magnetic fields.

We can take a more formal look by recalling Eq. 24-18, which defines the potential difference between two points i and f in an electric field $\vec{E}$ in terms of an integration between those points:

$$V_f - V_i = -\int_i^f \vec{E} \cdot d\vec{s}. \tag{30-21}$$

In Chapter 24 we had not yet encountered Faraday's law of induction; so the electric fields involved in the derivation of Eq. 24-18 were those due to static charges. If i and f in Eq. 30-21 are the same point, the path connecting them is a closed loop, V_i and V_f are identical, and Eq. 30-21 reduces to

$$\oint \vec{E} \cdot d\vec{s} = 0. \tag{30-22}$$

However, when a changing magnetic flux is present, this integral is *not* zero but is $-d\Phi_B/dt$, as Eq. 30-20 asserts. Thus, assigning electric potential to an induced electric field leads us to a contradiction. We must conclude that electric potential has no meaning for electric fields associated with induction.

✓ **Checkpoint 4**

The figure shows five lettered regions in which a uniform magnetic field extends either directly out of the page or into the page, with the direction indicated only for region *a*. The field is increasing in magnitude at the same steady rate in all five regions; the regions are identical in area. Also shown are four numbered paths along which $\oint \vec{E} \cdot d\vec{s}$ has the magnitudes given below in terms of a quantity "mag." Determine whether the magnetic field is directed into or out of the page for regions *b* through *e*.

Path	1	2	3	4
$\oint \vec{E} \cdot d\vec{s}$	mag	2(mag)	3(mag)	0

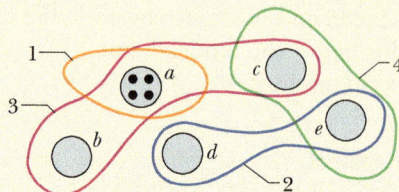

Sample Problem 30.04 **Induced electric field due to changing *B* field, inside and outside**

In Fig. 30-11*b*, take $R = 8.5$ cm and $dB/dt = 0.13$ T/s.

(a) Find an expression for the magnitude E of the induced electric field at points within the magnetic field, at radius r from the center of the magnetic field. Evaluate the expression for $r = 5.2$ cm.

KEY IDEA

An electric field is induced by the changing magnetic field, according to Faraday's law.

Calculations: To calculate the field magnitude E, we apply Faraday's law in the form of Eq. 30-20. We use a circular path of integration with radius $r \leq R$ because we want E for points within the magnetic field. We assume from the symmetry that $\vec{E}$ in Fig. 30-11*b* is tangent to the circular path at all points. The path vector $d\vec{s}$ is also always tangent to the circular path; so the dot product $\vec{E} \cdot d\vec{s}$ in Eq. 30-20 must have the magnitude $E\,ds$ at all points on the path. We can also assume from the symmetry that E has the same value at all points along the circular path. Then the left side of Eq. 30-20 becomes

$$\oint \vec{E} \cdot d\vec{s} = \oint E\,ds = E \oint ds = E(2\pi r). \quad (30\text{-}23)$$

(The integral $\oint ds$ is the circumference $2\pi r$ of the circular path.)

Next, we need to evaluate the right side of Eq. 30-20. Because $\vec{B}$ is uniform over the area A encircled by the path of integration and is directed perpendicular to that area, the magnetic flux is given by Eq. 30-2:

$$\Phi_B = BA = B(\pi r^2). \quad (30\text{-}24)$$

Substituting this and Eq. 30-23 into Eq. 30-20 and dropping

the minus sign, we find that

$$E(2\pi r) = (\pi r^2)\frac{dB}{dt}$$

or

$$E = \frac{r}{2}\frac{dB}{dt}. \quad \text{(Answer)} \quad (30\text{-}25)$$

Equation 30-25 gives the magnitude of the electric field at any point for which $r \leq R$ (that is, within the magnetic field). Substituting given values yields, for the magnitude of $\vec{E}$ at $r = 5.2$ cm,

$$E = \frac{(5.2 \times 10^{-2}\ \text{m})}{2}(0.13\ \text{T/s})$$

$$= 0.0034\ \text{V/m} = 3.4\ \text{mV/m}. \quad \text{(Answer)}$$

(b) Find an expression for the magnitude E of the induced electric field at points that are outside the magnetic field, at radius r from the center of the magnetic field. Evaluate the expression for $r = 12.5$ cm.

KEY IDEAS

Here again an electric field is induced by the changing magnetic field, according to Faraday's law, except that now we use a circular path of integration with radius $r \geq R$ because we want to evaluate E for points outside the magnetic field. Proceeding as in (a), we again obtain Eq. 30-23. However, we do not then obtain Eq. 30-24 because the new path of integration is now outside the magnetic field, and so the magnetic flux encircled by the new path is only that in the area πR^2 of the magnetic field region.

Calculations: We can now write

$$\Phi_B = BA = B(\pi R^2). \quad (30\text{-}26)$$

Substituting this and Eq. 30-23 into Eq. 30-20 (without the minus sign) and solving for E yield

$$E = \frac{R^2}{2r}\frac{dB}{dt}.\qquad\text{(Answer)}\quad(30\text{-}27)$$

Because E is not zero here, we know that an electric field is induced even at points that are outside the changing magnetic field, an important result that (as you will see in Module 31-6) makes transformers possible.

With the given data, Eq. 30-27 yields the magnitude of $\vec{E}$ at $r = 12.5$ cm:

$$E = \frac{(8.5 \times 10^{-2}\,\text{m})^2}{(2)(12.5 \times 10^{-2}\,\text{m})}\,(0.13\,\text{T/s})$$

$$= 3.8 \times 10^{-3}\,\text{V/m} = 3.8\,\text{mV/m}.\qquad\text{(Answer)}$$

Equations 30-25 and 30-27 give the same result for $r = R$. Figure 30-12 shows a plot of $E(r)$. Note that the inside and outside plots meet at $r = R$.

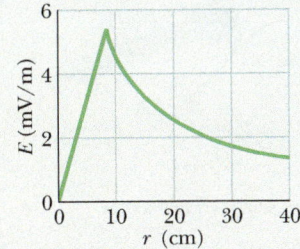

Figure 30-12 A plot of the induced electric field $E(r)$.

WILEY PLUS Additional examples, video, and practice available at *WileyPLUS*

30-4 INDUCTORS AND INDUCTANCE

Learning Objectives

After reading this module, you should be able to . . .

30.19 Identify an inductor.

30.20 For an inductor, apply the relationship between inductance L, total flux $N\Phi$, and current i.

30.21 For a solenoid, apply the relationship between the inductance per unit length L/l, the area A of each turn, and the number of turns per unit length n.

Key Ideas

● An inductor is a device that can be used to produce a known magnetic field in a specified region. If a current i is established through each of the N windings of an inductor, a magnetic flux Φ_B links those windings. The inductance L of the inductor is

$$L = \frac{N\Phi_B}{i}\qquad\text{(inductance defined)}.$$

● The SI unit of inductance is the henry (H), where 1 henry = $1\,\text{H} = 1\,\text{T}\cdot\text{m}^2/\text{A}$.

● The inductance per unit length near the middle of a long solenoid of cross-sectional area A and n turns per unit length is

$$\frac{L}{l} = \mu_0 n^2 A\qquad\text{(solenoid)}.$$

Inductors and Inductance

We found in Chapter 25 that a capacitor can be used to produce a desired electric field. We considered the parallel-plate arrangement as a basic type of capacitor. Similarly, an **inductor** (symbol ⚬⚬⚬⚬) can be used to produce a desired magnetic field. We shall consider a long solenoid (more specifically, a short length near the middle of a long solenoid, to avoid any fringing effects) as our basic type of inductor.

If we establish a current i in the windings (turns) of the solenoid we are taking as our inductor, the current produces a magnetic flux Φ_B through the central region of the inductor. The **inductance** of the inductor is then defined in terms of that flux as

$$L = \frac{N\Phi_B}{i}\qquad\text{(inductance defined)},\qquad(30\text{-}28)$$

The Royal Institution/Bridgeman Art Library/NY

The crude inductors with which Michael Faraday discovered the law of induction. In those days amenities such as insulated wire were not commercially available. It is said that Faraday insulated his wires by wrapping them with strips cut from one of his wife's petticoats.

in which N is the number of turns. The windings of the inductor are said to be *linked* by the shared flux, and the product $N\Phi_B$ is called the *magnetic flux linkage*. The inductance L is thus a measure of the flux linkage produced by the inductor per unit of current.

Because the SI unit of magnetic flux is the tesla–square meter, the SI unit of inductance is the tesla–square meter per ampere (T·m²/A). We call this the **henry** (H), after American physicist Joseph Henry, the codiscoverer of the law of induction and a contemporary of Faraday. Thus,

$$1 \text{ henry} = 1 \text{ H} = 1 \text{ T·m}^2/\text{A}. \tag{30-29}$$

Through the rest of this chapter we assume that all inductors, no matter what their geometric arrangement, have no magnetic materials such as iron in their vicinity. Such materials would distort the magnetic field of an inductor.

Inductance of a Solenoid

Consider a long solenoid of cross-sectional area A. What is the inductance per unit length near its middle? To use the defining equation for inductance (Eq. 30-28), we must calculate the flux linkage set up by a given current in the solenoid windings. Consider a length l near the middle of this solenoid. The flux linkage there is

$$N\Phi_B = (nl)(BA),$$

in which n is the number of turns per unit length of the solenoid and B is the magnitude of the magnetic field within the solenoid.

The magnitude B is given by Eq. 29-23,

$$B = \mu_0 in,$$

and so from Eq. 30-28,

$$L = \frac{N\Phi_B}{i} = \frac{(nl)(BA)}{i} = \frac{(nl)(\mu_0 in)(A)}{i}$$
$$= \mu_0 n^2 l A. \tag{30-30}$$

Thus, the inductance per unit length near the center of a long solenoid is

$$\frac{L}{l} = \mu_0 n^2 A \quad \text{(solenoid)}. \tag{30-31}$$

Inductance—like capacitance—depends only on the geometry of the device. The dependence on the square of the number of turns per unit length is to be expected. If you, say, triple n, you not only triple the number of turns (N) but you also triple the flux ($\Phi_B = BA = \mu_0 in A$) through each turn, multiplying the flux linkage $N\Phi_B$ and thus the inductance L by a factor of 9.

If the solenoid is very much longer than its radius, then Eq. 30-30 gives its inductance to a good approximation. This approximation neglects the spreading of the magnetic field lines near the ends of the solenoid, just as the parallel-plate capacitor formula ($C = \varepsilon_0 A/d$) neglects the fringing of the electric field lines near the edges of the capacitor plates.

From Eq. 30-30, and recalling that n is a number per unit length, we can see that an inductance can be written as a product of the permeability constant μ_0 and a quantity with the dimensions of a length. This means that μ_0 can be expressed in the unit henry per meter:

$$\mu_0 = 4\pi \times 10^{-7} \text{ T·m/A}$$
$$= 4\pi \times 10^{-7} \text{ H/m}. \tag{30-32}$$

The latter is the more common unit for the permeability constant.

30-5 SELF-INDUCTION

Learning Objectives

After reading this module, you should be able to . . .

30.22 Identify that an induced emf appears in a coil when the current through the coil is changing.

30.23 Apply the relationship between the induced emf in a coil, the coil's inductance L, and the rate di/dt at which the current is changing.

30.24 When an emf is induced in a coil because the current in the coil is changing, determine the direction of the emf by using Lenz's law to show that the emf always opposes the change in the current, attempting to maintain the initial current.

Key Ideas

● If a current i in a coil changes with time, an emf is induced in the coil. This self-induced emf is

$$\mathscr{E}_L = -L\frac{di}{dt}.$$

● The direction of $\mathscr{E}_L$ is found from Lenz's law: The self-induced emf acts to oppose the change that produces it.

Self-Induction

If two coils—which we can now call inductors—are near each other, a current i in one coil produces a magnetic flux Φ_B through the second coil. We have seen that if we change this flux by changing the current, an induced emf appears in the second coil according to Faraday's law. An induced emf appears in the first coil as well.

An induced emf $\mathscr{E}_L$ appears in any coil in which the current is changing.

This process (see Fig. 30-13) is called **self-induction,** and the emf that appears is called a **self-induced emf.** It obeys Faraday's law of induction just as other induced emfs do.

For any inductor, Eq. 30-28 tells us that

$$N\Phi_B = Li. \tag{30-33}$$

Faraday's law tells us that

$$\mathscr{E}_L = -\frac{d(N\Phi_B)}{dt}. \tag{30-34}$$

By combining Eqs. 30-33 and 30-34 we can write

$$\mathscr{E}_L = -L\frac{di}{dt} \quad \text{(self-induced emf).} \tag{30-35}$$

Thus, in any inductor (such as a coil, a solenoid, or a toroid) a self-induced emf appears whenever the current changes with time. The magnitude of the current has no influence on the magnitude of the induced emf; only the rate of change of the current counts.

Direction. You can find the *direction* of a self-induced emf from Lenz's law. The minus sign in Eq. 30-35 indicates that—as the law states—the self-induced emf $\mathscr{E}_L$ has the orientation such that it opposes the change in current i. We can drop the minus sign when we want only the magnitude of $\mathscr{E}_L$.

Suppose that you set up a current i in a coil and arrange to have the current increase with time at a rate di/dt. In the language of Lenz's law, this increase in the current in the coil is the "change" that the self-induction must oppose. Thus, a self-induced emf must appear in the coil, pointing so as to oppose the increase in the current, trying (but failing) to maintain the initial condition, as

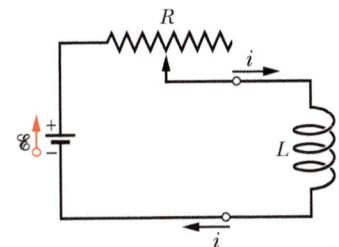

Figure 30-13 If the current in a coil is changed by varying the contact position on a variable resistor, a self-induced emf $\mathscr{E}_L$ will appear in the coil *while the current is changing.*

(a)

(b)

The changing current changes the flux, which creates an emf that opposes the change.

Figure 30-14 (a) The current i is increasing, and the self-induced emf $\mathcal{E}_L$ appears along the coil in a direction such that it opposes the increase. The arrow representing $\mathcal{E}_L$ can be drawn along a turn of the coil or alongside the coil. Both are shown. (b) The current i is decreasing, and the self-induced emf appears in a direction such that it opposes the decrease.

shown in Fig. 30-14a. If, instead, the current decreases with time, the self-induced emf must point in a direction that tends to oppose the decrease (Fig. 30-14b), again trying to maintain the initial condition.

Electric Potential. In Module 30-3 we saw that we cannot define an electric potential for an electric field (and thus for an emf) that is induced by a changing magnetic flux. This means that when a self-induced emf is produced in the inductor of Fig. 30-13, we cannot define an electric potential within the inductor itself, where the flux is changing. However, potentials can still be defined at points of the circuit that are not within the inductor—points where the electric fields are due to charge distributions and their associated electric potentials.

Moreover, we can define a self-induced potential difference V_L across an inductor (between its terminals, which we assume to be outside the region of changing flux). For an *ideal inductor* (its wire has negligible resistance), the magnitude of V_L is equal to the magnitude of the self-induced emf $\mathcal{E}_L$.

If, instead, the wire in the inductor has resistance r, we mentally separate the inductor into a resistance r (which we take to be outside the region of changing flux) and an ideal inductor of self-induced emf $\mathcal{E}_L$. As with a real battery of emf $\mathcal{E}$ and internal resistance r, the potential difference across the terminals of a real inductor then differs from the emf. Unless otherwise indicated, we assume here that inductors are ideal.

✓ **Checkpoint 5**

The figure shows an emf $\mathcal{E}_L$ induced in a coil. Which of the following can describe the current through the coil: (a) constant and rightward, (b) constant and leftward, (c) increasing and rightward, (d) decreasing and rightward, (e) increasing and leftward, (f) decreasing and leftward?

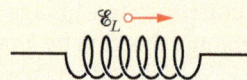

30-6 *RL* CIRCUITS

Learning Objectives

After reading this module, you should be able to . . .

30.25 Sketch a schematic diagram of an *RL* circuit in which the current is rising.

30.26 Write a loop equation (a differential equation) for an *RL* circuit in which the current is rising.

30.27 For an *RL* circuit in which the current is rising, apply the equation i(t) for the current as a function of time.

30.28 For an *RL* circuit in which the current is rising, find equations for the potential difference V across the resistor, the rate di/dt at which the current changes, and the emf of the inductor, as functions of time.

30.29 Calculate an inductive time constant τ_L.

30.30 Sketch a schematic diagram of an *RL* circuit in which the current is decaying.

30.31 Write a loop equation (a differential equation) for an *RL* circuit in which the current is decaying.

30.32 For an *RL* circuit in which the current is decaying, apply the equation i(t) for the current as a function of time.

30.33 From an equation for decaying current in an *RL* circuit, find equations for the potential difference V across the resistor, the rate di/dt at which current is changing, and the emf of the inductor, as functions of time.

30.34 For an *RL* circuit, identify the current through the inductor and the emf across it just as current in the circuit begins to change (the initial condition) and a long time later when equilibrium is reached (the final condition).

Key Ideas

● If a constant emf $\mathcal{E}$ is introduced into a single-loop circuit containing a resistance R and an inductance L, the current rises to an equilibrium value of $\mathcal{E}/R$ according to

$$i = \frac{\mathcal{E}}{R}\left(1 - e^{-t/\tau_L}\right) \quad \text{(rise of current)}.$$

Here τ_L (= L/R) governs the rate of rise of the current and is called the inductive time constant of the circuit.

● When the source of constant emf is removed, the current decays from a value i_0 according to

$$i = i_0 e^{-t/\tau_L} \quad \text{(decay of current)}.$$

RL Circuits

In Module 27-4 we saw that if we suddenly introduce an emf $\mathscr{E}$ into a single-loop circuit containing a resistor R and a capacitor C, the charge on the capacitor does not build up immediately to its final equilibrium value $C\mathscr{E}$ but approaches it in an exponential fashion:

$$q = C\mathscr{E}(1 - e^{-t/\tau_C}). \quad (30\text{-}36)$$

The rate at which the charge builds up is determined by the capacitive time constant τ_C, defined in Eq. 27-36 as

$$\tau_C = RC. \quad (30\text{-}37)$$

If we suddenly remove the emf from this same circuit, the charge does not immediately fall to zero but approaches zero in an exponential fashion:

$$q = q_0 e^{-t/\tau_C}. \quad (30\text{-}38)$$

The time constant τ_C describes the fall of the charge as well as its rise.

An analogous slowing of the rise (or fall) of the current occurs if we introduce an emf $\mathscr{E}$ into (or remove it from) a single-loop circuit containing a resistor R and an inductor L. When the switch S in Fig. 30-15 is closed on a, for example, the current in the resistor starts to rise. If the inductor were not present, the current would rise rapidly to a steady value $\mathscr{E}/R$. Because of the inductor, however, a self-induced emf $\mathscr{E}_L$ appears in the circuit; from Lenz's law, this emf opposes the rise of the current, which means that it opposes the battery emf $\mathscr{E}$ in polarity. Thus, the current in the resistor responds to the difference between two emfs, a constant $\mathscr{E}$ due to the battery and a variable $\mathscr{E}_L \ (= -L\ di/dt)$ due to self-induction. As long as this $\mathscr{E}_L$ is present, the current will be less than $\mathscr{E}/R$.

As time goes on, the rate at which the current increases becomes less rapid and the magnitude of the self-induced emf, which is proportional to di/dt, becomes smaller. Thus, the current in the circuit approaches $\mathscr{E}/R$ asymptotically.

We can generalize these results as follows:

> Initially, an inductor acts to oppose changes in the current through it. A long time later, it acts like ordinary connecting wire.

Now let us analyze the situation quantitatively. With the switch S in Fig. 30-15 thrown to a, the circuit is equivalent to that of Fig. 30-16. Let us apply the loop rule, starting at point x in this figure and moving clockwise around the loop along with current i.

1. *Resistor.* Because we move through the resistor in the direction of current i, the electric potential decreases by iR. Thus, as we move from point x to point y, we encounter a potential change of $-iR$.

2. *Inductor.* Because current i is changing, there is a self-induced emf $\mathscr{E}_L$ in the inductor. The magnitude of $\mathscr{E}_L$ is given by Eq. 30-35 as $L\ di/dt$. The direction of $\mathscr{E}_L$ is upward in Fig. 30-16 because current i is downward through the inductor *and* increasing. Thus, as we move from point y to point z, opposite the direction of $\mathscr{E}_L$, we encounter a potential change of $-L\ di/dt$.

3. *Battery.* As we move from point z back to starting point x, we encounter a potential change of $+\mathscr{E}$ due to the battery's emf.

Thus, the loop rule gives us

$$-iR - L\frac{di}{dt} + \mathscr{E} = 0$$

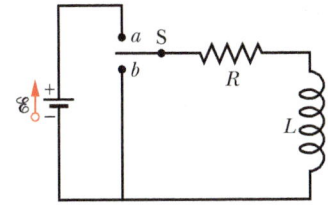

Figure 30-15 An *RL* circuit. When switch S is closed on a, the current rises and approaches a limiting value $\mathscr{E}/R$.

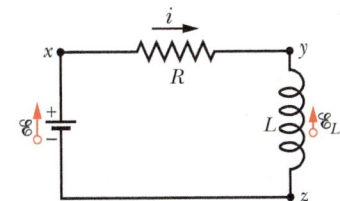

Figure 30-16 The circuit of Fig. 30-15 with the switch closed on a. We apply the loop rule for the circuit clockwise, starting at x.

or
$$L\frac{di}{dt} + Ri = \mathscr{E} \quad \text{(RL circuit).} \tag{30-39}$$

Equation 30-39 is a differential equation involving the variable i and its first derivative di/dt. To solve it, we seek the function $i(t)$ such that when $i(t)$ and its first derivative are substituted in Eq. 30-39, the equation is satisfied and the initial condition $i(0) = 0$ is satisfied.

Equation 30-39 and its initial condition are of exactly the form of Eq. 27-32 for an RC circuit, with i replacing q, L replacing R, and R replacing $1/C$. The solution of Eq. 30-39 must then be of exactly the form of Eq. 27-33 with the same replacements. That solution is

$$i = \frac{\mathscr{E}}{R}(1 - e^{-Rt/L}), \tag{30-40}$$

which we can rewrite as

$$i = \frac{\mathscr{E}}{R}(1 - e^{-t/\tau_L}) \quad \text{(rise of current).} \tag{30-41}$$

Here τ_L, the **inductive time constant,** is given by

$$\tau_L = \frac{L}{R} \quad \text{(time constant).} \tag{30-42}$$

Let's examine Eq. 30-41 for just after the switch is closed (at time $t = 0$) and for a time long after the switch is closed ($t \rightarrow \infty$). If we substitute $t = 0$ into Eq. 30-41, the exponential becomes $e^{-0} = 1$. Thus, Eq. 30-41 tells us that the current is initially $i = 0$, as we expected. Next, if we let t go to ∞, then the exponential goes to $e^{-\infty} = 0$. Thus, Eq. 30-41 tells us that the current goes to its equilibrium value of $\mathscr{E}/R$.

We can also examine the potential differences in the circuit. For example, Fig. 30-17 shows how the potential differences V_R ($= iR$) across the resistor and V_L ($= L\,di/dt$) across the inductor vary with time for particular values of $\mathscr{E}$, L, and R. Compare this figure carefully with the corresponding figure for an RC circuit (Fig. 27-16).

The resistor's potential difference turns on. The inductor's potential difference turns off.

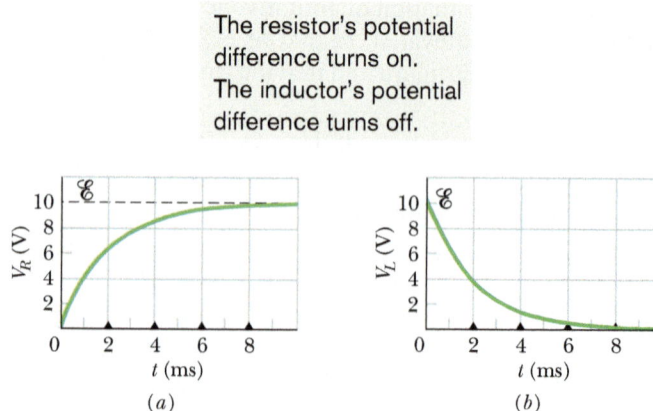

(a)

(b)

Figure 30-17 The variation with time of (a) V_R, the potential difference across the resistor in the circuit of Fig. 30-16, and (b) V_L, the potential difference across the inductor in that circuit. The small triangles represent successive intervals of one inductive time constant $\tau_L = L/R$. The figure is plotted for $R = 2000\ \Omega$, $L = 4.0$ H, and $\mathscr{E} = 10$ V.

To show that the quantity τ_L $(= L/R)$ has the dimension of time (as it must, because the argument of the exponential function in Eq. 30-41 must be dimensionless), we convert from henries per ohm as follows:

$$1\frac{H}{\Omega} = 1\frac{H}{\Omega}\left(\frac{1\,V \cdot s}{1\,H \cdot A}\right)\left(\frac{1\,\Omega \cdot A}{1\,V}\right) = 1\,s.$$

The first quantity in parentheses is a conversion factor based on Eq. 30-35, and the second one is a conversion factor based on the relation $V = iR$.

Time Constant. The physical significance of the time constant follows from Eq. 30-41. If we put $t = \tau_L = L/R$ in this equation, it reduces to

$$i = \frac{\mathscr{E}}{R}(1 - e^{-1}) = 0.63\,\frac{\mathscr{E}}{R}. \tag{30-43}$$

Thus, the time constant τ_L is the time it takes the current in the circuit to reach about 63% of its final equilibrium value $\mathscr{E}/R$. Since the potential difference V_R across the resistor is proportional to the current i, a graph of the increasing current versus time has the same shape as that of V_R in Fig. 30-17*a*.

Current Decay. If the switch S in Fig. 30-15 is closed on *a* long enough for the equilibrium current $\mathscr{E}/R$ to be established and then is thrown to *b*, the effect will be to remove the battery from the circuit. (The connection to *b* must actually be made an instant before the connection to *a* is broken. A switch that does this is called a *make-before-break* switch.) With the battery gone, the current through the resistor will decrease. However, it cannot drop immediately to zero but must decay to zero over time. The differential equation that governs the decay can be found by putting $\mathscr{E} = 0$ in Eq. 30-39:

$$L\frac{di}{dt} + iR = 0. \tag{30-44}$$

By analogy with Eqs. 27-38 and 27-39, the solution of this differential equation that satisfies the initial condition $i(0) = i_0 = \mathscr{E}/R$ is

$$i = \frac{\mathscr{E}}{R}e^{-t/\tau_L} = i_0e^{-t/\tau_L} \quad \text{(decay of current)}. \tag{30-45}$$

We see that both current rise (Eq. 30-41) and current decay (Eq. 30-45) in an *RL* circuit are governed by the same inductive time constant, τ_L.

We have used i_0 in Eq. 30-45 to represent the current at time $t = 0$. In our case that happened to be $\mathscr{E}/R$, but it could be any other initial value.

✔ **Checkpoint 6**

The figure shows three circuits with identical batteries, inductors, and resistors. Rank the circuits according to the current through the battery (a) just after the switch is closed and (b) a long time later, greatest first. (If you have trouble here, work through the next sample problem and then try again.)

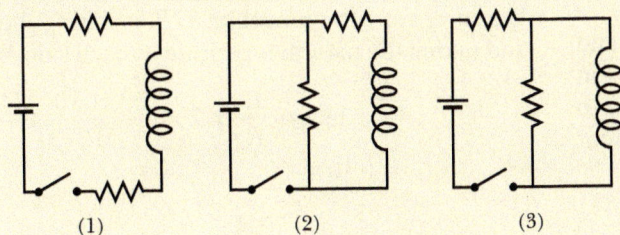

(1)   (2)   (3)

Sample Problem 30.05 *RL* circuit, immediately after switching and after a long time

Figure 30-18a shows a circuit that contains three identical resistors with resistance $R = 9.0\ \Omega$, two identical inductors with inductance $L = 2.0$ mH, and an ideal battery with emf $\mathscr{E} = 18$ V.

(a) What is the current i through the battery just after the switch is closed?

KEY IDEA

Just after the switch is closed, the inductor acts to oppose a change in the current through it.

Calculations: Because the current through each inductor is zero before the switch is closed, it will also be zero just afterward. Thus, immediately after the switch is closed, the inductors act as broken wires, as indicated in Fig. 30-18b. We then have a single-loop circuit for which the loop rule gives us

$$\mathscr{E} - iR = 0.$$

Substituting given data, we find that

$$i = \frac{\mathscr{E}}{R} = \frac{18\text{ V}}{9.0\ \Omega} = 2.0\text{ A.} \qquad \text{(Answer)}$$

(b) What is the current i through the battery long after the switch has been closed?

KEY IDEA

Long after the switch has been closed, the currents in the circuit have reached their equilibrium values, and the inductors act as simple connecting wires, as indicated in Fig. 30-18c.

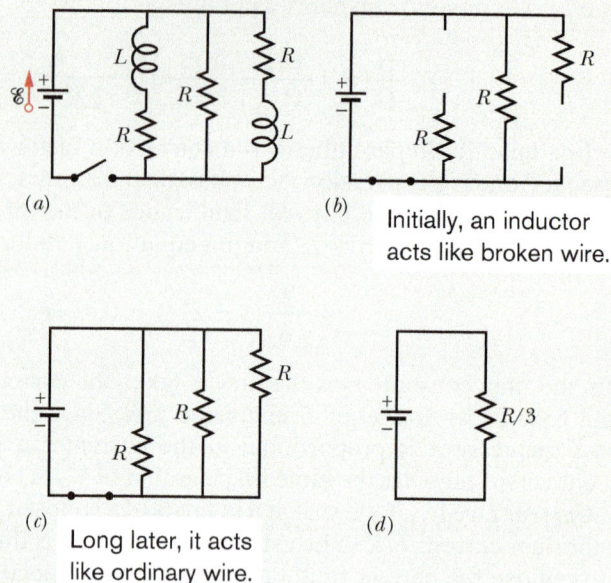

(a)   (b)

Initially, an inductor acts like broken wire.

(c)   (d)

Long later, it acts like ordinary wire.

Figure 30-18 (a) A multiloop *RL* circuit with an open switch. (b) The equivalent circuit just after the switch has been closed. (c) The equivalent circuit a long time later. (d) The single-loop circuit that is equivalent to circuit (c).

Calculations: We now have a circuit with three identical resistors in parallel; from Eq. 27-23, their equivalent resistance is $R_{eq} = R/3 = (9.0\ \Omega)/3 = 3.0\ \Omega$. The equivalent circuit shown in Fig. 30-18d then yields the loop equation $\mathscr{E} - iR_{eq} = 0$, or

$$i = \frac{\mathscr{E}}{R_{eq}} = \frac{18\text{ V}}{3.0\ \Omega} = 6.0\text{ A.} \qquad \text{(Answer)}$$

Sample Problem 30.06 *RL* circuit, current during the transition

A solenoid has an inductance of 53 mH and a resistance of $0.37\ \Omega$. If the solenoid is connected to a battery, how long will the current take to reach half its final equilibrium value? (This is a *real solenoid* because we are considering its small, but nonzero, internal resistance.)

KEY IDEA

We can mentally separate the solenoid into a resistance and an inductance that are wired in series with a battery, as in Fig. 30-16. Then application of the loop rule leads to Eq. 30-39, which has the solution of Eq. 30-41 for the current i in the circuit.

Calculations: According to that solution, current i increases exponentially from zero to its final equilibrium value of $\mathscr{E}/R$. Let t_0 be the time that current i takes to reach half its equilibrium value. Then Eq. 30-41 gives us

$$\frac{1}{2}\frac{\mathscr{E}}{R} = \frac{\mathscr{E}}{R}(1 - e^{-t_0/\tau_L}).$$

We solve for t_0 by canceling $\mathscr{E}/R$, isolating the exponential, and taking the natural logarithm of each side. We find

$$t_0 = \tau_L \ln 2 = \frac{L}{R}\ln 2 = \frac{53 \times 10^{-3}\text{ H}}{0.37\ \Omega}\ln 2$$

$$= 0.10\text{ s.} \qquad \text{(Answer)}$$

30-7 ENERGY STORED IN A MAGNETIC FIELD

Learning Objectives

After reading this module, you should be able to . . .

30.35 Describe the derivation of the equation for the magnetic field energy of an inductor in an *RL* circuit with a constant emf source.

30.36 For an inductor in an *RL* circuit, apply the relationship between the magnetic field energy U, the inductance L, and the current i.

Key Idea

● If an inductor L carries a current i, the inductor's magnetic field stores an energy given by

$$U_B = \tfrac{1}{2}Li^2 \quad \text{(magnetic energy)}.$$

Energy Stored in a Magnetic Field

When we pull two charged particles of opposite signs away from each other, we say that the resulting electric potential energy is stored in the electric field of the particles. We get it back from the field by letting the particles move closer together again. In the same way we say energy is stored in a magnetic field, but now we deal with current instead of electric charges.

To derive a quantitative expression for that stored energy, consider again Fig. 30-16, which shows a source of emf $\mathscr{E}$ connected to a resistor R and an inductor L. Equation 30-39, restated here for convenience,

$$\mathscr{E} = L\frac{di}{dt} + iR, \tag{30-46}$$

is the differential equation that describes the growth of current in this circuit. Recall that this equation follows immediately from the loop rule and that the loop rule in turn is an expression of the principle of conservation of energy for single-loop circuits. If we multiply each side of Eq. 30-46 by i, we obtain

$$\mathscr{E}i = Li\frac{di}{dt} + i^2R, \tag{30-47}$$

which has the following physical interpretation in terms of the work done by the battery and the resulting energy transfers:

1. If a differential amount of charge dq passes through the battery of emf $\mathscr{E}$ in Fig. 30-16 in time dt, the battery does work on it in the amount $\mathscr{E}\,dq$. The rate at which the battery does work is $(\mathscr{E}\,dq)/dt$, or $\mathscr{E}i$. Thus, the left side of Eq. 30-47 represents the rate at which the emf device delivers energy to the rest of the circuit.

2. The rightmost term in Eq. 30-47 represents the rate at which energy appears as thermal energy in the resistor.

3. Energy that is delivered to the circuit but does not appear as thermal energy must, by the conservation-of-energy hypothesis, be stored in the magnetic field of the inductor. Because Eq. 30-47 represents the principle of conservation of energy for *RL* circuits, the middle term must represent the rate dU_B/dt at which magnetic potential energy U_B is stored in the magnetic field.

Thus

$$\frac{dU_B}{dt} = Li\frac{di}{dt}. \tag{30-48}$$

We can write this as

$$dU_B = Li\,di.$$

Integrating yields

$$\int_0^{U_B} dU_B = \int_0^i Li\,di$$

or $$U_B = \tfrac{1}{2}Li^2 \quad \text{(magnetic energy)}, \tag{30-49}$$

which represents the total energy stored by an inductor L carrying a current i. Note the similarity in form between this expression for the energy stored in a magnetic field and the expression for the energy stored in an electric field by a capacitor with capacitance C and charge q; namely,

$$U_E = \frac{q^2}{2C}. \tag{30-50}$$

(The variable i^2 corresponds to q^2, and the constant L corresponds to $1/C$.)

Sample Problem 30.07 Energy stored in a magnetic field

A coil has an inductance of 53 mH and a resistance of 0.35 Ω.

(a) If a 12 V emf is applied across the coil, how much energy is stored in the magnetic field after the current has built up to its equilibrium value?

KEY IDEA

The energy stored in the magnetic field of a coil at any time depends on the current through the coil at that time, according to Eq. 30-49 ($U_B = \tfrac{1}{2}Li^2$).

Calculations: Thus, to find the energy $U_{B\infty}$ stored at equilibrium, we must first find the equilibrium current. From Eq. 30-41, the equilibrium current is

$$i_\infty = \frac{\mathscr{E}}{R} = \frac{12\ \text{V}}{0.35\ \Omega} = 34.3\ \text{A}. \tag{30-51}$$

Then substitution yields

$$U_{B\infty} = \tfrac{1}{2}Li_\infty^2 = (\tfrac{1}{2})(53 \times 10^{-3}\ \text{H})(34.3\ \text{A})^2$$

$$= 31\ \text{J}. \tag{Answer}$$

(b) After how many time constants will half this equilibrium energy be stored in the magnetic field?

Calculations: Now we are being asked: At what time t will the relation

$$U_B = \tfrac{1}{2}U_{B\infty}$$

be satisfied? Using Eq. 30-49 twice allows us to rewrite this energy condition as

$$\tfrac{1}{2}Li^2 = (\tfrac{1}{2})\tfrac{1}{2}Li_\infty^2$$

or $$i = \left(\frac{1}{\sqrt{2}}\right)i_\infty. \tag{30-52}$$

This equation tells us that, as the current increases from its initial value of 0 to its final value of i_∞, the magnetic field will have half its final stored energy when the current has increased to this value. In general, we know that i is given by Eq. 30-41, and here i_∞ (see Eq. 30-51) is $\mathscr{E}/R$; so Eq. 30-52 becomes

$$\frac{\mathscr{E}}{R}(1 - e^{-t/\tau_L}) = \frac{\mathscr{E}}{\sqrt{2}R}.$$

By canceling $\mathscr{E}/R$ and rearranging, we can write this as

$$e^{-t/\tau_L} = 1 - \frac{1}{\sqrt{2}} = 0.293,$$

which yields

$$\frac{t}{\tau_L} = -\ln 0.293 = 1.23$$

or $$t \approx 1.2\tau_L. \tag{Answer}$$

Thus, the energy stored in the magnetic field of the coil by the current will reach half its equilibrium value 1.2 time constants after the emf is applied.

PLUS Additional examples, video, and practice available at *WileyPLUS*

30-8 ENERGY DENSITY OF A MAGNETIC FIELD

Learning Objectives

After reading this module, you should be able to . . .

30.37 Identify that energy is associated with any magnetic field.

30.38 Apply the relationship between energy density u_B of a magnetic field and the magnetic field magnitude B.

Key Idea

- If B is the magnitude of a magnetic field at any point (in an inductor or anywhere else), the density of stored magnetic energy at that point is

$$u_B = \frac{B^2}{2\mu_0} \quad \text{(magnetic energy density)}.$$

Energy Density of a Magnetic Field

Consider a length l near the middle of a long solenoid of cross-sectional area A carrying current i; the volume associated with this length is Al. The energy U_B stored by the length l of the solenoid must lie entirely within this volume because the magnetic field outside such a solenoid is approximately zero. Moreover, the stored energy must be uniformly distributed within the solenoid because the magnetic field is (approximately) uniform everywhere inside.

Thus, the energy stored per unit volume of the field is

$$u_B = \frac{U_B}{Al}$$

or, since

$$U_B = \tfrac{1}{2}Li^2,$$

we have

$$u_B = \frac{Li^2}{2Al} = \frac{L}{l}\frac{i^2}{2A}. \tag{30-53}$$

Here L is the inductance of length l of the solenoid.

Substituting for L/l from Eq. 30-31, we find

$$u_B = \tfrac{1}{2}\mu_0 n^2 i^2, \tag{30-54}$$

where n is the number of turns per unit length. From Eq. 29-23 ($B = \mu_0 in$) we can write this *energy density* as

$$u_B = \frac{B^2}{2\mu_0} \quad \text{(magnetic energy density)}. \tag{30-55}$$

This equation gives the density of stored energy at any point where the magnitude of the magnetic field is B. Even though we derived it by considering the special case of a solenoid, Eq. 30-55 holds for all magnetic fields, no matter how they are generated. The equation is comparable to Eq. 25-25,

$$u_E = \tfrac{1}{2}\varepsilon_0 E^2, \tag{30-56}$$

which gives the energy density (in a vacuum) at any point in an electric field. Note that both u_B and u_E are proportional to the square of the appropriate field magnitude, B or E.

☑ **Checkpoint 7**

The table lists the number of turns per unit length, current, and cross-sectional area for three solenoids. Rank the solenoids according to the magnetic energy density within them, greatest first.

Solenoid	Turns per Unit Length	Current	Area
a	$2n_1$	i_1	$2A_1$
b	n_1	$2i_1$	A_1
c	n_1	i_1	$6A_1$

30-9 MUTUAL INDUCTION

Learning Objectives

After reading this module, you should be able to . . .

30.39 Describe the mutual induction of two coils and sketch the arrangement.

30.40 Calculate the mutual inductance of one coil with respect to a second coil (or some second current that is changing).

30.41 Calculate the emf induced in one coil by a second coil in terms of the mutual inductance and the rate of change of the current in the second coil.

Key Idea

● If coils 1 and 2 are near each other, a changing current in either coil can induce an emf in the other. This mutual induction is described by

$$\mathscr{E}_2 = -M \frac{di_1}{dt}$$

and

$$\mathscr{E}_1 = -M \frac{di_2}{dt},$$

where M (measured in henries) is the mutual inductance.

Mutual Induction

In this section we return to the case of two interacting coils, which we first discussed in Module 30-1, and we treat it in a somewhat more formal manner. We saw earlier that if two coils are close together as in Fig. 30-2, a steady current i in one coil will set up a magnetic flux Φ through the other coil (*linking* the other coil). If we change i with time, an emf $\mathscr{E}$ given by Faraday's law appears in the second coil; we called this process *induction*. We could better have called it **mutual induction,** to suggest the mutual interaction of the two coils and to distinguish it from *self-induction*, in which only one coil is involved.

Let us look a little more quantitatively at mutual induction. Figure 30-19a shows two circular close-packed coils near each other and sharing a common central axis. With the variable resistor set at a particular resistance R, the battery produces a steady current i_1 in coil 1. This current creates a magnetic field represented by the lines of $\vec{B}_1$ in the figure. Coil 2 is connected to a sensitive meter but contains no battery; a magnetic flux Φ_{21} (the flux through coil 2 associated with the current in coil 1) links the N_2 turns of coil 2.

We define the mutual inductance M_{21} of coil 2 with respect to coil 1 as

$$M_{21} = \frac{N_2 \Phi_{21}}{i_1}, \tag{30-57}$$

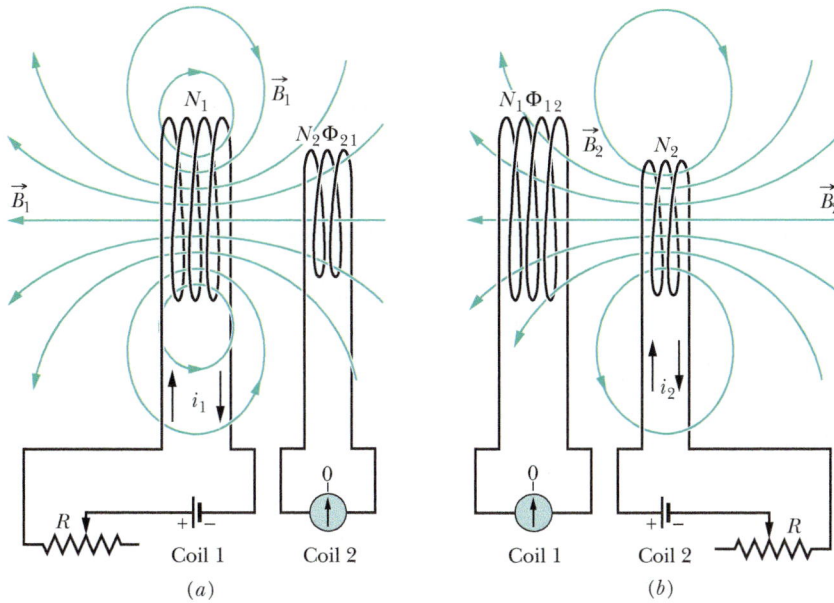

Figure 30-19 Mutual induction. (*a*) The magnetic field $\vec{B}_1$ produced by current i_1 in coil 1 extends through coil 2. If i_1 is varied (by varying resistance R), an emf is induced in coil 2 and current registers on the meter connected to coil 2. (*b*) The roles of the coils interchanged.

which has the same form as Eq. 30-28,

$$L = N\Phi/i, \tag{30-58}$$

the definition of inductance. We can recast Eq. 30-57 as

$$M_{21}i_1 = N_2\Phi_{21}. \tag{30-59}$$

If we cause i_1 to vary with time by varying R, we have

$$M_{21}\frac{di_1}{dt} = N_2\frac{d\Phi_{21}}{dt}. \tag{30-60}$$

The right side of this equation is, according to Faraday's law, just the magnitude of the emf $\mathscr{E}_2$ appearing in coil 2 due to the changing current in coil 1. Thus, with a minus sign to indicate direction,

$$\mathscr{E}_2 = -M_{21}\frac{di_1}{dt}, \tag{30-61}$$

which you should compare with Eq. 30-35 for self-induction ($\mathscr{E} = -L\,di/dt$).

Interchange. Let us now interchange the roles of coils 1 and 2, as in Fig. 30-19*b*; that is, we set up a current i_2 in coil 2 by means of a battery, and this produces a magnetic flux Φ_{12} that links coil 1. If we change i_2 with time by varying R, we then have, by the argument given above,

$$\mathscr{E}_1 = -M_{12}\frac{di_2}{dt}. \tag{30-62}$$

Thus, we see that the emf induced in either coil is proportional to the rate of change of current in the other coil. The proportionality constants M_{21} and M_{12} seem to be different. However, they turn out to be the same, although we cannot prove that fact here. Thus, we have

$$M_{21} = M_{12} = M, \tag{30-63}$$

and we can rewrite Eqs. 30-61 and 30-62 as

$$\mathscr{E}_2 = -M\frac{di_1}{dt} \tag{30-64}$$

and

$$\mathscr{E}_1 = -M\frac{di_2}{dt}. \tag{30-65}$$

Sample Problem 30.08 Mutual inductance of two parallel coils

Figure 30-20 shows two circular close-packed coils, the smaller (radius R_2, with N_2 turns) being coaxial with the larger (radius R_1, with N_1 turns) and in the same plane.

(a) Derive an expression for the mutual inductance M for this arrangement of these two coils, assuming that $R_1 \gg R_2$.

KEY IDEA

The mutual inductance M for these coils is the ratio of the flux linkage ($N\Phi$) through one coil to the current i in the other coil, which produces that flux linkage. Thus, we need to assume that currents exist in the coils; then we need to calculate the flux linkage in one of the coils.

Calculations: The magnetic field through the larger coil due to the smaller coil is nonuniform in both magnitude and direction; so the flux through the larger coil due to the smaller coil is nonuniform and difficult to calculate. However, the smaller coil is small enough for us to assume that the magnetic field through it due to the larger coil is approximately uniform. Thus, the flux through it due to the larger coil is also approximately uniform. Hence, to find M we shall assume a current i_1 in the larger coil and calculate the flux linkage $N_2\Phi_{21}$ in the smaller coil:

$$M = \frac{N_2\Phi_{21}}{i_1}. \qquad (30\text{-}66)$$

The flux Φ_{21} through each turn of the smaller coil is, from Eq. 30-2,

$$\Phi_{21} = B_1 A_2,$$

where B_1 is the magnitude of the magnetic field at points within the small coil due to the larger coil and A_2 ($= \pi R_2^2$) is the area enclosed by the turn. Thus, the flux linkage in the smaller coil (with its N_2 turns) is

$$N_2\Phi_{21} = N_2 B_1 A_2. \qquad (30\text{-}67)$$

To find B_1 at points within the smaller coil, we can use Eq. 29-26,

$$B(z) = \frac{\mu_0 i R^2}{2(R^2 + z^2)^{3/2}},$$

with z set to 0 because the smaller coil is in the plane of the larger coil. That equation tells us that each turn of the larger coil produces a magnetic field of magnitude $\mu_0 i_1/2R_1$ at points within the smaller coil. Thus, the larger coil (with its N_1 turns) produces a total magnetic field of magnitude

$$B_1 = N_1 \frac{\mu_0 i_1}{2R_1} \qquad (30\text{-}68)$$

at points within the smaller coil.

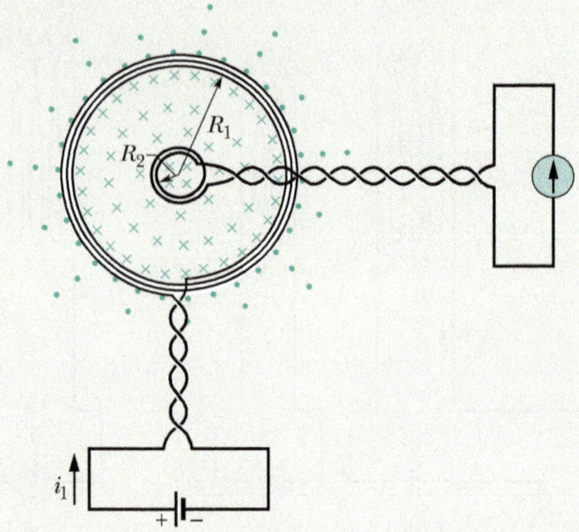

Figure 30-20 A small coil is located at the center of a large coil. The mutual inductance of the coils can be determined by sending current i_1 through the large coil.

Substituting Eq. 30-68 for B_1 and πR_2^2 for A_2 in Eq. 30-67 yields

$$N_2\Phi_{21} = \frac{\pi\mu_0 N_1 N_2 R_2^2 i_1}{2R_1}.$$

Substituting this result into Eq. 30-66, we find

$$M = \frac{N_2\Phi_{21}}{i_1} = \frac{\pi\mu_0 N_1 N_2 R_2^2}{2R_1}. \qquad \text{(Answer)} \quad (30\text{-}69)$$

(b) What is the value of M for $N_1 = N_2 = 1200$ turns, $R_2 = 1.1$ cm, and $R_1 = 15$ cm?

Calculations: Equation 30-69 yields

$$M = \frac{(\pi)(4\pi \times 10^{-7}\ \text{H/m})(1200)(1200)(0.011\ \text{m})^2}{(2)(0.15\ \text{m})}$$

$$= 2.29 \times 10^{-3}\ \text{H} \approx 2.3\ \text{mH}. \qquad \text{(Answer)}$$

Consider the situation if we reverse the roles of the two coils—that is, if we produce a current i_2 in the smaller coil and try to calculate M from Eq. 30-57 in the form

$$M = \frac{N_1\Phi_{12}}{i_2}.$$

The calculation of Φ_{12} (the nonuniform flux of the smaller coil's magnetic field encompassed by the larger coil) is not simple. If we were to do the calculation numerically using a computer, we would find M to be 2.3 mH, as above! This emphasizes that Eq. 30-63 ($M_{21} = M_{12} = M$) is not obvious.

Review & Summary

Magnetic Flux   The *magnetic flux* Φ_B through an area A in a magnetic field $\vec{B}$ is defined as

$$\Phi_B = \int \vec{B} \cdot d\vec{A}, \tag{30-1}$$

where the integral is taken over the area. The SI unit of magnetic flux is the weber, where 1 Wb = 1 T·m². If $\vec{B}$ is perpendicular to the area and uniform over it, Eq. 30-1 becomes

$$\Phi_B = BA \quad (\vec{B} \perp A, \vec{B}\,\text{uniform}). \tag{30-2}$$

Faraday's Law of Induction   If the magnetic flux Φ_B through an area bounded by a closed conducting loop changes with time, a current and an emf are produced in the loop; this process is called *induction*. The induced emf is

$$\mathscr{E} = -\frac{d\Phi_B}{dt} \quad \text{(Faraday's law)}. \tag{30-4}$$

If the loop is replaced by a closely packed coil of N turns, the induced emf is

$$\mathscr{E} = -N\frac{d\Phi_B}{dt}. \tag{30-5}$$

Lenz's Law   An induced current has a direction such that the magnetic field *due to the current* opposes the change in the magnetic flux that induces the current. The induced emf has the same direction as the induced current.

Emf and the Induced Electric Field   An emf is induced by a changing magnetic flux even if the loop through which the flux is changing is not a physical conductor but an imaginary line. The changing magnetic field induces an electric field $\vec{E}$ at every point of such a loop; the induced emf is related to $\vec{E}$ by

$$\mathscr{E} = \oint \vec{E} \cdot d\vec{s}, \tag{30-19}$$

where the integration is taken around the loop. From Eq. 30-19 we can write Faraday's law in its most general form,

$$\oint \vec{E} \cdot d\vec{s} = -\frac{d\Phi_B}{dt} \quad \text{(Faraday's law)}. \tag{30-20}$$

A changing magnetic field induces an electric field $\vec{E}$.

Inductors   An **inductor** is a device that can be used to produce a known magnetic field in a specified region. If a current i is established through each of the N windings of an inductor, a magnetic flux Φ_B links those windings. The **inductance** L of the inductor is

$$L = \frac{N\Phi_B}{i} \quad \text{(inductance defined)}. \tag{30-28}$$

The SI unit of inductance is the **henry** (H), where 1 henry = 1 H = 1 T·m²/A. The inductance per unit length near the middle of a long solenoid of cross-sectional area A and n turns per unit length is

$$\frac{L}{l} = \mu_0 n^2 A \quad \text{(solenoid)}. \tag{30-31}$$

Self-Induction   If a current i in a coil changes with time, an emf is induced in the coil. This self-induced emf is

$$\mathscr{E}_L = -L\frac{di}{dt}. \tag{30-35}$$

The direction of $\mathscr{E}_L$ is found from Lenz's law: The self-induced emf acts to oppose the change that produces it.

Series *RL* Circuits   If a constant emf $\mathscr{E}$ is introduced into a single-loop circuit containing a resistance R and an inductance L, the current rises to an equilibrium value of $\mathscr{E}/R$:

$$i = \frac{\mathscr{E}}{R}\left(1 - e^{-t/\tau_L}\right) \quad \text{(rise of current)}. \tag{30-41}$$

Here $\tau_L \,(= L/R)$ is the **inductive time constant**. When the source of constant emf is removed, the current decays from a value i_0 according to

$$i = i_0 e^{-t/\tau_L} \quad \text{(decay of current)}. \tag{30-45}$$

Magnetic Energy   If an inductor L carries a current i, the inductor's magnetic field stores an energy given by

$$U_B = \tfrac{1}{2}Li^2 \quad \text{(magnetic energy)}. \tag{30-49}$$

If B is the magnitude of a magnetic field at any point (in an inductor or anywhere else), the density of stored magnetic energy at that point is

$$u_B = \frac{B^2}{2\mu_0} \quad \text{(magnetic energy density)}. \tag{30-55}$$

Mutual Induction   If coils 1 and 2 are near each other, a changing current in either coil can induce an emf in the other. This mutual induction is described by

$$\mathscr{E}_2 = -M\frac{di_1}{dt} \tag{30-64}$$

and

$$\mathscr{E}_1 = -M\frac{di_2}{dt}, \tag{30-65}$$

where M (measured in henries) is the mutual inductance.

Questions

1   If the circular conductor in Fig. 30-21 undergoes thermal expansion while it is in a uniform magnetic field, a current is induced clockwise around it. Is the magnetic field directed into or out of the page?

2   The wire loop in Fig. 30-22a is subjected, in turn, to six uniform magnetic fields, each directed parallel to the z

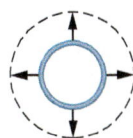

Figure 30-21 Question 1.

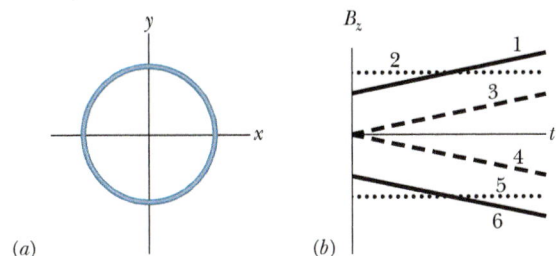

(a)   (b)

Figure 30-22 Question 2.

axis, which is directed out of the plane of the figure. Figure 30-22b gives the z components B_z of the fields versus time t. (Plots 1 and 3 are parallel; so are plots 4 and 6. Plots 2 and 5 are parallel to the time axis.) Rank the six plots according to the emf induced in the loop, greatest clockwise emf first, greatest counterclockwise emf last.

3 In Fig. 30-23, a long straight wire with current i passes (without touching) three rectangular wire loops with edge lengths L, $1.5L$, and $2L$. The loops are widely spaced (so as not to affect one another). Loops 1 and 3 are symmetric about the long wire. Rank the loops according to the size of the current induced in them if current i is (a) constant and (b) increasing, greatest first.

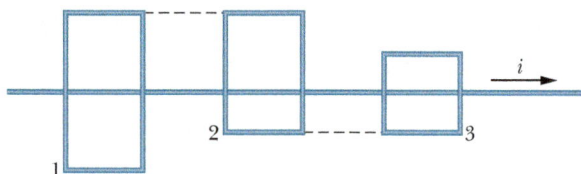

Figure 30-23 Question 3.

4 Figure 30-24 shows two circuits in which a conducting bar is slid at the same speed v through the same uniform magnetic field and along a **U**-shaped wire. The parallel lengths of the wire are separated by $2L$ in circuit 1 and by L in circuit 2. The current induced in circuit 1 is counterclockwise. (a) Is the magnetic field into or out of the page? (b) Is the current induced in circuit 2 clockwise or counterclockwise? (c) Is the emf induced in circuit 1 larger than, smaller than, or the same as that in circuit 2?

Figure 30-24 Question 4.

5 Figure 30-25 shows a circular region in which a decreasing uniform magnetic field is directed out of the page, as well as four concentric circular paths. Rank the paths according to the magnitude of $\oint \vec{E} \cdot d\vec{s}$ evaluated along them, greatest first.

Figure 30-25 Question 5.

6 In Fig. 30-26, a wire loop has been bent so that it has three segments: segment bc (a quarter-circle), ac (a square corner), and ab (straight). Here are three choices for a magnetic field through the loop:

(1) $\vec{B}_1 = 3\hat{i} + 7\hat{j} - 5t\hat{k}$,
(2) $\vec{B}_2 = 5t\hat{i} - 4\hat{j} - 15\hat{k}$,
(3) $\vec{B}_3 = 2\hat{i} - 5t\hat{j} - 12\hat{k}$,

where $\vec{B}$ is in milliteslas and t is in seconds. Without written calcula-

tion, rank the choices according to (a) the work done per unit charge in setting up the induced current and (b) that induced current, greatest first. (c) For each choice, what is the direction of the induced current in the figure?

7 Figure 30-27 shows a circuit with two identical resistors and an ideal inductor. Is the current through the central resistor more than, less than, or the same as that through the other resistor (a) just after the closing of switch S, (b) a long time after that, (c) just after S is reopened a long time later, and (d) a long time after that?

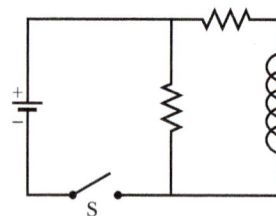

Figure 30-26 Question 6.

Figure 30-27 Question 7.

8 The switch in the circuit of Fig. 30-15 has been closed on a for a very long time when it is then thrown to b. The resulting current through the inductor is indicated in Fig. 30-28 for four sets of values for the resistance R and inductance L: (1) R_0 and L_0, (2) $2R_0$ and L_0, (3) R_0 and $2L_0$, (4) $2R_0$ and $2L_0$. Which set goes with which curve?

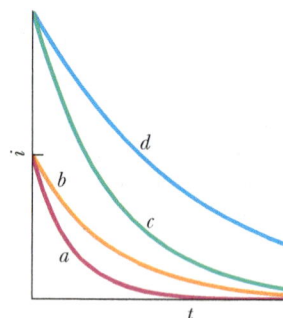

9 Figure 30-29 shows three circuits with identical batteries, inductors, and resistors. Rank the circuits, greatest first, according to the current through the resistor labeled R (a) long after the switch is closed, (b) just after the switch is reopened a long time later, and (c) long after it is reopened.

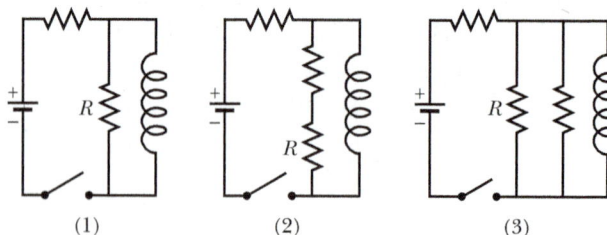

Figure 30-28 Question 8.

Figure 30-29 Question 9.

10 Figure 30-30 gives the variation with time of the potential difference V_R across a resistor in three circuits wired as shown in Fig. 30-16. The circuits contain the same resistance R and emf $\mathscr{E}$ but differ in the inductance L. Rank the circuits according to the value of L, greatest first.

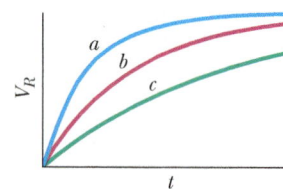

Figure 30-30 Question 10.

11 Figure 30-31 shows three situations in which a wire loop lies partially in a magnetic field. The magnitude of the field is either increasing or decreasing, as indicated. In each situation, a battery is part of the loop. In which situations are the induced emf and the battery emf in the same direction along the loop?

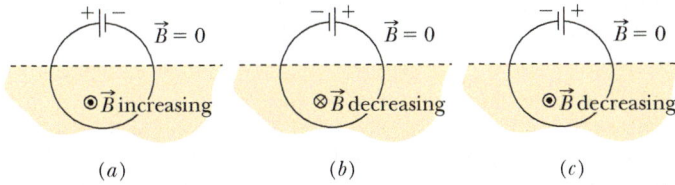

Figure 30-31 Question 11.

12 Figure 30-32 gives four situations in which we pull rectangular wire loops out of identical magnetic fields (directed into the page) at the same constant speed. The loops have edge lengths of either L or $2L$, as drawn. Rank the situations according to (a) the magnitude of the force required of us and (b) the rate at which energy is transferred from us to thermal energy of the loop, greatest first.

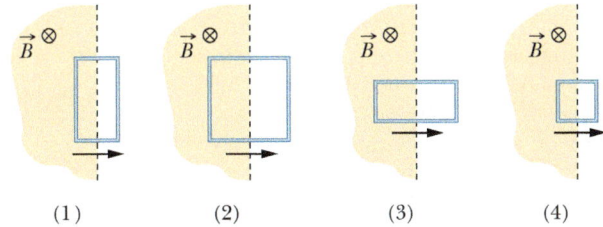

Figure 30-32 Question 12.

Problems

GO Tutoring problem available (at instructor's discretion) in *WileyPLUS* and WebAssign
SSM Worked-out solution available in Student Solutions Manual
•–••• Number of dots indicates level of problem difficulty
WWW Worked-out solution is at http://www.wiley.com/college/halliday
ILW Interactive solution is at
Additional information available in *The Flying Circus of Physics* and at flyingcircusofphysics.com

Module 30-1 Faraday's Law and Lenz's Law

•1 In Fig. 30-33, a circular loop of wire 10 cm in diameter (seen edge-on) is placed with its normal $\vec{N}$ at an angle $\theta = 30°$ with the direction of a uniform magnetic field $\vec{B}$ of magnitude 0.50 T. The loop is then rotated such that $\vec{N}$ rotates in a cone about the field direction at the rate 100 rev/min; angle θ remains unchanged during the process. What is the emf induced in the loop?

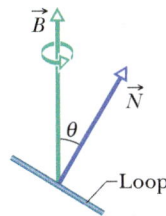

Figure 30-33 Problem 1.

•2 A certain elastic conducting material is stretched into a circular loop of 12.0 cm radius. It is placed with its plane perpendicular to a uniform 0.800 T magnetic field. When released, the radius of the loop starts to shrink at an instantaneous rate of 75.0 cm/s. What emf is induced in the loop at that instant?

•3 SSM WWW In Fig. 30-34, a 120-turn coil of radius 1.8 cm and resistance 5.3 Ω is coaxial with a solenoid of 220 turns/cm and diameter 3.2 cm. The solenoid current drops from 1.5 A to zero in time interval Δt = 25 ms. What current is induced in the coil during Δt?

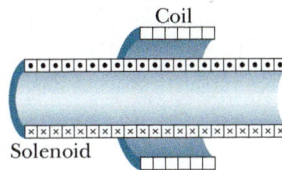

Figure 30-34 Problem 3.

•4 A wire loop of radius 12 cm and resistance 8.5 Ω is located in a uniform magnetic field $\vec{B}$ that changes in magnitude as given in Fig. 30-35. The vertical axis scale is set by $B_s = 0.50$ T, and the horizontal axis scale is set by $t_s = 6.00$ s. The loop's plane is perpendicular to $\vec{B}$. What emf is induced in the loop during time intervals (a) 0 to 2.0 s, (b) 2.0 s to 4.0 s, and (c) 4.0 s to 6.0 s?

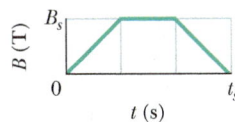

Figure 30-35 Problem 4.

•5 In Fig. 30-36, a wire forms a closed circular loop, of radius $R = 2.0$ m and resistance 4.0 Ω. The circle is centered on a long straight wire; at time $t = 0$, the current in the long straight wire is 5.0 A rightward. Thereafter, the current changes according to $i = 5.0$ A − (2.0 A/s²)t^2. (The straight wire is insulated; so there is no electrical contact between it and the wire of the loop.) What is the magnitude of the current induced in the loop at times $t > 0$?

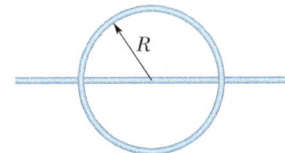

Figure 30-36 Problem 5.

•6 Figure 30-37a shows a circuit consisting of an ideal battery with emf $\mathscr{E} = 6.00$ μV, a resistance R, and a small wire loop of area 5.0 cm². For the time interval $t = 10$ s to $t = 20$ s, an external magnetic field is set up throughout the loop. The field is uniform, its direction is into the page in Fig. 30-37a, and the field magnitude is given by $B = at$, where B is in teslas, a is a constant, and t is in seconds. Figure 30-37b gives the current i in the circuit before, during, and after the external field is set up. The vertical axis scale is set by $i_s = 2.0$ mA. Find the constant a in the equation for the field magnitude.

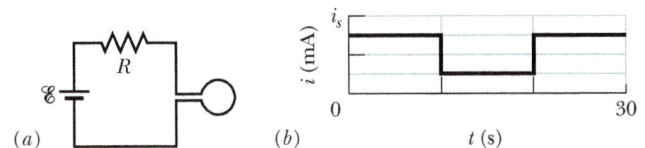

Figure 30-37 Problem 6.

•7 In Fig. 30-38, the magnetic flux through the loop increases according to the relation $\Phi_B = 6.0t^2 + 7.0t$, where Φ_B is in milliwebers and t is in seconds. (a) What is the magnitude of the emf induced in the loop when $t = 2.0$ s? (b) Is the direction of the current through R to the right or left?

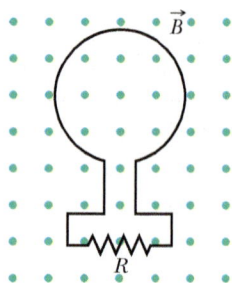

Figure 30-38 Problem 7.

•8 A uniform magnetic field $\vec{B}$ is perpendicular to the plane of a circular loop of diameter 10 cm formed from wire of diameter 2.5 mm and resistivity $1.69 \times 10^{-8}\,\Omega\cdot$m. At what rate must the magnitude of $\vec{B}$ change to induce a 10 A current in the loop?

•9 A small loop of area 6.8 mm^2 is placed inside a long solenoid that has 854 turns/cm and carries a sinusoidally varying current i of amplitude 1.28 A and angular frequency 212 rad/s. The central axes of the loop and solenoid coincide. What is the amplitude of the emf induced in the loop?

••10 Figure 30-39 shows a closed loop of wire that consists of a pair of equal semicircles, of radius 3.7 cm, lying in mutually perpendicular planes. The loop was formed by folding a flat circular loop along a diameter until the two halves became perpendicular to each other. A uniform magnetic field $\vec{B}$ of magnitude 76 mT is directed perpendicular to the fold diameter and makes equal angles (of 45°) with the planes of the

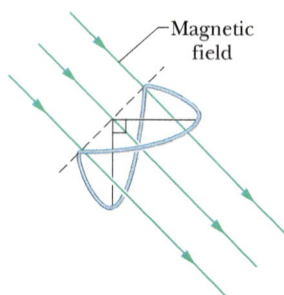

Figure 30-39 Problem 10.

semicircles. The magnetic field is reduced to zero at a uniform rate during a time interval of 4.5 ms. During this interval, what are the (a) magnitude and (b) direction (clockwise or counterclockwise when viewed along the direction of $\vec{B}$) of the emf induced in the loop?

••11 A rectangular coil of N turns and of length a and width b is rotated at frequency f in a uniform magnetic field $\vec{B}$, as indicated in Fig. 30-40. The coil is connected to co-rotating cylinders, against which metal brushes slide to make contact. (a) Show that the emf induced in the coil is given (as a function of time t) by

$$\mathscr{E} = 2\pi f NabB \sin(2\pi ft) = \mathscr{E}_0 \sin(2\pi ft).$$

This is the principle of the commercial alternating-current generator. (b) What value of Nab gives an emf with $\mathscr{E}_0 = 150$ V when the loop is rotated at 60.0 rev/s in a uniform magnetic field of 0.500 T?

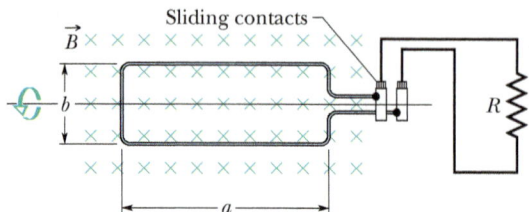

Figure 30-40 Problem 11.

••12 In Fig. 30-41, a wire loop of lengths $L = 40.0$ cm and $W = 25.0$ cm lies in a magnetic field $\vec{B}$. What are the (a) magnitude $\mathscr{E}$ and (b) direction (clockwise or counterclockwise—or "none" if $\mathscr{E} = 0$)

of the emf induced in the loop if $\vec{B} = (4.00 \times 10^{-2}\,\text{T/m})y\hat{k}$? What are (c) $\mathscr{E}$ and (d) the direction if $\vec{B} = (6.00 \times 10^{-2}\,\text{T/s})t\hat{k}$? What are (e) $\mathscr{E}$ and (f) the direction if $\vec{B} = (8.00 \times 10^{-2}\,\text{T/m}\cdot\text{s})yt\hat{k}$? What are (g) $\mathscr{E}$ and (h) the direction if $\vec{B} = (3.00 \times 10^{-2}\,\text{T/m}\cdot\text{s})xt\hat{j}$? What are (i) $\mathscr{E}$ and (j) the direction if $\vec{B} = (5.00 \times 10^{-2}\,\text{T/m}\cdot\text{s})yt\hat{i}$?

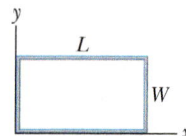

Figure 30-41 Problem 12.

••13 **ILW** One hundred turns of (insulated) copper wire are wrapped around a wooden cylindrical core of cross-sectional area 1.20×10^{-3} m^2. The two ends of the wire are connected to a resistor. The total resistance in the circuit is 13.0 Ω. If an externally applied uniform longitudinal magnetic field in the core changes from 1.60 T in one direction to 1.60 T in the opposite direction, how much charge flows through a point in the circuit during the change?

••14 **GO** In Fig. 30-42a, a uniform magnetic field $\vec{B}$ increases in magnitude with time t as given by Fig. 30-42b, where the vertical axis scale is set by $B_s = 9.0$ mT and the horizontal scale is set by $t_s = 3.0$ s. A circular conducting loop of area 8.0×10^{-4} m^2 lies in the field, in the plane of the page. The amount of charge q passing point A on the loop is given in Fig. 30-42c as a function of t, with the vertical axis scale set by $q_s = 6.0$ mC and the horizontal axis scale again set by $t_s = 3.0$ s. What is the loop's resistance?

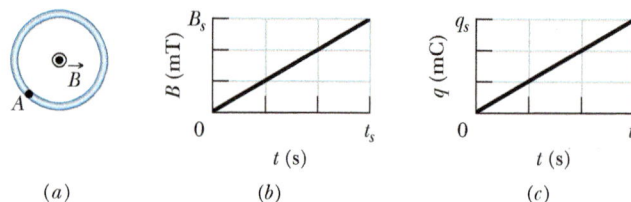

Figure 30-42 Problem 14.

••15 **GO** A square wire loop with 2.00 m sides is perpendicular to a uniform magnetic field, with half the area of the loop in the field as shown in Fig. 30-43. The loop contains an ideal battery with emf $\mathscr{E} = 20.0$ V. If the magnitude of the field varies with time according to $B = 0.0420 - 0.870t$, with B in teslas and t in seconds, what are (a) the net emf in the circuit and (b) the direction of the (net) current around the loop?

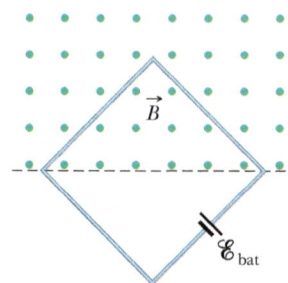

Figure 30-43 Problem 15.

••16 **GO** Figure 30-44a shows a wire that forms a rectangle ($W = 20$ cm, $H = 30$ cm) and has a resistance of 5.0 mΩ. Its

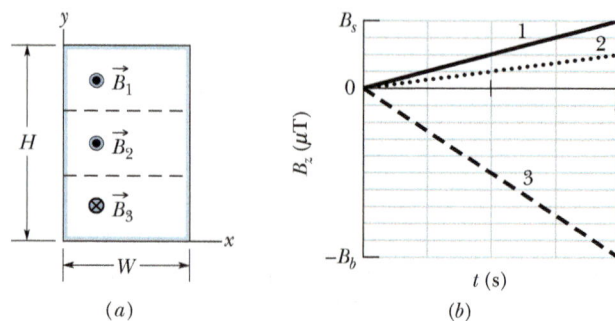

Figure 30-44 Problem 16.

interior is split into three equal areas, with magnetic fields $\vec{B}_1$, $\vec{B}_2$, and $\vec{B}_3$. The fields are uniform within each region and directly out of or into the page as indicated. Figure 30-44b gives the change in the z components B_z of the three fields with time t; the vertical axis scale is set by $B_s = 4.0\ \mu T$ and $B_b = -2.5B_s$, and the horizontal axis scale is set by $t_s = 2.0$ s. What are the (a) magnitude and (b) direction of the current induced in the wire?

••17 A small circular loop of area 2.00 cm² is placed in the plane of, and concentric with, a large circular loop of radius 1.00 m. The current in the large loop is changed at a constant rate from 200 A to −200 A (a change in direction) in a time of 1.00 s, starting at $t = 0$. What is the magnitude of the magnetic field $\vec{B}$ at the center of the small loop due to the current in the large loop at (a) $t = 0$, (b) $t = 0.500$ s, and (c) $t = 1.00$ s? (d) From $t = 0$ to $t = 1.00$ s, is $\vec{B}$ reversed? Because the inner loop is small, assume $\vec{B}$ is uniform over its area. (e) What emf is induced in the small loop at $t = 0.500$ s?

••18 In Fig. 30-45, two straight conducting rails form a right angle. A conducting bar in contact with the rails starts at the vertex at time $t = 0$ and moves with a constant velocity of 5.20 m/s along them. A magnetic field with $B = 0.350$ T is directed out of the page. Calculate (a) the flux through the triangle formed by the rails and bar at $t = 3.00$ s and (b) the emf around the triangle at that time. (c) If the emf is $\mathcal{E} = at^n$, where a and n are constants, what is the value of n?

Figure 30-45 Problem 18.

••19 ILW An electric generator contains a coil of 100 turns of wire, each forming a rectangular loop 50.0 cm by 30.0 cm. The coil is placed entirely in a uniform magnetic field with magnitude $B = 3.50$ T and with $\vec{B}$ initially perpendicular to the coil's plane. What is the maximum value of the emf produced when the coil is spun at 1000 rev/min about an axis perpendicular to $\vec{B}$?

••20 At a certain place, Earth's magnetic field has magnitude $B = 0.590$ gauss and is inclined downward at an angle of 70.0° to the horizontal. A flat horizontal circular coil of wire with a radius of 10.0 cm has 1000 turns and a total resistance of 85.0 Ω. It is connected in series to a meter with 140 Ω resistance. The coil is flipped through a half-revolution about a diameter, so that it is again horizontal. How much charge flows through the meter during the flip?

••21 In Fig. 30-46, a stiff wire bent into a semicircle of radius $a = 2.0$ cm is rotated at constant angular speed 40 rev/s in a uniform 20 mT magnetic field. What are the (a) frequency and (b) amplitude of the emf induced in the loop?

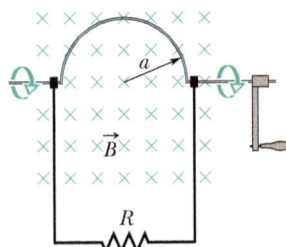

Figure 30-46 Problem 21.

••22 A rectangular loop (area = 0.15 m²) turns in a uniform magnetic field, $B = 0.20$ T. When the angle between the field and the normal to the plane of the loop is $\pi/2$ rad and increasing at 0.60 rad/s, what emf is induced in the loop?

••23 SSM Figure 30-47 shows two parallel loops of wire having a common axis. The smaller loop (radius r) is above the larger loop (radius R)

Figure 30-47 Problem 23.

by a distance $x \gg R$. Consequently, the magnetic field due to the counterclockwise current i in the larger loop is nearly uniform throughout the smaller loop. Suppose that x is increasing at the constant rate $dx/dt = v$. (a) Find an expression for the magnetic flux through the area of the smaller loop as a function of x. (Hint: See Eq. 29-27.) In the smaller loop, find (b) an expression for the induced emf and (c) the direction of the induced current.

••24 A wire is bent into three circular segments, each of radius $r = 10$ cm, as shown in Fig. 30-48. Each segment is a quadrant of a circle, ab lying in the xy plane, bc lying in the yz plane, and ca lying in the zx plane. (a) If a uniform magnetic field $\vec{B}$ points in the positive x direction, what is the magnitude of the emf developed in the wire when B increases at the rate of 3.0 mT/s? (b) What is the direction of the current in segment bc?

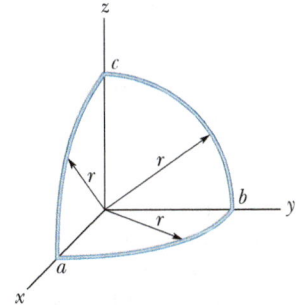

Figure 30-48 Problem 24.

•••25 GO Two long, parallel copper wires of diameter 2.5 mm carry currents of 10 A in opposite directions. (a) Assuming that their central axes are 20 mm apart, calculate the magnetic flux per meter of wire that exists in the space between those axes. (b) What percentage of this flux lies inside the wires? (c) Repeat part (a) for parallel currents.

•••26 GO For the wire arrangement in Fig. 30-49, $a = 12.0$ cm and $b = 16.0$ cm. The current in the long straight wire is $i = 4.50t^2 - 10.0t$, where i is in amperes and t is in seconds. (a) Find the emf in the square loop at $t = 3.00$ s. (b) What is the direction of the induced current in the loop?

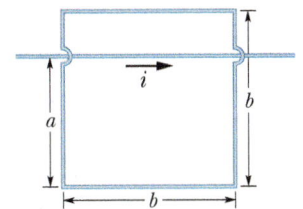

Figure 30-49 Problem 26.

•••27 ILW As seen in Fig. 30-50, a square loop of wire has sides of length 2.0 cm. A magnetic field is directed out of the page; its magnitude is given by $B = 4.0t^2y$, where B is in teslas, t is in seconds, and y is in meters. At $t = 2.5$ s, what are the (a) magnitude and (b) direction of the emf induced in the loop?

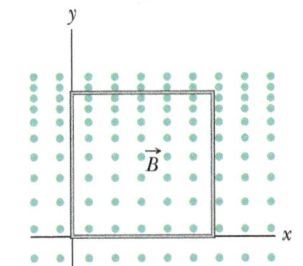

Figure 30-50 Problem 27.

•••28 GO In Fig. 30-51, a rectangular loop of wire with length $a = 2.2$ cm, width $b = 0.80$ cm, and resistance $R = 0.40$ mΩ is placed near an infinitely long wire carrying current $i = 4.7$ A. The loop is then moved away from the wire at constant speed $v = 3.2$ mm/s. When the center of the loop is at distance $r = 1.5b$, what are the (a) magnitude of the magnetic flux through the loop and (b) the current induced in the loop?

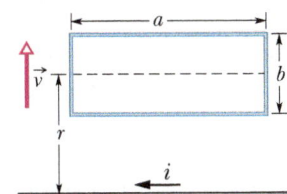

Figure 30-51 Problem 28.

Module 30-2 Induction and Energy Transfers

•29 In Fig. 30-52, a metal rod is forced to move with constant velocity $\vec{v}$ along two parallel metal rails, connected with a strip of metal at one end. A magnetic field of magnitude $B = 0.350$ T points out of the page. (a) If the rails are separated by $L = 25.0$ cm and the speed of the rod is 55.0 cm/s, what emf is generated?

Figure 30-52
Problems 29 and 35.

(b) If the rod has a resistance of 18.0 Ω and the rails and connector have negligible resistance, what is the current in the rod? (c) At what rate is energy being transferred to thermal energy?

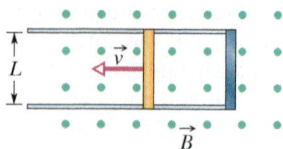

•30 In Fig. 30-53a, a circular loop of wire is concentric with a solenoid and lies in a plane perpendicular to the solenoid's central axis. The loop has radius 6.00 cm. The solenoid has radius 2.00 cm, consists of 8000 turns/m, and has a current i_{sol} varying with time t as given in Fig. 30-53b, where the vertical axis scale is set by $i_s = 1.00$ A and the horizontal axis scale is set by $t_s = 2.0$ s. Figure 30-53c shows, as a function of time, the energy E_{th} that is transferred to thermal energy of the loop; the vertical axis scale is set by $E_s = 100.0$ nJ. What is the loop's resistance?

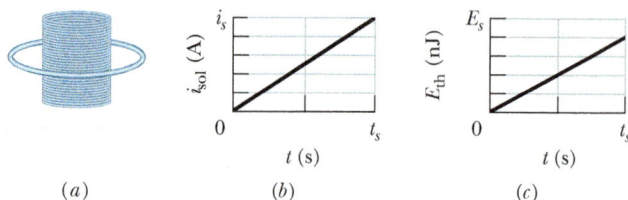

Figure 30-53 Problem 30.

•31 SSM ILW If 50.0 cm of copper wire (diameter = 1.00 mm) is formed into a circular loop and placed perpendicular to a uniform magnetic field that is increasing at the constant rate of 10.0 mT/s, at what rate is thermal energy generated in the loop?

•32 A loop antenna of area 2.00 cm² and resistance 5.21 μΩ is perpendicular to a uniform magnetic field of magnitude 17.0 μT. The field magnitude drops to zero in 2.96 ms. How much thermal energy is produced in the loop by the change in field?

••33 GO Figure 30-54 shows a rod of length $L = 10.0$ cm that is forced to move at constant speed $v = 5.00$ m/s along horizontal rails. The rod, rails, and connecting strip at the right form a conducting loop. The rod has resistance 0.400 Ω; the rest of the loop has negligible resistance. A current $i = 100$ A through the long straight wire at distance $a = 10.0$ mm from the loop sets up a (nonuniform) magnetic field through the loop. Find the (a) emf and (b) current induced in the loop. (c) At what rate is thermal energy generated in the rod? (d) What is the magnitude of the force that must be applied to the rod to make it move at constant speed? (e) At what rate does this force do work on the rod?

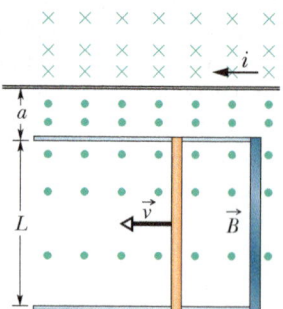

Figure 30-54 Problem 33.

••34 In Fig. 30-55, a long rectangular conducting loop, of width L, resistance R, and mass m, is hung in a horizontal, uniform magnetic field $\vec{B}$ that is directed into the page and that exists only above line aa. The loop is then dropped; during its fall, it accelerates until it reaches a certain terminal speed v_t. Ignoring air drag, find an expression for v_t.

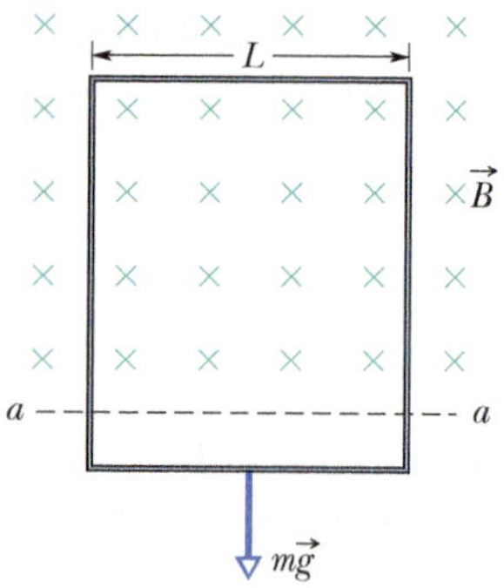

Figure 30-55 Problem 34.

••35 The conducting rod shown in Fig. 30-52 has length L and is being pulled along horizontal, frictionless conducting rails at a constant velocity $\vec{v}$. The rails are connected at one end with a metal strip. A uniform magnetic field $\vec{B}$, directed out of the page, fills the region in which the rod moves. Assume that $L = 10$ cm, $v = 5.0$ m/s, and $B = 1.2$ T. What are the (a) magnitude and (b) direction (up or down the page) of the emf induced in the rod? What are the (c) size and (d) direction of the current in the conducting loop? Assume that the resistance of the rod is 0.40 Ω and that the resistance of the rails and metal strip is negligibly small. (e) At what rate is thermal energy being generated in the rod? (f) What external force on the rod is needed to maintain $\vec{v}$? (g) At what rate does this force do work on the rod?

Module 30-3 Induced Electric Fields

•36 Figure 30-56 shows two circular regions R_1 and R_2 with radii $r_1 = 20.0$ cm and $r_2 = 30.0$ cm. In R_1 there is a uniform magnetic field of magnitude $B_1 = 50.0$ mT directed into the page, and in R_2 there is a uniform magnetic field of magnitude $B_2 = 75.0$ mT directed out of the page (ignore fringing). Both fields are decreasing at the rate of 8.50 mT/s. Calculate $\oint \vec{E} \cdot d\vec{s}$ for (a) path 1, (b) path 2, and (c) path 3.

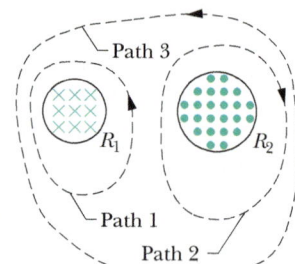

Figure 30-56 Problem 36.

•37 SSM ILW A long solenoid has a diameter of 12.0 cm. When a current i exists in its windings, a uniform magnetic field of magnitude $B = 30.0$ mT is produced in its interior. By decreasing i, the field is caused to decrease at the rate of 6.50 mT/s. Calculate the magnitude of the induced electric field (a) 2.20 cm and (b) 8.20 cm from the axis of the solenoid.

••38 GO A circular region in an xy plane is penetrated by a uniform magnetic field in the positive direction of the z axis. The field's magnitude B (in teslas) increases with time t (in seconds) according to $B = at$, where a is a constant. The magnitude E of the electric field set up by that increase in the magnetic field is given by Fig. 30-57 versus radial distance r; the vertical axis scale is set by $E_s = 300$ μN/C, and the horizontal axis scale is set by $r_s = 4.00$ cm. Find a.

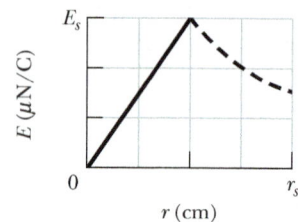

Figure 30-57 Problem 38.

••39 The magnetic field of a cylindrical magnet that has a pole-face diameter of 3.3 cm can be varied sinusoidally between 29.6 T and 30.0 T at a frequency of 15 Hz. (The current in a wire wrapped around a permanent magnet is varied to give this variation in the net field.) At a radial distance of 1.6 cm, what is the amplitude of the electric field induced by the variation?

Module 30-4   Inductors and Inductance

•40 The inductance of a closely packed coil of 400 turns is 8.0 mH. Calculate the magnetic flux through the coil when the current is 5.0 mA.

•41 A circular coil has a 10.0 cm radius and consists of 30.0 closely wound turns of wire. An externally produced magnetic field of magnitude 2.60 mT is perpendicular to the coil. (a) If no current is in the coil, what magnetic flux links its turns? (b) When the current in the coil is 3.80 A in a certain direction, the net flux through the coil is found to vanish. What is the inductance of the coil?

••42 Figure 30-58 shows a copper strip of width $W = 16.0$ cm that has been bent to form a shape that consists of a tube of radius $R = 1.8$ cm plus two parallel flat extensions. Current $i = 35$ mA is distributed uniformly across the width so that the tube is effectively a one-turn solenoid. Assume that the magnetic field outside the tube is negligible and the field inside the tube is uniform. What are (a) the magnetic field magnitude inside the tube and (b) the inductance of the tube (excluding the flat extensions)?

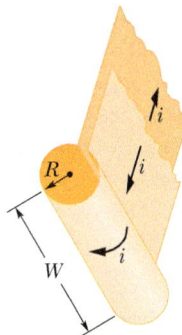

Figure 30-58 Problem 42.

••43 GO Two identical long wires of radius $a = 1.53$ mm are parallel and carry identical currents in opposite directions. Their center-to-center separation is $d = 14.2$ cm. Neglect the flux within the wires but consider the flux in the region between the wires. What is the inductance per unit length of the wires?

Module 30-5   Self-Induction

•44 A 12 H inductor carries a current of 2.0 A. At what rate must the current be changed to produce a 60 V emf in the inductor?

•45 At a given instant the current and self-induced emf in an inductor are directed as indicated in Fig. 30-59. (a) Is the current increasing or decreasing? (b) The induced emf is 17 V, and the rate of change of the current is 25 kA/s; find the inductance.

Figure 30-59 Problem 45.

••46 The current i through a 4.6 H inductor varies with time t as shown by the graph of Fig. 30-60, where the vertical axis scale is set by $i_s = 8.0$ A and the horizontal axis scale is set by $t_s = 6.0$ ms. The inductor has a resistance of 12 Ω. Find the magnitude of the induced emf $\mathscr{E}$ during time intervals (a) 0 to 2 ms, (b) 2 ms to 5 ms, and (c) 5 ms to 6 ms. (Ignore the behavior at the ends of the intervals.)

Figure 30-60 Problem 46.

••47 *Inductors in series.* Two inductors L_1 and L_2 are connected in series and are separated by a large distance so that the magnetic field of one cannot affect the other. (a) Show that the equivalent inductance is given by

$$L_{eq} = L_1 + L_2.$$

(*Hint:* Review the derivations for resistors in series and capacitors in series. Which is similar here?) (b) What is the generalization of (a) for N inductors in series?

••48 *Inductors in parallel.* Two inductors L_1 and L_2 are connected in parallel and separated by a large distance so that the magnetic field of one cannot affect the other. (a) Show that the equivalent inductance is given by

$$\frac{1}{L_{eq}} = \frac{1}{L_1} + \frac{1}{L_2}.$$

(*Hint:* Review the derivations for resistors in parallel and capacitors in parallel. Which is similar here?) (b) What is the generalization of (a) for N inductors in parallel?

••49 The inductor arrangement of Fig. 30-61, with $L_1 = 30.0$ mH, $L_2 = 50.0$ mH, $L_3 = 20.0$ mH, and $L_4 = 15.0$ mH, is to be connected to a varying current source. What is the equivalent inductance of the arrangement? (First see Problems 47 and 48.)

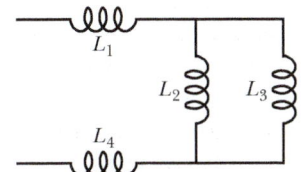

Figure 30-61 Problem 49.

Module 30-6   RL Circuits

•50 The current in an RL circuit builds up to one-third of its steady-state value in 5.00 s. Find the inductive time constant.

•51 ILW The current in an RL circuit drops from 1.0 A to 10 mA in the first second following removal of the battery from the circuit. If L is 10 H, find the resistance R in the circuit.

•52 The switch in Fig. 30-15 is closed on a at time $t = 0$. What is the ratio $\mathscr{E}_L/\mathscr{E}$ of the inductor's self-induced emf to the battery's emf (a) just after $t = 0$ and (b) at $t = 2.00\tau_L$? (c) At what multiple of τ_L will $\mathscr{E}_L/\mathscr{E} = 0.500$?

•53 SSM A solenoid having an inductance of 6.30 μH is connected in series with a 1.20 kΩ resistor. (a) If a 14.0 V battery is connected across the pair, how long will it take for the current through the resistor to reach 80.0% of its final value? (b) What is the current through the resistor at time $t = 1.0\tau_L$?

•54 In Fig. 30-62, $\mathscr{E} = 100$ V, $R_1 = 10.0$ Ω, $R_2 = 20.0$ Ω, $R_3 = 30.0$ Ω, and $L = 2.00$ H. Immediately after switch S is closed, what are (a) i_1 and (b) i_2? (Let currents in the indicated directions have positive values and currents in the opposite directions have negative values.) A long time later, what are (c) i_1 and (d) i_2? The switch is then reopened. Just then, what are (e) i_1 and (f) i_2? A long time later, what are (g) i_1 and (h) i_2?

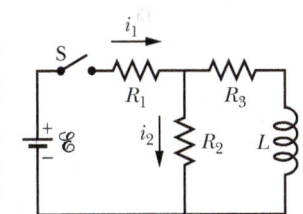

Figure 30-62 Problem 54.

•55 SSM A battery is connected to a series RL circuit at time $t = 0$. At what multiple of τ_L will the current be 0.100% less than its equilibrium value?

•56 In Fig. 30-63, the inductor has 25 turns and the ideal battery has an emf of 16 V. Figure 30-64 gives the magnetic flux Φ through each turn versus the current i through the inductor. The vertical

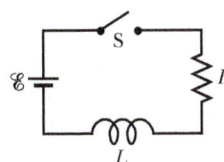

Figure 30-63 Problems 56, 80, 83, and 93.

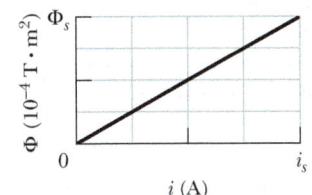

Figure 30-64 Problem 56.

axis scale is set by $\Phi_s = 4.0 \times 10^{-4}$ T·m², and the horizontal axis scale is set by $i_s = 2.00$ A. If switch S is closed at time $t = 0$, at what rate di/dt will the current be changing at $t = 1.5\tau_L$?

••**57** GO In Fig. 30-65, $R = 15\ \Omega$, $L = 5.0$ H, the ideal battery has $\mathscr{E} = 10$ V, and the fuse in the upper branch is an ideal 3.0 A fuse. It has zero resistance as long as the current through it remains less than 3.0 A. If the current reaches 3.0 A, the fuse "blows" and thereafter has infinite resistance. Switch S is closed

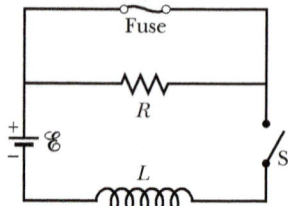

Figure 30-65 Problem 57.

at time $t = 0$. (a) When does the fuse blow? (*Hint:* Equation 30-41 does not apply. Rethink Eq. 30-39.) (b) Sketch a graph of the current i through the inductor as a function of time. Mark the time at which the fuse blows.

••**58** GO Suppose the emf of the battery in the circuit shown in Fig. 30-16 varies with time t so that the current is given by $i(t) = 3.0 + 5.0t$, where i is in amperes and t is in seconds. Take $R = 4.0\ \Omega$ and $L = 6.0$ H, and find an expression for the battery emf as a function of t. (*Hint:* Apply the loop rule.)

•••**59** SSM WWW In Fig. 30-66, after switch S is closed at time $t = 0$, the emf of the source is automatically adjusted to maintain a constant current i through S. (a) Find the current through the inductor as a function of time. (b) At what time is the current through the resistor equal to the current through the inductor?

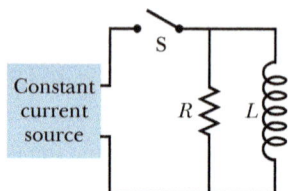

Figure 30-66 Problem 59.

•••**60** A wooden toroidal core with a square cross section has an inner radius of 10 cm and an outer radius of 12 cm. It is wound with one layer of wire (of diameter 1.0 mm and resistance per meter 0.020 Ω/m). What are (a) the inductance and (b) the inductive time constant of the resulting toroid? Ignore the thickness of the insulation on the wire.

Module 30-7 Energy Stored in a Magnetic Field

•**61** SSM A coil is connected in series with a 10.0 kΩ resistor. An ideal 50.0 V battery is applied across the two devices, and the current reaches a value of 2.00 mA after 5.00 ms. (a) Find the inductance of the coil. (b) How much energy is stored in the coil at this same moment?

•**62** A coil with an inductance of 2.0 H and a resistance of 10 Ω is suddenly connected to an ideal battery with $\mathscr{E} = 100$ V. At 0.10 s after the connection is made, what is the rate at which (a) energy is being stored in the magnetic field, (b) thermal energy is appearing in the resistance, and (c) energy is being delivered by the battery?

•**63** ILW At $t = 0$, a battery is connected to a series arrangement of a resistor and an inductor. If the inductive time constant is 37.0 ms, at what time is the rate at which energy is dissipated in the resistor equal to the rate at which energy is stored in the inductor's magnetic field?

•**64** At $t = 0$, a battery is connected to a series arrangement of a resistor and an inductor. At what multiple of the inductive time constant will the energy stored in the inductor's magnetic field be 0.500 its steady-state value?

••**65** GO For the circuit of Fig. 30-16, assume that $\mathscr{E} = 10.0$ V, $R = 6.70\ \Omega$, and $L = 5.50$ H. The ideal battery is connected at time $t = 0$.

(a) How much energy is delivered by the battery during the first 2.00 s? (b) How much of this energy is stored in the magnetic field of the inductor? (c) How much of this energy is dissipated in the resistor?

Module 30-8 Energy Density of a Magnetic Field

•**66** A circular loop of wire 50 mm in radius carries a current of 100 A. Find the (a) magnetic field strength and (b) energy density at the center of the loop.

•**67** SSM A solenoid that is 85.0 cm long has a cross-sectional area of 17.0 cm². There are 950 turns of wire carrying a current of 6.60 A. (a) Calculate the energy density of the magnetic field inside the solenoid. (b) Find the total energy stored in the magnetic field there (neglect end effects).

•**68** A toroidal inductor with an inductance of 90.0 mH encloses a volume of 0.0200 m³. If the average energy density in the toroid is 70.0 J/m³, what is the current through the inductor?

•**69** ILW What must be the magnitude of a uniform electric field if it is to have the same energy density as that possessed by a 0.50 T magnetic field?

••**70** GO Figure 30-67a shows, in cross section, two wires that are straight, parallel, and very long. The ratio i_1/i_2 of the current carried by wire 1 to that carried by wire 2 is 1/3. Wire 1 is fixed in place. Wire 2 can be moved along the positive side of the x axis so as to change the magnetic energy density u_B set up by the two currents at the origin. Figure 30-67b gives u_B as a function of the position x of wire 2. The curve has an asymptote of $u_B = 1.96$ nJ/m³ as $x \to \infty$, and the horizontal axis scale is set by $x_s = 60.0$ cm. What is the value of (a) i_1 and (b) i_2?

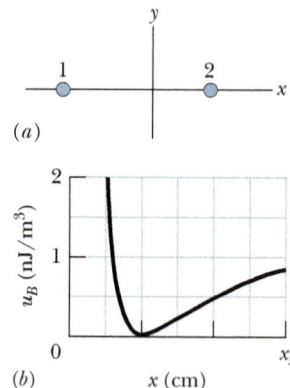

Figure 30-67 Problem 70.

••**71** A length of copper wire carries a current of 10 A uniformly distributed through its cross section. Calculate the energy density of (a) the magnetic field and (b) the electric field at the surface of the wire. The wire diameter is 2.5 mm, and its resistance per unit length is 3.3 Ω/km.

Module 30-9 Mutual Induction

•**72** Coil 1 has $L_1 = 25$ mH and $N_1 = 100$ turns. Coil 2 has $L_2 = 40$ mH and $N_2 = 200$ turns. The coils are fixed in place; their mutual inductance M is 3.0 mH. A 6.0 mA current in coil 1 is changing at the rate of 4.0 A/s. (a) What magnetic flux Φ_{12} links coil 1, and (b) what self-induced emf appears in that coil? (c) What magnetic flux Φ_{21} links coil 2, and (d) what mutually induced emf appears in that coil?

•**73** SSM Two coils are at fixed locations. When coil 1 has no current and the current in coil 2 increases at the rate 15.0 A/s, the emf in coil 1 is 25.0 mV. (a) What is their mutual inductance? (b) When coil 2 has no current and coil 1 has a current of 3.60 A, what is the flux linkage in coil 2?

•**74** Two solenoids are part of the spark coil of an automobile. When the current in one solenoid falls from 6.0 A to zero in 2.5 ms, an emf of 30 kV is induced in the other solenoid. What is the mutual inductance M of the solenoids?

••75 **ILW** A rectangular loop of N closely packed turns is positioned near a long straight wire as shown in Fig. 30-68. What is the mutual inductance M for the loop–wire combination if $N = 100$, $a = 1.0$ cm, $b = 8.0$ cm, and $l = 30$ cm?

Figure 30-68 Problem 75.

••76 A coil C of N turns is placed around a long solenoid S of radius R and n turns per unit length, as in Fig. 30-69. (a) Show that the mutual inductance for the coil–solenoid combination is given by $M = \mu_0 \pi R^2 nN$. (b) Explain why M does not depend on the shape, size, or possible lack of close packing of the coil.

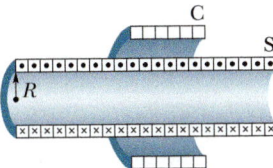

Figure 30-69 Problem 76.

••77 **SSM** Two coils connected as shown in Fig. 30-70 separately have inductances L_1 and L_2. Their mutual inductance is M. (a) Show that this combination can be replaced by a single coil of equivalent inductance given by

$$L_{eq} = L_1 + L_2 + 2M.$$

(b) How could the coils in Fig. 30-70 be reconnected to yield an equivalent inductance of

$$L_{eq} = L_1 + L_2 - 2M?$$

(This problem is an extension of Problem 47, but the requirement that the coils be far apart has been removed.)

Figure 30-70 Problem 77.

Additional Problems

78 At time $t = 0$, a 12.0 V potential difference is suddenly applied to the leads of a coil of inductance 23.0 mH and a certain resistance R. At time $t = 0.150$ ms, the current through the inductor is changing at the rate of 280 A/s. Evaluate R.

79 **SSM** In Fig. 30-71, the battery is ideal and $\mathscr{E} = 10$ V, $R_1 = 5.0$ Ω, $R_2 = 10$ Ω, and $L = 5.0$ H. Switch S is closed at time $t = 0$. Just afterwards, what are (a) i_1, (b) i_2, (c) the current i_S through the switch, (d) the potential difference V_2 across resistor 2, (e) the potential difference V_L across the inductor, and (f) the rate of change di_2/dt? A long time later, what are (g) i_1, (h) i_2, (i) i_S, (j) V_2, (k) V_L, and (l) di_2/dt?

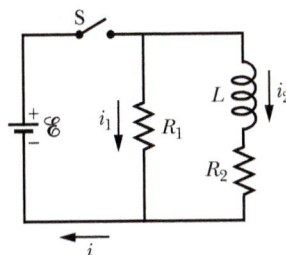

Figure 30-71 Problem 79.

80 In Fig. 30-63, $R = 4.0$ kΩ, $L = 8.0$ μH, and the ideal battery has $\mathscr{E} = 20$ V. How long after switch S is closed is the current 2.0 mA?

81 **SSM** Figure 30-72a shows a rectangular conducting loop of resistance $R = 0.020$ Ω, height $H = 1.5$ cm, and length $D = 2.5$ cm being pulled at constant speed $v = 40$ cm/s through two regions of uniform magnetic field. Figure 30-72b gives the current i induced in the loop as a function of the position x of the right side of the loop. The vertical axis scale is set by $i_s = 3.0$ μA. For example, a current equal to i_s is induced clockwise as the loop enters region 1. What are the (a) magnitude and (b) direction (into or out of the page) of the magnetic field in region 1? What are the (c) magnitude and (d) direction of the magnetic field in region 2?

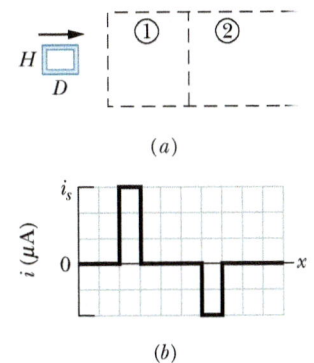

Figure 30-72 Problem 81.

82 A uniform magnetic field $\vec{B}$ is perpendicular to the plane of a circular wire loop of radius r. The magnitude of the field varies with time according to $B = B_0 e^{-t/\tau}$, where B_0 and τ are constants. Find an expression for the emf in the loop as a function of time.

83 Switch S in Fig. 30-63 is closed at time $t = 0$, initiating the buildup of current in the 15.0 mH inductor and the 20.0 Ω resistor. At what time is the emf across the inductor equal to the potential difference across the resistor?

84 **GO** Figure 30-73a shows two concentric circular regions in which uniform magnetic fields can change. Region 1, with radius $r_1 = 1.0$ cm, has an outward magnetic field $\vec{B}_1$ that is increasing in magnitude. Region 2, with radius $r_2 = 2.0$ cm, has an outward magnetic field $\vec{B}_2$ that may also be changing. Imagine that a conducting ring of radius R is centered on the two regions and then the emf $\mathscr{E}$ around the ring is determined. Figure 30-73b gives emf $\mathscr{E}$ as a function of the square R^2 of the ring's radius, to the outer edge of region 2. The vertical axis scale is set by $\mathscr{E}_s = 20.0$ nV. What are the rates (a) dB_1/dt and (b) dB_2/dt? (c) Is the magnitude of $\vec{B}_2$ increasing, decreasing, or remaining constant?

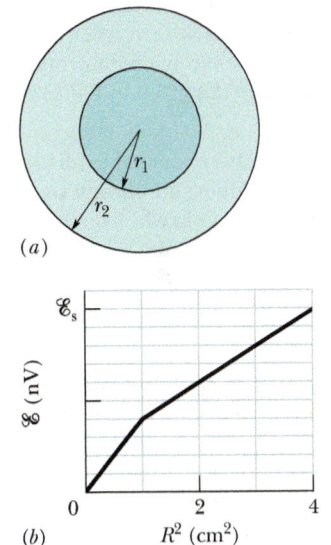

Figure 30-73 Problem 84.

85 **SSM** Figure 30-74 shows a uniform magnetic field $\vec{B}$ confined to a cylindrical volume of radius R. The magnitude of $\vec{B}$ is decreasing at a constant rate of 10 mT/s. In unit-vector notation, what is the initial acceleration of an electron released at (a) point a (radial distance $r = 5.0$ cm), (b) point b ($r = 0$), and (c) point c ($r = 5.0$ cm)?

Figure 30-74 Problem 85.

86 **GO** In Fig. 30-75a, switch S has been closed on A long enough to establish a steady current in the inductor of inductance

$L_1 = 5.00$ mH and the resistor of resistance $R_1 = 25.0$ Ω. Similarly, in Fig. 30-75b, switch S has been closed on A long enough to establish a steady current in the inductor of inductance $L_2 = 3.00$ mH and the resistor of resistance $R_2 = 30.0$ Ω. The ratio Φ_{02}/Φ_{01} of the magnetic flux through a turn in inductor 2 to that in inductor 1 is 1.50. At time $t = 0$, the two switches are closed on B. At what time t is the flux through a turn in the two inductors equal?

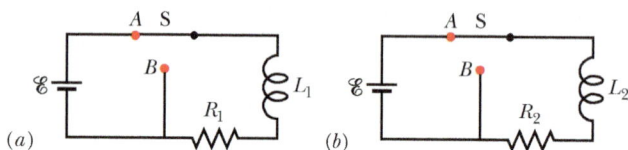

Figure 30-75 Problem 86.

87 SSM A square wire loop 20 cm on a side, with resistance 20 mΩ, has its plane normal to a uniform magnetic field of magnitude $B = 2.0$ T. If you pull two opposite sides of the loop away from each other, the other two sides automatically draw toward each other, reducing the area enclosed by the loop. If the area is reduced to zero in time $\Delta t = 0.20$ s, what are (a) the average emf and (b) the average current induced in the loop during Δt?

88 A coil with 150 turns has a magnetic flux of 50.0 nT·m² through each turn when the current is 2.00 mA. (a) What is the inductance of the coil? What are the (b) inductance and (c) flux through each turn when the current is increased to 4.00 mA? (d) What is the maximum emf $\mathscr{E}$ across the coil when the current through it is given by $i = (3.00$ mA$)\cos(377t)$, with t in seconds?

89 A coil with an inductance of 2.0 H and a resistance of 10 Ω is suddenly connected to an ideal battery with $\mathscr{E} = 100$ V. (a) What is the equilibrium current? (b) How much energy is stored in the magnetic field when this current exists in the coil?

90 How long would it take, following the removal of the battery, for the potential difference across the resistor in an RL circuit (with $L = 2.00$ H, $R = 3.00$ Ω) to decay to 10.0% of its initial value?

91 SSM In the circuit of Fig. 30-76, $R_1 = 20$ kΩ, $R_2 = 20$ Ω, $L = 50$ mH, and the ideal battery has $\mathscr{E} = 40$ V. Switch S has been open for a long time when it is closed at time $t = 0$. Just after the switch is closed, what are (a) the current i_{bat} through the battery and (b) the rate di_{bat}/dt? At $t = 3.0$ μs, what are (c) i_{bat} and (d) di_{bat}/dt? A long time later, what are (e) i_{bat} and (f) di_{bat}/dt?

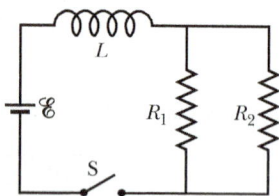

Figure 30-76 Problem 91.

92 The flux linkage through a certain coil of 0.75 Ω resistance would be 26 mWb if there were a current of 5.5 A in it. (a) Calculate the inductance of the coil. (b) If a 6.0 V ideal battery were suddenly connected across the coil, how long would it take for the current to rise from 0 to 2.5 A?

93 In Fig. 30-63, a 12.0 V ideal battery, a 20.0 Ω resistor, and an inductor are connected by a switch at time $t = 0$. At what rate is the battery transferring energy to the inductor's field at $t = 1.61\tau_L$?

94 A long cylindrical solenoid with 100 turns/cm has a radius of 1.6 cm. Assume that the magnetic field it produces is parallel to its axis and is uniform in its interior. (a) What is its inductance per

meter of length? (b) If the current changes at the rate of 13 A/s, what emf is induced per meter?

95 In Fig. 30-77, $R_1 = 8.0$ Ω, $R_2 = 10$ Ω, $L_1 = 0.30$ H, $L_2 = 0.20$ H, and the ideal battery has $\mathscr{E} = 6.0$ V. (a) Just after switch S is closed, at what rate is the current in inductor 1 changing? (b) When the circuit is in the steady state, what is the current in inductor 1?

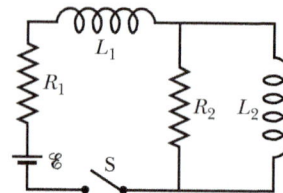

Figure 30-77 Problem 95.

96 A square loop of wire is held in a uniform 0.24 T magnetic field directed perpendicular to the plane of the loop. The length of each side of the square is decreasing at a constant rate of 5.0 cm/s. What emf is induced in the loop when the length is 12 cm?

97 At time $t = 0$, a 45 V potential difference is suddenly applied to the leads of a coil with inductance $L = 50$ mH and resistance $R = 180$ Ω. At what rate is the current through the coil increasing at $t = 1.2$ ms?

98 The inductance of a closely wound coil is such that an emf of 3.00 mV is induced when the current changes at the rate of 5.00 A/s. A steady current of 8.00 A produces a magnetic flux of 40.0 μWb through each turn. (a) Calculate the inductance of the coil. (b) How many turns does the coil have?

99 The magnetic field in the interstellar space of our galaxy has a magnitude of about 10^{-10} T. How much energy is stored in this field in a cube 10 light-years on edge? (For scale, note that the nearest star is 4.3 light-years distant and the radius of the galaxy is about 8×10^4 light-years.)

100 Figure 30-78 shows a wire that has been bent into a circular arc of radius $r = 24.0$ cm, centered at O. A straight wire OP can be rotated about O and makes sliding contact with the arc at P. Another straight wire OQ completes the conducting loop. The three wires have cross-sectional area 1.20 mm² and resistivity 1.70×10^{-8} $\Omega \cdot$m, and the apparatus lies in a uniform magnetic field of magnitude $B = 0.150$ T directed out of the figure. Wire OP begins from rest at angle $\theta = 0$ and has constant angular acceleration of 12 rad/s². As functions of θ (in rad), find (a) the loop's resistance and (b) the magnetic flux through the loop. (c) For what θ is the induced current maximum and (d) what is that maximum?

Figure 30-78 Problem 100.

101 A toroid has a 5.00 cm square cross section, an inside radius of 15.0 cm, 500 turns of wire, and a current of 0.800 A. What is the magnetic flux through the cross section?

Electromagnetic Oscillations and Alternating Current

31-1 *LC* OSCILLATIONS

Learning Objectives

After reading this module, you should be able to . . .

31.01 Sketch an *LC* oscillator and explain which quantities oscillate and what constitutes one period of the oscillation.

31.02 For an *LC* oscillator, sketch graphs of the potential difference across the capacitor and the current through the inductor as functions of time, and indicate the period *T* on each graph.

31.03 Explain the analogy between a block–spring oscillator and an *LC* oscillator.

31.04 For an *LC* oscillator, apply the relationships between the angular frequency ω (and the related frequency *f* and period *T*) and the values of the inductance and capacitance.

31.05 Starting with the energy of a block–spring system, explain the derivation of the differential equation for charge *q* in an *LC* oscillator and then identify the solution for *q(t)*.

31.06 For an *LC* oscillator, calculate the charge *q* on the capacitor for any given time and identify the amplitude *Q* of the charge oscillations.

31.07 Starting from the equation giving the charge *q(t)* on the capacitor in an *LC* oscillator, find the current *i(t)* in the inductor as a function of time.

31.08 For an *LC* oscillator, calculate the current *i* in the inductor for any given time and identify the amplitude *I* of the current oscillations.

31.09 For an *LC* oscillator, apply the relationship between the charge amplitude *Q*, the current amplitude *I*, and the angular frequency ω.

31.10 From the expressions for the charge *q* and the current *i* in an *LC* oscillator, find the magnetic field energy $U_B(t)$ and the electric field energy $U_E(t)$ and the total energy.

31.11 For an *LC* oscillator, sketch graphs of the magnetic field energy $U_B(t)$, the electric field energy $U_E(t)$, and the total energy, all as functions of time.

31.12 Calculate the maximum values of the magnetic field energy U_B and the electric field energy U_E and also calculate the total energy.

Key Ideas

● In an oscillating *LC* circuit, energy is shuttled periodically between the electric field of the capacitor and the magnetic field of the inductor; instantaneous values of the two forms of energy are

$$U_E = \frac{q^2}{2C} \quad \text{and} \quad U_B = \frac{Li^2}{2},$$

where *q* is the instantaneous charge on the capacitor and *i* is the instantaneous current through the inductor.

● The total energy $U\ (= U_E + U_B)$ remains constant.

● The principle of conservation of energy leads to

$$L\frac{d^2q}{dt^2} + \frac{1}{C}q = 0 \quad (LC \text{ oscillations})$$

as the differential equation of *LC* oscillations (with no resistance).

● The solution of this differential equation is

$$q = Q\cos(\omega t + \phi) \quad \text{(charge)},$$

in which *Q* is the charge amplitude (maximum charge on the capacitor) and the angular frequency ω of the oscillations is

$$\omega = \frac{1}{\sqrt{LC}}.$$

● The phase constant ϕ is determined by the initial conditions (at $t = 0$) of the system.

● The current *i* in the system at any time *t* is

$$i = -\omega Q\sin(\omega t + \phi) \quad \text{(current)},$$

in which ωQ is the current amplitude *I*.

What Is Physics?

We have explored the basic physics of electric and magnetic fields and how energy can be stored in capacitors and inductors. We next turn to the associated applied physics, in which the energy stored in one location can be transferred to another location so that it can be put to use. For example, energy produced at a power plant can show up at your home to run a computer. The total worth of this applied physics is now so high that its estimation is almost impossible. Indeed, modern civilization would be impossible without this applied physics.

In most parts of the world, electrical energy is transferred not as a direct current but as a sinusoidally oscillating current (alternating current, or ac). The challenge to both physicists and engineers is to design ac systems that transfer energy efficiently and to build appliances that make use of that energy. Our first step here is to study the oscillations in a circuit with inductance L and capacitance C.

LC Oscillations, Qualitatively

Of the three circuit elements, resistance R, capacitance C, and inductance L, we have so far discussed the series combinations RC (in Module 27-4) and RL (in Module 30-6). In these two kinds of circuit we found that the charge, current, and potential difference grow and decay exponentially. The time scale of the growth or decay is given by a *time constant* τ, which is either capacitive or inductive.

We now examine the remaining two-element circuit combination LC. You will see that in this case the charge, current, and potential difference do not decay exponentially with time but vary sinusoidally (with period T and angular frequency ω). The resulting oscillations of the capacitor's electric field and the inductor's magnetic field are said to be **electromagnetic oscillations.** Such a circuit is said to oscillate.

Parts *a* through *h* of Fig. 31-1 show succeeding stages of the oscillations in a simple LC circuit. From Eq. 25-21, the energy stored in the electric field of the

Figure 31-1 Eight stages in a single cycle of oscillation of a resistanceless LC circuit. The bar graphs by each figure show the stored magnetic and electrical energies. The magnetic field lines of the inductor and the electric field lines of the capacitor are shown. (*a*) Capacitor with maximum charge, no current. (*b*) Capacitor discharging, current increasing. (*c*) Capacitor fully discharged, current maximum. (*d*) Capacitor charging but with polarity opposite that in (*a*), current decreasing. (*e*) Capacitor with maximum charge having polarity opposite that in (*a*), no current. (*f*) Capacitor discharging, current increasing with direction opposite that in (*b*). (*g*) Capacitor fully discharged, current maximum. (*h*) Capacitor charging, current decreasing.

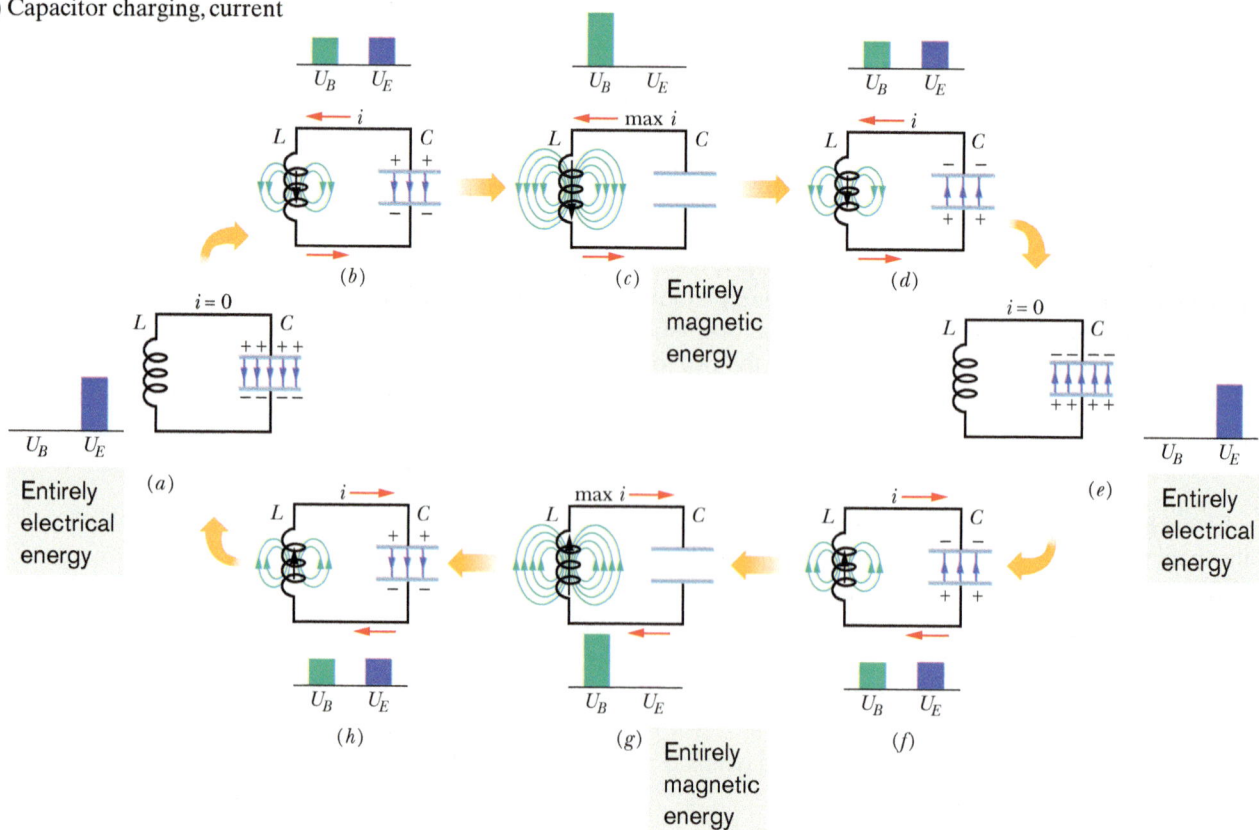

capacitor at any time is

$$U_E = \frac{q^2}{2C},$$

(31-1)

where q is the charge on the capacitor at that time. From Eq. 30-49, the energy stored in the magnetic field of the inductor at any time is

$$U_B = \frac{Li^2}{2},$$

(31-2)

where i is the current through the inductor at that time.

We now adopt the convention of representing *instantaneous values* of the electrical quantities of a sinusoidally oscillating circuit with small letters, such as q, and the *amplitudes* of those quantities with capital letters, such as Q. With this convention in mind, let us assume that initially the charge q on the capacitor in Fig. 31-1 is at its maximum value Q and that the current i through the inductor is zero. This initial state of the circuit is shown in Fig. 31-1a. The bar graphs for energy included there indicate that at this instant, with zero current through the inductor and maximum charge on the capacitor, the energy U_B of the magnetic field is zero and the energy U_E of the electric field is a maximum. As the circuit oscillates, energy shifts back and forth from one type of stored energy to the other, but the total amount is conserved.

The capacitor now starts to discharge through the inductor, positive charge carriers moving counterclockwise, as shown in Fig. 31-1b. This means that a current i, given by dq/dt and pointing down in the inductor, is established. As the capacitor's charge decreases, the energy stored in the electric field within the capacitor also decreases. This energy is transferred to the magnetic field that appears around the inductor because of the current i that is building up there. Thus, the electric field decreases and the magnetic field builds up as energy is transferred from the electric field to the magnetic field.

The capacitor eventually loses all its charge (Fig. 31-1c) and thus also loses its electric field and the energy stored in that field. The energy has then been fully transferred to the magnetic field of the inductor. The magnetic field is then at its maximum magnitude, and the current through the inductor is then at its maximum value I.

Although the charge on the capacitor is now zero, the counterclockwise current must continue because the inductor does not allow it to change suddenly to zero. The current continues to transfer positive charge from the top plate to the bottom plate through the circuit (Fig. 31-1d). Energy now flows from the inductor back to the capacitor as the electric field within the capacitor builds up again. The current gradually decreases during this energy transfer. When, eventually, the energy has been transferred completely back to the capacitor (Fig. 31-1e), the current has decreased to zero (momentarily). The situation of Fig. 31-1e is like the initial situation, except that the capacitor is now charged oppositely.

The capacitor then starts to discharge again but now with a clockwise current (Fig. 31-1f). Reasoning as before, we see that the clockwise current builds to a maximum (Fig. 31-1g) and then decreases (Fig. 31-1h), until the circuit eventually returns to its initial situation (Fig. 31-1a). The process then repeats at some frequency f and thus at an angular frequency $\omega = 2\pi f$. In the ideal *LC* circuit with no resistance, there are no energy transfers other than that between the electric field of the capacitor and the magnetic field of the inductor. Because of the conservation of energy, the oscillations continue indefinitely. The oscillations need not begin with the energy all in the electric field; the initial situation could be any other stage of the oscillation.

Figure 31-2 (*a*) The potential difference across the capacitor in the circuit of Fig. 31-1 as a function of time. This quantity is proportional to the charge on the capacitor. (*b*) A potential proportional to the current in the circuit of Fig. 31-1. The letters refer to the correspondingly labeled oscillation stages in Fig. 31-1.

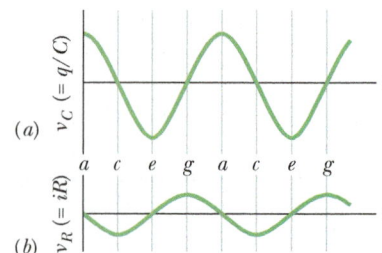

To determine the charge q on the capacitor as a function of time, we can put in a voltmeter to measure the time-varying potential difference (or *voltage*) v_C that exists across the capacitor C. From Eq. 25-1 we can write

$$v_C = \left(\frac{1}{C}\right) q,$$

which allows us to find q. To measure the current, we can connect a small resistance R in series with the capacitor and inductor and measure the time-varying potential difference v_R across it; v_R is proportional to i through the relation

$$v_R = iR.$$

We assume here that R is so small that its effect on the behavior of the circuit is negligible. The variations in time of v_C and v_R, and thus of q and i, are shown in Fig. 31-2. All four quantities vary sinusoidally.

In an actual LC circuit, the oscillations will not continue indefinitely because there is always some resistance present that will drain energy from the electric and magnetic fields and dissipate it as thermal energy (the circuit may become warmer). The oscillations, once started, will die away as Fig. 31-3 suggests. Compare this figure with Fig. 15-17, which shows the decay of mechanical oscillations caused by frictional damping in a block–spring system.

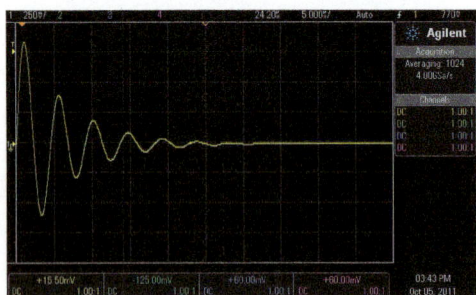

Courtesy Agilent Technologies

Figure 31-3 An oscilloscope trace showing how the oscillations in an RLC circuit actually die away because energy is dissipated in the resistor as thermal energy.

✅ Checkpoint 1

A charged capacitor and an inductor are connected in series at time $t = 0$. In terms of the period T of the resulting oscillations, determine how much later the following reach their maximum value: (a) the charge on the capacitor; (b) the voltage across the capacitor, with its original polarity; (c) the energy stored in the electric field; and (d) the current.

The Electrical–Mechanical Analogy

Let us look a little closer at the analogy between the oscillating LC system of Fig. 31-1 and an oscillating block–spring system. Two kinds of energy are involved in the block–spring system. One is potential energy of the compressed or extended spring; the other is kinetic energy of the moving block. These two energies are given by the formulas in the first energy column in Table 31-1.

Table 31-1 Comparison of the Energy in Two Oscillating Systems

Block–Spring System		LC Oscillator	
Element	Energy	Element	Energy
Spring	Potential, $\frac{1}{2}kx^2$	Capacitor	Electrical, $\frac{1}{2}(1/C)q^2$
Block	Kinetic, $\frac{1}{2}mv^2$	Inductor	Magnetic, $\frac{1}{2}Li^2$
	$v = dx/dt$		$i = dq/dt$

The table also shows, in the second energy column, the two kinds of energy involved in *LC* oscillations. By looking across the table, we can see an analogy between the forms of the two pairs of energies—the mechanical energies of the block–spring system and the electromagnetic energies of the *LC* oscillator. The equations for v and i at the bottom of the table help us see the details of the analogy. They tell us that q corresponds to x and i corresponds to v (in both equations, the former is differentiated to obtain the latter). These correspondences then suggest that, in the energy expressions, $1/C$ corresponds to k and L corresponds to m. Thus,

$$q \text{ corresponds to } x, \qquad 1/C \text{ corresponds to } k,$$
$$i \text{ corresponds to } v, \quad \text{and} \quad L \text{ corresponds to } m.$$

These correspondences suggest that in an *LC* oscillator, the capacitor is mathematically like the spring in a block–spring system and the inductor is like the block.

In Module 15-1 we saw that the angular frequency of oscillation of a (frictionless) block–spring system is

$$\omega = \sqrt{\frac{k}{m}} \quad \text{(block–spring system)}. \tag{31-3}$$

The correspondences listed above suggest that to find the angular frequency of oscillation for an ideal (resistanceless) *LC* circuit, k should be replaced by $1/C$ and m by L, yielding

$$\omega = \frac{1}{\sqrt{LC}} \quad \text{(LC circuit)}. \tag{31-4}$$

LC Oscillations, Quantitatively

Here we want to show explicitly that Eq. 31-4 for the angular frequency of *LC* oscillations is correct. At the same time, we want to examine even more closely the analogy between *LC* oscillations and block–spring oscillations. We start by extending somewhat our earlier treatment of the mechanical block–spring oscillator.

The Block-Spring Oscillator

We analyzed block–spring oscillations in Chapter 15 in terms of energy transfers and did not—at that early stage—derive the fundamental differential equation that governs those oscillations. We do so now.

We can write, for the total energy U of a block–spring oscillator at any instant,

$$U = U_b + U_s = \tfrac{1}{2}mv^2 + \tfrac{1}{2}kx^2, \tag{31-5}$$

where U_b and U_s are, respectively, the kinetic energy of the moving block and the potential energy of the stretched or compressed spring. If there is no friction—which we assume—the total energy U remains constant with time, even though v and x vary. In more formal language, $dU/dt = 0$. This leads to

$$\frac{dU}{dt} = \frac{d}{dt}\left(\tfrac{1}{2}mv^2 + \tfrac{1}{2}kx^2\right) = mv\frac{dv}{dt} + kx\frac{dx}{dt} = 0. \tag{31-6}$$

Substituting $v = dx/dt$ and $dv/dt = d^2x/dt^2$, we find

$$m\frac{d^2x}{dt^2} + kx = 0 \quad \text{(block–spring oscillations)}. \tag{31-7}$$

Equation 31-7 is the fundamental *differential equation* that governs the frictionless block–spring oscillations.

The general solution to Eq. 31-7 is (as we saw in Eq. 15-3)

$$x = X\cos(\omega t + \phi) \quad \text{(displacement)}, \tag{31-8}$$

in which X is the amplitude of the mechanical oscillations (x_m in Chapter 15), ω is the angular frequency of the oscillations, and ϕ is a phase constant.

The *LC* Oscillator

Now let us analyze the oscillations of a resistanceless *LC* circuit, proceeding exactly as we just did for the block–spring oscillator. The total energy U present at any instant in an oscillating *LC* circuit is given by

$$U = U_B + U_E = \frac{Li^2}{2} + \frac{q^2}{2C}, \tag{31-9}$$

in which U_B is the energy stored in the magnetic field of the inductor and U_E is the energy stored in the electric field of the capacitor. Since we have assumed the circuit resistance to be zero, no energy is transferred to thermal energy and U remains constant with time. In more formal language, dU/dt must be zero. This leads to

$$\frac{dU}{dt} = \frac{d}{dt}\left(\frac{Li^2}{2} + \frac{q^2}{2C}\right) = Li\frac{di}{dt} + \frac{q}{C}\frac{dq}{dt} = 0. \tag{31-10}$$

However, $i = dq/dt$ and $di/dt = d^2q/dt^2$. With these substitutions, Eq. 31-10 becomes

$$L\frac{d^2q}{dt^2} + \frac{1}{C}q = 0 \quad \text{(}LC\text{ oscillations).} \tag{31-11}$$

This is the *differential equation* that describes the oscillations of a resistanceless *LC* circuit. Equations 31-11 and 31-7 are exactly of the same mathematical form.

Charge and Current Oscillations

Since the differential equations are mathematically identical, their solutions must also be mathematically identical. Because q corresponds to x, we can write the general solution of Eq. 31-11, by analogy to Eq. 31-8, as

$$q = Q\cos(\omega t + \phi) \quad \text{(charge),} \tag{31-12}$$

where Q is the amplitude of the charge variations, ω is the angular frequency of the electromagnetic oscillations, and ϕ is the phase constant. Taking the first derivative of Eq. 31-12 with respect to time gives us the current:

$$i = \frac{dq}{dt} = -\omega Q\sin(\omega t + \phi) \quad \text{(current).} \tag{31-13}$$

The amplitude I of this sinusoidally varying current is

$$I = \omega Q, \tag{31-14}$$

and so we can rewrite Eq. 31-13 as

$$i = -I\sin(\omega t + \phi). \tag{31-15}$$

Angular Frequencies

We can test whether Eq. 31-12 is a solution of Eq. 31-11 by substituting Eq. 31-12 and its second derivative with respect to time into Eq. 31-11. The first derivative of Eq. 31-12 is Eq. 31-13. The second derivative is then

$$\frac{d^2q}{dt^2} = -\omega^2 Q\cos(\omega t + \phi).$$

Substituting for q and d^2q/dt^2 in Eq. 31-11, we obtain

$$-L\omega^2 Q\cos(\omega t + \phi) + \frac{1}{C}Q\cos(\omega t + \phi) = 0.$$

Canceling $Q\cos(\omega t + \phi)$ and rearranging lead to

$$\omega = \frac{1}{\sqrt{LC}}.$$

Thus, Eq. 31-12 is indeed a solution of Eq. 31-11 if ω has the constant value $1/\sqrt{LC}$. Note that this expression for ω is exactly that given by Eq. 31-4.

The phase constant ϕ in Eq. 31-12 is determined by the conditions that exist at any certain time—say, $t = 0$. If the conditions yield $\phi = 0$ at $t = 0$, Eq. 31-12 requires that $q = Q$ and Eq. 31-13 requires that $i = 0$; these are the initial conditions represented by Fig. 31-1a.

Electrical and Magnetic Energy Oscillations

The electrical energy stored in the *LC* circuit at time t is, from Eqs. 31-1 and 31-12,

$$U_E = \frac{q^2}{2C} = \frac{Q^2}{2C}\cos^2(\omega t + \phi). \tag{31-16}$$

The magnetic energy is, from Eqs. 31-2 and 31-13,

$$U_B = \tfrac{1}{2}Li^2 = \tfrac{1}{2}L\omega^2Q^2\sin^2(\omega t + \phi).$$

Substituting for ω from Eq. 31-4 then gives us

$$U_B = \frac{Q^2}{2C}\sin^2(\omega t + \phi). \tag{31-17}$$

Figure 31-4 shows plots of $U_E(t)$ and $U_B(t)$ for the case of $\phi = 0$. Note that

1. The maximum values of U_E and U_B are both $Q^2/2C$.
2. At any instant the sum of U_E and U_B is equal to $Q^2/2C$, a constant.
3. When U_E is maximum, U_B is zero, and conversely.

The electrical and magnetic energies vary but the total is constant.

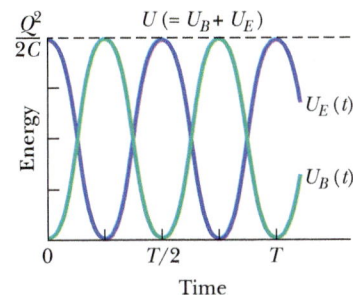

Figure 31-4 The stored magnetic energy and electrical energy in the circuit of Fig. 31-1 as a function of time. Note that their sum remains constant. T is the period of oscillation.

> ### ✓ Checkpoint 2
>
> A capacitor in an *LC* oscillator has a maximum potential difference of 17 V and a maximum energy of 160 μJ. When the capacitor has a potential difference of 5 V and an energy of 10 μJ, what are (a) the emf across the inductor and (b) the energy stored in the magnetic field?

Sample Problem 31.01 *LC* oscillator: potential change, rate of current change

A 1.5 μF capacitor is charged to 57 V by a battery, which is then removed. At time $t = 0$, a 12 mH coil is connected in series with the capacitor to form an *LC* oscillator (Fig. 31-1).

(a) What is the potential difference $v_L(t)$ across the inductor as a function of time?

KEY IDEAS

(1) The current and potential differences of the circuit (both the potential difference of the capacitor and the potential difference of the coil) undergo sinusoidal oscillations. (2) We can still apply the loop rule to these oscillating potential differences, just as we did for the nonoscillating circuits of Chapter 27.

Calculations: At any time t during the oscillations, the loop rule and Fig. 31-1 give us

$$v_L(t) = v_C(t); \tag{31-18}$$

that is, the potential difference v_L across the inductor must always be equal to the potential difference v_C across the capacitor, so that the net potential difference around the circuit is zero. Thus, we will find $v_L(t)$ if we can find $v_C(t)$, and we can find $v_C(t)$ from $q(t)$ with Eq. 25-1 ($q = CV$).

Because the potential difference $v_C(t)$ is maximum when the oscillations begin at time $t = 0$, the charge q on the capacitor must also be maximum then. Thus, phase constant ϕ must be zero; so Eq. 31-12 gives us

$$q = Q\cos\omega t. \tag{31-19}$$

(Note that this cosine function does indeed yield maximum q ($= Q$) when $t = 0$.) To get the potential difference $v_C(t)$, we divide both sides of Eq. 31-19 by C to write

$$\frac{q}{C} = \frac{Q}{C} \cos \omega t,$$

and then use Eq. 25-1 to write

$$v_C = V_C \cos \omega t. \qquad (31\text{-}20)$$

Here, V_C is the amplitude of the oscillations in the potential difference v_C across the capacitor.

Next, substituting $v_C = v_L$ from Eq. 31-18, we find

$$v_L = V_C \cos \omega t. \qquad (31\text{-}21)$$

We can evaluate the right side of this equation by first noting that the amplitude V_C is equal to the initial (maximum) potential difference of 57 V across the capacitor. Then we find ω with Eq. 31-4:

$$\omega = \frac{1}{\sqrt{LC}} = \frac{1}{[(0.012\ \text{H})(1.5 \times 10^{-6}\ \text{F})]^{0.5}}$$

$$= 7454\ \text{rad/s} \approx 7500\ \text{rad/s}.$$

Thus, Eq. 31-21 becomes

$$v_L = (57\ \text{V}) \cos(7500\ \text{rad/s})t. \qquad \text{(Answer)}$$

(b) What is the maximum rate $(di/dt)_{max}$ at which the current i changes in the circuit?

KEY IDEA

With the charge on the capacitor oscillating as in Eq. 31-12, the current is in the form of Eq. 31-13. Because $\phi = 0$, that equation gives us

$$i = -\omega Q \sin \omega t.$$

Calculations: Taking the derivative, we have

$$\frac{di}{dt} = \frac{d}{dt}(-\omega Q \sin \omega t) = -\omega^2 Q \cos \omega t.$$

We can simplify this equation by substituting CV_C for Q (because we know C and V_C but not Q) and $1/\sqrt{LC}$ for ω according to Eq. 31-4. We get

$$\frac{di}{dt} = -\frac{1}{LC}CV_C \cos \omega t = -\frac{V_C}{L} \cos \omega t.$$

This tells us that the current changes at a varying (sinusoidal) rate, with its maximum rate of change being

$$\frac{V_C}{L} = \frac{57\ \text{V}}{0.012\ \text{H}} = 4750\ \text{A/s} \approx 4800\ \text{A/s}. \qquad \text{(Answer)}$$

WILEY PLUS Additional examples, video, and practice available at *WileyPLUS*

31-2 DAMPED OSCILLATIONS IN AN *RLC* CIRCUIT

Learning Objectives

After reading this module, you should be able to . . .

31.13 Draw the schematic of a damped RLC circuit and explain why the oscillations are damped.

31.14 Starting with the expressions for the field energies and the rate of energy loss in a damped RLC circuit, write the differential equation for the charge q on the capacitor.

31.15 For a damped RLC circuit, apply the expression for charge $q(t)$.

31.16 Identify that in a damped RLC circuit, the charge amplitude and the amplitude of the electric field energy decrease exponentially with time.

31.17 Apply the relationship between the angular frequency ω' of a given damped RLC oscillator and the angular frequency ω of the circuit if R is removed.

31.18 For a damped RLC circuit, apply the expression for the electric field energy U_E as a function of time.

Key Ideas

● Oscillations in an LC circuit are damped when a dissipative element R is also present in the circuit. Then

$$L\frac{d^2q}{dt^2} + R\frac{dq}{dt} + \frac{1}{C}q = 0 \quad (RLC\ \text{circuit}).$$

● The solution of this differential equation is

$$q = Qe^{-Rt/2L} \cos(\omega't + \phi),$$

where $\omega' = \sqrt{\omega^2 - (R/2L)^2}$.

We consider only situations with small R and thus small damping; then $\omega' \approx \omega$.

Damped Oscillations in an *RLC* Circuit

A circuit containing resistance, inductance, and capacitance is called an *RLC circuit.* We shall here discuss only *series RLC circuits* like that shown in Fig. 31-5. With a resistance R present, the total *electromagnetic energy U* of the circuit (the sum of the electrical energy and magnetic energy) is no longer constant; instead, it decreases with time as energy is transferred to thermal energy in the resistance. Because of this loss of energy, the oscillations of charge, current, and potential difference continuously decrease in amplitude, and the oscillations are said to be *damped,* just as with the damped block–spring oscillator of Module 15-5.

To analyze the oscillations of this circuit, we write an equation for the total electromagnetic energy U in the circuit at any instant. Because the resistance does not store electromagnetic energy, we can use Eq. 31-9:

$$U = U_B + U_E = \frac{Li^2}{2} + \frac{q^2}{2C}. \tag{31-22}$$

Now, however, this total energy decreases as energy is transferred to thermal energy. The rate of that transfer is, from Eq. 26-27,

$$\frac{dU}{dt} = -i^2R, \tag{31-23}$$

where the minus sign indicates that U decreases. By differentiating Eq. 31-22 with respect to time and then substituting the result in Eq. 31-23, we obtain

$$\frac{dU}{dt} = Li\frac{di}{dt} + \frac{q}{C}\frac{dq}{dt} = -i^2R.$$

Substituting dq/dt for i and d^2q/dt^2 for di/dt, we obtain

$$L\frac{d^2q}{dt^2} + R\frac{dq}{dt} + \frac{1}{C}q = 0 \quad \text{(RLC circuit)}, \tag{31-24}$$

which is the differential equation for damped oscillations in an *RLC* circuit.

Charge Decay. The solution to Eq. 31-24 is

$$q = Qe^{-Rt/2L}\cos(\omega't + \phi), \tag{31-25}$$

in which

$$\omega' = \sqrt{\omega^2 - (R/2L)^2}, \tag{31-26}$$

where $\omega = 1/\sqrt{LC}$, as with an undamped oscillator. Equation 31-25 tells us how the charge on the capacitor oscillates in a damped *RLC* circuit; that equation is the electromagnetic counterpart of Eq. 15-42, which gives the displacement of a damped block–spring oscillator.

Equation 31-25 describes a sinusoidal oscillation (the cosine function) with an *exponentially decaying amplitude* $Qe^{-Rt/2L}$ (the factor that multiplies the cosine). The angular frequency ω' of the damped oscillations is always less than the angular frequency ω of the undamped oscillations; however, we shall here consider only situations in which R is small enough for us to replace ω' with ω.

Energy Decay. Let us next find an expression for the total electromagnetic energy U of the circuit as a function of time. One way to do so is to monitor the energy of the electric field in the capacitor, which is given by Eq. 31-1 ($U_E = q^2/2C$). By substituting Eq. 31-25 into Eq. 31-1, we obtain

$$U_E = \frac{q^2}{2C} = \frac{[Qe^{-Rt/2L}\cos(\omega't + \phi)]^2}{2C} = \frac{Q^2}{2C}e^{-Rt/L}\cos^2(\omega't + \phi). \tag{31-27}$$

Thus, the energy of the electric field oscillates according to a cosine-squared term, and the amplitude of that oscillation decreases exponentially with time.

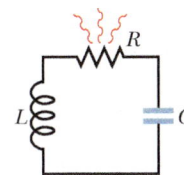

Figure 31-5 A series *RLC* circuit. As the charge contained in the circuit oscillates back and forth through the resistance, electromagnetic energy is dissipated as thermal energy, damping (decreasing the amplitude of) the oscillations.

Sample Problem 31.02 Damped *RLC* circuit: charge amplitude

A series *RLC* circuit has inductance $L = 12$ mH, capacitance $C = 1.6$ μF, and resistance $R = 1.5$ Ω and begins to oscillate at time $t = 0$.

(a) At what time t will the amplitude of the charge oscillations in the circuit be 50% of its initial value? (Note that we do not know that initial value.)

KEY IDEA

The amplitude of the charge oscillations decreases exponentially with time t: According to Eq. 31-25, the charge amplitude at any time t is $Qe^{-Rt/2L}$, in which Q is the amplitude at time $t = 0$.

Calculations: We want the time when the charge amplitude has decreased to $0.50Q$— that is, when

$$Qe^{-Rt/2L} = 0.50Q.$$

We can now cancel Q (which also means that we can answer the question without knowing the initial charge). Taking the natural logarithms of both sides (to eliminate the exponential function), we have

$$-\frac{Rt}{2L} = \ln 0.50.$$

Solving for t and then substituting given data yield

$$t = -\frac{2L}{R} \ln 0.50 = -\frac{(2)(12 \times 10^{-3}\text{ H})(\ln 0.50)}{1.5\ \Omega}$$

$$= 0.0111\text{ s} \approx 11\text{ ms}. \qquad \text{(Answer)}$$

(b) How many oscillations are completed within this time?

KEY IDEA

The time for one complete oscillation is the period $T = 2\pi/\omega$, where the angular frequency for *LC* oscillations is given by Eq. 31-4 ($\omega = 1/\sqrt{LC}$).

Calculation: In the time interval $\Delta t = 0.0111$ s, the number of complete oscillations is

$$\frac{\Delta t}{T} = \frac{\Delta t}{2\pi\sqrt{LC}}$$

$$= \frac{0.0111\text{ s}}{2\pi[(12 \times 10^{-3}\text{ H})(1.6 \times 10^{-6}\text{ F})]^{1/2}} \approx 13.$$

$$\text{(Answer)}$$

Thus, the amplitude decays by 50% in about 13 complete oscillations. This damping is less severe than that shown in Fig. 31-3, where the amplitude decays by a little more than 50% in one oscillation.

PLUS Additional examples, video, and practice available at *WileyPLUS*

31-3 FORCED OSCILLATIONS OF THREE SIMPLE CIRCUITS

Learning Objectives

After reading this module, you should be able to . . .

31.19 Distinguish alternating current from direct current.

31.20 For an ac generator, write the emf as a function of time, identifying the emf amplitude and driving angular frequency.

31.21 For an ac generator, write the current as a function of time, identifying its amplitude and its phase constant with respect to the emf.

31.22 Draw a schematic diagram of a (series) *RLC* circuit that is driven by a generator.

31.23 Distinguish driving angular frequency ω_d from natural angular frequency ω.

31.24 In a driven (series) *RLC* circuit, identify the conditions for resonance and the effect of resonance on the current amplitude.

31.25 For each of the three basic circuits (purely resistive load, purely capacitive load, and purely inductive load),

draw the circuit and sketch graphs and phasor diagrams for voltage $v(t)$ and current $i(t)$.

31.26 For the three basic circuits, apply equations for voltage $v(t)$ and current $i(t)$.

31.27 On a phasor diagram for each of the basic circuits, identify angular speed, amplitude, projection on the vertical axis, and rotation angle.

31.28 For each basic circuit, identify the phase constant, and interpret it in terms of the relative orientations of the current phasor and voltage phasor and also in terms of leading and lagging.

31.29 Apply the mnemonic "*ELI* positively is the *ICE* man."

31.30 For each basic circuit, apply the relationships between the voltage amplitude V and the current amplitude I.

31.31 Calculate capacitive reactance X_C and inductive reactance X_L.

Key Ideas

● A series RLC circuit may be set into forced oscillation at a driving angular frequency ω_d by an external alternating emf

$$\mathscr{E} = \mathscr{E}_m \sin \omega_d t.$$

● The current driven in the circuit is

$$i = I \sin(\omega_d t - \phi),$$

where ϕ is the phase constant of the current.

● The alternating potential difference across a resistor has

amplitude $V_R = IR$; the current is in phase with the potential difference.

● For a capacitor, $V_C = IX_C$, in which $X_C = 1/\omega_d C$ is the capacitive reactance; the current here leads the potential difference by 90° ($\phi = -90° = -\pi/2$ rad).

● For an inductor, $V_L = IX_L$, in which $X_L = \omega_d L$ is the inductive reactance; the current here lags the potential difference by 90° ($\phi = +90° = +\pi/2$ rad).

Alternating Current

The oscillations in an RLC circuit will not damp out if an external emf device supplies enough energy to make up for the energy dissipated as thermal energy in the resistance R. Circuits in homes, offices, and factories, including countless RLC circuits, receive such energy from local power companies. In most countries the energy is supplied via oscillating emfs and currents—the current is said to be an **alternating current,** or **ac** for short. (The nonoscillating current from a battery is said to be a **direct current,** or **dc**.) These oscillating emfs and currents vary sinusoidally with time, reversing direction (in North America) 120 times per second and thus having frequency $f = 60$ Hz.

Electron Oscillations. At first sight this may seem to be a strange arrangement. We have seen that the drift speed of the conduction electrons in household wiring may typically be 4×10^{-5} m/s. If we now reverse their direction every $\frac{1}{120}$ s, such electrons can move only about 3×10^{-7} m in a half-cycle. At this rate, a typical electron can drift past no more than about 10 atoms in the wiring before it is required to reverse its direction. How, you may wonder, can the electron ever get anywhere?

Although this question may be worrisome, it is a needless concern. The conduction electrons do not have to "get anywhere." When we say that the current in a wire is one ampere, we mean that charge passes through any plane cutting across that wire at the rate of one coulomb per second. The speed at which the charge carriers cross that plane does not matter directly; one ampere may correspond to many charge carriers moving very slowly or to a few moving very rapidly. Furthermore, the signal to the electrons to reverse directions—which originates in the alternating emf provided by the power company's generator—is propagated along the conductor at a speed close to that of light. All electrons, no matter where they are located, get their reversal instructions at about the same instant. Finally, we note that for many devices, such as lightbulbs and toasters, the direction of motion is unimportant as long as the electrons do move so as to transfer energy to the device via collisions with atoms in the device.

Why ac? The basic advantage of alternating current is this: *As the current alternates, so does the magnetic field that surrounds the conductor.* This makes possible the use of Faraday's law of induction, which, among other things, means that we can step up (increase) or step down (decrease) the magnitude of an alternating potential difference at will, using a device called a transformer, as we shall discuss later. Moreover, alternating current is more readily adaptable to rotating machinery such as generators and motors than is (nonalternating) direct current.

Emf and Current. Figure 31-6 shows a simple model of an ac generator. As the conducting loop is forced to rotate through the external magnetic field $\vec{B}$, a sinusoidally oscillating emf $\mathscr{E}$ is induced in the loop:

$$\mathscr{E} = \mathscr{E}_m \sin \omega_d t. \tag{31-28}$$

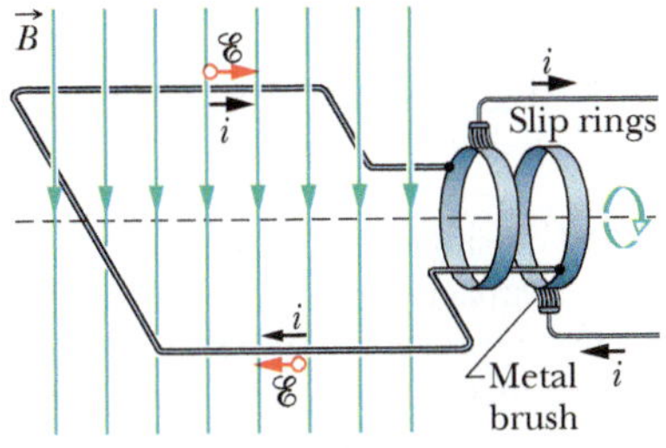

Figure 31-6 The basic mechanism of an alternating-current generator is a conducting loop rotated in an external magnetic field. In practice, the alternating emf induced in a coil of many turns of wire is made accessible by means of slip rings attached to the rotating loop. Each ring is connected to one end of the loop wire and is electrically connected to the rest of the generator circuit by a conducting brush against which the ring slips as the loop (and it) rotates.

The *angular frequency* ω_d of the emf is equal to the angular speed with which the loop rotates in the magnetic field, the *phase* of the emf is $\omega_d t$, and the *amplitude* of the emf is $\mathscr{E}_m$ (where the subscript stands for maximum). When the rotating loop is part of a closed conducting path, this emf produces (*drives*) a sinusoidal (alternating) current along the path with the same angular frequency ω_d, which then is called the **driving angular frequency.** We can write the current as

$$i = I \sin(\omega_d t - \phi), \tag{31-29}$$

in which I is the amplitude of the driven current. (The phase $\omega_d t - \phi$ of the current is traditionally written with a minus sign instead of as $\omega_d t + \phi$.) We include a phase constant ϕ in Eq. 31-29 because the current i may not be in phase with the emf $\mathscr{E}$. (As you will see, the phase constant depends on the circuit to which the generator is connected.) We can also write the current i in terms of the **driving frequency** f_d of the emf, by substituting $2\pi f_d$ for ω_d in Eq. 31-29.

Forced Oscillations

We have seen that once started, the charge, potential difference, and current in both undamped *LC* circuits and damped *RLC* circuits (with small enough *R*) oscillate at angular frequency $\omega = 1/\sqrt{LC}$. Such oscillations are said to be *free oscillations* (free of any external emf), and the angular frequency ω is said to be the circuit's **natural angular frequency.**

When the external alternating emf of Eq. 31-28 is connected to an *RLC* circuit, the oscillations of charge, potential difference, and current are said to be *driven oscillations* or *forced oscillations.* These oscillations always occur at the driving angular frequency ω_d:

> Whatever the natural angular frequency ω of a circuit may be, forced oscillations of charge, current, and potential difference in the circuit always occur at the driving angular frequency ω_d.

However, as you will see in Module 31-4, the amplitudes of the oscillations very much depend on how close ω_d is to ω. When the two angular frequencies match—a condition known as **resonance**—the amplitude I of the current in the circuit is maximum.

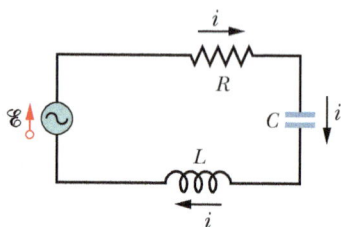

Three Simple Circuits

Later in this chapter, we shall connect an external alternating emf device to a series *RLC* circuit as in Fig. 31-7. We shall then find expressions for the amplitude I and phase constant ϕ of the sinusoidally oscillating current in terms of the amplitude $\mathscr{E}_m$ and angular frequency ω_d of the external emf. First, let's consider three simpler circuits, each having an external emf and only one other circuit element: *R*, *C*, or *L*. We start with a resistive element (a purely *resistive load*).

A Resistive Load

Figure 31-8 shows a circuit containing a resistance element of value *R* and an ac generator with the alternating emf of Eq. 31-28. By the loop rule, we have

$$\mathscr{E} - v_R = 0.$$

With Eq. 31-28, this gives us

$$v_R = \mathscr{E}_m \sin \omega_d t.$$

Because the amplitude V_R of the alternating potential difference (or voltage) across the resistance is equal to the amplitude $\mathscr{E}_m$ of the alternating emf, we can

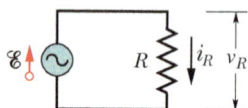

Figure 31-7 A single-loop circuit containing a resistor, a capacitor, and an inductor. A generator, represented by a sine wave in a circle, produces an alternating emf that establishes an alternating current; the directions of the emf and current are indicated here at only one instant.

Figure 31-8 A resistor is connected across an alternating-current generator.

write this as

$$v_R = V_R \sin \omega_d t. \qquad (31\text{-}30)$$

From the definition of resistance ($R = V/i$), we can now write the current i_R in the resistance as

$$i_R = \frac{v_R}{R} = \frac{V_R}{R} \sin \omega_d t. \qquad (31\text{-}31)$$

From Eq. 31-29, we can also write this current as

$$i_R = I_R \sin(\omega_d t - \phi), \qquad (31\text{-}32)$$

where I_R is the amplitude of the current i_R in the resistance. Comparing Eqs. 31-31 and 31-32, we see that for a purely resistive load the phase constant $\phi = 0°$. We also see that the voltage amplitude and current amplitude are related by

$$V_R = I_R R \quad \text{(resistor).} \qquad (31\text{-}33)$$

Although we found this relation for the circuit of Fig. 31-8, it applies to any resistance in any ac circuit.

By comparing Eqs. 31-30 and 31-31, we see that the time-varying quantities v_R and i_R are both functions of $\sin \omega_d t$ with $\phi = 0°$. Thus, these two quantities are *in phase*, which means that their corresponding maxima (and minima) occur at the same times. Figure 31-9a, which is a plot of $v_R(t)$ and $i_R(t)$, illustrates this fact. Note that v_R and i_R do not decay here because the generator supplies energy to the circuit to make up for the energy dissipated in R.

The time-varying quantities v_R and i_R can also be represented geometrically by *phasors*. Recall from Module 16-6 that phasors are vectors that rotate around an origin. Those that represent the voltage across and current in the resistor of Fig. 31-8 are shown in Fig. 31-9b at an arbitrary time t. Such phasors have the following properties:

Angular speed: Both phasors rotate counterclockwise about the origin with an angular speed equal to the angular frequency ω_d of v_R and i_R.

Length: The length of each phasor represents the amplitude of the alternating quantity: V_R for the voltage and I_R for the current.

Projection: The projection of each phasor on the *vertical* axis represents the value of the alternating quantity at time t: v_R for the voltage and i_R for the current.

Rotation angle: The rotation angle of each phasor is equal to the phase of the

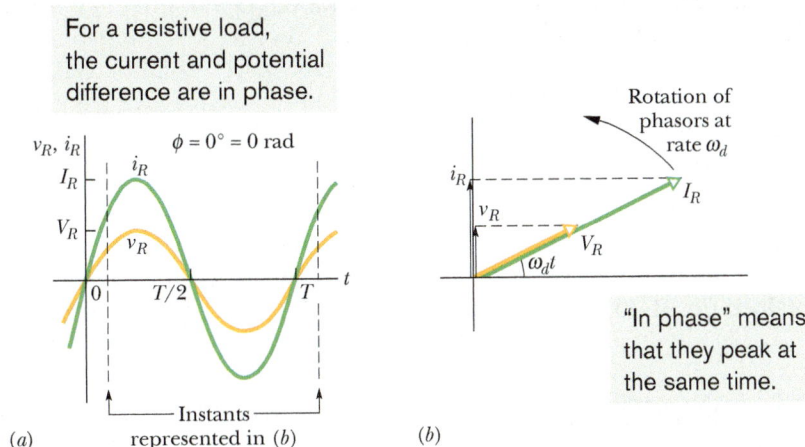

For a resistive load, the current and potential difference are in phase.

"In phase" means that they peak at the same time.

(a) Instants represented in (b)   (b)

Figure 31-9 (a) The current i_R and the potential difference v_R across the resistor are plotted on the same graph, both versus time t. They are in phase and complete one cycle in one period T. (b) A phasor diagram shows the same thing as (a).

alternating quantity at time t. In Fig. 31-9b, the voltage and current are in phase; so their phasors always have the same phase $\omega_d t$ and the same rotation angle, and thus they rotate together.

Mentally follow the rotation. Can you see that when the phasors have rotated so that $\omega_d t = 90°$ (they point vertically upward), they indicate that just then $v_R = V_R$ and $i_R = I_R$? Equations 31-30 and 31-32 give the same results.

✓ Checkpoint 3

If we increase the driving frequency in a circuit with a purely resistive load, do (a) amplitude V_R and (b) amplitude I_R increase, decrease, or remain the same?

Sample Problem 31.03 Purely resistive load: potential difference and current

In Fig. 31-8, resistance R is 200 Ω and the sinusoidal alternating emf device operates at amplitude $\mathcal{E}_m = 36.0$ V and frequency $f_d = 60.0$ Hz.

(a) What is the potential difference $v_R(t)$ across the resistance as a function of time t, and what is the amplitude V_R of $v_R(t)$?

KEY IDEA

In a circuit with a purely resistive load, the potential difference $v_R(t)$ across the resistance is always equal to the potential difference $\mathcal{E}(t)$ across the emf device.

Calculations: For our situation, $v_R(t) = \mathcal{E}(t)$ and $V_R = \mathcal{E}_m$. Since $\mathcal{E}_m$ is given, we can write

$$V_R = \mathcal{E}_m = 36.0 \text{ V}. \quad \text{(Answer)}$$

To find $v_R(t)$, we use Eq. 31-28 to write

$$v_R(t) = \mathcal{E}(t) = \mathcal{E}_m \sin \omega_d t \quad (31\text{-}34)$$

and then substitute $\mathcal{E}_m = 36.0$ V and

$$\omega_d = 2\pi f_d = 2\pi(60 \text{ Hz}) = 120\pi$$

to obtain

$$v_R = (36.0 \text{ V}) \sin(120\pi t). \quad \text{(Answer)}$$

We can leave the argument of the sine in this form for convenience, or we can write it as $(377 \text{ rad/s})t$ or as $(377 \text{ s}^{-1})t$.

(b) What are the current $i_R(t)$ in the resistance and the amplitude I_R of $i_R(t)$?

KEY IDEA

In an ac circuit with a purely resistive load, the alternating current $i_R(t)$ in the resistance is *in phase* with the alternating potential difference $v_R(t)$ across the resistance; that is, the phase constant ϕ for the current is zero.

Calculations: Here we can write Eq. 31-29 as

$$i_R = I_R \sin(\omega_d t - \phi) = I_R \sin \omega_d t. \quad (31\text{-}35)$$

From Eq. 31-33, the amplitude I_R is

$$I_R = \frac{V_R}{R} = \frac{36.0 \text{ V}}{200 \text{ Ω}} = 0.180 \text{ A}. \quad \text{(Answer)}$$

Substituting this and $\omega_d = 2\pi f_d = 120\pi$ into Eq. 31-35, we have

$$i_R = (0.180 \text{ A}) \sin(120\pi t). \quad \text{(Answer)}$$

PLUS Additional examples, video, and practice available at *WileyPLUS*

A Capacitive Load

Figure 31-10 shows a circuit containing a capacitance and a generator with the alternating emf of Eq. 31-28. Using the loop rule and proceeding as we did when we obtained Eq. 31-30, we find that the potential difference across the capacitor is

$$v_C = V_C \sin \omega_d t, \quad (31\text{-}36)$$

where V_C is the amplitude of the alternating voltage across the capacitor. From the definition of capacitance we can also write

$$q_C = C v_C = C V_C \sin \omega_d t. \quad (31\text{-}37)$$

Our concern, however, is with the current rather than the charge. Thus, we differ-

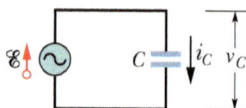

Figure 31-10 A capacitor is connected across an alternating-current generator.

entiate Eq. 31-37 to find

$$i_C = \frac{dq_C}{dt} = \omega_d CV_C \cos \omega_d t. \qquad (31\text{-}38)$$

We now modify Eq. 31-38 in two ways. First, for reasons of symmetry of notation, we introduce the quantity X_C, called the **capacitive reactance** of a capacitor, defined as

$$X_C = \frac{1}{\omega_d C} \quad \text{(capacitive reactance).} \qquad (31\text{-}39)$$

Its value depends not only on the capacitance but also on the driving angular frequency ω_d. We know from the definition of the capacitive time constant ($\tau = RC$) that the SI unit for C can be expressed as seconds per ohm. Applying this to Eq. 31-39 shows that the SI unit of X_C is the *ohm*, just as for resistance R.

Second, we replace $\cos \omega_d t$ in Eq. 31-38 with a phase-shifted sine:

$$\cos \omega_d t = \sin(\omega_d t + 90°).$$

You can verify this identity by shifting a sine curve 90° in the negative direction.

With these two modifications, Eq. 31-38 becomes

$$i_C = \left(\frac{V_C}{X_C}\right) \sin(\omega_d t + 90°). \qquad (31\text{-}40)$$

From Eq. 31-29, we can also write the current i_C in the capacitor of Fig. 31-10 as

$$i_C = I_C \sin(\omega_d t - \phi), \qquad (31\text{-}41)$$

where I_C is the amplitude of i_C. Comparing Eqs. 31-40 and 31-41, we see that for a purely capacitive load the phase constant ϕ for the current is −90°. We also see that the voltage amplitude and current amplitude are related by

$$V_C = I_C X_C \quad \text{(capacitor).} \qquad (31\text{-}42)$$

Although we found this relation for the circuit of Fig. 31-10, it applies to any capacitance in any ac circuit.

Comparison of Eqs. 31-36 and 31-40, or inspection of Fig. 31-11*a*, shows that the quantities v_C and i_C are 90°, $\pi/2$ rad, or one-quarter cycle, out of phase. Furthermore, we see that i_C *leads* v_C, which means that, if you monitored the current i_C and the potential difference v_C in the circuit of Fig. 31-10, you would find that i_C reaches its maximum *before* v_C does, by one-quarter cycle.

For a capacitive load, the current leads the potential difference by 90°.

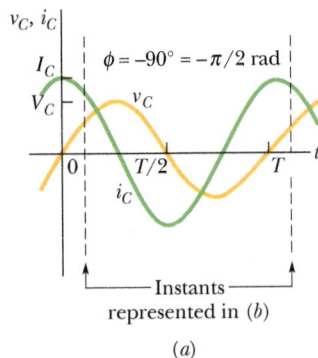

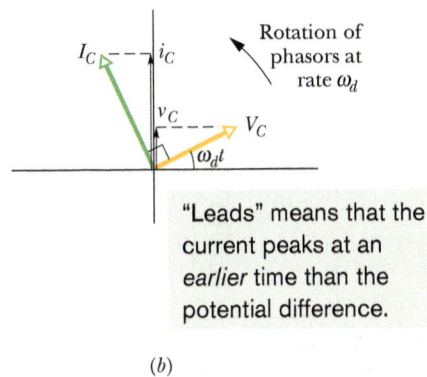

Rotation of phasors at rate ω_d

"Leads" means that the current peaks at an *earlier* time than the potential difference.

Figure 31-11 (*a*) The current in the capacitor leads the voltage by 90° (= $\pi/2$ rad). (*b*) A phasor diagram shows the same thing.

This relation between i_C and v_C is illustrated by the phasor diagram of Fig. 31-11b. As the phasors representing these two quantities rotate counterclockwise together, the phasor labeled I_C does indeed lead that labeled V_C, and by an angle of 90°; that is, the phasor I_C coincides with the vertical axis one-quarter cycle before the phasor V_C does. Be sure to convince yourself that the phasor diagram of Fig. 31-11b is consistent with Eqs. 31-36 and 31-40.

✓ Checkpoint 4

The figure shows, in (a), a sine curve $S(t) = \sin(\omega_d t)$ and three other sinusoidal curves $A(t)$, $B(t)$, and $C(t)$, each of the form $\sin(\omega_d t - \phi)$. (a) Rank the three other curves according to the value of ϕ, most positive first and most negative last. (b) Which curve corresponds to which phasor in (b) of the figure? (c) Which curve leads the others?

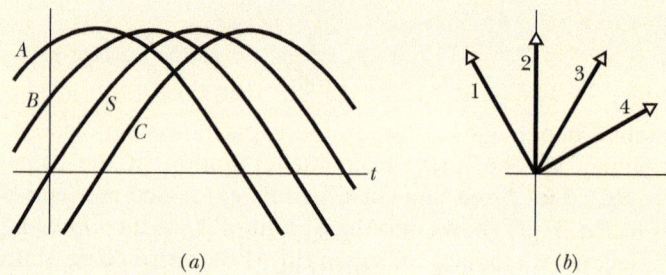

(a) (b)

Sample Problem 31.04   Purely capacitive load: potential difference and current

In Fig. 31-10, capacitance C is 15.0 μF and the sinusoidal alternating emf device operates at amplitude $\mathscr{E}_m = 36.0$ V and frequency $f_d = 60.0$ Hz.

(a) What are the potential difference $v_C(t)$ across the capacitance and the amplitude V_C of $v_C(t)$?

KEY IDEA

In a circuit with a purely capacitive load, the potential difference $v_C(t)$ across the capacitance is always equal to the potential difference $\mathscr{E}(t)$ across the emf device.

Calculations: Here we have $v_C(t) = \mathscr{E}(t)$ and $V_C = \mathscr{E}_m$. Since $\mathscr{E}_m$ is given, we have

$$V_C = \mathscr{E}_m = 36.0 \text{ V.} \qquad \text{(Answer)}$$

To find $v_C(t)$, we use Eq. 31-28 to write

$$v_C(t) = \mathscr{E}(t) = \mathscr{E}_m \sin \omega_d t. \qquad (31\text{-}43)$$

Then, substituting $\mathscr{E}_m = 36.0$ V and $\omega_d = 2\pi f_d = 120\pi$ into Eq. 31-43, we have

$$v_C = (36.0 \text{ V}) \sin(120\pi t). \qquad \text{(Answer)}$$

(b) What are the current $i_C(t)$ in the circuit as a function of time and the amplitude I_C of $i_C(t)$?

KEY IDEA

In an ac circuit with a purely capacitive load, the alternating current $i_C(t)$ in the capacitance leads the alternating potential difference $v_C(t)$ by 90°; that is, the phase constant ϕ for the current is −90°, or −$\pi/2$ rad.

Calculations: Thus, we can write Eq. 31-29 as

$$i_C = I_C \sin(\omega_d t - \phi) = I_C \sin(\omega_d t + \pi/2). \qquad (31\text{-}44)$$

We can find the amplitude I_C from Eq. 31-42 ($V_C = I_C X_C$) if we first find the capacitive reactance X_C. From Eq. 31-39 ($X_C = 1/\omega_d C$), with $\omega_d = 2\pi f_d$, we can write

$$X_C = \frac{1}{2\pi f_d C} = \frac{1}{(2\pi)(60.0 \text{ Hz})(15.0 \times 10^{-6} \text{ F})}$$

$$= 177 \ \Omega.$$

Then Eq. 31-42 tells us that the current amplitude is

$$I_C = \frac{V_C}{X_C} = \frac{36.0 \text{ V}}{177 \ \Omega} = 0.203 \text{ A.} \qquad \text{(Answer)}$$

Substituting this and $\omega_d = 2\pi f_d = 120\pi$ into Eq. 31-44, we have

$$i_C = (0.203 \text{ A}) \sin(120\pi t + \pi/2). \qquad \text{(Answer)}$$

PLUS Additional examples, video, and practice available at *WileyPLUS*

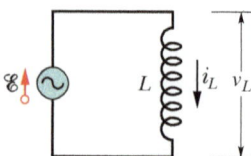

Figure 31-12 An inductor is connected across an alternating-current generator.

An Inductive Load

Figure 31-12 shows a circuit containing an inductance and a generator with the alternating emf of Eq. 31-28. Using the loop rule and proceeding as we did to obtain Eq. 31-30, we find that the potential difference across the inductance is

$$v_L = V_L \sin \omega_d t, \qquad (31\text{-}45)$$

where V_L is the amplitude of v_L. From Eq. 30-35 ($\mathscr{E}_L = -L\, di/dt$), we can write the potential difference across an inductance L in which the current is changing at the rate di_L/dt as

$$v_L = L\frac{di_L}{dt}. \qquad (31\text{-}46)$$

If we combine Eqs. 31-45 and 31-46, we have

$$\frac{di_L}{dt} = \frac{V_L}{L}\sin\omega_d t. \qquad (31\text{-}47)$$

Our concern, however, is with the current, so we integrate:

$$i_L = \int di_L = \frac{V_L}{L}\int \sin\omega_d t\, dt = -\left(\frac{V_L}{\omega_d L}\right)\cos\omega_d t. \qquad (31\text{-}48)$$

We now modify this equation in two ways. First, for reasons of symmetry of notation, we introduce the quantity X_L, called the **inductive reactance** of an inductor, which is defined as

$$X_L = \omega_d L \quad \text{(inductive reactance).} \qquad (31\text{-}49)$$

The value of X_L depends on the driving angular frequency ω_d. The unit of the inductive time constant τ_L indicates that the SI unit of X_L is the *ohm*, just as it is for X_C and for R.

Second, we replace $-\cos\omega_d t$ in Eq. 31-48 with a phase-shifted sine:

$$-\cos\omega_d t = \sin(\omega_d t - 90°).$$

You can verify this identity by shifting a sine curve 90° in the positive direction. With these two changes, Eq. 31-48 becomes

$$i_L = \left(\frac{V_L}{X_L}\right)\sin(\omega_d t - 90°). \qquad (31\text{-}50)$$

From Eq. 31-29, we can also write this current in the inductance as

$$i_L = I_L\sin(\omega_d t - \phi), \qquad (31\text{-}51)$$

where I_L is the amplitude of the current i_L. Comparing Eqs. 31-50 and 31-51, we see that for a purely inductive load the phase constant ϕ for the current is $+90°$. We also see that the voltage amplitude and current amplitude are related by

$$V_L = I_L X_L \quad \text{(inductor).} \qquad (31\text{-}52)$$

Although we found this relation for the circuit of Fig. 31-12, it applies to any inductance in any ac circuit.

Comparison of Eqs. 31-45 and 31-50, or inspection of Fig. 31-13a, shows that the quantities i_L and v_L are 90° out of phase. In this case, however, i_L *lags* v_L; that is, monitoring the current i_L and the potential difference v_L in the circuit of Fig. 31-12 shows that i_L reaches its maximum value *after* v_L does, by one-quarter cycle. The phasor diagram of Fig. 31-13b also contains this information. As the phasors rotate counterclockwise in the figure, the phasor labeled I_L does indeed lag that labeled V_L, and by an angle of 90°. Be sure to convince yourself that Fig. 31-13b represents Eqs. 31-45 and 31-50.

For an inductive load, the current lags the potential difference by 90°.

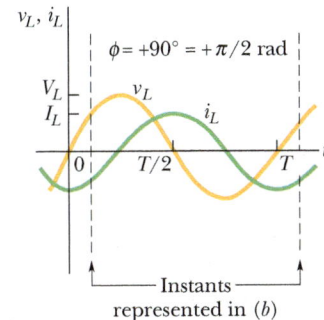

(a)

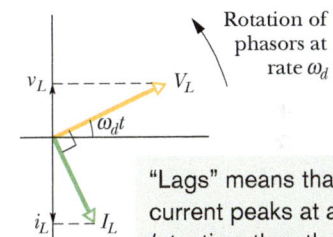

"Lags" means that the current peaks at a *later* time than the potential difference.

(b)

Figure 31-13 (a) The current in the inductor lags the voltage by 90° (= $\pi/2$ rad). (b) A phasor diagram shows the same thing.

☑ **Checkpoint 5**

If we increase the driving frequency in a circuit with a purely capacitive load, do (a) amplitude V_C and (b) amplitude I_C increase, decrease, or remain the same? If, instead, the circuit has a purely inductive load, do (c) amplitude V_L and (d) amplitude I_L increase, decrease, or remain the same?

Problem-Solving Tactics

Leading and Lagging in AC Circuits: Table 31-2 summarizes the relations between the current i and the voltage v for each of the three kinds of circuit elements we have considered. When an applied alternating voltage produces an alternating current in these elements, the current is always in phase with the voltage across a resistor, always leads the voltage across a capacitor, and always lags the voltage across an inductor.

Many students remember these results with the mnemonic "*ELI* the *ICE* man." *ELI* contains the letter L (for inductor), and in it the letter I (for current) comes *after* the letter E (for emf or voltage). Thus, for an inductor, the current *lags* (comes after) the voltage. Similarly, *ICE* (which contains a C for capacitor) means that the current *leads* (comes before) the voltage. You might also use the modified mnemonic "*ELI positively* is the *ICE* man" to remember that the phase constant ϕ is positive for an inductor.

If you have difficulty in remembering whether X_C is equal to $\omega_d C$ (wrong) or $1/\omega_d C$ (right), try remembering that C is in the "cellar"—that is, in the denominator.

Table 31-2 Phase and Amplitude Relations for Alternating Currents and Voltages

Circuit Element	Symbol	Resistance or Reactance	Phase of the Current	Phase Constant (or Angle) ϕ	Amplitude Relation
Resistor	R	R	In phase with v_R	$0° (= 0$ rad)	$V_R = I_R R$
Capacitor	C	$X_C = 1/\omega_d C$	Leads v_C by 90° $(= \pi/2$ rad)	$-90° (= -\pi/2$ rad)	$V_C = I_C X_C$
Inductor	L	$X_L = \omega_d L$	Lags v_L by 90° $(= \pi/2$ rad)	$+90° (= +\pi/2$ rad)	$V_L = I_L X_L$

Sample Problem 31.05 Purely inductive load: potential difference and current

In Fig. 31-12, inductance L is 230 mH and the sinusoidal alternating emf device operates at amplitude $\mathcal{E}_m = 36.0$ V and frequency $f_d = 60.0$ Hz.

(a) What are the potential difference $v_L(t)$ across the inductance and the amplitude V_L of $v_L(t)$?

KEY IDEA

In a circuit with a purely inductive load, the potential difference $v_L(t)$ across the inductance is always equal to the potential difference $\mathcal{E}(t)$ across the emf device.

Calculations: Here we have $v_L(t) = \mathcal{E}(t)$ and $V_L = \mathcal{E}_m$. Since $\mathcal{E}_m$ is given, we know that

$$V_L = \mathcal{E}_m = 36.0 \text{ V.} \quad \text{(Answer)}$$

To find $v_L(t)$, we use Eq. 31-28 to write

$$v_L(t) = \mathcal{E}(t) = \mathcal{E}_m \sin \omega_d t. \quad (31\text{-}53)$$

Then, substituting $\mathcal{E}_m = 36.0$ V and $\omega_d = 2\pi f_d = 120\pi$ into Eq. 31-53, we have

$$v_L = (36.0 \text{ V}) \sin(120\pi t). \quad \text{(Answer)}$$

(b) What are the current $i_L(t)$ in the circuit as a function of time and the amplitude I_L of $i_L(t)$?

KEY IDEA

In an ac circuit with a purely inductive load, the alternating current $i_L(t)$ in the inductance lags the alternating potential difference $v_L(t)$ by 90°. (In the mnemonic of the problem-solving tactic, this circuit is "positively an *ELI* circuit," which tells us that the emf E leads the current I and that ϕ is *positive*.)

Calculations: Because the phase constant ϕ for the current is +90°, or $+\pi/2$ rad, we can write Eq. 31-29 as

$$i_L = I_L \sin(\omega_d t - \phi) = I_L \sin(\omega_d t - \pi/2). \quad (31\text{-}54)$$

We can find the amplitude I_L from Eq. 31-52 ($V_L = I_L X_L$) if we first find the inductive reactance X_L. From Eq. 31-49 ($X_L = \omega_d L$), with $\omega_d = 2\pi f_d$, we can write

$$X_L = 2\pi f_d L = (2\pi)(60.0 \text{ Hz})(230 \times 10^{-3} \text{ H})$$
$$= 86.7 \ \Omega.$$

Then Eq. 31-52 tells us that the current amplitude is

$$I_L = \frac{V_L}{X_L} = \frac{36.0 \text{ V}}{86.7 \ \Omega} = 0.415 \text{ A.} \quad \text{(Answer)}$$

Substituting this and $\omega_d = 2\pi f_d = 120\pi$ into Eq. 31-54, we have

$$i_L = (0.415 \text{ A}) \sin(120\pi t - \pi/2). \quad \text{(Answer)}$$

31-4 THE SERIES *RLC* CIRCUIT

Learning Objectives

After reading this module, you should be able to . . .

31.32 Draw the schematic diagram of a series *RLC* circuit.

31.33 Identify the conditions for a mainly inductive circuit, a mainly capacitive circuit, and a resonant circuit.

31.34 For a mainly inductive circuit, a mainly capacitive circuit, and a resonant circuit, sketch graphs for voltage $v(t)$ and current $i(t)$ and sketch phasor diagrams, indicating leading, lagging, or resonance.

31.35 Calculate impedance Z.

31.36 Apply the relationship between current amplitude I, impedance Z, and emf amplitude $\mathscr{E}_m$.

31.37 Apply the relationships between phase constant ϕ and voltage amplitudes V_L and V_C, and also between

phase constant ϕ, resistance R, and reactances X_L and X_C.

31.38 Identify the values of the phase constant ϕ corresponding to a mainly inductive circuit, a mainly capacitive circuit, and a resonant circuit.

31.39 For resonance, apply the relationship between the driving angular frequency ω_d, the natural angular frequency ω, the inductance L, and the capacitance C.

31.40 Sketch a graph of current amplitude versus the ratio ω_d/ω, identifying the portions corresponding to a mainly inductive circuit, a mainly capacitive circuit, and a resonant circuit and indicating what happens to the curve for an increase in the resistance.

Key Ideas

● For a series *RLC* circuit with an external emf given by

$$\mathscr{E} = \mathscr{E}_m \sin \omega_d t,$$

and current given by

$$i = I \sin(\omega_d t - \phi),$$

the current amplitude is given by

$$I = \frac{\mathscr{E}_m}{\sqrt{R^2 + (X_L - X_C)^2}}$$

$$= \frac{\mathscr{E}_m}{\sqrt{R^2 + (\omega_d L - 1/\omega_d C)^2}} \quad \text{(current amplitude)}.$$

● The phase constant is given by

$$\tan \phi = \frac{X_L - X_C}{R} \quad \text{(phase constant)}.$$

● The impedance Z of the circuit is

$$Z = \sqrt{R^2 + (X_L - X_C)^2} \quad \text{(impedance)}.$$

● We relate the current amplitude and the impedance with

$$I = \mathscr{E}_m/Z.$$

● The current amplitude I is maximum ($I = \mathscr{E}_m/R$) when the driving angular frequency ω_d equals the natural angular frequency ω of the circuit, a condition known as resonance. Then $X_C = X_L$, $\phi = 0$, and the current is in phase with the emf.

The Series *RLC* Circuit

We are now ready to apply the alternating emf of Eq. 31-28,

$$\mathscr{E} = \mathscr{E}_m \sin \omega_d t \quad \text{(applied emf)}, \tag{31-55}$$

to the full *RLC* circuit of Fig. 31-7. Because R, L, and C are in series, the same current

$$i = I \sin(\omega_d t - \phi) \tag{31-56}$$

is driven in all three of them. We wish to find the current amplitude I and the phase constant ϕ and to investigate how these quantities depend on the driving angular frequency ω_d. The solution is simplified by the use of phasor diagrams as introduced for the three basic circuits of Module 31-3: capacitive load, inductive load, and resistive load. In particular we shall make use of how the voltage phasor is related to the current phasor for each of those basic circuits. We shall find that series *RLC* circuits can be separated into three types: mainly capacitive circuits, mainly inductive circuits, and circuits that are in resonance.

The Current Amplitude

We start with Fig. 31-14*a*, which shows the phasor representing the current of Eq. 31-56 at an arbitrary time *t*. The length of the phasor is the current amplitude *I*, the projection of the phasor on the vertical axis is the current *i* at time *t*, and the angle of rotation of the phasor is the phase $\omega_d t - \phi$ of the current at time *t*.

Figure 31-14*b* shows the phasors representing the voltages across *R*, *L*, and *C* at the same time *t*. Each phasor is oriented relative to the angle of rotation of current phasor *I* in Fig. 31-14*a*, based on the information in Table 31-2:

Resistor: Here current and voltage are in phase; so the angle of rotation of voltage phasor V_R is the same as that of phasor *I*.

Capacitor: Here current leads voltage by 90°; so the angle of rotation of voltage phasor V_C is 90° less than that of phasor *I*.

Inductor: Here current lags voltage by 90°; so the angle of rotation of voltage phasor v_L is 90° greater than that of phasor *I*.

Figure 31-14*b* also shows the instantaneous voltages v_R, v_C, and v_L across *R*, *C*, and *L* at time *t*; those voltages are the projections of the corresponding phasors on the vertical axis of the figure.

Figure 31-14*c* shows the phasor representing the applied emf of Eq. 31-55. The length of the phasor is the emf amplitude $\mathscr{E}_m$, the projection of the phasor on the vertical axis is the emf $\mathscr{E}$ at time *t*, and the angle of rotation of the phasor is the phase $\omega_d t$ of the emf at time *t*.

From the loop rule we know that at any instant the sum of the voltages v_R, v_C, and v_L is equal to the applied emf $\mathscr{E}$:

$$\mathscr{E} = v_R + v_C + v_L. \tag{31-57}$$

Thus, at time *t* the projection $\mathscr{E}$ in Fig. 31-14*c* is equal to the algebraic sum of the projections v_R, v_C, and v_L in Fig. 31-14*b*. In fact, as the phasors rotate together, this equality always holds. This means that phasor $\mathscr{E}_m$ in Fig. 31-14*c* must be equal to the vector sum of the three voltage phasors V_R, V_C, and V_L in Fig. 31-14*b*.

That requirement is indicated in Fig. 31-14*d*, where phasor $\mathscr{E}_m$ is drawn as the sum of phasors V_R, V_L, and V_C. Because phasors V_L and V_C have opposite directions in the figure, we simplify the vector sum by first combining V_L and V_C to form the single phasor $V_L - V_C$. Then we combine that single phasor with V_R to find the net phasor. Again, the net phasor must coincide with phasor $\mathscr{E}_m$, as shown.

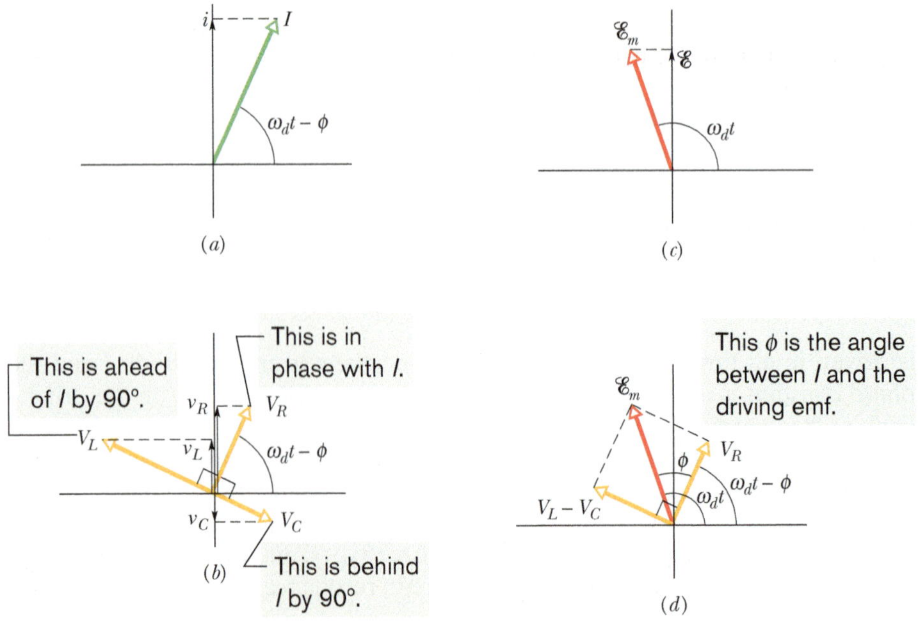

Figure 31-14 (*a*) A phasor representing the alternating current in the driven *RLC* circuit of Fig. 31-7 at time *t*. The amplitude *I*, the instantaneous value *i*, and the phase ($\omega_d t - \phi$) are shown. (*b*) Phasors representing the voltages across the inductor, resistor, and capacitor, oriented with respect to the current phasor in (*a*). (*c*) A phasor representing the alternating emf that drives the current of (*a*). (*d*) The emf phasor is equal to the vector sum of the three voltage phasors of (*b*). Here, voltage phasors V_L and V_C have been added vectorially to yield their net phasor ($V_L - V_C$).

Both triangles in Fig. 31-14d are right triangles. Applying the Pythagorean theorem to either one yields

$$\mathscr{E}_m^2 = V_R^2 + (V_L - V_C)^2. \tag{31-58}$$

From the voltage amplitude information displayed in the rightmost column of Table 31-2, we can rewrite this as

$$\mathscr{E}_m^2 = (IR)^2 + (IX_L - IX_C)^2, \tag{31-59}$$

and then rearrange it to the form

$$I = \frac{\mathscr{E}_m}{\sqrt{R^2 + (X_L - X_C)^2}}. \tag{31-60}$$

The denominator in Eq. 31-60 is called the **impedance** Z of the circuit for the driving angular frequency ω_d:

$$Z = \sqrt{R^2 + (X_L - X_C)^2} \quad \text{(impedance defined).} \tag{31-61}$$

We can then write Eq. 31-60 as

$$I = \frac{\mathscr{E}_m}{Z}. \tag{31-62}$$

If we substitute for X_C and X_L from Eqs. 31-39 and 31-49, we can write Eq. 31-60 more explicitly as

$$I = \frac{\mathscr{E}_m}{\sqrt{R^2 + (\omega_d L - 1/\omega_d C)^2}} \quad \text{(current amplitude).} \tag{31-63}$$

We have now accomplished half our goal: We have obtained an expression for the current amplitude I in terms of the sinusoidal driving emf and the circuit elements in a series *RLC* circuit.

The value of I depends on the difference between $\omega_d L$ and $1/\omega_d C$ in Eq. 31-63 or, equivalently, the difference between X_L and X_C in Eq. 31-60. In either equation, it does not matter which of the two quantities is greater because the difference is always squared.

The current that we have been describing in this module is the *steady-state current* that occurs after the alternating emf has been applied for some time. When the emf is first applied to a circuit, a brief *transient current* occurs. Its duration (before settling down into the steady-state current) is determined by the time constants $\tau_L = L/R$ and $\tau_C = RC$ as the inductive and capacitive elements "turn on." This transient current can, for example, destroy a motor on start-up if it is not properly taken into account in the motor's circuit design.

The Phase Constant

From the right-hand phasor triangle in Fig. 31-14d and from Table 31-2 we can write

$$\tan \phi = \frac{V_L - V_C}{V_R} = \frac{IX_L - IX_C}{IR}, \tag{31-64}$$

which gives us

$$\tan \phi = \frac{X_L - X_C}{R} \quad \text{(phase constant).} \tag{31-65}$$

This is the other half of our goal: an equation for the phase constant ϕ in the sinusoidally driven series *RLC* circuit of Fig. 31-7. In essence, it gives us three different results for the phase constant, depending on the relative values of the reactances X_L and X_C:

$X_L > X_C$: The circuit is said to be *more inductive than capacitive*. Equation 31-65 tells us that ϕ is positive for such a circuit, which means that phasor I rotates behind phasor $\mathscr{E}_m$ (Fig. 31-15a). A plot of $\mathscr{E}$ and i versus time is like that in Fig. 31-15b. (Figures 31-14c and d were drawn assuming $X_L > X_C$.)

$X_C > X_L$: The circuit is said to be *more capacitive than inductive*. Equation 31-65 tells us that ϕ is negative for such a circuit, which means that phasor I rotates ahead of phasor $\mathscr{E}_m$ (Fig. 31-15c). A plot of $\mathscr{E}$ and i versus time is like that in Fig. 31-15d.

$X_C = X_L$: The circuit is said to be in *resonance*, a state that is discussed next. Equation 31-65 tells us that $\phi = 0°$ for such a circuit, which means that phasors $\mathscr{E}_m$ and I rotate together (Fig. 31-15e). A plot of $\mathscr{E}$ and i versus time is like that in Fig. 31-15f.

As illustration, let us reconsider two extreme circuits: In the *purely inductive circuit* of Fig. 31-12, where X_L is nonzero and $X_C = R = 0$, Eq. 31-65 tells us that the circuit's phase constant is $\phi = +90°$ (the greatest value of ϕ), consistent with Fig. 31-13b. In the *purely capacitive circuit* of Fig. 31-10, where X_C is nonzero and $X_L = R = 0$, Eq. 31-65 tells us that the circuit's phase constant is $\phi = -90°$ (the least value of ϕ), consistent with Fig. 31-11b.

Resonance

Equation 31-63 gives the current amplitude I in an RLC circuit as a function of the driving angular frequency ω_d of the external alternating emf. For a given resistance R, that amplitude is a maximum when the quantity $\omega_d L - 1/\omega_d C$ in the

Figure 31-15 Phasor diagrams and graphs of the alternating emf $\mathscr{E}$ and current i for the driven RLC circuit of Fig. 31-7. In the phasor diagram of (a) and the graph of (b), the current i lags the driving emf $\mathscr{E}$ and the current's phase constant ϕ is positive. In (c) and (d), the current i leads the driving emf $\mathscr{E}$ and its phase constant ϕ is negative. In (e) and (f), the current i is in phase with the driving emf $\mathscr{E}$ and its phase constant ϕ is zero.

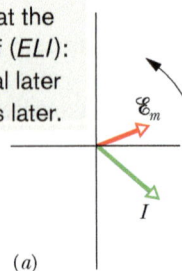

Positive ϕ means that the current lags the emf (*ELI*): the phasor is vertical later and the curve peaks later.

(a)

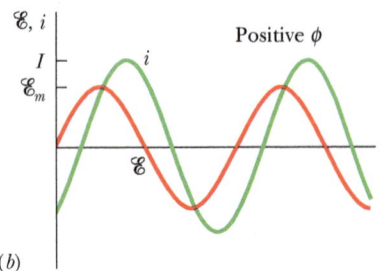

(b)

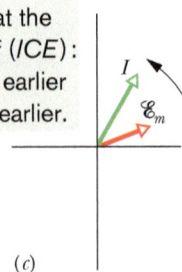

Negative ϕ means that the current leads the emf (*ICE*): the phasor is vertical earlier and the curve peaks earlier.

(c)

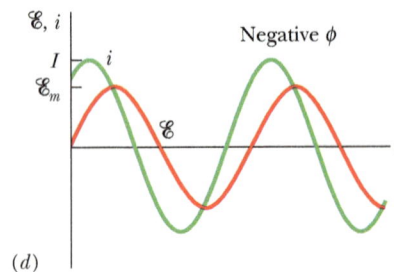

(d)

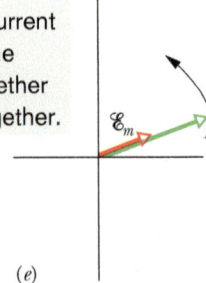

Zero ϕ means that the current and emf are in phase: the phasors are vertical together and the curves peak together.

(e)

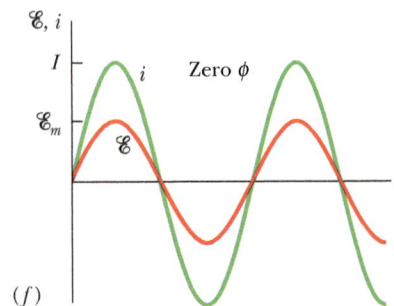

(f)

Driving ω_d equal to natural ω
- high current amplitude
- circuit is in resonance
- equally capacitive and inductive
- X_C equals X_L
- current and emf in phase
- zero ϕ

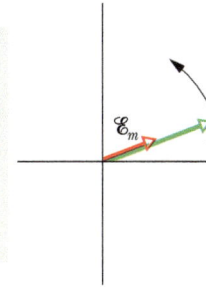

Figure 31-16 *Resonance curves* for the driven *RLC* circuit of Fig. 31-7 with $L = 100\ \mu\text{H}$, $C = 100\ \text{pF}$, and three values of R. The current amplitude I of the alternating current depends on how close the driving angular frequency ω_d is to the natural angular frequency ω. The horizontal arrow on each curve measures the curve's *half-width*, which is the width at the half-maximum level and is a measure of the sharpness of the resonance. To the left of $\omega_d/\omega = 1.00$, the circuit is mainly capacitive, with $X_C > X_L$; to the right, it is mainly inductive, with $X_L > X_C$.

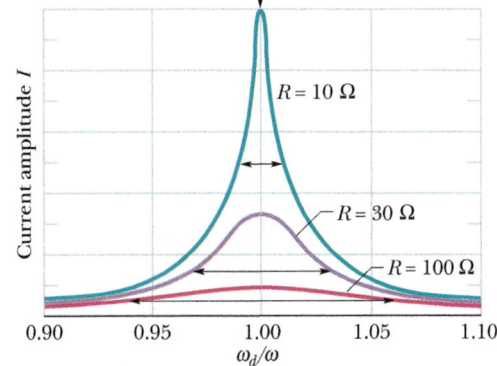

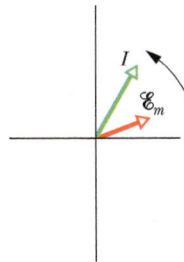

Low driving ω_d
- low current amplitude
- *ICE* side of the curve
- more capacitive
- X_C is greater
- current leads emf
- negative ϕ

High driving ω_d
- low current amplitude
- *ELI* side of the curve
- more inductive
- X_L is greater
- current lags emf
- positive ϕ

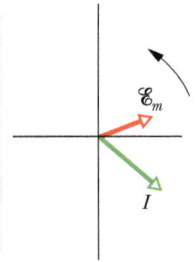

denominator is zero—that is, when

$$\omega_d L = \frac{1}{\omega_d C}$$

or
$$\omega_d = \frac{1}{\sqrt{LC}} \quad \text{(maximum } I\text{)}. \qquad (31\text{-}66)$$

Because the natural angular frequency ω of the *RLC* circuit is also equal to $1/\sqrt{LC}$, the maximum value of I occurs when the driving angular frequency matches the natural angular frequency—that is, at resonance. Thus, in an *RLC* circuit, resonance and maximum current amplitude I occur when

$$\omega_d = \omega = \frac{1}{\sqrt{LC}} \quad \text{(resonance)}. \qquad (31\text{-}67)$$

Resonance Curves. Figure 31-16 shows three *resonance curves* for sinusoidally driven oscillations in three series *RLC* circuits differing only in R. Each curve peaks at its maximum current amplitude I when the ratio ω_d/ω is 1.00, but the maximum value of I decreases with increasing R. (The maximum I is always $\mathcal{E}_m/R$; to see why, combine Eqs. 31-61 and 31-62.) In addition, the curves increase in width (measured in Fig. 31-16 at half the maximum value of I) with increasing R.

To make physical sense of Fig. 31-16, consider how the reactances X_L and X_C change as we increase the driving angular frequency ω_d, starting with a value much less than the natural frequency ω. For small ω_d, reactance X_L ($= \omega_d L$) is small and reactance X_C ($= 1/\omega_d C$) is large. Thus, the circuit is mainly capacitive and the impedance is dominated by the large X_C, which keeps the current low.

As we increase ω_d, reactance X_C remains dominant but decreases while reactance X_L increases. The decrease in X_C decreases the impedance, allowing the current to increase, as we see on the left side of any resonance curve in Fig. 31-16. When the increasing X_L and the decreasing X_C reach equal values, the current is greatest and the circuit is in resonance, with $\omega_d = \omega$.

As we continue to increase ω_d, the increasing reactance X_L becomes progressively more dominant over the decreasing reactance X_C. The impedance increases because of X_L and the current decreases, as on the right side of any resonance curve in Fig. 31-16. In summary, then: The low-angular-frequency side of a resonance curve is dominated by the capacitor's reactance, the high-angular-frequency side is dominated by the inductor's reactance, and resonance occurs in the middle.

✅ **Checkpoint 6**

Here are the capacitive reactance and inductive reactance, respectively, for three sinusoidally driven series *RLC* circuits: (1) 50 Ω, 100 Ω; (2) 100 Ω, 50 Ω; (3) 50 Ω, 50 Ω. (a) For each, does the current lead or lag the applied emf, or are the two in phase? (b) Which circuit is in resonance?

Sample Problem 31.06 Current amplitude, impedance, and phase constant

In Fig. 31-7, let $R = 200$ Ω, $C = 15.0$ μF, $L = 230$ mH, $f_d = 60.0$ Hz, and $\mathcal{E}_m = 36.0$ V. (These parameters are those used in the earlier sample problems.)

(a) What is the current amplitude I?

KEY IDEA

The current amplitude I depends on the amplitude $\mathcal{E}_m$ of the driving emf and on the impedance Z of the circuit, according to Eq. 31-62 ($I = \mathcal{E}_m/Z$).

Calculations: So, we need to find Z, which depends on resistance R, capacitive reactance X_C, and inductive reactance X_L. The circuit's resistance is the given resistance R. Its capacitive reactance is due to the given capacitance and, from an earlier sample problem, $X_C = 177$ Ω. Its inductive reactance is due to the given inductance and, from another sample problem, $X_L = 86.7$ Ω. Thus, the circuit's impedance is

$$Z = \sqrt{R^2 + (X_L - X_C)^2}$$
$$= \sqrt{(200\ \Omega)^2 + (86.7\ \Omega - 177\ \Omega)^2}$$
$$= 219\ \Omega.$$

We then find

$$I = \frac{\mathcal{E}_m}{Z} = \frac{36.0\ \text{V}}{219\ \Omega} = 0.164\ \text{A.} \qquad \text{(Answer)}$$

(b) What is the phase constant ϕ of the current in the circuit relative to the driving emf?

KEY IDEA

The phase constant depends on the inductive reactance, the capacitive reactance, and the resistance of the circuit, according to Eq. 31-65.

Calculation: Solving Eq. 31-65 for ϕ leads to

$$\phi = \tan^{-1} \frac{X_L - X_C}{R} = \tan^{-1} \frac{86.7\ \Omega - 177\ \Omega}{200\ \Omega}$$
$$= -24.3° = -0.424\ \text{rad.} \qquad \text{(Answer)}$$

The negative phase constant is consistent with the fact that the load is mainly capacitive; that is, $X_C > X_L$. In the common mnemonic for driven series *RLC* circuits, this circuit is an *ICE* circuit—the current *leads* the driving emf.

WILEY PLUS Additional examples, video, and practice available at *WileyPLUS*

31-5 POWER IN ALTERNATING-CURRENT CIRCUITS

Learning Objectives

After reading this module, you should be able to . . .

31.41 For the current, voltage, and emf in an ac circuit, apply the relationship between the rms values and the amplitudes.

31.42 For an alternating emf connected across a capacitor, an inductor, or a resistor, sketch graphs of the sinusoidal variation of the current and voltage and indicate the peak and rms values.

31.43 Apply the relationship between average power P_{avg}, rms current I_{rms}, and resistance R.

31.44 In a driven RLC circuit, calculate the power of each element.

31.45 For a driven RLC circuit in steady state, explain what happens to (a) the value of the average stored energy with time and (b) the energy that the generator puts into the circuit.

31.46 Apply the relationship between the power factor cos ϕ, the resistance R, and the impedance Z.

31.47 Apply the relationship between the average power P_{avg}, the rms emf $\mathcal{E}_{rms}$, the rms current I_{rms}, and the power factor cos ϕ.

31.48 Identify what power factor is required in order to maximize the rate at which energy is supplied to a resistive load.

Key Ideas

● In a series RLC circuit, the average power P_{avg} of the generator is equal to the production rate of thermal energy in the resistor:

$$P_{avg} = I_{rms}^2 R = \mathcal{E}_{rms} I_{rms} \cos \phi.$$

● The abbreviation rms stands for root-mean-square; the rms quantities are related to the maximum quantities by $I_{rms} = I/\sqrt{2}$, $V_{rms} = V/\sqrt{2}$, and $\mathcal{E}_{rms} = \mathcal{E}_m/\sqrt{2}$. The term cos ϕ is called the power factor of the circuit.

Power in Alternating-Current Circuits

In the RLC circuit of Fig. 31-7, the source of energy is the alternating-current generator. Some of the energy that it provides is stored in the electric field in the capacitor, some is stored in the magnetic field in the inductor, and some is dissipated as thermal energy in the resistor. In steady-state operation, the average stored energy remains constant. The net transfer of energy is thus from the generator to the resistor, where energy is dissipated.

The instantaneous rate at which energy is dissipated in the resistor can be written, with the help of Eqs. 26-27 and 31-29, as

$$P = i^2R = [I \sin(\omega_d t - \phi)]^2 R = I^2 R \sin^2(\omega_d t - \phi). \qquad (31\text{-}68)$$

The *average* rate at which energy is dissipated in the resistor, however, is the average of Eq. 31-68 over time. Over one complete cycle, the average value of sin θ, where θ is any variable, is zero (Fig. 31-17a) but the average value of sin^2 θ is $\frac{1}{2}$ (Fig. 31-17b). (Note in Fig. 31-17b how the shaded areas under the curve but above the horizontal line marked $+\frac{1}{2}$ exactly fill in the unshaded spaces below that line.) Thus, we can write, from Eq. 31-68,

$$P_{avg} = \frac{I^2 R}{2} = \left(\frac{I}{\sqrt{2}}\right)^2 R. \qquad (31\text{-}69)$$

The quantity $I/\sqrt{2}$ is called the **root-mean-square**, or **rms**, value of the current i:

$$I_{rms} = \frac{I}{\sqrt{2}} \quad \text{(rms current)}. \qquad (31\text{-}70)$$

We can now rewrite Eq. 31-69 as

$$P_{avg} = I_{rms}^2 R \quad \text{(average power)}. \qquad (31\text{-}71)$$

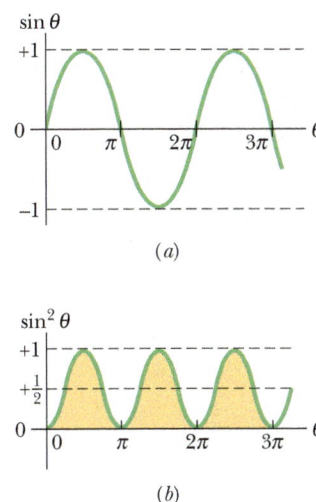

(a)

(b)

Figure 31-17 (a) A plot of sin θ versus θ. The average value over one cycle is zero. (b) A plot of sin^2 θ versus θ. The average value over one cycle is $\frac{1}{2}$.

Equation 31-71 has the same mathematical form as Eq. 26-27 ($P = i^2R$); the message here is that if we switch to the rms current, we can compute the average rate of energy dissipation for alternating-current circuits just as for direct-current circuits.

We can also define rms values of voltages and emfs for alternating-current circuits:

$$V_{rms} = \frac{V}{\sqrt{2}} \quad \text{and} \quad \mathscr{E}_{rms} = \frac{\mathscr{E}_m}{\sqrt{2}} \quad \text{(rms voltage; rms emf).} \qquad (31\text{-}72)$$

Alternating-current instruments, such as ammeters and voltmeters, are usually calibrated to read I_{rms}, V_{rms}, and $\mathscr{E}_{rms}$. Thus, if you plug an alternating-current voltmeter into a household electrical outlet and it reads 120 V, that is an rms voltage. The *maximum* value of the potential difference at the outlet is $\sqrt{2} \times$ (120 V), or 170 V. Generally scientists and engineers report rms values instead of maximum values.

Because the proportionality factor $1/\sqrt{2}$ in Eqs. 31-70 and 31-72 is the same for all three variables, we can write Eqs. 31-62 and 31-60 as

$$I_{rms} = \frac{\mathscr{E}_{rms}}{Z} = \frac{\mathscr{E}_{rms}}{\sqrt{R^2 + (X_L - X_C)^2}}, \qquad (31\text{-}73)$$

and, indeed, this is the form that we almost always use.

We can use the relationship $I_{rms} = \mathscr{E}_{rms}/Z$ to recast Eq. 31-71 in a useful equivalent way. We write

$$P_{avg} = \frac{\mathscr{E}_{rms}}{Z} I_{rms}R = \mathscr{E}_{rms}I_{rms} \frac{R}{Z}. \qquad (31\text{-}74)$$

From Fig. 31-14d, Table 31-2, and Eq. 31-62, however, we see that R/Z is just the cosine of the phase constant ϕ:

$$\cos \phi = \frac{V_R}{\mathscr{E}_m} = \frac{IR}{IZ} = \frac{R}{Z}. \qquad (31\text{-}75)$$

Equation 31-74 then becomes

$$P_{avg} = \mathscr{E}_{rms}I_{rms} \cos \phi \quad \text{(average power),} \qquad (31\text{-}76)$$

in which the term $\cos \phi$ is called the **power factor.** Because $\cos \phi = \cos(-\phi)$, Eq. 31-76 is independent of the sign of the phase constant ϕ.

To maximize the rate at which energy is supplied to a resistive load in an *RLC* circuit, we should keep the power factor $\cos \phi$ as close to unity as possible. This is equivalent to keeping the phase constant ϕ in Eq. 31-29 as close to zero as possible. If, for example, the circuit is highly inductive, it can be made less so by putting more capacitance in the circuit, connected in series. (Recall that putting an additional capacitance into a series of capacitances decreases the equivalent capacitance C_{eq} of the series.) Thus, the resulting decrease in C_{eq} in the circuit reduces the phase constant and increases the power factor in Eq. 31-76. Power companies place series-connected capacitors throughout their transmission systems to get these results.

✓ Checkpoint 7

(a) If the current in a sinusoidally driven series *RLC* circuit leads the emf, would we increase or decrease the capacitance to increase the rate at which energy is supplied to the resistance? (b) Would this change bring the resonant angular frequency of the circuit closer to the angular frequency of the emf or put it farther away?

Sample Problem 31.07 Driven *RLC* circuit: power factor and average power

A series *RLC* circuit, driven with $\mathscr{E}_{rms} = 120$ V at frequency $f_d = 60.0$ Hz, contains a resistance $R = 200\ \Omega$, an inductance with inductive reactance $X_L = 80.0\ \Omega$, and a capacitance with capacitive reactance $X_C = 150\ \Omega$.

(a) What are the power factor $\cos\phi$ and phase constant ϕ of the circuit?

KEY IDEA

The power factor $\cos\phi$ can be found from the resistance R and impedance Z via Eq. 31-75 ($\cos\phi = R/Z$).

Calculations: To calculate Z, we use Eq. 31-61:

$$Z = \sqrt{R^2 + (X_L - X_C)^2}$$
$$= \sqrt{(200\ \Omega)^2 + (80.0\ \Omega - 150\ \Omega)^2} = 211.90\ \Omega.$$

Equation 31-75 then gives us

$$\cos\phi = \frac{R}{Z} = \frac{200\ \Omega}{211.90\ \Omega} = 0.9438 \approx 0.944. \quad \text{(Answer)}$$

Taking the inverse cosine then yields

$$\phi = \cos^{-1} 0.944 = \pm 19.3°.$$

The inverse cosine on a calculator gives only the positive answer here, but both $+19.3°$ and $-19.3°$ have a cosine of 0.944. To determine which sign is correct, we must consider whether the current leads or lags the driving emf. Because $X_C > X_L$, this circuit is mainly capacitive, with the current leading the emf. Thus, ϕ must be negative:

$$\phi = -19.3°. \quad \text{(Answer)}$$

We could, instead, have found ϕ with Eq. 31-65. A calculator would then have given us the answer with the minus sign.

(b) What is the average rate P_{avg} at which energy is dissipated in the resistance?

KEY IDEAS

There are two ways and two ideas to use: (1) Because the circuit is assumed to be in steady-state operation, the rate at which energy is dissipated in the resistance is equal to the rate at which energy is supplied to the circuit, as given by Eq. 31-76 ($P_{avg} = \mathscr{E}_{rms}I_{rms}\cos\phi$). (2) The rate at which energy is dissipated in a resistance R depends on the square of the rms current I_{rms} through it, according to Eq. 31-71 ($P_{avg} = I_{rms}^2 R$).

First way: We are given the rms driving emf $\mathscr{E}_{rms}$ and we already know $\cos\phi$ from part (a). The rms current I_{rms} is

determined by the rms value of the driving emf and the circuit's impedance Z (which we know), according to Eq. 31-73:

$$I_{rms} = \frac{\mathscr{E}_{rms}}{Z}.$$

Substituting this into Eq. 31-76 then leads to

$$P_{avg} = \mathscr{E}_{rms}I_{rms}\cos\phi = \frac{\mathscr{E}_{rms}^2}{Z}\cos\phi$$
$$= \frac{(120\ \text{V})^2}{211.90\ \Omega}(0.9438) = 64.1\ \text{W}. \quad \text{(Answer)}$$

Second way: Instead, we can write

$$P_{avg} = I_{rms}^2 R = \frac{\mathscr{E}_{rms}^2}{Z^2} R$$
$$= \frac{(120\ \text{V})^2}{(211.90\ \Omega)^2}(200\ \Omega) = 64.1\ \text{W}. \quad \text{(Answer)}$$

(c) What new capacitance C_{new} is needed to maximize P_{avg} if the other parameters of the circuit are not changed?

KEY IDEAS

(1) The average rate P_{avg} at which energy is supplied and dissipated is maximized if the circuit is brought into resonance with the driving emf. (2) Resonance occurs when $X_C = X_L$.

Calculations: From the given data, we have $X_C > X_L$. Thus, we must decrease X_C to reach resonance. From Eq. 31-39 ($X_C = 1/\omega_d C$), we see that this means we must increase C to the new value C_{new}.

Using Eq. 31-39, we can write the resonance condition $X_C = X_L$ as

$$\frac{1}{\omega_d C_{new}} = X_L.$$

Substituting $2\pi f_d$ for ω_d (because we are given f_d and not ω_d) and then solving for C_{new}, we find

$$C_{new} = \frac{1}{2\pi f_d X_L} = \frac{1}{(2\pi)(60\ \text{Hz})(80.0\ \Omega)}$$
$$= 3.32 \times 10^{-5}\ \text{F} = 33.2\ \mu\text{F}. \quad \text{(Answer)}$$

Following the procedure of part (b), you can show that with C_{new}, the average power of energy dissipation P_{avg} would then be at its maximum value of

$$P_{avg,max} = 72.0\ \text{W}.$$

PLUS Additional examples, video, and practice available at *WileyPLUS*

31-6 TRANSFORMERS

Learning Objectives

After reading this module, you should be able to . . .

31.49 For power transmission lines, identify why the transmission should be at low current and high voltage.

31.50 Identify the role of transformers at the two ends of a transmission line.

31.51 Calculate the energy dissipation in a transmission line.

31.52 Identify a transformer's primary and secondary.

31.53 Apply the relationship between the voltage and number of turns on the two sides of a transformer.

31.54 Distinguish between a step-down transformer and a step-up transformer.

31.55 Apply the relationship between the current and number of turns on the two sides of a transformer.

31.56 Apply the relationship between the power into and out of an ideal transformer.

31.57 Identify the equivalent resistance as seen from the primary side of a transformer.

31.58 Apply the relationship between the equivalent resistance and the actual resistance.

31.59 Explain the role of a transformer in impedance matching.

Key Ideas

● A transformer (assumed to be ideal) is an iron core on which are wound a primary coil of N_p turns and a secondary coil of N_s turns. If the primary coil is connected across an alternating-current generator, the primary and secondary voltages are related by

$$V_s = V_p \frac{N_s}{N_p} \quad \text{(transformation of voltage)}.$$

● The currents through the coils are related by

$$I_s = I_p \frac{N_p}{N_s} \quad \text{(transformation of currents)}.$$

● The equivalent resistance of the secondary circuit, as seen by the generator, is

$$R_{eq} = \left(\frac{N_p}{N_s}\right)^2 R,$$

where R is the resistive load in the secondary circuit. The ratio N_p/N_s is called the transformer's turns ratio.

Transformers

Energy Transmission Requirements

When an ac circuit has only a resistive load, the power factor in Eq. 31-76 is $\cos 0° = 1$ and the applied rms emf $\mathscr{E}_{rms}$ is equal to the rms voltage V_{rms} across the load. Thus, with an rms current I_{rms} in the load, energy is supplied and dissipated at the average rate of

$$P_{avg} = \mathscr{E}I = IV. \tag{31-77}$$

(In Eq. 31-77 and the rest of this module, we follow conventional practice and drop the subscripts identifying rms quantities. Engineers and scientists assume that all time-varying currents and voltages are reported as rms values; that is what the meters read.) Equation 31-77 tells us that, to satisfy a given power requirement, we have a range of choices for I and V, provided only that the product IV is as required.

In electrical power distribution systems it is desirable for reasons of safety and for efficient equipment design to deal with relatively low voltages at both the generating end (the electrical power plant) and the receiving end (the home or factory). Nobody wants an electric toaster to operate at, say, 10 kV. However, in the transmission of electrical energy from the generating plant to the consumer, we want the lowest practical current (hence the largest practical voltage) to minimize I^2R losses (often called *ohmic losses*) in the transmission line.

As an example, consider the 735 kV line used to transmit electrical energy from the La Grande 2 hydroelectric plant in Quebec to Montreal, 1000 km away. Suppose that the current is 500 A and the power factor is close to unity. Then from Eq. 31-77, energy is supplied at the average rate

$$P_{avg} = \mathscr{E}I = (7.35 \times 10^5 \text{ V})(500 \text{ A}) = 368 \text{ MW}.$$

The resistance of the transmission line is about 0.220 Ω/km; thus, there is a total resistance of about 220 Ω for the 1000 km stretch. Energy is dissipated due to that resistance at a rate of about

$$P_{avg} = I^2R = (500\ A)^2(220\ \Omega) = 55.0\ MW,$$

which is nearly 15% of the supply rate.

Imagine what would happen if we doubled the current and halved the voltage. Energy would be supplied by the plant at the same average rate of 368 MW as previously, but now energy would be dissipated at the rate of about

$$P_{avg} = I^2R = (1000\ A)^2(220\ \Omega) = 220\ MW,$$

which is *almost 60% of the supply rate*. Hence the general energy transmission rule: Transmit at the highest possible voltage and the lowest possible current.

The Ideal Transformer

The transmission rule leads to a fundamental mismatch between the requirement for efficient high-voltage transmission and the need for safe low-voltage generation and consumption. We need a device with which we can raise (for transmission) and lower (for use) the ac voltage in a circuit, keeping the product current × voltage essentially constant. The **transformer** is such a device. It has no moving parts, operates by Faraday's law of induction, and has no simple direct-current counterpart.

The *ideal transformer* in Fig. 31-18 consists of two coils, with different numbers of turns, wound around an iron core. (The coils are insulated from the core.) In use, the primary winding, of N_p turns, is connected to an alternating-current generator whose emf $\mathcal{E}$ at any time t is given by

$$\mathcal{E} = \mathcal{E}_m \sin \omega t. \qquad (31\text{-}78)$$

The secondary winding, of N_s turns, is connected to load resistance R, but its circuit is an open circuit as long as switch S is open (which we assume for the present). Thus, there can be no current through the secondary coil. We assume further for this ideal transformer that the resistances of the primary and secondary windings are negligible. Well-designed, high-capacity transformers can have energy losses as low as 1%; so our assumptions are reasonable.

For the assumed conditions, the primary winding (or *primary*) is a pure inductance and the primary circuit is like that in Fig. 31-12. Thus, the (very small) primary current, also called the *magnetizing current* I_{mag}, lags the primary voltage V_p by 90°; the primary's power factor (= cos ϕ in Eq. 31-76) is zero; so no power is delivered from the generator to the transformer.

However, the small sinusoidally changing primary current I_{mag} produces a sinusoidally changing magnetic flux Φ_B in the iron core. The core acts to strengthen the flux and to bring it through the secondary winding (or *secondary*). Because Φ_B varies, it induces an emf $\mathcal{E}_{turn}$ (= $d\Phi_B/dt$) in each turn of the secondary. In fact, this emf per turn $\mathcal{E}_{turn}$ is the same in the primary and the secondary. Across the primary, the voltage V_p is the product of $\mathcal{E}_{turn}$ and the number of turns N_p; that is, $V_p = \mathcal{E}_{turn}N_p$. Similarly, across the secondary the voltage is $V_s = \mathcal{E}_{turn}N_s$. Thus, we can write

$$\mathcal{E}_{turn} = \frac{V_p}{N_p} = \frac{V_s}{N_s},$$

or $\qquad\qquad V_s = V_p \dfrac{N_s}{N_p} \qquad$ (transformation of voltage). $\qquad (31\text{-}79)$

If $N_s > N_p$, the device is a *step-up transformer* because it steps the primary's voltage V_p up to a higher voltage V_s. Similarly, if $N_s < N_p$, it is a *step-down transformer*.

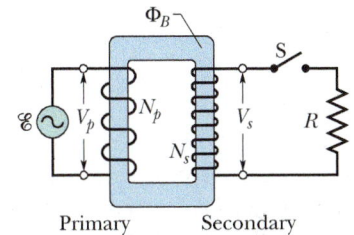

Figure 31-18 An ideal transformer (two coils wound on an iron core) in a basic transformer circuit. An ac generator produces current in the coil at the left (the *primary*). The coil at the right (the *secondary*) is connected to the resistive load R when switch S is closed.

With switch S open, no energy is transferred from the generator to the rest of the circuit, but when we close S to connect the secondary to the resistive load R, energy *is* transferred. (In general, the load would also contain inductive and capacitive elements, but here we consider just resistance R.) Here is the process:

1. An alternating current I_s appears in the secondary circuit, with corresponding energy dissipation rate $I_s^2 R$ ($= V_s^2/R$) in the resistive load.

2. This current produces its own alternating magnetic flux in the iron core, and this flux induces an opposing emf in the primary windings.

3. The voltage V_p of the primary, however, cannot change in response to this opposing emf because it must always be equal to the emf $\mathscr{E}$ that is provided by the generator; closing switch S cannot change this fact.

4. To maintain V_p, the generator now produces (in addition to I_{mag}) an alternating current I_p in the primary circuit; the magnitude and phase constant of I_p are just those required for the emf induced by I_p in the primary to exactly cancel the emf induced there by I_s. Because the phase constant of I_p is not 90° like that of I_{mag}, this current I_p can transfer energy to the primary.

Energy Transfers. We want to relate I_s to I_p. However, rather than analyze the foregoing complex process in detail, let us just apply the principle of conservation of energy. The rate at which the generator transfers energy to the primary is equal to $I_p V_p$. The rate at which the primary then transfers energy to the secondary (via the alternating magnetic field linking the two coils) is $I_s V_s$. Because we assume that no energy is lost along the way, conservation of energy requires that

$$I_p V_p = I_s V_s.$$

Substituting for V_s from Eq. 31-79, we find that

$$I_s = I_p \frac{N_p}{N_s} \quad \text{(transformation of currents).} \tag{31-80}$$

This equation tells us that the current I_s in the secondary can differ from the current I_p in the primary, depending on the *turns ratio* N_p/N_s.

Current I_p appears in the primary circuit because of the resistive load R in the secondary circuit. To find I_p, we substitute $I_s = V_s/R$ into Eq. 31-80 and then we substitute for V_s from Eq. 31-79. We find

$$I_p = \frac{1}{R} \left(\frac{N_s}{N_p} \right)^2 V_p. \tag{31-81}$$

This equation has the form $I_p = V_p/R_{\text{eq}}$, where equivalent resistance R_{eq} is

$$R_{\text{eq}} = \left(\frac{N_p}{N_s} \right)^2 R. \tag{31-82}$$

This R_{eq} is the value of the load resistance as "seen" by the generator; the generator produces the current I_p and voltage V_p as if the generator were connected to a resistance R_{eq}.

Impedance Matching

Equation 31-82 suggests still another function for the transformer. For maximum transfer of energy from an emf device to a resistive load, the resistance of the emf device must equal the resistance of the load. The same relation holds for ac circuits except that the *impedance* (rather than just the resistance) of the generator must equal that of the load. Often this condition is not met. For example, in a music-playing system, the amplifier has high impedance and the speaker set has low impedance. We can match the impedances of the two devices by coupling them through a transformer that has a suitable turns ratio N_p/N_s.

☑ **Checkpoint 8**

An alternating-current emf device in a certain circuit has a smaller resistance than that of the resistive load in the circuit; to increase the transfer of energy from the device to the load, a transformer will be connected between the two. (a) Should N_s be greater than or less than N_p? (b) Will that make it a step-up or step-down transformer?

Sample Problem 31.08   Transformer: turns ratio, average power, rms currents

A transformer on a utility pole operates at $V_p = 8.5$ kV on the primary side and supplies electrical energy to a number of nearby houses at $V_s = 120$ V, both quantities being rms values. Assume an ideal step-down transformer, a purely resistive load, and a power factor of unity.

(a) What is the turns ratio N_p/N_s of the transformer?

KEY IDEA

The turns ratio N_p/N_s is related to the (given) rms primary and secondary voltages via Eq. 31-79 ($V_s = V_p N_s/N_p$).

Calculation: We can write Eq. 31-79 as

$$\frac{V_s}{V_p} = \frac{N_s}{N_p}. \tag{31-83}$$

(Note that the right side of this equation is the *inverse* of the turns ratio.) Inverting both sides of Eq. 31-83 gives us

$$\frac{N_p}{N_s} = \frac{V_p}{V_s} = \frac{8.5 \times 10^3 \text{ V}}{120 \text{ V}} = 70.83 \approx 71. \quad \text{(Answer)}$$

(b) The average rate of energy consumption (or dissipation) in the houses served by the transformer is 78 kW. What are the rms currents in the primary and secondary of the transformer?

KEY IDEA

For a purely resistive load, the power factor $\cos \phi$ is unity; thus, the average rate at which energy is supplied and dissipated is given by Eq. 31-77 ($P_{avg} = \mathcal{E}I = IV$).

Calculations: In the primary circuit, with $V_p = 8.5$ kV,

Eq. 31-77 yields

$$I_p = \frac{P_{avg}}{V_p} = \frac{78 \times 10^3 \text{ W}}{8.5 \times 10^3 \text{ V}} = 9.176 \text{ A} \approx 9.2 \text{ A}. \quad \text{(Answer)}$$

Similarly, in the secondary circuit,

$$I_s = \frac{P_{avg}}{V_s} = \frac{78 \times 10^3 \text{ W}}{120 \text{ V}} = 650 \text{ A}. \quad \text{(Answer)}$$

You can check that $I_s = I_p(N_p/N_s)$ as required by Eq. 31-80.

(c) What is the resistive load R_s in the secondary circuit? What is the corresponding resistive load R_p in the primary circuit?

One way: We can use $V = IR$ to relate the resistive load to the rms voltage and current. For the secondary circuit, we find

$$R_s = \frac{V_s}{I_s} = \frac{120 \text{ V}}{650 \text{ A}} = 0.1846 \ \Omega \approx 0.18 \ \Omega. \quad \text{(Answer)}$$

Similarly, for the primary circuit we find

$$R_p = \frac{V_p}{I_p} = \frac{8.5 \times 10^3 \text{ V}}{9.176 \text{ A}} = 926 \ \Omega \approx 930 \ \Omega. \quad \text{(Answer)}$$

Second way: We use the fact that R_p equals the equivalent resistive load "seen" from the primary side of the transformer, which is a resistance modified by the turns ratio and given by Eq. 31-82 ($R_{eq} = (N_p/N_s)^2 R$). If we substitute R_p for R_{eq} and R_s for R, that equation yields

$$R_p = \left(\frac{N_p}{N_s}\right)^2 R_s = (70.83)^2 (0.1846 \ \Omega)$$

$$= 926 \ \Omega \approx 930 \ \Omega. \quad \text{(Answer)}$$

WILEY PLUS   Additional examples, video, and practice available at *WileyPLUS*

Review & Summary

LC Energy Transfers   In an oscillating LC circuit, energy is shuttled periodically between the electric field of the capacitor and the magnetic field of the inductor; instantaneous values of the two forms of energy are

$$U_E = \frac{q^2}{2C} \quad \text{and} \quad U_B = \frac{Li^2}{2}, \tag{31-1, 31-2}$$

where q is the instantaneous charge on the capacitor and i is the

instantaneous current through the inductor. The total energy $U \ (= U_E + U_B)$ remains constant.

LC Charge and Current Oscillations   The principle of conservation of energy leads to

$$L\frac{d^2q}{dt^2} + \frac{1}{C}q = 0 \quad (LC \text{ oscillations}) \tag{31-11}$$

as the differential equation of LC oscillations (with no resistance). The solution of Eq. 31-11 is

$$q = Q \cos(\omega t + \phi) \quad \text{(charge)}, \qquad (31\text{-}12)$$

in which Q is the *charge amplitude* (maximum charge on the capacitor) and the angular frequency ω of the oscillations is

$$\omega = \frac{1}{\sqrt{LC}}. \qquad (31\text{-}4)$$

The phase constant ϕ in Eq. 31-12 is determined by the initial conditions (at $t = 0$) of the system.

The current i in the system at any time t is

$$i = -\omega Q \sin(\omega t + \phi) \quad \text{(current)}, \qquad (31\text{-}13)$$

in which ωQ is the *current amplitude* I.

Damped Oscillations
Oscillations in an LC circuit are damped when a dissipative element R is also present in the circuit. Then

$$L\frac{d^2q}{dt^2} + R\frac{dq}{dt} + \frac{1}{C}q = 0 \quad (RLC \text{ circuit}). \qquad (31\text{-}24)$$

The solution of this differential equation is

$$q = Qe^{-Rt/2L}\cos(\omega' t + \phi), \qquad (31\text{-}25)$$

where

$$\omega' = \sqrt{\omega^2 - (R/2L)^2}. \qquad (31\text{-}26)$$

We consider only situations with small R and thus small damping; then $\omega' \approx \omega$.

Alternating Currents; Forced Oscillations
A series RLC circuit may be set into *forced oscillation* at a *driving angular frequency* ω_d by an external alternating emf

$$\mathscr{E} = \mathscr{E}_m \sin \omega_d t. \qquad (31\text{-}28)$$

The current driven in the circuit is

$$i = I \sin(\omega_d t - \phi), \qquad (31\text{-}29)$$

where ϕ is the phase constant of the current.

Resonance
The current amplitude I in a series RLC circuit driven by a sinusoidal external emf is a maximum ($I = \mathscr{E}_m/R$) when the driving angular frequency ω_d equals the natural angular frequency ω of the circuit (that is, at *resonance*). Then $X_C = X_L$, $\phi = 0$, and the current is in phase with the emf.

Single Circuit Elements
The alternating potential difference across a resistor has amplitude $V_R = IR$; the current is in phase with the potential difference.

For a *capacitor*, $V_C = IX_C$, in which $X_C = 1/\omega_d C$ is the **capacitive reactance;** the current here leads the potential difference by 90° ($\phi = -90° = -\pi/2$ rad).

For an *inductor*, $V_L = IX_L$, in which $X_L = \omega_d L$ is the **inductive reactance;** the current here lags the potential difference by 90° ($\phi = +90° = +\pi/2$ rad).

Series *RLC* Circuits
For a series RLC circuit with an alternating external emf given by Eq. 31-28 and a resulting alternating current given by Eq. 31-29,

$$I = \frac{\mathscr{E}_m}{\sqrt{R^2 + (X_L - X_C)^2}}$$

$$= \frac{\mathscr{E}_m}{\sqrt{R^2 + (\omega_d L - 1/\omega_d C)^2}} \quad \text{(current amplitude)} \quad (31\text{-}60, 31\text{-}63)$$

and

$$\tan \phi = \frac{X_L - X_C}{R} \quad \text{(phase constant)}. \qquad (31\text{-}65)$$

Defining the impedance Z of the circuit as

$$Z = \sqrt{R^2 + (X_L - X_C)^2} \quad \text{(impedance)} \qquad (31\text{-}61)$$

allows us to write Eq. 31-60 as $I = \mathscr{E}_m/Z$.

Power
In a series RLC circuit, the **average power** P_{avg} of the generator is equal to the production rate of thermal energy in the resistor:

$$P_{\text{avg}} = I_{\text{rms}}^2 R = \mathscr{E}_{\text{rms}} I_{\text{rms}} \cos \phi. \qquad (31\text{-}71, 31\text{-}76)$$

Here rms stands for **root-mean-square;** the rms quantities are related to the maximum quantities by $I_{\text{rms}} = I/\sqrt{2}$, $V_{\text{rms}} = V/\sqrt{2}$, and $\mathscr{E}_{\text{rms}} = \mathscr{E}_m/\sqrt{2}$. The term $\cos \phi$ is called the **power factor** of the circuit.

Transformers
A *transformer* (assumed to be ideal) is an iron core on which are wound a primary coil of N_p turns and a secondary coil of N_s turns. If the primary coil is connected across an alternating-current generator, the primary and secondary voltages are related by

$$V_s = V_p \frac{N_s}{N_p} \quad \text{(transformation of voltage)}. \qquad (31\text{-}79)$$

The currents through the coils are related by

$$I_s = I_p \frac{N_p}{N_s} \quad \text{(transformation of currents)}, \qquad (31\text{-}80)$$

and the equivalent resistance of the secondary circuit, as seen by the generator, is

$$R_{\text{eq}} = \left(\frac{N_p}{N_s}\right)^2 R, \qquad (31\text{-}82)$$

where R is the resistive load in the secondary circuit. The ratio N_p/N_s is called the transformer's *turns ratio*.

Questions

1 Figure 31-19 shows three oscillating LC circuits with identical inductors and capacitors. At a particular time, the charges on the capacitor plates (and thus the electric fields between the plates) are all at their maximum values. Rank the circuits according to the time taken to fully discharge the capacitors during the oscillations, greatest first.

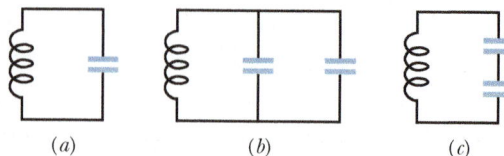

(a)   (b)   (c)

Figure 31-19 Question 1.

2 Figure 31-20 shows graphs of capacitor voltage v_C for LC circuits 1 and 2, which contain identical capacitances and have the same maximum charge Q. Are (a) the inductance L and (b) the maximum current I in circuit 1 greater than, less than, or the same as those in circuit 2?

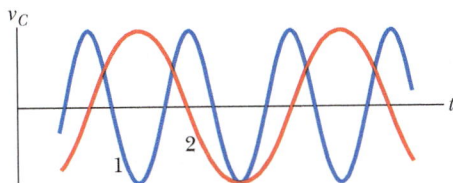

Figure 31-20 Question 2.

3 A charged capacitor and an inductor are connected at time $t = 0$. In terms of the period T of the resulting oscillations, what is the first later time at which the following reach a maximum: (a) U_B, (b) the magnetic flux through the inductor, (c) di/dt, and (d) the emf of the inductor?

4 What values of phase constant ϕ in Eq. 31-12 allow situations $(a), (c), (e),$ and (g) of Fig. 31-1 to occur at $t = 0$?

5 Curve a in Fig. 31-21 gives the impedance Z of a driven RC circuit versus the driving angular frequency ω_d. The other two curves are similar but for different values of resistance R and capacitance C. Rank the three curves according to the corresponding value of R, greatest first.

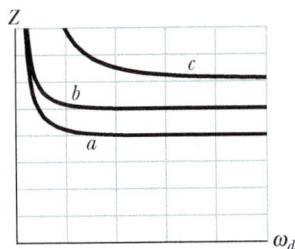

Figure 31-21 Question 5.

6 Charges on the capacitors in three oscillating LC circuits vary as: (1) $q = 2 \cos 4t$, (2) $q = 4 \cos t$, (3) $q = 3 \cos 4t$ (with q in coulombs and t in seconds). Rank the circuits according to (a) the current amplitude and (b) the period, greatest first.

7 An alternating emf source with a certain emf amplitude is connected, in turn, to a resistor, a capacitor, and then an inductor. Once connected to one of the devices, the driving frequency f_d is varied and the amplitude I of the resulting current through the device is measured and plotted. Which of the three plots in Fig. 31-22 corresponds to which of the three devices?

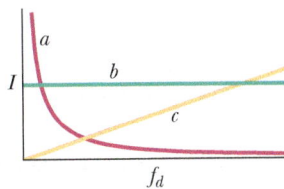

Figure 31-22 Question 7.

8 The values of the phase constant ϕ for four sinusoidally driven series RLC circuits are (1) $-15°$, (2) $+35°$, (3) $\pi/3$ rad, and (4) $-\pi/6$ rad. (a) In which is the load primarily capacitive? (b) In which does the current lag the alternating emf?

9 Figure 31-23 shows the current i and driving emf $\mathcal{E}$ for a series RLC circuit. (a) Is the phase constant positive or negative? (b) To increase the rate at which energy is transferred to the resistive load, should L be increased or decreased? (c) Should, instead, C be increased or decreased?

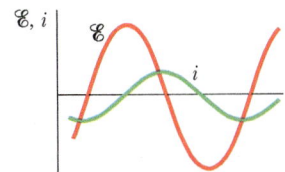

Figure 31-23 Question 9.

10 Figure 31-24 shows three situations like those of Fig. 31-15. Is the driving angular frequency greater than, less than, or equal to the resonant angular frequency of the circuit in (a) situation 1, (b) situation 2, and (c) situation 3?

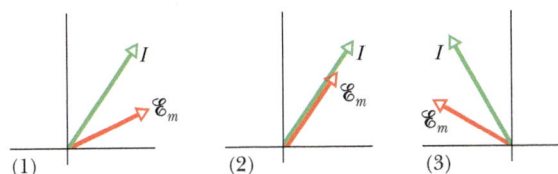

Figure 31-24 Question 10.

11 Figure 31-25 shows the current i and driving emf $\mathcal{E}$ for a series RLC circuit. Relative to the emf curve, does the current curve shift leftward or rightward and does the amplitude of that curve increase or decrease if we slightly increase (a) L, (b) C, and (c) ω_d?

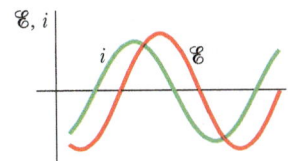

Figure 31-25 Questions 11 and 12.

12 Figure 31-25 shows the current i and driving emf $\mathcal{E}$ for a series RLC circuit. (a) Does the current lead or lag the emf? (b) Is the circuit's load mainly capacitive or mainly inductive? (c) Is the angular frequency ω_d of the emf greater than or less than the natural angular frequency ω?

13 (a) Does the phasor diagram of Fig. 31-26 correspond to an alternating emf source connected to a resistor, a capacitor, or an inductor? (b) If the angular speed of the phasors is increased, does the current phasor length increase or decrease when the scale of the diagram is maintained?

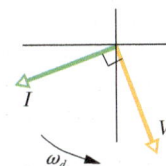

Figure 31-26 Question 13.

Problems

GO Tutoring problem available (at instructor's discretion) in *WileyPLUS* and WebAssign

SSM Worked-out solution available in Student Solutions Manual

• – ••• Number of dots indicates level of problem difficulty

Additional information available in *The Flying Circus of Physics* and at flyingcircusofphysics.com

WWW Worked-out solution is at

ILW Interactive solution is at   http://www.wiley.com/college/halliday

Module 31-1   *LC Oscillations*

•1 An oscillating LC circuit consists of a 75.0 mH inductor and a 3.60 μF capacitor. If the maximum charge on the capacitor is 2.90 μC, what are (a) the total energy in the circuit and (b) the maximum current?

•2 The frequency of oscillation of a certain LC circuit is 200 kHz. At time $t = 0$, plate A of the capacitor has maximum positive charge. At what earliest time $t > 0$ will (a) plate A again have maximum positive charge, (b) the other plate of the capacitor have maximum positive charge, and (c) the inductor have maximum magnetic field?

•3   In a certain oscillating LC circuit, the total energy is converted from electrical energy in the capacitor to magnetic energy in the inductor in 1.50 μs. What are (a) the period of oscillation and (b) the frequency of oscillation? (c) How long after the magnetic energy is a maximum will it be a maximum again?

•4   What is the capacitance of an oscillating LC circuit if the maximum charge on the capacitor is 1.60 μC and the total energy is 140 μJ?

•5   In an oscillating LC circuit, $L = 1.10$ mH and $C = 4.00$ μF. The maximum charge on the capacitor is 3.00 μC. Find the maximum current.

•6   A 0.50 kg body oscillates in SHM on a spring that, when extended 2.0 mm from its equilibrium position, has an 8.0 N restoring force. What are (a) the angular frequency of oscillation, (b) the period of oscillation, and (c) the capacitance of an LC circuit with the same period if L is 5.0 H?

•7 SSM   The energy in an oscillating LC circuit containing a 1.25 H inductor is 5.70 μJ. The maximum charge on the capacitor is 175 μC. For a mechanical system with the same period, find the (a) mass, (b) spring constant, (c) maximum displacement, and (d) maximum speed.

•8   A single loop consists of inductors ($L_1, L_2, \ldots$), capacitors ($C_1, C_2, \ldots$), and resistors ($R_1, R_2, \ldots$) connected in series as shown, for example, in Fig. 31-27a. Show that regardless of the sequence of these circuit elements in the loop, the behavior of this circuit is identical to that of the simple LC circuit shown in Fig. 31-27b. (*Hint:* Consider the loop rule and see Problem 47 in Chapter 30.)

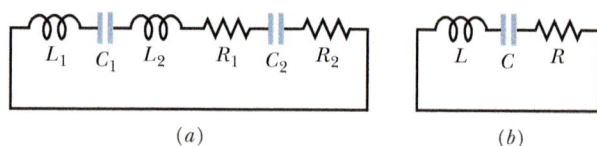

Figure 31-27 Problem 8.

•9 ILW   In an oscillating LC circuit with $L = 50$ mH and $C = 4.0$ μF, the current is initially a maximum. How long will it take before the capacitor is fully charged for the first time?

•10 LC oscillators have been used in circuits connected to loudspeakers to create some of the sounds of electronic music. What inductance must be used with a 6.7 μF capacitor to produce a frequency of 10 kHz, which is near the middle of the audible range of frequencies?

••11 SSM WWW   A variable capacitor with a range from 10 to 365 pF is used with a coil to form a variable-frequency LC circuit to tune the input to a radio. (a) What is the ratio of maximum frequency to minimum frequency that can be obtained with such a capacitor? If this circuit is to obtain frequencies from 0.54 MHz to 1.60 MHz, the ratio computed in (a) is too large. By adding a capacitor in parallel to the variable capacitor, this range can be adjusted. To obtain the desired frequency range, (b) what capacitance should be added and (c) what inductance should the coil have?

••12   In an oscillating LC circuit, when 75.0% of the total energy is stored in the inductor's magnetic field, (a) what multiple of the maximum charge is on the capacitor and (b) what multiple of the maximum current is in the inductor?

••13   In an oscillating LC circuit, $L = 3.00$ mH and $C = 2.70$ μF. At $t = 0$ the charge on the capacitor is zero and the current is 2.00 A. (a) What is the maximum charge that will appear on the capacitor? (b) At what earliest time $t > 0$ is the rate at which energy is stored in the capacitor greatest, and (c) what is that greatest rate?

••14   To construct an oscillating LC system, you can choose from a 10 mH inductor, a 5.0 μF capacitor, and a 2.0 μF capacitor. What are the (a) smallest, (b) second smallest, (c) second largest, and (d) largest oscillation frequency that can be set up by these elements in various combinations?

••15 ILW   An oscillating LC circuit consisting of a 1.0 nF capacitor and a 3.0 mH coil has a maximum voltage of 3.0 V. What are (a) the maximum charge on the capacitor, (b) the maximum current through the circuit, and (c) the maximum energy stored in the magnetic field of the coil?

••16   An inductor is connected across a capacitor whose capacitance can be varied by turning a knob. We wish to make the frequency of oscillation of this LC circuit vary linearly with the angle of rotation of the knob, going from 2×10^5 to 4×10^5 Hz as the knob turns through 180°. If $L = 1.0$ mH, plot the required capacitance C as a function of the angle of rotation of the knob.

••17 GO ILW   In Fig. 31-28, $R = 14.0$ Ω, $C = 6.20$ μF, and $L = 54.0$ mH, and the ideal battery has emf $\mathscr{E} = 34.0$ V. The switch is kept at a for a long time and then thrown to position b. What are the (a) frequency and (b) current amplitude of the resulting oscillations?

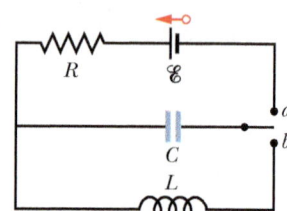

Figure 31-28 Problem 17.

••18   An oscillating LC circuit has a current amplitude of 7.50 mA, a potential amplitude of 250 mV, and a capacitance of 220 nF. What are (a) the period of oscillation, (b) the maximum energy stored in the capacitor, (c) the maximum energy stored in the inductor, (d) the maximum rate at which the current changes, and (e) the maximum rate at which the inductor gains energy?

••19   Using the loop rule, derive the differential equation for an LC circuit (Eq. 31-11).

••20 GO   In an oscillating LC circuit in which $C = 4.00$ μF, the maximum potential difference across the capacitor during the oscillations is 1.50 V and the maximum current through the inductor is 50.0 mA. What are (a) the inductance L and (b) the frequency of the oscillations? (c) How much time is required for the charge on the capacitor to rise from zero to its maximum value?

••21 ILW   In an oscillating LC circuit with $C = 64.0$ μF, the current is given by $i = (1.60) \sin(2500t + 0.680)$, where t is in seconds, i in amperes, and the phase constant in radians. (a) How soon after $t = 0$ will the current reach its maximum value? What are (b) the inductance L and (c) the total energy?

••22   A series circuit containing inductance L_1 and capacitance C_1 oscillates at angular frequency ω. A second series circuit, containing inductance L_2 and capacitance C_2, oscillates at the same angular frequency. In terms of ω, what is the angular frequency of oscillation of a series circuit containing all four of these elements? Neglect resistance. (*Hint:* Use the formulas for equivalent capacitance and equivalent inductance; see Module 25-3 and Problem 47 in Chapter 30.)

••23 GO In an oscillating LC circuit, $L = 25.0$ mH and $C = 7.80$ μF. At time $t = 0$ the current is 9.20 mA, the charge on the capacitor is 3.80 μC, and the capacitor is charging. What are (a) the total energy in the circuit, (b) the maximum charge on the capacitor, and (c) the maximum current? (d) If the charge on the capacitor is given by $q = Q\cos(\omega t + \phi)$, what is the phase angle ϕ? (e) Suppose the data are the same, except that the capacitor is discharging at $t = 0$. What then is ϕ?

Module 31-2 Damped Oscillations in an *RLC* Circuit

••24 GO A single-loop circuit consists of a 7.20 Ω resistor, a 12.0 H inductor, and a 3.20 μF capacitor. Initially the capacitor has a charge of 6.20 μC and the current is zero. Calculate the charge on the capacitor N complete cycles later for (a) $N = 5$, (b) $N = 10$, and (c) $N = 100$.

••25 ILW What resistance R should be connected in series with an inductance $L = 220$ mH and capacitance $C = 12.0$ μF for the maximum charge on the capacitor to decay to 99.0% of its initial value in 50.0 cycles? (Assume $\omega' \approx \omega$.)

••26 GO In an oscillating series RLC circuit, find the time required for the maximum energy present in the capacitor during an oscillation to fall to half its initial value. Assume $q = Q$ at $t = 0$.

•••27 SSM In an oscillating series RLC circuit, show that $\Delta U/U$, the fraction of the energy lost per cycle of oscillation, is given to a close approximation by $2\pi R/\omega L$. The quantity $\omega L/R$ is often called the Q of the circuit (for *quality*). A high-Q circuit has low resistance and a low fractional energy loss $(= 2\pi/Q)$ per cycle.

Module 31-3 Forced Oscillations of Three Simple Circuits

•28 A 1.50 μF capacitor is connected as in Fig. 31-10 to an ac generator with $\mathcal{E}_m = 30.0$ V. What is the amplitude of the resulting alternating current if the frequency of the emf is (a) 1.00 kHz and (b) 8.00 kHz?

•29 ILW A 50.0 mH inductor is connected as in Fig. 31-12 to an ac generator with $\mathcal{E}_m = 30.0$ V. What is the amplitude of the resulting alternating current if the frequency of the emf is (a) 1.00 kHz and (b) 8.00 kHz?

•30 A 50.0 Ω resistor is connected as in Fig. 31-8 to an ac generator with $\mathcal{E}_m = 30.0$ V. What is the amplitude of the resulting alternating current if the frequency of the emf is (a) 1.00 kHz and (b) 8.00 kHz?

•31 (a) At what frequency would a 6.0 mH inductor and a 10 μF capacitor have the same reactance? (b) What would the reactance be? (c) Show that this frequency would be the natural frequency of an oscillating circuit with the same L and C.

••32 GO An ac generator has emf $\mathcal{E} = \mathcal{E}_m \sin \omega_d t$, with $\mathcal{E}_m = 25.0$ V and $\omega_d = 377$ rad/s. It is connected to a 12.7 H inductor. (a) What is the maximum value of the current? (b) When the current is a maximum, what is the emf of the generator? (c) When the emf of the generator is -12.5 V and increasing in magnitude, what is the current?

••33 SSM An ac generator has emf $\mathcal{E} = \mathcal{E}_m \sin(\omega_d t - \pi/4)$, where $\mathcal{E}_m = 30.0$ V and $\omega_d = 350$ rad/s. The current produced in a connected circuit is $i(t) = I \sin(\omega_d t - 3\pi/4)$, where $I = 620$ mA. At what time after $t = 0$ does (a) the generator emf first reach a maximum and (b) the current first reach a maximum? (c) The circuit contains a single element other than the generator. Is it a capacitor, an inductor, or a resistor? Justify your answer. (d) What is

the value of the capacitance, inductance, or resistance, as the case may be?

••34 GO An ac generator with emf $\mathcal{E} = \mathcal{E}_m \sin \omega_d t$, where $\mathcal{E}_m = 25.0$ V and $\omega_d = 377$ rad/s, is connected to a 4.15 μF capacitor. (a) What is the maximum value of the current? (b) When the current is a maximum, what is the emf of the generator? (c) When the emf of the generator is -12.5 V and increasing in magnitude, what is the current?

Module 31-4 The Series *RLC* Circuit

•35 ILW A coil of inductance 88 mH and unknown resistance and a 0.94 μF capacitor are connected in series with an alternating emf of frequency 930 Hz. If the phase constant between the applied voltage and the current is 75°, what is the resistance of the coil?

•36 An alternating source with a variable frequency, a capacitor with capacitance C, and a resistor with resistance R are connected in series. Figure 31-29 gives the impedance Z of the circuit versus the driving angular frequency ω_d; the curve reaches an asymptote of 500 Ω, and the horizontal scale is set by $\omega_{ds} = 300$ rad/s. The figure also gives the reactance X_C for the capacitor versus ω_d. What are (a) R and (b) C?

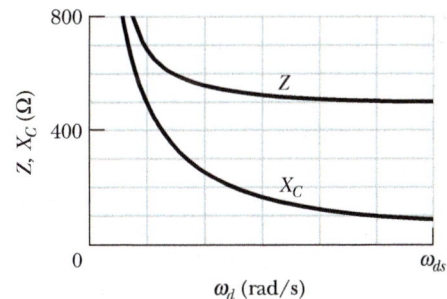

Figure 31-29 Problem 36.

•37 An electric motor has an effective resistance of 32.0 Ω and an inductive reactance of 45.0 Ω when working under load. The voltage amplitude across the alternating source is 420 V. Calculate the current amplitude.

•38 The current amplitude I versus driving angular frequency ω_d for a driven RLC circuit is given in Fig. 31-30, where the vertical axis scale is set by $I_s = 4.00$ A. The inductance is 200 μH, and the emf amplitude is 8.0 V. What are (a) C and (b) R?

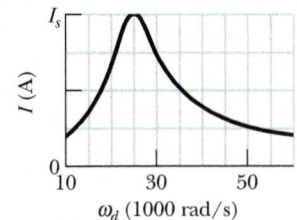

Figure 31-30 Problem 38.

•39 Remove the inductor from the circuit in Fig. 31-7 and set $R = 200$ Ω, $C = 15.0$ μF, $f_d = 60.0$ Hz, and $\mathcal{E}_m = 36.0$ V. What are (a) Z, (b) ϕ, and (c) I? (d) Draw a phasor diagram.

•40 An alternating source drives a series RLC circuit with an emf amplitude of 6.00 V, at a phase angle of $+30.0°$. When the potential difference across the capacitor reaches its maximum positive value of $+5.00$ V, what is the potential difference across the inductor (sign included)?

•41 SSM In Fig. 31-7, set $R = 200$ Ω, $C = 70.0$ μF, $L = 230$ mH, $f_d = 60.0$ Hz, and $\mathcal{E}_m = 36.0$ V. What are (a) Z, (b) ϕ, and (c) I? (d) Draw a phasor diagram.

•42 An alternating source with a variable frequency, an inductor with inductance L, and a resistor with resistance R are connected in series. Figure 31-31 gives the impedance Z of the circuit versus the driving angular frequency ω_d, with the horizontal axis scale set by $\omega_{ds} = 1600$ rad/s. The figure also gives the reactance X_L for the inductor versus ω_d. What are (a) R and (b) L?

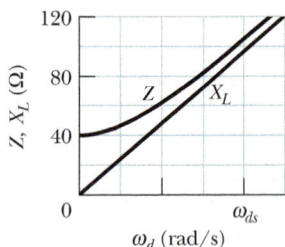

Figure 31-31 Problem 42.

•43 Remove the capacitor from the circuit in Fig. 31-7 and set $R = 200\ \Omega$, $L = 230$ mH, $f_d = 60.0$ Hz, and $\mathscr{E}_m = 36.0$ V. What are (a) Z, (b) ϕ, and (c) I? (d) Draw a phasor diagram.

••44 GO An ac generator with emf amplitude $\mathscr{E}_m = 220$ V and operating at frequency 400 Hz causes oscillations in a series RLC circuit having $R = 220\ \Omega$, $L = 150$ mH, and $C = 24.0\ \mu$F. Find (a) the capacitive reactance X_C, (b) the impedance Z, and (c) the current amplitude I. A second capacitor of the same capacitance is then connected in series with the other components. Determine whether the values of (d) X_C, (e) Z, and (f) I increase, decrease, or remain the same.

••45 GO ILW (a) In an RLC circuit, can the amplitude of the voltage across an inductor be greater than the amplitude of the generator emf? (b) Consider an RLC circuit with emf amplitude $\mathscr{E}_m = 10$ V, resistance $R = 10\ \Omega$, inductance $L = 1.0$ H, and capacitance $C = 1.0\ \mu$F. Find the amplitude of the voltage across the inductor at resonance.

••46 GO An alternating emf source with a variable frequency f_d is connected in series with a 50.0 Ω resistor and a 20.0 μF capacitor. The emf amplitude is 12.0 V. (a) Draw a phasor diagram for phasor V_R (the potential across the resistor) and phasor V_C (the potential across the capacitor). (b) At what driving frequency f_d do the two phasors have the same length? At that driving frequency, what are (c) the phase angle in degrees, (d) the angular speed at which the phasors rotate, and (e) the current amplitude?

••47 SSM WWW An RLC circuit such as that of Fig. 31-7 has $R = 5.00\ \Omega$, $C = 20.0\ \mu$F, $L = 1.00$ H, and $\mathscr{E}_m = 30.0$ V. (a) At what angular frequency ω_d will the current amplitude have its maximum value, as in the resonance curves of Fig. 31-16? (b) What is this maximum value? At what (c) lower angular frequency ω_{d1} and (d) higher angular frequency ω_{d2} will the current amplitude be half this maximum value? (e) For the resonance curve for this circuit, what is the fractional half-width $(\omega_{d1} - \omega_{d2})/\omega$?

••48 GO Figure 31-32 shows a driven RLC circuit that contains two identical capacitors and two switches. The emf amplitude is set at 12.0 V, and the driving frequency is set at 60.0 Hz. With both switches open, the current leads the emf by 30.9°. With switch S_1 closed and switch S_2 still open, the emf leads the current by 15.0°. With both switches closed, the current amplitude is 447 mA. What are (a) R, (b) C, and (c) L?

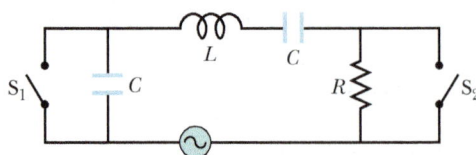

Figure 31-32 Problem 48.

••49 GO In Fig. 31-33, a generator with an adjustable frequency of oscillation is connected to resistance $R = 100\ \Omega$, inductances $L_1 = 1.70$ mH and $L_2 = 2.30$ mH, and capacitances $C_1 = 4.00$ μF, $C_2 = 2.50\ \mu$F, and $C_3 = 3.50$ μF. (a) What is the resonant frequency of the circuit? (*Hint:* See Problem 47 in Chapter 30.) What happens to the resonant frequency if (b) R is increased, (c) L_1 is increased, and (d) C_3 is removed from the circuit?

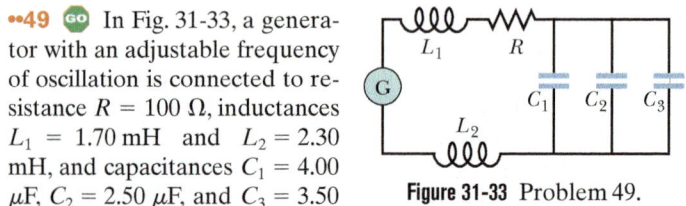

Figure 31-33 Problem 49.

••50 An alternating emf source with a variable frequency f_d is connected in series with an 80.0 Ω resistor and a 40.0 mH inductor. The emf amplitude is 6.00 V. (a) Draw a phasor diagram for phasor V_R (the potential across the resistor) and phasor V_L (the potential across the inductor). (b) At what driving frequency f_d do the two phasors have the same length? At that driving frequency, what are (c) the phase angle in degrees, (d) the angular speed at which the phasors rotate, and (e) the current amplitude?

••51 SSM The fractional half-width $\Delta\omega_d$ of a resonance curve, such as the ones in Fig. 31-16, is the width of the curve at half the maximum value of I. Show that $\Delta\omega_d/\omega = R(3C/L)^{1/2}$, where ω is the angular frequency at resonance. Note that the ratio $\Delta\omega_d/\omega$ increases with R, as Fig. 31-16 shows.

Module 31-5 Power in Alternating-Current Circuits

•52 An ac voltmeter with large impedance is connected in turn across the inductor, the capacitor, and the resistor in a series circuit having an alternating emf of 100 V (rms); the meter gives the same reading in volts in each case. What is this reading?

•53 SSM An air conditioner connected to a 120 V rms ac line is equivalent to a 12.0 Ω resistance and a 1.30 Ω inductive reactance in series. Calculate (a) the impedance of the air conditioner and (b) the average rate at which energy is supplied to the appliance.

•54 What is the maximum value of an ac voltage whose rms value is 100 V?

•55 What direct current will produce the same amount of thermal energy, in a particular resistor, as an alternating current that has a maximum value of 2.60 A?

••56 A typical light dimmer used to dim the stage lights in a theater consists of a variable inductor L (whose inductance is adjustable between zero and L_{max}) connected in series with a lightbulb B, as shown in Fig. 31-34. The electrical supply is 120 V (rms) at 60.0 Hz; the lightbulb is rated at 120 V, 1000 W. (a) What L_{max} is required if the rate of energy dissipation in the lightbulb is to be varied by a factor of 5 from its upper limit of 1000 W? Assume that the resistance of the lightbulb is independent of its temperature. (b) Could one use a variable resistor (adjustable between zero and R_{max}) instead of an inductor? (c) If so, what R_{max} is required? (d) Why isn't this done?

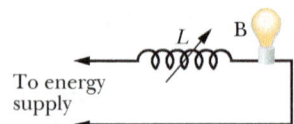

Figure 31-34 Problem 56.

••57 In an RLC circuit such as that of Fig. 31-7 assume that $R = 5.00\ \Omega$, $L = 60.0$ mH, $f_d = 60.0$ Hz, and $\mathscr{E}_m = 30.0$ V. For what values of the capacitance would the average rate at which energy is dissipated in the resistance be (a) a maximum and (b) a minimum? What are (c) the maximum dissipation rate and the corresponding

(d) phase angle and (e) power factor? What are (f) the minimum dissipation rate and the corresponding (g) phase angle and (h) power factor?

••58 For Fig. 31-35, show that the average rate at which energy is dissipated in resistance R is a maximum when R is equal to the internal resistance r of the ac generator. (In the text discussion we tacitly assumed that $r = 0$.)

Figure 31-35 Problems 58 and 66.

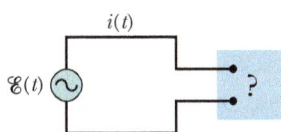

••59 **GO** In Fig. 31-7, $R = 15.0\ \Omega$, $C = 4.70\ \mu F$, and $L = 25.0$ mH. The generator provides an emf with rms voltage 75.0 V and frequency 550 Hz. (a) What is the rms current? What is the rms voltage across (b) R, (c) C, (d) L, (e) C and L together, and (f) R, C, and L together? At what average rate is energy dissipated by (g) R, (h) C, and (i) L?

••60 **GO** In a series oscillating RLC circuit, $R = 16.0\ \Omega$, $C = 31.2\ \mu F$, $L = 9.20$ mH, and $\mathscr{E}_m = \mathscr{E}_m \sin \omega_d t$ with $\mathscr{E}_m = 45.0$ V and $\omega_d = 3000$ rad/s. For time $t = 0.442$ ms find (a) the rate P_g at which energy is being supplied by the generator, (b) the rate P_C at which the energy in the capacitor is changing, (c) the rate P_L at which the energy in the inductor is changing, and (d) the rate P_R at which energy is being dissipated in the resistor. (e) Is the sum of P_C, P_L, and P_R greater than, less than, or equal to P_g?

••61 **SSM** **WWW** Figure 31-36 shows an ac generator connected to a "black box" through a pair of terminals. The box contains an RLC circuit, possibly even a multiloop circuit, whose elements and connections we do not know. Measurements outside the box reveal that

Figure 31-36 Problem 61.

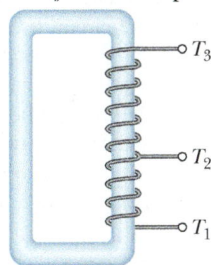

$$\mathscr{E}(t) = (75.0\ \text{V}) \sin \omega_d t$$

and

$$i(t) = (1.20\ \text{A}) \sin(\omega_d t + 42.0°).$$

(a) What is the power factor? (b) Does the current lead or lag the emf? (c) Is the circuit in the box largely inductive or largely capacitive? (d) Is the circuit in the box in resonance? (e) Must there be a capacitor in the box? (f) An inductor? (g) A resistor? (h) At what average rate is energy delivered to the box by the generator? (i) Why don't you need to know ω_d to answer all these questions?

Module 31-6 Transformers

•62 A generator supplies 100 V to a transformer's primary coil, which has 50 turns. If the secondary coil has 500 turns, what is the secondary voltage?

•63 **SSM** **ILW** A transformer has 500 primary turns and 10 secondary turns. (a) If V_p is 120 V (rms), what is V_s with an open circuit? If the secondary now has a resistive load of 15 Ω, what is the current in the (b) primary and (c) secondary?

•64 Figure 31-37 shows an "autotransformer." It consists of a single coil (with an iron core). Three taps T_i are provided. Between taps T_1 and T_2 there are 200 turns, and between taps T_2 and T_3 there are 800 turns. Any two taps can be chosen as the primary terminals, and any two taps can be chosen as the secondary terminals. For

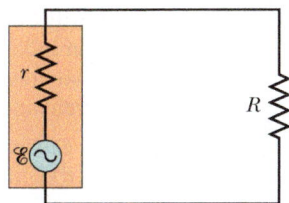

Figure 31-37 Problem 64.

choices producing a step-up transformer, what are the (a) smallest, (b) second smallest, and (c) largest values of the ratio V_s/V_p? For a step-down transformer, what are the (d) smallest, (e) second smallest, and (f) largest values of V_s/V_p?

••65 An ac generator provides emf to a resistive load in a remote factory over a two-cable transmission line. At the factory a step-down transformer reduces the voltage from its (rms) transmission value V_t to a much lower value that is safe and convenient for use in the factory. The transmission line resistance is 0.30 Ω/cable, and the power of the generator is 250 kW. If $V_t = 80$ kV, what are (a) the voltage decrease ΔV along the transmission line and (b) the rate P_d at which energy is dissipated in the line as thermal energy? If $V_t = 8.0$ kV, what are (c) ΔV and (d) P_d? If $V_t = 0.80$ kV, what are (e) ΔV and (f) P_d?

Additional Problems

66 In Fig. 31-35, let the rectangular box on the left represent the (high-impedance) output of an audio amplifier, with $r = 1000\ \Omega$. Let $R = 10\ \Omega$ represent the (low-impedance) coil of a loudspeaker. For maximum transfer of energy to the load R we must have $R = r$, and that is not true in this case. However, a transformer can be used to "transform" resistances, making them behave electrically as if they were larger or smaller than they actually are. (a) Sketch the primary and secondary coils of a transformer that can be introduced between the amplifier and the speaker in Fig. 31-35 to match the impedances. (b) What must be the turns ratio?

67 **GO** An ac generator produces emf $\mathscr{E} = \mathscr{E}_m \sin(\omega_d t - \pi/4)$, where $\mathscr{E}_m = 30.0$ V and $\omega_d = 350$ rad/s. The current in the circuit attached to the generator is $i(t) = I \sin(\omega_d t + \pi/4)$, where $I = 620$ mA. (a) At what time after $t = 0$ does the generator emf first reach a maximum? (b) At what time after $t = 0$ does the current first reach a maximum? (c) The circuit contains a single element other than the generator. Is it a capacitor, an inductor, or a resistor? Justify your answer. (d) What is the value of the capacitance, inductance, or resistance, as the case may be?

68 A series RLC circuit is driven by a generator at a frequency of 2000 Hz and an emf amplitude of 170 V. The inductance is 60.0 mH, the capacitance is 0.400 μF, and the resistance is 200 Ω. (a) What is the phase constant in radians? (b) What is the current amplitude?

69 A generator of frequency 3000 Hz drives a series RLC circuit with an emf amplitude of 120 V. The resistance is 40.0 Ω, the capacitance is 1.60 μF, and the inductance is 850 μH. What are (a) the phase constant in radians and (b) the current amplitude? (c) Is the circuit capacitive, inductive, or in resonance?

70 A 45.0 mH inductor has a reactance of 1.30 kΩ. (a) What is its operating frequency? (b) What is the capacitance of a capacitor with the same reactance at that frequency? If the frequency is doubled, what is the new reactance of (c) the inductor and (d) the capacitor?

71 An RLC circuit is driven by a generator with an emf amplitude of 80.0 V and a current amplitude of 1.25 A. The current leads the emf by 0.650 rad. What are the (a) impedance and (b) resistance of the circuit? (c) Is the circuit inductive, capacitive, or in resonance?

72 A series RLC circuit is driven in such a way that the maximum voltage across the inductor is 1.50 times the maximum voltage across the capacitor and 2.00 times the maximum voltage across the resistor. (a) What is ϕ for the circuit? (b) Is the circuit inductive, capacitive, or in resonance? The resistance is 49.9 Ω, and the current amplitude is 200 mA. (c) What is the amplitude of the driving emf?

73 A capacitor of capacitance 158 μF and an inductor form an LC circuit that oscillates at 8.15 kHz, with a current amplitude of 4.21 mA. What are (a) the inductance, (b) the total energy in the circuit, and (c) the maximum charge on the capacitor?

74 An oscillating LC circuit has an inductance of 3.00 mH and a capacitance of 10.0 μF. Calculate the (a) angular frequency and (b) period of the oscillation. (c) At time $t = 0$, the capacitor is charged to 200 μC and the current is zero. Roughly sketch the charge on the capacitor as a function of time.

75 For a certain driven series RLC circuit, the maximum generator emf is 125 V and the maximum current is 3.20 A. If the current leads the generator emf by 0.982 rad, what are the (a) impedance and (b) resistance of the circuit? (c) Is the circuit predominantly capacitive or inductive?

76 A 1.50 μF capacitor has a capacitive reactance of 12.0 Ω. (a) What must be its operating frequency? (b) What will be the capacitive reactance if the frequency is doubled?

77 SSM In Fig. 31-38, a three-phase generator G produces electrical power that is transmitted by means of three wires. The electric potentials (each relative to a common reference level) are $V_1 = A \sin \omega_d t$ for wire 1, $V_2 = A \sin(\omega_d t - 120°)$ for wire 2, and $V_3 = A \sin(\omega_d t - 240°)$ for wire 3. Some types of industrial equipment (for example, motors) have three terminals and are designed to be connected directly to these three wires. To use a more conventional two-terminal device (for example, a lightbulb), one connects it to any two of the three wires. Show that the potential difference between *any two* of the wires (a) oscillates sinusoidally with angular frequency ω_d and (b) has an amplitude of $A\sqrt{3}$.

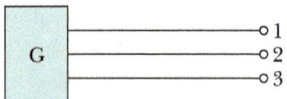
Three-wire transmission line
Figure 31-38 Problem 77.

78 An electric motor connected to a 120 V, 60.0 Hz ac outlet does mechanical work at the rate of 0.100 hp (1 hp = 746 W). (a) If the motor draws an rms current of 0.650 A, what is its effective resistance, relative to power transfer? (b) Is this the same as the resistance of the motor's coils, as measured with an ohmmeter with the motor disconnected from the outlet?

79 SSM (a) In an oscillating LC circuit, in terms of the maximum charge Q on the capacitor, what is the charge there when the energy in the electric field is 50.0% of that in the magnetic field? (b) What fraction of a period must elapse following the time the capacitor is fully charged for this condition to occur?

80 A series RLC circuit is driven by an alternating source at a frequency of 400 Hz and an emf amplitude of 90.0 V. The resistance is 20.0 Ω, the capacitance is 12.1 μF, and the inductance is 24.2 mH. What is the rms potential difference across (a) the resistor, (b) the capacitor, and (c) the inductor? (d) What is the average rate at which energy is dissipated?

81 SSM In a certain series RLC circuit being driven at a frequency of 60.0 Hz, the maximum voltage across the inductor is 2.00 times the maximum voltage across the resistor and 2.00 times the maximum voltage across the capacitor. (a) By what angle does the current lag the generator emf? (b) If the maximum generator emf is 30.0 V, what should be the resistance of the circuit to obtain a maximum current of 300 mA?

82 A 1.50 mH inductor in an oscillating LC circuit stores a maximum energy of 10.0 μJ. What is the maximum current?

83 A generator with an adjustable frequency of oscillation is wired in series to an inductor of $L = 2.50$ mH and a capacitor of $C = 3.00$ μF. At what frequency does the generator produce the largest possible current amplitude in the circuit?

84 A series RLC circuit has a resonant frequency of 6.00 kHz. When it is driven at 8.00 kHz, it has an impedance of 1.00 kΩ and a phase constant of 45°. What are (a) R, (b) L, and (c) C for this circuit?

85 SSM An LC circuit oscillates at a frequency of 10.4 kHz. (a) If the capacitance is 340 μF, what is the inductance? (b) If the maximum current is 7.20 mA, what is the total energy in the circuit? (c) What is the maximum charge on the capacitor?

86 When under load and operating at an rms voltage of 220 V, a certain electric motor draws an rms current of 3.00 A. It has a resistance of 24.0 Ω and no capacitive reactance. What is its inductive reactance?

87 The ac generator in Fig. 31-39 supplies 120 V at 60.0 Hz. With the switch open as in the diagram, the current leads the generator emf by 20.0°. With the switch in position 1, the current lags the generator emf by 10.0°. When the switch is in position 2, the current amplitude is 2.00 A. What are (a) R, (b) L, and (c) C?

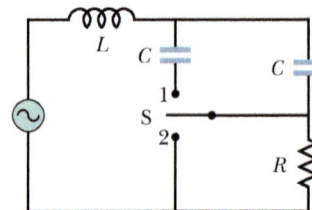
Figure 31-39 Problem 87.

88 In an oscillating LC circuit, $L = 8.00$ mH and $C = 1.40$ μF. At time $t = 0$, the current is maximum at 12.0 mA. (a) What is the maximum charge on the capacitor during the oscillations? (b) At what earliest time $t > 0$ is the rate of change of energy in the capacitor maximum? (c) What is that maximum rate of change?

89 SSM For a sinusoidally driven series RLC circuit, show that over one complete cycle with period T (a) the energy stored in the capacitor does not change; (b) the energy stored in the inductor does not change; (c) the driving emf device supplies energy $(\frac{1}{2}T)\mathscr{E}_m I \cos \phi$; and (d) the resistor dissipates energy $(\frac{1}{2}T)RI^2$. (e) Show that the quantities found in (c) and (d) are equal.

90 What capacitance would you connect across a 1.30 mH inductor to make the resulting oscillator resonate at 3.50 kHz?

91 A series circuit with resistor–inductor–capacitor combination R_1, L_1, C_1 has the same resonant frequency as a second circuit with a different combination R_2, L_2, C_2. You now connect the two combinations in series. Show that this new circuit has the same resonant frequency as the separate circuits.

92 Consider the circuit shown in Fig. 31-40. With switch S_1 closed and the other two switches open, the circuit has a time constant τ_C. With switch S_2 closed and the other two switches open, the circuit has a time constant τ_L. With switch S_3 closed and the other two switches open, the circuit oscillates with a period T. Show that $T = 2\pi\sqrt{\tau_C \tau_L}$.

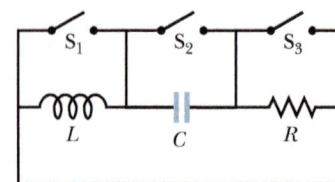
Figure 31-40 Problem 92.

93 When the generator emf in Sample Problem 31.07 is a maximum, what is the voltage across (a) the generator, (b) the resistance, (c) the capacitance, and (d) the inductance? (e) By summing these with appropriate signs, verify that the loop rule is satisfied.

Maxwell's Equations; Magnetism of Matter

32-1 GAUSS' LAW FOR MAGNETIC FIELDS

Learning Objectives

After reading this module, you should be able to . . .

32.01 Identify that the simplest magnetic structure is a magnetic dipole.

32.02 Calculate the magnetic flux Φ through a surface by integrating the dot product of the magnetic field vector

$\vec{B}$ and the area vector $d\vec{A}$ (for patch elements) over the surface.

32.03 Identify that the net magnetic flux through a Gaussian surface (which is a closed surface) is zero.

Key Idea

● The simplest magnetic structures are magnetic dipoles. Magnetic monopoles do not exist (as far as we know). Gauss' law for magnetic fields,

$$\Phi_B = \oint \vec{B} \cdot d\vec{A} = 0,$$

states that the net magnetic flux through any (closed) Gaussian surface is zero. It implies that magnetic monopoles do not exist.

What Is Physics?

This chapter reveals some of the breadth of physics because it ranges from the basic science of electric and magnetic fields to the applied science and engineering of magnetic materials. First, we conclude our basic discussion of electric and magnetic fields, finding that most of the physics principles in the last 11 chapters can be summarized in only *four* equations, known as Maxwell's equations.

Second, we examine the science and engineering of magnetic materials. The careers of many scientists and engineers are focused on understanding why some materials are magnetic and others are not and on how existing magnetic materials can be improved. These researchers wonder why Earth has a magnetic field but you do not. They find countless applications for inexpensive magnetic materials in cars, kitchens, offices, and hospitals, and magnetic materials often show up in unexpected ways. For example, if you have a tattoo (Fig. 32-1) and undergo an MRI (magnetic resonance imaging) scan, the large magnetic field used in the scan may noticeably tug on your tattooed skin because some tattoo inks contain magnetic particles. In another example, some breakfast cereals are advertised as being "iron fortified" because they contain small bits of iron for you to ingest. Because these iron bits are magnetic, you can collect them by passing a magnet over a slurry of water and cereal.

Our first step here is to revisit Gauss' law, but this time for magnetic fields.

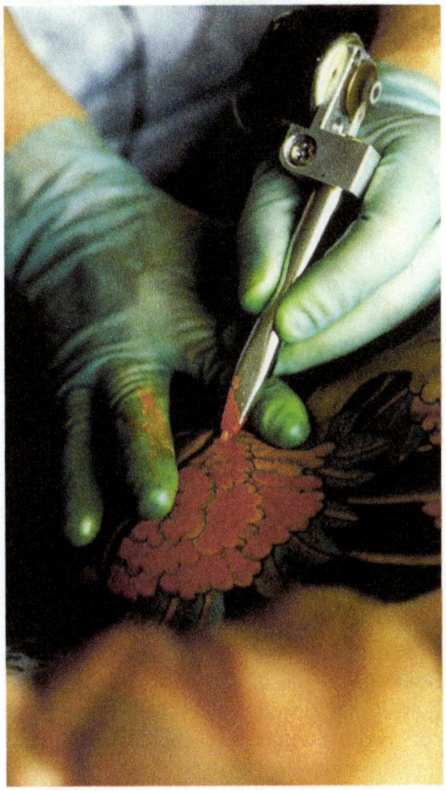

Oliver Strewe/Getty Images, Inc.

Figure 32-1 Some of the inks used for tattoos contain magnetic particles.

Gauss' Law for Magnetic Fields

Figure 32-2 shows iron powder that has been sprinkled onto a transparent sheet placed above a bar magnet. The powder grains, trying to align themselves with the magnet's magnetic field, have fallen into a pattern that reveals the field. One end of the magnet is a *source* of the field (the field lines diverge from it) and the other end is a *sink* of the field (the field lines converge toward it). By convention, we call the source the *north pole* of the magnet and the sink the *south pole,* and we say that the magnet, with its two poles, is an example of a **magnetic dipole.**

Suppose we break a bar magnet into pieces the way we can break a piece of chalk (Fig. 32-3). We should, it seems, be able to isolate a single magnetic pole, called a *magnetic monopole.* However, we cannot—not even if we break the magnet down to its individual atoms and then to its electrons and nuclei. Each fragment has a north pole and a south pole. Thus:

The simplest magnetic structure that can exist is a magnetic dipole. Magnetic monopoles do not exist (as far as we know).

Gauss' law for magnetic fields is a formal way of saying that magnetic monopoles do not exist. The law asserts that the net magnetic flux Φ_B through any closed Gaussian surface is zero:

$$\Phi_B = \oint \vec{B} \cdot d\vec{A} = 0 \quad \text{(Gauss' law for magnetic fields).} \qquad (32\text{-}1)$$

Contrast this with Gauss' law for electric fields,

$$\Phi_E = \oint \vec{E} \cdot d\vec{A} = \frac{q_{\text{enc}}}{\varepsilon_0} \quad \text{(Gauss' law for electric fields).}$$

In both equations, the integral is taken over a *closed* Gaussian surface. Gauss' law for electric fields says that this integral (the net electric flux through the surface) is proportional to the net electric charge q_{enc} enclosed by the surface. Gauss' law for magnetic fields says that there can be no net magnetic flux through the surface because there can be no net "magnetic charge" (individual magnetic poles) enclosed by the surface. The simplest magnetic structure that can exist and thus be enclosed by a Gaussian surface is a dipole, which consists of both a source and a sink for the field lines. Thus, there must always be as much magnetic flux into the surface as out of it, and the net magnetic flux must always be zero.

Gauss' law for magnetic fields holds for structures more complicated than a magnetic dipole, and it holds even if the Gaussian surface does not enclose the entire structure. Gaussian surface II near the bar magnet of Fig. 32-4 encloses no poles, and we can easily conclude that the net magnetic flux through it is zero. Gaussian surface I is more difficult. It may seem to enclose only the north pole of the magnet because it encloses the label N and not the label S. However, a south pole must be associated with the lower boundary of the surface because magnetic field lines enter the surface there. (The enclosed section is like one piece of the broken bar magnet in Fig. 32-3.) Thus, Gaussian surface I encloses a magnetic dipole, and the net flux through the surface is zero.

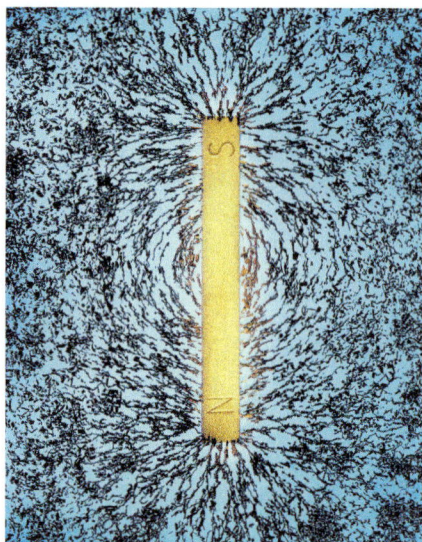

Richard Megna/Fundamental Photographs

Figure 32-2 A bar magnet is a magnetic dipole. The iron filings suggest the magnetic field lines. (Colored light fills the background.)

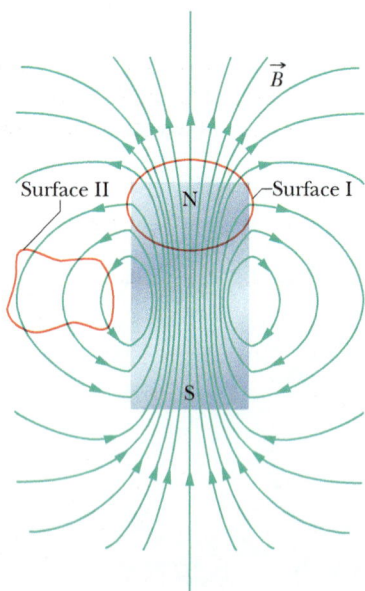

Figure 32-4 The field lines for the magnetic field $\vec{B}$ of a short bar magnet. The red curves represent cross sections of closed, three-dimensional Gaussian surfaces.

Figure 32-3 If you break a magnet, each fragment becomes a separate magnet, with its own north and south poles.

✓ Checkpoint 1

The figure here shows four closed surfaces with flat top and bottom faces and curved sides. The table gives the areas A of the faces and the magnitudes B of the uniform and perpendicular magnetic fields through those faces; the units of A and B are arbitrary but consistent. Rank the surfaces according to the magnitudes of the magnetic flux through their curved sides, greatest first.

Surface	A_{top}	B_{top}	A_{bot}	B_{bot}
a	2	6, outward	4	3, inward
b	2	1, inward	4	2, inward
c	2	6, inward	2	8, outward
d	2	3, outward	3	2, outward

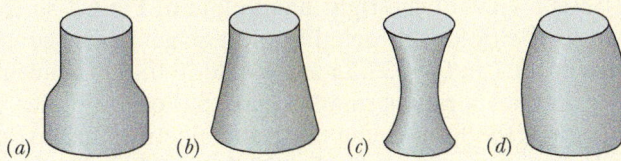

(a) (b) (c) (d)

32-2 INDUCED MAGNETIC FIELDS

Learning Objectives

After reading this module, you should be able to . . .

32.04 Identify that a changing electric flux induces a magnetic field.

32.05 Apply Maxwell's law of induction to relate the magnetic field induced around a closed loop to the rate of change of electric flux encircled by the loop.

32.06 Draw the field lines for an induced magnetic field inside a capacitor with parallel circular plates that are being charged, indicating the orientations of the vectors for the electric field and the magnetic field.

32.07 For the general situation in which magnetic fields can be induced, apply the Ampere–Maxwell (combined) law.

Key Ideas

● A changing electric flux induces a magnetic field $\vec{B}$. Maxwell's law,

$$\oint \vec{B} \cdot d\vec{s} = \mu_0 \varepsilon_0 \frac{d\Phi_E}{dt} \quad \text{(Maxwell's law of induction),}$$

relates the magnetic field induced along a closed loop to the changing electric flux Φ_E through the loop.

● Ampere's law, $\oint \vec{B} \cdot d\vec{s} = \mu_0 i_{enc}$, gives the magnetic field generated by a current i_{enc} encircled by a closed loop. Maxwell's law and Ampere's law can be written as the single equation

$$\oint \vec{B} \cdot d\vec{s} = \mu_0 \varepsilon_0 \frac{d\Phi_E}{dt} + \mu_0 i_{enc} \quad \text{(Ampere–Maxwell law).}$$

Induced Magnetic Fields

In Chapter 30 you saw that a changing magnetic flux induces an electric field, and we ended up with Faraday's law of induction in the form

$$\oint \vec{E} \cdot d\vec{s} = -\frac{d\Phi_B}{dt} \quad \text{(Faraday's law of induction).} \quad (32\text{-}2)$$

Here $\vec{E}$ is the electric field induced along a closed loop by the changing magnetic flux Φ_B encircled by that loop. Because symmetry is often so powerful in physics, we should be tempted to ask whether induction can occur in the opposite sense; that is, can a changing electric flux induce a magnetic field?

The answer is that it can; furthermore, the equation governing the induction of a magnetic field is almost symmetric with Eq. 32-2. We often call it Maxwell's

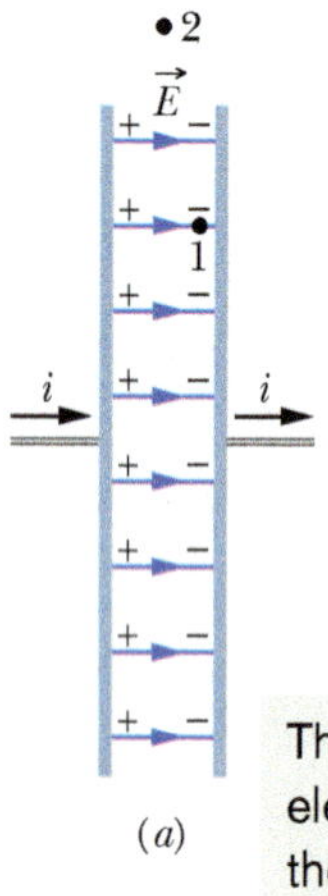

(a)

The changing of the electric field between the plates creates a magnetic field.

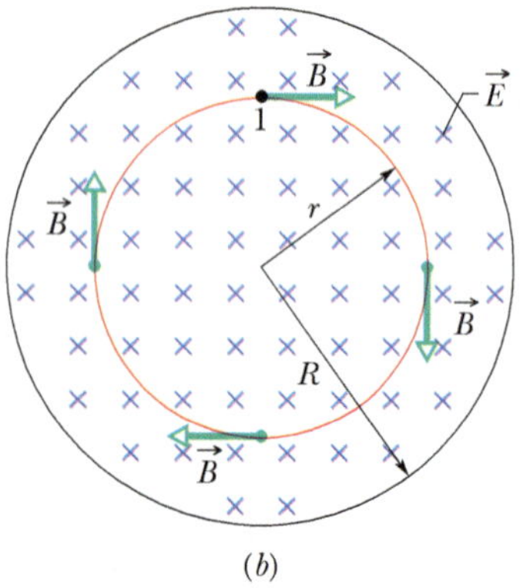

(b)

Figure 32-5 (a) A circular parallel-plate capacitor, shown in side view, is being charged by a constant current i. (b) A view from within the capacitor, looking toward the plate at the right in (a). The electric field $\vec{E}$ is uniform, is directed into the page (toward the plate), and grows in magnitude as the charge on the capacitor increases. The magnetic field $\vec{B}$ induced by this changing electric field is shown at four points on a circle with a radius r less than the plate radius R.

Figure 32-6 A uniform magnetic field $\vec{B}$ in a circular region. The field, directed into the page, is increasing in magnitude. The electric field $\vec{E}$ induced by the changing magnetic field is shown at four points on a circle concentric with the circular region. Compare this situation with that of Fig. 32-5b.

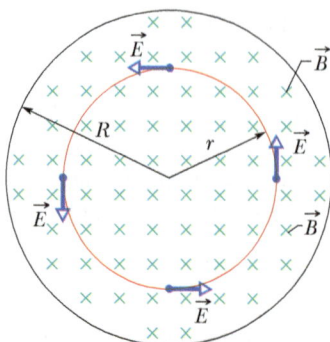

law of induction after James Clerk Maxwell, and we write it as

$$\oint \vec{B} \cdot d\vec{s} = \mu_0\varepsilon_0 \frac{d\Phi_E}{dt} \quad \text{(Maxwell's law of induction).} \quad (32\text{-}3)$$

Here $\vec{B}$ is the magnetic field induced along a closed loop by the changing electric flux Φ_E in the region encircled by that loop.

Charging a Capacitor. As an example of this sort of induction, we consider the charging of a parallel-plate capacitor with circular plates. (Although we shall focus on this arrangement, a changing electric flux will always induce a magnetic field whenever it occurs.) We assume that the charge on our capacitor (Fig. 32-5a) is being increased at a steady rate by a constant current i in the connecting wires. Then the electric field magnitude between the plates must also be increasing at a steady rate.

Figure 32-5b is a view of the right-hand plate of Fig. 32-5a from between the plates. The electric field is directed into the page. Let us consider a circular loop through point 1 in Figs. 32-5a and b, a loop that is concentric with the capacitor plates and has a radius smaller than that of the plates. Because the electric field through the loop is changing, the electric flux through the loop must also be changing. According to Eq. 32-3, this changing electric flux induces a magnetic field around the loop.

Experiment proves that a magnetic field $\vec{B}$ is indeed induced around such a loop, directed as shown. This magnetic field has the same magnitude at every point around the loop and thus has circular symmetry about the *central axis* of the capacitor plates (the axis extending from one plate center to the other).

If we now consider a larger loop—say, through point 2 outside the plates in Figs. 32-5a and b—we find that a magnetic field is induced around that loop as well. Thus, while the electric field is changing, magnetic fields are induced between the plates, both inside and outside the gap. When the electric field stops changing, these induced magnetic fields disappear.

Although Eq. 32-3 is similar to Eq. 32-2, the equations differ in two ways. First, Eq. 32-3 has the two extra symbols μ_0 and ε_0, but they appear only because we employ SI units. Second, Eq. 32-3 lacks the minus sign of Eq. 32-2, meaning that the induced electric field $\vec{E}$ and the induced magnetic field $\vec{B}$ have opposite directions when they are produced in otherwise similar situations. To see this opposition, examine Fig. 32-6, in which an increasing magnetic field $\vec{B}$, directed into the page, induces an electric field $\vec{E}$. The induced field $\vec{E}$ is counterclockwise, opposite the induced magnetic field $\vec{B}$ in Fig. 32-5b.

Ampere–Maxwell Law

Now recall that the left side of Eq. 32-3, the integral of the dot product $\vec{B} \cdot d\vec{s}$ around a closed loop, appears in another equation—namely, Ampere's law:

$$\oint \vec{B} \cdot d\vec{s} = \mu_0 i_{\text{enc}} \quad \text{(Ampere's law),} \quad (32\text{-}4)$$

The induced $\vec{E}$ direction here is opposite the induced $\vec{B}$ direction in the preceding figure.

where i_{enc} is the current encircled by the closed loop. Thus, our two equations that specify the magnetic field $\vec{B}$ produced by means other than a magnetic material (that is, by a current and by a changing electric field) give the field in exactly the same form. We can combine the two equations into the single equation

$$\oint \vec{B} \cdot d\vec{s} = \mu_0 \varepsilon_0 \frac{d\Phi_E}{dt} + \mu_0 i_{enc} \quad \text{(Ampere–Maxwell law).} \quad (32\text{-}5)$$

When there is a current but no change in electric flux (such as with a wire carrying a constant current), the first term on the right side of Eq. 32-5 is zero, and so Eq. 32-5 reduces to Eq. 32-4, Ampere's law. When there is a change in electric flux but no current (such as inside or outside the gap of a charging capacitor), the second term on the right side of Eq. 32-5 is zero, and so Eq. 32-5 reduces to Eq. 32-3, Maxwell's law of induction.

✓ Checkpoint 2

The figure shows graphs of the electric field magnitude E versus time t for four uniform electric fields, all contained within identical circular regions as in Fig. 32-5b. Rank the fields according to the magnitudes of the magnetic fields they induce at the edge of the region, greatest first.

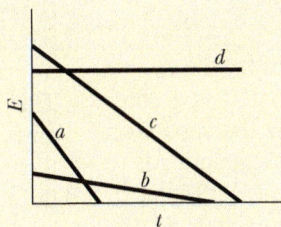

Sample Problem 32.01 Magnetic field induced by changing electric field

A parallel-plate capacitor with circular plates of radius R is being charged as in Fig. 32-5a.

(a) Derive an expression for the magnetic field at radius r for the case $r \leq R$.

KEY IDEAS

A magnetic field can be set up by a current and by induction due to a changing electric flux; both effects are included in Eq. 32-5. There is no current between the capacitor plates of Fig. 32-5, but the electric flux there is changing. Thus, Eq. 32-5 reduces to

$$\oint \vec{B} \cdot d\vec{s} = \mu_0 \varepsilon_0 \frac{d\Phi_E}{dt}. \quad (32\text{-}6)$$

We shall separately evaluate the left and right sides of this equation.

Left side of Eq. 32-6: We choose a circular Amperian loop with a radius $r \leq R$ as shown in Fig. 32-5b because we want to evaluate the magnetic field for $r \leq R$—that is, inside the capacitor. The magnetic field $\vec{B}$ at all points along the loop is tangent to the loop, as is the path element $d\vec{s}$. Thus, $\vec{B}$ and $d\vec{s}$ are either parallel or antiparallel at each point of the loop. For simplicity, assume they are parallel (the choice does not alter our outcome here). Then

$$\oint \vec{B} \cdot d\vec{s} = \oint B \, ds \cos 0° = \oint B \, ds.$$

Due to the circular symmetry of the plates, we can also assume that $\vec{B}$ has the same magnitude at every point around the loop. Thus, B can be taken outside the integral on the right side of the above equation. The integral that remains is $\oint ds$, which simply gives the circumference $2\pi r$ of the loop. The left side of Eq. 32-6 is then $(B)(2\pi r)$.

Right side of Eq. 32-6: We assume that the electric field $\vec{E}$ is uniform between the capacitor plates and directed perpendicular to the plates. Then the electric flux Φ_E through the Amperian loop is EA, where A is the area encircled by the loop within the electric field. Thus, the right side of Eq. 32-6 is $\mu_0 \varepsilon_0 \, d(EA)/dt$.

Combining results: Substituting our results for the left and right sides into Eq. 32-6, we get

$$(B)(2\pi r) = \mu_0 \varepsilon_0 \frac{d(EA)}{dt}.$$

Because A is a constant, we write $d(EA)$ as $A \, dE$; so we have

$$(B)(2\pi r) = \mu_0 \varepsilon_0 A \frac{dE}{dt}. \quad (32\text{-}7)$$

The area A that is encircled by the Amperian loop within the electric field is the *full* area πr^2 of the loop because the loop's radius r is less than (or equal to) the plate radius R. Substituting πr^2 for A in Eq. 32-7 leads to, for $r \leq R$,

$$B = \frac{\mu_0 \varepsilon_0 r}{2} \frac{dE}{dt}. \quad \text{(Answer)} \quad (32\text{-}8)$$

This equation tells us that, inside the capacitor, B increases linearly with increased radial distance r, from 0 at the central axis to a maximum value at plate radius R.

(b) Evaluate the field magnitude B for $r = R/5 = 11.0$ mm and $dE/dt = 1.50 \times 10^{12}$ V/m·s.

Calculation: From the answer to (a), we have

$$B = \frac{1}{2} \mu_0 \varepsilon_0 r \frac{dE}{dt}$$

$$= \tfrac{1}{2}(4\pi \times 10^{-7} \text{ T·m/A})(8.85 \times 10^{-12} \text{ C}^2/\text{N·m}^2)$$
$$\times (11.0 \times 10^{-3} \text{ m})(1.50 \times 10^{12} \text{ V/m·s})$$
$$= 9.18 \times 10^{-8} \text{ T.} \qquad \text{(Answer)}$$

(c) Derive an expression for the induced magnetic field for the case $r \geq R$.

Calculation: Our procedure is the same as in (a) except we now use an Amperian loop with a radius r that is greater than the plate radius R, to evaluate B outside the capacitor. Evaluating the left and right sides of Eq. 32-6 again leads to Eq. 32-7. However, we then need this subtle point: The electric field exists only between the plates, not outside the plates. Thus, the area A that is encircled by the Amperian

loop in the electric field is *not* the full area πr^2 of the loop. Rather, A is only the plate area πR^2.

Substituting πR^2 for A in Eq. 32-7 and solving the result for B give us, for $r \geq R$,

$$B = \frac{\mu_0 \varepsilon_0 R^2}{2r} \frac{dE}{dt}. \qquad \text{(Answer)} \quad (32\text{-}9)$$

This equation tells us that, outside the capacitor, B decreases with increased radial distance r, from a maximum value at the plate edges (where $r = R$). By substituting $r = R$ into Eqs. 32-8 and 32-9, you can show that these equations are consistent; that is, they give the same maximum value of B at the plate radius.

The magnitude of the induced magnetic field calculated in (b) is so small that it can scarcely be measured with simple apparatus. This is in sharp contrast to the magnitudes of induced electric fields (Faraday's law), which can be measured easily. This experimental difference exists partly because induced emfs can easily be multiplied by using a coil of many turns. No technique of comparable simplicity exists for multiplying induced magnetic fields. In any case, the experiment suggested by this sample problem has been done, and the presence of the induced magnetic fields has been verified quantitatively.

WILEY PLUS Additional examples, video, and practice available at *WileyPLUS*

32-3 DISPLACEMENT CURRENT

Learning Objectives

After reading this module, you should be able to . . .

32.08 Identify that in the Ampere–Maxwell law, the contribution to the induced magnetic field by the changing electric flux can be attributed to a fictitious current ("displacement current") to simplify the expression.

32.09 Identify that in a capacitor that is being charged or discharged, a displacement current is said to be spread uniformly over the plate area, from one plate to the other.

32.10 Apply the relationship between the rate of change of an electric flux and the associated displacement current.

32.11 For a charging or discharging capacitor, relate the amount of displacement current to the amount of actual

current and identify that the displacement current exists only when the electric field within the capacitor is changing.

32.12 Mimic the equations for the magnetic field inside and outside a wire with real current to write (and apply) the equations for the magnetic field inside and outside a region of displacement current.

32.13 Apply the Ampere–Maxwell law to calculate the magnetic field of a real current and a displacement current.

32.14 For a charging or discharging capacitor with parallel circular plates, draw the magnetic field lines due to the displacement current.

32.15 List Maxwell's equations and the purpose of each.

Key Ideas

● We define the fictitious displacement current due to a changing electric field as

$$i_d = \varepsilon_0 \frac{d\Phi_E}{dt}.$$

● The Ampere–Maxwell law then becomes

$$\oint \vec{B} \cdot d\vec{s} = \mu_0 i_{d,\text{enc}} + \mu_0 i_{\text{enc}} \quad \text{(Ampere–Maxwell law),}$$

where $i_{d,\text{enc}}$ is the displacement current encircled by the integration loop.

● The idea of a displacement current allows us to retain the notion of continuity of current through a capacitor. However, displacement current is *not* a transfer of charge.

● Maxwell's equations, displayed in Table 32-1, summarize electromagnetism and form its foundation, including optics.

Displacement Current

If you compare the two terms on the right side of Eq. 32-5, you will see that the product $\varepsilon_0(d\Phi_E/dt)$ must have the dimension of a current. In fact, that product has been treated as being a fictitious current called the **displacement current** i_d:

$$i_d = \varepsilon_0 \frac{d\Phi_E}{dt} \quad \text{(displacement current).} \quad (32\text{-}10)$$

"Displacement" is poorly chosen in that nothing is being displaced, but we are stuck with the word. Nevertheless, we can now rewrite Eq. 32-5 as

$$\oint \vec{B} \cdot d\vec{s} = \mu_0 i_{d,\text{enc}} + \mu_0 i_{\text{enc}} \quad \text{(Ampere–Maxwell law),} \quad (32\text{-}11)$$

in which $i_{d,\text{enc}}$ is the displacement current that is encircled by the integration loop.

Let us again focus on a charging capacitor with circular plates, as in Fig. 32-7a. The real current i that is charging the plates changes the electric field $\vec{E}$ between the plates. The fictitious displacement current i_d between the plates is associated with that changing field $\vec{E}$. Let us relate these two currents.

The charge q on the plates at any time is related to the magnitude E of the field between the plates at that time and the plate area A by Eq. 25-4:

$$q = \varepsilon_0 A E. \quad (32\text{-}12)$$

To get the real current i, we differentiate Eq. 32-12 with respect to time, finding

$$\frac{dq}{dt} = i = \varepsilon_0 A \frac{dE}{dt}. \quad (32\text{-}13)$$

To get the displacement current i_d, we can use Eq. 32-10. Assuming that the electric field $\vec{E}$ between the two plates is uniform (we neglect any fringing), we can

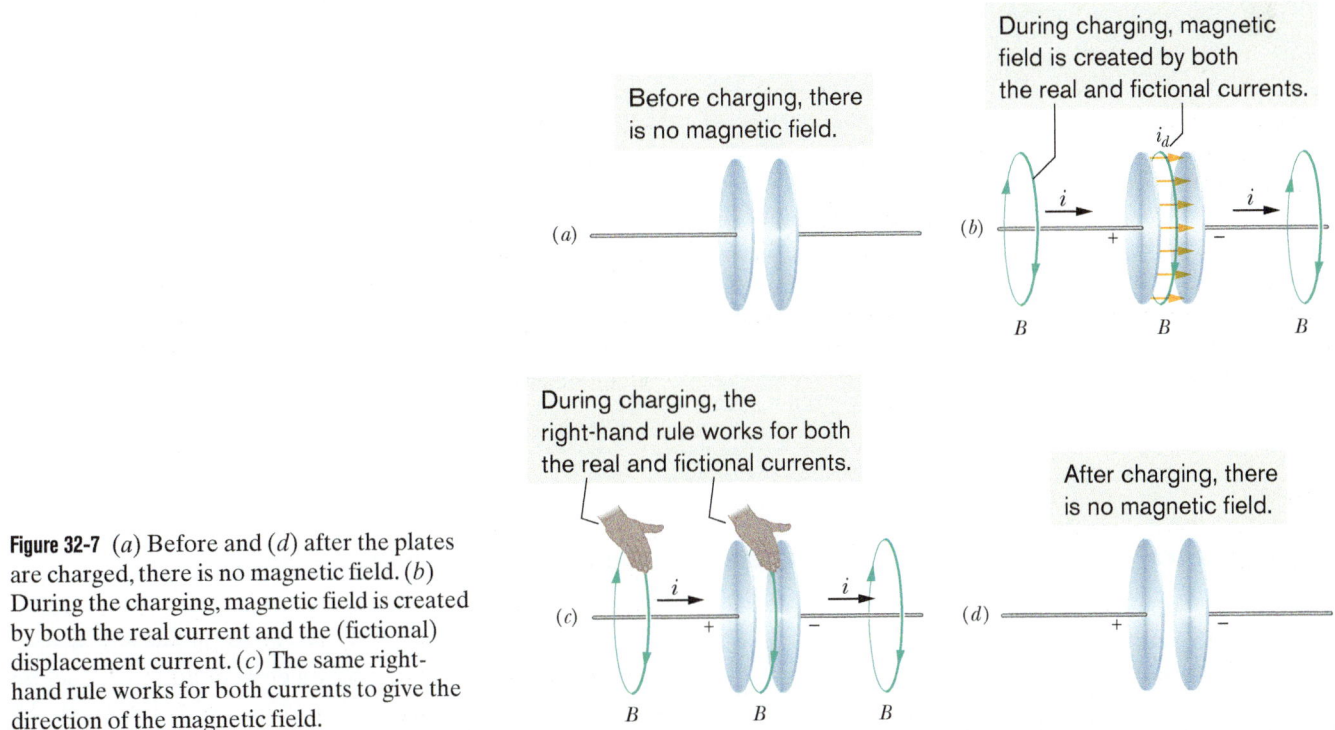

Figure 32-7 (a) Before and (d) after the plates are charged, there is no magnetic field. (b) During the charging, magnetic field is created by both the real current and the (fictional) displacement current. (c) The same right-hand rule works for both currents to give the direction of the magnetic field.

replace the electric flux Φ_E in that equation with EA. Then Eq. 32-10 becomes

$$i_d = \varepsilon_0 \frac{d\Phi_E}{dt} = \varepsilon_0 \frac{d(EA)}{dt} = \varepsilon_0 A \frac{dE}{dt}. \qquad (32\text{-}14)$$

Same Value. Comparing Eqs. 32-13 and 32-14, we see that the real current i charging the capacitor and the fictitious displacement current i_d between the plates have the same value:

$$i_d = i \quad \text{(displacement current in a capacitor).} \qquad (32\text{-}15)$$

Thus, we can consider the fictitious displacement current i_d to be simply a continuation of the real current i from one plate, across the capacitor gap, to the other plate. Because the electric field is uniformly spread over the plates, the same is true of this fictitious displacement current i_d, as suggested by the spread of current arrows in Fig. 32-7b. Although no charge actually moves across the gap between the plates, the idea of the fictitious current i_d can help us to quickly find the direction and magnitude of an induced magnetic field, as follows.

Finding the Induced Magnetic Field

In Chapter 29 we found the direction of the magnetic field produced by a real current i by using the right-hand rule of Fig. 29-5. We can apply the same rule to find the direction of an induced magnetic field produced by a fictitious displacement current i_d, as is shown in the center of Fig. 32-7c for a capacitor.

We can also use i_d to find the magnitude of the magnetic field induced by a charging capacitor with parallel circular plates of radius R. We simply consider the space between the plates to be an imaginary circular wire of radius R carrying the imaginary current i_d. Then, from Eq. 29-20, the magnitude of the magnetic field at a point inside the capacitor at radius r from the center is

$$B = \left(\frac{\mu_0 i_d}{2\pi R^2}\right)r \quad \text{(inside a circular capacitor).} \qquad (32\text{-}16)$$

Similarly, from Eq. 29-17, the magnitude of the magnetic field at a point outside the capacitor at radius r is

$$B = \frac{\mu_0 i_d}{2\pi r} \quad \text{(outside a circular capacitor).} \qquad (32\text{-}17)$$

✓ **Checkpoint 3**

The figure is a view of one plate of a parallel-plate capacitor from within the capacitor. The dashed lines show four integration paths (path b follows the edge of the plate). Rank the paths according to the magnitude of $\oint \vec{B} \cdot d\vec{s}$ along the paths during the discharging of the capacitor, greatest first.

Sample Problem 32.02 Treating a changing electric field as a displacement current

A circular parallel-plate capacitor with plate radius R is being charged with a current i.

(a) Between the plates, what is the magnitude of $\oint \vec{B} \cdot d\vec{s}$, in terms of μ_0 and i, at a radius $r = R/5$ from their center?

KEY IDEA

A magnetic field can be set up by a current and by induction due to a changing electric flux (Eq. 32-5). Between the plates in Fig. 32-5, the current is zero and we can account for the changing electric flux with a fictitious displacement current i_d. Then integral $\oint \vec{B} \cdot d\vec{s}$ is given by Eq. 32-11, but because there is no real current i between the capacitor plates, the equation reduces to

$$\oint \vec{B} \cdot d\vec{s} = \mu_0 i_{d,\text{enc}}. \qquad (32\text{-}18)$$

Calculations: Because we want to evaluate $\oint \vec{B} \cdot d\vec{s}$ at radius $r = R/5$ (within the capacitor), the integration loop encircles only a portion $i_{d,\text{enc}}$ of the total displacement current i_d. Let's assume that i_d is uniformly spread over the full plate area. Then the portion of the displacement current encircled by the loop is proportional to the area encircled by the loop:

$$\frac{\left(\begin{array}{c}\text{encircled displacement}\\\text{current } i_{d,\text{enc}}\end{array}\right)}{\left(\begin{array}{c}\text{total displacement}\\\text{current } i_d\end{array}\right)} = \frac{\text{encircled area } \pi r^2}{\text{full plate area } \pi R^2}.$$

This gives us

$$i_{d,\text{enc}} = i_d \frac{\pi r^2}{\pi R^2}.$$

Substituting this into Eq. 32-18, we obtain

$$\oint \vec{B} \cdot d\vec{s} = \mu_0 i_d \frac{\pi r^2}{\pi R^2}. \qquad (32\text{-}19)$$

Now substituting $i_d = i$ (from Eq. 32-15) and $r = R/5$ into Eq. 32-19 leads to

$$\oint \vec{B} \cdot d\vec{s} = \mu_0 i \frac{(R/5)^2}{R^2} = \frac{\mu_0 i}{25}. \qquad \text{(Answer)}$$

WILEY PLUS Additional examples, video, and practice available at *WileyPLUS*

(b) In terms of the maximum induced magnetic field, what is the magnitude of the magnetic field induced at $r = R/5$, inside the capacitor?

KEY IDEA

Because the capacitor has parallel circular plates, we can treat the space between the plates as an imaginary wire of radius R carrying the imaginary current i_d. Then we can use Eq. 32-16 to find the induced magnetic field magnitude B at any point inside the capacitor.

Calculations: At $r = R/5$, Eq. 32-16 yields

$$B = \left(\frac{\mu_0 i_d}{2\pi R^2}\right)r = \frac{\mu_0 i_d(R/5)}{2\pi R^2} = \frac{\mu_0 i_d}{10\pi R}. \qquad (32\text{-}20)$$

From Eq. 32-16, the maximum field magnitude $B_{\max}$ within the capacitor occurs at $r = R$. It is

$$B_{\max} = \left(\frac{\mu_0 i_d}{2\pi R^2}\right)R = \frac{\mu_0 i_d}{2\pi R}. \qquad (32\text{-}21)$$

Dividing Eq. 32-20 by Eq. 32-21 and rearranging the result, we find that the field magnitude at $r = R/5$ is

$$B = \tfrac{1}{5}B_{\max}. \qquad \text{(Answer)}$$

We should be able to obtain this result with a little reasoning and less work. Equation 32-16 tells us that inside the capacitor, B increases linearly with r. Therefore, a point $\tfrac{1}{5}$ the distance out to the full radius R of the plates, where $B_{\max}$ occurs, should have a field B that is $\tfrac{1}{5}B_{\max}$.

Maxwell's Equations

Equation 32-5 is the last of the four fundamental equations of electromagnetism, called *Maxwell's equations* and displayed in Table 32-1. These four equations

Table 32-1 Maxwell's Equations[a]

Name	Equation	
Gauss' law for electricity	$\oint \vec{E} \cdot d\vec{A} = q_{\text{enc}}/\varepsilon_0$	Relates net electric flux to net enclosed electric charge
Gauss' law for magnetism	$\oint \vec{B} \cdot d\vec{A} = 0$	Relates net magnetic flux to net enclosed magnetic charge
Faraday's law	$\oint \vec{E} \cdot d\vec{s} = -\dfrac{d\Phi_B}{dt}$	Relates induced electric field to changing magnetic flux
Ampere–Maxwell law	$\oint \vec{B} \cdot d\vec{s} = \mu_0\varepsilon_0\dfrac{d\Phi_E}{dt} + \mu_0 i_{\text{enc}}$	Relates induced magnetic field to changing electric flux and to current

[a]Written on the assumption that no dielectric or magnetic materials are present.

explain a diverse range of phenomena, from why a compass needle points north to why a car starts when you turn the ignition key. They are the basis for the functioning of such electromagnetic devices as electric motors, television transmitters and receivers, telephones, scanners, radar, and microwave ovens.

Maxwell's equations are the basis from which many of the equations you have seen since Chapter 21 can be derived. They are also the basis of many of the equations you will see in Chapters 33 through 36 concerning optics.

32-4 MAGNETS

Learning Objectives

After reading this module, you should be able to . . .

32.16 Identify lodestones.

32.17 In Earth's magnetic field, identify that the field is approximately that of a dipole and also identify in

which hemisphere the north geomagnetic pole is located.

32.18 Identify field declination and field inclination.

Key Ideas

● Earth is approximately a magnetic dipole with a dipole axis somewhat off the rotation axis and with the south pole in the Northern Hemisphere.

● The local field direction is given by the field declination (the angle left or right from geographic north) and the field inclination (the angle up or down from the horizontal).

Magnets

The first known magnets were *lodestones*, which are stones that have been *magnetized* (made magnetic) naturally. When the ancient Greeks and ancient Chinese discovered these rare stones, they were amused by the stones' ability to attract metal over a short distance, as if by magic. Only much later did they learn to use lodestones (and artificially magnetized pieces of iron) in compasses to determine direction.

Today, magnets and magnetic materials are ubiquitous. Their magnetic properties can be traced to their atoms and electrons. In fact, the inexpensive magnet you might use to hold a note on the refrigerator door is a direct result of the quantum physics taking place in the atomic and subatomic material within the magnet. Before we explore some of this physics, let's briefly discuss the largest magnet we commonly use—namely, Earth itself.

The Magnetism of Earth

Earth is a huge magnet; for points near Earth's surface, its magnetic field can be approximated as the field of a huge bar magnet—a magnetic dipole—that straddles the center of the planet. Figure 32-8 is an idealized symmetric depiction of the dipole field, without the distortion caused by passing charged particles from the Sun.

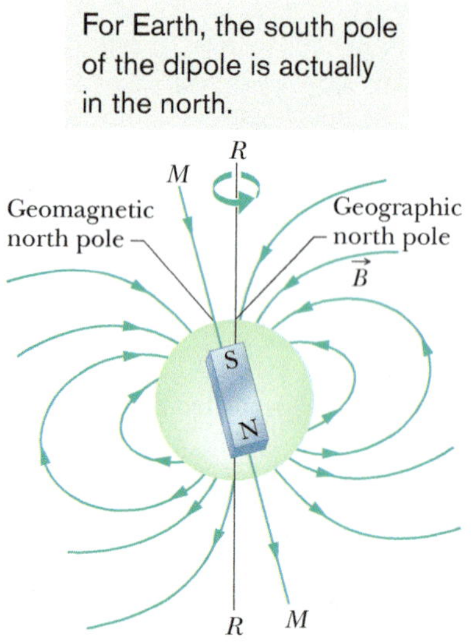

Because Earth's magnetic field is that of a magnetic dipole, a magnetic dipole moment $\vec{\mu}$ is associated with the field. For the idealized field of Fig. 32-8, the magnitude of $\vec{\mu}$ is 8.0×10^{22} J/T and the direction of $\vec{\mu}$ makes an angle of $11.5°$ with the rotation axis (RR) of Earth. The *dipole axis* (MM in Fig. 32-8) lies along $\vec{\mu}$ and intersects Earth's surface at the *geomagnetic north pole* off the northwest coast of Greenland and the *geomagnetic south pole* in Antarctica. The lines of the magnetic field $\vec{B}$ generally emerge in the Southern Hemisphere and reenter Earth in the Northern Hemisphere. Thus, the magnetic pole that is in Earth's Northern Hemisphere and known as a "north magnetic pole" *is really the south pole of Earth's magnetic dipole.*

For Earth, the south pole of the dipole is actually in the north.

Figure 32-8 Earth's magnetic field represented as a dipole field. The dipole axis MM makes an angle of $11.5°$ with Earth's rotational axis RR. The south pole of the dipole is in Earth's Northern Hemisphere.

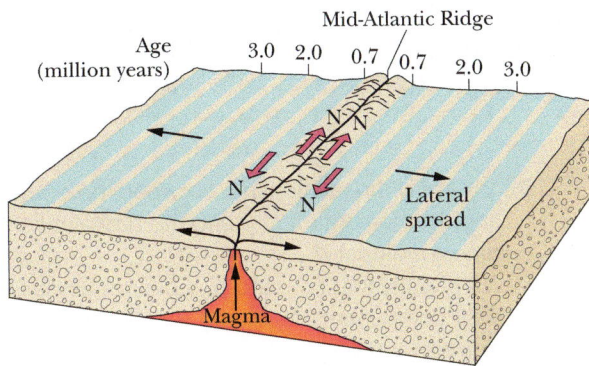

Figure 32-9 A magnetic profile of the seafloor on either side of the Mid-Atlantic Ridge. The seafloor, extruded through the ridge and spreading out as part of the tectonic drift system, displays a record of the past magnetic history of Earth's core. The direction of the magnetic field produced by the core reverses about every million years.

The direction of the magnetic field at any location on Earth's surface is commonly specified in terms of two angles. The **field declination** is the angle (left or right) between geographic north (which is toward 90° latitude) and the horizontal component of the field. The **field inclination** is the angle (up or down) between a horizontal plane and the field's direction.

Measurement. *Magnetometers* measure these angles and determine the field with much precision. However, you can do reasonably well with just a *compass* and a *dip meter.* A compass is simply a needle-shaped magnet that is mounted so it can rotate freely about a vertical axis. When it is held in a horizontal plane, the north-pole end of the needle points, generally, toward the geomagnetic north pole (really a south magnetic pole, remember). The angle between the needle and geographic north is the field declination. A dip meter is a similar magnet that can rotate freely about a horizontal axis. When its vertical plane of rotation is aligned with the direction of the compass, the angle between the meter's needle and the horizontal is the field inclination.

At any point on Earth's surface, the measured magnetic field may differ appreciably, in both magnitude and direction, from the idealized dipole field of Fig. 32-8. In fact, the point where the field is actually perpendicular to Earth's surface and inward is not located at the geomagnetic north pole off Greenland as we would expect; instead, this so-called *dip north pole* is located in the Queen Elizabeth Islands in northern Canada, far from Greenland.

In addition, the field observed at any location on the surface of Earth varies with time, by measurable amounts over a period of a few years and by substantial amounts over, say, 100 years. For example, between 1580 and 1820 the direction indicated by compass needles in London changed by 35°.

In spite of these local variations, the average dipole field changes only slowly over such relatively short time periods. Variations over longer periods can be studied by measuring the weak magnetism of the ocean floor on either side of the Mid-Atlantic Ridge (Fig. 32-9). This floor has been formed by molten magma that oozed up through the ridge from Earth's interior, solidified, and was pulled away from the ridge (by the drift of tectonic plates) at the rate of a few centimeters per year. As the magma solidified, it became weakly magnetized with its magnetic field in the direction of Earth's magnetic field at the time of solidification. Study of this solidified magma across the ocean floor reveals that Earth's field has reversed its *polarity* (directions of the north pole and south pole) about every million years. Theories explaining the reversals are still in preliminary stages. In fact, the mechanism that produces Earth's magnetic field is only vaguely understood.

32-5 MAGNETISM AND ELECTRONS

Learning Objectives

After reading this module, you should be able to . . .

32.19 Identify that a spin angular momentum $\vec{S}$ (usually simply called spin) and a spin magnetic dipole moment $\vec{\mu}_s$ are intrinsic properties of electrons (and also protons and neutrons).

32.20 Apply the relationship between the spin vector $\vec{S}$ and the spin magnetic dipole moment vector $\vec{\mu}_s$.

32.21 Identify that $\vec{S}$ and $\vec{\mu}_s$ cannot be observed (measured); only their components on an axis of measurement (usually called the z axis) can be observed.

32.22 Identify that the observed components S_z and $\mu_{s,z}$ are quantized and explain what that means.

32.23 Apply the relationship between the component S_z and the spin magnetic quantum number m_s, specifying the allowed values of m_s.

32.24 Distinguish spin up from spin down for the spin orientation of an electron.

32.25 Determine the z components $\mu_{s,z}$ of the spin magnetic dipole moment, both as a value and in terms of the Bohr magneton μ_B.

32.26 If an electron is in an external magnetic field, determine the orientation energy U of its spin magnetic dipole moment $\vec{\mu}_s$.

32.27 Identify that an electron in an atom has an orbital angular momentum $\vec{L}_{orb}$ and an orbital magnetic dipole moment $\vec{\mu}_{orb}$.

32.28 Apply the relationship between the orbital angular momentum $\vec{L}_{orb}$ and the orbital magnetic dipole moment $\vec{\mu}_{orb}$.

32.29 Identity that $\vec{L}_{orb}$ and $\vec{\mu}_{orb}$ cannot be observed but their components $L_{orb,z}$ and $\mu_{orb,z}$ on a z (measurement) axis can.

32.30 Apply the relationship between the component $L_{orb,z}$ of the orbital angular momentum and the orbital magnetic quantum number m_ℓ, specifying the allowed values of m_ℓ.

32.31 Determine the z components $\mu_{orb,z}$ of the orbital magnetic dipole moment, both as a value and in terms of the Bohr magneton μ_B.

32.32 If an atom is in an external magnetic field, determine the orientation energy U of the orbital magnetic dipole moment $\vec{\mu}_{orb}$.

32.33 Calculate the magnitude of the magnetic moment of a charged particle moving in a circle or a ring of uniform charge rotating like a merry-go-round at a constant angular speed around a central axis.

32.34 Explain the classical loop model for an orbiting electron and the forces on such a loop in a nonuniform magnetic field.

32.35 Distinguish diamagnetism, paramagnetism, and ferromagnetism.

Key Ideas

● An electron has an intrinsic angular momentum called *spin angular momentum* (or *spin*) $\vec{S}$, with which an intrinsic *spin magnetic dipole moment* $\vec{\mu}_s$ is associated:

$$\vec{\mu}_s = -\frac{e}{m}\,\vec{S}.$$

● For a measurement along a z axis, the component S_z can have only the values given by

$$S_z = m_s\frac{h}{2\pi}, \qquad \text{for } m_s = \pm\tfrac{1}{2},$$

where $h\,(= 6.63 \times 10^{-34}\text{ J·s})$ is the Planck constant.

● Similarly,

$$\mu_{s,z} = \pm\frac{eh}{4\pi m} = \pm\mu_B,$$

where μ_B is the Bohr magneton:

$$\mu_B = \frac{eh}{4\pi m} = 9.27 \times 10^{-24}\text{ J/T}.$$

● The energy U associated with the orientation of the spin magnetic dipole moment in an external magnetic field $\vec{B}_{ext}$ is

$$U = -\vec{\mu}_s \cdot \vec{B}_{ext} = -\mu_{s,z}B_{ext}.$$

● An electron in an atom has an additional angular momentum called its orbital angular momentum $\vec{L}_{orb}$, with which an orbital magnetic dipole moment $\vec{\mu}_{orb}$ is associated:

$$\vec{\mu}_{orb} = -\frac{e}{2m}\,\vec{L}_{orb}.$$

● Orbital angular momentum is quantized and can have only measured values given by

$$L_{orb,z} = m_\ell\frac{h}{2\pi},$$

$$\text{for } m_\ell = 0,\ \pm1,\ \pm2,\ \dots,\ \pm\,(\text{limit}).$$

● The associated magnetic dipole moment is given by

$$\mu_{orb,z} = -m_\ell\frac{eh}{4\pi m} = -m_\ell\mu_B.$$

● The energy U associated with the orientation of the orbital magnetic dipole moment in an external magnetic field $\vec{B}_{ext}$ is

$$U = -\vec{\mu}_{orb} \cdot \vec{B}_{ext} = -\mu_{orb,z}B_{ext}.$$

Magnetism and Electrons

Magnetic materials, from lodestones to tattoos, are magnetic because of the electrons within them. We have already seen one way in which electrons can generate a magnetic field: Send them through a wire as an electric current, and their motion produces a magnetic field around the wire. There are two more ways, each involving a magnetic dipole moment that produces a magnetic field in the surrounding space. However, their explanation requires quantum physics that is beyond the physics presented in this book, and so here we shall only outline the results.

Spin Magnetic Dipole Moment

An electron has an intrinsic angular momentum called its **spin angular momentum** (or just **spin**) $\vec{S}$; associated with this spin is an intrinsic **spin magnetic dipole moment** $\vec{\mu}_s$. (By *intrinsic*, we mean that $\vec{S}$ and $\vec{\mu}_s$ are basic characteristics of an electron, like its mass and electric charge.) Vectors $\vec{S}$ and $\vec{\mu}_s$ are related by

$$\vec{\mu}_s = -\frac{e}{m}\vec{S}, \tag{32-22}$$

in which e is the elementary charge (1.60×10^{-19} C) and m is the mass of an electron (9.11×10^{-31} kg). The minus sign means that $\vec{\mu}_s$ and $\vec{S}$ are oppositely directed.

Spin $\vec{S}$ is different from the angular momenta of Chapter 11 in two respects:

1. Spin $\vec{S}$ itself cannot be measured. However, its component along any axis can be measured.

2. A measured component of $\vec{S}$ is *quantized*, which is a general term that means it is restricted to certain values. A measured component of $\vec{S}$ can have only two values, which differ only in sign.

Let us assume that the component of spin $\vec{S}$ is measured along the z axis of a coordinate system. Then the measured component S_z can have only the two values given by

$$S_z = m_s \frac{h}{2\pi}, \quad \text{for } m_s = \pm\tfrac{1}{2}, \tag{32-23}$$

where m_s is called the *spin magnetic quantum number* and h ($= 6.63 \times 10^{-34}$ J·s) is the Planck constant, the ubiquitous constant of quantum physics. The signs given in Eq. 32-23 have to do with the direction of S_z along the z axis. When S_z is parallel to the z axis, m_s is $+\tfrac{1}{2}$ and the electron is said to be *spin up*. When S_z is antiparallel to the z axis, m_s is $-\tfrac{1}{2}$ and the electron is said to be *spin down*.

The spin magnetic dipole moment $\vec{\mu}_s$ of an electron also cannot be measured; only its component along any axis can be measured, and that component too is quantized, with two possible values of the same magnitude but different signs. We can relate the component $\mu_{s,z}$ measured on the z axis to S_z by rewriting Eq. 32-22 in component form for the z axis as

$$\mu_{s,z} = -\frac{e}{m}S_z.$$

Substituting for S_z from Eq. 32-23 then gives us

$$\mu_{s,z} = \pm\frac{eh}{4\pi m}, \tag{32-24}$$

where the plus and minus signs correspond to $\mu_{s,z}$ being parallel and antiparallel to the z axis, respectively. The quantity on the right is the *Bohr magneton* μ_B:

$$\mu_\mathrm{B} = \frac{eh}{4\pi m} = 9.27 \times 10^{-24} \text{ J/T} \quad \text{(Bohr magneton).} \tag{32-25}$$

For an electron, the spin is opposite the magnetic dipole moment.

Figure 32-10 The spin $\vec{S}$, spin magnetic dipole moment $\vec{\mu}_s$, and magnetic dipole field $\vec{B}$ of an electron represented as a microscopic sphere.

Spin magnetic dipole moments of electrons and other elementary particles can be expressed in terms of μ_B. For an electron, the magnitude of the measured z component of $\vec{\mu}_s$ is

$$|\mu_{s,z}| = 1\mu_B. \tag{32-26}$$

(The quantum physics of the electron, called *quantum electrodynamics*, or QED, reveals that $\mu_{s,z}$ is actually slightly greater than $1\mu_B$, but we shall neglect that fact.)

Energy. When an electron is placed in an external magnetic field $\vec{B}_{ext}$, an energy U can be associated with the orientation of the electron's spin magnetic dipole moment $\vec{\mu}_s$ just as an energy can be associated with the orientation of the magnetic dipole moment $\vec{\mu}$ of a current loop placed in $\vec{B}_{ext}$. From Eq. 28-38, the orientation energy for the electron is

$$U = -\vec{\mu}_s \cdot \vec{B}_{ext} = -\mu_{s,z} B_{ext}, \tag{32-27}$$

where the z axis is taken to be in the direction of $\vec{B}_{ext}$.

If we imagine an electron to be a microscopic sphere (which it is not), we can represent the spin $\vec{S}$, the spin magnetic dipole moment $\vec{\mu}_s$, and the associated magnetic dipole field as in Fig. 32-10. Although we use the word "spin" here, electrons do not spin like tops. How, then, can something have angular momentum without actually rotating? Again, we would need quantum physics to provide the answer.

Protons and neutrons also have an intrinsic angular momentum called spin and an associated intrinsic spin magnetic dipole moment. For a proton those two vectors have the same direction, and for a neutron they have opposite directions. We shall not examine the contributions of these dipole moments to the magnetic fields of atoms because they are about a thousand times smaller than that due to an electron.

✓ Checkpoint 4

The figure here shows the spin orientations of two particles in an external magnetic field $\vec{B}_{ext}$. (a) If the particles are electrons, which spin orientation is at lower energy? (b) If, instead, the particles are protons, which spin orientation is at lower energy?

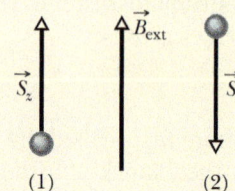

(1)　　　(2)

Orbital Magnetic Dipole Moment

When it is in an atom, an electron has an additional angular momentum called its **orbital angular momentum** $\vec{L}_{orb}$. Associated with $\vec{L}_{orb}$ is an **orbital magnetic dipole moment** $\vec{\mu}_{orb}$; the two are related by

$$\vec{\mu}_{orb} = -\frac{e}{2m} \vec{L}_{orb}. \tag{32-28}$$

The minus sign means that $\vec{\mu}_{orb}$ and $\vec{L}_{orb}$ have opposite directions.

Orbital angular momentum $\vec{L}_{orb}$ cannot be measured; only its component along any axis can be measured, and that component is quantized. The component along, say, a z axis can have only the values given by

$$L_{orb,z} = m_\ell \frac{h}{2\pi}, \qquad \text{for } m_\ell = 0, \pm 1, \pm 2, \ldots, \pm (\text{limit}), \tag{32-29}$$

in which m_ℓ is called the *orbital magnetic quantum number* and "limit" refers to some largest allowed integer value for m_ℓ. The signs in Eq. 32-29 have to do with the direction of $L_{orb,z}$ along the z axis.

The orbital magnetic dipole moment $\vec{\mu}_{orb}$ of an electron also cannot itself be measured; only its component along an axis can be measured, and that component is quantized. By writing Eq. 32-28 for a component along the same z axis as above and then substituting for $L_{orb,z}$ from Eq. 32-29, we can write the z component $\mu_{orb,z}$ of the orbital magnetic dipole moment as

$$\mu_{orb,z} = -m_\ell \frac{eh}{4\pi m} \qquad (32\text{-}30)$$

and, in terms of the Bohr magneton, as

$$\mu_{orb,z} = -m_\ell \mu_B. \qquad (32\text{-}31)$$

When an atom is placed in an external magnetic field $\vec{B}_{ext}$, an energy U can be associated with the orientation of the orbital magnetic dipole moment of each electron in the atom. Its value is

$$U = -\vec{\mu}_{orb} \cdot \vec{B}_{ext} = -\mu_{orb,z}B_{ext}, \qquad (32\text{-}32)$$

where the z axis is taken in the direction of $\vec{B}_{ext}$.

Although we have used the words "orbit" and "orbital" here, electrons do not orbit the nucleus of an atom like planets orbiting the Sun. How can an electron have an orbital angular momentum without orbiting in the common meaning of the term? Once again, this can be explained only with quantum physics.

Loop Model for Electron Orbits

We can obtain Eq. 32-28 with the nonquantum derivation that follows, in which we assume that an electron moves along a circular path with a radius that is much larger than an atomic radius (hence the name "loop model"). However, the derivation does not apply to an electron within an atom (for which we need quantum physics).

We imagine an electron moving at constant speed v in a circular path of radius r, counterclockwise as shown in Fig. 32-11. The motion of the negative charge of the electron is equivalent to a conventional current i (of positive charge) that is clockwise, as also shown in Fig. 32-11. The magnitude of the orbital magnetic dipole moment of such a *current loop* is obtained from Eq. 28-35 with $N = 1$:

$$\mu_{orb} = iA, \qquad (32\text{-}33)$$

where A is the area enclosed by the loop. The direction of this magnetic dipole moment is, from the right-hand rule of Fig. 29-21, downward in Fig. 32-11.

To evaluate Eq. 32-33, we need the current i. Current is, generally, the rate at which charge passes some point in a circuit. Here, the charge of magnitude e takes a time $T = 2\pi r/v$ to circle from any point back through that point, so

$$i = \frac{\text{charge}}{\text{time}} = \frac{e}{2\pi r/v}. \qquad (32\text{-}34)$$

Substituting this and the area $A = \pi r^2$ of the loop into Eq. 32-33 gives us

$$\mu_{orb} = \frac{e}{2\pi r/v}\pi r^2 = \frac{evr}{2}. \qquad (32\text{-}35)$$

To find the electron's orbital angular momentum $\vec{L}_{orb}$, we use Eq. 11-18, $\vec{\ell} = m(\vec{r} \times \vec{v})$. Because $\vec{r}$ and $\vec{v}$ are perpendicular, $\vec{L}_{orb}$ has the magnitude

$$L_{orb} = mrv \sin 90° = mrv. \qquad (32\text{-}36)$$

The vector $\vec{L}_{orb}$ is directed upward in Fig. 32-11 (see Fig. 11-12). Combining

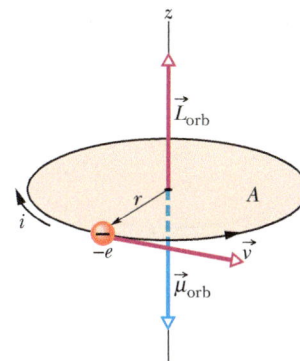

Figure 32-11 An electron moving at constant speed v in a circular path of radius r that encloses an area A. The electron has an orbital angular momentum $\vec{L}_{orb}$ and an associated orbital magnetic dipole moment $\vec{\mu}_{orb}$. A clockwise current i (of positive charge) is equivalent to the counterclockwise circulation of the negatively charged electron.

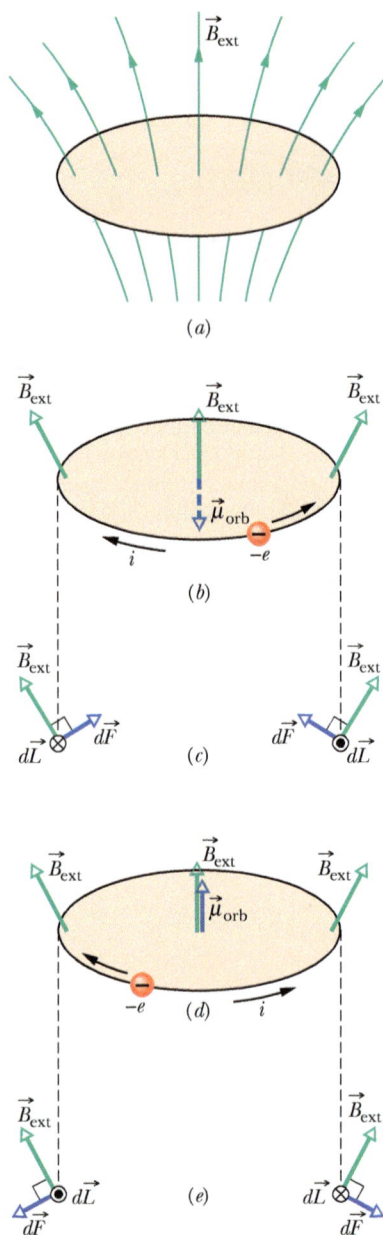

Figure 32-12 (a) A loop model for an electron orbiting in an atom while in a nonuniform magnetic field $\vec{B}_{ext}$. (b) Charge $-e$ moves counterclockwise; the associated conventional current i is clockwise. (c) The magnetic forces $d\vec{F}$ on the left and right sides of the loop, as seen from the plane of the loop. The net force on the loop is upward. (d) Charge $-e$ now moves clockwise. (e) The net force on the loop is now downward.

Eqs. 32-35 and 32-36, generalizing to a vector formulation, and indicating the opposite directions of the vectors with a minus sign yield

$$\vec{\mu}_{orb} = -\frac{e}{2m}\vec{L}_{orb},$$

which is Eq. 32-28. Thus, by "classical" (nonquantum) analysis we have obtained the same result, in both magnitude and direction, given by quantum physics. You might wonder, seeing as this derivation gives the correct result for an electron within an atom, why the derivation is invalid for that situation. The answer is that this line of reasoning yields other results that are contradicted by experiments.

Loop Model in a Nonuniform Field

We continue to consider an electron orbit as a current loop, as we did in Fig. 32-11. Now, however, we draw the loop in a nonuniform magnetic field $\vec{B}_{ext}$ as shown in Fig. 32-12a. (This field could be the diverging field near the north pole of the magnet in Fig. 32-4.) We make this change to prepare for the next several modules, in which we shall discuss the forces that act on magnetic materials when the materials are placed in a nonuniform magnetic field. We shall discuss these forces by assuming that the electron orbits in the materials are tiny current loops like that in Fig. 32-12a.

Here we assume that the magnetic field vectors all around the electron's circular path have the same magnitude and form the same angle with the vertical, as shown in Figs. 32-12b and d. We also assume that all the electrons in an atom move either counterclockwise (Fig. 32-12b) or clockwise (Fig. 32-12d). The associated conventional current i around the current loop and the orbital magnetic dipole moment $\vec{\mu}_{orb}$ produced by i are shown for each direction of motion.

Figures 32-12c and e show diametrically opposite views of a length element $d\vec{L}$ of the loop that has the same direction as i, as seen from the plane of the orbit. Also shown are the field $\vec{B}_{ext}$ and the resulting magnetic force $d\vec{F}$ on $d\vec{L}$. Recall that a current along an element $d\vec{L}$ in a magnetic field $\vec{B}_{ext}$ experiences a magnetic force $d\vec{F}$ as given by Eq. 28-28:

$$d\vec{F} = i\, d\vec{L} \times \vec{B}_{ext}. \qquad (32\text{-}37)$$

On the left side of Fig. 32-12c, Eq. 32-37 tells us that the force $d\vec{F}$ is directed upward and rightward. On the right side, the force $d\vec{F}$ is just as large and is directed upward and leftward. Because their angles are the same, the horizontal components of these two forces cancel and the vertical components add. The same is true at any other two symmetric points on the loop. Thus, the net force on the current loop of Fig. 32-12b must be upward. The same reasoning leads to a downward net force on the loop in Fig. 32-12d. We shall use these two results shortly when we examine the behavior of magnetic materials in nonuniform magnetic fields.

Magnetic Materials

Each electron in an atom has an orbital magnetic dipole moment and a spin magnetic dipole moment that combine vectorially. The resultant of these two vector quantities combines vectorially with similar resultants for all other electrons in the atom, and the resultant for each atom combines with those for all the other atoms in a sample of a material. If the combination of all these magnetic dipole moments produces a magnetic field, then the material is magnetic. There are three general types of magnetism: diamagnetism, paramagnetism, and ferromagnetism.

1. *Diamagnetism* is exhibited by all common materials but is so feeble that it is masked if the material also exhibits magnetism of either of the other two types. In diamagnetism, weak magnetic dipole moments are produced in the atoms of the material when the material is placed in an external magnetic field $\vec{B}_{ext}$; the combination of all those induced dipole moments gives the material as a whole only a feeble net magnetic field. The dipole moments and thus their net field disappear when $\vec{B}_{ext}$ is removed. The term *diamagnetic material* usually refers to materials that exhibit only diamagnetism.

2. *Paramagnetism* is exhibited by materials containing transition elements, rare earth elements, and actinide elements (see Appendix G). Each atom of such a material has a permanent resultant magnetic dipole moment, but the moments are randomly oriented in the material and the material as a whole lacks a net magnetic field. However, an external magnetic field $\vec{B}_{ext}$ can partially align the atomic magnetic dipole moments to give the material a net magnetic field. The alignment and thus its field disappear when $\vec{B}_{ext}$ is removed. The term *paramagnetic material* usually refers to materials that exhibit primarily paramagnetism.

3. *Ferromagnetism* is a property of iron, nickel, and certain other elements (and of compounds and alloys of these elements). Some of the electrons in these materials have their resultant magnetic dipole moments aligned, which produces regions with strong magnetic dipole moments. An external field $\vec{B}_{ext}$ can then align the magnetic moments of such regions, producing a strong magnetic field for a sample of the material; the field partially persists when $\vec{B}_{ext}$ is removed. We usually use the terms *ferromagnetic material* and *magnetic material* to refer to materials that exhibit primarily ferromagnetism.

The next three modules explore these three types of magnetism.

32-6 DIAMAGNETISM

Learning Objectives

After reading this module, you should be able to . . .

32.36 For a diamagnetic sample placed in an external magnetic field, identify that the field produces a magnetic dipole moment in the sample, and identify the relative orientations of that moment and the field.

32.37 For a diamagnetic sample in a nonuniform magnetic field, describe the force on the sample and the resulting motion.

Key Ideas

● Diamagnetic materials exhibit magnetism only when placed in an external magnetic field; there they form magnetic dipoles directed opposite the external field.

● In a nonuniform field, diamagnetic materials are repelled from the region of greater magnetic field.

Diamagnetism

We cannot yet discuss the quantum physical explanation of diamagnetism, but we can provide a classical explanation with the loop model of Figs. 32-11 and 32-12. To begin, we assume that in an atom of a diamagnetic material each electron can orbit only clockwise as in Fig. 32-12d or counterclockwise as in Fig. 32-12b. To account for the lack of magnetism in the absence of an external magnetic field $\vec{B}_{ext}$, we assume the atom lacks a net magnetic dipole moment. This implies that before $\vec{B}_{ext}$ is applied, the number of electrons orbiting in one direction is the same as that orbiting in the opposite direction, with the result that the net upward magnetic dipole moment of the atom equals the net downward magnetic dipole moment.

Courtesy A.K. Geim, University of Manchester, UK

Figure 32-13 An overhead view of a frog that is being levitated in a magnetic field produced by current in a vertical solenoid below the frog.

Now let's turn on the nonuniform field $\vec{B}_{ext}$ of Fig. 32-12a, in which $\vec{B}_{ext}$ is directed upward but is diverging (the magnetic field lines are diverging). We could do this by increasing the current through an electromagnet or by moving the north pole of a bar magnet closer to, and below, the orbits. As the magnitude of $\vec{B}_{ext}$ increases from zero to its final maximum, steady-state value, a clockwise electric field is induced around each electron's orbital loop according to Faraday's law and Lenz's law. Let us see how this induced electric field affects the orbiting electrons in Figs. 32-12b and d.

In Fig. 32-12b, the counterclockwise electron is accelerated by the clockwise electric field. Thus, as the magnetic field $\vec{B}_{ext}$ increases to its maximum value, the electron speed increases to a maximum value. This means that the associated conventional current i and the downward magnetic dipole moment $\vec{\mu}$ due to i also *increase*.

In Fig. 32-12d, the clockwise electron is decelerated by the clockwise electric field. Thus, here, the electron speed, the associated current i, and the upward magnetic dipole moment $\vec{\mu}$ due to i all *decrease*. By turning on field $\vec{B}_{ext}$, we have given the atom a *net* magnetic dipole moment that is downward. This would also be so if the magnetic field were uniform.

Force. The nonuniformity of field $\vec{B}_{ext}$ also affects the atom. Because the current i in Fig. 32-12b increases, the upward magnetic forces $d\vec{F}$ in Fig. 32-12c also increase, as does the net upward force on the current loop. Because current i in Fig. 32-12d decreases, the downward magnetic forces $d\vec{F}$ in Fig. 32-12e also decrease, as does the net downward force on the current loop. Thus, by turning on the *nonuniform* field $\vec{B}_{ext}$, we have produced a net force on the atom; moreover, that force is directed *away* from the region of greater magnetic field.

We have argued with fictitious electron orbits (current loops), but we have ended up with exactly what happens to a diamagnetic material: If we apply the magnetic field of Fig. 32-12, the material develops a downward magnetic dipole moment and experiences an upward force. When the field is removed, both the dipole moment and the force disappear. The external field need not be positioned as shown in Fig. 32-12; similar arguments can be made for other orientations of $\vec{B}_{ext}$. In general,

> A diamagnetic material placed in an external magnetic field $\vec{B}_{ext}$ develops a magnetic dipole moment directed opposite $\vec{B}_{ext}$. If the field is nonuniform, the diamagnetic material is repelled *from* a region of greater magnetic field *toward* a region of lesser field.

The frog in Fig. 32-13 is diamagnetic (as is any other animal). When the frog was placed in the diverging magnetic field near the top end of a vertical current-carrying solenoid, every atom in the frog was repelled upward, away from the region of stronger magnetic field at that end of the solenoid. The frog moved upward into weaker and weaker magnetic field until the upward magnetic force balanced the gravitational force on it, and there it hung in midair. The frog is not in discomfort because *every* atom is subject to the same forces and thus there is no force variation within the frog. The sensation is similar to the "weightless" situation of floating in water, which frogs like very much. If we went to the expense of building a much larger solenoid, we could similarly levitate a person in midair due to the person's diamagnetism.

✓ **Checkpoint 5**

The figure shows two diamagnetic spheres located near the south pole of a bar magnet. Are (a) the magnetic forces on the spheres and (b) the magnetic dipole moments of the spheres directed toward or away from the bar magnet? (c) Is the magnetic force on sphere 1 greater than, less than, or equal to that on sphere 2?

32-7 PARAMAGNETISM

Learning Objectives

After reading this module, you should be able to . . .

32.38 For a paramagnetic sample placed in an external magnetic field, identify the relative orientations of the field and the sample's magnetic dipole moment.

32.39 For a paramagnetic sample in a nonuniform magnetic field, describe the force on the sample and the resulting motion.

32.40 Apply the relationship between a sample's magnetization M, its measured magnetic moment, and its volume.

32.41 Apply Curie's law to relate a sample's magnetization M

to its temperature T, its Curie constant C, and the magnitude B of the external field.

32.42 Given a magnetization curve for a paramagnetic sample, relate the extent of the magnetization for a given magnetic field and temperature.

32.43 For a paramagnetic sample at a given temperature and in a given magnetic field, compare the energy associated with the dipole orientations and the thermal motion.

Key Ideas

● Paramagnetic materials have atoms with a permanent magnetic dipole moment but the moments are randomly oriented, with no net moment, unless the material is in an external magnetic field $\vec{B}_{ext}$, where the dipoles tend to align with that field.

● The extent of alignment within a volume V is measured as the magnetization M, given by

$$M = \frac{\text{measured magnetic moment}}{V}.$$

● Complete alignment (saturation) of all N dipoles in the volume gives a maximum value $M_{max} = N\mu/V$.

● At low values of the ratio B_{ext}/T,

$$M = C\frac{B_{ext}}{T} \quad \text{(Curie's law)},$$

where T is the temperature (in kelvins) and C is a material's Curie constant.

● In a nonuniform external field, a paramagnetic material is attracted to the region of greater magnetic field.

Paramagnetism

In paramagnetic materials, the spin and orbital magnetic dipole moments of the electrons in each atom do not cancel but add vectorially to give the atom a net (and permanent) magnetic dipole moment $\vec{\mu}$. In the absence of an external magnetic field, these atomic dipole moments are randomly oriented, and the net magnetic dipole moment of the material is zero. However, if a sample of the material is placed in an external magnetic field $\vec{B}_{ext}$, the magnetic dipole moments tend to line up with the field, which gives the sample a net magnetic dipole moment. This alignment with the external field is the opposite of what we saw with diamagnetic materials.

★ A paramagnetic material placed in an external magnetic field $\vec{B}_{ext}$ develops a magnetic dipole moment in the direction of $\vec{B}_{ext}$. If the field is nonuniform, the paramagnetic material is attracted *toward* a region of greater magnetic field *from* a region of lesser field.

A paramagnetic sample with N atoms would have a magnetic dipole moment of magnitude $N\mu$ if alignment of its atomic dipoles were complete. However, random collisions of atoms due to their thermal agitation transfer energy among the atoms, disrupting their alignment and thus reducing the sample's magnetic dipole moment.

Thermal Agitation. The importance of thermal agitation may be measured by comparing two energies. One, given by Eq. 19-24, is the mean translational kinetic energy $K\,(=\frac{3}{2}kT)$ of an atom at temperature T, where k is the Boltzmann constant $(1.38 \times 10^{-23}$ J/K) and T is in kelvins (not Celsius degrees). The other,

Richard Megna/Fundamental Photographs
Liquid oxygen is suspended between the two pole faces of a magnet because the liquid is paramagnetic and is magnetically attracted to the magnet.

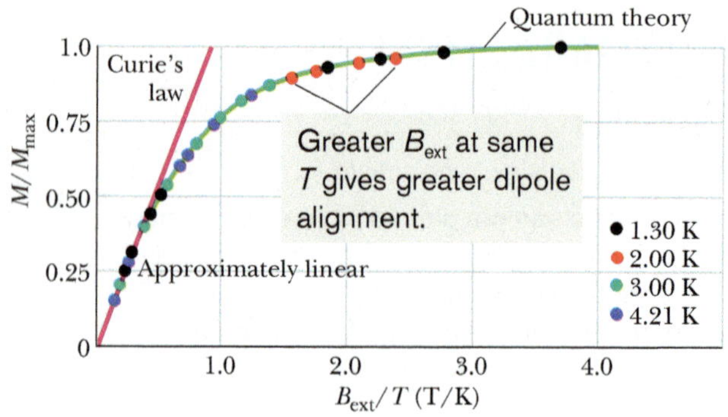

Figure 32-14 A *magnetization curve* for potassium chromium sulfate, a paramagnetic salt. The ratio of magnetization M of the salt to the maximum possible magnetization M_{max} is plotted versus the ratio of the applied magnetic field magnitude B_{ext} to the temperature T. Curie's law fits the data at the left; quantum theory fits all the data. Based on measurements by W. E. Henry.

derived from Eq. 28-38, is the difference in energy ΔU_B ($= 2\mu B_{ext}$) between parallel alignment and antiparallel alignment of the magnetic dipole moment of an atom and the external field. (The lower energy state is $-\mu B_{ext}$ and the higher energy state is $+\mu B_{ext}$.) As we shall show below, $K \gg \Delta U_B$, even for ordinary temperatures and field magnitudes. Thus, energy transfers during collisions among atoms can significantly disrupt the alignment of the atomic dipole moments, keeping the magnetic dipole moment of a sample much less than $N\mu$.

Magnetization. We can express the extent to which a given paramagnetic sample is magnetized by finding the ratio of its magnetic dipole moment to its volume V. This vector quantity, the magnetic dipole moment per unit volume, is the **magnetization** $\vec{M}$ of the sample, and its magnitude is

$$M = \frac{\text{measured magnetic moment}}{V}. \tag{32-38}$$

The unit of $\vec{M}$ is the ampere–square meter per cubic meter, or ampere per meter (A/m). Complete alignment of the atomic dipole moments, called *saturation* of the sample, corresponds to the maximum value $M_{max} = N\mu/V$.

In 1895 Pierre Curie discovered experimentally that the magnetization of a paramagnetic sample is directly proportional to the magnitude of the external magnetic field $\vec{B}_{ext}$ and inversely proportional to the temperature T in kelvins:

$$M = C \frac{B_{ext}}{T}. \tag{32-39}$$

Equation 32-39 is known as *Curie's law*, and C is called the *Curie constant*. Curie's law is reasonable in that increasing B_{ext} tends to align the atomic dipole moments in a sample and thus to increase M, whereas increasing T tends to disrupt the alignment via thermal agitation and thus to decrease M. However, the law is actually an approximation that is valid only when the ratio B_{ext}/T is not too large.

Figure 32-14 shows the ratio M/M_{max} as a function of B_{ext}/T for a sample of the salt potassium chromium sulfate, in which chromium ions are the paramagnetic substance. The plot is called a *magnetization curve*. The straight line for Curie's law fits the experimental data at the left, for B_{ext}/T below about 0.5 T/K. The curve that fits all the data points is based on quantum physics. The data on the right side, near saturation, are very difficult to obtain because they require very strong magnetic fields (about 100 000 times Earth's field), even at very low temperatures.

✓ Checkpoint 6

The figure here shows two paramagnetic spheres located near the south pole of a bar magnet. Are (a) the magnetic forces on the spheres and (b) the magnetic dipole moments of the spheres directed toward or away from the bar magnet? (c) Is the magnetic force on sphere 1 greater than, less than, or equal to that on sphere 2?

Sample Problem 32.03 Orientation energy of a paramagnetic gas in a magnetic field

A paramagnetic gas at room temperature ($T = 300$ K) is placed in an external uniform magnetic field of magnitude $B = 1.5$ T; the atoms of the gas have magnetic dipole moment $\mu = 1.0\mu_B$. Calculate the mean translational kinetic energy K of an atom of the gas and the energy difference ΔU_B between parallel alignment and antiparallel alignment of the atom's magnetic dipole moment with the external field.

KEY IDEAS

(1) The mean translational kinetic energy K of an atom in a gas depends on the temperature of the gas. (2) The energy U_B of a magnetic dipole $\vec{\mu}$ in an external magnetic field $\vec{B}$ depends on the angle θ between the directions of $\vec{\mu}$ and $\vec{B}$.

Calculations: From Eq. 19-24, we have

$$K = \tfrac{3}{2}kT = \tfrac{3}{2}(1.38 \times 10^{-23} \text{ J/K})(300 \text{ K})$$
$$= 6.2 \times 10^{-21} \text{ J} = 0.039 \text{ eV}. \quad \text{(Answer)}$$

From Eq. 28-38 ($U_B = -\vec{\mu} \cdot \vec{B}$), we can write the difference ΔU_B between parallel alignment ($\theta = 0°$) and antiparallel alignment ($\theta = 180°$) as

$$\Delta U_B = -\mu B \cos 180° - (-\mu B \cos 0°) = 2\mu B$$
$$= 2\mu_B B = 2(9.27 \times 10^{-24} \text{ J/T})(1.5 \text{ T})$$
$$= 2.8 \times 10^{-23} \text{ J} = 0.000\,17 \text{ eV}. \quad \text{(Answer)}$$

Here K is about 230 times ΔU_B; so energy exchanges among the atoms during their collisions with one another can easily reorient any magnetic dipole moments that might be aligned with the external magnetic field. That is, as soon as a magnetic dipole moment happens to become aligned with the external field, in the dipole's low energy state, chances are very good that a neighboring atom will hit the atom, transferring enough energy to put the dipole in a higher energy state. Thus, the magnetic dipole moment exhibited by the paramagnetic gas must be due to fleeting partial alignments of the atomic dipole moments.

PLUS Additional examples, video, and practice available at *WileyPLUS*

32-8 FERROMAGNETISM

Learning Objectives

After reading this module, you should be able to . . .

32.44 Identify that ferromagnetism is due to a quantum mechanical interaction called exchange coupling.
32.45 Explain why ferromagnetism disappears when the temperature exceeds the material's Curie temperature.
32.46 Apply the relationship between the magnetization of a ferromagnetic sample and the magnetic moment of its atoms.
32.47 For a ferromagnetic sample at a given temperature and in a given magnetic field, compare the energy associated with the dipole orientations and the thermal motion.
32.48 Describe and sketch a Rowland ring.
32.49 Identify magnetic domains.
32.50 For a ferromagnetic sample placed in an external magnetic field, identify the relative orientations of the field and the magnetic dipole moment.
32.51 Identify the motion of a ferromagnetic sample in a nonuniform field.
32.52 For a ferromagnetic object placed in a uniform magnetic field, calculate the torque and orientation energy.
32.53 Explain hysteresis and a hysteresis loop.
32.54 Identify the origin of lodestones.

Key Ideas

● The magnetic dipole moments in a ferromagnetic material can be aligned by an external magnetic field and then, after the external field is removed, remain partially aligned in regions (domains).

● Alignment is eliminated at temperatures above a material's Curie temperature.
● In a nonuniform external field, a ferromagnetic material is attracted to the region of greater magnetic field.

Ferromagnetism

When we speak of magnetism in everyday conversation, we almost always have a mental picture of a bar magnet or a disk magnet (probably clinging to a refrigerator door). That is, we picture a ferromagnetic material having strong,

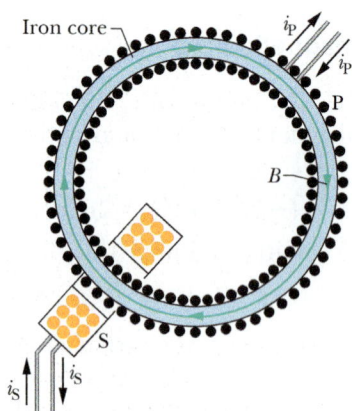

Figure 32-15 A Rowland ring. A primary coil P has a core made of the ferromagnetic material to be studied (here iron). The core is magnetized by a current i_P sent through coil P. (The turns of the coil are represented by dots.) The extent to which the core is magnetized determines the total magnetic field $\vec{B}$ within coil P. Field $\vec{B}$ can be measured by means of a secondary coil S.

permanent magnetism, and not a diamagnetic or paramagnetic material having weak, temporary magnetism.

Iron, cobalt, nickel, gadolinium, dysprosium, and alloys containing these elements exhibit ferromagnetism because of a quantum physical effect called *exchange coupling* in which the electron spins of one atom interact with those of neighboring atoms. The result is alignment of the magnetic dipole moments of the atoms, in spite of the randomizing tendency of atomic collisions due to thermal agitation. This persistent alignment is what gives ferromagnetic materials their permanent magnetism.

Thermal Agitation. If the temperature of a ferromagnetic material is raised above a certain critical value, called the *Curie temperature*, the exchange coupling ceases to be effective. Most such materials then become simply paramagnetic; that is, the dipoles still tend to align with an external field but much more weakly, and thermal agitation can now more easily disrupt the alignment. The Curie temperature for iron is 1043 K ($= 770°C$).

Measurement. The magnetization of a ferromagnetic material such as iron can be studied with an arrangement called a *Rowland ring* (Fig. 32-15). The material is formed into a thin toroidal core of circular cross section. A primary coil P having n turns per unit length is wrapped around the core and carries current i_P. (The coil is essentially a long solenoid bent into a circle.) If the iron core were not present, the magnitude of the magnetic field inside the coil would be, from Eq. 29-23,

$$B_0 = \mu_0 i_P n. \qquad (32\text{-}40)$$

However, with the iron core present, the magnetic field $\vec{B}$ inside the coil is greater than $\vec{B}_0$, usually by a large amount. We can write the magnitude of this field as

$$B = B_0 + B_M, \qquad (32\text{-}41)$$

where B_M is the magnitude of the magnetic field contributed by the iron core. This contribution results from the alignment of the atomic dipole moments within the iron, due to exchange coupling and to the applied magnetic field B_0, and is proportional to the magnetization M of the iron. That is, the contribution B_M is proportional to the magnetic dipole moment per unit volume of the iron. To determine B_M we use a secondary coil S to measure B, compute B_0 with Eq. 32-40, and subtract as suggested by Eq. 32-41.

Figure 32-16 shows a magnetization curve for a ferromagnetic material in a Rowland ring: The ratio $B_M/B_{M,max}$, where $B_{M,max}$ is the maximum possible value of B_M, corresponding to saturation, is plotted versus B_0. The curve is like Fig. 32-14, the magnetization curve for a paramagnetic substance: Both curves show the extent to which an applied magnetic field can align the atomic dipole moments of a material.

For the ferromagnetic core yielding Fig. 32-16, the alignment of the dipole moments is about 70% complete for $B_0 \approx 1 \times 10^{-3}$ T. If B_0 were increased to 1 T, the alignment would be almost complete (but $B_0 = 1$ T, and thus almost complete saturation, is quite difficult to obtain).

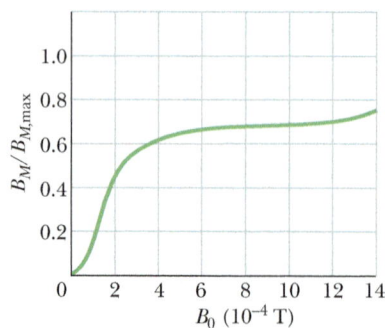

Figure 32-16 A magnetization curve for a ferromagnetic core material in the Rowland ring of Fig. 32-15. On the vertical axis, 1.0 corresponds to complete alignment (saturation) of the atomic dipoles within the material.

Magnetic Domains

Exchange coupling produces strong alignment of adjacent atomic dipoles in a ferromagnetic material at a temperature below the Curie temperature. Why, then, isn't the material naturally at saturation even when there is no applied magnetic field B_0? Why isn't every piece of iron a naturally strong magnet?

To understand this, consider a specimen of a ferromagnetic material such as iron that is in the form of a single crystal; that is, the arrangement of the atoms that make it up—its crystal lattice—extends with unbroken regularity throughout the volume of the specimen. Such a crystal will, in its normal state, be made up of a number of *magnetic domains*. These are regions of the crystal throughout which the alignment of the atomic dipoles is essentially perfect. The domains, however, are not all aligned. For the crystal as a whole, the domains are so oriented that they largely cancel with one another as far as their external magnetic effects are concerned.

Figure 32-17 is a magnified photograph of such an assembly of domains in a single crystal of nickel. It was made by sprinkling a colloidal suspension of finely powdered iron oxide on the surface of the crystal. The domain boundaries, which are thin regions in which the alignment of the elementary dipoles changes from a certain orientation in one of the domains forming the boundary to a different orientation in the other domain, are the sites of intense, but highly localized and nonuniform, magnetic fields. The suspended colloidal particles are attracted to these boundaries and show up as the white lines (not all the domain boundaries are apparent in Fig. 32-17). Although the atomic dipoles in each domain are completely aligned as shown by the arrows, the crystal as a whole may have only a very small resultant magnetic moment.

Actually, a piece of iron as we ordinarily find it is not a single crystal but an assembly of many tiny crystals, randomly arranged; we call it a *polycrystalline solid.* Each tiny crystal, however, has its array of variously oriented domains, just as in Fig. 32-17. If we magnetize such a specimen by placing it in an external magnetic field of gradually increasing strength, we produce two effects; together they produce a magnetization curve of the shape shown in Fig. 32-16. One effect is a growth in size of the domains that are oriented along the external field at the expense of those that are not. The second effect is a shift of the orientation of the dipoles within a domain, as a unit, to become closer to the field direction.

Exchange coupling and domain shifting give us the following result:

> ⭐ A ferromagnetic material placed in an external magnetic field $\vec{B}_{ext}$ develops a strong magnetic dipole moment in the direction of $\vec{B}_{ext}$. If the field is nonuniform, the ferromagnetic material is attracted *toward* a region of greater magnetic field *from* a region of lesser field.

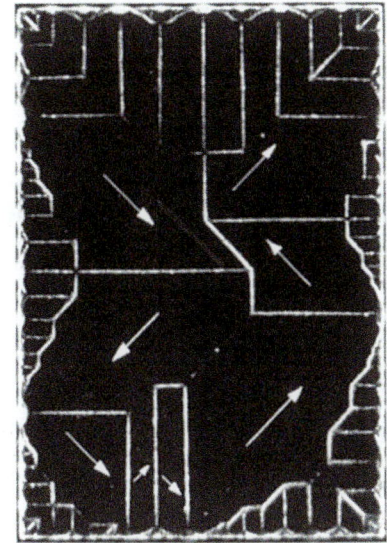

Courtesy Ralph W. DeBlois

Figure 32-17 A photograph of domain patterns within a single crystal of nickel; white lines reveal the boundaries of the domains. The white arrows superimposed on the photograph show the orientations of the magnetic dipoles within the domains and thus the orientations of the net magnetic dipoles of the domains. The crystal as a whole is unmagnetized if the net magnetic field (the vector sum over all the domains) is zero.

Hysteresis

Magnetization curves for ferromagnetic materials are not retraced as we increase and then decrease the external magnetic field B_0. Figure 32-18 is a plot of B_M versus B_0 during the following operations with a Rowland ring: (1) Starting with the iron unmagnetized (point *a*), increase the current in the toroid until B_0 ($= \mu_0 in$) has the value corresponding to point *b*; (2) reduce the current in the toroid winding (and thus B_0) back to zero (point *c*); (3) reverse the toroid current and increase it in magnitude until B_0 has the value corresponding to point *d*; (4) reduce the current to zero again (point *e*); (5) reverse the current once more until point *b* is reached again.

The lack of retraceability shown in Fig. 32-18 is called **hysteresis,** and the curve *bcdeb* is called a *hysteresis loop.* Note that at points *c* and *e* the iron core is magnetized, even though there is no current in the toroid windings; this is the familiar phenomenon of permanent magnetism.

Hysteresis can be understood through the concept of magnetic domains. Evidently the motions of the domain boundaries and the reorientations of the domain directions are not totally reversible. When the applied magnetic field B_0 is increased and then decreased back to its initial value, the domains do not return completely to their original configuration but retain some "memory" of their alignment after the initial increase. This memory of magnetic materials is essential for the magnetic storage of information.

This memory of the alignment of domains can also occur naturally. When lightning sends currents along multiple tortuous paths through the ground, the currents produce intense magnetic fields that can suddenly magnetize any

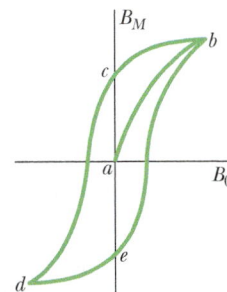

Figure 32-18 A magnetization curve (*ab*) for a ferromagnetic specimen and an associated hysteresis loop (*bcdeb*).

ferromagnetic material in nearby rock. Because of hysteresis, such rock material retains some of that magnetization after the lightning strike (after the currents disappear). Pieces of the rock—later exposed, broken, and loosened by weathering—are then lodestones.

Sample Problem 32.04   Magnetic dipole moment of a compass needle

A compass needle made of pure iron (density 7900 kg/m³) has a length L of 3.0 cm, a width of 1.0 mm, and a thickness of 0.50 mm. The magnitude of the magnetic dipole moment of an iron atom is $\mu_{\text{Fe}} = 2.1 \times 10^{-23}$ J/T. If the magnetization of the needle is equivalent to the alignment of 10% of the atoms in the needle, what is the magnitude of the needle's magnetic dipole moment $\vec{\mu}$?

KEY IDEAS

(1) Alignment of all N atoms in the needle would give a magnitude of $N\mu_{\text{Fe}}$ for the needle's magnetic dipole moment $\vec{\mu}$. However, the needle has only 10% alignment (the random orientation of the rest does not give any net contribution to $\vec{\mu}$). Thus,

$$\mu = 0.10 N \mu_{\text{Fe}}. \tag{32-42}$$

(2) We can find the number of atoms N in the needle from the needle's mass:

$$N = \frac{\text{needle's mass}}{\text{iron's atomic mass}}. \tag{32-43}$$

Finding N: Iron's atomic mass is not listed in Appendix F, but its molar mass M is. Thus, we write

$$\text{iron's atomic mass} = \frac{\text{iron's molar mass } M}{\text{Avogadro's number } N_{\text{A}}}. \tag{32-44}$$

Next, we can rewrite Eq. 32-43 in terms of the needle's mass m, the molar mass M, and Avogadro's number N_{A}:

$$N = \frac{m N_{\text{A}}}{M}. \tag{32-45}$$

The needle's mass m is the product of its density and its volume. The volume works out to be 1.5×10^{-8} m³; so

$$\begin{aligned} \text{needle's mass } m &= (\text{needle's density})(\text{needle's volume}) \\ &= (7900 \text{ kg/m}^3)(1.5 \times 10^{-8} \text{ m}^3) \\ &= 1.185 \times 10^{-4} \text{ kg}. \end{aligned}$$

Substituting into Eq. 32-45 with this value for m, and also 55.847 g/mol ($= 0.055\,847$ kg/mol) for M and 6.02×10^{23} for N_{A}, we find

$$N = \frac{(1.185 \times 10^{-4} \text{ kg})(6.02 \times 10^{23})}{0.055\,847 \text{ kg/mol}}$$

$$= 1.2774 \times 10^{21}.$$

Finding μ: Substituting our value of N and the value of μ_{Fe} into Eq. 32-42 then yields

$$\mu = (0.10)(1.2774 \times 10^{21})(2.1 \times 10^{-23} \text{ J/T})$$

$$= 2.682 \times 10^{-3} \text{ J/T} \approx 2.7 \times 10^{-3} \text{ J/T}. \quad \text{(Answer)}$$

WILEY PLUS Additional examples, video, and practice available at *WileyPLUS*

Review & Summary

Gauss' Law for Magnetic Fields The simplest magnetic structures are magnetic dipoles. Magnetic monopoles do not exist (as far as we know). **Gauss' law** for magnetic fields,

$$\Phi_B = \oint \vec{B} \cdot d\vec{A} = 0, \tag{32-1}$$

states that the net magnetic flux through any (closed) Gaussian surface is zero. It implies that magnetic monopoles do not exist.

Maxwell's Extension of Ampere's Law A changing electric flux induces a magnetic field $\vec{B}$. Maxwell's law,

$$\oint \vec{B} \cdot d\vec{s} = \mu_0 \varepsilon_0 \frac{d\Phi_E}{dt} \quad \text{(Maxwell's law of induction)}, \tag{32-3}$$

relates the magnetic field induced along a closed loop to the changing electric flux Φ_E through the loop. Ampere's law,

$\oint \vec{B} \cdot d\vec{s} = \mu_0 i_{\text{enc}}$ (Eq. 32-4), gives the magnetic field generated by a current i_{enc} encircled by a closed loop. Maxwell's law and Ampere's law can be written as the single equation

$$\oint \vec{B} \cdot d\vec{s} = \mu_0 \varepsilon_0 \frac{d\Phi_E}{dt} + \mu_0 i_{\text{enc}} \quad \text{(Ampere–Maxwell law)}. \tag{32-5}$$

Displacement Current We define the fictitious *displacement current* due to a changing electric field as

$$i_d = \varepsilon_0 \frac{d\Phi_E}{dt}. \tag{32-10}$$

Equation 32-5 then becomes

$$\oint \vec{B} \cdot d\vec{s} = \mu_0 i_{d,\text{enc}} + \mu_0 i_{\text{enc}} \quad \text{(Ampere–Maxwell law)}, \tag{32-11}$$

where $i_{d,\text{enc}}$ is the displacement current encircled by the integration loop. The idea of a displacement current allows us to retain the notion of continuity of current through a capacitor. However, displacement current is *not* a transfer of charge.

Maxwell's Equations Maxwell's equations, displayed in Table 32-1, summarize electromagnetism and form its foundation, including optics.

Earth's Magnetic Field Earth's magnetic field can be approximated as being that of a magnetic dipole whose dipole moment makes an angle of 11.5° with Earth's rotation axis, and with the south pole of the dipole in the Northern Hemisphere. The direction of the local magnetic field at any point on Earth's surface is given by the *field declination* (the angle left or right from geographic north) and the *field inclination* (the angle up or down from the horizontal).

Spin Magnetic Dipole Moment An electron has an intrinsic angular momentum called *spin angular momentum* (or *spin*) $\vec{S}$, with which an intrinsic *spin magnetic dipole moment* $\vec{\mu}_s$ is associated:

$$\vec{\mu}_s = -\frac{e}{m}\,\vec{S}. \tag{32-22}$$

For a measurement along a z axis, the component S_z can have only the values given by

$$S_z = m_s \frac{h}{2\pi}, \qquad \text{for } m_s = \pm\tfrac{1}{2}, \tag{32-23}$$

where h ($= 6.63 \times 10^{-34}$ J·s) is the Planck constant. Similarly,

$$\mu_{s,z} = \pm\frac{eh}{4\pi m} = \pm\mu_\mathrm{B}, \tag{32-24, 32-26}$$

where μ_B is the *Bohr magneton*:

$$\mu_\mathrm{B} = \frac{eh}{4\pi m} = 9.27 \times 10^{-24} \text{ J/T}. \tag{32-25}$$

The energy U associated with the orientation of the spin magnetic dipole moment in an external magnetic field $\vec{B}_{\text{ext}}$ is

$$U = -\vec{\mu}_s \cdot \vec{B}_{\text{ext}} = -\mu_{s,z}B_{\text{ext}}. \tag{32-27}$$

Orbital Magnetic Dipole Moment An electron in an atom has an additional angular momentum called its *orbital angular momentum* $\vec{L}_{\text{orb}}$, with which an *orbital magnetic dipole moment* $\vec{\mu}_{\text{orb}}$ is associated:

$$\vec{\mu}_{\text{orb}} = -\frac{e}{2m}\,\vec{L}_{\text{orb}}. \tag{32-28}$$

Orbital angular momentum is quantized and can have only measured values given by

$$L_{\text{orb},z} = m_\ell \frac{h}{2\pi},$$

$$\text{for } m_\ell = 0, \pm1, \pm2, \ldots, \pm\text{(limit)}. \tag{32-29}$$

The associated magnetic dipole moment is given by

$$\mu_{\text{orb},z} = -m_\ell \frac{eh}{4\pi m} = -m_\ell \mu_\mathrm{B}. \tag{32-30, 32-31}$$

The energy U associated with the orientation of the orbital magnetic dipole moment in an external magnetic field $\vec{B}_{\text{ext}}$ is

$$U = -\vec{\mu}_{\text{orb}} \cdot \vec{B}_{\text{ext}} = -\mu_{\text{orb},z}B_{\text{ext}}. \tag{32-32}$$

Diamagnetism *Diamagnetic materials* exhibit magnetism only when placed in an external magnetic field; there they form magnetic dipoles directed opposite the external field. In a nonuniform field, they are repelled from the region of greater magnetic field.

Paramagnetism *Paramagnetic materials* have atoms with a permanent magnetic dipole moment but the moments are randomly oriented unless the material is in an external magnetic field $\vec{B}_{\text{ext}}$, where the dipoles tend to align with the external field. The extent of alignment within a volume V is measured as the *magnetization M*, given by

$$M = \frac{\text{measured magnetic moment}}{V}. \tag{32-38}$$

Complete alignment (*saturation*) of all N dipoles in the volume gives a maximum value $M_{\max} = N\mu/V$. At low values of the ratio B_{ext}/T,

$$M = C\frac{B_{\text{ext}}}{T} \qquad \text{(Curie's law)}, \tag{32-39}$$

where T is the temperature (kelvins) and C is a material's *Curie constant*.

In a nonuniform external field, a paramagnetic material is attracted to the region of greater magnetic field.

Ferromagnetism The magnetic dipole moments in a *ferromagnetic material* can be aligned by an external magnetic field and then, after the external field is removed, remain partially aligned in regions (*domains*). Alignment is eliminated at temperatures above a material's *Curie temperature*. In a nonuniform external field, a ferromagnetic material is attracted to the region of greater magnetic field.

Questions

1 Figure 32-19a shows a capacitor, with circular plates, that is being charged. Point a (near one of the connecting wires) and point b (inside the capacitor gap) are equidistant from the central axis, as are point c (not so near the wire) and point d (between the plates but outside the gap). In Fig. 32-19b, one curve gives the variation with distance r of the magnitude of the magnetic field inside and outside the wire. The other curve gives the variation with distance r of the magnitude of the magnetic field inside and outside the gap. The two curves partially overlap. Which of the three points on the curves correspond to which of the four points of Fig. 32-19a?

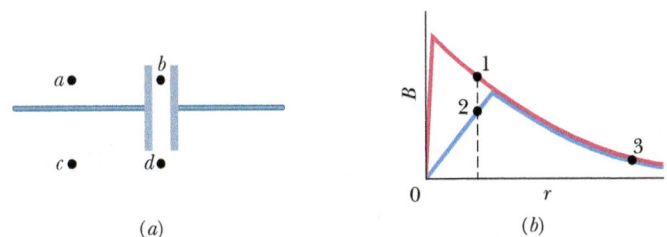

Figure 32-19 Question 1.

2 Figure 32-20 shows a parallel-plate capacitor and the current in the connecting wires that is discharging the capacitor. Are the directions of (a) electric field $\vec{E}$ and (b) displacement current i_d leftward or rightward between the plates? (c) Is the magnetic field at point P into or out of the page?

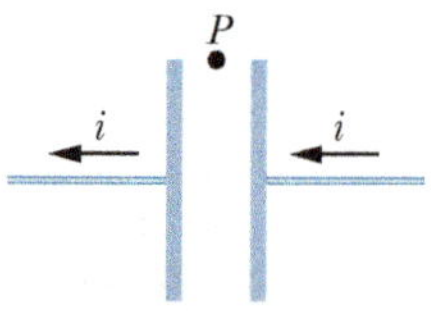

Figure 32-20 Question 2.

3 Figure 32-21 shows, in two situations, an electric field vector $\vec{E}$ and an induced magnetic field line. In each, is the magnitude of $\vec{E}$ increasing or decreasing?

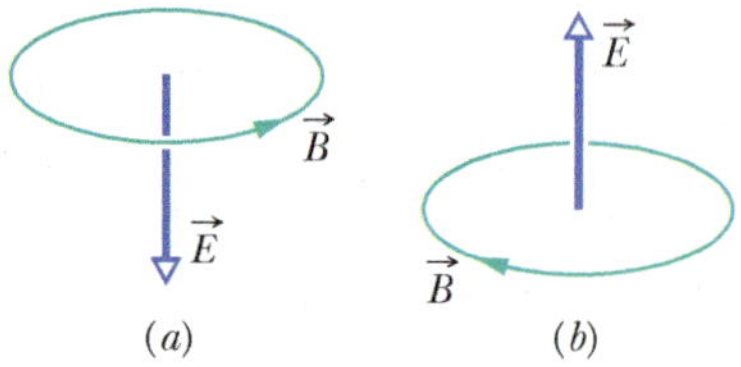

Figure 32-21 Question 3.

4 Figure 32-22a shows a pair of opposite spin orientations for an electron in an external magnetic field $\vec{B}_{ext}$. Figure 32-22b gives three choices for the graph of the energies associated with those orientations as a function of the magnitude of $\vec{B}_{ext}$. Choices b and c consist of intersecting lines, choice a of parallel lines. Which is the correct choice?

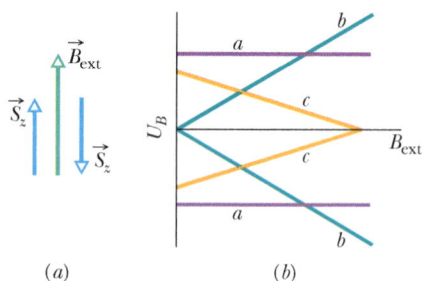

Figure 32-22 Question 4.

5 An electron in an external magnetic field $\vec{B}_{ext}$ has its spin angular momentum S_z antiparallel to $\vec{B}_{ext}$. If the electron undergoes a *spin-flip* so that S_z is then parallel with $\vec{B}_{ext}$, must energy be supplied to or lost by the electron?

6 Does the magnitude of the net force on the current loop of Figs. 32-12a and b increase, decrease, or remain the same if we increase (a) the magnitude of $\vec{B}_{ext}$ and (b) the divergence of $\vec{B}_{ext}$?

7 Figure 32-23 shows a face-on view of one of the two square plates of a parallel-plate capacitor, as well as four loops that are located between the plates. The capacitor is being discharged. (a) Neglecting fringing of the magnetic field, rank the loops according to the magnitude of $\oint \vec{B} \cdot d\vec{s}$ along them, greatest first. (b) Along which loop, if any, is the angle between the directions of $\vec{B}$ and $d\vec{s}$ constant

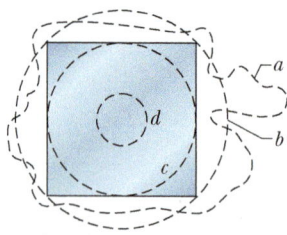

Figure 32-23 Question 7.

(so that their dot product can easily be evaluated)? (c) Along which loop, if any, is B constant (so that B can be brought in front of the integral sign in Eq. 32-3)?

8 Figure 32-24 shows three loop models of an electron orbiting counterclockwise within a magnetic field. The fields are nonuniform for models 1 and 2 and uniform for model 3. For each model, are (a) the magnetic dipole moment of the loop and (b) the magnetic force on the loop directed up, directed down, or zero?

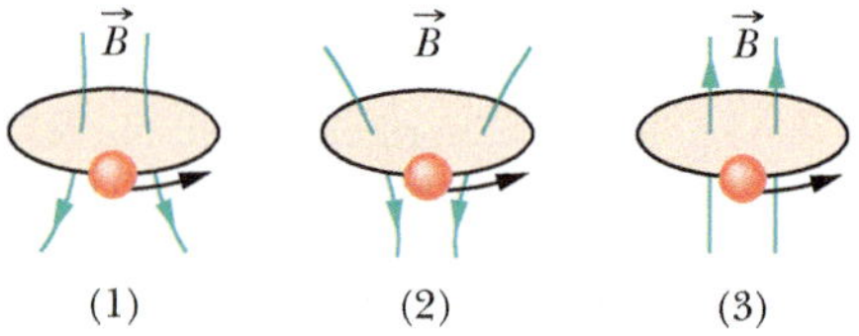

Figure 32-24 Questions 8, 9, and 10.

9 Replace the current loops of Question 8 and Fig. 32-24 with diamagnetic spheres. For each field, are (a) the magnetic dipole moment of the sphere and (b) the magnetic force on the sphere directed up, directed down, or zero?

10 Replace the current loops of Question 8 and Fig. 32-24 with paramagnetic spheres. For each field, are (a) the magnetic dipole moment of the sphere and (b) the magnetic force on the sphere directed up, directed down, or zero?

11 Figure 32-25 represents three rectangular samples of a ferromagnetic material in which the magnetic dipoles of the domains have been directed out of the page (encircled dot) by a very strong applied field B_0. In each sample, an island domain still has its magnetic field directed into the page (encircled $\times$). Sample 1 is one (pure) crystal. The other samples contain impurities collected along lines; domains cannot easily spread across such lines.

The applied field is now to be reversed and its magnitude kept moderate. The change causes the island domain to grow. (a) Rank the three samples according to the success of that growth, greatest growth first. Ferromagnetic materials in which the magnetic dipoles are easily changed are said to be *magnetically soft*; when the changes are difficult, requiring strong applied fields, the materials are said to be *magnetically hard*. (b) Of the three samples, which is the most magnetically hard?

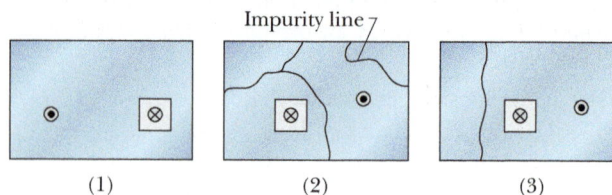

Figure 32-25 Question 11.

12 Figure 32-26 shows four steel bars; three are permanent magnets. One of the poles is indicated. Through experiment we find that ends a and d attract each other, ends c and f repel, ends e and h attract, and ends a and h attract. (a) Which ends are north poles? (b) Which bar is not a magnet?

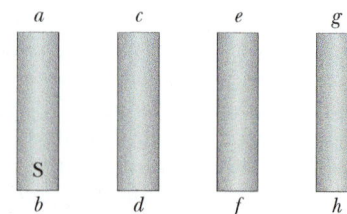

Figure 32-26 Question 12.

Problems

Module 32-1 Gauss' Law for Magnetic Fields

•1 The magnetic flux through each of five faces of a die (singular of "dice") is given by $\Phi_B = \pm N$ Wb, where N ($= 1$ to 5) is the number of spots on the face. The flux is positive (outward) for N even and negative (inward) for N odd. What is the flux through the sixth face of the die?

•2 Figure 32-27 shows a closed surface. Along the flat top face, which has a radius of 2.0 cm, a perpendicular magnetic field $\vec{B}$ of magnitude 0.30 T is directed outward. Along the flat bottom face, a magnetic flux of 0.70 mWb is directed outward. What are the (a) magnitude and (b) direction (inward or outward) of the magnetic flux through the curved part of the surface?

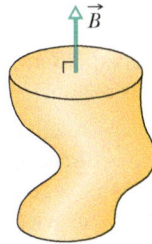

Figure 32-27
Problem 2.

••3 SSM ILW A Gaussian surface in the shape of a right circular cylinder with end caps has a radius of 12.0 cm and a length of 80.0 cm. Through one end there is an inward magnetic flux of 25.0 μWb. At the other end there is a uniform magnetic field of 1.60 mT, normal to the surface and directed outward. What are the (a) magnitude and (b) direction (inward or outward) of the net magnetic flux through the curved surface?

•••4 GO Two wires, parallel to a z axis and a distance $4r$ apart, carry equal currents i in opposite directions, as shown in Fig. 32-28. A circular cylinder of radius r and length L has its axis on the z axis, midway between the wires. Use Gauss' law for magnetism to derive an expression for the net outward magnetic flux through the half of the cylindrical surface above the x axis. (*Hint:* Find the flux through the portion of the xz plane that lies within the cylinder.)

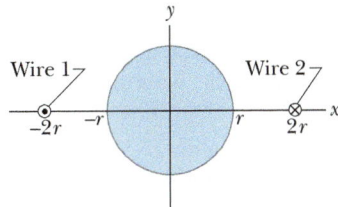

Figure 32-28 Problem 4.

Module 32-2 Induced Magnetic Fields

•5 SSM The induced magnetic field at radial distance 6.0 mm from the central axis of a circular parallel-plate capacitor is 2.0×10^{-7} T. The plates have radius 3.0 mm. At what rate dE/dt is the electric field between the plates changing?

•6 A capacitor with square plates of edge length L is being discharged by a current of 0.75 A. Figure 32-29 is a head-on view of one of the plates from inside the capacitor. A dashed rectangular path is shown. If $L = 12$ cm, $W = 4.0$ cm, and $H = 2.0$ cm, what is the value of $\oint \vec{B} \cdot d\vec{s}$ around the dashed path?

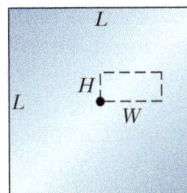

Figure 32-29
Problem 6.

••7 GO *Uniform electric flux.* Figure 32-30 shows a circular region of radius $R = 3.00$ cm in which a uniform electric flux is directed out of the plane of the page. The total electric flux through the region is given by $\Phi_E = (3.00 \text{ mV}\cdot\text{m/s})t$, where t is in seconds. What is the magnitude of the magnetic field that is induced at radial distances (a) 2.00 cm and (b) 5.00 cm?

••8 GO *Nonuniform electric flux.* Figure 32-30 shows a circular region of radius $R = 3.00$ cm in which an electric flux is directed out of the plane of the page. The flux encircled by a concentric circle of radius r is given by $\Phi_{E,\text{enc}} = (0.600 \text{ V}\cdot\text{m/s})(r/R)t$, where $r \le R$ and t is in seconds. What is the magnitude of the induced magnetic field at radial distances (a) 2.00 cm and (b) 5.00 cm?

Figure 32-30
Problems 7 to 10
and 19 to 22.

••9 GO *Uniform electric field.* In Fig. 32-30, a uniform electric field is directed out of the page within a circular region of radius $R = 3.00$ cm. The field magnitude is given by $E = (4.50 \times 10^{-3} \text{ V/m}\cdot\text{s})t$, where t is in seconds. What is the magnitude of the induced magnetic field at radial distances (a) 2.00 cm and (b) 5.00 cm?

••10 GO *Nonuniform electric field.* In Fig. 32-30, an electric field is directed out of the page within a circular region of radius $R = 3.00$ cm. The field magnitude is $E = (0.500 \text{ V/m}\cdot\text{s})(1 - r/R)t$, where t is in seconds and r is the radial distance ($r \le R$). What is the magnitude of the induced magnetic field at radial distances (a) 2.00 cm and (b) 5.00 cm?

••11 Suppose that a parallel-plate capacitor has circular plates with radius $R = 30$ mm and a plate separation of 5.0 mm. Suppose also that a sinusoidal potential difference with a maximum value of 150 V and a frequency of 60 Hz is applied across the plates; that is,

$$V = (150 \text{ V}) \sin[2\pi(60 \text{ Hz})t].$$

(a) Find $B_{\max}(R)$, the maximum value of the induced magnetic field that occurs at $r = R$. (b) Plot $B_{\max}(r)$ for $0 < r < 10$ cm.

••12 GO A parallel-plate capacitor with circular plates of radius 40 mm is being discharged by a current of 6.0 A. At what radius (a) inside and (b) outside the capacitor gap is the magnitude of the induced magnetic field equal to 75% of its maximum value? (c) What is that maximum value?

Module 32-3 Displacement Current

•13 At what rate must the potential difference between the plates of a parallel-plate capacitor with a 2.0 μF capacitance be changed to produce a displacement current of 1.5 A?

•14 A parallel-plate capacitor with circular plates of radius R is being charged. Show that the magnitude of the current density of the displacement current is $J_d = \varepsilon_0(dE/dt)$ for $r \le R$.

•15 SSM Prove that the displacement current in a parallel-plate capacitor of capacitance C can be written as $i_d = C(dV/dt)$, where V is the potential difference between the plates.

•16 A parallel-plate capacitor with circular plates of radius 0.10 m is being discharged. A circular loop of radius 0.20 m is concentric with the capacitor and halfway between the plates. The displacement current through the loop is 2.0 A. At what rate is the electric field between the plates changing?

••17 GO A silver wire has resistivity $\rho = 1.62 \times 10^{-8}$ $\Omega \cdot$m and a cross-sectional area of 5.00 mm^2. The current in the wire is uniform and changing at the rate of 2000 A/s when the current is 100 A. (a) What is the magnitude of the (uniform) electric field in the wire when the current in the wire is 100 A? (b) What is the displacement current in the wire at that time? (c) What is the ratio of the magnitude of the magnetic field due to the displacement current to that due to the current at a distance r from the wire?

••18 GO The circuit in Fig. 32-31 consists of switch S, a 12.0 V ideal battery, a 20.0 MΩ resistor, and an air-filled capacitor. The capacitor has parallel circular plates of radius 5.00 cm, separated by 3.00 mm. At time $t = 0$, switch S is closed to begin charging the capacitor. The electric field between the plates is uniform. At $t = 250$ μs, what is the magnitude of the magnetic field within the capacitor, at radial distance 3.00 cm?

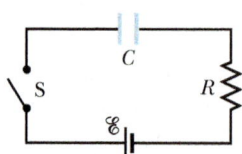

Figure 32-31 Problem 18.

••19 *Uniform displacement-current density.* Figure 32-30 shows a circular region of radius $R = 3.00$ cm in which a displacement current is directed out of the page. The displacement current has a uniform density of magnitude $J_d = 6.00$ A/m^2. What is the magnitude of the magnetic field due to the displacement current at radial distances (a) 2.00 cm and (b) 5.00 cm?

••20 *Uniform displacement current.* Figure 32-30 shows a circular region of radius $R = 3.00$ cm in which a uniform displacement current $i_d = 0.500$ A is out of the page. What is the magnitude of the magnetic field due to the displacement current at radial distances (a) 2.00 cm and (b) 5.00 cm?

••21 GO *Nonuniform displacement-current density.* Figure 32-30 shows a circular region of radius $R = 3.00$ cm in which a displacement current is directed out of the page. The magnitude of the density of this displacement current is $J_d = (4.00$ A/m$^2)(1 - r/R)$, where r is the radial distance ($r \le R$). What is the magnitude of the magnetic field due to the displacement current at (a) $r = 2.00$ cm and (b) $r = 5.00$ cm?

••22 GO *Nonuniform displacement current.* Figure 32-30 shows a circular region of radius $R = 3.00$ cm in which a displacement current i_d is directed out of the figure. The magnitude of the displacement current is $i_d = (3.00$ A$)(r/R)$, where r is the radial distance ($r \le R$) from the center. What is the magnitude of the magnetic field due to i_d at radial distances (a) 2.00 cm and (b) 5.00 cm?

••23 SSM ILW In Fig. 32-32, a parallel-plate capacitor has square plates of edge length $L = 1.0$ m. A current of 2.0 A charges the capacitor, producing a uniform electric field $\vec{E}$ between the plates, with $\vec{E}$ perpendicular to the plates. (a) What is the displacement current i_d through the region between the plates? (b) What is dE/dt in this region? (c) What is the displacement current encircled by the square dashed path of edge length $d = 0.50$ m? (d) What is the value of $\oint \vec{B} \cdot d\vec{s}$ around this square dashed path?

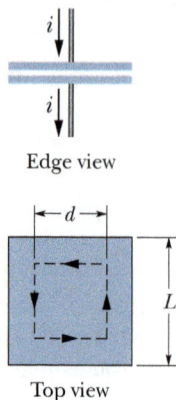

Edge view

Top view

Figure 32-32 Problem 23.

••24 The magnitude of the electric field between the two circular parallel plates in Fig. 32-33 is $E = (4.0 \times 10^5) - (6.0 \times 10^4 t)$, with E in volts per meter and t in seconds. At $t = 0$, $\vec{E}$ is upward. The plate area is 4.0×10^{-2} m^2. For $t \ge 0$, what are the (a) magnitude and (b) direction (up or down) of the displacement current between the plates and (c) is the direction of the induced magnetic field clockwise or counterclockwise in the figure?

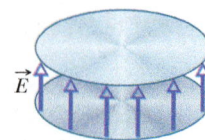

Figure 32-33 Problem 24.

••25 ILW As a parallel-plate capacitor with circular plates 20 cm in diameter is being charged, the current density of the displacement current in the region between the plates is uniform and has a magnitude of 20 A/m^2. (a) Calculate the magnitude B of the magnetic field at a distance $r = 50$ mm from the axis of symmetry of this region. (b) Calculate dE/dt in this region.

••26 A capacitor with parallel circular plates of radius $R = 1.20$ cm is discharging via a current of 12.0 A. Consider a loop of radius $R/3$ that is centered on the central axis between the plates. (a) How much displacement current is encircled by the loop? The maximum induced magnetic field has a magnitude of 12.0 mT. At what radius (b) inside and (c) outside the capacitor gap is the magnitude of the induced magnetic field 3.00 mT?

••27 ILW In Fig. 32-34, a uniform electric field $\vec{E}$ collapses. The vertical axis scale is set by $E_s = 6.0 \times 10^5$ N/C, and the horizontal axis scale is set by $t_s = 12.0$ μs. Calculate the magnitude of the displacement current through a 1.6 m^2 area perpendicular to the field during each of the time intervals a, b, and c shown on the graph. (Ignore the behavior at the ends of the intervals.)

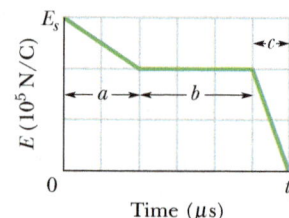

Figure 32-34 Problem 27.

••28 GO Figure 32-35a shows the current i that is produced in a wire of resistivity 1.62×10^{-8} $\Omega \cdot$m. The magnitude of the current versus time t is shown in Fig. 32-35b. The vertical axis scale is set by $i_s = 10.0$ A, and the horizontal axis scale is set by $t_s = 50.0$ ms. Point P is at radial distance 9.00 mm from the wire's center. Determine the magnitude of the magnetic field $\vec{B}_i$ at point P due to the actual current i in the wire at (a) $t = 20$ ms, (b) $t = 40$ ms, and (c) $t = 60$ ms. Next, assume that the electric field driving the current is confined to the wire. Then determine the magnitude of the magnetic field $\vec{B}_{id}$ at point P due to the displacement current i_d in the wire at (d) $t = 20$ ms, (e) $t = 40$ ms, and (f) $t = 60$ ms. At point P at $t = 20$ s, what is the direction (into or out of the page) of (g) $\vec{B}_i$ and (h) $\vec{B}_{id}$?

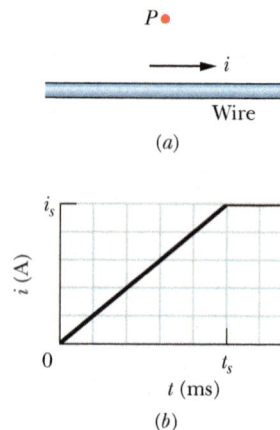

(a)

(b)

Figure 32-35 Problem 28.

•••29 In Fig. 32-36, a capacitor with circular plates of radius $R = 18.0$ cm

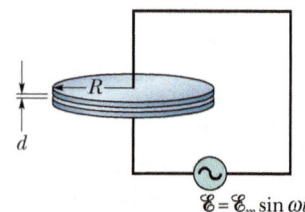

Figure 32-36 Problem 29.

is connected to a source of emf $\mathcal{E} = \mathcal{E}_m \sin \omega t$, where $\mathcal{E}_m = 220$ V and $\omega = 130$ rad/s. The maximum value of the displacement current is $i_d = 7.60$ μA. Neglect fringing of the electric field at the edges of the plates. (a) What is the maximum value of the current i in the circuit? (b) What is the maximum value of $d\Phi_E/dt$, where Φ_E is the electric flux through the region between the plates? (c) What is the separation d between the plates? (d) Find the maximum value of the magnitude of $\vec{B}$ between the plates at a distance $r = 11.0$ cm from the center.

Module 32-4   Magnets

•30 Assume the average value of the vertical component of Earth's magnetic field is 43 μT (downward) for all of Arizona, which has an area of 2.95×10^5 km². What then are the (a) magnitude and (b) direction (inward or outward) of the net magnetic flux through the rest of Earth's surface (the entire surface excluding Arizona)?

•31 In New Hampshire the average horizontal component of Earth's magnetic field in 1912 was 16 μT, and the average inclination or "dip" was 73°. What was the corresponding magnitude of Earth's magnetic field?

Module 32-5   Magnetism and Electrons

•32 Figure 32-37*a* is a one-axis graph along which two of the allowed energy values (*levels*) of an atom are plotted. When the atom is placed in a magnetic field of 0.500 T, the graph changes to that of Fig. 32-37*b* because of the energy associated with $\vec{\mu}_{orb} \cdot \vec{B}$. (We neglect $\vec{\mu}_s$.) Level E_1 is unchanged, but level E_2 splits into a (closely spaced) triplet of levels. What are the allowed values of m_ℓ associated with (a) energy level E_1 and (b) energy level E_2? (c) In joules, what amount of energy is represented by the spacing between the triplet levels?

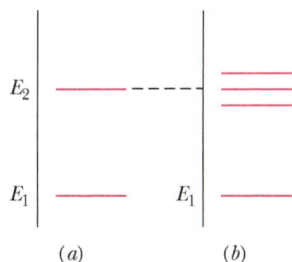

Figure 32-37 Problem 32.

•33 SSM WWW If an electron in an atom has an orbital angular momentum with $m = 0$, what are the components (a) $L_{orb,z}$ and (b) $\mu_{orb,z}$? If the atom is in an external magnetic field $\vec{B}$ that has magnitude 35 mT and is directed along the z axis, what are (c) the energy U_{orb} associated with $\vec{\mu}_{orb}$ and (d) the energy U_{spin} associated with $\vec{\mu}_s$? If, instead, the electron has $m = -3$, what are (e) $L_{orb,z}$, (f) $\mu_{orb,z}$, (g) U_{orb}, and (h) U_{spin}?

•34 What is the energy difference between parallel and antiparallel alignment of the z component of an electron's spin magnetic dipole moment with an external magnetic field of magnitude 0.25 T, directed parallel to the z axis?

•35 What is the measured component of the orbital magnetic dipole moment of an electron with (a) $m_\ell = 1$ and (b) $m_\ell = -2$?

•36 An electron is placed in a magnetic field $\vec{B}$ that is directed along a z axis. The energy difference between parallel and antiparallel alignments of the z component of the electron's spin magnetic moment with $\vec{B}$ is 6.00×10^{-25} J. What is the magnitude of $\vec{B}$?

Module 32-6   Diamagnetism

•37 Figure 32-38 shows a loop model (loop L) for a diamagnetic material. (a) Sketch the magnetic field lines within and about the material due to the bar magnet. What is

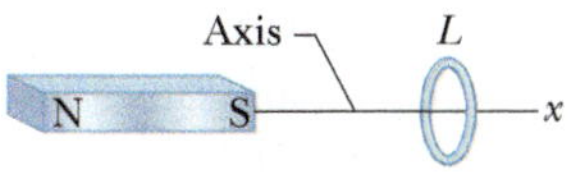

Figure 32-38
Problems 37 and 71.

the direction of (b) the loop's net magnetic dipole moment $\vec{\mu}$, (c) the conventional current i in the loop (clockwise or counterclockwise in the figure), and (d) the magnetic force on the loop?

•••38 Assume that an electron of mass m and charge magnitude e moves in a circular orbit of radius r about a nucleus. A uniform magnetic field $\vec{B}$ is then established perpendicular to the plane of the orbit. Assuming also that the radius of the orbit does not change and that the change in the speed of the electron due to field $\vec{B}$ is small, find an expression for the change in the orbital magnetic dipole moment of the electron due to the field.

Module 32-7   Paramagnetism

•39 A sample of the paramagnetic salt to which the magnetization curve of Fig. 32-14 applies is to be tested to see whether it obeys Curie's law. The sample is placed in a uniform 0.50 T magnetic field that remains constant throughout the experiment. The magnetization M is then measured at temperatures ranging from 10 to 300 K. Will it be found that Curie's law is valid under these conditions?

•40 A sample of the paramagnetic salt to which the magnetization curve of Fig. 32-14 applies is held at room temperature (300 K). At what applied magnetic field will the degree of magnetic saturation of the sample be (a) 50% and (b) 90%? (c) Are these fields attainable in the laboratory?

•41 SSM ILW A magnet in the form of a cylindrical rod has a length of 5.00 cm and a diameter of 1.00 cm. It has a uniform magnetization of 5.30×10^3 A/m. What is its magnetic dipole moment?

•42 A 0.50 T magnetic field is applied to a paramagnetic gas whose atoms have an intrinsic magnetic dipole moment of 1.0×10^{-23} J/T. At what temperature will the mean kinetic energy of translation of the atoms equal the energy required to reverse such a dipole end for end in this magnetic field?

••43 An electron with kinetic energy K_e travels in a circular path that is perpendicular to a uniform magnetic field, which is in the positive direction of a z axis. The electron's motion is subject only to the force due to the field. (a) Show that the magnetic dipole moment of the electron due to its orbital motion has magnitude $\mu = K_e/B$ and that it is in the direction opposite that of $\vec{B}$. What are the (b) magnitude and (c) direction of the magnetic dipole moment of a positive ion with kinetic energy K_i under the same circumstances? (d) An ionized gas consists of 5.3×10^{21} electrons/m³ and the same number density of ions. Take the average electron kinetic energy to be 6.2×10^{-20} J and the average ion kinetic energy to be 7.6×10^{-21} J. Calculate the magnetization of the gas when it is in a magnetic field of 1.2 T.

••44 Figure 32-39 gives the magnetization curve for a paramagnetic material. The vertical axis scale is set by $a = 0.15$, and the horizontal axis scale is set by $b = 0.2$ T/K. Let μ_{sam} be the measured net magnetic moment of a sample of the material and μ_{max} be the maximum possible net magnetic moment of that sample. According to Curie's law, what would be the ratio μ_{sam}/μ_{max} were the sample placed in a uniform magnetic field of magnitude 0.800 T, at a temperature of 2.00 K?

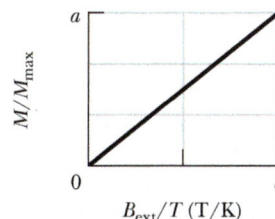

Figure 32-39 Problem 44.

•••45 SSM Consider a solid containing N atoms per unit volume, each atom having a magnetic dipole moment $\vec{\mu}$. Suppose the direction of $\vec{\mu}$ can be only parallel or antiparallel to an externally applied magnetic field $\vec{B}$ (this will be the case if $\vec{\mu}$ is due to the spin of a single electron). According to statistical mechanics, the probability of an atom being in a state with energy U is proportional to $e^{-U/kT}$, where T is the temperature and k is Boltzmann's constant. Thus, because energy U is $-\vec{\mu} \cdot \vec{B}$, the fraction of atoms whose dipole moment is parallel to $\vec{B}$ is proportional to $e^{\mu B/kT}$ and the fraction of atoms whose dipole moment is antiparallel to $\vec{B}$ is proportional to $e^{-\mu B/kT}$. (a) Show that the magnitude of the magnetization of this solid is $M = N\mu \tanh(\mu B/kT)$. Here tanh is the hyperbolic tangent function: $\tanh(x) = (e^x - e^{-x})/(e^x + e^{-x})$. (b) Show that the result given in (a) reduces to $M = N\mu^2 B/kT$ for $\mu B \ll kT$. (c) Show that the result of (a) reduces to $M = N\mu$ for $\mu B \gg kT$. (d) Show that both (b) and (c) agree qualitatively with Fig. 32-14.

Module 32-8 Ferromagnetism

••46 GO You place a magnetic compass on a horizontal surface, allow the needle to settle, and then give the compass a gentle wiggle to cause the needle to oscillate about its equilibrium position. The oscillation frequency is 0.312 Hz. Earth's magnetic field at the location of the compass has a horizontal component of 18.0 μT. The needle has a magnetic moment of 0.680 mJ/T. What is the needle's rotational inertia about its (vertical) axis of rotation?

••47 SSM ILW WWW The magnitude of the magnetic dipole moment of Earth is 8.0×10^{22} J/T. (a) If the origin of this magnetism were a magnetized iron sphere at the center of Earth, what would be its radius? (b) What fraction of the volume of Earth would such a sphere occupy? Assume complete alignment of the dipoles. The density of Earth's inner core is 14 g/cm³. The magnetic dipole moment of an iron atom is 2.1×10^{-23} J/T. (*Note:* Earth's inner core is in fact thought to be in both liquid and solid forms and partly iron, but a permanent magnet as the source of Earth's magnetism has been ruled out by several considerations. For one, the temperature is certainly above the Curie point.)

••48 The magnitude of the dipole moment associated with an atom of iron in an iron bar is 2.1×10^{-23} J/T. Assume that all the atoms in the bar, which is 5.0 cm long and has a cross-sectional area of 1.0 cm², have their dipole moments aligned. (a) What is the dipole moment of the bar? (b) What torque must be exerted to hold this magnet perpendicular to an external field of magnitude 1.5 T? (The density of iron is 7.9 g/cm³.)

••49 SSM The exchange coupling mentioned in Module 32-8 as being responsible for ferromagnetism is *not* the mutual magnetic interaction between two elementary magnetic dipoles. To show this, calculate (a) the magnitude of the magnetic field a distance of 10 nm away, along the dipole axis, from an atom with magnetic dipole moment 1.5×10^{-23} J/T (cobalt), and (b) the minimum energy required to turn a second identical dipole end for end in this field. (c) By comparing the latter with the mean translational kinetic energy of 0.040 eV, what can you conclude?

••50 A magnetic rod with length 6.00 cm, radius 3.00 mm, and (uniform) magnetization 2.70×10^3 A/m can turn about its center like a compass needle. It is placed in a uniform magnetic field $\vec{B}$ of magnitude 35.0 mT, such that the directions of its dipole moment and $\vec{B}$ make an angle of 68.0°. (a) What is the magnitude of the torque on the rod due to $\vec{B}$? (b) What is the change in the orientation energy of the rod if the angle changes to 34.0°?

••51 The saturation magnetization M_{max} of the ferromagnetic metal nickel is 4.70×10^5 A/m. Calculate the magnetic dipole moment of a single nickel atom. (The density of nickel is 8.90 g/cm³, and its molar mass is 58.71 g/mol.)

••52 Measurements in mines and boreholes indicate that Earth's interior temperature increases with depth at the average rate of 30 C°/km. Assuming a surface temperature of 10°C, at what depth does iron cease to be ferromagnetic? (The Curie temperature of iron varies very little with pressure.)

••53 A Rowland ring is formed of ferromagnetic material. It is circular in cross section, with an inner radius of 5.0 cm and an outer radius of 6.0 cm, and is wound with 400 turns of wire. (a) What current must be set up in the windings to attain a toroidal field of magnitude $B_0 = 0.20$ mT? (b) A secondary coil wound around the toroid has 50 turns and resistance 8.0 Ω. If, for this value of B_0, we have $B_M = 800B_0$, how much charge moves through the secondary coil when the current in the toroid windings is turned on?

Additional Problems

54 Using the approximations given in Problem 61, find (a) the altitude above Earth's surface where the magnitude of its magnetic field is 50.0% of the surface value at the same latitude; (b) the maximum magnitude of the magnetic field at the core–mantle boundary, 2900 km below Earth's surface; and the (c) magnitude and (d) inclination of Earth's magnetic field at the north geographic pole. (e) Suggest why the values you calculated for (c) and (d) differ from measured values.

55 Earth has a magnetic dipole moment of 8.0×10^{22} J/T. (a) What current would have to be produced in a single turn of wire extending around Earth at its geomagnetic equator if we wished to set up such a dipole? Could such an arrangement be used to cancel out Earth's magnetism (b) at points in space well above Earth's surface or (c) on Earth's surface?

56 A charge q is distributed uniformly around a thin ring of radius r. The ring is rotating about an axis through its center and perpendicular to its plane, at an angular speed ω. (a) Show that the magnetic moment due to the rotating charge has magnitude $\mu = \frac{1}{2}q\omega r^2$. (b) What is the direction of this magnetic moment if the charge is positive?

57 A magnetic compass has its needle, of mass 0.050 kg and length 4.0 cm, aligned with the horizontal component of Earth's magnetic field at a place where that component has the value $B_h = 16$ μT. After the compass is given a momentary gentle shake, the needle oscillates with angular frequency $\omega = 45$ rad/s. Assuming that the needle is a uniform thin rod mounted at its center, find the magnitude of its magnetic dipole moment.

58 The capacitor in Fig. 32-7 is being charged with a 2.50 A current. The wire radius is 1.50 mm, and the plate radius is 2.00 cm. Assume that the current i in the wire and the displacement current i_d in the capacitor gap are both uniformly distributed. What is the magnitude of the magnetic field due to i at the following radial distances from the wire's center: (a) 1.00 mm (inside the wire), (b) 3.00 mm (outside the wire), and (c) 2.20 cm (outside the wire)? What is the magnitude of the magnetic field due to i_d at the following radial distances from the central axis between the plates: (d) 1.00 mm (inside the gap), (e) 3.00 mm (inside the gap), and (f) 2.20 cm (outside the gap)? (g) Explain why the fields at the two smaller radii are so different for the wire and the gap but the fields at the largest radius are not.

59 A parallel-plate capacitor with circular plates of radius $R = 16$ mm and gap width $d = 5.0$ mm has a uniform electric field between the plates. Starting at time $t = 0$, the potential difference between the two plates is $V = (100 \text{ V})e^{-t/\tau}$, where the time constant $\tau = 12$ ms. At radial distance $r = 0.80R$ from the central axis, what is the magnetic field magnitude (a) as a function of time for $t \geq 0$ and (b) at time $t = 3\tau$?

60 A magnetic flux of 7.0 mWb is directed outward through the flat bottom face of the closed surface shown in Fig. 32-40. Along the flat top face (which has a radius of 4.2 cm) there is a 0.40 T magnetic field $\vec{B}$ directed perpendicular to the face. What are the (a) magnitude and (b) direction (inward or outward) of the magnetic flux through the curved part of the surface?

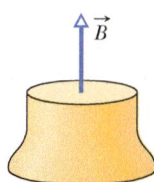

Figure 32-40 Problem 60.

61 SSM The magnetic field of Earth can be approximated as the magnetic field of a dipole. The horizontal and vertical components of this field at any distance r from Earth's center are given by

$$B_h = \frac{\mu_0\mu}{4\pi r^3}\cos\lambda_m, \qquad B_v = \frac{\mu_0\mu}{2\pi r^3}\sin\lambda_m,$$

where λ_m is the *magnetic latitude* (this type of latitude is measured from the geomagnetic equator toward the north or south geomagnetic pole). Assume that Earth's magnetic dipole moment has magnitude $\mu = 8.00 \times 10^{22}$ A·m². (a) Show that the magnitude of Earth's field at latitude λ_m is given by

$$B = \frac{\mu_0\mu}{4\pi r^3}\sqrt{1 + 3\sin^2\lambda_m}.$$

(b) Show that the inclination ϕ_i of the magnetic field is related to the magnetic latitude λ_m by $\tan\phi_i = 2\tan\lambda_m$.

62 Use the results displayed in Problem 61 to predict the (a) magnitude and (b) inclination of Earth's magnetic field at the geomagnetic equator, the (c) magnitude and (d) inclination at geomagnetic latitude 60.0°, and the (e) magnitude and (f) inclination at the north geomagnetic pole.

63 A parallel-plate capacitor with circular plates of radius 55.0 mm is being charged. At what radius (a) inside and (b) outside the capacitor gap is the magnitude of the induced magnetic field equal to 50.0% of its maximum value?

64 A sample of the paramagnetic salt to which the magnetization curve of Fig. 32-14 applies is immersed in a uniform magnetic field of 2.0 T. At what temperature will the degree of magnetic saturation of the sample be (a) 50% and (b) 90%?

65 A parallel-plate capacitor with circular plates of radius R is being discharged. The displacement current through a central circular area, parallel to the plates and with radius $R/2$, is 2.0 A. What is the discharging current?

66 Figure 32-41 gives the variation of an electric field that is perpendicular to a circular area of 2.0 m². During the time period shown, what is the greatest displacement current through the area?

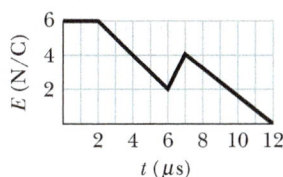

Figure 32-41 Problem 66.

67 In Fig. 32-42, a parallel-plate capacitor is being discharged by a current $i = 5.0$ A. The plates are square with edge length $L = 8.0$ mm. (a) What is the rate at which the electric field between the plates is changing? (b) What is the value of $\oint \vec{B}\cdot d\vec{s}$ around the dashed path, where $H = 2.0$ mm and $W = 3.0$ mm?

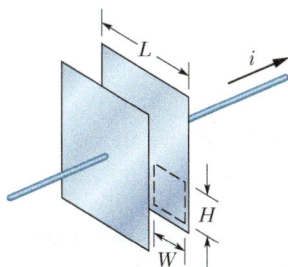

Figure 32-42 Problem 67.

68 What is the measured component of the orbital magnetic dipole moment of an electron with the values (a) $m_\ell = 3$ and (b) $m_\ell = -4$?

69 In Fig. 32-43, a bar magnet lies near a paper cylinder. (a) Sketch the magnetic field lines that pass through the surface of the cylinder. (b) What is the sign of $\vec{B}\cdot d\vec{A}$ for every area $d\vec{A}$ on the surface? (c) Does this contradict Gauss' law for magnetism? Explain.

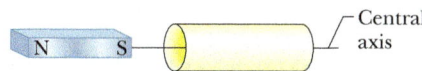

Figure 32-43 Problem 69.

70 In the lowest energy state of the hydrogen atom, the most probable distance of the single electron from the central proton (the nucleus) is 5.2×10^{-11} m. (a) Compute the magnitude of the proton's electric field at that distance. The component $\mu_{s,z}$ of the proton's spin magnetic dipole moment measured on a z axis is 1.4×10^{-26} J/T. (b) Compute the magnitude of the proton's magnetic field at the distance 5.2×10^{-11} m on the z axis. (*Hint:* Use Eq. 29-27.) (c) What is the ratio of the spin magnetic dipole moment of the electron to that of the proton?

71 Figure 32-38 shows a loop model (loop L) for a paramagnetic material. (a) Sketch the field lines through and about the material due to the magnet. What is the direction of (b) the loop's net magnetic dipole moment $\vec{\mu}$, (c) the conventional current i in the loop (clockwise or counterclockwise in the figure), and (d) the magnetic force acting on the loop?

72 Two plates (as in Fig. 32-7) are being discharged by a constant current. Each plate has a radius of 4.00 cm. During the discharging, at a point between the plates at radial distance 2.00 cm from the central axis, the magnetic field has a magnitude of 12.5 nT. (a) What is the magnitude of the magnetic field at radial distance 6.00 cm? (b) What is the current in the wires attached to the plates?

73 SSM If an electron in an atom has orbital angular momentum with m_ℓ values limited by ±3, how many values of (a) $L_{orb,z}$ and (b) $\mu_{orb,z}$ can the electron have? In terms of h, m, and e, what is the greatest allowed magnitude for (c) $L_{orb,z}$ and (d) $\mu_{orb,z}$? (e) What is the greatest allowed magnitude for the z component of the electron's *net* angular momentum (orbital plus spin)? (f) How many values (signs included) are allowed for the z component of its net angular momentum?

74 A parallel-plate capacitor with circular plates is being charged. Consider a circular loop centered on the central axis and located between the plates. If the loop radius of 3.00 cm is greater than the plate radius, what is the displacement current between the plates when the magnetic field along the loop has magnitude $2.00\ \mu$T?

75 Suppose that ±4 are the limits to the values of m_ℓ for an electron in an atom. (a) How many different values of the electron's $\mu_{orb,z}$ are possible? (b) What is the greatest magnitude of those possible values? Next, if the atom is in a magnetic field of magnitude 0.250 T, in the positive direction of the z axis, what are (c) the maximum energy and (d) the minimum energy associated with those possible values of $\mu_{orb,z}$?

76 What are the measured components of the orbital magnetic dipole moment of an electron with (a) $m_\ell = 3$ and (b) $m_\ell = -4$?

CHAPTER 33

Electromagnetic Waves

33-1 ELECTROMAGNETIC WAVES

Learning Objectives

After reading this module, you should be able to . . .

33.01 In the electromagnetic spectrum, identify the relative wavelengths (longer or shorter) of AM radio, FM radio, television, infrared light, visible light, ultraviolet light, x rays, and gamma rays.

33.02 Describe the transmission of an electromagnetic wave by an LC oscillator and an antenna.

33.03 For a transmitter with an LC oscillator, apply the relationships between the oscillator's inductance L, capacitance C, and angular frequency ω, and the emitted wave's frequency f and wavelength λ.

33.04 Identify the speed of an electromagnetic wave in vacuum (and approximately in air).

33.05 Identify that electromagnetic waves do not require a medium and can travel through vacuum.

33.06 Apply the relationship between the speed of an electromagnetic wave, the straight-line distance traveled by the wave, and the time required for the travel.

33.07 Apply the relationships between an electromagnetic

wave's frequency f, wavelength λ, period T, angular frequency ω, and speed c.

33.08 Identify that an electromagnetic wave consists of an electric component and a magnetic component that are (a) perpendicular to the direction of travel, (b) perpendicular to each other, and (c) sinusoidal waves with the same frequency and phase.

33.09 Apply the sinusoidal equations for the electric and magnetic components of an EM wave, written as functions of position and time.

33.10 Apply the relationship between the speed of light c, the permittivity constant ε_0, and the permeability constant μ_0.

33.11 For any instant and position, apply the relationship between the electric field magnitude E, the magnetic field magnitude B, and the speed of light c.

33.12 Describe the derivation of the relationship between the speed of light c and the ratio of the electric field amplitude E to the magnetic field amplitude B.

Key Ideas

● An electromagnetic wave consists of oscillating electric and magnetic fields.

● The various possible frequencies of electromagnetic waves form a spectrum, a small part of which is visible light.

● An electromagnetic wave traveling along an x axis has an electric field $\vec{E}$ and a magnetic field $\vec{B}$ with magnitudes that depend on x and t:

$$E = E_m \sin(kx - \omega t)$$

and

$$B = B_m \sin(kx - \omega t),$$

where E_m and B_m are the amplitudes of $\vec{E}$ and $\vec{B}$. The electric field induces the magnetic field and vice versa.

● The speed of any electromagnetic wave in vacuum is c, which can be written as

$$c = \frac{E}{B} = \frac{1}{\sqrt{\mu_0 \varepsilon_0}},$$

where E and B are the simultaneous magnitudes of the fields.

What Is Physics?

The information age in which we live is based almost entirely on the physics of electromagnetic waves. Like it or not, we are now globally connected by television, telephones, and the web. And like it or not, we are constantly immersed in those signals because of television, radio, and telephone transmitters.

Much of this global interconnection of information processors was not imagined by even the most visionary engineers of 40 years ago. The challenge for today's engineers is trying to envision what the global interconnection will be like 40 years from now. The starting point in meeting that challenge is understanding the basic physics of electromagnetic waves, which come in so many different types that they are poetically said to form *Maxwell's rainbow.*

Maxwell's Rainbow

The crowning achievement of James Clerk Maxwell (see Chapter 32) was to show that a beam of light is a traveling wave of electric and magnetic fields—an **electromagnetic wave**—and thus that optics, the study of visible light, is a branch of electromagnetism. In this chapter we move from one to the other: we conclude our discussion of strictly electrical and magnetic phenomena, and we build a foundation for optics.

In Maxwell's time (the mid 1800s), the visible, infrared, and ultraviolet forms of light were the only electromagnetic waves known. Spurred on by Maxwell's work, however, Heinrich Hertz discovered what we now call radio waves and verified that they move through the laboratory at the same speed as visible light, indicating that they have the same basic nature as visible light.

As Fig. 33-1 shows, we now know a wide *spectrum* (or range) of electromagnetic waves: Maxwell's rainbow. Consider the extent to which we are immersed in electromagnetic waves throughout this spectrum. The Sun, whose radiations define the environment in which we as a species have evolved and adapted, is the dominant source. We are also crisscrossed by radio and television signals. Microwaves from radar systems and from telephone relay systems may reach us. There are electromagnetic waves from lightbulbs, from the heated engine blocks of automobiles, from x-ray machines, from lightning flashes, and from buried radioactive materials. Beyond this, radiation reaches us from stars and other objects in our galaxy and from other galaxies. Electromagnetic waves also travel in the other direction. Television signals, transmitted from Earth since about 1950, have now taken news about us (along with episodes of *I Love Lucy,* albeit *very* faintly) to whatever technically sophisticated inhabitants there may be on whatever planets may encircle the nearest 400 or so stars.

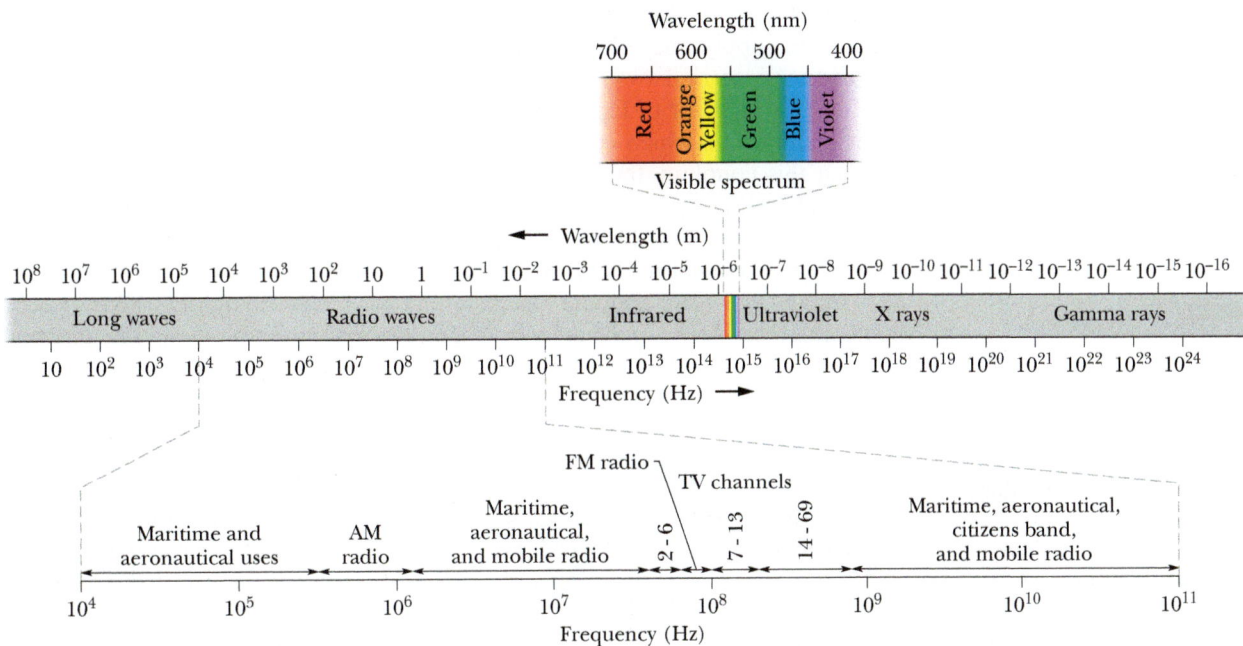

Figure 33-1 The electromagnetic spectrum.

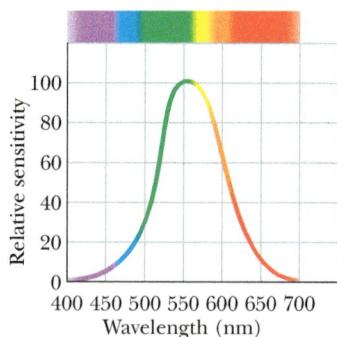

Figure 33-2 The relative sensitivity of the average human eye to electromagnetic waves at different wavelengths. This portion of the electromagnetic spectrum to which the eye is sensitive is called *visible light*.

In the wavelength scale in Fig. 33-1 (and similarly the corresponding frequency scale), each scale marker represents a change in wavelength (and correspondingly in frequency) by a factor of 10. The scale is open-ended; the wavelengths of electromagnetic waves have no inherent upper or lower bound.

Certain regions of the electromagnetic spectrum in Fig. 33-1 are identified by familiar labels, such as *x rays* and *radio waves*. These labels denote roughly defined wavelength ranges within which certain kinds of sources and detectors of electromagnetic waves are in common use. Other regions of Fig. 33-1, such as those labeled TV channels and AM radio, represent specific wavelength bands assigned by law for certain commercial or other purposes. There are no gaps in the electromagnetic spectrum—and all electromagnetic waves, no matter where they lie in the spectrum, travel through *free space* (vacuum) with the same speed c.

The visible region of the spectrum is of course of particular interest to us. Figure 33-2 shows the relative sensitivity of the human eye to light of various wavelengths. The center of the visible region is about 555 nm, which produces the sensation that we call yellow-green.

The limits of this visible spectrum are not well defined because the eye sensitivity curve approaches the zero-sensitivity line asymptotically at both long and short wavelengths. If we take the limits, arbitrarily, as the wavelengths at which eye sensitivity has dropped to 1% of its maximum value, these limits are about 430 and 690 nm; however, the eye can detect electromagnetic waves somewhat beyond these limits if they are intense enough.

The Traveling Electromagnetic Wave, Qualitatively

Some electromagnetic waves, including x rays, gamma rays, and visible light, are *radiated* (emitted) from sources that are of atomic or nuclear size, where quantum physics rules. Here we discuss how other electromagnetic waves are generated. To simplify matters, we restrict ourselves to that region of the spectrum (wavelength $\lambda \approx 1$ m) in which the source of the *radiation* (the emitted waves) is both macroscopic and of manageable dimensions.

Figure 33-3 shows, in broad outline, the generation of such waves. At its heart is an *LC oscillator*, which establishes an angular frequency $\omega\,(= 1/\sqrt{LC})$. Charges and currents in this circuit vary sinusoidally at this frequency, as depicted in Fig. 31-1. An external source—possibly an ac generator—must be included to supply energy to compensate both for thermal losses in the circuit and for energy carried away by the radiated electromagnetic wave.

The *LC* oscillator of Fig. 33-3 is coupled by a transformer and a transmission line to an *antenna*, which consists essentially of two thin, solid, conducting rods. Through this coupling, the sinusoidally varying current in the oscillator causes charge to oscillate sinusoidally along the rods of the antenna at the angular frequency ω of the *LC* oscillator. The current in the rods associated with this movement of charge also varies sinusoidally, in magnitude and direction, at angular frequency ω. The antenna has the effect of an electric dipole whose electric dipole moment varies sinusoidally in magnitude and direction along the antenna.

Figure 33-3 An arrangement for generating a traveling electromagnetic wave in the shortwave radio region of the spectrum: an *LC* oscillator produces a sinusoidal current in the antenna, which generates the wave. *P* is a distant point at which a detector can monitor the wave traveling past it.

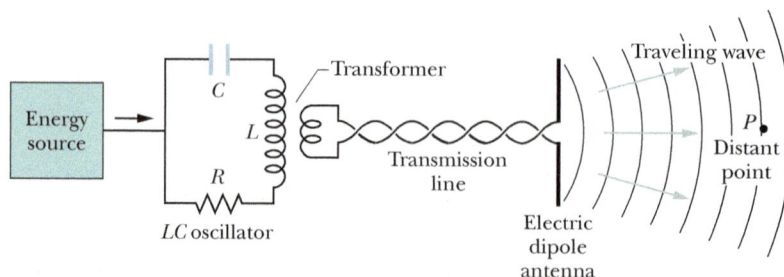

Because the dipole moment varies in magnitude and direction, the electric field produced by the dipole varies in magnitude and direction. Also, because the current varies, the magnetic field produced by the current varies in magnitude and direction. However, the changes in the electric and magnetic fields do not happen everywhere instantaneously; rather, the changes travel outward from the antenna at the speed of light c. Together the changing fields form an electromagnetic wave that travels away from the antenna at speed c. The angular frequency of this wave is ω, the same as that of the LC oscillator.

Electromagnetic Wave. Figure 33-4 shows how the electric field $\vec{E}$ and the magnetic field $\vec{B}$ change with time as one wavelength of the wave sweeps past the distant point P of Fig. 33-3; in each part of Fig. 33-4, the wave is traveling directly out of the page. (We choose a distant point so that the curvature of the waves suggested in Fig. 33-3 is small enough to neglect. At such points, the wave is said to be a *plane wave*, and discussion of the wave is much simplified.) Note several key features in Fig. 33-4; they are present regardless of how the wave is created:

1. The electric and magnetic fields $\vec{E}$ and $\vec{B}$ are always perpendicular to the direction in which the wave is traveling. Thus, the wave is a *transverse wave*, as discussed in Chapter 16.

2. The electric field is always perpendicular to the magnetic field.

3. The cross product $\vec{E} \times \vec{B}$ always gives the direction in which the wave travels.

4. The fields always vary sinusoidally, just like the transverse waves discussed in Chapter 16. Moreover, the fields vary with the same frequency and *in phase* (in step) with each other.

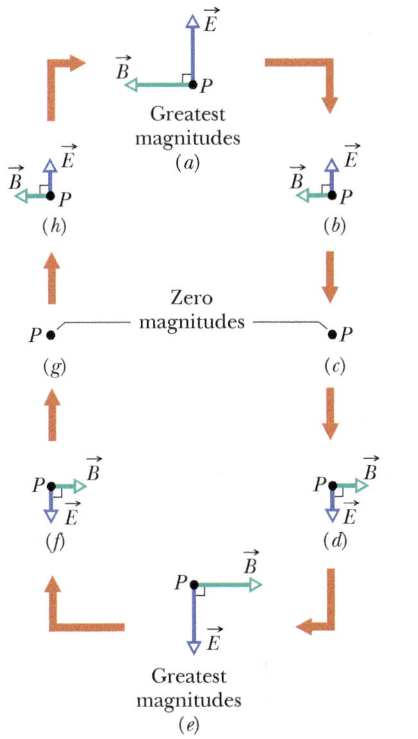

In keeping with these features, we can assume that the electromagnetic wave is traveling toward P in the positive direction of an x axis, that the electric field in Fig. 33-4 is oscillating parallel to the y axis, and that the magnetic field is then oscillating parallel to the z axis (using a right-handed coordinate system, of course). Then we can write the electric and magnetic fields as sinusoidal functions of position x (along the path of the wave) and time t:

$$E = E_m \sin(kx - \omega t), \tag{33-1}$$

$$B = B_m \sin(kx - \omega t), \tag{33-2}$$

in which E_m and B_m are the amplitudes of the fields and, as in Chapter 16, ω and k are the angular frequency and angular wave number of the wave, respectively. From these equations, we note that not only do the two fields form the electromagnetic wave but each also forms its own wave. Equation 33-1 gives the *electric wave component* of the electromagnetic wave, and Eq. 33-2 gives the *magnetic wave component*. As we shall discuss below, these two wave components cannot exist independently.

Wave Speed. From Eq. 16-13, we know that the speed of the wave is ω/k. However, because this is an electromagnetic wave, its speed (in vacuum) is given the symbol c rather than v. In the next section you will see that c has the value

$$c = \frac{1}{\sqrt{\mu_0 \varepsilon_0}} \quad \text{(wave speed)}, \tag{33-3}$$

which is about 3.0×10^8 m/s. In other words,

All electromagnetic waves, including visible light, have the same speed c in vacuum.

You will also see that the wave speed c and the amplitudes of the electric and

Figure 33-4 (a)–(h) The variation in the electric field $\vec{E}$ and the magnetic field $\vec{B}$ at the distant point P of Fig. 33-3 as one wavelength of the electromagnetic wave travels past it. In this perspective, the wave is traveling directly out of the page. The two fields vary sinusoidally in magnitude and direction. Note that they are always perpendicular to each other and to the wave's direction of travel.

magnetic fields are related by

$$\frac{E_m}{B_m} = c \quad \text{(amplitude ratio).} \tag{33-4}$$

If we divide Eq. 33-1 by Eq. 33-2 and then substitute with Eq. 33-4, we find that the magnitudes of the fields at every instant and at any point are related by

$$\frac{E}{B} = c \quad \text{(magnitude ratio).} \tag{33-5}$$

Rays and Wavefronts. We can represent the electromagnetic wave as in Fig. 33-5a, with a *ray* (a directed line showing the wave's direction of travel) or with *wavefronts* (imaginary surfaces over which the wave has the same magnitude of electric field), or both. The two wavefronts shown in Fig. 33-5a are separated by one wavelength λ ($= 2\pi/k$) of the wave. (Waves traveling in approximately the same direction form a *beam*, such as a laser beam, which can also be represented with a ray.)

Drawing the Wave. We can also represent the wave as in Fig. 33-5b, which shows the electric and magnetic field vectors in a "snapshot" of the wave at a certain instant. The curves through the tips of the vectors represent the sinusoidal oscillations given by Eqs. 33-1 and 33-2; the wave components $\vec{E}$ and $\vec{B}$ are in phase, perpendicular to each other, and perpendicular to the wave's direction of travel.

Interpretation of Fig. 33-5b requires some care. The similar drawings for a transverse wave on a taut string that we discussed in Chapter 16 represented the up and down displacement of sections of the string as the wave passed (*something actually moved*). Figure 33-5b is more abstract. At the instant shown, the electric and magnetic fields each have a certain magnitude and direction (but always perpendicular to the *x* axis) at each point along the *x* axis. We choose to represent these vector quantities with a pair of arrows for each point, and so we must draw arrows of different lengths for different points, all directed away from the *x* axis, like thorns on a rose stem. However, the arrows represent field values only at points that are on the *x* axis. Neither the arrows nor the sinusoidal curves represent a sideways motion of anything, nor do the arrows connect points on the *x* axis with points off the axis.

Feedback. Drawings like Fig. 33-5 help us visualize what is actually a very complicated situation. First consider the magnetic field. Because it varies sinusoidally, it induces (via Faraday's law of induction) a perpendicular electric field that also varies sinusoidally. However, because that electric field is varying sinusoidally, it induces (via Maxwell's law of induction) a perpendicular magnetic field that also varies sinusoidally. And so on. The two fields continuously create each other via induction, and the resulting sinusoidal variations in the fields travel as a wave—the electromagnetic wave. Without this amazing result, we could not see; indeed, because we need electromagnetic waves

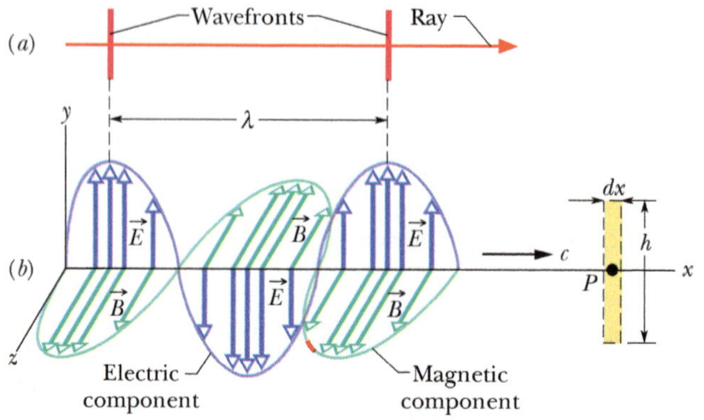

Figure 33-5 (*a*) An electromagnetic wave represented with a ray and two wavefronts; the wavefronts are separated by one wavelength λ. (*b*) The same wave represented in a "snapshot" of its electric field $\vec{E}$ and magnetic field $\vec{B}$ at points on the *x* axis, along which the wave travels at speed *c*. As it travels past point *P*, the fields vary as shown in Fig. 33-4. The electric component of the wave consists of only the electric fields; the magnetic component consists of only the magnetic fields. The dashed rectangle at *P* is used in Fig. 33-6.

from the Sun to maintain Earth's temperature, without this result we could not even exist.

A Most Curious Wave

The waves we discussed in Chapters 16 and 17 require a *medium* (some material) through which or along which to travel. We had waves traveling along a string, through Earth, and through the air. However, an electromagnetic wave (let's use the term *light wave* or *light*) is curiously different in that it requires no medium for its travel. It can, indeed, travel through a medium such as air or glass, but it can also travel through the vacuum of space between a star and us.

Once the special theory of relativity became accepted, long after Einstein published it in 1905, the speed of light waves was realized to be special. One reason is that light has the same speed regardless of the frame of reference from which it is measured. If you send a beam of light along an axis and ask several observers to measure its speed while they move at different speeds along that axis, either in the direction of the light or opposite it, they will all measure the *same speed* for the light. This result is an amazing one and quite different from what would have been found if those observers had measured the speed of any other type of wave; for other waves, the speed of the observers relative to the wave would have affected their measurements.

The meter has now been defined so that the speed of light (any electromagnetic wave) in vacuum has the exact value

$$c = 299\ 792\ 458 \text{ m/s,}$$

which can be used as a standard. In fact, if you now measure the travel time of a pulse of light from one point to another, you are not really measuring the speed of the light but rather the distance between those two points.

The Traveling Electromagnetic Wave, Quantitatively

We shall now derive Eqs. 33-3 and 33-4 and, even more important, explore the dual induction of electric and magnetic fields that gives us light.

Equation 33-4 and the Induced Electric Field

The dashed rectangle of dimensions dx and h in Fig. 33-6 is fixed at point P on the x axis and in the xy plane (it is shown on the right in Fig. 33-5b). As the electromagnetic wave moves rightward past the rectangle, the magnetic flux Φ_B through the rectangle changes and—according to Faraday's law of induction—induced electric fields appear throughout the region of the rectangle. We take $\vec{E}$ and $\vec{E} + d\vec{E}$ to be the induced fields along the two long sides of the rectangle. These induced electric fields are, in fact, the electrical component of the electromagnetic wave.

Note the small red portion of the magnetic field component curve far from the y axis in Fig. 33-5b. Let's consider the induced electric fields at the instant when this red portion of the magnetic component is passing through the rectangle. Just then, the magnetic field through the rectangle points in the positive z direction and is decreasing in magnitude (the magnitude was greater just before the red section arrived). Because the magnetic field is decreasing, the magnetic flux Φ_B through the rectangle is also decreasing. According to Faraday's law, this change in flux is opposed by induced electric fields, which produce a magnetic field $\vec{B}$ in the positive z direction.

According to Lenz's law, this in turn means that if we imagine the boundary of the rectangle to be a conducting loop, a counterclockwise induced current would have to appear in it. There is, of course, no conducting loop; but this analysis shows that the induced electric field vectors $\vec{E}$ and $\vec{E} + d\vec{E}$ are indeed

The oscillating magnetic field induces an oscillating and perpendicular electric field.

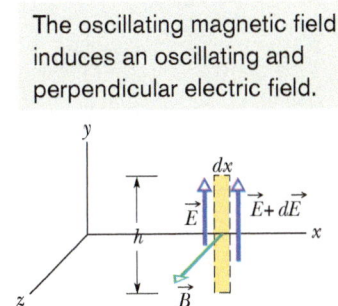

Figure 33-6 As the electromagnetic wave travels rightward past point P in Fig. 33-5b, the sinusoidal variation of the magnetic field $\vec{B}$ through a rectangle centered at P induces electric fields along the rectangle. At the instant shown, $\vec{B}$ is decreasing in magnitude and the induced electric field is therefore greater in magnitude on the right side of the rectangle than on the left.

oriented as shown in Fig. 33-6, with the magnitude of $\vec{E} + d\vec{E}$ greater than that of $\vec{E}$. Otherwise, the net induced electric field would not act counterclockwise around the rectangle.

Faraday's Law. Let us now apply Faraday's law of induction,

$$\oint \vec{E} \cdot d\vec{s} = -\frac{d\Phi_B}{dt}, \tag{33-6}$$

counterclockwise around the rectangle of Fig. 33-6. There is no contribution to the integral from the top or bottom of the rectangle because $\vec{E}$ and $d\vec{s}$ are perpendicular to each other there. The integral then has the value

$$\oint \vec{E} \cdot d\vec{s} = (E + dE)h - Eh = h\,dE. \tag{33-7}$$

The flux Φ_B through this rectangle is

$$\Phi_B = (B)(h\,dx), \tag{33-8}$$

where B is the average magnitude of $\vec{B}$ within the rectangle and $h\,dx$ is the area of the rectangle. Differentiating Eq. 33-8 with respect to t gives

$$\frac{d\Phi_B}{dt} = h\,dx\,\frac{dB}{dt}. \tag{33-9}$$

If we substitute Eqs. 33-7 and 33-9 into Eq. 33-6, we find

$$h\,dE = -h\,dx\,\frac{dB}{dt}$$

or

$$\frac{dE}{dx} = -\frac{dB}{dt}. \tag{33-10}$$

Actually, both B and E are functions of *two* variables, coordinate x and time t, as Eqs. 33-1 and 33-2 show. However, in evaluating dE/dx, we must assume that t is constant because Fig. 33-6 is an "instantaneous snapshot." Also, in evaluating dB/dt we must assume that x is constant (a particular value) because we are dealing with the time rate of change of B at a particular place, the point P shown in Fig. 33-5b. The derivatives under these circumstances are *partial derivatives,* and Eq. 33-10 must be written

$$\frac{\partial E}{\partial x} = -\frac{\partial B}{\partial t}. \tag{33-11}$$

The minus sign in this equation is appropriate and necessary because, although magnitude E is increasing with x at the site of the rectangle in Fig. 33-6, magnitude B is decreasing with t.

From Eq. 33-1 we have

$$\frac{\partial E}{\partial x} = kE_m \cos(kx - \omega t)$$

and from Eq. 33-2

$$\frac{\partial B}{\partial t} = -\omega B_m \cos(kx - \omega t).$$

Then Eq. 33-11 reduces to

$$kE_m \cos(kx - \omega t) = \omega B_m \cos(kx - \omega t). \tag{33-12}$$

The ratio ω/k for a traveling wave is its speed, which we are calling c. Equation 33-12 then becomes

$$\frac{E_m}{B_m} = c \quad \text{(amplitude ratio)}, \tag{33-13}$$

which is just Eq. 33-4.

Equation 33-3 and the Induced Magnetic Field

Figure 33-7 shows another dashed rectangle at point P of Fig. 33-5b; this one is in the xz plane. As the electromagnetic wave moves rightward past this new rectangle, the electric flux Φ_E through the rectangle changes and—according to Maxwell's law of induction—induced magnetic fields appear throughout the region of the rectangle. These induced magnetic fields are, in fact, the magnetic component of the electromagnetic wave.

We see from Fig. 33-5b that at the instant chosen for the magnetic field represented in Fig. 33-6, marked in red on the magnetic component curve, the electric field through the rectangle of Fig. 33-7 is directed as shown. Recall that at the chosen instant, the magnetic field in Fig. 33-6 is decreasing. Because the two fields are in phase, the electric field in Fig. 33-7 must also be decreasing, and so must the electric flux Φ_E through the rectangle. By applying the same reasoning we applied to Fig. 33-6, we see that the changing flux Φ_E will induce a magnetic field with vectors $\vec{B}$ and $\vec{B} + d\vec{B}$ oriented as shown in Fig. 33-7, where field $\vec{B} + d\vec{B}$ is greater than field $\vec{B}$.

Maxwell's Law. Let us apply Maxwell's law of induction,

$$\oint \vec{B} \cdot d\vec{s} = \mu_0 \varepsilon_0 \frac{d\Phi_E}{dt}, \tag{33-14}$$

by proceeding counterclockwise around the dashed rectangle of Fig. 33-7. Only the long sides of the rectangle contribute to the integral because the dot product along the short sides is zero. Thus, we can write

$$\oint \vec{B} \cdot d\vec{s} = -(B + dB)h + Bh = -h\,dB. \tag{33-15}$$

The flux Φ_E through the rectangle is

$$\Phi_E = (E)(h\,dx), \tag{33-16}$$

where E is the average magnitude of $\vec{E}$ within the rectangle. Differentiating Eq. 33-16 with respect to t gives

$$\frac{d\Phi_E}{dt} = h\,dx\,\frac{dE}{dt}.$$

If we substitute this and Eq. 33-15 into Eq. 33-14, we find

$$-h\,dB = \mu_0\varepsilon_0\left(h\,dx\,\frac{dE}{dt}\right)$$

or, changing to partial-derivative notation as we did for Eq. 33-11,

$$-\frac{\partial B}{\partial x} = \mu_0\varepsilon_0\frac{\partial E}{\partial t}. \tag{33-17}$$

Again, the minus sign in this equation is necessary because, although B is increasing with x at point P in the rectangle in Fig. 33-7, E is decreasing with t.

Evaluating Eq. 33-17 by using Eqs. 33-1 and 33-2 leads to

$$-kB_m\cos(kx - \omega t) = -\mu_0\varepsilon_0\omega E_m\cos(kx - \omega t),$$

which we can write as

$$\frac{E_m}{B_m} = \frac{1}{\mu_0\varepsilon_0(\omega/k)} = \frac{1}{\mu_0\varepsilon_0 c}.$$

Combining this with Eq. 33-13 leads at once to

$$c = \frac{1}{\sqrt{\mu_0\varepsilon_0}} \quad \text{(wave speed)}, \tag{33-18}$$

which is exactly Eq. 33-3.

The oscillating electric field induces an oscillating and perpendicular magnetic field.

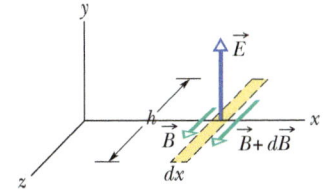

Figure 33-7 The sinusoidal variation of the electric field through this rectangle, located (but not shown) at point P in Fig. 33-5b, induces magnetic fields along the rectangle. The instant shown is that of Fig. 33-6: $\vec{E}$ is decreasing in magnitude, and the magnitude of the induced magnetic field is greater on the right side of the rectangle than on the left.

✓ **Checkpoint 1**

The magnetic field $\vec{B}$ through the rectangle of Fig. 33-6 is shown at a different instant in part 1 of the figure here; $\vec{B}$ is directed in the xz plane, parallel to the z axis, and its magnitude is increasing. (a) Complete part 1 by drawing the induced electric fields, indicating both directions and relative magnitudes (as in Fig. 33-6). (b) For the same instant, complete part 2 of the figure by drawing the electric field of the electromagnetic wave. Also draw the induced magnetic fields, indicating both directions and relative magnitudes (as in Fig. 33-7).

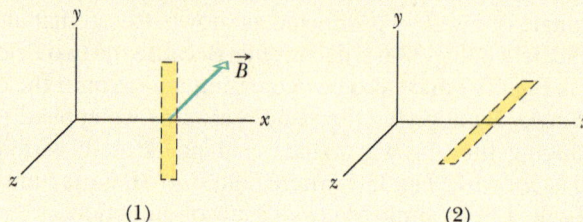

(1)   (2)

33-2 ENERGY TRANSPORT AND THE POYNTING VECTOR

Learning Objectives

After reading this module, you should be able to . . .

33.13 Identify that an electromagnetic wave transports energy.

33.14 For a target, identify that an EM wave's rate of energy transport per unit area is given by the Poynting vector $\vec{S}$, which is related to the cross product of the electric field $\vec{E}$ and magnetic field $\vec{B}$.

33.15 Determine the direction of travel (and thus energy transport) of an electromagnetic wave by applying the cross product for the corresponding Poynting vector.

33.16 Calculate the instantaneous rate S of energy flow of an EM wave in terms of the instantaneous electric field magnitude E.

33.17 For the electric field component of an electromagnetic wave, relate the rms value E_{rms} to the amplitude E_m.

33.18 Identify an EM wave's intensity I in terms of energy transport.

33.19 Apply the relationships between an EM wave's intensity I and the electric field's rms value E_{rms} and amplitude E_m.

33.20 Apply the relationship between average power P_{avg}, energy transfer ΔE, and the time Δt taken by that transfer, and apply the relationship between the instantaneous power P and the rate of energy transfer dE/dt.

33.21 Identify an isotropic point source of light.

33.22 For an isotropic point source of light, apply the relationship between the emission power P, the distance r to a point of measurement, and the intensity I at that point.

33.23 In terms of energy conservation, explain why the intensity from an isotropic point source of light decreases as $1/r^2$.

Key Ideas

● The rate per unit area at which energy is transported via an electromagnetic wave is given by the Poynting vector $\vec{S}$:

$$\vec{S} = \frac{1}{\mu_0} \vec{E} \times \vec{B}.$$

The direction of $\vec{S}$ (and thus of the wave's travel and the energy transport) is perpendicular to the directions of both $\vec{E}$ and $\vec{B}$.

● The time-averaged rate per unit area at which energy is transported is S_{avg}, which is called the intensity I of

the wave:

$$I = \frac{1}{c\mu_0} E_{rms}^2,$$

in which $E_{rms} = E_m/\sqrt{2}$.

● A point source of electromagnetic waves emits the waves isotropically—that is, with equal intensity in all directions. The intensity of the waves at distance r from a point source of power P_s is

$$I = \frac{P_s}{4\pi r^2}.$$

Energy Transport and the Poynting Vector

All sunbathers know that an electromagnetic wave can transport energy and deliver it to a body on which the wave falls. The rate of energy transport per unit area in such a wave is described by a vector $\vec{S}$, called the **Poynting vector** after physicist John Henry Poynting (1852–1914), who first discussed its properties. This vector is defined as

$$\vec{S} = \frac{1}{\mu_0}\vec{E} \times \vec{B} \quad \text{(Poynting vector).} \tag{33-19}$$

Its magnitude S is related to the rate at which energy is transported by a wave across a unit area at any instant (inst):

$$S = \left(\frac{\text{energy/time}}{\text{area}}\right)_{\text{inst}} = \left(\frac{\text{power}}{\text{area}}\right)_{\text{inst}}. \tag{33-20}$$

From this we can see that the SI unit for $\vec{S}$ is the watt per square meter (W/m²).

⭐ The direction of the Poynting vector $\vec{S}$ of an electromagnetic wave at any point gives the wave's direction of travel and the direction of energy transport at that point.

Because $\vec{E}$ and $\vec{B}$ are perpendicular to each other in an electromagnetic wave, the magnitude of $\vec{E} \times \vec{B}$ is EB. Then the magnitude of $\vec{S}$ is

$$S = \frac{1}{\mu_0}EB, \tag{33-21}$$

in which S, E, and B are instantaneous values. The magnitudes E and B are so closely coupled to each other that we need to deal with only one of them; we choose E, largely because most instruments for detecting electromagnetic waves deal with the electric component of the wave rather than the magnetic component. Using $B = E/c$ from Eq. 33-5, we can rewrite Eq. 33-21 in terms of just the electric component as

$$S = \frac{1}{c\mu_0}E^2 \quad \text{(instantaneous energy flow rate).} \tag{33-22}$$

Intensity. By substituting $E = E_m \sin(kx - \omega t)$ into Eq. 33-22, we could obtain an equation for the energy transport rate as a function of time. More useful in practice, however, is the average energy transported over time; for that, we need to find the time-averaged value of S, written S_{avg} and also called the **intensity** I of the wave. Thus from Eq. 33-20, the intensity I is

$$I = S_{\text{avg}} = \left(\frac{\text{energy/time}}{\text{area}}\right)_{\text{avg}} = \left(\frac{\text{power}}{\text{area}}\right)_{\text{avg}}. \tag{33-23}$$

From Eq. 33-22, we find

$$I = S_{\text{avg}} = \frac{1}{c\mu_0}[E^2]_{\text{avg}} = \frac{1}{c\mu_0}[E_m^2 \sin^2(kx - \omega t)]_{\text{avg}}. \tag{33-24}$$

Over a full cycle, the average value of $\sin^2\theta$, for any angular variable θ, is $\frac{1}{2}$ (see Fig. 31-17). In addition, we define a new quantity E_{rms}, the *root-mean-square* value of the electric field, as

$$E_{\text{rms}} = \frac{E_m}{\sqrt{2}}. \tag{33-25}$$

The energy emitted by light source S must pass through the sphere of radius r.

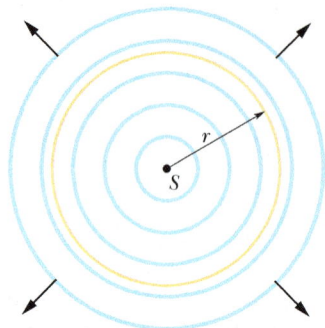

Figure 33-8 A point source S emits electromagnetic waves uniformly in all directions. The spherical wavefronts pass through an imaginary sphere of radius r that is centered on S.

We can then rewrite Eq. 33-24 as

$$I = \frac{1}{c\mu_0} E_{rms}^2. \qquad (33\text{-}26)$$

Because $E = cB$ and c is such a very large number, you might conclude that the energy associated with the electric field is much greater than that associated with the magnetic field. That conclusion is incorrect; the two energies are exactly equal. To show this, we start with Eq. 25-25, which gives the energy density $u \left(= \frac{1}{2}\varepsilon_0 E^2 \right)$ within an electric field, and substitute cB for E; then we can write

$$u_E = \tfrac{1}{2}\varepsilon_0 E^2 = \tfrac{1}{2}\varepsilon_0 (cB)^2.$$

If we now substitute for c with Eq. 33-3, we get

$$u_E = \tfrac{1}{2}\varepsilon_0 \frac{1}{\mu_0 \varepsilon_0} B^2 = \frac{B^2}{2\mu_0}.$$

However, Eq. 30-55 tells us that $B^2/2\mu_0$ is the energy density u_B of a magnetic field $\vec{B}$; so we see that $u_E = u_B$ everywhere along an electromagnetic wave.

Variation of Intensity with Distance

How intensity varies with distance from a real source of electromagnetic radiation is often complex—especially when the source (like a searchlight at a movie premier) beams the radiation in a particular direction. However, in some situations we can assume that the source is a *point source* that emits the light *isotropically*—that is, with equal intensity in all directions. The spherical wavefronts spreading from such an isotropic point source S at a particular instant are shown in cross section in Fig. 33-8.

Let us assume that the energy of the waves is conserved as they spread from this source. Let us also center an imaginary sphere of radius r on the source, as shown in Fig. 33-8. All the energy emitted by the source must pass through the sphere. Thus, the rate at which energy passes through the sphere via the radiation must equal the rate at which energy is emitted by the source—that is, the source power P_s. The intensity I (power per unit area) measured at the sphere must then be, from Eq. 33-23,

$$I = \frac{\text{power}}{\text{area}} = \frac{P_s}{4\pi r^2}, \qquad (33\text{-}27)$$

where $4\pi r^2$ is the area of the sphere. Equation 33-27 tells us that the intensity of the electromagnetic radiation from an isotropic point source decreases with the square of the distance r from the source.

✓ **Checkpoint 2**

The figure here gives the electric field of an electromagnetic wave at a certain point and a certain instant. The wave is transporting energy in the negative z direction. What is the direction of the magnetic field of the wave at that point and instant?

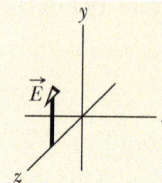

Sample Problem 33.01   **Light wave: rms values of the electric and magnetic fields**

When you look at the North Star (Polaris), you intercept light from a star at a distance of 431 ly and emitting energy at a rate of 2.2×10^3 times that of our Sun ($P_{sun} = 3.90 \times$ 10^{26} W). Neglecting any atmospheric absorption, find the rms values of the electric and magnetic fields when the starlight reaches you.

KEY IDEAS

1. The rms value E_{rms} of the electric field in light is related to the intensity I of the light via Eq. 33-26 ($I = E_{rms}^2/c\mu_0$).

2. Because the source is so far away and emits light with equal intensity in all directions, the intensity I at any distance r from the source is related to the source's power P_s via Eq. 33-27 ($I = P_s/4\pi r^2$).

3. The magnitudes of the electric field and magnetic field of an electromagnetic wave at any instant and at any point in the wave are related by the speed of light c according to Eq. 33-5 ($E/B = c$). Thus, the rms values of those fields are also related by Eq. 33-5.

Electric field: Putting the first two ideas together gives us

$$I = \frac{P_s}{4\pi r^2} = \frac{E_{rms}^2}{c\mu_0}$$

and

$$E_{rms} = \sqrt{\frac{P_s c\mu_0}{4\pi r^2}}.$$

By substituting $P_s = (2.2 \times 10^3)(3.90 \times 10^{26}$ W$)$, $r = 431$ ly $= 4.08 \times 10^{18}$ m, and values for the constants, we find

$$E_{rms} = 1.24 \times 10^{-3} \text{ V/m} \approx 1.2 \text{ mV/m}. \qquad \text{(Answer)}$$

Magnetic field: From Eq. 33-5, we write

$$B_{rms} = \frac{E_{rms}}{c} = \frac{1.24 \times 10^{-3} \text{ V/m}}{3.00 \times 10^8 \text{ m/s}}$$

$$= 4.1 \times 10^{-12} \text{ T} = 4.1 \text{ pT}.$$

Cannot compare the fields: Note that E_{rms} ($= 1.2$ mV/m) is small as judged by ordinary laboratory standards, but B_{rms} ($= 4.1$ pT) is quite small. This difference helps to explain why most instruments used for the detection and measurement of electromagnetic waves are designed to respond to the electric component. It is wrong, however, to say that the electric component of an electromagnetic wave is "stronger" than the magnetic component. You cannot compare quantities that are measured in different units. However, these electric and magnetic components are on an equal basis because their average energies, which *can* be compared, are equal.

PLUS Additional examples, video, and practice available at *WileyPLUS*

33-3 RADIATION PRESSURE

Learning Objectives

After reading this module, you should be able to . . .

33.24 Distinguish between force and pressure.

33.25 Identify that an electromagnetic wave transports momentum and can exert a force and a pressure on a target.

33.26 For a uniform electromagnetic beam that is perpendicular to a target area, apply the relationships between that

area, the wave's intensity, and the force on the target, for both total absorption and total backward reflection.

33.27 For a uniform electromagnetic beam that is perpendicular to a target area, apply the relationships between the wave's intensity and the pressure on the target, for both total absorption and total backward reflection.

Key Ideas

● When a surface intercepts electromagnetic radiation, a force and a pressure are exerted on the surface.

● If the radiation is totally absorbed by the surface, the force is

$$F = \frac{IA}{c} \quad \text{(total absorption)},$$

in which I is the intensity of the radiation and A is the area of the surface perpendicular to the path of the radiation.

● If the radiation is totally reflected back along its original

path, the force is

$$F = \frac{2IA}{c} \quad \text{(total reflection back along path)}.$$

● The radiation pressure p_r is the force per unit area:

$$p_r = \frac{I}{c} \quad \text{(total absorption)}$$

and

$$p_r = \frac{2I}{c} \quad \text{(total reflection back along path)}.$$

Radiation Pressure

Electromagnetic waves have linear momentum and thus can exert a pressure on an object when shining on it. However, the pressure must be very small because, for example, you do not feel a punch during a camera flash.

To find an expression for the pressure, let us shine a beam of electromagnetic radiation—light, for example—on an object for a time interval Δt. Further, let us assume that the object is free to move and that the radiation is entirely **absorbed** (taken up) by the object. This means that during the interval Δt, the object gains an energy ΔU from the radiation. Maxwell showed that the object also gains linear momentum. The magnitude Δp of the momentum change of the object is related to the energy change ΔU by

$$\Delta p = \frac{\Delta U}{c} \quad \text{(total absorption)}, \tag{33-28}$$

where c is the speed of light. The direction of the momentum change of the object is the direction of the *incident* (incoming) beam that the object absorbs.

Instead of being absorbed, the radiation can be **reflected** by the object; that is, the radiation can be sent off in a new direction as if it bounced off the object. If the radiation is entirely reflected back along its original path, the magnitude of the momentum change of the object is twice that given above, or

$$\Delta p = \frac{2 \, \Delta U}{c} \quad \text{(total reflection back along path)}. \tag{33-29}$$

In the same way, an object undergoes twice as much momentum change when a perfectly elastic tennis ball is bounced from it as when it is struck by a perfectly inelastic ball (a lump of wet putty, say) of the same mass and velocity. If the incident radiation is partly absorbed and partly reflected, the momentum change of the object is between $\Delta U/c$ and $2 \, \Delta U/c$.

Force. From Newton's second law in its linear momentum form (Module 9-3), we know that a change in momentum is related to a force by

$$F = \frac{\Delta p}{\Delta t}. \tag{33-30}$$

To find expressions for the force exerted by radiation in terms of the intensity I of the radiation, we first note that intensity is

$$I = \frac{\text{power}}{\text{area}} = \frac{\text{energy/time}}{\text{area}}.$$

Next, suppose that a flat surface of area A, perpendicular to the path of the radiation, intercepts the radiation. In time interval Δt, the energy intercepted by area A is

$$\Delta U = IA \, \Delta t. \tag{33-31}$$

If the energy is completely absorbed, then Eq. 33-28 tells us that $\Delta p = IA \, \Delta t/c$, and, from Eq. 33-30, the magnitude of the force on the area A is

$$F = \frac{IA}{c} \quad \text{(total absorption)}. \tag{33-32}$$

Similarly, if the radiation is totally reflected back along its original path, Eq. 33-29 tells us that $\Delta p = 2IA \, \Delta t/c$ and, from Eq. 33-30,

$$F = \frac{2IA}{c} \quad \text{(total reflection back along path)}. \tag{33-33}$$

If the radiation is partly absorbed and partly reflected, the magnitude of the force on area A is between the values of IA/c and $2IA/c$.

Pressure. The force per unit area on an object due to radiation is the radiation pressure p_r. We can find it for the situations of Eqs. 33-32 and 33-33 by dividing both sides of each equation by A. We obtain

$$p_r = \frac{I}{c} \quad \text{(total absorption)} \tag{33-34}$$

and

$$p_r = \frac{2I}{c} \quad \text{(total reflection back along path).}\qquad (33\text{-}35)$$

Be careful not to confuse the symbol p_r for radiation pressure with the symbol p for momentum. Just as with fluid pressure in Chapter 14, the SI unit of radiation pressure is the newton per square meter (N/m^2), which is called the pascal (Pa).

The development of laser technology has permitted researchers to achieve radiation pressures much greater than, say, that due to a camera flashlamp. This comes about because a beam of laser light—unlike a beam of light from a small lamp filament—can be focused to a tiny spot. This permits the delivery of great amounts of energy to small objects placed at that spot.

✓ Checkpoint 3

Light of uniform intensity shines perpendicularly on a totally absorbing surface, fully illuminating the surface. If the area of the surface is decreased, do (a) the radiation pressure and (b) the radiation force on the surface increase, decrease, or stay the same?

33-4 POLARIZATION

Learning Objectives

After reading this module, you should be able to . . .

33.28 Distinguish between polarized light and unpolarized light.

33.29 For a light beam headed toward you, sketch representations of polarized light and unpolarized light.

33.30 When a beam is sent into a polarizing sheet, explain the function of the sheet in terms of its polarizing direction (or axis) and the electric field component that is absorbed and the component that is transmitted.

33.31 For light that emerges from a polarizing sheet, identify its polarization relative to the sheet's polarizing direction.

33.32 For a light beam incident perpendicularly on a polarizing sheet, apply the one-half rule and the cosine-squared rule, distinguishing their uses.

33.33 Distinguish between a polarizer and an analyzer.

33.34 Explain what is meant if two sheets are crossed.

33.35 When a beam is sent into a system of polarizing sheets, work through the sheets one by one, finding the transmitted intensity and polarization.

Key Ideas

● Electromagnetic waves are polarized if their electric field vectors are all in a single plane, called the plane of oscillation. Light waves from common sources are not polarized; that is, they are unpolarized, or polarized randomly.

● When a polarizing sheet is placed in the path of light, only electric field components of the light parallel to the sheet's polarizing direction are transmitted by the sheet; components perpendicular to the polarizing direction are absorbed. The light that emerges from a polarizing sheet is polarized parallel to the polarizing direction of the sheet.

● If the original light is initially unpolarized, the transmitted intensity I is half the original intensity I_0:

$$I = \tfrac{1}{2}I_0.$$

● If the original light is initially polarized, the transmitted intensity depends on the angle θ between the polarization direction of the original light and the polarizing direction of the sheet:

$$I = I_0 \cos^2 \theta.$$

Polarization

VHF (very high frequency) television antennas in England are oriented vertically, but those in North America are horizontal. The difference is due to the direction of oscillation of the electromagnetic waves carrying the TV signal. In England, the transmitting equipment is designed to produce waves that are **polarized** vertically; that is, their electric field oscillates vertically. Thus, for the

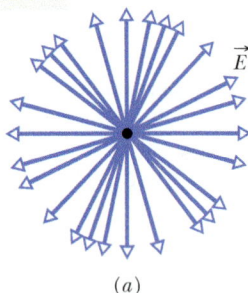

Figure 33-9 (*a*) The plane of oscillation of a polarized electromagnetic wave. (*b*) To represent the polarization, we view the plane of oscillation head-on and indicate the directions of the oscillating electric field with a double arrow.

electric field of the incident television waves to drive a current along an antenna (and provide a signal to a television set), the antenna must be vertical. In North America, the waves are polarized horizontally.

Figure 33-9*a* shows an electromagnetic wave with its electric field oscillating parallel to the vertical *y* axis. The plane containing the $\vec{E}$ vectors is called the **plane of oscillation** of the wave (hence, the wave is said to be *plane-polarized* in the *y* direction). We can represent the wave's *polarization* (state of being polarized) by showing the directions of the electric field oscillations in a head-on view of the plane of oscillation, as in Fig. 33-9*b*. The vertical double arrow in that figure indicates that as the wave travels past us, its electric field oscillates vertically—it continuously changes between being directed up and down the *y* axis.

Polarized Light

The electromagnetic waves emitted by a television station all have the same polarization, but the electromagnetic waves emitted by any common source of light (such as the Sun or a bulb) are **polarized randomly,** or **unpolarized** (the two terms mean the same thing). That is, the electric field at any given point is always perpendicular to the direction of travel of the waves but changes directions randomly. Thus, if we try to represent a head-on view of the oscillations over some time period, we do not have a simple drawing with a single double arrow like that of Fig. 33-9*b*; instead we have a mess of double arrows like that in Fig. 33-10*a*.

In principle, we can simplify the mess by resolving each electric field of Fig. 33-10*a* into *y* and *z* components. Then as the wave travels past us, the net *y* component oscillates parallel to the *y* axis and the net *z* component oscillates parallel to the *z* axis. We can then represent the unpolarized light with a pair of double arrows as shown in Fig. 33-10*b*. The double arrow along the *y* axis represents the oscillations of the net *y* component of the electric field. The double arrow along the *z* axis represents the oscillations of the net *z* component of the electric field. In doing all this, we effectively change unpolarized light into the superposition of two polarized waves whose planes of oscillation are perpendicular to each other—one plane contains the *y* axis and the other contains the *z* axis. One reason to make this change is that drawing Fig. 33-10*b* is a lot easier than drawing Fig. 33-10*a*.

We can draw similar figures to represent light that is **partially polarized** (its field oscillations are not completely random as in Fig. 33-10*a*, nor are they parallel to a single axis as in Fig. 33-9*b*). For this situation, we draw one of the double arrows in a perpendicular pair of double arrows longer than the other one.

Polarizing Direction. We can transform unpolarized visible light into polarized light by sending it through a *polarizing sheet*, as is shown in Fig. 33-11. Such sheets, commercially known as Polaroids or Polaroid filters, were invented in 1932 by Edwin Land while he was an undergraduate student. A polarizing sheet consists of certain long molecules embedded in plastic. When the sheet is manu-

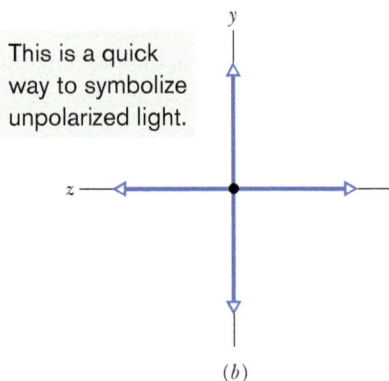

Figure 33-10 (*a*) Unpolarized light consists of waves with randomly directed electric fields. Here the waves are all traveling along the same axis, directly out of the page, and all have the same amplitude *E*. (*b*) A second way of representing unpolarized light—the light is the superposition of two polarized waves whose planes of oscillation are perpendicular to each other.

factured, it is stretched to align the molecules in parallel rows, like rows in a plowed field. When light is then sent through the sheet, electric field components along one direction pass through the sheet, while components perpendicular to that direction are absorbed by the molecules and disappear.

We shall not dwell on the molecules but, instead, shall assign to the sheet a *polarizing direction*, along which electric field components are passed:

⭐ An electric field component parallel to the polarizing direction is passed (*transmitted*) by a polarizing sheet; a component perpendicular to it is absorbed.

Thus, the electric field of the light emerging from the sheet consists of only the components that are parallel to the polarizing direction of the sheet; hence the light is polarized in that direction. In Fig. 33-11, the vertical electric field components are transmitted by the sheet; the horizontal components are absorbed. The transmitted waves are then vertically polarized.

Intensity of Transmitted Polarized Light

We now consider the intensity of light transmitted by a polarizing sheet. We start with unpolarized light, whose electric field oscillations we can resolve into y and z components as represented in Fig. 33-10b. Further, we can arrange for the y axis to be parallel to the polarizing direction of the sheet. Then only the y components of the light's electric field are passed by the sheet; the z components are absorbed. As suggested by Fig. 33-10b, if the original waves are randomly oriented, the sum of the y components and the sum of the z components are equal. When the z components are absorbed, half the intensity I_0 of the original light is lost. The intensity I of the emerging polarized light is then

$$I = \tfrac{1}{2}I_0 \quad \text{(one-half rule)}. \tag{33-36}$$

Let us call this the *one-half rule;* we can use it *only* when the light reaching a polarizing sheet is unpolarized.

Suppose now that the light reaching a polarizing sheet is already polarized. Figure 33-12 shows a polarizing sheet in the plane of the page and the electric field $\vec{E}$ of such a polarized light wave traveling toward the sheet (and thus prior to any absorption). We can resolve $\vec{E}$ into two components relative to the polarizing direction of the sheet: parallel component E_y is transmitted by the sheet, and perpendicular component E_z is absorbed. Since θ is the angle between $\vec{E}$ and the polarizing direction of the sheet, the transmitted parallel component is

$$E_y = E \cos \theta. \tag{33-37}$$

Recall that the intensity of an electromagnetic wave (such as our light wave) is proportional to the square of the electric field's magnitude (Eq. 33-26, $I = E_{\text{rms}}^2/c\mu_0$). In our present case then, the intensity I of the emerging wave is proportional to E_y^2 and the intensity I_0 of the original wave is proportional to E^2. Hence, from Eq. 33-37 we can write $I/I_0 = \cos^2 \theta$, or

$$I = I_0 \cos^2 \theta \quad \text{(cosine-squared rule)}. \tag{33-38}$$

Let us call this the *cosine-squared rule;* we can use it *only* when the light reaching a polarizing sheet is already polarized. Then the transmitted intensity I is a maximum and is equal to the original intensity I_0 when the original wave is polarized parallel to the polarizing direction of the sheet (when θ in Eq. 33-38 is 0° or 180°). The transmitted intensity is zero when the original wave is polarized perpendicular to the polarizing direction of the sheet (when θ is 90°).

The sheet's polarizing axis is vertical, so only vertically polarized light emerges.

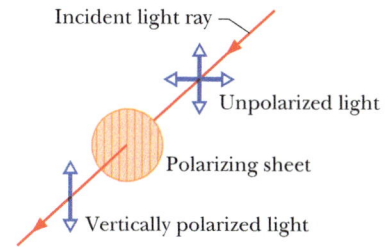

Figure 33-11 Unpolarized light becomes polarized when it is sent through a polarizing sheet. Its direction of polarization is then parallel to the polarizing direction of the sheet, which is represented here by the vertical lines drawn in the sheet.

The sheet's polarizing axis is vertical, so only vertical components of the electric fields pass.

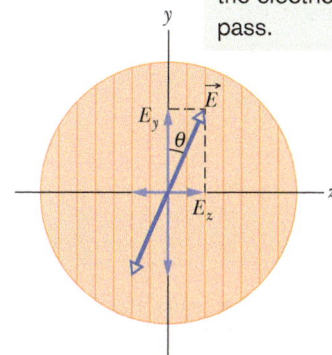

Figure 33-12 Polarized light approaching a polarizing sheet. The electric field $\vec{E}$ of the light can be resolved into components E_y (parallel to the polarizing direction of the sheet) and E_z (perpendicular to that direction). Component E_y will be transmitted by the sheet; component E_z will be absorbed.

Polarizing direction

This light is vertically polarized.

The sheet's polarizing axis is tilted, so only a fraction of the intensity passes.

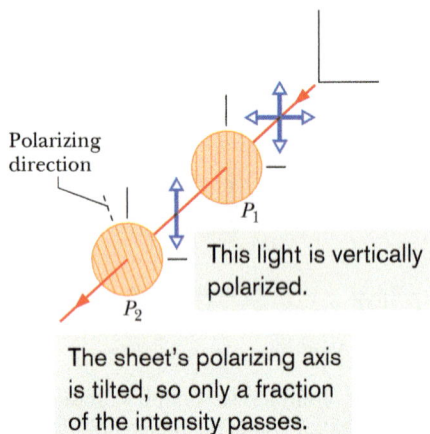

Figure 33-13 The light transmitted by polarizing sheet P_1 is vertically polarized, as represented by the vertical double arrow. The amount of that light that is then transmitted by polarizing sheet P_2 depends on the angle between the polarization direction of that light and the polarizing direction of P_2 (indicated by the lines drawn in the sheet and by the dashed line).

Figure 33-14 (a) Overlapping polarizing sheets transmit light fairly well when their polarizing directions have the same orientation, but (b) they block most of the light when they are crossed.

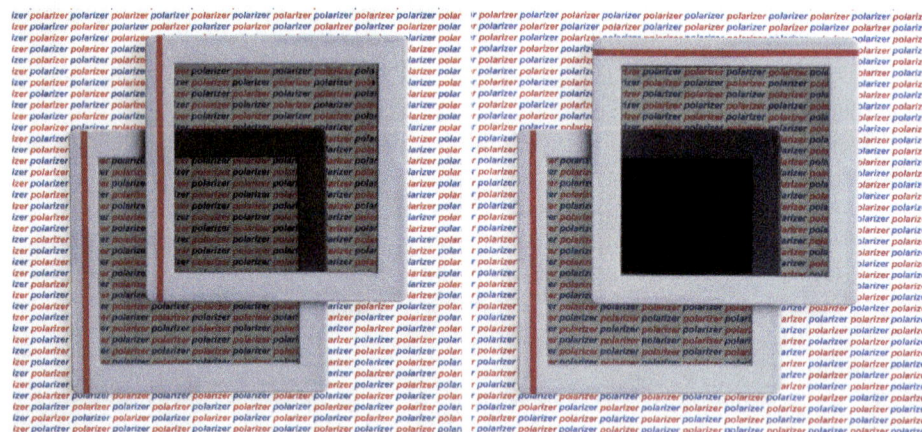

Two Polarizing Sheets. Figure 33-13 shows an arrangement in which initially unpolarized light is sent through two polarizing sheets P_1 and P_2. (Often, the first sheet is called the *polarizer,* and the second the *analyzer.*) Because the polarizing direction of P_1 is vertical, the light transmitted by P_1 to P_2 is polarized vertically. If the polarizing direction of P_2 is also vertical, then all the light transmitted by P_1 is transmitted by P_2. If the polarizing direction of P_2 is horizontal, none of the light transmitted by P_1 is transmitted by P_2. We reach the same conclusions by considering only the *relative* orientations of the two sheets: If their polarizing directions are parallel, all the light passed by the first sheet is passed by the second sheet (Fig. 33-14a). If those directions are perpendicular (the sheets are said to be *crossed*), no light is passed by the second sheet (Fig. 33-14b). Finally, if the two polarizing directions of Fig. 33-13 make an angle between 0° and 90°, some of the light transmitted by P_1 will be transmitted by P_2, as set by Eq. 33-38.

Other Means. Light can be polarized by means other than polarizing sheets, such as by reflection (discussed in Module 33-7) and by scattering from atoms or molecules. In *scattering,* light that is intercepted by an object, such as a molecule, is sent off in many, perhaps random, directions. An example is the scattering of sunlight by molecules in the atmosphere, which gives the sky its general glow.

Although direct sunlight is unpolarized, light from much of the sky is at least partially polarized by such scattering. Bees use the polarization of sky light in navigating to and from their hives. Similarly, the Vikings used it to navigate across the North Sea when the daytime Sun was below the horizon (because of the high latitude of the North Sea). These early seafarers had discovered certain crystals (now called cordierite) that changed color when rotated in polarized light. By looking at the sky through such a crystal while rotating it about their line of sight, they could locate the hidden Sun and thus determine which way was south.

Richard Megna/Fundamental Photographs

(a)   (b)

✓ **Checkpoint 4**

The figure shows four pairs of polarizing sheets, seen face-on. Each pair is mounted in the path of initially unpolarized light. The polarizing direction of each sheet (indicated by the dashed line) is referenced to either a horizontal x axis or a vertical y axis. Rank the pairs according to the fraction of the initial intensity that they pass, greatest first.

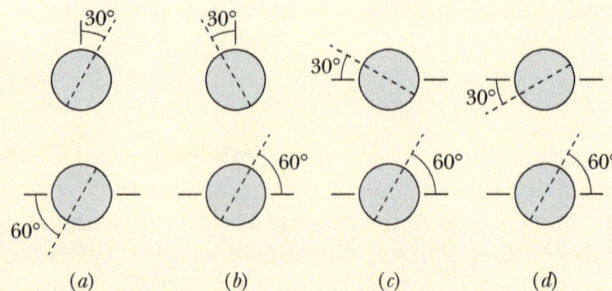

(a)   (b)   (c)   (d)

Sample Problem 33.02 Polarization and intensity with three polarizing sheets

Figure 33-15a, drawn in perspective, shows a system of three polarizing sheets in the path of initially unpolarized light. The polarizing direction of the first sheet is parallel to the y axis, that of the second sheet is at an angle of 60° counterclockwise from the y axis, and that of the third sheet is parallel to the x axis. What fraction of the initial intensity I_0 of the light emerges from the three-sheet system, and in which direction is that emerging light polarized?

KEY IDEAS

1. We work through the system sheet by sheet, from the first one encountered by the light to the last one.

2. To find the intensity transmitted by any sheet, we apply either the one-half rule or the cosine-squared rule, depending on whether the light reaching the sheet is unpolarized or already polarized.

3. The light that is transmitted by a polarizing sheet is always polarized parallel to the polarizing direction of the sheet.

First sheet: The original light wave is represented in Fig. 33-15b, using the head-on, double-arrow representation of Fig. 33-10b. Because the light is initially unpolarized, the intensity I_1 of the light transmitted by the first sheet is given by the one-half rule (Eq. 33-36):

$$I_1 = \tfrac{1}{2}I_0.$$

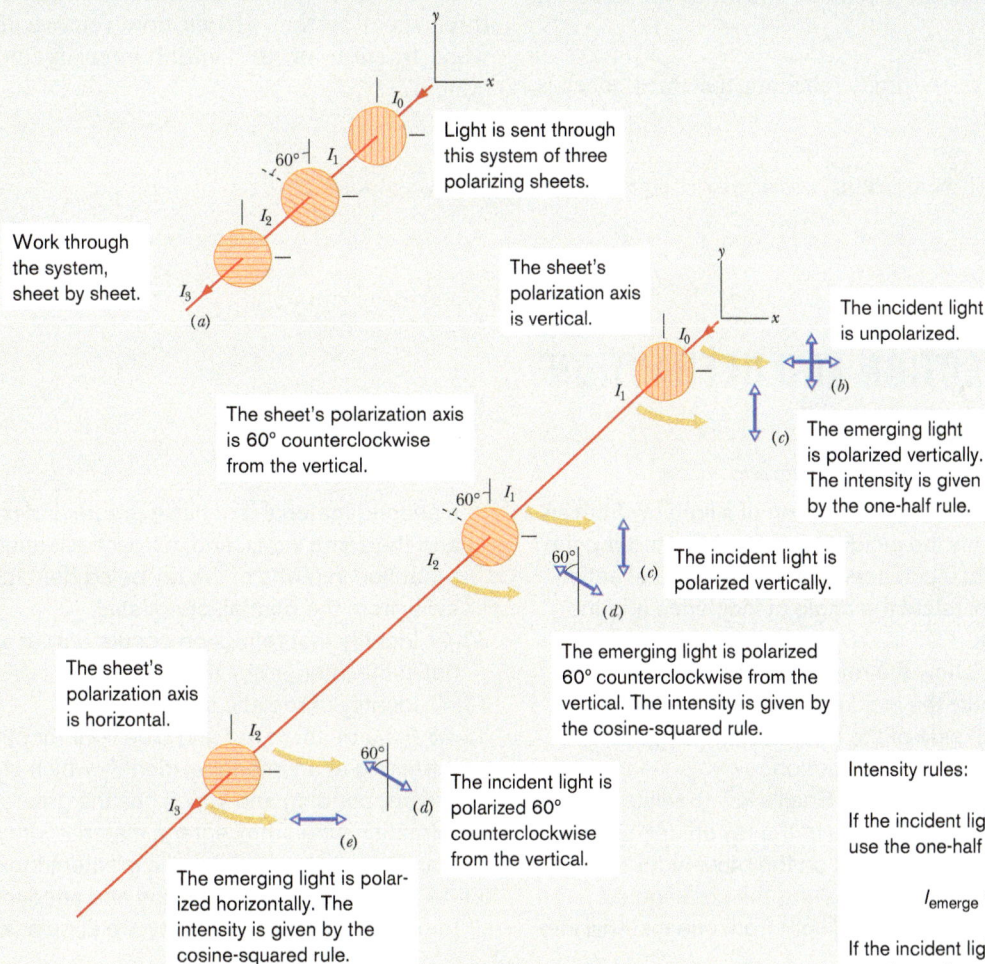

Light is sent through this system of three polarizing sheets.

Work through the system, sheet by sheet.

(a)

The sheet's polarization axis is 60° counterclockwise from the vertical.

The sheet's polarization axis is vertical.

The incident light is unpolarized.

(b)

The emerging light is polarized vertically. The intensity is given by the one-half rule.

(c)

The incident light is polarized vertically.

(d)

The emerging light is polarized 60° counterclockwise from the vertical. The intensity is given by the cosine-squared rule.

The sheet's polarization axis is horizontal.

The incident light is polarized 60° counterclockwise from the vertical.

(d)

(e)

The emerging light is polarized horizontally. The intensity is given by the cosine-squared rule.

Intensity rules:

If the incident light is unpolarized, use the one-half rule:

$$I_{emerge} = 0.5I_{incident}.$$

If the incident light is already polarized, use the cosine-squared rule:

$$I_{emerge} = I_{incident}(\cos\theta)^2,$$

but be sure to insert the angle between the polarization of the incident light and the polarization axis of the sheet.

Figure 33-15 (a) Initially unpolarized light of intensity I_0 is sent into a system of three polarizing sheets. The intensities I_1, I_2, and I_3 of the light transmitted by the sheets are labeled. Shown also are the polarizations, from head-on views, of (b) the initial light and the light transmitted by (c) the first sheet, (d) the second sheet, and (e) the third sheet.

Because the polarizing direction of the first sheet is parallel to the y axis, the polarization of the light transmitted by it is also, as shown in the head-on view of Fig. 33-15c.

Second sheet: Because the light reaching the second sheet is polarized, the intensity I_2 of the light transmitted by that sheet is given by the cosine-squared rule (Eq. 33-38). The angle θ in the rule is the angle between the polarization direction of the entering light (parallel to the y axis) and the polarizing direction of the second sheet (60° counterclockwise from the y axis), and so θ is 60°. (The larger angle between the two directions, namely 120°, can also be used.) We have

$$I_2 = I_1 \cos^2 60°.$$

The polarization of this transmitted light is parallel to the polarizing direction of the sheet transmitting it—that is, 60° counterclockwise from the y axis, as shown in the head-on view of Fig. 33-15d.

Third sheet: Because the light reaching the third sheet is polarized, the intensity I_3 of the light transmitted by that sheet is given by the cosine-squared rule. The angle θ is now the angle between the polarization direction of the entering light (Fig. 33-15d) and the polarizing direction of the third sheet (parallel to the x axis), and so $\theta = 30°$. Thus,

$$I_3 = I_2 \cos^2 30°.$$

This final transmitted light is polarized parallel to the x axis (Fig. 33-15e). We find its intensity by substituting first for I_2 and then for I_1 in the equation above:

$$I_3 = I_2 \cos^2 30° = (I_1 \cos^2 60°) \cos^2 30°$$
$$= (\tfrac{1}{2}I_0) \cos^2 60° \cos^2 30° = 0.094 I_0.$$

Thus, $$\frac{I_3}{I_0} = 0.094. \qquad \text{(Answer)}$$

That is to say, 9.4% of the initial intensity emerges from the three-sheet system. (If we now remove the second sheet, what fraction of the initial intensity emerges from the system?)

WILEY PLUS Additional examples, video, and practice available at *WileyPLUS*

33-5 REFLECTION AND REFRACTION

Learning Objectives

After reading this module, you should be able to . . .

33.36 With a sketch, show the reflection of a light ray from an interface and identify the incident ray, the reflected ray, the normal, the angle of incidence, and the angle of reflection.

33.37 For a reflection, relate the angle of incidence and the angle of reflection.

33.38 With a sketch, show the refraction of a light ray at an interface and identify the incident ray, the refracted ray, the normal on each side of the interface, the angle of incidence, and the angle of refraction.

33.39 For refraction of light, apply Snell's law to relate the index of refraction and the angle of the ray on one side of the interface to those quantities on the other side.

33.40 In a sketch and using a line along the undeflected direction, show the refraction of light from one material into

a second material that has a greater index, a smaller index, and the same index, and, for each situation, describe the refraction in terms of the ray being bent toward the normal, away from the normal, or not at all.

33.41 Identify that refraction occurs only at an interface and not in the interior of a material.

33.42 Identify chromatic dispersion.

33.43 For a beam of red and blue light (or other colors) refracting at an interface, identify which color has the greater bending and which has the greater angle of refraction when they enter a material with a lower index than the initial material and a greater index.

33.44 Describe how the primary and secondary rainbows are formed and explain why they are circular arcs.

Key Ideas

● Geometrical optics is an approximate treatment of light in which light waves are represented as straight-line rays.

● When a light ray encounters a boundary between two transparent media, a reflected ray and a refracted ray generally appear. Both rays remain in the plane of incidence. The angle of reflection is equal to the angle of incidence, and

the angle of refraction is related to the angle of incidence by Snell's law,

$$n_2 \sin \theta_2 = n_1 \sin \theta_1 \quad \text{(refraction)},$$

where n_1 and n_2 are the indexes of refraction of the media in which the incident and refracted rays travel.

Reflection and Refraction

Although a light wave spreads as it moves away from its source, we can often approximate its travel as being in a straight line; we did so for the light wave in Fig. 33-5a. The study of the properties of light waves under that approximation is called *geometrical optics*. For the rest of this chapter and all of Chapter 34, we shall discuss the geometrical optics of visible light.

The photograph in Fig. 33-16a shows an example of light waves traveling in approximately straight lines. A narrow beam of light (the *incident* beam), angled downward from the left and traveling through air, encounters a *plane* (flat) water surface. Part of the light is **reflected** by the surface, forming a beam directed upward toward the right, traveling as if the original beam had bounced from the surface. The rest of the light travels through the surface and into the water, forming a beam directed downward to the right. Because light can travel through it, the water is said to be *transparent;* that is, we can see through it. (In this chapter we shall consider only transparent materials and not opaque materials, through which light cannot travel.)

The travel of light through a surface (or *interface*) that separates two media is called **refraction,** and the light is said to be *refracted.* Unless an incident beam of light is perpendicular to the surface, refraction changes the light's direction of travel. For this reason, the beam is said to be "bent" by the refraction. Note in Fig. 33-16a that the bending occurs only at the surface; within the water, the light travels in a straight line.

In Figure 33-16b, the beams of light in the photograph are represented with an *incident ray,* a *reflected ray,* and a *refracted ray* (and wavefronts). Each ray is oriented with respect to a line, called the *normal,* that is perpendicular to the surface at the point of reflection and refraction. In Fig. 33-16b, the **angle of incidence** is θ_1, the **angle of reflection** is θ_1', and the **angle of refraction** is θ_2, all measured *relative to the normal.* The plane containing the incident ray and the normal is the *plane of incidence,* which is in the plane of the page in Fig. 33-16b.

Experiment shows that reflection and refraction are governed by two laws:

Law of reflection: A reflected ray lies in the plane of incidence and has an angle of reflection equal to the angle of incidence (both relative to the normal). In Fig. 33-16b, this means that

$$\theta_1' = \theta_1 \quad \text{(reflection).} \tag{33-39}$$

(We shall now usually drop the prime on the angle of reflection.)

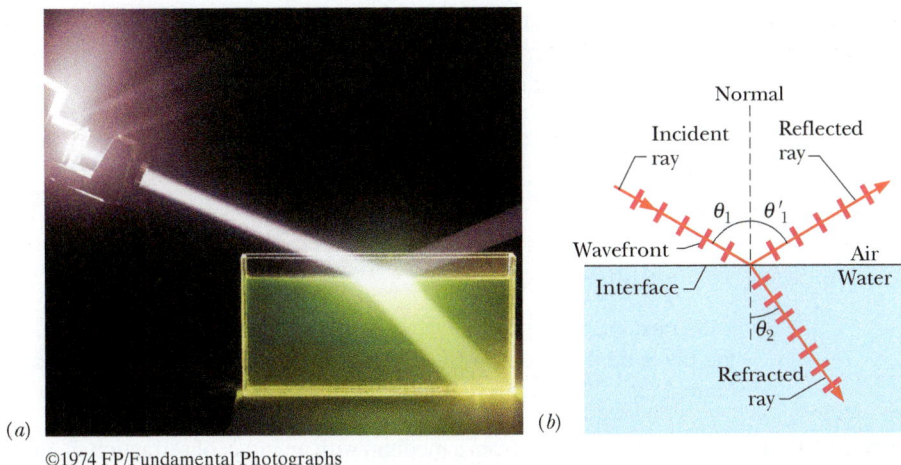

(a)

©1974 FP/Fundamental Photographs

(b)

Figure 33-16 (a) A photograph showing an incident beam of light reflected and refracted by a horizontal water surface. (b) A ray representation of (a). The angles of incidence (θ_1), reflection (θ_1'), and refraction (θ_2) are marked.

Table 33-1 Some Indexes of Refraction[a]

Medium	Index	Medium	Index
Vacuum	Exactly 1	Typical crown glass	1.52
Air (STP)[b]	1.00029	Sodium chloride	1.54
Water (20°C)	1.33	Polystyrene	1.55
Acetone	1.36	Carbon disulfide	1.63
Ethyl alcohol	1.36	Heavy flint glass	1.65
Sugar solution (30%)	1.38	Sapphire	1.77
Fused quartz	1.46	Heaviest flint glass	1.89
Sugar solution (80%)	1.49	Diamond	2.42

[a]For a wavelength of 589 nm (yellow sodium light).
[b]STP means "standard temperature (0°C) and pressure (1 atm)."

Law of refraction: A refracted ray lies in the plane of incidence and has an angle of refraction θ_2 that is related to the angle of incidence θ_1 by

$$n_2 \sin \theta_2 = n_1 \sin \theta_1 \quad \text{(refraction).} \quad (33\text{-}40)$$

Here each of the symbols n_1 and n_2 is a dimensionless constant, called the **index of refraction,** that is associated with a medium involved in the refraction. We derive this equation, called **Snell's law,** in Chapter 35. As we shall discuss there, the index of refraction of a medium is equal to c/v, where v is the speed of light in that medium and c is its speed in vacuum.

Table 33-1 gives the indexes of refraction of vacuum and some common substances. For vacuum, n is defined to be exactly 1; for air, n is very close to 1.0 (an approximation we shall often make). Nothing has an index of refraction below 1.

We can rearrange Eq. 33-40 as

$$\sin \theta_2 = \frac{n_1}{n_2} \sin \theta_1 \quad (33\text{-}41)$$

to compare the angle of refraction θ_2 with the angle of incidence θ_1. We can then see that the relative value of θ_2 depends on the relative values of n_2 and n_1:

1. If n_2 is equal to n_1, then θ_2 is equal to θ_1 and refraction does not bend the light beam, which continues in the *undeflected direction,* as in Fig. 33-17a.

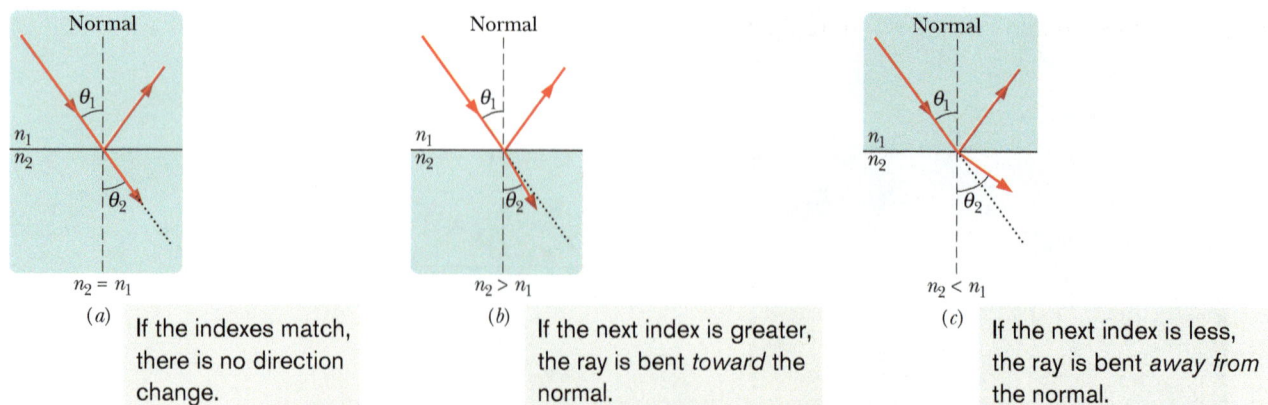

(a) If the indexes match, there is no direction change.

(b) If the next index is greater, the ray is bent *toward* the normal.

(c) If the next index is less, the ray is bent *away from* the normal.

Figure 33-17 Refraction of light traveling from a medium with an index of refraction n_1 into a medium with an index of refraction n_2. (a) The beam does not bend when $n_2 = n_1$; the refracted light then travels in the *undeflected direction* (the dotted line), which is the same as the direction of the incident beam. The beam bends (b) toward the normal when $n_2 > n_1$ and (c) away from the normal when $n_2 < n_1$.

2. If n_2 is greater than n_1, then θ_2 is less than θ_1. In this case, refraction bends the light beam away from the undeflected direction and toward the normal, as in Fig. 33-17b.

3. If n_2 is less than n_1, then θ_2 is greater than θ_1. In this case, refraction bends the light beam away from the undeflected direction and away from the normal, as in Fig. 33-17c.

Refraction *cannot* bend a beam so much that the refracted ray is on the same side of the normal as the incident ray.

Chromatic Dispersion

The index of refraction n encountered by light in any medium except vacuum depends on the wavelength of the light. The dependence of n on wavelength implies that when a light beam consists of rays of different wavelengths, the rays will be refracted at different angles by a surface; that is, the light will be spread out by the refraction. This spreading of light is called **chromatic dispersion,** in which "chromatic" refers to the colors associated with the individual wavelengths and "dispersion" refers to the spreading of the light according to its wavelengths or colors. The refractions of Figs. 33-16 and 33-17 do not show chromatic dispersion because the beams are *monochromatic* (of a single wavelength or color).

Generally, the index of refraction of a given medium is *greater* for a shorter wavelength (corresponding to, say, blue light) than for a longer wavelength (say, red light). As an example, Fig. 33-18 shows how the index of refraction of fused quartz depends on the wavelength of light. Such dependence means that when a beam made up of waves of both blue and red light is refracted through a surface, such as from air into quartz or vice versa, the blue *component* (the ray corresponding to the wave of blue light) bends more than the red component.

A beam of *white light* consists of components of all (or nearly all) the colors in the visible spectrum with approximately uniform intensities. When you see such a beam, you perceive white rather than the individual colors. In Fig. 33-19a, a beam of white light in air is incident on a glass surface. (Because the pages of this book are white, a beam of white light is represented with a gray ray here. Also, a beam of monochromatic light is generally represented with a red ray.) Of the refracted light in Fig. 33-19a, only the red and blue components are shown. Because the blue component is bent more than the red component, the angle of refraction θ_{2b} for the blue component is *smaller* than the angle of refraction θ_{2r} for the red component. (Remember, angles are measured relative to the normal.) In Fig. 33-19b, a ray of white light in glass is incident on a glass–air interface. Again, the blue component is bent more than the red component, but now θ_{2b} is greater than θ_{2r}.

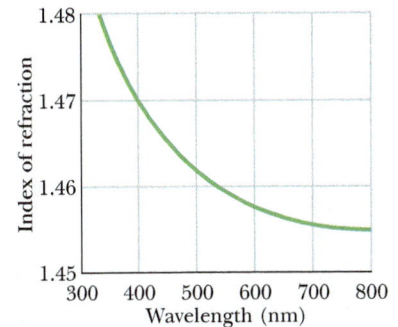

Figure 33-18 The index of refraction as a function of wavelength for fused quartz. The graph indicates that a beam of short-wavelength light, for which the index of refraction is higher, is bent more upon entering or leaving quartz than a beam of long-wavelength light.

Figure 33-19 Chromatic dispersion of white light. The blue component is bent more than the red component. (a) Passing from air to glass, the blue component ends up with the smaller angle of refraction. (b) Passing from glass to air, the blue component ends up with the greater angle of refraction. Each dotted line represents the direction in which the light would continue to travel if it were not bent by the refraction.

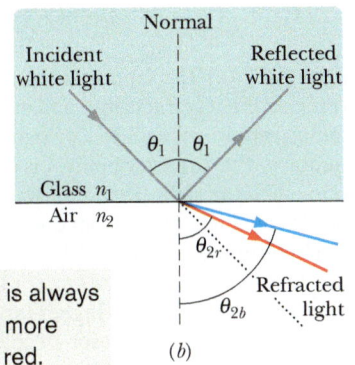

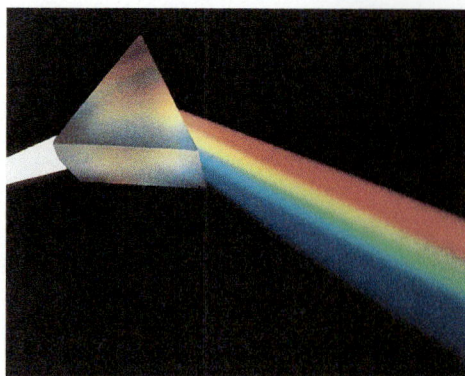

Courtesy Bausch & Lomb

(a)

White light

(b)

Figure 33-20 (a) A triangular prism separating white light into its component colors. (b) Chromatic dispersion occurs at the first surface and is increased at the second surface.

To increase the color separation, we can use a solid glass prism with a triangular cross section, as in Fig. 33-20a. The dispersion at the first surface (on the left in Figs. 33-20a, b) is then enhanced by the dispersion at the second surface.

Rainbows

The most charming example of chromatic dispersion is a rainbow. When sunlight (which consists of all visible colors) is intercepted by a falling raindrop, some of the light refracts into the drop, reflects once from the drop's inner surface, and then refracts out of the drop. Figure 33-21a shows the situation when the Sun is on the horizon at the left (and thus when the rays of sunlight are horizontal). The first refraction separates the sunlight into its component colors, and the second refraction increases the separation. (Only the red and blue rays are shown in the figure.) If many falling drops are brightly illuminated, you can see the separated colors they produce when the drops are at an angle of 42° from the direction of the *antisolar point A*, the point directly opposite the Sun in your view.

To locate the drops, face away from the Sun and point both arms directly away from the Sun, toward the shadow of your head. Then move your right arm directly up, directly rightward, or in any intermediate direction until the angle between your arms is 42°. If illuminated drops happen to be in the direction of your right arm, you see color in that direction.

Because any drop at an angle of 42° in any direction from A can contribute to the rainbow, the rainbow is always a 42° circular arc around A (Fig. 33-21b) and the top of a rainbow is never more than 42° above the horizon. When the Sun is above the horizon, the direction of A is below the horizon, and only a shorter, lower rainbow arc is possible (Fig. 33-21c).

Because rainbows formed in this way involve one reflection of light inside each drop, they are often called *primary rainbows*. A *secondary rainbow* involves two reflections inside a drop, as shown in Fig. 33-21d. Colors appear in the secondary rainbow at an angle of 52° from the direction of A. A secondary rainbow

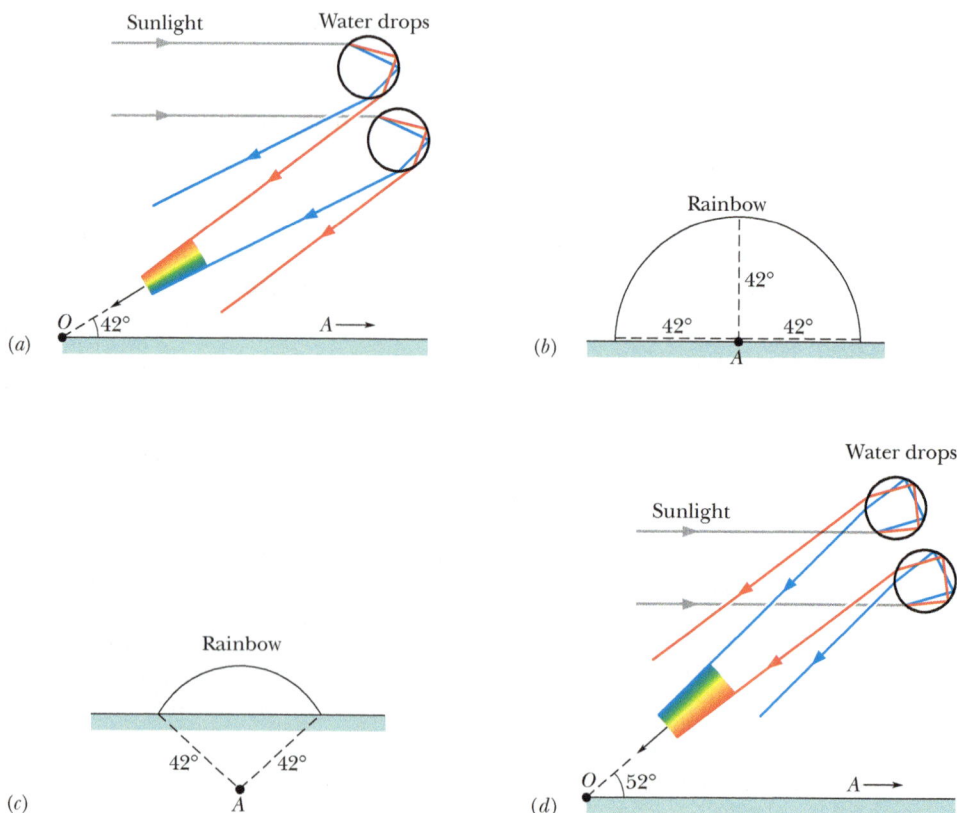

Figure 33-21 (a) The separation of colors when sunlight refracts into and out of falling raindrops leads to a primary rainbow. The antisolar point A is on the horizon at the right. The rainbow colors appear at an angle of 42° from the direction of A. (b) Drops at 42° from A in any direction can contribute to the rainbow. (c) The rainbow arc when the Sun is higher (and thus A is lower). (d) The separation of colors leading to a secondary rainbow.

is wider and dimmer than a primary rainbow and thus is more difficult to see. Also, the order of colors in a secondary rainbow is reversed from the order in a primary rainbow, as you can see by comparing parts *a* and *d* of Fig. 33-21.

Rainbows involving three or four reflections occur in the direction of the Sun and cannot be seen against the glare of sunshine in that part of the sky but have been photographed with special techniques.

Checkpoint 5

Which of the three drawings here (if any) show physically possible refraction?

(a)   (b)   (c)

Sample Problem 33.03   Reflection and refraction of a monochromatic beam

(a) In Fig. 33-22*a*, a beam of monochromatic light reflects and refracts at point *A* on the interface between material 1 with index of refraction $n_1 = 1.33$ and material 2 with index of refraction $n_2 = 1.77$. The incident beam makes an angle of 50° with the interface. What is the angle of reflection at point *A*? What is the angle of refraction there?

KEY IDEAS

(1) The angle of reflection is equal to the angle of incidence, and both angles are measured relative to the normal to the surface at the point of reflection. (2) When light reaches the interface between two materials with different indexes of refraction (call them n_1 and n_2), part of the light can be refracted by the interface according to Snell's law, Eq. 33-40:

$$n_2 \sin \theta_2 = n_1 \sin \theta_1, \qquad (33\text{-}42)$$

where both angles are measured relative to the normal at the point of refraction.

Calculations: In Fig. 33-22*a*, the normal at point *A* is drawn as a dashed line through the point. Note that the angle of incidence θ_1 is not the given 50° but is 90° − 50° = 40°. Thus, the angle of reflection is

$$\theta_1' = \theta_1 = 40°. \qquad \text{(Answer)}$$

The light that passes from material 1 into material 2 undergoes refraction at point *A* on the interface between the two materials. Again we measure angles between light rays and a normal, here at the point of refraction. Thus, in Fig. 33-22*a*, the angle of refraction is the angle marked θ_2. Solving Eq. 33-42 for θ_2 gives us

$$\theta_2 = \sin^{-1}\left(\frac{n_1}{n_2}\sin\theta_1\right) = \sin^{-1}\left(\frac{1.33}{1.77}\sin 40°\right)$$

$$= 28.88° \approx 29°. \qquad \text{(Answer)}$$

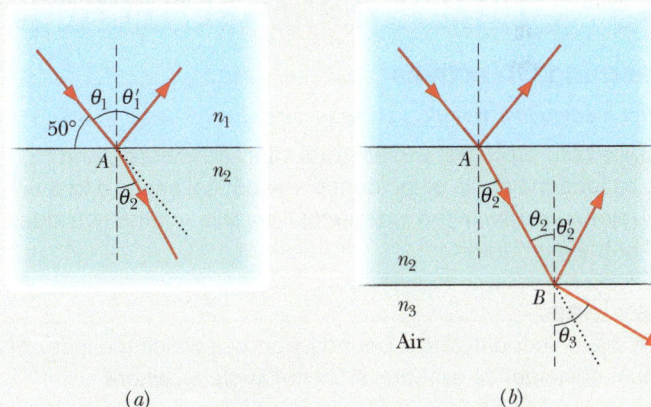

Figure 33-22 (*a*) Light reflects and refracts at point *A* on the interface between materials 1 and 2. (*b*) The light that passes through material 2 reflects and refracts at point *B* on the interface between materials 2 and 3 (air). Each dashed line is a normal. Each dotted line gives the incident direction of travel.

This result means that the beam swings toward the normal (it was at 40° to the normal and is now at 29°). The reason is that when the light travels across the interface, it moves into a material with a greater index of refraction. *Caution:* Note that the beam does *not* swing through the normal so that it appears on the left side of Fig. 33-22*a*.

(b) The light that enters material 2 at point *A* then reaches point *B* on the interface between material 2 and material 3, which is air, as shown in Fig. 33-22*b*. The interface through *B* is parallel to that through *A*. At *B*, some of the light reflects and the rest enters the air. What is the angle of reflection? What is the angle of refraction into the air?

Calculations: We first need to relate one of the angles at point B with a known angle at point A. Because the interface through point B is parallel to that through point A, the incident angle at B must be equal to the angle of refraction θ_2, as shown in Fig. 33-22b. Then for reflection, we again use the law of reflection. Thus, the angle of reflection at B is

$$\theta_2' = \theta_2 = 28.88° \approx 29°. \quad \text{(Answer)}$$

Next, the light that passes from material 2 into the air undergoes refraction at point B, with refraction angle θ_3. Thus, we again apply Snell's law of refraction, but this time

we write Eq. 33-40 as

$$n_3 \sin \theta_3 = n_2 \sin \theta_2. \quad (33\text{-}43)$$

Solving for θ_3 then leads to

$$\theta_3 = \sin^{-1}\left(\frac{n_2}{n_3}\sin \theta_2\right) = \sin^{-1}\left(\frac{1.77}{1.00}\sin 28.88°\right)$$

$$= 58.75° \approx 59°. \quad \text{(Answer)}$$

Thus, the beam swings away from the normal (it was at 29° to the normal and is now at 59°) because it moves into a material (air) with a lower index of refraction.

PLUS Additional examples, video, and practice available at *WileyPLUS*

33-6 TOTAL INTERNAL REFLECTION

Learning Objectives

After reading this module, you should be able to . . .

33.45 With sketches, explain total internal reflection and include the angle of incidence, the critical angle, and the relative values of the indexes of refraction on the two sides of the interface.

33.46 Identify the angle of refraction for incidence at a critical angle.

33.47 For a given pair of indexes of refraction, calculate the critical angle.

Key Idea

● A wave encountering a boundary across which the index of refraction decreases will experience total internal reflection if the angle of incidence exceeds a critical angle θ_c, where

$$\theta_c = \sin^{-1}\frac{n_2}{n_1} \quad \text{(critical angle)}.$$

Total Internal Reflection

Figure 33-23a shows rays of monochromatic light from a point source S in glass incident on the interface between the glass and air. For ray a, which is perpendicular to the interface, part of the light reflects at the interface and the rest travels through it with no change in direction.

For rays b through e, which have progressively larger angles of incidence at the interface, there are also both reflection and refraction at the interface. As the angle of incidence increases, the angle of refraction increases; for ray e it is 90°, which means that the refracted ray points directly along the interface. The angle of incidence giving this situation is called the **critical angle** θ_c. For angles of incidence larger than θ_c, such as for rays f and g, there is no refracted ray and *all* the light is reflected; this effect is called **total internal reflection** because all the light remains inside the glass.

To find θ_c, we use Eq. 33-40; we arbitrarily associate subscript 1 with the glass and subscript 2 with the air, and then we substitute θ_c for θ_1 and 90° for θ_2, which leads to

$$n_1 \sin \theta_c = n_2 \sin 90°, \quad (33\text{-}44)$$

(a)

(b)

Ken Kay/Fundamental Photographs

Figure 33-23 (*a*) Total internal reflection of light from a point source *S* in glass occurs for all angles of incidence greater than the critical angle θ_c. At the critical angle, the refracted ray points along the air–glass interface. (*b*) A source in a tank of water.

which gives us

$$\theta_c = \sin^{-1}\frac{n_2}{n_1} \quad \text{(critical angle).} \qquad (33\text{-}45)$$

Because the sine of an angle cannot exceed unity, n_2 cannot exceed n_1 in this equation. This restriction tells us that total internal reflection cannot occur when the incident light is in the medium of lower index of refraction. If source *S* were in the air in Fig. 33-23*a*, all its rays that are incident on the air–glass interface (including *f* and *g*) would be both reflected *and* refracted at the interface.

Total internal reflection has found many applications in medical technology. For example, a physician can view the interior of an artery of a patient by running two thin bundles of *optical fibers* through the chest wall and into an artery (Fig. 33-24). Light introduced at the outer end of one bundle undergoes repeated total internal reflection within the fibers so that, even though the bundle provides a curved path, most of the light ends up exiting the other end and illuminating the interior of the artery. Some of the light reflected from the interior then comes back up the second bundle in a similar way, to be detected and converted to an image on a monitor's screen for the physician to view. The physician can then perform a surgical procedure, such as the placement of a stent.

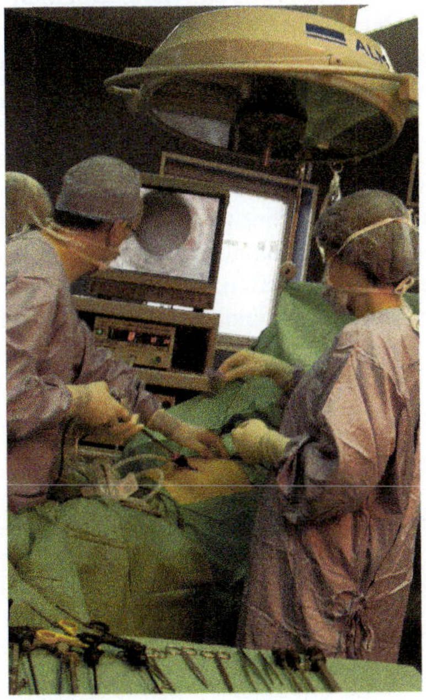

©Laurent/Phototake

Figure 33-24 An endoscope used to inspect an artery.

33-7 POLARIZATION BY REFLECTION

Learning Objectives

After reading this module, you should be able to . . .

33.48 With sketches, explain how unpolarized light can be converted to polarized light by reflection from an interface.

33.49 Identify Brewster's angle.

33.50 Apply the relationship between Brewster's angle and the indexes of refraction on the two sides of an interface.

33.51 Explain the function of polarizing sunglasses.

Key Idea

● A reflected wave will be fully polarized, with its $\vec{E}$ vectors perpendicular to the plane of incidence, if it strikes a boundary at the Brewster angle θ_B, where

$$\theta_B = \tan^{-1}\frac{n_2}{n_1} \quad \text{(Brewster angle).}$$

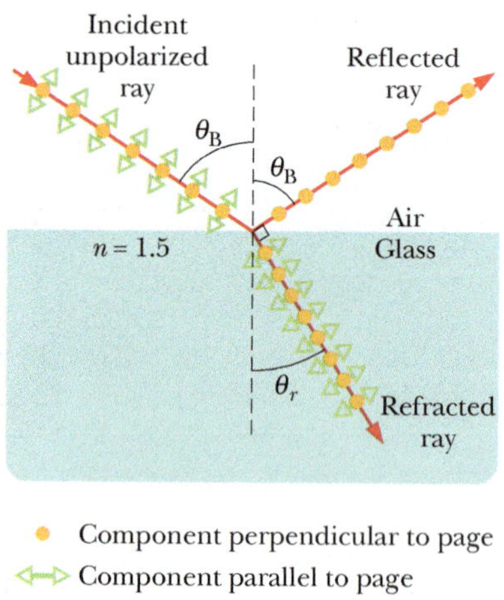

Figure 33-25 A ray of unpolarized light in air is incident on a glass surface at the Brewster angle θ_B. The electric fields along that ray have been resolved into components perpendicular to the page (the plane of incidence, reflection, and refraction) and components parallel to the page. The reflected light consists only of components perpendicular to the page and is thus polarized in that direction. The refracted light consists of the original components parallel to the page and weaker components perpendicular to the page; this light is partially polarized.

Polarization by Reflection

You can vary the glare you see in sunlight that has been reflected from, say, water by looking through a polarizing sheet (such as a polarizing sunglass lens) and then rotating the sheet's polarizing axis around your line of sight. You can do so because any light that is reflected from a surface is either fully or partially polarized by the reflection.

Figure 33-25 shows a ray of unpolarized light incident on a glass surface. Let us resolve the electric field vectors of the light into two components. The *perpendicular components* are perpendicular to the plane of incidence and thus also to the page in Fig. 33-25; these components are represented with dots (as if we see the tips of the vectors). The *parallel components* are parallel to the plane of incidence and the page; they are represented with double-headed arrows. Because the light is unpolarized, these two components are of equal magnitude.

In general, the reflected light also has both components but with unequal magnitudes. This means that the reflected light is partially polarized—the electric fields oscillating along one direction have greater amplitudes than those oscillating along other directions. However, when the light is incident at a particular incident angle, called the *Brewster angle* θ_B, the reflected light has only perpendicular components, as shown in Fig. 33-25. The reflected light is then fully polarized perpendicular to the plane of incidence. The parallel components of the incident light do not disappear but (along with perpendicular components) refract into the glass.

Polarizing Sunglasses. Glass, water, and the other dielectric materials discussed in Module 25-5 can partially and fully polarize light by reflection. When you intercept sunlight reflected from such a surface, you see a bright spot (the glare) on the surface where the reflection takes place. If the surface is horizontal as in Fig. 33-25, the reflected light is partially or fully polarized horizontally. To eliminate such glare from horizontal surfaces, the lenses in polarizing sunglasses are mounted with their polarizing direction vertical.

Brewster's Law

For light incident at the Brewster angle θ_B, we find experimentally that the reflected and refracted rays are perpendicular to each other. Because the reflected ray is reflected at the angle θ_B in Fig. 33-25 and the refracted ray is at an angle θ_r, we have

$$\theta_B + \theta_r = 90°. \quad (33\text{-}46)$$

These two angles can also be related with Eq. 33-40. Arbitrarily assigning subscript 1 in Eq. 33-40 to the material through which the incident and reflected rays travel, we have, from that equation,

$$n_1 \sin \theta_B = n_2 \sin \theta_r. \quad (33\text{-}47)$$

Combining these equations leads to

$$n_1 \sin \theta_B = n_2 \sin(90° - \theta_B) = n_2 \cos \theta_B, \quad (33\text{-}48)$$

which gives us

$$\theta_B = \tan^{-1} \frac{n_2}{n_1} \quad \text{(Brewster angle).} \quad (33\text{-}49)$$

(Note carefully that the subscripts in Eq. 33-49 are *not* arbitrary because of our decision as to their meanings.) If the incident and reflected rays travel *in air*, we can approximate n_1 as unity and let n represent n_2 in order to write Eq. 33-49 as

$$\theta_B = \tan^{-1} n \quad \text{(Brewster's law).} \quad (33\text{-}50)$$

This simplified version of Eq. 33-49 is known as **Brewster's law.** Like θ_B, it is named after Sir David Brewster, who found both experimentally in 1812.

Review & Summary

Electromagnetic Waves An electromagnetic wave consists of oscillating electric and magnetic fields. The various possible frequencies of electromagnetic waves form a *spectrum*, a small part of which is visible light. An electromagnetic wave traveling along an x axis has an electric field $\vec{E}$ and a magnetic field $\vec{B}$ with magnitudes that depend on x and t:

$$E = E_m \sin(kx - \omega t)$$

and
$$B = B_m \sin(kx - \omega t), \qquad (33\text{-}1, 33\text{-}2)$$

where E_m and B_m are the amplitudes of $\vec{E}$ and $\vec{B}$. The oscillating electric field induces the magnetic field, and the oscillating magnetic field induces the electric field. The speed of any electromagnetic wave in vacuum is c, which can be written as

$$c = \frac{E}{B} = \frac{1}{\sqrt{\mu_0 \varepsilon_0}}, \qquad (33\text{-}5, 33\text{-}3)$$

where E and B are the simultaneous (but nonzero) magnitudes of the two fields.

Energy Flow The rate per unit area at which energy is transported via an electromagnetic wave is given by the Poynting vector $\vec{S}$:

$$\vec{S} = \frac{1}{\mu_0} \vec{E} \times \vec{B}. \qquad (33\text{-}19)$$

The direction of $\vec{S}$ (and thus of the wave's travel and the energy transport) is perpendicular to the directions of both $\vec{E}$ and $\vec{B}$. The time-averaged rate per unit area at which energy is transported is S_{avg}, which is called the *intensity I* of the wave:

$$I = \frac{1}{c\mu_0} E_{\text{rms}}^2, \qquad (33\text{-}26)$$

in which $E_{\text{rms}} = E_m / \sqrt{2}$. A *point source* of electromagnetic waves emits the waves *isotropically*—that is, with equal intensity in all directions. The intensity of the waves at distance r from a point source of power P_s is

$$I = \frac{P_s}{4\pi r^2}. \qquad (33\text{-}27)$$

Radiation Pressure When a surface intercepts electromagnetic radiation, a force and a pressure are exerted on the surface. If the radiation is totally absorbed by the surface, the force is

$$F = \frac{IA}{c} \quad \text{(total absorption)}, \qquad (33\text{-}32)$$

in which I is the intensity of the radiation and A is the area of the surface perpendicular to the path of the radiation. If the radiation is totally reflected back along its original path, the force is

$$F = \frac{2IA}{c} \quad \text{(total reflection back along path)}. \qquad (33\text{-}33)$$

The radiation pressure p_r is the force per unit area:

$$p_r = \frac{I}{c} \quad \text{(total absorption)} \qquad (33\text{-}34)$$

and
$$p_r = \frac{2I}{c} \quad \text{(total reflection back along path)}. \qquad (33\text{-}35)$$

Polarization Electromagnetic waves are **polarized** if their electric field vectors are all in a single plane, called the *plane of oscillation*. From a head-on view, the field vectors oscillate parallel to a single axis perpendicular to the path taken by the waves. Light waves from common sources are not polarized; that is, they are **unpolarized**, or **polarized randomly**. From a head-on view, the vectors oscillate parallel to every possible axis that is perpendicular to the path taken by the waves.

Polarizing Sheets When a polarizing sheet is placed in the path of light, only electric field components of the light parallel to the sheet's **polarizing direction** are *transmitted* by the sheet; components perpendicular to the polarizing direction are absorbed. The light that emerges from a polarizing sheet is polarized parallel to the polarizing direction of the sheet.

If the original light is initially unpolarized, the transmitted intensity I is half the original intensity I_0:

$$I = \tfrac{1}{2} I_0. \qquad (33\text{-}36)$$

If the original light is initially polarized, the transmitted intensity depends on the angle θ between the polarization direction of the original light (the axis along which the fields oscillate) and the polarizing direction of the sheet:

$$I = I_0 \cos^2 \theta. \qquad (33\text{-}38)$$

Geometrical Optics *Geometrical optics* is an approximate treatment of light in which light waves are represented as straight-line rays.

Reflection and Refraction When a light ray encounters a boundary between two transparent media, a **reflected** ray and a **refracted** ray generally appear. Both rays remain in the plane of incidence. The **angle of reflection** is equal to the angle of incidence, and the **angle of refraction** is related to the angle of incidence by Snell's law,

$$n_2 \sin \theta_2 = n_1 \sin \theta_1 \quad \text{(refraction)}, \qquad (33\text{-}40)$$

where n_1 and n_2 are the indexes of refraction of the media in which the incident and refracted rays travel.

Total Internal Reflection A wave encountering a boundary across which the index of refraction decreases will experience **total internal reflection** if the angle of incidence exceeds a **critical angle** θ_c, where

$$\theta_c = \sin^{-1} \frac{n_2}{n_1} \quad \text{(critical angle)}. \qquad (33\text{-}45)$$

Polarization by Reflection A reflected wave will be fully **polarized**, with its $\vec{E}$ vectors perpendicular to the plane of incidence, if the incident, unpolarized wave strikes a boundary at the **Brewster angle** θ_B, where

$$\theta_B = \tan^{-1} \frac{n_2}{n_1} \quad \text{(Brewster angle)}. \qquad (33\text{-}49)$$

Questions

1 If the magnetic field of a light wave oscillates parallel to a y axis and is given by $B_y = B_m \sin(kz - \omega t)$, (a) in what direction does the wave travel and (b) parallel to which axis does the associated electric field oscillate?

2 Suppose we rotate the second sheet in Fig. 33-15a, starting with the polarization direction aligned with the y axis ($\theta = 0$) and ending with it aligned with the x axis ($\theta = 90°$). Which of the four curves in Fig. 33-26 best shows the intensity of the light through the three-sheet system during this 90° rotation?

Figure 33-26 Question 2.

3 (a) Figure 33-27 shows light reaching a polarizing sheet whose polarizing direction is parallel to a y axis. We shall rotate the sheet 40° clockwise about the light's indicated line of travel. During this rotation, does the fraction of the initial light intensity passed by the sheet increase, decrease, or remain the same if the light is (a) initially unpolarized, (b) initially polarized parallel to the x axis, and (c) initially polarized parallel to the y axis?

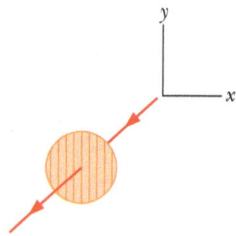

Figure 33-27 Question 3.

4 Figure 33-28 shows the electric and magnetic fields of an electromagnetic wave at a certain instant. Is the wave traveling into the page or out of the page?

Figure 33-28 Question 4.

5 In the arrangement of Fig. 33-15a, start with light that is initially polarized parallel to the x axis, and write the ratio of its final intensity I_3 to its initial intensity I_0 as $I_3/I_0 = A \cos^n \theta$. What are A, n, and θ if we rotate the polarizing direction of the first sheet (a) 60° counterclockwise and (b) 90° clockwise from what is shown?

6 In Fig. 33-29, unpolarized light is sent into a system of five polarizing sheets. Their polarizing directions, measured counterclockwise from the positive direction of the y axis, are the following: sheet 1, 35°; sheet 2, 0°; sheet 3, 0°; sheet 4, 110°; sheet 5, 45°. Sheet 3 is then rotated 180° counterclockwise about the light ray. During that rotation, at what angles (measured counterclockwise from the y axis) is the transmission of light through the system eliminated?

Figure 33-29 Question 6.

7 Figure 33-30 shows rays of monochromatic light propagating through three materials a, b, and c. Rank the materials according to the index of refraction, greatest first.

8 Figure 33-31 shows the multiple reflections of a light ray along a glass corridor where the walls are either parallel or perpendicular to one another. If the angle of incidence at point a is 30°, what are the angles of reflection of the light ray at points b, c, d, e, and f?

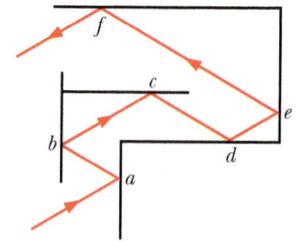

9 Figure 33-32 shows four long horizontal layers $A–D$ of different materials, with air above and below them. The index of refraction of each material is given. Rays of light are sent into the left end of each layer as shown. In which layer is there the possibility of totally trapping the light in that layer so that, after many reflections, all the light reaches the right end of the layer?

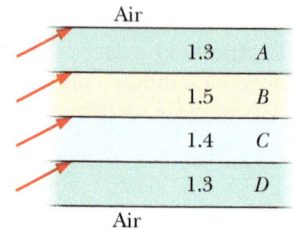

10 The leftmost block in Fig. 33-33 depicts total internal reflection for light inside a material with an index of refraction n_1 when air is outside the material. A light ray reaching point A from anywhere within the shaded region at the left (such as the ray shown) fully reflects at that point and ends up in the shaded region at the right. The other blocks show similar situations for two other materials. Rank the indexes of refraction of the three materials, greatest first.

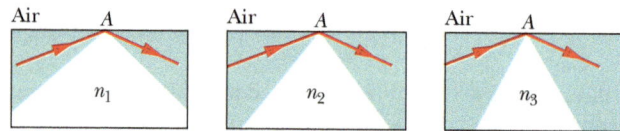

Figure 33-30 Question 7.

Figure 33-31 Question 8.

Figure 33-32 Question 9.

Figure 33-33 Question 10.

11 Each part of Fig. 33-34 shows light that refracts through an interface between two materials. The incident ray (shown gray in the figure) consists of red and blue light. The approximate index of refraction for visible light is indicated for each material. Which of the three parts show physically possible refraction? (*Hint:* First consider the refraction in general, regardless of the color, and then consider how red and blue light refract differently.)

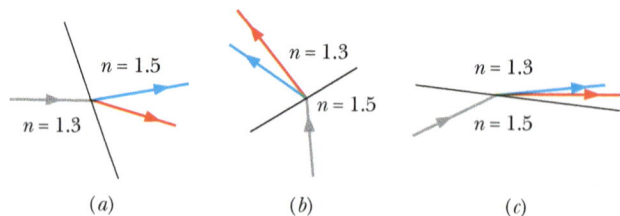

Figure 33-34 Question 11.

12 In Fig. 33-35, light travels from material a, through three layers of other materials with surfaces parallel to one another, and then back into another layer of material a. The refractions (but not the associated reflections) at the surfaces are shown. Rank the materials according to index of refraction, greatest first. (*Hint:* The parallel arrangement of the surfaces allows comparison.)

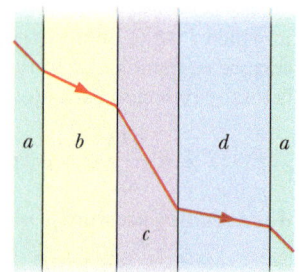

Figure 33-35 Question 12.

Problems

Module 33-1   Electromagnetic Waves

•1   A certain helium–neon laser emits red light in a narrow band of wavelengths centered at 632.8 nm and with a "wavelength width" (such as on the scale of Fig. 33-1) of 0.0100 nm. What is the corresponding "frequency width" for the emission?

•2   Project Seafarer was an ambitious program to construct an enormous antenna, buried underground on a site about 10 000 km^2 in area. Its purpose was to transmit signals to submarines while they were deeply submerged. If the effective wavelength were 1.0×10^4 Earth radii, what would be the (a) frequency and (b) period of the radiations emitted? Ordinarily, electromagnetic radiations do not penetrate very far into conductors such as seawater, and so normal signals cannot reach the submarines.

•3   From Fig. 33-2, approximate the (a) smaller and (b) larger wavelength at which the eye of a standard observer has half the eye's maximum sensitivity. What are the (c) wavelength, (d) frequency, and (e) period of the light at which the eye is the most sensitive?

•4   About how far apart must you hold your hands for them to be separated by 1.0 nano-light-second (the distance light travels in 1.0 ns)?

•5   **SSM**   What inductance must be connected to a 17 pF capacitor in an oscillator capable of generating 550 nm (i.e., visible) electromagnetic waves? Comment on your answer.

•6   What is the wavelength of the electromagnetic wave emitted by the oscillator–antenna system of Fig. 33-3 if $L = 0.253 \ \mu$H and $C = 25.0$ pF?

Module 33-2   Energy Transport and the Poynting Vector

•7   What is the intensity of a traveling plane electromagnetic wave if B_m is 1.0×10^{-4} T?

•8   Assume (unrealistically) that a TV station acts as a point source broadcasting isotropically at 1.0 MW. What is the intensity of the transmitted signal reaching Proxima Centauri, the star nearest our solar system, 4.3 ly away? (An alien civilization at that distance might be able to watch *X Files.*) A light-year (ly) is the distance light travels in one year.

•9   **ILW**   Some neodymium–glass lasers can provide 100 TW of power in 1.0 ns pulses at a wavelength of 0.26 μm. How much energy is contained in a single pulse?

•10   A plane electromagnetic wave has a maximum electric field magnitude of 3.20×10^{-4} V/m. Find the magnetic field amplitude.

•11   **ILW**   A plane electromagnetic wave traveling in the positive direction of an x axis in vacuum has components $E_x = E_y = 0$ and $E_z = (2.0 \text{ V/m}) \cos[(\pi \times 10^{15} \text{ s}^{-1})(t - x/c)]$. (a) What is the amplitude of the magnetic field component? (b) Parallel to which axis does the magnetic field oscillate? (c) When the electric field component is in the positive direction of the z axis at a certain point P, what is the direction of the magnetic field component there?

•12   In a plane radio wave the maximum value of the electric field component is 5.00 V/m. Calculate (a) the maximum value of the magnetic field component and (b) the wave intensity.

••13   Sunlight just outside Earth's atmosphere has an intensity of 1.40 kW/m^2. Calculate (a) E_m and (b) B_m for sunlight there, assuming it to be a plane wave.

••14   **GO**   An isotropic point source emits light at wavelength 500 nm, at the rate of 200 W. A light detector is positioned 400 m from the source. What is the maximum rate $\partial B/\partial t$ at which the magnetic component of the light changes with time at the detector's location?

••15   An airplane flying at a distance of 10 km from a radio transmitter receives a signal of intensity 10 μW/m^2. What is the amplitude of the (a) electric and (b) magnetic component of the signal at the airplane? (c) If the transmitter radiates uniformly over a hemisphere, what is the transmission power?

••16   Frank D. Drake, an investigator in the SETI (Search for Extra-Terrestrial Intelligence) program, once said that the large radio telescope in Arecibo, Puerto Rico (Fig. 33-36), "can detect a signal which lays down on the entire surface of the earth a power of only one picowatt." (a) What is the power that would be received by the Arecibo antenna for such a signal? The antenna diameter is 300 m. (b) What would be the power of an isotropic source at the center of our galaxy that could provide such a signal? The galactic center is 2.2×10^4 ly away. A light-year is the distance light travels in one year.

Courtesy SRI International, USRA, UMET

Figure 33-36   Problem 16. Radio telescope at Arecibo.

••17   The maximum electric field 10 m from an isotropic point source of light is 2.0 V/m. What are (a) the maximum value of the magnetic field and (b) the average intensity of the light there? (c) What is the power of the source?

••18   The intensity I of light from an isotropic point source is determined as a function of distance r from the source. Figure 33-37 gives

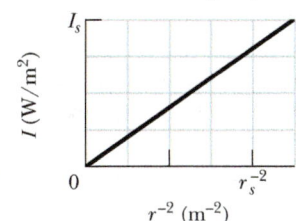

Figure 33-37   Problem 18.

intensity I versus the inverse square r^{-2} of that distance. The vertical axis scale is set by $I_s = 200$ W/m^2, and the horizontal axis scale is set by $r_s^{-2} = 8.0$ m^{-2}. What is the power of the source?

Module 33-3 Radiation Pressure

•19 SSM High-power lasers are used to compress a plasma (a gas of charged particles) by radiation pressure. A laser generating radiation pulses with peak power 1.5×10^3 MW is focused onto 1.0 mm^2 of high-electron-density plasma. Find the pressure exerted on the plasma if the plasma reflects all the light beams directly back along their paths.

•20 Radiation from the Sun reaching Earth (just outside the atmosphere) has an intensity of 1.4 kW/m^2. (a) Assuming that Earth (and its atmosphere) behaves like a flat disk perpendicular to the Sun's rays and that all the incident energy is absorbed, calculate the force on Earth due to radiation pressure. (b) For comparison, calculate the force due to the Sun's gravitational attraction.

•21 ILW What is the radiation pressure 1.5 m away from a 500 W lightbulb? Assume that the surface on which the pressure is exerted faces the bulb and is perfectly absorbing and that the bulb radiates uniformly in all directions.

•22 A black, totally absorbing piece of cardboard of area $A = 2.0$ cm^2 intercepts light with an intensity of 10 W/m^2 from a camera strobe light. What radiation pressure is produced on the cardboard by the light?

••23 Someone plans to float a small, totally absorbing sphere 0.500 m above an isotropic point source of light, so that the upward radiation force from the light matches the downward gravitational force on the sphere. The sphere's density is 19.0 g/cm^3, and its radius is 2.00 mm. (a) What power would be required of the light source? (b) Even if such a source were made, why would the support of the sphere be unstable?

••24 GO It has been proposed that a spaceship might be propelled in the solar system by radiation pressure, using a large sail made of foil. How large must the surface area of the sail be if the radiation force is to be equal in magnitude to the Sun's gravitational attraction? Assume that the mass of the ship + sail is 1500 kg, that the sail is perfectly reflecting, and that the sail is oriented perpendicular to the Sun's rays. See Appendix C for needed data. (With a larger sail, the ship is continuously driven away from the Sun.)

••25 SSM Prove, for a plane electromagnetic wave that is normally incident on a flat surface, that the radiation pressure on the surface is equal to the energy density in the incident beam. (This relation between pressure and energy density holds no matter what fraction of the incident energy is reflected.)

••26 In Fig. 33-38, a laser beam of power 4.60 W and diameter $D = 2.60$ mm is directed upward at one circular face (of diameter $d < 2.60$ mm) of a perfectly reflecting cylinder. The cylinder is levitated because the upward radiation force matches the downward gravitational force. If the cylinder's density is 1.20 g/cm^3, what is its height H?

••27 SSM WWW A plane electromagnetic wave, with wavelength 3.0 m, travels in vacuum in the positive direction of an x axis. The electric field, of amplitude 300 V/m, oscillates parallel to

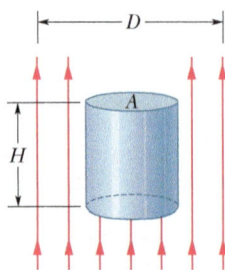

Figure 33-38
Problem 26.

the y axis. What are the (a) frequency, (b) angular frequency, and (c) angular wave number of the wave? (d) What is the amplitude of the magnetic field component? (e) Parallel to which axis does the magnetic field oscillate? (f) What is the time-averaged rate of energy flow in watts per square meter associated with this wave? The wave uniformly illuminates a surface of area 2.0 m^2. If the surface totally absorbs the wave, what are (g) the rate at which momentum is transferred to the surface and (h) the radiation pressure on the surface?

••28 The average intensity of the solar radiation that strikes normally on a surface just outside Earth's atmosphere is 1.4 kW/m^2. (a) What radiation pressure p_r is exerted on this surface, assuming complete absorption? (b) For comparison, find the ratio of p_r to Earth's sea-level atmospheric pressure, which is 1.0×10^5 Pa.

••29 SSM A small spaceship with a mass of only 1.5×10^3 kg (including an astronaut) is drifting in outer space with negligible gravitational forces acting on it. If the astronaut turns on a 10 kW laser beam, what speed will the ship attain in 1.0 day because of the momentum carried away by the beam?

••30 A small laser emits light at power 5.00 mW and wavelength 633 nm. The laser beam is focused (narrowed) until its diameter matches the 1266 nm diameter of a sphere placed in its path. The sphere is perfectly absorbing and has density 5.00×10^3 kg/m^3. What are (a) the beam intensity at the sphere's location, (b) the radiation pressure on the sphere, (c) the magnitude of the corresponding force, and (d) the magnitude of the acceleration that force alone would give the sphere?

•••31 GO As a comet swings around the Sun, ice on the comet's surface vaporizes, releasing trapped dust particles and ions. The ions, because they are electrically charged, are forced by the electrically charged *solar wind* into a straight *ion tail* that points radially away from the Sun (Fig. 33-39). The (electrically neutral) dust particles are pushed radially outward from the Sun by the radiation force on them from sunlight. Assume that the dust particles are spherical, have density 3.5×10^3 kg/m^3, and are totally absorbing. (a) What radius must a particle have in order to follow a straight path, like path 2 in the figure? (b) If its radius is larger, does its path curve away from the Sun (like path 1) or toward the Sun (like path 3)?

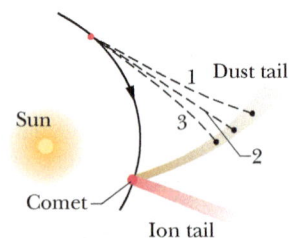

Figure 33-39 Problem 31.

Module 33-4 Polarization

•32 In Fig. 33-40, initially unpolarized light is sent into a system of three polarizing sheets whose polarizing directions make angles

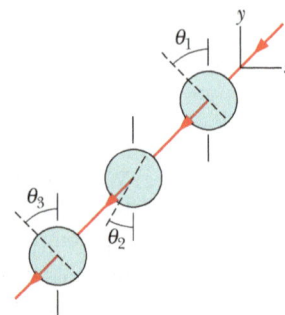

Figure 33-40 Problems 32 and 33.

of $\theta_1 = \theta_2 = \theta_3 = 50°$ with the direction of the y axis. What percentage of the initial intensity is transmitted by the system? (*Hint:* Be careful with the angles.)

•**33 SSM** In Fig. 33-40, initially unpolarized light is sent into a system of three polarizing sheets whose polarizing directions make angles of $\theta_1 = 40°$, $\theta_2 = 20°$, and $\theta_3 = 40°$ with the direction of the y axis. What percentage of the light's initial intensity is transmitted by the system? (*Hint:* Be careful with the angles.)

•**34 GO** In Fig. 33-41, a beam of unpolarized light, with intensity 43 W/m², is sent into a system of two polarizing sheets with polarizing directions at angles $\theta_1 = 70°$ and $\theta_2 = 90°$ to the y axis. What is the intensity of the light transmitted by the system?

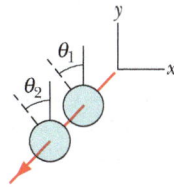

•**35 ILW** In Fig. 33-41, a beam of light, with intensity 43 W/m² and polarization parallel to a y axis, is sent into a system of two polarizing sheets with polarizing directions at angles of $\theta_1 = 70°$ and $\theta_2 = 90°$ to the y axis. What is the intensity of the light transmitted by the two-sheet system?

Figure 33-41 Problems 34, 35, and 42.

••**36** At a beach the light is generally partially polarized due to reflections off sand and water. At a particular beach on a particular day near sundown, the horizontal component of the electric field vector is 2.3 times the vertical component. A standing sunbather puts on polarizing sunglasses; the glasses eliminate the horizontal field component. (a) What fraction of the light intensity received before the glasses were put on now reaches the sunbather's eyes? (b) The sunbather, still wearing the glasses, lies on his side. What fraction of the light intensity received before the glasses were put on now reaches his eyes?

••**37 SSM WWW** We want to rotate the direction of polarization of a beam of polarized light through 90° by sending the beam through one or more polarizing sheets. (a) What is the minimum number of sheets required? (b) What is the minimum number of sheets required if the transmitted intensity is to be more than 60% of the original intensity?

••**38 GO** In Fig. 33-42, unpolarized light is sent into a system of three polarizing sheets. The angles θ_1, θ_2, and θ_3 of the polarizing directions are measured counterclockwise from the positive direction of the y axis (they are not drawn to scale). Angles θ_1 and θ_3 are fixed, but angle θ_2 can be varied. Figure 33-43 gives the intensity of the light emerging from sheet 3 as a function of θ_2. (The scale of the intensity axis is not indicated.) What percentage of the light's initial intensity is transmitted by the system when $\theta_2 = 30°$?

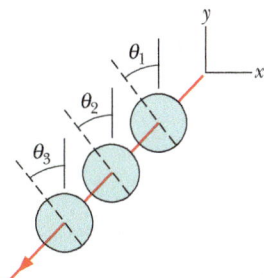

Figure 33-42 Problems 38, 40, and 44.

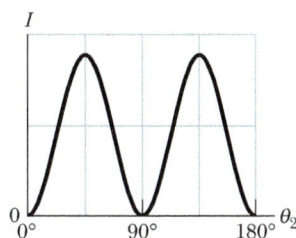

Figure 33-43 Problem 38.

••**39** Unpolarized light of intensity 10 mW/m² is sent into a polarizing sheet as in Fig. 33-11. What are (a) the amplitude of the electric field component of the transmitted light and (b) the radiation pressure on the sheet due to its absorbing some of the light?

••**40 GO** In Fig. 33-42, unpolarized light is sent into a system of three polarizing sheets. The angles θ_1, θ_2, and θ_3 of the polarizing directions are measured counterclockwise from the positive direction of the y axis (they are not drawn to scale). Angles θ_1 and θ_3 are fixed, but angle θ_2 can be varied. Figure 33-44 gives the intensity of the light emerging from sheet 3 as a function of θ_2. (The scale of the intensity axis is not indicated.) What percentage of the light's initial intensity is transmitted by the three-sheet system when $\theta_2 = 90°$?

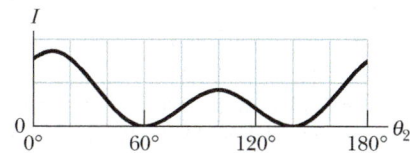

Figure 33-44 Problem 40.

••**41** A beam of polarized light is sent into a system of two polarizing sheets. Relative to the polarization direction of that incident light, the polarizing directions of the sheets are at angles θ for the first sheet and 90° for the second sheet. If 0.10 of the incident intensity is transmitted by the two sheets, what is θ?

••**42 GO** In Fig. 33-41, unpolarized light is sent into a system of two polarizing sheets. The angles θ_1 and θ_2 of the polarizing directions of the sheets are measured counterclockwise from the positive direction of the y axis (they are not drawn to scale in the figure). Angle θ_1 is fixed but angle θ_2 can be varied. Figure 33-45 gives the intensity of the light emerging from sheet 2 as a function of θ_2. (The scale of the intensity axis is not indicated.) What percentage of the light's initial intensity is transmitted by the two-sheet system when $\theta_2 = 90°$?

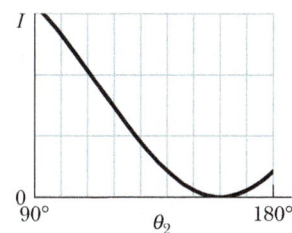

Figure 33-45 Problem 42.

••**43** A beam of partially polarized light can be considered to be a mixture of polarized and unpolarized light. Suppose we send such a beam through a polarizing filter and then rotate the filter through 360° while keeping it perpendicular to the beam. If the transmitted intensity varies by a factor of 5.0 during the rotation, what fraction of the intensity of the original beam is associated with the beam's polarized light?

••**44** In Fig. 33-42, unpolarized light is sent into a system of three polarizing sheets, which transmits 0.0500 of the initial light intensity. The polarizing directions of the first and third sheets are at angles $\theta_1 = 0°$ and $\theta_3 = 90°$. What are the (a) smaller and (b) larger possible values of angle θ_2 ($< 90°$) for the polarizing direction of sheet 2?

Module 33-5 Reflection and Refraction

•45 When the rectangular metal tank in Fig. 33-46 is filled to the top with an unknown liquid, observer O, with eyes level with the top of the tank, can just see corner E. A ray that refracts toward O at the top surface of the liquid is shown. If $D = 85.0$ cm and $L = 1.10$ m, what is the index of refraction of the liquid?

Figure 33-46 Problem 45.

•46 In Fig. 33-47a, a light ray in an underlying material is incident at angle θ_1 on a boundary with water, and some of the light refracts into the water. There are two choices of underlying material. For each, the angle of refraction θ_2 versus the incident angle θ_1 is given in Fig. 33-47b. The horizontal axis scale is set by $\theta_{1s} = 90°$. Without calculation, determine whether the index of refraction of (a) material 1 and (b) material 2 is greater or less than the index of water ($n = 1.33$). What is the index of refraction of (c) material 1 and (d) material 2?

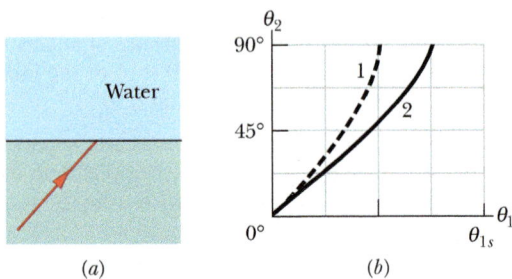

(a)   (b)

Figure 33-47 Problem 46.

•47 Light in vacuum is incident on the surface of a glass slab. In the vacuum the beam makes an angle of 32.0° with the normal to the surface, while in the glass it makes an angle of 21.0° with the normal. What is the index of refraction of the glass?

•48 In Fig. 33-48a, a light ray in water is incident at angle θ_1 on a boundary with an underlying material, into which some of the light refracts. There are two choices of underlying material. For each, the angle of refraction θ_2 versus the incident angle θ_1 is given in Fig. 33-48b. The vertical axis scale is set by $\theta_{2s} = 90°$. Without calculation, determine whether the index of refraction of (a) material 1 and (b) material 2 is greater or less than the index of water ($n = 1.33$). What is the index of refraction of (c) material 1 and (d) material 2?

(a)   (b)

Figure 33-48 Problem 48.

•49 Figure 33-49 shows light reflecting from two perpendicular reflecting surfaces A and B. Find the angle between the incoming ray i and the outgoing ray r'.

Figure 33-49 Problem 49.

••50 In Fig. 33-50a, a beam of light in material 1 is incident on a boundary at an angle $\theta_1 = 40°$. Some of the light travels through material 2, and then some of it emerges into material 3. The two boundaries between the three materials are parallel. The final direction of the beam depends, in part, on the index of refraction n_3 of the third material. Figure 33-50b gives the angle of refraction θ_3 in that material versus n_3 for a range of possible n_3 values. The vertical axis scale is set by $\theta_{3a} = 30.0°$ and $\theta_{3b} = 50.0°$. (a) What is the index of refraction of material 1, or is the index impossible to calculate without more information? (b) What is the index of refraction of material 2, or is the index impossible to calculate without more information? (c) If θ_1 is changed to 70° and the index of refraction of material 3 is 2.4, what is θ_3?

(a)   (b)

Figure 33-50 Problem 50.

••51 GO In Fig. 33-51, light is incident at angle $\theta_1 = 40.1°$ on a boundary between two transparent materials. Some of the light travels down through the next three layers of transparent materials, while some of it reflects upward and then escapes into the air. If $n_1 = 1.30$, $n_2 = 1.40$, $n_3 = 1.32$, and $n_4 = 1.45$, what is the value of (a) θ_5 in the air and (b) θ_4 in the bottom material?

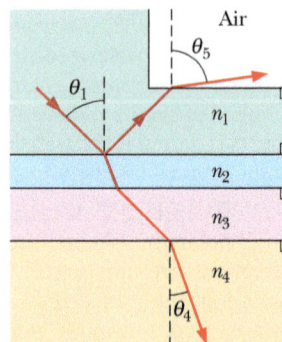

••52 In Fig. 33-52a, a beam of light in material 1 is incident on a boundary at an angle of $\theta_1 = 30°$. The extent of refraction of the light into material 2 depends, in part, on the index of refraction n_2 of material 2. Figure 33-52b gives the angle of refraction θ_2 versus n_2 for a range of possible n_2 values. The vertical axis scale is set by $\theta_{2a} = 20.0°$ and $\theta_{2b} = 40.0°$. (a) What is the index of refraction of

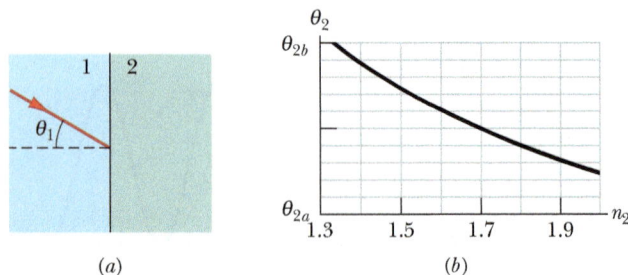

Figure 33-51 Problem 51.

(a)   (b)

Figure 33-52 Problem 52.

material 1? (b) If the incident angle is changed to 60° and material 2 has $n_2 = 2.4$, then what is angle θ_2?

••53 **SSM** **WWW** **ILW** In Fig. 33-53, a ray is incident on one face of a triangular glass prism in air. The angle of incidence θ is chosen so that the emerging ray also makes the same angle θ with the normal to the other face. Show that the index of refraction n of the glass prism is given by

$$n = \frac{\sin\frac{1}{2}(\psi + \phi)}{\sin\frac{1}{2}\phi},$$

where ϕ is the vertex angle of the prism and ψ is the *deviation angle,* the total angle through which the beam is turned in passing through the prism. (Under these conditions the deviation angle ψ has the smallest possible value, which is called the *angle of minimum deviation.*)

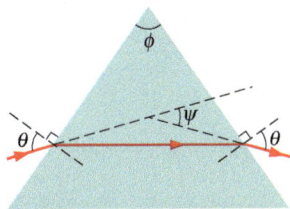

Figure 33-53 Problems 53 and 64.

••54 *Dispersion in a window pane.* In Fig. 33-54, a beam of white light is incident at angle $\theta = 50°$ on a common window pane (shown in cross section). For the pane's type of glass, the index of refraction for visible light ranges from 1.524 at the blue end of the spectrum to 1.509 at the red end. The two sides of the pane are parallel. What is the angular spread of the colors in the beam (a) when the light enters the pane and (b) when it emerges from the opposite side? (*Hint:* When you look at an object through a window pane, are the colors in the light from the object dispersed as shown in, say, Fig. 33-20?)

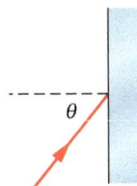

Figure 33-54 Problem 54.

••55 **GO** **SSM** In Fig. 33-55, a 2.00-m-long vertical pole extends from the bottom of a swimming pool to a point 50.0 cm above the water. Sunlight is incident at angle $\theta = 55.0°$. What is the length of the shadow of the pole on the level bottom of the pool?

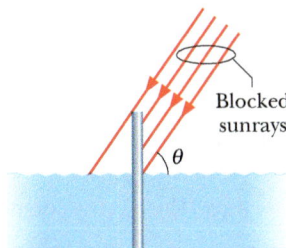

Figure 33-55 Problem 55.

••56 *Rainbows from square drops.* Suppose that, on some surreal world, raindrops had a square cross section and always fell with one face horizontal. Figure 33-56 shows such a falling drop, with a white beam of sunlight incident at $\theta = 70.0°$ at point P. The part of the light that enters the drop then travels to point A, where some of it refracts out into the air and the rest reflects. That reflected light then travels to point B, where again some of the light refracts out into the air and the rest reflects. What is the difference in the angles of the red light ($n = 1.331$) and the blue light ($n = 1.343$) that emerge at

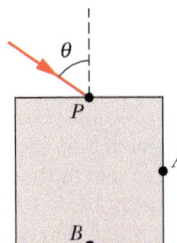

Figure 33-56 Problem 56.

(a) point A and (b) point B? (This angular difference in the light emerging at, say, point A would be the rainbow's angular width.)

Module 33-6   Total Internal Reflection

•57 A point source of light is 80.0 cm below the surface of a body of water. Find the diameter of the circle at the surface through which light emerges from the water.

•58 The index of refraction of benzene is 1.8. What is the critical angle for a light ray traveling in benzene toward a flat layer of air above the benzene?

••59 **SSM** **ILW** In Fig. 33-57, a ray of light is perpendicular to the face ab of a glass prism ($n = 1.52$). Find the largest value for the angle ϕ so that the ray is totally reflected at face ac if the prism is immersed (a) in air and (b) in water.

Figure 33-57 Problem 59.

••60 In Fig. 33-58, light from ray A refracts from material 1 ($n_1 = 1.60$) into a thin layer of material 2 ($n_2 = 1.80$), crosses that layer, and is then incident at the critical angle on the interface between materials 2 and 3 ($n_3 = 1.30$). (a) What is the value of incident angle θ_A? (b) If θ_A is decreased, does part of the light refract into material 3?

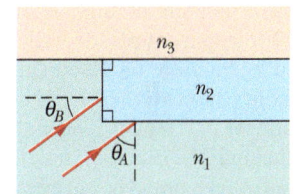

Figure 33-58 Problem 60.

Light from ray B refracts from material 1 into the thin layer, crosses that layer, and is then incident at the critical angle on the interface between materials 2 and 3. (c) What is the value of incident angle θ_B? (d) If θ_B is decreased, does part of the light refract into material 3?

••61 **GO** In Fig. 33-59, light initially in material 1 refracts into material 2, crosses that material, and is then incident at the critical angle on the interface between materials 2 and 3. The indexes of refraction are $n_1 = 1.60$, $n_2 = 1.40$, and $n_3 = 1.20$. (a) What is angle θ? (b) If θ is increased, is there refraction of light into material 3?

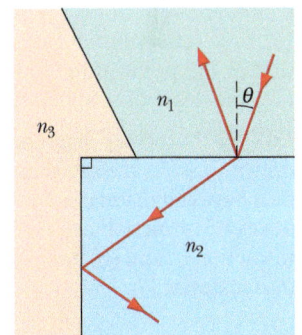

Figure 33-59 Problem 61.

••62 **GO** A catfish is 2.00 m below the surface of a smooth lake. (a) What is the diameter of the circle on the surface through which the fish can see the world outside the water? (b) If the fish descends, does the diameter of the circle increase, decrease, or remain the same?

••63 In Fig. 33-60, light enters a 90° triangular prism at point P with incident angle θ, and then some of it refracts at point Q with an angle of refraction of 90°. (a) What is the index of refraction of the prism in terms of θ? (b) What, numerically, is the maximum value that the index of refraction can have? Does light emerge at Q if the incident angle at P is (c) increased slightly and (d) decreased slightly?

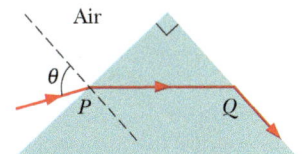

Figure 33-60 Problem 63.

••64 Suppose the prism of Fig. 33-53 has apex angle $\phi = 60.0°$ and index of refraction $n = 1.60$. (a) What is the smallest angle of incidence θ for which a ray can enter the left face of the prism and exit the right face? (b) What angle of incidence θ is required for the ray to exit the prism with an identical angle θ for its refraction, as it does in Fig. 33-53?

••65 GO Figure 33-61 depicts a simplistic optical fiber: a plastic core ($n_1 = 1.58$) is surrounded by a plastic sheath ($n_2 = 1.53$). A light ray is incident on one end of the fiber at angle θ. The ray is to undergo total internal reflection at point A, where it encounters the core–sheath boundary. (Thus there is no loss of light through that boundary.) What is the maximum value of θ that allows total internal reflection at A?

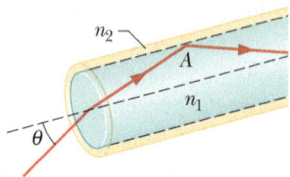

Figure 33-61 Problem 65.

••66 GO In Fig. 33-62, a light ray in air is incident at angle θ_1 on a block of transparent plastic with an index of refraction of 1.56. The dimensions indicated are $H = 2.00$ cm and $W = 3.00$ cm. The light passes through the block to one of its sides and there undergoes reflection (inside the block) and possibly refraction (out into the air). This is the point of *first reflection*. The reflected light then passes through

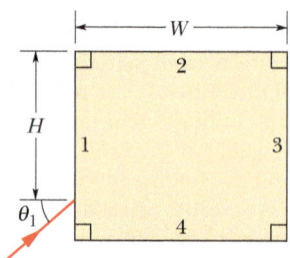

Figure 33-62 Problem 66.

the block to another of its sides—a point of *second reflection*. If $\theta_1 = 40°$, on which side is the point of (a) first reflection and (b) second reflection? If there is refraction at the point of (c) first reflection and (d) second reflection, give the angle of refraction; if not, answer "none." If $\theta_1 = 70°$, on which side is the point of (e) first reflection and (f) second reflection? If there is refraction at the point of (g) first reflection and (h) second reflection, give the angle of refraction; if not, answer "none."

••67 GO In the ray diagram of Fig. 33-63, where the angles are not drawn to scale, the ray is incident at the critical angle on the interface between materials 2 and 3. Angle $\phi = 60.0°$, and two of the indexes of refraction are $n_1 = 1.70$ and $n_2 = 1.60$. Find (a) index of refraction n_3 and (b) angle θ. (c) If θ is decreased, does light refract into material 3?

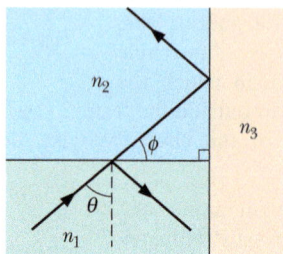

Figure 33-63 Problem 67.

Module 33-7 Polarization by Reflection

•68 (a) At what angle of incidence will the light reflected from water be completely polarized? (b) Does this angle depend on the wavelength of the light?

•69 SSM Light that is traveling in water (with an index of refraction of 1.33) is incident on a plate of glass (with index of refraction 1.53). At what angle of incidence does the reflected light end up fully polarized?

••70 In Fig. 33-64, a light ray in air is incident on a flat layer of material 2 that has an index of refraction $n_2 = 1.5$. Beneath material 2 is material 3 with an index of refraction n_3. The ray is incident on the air–material 2 interface at the Brewster angle for that interface. The ray of light refracted into material 3 happens to be incident on the material 2–material 3 interface at the Brewster angle for that interface. What is the value of n_3?

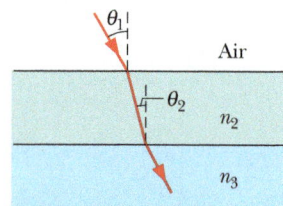

Figure 33-64 Problem 70.

Additional Problems

71 SSM (a) How long does it take a radio signal to travel 150 km from a transmitter to a receiving antenna? (b) We see a full Moon by reflected sunlight. How much earlier did the light that enters our eye leave the Sun? The Earth–Moon and Earth–Sun distances are 3.8×10^5 km and 1.5×10^8 km, respectively. (c) What is the round-trip travel time for light between Earth and a spaceship orbiting Saturn, 1.3×10^9 km distant? (d) The Crab nebula, which is about 6500 light-years (ly) distant, is thought to be the result of a supernova explosion recorded by Chinese astronomers in A.D. 1054. In approximately what year did the explosion actually occur? (When we look into the night sky, we are effectively looking back in time.)

72 An electromagnetic wave with frequency 4.00×10^{14} Hz travels through vacuum in the positive direction of an x axis. The wave has its electric field oscillating parallel to the y axis, with an amplitude E_m. At time $t = 0$, the electric field at point P on the x axis has a value of $+E_m/4$ and is decreasing with time. What is the distance along the x axis from point P to the first point with $E = 0$ if we search in (a) the negative direction and (b) the positive direction of the x axis?

73 SSM The electric component of a beam of polarized light is

$$E_y = (5.00 \text{ V/m}) \sin[(1.00 \times 10^6 \text{ m}^{-1})z + \omega t].$$

(a) Write an expression for the magnetic field component of the wave, including a value for ω. What are the (b) wavelength, (c) period, and (d) intensity of this light? (e) Parallel to which axis does the magnetic field oscillate? (f) In which region of the electromagnetic spectrum is this wave?

74 A particle in the solar system is under the combined influence of the Sun's gravitational attraction and the radiation force due to the Sun's rays. Assume that the particle is a sphere of density 1.0×10^3 kg/m³ and that all the incident light is absorbed. (a) Show that, if its radius is less than some critical radius R, the particle will be blown out of the solar system. (b) Calculate the critical radius.

75 **SSM** In Fig. 33-65, a light ray enters a glass slab at point A at incident angle $\theta_1 = 45.0°$ and then undergoes total internal reflection at point B. (The reflection at A is not shown.) What minimum value for the index of refraction of the glass can be inferred from this information?

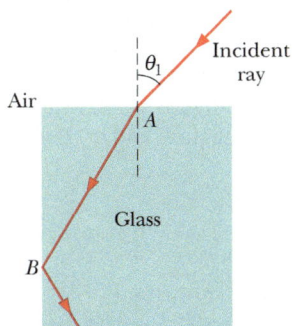

76 **GO** In Fig. 33-66, unpolarized light with an intensity of 25 W/m^2 is sent into a system of four polarizing sheets with polarizing directions at angles $\theta_1 = 40°$, $\theta_2 = 20°$, $\theta_3 = 20°$, and $\theta_4 = 30°$. What is the intensity of the light that emerges from the system?

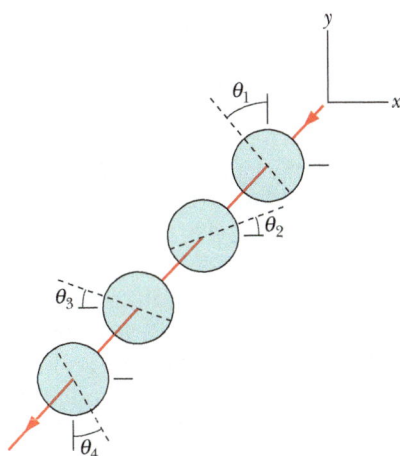

Figure 33-65 Problem 75.

Figure 33-66 Problem 76.

77 Rainbow. Figure 33-67 shows a light ray entering and then leaving a falling, spherical raindrop after one internal reflection (see Fig. 33-21a). The final direction of travel is deviated (turned) from the initial direction of travel by angular deviation θ_{dev}. (a) Show that θ_{dev} is

$$\theta_{dev} = 180° + 2\theta_i - 4\theta_r,$$

where θ_i is the angle of incidence of the ray on the drop and θ_r is the angle of refraction of the ray within the drop. (b) Using Snell's law, substitute for θ_r in terms of θ_i and the index of refraction n of the water. Then, on a graphing calculator or with a computer graphing package, graph θ_{dev} versus θ_i for the range of possible θ_i values and for $n = 1.331$ for red light (at one end of the visible spectrum) and $n = 1.333$ for blue light (at the other end).

The red-light curve and the blue-light curve have different minima, which means that there is a different *angle of minimum deviation* for each color. The light of any given color that leaves the drop at that color's angle of minimum deviation is especially bright because rays bunch up at that angle. Thus, the bright red light leaves the drop at one angle and the bright blue light leaves it at another angle.

Determine the angle of minimum deviation from the θ_{dev} curve

Figure 33-67 Problem 77.

for (c) red light and (d) blue light. (e) If these colors form the inner and outer edges of a rainbow (Fig. 33-21a), what is the angular width of the rainbow?

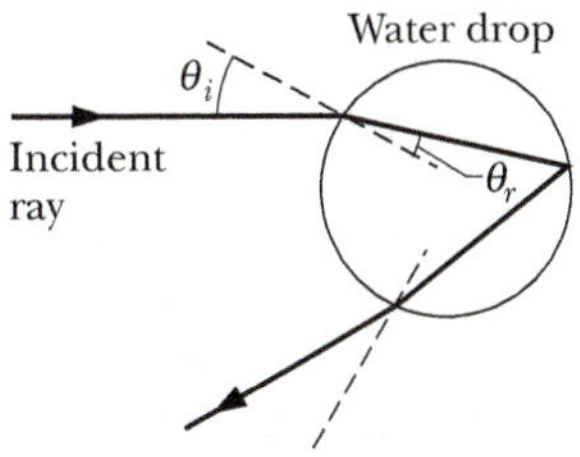

78 The *primary rainbow* described in Problem 77 is the type commonly seen in regions where rainbows appear. It is produced by light reflecting once inside the drops. Rarer is the *secondary rainbow* described in Module 33-5, produced by light reflecting twice inside the drops (Fig. 33-68a). (a) Show that the angular deviation of light entering and then leaving a spherical water drop is

$$\theta_{dev} = (180°)k + 2\theta_i - 2(k+1)\theta_r,$$

where k is the number of internal reflections. Using the procedure of Problem 77, find the angle of minimum deviation for (b) red light and (c) blue light in a secondary rainbow. (d) What is the angular width of that rainbow (Fig. 33-21d)?

The *tertiary rainbow* depends on three internal reflections (Fig. 33-68b). It probably occurs but, as noted in Module 33-5, cannot be seen with the eye because it is very faint and lies in the bright sky surrounding the Sun. What is the angle of minimum deviation for (e) the red light and (f) the blue light in this rainbow? (g) What is the rainbow's angular width?

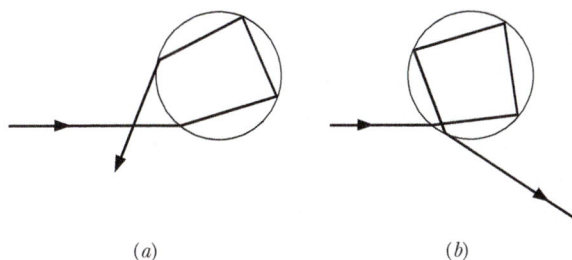

(a)   (b)

Figure 33-68 Problem 78.

79 **SSM** (a) Prove that a ray of light incident on the surface of a sheet of plate glass of thickness t emerges from the opposite face parallel to its initial direction but displaced sideways, as in Fig. 33-69. (b) Show that, for small angles of incidence θ, this displacement is given by

$$x = t\theta \frac{n-1}{n},$$

where n is the index of refraction of the glass and θ is measured in radians.

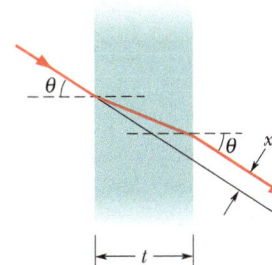

Figure 33-69 Problem 79.

80 An electromagnetic wave is traveling in the negative direction of a y axis. At a particular position and time, the electric field is directed along the positive direction of the z axis and has a magnitude of 100 V/m. What are the (a) magnitude and (b) direction of the corresponding magnetic field?

81 The magnetic component of a polarized wave of light is

$$B_x = (4.0 \times 10^{-6} \text{ T}) \sin[(1.57 \times 10^7 \text{ m}^{-1})y + \omega t].$$

(a) Parallel to which axis is the light polarized? What are the (b) frequency and (c) intensity of the light?

82 In Fig. 33-70, unpolarized light is sent into the system of three polarizing sheets, where the polarizing directions of the first and third sheets are at angles $\theta_1 = 30°$ (counterclockwise) and $\theta_3 = 30°$ (clockwise). What fraction of the initial light intensity emerges from the system?

Figure 33-70 Problem 82.

83 **SSM** A ray of white light traveling through fused quartz is incident at a quartz–air interface at angle θ_1. Assume that the index of refraction of quartz is $n = 1.456$ at the red end of the visible range and $n = 1.470$ at the blue end. If θ_1 is (a) 42.00°, (b) 43.10°, and (c) 44.00°, is the refracted light white, white dominated by the red end of the visible range, or white dominated by the blue end of the visible range, or is there no refracted light?

84 Three polarizing sheets are stacked. The first and third are crossed; the one between has its polarizing direction at 45.0° to the polarizing directions of the other two. What fraction of the intensity of an originally unpolarized beam is transmitted by the stack?

85 In a region of space where gravitational forces can be neglected, a sphere is accelerated by a uniform light beam of intensity 6.0 mW/m². The sphere is totally absorbing and has a radius of 2.0 μm and a uniform density of 5.0×10^3 kg/m³. What is the magnitude of the sphere's acceleration due to the light?

86 An unpolarized beam of light is sent into a stack of four polarizing sheets, oriented so that the angle between the polarizing directions of adjacent sheets is 30°. What fraction of the incident intensity is transmitted by the system?

87 **SSM** During a test, a NATO surveillance radar system, operating at 12 GHz at 180 kW of power, attempts to detect an incoming stealth aircraft at 90 km. Assume that the radar beam is emitted uniformly over a hemisphere. (a) What is the intensity of the beam when the beam reaches the aircraft's location? The aircraft reflects radar waves as though it has a cross-sectional area of only 0.22 m². (b) What is the power of the aircraft's reflection? Assume that the beam is reflected uniformly over a hemisphere. Back at the radar site, what are (c) the intensity, (d) the maximum value of the electric field vector, and (e) the rms value of the magnetic field of the reflected radar beam?

88 The magnetic component of an electromagnetic wave in vacuum has an amplitude of 85.8 nT and an angular wave number of 4.00 m⁻¹. What are (a) the frequency of the wave, (b) the rms value of the electric component, and (c) the intensity of the light?

89 Calculate the (a) upper and (b) lower limit of the Brewster angle for white light incident on fused quartz. Assume that the wavelength limits of the light are 400 and 700 nm.

90 In Fig. 33-71, two light rays pass from air through five layers of transparent plastic and then back into air. The layers have parallel interfaces and unknown thicknesses; their indexes of refraction are $n_1 = 1.7$, $n_2 = 1.6$, $n_3 = 1.5$, $n_4 = 1.4$, and $n_5 = 1.6$. Ray b is incident at angle $\theta_b = 20°$. Relative to a normal at the last interface, at what angle do (a) ray a and (b) ray b emerge? (*Hint:* Solving the problem algebraically can save time.) If the air at the left and right sides in the figure were, instead, glass with index of refraction 1.5, at what angle would (c) ray a and (d) ray b emerge?

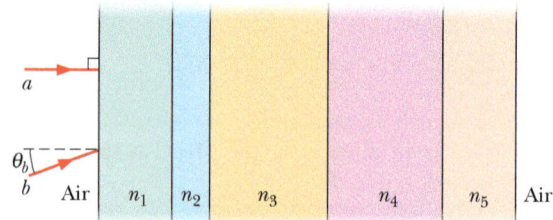

Figure 33-71 Problem 90.

91 A helium–neon laser, radiating at 632.8 nm, has a power output of 3.0 mW. The beam diverges (spreads) at angle $\theta = 0.17$ mrad (Fig. 33-72). (a) What is the intensity of the beam 40 m from the laser? (b) What is the power of a point source providing that intensity at that distance?

Figure 33-72 Problem 91.

92 In about A.D. 150, Claudius Ptolemy gave the following measured values for the angle of incidence θ_1 and the angle of refraction θ_2 for a light beam passing from air to water:

θ_1	θ_2	θ_1	θ_2
10°	8°	50°	35°
20°	15°30′	60°	40°30′
30°	22°30′	70°	45°30′
40°	29°	80°	50°

Assuming these data are consistent with the law of refraction, use them to find the index of refraction of water. These data are interesting as perhaps the oldest recorded physical measurements.

93 A beam of initially unpolarized light is sent through two polarizing sheets placed one on top of the other. What must be the angle between the polarizing directions of the sheets if the intensity of the transmitted light is to be one-third the incident intensity?

94 In Fig. 33-73, a long, straight copper wire (diameter 2.50 mm and resistance 1.00 Ω per 300 m) carries a uniform current of 25.0 A in the positive x direction. For point P on the wire's surface, calculate the magnitudes of (a) the electric field $\vec{E}$, (b) the magnetic field $\vec{B}$, and (c) the Poynting vector $\vec{S}$, and (d) determine the direction of $\vec{S}$.

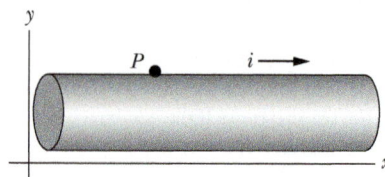

Figure 33-73 Problem 94.

95 Figure 33-74 shows a cylindrical resistor of length l, radius a, and resistivity ρ, carrying current i. (a) Show that the Poynting vector $\vec{S}$ at the surface of the resistor is everywhere directed normal to the surface, as shown. (b) Show that the rate P at which energy flows into the resistor through its cylindrical surface, calculated by integrating the Poynting vector over this surface, is equal to the rate at which thermal energy is produced:

$$\int \vec{S} \cdot d\vec{A} = i^2 R,$$

where $d\vec{A}$ is an element of area on the cylindrical surface and R is the resistance.

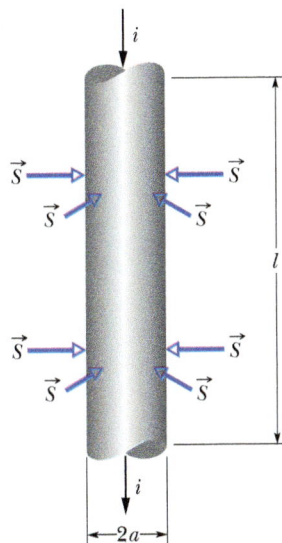

Figure 33-74 Problem 95.

96 A thin, totally absorbing sheet of mass m, face area A, and specific heat c_s is fully illuminated by a perpendicular beam of a plane electromagnetic wave. The magnitude of the maximum electric field of the wave is E_m. What is the rate dT/dt at which the sheet's temperature increases due to the absorption of the wave?

97 Two polarizing sheets, one directly above the other, transmit $p\%$ of the initially unpolarized light that is perpendicularly incident on the top sheet. What is the angle between the polarizing directions of the two sheets?

98 A laser beam of intensity I reflects from a flat, totally reflecting surface of area A, with a normal at angle θ with the beam. Write an expression for the beam's radiation pressure $p_r(\theta)$ on the surface in terms of the beam's pressure $p_{r\perp}$ when $\theta = 0°$.

99 A beam of intensity I reflects from a long, totally reflecting cylinder of radius R; the beam is perpendicular to the central axis of the cylinder and has a diameter larger than $2R$. What is the beam's force per unit length on the cylinder?

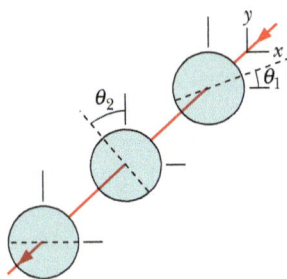

100 In Fig. 33-75, unpolarized light is sent into a system of three polarizing sheets, where the polarizing directions of the first and second sheets are at angles $\theta_1 = 20°$ and $\theta_2 = 40°$. What fraction of the initial light intensity emerges from the system?

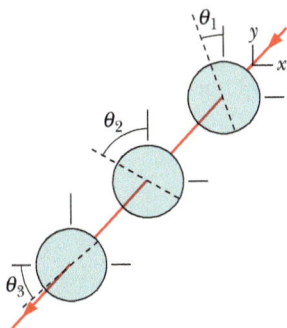

Figure 33-75 Problem 100.

101 In Fig. 33-76, unpolarized light is sent into a system of three polarizing sheets with polarizing directions at angles $\theta_1 = 20°$, $\theta_2 = 60°$, and $\theta_3 = 40°$. What fraction of the initial light intensity emerges from the system?

102 A square, perfectly reflecting surface is oriented in space to be perpendicular to the light rays from the Sun. The surface has an

Figure 33-76 Problem 101.

edge length of 2.0 m and is located 3.0×10^{11} m from the Sun's center. What is the radiation force on the surface from the light rays?

103 The rms value of the electric field in a certain light wave is 0.200 V/m. What is the amplitude of the associated magnetic field?

104 In Fig. 33-77, an albatross glides at a constant 15 m/s horizontally above level ground, moving in a vertical plane that contains the Sun. It glides toward a wall of height $h = 2.0$ m, which it will just barely clear. At that time of day, the angle of the Sun relative to the ground is $\theta = 30°$. At what speed does the shadow of the albatross move (a) across the level ground and then (b) up the wall? Suppose that later a hawk happens to glide along the same path, also at 15 m/s. You see that when its shadow reaches the wall, the speed of the shadow noticeably increases. (c) Is the Sun now higher or lower in the sky than when the albatross flew by earlier? (d) If the speed of the hawk's shadow on the wall is 45 m/s, what is the angle θ of the Sun just then?

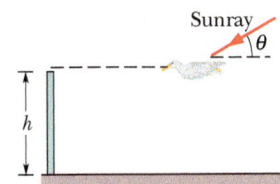

Figure 33-77 Problem 104.

105 The magnetic component of a polarized wave of light is given by $B_x = (4.00 \ \mu T) \sin [ky + (2.00 \times 10^{15} \ s^{-1})t]$. (a) In which direction does the wave travel, (b) parallel to which axis is it polarized, and (c) what is its intensity? (d) Write an expression for the electric field of the wave, including a value for the angular wave number. (e) What is the wavelength? (f) In which region of the electromagnetic spectrum is this electromagnetic wave?

106 In Fig. 33-78, where $n_1 = 1.70$, $n_2 = 1.50$, and $n_3 = 1.30$, light refracts from material 1 into material 2. If it is incident at point A at the critical angle for the interface between materials 2 and 3, what are (a) the angle of refraction at point B and (b) the initial angle θ? If, instead, light is incident at B at the critical angle for the interface between materials 2 and 3, what are (c) the angle of refraction at point A and (d) the initial angle θ? If, instead of all that, light is incident at point A at Brewster's angle for the interface between materials 2 and 3, what are (e) the angle of refraction at point B and (f) the initial angle θ?

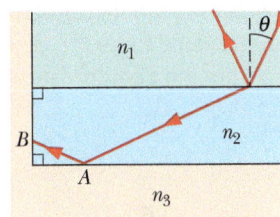

Figure 33-78 Problem 106.

107 When red light in vacuum is incident at the Brewster angle on a certain glass slab, the angle of refraction is 32.0°. What are (a) the index of refraction of the glass and (b) the Brewster angle?

108 Start from Eqs. 33-11 and 33-17 and show that $E(x, t)$ and $B(x, t)$, the electric and magnetic field components of a plane traveling electromagnetic wave, must satisfy the "wave equations"

$$\frac{\partial^2 E}{\partial t^2} = c^2 \frac{\partial^2 E}{\partial x^2} \quad \text{and} \quad \frac{\partial^2 B}{\partial t^2} = c^2 \frac{\partial^2 B}{\partial x^2}.$$

109 SSM (a) Show that Eqs. 33-1 land 33-2 satisfy the wave equations displayed in Problem 108. (b) Show that any expressions of the form $E = E_m f(kx \pm \omega t)$ and $B = B_m f(kx \pm \omega t)$, where $f(kx \pm \omega t)$ denotes an arbitrary function, also satisfy these wave equations.

110 A point source of light emits isotropically with a power of 200 W. What is the force due to the light on a totally absorbing sphere of radius 2.0 cm at a distance of 20 m from the source?

Images

34-1 IMAGES AND PLANE MIRRORS

Learning Objectives

After reading this module, you should be able to . . .

34.01 Distinguish virtual images from real images.

34.02 Explain the common roadway mirage.

34.03 Sketch a ray diagram for the reflection of a point source of light by a plane mirror, indicating the object distance and image distance.

34.04 Using the proper algebraic sign, relate the object distance p to the image distance i.

34.05 Give an example of the apparent hallway that you can see in a mirror maze based on equilateral triangles.

Key Ideas

● An image is a reproduction of an object via light. If the image can form on a surface, it is a real image and can exist even if no observer is present. If the image requires the visual system of an observer, it is a virtual image.

● A plane (flat) mirror can form a virtual image of a light source (said to be the object) by redirecting light rays emerging from the source. The image can be seen where backward extensions of reflected rays pass through one another. The object's distance p from the mirror is related to the (apparent) image distance i from the mirror by

$$i = -p \quad \text{(plane mirror).}$$

Object distance p is a positive quantity. Image distance i for a virtual image is a negative quantity.

What Is Physics?

One goal of physics is to discover the basic laws governing light, such as the law of refraction. A broader goal is to put those laws to use, and perhaps the most important use is the production of images. The first photographic images, made in 1824, were only novelties, but our world now thrives on images. Huge industries are based on the production of images on television, computer, and theater screens. Images from satellites guide military strategists during times of conflict and environmental strategists during times of blight. Camera surveillance can make a subway system more secure, but it can also invade the privacy of unsuspecting citizens. Physiologists and medical engineers are still puzzled by how images are produced by the human eye and the visual cortex of the brain, but they have managed to create mental images in some sightless people by electrical stimulation of the brain's visual cortex.

Our first step in this chapter is to define and classify images. Then we examine several basic ways in which they can be produced.

Two Types of Image

For you to see, say, a penguin, your eye must intercept some of the light rays spreading from the penguin and then redirect them onto the retina at the rear of the eye. Your visual system, starting with the retina and ending with the visual cortex at the rear of your brain, automatically and subconsciously processes the information provided by the light. That system identifies edges, orientations, textures, shapes, and colors and then rapidly brings to your consciousness an **image** (a reproduction derived from light) of the penguin; you perceive and recognize the penguin as being in the direction from which the light rays came and at the proper distance.

Your visual system goes through this processing and recognition even if the light rays do not come directly from the penguin, but instead reflect toward you from a mirror or refract through the lenses in a pair of binoculars. However, you now see the penguin in the direction from which the light rays came after they reflected or refracted, and the distance you perceive may be quite different from the penguin's true distance.

For example, if the light rays have been reflected toward you from a standard flat mirror, the penguin appears to be behind the mirror because the rays you intercept come from that direction. Of course, the penguin is not back there. This type of image, which is called a **virtual image,** truly exists only within the brain but nevertheless is *said* to exist at the perceived location.

A **real image** differs in that it can be formed on a surface, such as a card or a movie screen. You can see a real image (otherwise movie theaters would be empty), but the existence of the image does not depend on your seeing it and it is present even if you are not. Before we discuss real and virtual images in detail, let's examine a natural virtual image.

A Common Mirage

A common example of a virtual image is a pool of water that appears to lie on the road some distance ahead of you on a sunny day, but that you can never reach. The pool is a *mirage* (a type of illusion), formed by light rays coming from the low section of the sky in front of you (Fig. 34-1a). As the rays approach the road, they travel through progressively warmer air that has been heated by the road, which is usually relatively warm. With an increase in air temperature, the density of the air—and hence the index of refraction of the air—decreases slightly. Thus, as the rays descend, encountering progressively smaller indexes of refraction, they continuously bend toward the horizontal (Fig. 34-1b).

Once a ray is horizontal, somewhat above the road's surface, it still bends because the lower portion of each associated wavefront is in slightly warmer air and is moving slightly faster than the upper portion of the wavefront (Fig. 34-1c). This nonuniform motion of the wavefronts bends the ray upward. As the ray then ascends, it continues to bend upward through progressively greater indexes of refraction (Fig. 34-1d).

If you intercept some of this light, your visual system automatically infers that it originated along a backward extension of the rays you have intercepted and, to make sense of the light, assumes that it came from the road surface. If the light happens to be bluish from blue sky, the mirage appears bluish, like water. Because the air is probably turbulent due to the heating, the mirage shimmies, as if water waves were present. The bluish coloring and the shimmy enhance the illusion of a pool of water, but you are actually seeing a virtual image of a low section of the sky. As you travel toward the illusionary pool, you no longer intercept the shallow refracted rays and the illusion disappears.

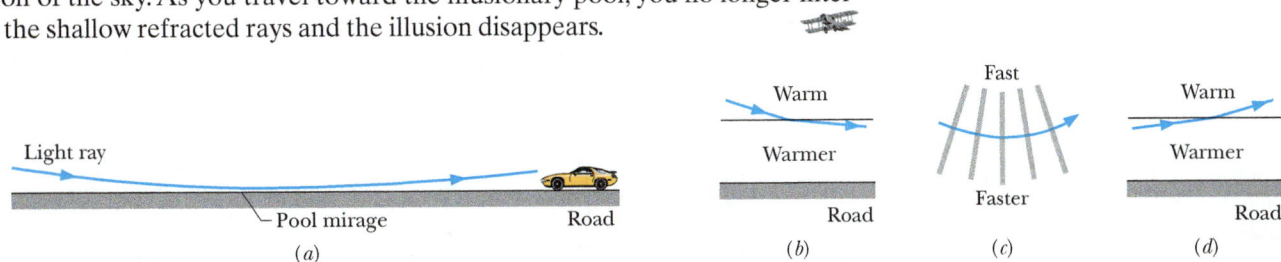

Figure 34-1 (a) A ray from a low section of the sky refracts through air that is heated by a road (without reaching the road). An observer who intercepts the light perceives it to be from a pool of water on the road. (b) Bending (exaggerated) of a light ray descending across an imaginary boundary from warm air to warmer air. (c) Shifting of wavefronts and associated bending of a ray, which occur because the lower ends of wavefronts move faster in warmer air. (d) Bending of a ray ascending across an imaginary boundary to warm air from warmer air.

In a plane mirror the light seems to come from an object on the other side.

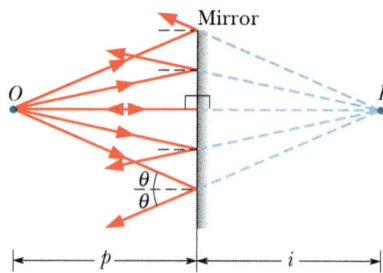

Mirror

Figure 34-2 A point source of light O, called the *object*, is a perpendicular distance p in front of a plane mirror. Light rays reaching the mirror from O reflect from the mirror. If your eye intercepts some of the reflected rays, you perceive a point source of light I to be behind the mirror, at a perpendicular distance i. The perceived source I is a virtual image of object O.

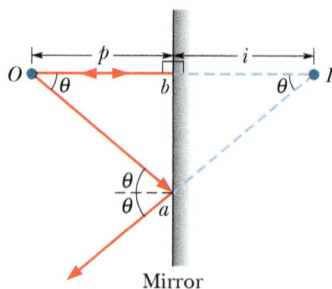

Figure 34-3 Two rays from Fig. 34-2. Ray Oa makes an arbitrary angle θ with the normal to the mirror surface. Ray Ob is perpendicular to the mirror.

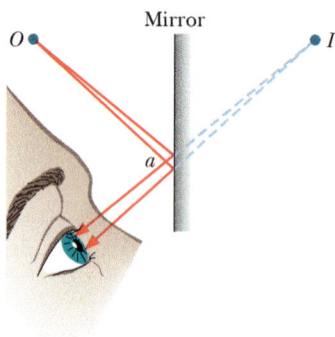

Mirror

Figure 34-4 A "pencil" of rays from O enters the eye after reflection at the mirror. Only a small portion of the mirror near a is involved in this reflection. The light appears to originate at point I behind the mirror.

Plane Mirrors

A **mirror** is a surface that can reflect a beam of light in one direction instead of either scattering it widely in many directions or absorbing it. A shiny metal surface acts as a mirror; a concrete wall does not. In this module we examine the images that a **plane mirror** (a flat reflecting surface) can produce.

Figure 34-2 shows a point source of light O, which we shall call the *object*, at a perpendicular distance p in front of a plane mirror. The light that is incident on the mirror is represented with rays spreading from O. The reflection of that light is represented with reflected rays spreading from the mirror. If we extend the reflected rays backward (behind the mirror), we find that the extensions intersect at a point that is a perpendicular distance i behind the mirror.

If you look into the mirror of Fig. 34-2, your eyes intercept some of the reflected light. To make sense of what you see, you perceive a point source of light located at the point of intersection of the extensions. This point source is the image I of object O. It is called a *point image* because it is a point, and it is a virtual image because the rays do not actually pass through it. (As you will see, rays *do* pass through a point of intersection for a real image.)

Ray Tracing. Figure 34-3 shows two rays selected from the many rays in Fig. 34-2. One reaches the mirror at point b, perpendicularly. The other reaches it at an arbitrary point a, with an angle of incidence θ. The extensions of the two reflected rays are also shown. The right triangles $aOba$ and $aIba$ have a common side and three equal angles and are thus congruent (equal in size); so their horizontal sides have the same length. That is,

$$Ib = Ob, \qquad (34\text{-}1)$$

where Ib and Ob are the distances from the mirror to the image and the object, respectively. Equation 34-1 tells us that the image is as far behind the mirror as the object is in front of it. By convention (that is, to get our equations to work out), *object distances p* are taken to be positive quantities and *image distances i* for virtual images (as here) are taken to be negative quantities. Thus, Eq. 34-1 can be written as $|i| = p$ or as

$$i = -p \quad \text{(plane mirror)}. \qquad (34\text{-}2)$$

Only rays that are fairly close together can enter the eye after reflection at a mirror. For the eye position shown in Fig. 34-4, only a small portion of the mirror near point a (a portion smaller than the pupil of the eye) is useful in forming the image. To find this portion, close one eye and look at the mirror image of a small object such as the tip of a pencil. Then move your fingertip over the mirror surface until you cannot see the image. Only that small portion of the mirror under your fingertip produced the image.

Extended Objects

In Fig. 34-5, an extended object O, represented by an upright arrow, is at perpendicular distance p in front of a plane mirror. Each small portion of the

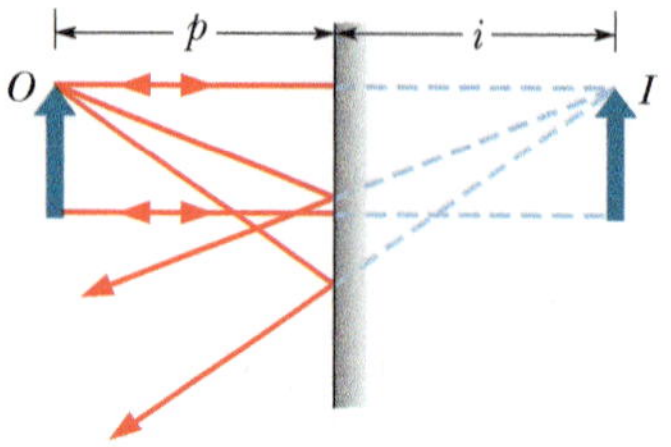

In a plane mirror the image is just as far from the mirror as the object.

Figure 34-5 An extended object O and its virtual image I in a plane mirror.

Figure 34-6 A maze of mirrors.

Courtesy Adrian Fisher, www.mazemaker.com

object that faces the mirror acts like the point source O of Figs. 34-2 and 34-3. If you intercept the light reflected by the mirror, you perceive a virtual image I that is a composite of the virtual point images of all those portions of the object. This virtual image seems to be at (negative) distance i behind the mirror, with i and p related by Eq. 34-2.

We can also locate the image of an extended object as we did for a point object in Fig. 34-2: we draw some of the rays that reach the mirror from the top of the object, draw the corresponding reflected rays, and then extend those reflected rays behind the mirror until they intersect to form an image of the top of the object. We then do the same for rays from the bottom of the object. As shown in Fig. 34-5, we find that virtual image I has the same orientation and *height* (measured parallel to the mirror) as object O.

Mirror Maze

In a mirror maze (Fig. 34-6), each wall is covered, floor to ceiling, with a mirror. Walk through such a maze and what you see in most directions is a confusing montage of reflections. In some directions, however, you see a hallway that seems to offer a path through the maze. Take these hallways, though, and you soon learn, after smacking into mirror after mirror, that the hallways are largely an illusion.

Figure 34-7a is an overhead view of a simple mirror maze in which differently painted floor sections form equilateral triangles (60° angles) and walls are covered with vertical mirrors. You look into the maze while standing at point O at the middle of the maze entrance. In most directions, you see a confusing jumble of images. However, you see something curious in the direction of the ray shown in Fig. 34-7a. That ray leaves the middle of mirror B and reflects to you at the middle of mirror A. (The reflection obeys the law of reflection, with the angle of incidence and the angle of reflection both equal to 30°.)

To make sense of the origin of the ray reaching you, your brain automatically extends the ray backward. It appears to originate at a point lying *behind* mirror A. That is, you perceive a virtual image of B behind A, at a distance equal to the actual distance between A and B (Fig. 34-7b). Thus, when you face into the maze in this direction, you see B along an apparent straight hallway consisting of four triangular floor sections.

This story is incomplete, however, because the ray reaching you does not *originate* at mirror B—it only reflects there. To find the origin, we continue to apply the law of reflection as we work backwards, reflection by reflection on the mirrors (Fig. 34-7c). We finally come to the origin of the ray: you! What you see when you look along the apparent hallway is a virtual image of yourself, at a distance of nine triangular floor sections from you (Fig. 34-7d).

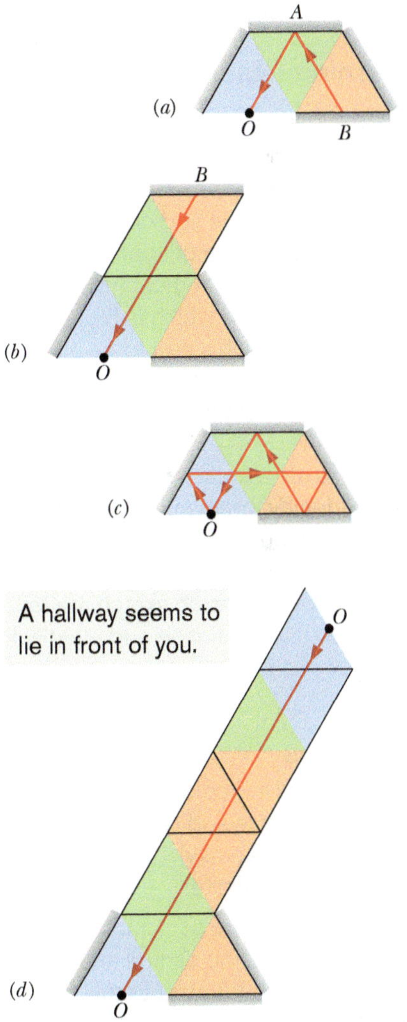

A hallway seems to lie in front of you.

Figure 34-7 (a) Overhead view of a mirror maze. A ray from mirror B reaches you at O by reflecting from mirror A. (b) Mirror B appears to be behind A. (c) The ray reaching you comes from you. (d) You see a virtual image of yourself at the end of an apparent hallway. (Can you find a second apparent hallway extending away from point O?)

☑ **Checkpoint 1**

In the figure you are in a system of two vertical parallel mirrors A and B separated by distance d. A grinning gargoyle is perched at point O, a distance $0.2d$ from mirror A. Each mirror produces a *first* (least deep) image of the gargoyle. Then each mirror produces a *second* image with the object being the first image in the opposite mirror. Then each mirror produces a *third* image with the object being the second image in the opposite mirror, and so on—you might see hundreds of grinning gargoyle images. How deep behind mirror A are the first, second, and third images in mirror A?

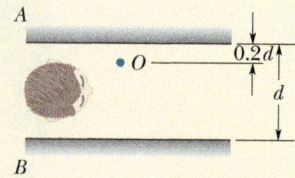

34-2 SPHERICAL MIRRORS

Learning Objectives

After reading this module, you should be able to . . .

34.06 Distinguish a concave spherical mirror from a convex spherical mirror.

34.07 For concave and convex mirrors, sketch a ray diagram for the reflection of light rays that are initially parallel to the central axis, indicating how they form the focal points, and identifying which is real and which is virtual.

34.08 Distinguish a real focal point from a virtual focal point, identify which corresponds to which type of mirror, and identify the algebraic sign associated with each focal length.

34.09 Relate a focal length of a spherical mirror to the radius.

34.10 Identify the terms "inside the focal point" and "outside the focal point."

34.11 For an object (a) inside and (b) outside the focal point of a concave mirror, sketch the reflections of at least two rays to find the image and identify the type and orientation of the image.

34.12 For a concave mirror, distinguish the locations and orientations of a real image and a virtual image.

34.13 For an object in front of a convex mirror, sketch the reflections of at least two rays to find the image and identify the type and orientation of the image.

34.14 Identify which type of mirror can produce both real and virtual images and which type can produce only virtual images.

34.15 Identify the algebraic signs of the image distance i for real images and virtual images.

34.16 For convex, concave, and plane mirrors, apply the relationship between the focal length f, object distance p, and image distance i.

34.17 Apply the relationships between lateral magnification m, image height h', object height h, image distance i, and object distance p.

Key Ideas

● A spherical mirror is in the shape of a small section of a spherical surface and can be concave (the radius of curvature r is a positive quantity), convex (r is a negative quantity), or plane (flat, r is infinite).

● If parallel rays are sent into a (spherical) concave mirror parallel to the central axis, the reflected rays pass through a common point (a real focus F) at a distance f (a positive quantity) from the mirror. If they are sent toward a (spherical) convex mirror, backward extensions of the reflected rays pass through a common point (a virtual focus F) at a distance f (a negative quantity) from the mirror.

● A concave mirror can form a real image (if the object is outside the focal point) or a virtual image (if the object is inside the focal point).

● A convex mirror can form only a virtual image.

● The mirror equation relates an object distance p, the mirror's focal length f and radius of curvature r, and the image distance i:

$$\frac{1}{p} + \frac{1}{i} = \frac{1}{f} = \frac{2}{r}.$$

● The magnitude of the lateral magnification m of an object is the ratio of the image height h' to object height h,

$$|m| = \frac{h'}{h},$$

and is related to the object distance p and image distance i by

$$m = -\frac{i}{p}.$$

Spherical Mirrors

We turn now from images produced by plane mirrors to images produced by mirrors with curved surfaces. In particular, we consider spherical mirrors, which are simply mirrors in the shape of a small section of the surface of a sphere. A plane mirror is in fact a spherical mirror with an infinitely large *radius of curvature* and thus an approximately flat surface.

Making a Spherical Mirror

We start with the plane mirror of Fig. 34-8a, which faces leftward toward an object *O* that is shown and an observer that is not shown. We make a **concave mirror** by curving the mirror's surface so it is *concave* ("caved in") as in Fig. 34-8b. Curving the surface in this way changes several characteristics of the mirror and the image it produces of the object:

1. The *center of curvature C* (the center of the sphere of which the mirror's surface is part) was infinitely far from the plane mirror; it is now closer but still in front of the concave mirror.

2. The *field of view*—the extent of the scene that is reflected to the observer—was wide; it is now smaller.

3. The image of the object was as far behind the plane mirror as the object was in front; the image is farther behind the concave mirror; that is, |i| is greater.

4. The height of the image was equal to the height of the object; the height of the image is now greater. This feature is why many makeup mirrors and shaving mirrors are concave—they produce a larger image of a face.

We can make a **convex mirror** by curving a plane mirror so its surface is *convex* ("flexed out") as in Fig. 34-8c. Curving the surface in this way (1) moves the center of curvature C to *behind* the mirror and (2) *increases* the field of view. It also (3) moves the image of the object *closer* to the mirror and (4) *shrinks* it. Store surveillance mirrors are usually convex to take advantage of the increase in the field of view—more of the store can then be seen with a single mirror.

Focal Points of Spherical Mirrors

For a plane mirror, the magnitude of the image distance *i* is always equal to the object distance *p*. Before we can determine how these two distances are related for a spherical mirror, we must consider the reflection of light from an object *O* located an effectively infinite distance in front of a spherical mirror, on the mirror's *central axis*. That axis extends through the center of curvature C and the center c of the mirror. Because of the great distance between the object and the mirror, the light waves spreading from the object are plane waves when they reach the mirror along the central axis. This means that the rays representing the light waves are all parallel to the central axis when they reach the mirror.

Forming a Focus. When these parallel rays reach a concave mirror like that of Fig. 34-9a, those near the central axis are reflected through a common point *F*; two of these reflected rays are shown in the figure. If we placed a (small) card at *F*, a point image of the infinitely distant object *O* would appear on the card. (This would occur for any infinitely distant object.) Point *F* is called the **focal point** (or **focus**) of the mirror, and its distance from the center of the mirror c is the **focal length** *f* of the mirror.

If we now substitute a convex mirror for the concave mirror, we find that the parallel rays are no longer reflected through a common point. Instead, they diverge as shown in Fig. 34-9b. However, if your eye intercepts some of the reflected light, you perceive the light as originating from a point source behind the mirror. This perceived source is located where extensions of the reflected rays pass through a common point (*F* in Fig. 34-9b). That point is the focal point (or

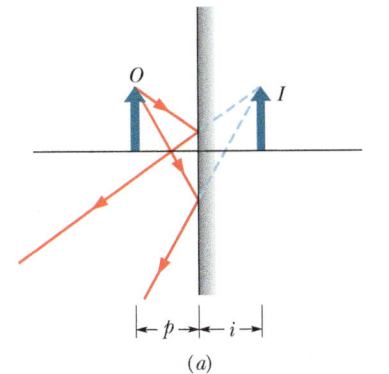

(a)

Bending the mirror this way shifts the image away.

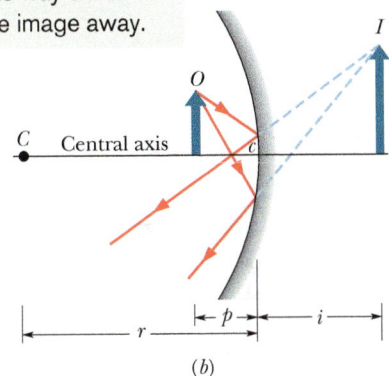

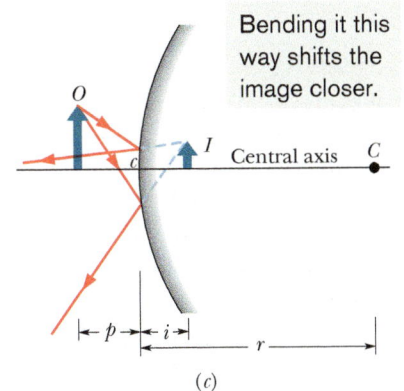

(b)

Bending it this way shifts the image closer.

(c)

Figure 34-8 (a) An object *O* forms a virtual image *I* in a plane mirror. (b) If the mirror is bent so that it becomes *concave*, the image moves farther away and becomes larger. (c) If the plane mirror is bent so that it becomes *convex*, the image moves closer and becomes smaller.

To find the focus, send in rays parallel to the central axis.

If you intercept the reflections, they seem to come from this point.

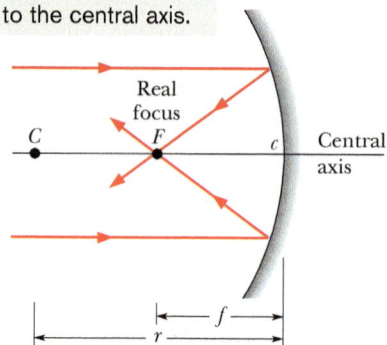
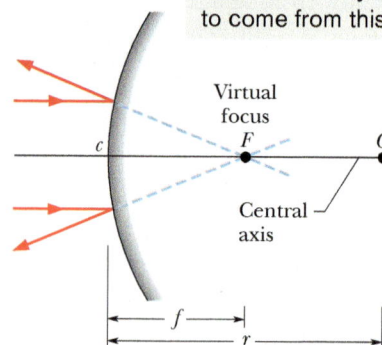

(a)   (b)

Figure 34-9 (a) In a concave mirror, incident parallel light rays are brought to a real focus at F, on the same side of the mirror as the incident light rays. (b) In a convex mirror, incident parallel light rays seem to diverge from a virtual focus at F, on the side of the mirror opposite the light rays.

focus) F of the convex mirror, and its distance from the mirror surface is the focal length f of the mirror. If we placed a card at this focal point, an image of object O would *not* appear on the card; so this focal point is not like that of a concave mirror.

Two Types. To distinguish the actual focal point of a concave mirror from the perceived focal point of a convex mirror, the former is said to be a *real focal point* and the latter is said to be a *virtual focal point*. Moreover, the focal length f of a concave mirror is taken to be a positive quantity, and that of a convex mirror a negative quantity. For mirrors of both types, the focal length f is related to the radius of curvature r of the mirror by

$$f = \tfrac{1}{2}r \quad \text{(spherical mirror)}, \tag{34-3}$$

where r is positive for a concave mirror and negative for a convex mirror.

Images from Spherical Mirrors

Inside. With the focal point of a spherical mirror defined, we can find the relation between image distance i and object distance p for concave and convex spherical mirrors. We begin by placing the object O *inside the focal point* of the concave mirror—that is, between the mirror and its focal point F (Fig. 34-10a).

Changing the location of the object relative to F changes the image.

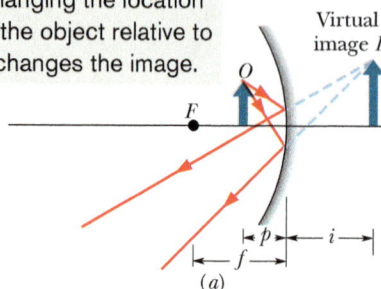
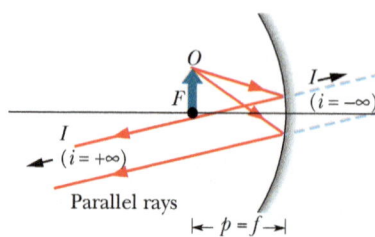
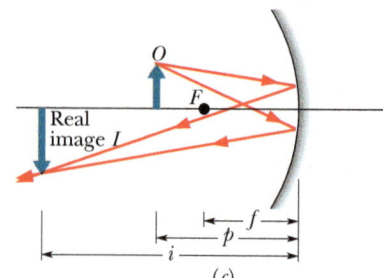

(a)   (b)   (c)

Figure 34-10 (a) An object O inside the focal point of a concave mirror, and its virtual image I. (b) The object at the focal point F. (c) The object outside the focal point, and its real image I.

An observer can then see a virtual image of O in the mirror: The image appears to be behind the mirror, and it has the same orientation as the object.

If we now move the object away from the mirror until it is at the focal point, the image moves farther and farther back from the mirror until, when the object is at the focal point, the image is at infinity (Fig. 34-10*b*). The image is then ambiguous and imperceptible because neither the rays reflected by the mirror nor the ray extensions behind the mirror cross to form an image of O.

Outside. If we next move the object *outside the focal point*—that is, farther away from the mirror than the focal point—the rays reflected by the mirror converge to form an *inverted* image of object O (Fig. 34-10*c*) in front of the mirror. That image moves in from infinity as we move the object farther outside F. If you were to hold a card at the position of the image, the image would show up on the card—the image is said to be *focused* on the card by the mirror. (The verb "focus," which in this context means to produce an image, differs from the noun "focus," which is another name for the focal point.) Because this image can actually appear on a surface, it is a real image—the rays actually intersect to create the image, regardless of whether an observer is present. The image distance i of a real image is a positive quantity, in contrast to that for a virtual image. We can now generalize about the location of images from spherical mirrors:

> Real images form on the side of a mirror where the object is, and virtual images form on the opposite side.

Main Equation. As we shall prove in Module 34-6, when light rays from an object make only small angles with the central axis of a spherical mirror, a simple equation relates the object distance p, the image distance i, and the focal length f:

$$\frac{1}{p} + \frac{1}{i} = \frac{1}{f} \quad \text{(spherical mirror).} \qquad (34\text{-}4)$$

We assume such small angles in figures such as Fig. 34-10, but for clarity the rays are drawn with exaggerated angles. With that assumption, Eq. 34-4 applies to any concave, convex, or plane mirror. For a convex or plane mirror, only a virtual image can be formed, regardless of the object's location on the central axis. As shown in the example of a convex mirror in Fig. 34-8*c*, the image is always on the opposite side of the mirror from the object and has the same orientation as the object.

Magnification. The size of an object or image, as measured *perpendicular* to the mirror's central axis, is called the object or image *height*. Let h represent the height of the object, and h' the height of the image. Then the ratio h'/h is called the **lateral magnification** m produced by the mirror. However, by convention, the lateral magnification always includes a plus sign when the image orientation is that of the object and a minus sign when the image orientation is opposite that of the object. For this reason, we write the formula for m as

$$|m| = \frac{h'}{h} \quad \text{(lateral magnification).} \qquad (34\text{-}5)$$

We shall soon prove that the lateral magnification can also be written as

$$m = -\frac{i}{p} \quad \text{(lateral magnification).} \qquad (34\text{-}6)$$

For a plane mirror, for which $i = -p$, we have $m = +1$. The magnification of 1 means that the image is the same size as the object. The plus sign means that

Table 34-1 Your Organizing Table for Mirrors

Mirror Type	Object Location	Image			Sign			
		Location	Type	Orientation	of f	of r	of i	of m
Plane	Anywhere							
Concave	Inside F							
	Outside F							
Convex	Anywhere							

Figure 34-11 (a, b) Four rays that may be drawn to find the image formed by a concave mirror. For the object position shown, the image is real, inverted, and smaller than the object. (c, d) Four similar rays for the case of a convex mirror. For a convex mirror, the image is always virtual, oriented like the object, and smaller than the object. [In (c), ray 2 is initially directed toward focal point F. In (d), ray 3 is initially directed toward center of curvature C.]

the image and the object have the same orientation. For the concave mirror of Fig. 34-10c, $m \approx -1.5$.

Organizing Table. Equations 34-3 through 34-6 hold for all plane mirrors, concave spherical mirrors, and convex spherical mirrors. In addition to those equations, you have been asked to absorb a lot of information about these mirrors, and you should organize it for yourself by filling in Table 34-1. Under Image Location, note whether the image is on the *same* side of the mirror as the object or on the *opposite* side. Under Image Type, note whether the image is *real* or *virtual*. Under Image Orientation, note whether the image has the *same* orientation as the object or is *inverted*. Under Sign, give the sign of the quantity or fill in ± if the sign is ambiguous. You will need this organization to tackle homework or a test.

Locating Images by Drawing Rays

Figures 34-11a and b show an object O in front of a concave mirror. We can graphically locate the image of any off-axis point of the object by drawing a *ray diagram* with any two of four special rays through the point:

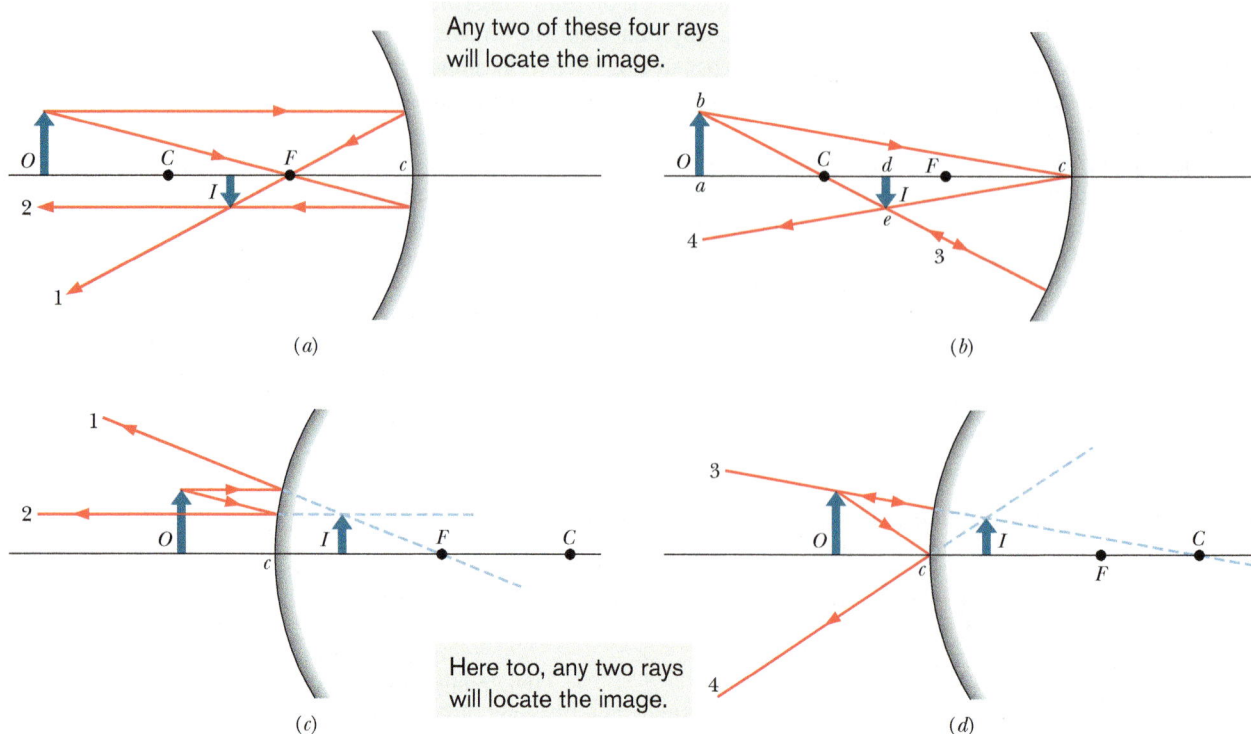

(a)

(b)

(c)

(d)

Any two of these four rays will locate the image.

Here too, any two rays will locate the image.

1. A ray that is initially parallel to the central axis reflects through the focal point F (ray 1 in Fig. 34-11a).

2. A ray that reflects from the mirror after passing through the focal point emerges parallel to the central axis (ray 2 in Fig. 34-11a).

3. A ray that reflects from the mirror after passing through the center of curvature C returns along itself (ray 3 in Fig. 34-11b).

4. A ray that reflects from the mirror at point c is reflected symmetrically about that axis (ray 4 in Fig. 34-11b).

The image of the point is at the intersection of the two special rays you choose. The image of the object can then be found by locating the images of two or more of its off-axis points (say, the point most off axis) and then sketching in the rest of the image. You need to modify the descriptions of the rays slightly to apply them to convex mirrors, as in Figs. 34-11c and d.

Proof of Equation 34-6

We are now in a position to derive Eq. 34-6 ($m = -i/p$), the equation for the lateral magnification of an object reflected in a mirror. Consider ray 4 in Fig. 34-11b. It is reflected at point c so that the incident and reflected rays make equal angles with the axis of the mirror at that point.

The two right triangles abc and dec in the figure are similar (have the same set of angles); so we can write

$$\frac{de}{ab} = \frac{cd}{ca}.$$

The quantity on the left (apart from the question of sign) is the lateral magnification m produced by the mirror. Because we indicate an inverted image as a *negative* magnification, we symbolize this as $-m$. However, $cd = i$ and $ca = p$; so we have

$$m = -\frac{i}{p} \quad \text{(magnification)}, \tag{34-7}$$

which is the relation we set out to prove.

✓ Checkpoint 2

A Central American vampire bat, dozing on the central axis of a spherical mirror, is magnified by $m = -4$. Is its image (a) real or virtual, (b) inverted or of the same orientation as the bat, and (c) on the same side of the mirror as the bat or on the opposite side?

Sample Problem 34.01 Image produced by a spherical mirror

A tarantula of height h sits cautiously before a spherical mirror whose focal length has absolute value $|f| = 40$ cm. The image of the tarantula produced by the mirror has the same orientation as the tarantula and has height $h' = 0.20h$.

(a) Is the image real or virtual, and is it on the same side of the mirror as the tarantula or the opposite side?

Reasoning: Because the image has the same orientation as the tarantula (the object), it must be virtual and on the opposite side of the mirror. (You can easily see this result if you have filled out Table 34-1.)

(b) Is the mirror concave or convex, and what is its focal length f, sign included?

KEY IDEA

We *cannot* tell the type of mirror from the type of image because both types of mirror can produce virtual images. Similarly, we cannot tell the type of mirror from the sign of the focal length f, as obtained from Eq. 34-3 or Eq. 34-4, because we lack enough information to use either equation. However, we can make use of the magnification information.

Calculations: From the given information, we know that the ratio of image height h' to object height h is 0.20. Thus, from Eq. 34-5 we have

$$|m| = \frac{h'}{h} = 0.20.$$

Because the object and image have the same orientation, we know that m must be positive: $m = +0.20$. Substituting this into Eq. 34-6 and solving for, say, i gives us

$$i = -0.20p,$$

which does not appear to be of help in finding f. However, it is helpful if we substitute it into Eq. 34-4. That equation gives us

$$\frac{1}{f} = \frac{1}{i} + \frac{1}{p} = \frac{1}{-0.20p} + \frac{1}{p} = \frac{1}{p}(-5 + 1),$$

from which we find

$$f = -p/4.$$

Now we have it: Because p is positive, f must be negative, which means that the mirror is convex with

$$f = -40 \text{ cm.} \qquad \text{(Answer)}$$

WILEY PLUS Additional examples, video, and practice available at *WileyPLUS*

34-3 SPHERICAL REFRACTING SURFACES

Learning Objectives

After reading this module, you should be able to . . .

34.18 Identify that the refraction of rays by a spherical surface can produce real images and virtual images of an object, depending on the indexes of refraction on the two sides, the surface's radius of curvature r, and whether the object faces a concave or convex surface.

34.19 For a point object on the central axis of a spherical refracting surface, sketch the refraction of a ray in the six general arrangements and identify whether the image is real or virtual.

34.20 For a spherical refracting surface, identify what type of image appears on the same side as the object and what type appears on the opposite side.

34.21 For a spherical refracting surface, apply the relationship between the two indexes of refraction, the object distance p, the image distance i, and the radius of curvature r.

34.22 Identify the algebraic signs of the radius r for an object facing a concave refracting surface and a convex refracting surface.

Key Ideas

● A single spherical surface that refracts light can form an image.

● The object distance p, the image distance i, and the radius of curvature r of the surface are related by

$$\frac{n_1}{p} + \frac{n_2}{i} = \frac{n_2 - n_1}{r},$$

where n_1 is the index of refraction of the material where the object is located and n_2 is the index of refraction on the other side of the surface.

● If the surface faced by the object is convex, r is positive, and if it is concave, r is negative.

● Images on the object's side of the surface are virtual, and images on the opposite side are real.

Spherical Refracting Surfaces

We now turn from images formed by reflection to images formed by refraction through surfaces of transparent materials, such as glass. We shall consider only spherical surfaces, with radius of curvature r and center of curvature C. The light will be emitted by a point object O in a medium with index of refraction n_1; it will refract through a spherical surface into a medium of index of refraction n_2.

Our concern is whether the light rays, after refracting through the surface, form a real image (no observer necessary) or a virtual image (assuming that an observer intercepts the rays). The answer depends on the relative values of n_1 and n_2 and on the geometry of the situation.

Six possible results are shown in Fig. 34-12. In each part of the figure, the medium with the greater index of refraction is shaded, and object O is always in the medium with index of refraction n_1, to the left of the refracting surface. In each part, a representative ray is shown refracting through the surface. (That ray and a ray along the central axis suffice to determine the position of the image in each case.)

At the point of refraction of each ray, the normal to the refracting surface is a radial line through the center of curvature C. Because of the refraction, the ray bends toward the normal if it is entering a medium of greater index of refraction and away from the normal if it is entering a medium of lesser index of refraction. If the bending sends the ray toward the central axis, that ray and others (undrawn) form a real image on that axis. If the bending sends the ray away from the central axis, the ray cannot form a real image; however, backward extensions of it and other refracted rays can form a virtual image, provided (as with mirrors) some of those rays are intercepted by an observer.

Real images I are formed (at image distance i) in parts a and b of Fig. 34-12, where the refraction directs the ray *toward* the central axis. Virtual images are formed in parts c and d, where the refraction directs the ray *away* from the central axis. Note, in these four parts, that real images are formed when the object is relatively far from the refracting surface and virtual images are formed when the object is nearer the refracting surface. In the final situations (Figs. 34-12*e* and *f*), refraction always directs the ray away from the central axis and virtual images are always formed, regardless of the object distance.

Note the following major difference from reflected images:

⭐ Real images form on the side of a refracting surface that is opposite the object, and virtual images form on the same side as the object.

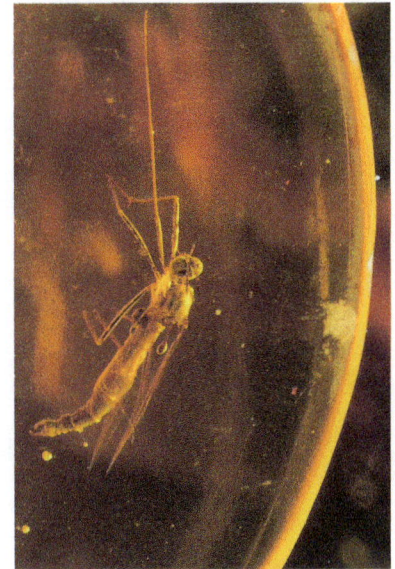

Dr. Paul A. Zahl/Photo Researchers, Inc.

This insect has been entombed in amber for about 25 million years. Because we view the insect through a curved refracting surface, the location of the image we see does not coincide with the location of the insect (see Fig. 34-12*d*).

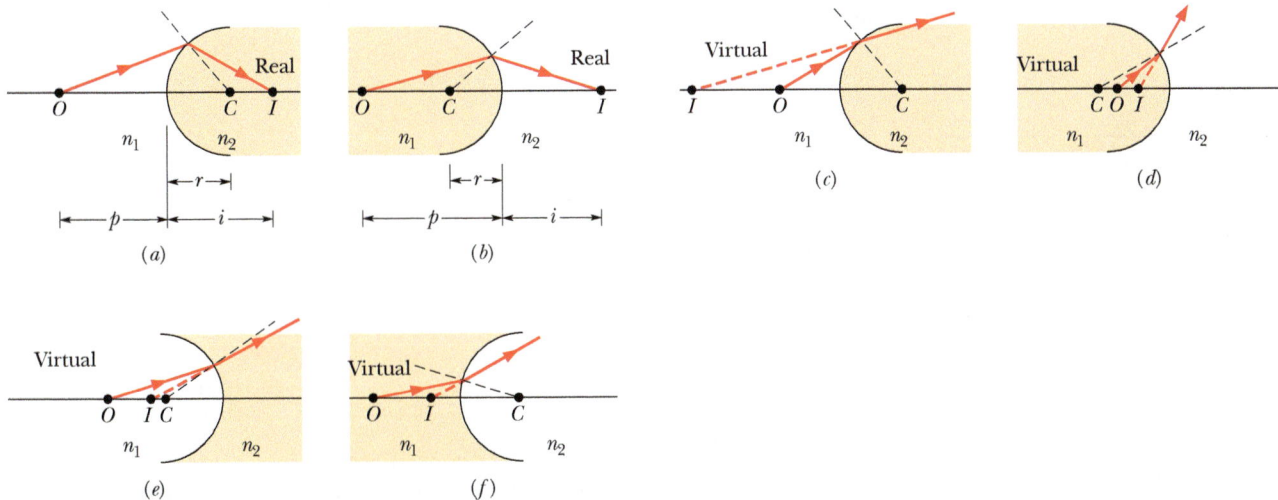

(a) (b) (c) (d)

(e) (f)

Figure 34-12 Six possible ways in which an image can be formed by refraction through a spherical surface of radius r and center of curvature C. The surface separates a medium with index of refraction n_1 from a medium with index of refraction n_2. The point object O is always in the medium with n_1, to the left of the surface. The material with the lesser index of refraction is unshaded (think of it as being air, and the other material as being glass). Real images are formed in (a) and (b); virtual images are formed in the other four situations.

In Module 34-6, we shall show that (for light rays making only small angles with the central axis)

$$\frac{n_1}{p} + \frac{n_2}{i} = \frac{n_2 - n_1}{r}. \tag{34-8}$$

Just as with mirrors, the object distance p is positive, and the image distance i is positive for a real image and negative for a virtual image. However, to keep all the signs correct in Eq. 34-8, we must use the following rule for the sign of the radius of curvature r:

> When the object faces a convex refracting surface, the radius of curvature r is positive. When it faces a concave surface, r is negative.

Be careful: This is just the reverse of the sign convention we have for mirrors, which can be a slippery point in the heat of an exam.

✓ Checkpoint 3

A bee is hovering in front of the concave spherical refracting surface of a glass sculpture. (a) Which part of Fig. 34-12 is like this situation? (b) Is the image produced by the surface real or virtual, and (c) is it on the same side as the bee or the opposite side?

Sample Problem 34.02 Image produced by a refracting surface

A Jurassic mosquito is discovered embedded in a chunk of amber, which has index of refraction 1.6. One surface of the amber is spherically convex with radius of curvature 3.0 mm (Fig. 34-13). The mosquito's head happens to be on the central axis of that surface and, when viewed along the axis, appears to be buried 5.0 mm into the amber. How deep is it really?

KEY IDEAS

The head appears to be 5.0 mm into the amber only because the light rays that the observer intercepts are bent by refraction at the convex amber surface. The image distance i differs from the object distance p according to Eq. 34-8. To use that equation to find the object distance, we first note:

1. Because the object (the head) and its image are on the same side of the refracting surface, the image must be virtual and so $i = -5.0$ mm.
2. Because the object is always taken to be in the medium of index of refraction n_1, we must have $n_1 = 1.6$ and $n_2 = 1.0$.
3. Because the *object* faces a concave refracting surface, the radius of curvature r is negative, and so $r = -3.0$ mm.

Calculations: Making these substitutions in Eq. 34-8,

$$\frac{n_1}{p} + \frac{n_2}{i} = \frac{n_2 - n_1}{r},$$

yields

$$\frac{1.6}{p} + \frac{1.0}{-5.0 \text{ mm}} = \frac{1.0 - 1.6}{-3.0 \text{ mm}}$$

and
$$p = 4.0 \text{ mm.} \qquad \text{(Answer)}$$

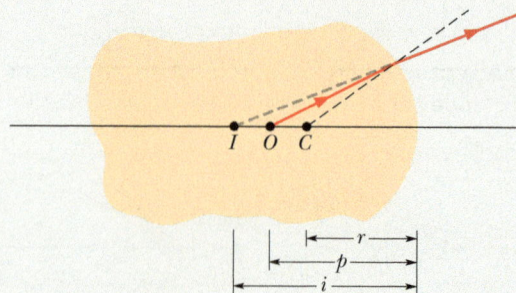

Figure 34-13 A piece of amber with a mosquito from the Jurassic period, with the head buried at point O. The spherical refracting surface at the right end, with center of curvature C, provides an image I to an observer intercepting rays from the object at O.

WILEY **PLUS** Additional examples, video, and practice available at *WileyPLUS*

34-4 THIN LENSES

Learning Objectives

After reading this module, you should be able to . . .

34.23 Distinguish converging lenses from diverging lenses.

34.24 For converging and diverging lenses, sketch a ray diagram for rays initially parallel to the central axis, indicating how they form focal points, and identifying which is real and which is virtual.

34.25 Distinguish a real focal point from a virtual focal point, identify which corresponds to which type of lens and under which circumstances, and identify the algebraic sign associated with each focal length.

34.26 For an object (a) inside and (b) outside the focal point of a converging lens, sketch at least two rays to find the image and identify the type and orientation of the image.

34.27 For a converging lens, distinguish the locations and orientations of a real image and a virtual image.

34.28 For an object in front of a diverging lens, sketch at least two rays to find the image and identify the type and orientation of the image.

34.29 Identify which type of lens can produce both real

and virtual images and which type can produce only virtual images.

34.30 Identify the algebraic sign of the image distance i for a real image and for a virtual image.

34.31 For converging and diverging lenses, apply the relationship between the focal length f, object distance p, and image distance i.

34.32 Apply the relationships between lateral magnification m, image height h', object height h, image distance i, and object distance p.

34.33 Apply the lens maker's equation to relate a focal length to the index of refraction of a lens (assumed to be in air) and the radii of curvature of the two sides of the lens.

34.34 For a multiple-lens system with the object in front of lens 1, find the image produced by lens 1 and then use it as the object for lens 2, and so on.

34.35 For a multiple-lens system, determine the overall magnification (of the final image) from the magnifications produced by each lens.

Key Ideas

● This module primarily considers thin lenses with symmetric, spherical surfaces.

● If parallel rays are sent through a converging lens parallel to the central axis, the refracted rays pass through a common point (a real focus F) at a focal distance f (a positive quantity) from the lens. If they are sent through a diverging lens, backward extensions of the refracted rays pass through a common point (a virtual focus F) at a focal distance f (a negative quantity) from the lens.

● A converging lens can form a real image (if the object is outside the focal point) or a virtual image (if the object is inside the focal point).

● A diverging lens can form only a virtual image.

● For an object in front of a lens, object distance p and image distance i are related to the lens's focal length f, index

of refraction n, and radii of curvature r_1 and r_2 by

$$\frac{1}{p} + \frac{1}{i} = \frac{1}{f} = (n - 1)\left(\frac{1}{r_1} - \frac{1}{r_2}\right).$$

● The magnitude of the lateral magnification m of an object is the ratio of the image height h' to object height h,

$$|m| = \frac{h'}{h},$$

and is related to the object distance p and image distance i by

$$m = -\frac{i}{p}.$$

● For a system of lenses with a common central axis, the image produced by the first lens acts as the object for the second lens, and so on, and the overall magnification is the product of the individual magnifications.

Thin Lenses

A **lens** is a transparent object with two refracting surfaces whose central axes coincide. The common central axis is the central axis of the lens. When a lens is surrounded by air, light refracts from the air into the lens, crosses through the lens, and then refracts back into the air. Each refraction can change the direction of travel of the light.

A lens that causes light rays initially parallel to the central axis to converge is (reasonably) called a **converging lens.** If, instead, it causes such rays to diverge, the lens is a **diverging lens.** When an object is placed in front of a lens of either type, light rays from the object that refract into and out of the lens can produce an image of the object.

Courtesy Matthew G. Wheeler

A fire is being started by focusing sunlight onto newspaper by means of a converging lens made of clear ice. The lens was made by melting both sides of a flat piece of ice into a convex shape in the shallow vessel (which has a curved bottom).

Lens Equations. We shall consider only the special case of a **thin lens**—that is, a lens in which the thickest part is thin relative to the object distance p, the image distance i, and the radii of curvature r_1 and r_2 of the two surfaces of the lens. We shall also consider only light rays that make small angles with the central axis (they are exaggerated in the figures here). In Module 34-6 we shall prove that for such rays, a thin lens has a focal length f. Moreover, i and p are related to each other by

$$\frac{1}{f} = \frac{1}{p} + \frac{1}{i} \quad \text{(thin lens)}, \tag{34-9}$$

which is the same as we had for mirrors. We shall also prove that when a thin lens with index of refraction n is surrounded by air, this focal length f is given by

$$\frac{1}{f} = (n-1)\left(\frac{1}{r_1} - \frac{1}{r_2}\right) \quad \text{(thin lens in air)}, \tag{34-10}$$

which is often called the *lens maker's equation*. Here r_1 is the radius of curvature of the lens surface nearer the object and r_2 is that of the other surface. The signs of these radii are found with the rules in Module 34-3 for the radii of spherical refracting surfaces. If the lens is surrounded by some medium other than air (say, corn oil) with index of refraction n_{medium}, we replace n in Eq. 34-10 with n/n_{medium}. Keep in mind the basis of Eqs. 34-9 and 34-10:

> A lens can produce an image of an object only because the lens can bend light rays, but it can bend light rays only if its index of refraction differs from that of the surrounding medium.

Forming a Focus. Figure 34-14a shows a thin lens with convex refracting surfaces, or *sides*. When rays that are parallel to the central axis of the lens are sent through the lens, they refract twice, as is shown enlarged in Fig. 34-14b. This

Figure 34-14 (a) Rays initially parallel to the central axis of a converging lens are made to converge to a real focal point F_2 by the lens. The lens is thinner than drawn, with a width like that of the vertical line through it. We shall consider all the bending of rays as occurring at this central line. (b) An enlargement of the top part of the lens of (a); normals to the surfaces are shown dashed. Note that both refractions bend the ray downward, toward the central axis. (c) The same initially parallel rays are made to diverge by a diverging lens. Extensions of the diverging rays pass through a virtual focal point F_2. (d) An enlargement of the top part of the lens of (c). Note that both refractions bend the ray upward, away from the central axis.

To find the focus, send in rays parallel to the central axis.

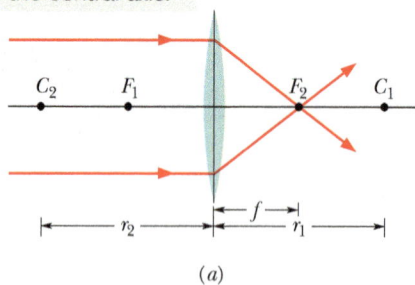

The bending occurs only at the surfaces.

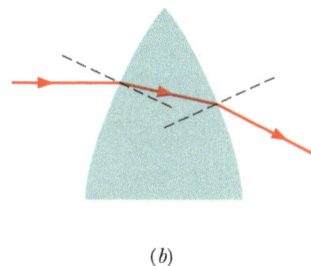

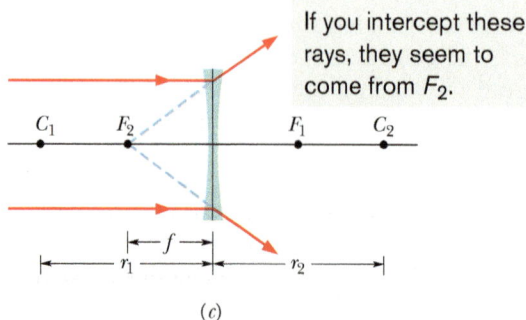

(a)

(b)

If you intercept these rays, they seem to come from F_2.

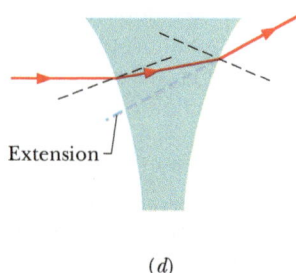

Extension

(c)

(d)

double refraction causes the rays to converge and pass through a common point F_2 at a distance f from the center of the lens. Hence, this lens is a converging lens; further, a *real* focal point (or focus) exists at F_2 (because the rays really do pass through it), and the associated focal length is f. When rays parallel to the central axis are sent in the opposite direction through the lens, we find another real focal point at F_1 on the other side of the lens. For a thin lens, these two focal points are equidistant from the lens.

Signs, Signs, Signs. Because the focal points of a converging lens are real, we take the associated focal lengths f to be positive, just as we do with a real focus of a concave mirror. However, signs in optics can be tricky; so we had better check this in Eq. 34-10. The left side of that equation is positive if f is positive; how about the right side? We examine it term by term. Because the index of refraction n of glass or any other material is greater than 1, the term $(n-1)$ must be positive. Because the source of the light (which is the object) is at the left and faces the convex left side of the lens, the radius of curvature r_1 of that side must be positive according to the sign rule for refracting surfaces. Similarly, because the object faces a concave right side of the lens, the radius of curvature r_2 of that side must be negative according to that rule. Thus, the term $(1/r_1 - 1/r_2)$ is positive, the whole right side of Eq. 34-10 is positive, and all the signs are consistent.

Figure 34-14c shows a thin lens with concave sides. When rays that are parallel to the central axis of the lens are sent through this lens, they refract twice, as is shown enlarged in Fig. 34-14d; these rays *diverge*, never passing through any common point, and so this lens is a diverging lens. However, extensions of the rays do pass through a common point F_2 at a distance f from the center of the lens. Hence, the lens has a *virtual* focal point at F_2. (If your eye intercepts some of the diverging rays, you perceive a bright spot to be at F_2, as if it is the source of the light.) Another virtual focus exists on the opposite side of the lens at F_1, symmetrically placed if the lens is thin. Because the focal points of a diverging lens are virtual, we take the focal length f to be negative.

Images from Thin Lenses

We now consider the types of image formed by converging and diverging lenses. Figure 34-15a shows an object O outside the focal point F_1 of a converging lens. The two rays drawn in the figure show that the lens forms a real, inverted image I of the object on the side of the lens opposite the object.

When the object is placed inside the focal point F_1, as in Fig. 34-15b, the lens forms a virtual image I on the same side of the lens as the object and with the same orientation. Hence, a converging lens can form either a real image or a virtual image, depending on whether the object is outside or inside the focal point, respectively.

Figure 34-15c shows an object O in front of a diverging lens. Regardless of the object distance (regardless of whether O is inside or outside the virtual focal point), this lens produces a virtual image that is on the same side of the lens as the object and has the same orientation.

As with mirrors, we take the image distance i to be positive when the image is real and negative when the image is virtual. However, the locations of real and virtual images from lenses are the reverse of those from mirrors:

Real images form on the side of a lens that is opposite the object, and virtual images form on the side where the object is.

The lateral magnification m produced by converging and diverging lenses is given by Eqs. 34-5 and 34-6, the same as for mirrors.

You have been asked to absorb a lot of information in this module, and you should organize it for yourself by filling in Table 34-2 for thin *symmetric lenses* (both

Converging lenses can give either type of image.

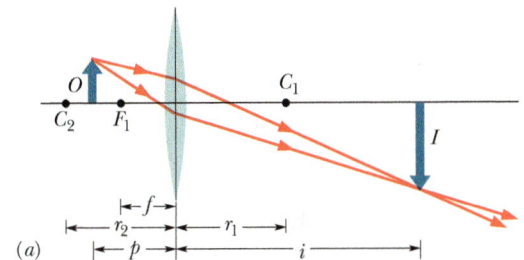

Diverging lenses can give only virtual images.

Figure 34-15 (a) A real, inverted image I is formed by a converging lens when the object O is outside the focal point F_1. (b) The image I is virtual and has the same orientation as O when O is inside the focal point. (c) A diverging lens forms a virtual image I, with the same orientation as the object O, whether O is inside or outside the focal point of the lens.

Table 34-2 Your Organizing Table for Thin Lenses

Lens Type	Object Location	Image			Sign		
		Location	Type	Orientation	of f	of i	of m
Converging	Inside F						
	Outside F						
Diverging	Anywhere						

sides are convex or both sides are concave). Under Image Location note whether the image is on the *same* side of the lens as the object or on the *opposite* side. Under Image Type note whether the image is *real* or *virtual.* Under Image Orientation note whether the image has the *same* orientation as the object or is *inverted.*

Locating Images of Extended Objects by Drawing Rays

Figure 34-16a shows an object O outside focal point F_1 of a converging lens. We can graphically locate the image of any off-axis point on such an object (such as the tip of the arrow in Fig. 34-16a) by drawing a ray diagram with any two of three special rays through the point. These special rays, chosen from all those that pass through the lens to form the image, are the following:

1. A ray that is initially parallel to the central axis of the lens will pass through focal point F_2 (ray 1 in Fig. 34-16a).

2. A ray that initially passes through focal point F_1 will emerge from the lens parallel to the central axis (ray 2 in Fig. 34-16a).

3. A ray that is initially directed toward the center of the lens will emerge from the lens with no change in its direction (ray 3 in Fig. 34-16a) because the ray encounters the two sides of the lens where they are almost parallel.

The image of the point is located where the rays intersect on the far side of the lens. The image of the object is found by locating the images of two or more of its points.

Figure 34-16b shows how the extensions of the three special rays can be used to locate the image of an object placed inside focal point F_1 of a converging lens. Note that the description of ray 2 requires modification (it is now a ray whose backward extension passes through F_1).

You need to modify the descriptions of rays 1 and 2 to use them to locate an image placed (anywhere) in front of a diverging lens. In Fig. 34-16c, for example, we find the point where ray 3 intersects the backward extensions of rays 1 and 2.

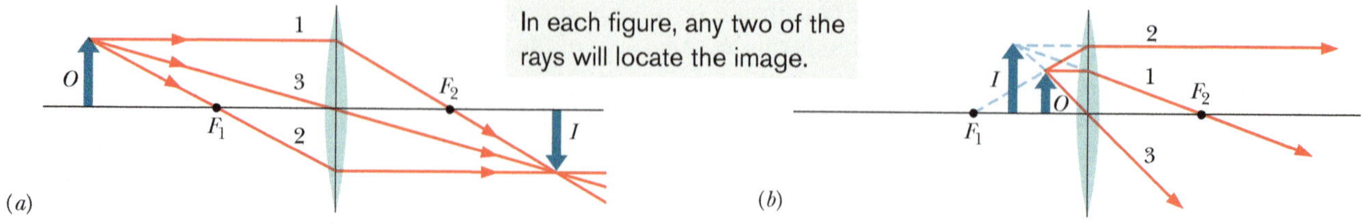

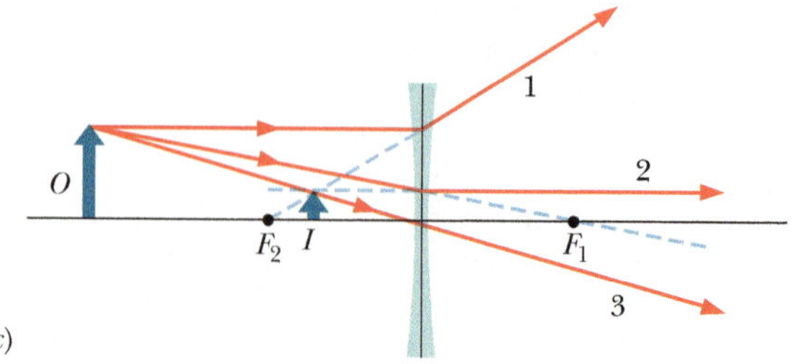

In each figure, any two of the rays will locate the image.

(a)

(b)

Figure 34-16 Three special rays allow us to locate an image formed by a thin lens whether the object O is (a) outside or (b) inside the focal point of a converging lens, or (c) anywhere in front of a diverging lens.

(c)

Two-Lens Systems

Here we consider an object sitting in front of a system of two lenses whose central axes coincide. Some of the possible two-lens systems are sketched in Fig. 34-17, but the figures are not drawn to scale. In each, the object sits to the left of lens 1 but can be inside or outside the focal point of the lens. Although tracing the light rays through any such two-lens system can be challenging, we can use the following simple two-step solution:

Step 1 Neglecting lens 2, use Eq. 34-9 to locate the image I_1 produced by lens 1. Determine whether the image is on the left or right side of the lens, whether it is real or virtual, and whether it has the same orientation as the object. Roughly sketch I_1. The top part of Fig. 34-17a gives an example.

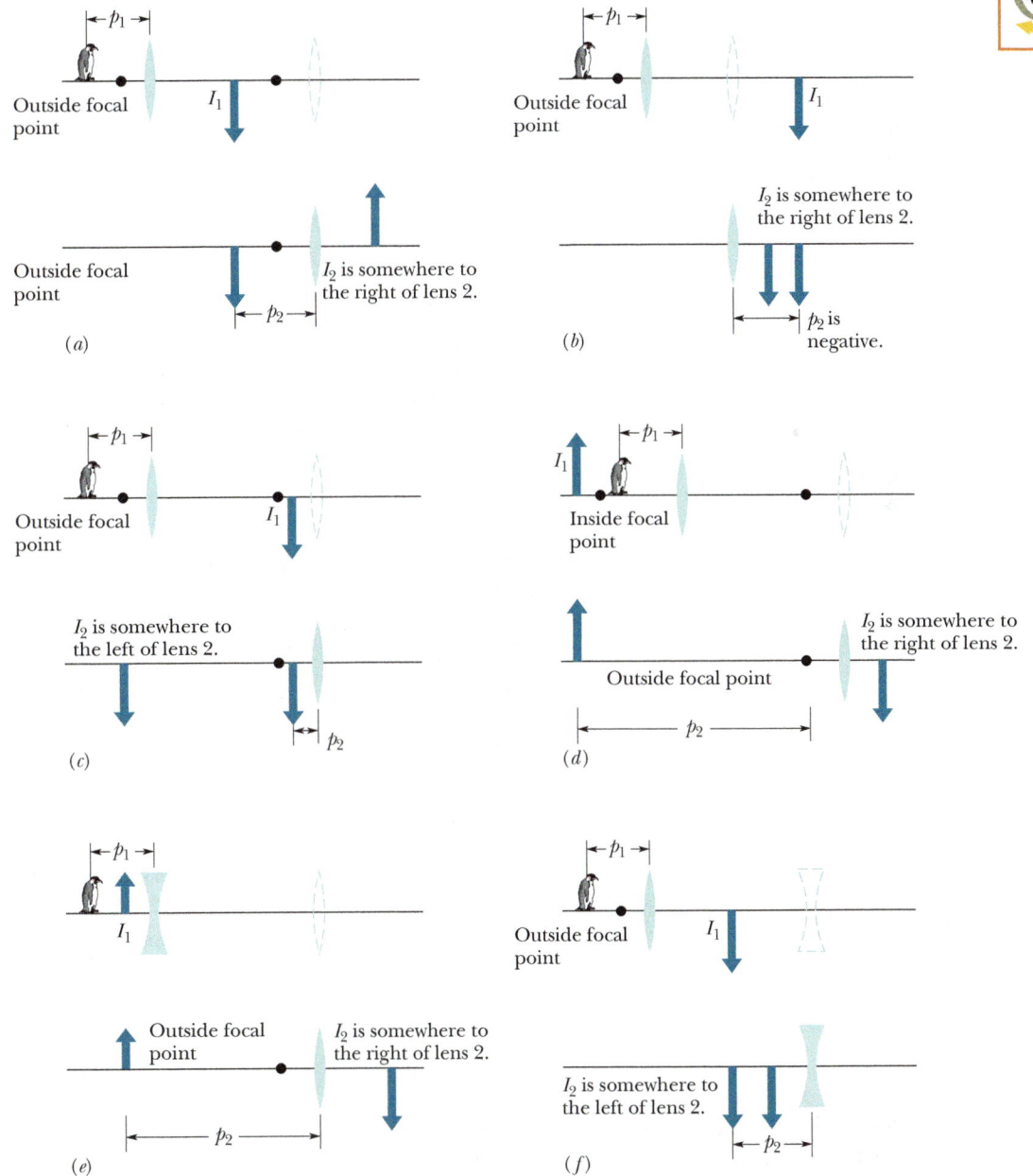

Figure 34-17 Several sketches (not to scale) of a two-lens system in which an object sits to the left of lens 1. In step 1 of the solution, we consider lens 1 and ignore lens 2 (shown in dashes). In step 2, we consider lens 2 and ignore lens 1 (no longer shown). We want to find the final image, that is, the image produced by lens 2.

Step 2 Neglecting lens 1, treat I_1 as though it is the *object* for lens 2. Use Eq. 34-9 to locate the image I_2 produced by lens 2. This is the final image of the system. Determine whether the image is on the left or right side of the lens, whether it is real or virtual, and whether it has the same orientation as the object for lens 2. Roughly sketch I_2. The bottom part of Fig. 34-17a gives an example.

Thus we treat the two-lens system with two single-lens calculations, using the normal decisions and rules for a single lens. The only exception to the procedure occurs if I_1 lies to the right of lens 2 (past lens 2). We still treat it as the object for lens 2, but we take the object distance p_2 as a *negative* number when we use Eq. 34-9 to find I_2. Then, as in our other examples, if the image distance i_2 is positive, the image is real and on the right side of the lens. An example is sketched in Fig. 34-17b.

This same step-by-step analysis can be applied for any number of lenses. It can also be applied if a mirror is substituted for lens 2. The *overall* (or *net*) lateral magnification M of a system of lenses (or lenses and a mirror) is the product of the individual lateral magnifications as given by Eq. 34-7 ($m = -i/p$). Thus, for a two-lens system, we have

$$M = m_1 m_2. \tag{34-11}$$

If M is positive, the final image has the same orientation as the object (the one in front of lens 1). If M is negative, the final image is inverted from the object. In the situation where p_2 is negative, such as in Fig. 34-17b, determining the orientation of the final image is probably easiest by examining the sign of M.

✓ **Checkpoint 4**

A thin symmetric lens provides an image of a fingerprint with a magnification of +0.2 when the fingerprint is 1.0 cm farther from the lens than the focal point of the lens. What are the (a) type and (b) orientation of the image, and (c) what is the type of lens?

Sample Problem 34.03 Image produced by a thin symmetric lens

A praying mantis preys along the central axis of a thin symmetric lens, 20 cm from the lens. The lateral magnification of the mantis provided by the lens is $m = -0.25$, and the index of refraction of the lens material is 1.65.

(a) Determine the type of image produced by the lens, the type of lens, whether the object (mantis) is inside or outside the focal point, on which side of the lens the image appears, and whether the image is inverted.

Reasoning: We can tell a lot about the lens and the image from the given value of m. From it and Eq. 34-6 ($m = -i/p$), we see that

$$i = -mp = 0.25p.$$

Even without finishing the calculation, we can answer the questions. Because p is positive, i here must be positive. That means we have a real image, which means we have a converging lens (the only lens that can produce a real image).

The object must be outside the focal point (the only way a real image can be produced). Also, the image is inverted and on the side of the lens opposite the object. (That is how a converging lens makes a real image.)

(b) What are the two radii of curvature of the lens?

KEY IDEAS

1. Because the lens is symmetric, r_1 (for the surface nearer the object) and r_2 have the same magnitude r.
2. Because the lens is a converging lens, the object faces a convex surface on the nearer side and so $r_1 = +r$. Similarly, it faces a concave surface on the farther side; so $r_2 = -r$.
3. We can relate these radii of curvature to the focal length f via the lens maker's equation, Eq. 34-10 (our only equation involving the radii of curvature of a lens).
4. We can relate f to the object distance p and image distance i via Eq. 34-9.

Calculations: We know p, but we do not know i. Thus, our starting point is to finish the calculation for i in part (a); we obtain

$$i = (0.25)(20 \text{ cm}) = 5.0 \text{ cm}.$$

Now Eq. 34-9 gives us

$$\frac{1}{f} = \frac{1}{p} + \frac{1}{i} = \frac{1}{20 \text{ cm}} + \frac{1}{5.0 \text{ cm}},$$

from which we find $f = 4.0$ cm.

Equation 34-10 then gives us

$$\frac{1}{f} = (n-1)\left(\frac{1}{r_1} - \frac{1}{r_2}\right) = (n-1)\left(\frac{1}{+r} - \frac{1}{-r}\right)$$

or, with known values inserted,

$$\frac{1}{4.0 \text{ cm}} = (1.65 - 1)\frac{2}{r},$$

which yields

$$r = (0.65)(2)(4.0 \text{ cm}) = 5.2 \text{ cm}. \qquad \text{(Answer)}$$

Sample Problem 34.04   Image produced by a system of two thin lenses

Figure 34-18a shows a jalapeño seed O_1 that is placed in front of two thin symmetrical coaxial lenses 1 and 2, with focal lengths $f_1 = +24$ cm and $f_2 = +9.0$ cm, respectively, and with lens separation $L = 10$ cm. The seed is 6.0 cm from lens 1. Where does the system of two lenses produce an image of the seed?

KEY IDEA

We could locate the image produced by the system of lenses by tracing light rays from the seed through the two lenses. However, we can, instead, calculate the location of that image by working through the system in steps, lens by lens. We begin with the lens closer to the seed. The image we seek is the final one—that is, image I_2 produced by lens 2.

Lens 1: Ignoring lens 2, we locate the image I_1 produced by lens 1 by applying Eq. 34-9 to lens 1 alone:

$$\frac{1}{p_1} + \frac{1}{i_1} = \frac{1}{f_1}.$$

The object O_1 for lens 1 is the seed, which is 6.0 cm from the lens; thus, we substitute $p_1 = +6.0$ cm. Also substituting the given value of f_1, we then have

$$\frac{1}{+6.0 \text{ cm}} + \frac{1}{i_1} = \frac{1}{+24 \text{ cm}},$$

which yields $i_1 = -8.0$ cm.

This tells us that image I_1 is 8.0 cm from lens 1 and virtual. (We could have guessed that it is virtual by noting that the seed is inside the focal point of lens 1, that is, between the lens and its focal point.) Because I_1 is virtual, it is on the same side of the lens as object O_1 and has the same orientation as the seed, as shown in Fig. 34-18b.

Lens 2: In the second step of our solution, we treat image I_1 as an object O_2 for the second lens and now ignore lens 1. We first note that this object O_2 is outside the focal point

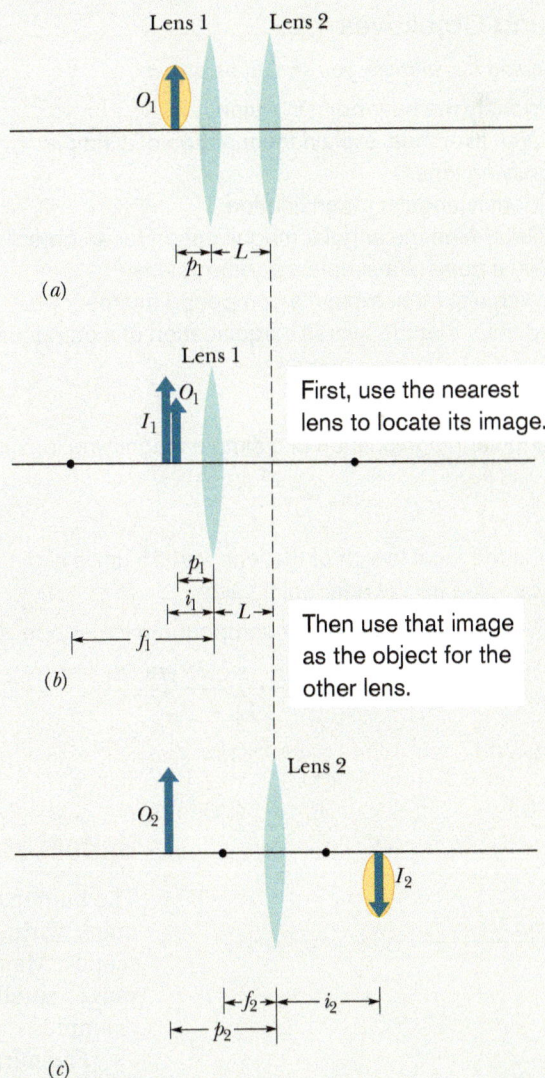

First, use the nearest lens to locate its image.

Then use that image as the object for the other lens.

Figure 34-18 (a) Seed O_1 is distance p_1 from a two-lens system with lens separation L. We use the arrow to orient the seed. (b) The image I_1 produced by lens 1 alone. (c) Image I_1 acts as object O_2 for lens 2 alone, which produces the final image I_2.

of lens 2. So the image I_2 produced by lens 2 must be real, inverted, and on the side of the lens opposite O_2. Let us see.

The distance p_2 between this object O_2 and lens 2 is, from Fig. 34-18c,

$$p_2 = L + |i_1| = 10 \text{ cm} + 8.0 \text{ cm} = 18 \text{ cm}.$$

Then Eq. 34-9, now written for lens 2, yields

$$\frac{1}{+18 \text{ cm}} + \frac{1}{i_2} = \frac{1}{+9.0 \text{ cm}}.$$

Hence, $i_2 = +18 \text{ cm}.$ (Answer)

The plus sign confirms our guess: Image I_2 produced by lens 2 is real, inverted, and on the side of lens 2 opposite O_2, as shown in Fig. 34-18c. Thus, the image would appear on a card placed at its location.

PLUS Additional examples, video, and practice available at *WileyPLUS*

34-5 OPTICAL INSTRUMENTS

Learning Objectives

After reading this module, you should be able to . . .

34.36 Identify the near point in vision.

34.37 With sketches, explain the function of a simple magnifying lens.

34.38 Identify angular magnification.

34.39 Determine the angular magnification for an object at the focal point of a simple magnifying lens.

34.40 With a sketch, explain a compound microscope.

34.41 Identify that the overall magnification of a compound microscope is due to the lateral magnification by the objective and the angular magnification by the eyepiece.

34.42 Calculate the overall magnification of a compound microscope.

34.43 With a sketch, explain a refracting telescope.

34.44 Calculate the angular magnification of a refracting telescope.

Key Ideas

● The angular magnification of a simple magnifying lens is

$$m_\theta = \frac{25 \text{ cm}}{f},$$

where f is the focal length of the lens and 25 cm is a reference value for the near point value.

● The overall magnification of a compound microscope is

$$M = m m_\theta = -\frac{s}{f_{ob}} \frac{25 \text{ cm}}{f_{ey}},$$

where m is the lateral magnification of the objective, m_θ is the angular magnification of the eyepiece, s is the tube length, f_{ob} is the focal length of the objective, and f_{ey} is the focal length of the eyepiece.

● The angular magnification of a refracting telescope is

$$m_\theta = -\frac{f_{ob}}{f_{ey}}.$$

Optical Instruments

The human eye is a remarkably effective organ, but its range can be extended in many ways by optical instruments such as eyeglasses, microscopes, and telescopes. Many such devices extend the scope of our vision beyond the visible range; satellite-borne infrared cameras and x-ray microscopes are just two examples.

The mirror and thin-lens formulas can be applied only as approximations to most sophisticated optical instruments. The lenses in typical laboratory microscopes are by no means "thin." In most optical instruments the lenses are compound lenses; that is, they are made of several components, the interfaces rarely being exactly spherical. Now we discuss three optical instruments, assuming, for simplicity, that the thin-lens formulas apply.

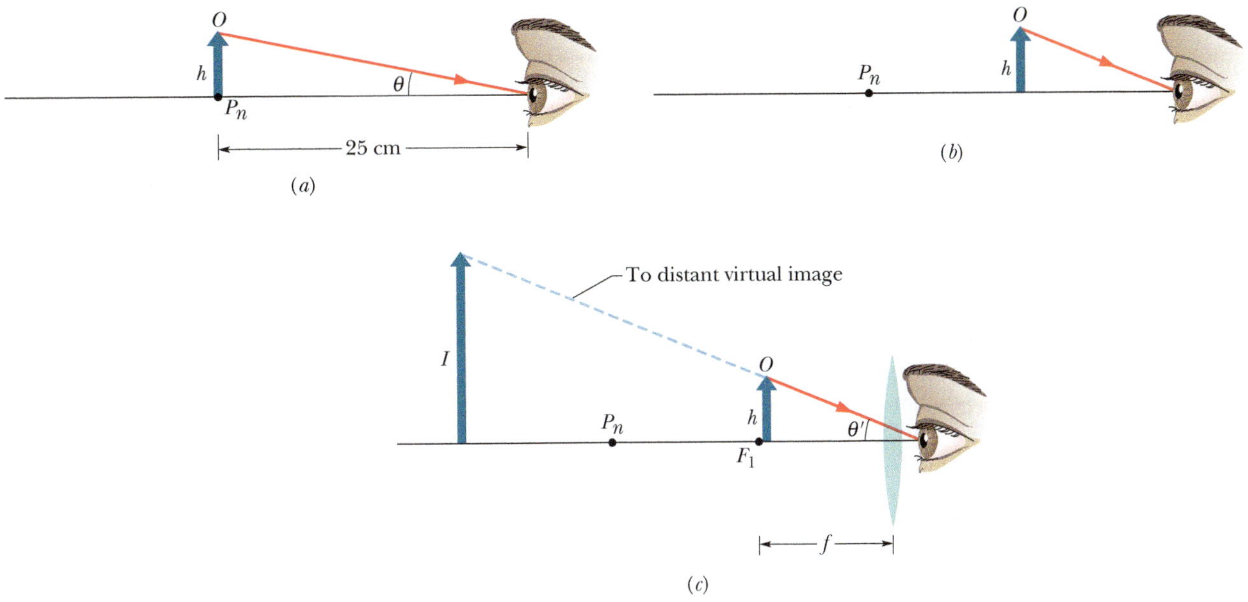

Figure 34-19 (*a*) An object *O* of height *h* placed at the near point of a human eye occupies angle *θ* in the eye's view. (*b*) The object is moved closer to increase the angle, but now the observer cannot bring the object into focus. (*c*) A converging lens is placed between the object and the eye, with the object just inside the focal point F_1 of the lens. The image produced by the lens is then far enough away to be focused by the eye, and the image occupies a larger angle *θ′* than object *O* does in (*a*).

Simple Magnifying Lens

The normal human eye can focus a sharp image of an object on the retina (at the rear of the eye) if the object is located anywhere from infinity to a certain point called the *near point* P_n. If you move the object closer to the eye than the near point, the perceived retinal image becomes fuzzy. The location of the near point normally varies with age, generally moving away from the person. To find your own near point, remove your glasses or contacts if you wear any, close one eye, and then bring this page closer to your open eye until it becomes indistinct. In what follows, we take the near point to be 25 cm from the eye, a bit more than the typical value for 20-year-olds.

Figure 34-19*a* shows an object *O* placed at the near point P_n of an eye. The size of the image of the object produced on the retina depends on the angle *θ* that the object occupies in the field of view from that eye. By moving the object closer to the eye, as in Fig. 34-19*b*, you can increase the angle and, hence, the possibility of distinguishing details of the object. However, because the object is then closer than the near point, it is no longer *in focus;* that is, the image is no longer clear.

You can restore the clarity by looking at *O* through a converging lens, placed so that *O* is just inside the focal point F_1 of the lens, which is at focal length *f* (Fig. 34-19*c*). What you then see is the virtual image of *O* produced by the lens. That image is farther away than the near point; thus, the eye can see it clearly.

Moreover, the angle *θ′* occupied by the virtual image is larger than the largest angle *θ* that the object alone can occupy and still be seen clearly. The *angular magnification* m_θ (not to be confused with lateral magnification *m*) of what is seen is

$$m_\theta = \theta'/\theta.$$

In words, the angular magnification of a simple magnifying lens is a comparison of the angle occupied by the image the lens produces with the angle occupied by the object when the object is moved to the near point of the viewer.

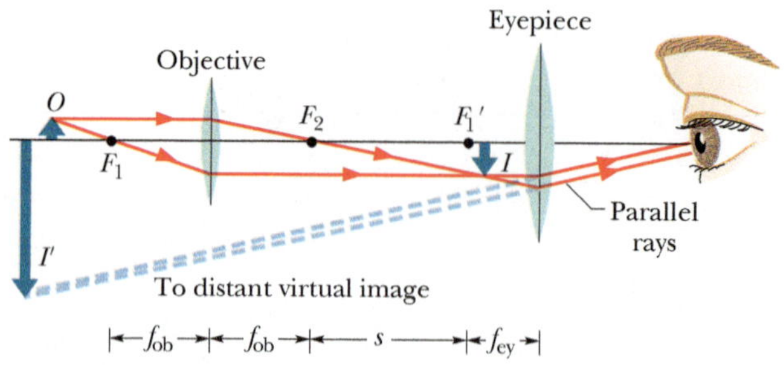

Figure 34-20 A thin-lens representation of a compound microscope (not to scale). The objective produces a real image I of object O just inside the focal point F_1' of the eyepiece. Image I then acts as an object for the eyepiece, which produces a virtual final image I' that is seen by the observer. The objective has focal length f_{ob}; the eyepiece has focal length f_{ey}; and s is the tube length.

From Fig. 34-19, assuming that O is at the focal point of the lens, and approximating $\tan \theta$ as θ and $\tan \theta'$ as θ' for small angles, we have

$$\theta \approx h/25 \text{ cm} \quad \text{and} \quad \theta' \approx h/f.$$

We then find that

$$m_\theta \approx \frac{25 \text{ cm}}{f} \quad \text{(simple magnifier)}. \tag{34-12}$$

Compound Microscope

Figure 34-20 shows a thin-lens version of a compound microscope. The instrument consists of an *objective* (the front lens) of focal length f_{ob} and an *eyepiece* (the lens near the eye) of focal length f_{ey}. It is used for viewing small objects that are very close to the objective.

The object O to be viewed is placed just outside the first focal point F_1 of the objective, close enough to F_1 that we can approximate its distance p from the lens as being f_{ob}. The separation between the lenses is then adjusted so that the enlarged, inverted, real image I produced by the objective is located just inside the first focal point F_1' of the eyepiece. The *tube length s* shown in Fig. 34-20 is actually large relative to f_{ob}, and therefore we can approximate the distance i between the objective and the image I as being length s.

From Eq. 34-6, and using our approximations for p and i, we can write the lateral magnification produced by the objective as

$$m = -\frac{i}{p} = -\frac{s}{f_{ob}}. \tag{34-13}$$

Because the image I is located just inside the focal point F_1' of the eyepiece, the eyepiece acts as a simple magnifying lens, and an observer sees a final (virtual, inverted) image I' through it. The overall magnification of the instrument is the product of the lateral magnification m produced by the objective, given by Eq. 34-13, and the angular magnification m_θ produced by the eyepiece, given by Eq. 34-12; that is,

$$M = mm_\theta = -\frac{s}{f_{ob}} \frac{25 \text{ cm}}{f_{ey}} \quad \text{(microscope)}. \tag{34-14}$$

Refracting Telescope

Telescopes come in a variety of forms. The form we describe here is the simple refracting telescope that consists of an objective and an eyepiece; both are represented in Fig. 34-21 with simple lenses, although in practice, as is also true for most microscopes, each lens is actually a compound lens system.

The lens arrangements for telescopes and for microscopes are similar, but telescopes are designed to view large objects, such as galaxies, stars, and planets, at large distances, whereas microscopes are designed for just the opposite purpose. This difference requires that in the telescope of Fig. 34-21 the second focal point of the objective F_2 coincide with the first focal point of the eyepiece F_1', whereas in the microscope of Fig. 34-20 these points are separated by the tube length s.

In Fig. 34-21a, parallel rays from a distant object strike the objective, making an angle θ_{ob} with the telescope axis and forming a real, inverted image I at the common focal point F_2, F_1'. This image I acts as an object for the eyepiece, through which an observer sees a distant (still inverted) virtual image I'. The rays defining the image make an angle θ_{ey} with the telescope axis.

The angular magnification m_θ of the telescope is θ_{ey}/θ_{ob}. From Fig. 34-21b, for rays close to the central axis, we can write $\theta_{ob} = h'/f_{ob}$ and $\theta_{ey} \approx h'/f_{ey}$, which gives us

$$m_\theta = -\frac{f_{ob}}{f_{ey}} \quad \text{(telescope)}, \tag{34-15}$$

where the minus sign indicates that I' is inverted. In words, the angular magnification of a telescope is a comparison of the angle occupied by the image the telescope produces with the angle occupied by the distant object as seen without the telescope.

Magnification is only one of the design factors for an astronomical telescope and is indeed easily achieved. A good telescope needs *light-gathering power,* which determines how bright the image is. This is important for viewing faint objects such as distant galaxies and is accomplished by making the objective diameter as large as possible. A telescope also needs *resolving power,* which is the ability to distinguish between two distant objects (stars, say) whose angular separation is small. *Field of view* is another important design parameter. A telescope designed to look at galaxies (which occupy a tiny field of view) is much different from one designed to track meteors (which move over a wide field of view).

The telescope designer must also take into account the difference between real lenses and the ideal thin lenses we have discussed. A real lens with spherical surfaces does not form sharp images, a flaw called *spherical aberration.* Also, because refraction by the two surfaces of a real lens depends on wavelength, a real lens does not focus light of different wavelengths to the same point, a flaw called *chromatic aberration.*

This brief discussion by no means exhausts the design parameters of astronomical telescopes—many others are involved. We could make a similar listing for any other high-performance optical instrument.

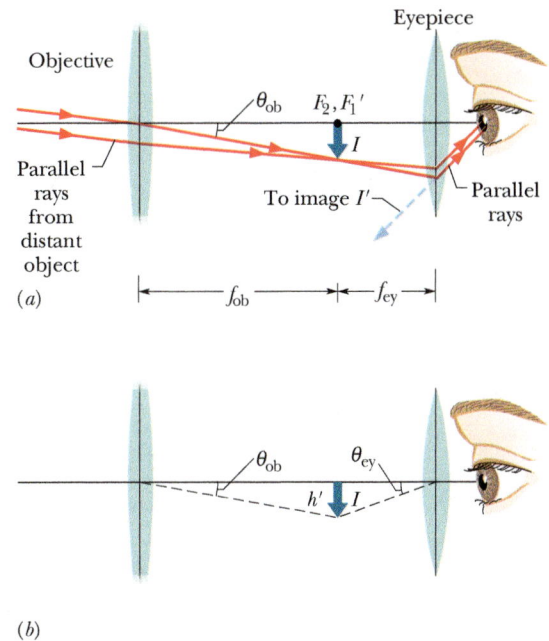

Figure 34-21 (*a*) A thin-lens representation of a refracting telescope. From rays that are approximately parallel when they reach the objective, the objective produces a real image I of a distant source of light (the object). (One end of the object is assumed to lie on the central axis.) Image I, formed at the common focal points F_2 and F_1', acts as an object for the eyepiece, which produces a virtual final image I' at a great distance from the observer. The objective has focal length f_{ob}; the eyepiece has focal length f_{ey}. (*b*) Image I has height h' and takes up angle θ_{ob} measured from the objective and angle θ_{ey} measured from the eyepiece.

34-6 THREE PROOFS

The Spherical Mirror Formula (Eq. 34-4)

Figure 34-22 shows a point object O placed on the central axis of a concave spherical mirror, outside its center of curvature C. A ray from O that makes an angle α with the axis intersects the axis at I after reflection from the mirror at a. A ray that leaves O along the axis is reflected back along itself at c and also passes through I. Thus, because both rays pass through that common point, I is the image of O; it is a *real* image because light actually passes through it. Let us find the image distance i.

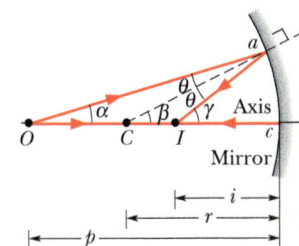

Figure 34-22 A concave spherical mirror forms a real point image I by reflecting light rays from a point object O.

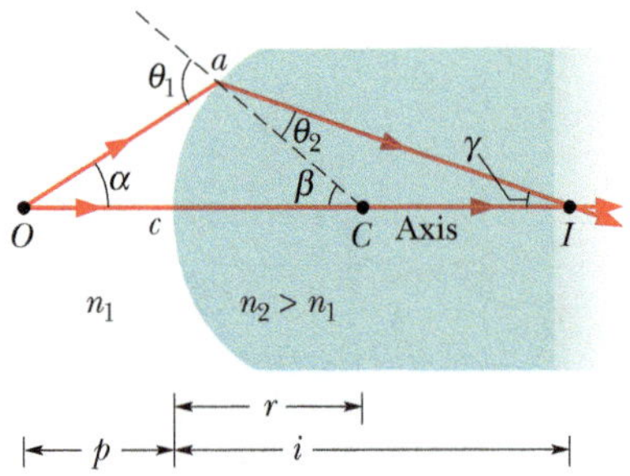

Figure 34-23 A real point image I of a point object O is formed by refraction at a spherical convex surface between two media.

A trigonometry theorem that is useful here tells us that an exterior angle of a triangle is equal to the sum of the two opposite interior angles. Applying this to triangles OaC and OaI in Fig. 34-22 yields

$$\beta = \alpha + \theta \quad \text{and} \quad \gamma = \alpha + 2\theta.$$

If we eliminate θ between these two equations, we find

$$\alpha + \gamma = 2\beta. \tag{34-16}$$

We can write angles α, β, and γ, in radian measure, as

$$\alpha \approx \frac{\widehat{ac}}{cO} = \frac{\widehat{ac}}{p}, \qquad \beta = \frac{\widehat{ac}}{cC} = \frac{\widehat{ac}}{r},$$

and

$$\gamma \approx \frac{\widehat{ac}}{cI} = \frac{\widehat{ac}}{i}, \tag{34-17}$$

where the overhead symbol means "arc." Only the equation for β is exact, because the center of curvature of $\widehat{ac}$ is at C. However, the equations for α and γ are approximately correct if these angles are small enough (that is, for rays close to the central axis). Substituting Eqs. 34-17 into Eq. 34-16, using Eq. 34-3 to replace r with $2f$, and canceling $\widehat{ac}$ lead exactly to Eq. 34-4, the relation that we set out to prove.

The Refracting Surface Formula (Eq. 34-8)

The incident ray from point object O in Fig. 34-23 that falls on point a of a spherical refracting surface is refracted there according to Eq. 33-40,

$$n_1 \sin \theta_1 = n_2 \sin \theta_2.$$

If α is small, θ_1 and θ_2 will also be small and we can replace the sines of these angles with the angles themselves. Thus, the equation above becomes

$$n_1 \theta_1 \approx n_2 \theta_2. \tag{34-18}$$

We again use the fact that an exterior angle of a triangle is equal to the sum of the two opposite interior angles. Applying this to triangles COa and ICa yields

$$\theta_1 = \alpha + \beta \quad \text{and} \quad \beta = \theta_2 + \gamma. \tag{34-19}$$

If we use Eqs. 34-19 to eliminate θ_1 and θ_2 from Eq. 34-18, we find

$$n_1 \alpha + n_2 \gamma = (n_2 - n_1)\beta. \tag{34-20}$$

In radian measure the angles α, β, and γ are

$$\alpha \approx \frac{\widehat{ac}}{p}; \qquad \beta = \frac{\widehat{ac}}{r}; \qquad \gamma \approx \frac{\widehat{ac}}{i}. \tag{34-21}$$

Only the second of these equations is exact. The other two are approximate because I and O are not the centers of circles of which $\widehat{ac}$ is a part. However, for α small enough (for rays close to the axis), the inaccuracies in Eqs. 34-21 are small. Substituting Eqs. 34-21 into Eq. 34-20 leads directly to Eq. 34-8, as we wanted.

The Thin-Lens Formulas (Eqs. 34-9 and 34-10)

Our plan is to consider each lens surface as a separate refracting surface, and to use the image formed by the first surface as the object for the second.

We start with the thick glass "lens" of length L in Fig. 34-24a whose left and right refracting surfaces are ground to radii r' and r''. A point object O' is placed near the left surface as shown. A ray leaving O' along the central axis is not deflected on entering or leaving the lens.

A second ray leaving O' at an angle α with the central axis intersects the left surface at point a', is refracted, and intersects the second (right) surface at point a''.

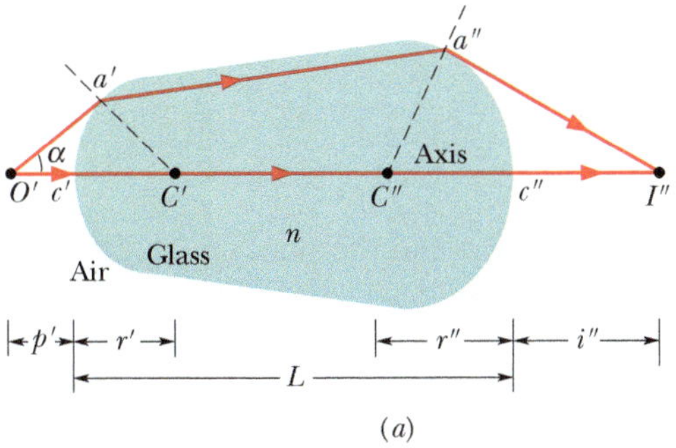

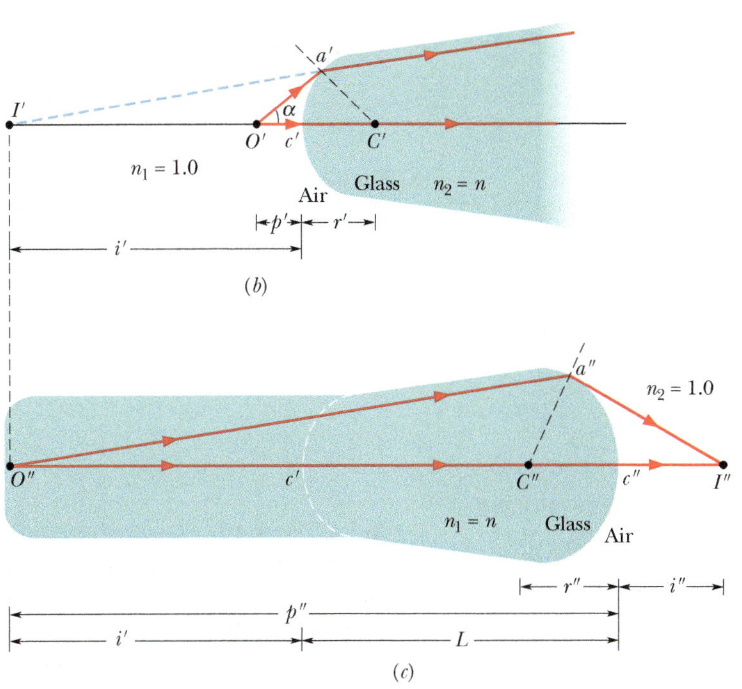

Figure 34-24 (a) Two rays from point object O' form a real image I'' after refracting through two spherical surfaces of a lens. The object faces a convex surface at the left side of the lens and a concave surface at the right side. The ray traveling through points a' and a'' is actually close to the central axis through the lens. (b) The left side and (c) the right side of the lens in (a), shown separately.

The ray is again refracted and crosses the axis at I'', which, being the intersection of two rays from O', is the image of point O', formed after refraction at two surfaces.

Figure 34-24b shows that the first (left) surface also forms a virtual image of O' at I'. To locate I', we use Eq. 34-8,

$$\frac{n_1}{p} + \frac{n_2}{i} = \frac{n_2 - n_1}{r}.$$

Putting $n_1 = 1$ for air and $n_2 = n$ for lens glass and bearing in mind that the (virtual) image distance is negative (that is, $i = -i'$ in Fig. 34-24b), we obtain

$$\frac{1}{p'} - \frac{n}{i'} = \frac{n-1}{r'}. \tag{34-22}$$

(Because the minus sign is explicit, i' will be a positive number.)

Figure 34-24c shows the second surface again. Unless an observer at point a'' were aware of the existence of the first surface, the observer would think that the light striking that point originated at point I' in Fig. 34-24b and that the region to the left of the surface was filled with glass as indicated. Thus, the (virtual) image I' formed by the first surface serves as a real object O'' for the second surface. The distance of this object from the second surface is

$$p'' = i' + L. \tag{34-23}$$

To apply Eq. 34-8 to the second surface, we must insert $n_1 = n$ and $n_2 = 1$ because the object now is effectively imbedded in glass. If we substitute with Eq. 34-23, then Eq. 34-8 becomes

$$\frac{n}{i' + L} + \frac{1}{i''} = \frac{1-n}{r''}. \tag{34-24}$$

Let us now assume that the thickness L of the "lens" in Fig. 34-24a is so small that we can neglect it in comparison with our other linear quantities (such as p', i', p'', i'', r', and r''). In all that follows we make this *thin-lens approximation*. Putting $L = 0$ in Eq. 34-24 and rearranging the right side lead to

$$\frac{n}{i'} + \frac{1}{i''} = -\frac{n-1}{r''}. \tag{34-25}$$

Adding Eqs. 34-22 and 34-25 leads to

$$\frac{1}{p'} + \frac{1}{i''} = (n-1)\left(\frac{1}{r'} - \frac{1}{r''}\right).$$

Finally, calling the original object distance simply p and the final image distance simply i leads to

$$\frac{1}{p} + \frac{1}{i} = (n-1)\left(\frac{1}{r'} - \frac{1}{r''}\right), \tag{34-26}$$

which, with a small change in notation, is Eqs. 34-9 and 34-10.

Review & Summary

Real and Virtual Images An *image* is a reproduction of an object via light. If the image can form on a surface, it is a *real image* and can exist even if no observer is present. If the image requires the visual system of an observer, it is a *virtual image*.

Image Formation *Spherical mirrors, spherical refracting surfaces,* and *thin lenses* can form images of a source of light—the object—by redirecting rays emerging from the source. The image occurs where the redirected rays cross (forming a real image) or where backward extensions of those rays cross (forming a virtual image). If the rays are sufficiently close to the *central axis* through the spherical mirror, refracting surface, or thin lens, we have the following relations between the *object distance p* (which is positive) and the *image distance i* (which is positive for real images and negative for virtual images):

1. **Spherical Mirror:**

$$\frac{1}{p} + \frac{1}{i} = \frac{1}{f} = \frac{2}{r}, \tag{34-4, 34-3}$$

where f is the mirror's focal length and r is its radius of curvature. A *plane mirror* is a special case for which $r \to \infty$, so that $p = -i$. Real images form on the side of a mirror where the object is located, and virtual images form on the opposite side.

2. **Spherical Refracting Surface:**

$$\frac{n_1}{p} + \frac{n_2}{i} = \frac{n_2 - n_1}{r} \quad \text{(single surface)}, \tag{34-8}$$

where n_1 is the index of refraction of the material where the object is located, n_2 is the index of refraction of the material on the other side of the refracting surface, and r is the radius of curvature of the surface. When the object faces a convex refracting surface, the radius r is positive. When it faces a concave surface, r is negative. Real images form on the side of a refracting surface that is opposite the object, and virtual images form on the same side as the object.

3. **Thin Lens:**

$$\frac{1}{p} + \frac{1}{i} = \frac{1}{f} = (n-1)\left(\frac{1}{r_1} - \frac{1}{r_2}\right), \tag{34-9, 34-10}$$

where f is the lens's focal length, n is the index of refraction of the lens material, and r_1 and r_2 are the radii of curvature of the two sides of the lens, which are spherical surfaces. A convex lens surface that

faces the object has a positive radius of curvature; a concave lens surface that faces the object has a negative radius of curvature. Real images form on the side of a lens that is opposite the object, and virtual images form on the same side as the object.

Lateral Magnification The *lateral magnification m* produced by a spherical mirror or a thin lens is

$$m = -\frac{i}{p}. \tag{34-6}$$

The magnitude of m is given by

$$|m| = \frac{h'}{h}, \tag{34-5}$$

where h and h' are the heights (measured perpendicular to the central axis) of the object and image, respectively.

Optical Instruments Three optical instruments that extend human vision are:

1. The *simple magnifying lens,* which produces an *angular magnification* m_θ given by

$$m_\theta = \frac{25 \text{ cm}}{f}, \tag{34-12}$$

where f is the focal length of the magnifying lens. The distance of 25 cm is a traditionally chosen value that is a bit more than the typical near point for someone 20 years old.

2. The *compound microscope,* which produces an *overall magnification M* given by

$$M = mm_\theta = -\frac{s}{f_{ob}}\frac{25 \text{ cm}}{f_{ey}}, \tag{34-14}$$

where m is the lateral magnification produced by the objective, m_θ is the angular magnification produced by the eyepiece, s is the tube length, and f_{ob} and f_{ey} are the focal lengths of the objective and eyepiece, respectively.

3. The *refracting telescope,* which produces an *angular magnification* m_θ given by

$$m_\theta = -\frac{f_{ob}}{f_{ey}}. \tag{34-15}$$

Questions

1 Figure 34-25 shows a fish and a fish stalker in water. (a) Does the stalker see the fish in the general region of point a or point b? (b) Does the fish see the (wild) eyes of the stalker in the general region of point c or point d?

Figure 34-25 Question 1.

2 In Fig. 34-26, stick figure O stands in front of a spherical mirror that is mounted within the boxed region; the central axis through the mirror is shown. The four stick figures I_1 to I_4 suggest general locations and orientations for the images that might be produced by the mirror. (The figures are only sketched in; neither their heights nor their distances from the mirror are drawn to scale.) (a) Which of the stick figures could not possibly represent images? Of the possible images, (b) which would be due to a concave mirror, (c) which would be due to a convex mirror, (d) which would be virtual, and (e) which would involve negative magnification?

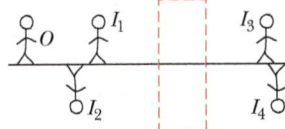

Figure 34-26
Questions 2 and 10.

3 Figure 34-27 is an overhead view of a mirror maze based on floor sections that are equilateral triangles. Every wall within the maze is mirrored. If you stand at entrance x, (a) which of the maze monsters a, b, and c hiding in the maze can you see along the virtual hallways extending from entrance x; (b) how many times does each visible monster appear in a hallway; and (c) what is at the far end of a hallway?

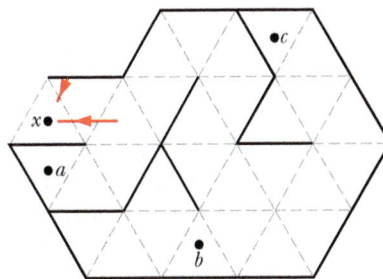

Figure 34-27 Question 3.

4 A penguin waddles along the central axis of a concave mirror, from the focal point to an effectively infinite distance. (a) How does its image move? (b) Does the height of its image increase continuously, decrease continuously, or change in some more complicated manner?

5 When a *T. rex* pursues a jeep in the movie *Jurassic Park*, we see a reflected image of the *T. rex* via a side-view mirror, on which is printed the (then darkly humorous) warning: "Objects in mirror are closer than they appear." Is the mirror flat, convex, or concave?

6 An object is placed against the center of a concave mirror and then moved along the central axis until it is 5.0 m from the mirror. During the motion, the distance $|i|$ between the mirror and the image it produces is measured. The procedure is then repeated with a convex mirror and a plane mirror. Figure 34-28 gives the results versus object distance p. Which curve corresponds to which mirror? (Curve 1 has two segments.)

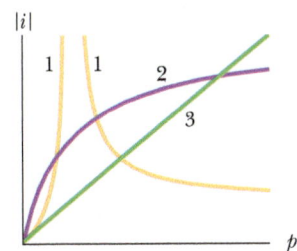

Figure 34-28 Questions 6 and 8.

7 The table details six variations of the basic arrangement of two thin lenses represented in Fig. 34-29. (The points labeled F_1 and F_2 are the focal points of lenses 1 and 2.) An object is distance p_1 to the left of lens 1, as in Fig. 34-18. (a) For which variations can we tell, *without calculation,* whether the final image (that due to lens 2) is to the left or right of lens 2 and whether it has the same orientation as the object? (b) For those "easy" variations, give the image location as "left" or "right" and the orientation as "same" or "inverted."

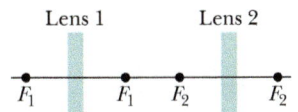

Figure 34-29 Question 7.

Variation	Lens 1	Lens 2	
1	Converging	Converging	$p_1 < \|f_1\|$
2	Converging	Converging	$p_1 > \|f_1\|$
3	Diverging	Converging	$p_1 < \|f_1\|$
4	Diverging	Converging	$p_1 > \|f_1\|$
5	Diverging	Diverging	$p_1 < \|f_1\|$
6	Diverging	Diverging	$p_1 > \|f_1\|$

8 An object is placed against the center of a converging lens and then moved along the central axis until it is 5.0 m from the lens. During the motion, the distance $|i|$ between the lens and the image it produces is measured. The procedure is then repeated with a diverging lens. Which of the curves in Fig. 34-28 best gives $|i|$ versus the object distance p for these lenses? (Curve 1 consists of two segments. Curve 3 is straight.)

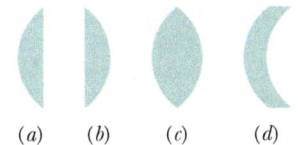

9 Figure 34-30 shows four thin lenses, all of the same material, with sides that either are flat or have a radius of curvature of magnitude 10 cm. Without written calculation, rank the lenses according to the magnitude of the focal length, greatest first.

Figure 34-30 Question 9.

10 In Fig. 34-26, stick figure O stands in front of a thin, symmetric lens that is mounted within the boxed region; the central axis through the lens is shown. The four stick figures I_1 to I_4 suggest general locations and orientations for the images that might be produced by the lens. (The figures are only sketched in; neither their height nor their distance from the lens is drawn to scale.) (a) Which of the stick figures could not possibly represent images? Of the possible images, (b) which would be due to a converging lens, (c) which would be due to a diverging lens, (d) which would be virtual, and (e) which would involve negative magnification?

11 Figure 34-31 shows a coordinate system in front of a flat mirror, with the x axis perpendicular to the mirror. Draw the image of the system in the mirror. (a) Which axis is reversed by the reflection? (b) If you face a mirror, is your image inverted (top for bottom)? (c) Is it reversed left and right (as commonly believed)? (d) What then is reversed?

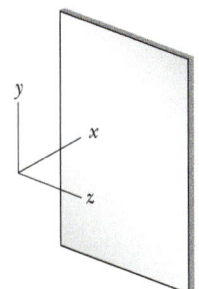

Figure 34-31
Question 11.

Problems

Module 34-1 Images and Plane Mirrors

•1 You look through a camera toward an image of a hummingbird in a plane mirror. The camera is 4.30 m in front of the mirror. The bird is at camera level, 5.00 m to your right and 3.30 m from the mirror. What is the distance between the camera and the apparent position of the bird's image in the mirror?

•2 **ILW** A moth at about eye level is 10 cm in front of a plane mirror; you are behind the moth, 30 cm from the mirror. What is the distance between your eyes and the apparent position of the moth's image in the mirror?

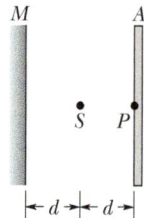

Figure 34-32 Problem 3.

••3 In Fig. 34-32, an isotropic point source of light S is positioned at distance d from a viewing screen A and the light intensity I_P at point P (level with S) is measured. Then a plane mirror M is placed behind S at distance d. By how much is I_P multiplied by the presence of the mirror?

••4 Figure 34-33 shows an overhead view of a corridor with a plane mirror M mounted at one end. A burglar B sneaks along the corridor directly toward the center of the mirror. If $d = 3.0$ m, how far from the mirror will she be when the security guard S can first see her in the mirror?

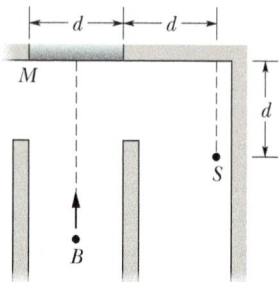

Figure 34-33 Problem 4.

•••5 **SSM** **WWW** Figure 34-34 shows a small lightbulb suspended at distance $d_1 = 250$ cm above the surface of the water in a swimming pool where the water depth is $d_2 = 200$ cm. The bottom of the pool is a large mirror. How far below the mir-

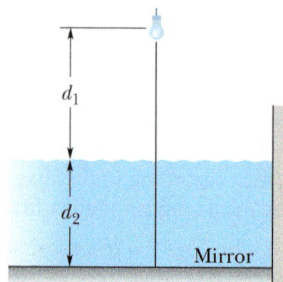

Figure 34-34 Problem 5.

ror surface is the image of the bulb? (*Hint:* Assume that the rays are close to a vertical axis through the bulb, and use the small-angle approximation in which $\sin \theta \approx \tan \theta \approx \theta$.)

Module 34-2 Spherical Mirrors

•6 An object is moved along the central axis of a spherical mirror while the lateral magnification m of it is measured. Figure 34-35 gives m versus object distance p for the range $p_a = 2.0$ cm to $p_b = 8.0$ cm. What is m for $p = 14.0$ cm?

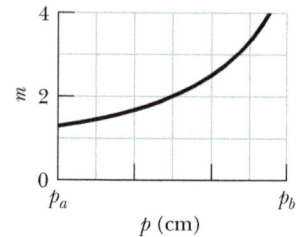

Figure 34-35 Problem 6.

•7 A concave shaving mirror has a radius of curvature of 35.0 cm. It is positioned so that the (upright) image of a man's face is 2.50 times the size of the face. How far is the mirror from the face?

•8 An object is placed against the center of a spherical mirror and then moved 70 cm from it along the central axis as the image distance i is measured. Figure 34-36 gives i versus object distance p out to $p_s = 40$ cm. What is i for $p = 70$ cm?

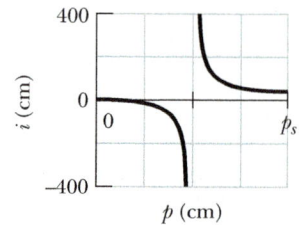

Figure 34-36 Problem 8.

••9 through 16 **GO** 12 **SSM** 9, 11, 13 *Spherical mirrors.* Object O stands on the central axis of a spherical mirror. For this situation, each problem in Table 34-3 gives object distance p_s (centimeters), the type of mirror, and then the distance (centimeters, without proper sign) between the focal point and the mirror. Find (a) the radius of curvature r (including sign), (b) the image distance i, and (c) the lateral magnification m. Also, determine whether the image is (d) real (R) or virtual (V), (e) inverted (I) from object O or noninverted (NI), and (f) on the *same* side of the mirror as O or on the *opposite* side.

••17 through 29 **GO** 22 **SSM** 23, 29 *More mirrors.* Object O stands on the central axis of a spherical or plane mirror. For this sit-

Table 34-3 Problems 9 through 16: Spherical Mirrors. See the setup for these problems.

	p	Mirror	(a) r	(b) i	(c) m	(d) R/V	(e) I/NI	(f) Side
9	+18	Concave, 12						
10	+15	Concave, 10						
11	+8.0	Convex, 10						
12	+24	Concave, 36						
13	+12	Concave, 18						
14	+22	Convex, 35						
15	+10	Convex, 8.0						
16	+17	Convex, 14						

Table 34-4 Problems 17 through 29: More Mirrors. See the setup for these problems.

	(a) Type	(b) f	(c) r	(d) p	(e) i	(f) m	(g) R/V	(h) I/NI	(i) Side
17	Concave	20		+10					
18				+24		0.50		I	
19			−40		−10				
20				+40		−0.70			
21		+20		+30					
22		20				+0.10			
23		30				+0.20			
24				+60		−0.50			
25				+30		0.40		I	
26		20		+60					Same
27		−30			−15				
28				+10		+1.0			
29	Convex		40		4.0				

uation, each problem in Table 34-4 refers to (a) the type of mirror, (b) the focal distance f, (c) the radius of curvature r, (d) the object distance p, (e) the image distance i, and (f) the lateral magnification m. (All distances are in centimeters.) It also refers to whether (g) the image is real (R) or virtual (V), (h) inverted (I) or noninverted (NI) from O, and (i) on the *same* side of the mirror as object O or on the *opposite* side. Fill in the missing information. Where only a sign is missing, answer with the sign.

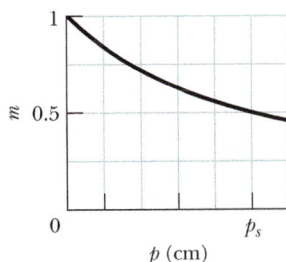

Figure 34-37 Problem 30.

••**30** Figure 34-37 gives the lateral magnification m of an object versus the object distance p from a spherical mirror as the object is moved along the mirror's central axis through a range of values for p. The horizontal scale is set by $p_s = 10.0$ cm. What is the magnification of the object when the object is 21 cm from the mirror?

••**31** (a) A luminous point is moving at speed v_O toward a spherical mirror with radius of curvature r, along the central axis of the mirror. Show that the image of this point is moving at speed

$$v_I = -\left(\frac{r}{2p-r}\right)^2 v_O,$$

where p is the distance of the luminous point from the mirror at any given time. Now assume the mirror is concave, with $r = 15$ cm,

and let $v_O = 5.0$ cm/s. Find v_I when (b) $p = 30$ cm (far outside the focal point), (c) $p = 8.0$ cm (just outside the focal point), and (d) $p = 10$ mm (very near the mirror).

Module 34-3   Spherical Refracting Surfaces

••**32 through 38** 37, 38 SSM 33, 35 *Spherical refracting surfaces.* An object O stands on the central axis of a spherical refracting surface. For this situation, each problem in Table 34-5 refers to the index of refraction n_1 where the object is located, (a) the index of refraction n_2 on the other side of the refracting surface, (b) the object distance p, (c) the radius of curvature r of the surface, and (d) the image distance i. (All distances are in centimeters.) Fill in the missing information, including whether the image is (e) real (R) or virtual (V) and (f) on the *same* side of the surface as object O or on the *opposite* side.

••**39** In Fig. 34-38, a beam of parallel light rays from a laser is incident on a solid transparent sphere of index of refraction n. (a) If a point image is produced at the back of the sphere, what is the index of refraction of the sphere? (b) What index of refraction, if any, will produce a point image at the center of the sphere?

Figure 34-38 Problem 39.

••**40** A glass sphere has radius $R = 5.0$ cm and index of refraction 1.6. A paperweight is constructed by slicing through the sphere along a plane that is 2.0 cm

Table 34-5 Problems 32 through 38: Spherical Refracting Surfaces. See the setup for these problems.

	n_1	(a) n_2	(b) p	(c) r	(d) i	(e) R/V	(f) Side
32	1.0	1.5	+10	+30			
33	1.0	1.5	+10		−13		
34	1.5		+100	−30	+600		
35	1.5	1.0	+70	+30			
36	1.5	1.0		−30	−7.5		
37	1.5	1.0	+10		−6.0		
38	1.0	1.5		+30	+600		

from the center of the sphere, leaving height $h = 3.0$ cm. The paperweight is placed on a table and viewed from directly above by an observer who is distance $d = 8.0$ cm from the tabletop (Fig. 34-39). When viewed through the paperweight, how far away does the tabletop appear to be to the observer?

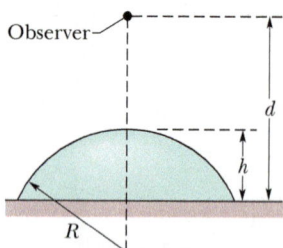

Figure 34-39 Problem 40.

Module 34-4 Thin Lenses

•41 A lens is made of glass having an index of refraction of 1.5. One side of the lens is flat, and the other is convex with a radius of curvature of 20 cm. (a) Find the focal length of the lens. (b) If an object is placed 40 cm in front of the lens, where is the image?

•42 Figure 34-40 gives the lateral magnification m of an object versus the object distance p from a lens as the object is moved along the central axis of the lens through a range of values for p out to $p_s = 20.0$ cm. What is the magnification of the object when the object is 35 cm from the lens?

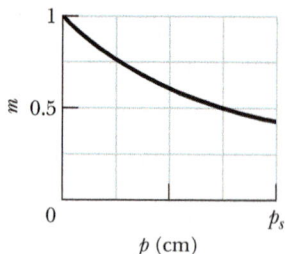

Figure 34-40 Problem 42.

•43 A movie camera with a (single) lens of focal length 75 mm takes a picture of a person standing 27 m away. If the person is 180 cm tall, what is the height of the image on the film?

•44 An object is placed against the center of a thin lens and then moved away from it along the central axis as the image distance i is measured. Figure 34-41 gives i versus object distance p out to $p_s = 60$ cm. What is the image distance when $p = 100$ cm?

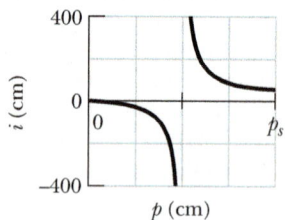

Figure 34-41 Problem 44.

•45 You produce an image of the Sun on a screen, using a thin lens whose focal length is 20.0 cm. What is the diameter of the image? (See Appendix C for needed data on the Sun.)

•46 An object is placed against the center of a thin lens and then moved 70 cm from it along the central axis as the image distance i

is measured. Figure 34-42 gives i versus object distance p out to $p_s = 40$ cm. What is the image distance when $p = 70$ cm?

•47 SSM WWW A double-convex lens is to be made of glass with an index of refraction of 1.5. One surface is to have twice the radius of curvature of the other and the focal length is to be 60 mm. What is the (a) smaller and (b) larger radius?

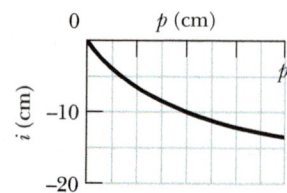

Figure 34-42 Problem 46.

•48 An object is moved along the central axis of a thin lens while the lateral magnification m is measured. Figure 34-43 gives m versus object distance p out to $p_s = 8.0$ cm. What is the magnification of the object when the object is 14.0 cm from the lens?

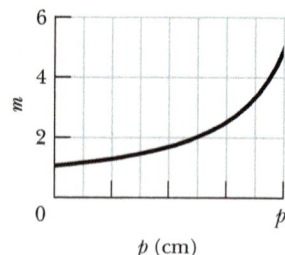

Figure 34-43 Problem 48.

•49 SSM An illuminated slide is held 44 cm from a screen. How far from the slide must a lens of focal length 11 cm be placed (between the slide and the screen) to form an image of the slide's picture on the screen?

••50 through 57 GO 55, 57 SSM 53 *Thin lenses.* Object O stands on the central axis of a thin symmetric lens. For this situation, each problem in Table 34-6 gives object distance p (centimeters), the type of lens (C stands for converging and D for diverging), and then the distance (centimeters, without proper sign) between a focal point and the lens. Find (a) the image distance i and (b) the lateral magnification m of the object, including signs. Also, determine whether the image is (c) real (R) or virtual (V), (d) inverted (I) from object O or noninverted (NI), and (e) on the *same* side of the lens as object O or on the *opposite* side.

••58 through 67 GO 61 SSM 59 *Lenses with given radii.* Object O stands in front of a thin lens, on the central axis. For this

Table 34-6 Problems 50 through 57: Thin Lenses. See the setup for these problems.

	p	Lens	(a) i	(b) m	(c) R/V	(d) I/NI	(e) Side
50	+16	C, 4.0					
51	+12	C, 16					
52	+25	C, 35					
53	+8.0	D, 12					
54	+10	D, 6.0					
55	+22	D, 14					
56	+12	D, 31					
57	+45	C, 20					

Table 34-7 Problems 58 through 67: Lenses with Given Radii. See the setup for these problems.

	p	n	r_1	r_2	(a) i	(b) m	(c) R/V	(d) I/NI	(e) Side
58	+29	1.65	+35	∞					
59	+75	1.55	+30	−42					
60	+6.0	1.70	+10	−12					
61	+24	1.50	−15	−25					
62	+10	1.50	+30	−30					
63	+35	1.70	+42	+33					
64	+10	1.50	−30	−60					
65	+10	1.50	−30	+30					
66	+18	1.60	−27	+24					
67	+60	1.50	+35	−35					

situation, each problem in Table 34-7 gives object distance p, index of refraction n of the lens, radius r_1 of the nearer lens surface, and radius r_2 of the farther lens surface. (All distances are in centimeters.) Find (a) the image distance i and (b) the lateral magnification m of the object, including signs. Also, determine whether the image is (c) real (R) or virtual (V), (d) inverted (I) from object O or noninverted (NI), and (e) on the *same* side of the lens as object O or on the *opposite* side.

··68 In Fig. 34-44, a real inverted image I of an object O is formed by a particular lens (not shown); the object–image separation is $d = 40.0$ cm, measured along the central axis of the lens. The image is just half the size of the object. (a) What kind of lens must be used to produce this image? (b) How far from the object must the lens be placed? (c) What is the focal length of the lens?

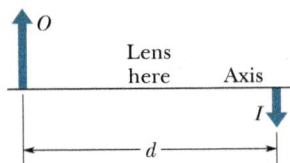
Figure 34-44 Problem 68.

··69 through 79 GO 76, 78 SSM 75, 77 *More lenses.* Object O stands on the central axis of a thin symmetric lens. For this situation, each problem in Table 34-8 refers to (a) the lens type, con-

verging (C) or diverging (D), (b) the focal distance f, (c) the object distance p, (d) the image distance i, and (e) the lateral magnification m. (All distances are in centimeters.) It also refers to whether (f) the image is real (R) or virtual (V), (g) inverted (I) or noninverted (NI) from O, and (h) on the *same* side of the lens as O or on the *opposite* side. Fill in the missing information, including the value of m when only an inequality is given. Where only a sign is missing, answer with the sign.

··80 through 87 GO 80, 87 SSM WWW 83 *Two-lens systems.* In Fig. 34-45, stick figure O (the object) stands on the common central axis of two thin, symmetric lenses, which are mounted in the boxed regions. Lens 1 is mounted within the boxed region closer to O, which is at object distance p_1. Lens 2 is mounted within the farther

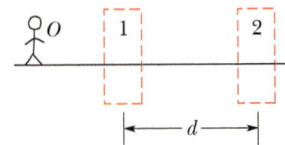
Figure 34-45 Problems 80 through 87.

Table 34-8 Problems 69 through 79: More Lenses. See the setup for these problems.

	(a) Type	(b) f	(c) p	(d) i	(e) m	(f) R/V	(g) I/NI	(h) Side
69		+10	+5.0					
70		20	+8.0		<1.0		NI	
71			+16		+0.25			
72			+16		−0.25			
73			+10		−0.50			
74	C	10	+20					
75		10	+5.0		<1.0			Same
76		10	+5.0		>1.0			
77			+16		+1.25			
78			+10		0.50		NI	
79		20	+8.0		>1.0			

Table 34-9 Problems 80 through 87: Two-Lens Systems. See the setup for these problems.

	p_1	Lens 1	d	Lens 2	(a) i_2	(b) M	(c) R/V	(d) I/NI	(e) Side
80	+10	C, 15	10	C, 8.0					
81	+12	C, 8.0	32	C, 6.0					
82	+8.0	D, 6.0	12	C, 6.0					
83	+20	C, 9.0	8.0	C, 5.0					
84	+15	C, 12	67	C, 10					
85	+4.0	C, 6.0	8.0	D, 6.0					
86	+12	C, 8.0	30	D, 8.0					
87	+20	D, 12	10	D, 8.0					

boxed region, at distance d. Each problem in Table 34-9 refers to a different combination of lenses and different values for distances, which are given in centimeters. The type of lens is indicated by C for converging and D for diverging; the number after C or D is the distance between a lens and either of its focal points (the proper sign of the focal distance is not indicated).

Find (a) the image distance i_2 for the image produced by lens 2 (the final image produced by the system) and (b) the overall lateral magnification M for the system, including signs. Also, determine whether the final image is (c) real (R) or virtual (V), (d) inverted (I) from object O or noninverted (NI), and (e) on the *same* side of lens 2 as object O or on the *opposite* side.

Module 34-5 Optical Instruments

•88 If the angular magnification of an astronomical telescope is 36 and the diameter of the objective is 75 mm, what is the minimum diameter of the eyepiece required to collect all the light entering the objective from a distant point source on the telescope axis?

•89 SSM In a microscope of the type shown in Fig. 34-20, the focal length of the objective is 4.00 cm, and that of the eyepiece is 8.00 cm. The distance between the lenses is 25.0 cm. (a) What is the tube length s? (b) If image I in Fig. 34-20 is to be just inside focal point F'_1, how far from the objective should the object be? What then are (c) the lateral magnification m of the objective, (d) the angular magnification m_θ of the eyepiece, and (e) the overall magnification M of the microscope?

••90 Figure 34-46a shows the basic structure of an old film camera. A lens can be moved forward or back to produce an image on film at the back of the camera. For a certain camera, with the distance i between the lens and the film set at $f = 5.0$ cm, parallel light rays from a very distant object O converge to a point image on the film, as shown. The object is now brought closer, to a distance of $p = 100$ cm, and the lens–film distance is adjusted so that an inverted real image forms on the film (Fig. 34-46b). (a) What is the lens–film distance i now? (b) By how much was distance i changed?

••91 SSM Figure 34-47a shows the basic structure of a human eye. Light refracts into the eye through the cornea and is then further redirected by a lens whose shape (and thus ability to focus the light) is controlled by muscles. We can treat the cornea and eye lens as a single effective thin lens (Fig. 34-47b). A "normal" eye can focus parallel light rays from a distant object O to a point on the retina at the back of the eye, where processing of the visual information begins. As an object is brought close to the eye, however, the muscles must change the shape of the lens so that rays form an inverted real image on the retina (Fig. 34-47c). (a) Suppose that for the parallel rays of Figs. 34-47a and b, the focal length f of the effective thin lens of the eye is 2.50 cm. For an object at distance $p = 40.0$ cm, what focal length f' of the effective lens is required for the object to be seen clearly? (b) Must the eye muscles increase or decrease the radii of curvature of the eye lens to give focal length f'?

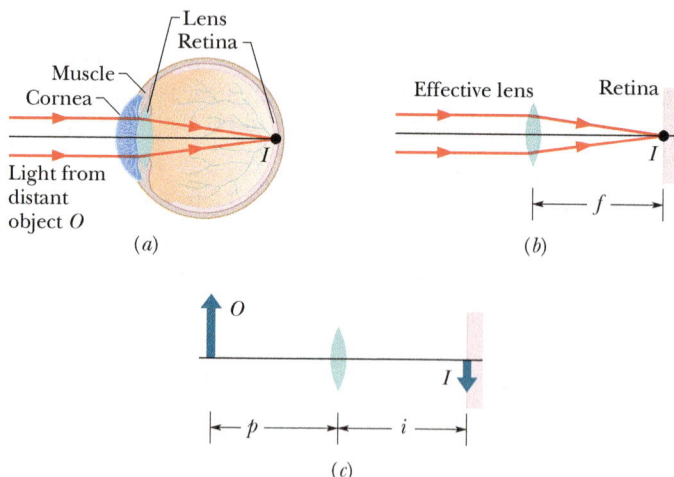

(a)

(b)

(c)

Figure 34-47 Problem 91.

••92 An object is 10.0 mm from the objective of a certain compound microscope. The lenses are 300 mm apart, and the intermediate image is 50.0 mm from the eyepiece. What overall magnification is produced by the instrument?

••93 Someone with a near point P_n of 25 cm views a thimble through a simple magnifying lens of focal length 10 cm by placing

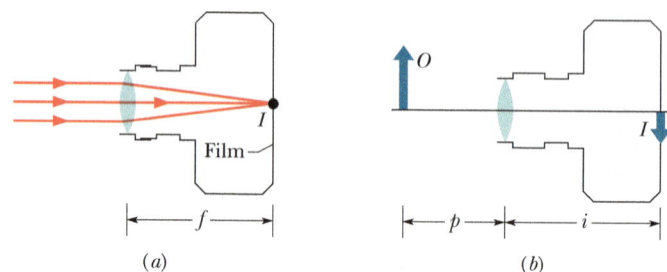

(a)

(b)

Figure 34-46 Problem 90.

the lens near his eye. What is the angular magnification of the thimble if it is positioned so that its image appears at (a) P_n and (b) infinity?

Additional Problems

94 An object is placed against the center of a spherical mirror and then moved 70 cm from it along the central axis as the image distance i is measured. Figure 34-48 gives i versus object distance p out to $p_s = 40$ cm. What is the image distance when the object is 70 cm from the mirror?

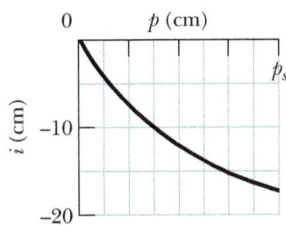

Figure 34-48 Problem 94.

95 through 100 🔵GO 95, 96, 99 *Three-lens systems.* In Fig. 34-49, stick figure O (the object) stands on the common central axis of three thin, symmetric lenses, which are mounted in the boxed regions. Lens 1 is mounted within the boxed region closest to O, which is at object distance p_1. Lens 2 is mounted within the middle boxed region, at distance d_{12} from lens 1. Lens 3 is mounted in the farthest boxed region, at distance d_{23} from lens 2. Each problem in Table 34-10 refers to a different combination of lenses and different values for distances, which are given in centimeters. The type of lens is indicated by C for converging and D for diverging; the number after C or D is the distance between a lens and either of the focal points (the proper sign of the focal distance is not indicated).

Find (a) the image distance i_3 for the (final) image produced by lens 3 (the final image produced by the system) and (b) the overall lateral magnification M for the system, including signs. Also, determine whether the final image is (c) real (R) or virtual (V), (d) inverted (I) from object O or noninverted (NI), and (e) on the same side of lens 3 as object O or on the opposite side.

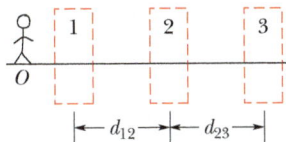

Figure 34-49 Problems 95 through 100.

101 SSM The formula $1/p + 1/i = 1/f$ is called the *Gaussian* form of the thin-lens formula. Another form of this formula, the *Newtonian* form, is obtained by considering the distance x from the object to the first focal point and the distance x' from the second focal point to the image. Show that $xx' = f^2$ is the Newtonian form of the thin-lens formula.

102 Figure 34-50a is an overhead view of two vertical plane mirrors with an object O placed between them. If you look into the

mirrors, you see multiple images of O. You can find them by drawing the reflection in each mirror of the angular region between the mirrors, as is done in Fig. 34-50b for the left-hand mirror. Then draw the reflection of the reflection. Continue this on the left and on the right until the reflections meet or overlap at the rear of the mirrors. Then you can count the number of images of O. How many images of O would you see if θ is (a) 90°, (b) 45°, and (c) 60°? If $\theta = 120°$, determine the (d) smallest and (e) largest number of images that can be seen, depending on your perspective and the location of O. (f) In each situation, draw the image locations and orientations as in Fig. 34-50b.

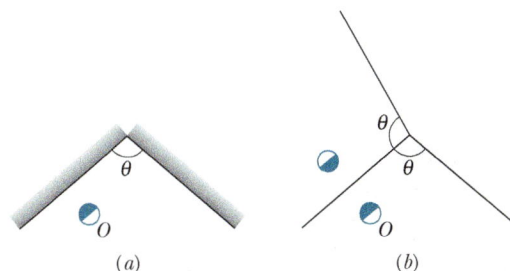

Figure 34-50 Problem 102.

103 SSM Two thin lenses of focal lengths f_1 and f_2 are in contact and share the same central axis. Show that, in image formation, they are equivalent to a single thin lens for which the focal length is $f = f_1 f_2/(f_1 + f_2)$.

104 Two plane mirrors are placed parallel to each other and 40 cm apart. An object is placed 10 cm from one mirror. Determine the (a) smallest, (b) second smallest, (c) third smallest (occurs twice), and (d) fourth smallest distance between the object and image of the object.

105 In Fig. 34-51, a box is somewhere at the left, on the central axis of the thin converging lens. The image I_m of the box produced by the plane mirror is 4.00 cm "inside" the mirror. The lens–mirror separation is 10.0 cm, and the focal length of the lens is 2.00 cm. (a) What is the distance between the box and the lens? Light reflected by the mirror travels back through the lens, which produces a final image of the box. (b) What is the distance between the lens and that final image?

Figure 34-51 Problem 105.

Table 34-10 Problems 95 through 100: Three-Lens Systems. See the setup for these problems.

	p_1	Lens 1	d_{12}	Lens 2	d_{23}	Lens 3	(a) i_3	(b) M	(c) R/V	(d) I/NI	(e) Side
95	+12	C, 8.0	28	C, 6.0	8.0	C, 6.0					
96	+4.0	D, 6.0	9.6	C, 6.0	14	C, 4.0					
97	+18	C, 6.0	15	C, 3.0	11	C, 3.0					
98	+2.0	C, 6.0	15	C, 6.0	19	C, 5.0					
99	+8.0	D, 8.0	8.0	D, 16	5.1	C, 8.0					
100	+4.0	C, 6.0	8.0	D, 4.0	5.7	D, 12					

106 In Fig. 34-52, an object is placed in front of a converging lens at a distance equal to twice the focal length f_1 of the lens. On the other side of the lens is a concave mirror of focal length f_2 separated from the lens by a distance $2(f_1 + f_2)$. Light from the object passes rightward through the lens, reflects from the mirror, passes leftward through the lens, and forms a final image of the object. What are (a) the distance between the lens and that final image and (b) the overall lateral magnification M of the object? Is the image (c) real or virtual (if it is virtual, it requires someone looking through the lens toward the mirror), (d) to the left or right of the lens, and (e) inverted or noninverted relative to the object?

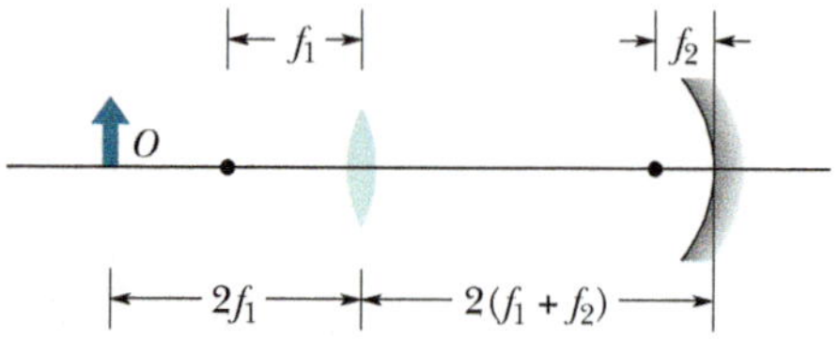

Figure 34-52 Problem 106.

107 SSM A fruit fly of height H sits in front of lens 1 on the central axis through the lens. The lens forms an image of the fly at a distance $d = 20$ cm from the fly; the image has the fly's orientation and height $H_I = 2.0H$. What are (a) the focal length f_1 of the lens and (b) the object distance p_1 of the fly? The fly then leaves lens 1 and sits in front of lens 2, which also forms an image at $d = 20$ cm that has the same orientation as the fly, but now $H_I = 0.50H$. What are (c) f_2 and (d) p_2?

108 You grind the lenses shown in Fig. 34-53 from flat glass disks ($n = 1.5$) using a machine that can grind a radius of curvature of either 40 cm or 60 cm. In a lens where either radius is appropriate, you select the 40 cm radius. Then you hold each lens in sunshine to form an image of the Sun. What are the (a) focal length f and (b) image type (real or virtual) for (bi-convex) lens 1, (c) f and (d) image type for (plane-convex) lens 2, (e) f and (f) image type for (meniscus convex) lens 3, (g) f and (h) image type for (bi-concave) lens 4, (i) f and (j) image type for (plane-concave) lens 5, and (k) f and (l) image type for (meniscus concave) lens 6?

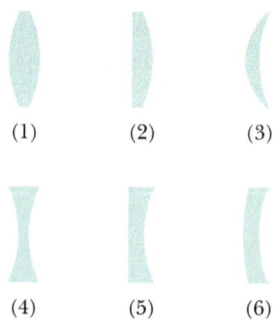

Figure 34-53 Problem 108.

109 In Fig. 34-54, a fish watcher at point P watches a fish through a glass wall of a fish tank. The watcher is level with the fish; the index of refraction of the glass is 8/5, and that of the water is 4/3. The distances are $d_1 = 8.0$ cm, $d_2 = 3.0$ cm, and $d_3 = 6.8$ cm. (a) To the fish, how far away does the watcher appear to be? (Hint: The watcher is the object. Light from that object passes through the wall's outside surface, which acts as a refracting surface. Find the image produced by that surface. Then treat that image as an object whose light passes through the wall's inside surface, which acts as another refracting surface.) (b) To the watcher, how far away does the fish appear to be?

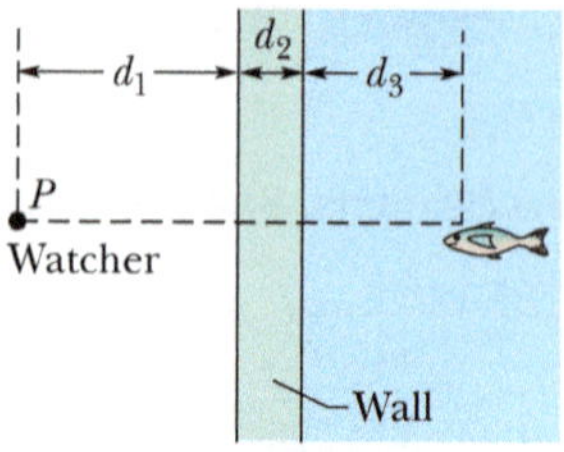

Figure 34-54 Problem 109.

110 A goldfish in a spherical fish bowl of radius R is at the level of the center C of the bowl and at distance $R/2$ from the glass (Fig. 34-55). What magnification of the fish is produced by the water in the bowl for a viewer looking along a line that includes the fish and the center, with the fish on the near side of the center? The index of refraction of the water is 1.33. Neglect the glass wall of the bowl. Assume the viewer looks with one eye. (Hint: Equation 34-5 holds, but Eq. 34-6 does not. You need to work with a ray diagram of the situation and assume that the rays are close to the observer's line of sight—that is, they deviate from that line by only small angles.)

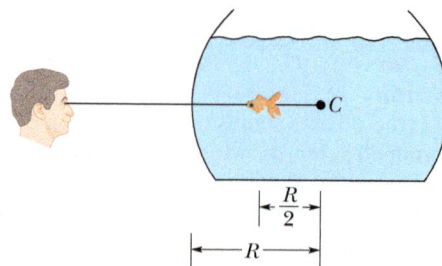

Figure 34-55 Problem 110.

111 Figure 34-56 shows a beam expander made with two coaxial converging lenses of focal lengths f_1 and f_2 and separation $d = f_1 + f_2$. The device can expand a laser beam while keeping the light rays in the beam parallel to the central axis through the lenses. Suppose a uniform laser beam of width $W_i = 2.5$ mm and intensity $I_i = 9.0$ kW/m^2 enters a beam expander for which $f_1 = 12.5$ cm and $f_2 = 30.0$ cm. What are (a) W_f and (b) I_f of the beam leaving the expander? (c) What value of d is needed for the beam expander if lens 1 is replaced with a diverging lens of focal length $f_1 = -26.0$ cm?

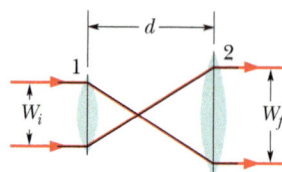

Figure 34-56 Problem 111.

112 You look down at a coin that lies at the bottom of a pool of liquid of depth d and index of refraction n (Fig. 34-57). Because you view with two eyes, which intercept different rays of light from the coin, you perceive the coin to be where extensions of the intercepted rays cross, at depth d_a instead of d. Assuming that the intercepted rays in Fig. 34-57 are close to a vertical axis through the coin, show that $d_a = d/n$. (Hint: Use the small-angle approximation $\sin \theta \approx \tan \theta \approx \theta$.)

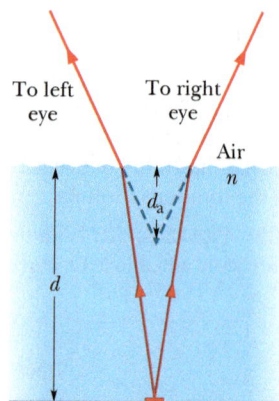

Figure 34-57 Problem 112.

113 A pinhole camera has the hole a distance 12 cm from the film plane, which is a rectangle of height 8.0 cm and width 6.0 cm. How far from a painting of dimensions 50 cm by 50 cm should the camera be placed so as to get the largest complete image possible on the film plane?

114 Light travels from point A to point B via reflection at point O on the surface of a mirror. Without using calculus, show that length AOB is a minimum when the angle of incidence θ is equal to the angle of reflection ϕ. (Hint: Consider the image of A in the mirror.)

115 A point object is 10 cm away from a plane mirror, and the eye of an observer (with pupil diameter 5.0 mm) is 20 cm away. Assuming the eye and the object to be on the same line perpendicular to the mirror surface, find the area of the mirror used in observing the reflection of the point. (*Hint:* Adapt Fig. 34-4.)

116 Show that the distance between an object and its real image formed by a thin converging lens is always greater than or equal to four times the focal length of the lens.

117 A luminous object and a screen are a fixed distance D apart. (a) Show that a converging lens of focal length f, placed between object and screen, will form a real image on the screen for two lens positions that are separated by a distance $d = \sqrt{D(D - 4f)}$. (b) Show that

$$\left(\frac{D - d}{D + d}\right)^2$$

gives the ratio of the two image sizes for these two positions of the lens.

118 An eraser of height 1.0 cm is placed 10.0 cm in front of a two-lens system. Lens 1 (nearer the eraser) has focal length $f_1 = -15$ cm, lens 2 has $f_2 = 12$ cm, and the lens separation is $d = 12$ cm. For the image produced by lens 2, what are (a) the image distance i_2 (including sign), (b) the image height, (c) the image type (real or virtual), and (d) the image orientation (inverted relative to the eraser or not inverted)?

119 A peanut is placed 40 cm in front of a two-lens system: lens 1 (nearer the peanut) has focal length $f_1 = +20$ cm, lens 2 has $f_2 = -15$ cm, and the lens separation is $d = 10$ cm. For the image produced by lens 2, what are (a) the image distance i_2 (including sign), (b) the image orientation (inverted relative to the peanut or not inverted), and (c) the image type (real or virtual)? (d) What is the net lateral magnification?

120 A coin is placed 20 cm in front of a two-lens system. Lens 1 (nearer the coin) has focal length $f_1 = +10$ cm, lens 2 has $f_2 = +12.5$ cm, and the lens separation is $d = 30$ cm. For the image produced by lens 2, what are (a) the image distance i_2 (including sign), (b) the overall lateral magnification, (c) the image type (real or virtual), and (d) the image orientation (inverted relative to the coin or not inverted)?

121 An object is 20 cm to the left of a thin diverging lens that has a 30 cm focal length. (a) What is the image distance i? (b) Draw a ray diagram showing the image position.

122 In Fig 34-58 a pinecone is at distance $p_1 = 1.0$ m in front of a lens of focal length $f_1 = 0.50$ m; a flat mirror is at distance $d = 2.0$ m behind the lens. Light from the pinecone passes rightward through the lens, reflects from the mirror, passes leftward through the lens, and forms a final image of the pinecone. What are (a) the distance between the lens and that image and (b) the overall lateral magnification of the pinecone? Is the image (c) real or virtual (if it is virtual, it requires someone looking through the lens toward the mirror), (d) to the left or right of the lens, and (e) inverted relative to the pinecone or not inverted?

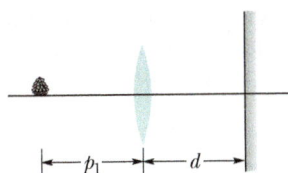

Figure 34-58 Problem 122.

123 One end of a long glass rod ($n = 1.5$) is a convex surface of radius 6.0 cm. An object is located in air along the axis of the rod, at a distance of 10 cm from the convex end. (a) How far apart are the

object and the image formed by the glass rod? (b) Within what range of distances from the end of the rod must the object be located in order to produce a virtual image?

124 A short straight object of length L lies along the central axis of a spherical mirror, a distance p from the mirror. (a) Show that its image in the mirror has a length L', where

$$L' = L\left(\frac{f}{p - f}\right)^2.$$

(*Hint:* Locate the two ends of the object.) (b) Show that the *longitudinal magnification* $m'(= L'/L)$ is equal to m^2, where m is the lateral magnification.

125 Prove that if a plane mirror is rotated through an angle α, the reflected beam is rotated through an angle 2α. Show that this result is reasonable for $\alpha = 45°$.

126 An object is 30.0 cm from a spherical mirror, along the mirror's central axis. The mirror produces an inverted image with a lateral magnification of absolute value 0.500. What is the focal length of the mirror?

127 A concave mirror has a radius of curvature of 24 cm. How far is an object from the mirror if the image formed is (a) virtual and 3.0 times the size of the object, (b) real and 3.0 times the size of the object, and (c) real and 1/3 the size of the object?

128 A pepper seed is placed in front of a lens. The lateral magnification of the seed is +0.300. The absolute value of the lens's focal length is 40.0 cm. How far from the lens is the image?

129 The equation $1/p + 1/i = 2/r$ for spherical mirrors is an approximation that is valid if the image is formed by rays that make only small angles with the central axis. In reality, many of the angles are large, which smears the image a little. You can determine how much. Refer to Fig. 34-22 and consider a ray that leaves a point source (the object) on the central axis and that makes an angle α with that axis.

First, find the point of intersection of the ray with the mirror. If the coordinates of this intersection point are x and y and the origin is placed at the center of curvature, then $y = (x + p - r)\tan \alpha$ and $x^2 + y^2 = r^2$, where p is the object distance and r is the mirror's radius of curvature. Next, use $\tan \beta = y/x$ to find the angle β at the point of intersection, and then use $\alpha + \gamma = 2\beta$ to find the value of γ. Finally, use the relation $\tan \gamma = y/(x + i - r)$ to find the distance i of the image.

(a) Suppose $r = 12$ cm and $p = 20$ cm. For each of the following values of α, find the position of the image — that is, the position of the point where the reflected ray crosses the central axis: 0.500, 0.100, 0.0100 rad. Compare the results with those obtained with the equation $1/p + 1/i = 2/r$. (b) Repeat the calculations for $p = 4.00$ cm.

130 A small cup of green tea is positioned on the central axis of a spherical mirror. The lateral magnification of the cup is +0.250, and the distance between the mirror and its focal point is 2.00 cm. (a) What is the distance between the mirror and the image it produces? (b) Is the focal length positive or negative? (c) Is the image real or virtual?

131 A 20-mm-thick layer of water ($n = 1.33$) floats on a 40-mm-thick layer of carbon tetrachloride ($n = 1.46$) in a tank. A coin lies at the bottom of the tank. At what depth below the top water surface do you perceive the coin? (*Hint:* Use the result and assumptions of Problem 112 and work with a ray diagram.)

132 A millipede sits 1.0 m in front of the nearest part of the surface of a shiny sphere of diameter 0.70 m. (a) How far from the surface does the millipede's image appear? (b) If the millipede's height is 2.0 mm, what is the image height? (c) Is the image inverted?

133 (a) Show that if the object O in Fig. 34-19c is moved from focal point F_1 toward the observer's eye, the image moves in from infinity and the angle θ' (and thus the angular magnification m_θ) increases. (b) If you continue this process, where is the image when m_θ has its maximum usable value? (You can then still increase m_θ, but the image will no longer be clear.) (c) Show that the maximum usable value of m_θ is $1 + (25\text{ cm})/f$. (d) Show that in this situation the angular magnification is equal to the lateral magnification.

134 Isaac Newton, having convinced himself (erroneously as it turned out) that chromatic aberration is an inherent property of refracting telescopes, invented the reflecting telescope, shown schematically in Fig. 34-59. He presented his second model of this telescope, with a magnifying power of 38, to the Royal Society (of London), which still has it. In Fig. 34-59 incident light falls, closely parallel to the telescope axis, on the objective mirror M. After reflection from small mirror M' (the figure is not to scale), the rays form a real, inverted image in the *focal plane* (the plane perpendicular to the line of sight, at focal point F). This image is then viewed through an eyepiece. (a) Show that the angular magnification m_θ for the device is given by Eq. 34-15:

$$m_\theta = -f_{ob}/f_{ey},$$

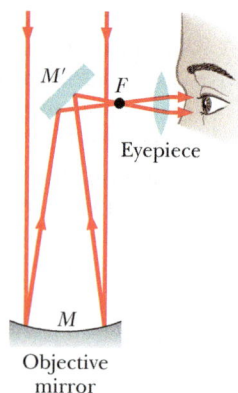

Figure 34-59 Problem 134.

where f_{ob} is the focal length of the objective mirror and f_{ey} is that of the eyepiece. (b) The 200 in. mirror in the reflecting telescope at Mt. Palomar in California has a focal length of 16.8 m. Estimate the size of the image formed by this mirror when the object is a meter stick 2.0 km away. Assume parallel incident rays. (c) The mirror of a different reflecting astronomical telescope has an effective radius of curvature of 10 m ("effective" because such mirrors are ground to a parabolic rather than a spherical shape, to eliminate spherical aberration defects). To give an angular magnification of 200, what must be the focal length of the eyepiece?

135 A narrow beam of parallel light rays is incident on a glass sphere from the left, directed toward the center of the sphere. (The

sphere is a lens but certainly not a *thin* lens.) Approximate the angle of incidence of the rays as 0°, and assume that the index of refraction of the glass is $n < 2.0$. (a) In terms of n and the sphere radius r, what is the distance between the image produced by the sphere and the right side of the sphere? (b) Is the image to the left or right of that side? (*Hint:* Apply Eq. 34-8 to locate the image that is produced by refraction at the left side of the sphere; then use that image as the object for refraction at the right side of the sphere to locate the final image. In the second refraction, is the object distance p positive or negative?)

136 A *corner reflector*, much used in optical, microwave, and other applications, consists of three plane mirrors fastened together to form the corner of a cube. Show that after three reflections, an incident ray is returned with its direction exactly reversed.

137 A cheese enchilada is 4.00 cm in front of a converging lens. The magnification of the enchilada is -2.00. What is the focal length of the lens?

138 A grasshopper hops to a point on the central axis of a spherical mirror. The absolute magnitude of the mirror's focal length is 40.0 cm, and the lateral magnification of the image produced by the mirror is $+0.200$. (a) Is the mirror convex or concave? (b) How far from the mirror is the grasshopper?

139 In Fig. 34-60, a sand grain is 3.00 cm from thin lens 1, on the central axis through the two symmetric lenses. The distance between focal point and lens is 4.00 cm for both lenses; the

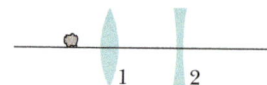

Figure 34-60 Problem 139.

lenses are separated by 8.00 cm. (a) What is the distance between lens 2 and the image it produces of the sand grain? Is that image (b) to the left or right of lens 2, (c) real or virtual, and (d) inverted relative to the sand grain or not inverted?

140 Suppose the farthest distance a person can see without visual aid is 50 cm. (a) What is the focal length of the corrective lens that will allow the person to see very far away? (b) Is the lens converging or diverging? (c) The *power P* of a lens (in *diopters*) is equal to $1/f$, where f is in meters. What is P for the lens?

141 A simple magnifier of focal length f is placed near the eye of someone whose near point P_n is 25 cm. An object is positioned so that its image in the magnifier appears at P_n. (a) What is the angular magnification of the magnifier? (b) What is the angular magnification if the object is moved so that its image appears at infinity? For $f = 10$ cm, evaluate the angular magnifications of (c) the situation in (a) and (d) the situation in (b). (Viewing an image at P_n requires effort by muscles in the eye, whereas viewing an image at infinity requires no such effort for many people.)

Interference

35-1 LIGHT AS A WAVE

Learning Objectives

After reading this module, you should be able to . . .

35.01 Using a sketch, explain Huygens' principle.

35.02 With a few simple sketches, explain refraction in terms of the gradual change in the speed of a wavefront as it passes through an interface at an angle to the normal.

35.03 Apply the relationship between the speed of light in vacuum c, the speed of light in a material v, and the index of refraction of the material n.

35.04 Apply the relationship between a distance L in a material, the speed of light in that material, and the time required for a pulse of the light to travel through L.

35.05 Apply Snell's law of refraction.

35.06 When light refracts through an interface, identify that the frequency does not change but the wavelength and effective speed do.

35.07 Apply the relationship between the wavelength in vacuum λ, the wavelength λ_n in a material (the internal wavelength), and the index of refraction n of the material.

35.08 For light in a certain length of a material, calculate the number of internal wavelengths that fit into the length.

35.09 If two light waves travel through different materials with different indexes of refraction and then reach a common point, determine their phase difference and interpret the resulting interference in terms of maximum brightness, intermediate brightness, and darkness.

35.10 Apply the learning objectives of Module 17-3 (sound waves there, light waves here) to find the phase difference and interference of two waves that reach a common point after traveling paths of different lengths.

35.11 Given the initial phase difference between two waves with the same wavelength, determine their phase difference after they travel through different path lengths and through different indexes of refraction.

35.12 Identify that rainbows are examples of optical interference.

Key Ideas

● The three-dimensional transmission of waves, including light, may often be predicted by Huygens' principle, which states that all points on a wavefront serve as point sources of spherical secondary wavelets. After a time t, the new position of the wavefront will be that of a surface tangent to these secondary wavelets.

● The law of refraction can be derived from Huygens' principle by assuming that the index of refraction of any medium is $n = c/v$, in which v is the speed of light in the medium and c is the speed of light in vacuum.

● The wavelength λ_n of light in a medium depends on the index of refraction n of the medium:

$$\lambda_n = \frac{\lambda}{n},$$

in which λ is the wavelength in vacuum.

● Because of this dependency, the phase difference between two waves can change if they pass through different materials with different indexes of refraction.

What Is Physics?

One of the major goals of physics is to understand the nature of light. This goal has been difficult to achieve (and has not yet fully been achieved) because light is complicated. However, this complication means that light offers many opportunities for applications, and some of the richest opportunities involve the interference of light waves—**optical interference.**

Nature has long used optical interference for coloring. For example, the wings of a *Morpho* butterfly are a dull, uninspiring brown, as can be seen on the bottom wing surface, but the brown is hidden on the top surface by an arresting blue due to the interference of light reflecting from that surface (Fig. 35-1). Moreover, the top surface is color-shifting; if you change your perspective or if the wing moves, the tint of the color changes. Similar color shifting is used in the inks on many currencies to thwart counterfeiters, whose copy machines can duplicate color from only one perspective and therefore cannot duplicate any shift in color caused by a change in perspective.

To understand the basic physics of optical interference, we must largely abandon the simplicity of geometrical optics (in which we describe light as rays) and return to the wave nature of light.

Light as a Wave

The first convincing wave theory for light was in 1678 by Dutch physicist Christian Huygens. Mathematically simpler than the electromagnetic theory of Maxwell, it nicely explained reflection and refraction in terms of waves and gave physical meaning to the index of refraction.

Huygens' wave theory is based on a geometrical construction that allows us to tell where a given wavefront will be at any time in the future if we know its present position. **Huygens' principle** is:

> All points on a wavefront serve as point sources of spherical secondary wavelets. After a time t, the new position of the wavefront will be that of a surface tangent to these secondary wavelets.

Here is a simple example. At the left in Fig. 35-2, the present location of a wavefront of a plane wave traveling to the right in vacuum is represented by plane ab, perpendicular to the page. Where will the wavefront be at time Δt later? We let several points on plane ab (the dots) serve as sources of spherical secondary wavelets that are emitted at $t = 0$. At time Δt, the radius of all these spherical wavelets will have grown to $c\,\Delta t$, where c is the speed of light in vacuum. We draw plane de tangent to these wavelets at time Δt. This plane represents the wavefront of the plane wave at time Δt; it is parallel to plane ab and a perpendicular distance $c\,\Delta t$ from it.

Figure 35-2 The propagation of a plane wave in vacuum, as portrayed by Huygens' principle.

The Law of Refraction

We now use Huygens' principle to derive the law of refraction, Eq. 33-40 (Snell's law). Figure 35-3 shows three stages in the refraction of several wavefronts at a flat interface between air (medium 1) and glass (medium 2). We arbitrarily choose the wavefronts in the incident light beam to be separated by λ_1, the wavelength in medium 1. Let the speed of light in air be v_1 and that in glass be v_2. We assume that $v_2 < v_1$, which happens to be true.

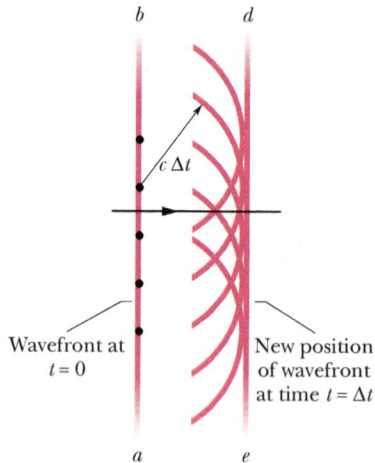

Figure 35-1 The blue of the top surface of a *Morpho* butterfly wing is due to optical interference and shifts in color as your viewing perspective changes.

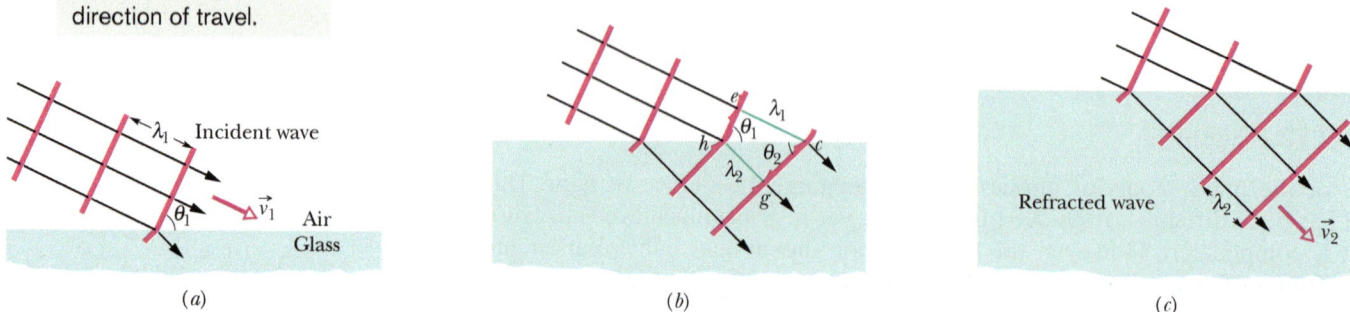

Figure 35-3 The refraction of a plane wave at an air–glass interface, as portrayed by Huygens' principle. The wavelength in glass is smaller than that in air. For simplicity, the reflected wave is not shown. Parts (a) through (c) represent three successive stages of the refraction.

Angle θ_1 in Fig. 35-3a is the angle between the wavefront and the interface; it has the same value as the angle between the *normal* to the wavefront (that is, the incident ray) and the *normal* to the interface. Thus, θ_1 is the angle of incidence.

As the wave moves into the glass, a Huygens wavelet at point *e* in Fig. 35-3b will expand to pass through point *c*, at a distance of λ_1 from point *e*. The time interval required for this expansion is that distance divided by the speed of the wavelet, or λ_1/v_1. Now note that in this same time interval, a Huygens wavelet at point *h* will expand to pass through point *g*, at the reduced speed v_2 and with wavelength λ_2. Thus, this time interval must also be equal to λ_2/v_2. By equating these times of travel, we obtain the relation

$$\frac{\lambda_1}{\lambda_2} = \frac{v_1}{v_2}, \tag{35-1}$$

which shows that the wavelengths of light in two media are proportional to the speeds of light in those media.

By Huygens' principle, the refracted wavefront must be tangent to an arc of radius λ_2 centered on *h*, say at point *g*. The refracted wavefront must also be tangent to an arc of radius λ_1 centered on *e*, say at *c*. Then the refracted wavefront must be oriented as shown. Note that θ_2, the angle between the refracted wavefront and the interface, is actually the angle of refraction.

For the right triangles *hce* and *hcg* in Fig. 35-3b we may write

$$\sin \theta_1 = \frac{\lambda_1}{hc} \quad \text{(for triangle } hce\text{)}$$

and

$$\sin \theta_2 = \frac{\lambda_2}{hc} \quad \text{(for triangle } hcg\text{)}.$$

Dividing the first of these two equations by the second and using Eq. 35-1, we find

$$\frac{\sin \theta_1}{\sin \theta_2} = \frac{\lambda_1}{\lambda_2} = \frac{v_1}{v_2}. \tag{35-2}$$

We can define the **index of refraction** n for each medium as the ratio of the speed of light in vacuum to the speed of light v in the medium. Thus,

$$n = \frac{c}{v} \quad \text{(index of refraction).} \tag{35-3}$$

In particular, for our two media, we have

$$n_1 = \frac{c}{v_1} \quad \text{and} \quad n_2 = \frac{c}{v_2}.$$

We can now rewrite Eq. 35-2 as

$$\frac{\sin \theta_1}{\sin \theta_2} = \frac{c/n_1}{c/n_2} = \frac{n_2}{n_1}$$

or

$$n_1 \sin \theta_1 = n_2 \sin \theta_2 \quad \text{(law of refraction),} \tag{35-4}$$

as introduced in Chapter 33.

✔ Checkpoint 1

The figure shows a monochromatic ray of light traveling across parallel interfaces, from an original material *a*, through layers of materials *b* and *c*, and then back into material *a*. Rank the materials according to the speed of light in them, greatest first.

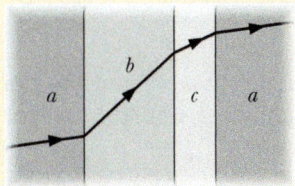

Wavelength and Index of Refraction

We have now seen that the wavelength of light changes when the speed of the light changes, as happens when light crosses an interface from one medium into another. Further, the speed of light in any medium depends on the index of refraction of the medium, according to Eq. 35-3. Thus, the wavelength of light in any medium depends on the index of refraction of the medium. Let a certain monochromatic light have wavelength λ and speed c in vacuum and wavelength λ_n and speed v in a medium with an index of refraction n. Now we can rewrite Eq. 35-1 as

$$\lambda_n = \lambda \frac{v}{c}. \tag{35-5}$$

Using Eq. 35-3 to substitute $1/n$ for v/c then yields

$$\lambda_n = \frac{\lambda}{n}. \tag{35-6}$$

This equation relates the wavelength of light in any medium to its wavelength in vacuum: A greater index of refraction means a smaller wavelength.

Next, let f_n represent the frequency of the light in a medium with index of refraction n. Then from the general relation of Eq. 16-13 ($v = \lambda f$), we can write

$$f_n = \frac{v}{\lambda_n}.$$

Substituting Eqs. 35-3 and 35-6 then gives us

$$f_n = \frac{c/n}{\lambda/n} = \frac{c}{\lambda} = f,$$

where f is the frequency of the light in vacuum. Thus, although the speed and wavelength of light in the medium are different from what they are in vacuum, *the frequency of the light in the medium is the same as it is in vacuum.*

Phase Difference. The fact that the wavelength of light depends on the index of refraction via Eq. 35-6 is important in certain situations involving the interference of light waves. For example, in Fig. 35-4, the *waves of the rays* (that is, the waves represented by the rays) have identical wavelengths λ and are initially in phase in air ($n \approx 1$). One of the waves travels through medium 1 of index of refraction n_1 and length L. The other travels through medium 2 of index of refraction n_2 and the same length L. When the waves leave the two media, they will have the same wavelength—their wavelength λ in air. However, because their wavelengths differed in the two media, the two waves may no longer be in phase.

The difference in indexes causes a phase shift between the rays.

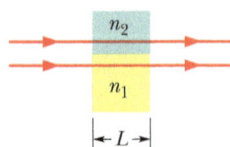

Figure 35-4 Two light rays travel through two media having different indexes of refraction.

⭐ The phase difference between two light waves can change if the waves travel through different materials having different indexes of refraction.

As we shall discuss soon, this change in the phase difference can determine how the light waves will interfere if they reach some common point.

To find their new phase difference in terms of wavelengths, we first count the number N_1 of wavelengths there are in the length L of medium 1. From Eq. 35-6, the wavelength in medium 1 is $\lambda_{n1} = \lambda/n_1$; so

$$N_1 = \frac{L}{\lambda_{n1}} = \frac{Ln_1}{\lambda}. \tag{35-7}$$

Similarly, we count the number N_2 of wavelengths there are in the length L of medium 2, where the wavelength is $\lambda_{n2} = \lambda/n_2$:

$$N_2 = \frac{L}{\lambda_{n2}} = \frac{Ln_2}{\lambda}. \tag{35-8}$$

To find the new phase difference between the waves, we subtract the smaller of N_1 and N_2 from the larger. Assuming $n_2 > n_1$, we obtain

$$N_2 - N_1 = \frac{Ln_2}{\lambda} - \frac{Ln_1}{\lambda} = \frac{L}{\lambda}(n_2 - n_1). \qquad (35\text{-}9)$$

Suppose Eq. 35-9 tells us that the waves now have a phase difference of 45.6 wavelengths. That is equivalent to taking the initially in-phase waves and shifting one of them by 45.6 wavelengths. However, a shift of an integer number of wavelengths (such as 45) would put the waves back in phase; so it is only the decimal fraction (here, 0.6) that is important. A phase difference of 45.6 wavelengths is equivalent to an *effective phase difference* of 0.6 wavelength.

A phase difference of 0.5 wavelength puts two waves exactly out of phase. If the waves had equal amplitudes and were to reach some common point, they would then undergo fully destructive interference, producing darkness at that point. With a phase difference of 0.0 or 1.0 wavelength, they would, instead, undergo fully constructive interference, resulting in brightness at the common point. Our phase difference of 0.6 wavelength is an intermediate situation but closer to fully destructive interference, and the waves would produce a dimly illuminated common point.

We can also express phase difference in terms of radians and degrees, as we have done already. A phase difference of one wavelength is equivalent to phase differences of 2π rad and $360°$.

Path Length Difference. As we discussed with sound waves in Module 17-3, two waves that begin with some initial phase difference can end up with a different phase difference if they travel through paths with different lengths before coming back together. The key for the waves (whatever their type might be) is the path length difference ΔL, or more to the point, how ΔL compares to the wavelength λ of the waves. From Eqs. 17-23 and 17-24, we know that, for light waves, fully constructive interference (maximum brightness) occurs when

$$\frac{\Delta L}{\lambda} = 0, 1, 2, \ldots \quad \text{(fully constructive interference)}, \qquad (35\text{-}10)$$

and that fully destructive interference (darkness) occurs when

$$\frac{\Delta L}{\lambda} = 0.5, 1.5, 2.5, \ldots \quad \text{(fully destructive interference)}. \qquad (35\text{-}11)$$

Intermediate values correspond to intermediate interference and thus also illumination.

Rainbows and Optical Interference

In Module 33-5, we discussed how the colors of sunlight are separated into a rainbow when sunlight travels through falling raindrops. We dealt with a simplified situation in which a single ray of white light entered a drop. Actually, light waves pass into a drop along the entire side that faces the Sun. Here we cannot discuss the details of how these waves travel through the drop and then emerge, but we can see that different parts of an incoming wave will travel different paths within the drop. That means waves will emerge from the drop with different phases. Thus, we can see that at some angles the emerging light will be in phase and give constructive interference. The rainbow is the result of such constructive interference. For example, the red of the rainbow appears because waves of red light emerge in phase from each raindrop in the direction in which you see that part of the rainbow. The light waves that emerge in other directions from each raindrop have a range of different phases because they take a

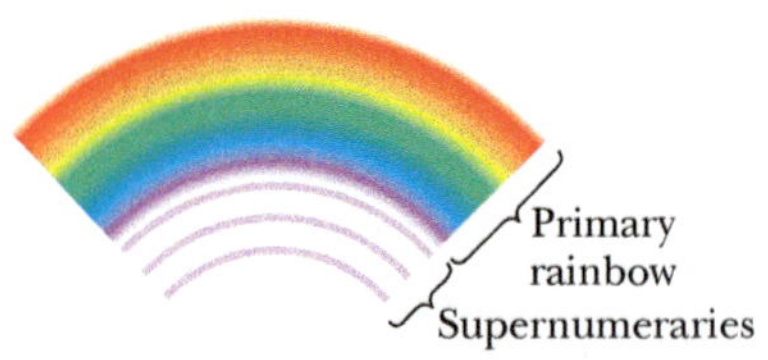

Figure 35-5 A primary rainbow and the faint supernumeraries below it are due to optical interference.

range of different paths through each drop. This light is neither bright nor colorful, and so you do not notice it.

If you are lucky and look carefully below a primary rainbow, you can see dimmer colored arcs called *supernumeraries* (Fig. 35-5). Like the main arcs of the rainbow, the supernumeraries are due to waves that emerge from each drop approximately in phase with one another to give constructive interference. If you are very lucky and look very carefully above a secondary rainbow, you might see even more (but even dimmer) supernumeraries. Keep in mind that both types of rainbows and both sets of supernumeraries are naturally occurring examples of optical interference and naturally occurring evidence that light consists of waves.

✓ Checkpoint 2

The light waves of the rays in Fig. 35-4 have the same wavelength and amplitude and are initially in phase. (a) If 7.60 wavelengths fit within the length of the top material and 5.50 wavelengths fit within that of the bottom material, which material has the greater index of refraction? (b) If the rays are angled slightly so that they meet at the same point on a distant screen, will the interference there result in the brightest possible illumination, bright intermediate illumination, dark intermediate illumination, or darkness?

Sample Problem 35.01 Phase difference of two waves due to difference in refractive indexes

In Fig. 35-4, the two light waves that are represented by the rays have wavelength 550.0 nm before entering media 1 and 2. They also have equal amplitudes and are in phase. Medium 1 is now just air, and medium 2 is a transparent plastic layer of index of refraction 1.600 and thickness 2.600 μm.

(a) What is the phase difference of the emerging waves in wavelengths, radians, and degrees? What is their effective phase difference (in wavelengths)?

KEY IDEA

The phase difference of two light waves can change if they travel through different media, with different indexes of refraction. The reason is that their wavelengths are different in the different media. We can calculate the change in phase difference by counting the number of wavelengths that fits into each medium and then subtracting those numbers.

Calculations: When the path lengths of the waves in the two media are identical, Eq. 35-9 gives the result of the subtraction. Here we have $n_1 = 1.000$ (for the air), $n_2 = 1.600$, $L = 2.600$ μm, and $\lambda = 550.0$ nm. Thus, Eq. 35-9 yields

$$N_2 - N_1 = \frac{L}{\lambda}(n_2 - n_1)$$

$$= \frac{2.600 \times 10^{-6}\text{ m}}{5.500 \times 10^{-7}\text{ m}}(1.600 - 1.000)$$

$$= 2.84. \qquad \text{(Answer)}$$

Thus, the phase difference of the emerging waves is 2.84 wavelengths. Because 1.0 wavelength is equivalent to 2π rad and 360°, you can show that this phase difference is equivalent to

$$\text{phase difference} = 17.8\text{ rad} \approx 1020°. \qquad \text{(Answer)}$$

The effective phase difference is the decimal part of the actual phase difference *expressed in wavelengths*. Thus, we have

$$\text{effective phase difference} = 0.84\text{ wavelength.} \qquad \text{(Answer)}$$

You can show that this is equivalent to 5.3 rad and about 300°. *Caution:* We do *not* find the effective phase difference by taking the decimal part of the actual phase difference as expressed in radians or degrees. For example, we do *not* take 0.8 rad from the actual phase difference of 17.8 rad.

(b) If the waves reached the same point on a distant screen, what type of interference would they produce?

Reasoning: We need to compare the effective phase difference of the waves with the phase differences that give the extreme types of interference. Here the effective phase difference of 0.84 wavelength is between 0.5 wavelength (for fully destructive interference, or the darkest possible result) and 1.0 wavelength (for fully constructive interference, or the brightest possible result), but closer to 1.0 wavelength. Thus, the waves would produce intermediate interference that is closer to fully constructive interference—they would produce a relatively bright spot.

PLUS Additional examples, video, and practice available at *WileyPLUS*

35-2 YOUNG'S INTERFERENCE EXPERIMENT

Learning Objectives

After reading this module, you should be able to . . .

35.13 Describe the diffraction of light by a narrow slit and the effect of narrowing the slit.

35.14 With sketches, describe the production of the interference pattern in a double-slit interference experiment using monochromatic light.

35.15 Identify that the phase difference between two waves can change if the waves travel along paths of different lengths, as in the case of Young's experiment.

35.16 In a double-slit experiment, apply the relationship between the path length difference ΔL and the wavelength λ, and then interpret the result in terms of interference (maximum brightness, intermediate brightness, and darkness).

35.17 For a given point in a double-slit interference pattern, express the path length difference ΔL of the rays reaching that point in terms of the slit separation d and the angle θ to that point.

35.18 In a Young's experiment, apply the relationships between the slit separation d, the light wavelength λ, and the

angles θ to the minima (dark fringes) and to the maxima (bright fringes) in the interference pattern.

35.19 Sketch the double-slit interference pattern, identifying what lies at the center and what the various bright and dark fringes are called (such as "first side maximum" and "third order").

35.20 Apply the relationship between the distance D between a double-slit screen and a viewing screen, the angle θ to a point in the interference pattern, and the distance y to that point from the pattern's center.

35.21 For a double-slit interference pattern, identify the effects of changing d or λ and also identify what determines the angular limit to the pattern.

35.22 For a transparent material placed over one slit in a Young's experiment, determine the thickness or index of refraction required to shift a given fringe to the center of the interference pattern.

Key Ideas

● In Young's interference experiment, light passing through a single slit falls on two slits in a screen. The light leaving these slits flares out (by diffraction), and interference occurs in the region beyond the screen. A fringe pattern, due to the interference, forms on a viewing screen.

● The conditions for maximum and minimum intensity are

$$d \sin \theta = m\lambda, \quad \text{for } m = 0, 1, 2, \ldots \quad \text{(maxima—bright fringes)},$$

$$d \sin \theta = (m + \tfrac{1}{2})\lambda, \quad \text{for } m = 0, 1, 2, \ldots \quad \text{(minima—dark fringes)},$$

where θ is the angle the light path makes with a central axis and d is the slit separation.

Diffraction

In this module we shall discuss the experiment that first proved that light is a wave. To prepare for that discussion, we must introduce the idea of **diffraction** of waves, a phenomenon that we explore much more fully in Chapter 36. Its essence is this: If a wave encounters a barrier that has an opening of dimensions similar to the wavelength, the part of the wave that passes through the opening will flare (spread) out—will *diffract*—into the region beyond the barrier. The flaring is consistent with the spreading of wavelets in the Huygens construction of Fig. 35-2. Diffraction occurs for waves of all types, not just light waves; Fig. 35-6 shows the diffraction of water waves traveling across the surface of water in a shallow tank. Similar diffraction of ocean waves through openings in a barrier can actually increase the erosion of a beach the barrier is intended to protect.

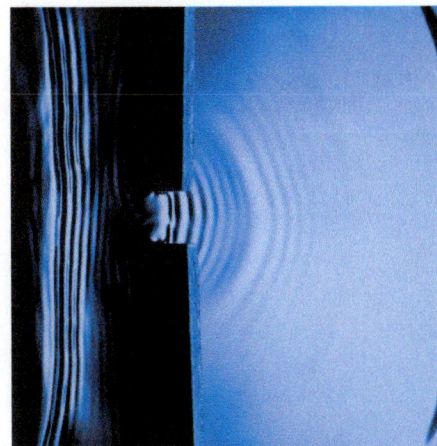

Figure 35-6 Waves produced by an oscillating paddle at the left flare out through an opening in a barrier along the water surface.

George Resch/Fundamental Photographs

A wave passing through a slit flares (diffracts).

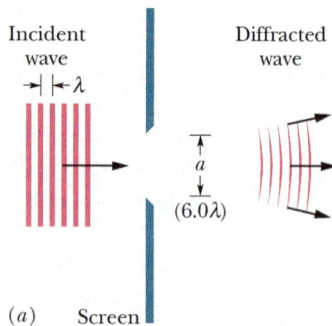

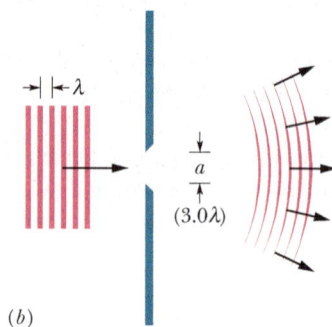

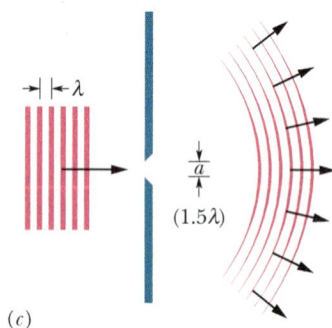

(a)   Screen

(b)

(c)

Figure 35-7 Diffraction represented schematically. For a given wavelength λ, the diffraction is more pronounced the smaller the slit width a. The figures show the cases for (a) slit width $a = 6.0\lambda$, (b) slit width $a = 3.0\lambda$, and (c) slit width $a = 1.5\lambda$. In all three cases, the screen and the length of the slit extend well into and out of the page, perpendicular to it.

Figure 35-7a shows the situation schematically for an incident plane wave of wavelength λ encountering a slit that has width $a = 6.0\lambda$ and extends into and out of the page. The part of the wave that passes through the slit flares out on the far side. Figures 35-7b (with $a = 3.0\lambda$) and 35-7c ($a = 1.5\lambda$) illustrate the main feature of diffraction: the narrower the slit, the greater the diffraction.

Diffraction limits geometrical optics, in which we represent an electromagnetic wave with a ray. If we actually try to form a ray by sending light through a narrow slit, or through a series of narrow slits, diffraction will always defeat our effort because it always causes the light to spread. Indeed, the narrower we make the slits (in the hope of producing a narrower beam), the greater the spreading is. Thus, geometrical optics holds only when slits or other apertures that might be located in the path of light do not have dimensions comparable to or smaller than the wavelength of the light.

Young's Interference Experiment

In 1801, Thomas Young experimentally proved that light is a wave, contrary to what most other scientists then thought. He did so by demonstrating that light undergoes interference, as do water waves, sound waves, and waves of all other types. In addition, he was able to measure the average wavelength of sunlight; his value, 570 nm, is impressively close to the modern accepted value of 555 nm. We shall here examine Young's experiment as an example of the interference of light waves.

Figure 35-8 gives the basic arrangement of Young's experiment. Light from a distant monochromatic source illuminates slit S_0 in screen A. The emerging light then spreads via diffraction to illuminate two slits S_1 and S_2 in screen B. Diffraction of the light by these two slits sends overlapping circular waves into

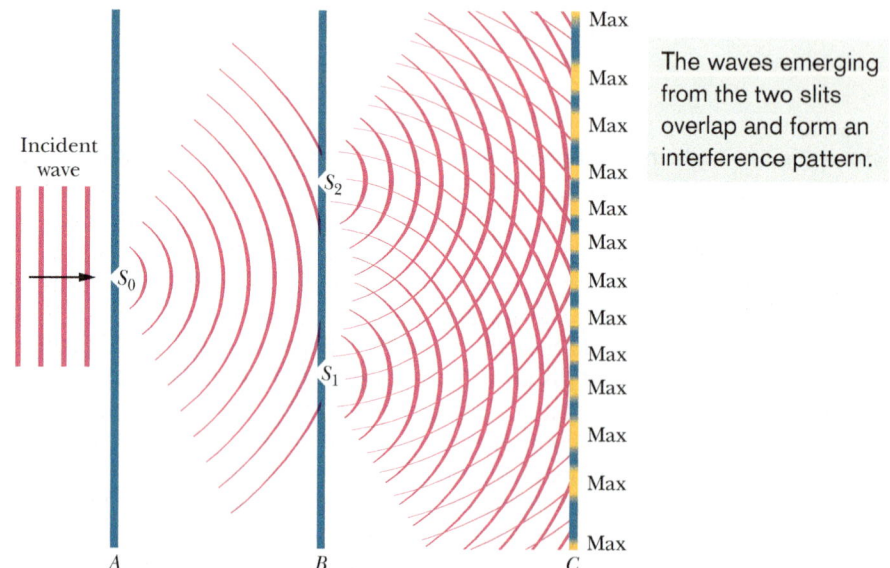

The waves emerging from the two slits overlap and form an interference pattern.

Figure 35-8 In Young's interference experiment, incident monochromatic light is diffracted by slit S_0, which then acts as a point source of light that emits semicircular wavefronts. As that light reaches screen B, it is diffracted by slits S_1 and S_2, which then act as two point sources of light. The light waves traveling from slits S_1 and S_2 overlap and undergo interference, forming an interference pattern of maxima and minima on viewing screen C. This figure is a cross section; the screens, slits, and interference pattern extend into and out of the page. Between screens B and C, the semicircular wavefronts centered on S_2 depict the waves that would be there if only S_2 were open. Similarly, those centered on S_1 depict waves that would be there if only S_1 were open.

the region beyond screen *B*, where the waves from one slit interfere with the waves from the other slit.

The "snapshot" of Fig. 35-8 depicts the interference of the overlapping waves. However, we cannot see evidence for the interference except where a viewing screen *C* intercepts the light. Where it does so, points of interference maxima form visible bright rows—called *bright bands, bright fringes,* or (loosely speaking) *maxima*—that extend across the screen (into and out of the page in Fig. 35-8). Dark regions—called *dark bands, dark fringes,* or (loosely speaking) *minima*—result from fully destructive interference and are visible between adjacent pairs of bright fringes. (*Maxima* and *minima* more properly refer to the center of a band.) The pattern of bright and dark fringes on the screen is called an **interference pattern.** Figure 35-9 is a photograph of part of the interference pattern that would be seen by an observer standing to the left of screen *C* in the arrangement of Fig. 35-8.

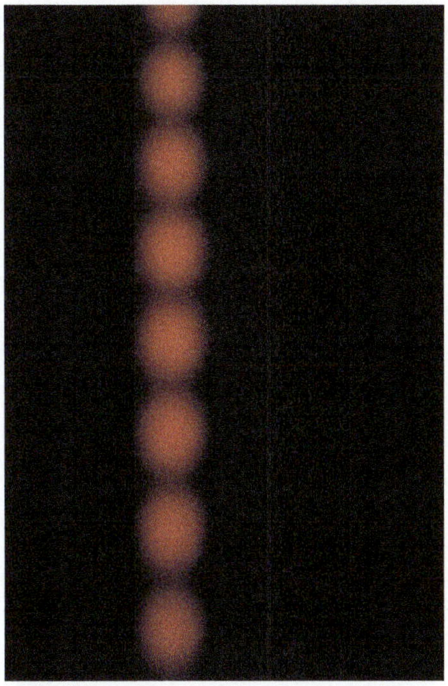

Locating the Fringes

Light waves produce fringes in a *Young's double-slit interference experiment,* as it is called, but what exactly determines the locations of the fringes? To answer, we shall use the arrangement in Fig. 35-10a. There, a plane wave of monochromatic light is incident on two slits S_1 and S_2 in screen *B*; the light diffracts through the slits and produces an interference pattern on screen *C*. We draw a central axis from the point halfway between the slits to screen *C* as a reference. We then pick, for discussion, an arbitrary point *P* on the screen, at angle θ to the central axis. This point intercepts the wave of ray r_1 from the bottom slit and the wave of ray r_2 from the top slit.

    Path Length Difference. These waves are in phase when they pass through the two slits because there they are just portions of the same incident wave. However, once they have passed the slits, the two waves must travel different distances to reach *P*. We saw a similar situation in Module 17-3 with sound waves and concluded that

> ★ The phase difference between two waves can change if the waves travel paths of different lengths.

The change in phase difference is due to the *path length difference* ΔL in the paths taken by the waves. Consider two waves initially exactly in phase, traveling along paths with a path length difference ΔL, and then passing through some common point. When ΔL is zero or an integer number of wavelengths, the waves arrive at the common point exactly in phase and they interfere fully constructively there. If that is true for the waves of rays r_1 and r_2 in Fig. 35-10, then

Courtesy Jearl Walker

Figure 35-9 A photograph of the interference pattern produced by the arrangement shown in Fig. 35-8, but with short slits. (The photograph is a front view of part of screen *C*.) The alternating maxima and minima are called *interference fringes* (because they resemble the decorative fringe sometimes used on clothing and rugs).

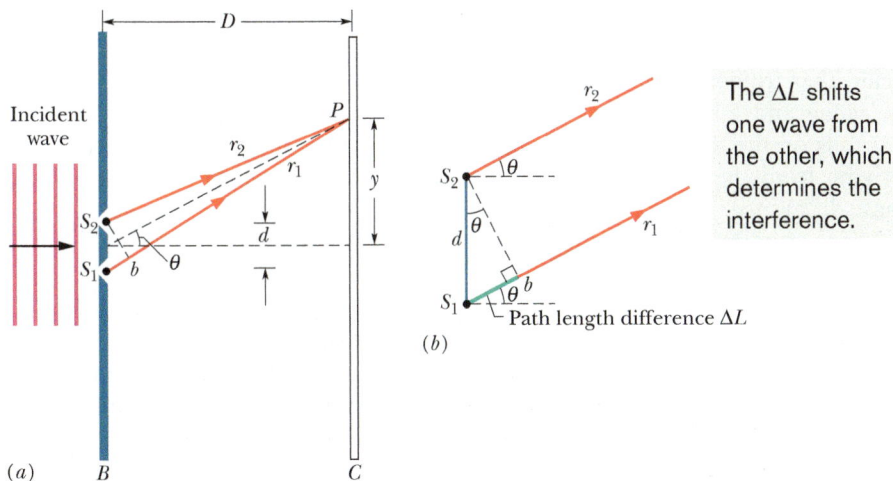

The ΔL shifts one wave from the other, which determines the interference.

Figure 35-10 (*a*) Waves from slits S_1 and S_2 (which extend into and out of the page) combine at *P*, an arbitrary point on screen *C* at distance *y* from the central axis. The angle θ serves as a convenient locator for *P*. (*b*) For $D \gg d$, we can approximate rays r_1 and r_2 as being parallel, at angle θ to the central axis.

point P is part of a bright fringe. When, instead, ΔL is an odd multiple of half a wavelength, the waves arrive at the common point exactly out of phase and they interfere fully destructively there. If that is true for the waves of rays r_1 and r_2, then point P is part of a dark fringe. (And, of course, we can have intermediate situations of interference and thus intermediate illumination at P.) Thus,

> What appears at each point on the viewing screen in a Young's double-slit interference experiment is determined by the path length difference ΔL of the rays reaching that point.

Angle. We can specify where each bright fringe and each dark fringe is located on the screen by giving the angle θ from the central axis to that fringe. To find θ, we must relate it to ΔL. We start with Fig. 35-10a by finding a point b along ray r_1 such that the path length from b to P equals the path length from S_2 to P. Then the path length difference ΔL between the two rays is the distance from S_1 to b.

The relation between this S_1-to-b distance and θ is complicated, but we can simplify it considerably if we arrange for the distance D from the slits to the screen to be much greater than the slit separation d. Then we can approximate rays r_1 and r_2 as being parallel to each other and at angle θ to the central axis (Fig. 35-10b). We can also approximate the triangle formed by S_1, S_2, and b as being a right triangle, and approximate the angle inside that triangle at S_2 as being θ. Then, for that triangle, $\sin \theta = \Delta L/d$ and thus

$$\Delta L = d \sin \theta \quad \text{(path length difference)}. \tag{35-12}$$

For a bright fringe, we saw that ΔL must be either zero or an integer number of wavelengths. Using Eq. 35-12, we can write this requirement as

$$\Delta L = d \sin \theta = (\text{integer})(\lambda), \tag{35-13}$$

or as

$$d \sin \theta = m\lambda, \quad \text{for } m = 0, 1, 2, \ldots \quad \text{(maxima—bright fringes)}. \tag{35-14}$$

For a dark fringe, ΔL must be an odd multiple of half a wavelength. Again using Eq. 35-12, we can write this requirement as

$$\Delta L = d \sin \theta = (\text{odd number})(\tfrac{1}{2}\lambda), \tag{35-15}$$

or as

$$d \sin \theta = (m + \tfrac{1}{2})\lambda, \quad \text{for } m = 0, 1, 2, \ldots \quad \text{(minima—dark fringes)}. \tag{35-16}$$

With Eqs. 35-14 and 35-16, we can find the angle θ to any fringe and thus locate that fringe; further, we can use the values of m to label the fringes. For the value and label $m = 0$, Eq. 35-14 tells us that a bright fringe is at $\theta = 0$ and thus on the central axis. This *central maximum* is the point at which waves arriving from the two slits have a path length difference $\Delta L = 0$, hence zero phase difference.

For, say, $m = 2$, Eq. 35-14 tells us that *bright* fringes are at the angle

$$\theta = \sin^{-1}\left(\frac{2\lambda}{d}\right)$$

above and below the central axis. Waves from the two slits arrive at these two fringes with $\Delta L = 2\lambda$ and with a phase difference of two wavelengths. These fringes are said to be the *second-order bright fringes* (meaning $m = 2$) or the *second side maxima* (the second maxima to the side of the central maximum), or they are described as being the second bright fringes from the central maximum.

For $m = 1$, Eq. 35-16 tells us that *dark* fringes are at the angle

$$\theta = \sin^{-1}\left(\frac{1.5\lambda}{d}\right)$$

above and below the central axis. Waves from the two slits arrive at these two fringes with $\Delta L = 1.5\lambda$ and with a phase difference, in wavelengths, of 1.5. These fringes are called the *second-order dark fringes* or *second minima* because they are the second dark fringes to the side of the central axis. (The first dark fringes, or first minima, are at locations for which $m = 0$ in Eq. 35-16.)

Nearby Screen. We derived Eqs. 35-14 and 35-16 for the situation $D \gg d$. However, they also apply if we place a converging lens between the slits and the viewing screen and then move the viewing screen closer to the slits, to the focal point of the lens. (The screen is then said to be in the *focal plane* of the lens; that is, it is in the plane perpendicular to the central axis at the focal point.) One property of a converging lens is that it focuses all rays that are parallel to one another to the same point on its focal plane. Thus, the rays that now arrive at any point on the screen (in the focal plane) were exactly parallel (rather than approximately) when they left the slits. They are like the initially parallel rays in Fig. 34-14a that are directed to a point (the focal point) by a lens.

✓ Checkpoint 3

In Fig. 35-10a, what are ΔL (as a multiple of the wavelength) and the phase difference (in wavelengths) for the two rays if point P is (a) a third side maximum and (b) a third minimum?

Sample Problem 35.02   Double-slit interference pattern

What is the distance on screen C in Fig. 35-10a between adjacent maxima near the center of the interference pattern? The wavelength λ of the light is 546 nm, the slit separation d is 0.12 mm, and the slit–screen separation D is 55 cm. Assume that θ in Fig. 35-10 is small enough to permit use of the approximations $\sin \theta \approx \tan \theta \approx \theta$, in which θ is expressed in radian measure.

KEY IDEAS

(1) First, let us pick a maximum with a low value of m to ensure that it is near the center of the pattern. Then, from the geometry of Fig. 35-10a, the maximum's vertical distance y_m from the center of the pattern is related to its angle θ from the central axis by

$$\tan \theta \approx \theta = \frac{y_m}{D}.$$

(2) From Eq. 35-14, this angle θ for the mth maximum is given by

$$\sin \theta \approx \theta = \frac{m\lambda}{d}.$$

Calculations: If we equate our two expressions for angle θ and then solve for y_m, we find

$$y_m = \frac{m\lambda D}{d}. \tag{35-17}$$

For the next maximum as we move away from the pattern's center, we have

$$y_{m+1} = \frac{(m + 1)\lambda D}{d}. \tag{35-18}$$

We find the distance between these adjacent maxima by subtracting Eq. 35-17 from Eq. 35-18:

$$\Delta y = y_{m+1} - y_m = \frac{\lambda D}{d}$$

$$= \frac{(546 \times 10^{-9} \text{ m})(55 \times 10^{-2} \text{ m})}{0.12 \times 10^{-3} \text{ m}}$$

$$= 2.50 \times 10^{-3} \text{ m} \approx 2.5 \text{ mm}. \qquad \text{(Answer)}$$

As long as d and θ in Fig. 35-10a are small, the separation of the interference fringes is independent of m; that is, the fringes are evenly spaced.

PLUS Additional examples, video, and practice available at *WileyPLUS*

Sample Problem 35.03 Double-slit interference pattern with plastic over one slit

A double-slit interference pattern is produced on a screen, as in Fig. 35-10; the light is monochromatic at a wavelength of 600 nm. A strip of transparent plastic with index of refraction $n = 1.50$ is to be placed over one of the slits. Its presence changes the interference between light waves from the two slits, causing the interference pattern to be shifted across the screen from the original pattern. Figure 35-11a shows the original locations of the central bright fringe ($m = 0$) and the first bright fringes ($m = 1$) above and below the central fringe. The purpose of the plastic is to shift the pattern upward so that the lower $m = 1$ bright fringe is shifted to the center of the pattern. Should the plastic be placed over the top slit (as arbitrarily drawn in Fig. 35-11b) or the bottom slit, and what thickness L should it have?

KEY IDEA

The interference at a point on the screen depends on the phase difference of the light rays arriving from the two slits. The light rays are in phase at the slits because they derive from the same wave, but their relative phase can shift on the way to the screen due to (1) a difference in the length of the paths they follow and (2) a difference in the number of their internal wavelengths λ_n in the materials through which they pass. The first condition applies to any off-center point, and the second condition applies when the plastic covers one of the slits.

Path length difference: Figure 35-11a shows rays r_1 and r_2 along which waves from the two slits travel to reach the lower $m = 1$ bright fringe. Those waves start in phase at the slits but arrive at the fringe with a phase difference of exactly 1 wavelength. To remind ourselves of this main characteristic of the fringe, let us call it the 1λ fringe. The one-wavelength phase difference is due to the one-wavelength path length difference between the rays reaching the fringe; that is, there is exactly one more wavelength along ray r_2 than along r_1.

Figure 35-11b shows the 1λ fringe shifted up to the center of the pattern with the plastic strip over the top slit (we still do not know whether the plastic should be there or over the bottom slit). The figure also shows the new orientations of rays r_1 and r_2 to reach that fringe. There still must be one more wavelength along r_2 than along r_1 (because they still produce the 1λ fringe), but now the path length difference between those rays is zero, as we can tell from the geometry of Fig. 35-11b. However, r_2 now passes through the plastic.

Internal wavelength: The wavelength λ_n of light in a material with index of refraction n is smaller than the wavelength in vacuum, as given by Eq. 35-6 ($\lambda_n = \lambda/n$). Here, this means that the wavelength of the light is smaller in the plastic than in the air. Thus, the ray that passes through the plastic will have more wavelengths along it than the ray that passes through only air — so we do get the one extra wavelength we need along ray r_2 by placing the plastic over the top slit, as drawn in Fig. 35-11b.

Thickness: To determine the required thickness L of the plastic, we first note that the waves are initially in phase and travel equal distances L through different materials (plastic and air). Because we know the phase difference and require L, we use Eq. 35-9,

$$N_2 - N_1 = \frac{L}{\lambda}(n_2 - n_1). \tag{35-19}$$

We know that $N_2 - N_1$ is 1 for a phase difference of one wavelength, n_2 is 1.50 for the plastic in front of the top slit, n_1 is 1.00 for the air in front of the bottom slit, and λ is 600×10^{-9} m. Then Eq. 35-19 tells us that, to shift the lower $m = 1$ bright fringe up to the center of the interference pattern, the plastic must have the thickness

$$L = \frac{\lambda(N_2 - N_1)}{n_2 - n_1} = \frac{(600 \times 10^{-9}\ \text{m})(1)}{1.50 - 1.00}$$

$$= 1.2 \times 10^{-6}\ \text{m}. \qquad \text{(Answer)}$$

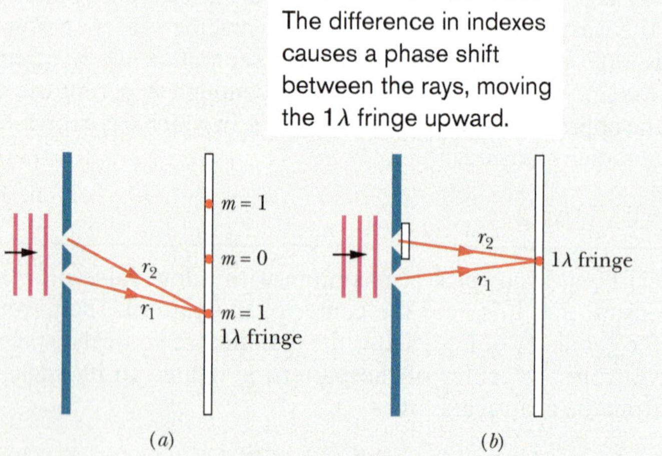

> The difference in indexes causes a phase shift between the rays, moving the 1λ fringe upward.

Figure 35-11 (a) Arrangement for two-slit interference (not to scale). The locations of three bright fringes (or maxima) are indicated. (b) A strip of plastic covers the top slit. We want the 1λ fringe to be at the center of the pattern.

35-3 INTERFERENCE AND DOUBLE-SLIT INTENSITY

Learning Objectives

After reading this module, you should be able to . . .

35.23 Distinguish between coherent and incoherent light.

35.24 For two light waves arriving at a common point, write expressions for their electric field components as functions of time and a phase constant.

35.25 Identify that the phase difference between two waves determines their interference.

35.26 For a point in a double-slit interference pattern, calculate the intensity in terms of the phase difference of

the arriving waves and relate that phase difference to the angle θ locating that point in the pattern.

35.27 Use a phasor diagram to find the resultant wave (amplitude and phase constant) of two or more light waves arriving at a common point and use that result to determine the intensity.

35.28 Apply the relationship between a light wave's angular frequency ω and the angular speed ω of the phasor representing the wave.

Key Ideas

● If two light waves that meet at a point are to interfere perceptibly, the phase difference between them must remain constant with time; that is, the waves must be coherent. When two coherent waves meet, the resulting intensity may be found by using phasors.

● In Young's interference experiment, two waves, each with intensity I_0, yield a resultant wave of intensity I at the viewing screen, with

$$I = 4I_0 \cos^2 \tfrac{1}{2}\phi, \qquad \text{where } \phi = \frac{2\pi d}{\lambda} \sin \theta.$$

Coherence

For the interference pattern to appear on viewing screen C in Fig. 35-8, the light waves reaching any point P on the screen must have a phase difference that does not vary in time. That is the case in Fig. 35-8 because the waves passing through slits S_1 and S_2 are portions of the single light wave that illuminates the slits. Because the phase difference remains constant, the light from slits S_1 and S_2 is said to be completely **coherent.**

Sunlight and Fingernails. Direct sunlight is partially coherent; that is, sunlight waves intercepted at two points have a constant phase difference only if the points are very close. If you look closely at your fingernail in bright sunlight, you can see a faint interference pattern called *speckle* that causes the nail to appear to be covered with specks. You see this effect because light waves scattering from very close points on the nail are sufficiently coherent to interfere with one another at your eye. The slits in a double-slit experiment, however, are not close enough, and in direct sunlight, the light at the slits would be **incoherent.** To get coherent light, we would have to send the sunlight through a single slit as in Fig. 35-8; because that single slit is small, light that passes through it is coherent. In addition, the smallness of the slit causes the coherent light to spread via diffraction to illuminate both slits in the double-slit experiment.

Incoherent Sources. If we replace the double slits with two similar but independent monochromatic light sources, such as two fine incandescent wires, the phase difference between the waves emitted by the sources varies rapidly and randomly. (This occurs because the light is emitted by vast numbers of atoms in the wires, acting randomly and independently for extremely short times—of the order of nanoseconds.) As a result, at any given point on the viewing screen, the interference between the waves from the two sources varies rapidly and randomly between fully constructive and fully destructive. The eye (and most common optical detectors) cannot follow such changes, and no interference pattern can be seen. The fringes disappear, and the screen is seen as being uniformly illuminated.

Coherent Source. A *laser* differs from common light sources in that its atoms emit light in a cooperative manner, thereby making the light coherent. Moreover, the light is almost monochromatic, is emitted in a thin beam with little spreading, and can be focused to a width that almost matches the wavelength of the light.

Intensity in Double-Slit Interference

Equations 35-14 and 35-16 tell us how to locate the maxima and minima of the double-slit interference pattern on screen C of Fig. 35-10 as a function of the angle θ in that figure. Here we wish to derive an expression for the intensity I of the fringes as a function of θ.

The light leaving the slits is in phase. However, let us assume that the light waves from the two slits are not in phase when they arrive at point P. Instead, the electric field components of those waves at point P are not in phase and vary with time as

$$E_1 = E_0 \sin \omega t \tag{35-20}$$

and
$$E_2 = E_0 \sin(\omega t + \phi), \tag{35-21}$$

where ω is the angular frequency of the waves and ϕ is the phase constant of wave E_2. Note that the two waves have the same amplitude E_0 and a phase difference of ϕ. Because that phase difference does not vary, the waves are coherent. We shall show that these two waves will combine at P to produce an intensity I given by

$$I = 4I_0 \cos^2 \tfrac{1}{2}\phi, \tag{35-22}$$

and that

$$\phi = \frac{2\pi d}{\lambda} \sin \theta. \tag{35-23}$$

In Eq. 35-22, I_0 is the intensity of the light that arrives on the screen from one slit when the other slit is temporarily covered. We assume that the slits are so narrow in comparison to the wavelength that this single-slit intensity is essentially uniform over the region of the screen in which we wish to examine the fringes.

Equations 35-22 and 35-23, which together tell us how the intensity I of the fringe pattern varies with the angle θ in Fig. 35-10, necessarily contain information about the location of the maxima and minima. Let us see if we can extract that information to find equations about those locations.

Maxima. Study of Eq. 35-22 shows that intensity maxima will occur when

$$\tfrac{1}{2}\phi = m\pi, \qquad \text{for } m = 0, 1, 2, \ldots. \tag{35-24}$$

If we put this result into Eq. 35-23, we find

$$2m\pi = \frac{2\pi d}{\lambda} \sin \theta, \qquad \text{for } m = 0, 1, 2, \ldots$$

or
$$d \sin \theta = m\lambda, \qquad \text{for } m = 0, 1, 2, \ldots \quad \text{(maxima)}, \tag{35-25}$$

which is exactly Eq. 35-14, the expression that we derived earlier for the locations of the maxima.

Minima. The minima in the fringe pattern occur when

$$\tfrac{1}{2}\phi = (m + \tfrac{1}{2})\pi, \qquad \text{for } m = 0, 1, 2, \ldots. \tag{35-26}$$

If we combine this relation with Eq. 35-23, we are led at once to

$$d \sin \theta = (m + \tfrac{1}{2})\lambda, \qquad \text{for } m = 0, 1, 2, \ldots \quad \text{(minima)}, \tag{35-27}$$

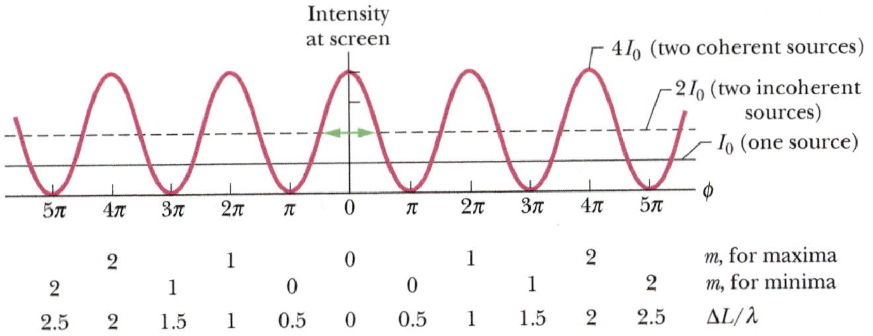

Figure 35-12 A plot of Eq. 35-22, showing the intensity of a double-slit interference pattern as a function of the phase difference between the waves when they arrive from the two slits. I_0 is the (uniform) intensity that would appear on the screen if one slit were covered. The average intensity of the fringe pattern is $2I_0$, and the *maximum* intensity (for coherent light) is $4I_0$.

which is just Eq. 35-16, the expression we derived earlier for the locations of the fringe minima.

Figure 35-12, which is a plot of Eq. 35-22, shows the intensity of double-slit interference patterns as a function of the phase difference ϕ between the waves at the screen. The horizontal solid line is I_0, the (uniform) intensity on the screen when one of the slits is covered up. Note in Eq. 35-22 and the graph that the intensity I varies from zero at the fringe minima to $4I_0$ at the fringe maxima.

If the waves from the two sources (slits) were *incoherent*, so that no enduring phase relation existed between them, there would be no fringe pattern and the intensity would have the uniform value $2I_0$ for all points on the screen; the horizontal dashed line in Fig. 35-12 shows this uniform value.

Interference cannot create or destroy energy but merely redistributes it over the screen. Thus, the *average* intensity on the screen must be the same $2I_0$ regardless of whether the sources are coherent. This follows at once from Eq. 35-22; if we substitute $\frac{1}{2}$, the average value of the cosine-squared function, this equation reduces to $I_{avg} = 2I_0$.

Proof of Eqs. 35-22 and 35-23

We shall combine the electric field components E_1 and E_2, given by Eqs. 35-20 and 35-21, respectively, by the method of phasors as is discussed in Module 16-6. In Fig. 35-13a, the waves with components E_1 and E_2 are represented by phasors of magnitude E_0 that rotate around the origin at angular speed ω. The values of E_1 and E_2 at any time are the projections of the corresponding phasors on the vertical axis. Figure 35-13a shows the phasors and their projections at an arbitrary time t. Consistent with Eqs. 35-20 and 35-21, the phasor for E_1 has a rotation angle ωt and the phasor for E_2 has a rotation angle $\omega t + \phi$ (it is phase-shifted ahead of E_1). As each phasor rotates, its projection on the vertical axis varies with time in the same way that the sinusoidal functions of Eqs. 35-20 and 35-21 vary with time.

To combine the field components E_1 and E_2 at any point P in Fig. 35-10, we add their phasors vectorially, as shown in Fig. 35-13b. The magnitude of the vector sum is the amplitude E of the resultant wave at point P, and that wave has a certain phase constant β. To find the amplitude E in Fig. 35-13b, we first note that the two angles marked β are equal because they are opposite equal-length sides of a triangle. From the theorem (for triangles) that an exterior angle (here ϕ, as shown in Fig. 35-13b) is equal to the sum of the two opposite interior angles (here that sum is $\beta + \beta$), we see that $\beta = \frac{1}{2}\phi$. Thus, we have

$$E = 2(E_0 \cos \beta)$$

$$= 2E_0 \cos \tfrac{1}{2}\phi. \qquad (35\text{-}28)$$

Phasors that represent waves can be added to find the net wave.

(a)

(b)

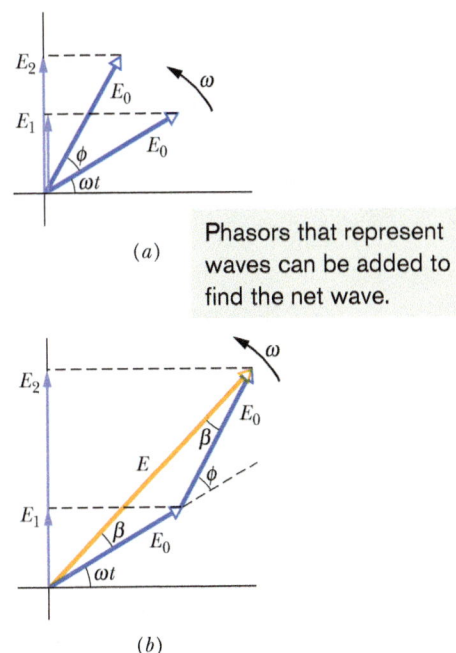

Figure 35-13 (a) Phasors representing, at time t, the electric field components given by Eqs. 35-20 and 35-21. Both phasors have magnitude E_0 and rotate with angular speed ω. Their phase difference is ϕ. (b) Vector addition of the two phasors gives the phasor representing the resultant wave, with amplitude E and phase constant β.

If we square each side of this relation, we obtain

$$E^2 = 4E_0^2 \cos^2 \tfrac{1}{2}\phi. \qquad (35\text{-}29)$$

Intensity. Now, from Eq. 33-24, we know that the intensity of an electromagnetic wave is proportional to the square of its amplitude. Therefore, the waves we are combining in Fig. 35-13b, whose amplitudes are E_0, each has an intensity I_0 that is proportional to E_0^2, and the resultant wave, with amplitude E, has an intensity I that is proportional to E^2. Thus,

$$\frac{I}{I_0} = \frac{E^2}{E_0^2}.$$

Substituting Eq. 35-29 into this equation and rearranging then yield

$$I = 4I_0 \cos^2 \tfrac{1}{2}\phi,$$

which is Eq. 35-22, which we set out to prove.

We still must prove Eq. 35-23, which relates the phase difference ϕ between the waves arriving at any point P on the screen of Fig. 35-10 to the angle θ that serves as a locator of that point.

The phase difference ϕ in Eq. 35-21 is associated with the path length difference S_1b in Fig. 35-10b. If S_1b is $\tfrac{1}{2}\lambda$, then ϕ is π; if S_1b is λ, then ϕ is 2π, and so on. This suggests

$$\left(\begin{array}{c}\text{phase}\\\text{difference}\end{array}\right) = \frac{2\pi}{\lambda}\left(\begin{array}{c}\text{path length}\\\text{difference}\end{array}\right). \qquad (35\text{-}30)$$

The path length difference S_1b in Fig. 35-10b is $d \sin \theta$ (a leg of the right triangle); so Eq. 35-30 for the phase difference between the two waves arriving at point P on the screen becomes

$$\phi = \frac{2\pi d}{\lambda} \sin \theta,$$

which is Eq. 35-23, the other equation that we set out to prove to relate ϕ to the angle θ that locates P.

Combining More Than Two Waves

In a more general case, we might want to find the resultant of more than two sinusoidally varying waves at a point. Whatever the number of waves is, our general procedure is this:

1. Construct a series of phasors representing the waves to be combined. Draw them end to end, maintaining the proper phase relations between adjacent phasors.

2. Construct the vector sum of this array. The length of this vector sum gives the amplitude of the resultant phasor. The angle between the vector sum and the first phasor is the phase of the resultant with respect to this first phasor. The projection of this vector-sum phasor on the vertical axis gives the time variation of the resultant wave.

✓ Checkpoint 4

Each of four pairs of light waves arrives at a certain point on a screen. The waves have the same wavelength. At the arrival point, their amplitudes and phase differences are (a) $2E_0, 6E_0$, and π rad; (b) $3E_0, 5E_0$, and π rad; (c) $9E_0, 7E_0$, and 3π rad; (d) $2E_0, 2E_0$, and 0 rad. Rank the four pairs according to the intensity of the light at the arrival point, greatest first. (*Hint:* Draw phasors.)

Sample Problem 35.04   Combining three light waves by using phasors

Three light waves combine at a certain point where their electric field components are

$$E_1 = E_0 \sin \omega t,$$
$$E_2 = E_0 \sin(\omega t + 60°),$$
$$E_3 = E_0 \sin(\omega t - 30°).$$

Find their resultant component $E(t)$ at that point.

KEY IDEA

The resultant wave is

$$E(t) = E_1(t) + E_2(t) + E_3(t).$$

We can use the method of phasors to find this sum, and we are free to evaluate the phasors at any time t.

Calculations: To simplify the solution, we choose $t = 0$, for which the phasors representing the three waves are shown in Fig. 35-14. We can add these three phasors either directly on a vector-capable calculator or by components. For the component approach, we first write the sum of their horizontal components as

$$\sum E_h = E_0 \cos 0 + E_0 \cos 60° + E_0 \cos(-30°) = 2.37E_0.$$

The sum of their vertical components, which is the value of E at $t = 0$, is

$$\sum E_v = E_0 \sin 0 + E_0 \sin 60° + E_0 \sin(-30°) = 0.366E_0.$$

The resultant wave $E(t)$ thus has an amplitude E_R of

$$E_R = \sqrt{(2.37E_0)^2 + (0.366E_0)^2} = 2.4E_0,$$

and a phase angle β relative to the phasor representing E_1 of

$$\beta = \tan^{-1}\left(\frac{0.366E_0}{2.37E_0}\right) = 8.8°.$$

We can now write, for the resultant wave $E(t)$,

$$E = E_R \sin(\omega t + \beta)$$
$$= 2.4E_0 \sin(\omega t + 8.8°). \qquad \text{(Answer)}$$

Be careful to interpret the angle β correctly in Fig. 35-14: It is the constant angle between E_R and the phasor representing E_1 as the four phasors rotate as a single unit around the origin. The angle between E_R and the horizontal axis in Fig. 35-14 does *not* remain equal to β.

Phasors that represent waves can be added to find the net wave.

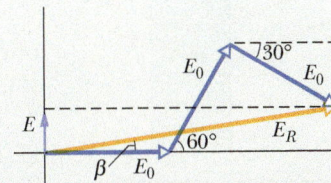

Figure 35-14 Three phasors, representing waves with equal amplitudes E_0 and with phase constants 0°, 60°, and −30°, shown at time $t = 0$. The phasors combine to give a resultant phasor with magnitude E_R, at angle β.

WILEY **PLUS**   Additional examples, video, and practice available at *WileyPLUS*

35-4 INTERFERENCE FROM THIN FILMS

Learning Objectives

After reading this module, you should be able to . . .

35.29 Sketch the setup for thin-film interference, showing the incident ray and reflected rays (perpendicular to the film but drawn slightly slanted for clarity) and identifying the thickness and the three indexes of refraction.

35.30 Identify the condition in which a reflection can result in a phase shift, and give the value of that phase shift.

35.31 Identify the three factors that determine the interference of the reflected waves: reflection shifts, path length difference, and internal wavelength (set by the film's index of refraction).

35.32 For a thin film, use the reflection shifts and the desired result (the *reflected* waves are in phase or out of phase, or

the *transmitted* waves are in phase or out of phase) to determine and then apply the necessary equation relating the thickness L, the wavelength λ (measured in air), and the index of refraction n of the film.

35.33 For a very thin film in air (with thickness much less than the wavelength of visible light), explain why the film is always dark.

35.34 At each end of a thin film in the form of a wedge, determine and then apply the necessary equation relating the thickness L, the wavelength λ (measured in air), and the index of refraction n of the film, and then count the number of bright bands and dark bands across the film.

Key Ideas

● When light is incident on a thin transparent film, the light waves reflected from the front and back surfaces interfere. For near-normal incidence, the wavelength conditions for maximum and minimum intensity of the light reflected from a *film in air* are

$$2L = (m + \tfrac{1}{2})\frac{\lambda}{n_2}, \qquad \text{for } m = 0, 1, 2, \ldots$$

(maxima—bright film in air),

$$2L = m\frac{\lambda}{n_2}, \qquad \text{for } m = 0, 1, 2, \ldots$$

(minima—dark film in air),

where n_2 is the index of refraction of the film, L is its thickness, and λ is the wavelength of the light in air.

● If a film is sandwiched between media other than air, these equations for bright and dark films may be interchanged, depending on the relative indexes of refraction.

● If the light incident at an interface between media with different indexes of refraction is in the medium with the smaller index of refraction, the reflection causes a phase change of π rad, or half a wavelength, in the reflected wave. Otherwise, there is no phase change due to the reflection. Refraction causes no phase shift.

The interference depends on the reflections and the path lengths.

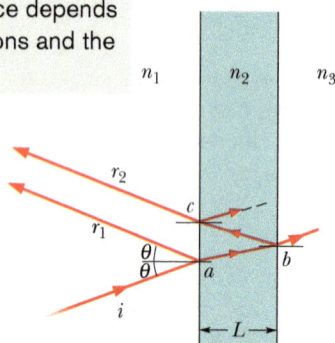

Figure 35-15 Light waves, represented with ray *i*, are incident on a thin film of thickness L and index of refraction n_2. Rays r_1 and r_2 represent light waves that have been reflected by the front and back surfaces of the film, respectively. (All three rays are actually nearly perpendicular to the film.) The interference of the waves of r_1 and r_2 with each other depends on their phase difference. The index of refraction n_1 of the medium at the left can differ from the index of refraction n_3 of the medium at the right, but for now we assume that both media are air, with $n_1 = n_3 = 1.0$, which is less than n_2.

Interference from Thin Films

The colors on a sunlit soap bubble or an oil slick are caused by the interference of light waves reflected from the front and back surfaces of a thin transparent film. The thickness of the soap or oil film is typically of the order of magnitude of the wavelength of the (visible) light involved. (Greater thicknesses spoil the coherence of the light needed to produce the colors due to interference.)

Figure 35-15 shows a thin transparent film of uniform thickness L and index of refraction n_2, illuminated by bright light of wavelength λ from a distant point source. For now, we assume that air lies on both sides of the film and thus that $n_1 = n_3$ in Fig. 35-15. For simplicity, we also assume that the light rays are almost perpendicular to the film ($\theta \approx 0$). We are interested in whether the film is bright or dark to an observer viewing it almost perpendicularly. (Since the film is brightly illuminated, how could it possibly be dark? You will see.)

The incident light, represented by ray *i*, intercepts the front (left) surface of the film at point *a* and undergoes both reflection and refraction there. The reflected ray r_1 is intercepted by the observer's eye. The refracted light crosses the film to point *b* on the back surface, where it undergoes both reflection and refraction. The light reflected at *b* crosses back through the film to point *c*, where it undergoes both reflection and refraction. The light refracted at *c*, represented by ray r_2, is intercepted by the observer's eye.

If the light waves of rays r_1 and r_2 are exactly in phase at the eye, they produce an interference maximum and region *ac* on the film is bright to the observer. If they are exactly out of phase, they produce an interference minimum and region *ac* is dark to the observer, *even though it is illuminated.* If there is some intermediate phase difference, there are intermediate interference and brightness.

The Key. Thus, the key to what the observer sees is the phase difference between the waves of rays r_1 and r_2. Both rays are derived from the same ray *i*, but the path involved in producing r_2 involves light traveling twice across the film (*a* to *b*, and then *b* to *c*), whereas the path involved in producing r_1 involves no travel through the film. Because θ is about zero, we approximate the path length difference between the waves of r_1 and r_2 as $2L$. However, to find the phase difference between the waves, we cannot just find the number of wavelengths λ that is equivalent to a path length difference of $2L$. This simple approach is impossible for two reasons: (1) the path length difference occurs in a medium other than air, and (2) reflections are involved, which can change the phase.

⭐ The phase difference between two waves can change if one or both are reflected.

Let's next discuss changes in phase that are caused by reflections.

Reflection Phase Shifts

Refraction at an interface never causes a phase change—but reflection can, depending on the indexes of refraction on the two sides of the interface. Figure 35-16 shows what happens when reflection causes a phase change, using as an example pulses on a denser string (along which pulse travel is relatively slow) and a lighter string (along which pulse travel is relatively fast).

When a pulse traveling relatively slowly along the denser string in Fig. 35-16a reaches the interface with the lighter string, the pulse is partially transmitted and partially reflected, with no change in orientation. For light, this situation corresponds to the incident wave traveling in the medium of greater index of refraction n (recall that greater n means slower speed). In that case, the wave that is reflected at the interface does not undergo a change in phase; that is, its *reflection phase shift* is zero.

When a pulse traveling more quickly along the lighter string in Fig. 35-16b reaches the interface with the denser string, the pulse is again partially transmitted and partially reflected. The transmitted pulse again has the same orientation as the incident pulse, but now the reflected pulse is inverted. For a sinusoidal wave, such an inversion involves a phase change of π rad, or half a wavelength. For light, this situation corresponds to the incident wave traveling in the medium of lesser index of refraction (with greater speed). In that case, the wave that is reflected at the interface undergoes a phase shift of π rad, or half a wavelength.

We can summarize these results for light in terms of the index of refraction of the medium off which (or from which) the light reflects:

Reflection	Reflection phase shift
Off lower index	0
Off higher index	0.5 wavelength

This might be remembered as "higher means half."

Equations for Thin-Film Interference

In this chapter we have now seen three ways in which the phase difference between two waves can change:

1. by reflection
2. by the waves traveling along paths of different lengths
3. by the waves traveling through media of different indexes of refraction

When light reflects from a thin film, producing the waves of rays r_1 and r_2 shown in Fig. 35-15, all three ways are involved. Let us consider them one by one.

Reflection Shift. We first reexamine the two reflections in Fig. 35-15. At point a on the front interface, the incident wave (in air) reflects from the medium having the higher of the two indexes of refraction; so the wave of reflected ray r_1 has its phase shifted by 0.5 wavelength. At point b on the back interface, the incident wave reflects from the medium (air) having the lower of the two indexes of refraction; so the wave reflected there is not shifted in phase by the reflection, and thus neither is the portion of it that exits the film as ray r_2. We can organize this information with the first line in Table 35-1, which refers to the simplified drawing in Fig. 35-17 for a thin film in air. So far, as a result of the reflection phase shifts, the waves of r_1 and r_2 have a phase difference of 0.5 wavelength and thus are exactly out of phase.

Path Length Difference. Now we must consider the path length difference $2L$ that occurs because the wave of ray r_2 crosses the film twice. (This difference

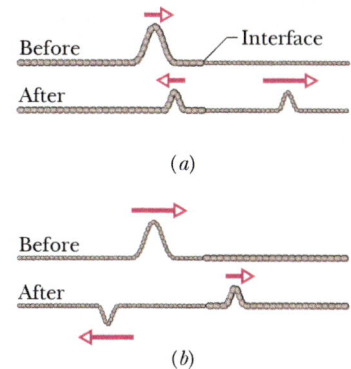

Figure 35-16 Phase changes when a pulse is reflected at the interface between two stretched strings of different linear densities. The wave speed is greater in the lighter string. (a) The incident pulse is in the denser string. (b) The incident pulse is in the lighter string. Only here is there a phase change, and only in the reflected wave.

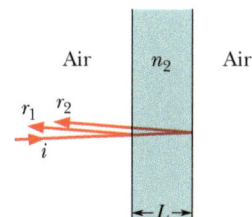

Figure 35-17 Reflections from a thin film in air.

Table 35-1 An Organizing Table for
Thin-Film Interference in Air (Fig. 35-17)[a]

	r_1	r_2
Reflection phase shifts	0.5 wavelength	0
Path length difference	$2L$	
Index in which path length difference occurs	n_2	
In phase[a]:	$2L = \dfrac{\text{odd number}}{2} \times \dfrac{\lambda}{n_2}$	
Out of phase[a]:	$2L = \text{integer} \times \dfrac{\lambda}{n_2}$	

[a]Valid for $n_2 > n_1$ and $n_2 > n_3$.

$2L$ is shown on the second line in Table 35-1.) If the waves of r_1 and r_2 are to be exactly in phase so that they produce fully constructive interference, the path length $2L$ must cause an additional phase difference of 0.5, 1.5, 2.5, ... wavelengths. Only then will the net phase difference be an integer number of wavelengths. Thus, for a bright film, we must have

$$2L = \frac{\text{odd number}}{2} \times \text{wavelength} \quad \text{(in-phase waves).} \quad (35\text{-}31)$$

The wavelength we need here is the wavelength λ_{n2} of the light in the medium containing path length $2L$—that is, in the medium with index of refraction n_2. Thus, we can rewrite Eq. 35-31 as

$$2L = \frac{\text{odd number}}{2} \times \lambda_{n2} \quad \text{(in-phase waves).} \quad (35\text{-}32)$$

If, instead, the waves are to be exactly out of phase so that there is fully destructive interference, the path length $2L$ must cause either no additional phase difference or a phase difference of 1, 2, 3, ... wavelengths. Only then will the net phase difference be an odd number of half-wavelengths. For a dark film, we must have

$$2L = \text{integer} \times \text{wavelength} \quad \text{(out-of-phase waves).} \quad (35\text{-}33)$$

where, again, the wavelength is the wavelength λ_{n2} in the medium containing $2L$. Thus, this time we have

$$2L = \text{integer} \times \lambda_{n2} \quad \text{(out-of-phase waves).} \quad (35\text{-}34)$$

Now we can use Eq. 35-6 ($\lambda_n = \lambda/n$) to write the wavelength of the wave of ray r_2 inside the film as

$$\lambda_{n2} = \frac{\lambda}{n_2}, \quad (35\text{-}35)$$

where λ is the wavelength of the incident light in vacuum (and approximately also in air). Substituting Eq. 35-35 into Eq. 35-32 and replacing "odd number/2" with $(m + \frac{1}{2})$ give us

$$2L = (m + \tfrac{1}{2})\frac{\lambda}{n_2}, \quad \text{for } m = 0, 1, 2, \ldots \quad \text{(maxima—bright film in air).} \quad (35\text{-}36)$$

Similarly, with m replacing "integer," Eq. 35-34 yields

$$2L = m\frac{\lambda}{n_2}, \quad \text{for } m = 0, 1, 2, \ldots \quad \text{(minima—dark film in air).} \quad (35\text{-}37)$$

For a given film thickness L, Eqs. 35-36 and 35-37 tell us the wavelengths of light for which the film appears bright and dark, respectively, one wavelength for each value of m. Intermediate wavelengths give intermediate brightnesses. For a given wavelength λ, Eqs. 35-36 and 35-37 tell us the thicknesses of the films that appear bright and dark in that light, respectively, one thickness for each value of m. Intermediate thicknesses give intermediate brightnesses.

Heads Up. (1) For a thin film surrounded by air, Eq. 35-36 corresponds to bright reflections and Eq. 35-37 corresponds to no reflections. For transmissions, the roles of the equations are reversed (after all, if the light is brightly reflected, then it is not transmitted, and vice versa). (2) If we have a different set of values of the indexes of refraction, the roles of the equations may be reversed. For any given set of indexes, you must go through the thought process behind Table 35-1 and, in particular, determine the reflection shifts to see which equation applies to bright reflections and which applies to no reflections. (3) The index of refraction in the equations is that of the thin film, where the path length difference occurs.

Film Thickness Much Less Than λ

A special situation arises when a film is so thin that L is much less than λ, say, $L < 0.1\lambda$. Then the path length difference $2L$ can be neglected, and the phase difference between r_1 and r_2 is due *only* to reflection phase shifts. If the film of Fig. 35-17, where the reflections cause a phase difference of 0.5 wavelength, has thickness $L < 0.1\lambda$, then r_1 and r_2 are exactly out of phase, and thus the film is dark, regardless of the wavelength and intensity of the light. This special situation corresponds to $m = 0$ in Eq. 35-37. We shall count *any* thickness $L < 0.1\lambda$ as being the least thickness specified by Eq. 35-37 to make the film of Fig. 35-17 dark. (Every such thickness will correspond to $m = 0$.) The next greater thickness that will make the film dark is that corresponding to $m = 1$.

In Fig. 35-18, bright white light illuminates a vertical soap film whose thickness increases from top to bottom. However, the top portion is so thin that it is dark. In the (somewhat thicker) middle we see fringes, or bands, whose color depends primarily on the wavelength at which reflected light undergoes fully constructive interference for a particular thickness. Toward the (thickest) bottom the fringes become progressively narrower and the colors begin to overlap and fade.

Richard Megna/Fundamental Photographs

Figure 35-18 The reflection of light from a soapy water film spanning a vertical loop. The top portion is so thin (due to gravitational slumping) that the light reflected there undergoes destructive interference, making that portion dark. Colored interference fringes, or bands, decorate the rest of the film but are marred by circulation of liquid within the film as the liquid is gradually pulled downward by gravitation.

✓ **Checkpoint 5**

The figure shows four situations in which light reflects perpendicularly from a thin film of thickness L,

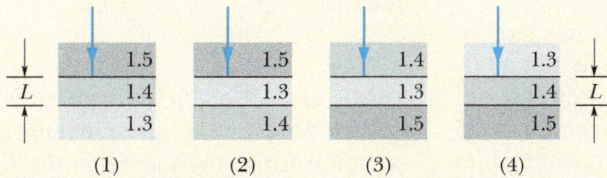

with indexes of refraction as given. (a) For which situations does reflection at the film interfaces cause a zero phase difference for the two reflected rays? (b) For which situations will the film be dark if the path length difference $2L$ causes a phase difference of 0.5 wavelength?

Sample Problem 35.05   Thin-film interference of a water film in air

White light, with a uniform intensity across the visible wavelength range of 400 to 690 nm, is perpendicularly incident on a water film, of index of refraction $n_2 = 1.33$ and thickness $L = 320$ nm, that is suspended in air. At what wavelength λ is the light reflected by the film brightest to an observer?

KEY IDEA

The reflected light from the film is brightest at the wavelengths λ for which the reflected rays are in phase with one another. The equation relating these wavelengths λ to the given film thickness L and film index of refraction n_2 is either Eq. 35-36 or Eq. 35-37, depending on the reflection phase shifts for this particular film.

Calculations: To determine which equation is needed, we should fill out an organizing table like Table 35-1. However, because there is air on both sides of the water film, the situation here is exactly like that in Fig. 35-17, and thus the table would be exactly like Table 35-1. Then from Table 35-1, we

see that the reflected rays are in phase (and thus the film is brightest) when

$$2L = \frac{\text{odd number}}{2} \times \frac{\lambda}{n_2},$$

which leads to Eq. 35-36:

$$2L = \left(m + \tfrac{1}{2}\right)\frac{\lambda}{n_2}.$$

Solving for λ and substituting for L and n_2, we find

$$\lambda = \frac{2n_2 L}{m + \tfrac{1}{2}} = \frac{(2)(1.33)(320 \text{ nm})}{m + \tfrac{1}{2}} = \frac{851 \text{ nm}}{m + \tfrac{1}{2}}.$$

For $m = 0$, this gives us $\lambda = 1700$ nm, which is in the infrared region. For $m = 1$, we find $\lambda = 567$ nm, which is yellow-green light, near the middle of the visible spectrum. For $m = 2$, $\lambda = 340$ nm, which is in the ultraviolet region. Thus, the wavelength at which the light seen by the observer is brightest is

$$\lambda = 567 \text{ nm.} \qquad \text{(Answer)}$$

WILEY PLUS   Additional examples, video, and practice available at *WileyPLUS*

Sample Problem 35.06 Thin-film interference of a coating on a glass lens

In Fig. 35-19, a glass lens is coated on one side with a thin film of magnesium fluoride (MgF_2) to reduce reflection from the lens surface. The index of refraction of MgF_2 is 1.38; that of the glass is 1.50. What is the least coating thickness that eliminates (via interference) the reflections at the middle of the visible spectrum ($\lambda = 550$ nm)? Assume that the light is approximately perpendicular to the lens surface.

KEY IDEA

Reflection is eliminated if the film thickness L is such that light waves reflected from the two film interfaces are exactly out of phase. The equation relating L to the given wavelength λ and the index of refraction n_2 of the thin film is either Eq. 35-36 or Eq. 35-37, depending on the reflection phase shifts at the interfaces.

Calculations: To determine which equation is needed, we fill out an organizing table like Table 35-1. At the first interface, the incident light is in air, which has a lesser index of refraction than the MgF_2 (the thin film). Thus, we fill in 0.5 wavelength under r_1 in our organizing table (meaning that the waves of ray r_1 are shifted by 0.5λ at the first interface). At the second interface, the incident light is in the MgF_2, which has a lesser index of refraction than the glass on the other side of the interface. Thus, we fill in 0.5 wavelength under r_2 in our table.

   Because both reflections cause the same phase shift, they tend to put the waves of r_1 and r_2 in phase. Since we want those waves to be *out of phase*, their path length difference $2L$ must be an odd number of half-wavelengths:

$$2L = \frac{\text{odd number}}{2} \times \frac{\lambda}{n_2}.$$

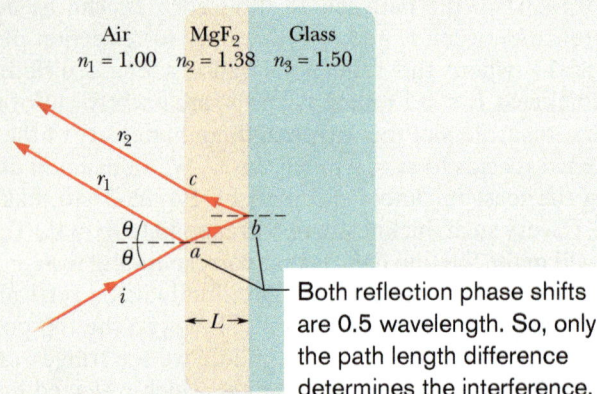

Air MgF_2 Glass
$n_1 = 1.00$ $n_2 = 1.38$ $n_3 = 1.50$

Both reflection phase shifts are 0.5 wavelength. So, only the path length difference determines the interference.

Figure 35-19 Unwanted reflections from glass can be suppressed (at a chosen wavelength) by coating the glass with a thin transparent film of magnesium fluoride of the properly chosen thickness.

This leads to Eq. 35-36 (for a bright film sandwiched in air but for a dark film in the arrangement here). Solving that equation for L then gives us the film thicknesses that will eliminate reflection from the lens and coating:

$$L = \left(m + \tfrac{1}{2}\right)\frac{\lambda}{2n_2}, \qquad \text{for } m = 0, 1, 2, \ldots . \quad (35\text{-}38)$$

We want the least thickness for the coating—that is, the smallest value of L. Thus, we choose $m = 0$, the smallest possible value of m. Substituting it and the given data in Eq. 35-38, we obtain

$$L = \frac{\lambda}{4n_2} = \frac{550 \text{ nm}}{(4)(1.38)} = 99.6 \text{ nm.} \qquad \text{(Answer)}$$

Sample Problem 35.07 Thin-film interference of a transparent wedge

Figure 35-20a shows a transparent plastic block with a thin wedge of air at the right. (The wedge thickness is exaggerated in the figure.) A broad beam of red light, with wavelength $\lambda = 632.8$ nm, is directed downward through the top of the block (at an incidence angle of 0°). Some of the light that passes into the plastic is reflected back up from the top and bottom surfaces of the wedge, which acts as a thin film (of air) with a thickness that varies uniformly and gradually from L_L at the left-hand end to L_R at the right-hand end. (The plastic layers above and below the wedge of air are too thick to act as thin films.) An observer looking down on the block sees an interference pattern consisting of six dark fringes and five bright red fringes along the wedge. What is the change in thickness ΔL ($= L_R - L_L$) along the wedge?

KEY IDEAS

(1) The brightness at any point along the left–right length of the air wedge is due to the interference of the waves reflected at the top and bottom interfaces of the wedge. (2) The variation of brightness in the pattern of bright and dark fringes is due to the variation in the thickness of the wedge. In some regions, the thickness puts the reflected waves in phase and thus produces a bright reflection (a bright red fringe). In other regions, the thickness puts the reflected waves out of phase and thus produces no reflection (a dark fringe).

Organizing the reflections: Because the observer sees more dark fringes than bright fringes, we can assume that a dark fringe is produced at both the left and right ends of

Overhead incident light

Side view
(a)

L_L L_R

(c)

L n_1
n_2
n_3

r_1
r_2
i

$m_L + 2$ $m_L + 4$

Overhead view
(b)

m_L $m_L + 1$ $m_L + 3$ $m_L + 5$

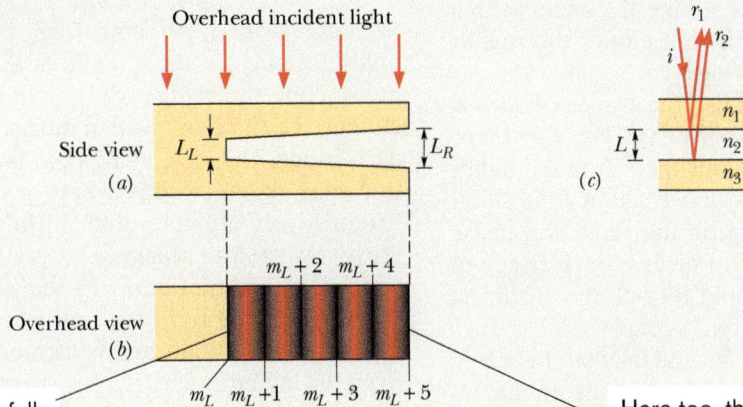

This dark fringe is due to fully destructive interference. So, the reflected rays must be *out* of phase.

Here too, the dark fringe means that the reflected waves are *out* of phase.

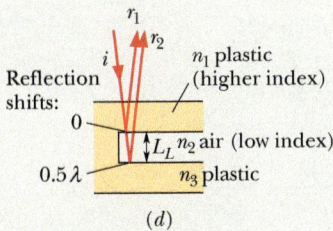

r_1
r_2
i

Reflection shifts:
0

n_1 plastic (higher index)

L_L n_2 air (low index)
n_3 plastic

0.5λ

(d)

The path length difference (down and back up) is 2L.

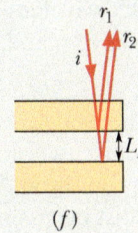

r_1
r_2
i

L_R

(f)

The path length difference is 2L here too but the L is larger.

Total reflection shift = 0.5 wavelength. So, the reflections put the waves out of phase.

(e) **Organizing Table**

	r_1	r_2
Reflection phase shifts	0	0.5 wavelength
Path length difference	2L	

Here again, the waves are already out of phase by the reflection shifts. So, the path length difference must be 2L = (integer)λ/n_2, but with the larger L.

We want the reflected waves to be out of phase. They already are out of phase because of the reflection shifts. So, we don't want the path length difference 2L to change that. Thus, $2L = $ (integer)λ/n_2.

Figure 35-20 (a) Red light is incident on a thin, air-filled wedge in the side of a transparent plastic block. The thickness of the wedge is L_L at the left end and L_R at the right end. (b) The view from above the block: an interference pattern of six dark fringes and five bright red fringes lies over the region of the wedge. (c) A representation of the incident ray i, reflected rays r_1 and r_2, and thickness L of the wedge anywhere along the length of the wedge. The reflection rays at the (d) left and (f) right ends of the wedge and (e) their organizing table.

the wedge. Thus, the interference pattern is that shown in Fig. 35-20b.

We can represent the reflection of light at the top and bottom interfaces of the wedge, at any point along its length, with Fig. 35-20c, in which L is the wedge thickness at that point. Let us apply this figure to the left end of the wedge, where the reflections give a dark fringe.

We know that, for a dark fringe, the waves of rays r_1 and r_2 in Fig. 35-20d must be out of phase. We also know that the equation relating the film thickness L to the light's wavelength λ and the film's index of refraction n_2 is either Eq. 35-36 or Eq. 35-37, depending on the reflection phase shifts. To determine which equation gives a dark fringe at the left end of the wedge, we should fill out an organizing table like Table 35-1, as shown in Fig. 35-20e.

At the top interface of the wedge, the incident light is in the plastic, which has a greater n than the air beneath that interface. So, we fill in 0 under r_1 in our organizing table. At the bottom interface of the wedge, the incident light is in air, which has a lesser n than the plastic beneath that interface. So we fill in 0.5 wavelength under r_2. So, the phase difference due to the reflection shifts is 0.5 wavelength. Thus the reflections alone tend to put the waves of r_1 and r_2 out of phase.

Reflections at left end (Fig. 35-20d): Because we see a dark fringe at the left end of the wedge, which the reflection phase shifts alone would produce, we don't want the path length difference to alter that condition. So, the path length difference $2L$ at the left end must be given by

$$2L = \text{integer} \times \frac{\lambda}{n_2},$$

which leads to Eq. 35-37:

$$2L = m\frac{\lambda}{n_2}, \qquad \text{for } m = 0, 1, 2, \ldots. \qquad (35\text{-}39)$$

Reflections at right end (Fig. 35-20f): Equation 35-39 holds not only for the left end of the wedge but also for any point along the wedge where a dark fringe is observed, including the right end, with a different integer value of m for each fringe. The least value of m is associated with the least thickness of the wedge where a dark fringe is observed. Progressively greater values of m are associated with progressively greater thicknesses of the wedge where a dark fringe is observed. Let m_L be the value at the left end. Then the value at the right end must be $m_L + 5$ because, from Fig. 35-20b, the right end is located at the fifth dark fringe from the left end.

Thickness difference: To find ΔL, we first solve Eq. 35-39 twice—once for the thickness L_L at the left end and once for the thickness L_R at the right end:

$$L_L = (m_L)\frac{\lambda}{2n_2}, \qquad L_R = (m_L + 5)\frac{\lambda}{2n_2}. \qquad (35\text{-}40)$$

We can now subtract L_L from L_R and substitute $n_2 = 1.00$ for the air within the wedge and $\lambda = 632.8 \times 10^{-9}$ m:

$$\Delta L = L_R - L_L = \frac{(m_L + 5)\lambda}{2n_2} - \frac{m_L\lambda}{2n_2} = \frac{5}{2}\frac{\lambda}{n_2}$$

$$= 1.58 \times 10^{-6} \text{ m.} \qquad \text{(Answer)}$$

WILEY PLUS Additional examples, video, and practice available at *WileyPLUS*

35-5 MICHELSON'S INTERFEROMETER

Learning Objectives

After reading this module, you should be able to . . .

35.35 With a sketch, explain how an interferometer works.

35.36 When a transparent material is inserted into one of the beams in an interferometer, apply the relationship between the phase change of the light (in terms of

wavelength) and the material's thickness and index of refraction.

35.37 For an interferometer, apply the relationship between the distance a mirror is moved and the resulting fringe shift in the interference pattern.

Key Ideas

● In Michelson's interferometer, a light wave is split into two beams that then recombine after traveling along different paths.

● The interference pattern they produce depends on the difference in the lengths of those paths and the indexes of refraction along the paths.

● If a transparent material of index n and thickness L is in one path, the phase difference (in terms of wavelength) in the recombining beams is equal to

$$\text{phase difference} = \frac{2L}{\lambda}(n - 1),$$

where λ is the wavelength of the light.

Michelson's Interferometer

An **interferometer** is a device that can be used to measure lengths or changes in length with great accuracy by means of interference fringes. We describe the form originally devised and built by A. A. Michelson in 1881.

Consider light that leaves point P on extended source S in Fig. 35-21 and encounters *beam splitter M*. A beam splitter is a mirror that transmits half the incident light and reflects the other half. In the figure we have assumed, for convenience, that this mirror possesses negligible thickness. At M the light thus divides into two waves. One proceeds by transmission toward mirror M_1 at the end of one arm of the instrument; the other proceeds by reflection toward mirror M_2 at the end of the other arm. The waves are entirely reflected at these mirrors and are sent back along their directions of incidence, each wave eventually entering telescope T. What the observer sees is a pattern of curved or approximately straight interference fringes; in the latter case the fringes resemble the stripes on a zebra.

Mirror Shift. The path length difference for the two waves when they recombine at the telescope is $2d_2 - 2d_1$, and anything that changes this path length difference will cause a change in the phase difference between these two waves at the eye. As an example, if mirror M_2 is moved by a distance $\frac{1}{2}\lambda$, the path length difference is changed by λ and the fringe pattern is shifted by one fringe (as if each dark stripe on a zebra had moved to where the adjacent dark stripe had been). Similarly, moving mirror M_2 by $\frac{1}{4}\lambda$ causes a shift by half a fringe (each dark zebra stripe shifts to where the adjacent white stripe had been).

Insertion. A shift in the fringe pattern can also be caused by the insertion of a thin transparent material into the optical path of one of the mirrors—say, M_1. If the material has thickness L and index of refraction n, then the number of wavelengths along the light's to-and-fro path through the material is, from Eq. 35-7,

$$N_m = \frac{2L}{\lambda_n} = \frac{2Ln}{\lambda}. \tag{35-41}$$

The number of wavelengths in the same thickness $2L$ of air before the insertion of the material is

$$N_a = \frac{2L}{\lambda}. \tag{35-42}$$

When the material is inserted, the light returned from mirror M_1 undergoes a phase change (in terms of wavelengths) of

$$N_m - N_a = \frac{2Ln}{\lambda} - \frac{2L}{\lambda} = \frac{2L}{\lambda}(n-1). \tag{35-43}$$

For each phase change of one wavelength, the fringe pattern is shifted by one fringe. Thus, by counting the number of fringes through which the material causes the pattern to shift, and substituting that number for $N_m - N_a$ in Eq. 35-43, you can determine the thickness L of the material in terms of λ.

Standard of Length. By such techniques the lengths of objects can be expressed in terms of the wavelengths of light. In Michelson's day, the standard of length—the meter—was the distance between two fine scratches on a certain metal bar preserved at Sèvres, near Paris. Michelson showed, using his interferometer, that the standard meter was equivalent to 1 553 163.5 wavelengths of a certain monochromatic red light emitted from a light source containing cadmium. For this careful measurement, Michelson received the 1907 Nobel Prize in physics. His work laid the foundation for the eventual abandonment (in 1961) of the meter bar as a standard of length and for the redefinition of the meter in terms of the wavelength of light. By 1983, even this wavelength standard was not precise enough to meet the growing technical needs, and it was replaced with a new standard based on a defined value for the speed of light.

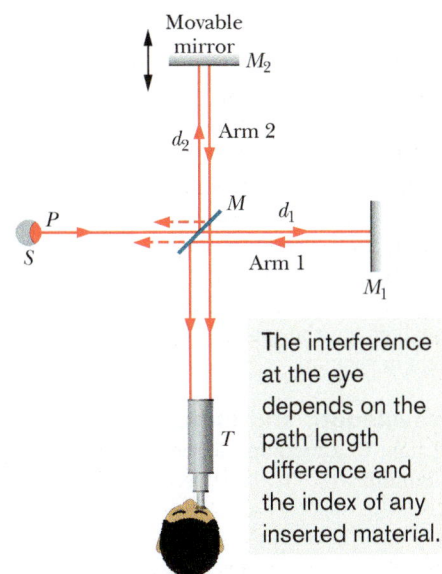

Figure 35-21 Michelson's interferometer, showing the path of light originating at point P of an extended source S. Mirror M splits the light into two beams, which reflect from mirrors M_1 and M_2 back to M and then to telescope T. In the telescope an observer sees a pattern of interference fringes.

Review & Summary

Huygens' Principle The three-dimensional transmission of waves, including light, may often be predicted by *Huygens' principle,* which states that all points on a wavefront serve as point sources of spherical secondary wavelets. After a time t, the new position of the wavefront will be that of a surface tangent to these secondary wavelets.

The law of refraction can be derived from Huygens' principle by assuming that the index of refraction of any medium is $n = c/v$, in which v is the speed of light in the medium and c is the speed of light in vacuum.

Wavelength and Index of Refraction The wavelength λ_n of light in a medium depends on the index of refraction n of the medium:

$$\lambda_n = \frac{\lambda}{n}, \qquad (35\text{-}6)$$

in which λ is the wavelength in vacuum. Because of this dependency, the phase difference between two waves can change if they pass through different materials with different indexes of refraction.

Young's Experiment In **Young's interference experiment,** light passing through a single slit falls on two slits in a screen. The light leaving these slits flares out (by diffraction), and interference occurs in the region beyond the screen. A fringe pattern, due to the interference, forms on a viewing screen.

The light intensity at any point on the viewing screen depends in part on the difference in the path lengths from the slits to that point. If this difference is an integer number of wavelengths, the waves interfere constructively and an intensity maximum results. If it is an odd number of half-wavelengths, there is destructive interference and an intensity minimum occurs. The conditions for maximum and minimum intensity are

$$d \sin \theta = m\lambda, \qquad \text{for } m = 0, 1, 2, \dots$$
$$\text{(maxima—bright fringes)}, \qquad (35\text{-}14)$$

$$d \sin \theta = (m + \tfrac{1}{2})\lambda, \qquad \text{for } m = 0, 1, 2, \dots$$
$$\text{(minima—dark fringes)}, \qquad (35\text{-}16)$$

where θ is the angle the light path makes with a central axis and d is the slit separation.

Coherence If two light waves that meet at a point are to interfere perceptibly, the phase difference between them must remain constant with time; that is, the waves must be **coherent.** When two coherent waves meet, the resulting intensity may be found by using phasors.

Intensity in Two-Slit Interference In Young's interference experiment, two waves, each with intensity I_0, yield a resultant wave of intensity I at the viewing screen, with

$$I = 4I_0 \cos^2 \tfrac{1}{2}\phi, \qquad \text{where } \phi = \frac{2\pi d}{\lambda} \sin \theta. \qquad (35\text{-}22, 35\text{-}23)$$

Equations 35-14 and 35-16, which identify the positions of the fringe maxima and minima, are contained within this relation.

Thin-Film Interference When light is incident on a thin transparent film, the light waves reflected from the front and back surfaces interfere. For near-normal incidence, the wavelength conditions for maximum and minimum intensity of the light reflected from a *film in air* are

$$2L = (m + \tfrac{1}{2}) \frac{\lambda}{n_2}, \qquad \text{for } m = 0, 1, 2, \dots$$
$$\text{(maxima—bright film in air)}, \qquad (35\text{-}36)$$

$$2L = m \frac{\lambda}{n_2}, \qquad \text{for } m = 0, 1, 2, \dots$$
$$\text{(minima—dark film in air)}, \qquad (35\text{-}37)$$

where n_2 is the index of refraction of the film, L is its thickness, and λ is the wavelength of the light in air.

If the light incident at an interface between media with different indexes of refraction is in the medium with the smaller index of refraction, the reflection causes a phase change of π rad, or half a wavelength, in the reflected wave. Otherwise, there is no phase change due to the reflection. Refraction causes no phase shift.

The Michelson Interferometer In *Michelson's interferometer* a light wave is split into two beams that, after traversing paths of different lengths, are recombined so they interfere and form a fringe pattern. Varying the path length of one of the beams allows distances to be accurately expressed in terms of wavelengths of light, by counting the number of fringes through which the pattern shifts because of the change.

Questions

1 Does the spacing between fringes in a two-slit interference pattern increase, decrease, or stay the same if (a) the slit separation is increased, (b) the color of the light is switched from red to blue, and (c) the whole apparatus is submerged in cooking sherry? (d) If the slits are illuminated with white light, then at any side maximum, does the blue component or the red component peak closer to the central maximum?

2 (a) If you move from one bright fringe in a two-slit interference pattern to the next one farther out, (b) does the path length difference ΔL increase or decrease and (c) by how much does it change, in wavelengths λ?

3 Figure 35-22 shows two light rays that are initially exactly in phase and that reflect from several glass surfaces. Neglect the slight slant in the path of the light in the second arrangement. (a) What is the path length difference of the rays? In wavelengths λ, (b) what should that path length difference equal if the rays are to be exactly out of phase when they emerge, and (c) what is the smallest value of d that will allow that final phase difference?

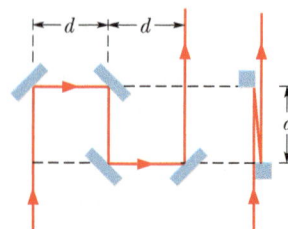

Figure 35-22 Question 3.

4 In Fig. 35-23, three pulses of light—a, b, and c—of the same wavelength are sent through layers

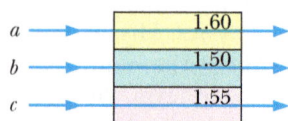

Figure 35-23 Question 4.

of plastic having the given indexes of refraction and along the paths indicated. Rank the pulses according to their travel time through the plastic layers, greatest first.

5 Is there an interference maximum, a minimum, an intermediate state closer to a maximum, or an intermediate state closer to a minimum at point P in Fig. 35-10 if the path length difference of the two rays is (a) 2.2λ, (b) 3.5λ, (c) 1.8λ, and (d) 1.0λ? For each situation, give the value of m associated with the maximum or minimum involved.

6 Figure 35-24a gives intensity I versus position x on the viewing screen for the central portion of a two-slit interference pattern. The other parts of the figure give phasor diagrams for the electric field components of the waves arriving at the screen from the two slits (as in Fig. 35-13a). Which numbered points on the screen best correspond to which phasor diagram?

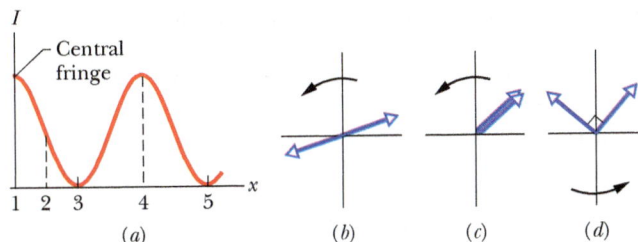

Figure 35-24 Question 6.

7 Figure 35-25 shows two sources S_1 and S_2 that emit radio waves of wavelength λ in all directions. The sources are exactly in phase and are separated by a distance equal to 1.5λ. The vertical broken line is the perpendicular bisector of the distance between the sources. (a) If we start at the indicated start point and travel along path 1, does the interference produce a maximum all along the path, a minimum all along the path, or alternating maxima and minima? Repeat for (b) path 2 (along an axis through the sources) and (c) path 3 (along a perpendicular to that axis).

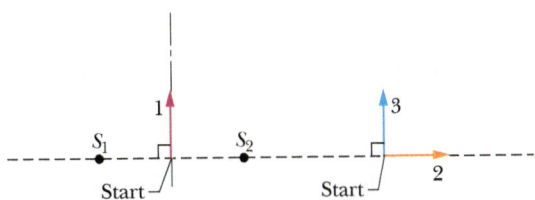

Figure 35-25 Question 7.

8 Figure 35-26 shows two rays of light, of wavelength 600 nm, that reflect from glass surfaces separated by 150 nm. The rays are initially in phase. (a) What is the path length difference of the rays? (b) When they have cleared the reflection region, are the rays exactly in phase, exactly out of phase, or in some intermediate state?

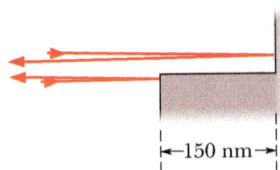

Figure 35-26 Question 8.

9 Light travels along the length of a 1500-nm-long nanostructure. When a peak of the wave is at one end of the nanostructure, is there a peak or a valley at the other end if the wavelength is (a) 500 nm and (b) 1000 nm?

10 Figure 35-27a shows the cross section of a vertical thin film whose width increases downward because gravitation causes slumping. Figure 35-27b is a face-on view of the film, showing four bright (red) interference fringes that result when the film is illuminated with a perpendicular beam of red light. Points in the cross section corresponding to the bright fringes are labeled. In terms of the wavelength of the light inside the film, what is the difference in film thickness between (a) points a and b and (b) points b and d?

Figure 35-27 Question 10.

11 Figure 35-28 shows four situations in which light reflects perpendicularly from a thin film of thickness L sandwiched between much thicker materials. The indexes of refraction are given. In which situations does Eq. 35-36 correspond to the reflections yielding maxima (that is, a bright film)?

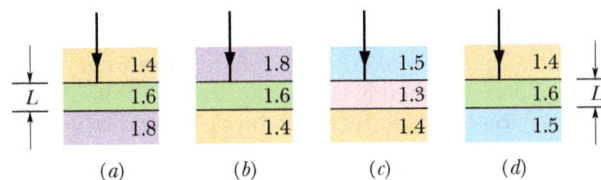

Figure 35-28 Question 11.

12 Figure 35-29 shows the transmission of light through a thin film in air by a perpendicular beam (tilted in the figure for clarity). (a) Did ray r_3 undergo a phase shift due to reflection? (b) In wavelengths, what is the reflection phase shift for ray r_4? (c) If the film thickness is L, what is the path length difference between rays r_3 and r_4?

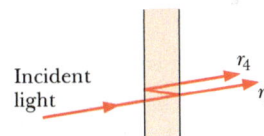

Figure 35-29 Question 12.

13 Figure 35-30 shows three situations in which two rays of sunlight penetrate slightly into and then scatter out of lunar soil. Assume that the rays are initially in phase. In which situation are the associated waves most likely to end up in phase? (Just as the Moon becomes full, its brightness suddenly peaks, becoming 25% greater than its brightness on the nights before and after, because at full Moon we intercept light waves that are scattered by lunar soil back toward the Sun and undergo constructive interference at our eyes. Before astronauts first landed on the Moon, NASA was concerned that backscatter of sunlight from the soil might blind the lunar astronauts if they did not have proper viewing shields on their helmets.)

Figure 35-30 Question 13.

Problems

GO Tutoring problem available (at instructor's discretion) in *WileyPLUS* and WebAssign

SSM Worked-out solution available in Student Solutions Manual

• – ••• Number of dots indicates level of problem difficulty

WWW Worked-out solution is at

ILW Interactive solution is at http://www.wiley.com/college/halliday

Additional information available in *The Flying Circus of Physics* and at flyingcircusofphysics.com

Module 35-1 Light as a Wave

•1 In Fig. 35-31, a light wave along ray r_1 reflects once from a mirror and a light wave along ray r_2 reflects twice from that same mirror and once from a tiny mirror at distance L from the bigger mirror. (Neglect the slight tilt of the rays.) The waves have wavelength 620 nm and are initially in phase. (a) What is the smallest value of L that puts the final light waves exactly out of phase? (b) With the tiny mirror initially at that value of L, how far must it be moved away from the bigger mirror to again put the final waves out of phase?

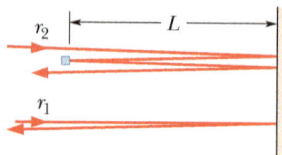

Figure 35-31 Problems 1 and 2.

•2 In Fig. 35-31, a light wave along ray r_1 reflects once from a mirror and a light wave along ray r_2 reflects twice from that same mirror and once from a tiny mirror at distance L from the bigger mirror. (Neglect the slight tilt of the rays.) The waves have wavelength λ and are initially exactly out of phase. What are the (a) smallest, (b) second smallest, and (c) third smallest values of L/λ that result in the final waves being exactly in phase?

•3 **SSM** In Fig. 35-4, assume that two waves of light in air, of wavelength 400 nm, are initially in phase. One travels through a glass layer of index of refraction $n_1 = 1.60$ and thickness L. The other travels through an equally thick plastic layer of index of refraction $n_2 = 1.50$. (a) What is the smallest value L should have if the waves are to end up with a phase difference of 5.65 rad? (b) If the waves arrive at some common point with the same amplitude, is their interference fully constructive, fully destructive, intermediate but closer to fully constructive, or intermediate but closer to fully destructive?

•4 In Fig. 35-32a, a beam of light in material 1 is incident on a boundary at an angle of 30°. The extent to which the light is bent due to refraction depends, in part, on the index of refraction n_2 of material 2. Figure 35-32b gives the angle of refraction θ_2 versus n_2 for a range of possible n_2 values, from $n_a = 1.30$ to $n_b = 1.90$. What is the speed of light in material 1?

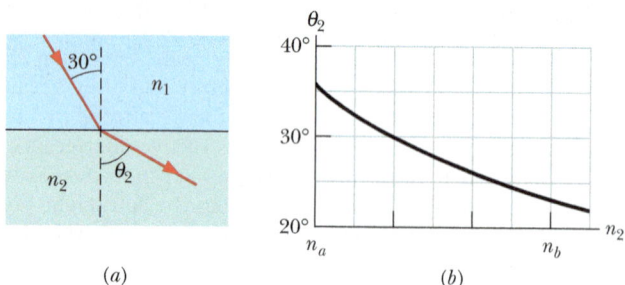

(a)

(b)

Figure 35-32 Problem 4.

•5 How much faster, in meters per second, does light travel in sapphire than in diamond? See Table 33-1.

•6 The wavelength of yellow sodium light in air is 589 nm. (a) What is its frequency? (b) What is its wavelength in glass whose index of refraction is 1.52? (c) From the results of (a) and (b), find its speed in this glass.

•7 The speed of yellow light (from a sodium lamp) in a certain liquid is measured to be 1.92×10^8 m/s. What is the index of refraction of this liquid for the light?

•8 In Fig. 35-33, two light pulses are sent through layers of plastic with thicknesses of either L or $2L$ as shown and indexes of refraction $n_1 = 1.55$, $n_2 = 1.70$, $n_3 = 1.60$, $n_4 = 1.45$, $n_5 = 1.59$, $n_6 = 1.65$, and $n_7 = 1.50$. (a) Which pulse travels through the plastic in less time? (b) What multiple of L/c gives the difference in the traversal times of the pulses?

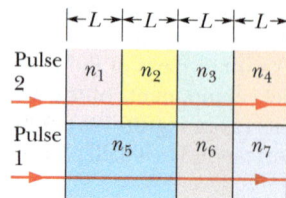

Figure 35-33 Problem 8.

••9 In Fig. 35-4, assume that the two light waves, of wavelength 620 nm in air, are initially out of phase by π rad. The indexes of refraction of the media are $n_1 = 1.45$ and $n_2 = 1.65$. What are the (a) smallest and (b) second smallest value of L that will put the waves exactly in phase once they pass through the two media?

••10 In Fig. 35-34, a light ray is incident at angle $\theta_1 = 50°$ on a series of five transparent layers with parallel boundaries. For layers 1 and 3, $L_1 = 20$ μm, $L_3 = 25$ μm, $n_1 = 1.6$, and $n_3 = 1.45$. (a) At what angle does the light emerge back into air at the right? (b) How much time does the light take to travel through layer 3?

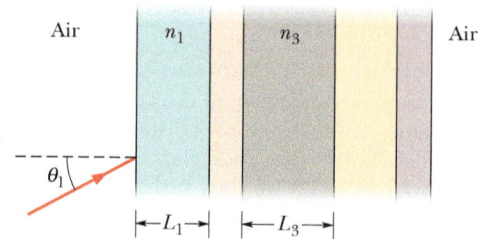

Figure 35-34 Problem 10.

••11 Suppose that the two waves in Fig. 35-4 have wavelength $\lambda = 500$ nm in air. What multiple of λ gives their phase difference when they emerge if (a) $n_1 = 1.50$, $n_2 = 1.60$, and $L = 8.50$ μm; (b) $n_1 = 1.62$, $n_2 = 1.72$, and $L = 8.50$ μm; and (c) $n_1 = 1.59$, $n_2 = 1.79$, and $L = 3.25$ μm? (d) Suppose that in each of these three situations the waves arrive at a common point (with the same amplitude) after emerging. Rank the situations according to the brightness the waves produce at the common point.

••12 In Fig. 35-35, two light rays go through different paths by reflecting from the various flat surfaces shown. The light waves have a wavelength of 420.0 nm and are initially in phase. What are the (a) smallest and (b) second smallest value of distance L that will put the waves exactly out of phase as they emerge from the region?

••13 **GO** **ILW** Two waves of light in air, of wavelength $\lambda = 600.0$ nm, are initially in phase. They then

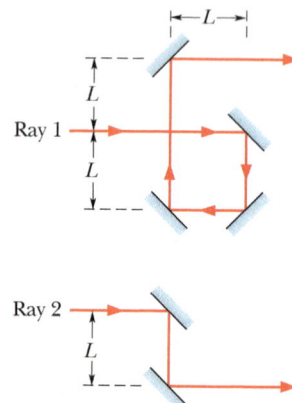

Figure 35-35 Problems 12 and 98.

both travel through a layer of plastic as shown in Fig. 35-36, with $L_1 = 4.00 \ \mu m$, $L_2 = 3.50 \ \mu m$, $n_1 = 1.40$, and $n_2 = 1.60$. (a) What multiple of λ gives their phase difference after they both have emerged from the layers? (b) If the waves later arrive at some common point with the same amplitude, is their interference fully constructive, fully destructive, intermediate but closer to fully constructive, or intermediate but closer to fully destructive?

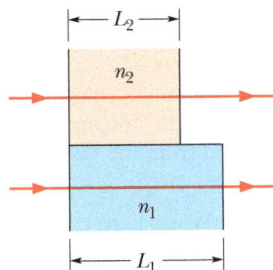

Figure 35-36 Problem 13.

Module 35-2 Young's Interference Experiment

•**14** In a double-slit arrangement the slits are separated by a distance equal to 100 times the wavelength of the light passing through the slits. (a) What is the angular separation in radians between the central maximum and an adjacent maximum? (b) What is the distance between these maxima on a screen 50.0 cm from the slits?

•**15** **SSM** A double-slit arrangement produces interference fringes for sodium light ($\lambda = 589$ nm) that have an angular separation of 3.50×10^{-3} rad. For what wavelength would the angular separation be 10.0% greater?

•**16** A double-slit arrangement produces interference fringes for sodium light ($\lambda = 589$ nm) that are 0.20° apart. What is the angular separation if the arrangement is immersed in water ($n = 1.33$)?

•**17** **GO** **SSM** In Fig. 35-37, two radio-frequency point sources S_1 and S_2, separated by distance $d = 2.0$ m, are radiating in phase with $\lambda = 0.50$ m. A detector moves in a large circular path around the two sources in a plane containing them. How many maxima does it detect?

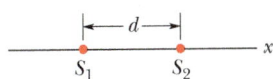

Figure 35-37 Problems 17 and 22.

•**18** In the two-slit experiment of Fig. 35-10, let angle θ be 20.0°, the slit separation be 4.24 μm, and the wavelength be $\lambda = 500$ nm. (a) What multiple of λ gives the phase difference between the waves of rays r_1 and r_2 when they arrive at point P on the distant screen? (b) What is the phase difference in radians? (c) Determine where in the interference pattern point P lies by giving the maximum or minimum on which it lies, or the maximum and minimum between which it lies.

•**19** **SSM** **ILW** Suppose that Young's experiment is performed with blue-green light of wavelength 500 nm. The slits are 1.20 mm apart, and the viewing screen is 5.40 m from the slits. How far apart are the bright fringes near the center of the interference pattern?

•**20** Monochromatic green light, of wavelength 550 nm, illuminates two parallel narrow slits 7.70 μm apart. Calculate the angular deviation (θ in Fig. 35-10) of the third-order ($m = 3$) bright fringe (a) in radians and (b) in degrees.

••**21** In a double-slit experiment, the distance between slits is 5.0 mm and the slits are 1.0 m from the screen. Two interference patterns can be seen on the screen: one due to light of wavelength 480 nm, and the other due to light of wavelength 600 nm. What is the separation on the screen between the third-order ($m = 3$) bright fringes of the two interference patterns?

••**22** In Fig. 35-37, two isotropic point sources S_1 and S_2 emit identical light waves in phase at wavelength λ. The sources lie at separation d on an x axis, and a light detector is moved in a circle of large radius around the midpoint between them. It detects 30 points of zero intensity, including two on the x axis, one of them to the left of the sources and the other to the right of the sources. What is the value of d/λ?

••**23** **GO** In Fig. 35-38, sources A and B emit long-range radio waves of wavelength 400 m, with the phase of the emission from A ahead of that from source B by 90°. The distance r_A from A to detector D is greater than the corresponding distance r_B by 100 m. What is the phase difference of the waves at D?

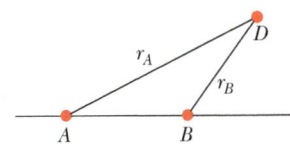

Figure 35-38 Problem 23.

••**24** In Fig. 35-39, two isotropic point sources S_1 and S_2 emit light in phase at wavelength λ and at the same amplitude. The sources are separated by distance $2d = 6.00\lambda$. They lie on an axis that is parallel to an x axis, which runs along a viewing screen at distance $D = 20.0\lambda$. The origin lies on the perpendicular bisector between the sources. The figure shows two rays reaching point P on the screen, at position x_P. (a) At what value of x_P do the rays have the minimum possible phase difference? (b) What multiple of λ gives that minimum phase difference? (c) At what value of x_P do the rays have the maximum possible phase difference? What multiple of λ gives (d) that maximum phase difference and (e) the phase difference when $x_P = 6.00\lambda$? (f) When $x_P = 6.00\lambda$, is the resulting intensity at point P maximum, minimum, intermediate but closer to maximum, or intermediate but closer to minimum?

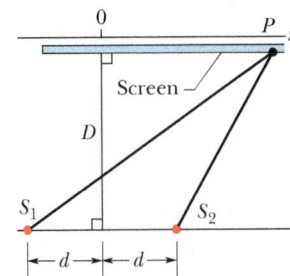

Figure 35-39 Problem 24.

••**25** **GO** In Fig. 35-40, two isotropic point sources of light (S_1 and S_2) are separated by distance 2.70 μm along a y axis and emit in phase at wavelength 900 nm and at the same amplitude. A light detector is located at point P at coordinate x_P on the x axis. What is the greatest value of x_P at which the detected light is minimum due to destructive interference?

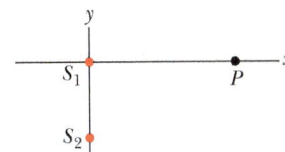

Figure 35-40 Problems 25 and 28.

••**26** In a double-slit experiment, the fourth-order maximum for a wavelength of 450 nm occurs at an angle of $\theta = 90°$. (a) What range of wavelengths in the visible range (400 nm to 700 nm) are not present in the third-order maxima? To eliminate all visible light in the fourth-order maximum, (b) should the slit separation be increased or decreased and (c) what least change is needed?

•••**27** A thin flake of mica ($n = 1.58$) is used to cover one slit of a double-slit interference arrangement. The central point on the viewing screen is now occupied by what had been the seventh bright side fringe ($m = 7$). If $\lambda = 550$ nm, what is the thickness of the mica?

•••**28** **GO** Figure 35-40 shows two isotropic point sources of light (S_1 and S_2) that emit in phase at wavelength 400 nm and at the same amplitude. A detection point P is shown on an x axis that extends through source S_1. The phase difference ϕ between the light arriving at point P from the two sources is to be measured as P is moved along the x axis from $x = 0$ out to $x = +\infty$. The results out to $x_s = 10 \times 10^{-7}$ m are given in Fig. 35-41. On the way out to

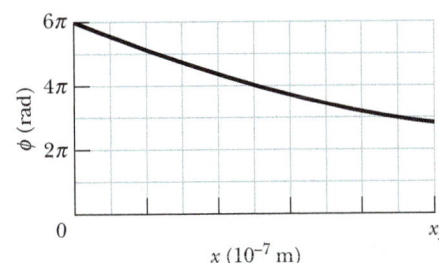

Figure 35-41 Problem 28.

$+\infty$, what is the greatest value of x at which the light arriving at P from S_1 is exactly out of phase with the light arriving at P from S_2?

Module 35-3 Interference and Double-Slit Intensity

•29 SSM Two waves of the same frequency have amplitudes 1.00 and 2.00. They interfere at a point where their phase difference is 60.0°. What is the resultant amplitude?

•30 Find the sum y of the following quantities:

$$y_1 = 10 \sin \omega t \quad \text{and} \quad y_2 = 8.0 \sin(\omega t + 30°).$$

••31 ILW Add the quantities $y_1 = 10 \sin \omega t$, $y_2 = 15 \sin(\omega t + 30°)$, and $y_3 = 5.0 \sin(\omega t - 45°)$ using the phasor method.

••32 GO In the double-slit experiment of Fig. 35-10, the electric fields of the waves arriving at point P are given by

$$E_1 = (2.00 \ \mu\text{V/m}) \sin[(1.26 \times 10^{15})t]$$

$$E_2 = (2.00 \ \mu\text{V/m}) \sin[(1.26 \times 10^{15})t + 39.6 \text{ rad}],$$

where time t is in seconds. (a) What is the amplitude of the resultant electric field at point P? (b) What is the ratio of the intensity I_P at point P to the intensity I_{cen} at the center of the interference pattern? (c) Describe where point P is in the interference pattern by giving the maximum or minimum on which it lies, or the maximum and minimum between which it lies. In a phasor diagram of the electric fields, (d) at what rate would the phasors rotate around the origin and (e) what is the angle between the phasors?

••33 GO Three electromagnetic waves travel through a certain point P along an x axis. They are polarized parallel to a y axis, with the following variations in their amplitudes. Find their resultant at P.

$$E_1 = (10.0 \ \mu\text{V/m}) \sin[(2.0 \times 10^{14} \text{ rad/s})t]$$
$$E_2 = (5.00 \ \mu\text{V/m}) \sin[(2.0 \times 10^{14} \text{ rad/s})t + 45.0°]$$
$$E_3 = (5.00 \ \mu\text{V/m}) \sin[(2.0 \times 10^{14} \text{ rad/s})t - 45.0°]$$

••34 In the double-slit experiment of Fig. 35-10, the viewing screen is at distance $D = 4.00$ m, point P lies at distance $y = 20.5$ cm from the center of the pattern, the slit separation d is 4.50 μm, and the wavelength λ is 580 nm. (a) Determine where point P is in the interference pattern by giving the maximum or minimum on which it lies, or the maximum and minimum between which it lies. (b) What is the ratio of the intensity I_P at point P to the intensity I_{cen} at the center of the pattern?

Module 35-4 Interference from Thin Films

•35 SSM We wish to coat flat glass ($n = 1.50$) with a transparent material ($n = 1.25$) so that reflection of light at wavelength 600 nm is eliminated by interference. What minimum thickness can the coating have to do this?

•36 A 600-nm-thick soap film ($n = 1.40$) in air is illuminated with white light in a direction perpendicular to the film. For how many different wavelengths in the 300 to 700 nm range is there (a) fully constructive interference and (b) fully destructive interference in the reflected light?

•37 The rhinestones in costume jewelry are glass with index of refraction 1.50. To make them more reflective, they are often coated with a layer of silicon monoxide of index of refraction 2.00. What is the minimum coating thickness needed to ensure that light of wavelength 560 nm and of perpendicular incidence will be reflected from the two surfaces of the coating with fully constructive interference?

•38 White light is sent downward onto a horizontal thin film that is sandwiched between two materials. The indexes of refraction are 1.80 for the top material, 1.70 for the thin film, and 1.50 for the bottom material. The film thickness is 5.00×10^{-7} m. Of the visible wavelengths (400 to 700 nm) that result in fully constructive interference at an observer above the film, which is the (a) longer and (b) shorter wavelength? The materials and film are then heated so that the film thickness increases. (c) Does the light resulting in fully constructive interference shift toward longer or shorter wavelengths?

•39 ILW Light of wavelength 624 nm is incident perpendicularly on a soap film ($n = 1.33$) suspended in air. What are the (a) least and (b) second least thicknesses of the film for which the reflections from the film undergo fully constructive interference?

••40 A thin film of acetone ($n = 1.25$) coats a thick glass plate ($n = 1.50$). White light is incident normal to the film. In the reflections, fully destructive interference occurs at 600 nm and fully constructive interference at 700 nm. Calculate the thickness of the acetone film.

••41 through 52 GO 43, 51 SSM 47, 51
Reflection by thin layers. In Fig. 35-42, light is incident perpendicularly on a thin layer of material 2 that lies between (thicker) materials 1 and 3. (The rays are tilted only for clarity.) The waves of rays r_1 and r_2 interfere, and here we consider the type of interference to be either maximum (max) or minimum (min). For this situation, each problem in Table 35-2 refers to the indexes of refraction n_1, n_2, and n_3, the type of interference, the thin-layer thickness L in nanometers, and the wavelength λ in nanometers of the light as measured in air. Where λ is missing, give the wavelength that is in the visible range. Where L is missing, give the second least thickness or the third least thickness as indicated.

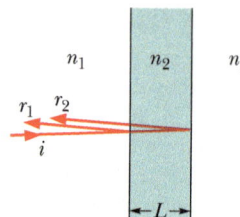

Figure 35-42 Problems 41 through 52.

Table 35-2 Problems 41 through 52: Reflection by Thin Layers. See the setup for these problems.

	n_1	n_2	n_3	Type	L	λ
41	1.68	1.59	1.50	min	2nd	342
42	1.55	1.60	1.33	max	285	
43	1.60	1.40	1.80	min	200	
44	1.50	1.34	1.42	max	2nd	587
45	1.55	1.60	1.33	max	3rd	612
46	1.68	1.59	1.50	min	415	
47	1.50	1.34	1.42	min	380	
48	1.60	1.40	1.80	max	2nd	632
49	1.32	1.75	1.39	max	3rd	382
50	1.40	1.46	1.75	min	2nd	482
51	1.40	1.46	1.75	min	210	
52	1.32	1.75	1.39	max	325	

••53 The reflection of perpendicularly incident white light by a soap film in air has an interference maximum at 600 nm and a minimum at 450 nm, with no minimum in between. If $n = 1.33$ for the film, what is the film thickness, assumed uniform?

••54 A plane wave of monochromatic light is incident normally on a uniform thin film of oil that covers a glass plate. The wavelength of the source can be varied continuously. Fully destructive interference of the reflected light is observed for wavelengths of 500 and 700 nm and for no wavelengths in between. If the index of refraction of the oil is 1.30 and that of the glass is 1.50, find the thickness of the oil film.

••55 **SSM** **WWW** A disabled tanker leaks kerosene ($n = 1.20$) into the Persian Gulf, creating a large slick on top of the water ($n = 1.30$). (a) If you are looking straight down from an airplane, while the Sun is overhead, at a region of the slick where its thickness is 460 nm, for which wavelength(s) of visible light is the reflection brightest because of constructive interference? (b) If you are scuba diving directly under this same region of the slick, for which wavelength(s) of visible light is the transmitted intensity strongest?

••56 A thin film, with a thickness of 272.7 nm and with air on both sides, is illuminated with a beam of white light. The beam is perpendicular to the film and consists of the full range of wavelengths for the visible spectrum. In the light reflected by the film, light with a wavelength of 600.0 nm undergoes fully constructive interference. At what wavelength does the reflected light undergo fully destructive interference? (*Hint:* You must make a reasonable assumption about the index of refraction.)

••57 through 68 **GO** 64, 65 **SSM** 59 *Transmission through thin layers.* In Fig. 35-43, light is incident perpendicularly on a thin layer of material 2 that lies between (thicker) materials 1 and 3. (The rays are tilted only for clarity.) Part of the light ends up in material 3 as ray r_3 (the light does not reflect inside material 2) and r_4 (the light reflects twice inside material 2). The waves of r_3 and r_4 interfere, and here we consider the type of interference to be either maximum (max) or minimum (min). For this situation, each problem in Table 35-3 refers to the indexes of refraction n_1, n_2, and n_3, the type

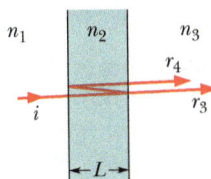

Figure 35-43
Problems 57
through 68.

of interference, the thin-layer thickness L in nanometers, and the wavelength λ in nanometers of the light as measured in air. Where λ is missing, give the wavelength that is in the visible range. Where L is missing, give the second least thickness or the third least thickness as indicated.

••69 **GO** In Fig. 35-44, a broad beam of light of wavelength 630 nm is incident at 90° on a thin, wedge-shaped film with index of refraction 1.50. Transmission gives 10 bright and 9 dark fringes along the film's length. What is the left-to-right change in film thickness?

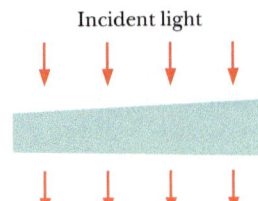

Incident light

Figure 35-44 Problem 69.

••70 **GO** In Fig. 35-45, a broad beam of light of wavelength 620 nm is sent directly downward through the top plate of a pair of glass plates touching at the left end. The air between the plates acts as a thin film, and an interference pattern can be seen from above the plates. Initially, a dark fringe lies at the left end, a bright fringe lies at the right end, and nine dark fringes lie between those two end fringes. The plates are then very gradually squeezed together at a constant rate to decrease the angle between them. As a result, the fringe at the right side changes between being bright to being dark every 15.0 s. (a) At what rate is the spacing between the plates at the right end being changed? (b) By how much has the spacing there changed when both left and right ends have a dark fringe and there are five dark fringes between them?

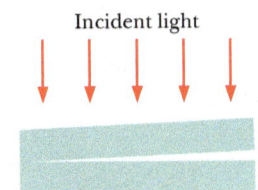

Incident light

Figure 35-45
Problems 70–74.

••71 In Fig. 35-45, two microscope slides touch at one end and are separated at the other end. When light of wavelength 500 nm shines vertically down on the slides, an overhead observer sees an interference pattern on the slides with the dark fringes separated by 1.2 mm. What is the angle between the slides?

••72 In Fig. 35-45, a broad beam of monochromatic light is directed perpendicularly through two glass plates that are held together at one end to create a wedge of air between them. An observer intercepting light reflected from the wedge of air, which acts as a thin film, sees 4001 dark fringes along the length of the wedge. When the air between the plates is evacuated, only 4000 dark fringes are seen. Calculate to six significant figures the index of refraction of air from these data.

••73 **SSM** In Fig. 35-45, a broad beam of light of wavelength 683 nm is sent directly downward through the top plate of a pair of glass plates. The plates are 120 mm long, touch at the left end, and are separated by 48.0 μm at the right end. The air between the plates acts as a thin film. How many bright fringes will be seen by an observer looking down through the top plate?

••74 **GO** Two rectangular glass plates ($n = 1.60$) are in contact along one edge and are separated along the opposite edge (Fig. 35-45). Light with a wavelength of 600

Table 35-3 Problems 57 through 68: Transmission Through Thin Layers. See the setup for these problems.

	n_1	n_2	n_3	Type	L	λ
57	1.55	1.60	1.33	min	285	
58	1.32	1.75	1.39	min	3rd	382
59	1.68	1.59	1.50	max	415	
60	1.50	1.34	1.42	max	380	
61	1.32	1.75	1.39	min	325	
62	1.68	1.59	1.50	max	2nd	342
63	1.40	1.46	1.75	max	2nd	482
64	1.40	1.46	1.75	max	210	
65	1.60	1.40	1.80	min	2nd	632
66	1.60	1.40	1.80	max	200	
67	1.50	1.34	1.42	min	2nd	587
68	1.55	1.60	1.33	min	3rd	612

nm is incident perpendicularly onto the top plate. The air between the plates acts as a thin film. Nine dark fringes and eight bright fringes are observed from above the top plate. If the distance between the two plates along the separated edges is increased by 600 nm, how many dark fringes will there then be across the top plate?

••75 SSM ILW Figure 35-46a shows a lens with radius of curvature R lying on a flat glass plate and illuminated from above by light with wavelength λ. Figure 35-46b (a photograph taken from above the lens) shows that circular interference fringes (known as *Newton's rings*) appear, associated with the variable thickness d of the air film between the lens and the plate. Find the radii r of the interference maxima assuming $r/R \ll 1$.

Figure 35-46 Problems 75–77.

(b) Courtesy Bausch & Lomb

••76 The lens in a Newton's rings experiment (see Problem 75) has diameter 20 mm and radius of curvature $R = 5.0$ m. For $\lambda = 589$ nm in air, how many bright rings are produced with the setup (a) in air and (b) immersed in water ($n = 1.33$)?

••77 A Newton's rings apparatus is to be used to determine the radius of curvature of a lens (see Fig. 35-46 and Problem 75). The radii of the nth and $(n + 20)$th bright rings are found to be 0.162 and 0.368 cm, respectively, in light of wavelength 546 nm. Calculate the radius of curvature of the lower surface of the lens.

•••78 A thin film of liquid is held in a horizontal circular ring, with air on both sides of the film. A beam of light at wavelength 550 nm is directed perpendicularly onto the film, and the intensity I of its reflection is monitored. Figure 35-47 gives intensity I as a function of time t; the horizontal scale is set by $t_s = 20.0$ s. The intensity changes because of evaporation from the two sides of the film. Assume that the film is flat and has parallel sides, a radius of 1.80 cm, and an index of refraction of 1.40. Also assume that the film's volume decreases at a constant rate. Find that rate.

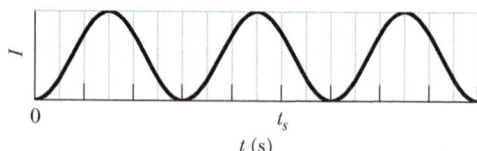

Figure 35-47 Problem 78.

Module 35-5 Michelson's Interferometer

•79 If mirror M_2 in a Michelson interferometer (Fig. 35-21) is moved through 0.233 mm, a shift of 792 bright fringes occurs. What is the wavelength of the light producing the fringe pattern?

•80 A thin film with index of refraction $n = 1.40$ is placed in one arm of a Michelson interferometer, perpendicular to the optical path. If this causes a shift of 7.0 bright fringes of the pattern produced by light of wavelength 589 nm, what is the film thickness?

••81 SSM WWW In Fig. 35-48, an airtight chamber of length $d = 5.0$ cm is placed in one of the arms of a Michelson interferometer. (The glass window on each end of the chamber has negligible thickness.) Light of wavelength $\lambda = 500$ nm is used. Evacuating the air from the chamber causes a shift of 60 bright fringes. From these data and to six significant figures, find the index of refraction of air at atmospheric pressure.

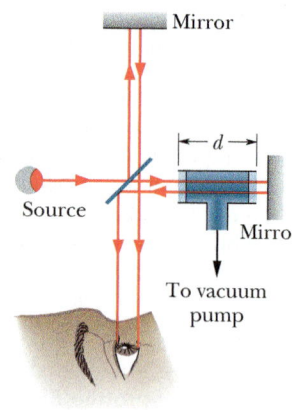

Figure 35-48 Problem 81.

••82 The element sodium can emit light at two wavelengths, $\lambda_1 = 588.9950$ nm and $\lambda_2 = 589.5924$ nm. Light from sodium is being used in a Michelson interferometer (Fig. 35-21). Through what distance must mirror M_2 be moved if the shift in the fringe pattern for one wavelength is to be 1.00 fringe more than the shift in the fringe pattern for the other wavelength?

Additional Problems

83 GO Two light rays, initially in phase and with a wavelength of 500 nm, go through different paths by reflecting from the various mirrors shown in Fig. 35-49. (Such a reflection does not itself produce a phase shift.) (a) What least value of distance d will put the rays exactly out of phase when they emerge from the region? (Ignore the slight tilt of the path for ray 2.) (b) Repeat the question assuming that the entire apparatus is immersed in a protein solution with an index of refraction of 1.38.

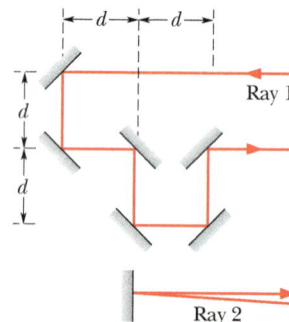

Figure 35-49 Problem 83.

84 GO In Figure 35-50, two isotropic point sources S_1 and S_2 emit light in phase at wavelength λ and at the same amplitude. The sources are separated by distance $d = 6.00\lambda$ on an x axis. A viewing screen is at distance $D = 20.0\lambda$ from S_2 and parallel to the y axis. The figure shows two rays reaching point P on the screen, at height y_P. (a) At what value of y_P do the rays have the minimum possible phase difference? (b) What multiple of λ gives that minimum phase difference? (c) At what value of y_P do the rays have the maximum possible phase difference? What multiple of λ gives (d) that maximum phase difference and (e) the phase difference when $y_P = d$? (f) When $y_P = d$, is the resulting intensity at point P maximum, mini-

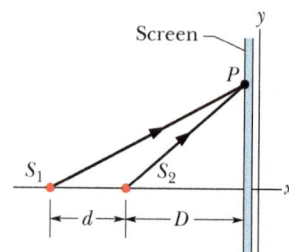

Figure 35-50 Problem 84.

mum, intermediate but closer to maximum, or intermediate but closer to minimum?

85 SSM A double-slit arrangement produces bright interference fringes for sodium light (a distinct yellow light at a wavelength of $\lambda = 589$ nm). The fringes are angularly separated by $0.30°$ near the center of the pattern. What is the angular fringe separation if the entire arrangement is immersed in water, which has an index of refraction of 1.33?

86 GO In Fig. 35-51a, the waves along rays 1 and 2 are initially in phase, with the same wavelength λ in air. Ray 2 goes through a material with length L and index of refraction n. The rays are then reflected by mirrors to a common point P on a screen. Suppose that we can vary n from $n = 1.0$ to $n = 2.5$. Suppose also that, from $n = 1.0$ to $n = n_s = 1.5$, the intensity I of the light at point P varies with n as given in Fig. 35-51b. At what values of n greater than 1.4 is intensity I (a) maximum and (b) zero? (c) What multiple of λ gives the phase difference between the rays at point P when $n = 2.0$?

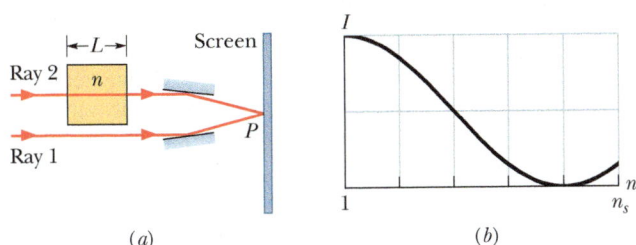

Figure 35-51 Problems 86 and 87.

87 SSM In Fig. 35-51a, the waves along rays 1 and 2 are initially in phase, with the same wavelength λ in air. Ray 2 goes through a material with length L and index of refraction n. The rays are then reflected by mirrors to a common point P on a screen. Suppose that we can vary L from 0 to 2400 nm. Suppose also that, from $L = 0$ to $L_s = 900$ nm, the intensity I of the light at point P varies with L as given in Fig. 35-52. At what values of L greater than L_s is intensity I (a) maximum and (b) zero? (c) What multiple of λ gives the phase difference between ray 1 and ray 2 at common point P when $L = 1200$ nm?

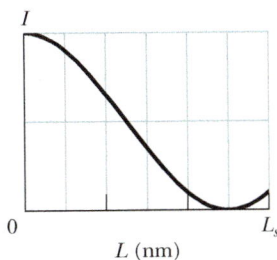

Figure 35-52 Problem 87.

88 Light of wavelength 700.0 nm is sent along a route of length 2000 nm. The route is then filled with a medium having an index of refraction of 1.400. In degrees, by how much does the medium phase-shift the light? Give (a) the full shift and (b) the equivalent shift that has a value less than $360°$.

89 SSM In Fig. 35-53, a microwave transmitter at height a above the water level of a wide lake transmits microwaves of wavelength λ toward a receiver on the opposite shore, a distance x above the water level. The microwaves reflecting from the water interfere with the microwaves arriving directly from the transmitter.

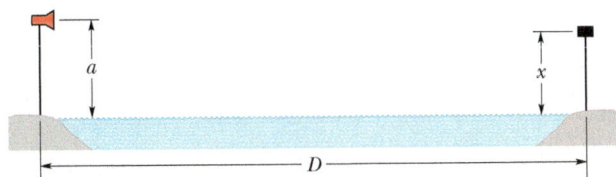

Figure 35-53 Problem 89.

Assuming that the lake width D is much greater than a and x, and that $\lambda \geq a$, find an expression that gives the values of x for which the signal at the receiver is maximum. (*Hint:* Does the reflection cause a phase change?)

90 In Fig. 35-54, two isotropic point sources S_1 and S_2 emit light at wavelength $\lambda = 400$ nm. Source S_1 is located at $y = 640$ nm; source S_2 is located at $y = -640$ nm. At point P_1 (at $x = 720$ nm), the wave from S_2 arrives ahead of the wave from S_1 by a phase difference of 0.600π rad. (a) What multiple of λ gives the phase difference between the waves from the two sources as the waves arrive at point P_2, which is located at $y = 720$ nm? (The figure is not drawn to scale.) (b) If the waves arrive at P_2 with equal amplitudes, is the interference there fully constructive, fully destructive, intermediate but closer to fully constructive, or intermediate but closer to fully destructive?

Figure 35-54
Problem 90.

91 Ocean waves moving at a speed of 4.0 m/s are approaching a beach at angle $\theta_1 = 30°$ to the normal, as shown from above in Fig. 35-55. Suppose the water depth changes abruptly at a certain distance from the beach and the wave speed there drops to 3.0 m/s. (a) Close to the beach, what is the angle θ_2 between the direction of wave motion and the normal? (Assume the same law of refraction as for light.) (b) Explain why most waves come in normal to a shore even though at large distances they approach at a variety of angles.

Figure 35-55 Problem 91.

92 Figure 35-56a shows two light rays that are initially in phase as they travel upward through a block of plastic, with wavelength 400 nm as measured in air. Light ray r_1 exits directly into air. However, before light ray r_2 exits into air, it travels through a liquid in a hollow cylinder within the plastic. Initially the height L_{liq} of the liquid is 40.0 μm, but then the liquid begins to evaporate. Let ϕ be the phase difference between rays r_1 and r_2 once they both exit into the air. Figure 35-56b shows ϕ versus the liquid's height L_{liq} until the liquid disappears, with ϕ given in terms of wavelength and the horizontal scale set by $L_s = 40.00$ μm. What are (a) the index of refraction of the plastic and (b) the index of refraction of the liquid?

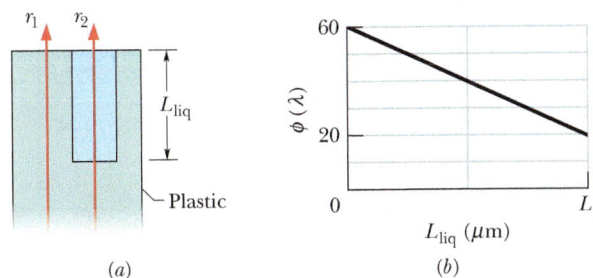

Figure 35-56 Problem 92.

93 SSM If the distance between the first and tenth minima of a double-slit pattern is 18.0 mm and the slits are separated by 0.150 mm with the screen 50.0 cm from the slits, what is the wavelength of the light used?

94 Figure 35-57 shows an optical fiber in which a central plastic core of index of refraction $n_1 = 1.58$ is surrounded by a plastic sheath of index of refraction $n_2 = 1.53$. Light can travel along different paths within the central

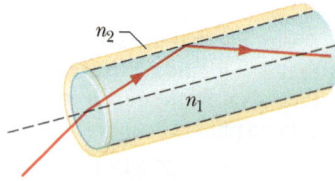

Figure 35-57 Problem 94.

core, leading to different travel times through the fiber. This causes an initially short pulse of light to spread as it travels along the fiber, resulting in information loss. Consider light that travels directly along the central axis of the fiber and light that is repeatedly reflected at the critical angle along the core–sheath interface, reflecting from side to side as it travels down the central core. If the fiber length is 300 m, what is the difference in the travel times along these two routes?

95 SSM Two parallel slits are illuminated with monochromatic light of wavelength 500 nm. An interference pattern is formed on a screen some distance from the slits, and the fourth dark band is located 1.68 cm from the central bright band on the screen. (a) What is the path length difference corresponding to the fourth dark band? (b) What is the distance on the screen between the central bright band and the first bright band on either side of the central band? (*Hint:* The angle to the fourth dark band and the angle to the first bright band are small enough that $\tan \theta \approx \sin \theta$.)

96 A camera lens with index of refraction greater than 1.30 is coated with a thin transparent film of index of refraction 1.25 to eliminate by interference the reflection of light at wavelength λ that is incident perpendicularly on the lens. What multiple of λ gives the minimum film thickness needed?

97 SSM Light of wavelength λ is used in a Michelson interferometer. Let x be the position of the movable mirror, with $x = 0$ when the arms have equal lengths $d_2 = d_1$. Write an expression for the intensity of the observed light as a function of x, letting I_m be the maximum intensity.

98 In two experiments, light is to be sent along the two paths shown in Fig. 35-35 by reflecting it from the various flat surfaces shown. In the first experiment, rays 1 and 2 are initially in phase and have a wavelength of 620.0 nm. In the second experiment, rays 1 and 2 are initially in phase and have a wavelength of 496.0 nm. What least value of distance L is required such that the 620.0 nm waves emerge from the region exactly in phase but the 496.0 nm waves emerge exactly out of phase?

99 Figure 35-58 shows the design of a Texas arcade game. Four laser pistols are pointed toward the center of an array of plastic

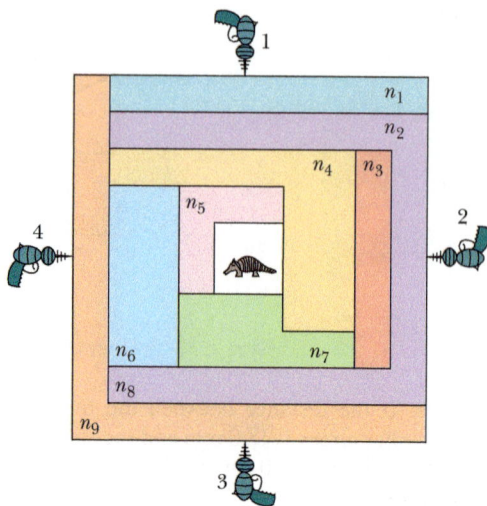

Figure 35-58
Problem 99.

layers where a clay armadillo is the target. The indexes of refraction of the layers are $n_1 = 1.55$, $n_2 = 1.70$, $n_3 = 1.45$, $n_4 = 1.60$, $n_5 = 1.45$, $n_6 = 1.61$, $n_7 = 1.59$, $n_8 = 1.70$, and $n_9 = 1.60$. The layer thicknesses are either 2.00 mm or 4.00 mm, as drawn. What is the travel time through the layers for the laser burst from (a) pistol 1, (b) pistol 2, (c) pistol 3, and (d) pistol 4? (e) If the pistols are fired simultaneously, which laser burst hits the target first?

100 A thin film suspended in air is 0.410 μm thick and is illuminated with white light incident perpendicularly on its surface. The index of refraction of the film is 1.50. At what wavelength will visible light that is reflected from the two surfaces of the film undergo fully constructive interference?

101 Find the slit separation of a double-slit arrangement that will produce interference fringes 0.018 rad apart on a distant screen when the light has wavelength $\lambda = 589$ nm.

102 In a phasor diagram for any point on the viewing screen for the two-slit experiment in Fig. 35-10, the resultant-wave phasor rotates 60.0° in 2.50×10^{-16} s. What is the wavelength?

103 In Fig. 35-59, an oil drop ($n = 1.20$) floats on the surface of water ($n = 1.33$) and is viewed from overhead when illuminated by sunlight shining vertically downward and reflected vertically upward. (a) Are the outer (thinnest) regions of the drop bright or dark? The oil film displays several spectra of colors. (b) Move from the rim inward to the third blue band and, using a wavelength of 475 nm for blue light, determine the film thickness there. (c) If the oil thickness increases, why do the colors gradually fade and then disappear?

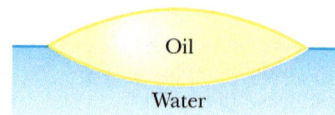

Figure 35-59 Problem 103.

104 *Lloyd's Mirror.* In Fig. 35-60, monochromatic light of wavelength λ diffracts through a narrow slit S in an otherwise opaque screen. On the other side, a plane mirror is perpendicular to the screen and a distance h from the slit. A viewing screen A is a distance much greater than h. (Because it sits in a plane through the focal point of the lens, screen A is effectively very distant. The lens

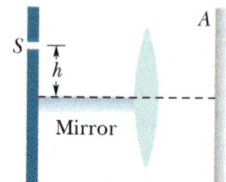

Figure 35-60
Problem 104.

plays no other role in the experiment and can otherwise be neglected.) Light that travels from the slit directly to A interferes with light from the slit that reflects from the mirror to A. The reflection causes a half-wavelength phase shift. (a) Is the fringe that corresponds to a zero path length difference bright or dark? Find expressions (like Eqs. 35-14 and 35-16) that locate (b) the bright fringes and (c) the dark fringes in the interference pattern. (*Hint:* Consider the image of S produced by the mirror as seen from a point on the viewing screen, and then consider Young's two-slit interference.)

105 The two point sources in Fig. 35-61 emit coherent waves. Show that all curves (such as the one shown), over which the phase difference for rays r_1 and r_2 is a constant, are hyperbolas. (*Hint:* A constant phase difference implies a constant difference in length between r_1 and r_2.)

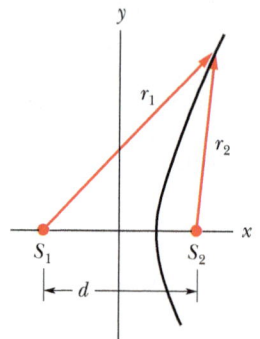

Figure 35-61
Problem 105.

Diffraction

36-1 SINGLE-SLIT DIFFRACTION

Learning Objectives

After reading this module, you should be able to . . .

36.01 Describe the diffraction of light waves by a narrow opening and an edge, and also describe the resulting interference pattern.

36.02 Describe an experiment that demonstrates the Fresnel bright spot.

36.03 With a sketch, describe the arrangement for a single-slit diffraction experiment.

36.04 With a sketch, explain how splitting a slit width into equal zones leads to the equations giving the angles to the minima in the diffraction pattern.

36.05 Apply the relationships between width a of a thin,

rectangular slit or object, the wavelength λ, the angle θ to any of the minima in the diffraction pattern, the distance to a viewing screen, and the distance between a minimum and the center of the pattern.

36.06 Sketch the diffraction pattern for monochromatic light, identifying what lies at the center and what the various bright and dark fringes are called (such as "first minimum").

36.07 Identify what happens to a diffraction pattern when the wavelength of the light or the width of the diffracting aperture or object is varied.

Key Ideas

● When waves encounter an edge, an obstacle, or an aperture the size of which is comparable to the wavelength of the waves, those waves spread out as they travel and, as a result, undergo interference. This type of interference is called diffraction.

● Waves passing through a long narrow slit of width a produce, on a viewing screen, a single-slit diffraction

pattern that includes a central maximum (bright fringe) and other maxima. They are separated by minima that are located relative to the central axis by angles θ:

$$a \sin \theta = m\lambda, \quad \text{for } m = 1, 2, 3, \ldots \quad \text{(minima)}.$$

● The maxima are located approximately halfway between minima.

What Is Physics?

One focus of physics in the study of light is to understand and put to use the diffraction of light as it passes through a narrow slit or (as we shall discuss) past either a narrow obstacle or an edge. We touched on this phenomenon in Chapter 35 when we looked at how light flared—diffracted—through the slits in Young's experiment. Diffraction through a given slit is more complicated than simple flaring, however, because the light also interferes with itself and produces an interference pattern. It is because of such complications that light is rich with application opportunities. Even though the diffraction of light as it passes through a slit or past an obstacle seems awfully academic, countless engineers and scientists make their living using this physics, and the total worth of diffraction applications worldwide is probably incalculable.

Before we can discuss some of these applications, we first must discuss why diffraction is due to the wave nature of light.

Diffraction and the Wave Theory of Light

In Chapter 35 we defined diffraction rather loosely as the flaring of light as it emerges from a narrow slit. More than just flaring occurs, however, because the

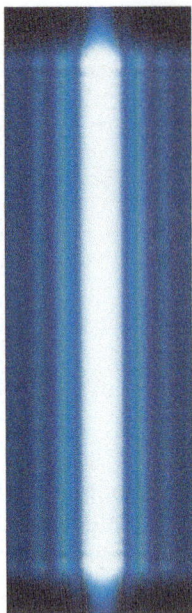

Ken Kay/Fundamental
Photographs

Figure 36-1 This diffraction pattern appeared on a viewing screen when light that had passed through a narrow vertical slit reached the screen. Diffraction caused the light to flare out perpendicular to the long sides of the slit. That flaring produced an interference pattern consisting of a broad central maximum plus less intense and narrower secondary (or side) maxima, with minima between them.

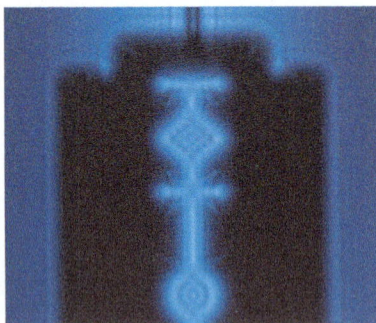

Ken Kay/Fundamental Photographs

Figure 36-2 The diffraction pattern produced by a razor blade in monochromatic light. Note the lines of alternating maximum and minimum intensity.

light produces an interference pattern called a **diffraction pattern.** For example, when monochromatic light from a distant source (or a laser) passes through a narrow slit and is then intercepted by a viewing screen, the light produces on the screen a diffraction pattern like that in Fig. 36-1. This pattern consists of a broad and intense (very bright) central maximum plus a number of narrower and less intense maxima (called **secondary** or **side** maxima) to both sides. In between the maxima are minima. Light flares into those dark regions, but the light waves cancel out one another.

Such a pattern would be totally unexpected in geometrical optics: If light traveled in straight lines as rays, then the slit would allow some of those rays through to form a sharp rendition of the slit on the viewing screen instead of a pattern of bright and dark bands as we see in Fig. 36-1. As in Chapter 35, we must conclude that geometrical optics is only an approximation.

Edges. Diffraction is not limited to situations in which light passes through a narrow opening (such as a slit or pinhole). It also occurs when light passes an edge, such as the edges of the razor blade whose diffraction pattern is shown in Fig. 36-2. Note the lines of maxima and minima that run approximately parallel to the edges, at both the inside edges of the blade and the outside edges. As the light passes, say, the vertical edge at the left, it flares left and right and undergoes interference, producing the pattern along the left edge. The rightmost portion of that pattern actually lies behind the blade, within what would be the blade's shadow if geometrical optics prevailed.

Floaters. You encounter a common example of diffraction when you look at a clear blue sky and see tiny specks and hairlike structures floating in your view. These *floaters*, as they are called, are produced when light passes the edges of tiny deposits in the vitreous humor, the transparent material filling most of the eyeball. What you are seeing when a floater is in your field of vision is the diffraction pattern produced on the retina by one of these deposits. If you sight through a pinhole in a piece of cardboard so as to make the light entering your eye approximately a plane wave, you can distinguish individual maxima and minima in the patterns.

Cheerleaders. Diffraction is a wave effect. That is, it occurs because light is a wave and it occurs with other types of waves as well. For example, you have probably seen diffraction in action at football games. When a cheerleader near the playing field yells up at several thousand noisy fans, the yell can hardly be heard because the sound waves diffract when they pass through the narrow opening of the cheerleader's mouth. This flaring leaves little of the waves traveling toward the fans in front of the cheerleader. To offset the diffraction, the cheerleader can yell through a megaphone. The sound waves then emerge from the much wider opening at the end of the megaphone. The flaring is thus reduced, and much more of the sound reaches the fans in front of the cheerleader.

The Fresnel Bright Spot

Diffraction finds a ready explanation in the wave theory of light. However, this theory, originally advanced in the late 1600s by Huygens and used 123 years later by Young to explain double-slit interference, was very slow in being adopted, largely because it ran counter to Newton's theory that light was a stream of particles.

Newton's view was the prevailing view in French scientific circles of the early 19th century, when Augustin Fresnel was a young military engineer. Fresnel, who believed in the wave theory of light, submitted a paper to the French Academy of Sciences describing his experiments with light and his wave-theory explanations of them.

In 1819, the Academy, dominated by supporters of Newton and thinking to challenge the wave point of view, organized a prize competition for an essay on the subject of diffraction. Fresnel won. The Newtonians, however, were not swayed. One of them, S. D. Poisson, pointed out the "strange result" that if Fresnel's theories were correct, then light waves should flare into the shadow region of a sphere as they pass the edge of the sphere, producing a bright spot at the center of the shadow. The prize committee arranged a test of Poisson's prediction and discovered that the predicted *Fresnel bright spot,* as we call it today, was indeed there (Fig. 36-3). Nothing builds confidence in a theory so much as having one of its unexpected and counterintuitive predictions verified by experiment.

Diffraction by a Single Slit: Locating the Minima

Let us now examine the diffraction pattern of plane waves of light of wavelength λ that are diffracted by a single long, narrow slit of width a in an otherwise opaque screen B, as shown in cross section in Fig. 36-4. (In that figure, the slit's length extends into and out of the page, and the incoming wavefronts are parallel to screen B.) When the diffracted light reaches viewing screen C, waves from different points within the slit undergo interference and produce a diffraction pattern of bright and dark fringes (interference maxima and minima) on the screen. To locate the fringes, we shall use a procedure somewhat similar to the one we used to locate the fringes in a two-slit interference pattern. However, diffraction is more mathematically challenging, and here we shall be able to find equations for only the dark fringes.

Before we do that, however, we can justify the central bright fringe seen in Fig. 36-1 by noting that the Huygens wavelets from all points in the slit travel about the same distance to reach the center of the pattern and thus are in phase there. As for the other bright fringes, we can say only that they are approximately halfway between adjacent dark fringes.

Pairings. To find the dark fringes, we shall use a clever (and simplifying) strategy that involves pairing up all the rays coming through the slit and then finding what conditions cause the wavelets of the rays in each pair to cancel each other. We apply this strategy in Fig. 36-4 to locate the first dark fringe, at point P_1. First, we mentally divide the slit into two *zones* of equal widths $a/2$. Then we extend to P_1 a light ray r_1 from the top point of the top zone and a light ray r_2 from the top point of the bottom zone. We want the wavelets along these two rays to cancel each other when they arrive at P_1. Then any similar pairing of rays from the two zones will give cancellation. A central axis is drawn from the center of the slit to screen C, and P_1 is located at an angle θ to that axis.

Path Length Difference. The wavelets of the pair of rays r_1 and r_2 are in phase within the slit because they originate from the same wavefront passing through the slit, along the width of the slit. However, to produce the first dark fringe they must be out of phase by $\lambda/2$ when they reach P_1; this phase difference is due to their path length difference, with the path traveled by the wavelet of r_2 to reach P_1 being longer than the path traveled by the wavelet of r_1. To display this path length difference, we find a point b on ray r_2 such that the path length from b to P_1 matches the path length of ray r_1. Then the path length difference between the two rays is the distance from the center of the slit to b.

When viewing screen C is near screen B, as in Fig. 36-4, the diffraction pattern on C is difficult to describe mathematically. However, we can simplify the mathematics considerably if we arrange for the screen separation D to be much larger than the slit width a. Then, as in Fig. 36-5, we can approximate rays r_1 and r_2

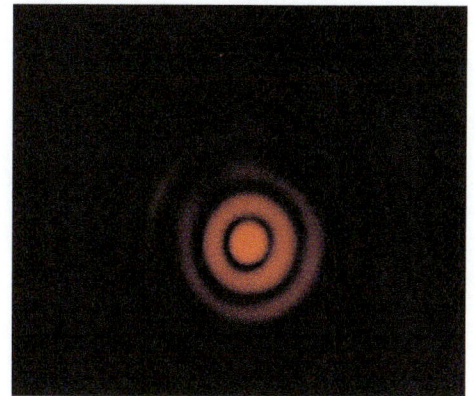

Courtesy Jearl Walker

Figure 36-3 A photograph of the diffraction pattern of a disk. Note the concentric diffraction rings and the Fresnel bright spot at the center of the pattern. This experiment is essentially identical to that arranged by the committee testing Fresnel's theories, because both the sphere they used and the disk used here have a cross section with a circular edge.

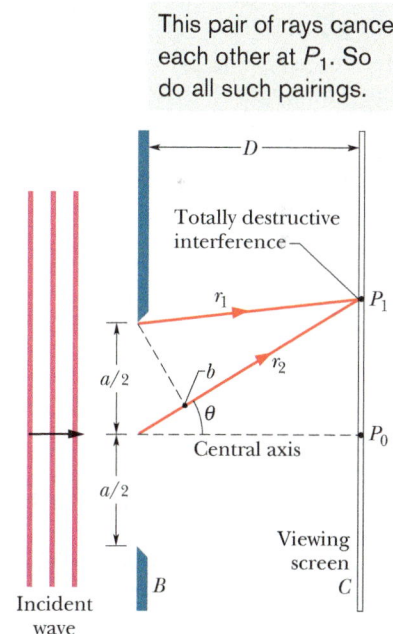

This pair of rays cancel each other at P_1. So do all such pairings.

Figure 36-4 Waves from the top points of two zones of width $a/2$ undergo fully destructive interference at point P_1 on viewing screen C.

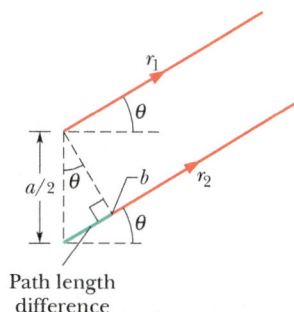

Figure 36-5 For $D \gg a$, we can approximate rays r_1 and r_2 as being parallel, at angle θ to the central axis.

This path length difference shifts one wave from the other, which determines the interference.

Path length difference

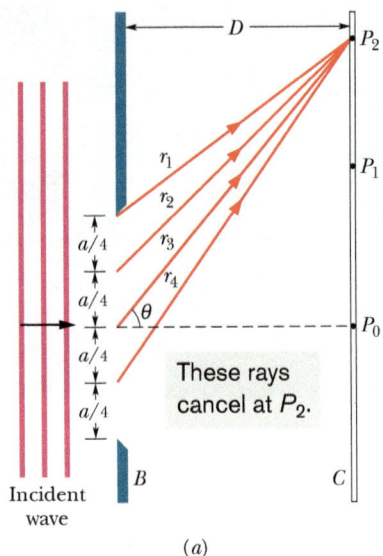

(a)

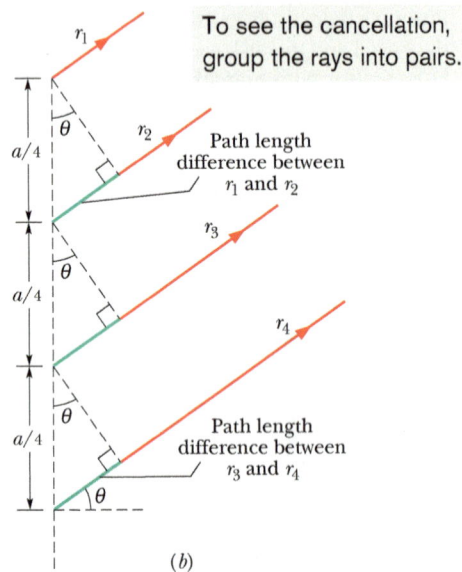

To see the cancellation, group the rays into pairs.

(b)

Figure 36-6 (a) Waves from the top points of four zones of width $a/4$ undergo fully destructive interference at point P_2. (b) For $D \gg a$, we can approximate rays $r_1, r_2, r_3,$ and r_4 as being parallel, at angle θ to the central axis.

as being parallel, at angle θ to the central axis. We can also approximate the triangle formed by point b, the top point of the slit, and the center point of the slit as being a right triangle, and one of the angles inside that triangle as being θ. The path length difference between rays r_1 and r_2 (which is still the distance from the center of the slit to point b) is then equal to $(a/2) \sin \theta$.

First Minimum. We can repeat this analysis for any other pair of rays originating at corresponding points in the two zones (say, at the midpoints of the zones) and extending to point P_1. Each such pair of rays has the same path length difference $(a/2) \sin \theta$. Setting this common path length difference equal to $\lambda/2$ (our condition for the first dark fringe), we have

$$\frac{a}{2} \sin \theta = \frac{\lambda}{2},$$

which gives us

$$a \sin \theta = \lambda \qquad \text{(first minimum).} \tag{36-1}$$

Given slit width a and wavelength λ, Eq. 36-1 tells us the angle θ of the first dark fringe above and (by symmetry) below the central axis.

Narrowing the Slit. Note that if we begin with $a > \lambda$ and then narrow the slit while holding the wavelength constant, we increase the angle at which the first dark fringes appear; that is, the extent of the diffraction (the extent of the flaring and the width of the pattern) is *greater* for a *narrower* slit. When we have reduced the slit width to the wavelength (that is, $a = \lambda$), the angle of the first dark fringes is 90°. Since the first dark fringes mark the two edges of the central bright fringe, that bright fringe must then cover the entire viewing screen.

Second Minimum. We find the second dark fringes above and below the central axis as we found the first dark fringes, except that we now divide the slit into *four* zones of equal widths $a/4$, as shown in Fig. 36-6a. We then extend rays r_1, $r_2, r_3,$ and r_4 from the top points of the zones to point P_2, the location of the second dark fringe above the central axis. To produce that fringe, the path length difference between r_1 and r_2, that between r_2 and r_3, and that between r_3 and r_4 must all be equal to $\lambda/2$.

For $D \gg a$, we can approximate these four rays as being parallel, at angle θ to the central axis. To display their path length differences, we extend a perpendicular line through each adjacent pair of rays, as shown in Fig. 36-6b, to form a series of right triangles, each of which has a path length difference as one side. We see from the top triangle that the path length difference between r_1 and r_2 is $(a/4) \sin \theta$. Similarly, from the bottom triangle, the path length difference between r_3 and r_4 is also $(a/4) \sin \theta$. In fact, the path length difference for any two rays that originate at corresponding points in two adjacent zones is $(a/4) \sin \theta$. Since in each such case the path length difference is equal to $\lambda/2$, we have

$$\frac{a}{4} \sin \theta = \frac{\lambda}{2},$$

which gives us

$$a \sin \theta = 2\lambda \qquad \text{(second minimum).} \tag{36-2}$$

All Minima. We could now continue to locate dark fringes in the diffraction pattern by splitting up the slit into more zones of equal width. We would always choose an even number of zones so that the zones (and their waves) could be paired as we have been doing. We would find that the dark fringes above and below the central axis can be located with the general equation

$$a \sin \theta = m\lambda, \qquad \text{for } m = 1, 2, 3, \dots \quad \text{(minima—dark fringes).} \tag{36-3}$$

You can remember this result in the following way. Draw a triangle like the one in Fig. 36-5, but for the full slit width a, and note that the path length difference between the top and bottom rays equals $a \sin \theta$. Thus, Eq. 36-3 says:

In a single-slit diffraction experiment, dark fringes are produced where the path length differences ($a \sin \theta$) between the top and bottom rays are equal to $\lambda, 2\lambda, 3\lambda, \ldots$.

This may seem to be wrong because the waves of those two particular rays will be exactly in phase with each other when their path length difference is an integer number of wavelengths. However, they each will still be part of a pair of waves that are exactly out of phase with each other; thus, *each* wave will be canceled by some other wave, resulting in darkness. (Two light waves that are exactly out of phase will always cancel each other, giving a net wave of zero, even if they happen to be exactly in phase with other light waves.)

Using a Lens. Equations 36-1, 36-2, and 36-3 are derived for the case of $D \gg a$. However, they also apply if we place a converging lens between the slit and the viewing screen and then move the screen in so that it coincides with the focal plane of the lens. The lens ensures that rays which now reach any point on the screen are *exactly* parallel (rather than approximately) back at the slit. They are like the initially parallel rays of Fig. 34-14a that are directed to the focal point by a converging lens.

Checkpoint 1

We produce a diffraction pattern on a viewing screen by means of a long narrow slit illuminated by blue light. Does the pattern expand away from the bright center (the maxima and minima shift away from the center) or contract toward it if we (a) switch to yellow light or (b) decrease the slit width?

Sample Problem 36.01 Single-slit diffraction pattern with white light

A slit of width a is illuminated by white light.

(a) For what value of a will the first minimum for red light of wavelength $\lambda = 650$ nm appear at $\theta = 15°$?

KEY IDEA

Diffraction occurs separately for each wavelength in the range of wavelengths passing through the slit, with the locations of the minima for each wavelength given by Eq. 36-3 ($a \sin \theta = m\lambda$).

Calculation: When we set $m = 1$ (for the first minimum) and substitute the given values of θ and λ, Eq. 36-3 yields

$$a = \frac{m\lambda}{\sin \theta} = \frac{(1)(650 \text{ nm})}{\sin 15°}$$

$$= 2511 \text{ nm} \approx 2.5 \ \mu\text{m}. \qquad \text{(Answer)}$$

For the incident light to flare out that much ($\pm 15°$ to the first minima) the slit has to be very fine indeed—in this case, a mere four times the wavelength. For comparison, note that a fine human hair may be about $100 \ \mu\text{m}$ in diameter.

(b) What is the wavelength λ' of the light whose first side diffraction maximum is at $15°$, thus coinciding with the first minimum for the red light?

KEY IDEA

The first side maximum for any wavelength is about halfway between the first and second minima for that wavelength.

Calculations: Those first and second minima can be located with Eq. 36-3 by setting $m = 1$ and $m = 2$, respectively. Thus, the first side maximum can be located *approximately* by setting $m = 1.5$. Then Eq. 36-3 becomes

$$a \sin \theta = 1.5\lambda'.$$

Solving for λ' and substituting known data yield

$$\lambda' = \frac{a \sin \theta}{1.5} = \frac{(2511 \text{ nm})(\sin 15°)}{1.5}$$

$$= 430 \text{ nm}. \qquad \text{(Answer)}$$

Light of this wavelength is violet (far blue, near the short-wavelength limit of the human range of visible light). From the two equations we used, can you see that the first side maximum for light of wavelength 430 nm will always coincide with the first minimum for light of wavelength 650 nm, no matter what the slit width is? However, the angle θ at which this overlap occurs does depend on slit width. If the slit is relatively narrow, the angle will be relatively large, and conversely.

PLUS Additional examples, video, and practice available at *WileyPLUS*

36-2 INTENSITY IN SINGLE-SLIT DIFFRACTION

Learning Objectives

After reading this module, you should be able to . . .

36.08 Divide a thin slit into multiple zones of equal width and write an expression for the phase difference of the wavelets from adjacent zones in terms of the angle θ to a point on the viewing screen.

36.09 For single-slit diffraction, draw phasor diagrams for the central maximum and several of the minima and maxima off to one side, indicating the phase difference between adjacent phasors, explaining how the net electric field is calculated, and

identifying the corresponding part of the diffraction pattern.

36.10 Describe a diffraction pattern in terms of the net electric field at points in the pattern.

36.11 Evaluate α, the convenient connection between angle θ to a point in a diffraction pattern and the intensity I at that point.

36.12 For a given point in a diffraction pattern, at a given angle, calculate the intensity I in terms of the intensity I_m at the center of the pattern.

Key Idea

● The intensity of the diffraction pattern at any given angle θ is

$$I(\theta) = I_m \left(\frac{\sin \alpha}{\alpha} \right)^2,$$

where I_m is the intensity at the center of the pattern and

$$\alpha = \frac{\pi a}{\lambda} \sin \theta.$$

Intensity in Single-Slit Diffraction, Qualitatively

In Module 36-1 we saw how to find the positions of the minima and the maxima in a single-slit diffraction pattern. Now we turn to a more general problem: find an expression for the intensity I of the pattern as a function of θ, the angular position of a point on a viewing screen.

To do this, we divide the slit of Fig. 36-4 into N zones of equal widths Δx small enough that we can assume each zone acts as a source of Huygens wavelets. We wish to superimpose the wavelets arriving at an arbitrary point P on the viewing screen, at angle θ to the central axis, so that we can determine the amplitude E_θ of the electric component of the resultant wave at P. The intensity of the light at P is then proportional to the square of that amplitude.

To find E_θ, we need the phase relationships among the arriving wavelets. The point here is that in general they have different phases because they travel different distances to reach P. The phase difference between wavelets from adjacent zones is given by

$$\left(\begin{array}{c} \text{phase} \\ \text{difference} \end{array} \right) = \left(\frac{2\pi}{\lambda} \right) \left(\begin{array}{c} \text{path length} \\ \text{difference} \end{array} \right).$$

For point P at angle θ, the path length difference between wavelets from adjacent zones is $\Delta x \sin \theta$. Thus, we can write the phase difference $\Delta\phi$ between wavelets from adjacent zones as

$$\Delta\phi = \left(\frac{2\pi}{\lambda} \right) (\Delta x \sin \theta). \tag{36-4}$$

We assume that the wavelets arriving at P all have the same amplitude ΔE. To find the amplitude E_θ of the resultant wave at P, we add the amplitudes ΔE via phasors. To do this, we construct a diagram of N phasors, one corresponding to the wavelet from each zone in the slit.

Central Maximum. For point P_0 at $\theta = 0$ on the central axis of Fig. 36-4, Eq. 36-4 tells us that the phase difference $\Delta\phi$ between the wavelets is zero; that is, the wavelets all arrive in phase. Figure 36-7a is the corresponding phasor diagram; adjacent phasors represent wavelets from adjacent zones and are arranged head to tail. Because there is zero phase difference between the wavelets, there is zero angle between each pair of adjacent phasors. The amplitude E_θ of the net

Here, with an even larger phase difference, they add to give a small amplitude and thus a small intensity.

E_θ

(d)

The last phasor is out of phase with the first phasor by 2π rad (full circle).

Here, with a larger phase difference, the phasors add to give zero amplitude and thus a minimum in the pattern.

$E_\theta = 0$

(c)

E_θ

Here the phasors have a small phase difference and add to give a smaller amplitude and thus less intensity in the pattern.

(b)

$E_\theta \ (= E_m)$

ΔE

Phasor for top ray

Phasor for bottom ray

(a)   The phasors from the 18 zones in the slit are in phase and add to give a maximum amplitude and thus the central maximum in the diffraction pattern.

I

θ

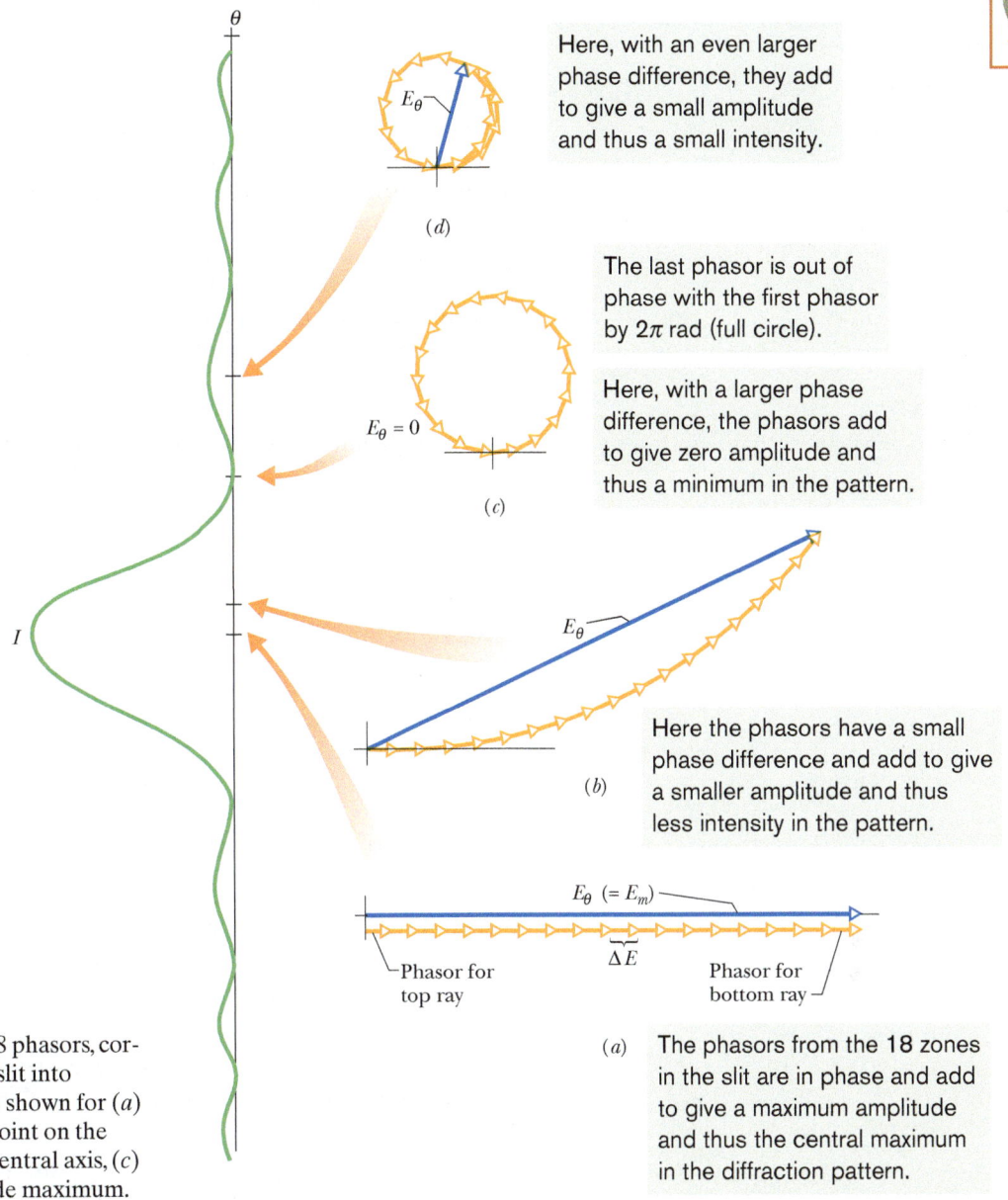

Figure 36-7 Phasor diagrams for $N = 18$ phasors, corresponding to the division of a single slit into 18 zones. Resultant amplitudes E_θ are shown for (a) the central maximum at $\theta = 0$, (b) a point on the screen lying at a small angle θ to the central axis, (c) the first minimum, and (d) the first side maximum.

wave at P_0 is the vector sum of these phasors. This arrangement of the phasors turns out to be the one that gives the greatest value for the amplitude E_θ. We call this value E_m; that is, E_m is the value of E_θ for $\theta = 0$.

We next consider a point P that is at a small angle θ to the central axis. Equation 36-4 now tells us that the phase difference $\Delta\phi$ between wavelets from adjacent zones is no longer zero. Figure 36-7b shows the corresponding phasor diagram; as before, the phasors are arranged head to tail, but now there is an angle $\Delta\phi$ between adjacent phasors. The amplitude E_θ at this new point is still the vector sum of the phasors, but it is smaller than that in Fig. 36-7a, which means that the intensity of the light is less at this new point P than at P_0.

First Minimun. If we continue to increase θ, the angle $\Delta\phi$ between adjacent phasors increases, and eventually the chain of phasors curls completely around so that the head of the last phasor just reaches the tail of the first phasor (Fig. 36-7c). The amplitude E_θ is now zero, which means that the intensity of the light is also zero. We have reached the first minimum, or dark fringe, in the diffraction pattern. The first and last phasors now have a phase difference of 2π rad, which means that the path length difference between the top and bottom rays through the slit equals one wavelength. Recall that this is the condition we determined for the first diffraction minimum.

First Side Maximum. As we continue to increase θ, the angle $\Delta\phi$ between adjacent phasors continues to increase, the chain of phasors begins to wrap back on itself, and the resulting coil begins to shrink. Amplitude E_θ now increases until it reaches a maximum value in the arrangement shown in Fig. 36-7d. This arrangement corresponds to the first side maximum in the diffraction pattern.

Second Minimum. If we increase θ a bit more, the resulting shrinkage of the coil decreases E_θ, which means that the intensity also decreases. When θ is increased enough, the head of the last phasor again meets the tail of the first phasor. We have then reached the second minimum.

We could continue this qualitative method of determining the maxima and minima of the diffraction pattern but, instead, we shall now turn to a quantitative method.

✓ **Checkpoint 2**

The figures represent, in smoother form (with more phasors) than Fig. 36-7, the phasor diagrams for two points of a diffraction pattern that are on opposite sides of a certain diffraction maximum. (a) Which maximum is it? (b) What is the approximate value of m (in Eq. 36-3) that corresponds to this maximum?

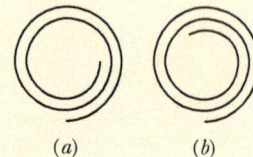

(a) (b)

Intensity in Single-Slit Diffraction, Quantitatively

Equation 36-3 tells us how to locate the minima of the single-slit diffraction pattern on screen C of Fig. 36-4 as a function of the angle θ in that figure. Here we wish to derive an expression for the intensity $I(\theta)$ of the pattern as a function of θ. We state, and shall prove below, that the intensity is given by

$$I(\theta) = I_m \left(\frac{\sin\alpha}{\alpha} \right)^2, \tag{36-5}$$

where

$$\alpha = \tfrac{1}{2}\phi = \frac{\pi a}{\lambda} \sin\theta. \tag{36-6}$$

The symbol α is just a convenient connection between the angle θ that locates a point on the viewing screen and the light intensity $I(\theta)$ at that point. The intensity I_m is the greatest value of the intensities $I(\theta)$ in the pattern and occurs at the central maximum (where $\theta = 0$), and ϕ is the phase difference (in radians) between the top and bottom rays from the slit of width a.

Study of Eq. 36-5 shows that intensity minima will occur where

$$\alpha = m\pi, \quad \text{for } m = 1, 2, 3, \ldots. \tag{36-7}$$

If we put this result into Eq. 36-6, we find

$$m\pi = \frac{\pi a}{\lambda} \sin\theta, \quad \text{for } m = 1, 2, 3, \ldots,$$

or

$$a \sin\theta = m\lambda, \quad \text{for } m = 1, 2, 3, \ldots \quad \text{(minima—dark fringes)}, \tag{36-8}$$

which is exactly Eq. 36-3, the expression that we derived earlier for the location of the minima.

Plots. Figure 36-8 shows plots of the intensity of a single-slit diffraction pattern, calculated with Eqs. 36-5 and 36-6 for three slit widths: $a = \lambda$, $a = 5\lambda$, and $a = 10\lambda$. Note that as the slit width increases (relative to the wavelength), the width of the *central diffraction maximum* (the central hill-like region of the graphs) decreases; that is, the light undergoes less flaring by the slit. The secondary maxima also decrease in width (and become weaker). In the limit of slit width a being much greater than wavelength λ, the secondary maxima due to the slit disappear; we then no longer have single-slit diffraction (but we still have diffraction due to the edges of the wide slit, like that produced by the edges of the razor blade in Fig. 36-2).

Proof of Eqs. 36-5 and 36-6

To find an expression for the intensity at a point in the diffraction pattern, we need to divide the slit into many zones and then add the phasors corresponding to those zones, as we did in Fig. 36-7. The arc of phasors in Fig. 36-9 represents the wavelets that reach an arbitrary point P on the viewing screen of Fig. 36-4, corresponding to a particular small angle θ. The amplitude E_θ of the resultant wave at P is the vector sum of these phasors. If we divide the slit of Fig. 36-4 into infinitesimal zones of width Δx, the arc of phasors in Fig. 36-9 approaches the arc of a circle; we call its radius R as indicated in that figure. The length of the arc must be E_m, the amplitude at the center of the diffraction pattern, because if we straightened out the arc we would have the phasor arrangement of Fig. 36-7a (shown lightly in Fig. 36-9).

The angle ϕ in the lower part of Fig. 36-9 is the difference in phase between the infinitesimal vectors at the left and right ends of arc E_m. From the geometry, ϕ is also the angle between the two radii marked R in Fig. 36-9. The dashed line in that figure, which bisects ϕ, then forms two congruent right triangles. From either triangle we can write

$$\sin \tfrac{1}{2}\phi = \frac{E_\theta}{2R}. \tag{36-9}$$

In radian measure, ϕ is (with E_m considered to be a circular arc)

$$\phi = \frac{E_m}{R}.$$

Solving this equation for R and substituting in Eq. 36-9 lead to

$$E_\theta = \frac{E_m}{\tfrac{1}{2}\phi} \sin \tfrac{1}{2}\phi. \tag{36-10}$$

Intensity. In Module 33-2 we saw that the intensity of an electromagnetic wave is proportional to the square of the amplitude of its electric field. Here, this means that the maximum intensity I_m (at the center of the pattern) is proportional to E_m^2 and the intensity $I(\theta)$ at angle θ is proportional to E_θ^2. Thus,

$$\frac{I(\theta)}{I_m} = \frac{E_\theta^2}{E_m^2}. \tag{36-11}$$

Substituting for E_θ with Eq. 36-10 and then substituting $\alpha = \tfrac{1}{2}\phi$, we are led to Eq. 36-5 for the intensity as a function of θ:

$$I(\theta) = I_m \left(\frac{\sin \alpha}{\alpha} \right)^2.$$

The second equation we wish to prove relates α to θ. The phase difference ϕ between the rays from the top and bottom of the entire slit may be related to a path length difference with Eq. 36-4; it tells us that

$$\phi = \left(\frac{2\pi}{\lambda} \right) (a \sin \theta),$$

where a is the sum of the widths Δx of the infinitesimal zones. However, $\phi = 2\alpha$, so this equation reduces to Eq. 36-6.

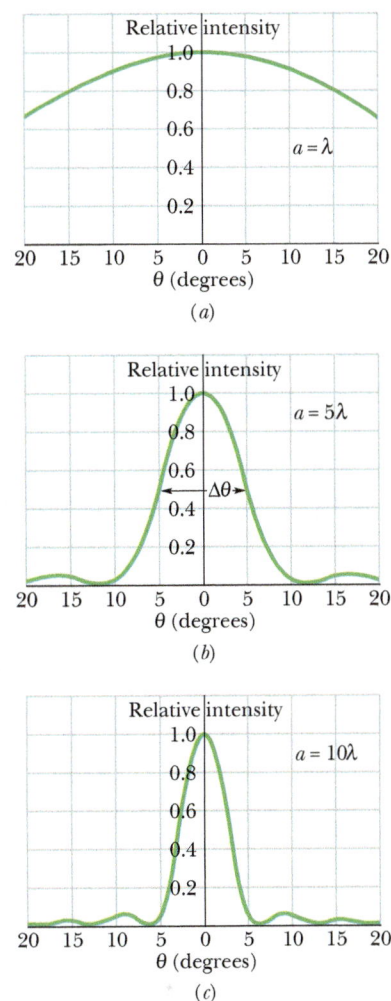

(a)

(b)

(c)

Figure 36-8 The relative intensity in single-slit diffraction for three values of the ratio a/λ. The wider the slit is, the narrower is the central diffraction maximum.

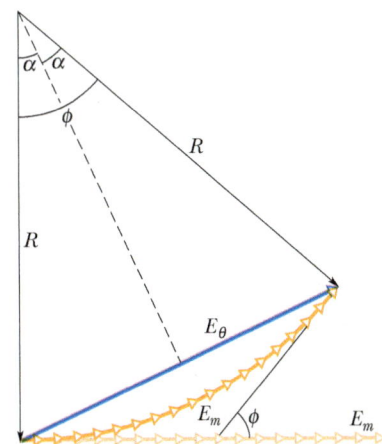

Figure 36-9 A construction used to calculate the intensity in single-slit diffraction. The situation shown corresponds to that of Fig. 36-7b.

✅ **Checkpoint 3**

Two wavelengths, 650 and 430 nm, are used separately in a single-slit diffraction experiment. The figure shows the results as graphs of intensity I versus angle θ for the two diffraction patterns. If both wavelengths are then used simultaneously, what color will be seen in the combined diffraction pattern at (a) angle A and (b) angle B?

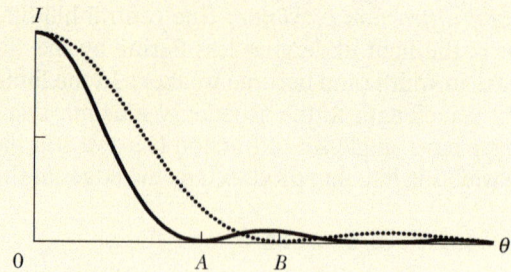

Sample Problem 36.02 Intensities of the maxima in a single-slit interference pattern

Find the intensities of the first three secondary maxima (side maxima) in the single-slit diffraction pattern of Fig. 36-1, measured as a percentage of the intensity of the central maximum.

KEY IDEAS

The secondary maxima lie approximately halfway between the minima, whose angular locations are given by Eq. 36-7 ($\alpha = m\pi$). The locations of the secondary maxima are then given (approximately) by

$$a = (m + \tfrac{1}{2})\pi, \quad \text{for } m = 1, 2, 3, \ldots,$$

with α in radian measure. We can relate the intensity I at any point in the diffraction pattern to the intensity I_m of the central maximum via Eq. 36-5.

Calculations: Substituting the approximate values of α for the secondary maxima into Eq. 36-5 to obtain the relative

intensities at those maxima, we get

$$\frac{I}{I_m} = \left(\frac{\sin \alpha}{\alpha}\right)^2 = \left(\frac{\sin(m + \tfrac{1}{2})\pi}{(m + \tfrac{1}{2})\pi}\right)^2, \quad \text{for } m = 1, 2, 3, \ldots.$$

The first of the secondary maxima occurs for $m = 1$, and its relative intensity is

$$\frac{I_1}{I_m} = \left(\frac{\sin(1 + \tfrac{1}{2})\pi}{(1 + \tfrac{1}{2})\pi}\right)^2 = \left(\frac{\sin 1.5\pi}{1.5\pi}\right)^2$$

$$= 4.50 \times 10^{-2} \approx 4.5\%. \quad \text{(Answer)}$$

For $m = 2$ and $m = 3$ we find that

$$\frac{I_2}{I_m} = 1.6\% \quad \text{and} \quad \frac{I_3}{I_m} = 0.83\%. \quad \text{(Answer)}$$

As you can see from these results, successive secondary maxima decrease rapidly in intensity. Figure 36-1 was deliberately overexposed to reveal them.

WILEY PLUS Additional examples, video, and practice available at *WileyPLUS*

36-3 DIFFRACTION BY A CIRCULAR APERTURE

Learning Objectives

After reading this module, you should be able to . . .

36.13 Describe and sketch the diffraction pattern from a small circular aperture or obstacle.

36.14 For diffraction by a small circular aperture or obstacle, apply the relationships between the angle θ to the first minimum, the wavelength λ of the light, the diameter d of the aperture, the distance D to a viewing screen, and the distance y between the minimum and the center of the diffraction pattern.

36.15 By discussing the diffraction patterns of point objects,

explain how diffraction limits visual resolution of objects.

36.16 Identify that Rayleigh's criterion for resolvability gives the (approximate) angle at which two point objects are just barely resolvable.

36.17 Apply the relationships between the angle θ_R in Rayleigh's criterion, the wavelength λ of the light, the diameter d of the aperture (for example, the diameter of the pupil of an eye), the angle θ subtended by two distant point objects, and the distance L to those objects.

Key Ideas

● Diffraction by a circular aperture or a lens with diameter d produces a central maximum and concentric maxima and minima, with the first minimum at an angle θ given by

$$\sin \theta = 1.22 \frac{\lambda}{d} \quad \text{(first minimum—circular aperture).}$$

● Rayleigh's criterion suggests that two objects are on the verge of resolvability if the central diffraction maximum of one is at the first minimum of the other. Their angular separation can then be no less than

$$\theta_R = 1.22 \frac{\lambda}{d} \quad \text{(Rayleigh's criterion),}$$

in which d is the diameter of the aperture through which the light passes.

Diffraction by a Circular Aperture

Here we consider diffraction by a circular aperture—that is, a circular opening, such as a circular lens, through which light can pass. Figure 36-10 shows the image formed by light from a laser that was directed onto a circular aperture with a very small diameter. This image is not a point, as geometrical optics would suggest, but a circular disk surrounded by several progressively fainter secondary rings. Comparison with Fig. 36-1 leaves little doubt that we are dealing with a diffraction phenomenon. Here, however, the aperture is a circle of diameter d rather than a rectangular slit.

The (complex) analysis of such patterns shows that the first minimum for the diffraction pattern of a circular aperture of diameter d is located by

$$\sin \theta = 1.22 \frac{\lambda}{d} \quad \text{(first minimum—circular aperture).} \tag{36-12}$$

The angle θ here is the angle from the central axis to any point on that (circular) minimum. Compare this with Eq. 36-1,

$$\sin \theta = \frac{\lambda}{a} \quad \text{(first minimum—single slit),} \tag{36-13}$$

which locates the first minimum for a long narrow slit of width a. The main difference is the factor 1.22, which enters because of the circular shape of the aperture.

Courtesy Jearl Walker

Figure 36-10 The diffraction pattern of a circular aperture. Note the central maximum and the circular secondary maxima. The figure has been overexposed to bring out these secondary maxima, which are much less intense than the central maximum.

Resolvability

The fact that lens images are diffraction patterns is important when we wish to *resolve* (distinguish) two distant point objects whose angular separation is small. Figure 36-11 shows, in three different cases, the visual appearance and corresponding intensity pattern for two distant point objects (stars, say) with small

Figure 36-11 At the top, the images of two point sources (stars) formed by a converging lens. At the bottom, representations of the image intensities. In (*a*) the angular separation of the sources is too small for them to be distinguished, in (*b*) they can be marginally distinguished, and in (*c*) they are clearly distinguished. Rayleigh's criterion is satisfied in (*b*), with the central maximum of one diffraction pattern coinciding with the first minimum of the other.

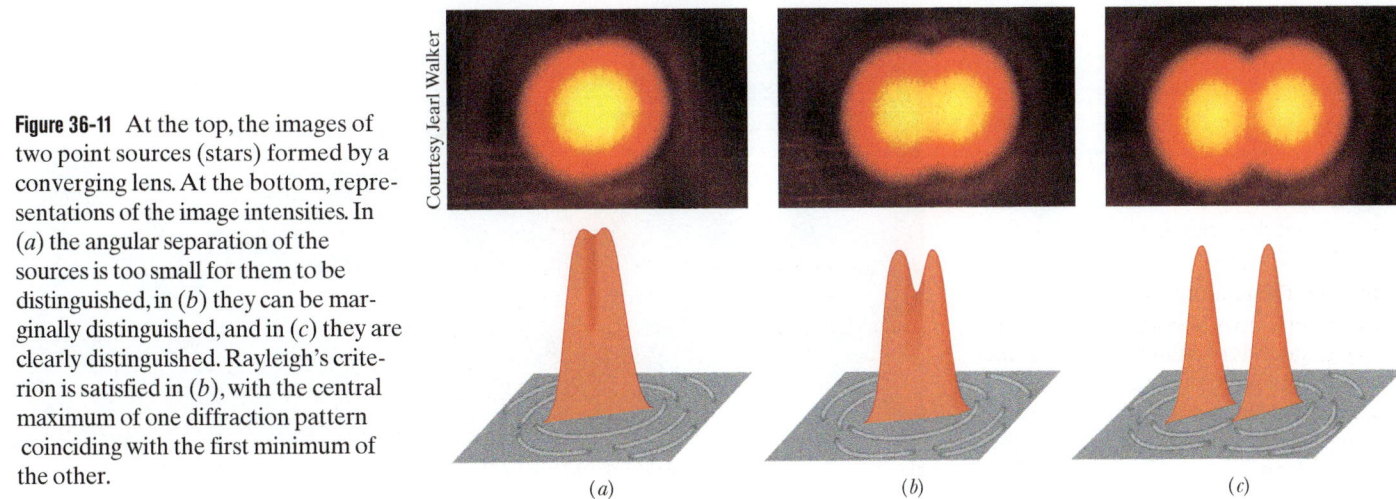

Courtesy Jearl Walker

(*a*)   (*b*)   (*c*)

angular separation. In Figure 36-11a, the objects are not resolved because of diffraction; that is, their diffraction patterns (mainly their central maxima) overlap so much that the two objects cannot be distinguished from a single point object. In Fig. 36-11b the objects are barely resolved, and in Fig. 36-11c they are fully resolved.

In Fig. 36-11b the angular separation of the two point sources is such that the central maximum of the diffraction pattern of one source is centered on the first minimum of the diffraction pattern of the other, a condition called **Rayleigh's criterion** for resolvability. From Eq. 36-12, two objects that are barely resolvable by this criterion must have an angular separation θ_R of

$$\theta_R = \sin^{-1} \frac{1.22\lambda}{d}.$$

Since the angles are small, we can replace $\sin \theta_R$ with θ_R expressed in radians:

$$\theta_R = 1.22 \frac{\lambda}{d} \quad \text{(Rayleigh's criterion)}. \tag{36-14}$$

Human Vision. Applying Rayleigh's criterion for resolvability to human vision is only an approximation because visual resolvability depends on many factors, such as the relative brightness of the sources and their surroundings, turbulence in the air between the sources and the observer, and the functioning of the observer's visual system. Experimental results show that the least angular separation that can actually be resolved by a person is generally somewhat greater than the value given by Eq. 36-14. However, for calculations here, we shall take Eq. 36-14 as being a precise criterion: If the angular separation θ between the sources is greater than θ_R, we can visually resolve the sources; if it is less, we cannot.

Pointillism. Rayleigh's criterion can explain the arresting illusions of color in the style of painting known as pointillism (Fig. 36-12). In this style, a painting is made not with brush strokes in the usual sense but rather with a myriad of small colored dots. One fascinating aspect of a pointillistic painting is that when you change your distance from it, the colors shift in subtle, almost subconscious ways. This color shifting has to do with whether you can resolve the colored dots. When you stand close enough to the painting, the angular separations θ of adjacent dots are greater than θ_R and thus the dots can be seen individually. Their colors are the true colors of the paints used. However, when

Figure 36-12 The pointillistic painting *The Seine at Herblay* by Maximilien Luce consists of thousands of colored dots. With the viewer very close to the canvas, the dots and their true colors are visible. At normal viewing distances, the dots are irresolvable and thus blend.

Maximilien Luce, *The Seine at Herblay*, 1890. Musée d'Orsay, Paris, France. Photo by Erich Lessing/Art Resource

you stand far enough from the painting, the angular separations θ are less than θ_R and the dots cannot be seen individually. The resulting blend of colors coming into your eye from any group of dots can then cause your brain to "make up" a color for that group—a color that may not actually exist in the group. In this way, a pointillistic painter uses your visual system to create the colors of the art.

When we wish to use a lens instead of our visual system to resolve objects of small angular separation, it is desirable to make the diffraction pattern as small as possible. According to Eq. 36-14, this can be done either by increasing the lens diameter or by using light of a shorter wavelength. For this reason ultraviolet light is often used with microscopes because its wavelength is shorter than a visible light wavelength.

✔ Checkpoint 4

Suppose that you can barely resolve two red dots because of diffraction by the pupil of your eye. If we increase the general illumination around you so that the pupil decreases in diameter, does the resolvability of the dots improve or diminish? Consider only diffraction. (You might experiment to check your answer.)

Sample Problem 36.03   Pointillistic paintings use the diffraction of your eye

Figure 36-13a is a representation of the colored dots on a pointillistic painting. Assume that the average center-to-center separation of the dots is $D = 2.0$ mm. Also assume that the diameter of the pupil of your eye is $d = 1.5$ mm and that the least angular separation between dots you can resolve is set only by Rayleigh's criterion. What is the least viewing distance from which you cannot distinguish any dots on the painting?

KEY IDEA

Consider any two adjacent dots that you can distinguish when you are close to the painting. As you move away, you continue to distinguish the dots until their angular separation θ (in your view) has decreased to the angle given by Rayleigh's criterion:

$$\theta_R = 1.22 \frac{\lambda}{d}. \qquad (36\text{-}15)$$

Calculations: Figure 36-13b shows, from the side, the angular separation θ of the dots, their center-to-center separation D, and your distance L from them. Because D/L is small, angle θ is also small and we can make the approximation

$$\theta = \frac{D}{L}. \qquad (36\text{-}16)$$

Setting θ of Eq. 36-16 equal to θ_R of Eq. 36-15 and solving for L, we then have

$$L = \frac{Dd}{1.22\lambda}. \qquad (36\text{-}17)$$

Equation 36-17 tells us that L is larger for smaller λ. Thus, as you move away from the painting, adjacent red dots (long wavelengths) become indistinguishable before adjacent blue dots do. To find the least distance L at which *no* colored dots are distinguishable, we substitute $\lambda = 400$ nm (blue or violet light) into Eq. 36-17:

$$L = \frac{(2.0 \times 10^{-3}\,\text{m})(1.5 \times 10^{-3}\,\text{m})}{(1.22)(400 \times 10^{-9}\,\text{m})} = 6.1\ \text{m.}\quad\text{(Answer)}$$

At this or a greater distance, the color you perceive at any given spot on the painting is a blended color that may not actually exist there.

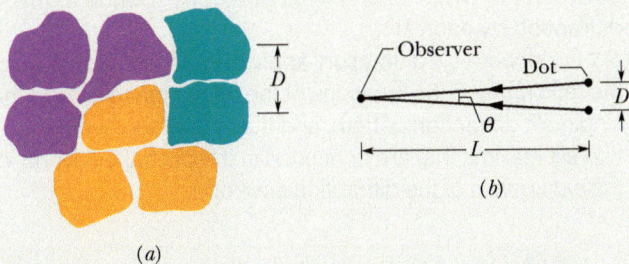

Figure 36-13 (a) Representation of some dots on a pointillistic painting, showing an average center-to-center separation D. (b) The arrangement of separation D between two dots, their angular separation θ, and the viewing distance L.

PLUS Additional examples, video, and practice available at *WileyPLUS*

Sample Problem 36.04 Rayleigh's criterion for resolving two distant objects

A circular converging lens, with diameter $d = 32$ mm and focal length $f = 24$ cm, forms images of distant point objects in the focal plane of the lens. The wavelength is $\lambda = 550$ nm.

(a) Considering diffraction by the lens, what angular separation must two distant point objects have to satisfy Rayleigh's criterion?

KEY IDEA

Figure 36-14 shows two distant point objects P_1 and P_2, the lens, and a viewing screen in the focal plane of the lens. It also shows, on the right, plots of light intensity I versus position on the screen for the central maxima of the images formed by the lens. Note that the angular separation θ_o of the objects equals the angular separation θ_i of the images. Thus, if the images are to satisfy Rayleigh's criterion, these separations must be given by Eq. 36-14 (for small angles).

Calculations: From Eq. 36-14, we obtain

$$\theta_o = \theta_i = \theta_R = 1.22 \frac{\lambda}{d}$$

$$= \frac{(1.22)(550 \times 10^{-9}\,\text{m})}{32 \times 10^{-3}\,\text{m}} = 2.1 \times 10^{-5}\,\text{rad.} \quad \text{(Answer)}$$

Each central maximum in the two intensity curves of Fig. 36-14 is centered on the first minimum of the other curve.

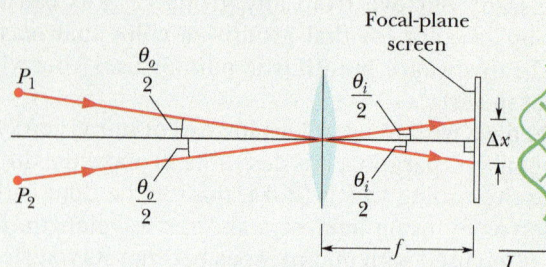

Figure 36-14 Light from two distant point objects P_1 and P_2 passes through a converging lens and forms images on a viewing screen in the focal plane of the lens. Only one representative ray from each object is shown. The images are not points but diffraction patterns, with intensities approximately as plotted at the right.

(b) What is the separation Δx of the centers of the *images* in the focal plane? (That is, what is the separation of the *central* peaks in the two intensity-versus-position curves?)

Calculations: From either triangle between the lens and the screen in Fig. 36-14, we see that $\tan \theta_i/2 = \Delta x/2f$. Rearranging this equation and making the approximation $\tan \theta \approx \theta$, we find

$$\Delta x = f\theta_i, \quad (36\text{-}18)$$

where θ_i is in radian measure. We then find

$$\Delta x = (0.24\,\text{m})(2.1 \times 10^{-5}\,\text{rad}) = 5.0\,\mu\text{m}. \quad \text{(Answer)}$$

WILEY PLUS Additional examples, video, and practice available at *WileyPLUS*

36-4 DIFFRACTION BY A DOUBLE SLIT

Learning Objectives

After reading this module, you should be able to . . .

36.18 In a sketch of a double-slit experiment, explain how the diffraction through each slit modifies the two-slit interference pattern, and identify the diffraction envelope, the central peak, and the side peaks of that envelope.

36.19 For a given point in a double-slit diffraction pattern, calculate the intensity I in terms of the intensity I_m at the center of the pattern.

36.20 In the intensity equation for a double-slit diffraction

pattern, identify what part corresponds to the interference between the two slits and what part corresponds to the diffraction by each slit.

36.21 For double-slit diffraction, apply the relationship between the ratio d/a and the locations of the diffraction minima in the single-slit diffraction pattern, and then count the number of two-slit maxima that are contained in the central peak and in the side peaks of the diffraction envelope.

Key Ideas

● Waves passing through two slits produce a combination of double-slit interference and diffraction by each slit.

● For identical slits with width a and center-to-center separation d, the intensity in the pattern varies with the angle θ from the central axis as

$$I(\theta) = I_m(\cos^2 \beta)\left(\frac{\sin \alpha}{\alpha}\right)^2 \quad \text{(double slit)},$$

where I_m is the intensity at the center of the pattern,

$$\beta = \left(\frac{\pi d}{\lambda}\right) \sin\theta,$$

and

$$\alpha = \left(\frac{\pi a}{\lambda}\right) \sin\theta.$$

Diffraction by a Double Slit

In the double-slit experiments of Chapter 35, we implicitly assumed that the slits were much narrower than the wavelength of the light illuminating them; that is, $a \ll \lambda$. For such narrow slits, the central maximum of the diffraction pattern of either slit covers the entire viewing screen. Moreover, the interference of light from the two slits produces bright fringes with approximately the same intensity (Fig. 35-12).

In practice with visible light, however, the condition $a \ll \lambda$ is often not met. For relatively wide slits, the interference of light from two slits produces bright fringes that do not all have the same intensity. That is, the intensities of the fringes produced by double-slit interference (as discussed in Chapter 35) are modified by diffraction of the light passing through each slit (as discussed in this chapter).

Plots. As an example, the intensity plot of Fig. 36-15a suggests the double-slit interference pattern that would occur if the slits were infinitely narrow (and thus $a \ll \lambda$); all the bright interference fringes would have the same intensity. The intensity plot of Fig. 36-15b is that for diffraction by a single actual slit; the diffraction pattern has a broad central maximum and weaker secondary maxima at $\pm 17°$. The plot of Fig. 36-15c suggests the interference pattern for two actual slits. That plot was constructed by using the curve of Fig. 36-15b as an *envelope* on the intensity plot in Fig. 36-15a. The positions of the fringes are not changed; only the intensities are affected.

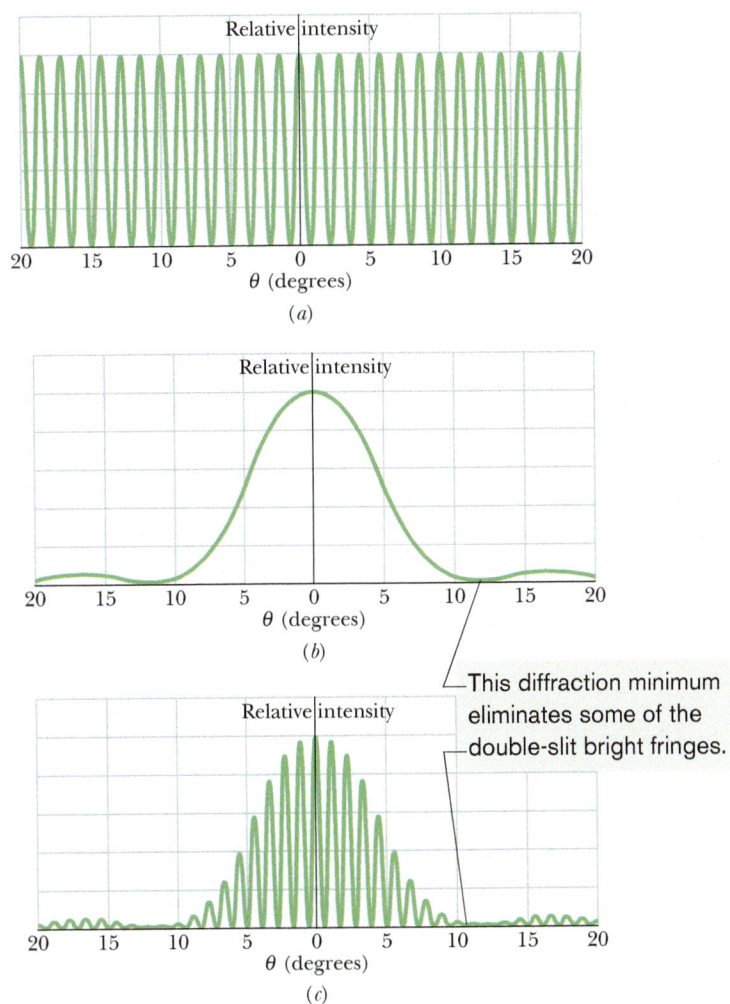

This diffraction minimum eliminates some of the double-slit bright fringes.

Figure 36-15 (*a*) The intensity plot to be expected in a double-slit interference experiment with vanishingly narrow slits. (*b*) The intensity plot for diffraction by a typical slit of width *a* (not vanishingly narrow). (*c*) The intensity plot to be expected for two slits of width *a*. The curve of (*b*) acts as an envelope, limiting the intensity of the double-slit fringes in (*a*). Note that the first minima of the diffraction pattern of (*b*) eliminate the double-slit fringes that would occur near 12° in (*c*).

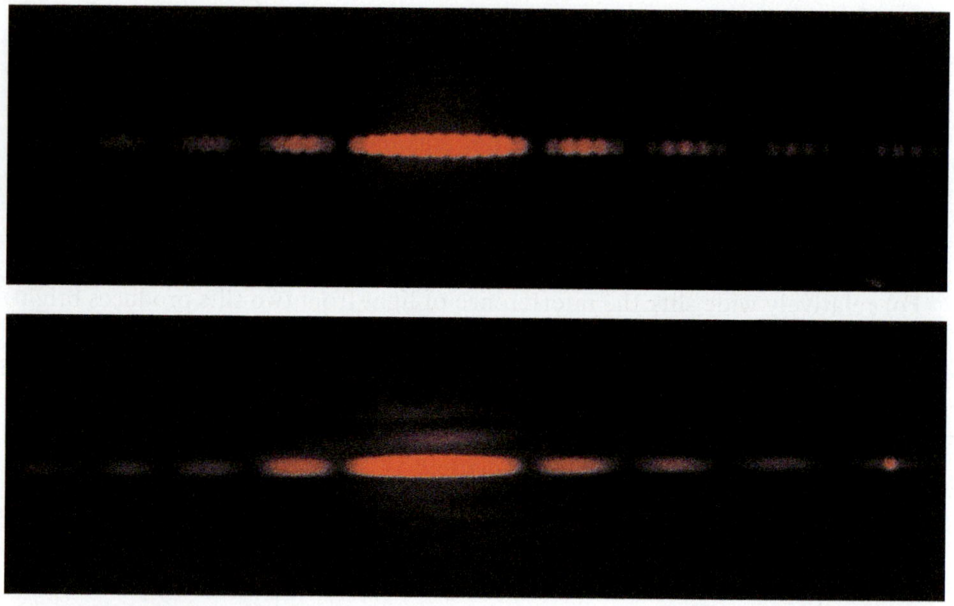

Figure 36-16 (*a*) Interference fringes for an actual double-slit system; compare with Fig. 36-15*c*. (*b*) The diffraction pattern of a single slit; compare with Fig. 36-15*b*.

Courtesy Jearl Walker

Photos. Figure 36-16*a* shows an actual pattern in which both double-slit interference and diffraction are evident. If one slit is covered, the single-slit diffraction pattern of Fig. 36-16*b* results. Note the correspondence between Figs. 36-16*a* and 36-15*c*, and between Figs. 36-16*b* and 36-15*b*. In comparing these figures, bear in mind that Fig. 36-16 has been deliberately overexposed to bring out the faint secondary maxima and that several secondary maxima (rather than one) are shown.

Intensity. With diffraction effects taken into account, the intensity of a double-slit interference pattern is given by

$$I(\theta) = I_m(\cos^2 \beta)\left(\frac{\sin \alpha}{\alpha}\right)^2 \quad \text{(double slit)}, \tag{36-19}$$

in which

$$\beta = \frac{\pi d}{\lambda}\sin\theta \tag{36-20}$$

and

$$\alpha = \frac{\pi a}{\lambda}\sin\theta. \tag{36-21}$$

Here *d* is the distance between the centers of the slits and *a* is the slit width. Note carefully that the right side of Eq. 36-19 is the product of I_m and two factors. (1) The *interference factor* $\cos^2 \beta$ is due to the interference between two slits with slit separation *d* (as given by Eqs. 35-22 and 35-23). (2) The *diffraction factor* $[(\sin \alpha)/\alpha]^2$ is due to diffraction by a single slit of width *a* (as given by Eqs. 36-5 and 36-6).

Let us check these factors. If we let $a \to 0$ in Eq. 36-21, for example, then $\alpha \to 0$ and $(\sin \alpha)/\alpha \to 1$. Equation 36-19 then reduces, as it must, to an equation describing the interference pattern for a pair of vanishingly narrow slits with slit separation *d*. Similarly, putting $d = 0$ in Eq. 36-20 is equivalent physically to causing the two slits to merge into a single slit of width *a*. Then Eq. 36-20 yields $\beta = 0$ and $\cos^2 \beta = 1$. In this case Eq. 36-19 reduces, as it must, to an equation describing the diffraction pattern for a single slit of width *a*.

Language. The double-slit pattern described by Eq. 36-19 and displayed in Fig. 36-16*a* combines interference and diffraction in an intimate way. Both are superposition effects, in that they result from the combining of waves with different phases at a given point. If the combining waves originate from a small number of elementary coherent sources—as in a double-slit experiment with $a \ll \lambda$—we call the process *interference.* If the combining waves originate in a single wavefront—as in a single-slit experiment—we call the process *diffraction.* This distinction between interference and diffraction (which is somewhat arbitrary and not always adhered to) is a convenient one, but we should not forget that both are superposition effects and usually both are present simultaneously (as in Fig. 36-16a).

Sample Problem 36.05 Double-slit experiment with diffraction of each slit included

In a double-slit experiment, the wavelength λ of the light source is 405 nm, the slit separation d is 19.44 μm, and the slit width a is 4.050 μm. Consider the interference of the light from the two slits and also the diffraction of the light through each slit.

(a) How many bright interference fringes are within the central peak of the diffraction envelope?

KEY IDEAS

We first analyze the two basic mechanisms responsible for the optical pattern produced in the experiment:

1. *Single-slit diffraction:* The limits of the central peak are the first minima in the diffraction pattern due to either slit individually. (See Fig. 36-15.) The angular locations of those minima are given by Eq. 36-3 ($a \sin \theta = m\lambda$). Here let us rewrite this equation as $a \sin \theta = m_1\lambda$, with the subscript 1 referring to the one-slit diffraction. For the first minima in the diffraction pattern, we substitute $m_1 = 1$, obtaining

$$a \sin \theta = \lambda. \qquad (36\text{-}22)$$

2. *Double-slit interference:* The angular locations of the bright fringes of the double-slit interference pattern are given by Eq. 35-14, which we can write as

$$d \sin \theta = m_2\lambda, \qquad \text{for } m_2 = 0, 1, 2, \ldots. \qquad (36\text{-}23)$$

Here the subscript 2 refers to the double-slit interference.

Calculations: We can locate the first diffraction minimum within the double-slit fringe pattern by dividing Eq. 36-23 by Eq. 36-22 and solving for m_2. By doing so and then substituting the given data, we obtain

$$m_2 = \frac{d}{a} = \frac{19.44 \ \mu\text{m}}{4.050 \ \mu\text{m}} = 4.8.$$

This tells us that the bright interference fringe for $m_2 = 4$ fits into the central peak of the one-slit diffraction pattern, but the fringe for $m_2 = 5$ does not fit. Within the central diffraction peak we have the central bright fringe ($m_2 = 0$), and four bright fringes (up to $m_2 = 4$) on each side of it. Thus, a total of nine bright fringes of the double-slit interference pattern are within the central peak of the diffraction

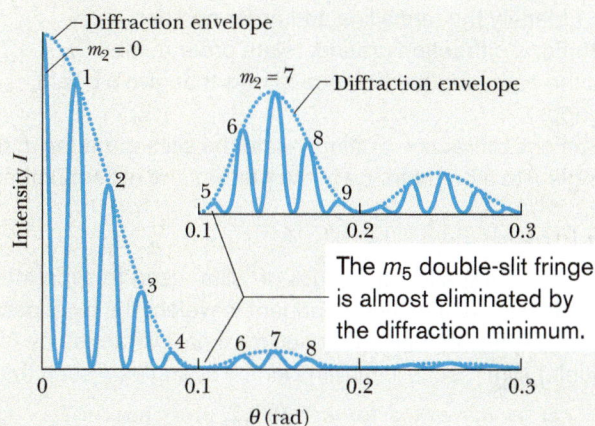

Figure 36-17 One side of the intensity plot for a two-slit interference experiment. The inset shows (vertically expanded) the plot within the first and second side peaks of the diffraction envelope.

envelope. The bright fringes to one side of the central bright fringe are shown in Fig. 36-17.

(b) How many bright fringes are within either of the first side peaks of the diffraction envelope?

KEY IDEA

The outer limits of the first side diffraction peaks are the second diffraction minima, each of which is at the angle θ given by $a \sin \theta = m_1\lambda$ with $m_1 = 2$:

$$a \sin \theta = 2\lambda. \qquad (36\text{-}24)$$

Calculation: Dividing Eq. 36-23 by Eq. 36-24, we find

$$m_2 = \frac{2d}{a} = \frac{(2)(19.44 \ \mu\text{m})}{4.050 \ \mu\text{m}} = 9.6.$$

This tells us that the second diffraction minimum occurs just before the bright interference fringe for $m_2 = 10$ in Eq. 36-23. Within either first side diffraction peak we have the fringes from $m_2 = 5$ to $m_2 = 9$, for a total of five bright fringes of the double-slit interference pattern (shown in the inset of Fig. 36-17). However, if the $m_2 = 5$ bright fringe, which is almost eliminated by the first diffraction minimum, is considered too dim to count, then only four bright fringes are in the first side diffraction peak.

36-5 DIFFRACTION GRATINGS

Learning Objectives

After reading this module, you should be able to . . .

36.22 Describe a diffraction grating and sketch the interference pattern it produces in monochromatic light.

36.23 Distinguish the interference patterns of a diffraction grating and a double-slit arrangement.

36.24 Identify the terms line and order number.

36.25 For a diffraction grating, relate order number m to the path length difference of rays that give a bright fringe.

36.26 For a diffraction grating, relate the slit separation d, the angle θ to a bright fringe in the pattern, the order number

m of that fringe, and the wavelength λ of the light.

36.27 Identify the reason why there is a maximum order number for a given diffraction grating.

36.28 Explain the derivation of the equation for a line's half-width in a diffraction-grating pattern.

36.29 Calculate the half-width of a line at a given angle in a diffraction-grating pattern.

36.30 Explain the advantage of increasing the number of slits in a diffraction grating.

36.31 Explain how a grating spectroscope works.

Key Idea

● A diffraction grating is a series of "slits" used to separate an incident wave into its component wavelengths by separating and displaying their diffraction maxima. Diffraction by N (multiple) slits results in maxima (lines) at angles θ such that

$$d \sin \theta = m\lambda, \quad \text{for } m = 0, 1, 2, \ldots \quad \text{(maxima)}.$$

● A line's half-width is the angle from its center to the point where it disappears into the darkness and is given by

$$\Delta\theta_{hw} = \frac{\lambda}{Nd \cos \theta} \quad \text{(half-width)}.$$

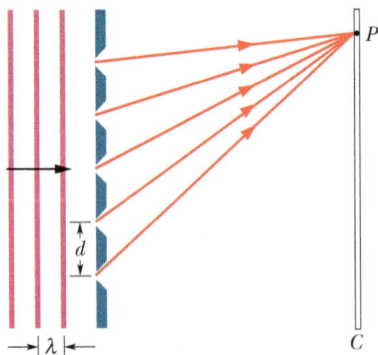

Figure 36-18 An idealized diffraction grating, consisting of only five rulings, that produces an interference pattern on a distant viewing screen C.

Diffraction Gratings

One of the most useful tools in the study of light and of objects that emit and absorb light is the **diffraction grating.** This device is somewhat like the double-slit arrangement of Fig. 35-10 but has a much greater number N of slits, often called *rulings,* perhaps as many as several thousand per millimeter. An idealized grating consisting of only five slits is represented in Fig. 36-18. When monochromatic light is sent through the slits, it forms narrow interference fringes that can be analyzed to determine the wavelength of the light. (Diffraction gratings can also be opaque surfaces with narrow parallel grooves arranged like the slits in Fig. 36-18. Light then scatters back from the grooves to form interference fringes rather than being transmitted through open slits.)

Pattern. With monochromatic light incident on a diffraction grating, if we gradually increase the number of slits from two to a large number N, the intensity plot changes from the typical double-slit plot of Fig. 36-15c to a much more complicated one and then eventually to a simple graph like that shown in Fig. 36-19a. The pattern you would see on a viewing screen using monochromatic red light from,

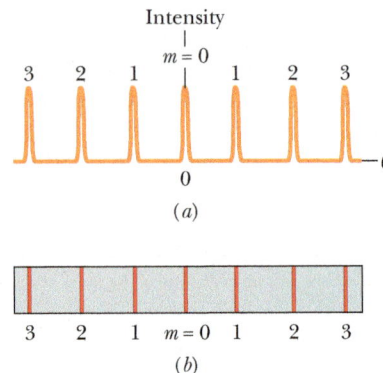

Figure 36-19 (*a*) The intensity plot produced by a diffraction grating with a great many rulings consists of narrow peaks, here labeled with their order numbers m. (*b*) The corresponding bright fringes seen on the screen are called lines and are here also labeled with order numbers m.

say, a helium–neon laser is shown in Fig. 36-19b. The maxima are now very narrow (and so are called *lines*); they are separated by relatively wide dark regions.

Equation. We use a familiar procedure to find the locations of the bright lines on the viewing screen. We first assume that the screen is far enough from the grating so that the rays reaching a particular point *P* on the screen are approximately parallel when they leave the grating (Fig. 36-20). Then we apply to each pair of adjacent rulings the same reasoning we used for double-slit interference. The separation *d* between rulings is called the *grating spacing*. (If *N* rulings occupy a total width *w*, then $d = w/N$.) The path length difference between adjacent rays is again $d \sin \theta$ (Fig. 36-20), where θ is the angle from the central axis of the grating (and of the diffraction pattern) to point *P*. A line will be located at *P* if the path length difference between adjacent rays is an integer number of wavelengths:

$$d \sin \theta = m\lambda, \quad \text{for } m = 0, 1, 2, \ldots \quad \text{(maxima—lines)}, \qquad (36\text{-}25)$$

where λ is the wavelength of the light. Each integer *m* represents a different line; hence these integers can be used to label the lines, as in Fig. 36-19. The integers are then called the *order numbers,* and the lines are called the zeroth-order line (the central line, with $m = 0$), the first-order line ($m = 1$), the second-order line ($m = 2$), and so on.

Determining Wavelength. If we rewrite Eq. 36-25 as $\theta = \sin^{-1}(m\lambda/d)$, we see that, for a given diffraction grating, the angle from the central axis to any line (say, the third-order line) depends on the wavelength of the light being used. Thus, when light of an unknown wavelength is sent through a diffraction grating, measurements of the angles to the higher-order lines can be used in Eq. 36-25 to determine the wavelength. Even light of several unknown wavelengths can be distinguished and identified in this way. We cannot do that with the double-slit arrangement of Module 35-2, even though the same equation and wavelength dependence apply there. In double-slit interference, the bright fringes due to different wavelengths overlap too much to be distinguished.

Width of the Lines

A grating's ability to resolve (separate) lines of different wavelengths depends on the width of the lines. We shall here derive an expression for the *half-width* of the central line (the line for which $m = 0$) and then state an expression for the half-widths of the higher-order lines. We define the **half-width** of the central line as being the angle $\Delta\theta_{hw}$ from the center of the line at $\theta = 0$ outward to where the line effectively ends and darkness effectively begins with the first minimum (Fig. 36-21). At such a minimum, the *N* rays from the *N* slits of the grating cancel one another. (The actual width of the central line is, of course, $2(\Delta\theta_{hw})$, but line widths are usually compared via half-widths.)

In Module 36-1 we were also concerned with the cancellation of a great many rays, there due to diffraction through a single slit. We obtained Eq. 36-3, which, because of the similarity of the two situations, we can use to find the first minimum here. It tells us that the first minimum occurs where the path length difference between the top and bottom rays equals λ. For single-slit diffraction, this difference is $a \sin \theta$. For a grating of *N* rulings, each separated from the next by distance *d*, the distance between the top and bottom rulings is *Nd* (Fig. 36-22), and so the path length difference between the top and bottom rays here is $Nd \sin \Delta\theta_{hw}$. Thus, the first minimum occurs where

$$Nd \sin \Delta\theta_{hw} = \lambda. \qquad (36\text{-}26)$$

Because $\Delta\theta_{hw}$ is small, $\sin \Delta\theta_{hw} = \Delta\theta_{hw}$ (in radian measure). Substituting this in Eq. 36-26 gives the half-width of the central line as

$$\Delta\theta_{hw} = \frac{\lambda}{Nd} \quad \text{(half-width of central line)}. \qquad (36\text{-}27)$$

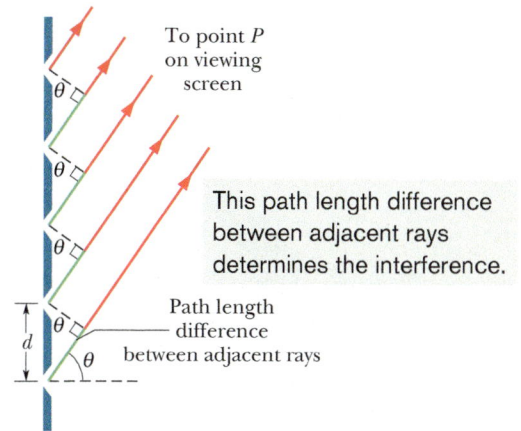

Figure 36-20 The rays from the rulings in a diffraction grating to a distant point *P* are approximately parallel. The path length difference between each two adjacent rays is $d \sin \theta$, where θ is measured as shown. (The rulings extend into and out of the page.)

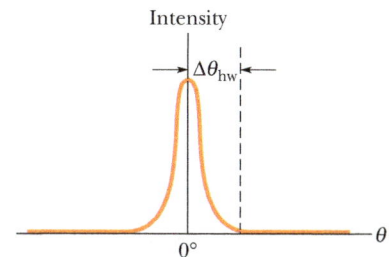

Figure 36-21 The half-width $\Delta\theta_{hw}$ of the central line is measured from the center of that line to the adjacent minimum on a plot of *I* versus θ like Fig. 36-19a.

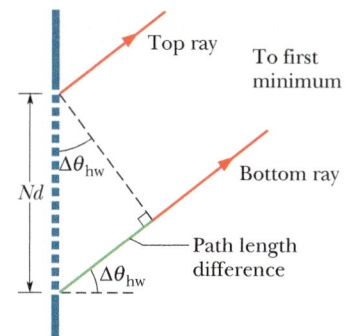

Figure 36-22 The top and bottom rulings of a diffraction grating of *N* rulings are separated by *Nd*. The top and bottom rays passing through these rulings have a path length difference of $Nd \sin \Delta\theta_{hw}$, where $\Delta\theta_{hw}$ is the angle to the first minimum. (The angle is here greatly exaggerated for clarity.)

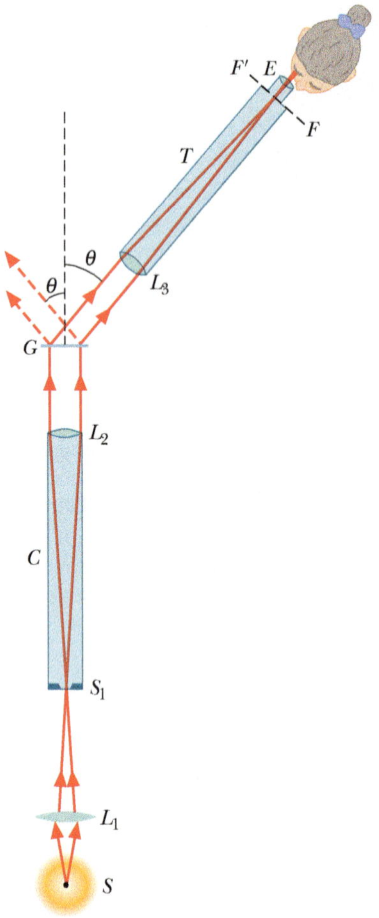

We state without proof that the half-width of any other line depends on its location relative to the central axis and is

$$\Delta\theta_{hw} = \frac{\lambda}{Nd\cos\theta} \quad \text{(half-width of line at } \theta\text{).}$$ (36-28)

Note that for light of a given wavelength λ and a given ruling separation d, the widths of the lines decrease with an increase in the number N of rulings. Thus, of two diffraction gratings, the grating with the larger value of N is better able to distinguish between wavelengths because its diffraction lines are narrower and so produce less overlap.

Grating Spectroscope

Diffraction gratings are widely used to determine the wavelengths that are emitted by sources of light ranging from lamps to stars. Figure 36-23 shows a simple *grating spectroscope* in which a grating is used for this purpose. Light from source S is focused by lens L_1 on a vertical slit S_1 placed in the focal plane of lens L_2. The light emerging from tube C (called a *collimator*) is a plane wave and is incident perpendicularly on grating G, where it is diffracted into a diffraction pattern, with the $m = 0$ order diffracted at angle $\theta = 0$ along the central axis of the grating.

We can view the diffraction pattern that would appear on a viewing screen at any angle θ simply by orienting telescope T in Fig. 36-23 to that angle. Lens L_3 of the telescope then focuses the light diffracted at angle θ (and at slightly smaller and larger angles) onto a focal plane FF' within the telescope. When we look through eyepiece E, we see a magnified view of this focused image.

By changing the angle θ of the telescope, we can examine the entire diffraction pattern. For any order number other than $m = 0$, the original light is spread out according to wavelength (or color) so that we can determine, with Eq. 36-25, just what wavelengths are being emitted by the source. If the source emits discrete wavelengths, what we see as we rotate the telescope horizontally through the angles corresponding to an order m is a vertical line of color for each wavelength, with the shorter-wavelength line at a smaller angle θ than the longer-wavelength line.

Hydrogen. For example, the light emitted by a hydrogen lamp, which contains hydrogen gas, has four discrete wavelengths in the visible range. If our eyes intercept this light directly, it appears to be white. If, instead, we view it through a grating spectroscope, we can distinguish, in several orders, the lines of the four colors corresponding to these visible wavelengths. (Such lines are called *emission lines.*) Four orders are represented in Fig. 36-24. In the central order ($m = 0$), the lines corresponding to all four wavelengths are superimposed, giving a single white line at $\theta = 0$. The colors are separated in the higher orders.

The third order is not shown in Fig. 36-24 for the sake of clarity; it actually overlaps the second and fourth orders. The fourth-order red line is missing because it is not formed by the grating used here. That is, when we attempt to

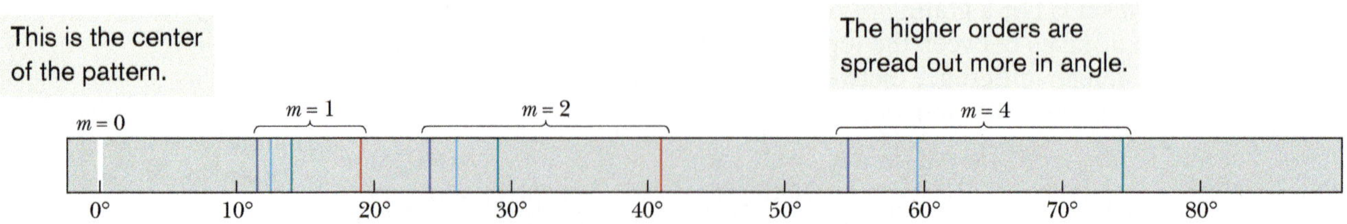

Figure 36-23 A simple type of grating spectroscope used to analyze the wavelengths of light emitted by source S.

This is the center of the pattern.

The higher orders are spread out more in angle.

Figure 36-24 The zeroth, first, second, and fourth orders of the visible emission lines from hydrogen. Note that the lines are farther apart at greater angles. (They are also dimmer and wider, although that is not shown here.)

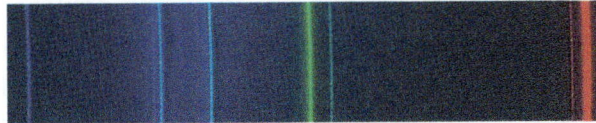

Figure 36-25 The visible emission lines of cadmium, as seen through a grating spectroscope.

Department of Physics, Imperial College/Science Photo Library/ Photo Researchers, Inc.

solve Eq. 36-25 for the angle θ for the red wavelength when $m = 4$, we find that $\sin \theta$ is greater than unity, which is not possible. The fourth order is then said to be *incomplete* for this grating; it might not be incomplete for a grating with greater spacing d, which will spread the lines less than in Fig. 36-24. Figure 36-25 is a photograph of the visible emission lines produced by cadmium.

✓ Checkpoint 5

The figure shows lines of different orders produced by a diffraction grating in monochromatic red light. (a) Is the center of the pattern to the left or right? (b) In monochromatic green light, are the half-widths of the lines produced in the same orders greater than, less than, or the same as the half-widths of the lines shown?

36-6 GRATINGS: DISPERSION AND RESOLVING POWER

Learning Objectives

After reading this module, you should be able to . . .

36.32 Identify dispersion as the spreading apart of the diffraction lines associated with different wavelengths.

36.33 Apply the relationships between dispersion D, wavelength difference $\Delta\lambda$, angular separation $\Delta\theta$, slit separation d, order number m, and the angle θ corresponding to the order number.

36.34 Identify the effect on the dispersion of a diffraction

grating if the slit separation is varied.

36.35 Identify that for us to resolve lines, a diffraction grating must make them distinguishable.

36.36 Apply the relationship between resolving power R, wavelength difference $\Delta\lambda$, average wavelength λ_{avg}, number of rulings N, and order number m.

36.37 Identify the effect on the resolving power R if the number of slits N is increased.

Key Ideas

● The dispersion D of a diffraction grating is a measure of the angular separation $\Delta\theta$ of the lines it produces for two wavelengths differing by $\Delta\lambda$. For order number m, at angle θ, the dispersion is given by

$$D = \frac{\Delta\theta}{\Delta\lambda} = \frac{m}{d\cos\theta} \quad \text{(dispersion)}.$$

● The resolving power R of a diffraction grating is a measure of its ability to make the emission lines of two close wavelengths distinguishable. For two wavelengths differing by $\Delta\lambda$ and with an average value of λ_{avg}, the resolving power is given by

$$R = \frac{\lambda_{avg}}{\Delta\lambda} = Nm \quad \text{(resolving power)}.$$

Gratings: Dispersion and Resolving Power

Dispersion

To be useful in distinguishing wavelengths that are close to each other (as in a grating spectroscope), a grating must spread apart the diffraction lines associated with the various wavelengths. This spreading, called **dispersion,** is defined as

$$D = \frac{\Delta\theta}{\Delta\lambda} \quad \text{(dispersion defined)}. \qquad (36\text{-}29)$$

Kristen Brochmann/Fundamental Photographs

The fine rulings, each 0.5 μm wide, on a compact disc function as a diffraction grating. When a small source of white light illuminates a disc, the diffracted light forms colored "lanes" that are the composite of the diffraction patterns from the rulings.

Here $\Delta\theta$ is the angular separation of two lines whose wavelengths differ by $\Delta\lambda$. The greater D is, the greater is the distance between two emission lines whose wavelengths differ by $\Delta\lambda$. We show below that the dispersion of a grating at angle θ is given by

$$D = \frac{m}{d\cos\theta} \quad \text{(dispersion of a grating).} \tag{36-30}$$

Thus, to achieve higher dispersion we must use a grating of smaller grating spacing d and work in a higher-order m. Note that the dispersion does not depend on the number of rulings N in the grating. The SI unit for D is the degree per meter or the radian per meter.

Resolving Power

To *resolve* lines whose wavelengths are close together (that is, to make the lines distinguishable), the line should also be as narrow as possible. Expressed otherwise, the grating should have a high **resolving power** R, defined as

$$R = \frac{\lambda_{\text{avg}}}{\Delta\lambda} \quad \text{(resolving power defined).} \tag{36-31}$$

Here λ_{avg} is the mean wavelength of two emission lines that can barely be recognized as separate, and $\Delta\lambda$ is the wavelength difference between them. The greater R is, the closer two emission lines can be and still be resolved. We shall show below that the resolving power of a grating is given by the simple expression

$$R = Nm \quad \text{(resolving power of a grating).} \tag{36-32}$$

To achieve high resolving power, we must use many rulings (large N).

Proof of Eq. 36-30

Let us start with Eq. 36-25, the expression for the locations of the lines in the diffraction pattern of a grating:

$$d\sin\theta = m\lambda.$$

Let us regard θ and λ as variables and take differentials of this equation. We find

$$d(\cos\theta)\,d\theta = m\,d\lambda.$$

For small enough angles, we can write these differentials as small differences, obtaining

$$d(\cos\theta)\,\Delta\theta = m\,\Delta\lambda \tag{36-33}$$

or

$$\frac{\Delta\theta}{\Delta\lambda} = \frac{m}{d\cos\theta}.$$

The ratio on the left is simply D (see Eq. 36-29), and so we have indeed derived Eq. 36-30.

Proof of Eq. 36-32

We start with Eq. 36-33, which was derived from Eq. 36-25, the expression for the locations of the lines in the diffraction pattern formed by a grating. Here $\Delta\lambda$ is the small wavelength difference between two waves that are diffracted by the grating, and $\Delta\theta$ is the angular separation between them in the diffraction pattern. If $\Delta\theta$ is to be the smallest angle that will permit the two lines to be resolved, it must (by Rayleigh's criterion) be equal to the half-width of each line, which is given by Eq. 36-28:

$$\Delta\theta_{\text{hw}} = \frac{\lambda}{Nd\cos\theta}.$$

Table 36-1 Three Gratings[a]

Grating	N	d (nm)	θ	D (°/μm)	R
A	10 000	2540	13.4°	23.2	10 000
B	20 000	2540	13.4°	23.2	20 000
C	10 000	1360	25.5°	46.3	10 000

[a]Data are for $\lambda = 589$ nm and $m = 1$.

If we substitute $\Delta\theta_{hw}$ as given here for $\Delta\theta$ in Eq. 36-33, we find that

$$\frac{\lambda}{N} = m\,\Delta\lambda,$$

from which it readily follows that

$$R = \frac{\lambda}{\Delta\lambda} = Nm.$$

This is Eq. 36-32, which we set out to derive.

Dispersion and Resolving Power Compared

The resolving power of a grating must not be confused with its dispersion. Table 36-1 shows the characteristics of three gratings, all illuminated with light of wavelength $\lambda = 589$ nm, whose diffracted light is viewed in the first order ($m = 1$ in Eq. 36-25). You should verify that the values of D and R as given in the table can be calculated with Eqs. 36-30 and 36-32, respectively. (In the calculations for D, you will need to convert radians per meter to degrees per micrometer.)

For the conditions noted in Table 36-1, gratings A and B have the same *dispersion D* and A and C have the same *resolving power R*.

Figure 36-26 shows the intensity patterns (also called *line shapes*) that would be produced by these gratings for two lines of wavelengths λ_1 and λ_2, in the vicinity of $\lambda = 589$ nm. Grating B, with the higher resolving power, produces narrower lines and thus is capable of distinguishing lines that are much closer together in wavelength than those in the figure. Grating C, with the higher dispersion, produces the greater angular separation between the lines.

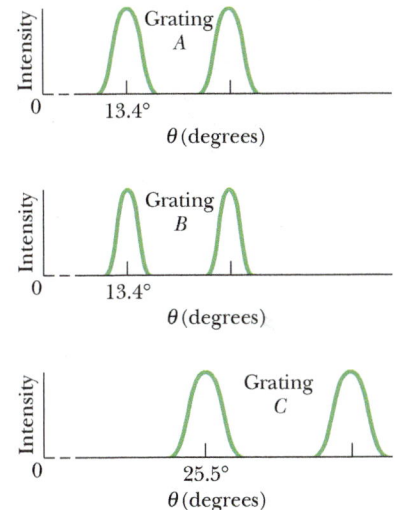

Figure 36-26 The intensity patterns for light of two wavelengths sent through the gratings of Table 36-1. Grating B has the highest resolving power, and grating C the highest dispersion.

Sample Problem 36.06 Dispersion and resolving power of a diffraction grating

A diffraction grating has 1.26×10^4 rulings uniformly spaced over width $w = 25.4$ mm. It is illuminated at normal incidence by yellow light from a sodium vapor lamp. This light contains two closely spaced emission lines (known as the sodium doublet) of wavelengths 589.00 nm and 589.59 nm.

(a) At what angle does the first-order maximum occur (on either side of the center of the diffraction pattern) for the wavelength of 589.00 nm?

KEY IDEA

The maxima produced by the diffraction grating can be determined with Eq. 36-25 ($d \sin\theta = m\lambda$).

Calculations: The grating spacing d is

$$d = \frac{w}{N} = \frac{25.4 \times 10^{-3}\text{ m}}{1.26 \times 10^4}$$
$$= 2.016 \times 10^{-6}\text{ m} = 2016\text{ nm}.$$

The first-order maximum corresponds to $m = 1$. Substituting these values for d and m into Eq. 36-25 leads to

$$\theta = \sin^{-1}\frac{m\lambda}{d} = \sin^{-1}\frac{(1)(589.00\text{ nm})}{2016\text{ nm}}$$
$$= 16.99° \approx 17.0°. \qquad \text{(Answer)}$$

(b) Using the dispersion of the grating, calculate the angular separation between the two lines in the first order.

KEY IDEAS

(1) The angular separation $\Delta\theta$ between the two lines in the first order depends on their wavelength difference $\Delta\lambda$ and the dispersion D of the grating, according to Eq. 36-29 ($D = \Delta\theta/\Delta\lambda$). (2) The dispersion D depends on the angle θ at which it is to be evaluated.

Calculations: We can assume that, in the first order, the two sodium lines occur close enough to each other for us to evaluate D at the angle $\theta = 16.99°$ we found in part (a) for one of those lines. Then Eq. 36-30 gives the dispersion as

$$D = \frac{m}{d\cos\theta} = \frac{1}{(2016\text{ nm})(\cos 16.99°)}$$

$$= 5.187 \times 10^{-4}\text{ rad/nm}.$$

From Eq. 36-29 and with $\Delta\lambda$ in nanometers, we then have

$$\Delta\theta = D\,\Delta\lambda = (5.187 \times 10^{-4}\text{ rad/nm})(589.59 - 589.00)$$

$$= 3.06 \times 10^{-4}\text{ rad} = 0.0175°. \qquad \text{(Answer)}$$

You can show that this result depends on the grating spacing d but not on the number of rulings there are in the grating.

(c) What is the least number of rulings a grating can have and still be able to resolve the sodium doublet in the first order?

KEY IDEAS

(1) The resolving power of a grating in any order m is physically set by the number of rulings N in the grating according to Eq. 36-32 ($R = Nm$). (2) The smallest wavelength difference $\Delta\lambda$ that can be resolved depends on the average wavelength involved and on the resolving power R of the grating, according to Eq. 36-31 ($R = \lambda_{avg}/\Delta\lambda$).

Calculation: For the sodium doublet to be barely resolved, $\Delta\lambda$ must be their wavelength separation of 0.59 nm, and λ_{avg} must be their average wavelength of 589.30 nm. Thus, we find that the smallest number of rulings for a grating to resolve the sodium doublet is

$$N = \frac{R}{m} = \frac{\lambda_{avg}}{m\,\Delta\lambda}$$

$$= \frac{589.30\text{ nm}}{(1)(0.59\text{ nm})} = 999\text{ rulings}. \qquad \text{(Answer)}$$

WILEY PLUS Additional examples, video, and practice available at *WileyPLUS*

36-7 X-RAY DIFFRACTION

Learning Objectives

After reading this module, you should be able to . . .

36.38 Identify approximately where x rays are located in the electromagnetic spectrum.
36.39 Define a unit cell.
36.40 Define reflecting planes (or crystal planes) and interplanar spacing.
36.41 Sketch two rays that scatter from adjacent planes, showing the angle that is used in calculations.

36.42 For the intensity maxima in x-ray scattering by a crystal, apply the relationship between the interplanar spacing d, the angle θ of scattering, the order number m, and the wavelength λ of the x rays.
36.43 Given a drawing of a unit cell, demonstrate how an interplanar spacing can be determined.

Key Ideas

● If x rays are directed toward a crystal structure, they undergo Bragg scattering, which is easiest to visualize if the crystal atoms are considered to be in parallel planes.

● For x rays of wavelength λ scattering from crystal planes

with separation d, the angles θ at which the scattered intensity is maximum are given by

$$2d\sin\theta = m\lambda, \qquad \text{for } m = 1, 2, 3, \dots \quad \text{(Bragg's law)}.$$

X-Ray Diffraction

X rays are electromagnetic radiation whose wavelengths are of the order of 1 Å ($= 10^{-10}$ m). Compare this with a wavelength of 550 nm ($= 5.5 \times 10^{-7}$ m) at the center of the visible spectrum. Figure 36-27 shows that x rays are produced when electrons escaping from a heated filament F are accelerated by a potential difference V and strike a metal target T.

A standard optical diffraction grating cannot be used to discriminate between different wavelengths in the x-ray wavelength range. For $\lambda = 1 \text{ Å}$ ($= 0.1$ nm) and $d = 3000$ nm, for example, Eq. 36-25 shows that the first-order maximum occurs at

$$\theta = \sin^{-1} \frac{m\lambda}{d} = \sin^{-1} \frac{(1)(0.1 \text{ nm})}{3000 \text{ nm}} = 0.0019°.$$

This is too close to the central maximum to be practical. A grating with $d \approx \lambda$ is desirable, but, because x-ray wavelengths are about equal to atomic diameters, such gratings cannot be constructed mechanically.

In 1912, it occurred to German physicist Max von Laue that a crystalline solid, which consists of a regular array of atoms, might form a natural three-dimensional "diffraction grating" for x rays. The idea is that, in a crystal such as sodium chloride (NaCl), a basic unit of atoms (called the *unit cell*) repeats itself throughout the array. Figure 36-28*a* represents a section through a crystal of NaCl and identifies this basic unit. The unit cell is a cube measuring a_0 on each side.

When an x-ray beam enters a crystal such as NaCl, x rays are *scattered*—that is, redirected—in all directions by the crystal structure. In some directions the scattered waves undergo destructive interference, resulting in intensity minima; in other directions the interference is constructive, resulting in intensity maxima. This process of scattering and interference is a form of diffraction.

Fictional Planes. Although the process of diffraction of x rays by a crystal is complicated, the maxima turn out to be in directions *as if* the x rays were

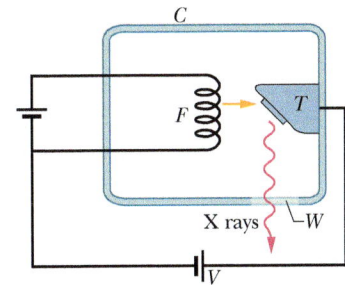

Figure 36-27 X rays are generated when electrons leaving heated filament F are accelerated through a potential difference V and strike a metal target T. The "window" W in the evacuated chamber C is transparent to x rays.

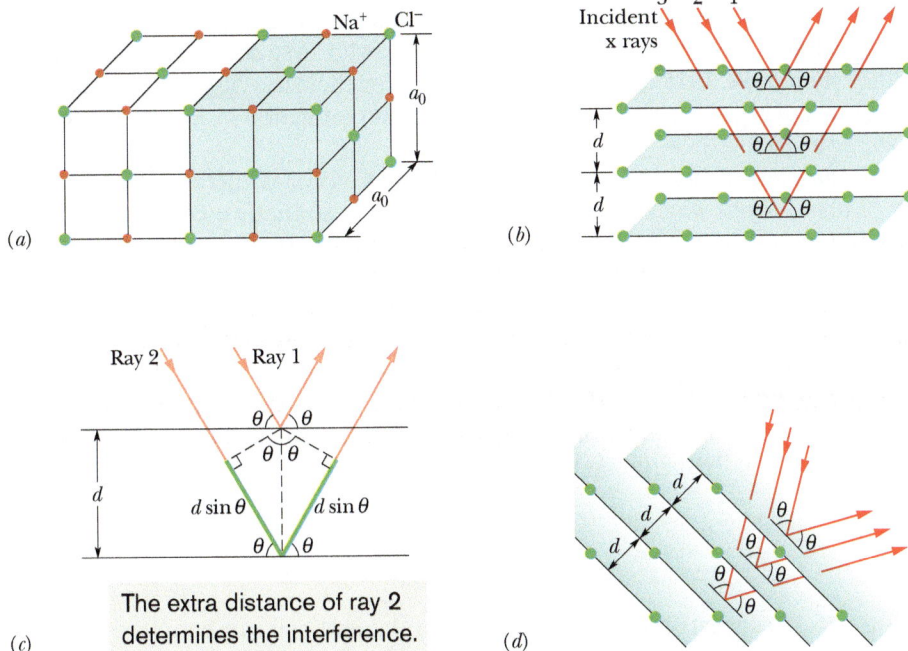

Figure 36-28 (*a*) The cubic structure of NaCl, showing the sodium and chlorine ions and a unit cell (shaded). (*b*) Incident x rays undergo diffraction by the structure of (*a*). The x rays are diffracted as if they were reflected by a family of parallel planes, with angles measured relative to the planes (not relative to a normal as in optics). (*c*) The path length difference between waves effectively reflected by two adjacent planes is $2d \sin \theta$. (*d*) A different orientation of the incident x rays relative to the structure. A different family of parallel planes now effectively reflects the x rays.

reflected by a family of parallel *reflecting planes* (or *crystal planes*) that extend through the atoms within the crystal and that contain regular arrays of the atoms. (The x rays are not actually reflected; we use these fictional planes only to simplify the analysis of the actual diffraction process.)

Figure 36-28b shows three reflecting planes (part of a family containing many parallel planes) with *interplanar spacing d,* from which the incident rays shown are said to reflect. Rays 1, 2, and 3 reflect from the first, second, and third planes, respectively. At each reflection the angle of incidence and the angle of reflection are represented with θ. Contrary to the custom in optics, these angles are defined relative to the *surface* of the reflecting plane rather than a normal to that surface. For the situation of Fig. 36-28b, the interplanar spacing happens to be equal to the unit cell dimension a_0.

Figure 36-28c shows an edge-on view of reflection from an adjacent pair of planes. The waves of rays 1 and 2 arrive at the crystal in phase. After they are reflected, they must again be in phase because the reflections and the reflecting planes have been defined solely to explain the intensity maxima in the diffraction of x rays by a crystal. Unlike light rays, the x rays do not refract upon entering the crystal; moreover, we do not define an index of refraction for this situation. Thus, the relative phase between the waves of rays 1 and 2 as they leave the crystal is set solely by their path length difference. For these rays to be in phase, the path length difference must be equal to an integer multiple of the wavelength λ of the x rays.

Diffraction Equation. By drawing the dashed perpendiculars in Fig. 36-28c, we find that the path length difference is $2d \sin \theta$. In fact, this is true for any pair of adjacent planes in the family of planes represented in Fig. 36-28b. Thus, we have, as the criterion for intensity maxima for x-ray diffraction,

$$2d \sin \theta = m\lambda, \qquad \text{for } m = 1, 2, 3, \ldots \quad \text{(Bragg's law)}, \qquad (36\text{-}34)$$

where m is the order number of an intensity maximum. Equation 36-34 is called **Bragg's law** after British physicist W. L. Bragg, who first derived it. (He and his father shared the 1915 Nobel Prize in physics for their use of x rays to study the structures of crystals.) The angle of incidence and reflection in Eq. 36-34 is called a *Bragg angle*.

Regardless of the angle at which x rays enter a crystal, there is always a family of planes from which they can be said to reflect so that we can apply Bragg's law. In Fig. 36-28d, notice that the crystal structure has the same orientation as it does in Fig. 36-28a, but the angle at which the beam enters the structure differs from that shown in Fig. 36-28b. This new angle requires a new family of reflecting planes, with a different interplanar spacing d and different Bragg angle θ, in order to explain the x-ray diffraction via Bragg's law.

Determining a Unit Cell. Figure 36-29 shows how the interplanar spacing d can be related to the unit cell dimension a_0. For the particular family of planes shown there, the Pythagorean theorem gives

$$5d = \sqrt{\tfrac{5}{4}a_0^2},$$

or

$$d = \frac{a_0}{\sqrt{20}} = 0.2236a_0. \qquad (36\text{-}35)$$

Figure 36-29 suggests how the dimensions of the unit cell can be found once the interplanar spacing has been measured by means of x-ray diffraction.

X-ray diffraction is a powerful tool for studying both x-ray spectra and the arrangement of atoms in crystals. To study spectra, a particular set of crystal planes, having a known spacing d, is chosen. These planes effectively reflect different wavelengths at different angles. A detector that can discriminate one angle from another can then be used to determine the wavelength of radiation reaching it. The crystal itself can be studied with a monochromatic x-ray beam, to determine not only the spacing of various crystal planes but also the structure of the unit cell.

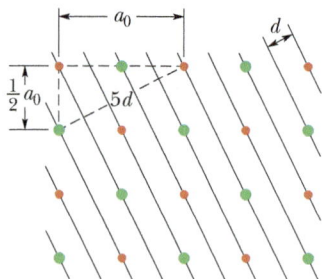

Figure 36-29 A family of planes through the structure of Fig. 36-28a, and a way to relate the edge length a_0 of a unit cell to the interplanar spacing d.

Review & Summary

Diffraction When waves encounter an edge, an obstacle, or an aperture the size of which is comparable to the wavelength of the waves, those waves spread out as they travel and, as a result, undergo interference. This is called **diffraction.**

Single-Slit Diffraction Waves passing through a long narrow slit of width a produce, on a viewing screen, a **single-slit diffraction pattern** that includes a central maximum and other maxima, separated by minima located at angles θ to the central axis that satisfy

$$a \sin \theta = m\lambda, \quad \text{for } m = 1, 2, 3, \ldots \text{ (minima).} \quad (36\text{-}3)$$

The intensity of the diffraction pattern at any given angle θ is

$$I(\theta) = I_m \left(\frac{\sin \alpha}{\alpha} \right)^2, \quad \text{where } \alpha = \frac{\pi a}{\lambda} \sin \theta \quad (36\text{-}5, 36\text{-}6)$$

and I_m is the intensity at the center of the pattern.

Circular-Aperture Diffraction Diffraction by a circular aperture or a lens with diameter d produces a central maximum and concentric maxima and minima, with the first minimum at an angle θ given by

$$\sin \theta = 1.22 \frac{\lambda}{d} \quad \text{(first minimum—circular aperture).} \quad (36\text{-}12)$$

Rayleigh's Criterion *Rayleigh's criterion* suggests that two objects are on the verge of resolvability if the central diffraction maximum of one is at the first minimum of the other. Their angular separation can then be no less than

$$\theta_R = 1.22 \frac{\lambda}{d} \quad \text{(Rayleigh's criterion),} \quad (36\text{-}14)$$

in which d is the diameter of the aperture through which the light passes.

Double-Slit Diffraction Waves passing through two slits, each of width a, whose centers are a distance d apart, display diffraction patterns whose intensity I at angle θ is

$$I(\theta) = I_m (\cos^2 \beta) \left(\frac{\sin \alpha}{\alpha} \right)^2 \quad \text{(double slit),} \quad (36\text{-}19)$$

with $\beta = (\pi d/\lambda) \sin \theta$ and α as for single-slit diffraction.

Diffraction Gratings A *diffraction grating* is a series of "slits" used to separate an incident wave into its component wavelengths by separating and displaying their diffraction maxima. Diffraction by N (multiple) slits results in maxima (lines) at angles θ such that

$$d \sin \theta = m\lambda, \quad \text{for } m = 0, 1, 2, \ldots \text{ (maxima),} \quad (36\text{-}25)$$

with the **half-widths** of the lines given by

$$\Delta\theta_{\text{hw}} = \frac{\lambda}{Nd \cos \theta} \quad \text{(half-widths).} \quad (36\text{-}28)$$

The dispersion D and resolving power R are given by

$$D = \frac{\Delta\theta}{\Delta\lambda} = \frac{m}{d \cos \theta} \quad (36\text{-}29, 36\text{-}30)$$

and

$$R = \frac{\lambda_{\text{avg}}}{\Delta\lambda} = Nm. \quad (36\text{-}31, 36\text{-}32)$$

X-Ray Diffraction The regular array of atoms in a crystal is a three-dimensional diffraction grating for short-wavelength waves such as x rays. For analysis purposes, the atoms can be visualized as being arranged in planes with characteristic interplanar spacing d. Diffraction maxima (due to constructive interference) occur if the incident direction of the wave, measured from the surfaces of these planes, and the wavelength λ of the radiation satisfy **Bragg's law:**

$$2d \sin \theta = m\lambda, \quad \text{for } m = 1, 2, 3, \ldots \text{ (Bragg's law).} \quad (36\text{-}34)$$

Questions

1 You are conducting a single-slit diffraction experiment with light of wavelength λ. What appears, on a distant viewing screen, at a point at which the top and bottom rays through the slit have a path length difference equal to (a) 5λ and (b) 4.5λ?

2 In a single-slit diffraction experiment, the top and bottom rays through the slit arrive at a certain point on the viewing screen with a path length difference of 4.0 wavelengths. In a phasor representation like those in Fig 36-7, how many overlapping circles does the chain of phasors make?

3 For three experiments, Fig. 36-30 gives the parameter β of Eq. 36-20 versus angle θ for two-slit interference using light of wavelength 500 nm. The slit separations in the three experiments differ. Rank the experiments according to (a) the slit separations and (b) the total number of two-slit interference maxima in the pattern, greatest first.

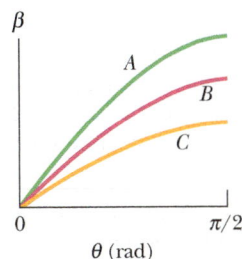

Figure 36-30 Question 3.

4 For three experiments, Fig. 36-31 gives α versus angle θ in one-slit diffraction using light of wavelength 500 nm. Rank the experiments according to (a) the slit widths and (b) the total number of diffraction minima in the pattern, greatest first.

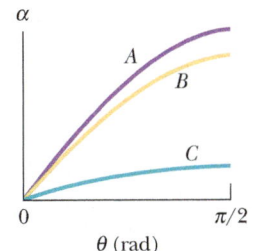

Figure 36-31 Question 4.

5 Figure 36-32 shows four choices for the rectangular opening of a source of either sound waves or light waves. The sides have lengths of either L or $2L$, with L being 3.0 times the wavelength of the waves. Rank the openings according to the extent of (a) left–right spreading and (b) up–down spreading of the waves due to diffraction, greatest first.

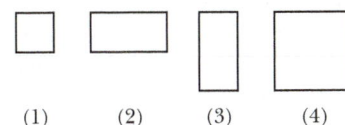

(1)   (2)   (3)   (4)

Figure 36-32 Question 5.

6 Light of frequency f illuminating a long narrow slit produces a diffraction pattern. (a) If we switch to light of frequency $1.3f$, does the pattern expand away from the center or contract toward the center? (b) Does the pattern expand or contract if, instead, we submerge the equipment in clear corn syrup?

7 At night many people see rings (called *entoptic halos*) surrounding bright outdoor lamps in otherwise dark surroundings. The rings are the first of the side maxima in diffraction patterns produced by structures that are thought to be within the cornea (or possibly the lens) of the observer's eye. (The central maxima of such patterns overlap the lamp.) (a) Would a particular ring become smaller or larger if the lamp were switched from blue to red light? (b) If a lamp emits white light, is blue or red on the outside edge of the ring?

8 (a) For a given diffraction grating, does the smallest difference $\Delta\lambda$ in two wavelengths that can be resolved increase, decrease, or remain the same as the wavelength increases? (b) For a given wavelength region (say, around 500 nm), is $\Delta\lambda$ greater in the first order or in the third order?

9 Figure 36-33 shows a red line and a green line of the same order in the pattern produced by a diffraction grating. If we increased the number of rulings in the grating—say, by removing tape that had covered the outer half of the rulings—would (a) the half-widths of the lines and (b) the separation of the lines increase, decrease, or remain the same? (c) Would the lines shift to the right, shift to the left, or remain in place?

Figure 36-33 Questions 9 and 10.

10 For the situation of Question 9 and Fig. 36-33, if instead we increased the grating spacing, would (a) the half-widths of the lines and (b) the separation of the lines increase, decrease, or remain the same? (c) Would the lines shift to the right, shift to the left, or remain in place?

11 (a) Figure 36-34a shows the lines produced by diffraction gratings A and B using light of the same wavelength; the lines are of the same order and appear at the same angles θ. Which grating has the greater number of rulings? (b) Figure 36-34b shows lines of two orders produced by a single diffraction grating using light of two wavelengths, both in the red region of the spectrum. Which lines, the left pair or right pair, are in the order with greater m? Is the center of the diffraction pattern located to the left or to the right in (c) Fig. 36-34a and (d) Fig. 36-34b?

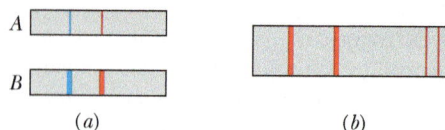

Figure 36-34 Question 11.

12 Figure 36-35 shows the bright fringes that lie within the central diffraction envelope in two double-slit diffraction experiments using the same wavelength of light. Are (a) the slit width a, (b) the slit separation d, and (c) the ratio d/a in experiment B greater than, less than, or the same as those quantities in experiment A?

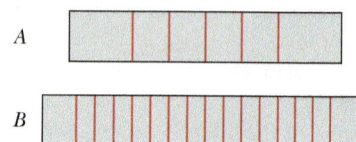

Figure 36-35 Question 12.

13 In three arrangements you view two closely spaced small objects that are the same large distance from you. The angles that the objects occupy in your field of view and their distances from you are the following: (1) 2ϕ and R; (2) 2ϕ and $2R$; (3) $\phi/2$ and $R/2$. (a) Rank the arrangements according to the separation between the objects, greatest first. If you can just barely resolve the two objects in arrangement 2, can you resolve them in (b) arrangement 1 and (c) arrangement 3?

14 For a certain diffraction grating, the ratio λ/a of wavelength to ruling spacing is 1/3.5. Without written calculation or use of a calculator, determine which of the orders beyond the zeroth order appear in the diffraction pattern.

Problems

GO Tutoring problem available (at instructor's discretion) in *WileyPLUS* and WebAssign

SSM Worked-out solution available in Student Solutions Manual

• – ••• Number of dots indicates level of problem difficulty

WWW Worked-out solution is at

ILW Interactive solution is at http://www.wiley.com/college/halliday

Additional information available in *The Flying Circus of Physics* and at flyingcircusofphysics.com

Module 36-1 Single-Slit Diffraction

•1 **GO** The distance between the first and fifth minima of a single-slit diffraction pattern is 0.35 mm with the screen 40 cm away from the slit, when light of wavelength 550 nm is used. (a) Find the slit width. (b) Calculate the angle θ of the first diffraction minimum.

•2 What must be the ratio of the slit width to the wavelength for a single slit to have the first diffraction minimum at $\theta = 45.0°$?

•3 A plane wave of wavelength 590 nm is incident on a slit with a width of $a = 0.40$ mm. A thin converging lens of focal length +70 cm is placed between the slit and a viewing screen and focuses the light on the screen. (a) How far is the screen from the lens? (b) What is the distance on the screen from the center of the diffraction pattern to the first minimum?

•4 In conventional television, signals are broadcast from towers to home receivers. Even when a receiver is not in direct view of a tower because of a hill or building, it can still intercept a signal if the signal diffracts enough around the obstacle, into the obstacle's "shadow region." Previously, television signals had a wavelength of about 50 cm, but digital television signals that are transmitted from towers have a wavelength of about 10 mm. (a) Did this change in wavelength increase or decrease the diffraction of the signals into the shadow regions of obstacles? Assume that a signal passes through an opening of 5.0 m width between two adjacent buildings. What is the angular spread of the central diffraction maximum (out to the first minima) for wavelengths of (b) 50 cm and (c) 10 mm?

•5 A single slit is illuminated by light of wavelengths λ_a and λ_b, chosen so that the first diffraction minimum of the λ_a component coincides with the second minimum of the λ_b component. (a) If $\lambda_b = 350$ nm, what is λ_a? For what order number m_b (if any) does a minimum of the λ_b component coincide with the minimum of the λ_a component in the order number (b) $m_a = 2$ and (c) $m_a = 3$?

•6 Monochromatic light of wavelength 441 nm is incident on a narrow slit. On a screen 2.00 m away, the distance between the second diffraction minimum and the central maximum is 1.50 cm. (a) Calculate the angle of diffraction θ of the second minimum. (b) Find the width of the slit.

•7 Light of wavelength 633 nm is incident on a narrow slit. The angle between the first diffraction minimum on one side of the central maximum and the first minimum on the other side is 1.20°. What is the width of the slit?

••8 Sound waves with frequency 3000 Hz and speed 343 m/s diffract through the rectangular opening of a speaker cabinet and into a large auditorium of length $d = 100$ m. The opening, which has a horizontal width of 30.0 cm, faces a wall 100 m away (Fig. 36-36). Along that wall, how far from the central axis will a listener be at the first diffraction minimum and thus have difficulty hearing the sound? (Neglect reflections.)

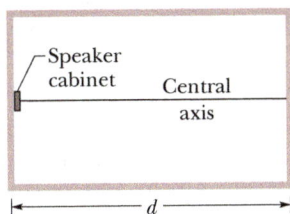

Figure 36-36 Problem 8.

••9 SSM ILW A slit 1.00 mm wide is illuminated by light of wavelength 589 nm. We see a diffraction pattern on a screen 3.00 m away. What is the distance between the first two diffraction minima on the same side of the central diffraction maximum?

••10 GO Manufacturers of wire (and other objects of small dimension) sometimes use a laser to continually monitor the thickness of the product. The wire intercepts the laser beam, producing a diffraction pattern like that of a single slit of the same width as the wire diameter (Fig. 36-37). Suppose a helium–neon laser, of wavelength 632.8 nm, illuminates a wire, and the diffraction pattern appears on a screen at distance $L = 2.60$ m. If the desired wire diameter is 1.37 mm, what is the observed distance between the two tenth-order minima (one on each side of the central maximum)?

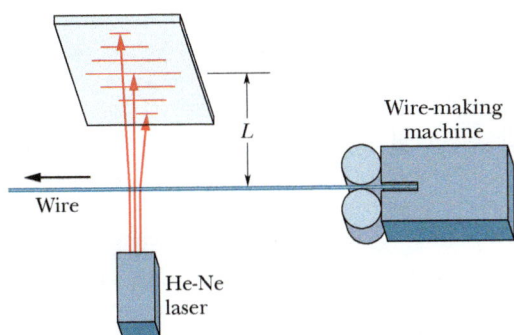

Figure 36-37 Problem 10.

Module 36-2   Intensity in Single-Slit Diffraction

•11 A 0.10-mm-wide slit is illuminated by light of wavelength 589 nm. Consider a point P on a viewing screen on which the diffraction pattern of the slit is viewed; the point is at 30° from the central axis of the slit. What is the phase difference between the Huygens wavelets arriving at point P from the top and midpoint of the slit? (*Hint:* See Eq. 36-4.)

•12 Figure 36-38 gives α versus the sine of the angle θ in a single-slit diffraction experiment using light of wavelength 610 nm. The vertical axis scale is set by $\alpha_s = 12$ rad. What are (a) the slit width, (b) the total number of diffraction minima in the pattern (count them on both sides of the center of the diffraction pattern), (c) the least angle for a minimum, and (d) the greatest angle for a minimum?

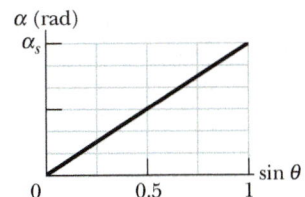

Figure 36-38 Problem 12.

•13 Monochromatic light with wavelength 538 nm is incident on a slit with width 0.025 mm. The distance from the slit to a screen is 3.5 m. Consider a point on the screen 1.1 cm from the central maximum. Calculate (a) θ for that point, (b) α, and (c) the ratio of the intensity at that point to the intensity at the central maximum.

•14 In the single-slit diffraction experiment of Fig. 36-4, let the wavelength of the light be 500 nm, the slit width be 6.00 μm, and the viewing screen be at distance $D = 3.00$ m. Let a y axis extend upward along the viewing screen, with its origin at the center of the diffraction pattern. Also let I_P represent the intensity of the diffracted light at point P at $y = 15.0$ cm. (a) What is the ratio of I_P to the intensity I_m at the center of the pattern? (b) Determine where point P is in the diffraction pattern by giving the maximum and minimum between which it lies, or the two minima between which it lies.

••15 SSM WWW The full width at half-maximum (FWHM) of a central diffraction maximum is defined as the angle between the two points in the pattern where the intensity is one-half that at the center of the pattern. (See Fig. 36-8b.) (a) Show that the intensity drops to one-half the maximum value when $\sin^2 \alpha = \alpha^2/2$. (b) Verify that $\alpha = 1.39$ rad (about 80°) is a solution to the transcendental equation of (a). (c) Show that the FWHM is $\Delta\theta = 2 \sin^{-1}(0.443\lambda/a)$, where a is the slit width. Calculate the FWHM of the central maximum for slit width (d) 1.00λ, (e) 5.00λ, and (f) 10.0λ.

••16 *Babinet's principle.* A monochromatic beam of parallel light is incident on a "collimating" hole of diameter $x \gg \lambda$. Point P lies in the geometrical shadow region on a *distant* screen (Fig. 36-39a). Two diffracting objects, shown in Fig. 36-39b, are placed in turn over the collimating hole. Object A is an opaque circle with a hole in it, and B is the "photographic negative" of A. Using superposition concepts, show that the intensity at P is identical for the two diffracting objects A and B.

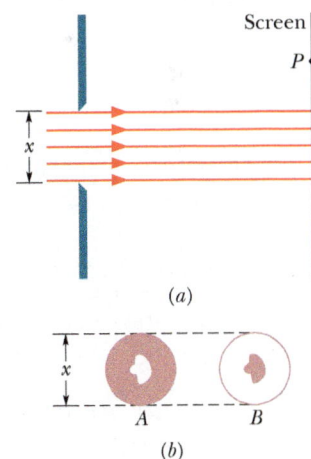

Figure 36-39 Problem 16.

••17 (a) Show that the values of α at which intensity maxima for single-slit diffraction occur can be found exactly by differentiating Eq. 36-5 with respect to α and equating the result to zero, obtaining the condition $\tan \alpha = \alpha$. To find values of α satisfying this relation, plot the curve $y = \tan \alpha$ and the straight line $y = \alpha$ and then find their intersections, or use a calculator to find an appropriate value of α by trial and error. Next, from $\alpha = (m + \frac{1}{2})\pi$, determine the values of m associated with the maxima in the single-slit pattern. (These m values are *not* integers because secondary maxima do not lie exactly halfway between minima.) What are the (b) smallest α and (c) associated m, the (d) second smallest α and (e) associated m, and the (f) third smallest α and (g) associated m?

Module 36-3 **Diffraction by a Circular Aperture**

•18 The wall of a large room is covered with acoustic tile in which small holes are drilled 5.0 mm from center to center. How far can a person be from such a tile and still distinguish the individual holes, assuming ideal conditions, the pupil diameter of the observer's eye to be 4.0 mm, and the wavelength of the room light to be 550 nm?

•19 (a) How far from grains of red sand must you be to position yourself just at the limit of resolving the grains if your pupil diameter is 1.5 mm, the grains are spherical with radius 50 μm, and the light from the grains has wavelength 650 nm? (b) If the grains were blue and the light from them had wavelength 400 nm, would the answer to (a) be larger or smaller?

•20 The radar system of a navy cruiser transmits at a wavelength of 1.6 cm, from a circular antenna with a diameter of 2.3 m. At a range of 6.2 km, what is the smallest distance that two speedboats can be from each other and still be resolved as two separate objects by the radar system?

•21 SSM WWW Estimate the linear separation of two objects on Mars that can just be resolved under ideal conditions by an observer on Earth (a) using the naked eye and (b) using the 200 in. (= 5.1 m) Mount Palomar telescope. Use the following data: distance to Mars = 8.0×10^7 km, diameter of pupil = 5.0 mm, wavelength of light = 550 nm.

•22 Assume that Rayleigh's criterion gives the limit of resolution of an astronaut's eye looking down on Earth's surface from a typical space shuttle altitude of 400 km. (a) Under that idealized assumption, estimate the smallest linear width on Earth's surface that the astronaut can resolve. Take the astronaut's pupil diameter to be 5 mm and the wavelength of visible light to be 550 nm. (b) Can the astronaut resolve the Great Wall of China (Fig. 36-40), which is more than 3000 km long, 5 to 10 m thick at its base, 4 m thick at its top, and 8 m in height? (c) Would the astronaut be able to resolve any unmistakable sign of intelligent life on Earth's surface?

©AP/Wide World Photos

Figure 36-40 Problem 22. The Great Wall of China.

•23 SSM The two headlights of an approaching automobile are 1.4 m apart. At what (a) angular separation and (b) maximum distance will the eye resolve them? Assume that the pupil diameter is 5.0 mm, and use a wavelength of 550 nm for the light. Also assume that diffraction effects alone limit the resolution so that Rayleigh's criterion can be applied.

•24 *Entoptic halos.* If someone looks at a bright outdoor lamp in otherwise dark surroundings, the lamp appears to be surrounded by bright and dark rings (hence *halos*) that are actually a circular diffraction pattern as in Fig. 36-10, with the central maximum overlapping the direct light from the lamp. The diffraction is produced by structures within the cornea or lens of the eye (hence *entoptic*). If the lamp is monochromatic at wavelength 550 nm and the first dark ring subtends angular diameter 2.5° in the observer's view, what is the (linear) diameter of the structure producing the diffraction?

•25 ILW Find the separation of two points on the Moon's surface that can just be resolved by the 200 in. (= 5.1 m) telescope at Mount Palomar, assuming that this separation is determined by diffraction effects. The distance from Earth to the Moon is 3.8×10^5 km. Assume a wavelength of 550 nm for the light.

•26 The telescopes on some commercial surveillance satellites can resolve objects on the ground as small as 85 cm across (see Google Earth), and the telescopes on military surveillance satellites reportedly can resolve objects as small as 10 cm across. Assume first that object resolution is determined entirely by Rayleigh's criterion and is not degraded by turbulence in the atmosphere. Also assume that the satellites are at a typical altitude of 400 km and that the wavelength of visible light is 550 nm. What would be the required diameter of the telescope aperture for (a) 85 cm resolution and (b) 10 cm resolution? (c) Now, considering that turbulence is certain to degrade resolution and that the aperture diameter of the Hubble Space Telescope is 2.4 m, what can you say about the answer to (b) and about how the military surveillance resolutions are accomplished?

•27 If Superman really had x-ray vision at 0.10 nm wavelength and a 4.0 mm pupil diameter, at what maximum altitude could he distinguish villains from heroes, assuming that he needs to resolve points separated by 5.0 cm to do this?

••28 GO The wings of tiger beetles (Fig. 36-41) are colored by interference due to thin cuticle-like layers. In addition, these layers are arranged in patches that are 60 μm across and produce different colors. The color you see is a pointillistic mixture of thin-film interference colors that varies with perspective. Approximately

Kjell B. Sandved/Bruce Coleman, Inc./Photoshot Holdings Ltd.

Figure 36-41 Problem 28. Tiger beetles are colored by pointillistic mixtures of thin-film interference colors.

what viewing distance from a wing puts you at the limit of resolving the different colored patches according to Rayleigh's criterion? Use 550 nm as the wavelength of light and 3.00 mm as the diameter of your pupil.

••29 (a) What is the angular separation of two stars if their images are barely resolved by the Thaw refracting telescope at the Allegheny Observatory in Pittsburgh? The lens diameter is 76 cm and its focal length is 14 m. Assume $\lambda = 550$ nm. (b) Find the distance between these barely resolved stars if each of them is 10 light-years distant from Earth. (c) For the image of a single star in this telescope, find the diameter of the first dark ring in the diffraction pattern, as measured on a photographic plate placed at the focal plane of the telescope lens. Assume that the structure of the image is associated entirely with diffraction at the lens aperture and not with lens "errors."

••30 **GO** ✈ *Floaters.* The floaters you see when viewing a bright, featureless background are diffraction patterns of defects in the vitreous humor that fills most of your eye. Sighting through a pinhole sharpens the diffraction pattern. If you also view a small circular dot, you can approximate the defect's size. Assume that the defect diffracts light as a circular aperture does. Adjust the dot's distance L from your eye (or eye lens) until the dot and the circle of the first minimum in the diffraction pattern appear to have the same size in your view. That is, until they have the same diameter D' on the retina at distance $L' = 2.0$ cm from the front of the eye, as suggested in Fig. 36-42*a*, where the angles on the two sides of the eye lens are equal. Assume that the wavelength of visible light is $\lambda = 550$ nm. If the dot has diameter $D = 2.0$ mm and is distance $L = 45.0$ cm from the eye and the defect is $x = 6.0$ mm in front of the retina (Fig. 36-42*b*), what is the diameter of the defect?

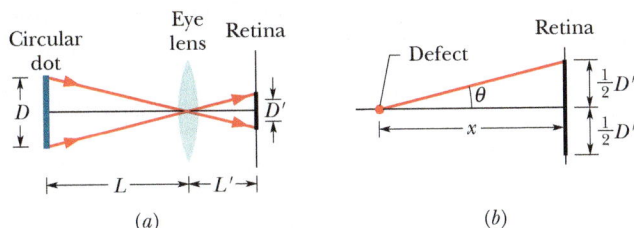

Figure 36-42 Problem 30.

••31 **SSM** Millimeter-wave radar generates a narrower beam than conventional microwave radar, making it less vulnerable to anti-radar missiles than conventional radar. (a) Calculate the angular width 2θ of the central maximum, from first minimum to first minimum, produced by a 220 GHz radar beam emitted by a 55.0-cm-diameter circular antenna. (The frequency is chosen to coincide with a low-absorption atmospheric "window.") (b) What is 2θ for a more conventional circular antenna that has a diameter of 2.3 m and emits at wavelength 1.6 cm?

••32 (a) A circular diaphragm 60 cm in diameter oscillates at a frequency of 25 kHz as an underwater source of sound used for submarine detection. Far from the source, the sound intensity is distributed as the diffraction pattern of a circular hole whose diameter equals that of the diaphragm. Take the speed of sound in water to be 1450 m/s and find the angle between the normal to the diaphragm and a line from the diaphragm to the first minimum. (b) Is there such a minimum for a source having an (audible) frequency of 1.0 kHz?

••33 **GO** Nuclear-pumped x-ray lasers are seen as a possible weapon to destroy ICBM booster rockets at ranges up to 2000 km.

One limitation on such a device is the spreading of the beam due to diffraction, with resulting dilution of beam intensity. Consider such a laser operating at a wavelength of 1.40 nm. The element that emits light is the end of a wire with diameter 0.200 mm. (a) Calculate the diameter of the central beam at a target 2000 km away from the beam source. (b) What is the ratio of the beam intensity at the target to that at the end of the wire? (The laser is fired from space, so neglect any atmospheric absorption.)

•••34 **GO** ✈ A circular obstacle produces the same diffraction pattern as a circular hole of the same diameter (except very near $\theta = 0$). Airborne water drops are examples of such obstacles. When you see the Moon through suspended water drops, such as in a fog, you intercept the diffraction pattern from many drops. The composite of the central diffraction maxima of those drops forms a white region that surrounds the Moon and may obscure it. Figure 36-43 is a photograph in which the Moon is obscured. There are two faint, colored rings around the Moon (the larger one may be too faint to be seen in your copy of the photograph). The smaller ring is on the outer edge of the central maxima from the drops; the somewhat larger ring is on the outer edge of the smallest of the secondary maxima from the drops (see Fig. 36-10). The color is visible because the rings are adjacent to the diffraction minima (dark rings) in the patterns. (Colors in other parts of the pattern overlap too much to be visible.)

(a) What is the color of these rings on the outer edges of the diffraction maxima? (b) The colored ring around the central maxima in Fig. 36-43 has an angular diameter that is 1.35 times the angular diameter of the Moon, which is 0.50°. Assume that the drops all have about the same diameter. Approximately what is that diameter?

Pekka Parvianen/Photo Researchers, Inc.

Figure 36-43 Problem 34. The corona around the Moon is a composite of the diffraction patterns of airborne water drops.

Module 36-4 Diffraction by a Double Slit

•35 Suppose that the central diffraction envelope of a double-slit diffraction pattern contains 11 bright fringes and the first diffraction minima eliminate (are coincident with) bright fringes. How many bright fringes lie between the first and second minima of the diffraction envelope?

•36 A beam of light of a single wavelength is incident perpendicularly on a double-slit arrangement, as in Fig. 35-10. The slit widths are each 46 μm and the slit separation is 0.30 mm. How many complete bright fringes appear between the two first-order minima of the diffraction pattern?

•37 In a double-slit experiment, the slit separation d is 2.00 times the slit width w. How many bright interference fringes are in the central diffraction envelope?

•38 In a certain two-slit interference pattern, 10 bright fringes lie within the second side peak of the diffraction envelope and diffraction minima coincide with two-slit interference maxima. What is the ratio of the slit separation to the slit width?

••39 Light of wavelength 440 nm passes through a double slit, yielding a diffraction pattern whose graph of intensity I versus angular position θ is shown in Fig. 36-44. Calculate (a) the slit width and (b) the slit separation. (c) Verify the displayed intensities of the $m = 1$ and $m = 2$ interference fringes.

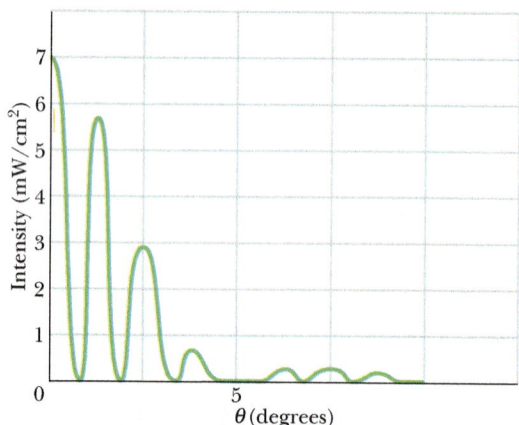

Figure 36-44 Problem 39.

••40 🟢 Figure 36-45 gives the parameter β of Eq. 36-20 versus the sine of the angle θ in a two-slit interference experiment using light of wavelength 435 nm. The vertical axis scale is set by $\beta_s = 80.0$ rad. What are (a) the slit separation, (b) the total number of interference maxima (count them on both sides of the pattern's center), (c) the smallest angle for a maxima, and (d) the greatest angle for a minimum? Assume that none of the interference maxima are completely eliminated by a diffraction minimum.

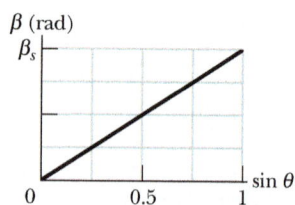

Figure 36-45 Problem 40.

••41 🟢 In the two-slit interference experiment of Fig. 35-10, the slit widths are each 12.0 μm, their separation is 24.0 μm, the wavelength is 600 nm, and the viewing screen is at a distance of 4.00 m. Let I_P represent the intensity at point P on the screen, at height $y = 70.0$ cm. (a) What is the ratio of I_P to the intensity I_m at the center of the pattern? (b) Determine where P is in the two-slit interference pattern by giving the maximum or minimum on which it lies or the maximum and minimum between which it lies. (c) In the same way, for the diffraction that occurs, determine where point P is in the diffraction pattern.

••42 🟢 (a) In a double-slit experiment, what largest ratio of d to a causes diffraction to eliminate the fourth bright side fringe? (b) What other bright fringes are also eliminated? (c) How many other ratios of d to a cause the diffraction to (exactly) eliminate that bright fringe?

••43 SSM WWW (a) How many bright fringes appear between

the first diffraction-envelope minima to either side of the central maximum in a double-slit pattern if $\lambda = 550$ nm, $d = 0.150$ mm, and $a = 30.0$ μm? (b) What is the ratio of the intensity of the third bright fringe to the intensity of the central fringe?

Module 36-5 Diffraction Gratings

•44 ✈ Perhaps to confuse a predator, some tropical gyrinid beetles (whirligig beetles) are colored by optical interference that is due to scales whose alignment forms a diffraction grating (which scatters light instead of transmitting it). When the incident light rays are perpendicular to the grating, the angle between the first-order maxima (on opposite sides of the zeroth-order maximum) is about 26° in light with a wavelength of 550 nm. What is the grating spacing of the beetle?

•45 A diffraction grating 20.0 mm wide has 6000 rulings. Light of wavelength 589 nm is incident perpendicularly on the grating. What are the (a) largest, (b) second largest, and (c) third largest values of θ at which maxima appear on a distant viewing screen?

•46 Visible light is incident perpendicularly on a grating with 315 rulings/mm. What is the longest wavelength that can be seen in the fifth-order diffraction?

•47 SSM ILW A grating has 400 lines/mm. How many orders of the entire visible spectrum (400–700 nm) can it produce in a diffraction experiment, in addition to the $m = 0$ order?

••48 A diffraction grating is made up of slits of width 300 nm with separation 900 nm. The grating is illuminated by monochromatic plane waves of wavelength $\lambda = 600$ nm at normal incidence. (a) How many maxima are there in the full diffraction pattern? (b) What is the angular width of a spectral line observed in the first order if the grating has 1000 slits?

••49 SSM WWW Light of wavelength 600 nm is incident normally on a diffraction grating. Two adjacent maxima occur at angles given by $\sin \theta = 0.2$ and $\sin \theta = 0.3$. The fourth-order maxima are missing. (a) What is the separation between adjacent slits? (b) What is the smallest slit width this grating can have? For that slit width, what are the (c) largest, (d) second largest, and (e) third largest values of the order number m of the maxima produced by the grating?

••50 With light from a gaseous discharge tube incident normally on a grating with slit separation 1.73 μm, sharp maxima of green light are experimentally found at angles $\theta = \pm 17.6°, 37.3°, -37.1°, 65.2°$, and $-65.0°$. Compute the wavelength of the green light that best fits these data.

••51 🟢 A diffraction grating having 180 lines/mm is illuminated with a light signal containing only two wavelengths, $\lambda_1 = 400$ nm and $\lambda_2 = 500$ nm. The signal is incident perpendicularly on the grating. (a) What is the angular separation between the second-order maxima of these two wavelengths? (b) What is the smallest angle at which two of the resulting maxima are superimposed? (c) What is the highest order for which maxima for both wavelengths are present in the diffraction pattern?

••52 🟢 A beam of light consisting of wavelengths from 460.0 nm to 640.0 nm is directed perpendicularly onto a diffraction grating with 160 lines/mm. (a) What is the lowest order that is overlapped by another order? (b) What is the highest order for which the complete wavelength range of the beam is present? In that highest order, at what angle does the light at wavelength (c) 460.0 nm and (d) 640.0 nm appear? (e) What is the greatest angle at which the light at wavelength 460.0 nm appears?

••53 🟢 A grating has 350 rulings/mm and is illuminated at normal incidence by white light. A spectrum is formed on a screen 30.0 cm from the grating. If a hole 10.0 mm square is cut in the screen, its inner edge being 50.0 mm from the central maximum and parallel to it, what are the (a) shortest and (b) longest wavelengths of the light that passes through the hole?

••54 Derive this expression for the intensity pattern for a three-slit "grating":

$$I = \tfrac{1}{9}I_m(1 + 4\cos\phi + 4\cos^2\phi),$$

where $\phi = (2\pi d \sin\theta)/\lambda$ and $a \ll \lambda$.

Module 36-6   Gratings: Dispersion and Resolving Power

•55 SSM ILW A source containing a mixture of hydrogen and deuterium atoms emits red light at two wavelengths whose mean is 656.3 nm and whose separation is 0.180 nm. Find the minimum number of lines needed in a diffraction grating that can resolve these lines in the first order.

•56 (a) How many rulings must a 4.00-cm-wide diffraction grating have to resolve the wavelengths 415.496 and 415.487 nm in the second order? (b) At what angle are the second-order maxima found?

•57 Light at wavelength 589 nm from a sodium lamp is incident perpendicularly on a grating with 40 000 rulings over width 76 nm. What are the first-order (a) dispersion D and (b) resolving power R, the second-order (c) D and (d) R, and the third-order (e) D and (f) R?

•58 A grating has 600 rulings/mm and is 5.0 mm wide. (a) What is the smallest wavelength interval it can resolve in the third order at $\lambda = 500$ nm? (b) How many higher orders of maxima can be seen?

•59 A diffraction grating with a width of 2.0 cm contains 1000 lines/cm across that width. For an incident wavelength of 600 nm, what is the smallest wavelength difference this grating can resolve in the second order?

•60 The D line in the spectrum of sodium is a doublet with wavelengths 589.0 and 589.6 nm. Calculate the minimum number of lines needed in a grating that will resolve this doublet in the second-order spectrum.

•61 With a particular grating the sodium doublet (589.00 nm and 589.59 nm) is viewed in the third order at 10° to the normal and is barely resolved. Find (a) the grating spacing and (b) the total width of the rulings.

••62 A diffraction grating illuminated by monochromatic light normal to the grating produces a certain line at angle θ. (a) What is the product of that line's half-width and the grating's resolving power? (b) Evaluate that product for the first order of a grating of slit separation 900 nm in light of wavelength 600 nm.

••63 Assume that the limits of the visible spectrum are arbitrarily chosen as 430 and 680 nm. Calculate the number of rulings per millimeter of a grating that will spread the first-order spectrum through an angle of 20.0°.

Module 36-7   X-Ray Diffraction

•64 What is the smallest Bragg angle for x rays of wavelength 30 pm to reflect from reflecting planes spaced 0.30 nm apart in a calcite crystal?

•65 An x-ray beam of wavelength A undergoes first-order reflection (Bragg law diffraction) from a crystal when its angle of incidence to a crystal face is 23°, and an x-ray beam of wavelength 97 pm undergoes third-order reflection when its angle of incidence to that face is 60°. Assuming that the two beams reflect from the same family of reflecting planes, find (a) the interplanar spacing and (b) the wavelength A.

•66 An x-ray beam of a certain wavelength is incident on an NaCl crystal, at 30.0° to a certain family of reflecting planes of spacing 39.8 pm. If the reflection from those planes is of the first order, what is the wavelength of the x rays?

•67 Figure 36-46 is a graph of intensity versus angular position θ for the diffraction of an x-ray beam by a crystal. The horizontal scale is set by $\theta_s = 2.00°$. The beam consists of two wavelengths, and the spacing between the reflecting planes is 0.94 nm. What are the (a) shorter and (b) longer wavelengths in the beam?

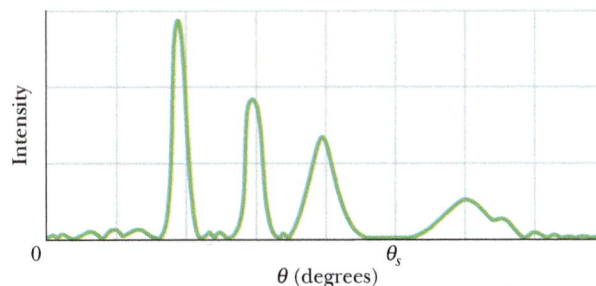

Figure 36-46 Problem 67.

•68 If first-order reflection occurs in a crystal at Bragg angle 3.4°, at what Bragg angle does second-order reflection occur from the same family of reflecting planes?

•69 X rays of wavelength 0.12 nm are found to undergo second-order reflection at a Bragg angle of 28° from a lithium fluoride crystal. What is the interplanar spacing of the reflecting planes in the crystal?

••70 GO In Fig. 36-47, first-order reflection from the reflection planes shown occurs when an x-ray beam of wavelength 0.260 nm makes an angle $\theta = 63.8°$ with the top face of the crystal. What is the unit cell size a_0?

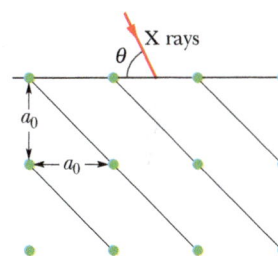

Figure 36-47 Problem 70.

••71 WWW In Fig. 36-48, let a beam of x rays of wavelength 0.125 nm be incident on an NaCl crystal at angle $\theta = 45.0°$ to the top face of the crystal and a family of reflecting planes. Let the reflecting planes have separation $d = 0.252$ nm. The crystal is turned through angle ϕ around an axis perpendicular to the plane of the page until these reflecting planes give diffraction maxima. What are the (a) smaller and (b) larger value of ϕ if the crystal is turned clockwise and the (c) smaller and (d) larger value of ϕ if it is turned counterclockwise?

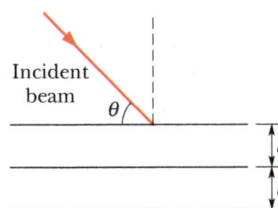

Figure 36-48 Problems 71 and 72.

••72 In Fig. 36-48, an x-ray beam of wavelengths from 95.0 to 140 pm is incident at $\theta = 45.0°$ to a family of reflecting planes with spacing $d = 275$ pm. What are the (a) longest wavelength λ and (b) associated order number m and the (c) shortest λ and (d) associated m of the intensity maxima in the diffraction of the beam?

••73 Consider a two-dimensional square crystal structure, such as one side of the structure shown in Fig. 36-28a. The largest interplanar spacing of reflecting planes is the unit cell size a_0. Calculate and sketch the (a) second largest, (b) third largest, (c) fourth largest, (d) fifth largest, and (e) sixth largest interplanar spacing. (f) Show that your results in (a) through (e) are consistent with the general formula

$$d = \frac{a_0}{\sqrt{h^2 + k^2}},$$

where h and k are relatively prime integers (they have no common factor other than unity).

Additional Problems

74 An astronaut in a space shuttle claims she can just barely resolve two point sources on Earth's surface, 160 km below. Calculate their (a) angular and (b) linear separation, assuming ideal conditions. Take $\lambda = 540$ nm and the pupil diameter of the astronaut's eye to be 5.0 mm.

75 SSM Visible light is incident perpendicularly on a diffraction grating of 200 rulings/mm. What are the (a) longest, (b) second longest, and (c) third longest wavelengths that can be associated with an intensity maximum at $\theta = 30.0°$?

76 A beam of light consists of two wavelengths, 590.159 nm and 590.220 nm, that are to be resolved with a diffraction grating. If the grating has lines across a width of 3.80 cm, what is the minimum number of lines required for the two wavelengths to be resolved in the second order?

77 SSM In a single-slit diffraction experiment, there is a minimum of intensity for orange light ($\lambda = 600$ nm) and a minimum of intensity for blue-green light ($\lambda = 500$ nm) at the same angle of 1.00 mrad. For what minimum slit width is this possible?

78 GO A double-slit system with individual slit widths of 0.030 mm and a slit separation of 0.18 mm is illuminated with 500 nm light directed perpendicular to the plane of the slits. What is the total number of complete bright fringes appearing between the two first-order minima of the diffraction pattern? (Do not count the fringes that coincide with the minima of the diffraction pattern.)

79 SSM A diffraction grating has resolving power $R = \lambda_{avg}/\Delta\lambda = Nm$. (a) Show that the corresponding frequency range Δf that can just be resolved is given by $\Delta f = c/Nm\lambda$. (b) From Fig. 36-22, show that the times required for light to travel along the ray at the bottom of the figure and the ray at the top differ by $\Delta t = (Nd/c) \sin \theta$. (c) Show that $(\Delta f)(\Delta t) = 1$, this relation being independent of the various grating parameters. Assume $N \gg 1$.

80 The pupil of a person's eye has a diameter of 5.00 mm. According to Rayleigh's criterion, what distance apart must two small objects be if their images are just barely resolved when they are 250 mm from the eye? Assume they are illuminated with light of wavelength 500 nm.

81 Light is incident on a grating at an angle ψ as shown in Fig. 36-49.

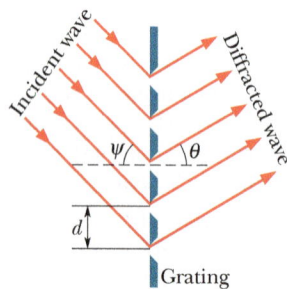

Figure 36-49 Problem 81.

Show that bright fringes occur at angles θ that satisfy the equation

$$d(\sin \psi + \sin \theta) = m\lambda, \quad \text{for } m = 0, 1, 2, \ldots.$$

(Compare this equation with Eq. 36-25.) Only the special case $\psi = 0$ has been treated in this chapter.

82 A grating with $d = 1.50$ μm is illuminated at various angles of incidence by light of wavelength 600 nm. Plot, as a function of the angle of incidence (0 to 90°), the angular deviation of the first-order maximum from the incident direction. (See Problem 81.)

83 SSM In two-slit interference, if the slit separation is 14 μm and the slit widths are each 2.0 μm, (a) how many two-slit maxima are in the central peak of the diffraction envelope and (b) how many are in either of the first side peak of the diffraction envelope?

84 GO In a two-slit interference pattern, what is the ratio of slit separation to slit width if there are 17 bright fringes within the central diffraction envelope and the diffraction minima coincide with two-slit interference maxima?

85 A beam of light with a narrow wavelength range centered on 450 nm is incident perpendicularly on a diffraction grating with a width of 1.80 cm and a line density of 1400 lines/cm across that width. For this light, what is the smallest wavelength difference this grating can resolve in the third order?

86 If you look at something 40 m from you, what is the smallest length (perpendicular to your line of sight) that you can resolve, according to Rayleigh's criterion? Assume the pupil of your eye has a diameter of 4.00 mm, and use 500 nm as the wavelength of the light reaching you.

87 Two yellow flowers are separated by 60 cm along a line perpendicular to your line of sight to the flowers. How far are you from the flowers when they are at the limit of resolution according to the Rayleigh criterion? Assume the light from the flowers has a single wavelength of 550 nm and that your pupil has a diameter of 5.5 mm.

88 In a single-slit diffraction experiment, what must be the ratio of the slit width to the wavelength if the second diffraction minima are to occur at an angle of 37.0° from the center of the diffraction pattern on a viewing screen?

89 A diffraction grating 3.00 cm wide produces the second order at 33.0° with light of wavelength 600 nm. What is the total number of lines on the grating?

90 A single-slit diffraction experiment is set up with light of wavelength 420 nm, incident perpendicularly on a slit of width 5.10 μm. The viewing screen is 3.20 m distant. On the screen, what is the distance between the center of the diffraction pattern and the second diffraction minimum?

91 A diffraction grating has 8900 slits across 1.20 cm. If light with a wavelength of 500 nm is sent through it, how many orders (maxima) lie to one side of the central maximum?

92 In an experiment to monitor the Moon's surface with a light beam, pulsed radiation from a ruby laser ($\lambda = 0.69$ μm) was directed to the Moon through a reflecting telescope with a mirror radius of 1.3 m. A reflector on the Moon behaved like a circular flat mirror with radius 10 cm, reflecting the light directly back toward the telescope on Earth. The reflected light was then detected after being brought to a focus by this telescope. Approximately what fraction of the original light energy was picked up by the detector? Assume that for each direction of travel all the energy is in the central diffraction peak.

93 In June 1985, a laser beam was sent out from the Air Force Optical Station on Maui, Hawaii, and reflected back from the shuttle *Discovery* as it sped by 354 km overhead. The diameter of the central maximum of the beam at the shuttle position was said to be 9.1 m, and the beam wavelength was 500 nm. What is the effective diameter of the laser aperture at the Maui ground station? (*Hint:* A laser beam spreads only because of diffraction; assume a circular exit aperture.)

94 A diffraction grating 1.00 cm wide has 10 000 parallel slits. Monochromatic light that is incident normally is diffracted through 30° in the first order. What is the wavelength of the light?

95 SSM If you double the width of a single slit, the intensity of the central maximum of the diffraction pattern increases by a factor of 4, even though the energy passing through the slit only doubles. Explain this quantitatively.

96 When monochromatic light is incident on a slit 22.0 μm wide, the first diffraction minimum lies at 1.80° from the direction of the incident light. What is the wavelength?

97 A spy satellite orbiting at 160 km above Earth's surface has a lens with a focal length of 3.6 m and can resolve objects on the ground as small as 30 cm. For example, it can easily measure the size of an aircraft's air intake port. What is the effective diameter of the lens as determined by diffraction consideration alone? Assume $\lambda = 550$ nm.

98 Suppose that two points are separated by 2.0 cm. If they are viewed by an eye with a pupil opening of 5.0 mm, what distance from the viewer puts them at the Rayleigh limit of resolution? Assume a light wavelength of 500 nm.

99 A diffraction grating has 200 lines/mm. Light consisting of a continuous range of wavelengths between 550 nm and 700 nm is incident perpendicularly on the grating. (a) What is the lowest order that is overlapped by another order? (b) What is the highest order for which the complete spectrum is present?

100 A diffraction grating has 200 rulings/mm, and it produces an intensity maximum at $\theta = 30.0°$. (a) What are the possible wavelengths of the incident visible light? (b) To what colors do they correspond?

101 SSM Show that the dispersion of a grating is $D = (\tan \theta)/\lambda$.

102 Monochromatic light (wavelength = 450 nm) is incident perpendicularly on a single slit (width = 0.40 mm). A screen is placed parallel to the slit plane, and on it the distance between the two minima on either side of the central maximum is 1.8 mm. (a) What is the distance from the slit to the screen? (*Hint:* The angle to either minimum is small enough that $\sin \theta \approx \tan \theta$.) (b) What is the distance on the screen between the first minimum and the third minimum on the same side of the central maximum?

103 Light containing a mixture of two wavelengths, 500 and 600 nm, is incident normally on a diffraction grating. It is desired (1) that the first and second maxima for each wavelength appear at $\theta \leq 30°$, (2) that the dispersion be as high as possible, and (3) that the third order for the 600 nm light be a missing order. (a) What should be the slit separation? (b) What is the smallest individual slit width that can be used? (c) For the values calculated in (a) and (b) and the light of wavelength 600 nm, what is the largest order of maxima produced by the grating?

104 A beam of x rays with wavelengths ranging from 0.120 nm to 0.0700 nm scatters from a family of reflecting planes in a crystal. The plane separation is 0.250 nm. It is observed that scattered beams are produced for 0.100 nm and 0.0750 nm. What is the angle between the incident and scattered beams?

105 Show that a grating made up of alternately transparent and opaque strips of equal width eliminates all the even orders of maxima (except $m = 0$).

106 Light of wavelength 500 nm diffracts through a slit of width 2.00 μm and onto a screen that is 2.00 m away. On the screen, what is the distance between the center of the diffraction pattern and the third diffraction minimum?

107 If, in a two-slit interference pattern, there are 8 bright fringes within the first side peak of the diffraction envelope and diffraction minima coincide with two-slit interference maxima, then what is the ratio of slit separation to slit width?

108 White light (consisting of wavelengths from 400 nm to 700 nm) is normally incident on a grating. Show that, no matter what the value of the grating spacing d, the second order and third order overlap.

109 If we make $d = a$ in Fig. 36-50, the two slits coalesce into a single slit of width $2a$. Show that Eq. 36-19 reduces to give the diffraction pattern for such a slit.

110 Derive Eq. 36-28, the expression for the half-width of the lines in a grating's diffraction pattern.

111 Prove that it is not possible to determine both wavelength of incident radiation and spacing of reflecting planes in a crystal by measuring the Bragg angles for several orders.

112 How many orders of the entire visible spectrum (400–700 nm) can be produced by a grating of 500 lines/mm?

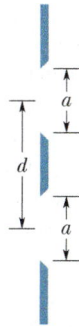

Figure 36-50
Problem 109.

113 An acoustic double-slit system (of slit separation d and slit width a) is driven by two loudspeakers as shown in Fig. 36-51. By use of a variable delay line, the phase of one of the speakers may be varied relative to the other speaker. Describe in detail what changes occur in the double-slit diffraction pattern at large distances as the phase difference between the speakers is varied from zero to 2π. Take both interference and diffraction effects into account.

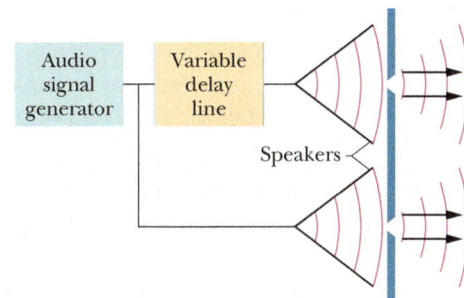

Figure 36-51 Problem 113.

114 Two emission lines have wavelengths λ and $\lambda + \Delta\lambda$, respectively, where $\Delta\lambda \ll \lambda$. Show that their angular separation $\Delta\theta$ in a grating spectrometer is given approximately by

$$\Delta\theta = \frac{\Delta\lambda}{\sqrt{(d/m)^2 - \lambda^2}},$$

where d is the slit separation and m is the order at which the lines are observed. Note that the angular separation is greater in the higher orders than the lower orders.

Relativity

37-1 SIMULTANEITY AND TIME DILATION

Learning Objectives

After reading this module, you should be able to . . .

37.01 Identify the two postulates of (special) relativity and the type of frames to which they apply.

37.02 Identify the speed of light as the ultimate speed and give its approximate value.

37.03 Explain how the space and time coordinates of an event can be measured with a three-dimensional array of clocks and measuring rods and how that eliminates the need of a signal's travel time to an observer.

37.04 Identify that the relativity of space and time has to do with transferring measurements *between* two inertial frames with relative motion but we still use classical kinematics and Newtonian mechanics within a frame.

37.05 Identify that for reference frames with relative motion,

simultaneous events in one of the frames will generally not be simultaneous in the other frame.

37.06 Explain what is meant by the entanglement of the spatial and temporal separations between two events.

37.07 Identify the conditions in which a temporal separation of two events is a proper time.

37.08 Identify that if the temporal separation of two events is a proper time as measured in one frame, that separation is greater (dilated) as measured in another frame.

37.09 Apply the relationship between proper time Δt_0, dilated time Δt, and the relative speed v between two frames.

37.10 Apply the relationships between the relative speed v, the speed parameter β, and the Lorentz factor γ.

Key Ideas

● Einstein's special theory of relativity is based on two postulates: (1) The laws of physics are the same for observers in all inertial reference frames. (2) The speed of light in vacuum has the same value c in all directions and in all inertial reference frames.

● Three space coordinates and one time coordinate specify an event. One task of special relativity is to relate these coordinates as assigned by two observers who are in uniform motion with respect to each other.

● If two observers are in relative motion, they generally will not agree as to whether two events are simultaneous.

● If two successive events occur at the same place in an inertial reference frame, the time interval Δt_0 between them, measured on a single clock where they occur, is the proper time between them. Observers in frames moving relative to that frame will always measure a *larger* value Δt for the time interval, an effect known as time dilation.

● If the relative speed between the two frames is v, then

$$\Delta t = \frac{\Delta t_0}{\sqrt{1 - (v/c)^2}} = \frac{\Delta t_0}{\sqrt{1 - \beta^2}} = \gamma \,\Delta t_0,$$

where $\beta = v/c$ is the speed parameter and $\gamma = 1/\sqrt{1 - \beta^2}$ is the Lorentz factor.

What Is Physics?

One principal subject of physics is **relativity,** the field of study that measures events (things that happen): where and when they happen, and by how much any two events are separated in space and in time. In addition, relativity has to do with transforming such measurements (and also measurements of energy and momentum) between reference frames that move relative to each other. (Hence the name *relativity*.)

Transformations and moving reference frames, such as those we discussed in Modules 4-6 and 4-7, were well understood and quite routine to physicists in 1905.

Then Albert Einstein (Fig. 37-1) published his **special theory of relativity.** The adjective *special* means that the theory deals only with **inertial reference frames,** which are frames in which Newton's laws are valid. (Einstein's *general theory of relativity* treats the more challenging situation in which reference frames can undergo gravitational acceleration; in this chapter the term *relativity* implies only inertial reference frames.)

Starting with two deceivingly simple postulates, Einstein stunned the scientific world by showing that the old ideas about relativity were wrong, even though everyone was so accustomed to them that they seemed to be unquestionable common sense. This supposed common sense, however, was derived only from experience with things that move rather slowly. Einstein's relativity, which turns out to be correct for all physically possible speeds, predicted many effects that were, at first study, bizarre because no one had ever experienced them.

Entangled. In particular, Einstein demonstrated that space and time are entangled; that is, the time between two events depends on how far apart they occur, and vice versa. Also, the entanglement is different for observers who move relative to each other. One result is that time does not pass at a fixed rate, as if it were ticked off with mechanical regularity on some master grandfather clock that controls the universe. Rather, that rate is adjustable: Relative motion can change the rate at which time passes. Prior to 1905, no one but a few daydreamers would have thought that. Now, engineers and scientists take it for granted because their experience with special relativity has reshaped their common sense. For example, any engineer involved with the Global Positioning System of the NAVSTAR satellites must routinely use relativity (both special relativity and general relativity) to determine the rate at which time passes on the satellites because that rate differs from the rate on Earth's surface. If the engineers failed to take relativity into account, GPS would become almost useless in less than one day.

Special relativity has the reputation of being difficult. It is not difficult mathematically, at least not here. However, it is difficult in that we must be very careful about *who* measures *what* about an event and just *how* that measurement is made — and it can be difficult because it can contradict routine experience.

© Corbis-Bettmann

Figure 37-1 Einstein posing for a photograph as fame began to accumulate.

The Postulates

We now examine the two postulates of relativity, on which Einstein's theory is based:

1. The Relativity Postulate: The laws of physics are the same for observers in all inertial reference frames. No one frame is preferred over any other.

Galileo assumed that the laws of *mechanics* were the same in all inertial reference frames. Einstein extended that idea to include *all* the laws of physics, especially those of electromagnetism and optics. This postulate does *not* say that the measured values of all physical quantities are the same for all inertial observers; most are not the same. It is the *laws of physics,* which relate these measurements to one another, that are the same.

2. The Speed of Light Postulate: The speed of light in vacuum has the same value c in all directions and in all inertial reference frames.

We can also phrase this postulate to say that there is in nature an *ultimate speed c,* the same in all directions and in all inertial reference frames. Light happens to travel at this ultimate speed. However, no entity that carries energy or information can exceed this limit. Moreover, no particle that has mass can actually reach speed c, no matter how much or for how long that particle is accelerated. (Alas,

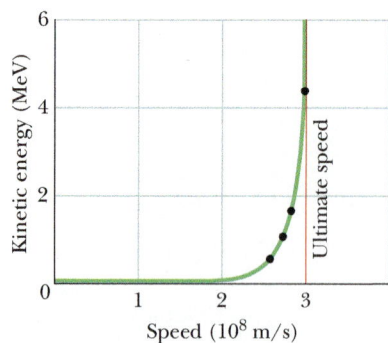

Figure 37-2 The dots show measured values of the kinetic energy of an electron plotted against its measured speed. No matter how much energy is given to an electron (or to any other particle having mass), its speed can never equal or exceed the ultimate limiting speed c. (The plotted curve through the dots shows the predictions of Einstein's special theory of relativity.)

the faster-than-light warp drive used in many science fiction stories appears to be impossible.)

Both postulates have been exhaustively tested, and no exceptions have ever been found.

The Ultimate Speed

The existence of a limit to the speed of accelerated electrons was shown in a 1964 experiment by W. Bertozzi, who accelerated electrons to various measured speeds and—by an independent method—measured their kinetic energies. He found that as the force on a very fast electron is increased, the electron's measured kinetic energy increases toward very large values but its speed does not increase appreciably (Fig. 37-2). Electrons have been accelerated in laboratories to at least 0.999 999 999 95 times the speed of light but—close though it may be—that speed is still less than the ultimate speed c.

This ultimate speed has been defined to be exactly

$$c = 299\ 792\ 458 \text{ m/s.} \tag{37-1}$$

Caution: So far in this book we have (appropriately) approximated c as 3.0×10^8 m/s, but in this chapter we shall often use the exact value. You might want to store the exact value in your calculator's memory (if it is not there already), to be called up when needed.

Testing the Speed of Light Postulate

If the speed of light is the same in all inertial reference frames, then the speed of light emitted by a source moving relative to, say, a laboratory should be the same as the speed of light that is emitted by a source at rest in the laboratory. This claim has been tested directly, in an experiment of high precision. The "light source" was the *neutral pion* (symbol π^0), an unstable, short-lived particle that can be produced by collisions in a particle accelerator. It decays (transforms) into two gamma rays by the process

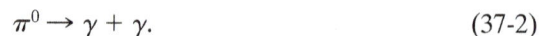

$$\pi^0 \rightarrow \gamma + \gamma. \tag{37-2}$$

Gamma rays are part of the electromagnetic spectrum (at very high frequencies) and so obey the speed of light postulate, just as visible light does. (In this chapter we shall use the term light for any type of electromagnetic wave, visible or not.)

In 1964, physicists at CERN, the European particle-physics laboratory near Geneva, generated a beam of pions moving at a speed of 0.999 75c with respect to the laboratory. The experimenters then measured the speed of the gamma rays emitted from these very rapidly moving sources. They found that the speed of the light emitted by the pions was the same as it would be if the pions were at rest in the laboratory, namely c.

Measuring an Event

An **event** is something that happens, and every event can be assigned three space coordinates and one time coordinate. Among many possible events are (1) the turning on or off of a tiny lightbulb, (2) the collision of two particles, (3) the passage of a pulse of light through a specified point, (4) an explosion, and (5) the sweeping of the hand of a clock past a marker on the rim of the clock. A certain observer, fixed in a certain inertial reference frame, might, for example, assign to an event A the coordinates given in Table 37-1. Because space and time are entangled with each other in relativity, we can describe these coordinates collectively as *spacetime* coordinates. The coordinate system itself is part of the reference frame of the observer.

A given event may be recorded by any number of observers, each in a different inertial reference frame. In general, different observers will assign differ-

Table 37-1 Record of Event A

Coordinate	Value
x	3.58 m
y	1.29 m
z	0 m
t	34.5 s

ent spacetime coordinates to the same event. Note that an event does not "belong" to any particular inertial reference frame. An event is just something that happens, and anyone in any reference frame may detect it and assign space-time coordinates to it.

Travel Times. Making such an assignment can be complicated by a practical problem. For example, suppose a balloon bursts 1 km to your right while a fire-cracker pops 2 km to your left, both at 9:00 A.M. However, you do not detect either event precisely at 9:00 A.M. because at that instant light from the events has not yet reached you. Because light from the firecracker pop has farther to go, it arrives at your eyes later than does light from the balloon burst, and thus the pop will seem to have occurred later than the burst. To sort out the actual times and to assign 9:00 A.M. as the happening time for both events, you must calculate the travel times of the light and then subtract these times from the arrival times.

This procedure can be very messy in more challenging situations, and we need an easier procedure that automatically eliminates any concern about the travel time from an event to an observer. To set up such a procedure, we shall construct an imaginary array of measuring rods and clocks throughout the observer's inertial frame (the array moves rigidly with the observer). This construction may seem contrived, but it spares us much confusion and calculation and allows us to find the coordinates, as follows.

1. **The Space Coordinates.** We imagine the observer's coordinate system fitted with a close-packed, three-dimensional array of measuring rods, one set of rods parallel to each of the three coordinate axes. These rods provide a way to determine coordinates along the axes. Thus, if the event is, say, the turning on of a small lightbulb, the observer, in order to locate the position of the event, need only read the three space coordinates at the bulb's location.

2. **The Time Coordinate.** For the time coordinate, we imagine that every point of intersection in the array of measuring rods includes a tiny clock, which the observer can read because the clock is illuminated by the light generated by the event. Figure 37-3 suggests one plane in the "jungle gym" of clocks and measuring rods we have described.

   The array of clocks must be synchronized properly. It is not enough to assemble a set of identical clocks, set them all to the same time, and then move them to their assigned positions. We do not know, for example, whether moving the clocks will change their rates. (Actually, it will.) We must put the clocks in place and *then* synchronize them.

   If we had a method of transmitting signals at infinite speed, synchroniza-tion would be a simple matter. However, no known signal has this property. We therefore choose light (any part of the electromagnetic spectrum) to send out our synchronizing signals because, in vacuum, light travels at the greatest possible speed, the limiting speed c.

   Here is one of many ways in which an observer might synchronize an array of clocks using light signals: The observer enlists the help of a great num-ber of temporary helpers, one for each clock. The observer then stands at a point selected as the origin and sends out a pulse of light when the origin clock reads $t = 0$. When the light pulse reaches the location of a helper, that helper sets the clock there to read $t = r/c$, where r is the distance between the helper and the origin. The clocks are then synchronized.

3. **The Spacetime Coordinates.** The observer can now assign spacetime coordinates to an event by simply recording the time on the clock nearest the event and the position as measured on the nearest measuring rods. If there are two events, the observer computes their separation in time as the difference in the times on clocks near each and their separation in space from the differ-ences in coordinates on rods near each. We thus avoid the practical problem of calculating the travel times of the signals to the observer from the events.

We use this array to assign spacetime coordinates.

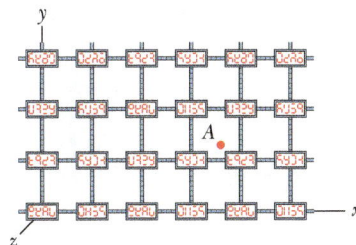

Figure 37-3 One section of a three-dimensional array of clocks and measuring rods by which an observer can assign spacetime coordinates to an event, such as a flash of light at point A. The event's space coordinates are approximately $x = 3.6$ rod lengths, $y = 1.3$ rod lengths, and $z = 0$. The time coordinate is whatever time appears on the clock closest to A at the instant of the flash.

The Relativity of Simultaneity

Suppose that one observer (Sam) notes that two independent events (event Red and event Blue) occur at the same time. Suppose also that another observer (Sally), who is moving at a constant velocity $\vec{v}$ with respect to Sam, also records these same two events. Will Sally also find that they occur at the same time?

The answer is that in general she will not:

> If two observers are in relative motion, they will not, in general, agree as to whether two events are simultaneous. If one observer finds them to be simultaneous, the other generally will not.

We cannot say that one observer is right and the other wrong. Their observations are equally valid, and there is no reason to favor one over the other.

The realization that two contradictory statements about the same natural events can be correct is a seemingly strange outcome of Einstein's theory. However, in Chapter 17 we saw another way in which motion can affect measurement without balking at the contradictory results: In the Doppler effect, the frequency an observer measures for a sound wave depends on the relative motion of observer and source. Thus, two observers moving relative to each other can measure different frequencies for the same wave, and both measurements are correct.

We conclude the following:

> Simultaneity is not an absolute concept but rather a relative one, depending on the motion of the observer.

If the relative speed of the observers is very much less than the speed of light, then measured departures from simultaneity are so small that they are not noticeable. Such is the case for all our experiences of daily living; that is why the relativity of simultaneity is unfamiliar.

A Closer Look at Simultaneity

Let us clarify the relativity of simultaneity with an example based on the postulates of relativity, no clocks or measuring rods being directly involved. Figure 37-4 shows two long spaceships (the SS *Sally* and the SS *Sam*), which can serve as inertial reference frames for observers Sally and Sam. The two observers are stationed at the midpoints of their ships. The ships are separating along a common x axis, the relative velocity of *Sally* with respect to *Sam* being $\vec{v}$. Figure 37-4a shows the ships with the two observer stations momentarily aligned opposite each other.

Two large meteorites strike the ships, one setting off a red flare (event Red) and the other a blue flare (event Blue), not necessarily simultaneously. Each event leaves a permanent mark on each ship, at positions RR' and BB'.

Let us suppose that the expanding wavefronts from the two events happen to reach Sam at the same time, as Fig. 37-4b shows. Let us further suppose that, after the episode, Sam finds, by measuring the marks on his spaceship, that he was indeed stationed exactly halfway between the markers B and R on his ship when the two events occurred. He will say:

Sam   Light from event Red and light from event Blue reached me at the same time. From the marks on my spaceship, I find that I was standing halfway between the two sources. Therefore, event Red and event Blue were simultaneous events.

As study of Fig. 37-4 shows, Sally and the expanding wavefront from event Red are moving *toward* each other, while she and the expanding wavefront from event Blue are moving in the *same direction*. Thus, the wavefront from event Red will reach Sally *before* the wavefront from event Blue does. She will say:

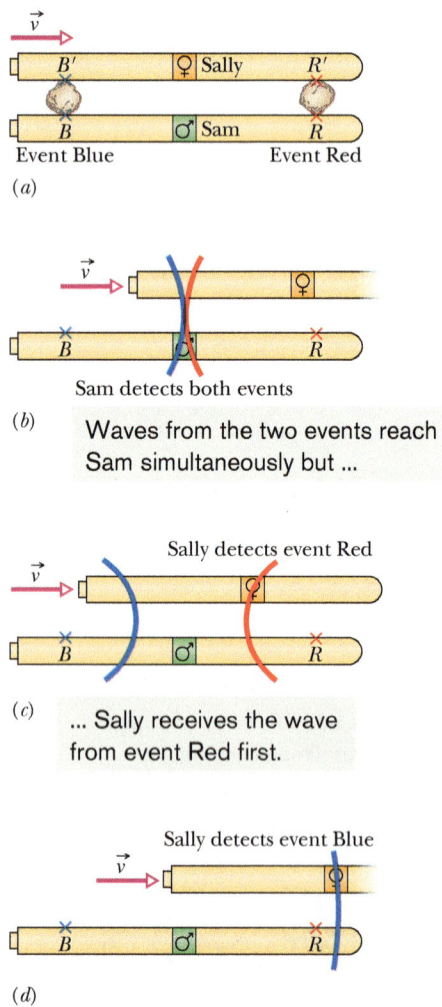

Figure 37-4 The spaceships of Sally and Sam and the occurrences of events from Sam's view. Sally's ship moves rightward with velocity $\vec{v}$. (a) Event Red occurs at positions RR' and event Blue occurs at positions BB'; each event sends out a wave of light. (b) Sam simultaneously detects the waves from event Red and event Blue. (c) Sally detects the wave from event Red. (d) Sally detects the wave from event Blue.

Sally   Light from event Red reached me before light from event Blue did. From the marks on my spaceship, I found that I too was standing halfway between the two sources. Therefore, the events were not simultaneous; event Red occurred first, followed by event Blue.

These reports do not agree. Nevertheless, *both* observers are correct.

Note carefully that there is only one wavefront expanding from the site of each event and that *this wavefront travels with the same speed c in both reference frames,* exactly as the speed of light postulate requires.

It *might* have happened that the meteorites struck the ships in such a way that the two hits appeared to Sally to be simultaneous. If that had been the case, then Sam would have declared them not to be simultaneous.

The Relativity of Time

If observers who move relative to each other measure the time interval (or *temporal separation*) between two events, they generally will find different results. Why? Because the spatial separation of the events can affect the time intervals measured by the observers.

The time interval between two events depends on how far apart they occur in both space and time; that is, their spatial and temporal separations are entangled.

In this module we discuss this entanglement by means of an example; however, the example is restricted in a crucial way: *To one of two observers, the two events occur at the same location.* We shall not get to more general examples until Module 37-3.

Figure 37-5a shows the basics of an experiment Sally conducts while she and her equipment—a light source, a mirror, and a clock—ride in a train moving with constant velocity $\vec{v}$ relative to a station. A pulse of light leaves the light source B (event 1), travels vertically upward, is reflected vertically downward by the mirror, and then is detected back at the source (event 2). Sally measures a certain time interval Δt_0 between the two events, related to the distance D from

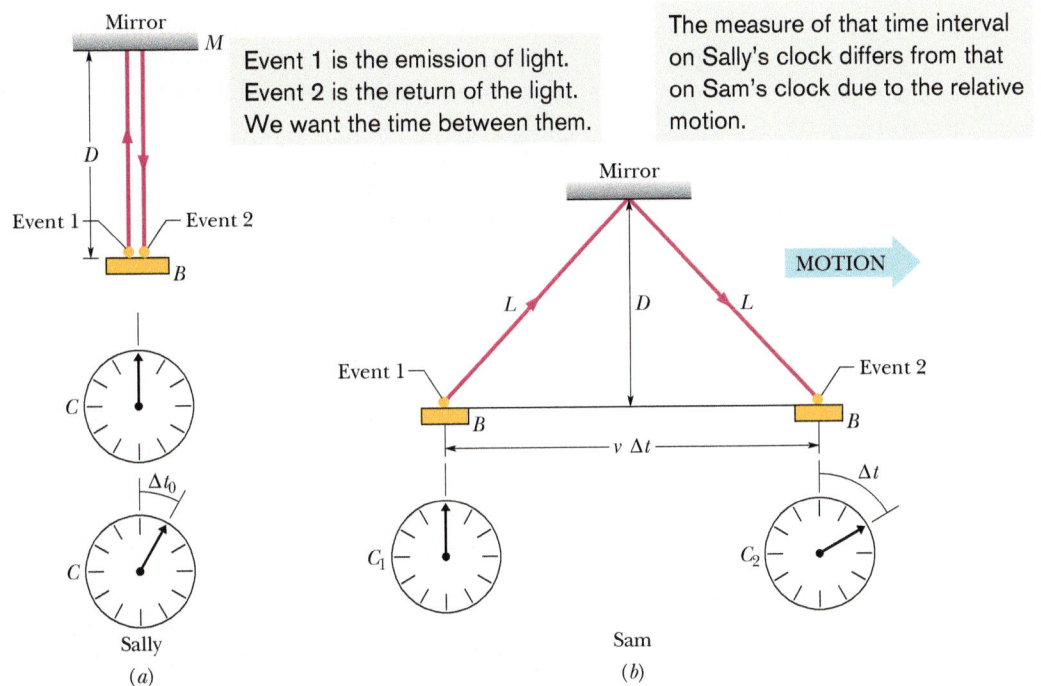

Event 1 is the emission of light.
Event 2 is the return of the light.
We want the time between them.

The measure of that time interval on Sally's clock differs from that on Sam's clock due to the relative motion.

Figure 37-5 (*a*) Sally, on the train, measures the time interval Δt_0 between events 1 and 2 using a single clock C on the train. That clock is shown twice: first for event 1 and then for event 2. (*b*) Sam, watching from the station as the events occur, requires two synchronized clocks, C_1 at event 1 and C_2 at event 2, to measure the time interval between the two events; his measured time interval is Δt.

source to mirror by

$$\Delta t_0 = \frac{2D}{c} \quad \text{(Sally)}. \tag{37-3}$$

The two events occur at the same location in Sally's reference frame, and she needs only one clock C at that location to measure the time interval. Clock C is shown twice in Fig. 37-5a, at the beginning and end of the interval.

Consider now how these same two events are measured by Sam, who is standing on the station platform as the train passes. Because the equipment moves with the train during the travel time of the light, Sam sees the path of the light as shown in Fig. 37-5b. For him, the two events occur at different places in his reference frame, and so to measure the time interval between events, Sam must use *two* synchronized clocks, C_1 and C_2, one at each event. According to Einstein's speed of light postulate, the light travels at the same speed c for Sam as for Sally. Now, however, the light travels distance $2L$ between events 1 and 2. The time interval measured by Sam between the two events is

$$\Delta t = \frac{2L}{c} \quad \text{(Sam)}, \tag{37-4}$$

in which
$$L = \sqrt{(\tfrac{1}{2}v\,\Delta t)^2 + D^2}. \tag{37-5}$$

From Eq. 37-3, we can write this as

$$L = \sqrt{(\tfrac{1}{2}v\,\Delta t)^2 + (\tfrac{1}{2}c\,\Delta t_0)^2}. \tag{37-6}$$

If we eliminate L between Eqs. 37-4 and 37-6 and solve for Δt, we find

$$\Delta t = \frac{\Delta t_0}{\sqrt{1 - (v/c)^2}}. \tag{37-7}$$

Equation 37-7 tells us how Sam's measured interval Δt between the events compares with Sally's interval Δt_0. Because v must be less than c, the denominator in Eq. 37-7 must be less than unity. Thus, Δt must be greater than Δt_0: Sam measures a *greater* time interval between the two events than does Sally. Sam and Sally have measured the time interval between the *same* two events, but the relative motion between Sam and Sally made their measurements *different*. We conclude that relative motion can change the *rate* at which time passes between two events; the key to this effect is the fact that the speed of light is the same for both observers.

We distinguish between the measurements of Sam and Sally in this way:

> When two events occur at the same location in an inertial reference frame, the time interval between them, measured in that frame, is called the **proper time interval** or the **proper time.** Measurements of the same time interval from any other inertial reference frame are always greater.

Thus, Sally measures a proper time interval, and Sam measures a greater time interval. (The term *proper* is unfortunate in that it implies that any other measurement is improper or nonreal. That is just not so.) The amount by which a measured time interval is greater than the corresponding proper time interval is called **time dilation.** (To dilate is to expand or stretch; here the time interval is expanded or stretched.)

Often the dimensionless ratio v/c in Eq. 37-7 is replaced with β, called the **speed parameter,** and the dimensionless inverse square root in Eq. 37-7 is often replaced with γ, called the **Lorentz factor:**

$$\gamma = \frac{1}{\sqrt{1 - \beta^2}} = \frac{1}{\sqrt{1 - (v/c)^2}}. \tag{37-8}$$

With these replacements, we can rewrite Eq. 37-7 as

$$\Delta t = \gamma \, \Delta t_0 \quad \text{(time dilation)}. \tag{37-9}$$

The speed parameter β is always less than unity, and, provided v is not zero, γ is always greater than unity. However, the difference between γ and 1 is not significant unless $v > 0.1c$. Thus, in general, "old relativity" works well enough for $v < 0.1c$, but we must use special relativity for greater values of v. As shown in Fig. 37-6, γ increases rapidly in magnitude as β approaches 1 (as v approaches c). Therefore, the greater the relative speed between Sally and Sam is, the greater will be the time interval measured by Sam, until at a great enough speed, the interval takes "forever."

You might wonder what Sally says about Sam's having measured a greater time interval than she did. His measurement comes as no surprise to her, because to her, he failed to synchronize his clocks C_1 and C_2 in spite of his insistence that he did. Recall that observers in relative motion generally do not agree about simultaneity. Here, Sam insists that his two clocks simultaneously read the same time when event 1 occurred. To Sally, however, Sam's clock C_2 was erroneously set ahead during the synchronization process. Thus, when Sam read the time of event 2 on it, to Sally he was reading off a time that was too large, and that is why the time interval he measured between the two events was greater than the interval she measured.

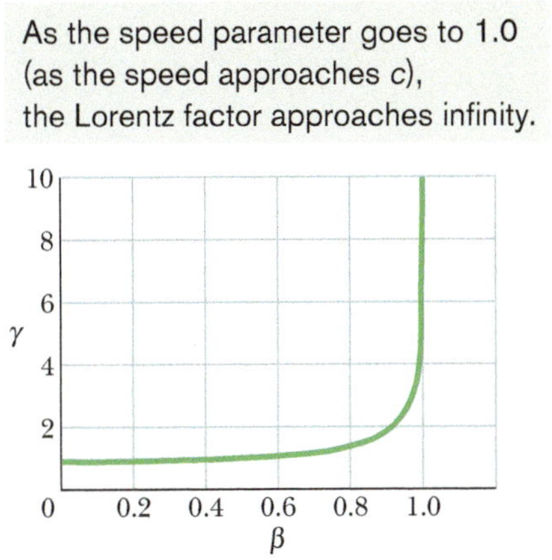

Two Tests of Time Dilation

1. **Microscopic Clocks.** Subatomic particles called *muons* are unstable; that is, when a muon is produced, it lasts for only a short time before it *decays* (transforms into particles of other types). The *lifetime* of a muon is the time interval between its production (event 1) and its decay (event 2). When muons are stationary and their lifetimes are measured with stationary clocks (say, in a laboratory), their average lifetime is 2.200 μs. This is a proper time interval because, for each muon, events 1 and 2 occur at the same location in the reference frame of the muon—namely, at the muon itself. We can represent this proper time interval with Δt_0; moreover, we can call the reference frame in which it is measured the *rest frame* of the muon.

If, instead, the muons are moving, say, through a laboratory, then measurements of their lifetimes made with the laboratory clocks should yield a greater average lifetime (a dilated average lifetime). To check this conclusion, measurements were made of the average lifetime of muons moving with a speed of 0.9994c relative to laboratory clocks. From Eq. 37-8, with $\beta = 0.9994$, the Lorentz factor for this speed is

$$\gamma = \frac{1}{\sqrt{1-\beta^2}} = \frac{1}{\sqrt{1-(0.9994)^2}} = 28.87.$$

Equation 37-9 then yields, for the average dilated lifetime,

$$\Delta t = \gamma \, \Delta t_0 = (28.87)(2.200 \ \mu\text{s}) = 63.51 \ \mu\text{s}.$$

The actual measured value matched this result within experimental error.

2. **Macroscopic Clocks.** In October 1971, Joseph Hafele and Richard Keating carried out what must have been a grueling experiment. They flew four portable atomic clocks twice around the world on commercial airlines, once in each direction. Their purpose was "to test Einstein's theory of relativity with macroscopic clocks." As we have just seen, the time dilation predictions of Einstein's theory have been confirmed on a microscopic scale, but there is great comfort in seeing a confirmation made with an actual clock. Such macroscopic measurements became possible only because of the very high precision of modern atomic clocks. Hafele and Keating verified the predictions of the theory to within 10%. (Einstein's *general* theory of relativity, which predicts

As the speed parameter goes to 1.0 (as the speed approaches c), the Lorentz factor approaches infinity.

Figure 37-6 A plot of the Lorentz factor γ as a function of the speed parameter $\beta \, (= v/c)$.

that the rate at which time passes on a clock is influenced by the gravitational force on the clock, also plays a role in this experiment.)

A few years later, physicists at the University of Maryland flew an atomic clock round and round over Chesapeake Bay for flights lasting 15 h and succeeded in checking the time dilation prediction to better than 1%. Today, when atomic clocks are transported from one place to another for calibration or other purposes, the time dilation caused by their motion is always taken into account.

✓ Checkpoint 1

Standing beside railroad tracks, we are suddenly startled by a relativistic boxcar traveling past us as shown in the figure. Inside, a well-equipped hobo fires a laser pulse from the front of the boxcar to its rear. (a) Is our measurement of the speed of the pulse greater than, less than, or the same as that measured by the hobo? (b) Is his measurement of the flight time of the pulse a proper time? (c) Are his measurement and our measurement of the flight time related by Eq. 37-9?

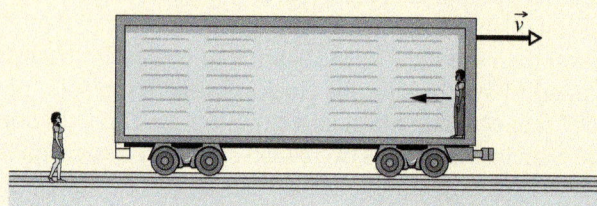

Sample Problem 37.01 Time dilation for a space traveler who returns to Earth

Your starship passes Earth with a relative speed of 0.9990c. After traveling 10.0 y (your time), you stop at lookout post LP13, turn, and then travel back to Earth with the same relative speed. The trip back takes another 10.0 y (your time). How long does the round trip take according to measurements made on Earth? (Neglect any effects due to the accelerations involved with stopping, turning, and getting back up to speed.)

KEY IDEAS

We begin by analyzing the outward trip:

1. This problem involves measurements made from two (inertial) reference frames, one attached to Earth and the other (your reference frame) attached to your ship.
2. The outward trip involves two events: the start of the trip at Earth and the end of the trip at LP13.
3. Your measurement of 10.0 y for the outward trip is the proper time Δt_0 between those two events, because the events occur at the same location in your reference frame—namely, on your ship.

4. The Earth-frame measurement of the time interval Δt for the outward trip must be greater than Δt_0, according to Eq. 37-9 ($\Delta t = \gamma \Delta t_0$) for time dilation.

Calculations: Using Eq. 37-8 to substitute for γ in Eq. 37-9, we find

$$\Delta t = \frac{\Delta t_0}{\sqrt{1 - (v/c)^2}}$$
$$= \frac{10.0 \text{ y}}{\sqrt{1 - (0.9990c/c)^2}} = (22.37)(10.0 \text{ y}) = 224 \text{ y}.$$

On the return trip, we have the same situation and the same data. Thus, the round trip requires 20 y of your time but

$$\Delta t_{\text{total}} = (2)(224 \text{ y}) = 448 \text{ y} \quad \text{(Answer)}$$

of Earth time. In other words, you have aged 20 y while the Earth has aged 448 y. Although you cannot travel into the past (as far as we know), you can travel into the future of, say, Earth, by using high-speed relative motion to adjust the rate at which time passes.

Sample Problem 37.02 Time dilation and travel distance for a relativistic particle

The elementary particle known as the *positive kaon* (K⁺) is unstable in that it can *decay* (transform) into other particles. Although the decay occurs randomly, we find that, on average, a positive kaon has a lifetime of 0.1237 μs when stationary—that is, when the lifetime is measured in the rest frame of the kaon. If a positive kaon has a speed of 0.990c relative to a laboratory reference frame when the kaon is produced, how far can it travel in that frame during its lifetime according to *classical physics* (which is a reasonable approximation for speeds much less than c)

and according to special relativity (which is correct for all physically possible speeds)?

KEY IDEAS

1. We have two (inertial) reference frames, one attached to the kaon and the other attached to the laboratory.
2. This problem also involves two events: the start of the kaon's travel (when the kaon is produced) and the end of that travel (at the end of the kaon's lifetime).

3. The distance traveled by the kaon between those two events is related to its speed v and the time interval for the travel by

$$v = \frac{\text{distance}}{\text{time interval}}. \qquad (37\text{-}10)$$

With these ideas in mind, let us solve for the distance first with classical physics and then with special relativity.

Classical physics: In classical physics we would find the same distance and time interval (in Eq. 37-10) whether we measured them from the kaon frame or from the laboratory frame. Thus, we need not be careful about the frame in which the measurements are made. To find the kaon's travel distance d_{cp} according to classical physics, we first rewrite Eq. 37-10 as

$$d_{cp} = v \, \Delta t, \qquad (37\text{-}11)$$

where Δt is the time interval between the two events in either frame. Then, substituting $0.990c$ for v and $0.1237\ \mu s$ for Δt in Eq. 37-11, we find

$$d_{cp} = (0.990c) \, \Delta t$$

$$= (0.990)(299\ 792\ 458\ \text{m/s})(0.1237 \times 10^{-6}\ \text{s})$$

$$= 36.7\ \text{m.} \qquad \text{(Answer)}$$

This is how far the kaon would travel if classical physics were correct at speeds close to c.

Special relativity: In special relativity we must be very careful that both the distance and the time interval in Eq. 37-10 are measured in the *same* reference frame—especially when the speed is close to c, as here. Thus, to find the actual travel distance d_{sr} of the kaon *as measured from the laboratory frame* and according to special relativity, we rewrite Eq. 37-10 as

$$d_{sr} = v \, \Delta t, \qquad (37\text{-}12)$$

where Δt is the time interval between the two events *as measured from the laboratory frame.*

Before we can evaluate d_{sr} in Eq. 37-12, we must find Δt. The $0.1237\ \mu s$ time interval is a proper time because the two events occur at the same location in the kaon frame—namely, at the kaon itself. Therefore, let Δt_0 represent this proper time interval. Then we can use Eq. 37-9 ($\Delta t = \gamma \, \Delta t_0$) for time dilation to find the time interval Δt as measured from the laboratory frame. Using Eq. 37-8 to substitute for γ in Eq. 37-9 leads to

$$\Delta t = \frac{\Delta t_0}{\sqrt{1 - (v/c)^2}} = \frac{0.1237 \times 10^{-6}\ \text{s}}{\sqrt{1 - (0.990c/c)^2}} = 8.769 \times 10^{-7}\ \text{s.}$$

This is about seven times longer than the kaon's proper lifetime. That is, the kaon's lifetime is about seven times longer in the laboratory frame than in its own frame—the kaon's lifetime is dilated. We can now evaluate Eq. 37-12 for the travel distance d_{sr} in the laboratory frame as

$$d_{sr} = v \, \Delta t = (0.990c) \, \Delta t$$

$$= (0.990)(299\ 792\ 458\ \text{m/s})(8.769 \times 10^{-7}\ \text{s})$$

$$= 260\ \text{m.} \qquad \text{(Answer)}$$

This is about seven times d_{cp}. Experiments like the one outlined here, which verify special relativity, became routine in physics laboratories decades ago. The engineering design and the construction of any scientific or medical facility that employs high-speed particles must take relativity into account.

37-2 THE RELATIVITY OF LENGTH

Learning Objectives

After reading this module, you should be able to . . .

37.11 Identify that because spatial and temporal separations are entangled, measurements of the lengths of objects may be different in two frames with relative motion.

37.12 Identify the condition in which a measured length is a proper length.

37.13 Identify that if a length is a proper length as measured in one frame, the length is less (contracted) as measured in another frame that is in relative motion *parallel* to the length.

37.14 Apply the relationship between contracted length L, proper length L_0, and the relative speed v between two frames.

Key Ideas

● The length L_0 of an object measured by an observer in an inertial reference frame in which the object is at rest is called its proper length. Observers in frames moving relative to that frame and parallel to that length will always measure a shorter length, an effect known as length contraction.

● If the relative speed between frames is v, the contracted length L and the proper length L_0 are related by

$$L = L_0 \sqrt{1 - \beta^2} = \frac{L_0}{\gamma},$$

where $\beta = v/c$ is the speed parameter and $\gamma = 1/\sqrt{1 - \beta^2}$ is the Lorentz factor.

The Relativity of Length

If you want to measure the length of a rod that is at rest with respect to you, you can—at your leisure—note the positions of its end points on a long stationary scale and subtract one reading from the other. If the rod is moving, however, you must note the positions of the end points *simultaneously* (in your reference frame) or your measurement cannot be called a length. Figure 37-7 suggests the difficulty of trying to measure the length of a moving penguin by locating its front and back at different times. Because simultaneity is relative and it enters into length measurements, length should also be a relative quantity. It is.

Let L_0 be the length of a rod that you measure when the rod is stationary (meaning you and it are in the same reference frame, the rod's rest frame). If, instead, there is relative motion at speed v between you and the rod *along the length of the rod,* then with simultaneous measurements you obtain a length L given by

$$L = L_0 \sqrt{1 - \beta^2} = \frac{L_0}{\gamma} \quad \text{(length contraction)}. \quad (37\text{-}13)$$

Because the Lorentz factor γ is always greater than unity if there is relative motion, L is less than L_0. The relative motion causes a *length contraction,* and L is called a *contracted length.* A greater speed v results in a greater contraction.

> The length L_0 of an object measured in the rest frame of the object is its **proper length** or **rest length.** Measurements of the length from any reference frame that is in relative motion *parallel* to that length are always less than the proper length.

Be careful: Length contraction occurs only along the direction of relative motion. Also, the length that is measured does not have to be that of an object like a rod or a circle. Instead, it can be the length (or distance) between two objects in the same rest frame—for example, the Sun and a nearby star (which are, at least approximately, at rest relative to each other).

Does a moving object *really* shrink? Reality is based on observations and measurements; if the results are always consistent and if no error can be determined, then what is observed and measured is real. In that sense, the object really does shrink. However, a more precise statement is that the object *is really measured* to shrink—motion affects that measurement and thus reality.

When you measure a contracted length for, say, a rod, what does an observer moving with the rod say of your measurement? To that observer, you did not locate the two ends of the rod simultaneously. (Recall that observers in motion relative to each other do not agree about simultaneity.) To the observer, you first located the rod's front end and then, slightly later, its rear end, and that is why you measured a length that is less than the proper length.

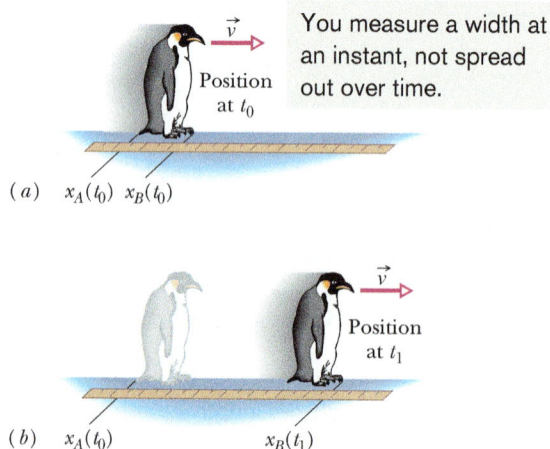

You measure a width at an instant, not spread out over time.

(a) $x_A(t_0)$ $x_B(t_0)$

(b) $x_A(t_0)$ $x_B(t_1)$

Figure 37-7 If you want to measure the front-to-back length of a penguin while it is moving, you must mark the positions of its front and back simultaneously (in your reference frame), as in (a), rather than at different times, as in (b).

Proof of Eq. 37-13

Length contraction is a direct consequence of time dilation. Consider once more our two observers. This time, both Sally, seated on a train moving through a station, and Sam, again on the station platform, want to measure the length of the platform. Sam, using a tape measure, finds the length to be L_0, a proper length because the platform is at rest with respect to him. Sam also notes that Sally, on the train, moves through this length in a time $\Delta t = L_0/v$, where v is the speed of the train; that is,

$$L_0 = v\,\Delta t \quad \text{(Sam)}. \quad (37\text{-}14)$$

This time interval Δt is not a proper time interval because the two events that define it (Sally passes the back of the platform and Sally passes the front of the platform) occur at two different places, and therefore Sam must use two synchronized clocks to measure the time interval Δt.

For Sally, however, the platform is moving past her. She finds that the two events measured by Sam occur *at the same place* in her reference frame. She can time them with a single stationary clock, and so the interval Δt_0 that she measures is a proper time interval. To her, the length L of the platform is given by

$$L = v\,\Delta t_0 \qquad \text{(Sally)}. \qquad (37\text{-}15)$$

If we divide Eq. 37-15 by Eq. 37-14 and apply Eq. 37-9, the time dilation equation, we have

$$\frac{L}{L_0} = \frac{v\,\Delta t_0}{v\,\Delta t} = \frac{1}{\gamma},$$

or

$$L = \frac{L_0}{\gamma}, \qquad (37\text{-}16)$$

which is Eq. 37-13, the length contraction equation.

Sample Problem 37.03 Time dilation and length contraction as seen from each frame

In Fig. 37-8, Sally (at point A) and Sam's spaceship (of proper length $L_0 = 230$ m) pass each other with constant relative speed v. Sally measures a time interval of 3.57 μs for the ship to pass her (from the passage of point B in Fig. 37-8a to the passage of point C in Fig. 37-8b). In terms of c, what is the relative speed v between Sally and the ship?

KEY IDEAS

Let's assume that speed v is near c. Then:

1. This problem involves measurements made from two (inertial) reference frames, one attached to Sally and the other attached to Sam and his spaceship.

2. This problem also involves two events: the first is the passage of point B past Sally (Fig. 37-8a) and the second is the passage of point C past her (Fig. 37-8b).

3. From either reference frame, the other reference frame passes at speed v and moves a certain distance in the time interval between the two events:

$$v = \frac{\text{distance}}{\text{time interval}}. \qquad (37\text{-}17)$$

Because speed v is assumed to be near the speed of light, we must be careful that the distance and the time interval in Eq. 37-17 are measured in the *same* reference frame.

Calculations: We are free to use either frame for the measurements. Because we know that the time interval Δt between the two events measured from Sally's frame is 3.57 μs, let us also use the distance L between the two events measured from her frame. Equation 37-17 then becomes

$$v = \frac{L}{\Delta t}. \qquad (37\text{-}18)$$

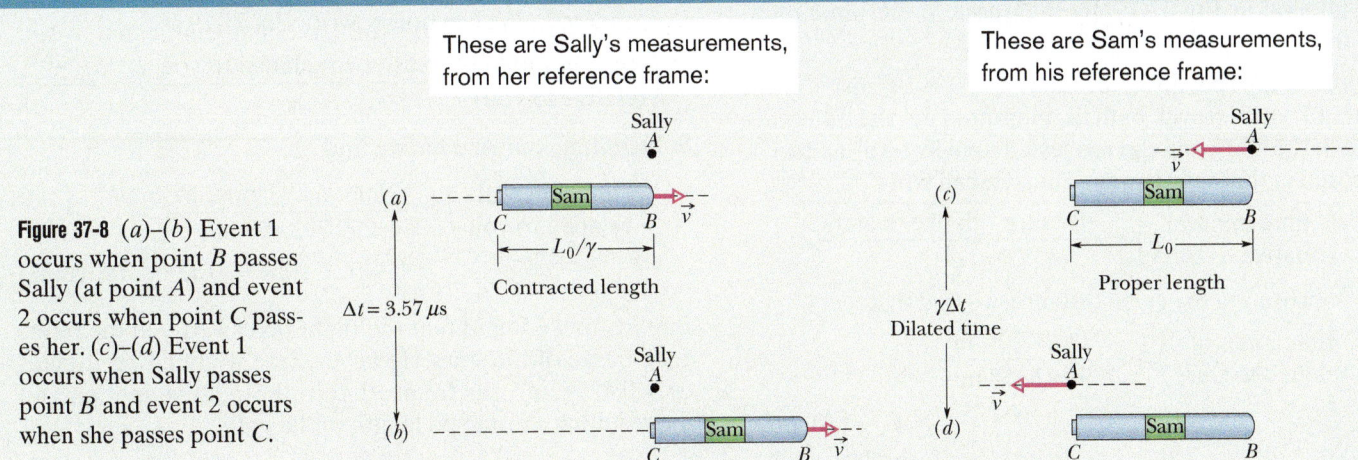

These are Sally's measurements, from her reference frame:

These are Sam's measurements, from his reference frame:

Figure 37-8 (a)–(b) Event 1 occurs when point B passes Sally (at point A) and event 2 occurs when point C passes her. (c)–(d) Event 1 occurs when Sally passes point B and event 2 occurs when she passes point C.

$\Delta t = 3.57\ \mu s$

Contracted length

$\gamma \Delta t$
Dilated time

Proper length

We do not know L, but we can relate it to the given L_0: The distance between the two events as measured from Sam's frame is the ship's proper length L_0. Thus, the distance L measured from Sally's frame must be less than L_0, as given by Eq. 37-13 ($L = L_0/\gamma$) for length contraction. Substituting L_0/γ for L in Eq. 37-18 and then substituting Eq. 37-8 for γ, we find

$$v = \frac{L_0/\gamma}{\Delta t} = \frac{L_0\sqrt{(1 - (v/c)^2}}{\Delta t}.$$

Solving this equation for v (notice that it is on the left and also buried in the Lorentz factor) leads us to

$$v = \frac{L_0 c}{\sqrt{(c\,\Delta t)^2 + L_0^2}}$$

$$= \frac{(230\text{ m})c}{\sqrt{(299\ 792\ 458\text{ m/s})^2(3.57 \times 10^{-6}\text{ s})^2 + (230\text{ m})^2}}$$

$$= 0.210c. \qquad \text{(Answer)}$$

Note that only the relative motion of Sally and Sam

matters here; whether either is stationary relative to, say, a space station is irrelevant. In Figs. 37-8a and b we took Sally to be stationary, but we could instead have taken the ship to be stationary, with Sally moving to the left past it. Event 1 is again when Sally and point B are aligned (Fig. 37-8c), and event 2 is again when Sally and point C are aligned (Fig. 37-8d). However, we are now using Sam's measurements. So the length between the two events in *his* frame is the proper length L_0 of the ship and the time interval between them is not Sally's measurement Δt but a dilated time interval $\gamma\,\Delta t$.

Substituting Sam's measurements into Eq. 37-17, we have

$$v = \frac{L_0}{\gamma\,\Delta t},$$

which is exactly what we found using Sally's measurements. Thus, we get the same result of $v = 0.210c$ with either set of measurements, *but we must be careful not to mix the measurements from the two frames.*

Sample Problem 37.04 Time dilation and length contraction in outrunning a supernova

Caught by surprise near a supernova, you race away from the explosion in your spaceship, hoping to outrun the high-speed material ejected toward you. Your Lorentz factor γ relative to the inertial reference frame of the local stars is 22.4.

(a) To reach a safe distance, you figure you need to cover 9.00×10^{16} m as measured in the reference frame of the local stars. How long will the flight take, as measured in that frame?

KEY IDEAS

From Chapter 2, for constant speed, we know that

$$\text{speed} = \frac{\text{distance}}{\text{time interval}}. \qquad (37\text{-}19)$$

From Fig. 37-6, we see that because your Lorentz factor γ relative to the stars is 22.4 (large), your relative speed v is almost c—so close that we can approximate it as c. Then for speed $v \approx c$, we must be careful that the distance and the time interval in Eq. 37-19 are measured in the *same* reference frame.

Calculations: The given distance (9.00×10^{16} m) for the length of your travel path is measured in the reference frame of the stars, and the requested time interval Δt is to be measured in that same frame. Thus, we can write

$$\left(\begin{array}{c}\text{time interval}\\\text{relative to stars}\end{array}\right) = \frac{\text{distance relative to stars}}{c}.$$

Then substituting the given distance, we find that

$$\left(\begin{array}{c}\text{time interval}\\\text{relative to stars}\end{array}\right) = \frac{9.00 \times 10^{16}\text{ m}}{299\ 792\ 458\text{ m/s}}$$

$$= 3.00 \times 10^8\text{ s} = 9.51\text{ y}. \quad \text{(Answer)}$$

(b) How long does that trip take according to you (in your reference frame)?

KEY IDEAS

1. We now want the time interval measured in a different reference frame—namely, yours. Thus, we need to transform the data given in the reference frame of the stars to your frame.

2. The given path length of 9.00×10^{16} m, measured in the reference frame of the stars, is a proper length L_0, because the two ends of the path are at rest in that frame. As observed from your reference frame, the stars' reference frame and those two ends of the path race past you at a relative speed of $v \approx c$.

3. You measure a contracted length L_0/γ for the path, not the proper length L_0.

Calculations: We can now rewrite Eq. 37-19 as

$$\left(\begin{array}{c}\text{time interval}\\\text{relative to you}\end{array}\right) = \frac{\text{distance relative to you}}{c} = \frac{L_0/\gamma}{c}.$$

Substituting known data, we find

$$\left(\begin{array}{c}\text{time interval}\\\text{relative to you}\end{array}\right) = \frac{(9.00 \times 10^{16}\text{ m})/22.4}{299\ 792\ 458\text{ m/s}}$$

$$= 1.340 \times 10^7\text{ s} = 0.425\text{ y}. \quad \text{(Answer)}$$

In part (a) we found that the flight takes 9.51 y in the reference frame of the stars. However, here we find that it takes only 0.425 y in your frame, due to the relative motion and the resulting contracted length of the path.

PLUS Additional examples, video, and practice available at *WileyPLUS*

37-3 THE LORENTZ TRANSFORMATION

Learning Objectives

After reading this module, you should be able to . . .

37.15 For frames with relative motion, apply the Galilean transformation to transform an event's position from one frame to the other.

37.16 Identify that a Galilean transformation is approximately correct for slow relative speeds but the Lorentz transformations are the correct transformations for any physically possible speed.

37.17 Apply the Lorentz transformations for the spatial and temporal separations of two events as measured in two frames with a relative speed v.

37.18 From the Lorentz transformations, derive the equations for time dilation and length contraction.

37.19 From the Lorentz transformations show that if two events are simultaneous but spatially separated in one frame, they cannot be simultaneous in another frame with relative motion.

Key Idea

● The Lorentz transformation equations relate the spacetime coordinates of a single event as seen by observers in two inertial frames, S and S', where S' is moving relative to S with velocity v in the positive x and x' direction. The four coordinates are related by

$$x' = \gamma(x - vt),$$
$$y' = y,$$
$$z' = z,$$
$$t' = \gamma(t - vx/c^2).$$

The Lorentz Transformation

Figure 37-9 shows inertial reference frame S' moving with speed v relative to frame S, in the common positive direction of their horizontal axes (marked x and x'). An observer in S reports spacetime coordinates x, y, z, t for an event, and an observer in S' reports x', y', z', t' for the same event. How are these sets of numbers related? We claim at once (although it requires proof) that the y and z coordinates, which are perpendicular to the motion, are not affected by the motion; that is, $y = y'$ and $z = z'$. Our interest then reduces to the relation between x and x' and that between t and t'.

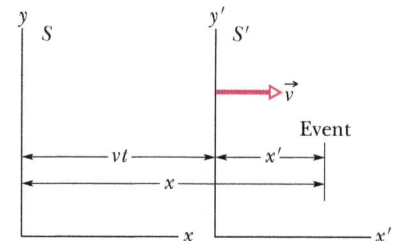

Figure 37-9 Two inertial reference frames: frame S' has velocity $\vec{v}$ relative to frame S.

The Galilean Transformation Equations

Prior to Einstein's publication of his special theory of relativity, the four coordinates of interest were assumed to be related by the *Galilean transformation equations:*

$$\begin{aligned} x' &= x - vt \\ t' &= t \end{aligned} \qquad \begin{array}{l}\text{(Galilean transformation equations;} \\ \text{approximately valid at low speeds).}\end{array} \qquad (37\text{-}20)$$

(These equations are written with the assumption that $t = t' = 0$ when the origins of S and S' coincide.) You can verify the first equation with Fig. 37-9. The second equation effectively claims that time passes at the same rate for observers in both reference frames. That would have been so obviously true to a scientist prior to Einstein that it would not even have been mentioned. When speed v is small compared to c, Eqs. 37-20 generally work well.

The Lorentz Transformation Equations

Equations 37-20 work well when speed v is small compared to c, but they are actually incorrect for any speed and are very wrong when v is greater than about $0.10c$. The equations that are correct for any physically possible speed are called the **Lorentz transformation equations*** (or simply the Lorentz transformations).

---

*You may wonder why we do not call these the *Einstein transformation equations* (and why not the *Einstein factor* for γ). H. A. Lorentz actually derived these equations before Einstein did, but as the great Dutch physicist graciously conceded, he did not take the further bold step of interpreting these equations as describing the true nature of space and time. It is this interpretation, first made by Einstein, that is at the heart of relativity.

We can derive them from the postulates of relativity, but here we shall instead first examine them and then justify them by showing them to be consistent with our results for simultaneity, time dilation, and length contraction. Assuming that $t = t' = 0$ when the origins of S and S' coincide in Fig. 37-9 (event 1), then the spatial and temporal coordinates of any other event are given by

$$x' = \gamma(x - vt),$$
$$y' = y,$$
$$z' = z,$$
$$t' = \gamma(t - vx/c^2)$$

(Lorentz transformation equations; valid at all physically possible speeds). (37-21)

Note that the spatial values x and the temporal values t are bound together in the first and last equations. This entanglement of space and time was a prime message of Einstein's theory, a message that was long rejected by many of his contemporaries.

It is a formal requirement of relativistic equations that they should reduce to familiar classical equations if we let c approach infinity. That is, if the speed of light were infinitely great, *all* finite speeds would be "low" and classical equations would never fail. If we let $c \rightarrow \infty$ in Eqs. 37-21, $\gamma \rightarrow 1$ and these equations reduce—as we expect—to the Galilean equations (Eqs. 37-20). You should check this.

Equations 37-21 are written in a form that is useful if we are given x and t and wish to find x' and t'. We may wish to go the other way, however. In that case we simply solve Eqs. 37-21 for x and t, obtaining

$$x = \gamma(x' + vt') \quad \text{and} \quad t = \gamma(t' + vx'/c^2). \quad (37\text{-}22)$$

Comparison shows that, starting from either Eqs. 37-21 or Eqs. 37-22, you can find the other set by interchanging primed and unprimed quantities and reversing the sign of the relative velocity v. (For example, if the S' frame has a positive velocity relative to an observer in the S frame as in Fig. 37-9, then the S frame has a *negative* velocity relative to an observer in the S' frame.)

Equations 37-21 relate the coordinates of a second event when the first event is the passing of the origins of S and S' at $t = t' = 0$. However, in general we do not want to restrict the first event to being such a passage. So, let's rewrite the Lorentz transformations in terms of any pair of events 1 and 2, with spatial and temporal separations

$$\Delta x = x_2 - x_1 \quad \text{and} \quad \Delta t = t_2 - t_1,$$

as measured by an observer in S, and

$$\Delta x' = x'_2 - x'_1 \quad \text{and} \quad \Delta t' = t'_2 - t'_1,$$

as measured by an observer in S'. Table 37-2 displays the Lorentz equations in difference form, suitable for analyzing pairs of events. The equations in the table were derived by simply substituting differences (such as Δx and $\Delta x'$) for the four variables in Eqs. 37-21 and 37-22.

Be careful: When substituting values for these differences, you must be consistent and not mix the values for the first event with those for the second event. Also, if, say, Δx is a negative quantity, you must be certain to include the minus sign in a substitution.

Table 37-2 The Lorentz Transformation Equations for Pairs of Events

1. $\Delta x = \gamma(\Delta x' + v\,\Delta t')$	1'. $\Delta x' = \gamma(\Delta x - v\,\Delta t)$
2. $\Delta t = \gamma(\Delta t' + v\,\Delta x'/c^2)$	2'. $\Delta t' = \gamma(\Delta t - v\,\Delta x/c^2)$

$$\gamma = \frac{1}{\sqrt{1 - (v/c)^2}} = \frac{1}{\sqrt{1 - \beta^2}}$$

Frame S' moves at velocity v relative to frame S.

✓ **Checkpoint 2**

In Fig. 37-9, frame S' has velocity $0.90c$ relative to frame S. An observer in frame S' measures two events as occurring at the following spacetime coordinates: event Yellow at $(5.0 \text{ m}, 20 \text{ ns})$ and event Green at $(-2.0 \text{ m}, 45 \text{ ns})$. An observer in frame S wants to find the temporal separation $\Delta t_{GY} = t_G - t_Y$ between the events. (a) Which equation in Table 37-2 should be used? (b) Should $+0.90c$ or $-0.90c$ be substituted for v in the parentheses on the equation's right side and in the Lorentz factor γ? What value should be substituted into the (c) first and (d) second term in the parentheses?

Some Consequences of the Lorentz Equations

Here we use the equations of Table 37-2 to affirm some of the conclusions that we reached earlier by arguments based directly on the postulates.

Simultaneity

Consider Eq. 2 of Table 37-2,

$$\Delta t = \gamma \left(\Delta t' + \frac{v \, \Delta x'}{c^2} \right). \tag{37-23}$$

If two events occur at different places in reference frame S' of Fig. 37-9, then $\Delta x'$ in this equation is not zero. It follows that even if the events are simultaneous in S' (thus $\Delta t' = 0$), they will not be simultaneous in frame S. (This is in accord with our conclusion in Module 37-1.) The time interval between the events in S will be

$$\Delta t = \gamma \frac{v \, \Delta x'}{c^2} \quad \text{(simultaneous events in } S'\text{).}$$

Thus, the spatial separation $\Delta x'$ guarantees a temporal separation Δt.

Time Dilation

Suppose now that two events occur at the same place in S' (thus $\Delta x' = 0$) but at different times (thus $\Delta t' \neq 0$). Equation 37-23 then reduces to

$$\Delta t = \gamma \, \Delta t' \quad \text{(events in same place in } S'\text{).} \tag{37-24}$$

This confirms time dilation between frames S and S'. Moreover, because the two events occur at the same place in S', the time interval $\Delta t'$ between them can be measured with a single clock, located at that place. Under these conditions, the measured interval is a proper time interval, and we can label it Δt_0 as we have previously labeled proper times. Thus, with that label Eq. 37-24 becomes

$$\Delta t = \gamma \, \Delta t_0 \quad \text{(time dilation),}$$

which is exactly Eq. 37-9, the time dilation equation. Thus, time dilation is a special case of the more general Lorentz equations.

Length Contraction

Consider Eq. 1' of Table 37-2,

$$\Delta x' = \gamma (\Delta x - v \, \Delta t). \tag{37-25}$$

If a rod lies parallel to the x and x' axes of Fig. 37-9 and is at rest in reference frame S', an observer in S' can measure its length at leisure. One way to do so is by subtracting the coordinates of the end points of the rod. The value of $\Delta x'$ that is obtained will be the proper length L_0 of the rod because the measurements are made in a frame where the rod is at rest.

Suppose the rod is moving in frame S. This means that Δx can be identified as the length L of the rod in frame S only if the coordinates of the rod's end points are measured *simultaneously*—that is, if $\Delta t = 0$. If we put $\Delta x' = L_0$, $\Delta x = L$, and $\Delta t = 0$ in Eq. 37-25, we find

$$L = \frac{L_0}{\gamma} \qquad \text{(length contraction)}, \qquad (37\text{-}26)$$

which is exactly Eq. 37-13, the length contraction equation. Thus, length contraction is a special case of the more general Lorentz equations.

Sample Problem 37.05 Lorentz transformations and reversing the sequence of events

An Earth starship has been sent to check an Earth outpost on the planet P1407, whose moon houses a battle group of the often hostile Reptulians. As the ship follows a straight-line course first past the planet and then past the moon, it detects a high-energy microwave burst at the Reptulian moon base and then, 1.10 s later, an explosion at the Earth outpost, which is 4.00×10^8 m from the Reptulian base as measured from the ship's reference frame. The Reptulians have obviously attacked the Earth outpost, and so the starship begins to prepare for a confrontation with them.

(a) The speed of the ship relative to the planet and its moon is 0.980c. What are the distance and time interval between the burst and the explosion as measured in the planet–moon frame (and thus according to the occupants of the stations)?

KEY IDEAS

1. This problem involves measurements made from two reference frames, the planet–moon frame and the starship frame.

2. We have two events: the burst and the explosion.

3. We need to transform the given data as measured in the starship frame to the corresponding data as measured in the planet–moon frame.

Starship frame: Before we get to the transformation, we need to carefully choose our notation. We begin with a sketch of the situation as shown in Fig. 37-10. There, we have chosen the ship's frame S to be stationary and the planet–moon frame S' to be moving with positive velocity (rightward). (This is an arbitrary choice; we could, instead, have chosen the planet–moon frame to be stationary. Then we would redraw $\vec{v}$ in Fig. 37-10 as being attached to the S frame and indicating leftward motion; v would then be a negative quantity. The results would be the same.) Let subscripts e and b represent the explosion and burst, respectively. Then the given data, all in the unprimed (starship) reference frame, are

$$\Delta x = x_e - x_b = +4.00 \times 10^8 \text{ m}$$

and $\qquad \Delta t = t_e - t_b = +1.10 \text{ s}.$

Here, Δx is a positive quantity because in Fig. 37-10, the co-ordinate x_e for the explosion is greater than the coordinate x_b

The relative motion alters the time intervals between events and maybe even their sequence.

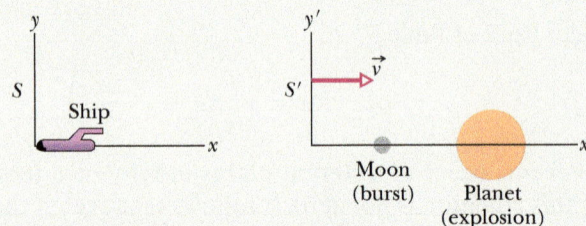

Figure 37-10 A planet and its moon in reference frame S' move rightward with speed v relative to a starship in reference frame S.

for the burst; Δt is also a positive quantity because the time t_e of the explosion is greater (later) than the time t_b of the burst.

Planet–moon frame: We seek $\Delta x'$ and $\Delta t'$, which we shall get by transforming the given S-frame data to the planet–moon frame S'. Because we are considering a pair of events, we choose transformation equations from Table 37-2—namely, Eqs. 1' and 2':

$$\Delta x' = \gamma(\Delta x - v\,\Delta t) \qquad (37\text{-}27)$$

and $\qquad \Delta t' = \gamma\left(\Delta t - \dfrac{v\,\Delta x}{c^2}\right). \qquad (37\text{-}28)$

Here, $v = +0.980c$ and the Lorentz factor is

$$\gamma = \frac{1}{\sqrt{1 - (v/c)^2}} = \frac{1}{\sqrt{1 - (+0.980c/c)^2}} = 5.0252.$$

Equation 37-27 then becomes

$$\Delta x' = (5.0252)[4.00 \times 10^8 \text{ m} - (+0.980\,c)(1.10 \text{ s})]$$

$$= 3.86 \times 10^8 \text{ m}, \qquad \text{(Answer)}$$

and Eq. 37-28 becomes

$$\Delta t' = (5.0252)\left[(1.10 \text{ s}) - \frac{(+0.980c)(4.00 \times 10^8 \text{m})}{c^2}\right]$$

$$= -1.04 \text{ s}. \qquad \text{(Answer)}$$

(b) What is the meaning of the minus sign in the value for $\Delta t'$?

Reasoning: We must be consistent with the notation we set up in part (a). Recall how we originally defined the time interval between burst and explosion: $\Delta t = t_e - t_b = +1.10$ s. To be consistent with that choice of notation, our definition of $\Delta t'$ must be $t'_e - t'_b$; thus, we have found that

$$\Delta t' = t'_e - t'_b = -1.04 \text{ s.}$$

The minus sign here tells us that $t'_b > t'_e$; that is, in the planet–moon reference frame, the burst occurred 1.04 s *after* the explosion, not 1.10 s *before* the explosion as detected in the ship frame.

(c) Did the burst cause the explosion, or vice versa?

KEY IDEA

The sequence of events measured in the planet–moon reference frame is the reverse of that measured in the ship frame. In either situation, if there is a causal relationship between the two events, information must travel from the location of one event to the location of the other to cause it.

Checking the speed: Let us check the required speed of the information. In the ship frame, this speed is

$$v_{info} = \frac{\Delta x}{\Delta t} = \frac{4.00 \times 10^8 \text{ m}}{1.10 \text{ s}} = 3.64 \times 10^8 \text{ m/s,}$$

but that speed is impossible because it exceeds *c*. In the planet–moon frame, the speed comes out to be 3.70×10^8 m/s, also impossible. Therefore, neither event could possibly have caused the other event; that is, they are *unrelated* events. Thus, the starship should stand down and not confront the Reptulians.

WILEY PLUS Additional examples, video, and practice available at *WileyPLUS*

37-4 THE RELATIVITY OF VELOCITIES

Learning Objectives

After reading this module, you should be able to . . .

37.20 With a sketch, explain the arrangement in which a particle's velocity is to be measured relative to two frames that have relative motion.

37.21 Apply the relationship for a relativistic velocity transformation between two frames with relative motion.

Key Idea

● When a particle is moving with speed u' in the positive x' direction in an inertial reference frame S' that itself is moving with speed v parallel to the x direction of a second inertial frame S, the speed u of the particle as measured in S is

$$u = \frac{u' + v}{1 + u'v/c^2} \quad \text{(relativistic velocity).}$$

The Relativity of Velocities

Here we wish to use the Lorentz transformation equations to compare the velocities that two observers in different inertial reference frames S and S' would measure for the same moving particle. Let S' move with velocity v relative to S.

Suppose that the particle, moving with constant velocity parallel to the x and x' axes in Fig. 37-11, sends out two signals as it moves. Each observer measures the space interval and the time interval between these two events. These four measurements are related by Eqs. 1 and 2 of Table 37-2,

$$\Delta x = \gamma(\Delta x' + v\,\Delta t')$$

and

$$\Delta t = \gamma\left(\Delta t' + \frac{v\,\Delta x'}{c^2}\right).$$

If we divide the first of these equations by the second, we find

$$\frac{\Delta x}{\Delta t} = \frac{\Delta x' + v\,\Delta t'}{\Delta t' + v\,\Delta x'/c^2}.$$

The speed of the moving particle depends on the frame.

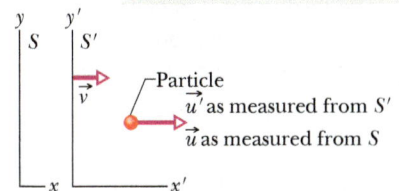

Figure 37-11 Reference frame S' moves with velocity $\vec{v}$ relative to frame S. A particle has velocity $\vec{u}'$ relative to reference frame S' and velocity $\vec{u}$ relative to reference frame S.

Dividing the numerator and denominator of the right side by $\Delta t'$, we find

$$\frac{\Delta x}{\Delta t} = \frac{\Delta x'/\Delta t' + v}{1 + v(\Delta x'/\Delta t')/c^2}.$$

However, in the differential limit, $\Delta x/\Delta t$ is u, the velocity of the particle as measured in S, and $\Delta x'/\Delta t'$ is u', the velocity of the particle as measured in S'. Then we have, finally,

$$u = \frac{u' + v}{1 + u'v/c^2} \quad \text{(relativistic velocity transformation)} \tag{37-29}$$

as the relativistic velocity transformation equation. (*Caution:* Be careful to substitute the correct signs for the velocities.) Equation 37-29 reduces to the classical, or Galilean, velocity transformation equation,

$$u = u' + v \quad \text{(classical velocity transformation)}, \tag{37-30}$$

when we apply the formal test of letting $c \rightarrow \infty$. In other words, Eq. 37-29 is correct for all physically possible speeds, but Eq. 37-30 is approximately correct for speeds much less than c.

37-5 DOPPLER EFFECT FOR LIGHT

Learning Objectives

After reading this module, you should be able to . . .

37.22 Identify that the frequency of light as measured in a frame attached to the light source (the rest frame) is the proper frequency.

37.23 For source–detector separations increasing and decreasing, identify whether the detected frequency is shifted up or down from the proper frequency, identify that the shift increases with an increase in relative speed, and apply the terms blue shift and red shift.

37.24 Identify radial speed.

37.25 For source–detector separations increasing and decreasing, apply the relationships between proper frequency f_0, detected frequency f, and radial speed v.

37.26 Convert between equations for frequency shift and wavelength shift.

37.27 When a radial speed is much less than light speed, apply the approximation relating wavelength shift $\Delta\lambda$, proper wavelength λ_0, and radial speed v.

37.28 Identify that for light (not sound) there is a shift in the frequency even when the velocity of the source is perpendicular to the line between the source and the detector, an effect due to time dilation.

37.29 Apply the relationship for the transverse Doppler effect by relating detected frequency f, proper frequency f_0, and relative speed v.

Key Ideas

● When a light source and a light detector move relative to each other, the wavelength of the light as measured in the rest frame of the source is the proper wavelength λ_0. The detected wavelength λ is either longer (a red shift) or shorter (a blue shift) depending on whether the source–detector separation is increasing or decreasing.

● When the separation is increasing, the wavelengths are related by

$$\lambda = \lambda_0 \sqrt{\frac{1 + \beta}{1 - \beta}} \quad \text{(source and detector separating)},$$

where $\beta = v/c$ and v is the relative radial speed (along a line through the source and detector). If the separation is

decreasing, the signs in front of the β symbols are reversed.

● For speeds much less than c, the magnitude of the Doppler wavelength shift $\Delta\lambda = \lambda - \lambda_0$ is approximately related to v by

$$v = \frac{|\Delta\lambda|}{\lambda_0} c \quad (v \ll c).$$

● If the relative motion of the light source is perpendicular to a line through the source and detector, the detected frequency f is related to the proper frequency f_0 by

$$f = f_0 \sqrt{1 - \beta^2}.$$

This transverse Doppler effect is due to time dilation.

Doppler Effect for Light

In Module 17-7 we discussed the Doppler effect (a shift in detected frequency) for sound waves, finding that the effect depends on the source and detector velocities relative to the air. That is not the situation with light waves, which require no medium (they can even travel through vacuum). The Doppler effect for light waves depends on only the relative velocity $\vec{v}$ between source and detector, as measured from the reference frame of either. Let f_0 represent the **proper frequency** of the source—that is, the frequency that is measured by an observer in the rest frame of the source. Let f represent the frequency detected by an observer moving with velocity $\vec{v}$ relative to that rest frame. Then, when the direction of $\vec{v}$ is directly away from the source,

$$f = f_0 \sqrt{\frac{1-\beta}{1+\beta}} \quad \text{(source and detector separating),} \tag{37-31}$$

where $\beta = v/c$.

Because measurements involving light are usually done in wavelengths rather than frequencies, let's rewrite Eq. 37-31 by replacing f with c/λ and f_0 with c/λ_0, where λ is the measured wavelength and λ_0 is the **proper wavelength** (the wavelength associated with f_0). After canceling c from both sides, we then have

$$\lambda = \lambda_0 \sqrt{\frac{1+\beta}{1-\beta}} \quad \text{(source and detector separating).} \tag{37-32}$$

When the direction of $\vec{v}$ is directly toward the source, we must change the signs in front of the β symbols in Eqs. 37-31 and 37-32.

For an increasing separation, we can see from Eq. 37-32 (with an addition in the numerator and a subtraction in the denominator) that the measured wavelength is greater than the proper wavelength. Such a Doppler shift is described as being a *red shift*, where *red* does not mean the measured wavelength is red or even visible. The term merely serves as a memory device because red is at the *long*-wavelength end of the visible spectrum. Thus λ is longer than λ_0. Similarly, for a decreasing separation, λ is shorter than λ_0, and the Doppler shift is described as being a *blue shift*.

Low-Speed Doppler Effect

For low speeds ($\beta \ll 1$), Eq. 37-31 can be expanded in a power series in β and approximated as

$$f = f_0(1 - \beta + \tfrac{1}{2}\beta^2) \quad \text{(source and detector separating, } \beta \ll 1). \tag{37-33}$$

The corresponding low-speed equation for the Doppler effect with sound waves (or any waves except light waves) has the same first two terms but a different coefficient in the third term. Thus, the relativistic effect for low-speed light sources and detectors shows up only with the β^2 term.

A police radar unit employs the Doppler effect with microwaves to measure the speed v of a car. A source in the radar unit emits a microwave beam at a certain (proper) frequency f_0 along the road. A car that is moving toward the unit intercepts that beam but at a frequency that is shifted upward by the Doppler effect due to the car's motion toward the radar unit. The car reflects the beam back toward the radar unit. Because the car is moving toward the radar unit, the detector in the unit intercepts a reflected beam that is further shifted up in frequency. The unit compares that detected frequency with f_0 and computes the speed v of the car.

Astronomical Doppler Effect

In astronomical observations of stars, galaxies, and other sources of light, we can determine how fast the sources are moving, either directly away from us or directly toward us, by measuring the *Doppler shift* of the light that reaches us. If a certain star were at rest relative to us, we would detect light from it with a certain proper frequency f_0. However, if the star is moving either directly away from us or directly toward us, the light we detect has a frequency f that is shifted from f_0 by the Doppler effect. This Doppler shift is due only to the *radial* motion of the star (its motion directly toward us or away from us), and the speed we can determine by measuring this Doppler shift is only the *radial speed* v of the star—that is, only the radial component of the star's velocity relative to us.

Suppose a star (or any other light source) moves away from us with a radial speed v that is low enough (β is small enough) for us to neglect the β^2 term in Eq. 37-33. Then we have

$$f = f_0(1 - \beta). \tag{37-34}$$

Because astronomical measurements involving light are usually done in wavelengths rather than frequencies, let's rewrite Eq. 37-34 as

$$\frac{c}{\lambda} = \frac{c}{\lambda_0}(1 - \beta),$$

or
$$\lambda = \lambda_0(1 - \beta)^{-1}.$$

Because we assume β is small, we can expand $(1 - \beta)^{-1}$ in a power series. Doing so and retaining only the first power of β, we have

$$\lambda = \lambda_0(1 + \beta),$$

or
$$\beta = \frac{\lambda - \lambda_0}{\lambda_0}. \tag{37-35}$$

Replacing β with v/c and $\lambda - \lambda_0$ with $|\Delta\lambda|$ leads to

$$v = \frac{|\Delta\lambda|}{\lambda_0}\, c \quad \text{(radial speed of light source, } v \ll c\text{)}. \tag{37-36}$$

The difference $\Delta\lambda$ is the *wavelength Doppler shift* of the light source. We enclose it with an absolute sign so that we always have a magnitude of the shift. Equation 37-36 is an approximation that can be applied whether the light source is moving toward or away from us but only when $v \ll c$.

✓ Checkpoint 3

The figure shows a source that emits light of proper frequency f_0 while moving directly toward the right with speed $c/4$ as measured from reference frame S. The figure also shows a light detector, which measures a frequency $f > f_0$ for the emitted light. (a) Is the detector moving toward the left or the right? (b) Is the speed of the detector as measured from reference frame S more than $c/4$, less than $c/4$, or equal to $c/4$?

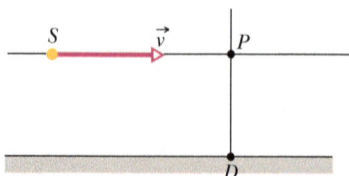

Transverse Doppler Effect

So far, we have discussed the Doppler effect, here and in Chapter 17, only for situations in which the source and the detector move either directly toward or directly away from each other. Figure 37-12 shows a different arrangement, in which a source S moves past a detector D. When S reaches point P, the velocity of S is perpendicular to the line joining P and D, and at that instant S is moving neither toward nor away from D. If the source is emitting sound waves of frequency f_0, D detects that frequency (with no Doppler effect) when it intercepts the

Figure 37-12 A light source S travels with velocity $\vec{v}$ past a detector at D. The special theory of relativity predicts a transverse Doppler effect as the source passes through point P, where the direction of travel is perpendicular to the line extending through D. Classical theory predicts no such effect.

waves that were emitted at point P. However, if the source is emitting light waves, there is still a Doppler effect, called the **transverse Doppler effect.** In this situation, the detected frequency of the light emitted when the source is at point P is

$$f = f_0 \sqrt{1 - \beta^2} \quad \text{(transverse Doppler effect).} \tag{37-37}$$

For low speeds ($\beta \ll 1$), Eq. 37-37 can be expanded in a power series in β and approximated as

$$f = f_0(1 - \tfrac{1}{2}\beta^2) \quad \text{(low speeds).} \tag{37-38}$$

Here the first term is what we would expect for sound waves, and again the relativistic effect for low-speed light sources and detectors appears with the β^2 term.

In principle, a police radar unit can determine the speed of a car even when the path of the radar beam is perpendicular (transverse) to the path of the car. However, Eq. 37-38 tells us that because β is small even for a fast car, the relativistic term $\beta^2/2$ in the transverse Doppler effect is extremely small. Thus, $f \approx f_0$ and the radar unit computes a speed of zero.

The transverse Doppler effect is really another test of time dilation. If we rewrite Eq. 37-37 in terms of the period T of oscillation of the emitted light wave instead of the frequency, we have, because $T = 1/f$,

$$T = \frac{T_0}{\sqrt{1 - \beta^2}} = \gamma T_0, \tag{37-39}$$

in which T_0 ($= 1/f_0$) is the **proper period** of the source. As comparison with Eq. 37-9 shows, Eq. 37-39 is simply the time dilation formula.

37-6 MOMENTUM AND ENERGY

Learning Objectives

After reading this module, you should be able to . . .

37.30 Identify that the classical expressions for momentum and kinetic energy are approximately correct for slow speeds whereas the relativistic expressions are correct for any physically possible speed.

37.31 Apply the relationship between momentum, mass, and relative speed.

37.32 Identify that an object has a mass energy (or rest energy) associated with its mass.

37.33 Apply the relationships between total energy, rest energy, kinetic energy, momentum, mass, speed, the speed parameter, and the Lorentz factor.

37.34 Sketch a graph of kinetic energy versus the ratio v/c (of speed to light speed) for both classical and relativistic expressions of kinetic energy.

37.35 Apply the work–kinetic energy theorem to relate work by an applied force and the resulting change in kinetic energy.

37.36 For a reaction, apply the relationship between the Q value and the change in the mass energy.

37.37 For a reaction, identify the correlation between the algebraic sign of Q and whether energy is released or absorbed by the reaction.

Key Ideas

● The following definitions of linear momentum $\vec{p}$, kinetic energy K, and total energy E for a particle of mass m are valid at any physically possible speed:

$$\vec{p} = \gamma m \vec{v} \quad \text{(momentum),}$$
$$E = mc^2 + K = \gamma mc^2 \quad \text{(total energy),}$$
$$K = mc^2(\gamma - 1) \quad \text{(kinetic energy).}$$

Here γ is the Lorentz factor for the particle's motion, and mc^2 is the *mass energy*, or *rest energy*, associated with the mass of the particle.

● These equations lead to the relationships

$$(pc)^2 = K^2 + 2Kmc^2$$

and
$$E^2 = (pc)^2 + (mc^2)^2.$$

● When a system of particles undergoes a chemical or nuclear reaction, the Q of the reaction is the negative of the change in the system's total mass energy:

$$Q = M_i c^2 - M_f c^2 = -\Delta M c^2,$$

where M_i is the system's total mass before the reaction and M_f is its total mass after the reaction.

A New Look at Momentum

Suppose that a number of observers, each in a different inertial reference frame, watch an isolated collision between two particles. In classical mechanics, we have seen that—even though the observers measure different velocities for the colliding particles—they all find that the law of conservation of momentum holds. That is, they find that the total momentum of the system of particles after the collision is the same as it was before the collision.

How is this situation affected by relativity? We find that if we continue to define the momentum $\vec{p}$ of a particle as $m\vec{v}$, the product of its mass and its velocity, total momentum is *not* conserved for the observers in different inertial frames. So, we need to redefine momentum in order to save that conservation law.

Consider a particle moving with constant speed v in the positive direction of an x axis. Classically, its momentum has magnitude

$$p = mv = m\,\frac{\Delta x}{\Delta t} \quad \text{(classical momentum),} \tag{37-40}$$

in which Δx is the distance it travels in time Δt. To find a relativistic expression for momentum, we start with the new definition

$$p = m\,\frac{\Delta x}{\Delta t_0}.$$

Here, as before, Δx is the distance traveled by a moving particle as viewed by an observer watching that particle. However, Δt_0 is the time required to travel that distance, measured not by the observer watching the moving particle but by an observer moving with the particle. The particle is at rest with respect to this second observer; thus that measured time is a proper time.

Using the time dilation formula, $\Delta t = \gamma\,\Delta t_0$ (Eq. 37-9), we can then write

$$p = m\,\frac{\Delta x}{\Delta t_0} = m\,\frac{\Delta x}{\Delta t}\,\frac{\Delta t}{\Delta t_0} = m\,\frac{\Delta x}{\Delta t}\,\gamma.$$

However, since $\Delta x/\Delta t$ is just the particle velocity v, we have

$$p = \gamma m v \quad \text{(momentum).} \tag{37-41}$$

Note that this differs from the classical definition of Eq. 37-40 only by the Lorentz factor γ. However, that difference is important: Unlike classical momentum, relativistic momentum approaches an infinite value as v approaches c.

We can generalize the definition of Eq. 37-41 to vector form as

$$\vec{p} = \gamma m \vec{v} \quad \text{(momentum).} \tag{37-42}$$

This equation gives the correct definition of momentum for all physically possible speeds. For a speed much less than c, it reduces to the classical definition of momentum ($\vec{p} = m\vec{v}$).

A New Look at Energy

Mass Energy

The science of chemistry was initially developed with the assumption that in chemical reactions, energy and mass are conserved separately. In 1905, Einstein showed that as a consequence of his theory of special relativity, mass can be considered to be another form of energy. Thus, the law of conservation of energy is really the law of conservation of mass–energy.

In a *chemical reaction* (a process in which atoms or molecules interact), the amount of mass that is transferred into other forms of energy (or vice versa) is such a tiny fraction of the total mass involved that there is no hope of measuring the mass change with even the best laboratory balances. Mass and energy truly *seem* to be separately conserved. However, in a *nuclear reaction* (in which nuclei or fundamental particles interact), the energy released is often about a million times greater than in a chemical reaction, and the change in mass can easily be measured.

An object's mass m and the equivalent energy E_0 are related by

$$E_0 = mc^2, \tag{37-43}$$

which, without the subscript 0, is the best-known science equation of all time. This energy that is associated with the mass of an object is called **mass energy** or **rest energy.** The second name suggests that E_0 is an energy that the object has even when it is at rest, simply because it has mass. (If you continue your study of physics beyond this book, you will see more refined discussions of the relation between mass and energy. You might even encounter disagreements about just what that relation is and means.)

Table 37-3 shows the (approximate) mass energy, or rest energy, of a few objects. The mass energy of, say, a U.S. penny is enormous; the equivalent amount of electrical energy would cost well over a million dollars. On the other hand, the entire annual U.S. electrical energy production corresponds to a mass of only a few hundred kilograms of matter (stones, burritos, or anything else).

In practice, SI units are rarely used with Eq. 37-43 because they are too large to be convenient. Masses are usually measured in atomic mass units, where

$$1\ u = 1.660\ 538\ 86 \times 10^{-27}\ \text{kg}, \tag{37-44}$$

and energies are usually measured in electron-volts or multiples of it, where

$$1\ \text{eV} = 1.602\ 176\ 462 \times 10^{-19}\ \text{J}. \tag{37-45}$$

In the units of Eqs. 37-44 and 37-45, the multiplying constant c^2 has the values

$$c^2 = 9.314\ 940\ 13 \times 10^8\ \text{eV/u} = 9.314\ 940\ 13 \times 10^5\ \text{keV/u}$$
$$= 931.494\ 013\ \text{MeV/u}. \tag{37-46}$$

Total Energy

Equation 37-43 gives, for any object, the mass energy E_0 that is associated with the object's mass m, regardless of whether the object is at rest or moving. If the object is moving, it has additional energy in the form of kinetic energy K. If we assume that the object's potential energy is zero, then its total energy E is the sum of its mass energy and its kinetic energy:

$$E = E_0 + K = mc^2 + K. \tag{37-47}$$

Although we shall not prove it, the total energy E can also be written as

$$E = \gamma mc^2, \tag{37-48}$$

where γ is the Lorentz factor for the object's motion.

Table 37-3 The Energy Equivalents of a Few Objects

Object	Mass (kg)	Energy Equivalent	
Electron	$\approx 9.11 \times 10^{-31}$	$\approx 8.19 \times 10^{-14}$ J	(≈ 511 keV)
Proton	$\approx 1.67 \times 10^{-27}$	$\approx 1.50 \times 10^{-10}$ J	(≈ 938 MeV)
Uranium atom	$\approx 3.95 \times 10^{-25}$	$\approx 3.55 \times 10^{-8}$ J	(≈ 225 GeV)
Dust particle	$\approx 1 \times 10^{-13}$	$\approx 1 \times 10^4$ J	(≈ 2 kcal)
U.S. penny	$\approx 3.1 \times 10^{-3}$	$\approx 2.8 \times 10^{14}$ J	(≈ 78 GW·h)

Since Chapter 7, we have discussed many examples involving changes in the total energy of a particle or a system of particles. However, we did not include mass energy in the discussions because the changes in mass energy were either zero or small enough to be neglected. The law of conservation of total energy still applies when changes in mass energy are significant. Thus, regardless of what happens to the mass energy, the following statement from Module 8-5 is still true:

The total energy E of an *isolated system* cannot change.

For example, if the total mass energy of two interacting particles in an isolated system decreases, some other type of energy in the system must increase because the total energy cannot change.

Q Value. In a system undergoing a chemical or nuclear reaction, a change in the total mass energy of the system due to the reaction is often given as a Q value. The Q value for a reaction is obtained from the relation

$$\begin{pmatrix} \text{system's initial} \\ \text{total mass energy} \end{pmatrix} = \begin{pmatrix} \text{system's final} \\ \text{total mass energy} \end{pmatrix} + Q$$

or
$$E_{0i} = E_{0f} + Q. \tag{37-49}$$

Using Eq. 37-43 ($E_0 = mc^2$), we can rewrite this in terms of the initial *total* mass M_i and the final *total* mass M_f as

$$M_i c^2 = M_f c^2 + Q$$

or
$$Q = M_i c^2 - M_f c^2 = -\Delta M c^2, \tag{37-50}$$

where the change in mass due to the reaction is $\Delta M = M_f - M_i$.

If a reaction results in the transfer of energy from mass energy to, say, kinetic energy of the reaction products, the system's total mass energy E_0 (and total mass M) decreases and Q is positive. If, instead, a reaction requires that energy be transferred to mass energy, the system's total mass energy E_0 (and its total mass M) increases and Q is negative.

For example, suppose two hydrogen nuclei undergo a *fusion reaction* in which they join together to form a single nucleus and release two particles:

$$^1\text{H} + {}^1\text{H} \rightarrow {}^2\text{H} + \text{e}^+ + \nu,$$

where ^{2}H is another type of hydrogen nucleus (with a neutron in addition to the proton), e$^+$ is a positron, and ν is a neutrino. The total mass energy (and total mass) of the resultant single nucleus and two released particles is less than the total mass energy (and total mass) of the initial hydrogen nuclei. Thus, the Q of the fusion reaction is positive, and energy is said to be *released* (transferred from mass energy) by the reaction. This release is important to you because the fusion of hydrogen nuclei in the Sun is one part of the process that results in sunshine on Earth and makes life here possible.

Kinetic Energy

In Chapter 7 we defined the kinetic energy K of an object of mass m moving at speed v well below c to be

$$K = \tfrac{1}{2}mv^2. \tag{37-51}$$

However, this classical equation is only an approximation that is good enough when the speed is well below the speed of light.

Let us now find an expression for kinetic energy that is correct for *all* physically possible speeds, including speeds close to c. Solving Eq. 37-47 for K and then substituting for E from Eq. 37-48 lead to

$$K = E - mc^2 = \gamma mc^2 - mc^2$$

$$= mc^2(\gamma - 1) \quad \text{(kinetic energy)}, \qquad (37\text{-}52)$$

where $\gamma \,(= 1/\sqrt{1 - (v/c)^2})$ is the Lorentz factor for the object's motion.

Figure 37-13 shows plots of the kinetic energy of an electron as calculated with the correct definition (Eq. 37-52) and the classical approximation (Eq. 37-51), both as functions of v/c. Note that on the left side of the graph the two plots coincide; this is the part of the graph—at lower speeds—where we have calculated kinetic energies so far in this book. This part of the graph tells us that we have been justified in calculating kinetic energy with the classical expression of Eq. 37-51. However, on the right side of the graph—at speeds near c—the two plots differ significantly. As v/c approaches 1.0, the plot for the classical definition of kinetic energy increases only moderately while the plot for the correct definition of kinetic energy increases dramatically, approaching an infinite value as v/c approaches 1.0. Thus, when an object's speed v is near c, we *must* use Eq. 37-52 to calculate its kinetic energy.

Work. Figure 37-13 also tells us something about the work we must do on an object to increase its speed by, say, 1%. The required work W is equal to the resulting change ΔK in the object's kinetic energy. If the change is to occur on the low-speed, left side of Fig. 37-13, the required work might be modest. However, if the change is to occur on the high-speed, right side of Fig. 37-13, the required work could be enormous because the kinetic energy K increases so rapidly there with an increase in speed v. To increase an object's speed to c would require, in principle, an infinite amount of energy; thus, doing so is impossible.

The kinetic energies of electrons, protons, and other particles are often stated with the unit electron-volt or one of its multiples used as an adjective. For example, an electron with a kinetic energy of 20 MeV may be described as a 20 MeV electron.

As v/c approaches 1.0, the actual kinetic energy approaches infinity.

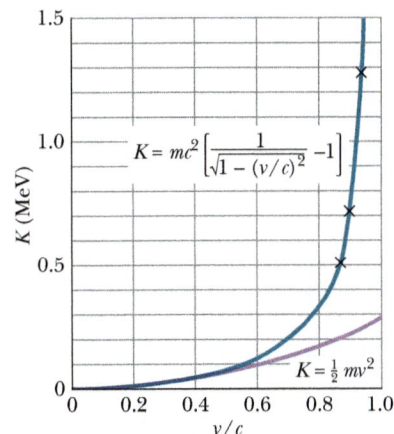

Figure 37-13 The relativistic (Eq. 37-52) and classical (Eq. 37-51) equations for the kinetic energy of an electron, plotted as a function of v/c, where v is the speed of the electron and c is the speed of light. Note that the two curves blend together at low speeds and diverge widely at high speeds. Experimental data (at the $\times$ marks) show that at high speeds the relativistic curve agrees with experiment but the classical curve does not.

Momentum and Kinetic Energy

In classical mechanics, the momentum p of a particle is mv and its kinetic energy K is $\frac{1}{2}mv^2$. If we eliminate v between these two expressions, we find a direct relation between momentum and kinetic energy:

$$p^2 = 2Km \quad \text{(classical)}. \qquad (37\text{-}53)$$

We can find a similar connection in relativity by eliminating v between the relativistic definition of momentum (Eq. 37-41) and the relativistic definition of kinetic energy (Eq. 37-52). Doing so leads, after some algebra, to

$$(pc)^2 = K^2 + 2Kmc^2. \qquad (37\text{-}54)$$

With the aid of Eq. 37-47, we can transform Eq. 37-54 into a relation between the momentum p and the total energy E of a particle:

$$E^2 = (pc)^2 + (mc^2)^2. \qquad (37\text{-}55)$$

The right triangle of Fig. 37-14 can help you keep these useful relations in mind. You can also show that, in that triangle,

$$\sin \theta = \beta \quad \text{and} \quad \cos \theta = 1/\gamma. \qquad (37\text{-}56)$$

With Eq. 37-55 we can see that the product pc must have the same unit as energy E; thus, we can express the unit of momentum p as an energy unit divided by c, usually as MeV/c or GeV/c in fundamental particle physics.

This might help you to remember the relations.

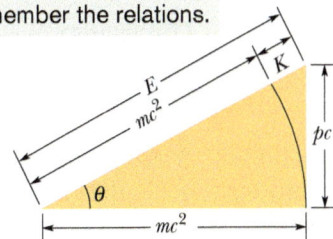

Figure 37-14 A useful memory diagram for the relativistic relations among the total energy E, the rest energy or mass energy mc^2, the kinetic energy K, and the momentum magnitude p.

✅ **Checkpoint 4**

Are (a) the kinetic energy and (b) the total energy of a 1 GeV electron more than, less than, or equal to those of a 1 GeV proton?

Sample Problem 37.06 Energy and momentum of a relativistic electron

(a) What is the total energy E of a 2.53 MeV electron?

KEY IDEA

From Eq. 37-47, the total energy E is the sum of the electron's mass energy (or rest energy) mc^2 and its kinetic energy:

$$E = mc^2 + K. \qquad (37\text{-}57)$$

Calculations: The adjective "2.53 MeV" in the problem statement means that the electron's kinetic energy is 2.53 MeV. To evaluate the electron's mass energy mc^2, we substitute the electron's mass m from Appendix B, obtaining

$$mc^2 = (9.109 \times 10^{-31} \text{ kg})(299\ 792\ 458 \text{ m/s})^2$$

$$= 8.187 \times 10^{-14} \text{ J}.$$

Then dividing this result by 1.602×10^{-13} J/MeV gives us 0.511 MeV as the electron's mass energy (confirming the value in Table 37-3). Equation 37-57 then yields

$$E = 0.511 \text{ MeV} + 2.53 \text{ MeV} = 3.04 \text{ MeV}. \quad \text{(Answer)}$$

(b) What is the magnitude p of the electron's momentum, in the unit MeV/c? (Note that c is the symbol for the speed of light and not itself a unit.)

KEY IDEA

We can find p from the total energy E and the mass energy mc^2 via Eq. 37-55,

$$E^2 = (pc)^2 + (mc^2)^2.$$

Calculations: Solving for pc gives us

$$pc = \sqrt{E^2 - (mc^2)^2}$$

$$= \sqrt{(3.04 \text{ MeV})^2 - (0.511 \text{ MeV})^2} = 3.00 \text{ MeV}.$$

Finally, dividing both sides by c we find

$$p = 3.00 \text{ MeV}/c. \qquad \text{(Answer)}$$

Sample Problem 37.07 Energy and an astounding discrepancy in travel time

The most energetic proton ever detected in the cosmic rays coming to Earth from space had an astounding kinetic energy of 3.0×10^{20} eV (enough energy to warm a teaspoon of water by a few degrees).

(a) What were the proton's Lorentz factor γ and speed v (both relative to the ground-based detector)?

KEY IDEAS

(1) The proton's Lorentz factor γ relates its total energy E to its mass energy mc^2 via Eq. 37-48 ($E = \gamma mc^2$). (2) The proton's total energy is the sum of its mass energy mc^2 and its (given) kinetic energy K.

Calculations: Putting these ideas together we have

$$\gamma = \frac{E}{mc^2} = \frac{mc^2 + K}{mc^2} = 1 + \frac{K}{mc^2}. \qquad (37\text{-}58)$$

From Table 37-3, the proton's mass energy mc^2 is 938 MeV. Substituting this and the given kinetic energy into Eq. 37-58, we obtain

$$\gamma = 1 + \frac{3.0 \times 10^{20} \text{ eV}}{938 \times 10^6 \text{ eV}}$$

$$= 3.198 \times 10^{11} \approx 3.2 \times 10^{11}. \qquad \text{(Answer)}$$

This computed value for γ is so large that we cannot use the definition of γ (Eq. 37-8) to find v. Try it; your calculator will tell you that β is effectively equal to 1 and thus that v is effectively equal to c. Actually, v is almost c, but we want a more accurate answer, which we can obtain by first solving

Eq. 37-8 for $1 - \beta$. To begin we write

$$\gamma = \frac{1}{\sqrt{1 - \beta^2}} = \frac{1}{\sqrt{(1 - \beta)(1 + \beta)}} \approx \frac{1}{\sqrt{2(1 - \beta)}},$$

where we have used the fact that β is so close to unity that $1 + \beta$ is very close to 2. (We can round off the sum of two very close numbers but not their difference.) The velocity we seek is contained in the $1 - \beta$ term. Solving for $1 - \beta$ then yields

$$1 - \beta = \frac{1}{2\gamma^2} = \frac{1}{(2)(3.198 \times 10^{11})^2}$$

$$= 4.9 \times 10^{-24} \approx 5 \times 10^{-24}.$$

Thus, $$\beta = 1 - 5 \times 10^{-24}$$

and, since $v = \beta c$,

$$v \approx 0.999\ 999\ 999\ 999\ 999\ 999\ 999\ 995c. \quad \text{(Answer)}$$

(b) Suppose that the proton travels along a diameter of the Milky Way galaxy (9.8×10^4 ly). Approximately how long does the proton take to travel that diameter as measured from the common reference frame of Earth and the Galaxy?

Reasoning: We just saw that this *ultrarelativistic* proton is traveling at a speed barely less than c. By the definition of light-year, light takes 1 y to travel a distance of 1 ly, and so light should take 9.8×10^4 y to travel 9.8×10^4 ly, and this proton should take almost the same time. Thus, from our Earth–Milky Way reference frame, the proton's trip takes

$$\Delta t = 9.8 \times 10^4 \text{ y}. \qquad \text{(Answer)}$$

(c) How long does the trip take as measured in the reference frame of the proton?

KEY IDEAS

1. This problem involves measurements made from two (inertial) reference frames: one is the Earth–Milky Way frame and the other is attached to the proton.

2. This problem also involves two events: the first is when the proton passes one end of the diameter along the Galaxy, and the second is when it passes the opposite end.

3. The time interval between those two events as measured in the proton's reference frame is the proper time interval Δt_0 because the events occur at the same location in that frame—namely, at the proton itself.

4. We can find the proper time interval Δt_0 from the time

interval Δt measured in the Earth–Milky Way frame by using Eq. 37-9 ($\Delta t = \gamma \Delta t_0$) for time dilation. (Note that we can use that equation because one of the time measures *is* a proper time. However, we get the same relation if we use a Lorentz transformation.)

Calculation: Solving Eq. 37-9 for Δt_0 and substituting γ from (a) and Δt from (b), we find

$$\Delta t_0 = \frac{\Delta t}{\gamma} = \frac{9.8 \times 10^4 \text{ y}}{3.198 \times 10^{11}}$$

$$= 3.06 \times 10^{-7} \text{ y} = 9.7 \text{ s.} \qquad \text{(Answer)}$$

In our frame, the trip takes 98 000 y. In the proton's frame, it takes 9.7 s! As promised at the start of this chapter, relative motion can alter the rate at which time passes, and we have here an extreme example.

WILEY PLUS Additional examples, video, and practice available at *WileyPLUS*

Review & Summary

The Postulates Einstein's **special theory of relativity** is based on two postulates:

1. The laws of physics are the same for observers in all inertial reference frames. No one frame is preferred over any other.

2. The speed of light in vacuum has the same value c in all directions and in all inertial reference frames.

The speed of light c in vacuum is an ultimate speed that cannot be exceeded by any entity carrying energy or information.

Coordinates of an Event Three space coordinates and one time coordinate specify an **event.** One task of special relativity is to relate these coordinates as assigned by two observers who are in uniform motion with respect to each other.

Simultaneous Events If two observers are in relative motion, they will not, in general, agree as to whether two events are simultaneous.

Time Dilation If two successive events occur at the same place in an inertial reference frame, the time interval Δt_0 between them, measured on a single clock where they occur, is the **proper time** between the events. *Observers in frames moving relative to that frame will measure a larger value for this interval.* For an observer moving with relative speed v, the measured time interval is

$$\Delta t = \frac{\Delta t_0}{\sqrt{1-(v/c)^2}} = \frac{\Delta t_0}{\sqrt{1-\beta^2}}$$

$$= \gamma \Delta t_0 \quad \text{(time dilation).} \qquad \text{(37-7 to 37-9)}$$

Here $\beta = v/c$ is the **speed parameter** and $\gamma = 1/\sqrt{1-\beta^2}$ is the

Lorentz factor. An important result of time dilation is that moving clocks run slow as measured by an observer at rest.

Length Contraction The length L_0 of an object measured by an observer in an inertial reference frame in which the object is at rest is called its **proper length.** *Observers in frames moving relative to that frame and parallel to that length will measure a shorter length.* For an observer moving with relative speed v, the measured length is

$$L = L_0\sqrt{1-\beta^2} = \frac{L_0}{\gamma} \quad \text{(length contraction).} \qquad (37\text{-}13)$$

The Lorentz Transformation The *Lorentz transformation* equations relate the spacetime coordinates of a single event as seen by observers in two inertial frames, S and S', where S' is moving relative to S with velocity v in the positive x and x' direction. The four coordinates are related by

$$
\begin{aligned}
x' &= \gamma(x - vt), \\
y' &= y, \\
z' &= z, \\
t' &= \gamma(t - vx/c^2).
\end{aligned}
\qquad (37\text{-}21)
$$

Relativity of Velocities When a particle is moving with speed u' in the positive x' direction in an inertial reference frame S' that itself is moving with speed v parallel to the x direction of a second inertial frame S, the speed u of the particle as measured in S is

$$u = \frac{u' + v}{1 + u'v/c^2} \quad \text{(relativistic velocity).} \qquad (37\text{-}29)$$

Relativistic Doppler Effect When a light source and a light detector move directly relative to each other, the wavelength of the light as measured in the rest frame of the source is the *proper wavelength* λ_0. The detected wavelength λ is either longer (a *red shift*) or shorter (a *blue shift*) depending on whether the source–detector separation is increasing or decreasing. When the separation is increasing, the wavelengths are related by

$$\lambda = \lambda_0 \sqrt{\frac{1 + \beta}{1 - \beta}} \quad \text{(source and detector separating),} \quad (37\text{-}32)$$

where $\beta = v/c$ and v is the relative radial speed (along a line connecting the source and detector). If the separation is decreasing, the signs in front of the β symbols are reversed. For speeds much less than c, the magnitude of the Doppler wavelength shift ($\Delta\lambda = \lambda - \lambda_0$) is approximately related to v by

$$v = \frac{|\Delta\lambda|}{\lambda_0} c \quad (v \ll c). \quad (37\text{-}36)$$

Transverse Doppler Effect   If the relative motion of the light source is perpendicular to a line joining the source and detector, the detected frequency f is related to the proper frequency f_0 by

$$f = f_0 \sqrt{1 - \beta^2}. \quad (37\text{-}37)$$

Momentum and Energy   The following definitions of linear momentum $\vec{p}$, kinetic energy K, and total energy E for a particle of mass m are valid at any physically possible speed:

$$\vec{p} = \gamma m \vec{v} \quad \text{(momentum),} \quad (37\text{-}42)$$

$$E = mc^2 + K = \gamma mc^2 \quad \text{(total energy),} \quad (37\text{-}47, 37\text{-}48)$$

$$K = mc^2(\gamma - 1) \quad \text{(kinetic energy).} \quad (37\text{-}52)$$

Here γ is the Lorentz factor for the particle's motion, and mc^2 is the *mass energy*, or *rest energy*, associated with the mass of the particle. These equations lead to the relationships

$$(pc)^2 = K^2 + 2Kmc^2 \quad (37\text{-}54)$$

and

$$E^2 = (pc)^2 + (mc^2)^2. \quad (37\text{-}55)$$

When a system of particles undergoes a chemical or nuclear reaction, the Q of the reaction is the negative of the change in the system's total mass energy:

$$Q = M_i c^2 - M_f c^2 = -\Delta M\, c^2, \quad (37\text{-}50)$$

where M_i is the system's total mass before the reaction and M_f is its total mass after the reaction.

Questions

1   A rod is to move at constant speed v along the x axis of reference frame S, with the rod's length parallel to that axis. An observer in frame S is to measure the length L of the rod. Which of the curves in Fig. 37-15 best gives length L (vertical axis of the graph) versus speed parameter β?

2   Figure 37-16 shows a ship (attached to reference frame S') passing us (standing in reference frame S). A proton is fired at nearly the speed of light along the length of the ship, from the front to the rear. (a) Is the spatial separation $\Delta x'$ between the point at which the proton is fired and the point at which it hits the ship's rear wall a positive or negative quantity? (b) Is the temporal separation $\Delta t'$ between those events a positive or negative quantity?

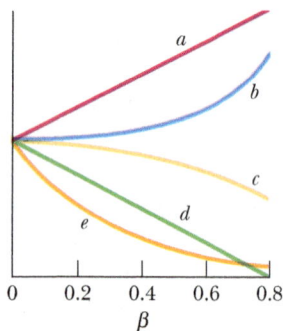

Figure 37-15
Questions 1 and 3.

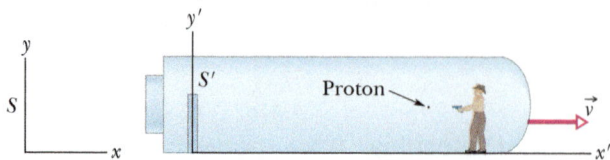

Figure 37-16   Question 2 and Problem 68.

3   Reference frame S' is to pass reference frame S at speed v along the common direction of the x' and x axes, as in Fig. 37-9. An observer who rides along with frame S' is to count off 25 s on his wristwatch. The corresponding time interval Δt is to be measured by an observer in frame S. Which of the curves in Fig. 37-15 best

gives Δt (vertical axis of the graph) versus speed parameter β?

4   Figure 37-17 shows two clocks in stationary frame S' (they are synchronized in that frame) and one clock in moving frame S. Clocks C_1 and C_1' read zero when they pass each other. When clocks C_1 and C_2' pass each other, (a) which clock has the smaller reading and (b) which clock measures a proper time?

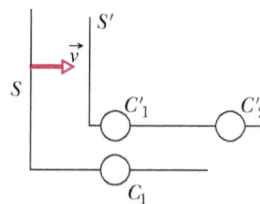

Figure 37-17   Question 4.

5   Figure 37-18 shows two clocks in stationary frame S (they are synchronized in that frame) and one clock in moving frame S'. Clocks C_1 and C_1' read zero when they pass each other. When clocks C_1' and C_2 pass each other, (a) which clock has the smaller reading and (b) which clock measures a proper time?

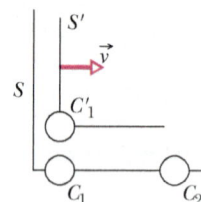

Figure 37-18
Question 5.

6   Sam leaves Venus in a spaceship headed to Mars and passes Sally, who is on Earth, with a relative speed of $0.5c$. (a) Each measures the Venus–Mars voyage time. Who measures a proper time: Sam, Sally, or neither? (b) On the way, Sam sends a pulse of light to Mars. Each measures the travel time of the pulse. Who measures a proper time: Sam, Sally, or neither?

7   The plane of clocks and measuring rods in Fig. 37-19 is like that in Fig. 37-3. The clocks along the x axis are separated (center to center) by 1

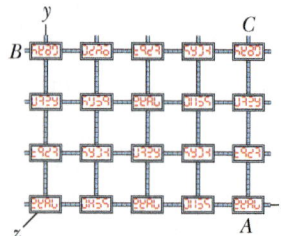

Figure 37-19   Question 7.

light-second, as are the clocks along the y axis, and all the clocks are synchronized via the procedure described in Module 37-1. When the initial synchronizing signal of $t = 0$ from the origin reaches (a) clock A, (b) clock B, and (c) clock C, what initial time is then set on those clocks? An event occurs at clock A when it reads 10 s. (d) How long does the signal of that event take to travel to an observer stationed at the origin? (e) What time does that observer assign to the event?

8 The rest energy and total energy, respectively, of three particles, expressed in terms of a basic amount A are (1) $A, 2A$; (2) $A, 3A$; (3) $3A$, $4A$. Without written calculation, rank the particles according to their (a) mass, (b) kinetic energy, (c) Lorentz factor, and (d) speed, greatest first.

9 Figure 37-20 shows the triangle of Fig 37-14 for six particles; the slanted lines 2 and 4 have the same length. Rank the particles according to (a) mass, (b) momentum magnitude, and (c) Lorentz factor, greatest first. (d) Identify which two particles have the same total energy. (e) Rank the three lowest-mass particles according to kinetic energy, greatest first.

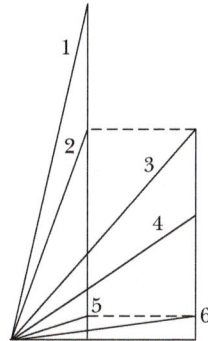

Figure 37-20
Question 9.

10 While on board a starship, you intercept signals from four shuttle craft that are moving either directly toward or directly

away from you. The signals have the same proper frequency f_0. The speed and direction (both relative to you) of the shuttle craft are (a) $0.3c$ toward, (b) $0.6c$ toward, (c) $0.3c$ away, and (d) $0.6c$ away. Rank the shuttle craft according to the frequency you receive, greatest first.

11 Figure 37-21 shows one of four star cruisers that are in a race. As each cruiser passes the starting line, a shuttle craft leaves the cruiser and races toward the finish line. You, judging the race, are stationary relative to the starting and finish lines. The speeds v_c of the cruisers relative to you and the speeds v_s of the shuttle craft relative to their respective starships are, in that order, (1) $0.70c, 0.40c$; (2) $0.40c, 0.70c$; (3) $0.20c, 0.90c$; (4) $0.50c, 0.60c$. (a) Rank the shuttle craft according to their speeds relative to you, greatest first. (b) Rank the shuttle craft according to the distances their pilots measure from the starting line to the finish line, greatest first. (c) Each starship sends a signal to its shuttle craft at a certain frequency f_0 as measured on board the starship. Rank the shuttle craft according to the frequencies they detect, greatest first.

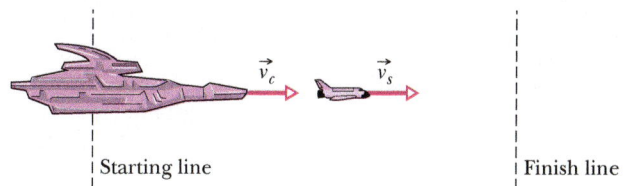

Figure 37-21 Question 11.

Problems

Module 37-1 Simultaneity and Time Dilation

•1 The mean lifetime of stationary muons is measured to be 2.2000 μs. The mean lifetime of high-speed muons in a burst of cosmic rays observed from Earth is measured to be 16.000 μs. To five significant figures, what is the speed parameter β of these cosmic-ray muons relative to Earth?

•2 To eight significant figures, what is speed parameter β if the Lorentz factor γ is (a) 1.010 000 0, (b) 10.000 000, (c) 100.000 00, and (d) 1000.000 0?

••3 You wish to make a round trip from Earth in a spaceship, traveling at constant speed in a straight line for exactly 6 months (as you measure the time interval) and then returning at the same constant speed. You wish further, on your return, to find Earth as it will be exactly 1000 years in the future. (a) To eight significant figures, at what speed parameter β must you travel? (b) Does it matter whether you travel in a straight line on your journey?

••4 *(Come) back to the future.* Suppose that a father is 20.00 y older than his daughter. He wants to travel outward from Earth for 2.000 y and then back for another 2.000 y (both intervals as he measures them) such that he is then 20.00 y *younger* than his daughter. What constant speed parameter β (relative to Earth) is required?

••5 **ILW** An unstable high-energy particle enters a detector and leaves a track of length 1.05 mm before it decays. Its speed relative to the detector was $0.992c$. What is its proper lifetime? That is, how

long would the particle have lasted before decay had it been at rest with respect to the detector?

••6 **GO** Reference frame S' is to pass reference frame S at speed v along the common direction of the x' and x axes, as in Fig. 37-9. An observer who rides along with frame S' is to count off a certain time interval on his wristwatch. The corresponding time interval Δt is to be measured by an observer in frame S. Figure 37-22 gives Δt versus speed parameter β for a range of values for β. The vertical axis scale is set by $\Delta t_a = 14.0$ s. What is interval Δt if $v = 0.98c$?

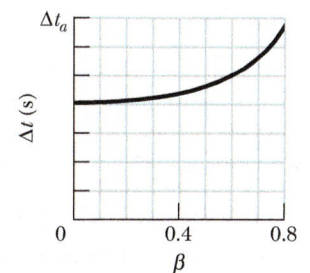

Figure 37-22 Problem 6.

••7 The premise of the *Planet of the Apes* movies and book is that hibernating astronauts travel far into Earth's future, to a time when human civilization has been replaced by an ape civilization. Considering only special relativity, determine how far into Earth's future the astronauts would travel if they slept for 120 y while traveling relative to Earth with a speed of $0.9990c$, first outward from Earth and then back again.

Module 37-2 The Relativity of Length

•8 An electron of $\beta = 0.999\ 987$ moves along the axis of an evacuated tube that has a length of 3.00 m as measured by a laboratory observer S at rest relative to the tube. An observer S' who is at rest relative to the electron, however, would see this tube moving with speed $v (= \beta c)$. What length would observer S' measure for the tube?

•9 SSM A spaceship of rest length 130 m races past a timing station at a speed of $0.740c$. (a) What is the length of the spaceship as measured by the timing station? (b) What time interval will the station clock record between the passage of the front and back ends of the ship?

•10 A meter stick in frame S' makes an angle of $30°$ with the x' axis. If that frame moves parallel to the x axis of frame S with speed $0.90c$ relative to frame S, what is the length of the stick as measured from S?

•11 A rod lies parallel to the x axis of reference frame S, moving along this axis at a speed of $0.630c$. Its rest length is 1.70 m. What will be its measured length in frame S?

••12 The length of a spaceship is measured to be exactly half its rest length. (a) To three significant figures, what is the speed parameter β of the spaceship relative to the observer's frame? (b) By what factor do the spaceship's clocks run slow relative to clocks in the observer's frame?

••13 GO A space traveler takes off from Earth and moves at speed $0.9900c$ toward the star Vega, which is 26.00 ly distant. How much time will have elapsed by Earth clocks (a) when the traveler reaches Vega and (b) when Earth observers receive word from the traveler that she has arrived? (c) How much older will Earth observers calculate the traveler to be (measured from her frame) when she reaches Vega than she was when she started the trip?

••14 GO A rod is to move at constant speed v along the x axis of reference frame S, with the rod's length parallel to that axis. An observer in frame S is to measure the length L of the rod. Figure 37-23 gives length L versus speed parameter β for a range of values for β. The vertical axis scale is set by $L_a = 1.00$ m. What is L if $v = 0.95c$?

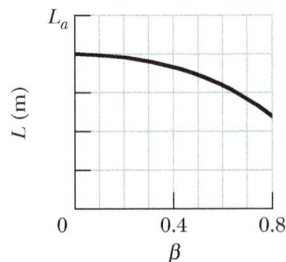

Figure 37-23 Problem 14.

••15 GO The center of our Milky Way galaxy is about 23 000 ly away. (a) To eight significant figures, at what constant speed parameter would you need to travel exactly 23 000 ly (measured in the Galaxy frame) in exactly 30 y (measured in your frame)? (b) Measured in your frame and in light-years, what length of the Galaxy would pass by you during the trip?

Module 37-3 The Lorentz Transformation

•16 Observer S reports that an event occurred on the x axis of his reference frame at $x = 3.00 \times 10^8$ m at time $t = 2.50$ s. Observer S' and her frame are moving in the positive direction of the x axis at a speed of $0.400c$. Further, $x = x' = 0$ at $t = t' = 0$. What are the (a) spatial and (b) temporal coordinate of the event according to S'? If S' were, instead, moving in the *negative* direction of the x axis, what would be the (c) spatial and (d) temporal coordinate of the event according to S'?

•17 SSM WWW In Fig. 37-9, the origins of the two frames coincide at $t = t' = 0$ and the relative speed is $0.950c$. Two micrometeorites collide at coordinates $x = 100$ km and $t = 200$ μs according to an observer in frame S. What are the (a) spatial and (b) temporal coordinate of the collision according to an observer in frame S'?

•18 Inertial frame S' moves at a speed of $0.60c$ with respect to frame S (Fig. 37-9). Further, $x = x' = 0$ at $t = t' = 0$. Two events are recorded. In frame S, event 1 occurs at the origin at $t = 0$ and event 2 occurs on the x axis at $x = 3.0$ km at $t = 4.0$ μs. According to observer S', what is the time of (a) event 1 and (b) event 2? (c) Do the two observers see the same sequence or the reverse?

•19 An experimenter arranges to trigger two flashbulbs simultaneously, producing a big flash located at the origin of his reference frame and a small flash at $x = 30.0$ km. An observer moving at a speed of $0.250c$ in the positive direction of x also views the flashes. (a) What is the time interval between them according to her? (b) Which flash does she say occurs first?

••20 GO As in Fig. 37-9, reference frame S' passes reference frame S with a certain velocity. Events 1 and 2 are to have a certain temporal separation $\Delta t'$ according to the S' observer. However, their spatial separation $\Delta x'$ according to that observer has not been set yet. Figure 37-24 gives their temporal separation Δt according to the S observer as a function of $\Delta x'$ for a range of $\Delta x'$ values. The vertical axis scale is set by $\Delta t_a = 6.00$ μs. What is $\Delta t'$?

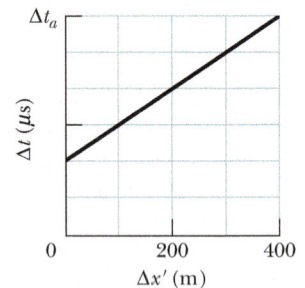

Figure 37-24 Problem 20.

••21 *Relativistic reversal of events.* Figures 37-25a and b show the (usual) situation in which a primed reference frame passes an unprimed reference frame, in the common positive direction of the x and x' axes, at a constant relative velocity of magnitude v. We are at rest in the unprimed frame; Bullwinkle, an astute student of relativity in spite of his cartoon upbringing, is at rest in the primed frame. The figures also indicate events A and B that occur at the following spacetime coordinates as measured in our unprimed frame and in Bullwinkle's primed frame:

Event	Unprimed	Primed
A	(x_A, t_A)	(x'_A, t'_A)
B	(x_B, t_B)	(x'_B, t'_B)

In our frame, event A occurs before event B, with temporal separation $\Delta t = t_B - t_A = 1.00$ μs and spatial separation $\Delta x = x_B - x_A = 400$ m. Let $\Delta t'$ be the temporal separation of the events according to Bullwinkle. (a) Find an expression for $\Delta t'$ in terms of the speed parameter $\beta (= v/c)$ and the given data. Graph $\Delta t'$ versus β for the following two ranges of β:

(b) 0 to 0.01 (v is low, from 0 to $0.01c$)

(c) 0.1 to 1 (v is high, from $0.1c$ to the limit c)

(d) At what value of β is $\Delta t' = 0$? For what range of β is the

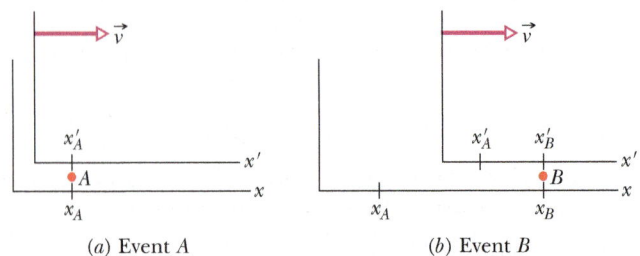

(a) Event A (b) Event B

Figure 37-25 Problems 21, 22, 60, and 61.

sequence of events *A* and *B* according to Bullwinkle (e) the same as ours and (f) the reverse of ours? (g) Can event *A* cause event *B*, or vice versa? Explain.

••22 For the passing reference frames in Fig. 37-25, events *A* and *B* occur at the following spacetime coordinates: according to the unprimed frame, (x_A, t_A) and (x_B, t_B); according to the primed frame, (x'_A, t'_A) and (x'_B, t'_B). In the unprimed frame, $\Delta t = t_B - t_A = 1.00 \ \mu s$ and $\Delta x = x_B - x_A = 400$ m. (a) Find an expression for $\Delta x'$ in terms of the speed parameter β and the given data. Graph $\Delta x'$ versus β for two ranges of β: (b) 0 to 0.01 and (c) 0.1 to 1. (d) At what value of β is $\Delta x'$ minimum, and (e) what is that minimum?

••23 ILW A clock moves along an *x* axis at a speed of 0.600*c* and reads zero as it passes the origin of the axis. (a) Calculate the clock's Lorentz factor. (b) What time does the clock read as it passes *x* = 180 m?

••24 Bullwinkle in reference frame *S′* passes you in reference frame *S* along the common direction of the *x′* and *x* axes, as in Fig. 37-9. He carries three meter sticks: meter stick 1 is parallel to the *x′* axis, meter stick 2 is parallel to the *y′* axis, and meter stick 3 is parallel to the *z′* axis. On his wristwatch he counts off 15.0 s, which takes 30.0 s according to you. Two events occur during his passage. According to you, event 1 occurs at $x_1 = 33.0$ m and $t_1 = 22.0$ ns, and event 2 occurs at $x_2 = 53.0$ m and $t_2 = 62.0$ ns. According to your measurements, what is the length of (a) meter stick 1, (b) meter stick 2, and (c) meter stick 3? According to Bullwinkle, what are (d) the spatial separation and (e) the temporal separation between events 1 and 2, and (f) which event occurs first?

••25 In Fig. 37-9, observer *S* detects two flashes of light. A big flash occurs at $x_1 = 1200$ m and, 5.00 μs later, a small flash occurs at $x_2 = 480$ m. As detected by observer *S′*, the two flashes occur at a single coordinate *x′*. (a) What is the speed parameter of *S′*, and (b) is *S′* moving in the positive or negative direction of the *x* axis? To *S′*, (c) which flash occurs first and (d) what is the time interval between the flashes?

••26 In Fig. 37-9, observer *S* detects two flashes of light. A big flash occurs at $x_1 = 1200$ m and, slightly later, a small flash occurs at $x_2 = 480$ m. The time interval between the flashes is $\Delta t = t_2 - t_1$. What is the smallest value of Δt for which observer *S′* will determine that the two flashes occur at the same *x′* coordinate?

Module 37-4 The Relativity of Velocities

•27 SSM A particle moves along the *x′* axis of frame *S′* with velocity 0.40*c*. Frame *S′* moves with velocity 0.60*c* with respect to frame *S*. What is the velocity of the particle with respect to frame *S*?

•28 In Fig. 37-11, frame *S′* moves relative to frame *S* with velocity $0.62c\hat{i}$ while a particle moves parallel to the common *x* and *x′* axes. An observer attached to frame *S′* measures the particle's velocity to be $0.47c\hat{i}$. In terms of *c*, what is the particle's velocity as measured by an observer attached to frame *S* according to the (a) relativistic and (b) classical velocity transformation? Suppose, instead, that the *S′* measure of the particle's velocity is $-0.47c\hat{i}$. What velocity does the observer in *S* now measure according to the (c) relativistic and (d) classical velocity transformation?

•29 Galaxy A is reported to be receding from us with a speed of 0.35*c*. Galaxy B, located in precisely the opposite direction, is also found to be receding from us at this same speed. What multiple of *c* gives the recessional speed an observer on Galaxy A would find for (a) our galaxy and (b) Galaxy B?

•30 Stellar system Q_1 moves away from us at a speed of 0.800*c*. Stellar system Q_2, which lies in the same direction in space but is closer to us, moves away from us at speed 0.400*c*. What multiple of *c* gives the speed of Q_2 as measured by an observer in the reference frame of Q_1?

••31 SSM WWW ILW A spaceship whose rest length is 350 m has a speed of 0.82*c* with respect to a certain reference frame. A micrometeorite, also with a speed of 0.82*c* in this frame, passes the spaceship on an antiparallel track. How long does it take this object to pass the ship as measured on the ship?

••32 GO In Fig. 37-26*a*, particle *P* is to move parallel to the *x* and *x′* axes of reference frames *S* and *S′*, at a certain velocity relative to frame *S*. Frame *S′* is to move parallel to the *x* axis of frame *S* at velocity *v*. Figure 37-26*b* gives the velocity *u′* of the particle relative to frame *S′* for a range of values for *v*. The vertical axis scale is set by $u'_a = 0.800c$. What value will *u′* have if (a) *v* = 0.90*c* and (b) *v* → *c*?

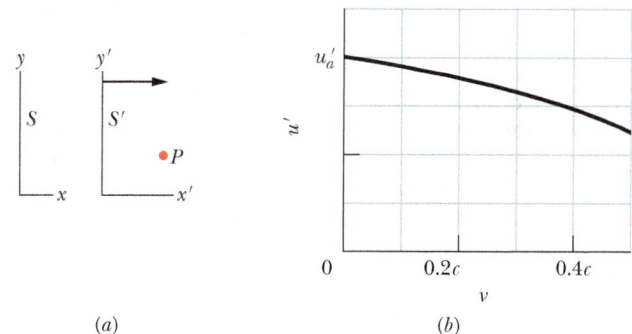

Figure 37-26 Problem 32.

••33 GO An armada of spaceships that is 1.00 ly long (as measured in its rest frame) moves with speed 0.800*c* relative to a ground station in frame *S*. A messenger travels from the rear of the armada to the front with a speed of 0.950*c* relative to *S*. How long does the trip take as measured (a) in the rest frame of the messenger, (b) in the rest frame of the armada, and (c) by an observer in the ground frame *S*?

Module 37-5 Doppler Effect for Light

•34 A sodium light source moves in a horizontal circle at a constant speed of 0.100*c* while emitting light at the proper wavelength of $\lambda_0 = 589.00$ nm. Wavelength λ is measured for that light by a detector fixed at the center of the circle. What is the wavelength shift $\lambda - \lambda_0$?

•35 SSM A spaceship, moving away from Earth at a speed of 0.900*c*, reports back by transmitting at a frequency (measured in the spaceship frame) of 100 MHz. To what frequency must Earth receivers be tuned to receive the report?

•36 Certain wavelengths in the light from a galaxy in the constellation Virgo are observed to be 0.4% longer than the corresponding light from Earth sources. (a) What is the radial speed of this galaxy with respect to Earth? (b) Is the galaxy approaching or receding from Earth?

•37 Assuming that Eq. 37-36 holds, find how fast you would have to go through a red light to have it appear green. Take 620 nm as the wavelength of red light and 540 nm as the wavelength of green light.

•38 Figure 37-27 is a graph of intensity versus wavelength for light reaching Earth from galaxy NGC 7319, which is about 3×10^8 light-years away. The most intense light is emitted by the oxygen in NGC 7319. In a laboratory that emission is at wavelength $\lambda = 513$ nm, but in the light from NGC 7319 it has been shifted to 525 nm due to the Doppler effect (all the emissions from NGC 7319 have been shifted). (a) What is the radial speed of NGC 7319 relative to Earth? (b) Is the relative motion toward or away from our planet?

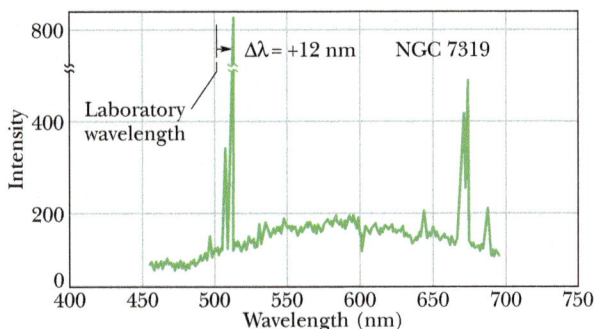

Figure 37-27 Problem 38.

••39 **SSM** A spaceship is moving away from Earth at speed $0.20c$. A source on the rear of the ship emits light at wavelength 450 nm according to someone on the ship. What (a) wavelength and (b) color (blue, green, yellow, or red) are detected by someone on Earth watching the ship?

Module 37-6 Momentum and Energy

•40 How much work must be done to increase the speed of an electron from rest to (a) $0.500c$, (b) $0.990c$, and (c) $0.9990c$?

•41 **SSM** **WWW** The mass of an electron is $9.109\ 381\ 88 \times 10^{-31}$ kg. To six significant figures, find (a) γ and (b) β for an electron with kinetic energy $K = 100.000$ MeV.

•42 What is the minimum energy that is required to break a nucleus of ^{12}C (of mass 11.996 71 u) into three nuclei of ^{4}He (of mass 4.001 51 u each)?

•43 How much work must be done to increase the speed of an electron (a) from $0.18c$ to $0.19c$ and (b) from $0.98c$ to $0.99c$? Note that the speed increase is $0.01c$ in both cases.

•44 In the reaction $p + {}^{19}F \rightarrow \alpha + {}^{16}O$, the masses are

$$m(p) = 1.007825\text{ u}, \qquad m(\alpha) = 4.002603\text{ u},$$
$$m(F) = 18.998405\text{ u}, \qquad m(O) = 15.994915\text{ u}.$$

Calculate the Q of the reaction from these data.

••45 In a high-energy collision between a cosmic-ray particle and a particle near the top of Earth's atmosphere, 120 km above sea level, a pion is created. The pion has a total energy E of 1.35×10^5 MeV and is traveling vertically downward. In the pion's rest frame, the pion decays 35.0 ns after its creation. At what altitude above sea level, as measured from Earth's reference frame, does the decay occur? The rest energy of a pion is 139.6 MeV.

••46 (a) If m is a particle's mass, p is its momentum magnitude, and K is its kinetic energy, show that

$$m = \frac{(pc)^2 - K^2}{2Kc^2}.$$

(b) For low particle speeds, show that the right side of the equation reduces to m. (c) If a particle has $K = 55.0$ MeV when $p =$ 121 MeV/c, what is the ratio m/m_e of its mass to the electron mass?

••47 **SSM** A 5.00-grain aspirin tablet has a mass of 320 mg. For how many kilometers would the energy equivalent of this mass power an automobile? Assume 12.75 km/L and a heat of combustion of 3.65×10^7 J/L for the gasoline used in the automobile.

••48 **GO** The mass of a muon is 207 times the electron mass; the average lifetime of muons at rest is 2.20 μs. In a certain experiment, muons moving through a laboratory are measured to have an average lifetime of 6.90 μs. For the moving muons, what are (a) β, (b) K, and (c) p (in MeV/c)?

••49 **GO** As you read this page (on paper or monitor screen), a cosmic ray proton passes along the left–right width of the page with relative speed v and a total energy of 14.24 nJ. According to your measurements, that left–right width is 21.0 cm. (a) What is the width according to the proton's reference frame? How much time did the passage take according to (b) your frame and (c) the proton's frame?

••50 To four significant figures, find the following when the kinetic energy is 10.00 MeV: (a) γ and (b) β for an electron ($E_0 = 0.510\ 998$ MeV), (c) γ and (d) β for a proton ($E_0 = 938.272$ MeV), and (e) γ and (f) β for an α particle ($E_0 = 3727.40$ MeV).

••51 **ILW** What must be the momentum of a particle with mass m so that the total energy of the particle is 3.00 times its rest energy?

••52 Apply the binomial theorem (Appendix E) to the last part of Eq. 37-52 for the kinetic energy of a particle. (a) Retain the first two terms of the expansion to show the kinetic energy in the form

$$K = (\text{first term}) + (\text{second term}).$$

The first term is the classical expression for kinetic energy. The second term is the first-order correction to the classical expression. Assume the particle is an electron. If its speed v is $c/20$, what is the value of (b) the classical expression and (c) the first-order correction? If the electron's speed is $0.80c$, what is the value of (d) the classical expression and (e) the first-order correction? (f) At what speed parameter β does the first-order correction become 10% or greater of the classical expression?

••53 In Module 28-4, we showed that a particle of charge q and mass m will move in a circle of radius $r = mv/|q|B$ when its velocity $\vec{v}$ is perpendicular to a uniform magnetic field $\vec{B}$. We also found that the period T of the motion is independent of speed v. These two results are approximately correct if $v \ll c$. For relativistic speeds, we must use the correct equation for the radius:

$$r = \frac{p}{|q|B} = \frac{\gamma m v}{|q|B}.$$

(a) Using this equation and the definition of period ($T = 2\pi r/v$), find the correct expression for the period. (b) Is T independent of v? If a 10.0 MeV electron moves in a circular path in a uniform magnetic field of magnitude 2.20 T, what are (c) the radius according to Chapter 28, (d) the correct radius, (e) the period according to Chapter 28, and (f) the correct period?

••54 **GO** What is β for a particle with (a) $K = 2.00E_0$ and (b) $E = 2.00E_0$?

••55 A certain particle of mass m has momentum of magnitude mc. What are (a) β, (b) γ, and (c) the ratio K/E_0?

••56 (a) The energy released in the explosion of 1.00 mol of TNT is 3.40 MJ. The molar mass of TNT is 0.227 kg/mol. What weight of TNT is needed for an explosive release of 1.80×10^{14} J? (b) Can you carry that weight in a backpack, or is a truck or train required? (c) Suppose that in an explosion of a fission bomb, 0.080% of the fissionable mass is converted to released energy. What weight of fissionable material is needed for an explosive release of 1.80×10^{14} J? (d) Can you carry that weight in a backpack, or is a truck or train required?

••57 Quasars are thought to be the nuclei of active galaxies in the early stages of their formation. A typical quasar radiates energy at the rate of 10^{41} W. At what rate is the mass of this quasar being reduced to supply this energy? Express your answer in solar mass units per year, where one solar mass unit (1 smu $= 2.0 \times 10^{30}$ kg) is the mass of our Sun.

••58 The mass of an electron is $9.109\,381\,88 \times 10^{-31}$ kg. To eight significant figures, find the following for the given electron kinetic energy: (a) γ and (b) β for $K = 1.000\,000\,0$ keV, (c) γ and (d) β for $K = 1.000\,000\,0$ MeV, and then (e) γ and (f) β for $K = 1.000\,000\,0$ GeV.

•••59 GO An alpha particle with kinetic energy 7.70 MeV collides with an ^{14}N nucleus at rest, and the two transform into an ^{17}O nucleus and a proton. The proton is emitted at 90° to the direction of the incident alpha particle and has a kinetic energy of 4.44 MeV. The masses of the various particles are alpha particle, 4.00260 u; ^{14}N, 14.00307 u; proton, 1.007825 u; and ^{17}O, 16.99914 u. In MeV, what are (a) the kinetic energy of the oxygen nucleus and (b) the Q of the reaction? (*Hint:* The speeds of the particles are much less than c.)

Additional Problems

60 *Temporal separation between two events.* Events A and B occur with the following spacetime coordinates in the reference frames of Fig. 37-25: according to the unprimed frame, (x_A, t_A) and (x_B, t_B); according to the primed frame, (x'_A, t'_A) and (x'_B, t'_B). In the unprimed frame, $\Delta t = t_B - t_A = 1.00$ μs and $\Delta x = x_B - x_A = 240$ m. (a) Find an expression for $\Delta t'$ in terms of the speed parameter β and the given data. Graph $\Delta t'$ versus β for the following two ranges of β: (b) 0 to 0.01 and (c) 0.1 to 1. (d) At what value of β is $\Delta t'$ minimum and (e) what is that minimum? (f) Can one of these events cause the other? Explain.

61 *Spatial separation between two events.* For the passing reference frames of Fig. 37-25, events A and B occur with the following spacetime coordinates: according to the unprimed frame, (x_A, t_A) and (x_B, t_B); according to the primed frame, (x'_A, t'_A) and (x'_B, t'_B). In the unprimed frame, $\Delta t = t_B - t_A = 1.00$ μs and $\Delta x = x_B - x_A = 240$ m. (a) Find an expression for $\Delta x'$ in terms of the speed parameter β and the given data. Graph $\Delta x'$ versus β for two ranges of β: (b) 0 to 0.01 and (c) 0.1 to 1. (d) At what value of β is $\Delta x' = 0$?

62 GO In Fig. 37-28a, particle P is to move parallel to the x and x' axes of reference frames S and S', at a certain velocity relative to frame S. Frame S' is to move parallel to the x axis of frame S at

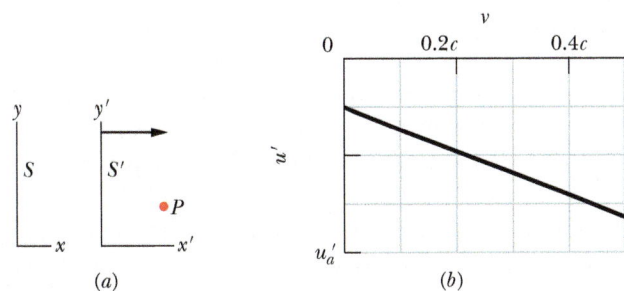

Figure 37-28 Problem 62.

velocity v. Figure 37-28b gives the velocity u' of the particle relative to frame S' for a range of values for v. The vertical axis scale is set by $u'_a = -0.800c$. What value will u' have if (a) $v = 0.80c$ and (b) $v \rightarrow c$?

63 GO *Superluminal jets.* Figure 37-29a shows the path taken by a knot in a jet of ionized gas that has been expelled from a galaxy. The knot travels at constant velocity $\vec{v}$ at angle θ from the direction of Earth. The knot occasionally emits a burst of light, which is eventually detected on Earth. Two bursts are indicated in Fig. 37-29a, separated by time t as measured in a stationary frame near the bursts. The bursts are shown in Fig. 37-29b as if they were photographed on the same piece of film, first when light from burst 1 arrived on Earth and then later when light from burst 2 arrived. The apparent distance D_{app} traveled by the knot between the two bursts is the distance across an Earth-observer's view of the knot's path. The apparent time T_{app} between the bursts is the difference in the arrival times of the light from them. The apparent speed of the knot is then $V_{app} = D_{app}/T_{app}$. In terms of v, t, and θ, what are (a) D_{app} and (b) T_{app}? (c) Evaluate V_{app} for $v = 0.980c$ and $\theta = 30.0°$. When superluminal (faster than light) jets were first observed, they seemed to defy special relativity—at least until the correct geometry (Fig. 37-29a) was understood.

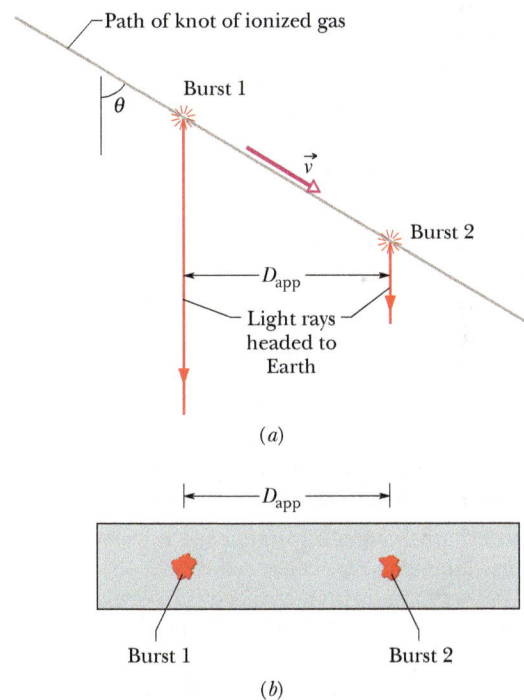

Figure 37-29 Problem 63.

64 GO Reference frame S' passes reference frame S with a certain velocity as in Fig. 37-9. Events 1 and 2 are to have a certain spatial separation $\Delta x'$ according to the S' observer. However, their temporal separation $\Delta t'$ according to that observer has not been set yet. Figure 37-30 gives their spatial separation Δx according to the S observer as a function of $\Delta t'$ for a range of $\Delta t'$ values. The vertical axis scale is set by $\Delta x_a = 10.0$ m. What is $\Delta x'$?

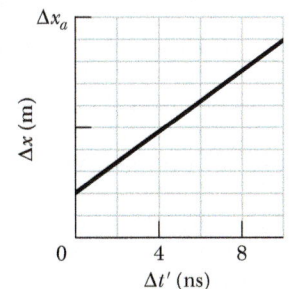

Figure 37-30 Problem 64.

65 *Another approach to velocity transformations.* In Fig. 37-31, reference frames B and C move past reference frame A in the common direction of their x axes. Represent the x components of the velocities of one frame relative to another with a two-letter subscript. For example, v_{AB} is the x component of the velocity of A relative to B. Similarly, represent the corresponding speed parameters with two-letter subscripts. For example, $\beta_{AB}\ (=v_{AB}/c)$ is the speed parameter corresponding to v_{AB}. (a) Show that

$$\beta_{AC} = \frac{\beta_{AB} + \beta_{BC}}{1 + \beta_{AB}\beta_{BC}}.$$

Let M_{AB} represent the ratio $(1 - \beta_{AB})/(1 + \beta_{AB})$, and let M_{BC} and M_{AC} represent similar ratios. (b) Show that the relation

$$M_{AC} = M_{AB}M_{BC}$$

is true by deriving the equation of part (a) from it.

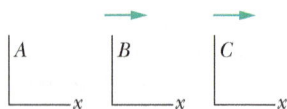

Figure 37-31 Problems 65, 66, and 67.

66 *Continuation of Problem 65.* Use the result of part (b) in Problem 65 for the motion along a single axis in the following situation. Frame A in Fig. 37-31 is attached to a particle that moves with velocity $+0.500c$ past frame B, which moves past frame C with a velocity of $+0.500c$. What are (a) M_{AC}, (b) β_{AC}, and (c) the velocity of the particle relative to frame C?

67 *Continuation of Problem 65.* Let reference frame C in Fig. 37-31 move past reference frame D (not shown). (a) Show that

$$M_{AD} = M_{AB}M_{BC}M_{CD}.$$

(b) Now put this general result to work: Three particles move parallel to a single axis on which an observer is stationed. Let plus and minus signs indicate the directions of motion along that axis. Particle A moves past particle B at $\beta_{AB} = +0.20$. Particle B moves past particle C at $\beta_{BC} = -0.40$. Particle C moves past observer D at $\beta_{CD} = +0.60$. What is the velocity of particle A relative to observer D? (The solution technique here is *much* faster than using Eq. 37-29.)

68 Figure 37-16 shows a ship (attached to reference frame S') passing us (standing in reference frame S) with velocity $\vec{v} = 0.950c\hat{i}$. A proton is fired at speed $0.980c$ relative to the ship from the front of the ship to the rear. The proper length of the ship is 760 m. What is the temporal separation between the time the proton is fired and the time it hits the rear wall of the ship according to (a) a passenger in the ship and (b) us? Suppose that, instead, the proton is fired from the rear to the front. What then is the temporal separation between the time it is fired and the time it hits the front wall according to (c) the passenger and (d) us?

69 *The car-in-the-garage problem.* Carman has just purchased the world's longest stretch limo, which has a proper length of $L_c = 30.5$ m. In Fig. 37-32a, it is shown parked in front of a garage with a proper length of $L_g = 6.00$ m. The garage has a front door (shown open) and a back door (shown closed). The limo is obviously longer than the garage. Still, Garageman, who owns the garage and knows something about relativistic length contraction, makes a bet with Carman that the limo can fit in the garage with both doors closed. Carman, who dropped his physics course before reaching special relativity, says such a thing, even in principle, is impossible.

To analyze Garageman's scheme, an x_c axis is attached to the limo, with $x_c = 0$ at the rear bumper, and an x_g axis is attached to the garage, with $x_g = 0$ at the (now open) front door. Then Carman is to drive the limo directly toward the front door at a velocity of $0.9980c$ (which is, of course, both technically and financially impossible). Carman is stationary in the x_c reference frame; Garageman is stationary in the x_g reference frame.

There are two events to consider. *Event 1:* When the rear bumper clears the front door, the front door is closed. Let the time of this event be zero to both Carman and Garageman: $t_{g1} = t_{c1} = 0$. The event occurs at $x_c = x_g = 0$. Figure 37-32b shows event 1 according to the x_g reference frame. *Event 2:* When the front bumper reaches the back door, that door opens. Figure 37-32c shows event 2 according to the x_g reference frame.

According to Garageman, (a) what is the length of the limo, and what are the spacetime coordinates (b) x_{g2} and (c) t_{g2} of event 2? (d) For how long is the limo temporarily "trapped" inside the garage with both doors shut? Now consider the situation from the x_c reference frame, in which the garage comes racing past the limo at a velocity of $-0.9980c$. According to Carman, (e) what is the length of the passing garage, what are the spacetime coordinates (f) x_{c2} and (g) t_{c2} of event 2, (h) is the limo ever in the garage with both doors shut, and (i) which event occurs first? (j) Sketch events 1 and 2 as seen by Carman. (k) Are the events causally related; that is, does one of them cause the other? (l) Finally, who wins the bet?

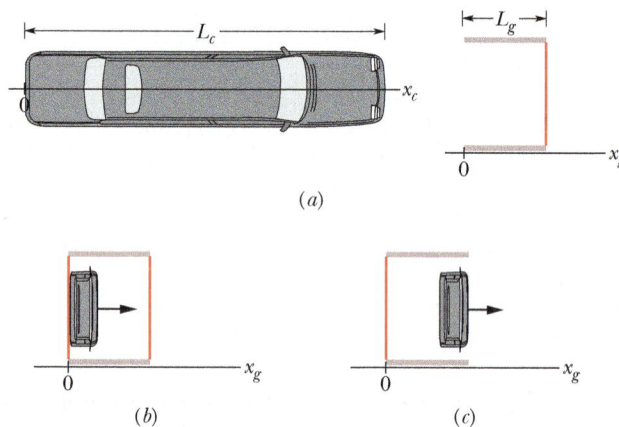

(a)

(b)

(c)

Figure 37-32 Problem 69.

70 An airplane has rest length 40.0 m and speed 630 m/s. To a ground observer, (a) by what fraction is its length contracted and (b) how long is needed for its clocks to be 1.00 μs slow.

71 SSM To circle Earth in low orbit, a satellite must have a speed of about 2.7×10^4 km/h. Suppose that two such satellites orbit Earth in opposite directions. (a) What is their relative speed as they pass, according to the classical Galilean velocity transformation equation? (b) What fractional error do you make in (a) by not using the (correct) relativistic transformation equation?

72 Find the speed parameter of a particle that takes 2.0 y longer than light to travel a distance of 6.0 ly.

73 SSM How much work is needed to accelerate a proton from a speed of $0.9850c$ to a speed of $0.9860c$?

74 A pion is created in the higher reaches of Earth's atmosphere when an incoming high-energy cosmic-ray particle collides with an atomic nucleus. A pion so formed descends toward Earth with a speed of $0.99c$. In a reference frame in which they are at rest, pions decay with an average life of 26 ns. As measured in a frame fixed with respect to Earth, how far (on the average) will such a pion move through the atmosphere before it decays?

75 **SSM** If we intercept an electron having total energy 1533 MeV that came from Vega, which is 26 ly from us, how far in light-years was the trip in the rest frame of the electron?

76 The total energy of a proton passing through a laboratory apparatus is 10.611 nJ. What is its speed parameter β? Use the proton mass given in Appendix B under "Best Value," not the commonly remembered rounded number.

77 A spaceship at rest in a certain reference frame S is given a speed increment of $0.50c$. Relative to its new rest frame, it is then given a further $0.50c$ increment. This process is continued until its speed with respect to its original frame S exceeds $0.999c$. How many increments does this process require?

78 In the red shift of radiation from a distant galaxy, a certain radiation, known to have a wavelength of 434 nm when observed in the laboratory, has a wavelength of 462 nm. (a) What is the radial speed of the galaxy relative to Earth? (b) Is the galaxy approaching or receding from Earth?

79 **SSM** What is the momentum in MeV/c of an electron with a kinetic energy of 2.00 MeV?

80 The radius of Earth is 6370 km, and its orbital speed about the Sun is 30 km/s. Suppose Earth moves past an observer at this speed. To the observer, by how much does Earth's diameter contract along the direction of motion?

81 A particle with mass m has speed $c/2$ relative to inertial frame S. The particle collides with an identical particle at rest relative to frame S. Relative to S, what is the speed of a frame S' in which the total momentum of these particles is zero? This frame is called the *center of momentum frame.*

82 An elementary particle produced in a laboratory experiment travels 0.230 mm through the lab at a relative speed of $0.960c$ before it decays (becomes another particle). (a) What is the proper lifetime of the particle? (b) What is the distance the particle travels as measured from its rest frame?

83 What are (a) K, (b) E, and (c) p (in GeV/c) for a proton moving at speed $0.990c$? What are (d) K, (e) E, and (f) p (in MeV/c) for an electron moving at speed $0.990c$?

84 A radar transmitter T is fixed to a reference frame S' that is moving to the right with speed v relative to reference frame S (Fig. 37-33). A mechanical timer (essentially a clock) in frame S', having a period τ_0 (measured in S'), causes transmitter T to emit timed radar pulses, which travel at the speed of light and are received by R, a receiver fixed in frame S. (a) What is the period τ of the timer as detected by observer A, who is fixed in frame S? (b) Show that at receiver R the time interval between pulses arriving from T is not τ or τ_0, but

$$\tau_R = \tau_0\sqrt{\frac{c + v}{c - v}}.$$

(c) Explain why receiver R and observer A, who are in the same

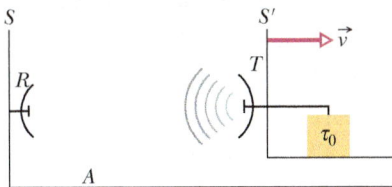
Figure 37-33 Problem 84.

reference frame, measure a different period for the transmitter. (*Hint:* A clock and a radar pulse are not the same thing.)

85 One cosmic-ray particle approaches Earth along Earth's north–south axis with a speed of $0.80c$ toward the geographic north pole, and another approaches with a speed of $0.60c$ toward the geographic south pole (Fig. 37-34). What is the relative speed of approach of one particle with respect to the other?

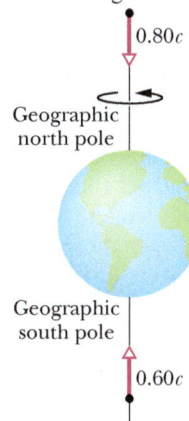
Figure 37-34 Problem 85.

86 (a) How much energy is released in the explosion of a fission bomb containing 3.0 kg of fissionable material? Assume that 0.10% of the mass is converted to released energy. (b) What mass of TNT would have to explode to provide the same energy release? Assume that each mole of TNT liberates 3.4 MJ of energy on exploding. The molecular mass of TNT is 0.227 kg/mol. (c) For the same mass of explosive, what is the ratio of the energy released in a nuclear explosion to that released in a TNT explosion?

87 (a) What potential difference would accelerate an electron to speed c according to classical physics? (b) With this potential difference, what speed would the electron actually attain?

88 A Foron cruiser moving directly toward a Reptulian scout ship fires a decoy toward the scout ship. Relative to the scout ship, the speed of the decoy is $0.980c$ and the speed of the Foron cruiser is $0.900c$. What is the speed of the decoy relative to the cruiser?

89 In Fig. 37-35, three spaceships are in a chase. Relative to an x axis in an inertial frame (say, Earth frame), their velocities are $v_A = 0.900c$, v_B, and $v_C = 0.800c$. (a) What value of v_B is required such that ships A and C approach ship B with the same speed relative to ship B, and (b) what is that relative speed?

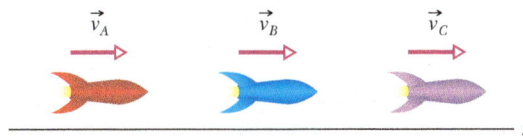
Figure 37-35 Problem 89.

90 Space cruisers A and B are moving parallel to the positive direction of an x axis. Cruiser A is faster, with a relative speed of $v = 0.900c$, and has a proper length of $L = 200$ m. According to the pilot of A, at the instant ($t = 0$) the tails of the cruisers are aligned, the noses are also. According to the pilot of B, how much later are the noses aligned?

91 In Fig. 37-36, two cruisers fly toward a space station. Relative to the station, cruiser A has speed $0.800c$. Relative to the station, what speed is required of cruiser B such that its pilot sees A and the station approach B at the same speed?

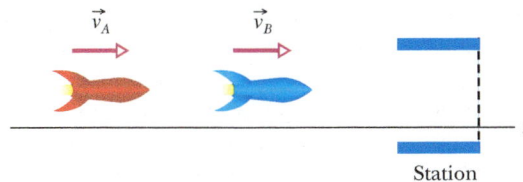
Figure 37-36 Problem 91.

92 A relativistic train of proper length 200 m approaches a tunnel of the same proper length, at a relative speed of 0.900c. A paint bomb in the engine room is set to explode (and cover everyone with blue paint) when the *front* of the train passes the *far* end of the tunnel (event FF). However, when the *rear* car passes the *near* end of the tunnel (event RN), a device in that car is set to send a signal to the engine room to deactivate the bomb. *Train view:* (a) What is the tunnel length? (b) Which event occurs first, FF or RN? (c) What is the time between those events? (d) Does the paint bomb explode? *Tunnel view:* (e) What is the train length? (f) Which event occurs first? (g) What is the time between those events? (h) Does the paint bomb explode? If your answers to (d) and (h) differ, you need to explain the paradox, because either the engine room is covered with blue paint or not; you cannot have it both ways. If your answers are the same, you need to explain why.

93 Particle A (with rest energy 200 MeV) is at rest in a lab frame when it decays to particle B (rest energy 100 MeV) and particle C (rest energy 50 MeV). What are the (a) total energy and (b) momentum of B and the (c) total energy and (d) momentum of C?

94 Figure 37-37 shows three situations in which a starship passes Earth (the dot) and then makes a round trip that brings it back past Earth, each at the given Lorentz factor. As measured in the rest frame of Earth, the round-trip distances are as follows: trip 1, $2D$; trip 2, $4D$; trip 3, $6D$. Neglecting any time needed for accelerations and in terms of D and c, find the travel times of (a) trip 1, (b) trip 2, and (c) trip 3 as measured from the rest frame of Earth. Next, find the travel times of (d) trip 1, (e) trip 2, and (f) trip 3 as measured from the rest frame of the starship. (*Hint:* For a large Lorentz factor, the relative speed is almost c.)

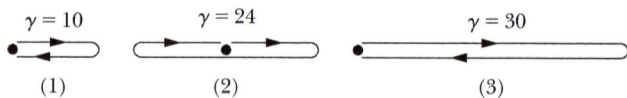

$\gamma = 10$ $\gamma = 24$ $\gamma = 30$

(1)   (2)   (3)

Figure 37-37 Problem 94.

95 Ionization measurements show that a particular lightweight nuclear particle carries a double charge ($= 2e$) and is moving with a speed of 0.710c. Its measured radius of curvature in a magnetic field of 1.00 T is 6.28 m. Find the mass of the particle and identify it. (*Hints:* Lightweight nuclear particles are made up of neutrons (which have no charge) and protons (charge $= +e$), in roughly equal numbers. Take the mass of each such particle to be 1.00 u. (See Problem 53.)

96 A 2.50 MeV electron moves perpendicularly to a magnetic field in a path with a 3.0 cm radius of curvature. What is the magnetic field B? (See Problem 53.)

97 A proton synchrotron accelerates protons to a kinetic energy of 500 GeV. At this energy, calculate (a) the Lorentz factor, (b) the speed parameter, and (c) the magnetic field for which the proton orbit has a radius of curvature of 750 m.

98 An astronaut exercising on a treadmill maintains a pulse rate of 150 per minute. If he exercises for 1.00 h as measured by a clock on his spaceship, with a stride length of 1.00 m/s, while the ship travels with a speed of 0.900c relative to a ground station, what are (a) the pulse rate and (b) the distance walked as measured by someone at the ground station?

99 A spaceship approaches Earth at a speed of 0.42c. A light on the front of the ship appears red (wavelength 650 nm) to passengers on the ship. What (a) wavelength and (b) color (blue, green, or yellow) would it appear to an observer on Earth?

100 Some of the familiar hydrogen lines appear in the spectrum of quasar 3C9, but they are shifted so far toward the red that their wavelengths are observed to be 3.0 times as long as those observed for hydrogen atoms at rest in the laboratory. (a) Show that the classical Doppler equation gives a relative velocity of recession greater than c for this situation. (b) Assuming that the relative motion of 3C9 and Earth is due entirely to the cosmological expansion of the universe, find the recession speed that is predicted by the relativistic Doppler equation.

101 In one year the United States consumption of electrical energy was about 2.2×10^{12} kW · h. (a) How much mass is equivalent to the consumed energy in that year? (b) Does it make any difference to your answer if this energy is generated in oil-burning, nuclear, or hydroelectric plants?

102 Quite apart from effects due to Earth's rotational and orbital motions, a laboratory reference frame is not strictly an inertial frame because a particle at rest there will not, in general, remain at rest; it will fall. Often, however, events happen so quickly that we can ignore the gravitational acceleration and treat the frame as inertial. Consider, for example, an electron of speed $v = 0.992c$, projected horizontally into a laboratory test chamber and moving through a distance of 20 cm. (a) How long would that take, and (b) how far would the electron fall during this interval? (c) What can you conclude about the suitability of the laboratory as an inertial frame in this case?

103 What is the speed parameter for the following speeds: (a) a typical rate of continental drift (1 in./y); (b) a typical drift speed for electrons in a current-carrying conductor (0.5 mm/s); (c) a highway speed limit of 55 mi/h; (d) the root-mean-square speed of a hydrogen molecule at room temperature; (e) a supersonic plane flying at Mach 2.5 (1200 km/h); (f) the escape speed of a projectile from the Earth's surface; (g) the speed of Earth in its orbit around the Sun; (h) a typical recession speed of a distant quasar due to the cosmological expansion (3.0×10^4 km/s)?

Photons and Matter Waves

38-1 THE PHOTON, THE QUANTUM OF LIGHT

Learning Objectives

After reading this module, you should be able to . . .

38.01 Explain the absorption and emission of light in terms of quantized energy and photons.

38.02 For photon absorption and emission, apply the relationships between energy, power, intensity, rate of photons, the Planck constant, the associated frequency, and the associated wavelength.

Key Ideas

● An electromagnetic wave (light) is quantized (allowed only in certain quantities), and the quanta are called photons.

● For light of frequency f and wavelength λ, the photon energy is
$$E = hf,$$
where h is the Planck constant.

What Is Physics?

One primary focus of physics is Einstein's theory of relativity, which took us into a world far beyond that of ordinary experience—the world of objects moving at speeds close to the speed of light. Among other surprises, Einstein's theory predicts that the rate at which a clock runs depends on how fast the clock is moving relative to the observer: the faster the motion, the slower the clock rate. This and other predictions of the theory have passed every experimental test devised thus far, and relativity theory has led us to a deeper and more satisfying view of the nature of space and time.

Now you are about to explore a second world that is outside ordinary experience—the subatomic world. You will encounter a new set of surprises that, though they may sometimes seem bizarre, have led physicists step by step to a deeper view of reality.

Quantum physics, as our new subject is called, answers such questions as: Why do the stars shine? Why do the elements exhibit the order that is so apparent in the periodic table? How do transistors and other microelectronic devices work? Why does copper conduct electricity but glass does not? In fact, scientists and engineers have applied quantum physics in almost every aspect of everyday life, from medical instrumentation to transportation systems to entertainment industries. Indeed, because quantum physics accounts for all of chemistry, including biochemistry, we need to understand it if we are to understand life itself.

Some of the predictions of quantum physics seem strange even to the physicists and philosophers who study its foundations. Still, experiment after experiment has proved the theory correct, and many have exposed even stranger aspects of the theory. The quantum world is an amusement park full of wonderful rides that are guaranteed to shake up the commonsense world view you have developed since childhood. We begin our exploration of that quantum park with the photon.

The Photon, the Quantum of Light

Quantum physics (which is also known as *quantum mechanics* and *quantum theory*) is largely the study of the microscopic world. In that world, many quantities are found only in certain minimum (*elementary*) amounts, or integer multiples of those elementary amounts; these quantities are then said to be *quantized*. The elementary amount that is associated with such a quantity is called the **quantum** of that quantity (*quanta* is the plural).

In a loose sense, U.S. currency is quantized because the coin of least value is the penny, or $0.01 coin, and the values of all other coins and bills are restricted to integer multiples of that least amount. In other words, the currency quantum is $0.01, and all greater amounts of currency are of the form $n($0.01)$, where n is always a positive integer. For example, you cannot hand someone $0.755 = 75.5($0.01)$.

In 1905, Einstein proposed that electromagnetic radiation (or simply *light*) is quantized and exists in elementary amounts (quanta) that we now call **photons**. This proposal should seem strange to you because we have just spent several chapters discussing the classical idea that light is a sinusoidal wave, with a wavelength λ, a frequency f, and a speed c such that

$$f = \frac{c}{\lambda}. \tag{38-1}$$

Furthermore, in Chapter 33 we discussed the classical light wave as being an interdependent combination of electric and magnetic fields, each oscillating at frequency f. How can this wave of oscillating fields consist of an elementary amount of something—the light quantum? What *is* a photon?

The concept of a light quantum, or a photon, turns out to be far more subtle and mysterious than Einstein imagined. Indeed, it is still very poorly understood. In this book, we shall discuss only some of the basic aspects of the photon concept, somewhat along the lines of Einstein's proposal. According to that proposal, the quantum of a light wave of frequency f has the energy

$$E = hf \quad \text{(photon energy)}. \tag{38-2}$$

Here h is the **Planck constant**, the constant we first met in Eq. 32-23, and which has the value

$$h = 6.63 \times 10^{-34} \, \text{J} \cdot \text{s} = 4.14 \times 10^{-15} \, \text{eV} \cdot \text{s}. \tag{38-3}$$

The smallest amount of energy a light wave of frequency f can have is hf, the energy of a single photon. If the wave has more energy, its total energy must be an integer multiple of hf. The light cannot have an energy of, say, $0.6hf$ or $75.5hf$.

Einstein further proposed that when light is absorbed or emitted by an object (matter), the absorption or emission event occurs in the atoms of the object. When light of frequency f is absorbed by an atom, the energy hf of one photon is transferred from the light to the atom. In this *absorption event*, the photon vanishes and the atom is said to absorb it. When light of frequency f is emitted by an atom, an amount of energy hf is transferred from the atom to the light. In this *emission event*, a photon suddenly appears and the atom is said to emit it. Thus, we can have *photon absorption* and *photon emission* by atoms in an object.

For an object consisting of many atoms, there can be many photon absorptions (such as with sunglasses) or photon emissions (such as with lamps). However, each absorption or emission event still involves the transfer of an amount of energy equal to that of a single photon of the light.

When we discussed the absorption or emission of light in previous chapters, our examples involved so much light that we had no need of quantum physics, and we got by with classical physics. However, in the late 20th century, technology became advanced enough that single-photon experiments could be conducted and put to practical use. Since then quantum physics has become part of standard engineering practice, especially in optical engineering.

☑ **Checkpoint 1**

Rank the following radiations according to their associated photon energies, greatest first: (a) yellow light from a sodium vapor lamp, (b) a gamma ray emitted by a radioactive nucleus, (c) a radio wave emitted by the antenna of a commercial radio station, (d) a microwave beam emitted by airport traffic control radar.

Sample Problem 38.01 Emission and absorption of light as photons

A sodium vapor lamp is placed at the center of a large sphere that absorbs all the light reaching it. The rate at which the lamp emits energy is 100 W; assume that the emission is entirely at a wavelength of 590 nm. At what rate are photons absorbed by the sphere?

KEY IDEAS

The light is emitted and absorbed as photons. We assume that all the light emitted by the lamp reaches (and thus is absorbed by) the sphere. So, the rate R at which photons are absorbed by the sphere is equal to the rate R_{emit} at which photons are emitted by the lamp.

Calculations: That rate is

$$R_{emit} = \frac{\text{rate of energy emission}}{\text{energy per emitted photon}} = \frac{P_{emit}}{E}.$$

Next, into this we can substitute from Eq. 38-2 ($E = hf$), Einstein's proposal about the energy E of each quantum of light (which we here call a photon in modern language). We can then write the absorption rate as

$$R = R_{emit} = \frac{P_{emit}}{hf}.$$

Using Eq. 38-1 ($f = c/\lambda$) to substitute for f and then entering known data, we obtain

$$R = \frac{P_{emit}\lambda}{hc}$$

$$= \frac{(100 \text{ W})(590 \times 10^{-9} \text{ m})}{(6.63 \times 10^{-34} \text{ J·s})(2.998 \times 10^8 \text{ m/s})}$$

$$= 2.97 \times 10^{20} \text{ photons/s.} \qquad \text{(Answer)}$$

WILEY PLUS Additional examples, video, and practice available at *WileyPLUS*

38-2 THE PHOTOELECTRIC EFFECT

Learning Objectives

After reading this module, you should be able to . . .

38.03 Make a simple and basic sketch of a photoelectric experiment, showing the incident light, the metal plate, the emitted electrons (photoelectrons), and the collector cup.

38.04 Explain the problems physicists had with the photoelectric effect prior to Einstein and the historical importance of Einstein's explanation of the effect.

38.05 Identify a stopping potential V_{stop} and relate it to the maximum kinetic energy K_{max} of escaping photoelectrons.

38.06 For a photoelectric setup, apply the relationships between the frequency and wavelength of the incident light, the maximum kinetic energy K_{max} of the photoelectrons, the work function Φ, and the stopping potential V_{stop}.

38.07 For a photoelectric setup, sketch a graph of the stopping potential V_{stop} versus the frequency of the light, identifying the cutoff frequency f_0 and relating the slope to the Planck constant h and the elementary charge e.

Key Ideas

● When light of high enough frequency illuminates a metal surface, electrons can gain enough energy to escape the metal by absorbing photons in the illumination, in what is called the photoelectric effect.

● The conservation of energy in such an absorption and escape is written as

$$hf = K_{max} + \Phi,$$

where hf is the energy of the absorbed photon, K_{max} is the kinetic energy of the most energetic of the escaping electrons, and Φ (called the work function) is the least energy required by an electron to escape the electric forces holding electrons in the metal.

● If $hf = \Phi$, electrons barely escape but have no kinetic energy and the frequency is called the cutoff frequency f_0.

● If $hf < \Phi$, electrons cannot escape.

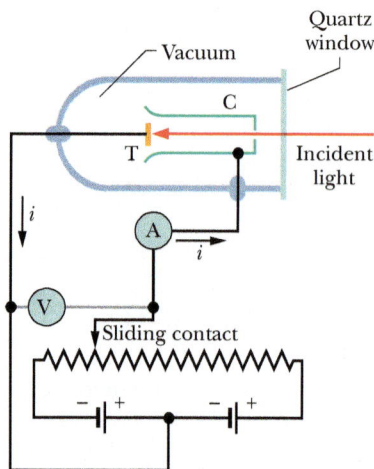

Figure 38-1 An apparatus used to study the photoelectric effect. The incident light shines on target T, ejecting electrons, which are collected by collector cup C. The electrons move in the circuit in a direction opposite the conventional current arrows. The batteries and the variable resistor are used to produce and adjust the electric potential difference between T and C.

The Photoelectric Effect

If you direct a beam of light of short enough wavelength onto a clean metal surface, the light will cause electrons to leave that surface (the light will *eject* the electrons from the surface). This **photoelectric effect** is used in many devices, including camcorders. Einstein's photon concept can explain it.

Let us analyze two basic photoelectric experiments, each using the apparatus of Fig. 38-1, in which light of frequency f is directed onto target T and ejects electrons from it. A potential difference V is maintained between target T and collector cup C to sweep up these electrons, said to be **photoelectrons.** This collection produces a **photoelectric current** i that is measured with meter A.

First Photoelectric Experiment

We adjust the potential difference V by moving the sliding contact in Fig. 38-1 so that collector C is slightly negative with respect to target T. This potential difference acts to slow down the ejected electrons. We then vary V until it reaches a certain value, called the **stopping potential** V_{stop}, at which point the reading of meter A has just dropped to zero. When $V = V_{stop}$, the most energetic ejected electrons are turned back just before reaching the collector. Then K_{max}, the kinetic energy of these most energetic electrons, is

$$K_{max} = eV_{stop}, \tag{38-4}$$

where e is the elementary charge.

Measurements show that for light of a given frequency, K_{max} *does not depend on the intensity of the light source.* Whether the source is dazzling bright or so feeble that you can scarcely detect it (or has some intermediate brightness), the maximum kinetic energy of the ejected electrons always has the same value.

This experimental result is a puzzle for classical physics. Classically, the incident light is a sinusoidally oscillating electromagnetic wave. An electron in the target should oscillate sinusoidally due to the oscillating electric force on it from the wave's electric field. If the amplitude of the electron's oscillation is great enough, the electron should break free of the target's surface—that is, be ejected from the target. Thus, if we increase the amplitude of the wave and its oscillating electric field, the electron should get a more energetic "kick" as it is being ejected. *However, that is not what happens.* For a given frequency, intense light beams and feeble light beams give exactly the same maximum kick to ejected electrons.

The actual result follows naturally if we think in terms of photons. Now the energy that can be transferred from the incident light to an electron in the target is that of a single photon. Increasing the light intensity increases the *number* of photons in the light, but the photon energy, given by Eq. 38-2 ($E = hf$), is unchanged because the frequency is unchanged. Thus, the energy transferred to the kinetic energy of an electron is also unchanged.

Second Photoelectric Experiment

Now we vary the frequency f of the incident light and measure the associated stopping potential V_{stop}. Figure 38-2 is a plot of V_{stop} versus f. Note that the photoelectric effect does not occur if the frequency is below a certain **cutoff frequency** f_0 or, equivalently, if the wavelength is greater than the corresponding **cutoff wavelength** $\lambda_0 = c/f_0$. This is so *no matter how intense the incident light is.*

This is another puzzle for classical physics. If you view light as an electromagnetic wave, you must expect that no matter how low the frequency, electrons can always be ejected by light if you supply them with enough energy—that is, if you use a light source that is bright enough. *That is not what happens.* For light below the cutoff frequency f_0, the photoelectric effect does not occur, no matter how bright the light source.

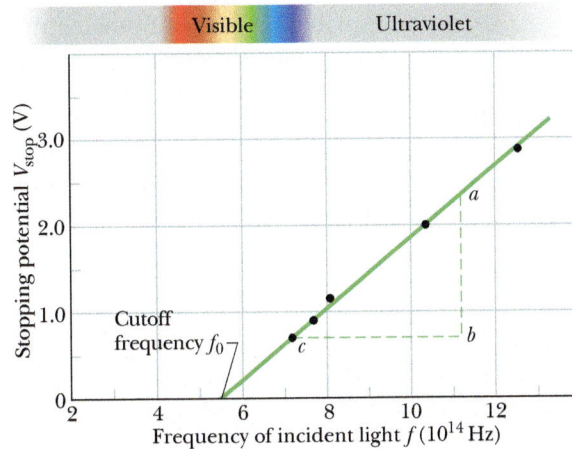

Electrons can escape only if the light frequency exceeds a certain value.

The escaping electron's kinetic energy is greater for a greater light frequency.

Figure 38-2 The stopping potential V_{stop} as a function of the frequency f of the incident light for a sodium target T in the apparatus of Fig. 38-1. (Data reported by R. A. Millikan in 1916.)

The existence of a cutoff frequency is, however, just what we should expect if the energy is transferred via photons. The electrons within the target are held there by electric forces. (If they weren't, they would drip out of the target due to the gravitational force on them.) To just escape from the target, an electron must pick up a certain minimum energy Φ, where Φ is a property of the target material called its **work function.** If the energy hf transferred to an electron by a photon exceeds the work function of the material (if $hf > \Phi$), the electron can escape the target. If the energy transferred does not exceed the work function (that is, if $hf < \Phi$), the electron cannot escape. This is what Fig. 38-2 shows.

The Photoelectric Equation

Einstein summed up the results of such photoelectric experiments in the equation

$$hf = K_{max} + \Phi \quad \text{(photoelectric equation).} \tag{38-5}$$

This is a statement of the conservation of energy for a single photon absorption by a target with work function Φ. Energy equal to the photon's energy hf is transferred to a single electron in the material of the target. If the electron is to escape from the target, it must pick up energy at least equal to Φ. Any additional energy $(hf - \Phi)$ that the electron acquires from the photon appears as kinetic energy K of the electron. In the most favorable circumstance, the electron can escape through the surface without losing any of this kinetic energy in the process; it then appears outside the target with the maximum possible kinetic energy K_{max}.

Let us rewrite Eq. 38-5 by substituting for K_{max} from Eq. 38-4 ($K_{max} = eV_{stop}$). After a little rearranging we get

$$V_{stop} = \left(\frac{h}{e}\right)f - \frac{\Phi}{e}. \tag{38-6}$$

The ratios h/e and Φ/e are constants, and so we would expect a plot of the measured stopping potential V_{stop} versus the frequency f of the light to be a straight line, as it is in Fig. 38-2. Further, the slope of that straight line should be h/e. As a check, we measure ab and bc in Fig. 38-2 and write

$$\frac{h}{e} = \frac{ab}{bc} = \frac{2.35 \text{ V} - 0.72 \text{ V}}{(11.2 \times 10^{14} - 7.2 \times 10^{14}) \text{ Hz}}$$
$$= 4.1 \times 10^{-15} \text{ V} \cdot \text{s}.$$

Multiplying this result by the elementary charge e, we find

$$h = (4.1 \times 10^{-15}\,\text{V}\cdot\text{s})(1.6 \times 10^{-19}\,\text{C}) = 6.6 \times 10^{-34}\,\text{J}\cdot\text{s},$$

which agrees with values measured by many other methods.

An aside: An explanation of the photoelectric effect certainly requires quantum physics. For many years, Einstein's explanation was also a compelling argument for the existence of photons. However, in 1969 an alternative explanation for the effect was found that used quantum physics but did not need the concept of photons. As shown in countless other experiments, light *is* in fact quantized as photons, but Einstein's explanation of the photoelectric effect is not the best argument for that fact.

Checkpoint 2

The figure shows data like those of Fig. 38-2 for targets of cesium, potassium, sodium, and lithium. The plots are parallel. (a) Rank the targets according to their work functions, greatest first. (b) Rank the plots according to the value of h they yield, greatest first.

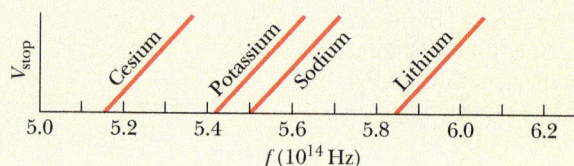

Sample Problem 38.02 Photoelectric effect and work function

Find the work function Φ of sodium from Fig. 38-2.

KEY IDEAS

We can find the work function Φ from the cutoff frequency f_0 (which we can measure on the plot). The reasoning is this: At the cutoff frequency, the kinetic energy K_{max} in Eq. 38-5 is zero. Thus, all the energy hf that is transferred from a photon to an electron goes into the electron's escape, which requires an energy of Φ.

Calculations: From that last idea, Eq. 38-5 then gives us, with $f = f_0$,

$$hf_0 = 0 + \Phi = \Phi.$$

In Fig. 38-2, the cutoff frequency f_0 is the frequency at which the plotted line intercepts the horizontal frequency axis, about 5.5×10^{14} Hz. We then have

$$\Phi = hf_0 = (6.63 \times 10^{-34}\,\text{J}\cdot\text{s})(5.5 \times 10^{14}\,\text{Hz})$$

$$= 3.6 \times 10^{-19}\,\text{J} = 2.3\,\text{eV}. \qquad \text{(Answer)}$$

WILEY PLUS Additional examples, video, and practice available at *WileyPLUS*

38-3 PHOTONS, MOMENTUM, COMPTON SCATTERING, LIGHT INTERFERENCE

Learning Objectives

After reading this module, you should be able to . . .

38.08 For a photon, apply the relationships between momentum, energy, frequency, and wavelength.

38.09 With sketches, describe the basics of a Compton scattering experiment.

38.10 Identify the historic importance of Compton scattering.

38.11 For an increase in the Compton-scattering angle ϕ, identify whether these quantities of the scattered x ray increase or decrease: kinetic energy, momentum, wavelength.

38.12 For Compton scattering, describe how the conserva-

tions of momentum and kinetic energy lead to the equation giving the wavelength shift $\Delta\lambda$.

38.13 For Compton scattering, apply the relationships between the wavelengths of the incident and scattered x rays, the wavelength shift $\Delta\lambda$, the angle ϕ of photon scattering, and the electron's final energy and momentum (both magnitude and angle).

38.14 In terms of photons, explain the double-slit experiment in the standard version, the single-photon version, and the single-photon, wide-angle version.

Key Ideas

● Although it is massless, a photon has momentum, which is related to its energy E, frequency f, and wavelength by

$$p = \frac{hf}{c} = \frac{h}{\lambda}.$$

● In Compton scattering, x rays scatter as particles (as photons) from loosely bound electrons in a target.

● In the scattering, an x-ray photon loses energy and momentum to the target electron.

● The resulting increase (Compton shift) in the photon wavelength is

$$\Delta\lambda = \frac{h}{mc}(1 - \cos\phi),$$

where m is the mass of the target electron and ϕ is the angle at which the photon is scattered from its initial travel direction.

● Photons: When light interacts with matter, the interaction is particle-like, occurring at a point and transferring energy and momentum.

● Wave: When a single photon is emitted by a source, we interpret its travel as being that of a probability wave.

● Wave: When many photons are emitted or absorbed by matter, we interpret the combined light as a classical electromagnetic wave.

Photons Have Momentum

In 1916, Einstein extended his concept of light quanta (photons) by proposing that a quantum of light has linear momentum. For a photon with energy hf, the magnitude of that momentum is

$$p = \frac{hf}{c} = \frac{h}{\lambda} \quad \text{(photon momentum)}, \qquad (38\text{-}7)$$

where we have substituted for f from Eq. 38-1 ($f = c/\lambda$). Thus, when a photon interacts with matter, energy *and* momentum are transferred, *as if* there were a collision between the photon and matter in the classical sense (as in Chapter 9).

In 1923, Arthur Compton at Washington University in St. Louis showed that both momentum and energy are transferred via photons. He directed a beam of x rays of wavelength λ onto a target made of carbon, as shown in Fig. 38-3. An x ray is a form of electromagnetic radiation, at high frequency and thus small wavelength. Compton measured the wavelengths and intensities of the x rays that were scattered in various directions from his carbon target.

Figure 38-4 shows his results. Although there is only a single wavelength ($\lambda = 71.1$ pm) in the incident x-ray beam, we see that the scattered x rays contain a range of wavelengths with two prominent intensity peaks. One peak is centered about the incident wavelength λ, the other about a wavelength λ' that is longer than λ by an amount $\Delta\lambda$, which is called the **Compton shift.** The value of the Compton shift varies with the angle at which the scattered x rays are detected and is greater for a greater angle.

Figure 38-4 is still another puzzle for classical physics. Classically, the incident x-ray beam is a sinusoidally oscillating electromagnetic wave. An electron in the

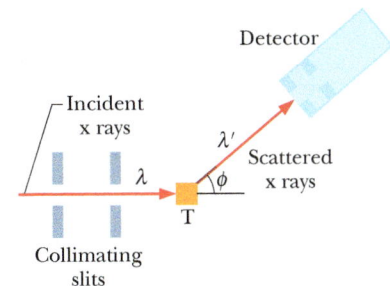

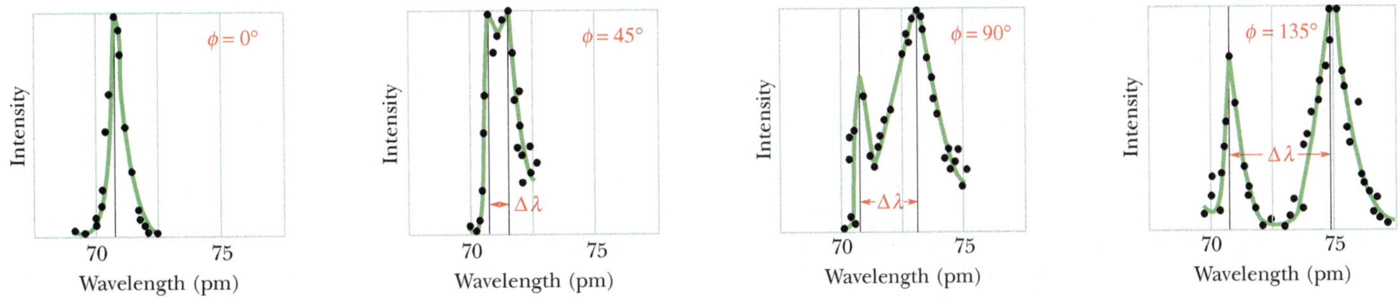

Figure 38-3 Compton's apparatus. A beam of x rays of wavelength $\lambda = 71.1$ pm is directed onto a carbon target T. The x rays scattered from the target are observed at various angles ϕ to the direction of the incident beam. The detector measures both the intensity of the scattered x rays and their wavelength.

Figure 38-4 Compton's results for four values of the scattering angle ϕ. Note that the Compton shift $\Delta\lambda$ increases as the scattering angle increases.

carbon target should oscillate sinusoidally due to the oscillating electric force on it from the wave's electric field. Further, the electron should oscillate at the same frequency as the wave and should send out waves *at this same frequency,* as if it were a tiny transmitting antenna. Thus, the x rays scattered by the electron should have the same frequency, and the same wavelength, as the x rays in the incident beam — but they don't.

Compton interpreted the scattering of x rays from carbon in terms of energy and momentum transfers, via photons, between the incident x-ray beam and loosely bound electrons in the carbon target. Let's see how this quantum physics interpretation leads to an understanding of Compton's results.

Suppose a single photon (of energy $E = hf$) is associated with the interaction between the incident x-ray beam and a stationary electron. In general, the direction of travel of the x ray will change (the x ray is scattered), and the electron will recoil, which means that the electron has obtained some kinetic energy. Energy is conserved in this isolated interaction. Thus, the energy of the scattered photon ($E' = hf'$) must be less than that of the incident photon. The scattered x rays must then have a lower frequency f' and thus a longer wavelength λ' than the incident x rays, just as Compton's experimental results in Fig. 38-4 show.

For the quantitative part, we first apply the law of conservation of energy. Figure 38-5 suggests a "collision" between an x ray and an initially stationary free electron in the target. As a result of the collision, an x ray of wavelength λ' moves off at an angle ϕ and the electron moves off at an angle θ, as shown. Conservation of energy then gives us

$$hf = hf' + K,$$

in which hf is the energy of the incident x-ray photon, hf' is the energy of the scattered x-ray photon, and K is the kinetic energy of the recoiling electron. Because the electron may recoil with a speed comparable to that of light, we must use the relativistic expression of Eq. 37-52,

$$K = mc^2(\gamma - 1),$$

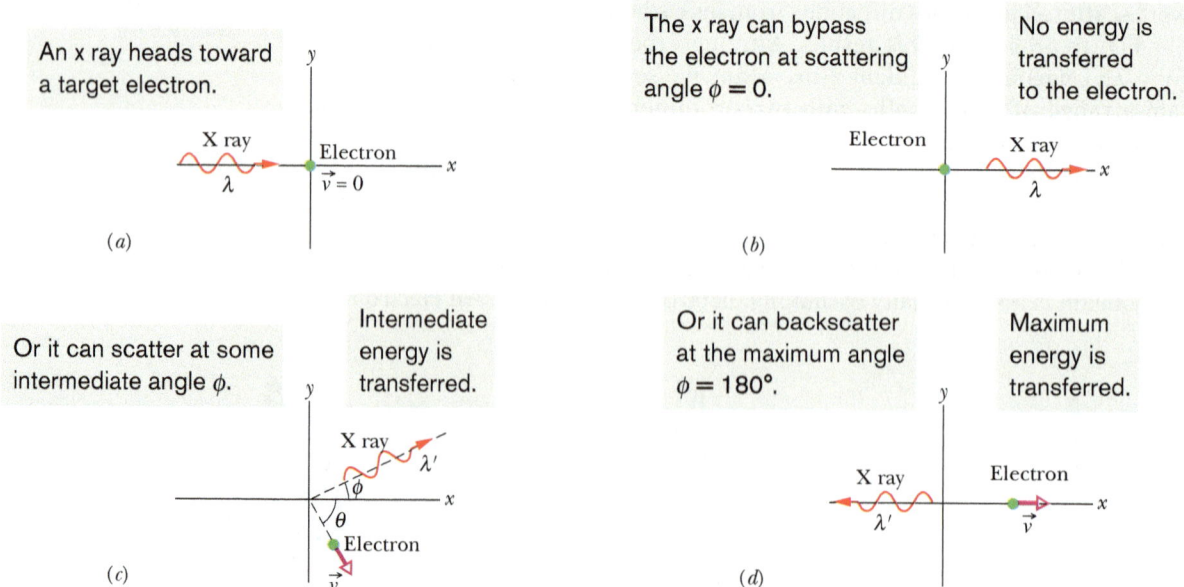

Figure 38-5 (a) An x ray approaches a stationary electron. The x ray can (b) bypass the electron (forward scatter) with no energy or momentum transfer, (c) scatter at some intermediate angle with an intermediate energy and momentum transfer, or (d) backscatter with the maximum energy and momentum transfer.

for the electron's kinetic energy. Here m is the electron's mass and γ is the Lorentz factor

$$\gamma = \frac{1}{\sqrt{1 - (v/c)^2}}.$$

Substituting for K in the conservation of energy equation yields

$$hf = hf' + mc^2(\gamma - 1).$$

Substituting c/λ for f and c/λ' for f' then leads to the new energy conservation equation

$$\frac{h}{\lambda} = \frac{h}{\lambda'} + mc(\gamma - 1). \tag{38-8}$$

Next we apply the law of conservation of momentum to the x-ray–electron collision of Fig. 38-5. From Eq. 38-7 ($p = h/\lambda$), the magnitude of the momentum of the incident photon is h/λ, and that of the scattered photon is h/λ'. From Eq. 37-41, the magnitude for the recoiling electron's momentum is $p = \gamma mv$. Because we have a two-dimensional situation, we write separate equations for the conservation of momentum along the x and y axes, obtaining

$$\frac{h}{\lambda} = \frac{h}{\lambda'} \cos \phi + \gamma mv \cos \theta \quad (x \text{ axis}) \tag{38-9}$$

and

$$0 = \frac{h}{\lambda'} \sin \phi - \gamma mv \sin \theta \quad (y \text{ axis}). \tag{38-10}$$

We want to find $\Delta\lambda$ ($= \lambda' - \lambda$), the Compton shift of the scattered x rays. Of the five collision variables (λ, λ', v, ϕ, and θ) that appear in Eqs. 38-8, 38-9, and 38-10, we choose to eliminate v and θ, which deal only with the recoiling electron. Carrying out the algebra (it is somewhat complicated) leads to

$$\Delta\lambda = \frac{h}{mc} (1 - \cos \phi) \quad \text{(Compton shift).} \tag{38-11}$$

Equation 38-11 agrees exactly with Compton's experimental results.

The quantity h/mc in Eq. 38-11 is a constant called the **Compton wavelength.** Its value depends on the mass m of the particle from which the x rays scatter. Here that particle is a loosely bound electron, and thus we would substitute the mass of an electron for m to evaluate the *Compton wavelength for Compton scattering from an electron.*

A Loose End

The peak at the incident wavelength λ ($= 71.1$ pm) in Fig. 38-4 still needs to be explained. This peak arises not from interactions between x rays and the very loosely bound electrons in the target but from interactions between x rays and the electrons that are *tightly* bound to the carbon atoms making up the target. Effectively, each of these latter collisions occurs between an incident x ray and an entire carbon atom. If we substitute for m in Eq. 38-11 the mass of a carbon atom (which is about 22 000 times that of an electron), we see that $\Delta\lambda$ becomes about 22 000 times smaller than the Compton shift for an electron—too small to detect. Thus, the x rays scattered in these collisions have the same wavelength as the incident x rays and give us the unshifted peaks in Fig. 38-4.

✓ Checkpoint 3

Compare Compton scattering for x rays ($\lambda \approx 20$ pm) and visible light ($\lambda \approx 500$ nm) at a particular angle of scattering. Which has the greater (a) Compton shift, (b) fractional wavelength shift, (c) fractional energy loss, and (d) energy imparted to the electron?

Sample Problem 38.03 Compton scattering of light by electrons

X rays of wavelength $\lambda = 22$ pm (photon energy = 56 keV) are scattered from a carbon target, and the scattered rays are detected at 85° to the incident beam.

(a) What is the Compton shift of the scattered rays?

KEY IDEA

The Compton shift is the wavelength change of the x rays due to scattering from loosely bound electrons in a target. Further, that shift depends on the angle at which the scattered x rays are detected, according to Eq. 38-11. The shift is zero for forward scattering at angle $\phi = 0°$, and it is maximum for backscattering at angle $\phi = 180°$. Here we have an intermediate situation at angle $\phi = 85°$.

Calculation: Substituting 85° for that angle and 9.11×10^{-31} kg for the electron mass (because the scattering is from electrons) in Eq. 38-11 gives us

$$\Delta\lambda = \frac{h}{mc}(1 - \cos\phi)$$

$$= \frac{(6.63 \times 10^{-34}\ \text{J·s})(1 - \cos 85°)}{(9.11 \times 10^{-31}\ \text{kg})(3.00 \times 10^{8}\ \text{m/s})}$$

$$= 2.21 \times 10^{-12}\ \text{m} \approx 2.2\ \text{pm}. \qquad \text{(Answer)}$$

(b) What percentage of the initial x-ray photon energy is transferred to an electron in such scattering?

KEY IDEA

We need to find the *fractional energy loss* (let us call it *frac*) for photons that scatter from the electrons:

$$frac = \frac{\text{energy loss}}{\text{initial energy}} = \frac{E - E'}{E}.$$

Calculations: From Eq. 38-2 ($E = hf$), we can substitute for the initial energy E and the detected energy E' of the x rays in terms of frequencies. Then, from Eq. 38-1 ($f = c/\lambda$), we can substitute for those frequencies in terms of the wavelengths. We find

$$frac = \frac{hf - hf'}{hf} = \frac{c/\lambda - c/\lambda'}{c/\lambda} = \frac{\lambda' - \lambda}{\lambda'}$$

$$= \frac{\Delta\lambda}{\lambda + \Delta\lambda}.$$

Substitution of data yields

$$frac = \frac{2.21\ \text{pm}}{22\ \text{pm} + 2.21\ \text{pm}} = 0.091, \text{ or } 9.1\%. \quad \text{(Answer)}$$

Although the Compton shift $\Delta\lambda$ is independent of the wavelength λ of the incident x rays (see Eq. 38-11), our result here tells us that the *fractional* photon energy loss of the x rays does depend on λ, increasing as the wavelength of the incident radiation decreases.

WILEY **PLUS** Additional examples, video, and practice available at *WileyPLUS*

Light as a Probability Wave

A fundamental mystery in physics is how light can be a wave (which spreads out over a region) in classical physics but be emitted and absorbed as photons (which originate and vanish at points) in quantum physics. The double-slit experiment of Module 35-2 lies at the heart of this mystery. Let us discuss three versions of it.

The Standard Version

Figure 38-6 is a sketch of the original experiment carried out by Thomas Young in 1801 (see also Fig. 35-8). Light shines on screen *B*, which contains two narrow parallel slits. The light waves emerging from the two slits spread out by diffraction and overlap on screen *C* where, by interference, they form a pattern of alternating intensity maxima and minima. In Module 35-2 we took the existence of these interference fringes as compelling evidence for the wave nature of light.

Let us place a tiny photon detector D at one point in the plane of screen *C*. Let the detector be a photoelectric device that clicks when it absorbs a photon. We would find that the detector produces a series of clicks, randomly spaced in time, each click signaling the transfer of energy from the light wave to the screen via a photon absorption. If we moved the detector very slowly up or down as indicated by the black arrow in Fig. 38-6, we would find that the click rate increases and decreases, passing through alternate maxima and minima that correspond exactly to the maxima and minima of the interference fringes.

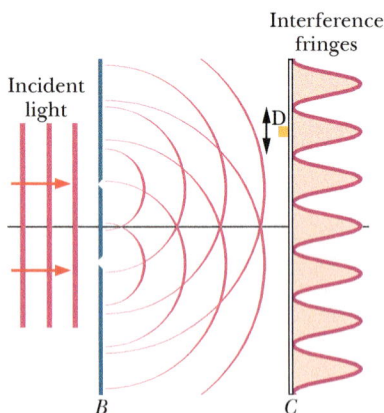

Figure 38-6 Light is directed onto screen *B*, which contains two parallel slits. Light emerging from these slits spreads out by diffraction. The two diffracted waves overlap at screen *C* and form a pattern of interference fringes. A small photon detector D in the plane of screen *C* generates a sharp click for each photon that it absorbs.

The point of this thought experiment is as follows. We cannot predict when a photon will be detected at any particular point on screen C; photons are detected at individual points at random times. We can, however, predict that the relative *probability* that a single photon will be detected at a particular point in a specified time interval is proportional to the light intensity at that point.

We know from Eq. 33-26 ($I = E_{rms}^2/c\mu_0$) in Module 33-2 that the intensity I of a light wave at any point is proportional to the square of E_m, the amplitude of the oscillating electric field vector of the wave at that point. Thus,

> The probability (per unit time interval) that a photon will be detected in any small volume centered on a given point in a light wave is proportional to the square of the amplitude of the wave's electric field vector at that point.

We now have a probabilistic description of a light wave, hence another way to view light. It is not only an electromagnetic wave but also a **probability wave.** That is, to every point in a light wave we can attach a numerical probability (per unit time interval) that a photon can be detected in any small volume centered on that point.

The Single-Photon Version

A single-photon version of the double-slit experiment was first carried out by G. I. Taylor in 1909 and has been repeated many times since. It differs from the standard version in that the light source in the Taylor experiment is so extremely feeble that it emits only one photon at a time, at random intervals. Astonishingly, interference fringes still build up on screen C if the experiment runs long enough (several months for Taylor's early experiment).

What explanation can we offer for the result of this single-photon double-slit experiment? Before we can even consider the result, we are compelled to ask questions like these: If the photons move through the apparatus one at a time, through which of the two slits in screen B does a given photon pass? How does a given photon even "know" that there is another slit present so that interference is a possibility? Can a single photon somehow pass through both slits and interfere with itself?

Bear in mind that the only thing we can know about photons is when light interacts with matter—we have no way of detecting them without an interaction with matter, such as with a detector or a screen. Thus, in the experiment of Fig. 38-6, all we can know is that photons originate at the light source and vanish at the screen. Between source and screen, we cannot know what the photon is or does. However, because an interference pattern eventually builds up on the screen, we can speculate that each photon travels from source to screen *as a wave* that fills up the space between source and screen and then vanishes in a photon absorption at some point on the screen, with a transfer of energy and momentum to the screen at that point.

We *cannot* predict where this transfer will occur (where a photon will be detected) for any given photon originating at the source. However, we *can* predict the probability that a transfer will occur at any given point on the screen. Transfers will tend to occur (and thus photons will tend to be absorbed) in the regions of the bright fringes in the interference pattern that builds up on the screen. Transfers will tend *not* to occur (and thus photons will tend *not* to be absorbed) in the regions of the dark fringes in the built-up pattern. Thus, we can say that the wave traveling from the source to the screen is a *probability wave*, which produces a pattern of "probability fringes" on the screen.

The Single-Photon, Wide-Angle Version

In the past, physicists tried to explain the single-photon double-slit experiment in terms of small packets of classical light waves that are individually sent toward the slits. They would define these small packets as photons. However, modern experiments invalidate this explanation and definition. One of these experiments, reported in 1992 by Ming Lai and Jean-Claude Diels of the University of New Mexico,

A *single* photon can take widely different paths and still interfere with itself.

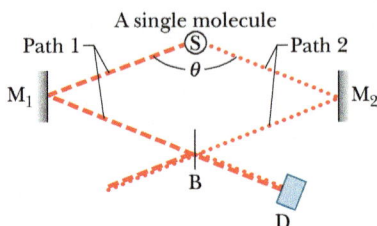

Figure 38-7 The light from a single photon emission in source S travels over two widely separated paths and interferes with itself at detector D after being recombined by beam splitter B. (Based on Ming Lai and Jean-Claude Diels, *Journal of the Optical Society of America B*, **9**, 2290–2294, December 1992.)

is depicted in Figure 38-7. Source S contains molecules that emit photons at well-separated times. Mirrors M_1 and M_2 are positioned to reflect light that the source emits along two distinct paths, 1 and 2, that are separated by an angle θ, which is close to 180°. This arrangement differs from the standard two-slit experiment, in which the angle between the paths of the light reaching two slits is very small.

After reflection from mirrors M_1 and M_2, the light waves traveling along paths 1 and 2 meet at beam splitter B, which transmits half the incident light and reflects the other half. On the right side of B in Fig. 38-7, the light wave traveling along path 2 and reflected by B combines with the light wave traveling along path 1 and transmitted by B. These two waves then interfere with each other at detector D (a *photomultiplier tube* that can detect individual photons).

The output of the detector is a randomly spaced series of electronic pulses, one for each detected photon. In the experiment, the beam splitter is moved slowly in a horizontal direction (in the reported experiment, a distance of only about 50 μm maximum), and the detector output is recorded on a chart recorder. Moving the beam splitter changes the lengths of paths 1 and 2, producing a phase shift between the light waves arriving at detector D. Interference maxima and minima appear in the detector's output signal.

This experiment is difficult to understand in traditional terms. For example, when a molecule in the source emits a single photon, does that photon travel along path 1 or path 2 in Fig. 38-7 (or along any other path)? Or can it move in both directions at once? To answer, we assume that when a molecule emits a photon, a probability wave radiates in all directions from it. The experiment samples this wave in two of those directions, chosen to be nearly opposite each other.

We see that we can interpret all three versions of the double-slit experiment if we assume that (1) light is generated in the source as photons, (2) light is absorbed in the detector as photons, and (3) light travels between source and detector as a probability wave.

38-4 THE BIRTH OF QUANTUM PHYSICS

Learning Objectives

After reading this module, you should be able to . . .

38.15 Identify an ideal blackbody radiator and its spectral radiancy $S(\lambda)$.

38.16 Identify the problem that physicists had with blackbody radiation prior to Planck's work, and explain how Planck and Einstein solved the problem.

38.17 Apply Planck's radiation law for a given wavelength and temperature.

38.18 For a narrow wavelength range and for a given wavelength and temperature, find the intensity in blackbody radiation.

38.19 Apply the relationship between intensity, power, and area.

38.20 Apply Wien's law to relate the surface temperature of an ideal blackbody radiator to the wavelength at which the spectral radiancy is maximum.

Key Ideas

● As a measure of the emission of thermal radiation by an ideal blackbody radiator, we define the spectral radiancy in terms of the emitted intensity per unit wavelength at a given wavelength λ:

$$S(\lambda) = \frac{\text{intensity}}{(\text{unit wavelength})}.$$

● The Planck radiation law, in which atomic oscillators produce the thermal radiation, is

$$S(\lambda) = \frac{2\pi c^2 h}{\lambda^5} \frac{1}{e^{hc/\lambda kT} - 1},$$

where h is the Planck constant, k is the Boltzmann constant, and T is the temperature of the radiating surface (in kelvins).

● Planck's law was the first suggestion that the energies of the atomic oscillators producing the radiation are quantized.

● Wien's law relates the temperature T of a blackbody radiator and the wavelength λ_{max} at which the spectral radiancy is maximum:

$$\lambda_{\text{max}} T = 2898 \ \mu\text{m} \cdot \text{K}.$$

The Birth of Quantum Physics

Now that we have seen how the photoelectric effect and Compton scattering propelled physicists into quantum physics, let's back up to the very beginning, when the idea of quantized energies gradually emerged out of experimental data. The story begins with what might seem mundane these days but which was a fixation point for physicists of 1900. The subject was the thermal radiation emitted by an ideal blackbody radiator—that is, a radiator whose emitted radiation depends only on its temperature and not on the material from which it is made, the nature of its surface, or anything other than temperature. In a nutshell here was the trouble: the experimental results differed wildly from the theoretical predictions and no one had a clue as to why.

Experimental Setup. We can make an ideal radiator by forming a cavity within a body and keeping the cavity walls at a uniform temperature. The atoms on the inner wall of the body oscillate (they have thermal energy), which causes them to emit electromagnetic waves, the thermal radiation. To sample that internal radiation, we drill a small hole through the wall so that some of the radiation can escape to be measured (but not enough to alter the radiation inside the cavity). We are interested in how the intensity of the radiation depends on wavelength.

That intensity distribution is handled by defining a **spectral radiancy** $S(\lambda)$ of the radiation emitted at given wavelength λ:

$$S(\lambda) = \frac{\text{intensity}}{\left(\begin{array}{c}\text{unit}\\\text{wavelength}\end{array}\right)} = \frac{\text{power}}{\left(\begin{array}{c}\text{unit area}\\\text{of emitter}\end{array}\right)\left(\begin{array}{c}\text{unit}\\\text{wavelength}\end{array}\right)}. \tag{38-12}$$

If we multiply $S(\lambda)$ by a narrow wavelength range $d\lambda$, we have the intensity (that is, the power per unit area of the hole in the wall) that is being emitted in the wavelength range λ to $\lambda + d\lambda$.

The solid curve in Fig. 38-8 shows the experimental results for a cavity with a wall temperature of 2000 K, for a range of wavelengths. Although such a radiator would glow brightly in a dark room, we can tell from the figure that only a small part of its radiated energy actually lies in the visible range (which is colorfully indicated). At that temperature, most of the radiated energy lies in the infrared region, with longer wavelengths.

Theory. The prediction of classical physics for the spectral radiancy, for a given temperature T in kelvins, is

$$S(\lambda) = \frac{2\pi c k T}{\lambda^4} \quad \text{(classical radiation law)}, \tag{38-13}$$

where k is the Boltzmann constant (Eq. 19-7) with the value

$$k = 1.38 \times 10^{-23}\,\text{J/K} = 8.62 \times 10^{-5}\,\text{eV/K}.$$

This classical result is plotted in Fig. 38-8 for $T = 2000$ K. Although the theoretical and experimental results agree well at long wavelengths (off the graph to the right), they are not even close in the short wavelength region. Indeed, the theoretical prediction does not even include a maximum as seen in the measured results and instead "blows up" up to infinity (which was quite disturbing, even embarrassing, to the physicists).

Planck's Solution. In 1900, Planck devised a formula for $S(\lambda)$ that neatly fitted the experimental results for all wavelengths and for all temperatures:

$$S(\lambda) = \frac{2\pi c^2 h}{\lambda^5}\frac{1}{e^{hc/\lambda kT} - 1} \quad \text{(Planck's radiation law)}. \tag{38-14}$$

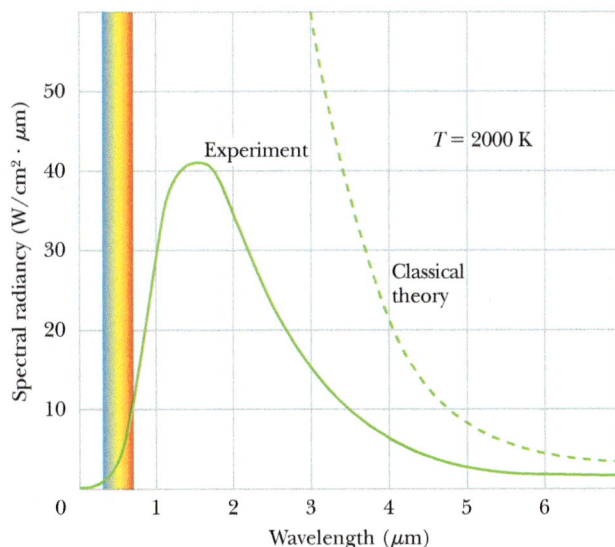

Figure 38-8 The solid curve shows the experimental spectral radiancy for a cavity at 2000 K. Note the failure of the classical theory, which is shown as a dashed curve. The range of visible wavelengths is indicated.

The key element in the equation lies in the argument of the exponential: hc/λ, which we can rewrite in a more suggestive form as hf. Equation 38-14 was the first use of the symbol h, and the appearance of hf suggests that the energies of the atomic oscillators in the cavity wall are quantized. However, Planck, with his training in classical physics, simply could not believe such a result in spite of the immediate success of his equation in fitting all experimental data.

Einstein's Solution. No one understood Eq. 38-14 for 17 years, but then Einstein explained it with a very simple model with two key ideas: (1) The energies of the cavity-wall atoms that are emitting the radiation are indeed quantized. (2) The energies of the radiation in the cavity are also quantized in the form of quanta (what we now call photons), each with energy $E = hf$. In his model he explained the processes by which atoms can emit and absorb photons and how the atoms can be in equilibrium with the emitted and absorbed light.

Maximum Value. The wavelength λ_{max} at which the $S(\lambda)$ is maximum (for a given temperature T) can be found by taking the first derivative of Eq. 38-14 with respect to λ, setting the derivative to zero, and then solving for the wavelength. The result is known as Wien's law:

$$\lambda_{\text{max}}T = 2898 \ \mu\text{m} \cdot \text{K} \qquad \text{(at maximum radiancy)}. \qquad (38\text{-}15)$$

For example, in Fig. 38-8 for which $T = 2000$ K, $\lambda_{\text{max}} = 1.5 \ \mu\text{m}$, which is greater than the long wavelength end of the visible spectrum and is in the infrared region, as shown. If we increase the temperature, λ_{max} decreases and the peak in Fig. 38-8 changes shape and shifts more into the visible range.

Radiated Power. If we integrate Eq. 38-14 over all wavelengths (for a given temperature), we find the power per unit area of a thermal radiator. If we then multiply by the total surface area A, we find the total radiated power P. We have already seen the result in Eq. 18-38 (with some changes in notation):

$$P = \sigma \varepsilon A T^4, \qquad (38\text{-}16)$$

where $\sigma \ (= 5.6704 \times 10^{-8} \ \text{W/m}^2 \cdot \text{K}^4)$ is the Stefan–Boltzmann constant and ε is the emissivity of the radiating surface ($\varepsilon = 1$ for an ideal blackbody radiator). Actually, integrating Eq. 38-14 over all wavelengths is difficult. However, for a given temperature T, wavelength λ, and wavelength range $\Delta\lambda$ that is small relative to λ, we can approximate the power in that range by simply evaluating $S(\lambda)A \ \Delta\lambda$.

38-5 ELECTRONS AND MATTER WAVES

Learning Objectives

After reading this module, you should be able to . . .

38.21 Identify that electrons (and protons and all other elementary particles) are matter waves.

38.22 For both relativistic and nonrelativistic particles, apply the relationships between the de Broglie wavelength, momentum, speed, and kinetic energy.

38.23 Describe the double-slit interference pattern obtained with particles such as electrons.

38.24 Apply the optical two-slit equations (Module 35-2) and diffraction equations (Module 36-1) to matter waves.

Key Ideas

● A moving particle such as an electron can be described as a matter wave.

● The wavelength associated with the matter wave is the particle's de Broglie wavelength $\lambda = h/p$, where p is the particle's momentum.

● Particle: When an electron interacts with matter, the interaction is particle-like, occurring at a point and transferring energy and momentum.

● Wave: When an electron is in transit, we interpret it as being a probability wave.

Electrons and Matter Waves

In 1924, French physicist Louis de Broglie made the following appeal to symmetry: A beam of light is a wave, but it transfers energy and momentum to matter only at points, via photons. Why can't a beam of particles have the same properties? That is, why can't we think of a moving electron—or any other particle—as a **matter wave** that transfers energy and momentum to other matter at points?

In particular, de Broglie suggested that Eq. 38-7 ($p = h/\lambda$) might apply not only to photons but also to electrons. We used that equation in Module 38-3 to assign a momentum p to a photon of light with wavelength λ. We now use it, in the form

$$\lambda = \frac{h}{p} \quad \text{(de Broglie wavelength)}, \tag{38-17}$$

to assign a wavelength λ to a particle with momentum of magnitude p. The wavelength calculated from Eq. 38-17 is called the **de Broglie wavelength** of the moving particle. De Broglie's prediction of the existence of matter waves was first verified experimentally in 1927, by C. J. Davisson and L. H. Germer of the Bell Telephone Laboratories and by George P. Thomson of the University of Aberdeen in Scotland.

Figure 38-9 shows photographic proof of matter waves in a more recent experiment. In the experiment, an interference pattern was built up when

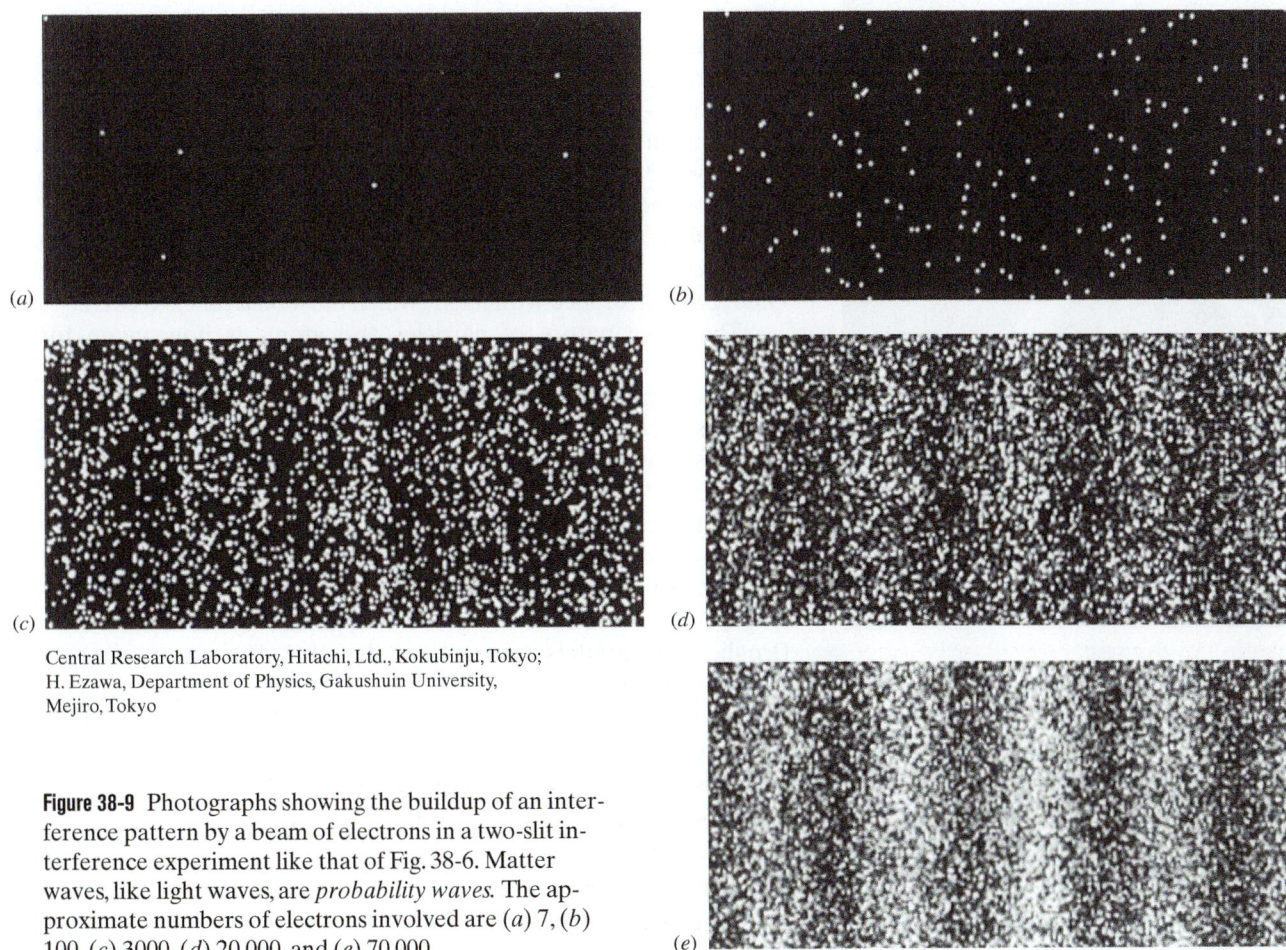

(a)

(b)

(c)

(d)

(e)

Central Research Laboratory, Hitachi, Ltd., Kokubinju, Tokyo;
H. Ezawa, Department of Physics, Gakushuin University,
Mejiro, Tokyo

Figure 38-9 Photographs showing the buildup of an interference pattern by a beam of electrons in a two-slit interference experiment like that of Fig. 38-6. Matter waves, like light waves, are *probability waves*. The approximate numbers of electrons involved are (a) 7, (b) 100, (c) 3000, (d) 20 000, and (e) 70 000.

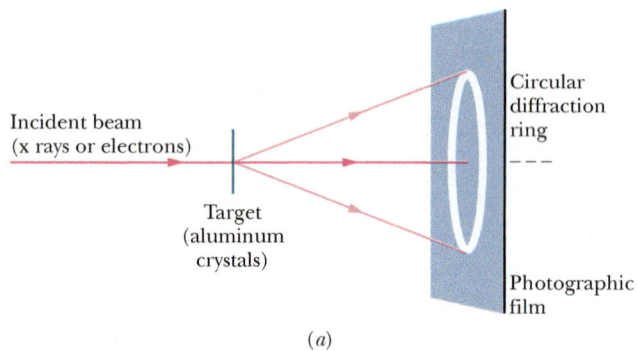

(a)

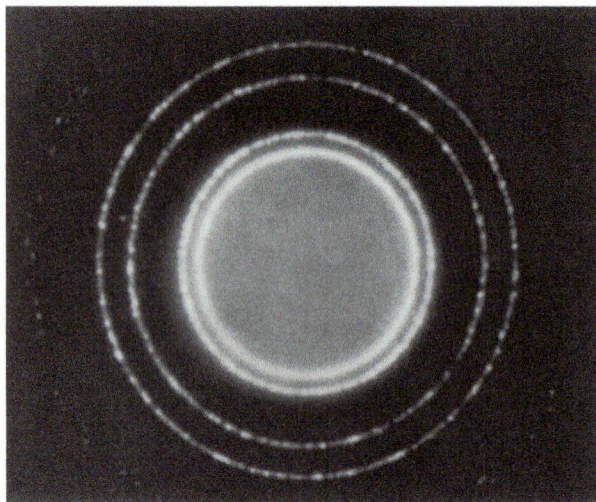

(b)

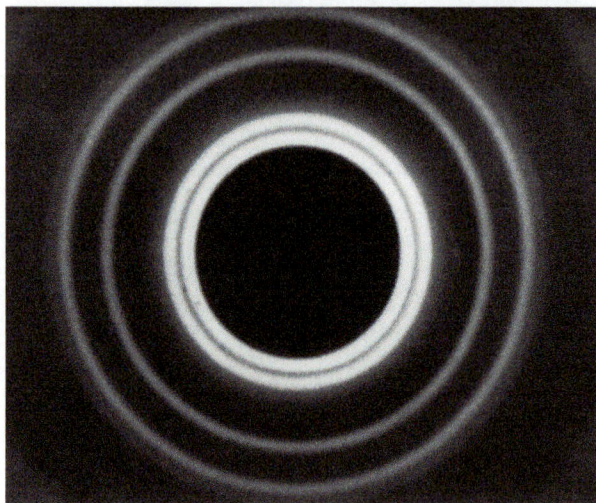

(c)

Parts (b) and (c) from PSSC film "Matter Waves," courtesy Education Development Center, Newton, Massachusetts

Figure 38-10 (a) An experimental arrangement used to demonstrate, by diffraction techniques, the wave-like character of the incident beam. Photographs of the diffraction patterns when the incident beam is (b) an x-ray beam (light wave) and (c) an electron beam (matter wave). Note that the two patterns are geometrically identical to each other.

electrons were sent, *one by one,* through a double-slit apparatus. The apparatus was like the ones we have previously used to demonstrate optical interference, except that the viewing screen was similar to an old-fashioned television screen. When an electron hit the screen, it caused a flash of light whose position was recorded.

The first several electrons (top two photos) revealed nothing interesting and seemingly hit the screen at random points. However, after many thousands of electrons were sent through the apparatus, a pattern appeared on the screen, revealing fringes where many electrons had hit the screen and fringes where few had hit the screen. The pattern is exactly what we would expect for wave interference. Thus, *each* electron passed through the apparatus as a matter wave—the portion of the matter wave that traveled through one slit interfered with the portion that traveled through the other slit. That interference then determined the probability that the electron would materialize at a given point on the screen, hitting the screen there. Many electrons materialized in regions corresponding to bright fringes in optical interference, and few electrons materialized in regions corresponding to dark fringes.

Similar interference has been demonstrated with protons, neutrons, and various atoms. In 1994, it was demonstrated with iodine molecules I_2, which are not only 500 000 times more massive than electrons but far more complex. In 1999, it was demonstrated with the even more complex *fullerenes* (or *buckyballs*) C_{60} and C_{70}. (Fullerenes are molecules of carbon atoms that are arranged in a structure resembling a soccer ball, 60 carbon atoms in C_{60} and 70 carbon atoms in C_{70}.) Apparently, such small objects as electrons, protons, atoms, and molecules travel as matter waves. However, as we consider larger and more complex objects, there must come a point at which we are no longer justified in considering the wave nature of an object. At that point, we are back in our familiar nonquantum world, with the physics of earlier chapters of this book. In short, an electron is a matter wave and can undergo interference with itself, but a cat is not a matter wave and cannot undergo interference with itself (which must be a relief to cats).

The wave nature of particles and atoms is now taken for granted in many scientific and engineering fields. For example, electron diffraction and neutron diffraction are used to study the atomic structures of solids and liquids, and electron diffraction is used to study the atomic features of surfaces on solids.

Figure 38-10a shows an arrangement that can be used to demonstrate the scattering of either x rays or electrons by crystals. A beam of one or the other is directed onto a target consisting of a layer of tiny aluminum crystals. The x rays have a certain wavelength λ. The electrons are given enough energy so that their de Broglie wavelength is the same wavelength λ. The scatter of x rays or electrons by the crystals produces a circular interference pattern on a photographic film. Figure 38-10b shows the pattern for the scatter of x rays, and Fig. 38-10c shows the pattern for the scatter of electrons. The patterns are the same—both x rays and electrons are waves.

Waves and Particles

Figures 38-9 and 38-10 are convincing evidence of the *wave* nature of matter, but we have countless experiments that suggest its *parti-*

cle nature. Figure 38-11, for example, shows the tracks of particles (rather than waves) revealed in a bubble chamber. When a charged particle passes through the liquid hydrogen that fills such a chamber, the particle causes the liquid to vaporize along the particle's path. A series of bubbles thus marks the path, which is usually curved due to a magnetic field set up perpendicular to the plane of the chamber.

In Fig. 38-11, a gamma ray left no track when it entered at the top because the ray is electrically neutral and thus caused no vapor bubbles as it passed through the liquid hydrogen. However, it collided with one of the hydrogen atoms, kicking an electron out of that atom; the curved path taken by the electron to the bottom of the photograph has been color coded green. Simultaneous with the collision, the gamma ray transformed into an electron and a positron in a pair production event (see Eq. 21-15). Those two particles then moved in tight spirals (color coded green for the electron and red for the positron) as they gradually lost energy in repeated collisions with hydrogen atoms. Surely these tracks are evidence of the particle nature of the electron and positron, but is there any evidence of waves in Fig. 38-11?

To simplify the situation, let us turn off the magnetic field so that the strings of bubbles will be straight. We can view each bubble as a detection point for the electron. Matter waves traveling between detection points such as *I* and *F* in Fig. 38-12 will explore all possible paths, a few of which are shown.

In general, for every path connecting *I* and *F* (except the straight-line path), there will be a neighboring path such that matter waves following the two paths cancel each other by interference. For the straight-line path joining *I* and *F*, matter waves traversing all neighboring paths reinforce the wave following the direct path. You can think of the bubbles that form the track as a series of detection points at which the matter wave undergoes constructive interference.

Lawrence Berkeley Laboratory/Science Photo Library/ Photo Researchers, Inc.

Figure 38-11 A bubble-chamber image showing where two electrons (paths color coded green) and one positron (red) moved after a gamma ray entered the chamber.

Figure 38-12 A few of the many paths that connect two particle detection points *I* and *F*. Only matter waves that follow paths close to the straight line between these points interfere constructively. For all other paths, the waves following any pair of neighboring paths interfere destructively.

Checkpoint 4

For an electron and a proton that have the same (a) kinetic energy, (b) momentum, or (c) speed, which particle has the shorter de Broglie wavelength?

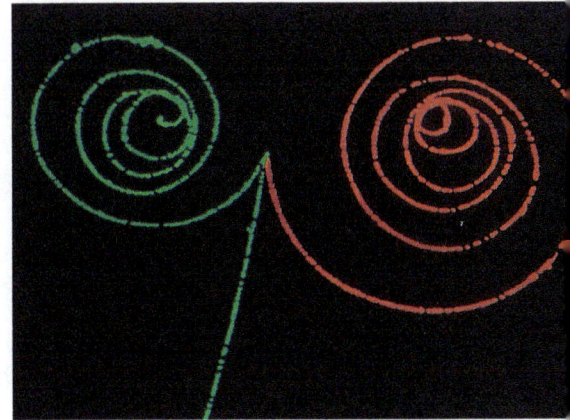

Sample Problem 38.04 de Broglie wavelength of an electron

What is the de Broglie wavelength of an electron with a kinetic energy of 120 eV?

KEY IDEAS

(1) We can find the electron's de Broglie wavelength λ from Eq. 38-17 ($\lambda = h/p$) if we first find the magnitude of its momentum p. (2) We find p from the given kinetic energy K of the electron. That kinetic energy is much less than the rest energy of an electron (0.511 MeV, from Table 37-3). Thus, we can get by with the classical approximations for momentum p ($= mv$) and kinetic energy K ($= \frac{1}{2}mv^2$).

Calculations: We are given the value of the kinetic energy. So, in order to use the de Broglie relation, we first solve the kinetic energy equation for v and then substitute into the momentum equation, finding

$$p = \sqrt{2mK}$$
$$= \sqrt{(2)(9.11 \times 10^{-31}\text{ kg})(120\text{ eV})(1.60 \times 10^{-19}\text{ J/eV})}$$
$$= 5.91 \times 10^{-24}\text{ kg·m/s.}$$

From Eq. 38-17 then

$$\lambda = \frac{h}{p}$$
$$= \frac{6.63 \times 10^{-34}\text{ J·s}}{5.91 \times 10^{-24}\text{ kg·m/s}}$$
$$= 1.12 \times 10^{-10}\text{ m} = 112\text{ pm.} \quad \text{(Answer)}$$

This wavelength associated with the electron is about the size of a typical atom. If we increase the electron's kinetic energy, the wavelength becomes even smaller.

PLUS Additional examples, video, and practice available at *WileyPLUS*

38-6 SCHRÖDINGER'S EQUATION

Learning Objectives

After reading this module, you should be able to . . .

38.25 Identify that matter waves are described by Schrödinger's equation.

38.26 For a nonrelativistic particle moving along an x axis, write the Schrödinger equation and its general solution for the spatial part of the wave function.

38.27 For a nonrelativistic particle, apply the relationships between angular wave number, energy, potential energy, kinetic energy, momentum, and de Broglie wavelength.

38.28 Given the spatial solution to the Schrödinger equation, write the full solution by including the time dependence.

38.29 Given a complex number, find the complex conjugate.

38.30 Given a wave function, calculate the probability density.

Key Ideas

● A matter wave (such as for an electron) is described by a wave function $\Psi(x, y, z, t)$, which can be separated into a space-dependent part $\psi(x, y, z)$ and a time-dependent part $e^{-i\omega t}$, where ω is the angular frequency associated with the wave.

● For a nonrelativistic particle of mass m traveling along an x axis, with energy E and potential energy U, the space-dependent part can be found by solving Schrödinger's equation,

$$\frac{d^2\psi}{dx^2} + k^2\psi = 0,$$

where k is the angular wave number, which is related to the de

Broglie wavelength λ, the momentum p, and the kinetic energy $E - U$ by

$$k = \frac{2\pi}{\lambda} = \frac{2\pi p}{h} = \frac{2\pi\sqrt{2m(E - U)}}{h}.$$

● A particle does not have a specific location until its location is actually measured.

● The probability of detecting a particle in a small volume centered on a given point is proportional to the probability density $|\psi|^2$ of the matter wave at that point.

Schrödinger's Equation

A simple traveling wave of any kind, be it a wave on a string, a sound wave, or a light wave, is described in terms of some quantity that varies in a wave-like fashion. For light waves, for example, this quantity is $\vec{E}(x, y, z, t)$, the electric field component of the wave. Its observed value at any point depends on the location of that point and on the time at which the observation is made.

What varying quantity should we use to describe a matter wave? We should expect this quantity, which we call the **wave function** $\Psi(x, y, z, t)$, to be more complicated than the corresponding quantity for a light wave because a matter wave, in addition to energy and momentum, transports mass and (often) electric charge. It turns out that Ψ, the uppercase Greek letter psi, usually represents a function that is complex in the mathematical sense; that is, we can always write its values in the form $a + ib$, in which a and b are real numbers and $i^2 = -1$.

In all the situations you will meet here, the space and time variables can be grouped separately and Ψ can be written in the form

$$\Psi(x, y, z, t) = \psi(x, y, z)\, e^{-i\omega t}, \tag{38-18}$$

where $\omega \, (= 2\pi f)$ is the angular frequency of the matter wave. Note that ψ, the lowercase Greek letter psi, represents only the space-dependent part of the complete, time-dependent wave function Ψ. We shall focus on ψ. Two questions arise: What is meant by the wave function? How do we find it?

What does the wave function mean? It has to do with the fact that a matter wave, like a light wave, is a probability wave. Suppose that a matter wave reaches a particle detector that is small; then the probability that a particle will be detected in a specified time interval is proportional to $|\psi|^2$, where $|\psi|$ is the absolute value of the wave function at the location of the detector. Although ψ

is usually a complex quantity, $|\psi|^2$ is always both real and positive. It is, then, $|\psi|^2$, which we call the **probability density,** and not ψ, that has *physical* meaning. Speaking loosely, the meaning is this:

> The probability of detecting a particle in a small volume centered on a given point in a matter wave is proportional to the value of $|\psi|^2$ at that point.

Because ψ is usually a complex quantity, we find the square of its absolute value by multiplying ψ by ψ^*, the *complex conjugate* of ψ. (To find ψ^* we replace the imaginary number i in ψ with $-i$, wherever it occurs.)

How do we find the wave function? Sound waves and waves on strings are described by the equations of Newtonian mechanics. Light waves are described by Maxwell's equations. Matter waves for nonrelativistic particles are described by **Schrödinger's equation,** advanced in 1926 by Austrian physicist Erwin Schrödinger.

Many of the situations that we shall discuss involve a particle traveling in the x direction through a region in which forces acting on the particle cause it to have a potential energy $U(x)$. In this special case, Schrödinger's equation reduces to

$$\frac{d^2\psi}{dx^2} + \frac{8\pi^2 m}{h^2}[E - U(x)]\psi = 0 \quad \begin{array}{l}\text{(Schrödinger's equation,}\\ \text{one-dimensional motion),}\end{array} \qquad (38\text{-}19)$$

in which E is the total mechanical energy of the moving particle. (We do *not* consider mass energy in this nonrelativistic equation.) We cannot derive Schrödinger's equation from more basic principles; it *is* the basic principle.

We can simplify the expression of Schrödinger's equation by rewriting the second term. First, note that $E - U(x)$ is the kinetic energy of the particle. Let's assume that the potential energy is uniform and constant (it might even be zero). Because the particle is nonrelativistic, we can write the kinetic energy classically in terms of speed v and then momentum p, and then we can introduce quantum theory by using the de Broglie wavelength:

$$E - U = \frac{1}{2}mv^2 = \frac{p^2}{2m} = \frac{1}{2m}\left(\frac{h}{\lambda}\right)^2. \qquad (38\text{-}20)$$

By putting 2π in both the numerator and denominator of the squared term, we can rewrite the kinetic energy in terms of the angular wave number $k = 2\pi/\lambda$:

$$E - U = \frac{1}{2m}\left(\frac{kh}{2\pi}\right)^2. \qquad (38\text{-}21)$$

Substituting this into Eq. 38-19 leads to

$$\frac{d^2\psi}{dx^2} + k^2\psi = 0 \quad \text{(Schrödinger's equation, uniform } U\text{)}, \qquad (38\text{-}22)$$

where, from Eq. 38-21, the angular wave number is

$$k = \frac{2\pi\sqrt{2m(E - U)}}{h} \quad \text{(angular wave number)}. \qquad (38\text{-}23)$$

The general solution of Eq. 38-22 is

$$\psi(x) = Ae^{ikx} + Be^{-ikx}, \qquad (38\text{-}24)$$

in which A and B are constants. You can show that this equation is indeed a solution of Eq. 38-22 by substituting it and its second derivative into that equation and noting that an identity results.

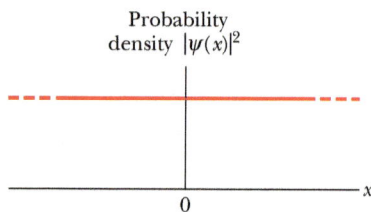

Probability
density $|\psi(x)|^2$

Figure 38-13 A plot of the probability density $|\psi|^2$ for a particle moving in the positive x direction with a uniform potential energy. Since $|\psi|^2$ has the same constant value for all values of x, the particle has the same probability of detection at all points along its path.

Equation 38-24 is the time-independent solution of Schrödinger's equation. We can assume it is the spatial part of the wave function at some initial time $t = 0$. Given values for E and U, we could determine the coefficients A and B to see how the wave function looks at $t = 0$. Then, if we wanted to see how the wave function evolves with time, we follow the guide of Eq. 38-18 and multiply Eq. 38-24 by the time dependence $e^{-i\omega t}$:

$$\Psi(x,t) = \psi(x)e^{-i\omega t} = (Ae^{ikx} + Be^{-ikx})e^{-i\omega t}$$
$$= Ae^{i(kx-\omega t)} + Be^{-i(kx+\omega t)}. \tag{38-25}$$

Here, however, we will not go that far.

Finding the Probability Density $|\psi|^2$

In Module 16-1 we saw that any function F of the form $F(kx \pm \omega t)$ represents a traveling wave. In Chapter 16, the functions were sinusoidal (sines and cosines); here they are exponentials. If we wanted, we could always switch between the two forms by using the Euler formula: For a general argument θ,

$$e^{i\theta} = \cos\theta + i\sin\theta \quad \text{and} \quad e^{-i\theta} = \cos\theta - i\sin\theta. \tag{38-26}$$

The first term on the right in Eq. 38-25 represents a wave traveling in the positive direction of x, and the second term represents a wave traveling in the negative direction of x. Let's evaluate the probability density $|\psi^2|$ for a particle with only positive motion. We eliminate the negative motion by setting B to zero, and then the solution at $t = 0$ becomes

$$\psi(x) = Ae^{ikx}. \tag{38-27}$$

To calculate the probability density, we take the square of the absolute value:

$$|\psi|^2 = |Ae^{ikx}|^2 = A^2|e^{ikx}|^2.$$

Because

$$|e^{ikx}|^2 = (e^{ikx})(e^{ikx})^* = e^{ikx}e^{-ikx} = e^{ikx-ikx} = e^0 = 1,$$

we get

$$|\psi|^2 = A^2(1)^2 = A^2.$$

Now here is the point: For the condition we have set up (uniform potential energy U, including $U = 0$ for a *free particle*), the probability density is a constant (the same value A^2) for any point along the x axis, as shown in the plot of Fig. 38-13. That means that if we make a measurement to locate the particle, the location could turn out to be at any x value. Thus, we cannot say that the particle is moving along the axis in a classical way as a car moves along a street. *In fact, the particle does not have a location until we measure it.*

38-7 HEISENBERG'S UNCERTAINTY PRINCIPLE

Learning Objective

After reading this module, you should be able to . . .

38.31 Apply the Heisenberg uncertainty principle for, say, an electron moving along the x axis and explain its meaning.

Key Idea

● The probabilistic nature of quantum physics places an important limitation on detecting a particle's position and momentum. That is, it is not possible to measure the position $\vec{r}$ and the momentum $\vec{p}$ of a particle simultaneously with unlimited precision. The uncertainties in the components of

these quantities are given by

$$\Delta x \cdot \Delta p_x \geq \hbar$$
$$\Delta y \cdot \Delta p_y \geq \hbar$$
$$\Delta z \cdot \Delta p_z \geq \hbar.$$

Heisenberg's Uncertainty Principle

Our inability to predict the position of a particle with a uniform electric potential energy, as indicated by Fig. 38-13, is our first example of **Heisenberg's uncertainty principle,** proposed in 1927 by German physicist Werner Heisenberg. It states that measured values cannot be assigned to the position $\vec{r}$ and the momentum $\vec{p}$ of a particle simultaneously with unlimited precision.

In terms of $\hbar = h/2\pi$ (called "h-bar"), the principle tells us

$$\Delta x \cdot \Delta p_x \geq \hbar$$
$$\Delta y \cdot \Delta p_y \geq \hbar \quad \text{(Heisenberg's uncertainty principle).} \qquad (38\text{-}28)$$
$$\Delta z \cdot \Delta p_z \geq \hbar$$

Here Δx and Δp_x represent the intrinsic uncertainties in the measurements of the x components of $\vec{r}$ and $\vec{p}$, with parallel meanings for the y and z terms. Even with the best measuring instruments, each product of a position uncertainty and a momentum uncertainty in Eq. 38-28 will be greater than $\hbar$, *never* less.

Here we shall not derive the uncertainty relationships but only apply them. They are due to the fact that electrons and other particles are matter waves and that repeated measurements of their positions and momenta involve probabilities, not certainties. In the statistics of such measurements, we can view, say, Δx and Δp_x as the spread (actually, the standard deviations) in the measurements.

We can also justify them with a physical (though highly simplified) argument: In earlier chapters we took for granted our ability to detect and measure location and motion, such as a car moving down a street or a pool ball rolling across a table. We could locate a moving object by watching it—that is, by intercepting light scattered by the object. That scattering did not alter the object's motion. In quantum physics, however, the act of detection in itself alters the location and motion. The more precisely we wish to determine the location of, say, an electron moving along an x axis (by using light or by any other means), the more we alter the electron's momentum and thus become less certain of the momentum. That is, by decreasing Δx, we necessarily increase Δp_x. Vice versa, if we determine the momentum very precisely (less Δp_x), we become less certain of where the electron will be located (we increase Δx).

That latter situation is what we found in Fig 38-13. We had an electron with a certain value of k, which, by the de Broglie relationship, means a certain momentum p_x. Thus, $\Delta p_x = 0$. By Eq. 38-28, that means that $\Delta x = \infty$. If we then set up an experiment to detect the electron, it could show up anywhere between $x = -\infty$ and $x = +\infty$.

You might push back on the argument: Couldn't we very precisely measure p_x and then next very precisely measure x wherever the electron happens to show up? Doesn't that mean that we have measured both p_x and x simultaneously and very precisely? No, the flaw is that although the first measurement can give us a precise value for p_x, the second measurement necessarily alters that value. Indeed, if the second measurement really does give us a precise value for x, we then have no idea what the value of p_x is.

Sample Problem 38.05   Uncertainty principle: position and momentum

Assume that an electron is moving along an x axis and that you measure its speed to be 2.05×10^6 m/s, which can be known with a precision of 0.50%. What is the minimum uncertainty (as allowed by the uncertainty principle in quantum theory) with which you can simultaneously measure the position of the electron along the x axis?

KEY IDEA

The minimum uncertainty allowed by quantum theory is given by Heisenberg's uncertainty principle in Eq. 38-28. We need only consider components along the x axis because we have motion only along that axis and want the uncertainty Δx in location along that axis. Since we want the minimum allowed uncertainty, we use the equality instead of the inequality in the x-axis part of Eq. 38-28, writing $\Delta x \cdot \Delta p_x = \hbar$.

Calculations: To evaluate the uncertainty Δp_x in the momentum, we must first evaluate the momentum component p_x. Because the electron's speed v_x is much less than the speed of light c, we can evaluate p_x with the classical expression for momentum instead of using a relativistic expression. We find

$$p_x = mv_x = (9.11 \times 10^{-31}\ \text{kg})(2.05 \times 10^6\ \text{m/s})$$
$$= 1.87 \times 10^{-24}\ \text{kg} \cdot \text{m/s}.$$

The uncertainty in the speed is given as 0.50% of the measured speed. Because p_x depends directly on speed,

the uncertainty Δp_x in the momentum must be 0.50% of the momentum:

$$\Delta p_x = (0.0050)p_x$$
$$= (0.0050)(1.87 \times 10^{-24}\ \text{kg} \cdot \text{m/s})$$
$$= 9.35 \times 10^{-27}\ \text{kg} \cdot \text{m/s}.$$

Then the uncertainty principle gives us

$$\Delta x = \frac{\hbar}{\Delta p_x} = \frac{(6.63 \times 10^{-34}\ \text{J} \cdot \text{s})/2\pi}{9.35 \times 10^{-27}\ \text{kg} \cdot \text{m/s}}$$
$$= 1.13 \times 10^{-8}\ \text{m} \approx 11\ \text{nm}, \qquad \text{(Answer)}$$

which is about 100 atomic diameters.

WILEY PLUS Additional examples, video, and practice available at *WileyPLUS*

38-8 REFLECTION FROM A POTENTIAL STEP

Learning Objectives

After reading this module, you should be able to . . .

38.32 Write the general wave function for Schrödinger's equation for an electron in a region of constant (including zero) potential energy.

38.33 With a sketch, identify a potential step for an electron, indicating the barrier height U_b.

38.34 For electron wave functions in two adjacent regions, determine the coefficients (probability amplitudes) by matching values and slopes at the boundary.

38.35 Determine the reflection and transmission coefficients for electrons incident on a potential step (or potential

energy step), where the incident electrons each have zero potential energy $U = 0$ and a mechanical energy E greater than the step height U_b.

38.36 Identify that because electrons are matter waves, they might reflect from a potential step even when they have more than enough energy to pass through the step.

38.37 Interpret the reflection and transmission coefficients in terms of the probability of an electron reflecting or passing through the boundary and also in terms of the average number of electrons out of the total number shot at the barrier.

Key Ideas

● A particle can reflect from a boundary at which its potential energy changes even when classically it would not reflect.

● The reflection coefficient R gives the probability of reflection of an individual particle at the boundary.

● For a beam of a great many particles, R gives the average fraction that will undergo reflection.

● The transmission coefficient T that gives the probability of transmission through the boundary is

$$T = 1 - R.$$

Can the electron be reflected by the region of negative potential?

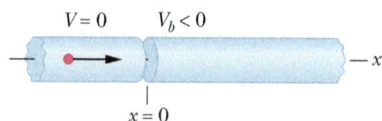

Figure 38-14 The elements of a tube in which an electron (the dot) approaches a region with a negative electric potential V_b.

Reflection from a Potential Step

Here is a quick taste of what you would see in more advanced quantum physics. In Fig. 38-14, we send a beam of a great many nonrelativistic electrons, each of total energy E, along an x axis through a narrow tube. Initially they are in region 1 where their potential energy is $U = 0$, but at $x = 0$ they encounter a region with a negative electric potential V_b. The transition is called a *potential step* or *potential energy step*. The step is said to have a *height* U_b, which is the potential energy an electron will have once it passes through the boundary at $x = 0$, as plotted in

Fig. 38-15 for potential energy as a function of position x. (Recall that $U = qV$. Here the potential V_b is negative, the electron's charge q is negative, and so the potential energy U_b is positive.)

Let's consider the situation where $E > U_b$. Classically, the electrons should all pass through the boundary—they certainly have enough energy. Indeed, we discussed such motion extensively in Chapters 22 through 24, where electrons moved into electric potentials and had changes in potential energy and kinetic energy. We simply conserved mechanical energy and noted that if the potential energy increases, the kinetic energy decreases by the same amount, and the speed thus also decreases. What we took for granted is that, because the electron energy E is greater than the potential energy U_b, all the electrons pass through the boundary. However, if we apply Schrödinger's equation, we find a big surprise—because electrons are matter waves, not tiny solid (classical) particles, some of them actually *reflect from the boundary*. Let's determine what fraction R of the incoming electrons reflect.

In region 1, where U is zero, Eq. 38-23 tells us that the angular wave number is

$$k = \frac{2\pi\sqrt{2mE}}{h} \tag{38-29}$$

and Eq. 38-24 tells us that the general space-dependent solution to Schrödinger's equation is

$$\psi_1(x) = Ae^{ikx} + Be^{-ikx} \quad \text{(region 1)}. \tag{38-30}$$

In region 2, where the potential energy is U_b, the angular wave number is

$$k_b = \frac{2\pi\sqrt{2m(E - U_b)}}{h}, \tag{38-31}$$

and the general solution, with this angular wave number, is

$$\psi_2(x) = Ce^{ik_bx} + De^{-ik_bx} \quad \text{(region 2)}. \tag{38-32}$$

We use coefficients C and D because they are not the same as the coefficients in region 1.

The terms with positive arguments in an exponential represent particles moving in the $+x$ direction; those with negative arguments represent particles moving in the $-x$ direction. However, because there is no electron source off to the right in Figs. 38-14 and 38-15, there can be no electrons moving to the left in region 2. So, we set $D = 0$, and the solution in region 2 is then simply

$$\psi_2(x) = Ce^{ik_bx} \quad \text{(region 2)}. \tag{38-33}$$

Next, we must make sure that our solutions are "well behaved" at the boundary. That is, they must be consistent with each other at $x = 0$, both in value and in slope. These conditions are said to be **boundary conditions**. We first substitute $x = 0$ into Eqs. 38-30 and 38-33 for the wave functions and then set the results equal to each other. This gives us our first boundary condition:

$$A + B = C \quad \text{(matching of values)}. \tag{38-34}$$

The functions have the same value at $x = 0$ provided the coefficients have this relationship.

Next, we take a derivative of Eq. 38-30 with respect to x and then substitute in $x = 0$. Then we take a derivative of Eq. 38-33 with respect to x and then substitute in $x = 0$. And then we set the two results equal to each other (one slope equal to the other slope at $x = 0$). We find

$$Ak - Bk = Ck_b \quad \text{(matching of slopes)}. \tag{38-35}$$

The slopes at $x = 0$ are equal provided that this relationship of coefficients and angular wave numbers is satisfied.

Classically, the electron has too much energy to be reflected by the potential step.

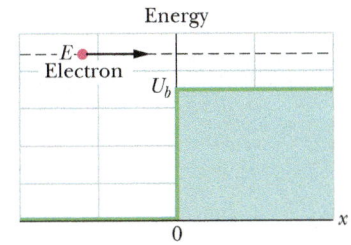

Figure 38-15 An energy diagram containing two plots for the situation of Fig. 38-14: (1) The electron's mechanical energy E is plotted. (2) The electron's electric potential energy U is plotted as a function of the electron's position x. The nonzero part of the plot (the potential step) has height U_b.

We want to find the probability that electrons reflect from the barrier. Recall that probability density is proportional to $|\psi|^2$. Here let's relate the probability density in the reflection (which is proportional to $|B|^2$) to the probability density in the incident beam (which is proportional to $|A|^2$) by defining a **reflection coefficient** R:

$$R = \frac{|B|^2}{|A|^2}. \tag{38-36}$$

This R gives the probability of reflection and thus is also the fraction of the incoming electrons that reflect. The **transmission coefficient** (the probability of transmission) is

$$T = 1 - R. \tag{38-37}$$

For example, suppose $R = 0.010$. Then if we send 10,000 electrons toward the barrier, we find that about 100 are reflected. However, we could never guess which 100 would be reflected. We have only the probability. The best we can say about any one electron is that it has a 1.0% chance of being reflected and a 99% chance of being transmitted. The wave nature of the electron does not allow us to be any more precise than that.

To evaluate R for any given values of E and U_b, we first solve Eqs. 38-34 and 38-35 for B in terms of A by eliminating C and then substitute the result into Eq. 38-36. Finally, using Eqs. 38-29 and 38-31, we substitute values for k and k_b. The surprise is that R is not simply zero (and T is not simply 1) as we assumed classically in earlier chapters.

38-9 TUNNELING THROUGH A POTENTIAL BARRIER

Learning Objectives

After reading this module, you should be able to . . .

38.38 With a sketch, identify a potential barrier for an electron, indicating the barrier height U_b and thickness L.
38.39 Identify the energy argument about what is classically required of a particle's energy if the particle is to pass through a potential barrier.
38.40 Identify the transmission coefficient for tunneling.
38.41 For tunneling, calculate the transmission coefficient T in terms of the particle's energy E and mass m and the barrier's height U_b and thickness L.

38.42 Interpret a transmission coefficient in terms of the probability of any one particle tunneling through a barrier and also in terms of the average fraction of many particles tunneling through the barrier.
38.43 In a tunneling setup, describe the probability density in front of the barrier, within the barrier, and then beyond the barrier.
38.44 Describe how a scanning tunneling microscope works.

Key Ideas

● A potential energy barrier is a region where a traveling particle will have an increased potential energy U_b.

● The particle can pass through the barrier if its total energy $E > U_b$.

● Classically, it cannot pass through it if $E < U_b$, but in quantum physics it can, an effect called tunneling.

● For a particle with mass m and a barrier of thickness L, the transmission coefficient is

$$T \approx e^{-2bL},$$

where

$$b = \sqrt{\frac{8\pi^2 m(U_b - E)}{h^2}}.$$

Tunneling Through a Potential Barrier

Let's replace the potential step of Fig. 38-14 with a **potential barrier** (or **potential energy barrier**), which is a region of thickness L (the *barrier thickness* or *length*) where the electric potential is V_b (< 0) and the barrier height is U_b ($= qV$), as shown in Fig. 38-16. To the right of the barrier is region 3 with $V = 0$. As before, we'll send a beam of nonrelativistic electrons toward the barrier, each with energy E. If we again consider $E > U_b$, we have a more complicated situation than our previous potential step because now electrons can possibly reflect from two boundaries, at $x = 0$ and $x = L$.

Instead of sorting that out, let's consider the situation where $E < U_b$—that is, where the mechanical energy is less than the potential energy that would be demanded of an electron in region 2. Such a demand would require that the electron's kinetic energy ($= E - U_b$) be negative in region 2, which is, of course, simply absurd because kinetic energies must always be positive (nothing in the expression $\frac{1}{2}mv^2$ can be negative). Therefore, region 2 is *classically* forbidden to an electron with $E < U_b$.

Tunneling. However, because an electron is a matter wave, it actually has a finite probability of leaking (or, better, *tunneling*) through the barrier and materializing on the other side. Once past the barrier, it again has its full mechanical energy E as though nothing (strange or otherwise) has happened in the region $0 \le x \le L$. Figure 38-17 shows the potential barrier and an approaching electron, with an energy less than the barrier height. We are interested in the probability of the electron appearing on the other side of the barrier. Thus, we want the transmission coefficient T.

To find an expression for T we would in principle follow the procedure for finding R for a potential step. We would solve Schrödinger's equation for the general solutions in each of three regions in Fig. 38-16. We would discard the region-3 solution for a wave traveling in the $-x$ direction (there is no electron source off to the right). Then we would determine the coefficients in terms of the coefficient A of the incident electrons by applying the boundary conditions—that is, by matching the values and slopes of the wave functions at the two boundaries. Finally, we would determine the relative probability density in region 3 in terms of the incident probability density. However, because all this requires a lot of mathematical manipulation, here we shall just examine the general results.

Figure 38-18 shows a plot of the probability densities in the three regions. The oscillating curve to the left of the barrier (for $x < 0$) is a combination of the incident matter wave and the reflected matter wave (which has a smaller amplitude than the incident wave). The oscillations occur because these two waves, traveling in opposite directions, interfere with each other, setting up a standing wave pattern.

Within the barrier (for $0 < x < L$) the probability density decreases exponentially with x. However, if L is small, the probability density is not quite zero at $x = L$.

To the right of the barrier (for $x > L$), the probability density plot describes a transmitted (through the barrier) wave with low but constant amplitude. Thus, the electron can be detected in this region but with a relatively small probability. (Compare this part of the figure with Fig. 38-13.)

As we did with a step potential, we can assign a transmission coefficient T to the incident matter wave and the barrier. This coefficient gives the probability with which an approaching electron will be transmitted through the barrier— that is, that tunneling will occur. As an example, if $T = 0.020$, then of every 1000 electrons fired at the barrier, 20 (on average) will tunnel through it and 980 will be reflected. The transmission coefficient T is approximately

$$T \approx e^{-2bL}, \tag{38-38}$$

in which

$$b = \sqrt{\frac{8\pi^2 m(U_b - E)}{h^2}}, \tag{38-39}$$

and e is the exponential function. Because of the exponential form of Eq. 38-38, the value of T is very sensitive to the three variables on which it depends: particle mass m, barrier thickness L, and energy difference $U_b - E$. (Because we do not include relativistic effects here, E does not include mass energy.)

Can the electron pass through the region of negative potential?

Figure 38-16 The elements of a narrow tube in which an electron (the dot) approaches a negative electric potential V_b in the region $x = 0$ to $x = L$.

Classically, the electron lacks the energy to pass through the barrier region.

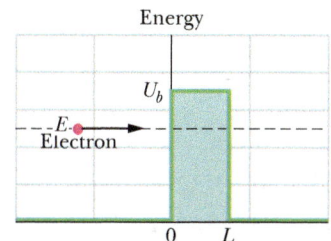

Figure 38-17 An energy diagram containing two plots for the situation of Fig. 38-16: (1) The electron's mechanical energy E is plotted when the electron is at any coordinate $x < 0$. (2) The electron's electric potential energy U is plotted as a function of the electron's position x, *assuming* that the electron can reach any value of x. The nonzero part of the plot (the potential barrier) has height U_b and thickness L.

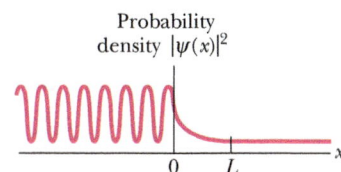

Figure 38-18 A plot of the probability density $|\psi|^2$ of the electron matter wave for the situation of Fig. 38-17. The value of $|\psi|^2$ is nonzero to the right of the potential barrier.

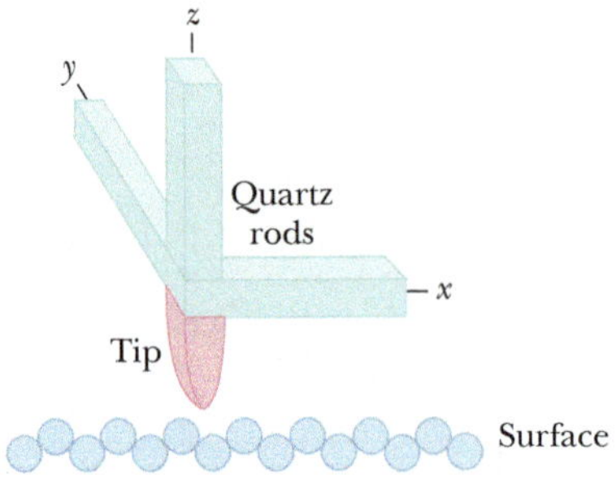

Figure 38-19 The essence of a scanning tunneling microscope (STM). Three quartz rods are used to scan a sharply pointed conducting tip across the surface of interest and to maintain a constant separation between tip and surface. The tip thus moves up and down to match the contours of the surface, and a record of its movement provides information for a computer to create an image of the surface.

Barrier tunneling finds many applications in technology, including the tunnel diode, in which a flow of electrons produced by tunneling can be rapidly turned on or off by electronically controlling the barrier height. The 1973 Nobel Prize in physics was shared by three "tunnelers," Leo Esaki (for tunneling in semiconductors), Ivar Giaever (for tunneling in superconductors), and Brian Josephson (for the Josephson junction, a rapid quantum switching device based on tunneling). The 1986 Nobel Prize was awarded to Gerd Binnig and Heinrich Rohrer for development of the scanning tunneling microscope.

✓ Checkpoint 5

Is the wavelength of the transmitted wave in Fig. 38-18 larger than, smaller than, or the same as that of the incident wave?

The Scanning Tunneling Microscope (STM)

The size of details that can be seen in an optical microscope is limited by the wavelength of the light the microscope uses (about 300 nm for ultraviolet light). The size of details that are required for images on the atomic scale is far smaller and thus requires much smaller wavelengths. The waves used are electron matter waves, but they do not scatter from the surface being examined the way waves do in an optical microscope. Instead, the images we see are created by electrons tunneling through potential barriers at the tip of a *scanning tunneling microscope* (STM).

Figure 38-19 shows the heart of the scanning tunneling microscope. A fine metallic tip, mounted at the intersection of three mutually perpendicular quartz rods, is placed close to the surface to be examined. A small potential difference, perhaps only 10 mV, is applied between tip and surface.

Crystalline quartz has an interesting property called *piezoelectricity:* When an electric potential difference is applied across a sample of crystalline quartz, the dimensions of the sample change slightly. This property is used to change the length of each of the three rods in Fig. 38-19, smoothly and by tiny amounts, so that the tip can be scanned back and forth over the surface (in the x and y directions) and also lowered or raised with respect to the surface (in the z direction).

The space between the surface and the tip forms a potential energy barrier, much like that plotted in Fig. 38-17. If the tip is close enough to the surface, electrons from the sample can tunnel through this barrier from the surface to the tip, forming a tunneling current.

In operation, an electronic feedback arrangement adjusts the vertical position of the tip to keep the tunneling current constant as the tip is scanned over the surface. This means that the tip–surface separation also remains constant during the scan. The output of the device is a video display of the varying vertical position of the tip, hence of the surface contour, as a function of the tip position in the xy plane.

An STM not only can provide an image of a static surface, it can also be used to manipulate atoms and molecules on a surface, such as was done in forming the *quantum corral* shown in Fig. 39-12 in the next chapter. In a process known as lateral manipulation, the STM probe is initially brought down near a molecule, close enough that the molecule is attracted to the probe without actually touching it. The probe is then moved across the background surface (such as copper), dragging the molecule with it until the molecule is in the desired location. Then the probe is backed up away from the molecule, weakening and then eliminating the attractive force on the molecule. Although the work requires very fine control, a design can eventually be formed. In Fig. 39-12, an STM probe has been used to move 48 iron atoms across a copper surface and into a circular corral 14 nm in diameter, in which electrons can be trapped.

Sample Problem 38.06 Barrier tunneling by matter wave

Suppose that the electron in Fig. 38-17, having a total energy E of 5.1 eV, approaches a barrier of height $U_b = 6.8$ eV and thickness $L = 750$ pm.

(a) What is the approximate probability that the electron will be transmitted through the barrier, to appear (and be detectable) on the other side of the barrier?

KEY IDEA

The probability we seek is the transmission coefficient T as given by Eq. 38-38 ($T \approx e^{-2bL}$), where

$$b = \sqrt{\frac{8\pi^2 m(U_b - E)}{h^2}}.$$

Calculations: The numerator of the fraction under the square-root sign is

$$(8\pi^2)(9.11 \times 10^{-31} \text{ kg})(6.8 \text{ eV} - 5.1 \text{ eV})$$
$$\times (1.60 \times 10^{-19} \text{ J/eV}) = 1.956 \times 10^{-47} \text{ J} \cdot \text{kg}.$$

Thus, $\quad b = \sqrt{\dfrac{1.956 \times 10^{-47} \text{ J} \cdot \text{kg}}{(6.63 \times 10^{-34} \text{ J} \cdot \text{s})^2}} = 6.67 \times 10^9 \text{ m}^{-1}.$

The (dimensionless) quantity $2bL$ is then

$$2bL = (2)(6.67 \times 10^9 \text{ m}^{-1})(750 \times 10^{-12} \text{ m}) = 10.0$$

and, from Eq. 38-38, the transmission coefficient is

$$T \approx e^{-2bL} = e^{-10.0} = 45 \times 10^{-6}. \qquad \text{(Answer)}$$

Thus, of every million electrons that strike the barrier, about 45 will tunnel through it, each appearing on the other side with its original total energy of 5.1 eV. (The transmission through the barrier does not alter an electron's energy or any other property.)

(b) What is the approximate probability that a proton with the same total energy of 5.1 eV will be transmitted through the barrier, to appear (and be detectable) on the other side of the barrier?

Reasoning: The transmission coefficient T (and thus the probability of transmission) depends on the mass of the particle. Indeed, because mass m is one of the factors in the exponent of e in the equation for T, the probability of transmission is very sensitive to the mass of the particle. This time, the mass is that of a proton (1.67×10^{-27} kg), which is significantly greater than that of the electron in (a). By substituting the proton's mass for the mass in (a) and then continuing as we did there, we find that $T \approx 10^{-186}$. Thus, although the probability that the proton will be transmitted is not exactly zero, it is barely more than zero. For even more massive particles with the same total energy of 5.1 eV, the probability of transmission is exponentially lower.

WILEY PLUS Additional examples, video, and practice available at *WileyPLUS*

Review & Summary

Light Quanta—Photons An electromagnetic wave (light) is quantized, and its quanta are called *photons*. For a light wave of frequency f and wavelength λ, the energy E and momentum magnitude p of a photon are

$$E = hf \quad \text{(photon energy)} \qquad (38\text{-}2)$$

and $\qquad p = \dfrac{hf}{c} = \dfrac{h}{\lambda} \quad \text{(photon momentum)}. \qquad (38\text{-}7)$

Photoelectric Effect When light of high enough frequency falls on a clean metal surface, electrons are emitted from the surface by photon–electron interactions within the metal. The governing relation is

$$hf = K_{max} + \Phi, \qquad (38\text{-}5)$$

in which hf is the photon energy, K_{max} is the kinetic energy of the most energetic emitted electrons, and Φ is the **work function** of the target material—that is, the minimum energy an electron must have if it is to emerge from the surface of the target. If hf is less than Φ, electrons are not emitted.

Compton Shift When x rays are scattered by loosely bound electrons in a target, some of the scattered x rays have a longer wavelength than do the incident x rays. This **Compton shift** (in wavelength) is given by

$$\Delta\lambda = \frac{h}{mc}(1 - \cos\phi), \qquad (38\text{-}11)$$

in which ϕ is the angle at which the x rays are scattered.

Light Waves and Photons When light interacts with matter, energy and momentum are transferred via photons. When light is in transit, however, we interpret the light wave as a **probability wave,** in which the probability (per unit time) that a photon can be detected is proportional to E_m^2, where E_m is the amplitude of the oscillating electric field of the light wave at the detector.

Ideal Blackbody Radiation As a measure of the emission of thermal radiation by an ideal blackbody radiator, we define the spectral radiancy $S(\lambda)$ in terms of the emitted intensity per unit wavelength at a given wavelength λ. For the Planck radiation law, in which atomic oscillators produce the thermal radiation, we have

$$S(\lambda) = \frac{2\pi c^2 h}{\lambda^5} \frac{1}{e^{hc/\lambda kT} - 1},$$ (38-14)

where h is the Planck constant, k is the Boltzmann constant, and T is the temperature of the radiating surface. Wien's law relates the temperature T of a blackbody radiator and the wavelength λ_{max} at which the spectral radiancy is maximum:

$$\lambda_{max} T = 2898 \ \mu m \cdot K.$$ (38-15)

Matter Waves A moving particle such as an electron or a proton can be described as a **matter wave;** its wavelength (called the **de Broglie wavelength**) is given by $\lambda = h/p$, where p is the magnitude of the particle's momentum.

The Wave Function A matter wave is described by its **wave function** $\Psi(x, y, z, t)$, which can be separated into a space-dependent part $\psi(x, y, z)$ and a time-dependent part $e^{-i\omega t}$. For a particle of mass m moving in the x direction with constant total energy E through a region in which its potential energy is $U(x)$, $\psi(x)$ can be found by solving the simplified **Schrödinger equation:**

$$\frac{d^2\psi}{dx^2} + \frac{8\pi^2 m}{h^2}[E - U(x)]\psi = 0.$$ (38-19)

A matter wave, like a light wave, is a probability wave in the sense that if a particle detector is inserted into the wave, the probability that the detector will register a particle during any specified time interval is proportional to $|\psi|^2$, a quantity called the **probability density.**

For a free particle—that is, a particle for which $U(x) = 0$—moving in the x direction, $|\psi|^2$ has a constant value for all positions along the x axis.

Heisenberg's Uncertainty Principle The probabilistic nature of quantum physics places an important limitation on detecting a particle's position and momentum. That is, it is not possible to measure the position $\vec{r}$ and the momentum $\vec{p}$ of a particle simultaneously with unlimited precision. The uncertainties in the components of these quantities are given by

$$\Delta x \cdot \Delta p_x \geq \hbar$$
$$\Delta y \cdot \Delta p_y \geq \hbar$$ (38-28)
$$\Delta z \cdot \Delta p_z \geq \hbar.$$

Potential Step This term defines a region where a particle's potential energy increases at the expense of its kinetic energy. According to classical physics, if a particle's initial kinetic energy exceeds the potential energy, it should never be reflected by the region. However, according to quantum physics, there is a reflection coefficient R that gives a finite probability of reflection. The probability of transmission is $T = 1 - R$.

Barrier Tunneling According to classical physics, an incident particle will be reflected from a potential energy barrier whose height is greater than the particle's kinetic energy. According to quantum physics, however, the particle has a finite probability of tunneling through such a barrier, appearing on the other side unchanged. The probability that a given particle of mass m and energy E will tunnel through a barrier of height U_b and thickness L is given by the transmission coefficient T:

$$T \approx e^{-2bL},$$ (38-38)

where

$$b = \sqrt{\frac{8\pi^2 m(U_b - E)}{h^2}}.$$ (38-39)

Questions

1 Photon A has twice the energy of photon B. (a) Is the momentum of A less than, equal to, or greater than that of B? (b) Is the wavelength of A less than, equal to, or greater than that of B?

2 In the photoelectric effect (for a given target and a given frequency of the incident light), which of these quantities, if any, depend on the intensity of the incident light beam: (a) the maximum kinetic energy of the electrons, (b) the maximum photoelectric current, (c) the stopping potential, (d) the cutoff frequency?

3 According to the figure for Checkpoint 2, is the maximum kinetic energy of the ejected electrons greater for a target made of sodium or of potassium for a given frequency of incident light?

4 Photoelectric effect: Figure 38-20 gives the stopping voltage V versus the wavelength λ of light for three different materials. Rank the materials according to their work function, greatest first.

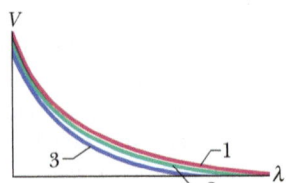

Figure 38-20 Question 4.

5 A metal plate is illuminated with light of a certain frequency. Which of the following determine whether or not electrons are ejected: (a) the intensity of the light, (b) how long the plate is exposed to the light, (c) the thermal conductivity of the plate, (d) the area of the plate, (e) the material of which the plate is made?

6 Let K be the kinetic energy that a stationary free electron gains when a photon scatters from it. We can plot K versus the angle ϕ at which the photon scatters; see curve 1 in Fig. 38-21. If we switch the target to be a stationary free proton, does the end point of the graph shift (a) upward as suggested by curve 2, (b) downward as suggested by curve 3, or (c) remain the same?

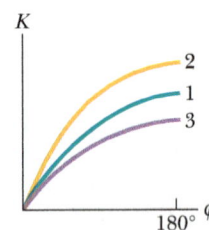

Figure 38-21 Question 6.

7 In a Compton-shift experiment, light (in the x-ray range) is scattered in the forward direction, at $\phi = 0$ in Fig. 38-3. What fraction of the light's energy does the electron acquire?

8 *Compton scattering.* Figure 38-22 gives the Compton shift $\Delta\lambda$ versus scattering angle ϕ for three different stationary target particles. Rank the particles according to their mass, greatest first.

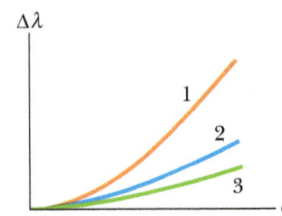

Figure 38-22 Question 8.

9 (a) If you double the kinetic energy of a nonrelativistic particle, how does its de Broglie wavelength change? (b) What if you double the speed of the particle?

10 Figure 38-23 shows an electron moving (a) opposite an electric field, (b) in the same direction as an electric field, (c) in the same direction as a magnetic field, and (d) perpendicular to a magnetic field. For each situation, is the de Broglie wavelength of the electron increasing, decreasing, or remaining the same?

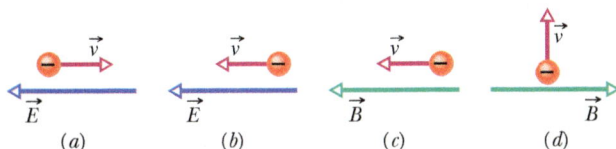

Figure 38-23 Question 10.

11 At the left in Fig. 38-18, why are the minima nonzero?

12 An electron and a proton have the same kinetic energy. Which has the greater de Broglie wavelength?

13 The following nonrelativistic particles all have the same kinetic energy. Rank them in order of their de Broglie wavelengths, greatest first: electron, alpha particle, neutron.

14 Figure 38-24 shows an electron moving through several re-

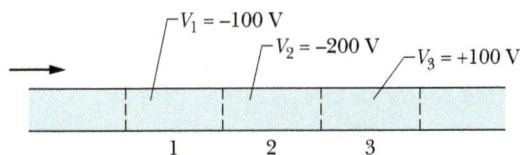

Figure 38-24 Question 14.

gions where uniform electric potentials V have been set up. Rank the three regions according to the de Broglie wavelength of the electron there, greatest first.

15 The table gives relative values for three situations for the barrier tunneling experiment of Figs. 38-16 and 38-17. Rank the situations according to the probability of the electron tunneling through the barrier, greatest first.

	Electron Energy	Barrier Height	Barrier Thickness
(a)	E	$5E$	L
(b)	E	$17E$	$L/2$
(c)	E	$2E$	$2L$

16 For three experiments, Fig. 38-25 gives the transmission coefficient T for electron tunneling through a potential barrier, plotted versus barrier thickness L. The de Broglie wavelengths of the electrons are identical in the three experiments. The only difference in the physical setups is the barrier heights U_b. Rank the three experiments according to U_b, greatest first.

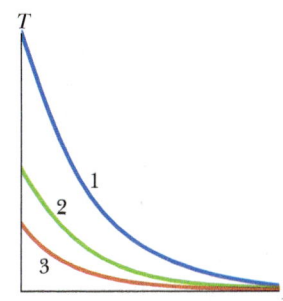

Figure 38-25 Question 16.

Problems

Module 38-1   The Photon, the Quantum of Light

•1 Monochromatic light (that is, light of a single wavelength) is to be absorbed by a sheet of photographic film and thus recorded on the film. Photon absorption will occur if the photon energy equals or exceeds 0.6 eV, the smallest amount of energy needed to dissociate an AgBr molecule in the film. (a) What is the greatest wavelength of light that can be recorded by the film? (b) In what region of the electromagnetic spectrum is this wavelength located?

•2 How fast must an electron move to have a kinetic energy equal to the photon energy of sodium light at wavelength 590 nm?

•3 At what rate does the Sun emit photons? For simplicity, assume that the Sun's entire emission at the rate of 3.9×10^{26} W is at the single wavelength of 550 nm.

•4 A helium–neon laser emits red light at wavelength $\lambda = 633$ nm in a beam of diameter 3.5 mm and at an energy-emission rate of 5.0 mW. A detector in the beam's path totally absorbs the beam. At what rate per unit area does the detector absorb photons?

•5 The meter was once defined as 1 650 763.73 wavelengths of the orange light emitted by a source containing krypton-86 atoms. What is the photon energy of that light?

•6 What is the photon energy for yellow light from a highway sodium lamp at a wavelength of 589 nm?

••7 A light detector (your eye) has an area of 2.00×10^{-6} m^2 and absorbs 80% of the incident light, which is at wavelength 500 nm. The detector faces an isotropic source, 3.00 m from the source. If the detector absorbs photons at the rate of exactly 4.000 s^{-1}, at what power does the emitter emit light?

••8 The beam emerging from a 1.5 W argon laser ($\lambda = 515$ nm) has a diameter d of 3.0 mm. The beam is focused by a lens system with an effective focal length f_L of 2.5 mm. The focused beam strikes a totally absorbing screen, where it forms a circular diffraction pattern whose central disk has a radius R given by $1.22 f_L \lambda/d$. It can be shown that 84% of the incident energy ends up within this central disk. At what rate are photons absorbed by the screen in the central disk of the diffraction pattern?

••9 **GO** A 100 W sodium lamp ($\lambda = 589$ nm) radiates energy uniformly in all directions. (a) At what rate are photons emitted by the lamp? (b) At what distance from the lamp will a totally absorbing screen absorb photons at the rate of 1.00 photon/cm^2·s? (c) What is the photon flux (photons per unit area per unit time) on a small screen 2.00 m from the lamp?

••10 A satellite in Earth orbit maintains a panel of solar cells of area 2.60 m^2 perpendicular to the direction of the Sun's light rays. The intensity of the light at the panel is 1.39 kW/m^2. (a) At what rate does solar energy arrive at the panel? (b) At what rate are solar photons absorbed by the panel? Assume that the solar radiation is monochromatic, with a wavelength of 550 nm, and that all the solar radiation striking the panel is absorbed. (c) How long would it take for a "mole of photons" to be absorbed by the panel?

••11 SSM WWW An ultraviolet lamp emits light of wavelength 400 nm at the rate of 400 W. An infrared lamp emits light of wavelength 700 nm, also at the rate of 400 W. (a) Which lamp emits photons at the greater rate and (b) what is that greater rate?

••12 Under ideal conditions, a visual sensation can occur in the human visual system if light of wavelength 550 nm is absorbed by the eye's retina at a rate as low as 100 photons per second. What is the corresponding rate at which energy is absorbed by the retina?

••13 A special kind of lightbulb emits monochromatic light of wavelength 630 nm. Electrical energy is supplied to it at the rate of 60 W, and the bulb is 93% efficient at converting that energy to light energy. How many photons are emitted by the bulb during its lifetime of 730 h?

••14 GO A light detector has an absorbing area of 2.00×10^{-6} m² and absorbs 50% of the incident light, which is at wavelength 600 nm. The detector faces an isotropic source, 12.0 m from the source. The energy E emitted by the source versus time t is given in Fig. 38-26 ($E_s = 7.2$ nJ, $t_s = 2.0$ s). At what rate are photons absorbed by the detector?

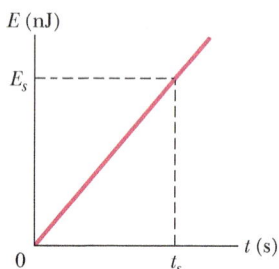

Figure 38-26 Problem 14.

Module 38-2   The Photoelectric Effect

•15 SSM Light strikes a sodium surface, causing photoelectric emission. The stopping potential for the ejected electrons is 5.0 V, and the work function of sodium is 2.2 eV. What is the wavelength of the incident light?

•16 Find the maximum kinetic energy of electrons ejected from a certain material if the material's work function is 2.3 eV and the frequency of the incident radiation is 3.0×10^{15} Hz.

•17 The work function of tungsten is 4.50 eV. Calculate the speed of the fastest electrons ejected from a tungsten surface when light whose photon energy is 5.80 eV shines on the surface.

•18 You wish to pick an element for a photocell that will operate via the photoelectric effect with visible light. Which of the following are suitable (work functions are in parentheses): tantalum (4.2 eV), tungsten (4.5 eV), aluminum (4.2 eV), barium (2.5 eV), lithium (2.3 eV)?

••19 (a) If the work function for a certain metal is 1.8 eV, what is the stopping potential for electrons ejected from the metal when light of wavelength 400 nm shines on the metal? (b) What is the maximum speed of the ejected electrons?

••20 Suppose the *fractional efficiency* of a cesium surface (with work function 1.80 eV) is 1.0×10^{-16}; that is, on average one electron is ejected for every 10^{16} photons that reach the surface. What would be the current of electrons ejected from such a surface if it were illuminated with 600 nm light from a 2.00 mW laser and all the ejected electrons took part in the charge flow?

••21 GO X rays with a wavelength of 71 pm are directed onto a gold foil and eject tightly bound electrons from the gold atoms. The

ejected electrons then move in circular paths of radius r in a region of uniform magnetic field $\vec{B}$. For the fastest of the ejected electrons, the product Br is equal to 1.88×10^{-4} T·m. Find (a) the maximum kinetic energy of those electrons and (b) the work done in removing them from the gold atoms.

••22 The wavelength associated with the cutoff frequency for silver is 325 nm. Find the maximum kinetic energy of electrons ejected from a silver surface by ultraviolet light of wavelength 254 nm.

••23 SSM Light of wavelength 200 nm shines on an aluminum surface; 4.20 eV is required to eject an electron. What is the kinetic energy of (a) the fastest and (b) the slowest ejected electrons? (c) What is the stopping potential for this situation? (d) What is the cutoff wavelength for aluminum?

••24 In a photoelectric experiment using a sodium surface, you find a stopping potential of 1.85 V for a wavelength of 300 nm and a stopping potential of 0.820 V for a wavelength of 400 nm. From these data find (a) a value for the Planck constant, (b) the work function Φ for sodium, and (c) the cutoff wavelength λ_0 for sodium.

••25 GO The stopping potential for electrons emitted from a surface illuminated by light of wavelength 491 nm is 0.710 V. When the incident wavelength is changed to a new value, the stopping potential is 1.43 V. (a) What is this new wavelength? (b) What is the work function for the surface?

••26 An orbiting satellite can become charged by the photoelectric effect when sunlight ejects electrons from its outer surface. Satellites must be designed to minimize such charging because it can ruin the sensitive microelectronics. Suppose a satellite is coated with platinum, a metal with a very large work function ($\Phi = 5.32$ eV). Find the longest wavelength of incident sunlight that can eject an electron from the platinum.

Module 38-3   Photons, Momentum, Compton Scattering, Light Interference

•27 SSM Light of wavelength 2.40 pm is directed onto a target containing free electrons. (a) Find the wavelength of light scattered at 30.0° from the incident direction. (b) Do the same for a scattering angle of 120°.

•28 (a) In MeV/c, what is the magnitude of the momentum associated with a photon having an energy equal to the electron rest energy? What are the (b) wavelength and (c) frequency of the corresponding radiation?

•29 What (a) frequency, (b) photon energy, and (c) photon momentum magnitude (in keV/c) are associated with x rays having wavelength 35.0 pm?

••30 What is the maximum wavelength shift for a Compton collision between a photon and a free *proton*?

••31 What percentage increase in wavelength leads to a 75% loss of photon energy in a photon–free electron collision?

••32 X rays of wavelength 0.0100 nm are directed in the positive direction of an x axis onto a target containing loosely bound electrons. For Compton scattering from one of those electrons, at an angle of 180°, what are (a) the Compton shift, (b) the corresponding change in photon energy, (c) the kinetic energy of the recoiling electron, and (d) the angle between the positive direction of the x axis and the electron's direction of motion?

••33 Calculate the percentage change in photon energy during a collision like that in Fig. 38-5 for $\phi = 90°$ and for radiation in

(a) the microwave range, with $\lambda = 3.0$ cm; (b) the visible range, with $\lambda = 500$ nm; (c) the x-ray range, with $\lambda = 25$ pm; and (d) the gamma-ray range, with a gamma photon energy of 1.0 MeV. (e) What are your conclusions about the feasibility of detecting the Compton shift in these various regions of the electromagnetic spectrum, judging solely by the criterion of energy loss in a single photon–electron encounter?

••**34** GO A photon undergoes Compton scattering off a stationary free electron. The photon scatters at 90.0° from its initial direction; its initial wavelength is 3.00×10^{-12} m. What is the electron's kinetic energy?

••**35** Calculate the Compton wavelength for (a) an electron and (b) a proton. What is the photon energy for an electromagnetic wave with a wavelength equal to the Compton wavelength of (c) the electron and (d) the proton?

••**36** Gamma rays of photon energy 0.511 MeV are directed onto an aluminum target and are scattered in various directions by loosely bound electrons there. (a) What is the wavelength of the incident gamma rays? (b) What is the wavelength of gamma rays scattered at 90.0° to the incident beam? (c) What is the photon energy of the rays scattered in this direction?

••**37** Consider a collision between an x-ray photon of initial energy 50.0 keV and an electron at rest, in which the photon is scattered backward and the electron is knocked forward. (a) What is the energy of the backscattered photon? (b) What is the kinetic energy of the electron?

••**38** Show that when a photon of energy E is scattered from a free electron at rest, the maximum kinetic energy of the recoiling electron is given by

$$K_{max} = \frac{E^2}{E + mc^2/2}.$$

••**39** Through what angle must a 200 keV photon be scattered by a free electron so that the photon loses 10% of its energy?

••**40** GO What is the maximum kinetic energy of electrons knocked out of a thin copper foil by Compton scattering of an incident beam of 17.5 keV x rays? Assume the work function is negligible.

••**41** What are (a) the Compton shift $\Delta\lambda$, (b) the fractional Compton shift $\Delta\lambda/\lambda$, and (c) the change ΔE in photon energy for light of wavelength $\lambda = 590$ nm scattering from a free, initially stationary electron if the scattering is at 90° to the direction of the incident beam? What are (d) $\Delta\lambda$, (e) $\Delta\lambda/\lambda$, and (f) ΔE for 90° scattering for photon energy 50.0 keV (x-ray range)?

Module 38-4 The Birth of Quantum Physics

•**42** The Sun is approximately an ideal blackbody radiator with a surface temperature of 5800 K. (a) Find the wavelength at which its spectral radiancy is maximum and (b) identify the type of electromagnetic wave corresponding to that wavelength. (See Fig. 33-1.) (c) As we shall discuss in Chapter 44, the universe is approximately an ideal blackbody radiator with radiation emitted when atoms first formed. Today the spectral radiancy of that radiation peaks at a wavelength of 1.06 mm (in the microwave region). What is the corresponding temperature of the universe?

•**43** Just after detonation, the fireball in a nuclear blast is approximately an ideal blackbody radiator with a surface temperature of about 1.0×10^7 K. (a) Find the wavelength at which the thermal radiation is maximum and (b) identify the type of electromagnetic wave corresponding to that wavelength. (See Fig. 33-1.) This radia-

tion is almost immediately absorbed by the surrounding air molecules, which produces another ideal blackbody radiator with a surface temperature of about 1.0×10^5 K. (c) Find the wavelength at which the thermal radiation is maximum and (d) identify the type of electromagnetic wave corresponding to that wavelength.

••**44** GO For the thermal radiation from an ideal blackbody radiator with a surface temperature of 2000 K, let I_c represent the intensity per unit wavelength according to the classical expression for the spectral radiancy and I_P represent the corresponding intensity per unit wavelength according to the Planck expression. What is the ratio I_c/I_P for a wavelength of (a) 400 nm (at the blue end of the visible spectrum) and (b) 200 μm (in the far infrared)? (c) Does the classical expression agree with the Planck expression in the shorter wavelength range or the longer wavelength range?

••**45** Assuming that your surface temperature is 98.6°F and that you are an ideal blackbody radiator (you are close), find (a) the wavelength at which your spectral radiancy is maximum, (b) the power at which you emit thermal radiation in a wavelength range of 1.00 nm at that wavelength, from a surface area of 4.00 cm², and (c) the corresponding rate at which you emit photons from that area. Using a wavelength of 500 nm (in the visible range), (d) recalculate the power and (e) the rate of photon emission. (As you have noticed, you do not visibly glow in the dark.)

Module 38-5 Electrons and Matter Waves

•**46** Calculate the de Broglie wavelength of (a) a 1.00 keV electron, (b) a 1.00 keV photon, and (c) a 1.00 keV neutron.

•**47** SSM In an old-fashioned television set, electrons are accelerated through a potential difference of 25.0 kV. What is the de Broglie wavelength of such electrons? (Relativity is not needed.)

••**48** The smallest dimension (*resolving power*) that can be resolved by an electron microscope is equal to the de Broglie wavelength of its electrons. What accelerating voltage would be required for the electrons to have the same resolving power as could be obtained using 100 keV gamma rays?

••**49** SSM WWW Singly charged sodium ions are accelerated through a potential difference of 300 V. (a) What is the momentum acquired by such an ion? (b) What is its de Broglie wavelength?

••**50** Electrons accelerated to an energy of 50 GeV have a de Broglie wavelength λ small enough for them to probe the structure within a target nucleus by scattering from the structure. Assume that the energy is so large that the extreme relativistic relation $p = E/c$ between momentum magnitude p and energy E applies. (In this extreme situation, the kinetic energy of an electron is much greater than its rest energy.) (a) What is λ? (b) If the target nucleus has radius $R = 5.0$ fm, what is the ratio R/λ?

••**51** SSM The wavelength of the yellow spectral emission line of sodium is 590 nm. At what kinetic energy would an electron have that wavelength as its de Broglie wavelength?

••**52** A stream of protons, each with a speed of $0.9900c$, are directed into a two-slit experiment where the slit separation is 4.00×10^{-9} m. A two-slit interference pattern is built up on the viewing screen. What is the angle between the center of the pattern and the second minimum (to either side of the center)?

••**53** What is the wavelength of (a) a photon with energy 1.00 eV, (b) an electron with energy 1.00 eV, (c) a photon of energy 1.00 GeV, and (d) an electron with energy 1.00 GeV?

••**54** An electron and a photon each have a wavelength of 0.20 nm.

What is the momentum (in kg·m/s) of the (a) electron and (b) photon? What is the energy (in eV) of the (c) electron and (d) photon?

••55 The highest achievable resolving power of a microscope is limited only by the wavelength used; that is, the smallest item that can be distinguished has dimensions about equal to the wavelength. Suppose one wishes to "see" inside an atom. Assuming the atom to have a diameter of 100 pm, this means that one must be able to resolve a width of, say, 10 pm. (a) If an electron microscope is used, what minimum electron energy is required? (b) If a light microscope is used, what minimum photon energy is required? (c) Which microscope seems more practical? Why?

••56 The existence of the atomic nucleus was discovered in 1911 by Ernest Rutherford, who properly interpreted some experiments in which a beam of alpha particles was scattered from a metal foil of atoms such as gold. (a) If the alpha particles had a kinetic energy of 7.5 MeV, what was their de Broglie wavelength? (b) Explain whether the wave nature of the incident alpha particles should have been taken into account in interpreting these experiments. The mass of an alpha particle is 4.00 u (atomic mass units), and its distance of closest approach to the nuclear center in these experiments was about 30 fm. (The wave nature of matter was not postulated until more than a decade after these crucial experiments were first performed.)

••57 A nonrelativistic particle is moving three times as fast as an electron. The ratio of the de Broglie wavelength of the particle to that of the electron is 1.813×10^{-4}. By calculating its mass, identify the particle.

••58 What are (a) the energy of a photon corresponding to wavelength 1.00 nm, (b) the kinetic energy of an electron with de Broglie wavelength 1.00 nm, (c) the energy of a photon corresponding to wavelength 1.00 fm, and (d) the kinetic energy of an electron with de Broglie wavelength 1.00 fm?

•••59 GO If the de Broglie wavelength of a proton is 100 fm, (a) what is the speed of the proton and (b) through what electric potential would the proton have to be accelerated to acquire this speed?

Module 38-6 Schrödinger's Equation

•60 Suppose we put $A = 0$ in Eq. 38-24 and relabeled B as ψ_0. (a) What would the resulting wave function then describe? (b) How, if at all, would Fig. 38-13 be altered?

•61 SSM The function $\psi(x)$ displayed in Eq. 38-27 can describe a free particle, for which the potential energy is $U(x) = 0$ in Schrödinger's equation (Eq. 38-19). Assume now that $U(x) = U_0 = $ a constant in that equation. Show that Eq. 38-27 is a solution of Schrödinger's equation, with

$$k = \frac{2\pi}{h}\sqrt{2m(E - U_0)}$$

giving the angular wave number k of the particle.

•62 Show that Eq. 38-24 is indeed a solution of Eq. 38-22 by substituting $\psi(x)$ and its second derivative into Eq. 38-22 and noting that an identity results.

•63 (a) Write the wave function $\psi(x)$ displayed in Eq. 38-27 in the form $\psi(x) = a + ib$, where a and b are real quantities. (Assume that ψ_0 is real.) (b) Write the time-dependent wave function $\Psi(x, t)$ that corresponds to $\psi(x)$ written in this form.

•64 SSM Show that the angular wave number k for a nonrela-

tivistic free particle of mass m can be written as

$$k = \frac{2\pi\sqrt{2mK}}{h},$$

in which K is the particle's kinetic energy.

•65 (a) Let $n = a + ib$ be a complex number, where a and b are real (positive or negative) numbers. Show that the product nn^* is always a positive real number. (b) Let $m = c + id$ be another complex number. Show that $|nm| = |n| |m|$.

••66 In Eq. 38-25 keep both terms, putting $A = B = \psi_0$. The equation then describes the superposition of two matter waves of equal amplitude, traveling in opposite directions. (Recall that this is the condition for a standing wave.) (a) Show that $|\Psi(x, t)|^2$ is then given by

$$|\Psi(x, t)|^2 = 2\psi_0^2[1 + \cos 2kx].$$

(b) Plot this function, and demonstrate that it describes the square of the amplitude of a standing matter wave. (c) Show that the nodes of this standing wave are located at

$$x = (2n + 1)\left(\frac{1}{4}\lambda\right), \quad \text{where } n = 0, 1, 2, 3, \ldots$$

and λ is the de Broglie wavelength of the particle. (d) Write a similar expression for the most probable locations of the particle.

Module 38-7 Heisenberg's Uncertainty Principle

•67 The uncertainty in the position of an electron along an x axis is given as 50 pm, which is about equal to the radius of a hydrogen atom. What is the least uncertainty in any simultaneous measurement of the momentum component p_x of this electron?

••68 You will find in Chapter 39 that electrons cannot move in definite orbits within atoms, like the planets in our solar system. To see why, let us try to "observe" such an orbiting electron by using a light microscope to measure the electron's presumed orbital position with a precision of, say, 10 pm (a typical atom has a radius of about 100 pm). The wavelength of the light used in the microscope must then be about 10 pm. (a) What would be the photon energy of this light? (b) How much energy would such a photon impart to an electron in a head-on collision? (c) What do these results tell you about the possibility of "viewing" an atomic electron at two or more points along its presumed orbital path? (*Hint:* The outer electrons of atoms are bound to the atom by energies of only a few electron-volts.)

••69 Figure 38-13 shows a case in which the momentum component p_x of a particle is fixed so that $\Delta p_x = 0$; then, from Heisenberg's uncertainty principle (Eq. 38-28), the position x of the particle is completely unknown. From the same principle it follows that the opposite is also true; that is, if the position of a particle is exactly known ($\Delta x = 0$), the uncertainty in its momentum is infinite.

Consider an intermediate case, in which the position of a particle is measured, not to infinite precision, but to within a distance of $\lambda/2\pi$, where λ is the particle's de Broglie wavelength. Show that the uncertainty in the (simultaneously measured) momentum component is then equal to the component itself; that is, $\Delta p_x = p$. Under these circumstances, would a measured momentum of zero surprise you? What about a measured momentum of $0.5p$? Of $2p$? Of $12p$?

Module 38-8 Reflection from a Potential Step

••70 An electron moves through a region of uniform electric potential of -200 V with a (total) energy of 500 eV. What are its (a) kinetic energy (in electron-volts), (b) momentum, (c) speed, (d) de Broglie wavelength, and (e) angular wave number?

••71 GO For the arrangement of Figs. 38-14 and 38-15, electrons in the incident beam in region 1 have energy $E = 800$ eV and the potential step has a height of $U_1 = 600$ eV. What is the angular wave number in (a) region 1 and (b) region 2? (c) What is the reflection coefficient? (d) If the incident beam sends 5.00×10^5 electrons against the potential step, approximately how many will be reflected?

••72 GO For the arrangement of Figs. 38-14 and 38-15, electrons in the incident beam in region 1 have a speed of 1.60×10^7 m/s and region 2 has an electric potential of $V_2 = -500$ V. What is the angular wave number in (a) region 1 and (b) region 2? (c) What is the reflection coefficient? (d) If the incident beam sends 3.00×10^9 electrons against the potential step, approximately how many will be reflected?

•••73 GO The current of a beam of electrons, each with a speed of 900 m/s, is 5.00 mA. At one point along its path, the beam encounters a potential step of height -1.25 μV. What is the current on the other side of the step boundary?

Module 38-9 Tunneling Through a Potential Barrier

••74 Consider a potential energy barrier like that of Fig. 38-17 but whose height U_b is 6.0 eV and whose thickness L is 0.70 nm. What is the energy of an incident electron whose transmission coefficient is 0.0010?

••75 A 3.0 MeV proton is incident on a potential energy barrier of thickness 10 fm and height 10 MeV. What are (a) the transmission coefficient T, (b) the kinetic energy K_t the proton will have on the other side of the barrier if it tunnels through the barrier, and (c) the kinetic energy K_r it will have if it reflects from the barrier? A 3.0 MeV deuteron (the same charge but twice the mass as a proton) is incident on the same barrier. What are (d) T, (e) K_t, and (f) K_r?

••76 GO (a) Suppose a beam of 5.0 eV protons strikes a potential energy barrier of height 6.0 eV and thickness 0.70 nm, at a rate equivalent to a current of 1000 A. How long would you have to wait—on average—for one proton to be transmitted? (b) How long would you have to wait if the beam consisted of electrons rather than protons?

••77 SSM WWW An electron with total energy $E = 5.1$ eV approaches a barrier of height $U_b = 6.8$ eV and thickness $L = 750$ pm. What percentage change in the transmission coefficient T occurs for a 1.0% change in (a) the barrier height, (b) the barrier thickness, and (c) the kinetic energy of the incident electron?

•••78 GO The current of a beam of electrons, each with a speed of 1.200×10^3 m/s, is 9.000 mA. At one point along its path, the beam encounters a potential barrier of height -4.719 μV and thickness 200.0 nm. What is the transmitted current?

Additional Problems

79 Figure 38-13 shows that because of Heisenberg's uncertainty principle, it is not possible to assign an x coordinate to the position of a free electron moving along an x axis. (a) Can you assign a y or a z coordinate? (*Hint:* The momentum of the electron has no y or z component.) (b) Describe the extent of the matter wave in three dimensions.

80 A spectral emission line is electromagnetic radiation that is emitted in a wavelength range narrow enough to be taken as a sin-

gle wavelength. One such emission line that is important in astronomy has a wavelength of 21 cm. What is the photon energy in the electromagnetic wave at that wavelength?

81 Using the classical equations for momentum and kinetic energy, show that an electron's de Broglie wavelength in nanometers can be written as $\lambda = 1.226/\sqrt{K}$, in which K is the electron's kinetic energy in electron-volts.

82 Derive Eq. 38-11, the equation for the Compton shift, from Eqs. 38-8, 38-9, and 38-10 by eliminating v and θ.

83 Neutrons in thermal equilibrium with matter have an average kinetic energy of $(3/2)kT$, where k is the Boltzmann constant and T, which may be taken to be 300 K, is the temperature of the environment of the neutrons. (a) What is the average kinetic energy of such a neutron? (b) What is the corresponding de Broglie wavelength?

84 Consider a balloon filled with helium gas at room temperature and atmospheric pressure. Calculate (a) the average de Broglie wavelength of the helium atoms and (b) the average distance between atoms under these conditions. The average kinetic energy of an atom is equal to $(3/2)kT$, where k is the Boltzmann constant. (c) Can the atoms be treated as particles under these conditions? Explain.

85 In about 1916, R. A. Millikan found the following stopping-potential data for lithium in his photoelectric experiments:

Wavelength (nm)	433.9	404.7	365.0	312.5	253.5
Stopping potential (V)	0.55	0.73	1.09	1.67	2.57

Use these data to make a plot like Fig. 38-2 (which is for sodium) and then use the plot to find (a) the Planck constant and (b) the work function for lithium.

86 Show that $|\psi|^2 = |\Psi|^2$, with ψ and Ψ related as in Eq. 38-14. That is, show that the probability density does not depend on the time variable.

87 Show that $\Delta E/E$, the fractional loss of energy of a photon during a collision with a particle of mass m, is given by

$$\frac{\Delta E}{E} = \frac{hf'}{mc^2}(1 - \cos \phi),$$

where E is the energy of the incident photon, f' is the frequency of the scattered photon, and ϕ is defined as in Fig. 38-5.

88 A bullet of mass 40 g travels at 1000 m/s. Although the bullet is clearly too large to be treated as a matter wave, determine what Eq. 38-17 predicts for the de Broglie wavelength of the bullet at that speed.

89 (a) The smallest amount of energy needed to eject an electron from metallic sodium is 2.28 eV. Does sodium show a photoelectric effect for red light, with $\lambda = 680$ nm? (That is, does the light cause electron emission?) (b) What is the cutoff wavelength for photoelectric emission from sodium? (c) To what color does that wavelength correspond?

90 SSM Imagine playing baseball in a universe (not ours!) where the Planck constant is 0.60 J·s and thus quantum physics affects macroscopic objects. What would be the uncertainty in the position of a 0.50 kg baseball that is moving at 20 m/s along an axis if the uncertainty in the speed is 1.0 m/s?

More About Matter Waves

39-1 ENERGIES OF A TRAPPED ELECTRON

Learning Objectives

After reading this module, you should be able to . . .

39.01 Identify the confinement principle: Confinement of a wave (including a matter wave) leads to the quantization of wavelengths and energy values.

39.02 Sketch a one-dimensional infinite potential well, indicating the length (or width) and the potential energy of the walls.

39.03 For an electron, apply the relationship between the de Broglie wavelength λ and the kinetic energy.

39.04 For an electron in a one-dimensional infinite potential well, apply the relationship between the de Broglie wavelength λ, the well's length, and the quantum number n.

39.05 For an electron in a one-dimensional infinite potential well, apply the relationship between the allowed energies E_n, the well length L, and the quantum number n.

39.06 Sketch an energy-level diagram for an electron in a one-dimensional infinite potential well, indicating the ground state and several excited states.

39.07 Identify that a trapped electron tends to be in its ground state, can be excited to a higher-energy state, and cannot exist between the allowed states.

39.08 Calculate the energy change required for an electron to move between states: a quantum jump up or down an energy-level diagram.

39.09 If a quantum jump involves light, identify that an upward jump requires the absorption of a photon (to increase the electron's energy) and a downward jump requires the emission of a photon (to reduce the electron's energy).

39.10 If a quantum jump involves light, apply the relationships between the energy change and the frequency and wavelength associated with the photon.

39.11 Identify the emission and absorption spectra of an electron in a one-dimensional infinite potential well.

Key Ideas

● Confinement of waves (string waves, matter waves—any type of wave) leads to quantization—that is, discrete states with certain energies. States with intermediate energies are not allowed.

● Because it is a matter wave, an electron confined to an infinite potential well can exist in only certain discrete states. If the well is one-dimensional with length L, the energies associated with these quantum states are

$$E_n = \left(\frac{h^2}{8mL^2}\right)n^2, \quad \text{for } n = 1, 2, 3, \ldots,$$

where m is the electron mass and n is a quantum number.

● The lowest energy is not zero but is given by $n = 1$.

● The electron can change (jump) from one quantum state to another only if its energy change is

$$\Delta E = E_{\text{high}} - E_{\text{low}},$$

where E_{high} is the higher energy and E_{low} is the lower energy.

● If the change is done by photon absorption or emission, the energy of the photon must be equal to the change in the electron's energy:

$$hf = \frac{hc}{\lambda} = \Delta E = E_{\text{high}} - E_{\text{low}},$$

where frequency f and wavelength λ are associated with the photon.

What Is Physics?

One of the long-standing goals of physics has been to understand the nature of atoms. Early in the 20th century nobody knew how the electrons in an atom are arranged, what their motions are, how atoms emit or absorb light, or even why atoms are stable. Without this knowledge it was not possible to understand how atoms combine to form molecules or stack up to form solids. As a consequence, the foundations of chemistry—including biochemistry, which underlies the nature of life itself—were more or less a mystery.

In 1926, all these questions and many others were answered with the development of quantum physics. Its basic premise is that moving electrons, protons, and particles of any kind are best viewed as matter waves, whose motions are governed by Schrödinger's equation. Although quantum theory also applies to larger objects, such as baseballs and planets, it yields the same results as Newtonian physics, which is easier to use and more intuitive.

Before we can apply quantum physics to the problem of atomic structure, we need to develop some insights by applying quantum ideas in a few simpler situations. Some of these situations may seem simplistic and unreal, but they allow us to discuss the basic principles of the quantum physics of atoms without having to deal with the often overwhelming complexity of atoms. Besides, with advances in nanotechnology, situations that were previously found only in textbooks are now being produced in laboratories and put to use in modern electronics and materials science applications. We are on the threshold of being able to use nanometer-scale constructions called *quantum corrals* and *quantum dots* to create "designer atoms" whose properties can be manipulated in the laboratory. For both natural atoms and these artificial ones, the starting point in our discussion is the wave nature of an electron.

String Waves and Matter Waves

In Chapter 16 we saw that waves of two kinds can be set up on a stretched string. If the string is so long that we can take it to be infinitely long, we can set up a *traveling wave* of essentially any frequency. However, if the stretched string has only a finite length, perhaps because it is rigidly clamped at both ends, we can set up only *standing waves* on it; further, these standing waves can have only discrete frequencies. In other words, confining the wave to a finite region of space leads to *quantization* of the motion — to the existence of discrete *states* for the wave, each state with a sharply defined frequency.

This observation applies to waves of all kinds, including matter waves. For matter waves, however, it is more convenient to deal with the energy E of the associated particle than with the frequency f of the wave. In all that follows we shall focus on the matter wave associated with an electron, but the results apply to any confined matter wave.

Consider the matter wave associated with an electron moving in the positive x direction and subject to no net force — a so-called *free particle*. The energy of such an electron can have any reasonable value, just as a wave traveling along a stretched string of infinite length can have any reasonable frequency.

Consider next the matter wave associated with an atomic electron, perhaps the *valence* (least tightly bound) electron. The electron — held within the atom by the attractive Coulomb force between it and the positively charged nucleus — is not a free particle. It can exist only in a set of discrete states, each having a discrete energy E. This sounds much like the discrete states and quantized frequencies that are available to a stretched string of finite length. For matter waves, then, as for all other kinds of waves, we may state a **confinement principle:**

Confinement of a wave leads to quantization — that is, to the existence of discrete states with discrete energies.

Energies of a Trapped Electron

One-Dimensional Traps

Here we examine the matter wave associated with a nonrelativistic electron confined to a limited region of space. We do so by analogy with standing waves on a string of finite length, stretched along an x axis and confined between rigid supports. Because the supports are rigid, the two ends of the string are nodes, or

An electron can be trapped in the V = 0 region.

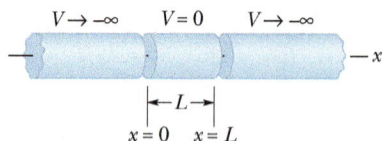

Figure 39-1 The elements of an idealized "trap" designed to confine an electron to the central cylinder. We take the semi-infinitely long end cylinders to be at an infinitely great negative potential and the central cylinder to be at zero potential.

An electron can be trapped in the U = 0 region.

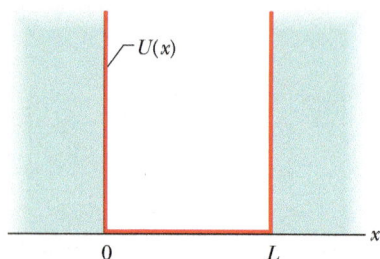

Figure 39-2 The electric potential energy $U(x)$ of an electron confined to the central cylinder of the idealized trap of Fig. 39-1. We see that $U = 0$ for $0 < x < L$, and $U \to \infty$ for $x < 0$ and $x > L$.

points at which the string is always at rest. There may be other nodes along the string, but these two must always be present, as Fig. 16-21 shows.

The states, or discrete standing wave patterns in which the string can oscillate, are those for which the length L of the string is equal to an integer number of half-wavelengths. That is, the string can occupy only states for which

$$L = \frac{n\lambda}{2}, \qquad \text{for } n = 1, 2, 3, \ldots . \tag{39-1}$$

Each value of n identifies a state of the oscillating string; using the language of quantum physics, we can call the integer n a **quantum number.**

For each state of the string permitted by Eq. 39-1, the transverse displacement of the string at any position x along the string is given by

$$y_n(x) = A \sin\left(\frac{n\pi}{L}x\right), \qquad \text{for } n = 1, 2, 3, \ldots, \tag{39-2}$$

in which the quantum number n identifies the oscillation pattern and A depends on the time at which you inspect the string. (Equation 39-2 is a short version of Eq. 16-60.) We see that for all values of n and for all times, there is a point of zero displacement (a node) at $x = 0$ and at $x = L$, as there must be. Figure 16-20 shows time exposures of such a stretched string for $n = 2, 3$, and 4.

Now let us turn our attention to matter waves. Our first problem is to physically confine an electron that is moving along the x axis so that it remains within a finite segment of that axis. Figure 39-1 shows a conceivable one-dimensional *electron trap.* It consists of two semi-infinitely long cylinders, each of which has an electric potential approaching $-\infty$; between them is a hollow cylinder of length L, which has an electric potential of zero. We put a single electron into this central cylinder to trap it.

The trap of Fig. 39-1 is easy to analyze but is not very practical. Single electrons *can,* however, be trapped in the laboratory with traps that are more complex in design but similar in concept. At the University of Washington, for example, a single electron has been held in a trap for months on end, permitting scientists to make extremely precise measurements of its properties.

Finding the Quantized Energies

Figure 39-2 shows the potential energy of the electron as a function of its position along the x axis of the idealized trap of Fig. 39-1. When the electron is in the central cylinder, its potential energy $U\ (= -eV)$ is zero because there the potential V is zero. If the electron could get outside this region, its potential energy would be positive and of infinite magnitude because there $V \to -\infty$. We call the potential energy pattern of Fig. 39-2 an **infinitely deep potential energy well** or, for short, an *infinite potential well.* It is a "well" because an electron placed in the central cylinder of Fig. 39-1 cannot escape from it. As the electron approaches either end of the cylinder, a force of essentially infinite magnitude reverses the electron's motion, thus trapping it. Because the electron can move along only a single axis, this trap can be called a *one-dimensional infinite potential well.*

Just like the standing wave in a length of stretched string, the matter wave describing the confined electron must have nodes at $x = 0$ and $x = L$. Moreover, Eq. 39-1 applies to such a matter wave if we interpret λ in that equation as the de Broglie wavelength associated with the moving electron.

The de Broglie wavelength λ is defined in Eq. 38-17 as $\lambda = h/p$, where p is the magnitude of the electron's momentum. Because the electron is nonrelativistic, this momentum magnitude p is related to the kinetic energy K by $p = \sqrt{2mK}$, where m is the mass of the electron. For an electron moving within the central cylinder of Fig. 39-1, where $U = 0$, the total (mechanical) energy E is equal to the kinetic energy. Hence, we can write the de Broglie wavelength of this electron as

$$\lambda = \frac{h}{p} = \frac{h}{\sqrt{2mE}}. \qquad (39\text{-}3)$$

If we substitute Eq. 39-3 into Eq. 39-1 and solve for the energy E, we find that E depends on n according to

$$E_n = \left(\frac{h^2}{8mL^2}\right)n^2, \qquad \text{for } n = 1, 2, 3, \ldots . \qquad (39\text{-}4)$$

The positive integer n here is the quantum number of the electron's quantum state in the trap.

    Equation 39-4 tells us something important: Because the electron is confined to the trap, it can have only the energies given by the equation. It *cannot* have an energy that is, say, halfway between the values for $n = 1$ and $n = 2$. Why this restriction? Because an electron is a matter wave. Were it, instead, a particle as assumed in classical physics, it could have *any* value of energy while it is confined to the trap.

    Figure 39-3 is a graph showing the lowest five allowed energy values for an electron in an infinite well with $L = 100$ pm (about the size of a typical atom). The values are called *energy levels*, and they are drawn in Fig. 39-3 as levels, or steps, on a ladder, in an *energy-level diagram*. Energy is plotted vertically; nothing is plotted horizontally.

    The quantum state with the lowest possible energy level E_1 allowed by Eq. 39-4, with quantum number $n = 1$, is called the *ground state* of the electron. The electron tends to be in this lowest energy state. All the quantum states with greater energies (corresponding to quantum numbers $n = 2$ or greater) are called *excited states* of the electron. The state with energy level E_2, for quantum number $n = 2$, is called the *first excited state* because it is the first of the excited states as we move up the energy-level diagram. The other states have similar names.

Energy Changes

A trapped electron tends to have the lowest allowed energy and thus to be in its ground state. It can be changed to an excited state (in which it has greater energy) only if an external source provides the additional energy that is required for the change. Let E_{low} be the initial energy of the electron and E_{high} be the greater energy in a state that is higher on its energy-level diagram. Then the amount of energy that is required for the electron's change of state is

$$\Delta E = E_{\text{high}} - E_{\text{low}}. \qquad (39\text{-}5)$$

    An electron that receives such energy is said to make a *quantum jump* (or *transition*), or to be *excited* from the lower-energy state to the higher-energy state. Figure 39-4a represents a quantum jump from the ground state (with energy level E_1) to the third excited state (with energy level E_4). As shown, the jump *must* be from one energy level to another, but it can bypass one or more intermediate energy levels.

    Photons. One way an electron can gain energy to make a quantum jump up to a greater energy level is to absorb a photon. However, this absorption and quantum jump can occur only if the following condition is met:

If a confined electron is to absorb a photon, the energy hf of the photon must equal the energy difference ΔE between the initial energy level of the electron and a higher level.

Thus, excitation by the absorption of light requires that

$$hf = \frac{hc}{\lambda} = \Delta E = E_{\text{high}} - E_{\text{low}}. \qquad (39\text{-}6)$$

These are the lowest five energy levels allowed the electron. (No intermediate levels are allowed.)

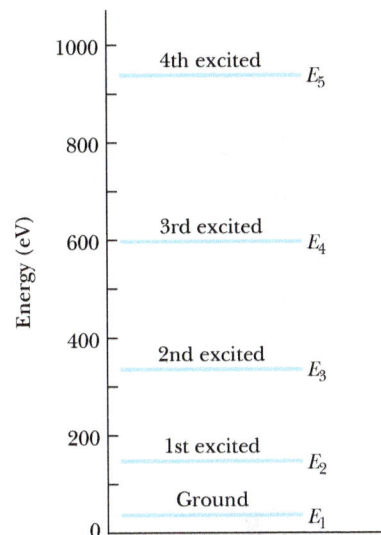

Figure 39-3 Several of the allowed energies for an electron confined to the infinite well of Fig. 39-2, with width $L = 100$ pm.

The electron is excited to a higher energy level.

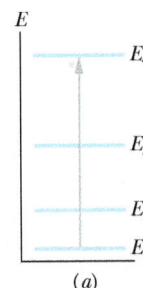

It can de-excite to a lower level in several ways (set by chance).

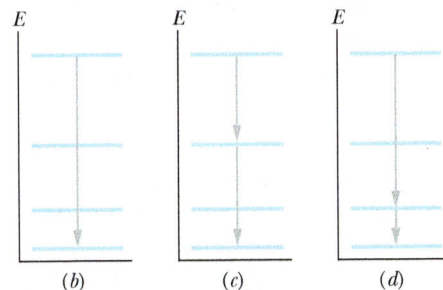

Figure 39-4 (a) Excitation of a trapped electron from the energy level of its ground state to the level of its third excited state. (b)–(d) Three of four possible ways the electron can de-excite to return to the energy level of its ground state. (Which way is not shown?)

When an electron reaches an excited state, it does not stay there but quickly *de-excites* by decreasing its energy. Figures 39-4b to d represent some of the possible quantum jumps down from the energy level of the third excited state. The electron can reach its ground-state level either with one direct quantum jump (Fig. 39-4b) or with shorter jumps via intermediate levels (Figs. 39-4c and d).

An electron can decrease its energy by emitting a photon but only this way:

> If a confined electron emits a photon, the energy hf of that photon must equal the energy difference ΔE between the initial energy level of the electron and a lower level.

Thus, Eq. 39-6 applies to both the absorption and the emission of light by a confined electron. That is, the absorbed or emitted light can have only certain values of hf and thus only certain values of frequency f and wavelength λ.

Aside: Although Eq. 39-6 and what we have discussed about photon absorption and emission can be applied to physical (real) electron traps, they actually cannot be applied to one-dimensional (unreal) electron traps. The reason involves the need to conserve angular momentum in a photon absorption or emission process. In this book, we shall neglect that need and use Eq. 39-6 even for one-dimensional traps.

✓ Checkpoint 1

Rank the following pairs of quantum states for an electron confined to an infinite well according to the energy differences between the states, greatest first: (a) $n = 3$ and $n = 1$, (b) $n = 5$ and $n = 4$, (c) $n = 4$ and $n = 3$.

Sample Problem 39.01   Energy levels in a 1D infinite potential well

An electron is confined to a one-dimensional, infinitely deep potential energy well of width $L = 100$ pm. (a) What is the smallest amount of energy the electron can have? (A trapped electron cannot have zero energy.)

KEY IDEA

Confinement of the electron (a matter wave) to the well leads to quantization of its energy. Because the well is infinitely deep, the allowed energies are given by Eq. 39-4 ($E_n = (h^2/8mL^2)n^2$), with the quantum number n a positive integer.

Lowest energy level: Here, the collection of constants in front of n^2 in Eq. 39-4 is evaluated as

$$\frac{h^2}{8mL^2} = \frac{(6.63 \times 10^{-34}\,\text{J·s})^2}{(8)(9.11 \times 10^{-31}\,\text{kg})(100 \times 10^{-12}\,\text{m})^2}$$

$$= 6.031 \times 10^{-18}\,\text{J}. \qquad (39\text{-}7)$$

The smallest amount of energy the electron can have corresponds to the lowest quantum number, which is $n = 1$ for the ground state of the electron. Thus, Eqs. 39-4 and 39-7 give us

$$E_1 = \left(\frac{h^2}{8mL^2}\right)n^2 = (6.031 \times 10^{-18}\,\text{J})(1^2)$$

$$\approx 6.03 \times 10^{-18}\,\text{J} = 37.7\,\text{eV}. \qquad \text{(Answer)}$$

(b) How much energy must be transferred to the electron if it is to make a quantum jump from its ground state to its second excited state?

KEY IDEA

First a caution: Note that, from Fig. 39-3, the *second* excited state corresponds to the *third* energy level, with quantum number $n = 3$. Then if the electron is to jump from the $n = 1$ level to the $n = 3$ level, the required change in its energy is, from Eq. 39-5,

$$\Delta E_{31} = E_3 - E_1. \qquad (39\text{-}8)$$

Upward jump: The energies E_3 and E_1 depend on the quantum number n, according to Eq. 39-4. Therefore, substituting that equation into Eq. 39-8 for energies E_3 and E_1 and using Eq. 39-7 lead to

$$\Delta E_{31} = \left(\frac{h^2}{8mL^2}\right)(3)^2 - \left(\frac{h^2}{8mL^2}\right)(1)^2$$

$$= \frac{h^2}{8mL^2}(3^2 - 1^2)$$

$$= (6.031 \times 10^{-18}\,\text{J})(8)$$

$$= 4.83 \times 10^{-17}\,\text{J} = 301\,\text{eV}. \qquad \text{(Answer)}$$

(c) If the electron gains the energy for the jump from energy level E_1 to energy level E_3 by absorbing light, what light wavelength is required?

KEY IDEAS

(1) If light is to transfer energy to the electron, the transfer must be by photon absorption. (2) The photon's energy must equal the energy difference ΔE between the initial energy

level of the electron and a higher level, according to Eq. 39-6 ($hf = \Delta E$). Otherwise, a photon *cannot* be absorbed.

Wavelength: Substituting c/λ for f, we can rewrite Eq. 39-6 as

$$\lambda = \frac{hc}{\Delta E}. \tag{39-9}$$

For the energy difference ΔE_{31} we found in (b), this equation gives us

$$\lambda = \frac{hc}{\Delta E_{31}}$$
$$= \frac{(6.63 \times 10^{-34}\ \text{J} \cdot \text{s})(2.998 \times 10^{8}\ \text{m/s})}{4.83 \times 10^{-17}\ \text{J}}$$
$$= 4.12 \times 10^{-9}\ \text{m}. \qquad \text{(Answer)}$$

(d) Once the electron has been excited to the second excited state, what wavelengths of light can it emit by de-excitation?

KEY IDEAS

1. The electron tends to de-excite, rather than remain in an excited state, until it reaches the ground state ($n = 1$).

2. If the electron is to de-excite, it must lose just enough energy to jump to a lower energy level.

3. If it is to lose energy by emitting light, then the loss of energy must be by emission of a photon.

Downward jumps: Starting in the second excited state (at the $n = 3$ level), the electron can reach the ground state ($n = 1$) by *either* making a quantum jump directly to the ground-state energy level (Fig. 39-5a) or by making two *separate* jumps by way of the $n = 2$ level (Figs. 39-5b and c).

The direct jump involves the same energy difference ΔE_{31} we found in (c). Then the wavelength is the same as we calculated in (c)—except now the wavelength is for light that is emitted, not absorbed. Thus, the electron can jump directly to the ground state by emitting light of wavelength

$$\lambda = 4.12 \times 10^{-9}\ \text{m}. \qquad \text{(Answer)}$$

Following the procedure of part (b), you can show that the energy differences for the jumps of Figs. 39-5b and c are

$$\Delta E_{32} = 3.016 \times 10^{-17}\ \text{J} \quad \text{and} \quad \Delta E_{21} = 1.809 \times 10^{-17}\ \text{J}.$$

From Eq. 39-9, we then find that the wavelength of the light emitted in the first of these jumps (from $n = 3$ to $n = 2$) is

$$\lambda = 6.60 \times 10^{-9}\ \text{m}, \qquad \text{(Answer)}$$

and the wavelength of the light emitted in the second of these jumps (from $n = 2$ to $n = 1$) is

$$\lambda = 1.10 \times 10^{-8}\ \text{m}. \qquad \text{(Answer)}$$

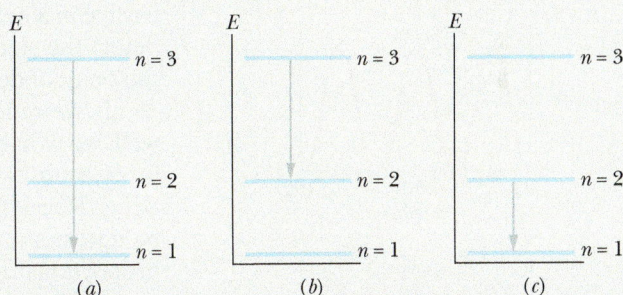

Figure 39-5 De-excitation from the second excited state to the ground state either directly (a) or via the first excited state (b, c).

WILEY PLUS Additional examples, video, and practice available at *WileyPLUS*

39-2 WAVE FUNCTIONS OF A TRAPPED ELECTRON

Learning Objectives

After reading this module, you should be able to . . .

39.12 For an electron trapped in a one-dimensional, infinite potential well, write its wave function in terms of coordinates inside the well and in terms of the quantum number n.

39.13 Identify probability density.

39.14 For an electron trapped in a one-dimensional, infinite potential well in a given state, write the probability density as a function of position inside the well, identify that the probability density is zero outside the well, and calculate the probability of detection between two given coordinates inside the well.

39.15 Identify the correspondence principle.

39.16 Normalize a given wave function and identify what that has to do with the probability of detection.

39.17 Identify that the lowest allowed energy (the zero-point energy) of a trapped electron is not zero.

Key Ideas

● The wave functions for an electron in an infinite, one-dimensional potential well with length L along an x axis are given by

$$\psi_n(x) = \sqrt{\frac{2}{L}} \sin\left(\frac{n\pi}{L}x\right), \qquad \text{for } n = 1, 2, 3, \ldots,$$

where n is the quantum number.

● The product $\psi_n^2(x)\, dx$ is the probability that the electron will be detected in the interval between coordinates x and $x + dx$.

● If the probability density of an electron is integrated over the entire x axis, the total probability must be 1:

$$\int_{-\infty}^{\infty} \psi_n^2(x)\, dx = 1.$$

The probability density must be zero at the infinite walls.

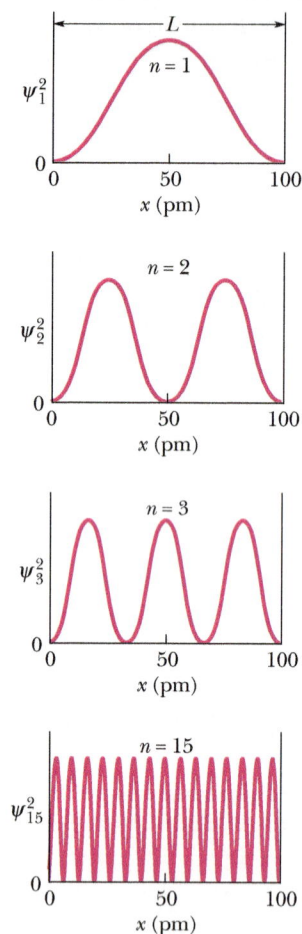

Figure 39-6 The probability density $\psi_n^2(x)$ for four states of an electron trapped in a one-dimensional infinite well; their quantum numbers are $n = 1, 2, 3,$ and 15. The electron is most likely to be found where $\psi_n^2(x)$ is greatest and least likely to be found where $\psi_n^2(x)$ is least.

Wave Functions of a Trapped Electron

If we solve Schrödinger's equation for an electron trapped in a one-dimensional infinite potential well of width L and impose the boundary condition that the solutions be zero at the infinite walls, we find that the wave functions for the electron are given by

$$\psi_n(x) = A \sin\left(\frac{n\pi}{L}x\right), \qquad \text{for } n = 1, 2, 3, \ldots, \qquad (39\text{-}10)$$

for $0 \le x \le L$ (the wave function is zero outside that range). We shall soon evaluate the amplitude constant A in this equation.

Note that the wave functions $\psi_n(x)$ have the same form as the displacement functions $y_n(x)$ for a standing wave on a string stretched between rigid supports (see Eq. 39-2). We can picture an electron trapped in a one-dimensional well between infinite-potential walls as being a standing matter wave.

Probability of Detection

The wave function $\psi_n(x)$ cannot be detected or directly measured in any way—we cannot simply look inside the well to see the wave the way we can see, say, a wave in a bathtub of water. All we can do is insert a probe of some kind to try to detect the electron. At the instant of detection, the electron would materialize at the point of detection, at some position along the x axis within the well.

If we repeated this detection procedure at many positions throughout the well, we would find that the probability of detecting the electron is related to the probe's position x in the well. In fact, they are related by the *probability density* $\psi_n^2(x)$. Recall from Module 38-6 that in general the probability that a particle can be detected in a specified infinitesimal volume centered on a specified point is proportional to $|\psi_n^2|$. Here, with the electron trapped in a one-dimensional well, we are concerned only with detection of the electron along the x axis. Thus, the probability density $\psi_n^2(x)$ here is a probability per unit length along the x axis. (We can omit the absolute value sign here because $\psi_n(x)$ in Eq. 39-10 is a real quantity, not a complex one.) The probability $p(x)$ that an electron can be detected at position x within the well is

$$\begin{pmatrix} \text{probability } p(x) \\ \text{of detection in width } dx \\ \text{centered on position } x \end{pmatrix} = \begin{pmatrix} \text{probability density } \psi_n^2(x) \\ \text{at position } x \end{pmatrix} (\text{width } dx),$$

or

$$p(x) = \psi_n^2(x)\, dx. \qquad (39\text{-}11)$$

From Eq. 39-10, we see that the probability density $\psi_n^2(x)$ is

$$\psi_n^2(x) = A^2 \sin^2\left(\frac{n\pi}{L}x\right), \qquad \text{for } n = 1, 2, 3, \ldots, \qquad (39\text{-}12)$$

for the range $0 \le x \le L$ (the probability density is zero outside that range). Figure 39-6 shows $\psi_n^2(x)$ for $n = 1, 2, 3,$ and 15 for an electron in an infinite well whose width L is 100 pm.

To find the probability that the electron can be detected in any finite section of the well—say, between point x_1 and point x_2—we must integrate $p(x)$ between those points. Thus, from Eqs. 39-11 and 39-12,

$$\begin{pmatrix} \text{probability of detection} \\ \text{between } x_1 \text{ and } x_2 \end{pmatrix} = \int_{x_1}^{x_2} p(x)$$

$$= \int_{x_1}^{x_2} A^2 \sin^2\left(\frac{n\pi}{L}x\right) dx. \qquad (39\text{-}13)$$

If the range Δx in which we search for the electron is much smaller than the well length L, then we can usually approximate the integral in Eq. 39-13 as being equal to the product $p(x) \Delta x$, with $p(x)$ evaluated in the center of Δx.

If classical physics prevailed, we would expect the trapped electron to be detectable with equal probabilities in all parts of the well. From Fig. 39-6 we see that it is not. For example, inspection of that figure or of Eq. 39-12 shows that for the state with $n = 2$, the electron is most likely to be detected near $x = 25$ pm and $x = 75$ pm. It can be detected with near-zero probability near $x = 0, x = 50$ pm, and $x = 100$ pm.

The case of $n = 15$ in Fig. 39-6 suggests that as n increases, the probability of detection becomes more and more uniform across the well. This result is an instance of a general principle called the **correspondence principle:**

At large enough quantum numbers, the predictions of quantum physics merge smoothly with those of classical physics.

This principle, first advanced by Danish physicist Niels Bohr, holds for all quantum predictions.

✅ **Checkpoint 2**

The figure shows three infinite potential wells of widths L, $2L$, and $3L$; each contains an electron in the state for which $n = 10$. Rank the wells according to (a) the number of maxima for the probability density of the electron and (b) the energy of the electron, greatest first.

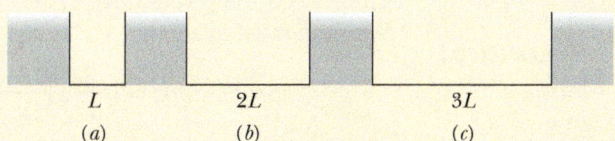

L $\qquad$ $2L$ $\qquad$ $3L$
(a) $\qquad$ (b) $\qquad$ (c)

Normalization

The product $\psi_n^2(x)\, dx$ gives the probability that an electron in an infinite well can be detected in the interval of the x axis that lies between x and $x + dx$. We know that the electron must be *somewhere* in the infinite well; so it must be true that

$$\int_{-\infty}^{+\infty} \psi_n^2(x)\,dx = 1 \quad \text{(normalization equation),} \tag{39-14}$$

because the probability 1 corresponds to certainty. Although the integral is taken over the entire x axis, only the region from $x = 0$ to $x = L$ makes any contribution to the probability. Graphically, the integral in Eq. 39-14 represents the area under each of the plots of Fig. 39-6. If we substitute $\psi_n^2(x)$ from Eq. 39-12 into Eq. 39-14, we find that $A = \sqrt{2/L}$. This process of using Eq. 39-14 to evaluate the amplitude of a wave function is called **normalizing** the wave function. The process applies to *all* one-dimensional wave functions.

Zero-Point Energy

Substituting $n = 1$ in Eq. 39-4 defines the state of lowest energy for an electron in an infinite potential well, the ground state. That is the state the confined electron will occupy unless energy is supplied to it to raise it to an excited state.

The question arises: Why can't we include $n = 0$ among the possibilities listed for n in Eq. 39-4? Putting $n = 0$ in this equation would indeed yield a ground-state energy of zero. However, putting $n = 0$ in Eq. 39-12 would also yield $\psi_n^2(x) = 0$ for all x, which we can interpret only to mean that there is no electron in the well. We know that there is; so $n = 0$ is not a possible quantum number.

It is an important conclusion of quantum physics that confined systems cannot exist in states with zero energy. They must always have a certain minimum energy called the **zero-point energy.**

We can make the zero-point energy as small as we like by making the infinite well wider—that is, by increasing L in Eq. 39-4 for $n = 1$. In the limit as $L \to \infty$, the zero-point energy $E_1 \to 0$. However, the electron is then a free particle, no longer confined in the x direction. Also, because the energy of a free particle is not quantized, that energy can have any value, including zero. Only a confined particle must have a finite zero-point energy and can never be at rest.

✓ Checkpoint 3

Each of the following particles is confined to an infinite well, and all four wells have the same width: (a) an electron, (b) a proton, (c) a deuteron, and (d) an alpha particle. Rank their zero-point energies, greatest first. The particles are listed in order of increasing mass.

Sample Problem 39.02 Detection probability in a 1D infinite potential well

A ground-state electron is trapped in the one-dimensional infinite potential well of Fig. 39-2, with width $L = 100$ pm.

(a) What is the probability that the electron can be detected in the left one-third of the well ($x_1 = 0$ to $x_2 = L/3$)?

KEY IDEAS

(1) If we probe the left one-third of the well, there is no guarantee that we will detect the electron. However, we can calculate the probability of detecting it with the integral of Eq. 39-13. (2) The probability very much depends on which state the electron is in—that is, the value of quantum number n.

Calculations: Because here the electron is in the ground state, we set $n = 1$ in Eq. 39-13. We also set the limits of integration as the positions $x_1 = 0$ and $x_2 = L/3$ and set the amplitude constant A as $\sqrt{2/L}$ (so that the wave function is normalized). We then see that

$$\begin{pmatrix} \text{probability of detection} \\ \text{in left one-third} \end{pmatrix} = \int_0^{L/3} \frac{2}{L} \sin^2\left(\frac{1\pi}{L}x\right) dx.$$

We could find this probability by substituting 100×10^{-12} m for L and then using a graphing calculator or a computer math package to evaluate the integral. Here, however, we shall evaluate the integral "by hand." First we switch to a new integration variable y:

$$y = \frac{\pi}{L}x \quad \text{and} \quad dx = \frac{L}{\pi} dy.$$

From the first of these equations, we find the new limits of integration to be $y_1 = 0$ for $x_1 = 0$ and $y_2 = \pi/3$ for $x_2 = L/3$. We then must evaluate

$$\text{probability} = \left(\frac{2}{L}\right)\left(\frac{L}{\pi}\right)\int_0^{\pi/3} (\sin^2 y)\, dy.$$

Using integral 11 in Appendix E, we then find

$$\text{probability} = \frac{2}{\pi}\left(\frac{y}{2} - \frac{\sin 2y}{4}\right)_0^{\pi/3} = 0.20.$$

Thus, we have

$$\begin{pmatrix} \text{probability of detection} \\ \text{in left one-third} \end{pmatrix} = 0.20. \qquad \text{(Answer)}$$

That is, if we repeatedly probe the left one-third of the well, then on average we can detect the electron with 20% of the probes.

(b) What is the probability that the electron can be detected in the middle one-third of the well?

Reasoning: We now know that the probability of detection in the left one-third of the well is 0.20. By symmetry, the probability of detection in the right one-third of the well is also 0.20. Because the electron is certainly in the well, the probability of detection in the entire well is 1. Thus, the probability of detection in the middle one-third of the well is

$$\begin{pmatrix} \text{probability of detection} \\ \text{in middle one-third} \end{pmatrix} = 1 - 0.20 - 0.20$$

$$= 0.60. \qquad \text{(Answer)}$$

Sample Problem 39.03 Normalizing wave functions in a 1D infinite potential well

Evaluate the amplitude constant A in Eq. 39-10 for an infinite potential well extending from $x = 0$ to $x = L$.

KEY IDEA

The wave functions of Eq. 39-10 must satisfy the normalization requirement of Eq. 39-14, which states that the probability that the electron can be detected somewhere along the x axis is 1.

Calculations: Substituting Eq. 39-10 into Eq. 39-14 and taking the constant A outside the integral yield

$$A^2 \int_0^L \sin^2\left(\frac{n\pi}{L}x\right) dx = 1. \qquad (39\text{-}15)$$

We have changed the limits of the integral from $-\infty$ and $+\infty$ to 0 and L because the "outside" wave function is zero.

We can simplify the indicated integration by changing the variable from x to the dimensionless variable y, where

$$y = \frac{n\pi}{L} x, \qquad (39\text{-}16)$$

hence

$$dx = \frac{L}{n\pi} dy.$$

When we change the variable, we must also change the integration limits (again). Equation 39-16 tells us that $y = 0$ when $x = 0$ and that $y = n\pi$ when $x = L$; thus 0 and $n\pi$ are our new limits. With all these substitutions, Eq. 39-15 becomes

$$A^2 \frac{L}{n\pi} \int_0^{n\pi} (\sin^2 y)\, dy = 1.$$

We can use integral 11 in Appendix E to evaluate the integral, obtaining the equation

$$\frac{A^2 L}{n\pi} \left[\frac{y}{2} - \frac{\sin 2y}{4} \right]_0^{n\pi} = 1.$$

Evaluating at the limits yields

$$\frac{A^2 L}{n\pi} \frac{n\pi}{2} = 1;$$

thus

$$A = \sqrt{\frac{2}{L}}. \qquad \text{(Answer)} \quad (39\text{-}17)$$

This result tells us that the dimension for A^2, and thus for $\psi_n^2(x)$, is an inverse length. This is appropriate because the probability density of Eq. 39-12 is a probability *per unit length*.

WILEY
PLUS   Additional examples, video, and practice available at *WileyPLUS*

39-3 AN ELECTRON IN A FINITE WELL

Learning Objectives

After reading this module, you should be able to . . .

39.18 Sketch a one-dimensional finite potential well, indicating the length and height.

39.19 For an electron trapped in a finite well with given energy levels, sketch the energy-level diagram, indicate the nonquantized region, and compare the energies and de Broglie wavelengths with those of an infinite well of the same length.

39.20 For an electron trapped in a finite well, explain (in principle) how the wave functions for the allowed states are determined.

39.21 For an electron trapped in a finite well with a given quantum number, sketch the probability density as a function of position across the well and into the walls.

39.22 Identify that a trapped electron can exist in only the allowed states and relate that energy of the state to the kinetic energy of the electron.

39.23 Calculate the energy that an electron must absorb or emit to move between the allowed states or between an allowed state and any value in the nonquantized region.

39.24 If a quantum jump involves light, apply the relationship between the energy change and the frequency and wavelength associated with the photon.

39.25 From a given allowed state in a finite well, calculate the minimum energy required for the electron to escape and the kinetic energy of the escaped electron if provided more than that minimal energy.

39.26 Identify the emission and absorption spectra of an electron in a one-dimensional infinite potential well, including escaping the trap and falling into the trap.

Key Ideas

● The wave function for an electron in a finite, one-dimensional potential well extends into the walls, where the wave function decreases exponentially with depth.

● Compared to the states in an infinite well of the same size, the states in a finite well have a limited number, longer de Broglie wavelengths, and lower energies.

An Electron in a Finite Well

A potential energy well of infinite depth is an idealization. Figure 39-7 shows a realizable potential energy well—one in which the potential energy of an electron outside the well is not infinitely great but has a finite positive value U_0,

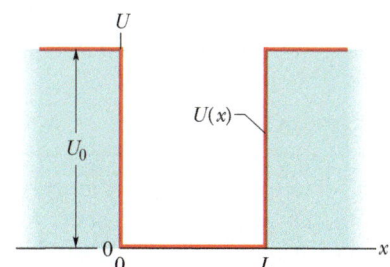

Figure 39-7 A *finite* potential energy well. The depth of the well is U_0 and its width is L. As in the infinite potential well of Fig. 39-2, the motion of the trapped electron is restricted to the x direction.

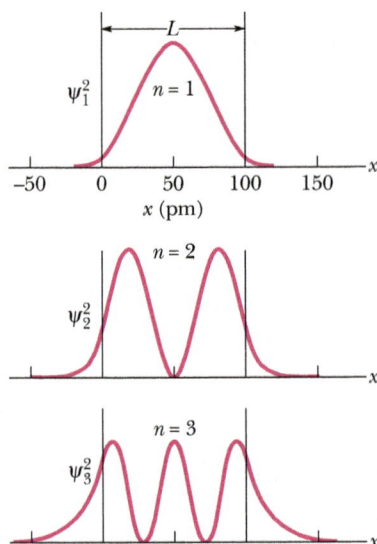

Figure 39-8 The first three probability densities $\psi_n^2(x)$ for an electron confined to a finite potential well of depth $U_0 = 450$ eV and width $L = 100$ pm. Only states $n = 1$, 2, 3, and 4 are allowed.

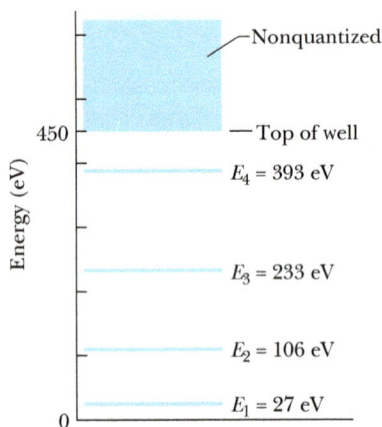

Figure 39-9 The energy-level diagram corresponding to the probability densities of Fig. 39-8. If an electron is trapped in the finite potential well, it can have only the energies corresponding to $n = 1, 2, 3,$ and 4. If it has an energy of 450 eV or greater, it is not trapped and its energy is not quantized.

called the **well depth.** The analogy between waves on a stretched string and matter waves fails us for wells of finite depth because we can no longer be sure that matter wave nodes exist at $x = 0$ and at $x = L$. (As we shall see, they don't.)

To find the wave functions describing the quantum states of an electron in the finite well of Fig. 39-7, we *must* resort to Schrödinger's equation, the basic equation of quantum physics. From Module 38-6 recall that, for motion in one dimension, we use Schrödinger's equation in the form of Eq. 38-19:

$$\frac{d^2\psi}{dx^2} + \frac{8\pi^2 m}{h^2}[E - U(x)]\psi = 0. \qquad (39\text{-}18)$$

Rather than attempting to solve this equation for the finite well, we simply state the results for particular numerical values of U_0 and L. Figure 39-8 shows three results as graphs of $\psi_n^2(x)$, the probability density, for a well with $U_0 = 450$ eV and $L = 100$ pm.

The probability density $\psi_n^2(x)$ for each graph in Fig. 39-8 satisfies Eq. 39-14, the normalization equation; so we know that the areas under all three probability density plots are numerically equal to 1.

If you compare Fig. 39-8 for a finite well with Fig. 39-6 for an infinite well, you will see one striking difference: For a finite well, the electron matter wave penetrates the walls of the well—into a region in which Newtonian mechanics says the electron cannot exist. This penetration should not be surprising because we saw in Module 38-9 that an electron can tunnel through a potential energy barrier. "Leaking" into the walls of a finite potential energy well is a similar phenomenon. From the plots of ψ^2 in Fig. 39-8, we see that the leakage is greater for greater values of quantum number n.

Because a matter wave *does* leak into the walls of a finite well, the wavelength λ for any given quantum state is greater when the electron is trapped in a finite well than when it is trapped in an infinite well of the same length L. Equation 39-3 ($\lambda = h/\sqrt{2mE}$) then tells us that the energy E for an electron in any given state is less in the finite well than in the infinite well.

That fact allows us to approximate the energy-level diagram for an electron trapped in a finite well. As an example, we can approximate the diagram for the finite well of Fig. 39-8, which has width $L = 100$ pm and depth $U_0 = 450$ eV. The energy-level diagram for an *infinite* well of that width is shown in Fig. 39-3. First we remove the portion of Fig. 39-3 above 450 eV. Then we shift the remaining four energy levels down, shifting the level for $n = 4$ the most because the wave leakage into the walls is greatest for $n = 4$. The result is approximately the energy-level diagram for the finite well. The actual diagram is Fig. 39-9.

In that figure, an electron with an energy greater than U_0 ($= 450$ eV) has too much energy to be trapped in the finite well. Thus, it is not confined, and its energy is not quantized; that is, its energy is not restricted to certain values. To reach this *nonquantized* portion of the energy-level diagram and thus to be free, a trapped electron must somehow obtain enough energy to have a mechanical energy of 450 eV or greater.

Sample Problem 39.04 Electron escaping from a finite potential well

Suppose a finite well with $U_0 = 450$ eV and $L = 100$ pm confines a single electron in its ground state.

(a) What wavelength of light is needed to barely free it with a single photon absorption?

KEY IDEA

For the electron to escape, it must receive enough energy to jump to the nonquantized energy region of Fig. 39-9 and end up with an energy of at least U_0 ($= 450$ eV).

Barely escaping: The electron is initially in its ground state, with an energy of $E_1 = 27$ eV. So, to barely become free, it must receive an energy of

$$U_0 - E_1 = 450 \text{ eV} - 27 \text{ eV} = 423 \text{ eV}.$$

Thus the photon must have this much energy. From Eq. 39-6 ($hf = E_{high} - E_{low}$), with c/λ substituted for f, we write

$$\frac{hc}{\lambda} = U_0 - E_1,$$

from which we find

$$\lambda = \frac{hc}{U_0 - E_1}$$

$$= \frac{(6.63 \times 10^{-34}\, \text{J} \cdot \text{s})(3.00 \times 10^{8}\, \text{m/s})}{(423\, \text{eV})(1.60 \times 10^{-19}\, \text{J/eV})}$$

$$= 2.94 \times 10^{-9}\, \text{m} = 2.94\, \text{nm.} \qquad \text{(Answer)}$$

Thus, if $\lambda = 2.94$ nm, the electron just barely escapes.

(b) Can the ground-state electron absorb light with $\lambda = 2.00$ nm? If so, what then is the electron's energy?

KEY IDEAS

1. In (a) we found that light of 2.94 nm will just barely free the electron from the potential well.
2. We are now considering light with a shorter wavelength of 2.00 nm and thus a greater energy per photon ($hf = hc/\lambda$).
3. Hence, the electron *can* absorb a photon of this light. The

energy transfer will not only free the electron but will also provide it with more kinetic energy. Further, because the electron is then no longer trapped, its energy is not quantized.

More than escaping: The energy transferred to the electron is the photon energy:

$$hf = h\frac{c}{\lambda} = \frac{(6.63 \times 10^{-34}\, \text{J} \cdot \text{s})(3.00 \times 10^{8}\, \text{m/s})}{2.00 \times 10^{-9}\, \text{m}}$$

$$= 9.95 \times 10^{-17}\, \text{J} = 622\, \text{eV.}$$

From (a), the energy required to just barely free the electron from the potential well is $U_0 - E_1$ (= 423 eV). The remainder of the 622 eV goes to kinetic energy. Thus, the kinetic energy of the freed electron is

$$K = hf - (U_0 - E_1)$$

$$= 622\, \text{eV} - 423\, \text{eV} = 199\, \text{eV.} \quad \text{(Answer)}$$

PLUS Additional examples, video, and practice available at *WileyPLUS*

39-4 TWO- AND THREE-DIMENSIONAL ELECTRON TRAPS

Learning Objectives

After reading this module, you should be able to . . .

39.27 Discuss nanocrystallites as being electron traps and explain how their threshold wavelength can determine their color.

39.28 Identify quantum dots and quantum corrals.

39.29 For a given state of an electron in an infinite potential well with two or three dimensions, write equations for the wave function and probability density and then calculate the probability of detection for a given range in the well.

39.30 For a given state of an electron in an infinite potential well with two or three dimensions, calculate the allowed

energies and draw an energy-level diagram, complete with labels for the quantum numbers, the ground state, and several excited states.

39.31 Identify degenerate states.

39.32 Calculate the energy that an electron must absorb or emit to move between the allowed states in a 2D or 3D trap.

39.33 If a quantum jump involves light, apply the relationships between the energy change and the frequency and wavelength associated with the photon.

Key Ideas

● The quantized energies for an electron trapped in a two-dimensional infinite potential well that forms a rectangular corral are

$$E_{nx,ny} = \frac{h^2}{8m}\left(\frac{n_x^2}{L_x^2} + \frac{n_y^2}{L_y^2}\right),$$

where n_x is a quantum number for well width L_x and n_y is a quantum number for well width L_y.

● The wave functions for an electron in a two-dimensional well are given by

$$\psi_{nx,ny} = \sqrt{\frac{2}{L_x}}\sin\left(\frac{n_x\pi}{L_x}x\right)\sqrt{\frac{2}{L_y}}\sin\left(\frac{n_y\pi}{L_y}y\right).$$

More Electron Traps

Here we discuss three types of artificial electron traps.

Nanocrystallites

Perhaps the most direct way to construct a potential energy well in the laboratory is to prepare a sample of a semiconducting material in the form of a powder

From *Scientific American*, January 1993, page 119. Reproduced with permission of Michael Steigerwald.

Figure 39-10 Two samples of powdered cadmium selenide, a semiconductor, differing only in the size of their granules. Each granule serves as an electron trap. The lower sample has the larger granules and consequently the smaller spacing between energy levels and the lower photon energy threshold for the absorption of light. Light not absorbed is scattered, causing the sample to scatter light of greater wavelength and appear red. The upper sample, because of its smaller granules, and consequently its larger level spacing and its larger energy threshold for absorption, appears yellow.

whose granules are small—in the nanometer range—and of uniform size. Each such granule—each **nanocrystallite**—acts as a potential well for the electrons trapped within it.

Equation 39-4 ($E = (h^2/8mL^2)n^2$) shows that we can increase the energy-level values of an electron trapped in an infinite well by reducing the width L of the well. This would also shift the photon energies that the well can absorb to higher values and thus shift the corresponding wavelengths to shorter values.

These general results are also true for a well formed by a nanocrystallite. A given nanocrystallite can absorb photons with an energy above a certain threshold energy E_t ($= hf_t$) and thus wavelengths below a corresponding threshold wavelength

$$\lambda_t = \frac{c}{f_t} = \frac{ch}{E_t}.$$

Light with any wavelength longer than λ_t is scattered by the nanocrystallite instead of being absorbed. The color we attribute to the nanocrystallite is then determined by the wavelength composition of the scattered light we intercept.

If we reduce the size of the nanocrystallite, the value of E_t is increased, the value of λ_t is decreased, and the light that is scattered to us changes in its wavelength composition. Thus, the color we attribute to the nanocrystallite changes. As an example, Fig. 39-10 shows two samples of the semiconductor cadmium selenide, each consisting of a powder of nanocrystallites of uniform size. The lower sample scatters light at the red end of the spectrum. The upper sample differs from the lower sample *only* in that the upper sample is composed of smaller nanocrystallites. For this reason its threshold energy E_t is greater and, from above, its threshold wavelength λ_t is shorter, in the green range of visible light. Thus, the sample now scatters both red and yellow. Because the yellow component happens to be brighter, the sample's color is now dominated by the yellow. The striking contrast in color between the two samples is compelling evidence of the quantization of the energies of trapped electrons and the dependence of these energies on the size of the electron trap.

Quantum Dots

The highly developed techniques used to fabricate computer chips can be used to construct, atom by atom, individual potential energy wells that behave, in many respects, like artificial atoms. These **quantum dots,** as they are usually called, have promising applications in electron optics and computer technology.

In one such arrangement, a "sandwich" is fabricated in which a thin layer of a semiconducting material, shown in purple in Fig. 39-11a, is deposited between two insulating layers, one of which is much thinner than the other. Metal end caps with conducting leads are added at both ends. The materials are chosen to ensure that the potential energy of an electron in the central layer is less than it is

Figure 39-11 A quantum dot, or "artificial atom." (*a*) A central semiconducting layer forms a potential energy well in which electrons are trapped. The lower insulating layer is thin enough to allow electrons to be added to or removed from the central layer by barrier tunneling if an appropriate voltage is applied between the leads. (*b*) A photograph of an actual quantum dot. The central purple band is the electron confinement region.

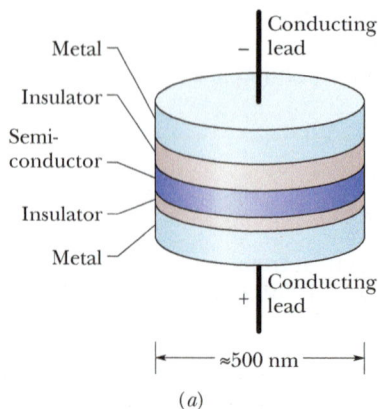

(*a*)

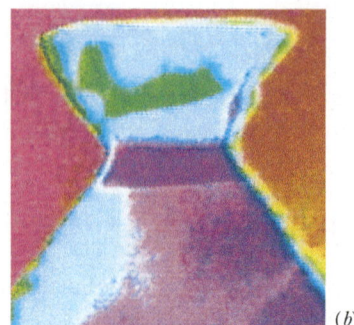

(*b*)

From *Scientific American*, September 1995, page 67. Image reproduced with permission of H. Temkin, Texas Tech University

in the two insulating layers, causing the central layer to act as a potential energy well. Figure 39-11*b* is a photograph of an actual quantum dot; the well in which individual electrons can be trapped is the purple region.

The lower (but not the upper) insulating layer in Fig. 39-11*a* is thin enough to permit electrons to tunnel through it if an appropriate potential difference is applied between the leads. In this way the number of electrons confined to the well can be controlled. The arrangement does indeed behave like an artificial atom with the property that the number of electrons it contains can be controlled. Quantum dots can be constructed in two-dimensional arrays that could well form the basis for computing systems of great speed and storage capacity.

Quantum Corrals

When a scanning tunneling microscope (described in Module 38-9) is in operation, its tip exerts a small force on isolated atoms that may be located on an otherwise smooth surface. By careful manipulation of the position of the tip, such isolated atoms can be "dragged" across the surface and deposited at another location. Using this technique, scientists at IBM's Almaden Research Center moved iron atoms across a carefully prepared copper surface, forming the atoms into a circle (Fig. 39-12), which they named a **quantum corral.** Each iron atom in the circle is nestled in a hollow in the copper surface, equidistant from three nearest-neighbor copper atoms. The corral was fabricated at a low temperature (about 4 K) to minimize the tendency of the iron atoms to move randomly about on the surface because of their thermal energy.

The ripples within the corral are due to matter waves associated with electrons that can move over the copper surface but are largely trapped in the potential well of the corral. The dimensions of the ripples are in excellent agreement with the predictions of quantum theory.

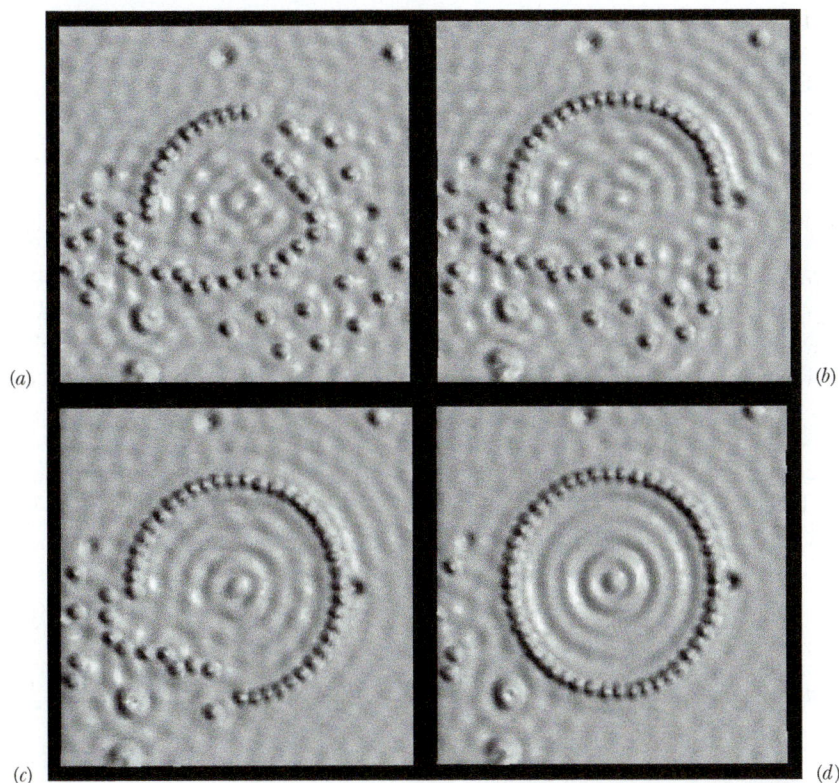

(a)   (b)

(c)   (d)

From M. F. Crommie, C. P. Lutz, D. M. Eigler, *Science*, 262: 218, 1993. Reprinted with permission from AAAS.

Figure 39-12 A quantum corral during four stages of construction. Note the appearance of ripples caused by electrons trapped in the corral when it is almost complete.

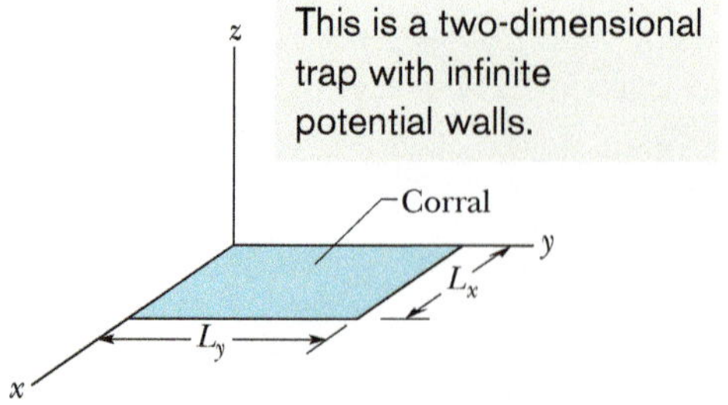

Figure 39-13 A rectangular corral—a two-dimensional version of the infinite potential well of Fig. 39-2—with widths L_x and L_y.

Two- and Three-Dimensional Electron Traps

In the next module, we shall discuss the hydrogen atom as being a three-dimensional finite potential well. As a warm-up for the hydrogen atom, let us extend our discussion of infinite potential wells to two and three dimensions.

Rectangular Corral

Figure 39-13 shows the rectangular area to which an electron can be confined by the two-dimensional version of Fig. 39-2—a two-dimensional infinite potential well of widths L_x and L_y that forms a rectangular corral. The corral might be on the surface of a body that somehow prevents the electron from moving parallel to the z axis and thus from leaving the surface. You have to imagine infinite potential energy functions (like $U(x)$ in Fig. 39-2) along each side of the corral, keeping the electron within the corral.

Solution of Schrödinger's equation for the rectangular corral of Fig. 39-13 shows that, for the electron to be trapped, its matter wave must fit into each of the two widths separately, just as the matter wave of a trapped electron must fit into a one-dimensional infinite well. This means the wave is separately quantized in width L_x and in width L_y. Let n_x be the quantum number for which the matter wave fits into width L_x, and let n_y be the quantum number for which the matter wave fits into width L_y. As with a one-dimensional potential well, these quantum numbers can be only positive integers. We can extend Eqs. 39-10 and 39-17 to write the normalized wave function as

$$\psi_{nx,ny} = \sqrt{\frac{2}{L_x}}\sin\left(\frac{n_x\pi}{L}x\right)\sqrt{\frac{2}{L_y}}\sin\left(\frac{n_y\pi}{L}y\right), \qquad (39\text{-}19)$$

The energy of the electron depends on both quantum numbers and is the sum of the energy the electron would have if it were confined along the x axis alone and the energy it would have if it were confined along the y axis alone. From Eq. 39-4, we can write this sum as

$$E_{nx,ny} = \left(\frac{h^2}{8mL_x^2}\right)n_x^2 + \left(\frac{h^2}{8mL_y^2}\right)n_y^2 = \frac{h^2}{8m}\left(\frac{n_x^2}{L_x^2} + \frac{n_y^2}{L_y^2}\right). \qquad (39\text{-}20)$$

Excitation of the electron by photon absorption and de-excitation of the electron by photon emission have the same requirements as for one-dimensional traps. Now, however, two quantum numbers (n_x and n_y) are involved. Because of that, different states might have the same energy; such states and their energy levels are said to be *degenerate*.

Rectangular Box

An electron can also be trapped in a three-dimensional infinite potential well—a *box*. If the box is rectangular as in Fig. 39-14, then Schrödinger's equation shows us that we can write the energy of the electron as

$$E_{nx,ny,nz} = \frac{h^2}{8m}\left(\frac{n_x^2}{L_x^2} + \frac{n_y^2}{L_y^2} + \frac{n_z^2}{L_z^2}\right). \qquad (39\text{-}21)$$

Figure 39-14 A rectangular box—a three-dimensional version of the infinite potential well of Fig. 39-2—with widths L_x, L_y, and L_z.

Here n_z is a third quantum number, for fitting the matter wave into width L_z.

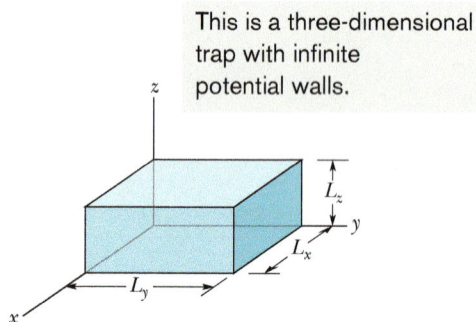

✓ **Checkpoint 4**

In the notation of Eq. 39-20, is $E_{0,0}$, $E_{1,0}$, $E_{0,1}$, or $E_{1,1}$ the ground-state energy of an electron in a (two-dimensional) rectangular corral?

Sample Problem 39.05   Energy levels in a 2D infinite potential well

An electron is trapped in a square corral that is a two-dimensional infinite potential well (Fig. 39-13) with widths $L_x = L_y$.

(a) Find the energies of the lowest five possible energy levels for this trapped electron, and construct the corresponding energy-level diagram.

KEY IDEA

Because the electron is trapped in a two-dimensional well that is rectangular, the electron's energy depends on two quantum numbers, n_x and n_y, according to Eq. 39-20.

Energy levels: Because the well here is square, we can let the widths be $L_x = L_y = L$. Then Eq. 39-20 simplifies to

$$E_{nx,ny} = \frac{h^2}{8mL^2}(n_x^2 + n_y^2). \qquad (39\text{-}22)$$

The lowest energy states correspond to low values of the quantum numbers n_x and n_y, which are the positive integers $1, 2, \ldots, \infty$. Substituting those integers for n_x and n_y in Eq. 39-22, starting with the lowest value 1, we can obtain the energy values as listed in Table 39-1. There we can see that several of the pairs of quantum numbers (n_x, n_y) give the same

These are the lowest five energy levels allowed the electron. Different quantum states may have the same energy.

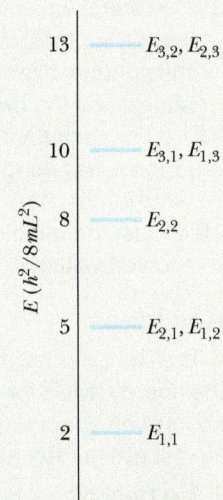

Figure 39-15 Energy-level diagram for an electron trapped in a square corral.

energy. For example, the (1, 2) and (2, 1) states both have an energy of $5(h^2/8mL^2)$. Each such pair is associated with degenerate energy levels. Note also that, perhaps surprisingly, the (4, 1) and (1, 4) states have less energy than the (3, 3) state.

From Table 39-1 (carefully keeping track of degenerate levels), we can construct the energy-level diagram of Fig. 39-15.

(b) As a multiple of $h^2/8mL^2$, what is the energy difference between the ground state and the third excited state?

Energy difference: From Fig. 39-15, we see that the ground state is the (1, 1) state, with an energy of $2(h^2/8mL^2)$. We also see that the third excited state (the third state up from the ground state in the energy-level diagram) is the degenerate (1, 3) and (3, 1) states, with an energy of $10(h^2/8mL^2)$. Thus, the difference ΔE between these two states is

$$\Delta E = 10\left(\frac{h^2}{8mL^2}\right) - 2\left(\frac{h^2}{8mL^2}\right) = 8\left(\frac{h^2}{8mL^2}\right).$$

(Answer)

Table 39-1 Energy Levels

n_x	n_y	Energy[a]	n_x	n_y	Energy[a]
1	3	10	2	4	20
3	1	10	4	2	20
2	2	8	3	3	18
1	2	5	1	4	17
2	1	5	4	1	17
1	1	2	2	3	13
			3	2	13

[a]In multiples of $h^2/8mL^2$.

WILEY PLUS   Additional examples, video, and practice available at *WileyPLUS*

39-5 THE HYDROGEN ATOM

Learning Objectives

After reading this module, you should be able to . . .

39.34 Identify Bohr's model of the hydrogen atom and explain how he derived the quantized radii and energies.

39.35 For a given quantum number n in the Bohr model, calculate the electron's orbital radius, kinetic energy, potential energy, total energy, orbital period, orbital frequency, momentum, and angular momentum.

39.36 Distinguish the Bohr and Schrödinger descriptions of

the hydrogen atom, including the discrepancy between the allowed angular momentum values.

39.37 For a hydrogen atom, apply the relationship between the quantized energies E_n and the quantum number n.

39.38 For a given jump in hydrogen, between quantized states or between a quantized state and a nonquantized state, calculate the change in energy and, if light is involved, the associated energy, frequency, wavelength, and momentum of the photon.

39.39 Sketch an energy-level diagram for hydrogen, identifying the ground state, several of the excited states, the nonquantized region, the Paschen series, the Balmer series, and the Lyman series (including the series limits).

39.40 For each transition series, identify the jumps giving the longest wavelength, the shortest wavelength for downward jumps, the series limit, and ionization.

39.41 List the quantum numbers for an atom and indicate the allowed values.

39.42 Given a normalized wave function for a state, find the radial probability density $P(r)$ and the probability of detecting the electron in a given range of radii.

39.43 For ground-state hydrogen, sketch a graph of the radial probability density versus radial distance and locate one Bohr radius a.

39.44 For a given normalized wave function for hydrogen, verify that it satisfies the Schrödinger equation.

39.45 Distinguish shell from subshell.

39.46 Explain a dot plot of a probability density.

Key Ideas

● The Bohr model of the hydrogen atom successfully derived the energy levels for the atom, to explain the emission/absorption spectrum of the atom, but it is incorrect in almost every other aspect.

● The Bohr model is a planetary model in which the electron orbits the central proton with an angular momentum L that is limited to values given by

$$L = n\hbar, \quad \text{for } n = 1, 2, 3, \ldots,$$

where n is a quantum number. The value $L = 0$ is incorrectly disallowed.

● Application of the Schrödinger equation gives the correct values of L and the quantized energies:

$$E_n = -\frac{me^4}{8\varepsilon_0^2 h^2}\frac{1}{n^2} = -\frac{13.60 \text{ eV}}{n^2}, \quad \text{for } n = 1, 2, 3, \ldots.$$

● The atom (or the electron in the atom) can change energy only by jumping between these allowed energies.

● If the jump is by photon absorption (the atom's energy increases) or photon emission (the atom's energy decreases), this restriction in energy changes leads to

$$\frac{1}{\lambda} = R\left(\frac{1}{n_{\text{low}}^2} - \frac{1}{n_{\text{high}}^2}\right),$$

for the wevelength of the light, where R is the Rydberg constant,

$$R = \frac{me^4}{8\varepsilon_0^2 h^3 c} = 1.097\,373 \times 10^7 \text{ m}^{-1}.$$

● The radial probability density $P(r)$ for a state of the hydrogen atom is defined so that $P(r)$ is the probability that the electron will be detected somewhere in the space between two spherical shells of radii r and $r + dr$ that are centered on the nucleus.

● Normalization requires that

$$\int_0^\infty P(r)\,dr = 1.$$

● The probability that the electron will be detected between any two given radii r_1 and r_2 is

$$(\text{probability of detection between } r_1 \text{ and } r_2) = \int_{r_1}^{r_2} P(r)\,dr.$$

The Hydrogen Atom Is an Electron Trap

We now move from artificial or fictitious electron traps to natural ones — atoms. In this chapter we focus on the simplest example, a hydrogen atom, which contains an electron that is trapped by the Coulomb force it experiences from the proton, which is the nucleus of the atom. Because the proton's mass is much greater than the electron's mass, we shall assume that the proton is fixed in place. So, we think of the atom as a fixed potential trap with the electron moving around inside it.

We have now discussed at length that confinement of an electron means that the electron's energy E is quantized and thus so is any change ΔE in its energy. In this module we want to calculate the quantized energies of the electron confined to a hydrogen atom. We shall, in principle at least, apply Schrödinger's equation to the trap, to find those energies and the associated wave functions. However, at the discretion of your instructor, let's take an historical aside to examine how the quantizing of atoms began, back when quantization was a revolutionary concept.

The Bohr Model of Hydrogen, a Lucky Break

By the early 1900s, scientists understood that matter came in tiny pieces called atoms and that an atom of hydrogen contained positive charge $+e$ at its center and negative charge $-e$ (an electron) outside that center. However, no one understood why the electrical attraction between the electron and the positive charge did not simply cause the two to collapse together.

Visible Wavelengths. One clue lay in the experimental fact that a hydrogen atom can emit and absorb only four wavelengths in the visible spectrum (656 nm, 486 nm, 434 nm, and 410 nm). Why did it not emit all wavelengths as, say, a hot blackbody radiator? In 1913, Niels Bohr had a remarkable idea that simultaneously explained not only the four visible wavelengths but also why the atom did not simply collapse. However, as successful as his theory was on those two counts, it turned out to be quite wrong in almost every other aspect of the atom and led to very little success in explaining atoms more complicated than hydrogen. Nevertheless, the Bohr model is historically important because it ushered in the quantum physics of atoms.

Assumptions. To build his model, Bohr made two bold (completely unjustified) assumptions: (1) The electron in a hydrogen atom orbits the nucleus in a circle much like Earth orbits the Sun (Fig. 39-16a). (2) The magnitude of the angular momentum $\vec{L}$ of the electron in its orbit is restricted (quantized) to the values

$$L = n\hbar, \quad \text{for } n = 1, 2, 3, \ldots, \tag{39-23}$$

where $\hbar$ (h-bar) is $h/2\pi$ and n is a positive integer (a quantum number). We are going to follow Bohr's relatively simple arguments to get an equation for the quantized energies of the hydrogen atom, but let's be explicit here: The electron is *not* simply a particle in a planetary orbit and Eq. 39-23 does *not* correctly give the angular momentum values. (For example, $L = 0$ is missing.)

Newton's Second Law. In the orbit picture of Fig. 39-16a, the electron is in uniform circular motion and thus experiences a centripetal force (Fig. 39-16b), which causes a centripetal acceleration. The force is the Coulomb force (Eq. 21-4) between the electron (with charge $-e$) and the proton (with charge $+e$), separated by the orbital radius r. The centripetal acceleration has the magnitude $a = v^2/r$ (Eq. 4-34), where v is the electron's speed. So, we can write Newton's second law for a radial axis as

$$F = ma$$

$$-\frac{1}{4\pi\varepsilon_0}\frac{|-e||e|}{r^2} = m\left(-\frac{v^2}{r}\right), \tag{39-24}$$

where m is the electron mass.

We next introduce quantization by using Bohr's assumption expressed in Eq. 39-23. From Eq. 11-19, the magnitude ℓ of the angular momentum of a particle of mass m and speed v moving in a circle of radius r is $\ell = rmv \sin\phi$, where ϕ (the angle between $\vec{r}$ and $\vec{v}$) is 90°. Replacing L in Eq. 39-23 with $rmv \sin 90°$ gives us

$$rmv = n\hbar,$$

or

$$v = \frac{n\hbar}{rm}. \tag{39-25}$$

Substituting this equation into Eq. 39-24, replacing $\hbar$ with $h/2\pi$, and rearranging,

Figure 39-16 (a) Circular orbit of an electron in the Bohr model of the hydrogen atom. (b) The Coulomb force $\vec{F}$ on the electron is directed radially inward toward the nucleus.

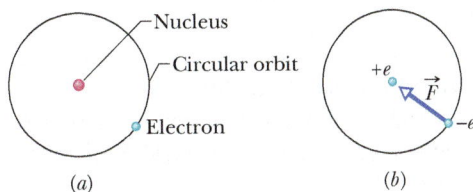

Bohr's model for hydrogen resembles the orbital model of a planet around a star.

we find

$$r = \frac{h^2\varepsilon_0}{\pi m e^2} n^2, \quad \text{for } n = 1, 2, 3, \ldots. \tag{39-26}$$

We can rewrite this as

$$r = an^2, \quad \text{for } n = 1, 2, 3, \ldots, \tag{39-27}$$

where

$$a = \frac{h^2\varepsilon_0}{\pi m e^2} = 5.291\,772 \times 10^{-11}\,\text{m} \approx 52.92\,\text{pm}. \tag{39-28}$$

These last three equations tell us that, in the *Bohr model of the hydrogen atom,* the electron's orbital radius r is quantized and the smallest possible orbital radius (for $n = 1$) is a, which is called the *Bohr radius.* According to the Bohr model, the electron cannot get any closer to the nucleus than orbital radius a, and that is why the attraction between electron and nucleus does not simply collapse them together.

Orbital Energy Is Quantized

Let's next find the energy of the hydrogen atom according to the Bohr model. The electron has kinetic energy $K = \frac{1}{2}mv^2$, and the electron–nucleus system has electric potential energy $U = q_1q_2/4\pi\varepsilon_0 r$ (Eq. 24-46). Again, let q_1 be the electron's charge $-e$ and q_2 be the nuclear charge $+e$. Then the mechanical energy is

$$E = K + U$$

$$= \tfrac{1}{2}mv^2 + \left(-\frac{1}{4\pi\varepsilon_0}\frac{e^2}{r}\right). \tag{39-29}$$

Solving Eq. 39-24 for mv^2 and substituting the result in Eq. 39-29 lead to

$$E = -\frac{1}{8\pi\varepsilon_0}\frac{e^2}{r}. \tag{39-30}$$

Next, replacing r with its equivalent from Eq. 39-26, we have

$$E_n = -\frac{me^4}{8\varepsilon_0^2 h^2}\frac{1}{n^2}, \quad \text{for } n = 1, 2, 3, \ldots, \tag{39-31}$$

where the subscript n on E signals that we have now quantized the energy.

From this equation, Bohr was able to calculate the visible wavelengths emitted and absorbed by hydrogen, but before we discuss how to go from the energy equation to the wavelengths, let's discuss the correct model of the hydrogen atom.

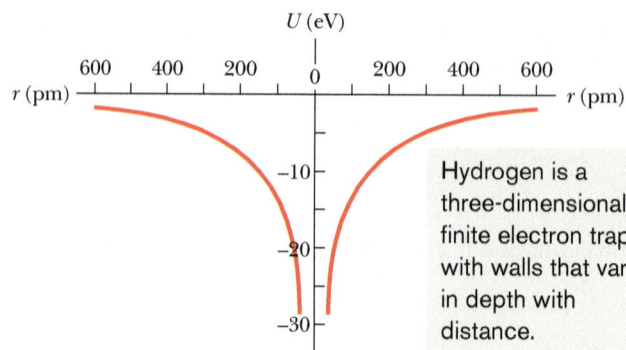

Figure 39-17 The potential energy U of a hydrogen atom as a function of the separation r between the electron and the central proton. The plot is shown twice (on the left and on the right) to suggest the three-dimensional spherically symmetric trap in which the electron is confined.

Schrödinger's Equation and the Hydrogen Atom

In Schrödinger's model of the hydrogen atom, the electron (charge $-e$) is in a potential energy trap due to its electrical attraction to the proton (charge $+e$) at the center of the atom. From Eq. 24-46, we write the potential energy function as

$$U(r) = \frac{-e^2}{4\pi\varepsilon_0 r}.$$ (39-32)

Because this well is three-dimensional, it is more complex than our previous one- and two-dimensional wells. Because this well is finite, it is more complex than the three-dimensional well of Fig. 39-14. Moreover, it does not have sharply defined walls. Rather, its walls vary in depth with radial distance r. Figure 39-17 is probably the best we can do in drawing the hydrogen potential well, but even that drawing takes much effort to interpret.

   To find the allowed energies and wave functions for an electron trapped in the potential well given by Eq. 39-32, we need to apply Schrödinger's equation. With some manipulation, we would find that we could separate the equation into three separate differential equations, two depending on angles and one depending on radial distance r. The solution of the latter equation requires a quantum number n and produces the energy values E_n of the electron:

$$E_n = -\frac{me^4}{8\varepsilon_0^2 h^2}\frac{1}{n^2}, \quad \text{for } n = 1, 2, 3, \ldots,$$ (39-33)

(This equation is exactly what Bohr found by using a very wrong planetary model of the atom.) Evaluating the constants in Eq. 39-33 gives us

$$E_n = -\frac{2.180 \times 10^{-18}\text{ J}}{n^2} = -\frac{13.61\text{ eV}}{n^2}, \quad \text{for } n = 1, 2, 3, \ldots.$$ (39-34)

This equation tells us that the energy E_n of the hydrogen atom is quantized; that is, E_n is restricted by its dependence on the quantum number n. Because the nucleus is assumed to be fixed in place and only the electron has motion, we can assign the energy values of Eq. 39-34 either to the atom as a whole or to the electron alone.

Energy Changes

The energy of a hydrogen atom (or, equivalently, of its electron) changes when the atom emits or absorbs light. As we have seen several times since Eq. 39-6, emission and absorption involve a quantum of light according to

$$hf = \Delta E = E_{\text{high}} - E_{\text{low}}.$$ (39-35)

   Let's make three changes to Eq. 39-35. On the left side, we substitute c/λ for f. On the right side, we use Eq. 39-33 twice to replace the energy terms. Then, with a simple rearrangement, we have

$$\frac{1}{\lambda} = -\frac{me^4}{8\varepsilon_0^2 h^3 c}\left(\frac{1}{n_{\text{high}}^2} - \frac{1}{n_{\text{low}}^2}\right).$$ (39-36)

We can rewrite this as

$$\frac{1}{\lambda} = R\left(\frac{1}{n_{\text{low}}^2} - \frac{1}{n_{\text{high}}^2}\right),$$ (39-37)

in which R is the *Rydberg constant*:

$$R = \frac{me^4}{8\varepsilon_0^2 h^3 c} = 1.097\,373 \times 10^7\text{ m}^{-1}.$$ (39-38)

For example, if we replace n_{low} with 2 in Eq. 39-36 and then restrict n_{high} to be 3, 4, 5, and 6, we generate the four visible wavelengths at which hydrogen can emit or absorb light: 656 nm, 486 nm, 434 nm, and 410 nm.

The Hydrogen Spectrum

Figure 39-18a shows the energy levels corresponding to various values of n in Eq. 39-34. The lowest level, for $n = 1$, is the ground state of hydrogen. Higher levels correspond to excited states, just as we saw for our simpler potential traps. Note several differences, however. (1) The energy levels now have negative values rather than the positive values we previously chose in, for instance, Figs. 39-3 and 39-9. (2) The levels now become progressively closer as we move to higher levels. (3) The energy for the greatest value of n—namely, $n = \infty$—is now $E_\infty = 0$. For any energy greater than $E_\infty = 0$, the electron and proton are not bound together (there is no hydrogen atom), and the $E > 0$ region in Fig. 39-18a is like the nonquantized region for the finite well of Fig. 39-9.

A hydrogen atom can jump between quantized energy levels by emitting or absorbing light at the wavelengths given by Eq. 39-36. Any such wavelength is often called a *line* because of the way it is detected with a spectroscope; thus, a hydrogen atom has *absorption lines* and *emission lines*. A collection of such lines, such as in those in the visible range, is called a **spectrum** of the hydrogen atom.

Series. The lines for hydrogen are said to be grouped into *series*, according to the level at which upward jumps start and downward jumps end. For example, the emission and absorption lines for all possible jumps up from the $n = 1$ level and down to the $n = 1$ level are said to be in the *Lyman series* (Fig. 39-18b), named after the person who first studied those lines. Further, we can say that the Lyman series has a *home-base level* of $n = 1$. Similarly, the *Balmer series* has a home-base level of $n = 2$ (Fig. 39-18c), and the *Paschen series* has a home-base level of $n = 3$ (Fig. 39-18d).

Some of the downward quantum jumps for these three series are shown in Fig. 39-18. Four lines in the Balmer series are in the visible range and are represented in Fig. 39-18c with arrows corresponding to their colors. The shortest of those arrows represents the shortest jump in the series, from the $n = 3$ level to the $n = 2$ level. Thus, that jump involves the smallest change in the electron's energy and the smallest amount of emitted photon energy for the series. The emitted light is red. The next jump in the series, from $n = 4$ to $n = 2$, is longer, the photon energy is greater, the wavelength of the emitted light is shorter, and the light is green. The third, fourth, and fifth arrows represent longer jumps and shorter wavelengths. For the fifth jump, the emitted light is in the ultraviolet range and thus is not visible.

The *series limit* of a series is the line produced by the jump between the home-base level and the highest energy level, which is the level with the limiting quantum number $n = \infty$. Thus, the series limit corresponds to the shortest wavelength in the series.

If a jump is upward into the nonquantized portion of Fig. 39-18, the electron's energy is no longer given by Eq. 39-34 because the electron is no longer trapped in the atom. That is, the hydrogen atom has been *ionized*, meaning that the electron has been removed to a distance so great that the Coulomb force on it from the nucleus is negligible. The atom can be ionized if it absorbs any wavelength shorter than the series limit. The free electron then has only kinetic energy $K \left(= \frac{1}{2} mv^2 \right.$, assuming a nonrelativistic situation).

Quantum Numbers for the Hydrogen Atom

Although the energies of the hydrogen atom states can be described by the single quantum number n, the wave functions describing these states require three quantum numbers, corresponding to the three dimensions in which the electron

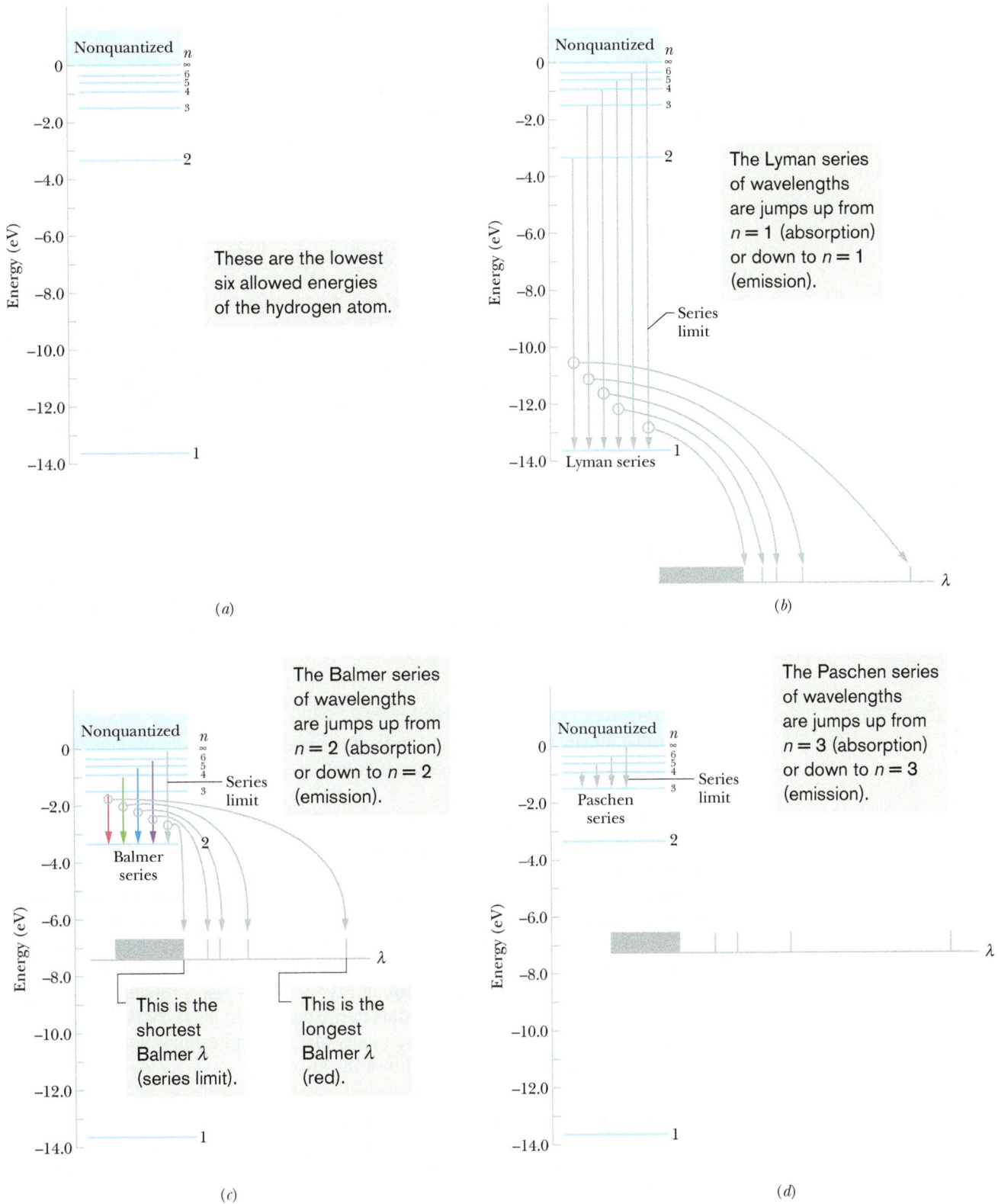

Figure 39-18 (*a*) An energy-level diagram for the hydrogen atom. Some of the transitions for (*b*) the Lyman series, (*c*) the Balmer series, and (*d*) the Paschen series. For each, the longest four wavelengths and the series-limit wavelength are plotted on a wavelength axis. Any wavelength shorter than the series-limit wavelength is allowed.

Table 39-2 Quantum Numbers for the Hydrogen Atom

Symbol	Name	Allowed Values
n	Principal quantum number	$1, 2, 3, \ldots$
ℓ	Orbital quantum number	$0, 1, 2, \ldots, n - 1$
m_ℓ	Orbital magnetic quantum number	$-\ell, -(\ell - 1), \ldots, +(\ell - 1), +\ell$

can move. The three quantum numbers, along with their names and the values that they may have, are shown in Table 39-2.

Each set of quantum numbers (n, ℓ, m_ℓ) identifies the wave function of a particular quantum state. The quantum number n, called the **principal quantum number,** appears in Eq. 39-34 for the energy of the state. The **orbital quantum number** ℓ is a measure of the magnitude of the angular momentum associated with the quantum state. The **orbital magnetic quantum number** m_ℓ is related to the orientation in space of this angular momentum vector. The restrictions on the values of the quantum numbers for the hydrogen atom, as listed in Table 39-2, are not arbitrary but come out of the solution to Schrödinger's equation. Note that for the ground state ($n = 1$), the restrictions require that $\ell = 0$ and $m_\ell = 0$. That is, the hydrogen atom in its ground state has zero angular momentum, which is not predicted by Eq. 39-23 in the Bohr model.

☑ **Checkpoint 5**

(a) A group of quantum states of the hydrogen atom has $n = 5$. How many values of ℓ are possible for states in this group? (b) A subgroup of hydrogen atom states in the $n = 5$ group has $\ell = 3$. How many values of m_ℓ are possible for states in this subgroup?

The Wave Function of the Hydrogen Atom's Ground State

The wave function for the ground state of the hydrogen atom, as obtained by solving the three-dimensional Schrödinger equation and normalizing the result, is

$$\psi(r) = \frac{1}{\sqrt{\pi}\, a^{3/2}}\, e^{-r/a} \quad \text{(ground state),} \tag{39-39}$$

where a ($= 5.291\ 772 \times 10^{-11}$ m) is the Bohr radius. This radius is loosely taken to be the effective radius of a hydrogen atom and turns out to be a convenient unit of length for other situations involving atomic dimensions.

As with other wave functions, $\psi(r)$ in Eq. 39-39 does not have physical meaning but $\psi^2(r)$ does, being the probability density—the probability per unit volume—that the electron can be detected. Specifically, $\psi^2(r)\, dV$ is the probability that the electron can be detected in any given (infinitesimal) volume element dV located at radius r from the center of the atom:

$$\begin{pmatrix} \text{probability of detection} \\ \text{in volume } dV \\ \text{at radius } r \end{pmatrix} = \begin{pmatrix} \text{volume probability} \\ \text{density } \psi^2(r) \\ \text{at radius } r \end{pmatrix} (\text{volume } dV). \tag{39-40}$$

Because $\psi^2(r)$ here depends only on r, it makes sense to choose, as a volume element dV, the volume between two concentric spherical shells whose radii are r and $r + dr$. That is, we take the volume element dV to be

$$dV = (4\pi r^2)\, dr, \tag{39-41}$$

in which $4\pi r^2$ is the surface area of the inner shell and dr is the radial distance between the two shells. Then, combining Eqs. 39-39, 39-40, and 39-41 gives us

$$\begin{pmatrix} \text{probability of detection} \\ \text{in volume } dV \\ \text{at radius } r \end{pmatrix} = \psi^2(r)\, dV = \frac{4}{a^3}\, e^{-2r/a} r^2\, dr. \qquad (39\text{-}42)$$

Describing the probability of detecting an electron is easier if we work with a **radial probability density** $P(r)$ instead of a volume probability density $\psi^2(r)$. This $P(r)$ is a linear probability density such that

$$\begin{pmatrix} \text{radial probability} \\ \text{density } P(r) \\ \text{at radius } r \end{pmatrix} \begin{pmatrix} \text{radial} \\ \text{width } dr \end{pmatrix} = \begin{pmatrix} \text{volume probability} \\ \text{density } \psi^2(r) \\ \text{at radius } r \end{pmatrix} (\text{volume } dV)$$

or

$$P(r)\, dr = \psi^2(r)\, dV. \qquad (39\text{-}43)$$

Substituting for $\psi^2(r)\, dV$ from Eq. 39-42, we obtain

$$P(r) = \frac{4}{a^3} r^2 e^{-2r/a} \quad \text{(radial probability density, hydrogen atom ground state).} \quad (39\text{-}44)$$

To find the probability of detecting the ground-state electron between any two radii r_1 and r_2 (that is, between a spherical shell of radius r_1 and another of radius r_2), we integrate Eq. 39-44 between those two radii:

$$\begin{pmatrix} \text{probability of detection} \\ \text{between } r_1 \text{ and } r_2 \end{pmatrix} = \int_{r_1}^{r_2} P(r)\, dr. \qquad (39\text{-}45)$$

If the radial range $\Delta r \; (= r_2 - r_1)$ in which we search for the electron is small enough such that $P(r)$ does not vary by much over the range, then we can usually approximate the integral in Eq. 39-45 as being equal to the product $P(r)\,\Delta r$, with $P(r)$ evaluated in the center of Δr.

Figure 39-19 is a plot of Eq. 39-44. The area under the plot is unity; that is,

$$\int_0^\infty P(r)\, dr = 1. \qquad (39\text{-}46)$$

This equation states that in a hydrogen atom, the electron must be *somewhere* in the space surrounding the nucleus.

The triangular marker on the horizontal axis of Fig. 39-19 is located one Bohr radius from the origin. The graph tells us that in the ground state of the hydrogen atom, the electron is most likely to be found at about this distance from the center of the atom.

Figure 39-19 conflicts sharply with the popular view that electrons in atoms follow well-defined orbits like planets moving around the Sun. *This popular view, however familiar, is incorrect.* Figure 39-19 shows us all that we can ever know about the location of the electron in the ground state of the hydrogen atom. The appropriate question is not "When will the electron arrive at such-and-such a point?" but "What are the odds that the electron will be detected in a small volume centered on such-and-such a point?" Figure 39-20, which we call a dot plot, suggests the probabilistic nature of the wave function: The density of dots represents the probability density of detection of the electron with the hydrogen atom in its ground state. Think of the atom in this state as a fuzzy ball with no sharply defined boundary and no hint of orbits.

It is not easy for a beginner to envision subatomic particles in this probabilistic way. The difficulty is our natural impulse to regard an electron as something like a tiny jelly bean, located at certain places at certain times and following a well-defined path. Electrons and other subatomic particles simply do not behave in this way.

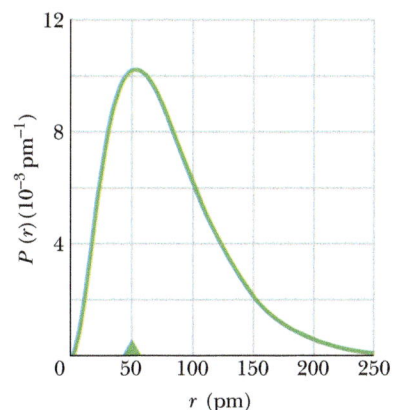

Figure 39-19 A plot of the radial probability density $P(r)$ for the ground state of the hydrogen atom. The triangular marker is located at one Bohr radius from the origin, and the origin represents the center of the atom.

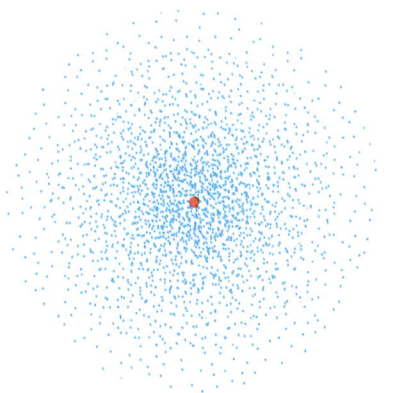

Figure 39-20 A "dot plot" showing the volume probability density $\psi^2(r)$—not the *radial* probability density $P(r)$—for the ground state of the hydrogen atom. The density of dots drops exponentially with increasing distance from the nucleus, which is represented here by a red spot.

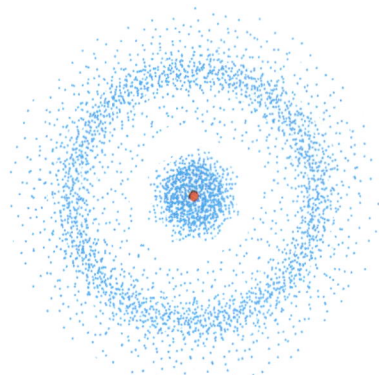

Figure 39-21 A dot plot showing the volume probability density $\psi^2(r)$ for the hydrogen atom in the quantum state with $n = 2$, $\ell = 0$, and $m_\ell = 0$. The plot has spherical symmetry about the central nucleus. The gap in the dot density pattern marks a spherical surface over which $\psi^2(r) = 0$.

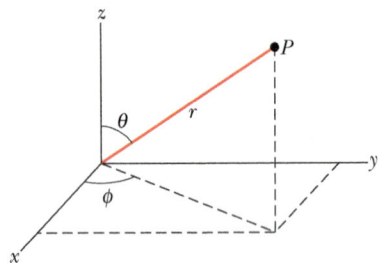

Figure 39-22 The relationship between the coordinates $x, y,$ and z of the rectangular coordinate system and the coordinates $r, \theta,$ and ϕ of the spherical coordinate system. The latter are more appropriate for analyzing situations involving spherical symmetry, such as the hydrogen atom.

Table 39-3 Quantum Numbers for Hydrogen Atom States with $n = 2$

n	ℓ	m_ℓ
2	0	0
2	1	+1
2	1	0
2	1	−1

The energy of the ground state, found by putting $n = 1$ in Eq. 39-34, is $E_1 = -13.60$ eV. The wave function of Eq. 39-39 results if you solve Schrödinger's equation with this value of the energy. Actually, you can find a solution of Schrödinger's equation for *any* value of the energy—say, $E = -11.6$ eV or -14.3 eV. This may suggest that the energies of the hydrogen atom states are not quantized—but we know that they are.

The puzzle was solved when physicists realized that such solutions of Schrödinger's equation are not physically acceptable because they yield increasingly large values as $r \to \infty$. These "wave functions" tell us that the electron is more likely to be found very far from the nucleus rather than closer to it, which makes no sense. We discard such solutions and accept only solutions that meet the boundary condition $\psi(r) \to 0$ as $r \to \infty$; that is, we agree to deal only with *confined* electrons. With this restriction, the solutions of Schrödinger's equation form a discrete set, with quantized energies given by Eq. 39-34.

Hydrogen Atom States with $n = 2$

According to the requirements of Table 39-2, there are four states of the hydrogen atom with $n = 2$; their quantum numbers are listed in Table 39-3. Consider first the state with $n = 2$ and $\ell = m_\ell = 0$; its probability density is represented by the dot plot of Fig. 39-21. Note that this plot, like the plot for the ground state shown in Fig. 39-20, is spherically symmetric. That is, in a spherical coordinate system like that defined in Fig. 39-22, the probability density is a function of the radial coordinate r only and is independent of the angular coordinates θ and ϕ.

It turns out that all quantum states with $\ell = 0$ have spherically symmetric wave functions. This is reasonable because the quantum number ℓ is a measure of the angular momentum associated with a given state. If $\ell = 0$, the angular momentum is also zero, which requires that the probability density representing the state have no preferred axis of symmetry.

Dot plots of ψ^2 for the three states with $n = 2$ and $\ell = 1$ are shown in Fig. 39-23. The probability densities for the states with $m_\ell = +1$ and $m_\ell = -1$ are

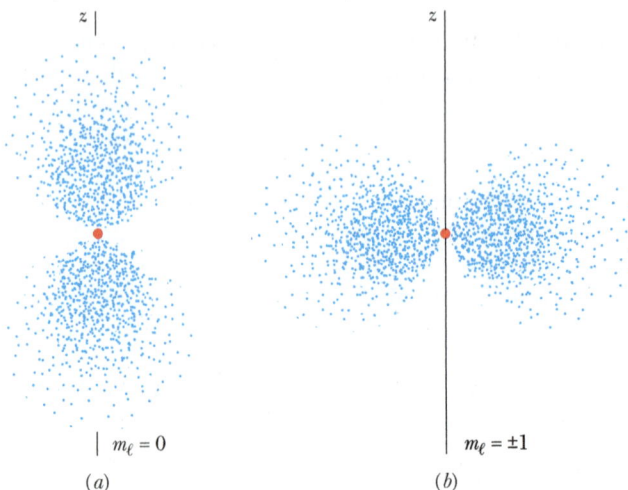

Figure 39-23 Dot plots of the volume probability density $\psi^2(r, \theta)$ for the hydrogen atom in states with $n = 2$ and $\ell = 1$. (a) Plot for $m_\ell = 0$. (b) Plot for $m_\ell = +1$ and $m_\ell = -1$. Both plots show that the probability density is symmetric about the z axis.

identical. Although these plots are symmetric about the z axis, they are *not* spherically symmetric. That is, the probability densities for these three states are functions of both r and the angular coordinate θ.

Here is a puzzle: What is there about the hydrogen atom that establishes the axis of symmetry that is so obvious in Fig. 39-23? The answer: *absolutely nothing.*

The solution to this puzzle comes about when we realize that all three states shown in Fig. 39-23 have the same energy. Recall that the energy of a state, given by Eq. 39-33, depends only on the principal quantum number n and is independent of ℓ and m_ℓ. In fact, for an *isolated* hydrogen atom there is no way to differentiate experimentally among the three states of Fig. 39-23.

If we add the volume probability densities for the three states for which $n = 2$ and $\ell = 1$, the combined probability density turns out to be spherically symmetrical, with no unique axis. One can, then, think of the electron as spending one-third of its time in each of the three states of Fig. 39-23, and one can think of the weighted sum of the three independent wave functions as defining a spherically symmetric **subshell** specified by the quantum numbers $n = 2$, $\ell = 1$. The individual states will display their separate existence only if we place the hydrogen atom in an external electric or magnetic field. The three states of the $n = 2$, $\ell = 1$ subshell will then have different energies, and the field direction will establish the necessary symmetry axis.

The $n = 2$, $\ell = 0$ state, whose volume probability density is shown in Fig. 39-21, *also* has the same energy as each of the three states of Fig. 39-23. We can view all four states whose quantum numbers are listed in Table 39-3 as forming a spherically symmetric **shell** specified by the single quantum number n. The importance of shells and subshells will become evident in Chapter 40, where we discuss atoms having more than one electron.

To round out our picture of the hydrogen atom, we display in Fig. 39-24 a dot plot of the *radial* probability density for a hydrogen atom state with a relatively high quantum number ($n = 45$) and the highest orbital quantum number that the restrictions of Table 39-2 permit ($\ell = n - 1 = 44$). The probability density forms a ring that is symmetrical about the z axis and lies very close to the xy plane. The mean radius of the ring is n^2a, where a is the Bohr radius. This mean radius is more than 2000 times the effective radius of the hydrogen atom in its ground state.

Figure 39-24 suggests the electron orbit of classical physics — it resembles the circular orbit of a planet around a star. Thus, we have another illustration of Bohr's correspondence principle — namely, that at large quantum numbers the predictions of quantum mechanics merge smoothly with those of classical physics. Imagine what a dot plot like that of Figure 39-24 would look like for *really* large values of n and ℓ — say, $n = 1000$ and $\ell = 999$.

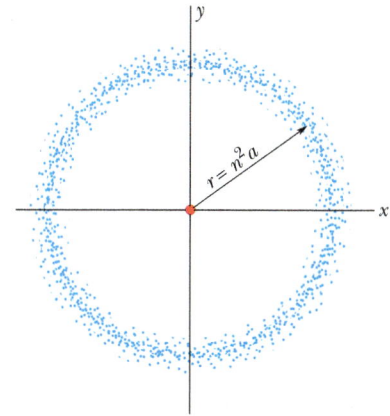

Figure 39-24 A dot plot of the radial probability density $P(r)$ for the hydrogen atom in a quantum state with a relatively large principal quantum number — namely, $n = 45$ — and angular momentum quantum number $\ell = n - 1 = 44$. The dots lie close to the xy plane, the ring of dots suggesting a classical electron orbit.

Sample Problem 39.06 Radial probability density for the electron in a hydrogen atom

Show that the radial probability density for the ground state of the hydrogen atom has a maximum at $r = a$.

KEY IDEAS

(1) The radial probability density for a ground-state hydrogen atom is given by Eq. 39-44,

$$P(r) = \frac{4}{a^3} r^2 e^{-2r/a}.$$

(2) To find the maximum (or minimum) of any function, we must differentiate the function and set the result equal to zero.

Calculation: If we differentiate $P(r)$ with respect to r, using derivative 7 of Appendix E and the chain rule for differentiating products, we get

$$\frac{dP}{dr} = \frac{4}{a^3} r^2 \left(\frac{-2}{a}\right) e^{-2r/a} + \frac{4}{a^3} 2r e^{-2r/a}$$

$$= \frac{8r}{a^3} e^{-2r/a} - \frac{8r^2}{a^4} e^{-2r/a}$$

$$= \frac{8}{a^4} r(a - r) e^{-2r/a}.$$

If we set the right side equal to zero, we obtain an equation that is true if $r = a$, so that the term $(a - r)$ in the middle of the equation is zero. In other words, dP/dr is equal to zero when $r = a$. (Note that we also have $dP/dr = 0$ at $r = 0$ and at $r = \infty$. However, these conditions correspond to a *minimum* in $P(r)$, as you can see in Fig. 39-19.)

Sample Problem 39.07   Probability of detection of the electron in a hydrogen atom

It can be shown that the probability $p(r)$ that the electron in the ground state of the hydrogen atom will be detected inside a sphere of radius r is given by

$$p(r) = 1 - e^{-2x}(1 + 2x + 2x^2),$$

in which x, a dimensionless quantity, is equal to r/a. Find r for $p(r) = 0.90$.

KEY IDEA

There is no guarantee of detecting the electron at any particular radial distance r from the center of the hydrogen atom. However, with the given function, we can calculate the probability that the electron will be detected *somewhere* within a sphere of radius r.

Calculation: We seek the radius of a sphere for which $p(r) = 0.90$. Substituting that value in the expression for $p(r)$, we have

$$0.90 = 1 - e^{-2x}(1 + 2x + 2x^2)$$

or

$$10e^{-2x}(1 + 2x + 2x^2) = 1.$$

We must find the value of x that satisfies this equality. It is not possible to solve explicitly for x, but an equation solver on a calculator yields $x = 2.66$. This means that the radius of a sphere within which the electron will be detected 90% of the time is $2.66a$. Mark this position on the horizontal axis of Fig. 39-19. The area under the curve from $r = 0$ to $r = 2.66a$ gives the probability of detection in that range and is 90% of the total area under the curve.

Sample Problem 39.08   Light emission from a hydrogen atom

(a) What is the wavelength of light for the least energetic photon emitted in the Lyman series of the hydrogen atom spectrum lines?

KEY IDEAS

(1) For any series, the transition that produces the least energetic photon is the transition between the home-base level that defines the series and the level immediately above it. (2) For the Lyman series, the home-base level is at $n = 1$ (Fig. 39-18b). Thus, the transition that produces the least energetic photon is the transition from the $n = 2$ level to the $n = 1$ level.

Calculations: From Eq. 39-34 the energy difference is

$$\Delta E = E_2 - E_1 = -(13.60 \text{ eV})\left(\frac{1}{2^2} - \frac{1}{1^2}\right) = 10.20 \text{ eV}.$$

Then from Eq. 39-6 ($\Delta E = hf$), with c/λ replacing f, we have

$$\lambda = \frac{hc}{\Delta E} = \frac{(6.63 \times 10^{-34} \text{ J} \cdot \text{s})(3.00 \times 10^8 \text{ m/s})}{(10.20 \text{ eV})(1.60 \times 10^{-19} \text{ J/eV})}$$

$$= 1.22 \times 10^{-7} \text{ m} = 122 \text{ nm}. \quad \text{(Answer)}$$

Light with this wavelength is in the ultraviolet range.

(b) What is the wavelength of the series limit for the Lyman series?

KEY IDEA

The series limit corresponds to a jump between the home-base level ($n = 1$ for the Lyman series) and the level at the limit $n = \infty$.

Calculations: Now that we have identified the values of n for the transition, we could proceed as in (a) to find the corresponding wavelength λ. Instead, let's use a more direct procedure. From Eq. 39-37, we find

$$\frac{1}{\lambda} = R\left(\frac{1}{n_{\text{low}}^2} - \frac{1}{n_{\text{high}}^2}\right)$$

$$= 1.097 373 \times 10^7 \text{ m}^{-1}\left(\frac{1}{1^2} - \frac{1}{\infty^2}\right),$$

which yields

$$\lambda = 9.11 \times 10^{-8} \text{ m} = 91.1 \text{ nm}. \quad \text{(Answer)}$$

Light with this wavelength is also in the ultraviolet range.

WILEY **PLUS**   Additional examples, video, and practice available at *WileyPLUS*

Review & Summary

Confinement Confinement of waves (string waves, matter waves—any type of wave) leads to quantization—that is, discrete states with certain energies. States with intermediate energies are not allowed.

Electron in an Infinite Potential Well Because it is a matter wave, an electron confined to an infinite potential well can exist in only certain discrete states. If the well is one-dimensional with length L, the energies associated with these quantum states are

$$E_n = \left(\frac{h^2}{8mL^2}\right)n^2, \quad \text{for } n = 1, 2, 3, \ldots, \quad (39\text{-}4)$$

where m is the electron mass and n is a *quantum number*. The lowest energy, said to be the *zero-point energy*, is not zero but is given by $n = 1$. The electron can change (jump) from one state to another only if its energy change is

$$\Delta E = E_{\text{high}} - E_{\text{low}}, \quad (39\text{-}5)$$

where E_{high} is the higher energy and E_{low} is the lower energy. If the change is done by photon absorption or emission, the energy of the photon must be equal to the change in the electron's energy:

$$hf = \frac{hc}{\lambda} = \Delta E = E_{\text{high}} - E_{\text{low}}, \quad (39\text{-}6)$$

where frequency f and wavelength λ are associated with the photon.

The wave functions for an electron in an infinite, one-dimensional potential well with length L along an x axis are given by

$$\psi_n(x) = \sqrt{\frac{2}{L}} \sin\left(\frac{n\pi}{L}x\right), \quad \text{for } n = 1, 2, 3, \ldots, \quad (39\text{-}10)$$

where n is the quantum number and the factor $\sqrt{2/L}$ comes from normalizing the wave function. The wave function $\psi_n(x)$ does not have physical meaning, but the probability density $\psi_n^2(x)$ does have physical meaning: The product $\psi_n^2(x)\,dx$ is the probability that the electron will be detected in the interval between x and $x + dx$. If the probability density of an electron is integrated over the entire x axis, the total probability must be 1, which means that the electron will be detected somewhere along the x axis:

$$\int_{-\infty}^{\infty} \psi_n^2(x)\,dx = 1. \quad (39\text{-}14)$$

Electron in a Finite Well The wave function for an electron in a finite, one-dimensional potential well extends into the walls. Compared to the states in an infinite well of the same size, the states in a finite well have a limited number, longer de Broglie wavelengths, and lower energies.

Two-Dimensional Electron Trap The quantized energies

for an electron trapped in a two-dimensional infinite potential well that forms a rectangular corral are

$$E_{nx,ny} = \frac{h^2}{8m}\left(\frac{n_x^2}{L_x^2} + \frac{n_y^2}{L_y^2}\right), \quad (39\text{-}20)$$

where n_x is a quantum number for which the electron's matter wave fits in well width L_x and n_y is a quantum number for which it fits in well width L_y. The wave functions for an electron in a two-dimensional well are given by

$$\psi_{nx,ny} = \sqrt{\frac{2}{L_x}}\sin\left(\frac{n_x\pi}{L_x}x\right)\sqrt{\frac{2}{L_y}}\sin\left(\frac{n_y\pi}{L_y}y\right). \quad (39\text{-}19)$$

The Hydrogen Atom The Bohr model of the hydrogen atom successfully derived the energy levels for the atom, to explain the emission/absorption spectrum of the atom, but it is incorrect in almost every other aspect. It is a planetary model in which the electron orbits the central proton with an angular momentum L that is limited to values given by

$$L = n\hbar, \quad \text{for } n = 1, 2, 3, \ldots, \quad (39\text{-}23)$$

where n is a quantum number. The equation is, however, incorrect. Application of the Schrödinger equation gives the correct values of L and the quantized energies:

$$E_n = -\frac{me^4}{8\varepsilon_0^2 h^2}\frac{1}{n^2} = -\frac{13.60\text{ eV}}{n^2}, \quad \text{for } n = 1, 2, 3, \ldots. \quad (39\text{-}34)$$

The atom (or, the electron in the atom) can change energy only by jumping between these allowed energies. If the jump is by photon absorption (the atom's energy increases) or photon emission (the atom's energy decreases), this restriction in energy changes leads to

$$\frac{1}{\lambda} = R\left(\frac{1}{n_{\text{low}}^2} - \frac{1}{n_{\text{high}}^2}\right), \quad (39\text{-}37)$$

for the wavelength of the light, where R is the Rydberg constant,

$$R = \frac{me^4}{8\varepsilon_0^2 h^3 c} = 1.097\,373 \times 10^7\text{ m}^{-1}. \quad (39\text{-}38)$$

The radial probability density $P(r)$ for a state of the hydrogen atom is defined so that $P(r)$ is the probability that the electron will be detected somewhere in the space between two spherical shells of radii r and $r + dr$ that are centered on the nucleus. The probability that the electron will be detected between any two given radii r_1 and r_2 is

$$(\text{probability of detection}) = \int_{r_1}^{r_2} P(r)\,dr. \quad (39\text{-}45)$$

Questions

1 Three electrons are trapped in three different one-dimensional infinite potential wells of widths (a) 50 pm, (b) 200 pm, and (c) 100 pm. Rank the electrons according to their ground-state energies, greatest first.

2 Is the ground-state energy of a proton trapped in a one-dimensional infinite potential well greater than, less than, or equal to that of an electron trapped in the same potential well?

3 An electron is trapped in a one-dimensional infinite potential well in a state with quantum number $n = 17$. How many points of (a) zero probability and (b) maximum probability does its matter wave have?

4 Figure 39-25 shows three infinite potential wells, each on an x axis. Without written calculation, determine the wave function ψ for a ground-state electron trapped in each well.

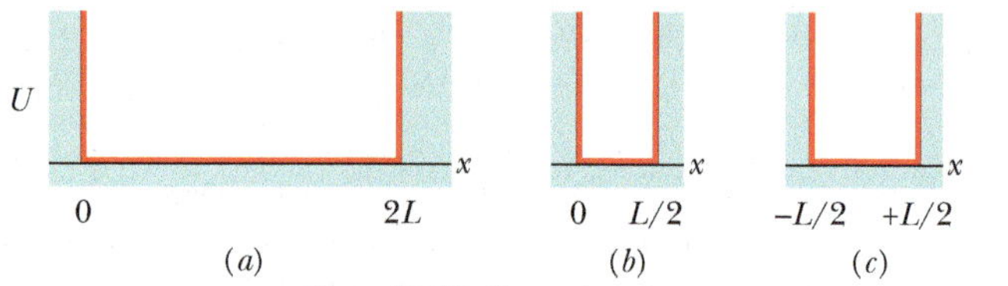

Figure 39-25 Question 4.

5 A proton and an electron are trapped in identical one-dimensional infinite potential wells; each particle is in its ground state. At the center of the wells, is the probability density for the proton greater than, less than, or equal to that of the electron?

6 If you double the width of a one-dimensional infinite potential well, (a) is the energy of the ground state of the trapped electron multiplied by 4, 2, $\frac{1}{2}$, $\frac{1}{4}$, or some other number? (b) Are the energies of the higher energy states multiplied by this factor or by some other factor, depending on their quantum number?

7 If you wanted to use the idealized trap of Fig. 39-1 to trap a positron, would you need to change (a) the geometry of the trap, (b) the electric potential of the central cylinder, or (c) the electric potentials of the two semi-infinite end cylinders? (A positron has the same mass as an electron but is positively charged.)

8 An electron is trapped in a finite potential well that is deep enough to allow the electron to exist in a state with $n = 4$. How many points of (a) zero probability and (b) maximum probability does its matter wave have within the well?

9 An electron that is trapped in a one-dimensional infinite potential well of width L is excited from the ground state to the first excited state. Does the excitation increase, decrease, or have no effect on the probability of detecting the electron in a small length of the x axis (a) at the center of the well and (b) near one of the well walls?

10 An electron, trapped in a finite potential energy well such as that of Fig. 39-7, is in its state of lowest energy. Are (a) its de Broglie wavelength, (b) the magnitude of its momentum, and (c) its energy greater than, the same as, or less than they would be if the potential well were infinite, as in Fig. 39-2?

11 From a visual inspection of Fig. 39-8, rank the quantum numbers of the three quantum states according to the de Broglie wavelength of the electron, greatest first.

12 You want to modify the finite potential well of Fig. 39-7 to allow its trapped electron to exist in more than four quantum states. Could you do so by making the well (a) wider or narrower, (b) deeper or shallower?

13 A hydrogen atom is in the third excited state. To what state (give the quantum number n) should it jump to (a) emit light with the longest possible wavelength, (b) emit light with the shortest possible wavelength, and (c) absorb light with the longest possible wavelength?

14 Figure 39-26 indicates the lowest energy levels (in electron-volts) for five situations in which an electron is trapped in a one-dimensional infinite potential well. In wells B, C, D, and E, the electron is in the ground state. We shall excite the electron in well A to the fourth excited state (at 25 eV). The electron can then de-excite to the ground state by emitting one or more photons, corresponding to one long jump or several short jumps. Which photon *emission* energies of this de-excitation match a photon *absorption* energy (from the ground state) of the other four electrons? Give the n values.

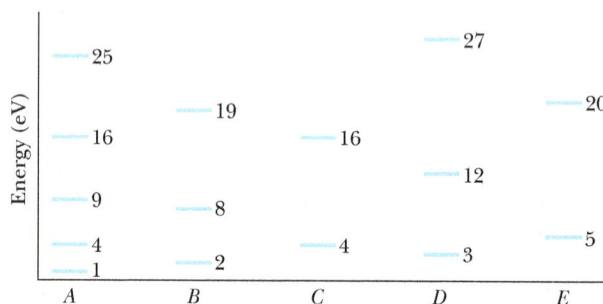

Figure 39-26 Question 14.

15 Table 39-4 lists the quantum numbers for five proposed hydrogen atom states. Which of them are not possible?

Table 39-4

	n	ℓ	m_ℓ
(a)	3	2	0
(b)	2	3	1
(c)	4	3	−4
(d)	5	5	0
(e)	5	3	−2

Problems

GO Tutoring problem available (at instructor's discretion) in *WileyPLUS* and WebAssign
SSM Worked-out solution available in Student Solutions Manual
• – ••• Number of dots indicates level of problem difficulty
Additional information available in *The Flying Circus of Physics* and at flyingcircusofphysics.com
WWW Worked-out solution is at
ILW Interactive solution is at
http://www.wiley.com/college/halliday

Module 39-1 Energies of a Trapped Electron

•1 An electron in a one-dimensional infinite potential well of length L has ground-state energy E_1. The length is changed to L' so that the new ground-state energy is $E'_1 = 0.500E_1$. What is the ratio L'/L?

•2 What is the ground-state energy of (a) an electron and (b) a proton if each is trapped in a one-dimensional infinite potential well that is 200 pm wide?

•3 The ground-state energy of an electron trapped in a one-dimensional infinite potential well is 2.6 eV. What will this quantity be if the width of the potential well is doubled?

•4 An electron, trapped in a one-dimensional infinite potential well 250 pm wide, is in its ground state. How much energy must it absorb if it is to jump up to the state with $n = 4$?

•5 What must be the width of a one-dimensional infinite potential well if an electron trapped in it in the $n = 3$ state is to have an energy of 4.7 eV?

•6 A proton is confined to a one-dimensional infinite potential well 100 pm wide. What is its ground-state energy?

•7 Consider an atomic nucleus to be equivalent to a one-dimensional infinite potential well with $L = 1.4 \times 10^{-14}$ m, a typical nuclear diameter. What would be the ground-state energy of an electron if it were trapped in such a potential well? (*Note:* Nuclei do not contain electrons.)

••8 GO An electron is trapped in a one-dimensional infinite well and is in its first excited state. Figure 39-27 indicates the five longest wavelengths of light that the electron could absorb in transitions from this initial state via a single photon absorption: $\lambda_a = 80.78$ nm, $\lambda_b = 33.66$ nm, $\lambda_c = 19.23$ nm, $\lambda_d = 12.62$ nm, and $\lambda_e = 8.98$ nm. What is the width of the potential well?

Figure 39-27 Problem 8.

••9 Suppose that an electron trapped in a one-dimensional infinite well of width 250 pm is excited from its first excited state to its third excited state. (a) What energy must be transferred to the electron for this quantum jump? The electron then de-excites back to its ground state by emitting light. In the various possible ways it can do this, what are the (b) shortest, (c) second shortest, (d) longest, and (e) second longest wavelengths that can be emitted? (f) Show the various possible ways on an energy-level diagram. If light of wavelength 29.4 nm happens to be emitted, what are the (g) longest and (h) shortest wavelength that can be emitted afterwards?

••10 An electron is trapped in a one-dimensional infinite potential well. For what (a) higher quantum number and (b) lower quantum number is the corresponding energy difference equal to the energy difference ΔE_{43} between the levels $n = 4$ and $n = 3$? (c) Show that no pair of adjacent levels has an energy difference equal to $2\Delta E_{43}$.

••11 An electron is trapped in a one-dimensional infinite potential well. For what (a) higher quantum number and (b) lower quantum number is the corresponding energy difference equal to the energy of the $n = 5$ level? (c) Show that no pair of adjacent levels has an energy difference equal to the energy of the $n = 6$ level.

••12 GO An electron is trapped in a one-dimensional infinite well of width 250 pm and is in its ground state. What are the (a) longest, (b) second longest, and (c) third longest wavelengths of light that can excite the electron from the ground state via a single photon absorption?

Module 39-2 Wave Functions of a Trapped Electron

••13 GO A one-dimensional infinite well of length 200 pm contains an electron in its third excited state. We position an electron-detector probe of width 2.00 pm so that it is centered on a point of maximum probability density. (a) What is the probability of detection by the probe? (b) If we insert the probe as described 1000 times, how many times should we expect the electron to materialize on the end of the probe (and thus be detected)?

••14 An electron is in a certain energy state in a one-dimensional, infinite potential well from $x = 0$ to $x = L = 200$ pm. The electron's probability density is zero at $x = 0.300L$, and $x = 0.400L$; it is not zero at intermediate values of x. The electron then jumps to the next lower energy level by emitting light. What is the change in the electron's energy?

••15 SSM WWW An electron is trapped in a one-dimensional infinite potential well that is 100 pm wide; the electron is in its ground state. What is the probability that you can detect the electron in an interval of width $\Delta x = 5.0$ pm centered at $x =$ (a) 25 pm, (b) 50 pm, and (c) 90 pm? (*Hint:* The interval Δx is so narrow that you can take the probability density to be constant within it.)

••16 A particle is confined to the one-dimensional infinite potential well of Fig. 39-2. If the particle is in its ground state, what is its probability of detection between (a) $x = 0$ and $x = 0.25L$, (b) $x = 0.75L$ and $x = L$, and (c) $x = 0.25L$ and $x = 0.75L$?

Module 39-3 An Electron in a Finite Well

•17 An electron in the $n = 2$ state in the finite potential well of Fig. 39-7 absorbs 400 eV of energy from an external source. Using the energy-level diagram of Fig. 39-9, determine the electron's kinetic energy after this absorption, assuming that the electron moves to a position for which $x > L$.

•18 Figure 39-9 gives the energy levels for an electron trapped in a finite potential energy well 450 eV deep. If the electron is in the $n = 3$ state, what is its kinetic energy?

••19 GO Figure 39-28a shows the energy-level diagram for a finite, one-dimensional energy well that contains an electron. The nonquantized region begins at $E_4 = 450.0$ eV. Figure 39-28b gives the absorption spectrum of the electron when it is in the ground state—it can absorb at the indicated wavelengths: $\lambda_a = 14.588$ nm and $\lambda_b = 4.8437$ nm and for any wavelength less than $\lambda_c = 2.9108$ nm. What is the energy of the first excited state?

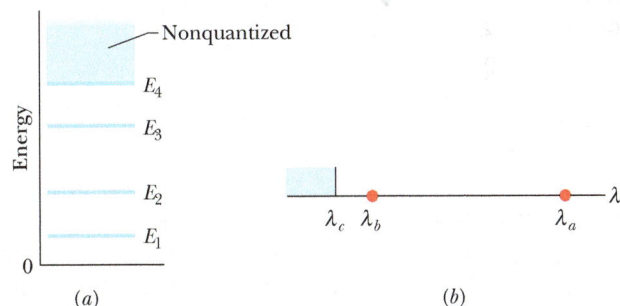

Figure 39-28 Problem 19.

••20 GO Figure 39-29a shows a thin tube in which a finite potential trap has been set up where $V_2 = 0$ V. An electron is shown traveling rightward toward the trap, in a region with a voltage of $V_1 =$

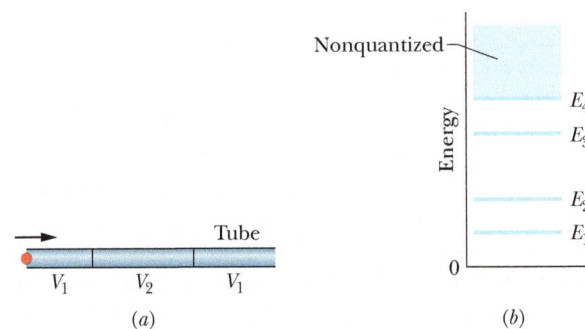

Figure 39-29 Problem 20.

−9.00 V, where it has a kinetic energy of 2.00 eV. When the electron enters the trap region, it can become trapped if it gets rid of enough energy by emitting a photon. The energy levels of the electron within the trap are $E_1 = 1.0$, $E_2 = 2.0$, and $E_3 = 4.0$ eV, and the nonquantized region begins at $E_4 = 9.0$ eV as shown in the energy-level diagram of Fig. 39-29b. What is the smallest energy (eV) such a photon can have?

••21 (a) Show that for the region $x > L$ in the finite potential well of Fig. 39-7, $\psi(x) = De^{2kx}$ is a solution of Schrödinger's equation in its one-dimensional form, where D is a constant and k is positive. (b) On what basis do we find this mathematically acceptable solution to be physically unacceptable?

Module 39-4 Two- and Three-Dimensional Electron Traps

•22 GO An electron is contained in the rectangular corral of Fig. 39-13, with widths $L_x = 800$ pm and $L_y = 1600$ pm. What is the electron's ground-state energy?

•23 An electron is contained in the rectangular box of Fig. 39-14, with widths $L_x = 800$ pm, $L_y = 1600$ pm, and $L_z = 390$ pm. What is the electron's ground-state energy?

••24 Figure 39-30 shows a two-dimensional, infinite-potential well lying in an xy plane that contains an electron. We probe for the electron along a line that bisects L_x and find three points at which the detection probability is maximum. Those points are separated by 2.00 nm. Then we probe along a line that bisects L_y and find five points at which the detection probability is maximum. Those points are separated by 3.00 nm. What is the energy of the electron?

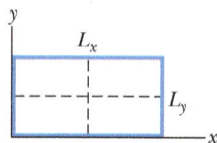

Figure 39-30 Problem 24.

••25 GO The two-dimensional, infinite corral of Fig. 39-31 is square, with edge length $L = 150$ pm. A square probe is centered at xy coordinates $(0.200L, 0.800L)$ and has an x width of 5.00 pm and a y width of 5.00 pm. What is the probability of detection if the electron is in the $E_{1,3}$ energy state?

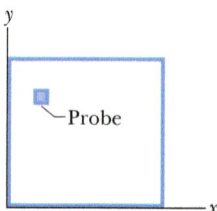

Figure 39-31 Problem 25.

••26 A rectangular corral of widths $L_x = L$ and $L_y = 2L$ holds an electron. What multiple of $h^2/8mL^2$, where m is the electron mass, gives (a) the energy of the electron's ground state, (b) the energy of its first excited state, (c) the energy of its lowest degenerate states, and (d) the difference between the energies of its second and third excited states?

••27 SSM WWW An electron (mass m) is contained in a rectangular corral of widths $L_x = L$ and $L_y = 2L$. (a) How many different frequencies of light could the electron emit or absorb if it makes a transition between a pair of the lowest five energy levels? What multiple of $h/8mL^2$ gives the (b) lowest, (c) second lowest, (d) third lowest, (e) highest, (f) second highest, and (g) third highest frequency?

••28 GO A cubical box of widths $L_x = L_y = L_z = L$ contains an electron. What multiple of $h^2/8mL^2$, where m is the electron mass, is (a) the energy of the electron's ground state, (b) the energy of its second excited state, and (c) the difference between the energies of its second and third excited states? How many degenerate states have the energy of (d) the first excited state and (e) the fifth excited state?

••29 An electron (mass m) is contained in a cubical box of widths $L_x = L_y = L_z$. (a) How many different frequencies of light could

the electron emit or absorb if it makes a transition between a pair of the lowest five energy levels? What multiple of $h/8mL^2$ gives the (b) lowest, (c) second lowest, (d) third lowest, (e) highest, (f) second highest, and (g) third highest frequency?

•••30 GO An electron is in the ground state in a two-dimensional, square, infinite potential well with edge lengths L. We will probe for it in a square of area 400 pm² that is centered at $x = L/8$ and $y = L/8$. The probability of detection turns out to be 4.5×10^{-8}. What is edge length L?

Module 39-5 The Hydrogen Atom

•31 SSM What is the ratio of the shortest wavelength of the Balmer series to the shortest wavelength of the Lyman series?

•32 An atom (not a hydrogen atom) absorbs a photon whose associated wavelength is 375 nm and then immediately emits a photon whose associated wavelength is 580 nm. How much net energy is absorbed by the atom in this process?

•33 What are the (a) energy, (b) magnitude of the momentum, and (c) wavelength of the photon emitted when a hydrogen atom undergoes a transition from a state with $n = 3$ to a state with $n = 1$?

•34 Calculate the radial probability density $P(r)$ for the hydrogen atom in its ground state at (a) $r = 0$, (b) $r = a$, and (c) $r = 2a$, where a is the Bohr radius.

•35 For the hydrogen atom in its ground state, calculate (a) the probability density $\psi^2(r)$ and (b) the radial probability density $P(r)$ for $r = a$, where a is the Bohr radius.

•36 (a) What is the energy E of the hydrogen-atom electron whose probability density is represented by the dot plot of Fig. 39-21? (b) What minimum energy is needed to remove this electron from the atom?

•37 SSM A neutron with a kinetic energy of 6.0 eV collides with a stationary hydrogen atom in its ground state. Explain why the collision must be elastic—that is, why kinetic energy must be conserved. (*Hint:* Show that the hydrogen atom cannot be excited as a result of the collision.)

•38 An atom (not a hydrogen atom) absorbs a photon whose associated frequency is 6.2×10^{14} Hz. By what amount does the energy of the atom increase?

••39 SSM Verify that Eq. 39-44, the radial probability density for the ground state of the hydrogen atom, is normalized. That is, verify that the following is true:

$$\int_0^\infty P(r)\, dr = 1$$

••40 What are the (a) wavelength range and (b) frequency range of the Lyman series? What are the (c) wavelength range and (d) frequency range of the Balmer series?

••41 What is the probability that an electron in the ground state of the hydrogen atom will be found between two spherical shells whose radii are r and $r + \Delta r$, (a) if $r = 0.500a$ and $\Delta r = 0.010a$ and (b) if $r = 1.00a$ and $\Delta r = 0.01a$, where a is the Bohr radius? (*Hint:* Δr is small enough to permit the radial probability density to be taken to be constant between r and $r + \Delta r$.)

••42 A hydrogen atom, initially at rest in the $n = 4$ quantum state, undergoes a transition to the ground state, emitting a photon in the process. What is the speed of the recoiling hydrogen atom? (*Hint:* This is similar to the explosions of Chapter 9.)

••43 In the ground state of the hydrogen atom, the electron has a total energy of -13.6 eV. What are (a) its kinetic energy and (b) its potential energy if the electron is one Bohr radius from the central nucleus?

••44 A hydrogen atom in a state having a *binding energy* (the energy required to remove an electron) of 0.85 eV makes a transition to a state with an *excitation energy* (the difference between the energy of the state and that of the ground state) of 10.2 eV. (a) What is the energy of the photon emitted as a result of the transition? What are the (b) higher quantum number and (c) lower quantum number of the transition producing this emission?

••45 SSM The wave functions for the three states with the dot plots shown in Fig. 39-23, which have $n = 2, \ell = 1$, and $m_\ell = 0, +1$, and -1, are

$$\psi_{210}(r, \theta) = (1/4\sqrt{2\pi})(a^{-3/2})(r/a)e^{-r/2a}\cos\theta,$$

$$\psi_{21+1}(r, \theta) = (1/8\sqrt{\pi})(a^{-3/2})(r/a)e^{-r/2a}(\sin\theta)e^{+i\phi},$$

$$\psi_{21-1}(r, \theta) = (1/8\sqrt{\pi})(a^{-3/2})(r/a)e^{-r/2a}(\sin\theta)e^{-i\phi},$$

in which the subscripts on $\psi(r, \theta)$ give the values of the quantum numbers n, ℓ, m_ℓ and the angles θ and ϕ are defined in Fig. 39-22. Note that the first wave function is real but the others, which involve the imaginary number i, are complex. Find the radial probability density $P(r)$ for (a) ψ_{210} and (b) ψ_{21+1} (same as for ψ_{21-1}). (c) Show that each $P(r)$ is consistent with the corresponding dot plot in Fig. 39-23. (d) Add the radial probability densities for ψ_{210}, ψ_{21+1}, and ψ_{21-1} and then show that the sum is spherically symmetric, depending only on r.

••46 Calculate the probability that the electron in the hydrogen atom, in its ground state, will be found between spherical shells whose radii are a and $2a$, where a is the Bohr radius.

••47 For what value of the principal quantum number n would the effective radius, as shown in a probability density dot plot for the hydrogen atom, be 1.0 mm? Assume that ℓ has its maximum value of $n - 1$. (*Hint:* See Fig. 39-24.)

••48 Light of wavelength 121.6 nm is emitted by a hydrogen atom. What are the (a) higher quantum number and (b) lower quantum number of the transition producing this emission? (c) What is the name of the series that includes the transition?

••49 How much work must be done to pull apart the electron and the proton that make up the hydrogen atom if the atom is initially in (a) its ground state and (b) the state with $n = 2$?

••50 Light of wavelength 102.6 nm is emitted by a hydrogen atom. What are the (a) higher quantum number and (b) lower quantum number of the transition producing this emission? (c) What is the name of the series that includes the transition?

••51 What is the probability that in the ground state of the hydrogen atom, the electron will be found at a radius greater than the Bohr radius?

••52 A hydrogen atom is excited from its ground state to the state with $n = 4$. (a) How much energy must be absorbed by the atom? Consider the photon energies that can be emitted by the atom as it de-excites to the ground state in the several possible ways. (b) How many different energies are possible; what are the (c) highest, (d) second highest, (e) third highest, (f) lowest, (g) second lowest, and (h) third lowest energies?

••53 SSM WWW Schrödinger's equation for states of the hy-

drogen atom for which the orbital quantum number ℓ is zero is

$$\frac{1}{r^2}\frac{d}{dr}\left(r^2\frac{d\psi}{dr}\right) + \frac{8\pi^2 m}{h^2}[E - U(r)]\psi = 0.$$

Verify that Eq. 39-39, which describes the ground state of the hydrogen atom, is a solution of this equation.

•••54 The wave function for the hydrogen-atom quantum state represented by the dot plot shown in Fig. 39-21, which has $n = 2$ and $\ell = m_\ell = 0$, is

$$\psi_{200}(r) = \frac{1}{4\sqrt{2\pi}}a^{-3/2}\left(2 - \frac{r}{a}\right)e^{-r/2a},$$

in which a is the Bohr radius and the subscript on $\psi(r)$ gives the values of the quantum numbers n, ℓ, m_ℓ. (a) Plot $\psi^2_{200}(r)$ and show that your plot is consistent with the dot plot of Fig. 39-21. (b) Show analytically that $\psi^2_{200}(r)$ has a maximum at $r = 4a$. (c) Find the radial probability density $P_{200}(r)$ for this state. (d) Show that

$$\int_0^\infty P_{200}(r)\, dr = 1$$

and thus that the expression above for the wave function $\psi_{200}(r)$ has been properly normalized.

•••55 The radial probability density for the ground state of the hydrogen atom is a maximum when $r = a$, where a is the Bohr radius. Show that the *average* value of r, defined as

$$r_{avg} = \int P(r)\, r\, dr,$$

has the value $1.5a$. In this expression for r_{avg}, each value of $P(r)$ is weighted with the value of r at which it occurs. Note that the average value of r is greater than the value of r for which $P(r)$ is a maximum.

Additional Problems

56 Let ΔE_{adj} be the energy difference between two adjacent energy levels for an electron trapped in a one-dimensional infinite potential well. Let E be the energy of either of the two levels. (a) Show that the ratio $\Delta E_{adj}/E$ approaches the value $2/n$ at large values of the quantum number n. As $n \to \infty$, does (b) ΔE_{adj}, (c) E, or (d) $\Delta E_{adj}/E$ approach zero? (e) What do these results mean in terms of the correspondence principle?

57 An electron is trapped in a one-dimensional infinite potential well. Show that the energy difference ΔE between its quantum levels n and $n + 2$ is $(h^2/2mL^2)(n + 1)$.

58 As Fig. 39-8 suggests, the probability density for an electron in the region $0 < x < L$ for the finite potential well of Fig. 39-7 is sinusoidal, being given by $\psi^2(x) = B \sin^2 kx$, in which B is a constant. (a) Show that the wave function $\psi(x)$ that may be found from this equation is a solution of Schrödinger's equation in its one-dimensional form. (b) Find an expression for k that makes this true.

59 SSM As Fig. 39-8 suggests, the probability density for the region $x > L$ in the finite potential well of Fig. 39-7 drops off exponentially according to $\psi^2(x) = Ce^{-2kx}$, where C is a constant. (a) Show that the wave function $\psi(x)$ that may be found from this equation is a solution of Schrödinger's equation in its one-dimensional form. (b) Find an expression for k for this to be true.

60 An electron is confined to a narrow evacuated tube of length 3.0 m; the tube functions as a one-dimensional infinite potential well. (a) What is the energy difference between the electron's ground state and its first excited state? (b) At what quantum number n would the energy difference between adjacent energy levels be 1.0 eV—which is measurable, unlike the result of (a)? At that quantum number, (c) what multiple of the electron's rest energy would give the electron's total energy and (d) would the electron be relativistic?

61 (a) Show that the terms in Schrödinger's equation (Eq. 39-18) have the same dimensions. (b) What is the common SI unit for each of these terms?

62 (a) What is the wavelength of light for the least energetic photon emitted in the Balmer series of the hydrogen atom spectrum lines? (b) What is the wavelength of the series limit?

63 (a) For a given value of the principal quantum number n for a hydrogen atom, how many values of the orbital quantum number ℓ are possible? (b) For a given value of ℓ, how many values of the orbital magnetic quantum number m_ℓ are possible? (c) For a given value of n, how many values of m_ℓ are possible?

64 Verify that the combined value of the constants appearing in Eq. 39-33 is 13.6 eV.

65 A diatomic gas molecule consists of two atoms of mass m separated by a fixed distance d rotating about an axis as indicated in Fig. 39-32. Assuming that its angular momentum is quantized as in the Bohr model for the hydrogen atom, find (a) the possible angular velocities and (b) the possible quantized rotational energies.

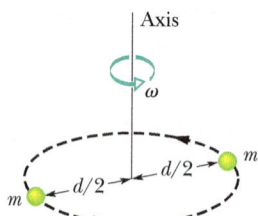

Figure 39-32 Problem 65.

66 In atoms there is a finite, though very small, probability that, at some instant, an orbital electron will actually be found inside the nucleus. In fact, some unstable nuclei use this occasional appearance of the electron to decay by *electron capture*. Assuming that the proton itself is a sphere of radius 1.1×10^{-15} m and that the wave function of the hydrogen atom's electron holds all the way to the proton's center, use the ground-state wave function to calculate the probability that the hydrogen atom's electron is inside its nucleus.

67 (a) What is the separation in energy between the lowest two energy levels for a container 20 cm on a side containing argon atoms? Assume, for simplicity, that the argon atoms are trapped in a one-dimensional well 20 cm wide. The molar mass of argon is 39.9 g/mol. (b) At 300 K, to the nearest power of ten, what is the ratio of the thermal energy of the atoms to this energy separation? (c) At what temperature does the thermal energy equal the energy separation?

68 A muon of charge $-e$ and mass $m = 207m_e$ (where m_e is the mass of an electron) orbits the nucleus of a singly ionized helium atom (He^+). Assuming that the Bohr model of the hydrogen atom can be applied to this muon–helium system, verify that the energy levels of the system are given by

$$E = -\frac{11.3 \text{ keV}}{n^2}.$$

69 From the energy-level diagram for hydrogen, explain the observation that the frequency of the second Lyman-series line is the sum of the frequencies of the first Lyman-series line and the first Balmer-series line. This is an example of the empirically discovered *Ritz combination principle*. Use the diagram to find some other valid combinations.

70 A hydrogen atom can be considered as having a central point-like proton of positive charge e and an electron of negative charge $-e$ that is distributed about the proton according to the volume charge density $\rho = A \exp(-2r/a_0)$. Here A is a constant, $a_0 = 0.53 \times 10^{-10}$ m, and r is the distance from the center of the atom. (a) Using the fact that the hydrogen is electrically neutral, find A. Then find the (b) magnitude and (c) direction of the atom's electric field at a_0.

71 An old model of a hydrogen atom has the charge $+e$ of the proton uniformly distributed over a sphere of radius a_0, with the electron of charge $-e$ and mass m at its center. (a) What would then be the force on the electron if it were displaced from the center by a distance $r \leq a_0$? (b) What would be the angular frequency of oscillation of the electron about the center of the atom once the electron was released?

72 In a simple model of a hydrogen atom, the single electron orbits the single proton (the nucleus) in a circular path. Calculate (a) the electric potential set up by the proton at the orbital radius of 52.9 pm, (b) the electric potential energy of the atom, and (c) the kinetic energy of the electron. (d) How much energy is required to ionize the atom (that is, to remove the electron to an infinite distance with no kinetic energy)? Give the energies in electron-volts.

73 Consider a conduction electron in a cubical crystal of a conducting material. Such an electron is free to move throughout the volume of the crystal but cannot escape to the outside. It is trapped in a three-dimensional infinite well. The electron can move in three dimensions, so that its total energy is given by

$$E = \frac{h^2}{8L^2m}(n_1^2 + n_2^2 + n_3^2),$$

in which $n_1, n_2,$ and n_3 are positive integer values. Calculate the energies of the lowest five distinct states for a conduction electron moving in a cubical crystal of edge length $L = 0.25$ μm.

All About Atoms

40-1 PROPERTIES OF ATOMS

Learning Objectives

After reading this module, you should be able to . . .

40.01 Discuss the pattern that is seen in a plot of ionization energies versus atomic number Z.

40.02 Identify that atoms have angular momentum and magnetism.

40.03 Explain the Einstein–de Haas experiment.

40.04 Identify the five quantum numbers of an electron in an atom and the allowed values of each.

40.05 Determine the number of electron states allowed in a given shell and subshell.

40.06 Identify that an electron in an atom has an orbital angular momentum $\vec{L}$ and an orbital magnetic dipole moment $\vec{\mu}_{orb}$.

40.07 Calculate magnitudes for orbital angular momentum $\vec{L}$ and orbital magnetic dipole moment $\vec{\mu}_{orb}$ in terms of the orbital quantum number ℓ.

40.08 Apply the relationship between orbital angular momentum $\vec{L}$ and orbital magnetic dipole moment $\vec{\mu}_{orb}$.

40.09 Identify that $\vec{L}$ and $\vec{\mu}_{orb}$ cannot be observed (measured) but a component on a measurement axis (usually called the z axis) can.

40.10 Calculate the z components L_z of an orbital angular momentum $\vec{L}$ using the orbital magnetic quantum number m_ℓ.

40.11 Calculate the z components $\mu_{orb,z}$ of an orbital magnetic dipole moment $\vec{\mu}_{orb}$ using the orbital magnetic quantum number m_ℓ and the Bohr magneton μ_B.

40.12 For a given orbital state or spin state, calculate the semiclassical angle θ.

40.13 Identify that a spin angular momentum $\vec{S}$ (usually simply called spin) and a spin magnetic dipole moment $\vec{\mu}_s$ are intrinsic properties of electrons (and also protons and neutrons).

40.14 Calculate magnitudes for spin angular momentum $\vec{S}$ and spin magnetic dipole moment $\vec{\mu}_s$ in terms of the spin quantum number s.

40.15 Apply the relationship between the spin angular momentum $\vec{S}$ and the spin magnetic dipole moment $\vec{\mu}_s$.

40.16 Identify that $\vec{S}$ and $\vec{\mu}_s$ cannot be observed (measured) but a component on a measurement axis can.

40.17 Calculate the z components S_z of the spin angular momentum $\vec{S}$ using the spin magnetic quantum number m_s.

40.18 Calculate the z components $\mu_{s,z}$ of the spin magnetic dipole moment $\vec{\mu}_s$ using the spin magnetic quantum number m_s and the Bohr magneton μ_B.

40.19 Identify the effective magnetic dipole moment of an atom.

Key Ideas

● Atoms have quantized energies and can make quantum jumps between them. If a jump between a higher energy and a lower energy involves the emission or absorption of a photon, the frequency associated with the light is given by

$$hf = E_{high} - E_{low}.$$

● States with the same value of quantum number n form a shell.

● States with the same values of quantum numbers n and ℓ form a subshell.

● The magnitude of the orbital angular momentum of an electron trapped in an atom has quantized values given by

$$L = \sqrt{\ell(\ell + 1)}\,\hbar, \quad \text{for } \ell = 0, 1, 2, \ldots, (n - 1),$$

where $\hbar$ is $h/2\pi$, ℓ is the orbital quantum number, and n is the electron's principal quantum number.

● The component L_z of the orbital angular momentum on a z axis is quantized and given by

$$L_z = m_\ell \hbar, \quad \text{for } m_\ell = 0, \pm1, \pm2, \ldots, \pm\ell,$$

where m_ℓ is the orbital magnetic quantum number.

● The magnitude μ_{orb} of the orbital magnetic moment of the electron is quantized with the values given by

$$\mu_{orb} = \frac{e}{2m}\sqrt{\ell(\ell + 1)}\,\hbar,$$

where m is the electron mass.

● The component $\mu_{orb,z}$ on a z axis is also quantized according to

$$\mu_{orb,z} = -\frac{e}{2m}m_\ell \hbar = -m_\ell \mu_B,$$

where μ_B is the Bohr magneton:

$$\mu_B = \frac{eh}{4\pi m} = \frac{e\hbar}{2m} = 9.274 \times 10^{-24} \text{ J/T}.$$

● Every electron, whether trapped or free, has an intrinsic spin angular momentum $\vec{S}$ with a magnitude that is quantized as

$$S = \sqrt{s(s + 1)}\,\hbar, \quad \text{for } s = \tfrac{1}{2},$$

where s is the spin quantum number. An electron is said to be a spin-$\frac{1}{2}$ particle.

● The component S_z on a z axis is also quantized according to

$$S_z = m_s\hbar, \quad \text{for } m_s = \pm s = \pm\tfrac{1}{2},$$

where m_s is the spin magnetic quantum number.

● Every electron, whether trapped or free, has an intrinsic spin magnetic dipole moment $\vec{\mu}_s$ with a magnitude that is

quantized as

$$\mu_s = \frac{e}{m}\sqrt{s(s+1)}\,\hbar, \quad \text{for } s = \tfrac{1}{2}.$$

● The component $\mu_{s,z}$ on a z axis is also quantized according to

$$\mu_{s,z} = -2m_s\mu_B, \quad \text{for } m_s = \pm\tfrac{1}{2}.$$

What Is Physics?

In this chapter we continue with a primary goal of physics—discovering and understanding the properties of atoms. About 100 years ago, researchers struggled to find experiments that would prove the existence of atoms. Now we take their existence for granted and even have photographs (scanning tunneling microscope images) of atoms. We can drag them around on surfaces, such as to make the quantum corral shown in the photograph of Fig. 39-12. We can even hold an individual atom indefinitely in a trap (Fig. 40-1) so as to study its properties when it is completely isolated from other atoms.

Some Properties of Atoms

You may think the details of atomic physics are remote from your daily life. However, consider how the following properties of atoms—so basic that we rarely think about them—affect the way we live in our world.

Atoms are stable. Essentially all the atoms that form our tangible world have existed without change for billions of years. What would the world be like if atoms continually changed into other forms, perhaps every few weeks or every few years?

Atoms combine with each other. They stick together to form stable molecules and stack up to form rigid solids. An atom is mostly empty space, but you can stand on a floor—made up of atoms—without falling through it.

These basic properties of atoms can be explained by quantum physics, as can the three less apparent properties that follow.

Atoms Are Put Together Systematically

Figure 40-2 shows an example of a repetitive property of the elements as a function of their position in the periodic table (Appendix G). The figure is a plot

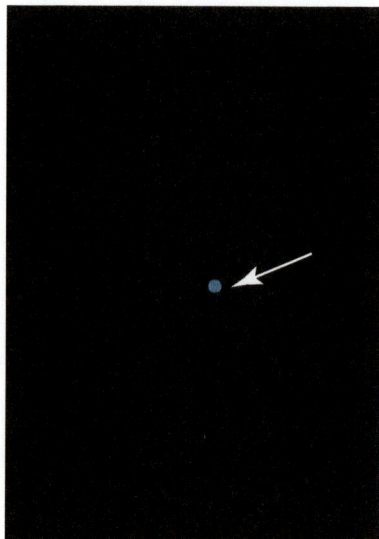

Courtesy Warren Nagourney

Figure 40-1 The blue dot is a photograph of the light emitted from a single barium ion held for a long time in a trap at the University of Washington. Special techniques caused the ion to emit light over and over again as it underwent transitions between the same pair of energy levels. The dot represents the cumulative emission of many photons.

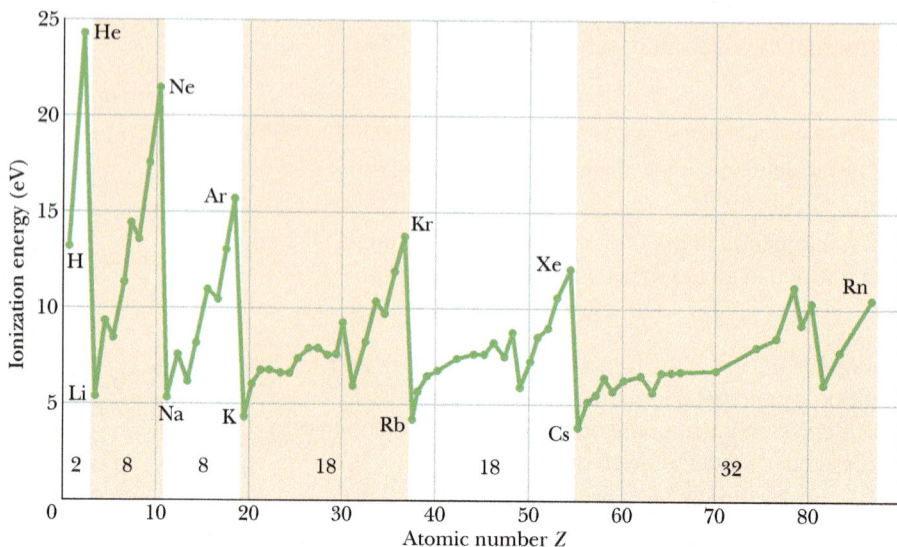

Figure 40-2 A plot of the ionization energies of the elements as a function of atomic number, showing the periodic repetition of properties through the six complete horizontal periods of the periodic table. The number of elements in each of these periods is indicated.

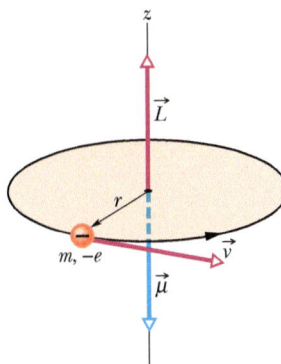

Figure 40-3 A classical model showing a particle of mass m and charge $-e$ moving with speed v in a circle of radius r. The moving particle has an angular momentum $\vec{L}$ given by $\vec{r} \times \vec{p}$, where $\vec{p}$ is its linear momentum $m\vec{v}$. The particle's motion is equivalent to a current loop that has an associated magnetic moment $\vec{\mu}$ that is directed opposite $\vec{L}$.

of the **ionization energy** of the elements; the energy required to remove the most loosely bound electron from a neutral atom is plotted as a function of the position in the periodic table of the element to which the atom belongs. The remarkable similarities in the chemical and physical properties of the elements in each vertical column of the periodic table are evidence enough that the atoms are constructed according to systematic rules.

The elements are arranged in the periodic table in six complete horizontal **periods** (and a seventh incomplete period): except for the first, each period starts at the left with a highly reactive alkali metal (lithium, sodium, potassium, and so on) and ends at the right with a chemically inert noble gas (neon, argon, krypton, and so on). Quantum physics accounts for the chemical properties of these elements. The numbers of elements in the six periods are

$$2, 8, 8, 18, 18, \text{ and } 32.$$

Quantum physics predicts these numbers.

Atoms Emit and Absorb Light

We have already seen that atoms can exist only in discrete quantum states, each state having a certain energy. An atom can make a transition from one state to another by emitting light (to jump to a lower energy level E_{low}) or by absorbing light (to jump to a higher energy level E_{high}). As we first discussed in Module 39-1, the light is emitted or absorbed as a photon with energy

$$hf = E_{\text{high}} - E_{\text{low}}. \qquad (40\text{-}1)$$

Thus, the problem of finding the frequencies of light emitted or absorbed by an atom reduces to the problem of finding the energies of the quantum states of that atom. Quantum physics allows us—in principle at least—to calculate these energies.

Atoms Have Angular Momentum and Magnetism

Figure 40-3 shows a negatively charged particle moving in a circular orbit around a fixed center. As we discussed in Module 32-5, the orbiting particle has both an angular momentum $\vec{L}$ and (because its path is equivalent to a tiny current loop) a magnetic dipole moment $\vec{\mu}$. As Fig. 40-3 shows, vectors $\vec{L}$ and $\vec{\mu}$ are both perpendicular to the plane of the orbit but, because the charge is negative, they point in opposite directions.

The model of Fig. 40-3 is strictly classical and does not accurately represent an electron in an atom. In quantum physics, the rigid orbit model has been replaced by the probability density model, best visualized as a dot plot. In quantum physics, however, it is still true that in general, each quantum state of an electron in an atom involves an angular momentum $\vec{L}$ and a magnetic dipole moment $\vec{\mu}$ that have opposite directions (those vector quantities are said to be *coupled*).

Figure 40-4 The Einstein–de Haas experimental setup. (*a*) Initially, the magnetic field in the iron cylinder is zero and the magnetic dipole moment vectors $\vec{\mu}$ of its atoms are randomly oriented. (*b*) When a magnetic field $\vec{B}$ is set up along the cylinder's axis, the magnetic dipole moment vectors line up parallel to $\vec{B}$ and the cylinder begins to rotate.

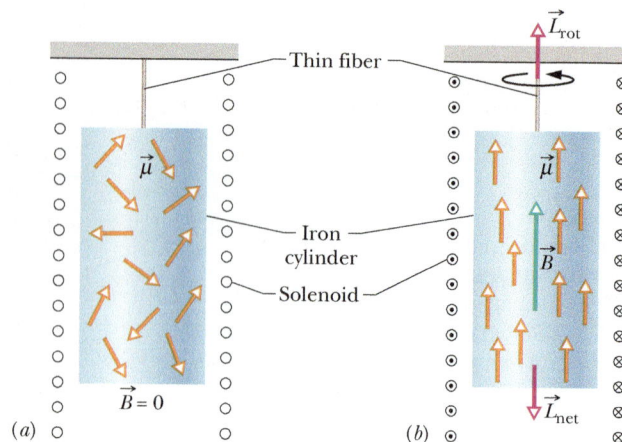

Aligning the magnetic moment vectors rotates the cylinder.

The Einstein–de Haas Experiment

In 1915, well before the discovery of quantum physics, Albert Einstein and Dutch physicist W. J. de Haas carried out a clever experiment designed to show that the angular momentum and magnetic moment of individual atoms are coupled.

Einstein and de Haas suspended an iron cylinder from a thin fiber, as shown in Fig. 40-4. A solenoid was placed around the cylinder but not touching it. Initially, the magnetic dipole moments $\vec{\mu}$ of the atoms of the cylinder point in random directions, and so their external magnetic effects cancel (Fig. 40-4*a*). However, when a current is switched on in the solenoid (Fig. 40-4*b*) so that a magnetic field $\vec{B}$ is set up parallel to the long axis of the cylinder, the magnetic dipole moments of the atoms of the cylinder reorient themselves, lining up with that field. If the angular momentum $\vec{L}$ of each atom is coupled to its magnetic moment $\vec{\mu}$, then this alignment of the atomic magnetic moments must cause an alignment of the atomic angular momenta opposite the magnetic field.

No external torques initially act on the cylinder; thus, its angular momentum must remain at its initial zero value. However, when $\vec{B}$ is turned on and the atomic angular momenta line up antiparallel to $\vec{B}$, they tend to give a net angular momentum $\vec{L}_{net}$ to the cylinder as a whole (directed downward in Fig. 40-4*b*). To maintain zero angular momentum, the cylinder begins to rotate around its central axis to produce an angular momentum $\vec{L}_{rot}$ in the opposite direction (upward in Fig. 40-4*b*).

The twisting of the fiber quickly produces a torque that momentarily stops the cylinder's rotation and then rotates the cylinder in the opposite direction as the twisting is undone. Thereafter, the fiber will twist and untwist as the cylinder oscillates about its initial orientation in angular simple harmonic motion.

Observation of the cylinder's rotation verified that the angular momentum and the magnetic dipole moment of an atom are coupled in opposite directions. Moreover, it dramatically demonstrated that the angular momenta associated with quantum states of atoms can result in *visible* rotation of an object of everyday size.

Angular Momentum, Magnetic Dipole Moments

Every quantum state of an electron in an atom has an associated orbital angular momentum and orbital magnetic dipole moment. Every electron, whether trapped in an atom or free, has a spin angular momentum and a spin magnetic dipole moment that are as intrinsic as its mass and charge. Let's next discuss these various quantities.

Orbital Angular Momentum

Classically, a moving particle has an angular momentum $\vec{L}$ with respect to any given reference point. In Chapter 11 we wrote this as the cross product

Table 40-1 Electron States for an Atom

Quantum Number	Symbol	Allowed Values	Related to
Principal	n	$1, 2, 3, \ldots$	Distance from the nucleus
Orbital	ℓ	$0, 1, 2, \ldots, (n-1)$	Orbital angular momentum
Orbital magnetic	m_ℓ	$0, \pm 1, \pm 2, \ldots, \pm \ell$	Orbital angular momentum (z component)
Spin	s	$\frac{1}{2}$	Spin angular momentum
Spin magnetic	m_s	$\pm\frac{1}{2}$	Spin angular momentum (z component)

$\vec{L} = \vec{r} \times \vec{p}$, where $\vec{r}$ is a position vector extending to the particle from the reference point and $\vec{p}$ is the particle's linear momentum ($m\vec{v}$). Although an electron in an atom is not a classical moving particle, it too has angular momentum given by $\vec{L} = \vec{r} \times \vec{p}$, with the reference point being the nucleus. However, unlike the classical particle, the electron's *orbital angular momentum* $\vec{L}$ is quantized. For the electron in a hydrogen atom, we can find the quantized (allowed) values by solving Schrödinger's equation. For that situation and any other, we can also find the quantized values by using the appropriate mathematics for a cross product in a quantum situation. (The mathematics is linear algebra, which you may have on your schedule of classes.) Either way we find that the allowed magnitudes of $\vec{L}$ are given by

$$L = \sqrt{\ell(\ell+1)}\,\hbar, \quad \text{for } \ell = 0, 1, 2, \ldots, (n-1), \tag{40-2}$$

where $\hbar$ is $h/2\pi$, ℓ is the orbital quantum number (introduced in Table 39-2, which is reproduced in Table 40-1), and n is the electron's principal quantum number.

The electron can have a definite value of L as given by one of the allowed states in Eq. 40-2, but it cannot have a definite direction for the vector $\vec{L}$. However, we can measure (detect) definite values of a component L_z along a chosen measurement axis (usually taken to be a z axis) as given by

$$L_z = m_\ell \hbar, \quad \text{for } m_\ell = 0, \pm 1, \pm 2, \ldots, \pm \ell, \tag{40-3}$$

where m_ℓ is the orbital magnetic quantum number (Table 40-1). However, if the electron has a definite value of L_z, it does not have definite values for L_x and L_y. We cannot get around this uncertainty by, say, first measuring L_z (getting a definite value) and then measuring L_x (getting a definite value) because the second measurement can change L_z and thus we no longer have a definite value for it. Also, we can never find $\vec{L}$ aligned with an axis because then it would have a definite direction and definite components along the other axes (namely, zero components).

A common way to depict the allowed values for L_z is shown in Fig. 40-5 for the situation in which $\ell = 2$. However, do not take the figure literally because it implies (incorrectly) that $\vec{L}$ has the definite direction of the drawn vector. Still, it allows us to relate the five possible z components to the full vector (which has a magnitude of $\hbar\sqrt{6}$) and to define the *semi-classical angle* θ given by

$$\cos\theta = \frac{L_z}{L}. \tag{40-4}$$

Orbital Magnetic Dipole Moment

Classically, an orbiting charged particle sets up the magnetic field of a magnetic dipole, as we discussed in Module 32-5. From Eq. 32-28, the dipole moment is related to the angular momentum of the classical particle by

$$\vec{\mu}_{\text{orb}} = -\frac{e}{2m}\vec{L}, \tag{40-5}$$

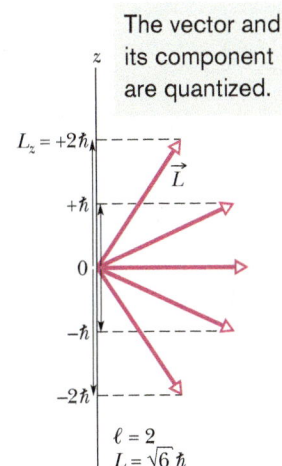

Figure 40-5 The allowed values of L_z for an electron in a quantum state with $\ell = 2$. For every orbital angular momentum vector $\vec{L}$ in the figure, there is a vector pointing in the opposite direction, representing the magnitude and direction of the orbital magnetic dipole moment $\vec{\mu}_{\text{orb}}$.

where m is the mass of the particle, here an electron. The minus sign means that the two vectors in Eq. 40-5 are in opposite directions, which is due to the fact that an electron is negatively charged.

An electron in an atom also has an orbital magnetic dipole moment given by Eq. 40-5, but $\vec{\mu}_{orb}$ is quantized. We find allowed values of the magnitude by substituting from Eq. 40-2:

$$\mu_{orb} = \frac{e}{2m} \sqrt{\ell(\ell + 1)} \, \hbar. \tag{40-6}$$

As with the angular momentum, $\vec{\mu}_{orb}$ can have a definite magnitude but does not have a definite direction. The best we can do is to measure its component on a z axis, and that component can have a definite value as given by

$$\mu_{orb,z} = -m_\ell \frac{e\hbar}{2m} = -m_\ell \mu_B, \tag{40-7}$$

where μ_B is the *Bohr magneton*:

$$\mu_B = \frac{eh}{4\pi m} = \frac{e\hbar}{2m} = 9.274 \times 10^{-24} \text{ J/T} \qquad \text{(Bohr magneton)}. \tag{40-8}$$

If the electron has a definite value of $\mu_{orb,z}$, it cannot have definite values of $\mu_{orb,x}$ and $\mu_{orb,y}$.

Spin Angular Momentum

Every electron, whether in an atom or free, has an intrinsic angular momentum that has no classical counterpart (it is *not* of the form $\vec{r} \times \vec{p}$). It is called *spin angular momentum* $\vec{S}$ (or simply *spin*), but the name is misleading because the electron is not spinning. Indeed there is nothing at all rotating in an electron, and yet the electron has angular momentum. The magnitude of $\vec{S}$ is quantized, with values restricted to

$$S = \sqrt{s(s + 1)} \, \hbar, \quad \text{for } s = \tfrac{1}{2}, \tag{40-9}$$

where s is the *spin quantum number*. For every electron, $s = \tfrac{1}{2}$ and the electron is said to be a spin-$\tfrac{1}{2}$ particle. (Protons and neutrons are also spin-$\tfrac{1}{2}$ particles.) The language here can be confusing, because both $\vec{S}$ and s are often referred to as spin.

As with the angular momentum associated with motion, this intrinsic angular momentum can have a definite magnitude but does not have a definite direction. The best we can do is to measure its component on a z axis, and that component can have only the definite values given by

$$S_z = m_s \hbar, \quad \text{for } m_s = \pm s = \pm\tfrac{1}{2}. \tag{40-10}$$

Here m_s is the *spin magnetic quantum number*, which can have only two values: $m_s = +s = +\tfrac{1}{2}$ (the electron is said to be *spin up*) and $m_s = -s = -\tfrac{1}{2}$ (the electron is said to be *spin down*). Also, if S_z has a definite value, then S_x and S_y do not. Figure 40-6 is another figure that you should not take literally but it serves to show the possible values of S_z.

The existence of electron spin was postulated on experimental evidence by two Dutch graduate students, George Uhlenbeck and Samuel Goudsmit, from their studies of atomic spectra. The theoretical basis for spin was provided a few years later by British physicist P. A. M. Dirac, who developed a relativistic quantum theory of the electron.

We have now seen the full set of quantum numbers for an electron, as listed in Table 40-1. If an electron is free, it has only its intrinsic quantum numbers s and m_s. If it is trapped in an atom, it has also has the quantum numbers n, ℓ, and m_ℓ.

Spin Magnetic Dipole Moment

As with the orbital angular momentum, a magnetic dipole moment is associated with the spin angular momentum:

$$\vec{\mu}_s = -\frac{e}{m}\vec{S},\qquad(40\text{-}11)$$

where the minus sign means that the two vectors are in opposite directions, which is due to the fact that an electron is negatively charged. This $\vec{\mu}_s$ is an intrinsic property of every electron. The vector $\vec{\mu}_s$ does not have a definite direction but it can have a definite magnitude, given by

$$\mu_s = \frac{e}{m}\sqrt{s(s+1)}\,\hbar.\qquad(40\text{-}12)$$

The vector can also have a definite component on a z axis, given by

$$\mu_{s,z} = -2m_s\mu_B,\qquad(40\text{-}13)$$

but that means that it cannot have a definite value of $\mu_{s,x}$ or $\mu_{s,y}$. Figure 40-6 shows the possible values of $\mu_{s,z}$. In the next module we shall discuss the early experimental evidence for the quantized nature in Eq. 40-13.

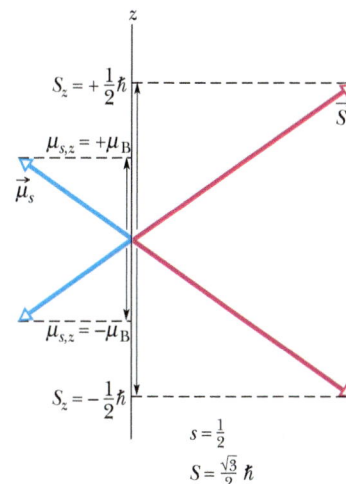

Figure 40-6 The allowed values of S_z and μ_z for an electron.

Shells and Subshells

As we discussed in Module 39-5, all states with the same n form a *shell*, and all states with the same value of n and ℓ form a *subshell*. As displayed in Table 40-1, for a given ℓ, there are $2\ell+1$ possible values of quantum number m_ℓ and, for each m_ℓ, there are two possible values for the quantum number m_s (spin up and spin down). Thus, there are $2(2\ell+1)$ states in a subshell. If we count all the states throughout a given shell with quantum number n, we find that the total number in the shell is $2n^2$.

Orbital and Spin Angular Momenta Combined

For an atom containing more than one electron, we define a total angular momentum $\vec{J}$, which is the vector sum of the angular momenta of the individual electrons—both their orbital and their spin angular momenta. Each element in the periodic table is defined by the number of protons in the nucleus of an atom of the element. This number of protons is defined as being the *atomic number* (or *charge number*) Z of the element. Because an electrically neutral atom contains equal numbers of protons and electrons, Z is also the number of electrons in the neutral atom, and we use this fact to indicate a $\vec{J}$ value for a neutral atom:

$$\vec{J} = (\vec{L}_1 + \vec{L}_2 + \vec{L}_3 + \cdots + \vec{L}_Z) + (\vec{S}_1 + \vec{S}_2 + \vec{S}_3 + \cdots + \vec{S}_Z).\quad(40\text{-}14)$$

Similarly, the total magnetic dipole moment of a multielectron atom is the vector sum of the magnetic dipole moments (both orbital and spin) of its individual electrons. However, because of the factor 2 in Eq. 40-13, the resultant magnetic dipole moment for the atom does not have the direction of vector $-\vec{J}$; instead, it makes a certain angle with that vector. The **effective magnetic dipole moment** $\vec{\mu}_{eff}$ for the atom is the component of the vector sum of the individual magnetic dipole moments in the direction of $-\vec{J}$ (Fig. 40-7). In typical atoms the orbital angular momenta and the spin angular momenta of most of the electrons sum vectorially to zero. Then $\vec{J}$ and $\vec{\mu}_{eff}$ of those atoms are due to a relatively small number of electrons, often only a single valence electron.

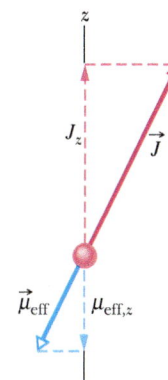

Figure 40-7 A classical model showing the total angular momentum vector $\vec{J}$ and the effective magnetic moment vector $\vec{\mu}_{eff}$.

✓ **Checkpoint 1**

An electron is in a quantum state for which the magnitude of the electron's orbital angular momentum $\vec{L}$ is $2\sqrt{3}\hbar$. How many projections of the electron's orbital magnetic dipole moment on a z axis are allowed?

40-2 THE STERN-GERLACH EXPERIMENT

Learning Objectives

After reading this module, you should be able to . . .

40.20 Sketch the Stern–Gerlach experiment and explain the type of atom required, the anticipated result, the actual result, and the importance of the experiment.

40.21 Apply the relationship between the magnetic field gradient and the force on an atom in a Stern–Gerlach experiment.

Key Ideas

● The Stern–Gerlach experiment demonstrated that the magnetic moment of silver atoms is quantized, experimental proof that magnetic moments at the atomic level are quantized.

● An atom with a magnetic dipole moment experiences a force in a nonuniform magnetic field. If the field changes at

the rate of dB/dz along a z axis, then the force is along the z axis and its magnitude is related to the component μ_z of the dipole moment:

$$F_z = \mu_z \frac{dB}{dz}.$$

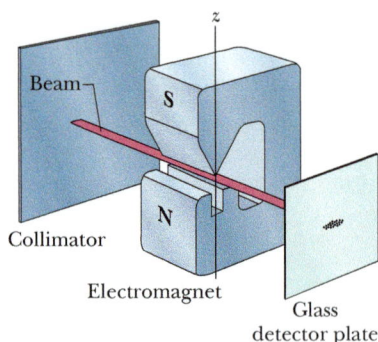

Figure 40-8 Apparatus used by Stern and Gerlach.

The Stern-Gerlach Experiment

In 1922, Otto Stern and Walther Gerlach at the University of Hamburg in Germany showed experimentally that the magnetic moment of silver atoms is quantized. In the Stern–Gerlach experiment, as it is now known, silver is vaporized in an oven, and some of the atoms in that vapor escape through a narrow slit in the oven wall and pass into an evacuated tube. Some of those escaping atoms then pass through a second narrow slit, to form a narrow beam of atoms (Fig. 40-8). (The atoms are said to be *collimated*—made into a beam—and the second slit is called a *collimator.*) The beam passes between the poles of an electromagnet and then lands on a glass detector plate where it forms a silver deposit.

When the electromagnet is off, the silver deposit is a narrow spot. However, when the electromagnet is turned on, the silver deposit should be spread vertically. The reason is that silver atoms are magnetic dipoles, and so vertical magnetic forces act on them as they pass through the vertical magnetic field of the electromagnet; these forces deflect them slightly up or down. Thus, by analyzing the silver deposit on the plate, we can determine what deflections the atoms underwent in the magnetic field. When Stern and Gerlach analyzed the pattern of silver on their detector plate, they found a surprise. However, before we discuss that surprise and its quantum implications, let us discuss the magnetic deflecting force acting on the silver atoms.

The Magnetic Deflecting Force on a Silver Atom

We have not previously discussed the type of magnetic force that deflects the silver atoms in a Stern–Gerlach experiment. It is *not* the magnetic deflecting force that acts on a moving charged particle, as given by Eq. 28-2 ($\vec{F} = q\vec{v} \times \vec{B}$). The reason is simple: A silver atom is electrically neutral (its net charge q is zero), and thus this type of magnetic force is also zero.

The type of magnetic force we seek is due to an interaction between the magnetic field $\vec{B}$ of the electromagnet and the magnetic dipole of the individual silver atom. We can derive an expression for the force in this interaction by starting with the energy U of the dipole in the magnetic field. Equation 28-38 tells us that

$$U = -\vec{\mu} \cdot \vec{B}, \qquad (40\text{-}15)$$

where $\vec{\mu}$ is the magnetic dipole moment of a silver atom. In Fig. 40-8, the positive direction of the z axis and the direction of $\vec{B}$ are vertically upward. Thus, we can write Eq. 40-15 in terms of the component μ_z of the atom's magnetic dipole moment along the direction of $\vec{B}$:

$$U = -\mu_z B. \qquad (40\text{-}16)$$

Then, using Eq. 8-22 ($F = -dU/dx$) for the z axis shown in Fig. 40-8, we obtain

$$F_z = -\frac{dU}{dz} = \mu_z \frac{dB}{dz}. \qquad (40\text{-}17)$$

This is what we sought—an equation for the magnetic force that deflects a silver atom as the atom passes through a magnetic field.

The term dB/dz in Eq. 40-17 is the *gradient* of the magnetic field along the z axis. If the magnetic field does not change along the z axis (as in a uniform magnetic field or no magnetic field), then $dB/dz = 0$ and a silver atom is not deflected as it moves between the magnet's poles. In the Stern–Gerlach experiment, the poles are designed to maximize the gradient dB/dz, so as to vertically deflect the silver atoms passing between the poles as much as possible, so that their deflections show up in the deposit on the glass plate.

According to classical physics, the components μ_z of silver atoms passing through the magnetic field in Fig. 40-8 should range in value from $-\mu$ (the dipole moment $\vec{\mu}$ is directed straight down the z axis) to $+\mu$ ($\vec{\mu}$ is directed straight up the z axis). Thus, from Eq. 40-17, there should be a range of forces on the atoms, and therefore a range of deflections of the atoms, from a greatest downward deflection to a greatest upward deflection. This means that we should expect the atoms to land along a vertical line on the glass plate, but they *don't*.

The Experimental Surprise

What Stern and Gerlach found was that the atoms formed two distinct spots on the glass plate, one spot above the point where they would have landed with no deflection and the other spot just as far below that point. The spots were initially too faint to be seen, but they became visible when Stern happened to breathe on the glass plate after smoking a cheap cigar. Sulfur in his breath (from the cigar) combined with the silver to produce a noticeably black silver sulfide.

This two-spot result can be seen in the plots of Fig. 40-9, which shows the outcome of a more recent version of the Stern–Gerlach experiment. In that version, a beam of cesium atoms (magnetic dipoles like the silver atoms in the original Stern–Gerlach experiment) was sent through a magnetic field with a large vertical gradient dB/dz. The field could be turned on and off, and a detector could be moved up and down through the beam.

When the field was turned off, the beam was, of course, undeflected and the detector recorded the central-peak pattern shown in Fig. 40-9. When the field was turned on, the original beam was split vertically by the magnetic field into two smaller beams, one beam higher than the previously undeflected beam and the other beam lower. As the detector moved vertically up through these two smaller beams, it recorded the two-peak pattern shown in Fig. 40-9.

The Meaning of the Results

In the original Stern–Gerlach experiment, two spots of silver were formed on the glass plate, not a vertical line of silver. This means that the component μ_z along $\vec{B}$ (and along z) could not have any value between $-\mu$ and $+\mu$ as classical physics predicts. Instead, μ_z is restricted to only two values, one for each spot on the glass. Thus, the original Stern–Gerlach experiment showed that μ_z is quantized, implying (correctly) that $\vec{\mu}$ is also. Moreover, because the angular momentum $\vec{L}$ of an atom is associated with $\vec{\mu}$, that angular momentum and its component L_z are also quantized.

With modern quantum theory, we can add to the explanation of the two-spot result in the Stern–Gerlach experiment. We now know that a silver atom consists

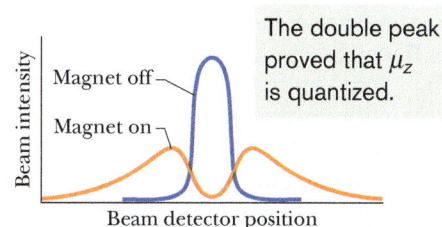

Figure 40-9 Results of a modern repetition of the Stern–Gerlach experiment. With the electromagnet turned off, there is only a single beam; with the electromagnet turned on, the original beam splits into two subbeams. The two subbeams correspond to parallel and antiparallel alignment of the magnetic moments of cesium atoms with the external magnetic field.

of many electrons, each with a spin magnetic moment and an orbital magnetic moment. We also know that all those moments vectorially cancel out *except* for a single electron, and the orbital dipole moment of that electron is zero. Thus, the combined dipole moment $\vec{\mu}$ of a silver atom is the *spin* magnetic dipole moment of that single electron. According to Eq. 40-13, this means that μ_z can have only two components along the z axis in Fig. 40-8. One component is for quantum number $m_s = +\frac{1}{2}$ (the single electron is spin up), and the other component is for quantum number $m_s = -\frac{1}{2}$ (the single electron is spin down). Substituting into Eq. 40-13 gives us

$$\mu_{s,z} = -2(+\tfrac{1}{2})\mu_B = -\mu_B \quad \text{and} \quad \mu_{s,z} = -2(-\tfrac{1}{2})\mu_B = +\mu_B. \quad (40\text{-}18)$$

Then substituting these expressions for μ_z in Eq. 40-17, we find that the force component F_z deflecting the silver atoms as they pass through the magnetic field can have only the two values

$$F_z = -\mu_B\left(\frac{dB}{dz}\right) \quad \text{and} \quad F_z = +\mu_B\left(\frac{dB}{dz}\right), \quad (40\text{-}19)$$

which result in the two spots of silver on the glass. Although no one knew about spin at the time, the Stern–Gerlach results were actually the first experimental evidence of electron spin.

Sample Problem 40.01 Beam separation in a Stern–Gerlach experiment

In the Stern–Gerlach experiment of Fig. 40-8, a beam of silver atoms passes through a magnetic field gradient dB/dz of magnitude 1.4 T/mm that is set up along the z axis. This region has a length w of 3.5 cm in the direction of the original beam. The speed of the atoms is 750 m/s. By what distance d have the atoms been deflected when they leave the region of the field gradient? The mass M of a silver atom is 1.8×10^{-25} kg.

KEY IDEAS

(1) The deflection of a silver atom in the beam is due to an interaction between the magnetic dipole of the atom and the magnetic field, because of the gradient dB/dz. The deflecting force is directed along the field gradient (along the z axis) and is given by Eqs. 40-19. Let us consider only deflection in the positive direction of z; thus, we shall use $F_z = \mu_B(dB/dz)$ from Eqs. 40-19.

(2) We assume the field gradient dB/dz has the same value throughout the region through which the silver atoms travel. Thus, force component F_z is constant in that region, and from Newton's second law, the acceleration a_z of an atom along the z axis due to F_z is also constant.

Calculations: Putting these ideas together, we write the acceleration as

$$a_z = \frac{F_z}{M} = \frac{\mu_B(dB/dz)}{M}.$$

Because this acceleration is constant, we can use Eq. 2-15 (from Table 2-1) to write the deflection d parallel to the z axis as

$$d = v_{0z}t + \tfrac{1}{2}a_z t^2 = 0t + \frac{1}{2}\left(\frac{\mu_B(dB/dz)}{M}\right)t^2. \quad (40\text{-}20)$$

Because the deflecting force on the atom acts perpendicular to the atom's original direction of travel, the component v of the atom's velocity along the original direction of travel is not changed by the force. Thus, the atom requires time $t = w/v$ to travel through length w in that direction. Substituting w/v for t into Eq. 40-20, we find

$$d = \frac{1}{2}\left(\frac{\mu_B(dB/dz)}{M}\right)\left(\frac{w}{v}\right)^2 = \frac{\mu_B(dB/dz)w^2}{2Mv^2}$$

$$= (9.27 \times 10^{-24} \text{ J/T})(1.4 \times 10^3 \text{ T/m})$$

$$\times \frac{(3.5 \times 10^{-2} \text{ m})^2}{(2)(1.8 \times 10^{-25} \text{ kg})(750 \text{ m/s})^2}$$

$$= 7.85 \times 10^{-5} \text{ m} \approx 0.08 \text{ mm}. \quad \text{(Answer)}$$

The separation between the two subbeams is twice this, or 0.16 mm. This separation is not large but is easily measured.

40-3 MAGNETIC RESONANCE

Learning Objectives

After reading this module, you should be able to . . .

40.22 For a proton in a magnetic field, sketch the field vector and the proton's magnetic moment vector for the lower energy state and the upper energy state and then include the labels of spin up and spin down.

40.23 For a proton in a magnetic field, calculate the energy difference between the two spin states and find the photon frequency and wavelength required for a transition between the states.

40.24 Explain the procedure of producing a nuclear magnetic resonance spectrum.

Key Ideas

● A proton has an intrinsic spin angular momentum $\vec{S}$ and an intrinsic magnetic dipole moment $\vec{\mu}$ that are in the same direction (because the proton is positively charged).

● The magnetic dipole moment $\vec{\mu}$ of a proton in a magnetic field $\vec{B}$ has two quantized components along the field axis: spin up (μ_z is in the direction $\vec{B}$ and spin down μ_z is in the opposite direction).

● Contrary to the situation with an electron, spin up is the lower energy orientation; the difference between the two orientations is $2\mu_z B$.

● The energy required of a photon to spin-flip the proton between the two orientations is

$$hf = 2\mu_z B.$$

● The field is the vector sum of an external field set up by equipment and an internal field set up by the atoms and nuclei surrounding the proton.

● Detection of spin-flips can lead to nuclear magnetic resonance spectra by which specific substances can be identified.

Magnetic Resonance

As we discussed briefly in Module 32-5, a proton has a spin magnetic dipole moment $\vec{\mu}$ that is associated with the proton's intrinsic spin angular momentum $\vec{S}$. The two vectors are said to be coupled together and, because the proton is positively charged, they are in the same direction. Suppose a proton is located in a magnetic field $\vec{B}$ that is directed along the positive direction of a z axis. Then $\vec{\mu}$ has two possible quantized components along that axis: the component can be $+\mu_z$ if the vector is in the direction of $\vec{B}$ (Fig. 40-10a) or $-\mu_z$ if it is opposite the direction of $\vec{B}$ (Fig. 40-10b).

From Eq. 28-38 ($U(\theta) = -\vec{\mu} \cdot \vec{B}$), recall that an energy is associated with the orientation of any magnetic dipole moment $\vec{\mu}$ located in an external magnetic field $\vec{B}$. Thus, energy is associated with the two orientations of Figs. 40-10a and b. The orientation in Fig. 40-10a is the lower-energy state ($-\mu_z B$) and is called the *spin-up state* because the proton's spin component S_z (not shown) is also aligned with $\vec{B}$. The orientation in Fig. 40-10b (the *spin-down state*) is the higher-energy state ($\mu_z B$). Thus, the energy difference between these two states is

$$\Delta E = \mu_z B - (-\mu_z B) = 2\mu_z B. \qquad (40\text{-}21)$$

If we place a sample of water in a magnetic field $\vec{B}$, the protons in the hydrogen portions of each water molecule tend to be in the lower-energy state. (We shall not consider the oxygen portions.) Any one of these protons can jump to the higher-energy state by absorbing a photon with an energy hf equal to ΔE. That is, the proton can jump by absorbing a photon of energy

$$hf = 2\mu_z B. \qquad (40\text{-}22)$$

Such absorption is called **magnetic resonance** or, as originally, **nuclear magnetic resonance** (NMR), and the consequent reversal of S_z is called *spin-flipping*.

In practice, the photons required for magnetic resonance have an associated frequency in the radio-frequency (RF) range and are provided by a small coil wrapped around the sample undergoing resonance. An electromagnetic oscillator called an *RF source* drives a sinusoidal current in the coil at frequency f. The electromagnetic (EM) field set up within the coil and sample also oscillates at

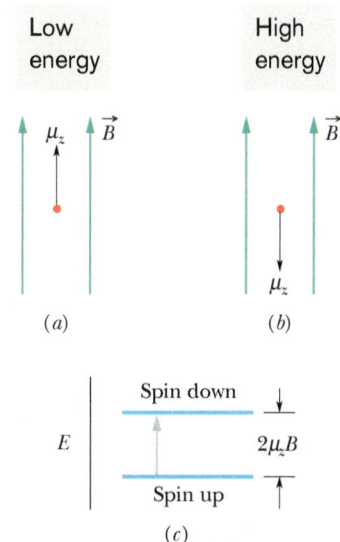

Figure 40-10 The z component of $\vec{\mu}$ for a proton in the (a) lower-energy (spin-up) and (b) higher-energy (spin-down) state. (c) An energy-level diagram for the states, showing the upward quantum jump the proton makes when its spin flips from up to down.

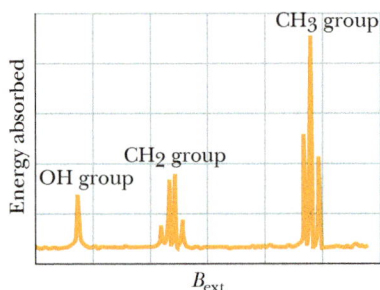

Figure 40-11 A nuclear magnetic resonance spectrum for ethanol, CH_3CH_2OH. The spectral lines represent the absorption of energy associated with spin-flips of protons. The three groups of lines correspond, as indicated, to protons in the OH group, the CH_2 group, and the CH_3 group of the ethanol molecule. Note that the two protons in the CH_2 group occupy four different local environments. The entire horizontal axis covers less than 10^{-4} T.

frequency f. If f meets the requirement of Eq. 40-22, the oscillating EM field can transfer a quantum of energy to a proton in the sample via a photon absorption, spin-flipping the proton.

The magnetic field magnitude B that appears in Eq. 40-22 is actually the magnitude of the net magnetic field $\vec{B}$ at the site where a given proton undergoes spin-flipping. That net field is the vector sum of the external field $\vec{B}_{ext}$ set up by the magnetic resonance equipment (primarily a large magnet) and the internal field $\vec{B}_{int}$ set up by the magnetic dipole moments of the atoms and nuclei near the given proton. For practical reasons we do not discuss here, magnetic resonance is usually detected by sweeping the magnitude B_{ext} through a range of values while the frequency f of the RF source is kept at a predetermined value and the energy of the RF source is monitored. A graph of the energy loss of the RF source versus B_{ext} shows a *resonance peak* when B_{ext} sweeps through the value at which spin-flipping occurs. Such a graph is called a *nuclear magnetic resonance spectrum,* or *NMR spectrum.*

Figure 40-11 shows the NMR spectrum of ethanol, which is a molecule consisting of three groups of atoms: CH_3, CH_2, and OH. Protons in each group can undergo magnetic resonance, but each group has its own unique magnetic-resonance value of B_{ext} because the groups lie in different internal fields $\vec{B}_{int}$ due to their arrangement within the CH_3CH_2OH molecule. Thus, the resonance peaks in the spectrum of Fig. 40-11 form a unique NMR signature by which ethanol can be indentified.

40-4 EXCLUSION PRINCIPLE AND MULTIPLE ELECTRONS IN A TRAP

Learning Objectives

After reading this module, you should be able to . . .

40.25 Identify the Pauli exclusion principle.
40.26 Explain the procedure for placing multiple electrons in traps of one, two, and three dimensions, including the need to obey the exclusion principle and to allow for

degenerate states, and explain the terms empty, partially occupied, and fully occupied.
40.27 For a system of multiple electrons in traps of one, two, and three dimensions, produce energy-level diagrams.

Key Idea

● Electrons in atoms and other traps obey the Pauli exclusion principle, which requires that no two electrons in a trap can have the same set of quantum numbers.

The Pauli Exclusion Principle

In Chapter 39 we considered a variety of electron traps, from fictional one-dimensional traps to the real three-dimensional trap of a hydrogen atom. In all those examples, we trapped only one electron. However, when we discuss traps containing two or more electrons (as we shall below), we must consider a principle that governs any particle whose spin quantum number s is not zero or an integer. This principle applies not only to electrons but also to protons and neutrons, all of which have $s = \frac{1}{2}$. The principle is known as the **Pauli exclusion principle** after Wolfgang Pauli, who formulated it in 1925. For electrons, it states that

> No two electrons confined to the same trap can have the same set of values for their quantum numbers.

As we shall discuss in Module 40-5, this principle means that no two electrons in an atom can have the same four values for the quantum numbers n, ℓ, m_ℓ, and m_s. All electrons have the same quantum number $s = \frac{1}{2}$. Thus, any two electrons in an atom must differ in at least one of these other quantum numbers. Were this not true, atoms would collapse, and thus you and the world could not exist.

Multiple Electrons in Rectangular Traps

To prepare for our discussion of multiple electrons in atoms, let us discuss two electrons confined to the rectangular traps of Chapter 39. However, here we shall also include the spin angular momenta. To do this, we assume that the traps are located in a uniform magnetic field. Then according to Eq. 40-10 ($S_z = m_s\hbar$), an electron can be either spin up with $m_s = \frac{1}{2}$ or spin down with $m_s = -\frac{1}{2}$. (We assume that the field is very weak so that the associated energy is negligible.)

As we confine the two electrons to one of the traps, we must keep the Pauli exclusion principle in mind; that is, the electrons cannot have the same set of values for their quantum numbers.

1. *One-dimensional trap.* In the one-dimensional trap of Fig. 39-2, fitting an electron wave to the trap's width L requires the single quantum number n. Therefore, any electron confined to the trap must have a certain value of n, and its quantum number m_s can be either $+\frac{1}{2}$ or $-\frac{1}{2}$. The two electrons could have different values of n, or they could have the same value of n if one of them is spin up and the other is spin down.

2. *Rectangular corral.* In the rectangular corral of Fig. 39-13, fitting an electron wave to the corral's widths L_x and L_y requires the two quantum numbers n_x and n_y. Thus, any electron confined to the trap must have certain values for those two quantum numbers, and its quantum number m_s can be either $+\frac{1}{2}$ or $-\frac{1}{2}$; so now there are three quantum numbers. According to the Pauli exclusion principle, two electrons confined to the trap must have different values for at least one of those three quantum numbers.

3. *Rectangular box.* In the rectangular box of Fig. 39-14, fitting an electron wave to the box's widths L_x, L_y, and L_z requires the three quantum numbers n_x, n_y, and n_z. Thus, any electron confined to the trap must have certain values for these three quantum numbers, and its quantum number m_s can be either $+\frac{1}{2}$ or $-\frac{1}{2}$; so now there are four quantum numbers. According to the Pauli exclusion principle, two electrons confined to the trap must have different values for at least one of those four quantum numbers.

Suppose we add more than two electrons, one by one, to a rectangular trap in the preceding list. The first electrons naturally go into the lowest possible energy level—they are said to *occupy* that level. However, eventually the Pauli exclusion principle disallows any more electrons from occupying that lowest energy level, and the next electron must occupy the next higher level. When an energy level cannot be occupied by more electrons because of the Pauli exclusion principle, we say that level is **full** or **fully occupied.** In contrast, a level that is not occupied by any electrons is **empty** or **unoccupied.** For intermediate situations, the level is **partially occupied.** The *electron configuration* of a system of trapped electrons is a listing or drawing either of the energy levels the electrons occupy or of the set of the quantum numbers of the electrons.

Finding the Total Energy

To find the energy of a system of two or more electrons confined to a trap, we assume that the electrons do not electrically interact with one another; that is, we shall neglect the electric potential energies of pairs of electrons. Then we can calculate the total energy for the system by calculating the energy of each electron (as in Chapter 39) and then summing those energies.

A good way to organize the energy values of a given system of electrons is with an energy-level diagram *for the system,* just as we did for a single electron in the traps of Chapter 39. The lowest level, with energy E_{gr}, corresponds to the ground state of the system. The next higher level, with energy E_{fe}, corresponds to the first excited state of the system. The next level, with energy E_{se}, corresponds to the second excited state of the system, and so on.

Sample Problem 40.02 Energy levels of multiple electrons in a 2D infinite potential well

Seven electrons are confined to a square corral (two-dimensional infinite potential well) with widths $L_x = L_y = L$ (Fig. 39-13). Assume that the electrons do not electrically interact with one another.

(a) What is the electron configuration for the ground state of the system of seven electrons?

One-electron diagram: We can determine the electron configuration of the system by placing the seven electrons in the corral one by one, to build up the system. Because we assume the electrons do not electrically interact with one another, we can use the energy-level diagram for a single trapped electron

in order to keep track of how we place the seven electrons in the corral. That *one-electron energy-level diagram* is given in Fig. 39-15 and partially reproduced here as Fig. 40-12a. Recall that the levels are labeled as $E_{nx,ny}$ for their associated energy. For example, the lowest level is for energy $E_{1,1}$, where quantum number n_x is 1 and quantum number n_y is 1.

Pauli principle: The trapped electrons must obey the Pauli exclusion principle; that is, no two electrons can have the same set of values for their quantum numbers n_x, n_y, and m_s. The first electron goes into energy level $E_{1,1}$ and can have $m_s = \frac{1}{2}$ or $m_s = -\frac{1}{2}$. We arbitrarily choose the latter

Figure 40-12 (a) Energy-level diagram for one electron in a square corral. (Energy E is in multiples of $h^2/8mL^2$.) A spin-down electron occupies the lowest level. (b) Two electrons (one spin down, the other spin up) occupy the lowest level of the one-electron energy-level diagram. (c) A third electron occupies the next energy level. (d) Four electrons can be put into the second level. (e) The system's ground-state configuration. (f) Three transitions to consider for the first excited state. (g) The system's lowest three total energies.

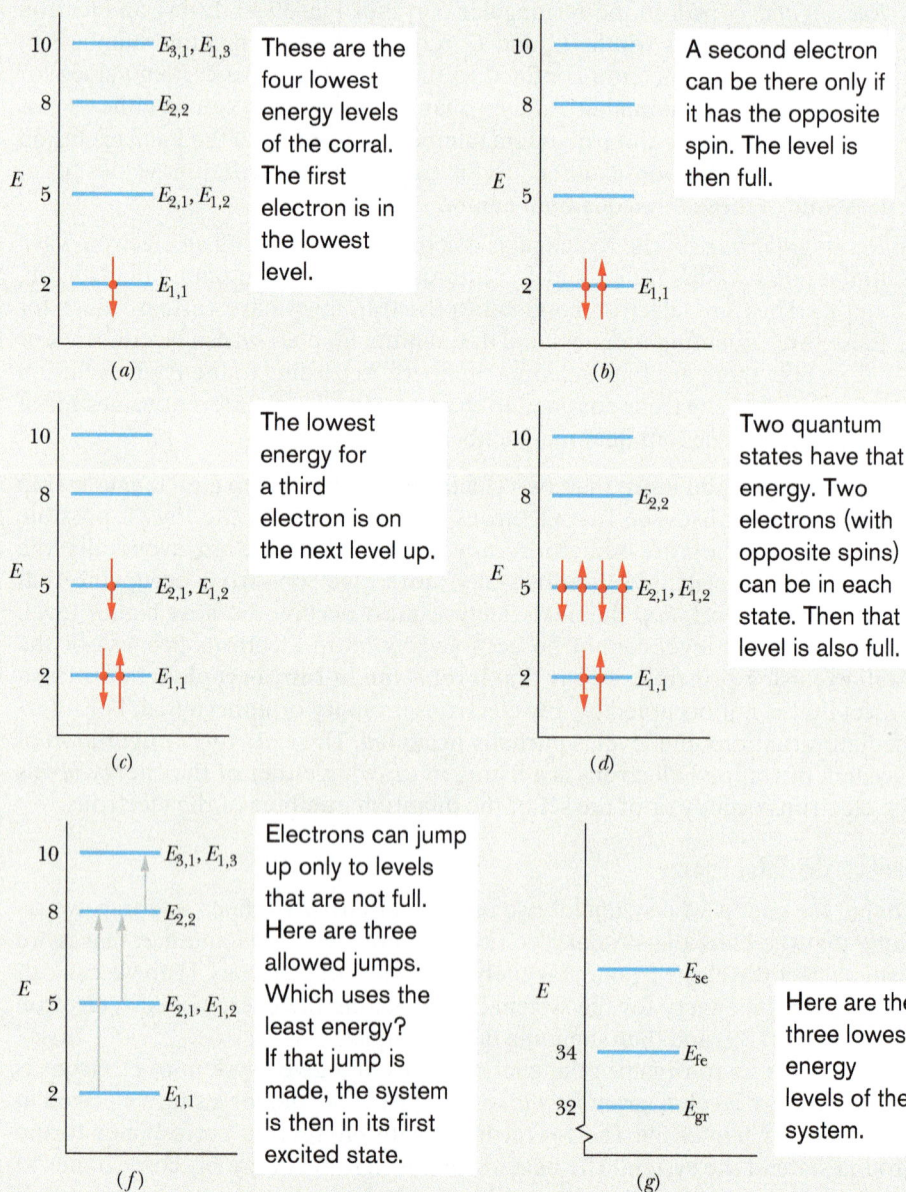

and draw a down arrow (to represent spin down) on the $E_{1,1}$ level in Fig. 40-12a. The second electron also goes into the $E_{1,1}$ level but must have $m_s = +\frac{1}{2}$ so that one of its quantum numbers differs from those of the first electron. We represent this second electron with an up arrow (for spin up) on the $E_{1,1}$ level in Fig. 40-12b.

Electrons, one by one: The level for energy $E_{1,1}$ is fully occupied, and thus the third electron cannot have that energy. Therefore, the third electron goes into the next higher level, which is for the equal energies $E_{2,1}$ and $E_{1,2}$ (the level is degenerate). This third electron can have quantum numbers n_x and n_y of either 1 and 2 or 2 and 1, respectively. It can also have a quantum number m_s of either $+\frac{1}{2}$ or $-\frac{1}{2}$. Let us arbitrarily assign it the quantum numbers $n_x = 2$, $n_y = 1$, and $m_s = -\frac{1}{2}$. We then represent it with a down arrow on the level for $E_{1,2}$ and $E_{2,1}$ in Fig. 40-12c.

You can show that the next three electrons can also go into the level for energies $E_{2,1}$ and $E_{1,2}$, provided that no set of three quantum numbers is completely duplicated. That level then contains four electrons (Fig. 40-12d), with quantum numbers (n_x, n_y, m_s) of

$$(2,1,-\tfrac{1}{2}), (2,1,+\tfrac{1}{2}), (1,2,-\tfrac{1}{2}), (1,2,+\tfrac{1}{2}),$$

and the level is fully occupied. Thus, the seventh electron goes into the next higher level, which is the $E_{2,2}$ level. Let us assume this electron is spin down, with $m_s = -\frac{1}{2}$.

Figure 40-12e shows all seven electrons on a one-electron energy-level diagram. We now have seven electrons in the corral, and they are in the configuration with the lowest energy that satisfies the Pauli exclusion principle. Thus, the ground-state configuration of the system is that shown in Fig. 40-12e and listed in Table 40-2.

(b) What is the total energy of the seven-electron system in its ground state, as a multiple of $h^2/8mL^2$?

KEY IDEA

The total energy E_{gr} is the sum of the energies of the individual electrons in the system's ground-state configuration.

Ground-state energy: The energy of each electron can be read from Table 39-1, which is partially reproduced in Table 40-2, or from Fig. 40-12e. Because there are two electrons in the first (lowest) level, four in the second level, and one in the third level, we have

$$E_{gr} = 2\left(2\frac{h^2}{8mL^2}\right) + 4\left(5\frac{h^2}{8mL^2}\right) + 1\left(8\frac{h^2}{8mL^2}\right)$$

$$= 32\frac{h^2}{8mL^2}. \qquad \text{(Answer)}$$

(c) How much energy must be transferred to the system for it to jump to its first excited state, and what is the energy of that state?

KEY IDEAS

1. If the system is to be excited, one of the seven electrons must make a quantum jump up the one-electron energy-level diagram of Fig. 40-12e.

2. If that jump is to occur, the energy change ΔE of the electron (and thus of the system) must be $\Delta E = E_{high} - E_{low}$ (Eq. 39-5), where E_{low} is the energy of the level where the jump begins and E_{high} is the energy of the level where the jump ends.

3. The Pauli exclusion principle must still apply; an electron *cannot* jump to a level that is fully occupied.

First-excited-state energy: Let us consider the three jumps shown in Fig. 40-12f; all are allowed by the Pauli exclusion principle because they are jumps to either empty or partially occupied states. In one of those possible jumps, an electron jumps from the $E_{1,1}$ level to the partially occupied $E_{2,2}$ level. The change in the energy is

$$\Delta E = E_{2,2} - E_{1,1} = 8\frac{h^2}{8mL^2} - 2\frac{h^2}{8mL^2} = 6\frac{h^2}{8mL^2}.$$

(We shall assume that the spin orientation of the electron making the jump can change as needed.)

In another of the possible jumps in Fig. 40-12f, an electron jumps from the degenerate level of $E_{2,1}$ and $E_{1,2}$ to the partially occupied $E_{2,2}$ level. The change in the energy is

$$\Delta E = E_{2,2} - E_{2,1} = 8\frac{h^2}{8mL^2} - 5\frac{h^2}{8mL^2} = 3\frac{h^2}{8mL^2}.$$

In the third possible jump in Fig. 40-12f, the electron in the $E_{2,2}$ level jumps to the unoccupied, degenerate level of $E_{1,3}$ and $E_{3,1}$. The change in energy is

$$\Delta E = E_{1,3} - E_{2,2} = 10\frac{h^2}{8mL^2} - 8\frac{h^2}{8mL^2} = 2\frac{h^2}{8mL^2}.$$

Table 40-2 Ground-State Configuration and Energies

n_x	n_y	m_s	Energy[a]
2	2	$-\frac{1}{2}$	8
2	1	$+\frac{1}{2}$	5
2	1	$-\frac{1}{2}$	5
1	2	$+\frac{1}{2}$	5
1	2	$-\frac{1}{2}$	5
1	1	$+\frac{1}{2}$	2
1	1	$-\frac{1}{2}$	2
			Total 32

[a]In multiples of $h^2/8mL^2$.

Of these three possible jumps, the one requiring the least energy change ΔE is the last one. We could consider even more possible jumps, but none would require less energy. Thus, for the system to jump from its ground state to its first excited state, the electron in the $E_{2,2}$ level must jump to the unoccupied, degenerate level of $E_{1,3}$ and $E_{3,1}$, and the required energy is

$$\Delta E = 2\frac{h^2}{8mL^2}. \qquad \text{(Answer)}$$

The energy E_{fe} of the first excited state of the system is then

$$E_{fe} = E_{gr} + \Delta E$$

$$= 32\frac{h^2}{8mL^2} + 2\frac{h^2}{8mL^2} = 34\frac{h^2}{8mL^2}. \qquad \text{(Answer)}$$

We can represent this energy and the energy E_{gr} for the ground state of the system on an energy-level diagram *for the system,* as shown in Fig. 40-12g.

WILEY PLUS Additional examples, video, and practice available at *WileyPLUS*

40-5 BUILDING THE PERIODIC TABLE

Learning Objectives

After reading this module, you should be able to . . .

40.28 Identify that all states in a subshell have the same energy that is determined primarily by quantum number n but to a lesser extent by quantum number ℓ.

40.29 Identify the labeling system for the orbital angular momentum quantum number.

40.30 Identify the procedure for filling up the shells and subshells in building up the periodic table for as long as the electron–electron interaction can be neglected.

40.31 Distinguish the noble gases from the other elements in terms of chemical interactions, net angular momentum, and ionization energy.

40.32 For a transition between two given atomic energy levels, for either emission or absorption of light, apply the relationship between the energy difference and the frequency and wavelength of the light.

Key Ideas

● In the periodic table, the elements are listed in order of increasing atomic number Z, where Z is the number of protons in the nucleus. For a neutral atom, Z is also the number of electrons.

● States with the same value of quantum number n form a shell.

● States with the same values of quantum numbers n and ℓ form a subshell.

● A closed shell and a closed subshell contain the maximum number of electrons as allowed by the Pauli exclusion principle. The net angular momentum and net magnetic moment of such closed structures are zero.

Building the Periodic Table

The four quantum numbers n, ℓ, m_ℓ, and m_s identify the quantum states of individual electrons in a multielectron atom. The wave functions for these states, however, are not the same as the wave functions for the corresponding states of the hydrogen atom because, in multielectron atoms, the potential energy associated with a given electron is determined not only by the charge and position of the atom's nucleus but also by the charges and positions of all the other electrons in the atom. Solutions of Schrödinger's equation for multielectron atoms can be carried out numerically—in principle at least—using a computer.

Shells and Subshells

As we discussed in Module 40-1, all states with the same n form a *shell*, and all states with the same value of n and ℓ form a *subshell*. For a given ℓ, there are $2\ell + 1$ possible values of quantum number m_ℓ and, for each m_ℓ, there are two possible values for the quantum number m_s (spin up and spin down). Thus, there are $2(2\ell + 1)$

states in a subshell. If we count all the states throughout a given shell with quantum number n, we find that the total number in the shell is $2n^2$. All states in a given subshell have about the same energy, which depends primarily on the value of n, but it also depends somewhat on the value of ℓ.

For the purpose of labeling subshells, the values of ℓ are represented by letters:

$$\ell = 0 \quad 1 \quad 2 \quad 3 \quad 4 \quad 5 \ldots$$
$$s \quad p \quad d \quad f \quad g \quad h \ldots$$

For example, the $n = 3, \ell = 2$ subshell would be labeled the $3d$ subshell.

When we assign electrons to states in a multielectron atom, we must be guided by the Pauli exclusion principle of Module 40-4; that is, no two electrons in an atom can have the same set of the quantum numbers n, ℓ, m_ℓ, and m_s. If this important principle did not hold, *all* the electrons in any atom could jump to the atom's lowest energy level, which would eliminate the chemistry of atoms and molecules, and thus also eliminate biochemistry and us. Let us examine the atoms of a few elements to see how the Pauli exclusion principle operates in the building up of the periodic table.

Neon

The neon atom has 10 electrons. Only two of them fit into the lowest-energy subshell, the $1s$ subshell. These two electrons both have $n = 1, \ell = 0$, and $m_\ell = 0$, but one has $m_s = +\frac{1}{2}$ and the other has $m_s = -\frac{1}{2}$. The $1s$ subshell contains $2[2(0) + 1] = 2$ states. Because this subshell then contains all the electrons permitted by the Pauli principle, it is said to be **closed.**

Two of the remaining eight electrons fill the next lowest energy subshell, the $2s$ subshell. The last six electrons just fill the $2p$ subshell, which, with $\ell = 1$, holds $2[2(1) + 1] = 6$ states.

In a closed subshell, all allowed z projections of the orbital angular momentum vector $\vec{L}$ are present and, as you can verify from Fig. 40-5, these projections cancel for the subshell as a whole; for every positive projection there is a corresponding negative projection of the same magnitude. Similarly, the z projections of the spin angular momenta also cancel. Thus, a closed subshell has no angular momentum and no magnetic moment of any kind. Furthermore, its probability density is spherically symmetric. Then neon with its three closed subshells ($1s$, $2s$, and $2p$) has no "loosely dangling electrons" to encourage chemical interaction with other atoms. Neon, like the other **noble gases** that form the right-hand column of the periodic table, is almost chemically inert.

Sodium

Next after neon in the periodic table comes sodium, with 11 electrons. Ten of them form a closed neon-like core, which, as we have seen, has zero angular momentum. The remaining electron is largely outside this inert core, in the $3s$ subshell—the next lowest energy subshell. Because this **valence electron** of sodium is in a state with $\ell = 0$ (that is, an s state using the lettering system above), the sodium atom's angular momentum and magnetic dipole moment must be due entirely to the spin of this single electron.

Sodium readily combines with other atoms that have a "vacancy" into which sodium's loosely bound valence electron can fit. Sodium, like the other **alkali metals** that form the left-hand column of the periodic table, is chemically active.

Chlorine

The chlorine atom, which has 17 electrons, has a closed 10-electron, neon-like core, with 7 electrons left over. Two of them fill the $3s$ subshell, leaving five to be assigned to the $3p$ subshell, which is the subshell next lowest in energy. This subshell, which has $\ell = 1$, can hold $2[2(1) + 1] = 6$ electrons, and so there is a vacancy, or a "hole," in this subshell.

Chlorine is receptive to interacting with other atoms that have a valence electron that might fill this hole. Sodium chloride (NaCl), for example, is a very stable compound. Chlorine, like the other **halogens** that form column VIIA of the periodic table, is chemically active.

Iron

The arrangement of the 26 electrons of the iron atom can be represented as follows:

$$\underbrace{1s^2 \quad 2s^2\ 2p^6 \quad 3s^2\ 3p^6}\ 3d^6 \quad 4s^2.$$

The subshells are listed in numerical order and, following convention, a superscript gives the number of electrons in each subshell. From Table 40-1 we can see that an s subshell ($\ell = 0$) can hold 2 electrons, a p subshell ($\ell = 1$) can hold 6, and a d subshell ($\ell = 2$) can hold 10. Thus, iron's first 18 electrons form the five filled subshells that are marked off by the bracket, leaving 8 electrons to be accounted for. Six of the eight go into the $3d$ subshell, and the remaining two go into the $4s$ subshell.

The reason the last two electrons do not also go into the $3d$ subshell (which can hold 10 electrons) is that the $3d^6\ 4s^2$ configuration results in a lower-energy state for the atom as a whole than would the $3d^8$ configuration. An iron atom with 8 electrons (rather than 6) in the $3d$ subshell would quickly make a transition to the $3d^6\ 4s^2$ configuration, emitting electromagnetic radiation in the process. The lesson here is that except for the simplest elements, the states may not be filled in what we might think of as their "logical" sequence.

40-6 X RAYS AND THE ORDERING OF THE ELEMENTS

Learning Objectives

After reading this module, you should be able to . . .

40.33 Identify where x rays are located in the electromagnetic spectrum.

40.34 Explain how x rays are produced in a laboratory or medical setting.

40.35 Distinguish between a continuous x-ray spectrum and a characteristic x-ray spectrum.

40.36 In a continuous x-ray spectrum, identify the cause of the cutoff wavelength λ_{min}.

40.37 Identify that in an electron–atom collision, energy and momentum are conserved.

40.38 Apply the relationship between a cutoff wavelength λ_{min} and the kinetic energy K_0 of the incident electrons.

40.39 Draw an energy-level diagram for holes and identify (with labels) the transitions that produce x rays.

40.40 For a given hole transition, calculate the wavelength of the emitted x ray.

40.41 Explain the importance of Moseley's work with regard to the periodic table.

40.42 Sketch a Moseley plot.

40.43 Describe the screening effect in a multielectron atom.

40.44 Apply the relationship between the frequency of the emitted K-alpha x rays and the atomic number Z of the atoms.

Key Ideas

● When a beam of high-energy electrons impact a target, the electrons can lose their energy by scattering from atoms and emitting a continuous spectrum of x rays.

● The shortest wavelength in the spectrum is the cutoff wavelength λ_{min}, which is emitted when an incident electron loses its full kinetic energy K_0 in a single collision:

$$\lambda_{min} = \frac{hc}{K_0}.$$

● The characteristic x-ray spectrum is produced when incident electrons eject low-lying electrons in the target atoms and electrons from upper levels jump down to the resulting holes, emitting light.

● A Moseley plot is a graph of the square root of the characteristic-emission frequencies $\sqrt{f}$ versus atomic number Z of the target atoms. The straight-line plot reveals that the position of an element in the periodic table is set by Z and not the atomic weight.

X Rays and the Ordering of the Elements

When a solid target, such as solid copper or tungsten, is bombarded with electrons whose kinetic energies are in the kiloelectron-volt range, electromagnetic radiation called **x rays** is emitted. Our concern here is what these rays can teach us about the atoms that absorb or emit them. Figure 40-13 shows the wavelength spectrum of the x rays produced when a beam of 35 keV electrons falls on a molybdenum target. We see a broad, continuous spectrum of radiation on which are superimposed two peaks of sharply defined wavelengths. The continuous spectrum and the peaks arise in different ways, which we next discuss separately.

The Continuous X-Ray Spectrum

Here we examine the continuous x-ray spectrum of Fig. 40-13, ignoring for the time being the two prominent peaks that rise from it. Consider an electron of initial kinetic energy K_0 that collides (interacts) with one of the target atoms, as in Fig. 40-14. The electron may lose an amount of energy ΔK, which will appear as the energy of an x-ray photon that is radiated away from the site of the collision. (Very little energy is transferred to the recoiling atom because of the relatively large mass of the atom; here we neglect that transfer.)

The scattered electron in Fig. 40-14, whose energy is now less than K_0, may have a second collision with a target atom, generating a second photon, with a different photon energy. This electron-scattering process can continue until the electron is approximately stationary. All the photons generated by these collisions form part of the continuous x-ray spectrum.

A prominent feature of that spectrum in Fig. 40-13 is the sharply defined **cutoff wavelength** λ_{min}, below which the continuous spectrum does not exist. This minimum wavelength corresponds to a collision in which an incident electron loses *all* its initial kinetic energy K_0 in a single head-on collision with a target atom. Essentially all this energy appears as the energy of a single photon, whose associated wavelength—the minimum possible x-ray wavelength—is found from

$$K_0 = hf = \frac{hc}{\lambda_{min}},$$

or
$$\lambda_{min} = \frac{hc}{K_0} \quad \text{(cutoff wavelength)}. \tag{40-23}$$

The cutoff wavelength is totally independent of the target material. If we were to switch from a molybdenum target to a copper target, for example, all features of the x-ray spectrum of Fig. 40-13 would change *except* the cutoff wavelength.

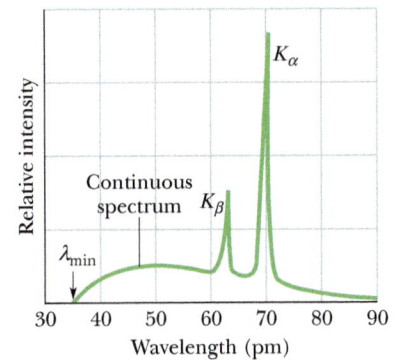

Figure 40-13 The distribution by wavelength of the x rays produced when 35 keV electrons strike a molybdenum target. The sharp peaks and the continuous spectrum from which they rise are produced by different mechanisms.

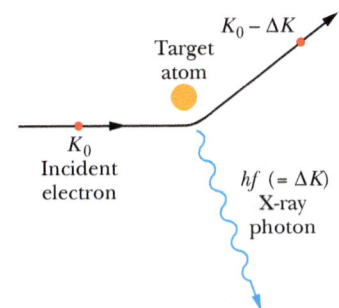

Figure 40-14 An electron of kinetic energy K_0 passing near an atom in the target may generate an x-ray photon, the electron losing part of its energy in the process. The continuous x-ray spectrum arises in this way.

✓ Checkpoint 2

Does the cutoff wavelength λ_{min} of the continuous x-ray spectrum increase, decrease, or remain the same if you (a) increase the kinetic energy of the electrons that strike the x-ray target, (b) allow the electrons to strike a thin foil rather than a thick block of the target material, (c) change the target to an element of higher atomic number?

The Characteristic X-Ray Spectrum

We now turn our attention to the two peaks of Fig. 40-13, labeled K_α and K_β. These (and other peaks that appear at wavelengths beyond the range displayed in Fig. 40-13) form the **characteristic x-ray spectrum** of the target material.

The peaks arise in a two-part process. (1) An energetic electron strikes an atom in the target and, while it is being scattered, the incident electron knocks out one of the atom's deep-lying (low n value) electrons. If the deep-lying elec-

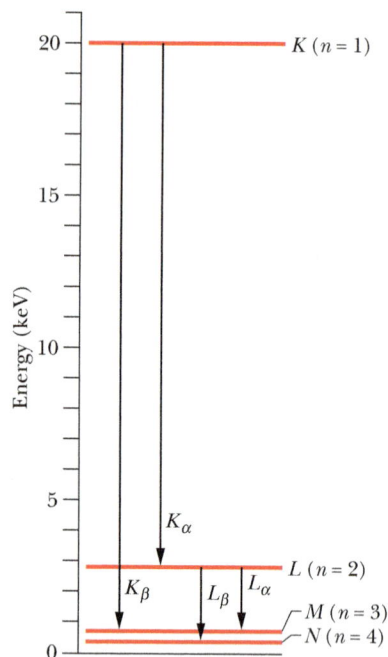

Figure 40-15 A simplified energy-level diagram for a molybdenum atom, showing the transitions (of holes rather than electrons) that give rise to some of the characteristic x rays of that element. Each horizontal line represents the energy of the atom with a hole (a missing electron) in the shell indicated.

tron is in the shell defined by $n = 1$ (called, for historical reasons, the K shell), there remains a vacancy, or *hole*, in this shell. (2) An electron in one of the shells with a higher energy jumps to the K shell, filling the hole in this shell. During this jump, the atom emits a characteristic x-ray photon. If the electron that fills the K-shell vacancy jumps from the shell with $n = 2$ (called the L shell), the emitted radiation is the K_α line of Fig. 40-13; if it jumps from the shell with $n = 3$ (called the M shell), it produces the K_β line, and so on. The hole left in either the L or M shell will be filled by an electron from still farther out in the atom.

In studying x rays, it is more convenient to keep track of where a hole is created deep in the atom's "electron cloud" than to record the changes in the quantum state of the electrons that jump to fill that hole. Figure 40-15 does exactly that; it is an energy-level diagram for molybdenum, the element to which Fig. 40-13 refers. The baseline ($E = 0$) represents the neutral atom in its ground state. The level marked K (at $E = 20$ keV) represents the energy of the molybdenum atom with a hole in its K shell, the level marked L (at $E = 2.7$ keV) represents the atom with a hole in its L shell, and so on.

The transitions marked K_α and K_β in Fig. 40-15 are the ones that produce the two x-ray peaks in Fig. 40-13. The K_α spectral line, for example, originates when an electron from the L shell fills a hole in the K shell. To state this transition in terms of what the arrows in Fig. 40-15 show, a hole originally in the K shell moves to the L shell.

Ordering the Elements

In 1913, British physicist H. G. J. Moseley generated characteristic x rays for as many elements as he could find—he found 38—by using them as targets for electron bombardment in an evacuated tube of his own design. By means of a trolley manipulated by strings, Moseley was able to move the individual targets into the path of an electron beam. He measured the wavelengths of the emitted x rays by the crystal diffraction method described in Module 36-7.

Moseley then sought (and found) regularities in these spectra as he moved from element to element in the periodic table. In particular, he noted that if, for a given spectral line such as K_α, he plotted for each element the square root of the frequency f against the position of the element in the periodic table, a straight line resulted. Figure 40-16 shows a portion of his extensive data. Moseley's conclusion was this:

> We have here a proof that there is in the atom a fundamental quantity, which increases by regular steps as we pass from one element to the next. This quantity can only be the charge on the central nucleus.

As a result of Moseley's work, the characteristic x-ray spectrum became the universally accepted signature of an element, permitting the solution of a number of

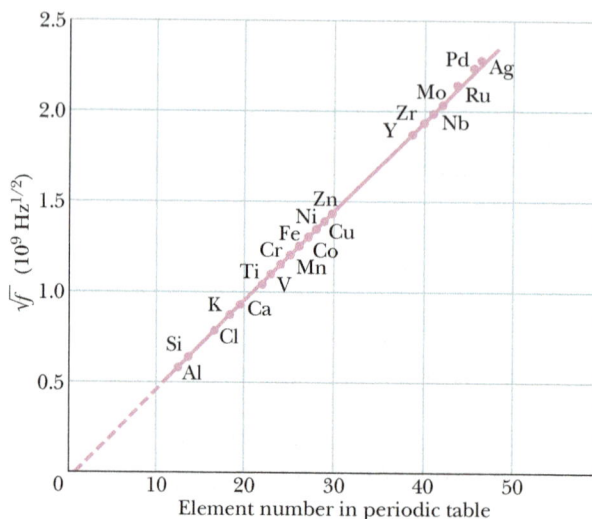

Figure 40-16 A Moseley plot of the K_α line of the characteristic x-ray spectra of 21 elements. The frequency is calculated from the measured wavelength.

periodic table puzzles. Prior to that time (1913), the positions of elements in the table were assigned in order of atomic *mass,* although it was necessary to invert this order for several pairs of elements because of compelling chemical evidence; Moseley showed that it is the nuclear charge (that is, atomic number Z) that is the real basis for ordering the elements.

In 1913 the periodic table had several empty squares, and a surprising number of claims for new elements had been advanced. The x-ray spectrum provided a conclusive test of such claims. The lanthanide elements, often called the rare earth elements, had been sorted out only imperfectly because their similar chemical properties made sorting difficult. Once Moseley's work was reported, these elements were properly organized.

It is not hard to see why the characteristic x-ray spectrum shows such impressive regularities from element to element whereas the optical spectrum in the visible and near-visible region does not: The key to the identity of an element is the charge on its nucleus. Gold, for example, is what it is because its atoms have a nuclear charge of $+79e$ (that is, $Z = 79$). An atom with one more elementary charge on its nucleus is mercury; with one fewer, it is platinum. The K electrons, which play such a large role in the production of the x-ray spectrum, lie very close to the nucleus and are thus sensitive probes of its charge. The optical spectrum, on the other hand, involves transitions of the outermost electrons, which are heavily screened from the nucleus by the remaining electrons of the atom and thus are *not* sensitive probes of nuclear charge.

Accounting for the Moseley Plot

Moseley's experimental data, of which the Moseley plot of Fig. 40-16 is but a part, can be used directly to assign the elements to their proper places in the periodic table. This can be done even if no theoretical basis for Moseley's results can be established. However, there is such a basis.

According to Eq. 39-33, the energy of the hydrogen atom is

$$E_n = -\frac{me^4}{8\varepsilon_0^2 h^2}\frac{1}{n^2} = -\frac{13.60\text{ eV}}{n^2}, \qquad \text{for } n = 1, 2, 3, \ldots. \quad (40\text{-}24)$$

Consider now one of the two innermost electrons in the K shell of a multielectron atom. Because of the presence of the other K-shell electron, our electron "sees" an effective nuclear charge of approximately $(Z - 1)e$, where e is the elementary charge and Z is the atomic number of the element. The factor e^4 in Eq. 40-24 is the product of e^2—the square of hydrogen's nuclear charge—and $(-e)^2$—the square of an electron's charge. For a multielectron atom, we can approximate the effective energy of the atom by replacing the factor e^4 in Eq. 40-24 with $(Z - 1)^2 e^2 \times (-e)^2$, or $e^4(Z - 1)^2$. That gives us

$$E_n = -\frac{(13.60\text{ eV})(Z - 1)^2}{n^2}. \quad (40\text{-}25)$$

We saw that the K_α x-ray photon (of energy hf) arises when an electron makes a transition from the L shell (with $n = 2$ and energy E_2) to the K shell (with $n = 1$ and energy E_1). Thus, using Eq. 40-25, we may write the energy change as

$$\begin{aligned}\Delta E &= E_2 - E_1\\ &= \frac{-(13.60\text{ eV})(Z - 1)^2}{2^2} - \frac{-(13.60\text{ eV})(Z - 1)^2}{1^2}\\ &= (10.2\text{ eV})(Z - 1)^2.\end{aligned}$$

Then the frequency f of the K_α line is

$$\begin{aligned}f = \frac{\Delta E}{h} &= \frac{(10.2\text{ eV})(Z - 1)^2}{(4.14 \times 10^{-15}\text{ eV}\cdot\text{s})}\\ &= (2.46 \times 10^{15}\text{ Hz})(Z - 1)^2. \quad (40\text{-}26)\end{aligned}$$

Taking the square root of both sides yields

$$\sqrt{f} = CZ - C, \qquad (40\text{-}27)$$

in which C is a constant ($= 4.96 \times 10^7 \text{ Hz}^{1/2}$). Equation 40-27 is the equation of a straight line. It shows that if we plot the square root of the frequency of the K_α x-ray spectral line against the atomic number Z, we should obtain a straight line. As Fig. 40-16 shows, that is exactly what Moseley found.

Sample Problem 40.03 Characteristic spectrum in x-ray production

A cobalt target is bombarded with electrons, and the wavelengths of its characteristic x-ray spectrum are measured. There is also a second, fainter characteristic spectrum, which is due to an impurity in the cobalt. The wavelengths of the K_α lines are 178.9 pm (cobalt) and 143.5 pm (impurity), and the proton number for cobalt is $Z_{\text{Co}} = 27$. Determine the impurity using only these data.

KEY IDEA

The wavelengths of the K_α lines for both the cobalt (Co) and the impurity (X) fall on a K_α Moseley plot, and Eq. 40-27 is the equation for that plot.

Calculations: Substituting c/λ for f in Eq. 40-27, we obtain

$$\sqrt{\frac{c}{\lambda_{\text{Co}}}} = CZ_{\text{Co}} - C \quad \text{and} \quad \sqrt{\frac{c}{\lambda_{\text{X}}}} = CZ_{\text{X}} - C.$$

Dividing the second equation by the first neatly eliminates C, yielding

$$\sqrt{\frac{\lambda_{\text{Co}}}{\lambda_{\text{X}}}} = \frac{Z_{\text{X}} - 1}{Z_{\text{Co}} - 1}.$$

Substituting the given data yields

$$\sqrt{\frac{178.9 \text{ pm}}{143.5 \text{ pm}}} = \frac{Z_{\text{X}} - 1}{27 - 1}.$$

Solving for the unknown, we find that

$$Z_{\text{X}} = 30.0. \qquad \text{(Answer)}$$

Thus, the number of protons in the impurity nucleus is 30, and a glance at the periodic table identifies the impurity as zinc. Note that with a larger value of Z than cobalt, zinc has a smaller value of the K_α line. This means that the energy associated with that jump must be greater in zinc than cobalt.

PLUS Additional examples, video, and practice available at *WileyPLUS*

40-7 LASERS

Learning Objectives

After reading this module, you should be able to . . .

40.45 Distinguish the light of a laser from the light of a common lightbulb.

40.46 Sketch energy-level diagrams for the three basic ways that light can interact with matter (atoms) and identify which is the basis of lasing.

40.47 Identify metastable states.

40.48 For two energy states, apply the relationship between the relative number of atoms in the higher state due to thermal agitation, the energy difference, and the temperature.

40.49 Identify population inversion, explain why it is required in a laser, and relate it to the lifetimes of the states.

40.50 Discuss how a helium–neon laser works, pointing out which gas lases and explaining why the other gas is required.

40.51 For stimulated emission, apply the relationships between energy change, frequency, and wavelength.

40.52 For stimulated emission, apply the relationships between energy, power, time, intensity, area, photon energy, and rate of photon emission.

Key Ideas

● In stimulated emission, an atom in an excited state can be induced to de-excite to a lower energy state by emitting a photon if an identical photon passes the atom.

● The light emitted in stimulated emission is in phase with and travels in the direction of the light causing the emission.

● A laser can emit light via stimulated emission provided that its atoms are in a population inversion. That is, for the pair of levels involved in the stimulated emission, more atoms must be in the upper level than the lower level so that there is more stimulated emission than just absorption.

Lasers and Laser Light

In the early 1960s, quantum physics made one of its many contributions to technology: the **laser.** Laser light, like the light from an ordinary lightbulb, is emitted when atoms make a transition from one quantum state to a lower one. However, in a lightbulb the emissions are random, both in time and direction, and in a laser they are coordinated so that the emissions are at the same time and in the same direction. As a result, laser light has the following characteristics:

1. *Laser light is highly monochromatic.* Light from an ordinary incandescent lightbulb is spread over a continuous range of wavelengths and is certainly not monochromatic. The radiation from a fluorescent neon sign is monochromatic, true, to about 1 part in 10^6, but the sharpness of definition of laser light can be many times greater, as much as 1 part in 10^{15}.

2. *Laser light is highly coherent.* Individual long waves (*wave trains*) for laser light can be several hundred kilometers long. When two separated beams that have traveled such distances over separate paths are recombined, they "remember" their common origin and are able to form a pattern of interference fringes. The corresponding *coherence length* for wave trains emitted by a lightbulb is typically less than a meter.

3. *Laser light is highly directional.* A laser beam spreads very little; it departs from strict parallelism only because of diffraction at the exit aperture of the laser. For example, a laser pulse used to measure the distance to the Moon generates a spot on the Moon's surface with a diameter of only a few kilometers. Light from an ordinary bulb can be made into an approximately parallel beam by a lens, but the beam divergence is much greater than for laser light. Each point on a lightbulb's filament forms its own separate beam, and the angular divergence of the overall composite beam is set by the size of the filament.

4. *Laser light can be sharply focused.* If two light beams transport the same amount of energy, the beam that can be focused to the smaller spot will have the greater intensity (power per unit area) at that spot. For laser light, the focused spot can be so small that an intensity of 10^{17} W/cm^2 is readily obtained. An oxyacetylene flame, by contrast, has an intensity of only about 10^3 W/cm^2.

Lasers Have Many Uses

The smallest lasers, used for voice and data transmission over optical fibers, have as their active medium a semiconducting crystal about the size of a pinhead. Small as they are, such lasers can generate about 200 mW of power. The largest lasers, used for nuclear fusion research and for astronomical and military applications, fill a large building. The largest such laser can generate brief pulses of laser light with a power level, during the pulse, of about 10^{14} W. This is a few hundred times greater than the total electrical power generating capacity of the United States. To avoid a brief national power blackout during a pulse, the energy required for each pulse is stored up at a steady rate during the relatively long interpulse interval.

Among the many uses of lasers are reading bar codes, manufacturing and reading compact discs and DVDs, performing surgery of many kinds (both as a surgical aid as in Fig. 40-17 and as a cutting and cauterizing tool), surveying, cutting cloth in the garment industry (several hundred layers at a time), welding auto bodies, and generating holograms.

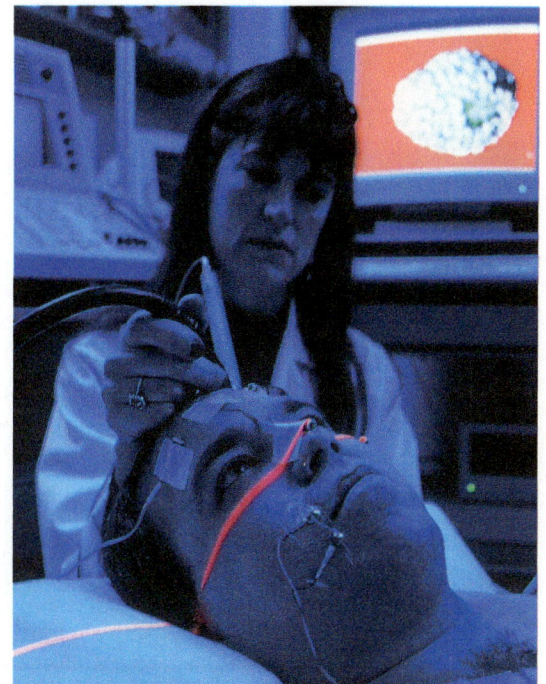

Figure 40-17 A patient's head is scanned and mapped by (red) laser light in preparation for brain surgery. During the surgery, the laser-derived image of the head will be superimposed on the model of the brain shown on the monitor, to guide the surgical team into the region shown in green (lower right) on the model displayed on the screen.

Sam Ogden/Photo Researchers, Inc.

How Lasers Work

Because the word "laser" is an acronym for "light amplification by the stimulated emission of radiation," you should not be surprised that stimulated emission is the key to laser operation. Einstein introduced this concept in 1917 in the paper where he explained the Planck formula for an ideal blackbody radiator (Eq. 38-14). Although the world had to wait until 1960 to see an operating laser, the groundwork for its development was put in place decades earlier.

Consider an isolated atom that can exist either in its state of lowest energy (its ground state), whose energy is E_0, or in a state of higher energy (an excited state), whose energy is E_x. Here are three processes by which the atom can move from one of these states to the other:

1. **Absorption.** Figure 40-18a shows the atom initially in its ground state. If the atom is placed in an electromagnetic field that is alternating at frequency f, the atom can absorb an amount of energy hf from that field and move to the higher-energy state. From the principle of conservation of energy we have

$$hf = E_x - E_0. \qquad (40\text{-}28)$$

   We call this process **absorption.**

2. **Spontaneous emission.** In Fig. 40-18b the atom is in its excited state and no external radiation is present. After a time, the atom will de-excite to its ground state, emitting a photon of energy hf in the process. We call this process **spontaneous emission**—*spontaneous* because the event is random and set by chance. The light from the filament of an ordinary lightbulb or any other common light source is generated in this way.

   Normally, the mean life of excited atoms before spontaneous emission occurs is about 10^{-8} s. However, for some excited states, this mean life is perhaps as much as 10^5 times longer. We call such long-lived states **metastable;** they play an important role in laser operation.

3. **Stimulated emission.** In Fig. 40-18c the atom is again in its excited state, but this time radiation with a frequency given by Eq. 40-28 is present. A photon of energy hf can stimulate the atom to move to its ground state, during which process the atom emits an additional photon, whose energy is also hf. We call this process **stimulated emission**—*stimulated* because the event is triggered by the external photon. The emitted photon is in every way identical to the stimulating photon. Thus, the waves associated with the photons have the same energy, phase, polarization, and direction of travel.

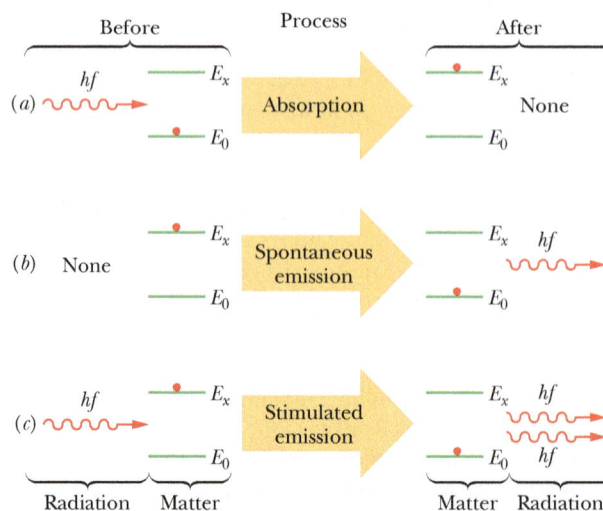

Figure 40-18 The interaction of radiation and matter in the processes of (a) absorption, (b) spontaneous emission, and (c) stimulated emission. An atom (matter) is represented by the red dot; the atom is in either a lower quantum state with energy E_0 or a higher quantum state with energy E_x. In (a) the atom absorbs a photon of energy hf from a passing light wave. In (b) it emits a light wave by emitting a photon of energy hf. In (c) a passing light wave with photon energy hf causes the atom to emit a photon of the same energy, increasing the energy of the light wave.

These are three ways that radiation (light) can interact with matter. The third way is the basis of lasing.

Figure 40-18c describes stimulated emission for a single atom. Suppose now that a sample contains a large number of atoms in thermal equilibrium at temperature *T*. Before any radiation is directed at the sample, a number N_0 of these atoms are in their ground state with energy E_0 and a number N_x are in a state of higher energy E_x. Ludwig Boltzmann showed that N_x is given in terms of N_0 by

$$N_x = N_0 e^{-(E_x - E_0)/kT}, \qquad (40\text{-}29)$$

in which *k* is Boltzmann's constant. This equation seems reasonable. The quantity *kT* is the mean kinetic energy of an atom at temperature *T*. The higher the temperature, the more atoms—on average—will have been "bumped up" by thermal agitation (that is, by atom–atom collisions) to the higher energy state E_x. Also, because $E_x > E_0$, Eq. 40-29 requires that $N_x < N_0$; that is, there will always be fewer atoms in the excited state than in the ground state. This is what we expect if the level populations N_0 and N_x are determined only by the action of thermal agitation. Figure 40-19a illustrates this situation.

If we now flood the atoms of Fig. 40-19a with photons of energy $E_x - E_0$, photons will disappear via absorption by ground-state atoms and photons will be generated largely via stimulated emission of excited-state atoms. Einstein showed that the probabilities per atom for these two processes are identical. Thus, because there are more atoms in the ground state, the *net* effect will be the absorption of photons.

To produce laser light, we must have more photons emitted than absorbed; that is, we must have a situation in which stimulated emission dominates. Thus, we need more atoms in the excited state than in the ground state, as in Fig. 40-19b. However, because such a **population inversion** is not consistent with thermal equilibrium, we must think up clever ways to set up and maintain one.

The Helium–Neon Gas Laser

Figure 40-20 shows a common type of laser developed in 1961 by Ali Javan and his coworkers. The glass discharge tube is filled with a 20 : 80 mixture of helium and neon gases, neon being the medium in which laser action occurs.

Figure 40-21 shows simplified energy-level diagrams for the two types of atoms. An electric current passed through the helium–neon gas mixture serves—through collisions between helium atoms and electrons of the current—to raise many helium

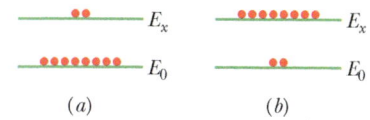

Figure 40-19 (a) The equilibrium distribution of atoms between the ground state E_0 and excited state E_x accounted for by thermal agitation. (b) An inverted population, obtained by special methods. Such a population inversion is essential for laser action.

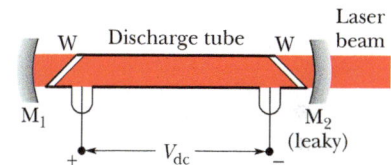

Figure 40-20 The elements of a helium–neon gas laser. An applied potential V_{dc} sends electrons through a discharge tube containing a mixture of helium gas and neon gas. Electrons collide with helium atoms, which then collide with neon atoms, which emit light along the length of the tube. The light passes through transparent windows W and reflects back and forth through the tube from mirrors M_1 and M_2 to cause more neon atom emissions. Some of the light leaks through mirror M_2 to form the laser beam.

Figure 40-21 Five essential energy levels for helium and neon atoms in a helium–neon gas laser. Laser action occurs between levels E_2 and E_1 of neon when more atoms are at the E_2 level than at the E_1 level.

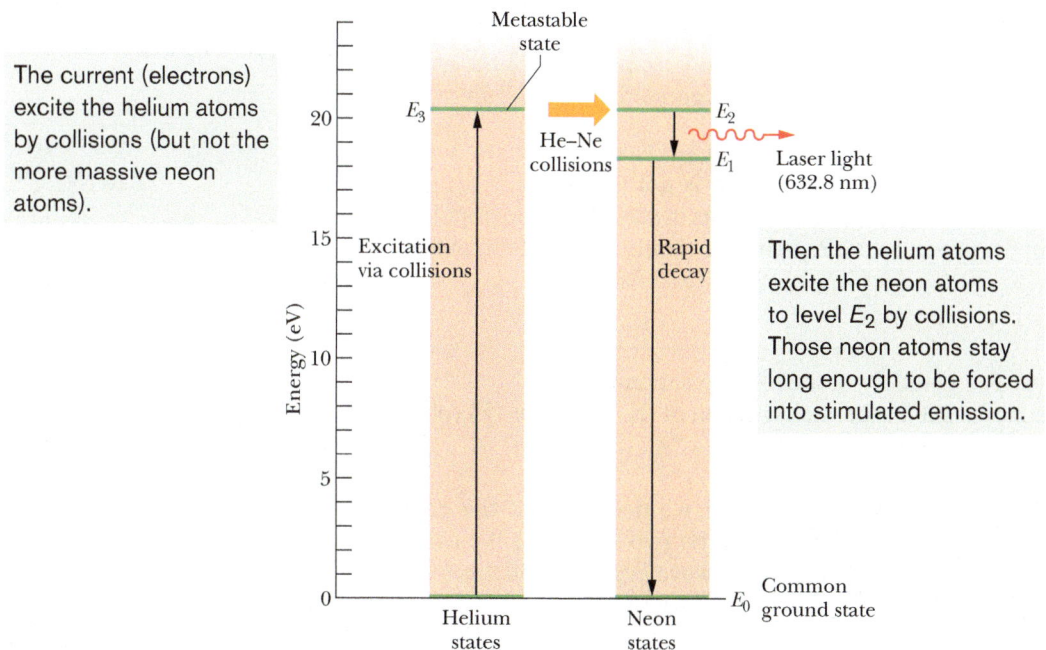

atoms to state E_3, which is metastable with a mean life of at least 1 μs. (The neon atoms are too massive to be excited by collisions with the (low-mass) electrons.)

The energy of helium state E_3 (20.61 eV) is very close to the energy of neon state E_2 (20.66 eV). Thus, when a metastable (E_3) helium atom and a ground-state (E_0) neon atom collide, the excitation energy of the helium atom is often transferred to the neon atom, which then moves to state E_2. In this manner, neon level E_2 (with a mean life of 170 ns) can become more heavily populated than neon level E_1 (which, with a mean life of only 10 ns, is almost empty).

This population inversion is relatively easy to set up because (1) initially there are essentially no neon atoms in state E_1, (2) the long mean life of helium level E_3 means that there is always a good chance that collisions will excite neon atoms to their E_2 level, and (3) once those neon atoms undergo stimulated emission and fall to their E_1 level, they almost immediately fall down to their ground state (via intermediate levels not shown) and are then ready to be re-excited by collisions.

Suppose now that a single photon is spontaneously emitted as a neon atom transfers from state E_2 to state E_1. Such a photon can trigger a stimulated emission event, which, in turn, can trigger other stimulated emission events. Through such a chain reaction, a coherent beam of laser light, moving parallel to the tube axis, can build up rapidly. This light, of wavelength 632.8 nm (red), moves through the discharge tube many times by successive reflections from mirrors M_1 and M_2 shown in Fig. 40-20, accumulating additional stimulated emission photons with each passage. M_1 is totally reflecting, but M_2 is slightly "leaky" so that a small fraction of the laser light escapes to form a useful external beam.

✓ Checkpoint 3

The wavelength of light from laser A (a helium–neon gas laser) is 632.8 nm; that from laser B (a carbon dioxide gas laser) is 10.6 μm; that from laser C (a gallium arsenide semiconductor laser) is 840 nm. Rank these lasers according to the energy interval between the two quantum states responsible for laser action, greatest first.

Sample Problem 40.04   Population inversion in a laser

In the helium–neon laser of Fig. 40-20, laser action occurs between two excited states of the neon atom. However, in many lasers, laser action (*lasing*) occurs between the ground state and an excited state, as suggested in Fig. 40-19b.

(a) Consider such a laser that emits at wavelength $\lambda = 550$ nm. If a population inversion is not generated, what is the ratio of the population of atoms in state E_x to the population in the ground state E_0, with the atoms at room temperature?

KEY IDEAS

(1) The naturally occurring population ratio N_x/N_0 of the two states is due to thermal agitation of the gas atoms (Eq. 40-29):

$$N_x/N_0 = e^{-(E_x - E_0)/kT}. \tag{40-30}$$

To find N_x/N_0 with Eq. 40-30, we need to find the energy separation $E_x - E_0$ between the two states. (2) We can obtain $E_x - E_0$ from the given wavelength of 550 nm for the lasing between those two states.

Calculation: The lasing wavelength gives us

$$E_x - E_0 = hf = \frac{hc}{\lambda}$$

$$= \frac{(6.63 \times 10^{-34} \text{ J·s})(3.00 \times 10^8 \text{ m/s})}{(550 \times 10^{-9} \text{ m})(1.60 \times 10^{-19} \text{ J/eV})}$$

$$= 2.26 \text{ eV}.$$

To solve Eq. 40-30, we also need the mean energy of thermal agitation kT for an atom at room temperature (assumed to be 300 K), which is

$$kT = (8.62 \times 10^{-5} \text{ eV/K})(300 \text{ K}) = 0.0259 \text{ eV},$$

in which k is Boltzmann's constant.

Substituting the last two results into Eq. 40-30 gives us the population ratio at room temperature:

$$N_x/N_0 = e^{-(2.26 \text{ eV})/(0.0259 \text{ eV})}$$

$$\approx 1.3 \times 10^{-38}. \qquad \text{(Answer)}$$

This is an extremely small number. It is not unreasonable, however. Atoms with a mean thermal agitation energy of only 0.0259 eV will not often impart an energy of 2.26 eV to another atom in a collision.

(b) For the conditions of (a), at what temperature would the ratio N_x/N_0 be 1/2?

Calculation: Now we want the temperature T such that thermal agitation has bumped enough neon atoms up to the higher-energy state to give $N_x/N_0 = 1/2$. Substituting that ratio into Eq. 40-30, taking the natural logarithm of both sides, and solving for T yield

$$T = \frac{E_x - E_0}{k(\ln 2)} = \frac{2.26 \text{ eV}}{(8.62 \times 10^{-5} \text{ eV/K})(\ln 2)}$$

$$= 38\,000 \text{ K.} \qquad \text{(Answer)}$$

This is much hotter than the surface of the Sun. Thus, it is clear that if we are to invert the populations of these two levels, some specific mechanism for bringing this about is needed—that is, we must "pump" the atoms. No temperature, however high, will naturally generate a population inversion by thermal agitation.

WILEY PLUS Additional examples, video, and practice available at *WileyPLUS*

Review & Summary

Some Properties of Atoms Atoms have quantized energies and can make quantum jumps between them. If a jump between a higher energy and a lower energy involves the emission or absorption of a photon, the frequency associated with the light is given by

$$hf = E_{\text{high}} - E_{\text{low}}. \qquad (40\text{-}1)$$

States with the same value of quantum number n form a shell. States with the same values of quantum numbers n and ℓ form a subshell.

Orbital Angular Momentum and Magnetic Dipole Moments The magnitude of the orbital angular momentum of an electron trapped in an atom has quantized values given by

$$L = \sqrt{\ell(\ell + 1)}\,\hbar, \qquad \text{for } \ell = 0, 1, 2, \dots, (n - 1), \qquad (40\text{-}2)$$

where $\hbar$ is $h/2\pi$, ℓ is the orbital magnetic quantum number, and n is the electron's principal quantum number. The component L_z of the orbital angular momentum on a z axis is quantized and given by

$$L_z = m_\ell \hbar, \qquad \text{for } m_\ell = 0, \pm 1, \pm 2, \dots, \pm \ell, \qquad (40\text{-}3)$$

where m_ℓ is the orbital magnetic quantum number. The magnitude μ_{orb} of the orbital magnetic moment of the electron is quantized with the values given by

$$\mu_{\text{orb}} = \frac{e}{2m}\sqrt{\ell(\ell + 1)}\,\hbar, \qquad (40\text{-}6)$$

where m is the electron mass. The component $\mu_{\text{orb},z}$ on a z axis is also quantized according to

$$\mu_{\text{orb},z} = -\frac{e}{2m}m_\ell \hbar = -m_\ell \mu_B, \qquad (40\text{-}7)$$

where μ_B is the Bohr magneton:

$$\mu_B = \frac{eh}{4\pi m} = \frac{e\hbar}{2m} = 9.274 \times 10^{-24} \text{ J/T}. \qquad (40\text{-}8)$$

Spin Angular Momentum and Magnetic Dipole Moment Every electron, whether trapped or free, has an intrinsic spin angular momentum $\vec{S}$ with a magnitude that is quantized as

$$S = \sqrt{s(s + 1)}\,\hbar, \qquad \text{for } s = \tfrac{1}{2}, \qquad (40\text{-}9)$$

where s is the spin quantum number. An electron is said to be a

spin-$\tfrac{1}{2}$ particle. The component S_z on a z axis is also quantized according to

$$S_z = m_s \hbar, \qquad \text{for } m_s = \pm s = \pm\tfrac{1}{2}, \qquad (40\text{-}10)$$

where m_s is the spin magnetic quantum number. Every electron, whether trapped or free, has an intrinsic spin magnetic dipole moment $\vec{\mu}_s$ with a magnitude that is quantized as

$$\mu_s = \frac{e}{m}\sqrt{s(s + 1)}\,\hbar, \qquad \text{for } s = \tfrac{1}{2}. \qquad (40\text{-}12)$$

The component $\mu_{s,z}$ on a z axis is also quantized according to

$$\mu_{s,z} = -2m_s \mu_B, \qquad \text{for } m_s = \pm\tfrac{1}{2}. \qquad (40\text{-}13)$$

Stern–Gerlach Experiment The Stern–Gerlach experiment demonstrated that the magnetic moment of silver atoms is quantized, experimental proof that magnetic moments at the atomic level are quantized. An atom with magnetic dipole moment experiences a force in a nonuniform magnetic field. If the field changes at the rate of dB/dz along a z axis, then the force is along the z axis and is related to the component μ_z of the dipole moment:

$$F_z = \mu_z \frac{dB}{dz}. \qquad (40\text{-}17)$$

A proton has an intrinsic spin angular momentum $\vec{S}$ and an intrinsic magnetic dipole moment $\vec{\mu}$ that are in the same direction.

Magnetic Resonance The magnetic dipole moment of a proton in a magnetic field $\vec{B}$ along a z axis has two quantized components on that axis: spin up (μ_z is in the direction $\vec{B}$) and spin down (μ_z is in the opposite direction). Contrary to the situation with an electron, spin up is the lower energy orientation; the difference between the two orientations is $2\mu_z B$. The energy required of a photon to spin-flip the proton between the two orientations is

$$hf = 2\mu_z B. \qquad (40\text{-}22)$$

The field is the vector sum of an external field set up by equipment and an internal field set up by the atoms and nuclei surrounding the proton. Detection of spin-flips can lead to nuclear magnetic resonance spectra by which specific substances can be identified.

Pauli Exclusion Principle Electrons in atoms and other traps obey the Pauli exclusion principle, which requires that no two electrons in a trap can have the same set of quantum numbers.

Building the Periodic Table In the periodic table, the elements are listed in order of increasing atomic number Z, where Z is the number of protons in the nucleus. For a neutral atom, Z is also the number of electrons. States with the same value of quantum number n form a shell. States with the same values of quantum numbers n and ℓ form a subshell. A closed shell and a closed subshell contain the maximum number of electrons as allowed by the Pauli exclusion principle. The net angular momentum and net magnetic moment of such closed structures is zero.

X Rays and the Numbering of the Elements When a beam of high-energy electrons impacts a target, the electrons can lose their energy by emitting x rays when they scatter from atoms in the target. The emission is over a range of wavelengths, said to be a continuous spectrum. The shortest wavelength in the spectrum is the cutoff wavelength λ_{min}, which is emitted when an incident electron loses its full kinetic energy K_0 in a single scattering event, with a single x-ray emission:

$$\lambda_{min} = \frac{hc}{K_0}.$$

The characteristic x-ray spectrum is produced when incident electrons eject low-lying electrons in the target atoms and electrons from upper levels jump down to the resulting holes, emitting light. A Moseley plot is a graph of the square root of the characteristic-emission frequencies $\sqrt{f}$ versus atomic number Z of the target atoms. The straight-line plot reveals that the position of an element in the periodic table is set by Z and not by the atomic weight.

Lasers In stimulated emission, an atom in an excited state can be induced to de-excite to a lower energy state by emitting a photon if an identical photon passes the atom. The light emitted in stimulated emission is in phase with and travels in the direction of the light causing the emission.

A laser can emit light via stimulated emission provided that its atoms are in population inversion. That is, for the pair of levels involved in the stimulated emission, more atoms must be in the upper level than the lower level so that there is more stimulated emission than just absorption.

Questions

1 How many (a) subshells and (b) electron states are in the $n = 2$ shell? How many (c) subshells and (d) electron states are in the $n = 5$ shell?

2 An electron in an atom of gold is in a state with $n = 4$. Which of these values of ℓ are possible for it: $-3, 0, 2, 3, 4, 5$?

3 Label these statements as true or false: (a) One (and only one) of these subshells cannot exist: $2p, 4f, 3d, 1p$. (b) The number of values of m_ℓ that are allowed depends only on ℓ and not on n. (c) There are four subshells with $n = 4$. (d) The smallest value of n for a given value of ℓ is $\ell + 1$. (e) All states with $\ell = 0$ also have $m_\ell = 0$. (f) There are n subshells for each value of n.

4 An atom of uranium has closed $6p$ and $7s$ subshells. Which subshell has the greater number of electrons?

5 An atom of silver has closed $3d$ and $4d$ subshells. Which subshell has the greater number of electrons, or do they have the same number?

6 From which atom of each of the following pairs is it easier to remove an electron: (a) krypton or bromine, (b) rubidium or cerium, (c) helium or hydrogen?

7 An electron in a mercury atom is in the $3d$ subshell. Which of the following m_ℓ values are possible for it: -3, $-1, 0, 1, 2$?

8 Figure 40-22 shows three points at which a spin-up electron can be placed in a nonuniform magnetic field (there is a gradient along the z axis). (a) Rank the three points according to the energy U of the electron's intrinsic magnetic dipole moment $\vec{\mu}_s$, most positive first. (b) What is the direction of the force on the electron due to the magnetic field if the spin-up electron is at point 2?

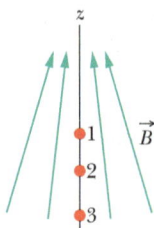

Figure 40-22
Question 8.

9 The K_α x-ray line for any element arises because of a transition between the K shell ($n = 1$) and the L shell ($n = 2$). Figure 40-13 shows this line (for a molybdenum target) occurring at a single wavelength. With higher resolution, however, the line splits into several wavelength components because the L shell does not have a unique energy. (a) How many components does the K_α line have? (b) Similarly, how many components does the K_β line have?

10 Consider the elements krypton and rubidium. (a) Which is more suitable for use in a Stern–Gerlach experiment of the kind described in connection with Fig. 40-8? (b) Which, if either, would not work at all?

11 On which quantum numbers does the energy of an electron depend in (a) a hydrogen atom and (b) a vanadium atom?

12 Which (if any) of the following are essential for laser action to occur between two energy levels of an atom? (a) There are more atoms in the upper level than in the lower. (b) The upper level is metastable. (c) The lower level is metastable. (d) The lower level is the ground state of the atom. (e) The lasing medium is a gas.

13 Figure 40-21 shows partial energy-level diagrams for the helium and neon atoms that are involved in the operation of a helium–neon laser. It is said that a helium atom in state E_3 can collide with a neon atom in its ground state and raise the neon atom to state E_2. The energy of helium state E_3 (20.61 eV) is close to, but not exactly equal to, the energy of neon state E_2 (20.66 eV). How can the energy transfer take place if these energies are not *exactly* equal?

14 The x-ray spectrum of Fig. 40-13 is for 35.0 keV electrons striking a molybdenum ($Z = 42$) target. If you substitute a silver ($Z = 47$) target for the molybdenum target, will (a) λ_{min}, (b) the wavelength for the K_α line, and (c) the wavelength for the K_β line increase, decrease, or remain unchanged?

Problems

Module 40-1   Properties of Atoms

•1   An electron in a hydrogen atom is in a state with $\ell = 5$. What is the minimum possible value of the semiclassical angle between $\vec{L}$ and L_z?

•2   How many electron states are there in a shell defined by the quantum number $n = 5$?

•3   (a) What is the magnitude of the orbital angular momentum in a state with $\ell = 3$? (b) What is the magnitude of its largest projection on an imposed z axis?

•4   How many electron states are there in the following shells: (a) $n = 4$, (b) $n = 1$, (c) $n = 3$, (d) $n = 2$?

•5   (a) How many ℓ values are associated with $n = 3$? (b) How many m_ℓ values are associated with $\ell = 1$?

•6   How many electron states are in these subshells: (a) $n = 4$, $\ell = 3$; (b) $n = 3, \ell = 1$; (c) $n = 4, \ell = 1$; (d) $n = 2, \ell = 0$?

•7   An electron in a multielectron atom has $m_\ell = +4$. For this electron, what are (a) the value of ℓ, (b) the smallest possible value of n, and (c) the number of possible values of m_s?

•8   In the subshell $\ell = 3$, (a) what is the greatest (most positive) m_ℓ value, (b) how many states are available with the greatest m_ℓ value, and (c) what is the total number of states available in the subshell?

••9 **SSM** **WWW** An electron is in a state with $\ell = 3$. (a) What multiple of $\hbar$ gives the magnitude of $\vec{L}$? (b) What multiple of μ_B gives the magnitude of $\vec{\mu}$? (c) What is the largest possible value of m_ℓ, (d) what multiple of $\hbar$ gives the corresponding value of L_z, and (e) what multiple of μ_B gives the corresponding value of $\mu_{orb,z}$? (f) What is the value of the semiclassical angle θ between the directions of L_z and $\vec{L}$? What is the value of angle θ for (g) the second largest possible value of m_ℓ and (h) the smallest (that is, most negative) possible value of m_ℓ?

••10   An electron is in a state with $n = 3$. What are (a) the number of possible values of ℓ, (b) the number of possible values of m_ℓ, (c) the number of possible values of m_s, (d) the number of states in the $n = 3$ shell, and (e) the number of subshells in the $n = 3$ shell?

••11 **SSM** If orbital angular momentum $\vec{L}$ is measured along, say, a z axis to obtain a value for L_z, show that

$$(L_x^2 + L_y^2)^{1/2} = [\ell(\ell + 1) - m_\ell^2]^{1/2}\hbar$$

is the most that can be said about the other two components of the orbital angular momentum.

•••12 **GO** A magnetic field is applied to a freely floating uniform iron sphere with radius $R = 2.00$ mm. The sphere initially had no net magnetic moment, but the field aligns 12% of the magnetic moments of the atoms (that is, 12% of the magnetic moments of the loosely bound electrons in the sphere, with one such electron per atom). The magnetic moment of those aligned electrons is the sphere's intrinsic magnetic moment $\vec{\mu}_s$. What is the sphere's resulting angular speed ω?

Module 40-2   The Stern–Gerlach Experiment

•13 **SSM** What is the acceleration of a silver atom as it passes through the deflecting magnet in the Stern–Gerlach experiment of Fig. 40-8 if the magnetic field gradient is 1.4 T/mm?

•14   Suppose that a hydrogen atom in its ground state moves 80 cm through and perpendicular to a vertical magnetic field that has a magnetic field gradient $dB/dz = 1.6 \times 10^2$ T/m. (a) What is the magnitude of force exerted by the field gradient on the atom due to the magnetic moment of the atom's electron, which we take to be 1 Bohr magneton? (b) What is the vertical displacement of the atom in the 80 cm of travel if its speed is 1.2×10^5 m/s?

•15   Calculate the (a) smaller and (b) larger value of the semiclassical angle between the electron spin angular momentum vector and the magnetic field in a Stern–Gerlach experiment. Bear in mind that the orbital angular momentum of the valence electron in the silver atom is zero.

•16   Assume that in the Stern–Gerlach experiment as described for neutral silver atoms, the magnetic field $\vec{B}$ has a magnitude of 0.50 T. (a) What is the energy difference between the magnetic moment orientations of the silver atoms in the two subbeams? (b) What is the frequency of the radiation that would induce a transition between these two states? (c) What is the wavelength of this radiation, and (d) to what part of the electromagnetic spectrum does it belong?

Module 40-3   Magnetic Resonance

•17   In an NMR experiment, the RF source oscillates at 34 MHz and magnetic resonance of the hydrogen atoms in the sample being investigated occurs when the external field $\vec{B}_{ext}$ has magnitude 0.78 T. Assume that $\vec{B}_{int}$ and $\vec{B}_{ext}$ are in the same direction and take the proton magnetic moment component μ_z to be 1.41×10^{-26} J/T. What is the magnitude of $\vec{B}_{int}$?

•18   A hydrogen atom in its ground state actually has two possible, closely spaced energy levels because the electron is in the magnetic field $\vec{B}$ of the proton (the nucleus). Accordingly, an energy is associated with the orientation of the electron's magnetic moment $\vec{\mu}$ relative to $\vec{B}$, and the electron is said to be either spin up (higher energy) or spin down (lower energy) in that field. If the electron is excited to the higher-energy level, it can de-excite by spin-flipping and emitting a photon. The wavelength associated with that photon is 21 cm. (Such a process occurs extensively in the Milky Way galaxy, and reception of the 21 cm radiation by radio telescopes reveals where hydrogen gas lies between stars.) What is the effective magnitude of $\vec{B}$ as experienced by the electron in the ground-state hydrogen atom?

•19   What is the wavelength associated with a photon that will induce a transition of an electron spin from parallel to antiparallel orientation in a magnetic field of magnitude 0.200 T? Assume that $\ell = 0$.

Module 40-4   Exclusion Principle and Multiple Electrons in a Trap

•20   A rectangular corral of widths $L_x = L$ and $L_y = 2L$ contains seven electrons. What multiple of $h^2/8mL^2$ gives the energy of the ground state of this system? Assume that the electrons do not interact with one another, and do not neglect spin.

•21 Seven electrons are trapped in a one-dimensional infinite potential well of width L. What multiple of $h^2/8mL^2$ gives the energy of the ground state of this system? Assume that the electrons do not interact with one another, and do not neglect spin.

•22 GO Figure 40-23 is an energy-level diagram for a fictitious infinite potential well that contains one electron. The number of degenerate states of the levels are indicated: "non" means nondegenerate (which includes the ground state of the electron), "double" means 2 states, and "triple" means 3 states. We put a total of 11 electrons in the well. If the electrostatic forces between the electrons can be neglected, what multiple of $h^2/8mL^2$ gives the energy of the first excited state of the 11-electron system?

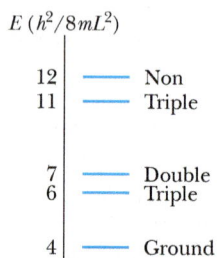

$E\ (h^2/8mL^2)$

12	—— Non
11	—— Triple
7	—— Double
6	—— Triple
4	—— Ground

Figure 40-23
Problem 22.

••23 GO SSM A cubical box of widths $L_x = L_y = L_z = L$ contains eight electrons. What multiple of $h^2/8mL^2$ gives the energy of the ground state of this system? Assume that the electrons do not interact with one another, and do not neglect spin.

••24 GO For Problem 20, what multiple of $h^2/8mL^2$ gives the energy of (a) the first excited state, (b) the second excited state, and (c) the third excited state of the system of seven electrons? (d) Construct an energy-level diagram for the lowest four energy levels.

••25 GO For the situation of Problem 21, what multiple of $h^2/8mL^2$ gives the energy of (a) the first excited state, (b) the second excited state, and (c) the third excited state of the system of seven electrons? (d) Construct an energy-level diagram for the lowest four energy levels of the system.

•••26 GO For the situation of Problem 23, what multiple of $h^2/8mL^2$ gives the energy of (a) the first excited state, (b) the second excited state, and (c) the third excited state of the system of eight electrons? (d) Construct an energy-level diagram for the lowest four energy levels of the system.

Module 40-5 Building the Periodic Table

•27 SSM WWW Two of the three electrons in a lithium atom have quantum numbers (n, ℓ, m_ℓ, m_s) of $(1, 0, 0, +\frac{1}{2})$ and $(1, 0, 0, -\frac{1}{2})$. What quantum numbers are possible for the third electron if the atom is (a) in the ground state and (b) in the first excited state?

•28 Show that the number of states with the same quantum number n is $2n^2$.

•29 GO A recently named element is darmstadtium (Ds), which has 110 electrons. Assume that you can put the 110 electrons into the atomic shells one by one and can neglect any electron–electron interaction. With the atom in ground state, what is the spectroscopic notation for the quantum number ℓ for the last electron?

•30 For a helium atom in its ground state, what are quantum numbers $(n, \ell, m_\ell, $ and $m_s)$ for the (a) spin-up electron and (b) spin-down electron?

•31 Consider the elements selenium ($Z = 34$), bromine ($Z = 35$), and krypton ($Z = 36$). In their part of the periodic table, the sub-

shells of the electronic states are filled in the sequence

$$1s\ 2s\ 2p\ 3s\ 3p\ 3d\ 4s\ 4p\ \ldots.$$

What are (a) the highest occupied subshell for selenium and (b) the number of electrons in it, (c) the highest occupied subshell for bromine and (d) the number of electrons in it, and (e) the highest occupied subshell for krypton and (f) the number of electrons in it?

•32 Suppose two electrons in an atom have quantum numbers $n = 2$ and $\ell = 1$. (a) How many states are possible for those two electrons? (Keep in mind that the electrons are indistinguishable.) (b) If the Pauli exclusion principle did not apply to the electrons, how many states would be possible?

Module 40-6 X Rays and the Ordering of the Elements

•33 Through what minimum potential difference must an electron in an x-ray tube be accelerated so that it can produce x rays with a wavelength of 0.100 nm?

••34 The wavelength of the K_α line from iron is 193 pm. What is the energy difference between the two states of the iron atom that give rise to this transition?

••35 SSM WWW In Fig. 40-13, the x rays shown are produced when 35.0 keV electrons strike a molybdenum ($Z = 42$) target. If the accelerating potential is maintained at this value but a silver ($Z = 47$) target is used instead, what values of (a) λ_{min}, (b) the wavelength of the K_α line, and (c) the wavelength of the K_β line result? The K, L, and M atomic x-ray levels for silver (compare Fig. 40-15) are 25.51, 3.56, and 0.53 keV.

••36 When electrons bombard a molybdenum target, they produce both continuous and characteristic x rays as shown in Fig. 40-13. In that figure the kinetic energy of the incident electrons is 35.0 keV. If the accelerating potential is increased to 50.0 keV, (a) what is the value of λ_{min}, and (b) do the wavelengths of the K_α and K_β lines increase, decrease, or remain the same?

••37 Show that a moving electron cannot spontaneously change into an x-ray photon in free space. A third body (atom or nucleus) must be present. Why is it needed? (*Hint:* Examine the conservation of energy and momentum.)

••38 Here are the K_α wavelengths of a few elements:

Element	λ (pm)	Element	λ (pm)
Ti	275	Co	179
V	250	Ni	166
Cr	229	Cu	154
Mn	210	Zn	143
Fe	193	Ga	134

Make a Moseley plot (like that in Fig. 40-16) from these data and verify that its slope agrees with the value given for C in Module 40-6.

••39 SSM Calculate the ratio of the wavelength of the K_α line for niobium (Nb) to that for gallium (Ga). Take needed data from the periodic table of Appendix G.

••40 (a) From Eq. 40-26, what is the ratio of the photon energies due to K_α transitions in two atoms whose atomic numbers are Z and Z'? (b) What is this ratio for uranium and aluminum? (c) For uranium and lithium?

••41 The binding energies of K-shell and L-shell electrons in copper are 8.979 and 0.951 keV, respectively. If a K_α x ray from copper is incident on a sodium chloride crystal and gives a first-order Bragg reflection at an angle of 74.1° measured relative to parallel planes of sodium atoms, what is the spacing between these parallel planes?

••42 From Fig. 40-13, calculate approximately the energy difference $E_L - E_M$ for molybdenum. Compare it with the value that may be obtained from Fig. 40-15.

••43 A tungsten ($Z = 74$) target is bombarded by electrons in an x-ray tube. The K, L, and M energy levels for tungsten (compare Fig. 40-15) have the energies 69.5, 11.3, and 2.30 keV, respectively. (a) What is the minimum value of the accelerating potential that will permit the production of the characteristic K_α and K_β lines of tungsten? (b) For this same accelerating potential, what is λ_{min}? What are the (c) K_α and (d) K_β wavelengths?

••44 A 20 keV electron is brought to rest by colliding twice with target nuclei as in Fig. 40-14. (Assume the nuclei remain stationary.) The wavelength associated with the photon emitted in the second collision is 130 pm greater than that associated with the photon emitted in the first collision. (a) What is the kinetic energy of the electron after the first collision? What are (b) the wavelength λ_1 and (c) the energy E_1 associated with the first photon? What are (d) λ_2 and (e) E_2 associated with the second photon?

••45 X rays are produced in an x-ray tube by electrons accelerated through an electric potential difference of 50.0 kV. Let K_0 be the kinetic energy of an electron at the end of the acceleration. The electron collides with a target nucleus (assume the nucleus remains stationary) and then has kinetic energy $K_1 = 0.500K_0$. (a) What wavelength is associated with the photon that is emitted? The electron collides with another target nucleus (assume it, too, remains stationary) and then has kinetic energy $K_2 = 0.500K_1$. (b) What wavelength is associated with the photon that is emitted?

•••46 Determine the constant C in Eq. 40-27 to five significant figures by finding C in terms of the fundamental constants in Eq. 40-24 and then using data from Appendix B to evaluate those constants. Using this value of C in Eq. 40-27, determine the theoretical energy E_{theory} of the K_α photon for the low-mass elements listed in the following table. The table includes the value (eV) of the measured energy E_{exp} of the K_α photon for each listed element. The percentage deviation between E_{theory} and E_{exp} can be calculated as

$$\text{percentage deviation} = \frac{E_{theory} - E_{exp}}{E_{exp}} 100.$$

What is the percentage deviation for (a) Li, (b) Be, (c) B, (d) C, (e) N, (f) O, (g) F, (h) Ne, (i) Na, and (j) Mg?

Li	54.3	O	524.9
Be	108.5	F	676.8
B	183.3	Ne	848.6
C	277	Na	1041
N	392.4	Mg	1254

(There is actually more than one K_α ray because of the splitting of the L energy level, but that effect is negligible for the elements listed here.)

Module 40-7 Lasers

•47 The active volume of a laser constructed of the semiconductor GaAlAs is only 200 μm^3 (smaller than a grain of sand), and yet the laser can continuously deliver 5.0 mW of power at a wavelength of 0.80 μm. At what rate does it generate photons?

•48 A high-powered laser beam ($\lambda = 600$ nm) with a beam diameter of 12 cm is aimed at the Moon, 3.8×10^5 km distant. The beam spreads only because of diffraction. The angular location of the edge of the central diffraction disk (see Eq. 36-12) is given by

$$\sin\theta = \frac{1.22\lambda}{d},$$

where d is the diameter of the beam aperture. What is the diameter of the central diffraction disk on the Moon's surface?

•49 Assume that lasers are available whose wavelengths can be precisely "tuned" to anywhere in the visible range—that is, in the range 450 nm $< \lambda <$ 650 nm. If every television channel occupies a bandwidth of 10 MHz, how many channels can be accommodated within this wavelength range?

•50 A hypothetical atom has only two atomic energy levels, separated by 3.2 eV. Suppose that at a certain altitude in the atmosphere of a star there are $6.1 \times 10^{13}/cm^3$ of these atoms in the higher-energy state and $2.5 \times 10^{15}/cm^3$ in the lower-energy state. What is the temperature of the star's atmosphere at that altitude?

•51 SSM A hypothetical atom has energy levels uniformly separated by 1.2 eV. At a temperature of 2000 K, what is the ratio of the number of atoms in the 13th excited state to the number in the 11th excited state?

•52 GO A laser emits at 424 nm in a single pulse that lasts 0.500 μs. The power of the pulse is 2.80 MW. If we assume that the atoms contributing to the pulse underwent stimulated emission only once during the 0.500 μs, how many atoms contributed?

•53 A helium–neon laser emits laser light at a wavelength of 632.8 nm and a power of 2.3 mW. At what rate are photons emitted by this device?

•54 A certain gas laser can emit light at wavelength 550 nm, which involves population inversion between ground state and an excited state. At room temperature, how many moles of neon are needed to put 10 atoms in that excited state by thermal agitation?

•55 A pulsed laser emits light at a wavelength of 694.4 nm. The pulse duration is 12 ps, and the energy per pulse is 0.150 J. (a) What is the length of the pulse? (b) How many photons are emitted in each pulse?

•56 A population inversion for two energy levels is often described by assigning a negative Kelvin temperature to the system. What negative temperature would describe a system in which the population of the upper energy level exceeds that of the lower level by 10% and the energy difference between the two levels is 2.26 eV?

••57 A hypothetical atom has two energy levels, with a transition wavelength between them of 580 nm. In a particular sample at 300 K, 4.0×10^{20} such atoms are in the state of lower energy. (a) How many atoms are in the upper state, assuming conditions of thermal equilibrium? (b) Suppose, instead, that 3.0×10^{20} of these atoms are "pumped" into the upper state by an external process, with 1.0×10^{20} atoms remaining in the lower state. What is the maximum energy that could be released by the atoms in a single laser pulse if each atom jumps once between those two states (either via absorption or via stimulated emission)?

••58 The mirrors in the laser of Fig. 40-20, which are separated by 8.0 cm, form an optical cavity in which standing waves of laser light can be set up. Each standing wave has an integral number n of half wavelengths in the 8.0 cm length, where n is large and the waves differ slightly in wavelength. Near $\lambda = 533$ nm, how far apart in wavelength are the standing waves?

••59 GO Figure 40-24 shows the energy levels of two types of atoms. Atoms A are in one tube, and atoms B are in another tube. The energies (relative to a ground-state energy of zero) are indicated; the average lifetime of atoms in each level is also indicated. All the atoms are initially pumped to levels higher than the levels shown in the figure. The atoms then drop down through the levels, and many become "stuck" on certain levels, leading to population inversion and lasing. The light emitted by A illuminates B and can cause stimulated emission of B. What is the energy per photon of that stimulated emission of B?

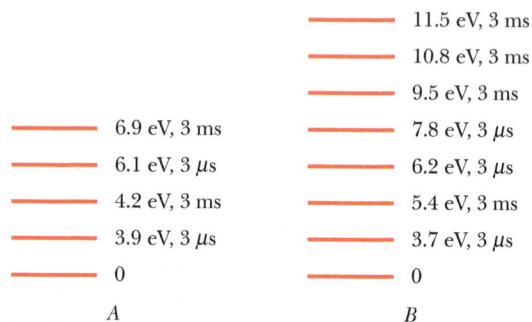

Figure 40-24 Problem 59.

••60 The beam from an argon laser (of wavelength 515 nm) has a diameter d of 3.00 mm and a continuous energy output rate of 5.00 W. The beam is focused onto a diffuse surface by a lens whose focal length f is 3.50 cm. A diffraction pattern such as that of Fig. 36-10 is formed, the radius of the central disk being given by

$$R = \frac{1.22 \, f\lambda}{d}$$

(see Eq. 36-12 and Fig. 36-14). The central disk can be shown to contain 84% of the incident power. (a) What is the radius of the central disk? (b) What is the average intensity (power per unit area) in the incident beam? (c) What is the average intensity in the central disk?

••61 The active medium in a particular laser that generates laser light at a wavelength of 694 nm is 6.00 cm long and 1.00 cm in diameter. (a) Treat the medium as an optical resonance cavity analogous to a closed organ pipe. How many standing-wave nodes are there along the laser axis? (b) By what amount Δf would the beam frequency have to shift to increase this number by one? (c) Show that Δf is just the inverse of the travel time of laser light for one round trip back and forth along the laser axis. (d) What is the corresponding fractional frequency shift $\Delta f/f$? The appropriate index of refraction of the lasing medium (a ruby crystal) is 1.75.

••62 GO Ruby lases at a wavelength of 694 nm. A certain ruby crystal has 4.00×10^{19} Cr ions (which are the atoms that lase). The lasing transition is between the first excited state and the ground state, and the output is a light pulse lasting 2.00 μs. As the pulse begins, 60.0% of the Cr ions are in the first excited state and the rest are in the ground state. What is the average power emitted during the pulse? (*Hint:* Don't just ignore the ground-state ions.)

Additional Problems

63 GO Figure 40-25 is an energy-level diagram for a fictitious three-dimensional infinite potential well that contains one electron. The number of degenerate states of the levels are indicated: "non" means nondegenerate (which includes the ground state) and "triple" means 3 states. If we put a total of 22 electrons in the well, what multiple of $h^2/8mL^2$ gives the energy of the ground state of the 22-electron system? Assume that the electrostatic forces between the electrons are negligible.

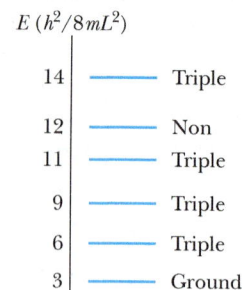

Figure 40-25
Problem 63.

64 *Martian CO_2 laser.* Where sunlight shines on the atmosphere of Mars, carbon dioxide molecules at an altitude of about 75 km undergo natural laser action. The energy levels involved in the action are shown in Fig. 40-26; population inversion occurs between energy levels E_2 and E_1. (a) What wavelength of sunlight excites the molecules in the lasing action? (b) At what wavelength does lasing occur? (c) In what region of the electromagnetic spectrum do the excitation and lasing wavelengths lie?

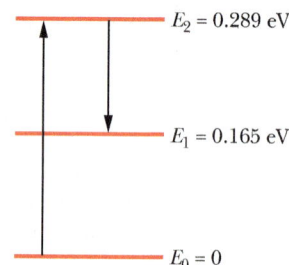

Figure 40-26 Problem 64.

65 GO Excited sodium atoms emit two closely spaced spectrum lines called the *sodium doublet* (Fig. 40-27) with wavelengths 588.995 nm and 589.592 nm. (a) What is the difference in energy between the two upper energy levels ($n = 3, \ell = 1$)? (b) This energy difference occurs because the electron's spin magnetic moment can be oriented either parallel or antiparallel to the internal magnetic field associated with the electron's orbital motion. Use your result in (a) to find the magnitude of this internal magnetic field.

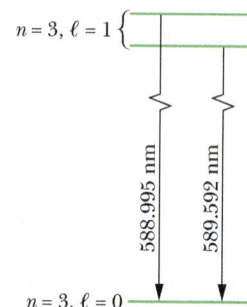

Figure 40-27 Problem 65.

66 *Comet stimulated emission.* When a comet approaches the Sun, the increased warmth evaporates water from the ice on the surface of the comet nucleus, producing a thin atmosphere of water vapor around the nucleus. Sunlight can then dissociate H_2O molecules in the vapor to H atoms and OH molecules. The sunlight can also excite the OH molecules to higher energy levels.

When the comet is still relatively far from the Sun, the sunlight causes equal excitation to the E_2 and E_1 levels (Fig. 40-28a). Hence, there is no population inversion between the two levels. However, as the comet approaches the Sun, the excitation to the E_1 level decreases and population inversion occurs. The reason has to do with one of the many wavelengths—said to be *Fraunhofer lines*—that are missing in sunlight because, as the light travels outward through the Sun's atmosphere, those particular wavelengths are absorbed by the atmosphere.

As a comet approaches the Sun, the Doppler effect due to the comet's speed relative to the Sun shifts the Fraunhofer lines in wavelength, apparently overlapping one of them with the wavelength required for excitation to the E_1 level in OH molecules. Population inversion then occurs in those molecules, and they radiate stimulated emission (Fig. 40-28b). For example, as comet Kouhoutek approached the Sun in December 1973 and January 1974, it radiated stimulated emission at about 1666 MHz during mid-January. (a) What was the energy difference $E_2 - E_1$ for that emission? (b) In what region of the electromagnetic spectrum was the emission?

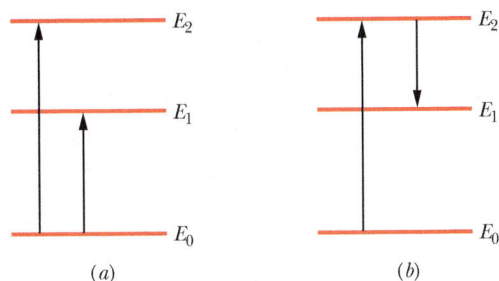

Figure 40-28 Problem 66.

67 Show that the cutoff wavelength (in picometers) in the continuous x-ray spectrum from any target is given by $\lambda_{min} = 1240/V$, where V is the potential difference (in kilovolts) through which the electrons are accelerated before they strike the target.

68 By measuring the go-and-return time for a laser pulse to travel from an Earth-bound observatory to a reflector on the Moon, it is possible to measure the separation between these bodies. (a) What is the predicted value of this time? (b) The separation can be measured to a precision of about 15 cm. To what uncertainty in travel time does this correspond? (c) If the laser beam forms a spot on the Moon 3 km in diameter, what is the angular divergence of the beam?

69 **SSM** Can an incoming intercontinental ballistic missile be destroyed by an intense laser beam? A beam of intensity 10^8 W/m² would probably burn into and destroy a nonspinning missile in 1 s. (a) If the laser had 5.0 MW power, 3.0 μm wavelength, and a 4.0 m beam diameter (a very powerful laser indeed), would it destroy a missile at a distance of 3000 km? (b) If the wavelength could be changed, what maximum value would work? Use the equation for the central diffraction maximum as given by Eq. 36-12 ($\sin \theta = 1.22\lambda/d$).

70 A molybdenum ($Z = 42$) target is bombarded with 35.0 keV electrons and the x-ray spectrum of Fig. 40-13 results. The K_β and K_α wavelengths are 63.0 and 71.0 pm, respectively. What photon energy corresponds to the (a) K_β and (b) K_α radiation? The two radiations are to be filtered through one of the substances in the following table such that the substance absorbs the K_β line more strongly than the K_α line. A substance will absorb radiation x_1 more strongly than it absorbs radiation x_2 if a photon of x_1 has enough energy to eject a K electron from an atom of the substance but a photon of x_2 does not. The table gives the ionization energy of the K electron in molybdenum and four other substances. Which substance in the table will serve (c) best and (d) second best as the filter?

	Zr	Nb	Mo	Tc	Ru
Z	40	40	42	43	44
E_K (keV)	18.00	18.99	20.00	21.04	22.12

71 An electron in a multielectron atom is known to have the quantum number $\ell = 3$. What are its possible n, m_ℓ, and m_s quantum numbers?

72 Show that if the 63 electrons in an atom of europium were assigned to shells according to the "logical" sequence of quantum numbers, this element would be chemically similar to sodium.

73 **SSM** Lasers can be used to generate pulses of light whose durations are as short as 10 fs. (a) How many wavelengths of light ($\lambda = 500$ nm) are contained in such a pulse? (b) In

$$\frac{10 \text{ fs}}{1 \text{ s}} = \frac{1 \text{ s}}{X},$$

what is the missing quantity X (in years)?

74 Show that $\hbar = 1.06 \times 10^{-34}$ J·s $= 6.59 \times 10^{-16}$ eV·s.

75 Suppose that the electron had no spin and that the Pauli exclusion principle still held. Which, if any, of the present noble gases would remain in that category?

76 (A correspondence principle problem.) Estimate (a) the quantum number ℓ for the orbital motion of Earth around the Sun and (b) the number of allowed orientations of the plane of Earth's orbit. (c) Find θ_{min}, the half-angle of the smallest cone that can be swept out by a perpendicular to Earth's orbit as Earth revolves around the Sun.

77 Knowing that the minimum x-ray wavelength produced by 40.0 keV electrons striking a target is 31.1 pm, determine the Planck constant h.

78 Consider an atom with two closely spaced excited states A and B. If the atom jumps to ground state from A or from B, it emits a wavelength of 500 nm or 510 nm, respectively. What is the energy difference between states A and B?

79 In 1911, Ernest Rutherford modeled an atom as being a point of positive charge Ze surrounded by a negative charge $-Ze$ uniformly distributed in a sphere of radius R centered at the point. At distance r within the sphere, the electric potential is

$$V = \frac{Ze}{4\pi\varepsilon_0}\left(\frac{1}{r} - \frac{3}{2R} + \frac{r^2}{2R^3}\right).$$

(a) From this formula, determine the magnitude of electric field for $0 \le r \le R$. What are the (b) electric field and (c) potential for $r \ge R$?

Conduction of Electricity in Solids

41-1 THE ELECTRICAL PROPERTIES OF METALS

Learning Objectives

After reading this module, you should be able to . . .

41.01 Identify the three basic properties of crystalline solids and sketch unit cells for them.

41.02 Distinguish insulators, metals, and semiconductors.

41.03 With sketches, explain the transition of an energy-level diagram for a single atom to an energy-band diagram for many atoms.

41.04 Draw a band–gap diagram for an insulator, indicating the filled and empty bands and explaining what prevents the electrons from participating in a current.

41.05 Draw a band–gap diagram for a metal, and explain what feature, in contrast to an insulator, allows electrons to participate in a current.

41.06 Identify the Fermi level, Fermi energy, and Fermi speed.

41.07 Distinguish monovalent atoms, bivalent atoms, and trivalent atoms.

41.08 For a conducting material, apply the relationships between the number density n of conduction electrons and the material's density, volume V, and molar mass M.

41.09 Identify that in a metal's partially filled band, thermal agitation can jump some of the conduction electrons to higher energy levels.

41.10 For a given energy level in a band, calculate the density of states $N(E)$ and identify that it is actually a double density (per volume and per energy).

41.11 Find the number of states per unit volume in a range ΔE at height E in a band by integrating $N(E)$ over that range or, if ΔE is small relative to E, by evaluating the product $N(E)\,\Delta E$.

41.12 For a given energy level, calculate the probability $P(E)$ that the level is occupied by electrons.

41.13 Identify that probability $P(E)$ is 0.5 at the Fermi level.

41.14 At a given energy level, calculate the density $N_o(E)$ of occupied states.

41.15 For a given range in energy levels, calculate the number of states and the number of occupied states.

41.16 Sketch graphs of the density of states $N(E)$, occupancy probability $P(E)$, and the density of occupied states $N_o(E)$, all versus height in a band.

41.17 Apply the relationship between the Fermi energy E_F and the number density of conduction electrons n.

Key Ideas

● Crystalline solids can be broadly divided into insulators, metals, and semiconductors.

● The quantized energy levels for a crystalline solid form bands that are separated by gaps.

● In a metal, the highest band that contains any electrons is only partially filled, and the highest filled level at a temperature of 0 K is called the Fermi level E_F.

● The electrons in the partially filled band are the conduction electrons, and their number density (number per unit volume) is

$$n = \frac{\text{material's density}}{M/N_A},$$

where M is the material's molar mass and N_A is Avogadro's number.

● The number density of states of the allowed energy levels per unit volume and per unit energy interval is

$$N(E) = \frac{8\sqrt{2}\,\pi m^{3/2}}{h^3}\,E^{1/2},$$

where m is the electron mass and E is the energy *in joules* at which $N(E)$ is to be evaluated.

● The occupancy probability $P(E)$ is the probability that a given available state will be occupied by an electron:

$$P(E) = \frac{1}{e^{(E-E_F)/kT} + 1}.$$

● The density of occupied states $N_o(E)$ is given by the product of the density of states function and the occupancy probability function:

$$N_o(E) = N(E)\,P(E).$$

● The Fermi energy E_F for a metal can be found by integrating $N_o(E)$ for temperature $T = 0$ K (absolute zero) from $E = 0$ to $E = E_F$. The result is

$$E_F = \left(\frac{3}{16\sqrt{2}\,\pi}\right)^{2/3} \frac{h^2}{m}\,n^{2/3} = \frac{0.121 h^2}{m}\,n^{2/3}.$$

What Is Physics?

A major question in physics, which underlies *solid-state* electronic devices, is this: What are the mechanisms by which a material conducts, or does not conduct, electricity? The answers are complex and poorly understood, largely because they involve the application of quantum physics to a tremendous number of particles and atoms grouped together and interacting. Let's start by characterizing conducting and nonconducting materials.

The Electrical Properties of Solids

We shall examine only **crystalline solids**—that is, solids whose atoms are arranged in a repetitive three-dimensional structure called a **lattice.** We shall not consider such solids as wood, plastic, glass, and rubber, whose atoms are not arranged in such repetitive patterns. Figure 41-1 shows the basic repetitive units (the **unit cells**) of the lattice structures of copper, our prototype of a metal, and silicon and diamond (carbon), our prototypes of a semiconductor and an insulator, respectively.

We can classify solids electrically according to three basic properties:

1. Their **resistivity** ρ at room temperature, with the SI unit ohm-meter $(\Omega \cdot m)$; resistivity is defined in Module 26-3.

2. Their **temperature coefficient of resistivity** α, defined as $\alpha = (1/\rho)(d\rho/dT)$ in Eq. 26-17 and having the SI unit inverse kelvin (K^{-1}). We can evaluate α for any solid by measuring ρ over a range of temperatures.

3. Their **number density of charge carriers** n. This quantity, the number of charge carriers per unit volume, can be found from measurements of the Hall effect, as discussed in Module 28-2, and has the SI unit inverse cubic meter (m^{-3}).

From measurements of resistivity, we find that there are some materials, **insulators**, that do not conduct electricity at all. These are materials with very high resistivity. Diamond, an excellent example, has a resistivity greater than that of copper by the enormous factor of about 10^{24}.

We can then use measurements of ρ, α, and n to divide most noninsulators, at least at low temperatures, into two categories: **metals** and **semiconductors.**

Semiconductors have a considerably greater resistivity ρ than metals.

Semiconductors have a temperature coefficient of resistivity α that is both high and negative. That is, the resistivity of a semiconductor *decreases* with temperature, whereas that of a metal *increases*.

Semiconductors have a considerably lower number density of charge carriers n than metals.

Table 41-1 shows values of these quantities for copper, our prototype metal, and silicon, our prototype semiconductor.

Now let's consider our central question: *What features make diamond an insulator, copper a metal, and silicon a semiconductor?*

(a)

(b)

Figure 41-1 (*a*) The unit cell for copper is a cube. There is one copper atom (darker) at each corner of the cube and one copper atom (lighter) at the center of each face of the cube. The arrangement is called *face-centered cubic.* (*b*) The unit cell for either silicon or the carbon atoms in diamond is also a cube, the atoms being arranged in what is called a *diamond lattice.* There is one atom (darkest) at each corner of the cube and one atom (lightest) at the center of each cube face; in addition, four atoms (medium color) lie within the cube. Every atom is bonded to its four nearest neighbors by a two-electron covalent bond (only the four atoms within the cube show all four nearest neighbors).

Table 41-1 Some Electrical Properties of Two Materials[a]

		Material	
Property	Unit	Copper	Silicon
Type of conductor		Metal	Semiconductor
Resistivity, ρ	$\Omega \cdot m$	2×10^{-8}	3×10^3
Temperature coefficient of resistivity, α	K^{-1}	$+4 \times 10^{-3}$	-70×10^{-3}
Number density of charge carriers, n	m^{-3}	9×10^{28}	1×10^{16}

[a]All values are for room temperature.

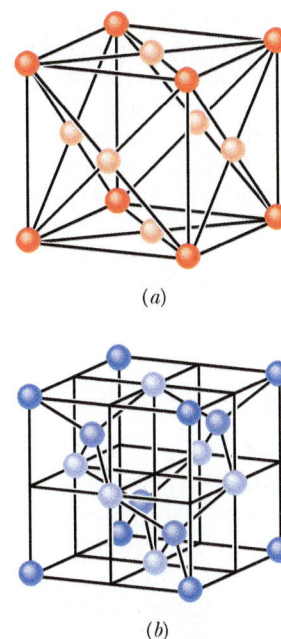

Figure 41-2 (*a*) Two copper atoms separated by a large distance; their electron distributions are represented by dot plots. (*b*) Each copper atom has 29 electrons distributed among a set of subshells. In the neutral atom in its ground state, all subshells up through the 3*d* level are filled, the 4*s* subshell contains one electron (it can hold two), and higher subshells are empty. For simplicity, the subshells are shown as being evenly spaced in energy.

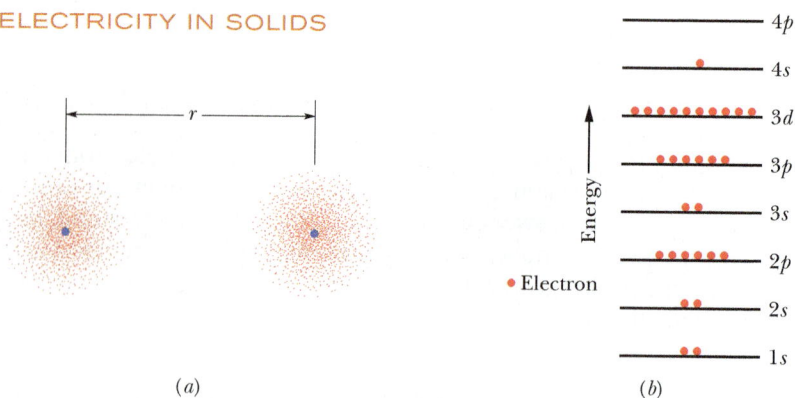

(*a*)　　(*b*)

Energy Levels in a Crystalline Solid

The distance between adjacent copper atoms in solid copper is 260 pm. Figure 41-2*a* shows two isolated copper atoms separated by a distance *r* that is much greater than that. As Fig. 41-2*b* shows, each of these isolated neutral atoms stacks up its 29 electrons in an array of discrete subshells as follows:

$$1s^2\ 2s^2\ 2p^6\ 3s^2\ 3p^6\ 3d^{10}\ 4s^1.$$

Here we use the shorthand notation of Module 40-5 to identify the subshells. Recall, for example, that the subshell with principal quantum number $n = 3$ and orbital quantum number $\ell = 1$ is called the 3*p* subshell; it can hold up to $2(2\ell + 1) = 6$ electrons; the number it actually contains is indicated by a numerical superscript. We see above that the first six subshells in copper are filled, but the (outermost) 4*s* subshell, which can hold two electrons, holds only one.

If we bring the atoms of Fig. 41-2*a* closer, their wave functions begin to overlap, starting with those of the outer electrons. We then have a single two-atom system with 58 electrons, not two independent atoms. The Pauli exclusion principle requires that each of these electrons occupy a different quantum state. In fact, 58 quantum states are available because each energy level of the isolated atom splits into *two* levels for the two-atom system.

If we bring up more atoms, we gradually assemble a lattice of solid copper. For *N* atoms, each level of an isolated copper atom must split into *N* levels in the solid. Thus, the individual energy levels of the solid form energy **bands,** adjacent bands being separated by an energy **gap,** with the gap representing a range of energies that no electron can possess. A typical band ranges over only a few electron-volts. Since *N* may be of the order of 10^{24}, the individual levels within a band are very close together indeed, and there are a vast number of levels.

Figure 41-3 suggests the band–gap structure of the energy levels in a generalized crystalline solid. Note that bands of lower energy are narrower than those of higher energy. This occurs because electrons that occupy the lower energy bands spend most of their time deep within the atom's electron cloud. The wave functions of these core electrons do not overlap as much as the wave functions of the outer electrons do. Hence the splitting of the lower energy levels (core electrons) is less than that of the higher energy levels (outer electrons).

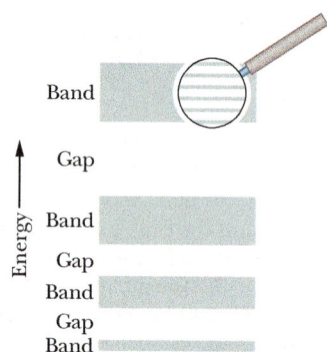

Figure 41-3 The band–gap pattern of energy levels for an idealized crystalline solid. As the magnified view suggests, each band consists of a very large number of very closely spaced energy levels. (In many solids, adjacent bands may overlap; for clarity, we have not shown this condition.)

Insulators

A solid is said to be an electrical insulator if no current exists within it when we apply a potential difference across it. For a current to exist, the kinetic energy of the average electron must increase. In other words, some electrons in the solid must move to a higher energy level. However, as Fig. 41-4 shows, in an insulator the highest band containing any electrons is fully occupied. Because the Pauli exclusion principle keeps electrons from moving to occupied levels, no electrons in the solid are allowed to move. Thus, the electrons in the filled band of an insulator have no place to go; they are in gridlock, like a child on a ladder filled with children.

In an insulator, electrons need a big energy jump.

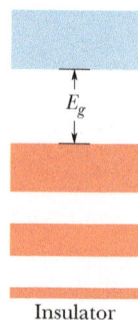

Figure 41-4 The band–gap pattern for an insulator; filled levels are shown in red and empty levels in blue.

There are plenty of unoccupied levels (or *vacant levels*) in the band above the filled band in Fig. 41-4. However, if an electron is to occupy one of those levels, it must acquire enough energy to jump across the substantial energy gap E_g that separates the two bands. In diamond, this gap is so wide (the energy needed to cross it is 5.5 eV, about 140 times the average thermal energy of a free particle at room temperature) that essentially no electron can jump across it. Diamond is thus an electrical insulator, and a very good one.

Sample Problem 41.01   Probability of electron excitation in an insulator

Approximately what is the probability that, at room temperature (300 K), an electron at the top of the highest filled band in diamond (an insulator) will jump the energy gap E_g in Fig. 41-4? For diamond, E_g is 5.5 eV.

KEY IDEA

In Chapter 40 we used Eq. 40-29,

$$\frac{N_x}{N_0} = e^{-(E_x - E_0)/kT}, \qquad (41\text{-}1)$$

to relate the population N_x of atoms at energy level E_x to the population N_0 at energy level E_0, where the atoms are part of a system at temperature T (measured in kelvins); k is the Boltzmann constant (8.62×10^{-5} eV/K). In this chapter we can use Eq. 41-1 to *approximate* the probability P that an electron in an insulator will jump the energy gap E_g in Fig. 41-4.

Calculations: We first set the energy difference $E_x - E_0$ to E_g. Then the probability P of the jump is approximately equal to the ratio N_x/N_0 of the number of electrons just above the energy gap to the number of electrons just below the gap.

For diamond, the exponent in Eq. 41-1 is

$$-\frac{E_g}{kT} = -\frac{5.5 \text{ eV}}{(8.62 \times 10^{-5} \text{ eV/K})(300 \text{ K})} = -213.$$

The required probability is then

$$P = \frac{N_x}{N_0} = e^{-(E_g/kT)} = e^{-213} \approx 3 \times 10^{-93}. \quad \text{(Answer)}$$

This result tells us that approximately 3 electrons out of 10^{93} electrons would jump across the energy gap. Because any diamond stone has fewer than 10^{23} electrons, we see that the probability of the jump is vanishingly small. No wonder diamond is such a good insulator.

WILEY PLUS Additional examples, video, and practice available at *WileyPLUS*

Metals

The feature that defines a metal is that, as Fig. 41-5 shows, the highest occupied energy level falls somewhere near the middle of an energy band. If we apply a potential difference across a metal, a current can exist because there are plenty of vacant levels at nearby higher energies into which electrons (the charge carriers in a metal) can jump. Thus, a metal can conduct electricity because electrons in its highest occupied band can easily move into higher energy levels.

In Module 26-4 we discussed the **free-electron model** of a metal, in which the **conduction electrons** are free to move throughout the volume of the sample like the molecules of a gas in a closed container. We used this model to derive an expression for the resistivity of a metal. Here we use the model to explain the behavior of the conduction electrons in the partially filled band of Fig. 41-5. However, we now assume the energies of these electrons to be quantized and the Pauli exclusion principle to hold.

Assuming that the electric potential energy U of a conduction electron is uniform throughout the lattice, let's set $U = 0$ so that the mechanical energy E is entirely kinetic. Then the level at the bottom of the partially filled band of Fig. 41-5 corresponds to $E = 0$. The highest occupied level in this band at absolute zero ($T = 0$ K) is called the **Fermi level,** and the energy corresponding to it is called the **Fermi energy** E_F; for copper, $E_F = 7.0$ eV.

The electron speed corresponding to the Fermi energy is called the **Fermi speed** v_F. For copper the Fermi speed is 1.6×10^6 m/s. Thus, all motion does *not* cease at absolute zero; at that temperature—and solely because of the Pauli ex-

In a conductor, electrons need only a small energy jump.

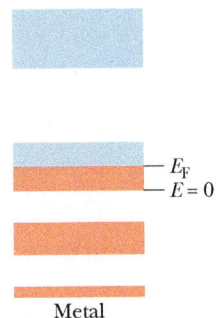

E_F
$E = 0$

Metal

Figure 41-5 The band–gap pattern for a metal. The highest filled level, called the Fermi level, lies near the middle of a band. Since vacant levels are available within that band, electrons in the band can easily change levels, and conduction can take place.

clusion principle—the conduction electrons are stacked up in the partially filled band of Fig. 41-5 with energies that range from zero to the Fermi energy.

How Many Conduction Electrons Are There?

If we could bring individual atoms together to form a sample of a metal, we would find that the conduction electrons in the metal are the *valence electrons* of the atoms (the electrons in the outermost occupied shells of the atoms). A *monovalent* atom contributes one such electron to the conduction electrons in a metal; a *bivalent* atom contributes two such electrons. Thus, the total number of conduction electrons is

$$\begin{pmatrix} \text{number of conduction} \\ \text{electrons in sample} \end{pmatrix} = \begin{pmatrix} \text{number of atoms} \\ \text{in sample} \end{pmatrix} \begin{pmatrix} \text{number of valence} \\ \text{electrons per atom} \end{pmatrix}. \quad (41\text{-}2)$$

(In this chapter, we shall write several equations largely in words because the symbols we have previously used for the quantities in them now represent other quantities.) The *number density n* of conduction electrons in a sample is the number of conduction electrons per unit volume:

$$n = \frac{\text{number of conduction electrons in sample}}{\text{sample volume } V}. \quad (41\text{-}3)$$

We can relate the number of atoms in a sample to various other properties of the sample and to the material making up the sample with the following:

$$\begin{pmatrix} \text{number of atoms} \\ \text{in sample} \end{pmatrix} = \frac{\text{sample mass } M_{\text{sam}}}{\text{atomic mass}} = \frac{\text{sample mass } M_{\text{sam}}}{(\text{molar mass } M)/N_A}$$

$$= \frac{(\text{material's density})(\text{sample volume } V)}{(\text{molar mass } M)/N_A}, \quad (41\text{-}4)$$

where the molar mass M is the mass of one mole of the material in the sample and N_A is Avogadro's number ($6.02 \times 10^{23} \text{ mol}^{-1}$).

Sample Problem 41.02 Number of conduction electrons in a metal

How many conduction electrons are in a cube of magnesium of volume $2.00 \times 10^{-6} \text{ m}^3$? Magnesium atoms are bivalent.

KEY IDEAS

1. Because magnesium atoms are bivalent, each magnesium atom contributes two conduction electrons.
2. The cube's number of conduction electrons is related to its number of magnesium atoms by Eq. 41-2.
3. We can find the number of atoms with Eq. 41-4 and known data about the cube's volume and magnesium's properties.

Calculations: We can write Eq. 41-4 as

$$\begin{pmatrix} \text{number} \\ \text{of atoms} \\ \text{in sample} \end{pmatrix} = \frac{(\text{density})(\text{sample volume } V)N_A}{\text{molar mass } M}.$$

Magnesium has density 1.738 g/cm³ (= 1.738×10^3 kg/m³)

and molar mass 24.312 g/mol (= 24.312×10^{-3} kg/mol) (see Appendix F). The numerator gives us

$$(1.738 \times 10^3 \text{ kg/m}^3)(2.00 \times 10^{-6} \text{ m}^3)$$
$$\times (6.02 \times 10^{23} \text{ atoms/mol}) = 2.0926 \times 10^{21} \text{ kg/mol}.$$

Thus, $\begin{pmatrix} \text{number of atoms} \\ \text{in sample} \end{pmatrix} = \dfrac{2.0926 \times 10^{21} \text{ kg/mol}}{24.312 \times 10^{-3} \text{ kg/mol}}$
$$= 8.61 \times 10^{22}.$$

Using this result and the fact that magnesium atoms are bivalent, we find that Eq. 41-2 yields

$$\begin{pmatrix} \text{number of} \\ \text{conduction electrons} \\ \text{in sample} \end{pmatrix}$$

$$= (8.61 \times 10^{22} \text{ atoms})\left(2 \frac{\text{electrons}}{\text{atom}} \right)$$
$$= 1.72 \times 10^{23} \text{ electrons.} \quad \text{(Answer)}$$

WILEY PLUS Additional examples, video, and practice available at *WileyPLUS*

Conductivity Above Absolute Zero

Our practical interest in the conduction of electricity in metals is at temperatures above absolute zero. What happens to the electron distribution of Fig. 41-5 at such higher temperatures? As we shall see, surprisingly little. Of the electrons in the partially filled band of Fig. 41-5, only those that are close to the Fermi energy find unoccupied levels above them, and only those electrons are free to be boosted to these higher levels by thermal agitation. Even at $T = 1000$ K (the copper would glow brightly in a dark room), the electron distribution among the available levels does not differ much from the distribution at $T = 0$ K.

Let us see why. The quantity kT, where k is the Boltzmann constant, is a convenient measure of the energy that may be given to a conduction electron by the random thermal motions of the lattice. At $T = 1000$ K, we have $kT = 0.086$ eV. No electron can hope to have its energy changed by more than a few times this relatively small amount by thermal agitation alone; so at best only those few conduction electrons whose energies are close to the Fermi energy are likely to jump to higher energy levels due to thermal agitation. Poetically stated, thermal agitation normally causes only ripples on the surface of the Fermi sea of electrons; the vast depths of that sea lie undisturbed.

How Many Quantum States Are There?

The ability of a metal to conduct electricity depends on how many quantum states are available to its electrons and what the energies of those states are. Thus, a question arises: What are the energies of the individual states in the partially filled band of Fig. 41-5? This question is too difficult to answer because we cannot possibly list the energies of so many states individually. We ask instead: How many states in a unit volume of a sample have energies in the energy range E to $E + dE$? We write this number as $N(E)\, dE$, where $N(E)$ is called the **density of states** at energy E. The conventional unit for $N(E)\, dE$ is states per cubic meter (states/m³, or simply m⁻³), and the conventional unit for $N(E)$ is states per cubic meter per electron-volt (m⁻³ eV⁻¹).

We can find an expression for the density of states by counting the number of standing electron matter waves that can fit into a box the size of the metal sample we are considering. This is analogous to counting the number of standing waves of sound that can exist in a closed organ pipe. Here the problem is three-dimensional (not one-dimensional) and the waves are matter waves (not sound waves). Such counting is covered in more advanced treatments of solid state physics; the result is

$$N(E) = \frac{8\sqrt{2}\,\pi m^{3/2}}{h^3}\, E^{1/2} \quad \text{(density of states, m}^{-3}\,\text{J}^{-1}\text{)}, \qquad (41\text{-}5)$$

where m ($= 9.109 \times 10^{-31}$ kg) is the electron mass, h ($= 6.626 \times 10^{-34}$ J·s) is the Planck constant, E is the energy in joules at which $N(E)$ is to be evaluated, and $N(E)$ is in states per cubic meter per joule (m⁻³ J⁻¹). To modify this equation so that the value of E is in electron-volts and the value of $N(E)$ is in states per cubic meter per electron-volt (m⁻³ eV⁻¹), multiply the right side of the equation by $e^{3/2}$, where e is the fundamental charge, 1.602×10^{-19} C. Figure 41-6 is a plot of such a modified version of Eq. 41-5. Note that nothing in Eq. 41-5 or Fig. 41-6 involves the shape, temperature, or composition of the sample.

The density of energy levels increases upward in a band.

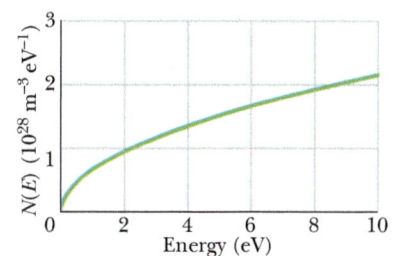

Figure 41-6 The density of states $N(E)$—that is, the number of electron energy levels per unit energy interval per unit volume—plotted as a function of electron energy. The density of states function simply counts the available states; it says nothing about whether these states are occupied by electrons.

✓ **Checkpoint 1**

Is the spacing between adjacent energy levels at $E = 4$ eV in copper larger than, the same as, or smaller than the spacing at $E = 6$ eV?

Sample Problem 41.03 **Number of states per electron volt in a metal**

(a) Using the data of Fig. 41-6, determine the number of states per electron-volt at 7 eV in a metal sample with a volume V of $2 \times 10^{-9} \, m^3$.

KEY IDEA

We can obtain the number of states per electron-volt at a given energy by using the density of states $N(E)$ at that energy and the sample's volume V.

Calculations: At an energy of 7 eV, we write

$$\begin{pmatrix} \text{number of states} \\ \text{per eV at 7 eV} \end{pmatrix} = \begin{pmatrix} \text{density of states} \\ N(E) \text{ at 7 eV} \end{pmatrix} \begin{pmatrix} \text{volume } V \\ \text{of sample} \end{pmatrix}.$$

From Fig. 41-6, we see that at an energy E of 7 eV, the density of states is about $1.8 \times 10^{28} \, m^{-3} \, eV^{-1}$. Thus,

$$\begin{pmatrix} \text{number of states} \\ \text{per eV at 7 eV} \end{pmatrix} = (1.8 \times 10^{28} \, m^{-3} \, eV^{-1})(2 \times 10^{-9} \, m^3)$$

$$= 3.6 \times 10^{19} \, eV^{-1}$$

$$\approx 4 \times 10^{19} \, eV^{-1}. \qquad \text{(Answer)}$$

(b) Next, determine the number of states N in the sample within a *small* energy range ΔE of 0.003 eV centered at 7 eV (the range is small relative to the energy level in the band).

Calculation: From Eq. 41-5 and Fig. 41-6, we know that the density of states is a function of energy E. However, for an energy range ΔE that is small relative to E, we can approximate the density of states (and thus the number of states per electron-volt) to be constant. Thus, at an energy of 7 eV, we find the number of states N in the energy range ΔE of 0.003 eV as

$$\begin{pmatrix} \text{number of states } N \\ \text{in range } \Delta E \text{ at 7 eV} \end{pmatrix} = \begin{pmatrix} \text{number of states} \\ \text{per eV at 7 eV} \end{pmatrix} \begin{pmatrix} \text{energy} \\ \text{range } \Delta E \end{pmatrix}$$

or

$$N = (3.6 \times 10^{19} \, eV^{-1})(0.003 \, eV)$$

$$= 1.1 \times 10^{17} \approx 1 \times 10^{17}. \qquad \text{(Answer)}$$

(When you are asked for the number of states in a certain energy range, first see if that range is small enough to allow this type of approximation.)

PLUS Additional examples, video, and practice available at *WileyPLUS*

The Occupancy Probability *P*(*E*)

If an energy level is available at energy E, what is the probability $P(E)$ that it is actually occupied by an electron? At $T = 0$ K, we know that all levels with energies below the Fermi energy are certainly occupied ($P(E) = 1$) and all higher levels are certainly not occupied ($P(E) = 0$). Figure 41-7a illustrates this situation. To find $P(E)$ at temperatures above absolute zero, we must use a set of quantum counting rules called **Fermi–Dirac statistics,** named for the physicists who introduced them. With these rules, the **occupancy probability** $P(E)$ is

$$P(E) = \frac{1}{e^{(E-E_F)/kT} + 1} \quad \text{(occupancy probability)}, \qquad (41\text{-}6)$$

in which E_F is the Fermi energy. Note that $P(E)$ depends not on the energy E of the level but only on the difference $E - E_F$, which may be positive or negative.

To see whether Eq. 41-6 describes Fig. 41-7a, we substitute $T = 0$ K in it. Then,

For $E < E_F$, the exponential term in Eq. 41-6 is $e^{-\infty}$, or zero; so $P(E) = 1$, in agreement with Fig. 41-7a.

For $E > E_F$, the exponential term is $e^{+\infty}$; so $P(E) = 0$, again in agreement with Fig. 41-7a.

Figure 41-7 The occupancy probability $P(E)$ is the probability that an energy level will be occupied by an electron. (a) At $T = 0$ K, $P(E)$ is unity for levels with energies E up to the Fermi energy E_F and zero for levels with higher energies. (b) At $T = 1000$ K, a few electrons whose energies were slightly less than the Fermi energy at $T = 0$ K move up to states with energies slightly greater than the Fermi energy. The dot on the curve shows that, for $E = E_F$, $P(E) = 0.5$.

The occupancy probability is high below the Fermi level.

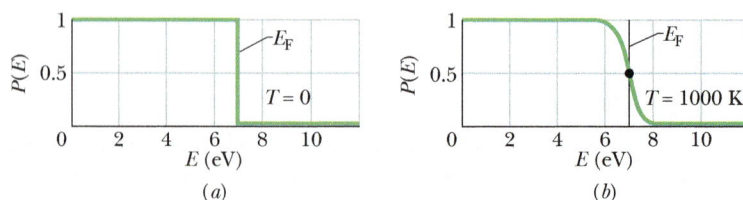

(a)

(b)

Figure 41-7b is a plot of $P(E)$ for $T = 1000$ K. Compared with Fig. 41-7a, it shows that, as stated above, changes in the distribution of electrons among the available states involve only states whose energies are near the Fermi energy E_F. Note that if $E = E_F$ (no matter what the temperature T), the exponential term in Eq. 41-6 is $e^0 = 1$ and $P(E) = 0.5$. This leads us to a more useful definition of the Fermi energy:

> The Fermi energy of a given material is the energy of a quantum state that has the probability 0.5 of being occupied by an electron.

Figures 41-7a and b are plotted for copper, which has a Fermi energy of 7.0 eV. Thus, for copper both at $T = 0$ K and at $T = 1000$ K, a state at energy $E = 7.0$ eV has a probability of 0.5 of being occupied.

Sample Problem 41.04 Probability of occupancy of an energy state in a metal

(a) What is the probability that a quantum state whose energy is 0.10 eV above the Fermi energy will be occupied? Assume a sample temperature of 800 K.

KEY IDEA

The occupancy probability of any state in a metal can be found from Fermi–Dirac statistics according to Eq. 41-6.

Calculations: Let's start with the exponent in Eq. 41-6:

$$\frac{E - E_F}{kT} = \frac{0.10 \text{ eV}}{(8.62 \times 10^{-5} \text{ eV/K})(800 \text{ K})} = 1.45.$$

Inserting this exponent into Eq. 41-6 yields

$$P(E) = \frac{1}{e^{1.45} + 1} = 0.19 \text{ or } 19\%. \qquad \text{(Answer)}$$

(b) What is the probability of occupancy for a state that is 0.10 eV *below* the Fermi energy?

Calculation: The Key Idea of part (a) applies here also except that now the state has an energy *below* the Fermi energy. Thus, the exponent in Eq. 41-6 has the same magnitude we found in part (a) but is negative, and that makes the denominator smaller. Equation 41-6 now yields

$$P(E) = \frac{1}{e^{-1.45} + 1} = 0.81 \text{ or } 81\%. \qquad \text{(Answer)}$$

For states below the Fermi energy, we are often more interested in the probability that the state is *not* occupied. This probability is just $1 - P(E)$, or 19%. Note that it is the same as the probability of occupancy in (a).

PLUS Additional examples, video, and practice available at *WileyPLUS*

How Many Occupied States Are There?

Equation 41-5 and Fig. 41-6 tell us how the available states are distributed in energy. The occupancy probability of Eq. 41-6 gives us the probability that any given state will actually be occupied by an electron. To find $N_o(E)$, the **density of occupied states,** we must multiply each available state by the corresponding value of the occupancy probability; that is,

$$\begin{pmatrix} \text{density of occupied states} \\ N_o(E) \text{ at energy } E \end{pmatrix} = \begin{pmatrix} \text{density of states} \\ N(E) \text{ at energy } E \end{pmatrix} \begin{pmatrix} \text{occupancy probability} \\ P(E) \text{ at energy } E \end{pmatrix}$$

or $\qquad\qquad N_o(E) = N(E) P(E) \quad$ (density of occupied states). $\qquad$ (41-7)

For copper at $T = 0$ K, Eq. 41-7 tells us to multiply, at each energy, the value of the density of states function (Eq. 41-6) by the value of the occupancy probability for absolute zero (Fig. 41-7a). The result is Fig. 41-8a. Figure 41-8b shows the density of occupied states at $T = 1000$ K.

The density of occupied states equals the product of ...

... the density of energy levels and ...

... the occupancy probability

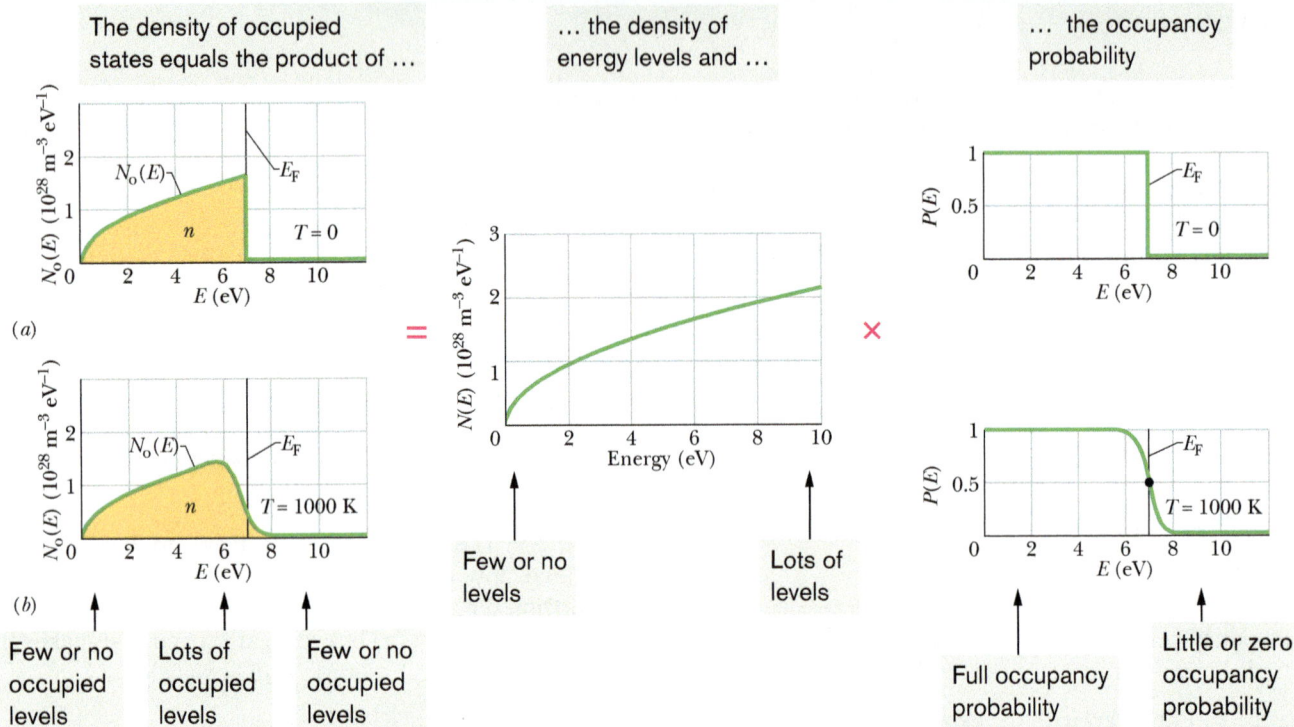

Figure 41-8 (a) The density of occupied states $N_o(E)$ for copper at absolute zero. The area under the curve is the number density of electrons n. Note that all states with energies up to the Fermi energy $E_F = 7$ eV are occupied, and all those with energies above the Fermi energy are vacant. (b) The same for copper at $T = 1000$ K. Note that only electrons with energies near the Fermi energy have been affected and redistributed.

Sample Problem 41.05   Number of occupied states in an energy range in a metal

A lump of copper (Fermi energy $= 7.0$ eV) has volume 2×10^{-9} m^3. How many occupied states per eV lie in a narrow energy range around 7.0 eV?

KEY IDEAS

(1) First we want the density of occupied states $N_o(E)$ as given by Eq. 41-7 ($N_o(E) = N(E) P(E)$). (2) Because we want to evaluate quantities for a narrow energy range around 7.0 eV (the Fermi energy for copper), the occupancy probability $P(E)$ is 0.50.

Calculations: From Fig. 41-6, we see that the density of states at 7 eV is about 1.8×10^{28} m^{-3} eV^{-1}. Thus, Eq. 41-7 tells us that the density of occupied states is

$$N_o(E) = N(E) P(E) = (1.8 \times 10^{28} \text{ m}^{-3} \text{ eV}^{-1})(0.50)$$
$$= 0.9 \times 10^{28} \text{ m}^{-3} \text{ eV}^{-1}.$$

Next, we write

$$\begin{pmatrix} \text{number of } occupied \\ \text{states per eV at 7 eV} \end{pmatrix} = \begin{pmatrix} \text{density of } occupied \\ \text{states } N_o(E) \text{ at 7 eV} \end{pmatrix}$$
$$\times \begin{pmatrix} \text{volume } V \\ \text{of sample} \end{pmatrix}.$$

Substituting for $N_o(E)$ and V gives us

$$\begin{pmatrix} \text{number of occupied} \\ \text{states per eV} \\ \text{at 7 eV} \end{pmatrix} = (0.9 \times 10^{28} \text{ m}^{-3} \text{ eV}^{-1})(2 \times 10^{-9} \text{ m}^3)$$
$$= 1.8 \times 10^{19} \text{ eV}^{-1}$$
$$\approx 2 \times 10^{19} \text{ eV}^{-1}. \qquad \text{(Answer)}$$

PLUS Additional examples, video, and practice available at *WileyPLUS*

Calculating the Fermi Energy

Suppose we add up (via integration) the number of occupied states per unit volume in Fig. 41-8a (for $T = 0$ K) at all energies between $E = 0$ and $E = E_F$. The result must equal n, the number of conduction electrons per unit volume for the metal, because at that temperature none of the energy states above the Fermi level are occupied. In equation form, we have

$$n = \int_0^{E_F} N_o(E) \, dE. \tag{41-8}$$

(Graphically, the integral here represents the area under the distribution curve of Fig. 41-8a.) Because $P(E) = 1$ for all energies below the Fermi energy when $T = 0$ K, Eq. 41-7 tells us we can replace $N_o(E)$ in Eq. 41-8 with $N(E)$ and then use Eq. 41-8 to find the Fermi energy E_F. If we substitute Eq. 41-5 into Eq. 41-8, we find that

$$n = \frac{8\sqrt{2}\pi m^{3/2}}{h^3} \int_0^{E_F} E^{1/2} \, dE = \frac{8\sqrt{2}\pi m^{3/2}}{h^3} \frac{2E_F^{3/2}}{3},$$

in which m is the electron mass. Solving for E_F now leads to

$$E_F = \left(\frac{3}{16\sqrt{2}\pi}\right)^{2/3} \frac{h^2}{m} n^{2/3} = \frac{0.121 h^2}{m} n^{2/3}. \tag{41-9}$$

Thus, when we know n, the number of conduction electrons per unit volume for a metal, we can find the Fermi energy for that metal.

41-2 SEMICONDUCTORS AND DOPING

Learning Objectives

After reading this module, you should be able to . . .

41.18 Sketch a band–gap diagram for a semiconductor, identifying the conduction and valence bands, conduction electrons, holes, and the energy gap.

41.19 Compare the energy gap of a semiconductor with that of an insulator.

41.20 Apply the relationship between a semiconductor's energy gap and the wavelength of light associated with a transition across the gap.

41.21 Sketch the lattice structure for pure silicon and doped silicon.

41.22 Identify holes, how they are produced, and how they move in an applied electric field.

41.23 For metals and semiconductors, compare the resistivity ρ and the temperature coefficient of resistivity α, and explain how the resistivity changes with temperature.

41.24 Explain the procedure for producing n-type semiconductors and p-type semiconductors.

41.25 Apply the relationship between the number of conduction electrons in a pure material and the number in the doped material.

41.26 Identify donors and acceptors and indicate where their energy levels lie in an energy-level diagram.

41.27 Identify majority carriers and minority carriers.

41.28 Explain the advantage of doping a semiconductor.

Key Ideas

● The band structure of a semiconductor is like that of an insulator except it has a much smaller gap width E_g, which can be jumped by thermally excited electrons.

● In silicon at room temperature, thermal agitation raises a few electrons to the conduction band, leaving an equal number of holes in the valence band. When the silicon is put under a potential difference, both electrons and holes serve as charge carriers.

● The number of electrons in the conduction band of silicon can be increased greatly by doping with small amounts of phosphorus, thus forming n-type material. The phosphorus atoms are said to be donor atoms.

● The number of holes in the valence band of silicon can be greatly increased by doping with small amounts of aluminum, thus forming p-type material. The aluminum atoms are said to be acceptor atoms.

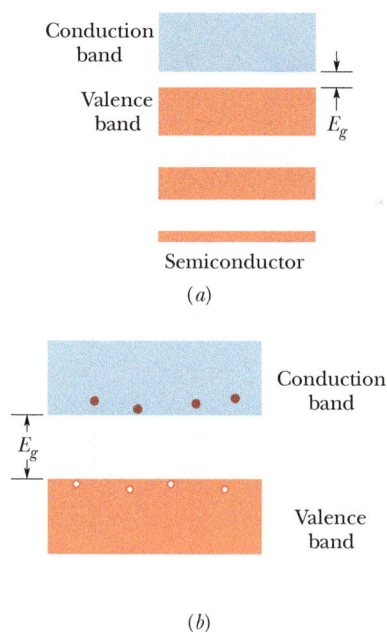

Figure 41-9 (a) The band–gap pattern for a semiconductor. It resembles that of an insulator (see Fig. 41-4) except that here the energy gap E_g is much smaller; thus electrons, because of their thermal agitation, have some reasonable probability of being able to jump the gap. (b) Thermal agitation has caused a few electrons to jump the gap from the valence band to the conduction band, leaving an equal number of holes in the valence band.

Semiconductors

If you compare Fig. 41-9a with Fig. 41-4, you can see that the band structure of a semiconductor is like that of an insulator. The main difference is that the semiconductor has a much smaller energy gap E_g between the top of the highest filled band (called the **valence band**) and the bottom of the vacant band just above it (called the **conduction band**). Thus, there is no doubt that silicon (E_g = 1.1 eV) is a semiconductor and diamond (E_g = 5.5 eV) is an insulator. In silicon—but not in diamond—there is a real possibility that thermal agitation at room temperature will cause electrons to jump the gap from valence to conduction band.

In Table 41-1 we compared three basic electrical properties of copper, our prototype metallic conductor, and silicon, our prototype semiconductor. Let us look again at that table, one row at a time, to see how a semiconductor differs from a metal.

Number Density of Charge Carriers *n*

The bottom row of Table 41-1 shows that copper has far more charge carriers per unit volume than silicon, by a factor of about 10^{13}. For copper, each atom contributes one electron, its single valence electron, to the conduction process. Charge carriers in silicon arise only because, at thermal equilibrium, thermal agitation causes a certain (very small) number of valence-band electrons to jump the energy gap into the conduction band, leaving an equal number of unoccupied energy states, called **holes,** in the valence band. Figure 41-9b shows the situation.

Both the electrons in the conduction band and the holes in the valence band serve as charge carriers. The holes do so by permitting a certain freedom of movement to the electrons remaining in the valence band, electrons that, in the absence of holes, would be gridlocked. If an electric field $\vec{E}$ is set up in a semiconductor, the electrons in the valence band, being negatively charged, tend to drift in the direction opposite $\vec{E}$. This causes the positions of the holes to drift in the direction of $\vec{E}$. In effect, the holes behave like moving particles of charge $+e$.

It may help to think of a row of cars parked bumper to bumper, with the lead car at one car's length from a barrier and the empty one-car-length distance being an available parking space. If the leading car moves forward to the barrier, it opens up a parking space behind it. The second car can then move up to fill that space, allowing the third car to move up, and so on. The motions of the many cars toward the barrier are most simply analyzed by focusing attention on the drift of the single "hole" (parking space) away from the barrier.

In semiconductors, conduction by holes is just as important as conduction by electrons. In thinking about hole conduction, we can assume that all unoccupied states in the valence band are occupied by particles of charge $+e$ and that all electrons in the valence band have been removed, so that these positive charge carriers can move freely throughout the band.

Resistivity *ρ*

Recall from Chapter 26 that the resistivity ρ of a material is $m/e^2 n\tau$, where m is the electron mass, e is the fundamental charge, n is the number of charge carriers per unit volume, and τ is the mean time between collisions of the charge carriers. Table 41-1 shows that, at room temperature, the resistivity of silicon is higher than that of copper by a factor of about 10^{11}. This vast difference can be accounted for by the vast difference in n. Other factors enter, but their effect on the resistivity is swamped by the enormous difference in n.

Temperature Coefficient of Resistivity *α*

Recall that α (see Eq. 26-17) is the fractional change in resistivity per unit change in temperature:

$$\alpha = \frac{1}{\rho}\frac{d\rho}{dT}.$$ (41-10)

The resistivity of copper *increases* with temperature (that is, $d\rho/dT > 0$) because collisions of copper's charge carriers occur more frequently at higher temperatures. Thus, α is *positive* for copper.

The collision frequency also increases with temperature for silicon. However, the resistivity of silicon actually *decreases* with temperature ($d\rho/dT < 0$) because the number of charge carriers n (electrons in the conduction band and holes in the valence band) increases so rapidly with temperature. (More electrons jump the gap from the valence band to the conduction band.) Thus, the fractional change α is *negative* for silicon.

Doped Semiconductors

The usefulness of semiconductors in technology can be greatly improved by introducing a small number of suitable replacement atoms (called impurities) into the semiconductor lattice—a process called **doping**. Typically, only about 1 silicon atom in 10^7 is replaced by a dopant atom in the doped semiconductor. Essentially all modern semiconducting devices are based on doped material. Such materials are of two types, called *n*-**type** and *p*-**type**; we discuss each in turn.

n-Type Semiconductors

The electrons in an isolated silicon atom are arranged in subshells according to the scheme

$$1s^2 \; 2s^2 \; 2p^6 \; 3s^2 \; 3p^2,$$

in which, as usual, the superscripts (which add to 14, the atomic number of silicon) represent the numbers of electrons in the specified subshells.

Figure 41-10a is a flattened-out representation of a portion of the lattice of pure silicon in which the portion has been projected onto a plane; compare the figure with Fig. 41-1b, which represents the unit cell of the lattice in three dimensions. Each silicon atom contributes its pair of 3s electrons and its pair of 3p electrons to form a rigid two-electron covalent bond with each of its four nearest neighbors. (A covalent bond is a link between two atoms in which the atoms share a pair of electrons.) The four atoms that lie within the unit cell in Fig. 41-1b show these four bonds.

The electrons that form the silicon–silicon bonds constitute the valence band of the silicon sample. If an electron is torn from one of these bonds so that it becomes free to wander throughout the lattice, we say that the electron has been raised from the valence band to the conduction band. The minimum energy required to do this is the gap energy E_g.

Because four of its electrons are involved in bonds, each silicon "atom" is actually an ion consisting of an inert neon-like electron cloud (containing 10 electrons) surrounding a nucleus whose charge is $+14e$, where 14 is the atomic number of silicon. The net charge of each of these ions is thus $+4e$, and the ions are said to have a *valence number* of 4.

In Fig. 41-10b the central silicon ion has been replaced by an atom of phosphorus (valence = 5). Four of the valence electrons of the phosphorus form bonds with the four surrounding silicon ions. The fifth ("extra") electron is only loosely bound to the phosphorus ion core. On an energy-band diagram, we usually say that such an electron occupies a localized energy state that lies within the energy gap, at an average energy interval E_d below the bottom of the conduction band; this is indicated in Fig. 41-11a. Because $E_d \ll E_g$, the energy required to excite electrons from *these* levels into the conduction band is much less than that required to excite silicon valence electrons into the conduction band.

The phosphorus atom is called a **donor** atom because it readily *donates* an electron to the conduction band. In fact, at room temperature virtually *all* the

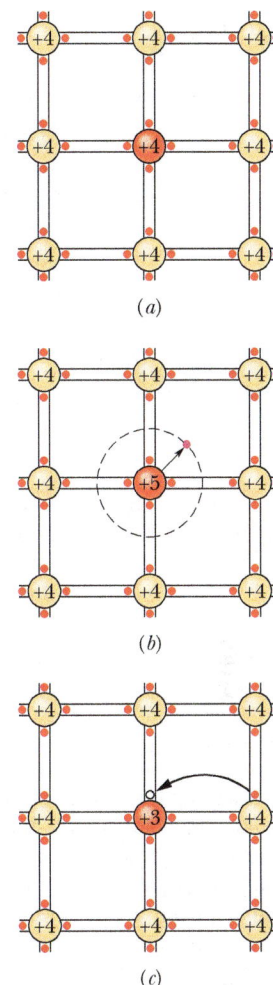

Figure 41-10 (a) A flattened-out representation of the lattice structure of pure silicon. Each silicon ion is coupled to its four nearest neighbors by a two-electron covalent bond (represented by a pair of red dots between two parallel black lines). The electrons belong to the bond—not to the individual atoms—and form the valence band of the sample. (b) One silicon atom is replaced by a phosphorus atom (valence = 5). The "extra" electron is only loosely bound to its ion core and may easily be elevated to the conduction band, where it is free to wander through the volume of the lattice. (c) One silicon atom is replaced by an aluminum atom (valence = 3). There is now a hole in one of the covalent bonds and thus in the valence band of the sample. The hole can easily migrate through the lattice as electrons from neighboring bonds move in to fill it. Here the hole migrates rightward.

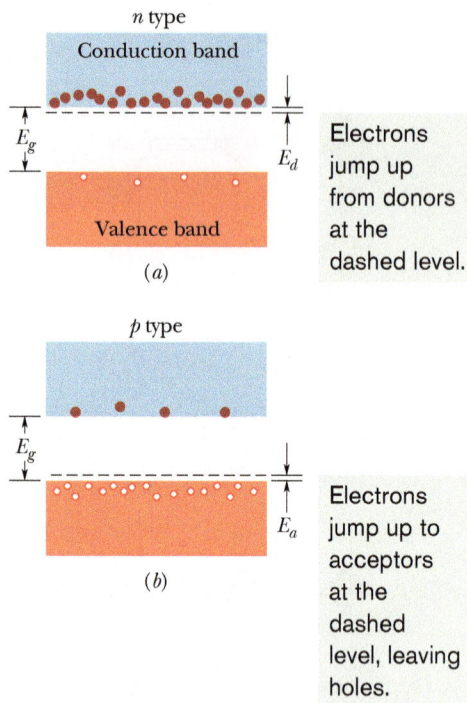

Electrons jump up from donors at the dashed level.

Electrons jump up to acceptors at the dashed level, leaving holes.

Figure 41-11 (*a*) In a doped *n*-type semiconductor, the energy levels of donor electrons lie a small interval E_d below the bottom of the conduction band. Because donor electrons can be easily excited to the conduction band, there are now many more electrons in that band. The valence band contains the same small number of holes as before the dopant was added. (*b*) In a doped *p*-type semiconductor, the acceptor levels lie a small energy interval E_a above the top of the valence band. There are now many more holes in the valence band. The conduction band contains the same small number of electrons as before the dopant was added. The ratio of majority carriers to minority carriers in both (*a*) and (*b*) is very much greater than is suggested by these diagrams.

electrons contributed by the donor atoms are in the conduction band. By adding donor atoms, it is possible to increase greatly the number of electrons in the conduction band, by a factor very much larger than Fig. 41-11*a* suggests.

Semiconductors doped with donor atoms are called **n-type semiconductors**; the *n* stands for *negative,* to imply that the negative charge carriers introduced into the conduction band greatly outnumber the positive charge carriers, which are the holes in the valence band. In *n*-type semiconductors, the electrons are called the **majority carriers** and the holes are called the **minority carriers.**

p-Type Semiconductors

Now consider Fig. 41-10*c*, in which one of the silicon atoms (valence = 4) has been replaced by an atom of aluminum (valence = 3). The aluminum atom can bond covalently with only three silicon atoms, and so there is now a "missing" electron (a hole) in one aluminum–silicon bond. With a small expenditure of energy, an electron can be torn from a neighboring silicon–silicon bond to fill this hole, thereby creating a hole in *that* bond. Similarly, an electron from some other bond can be moved to fill the newly created hole. In this way, the hole can migrate through the lattice.

The aluminum atom is called an **acceptor** atom because it readily *accepts* an electron from a neighboring bond—that is, from the valence band of silicon. As Fig. 41-11*b* suggests, this electron occupies a localized energy state that lies within the energy gap, at an average energy interval E_a above the top of the valence band. Because this energy interval E_a is small, valence electrons are easily bumped up to the acceptor level, leaving holes in the valence band. Thus, by adding acceptor atoms, it is possible to greatly increase the number of holes in the valence band, by a factor much larger than Fig. 41-11*b* suggests. In silicon at room temperature, virtually *all* the acceptor levels are occupied by electrons.

Semiconductors doped with acceptor atoms are called **p-type semiconductors**; the *p* stands for *positive* to imply that the holes introduced into the valence band, which behave like positive charge carriers, greatly outnumber the electrons in the conduction band. In *p*-type semiconductors, holes are the majority carriers and electrons are the minority carriers.

Table 41-2 summarizes the properties of a typical *n*-type and a typical *p*-type semiconductor. Note particularly that the donor and acceptor ion cores, although they are charged, are not charge *carriers* because they are fixed in place.

Table 41-2 Properties of Two Doped Semiconductors

Property	Type of Semiconductor	
	n	*p*
Matrix material	Silicon	Silicon
Matrix nuclear charge	$+14e$	$+14e$
Matrix energy gap	1.2 eV	1.2 eV
Dopant	Phosphorus	Aluminum
Type of dopant	Donor	Acceptor
Majority carriers	Electrons	Holes
Minority carriers	Holes	Electrons
Dopant energy gap	$E_d = 0.045$ eV	$E_a = 0.067$ eV
Dopant valence	5	3
Dopant nuclear charge	$+15e$	$+13e$
Dopant net ion charge	$+e$	$-e$

Sample Problem 41.06 Doping silicon with phosphorus

The number density n_0 of conduction electrons in pure silicon at room temperature is about 10^{16} m^{-3}. Assume that, by doping the silicon lattice with phosphorus, we want to increase this number by a factor of a million (10^6). What fraction of silicon atoms must we replace with phosphorus atoms? (Recall that at room temperature, thermal agitation is so effective that essentially every phosphorus atom donates its "extra" electron to the conduction band.)

Number of phosphorus atoms: Because each phosphorus atom contributes one conduction electron and because we want the total number density of conduction electrons to be $10^6 n_0$, the number density of phosphorus atoms n_P must be given by

$$10^6 n_0 = n_0 + n_P.$$

Then $n_P = 10^6 n_0 - n_0 \approx 10^6 n_0$

$$= (10^6)(10^{16} \text{ m}^{-3}) = 10^{22} \text{ m}^{-3}.$$

This tells us that we must add 10^{22} atoms of phosphorus per cubic meter of silicon.

Fraction of silicon atoms: We can find the number density n_{Si} of silicon atoms in pure silicon (before the doping) from Eq. 41-4, which we can write as

$$\left(\begin{array}{c}\text{number of atoms}\\\text{in sample}\end{array}\right)$$

$$= \frac{(\text{silicon density})(\text{sample volume } V)}{(\text{silicon molar mass } M_{Si})/N_A}.$$

Dividing both sides by the sample volume V to get the number density of silicon atoms n_{Si} on the left, we then have

$$n_{Si} = \frac{(\text{silicon density})N_A}{M_{Si}}.$$

Appendix F tells us that the density of silicon is 2.33 g/cm^3 ($= 2330$ kg/m^3) and the molar mass of silicon is 28.1 g/mol ($= 0.0281$ kg/mol). Thus, we have

$$n_{Si} = \frac{(2330 \text{ kg/m}^3)(6.02 \times 10^{23} \text{ atoms/mol})}{0.0281 \text{ kg/mol}}$$

$$= 5 \times 10^{28} \text{ atoms/m}^3 = 5 \times 10^{28} \text{ m}^{-3}.$$

The fraction we seek is approximately

$$\frac{n_P}{n_{Si}} = \frac{10^{22} \text{ m}^{-3}}{5 \times 10^{28} \text{ m}^{-3}} = \frac{1}{5 \times 10^6}. \qquad \text{(Answer)}$$

If we replace only *one silicon atom in five million* with a phosphorus atom, the number of electrons in the conduction band will be increased by a factor of a million.

How can such a tiny admixture of phosphorus have what seems to be such a big effect? The answer is that, although the effect is very significant, it is not "big." The number density of conduction electrons was 10^{16} m^{-3} before doping and 10^{22} m^{-3} after doping. For copper, however, the conduction-electron number density (given in Table 41-1) is about 10^{29} m^{-3}. Thus, even after doping, the number density of conduction electrons in silicon remains much less than that of a typical metal, such as copper, by a factor of about 10^7.

PLUS Additional examples, video, and practice available at *WileyPLUS*

41-3 THE *p-n* JUNCTION AND THE TRANSISTOR

Learning Objectives

After reading this module, you should be able to . . .

41.29 Describe a *p-n* junction and outline how it works.

41.30 Identify diffusion current, space charge, depletion zone, contact potential difference, and drift current.

41.31 Describe the functioning of a junction rectifier.

41.32 Distinguish forward bias and back bias.

41.33 Explain the general properties of a light-emitting diode, a photodiode, a junction laser, and a MOSFET.

Key Ideas

● A *p-n* junction is a single semiconducting crystal with one end doped to form *p*-type material and the other end doped to form *n*-type material. The two types meet at a junction plane.

● At thermal equilibrium, the following occur at the junction plane: (1) Majority carriers diffuse across the plane, producing a diffusion current I_{diff}. (2) Minority carriers are swept across the plane, forming a drift current I_{drift}. (3) A depletion zone forms at the plane. (4) A contact potential V_0 develops across the depletion zone.

● A *p-n* junction conducts electricity better for one direction of an applied potential difference (forward biased) than for the opposite direction (back biased), and thus the device can serve as a junction rectifier.

● A *p-n* junction made with certain materials can emit light when forward biased and thus can serve as a light-emitting diode (LED).

● A light-emitting *p-n* junction can also be made to emit stimulated emission and thus can serve as a laser.

The *p-n* Junction

A **p-n junction** (Fig. 41-12a), essential to most semiconductor devices, is a single semiconductor crystal that has been selectively doped so that one region is *n*-type material and the adjacent region is *p*-type material. Let's assume that the junction has been formed mechanically by jamming together a bar of *n*-type semiconductor and a bar of *p*-type semiconductor. Thus, the transition from one region to the other is perfectly sharp, occurring at a single **junction plane.**

Let us discuss the motions of electrons and holes just after the *n*-type bar and the *p*-type bar, both electrically neutral, have been jammed together to form the junction. We first examine the majority carriers, which are electrons in the *n*-type material and holes in the *p*-type material.

Motions of the Majority Carriers

If you burst a helium-filled balloon, helium atoms will diffuse (spread) outward into the surrounding air. This happens because there are very few helium atoms in normal air. In more formal language, there is a helium *density gradient* at the balloon–air interface (the number density of helium atoms varies across the interface); the helium atoms move so as to reduce the gradient.

In the same way, electrons on the *n* side of Fig. 41-12a that are close to the junction plane tend to diffuse across it (from right to left in the figure) and into the *p* side, where there are very few free electrons. Similarly, holes on the *p* side that are close to the junction plane tend to diffuse across that plane (from left to right) and into the *n* side, where there are very few holes. The motions of both the electrons and the holes contribute to a **diffusion current** I_{diff}, conventionally directed from left to right as indicated in Fig. 41-12d.

Recall that the *n*-side is studded throughout with positively charged donor ions, fixed firmly in their lattice sites. Normally, the excess positive charge of each of these ions is compensated electrically by one of the conduction-band electrons. When an *n*-side electron diffuses across the junction plane, however, the diffusion "uncovers" one of these donor ions, thus introducing a fixed positive charge near the junction plane on the *n* side. When the diffusing electron arrives on the *p* side, it quickly combines with an acceptor ion (which lacks one electron), thus introducing a fixed negative charge near the junction plane on the *p* side.

In this way electrons diffusing through the junction plane from right to left in Fig. 41-12a result in a buildup of **space charge** on each side of the junction plane, with positive charge on the *n* side and negative charge on the *p* side, as shown in Fig. 41-12b. Holes diffusing through the junction plane from left to right have exactly the same effect. (Take the time now to convince yourself of that.) The motions of both majority carriers—electrons and holes—contribute to the buildup of these two space charge regions, one positive and one negative. These two regions form a **depletion zone,** so named because it is relatively free of *mobile* charge carriers; its width is shown as d_0 in Fig. 41-12b.

The buildup of space charge generates an associated **contact potential difference** V_0 across the depletion zone, as Fig. 41-12c shows. This potential difference

Figure 41-12 (*a*) A *p-n* junction. (*b*) Motions of the majority charge carriers across the junction plane uncover a space charge associated with uncompensated donor ions (to the right of the plane) and acceptor ions (to the left). (*c*) Associated with the space charge is a contact potential difference V_0 across d_0. (*d*) The diffusion of majority carriers (both electrons and holes) across the junction plane produces a diffusion current I_{diff}. (In a real *p-n* junction, the boundaries of the depletion zone would not be sharp, as shown here, and the contact potential curve (*c*) would be smooth, with no sharp corners.)

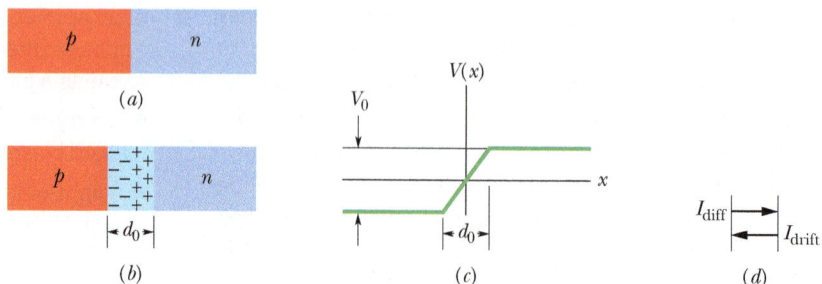

limits further diffusion of electrons and holes across the junction plane. Negative charges tend to avoid regions of low potential. Thus, an electron approaching the junction plane from the right in Fig. 41-12*b* is moving toward a region of low potential and would tend to turn back into the *n* side. Similarly, a positive charge (a hole) approaching the junction plane from the left is moving toward a region of high potential and would tend to turn back into the *p* side.

Motions of the Minority Carriers

As Fig. 41-11*a* shows, although the majority carriers in *n*-type material are electrons, there are a few holes. Likewise in *p*-type material (Fig. 41-11*b*), although the majority carriers are holes, there are also a few electrons. These few holes and electrons are the minority carriers in the corresponding materials.

Although the potential difference V_0 in Fig. 41-12*c* acts as a barrier for the majority carriers, it is a downhill trip for the minority carriers, be they electrons on the *p* side or holes on the *n* side. Positive charges (holes) tend to seek regions of low potential; negative charges (electrons) tend to seek regions of high potential. Thus, both types of minority carriers are *swept across* the junction plane by the contact potential difference and together constitute a **drift current** I_{drift} across the junction plane from right to left, as Fig. 41-12*d* indicates.

Thus, an isolated *p-n* junction is in an equilibrium state in which a contact potential difference V_0 exists between its ends. At equilibrium, the average diffusion current I_{diff} that moves through the junction plane from the *p* side to the *n* side is just balanced by an average drift current I_{drift} that moves in the opposite direction. These two currents cancel because the net current through the junction plane must be zero; otherwise charge would be transferred without limit from one end of the junction to the other.

> ### ✓ Checkpoint 2
>
> Which of the following five currents across the junction plane of Fig. 41-12*a* must be zero?
> (a) the net current due to holes, both majority and minority carriers included
> (b) the net current due to electrons, both majority and minority carriers included
> (c) the net current due to both holes and electrons, both majority and minority carriers included
> (d) the net current due to majority carriers, both holes and electrons included
> (e) the net current due to minority carriers, both holes and electrons included

The Junction Rectifier

Look now at Fig. 41-13. It shows that, if we place a potential difference across a *p-n* junction in one direction (here labeled + and "Forward bias"), there will be a current through the junction. However, if we reverse the direction of the potential difference, there will be approximately zero current through the junction.

One application of this property is the **junction rectifier,** whose symbol is shown in Fig. 41-14*b*; the arrowhead corresponds to the *p*-type end of the device and points in the allowed direction of conventional current. A sine wave input potential to the device (Fig. 41-14*a*) is transformed to a half-wave output potential (Fig. 41-14*c*) by the junction rectifier; that is, the rectifier acts as essentially a closed switch (zero resistance) for one polarity of the input potential and as essentially an open switch (infinite resistance) for the other. The average input voltage is zero, but the average output voltage is not. Thus, a junction rectifier can be used as part of an apparatus to convert an alternating potential difference into a constant potential difference, as for a power supply.

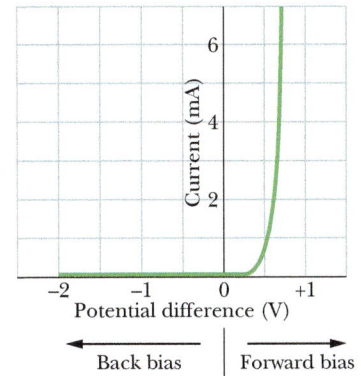

Figure 41-13 A current–voltage plot for a *p-n* junction, showing that the junction is highly conducting when forward-biased and essentially nonconducting when back-biased.

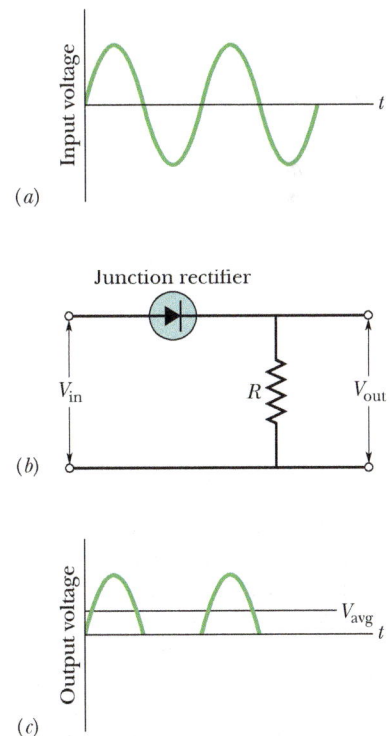

Figure 41-14 A *p-n* junction connected as a junction rectifier. The action of the circuit in (*b*) is to pass the positive half of the input wave form in (*a*) but to suppress the negative half. The average potential of the input wave form is zero; that of the output wave form in (*c*) has a positive value V_{avg}.

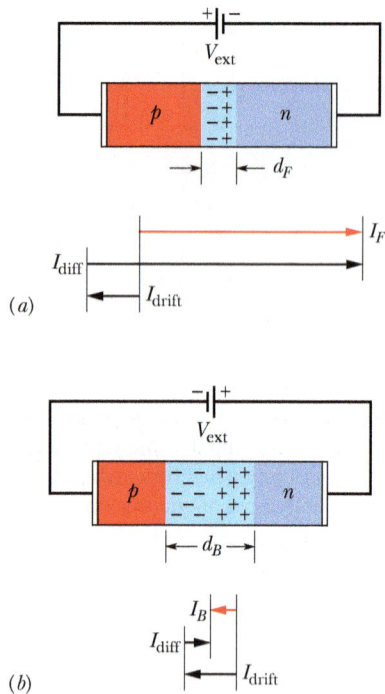

Figure 41-15 (*a*) The forward-bias connection of a *p-n* junction, showing the narrowed depletion zone and the large forward current I_F. (*b*) The back-bias connection, showing the widened depletion zone and the small back current I_B.

Figure 41-15 shows why a *p-n* junction operates as a junction rectifier. In Fig. 41-15*a*, a battery is connected across the junction with its positive terminal connected at the *p* side. In this **forward-bias connection,** the *p* side becomes more positive and the *n* side becomes more negative, thus *decreasing* the height of the potential barrier V_0 of Fig. 41-12*c*. More of the majority carriers can now surmount this smaller barrier; hence, the diffusion current I_{diff} increases markedly.

The minority carriers that form the drift current, however, sense no barrier; so the drift current I_{drift} is not affected by the external battery. The nice current balance that existed at zero bias (see Fig. 41-12*d*) is thus upset, and, as shown in Fig. 41-15*a*, a large net forward current I_F appears in the circuit.

Another effect of forward bias is to narrow the depletion zone, as a comparison of Fig. 41-12*b* and Fig. 41-15*a* shows. The depletion zone narrows because the reduced potential barrier associated with forward bias must be associated with a smaller space charge. Because the ions producing the space charge are fixed in their lattice sites, a reduction in their number can come about only through a reduction in the width of the depletion zone.

Because the depletion zone normally contains very few charge carriers, it is normally a region of high resistivity. However, when its width is substantially reduced by a forward bias, its resistance is also reduced substantially, as is consistent with the large forward current.

Figure 41-15*b* shows the **back-bias** connection, in which the negative terminal of the battery is connected at the *p*-type end of the *p-n* junction. Now the applied emf *increases* the contact potential difference, the diffusion current *decreases* substantially while the drift current remains unchanged, and a relatively *small* back current I_B results. The depletion zone *widens*, its *high* resistance being consistent with the *small* back current I_B.

The Light-Emitting Diode (LED)

Nowadays, we can hardly avoid the brightly colored "electronic" numbers that glow at us from cash registers and gasoline pumps, microwave ovens and alarm clocks, and we cannot seem to do without the invisible infrared beams that control elevator doors and operate television sets via remote control. In nearly all cases this light is emitted from a *p-n* junction operating as a **light-emitting diode** (LED). How can a *p-n* junction generate light?

Consider first a simple semiconductor. When an electron from the bottom of the conduction band falls into a hole at the top of the valence band, an energy E_g equal to the gap width is released. In silicon, germanium, and many other semiconductors, this energy is largely transformed into thermal energy of the vibrating lattice, and as a result, no light is emitted.

In some semiconductors, however, including gallium arsenide, the energy can be emitted as a photon of energy hf at wavelength

$$\lambda = \frac{c}{f} = \frac{c}{E_g/h} = \frac{hc}{E_g}. \tag{41-11}$$

To emit enough light to be useful as an LED, the material must have a suitably large number of electron–hole transitions. This condition is not satisfied by a pure semiconductor because, at room temperature, there are simply not enough electron–hole pairs. As Fig. 41-11 suggests, doping will not help. In doped *n*-type material the number of conduction electrons is greatly increased, but there are not enough holes for them to combine with; in doped *p*-type material there are plenty of holes but not enough electrons to combine with them. Thus, neither a pure semiconductor nor a doped semiconductor can provide enough electron–hole transitions to serve as a practical LED.

What we need is a semiconductor material with a very large number of electrons in the conduction band *and* a correspondingly large number of holes in the valence band. A device with this property can be fabricated by placing a strong forward bias on a heavily doped *p-n* junction, as in Fig. 41-16. In such an arrangement the current *I* through the device serves to inject electrons into the *n*-type material and to inject holes into the *p*-type material. If the doping is heavy enough and the current is great enough, the depletion zone can become very narrow, perhaps only a few micrometers wide. The result is a great number density of electrons in the *n*-type material facing a correspondingly great number density of holes in the *p*-type material, across the narrow depletion zone. With such great number densities so near each other, many electron–hole combinations occur, causing light to be emitted from that zone. Figure 41-17 shows the construction of an actual LED.

Commercial LEDs designed for the visible region are commonly based on gallium suitably doped with arsenic and phosphorus atoms. An arrangement in which 60% of the nongallium sites are occupied by arsenic ions and 40% by phosphorus ions results in a gap width E_g of about 1.8 eV, corresponding to red light. Other doping and transition-level arrangements make it possible to construct LEDs that emit light in essentially any desired region of the visible and near-visible spectra.

The Photodiode

Passing a current through a suitably arranged *p-n* junction can generate light. The reverse is also true; that is, shining light on a suitably arranged *p-n* junction can produce a current in a circuit that includes the junction. This is the basis for the **photodiode.**

When you click your television remote control, an LED in the device sends out a coded sequence of pulses of infrared light. The receiving device in your television set is an elaboration of the simple (two-terminal) photodiode that not only detects the infrared signals but also amplifies them and transforms them into electrical signals that change the channel or adjust the volume, among other tasks.

The Junction Laser

In the arrangement of Fig. 41-16, there are many electrons in the conduction band of the *n*-type material and many holes in the valence band of the *p*-type material. Thus, there is a **population inversion** for the electrons; that is, there are more electrons in higher energy levels than in lower energy levels. As we discussed in Module 40-7, this can lead to lasing.

When a single electron moves from the conduction band to the valence band, it can release its energy as a photon. This photon can stimulate a second electron to fall into the valence band, producing a second photon by stimulated emission. In this way, if the current through the junction is great enough, a chain reaction of stimulated emission events can occur and laser light can be generated. To bring this about, opposite faces of the *p-n* junction crystal must be flat and parallel, so that light can be reflected back and forth within the crystal. (Recall that in the helium–neon laser of Fig. 40-20, a pair of mirrors served this purpose.) Thus, a *p-n* junction can act as a **junction laser,** its light output being highly coherent and much more sharply defined in wavelength than light from an LED.

Junction lasers are built into compact disc (CD) players, where, by detecting reflections from the rotating disc, they are used to translate microscopic pits in the disc into sound. They are also much used in optical communication systems based on optical fibers. Figure 41-18 suggests their tiny scale. Junction lasers are usually designed to operate in the infrared region of the electromagnetic spectrum because optical fibers have two "windows" in that region (at $\lambda = 1.31$ and 1.55 μm) for which the energy absorption per unit length of the fiber is a minimum.

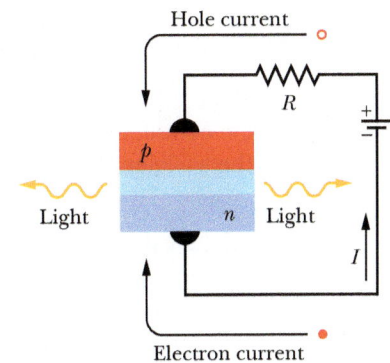

Figure 41-16 A forward-biased *p-n* junction, showing electrons being injected into the *n*-type material and holes into the *p*-type material. (Holes move in the conventional direction of the current *I*, equivalent to electrons moving in the opposite direction.) Light is emitted from the narrow depletion zone each time an electron and a hole combine across that zone.

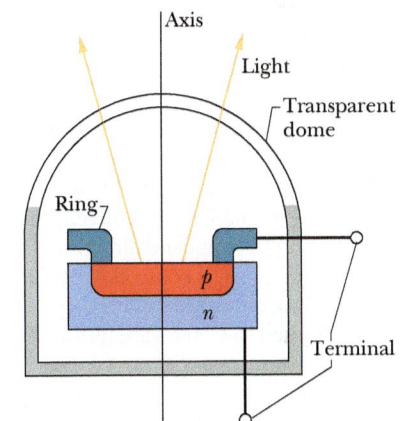

Figure 41-17 Cross section of an LED (the device has rotational symmetry about the central axis). The *p*-type material, which is thin enough to transmit light, is in the form of a circular disk. A connection is made to the *p*-type material through a circular metal ring that touches the disk at its periphery. The depletion zone between the *n*-type material and the *p*-type material is not shown.

Courtesy AT&T Archives and History Center, Warren, NJ

Figure 41-18 A junction laser developed at the AT&T Bell Laboratories. The cube at the right is a grain of salt.

Sample Problem 41.07 Light-emitting diode (LED)

An LED is constructed from a p-n junction based on a certain Ga-As-P semiconducting material whose energy gap is 1.9 eV. What is the wavelength of the emitted light?

Calculation: For jumps from the bottom of the conduction band to the top of the valence band, Eq. 41-11 tells us

$$\lambda = \frac{hc}{E_g} = \frac{(6.63 \times 10^{-34} \text{ J} \cdot \text{s})(3.00 \times 10^8 \text{ m/s})}{(1.9 \text{ eV})(1.60 \times 10^{-19} \text{ J/eV})}$$

$$= 6.5 \times 10^{-7} \text{ m} = 650 \text{ nm.} \qquad \text{(Answer)}$$

Light of this wavelength is red.

WILEY PLUS Additional examples, video, and practice available at *WileyPLUS*

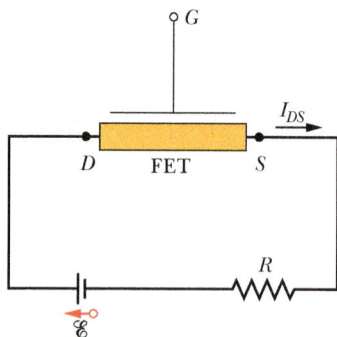

Figure 41-19 A circuit containing a generalized field-effect transistor through which electrons flow from the source terminal S to the drain terminal D. (The conventional current I_{DS} is in the opposite direction.) The magnitude of I_{DS} is controlled by the electric field set up within the FET by a potential applied to G, the gate terminal.

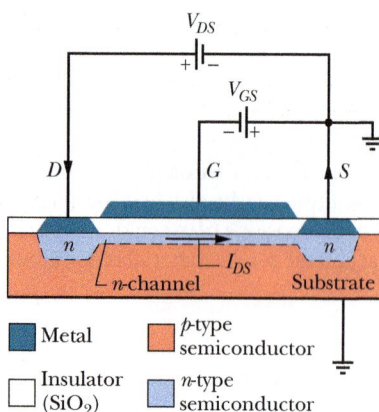

Figure 41-20 A particular type of field-effect transistor known as a MOSFET. The magnitude of the drain-to-source conventional current I_{DS} through the n channel is controlled by the potential difference V_{GS} applied between the source S and the gate G. A depletion zone that exists between the n-type material and the p-type substrate is not shown.

The Transistor

A **transistor** is a three-terminal semiconducting device that can be used to amplify input signals. Figure 41-19 shows a generalized **f**ield-**e**ffect **t**ransistor (FET); in it, the flow of electrons from terminal S (the *source*) leftward through the shaded region to terminal D (the *drain*) can be controlled by an electric field (hence field effect) set up within the device by a suitable electric potential applied to terminal G (the *gate*). Transistors are available in many types; we shall discuss only a particular FET called a MOSFET, or **m**etal-**o**xide-**s**emiconductor-**f**ield-**e**ffect **t**ransistor. The MOSFET has been described as the workhorse of the modern electronics industry.

For many applications the MOSFET is operated in only two states: with the drain-to-source current I_{DS} ON (gate open) or with it OFF (gate closed). The first of these can represent a 1 and the other a 0 in the binary arithmetic on which digital logic is based, and therefore MOSFETs can be used in digital logic circuits. Switching between the ON and OFF states can occur at high speed, so that binary logic data can be moved through MOSFET-based circuits very rapidly. MOSFETs about 500 nm in length—about the same as the wavelength of yellow light—are routinely fabricated for use in electronic devices of all kinds.

Figure 41-20 shows the basic structure of a MOSFET. A single crystal of silicon or other semiconductor is lightly doped to form p-type material that serves as the *substrate*. Embedded in this substrate, by heavily "overdoping" with n-type dopants, are two "islands" of n-type material, forming the drain D and the source S. The drain and source are connected by a thin channel of n-type material, called the **n channel**. A thin insulating layer of silicon dioxide (hence the O in MOSFET) is deposited on the crystal and penetrated by two metallic terminals (hence the M) at D and S, so that electrical contact can be made with the drain and the source. A thin metallic layer—the gate G—is deposited facing the n channel. Note that the gate makes no electrical contact with the transistor proper, being separated from it by the insulating oxide layer.

Consider first that the source and p-type substrate are grounded (at zero potential) and the gate is "floating"; that is, the gate is not connected to an external source of emf. Let a potential V_{DS} be applied between the drain and the source, such that the drain is positive. Electrons will then flow through the n channel from source to drain, and the conventional current I_{DS}, as shown in Fig. 41-20, will be from drain to source through the n channel.

Now let a potential V_{GS} be applied to the gate, making it negative with respect to the source. The negative gate sets up within the device an electric field (hence the "field effect") that tends to repel electrons from the n channel down into the substrate. This electron movement widens the (naturally occurring) depletion zone between the n channel and the substrate, at the expense of the n channel. The reduced width of the n channel, coupled with a reduction in the number of charge carriers in that channel, increases the resistance of that channel and thus decreases the current I_{DS}. With the proper value of V_{GS}, this current can be shut off completely; hence, by controlling V_{GS}, the MOSFET can be switched between its ON and OFF modes.

Charge carriers do not flow through the substrate because it (1) is lightly doped, (2) is not a good conductor, and (3) is separated from the n channel and the two n-type islands by an insulating depletion zone, not specifically shown in Fig. 41-20. Such a depletion zone always exists at a boundary between n-type material and p-type material, as Fig. 41-12b shows.

Computers and other electronic devices employ thousands (if not millions) of transistors and other electronic components, such as capacitors and resistors. These are not assembled as separate units but are crafted into a single semiconducting **chip,** forming an **integrated circuit** with millions of transistors and many other electronic components.

Review & Summary

Metals, Semiconductors, and Insulators   Three electrical properties that can be used to distinguish among crystalline solids are **resistivity** ρ, **temperature coefficient of resistivity** α, and **number density of charge carriers** n. Solids can be broadly divided into **insulators** (very high ρ), **metals** (low ρ, positive and low α, large n), and **semiconductors** (high ρ, negative and high α, small n).

Energy Levels and Gaps in a Crystalline Solid   An isolated atom can exist in only a discrete set of energy levels. As atoms come together to form a solid, the levels of the individual atoms merge to form the discrete energy **bands** of the solid. These energy bands are separated by energy **gaps,** each of which corresponds to a range of energies that no electron may possess.

Any energy band is made up of an enormous number of very closely spaced levels. The Pauli exclusion principle asserts that only one electron may occupy each of these levels.

Insulators   In an insulator, the highest band containing electrons is completely filled and is separated from the vacant band above it by an energy gap so large that electrons can essentially never become thermally agitated enough to jump across the gap.

Metals   In a metal, the highest band that contains any electrons is only partially filled. The energy of the highest filled level at a temperature of 0 K is called the **Fermi energy** E_F for the metal.

The electrons in the partially filled band are the **conduction electrons** and their number is

$$\begin{pmatrix} \text{number of conduction} \\ \text{electrons in sample} \end{pmatrix} = \begin{pmatrix} \text{number of atoms} \\ \text{in sample} \end{pmatrix}$$
$$\times \begin{pmatrix} \text{number of valence} \\ \text{electrons per atom} \end{pmatrix}. \quad (41\text{-}2)$$

The number of atoms in a sample is given by

$$\begin{pmatrix} \text{number of atoms} \\ \text{in sample} \end{pmatrix} = \frac{\text{sample mass } M_{\text{sam}}}{\text{atomic mass}}$$
$$= \frac{\text{sample mass } M_{\text{sam}}}{(\text{molar mass } M)/N_A}$$
$$= \frac{\begin{pmatrix} \text{material's} \\ \text{density} \end{pmatrix}\begin{pmatrix} \text{sample} \\ \text{volume } V \end{pmatrix}}{(\text{molar mass } M)/N_A}. \quad (41\text{-}4)$$

The number density n of the conduction electrons is

$$n = \frac{\text{number of conduction electrons in sample}}{\text{sample volume } V}. \quad (41\text{-}3)$$

The **density of states** function $N(E)$ is the number of available energy levels per unit volume of the sample and per unit energy interval and is given by

$$N(E) = \frac{8\sqrt{2}\pi m^{3/2}}{h^3}\, E^{1/2} \quad \text{(density of states, m}^{-3}\text{ J}^{-1}\text{),} \quad (41\text{-}5)$$

where m $(= 9.109 \times 10^{-31}$ kg) is the electron mass, h $(= 6.626 \times 10^{-34}$ J·s) is the Planck constant, and E is the energy in joules at which $N(E)$ is to be evaluated. To modify the equation so that the value of E is in eV and the value of $N(E)$ is in m^{-3} eV^{-1}, multiply the right side by $e^{3/2}$ (where $e = 1.602 \times 10^{-19}$ C).

The **occupancy probability** $P(E)$, the probability that a given available state will be occupied by an electron, is

$$P(E) = \frac{1}{e^{(E-E_F)/kT} + 1} \quad \text{(occupancy probability).} \quad (41\text{-}6)$$

The **density of occupied states** $N_o(E)$ is given by the product of the two quantities in Eqs. (41-5) and (41-6):

$$N_o(E) = N(E)\, P(E) \quad \text{(density of occupied states).} \quad (41\text{-}7)$$

The Fermi energy for a metal can be found by integrating $N_o(E)$ for $T = 0$ from $E = 0$ to $E = E_F$. The result is

$$E_F = \left(\frac{3}{16\sqrt{2}\pi}\right)^{2/3} \frac{h^2}{m}\, n^{2/3} = \frac{0.121 h^2}{m}\, n^{2/3}. \quad (41\text{-}9)$$

Semiconductors   The band structure of a semiconductor is like that of an insulator except that the gap width E_g is much smaller in the semiconductor. For silicon (a semiconductor) at room temperature, thermal agitation raises a few electrons to the **conduction band,** leaving an equal number of **holes** in the **valence band.** Both electrons and holes serve as charge carriers. The number of electrons in the conduction band of silicon can be increased greatly by doping with small amounts of phosphorus, thus forming **n-type material.** The number of holes in the valence band can be greatly increased by doping with aluminum, thus forming **p-type material.**

The p-n Junction A **p-n junction** is a single semiconducting crystal with one end doped to form *p*-type material and the other end doped to form *n*-type material, the two types meeting at a **junction plane.** At thermal equilibrium, the following occurs at that plane:

The **majority carriers** (electrons on the *n* side and holes on the *p* side) diffuse across the junction plane, producing a **diffusion current** I_{diff}.

The **minority carriers** (holes on the *n* side and electrons on the *p* side) are swept across the junction plane, forming a **drift current** I_{drift}. These two currents are equal in magnitude, making the net current zero.

A **depletion zone,** consisting largely of charged donor and acceptor ions, forms across the junction plane.

A **contact potential difference** V_0 develops across the depletion zone.

Applications of the p-n Junction When a potential difference is applied across a *p-n* junction, the device conducts electricity more readily for one polarity of the applied potential difference than for the other. Thus, a *p-n* junction can serve as a **junction rectifier.**

When a *p-n* junction is forward biased, it can emit light, hence can serve as a **light-emitting diode** (LED). The wavelength of the emitted light is given by

$$\lambda = \frac{c}{f} = \frac{hc}{E_g}. \tag{41-11}$$

A strongly forward-biased *p-n* junction with parallel end faces can operate as a **junction laser,** emitting light of a sharply defined wavelength.

Questions

1 On which of the following does the interval between adjacent energy levels in the highest occupied band of a metal depend: (a) the material of which the sample is made, (b) the size of the sample, (c) the position of the level in the band, (d) the temperature of the sample, (e) the Fermi energy of the metal?

2 Figure 41-1*a* shows 14 atoms that represent the unit cell of copper. However, because each of these atoms is shared with one or more adjoining unit cells, only a fraction of each atom belongs to the unit cell shown. What is the number of atoms per unit cell for copper? (To answer, count up the fractional atoms belonging to a single unit cell.)

3 Figure 41-1*b* shows 18 atoms that represent the unit cell of silicon. Fourteen of these atoms, however, are shared with one or more adjoining unit cells. What is the number of atoms per unit cell for silicon? (See Question 2.)

4 Figure 41-21 shows three labeled levels in a band and also the Fermi level for the material. The temperature is 0 K. Rank the three levels according to the probability of occupation, greatest first if the temperature is (a) 0 K and (b) 1000 K. (c) At the latter temperature, rank the levels according to the density of states $N(E)$ there, greatest first.

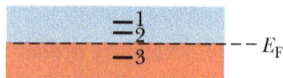

Figure 41-21 Question 4.

5 The occupancy probability at a certain energy E_1 in the valence band of a metal is 0.60 when the temperature is 300 K. Is E_1 above or below the Fermi energy?

6 An isolated atom of germanium has 32 electrons, arranged in subshells according to this scheme:

$$1s^2 \ 2s^2 \ 2p^6 \ 3s^2 \ 3p^6 \ 3d^{10} \ 4s^2 \ 4p^2.$$

This element has the same crystal structure as silicon and, like silicon, is a semiconductor. Which of these electrons form the valence band of crystalline germanium?

7 If the temperature of a piece of a metal is increased, does the probability of occupancy 0.1 eV above the Fermi level increase, decrease, or remain the same?

8 In the biased *p-n* junctions shown in Fig. 41-15, there is an electric field $\vec{E}$ in each of the two depletion zones, associated with the potential difference that exists across that zone. (a) Is the electric field vector directed from left to right in the figure or from right to left? (b) Is the magnitude of the field greater for forward bias or for back bias?

9 Consider a copper wire that is carrying, say, a few amperes of current. Is the drift speed v_d of the conduction electrons that form that current about equal to, much greater than, or much less than the Fermi speed v_F for copper (the speed associated with the Fermi energy for copper)?

10 In a silicon lattice, where should you look if you want to find (a) a conduction electron, (b) a valence electron, and (c) an electron associated with the 2*p* subshell of the isolated silicon atom?

11 The energy gaps E_g for the semiconductors silicon and germanium are, respectively, 1.12 and 0.67 eV. Which of the following statements, if any, are true? (a) Both substances have the same number density of charge carriers at room temperature. (b) At room temperature, germanium has a greater number density of charge carriers than silicon. (c) Both substances have a greater number density of conduction electrons than holes. (d) For each substance, the number density of electrons equals that of holes.

Problems

Module 41-1 The Electrical Properties of Metals

•**1** Show that Eq. 41-9 can be written as $E_F = An^{2/3}$, where the constant *A* has the value 3.65×10^{-19} m²·eV.

•**2** Calculate the density of states $N(E)$ for a metal at energy $E = 8.0$ eV and show that your result is consistent with the curve of Fig. 41-6.

•3 Copper, a monovalent metal, has molar mass 63.54 g/mol and density 8.96 g/cm^3. What is the number density n of conduction electrons in copper?

•4 A state 63 meV above the Fermi level has a probability of occupancy of 0.090. What is the probability of occupancy for a state 63 meV *below* the Fermi level?

•5 (a) Show that Eq. 41-5 can be written as $N(E) = CE^{1/2}$. (b) Evaluate C in terms of meters and electron-volts. (c) Calculate $N(E)$ for $E = 5.00$ eV.

•6 Use Eq. 41-9 to verify 7.0 eV as copper's Fermi energy.

•7 SSM What is the probability that a state 0.0620 eV above the Fermi energy will be occupied at (a) $T = 0$ K and (b) $T = 320$ K?

•8 What is the number density of conduction electrons in gold, which is a monovalent metal? Use the molar mass and density provided in Appendix F.

••9 SSM WWW Silver is a monovalent metal. Calculate (a) the number density of conduction electrons, (b) the Fermi energy, (c) the Fermi speed, and (d) the de Broglie wavelength corresponding to this electron speed. See Appendix F for the needed data on silver.

••10 Show that the probability $P(E)$ that an energy level having energy E is not occupied is

$$P(E) = \frac{1}{e^{-\Delta E/kT} + 1},$$

where $\Delta E = E - E_F$.

••11 Calculate $N_0(E)$, the density of occupied states, for copper at $T = 1000$ K for an energy E of (a) 4.00 eV, (b) 6.75 eV, (c) 7.00 eV, (d) 7.25 eV, and (e) 9.00 eV. Compare your results with the graph of Fig. 41-8b. The Fermi energy for copper is 7.00 eV.

••12 What is the probability that, at a temperature of $T = 300$ K, an electron will jump across the energy gap $E_g (= 5.5$ eV) in a diamond that has a mass equal to the mass of Earth? Use the molar mass of carbon in Appendix F; assume that in diamond there is one valence electron per carbon atom.

••13 GO The Fermi energy for copper is 7.00 eV. For copper at 1000 K, (a) find the energy of the energy level whose probability of being occupied by an electron is 0.900. For this energy, evaluate (b) the density of states $N(E)$ and (c) the density of occupied states $N_0(E)$.

••14 Assume that the total volume of a metal sample is the sum of the volume occupied by the metal ions making up the lattice and the (separate) volume occupied by the conduction electrons. The density and molar mass of sodium (a metal) are 971 kg/m^3 and 23.0 g/mol, respectively; assume the radius of the Na$^+$ ion is 98.0 pm. (a) What percent of the volume of a sample of metallic sodium is occupied by its conduction electrons? (b) Carry out the same calculation for copper, which has density, molar mass, and ionic radius of 8960 kg/m^3, 63.5 g/mol, and 135 pm, respectively. (c) For which of these metals do you think the conduction electrons behave more like a free-electron gas?

••15 SSM WWW In Eq. 41-6 let $E - E_F = \Delta E = 1.00$ eV. (a) At what temperature does the result of using this equation differ by 1.0% from the result of using the classical Boltzmann equation $P(E) = e^{-\Delta E/kT}$ (which is Eq. 41-1 with two changes in notation)? (b) At what temperature do the results from these two equations differ by 10%?

••16 Calculate the number density (number per unit volume) for

(a) molecules of oxygen gas at 0.0°C and 1.0 atm pressure and (b) conduction electrons in copper. (c) What is the ratio of the latter to the former? What is the average distance between (d) the oxygen molecules and (e) the conduction electrons, assuming this distance is the edge length of a cube with a volume equal to the available volume per particle (molecule or electron)?

••17 The Fermi energy of aluminum is 11.6 eV; its density and molar mass are 2.70 g/cm^3 and 27.0 g/mol, respectively. From these data, determine the number of conduction electrons per atom.

••18 GO A sample of a certain metal has a volume of 4.0×10^{-5} m^3. The metal has a density of 9.0 g/cm^3 and a molar mass of 60 g/mol. The atoms are bivalent. How many conduction electrons (or valence electrons) are in the sample?

••19 The Fermi energy for silver is 5.5 eV. At $T = 0$°C, what are the probabilities that states with the following energies are occupied: (a) 4.4 eV, (b) 5.4 eV, (c) 5.5 eV, (d) 5.6 eV, and (e) 6.4 eV? (f) At what temperature is the probability 0.16 that a state with energy $E = 5.6$ eV is occupied?

••20 GO What is the number of occupied states in the energy range of 0.0300 eV that is centered at a height of 6.10 eV in the valence band if the sample volume is 5.00×10^{-8} m^3, the Fermi level is 5.00 eV, and the temperature is 1500 K?

••21 At 1000 K, the fraction of the conduction electrons in a metal that have energies greater than the Fermi energy is equal to the area under the curve of Fig. 41-8b beyond E_F divided by the area under the entire curve. It is difficult to find these areas by direct integration. However, an approximation to this fraction at any temperature T is

$$frac = \frac{3kT}{2E_F}.$$

Note that $frac = 0$ for $T = 0$ K, just as we would expect. What is this fraction for copper at (a) 300 K and (b) 1000 K? For copper, $E_F = 7.0$ eV. (c) Check your answers by numerical integration using Eq. 41-7.

••22 At what temperature do 1.30% of the conduction electrons in lithium (a metal) have energies greater than the Fermi energy E_F, which is 4.70 eV? (See Problem 21.)

••23 Show that, at $T = 0$ K, the average energy E_{avg} of the conduction electrons in a metal is equal to $\frac{3}{5}E_F$. (*Hint:* By definition of average, $E_{avg} = (1/n) \int E N_0(E) dE$, where n is the number density of charge carriers.)

••24 GO A certain material has a molar mass of 20.0 g/mol, a Fermi energy of 5.00 eV, and 2 valence electrons per atom. What is the density (g/cm^3)?

••25 (a) Using the result of Problem 23 and 7.00 eV for copper's Fermi energy, determine how much energy would be released by the conduction electrons in a copper coin with mass 3.10 g if we could suddenly turn off the Pauli exclusion principle. (b) For how long would this amount of energy light a 100 W lamp? (*Note:* There is no way to turn off the Pauli principle!)

••26 At $T = 300$ K, how far above the Fermi energy is a state for which the probability of occupation by a conduction electron is 0.10?

••27 Zinc is a bivalent metal. Calculate (a) the number density of conduction electrons, (b) the Fermi energy, (c) the Fermi speed, and (d) the de Broglie wavelength corresponding to this electron speed. See Appendix F for the needed data on zinc.

••**28** GO What is the Fermi energy of gold (a monovalent metal with molar mass 197 g/mol and density 19.3 g/cm³)?

••**29** Use the result of Problem 23 to calculate the total translational kinetic energy of the conduction electrons in 1.00 cm³ of copper at $T = 0$ K.

•••**30** GO A certain metal has 1.70×10^{28} conduction electrons per cubic meter. A sample of that metal has a volume of 6.00×10^{-6} m³ and a temperature of 200 K. How many occupied states are in the energy range of 3.20×10^{-20} J that is centered on the energy 4.00×10^{-19} J? (*Caution:* Avoid round-off in the exponential.)

Module 41-2 Semiconductors and Doping

•**31** SSM (a) What maximum light wavelength will excite an electron in the valence band of diamond to the conduction band? The energy gap is 5.50 eV. (b) In what part of the electromagnetic spectrum does this wavelength lie?

••**32** The compound gallium arsenide is a commonly used semiconductor, having an energy gap E_g of 1.43 eV. Its crystal structure is like that of silicon, except that half the silicon atoms are replaced by gallium atoms and half by arsenic atoms. Draw a flattened-out sketch of the gallium arsenide lattice, following the pattern of Fig. 41-10a. What is the net charge of the (a) gallium and (b) arsenic ion core? (c) How many electrons per bond are there? (*Hint:* Consult the periodic table in Appendix G.)

••**33** The occupancy probability function (Eq. 41-6) can be applied to semiconductors as well as to metals. In semiconductors the Fermi energy is close to the midpoint of the gap between the valence band and the conduction band. For germanium, the gap width is 0.67 eV. What is the probability that (a) a state at the bottom of the conduction band is occupied and (b) a state at the top of the valence band is not occupied? Assume that $T = 290$ K. (*Note:* In a pure semiconductor, the Fermi energy lies symmetrically between the population of conduction electrons and the population of holes and thus is at the center of the gap. There need not be an available state at the location of the Fermi energy.)

••**34** In a simplified model of an undoped semiconductor, the actual distribution of energy states may be replaced by one in which there are N_v states in the valence band, all these states having the same energy E_v, and N_c states in the conduction band, all these states having the same energy E_c. The number of electrons in the conduction band equals the number of holes in the valence band. (a) Show that this last condition implies that

$$\frac{N_c}{\exp(\Delta E_c/kT) + 1} = \frac{N_v}{\exp(\Delta E_v/kT) + 1},$$

in which

$$\Delta E_c = E_c - E_F \quad \text{and} \quad \Delta E_v = -(E_v - E_F).$$

(b) If the Fermi level is in the gap between the two bands and its distance from each band is large relative to kT, then the exponentials dominate in the denominators. Under these conditions, show that

$$E_F = \frac{(E_c + E_v)}{2} + \frac{kT \ln(N_v/N_c)}{2}$$

and that, if $N_v \approx N_c$, the Fermi level for the undoped semiconductor is close to the gap's center.

••**35** SSM WWW What mass of phosphorus is needed to dope 1.0 g of silicon so that the number density of conduction electrons in the silicon is increased by a multiply factor of 10^6 from the 10^{16} m⁻³ in pure silicon.

••**36** GO A silicon sample is doped with atoms having donor states 0.110 eV below the bottom of the conduction band. (The energy gap in silicon is 1.11 eV.) If each of these donor states is occupied with a probability of 5.00×10^{-5} at $T = 300$ K, (a) is the Fermi level above or below the top of the silicon valence band and (b) how far above or below? (c) What then is the probability that a state at the bottom of the silicon conduction band is occupied?

••**37** GO Doping changes the Fermi energy of a semiconductor. Consider silicon, with a gap of 1.11 eV between the top of the valence band and the bottom of the conduction band. At 300 K the Fermi level of the pure material is nearly at the mid-point of the gap. Suppose that silicon is doped with donor atoms, each of which has a state 0.15 eV below the bottom of the silicon conduction band, and suppose further that doping raises the Fermi level to 0.11 eV below the bottom of that band (Fig. 41-22). For (a) pure and (b) doped silicon, calculate the probability that a state at the bottom of the silicon conduction band is occupied. (c) Calculate the probability that a state in the doped material (at the donor level) is occupied.

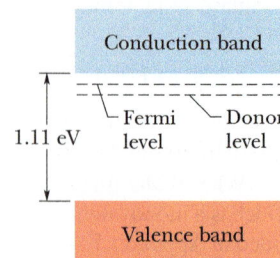

Figure 41-22 Problem 37.

••**38** Pure silicon at room temperature has an electron number density in the conduction band of about 5×10^{15} m⁻³ and an equal density of holes in the valence band. Suppose that one of every 10^7 silicon atoms is replaced by a phosphorus atom. (a) Which type will the doped semiconductor be, n or p? (b) What charge carrier number density will the phosphorus add? (c) What is the ratio of the charge carrier number density (electrons in the conduction band and holes in the valence band) in the doped silicon to that in pure silicon?

Module 41-3 The *p-n* Junction and the Transistor

•**39** SSM When a photon enters the depletion zone of a *p-n* junction, the photon can scatter from the valence electrons there, transferring part of its energy to each electron, which then jumps to the conduction band. Thus, the photon creates electron–hole pairs. For this reason, the junctions are often used as light detectors, especially in the x-ray and gamma-ray regions of the electromagnetic spectrum. Suppose a single 662 keV gamma-ray photon transfers its energy to electrons in multiple scattering events inside a semiconductor with an energy gap of 1.1 eV, until all the energy is transferred. Assuming that each electron jumps the gap from the top of the valence band to the bottom of the conduction band, find the number of electron–hole pairs created by the process.

•**40** For an ideal *p-n* junction rectifier with a sharp boundary between its two semiconducting sides, the current I is related to the potential difference V across the rectifier by

$$I = I_0(e^{eV/kT} - 1),$$

where I_0, which depends on the materials but not on I or V, is called the *reverse saturation current*. The potential difference V is positive if the rectifier is forward-biased and negative if it is back-biased. (a) Verify that this expression predicts the behavior of a junction rectifier by graphing I versus V from -0.12 V to $+0.12$ V. Take $T = 300$ K and $I_0 = 5.0$ nA. (b) For the same temperature, calculate the ratio of the current for a 0.50 V forward bias to the current for a 0.50 V back bias.

•41 In a particular crystal, the highest occupied band is full. The crystal is transparent to light of wavelengths longer than 295 nm but opaque at shorter wavelengths. Calculate, in electron-volts, the gap between the highest occupied band and the next higher (empty) band for this material.

•42 A potassium chloride crystal has an energy band gap of 7.6 eV above the topmost occupied band, which is full. Is this crystal opaque or transparent to light of wavelength 140 nm?

•43 A certain computer chip that is about the size of a postage stamp (2.54 cm × 2.22 cm) contains about 3.5 million transistors. If the transistors are square, what must be their *maximum* dimension? (*Note:* Devices other than transistors are also on the chip, and there must be room for the interconnections among the circuit elements. Transistors smaller than 0.7 μm are now commonly and inexpensively fabricated.)

•44 A silicon-based MOSFET has a square gate 0.50 μm on edge. The insulating silicon oxide layer that separates the gate from the p-type substrate is 0.20 μm thick and has a dielectric constant of 4.5. (a) What is the equivalent gate–substrate capacitance (treating the gate as one plate and the substrate as the other plate)? (b) Approximately how many elementary charges e appear in the gate when there is a gate–source potential difference of 1.0 V?

Additional Problems

45 SSM (a) Show that the slope dP/dE of Eq. 41-6 evaluated at $E = E_F$ is $-1/4kT$. (b) Show that the tangent line to the curve of Fig. 41-7b evaluated at $E = E_F$ intercepts the horizontal axis at $E = E_F + 2kT$.

46 Calculate $d\rho/dT$ at room temperature for (a) copper and (b) silicon, using data from Table 41-1.

47 (a) Find the angle θ between adjacent nearest-neighbor bonds in the silicon lattice. Recall that each silicon atom is bonded to four of its nearest neighbors. The four neighbors form a regular tetrahedron—a pyramid whose sides and base are equilateral triangles. (b) Find the bond length, given that the atoms at the corners of the tetrahedron are 388 pm apart.

48 Show that $P(E)$, the occupancy probability in Eq. 41-6, is symmetrical about the value of the Fermi energy; that is, show that

$$P(E_F + \Delta E) + P(E_F - \Delta E) = 1.$$

49 (a) Show that the density of states at the Fermi energy is given by

$$N(E_F) = \frac{(4)(3^{1/3})(\pi^{2/3})mn^{1/3}}{h^2}$$
$$= (4.11 \times 10^{18}\,\text{m}^{-2}\,\text{eV}^{-1})n^{1/3},$$

in which n is the number density of conduction electrons. (b) Calculate $N(E_F)$ for copper, which is a monovalent metal with molar mass 63.54 g/mol and density 8.96 g/cm³. (c) Verify your calculation with the curve of Fig. 41-6, recalling that $E_F = 7.0$ eV for copper.

50 Silver melts at 961°C. At the melting point, what fraction of the conduction electrons are in states with energies greater than the Fermi energy of 5.5 eV? (See Problem 21.)

51 The Fermi energy of copper is 7.0 eV. Verify that the corresponding Fermi speed is 1600 km/s.

52 Verify the numerical factor 0.121 in Eq. 41-9.

53 At what pressure, in atmospheres, would the number of molecules per unit volume in an ideal gas be equal to the number density of the conduction electrons in copper, with both gas and copper at temperature $T = 300$ K?

Nuclear Physics

42-1 DISCOVERING THE NUCLEUS

Learning Objectives

After reading this module, you should be able to . . .

42.01 Explain the general arrangement for Rutherford scattering and what was learned from it.

42.02 In a Rutherford scattering arrangement, apply the relationship between the projectile's initial kinetic energy and the distance of its closest approach to the target nucleus.

Key Ideas

● The positive charge of an atom is concentrated in the central nucleus rather than being spread through the volume of the atom. This structure was proposed in 1910 by Ernest Rutherford of England after he conducted experiments with what we now call Rutherford scattering. Alpha particles (positively charged particles consisting of two protons and two neutrons) are directed through a thin metal foil to be scattered by the (positive) nuclei within the atoms.

● The total energy (kinetic energy plus electric potential energy) of the system of alpha particle and target nucleus is conserved as the alpha particle approaches the nucleus.

What Is Physics?

We now turn to what lies at the center of an atom—the nucleus. For the last 90 years, a principal goal of physics has been to work out the quantum physics of nuclei, and, for almost as long, a principal goal of some types of engineering has been to apply that quantum physics with applications ranging from radiation therapy in the war on cancer to detectors of radon gas in basements.

Before we get to such applications and the quantum physics of nuclei, let's first discuss how physicists discovered that an atom has a nucleus. As obvious as that fact is today, it initially came as an incredible surprise.

Discovering the Nucleus

In the first years of the 20th century, not much was known about the structure of atoms beyond the fact that they contain electrons. The electron had been discovered (by J. J. Thomson) in 1897, and its mass was unknown in those early days. Thus, it was not possible even to say how many negatively charged electrons a given atom contained. Scientists reasoned that because atoms were electrically neutral, they must also contain some positive charge, but nobody knew what form this compensating positive charge took. One popular model was that the positive and negative charges were spread uniformly in a sphere.

In 1911 Ernest Rutherford proposed that the positive charge of the atom is densely concentrated at the center of the atom, forming its **nucleus,** and that, furthermore, the nucleus is responsible for most of the mass of the atom. Rutherford's proposal was no mere conjecture but was based firmly on the results of an experiment suggested by him and carried out by his collaborators, Hans Geiger (of Geiger counter fame) and Ernest Marsden, a 20-year-old student who had not yet earned his bachelor's degree.

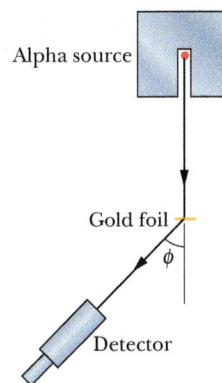

Figure 42-1 An arrangement (top view) used in Rutherford's laboratory in 1911–1913 to study the scattering of α particles by thin metal foils. The detector can be rotated to various values of the scattering angle ϕ. The alpha source was radon gas, a decay product of radium. With this simple "tabletop" apparatus, the atomic nucleus was discovered.

In Rutherford's day it was known that certain elements, called **radioactive,** transform into other elements spontaneously, emitting particles in the process. One such element is radon, which emits alpha (α) particles that have an energy of about 5.5 MeV. We now know that these particles are helium nuclei.

Rutherford's idea was to direct energetic alpha particles at a thin target foil and measure the extent to which they were deflected as they passed through the foil. Alpha particles, which are about 7300 times more massive than electrons, have a charge of $+2e$.

Figure 42-1 shows the experimental arrangement of Geiger and Marsden. Their alpha source was a thin-walled glass tube of radon gas. The experiment involves counting the number of alpha particles that are deflected through various scattering angles ϕ.

Figure 42-2 shows their results. Note especially that the vertical scale is logarithmic. We see that most of the particles are scattered through rather small angles, but—and this was the big surprise—a very small fraction of them are scattered through very large angles, approaching 180°. In Rutherford's words: "It was quite the most incredible event that ever happened to me in my life. It was almost as incredible as if you had fired a 15-inch shell at a piece of tissue paper and it [the shell] came back and hit you."

Why was Rutherford so surprised? At the time of these experiments, most physicists believed in the so-called plum pudding model of the atom, which had been advanced by J. J. Thomson. In this view the positive charge of the atom was thought to be spread out through the entire volume of the atom. The electrons (the "plums") were thought to vibrate about fixed points within this sphere of positive charge (the "pudding").

The maximum deflecting force that could act on an alpha particle as it passed through such a large positive sphere of charge would be far too small to deflect the alpha particle by even as much as 1°. (The expected deflection has been compared to what you would observe if you fired a bullet through a sack of snowballs.) The electrons in the atom would also have very little effect on the massive, energetic alpha particle. They would, in fact, be themselves strongly deflected, much as a swarm of gnats would be brushed aside by a stone thrown through them.

Rutherford saw that, to deflect the alpha particle backward, there must be a large force; this force could be provided if the positive charge, instead of being spread throughout the atom, were concentrated tightly at its center. Then the incoming alpha particle could get very close to the positive charge without penetrating it; such a close encounter would result in a large deflecting force.

Figure 42-3 shows possible paths taken by typical alpha particles as they pass through the atoms of the target foil. As we see, most are either undeflected or only slightly deflected, but a few (those whose incoming paths pass, by chance, very close to a nucleus) are deflected through large angles. From an analysis of the data, Rutherford concluded that the radius of the nucleus must be smaller than the radius of an atom by a factor of about 10^4. In other words, the atom is mostly empty space.

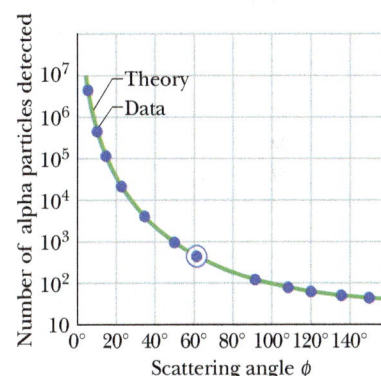

Figure 42-2 The dots are alpha-particle scattering data for a gold foil, obtained by Geiger and Marsden using the apparatus of Fig. 42-1. The solid curve is the theoretical prediction, based on the assumption that the atom has a small, massive, positively charged nucleus. The data have been adjusted to fit the theoretical curve at the experimental point that is enclosed in a circle.

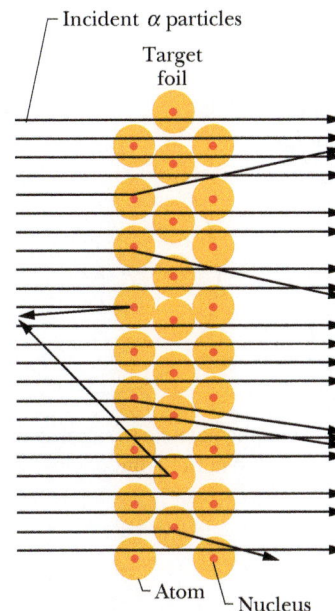

Figure 42-3 The angle through which an incident alpha particle is scattered depends on how close the particle's path lies to an atomic nucleus. Large deflections result only from very close encounters.

Sample Problem 42.01 Rutherford scattering of an alpha particle by a gold nucleus

An alpha particle with kinetic energy K_i = 5.30 MeV happens, by chance, to be headed directly toward the nucleus of a neutral gold atom (Fig. 42-4a). What is its *distance of closest approach d* (least center-to-center separation) to the nucleus? Assume that the atom remains stationary.

KEY IDEAS

(1) Throughout the motion, the total mechanical energy E of the particle–atom system is conserved. (2) In addition to the kinetic energy, that total energy includes electric potential energy U as given by Eq. 24-46 ($U = q_1q_2/4\pi\varepsilon_0 r$).

Calculations: The alpha particle has a charge of +2e because it contains two protons. The target nucleus has a charge of q_{Au} = +79e because it contains 79 protons. However, that nuclear charge is surrounded by an electron "cloud" with a charge of q_e = −79e, and thus the alpha particle initially "sees" a neutral atom with a net charge of q_{atom} = 0. The electric force on the particle is zero and the initial electric potential energy of the particle–atom system is $U_i = 0$.

Once the alpha particle enters the atom, we say that it passes through the electron cloud surrounding the nucleus.

That cloud then acts as a closed conducting spherical shell and, by Gauss' law, has no effect on the (now internal) charged alpha particle. Then the alpha particle "sees" only the nuclear charge q_{Au}. Because q_α and q_{Au} are both positively charged, a repulsive electric force acts on the alpha particle, slowing it, and the particle–atom system has a potential energy

$$U = \frac{1}{4\pi\varepsilon_0}\frac{q_\alpha q_{Au}}{r}$$

that depends on the center-to-center separation r of the incoming particle and the target nucleus (Fig. 42-4b).

As the repulsive force slows the alpha particle, energy is transferred from kinetic energy to electric potential energy. The transfer is complete when the alpha particle momentarily stops at the distance of closest approach d to the target nucleus (Fig. 42-4c). Just then the kinetic energy is $K_f = 0$ and the particle–atom system has the electric potential energy

$$U_f = \frac{1}{4\pi\varepsilon_0}\frac{q_\alpha q_{Au}}{d}.$$

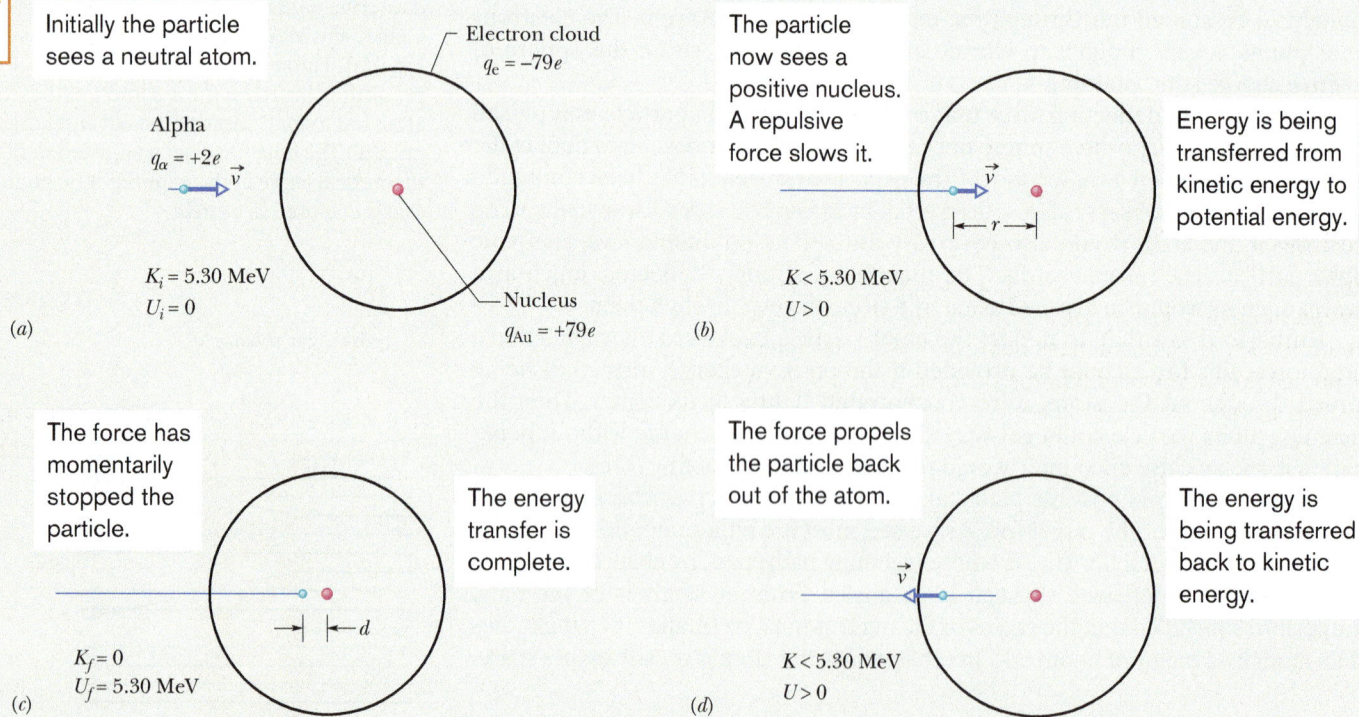

Initially the particle sees a neutral atom.

Electron cloud q_e = −79e

Alpha q_α = +2e

K_i = 5.30 MeV
$U_i = 0$
(a)

Nucleus q_{Au} = +79e

The particle now sees a positive nucleus. A repulsive force slows it.

$K < 5.30$ MeV
$U > 0$
(b)

Energy is being transferred from kinetic energy to potential energy.

The force has momentarily stopped the particle.

$K_f = 0$
U_f = 5.30 MeV
(c)

The energy transfer is complete.

The force propels the particle back out of the atom.

$K < 5.30$ MeV
$U > 0$
(d)

The energy is being transferred back to kinetic energy.

Figure 42-4 An alpha particle (a) approaches and (b) then enters a gold atom, headed toward the nucleus. The alpha particle (c) comes to a stop at the point of closest approach and (d) is propelled back out of the atom.

To find d, we conserve the total mechanical energy between the initial state i and this later state f, writing

$$K_i + U_i = K_f + U_f$$

and

$$K_i + 0 = 0 + \frac{1}{4\pi\varepsilon_0}\frac{q_\alpha q_{Au}}{d}.$$

(We are assuming that the alpha particle is not affected by the force holding the nucleus together, which acts over only a short distance.) Solving for d and then substituting for the charges and initial kinetic energy lead to

$$d = \frac{(2e)(79e)}{4\pi\varepsilon_0 K_\alpha}$$

$$= \frac{(2 \times 79)(1.60 \times 10^{-19}\,\text{C})^2}{4\pi\varepsilon_0(5.30\,\text{MeV})(1.60 \times 10^{-13}\,\text{J/MeV})}$$

$$= 4.29 \times 10^{-14}\,\text{m}. \qquad \text{(Answer)}$$

This distance is considerably larger than the sum of the radii of the gold nucleus and the alpha particle. Thus, this alpha particle reverses its motion (Fig. 42-4d) without ever actually "touching" the gold nucleus.

WILEY PLUS Additional examples, video, and practice available at *WileyPLUS*

42-2 SOME NUCLEAR PROPERTIES

Learning Objectives

After reading this module, you should be able to . . .

42.03 Identify nuclides, atomic number (or proton number), neutron number, mass number, nucleon, isotope, disintegration, neutron excess, isobar, zone of stable nuclei, and island of stability, and explain the symbols used for nuclei (such as ^{197}Au).

42.04 Sketch a graph of proton number versus neutron number and identify the approximate location of the stable nuclei, the proton-rich nuclei, and the neutron-rich nuclei.

42.05 For spherical nuclei, apply the relationship between radius and mass number and calculate the nuclear density.

42.06 Work with masses in atomic mass units, relate the mass number and the approximate nuclear mass, and convert between mass units and energy.

42.07 Calculate mass excess.

42.08 For a given nucleus, calculate the binding energy ΔE_{be} and the binding energy per nucleon ΔE_{ben}, and explain the meaning of each term.

42.09 Sketch a graph of the binding energy per nucleon versus mass number, indicating the nuclei that are the most tightly bound, those that can undergo fission with a release of energy, and those that can undergo fusion with a release of energy.

42.10 Identify the force that holds nucleons together.

Key Ideas

● Different types of nuclei are called nuclides. Each is characterized by an atomic number Z (the number of protons), a neutron number N, and a mass number A (the total number of nucleons—protons and neutrons). Thus, $A = Z + N$. A nuclide is represented with a symbol such as ^{197}Au or $^{197}_{79}\text{Au}$, where the chemical symbol carries a superscript with the value of A and (possibly) a subscript with the value of Z.

● Nuclides with the same atomic number but different neutron numbers are isotopes of one another.

● Nuclei have a mean radius r given by

$$r = r_0 A^{1/3}$$

where $r_0 \approx 1.2$ fm.

● Atomic masses are often reported in terms of mass excess

$$\Delta = M - A,$$

where M is the actual mass of an atom in atomic mass units and A is the mass number for that atom's nucleus.

● The binding energy of a nucleus is the difference

$$\Delta E_{be} = \Sigma(mc^2) - Mc^2,$$

where $\Sigma(mc^2)$ is the total mass energy of the individual protons and neutrons. The binding energy of a nucleus is the amount of energy needed to break the nucleus into its constituent parts (and is *not* an energy that resides in the nucleus).

● The binding energy per nucleon is

$$\Delta E_{ben} = \frac{\Delta E_{be}}{A}.$$

● The energy equivalent of one mass unit (u) is $931.494\,013$ MeV.

● A plot of the binding energy per nucleon ΔE_{ben} versus mass number A shows that middle-mass nuclides are the most stable and that energy can be released both by fission of high-mass nuclei and by fusion of low-mass nuclei.

Some Nuclear Properties

Table 42-1 shows some properties of a few atomic nuclei. When we are interested primarily in their properties as specific nuclear species (rather than as parts of atoms), we call these particles **nuclides.**

Some Nuclear Terminology

Nuclei are made up of protons and neutrons. The number of protons in a nucleus (called the **atomic number** or **proton number** of the nucleus) is represented by the symbol Z; the number of neutrons (the **neutron number**) is represented by the symbol N. The total number of neutrons and protons in a nucleus is called its **mass number** A; thus

$$A = Z + N. \tag{42-1}$$

Neutrons and protons, when considered collectively as members of a nucleus, are called **nucleons.**

We represent nuclides with symbols such as those displayed in the first column of Table 42-1. Consider ^{197}Au, for example. The superscript 197 is the mass number A. The chemical symbol Au tells us that this element is gold, whose atomic number is 79. Sometimes the atomic number is explicitly shown as a subscript, as in $^{197}_{79}$Au. From Eq. 42-1, the neutron number of this nuclide is the difference between the mass number and the atomic number, namely, $197 - 79$, or 118.

Nuclides with the same atomic number Z but different neutron numbers N are called **isotopes** of one another. The element gold has 36 isotopes, ranging from ^{173}Au to ^{204}Au. Only one of them (^{197}Au) is stable; the remaining 35 are radioactive. Such **radionuclides** undergo **decay** (or **disintegration**) by emitting a particle and thereby transforming to a different nuclide.

Organizing the Nuclides

The neutral atoms of all isotopes of an element (all with the same Z) have the same number of electrons and the same chemical properties, and they fit into the same box in the periodic table of the elements. The *nuclear* properties of the isotopes of a given element, however, are very different from one isotope to another. Thus, the periodic table is of limited use to the nuclear physicist, the nuclear chemist, or the nuclear engineer.

Table 42-1 **Some Properties of Selected Nuclides**

Nuclide	Z	N	A	Stability[a]	Mass[b] (u)	Spin[c]	Binding Energy (MeV/nucleon)
^{1}H	1	0	1	99.985%	1.007 825	$\frac{1}{2}$	—
^{7}Li	3	4	7	92.5%	7.016 004	$\frac{3}{2}$	5.60
^{31}P	15	16	31	100%	30.973 762	$\frac{1}{2}$	8.48
^{84}Kr	36	48	84	57.0%	83.911 507	0	8.72
^{120}Sn	50	70	120	32.4%	119.902 197	0	8.51
^{157}Gd	64	93	157	15.7%	156.923 957	$\frac{3}{2}$	8.21
^{197}Au	79	118	197	100%	196.966 552	$\frac{3}{2}$	7.91
^{227}Ac	89	138	227	21.8 y	227.027 747	$\frac{3}{2}$	7.65
^{239}Pu	94	145	239	24 100 y	239.052 157	$\frac{1}{2}$	7.56

[a]For stable nuclides, the **isotopic abundance** is given; this is the fraction of atoms of this type found in a typical sample of the element. For radioactive nuclides, the half-life is given.

[b]Following standard practice, the reported mass is that of the neutral atom, not that of the bare nucleus.

[c]Spin angular momentum in units of $\hbar$.

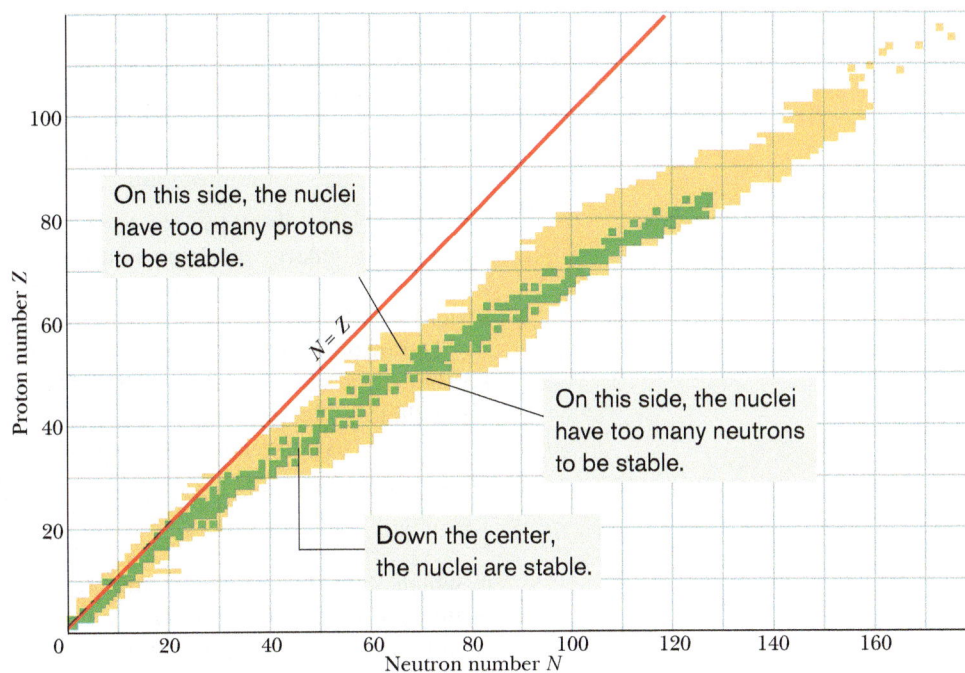

Figure 42-5 A plot of the known nuclides. The green shading identifies the band of stable nuclides, the beige shading the radionuclides. Low-mass, stable nuclides have essentially equal numbers of neutrons and protons, but more massive nuclides have an increasing excess of neutrons. The figure shows that there are no stable nuclides with $Z > 83$ (bismuth).

We organize the nuclides on a **nuclidic chart** like that in Fig. 42-5, in which a nuclide is represented by plotting its proton number against its neutron number. The stable nuclides in this figure are represented by the green, the radionuclides by the beige. As you can see, the radionuclides tend to lie on either side of—and at the upper end of—a well-defined band of stable nuclides. Note too that light stable nuclides tend to lie close to the line $N = Z$, which means that they have about the same numbers of neutrons and protons. Heavier nuclides, however, tend to have many more neutrons than protons. As an example, we saw that ^{197}Au has 118 neutrons and only 79 protons, a *neutron excess* of 39.

Nuclidic charts are available as wall charts, in which each small box on the chart is filled with data about the nuclide it represents. Figure 42-6 shows a section of such a chart, centered on ^{197}Au. Relative abundances (usually, as found on Earth) are shown for stable nuclides, and half-lives (a measure of decay rate) are shown for radionuclides. The sloping line points out a line of **isobars**—nuclides of the same mass number, $A = 198$ in this case.

In recent years, nuclides with atomic numbers as high as $Z = 118$ ($A = 294$) have been found in laboratory experiments (no elements with Z greater than 92 occur naturally). Although large nuclides generally should be highly unstable and last only a very brief time, certain supermassive nuclides are relatively stable, with fairly long lifetimes. These stable supermassive nuclides and other predicted ones form an *island of stability* at high values of Z and N on a nuclidic chart like Fig. 42-5.

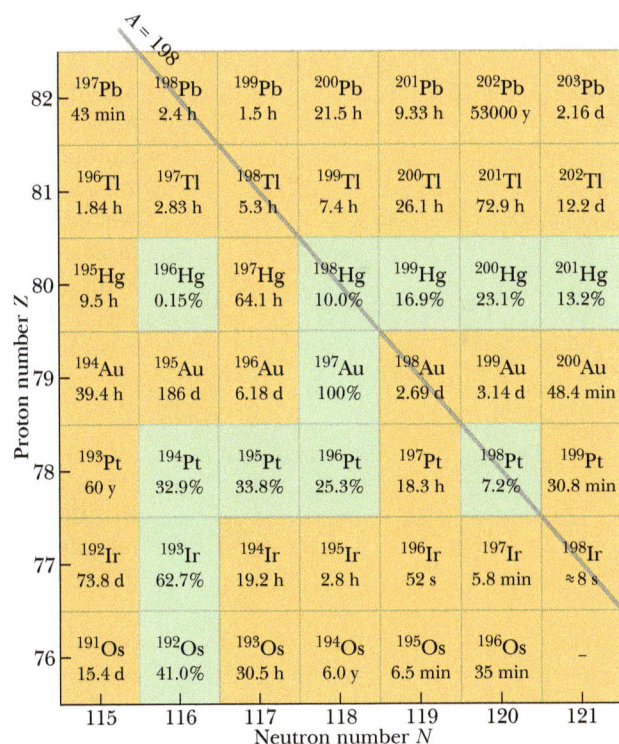

Figure 42-6 An enlarged and detailed section of the nuclidic chart of Fig. 42-5, centered on ^{197}Au. Green squares represent stable nuclides, for which relative isotopic abundances are given. Beige squares represent radionuclides, for which half-lives are given. Isobaric lines of constant mass number A slope as shown by the example line for $A = 198$.

✓ **Checkpoint 1**

Based on Fig. 42-5, which of the following nuclides do you conclude are not likely to be detected: ^{52}Fe ($Z = 26$), ^{90}As ($Z = 33$), ^{158}Nd ($Z = 60$), ^{175}Lu ($Z = 71$), ^{208}Pb ($Z = 82$)?

Nuclear Radii

A convenient unit for measuring distances on the scale of nuclei is the *femtometer* ($= 10^{-15}$ m). This unit is often called the *fermi;* the two names share the same abbreviation. Thus,

$$1 \text{ femtometer} = 1 \text{ fermi} = 1 \text{ fm} = 10^{-15} \text{ m}. \qquad (42\text{-}2)$$

We can learn about the size and structure of nuclei by bombarding them with a beam of high-energy electrons and observing how the nuclei deflect the incident electrons. The electrons must be energetic enough (at least 200 MeV) to have de Broglie wavelengths that are smaller than the nuclear structures they are to probe.

The nucleus, like the atom, is not a solid object with a well-defined surface. Furthermore, although most nuclides are spherical, some are notably ellipsoidal. Nevertheless, electron-scattering experiments (as well as experiments of other kinds) allow us to assign to each nuclide an effective radius given by

$$r = r_0 A^{1/3}, \qquad (42\text{-}3)$$

in which A is the mass number and $r_0 \approx 1.2$ fm. We see that the volume of a nucleus, which is proportional to r^3, is directly proportional to the mass number A and is independent of the separate values of Z and N. That is, we can treat most nuclei as being a sphere with a volume that depends on the number of nucleons, regardless of their type.

Equation 42-3 does not apply to *halo nuclides,* which are neutron-rich nuclides that were first produced in laboratories in the 1980s. These nuclides are larger than predicted by Eq. 42-3, because some of the neutrons form a *halo* around a spherical core of the protons and the rest of the neutrons. Lithium isotopes give an example. When a neutron is added to ^{8}Li to form ^{9}Li, neither of which are halo nuclides, the effective radius increases by about 4%. However, when two neutrons are added to ^{9}Li to form the neutron-rich isotope ^{11}Li (the largest of the lithium isotopes), they do not join that existing nucleus but instead form a halo around it, increasing the effective radius by about 30%. Apparently this halo configuration involves less energy than a core containing all 11 nucleons. (In this chapter we shall generally assume that Eq. 42-3 applies.)

Atomic Masses

Atomic masses are now measured to great precision, but usually nuclear masses are not directly measurable because stripping off all the electrons from an atom is difficult. As we briefly discussed in Module 37-6, atomic masses are often reported in *atomic mass units,* a system in which the atomic mass of neutral ^{12}C is defined to be exactly 12 u.

Precise atomic masses are available in tables on the web and are usually provided in homework problems. However, sometimes we need only an approximation of the mass of either a nucleus alone or a neutral atom. The mass number A of a nuclide gives such an approximate mass in atomic mass units. For example, the approximate mass of both the nucleus and the neutral atom for ^{197}Au is 197 u, which is close to the actual atomic mass of 196.966 552 u.

As we saw in Module 37-6,

$$1 \text{ u} = 1.660\,538\,86 \times 10^{-27} \text{ kg}. \qquad (42\text{-}4)$$

We also saw that if the total mass of the participants in a nuclear reaction changes by an amount Δm, there is an energy release or absorption given by Eq. 37-50 ($Q = -\Delta m\, c^2$). As we shall now see, nuclear energies are often reported in multiples of 1 MeV. Thus, a convenient conversion between mass units and energy units is provided by Eq. 37-46:

$$c^2 = 931.494\,013 \text{ MeV/u}. \qquad (42\text{-}5)$$

Scientists and engineers working with atomic masses often prefer to report the mass of an atom by means of the atom's *mass excess* Δ, defined as

$$\Delta = M - A \quad \text{(mass excess)}, \tag{42-6}$$

where M is the actual mass of the atom in atomic mass units and A is the mass number for that atom's nucleus.

Nuclear Binding Energies

The mass M of a nucleus is *less* than the total mass Σm of its individual protons and neutrons. That means that the mass energy Mc^2 of a nucleus is *less* than the total mass energy $\Sigma(mc^2)$ of its individual protons and neutrons. The difference between these two energies is called the **binding energy** of the nucleus:

$$\Delta E_{be} = \Sigma(mc^2) - Mc^2 \quad \text{(binding energy)}. \tag{42-7}$$

Caution: Binding energy is not an energy that resides in the nucleus. Rather, it is a *difference* in mass energy between a nucleus and its individual nucleons: If we were able to separate a nucleus into its nucleons, we would have to transfer a total energy equal to ΔE_{be} to those particles during the separating process. Although we cannot actually tear apart a nucleus in this way, the nuclear binding energy is still a convenient measure of how well a nucleus is held together, in the sense that it measures how difficult the nucleus would be to take apart.

A better measure is the **binding energy per nucleon** ΔE_{ben}, which is the ratio of the binding energy ΔE_{be} of a nucleus to the number A of nucleons in that nucleus:

$$\Delta E_{ben} = \frac{\Delta E_{be}}{A} \quad \text{(binding energy per nucleon)}. \tag{42-8}$$

We can think of the binding energy per nucleon as the average energy needed to separate a nucleus into its individual nucleons. *A greater binding energy per nucleon means a more tightly bound nucleus.*

Figure 42-7 is a plot of the binding energy per nucleon ΔE_{ben} versus mass number A for a large number of nuclei. Those high on the plot are very tightly

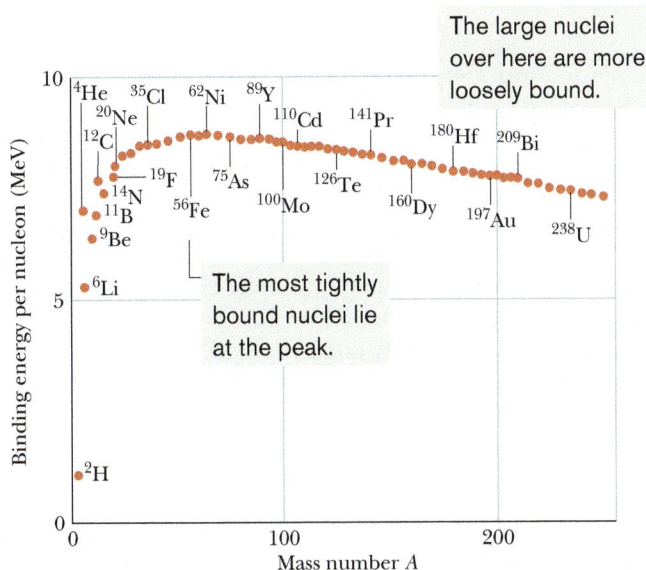

Figure 42-7 The binding energy per nucleon for some representative nuclides. The nickel nuclide ^{62}Ni has the highest binding energy per nucleon (about 8.794 60 MeV/nucleon) of any known stable nuclide. Note that the alpha particle (^{4}He) has a higher binding energy per nucleon than its neighbors in the periodic table and thus is also particularly stable.

bound; that is, we would have to supply a great amount of energy per nucleon to break apart one of those nuclei. The nuclei that are lower on the plot, at the left and right sides, are less tightly bound, and less energy per nucleon would be required to break them apart.

These simple statements about Fig. 42-7 have profound consequences. The nucleons in a nucleus on the right side of the plot would be more tightly bound if that nucleus were to split into two nuclei that lie near the top of the plot. Such a process, called **fission,** occurs naturally with large (high mass number A) nuclei such as uranium, which can undergo fission spontaneously (that is, without an external cause or source of energy). The process can also occur in nuclear weapons, in which many uranium or plutonium nuclei are made to fission all at once, to create an explosion.

The nucleons in any pair of nuclei on the left side of the plot would be more tightly bound if the pair were to combine to form a single nucleus that lies near the top of the plot. Such a process, called **fusion,** occurs naturally in stars. Were this not true, the Sun would not shine and thus life could not exist on Earth. As we shall discuss in the next chapter, fusion is also the basis of thermonuclear weapons (with an explosive release of energy) and anticipated power plants (with a sustained and controlled release of energy).

Nuclear Energy Levels

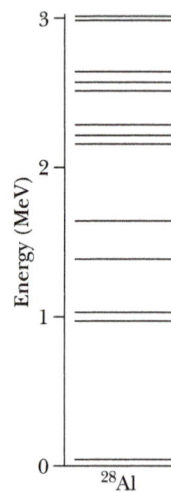

Figure 42-8 Energy levels for the nuclide ^{28}Al, deduced from nuclear reaction experiments.

The energy of nuclei, like that of atoms, is quantized. That is, nuclei can exist only in discrete quantum states, each with a well-defined energy. Figure 42-8 shows some of these energy levels for ^{28}Al, a typical low-mass nuclide. Note that the energy scale is in millions of electron-volts, rather than the electron-volts used for atoms. When a nucleus makes a transition from one level to a level of lower energy, the emitted photon is typically in the gamma-ray region of the electromagnetic spectrum.

Nuclear Spin and Magnetism

Many nuclides have an intrinsic *nuclear angular momentum*, or spin, and an associated intrinsic *nuclear magnetic moment*. Although nuclear angular momenta are roughly of the same magnitude as the angular momenta of atomic electrons, nuclear magnetic moments are much smaller than typical atomic magnetic moments.

The Nuclear Force

The force that controls the motions of atomic electrons is the familiar electromagnetic force. To bind the nucleus together, however, there must be a strong attractive nuclear force of a totally different kind, strong enough to overcome the repulsive force between the (positively charged) nuclear protons and to bind both protons and neutrons into the tiny nuclear volume. The nuclear force must also be of short range because its influence does not extend very far beyond the nuclear "surface."

The present view is that the nuclear force that binds neutrons and protons in the nucleus is not a fundamental force of nature but is a secondary, or "spillover," effect of the **strong force** that binds quarks together to form neutrons and protons. In much the same way, the attractive force between certain neutral molecules is a spillover effect of the Coulomb electric force that acts within each molecule to bind it together.

Sample Problem 42.02 Binding energy per nucleon

What is the binding energy per nucleon for ^{120}Sn?

KEY IDEAS

1. We can find the binding energy per nucleon ΔE_{ben} if we first find the binding energy ΔE_{be} and then divide by the number of nucleons A in the nucleus, according to Eq. 42-8 ($\Delta E_{ben} = \Delta E_{be}/A$).

2. We can find ΔE_{be} by finding the difference between the mass energy Mc^2 of the nucleus and the total mass energy $\Sigma(mc^2)$ of the individual nucleons that make up the nucleus, according to Eq. 42-7 ($\Delta E_{be} = \Sigma(mc^2) - Mc^2$).

Calculations: From Table 42-1, we see that a ^{120}Sn nucleus consists of 50 protons ($Z = 50$) and 70 neutrons ($N = A - Z = 120 - 50 = 70$). Thus, we need to imagine a ^{120}Sn nucleus being separated into its 50 protons and 70 neutrons,

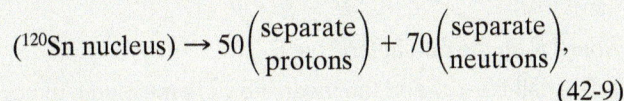

$$(^{120}\text{Sn nucleus}) \rightarrow 50\binom{\text{separate}}{\text{protons}} + 70\binom{\text{separate}}{\text{neutrons}},$$
$$(42\text{-}9)$$

and then compute the resulting change in mass energy.

For that computation, we need the masses of a ^{120}Sn nucleus, a proton, and a neutron. However, because the mass of a neutral atom (nucleus *plus* electrons) is much easier to measure than the mass of a bare nucleus, calculations of binding energies are traditionally done with atomic masses. Thus, let's modify Eq. 42-9 so that it has a neutral ^{120}Sn atom on the left side. To do that, we include 50 electrons on the left side (to match the 50 protons in the ^{120}Sn nucleus). We

must also add 50 electrons on the right side to balance Eq. 42-9. Those 50 electrons can be combined with the 50 protons, to form 50 neutral hydrogen atoms. We then have

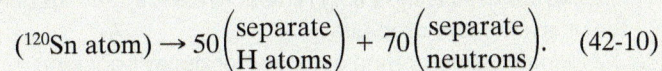

$$(^{120}\text{Sn atom}) \rightarrow 50\binom{\text{separate}}{\text{H atoms}} + 70\binom{\text{separate}}{\text{neutrons}}. \quad (42\text{-}10)$$

From the mass column of Table 42-1, the mass M_{Sn} of a ^{120}Sn atom is 119.902 197 u and the mass m_H of a hydrogen atom is 1.007 825 u; the mass m_n of a neutron is 1.008 665 u. Thus, Eq. 42-7 yields

$$\Delta E_{be} = \Sigma(mc^2) - Mc^2$$
$$= 50(m_H c^2) + 70(m_n c^2) - M_{Sn}c^2$$
$$= 50(1.007\ 825\ \text{u})c^2 + 70(1.008\ 665\ \text{u})c^2$$
$$\quad - (119.902\ 197\ \text{u})c^2$$
$$= (1.095\ 603\ \text{u})c^2$$
$$= (1.095\ 603\ \text{u})(931.494\ 013\ \text{MeV/u})$$
$$= 1020.5\ \text{MeV},$$

where Eq. 42-5 ($c^2 = 931.494\ 013$ MeV/u) provides an easy unit conversion. Note that using atomic masses instead of nuclear masses does not affect the result because the mass of the 50 electrons in the ^{120}Sn atom subtracts out from the mass of the electrons in the 50 hydrogen atoms.

Now Eq. 42-8 gives us the binding energy per nucleon as

$$\Delta E_{ben} = \frac{\Delta E_{be}}{A} = \frac{1020.5\ \text{MeV}}{120}$$
$$= 8.50\ \text{MeV/nucleon}. \qquad \text{(Answer)}$$

Sample Problem 42.03 Density of nuclear matter

We can think of all nuclides as made up of a neutron–proton mixture that we can call *nuclear matter*. What is the density of nuclear matter?

KEY IDEA

We can find the (average) density ρ of a nucleus by dividing its total mass by its volume.

Calculations: Let m represent the mass of a nucleon (either a proton or a neutron, because those particles have about the same mass). Then the mass of a nucleus containing A nucleons is Am. Next, we assume the nucleus is spherical with radius r. Then its volume is $\frac{4}{3}\pi r^3$, and we can write the density of the nucleus as

$$\rho = \frac{Am}{\frac{4}{3}\pi r^3}.$$

The radius r is given by Eq. 42-3 ($r = r_0 A^{1/3}$), where r_0 is 1.2 fm ($= 1.2 \times 10^{-15}$ m). Substituting for r then leads to

$$\rho = \frac{Am}{\frac{4}{3}\pi r_0^3 A} = \frac{m}{\frac{4}{3}\pi r_0^3}.$$

Note that A has canceled out; thus, this equation for density ρ applies to any nucleus that can be treated as spherical with a radius given by Eq. 42-3. Using 1.67×10^{-27} kg for the mass m of a nucleon, we then have

$$\rho = \frac{1.67 \times 10^{-27}\ \text{kg}}{\frac{4}{3}\pi(1.2 \times 10^{-15}\ \text{m})^3} \approx 2 \times 10^{17}\ \text{kg/m}^3. \quad \text{(Answer)}$$

This is about 2×10^{14} times the density of water and is the density of neutron stars, which contain only neutrons.

WILEY **PLUS** Additional examples, video, and practice available at *WileyPLUS*

42-3 RADIOACTIVE DECAY

Learning Objectives

After reading this module, you should be able to . . .

42.11 Explain what is meant by radioactive decay and identify that it is a random process.

42.12 Identify disintegration constant (or decay constant) λ.

42.13 Identify that, at any given instant, the rate dN/dt at which radioactive nuclei decay is proportional to the number N of them still present then.

42.14 Apply the relationship that gives the number N of radioactive nuclei as a function of time.

42.15 Apply the relationship that gives the decay rate R of radioactive nuclei as a function of time.

42.16 For any given time, apply the relationship between the decay rate R and the remaining number N of radioactive nuclei.

42.17 Identify activity.

42.18 Distinguish Becquerel (Bq), curie (Ci), and counts per unit time.

42.19 Distinguish half-life $T_{1/2}$ and mean life τ.

42.20 Apply the relationship between half-life $T_{1/2}$, mean life τ, and disintegration constant λ.

42.21 Identify that in any nuclear process, including radioactive decay, the charge and the number of nucleons are conserved.

Key Ideas

● Most nuclides spontaneously decay at a rate $R = dN/dt$ that is proportional to the number N of radioactive atoms present. The proportionality constant is the disintegration constant λ.

● The number of radioactive nuclei is given as a function of time by

$$N = N_0 e^{-\lambda t},$$

where N_0 is the number at time $t = 0$.

● The rate at which the nuclei decay is given as a function of time

by

$$R = R_0 e^{-\lambda t},$$

where R_0 is the rate at time $t = 0$.

● The half-life $T_{1/2}$ and the mean life τ are measures of how quickly radioactive nuclei decay and are related by

$$T_{1/2} = \frac{\ln 2}{\lambda} = \tau \ln 2.$$

Radioactive Decay

As Fig. 42-5 shows, most nuclides are radioactive. They each spontaneously (randomly) emit a particle and transform into a different nuclide. Thus these decays reveal that the laws for subatomic processes are statistical. For example, in a 1 mg sample of uranium metal, with 2.5×10^{18} atoms of the very long-lived radionuclide ^{238}U, only about 12 of the nuclei will decay in a given second by emitting an alpha particle and transforming into a nucleus of ^{234}Th. However,

There is absolutely no way to predict whether any given nucleus in a radioactive sample will be among the small number of nuclei that decay during any given second. All have the same chance.

Although we cannot predict which nuclei in a sample will decay, we can say that if a sample contains N radioactive nuclei, then the rate ($= -dN/dt$) at which nuclei will decay is proportional to N:

$$-\frac{dN}{dt} = \lambda N, \tag{42-11}$$

in which λ, the **disintegration constant** (or **decay constant**) has a characteristic value for every radionuclide. Its SI unit is the inverse second (s^{-1}).

To find N as a function of time t, we first rearrange Eq. 42-11 as

$$\frac{dN}{N} = -\lambda \, dt, \tag{42-12}$$

and then integrate both sides, obtaining

$$\int_{N_0}^{N} \frac{dN}{N} = -\lambda \int_{t_0}^{t} dt,$$

or

$$\ln N - \ln N_0 = -\lambda(t - t_0). \tag{42-13}$$

Here N_0 is the number of radioactive nuclei in the sample at some arbitrary initial time t_0. Setting $t_0 = 0$ and rearranging Eq. 42-13 give us

$$\ln \frac{N}{N_0} = -\lambda t. \tag{42-14}$$

Taking the exponential of both sides (the exponential function is the antifunction of the natural logarithm) leads to

$$\frac{N}{N_0} = e^{-\lambda t}$$

or
$$N = N_0 e^{-\lambda t} \quad \text{(radioactive decay)}, \tag{42-15}$$

in which N_0 is the number of radioactive nuclei in the sample at $t = 0$ and N is the number remaining at any subsequent time t. Note that lightbulbs (for one example) follow no such exponential decay law. If we life-test 1000 bulbs, we expect that they will all "decay" (that is, burn out) at more or less the same time. The decay of radionuclides follows quite a different law.

We are often more interested in the decay rate R ($= -dN/dt$) than in N itself. Differentiating Eq. 42-15, we find

$$R = -\frac{dN}{dt} = \lambda N_0 e^{-\lambda t}$$

or
$$R = R_0 e^{-\lambda t} \quad \text{(radioactive decay)}, \tag{42-16}$$

an alternative form of the law of radioactive decay (Eq. 42-15). Here R_0 is the decay rate at time $t = 0$ and R is the rate at any subsequent time t. We can now rewrite Eq. 42-11 in terms of the decay rate R of the sample as

$$R = \lambda N, \tag{42-17}$$

where R and the number of radioactive nuclei N that have not yet undergone decay must be evaluated at the same instant.

The total decay rate R of a sample of one or more radionuclides is called the **activity** of that sample. The SI unit for activity is the **becquerel,** named for Henri Becquerel, the discoverer of radioactivity:

$$1 \text{ becquerel} = 1 \text{ Bq} = 1 \text{ decay per second.}$$

An older unit, the **curie,** is still in common use:

$$1 \text{ curie} = 1 \text{ Ci} = 3.7 \times 10^{10} \text{ Bq.}$$

Often a radioactive sample will be placed near a detector that does not record all the disintegrations that occur in the sample. The reading of the detector under these circumstances is proportional to (and smaller than) the true activity of the sample. Such proportional activity measurements are reported not in becquerel units but simply in counts per unit time.

Lifetimes. There are two common time measures of how long any given type of radionuclides lasts. One measure is the **half-life** $T_{1/2}$ of a radionuclide, which is the time at which both N and R have been reduced to one-half their initial values. The other measure is the **mean** (or **average**) **life** τ, which is the time at which both N and R have been reduced to e^{-1} of their initial values.

To relate $T_{1/2}$ to the disintegration constant λ, we put $R = \frac{1}{2}R_0$ in Eq. 42-16 and substitute $T_{1/2}$ for t. We obtain

$$\tfrac{1}{2}R_0 = R_0 e^{-\lambda T_{1/2}}.$$

Taking the natural logarithm of both sides and solving for $T_{1/2}$, we find

$$T_{1/2} = \frac{\ln 2}{\lambda}.$$

Similarly, to relate τ to λ, we put $R = e^{-1}R_0$ in Eq. 42-16, substitute τ for t, and

solve for τ, finding

$$\tau = \frac{1}{\lambda}.$$

We summarize these results with the following:

$$T_{1/2} = \frac{\ln 2}{\lambda} = \tau \ln 2. \qquad (42\text{-}18)$$

✓ Checkpoint 2

The nuclide ^{131}I is radioactive, with a half-life of 8.04 days. At noon on January 1, the activity of a certain sample is 600 Bq. Using the concept of half-life, without written calculation, determine whether the activity at noon on January 24 will be a little less than 200 Bq, a little more than 200 Bq, a little less than 75 Bq, or a little more than 75 Bq.

Sample Problem 42.04 Finding the disintegration constant and half-life from a graph

The table that follows shows some measurements of the decay rate of a sample of ^{128}I, a radionuclide often used medically as a tracer to measure the rate at which iodine is absorbed by the thyroid gland.

Time (min)	R (counts/s)	Time (min)	R (counts/s)
4	392.2	132	10.9
36	161.4	164	4.56
68	65.5	196	1.86
100	26.8	218	1.00

Find the disintegration constant λ and the half-life $T_{1/2}$ for this radionuclide.

KEY IDEAS

The disintegration constant λ determines the exponential rate at which the decay rate R decreases with time t (as indicated by Eq. 42-16, $R = R_0 e^{-\lambda t}$). Therefore, we should be able to determine λ by plotting the measurements of R against the measurement times t. However, obtaining λ from a plot of R versus t is difficult because R decreases exponentially with t, according to Eq. 42-16. A neat solution is to transform Eq. 42-16 into a linear function of t, so that we can easily find λ. To do so, we take the natural logarithms of both sides of Eq. 42-16.

Calculations: We obtain

$$\ln R = \ln(R_0 e^{-\lambda t}) = \ln R_0 + \ln(e^{-\lambda t})$$
$$= \ln R_0 - \lambda t. \qquad (42\text{-}19)$$

Because Eq. 42-19 is of the form $y = b + mx$, with b and m constants, it is a linear equation giving the quantity $\ln R$ as a function of t. Thus, if we plot $\ln R$ (instead of R) versus t, we

should get a straight line. Further, the slope of the line should be equal to $-\lambda$.

Figure 42-9 shows a plot of $\ln R$ versus time t for the given measurements. The slope of the straight line that fits through the plotted points is

$$\text{slope} = \frac{0 - 6.2}{225 \text{ min} - 0} = -0.0276 \text{ min}^{-1}.$$

Thus, $-\lambda = -0.0276 \text{ min}^{-1}$

or $\lambda = 0.0276 \text{ min}^{-1} \approx 1.7 \text{ h}^{-1}.$ (Answer)

The time for the decay rate R to decrease by 1/2 is related to the disintegration constant λ via Eq. 42-18 ($T_{1/2} = (\ln 2)/\lambda$). From that equation, we find

$$T_{1/2} = \frac{\ln 2}{\lambda} = \frac{\ln 2}{0.0276 \text{ min}^{-1}} \approx 25 \text{ min}. \qquad \text{(Answer)}$$

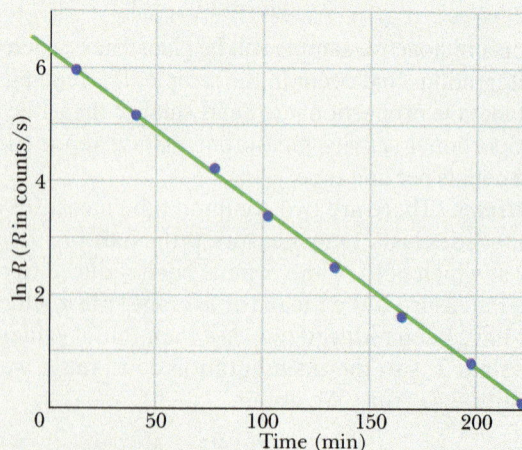

Figure 42-9 A semilogarithmic plot of the decay of a sample of ^{128}I, based on the data in the table.

Sample Problem 42.05 Radioactivity of the potassium in a banana

Of the 600 mg of potassium in a large banana, 0.0117% is radioactive ^{40}K, which has a half-life $T_{1/2}$ of 1.25×10^9 y. What is the activity of the banana?

KEY IDEAS

(1) We can relate the activity R to the disintegration constant λ with Eq. 42-17, but let's write it as $R = \lambda N_{40}$, where N_{40} is the number of ^{40}K nuclei (and thus atoms) in the banana. (2) We can relate the disintegration constant to the known half-life $T_{1/2}$ with Eq. 42-18 ($T_{1/2} = (\ln 2)/\lambda$).

Calculations: Combining Eqs. 42-18 and 42-17 yields

$$R = \frac{N_{40} \ln 2}{T_{1/2}}. \qquad (42\text{-}20)$$

We know that N_{40} is 0.0117% of the total number N of potassium atoms in the banana. We can find an expression for N by combining two equations that give the number of moles n of potassium in the banana. From Eq. 19-2, $n = N/N_A$, where N_A is Avogadro's number (6.02×10^{23} mol^{-1}). From Eq. 19-3, $n = M_{sam}/M$, where M_{sam} is the sample mass (here the given

600 mg of potassium) and M is the molar mass of potassium. Combining those two equations to eliminate n, we can write

$$N_{40} = (1.17 \times 10^{-4})\frac{M_{sam}N_A}{M}. \qquad (42\text{-}21)$$

From Appendix F, we see that the molar mass of potassium is 39.102 g/mol. Equation 42-21 then yields

$$N_{40} = (1.17 \times 10^{-4})\frac{(600 \times 10^{-3}\,g)(6.02 \times 10^{23}\,mol^{-1})}{39.102\,g/mol}$$
$$= 1.081 \times 10^{18}.$$

Substituting this value for N_{40} and the given half-life of 1.25×10^9 y for $T_{1/2}$ into Eq. 42-20 leads to

$$R = \frac{(1.081 \times 10^{18})(\ln 2)}{(1.25 \times 10^9\,y)(3.16 \times 10^7\,s/y)}$$
$$= 18.96\,\text{Bq} \approx 19.0\,\text{Bq.} \qquad \text{(Answer)}$$

This is about 0.51 nCi. Your body always has about 160 g of potassium. If you repeat our calculation here, you will find that the ^{40}K component of that everyday amount has an activity of 5.06×10^3 Bq (or 0.14 μCi). So, eating a banana adds less than 1% to the radiation your body receives daily from radioactive potassium.

PLUS Additional examples, video, and practice available at *WileyPLUS*

42-4 ALPHA DECAY

Learning Objectives

After reading this module, you should be able to . . .

42.22 Identify alpha particle and alpha decay.
42.23 For a given alpha decay, calculate the mass change and the Q of the reaction.
42.24 Determine the change in atomic number Z and mass

number A of a nucleus undergoing alpha decay.
42.25 In terms of the potential barrier, explain how an alpha particle can escape from a nucleus with less energy than the barrier height.

Key Idea

● Some nuclides decay by emitting an alpha particle (a helium nucleus, ^{4}He). Such decay is inhibited by a potential energy barrier that must be penetrated by tunneling.

Alpha Decay

When a nucleus undergoes **alpha decay,** it transforms to a different nuclide by emitting an alpha particle (a helium nucleus, ^{4}He). For example, when uranium ^{238}U undergoes alpha decay, it transforms to thorium ^{234}Th:

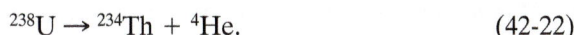

$$^{238}U \rightarrow {}^{234}Th + {}^4He. \qquad (42\text{-}22)$$

This alpha decay of ^{238}U can occur spontaneously (without an external source of energy) because the total mass of the decay products ^{234}Th and ^{4}He is less than the mass of the original ^{238}U. Thus, the total mass energy of the decay

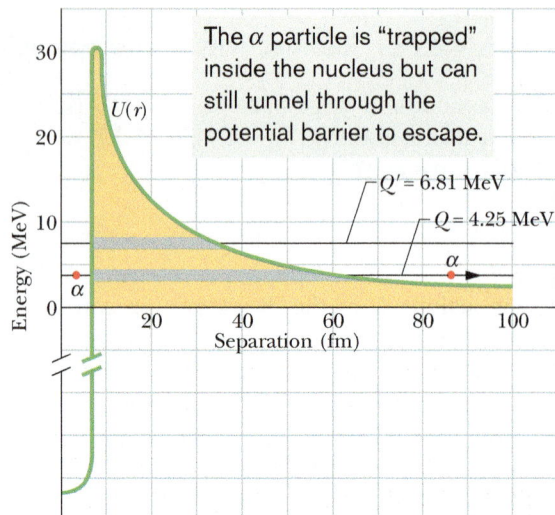

Figure 42-10 A potential energy function for the emission of an alpha particle by ^{238}U. The horizontal black line marked $Q = 4.25$ MeV shows the disintegration energy for the process. The thick gray portion of this line represents separations r that are classically forbidden to the alpha particle. The alpha particle is represented by a dot, both inside this potential energy barrier (at the left) and outside it (at the right), after the particle has tunneled through. The horizontal black line marked $Q' = 6.81$ MeV shows the disintegration energy for the alpha decay of ^{228}U. (Both isotopes have the same potential energy function because they have the same nuclear charge.)

products is less than the mass energy of the original nuclide. As defined by Eq. 37-50 ($Q = -\Delta M\ c^2$), in such a process the difference between the initial mass energy and the total final mass energy is called the Q of the process.

For a nuclear decay, we say that the difference in mass energy is the decay's *disintegration energy Q*. The Q for the decay in Eq. 42-22 is 4.25 MeV—that amount of energy is said to be released by the alpha decay of ^{238}U, with the energy transferred from mass energy to the kinetic energy of the two products.

The half-life of ^{238}U for this decay process is 4.5×10^9 y. Why so long? If ^{238}U can decay in this way, why doesn't every ^{238}U nuclide in a sample of ^{238}U atoms simply decay at once? To answer the questions, we must examine the process of alpha decay.

We choose a model in which the alpha particle is imagined to exist (already formed) inside the nucleus before it escapes from the nucleus. Figure 42-10 shows the approximate potential energy $U(r)$ of the system consisting of the alpha particle and the residual ^{234}Th nucleus, as a function of their separation r. This energy is a combination of (1) the potential energy associated with the (attractive) strong nuclear force that acts in the nuclear interior and (2) a Coulomb potential associated with the (repulsive) electric force that acts between the two particles before and after the decay has occurred.

The horizontal black line marked $Q = 4.25$ MeV shows the disintegration energy for the process. If we assume that this represents the total energy of the alpha particle during the decay process, then the part of the $U(r)$ curve above this line constitutes a potential energy barrier like that in Fig. 38-17. This barrier cannot be surmounted. If the alpha particle were able to be at some separation r within the barrier, its potential energy U would exceed its total energy E. This would mean, classically, that its kinetic energy K (which equals $E - U$) would be negative, an impossible situation.

Tunneling. We can see now why the alpha particle is not immediately emitted from the ^{238}U nucleus. That nucleus is surrounded by an impressive potential barrier, occupying—if you think of it in three dimensions—the volume lying between two spherical shells (of radii about 8 and 60 fm). This argument is so convincing that we now change our last question and ask: Since the particle seems permanently trapped inside the nucleus by the barrier, how can the ^{238}U nucleus *ever* emit an alpha particle? The answer is that, as you learned in Module 38-9, there is a finite probability that a particle can tunnel through an energy barrier that is classically insurmountable. In fact, alpha decay occurs as a result of barrier tunneling.

The very long half-life of ^{238}U tells us that the barrier is apparently not very "leaky." If we imagine that an already-formed alpha particle is rattling back and forth inside the nucleus, it would arrive at the inner surface of the barrier about 10^{38} times before it would succeed in tunneling through the barrier. This is about 10^{21} times per second for about 4×10^9 years (the age of Earth)! We, of course, are waiting on the outside, able to count only the alpha particles that *do* manage to escape without being able to tell what's going on inside the nucleus.

We can test this explanation of alpha decay by examining other alpha emitters. For an extreme contrast, consider the alpha decay of another uranium isotope, ^{228}U, which has a disintegration energy Q' of 6.81 MeV, about 60% higher than that of ^{238}U. (The value of Q' is also shown as a horizontal black line in Fig. 42-10.) Recall from Module 38-9 that the transmission coefficient of a barrier is very sensitive to small changes in the total energy of the particle seeking to penetrate it. Thus, we expect alpha decay to occur more readily for this nuclide than for ^{238}U. Indeed it does. As Table 42-2 shows, its half-life is only 9.1 min! An increase in Q by a factor of only 1.6 produces a decrease in half-life (that is, in the effectiveness of the barrier) by a factor of 3×10^{14}. This is sensitivity indeed.

Table 42-2 Two Alpha Emitters Compared

Radionuclide	Q	Half-Life
^{238}U	4.25 MeV	4.5×10^9 y
^{228}U	6.81 MeV	9.1 min

Sample Problem 42.06 *Q* value in an alpha decay, using masses

We are given the following atomic masses:

^{238}U	238.050 79 u	^{4}He	4.002 60 u
^{234}Th	234.043 63 u	^{1}H	1.007 83 u
^{237}Pa	237.051 21 u		

Here Pa is the symbol for the element protactinium ($Z = 91$).

(a) Calculate the energy released during the alpha decay of ^{238}U. The decay process is

$$^{238}\text{U} \rightarrow {}^{234}\text{Th} + {}^4\text{He}.$$

Note, incidentally, how nuclear charge is conserved in this equation: The atomic numbers of thorium (90) and helium (2) add up to the atomic number of uranium (92). The number of nucleons is also conserved: 238 = 234 + 4.

KEY IDEA

The energy released in the decay is the disintegration energy Q, which we can calculate from the change in mass ΔM due to the ^{238}U decay.

Calculation: To do this, we use Eq. 37-50,

$$Q = M_i c^2 - M_f c^2, \qquad (42\text{-}23)$$

where the initial mass M_i is that of ^{238}U and the final mass M_f is the sum of the ^{234}Th and ^{4}He masses. Using the atomic masses given in the problem statement, Eq. 42-23 becomes

$$Q = (238.050\ 79\ \text{u})c^2 - (234.043\ 63\ \text{u} + 4.002\ 60\ \text{u})c^2$$
$$= (0.004\ 56\ \text{u})c^2 = (0.004\ 56\ \text{u})(931.494\ 013\ \text{MeV/u})$$
$$= 4.25\ \text{MeV}. \qquad \text{(Answer)}$$

Note that using atomic masses instead of nuclear masses does not affect the result because the total mass of the electrons in the products subtracts out from the mass of the nucleons + electrons in the original ^{238}U.

(b) Show that ^{238}U cannot spontaneously emit a proton; that is, protons do not leak out of the nucleus in spite of the proton–proton repulsion within the nucleus.

Solution: If this happened, the decay process would be

$$^{238}\text{U} \rightarrow {}^{237}\text{Pa} + {}^1\text{H}.$$

(You should verify that both nuclear charge and the number of nucleons are conserved in this process.) Using the same Key Idea as in part (a) and proceeding as we did there, we would find that the mass of the two decay products

$$237.051\ 21\ \text{u} + 1.007\ 83\ \text{u}$$

would *exceed* the mass of ^{238}U by $\Delta m = 0.008\ 25$ u, with disintegration energy

$$Q = -7.68\ \text{MeV}.$$

The minus sign indicates that we must *add* 7.68 MeV to a ^{238}U nucleus before it will emit a proton; it will certainly not do so spontaneously.

PLUS Additional examples, video, and practice available at *WileyPLUS*

42-5 BETA DECAY

Learning Objectives

After reading this module, you should be able to . . .

42.26 Identify the two types of beta particles and the two types of beta decay.

42.27 Identify neutrino.

42.28 Explain why the beta particles in beta decays are emitted with a range of energies.

42.29 For a given beta decay, calculate the mass change and the Q of the reaction.

42.30 Determine the change in the atomic number Z of a nucleus undergoing a beta decay and identify that the mass number A does not change.

Key Ideas

● In beta decay, either an electron or a positron is emitted by a nucleus, along with a neutrino.

● The emitted particles share the available disintegration

energy. Sometimes the neutrino gets most of the energy and sometimes the electron or positron gets most of it.

Beta Decay

A nucleus that decays spontaneously by emitting an electron or a positron (a positively charged particle with the mass of an electron) is said to undergo **beta decay.** Like alpha decay, this is a spontaneous process, with a definite disintegration energy and half-life. Again like alpha decay, beta decay is a statistical process, governed by Eqs. 42-15 and 42-16. In *beta-minus* (β^-) decay, an electron is emitted by a nucleus, as in the decay

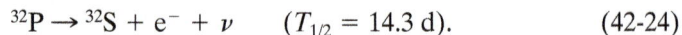

$$^{32}\text{P} \rightarrow {}^{32}\text{S} + e^- + \nu \qquad (T_{1/2} = 14.3 \text{ d}). \qquad (42\text{-}24)$$

In *beta-plus* (β^+) decay, a positron is emitted by a nucleus, as in the decay

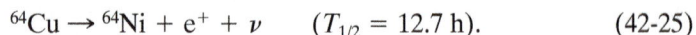

$$^{64}\text{Cu} \rightarrow {}^{64}\text{Ni} + e^+ + \nu \qquad (T_{1/2} = 12.7 \text{ h}). \qquad (42\text{-}25)$$

The symbol ν represents a **neutrino,** a neutral particle which has a very small mass, that is emitted from the nucleus along with the electron or positron during the decay process. Neutrinos interact only very weakly with matter and—for that reason—are so extremely difficult to detect that their presence long went unnoticed.*

Both charge and nucleon number are conserved in the above two processes. In the decay of Eq. 42-24, for example, we can write for charge conservation

$$(+15e) = (+16e) + (-e) + (0),$$

because ^{32}P has 15 protons, ^{32}S has 16 protons, and the neutrino ν has zero charge. Similarly, for nucleon conservation, we can write

$$(32) = (32) + (0) + (0),$$

because ^{32}P and ^{32}S each have 32 nucleons and neither the electron nor the neutrino is a nucleon.

It may seem surprising that nuclei can emit electrons, positrons, and neutrinos, since we have said that nuclei are made up of neutrons and protons only. However, we saw earlier that atoms emit photons, and we certainly do not say that atoms "contain" photons. We say that the photons are created during the emission process.

---

*Beta decay also includes *electron capture,* in which a nucleus decays by absorbing one of its atomic electrons, emitting a neutrino in the process. We do not consider that process here. Also, the neutral particle emitted in the decay process of Eq. 42-24 is actually an *antineutrino,* a distinction we shall not make in this introductory treatment.

It is the same with the electrons, positrons, and neutrinos emitted from nuclei during beta decay. They are created during the emission process. For beta-minus decay, a neutron transforms into a proton within the nucleus according to

$$n \rightarrow p + e^- + \nu. \quad (42\text{-}26)$$

For beta-plus decay, a proton transforms into a neutron via

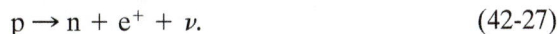

$$p \rightarrow n + e^+ + \nu. \quad (42\text{-}27)$$

These processes show why the mass number A of a nuclide undergoing beta decay does not change; one of its constituent nucleons simply changes its character according to Eq. 42-26 or 42-27.

In both alpha decay and beta decay, the same amount of energy is released in every individual decay of a particular radionuclide. In the alpha decay of a particular radionuclide, every emitted alpha particle has the same sharply defined kinetic energy. However, in the beta-minus decay of Eq. 42-26 with electron emission, the disintegration energy Q is shared—in varying proportions—between the emitted electron and neutrino. Sometimes the electron gets nearly all the energy, sometimes the neutrino does. In every case, however, the sum of the electron's energy and the neutrino's energy gives the same value Q. A similar sharing of energy, with a sum equal to Q, occurs in beta-plus decay (Eq. 42-27).

Thus, in beta decay the energy of the emitted electrons or positrons may range from near zero up to a certain maximum K_{max}. Figure 42-11 shows the distribution of positron energies for the beta decay of ^{64}Cu (see Eq. 42-25). The maximum positron energy K_{max} must equal the disintegration energy Q because the neutrino has approximately zero energy when the positron has K_{max}:

$$Q = K_{max}. \quad (42\text{-}28)$$

The Neutrino

Wolfgang Pauli first suggested the existence of neutrinos in 1930. His neutrino hypothesis not only permitted an understanding of the energy distribution of electrons or positrons in beta decay but also solved another early beta-decay puzzle involving "missing" angular momentum.

The neutrino is a truly elusive particle; the mean free path of an energetic neutrino in water has been calculated as no less than several thousand light-years. At the same time, neutrinos left over from the big bang that presumably marked the creation of the universe are the most abundant particles of physics. Billions of them pass through our bodies every second, leaving no trace.

In spite of their elusive character, neutrinos have been detected in the laboratory. This was first done in 1953 by F. Reines and C. L. Cowan, using neutrinos generated in a high-power nuclear reactor. (In 1995, Reines received a Nobel Prize for this work.) In spite of the difficulties of detection, experimental neutrino physics is now a well-developed branch of experimental physics, with avid practitioners at laboratories throughout the world.

The Sun emits neutrinos copiously from the nuclear furnace at its core, and at night these messengers from the center of the Sun come up at us from below, Earth being almost totally transparent to them. In February 1987, light from an exploding star in the Large Magellanic Cloud (a nearby galaxy) reached Earth after having traveled for 170 000 years. Enormous numbers of neutrinos were generated in this explosion, and about 10 of them were picked up by a sensitive neutrino detector in Japan; Fig. 42-12 shows a record of their passage.

Radioactivity and the Nuclidic Chart

We can increase the amount of information obtainable from the nuclidic chart of Fig. 42-5 by including a third axis showing the mass excess Δ expressed in the

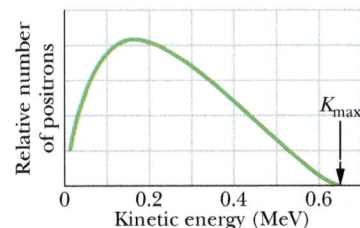

Figure 42-11 The distribution of the kinetic energies of positrons emitted in the beta decay of ^{64}Cu. The maximum kinetic energy of the distribution (K_{max}) is 0.653 MeV. In all ^{64}Cu decay events, this energy is shared between the positron and the neutrino, in varying proportions. The *most probable* energy for an emitted positron is about 0.15 MeV.

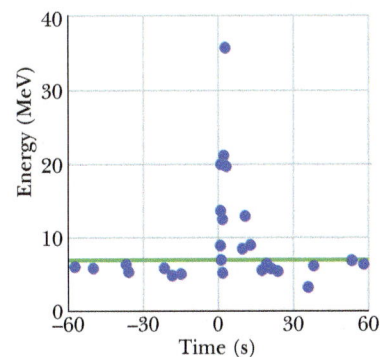

Figure 42-12 A burst of neutrinos from the supernova SN 1987A, which occurred at (relative) time 0, stands out from the usual *background* of neutrinos. (For neutrinos, 10 is a "burst.") The particles were detected by an elaborate detector housed deep in a mine in Japan. The supernova was visible only in the Southern Hemisphere; so the neutrinos had to penetrate Earth (a trifling barrier for them) to reach the detector.

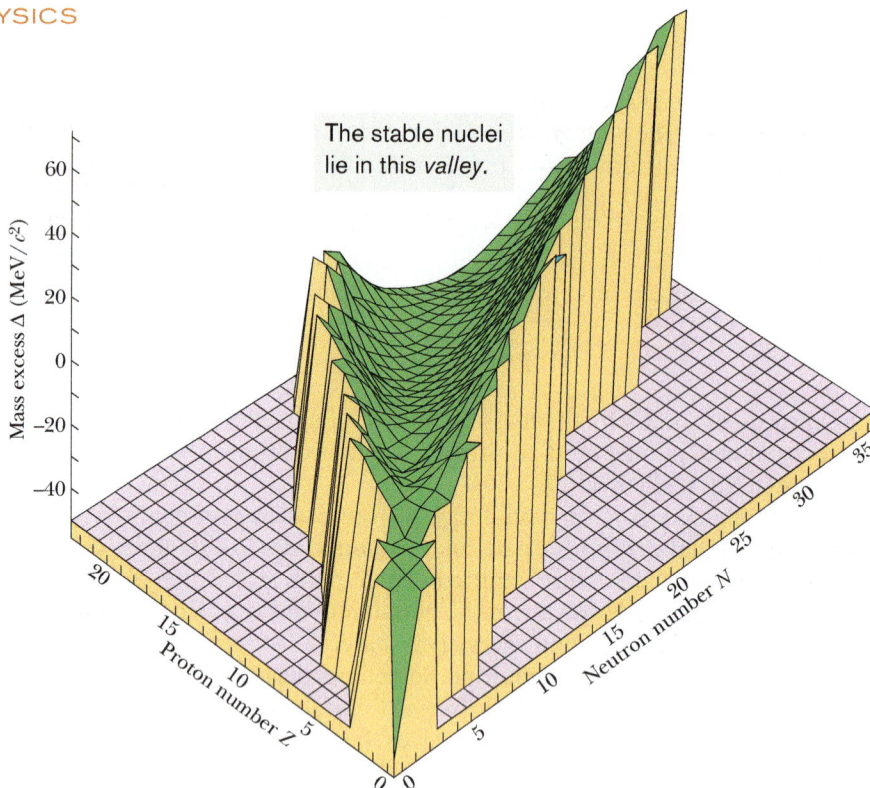

Figure 42-13 A portion of the valley of the nuclides, showing only the nuclides of low mass. Deuterium, tritium, and helium lie at the near end of the plot, with helium at the high point. The valley stretches away from us, with the plot stopping at about $Z = 22$ and $N = 35$. Nuclides with large values of A, which would be plotted much beyond the valley, can decay into the valley by repeated alpha emissions and by fission (splitting of a nuclide).

unit MeV/c^2. The inclusion of such an axis gives Fig. 42-13, which reveals the degree of nuclear stability of the nuclides. For the low-mass nuclides, we find a "valley of the nuclides," with the stability band of Fig. 42-5 running along its bottom. Nuclides on the proton-rich side of the valley decay into it by emitting positrons, and those on the neutron-rich side do so by emitting electrons.

✓ Checkpoint 3

^{238}U decays to ^{234}Th by the emission of an alpha particle. There follows a chain of further radioactive decays, either by alpha decay or by beta decay. Eventually a stable nuclide is reached and, after that, no further radioactive decay is possible. Which of the following stable nuclides is the end product of the ^{238}U radioactive decay chain: ^{206}Pb, ^{207}Pb, ^{208}Pb, or ^{209}Pb? (*Hint:* You can decide by considering the changes in mass number A for the two types of decay.)

Sample Problem 42.07   Q value in a beta decay, using masses

Calculate the disintegration energy Q for the beta decay of ^{32}P, as described by Eq. 42-24. The needed atomic masses are 31.973 91 u for ^{32}P and 31.972 07 u for ^{32}S.

KEY IDEA

The disintegration energy Q for the beta decay is the amount by which the mass energy is changed by the decay.

Calculations: Q is given by Eq. 37-50 ($Q = -\Delta M \, c^2$). However, we must be careful to distinguish between nuclear masses (which we do not know) and atomic masses (which we do know). Let the boldface symbols $\mathbf{m}_P$ and $\mathbf{m}_S$ represent the nuclear masses of ^{32}P and ^{32}S, and let the italic symbols

m_P and m_S represent their atomic masses. Then we can write the change in mass for the decay of Eq. 42-24 as

$$\Delta m = (\mathbf{m}_S + m_e) - \mathbf{m}_P,$$

in which m_e is the mass of the electron. If we add and subtract $15m_e$ on the right side of this equation, we obtain

$$\Delta m = (\mathbf{m}_S + 16m_e) - (\mathbf{m}_P + 15m_e).$$

The quantities in parentheses are the atomic masses of ^{32}S and ^{32}P; so

$$\Delta m = m_S - m_P.$$

We thus see that if we subtract only the atomic masses, the mass of the emitted electron is automatically taken into account. (This procedure will not work for positron emission.)

The disintegration energy for the ^{32}P decay is then

$$Q = -\Delta m\, c^2$$
$$= -(31.972\,07\ \text{u} - 31.973\,91\ \text{u})(931.494\,013\ \text{MeV/u})$$
$$= 1.71\ \text{MeV}. \qquad \text{(Answer)}$$

Experimentally, this calculated quantity proves to be equal to K_{max}, the maximum energy the emitted electrons can have. Although 1.71 MeV is released every time a ^{32}P nucleus decays, in essentially every case the electron carries away less energy than this. The neutrino gets all the rest, carrying it stealthily out of the laboratory.

WILEY PLUS Additional examples, video, and practice available at *WileyPLUS*

42-6 RADIOACTIVE DATING

Learning Objectives

After reading this module, you should be able to . . .

42.31 Apply the equations for radioactive decay to determine the age of rocks and archaeological materials.

42.32 Explain how radiocarbon dating can be used to date the age of biological samples.

Key Idea

● Naturally occurring radioactive nuclides provide a means for estimating the dates of historic and prehistoric events. For example, the ages of organic materials can often be found by measuring their ^{14}C content, and rock samples can be dated using the radioactive ^{40}K.

Radioactive Dating

If you know the half-life of a given radionuclide, you can in principle use the decay of that radionuclide as a clock to measure time intervals. The decay of very long-lived nuclides, for example, can be used to measure the age of rocks—that is, the time that has elapsed since they were formed. Such measurements for rocks from Earth and the Moon, and for meteorites, yield a consistent maximum age of about 4.5×10^9 y for these bodies.

The radionuclide ^{40}K, for example, decays to ^{40}Ar, a stable isotope of the noble gas argon. The half-life for this decay is 1.25×10^9 y. A measurement of the ratio of ^{40}K to ^{40}Ar, as found in the rock in question, can be used to calculate the age of that rock. Other long-lived decays, such as that of ^{235}U to ^{207}Pb (involving a number of intermediate stages of unstable nuclei), can be used to verify this calculation.

For measuring shorter time intervals, in the range of historical interest, radiocarbon dating has proved invaluable. The radionuclide ^{14}C (with $T_{1/2} = 5730$ y) is produced at a constant rate in the upper atmosphere as atmospheric nitrogen is bombarded by cosmic rays. This radiocarbon mixes with the carbon that is normally present in the atmosphere (as CO_2) so that there is about one atom of ^{14}C for every 10^{13} atoms of ordinary stable ^{12}C. Through biological activity such as photosynthesis and breathing, the atoms of atmospheric carbon trade places randomly, one atom at a time, with the atoms of carbon in every living thing, including broccoli, mushrooms, penguins, and humans. Eventually an exchange equilibrium is reached at which the carbon atoms of every living thing contain a fixed small fraction of the radioactive nuclide ^{14}C.

This equilibrium persists as long as the organism is alive. When the organism dies, the exchange with the atmosphere stops and the amount of radiocarbon trapped in the organism, since it is no longer being replenished, dwindles away with a half-life of 5730 y. By measuring the amount of radiocarbon per gram of organic matter, it is possible to measure the time that has elapsed since the organism died. Charcoal from ancient campfires, the Dead Sea scrolls (actually, the cloth used to plug the jars holding the scrolls), and many prehistoric artifacts have been dated in this way.

Top photo: George Rockwin/Bruce Coleman, Inc./Photoshot Holdings Ltd. Inset photo: www.BibleLandPictures.com/Alamy

A fragment of the Dead Sea scrolls and the caves from which the scrolls were recovered.

Sample Problem 42.08 Radioactive dating of a moon rock

In a Moon rock sample, the ratio of the number of (stable) ^{40}Ar atoms present to the number of (radioactive) ^{40}K atoms is 10.3. Assume that all the argon atoms were produced by the decay of potassium atoms, with a half-life of 1.25×10^9 y. How old is the rock?

KEY IDEAS

(1) If N_0 potassium atoms were present at the time the rock was formed by solidification from a molten form, the number of potassium atoms now remaining at the time of analysis is

$$N_K = N_0 e^{-\lambda t}, \qquad (42\text{-}29)$$

in which t is the age of the rock. (2) For every potassium atom that decays, an argon atom is produced. Thus, the number of argon atoms present at the time of the analysis is

$$N_{Ar} = N_0 - N_K. \qquad (42\text{-}30)$$

Calculations: We cannot measure N_0; so let's eliminate it from Eqs. 42-29 and 42-30. We find, after some algebra, that

$$\lambda t = \ln\left(1 + \frac{N_{Ar}}{N_K}\right), \qquad (42\text{-}31)$$

in which N_{Ar}/N_K *can* be measured. Solving for t and using Eq. 42-18 to replace λ with $(\ln 2)/T_{1/2}$ yield

$$t = \frac{T_{1/2}\ln(1 + N_{Ar}/N_K)}{\ln 2}$$
$$= \frac{(1.25 \times 10^9 \text{ y})[\ln(1 + 10.3)]}{\ln 2}$$
$$= 4.37 \times 10^9 \text{ y}. \qquad \text{(Answer)}$$

Lesser ages may be found for other lunar or terrestrial rock samples, but no substantially greater ones. Thus, the oldest rocks were formed soon after the solar system formed, and the solar system must be about 4 billion years old.

PLUS Additional examples, video, and practice available at *WileyPLUS*

42-7 MEASURING RADIATION DOSAGE

Learning Objectives

After reading this module, you should be able to . . .

42.33 Identify absorbed dose, dose equivalent, and the associated units.

42.34 Calculate absorbed dose and dose equivalent.

Key Ideas

● The Becquerel (1 Bq = 1 decay per second) measures the activity of a source.

● The amount of energy actually absorbed is measured in grays, with 1 Gy corresponding to 1 J/kg.

● The estimated biological effect of the absorbed energy is the dose equivalent and is measured in sieverts.

Measuring Radiation Dosage

The effect of radiation such as gamma rays, electrons, and alpha particles on living tissue (particularly our own) is a matter of public interest. Such radiation is found in nature in cosmic rays (from astronomical sources) and in the emissions by radioactive elements in Earth's crust. Radiation associated with some human activities, such as using x rays and radionuclides in medicine and in industry, also contributes.

Our task here is not to explore the various sources of radiation but simply to describe the units in which the properties and effects of such radiations are expressed. We have already discussed the *activity* of a radioactive source. There are two remaining quantities of interest.

1. *Absorbed Dose.* This is a measure of the radiation dose (as energy per unit mass) actually absorbed by a specific object, such as a patient's hand or chest. Its SI unit is the **gray** (Gy). An older unit, the **rad** (from **r**adiation **a**bsorbed

dose) is still in common use. The terms are defined and related as follows:

$$1 \text{ Gy} = 1 \text{ J/kg} = 100 \text{ rad.} \qquad (42\text{-}32)$$

A typical dose-related statement is: "A whole-body, short-term gamma-ray dose of 3 Gy (= 300 rad) will cause death in 50% of the population exposed to it." Thankfully, our present average absorbed dose per year, from sources of both natural and human origin, is only about 2 mGy (= 0.2 rad).

2. *Dose Equivalent.* Although different types of radiation (gamma rays and neutrons, say) may deliver the same amount of energy to the body, they do not have the same biological effect. The dose equivalent allows us to express the biological effect by multiplying the absorbed dose (in grays or rads) by a numerical **RBE** factor (from **r**elative **b**iological **e**ffectiveness). For x rays and electrons, for example, RBE = 1; for slow neutrons, RBE = 5; for alpha particles, RBE = 10; and so on. Personnel-monitoring devices such as film badges register the dose equivalent.

The SI unit of dose equivalent is the **sievert** (Sv). An earlier unit, the **rem,** is still in common use. Their relationship is

$$1 \text{ Sv} = 100 \text{ rem.} \qquad (42\text{-}33)$$

An example of the correct use of these terms is: "The recommendation of the National Council on Radiation Protection is that no individual who is (nonoccupationally) exposed to radiation should receive a dose equivalent greater than 5 mSv (= 0.5 rem) in any one year." This includes radiation of all kinds; of course the appropriate RBE factor must be used for each kind.

42-8 NUCLEAR MODELS

Learning Objectives

After reading this module, you should be able to . . .

42.35 Distinguish the collective model and the independent model, and explain the combined model.

42.36 Identify compound nucleus.

42.37 Identify magic numbers.

Key Ideas

● The collective model of nuclear structure assumes that nucleons collide constantly with one another and that relatively long-lived compound nuclei are formed when a projectile is captured. The formation and eventual decay of a compound nucleus are totally independent events.

● The independent particle model of nuclear structure assumes that each nucleon moves, essentially without collision, in a quantized state within the nucleus. The model predicts nucleon levels and magic nucleon numbers associated with closed shells of nucleons.

● The combined model assumes that extra nucleons occupy quantized states outside a central core of closed shells.

Nuclear Models

Nuclei are more complicated than atoms. For atoms, the basic force law (Coulomb's law) is simple in form and there is a natural force center, the nucleus. For nuclei, the force law is complicated and cannot, in fact, be written down explicitly in full detail. Furthermore, the nucleus—a jumble of protons and neutrons—has no natural force center to simplify the calculations.

In the absence of a comprehensive nuclear *theory,* we turn to the construction of nuclear *models.* A nuclear model is simply a way of looking at the nucleus that gives a physical insight into as wide a range of its properties as possible. The usefulness of a model is tested by its ability to provide predictions that can be verified experimentally in the laboratory.

Two models of the nucleus have proved useful. Although based on assumptions that seem flatly to exclude each other, each accounts very well for a selected group of nuclear properties. After describing them separately, we shall see how these two models may be combined to form a single coherent picture of the atomic nucleus.

The Collective Model

In the *collective model,* formulated by Niels Bohr, the nucleons, moving around within the nucleus at random, are imagined to interact strongly with each other, like the molecules in a drop of liquid. A given nucleon collides frequently with other nucleons in the nuclear interior, its mean free path as it moves about being substantially less than the nuclear radius.

The collective model permits us to correlate many facts about nuclear masses and binding energies; it is useful (as you will see later) in explaining nuclear fission. It is also useful for understanding a large class of nuclear reactions.

Consider, for example, a generalized nuclear reaction of the form

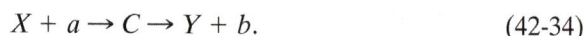

$$X + a \rightarrow C \rightarrow Y + b. \tag{42-34}$$

We imagine that projectile a enters target nucleus X, forming a **compound nucleus** C and conveying to it a certain amount of excitation energy. The projectile, perhaps a neutron, is at once caught up by the random motions that characterize the nuclear interior. It quickly loses its identity—so to speak—and the excitation energy it carried into the nucleus is quickly shared with all the other nucleons in C.

The quasi-stable state represented by C in Eq. 42-34 may have a mean life of 10^{-16} s before it decays to Y and b. By nuclear standards, this is a very long time, being about one million times longer than the time required for a nucleon with a few million electron-volts of energy to travel across a nucleus.

The central feature of this compound-nucleus concept is that the formation of the compound nucleus and its eventual decay are totally independent events. At the time of its decay, the compound nucleus has "forgotten" how it was formed. Hence, its mode of decay is not influenced by its mode of formation. As an example, Fig. 42-14 shows three possible ways in which the compound nucleus ^{20}Ne might be formed and three in which it might decay. Any of the three formation modes can lead to any of the three decay modes.

The Independent Particle Model

In the collective model, we assume that the nucleons move around at random and bump into one another frequently. The *independent particle model,* however, is based on just the opposite assumption—namely, that each nucleon remains in a well-defined quantum state within the nucleus and makes hardly any collisions at all! The nucleus, unlike the atom, has no fixed center of charge; we assume in this model that each nucleon moves in a potential well that is determined by the smeared-out (time-averaged) motions of all the other nucleons.

A nucleon in a nucleus, like an electron in an atom, has a set of quantum numbers that defines its state of motion. Also, nucleons obey the Pauli exclusion principle, just as electrons do; that is, no two nucleons in a nucleus may occupy the same quantum state at the same time. In this regard, the neutrons and the protons are treated separately, each particle type with its own set of quantum states.

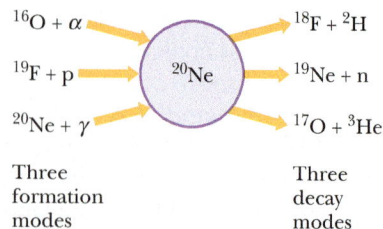

Figure 42-14 The formation modes and the decay modes of the compound nucleus ^{20}Ne.

The fact that nucleons obey the Pauli exclusion principle helps us to understand the relative stability of nucleon states. If two nucleons within the nucleus are to collide, the energy of each of them after the collision must correspond to the energy of an *unoccupied* state. If no such state is available, the collision simply cannot occur. Thus, any given nucleon experiencing repeated "frustrated collision opportunities" will maintain its state of motion long enough to give meaning to the statement that it exists in a quantum state with a well-defined energy.

In the atomic realm, the repetitions of physical and chemical properties that we find in the periodic table are associated with a property of atomic electrons—namely, they arrange themselves in shells that have a special stability when fully occupied. We can take the atomic numbers of the noble gases,

$$2, \ 10, \ 18, \ 36, \ 54, \ 86, \ldots,$$

as *magic electron numbers* that mark the completion (or closure) of such shells.

Nuclei also show such closed-shell effects, associated with certain **magic nucleon numbers:**

$$2, \ 8, \ 20, \ 28, 50, 82, 126, \ldots.$$

Any nuclide whose proton number Z or neutron number N has one of these values turns out to have a special stability that may be made apparent in a variety of ways.

Examples of "magic" nuclides are ^{18}O ($Z = 8$), ^{40}Ca ($Z = 20, N = 20$), ^{92}Mo ($N = 50$), and ^{208}Pb ($Z = 82, N = 126$). Both ^{40}Ca and ^{208}Pb are said to be "doubly magic" because they contain both filled shells of protons *and* filled shells of neutrons.

The magic number 2 shows up in the exceptional stability of the alpha particle (^4He), which, with $Z = N = 2$, is doubly magic. For example, on the binding energy curve of Fig. 42-7, the binding energy per nucleon for this nuclide stands well above those of its periodic-table neighbors hydrogen, lithium, and beryllium. The neutrons and protons making up the alpha particle are so tightly bound to one another, in fact, that it is impossible to add another proton or neutron to it; there is no stable nuclide with $A = 5$.

The central idea of a closed shell is that a single particle outside a closed shell can be relatively easily removed, but considerably more energy must be expended to remove a particle from the shell itself. The sodium atom, for example, has one (valence) electron outside a closed electron shell. Only about 5 eV is required to strip the valence electron away from a sodium atom; however, to remove a *second* electron (which must be plucked out of a closed shell) requires 22 eV. As a nuclear case, consider ^{121}Sb ($Z = 51$), which contains a single proton outside a closed shell of 50 protons. To remove this lone proton requires 5.8 MeV; to remove a *second* proton, however, requires an energy of 11 MeV. There is much additional experimental evidence that the nucleons in a nucleus form closed shells and that these shells exhibit stable properties.

We have seen that quantum theory can account beautifully for the magic electron numbers—that is, for the populations of the subshells into which atomic electrons are grouped. It turns out that, under certain assumptions, quantum theory can account equally well for the magic nucleon numbers! The 1963 Nobel Prize in physics was, in fact, awarded to Maria Mayer and Hans Jensen "for their discoveries concerning nuclear shell structure."

A Combined Model

Consider a nucleus in which a small number of neutrons (or protons) exist outside a core of closed shells that contains magic numbers of neutrons or protons. The outside nucleons occupy quantized states in a potential well established by the central core, thus preserving the central feature of the independent-particle model. These outside nucleons also interact with the core, deforming it and setting up "tidal wave" motions of rotation or vibration within it. These collective motions of the core preserve the central feature of the collective model. Such a model of nuclear structure thus succeeds in combining the seemingly irreconcilable points of view of the collective and independent-particle models. It has been remarkably successful in explaining observed nuclear properties.

Sample Problem 42.09 Lifetime of a compound nucleus made by neutron capture

Consider the neutron capture reaction

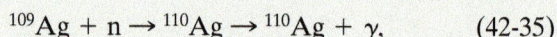

$$^{109}\text{Ag} + \text{n} \rightarrow {}^{110}\text{Ag} \rightarrow {}^{110}\text{Ag} + \gamma, \qquad (42\text{-}35)$$

in which a compound nucleus (^{110}Ag) is formed. Figure 42-15 shows the relative rate at which such events take place, plotted against the energy of the incoming neutron. Find the mean lifetime of this compound nucleus by using the uncertainty principle in the form

$$\Delta E \cdot \Delta t \approx \hbar. \qquad (42\text{-}36)$$

Here ΔE is a measure of the uncertainty with which the energy of a state can be defined. The quantity Δt is a measure of the time available to measure this energy. In fact, here Δt is just t_{avg}, the average life of the compound nucleus before it decays to its ground state.

Reasoning: We see that the relative reaction rate peaks sharply at a neutron energy of about 5.2 eV. This suggests that we are dealing with a single excited energy level of the compound nucleus ^{110}Ag. When the available energy (of the incoming neutron) just matches the energy of this level above the ^{110}Ag ground state, we have "resonance" and the reaction of Eq. 42-35 really "goes."

However, the resonance peak is not infinitely sharp but has an approximate half-width (ΔE in the figure) of about 0.20 eV. We can account for this resonance-peak width by saying that the excited level is not sharply defined in energy but has an energy uncertainty ΔE of about 0.20 eV.

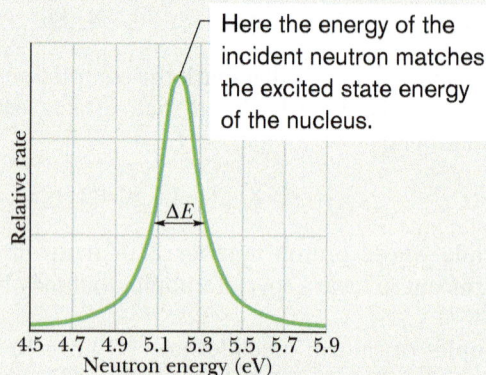

Here the energy of the incident neutron matches the excited state energy of the nucleus.

Figure 42-15 A plot of the relative number of reaction events of the type described by Eq. 42-35 as a function of the energy of the incident neutron. The half-width ΔE of the resonance peak is about 0.20 eV.

Calculation: Substituting that uncertainty of 0.20 eV into Eq. 42-36 gives us

$$\Delta t = t_{\text{avg}} \approx \frac{\hbar}{\Delta E} \approx \frac{(4.14 \times 10^{-15}\,\text{eV} \cdot \text{s})/2\pi}{0.20\,\text{eV}}$$

$$\approx 3 \times 10^{-15}\,\text{s}. \qquad \text{(Answer)}$$

This is several hundred times greater than the time a 5.2 eV neutron takes to cross the diameter of a ^{109}Ag nucleus. Therefore, the neutron is spending this time of 3×10^{-15} s *as part of* the nucleus.

WILEY PLUS Additional examples, video, and practice available at *WileyPLUS*

Review & Summary

The Nuclides Approximately 2000 **nuclides** are known to exist. Each is characterized by an **atomic number** Z (the number of protons), a **neutron number** N, and a **mass number** A (the total number of **nucleons**—protons and neutrons). Thus, $A = Z + N$. Nuclides with the same atomic number but different neutron numbers are **isotopes** of one another. Nuclei have a mean radius r given by

$$r = r_0 A^{1/3}, \qquad (42\text{-}3)$$

where $r_0 \approx 1.2$ fm.

Mass and Binding Energy Atomic masses are often re-

ported in terms of *mass excess*

$$\Delta = M - A \quad \text{(mass excess)}, \qquad (42\text{-}6)$$

where M is the actual mass of an atom in atomic mass units and A is the mass number for that atom's nucleus. The **binding energy** of a nucleus is the difference

$$\Delta E_{\text{be}} = \Sigma(mc^2) - Mc^2 \quad \text{(binding energy)}, \qquad (42\text{-}7)$$

where $\Sigma(mc^2)$ is the total mass energy of the *individual* protons and neutrons. The **binding energy per nucleon** is

$$\Delta E_{ben} = \frac{\Delta E_{be}}{A} \quad \text{(binding energy per nucleon).} \quad (42\text{-}8)$$

Mass–Energy Exchanges The energy equivalent of one mass unit (u) is 931.494 013 MeV. The binding energy curve shows that middle-mass nuclides are the most stable and that energy can be released both by fission of high-mass nuclei and by fusion of low-mass nuclei.

The Nuclear Force Nuclei are held together by an attractive force acting among the nucleons, part of the **strong force** acting between the quarks that make up the nucleons.

Radioactive Decay Most known nuclides are radioactive; they spontaneously decay at a rate R $(= -dN/dt)$ that is proportional to the number N of radioactive atoms present, the proportionality constant being the **disintegration constant** λ. This leads to the law of exponential decay:

$$N = N_0 e^{-\lambda t}, \qquad R = \lambda N = R_0 e^{-\lambda t}$$
$$\text{(radioactive decay).} \quad (42\text{-}15, 42\text{-}17, 42\text{-}16)$$

The **half-life** $T_{1/2} = (\ln 2)/\lambda$ of a radioactive nuclide is the time required for the decay rate R (or the number N) in a sample to drop to half its initial value.

Alpha Decay Some nuclides decay by emitting an alpha particle (a helium nucleus, ^{4}He). Such decay is inhibited by a potential energy barrier that cannot be penetrated according to classical physics but is subject to tunneling according to quantum physics. The barrier penetrability, and thus the half-life for alpha decay, is very sensitive to the energy of the emitted alpha particle.

Beta Decay In **beta decay** either an electron or a positron is emitted by a nucleus, along with a neutrino. The emitted particles share the available disintegration energy. The electrons and positrons emitted in beta decay have a continuous spectrum of energies from near zero up to a limit K_{max} $(= Q = -\Delta m c^2)$.

Radioactive Dating Naturally occurring radioactive nuclides provide a means for estimating the dates of historic and prehistoric events. For example, the ages of organic materials can often be found by measuring their ^{14}C content; rock samples can be dated using the radioactive isotope ^{40}K.

Radiation Dosage Three units are used to describe exposure to ionizing radiation. The **becquerel** (1 Bq = 1 decay per second) measures the **activity** of a source. The amount of energy actually absorbed is measured in **grays,** with 1 Gy corresponding to 1 J/kg. The estimated biological effect of the absorbed energy is measured in **sieverts;** a dose equivalent of 1 Sv causes the same biological effect regardless of the radiation type by which it was acquired.

Nuclear Models The **collective** model of nuclear structure assumes that nucleons collide constantly with one another and that relatively long-lived **compound nuclei** are formed when a projectile is captured. The formation and eventual decay of a compound nucleus are totally independent events.

The **independent particle** model of nuclear structure assumes that each nucleon moves, essentially without collisions, in a quantized state within the nucleus. The model predicts nucleon levels and **magic nucleon numbers** (2, 8, 20, 28, 50, 82, and 126) associated with closed shells of nucleons; nuclides with any of these numbers of neutrons or protons are particularly stable.

The **combined** model, in which extra nucleons occupy quantized states outside a central core of closed shells, is highly successful in predicting many nuclear properties.

Questions

1 The radionuclide ^{196}Ir decays by emitting an electron. (a) Into which square in Fig. 42-6 is it transformed? (b) Do further decays then occur?

2 Is the mass excess of an alpha particle (use a straightedge on Fig. 42-13) greater than or less than the particle's total binding energy (use the binding energy per nucleon from Fig. 42-7)?

3 At $t = 0$, a sample of radionuclide A has the same decay rate as a sample of radionuclide B has at $t = 30$ min. The disintegration constants are λ_A and λ_B, with $\lambda_A < \lambda_B$. Will the two samples ever have (simultaneously) the same decay rate? (*Hint:* Sketch a graph of their activities.)

4 A certain nuclide is said to be particularly stable. Does its binding energy per nucleon lie slightly above or slightly below the binding energy curve of Fig. 42-7?

5 Suppose the alpha particle in a Rutherford scattering experiment is replaced with a proton of the same initial kinetic energy and also headed directly toward the nucleus of the gold atom. (a) Will the distance from the center of the nucleus at which the proton stops be greater than, less than, or the same as that of the alpha particle? (b) If, instead, we switch the target to a nucleus with a larger value of Z, is the stopping distance of the alpha particle greater than, less than, or the same as with the gold target?

6 Figure 42-16 gives the activities of three radioactive samples versus time. Rank the samples according to their (a) half-life and (b) disintegration constant, greatest first. (*Hint:* For (a), use a straightedge on the graph.)

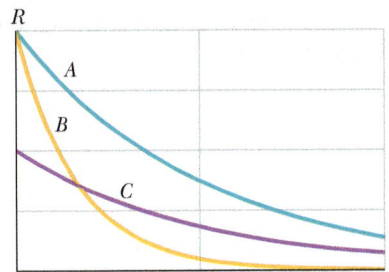

Figure 42-16 Question 6.

7 The nuclide ^{244}Pu $(Z = 94)$ is an alpha-emitter. Into which of the following nuclides does it decay: ^{240}Np $(Z = 93)$, ^{240}U $(Z = 92)$, ^{248}Cm $(Z = 96)$, or ^{244}Am $(Z = 95)$?

8 The radionuclide ^{49}Sc has a half-life of 57.0 min. At $t = 0$, the counting rate of a sample of it is 6000 counts/min above the general background activity, which is 30 counts/min. Without computation, determine whether the counting rate of the sample will be about equal to the background rate in 3 h, 7 h, 10 h, or a time much longer than 10 h.

9 At $t = 0$ we begin to observe two identical radioactive nuclei that have a half-life of 5 min. At $t = 1$ min, one of the nuclei decays. Does that event increase or decrease the chance that the second nucleus will decay in the next 4 min, or is there no effect on the second nucleus? (Are the events cause and effect, or random?)

10 Figure 42-17 shows the curve for the binding energy per nucleon ΔE_{ben} versus mass number A. Three isotopes are indicated. Rank them according to the energy required to remove a nucleon from the isotope, greatest first.

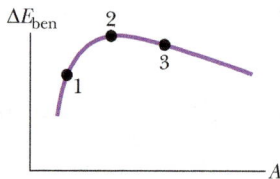

Figure 42-17 Question 10.

11 At $t = 0$, a sample of radionuclide A has twice the decay rate as a sample of radionuclide B. The disintegration constants are λ_A and λ_B,

with $\lambda_A > \lambda_B$. Will the two samples ever have (simultaneously) the same decay rate?

12 Figure 42-18 is a plot of mass number A versus charge number Z. The location of a certain nucleus is represented by a dot. Which of the arrows extending from the dot would best represent the transition were the nucleus to undergo (a) a β^- decay and (b) an α decay?

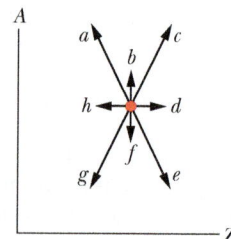

Figure 42-18 Question 12.

13 (a) Which of the following nuclides are magic: ^{122}Sn, ^{132}Sn, ^{98}Cd, ^{198}Au, ^{208}Pb? (b) Which, if any, are doubly magic?

14 If the mass of a radioactive sample is doubled, do (a) the activity of the sample and (b) the disintegration constant of the sample increase, decrease, or remain the same?

15 The magic nucleon numbers for nuclei are given in Module 42-8 as 2, 8, 20, 28, 50, 82, and 126. Are nuclides magic (that is, especially stable) when (a) only the mass number A, (b) only the atomic number Z, (c) only the neutron number N, or (d) either Z or N (or both) is equal to one of these numbers? Pick all correct phrases.

Problems

Module 42-1 Discovering the Nucleus

•1 A ^{7}Li nucleus with a kinetic energy of 3.00 MeV is sent toward a ^{232}Th nucleus. What is the least center-to-center separation between the two nuclei, assuming that the (more massive) ^{232}Th nucleus does not move?

•2 Calculate the distance of closest approach for a head-on collision between a 5.30 MeV alpha particle and a copper nucleus.

••3 A 10.2 MeV Li nucleus is shot directly at the center of a Ds nucleus. At what center-to-center distance does the Li momentarily stop, assuming the Ds does not move?

••4 **GO** In a Rutherford scattering experiment, assume that an incident alpha particle (radius 1.80 fm) is headed directly toward a target gold nucleus (radius 6.23 fm). What energy must the alpha particle have to just barely "touch" the gold nucleus?

••5 **GO** When an alpha particle collides elastically with a nucleus, the nucleus recoils. Suppose a 5.00 MeV alpha particle has a head-on elastic collision with a gold nucleus that is initially at rest. What is the kinetic energy of (a) the recoiling nucleus and (b) the rebounding alpha particle?

Module 42-2 Some Nuclear Properties

•6 The strong neutron excess (defined as $N - Z$) of high-mass nuclei is illustrated by noting that most high-mass nuclides could never fission into two stable nuclei without neutrons being left over. For example, consider the spontaneous fission of a ^{235}U nucleus into two stable *daughter nuclei* with atomic numbers 39 and 53. From Appendix F, determine the name of the (a) first and (b) second daughter nucleus. From Fig. 42-5, approximately how many

neutrons are in the (c) first and (d) second? (e) Approximately how many neutrons are left over?

•7 What is the nuclear mass density ρ_m of (a) the fairly low-mass nuclide ^{55}Mn and (b) the fairly high-mass nuclide ^{209}Bi? (c) Compare the two answers, with an explanation. What is the nuclear charge density ρ_q of (d) ^{55}Mn and (e) ^{209}Bi? (f) Compare the two answers, with an explanation.

•8 (a) Show that the mass M of an atom is given approximately by $M_{app} = A m_p$, where A is the mass number and m_p is the proton mass. For (b) ^{1}H, (c) ^{31}P, (d) ^{120}Sn, (e) ^{197}Au, and (f) ^{239}Pu, use Table 42-1 to find the percentage deviation between M_{app} and M:

$$\text{percentage deviation} = \frac{M_{app} - M}{M} \, 100.$$

(g) Is a value of M_{app} accurate enough to be used in a calculation of a nuclear binding energy?

•9 The nuclide ^{14}C contains (a) how many protons and (b) how many neutrons?

•10 What is the mass excess Δ_1 of ^{1}H (actual mass is 1.007 825 u) in (a) atomic mass units and (b) MeV/c^2? What is the mass excess Δ_n of a neutron (actual mass is 1.008 665 u) in (c) atomic mass units and (d) MeV/c^2? What is the mass excess Δ_{120} of ^{120}Sn (actual mass is 119.902 197 u) in (e) atomic mass units and (f) MeV/c^2?

•11 **SSM** Nuclear radii may be measured by scattering high-energy (high speed) electrons from nuclei. (a) What is the de Broglie wavelength for 200 MeV electrons? (b) Are these electrons suitable probes for this purpose?

•12 The electric potential energy of a uniform sphere of charge q and radius r is given by

$$U = \frac{3q^2}{20\pi\varepsilon_0 r}.$$

(a) Does the energy represent a tendency for the sphere to bind together or blow apart? The nuclide ^{239}Pu is spherical with radius 6.64 fm. For this nuclide, what are (b) the electric potential energy U according to the equation, (c) the electric potential energy per proton, and (d) the electric potential energy per nucleon? The binding energy per nucleon is 7.56 MeV. (e) Why is the nuclide bound so well when the answers to (c) and (d) are large and positive?

•13 A neutron star is a stellar object whose density is about that of nuclear matter, 2×10^{17} kg/m^3. Suppose that the Sun were to collapse and become such a star without losing any of its present mass. What would be its radius?

••14 GO What is the binding energy per nucleon of the americium isotope $^{244}_{95}$Am? Here are some atomic masses and the neutron mass.

$^{244}_{95}$Am	244.064 279 u	^{1}H	1.007 825 u
n	1.008 665 u		

••15 (a) Show that the energy associated with the strong force between nucleons in a nucleus is proportional to A, the mass number of the nucleus in question. (b) Show that the energy associated with the Coulomb force between protons in a nucleus is proportional to $Z(Z - 1)$. (c) Show that, as we move to larger and larger nuclei (see Fig. 42-5), the importance of the Coulomb force increases more rapidly than does that of the strong force.

••16 GO What is the binding energy per nucleon of the europium isotope $^{152}_{63}$Eu? Here are some atomic masses and the neutron mass.

$^{152}_{63}$Eu	151.921 742 u	^{1}H	1.007 825 u
n	1.008 665 u		

••17 Because the neutron has no charge, its mass must be found in some way other than by using a mass spectrometer. When a neutron and a proton meet (assume both to be almost stationary), they combine and form a deuteron, emitting a gamma ray whose energy is 2.2233 MeV. The masses of the proton and the deuteron are 1.007 276 467 u and 2.013 553 212 u, respectively. Find the mass of the neutron from these data.

••18 GO What is the binding energy per nucleon of the rutherfordium isotope $^{259}_{104}$Rf? Here are some atomic masses and the neutron mass.

$^{259}_{104}$Rf	259.105 63 u	^{1}H	1.007 825 u
n	1.008 665 u		

••19 A periodic table might list the average atomic mass of magnesium as being 24.312 u, which is the result of *weighting* the atomic masses of the magnesium isotopes according to their natural abundances on Earth. The three isotopes and their masses are ^{24}Mg (23.985 04 u), ^{25}Mg (24.985 84 u), and ^{26}Mg (25.982 59 u). The natural abundance of ^{24}Mg is 78.99% by mass (that is, 78.99% of the mass of a naturally occurring sample of magnesium is due to the presence of ^{24}Mg). What is the abundance of (a) ^{25}Mg and (b) ^{26}Mg?

••20 What is the binding energy per nucleon of ^{262}Bh? The mass of the atom is 262.1231 u.

••21 SSM WWW (a) Show that the total binding energy E_{be} of a given nuclide is

$$E_{be} = Z\Delta_H + N\Delta_n - \Delta,$$

where Δ_H is the mass excess of ^{1}H, Δ_n is the mass excess of a neutron, and Δ is the mass excess of the given nuclide. (b) Using this method, calculate the binding energy per nucleon for ^{197}Au. Compare your result with the value listed in Table 42-1. The needed mass excesses, rounded to three significant figures, are $\Delta_H = +7.29$ MeV, $\Delta_n = +8.07$ MeV, and $\Delta_{197} = -31.2$ MeV. Note the economy of calculation that results when mass excesses are used in place of the actual masses.

••22 GO An α particle (^{4}He nucleus) is to be taken apart in the following steps. Give the energy (work) required for each step: (a) remove a proton, (b) remove a neutron, and (c) separate the remaining proton and neutron. For an α particle, what are (d) the total binding energy and (e) the binding energy per nucleon? (f) Does either match an answer to (a), (b), or (c)? Here are some atomic masses and the neutron mass.

^{4}He	4.002 60 u	^{2}H	2.014 10 u
^{3}H	3.016 05 u	^{1}H	1.007 83 u
n	1.008 67 u		

••23 SSM Verify the binding energy per nucleon given in Table 42-1 for the plutonium isotope ^{239}Pu. The mass of the neutral atom is 239.052 16 u.

••24 A penny has a mass of 3.0 g. Calculate the energy that would be required to separate all the neutrons and protons in this coin from one another. For simplicity, assume that the penny is made entirely of ^{63}Cu atoms (of mass 62.929 60 u). The masses of the proton-plus-electron and the neutron are 1.007 83 u and 1.008 66 u, respectively.

Module 42-3 Radioactive Decay

•25 Cancer cells are more vulnerable to x and gamma radiation than are healthy cells. In the past, the standard source for radiation therapy was radioactive ^{60}Co, which decays, with a half-life of 5.27 y, into an excited nuclear state of ^{60}Ni. That nickel isotope then immediately emits two gamma-ray photons, each with an approximate energy of 1.2 MeV. How many radioactive ^{60}Co nuclei are present in a 6000 Ci source of the type used in hospitals? (Energetic particles from linear accelerators are now used in radiation therapy.)

•26 The half-life of a radioactive isotope is 140 d. How many days would it take for the decay rate of a sample of this isotope to fall to one-fourth of its initial value?

•27 A radioactive nuclide has a half-life of 30.0 y. What fraction of an initially pure sample of this nuclide will remain undecayed at the end of (a) 60.0 y and (b) 90.0 y?

•28 The plutonium isotope ^{239}Pu is produced as a by-product in nuclear reactors and hence is accumulating in our environment. It is radioactive, decaying with a half-life of 2.41×10^4 y. (a) How many nuclei of Pu constitute a chemically lethal dose of 2.00 mg? (b) What is the decay rate of this amount?

•29 SSM WWW A radioactive isotope of mercury, ^{197}Hg, decays to gold, ^{197}Au, with a disintegration constant of 0.0108 h^{-1}. (a) Calculate the half-life of the ^{197}Hg. What fraction of a sample will remain at the end of (b) three half-lives and (c) 10.0 days?

•30 The half-life of a particular radioactive isotope is 6.5 h. If there are initially 48×10^{19} atoms of this isotope, how many remain at the end of 26 h?

•31 Consider an initially pure 3.4 g sample of ^{67}Ga, an isotope that has a half-life of 78 h. (a) What is its initial decay rate? (b) What is its decay rate 48 h later?

•32 When aboveground nuclear tests were conducted, the explosions shot radioactive dust into the upper atmosphere. Global air circulations then spread the dust worldwide before it settled out on ground and water. One such test was conducted in October 1976. What fraction of the ^{90}Sr produced by that explosion still existed in October 2006? The half-life of ^{90}Sr is 29 y.

••33 The air in some caves includes a significant amount of radon gas, which can lead to lung cancer if breathed over a prolonged time. In British caves, the air in the cave with the greatest amount of the gas has an activity per volume of 1.55×10^5 Bq/m³. Suppose that you spend two full days exploring (and sleeping in) that cave. Approximately how many ^{222}Rn atoms would you take in and out of your lungs during your two-day stay? The radionuclide ^{222}Rn in radon gas has a half-life of 3.82 days. You need to estimate your lung capacity and average breathing rate.

••34 Calculate the mass of a sample of (initially pure) ^{40}K that has an initial decay rate of 1.70×10^5 disintegrations/s. The isotope has a half-life of 1.28×10^9 y.

••35 SSM A certain radionuclide is being manufactured in a cyclotron at a constant rate R. It is also decaying with disintegration constant λ. Assume that the production process has been going on for a time that is much longer than the half-life of the radionuclide. (a) Show that the number of radioactive nuclei present after such time remains constant and is given by $N = R/\lambda$. (b) Now show that this result holds no matter how many radioactive nuclei were present initially. The nuclide is said to be in *secular equilibrium* with its source; in this state its decay rate is just equal to its production rate.

••36 Plutonium isotope ^{239}Pu decays by alpha decay with a half-life of 24 100 y. How many milligrams of helium are produced by an initially pure 12.0 g sample of ^{239}Pu at the end of 20 000 y? (Consider only the helium produced directly by the plutonium and not by any by-products of the decay process.)

••37 The radionuclide ^{64}Cu has a half-life of 12.7 h. If a sample contains 5.50 g of initially pure ^{64}Cu at $t = 0$, how much of it will decay between $t = 14.0$ h and $t = 16.0$ h?

••38 A dose of 8.60 μCi of a radioactive isotope is injected into a patient. The isotope has a half-life of 3.0 h. How many of the isotope parents are injected?

••39 The radionuclide ^{56}Mn has a half-life of 2.58 h and is produced in a cyclotron by bombarding a manganese target with deuterons. The target contains only the stable manganese isotope ^{55}Mn, and the manganese–deuteron reaction that produces ^{56}Mn is

$$^{55}\text{Mn} + \text{d} \rightarrow {}^{56}\text{Mn} + \text{p}.$$

If the bombardment lasts much longer than the half-life of ^{56}Mn, the activity of the ^{56}Mn produced in the target reaches a final value of 8.88×10^{10} Bq. (a) At what rate is ^{56}Mn being produced? (b) How many ^{56}Mn nuclei are then in the target? (c) What is their total mass?

••40 A source contains two phosphorus radionuclides, ^{32}P ($T_{1/2} = 14.3$ d) and ^{33}P ($T_{1/2} = 25.3$ d). Initially, 10.0% of the decays come from ^{33}P. How long must one wait until 90.0% do so?

••41 A 1.00 g sample of samarium emits alpha particles at a rate of 120 particles/s. The responsible isotope is ^{147}Sm, whose natural abundance in bulk samarium is 15.0%. Calculate the half-life.

••42 What is the activity of a 20 ng sample of ^{92}Kr, which has a half-life of 1.84 s?

••43 GO A radioactive sample intended for irradiation of a hospital patient is prepared at a nearby laboratory. The sample has a half-life of 83.61 h. What should its initial activity be if its activity is to be 7.4×10^8 Bq when it is used to irradiate the patient 24 h later?

••44 GO Figure 42-19 shows the decay of parents in a radioactive sample. The axes are scaled by $N_s = 2.00 \times 10^6$ and $t_s = 10.0$ s. What is the activity of the sample at $t = 27.0$ s?

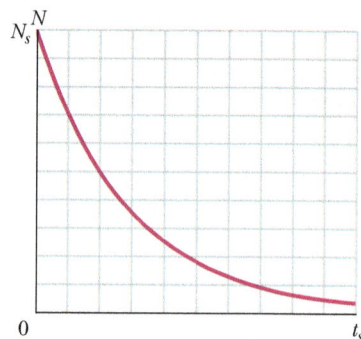

Figure 42-19 Problem 44.

••45 In 1992, Swiss police arrested two men who were attempting to smuggle osmium out of Eastern Europe for a clandestine sale. However, by error, the smugglers had picked up ^{137}Cs. Reportedly, each smuggler was carrying a 1.0 g sample of ^{137}Cs *in a pocket*! In (a) bequerels and (b) curies, what was the activity of each sample? The isotope ^{137}Cs has a half-life of 30.2 y. (The activities of radioisotopes commonly used in hospitals range up to a few millicuries.)

••46 The radioactive nuclide ^{99}Tc can be injected into a patient's bloodstream in order to monitor the blood flow, measure the blood volume, or find a tumor, among other goals. The nuclide is produced in a hospital by a "cow" containing ^{99}Mo, a radioactive nuclide that decays to ^{99}Tc with a half-life of 67 h. Once a day, the cow is "milked" for its ^{99}Tc, which is produced in an excited state by the ^{99}Mo; the ^{99}Tc de-excites to its lowest energy state by emitting a gamma-ray photon, which is recorded by detectors placed around the patient. The de-excitation has a half-life of 6.0 h. (a) By what process does ^{99}Mo decay to ^{99}Tc? (b) If a patient is injected with an 8.2×10^7 Bq sample of ^{99}Tc, how many gamma-ray photons are initially produced within the patient each second? (c) If the emission rate of gamma-ray photons from a small tumor that has collected ^{99}Tc is 38 per second at a certain time, how many excited-state ^{99}Tc are located in the tumor at that time?

••47 SSM After long effort, in 1902 Marie and Pierre Curie succeeded in separating from uranium ore the first substantial quantity of radium, one decigram of pure $RaCl_2$. The radium was the radioactive isotope ^{226}Ra, which has a half-life of 1600 y. (a) How many radium nuclei had the Curies isolated? (b) What was the decay rate of their sample, in disintegrations per second?

Module 42-4 Alpha Decay

•48 How much energy is released when a ^{238}U nucleus decays by emitting (a) an alpha particle and (b) a sequence of neutron, proton, neutron, proton? (c) Convince yourself both by reasoned argument and by direct calculation that the difference between these two numbers is just the total binding energy of the alpha particle. (d) Find that binding energy. Some needed atomic and particle masses are

^{238}U	238.050 79 u	^{234}Th	234.043 63 u
^{237}U	237.048 73 u	^{4}He	4.002 60 u
^{236}Pa	236.048 91 u	^{1}H	1.007 83 u
^{235}Pa	235.045 44 u	n	1.008 66 u

•49 SSM Generally, more massive nuclides tend to be more unstable to alpha decay. For example, the most stable isotope of uranium, ^{238}U, has an alpha decay half-life of 4.5×10^9 y. The most stable isotope of plutonium is ^{244}Pu with an 8.0×10^7 y half-life, and for curium we have ^{248}Cm and 3.4×10^5 y. When half of an original sample of ^{238}U has decayed, what fraction of the original sample of (a) plutonium and (b) curium is left?

••50 Large radionuclides emit an alpha particle rather than other combinations of nucleons because the alpha particle has such a stable, tightly bound structure. To confirm this statement, calculate the disintegration energies for these hypothetical decay processes and discuss the meaning of your findings:

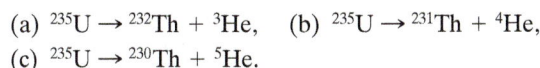

(a) ^{235}U $\rightarrow$ ^{232}Th + ^{3}He,   (b) ^{235}U $\rightarrow$ ^{231}Th + ^{4}He,

(c) ^{235}U $\rightarrow$ ^{230}Th + ^{5}He.

The needed atomic masses are

^{232}Th	232.0381 u	^{3}He	3.0160 u
^{231}Th	231.0363 u	^{4}He	4.0026 u
^{230}Th	230.0331 u	^{5}He	5.0122 u
^{235}U	235.0429 u		

••51 A ^{238}U nucleus emits a 4.196 MeV alpha particle. Calculate the disintegration energy Q for this process, taking the recoil energy of the residual ^{234}Th nucleus into account.

••52 Under certain rare circumstances, a nucleus can decay by emitting a particle more massive than an alpha particle. Consider the decays

$$^{223}\text{Ra} \rightarrow ^{209}\text{Pb} + ^{14}\text{C} \quad \text{and} \quad ^{223}\text{Ra} \rightarrow ^{219}\text{Rn} + ^{4}\text{He}.$$

Calculate the Q value for the (a) first and (b) second decay and determine that both are energetically possible. (c) The Coulomb barrier height for alpha-particle emission is 30.0 MeV. What is the barrier height for ^{14}C emission? (Be careful about the nuclear radii.) The needed atomic masses are

^{223}Ra	223.018 50 u	^{14}C	14.003 24 u
^{209}Pb	208.981 07 u	^{4}He	4.002 60 u
^{219}Rn	219.009 48 u		

Module 42-5   Beta Decay

•53 SSM The cesium isotope ^{137}Cs is present in the fallout from aboveground detonations of nuclear bombs. Because it decays with a slow (30.2 y) half-life into ^{137}Ba, releasing considerable energy in the process, it is of environmental concern. The atomic masses of the Cs and Ba are 136.9071 and 136.9058 u, respectively; calculate the total energy released in such a decay.

•54 Some radionuclides decay by capturing one of their own atomic electrons, a K-shell electron, say. An example is

$$^{49}\text{V} + \text{e}^- \rightarrow ^{49}\text{Ti} + \nu, \quad T_{1/2} = 331 \text{ d}.$$

Show that the disintegration energy Q for this process is given by

$$Q = (m_\text{V} - m_\text{Ti})c^2 - E_K,$$

where m_V and m_Ti are the atomic masses of ^{49}V and ^{49}Ti, respectively, and E_K is the binding energy of the vanadium K-shell electron. (*Hint:* Put $\mathbf{m}_\text{V}$ and $\mathbf{m}_\text{Ti}$ as the corresponding nuclear masses and then add in enough electrons to use the atomic masses.)

•55 A free neutron decays according to Eq. 42-26. If the neutron–hydrogen atom mass difference is 840 μu, what is the maximum kinetic energy K_max possible for the electron produced in a neutron decay?

•56 An electron is emitted from a middle-mass nuclide ($A = 150$, say) with a kinetic energy of 1.0 MeV. (a) What is its de Broglie wavelength? (b) Calculate the radius of the emitting nucleus. (c) Can such an electron be confined as a standing wave in a "box" of such dimensions? (d) Can you use these numbers to disprove the (abandoned) argument that electrons actually exist in nuclei?

••57 GO The radionuclide ^{11}C decays according to

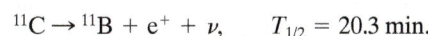

$$^{11}\text{C} \rightarrow ^{11}\text{B} + \text{e}^+ + \nu, \quad T_{1/2} = 20.3 \text{ min}.$$

The maximum energy of the emitted positrons is 0.960 MeV. (a) Show that the disintegration energy Q for this process is given by

$$Q = (m_\text{C} - m_\text{B} - 2m_\text{e})c^2,$$

where m_C and m_B are the atomic masses of ^{11}C and ^{11}B, respectively, and m_e is the mass of a positron. (b) Given the mass values $m_\text{C} = 11.011\,434$ u, $m_\text{B} = 11.009\,305$ u, and $m_\text{e} = 0.000\,548\,6$ u, calculate Q and compare it with the maximum energy of the emitted positron given above. (*Hint:* Let $\mathbf{m}_\text{C}$ and $\mathbf{m}_\text{B}$ be the nuclear masses and then add in enough electrons to use the atomic masses.)

••58 Two radioactive materials that alpha decay, ^{238}U and ^{232}Th, and one that beta decays, ^{40}K, are sufficiently abundant in granite to contribute significantly to the heating of Earth through the decay energy produced. The alpha-decay isotopes give rise to decay chains that stop when stable lead isotopes are formed. The isotope ^{40}K has a single beta decay. (Assume this is the only possible decay of that isotope.) Here is the information:

Parent	Decay Mode	Half-Life (y)	Stable End Point	Q (MeV)	f (ppm)
^{238}U	α	4.47×10^9	^{206}Pb	51.7	4
^{232}Th	α	1.41×10^{10}	^{208}Pb	42.7	13
^{40}K	β	1.28×10^9	^{40}Ca	1.31	4

In the table Q is the *total* energy released in the decay of one parent nucleus to the *final* stable end point and f is the abundance of the isotope in kilograms per kilogram of granite; ppm means parts per million. (a) Show that these materials produce energy as heat at the rate of 1.0×10^{-9} W for each kilogram of granite. (b) Assuming that there is 2.7×10^{22} kg of granite in a 20-km-thick spherical shell at the surface of Earth, estimate the power of this decay process over all of Earth. Compare this power with the total solar power intercepted by Earth, 1.7×10^{17} W.

•••59 SSM WWW The radionuclide ^{32}P decays to ^{32}S as described by Eq. 42-24. In a particular decay event, a 1.71 MeV electron is emitted, the maximum possible value. What is the kinetic energy of the recoiling ^{32}S atom in this event? (*Hint:* For the electron it is necessary to use the relativistic expressions for kinetic energy and linear momentum. The ^{32}S atom is nonrelativistic.)

Module 42-6 Radioactive Dating

•60 A 5.00 g charcoal sample from an ancient fire pit has a ^{14}C activity of 63.0 disintegrations/min. A living tree has a ^{14}C activity of 15.3 disintegrations/min per 1.00 g. The half-life of ^{14}C is 5730 y. How old is the charcoal sample?

•61 The isotope ^{238}U decays to ^{206}Pb with a half-life of 4.47×10^9 y. Although the decay occurs in many individual steps, the first step has by far the longest half-life; therefore, one can often consider the decay to go directly to lead. That is,

$$^{238}\text{U} \rightarrow {}^{206}\text{Pb} + \text{various decay products.}$$

A rock is found to contain 4.20 mg of ^{238}U and 2.135 mg of ^{206}Pb. Assume that the rock contained no lead at formation, so all the lead now present arose from the decay of uranium. How many atoms of (a) ^{238}U and (b) ^{206}Pb does the rock now contain? (c) How many atoms of ^{238}U did the rock contain at formation? (d) What is the age of the rock?

••62 A particular rock is thought to be 260 million years old. If it contains 3.70 mg of ^{238}U, how much ^{206}Pb should it contain? See Problem 61.

••63 GO A rock recovered from far underground is found to contain 0.86 mg of ^{238}U, 0.15 mg of ^{206}Pb, and 1.6 mg of ^{40}Ar. How much ^{40}K will it likely contain? Assume that ^{40}K decays to only ^{40}Ar with a half-life of 1.25×10^9 y. Also assume that ^{238}U has a half-life of 4.47×10^9 y.

•••64 GO The isotope ^{40}K can decay to either ^{40}Ca or ^{40}Ar; assume both decays have a half-life of 1.26×10^9 y. The ratio of the Ca produced to the Ar produced is 8.54/1 = 8.54. A sample originally had only ^{40}K. It now has equal amounts of ^{40}K and ^{40}Ar; that is, the ratio of K to Ar is 1/1 = 1. How old is the sample? (*Hint:* Work this like other radioactive-dating problems, except that this decay has two products.)

Module 42-7 Measuring Radiation Dosage

•65 SSM The nuclide ^{198}Au, with a half-life of 2.70 d, is used in cancer therapy. What mass of this nuclide is required to produce an activity of 250 Ci?

•66 A radiation detector records 8700 counts in 1.00 min. Assuming that the detector records all decays, what is the activity of the radiation source in (a) becquerels and (b) curies?

•67 An organic sample of mass 4.00 kg absorbs 2.00 mJ via slow neutron radiation (RBE = 5). What is the dose equivalent (mSv)?

••68 A 75 kg person receives a whole-body radiation dose of 2.4×10^{-4} Gy, delivered by alpha particles for which the RBE factor is 12. Calculate (a) the absorbed energy in joules and the dose equivalent in (b) sieverts and (c) rem.

••69 An 85 kg worker at a breeder reactor plant accidentally ingests 2.5 mg of ^{239}Pu dust. This isotope has a half-life of 24 100 y, decaying by alpha decay. The energy of the emitted alpha particles is 5.2 MeV, with an RBE factor of 13. Assume that the plutonium resides in the worker's body for 12 h (it is eliminated naturally by the digestive system rather than being absorbed by any of the internal organs) and that 95% of the emitted alpha particles are stopped within the body. Calculate (a) the number of plutonium atoms ingested, (b) the number that decay during the 12 h, (c) the

energy absorbed by the body, (d) the resulting physical dose in grays, and (e) the dose equivalent in sieverts.

Module 42-8 Nuclear Models

•70 A typical kinetic energy for a nucleon in a middle-mass nucleus may be taken as 5.00 MeV. To what effective nuclear temperature does this correspond, based on the assumptions of the collective model of nuclear structure?

•71 A measurement of the energy E of an intermediate nucleus must be made within the mean lifetime Δt of the nucleus and necessarily carries an uncertainty ΔE according to the uncertainty principle

$$\Delta E \cdot \Delta t = \hbar.$$

(a) What is the uncertainty ΔE in the energy for an intermediate nucleus if the nucleus has a mean lifetime of 10^{-22} s? (b) Is the nucleus a compound nucleus?

•72 In the following list of nuclides, identify (a) those with filled nucleon shells, (b) those with one nucleon outside a filled shell, and (c) those with one vacancy in an otherwise filled shell: ^{13}C, ^{18}O, ^{40}K, ^{49}Ti, ^{60}Ni, ^{91}Zr, ^{92}Mo, ^{121}Sb, ^{143}Nd, ^{144}Sm, ^{205}Tl, and ^{207}Pb.

••73 SSM Consider the three formation processes shown for the compound nucleus ^{20}Ne in Fig. 42-14. Here are some of the atomic and particle masses:

^{20}Ne	19.992 44 u	α	4.002 60 u
^{19}F	18.998 40 u	p	1.007 83 u
^{16}O	15.994 91 u		

What energy must (a) the alpha particle, (b) the proton, and (c) the γ-ray photon have to provide 25.0 MeV of excitation energy to the compound nucleus?

Additional Problems

74 GO In a certain rock, the ratio of lead atoms to uranium atoms is 0.300. Assume that uranium has a half-life of 4.47×10^9 y and that the rock had no lead atoms when it formed. How old is the rock?

75 SSM A certain stable nuclide, after absorbing a neutron, emits an electron, and the new nuclide splits spontaneously into two alpha particles. Identify the nuclide.

76 A typical chest x-ray radiation dose is 250 μSv, delivered by x rays with an RBE factor of 0.85. Assuming that the mass of the exposed tissue is one-half the patient's mass of 88 kg, calculate the energy absorbed in joules.

77 SSM How many years are needed to reduce the activity of ^{14}C to 0.020 of its original activity? The half-life of ^{14}C is 5730 y.

78 GO Radioactive element AA can decay to either element BB or element CC. The decay depends on chance, but the ratio of the resulting number of BB atoms to the resulting number of CC atoms is always 2/1. The decay has a half-life of 8.00 days. We start with a sample of pure AA. How long must we wait until the number of CC atoms is 1.50 times the number of AA atoms?

79 GO SSM One of the dangers of radioactive fallout from a nuclear bomb is its ^{90}Sr, which decays with a 29-year half-life. Because it has chemical properties much like those of calcium, the strontium, if ingested by a cow, becomes concentrated in the cow's milk. Some of the ^{90}Sr ends up in the bones of whoever drinks the milk. The energetic electrons emitted in the beta decay of ^{90}Sr damage the bone marrow and thus impair the production of red blood cells. A 1 megaton bomb produces approximately 400 g of ^{90}Sr. If the fallout spreads uniformly over a 2000 km^2 area, what ground area

would hold an amount of radioactivity equal to the "allowed" limit for one person, which is 74 000 counts/s?

80 Because of the 1986 explosion and fire in a reactor at the Chernobyl nuclear power plant in northern Ukraine, part of Ukraine is contaminated with ^{137}Cs, which undergoes beta-minus decay with a half-life of 30.2 y. In 1996, the total activity of this contamination over an area of 2.6×10^5 km^2 was estimated to be 1×10^{16} Bq. Assume that the ^{137}Cs is uniformly spread over that area and that the beta-decay electrons travel either directly upward or directly downward. How many beta-decay electrons would you intercept were you to lie on the ground in that area for 1 h (a) in 1996 and (b) today? (You need to estimate your cross-sectional area that intercepts those electrons.)

81 Figure 42-20 shows part of the decay scheme of ^{237}Np on a plot of mass number A versus proton number Z; five lines that represent either alpha decay or beta-minus decay connect dots that represent isotopes. What is the isotope at the end of the five decays (as marked with a question mark in Fig. 42-20)?

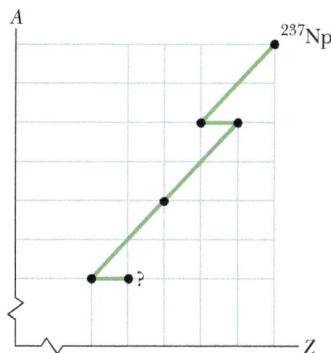

Figure 42-20 Problem 81.

82 After a brief neutron irradiation of silver, two isotopes are present: ^{108}Ag ($T_{1/2} = 2.42$ min) with an initial decay rate of 3.1×10^5/s, and ^{110}Ag ($T_{1/2} = 24.6$ s) with an initial decay rate of 4.1×10^6/s. Make a semilog plot similar to Fig. 42-9 showing the total combined decay rate of the two isotopes as a function of time from $t = 0$ until $t = 10$ min. We used Fig. 42-9 to illustrate the extraction of the half-life for simple (one isotope) decays. Given only your plot of total decay rate for the two-isotope system here, suggest a way to analyze it in order to find the half-lives of both isotopes.

83 Because a nucleon is confined to a nucleus, we can take the uncertainty in its position to be approximately the nuclear radius r. Use the uncertainty principle to determine the uncertainty Δp in the linear momentum of the nucleon. Using the approximation $p \approx \Delta p$ and the fact that the nucleon is nonrelativistic, calculate the kinetic energy of the nucleon in a nucleus with $A = 100$.

84 A radium source contains 1.00 mg of ^{226}Ra, which decays with a half-life of 1600 y to produce ^{222}Rn, a noble gas. This radon isotope in turn decays by alpha emission with a half-life of 3.82 d. If this process continues for a time much longer than the half-life of ^{222}Rn, the ^{222}Rn decay rate reaches a limiting value that matches the rate at which ^{222}Rn is being produced, which is approximately constant because of the relatively long half-life of ^{226}Ra. For the source under this limiting condition, what are (a) the activity of ^{226}Ra, (b) the activity of ^{222}Rn, and (c) the total mass of ^{222}Rn?

85 Make a nuclidic chart similar to Fig. 42-6 for the 25 nuclides $^{118-122}$Te, $^{117-121}$Sb, $^{116-120}$Sn, $^{115-119}$In, and $^{114-118}$Cd. Draw in and la-

bel (a) all isobaric (constant A) lines and (b) all lines of constant neutron excess, defined as $N - Z$.

86 GO A projectile alpha particle is headed directly toward a target aluminum nucleus. Both objects are assumed to be spheres. What energy is required of the alpha particle if it is to momentarily stop just as its "surface" touches the "surface" of the aluminum nucleus? Assume that the target nucleus remains stationary.

87 Consider a ^{238}U nucleus to be made up of an alpha particle (^{4}He) and a residual nucleus (^{234}Th). Plot the electrostatic potential energy $U(r)$, where r is the distance between these particles. Cover the approximate range 10 fm $< r <$ 100 fm and compare your plot with that of Fig. 42-10.

88 Characteristic nuclear time is a useful but loosely defined quantity, taken to be the time required for a nucleon with a few million electron-volts of kinetic energy to travel a distance equal to the diameter of a middle-mass nuclide. What is the order of magnitude of this quantity? Consider 5 MeV neutrons traversing a nuclear diameter of ^{197}Au; use Eq. 42-3.

89 What is the likely mass number of a spherical nucleus with a radius of 3.6 fm as measured by electron-scattering methods?

90 Using a nuclidic chart, write the symbols for (a) all stable isotopes with $Z = 60$, (b) all radioactive nuclides with $N = 60$, and (c) all nuclides with $A = 60$.

91 If the unit for atomic mass were defined so that the mass of ^{1}H were exactly 1.000 000 u, what would be the mass of (a) ^{12}C (actual mass 12.000 000 u) and (b) ^{238}U (actual mass 238.050 785 u)?

92 High-mass radionuclides, which may be either alpha or beta emitters, belong to one of four decay chains, depending on whether their mass number A is of the form $4n$, $4n + 1$, $4n + 2$, or $4n + 3$, where n is a positive integer. (a) Justify this statement and show that if a nuclide belongs to one of these families, all its decay products belong to the same family. Classify the following nuclides as to family: (b) ^{235}U, (c) ^{236}U, (d) ^{238}U, (e) ^{239}Pu, (f) ^{240}Pu, (g) ^{245}Cm, (h) ^{246}Cm, (i) ^{249}Cf, and (j) ^{253}Fm.

93 Find the disintegration energy Q for the decay of ^{49}V by K-electron capture (see Problem 54). The needed data are $m_V = 48.948\ 52$ u, $m_{Ti} = 48.947\ 87$ u, and $E_K = 5.47$ keV.

94 Locate the nuclides displayed in Table 42-1 on the nuclidic chart of Fig. 42-5. Verify that they lie in the stability zone.

95 The radionuclide ^{32}P ($T_{1/2} = 14.28$ d) is often used as a tracer to follow the course of biochemical reactions involving phosphorus. (a) If the counting rate in a particular experimental setup is initially 3050 counts/s, how much time will the rate take to fall to 170 counts/s? (b) A solution containing ^{32}P is fed to the root system of an experimental tomato plant, and the ^{32}P activity in a leaf is measured 3.48 days later. By what factor must this reading be multiplied to correct for the decay that has occurred since the experiment began?

96 At the end of World War II, Dutch authorities arrested Dutch artist Hans van Meegeren for treason because, during the war, he had sold a masterpiece painting to the Nazi Hermann Goering. The painting, *Christ and His Disciples at Emmaus* by Dutch master Johannes Vermeer (1632–1675), had been discovered in 1937 by van Meegeren, after it had been lost for almost 300 years. Soon after the discovery, art experts proclaimed that *Emmaus* was possibly the best Vermeer ever seen. Selling such a Dutch national treasure to the enemy was unthinkable treason.

However, shortly after being imprisoned, van Meegeren suddenly announced that he, not Vermeer, had painted *Emmaus*. He explained that he had carefully mimicked Vermeer's style, using a 300-year-old canvas and Vermeer's choice of pigments; he had then signed Vermeer's name to the work and baked the painting to give it an authentically old look.

Was van Meegeren lying to avoid a conviction of treason, hoping to be convicted of only the lesser crime of fraud? To art experts, *Emmaus* certainly looked like a Vermeer but, at the time of van Meegeren's trial in 1947, there was no scientific way to answer the question. However, in 1968 Bernard Keisch of Carnegie-Mellon University was able to answer the question with newly developed techniques of radioactive analysis.

Specifically, he analyzed a small sample of white lead-bearing pigment removed from *Emmaus*. This pigment is refined from lead ore, in which the lead is produced by a long radioactive decay series that starts with unstable ^{238}U and ends with stable ^{206}Pb. To follow the spirit of Keisch's analysis, focus on the following abbreviated portion of that decay series, in which intermediate, relatively short-lived radionuclides have been omitted:

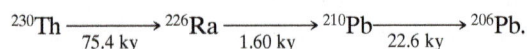

$$^{230}Th \xrightarrow[75.4\ ky]{} {}^{226}Ra \xrightarrow[1.60\ ky]{} {}^{210}Pb \xrightarrow[22.6\ ky]{} {}^{206}Pb.$$

The longer and more important half-lives in this portion of the decay series are indicated.

(a) Show that in a sample of lead ore, the rate at which the number of ^{210}Pb nuclei changes is given by

$$\frac{dN_{210}}{dt} = \lambda_{226}N_{226} - \lambda_{210}N_{210},$$

where N_{210} and N_{226} are the numbers of ^{210}Pb nuclei and ^{226}Ra nuclei in the sample and λ_{210} and λ_{226} are the corresponding disintegration constants.

Because the decay series has been active for billions of years and because the half-life of ^{210}Pb is much less than that of ^{226}Ra, the nuclides ^{226}Ra and ^{210}Pb are in *equilibrium*; that is, the numbers of these nuclides (and thus their concentrations) in the sample do not change. (b) What is the ratio R_{226}/R_{210} of the activities of these nuclides in the sample of lead ore? (c) What is the ratio N_{226}/N_{210} of their numbers?

When lead pigment is refined from the ore, most of the ^{226}Ra is eliminated. Assume that only 1.00% remains. Just after the pigment is produced, what are the ratios (d) R_{226}/R_{210} and (e) N_{226}/N_{210}?

Keisch realized that with time the ratio R_{226}/R_{210} of the pigment would gradually change from the value in freshly refined pigment back to the value in the ore, as equilibrium between the ^{210}Pb and the remaining ^{226}Ra is established in the pigment. If *Emmaus* were painted by Vermeer and the sample of pigment taken from it were 300 years old when examined in 1968, the ratio would be close to the answer of (b). If *Emmaus* were painted by van Meegeren in the 1930s and the sample were only about 30 years old, the ratio would be close to the answer of (d). Keisch found a ratio of 0.09. (f) Is *Emmaus* a Vermeer?

97 From data presented in the first few paragraphs of Module 42-3, find (a) the disintegration constant λ and (b) the half-life of ^{238}U.

Energy from the Nucleus

43-1 NUCLEAR FISSION

Learning Objectives

After reading this module, you should be able to . . .

43.01 Distinguish atomic and nuclear burning, noting that in both processes energy is produced because of a reduction of mass.

43.02 Define the fission process.

43.03 Describe the process of a thermal neutron causing a ^{235}U nucleus to undergo fission, and explain the role of the intermediate compound nucleus.

43.04 For the absorption of a thermal neutron, calculate the change in the system's mass and the energy put into the resulting oscillation of the intermediate compound nucleus.

43.05 For a given fission process, calculate the Q value in terms of the binding energy per nucleon.

43.06 Explain the Bohr–Wheeler model for nuclear fission, including the energy barrier.

43.07 Explain why thermal neutrons cannot cause ^{238}U to undergo fission.

43.08 Identify the approximate amount of energy (MeV) in the fission of any high-mass nucleus to two middle-mass nuclei.

43.09 Relate the rate at which nuclei fission and the rate at which energy is released.

Key Ideas

● Nuclear processes are about a million times more effective, per unit mass, than chemical processes in transforming mass into other forms of energy.

● If a thermal neutron is captured by a ^{235}U nucleus, the resulting ^{236}U can undergo fission, producing two intermediate-mass nuclei and one or more neutrons.

● The energy released in such a fission event is $Q \approx 200$ MeV.

● Fission can be understood in terms of the collective model, in which a nucleus is likened to a charged liquid drop carrying a certain excitation energy.

● A potential barrier must be tunneled through if fission is to occur. Fissionability depends on the relationship between the barrier height E_b and the excitation energy E_n transferred to the nucleus in the neutron capture.

What Is Physics?

Let's now turn to a central concern of physics and certain types of engineering: Can we get useful energy from nuclear sources, as people have done for thousands of years from atomic sources by burning materials like wood and coal? As you already know, the answer is yes, but there are major differences between the two energy sources. When we get energy from wood and coal by burning them, we are tinkering with atoms of carbon and oxygen, rearranging their outer *electrons* into more stable combinations. When we get energy from uranium in a nuclear reactor, we are again burning a fuel, but now we are tinkering with the uranium nucleus, rearranging its *nucleons* into more stable combinations.

Electrons are held in atoms by the electromagnetic Coulomb force, and it takes only a few electron-volts to pull one of them out. On the other hand, nucleons are held in nuclei by the strong force, and it takes a few *million* electron-volts to pull one of *them* out. This factor of a few million is reflected in the fact that we can extract a few million times more energy from a kilogram of uranium than we can from a kilogram of coal.

Table 43-1 Energy Released by 1 kg of Matter

Form of Matter	Process	Time[a]
Water	A 50 m waterfall	5 s
Coal	Burning	8 h
Enriched UO_2	Fission in a reactor	690 y
^{235}U	Complete fission	3×10^4 y
Hot deuterium gas	Complete fusion	3×10^4 y
Matter and antimatter	Complete annihilation	3×10^7 y

[a]This column shows the time interval for which the generated energy could power a 100 W lightbulb.

In both atomic and nuclear burning, the release of energy is accompanied by a decrease in mass, according to the equation $Q = -\Delta m \, c^2$. The central difference between burning uranium and burning coal is that, in the former case, a much larger fraction of the available mass (again, by a factor of a few million) is consumed.

The different processes that can be used for atomic or nuclear burning provide different levels of power, or rates at which the energy is delivered. In the nuclear case, we can burn a kilogram of uranium explosively in a bomb or slowly in a power reactor. In the atomic case, we might consider exploding a stick of dynamite or digesting a jelly doughnut.

Table 43-1 shows how much energy can be extracted from 1 kg of matter by doing various things to it. Instead of reporting the energy directly, the table shows how long the extracted energy could operate a 100 W lightbulb. Only processes in the first three rows of the table have actually been carried out; the remaining three represent theoretical limits that may not be attainable in practice. The bottom row, the total mutual annihilation of matter and antimatter, is an ultimate energy production goal. In that process, *all* the mass energy is transferred to other forms of energy.

The comparisons of Table 43-1 are computed on a per-unit-mass basis. Kilogram for kilogram, you get several million times more energy from uranium than you do from coal or from falling water. On the other hand, there is a lot of coal in Earth's crust, and water is easily backed up behind a dam.

Nuclear Fission: The Basic Process

In 1932 English physicist James Chadwick discovered the neutron. A few years later Enrico Fermi in Rome found that when various elements are bombarded by neutrons, new radioactive elements are produced. Fermi had predicted that the neutron, being uncharged, would be a useful nuclear projectile; unlike the proton or the alpha particle, it experiences no repulsive Coulomb force when it nears a nuclear surface. Even *thermal neutrons,* which are slowly moving neutrons in thermal equilibrium with the surrounding matter at room temperature, with a kinetic energy of only about 0.04 eV, are useful projectiles in nuclear studies.

In the late 1930s physicist Lise Meitner and chemists Otto Hahn and Fritz Strassmann, working in Berlin and following up on the work of Fermi and his co-workers, bombarded solutions of uranium salts with such thermal neutrons. They found that after the bombardment a number of new radionuclides were present. In 1939 one of the radionuclides produced in this way was positively identified, by repeated tests, as barium. But how, Hahn and Strassmann wondered, could this middle-mass element ($Z = 56$) be produced by bombarding uranium ($Z = 92$) with neutrons?

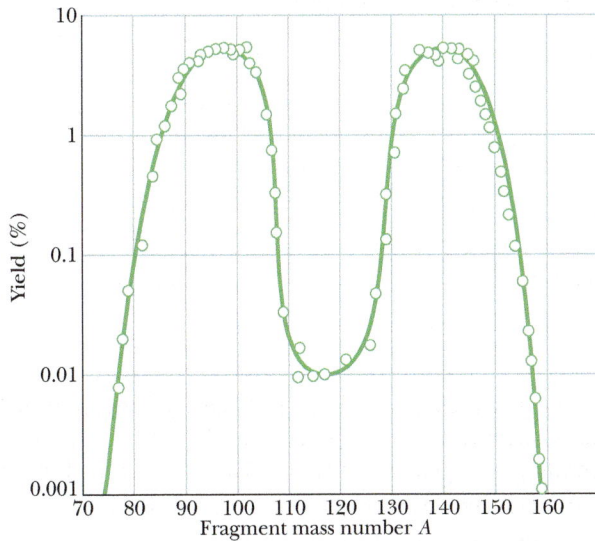

Figure 43-1 The distribution by mass number of the fragments that are found when many fission events of ^{235}U are examined. Note that the vertical scale is logarithmic.

The puzzle was solved within a few weeks by Meitner and her nephew Otto Frisch. They suggested the mechanism by which a uranium nucleus, having absorbed a thermal neutron, could split, with the release of energy, into two roughly equal parts, one of which might well be barium. Frisch named the process **fission.**

Meitner's central role in the discovery of fission was not fully recognized until recent historical research brought it to light. She did not share in the Nobel Prize in chemistry that was awarded to Otto Hahn in 1944. However, in 1997 Meitner was (finally) honored by having an element named after her: meitnerium (symbol Mt, $Z = 109$).

A Closer Look at Fission

Figure 43-1 shows the distribution by mass number of the fragments produced when ^{235}U is bombarded with thermal neutrons. The most probable mass numbers, occurring in about 7% of the events, are centered around $A \approx 95$ and $A \approx 140$. Curiously, the "double-peaked" character of Fig. 43-1 is still not understood.

In a typical ^{235}U fission event, a ^{235}U nucleus absorbs a thermal neutron, producing a compound nucleus ^{236}U in a highly excited state. It is *this* nucleus that actually undergoes fission, splitting into two fragments. These fragments — between them — rapidly emit two neutrons, leaving (in a typical case) ^{140}Xe ($Z = 54$) and ^{94}Sr ($Z = 38$) as fission fragments. Thus, the stepwise fission equation for this event is

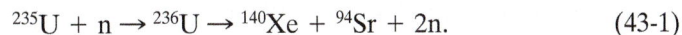

$$^{235}\text{U} + \text{n} \rightarrow {}^{236}\text{U} \rightarrow {}^{140}\text{Xe} + {}^{94}\text{Sr} + 2\text{n}. \qquad (43\text{-}1)$$

Note that during the formation and fission of the compound nucleus, there is conservation of the number of protons and of the number of neutrons involved in the process (and thus conservation of their total number and the net charge).

In Eq. 43-1, the fragments ^{140}Xe and ^{94}Sr are both highly unstable, undergoing beta decay (with the conversion of a neutron to a proton and the emission of an electron and a neutrino) until each reaches a stable end product. For xenon, the decay chain is

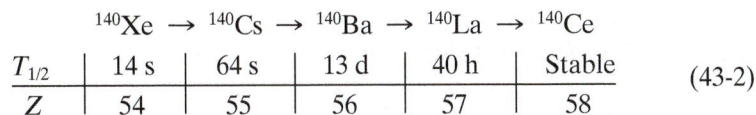

$$^{140}\text{Xe} \rightarrow {}^{140}\text{Cs} \rightarrow {}^{140}\text{Ba} \rightarrow {}^{140}\text{La} \rightarrow {}^{140}\text{Ce}$$

$T_{1/2}$	14 s	64 s	13 d	40 h	Stable
Z	54	55	56	57	58

$$(43\text{-}2)$$

For strontium, it is

$$^{94}\text{Sr} \;\rightarrow\; {}^{94}\text{Y} \;\rightarrow\; {}^{94}\text{Zr}$$

$T_{1/2}$	75 s	19 min	Stable
Z	38	39	40

(43-3)

As we should expect from Module 42-5, the mass numbers (140 and 94) of the fragments remain unchanged during these beta-decay processes and the atomic numbers (initially 54 and 38) increase by unity at each step.

Inspection of the stability band on the nuclidic chart of Fig. 42-5 shows why the fission fragments are unstable. The nuclide ^{236}U, which is the fissioning nucleus in the reaction of Eq. 43-1, has 92 protons and $236 - 92$, or 144, neutrons, for a neutron/proton ratio of about 1.6. The primary fragments formed immediately after the fission reaction have about this same neutron/proton ratio. However, stable nuclides in the middle-mass region have smaller neutron/proton ratios, in the range of 1.3 to 1.4. The primary fragments are thus *neutron rich* (they have too many neutrons) and will eject a few neutrons, two in the case of the reaction of Eq. 43-1. The fragments that remain are still too neutron rich to be stable. Beta decay offers a mechanism for getting rid of the excess neutrons—namely, by changing them into protons within the nucleus.

We can estimate the energy released by the fission of a high-mass nuclide by examining the total binding energy per nucleon ΔE_{ben} before and after the fission. The idea is that fission can occur because the total mass energy will decrease; that is, ΔE_{ben} will *increase* so that the products of the fission are *more* tightly bound. Thus, the energy Q released by the fission is

$$Q = \left(\begin{array}{c}\text{total final} \\ \text{binding energy}\end{array}\right) - \left(\begin{array}{c}\text{initial} \\ \text{binding energy}\end{array}\right). \qquad (43\text{-}4)$$

For our estimate, let us assume that fission transforms an initial high-mass nucleus to two middle-mass nuclei with the same number of nucleons. Then we have

$$Q = \left(\begin{array}{c}\text{final} \\ \Delta E_{\text{ben}}\end{array}\right)\left(\begin{array}{c}\text{final number} \\ \text{of nucleons}\end{array}\right) - \left(\begin{array}{c}\text{initial} \\ \Delta E_{\text{ben}}\end{array}\right)\left(\begin{array}{c}\text{initial number} \\ \text{of nucleons}\end{array}\right). \qquad (43\text{-}5)$$

From Fig. 42-7, we see that for a high-mass nuclide ($A \approx 240$), the binding energy per nucleon is about 7.6 MeV/nucleon. For middle-mass nuclides ($A \approx 120$), it is about 8.5 MeV/nucleon. Thus, the energy released by fission of a high-mass nuclide to two middle-mass nuclides is

$$Q = \left(8.5\,\frac{\text{MeV}}{\text{nucleon}}\right)(2\text{ nuclei})\left(120\,\frac{\text{nucleons}}{\text{nucleus}}\right)$$

$$- \left(7.6\,\frac{\text{MeV}}{\text{nucleon}}\right)(240\text{ nucleons}) \approx 200\text{ MeV}. \qquad (43\text{-}6)$$

✓ **Checkpoint 1**

A generic fission event is

$$^{235}\text{U} + \text{n} \rightarrow X + Y + 2\text{n}.$$

Which of the following pairs *cannot* represent X and Y: (a) ^{141}Xe and ^{93}Sr; (b) ^{139}Cs and ^{95}Rb; (c) ^{156}Nd and ^{79}Ge; (d) ^{121}In and ^{113}Ru?

A Model for Nuclear Fission

Soon after the discovery of fission, Niels Bohr and John Wheeler used the collective model of the nucleus (Module 42-8), based on the analogy between a nucleus and a charged liquid drop, to explain the main nuclear features. Figure 43-2 suggests how the fission process proceeds from this point of view. When a high-mass nucleus—let us say ^{235}U—absorbs a slow (thermal) neutron, as in Fig. 43-2a, that neutron falls into the potential well associated with the strong forces that act in the nuclear interior. The neutron's potential energy is then transformed into internal excitation energy of the nucleus, as Fig. 43-2b suggests. The amount of excitation energy that a slow neutron carries into a nucleus is equal to the binding energy E_n of the neutron in that nucleus, which is the change in mass energy of the neutron–nucleus system due to the neutron's capture.

Figures 43-2c and d show that the nucleus, behaving like an energetically oscillating charged liquid drop, will sooner or later develop a short "neck" and will begin to separate into two charged "globs." Two competing forces then act on the globs: Because they are positively charged, the electric force attempts to separate them. Because they hold protons and neutrons, the strong force attempts to pull them together. If the electric repulsion drives them far enough apart to break the neck, the two fragments, each still carrying some residual excitation energy, will fly apart (Figs. 43-2e and f). Fission has occurred.

This model gave a good qualitative picture of the fission process. What remained to be seen, however, was whether it could answer a hard question: Why are some high-mass nuclides (^{235}U and ^{239}Pu, say) readily fissionable by thermal neutrons when other, equally massive nuclides (^{238}U and ^{243}Am, say) are not?

The ^{235}U absorbs a slow neutron (with little kinetic energy), becoming ^{236}U.

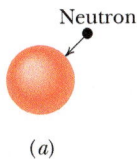

Energy is transferred from mass energy to energy of the oscillations caused by the absorption.

Both globs contain protons and are positively charged and thus they repel each other.

But the protons and neutrons also attract one another by the strong force that binds the nucleus.

Neutron

(a)

(b)

(c)

(d)

The strong force, however, decreases very quickly with distance between the globs.

So, if the globs move apart enough, the electric repulsion rips apart the nucleus.

This fission decreases the mass energy, thus releasing energy.

The two fragments eject neutrons, further reducing mass energy.

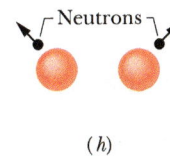

$\longleftarrow r \longrightarrow$

Neutrons

(e)

(f)

(g)

(h)

Figure 43-2 The stages of a typical fission process, according to the collective model of Bohr and Wheeler.

Figure 43-3 The potential energy at various stages in the fission process, as predicted from the collective model of Bohr and Wheeler. The Q of the reaction (about 200 MeV) and the fission barrier height E_b are both indicated.

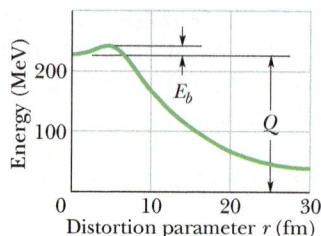

E_b is an energy barrier that must be overcome.

Q is the energy that would then be released.

Bohr and Wheeler were able to answer this question. Figure 43-3 shows a graph of the potential energy of the fissioning nucleus at various stages, derived from their model for the fission process. This energy is plotted against the *distortion parameter r*, which is a rough measure of the extent to which the oscillating nucleus departs from a spherical shape. When the fragments are far apart, this parameter is simply the distance between their centers (Fig. 43-2e).

The energy difference between the initial state ($r = 0$) and the final state ($r = \infty$) of the fissioning nucleus—that is, the disintegration energy Q—is labeled in Fig. 43-3. The central feature of that figure, however, is that the potential energy curve passes through a maximum at a certain value of r. Thus, there is a *potential barrier* of height E_b that must be surmounted (or tunneled through) before fission can occur. This reminds us of alpha decay (Fig. 42-10), which is also a process that is inhibited by a potential barrier.

We see then that fission will occur only if the absorbed neutron provides an excitation energy E_n great enough to overcome the barrier. This energy E_n need not be *quite* as great as the barrier height E_b because of the possibility of quantum-physics tunneling.

Table 43-2 shows, for four high-mass nuclides, this test of whether capture of a thermal neutron can cause fissioning. For each nuclide, the table shows both the barrier height E_b of the nucleus that is formed by the neutron capture and the excitation energy E_n due to the capture. The values of E_b are calculated from the theory of Bohr and Wheeler. The values of E_n are calculated from the change in mass energy due to the neutron capture.

For an example of the calculation of E_n, we can go to the first line in the table, which represents the neutron capture process

$$^{235}\text{U} + \text{n} \rightarrow {}^{236}\text{U}.$$

The masses involved are 235.043 922 u for ^{235}U, 1.008 665 u for the neutron, and 236.045 562 u for ^{236}U. It is easy to show that, because of the neutron capture, the mass decreases by 7.025×10^{-3} u. Thus, energy is transferred from mass energy to excitation energy E_n. Multiplying the change in mass by c^2 ($= 931.494\,013$ MeV/u) gives us $E_n = 6.5$ MeV, which is listed on the first line of the table.

The first and third results in Table 43-2 are historically profound because they are the reasons the two atomic bombs used in World War II contained ^{235}U (first bomb) and ^{239}Pu (second bomb). That is, for ^{235}U and ^{239}Pu, $E_n > E_b$. This means that fission by absorption of a thermal neutron is predicted to occur for these nuclides. For the other two nuclides in Table 43-2 (^{238}U and ^{243}Am), we have $E_n < E_b$; thus,

Table 43-2 Test of the Fissionability of Four Nuclides

Target Nuclide	Nuclide Being Fissioned	E_n (MeV)	E_b (MeV)	Fission by Thermal Neutrons?
^{235}U	^{236}U	6.5	5.2	Yes
^{238}U	^{239}U	4.8	5.7	No
^{239}Pu	^{240}Pu	6.4	4.8	Yes
^{243}Am	^{244}Am	5.5	5.8	No

Courtesy U.S. Department of Energy

Figure 43-4 This image has transfixed the world since World War II. When Robert Oppenheimer, the head of the scientific team that developed the atomic bomb, witnessed the first atomic explosion, he quoted from a sacred Hindu text: "Now I am become Death, the destroyer of worlds."

there is not enough energy from a thermal neutron for the excited nucleus to surmount the barrier or to tunnel through it effectively. Instead of fissioning, the nucleus gets rid of its excitation energy by emitting a gamma-ray photon.

The nuclides ^{238}U and ^{243}Am *can* be made to fission, however, if they absorb a substantially energetic (rather than a thermal) neutron. A ^{238}U nucleus, for example, might fission if it happens to absorb a neutron of at least 1.3 MeV in a so-called *fast fission* process ("fast" because the neutron is fast).

The two atomic bombs used in World War II depended on the ability of thermal neutrons to cause many high-mass nuclides in the cores of the bombs to fission nearly all at once. The process is initiated by a neutron emitter such as beryllium. After its emitted thermal neutrons cause the fission of the first set of ^{235}U, each fission releases more thermal neutrons, which cause more ^{235}U to fission and release thermal neutrons. This **chain reaction** would rapidly spread through the ^{235}U in the bomb, resulting in an explosive and devastating output of energy. Researchers knew that ^{235}U would work, but they had refined only enough for one bomb from uranium ore, which consists mainly of ^{238}U, which thermal neutrons will not fission. As the first bomb was being deployed, a ^{239}Pu bomb was tested successfully in New Mexico (Fig. 43-4), so the next deployed bomb contained ^{239}Pu rather than ^{235}U.

Sample Problem 43.01   Q value in a fission of uranium-235

Find the disintegration energy Q for the fission event of Eq. 43-1, taking into account the decay of the fission fragments as displayed in Eqs. 43-2 and 43-3. Some needed atomic and particle masses are

^{235}U	235.0439 u	^{140}Ce	139.9054 u
n	1.008 66 u	^{94}Zr	93.9063 u

KEY IDEAS

(1) The disintegration energy Q is the energy transferred from mass energy to kinetic energy of the decay products. (2) $Q = -\Delta m\, c^2$, where Δm is the change in mass.

Calculations: Because we are to include the decay of the fission fragments, we combine Eqs. 43-1, 43-2, and 43-3 to write the overall transformation as

$$^{235}\text{U} \rightarrow {}^{140}\text{Ce} + {}^{94}\text{Zr} + \text{n}. \qquad (43\text{-}7)$$

Only the single neutron appears here because the initiating neutron on the left side of Eq. 43-1 cancels one of the two

neutrons on the right of that equation. The mass difference for the reaction of Eq. 43-7 is

$$\Delta m = (139.9054\ \text{u} + 93.9063\ \text{u} + 1.008\ 66\ \text{u})$$
$$- (235.0439\ \text{u})$$
$$= -0.223\ 54\ \text{u},$$

and the corresponding disintegration energy is

$$Q = -\Delta m\, c^2 = -(-0.223\ 54\ \text{u})(931.494\ 013\ \text{MeV/u})$$
$$= 208\ \text{MeV}, \qquad \text{(Answer)}$$

which is in good agreement with our estimate of Eq. 43-6.

If the fission event takes place in a bulk solid, most of this disintegration energy, which first goes into kinetic energy of the decay products, appears eventually as an increase in the internal energy of that body, revealing itself as a rise in temperature. Five or six percent or so of the disintegration energy, however, is associated with neutrinos that are emitted during the beta decay of the primary fission fragments. This energy is carried out of the system and is lost.

PLUS Additional examples, video, and practice available at *WileyPLUS*

43-2 THE NUCLEAR REACTOR

Learning Objectives

After reading this module, you should be able to . . .

43.10 Define chain reaction.

43.11 Explain the neutron leakage problem, the neutron energy problem, and the neutron capture problem.

43.12 Identify the multiplication factor and apply it to relate the number of neutrons and power output after a given

number of cycles to the initial number of neutrons and power output.

43.13 Distinguish subcritical, critical, and supercritical.

43.14 Describe the control over the response time.

43.15 Give a general description of a complete generation.

Key Idea

● A nuclear reactor uses a controlled chain reaction of fission events to generate electrical power.

The Nuclear Reactor

For large-scale energy release due to fission, one fission event must trigger others, so that the process spreads throughout the nuclear fuel like flame through a log. The fact that more neutrons are produced in fission than are consumed raises the possibility of just such a chain reaction, with each neutron that is produced potentially triggering another fission. The reaction can be either rapid (as in a nuclear bomb) or controlled (as in a nuclear reactor).

Suppose that we wish to design a reactor based on the fission of ^{235}U by thermal neutrons. Natural uranium contains 0.7% of this isotope, the remaining 99.3% being ^{238}U, which is not fissionable by thermal neutrons. Let us give ourselves an edge by artificially *enriching* the uranium fuel so that it contains perhaps 3% ^{235}U. Three difficulties still stand in the way of a working reactor.

1. *The Neutron Leakage Problem.* Some of the neutrons produced by fission will leak out of the reactor and so not be part of the chain reaction. Leakage is a surface effect; its magnitude is proportional to the square of a typical reactor dimension (the surface area of a cube of edge length a is $6a^2$). Neutron production, however, occurs throughout the volume of the fuel and is thus proportional to the cube of a typical dimension (the volume of the same cube is a^3). We can make the fraction of neutrons lost by leakage as small as we wish by making the reactor core large enough, thereby reducing the surface-to-volume ratio ($= 6/a$ for a cube).

2. *The Neutron Energy Problem.* The neutrons produced by fission are fast, with kinetic energies of about 2 MeV. However, fission is induced most effectively by thermal neutrons. The fast neutrons can be slowed down by mixing the uranium fuel with a substance—called a **moderator**—that has two properties: It is effective in slowing down neutrons via elastic collisions, and it does not remove neutrons from the core by absorbing them so that they do not result in fission. Most power reactors in North America use water as a moderator; the hydrogen nuclei (protons) in the water are the effective component. We saw in Chapter 9 that if a moving particle has a head-on elastic collision with a stationary particle, the moving particle loses *all* its kinetic energy if the two particles have the same mass. Thus, protons form an effective moderator because they have approximately the same mass as the fast neutrons whose speed we wish to reduce.

3. *The Neutron Capture Problem.* As the fast (2 MeV) neutrons generated by fission are slowed down in the moderator to thermal energies (about 0.04 eV), they must pass through a critical energy interval (from 1 to 100 eV) in which they are particularly susceptible to nonfission capture by ^{238}U nuclei. Such *resonance capture,* which results in the emission of a gamma ray, removes the neutron from the fission chain. To minimize such nonfission capture, the uranium fuel and the moderator are not intimately mixed but rather are placed in different regions of the reactor volume.

In a typical reactor, the uranium fuel is in the form of uranium oxide pellets, which are inserted end to end into long, hollow metal tubes. The liquid moderator surrounds bundles of these **fuel rods,** forming the reactor **core.** This geometric arrangement increases the probability that a fast neutron, produced in a fuel rod, will find itself in the moderator when it passes through the critical energy interval. Once the neutron has reached thermal energies, it may *still* be captured in ways that do not result in fission (called *thermal capture*). However, it is much more likely that the thermal neutron will wander back into a fuel rod and produce a fission event.

Figure 43-5 shows the neutron balance in a typical power reactor operating at constant power. Let us trace a sample of 1000 thermal neutrons through one complete cycle, or *generation,* in the reactor core. They produce 1330 neutrons by fission in the ^{235}U fuel and 40 neutrons by fast fission in ^{238}U, which gives 370 neutrons more than the original 1000, all of them fast.

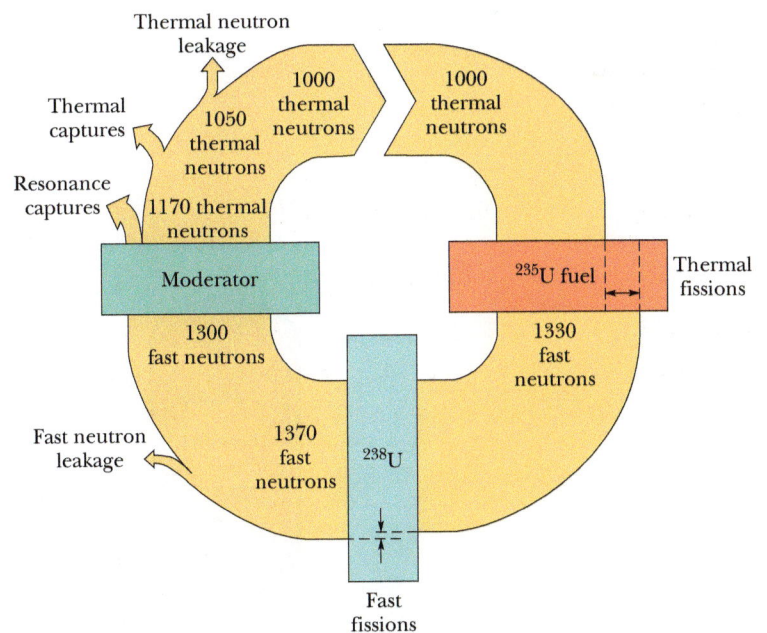

Figure 43-5 Neutron bookkeeping in a reactor. A generation of 1000 thermal neutrons interacts with the ^{235}U fuel, the ^{238}U matrix, and the moderator. They produce 1370 neutrons by fission, but 370 of these are lost by nonfission capture or by leakage, meaning that 1000 thermal neutrons are left to form the next generation. The figure is drawn for a reactor running at a steady power level.

When the reactor is operating at a steady power level, exactly the same number of neutrons (370) is then lost by leakage from the core and by nonfission capture, leaving 1000 thermal neutrons to start the next generation. In this cycle, of course, each of the 370 neutrons produced by fission events represents a deposit of energy in the reactor core, heating up the core.

The *multiplication factor k*—an important reactor parameter—is the ratio of the number of neutrons present at the conclusion of a particular generation to the number present at the beginning of that generation. In Fig. 43-5, the multiplication factor is 1000/1000, or exactly unity. For $k = 1$, the operation of the reactor is said to be exactly *critical,* which is what we wish it to be for steady-power operation. Reactors are actually designed so that they are inherently *supercritical* ($k > 1$); the multiplication factor is then adjusted to critical operation ($k = 1$) by inserting **control rods** into the reactor core. These rods, containing a material such as cadmium that absorbs neutrons readily, can be inserted farther to reduce the operating power level and withdrawn to increase the power level or to compensate for the tendency of reactors to go *subcritical* as (neutron-absorbing) fission products build up in the core during continued operation.

If you pulled out one of the control rods rapidly, how fast would the reactor power level increase? This *response time* is controlled by the fascinating circumstance that a small fraction of the neutrons generated by fission do not escape promptly from the newly formed fission fragments but are emitted from these fragments later, as the fragments decay by beta emission. Of the 370 "new" neutrons produced in Fig. 43-5, for example, perhaps 16 are delayed, being emitted from fragments following beta decays whose half-lives range from 0.2 to 55 s. These delayed neutrons are few in number, but they serve the essential purpose of slowing the reactor response time to match practical mechanical reaction times.

Figure 43-6 shows the broad outlines of an electrical power plant based on a *pressurized-water reactor* (PWR), a type in common use in North America. In such a reactor, water is used both as the moderator and as the heat transfer medium. In the *primary loop,* water is circulated through the reactor vessel and

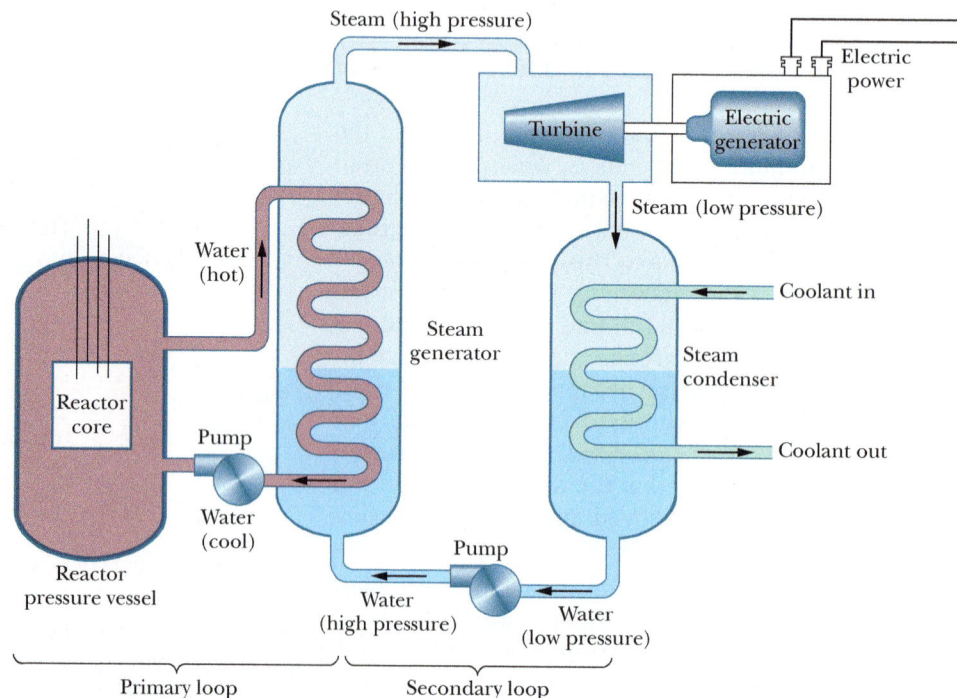

Figure 43-6 A simplified layout of a nuclear power plant, based on a pressurized-water reactor. Many features are omitted—among them the arrangement for cooling the reactor core in case of an emergency.

transfers energy at high temperature and pressure (possibly 600 K and 150 atm) from the hot reactor core to the steam generator, which is part of the *secondary loop*. In the steam generator, evaporation provides high-pressure steam to operate the turbine that drives the electric generator. To complete the secondary loop, low-pressure steam from the turbine is cooled and condensed to water and forced back into the steam generator by a pump. To give some idea of scale, a typical reactor vessel for a 1000 MW (electric) plant may be 12 m high and weigh 4 MN. Water flows through the primary loop at a rate of about 1 ML/min.

An unavoidable feature of reactor operation is the accumulation of radioactive wastes, including both fission products and heavy *transuranic* nuclides such as plutonium and americium. One measure of their radioactivity is the rate at which they release energy in thermal form. Figure 43-7 shows the thermal power generated by such wastes from one year's operation of a typical large nuclear plant. Note that both scales are logarithmic. Most "spent" fuel rods from power reactor operation are stored on site, immersed in water; permanent secure storage facilities for reactor waste have yet to be completed. Much weapons-derived radioactive waste accumulated during World War II and in subsequent years is also still in on-site storage.

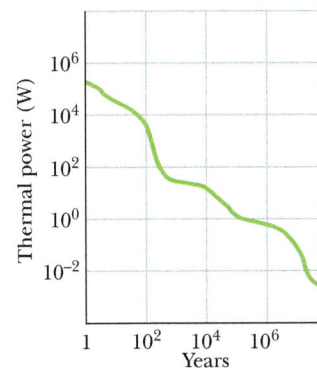

Figure 43-7 The thermal power released by the radioactive wastes from one year's operation of a typical large nuclear power plant, shown as a function of time. The curve is the superposition of the effects of many radionuclides, with a wide variety of half-lives. Note that both scales are logarithmic.

Sample Problem 43.02 Nuclear reactor: efficiency, fission rate, consumption rate

A large electric generating station is powered by a pressurized-water nuclear reactor. The thermal power produced in the reactor core is 3400 MW, and 1100 MW of electricity is generated by the station. The *fuel charge* is 8.60×10^4 kg of uranium, in the form of uranium oxide, distributed among 5.70×10^4 fuel rods. The uranium is enriched to 3.0% ^{235}U.

(a) What is the station's efficiency?

KEY IDEA

The efficiency for this power plant or any other energy device is given by this: Efficiency is the ratio of the output power (rate at which useful energy is provided) to the input power (rate at which energy must be supplied).

Calculation: Here the efficiency (eff) is

$$\text{eff} = \frac{\text{useful output}}{\text{energy input}} = \frac{1100 \text{ MW (electric)}}{3400 \text{ MW (thermal)}}$$

$$= 0.32, \text{ or } 32\%. \hspace{2cm} \text{(Answer)}$$

The efficiency—as for all power plants—is controlled by the second law of thermodynamics. To run this plant, energy at the rate of 3400 MW − 1100 MW, or 2300 MW, must be discharged as thermal energy to the environment.

(b) At what rate R do fission events occur in the reactor core?

KEY IDEAS

1. The fission events provide the input power P of 3400 MW ($= 3.4 \times 10^9$ J/s).

2. From Eq. 43-6, the energy Q released by each event is about 200 MeV.

Calculation: For steady-state operation (P is constant), we find

$$R = \frac{P}{Q} = \left(\frac{3.4 \times 10^9 \text{ J/s}}{200 \text{ MeV/fission}} \right) \left(\frac{1 \text{ MeV}}{1.60 \times 10^{-13} \text{ J}} \right)$$

$$= 1.06 \times 10^{20} \text{ fissions/s}$$

$$\approx 1.1 \times 10^{20} \text{ fissions/s}. \hspace{1.5cm} \text{(Answer)}$$

(c) At what rate (in kilograms per day) is the ^{235}U fuel disappearing? Assume conditions at start-up.

KEY IDEA

^{235}U disappears due to two processes: (1) the fission process with the rate calculated in part (b) and (2) the nonfission capture of neutrons at about one-fourth that rate.

Calculations: The total rate at which the number of atoms of ^{235}U decreases is

$$(1 + 0.25)(1.06 \times 10^{20} \text{ atoms/s}) = 1.33 \times 10^{20} \text{ atoms/s}.$$

We want the corresponding decrease in the mass of the ^{235}U fuel. We start with the mass of each ^{235}U atom. We cannot use the molar mass for uranium listed in Appendix F because that molar mass is for ^{238}U, the most common uranium isotope. Instead, we shall assume that the mass of each ^{235}U atom in atomic mass units is equal to the mass number A. Thus, the mass of each ^{235}U atom is 235 u ($= 3.90 \times 10^{-25}$ kg). Then the rate at which the ^{235}U fuel disappears is

$$\frac{dM}{dt} = (1.33 \times 10^{20} \text{ atoms/s})(3.90 \times 10^{-25} \text{ kg/atom})$$

$$= 5.19 \times 10^{-5} \text{ kg/s} \approx 4.5 \text{ kg/d}. \hspace{0.5cm} \text{(Answer)}$$

(d) At this rate of fuel consumption, how long would the fuel supply of ^{235}U last?

Calculation: At start-up, we know that the total mass of ^{235}U is 3.0% of the 8.60×10^4 kg of uranium oxide. So, the time T required to consume this total mass of ^{235}U at the steady rate of 4.5 kg/d is

$$T = \frac{(0.030)(8.60 \times 10^4 \text{ kg})}{4.5 \text{ kg/d}} \approx 570 \text{ d.} \quad \text{(Answer)}$$

In practice, the fuel rods must be replaced (usually in batches) before their ^{235}U content is entirely consumed.

(e) At what rate is mass being converted to other forms of energy by the fission of ^{235}U in the reactor core?

KEY IDEA

The conversion of mass energy to other forms of energy is linked only to the fissioning that produces the input power (3400 MW) and not to the nonfission capture of neutrons (although both these processes affect the rate at which ^{235}U is consumed).

Calculation: From Einstein's relation $E = mc^2$, we can write

$$\frac{dm}{dt} = \frac{dE/dt}{c^2} = \frac{3.4 \times 10^9 \text{ W}}{(3.00 \times 10^8 \text{ m/s})^2} \quad (43\text{-}8)$$

$$= 3.8 \times 10^{-8} \text{ kg/s} = 3.3 \text{ g/d.} \quad \text{(Answer)}$$

We see that the mass conversion rate is about the mass of one common coin per day, considerably less (by about three orders of magnitude) than the fuel consumption rate calculated in (c).

PLUS Additional examples, video, and practice available at *WileyPLUS*

43-3 A NATURAL NUCLEAR REACTOR

Learning Objectives

After reading this module, you should be able to . . .

43.16 Describe the evidence that a natural nuclear reactor operated in Gabon, West Africa, about 2 billion years ago.

43.17 Explain why a deposit of uranium ore could go critical in the past but not today.

Key Idea

● A natural nuclear reactor occurred in West Africa about two billion years ago.

A Natural Nuclear Reactor

On December 2, 1942, when their reactor first became operational (Fig. 43-8), Enrico Fermi and his associates had every right to assume that they had put into operation the first fission reactor that had ever existed on this planet. About 30 years later it was discovered that, if they did in fact think that, they were wrong.

Some two billion years ago, in a uranium deposit recently mined in Gabon, West Africa, a natural fission reactor apparently went into operation and ran for perhaps several hundred thousand years before shutting down. We can test whether this could actually have happened by considering two questions:

1. *Was There Enough Fuel?* The fuel for a uranium-based fission reactor must be the easily fissionable isotope ^{235}U, which, as noted earlier, constitutes only 0.72% of natural uranium. This isotopic ratio has been measured for terrestrial samples, in Moon rocks, and in meteorites; in all cases the abundance values are the same. The clue to the discovery in West Africa was that the uranium in that deposit was deficient in ^{235}U, some samples having abundances as low as 0.44%. Investigation led to the speculation that this deficit in ^{235}U could be accounted for if, at some earlier time, the ^{235}U was partially consumed by the operation of a natural fission reactor.

The serious problem remains that, with an isotopic abundance of only 0.72%, a reactor can be assembled (as Fermi and his team learned) only after thoughtful design and with scrupulous attention to detail. There seems no chance that a nuclear reactor could go critical "naturally."

However, things were different in the distant past. Both ^{235}U and ^{238}U are radioactive, with half-lives of 7.04×10^8 y and 44.7×10^8 y, respectively. Thus, the half-life of the readily fissionable ^{235}U is about 6.4 times shorter than that of ^{238}U. Because ^{235}U decays faster, there was more of it, relative to ^{238}U, in the past. Two billion years ago, in fact, this abundance was not 0.72%, as it is now, but 3.8%. This abundance happens to be just about the abundance to which natural uranium is artificially enriched to serve as fuel in modern power reactors.

With this readily fissionable fuel available, the presence of a natural reactor (provided certain other conditions are met) is less surprising. The fuel was there. Two billion years ago, incidentally, the highest order of life-form to have evolved was the blue-green alga.

Gary Sheehan, *Birth of the Atomic Age*, 1957. Reproduced courtesy Chicago Historical Society.

Figure 43-8 A painting of the first nuclear reactor, assembled during World War II on a squash court at the University of Chicago by a team headed by Enrico Fermi. This reactor was built of lumps of uranium embedded in blocks of graphite.

2. *What Is the Evidence?* The mere depletion of ^{235}U in an ore deposit does not prove the existence of a natural fission reactor. One looks for more convincing evidence.

If there was a reactor, there must now be fission products. Of the 30 or so elements whose stable isotopes are produced in a reactor, some must still remain. Study of their isotopic abundances could provide the evidence we need.

Of the several elements investigated, the case of neodymium is spectacularly convincing. Figure 43-9a shows the isotopic abundances of the seven stable neodymium isotopes as they are normally found in nature. Figure 43-9b shows these abundances as they appear among the ultimate stable fission products of the fission of ^{235}U. The clear differences are not surprising, considering the totally different origins of the two sets of isotopes. Note particularly that ^{142}Nd, the dominant isotope in the natural element, is absent from the fission products.

The big question is: What do the neodymium isotopes found in the uranium ore body in West Africa look like? If a natural reactor operated there, we would expect to find isotopes from *both* sources (that is, natural isotopes as well as fission-produced isotopes). Figure 43-9c shows the abundances after dual-source and other corrections have been made to the data. Comparison of Figs. 43-9b and c indicates that there was indeed a natural fission reactor at work.

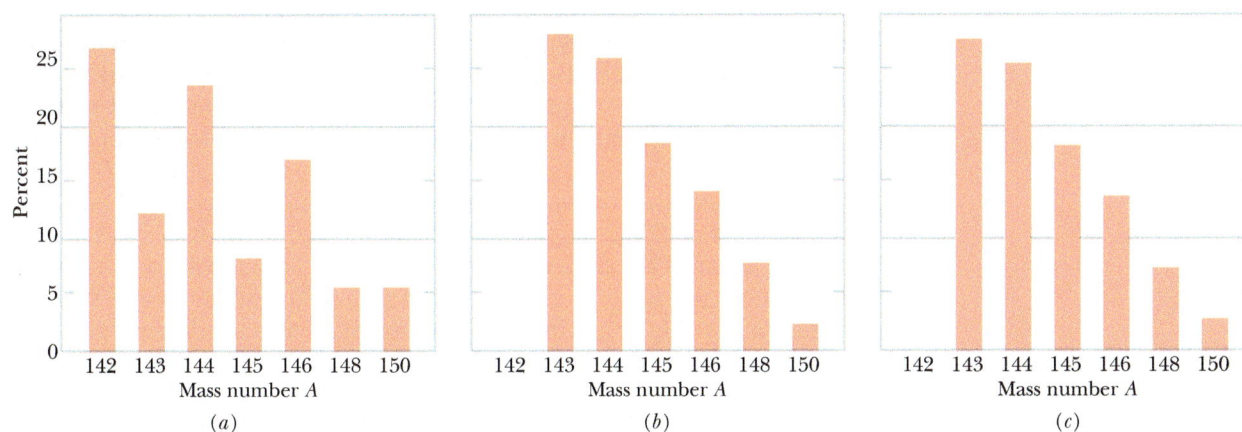

(a)   (b)   (c)

Figure 43-9 The distribution by mass number of the isotopes of neodymium as they occur in (a) natural terrestrial deposits of the ores of this element and (b) the spent fuel of a power reactor. (c) The distribution (after several corrections) found for neodymium from the uranium mine in Gabon, West Africa. Note that (b) and (c) are virtually identical and are quite different from (a).

43-4 THERMONUCLEAR FUSION: THE BASIC PROCESS

Learning Objectives

After reading this module, you should be able to . . .

43.18 Define thermonuclear fusion, explaining why the nuclei must be at a high temperature to fuse.

43.19 For nuclei, apply the relationship between their kinetic energy and their temperature.

43.20 Explain the two reasons why fusion of two nuclei can occur even when the kinetic energy associated with their most probable speed is insufficient to overcome their energy barrier.

Key Ideas

● The release of energy by fusion of two light nuclei is inhibited by their mutual Coulomb barrier (due to the electric repulsion between the two collections of protons).

● Fusion can occur in bulk matter only if the temperature is high enough (that is, if the particle energy is high enough) for appreciable barrier tunneling to occur.

Thermonuclear Fusion: The Basic Process

The binding energy curve of Fig. 42-7 shows that energy can be released if two light nuclei combine to form a single larger nucleus, a process called nuclear **fusion.** That process is hindered by the Coulomb repulsion that acts to prevent the two positively charged particles from getting close enough to be within range of their attractive nuclear forces and thus "fusing." The range of the nuclear force is short, hardly beyond the nuclear "surface," but the range of the repulsive Coulomb force is long and that force thus forms an energy barrier. The height of this *Coulomb barrier* depends on the charges and the radii of the two interacting nuclei. For two protons ($Z = 1$), the barrier height is 400 keV. For more highly charged particles, of course, the barrier is correspondingly higher.

To generate useful amounts of energy, nuclear fusion must occur in bulk matter. The best hope for bringing this about is to raise the temperature of the material until the particles have enough energy—due to their thermal motions alone—to penetrate the Coulomb barrier. We call this process **thermonuclear fusion.**

In thermonuclear studies, temperatures are reported in terms of the kinetic energy K of interacting particles via the relation

$$K = kT, \qquad (43\text{-}9)$$

in which K is the kinetic energy corresponding to the *most probable speed* of the interacting particles, k is the Boltzmann constant, and the temperature T is in kelvins. Thus, rather than saying, "The temperature at the center of the Sun is 1.5×10^7 K," it is more common to say, "The temperature at the center of the Sun is 1.3 keV."

Room temperature corresponds to $K \approx 0.03$ eV; a particle with only this amount of energy could not hope to overcome a barrier as high as, say, 400 keV. Even at the center of the Sun, where $kT = 1.3$ keV, the outlook for thermonuclear fusion does not seem promising at first glance. Yet we know that thermonuclear fusion not only occurs in the core of the Sun but is the dominant feature of that body and of all other stars.

The puzzle is solved when we realize two facts: (1) The energy calculated with Eq. 43-9 is that of the particles with the *most probable* speed, as defined in Module 19-6; there is a long tail of particles with much higher speeds and, correspondingly, much higher energies. (2) The barrier heights that we have calculated represent the *peaks* of the barriers. Barrier tunneling can occur at energies considerably below those peaks, as we saw with alpha decay in Module 42-4.

Figure 43-10 sums things up. The curve marked $n(K)$ in this figure is a Maxwell distribution curve for the protons in the Sun's core, drawn to correspond to the Sun's central temperature. This curve differs from the Maxwell distribution curve given in Fig. 19-8 in that here the curve is drawn in terms of energy and not of speed. Specifically, for any kinetic energy K, the expression $n(K)\, dK$ gives the probability that a proton will have a kinetic energy lying between the values K and $K + dK$. The value of kT in the core of the Sun is indicated by the vertical line in the figure; note that many of the Sun's core protons have energies greater than this value.

The curve marked $p(K)$ in Fig. 43-10 is the probability of barrier penetration by two colliding protons. The two curves in Fig. 43-10 suggest that there is a particular proton energy at which proton–proton fusion events occur at a maximum rate. At energies much above this value, the barrier is transparent enough but too few protons have these energies, and so the fusion reaction cannot be sustained. At energies much below this value, plenty of protons have these energies but the Coulomb barrier is too formidable.

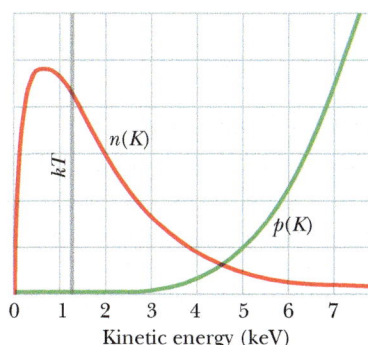

Figure 43-10 The curve marked $n(K)$ gives the number density per unit energy for protons at the center of the Sun. The curve marked $p(K)$ gives the probability of barrier penetration (and hence fusion) for proton–proton collisions at the Sun's core temperature. The vertical line marks the value of kT at this temperature. Note that the two curves are drawn to (separate) arbitrary vertical scales.

✓ **Checkpoint 2**

Which of these potential fusion reactions will *not* result in the net release of energy: (a) $^6Li + {}^6Li$, (b) $^4He + {}^4He$, (c) $^{12}C + {}^{12}C$, (d) $^{20}Ne + {}^{20}Ne$, (e) $^{35}Cl + {}^{35}Cl$, and (f) $^{14}N + {}^{35}Cl$? (*Hint:* Consult the curve of Fig. 42-7.)

Sample Problem 43.03   **Fusion in a gas of protons, and the required temperature**

Assume a proton is a sphere of radius $R \approx 1$ fm. Two protons are fired at each other with the same kinetic energy K.

(a) What must K be if the particles are brought to rest by their mutual Coulomb repulsion when they are just "touching" each other? We can take this value of K as a representative measure of the height of the Coulomb barrier.

KEY IDEAS

The mechanical energy E of the two-proton system is conserved as the protons move toward each other and momentarily stop. In particular, the initial mechanical energy E_i is equal to the mechanical energy E_f when they stop. The initial energy E_i consists only of the total kinetic energy $2K$ of the two protons. When the protons stop, energy E_f consists only of the electric potential energy U of the system, as given by Eq. 24-46 ($U = q_1 q_2 / 4\pi\varepsilon_0 r$).

Calculations: Here the distance r between the protons when they stop is their center-to-center distance $2R$, and their charges q_1 and q_2 are both e. Then we can write the conservation of energy $E_i = E_f$ as

$$2K = \frac{1}{4\pi\varepsilon_0} \frac{e^2}{2R}.$$

This yields, with known values,

$$
\begin{aligned}
K &= \frac{e^2}{16\pi\varepsilon_0 R} \\
&= \frac{(1.60 \times 10^{-19}\ \text{C})^2}{(16\pi)(8.85 \times 10^{-12}\ \text{F/m})(1 \times 10^{-15}\ \text{m})} \\
&= 5.75 \times 10^{-14}\ \text{J} = 360\ \text{keV} \approx 400\ \text{keV}. \quad \text{(Answer)}
\end{aligned}
$$

(b) At what temperature would a proton in a gas of protons have the average kinetic energy calculated in (a) and thus have energy equal to the height of the Coulomb barrier?

KEY IDEA

If we treat the proton gas as an ideal gas, then from Eq. 19-24, the average energy of the protons is $K_{avg} = \frac{3}{2}kT$, where k is the Boltzmann constant.

Calculation: Solving that equation for T and using the result of (a) yield

$$
T = \frac{2K_{avg}}{3k} = \frac{(2)(5.75 \times 10^{-14}\ \text{J})}{(3)(1.38 \times 10^{-23}\ \text{J/K})}
$$
$$
\approx 3 \times 10^9\ \text{K}. \quad \text{(Answer)}
$$

The temperature of the core of the Sun is only about 1.5×10^7 K; thus fusion in the Sun's core must involve protons whose energies are *far* above the average energy.

PLUS Additional examples, video, and practice available at *WileyPLUS*

43-5 THERMONUCLEAR FUSION IN THE SUN AND OTHER STARS

Learning Objectives

After reading this module, you should be able to . . .

43.21 Explain the proton–proton cycle for the Sun.

43.22 Explain the stages after the Sun has consumed its hydrogen.

43.23 Explain the probable source of the elements that are more massive than hydrogen and helium.

Key Ideas

● The Sun's energy arises mainly from the thermonuclear burning of hydrogen to form helium by the proton–proton cycle.

● Elements up to $A \approx 56$ (the peak of the binding energy curve) can be built up by other fusion processes once the hydrogen fuel supply of a star has been exhausted.

Thermonuclear Fusion in the Sun and Other Stars

The Sun has been radiating energy at the rate of 3.9×10^{26} W for several billion years. Where does all this energy come from? It does not come from chemical burning. (Even if the Sun were made of coal and had its own oxygen, burning the coal would last only 1000 y.) It also does not come from the Sun shrinking, transferring gravitational potential energy to thermal energy. (Its lifetime would be short by a factor of at least 500.) That leaves only thermonuclear fusion. The Sun, as you will see, burns not coal but hydrogen, and in a nuclear furnace, not an atomic or chemical one.

The fusion reaction in the Sun is a multistep process in which hydrogen is burned to form helium, hydrogen being the "fuel" and helium the "ashes." Figure 43-11 shows the **proton–proton** (p-p) **cycle** by which this occurs.

The p-p cycle starts with the collision of two protons (^{1}H + ^{1}H) to form a deuteron (^{2}H), with the simultaneous creation of a positron (e^+) and a neutrino (ν). The positron immediately annihilates with any nearby electron (e^-), their mass energy appearing as two gamma-ray photons (γ) as in Module 21-3.

A pair of such events is shown in the top row of Fig. 43-11. These events are actually extremely rare. In fact, only once in about 10^{26} proton–proton collisions is a deuteron formed; in the vast majority of cases, the two protons simply rebound elastically from each other. It is the slowness of this "bottleneck" process that regulates the rate of energy production and keeps the Sun from exploding. In spite of this slowness, there are so very many protons in the huge and dense volume of the Sun's core that deuterium is produced in just this way at the rate of 10^{12} kg/s.

Once a deuteron has been produced, it quickly collides with another proton and forms a ^{3}He nucleus, as the middle row of Fig. 43-11 shows. Two such ^{3}He nuclei may eventually (within 10^5 y; there is plenty of time) find each other, forming an alpha particle (^{4}He) and two protons, as the bottom row in the figure shows.

Overall, we see from Fig. 43-11 that the p-p cycle amounts to the combination of four protons and two electrons to form an alpha particle, two neutrinos, and

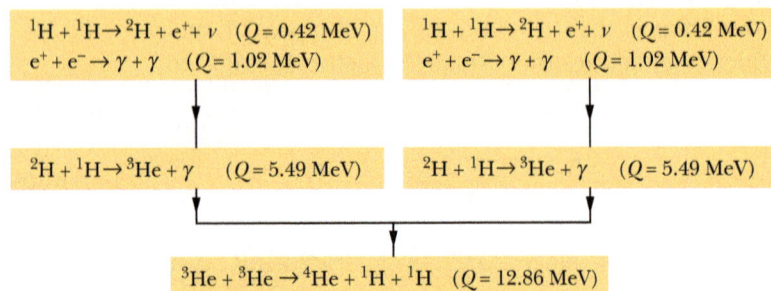

$$^1\text{H} + {^1\text{H}} \rightarrow {^2\text{H}} + e^+ + \nu \quad (Q = 0.42 \text{ MeV})$$
$$e^+ + e^- \rightarrow \gamma + \gamma \quad (Q = 1.02 \text{ MeV})$$

$$^1\text{H} + {^1\text{H}} \rightarrow {^2\text{H}} + e^+ + \nu \quad (Q = 0.42 \text{ MeV})$$
$$e^+ + e^- \rightarrow \gamma + \gamma \quad (Q = 1.02 \text{ MeV})$$

$$^2\text{H} + {^1\text{H}} \rightarrow {^3\text{He}} + \gamma \quad (Q = 5.49 \text{ MeV})$$

$$^2\text{H} + {^1\text{H}} \rightarrow {^3\text{He}} + \gamma \quad (Q = 5.49 \text{ MeV})$$

$$^3\text{He} + {^3\text{He}} \rightarrow {^4\text{He}} + {^1\text{H}} + {^1\text{H}} \quad (Q = 12.86 \text{ MeV})$$

Figure 43-11 The proton–proton mechanism that accounts for energy production in the Sun. In this process, protons fuse to form an alpha particle (^{4}He), with a net energy release of 26.7 MeV for each event.

six gamma-ray photons. That is,

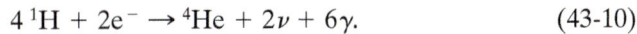

$$4 \, ^1\text{H} + 2e^- \rightarrow {}^4\text{He} + 2\nu + 6\gamma. \tag{43-10}$$

Let us now add two electrons to each side of Eq. 43-10, obtaining

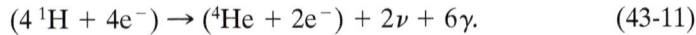

$$(4 \, ^1\text{H} + 4e^-) \rightarrow ({}^4\text{He} + 2e^-) + 2\nu + 6\gamma. \tag{43-11}$$

The quantities in the two sets of parentheses then represent *atoms* (not bare nuclei) of hydrogen and of helium. That allows us to compute the energy release in the overall reaction of Eq. 43-10 (and Eq. 43-11) as

$$\begin{aligned}
Q &= -\Delta m \, c^2 \\
&= -[4.002\,603 \text{ u} - (4)(1.007\,825 \text{ u})][931.5 \text{ MeV/u}] \\
&= 26.7 \text{ MeV},
\end{aligned}$$

in which 4.002 603 u is the mass of a helium atom and 1.007 825 u is the mass of a hydrogen atom. Neutrinos have a negligibly small mass, and gamma-ray photons have no mass; thus, they do not enter into the calculation of the disintegration energy.

This same value of Q follows (as it must) from adding up the Q values for the separate steps of the proton–proton cycle in Fig. 43-11. Thus,

$$\begin{aligned}
Q &= (2)(0.42 \text{ MeV}) + (2)(1.02 \text{ MeV}) + (2)(5.49 \text{ MeV}) + 12.86 \text{ MeV} \\
&= 26.7 \text{ MeV}.
\end{aligned}$$

About 0.5 MeV of this energy is carried out of the Sun by the two neutrinos indicated in Eqs. 43-10 and 43-11; the rest (= 26.2 MeV) is deposited in the core of the Sun as thermal energy. That thermal energy is then gradually transported to the Sun's surface, where it is radiated away from the Sun as electromagnetic waves, including visible light.

Hydrogen burning has been going on in the Sun for about 5×10^9 y, and calculations show that there is enough hydrogen left to keep the Sun going for about the same length of time into the future. In 5 billion years, however, the Sun's core, which by that time will be largely helium, will begin to cool and the Sun will start to collapse under its own gravity. This will raise the core temperature and cause the outer envelope to expand, turning the Sun into what is called a *red giant*.

If the core temperature increases to about 10^8 K again, energy can be produced through fusion once more—this time by burning helium to make carbon. As a star evolves further and becomes still hotter, other elements can be formed by other fusion reactions. However, elements more massive than those near the peak of the binding energy curve of Fig. 42-7 cannot be produced by further fusion processes.

Elements with mass numbers beyond the peak are thought to be formed by neutron capture during cataclysmic stellar explosions that we call *supernovas* (Fig. 43-12).

Figure 43-12 (*a*) The star known as Sanduleak, as it appeared until 1987. (*b*) We then began to intercept light from the star's supernova, designated SN1987a; the explosion was 100 million times brighter than our Sun and could be seen with the unaided eye even through it was outside our Galaxy.

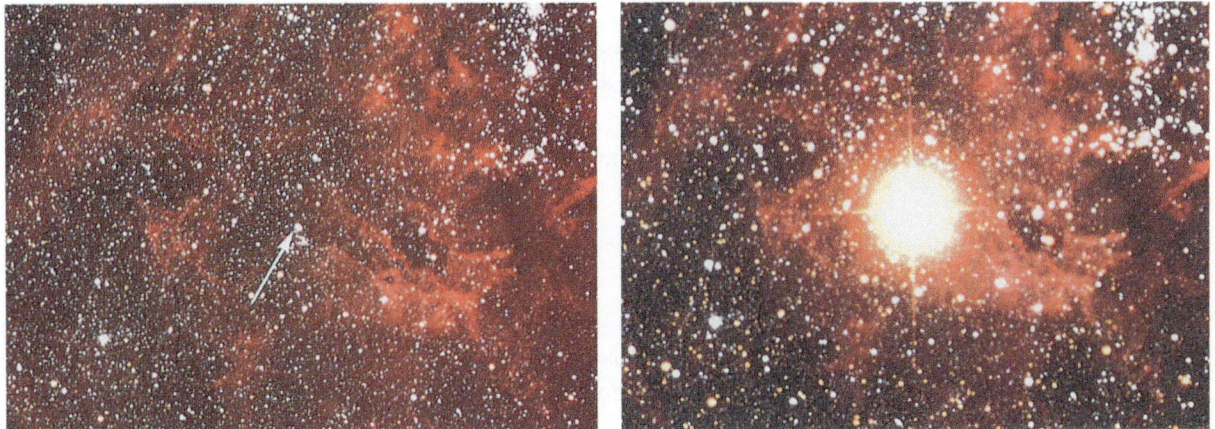

(*a*)   (*b*)   Courtesy Anglo Australian Telescope Board

In such an event the outer shell of the star is blown outward into space, where it mixes with the tenuous medium that fills the space between the stars. It is from this medium, continually enriched by debris from stellar explosions, that new stars form, by condensation under the influence of the gravitational force.

The abundance on Earth of elements heavier than hydrogen and helium suggests that our solar system has condensed out of interstellar material that contained the remnants of such explosions. Thus, all the elements around us—including those in our own bodies—were manufactured in the interiors of stars that no longer exist. As one scientist put it: "In truth, we are the children of the stars."

Sample Problem 43.04 **Consumption rate of hydrogen in the Sun**

At what rate dm/dt is hydrogen being consumed in the core of the Sun by the p-p cycle of Fig. 43-11?

KEY IDEA

The rate dE/dt at which energy is produced by hydrogen (proton) consumption within the Sun is equal to the rate P at which energy is radiated by the Sun:

$$P = \frac{dE}{dt}.$$

Calculations: To bring the mass consumption rate dm/dt into the power equation, we can rewrite it as

$$P = \frac{dE}{dt} = \frac{dE}{dm}\frac{dm}{dt} \approx \frac{\Delta E}{\Delta m}\frac{dm}{dt}, \qquad (43\text{-}12)$$

where ΔE is the energy produced when protons of mass Δm are consumed. From our discussion in this module, we know that 26.2 MeV ($= 4.20 \times 10^{-12}$ J) of thermal energy is produced when four protons are consumed. That is, $\Delta E = 4.20 \times 10^{-12}$ J for a mass consumption of $\Delta m = 4(1.67 \times 10^{-27}$ kg). Substituting these data into Eq. 43-12 and using the power P of the Sun given in Appendix C, we find that

$$\frac{dm}{dt} = \frac{\Delta m}{\Delta E}P = \frac{4(1.67 \times 10^{-27}\text{ kg})}{4.20 \times 10^{-12}\text{ J}}(3.90 \times 10^{26}\text{ W})$$

$$= 6.2 \times 10^{11}\text{ kg/s.} \qquad \text{(Answer)}$$

Thus, a huge amount of hydrogen is consumed by the Sun every second. However, you need not worry too much about the Sun running out of hydrogen, because its mass of 2×10^{30} kg will keep it burning for a long, long time.

WILEY PLUS Additional examples, video, and practice available at *WileyPLUS*

43-6 CONTROLLED THERMONUCLEAR FUSION

Learning Objectives

After reading this module, you should be able to . . .

43.24 Give the three requirements for a thermonuclear reactor.
43.25 Define Lawson's criterion.

43.26 Give general descriptions of the magnetic confinement approach and the inertial confinement approach.

Key Ideas

● Controlled thermonuclear fusion for energy generation has not yet been achieved. The d-d and d-t reactions are the most promising mechanisms.

● A successful fusion reactor must satisfy Lawson's criterion,

$$n\tau > 10^{20}\text{ s/m}^3,$$

and must have a suitably high plasma temperature T.

● In a tokamak, the plasma is confined by a magnetic field.

● In laser fusion, inertial confinement is used.

Controlled Thermonuclear Fusion

The first thermonuclear reaction on Earth occurred at Eniwetok Atoll on November 1, 1952, when the United States exploded a fusion device, generating an energy release equivalent to 10 million tons of TNT. The high temperatures and densities needed to initiate the reaction were provided by using a fission bomb as a trigger.

A sustained and controllable source of fusion power—a fusion reactor as part of, say, an electric generating plant—is considerably more difficult to achieve. That goal is nonetheless being pursued vigorously in many countries around the world, because many people look to the fusion reactor as the power source of the future, at least for the generation of electricity.

The p-p scheme displayed in Fig. 43-11 is not suitable for an Earth-bound fusion reactor because it is hopelessly slow. The process succeeds in the Sun only because of the enormous density of protons in the center of the Sun. The most attractive reactions for terrestrial use appear to be two deuteron–deuteron (d-d) reactions,

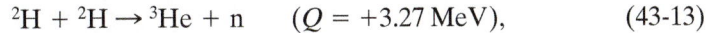

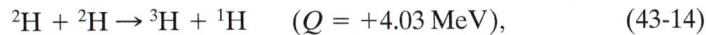

$$^2H + {}^2H \rightarrow {}^3He + n \qquad (Q = +3.27 \text{ MeV}), \qquad (43\text{-}13)$$

$$^2H + {}^2H \rightarrow {}^3H + {}^1H \qquad (Q = +4.03 \text{ MeV}), \qquad (43\text{-}14)$$

and the deuteron–triton (d-t) reaction

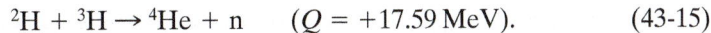

$$^2H + {}^3H \rightarrow {}^4He + n \qquad (Q = +17.59 \text{ MeV}). \qquad (43\text{-}15)$$

(The nucleus of the hydrogen isotope ^3H (tritium) is called the *triton* and has a half-life of 12.3 y.) Deuterium, the source of deuterons for these reactions, has an isotopic abundance of only 1 part in 6700 but is available in unlimited quantities as a component of seawater. Proponents of power from the nucleus have described our ultimate power choice—after we have burned up all our fossil fuels—as either "burning rocks" (fission of uranium extracted from ores) or "burning water" (fusion of deuterium extracted from water).

There are three requirements for a successful thermonuclear reactor:

1. *A High Particle Density n.*   The number density of interacting particles (the number of, say, deuterons per unit volume) must be great enough to ensure that the d-d collision rate is high enough. At the high temperatures required, the deuterium would be completely ionized, forming an electrically neutral **plasma** (ionized gas) of deuterons and electrons.

2. *A High Plasma Temperature T.*   The plasma must be hot. Otherwise the colliding deuterons will not be energetic enough to penetrate the Coulomb barrier that tends to keep them apart. A plasma ion temperature of 35 keV, corresponding to 4×10^8 K, has been achieved in the laboratory. This is about 30 times higher than the Sun's central temperature.

3. *A Long Confinement Time τ.*   A major problem is containing the hot plasma long enough to maintain it at a density and a temperature sufficiently high to ensure the fusion of enough of the fuel. Because it is clear that no solid container can withstand the high temperatures that are necessary, clever confining techniques are called for; we shall shortly discuss two of them.

It can be shown that, for the successful operation of a thermonuclear reactor using the d-t reaction, it is necessary to have

$$n\tau > 10^{20} \text{ s/m}^3. \qquad (43\text{-}16)$$

This condition, known as **Lawson's criterion,** tells us that we have a choice between confining a lot of particles for a short time or fewer particles for a longer time. Also, the plasma temperature must be high enough.

Two approaches to controlled nuclear power generation are currently under study. Although neither approach has yet been successful, both are being pursued because of their promise and because of the potential importance of controlled fusion to solving the world's energy problems.

Magnetic Confinement

One avenue to controlled fusion is to contain the fusing material in a very strong magnetic field—hence the name **magnetic confinement.** In one version of this approach, a suitably shaped magnetic field is used to confine the hot plasma in an evacuated doughnut-shaped chamber called a **tokamak** (the name is an abbreviation consisting of parts of three Russian words). The magnetic forces acting on

Figure 43-13 The small spheres on the quarter are deuterium–tritium fuel pellets, designed to be used in a laser fusion chamber.

Courtesy Los Alamos National Laboratory, New Mexico

the charged particles that make up the hot plasma keep the plasma from touching the walls of the chamber.

The plasma is heated by inducing a current in it and by bombarding it with an externally accelerated beam of particles. The first goal of this approach is to achieve **breakeven,** which occurs when the Lawson criterion is met or exceeded. The ultimate goal is **ignition,** which corresponds to a self-sustaining thermonuclear reaction and a net generation of energy.

Inertial Confinement

A second approach, called **inertial confinement,** involves "zapping" a solid fuel pellet from all sides with intense laser beams, evaporating some material from the surface of the pellet. This boiled-off material causes an inward-moving shock wave that compresses the core of the pellet, increasing both its particle density and its temperature. The process is called inertial confinement because (a) the fuel is *confined* to the pellet and (b) the particles do not escape from the heated pellet during the very short zapping interval because of their *inertia* (their mass).

Laser fusion, using the inertial confinement approach, is being investigated in many laboratories in the United States and elsewhere. At the Lawrence Livermore Laboratory, for example, deuterium–tritium fuel pellets, each smaller than a grain of sand (Fig. 43-13), are to be zapped by 10 synchronized high-power laser pulses symmetrically arranged around the pellet. The laser pulses are designed to deliver, in total, some 200 kJ of energy to each fuel pellet in less than a nanosecond. This is a delivered power of about 2×10^{14} W during the pulse, which is roughly 100 times the total installed (sustained) electrical power generating capacity of the world!

Sample Problem 43.05   Laser fusion: number of particles and Lawson's criterion

Suppose a fuel pellet in a laser fusion device contains equal numbers of deuterium and tritium atoms (and no other material). The density $d = 200 \text{ kg/m}^3$ of the pellet is increased by a factor of 10^3 by the action of the laser pulses.

(a) How many particles per unit volume (both deuterons and tritons) does the pellet contain in its compressed state? The molar mass M_d of deuterium atoms is 2.0×10^{-3} kg/mol, and the molar mass M_t of tritium atoms is 3.0×10^{-3} kg/mol.

KEY IDEA

For a system consisting of only one type of particle, we can write the (mass) density (the mass per unit volume) of the system in terms of the particle masses and number density (the number of particles per unit volume):

$$\begin{pmatrix} \text{density,} \\ \text{kg/m}^3 \end{pmatrix} = \begin{pmatrix} \text{number density,} \\ \text{m}^{-3} \end{pmatrix} \begin{pmatrix} \text{particle mass,} \\ \text{kg} \end{pmatrix}. \quad (43\text{-}17)$$

Let n be the total number of particles per unit volume in the compressed pellet. Then, because we know that the device contains equal numbers of deuterium and tritium atoms, the number of deuterium atoms per unit volume is $n/2$, and the number of tritium atoms per unit volume is also $n/2$.

Calculations: We can extend Eq. 43-17 to the system consisting of the two types of particles by writing the density $d*$ of the compressed pellet as the sum of the individual densities:

$$d* = \frac{n}{2} m_d + \frac{n}{2} m_t, \qquad (43\text{-}18)$$

where m_d and m_t are the masses of a deuterium atom and a tritium atom, respectively. We can replace those masses with the given molar masses by substituting

$$m_d = \frac{M_d}{N_A} \quad \text{and} \quad m_t = \frac{M_t}{N_A},$$

where N_A is Avogadro's number. After making those replacements and substituting $1000d$ for the compressed density $d*$, we solve Eq. 43-18 for the particle number density n to obtain

$$n = \frac{2000dN_A}{M_d + M_t},$$

which gives us

$$n = \frac{(2000)(200 \text{ kg/m}^3)(6.02 \times 10^{23} \text{ mol}^{-1})}{2.0 \times 10^{-3} \text{ kg/mol} + 3.0 \times 10^{-3} \text{ kg/mol}}$$

$$= 4.8 \times 10^{31} \text{ m}^{-3}. \qquad \text{(Answer)}$$

(b) According to Lawson's criterion, how long must the pellet maintain this particle density if breakeven operation is to take place at a suitably high temperature?

KEY IDEA

If breakeven operation is to occur, the compressed density must be maintained for a time period τ given by Eq. 43-16 ($n\tau > 10^{20}$ s/m^3).

Calculation: We can now write

$$\tau > \frac{10^{20} \text{ s/m}^3}{4.8 \times 10^{31} \text{ m}^{-3}} \approx 10^{-12} \text{ s}. \qquad \text{(Answer)}$$

WILEY PLUS Additional examples, video, and practice available at *WileyPLUS*

Review & Summary

Energy from the Nucleus Nuclear processes are about a million times more effective, per unit mass, than chemical processes in transforming mass into other forms of energy.

Nuclear Fission Equation 43-1 shows a **fission** of ^{236}U induced by thermal neutrons bombarding ^{235}U. Equations 43-2 and 43-3 show the beta-decay chains of the primary fragments. The energy released in such a fission event is $Q \approx 200$ MeV.

Fission can be understood in terms of the collective model, in which a nucleus is likened to a charged liquid drop carrying a certain excitation energy. A potential barrier must be tunneled through if fission is to occur. The ability of a nucleus to undergo fission depends on the relationship between the barrier height E_b and the excitation energy E_n.

The neutrons released during fission make possible a fission **chain reaction.** Figure 43-5 shows the neutron balance for one cycle of a typical reactor. Figure 43-6 suggests the layout of a complete nuclear power plant.

Nuclear Fusion The release of energy by the **fusion** of two light nuclei is inhibited by their mutual Coulomb barrier (due to

the electric repulsion between the two collections of protons). Fusion can occur in bulk matter only if the temperature is high enough (that is, if the particle energy is high enough) for appreciable barrier tunneling to occur.

The Sun's energy arises mainly from the thermonuclear burning of hydrogen to form helium by the **proton–proton cycle** outlined in Fig. 43-11. Elements up to $A \approx 56$ (the peak of the binding energy curve) can be built up by other fusion processes once the hydrogen fuel supply of a star has been exhausted. Fusion of more massive elements requires an input of energy and thus cannot be the source of a star's energy output.

Controlled Fusion Controlled **thermonuclear fusion** for energy generation has not yet been achieved. The d-d and d-t reactions are the most promising mechanisms. A successful fusion reactor must satisfy **Lawson's criterion,**

$$n\tau > 10^{20} \text{ s/m}^3, \qquad (43\text{-}16)$$

and must have a suitably high plasma temperature T.

In a **tokamak** the plasma is confined by a magnetic field. In **laser fusion** inertial confinement is used.

Questions

1 In the fission process

$$^{235}\text{U} + \text{n} \rightarrow {}^{132}\text{Sn} + \boxed{}\Box + 3\text{n},$$

what number goes in (a) the elevated box (the superscript) and (b) the descended box (the value of Z)?

2 If a fusion process requires an absorption of energy, does the average binding energy per nucleon increase or decrease?

3 Suppose a ^{238}U nucleus "swallows" a neutron and then decays not by fission but by beta-minus decay, in which it emits an electron and a neutrino. Which nuclide remains after this decay: ^{239}Pu, ^{238}Np, ^{239}Np, or ^{238}Pa?

4 Do the initial fragments formed by fission have more protons than neutrons, more neutrons than protons, or about the same number of each?

5 For the fission reaction

$$^{235}U + n \rightarrow X + Y + 2n,$$

rank the following possibilities for X (or Y), most likely first: ^{152}Nd, ^{140}I, ^{128}In, ^{115}Pd, ^{105}Mo. (*Hint:* See Fig. 43-1.)

6 To make the newly discovered, very large elements of the periodic table, researchers shoot a medium-size nucleus at a large nucleus. Sometimes a projectile nucleus and a target nucleus fuse to form one of the very large elements. In such a fusion, is the mass of the product greater than or less than the sum of the masses of the projectile and target nuclei?

7 If we split a nucleus into two smaller nuclei, with a release of energy, has the average binding energy per nucleon increased or decreased?

8 Which of these elements is *not* "cooked up" by thermonuclear fusion processes in stellar interiors: carbon, silicon, chromium, bromine?

9 Lawson's criterion for the d-t reaction (Eq. 43-16) is $n\tau > 10^{20}$ s/m^3. For the d-d reaction, do you expect the number on the right-hand side to be the same, smaller, or larger?

10 About 2% of the energy generated in the Sun's core by the p-p reaction is carried out of the Sun by neutrinos. Is the energy associated with this neutrino flux equal to, greater than, or less than the energy radiated from the Sun's surface as electromagnetic radiation?

11 A nuclear reactor is operating at a certain power level, with its multiplication factor k adjusted to unity. If the control rods are used to reduce the power output of the reactor to 25% of its former value, is the multiplication factor now a little less than unity, substantially less than unity, or still equal to unity?

12 Pick the most likely member of each pair to be one of the initial fragments formed by a fission event: (a) ^{93}Sr or ^{93}Ru, (b) ^{140}Gd or ^{140}I, (c) ^{155}Nd or ^{155}Lu. (*Hint:* See Fig. 42-5 and the periodic table, and consider the neutron abundance.)

Problems

Module 43-1 Nuclear Fission

•1 The isotope ^{235}U decays by alpha emission with a half-life of 7.0×10^8 y. It also decays (rarely) by spontaneous fission, and if the alpha decay did not occur, its half-life due to spontaneous fission alone would be 3.0×10^{17} y. (a) At what rate do spontaneous fission decays occur in 1.0 g of ^{235}U? (b) How many ^{235}U alpha-decay events are there for every spontaneous fission event?

•2 The nuclide ^{238}Np requires 4.2 MeV for fission. To remove a neutron from this nuclide requires an energy expenditure of 5.0 MeV. Is ^{237}Np fissionable by thermal neutrons?

•3 **GO** A thermal neutron (with approximately zero kinetic energy) is absorbed by a ^{238}U nucleus. How much energy is transferred from mass energy to the resulting oscillation of the nucleus? Here are some atomic masses and the neutron mass.

^{237}U	237.048 723 u	^{238}U	238.050 782 u
^{239}U	239.054 287 u	^{240}U	240.056 585 u
n	1.008 664 u		

•4 The fission properties of the plutonium isotope ^{239}Pu are very similar to those of ^{235}U. The average energy released per fission is 180 MeV. How much energy, in MeV, is released if all the atoms in 1.00 kg of pure ^{239}Pu undergo fission?

•5 During the Cold War, the Premier of the Soviet Union threatened the United States with 2.0 megaton ^{239}Pu warheads. (Each would have yielded the equivalent of an explosion of 2.0 megatons of TNT, where 1 megaton of TNT releases 2.6×10^{28} MeV of energy.) If the plutonium that actually fissioned had been 8.00% of the total mass of the plutonium in such a warhead, what was that total mass?

•6 (a)–(d) Complete the following table, which refers to the generalized fission reaction $^{235}U + n \rightarrow X + Y + bn$.

X	Y	b
^{140}Xe	(a)	1
^{139}I	(b)	2
(c)	^{100}Zr	2
^{141}Cs	^{92}Rb	(d)

•7 At what rate must ^{235}U nuclei undergo fission by neutron bombardment to generate energy at the rate of 1.0 W? Assume that $Q = 200$ MeV.

•8 (a) Calculate the disintegration energy Q for the fission of the molybdenum isotope ^{98}Mo into two equal parts. The masses you will need are 97.905 41 u for ^{98}Mo and 48.950 02 u for ^{49}Sc. (b) If Q turns out to be positive, discuss why this process does not occur spontaneously.

•9 (a) How many atoms are contained in 1.0 kg of pure ^{235}U? (b) How much energy, in joules, is released by the complete fissioning of 1.0 kg of ^{235}U? Assume $Q = 200$ MeV. (c) For how long would this energy light a 100 W lamp?

•10 **GO** Calculate the energy released in the fission reaction

$$^{235}U + n \rightarrow {}^{141}Cs + {}^{93}Rb + 2n.$$

Here are some atomic and particle masses.

^{235}U	235.043 92 u	^{93}Rb	92.921 57 u
^{141}Cs	140.919 63 u	n	1.008 66 u

•11 **GO** Calculate the disintegration energy Q for the fission of ^{52}Cr into two equal fragments. The masses you will need are

^{52}Cr	51.940 51 u	^{26}Mg	25.982 59 u.

••12 GO Consider the fission of ^{238}U by fast neutrons. In one fission event, no neutrons are emitted and the final stable end products, after the beta decay of the primary fission fragments, are ^{140}Ce and ^{99}Ru. (a) What is the total of the beta-decay events in the two beta-decay chains? (b) Calculate Q for this fission process. The relevant atomic and particle masses are

^{238}U	238.050 79 u	^{140}Ce	139.905 43 u
n	1.008 66 u	^{99}Ru	98.905 94 u

••13 GO Assume that immediately after the fission of ^{236}U according to Eq. 43-1, the resulting ^{140}Xe and ^{94}Sr nuclei are just touching at their surfaces. (a) Assuming the nuclei to be spherical, calculate the electric potential energy associated with the repulsion between the two fragments. (*Hint:* Use Eq. 42-3 to calculate the radii of the fragments.) (b) Compare this energy with the energy released in a typical fission event.

••14 A ^{236}U nucleus undergoes fission and breaks into two middle-mass fragments, ^{140}Xe and ^{96}Sr. (a) By what percentage does the surface area of the fission products differ from that of the original ^{236}U nucleus? (b) By what percentage does the volume change? (c) By what percentage does the electric potential energy change? The electric potential energy of a uniformly charged sphere of radius r and charge Q is given by

$$U = \frac{3}{5}\left(\frac{Q^2}{4\pi\varepsilon_0 r}\right).$$

••15 SSM A 66 kiloton atomic bomb is fueled with pure ^{235}U (Fig. 43-14), 4.0% of which actually undergoes fission. (a) What is the mass of the uranium in the bomb? (It is not 66 kilotons—that is the amount of released energy specified in terms of the mass of TNT required to produce the same amount of energy.) (b) How many primary fission fragments are produced? (c) How many fission neutrons generated are released to the environment? (On average, each fission produces 2.5 neutrons.)

Courtesy Martin Marietta Energy Systems/U.S. Department of Energy

Figure 43-14 Problem 15. A "button" of ^{235}U ready to be recast and machined for a warhead.

••16 In an atomic bomb, energy release is due to the uncontrolled fission of plutonium ^{239}Pu (or ^{235}U). The bomb's rating is the mag-nitude of the released energy, specified in terms of the mass of TNT required to produce the same energy release. One megaton of TNT releases 2.6×10^{28} MeV of energy. (a) Calculate the rating, in tons of TNT, of an atomic bomb containing 95.0 kg of ^{239}Pu, of which 2.5 kg actually undergoes fission. (See Problem 4.) (b) Why is the other 92.5 kg of ^{239}Pu needed if it does not fission?

••17 SSM WWW In a particular fission event in which ^{235}U is fissioned by slow neutrons, no neutron is emitted and one of the primary fission fragments is ^{83}Ge. (a) What is the other fragment? The disintegration energy is $Q = 170$ MeV. How much of this energy goes to (b) the ^{83}Ge fragment and (c) the other fragment? Just after the fission, what is the speed of (d) the ^{83}Ge fragment and (e) the other fragment?

Module 43-2   The Nuclear Reactor

•18 A 200 MW fission reactor consumes half its fuel in 3.00 y. How much ^{235}U did it contain initially? Assume that all the energy generated arises from the fission of ^{235}U and that this nuclide is consumed only by the fission process.

••19 The neutron generation time t_{gen} in a reactor is the average time needed for a fast neutron emitted in one fission event to be slowed to thermal energies by the moderator and then initiate another fission event. Suppose the power output of a reactor at time $t = 0$ is P_0. Show that the power output a time t later is $P(t)$, where $P(t) = P_0 k^{t/t_{gen}}$ and k is the multiplication factor. For constant power output, $k = 1$.

••20 A reactor operates at 400 MW with a neutron generation time (see Problem 19) of 30.0 ms. If its power increases for 5.00 min with a multiplication factor of 1.0003, what is the power output at the end of the 5.00 min?

••21 The thermal energy generated when radiation from radionuclides is absorbed in matter can serve as the basis for a small power source for use in satellites, remote weather stations, and other isolated locations. Such radionuclides are manufactured in abundance in nuclear reactors and may be separated chemically from the spent fuel. One suitable radionuclide is ^{238}Pu ($T_{1/2} = 87.7$ y), which is an alpha emitter with $Q = 5.50$ MeV. At what rate is thermal energy generated in 1.00 kg of this material?

••22 The neutron generation time t_{gen} (see Problem 19) in a particular reactor is 1.0 ms. If the reactor is operating at a power level of 500 MW, about how many free neutrons are present in the reactor at any moment?

••23 SSM WWW The neutron generation time (see Problem 19) of a particular reactor is 1.3 ms. The reactor is generating energy at the rate of 1200.0 MW. To perform certain maintenance checks, the power level must temporarily be reduced to 350.00 MW. It is desired that the transition to the reduced power level take 2.6000 s. To what (constant) value should the multiplication factor be set to effect the transition in the desired time?

••24 (See Problem 21.) Among the many fission products that may be extracted chemically from the spent fuel of a nuclear reactor is ^{90}Sr ($T_{1/2} = 29$ y). This isotope is produced in typical large reactors at the rate of about 18 kg/y. By its radioactivity, the isotope generates thermal energy at the rate of 0.93 W/g. (a) Calculate the effective disintegration energy Q_{eff} associated with the decay of a ^{90}Sr nucleus. (This energy Q_{eff} includes contributions from the decay of the ^{90}Sr daughter products in its decay chain but not from neutrinos, which escape totally from the sample.) (b) It is desired to construct a power source generating 150 W (electric power) to use in operating electronic equipment in an underwater acoustic beacon. If the power source is based on the thermal energy generated by ^{90}Sr and if the efficiency of the thermal–electric conversion process is 5.0%, how much ^{90}Sr is needed?

••25 SSM (a) A neutron of mass m_n and kinetic energy K makes a head-on elastic collision with a stationary atom of mass m. Show that the fractional kinetic energy loss of the neutron is given by

$$\frac{\Delta K}{K} = \frac{4m_n m}{(m + m_n)^2}.$$

Find $\Delta K/K$ for each of the following acting as the stationary atom: (b) hydrogen, (c) deuterium, (d) carbon, and (e) lead. (f) If $K = 1.00$ MeV initially, how many such head-on collisions would it take to reduce the neutron's kinetic energy to a thermal value (0.025 eV) if the stationary atoms it collides with are deuterium, a commonly used moderator? (In actual moderators, most collisions are not head-on.)

Module 43-3 A Natural Nuclear Reactor

•26 How long ago was the ratio ^{235}U/^{238}U in natural uranium deposits equal to 0.15?

•27 The natural fission reactor discussed in Module 43-3 is estimated to have generated 15 gigawatt-years of energy during its lifetime. (a) If the reactor lasted for 200 000 y, at what average power level did it operate? (b) How many kilograms of ^{235}U did it consume during its lifetime?

••28 Some uranium samples from the natural reactor site described in Module 43-3 were found to be slightly *enriched* in ^{235}U, rather than depleted. Account for this in terms of neutron absorption by the abundant isotope ^{238}U and the subsequent beta and alpha decay of its products.

••29 SSM The uranium ore mined today contains only 0.72% of fissionable ^{235}U, too little to make reactor fuel for thermal-neutron fission. For this reason, the mined ore must be enriched with ^{235}U. Both ^{235}U ($T_{1/2} = 7.0 \times 10^8$ y) and ^{238}U ($T_{1/2} = 4.5 \times 10^9$ y) are radioactive. How far back in time would natural uranium ore have been a practical reactor fuel, with a ^{235}U/^{238}U ratio of 3.0%?

Module 43-4 Thermonuclear Fusion: The Basic Process

•30 Verify that the fusion of 1.0 kg of deuterium by the reaction

$$^2\text{H} + {}^2\text{H} \rightarrow {}^3\text{He} + n \quad (Q = +3.27 \text{ MeV})$$

could keep a 100 W lamp burning for 2.5×10^4 y.

•31 SSM Calculate the height of the Coulomb barrier for the head-on collision of two deuterons, with effective radius 2.1 fm.

••32 For overcoming the Coulomb barrier for fusion, methods other than heating the fusible material have been suggested. For example, if you were to use two particle accelerators to accelerate two beams of deuterons directly toward each other so as to collide head-on, (a) what voltage would each accelerator require in order for the colliding deuterons to overcome the Coulomb barrier? (b) Why do you suppose this method is not presently used?

••33 Calculate the Coulomb barrier height for two ^{7}Li nuclei that are fired at each other with the same initial kinetic energy K. (*Hint:* Use Eq. 42-3 to calculate the radii of the nuclei.)

••34 In Fig. 43-10, the equation for $n(K)$, the number density per unit energy for particles, is

$$n(K) = 1.13n \frac{K^{1/2}}{(kT)^{3/2}} e^{-K/kT},$$

where n is the total particle number density. At the center of the

Sun, the temperature is 1.50×10^7 K and the average proton energy K_{avg} is 1.94 keV. Find the ratio of the proton number density at 5.00 keV to the number density at the average proton energy.

Module 43-5 Thermonuclear Fusion in the Sun and Other Stars

•35 Assume that the protons in a hot ball of protons each have a kinetic energy equal to kT, where k is the Boltzmann constant and T is the absolute temperature. If $T = 1 \times 10^7$ K, what (approximately) is the least separation any two protons can have?

•36 GO What is the Q of the following fusion process?

$$^2\text{H}_1 + {}^1\text{H}_1 \rightarrow {}^3\text{He}_2 + \text{photon}$$

Here are some atomic masses.

^{2}H$_1$	2.014 102 u	^{1}H$_1$	1.007 825 u
^{3}He$_2$	3.016 029 u		

•37 The Sun has mass 2.0×10^{30} kg and radiates energy at the rate 3.9×10^{26} W. (a) At what rate is its mass changing? (b) What fraction of its original mass has it lost in this way since it began to burn hydrogen, about 4.5×10^9 y ago?

•38 We have seen that Q for the overall proton–proton fusion cycle is 26.7 MeV. How can you relate this number to the Q values for the reactions that make up this cycle, as displayed in Fig. 43-11?

•39 GO Show that the energy released when three alpha particles fuse to form ^{12}C is 7.27 MeV. The atomic mass of ^{4}He is 4.0026 u, and that of ^{12}C is 12.0000 u.

••40 Calculate and compare the energy released by (a) the fusion of 1.0 kg of hydrogen deep within the Sun and (b) the fission of 1.0 kg of ^{235}U in a fission reactor.

••41 GO A star converts all its hydrogen to helium, achieving a 100% helium composition. Next it converts the helium to carbon via the triple-alpha process,

$$^4\text{He} + {}^4\text{He} + {}^4\text{He} \rightarrow {}^{12}\text{C} + 7.27 \text{ MeV}.$$

The mass of the star is 4.6×10^{32} kg, and it generates energy at the rate of 5.3×10^{30} W. How long will it take to convert all the helium to carbon at this rate?

••42 Verify the three Q values reported for the reactions given in Fig. 43-11. The needed atomic and particle masses are

^{1}H	1.007 825 u	^{4}He	4.002 603 u
^{2}H	2.014 102 u	$e^\pm$	0.000 548 6 u
^{3}He	3.016 029 u		

(*Hint:* Distinguish carefully between atomic and nuclear masses, and take the positrons properly into account.)

••43 Figure 43-15 shows an early proposal for a hydrogen bomb. The fusion fuel is deuterium, ^{2}H. The high temperature and particle density needed for fusion are provided by an atomic bomb "trig-

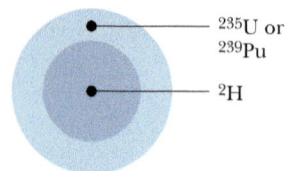

Figure 43-15 Problem 43.

ger" that involves a ^{235}U or ^{239}Pu fission fuel arranged to impress an imploding, compressive shock wave on the deuterium. The fusion reaction is

$$5\,^2\text{H} \rightarrow {}^3\text{He} + {}^4\text{He} + {}^1\text{H} + 2\text{n}.$$

(a) Calculate Q for the fusion reaction. For needed atomic masses, see Problem 42. (b) Calculate the rating (see Problem 16) of the fusion part of the bomb if it contains 500 kg of deuterium, 30.0% of which undergoes fusion.

••44 Assume that the core of the Sun has one-eighth of the Sun's mass and is compressed within a sphere whose radius is one-fourth of the solar radius. Assume further that the composition of the core is 35% hydrogen by mass and that essentially all the Sun's energy is generated there. If the Sun continues to burn hydrogen at the current rate of 6.2×10^{11} kg/s, how long will it be before the hydrogen is entirely consumed? The Sun's mass is 2.0×10^{30} kg.

••45 (a) Calculate the rate at which the Sun generates neutrinos. Assume that energy production is entirely by the proton–proton fusion cycle. (b) At what rate do solar neutrinos reach Earth?

••46 In certain stars the *carbon cycle* is more effective than the proton–proton cycle in generating energy. This carbon cycle is

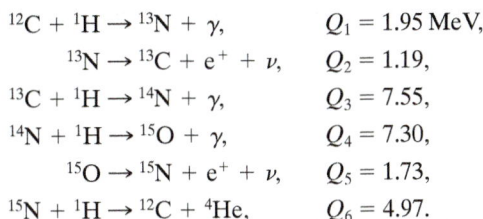

$$
\begin{array}{ll}
{}^{12}\text{C} + {}^1\text{H} \rightarrow {}^{13}\text{N} + \gamma, & Q_1 = 1.95\,\text{MeV}, \\
{}^{13}\text{N} \rightarrow {}^{13}\text{C} + e^+ + \nu, & Q_2 = 1.19, \\
{}^{13}\text{C} + {}^1\text{H} \rightarrow {}^{14}\text{N} + \gamma, & Q_3 = 7.55, \\
{}^{14}\text{N} + {}^1\text{H} \rightarrow {}^{15}\text{O} + \gamma, & Q_4 = 7.30, \\
{}^{15}\text{O} \rightarrow {}^{15}\text{N} + e^+ + \nu, & Q_5 = 1.73, \\
{}^{15}\text{N} + {}^1\text{H} \rightarrow {}^{12}\text{C} + {}^4\text{He}, & Q_6 = 4.97.
\end{array}
$$

(a) Show that this cycle is exactly equivalent in its overall effects to the proton–proton cycle of Fig. 43-11. (b) Verify that the two cycles, as expected, have the same Q value.

••47 SSM WWW Coal burns according to the reaction $C + O_2 \rightarrow CO_2$. The heat of combustion is 3.3×10^7 J/kg of atomic carbon consumed. (a) Express this in terms of energy per carbon atom. (b) Express it in terms of energy per kilogram of the initial reactants, carbon and oxygen. (c) Suppose that the Sun (mass = 2.0×10^{30} kg) were made of carbon and oxygen in combustible proportions and that it continued to radiate energy at its present rate of 3.9×10^{26} W. How long would the Sun last?

Module 43-6   Controlled Thermonuclear Fusion

•48 Verify the Q values reported in Eqs. 43-13, 43-14, and 43-15. The needed masses are

$$
\begin{array}{llll}
{}^1\text{H} & 1.007\,825\,\text{u} & {}^4\text{He} & 4.002\,603\,\text{u} \\
{}^2\text{H} & 2.014\,102\,\text{u} & \text{n} & 1.008\,665\,\text{u} \\
{}^3\text{H} & 3.016\,049\,\text{u} & &
\end{array}
$$

••49 Roughly 0.0150% of the mass of ordinary water is due to "heavy water," in which one of the two hydrogens in an H_2O molecule is replaced with deuterium, ^{2}H. How much average fusion power could be obtained if we "burned" all the ^{2}H in 1.00 liter of water in 1.00 day by somehow causing the deuterium to fuse via the reaction $^2\text{H} + {}^2\text{H} \rightarrow {}^3\text{He} + \text{n}$?

Additional Problems

50 The effective Q for the proton–proton cycle of Fig. 43-11 is 26.2 MeV. (a) Express this as energy per kilogram of hydrogen con-

sumed. (b) The power of the Sun is 3.9×10^{26} W. If its energy derives from the proton–proton cycle, at what rate is it losing hydrogen? (c) At what rate is it losing mass? (d) Account for the difference in the results for (b) and (c). (e) The mass of the Sun is 2.0×10^{30} kg. If it loses mass at the constant rate calculated in (c), how long will it take to lose 0.10% of its mass?

51 Many fear that nuclear power reactor technology will increase the likelihood of nuclear war because reactors can be used not only to produce electrical energy but also, as a by-product through neutron capture with inexpensive ^{238}U, to make ^{239}Pu, which is a "fuel" for nuclear bombs. What simple series of reactions involving neutron capture and beta decay would yield this plutonium isotope?

52 In the deuteron–triton fusion reaction of Eq. 43-15, what is the kinetic energy of (a) the alpha particle and (b) the neutron? Neglect the relatively small kinetic energies of the two combining particles.

53 Verify that, as stated in Module 43-1, neutrons in equilibrium with matter at room temperature, 300 K, have an average kinetic energy of about 0.04 eV.

54 Verify that, as reported in Table 43-1, fissioning of the ^{235}U in 1.0 kg of UO_2 (enriched so that ^{235}U is 3.0% of the total uranium) could keep a 100 W lamp burning for 690 y.

55 At the center of the Sun, the density of the gas is 1.5×10^5 kg/m^3 and the composition is essentially 35% hydrogen by mass and 65% helium by mass. (a) What is the number density of protons there? (b) What is the ratio of that proton density to the density of particles in an ideal gas at standard temperature (0°C) and pressure (1.01×10^5 Pa)?

56 Expressions for the Maxwell speed distribution for molecules in a gas are given in Chapter 19. (a) Show that the *most probable energy* is given by

$$K_p = \tfrac{1}{2}kT.$$

Verify this result with the energy distribution curve of Fig. 43-10, for which $T = 1.5 \times 10^7$ K. (b) Show that the *most probable speed* is given by

$$v_p = \sqrt{\frac{2kT}{m}}.$$

Find its value for protons at $T = 1.5 \times 10^7$ K. (c) Show that the *energy corresponding to the most probable speed* (which is not the same as the most probable energy) is

$$K_{v,p} = kT.$$

Locate this quantity on the curve of Fig. 43-10.

57 The uncompressed radius of the fuel pellet of Sample Problem 43.05 is 20 μm. Suppose that the compressed fuel pellet "burns" with an efficiency of 10%—that is, only 10% of the deuterons and 10% of the tritons participate in the fusion reaction of Eq. 43-15. (a) How much energy is released in each such microexplosion of a pellet? (b) To how much TNT is each such pellet equivalent? The heat of combustion of TNT is 4.6 MJ/kg. (c) If a fusion reactor is constructed on the basis of 100 microexplosions per second, what power would be generated? (Part of this power would be used to operate the lasers.)

58 Assume that a plasma temperature of 1×10^8 K is reached in a laser-fusion device. (a) What is the most probable speed of a deuteron at that temperature? (b) How far would such a deuteron move in a confinement time of 1×10^{-12} s?

Quarks, Leptons, and the Big Bang

44-1 GENERAL PROPERTIES OF ELEMENTARY PARTICLES

Learning Objectives

After reading this module, you should be able to . . .

44.01 Identify that a great many different elementary particles exist or can be created and that nearly all of them are unstable.

44.02 For the decay of an unstable particle, apply the same decay equations as used for the radioactive decay of nuclei.

44.03 Identify spin as the intrinsic angular momentum of a particle.

44.04 Distinguish fermions from bosons, and identify which are required to obey the Pauli exclusion principle.

44.05 Distinguish leptons and hadrons, and then identify the two types of hadrons.

44.06 Distinguish particle from antiparticle, and identify that if they meet, they undergo annihilation and are transformed into photons or into other elementary particles.

44.07 Distinguish the strong force and the weak force.

44.08 To see if a given process for elementary particles is physically possible, apply the conservation laws for charge, linear momentum, spin angular momentum, and energy (including mass energy).

Key Ideas

● The term fundamental particles refers to the basic building blocks of matter. We can divide the particles into several broad categories.

● The terms particles and antiparticles originally referred to common particles (such as the electrons, protons, and neutrons in your body) and their antiparticle counterparts

(the positrons, antiprotons, and antineutrons), but for most of the rarely detected particles, the distinction between particles and antiparticles is made largely to be consistent with experimental results.

● Fermions (such as the particles in your body) obey the Pauli exclusion principle; bosons do not.

What Is Physics?

Physicists often refer to the theories of relativity and quantum physics as "modern physics," to distinguish them from the theories of Newtonian mechanics and Maxwellian electromagnetism, which are lumped together as "classical physics." As the years go by, the word "modern" seems less and less appropriate for theories whose foundations were laid down in the opening years of the 20th century. After all, Einstein published his paper on the photoelectric effect and his first paper on special relativity in 1905, Bohr published his quantum model of the hydrogen atom in 1913, and Schrödinger published his matter wave equation in 1926. Nevertheless, the label of "modern physics" hangs on.

In this closing chapter we consider two lines of investigation that are truly "modern" but at the same time have the most ancient of roots. They center around two deceptively simple questions:

What is the universe made of?

How did the universe come to be the way it is?

Progress in answering these questions has been rapid in the last few decades.

Many new insights are based on experiments carried out with large particle accelerators. However, as they bang particles together at higher and higher energies using larger and larger accelerators, physicists come to realize that no conceivable Earth-bound accelerator can generate particles with energies great enough to test the ultimate theories of physics. There has been only one source of particles with these energies, and that was the universe itself within the first millisecond of its existence.

In this chapter you will encounter a host of new terms and a veritable flood of particles with names that you should not try to remember. If you are temporarily bewildered, you are sharing the bewilderment of the physicists who lived through these developments and who at times saw nothing but increasing complexity with little hope of understanding. If you stick with it, however, you will come to share the excitement physicists felt as marvelous new accelerators poured out new results, as the theorists put forth ideas each more daring than the last, and as clarity finally sprang from obscurity. The main message of this book is that, although we know a lot about the physics of the world, grand mysteries remain.

Particles, Particles, Particles

In the 1930s, there were many scientists who thought that the problem of the ultimate structure of matter was well on the way to being solved. The atom could be understood in terms of only three particles—the electron, the proton, and the neutron. Quantum physics accounted well for the structure of the atom and for radioactive alpha decay. The neutrino had been postulated and, although not yet observed, had been incorporated by Enrico Fermi into a successful theory of beta decay. There was hope that quantum theory applied to protons and neutrons would soon account for the structure of the nucleus. What else was there?

The euphoria did not last. The end of that same decade saw the beginning of a period of discovery of new particles that continues to this day. The new particles have names and symbols such as *muon* (μ), *pion* (π), *kaon* (K), and *sigma* (Σ). All the new particles are unstable; that is, they spontaneously transform into other types of particles according to the same functions of time that apply to unstable nuclei. Thus, if N_0 particles of any one type are present in a sample at time $t = 0$, the number N of those particles present at some later time t is given by Eq. 42-15,

$$N = N_0 e^{-\lambda t},\qquad(44\text{-}1)$$

the rate of decay R, from an initial value of R_0, is given by Eq. 42-16,

$$R = R_0 e^{-\lambda t},\qquad(44\text{-}2)$$

and the half-life $T_{1/2}$, decay constant λ, and mean life τ are related by Eq. 42-18,

$$T_{1/2} = \frac{\ln 2}{\lambda} = \tau \ln 2.\qquad(44\text{-}3)$$

The half-lives of the new particles range from about 10^{-6} s to 10^{-23} s. Indeed, some of the particles last so briefly that they cannot be detected directly but can only be inferred from indirect evidence.

These new particles have been commonly produced in head-on collisions between protons or electrons accelerated to high energies in accelerators at places like Brookhaven National Laboratory (on Long Island, New York), Fermilab (near Chicago), CERN (near Geneva, Switzerland), SLAC (at Stanford University in California), and DESY (near Hamburg, Germany). They have been discovered with particle detectors that have grown in sophistication until they rival the size and complexity of entire accelerators of only a few decades ago.

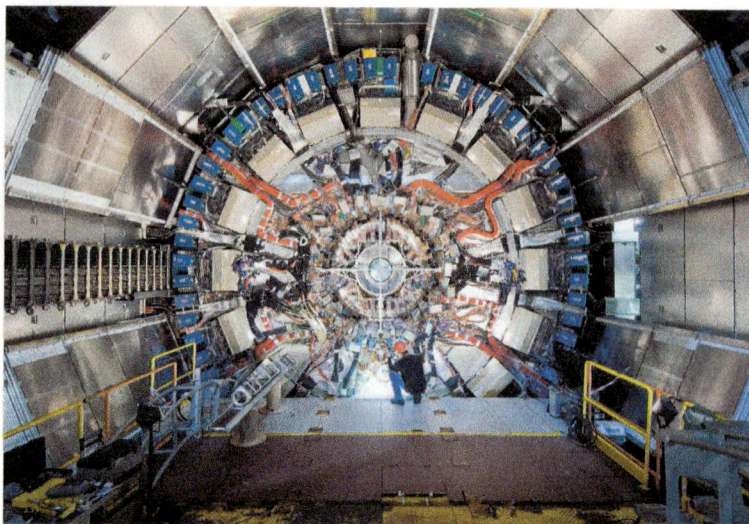

© CERN Geneva

One of the detectors at the Large Hadron Collider at CERN, where the Standard Model of the elementary particles is being put to the test.

Today there are several hundred known particles. Naming them has strained the resources of the Greek alphabet, and most are known only by an assigned number in a periodically issued compilation. To make sense of this array of particles, we look for simple physical criteria by which we can place the particles in categories. The result is known as the **Standard Model** of particles. Although this model is continuously challenged by theorists, it remains our best scheme of understanding all the particles discovered to date.

To explore the Standard Model, we make the following three rough cuts among the known particles: fermion or boson, hadron or lepton, particle or antiparticle? Let's now look at the categories one by one.

Fermion or Boson?

All particles have an intrinsic angular momentum called **spin**, as we discussed for electrons, protons, and neutrons in Module 32-5. Generalizing the notation of that section, we can write the component of spin $\vec{S}$ in any direction (assume the component to be along a z axis) as

$$S_z = m_s \hbar \qquad \text{for } m_s = s, s - 1, \ldots, -s, \qquad (44\text{-}4)$$

in which $\hbar$ is $h/2\pi$, m_s is the *spin magnetic quantum number*, and s is the *spin quantum number*. This last can have either positive half-integer values ($\frac{1}{2}, \frac{3}{2}, \ldots$) or nonnegative integer values ($0, 1, 2, \ldots$). For example, an electron has the value $s = \frac{1}{2}$. Hence the spin of an electron (measured along any direction, such as the z direction) can have the values

$$S_z = \tfrac{1}{2}\hbar \quad \text{(spin up)}$$

or $$S_z = -\tfrac{1}{2}\hbar \quad \text{(spin down)}.$$

Confusingly, the term *spin* is used in two ways: It properly means a particle's intrinsic angular momentum $\vec{S}$, but it is often used loosely to mean the particle's spin quantum number s. In the latter case, for example, an electron is said to be a spin-$\frac{1}{2}$ particle.

Particles with half-integer spin quantum numbers (like electrons) are called **fermions,** after Fermi, who (simultaneously with Paul Dirac) discovered the statistical laws that govern their behavior. Like electrons, protons and neutrons also have $s = \frac{1}{2}$ and are fermions.

Particles with zero or integer spin quantum numbers are called **bosons,** after Indian physicist Satyendra Nath Bose, who (simultaneously with Albert Einstein) discovered the governing statistical laws for *those* particles. Photons, which have $s = 1$, are bosons; you will soon meet other particles in this class.

This may seem a trivial way to classify particles, but it is very important for this reason:

Fermions obey the Pauli exclusion principle, which asserts that only a single particle can be assigned to a given quantum state. Bosons *do not* obey this principle. Any number of bosons can occupy a given quantum state.

We saw how important the Pauli exclusion principle is when we "built up" the atoms by assigning (spin-$\frac{1}{2}$) electrons to individual quantum states. Using that principle led to a full accounting of the structure and properties of atoms of different types and of solids such as metals and semiconductors.

Because bosons do *not* obey the Pauli principle, those particles tend to pile up in the quantum state of lowest energy. In 1995 a group in Boulder, Colorado, succeeded in producing a condensate of about 2000 rubidium-87 atoms—they are bosons—in a single quantum state of approximately zero energy.

For this to happen, the rubidium has to be a vapor with a temperature so low and a density so great that the de Broglie wavelengths of the individual atoms are greater than the average separation between the atoms. When this condition is met, the wave functions of the individual atoms overlap and the entire assembly becomes a single quantum system (one big atom) called a *Bose–Einstein condensate*. Figure 44-1 shows that, as the temperature of the rubidium vapor is lowered to about 1.70×10^{-7} K, the atoms do indeed "collapse" into a single sharply defined state corresponding to approximately zero speed.

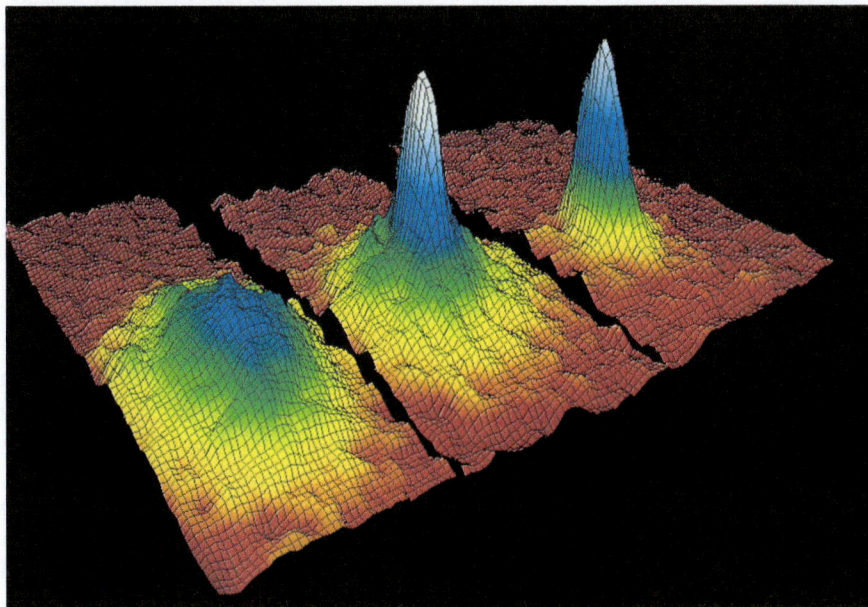

Courtesy Michael Mathews

(a)   (b)   (c)

Figure 44-1 Three plots of the particle speed distribution in a vapor of rubidium-87 atoms. The temperature of the vapor is successively reduced from plot (a) to plot (c). Plot (c) shows a sharp peak centered around zero speed; that is, all the atoms are in the same quantum state. The achievement of such a Bose–Einstein condensate, often called the Holy Grail of atomic physics, was finally recorded in 1995.

Hadron or Lepton?

We can also classify particles in terms of the four fundamental forces that act on them. The *gravitational force* acts on *all* particles, but its effects at the level of subatomic particles are so weak that we need not consider that force (at least not in today's research). The *electromagnetic force* acts on all *electrically charged* particles; its effects are well known, and we can take them into account when we need to; we largely ignore this force in this chapter.

We are left with the *strong force,* which is the force that binds nucleons together, and the *weak force,* which is involved in beta decay and similar processes. The weak force acts on all particles, the strong force only on some.

We can, then, roughly classify particles on the basis of whether the strong force acts on them. Particles on which the strong force acts are called **hadrons.** Particles on which the strong force does *not* act, leaving the weak force and the electromagnetic force as the dominant forces, are called **leptons.** Protons, neutrons, and pions are hadrons; electrons and neutrinos are leptons.

We can make a further distinction among the hadrons because some of them are bosons (we call them **mesons**); the pion is an example. The other hadrons are fermions (we call them **baryons**); the proton is an example.

Particle or Antiparticle?

In 1928 Dirac predicted that the electron e^- should have a positively charged counterpart of the same mass and spin. The counterpart, the *positron* e^+, was discovered in cosmic radiation in 1932 by Carl Anderson. Physicists then gradually realized that *every* particle has a corresponding **antiparticle.** The members of such pairs have the same mass and spin but opposite signs of electric charge (if they are charged) and opposite signs of quantum numbers that we have not yet discussed.

At first, *particle* was used to refer to the common particles such as electrons, protons, and neutrons, and *antiparticle* referred to their rarely detected counterparts. Later, for the less common particles, the assignment of *particle* and *antiparticle* was made so as to be consistent with certain conservation laws that we shall discuss later in this chapter. (Confusingly, both particles and antiparticles are sometimes called particles when no distinction is needed.) We often, but not always, represent an antiparticle by putting a bar over the symbol for the particle. Thus, p is the symbol for the proton, and p̄ (pronounced "p bar") is the symbol for the antiproton.

Annihilation. When a particle meets its antiparticle, the two can *annihilate* each other. That is, the particle and antiparticle disappear and their combined energies reappear in other forms. For an electron annihilating with a positron, this energy reappears as two gamma-ray photons:

$$e^- + e^+ \rightarrow \gamma + \gamma. \tag{44-5}$$

If the electron and positron are stationary when they annihilate, their total energy is their total mass energy, and that energy is then shared equally by the two photons. To conserve momentum and because photons cannot be stationary, the photons fly off in opposite directions.

Antihydrogen atoms (each with an antiproton and positron instead of a proton and electron in a hydrogen atom) are now being manufactured and studied at CERN. The Standard Model predicts that a transition in an antihydrogen atom (say, between the first excited state and the ground state) is identical to the same transition in a hydrogen atom. Thus, any difference in the transitions would clearly signal that the Standard Model is erroneous; no difference has yet been spotted.

An assembly of antiparticles, such as an antihydrogen atom, is often called *antimatter* to distinguish it from an assembly of common particles (*matter*). (The terms can easily be confusing when the word "matter" is used to describe anything that has mass.) We can speculate that future scientists and engineers may construct objects of antimatter. However, no evidence suggests that nature has already done this on an astronomical scale because all stars and galaxies appear to consist largely of matter and not antimatter. This is a perplexing observation because it means that when the universe began, some feature biased the conditions toward matter and away from antimatter. (For example, electrons are common but positrons are not.) This bias is still not well understood.

An Interlude

Before pressing on with the task of classifying the particles, let us step aside for a moment and capture some of the spirit of particle research by analyzing a typical particle event—namely, that shown in the bubble-chamber photograph of Fig. 44-2*a*.

The tracks in this figure consist of bubbles formed along the paths of electrically charged particles as they move through a chamber filled with liquid hydrogen. We can identify the particle that makes a particular track by—among other means—measuring the relative spacing between the bubbles. The chamber lies in a uniform magnetic field that deflects the tracks of positively

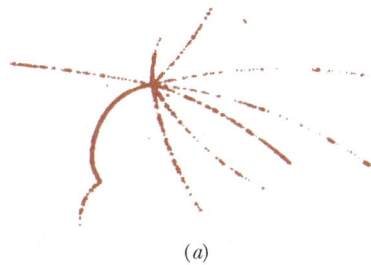

(*a*)

The moving antiproton collides with a stationary proton. The annihilation produces all the other particles.

The positive pion decays, producing a positive muon and an (unseen) neutrino.

The positive muon decays, producing an electron, a neutrino, and an antineutrino, all unseen.

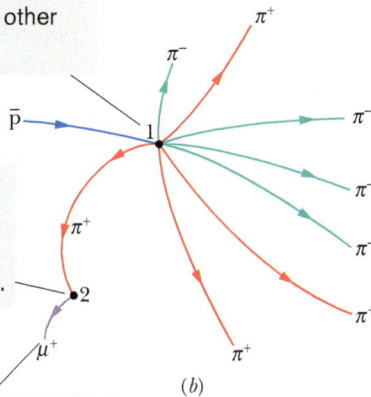

(*b*)

Here, clockwise curvature means negative charge and ...

... counterclockwise curvature means positive charge.

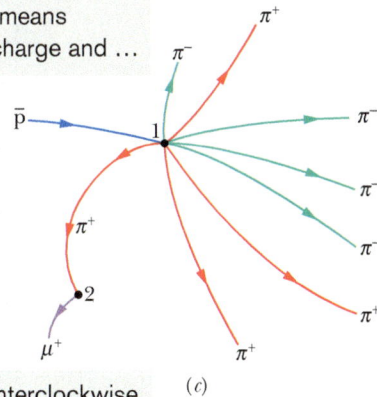

(*c*)

Part (*a*): Courtesy Lawrence Berkeley Laboratory

Figure 44-2 (*a*) A bubble-chamber photograph of a series of events initiated by an antiproton that enters the chamber from the left. (*b*) The tracks redrawn and labeled for clarity. (*c*) The tracks are curved because a magnetic field present in the chamber exerts a deflecting force on each moving charged particle.

Table 44-1 The Particles or Antiparticles Involved in the Event of Fig. 44-2

Particle	Symbol	Charge q	Mass (MeV/c^2)	Spin Quantum Number s	Identity	Mean Life (s)	Antiparticle
Neutrino	ν	0	$\approx 1 \times 10^{-7}$	$\frac{1}{2}$	Lepton	Stable	$\bar{\nu}$
Electron	e^-	-1	0.511	$\frac{1}{2}$	Lepton	Stable	e^+
Muon	μ^-	-1	105.7	$\frac{1}{2}$	Lepton	2.2×10^{-6}	μ^+
Pion	π^+	$+1$	139.6	0	Meson	2.6×10^{-8}	π^-
Proton	p	$+1$	938.3	$\frac{1}{2}$	Baryon	Stable	$\bar{p}$

charged particles counterclockwise and the tracks of negatively charged particles clockwise. By measuring the radius of curvature of a track, we can calculate the momentum of the particle that made it. Table 44-1 shows some properties of the particles and antiparticles that participated in the event of Fig. 44-2a, including those that did not make tracks. Following common practice, we express the masses of the particles listed in Table 44-1—and in all other tables in this chapter—in the unit MeV/c^2. The reason for this notation is that the rest energy of a particle is needed more often than its mass. Thus, the mass of a proton is shown in Table 44-1 to be 938.3 MeV/c^2. To find the proton's rest energy, multiply this mass by c^2 to obtain 938.3 MeV.

The general tools used for the analysis of photographs like Fig. 44-2a are the laws of conservation of energy, linear momentum, angular momentum, and electric charge, along with other conservation laws that we have not yet discussed. Figure 44-2a is actually one of a stereo pair of photographs so that, in practice, these analyses are carried out in three dimensions.

The event of Fig. 44-2a is triggered by an energetic antiproton ($\bar{p}$) that, generated in an accelerator at the Lawrence Berkeley Laboratory, enters the chamber from the left. There are three separate subevents; one occurs at point 1 in Fig. 44-2b, the second occurs at point 2, and the third occurs out of the frame of the figure. Let's examine each:

1. *Proton–Antiproton Annihilation.* At point 1 in Fig. 44-2b, the initiating antiproton (blue track) slams into a proton of the liquid hydrogen in the chamber, and the result is mutual annihilation. We can tell that annihilation occurred while the incoming antiproton was in flight because most of the particles generated in the encounter move in the forward direction—that is, toward the right in Fig. 44-2. From the principle of conservation of linear momentum, the incoming antiproton must have had a forward momentum when it underwent annihilation. Further, because the particles are charged and moving through a magnetic field, the curvature of the paths reveal whether the particles are negatively charged (like the incident antiproton) or positively charged (Fig. 44-2c).

   The total energy involved in the collision of the antiproton and the proton is the sum of the antiproton's kinetic energy and the two (identical) rest energies of those two particles (2×938.3 MeV, or 1876.6 MeV). This is enough energy to create a number of lighter particles and give them kinetic energy. In this case, the annihilation produces four positive pions (red tracks in Fig. 44-2b) and four negative pions (green tracks). (For simplicity, we assume that no gamma-ray photons, which would leave no tracks because they lack electric charge, are produced.) Thus we conclude that the annihilation process is

$$p + \bar{p} \rightarrow 4\pi^+ + 4\pi^-. \tag{44-6}$$

We see from Table 44-1 that the positive pions (π^+) are *particles* and the negative pions (π^-) are *antiparticles*. The reaction of Eq. 44-6 is a *strong interac-*

tion (it involves the strong force) because all the particles involved are hadrons.

Let us check whether electric charge is conserved in the reaction. To do so, we can write the electric charge of a particle as qe, in which q is a **charge quantum number.** Then determining whether electric charge is conserved in a process amounts to determining whether the initial net charge quantum number is equal to the final net charge quantum number. In the process of Eq. 44-6, the initial net charge number is $1 + (-1)$, or 0, and the final net charge number is $4(1) + 4(-1)$, or 0. Thus, charge *is* conserved.

For the energy balance, note from above that the energy available from the p-p̄ annihilation process is at least the sum of the proton and antiproton rest energies, 1876.6 MeV. The rest energy of a pion is 139.6 MeV, which means the rest energies of the eight pions amount to 8×139.6 MeV, or 1116.8 MeV. This leaves at least about 760 MeV to distribute among the eight pions as kinetic energy. Thus, the requirement of energy conservation is easily met.

2. *Pion Decay.* Pions are unstable particles and decay with a mean lifetime of 2.6×10^{-8} s. At point 2 in Fig. 44-2b, one of the positive pions comes to rest in the chamber and decays spontaneously into an antimuon μ^+ (purple track) and a neutrino ν:

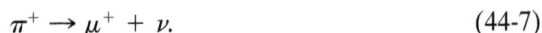

$$\pi^+ \rightarrow \mu^+ + \nu. \tag{44-7}$$

The neutrino, being uncharged, leaves no track. Both the antimuon and the neutrino are leptons; that is, they are particles on which the strong force does not act. Thus, the decay process of Eq. 44-7, which is governed by the weak force, is described as a *weak interaction*.

Let's consider the energies in the decay. From Table 44-1, the rest energy of an antimuon is 105.7 MeV and the rest energy of a neutrino is approximately 0. Because the pion is at rest when it decays, its energy is just its rest energy, 139.6 MeV. Thus, an energy of 139.6 MeV − 105.7 MeV, or 33.9 MeV, is available to share between the antimuon and the neutrino as kinetic energy.

Let us check whether spin angular momentum is conserved in the process of Eq. 44-7. This amounts to determining whether the net component S_z of spin angular momentum along some arbitrary z axis can be conserved by the process. The spin quantum numbers s of the particles in the process are 0 for the pion π^+ and $\frac{1}{2}$ for both the antimuon μ^+ and the neutrino ν. Thus, for π^+, the component S_z must be $0\hbar$, and for μ^+ and ν, it can be either $+\frac{1}{2}\hbar$ or $-\frac{1}{2}\hbar$.

The net component S_z is conserved by the process of Eq. 44-7 if there is *any* way in which the initial S_z ($= 0\hbar$) can be equal to the final net S_z. We see that if one of the products, either μ^+ or ν, has $S_z = +\frac{1}{2}\hbar$ and the other has $S_z = -\frac{1}{2}\hbar$, then their final net value is $0\hbar$. Thus, because S_z can be conserved, the decay process of Eq. 44-7 *can* occur.

From Eq. 44-7, we also see that the net charge is conserved by the process: before the process the net charge quantum number is $+1$, and after the process it is $+1 + 0 = +1$.

3. *Muon Decay.* Muons (whether μ^- or μ^+) are also unstable, decaying with a mean life of 2.2×10^{-6} s. Although the decay products are not shown in Fig. 44-2, the antimuon produced in the reaction of Eq. 44-7 comes to rest and decays spontaneously according to

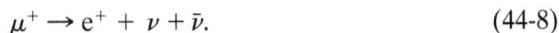

$$\mu^+ \rightarrow e^+ + \nu + \bar{\nu}. \tag{44-8}$$

The rest energy of the antimuon is 105.7 MeV, and that of the positron is only 0.511 MeV, leaving 105.2 MeV to be shared as kinetic energy among the three particles produced in the decay process of Eq. 44-8.

You may wonder: Why *two* neutrinos in Eq. 44-8? Why not just one, as in the pion decay in Eq. 44-7? One answer is that the spin quantum numbers of the antimuon, the positron, and the neutrino are each $\frac{1}{2}$; with only one neutrino, the net component S_z of spin angular momentum could not be conserved in the antimuon decay of Eq. 44-8. In Module 44-2 we shall discuss another reason.

Sample Problem 44.01 Momentum and kinetic energy in a pion decay

A stationary positive pion can decay according to

$$\pi^+ \rightarrow \mu^+ + \nu.$$

What is the kinetic energy of the antimuon μ^+? What is the kinetic energy of the neutrino?

KEY IDEA

The pion decay process must conserve both total energy and total linear momentum.

Energy conservation: Let us first write the conservation of total energy (rest energy mc^2 plus kinetic energy K) for the decay process as

$$m_\pi c^2 + K_\pi = m_\mu c^2 + K_\mu + m_\nu c^2 + K_\nu.$$

Because the pion was stationary, its kinetic energy K_π is zero. Then, using the masses listed for m_π, m_μ, and m_ν in Table 44-1, we find

$$K_\mu + K_\nu = m_\pi c^2 - m_\mu c^2 - m_\nu c^2$$

$$= 139.6 \text{ MeV} - 105.7 \text{ MeV} - 0$$

$$= 33.9 \text{ MeV}, \tag{44-9}$$

where we have approximated m_ν as zero.

Momentum conservation: We cannot solve Eq. 44-9 for either K_μ or K_ν separately, and so let us next apply the principle of conservation of linear momentum to the decay process. Because the pion is stationary when it decays, that principle requires that the muon and neutrino move in opposite directions after the decay. Assume that their motion is along an axis. Then, for components along that axis, we can write the conservation of linear momentum for the decay as

$$p_\pi = p_\mu + p_\nu,$$

which, with $p_\pi = 0$, gives us

$$p_\mu = -p_\nu. \tag{44-10}$$

Relating p and K: We want to relate these momenta p_μ and $-p_\nu$ to the kinetic energies K_μ and K_ν so that we can solve for the kinetic energies. Because we have no reason to believe that classical physics can be applied, we use Eq. 37-54, the momentum–kinetic energy relation from special relativity:

$$(pc)^2 = K^2 + 2Kmc^2. \tag{44-11}$$

From Eq. 44-10, we know that

$$(p_\mu c)^2 = (p_\nu c)^2. \tag{44-12}$$

Substituting from Eq. 44-11 for each side of Eq. 44-12 yields

$$K_\mu^2 + 2K_\mu m_\mu c^2 = K_\nu^2 + 2K_\nu m_\nu c^2.$$

Approximating the neutrino mass to be $m_\nu = 0$, substituting $K_\nu = 33.9 \text{ MeV} - K_\mu$ from Eq. 44-9, and then solving for K_μ, we find

$$K_\mu = \frac{(33.9 \text{ MeV})^2}{(2)(33.9 \text{ MeV} + m_\mu c^2)}$$

$$= \frac{(33.9 \text{ MeV})^2}{(2)(33.9 \text{ MeV} + 105.7 \text{ MeV})}$$

$$= 4.12 \text{ MeV}. \qquad \text{(Answer)}$$

The kinetic energy of the neutrino is then, from Eq. 44-9,

$$K_\nu = 33.9 \text{ MeV} - K_\mu = 33.9 \text{ MeV} - 4.12 \text{ MeV}$$

$$= 29.8 \text{ MeV}. \qquad \text{(Answer)}$$

We see that, although the magnitudes of the momenta of the two recoiling particles are the same, the neutrino gets the larger share (88%) of the kinetic energy.

Sample Problem 44.02 Q in a proton-pion reaction

The protons in the material filling a bubble chamber are bombarded with a beam of high-energy antiparticles known as negative pions. At collision points, a proton and a pion transform into a negative kaon and a positive sigma in this reaction:

$$\pi^- + p \rightarrow K^- + \Sigma^+.$$

The rest energies of these particles are

π^-	139.6 MeV	K^-	493.7 MeV
p	938.3 MeV	Σ^+	1189.4 MeV

What is the Q of the reaction?

KEY IDEA

The Q of a reaction is

$$Q = \left(\begin{array}{c}\text{initial total} \\ \text{mass energy}\end{array}\right) - \left(\begin{array}{c}\text{final total} \\ \text{mass energy}\end{array}\right).$$

Calculation: For the given reaction, we find

$$Q = (m_\pi c^2 + m_p c^2) - (m_K c^2 + m_\Sigma c^2)$$

$$= (139.6 \text{ MeV} + 938.3 \text{ MeV})$$

$$- (493.7 \text{ MeV} + 1189.4 \text{ MeV})$$

$$= -605 \text{ MeV}. \qquad \text{(Answer)}$$

The minus sign means that the reaction is *endothermic;* that is, the incoming pion (π^-) must have a kinetic energy greater than a certain threshold value if the reaction is to occur. The threshold energy is actually greater than 605 MeV because linear momentum must be conserved. (The incoming pion has momentum.) This means that the kaon (K^-) and the sigma (Σ^+) not only must be created but also must be given some kinetic energy. A relativistic calculation whose details are beyond our scope shows that the threshold energy for the reaction is 907 MeV.

WILEY PLUS Additional examples, video, and practice available at *WileyPLUS*

44-2 LEPTONS, HADRONS, AND STRANGENESS

Learning Objectives

After reading this module, you should be able to . . .

44.09 Identify that there are six leptons (with an antiparticle each) in three families, with a different type of neutrino in each family.

44.10 To see if a given process for elementary particles is physically possible, determine whether it conserves lepton number and whether it conserves the individual family lepton numbers.

44.11 Identify that there is a quantum number called baryon number associated with the baryons.

44.12 To see if a given process for elementary particles is physically possible, determine whether the process conserves baryon number.

44.13 Identify that there is a quantum number called strangeness associated with some of the baryons and mesons.

44.14 Identify that strangeness must be conserved in an interaction involving the strong force, but this conservation law can be broken for other interactions.

44.15 Describe the eightfold-way patterns.

Key Ideas

● We can classify particles and their antiparticles into two main types: leptons and hadrons. The latter consists of mesons and baryons.

● Three of the leptons (the electron, muon, and tau) have electric charge equal to $-1e$. There are also three uncharged neutrinos (also leptons), one corresponding to each of the charged leptons. The antiparticles for the charged leptons have positive charge.

● To explain the possible and impossible reactions of these particles, each is assigned a lepton quantum number, which must be conserved in a reaction.

● The leptons have half-integer spin quantum numbers and are thus fermions, which obey the Pauli exclusion principle.

● Baryons, including protons and neutrons, are hadrons with half-integer spin quantum numbers and thus are also fermions.

● Mesons are hadrons with integer spin quantum numbers and thus are bosons, which do not obey the Pauli exclusion principle.

● To explain the possible and impossible reactions of these particles, baryons are assigned a baryon quantum number, which must be conserved in a reaction.

● Baryons are also assigned a strangeness quantum number, but it is conserved only in reactions involving the strong force.

The Leptons

In this module, we discuss some of the particles of one of our classification schemes: lepton or hadron. We begin with the leptons, those particles on which the strong force does *not* act. So far, we have encountered the familiar electron and the neutrino that accompanies it in beta decay. The muon, whose decay is described in Eq. 44-8, is another member of this family. Physicists gradually learned that the neutrino that appears in Eq. 44-7, associated with the production of a muon, is *not the same particle* as the neutrino produced in beta decay, associated with the appearance of an electron. We call the former the **muon neutrino** (symbol ν_μ) and the latter the **electron neutrino** (symbol ν_e) when it is necessary to distinguish between them.

Table 44-2 The Leptons[a]

Family	Particle	Symbol	Mass (MeV/c^2)	Charge q	Antiparticle
Electron	Electron	e^-	0.511	−1	e^+
	Electron neutrino[b]	ν_e	$\approx 1 \times 10^{-7}$	0	$\bar{\nu}_e$
Muon	Muon	μ^-	105.7	−1	μ^+
	Muon neutrino[b]	ν_μ	$\approx 1 \times 10^{-7}$	0	$\bar{\nu}_\mu$
Tau	Tau	τ^-	1777	−1	τ^+
	Tau neutrino[b]	ν_τ	$\approx 1 \times 10^{-7}$	0	$\bar{\nu}_\tau$

[a]All leptons have spin quantum numbers of $\frac{1}{2}$ and are thus fermions.
[b]The neutrino masses have not been well determined. Also, because of neutrino oscillations, we might not be able to associate a particular mass with a particular neutrino.

These two types of neutrino are known to be different particles because, if a beam of muon neutrinos (produced from pion decay as in Eq. 44-7) strikes a solid target, *only muons*—and never electrons—are produced. On the other hand, if electron neutrinos (produced by the beta decay of fission products in a nuclear reactor) strike a solid target, *only electrons*—and never muons—are produced.

Another lepton, the **tau,** was discovered at SLAC in 1975; its discoverer, Martin Perl, shared the 1995 Nobel Prize in physics. The tau has its own associated neutrino, different still from the other two. Table 44-2 lists all the leptons (both particles and antiparticles); all have a spin quantum number s of $\frac{1}{2}$.

There are reasons for dividing the leptons into three families, each consisting of a particle (electron, muon, or tau), its associated neutrino, and the corresponding antiparticles. Furthermore, there are reasons to believe that there are *only* the three families of leptons shown in Table 44-2. Leptons have no internal structure and no measurable dimensions; they are believed to be truly pointlike fundamental particles when they interact with other particles or with electromagnetic waves.

The Conservation of Lepton Number

According to experiment, particle interactions involving leptons obey a conservation law for a quantum number called the **lepton number** L. Each (normal) particle in Table 44-2 is assigned $L = +1$, and each antiparticle is assigned $L = -1$. All other particles, which are not leptons, are assigned $L = 0$. Also according to experiment,

In all particle interactions, the net lepton number is conserved.

This experimental fact is called the law of **conservation of lepton number.** We do not know *why* the law must be obeyed; we only know that this conservation law is part of the way our universe works.

There are actually three types of lepton number, one for each lepton family: the electron lepton number L_e, the muon lepton number L_μ, and the tau lepton number L_τ. In nearly all observed interactions, these three quantum numbers are separately conserved. An important exception involves the neutrinos. For reasons that we cannot explore here, the fact that neutrinos are not massless means that they can "oscillate" between different types as they travel long distances. Such oscillations were proposed to explain why only about a third of the expected number of electron neutrinos arrive at Earth from the proton-proton fusion mechanism in the Sun (Fig. 43-11). The rest change on the way. The oscillations, then, mean that the individual family lepton numbers are not conserved for neutrinos. In this book we shall not consider such violations and shall always conserve the individual family lepton numbers.

Let's illustrate such conservation by reconsidering the antimuon decay process shown in Eq. 44-8, which we now write more fully as

$$\mu^+ \rightarrow e^+ + \nu_e + \bar{\nu}_\mu. \tag{44-13}$$

Consider this first in terms of the muon family of leptons. The μ^+ is an antiparticle (see Table 44-2) and thus has the muon lepton number $L_\mu = -1$. The two particles e^+ and ν_e do not belong to the muon family and thus have $L_\mu = 0$. This leaves $\bar{\nu}_\mu$ on the right which, being an antiparticle, also has the muon lepton number $L_\mu = -1$. Thus, both sides of Eq. 44-13 have the same net muon lepton number—namely, $L_\mu = -1$; if they did not, the μ^+ would not decay by this process.

No members of the electron family appear on the left in Eq. 44-13; so there the net electron lepton number must be $L_e = 0$. On the right side of Eq. 44-13, the positron, being an antiparticle (again see Table 44-2), has the electron lepton number $L_e = -1$. The electron neutrino ν_e, being a particle, has the electron number $L_e = +1$. Thus, the net electron lepton number for these two particles on the right in Eq. 44-13 is also zero; the electron lepton number is also conserved in the process.

Because no members of the tau family appear on either side of Eq. 44-13, we must have $L_\tau = 0$ on each side. Thus, each of the lepton quantum numbers L_μ, L_e, and L_τ remains unchanged during the decay process of Eq. 44-13, their constant values being $-1, 0,$ and 0, respectively.

✓ **Checkpoint 1**

(a) The π^+ meson decays by the process $\pi^+ \rightarrow \mu^+ + \nu$. To what lepton family does the neutrino ν belong? (b) Is this neutrino a particle or an antiparticle? (c) What is its lepton number?

The Hadrons

We are now ready to consider hadrons (baryons and mesons), those particles whose interactions are governed by the strong force. We start by adding another conservation law to our list: conservation of baryon number.

To develop this conservation law, let us consider the proton decay process

$$p \rightarrow e^+ + \nu_e. \tag{44-14}$$

This process *never* happens. We should be glad that it does not because otherwise all protons in the universe would gradually change into positrons, with disastrous consequences for us. Yet this decay process does not violate the conservation laws involving energy, linear momentum, or lepton number.

We account for the apparent stability of the proton—and for the absence of many other processes that might otherwise occur—by introducing a new quantum number, the **baryon number** B, and a new conservation law, the **conservation of baryon number:**

⭐ To every baryon we assign $B = +1$. To every antibaryon we assign $B = -1$. To all particles of other types we assign $B = 0$. A particle process cannot occur if it changes the net baryon number.

In the process of Eq. 44-14, the proton has a baryon number of $B = +1$ and the positron and neutrino both have a baryon number of $B = 0$. Thus, the process does not conserve baryon number and cannot occur.

✓ Checkpoint 2

This mode of decay for a neutron is *not* observed:

$$n \rightarrow p + e^-.$$

Which of the following conservation laws does this process violate: (a) energy, (b) angular momentum, (c) linear momentum, (d) charge, (e) lepton number, (f) baryon number? The masses are $m_n = 939.6 \text{ MeV}/c^2$, $m_p = 938.3 \text{ MeV}/c^2$, and $m_e = 0.511 \text{ MeV}/c^2$.

Still Another Conservation Law

Particles have intrinsic properties in addition to the ones we have listed so far: mass, charge, spin, lepton number, and baryon number. The first of these additional properties was discovered when researchers observed that certain new particles, such as the kaon (K) and the sigma (Σ), always seemed to be produced in pairs. It seemed impossible to produce only one of them at a time. Thus, if a beam of energetic pions interacts with the protons in a bubble chamber, the reaction

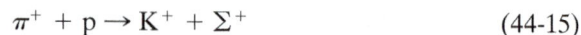

$$\pi^+ + p \rightarrow K^+ + \Sigma^+ \tag{44-15}$$

often occurs. The reaction

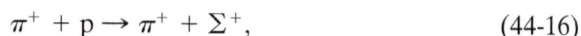

$$\pi^+ + p \rightarrow \pi^+ + \Sigma^+, \tag{44-16}$$

which violates no conservation law known in the early days of particle physics, never occurs.

It was eventually proposed (by Murray Gell-Mann in the United States and independently by K. Nishijima in Japan) that certain particles possess a new property, called **strangeness,** with its own quantum number S and its own conservation law. (Be careful not to confuse the symbol S here with the symbol for spin.) The name *strangeness* arises from the fact that, before the identities of these particles were pinned down, they were known as "strange particles," and the label stuck.

The proton, neutron, and pion have $S = 0$; that is, they are not "strange." It was proposed, however, that the K^+ particle has strangeness $S = +1$ and that Σ^+ has $S = -1$. In the reaction of Eq. 44-15, the net strangeness is initially zero and finally zero; thus, the reaction conserves strangeness. However, in the reaction shown in Eq. 44-16, the final net strangeness is -1; thus, that reaction does not conserve strangeness and cannot occur. Apparently, then, we must add one more conservation law to our list — the **conservation of strangeness:**

⭐ Strangeness is conserved in interactions involving the strong force.

Strange particles are produced only (rapidly) by strong interactions and only in pairs with a net strangeness of zero. They then decay (slowly) through weak interactions without conserving strangeness.

It may seem heavy-handed to invent a new property of particles just to account for a little puzzle like that posed by Eqs. 44-15 and 44-16. However, strangeness soon solved many other puzzles. Still, do not be misled by the whimsical name. Strangeness is no more mysterious a property of particles than is charge. Both are properties that particles may (or may not) have; each is described by an appropriate quantum number. Each obeys a conservation law. Still other properties of particles have been discovered and given even more whimsical names, such as *charm* and *bottomness,* but all are perfectly legitimate properties. Let us see, as an example, how the new property of strangeness "earns its keep" by leading us to uncover important regularities in the properties of the particles.

Table 44-3 Eight Spin-$\frac{1}{2}$ Baryons

Particle	Symbol	Mass (MeV/c^2)	Quantum Numbers Charge q	Strangeness S
Proton	p	938.3	+1	0
Neutron	n	939.6	0	0
Lambda	Λ^0	1115.6	0	−1
Sigma	Σ^+	1189.4	+1	−1
Sigma	Σ^0	1192.5	0	−1
Sigma	Σ^-	1197.3	−1	−1
Xi	Ξ^0	1314.9	0	−2
Xi	Ξ^-	1321.3	−1	−2

Table 44-4 Nine Spin-Zero Mesons[a]

Particle	Symbol	Mass (MeV/c^2)	Quantum Numbers Charge q	Strangeness S
Pion	π^0	135.0	0	0
Pion	π^+	139.6	+1	0
Pion	π^-	139.6	−1	0
Kaon	K^+	493.7	+1	+1
Kaon	K^-	493.7	−1	−1
Kaon	K^0	497.7	0	+1
Kaon	$\overline{K}^0$	497.7	0	−1
Eta	η	547.5	0	0
Eta prime	η'	957.8	0	0

[a]All mesons are bosons, having spins of 0, 1, 2, ... The ones listed here all have a spin of 0.

The Eightfold Way

There are eight baryons—the neutron and the proton among them—that have a spin quantum number of $\frac{1}{2}$. Table 44-3 shows some of their other properties. Figure 44-3a shows the fascinating pattern that emerges if we plot the strangeness of these baryons against their charge quantum number, using a sloping axis for the charge quantum numbers. Six of the eight form a hexagon with the two remaining baryons at its center.

Let us turn now from the hadrons called baryons to the hadrons called mesons. Nine with a spin of zero are listed in Table 44-4. If we plot them on a sloping strangeness–charge diagram, as in Fig. 44-3b, the same fascinating pattern emerges! These and related plots, called the **eightfold way** patterns,* were proposed independently in 1961 by Murray Gell-Mann at the California Institute of Technology and by Yuval Ne'eman at Imperial College, London. The two patterns of Fig. 44-3 are representative of a larger number of symmetrical patterns in which groups of baryons and mesons can be displayed.

The symmetry of the eightfold way pattern for the spin-$\frac{3}{2}$ baryons (not shown here) calls for ten particles arranged in a pattern like that of the tenpins in a bowling alley. However, when the pattern was first proposed, only nine such particles were known; the "headpin" was missing. In 1962, guided by theory and the symmetry of the pattern, Gell-Mann made a prediction in which he essentially said:

> There exists a spin-$\frac{3}{2}$ baryon with a charge of −1, a strangeness of −3, and a rest energy of about 1680 MeV. If you look for this omega minus particle (as I propose to call it), I think you will find it.

A team of physicists headed by Nicholas Samios of the Brookhaven National Laboratory took up the challenge and found the "missing" particle, confirming all its predicted properties. Nothing beats prompt experimental confirmation for building confidence in a theory!

The eightfold way patterns bear the same relationship to particle physics that the periodic table does to chemistry. In each case, there is a pattern of organization in which vacancies (missing particles or missing elements) stick out like sore thumbs, guiding experimenters in their searches. In the case of the periodic table, its very existence strongly suggests that the atoms of the elements are not fundamental particles but have an underlying structure.

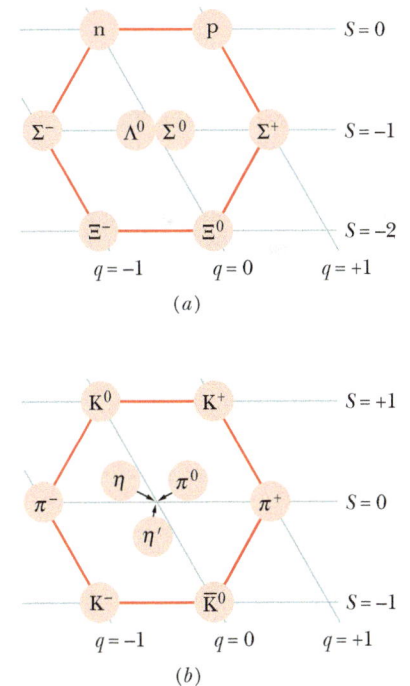

Figure 44-3 (a) The eightfold way pattern for the eight spin-$\frac{1}{2}$ baryons listed in Table 44-3. The particles are represented as disks on a strangeness–charge plot, using a sloping axis for the charge quantum number. (b) A similar pattern for the nine spin-zero mesons listed in Table 44-4.

---

*The name is a borrowing from Eastern mysticism. The "eight" refers to the eight quantum numbers (only a few of which we have defined here) that are involved in the symmetry-based theory that predicts the existence of the patterns.

Similarly, the eightfold way patterns strongly suggest that the mesons and the baryons must have an underlying structure, in terms of which their properties can be understood. That structure can be explained in terms of the *quark model,* which we now discuss.

Sample Problem 44.03   Proton decay: conservation of quantum numbers, energy, and momentum

Determine whether a stationary proton can decay according to the scheme

$$p \rightarrow \pi^0 + \pi^+.$$

Properties of the proton and the π^+ pion are listed in Table 44-1. The π^0 pion has zero charge, zero spin, and a mass energy of 135.0 MeV.

KEY IDEA

We need to see whether the proposed decay violates any of the conservation laws we have discussed.

Electric charge: We see that the net charge quantum number is initially +1 and finally 0 + 1, or +1. Thus, charge is conserved by the decay. Lepton number is also conserved, because none of the three particles is a lepton and thus each lepton number is zero.

Linear momentum: Because the proton is stationary, with zero linear momentum, the two pions must merely move in opposite directions with equal magnitudes of linear momentum (so that their total linear momentum is also zero) to conserve linear momentum. The fact that linear momentum *can* be conserved means that the process does not violate the conservation of linear momentum.

Energy: Is there energy for the decay? Because the proton is stationary, that question amounts to asking whether the proton's mass energy is sufficient to produce the mass

energies and kinetic energies of the pions. To answer, we evaluate the Q of the decay:

$$Q = \left(\begin{array}{c} \text{initial total} \\ \text{mass energy} \end{array} \right) - \left(\begin{array}{c} \text{final total} \\ \text{mass energy} \end{array} \right)$$
$$= m_p c^2 - (m_0 c^2 + m_+ c^2)$$
$$= 938.3 \text{ MeV} - (135.0 \text{ MeV} + 139.6 \text{ MeV})$$
$$= 663.7 \text{ MeV}.$$

The fact that Q is positive indicates that the initial mass energy exceeds the final mass energy. Thus, the proton *does* have enough mass energy to create the pair of pions.

Spin: Is spin angular momentum conserved by the decay? This amounts to determining whether the net component S_z of spin angular momentum along some arbitrary z axis can be conserved by the decay. The spin quantum numbers s of the particles in the process are $\frac{1}{2}$ for the proton and 0 for both pions. Thus, for the proton the component S_z can be either $+\frac{1}{2}\hbar$ or $-\frac{1}{2}\hbar$ and for each pion it is $0\hbar$. We see that there is no way that S_z can be conserved. Hence, spin angular momentum is not conserved, and the proposed decay of the proton cannot occur.

Baryon number: The decay also violates the conservation of baryon number: The proton has a baryon number of $B = +1$, and both pions have a baryon number of $B = 0$. Thus, nonconservation of baryon number is another reason the proposed decay cannot occur.

Sample Problem 44.04   Xi-minus decay: conservation of quantum numbers

A particle called xi-minus and having the symbol Ξ^- decays as follows:

$$\Xi^- \rightarrow \Lambda^0 + \pi^-.$$

The Λ^0 particle (called lambda-zero) and the π^- particle are both unstable. The following decay processes occur in *cascade* until only relatively stable products remain:

$$\Lambda^0 \rightarrow p + \pi^- \qquad \pi^- \rightarrow \mu^- + \bar{\nu}_\mu$$
$$\mu^- \rightarrow e^- + \nu_\mu + \bar{\nu}_e.$$

(a) Is the Ξ^- particle a lepton or a hadron? If the latter, is it a baryon or a meson?

KEY IDEAS

(1) Only three families of leptons exist (Table 44-2) and none include the Ξ^- particle. Thus, the Ξ^- must be a

hadron. (2) To answer the second question we need to determine the baryon number of the Ξ^- particle. If it is +1 or −1, then the Ξ^- is a baryon. If, instead, it is 0, then the Ξ^- is a meson.

Baryon number: To see, let us write the overall decay scheme, from the initial Ξ^- to the final relatively stable products, as

$$\Xi^- \rightarrow p + 2(e^- + \bar{\nu}_e) + 2(\nu_\mu + \bar{\nu}_\mu). \qquad (44\text{-}17)$$

On the right side, the proton has a baryon number of +1 and each electron and neutrino has a baryon number of 0. Thus, the net baryon number of the right side is +1. That must then be the baryon number of the lone Ξ^- particle on the left side. We conclude that the Ξ^- particle is a baryon.

(b) Does the decay process conserve the three lepton numbers?

KEY IDEA

Any process must separately conserve the net lepton number for each lepton family of Table 44-2.

Lepton number: Let us first consider the electron lepton number L_e, which is $+1$ for the electron e^-, -1 for the anti-electron neutrino $\bar{\nu}_e$, and 0 for the other particles in the overall decay of Eq. 44-17. We see that the net L_e is 0 before the decay and $2[+1 + (-1)] + 2(0 + 0) = 0$ after the decay. Thus, the net electron lepton number *is* conserved. You can similarly show that the net muon lepton number and the net tau lepton number are also conserved.

(c) What can you say about the spin of the Ξ^- particle?

KEY IDEA

The overall decay scheme of Eq. 44-17 must conserve the net spin component S_z.

Spin: We can determine the spin component S_z of the Ξ^- particle on the left side of Eq. 44-17 by considering the S_z components of the nine particles on the right side. All nine of those particles are spin-$\frac{1}{2}$ particles and thus can have S_z of either $+\frac{1}{2}\hbar$ or $-\frac{1}{2}\hbar$. No matter how we choose between those two possible values of S_z, the net S_z for those nine particles must be a *half-integer* times $\hbar$. Thus, the Ξ^- particle must have S_z of a *half-integer* times $\hbar$, and that means that its spin quantum number s must be a half-integer. (It is $\frac{1}{2}$.)

WILEY PLUS Additional examples, video, and practice available at *WileyPLUS*

44-3 QUARKS AND MESSENGER PARTICLES

Learning Objectives

After reading this module, you should be able to . . .

44.16 Identify that there are six quarks (with an antiparticle for each).

44.17 Identify that baryons contain three quarks (or antiquarks) and mesons contain a quark and an antiquark, and that many of these hadrons are excited states of the basic quark combinations.

44.18 For a given hadron, identify the quarks it contains, and vice versa.

44.19 Identify virtual particles.

44.20 Apply the relationship between the violation of energy by a virtual particle and the time interval allowed for that violation (an uncertainty principle written in terms of energy).

44.21 Identify the messenger particles for electromagnetic interactions, weak interactions, and strong interactions.

Key Ideas

● The six quarks (up, down, strange, charm, bottom, and top, in order of increasing mass) each have baryon number $+\frac{1}{3}$ and charge equal to either $+\frac{2}{3}$ or $-\frac{1}{3}$. The strange quark has strangeness -1, whereas the others all have strangeness 0. These four algebraic signs are reversed for the antiquarks.

● Leptons do not contain quarks and have no internal structure. Mesons contain one quark and one antiquark. Baryons contain three quarks or antiquarks. The quantum numbers of the quarks and antiquarks are assigned to be consistent with the quantum numbers of the mesons and baryons.

● Particles with electric charge interact through the electromagnetic force by exchanging virtual photons.

● Leptons can also interact with each other and with quarks through the weak force, via massive W and Z particles as messengers.

● Quarks primarily interact with each other through the color force, via gluons.

● The electromagnetic and weak forces are different manifestations of the same force, called the electroweak force.

The Quark Model

In 1964 Gell-Mann and George Zweig independently pointed out that the eightfold way patterns can be understood in a simple way if the mesons and the baryons are built up out of subunits that Gell-Mann called **quarks.** We deal first with three of them, called the *up quark* (symbol u), the *down quark* (symbol d), and the *strange quark* (symbol s). The names of the quarks, along with those assigned to three other quarks that we shall meet later, have no meaning other than

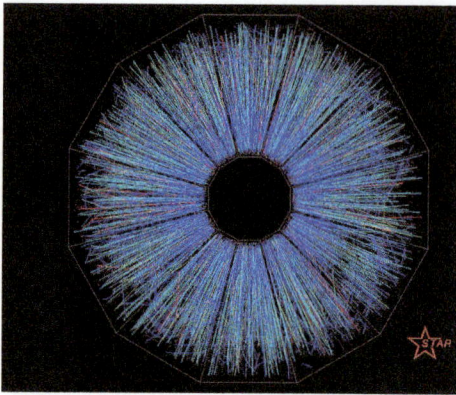

Courtesy Brookhaven National Laboratory

The violent head-on collision of two 30 GeV beams of gold atoms in the RHIC accelerator at the Brookhaven National Laboratory. In the moment of collision, a gas of individual quarks and gluons was created.

Table 44-5 The Quarks[a]

Particle	Symbol	Mass (MeV/c^2)	Quantum Numbers Charge q	Strangeness S	Baryon Number B	Antiparticle
Up	u	5	$+\frac{2}{3}$	0	$+\frac{1}{3}$	$\bar{u}$
Down	d	10	$-\frac{1}{3}$	0	$+\frac{1}{3}$	$\bar{d}$
Charm	c	1500	$+\frac{2}{3}$	0	$+\frac{1}{3}$	$\bar{c}$
Strange	s	200	$-\frac{1}{3}$	-1	$+\frac{1}{3}$	$\bar{s}$
Top	t	175 000	$+\frac{2}{3}$	0	$+\frac{1}{3}$	$\bar{t}$
Bottom	b	4300	$-\frac{1}{3}$	0	$+\frac{1}{3}$	$\bar{b}$

[a]All quarks (including antiquarks) have spin $\frac{1}{2}$ and thus are fermions. The quantum numbers q, S, and B for each antiquark are the negatives of those for the corresponding quark.

as convenient labels. Collectively, these names are called the *quark flavors*. We could just as well call them vanilla, chocolate, and strawberry instead of up, down, and strange. Some properties of the quarks are displayed in Table 44-5.

The fractional charge quantum numbers of the quarks may jar you a little. However, withhold judgment until you see how neatly these fractional charges account for the observed integer charges of the mesons and the baryons. In all normal situations, whether here on Earth or in an astronomical process, quarks are always bound up together in twos or threes (and perhaps more) for reasons that are still not well understood. Such requirements are our normal rule for quark combinations.

An exciting exception to the normal rule occurred in experiments at the RHIC particle collider at the Brookhaven National Laboratory. At the spot where two high-energy beams of gold nuclei collided head-on, the kinetic energy of the particles was so large that it matched the kinetic energy of particles that were present soon after the beginning of the universe (as we discuss in Module 44-4). The protons and neutrons of the gold nuclei were ripped apart to form a momentary gas of individual quarks. (The gas also contained gluons, the particles that normally hold quarks together.) These experiments at RHIC may be the first time that quarks have been set free of one another since the universe began.

(a)

(b)

Figure 44-4 (*a*) The quark compositions of the eight spin-$\frac{1}{2}$ baryons plotted in Fig. 44-3*a*. (Although the two central baryons share the same quark structure, they are different particles. The sigma is an excited state of the lambda, decaying into the lambda by emission of a gamma-ray photon.) (*b*) The quark compositions of the nine spin-zero mesons plotted in Fig. 44-3*b*.

Quarks and Baryons

Each baryon is a combination of three quarks; some of the combinations are given in Fig. 44-4*a*. With regard to baryon number, we see that any three quarks (each with $B = +\frac{1}{3}$) yield a proper baryon (with $B = +1$).

Charges also work out, as we can see from three examples. The proton has a quark composition of uud, and so its charge quantum number is

$$q(uud) = \tfrac{2}{3} + \tfrac{2}{3} + (-\tfrac{1}{3}) = +1.$$

The neutron has a quark composition of udd, and its charge quantum number is therefore

$$q(udd) = \tfrac{2}{3} + (-\tfrac{1}{3}) + (-\tfrac{1}{3}) = 0.$$

The Σ^- (sigma-minus) particle has a quark composition of dds, and its charge quantum number is therefore

$$q(dds) = -\tfrac{1}{3} + (-\tfrac{1}{3}) + (-\tfrac{1}{3}) = -1.$$

The strangeness quantum numbers work out as well. You can check this by using Table 44-3 for the Σ^- strangeness number and Table 44-5 for the strangeness numbers of the dds quarks.

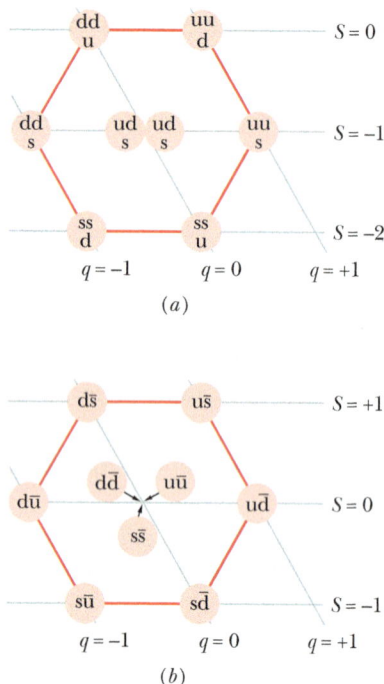

Note, however, that the mass of a proton, neutron, Σ^-, or any other baryon is *not* the sum of the masses of the constituent quarks. For example, the total mass of the three quarks in a proton is only 20 MeV/c^2, woefully less than the proton's mass of 938.3 MeV/c^2. Nearly all of the proton's mass is due to the internal energies of (1) the quark motion and (2) the fields that bind the quarks together. (Recall that mass is related to energy via Einstein's equation, which we can write as $m = E/c^2$.) Thus, because most of your mass is due to the protons and neutrons in your body, your mass (and therefore your weight on a bathroom scale) is primarily a measure of the energies of the quark motion and the quark-binding fields within you.

Quarks and Mesons

Mesons are quark–antiquark pairs; some of their compositions are given in Fig. 44-4b. The quark–antiquark model is consistent with the fact that mesons are not baryons; that is, mesons have a baryon number $B = 0$. The baryon number for a quark is $+\frac{1}{3}$ and for an antiquark is $-\frac{1}{3}$; thus, the combination of baryon numbers in a meson is zero.

Consider the meson π^+, which consists of an up quark u and an antidown quark $\bar{\text{d}}$. We see from Table 44-5 that the charge quantum number of the up quark is $+\frac{2}{3}$ and that of the antidown quark is $+\frac{1}{3}$ (the sign is opposite that of the down quark). This adds nicely to a charge quantum number of +1 for the π^+ meson; that is,

$$q(u\bar{\text{d}}) = \tfrac{2}{3} + \tfrac{1}{3} = +1.$$

All the charge and strangeness quantum numbers of Fig. 44-4b agree with those of Table 44-4 and Fig. 44-3b. Convince yourself that all possible up, down, and strange quark–antiquark combinations are used. Everything fits.

> ### ✓ Checkpoint 3
>
> Is a combination of a down quark (d) and an antiup quark ($\bar{\text{u}}$) called (a) a π^0 meson, (b) a proton, (c) a π^- meson, (d) a π^+ meson, or (e) a neutron?

A New Look at Beta Decay

Let us see how beta decay appears from the quark point of view. In Eq. 42-24, we presented a typical example of this process:

$$^{32}\text{P} \rightarrow {}^{32}\text{S} + \text{e}^- + \nu.$$

After the neutron was discovered and Fermi had worked out his theory of beta decay, physicists came to view the fundamental beta-decay process as the changing of a neutron into a proton inside the nucleus, according to the scheme

$$\text{n} \rightarrow \text{p} + \text{e}^- + \bar{\nu}_\text{e},$$

in which the neutrino is identified more completely. Today we look deeper and see that a neutron (udd) can change into a proton (uud) by changing a down quark into an up quark. We now view the fundamental beta-decay process as

$$\text{d} \rightarrow \text{u} + \text{e}^- + \bar{\nu}_\text{e}.$$

Thus, as we come to know more and more about the fundamental nature of matter, we can examine familiar processes at deeper and deeper levels. We see too that the quark model not only helps us to understand the structure of particles but also clarifies their interactions.

Still More Quarks

There are other particles and other eightfold way patterns that we have not discussed. To account for them, it turns out that we need to postulate three more quarks, the *charm quark* c, the *top quark* t, and the *bottom quark* b. Thus, a total of six quarks exist, as listed in Table 44-5.

Note that three quarks are exceptionally massive, the most massive of them (top) being almost 190 times more massive than a proton. To generate particles that contain such quarks, with such large mass energies, we must go to higher and higher energies, which is the reason that these three quarks were not discovered earlier.

The first particle containing a charm quark to be observed was the J/ψ meson, whose quark structure is $c\bar{c}$. It was discovered simultaneously and independently in 1974 by groups headed by Samuel Ting at the Brookhaven National Laboratory and Burton Richter at Stanford University.

The top quark defied all efforts to generate it in the laboratory until 1995, when its existence was finally demonstrated in the Tevatron, a large particle accelerator at Fermilab. In this accelerator, protons and antiprotons, each with an energy of 0.9 TeV (= 9×10^{11} eV), were made to collide at the centers of two large particle detectors. In a very few cases, the colliding particles generated a top–antitop ($t\bar{t}$) quark pair, which *very* quickly decays into particles that can be detected and thus can be used to infer the existence of the top–antitop pair.

Look back for a moment at Table 44-5 (the quark family) and Table 44-2 (the lepton family) and notice the neat symmetry of these two "six-packs" of particles, each dividing naturally into three corresponding two-particle families. In terms of what we know today, the quarks and the leptons seem to be truly fundamental particles having no internal structure.

Sample Problem 44.05   Quark composition of a xi-minus particle

The Ξ^- (xi-minus) particle is a baryon with a spin quantum number s of $\frac{1}{2}$, a charge quantum number q of -1, and a strangeness quantum number S of -2. Also, it does not contain a bottom quark. What combination of quarks makes up Ξ^-?

Reasoning: Because the Ξ^- is a baryon, it must consist of three quarks (not two as for a meson).

Let us next consider the strangeness $S = -2$ of the Ξ^-. Only the strange quark s and the antistrange quark $\bar{s}$ have nonzero values of strangeness (see Table 44-5). Further, because only the strange quark s has a *negative* value of strangeness, Ξ^- must contain that quark. In fact, for Ξ^- to have a strangeness of -2, it must contain two strange quarks.

To determine the third quark, call it x, we can consider the other known properties of Ξ^-. Its charge quantum number q is -1, and the charge quantum number q of each

strange quark is $-\frac{1}{3}$. Thus, the third quark x must have a charge quantum number of $-\frac{1}{3}$, so that we can have

$$q(\Xi^-) = q(ssx)$$
$$= -\tfrac{1}{3} + (-\tfrac{1}{3}) + (-\tfrac{1}{3}) = -1.$$

Besides the strange quark, the only quarks with $q = -\frac{1}{3}$ are the down quark d and bottom quark b. Because the problem statement ruled out a bottom quark, the third quark must be a down quark. This conclusion is also consistent with the baryon quantum numbers:

$$B(\Xi^-) = B(ssd)$$
$$= \tfrac{1}{3} + \tfrac{1}{3} + \tfrac{1}{3} = +1.$$

Thus, the quark composition of the Ξ^- particle is ssd.

PLUS Additional examples, video, and practice available at *WileyPLUS*

The Basic Forces and Messenger Particles

We turn now from cataloging the particles to considering the forces between them.

The Electromagnetic Force

At the atomic level, we say that two electrons exert electromagnetic forces on each other according to Coulomb's law. At a deeper level, this interaction is described by a highly successful theory called **quantum electrodynamics** (QED). From this point of view, we say that each electron senses the presence of the other by exchanging photons with it.

We cannot detect these photons because they are emitted by one electron and absorbed by the other a very short time later. Because of their undetectable existence, we call them **virtual photons.** Because they communicate between the two interacting charged particles, we sometimes call these photons *messenger particles.*

If a stationary electron emits a photon and remains itself unchanged, energy is not conserved. The principle of conservation of energy is saved, however, by an uncertainty principle written in the form

$$\Delta E \cdot \Delta t \approx \hbar. \qquad (44\text{-}18)$$

Here we interpret this relation to mean that you can "overdraw" an amount of energy ΔE, violating conservation of energy, *provided* you "return" it within an interval Δt given by $\hbar/\Delta E$ so that the violation cannot be detected. The virtual photons do just that. When, say, electron A emits a virtual photon, the overdraw in energy is quickly set right when that electron receives a virtual photon from electron B, and the violation is hidden by the inherent uncertainty.

The Weak Force

A theory of the weak force, which acts on all particles, was developed by analogy with the theory of the electromagnetic force. The messenger particles that transmit the weak force between particles, however, are not (massless) photons but massive particles, identified by the symbols W and Z. The theory was so successful that it revealed the electromagnetic force and the weak force as being different aspects of a single **electroweak force.** This accomplishment is a logical extension of the work of Maxwell, who revealed the electric and magnetic forces as being different aspects of a single *electromagnetic* force.

The electroweak theory was specific in predicting the properties of the messenger particles. In addition to the massless photon, the messenger of the electromagnetic interactions, the theory gives us three messengers for the weak interactions:

Particle	Charge	Mass
W	$\pm e$	$80.4 \text{ GeV}/c^2$
Z	0	$91.2 \text{ GeV}/c^2$

Recall that the proton mass is only $0.938 \text{ GeV}/c^2$; these are massive particles! The 1979 Nobel Prize in physics was awarded to Sheldon Glashow, Steven Weinberg, and Abdus Salam for their electroweak theory. The theory was confirmed in 1983 by Carlo Rubbia and his group at CERN, and the 1984 Nobel Prize in physics went to Rubbia and Simon van der Meer for this brilliant experimental work.

Some notion of the complexity of particle physics in this day and age can be found by looking at an earlier particle physics experiment that led to the Nobel Prize in physics — the discovery of the neutron. This vitally important discovery was a "tabletop" experiment, employing particles emitted by naturally occurring radioactive materials as projectiles; it was reported in 1932 under the title "Possible Existence of a Neutron," the single author being James Chadwick.

The discovery of the W and Z messenger particles in 1983, by contrast, was carried out at a large particle accelerator, about 7 km in circumference and operating in the range of several hundred billion electron-volts. The principal particle detector alone weighed 20 MN. The experiment employed more than 130 physicists from 12 institutions in 8 countries, along with a large support staff.

The Strong Force

A theory of the strong force—that is, the force that acts between quarks to bind hadrons together—has also been developed. The messenger particles in this case are called **gluons** and, like the photon, they are predicted to be massless. The theory assumes that each "flavor" of quark comes in three varieties that, for convenience, have been labeled *red, yellow,* and *blue.* Thus, there are three up quarks, one of each color, and so on. The antiquarks also come in three colors, which we call *antired, antiyellow,* and *antiblue.* You must not think that quarks are actually colored, like tiny jelly beans. The names are labels of convenience, but (for once) they do have a certain formal justification, as you will see.

The force acting between quarks is called a **color force** and the underlying theory, by analogy with quantum electrodynamics (QED), is called **quantum chromodynamics** (QCD). Apparently, quarks can be assembled only in combinations that are *color-neutral.*

There are two ways to bring about color neutrality. In the theory of actual colors, red + yellow + blue yields white, which is color-neutral, and we use the same scheme in dealing with quarks. Thus we can assemble three quarks to form a baryon, provided one is a yellow quark, one is a red quark, and one is a blue quark. Antired + antiyellow + antiblue is also white, so that we can assemble three antiquarks (of the proper anticolors) to form an antibaryon. Finally, red + antired, or yellow + antiyellow, or blue + antiblue also yields white. Thus, we can assemble a quark–antiquark combination to form a meson. The color-neutral rule does not permit any other combination of quarks, and none are observed.

The color force not only acts to bind together quarks as baryons and mesons, but it also acts between such particles, in which case it has traditionally been called the strong force. Hence, not only does the color force bind together quarks to form protons and neutrons, but it also binds together the protons and neutrons to form nuclei.

The Higgs Field and Particle

The Standard Model of the fundamental particles consists of the theory for the electroweak interactions and the theory for the strong interactions. A key success in the model has been to demonstrate the existence of the four messenger particles in the electroweak interactions: the photon, and the Z and W particles. However, a key puzzle has involved the masses of those particles. Why is the photon massless while the Z and W particles are extremely massive?

In the 1960s, Peter Higgs and, independently, Robert Brout and François Englert suggested that the mass discrepancy is due to a field (now called the *Higgs field*) that permeates all of space and thus is a property of the vacuum. Without this field, the four messenger particles would be massless and indistinguishable—they would be *symmetric.* The Brout–Englert–Higgs theory demonstrates how the field breaks that symmetry, producing the electroweak messengers with one being massless. It also explains why all other particles, except for the gluon, have mass. The quantum of that field is the **Higgs boson.** Because of its pivotal role for all particles and because the theory behind its existence is compelling (even beautiful), intense searches for the Higgs boson were conducted on the Tevatron at Brookhaven and the Large Hadron Collider at CERN. In 2012, tantalizing experimental evidence was announced for the Higgs boson, at a mass of 125 GeV/c^2.

Einstein's Dream

The unification of the fundamental forces of nature into a single force—which occupied Einstein's attention for much of his later life—is very much a current focus of research. We have seen that the weak force has been successfully combined with electromagnetism so that they may be jointly viewed as aspects of a single *electroweak force.* Theories that attempt to add the strong force to this combination—called *grand unification theories* (GUTs)—are being pursued actively. Theories that seek to complete the job by adding gravity—sometimes called *theories of everything* (TOE)—are at a speculative stage at this time. *String theory* (in which particles are tiny oscillating loops) is one approach.

44-4 COSMOLOGY

Learning Objectives

After reading this module, you should be able to . . .

44.22 Identify that the universe (all of spacetime) began with the big bang and has been expanding ever since.

44.23 Identify that all distant galaxies (and thus their stars, black holes, etc.), in all directions, are receding from us because of the expansion.

44.24 Apply Hubble's law to relate the recession speed v of a distant galaxy, its distance r from us, and the Hubble constant H.

44.25 Apply the Doppler equation for the red shift of light to relate the wavelength shift $\Delta\lambda$, the recession speed v, and the proper wavelength λ_0 of the emission.

44.26 Approximate the age of the universe using the Hubble constant.

44.27 Identify the cosmic background radiation and explain the importance of its detection.

44.28 Explain the evidence for the dark matter that apparently surrounds every galaxy.

44.29 Discuss the various stages of the universe from very soon after the big bang until atoms began to form.

44.30 Identify that the expansion of the universe is being accelerated by some unknown property dubbed dark energy.

44.31 Identify that the total energy of baryonic matter (protons and neutrons) is only a small part of the total energy of the universe.

Key Ideas

● The universe is expanding, which means that empty space is continuously appearing between us and any distant galaxy.

● The rate v at which a distance to a distant galaxy is increasing (the galaxy appears to be moving at speed v) is given by the Hubble law:

$$v = Hr,$$

where r is the current distance to the galaxy and H is the Hubble constant, which we take to be

$$H = 71.0 \text{ km/s} \cdot \text{Mpc} = 21.8 \text{ mm/s} \cdot \text{ly}.$$

● The expansion causes a red shift in the light we receive from distant galaxies. We can assume that the wavelength shift $\Delta\lambda$ is given (approximately) by the Doppler shift equa-

tion for light discussed in Module 37-5:

$$v = \frac{|\Delta\lambda|}{\lambda_0} c,$$

where λ_0 is the proper wavelength as measured in the frame of the light source (the galaxy).

● The expansion described by Hubble's law and the presence of ubiquitous background microwave radiation reveal that the universe began in a "big bang" 13.7 billion years ago.

● The rate of expansion is increasing due to a mysterious property of the vacuum called dark energy.

● Much of the energy of the universe is hidden in dark matter that apparently interacts with normal (baryonic) matter through the gravitational force.

A Pause for Reflection

Let us put what you have just learned in perspective. If all we are interested in is the structure of the world around us, we can get along nicely with the electron, the neutrino, the neutron, and the proton. As someone has said, we can operate "Spaceship Earth" quite well with just these particles. We can see a few of the more exotic particles by looking for them in the cosmic rays; however, to see most of them, we must build massive accelerators and look for them at great effort and expense.

The reason we must go to such effort is that—measured in energy terms— we live in a world of very low temperatures. Even at the center of the Sun, the value of kT is only about 1 keV. To produce the exotic particles, we must be able to accelerate protons or electrons to energies in the GeV and TeV range and higher.

Once upon a time the temperature everywhere *was* high enough to provide such energies. That time of extremely high temperatures occurred in the **big bang** beginning of the universe, when the universe (and both space and time) came into existence. Thus, one reason scientists study particles at high energies is to understand what the universe was like just after it began.

As we shall discuss shortly, *all* of space within the universe was initially tiny in extent, and the temperature of the particles within that space was incredibly high. With time, however, the universe expanded and cooled to lower temperatures, eventually to the size and temperature we see today.

Actually, the phrase "we see today" is complicated: When we look out into space, we are actually looking back in time because the light from the stars and galaxies has taken a long time to reach us. The most distant objects that we can detect are **quasars** (*quasi*stellar objects), which are the extremely bright cores of galaxies that are as much as 13×10^9 ly from us. Each such core contains a gigantic black hole; as material (gas and even stars) is pulled into one of those black holes, the material heats up and radiates a tremendous amount of light, enough for us to detect in spite of the huge distance. We therefore "see" a quasar not as it looks today but rather as it once was, when that light began its journey to us billions of years ago.

The Universe Is Expanding

As we saw in Module 37-5, it is possible to measure the relative speeds at which galaxies are approaching us or receding from us by measuring the shifts in the wavelength of the light they emit. If we look only at distant galaxies, beyond our immediate galactic neighbors, we find an astonishing fact: They are *all* moving away (receding) from us! In 1929 Edwin P. Hubble connected the recession speed v of a galaxy and its distance r from us—they are directly proportional:

$$v = Hr \quad \text{(Hubble's law)}, \tag{44-19}$$

in which H is called the **Hubble constant.** The value of H is usually measured in the unit kilometers per second-megaparsec (km/s · Mpc), where the megaparsec is a length unit commonly used in astrophysics and astronomy:

$$1 \text{ Mpc} = 3.084 \times 10^{19} \text{ km} = 3.260 \times 10^6 \text{ ly}. \tag{44-20}$$

The Hubble constant H has not had the same value since the universe began. Determining its current value is extremely difficult because doing so involves measurements of very distant galaxies. However, the Hubble constant is now known to be

$$H = 71.0 \text{ km/s} \cdot \text{Mpc} = 21.8 \text{ mm/s} \cdot \text{ly}. \tag{44-21}$$

We interpret the recession of the galaxies to mean that the universe is expanding, much as the raisins in what is to be a loaf of raisin bread grow farther apart as the dough expands. Observers on all other galaxies would find that distant galaxies were rushing away from them also, in accordance with Hubble's law. In keeping with our analogy, we can say that no raisin (galaxy) has a unique or preferred view.

Hubble's law is consistent with the hypothesis that the universe began with the big bang and has been expanding ever since. If we assume that the rate of expansion has been constant (that is, the value of H has been constant), then we can estimate the age T of the universe by using Eq. 44-19. Let us also assume that since the big bang, any given part of the universe (say, a galaxy) has been receding from our location at a speed v given by Eq. 44-19. Then the time required for the given part to recede a distance r is

$$T = \frac{r}{v} = \frac{r}{Hr} = \frac{1}{H} \quad \text{(estimated age of universe)}. \tag{44-22}$$

For the value of H in Eq. 44-21, T works out to be 13.8×10^9 y. Much more sophisticated studies of the expansion of the universe put T at 13.7×10^9 y.

Sample Problem 44.06   Using Hubble's law to relate distance and recessional speed

The wavelength shift in the light from a particular quasar indicates that the quasar has a recessional speed of 2.8×10^8 m/s (which is 93% of the speed of light). Approximately how far from us is the quasar?

KEY IDEA

We assume that the distance and speed are related by Hubble's law.

Calculation: From Eqs. 44-19 and 44-21, we find

$$r = \frac{v}{H} = \frac{2.8 \times 10^8 \text{ m/s}}{21.8 \text{ mm/s} \cdot \text{ly}} (1000 \text{ mm/m})$$
$$= 12.8 \times 10^9 \text{ ly}. \qquad \text{(Answer)}$$

This is only an approximation because the quasar has not always been receding from our location at the same speed v; that is, H has not had its current value throughout the time during which the universe has been expanding.

Sample Problem 44.07   Using Hubble's law to relate distance and Doppler shift

A particular emission line detected in the light from a galaxy has a detected wavelength $\lambda_{det} = 1.1\lambda$, where λ is the proper wavelength of the line. What is the galaxy's distance from us?

KEY IDEAS

(1) We assume that Hubble's law ($v = Hr$) applies to the recession of the galaxy. (2) We also assume that the astronomical Doppler shift of Eq. 37-36 ($v = c|\Delta\lambda|/\lambda$, for $v \ll c$) applies to the shift in wavelength due to the recession.

Calculations: We can then set the right side of these two equations equal to each other to write

$$Hr = \frac{c|\Delta\lambda|}{\lambda}, \qquad (44\text{-}23)$$

which leads us to

$$r = \frac{c|\Delta\lambda|}{H\lambda}. \qquad (44\text{-}24)$$

In this equation,

$$\Delta\lambda = \lambda_{det} - \lambda = 1.1\lambda - \lambda = 0.1\lambda.$$

Substituting this into Eq. 44-24 then gives us

$$r = \frac{c(0.1\lambda)}{H\lambda} = \frac{0.1c}{H}$$
$$= \frac{(0.1)(3.0 \times 10^8 \text{ m/s})}{21.8 \text{ mm/s} \cdot \text{ly}} (1000 \text{ mm/m})$$
$$= 1.4 \times 10^9 \text{ ly}. \qquad \text{(Answer)}$$

WILEY PLUS Additional examples, video, and practice available at *WileyPLUS*

The Cosmic Background Radiation

In 1965 Arno Penzias and Robert Wilson, of what was then the Bell Telephone Laboratories, were testing a sensitive microwave receiver used for communications research. They discovered a faint background "hiss" that remained unchanged in intensity no matter where their antenna was pointed. It soon became clear that Penzias and Wilson were observing a **cosmic background radiation,** generated in the early universe and filling all space almost uniformly. Currently this radiation has a maximum intensity at a wavelength of 1.1 mm, which lies in the microwave region of electromagnetic radiation (or light, for short). The wavelength distribution of this radiation matches the wavelength distribution of light that would be emitted by a laboratory enclosure with walls at a temperature of 2.7 K. Thus, for the cosmic background radiation, we say that the enclosure is the entire universe and that the universe is at an (average) temperature of 2.7 K. For their discovery of the cosmic background radiation, Penzias and Wilson were awarded the 1978 Nobel Prize in physics.

The cosmic background radiation is now known to be light that has been in flight across the universe since shortly after the universe began billions of years ago. When the universe was even younger, light could scarcely go any significant distance without being scattered by all the individual, high-speed particles along its path. If a light ray started from, say, point A, it would be scattered in so many directions that if you could have intercepted part of it, you would have not been

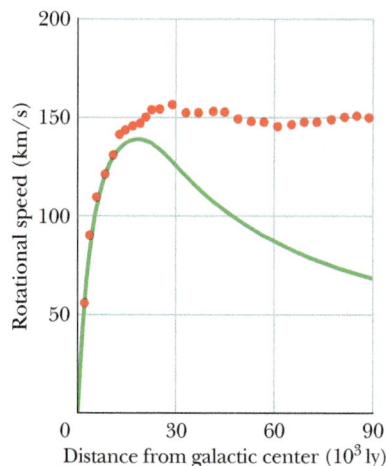

Figure 44-5 The rotational speed of stars in a typical galaxy as a function of their distance from the galactic center. The theoretical solid curve shows that if a galaxy contained only the mass that is visible, the observed rotational speed would drop off with distance at large distances. The dots are the experimental data, which show that the rotational speed is approximately constant at large distances.

able to tell that it originated at point A. However, after the particles began to form atoms, the scattering of light greatly decreased. A light ray from point A might then be able to travel for billions of years without being scattered. This light is the cosmic background radiation.

As soon as the nature of the radiation was recognized, researchers wondered, "Can we use this incoming radiation to distinguish the points at which it originated, so that we then can produce an image of the early universe, back when atoms first formed and light scattering largely ceased?" The answer is yes, and that image is coming up in a moment.

Dark Matter

At the Kitt Peak National Observatory in Arizona, Vera Rubin and her co-worker Kent Ford measured the rotational rates of a number of distant galaxies. They did so by measuring the Doppler shifts of bright clusters of stars located within each galaxy at various distances from the galactic center. As Fig. 44-5 shows, their results were surprising: The orbital speed of stars at the outer visible edge of the galaxy is about the same as that of stars close to the galactic center.

As the solid curve in Fig. 44-5 attests, that is not what we would expect to find if all the mass of the galaxy were represented by visible light. Nor is the pattern found by Rubin and Ford what we find in the solar system. For example, the orbital speed of Pluto (the "planet" most distant from the Sun) is only about one-tenth that of Mercury (the planet closest to the Sun).

The only explanation for the findings of Rubin and Ford that is consistent with Newtonian mechanics is that a typical galaxy contains much more matter than what we can actually see. In fact, the visible portion of a galaxy represents only about 5 to 10% of the total mass of the galaxy. In addition to these studies of galactic rotation, many other observations lead to the conclusion that the universe abounds in matter that we cannot see. This unseen matter is called **dark matter** because either it does not emit light or its light emission is too dim for us to detect.

Normal matter (such as stars, planets, dust, and molecules) is often called **baryonic matter** because its mass is primarily due to the combined mass of the protons and neutrons (baryons) it contains. (The much smaller mass of the electrons is neglected.) Some of the normal matter, such as burned-out stars and dim interstellar gas, is part of the dark matter in a galaxy.

However, according to various calculations, this dark normal matter is only a small part of the total dark matter. The rest is called **nonbaryonic dark matter** because it does not contain protons and neutrons. We know of only one member of this type of dark matter—the neutrinos. Although the mass of a neutrino is very small relative to the mass of a proton or neutron, the number of neutrinos in a galaxy is huge and thus the total mass of the neutrinos is large. Nevertheless, calculations indicate that not even the total mass of the neutrinos is enough to account for the total mass of the nonbaryonic dark matter. In spite of over a hundred years in which elementary particles have been detected and studied, the particles that make up the rest of this type of dark matter are undetected and their nature is unknown. Because we have no experience with them, they must interact only gravitationally with the common particles.

The Big Bang

In 1985, a physicist remarked at a scientific meeting:

> *It is as certain that the universe started with a big bang about 15 billion years ago as it is that the Earth goes around the Sun.*

This strong statement suggests the level of confidence in which the big bang theory, first advanced by Belgian physicist Georges Lemaître, is held by those who study these matters. However, you must not imagine that the big bang was like the explosion of some gigantic firecracker and that, in principle at least, you could have stood to one side and watched. There was no "one side" because the big bang represents the beginning of spacetime itself. From the point of view of our present universe, there is no position in space to which you can point and say, "The big bang happened there." It happened everywhere.

Moreover, there was no "before the big bang," because time *began* with that creation event. In this context, the word "before" loses its meaning. We can, however, conjecture about what went on during succeeding intervals of time after the big bang (Fig. 44-6).

$t \approx 10^{-43}$ s. This is the earliest time at which we can say anything meaningful about the development of the universe. It is at this moment that the concepts of space and time come to have their present meanings and the laws of physics as we know them become applicable. At this instant, the entire universe (that is, the *entire* spatial extent of the universe) is much smaller than a proton and its temperature is about 10^{32} K. Quantum fluctuations in the fabric of spacetime are the seeds that will eventually lead to the formation of galaxies, clusters of galaxies, and superclusters of galaxies.

$t \approx 10^{-34}$ s. By this moment the universe has undergone a tremendously rapid inflation, increasing in size by a factor of about 10^{30}, causing the formation of matter in a distribution set by the initial quantum fluctuations. The universe has become a hot soup of photons, quarks, and leptons at a temperature of about 10^{27} K, which is too hot for protons and neutrons to form.

$t \approx 10^{-4}$ s. Quarks can now combine to form protons and neutrons and their antiparticles. The universe has now cooled to such an extent by continued (but much slower) expansion that photons lack the energy needed to break up these new particles. Particles of matter and antimatter collide and annihilate each other. There is a slight excess of matter, which, failing to find annihilation partners, survives to form the world of matter that we know today.

$t \approx 1$ min. The universe has now cooled enough so that protons and neutrons, in colliding, can stick together to form the low-mass nuclei ^{2}H, ^{3}He, ^{4}He, and ^{7}Li. The predicted relative abundances of these nuclides are just what we observe in the universe today. Also, there is plenty of radiation present at $t \approx 1$ min,

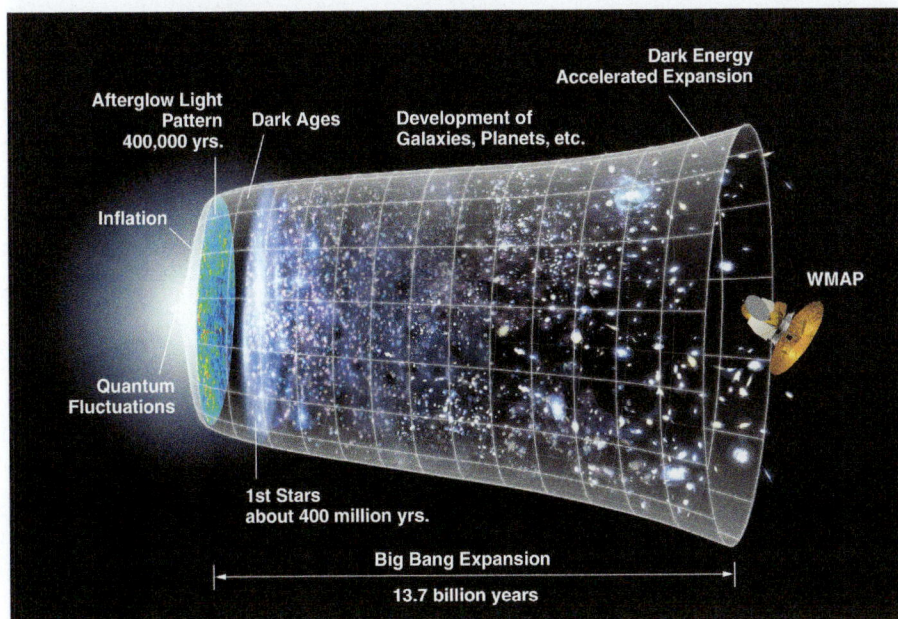

Courtesy NASA

Figure 44-6 An illustration of the universe from the initial quantum fluctuations just after $t = 0$ (at the left) to the current accelerated expansion, 13.7×10^9 y later (at the right). Don't take the illustration literally—there is *no* such "external view" of the universe because there is *no* exterior to the universe.

but this light cannot travel far before it interacts with a nucleus. Thus the universe is opaque.

$t \approx$ **379 000 y.** The temperature has now fallen to 2970 K, and electrons can stick to bare nuclei when the two collide, forming atoms. Because light does not interact appreciably with (uncharged) particles, such as neutral atoms, the light is now free to travel great distances. This radiation forms the cosmic background radiation that we discussed earlier. Atoms of hydrogen and helium, under the influence of gravity, begin to clump together, eventually starting the formation of galaxies and stars, but until then, the universe is relatively dark (Fig. 44-6).

Early measurements suggested that the cosmic background radiation is uniform in all directions, implying that 379 000 y after the big bang all matter in the universe was uniformly distributed. This finding was most puzzling because matter in the present universe is not uniformly distributed, but instead is collected in galaxies, clusters of galaxies, and superclusters of galactic clusters. There are also vast *voids* in which there is relatively little matter, and there are regions so crowded with matter that they are called *walls*. If the big bang theory of the beginning of the universe is even approximately correct, the seeds for this nonuniform distribution of matter must have been in place before the universe was 379 000 y old and now should show up as a nonuniform distribution of the microwave background radiation.

In 1992, measurements made by NASA's Cosmic Background Explorer (COBE) satellite revealed that the background radiation is, in fact, not perfectly uniform. In 2003, measurements by NASA's Wilkinson Microwave Anisotropy Probe (WMAP) greatly increased our resolution of this nonuniformity. The resulting image (Fig. 44-7) is effectively a color-coded photograph of the universe when it was only 379 000 y old. As you can see from the variations in the colors, large-scale collecting of matter had already begun. Thus, the big bang theory and the theory of inflation at $t \approx 10^{-34}$ s are on the right track.

The Accelerated Expansion of the Universe

Recall from Module 13-8 the statement that mass causes curvature of space. Now that we have seen that mass is a form of energy, as given by Einstein's equa-

Figure 44-7 This color-coded image is effectively a photograph of the universe when it was only 379 000 y old, which was about 13.7×10^9 y ago. This is what you would have seen then as you looked away in all directions (the view has been condensed to this oval). Patches of light from collections of atoms stretch across the "sky," but galaxies, stars, and planets have not yet formed.

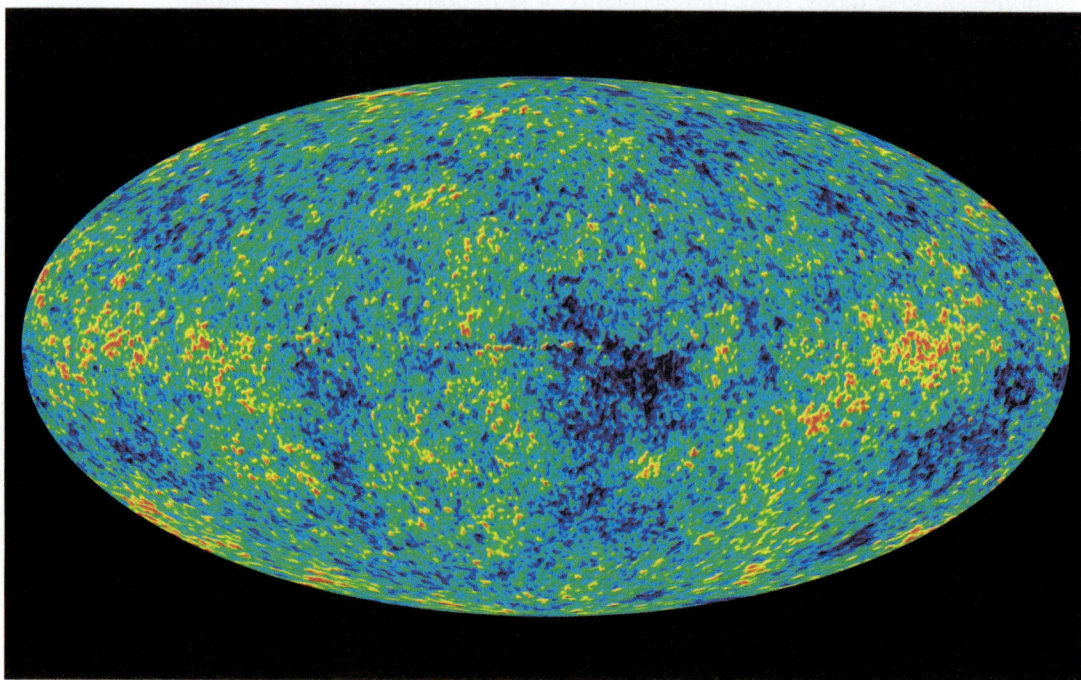

Courtesy WMAP Science Team/NASA

tion $E = mc^2$, we can generalize the statement: energy can cause curvature of space. This certainly happens to the space around the energy packed into a black hole and, more weakly, to the space around any other astronomical body, but is the space of the universe as a whole curved by the energy the universe contains?

The question was answered first by the 1992 COBE measurements of the cosmic background radiation. It was then answered more definitively by the 2003 WMAP measurements that produced the image in Fig. 44-7. The spots we see in that image are the original sources of the cosmic background radiation, and the angular distribution of the spots reveals the curvature of the universe through which the light has to travel to reach us. If adjacent spots subtend either more than 1° (Fig. 44-8a) or less than 1° (Fig. 44-8b) in the detector's view (or our view) into the universe, then the universe is curved. Analysis of the spot distribution in the WMAP image shows that the spots subtend about 1° (Fig. 44-8c), which means that the universe is *flat* (having no curvature). Thus, the initial curvature the universe presumably had when it began must have been flattened out by the rapid inflation the universe underwent at $t \approx 10^{-34}$ s.

This flatness poses a very difficult problem for physicists because it requires that the universe contain a certain amount of energy (as mass or otherwise). The trouble is that all estimations of the amount of energy in the universe (both in known forms and in the form of the unknown type of dark matter) fall dramatically short of the required amount.

One theory proposed about this missing energy gave it the gothic name of *dark energy* and predicted that it has the strange property of causing the expansion of the universe to accelerate. Until 1998, determining whether the expansion is, in fact, accelerating was very difficult because it requires measuring distances to very distant astronomical bodies where the acceleration might show up.

In 1998, however, advances in astronomical technology allowed astronomers to detect a certain type of supernovae at very great distances. More important, the astronomers could measure the duration of the burst of light from such a supernova. The duration reveals the brightness of the supernova that would be seen by an observer near the supernova. By measuring the brightness of the supernova as seen from Earth, astronomers could then determine the distance to the supernova. From the redshift of the light from the galaxy containing the supernova, astronomers could also determine how fast the galaxy is receding from us. Combining all this information, they could then calculate the expansion rate of the universe. The conclusion is that the expansion is indeed accelerating as predicted by the theory of dark energy (Fig. 44-6). However, we have no clue as to what this dark energy is.

Figure 44-9 gives our current state of knowledge about the energy in the universe. About 4% is associated with baryonic matter, which we understand fairly well. About 23% is associated with nonbaryonic dark matter, about which we have a few clues that might be fruitful. The rest, a whopping 73%, is associated with dark energy, about which we are clueless. There have been times in the history of physics, even in the 1990s, when pontiffs proclaimed that physics was nearly complete, that only details were left. In fact, we are nowhere near the end.

A Summing Up

In this closing paragraph, let's consider where we are headed as we accumulate knowledge about the universe more and more rapidly. What we have found is marvelous and profound, but it is also humbling in that each new step seems to reveal more clearly our own relative insignificance in the grand scheme of things. Thus, in roughly chronological order, we humans have come to realize that

Our Earth is not the center of the solar system.

Our Sun is but one star among many in our galaxy.

Our galaxy is but one of many, and our Sun is an insignificant star in it.

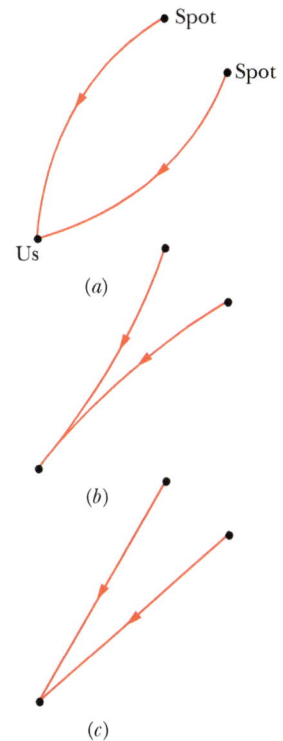

Figure 44-8 Light rays from two adjacent spots in our view of the cosmic background radiation would reach us at an angle (a) greater than 1° or (b) less than 1° if the space along the light-ray paths through the universe were curved. (c) An angle of 1° means that the space is not curved.

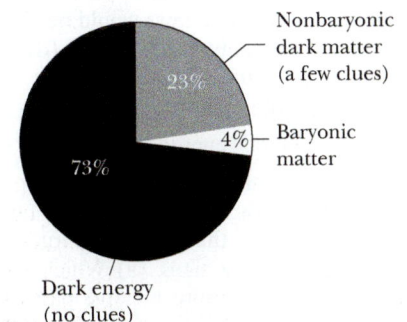

Figure 44-9 The distribution of energy (including mass) in the universe.

Our Earth has existed for perhaps only a third of the age of the universe and will surely disappear when our Sun burns up its fuel and becomes a red giant.

Our species has inhabited Earth for less than a million years—a blink in cosmological time.

Although our position in the universe may be insignificant, the laws of physics that we have discovered (uncovered?) seem to hold throughout the universe and—as far as we know—have held since the universe began and will continue to hold for all future time. At least, there is no evidence that other laws hold in other parts of the universe. Thus, until someone complains, we are entitled to stamp the laws of physics "Discovered on Earth." Much remains to be discovered. In the words of writer Eden Phillpotts, "*The universe is full of magical things, patiently waiting for our wits to grow sharper.*" That declaration allows us to answer one last time the question "What is physics?" that we have explored repeatedly in this book. Physics is the gateway to those magical things.

Review & Summary

Leptons and Quarks Current research supports the view that all matter is made of six kinds of **leptons** (Table 44-2), six kinds of **quarks** (Table 44-5), and 12 **antiparticles,** one corresponding to each lepton and each quark. All these particles have spin quantum numbers equal to $\frac{1}{2}$ and are thus **fermions** (particles with half-integer spin quantum numbers).

The Interactions Particles with electric charge interact through the electromagnetic force by exchanging **virtual photons.** Leptons can also interact with each other and with quarks through the **weak force,** via massive W and Z particles as messengers. In addition, quarks interact with each other through the **color force.** The electromagnetic and weak forces are different manifestations of the same force, called the **electroweak force.**

Leptons Three of the leptons (the **electron, muon,** and **tau**) have electric charge equal to $-1e$. There are also three uncharged **neutrinos** (also leptons), one corresponding to each of the charged leptons. The antiparticles for the charged leptons have positive charge.

Quarks The six quarks (up, down, strange, charm, bottom, and top, in order of increasing mass) each have baryon number $+\frac{1}{3}$ and charge equal to either $+\frac{2}{3}e$ or $-\frac{1}{3}e$. The strange quark has strange-

ness -1, whereas the others all have strangeness 0. These four algebraic signs are reversed for the antiquarks.

Hadrons: Baryons and Mesons Quarks combine into strongly interacting particles called **hadrons. Baryons** are hadrons with half-integer spin quantum numbers ($\frac{1}{2}$ or $\frac{3}{2}$). **Mesons** are hadrons with integer spin quantum numbers (0 or 1) and thus are **bosons.** Baryons are fermions. Mesons have baryon number equal to zero; baryons have baryon number equal to $+1$ or -1. **Quantum chromodynamics** predicts that the possible combinations of quarks are either a quark with an antiquark, three quarks, or three antiquarks (this prediction is consistent with experiment).

Expansion of the Universe Current evidence strongly suggests that the universe is expanding, with the distant galaxies moving away from us at a rate v given by **Hubble's law:**

$$v = Hr \quad \text{(Hubble's law)}. \quad (44\text{-}19)$$

Here we take H, the **Hubble constant,** to have the value

$$H = 71.0 \text{ km/s} \cdot \text{Mpc} = 21.8 \text{ mm/s} \cdot \text{ly}. \quad (44\text{-}21)$$

The expansion described by Hubble's law and the presence of ubiquitous background microwave radiation reveal that the universe began in a "big bang" 13.7 billion years ago.

Questions

1 An electron cannot decay into two neutrinos. Which of the following conservation laws would be violated if it did: (a) energy, (b) angular momentum, (c) charge, (d) lepton number, (e) linear momentum, (f) baryon number?

2 Which of the eight pions in Fig. 44-2b has the least kinetic energy?

3 Figure 44-10 shows the paths of two particles circling in a uniform magnetic field. The particles have the same magnitude of charge but opposite signs. (a) Which path corresponds to the more massive particle? (b) If the magnetic field is directed into the plane of the page, is the more massive particle positively or negatively charged?

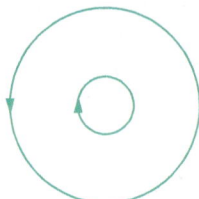

Figure 44-10
Question 3.

4 A proton has enough mass energy to decay into a shower made up of electrons, neutrinos, and their antiparticles. Which of the following conservation laws would necessarily be violated if it did: electron lepton number or baryon number?

5 A proton cannot decay into a neutron and a neutrino. Which of the following conservation laws would be violated if it did: (a) energy (assume the proton is stationary), (b) angular momentum, (c) charge, (d) lepton number, (e) linear momentum, (f) baryon number?

6 Does the proposed decay $\Lambda^0 \rightarrow p + K^-$ conserve (a) electric charge, (b) spin angular momentum, and (c) strangeness? (d) If the original particle is stationary, is there enough energy to create the decay products?

7 Not only particles such as electrons and protons but also entire atoms can be classified as fermions or bosons, depending on whether their overall spin quantum numbers are, respectively, half-integral or integral. Consider the helium isotopes ^{3}He and ^{4}He. Which of the following statements is correct? (a) Both are fermions. (b) Both are bosons. (c) ^{4}He is a fermion, and ^{3}He is a boson. (d) ^{3}He is a fermion, and ^{4}He is a boson. (The two helium electrons form a closed shell and play no role in this determination.)

8 Three cosmologists have each plotted a line on the Hubble-like graph of Fig. 44-11. If we calculate the corresponding age of the universe from the three plots, rank the plots according to that age, greatest first.

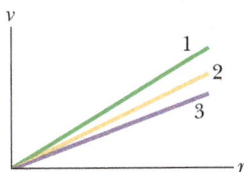

Figure 44-11
Question 8.

9 A Σ^+ particle has these quantum numbers: strangeness $S = -1$, charge $q = +1$, and spin $s = \frac{1}{2}$. Which of the following quark combinations produces it: (a) dds, (b) s$\bar{\text{s}}$, (c) uus, (d) ssu, or (e) uu$\bar{\text{s}}$?

10 As we have seen, the π^- meson has the quark structure d$\bar{\text{u}}$. Which of the following conservation laws would be violated if a π^- were formed, instead, from a d quark and a u quark: (a) energy, (b) angular momentum, (c) charge, (d) lepton number, (e) linear momentum, (f) baryon number?

11 Consider the neutrino whose symbol is $\bar{\nu}_\tau$. (a) Is it a quark, a lepton, a meson, or a baryon? (b) Is it a particle or an antiparticle? (c) Is it a boson or a fermion? (d) Is it stable against spontaneous decay?

Problems

GO Tutoring problem available (at instructor's discretion) in *WileyPLUS* and WebAssign
SSM Worked-out solution available in Student Solutions Manual
• – ••• Number of dots indicates level of problem difficulty
Additional information available in *The Flying Circus of Physics* and at flyingcircusofphysics.com
WWW Worked-out solution is at
ILW Interactive solution is at http://www.wiley.com/college/halliday

Module 44-1 General Properties of Elementary Particles

•**1** A positively charged pion decays by Eq. 44-7: $\pi^+ \to \mu^+ + \nu$. What must be the decay scheme of the negatively charged pion? (*Hint:* The π^- is the antiparticle of the π^+.)

•**2** Certain theories predict that the proton is unstable, with a half-life of about 10^{32} years. Assuming that this is true, calculate the number of proton decays you would expect to occur in one year in the water of an Olympic-sized swimming pool holding 4.32×10^5 L of water.

•**3** **GO** An electron and a positron undergo pair annihilation (Eq. 44-5). If they had approximately zero kinetic energy before the annihilation, what is the wavelength of each γ produced by the annihilation?

•**4** A neutral pion initially at rest decays into two gamma rays: $\pi^0 \to \gamma + \gamma$. Calculate the wavelength of the gamma rays. Why must they have the same wavelength?

•**5** An electron and a positron are separated by distance r. Find the ratio of the gravitational force to the electric force between them. From the result, what can you conclude concerning the forces acting between particles detected in a bubble chamber? (Should gravitational interactions be considered?)

••**6** (a) A stationary particle 1 decays into particles 2 and 3, which move off with equal but oppositely directed momenta. Show that the kinetic energy K_2 of particle 2 is given by

$$K_2 = \frac{1}{2E_1}[(E_1 - E_2)^2 - E_3^2],$$

where E_1, E_2, and E_3 are the rest energies of the particles. (b) A stationary positive pion π^+ (rest energy 139.6 MeV) can decay to an antimuon μ^+ (rest energy 105.7 MeV) and a neutrino ν (rest energy approximately 0). What is the resulting kinetic energy of the antimuon?

••**7** The rest energy of many short-lived particles cannot be measured directly but must be inferred from the measured momenta and known rest energies of the decay products. Consider the ρ^0 meson, which decays by the reaction $\rho^0 \to \pi^+ + \pi^-$. Calculate the rest energy of the ρ^0 meson given that the oppositely directed momenta of the created pions each have magnitude 358.3 MeV/c. See Table 44-4 for the rest energies of the pions.

••**8** **GO** A positive tau (τ^+, rest energy = 1777 MeV) is moving with 2200 MeV of kinetic energy in a circular path perpendicular to a uniform 1.20 T magnetic field. (a) Calculate the momentum of the tau in kilogram-meters per second. Relativistic effects must be considered. (b) Find the radius of the circular path.

••**9** **GO** Observations of neutrinos emitted by the supernova SN1987a (Fig. 43-12b) place an upper limit of 20 eV on the rest energy of the electron neutrino. If the rest energy of the electron neutrino were, in fact, 20 eV, what would be the speed difference between light and a 1.5 MeV electron neutrino?

••**10** **GO** A neutral pion has a rest energy of 135 MeV and a mean life of 8.3×10^{-17} s. If it is produced with an initial kinetic energy of 80 MeV and decays after one mean lifetime, what is the longest possible track this particle could leave in a bubble chamber? Use relativistic time dilation.

Module 44-2 Leptons, Hadrons, and Strangeness

•**11** **SSM** **WWW** Which conservation law is violated in each of these proposed decays? Assume that the initial particle is stationary and the decay products have zero orbital angular momentum. (a) $\mu^- \to e^- + \nu_\mu$; (b) $\mu^- \to e^+ + \nu_e + \bar{\nu}_\mu$; (c) $\mu^+ \to \pi^+ + \nu_\mu$.

•**12** The A_2^+ particle and its products decay according to the scheme

$$A_2^+ \to \rho^0 + \pi^+, \qquad \mu^+ \to e^+ + \nu + \bar{\nu},$$
$$\rho^0 \to \pi^+ + \pi^-, \qquad \pi^- \to \mu^- + \bar{\nu},$$
$$\pi^+ \to \mu^+ + \nu, \qquad \mu^- \to e^- + \nu + \bar{\nu}.$$

(a) What are the final stable decay products? From the evidence, (b) is the A_2^+ particle a fermion or a boson and (c) is it a meson or a baryon? (d) What is its baryon number?

•**13** Show that if, instead of plotting strangeness S versus charge q

for the spin-$\frac{1}{2}$ baryons in Fig. 44-3a and for the spin-zero mesons in Fig. 44-3b, we plot the quantity $Y = B + S$ versus the quantity $T_z = q - \frac{1}{2}(B + S)$, we get the hexagonal patterns without using sloping axes. (The quantity Y is called *hypercharge,* and T_z is related to a quantity called *isospin.*)

•14 Calculate the disintegration energy of the reactions (a) $\pi^+ + p \rightarrow \Sigma^+ + K^+$ and (b) $K^- + p \rightarrow \Lambda^0 + \pi^0$.

•15 Which conservation law is violated in each of these proposed reactions and decays? (Assume that the products have zero orbital angular momentum.) (a) $\Lambda^0 \rightarrow p + K^-$; (b) $\Omega^- \rightarrow \Sigma^- + \pi^0$ ($S = -3, q = -1, m = 1672$ MeV/c^2, and $m_s = \frac{3}{2}$ for Ω^-); (c) $K^- + p \rightarrow \Lambda^0 + \pi^+$.

•16 Does the proposed reaction

$$p + \bar{p} \rightarrow \Lambda^0 + \Sigma^+ + e^-$$

conserve (a) charge, (b) baryon number, (c) electron lepton number, (d) spin angular momentum, (e) strangeness, and (f) muon lepton number?

•17 Does the proposed decay process

$$\Xi^- \rightarrow \pi^- + n + K^- + p$$

conserve (a) charge, (b) baryon number, (c) spin angular momentum, and (d) strangeness?

•18 By examining strangeness, determine which of the following decays or reactions proceed via the strong interaction: (a) $K^0 \rightarrow \pi^+ + \pi^-$; (b) $\Lambda^0 + p \rightarrow \Sigma^+ + n$; (c) $\Lambda^0 \rightarrow p + \pi^-$; (d) $K^- + p \rightarrow \Lambda^0 + \pi^0$.

•19 The reaction $\pi^+ + p \rightarrow p + p + \bar{n}$ proceeds via the strong interaction. By applying the conservation laws, deduce the (a) charge quantum number, (b) baryon number, and (c) strangeness of the antineutron.

•20 There are 10 baryons with spin $\frac{3}{2}$. Their symbols and quantum numbers for charge q and strangeness S are as follows:

	q	S		q	S
Δ^-	-1	0	Σ^{*0}	0	-1
Δ^0	0	0	Σ^{*+}	$+1$	-1
Δ^+	$+1$	0	Ξ^{*-}	-1	-2
Δ^{++}	$+2$	0	Ξ^{*0}	0	-2
Σ^{*-}	-1	-1	Ω^-	-1	-3

Make a charge–strangeness plot for these baryons, using the sloping coordinate system of Fig. 44-3. Compare your plot with this figure.

••21 Use the conservation laws and Tables 44-3 and 44-4 to identify particle x in each of the following reactions, which proceed by means of the strong interaction: (a) $p + p \rightarrow p + \Lambda^0 + x$; (b) $p + \bar{p} \rightarrow n + x$; (c) $\pi^- + p \rightarrow \Xi^0 + K^0 + x$.

••22 **GO** A 220 MeV Σ^- particle decays: $\Sigma^- \rightarrow \pi^- + n$. Calculate the total kinetic energy of the decay products.

••23 **GO** Consider the decay $\Lambda^0 \rightarrow p + \pi^-$ with the Λ^0 at rest. (a) Calculate the disintegration energy. What is the kinetic energy of (b) the proton and (c) the pion? (*Hint:* See Problem 6.)

••24 The spin-$\frac{3}{2}$ Σ^{*0} baryon (see table in Problem 24) has a rest energy of 1385 MeV (with an intrinsic uncertainty ignored here); the spin-$\frac{1}{2}$ Σ^0 baryon has a rest energy of 1192.5 MeV. If each of these particles has a kinetic energy of 1000 MeV, (a) which is moving faster and (b) by how much?

Module 44-3 Quarks and Messenger Particles

•25 The quark makeups of the proton and neutron are uud and udd, respectively. What are the quark makeups of (a) the antiproton and (b) the antineutron?

•26 From Tables 44-3 and 44-5, determine the identity of the baryon formed from quarks (a) ddu, (b) uus, and (c) ssd. Check your answers against the baryon octet shown in Fig. 44-3a.

•27 What is the quark makeup of $\overline{K}^0$?

•28 What quark combination is needed to form (a) Λ^0 and (b) Ξ^0?

•29 Which hadron in Tables 44-3 and 44-4 corresponds to the quark bundles (a) ssu and (b) dds?

•30 **SSM** **WWW** Using the up, down, and strange quarks only, construct, if possible, a baryon (a) with $q = +1$ and strangeness $S = -2$ and (b) with $q = +2$ and strangeness $S = 0$.

Module 44-4 Cosmology

•31 In the laboratory, one of the lines of sodium is emitted at a wavelength of 590.0 nm. In the light from a particular galaxy, however, this line is seen at a wavelength of 602.0 nm. Calculate the distance to the galaxy, assuming that Hubble's law holds and that the Doppler shift of Eq. 37-36 applies.

•32 Because of the cosmological expansion, a particular emission from a distant galaxy has a wavelength that is 2.00 times the wavelength that emission would have in a laboratory. Assuming that Hubble's law holds and that we can apply Doppler-shift calculations, what was the distance (ly) to that galaxy when the light was emitted?

•33 What is the observed wavelength of the 656.3 nm (first Balmer) line of hydrogen emitted by a galaxy at a distance of 2.40×10^8 ly? Assume that the Doppler shift of Eq. 37-36 and Hubble's law apply.

•34 An object is 1.5×10^4 ly from us and does not have any motion relative to us except for the motion due to the expansion of the universe. If the space between us and it expands according to Hubble's law, with $H = 21.8$ mm/s · ly, (a) how much extra distance (meters) will be between us and the object by this time next year and (b) what is the speed of the object away from us?

•35 If Hubble's law can be extrapolated to very large distances, at what distance would the apparent recessional speed become equal to the speed of light?

•36 What would the mass of the Sun have to be if Pluto (the outermost "planet" most of the time) were to have the same orbital speed that Mercury (the innermost planet) has now? Use data from Appendix C, express your answer in terms of the Sun's current mass M_S, and assume circular orbits.

•37 The wavelength at which a thermal radiator at temperature T radiates electromagnetic waves most intensely is given by Wien's law: $\lambda_{max} = (2898 \ \mu m \cdot K)/T$. (a) Show that the energy E of a photon corresponding to that wavelength can be computed from

$$E = (4.28 \times 10^{-10} \text{ MeV/K})T.$$

(b) At what minimum temperature can this photon create an electron–positron pair (as discussed in Module 21-3)?

•38 Use Wien's law (see Problem 37) to answer the following questions: (a) The cosmic background radiation peaks in intensity at a wavelength of 1.1 mm. To what temperature does this correspond? (b) About 379 000 y after the big bang, the universe became transparent to electromagnetic radiation. Its temperature then was

2970 K. What was the wavelength at which the background radiation was then most intense?

••**39** Will the universe continue to expand forever? To attack this question, assume that the theory of dark energy is in error and that the recessional speed v of a galaxy a distance r from us is determined only by the gravitational interaction of the matter that lies inside a sphere of radius r centered on us. If the total mass inside this sphere is M, the escape speed v_e from the sphere is $v_e = \sqrt{2GM/r}$ (Eq. 13-28). (a) Show that to prevent unlimited expansion, the average density ρ inside the sphere must be at least equal to

$$\rho = \frac{3H^2}{8\pi G}.$$

(b) Evaluate this "critical density" numerically; express your answer in terms of hydrogen atoms per cubic meter. Measurements of the actual density are difficult and are complicated by the presence of dark matter.

••**40** Because the apparent recessional speeds of galaxies and quasars at great distances are close to the speed of light, the relativistic Doppler shift formula (Eq. 37-31) must be used. The shift is reported as fractional red shift $z = \Delta\lambda/\lambda_0$. (a) Show that, in terms of z, the recessional speed parameter $\beta = v/c$ is given by

$$\beta = \frac{z^2 + 2z}{z^2 + 2z + 2}.$$

(b) A quasar detected in 1987 has $z = 4.43$. Calculate its speed parameter. (c) Find the distance to the quasar, assuming that Hubble's law is valid to these distances.

••**41** [GO] An electron jumps from $n = 3$ to $n = 2$ in a hydrogen atom in a distant galaxy, emitting light. If we detect that light at a wavelength of 3.00 mm, by what multiplication factor has the wavelength, and thus the universe, expanded since the light was emitted?

••**42** Due to the presence everywhere of the cosmic background radiation, the minimum possible temperature of a gas in interstellar or intergalactic space is not 0 K but 2.7 K. This implies that a significant fraction of the molecules in space that can be in a low-level excited state may, in fact, be so. Subsequent de-excitation would lead to the emission of radiation that could be detected. Consider a (hypothetical) molecule with just one possible excited state. (a) What would the excitation energy have to be for 25% of the molecules to be in the excited state? (*Hint:* See Eq. 40-29.) (b) What would be the wavelength of the photon emitted in a transition back to the ground state?

••**43** SSM Suppose that the radius of the Sun were increased to 5.90×10^{12} m (the average radius of the orbit of Pluto), that the density of this expanded Sun were uniform, and that the planets revolved within this tenuous object. (a) Calculate Earth's orbital speed in this new configuration. (b) What is the ratio of the orbital speed calculated in (a) to Earth's present orbital speed of 29.8 km/s? Assume that the radius of Earth's orbit remains unchanged. (c) What would be Earth's new period of revolution? (The Sun's mass remains unchanged.)

••**44** Suppose that the matter (stars, gas, dust) of a particular galaxy, of total mass M, is distributed uniformly throughout a sphere of radius R. A star of mass m is revolving about the center of the galaxy in a circular orbit of radius $r < R$. (a) Show that the orbital speed v of the star is given by

$$v = r\sqrt{GM/R^3},$$

and therefore that the star's period T of revolution is

$$T = 2\pi\sqrt{R^3/GM},$$

independent of r. Ignore any resistive forces. (b) Next suppose that the galaxy's mass is concentrated near the galactic center, within a sphere of radius less than r. What expression then gives the star's orbital period?

Additional Problems

45 SSM There is no known meson with charge quantum number $q = +1$ and strangeness $S = -1$ or with $q = -1$ and $S = +1$. Explain why in terms of the quark model.

46 Figure 44-12 is a hypothetical plot of the recessional speeds v of galaxies against their distance r from us; the best-fit straight line through the data points is shown. From this plot determine the age of the universe, assuming that Hubble's law holds and that Hubble's constant has always had the same value.

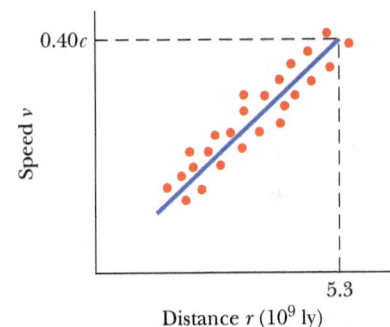

Figure 44-12 Problem 46.

47 SSM How much energy would be released if Earth were annihilated by collision with an anti-Earth?

48 *A particle game.* Figure 44-13 is a sketch of the tracks made by particles in a *fictional* cloud chamber experiment (with a uniform magnetic field directed perpendicular to the page), and Table 44-6 gives *fictional* quantum numbers associated with the particles making the tracks. Particle A entered the chamber at the lower left, leaving track 1 and decaying into three particles. Then the particle creating track 6 decayed into three other particles, and the particle creating

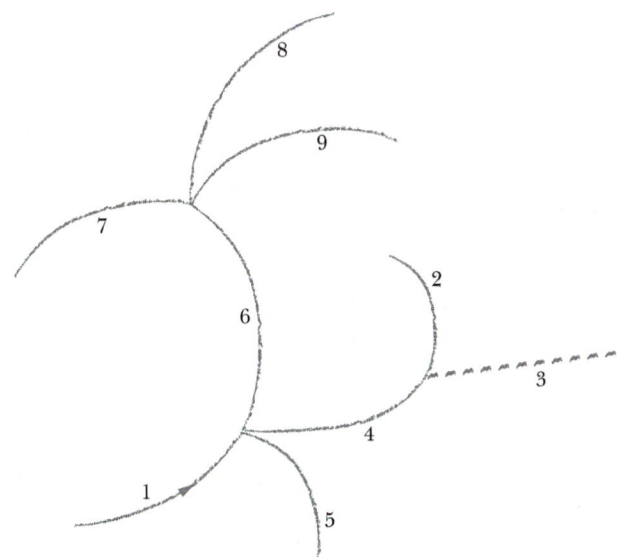

Figure 44-13 Problem 48.

Table 44-6 Problem 44-48

Particle	Charge	Whimsy	Seriousness	Cuteness
A	1	1	−2	−2
B	0	4	3	0
C	1	2	−3	−1
D	−1	−1	0	1
E	−1	0	−4	−2
F	1	0	0	0
G	−1	−1	1	−1
H	3	3	1	0
I	0	6	4	6
J	1	−6	−4	−6

track 4 decayed into two other particles, one of which was electrically uncharged—the path of that uncharged particle is represented by the dashed straight line because, being electrically neutral, it would not actually leave a track in a cloud chamber. The particle that created track 8 is known to have a seriousness quantum number of zero.

By conserving the fictional quantum numbers at each decay point and by noting the directions of curvature of the tracks, identify which particle goes with track (a) 1, (b) 2, (c) 3, (d) 4, (e) 5, (f) 6, (g) 7, (h) 8, and (i) 9. One of the listed particles is not formed; the others appear only once each.

49 Figure 44-14 shows part of the experimental arrangement in which antiprotons were discovered in the 1950s. A beam of 6.2 GeV protons emerged from a particle accelerator and collided with nuclei in a copper target. According to theoretical predictions at the time, collisions between protons in the beam and the protons and neutrons in those nuclei should produce antiprotons via the reactions

$$p + p \rightarrow p + p + p + \bar{p}$$

and

$$p + n \rightarrow p + n + p + \bar{p}.$$

However, even if these reactions did occur, they would be rare compared to the reactions

$$p + p \rightarrow p + p + \pi^+ + \pi^-$$

and

$$p + n \rightarrow p + n + \pi^+ + \pi^-.$$

Thus, most of the particles produced by the collisions between the 6.2 GeV protons and the copper target were pions.

To prove that antiprotons exist and were produced by some limited number of the collisions, particles leaving the target were sent into a series of magnetic fields and detectors as shown in Fig. 44-14. The first magnetic field (M1) curved the path of any charged particle passing through it; moreover, the field was arranged so that the only particles that emerged from it to reach the second magnetic field (Q1) had to be negatively charged (either a $\bar{p}$ or a π^-) and have a momentum of 1.19 GeV/c. Field Q1 was a special type of magnetic field (a *quadrapole field*) that focused the particles reaching it into a beam, allowing them to pass through a hole in thick shielding to a *scintillation counter* S1. The passage of a charged particle through the counter triggered a signal, with each signal indicating the passage of either a 1.19 GeV/c π^- or (presumably) a 1.19 GeV/c $\bar{p}$.

After being refocused by magnetic field Q2, the particles were directed by magnetic field M2 through a second scintillation

counter S2 and then through two *Cerenkov counters* C1 and C2. These latter detectors can be manufactured so that they send a signal only when the particle passing through them is moving with a speed that falls within a certain range. In the experiment, a particle with a speed greater than 0.79c would trigger C1 and a particle with a speed between 0.75c and 0.78c would trigger C2.

There were then two ways to distinguish the predicted rare antiprotons from the abundant negative pions. Both ways involved the fact that the speed of a 1.19 GeV/c $\bar{p}$ differs from that of a 1.19 GeV/c π^-: (1) According to calculations, a $\bar{p}$ would trigger one of the Cerenkov counters and a π^- would trigger the other. (2) The time interval Δt between signals from S1 and S2, which were separated by 12 m, would have one value for a $\bar{p}$ and another value for a π^-. Thus, if the correct Cerenkov counter were triggered and the time interval Δt had the correct value, the experiment would prove the existence of antiprotons.

What is the speed of (a) an antiproton with a momentum of 1.19 GeV/c and (b) a negative pion with that same momentum? (The speed of an antiproton through the Cerenkov detectors would actually be slightly less than calculated here because the antiproton would lose a little energy within the detectors.) Which Cerenkov detector was triggered by (c) an antiproton and (d) a negative pion? What time interval Δt indicated the passage of (e) an antiproton and (f) a negative pion? [Problem adapted from O. Chamberlain, E. Segrè, C. Wiegand, and T. Ypsilantis, "Observation of Antiprotons," *Physical Review*, Vol. 100, pp. 947–950 (1955).]

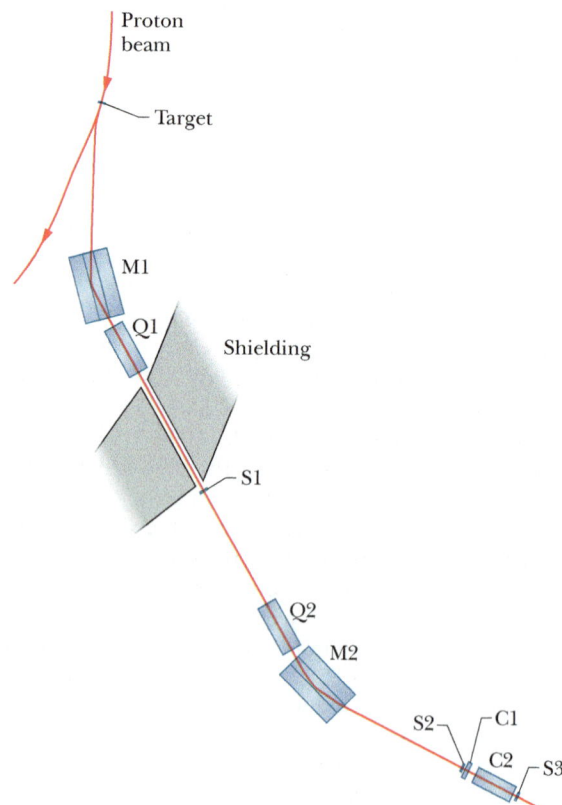

Figure 44-14 Problem 49.

50 Verify that the hypothetical proton decay scheme in Eq. 44-14 does not violate the conservation law of (a) charge, (b) energy, and (c) linear momentum. (d) How about angular momentum?

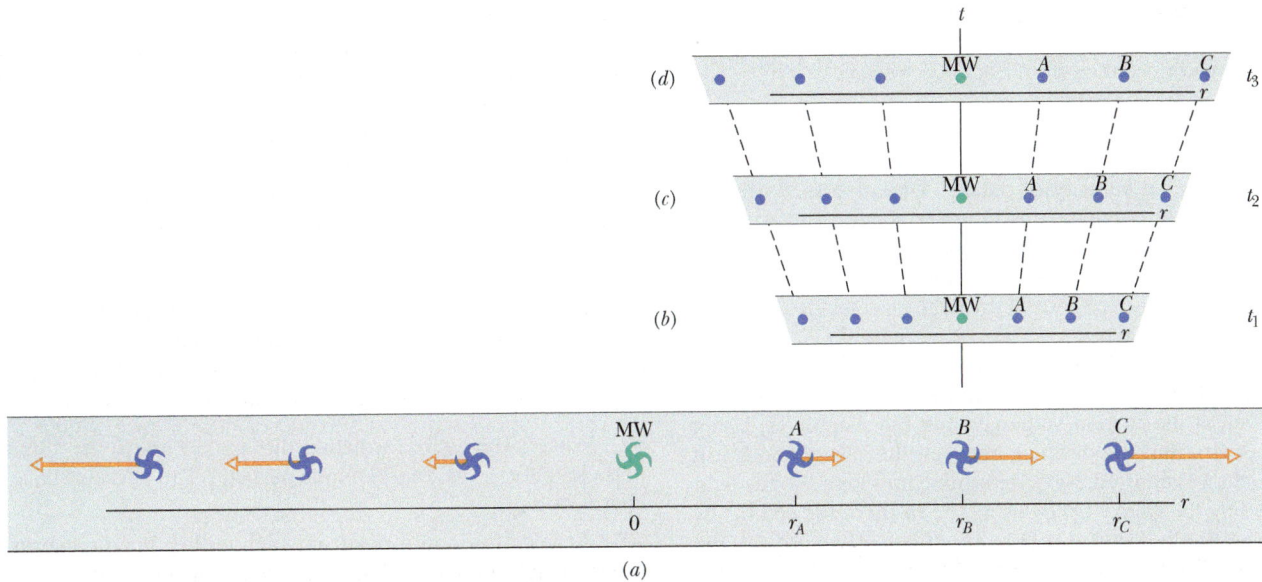

Figure 44-15 Problem 51.

51 SSM *Cosmological red shift.* The expansion of the universe is often represented with a drawing like Fig. 44-15a. In that figure, we are located at the symbol labeled MW (for the Milky Way galaxy), at the origin of an r axis that extends radially away from us in any direction. Other, very distant galaxies are also represented. Superimposed on their symbols are their velocity vectors as inferred from the red shift of the light reaching us from the galaxies. In accord with Hubble's law, the speed of each galaxy is proportional to its distance from us. Such drawings can be misleading because they imply (1) that the red shifts are due to the motions of galaxies relative to us, as they rush away from us through static (stationary) space, and (2) that we are at the center of all this motion.

Actually, the expansion of the universe and the increased separation of the galaxies are due not to an outward rush of the galaxies into pre-existing space but to an expansion of space itself throughout the universe. *Space is dynamic, not static.*

Figures 44-15b, c, and d show a different way of representing the universe and its expansion. Each part of the figure gives part of a one-dimensional section of the universe (along an r axis); the other two spatial dimensions of the universe are not shown. Each of the three parts of the figure shows the Milky Way and six other galaxies (represented by dots); the parts are positioned along a time axis, with time increasing upward. In part b, at the earliest time of the three parts, the Milky Way and the six other galaxies are represented as being relatively close to one another. As time progresses upward in the figures, space expands, causing the galaxies to move apart. Note that the figure parts are drawn relative to the Milky Way, and from that observation point all the other galaxies move away because of the expansion. However, there is nothing special about the Milky Way—the galaxies also move away from any other observation point we might have chosen.

Figures 44-16a and b focus on just the Milky Way galaxy and one of the other galaxies, galaxy A, at two particular times during the expansion. In part a, galaxy A is a distance r from the Milky Way and is emitting a light wave of wavelength λ. In part b, after a time interval Δt, that light wave is being detected at Earth. Let us represent the universe's expansion rate per unit length of space with α, which we assume to be constant during time interval Δt. Then during Δt, every unit length of space (say, every meter) expands by an amount $\alpha \Delta t$; hence, a distance r expands by $r\alpha \Delta t$. The light wave of Figs. 44.16a and b travels at speed c from galaxy A to Earth. (a) Show that

$$\Delta t = \frac{r}{c - r\alpha}.$$

The detected wavelength λ' of the light is greater than the emitted wavelength λ because space expanded during time interval Δt. This increase in wavelength is called the **cosmological red shift;** it is not a Doppler effect. (b) Show that the change in wavelength $\Delta\lambda$ $(=\lambda' - \lambda)$ is given by

$$\frac{\Delta\lambda}{\lambda} = \frac{r\alpha}{c - r\alpha}.$$

(c) Expand the right side of this equation using the binomial expansion (given in Appendix E). (d) If you retain only the first term

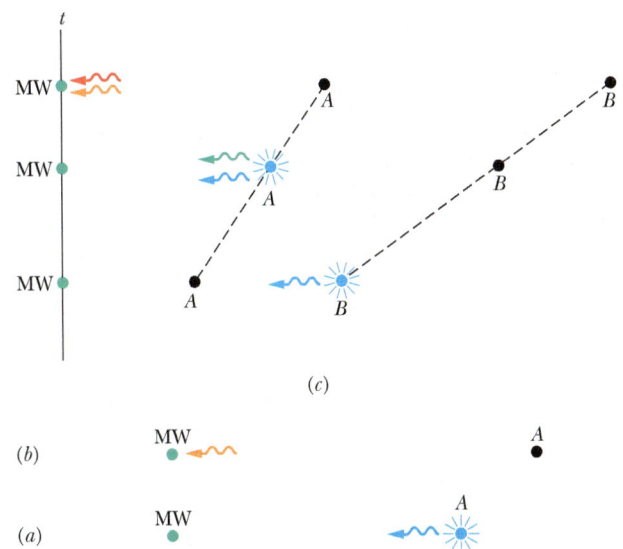

Figure 44-16 Problem 51.

of the expansion, what is the resulting equation for $\Delta\lambda/\lambda$?

If, instead, we assume that Fig. 44-15a applies and that $\Delta\lambda$ is due to a Doppler effect, then from Eq. 37-36 we have

$$\frac{\Delta\lambda}{\lambda} = \frac{v}{c},$$

where v is the radial velocity of galaxy A relative to Earth. (e) Using Hubble's law, compare this Doppler-effect result with the cosmological-expansion result of (d) and find a value for α. From this analysis you can see that the two results, derived with very different models about the red shift of the light we detect from distant galaxies, are compatible.

Suppose that the light we detect from galaxy A has a red shift of $\Delta\lambda/\lambda = 0.050$ and that the expansion rate of the universe has been constant at the current value given in the chapter. (f) Using the result of (b), find the distance between the galaxy and Earth when the light was emitted. Next, determine how long ago the light was emitted by the galaxy (g) by using the result of (a) and (h) by assuming that the red shift is a Doppler effect. (*Hint*: For (h), the time is just the distance at the time of emission divided by the speed of light, because if the red shift is just a Doppler effect, the distance

does not change during the light's travel to us. Here the two models about the red shift of the light differ in their results.) (i) At the time of detection, what is the distance between Earth and galaxy A? (We make the assumption that galaxy A still exists; if it ceased to exist, humans would not know about its death until the last light emitted by the galaxy reached Earth.)

Now suppose that the light we detect from galaxy B (Fig. 44-16c) has a red shift of $\Delta\lambda/\lambda = 0.080$. (j) Using the result of (b), find the distance between galaxy B and Earth when the light was emitted. (k) Using the result of (a), find how long ago the light was emitted by galaxy B. (l) When the light that we detect from galaxy A was emitted, what was the distance between galaxy A and galaxy B?

52   Calculate the difference in mass, in kilograms, between the muon and pion of Sample Problem 44.01.

53   What is the quark formation that makes up (a) the xi-minus particle and (b) the anti-xi-minus particle? The particles have no charm, bottom, or top.

54   An electron and a positron, each with a kinetic energy of 2.500 MeV, annihilate, creating two photons that travel away in opposite directions. What is the frequency of each photon?

THE INTERNATIONAL SYSTEM OF UNITS (SI)*

Table 1 The SI Base Units

Quantity	Name	Symbol	Definition
length	meter	m	"... the length of the path traveled by light in vacuum in 1/299,792,458 of a second." (1983)
mass	kilogram	kg	"... this prototype [a certain platinum–iridium cylinder] shall henceforth be considered to be the unit of mass." (1889)
time	second	s	"... the duration of 9,192,631,770 periods of the radiation corresponding to the transition between the two hyperfine levels of the ground state of the cesium-133 atom." (1967)
electric current	ampere	A	"... that constant current which, if maintained in two straight parallel conductors of infinite length, of negligible circular cross section, and placed 1 meter apart in vacuum, would produce between these conductors a force equal to 2×10^{-7} newton per meter of length." (1946)
thermodynamic temperature	kelvin	K	"... the fraction 1/273.16 of the thermodynamic temperature of the triple point of water." (1967)
amount of substance	mole	mol	"... the amount of substance of a system which contains as many elementary entities as there are atoms in 0.012 kilogram of carbon-12." (1971)
luminous intensity	candela	cd	"... the luminous intensity, in a given direction, of a source that emits monochromatic radiation of frequency 540×10^{12} hertz and that has a radiant intensity in that direction of 1/683 watt per steradian." (1979)

---

*Adapted from "The International System of Units (SI)," National Bureau of Standards Special Publication 330, 1972 edition. The definitions above were adopted by the General Conference of Weights and Measures, an international body, on the dates shown. In this book we do not use the candela.

Table 2 **Some SI Derived Units**

Quantity	Name of Unit	Symbol	
area	square meter	m^2	
volume	cubic meter	m^3	
frequency	hertz	Hz	s^{-1}
mass density (density)	kilogram per cubic meter	kg/m^3	
speed, velocity	meter per second	m/s	
angular velocity	radian per second	rad/s	
acceleration	meter per second per second	m/s^2	
angular acceleration	radian per second per second	rad/s^2	
force	newton	N	$kg \cdot m/s^2$
pressure	pascal	Pa	N/m^2
work, energy, quantity of heat	joule	J	$N \cdot m$
power	watt	W	J/s
quantity of electric charge	coulomb	C	$A \cdot s$
potential difference, electromotive force	volt	V	W/A
electric field strength	volt per meter (or newton per coulomb)	V/m	N/C
electric resistance	ohm	Ω	V/A
capacitance	farad	F	$A \cdot s/V$
magnetic flux	weber	Wb	$V \cdot s$
inductance	henry	H	$V \cdot s/A$
magnetic flux density	tesla	T	Wb/m^2
magnetic field strength	ampere per meter	A/m	
entropy	joule per kelvin	J/K	
specific heat	joule per kilogram kelvin	$J/(kg \cdot K)$	
thermal conductivity	watt per meter kelvin	$W/(m \cdot K)$	
radiant intensity	watt per steradian	W/sr	

Table 3 **The SI Supplementary Units**

Quantity	Name of Unit	Symbol
plane angle	radian	rad
solid angle	steradian	sr

SOME FUNDAMENTAL CONSTANTS OF PHYSICS*

Constant	Symbol	Computational Value	Best (1998) Value Value[a]	Best (1998) Value Uncertainty[b]
Speed of light in a vacuum	c	3.00×10^8 m/s	2.997 924 58	exact
Elementary charge	e	1.60×10^{-19} C	1.602 176 487	0.025
Gravitational constant	G	6.67×10^{-11} m³/s²·kg	6.674 28	100
Universal gas constant	R	8.31 J/mol·K	8.314 472	1.7
Avogadro constant	N_A	6.02×10^{23} mol⁻¹	6.022 141 79	0.050
Boltzmann constant	k	1.38×10^{-23} J/K	1.380 650 4	1.7
Stefan–Boltzmann constant	σ	5.67×10^{-8} W/m²·K⁴	5.670 400	7.0
Molar volume of ideal gas at STP[d]	V_m	2.27×10^{-2} m³/mol	2.271 098 1	1.7
Permittivity constant	ϵ_0	8.85×10^{-12} F/m	8.854 187 817 62	exact
Permeability constant	μ_0	1.26×10^{-6} H/m	1.256 637 061 43	exact
Planck constant	h	6.63×10^{-34} J·s	6.626 068 96	0.050
Electron mass[c]	m_e	9.11×10^{-31} kg	9.109 382 15	0.050
		5.49×10^{-4} u	5.485 799 094 3	4.2×10^{-4}
Proton mass[c]	m_p	1.67×10^{-27} kg	1.672 621 637	0.050
		1.0073 u	1.007 276 466 77	1.0×10^{-4}
Ratio of proton mass to electron mass	m_p/m_e	1840	1836.152 672 47	4.3×10^{-4}
Electron charge-to-mass ratio	e/m_e	1.76×10^{11} C/kg	1.758 820 150	0.025
Neutron mass[c]	m_n	1.68×10^{-27} kg	1.674 927 211	0.050
		1.0087 u	1.008 664 915 97	4.3×10^{-4}
Hydrogen atom mass[c]	m_{1_H}	1.0078 u	1.007 825 031 6	0.0005
Deuterium atom mass[c]	m_{2_H}	2.0136 u	2.013 553 212 724	3.9×10^{-5}
Helium atom mass[c]	$m_{4_{He}}$	4.0026 u	4.002 603 2	0.067
Muon mass	m_μ	1.88×10^{-28} kg	1.883 531 30	0.056
Electron magnetic moment	μ_e	9.28×10^{-24} J/T	9.284 763 77	0.025
Proton magnetic moment	μ_p	1.41×10^{-26} J/T	1.410 606 662	0.026
Bohr magneton	μ_B	9.27×10^{-24} J/T	9.274 009 15	0.025
Nuclear magneton	μ_N	5.05×10^{-27} J/T	5.050 783 24	0.025
Bohr radius	a	5.29×10^{-11} m	5.291 772 085 9	6.8×10^{-4}
Rydberg constant	R	1.10×10^7 m⁻¹	1.097 373 156 852 7	6.6×10^{-6}
Electron Compton wavelength	λ_C	2.43×10^{-12} m	2.426 310 217 5	0.0014

[a]Values given in this column should be given the same unit and power of 10 as the computational value.

[b]Parts per million.

[c]Masses given in u are in unified atomic mass units, where 1 u = $1.660\ 538\ 782 \times 10^{-27}$ kg.

[d]STP means standard temperature and pressure: 0°C and 1.0 atm (0.1 MPa).

*The values in this table were selected from the 1998 CODATA recommended values (www.physics.nist.gov).

SOME ASTRONOMICAL DATA

Some Distances from Earth

To the Moon*	3.82×10^8 m	To the center of our galaxy	2.2×10^{20} m
To the Sun*	1.50×10^{11} m	To the Andromeda Galaxy	2.1×10^{22} m
To the nearest star (Proxima Centauri)	4.04×10^{16} m	To the edge of the observable universe	$\sim 10^{26}$ m

*Mean distance.

The Sun, Earth, and the Moon

Property	Unit	Sun	Earth	Moon
Mass	kg	1.99×10^{30}	5.98×10^{24}	7.36×10^{22}
Mean radius	m	6.96×10^8	6.37×10^6	1.74×10^6
Mean density	kg/m^3	1410	5520	3340
Free-fall acceleration at the surface	m/s^2	274	9.81	1.67
Escape velocity	km/s	618	11.2	2.38
Period of rotation[a]	—	37 d at poles[b] 26 d at equator[b]	23 h 56 min	27.3 d
Radiation power[c]	W	3.90×10^{26}		

[a]Measured with respect to the distant stars.
[b]The Sun, a ball of gas, does not rotate as a rigid body.
[c]Just outside Earth's atmosphere solar energy is received, assuming normal incidence, at the rate of 1340 W/m^2.

Some Properties of the Planets

	Mercury	Venus	Earth	Mars	Jupiter	Saturn	Uranus	Neptune	Pluto[d]
Mean distance from Sun, 10^6 km	57.9	108	150	228	778	1430	2870	4500	5900
Period of revolution, y	0.241	0.615	1.00	1.88	11.9	29.5	84.0	165	248
Period of rotation,[a] d	58.7	-243^b	0.997	1.03	0.409	0.426	-0.451^b	0.658	6.39
Orbital speed, km/s	47.9	35.0	29.8	24.1	13.1	9.64	6.81	5.43	4.74
Inclination of axis to orbit	<28°	≈3°	23.4°	25.0°	3.08°	26.7°	97.9°	29.6°	57.5°
Inclination of orbit to Earth's orbit	7.00°	3.39°		1.85°	1.30°	2.49°	0.77°	1.77°	17.2°
Eccentricity of orbit	0.206	0.0068	0.0167	0.0934	0.0485	0.0556	0.0472	0.0086	0.250
Equatorial diameter, km	4880	12 100	12 800	6790	143 000	120 000	51 800	49 500	2300
Mass (Earth = 1)	0.0558	0.815	1.000	0.107	318	95.1	14.5	17.2	0.002
Density (water = 1)	5.60	5.20	5.52	3.95	1.31	0.704	1.21	1.67	2.03
Surface value of g,[c] m/s^2	3.78	8.60	9.78	3.72	22.9	9.05	7.77	11.0	0.5
Escape velocity,[c] km/s	4.3	10.3	11.2	5.0	59.5	35.6	21.2	23.6	1.3
Known satellites	0	0	1	2	67 + ring	62 + rings	27 + rings	13 + rings	4

[a]Measured with respect to the distant stars.
[b]Venus and Uranus rotate opposite their orbital motion.
[c]Gravitational acceleration measured at the planet's equator.
[d]Pluto is now classified as a dwarf planet.

CONVERSION FACTORS

Conversion factors may be read directly from these tables. For example, 1 degree = 2.778×10^{-3} revolutions, so $16.7° = 16.7 \times 2.778 \times 10^{-3}$ rev. The SI units are fully capitalized. Adapted in part from G. Shortley and D. Williams, *Elements of Physics,* 1971, Prentice-Hall, Englewood Cliffs, NJ.

Plane Angle

	°	′	″	RADIAN	rev
1 degree = 1	60	3600	1.745×10^{-2}	2.778×10^{-3}	
1 minute = 1.667×10^{-2}	1	60	2.909×10^{-4}	4.630×10^{-5}	
1 second = 2.778×10^{-4}	1.667×10^{-2}	1	4.848×10^{-6}	7.716×10^{-7}	
1 RADIAN = 57.30	3438	2.063×10^{5}	1	0.1592	
1 revolution = 360	2.16×10^{4}	1.296×10^{6}	6.283	1	

Solid Angle

1 sphere = 4π steradians = 12.57 steradians

Length

	cm	METER	km	in.	ft	mi
1 centimeter = 1	10^{-2}	10^{-5}	0.3937	3.281×10^{-2}	6.214×10^{-6}
1 METER = 100	1	10^{-3}	39.37	3.281	6.214×10^{-4}
1 kilometer = 10^{5}	1000	1	3.937×10^{4}	3281	0.6214
1 inch = 2.540	2.540×10^{-2}	2.540×10^{-5}	1	8.333×10^{-2}	1.578×10^{-5}
1 foot = 30.48	0.3048	3.048×10^{-4}	12	1	1.894×10^{-4}
1 mile = 1.609×10^{5}	1609	1.609	6.336×10^{4}	5280	1

1 angström = 10^{-10} m
1 nautical mile = 1852 m
= 1.151 miles = 6076 ft

1 fermi = 10^{-15} m
1 light-year = 9.461×10^{12} km
1 parsec = 3.084×10^{13} km

1 fathom = 6 ft
1 Bohr radius = 5.292×10^{-11} m
1 yard = 3 ft

1 rod = 16.5 ft
1 mil = 10^{-3} in.
1 nm = 10^{-9} m

Area

	METER^2	cm^2	ft^2	in.^{2}
1 SQUARE METER = 1	10^{4}	10.76	1550
1 square centimeter = 10^{-4}	1	1.076×10^{-3}	0.1550
1 square foot = 9.290×10^{-2}	929.0	1	144
1 square inch = 6.452×10^{-4}	6.452	6.944×10^{-3}	1

1 square mile = 2.788×10^{7} ft^2 = 640 acres
1 barn = 10^{-28} m^2

1 acre = 43 560 ft^2
1 hectare = 10^{4} m^2 = 2.471 acres

Volume

	METER^3	cm^3	L	ft^3	in.^{3}
1 CUBIC METER = 1		10^6	1000	35.31	6.102×10^4
1 cubic centimeter = 10^{-6}		1	1.000×10^{-3}	3.531×10^{-5}	6.102×10^{-2}
1 liter = 1.000×10^{-3}		1000	1	3.531×10^{-2}	61.02
1 cubic foot = 2.832×10^{-2}		2.832×10^4	28.32	1	1728
1 cubic inch = 1.639×10^{-5}		16.39	1.639×10^{-2}	5.787×10^{-4}	1

1 U.S. fluid gallon = 4 U.S. fluid quarts = 8 U.S. pints = 128 U.S. fluid ounces = 231 in.^{3}
1 British imperial gallon = 277.4 in.^{3} = 1.201 U.S. fluid gallons

Mass

Quantities in the colored areas are not mass units but are often used as such. For example, when we write 1 kg "=" 2.205 lb, this means that a kilogram is a *mass* that *weighs* 2.205 pounds at a location where g has the standard value of 9.80665 m/s^2.

	g	KILOGRAM	slug	u	oz	lb	ton
1 gram = 1		0.001	6.852×10^{-5}	6.022×10^{23}	3.527×10^{-2}	2.205×10^{-3}	1.102×10^{-6}
1 KILOGRAM = 1000		1	6.852×10^{-2}	6.022×10^{26}	35.27	2.205	1.102×10^{-3}
1 slug = 1.459×10^4		14.59	1	8.786×10^{27}	514.8	32.17	1.609×10^{-2}
1 atomic mass unit = 1.661×10^{-24}		1.661×10^{-27}	1.138×10^{-28}	1	5.857×10^{-26}	3.662×10^{-27}	1.830×10^{-30}
1 ounce = 28.35		2.835×10^{-2}	1.943×10^{-3}	1.718×10^{25}	1	6.250×10^{-2}	3.125×10^{-5}
1 pound = 453.6		0.4536	3.108×10^{-2}	2.732×10^{26}	16	1	0.0005
1 ton = 9.072×10^5		907.2	62.16	5.463×10^{29}	3.2×10^4	2000	1

1 metric ton = 1000 kg

Density

Quantities in the colored areas are weight densities and, as such, are dimensionally different from mass densities. See the note for the mass table.

	slug/ft^3	KILOGRAM/ METER^3	g/cm^3	lb/ft^3	lb/in.^{3}
1 slug per foot^3 = 1		515.4	0.5154	32.17	1.862×10^{-2}
1 KILOGRAM per METER^3 = 1.940×10^{-3}		1	0.001	6.243×10^{-2}	3.613×10^{-5}
1 gram per centimeter^3 = 1.940		1000	1	62.43	3.613×10^{-2}
1 pound per foot^3 = 3.108×10^{-2}		16.02	16.02×10^{-2}	1	5.787×10^{-4}
1 pound per inch^3 = 53.71		2.768×10^4	27.68	1728	1

Time

	y	d	h	min	SECOND
1 year = 1		365.25	8.766×10^3	5.259×10^5	3.156×10^7
1 day = 2.738×10^{-3}		1	24	1440	8.640×10^4
1 hour = 1.141×10^{-4}		4.167×10^{-2}	1	60	3600
1 minute = 1.901×10^{-6}		6.944×10^{-4}	1.667×10^{-2}	1	60
1 SECOND = 3.169×10^{-8}		1.157×10^{-5}	2.778×10^{-4}	1.667×10^{-2}	1

Speed

	ft/s	km/h	METER/SECOND	mi/h	cm/s
1 foot per second =	1	1.097	0.3048	0.6818	30.48
1 kilometer per hour =	0.9113	1	0.2778	0.6214	27.78
1 METER per SECOND =	3.281	3.6	1	2.237	100
1 mile per hour =	1.467	1.609	0.4470	1	44.70
1 centimeter per second =	3.281×10^{-2}	3.6×10^{-2}	0.01	2.237×10^{-2}	1

1 knot = 1 nautical mi/h = 1.688 ft/s   1 mi/min = 88.00 ft/s = 60.00 mi/h

Force

Force units in the colored areas are now little used. To clarify: 1 gram-force (= 1 gf) is the force of gravity that would act on an object whose mass is 1 gram at a location where g has the standard value of 9.80665 m/s^2.

	dyne	NEWTON	lb	pdl	gf	kgf
1 dyne =	1	10^{-5}	2.248×10^{-6}	7.233×10^{-5}	1.020×10^{-3}	1.020×10^{-6}
1 NEWTON =	10^5	1	0.2248	7.233	102.0	0.1020
1 pound =	4.448×10^5	4.448	1	32.17	453.6	0.4536
1 poundal =	1.383×10^4	0.1383	3.108×10^{-2}	1	14.10	1.410×10^2
1 gram-force =	980.7	9.807×10^{-3}	2.205×10^{-3}	7.093×10^{-2}	1	0.001
1 kilogram-force =	9.807×10^5	9.807	2.205	70.93	1000	1

1 ton = 2000 lb

Pressure

	atm	dyne/cm^2	inch of water	cm Hg	PASCAL	lb/in.^{2}	lb/ft^2
1 atmosphere =	1	1.013×10^6	406.8	76	1.013×10^5	14.70	2116
1 dyne per centimeter^2 =	9.869×10^{-7}	1	4.015×10^{-4}	7.501×10^{-5}	0.1	1.405×10^{-5}	2.089×10^{-3}
1 inch of water[a] at 4°C =	2.458×10^{-3}	2491	1	0.1868	249.1	3.613×10^{-2}	5.202
1 centimeter of mercury[a] at 0°C =	1.316×10^{-2}	1.333×10^4	5.353	1	1333	0.1934	27.85
1 PASCAL =	9.869×10^{-6}	10	4.015×10^{-3}	7.501×10^{-4}	1	1.450×10^{-4}	2.089×10^{-2}
1 pound per inch^2 =	6.805×10^{-2}	6.895×10^4	27.68	5.171	6.895×10^3	1	144
1 pound per foot^2 =	4.725×10^{-4}	478.8	0.1922	3.591×10^{-2}	47.88	6.944×10^{-3}	1

[a]Where the acceleration of gravity has the standard value of 9.80665 m/s^2.

1 bar = 10^6 dyne/cm^2 = 0.1 MPa   1 millibar = 10^3 dyne/cm^2 = 10^2 Pa   1 torr = 1 mm Hg

Energy, Work, Heat

Quantities in the colored areas are not energy units but are included for convenience. They arise from the relativistic mass–energy equivalence formula $E = mc^2$ and represent the energy released if a kilogram or unified atomic mass unit (u) is completely converted to energy (bottom two rows) or the mass that would be completely converted to one unit of energy (rightmost two columns).

	Btu	erg	ft·lb	hp·h	JOULE	cal	kW·h	eV	MeV	kg	u
1 British thermal unit = 1	1.055×10^{10}	777.9	3.929×10^{-4}	1055	252.0	2.930×10^{-4}	6.585×10^{21}	6.585×10^{15}	1.174×10^{-14}	7.070×10^{12}
1 erg = 9.481×10^{-11}	1	7.376×10^{-8}	3.725×10^{-14}	10^{-7}	2.389×10^{-8}	2.778×10^{-14}	6.242×10^{11}	6.242×10^{5}	1.113×10^{-24}	670.2
1 foot-pound = 1.285×10^{-3}	1.356×10^{7}	1	5.051×10^{-7}	1.356	0.3238	3.766×10^{-7}	8.464×10^{18}	8.464×10^{12}	1.509×10^{-17}	9.037×10^{9}
1 horsepower-hour = 2545	2.685×10^{13}	1.980×10^{6}	1	2.685×10^{6}	6.413×10^{5}	0.7457	1.676×10^{25}	1.676×10^{19}	2.988×10^{-11}	1.799×10^{16}
1 JOULE = 9.481×10^{-4}	10^{7}	0.7376	3.725×10^{-7}	1	0.2389	2.778×10^{-7}	6.242×10^{18}	6.242×10^{12}	1.113×10^{-17}	6.702×10^{9}
1 calorie = 3.968×10^{-3}	4.1868×10^{7}	3.088	1.560×10^{-6}	4.1868	1	1.163×10^{-6}	2.613×10^{19}	2.613×10^{13}	4.660×10^{-17}	2.806×10^{10}
1 kilowatt-hour = 3413	3.600×10^{13}	2.655×10^{6}	1.341	3.600×10^{6}	8.600×10^{5}	1	2.247×10^{25}	2.247×10^{19}	4.007×10^{-11}	2.413×10^{16}
1 electron-volt = 1.519×10^{-22}	1.602×10^{-12}	1.182×10^{-19}	5.967×10^{-26}	1.602×10^{-19}	3.827×10^{-20}	4.450×10^{-26}	1	10^{-6}	1.783×10^{-36}	1.074×10^{-9}
1 million electron-volts = 1.519×10^{-16}	1.602×10^{-6}	1.182×10^{-13}	5.967×10^{-20}	1.602×10^{-13}	3.827×10^{-14}	4.450×10^{-20}	10^{-6}	1	1.783×10^{-30}	1.074×10^{-3}
1 kilogram = 8.521×10^{13}	8.987×10^{23}	6.629×10^{16}	3.348×10^{10}	8.987×10^{16}	2.146×10^{16}	2.497×10^{10}	5.610×10^{35}	5.610×10^{29}	1	6.022×10^{26}
1 unified atomic mass unit = 1.415×10^{-13}	1.492×10^{-3}	1.101×10^{-10}	5.559×10^{-17}	1.492×10^{-10}	3.564×10^{-11}	4.146×10^{-17}	9.320×10^{8}	932.0	1.661×10^{-27}	1

Power

	Btu/h	ft·lb/s	hp	cal/s	kW	WATT
1 British thermal unit per hour = 1	0.2161	3.929×10^{-4}	6.998×10^{-2}	2.930×10^{-4}	0.2930
1 foot-pound per second = 4.628	1	1.818×10^{-3}	0.3239	1.356×10^{-3}	1.356
1 horsepower = 2545	550	1	178.1	0.7457	745.7
1 calorie per second = 14.29	3.088	5.615×10^{-3}	1	4.186×10^{-3}	4.186
1 kilowatt = 3413	737.6	1.341	238.9	1	1000
1 WATT = 3.413	0.7376	1.341×10^{-3}	0.2389	0.001	1

Magnetic Field

	gauss	TESLA	milligauss
1 gauss = 1	10^{-4}	1000
1 TESLA = 10^4	1	10^7
1 milligauss = 0.001	10^{-7}	1

Magnetic Flux

	maxwell	WEBER
1 maxwell = 1	10^{-8}
1 WEBER = 10^8	1

1 tesla = 1 weber/meter^2

MATHEMATICAL FORMULAS

Geometry

Circle of radius r: circumference $= 2\pi r$; area $= \pi r^2$.

Sphere of radius r: area $= 4\pi r^2$; volume $= \frac{4}{3}\pi r^3$.

Right circular cylinder of radius r and height h:
   area $= 2\pi r^2 + 2\pi rh$; volume $= \pi r^2 h$.

Triangle of base a and altitude h: area $= \frac{1}{2}ah$.

Quadratic Formula

If $ax^2 + bx + c = 0$, then $x = \dfrac{-b \pm \sqrt{b^2 - 4ac}}{2a}$.

Trigonometric Functions of Angle θ

$\sin\theta = \dfrac{y}{r}$ $\cos\theta = \dfrac{x}{r}$

$\tan\theta = \dfrac{y}{x}$ $\cot\theta = \dfrac{x}{y}$

$\sec\theta = \dfrac{r}{x}$ $\csc\theta = \dfrac{r}{y}$

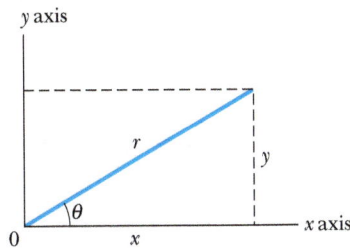

Pythagorean Theorem

In this right triangle,
   $a^2 + b^2 = c^2$

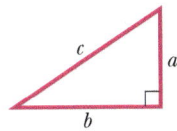

Triangles

Angles are A, B, C

Opposite sides are a, b, c

Angles $A + B + C = 180°$

$\dfrac{\sin A}{a} = \dfrac{\sin B}{b} = \dfrac{\sin C}{c}$

$c^2 = a^2 + b^2 - 2ab\cos C$

Exterior angle $D = A + C$

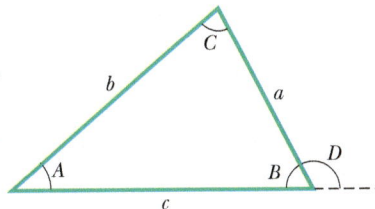

Mathematical Signs and Symbols

$=$ equals

$\approx$ equals approximately

$\sim$ is the order of magnitude of

$\neq$ is not equal to

$\equiv$ is identical to, is defined as

$>$ is greater than ($\gg$ is much greater than)

$<$ is less than ($\ll$ is much less than)

$\geq$ is greater than or equal to (or, is no less than)

$\leq$ is less than or equal to (or, is no more than)

$\pm$ plus or minus

$\propto$ is proportional to

Σ the sum of

x_{avg} the average value of x

Trigonometric Identities

$\sin(90° - \theta) = \cos\theta$

$\cos(90° - \theta) = \sin\theta$

$\sin\theta/\cos\theta = \tan\theta$

$\sin^2\theta + \cos^2\theta = 1$

$\sec^2\theta - \tan^2\theta = 1$

$\csc^2\theta - \cot^2\theta = 1$

$\sin 2\theta = 2\sin\theta\cos\theta$

$\cos 2\theta = \cos^2\theta - \sin^2\theta = 2\cos^2\theta - 1 = 1 - 2\sin^2\theta$

$\sin(\alpha \pm \beta) = \sin\alpha\cos\beta \pm \cos\alpha\sin\beta$

$\cos(\alpha \pm \beta) = \cos\alpha\cos\beta \mp \sin\alpha\sin\beta$

$\tan(\alpha \pm \beta) = \dfrac{\tan\alpha \pm \tan\beta}{1 \mp \tan\alpha\tan\beta}$

$\sin\alpha \pm \sin\beta = 2\sin\frac{1}{2}(\alpha \pm \beta)\cos\frac{1}{2}(\alpha \mp \beta)$

$\cos\alpha + \cos\beta = 2\cos\frac{1}{2}(\alpha + \beta)\cos\frac{1}{2}(\alpha - \beta)$

$\cos\alpha - \cos\beta = -2\sin\frac{1}{2}(\alpha + \beta)\sin\frac{1}{2}(\alpha - \beta)$

Binomial Theorem

$(1 + x)^n = 1 + \dfrac{nx}{1!} + \dfrac{n(n-1)x^2}{2!} + \cdots$ $(x^2 < 1)$

Exponential Expansion

$$e^x = 1 + x + \dfrac{x^2}{2!} + \dfrac{x^3}{3!} + \cdots$$

Logarithmic Expansion

$$\ln(1 + x) = x - \tfrac{1}{2}x^2 + \tfrac{1}{3}x^3 - \cdots \qquad (|x| < 1)$$

Trigonometric Expansions (θ in radians)

$$\sin \theta = \theta - \frac{\theta^3}{3!} + \frac{\theta^5}{5!} - \cdots$$

$$\cos \theta = 1 - \frac{\theta^2}{2!} + \frac{\theta^4}{4!} - \cdots$$

$$\tan \theta = \theta + \frac{\theta^3}{3} + \frac{2\theta^5}{15} + \cdots$$

Cramer's Rule

Two simultaneous equations in unknowns x and y,

$$a_1 x + b_1 y = c_1 \quad \text{and} \quad a_2 x + b_2 y = c_2,$$

have the solutions

$$x = \frac{\begin{vmatrix} c_1 & b_1 \\ c_2 & b_2 \end{vmatrix}}{\begin{vmatrix} a_1 & b_1 \\ a_2 & b_2 \end{vmatrix}} = \frac{c_1 b_2 - c_2 b_1}{a_1 b_2 - a_2 b_1}$$

and

$$y = \frac{\begin{vmatrix} a_1 & c_1 \\ a_2 & c_2 \end{vmatrix}}{\begin{vmatrix} a_1 & b_1 \\ a_2 & b_2 \end{vmatrix}} = \frac{a_1 c_2 - a_2 c_1}{a_1 b_2 - a_2 b_1}.$$

Products of Vectors

Let $\hat{i}$, $\hat{j}$, and $\hat{k}$ be unit vectors in the x, y, and z directions. Then

$$\hat{i} \cdot \hat{i} = \hat{j} \cdot \hat{j} = \hat{k} \cdot \hat{k} = 1, \quad \hat{i} \cdot \hat{j} = \hat{j} \cdot \hat{k} = \hat{k} \cdot \hat{i} = 0,$$

$$\hat{i} \times \hat{i} = \hat{j} \times \hat{j} = \hat{k} \times \hat{k} = 0,$$

$$\hat{i} \times \hat{j} = \hat{k}, \quad \hat{j} \times \hat{k} = \hat{i}, \quad \hat{k} \times \hat{i} = \hat{j}$$

Any vector $\vec{a}$ with components a_x, a_y, and a_z along the x, y, and z axes can be written as

$$\vec{a} = a_x \hat{i} + a_y \hat{j} + a_z \hat{k}.$$

Let $\vec{a}$, $\vec{b}$, and $\vec{c}$ be arbitrary vectors with magnitudes a, b, and c. Then

$$\vec{a} \times (\vec{b} + \vec{c}) = (\vec{a} \times \vec{b}) + (\vec{a} \times \vec{c})$$

$$(s\vec{a}) \times \vec{b} = \vec{a} \times (s\vec{b}) = s(\vec{a} \times \vec{b}) \qquad (s = \text{a scalar}).$$

Let θ be the smaller of the two angles between $\vec{a}$ and $\vec{b}$. Then

$$\vec{a} \cdot \vec{b} = \vec{b} \cdot \vec{a} = a_x b_x + a_y b_y + a_z b_z = ab \cos \theta$$

$$\vec{a} \times \vec{b} = -\vec{b} \times \vec{a} = \begin{vmatrix} \hat{i} & \hat{j} & \hat{k} \\ a_x & a_y & a_z \\ b_x & b_y & b_z \end{vmatrix}$$

$$= \hat{i} \begin{vmatrix} a_y & a_z \\ b_y & b_z \end{vmatrix} - \hat{j} \begin{vmatrix} a_x & a_z \\ b_x & b_z \end{vmatrix} + \hat{k} \begin{vmatrix} a_x & a_y \\ b_x & b_y \end{vmatrix}$$

$$= (a_y b_z - b_y a_z)\hat{i} + (a_z b_x - b_z a_x)\hat{j}$$

$$+ (a_x b_y - b_x a_y)\hat{k}$$

$$|\vec{a} \times \vec{b}| = ab \sin \theta$$

$$\vec{a} \cdot (\vec{b} \times \vec{c}) = \vec{b} \cdot (\vec{c} \times \vec{a}) = \vec{c} \cdot (\vec{a} \times \vec{b})$$

$$\vec{a} \times (\vec{b} \times \vec{c}) = (\vec{a} \cdot \vec{c})\vec{b} - (\vec{a} \cdot \vec{b})\vec{c}$$

Derivatives and Integrals

In what follows, the letters u and v stand for any functions of x, and a and m are constants. To each of the indefinite integrals should be added an arbitrary constant of integration. The *Handbook of Chemistry and Physics* (CRC Press Inc.) gives a more extensive tabulation.

1. $\dfrac{dx}{dx} = 1$

2. $\dfrac{d}{dx}(au) = a\dfrac{du}{dx}$

3. $\dfrac{d}{dx}(u + v) = \dfrac{du}{dx} + \dfrac{dv}{dx}$

4. $\dfrac{d}{dx}x^m = mx^{m-1}$

5. $\dfrac{d}{dx}\ln x = \dfrac{1}{x}$

6. $\dfrac{d}{dx}(uv) = u\dfrac{dv}{dx} + v\dfrac{du}{dx}$

7. $\dfrac{d}{dx}e^x = e^x$

8. $\dfrac{d}{dx}\sin x = \cos x$

9. $\dfrac{d}{dx}\cos x = -\sin x$

10. $\dfrac{d}{dx}\tan x = \sec^2 x$

11. $\dfrac{d}{dx}\cot x = -\csc^2 x$

12. $\dfrac{d}{dx}\sec x = \tan x \sec x$

13. $\dfrac{d}{dx}\csc x = -\cot x \csc x$

14. $\dfrac{d}{dx}e^u = e^u\dfrac{du}{dx}$

15. $\dfrac{d}{dx}\sin u = \cos u\dfrac{du}{dx}$

16. $\dfrac{d}{dx}\cos u = -\sin u\dfrac{du}{dx}$

1. $\displaystyle\int dx = x$

2. $\displaystyle\int au\,dx = a\int u\,dx$

3. $\displaystyle\int (u + v)\,dx = \int u\,dx + \int v\,dx$

4. $\displaystyle\int x^m\,dx = \dfrac{x^{m+1}}{m + 1}\ (m \neq -1)$

5. $\displaystyle\int \dfrac{dx}{x} = \ln |x|$

6. $\displaystyle\int u\dfrac{dv}{dx}\,dx = uv - \int v\dfrac{du}{dx}\,dx$

7. $\displaystyle\int e^x\,dx = e^x$

8. $\displaystyle\int \sin x\,dx = -\cos x$

9. $\displaystyle\int \cos x\,dx = \sin x$

10. $\displaystyle\int \tan x\,dx = \ln |\sec x|$

11. $\displaystyle\int \sin^2 x\,dx = \tfrac{1}{2}x - \tfrac{1}{4}\sin 2x$

12. $\displaystyle\int e^{-ax}\,dx = -\dfrac{1}{a}e^{-ax}$

13. $\displaystyle\int xe^{-ax}\,dx = -\dfrac{1}{a^2}(ax + 1)e^{-ax}$

14. $\displaystyle\int x^2 e^{-ax}\,dx = -\dfrac{1}{a^3}(a^2x^2 + 2ax + 2)e^{-ax}$

15. $\displaystyle\int_0^\infty x^n e^{-ax}\,dx = \dfrac{n!}{a^{n+1}}$

16. $\displaystyle\int_0^\infty x^{2n} e^{-ax^2}\,dx = \dfrac{1\cdot 3\cdot 5\,\cdots\,(2n-1)}{2^{n+1}a^n}\sqrt{\dfrac{\pi}{a}}$

17. $\displaystyle\int \dfrac{dx}{\sqrt{x^2 + a^2}} = \ln(x + \sqrt{x^2 + a^2})$

18. $\displaystyle\int \dfrac{x\,dx}{(x^2 + a^2)^{3/2}} = -\dfrac{1}{(x^2 + a^2)^{1/2}}$

19. $\displaystyle\int \dfrac{dx}{(x^2 + a^2)^{3/2}} = \dfrac{x}{a^2(x^2 + a^2)^{1/2}}$

20. $\displaystyle\int_0^\infty x^{2n+1} e^{-ax^2}\,dx = \dfrac{n!}{2a^{n+1}}\ (a > 0)$

21. $\displaystyle\int \dfrac{x\,dx}{x + d} = x - d\ln(x + d)$

PROPERTIES OF THE ELEMENTS

All physical properties are for a pressure of 1 atm unless otherwise specified.

Element	Symbol	Atomic Number Z	Molar Mass, g/mol	Density, g/cm^3 at 20°C	Melting Point, °C	Boiling Point, °C	Specific Heat, J/(g·°C) at 25°C
Actinium	Ac	89	(227)	10.06	1323	(3473)	0.092
Aluminum	Al	13	26.9815	2.699	660	2450	0.900
Americium	Am	95	(243)	13.67	1541	—	—
Antimony	Sb	51	121.75	6.691	630.5	1380	0.205
Argon	Ar	18	39.948	1.6626×10^{-3}	−189.4	−185.8	0.523
Arsenic	As	33	74.9216	5.78	817 (28 atm)	613	0.331
Astatine	At	85	(210)	—	(302)	—	—
Barium	Ba	56	137.34	3.594	729	1640	0.205
Berkelium	Bk	97	(247)	14.79	—	—	—
Beryllium	Be	4	9.0122	1.848	1287	2770	1.83
Bismuth	Bi	83	208.980	9.747	271.37	1560	0.122
Bohrium	Bh	107	262.12	—	—	—	—
Boron	B	5	10.811	2.34	2030	—	1.11
Bromine	Br	35	79.909	3.12 (liquid)	−7.2	58	0.293
Cadmium	Cd	48	112.40	8.65	321.03	765	0.226
Calcium	Ca	20	40.08	1.55	838	1440	0.624
Californium	Cf	98	(251)	—	—	—	—
Carbon	C	6	12.01115	2.26	3727	4830	0.691
Cerium	Ce	58	140.12	6.768	804	3470	0.188
Cesium	Cs	55	132.905	1.873	28.40	690	0.243
Chlorine	Cl	17	35.453	3.214×10^{-3} (0°C)	−101	−34.7	0.486
Chromium	Cr	24	51.996	7.19	1857	2665	0.448
Cobalt	Co	27	58.9332	8.85	1495	2900	0.423
Copernicium	Cn	112	(285)	—	—	—	—
Copper	Cu	29	63.54	8.96	1083.40	2595	0.385
Curium	Cm	96	(247)	13.3	—	—	—
Darmstadtium	Ds	110	(271)	—	—	—	—
Dubnium	Db	105	262.114	—	—	—	—
Dysprosium	Dy	66	162.50	8.55	1409	2330	0.172
Einsteinium	Es	99	(254)	—	—	—	—
Erbium	Er	68	167.26	9.15	1522	2630	0.167
Europium	Eu	63	151.96	5.243	817	1490	0.163
Fermium	Fm	100	(237)	—	—	—	—
Flerovium*	Fl	114	(289)	—	—	—	—
Fluorine	F	9	18.9984	1.696×10^{-3} (0°C)	−219.6	−188.2	0.753
Francium	Fr	87	(223)	—	(27)	—	—
Gadolinium	Gd	64	157.25	7.90	1312	2730	0.234
Gallium	Ga	31	69.72	5.907	29.75	2237	0.377
Germanium	Ge	32	72.59	5.323	937.25	2830	0.322
Gold	Au	79	196.967	19.32	1064.43	2970	0.131

Element	Symbol	Atomic Number Z	Molar Mass, g/mol	Density, g/cm³ at 20°C	Melting Point, °C	Boiling Point, °C	Specific Heat, J/(g·°C) at 25°C
Hafnium	Hf	72	178.49	13.31	2227	5400	0.144
Hassium	Hs	108	(265)	—	—	—	—
Helium	He	2	4.0026	0.1664×10^{-3}	−269.7	−268.9	5.23
Holmium	Ho	67	164.930	8.79	1470	2330	0.165
Hydrogen	H	1	1.00797	0.08375×10^{-3}	−259.19	−252.7	14.4
Indium	In	49	114.82	7.31	156.634	2000	0.233
Iodine	I	53	126.9044	4.93	113.7	183	0.218
Iridium	Ir	77	192.2	22.5	2447	(5300)	0.130
Iron	Fe	26	55.847	7.874	1536.5	3000	0.447
Krypton	Kr	36	83.80	3.488×10^{-3}	−157.37	−152	0.247
Lanthanum	La	57	138.91	6.189	920	3470	0.195
Lawrencium	Lr	103	(257)	—	—	—	—
Lead	Pb	82	207.19	11.35	327.45	1725	0.129
Lithium	Li	3	6.939	0.534	180.55	1300	3.58
Livermorium*	Lv	116	(293)	—	—	—	—
Lutetium	Lu	71	174.97	9.849	1663	1930	0.155
Magnesium	Mg	12	24.312	1.738	650	1107	1.03
Manganese	Mn	25	54.9380	7.44	1244	2150	0.481
Meitnerium	Mt	109	(266)	—	—	—	—
Mendelevium	Md	101	(256)	—	—	—	—
Mercury	Hg	80	200.59	13.55	−38.87	357	0.138
Molybdenum	Mo	42	95.94	10.22	2617	5560	0.251
Neodymium	Nd	60	144.24	7.007	1016	3180	0.188
Neon	Ne	10	20.183	0.8387×10^{-3}	−248.597	−246.0	1.03
Neptunium	Np	93	(237)	20.25	637	—	1.26
Nickel	Ni	28	58.71	8.902	1453	2730	0.444
Niobium	Nb	41	92.906	8.57	2468	4927	0.264
Nitrogen	N	7	14.0067	1.1649×10^{-3}	−210	−195.8	1.03
Nobelium	No	102	(255)	—	—	—	—
Osmium	Os	76	190.2	22.59	3027	5500	0.130
Oxygen	O	8	15.9994	1.3318×10^{-3}	−218.80	−183.0	0.913
Palladium	Pd	46	106.4	12.02	1552	3980	0.243
Phosphorus	P	15	30.9738	1.83	44.25	280	0.741
Platinum	Pt	78	195.09	21.45	1769	4530	0.134
Plutonium	Pu	94	(244)	19.8	640	3235	0.130
Polonium	Po	84	(210)	9.32	254	—	—
Potassium	K	19	39.102	0.862	63.20	760	0.758
Praseodymium	Pr	59	140.907	6.773	931	3020	0.197
Promethium	Pm	61	(145)	7.22	(1027)	—	—
Protactinium	Pa	91	(231)	15.37 (estimated)	(1230)	—	—
Radium	Ra	88	(226)	5.0	700	—	—
Radon	Rn	86	(222)	9.96×10^{-3} (0°C)	(−71)	−61.8	0.092
Rhenium	Re	75	186.2	21.02	3180	5900	0.134
Rhodium	Rh	45	102.905	12.41	1963	4500	0.243
Roentgenium	Rg	111	(280)	—	—	—	—
Rubidium	Rb	37	85.47	1.532	39.49	688	0.364
Ruthenium	Ru	44	101.107	12.37	2250	4900	0.239
Rutherfordium	Rf	104	261.11	—	—	—	—

Element	Symbol	Atomic Number Z	Molar Mass, g/mol	Density, g/cm³ at 20°C	Melting Point, °C	Boiling Point, °C	Specific Heat, J/(g·°C) at 25°C
Samarium	Sm	62	150.35	7.52	1072	1630	0.197
Scandium	Sc	21	44.956	2.99	1539	2730	0.569
Seaborgium	Sg	106	263.118	—	—	—	—
Selenium	Se	34	78.96	4.79	221	685	0.318
Silicon	Si	14	28.086	2.33	1412	2680	0.712
Silver	Ag	47	107.870	10.49	960.8	2210	0.234
Sodium	Na	11	22.9898	0.9712	97.85	892	1.23
Strontium	Sr	38	87.62	2.54	768	1380	0.737
Sulfur	S	16	32.064	2.07	119.0	444.6	0.707
Tantalum	Ta	73	180.948	16.6	3014	5425	0.138
Technetium	Tc	43	(99)	11.46	2200	—	0.209
Tellurium	Te	52	127.60	6.24	449.5	990	0.201
Terbium	Tb	65	158.924	8.229	1357	2530	0.180
Thallium	Tl	81	204.37	11.85	304	1457	0.130
Thorium	Th	90	(232)	11.72	1755	(3850)	0.117
Thulium	Tm	69	168.934	9.32	1545	1720	0.159
Tin	Sn	50	118.69	7.2984	231.868	2270	0.226
Titanium	Ti	22	47.90	4.54	1670	3260	0.523
Tungsten	W	74	183.85	19.3	3380	5930	0.134
Unnamed	Uut	113	(284)	—	—	—	—
Unnamed	Uup	115	(288)	—	—	—	—
Unnamed	Uus	117	—	—	—	—	—
Unnamed	Uuo	118	(294)	—	—	—	—
Uranium	U	92	(238)	18.95	1132	3818	0.117
Vanadium	V	23	50.942	6.11	1902	3400	0.490
Xenon	Xe	54	131.30	5.495×10^{-3}	−111.79	−108	0.159
Ytterbium	Yb	70	173.04	6.965	824	1530	0.155
Yttrium	Y	39	88.905	4.469	1526	3030	0.297
Zinc	Zn	30	65.37	7.133	419.58	906	0.389
Zirconium	Zr	40	91.22	6.506	1852	3580	0.276

The values in parentheses in the column of molar masses are the mass numbers of the longest-lived isotopes of those elements that are radioactive. Melting points and boiling points in parentheses are uncertain.

The data for gases are valid only when these are in their usual molecular state, such as H_2, He, O_2, Ne, etc. The specific heats of the gases are the values at constant pressure.

Source: Adapted from J. Emsley, *The Elements,* 3rd ed., 1998, Clarendon Press, Oxford. See also www.webelements.com for the latest values and newest elements.

*The names and symbols for elements 114 (Flerovium, Fl) and 116 (Livermorium, Lv) have been suggested but are not official.

PERIODIC TABLE OF THE ELEMENTS

	Metals
	Metalloids
	Nonmetals

THE HORIZONTAL PERIODS

Alkali metals

	IA	IIA					Transition metals								IIIA	IVA	VA	VIA	VIIA	Noble gases 0
1	1 H																			2 He
2	3 Li	4 Be													5 B	6 C	7 N	8 O	9 F	10 Ne
3	11 Na	12 Mg	IIIB	IVB	VB	VIB	VIIB		VIIIB		IB	IIB			13 Al	14 Si	15 P	16 S	17 Cl	18 Ar
4	19 K	20 Ca	21 Sc	22 Ti	23 V	24 Cr	25 Mn	26 Fe	27 Co	28 Ni	29 Cu	30 Zn	31 Ga	32 Ge	33 As	34 Se	35 Br	36 Kr
5	37 Rb	38 Sr	39 Y	40 Zr	41 Nb	42 Mo	43 Tc	44 Ru	45 Rh	46 Pd	47 Ag	48 Cd	49 In	50 Sn	51 Sb	52 Te	53 I	54 Xe
6	55 Cs	56 Ba	57-71 *	72 Hf	73 Ta	74 W	75 Re	76 Os	77 Ir	78 Pt	79 Au	80 Hg	81 Tl	82 Pb	83 Bi	84 Po	85 At	86 Rn
7	87 Fr	88 Ra	89-103 †	104 Rf	105 Db	106 Sg	107 Bh	108 Hs	109 Mt	110 Ds	111 Rg	112 Cn	113	114 Fl	115	116 Lv	117	118

Inner transition metals

Lanthanide series *

57 La	58 Ce	59 Pr	60 Nd	61 Pm	62 Sm	63 Eu	64 Gd	65 Tb	66 Dy	67 Ho	68 Er	69 Tm	70 Yb	71 Lu

Actinide series †

89 Ac	90 Th	91 Pa	92 U	93 Np	94 Pu	95 Am	96 Cm	97 Bk	98 Cf	99 Es	100 Fm	101 Md	102 No	103 Lr

Evidence for the discovery of elements 113 through 118 has been reported. See www.webelements.com for the latest information and newest elements. The names and symbols for elements 114 and 116 have been suggested but are not official.

To Checkpoints and Odd-Numbered Questions and Problems

Chapter 21

CP **1.** C and D attract; B and D attract **2.** (a) leftward;
(b) leftward; (c) leftward **3.** (a) a, c, b; (b) less than **4.** $-15e$
(net charge of $-30e$ is equally shared)
Q **1.** 3, 1, 2, 4 (zero) **3.** a and b **5.** $2kq^2/r^2$, up the page
7. b and c tie, then a (zero) **9.** (a) same; (b) less than; (c) cancel;
(d) add; (e) adding components; (f) positive direction of y;
(g) negative direction of y; (h) positive direction of x; (i) negative
direction of x **11.** (a) $+4e$; (b) $-2e$ upward; (c) $-3e$ upward;
(d) $-12e$ upward
P **1.** 0.500 **3.** 1.39 m **5.** 2.81 N **7.** -4.00 **9.** (a) $-1.00\ \mu$C;
(b) 3.00 μC **11.** (a) 0.17 N; (b) -0.046 N **13.** (a) -14 cm; (b) 0
15. (a) 35 N; (b) $-10°$; (c) -8.4 cm; (d) $+2.7$ cm **17.** (a) 1.60 N;
(b) 2.77 N **19.** (a) 3.00 cm; (b) 0; (c) -0.444 **21.** 3.8×10^{-8} C
23. (a) 0; (b) 12 cm; (c) 0; (d) 4.9×10^{-26} N **25.** 6.3×10^{11}
27. (a) 3.2×10^{-19} C; (b) 2 **29.** (a) -6.05 cm; (b) 6.05 cm
31. 122 mA **33.** 1.3×10^7 C **35.** (a) 0; (b) 1.9×10^{-9} N
37. (a) ^{9}B; (b) ^{13}N; (c) ^{12}C **39.** 1.31×10^{-22} N **41.** (a) 5.7×10^{13} C;
(b) cancels out; (c) 6.0×10^5 kg **43.** (b) 3.1 cm **45.** 0.19 MC
47. $-45\ \mu$C **49.** 3.8 N **51.** (a) 2.00×10^{10} electrons; (b) 1.33×10^{10}
electrons **53.** (a) 8.99×10^9 N; (b) 8.99 kN **55.** (a) 0.5; (b) 0.15;
(c) 0.85 **57.** 1.7×10^8 N **59.** -1.32×10^{13} C **61.** (a) $(0.829\ \text{N})\hat{\text{i}}$;
(b) $(-0.621\ \text{N})\hat{\text{j}}$ **63.** 2.2×10^{-6} kg **65.** 4.68×10^{-19} N
67. (a) 2.72L; (b) 0 **69.** (a) 5.1×10^2 N; (b) 7.7×10^{28} m/s^2
71. (a) 0; (b) 3.43×10^9 m/s^2 **73.** (a) 2.19×10^6 m/s;
(b) 1.09×10^6 m/s; (c) decrease **75.** 4.16×10^{42}

Chapter 22

CP **1.** (a) rightward; (b) leftward; (c) leftward; (d) rightward
(p and e have same charge magnitude, and p is farther)
2. (a) toward positive y; (b) toward positive x; (c) toward negative y
3. (a) leftward; (b) leftward; (c) decrease **4.** (a) all tie; (b) 1 and 3
tie, then 2 and 4 tie
Q **1.** a, b, c **3.** (a) yes; (b) toward; (c) no (the field vectors are not
along the same line); (d) cancel; (e) add; (f) adding components;
(g) toward negative y **5.** (a) to their left; (b) no **7.** (a) 4, 3, 1, 2;
(b) 3, then 1 and 4 tie, then 2 **9.** a, b, c **11.** e, b, then a and c tie,
then d (zero) **13.** a, b, c
P **3.** (a) 3.07×10^{21} N/C; (b) outward **5.** 56 pC **7.** $(1.02 \times 10^5\ \text{N/C})\hat{\text{j}}$ **9.** (a) 1.38×10^{-10} N/C; (b) $180°$ **11.** -30 cm
13. (a) 3.60×10^{-6} N/C; (b) 2.55×10^{-6} N/C; (c) 3.60×10^{-4} N/C;
(d) 7.09×10^{-7} N/C; (e) As the proton nears the disk, the forces on
it from electrons e_s more nearly cancel. **15.** (a) 160 N/C; (b) $45°$
17. (a) $-90°$; (b) $+2.0\ \mu$C; (c) $-1.6\ \mu$C **19.** (a) $qd/4\pi\varepsilon_0 r^3$; (b) $-90°$
23. 0.506 **25.** (a) 1.62×10^6 N/C; (b) $-45°$ **27.** (a) 23.8 N/C;
(b) $-90°$ **29.** 1.57 **31.** (a) -5.19×10^{-14} C/m; (b) 1.57×10^{-3} N/C;
(c) $-180°$; (d) 1.52×10^{-8} N/C; (e) 1.52×10^{-8} N/C **35.** 0.346 m
37. 28% **39.** $-5e$ **41.** (a) 1.5×10^3 N/C; (b) 2.4×10^{-16} N; (c) up;
(d) 1.6×10^{-26} N; (e) 1.5×10^{10} **43.** 3.51×10^{15} m/s^2
45. 6.6×10^{-15} N **47.** (a) 1.92×10^{12} m/s^2; (b) 1.96×10^5 m/s
49. (a) 0.245 N; (b) $-11.3°$; (c) 108 m; (d) -21.6 m **51.** 2.6×10^{-10} N;
(b) 3.1×10^{-8} N; (c) moves to stigma **53.** 27 μm **55.** (a) 2.7×10^6
m/s; (b) 1.0 kN/C **57.** (a) 9.30×10^{-15} C·m; (b) 2.05×10^{-11} J

59. 1.22×10^{-23} J **61.** $(1/2\pi)(pE/I)^{0.5}$ **63.** (a) 8.87×10^{-15} N;
(b) 120 **65.** $217°$ **67.** 61 N/C **69.** (a) 47 N/C; (b) 27 N/C
71. 38 N/C **73.** (a) -1.0 cm; (b) 0; (c) 10 pC **75.** $+1.00\ \mu$C
77. (a) 6.0 mm; (b) $180°$ **79.** 9:30 **81.** (a) -0.029 C; (b) repulsive
forces would explode the sphere **83.** (a) -1.49×10^{-26} J;
(b) $(-1.98 \times 10^{-26}\ \text{N·m})\hat{\text{k}}$; (c) 3.47×10^{-26} J **85.** (a) top row: 4, 8,
12; middle row: 5, 10, 14; bottom row: 7, 11, 16; (b) 1.63×10^{-19} C
87. (a) $(-1.80\ \text{N/C})\hat{\text{i}}$; (b) $(43.2\ \text{N/C})\hat{\text{i}}$; (c) $(-6.29\ \text{N/C})\hat{\text{i}}$

Chapter 23

CP **1.** (a) $+EA$; (b) $-EA$; (c) 0; (d) 0 **2.** (a) 2; (b) 3; (c) 1
3. (a) equal; (b) equal; (c) equal **4.** 3 and 4 tie, then 2, 1
Q **1.** (a) 8 N·m^2/C; (b) 0 **3.** all tie **5.** all tie **7.** a, c, then b and d
tie (zero) **9.** (a) 2, 1, 3; (b) all tie $(+4q)$ **11.** (a) impossible;
(b) $-3q_0$; (c) impossible
P **1.** -0.015 N·m^2/C **3.** (a) 0; (b) -3.92 N·m^2/C; (c) 0; (d) 0
5. 3.01 nN·m^2/C **7.** 2.0×10^5 N·m^2/C **9.** (a) 8.23 N·m^2/C;
(b) 72.9 pC; (c) 8.23 N·m^2/C; (d) 72.9 pC **11.** -1.70 nC
13. 3.54 μC **15.** (a) 0; (b) 0.0417 **17.** (a) 37 μC; (b) 4.1×10^6 N·m^2/C
19. (a) 4.5×10^{-7} C/m^2; (b) 5.1×10^4 N/C **21.** (a) -3.0×10^{-6} C;
(b) $+1.3 \times 10^{-5}$ C **23.** (a) 0.32 μC; (b) 0.14 μC **25.** 5.0 μC/m
27. 3.8×10^{-8} C/m^2 **29.** (a) 0.214 N/C; (b) inward; (c) 0.855 N/C;
(d) outward; (e) -3.40×10^{-12} C; (f) -3.40×10^{-12} C **31.** (a) 2.3×10^6 N/C; (b) outward; (c) 4.5×10^5 N/C; (d) inward **33.** (a) 0;
(b) 0; (c) $(-7.91 \times 10^{-11}\ \text{N/C})\hat{\text{i}}$ **35.** -1.5 **37.** (a) 5.3×10^7 N/C;
(b) 60 N/C **39.** 5.0 nC/m^2 **41.** 0.44 mm **43.** (a) 0; (b) 1.31 μN/C;
(c) 3.08 μN/C; (d) 3.08 μN/C **45.** (a) 2.50×10^4 N/C; (b) 1.35×10^4 N/C **47.** -7.5 nC **49.** (a) 0; (b) 56.2 mN/C; (c) 112 mN/C;
(d) 49.9 mN/C; (e) 0; (f) 0; (g) -5.00 fC; (h) 0 **51.** 1.79×10^{-11} C/m^2
53. (a) 7.78 fC; (b) 0; (c) 5.58 mN/C; (d) 22.3 mN/C **55.** $6K\varepsilon_0 r^3$
57. (a) 0; (b) 2.88×10^4 N/C; (c) 200 N/C **59.** (a) 5.4 N/C;
(b) 6.8 N/C **61.** (a) 0; (b) $q_a/4\pi\varepsilon_0 r^2$; (c) $(q_a + q_b)/4\pi\varepsilon_0 r^2$
63. -1.04 nC **65.** (a) 0.125; (b) 0.500 **67.** (a) $+2.0$ nC;
(b) -1.2 nC; (c) $+1.2$ nC; (d) $+0.80$ nC **69.** $(5.65 \times 10^4\ \text{N/C})\hat{\text{j}}$
71. (a) -2.53×10^{-2} N·m^2/C; (b) $+2.53 \times 10^{-2}$ N·m^2/C
75. 3.6 nC **77.** (a) $+4.0\ \mu$C; (b) $-4.0\ \mu$C **79.** (a) 693 kg/s;
(b) 693 kg/s; (c) 347 kg/s; (d) 347 kg/s; (e) 575 kg/s **81.** (a) 0.25R;
(b) 2.0R

Chapter 24

CP **1.** (a) negative; (b) increase; (c) positive; (d) higher
2. (a) rightward; (b) 1, 2, 3, 5: positive; 4, negative; (c) 3, then 1, 2,
and 5 tie, then 4 **3.** all tie **4.** a, c (zero), b **5.** (a) 2, then 1 and 3
tie; (b) 3; (c) accelerate leftward
Q **1.** $-4q/4\pi\varepsilon_0 d$ **3.** (a) 1 and 2; (b) none; (c) no; (d) 1 and 2, yes;
3 and 4, no **5.** (a) higher; (b) positive; (c) negative; (d) all tie
7. (a) 0; (b) 0; (c) 0; (d) all three quantities still 0 **9.** (a) 3 and 4 tie,
then 1 and 2 tie; (b) 1 and 2, increase; 3 and 4, decrease **11.** a, b, c
P **1.** (a) 3.0×10^5 C; (b) 3.6×10^6 J **3.** 2.8×10^5 **5.** 8.8 mm
7. -32.0 V **9.** (a) 1.87×10^{-21} J; (b) -11.7 mV **11.** (a) -0.268 mV;
(b) -0.681 mV **13.** (a) 3.3 nC; (b) 12 nC/m^2 **15.** (a) 0.54 mm;
(b) 790 V **17.** 0.562 mV **19.** (a) 6.0 cm; (b) -12.0 cm **21.** 16.3 μV
23. (a) 24.3 mV; (b) 0 **25.** (a) -2.30 V; (b) -1.78 V **27.** 13 kV

29. 32.4 mV   **31.** 47.1 μV   **33.** 18.6 mV   **35.** $(-12\ \text{V/m})\hat{i}+(12\ \text{V/m})\hat{j}$
37. 150 N/C   **39.** $(-4.0\times10^{-16}\ \text{N})\hat{i}+(1.6\times10^{-16}\ \text{N})\hat{j}$
41. (a) 0.90 J; (b) 4.5 J   **43.** -0.192 pJ   **45.** 2.5 km/s   **47.** 22 km/s
49. 0.32 km/s   **51.** (a) $+6.0\times10^4$ V; (b) -7.8×10^5 V; (c) 2.5 J;
(d) increase; (e) same; (f) same   **53.** (a) 0.225 J; (b) A 45.0 m/s^2,
B 22.5 m/s^2; (c) A 7.75 m/s, B 3.87 m/s   **55.** 1.6×10^{-9} m
57. (a) 3.0 J; (b) -8.5 m   **59.** (a) proton; (b) 65.3 km/s   **61.** (a) 12;
(b) 2   **63.** (a) -1.8×10^2 V; (b) 2.9 kV; (c) -8.9 kV
65. 2.5×10^{-8} C   **67.** (a) 12 kN/C; (b) 1.8 kV; (c) 5.8 cm
69. (a) 64 N/C; (b) 2.9 V; (c) 0   **71.** $p/2\pi\varepsilon_0 r^3$ **73.** (a) 3.6×10^5 V;
(b) no   **75.** 6.4×10^8 V   **77.** 2.90 kV   **79.** 7.0×10^5 m/s
81. (a) 1.8 cm; (b) 8.4×10^5 m/s; (c) 2.1×10^{-17} N; (d) positive;
(e) 1.6×10^{-17} N; (f) negative   **83.** (a) $+7.19\times10^{-10}$ V;
(b) $+2.30\times10^{-28}$ J; (c) $+2.43\times10^{-29}$ J   **85.** 2.30×10^{-28} J
87. 2.1 days   **89.** 2.30×10^{-22} J   **91.** 1.48×10^7 m/s   **93.** -1.92 MV
95. (a) $Q/4\pi\varepsilon_0 r$; (b) $(\rho/3\varepsilon_0)\ (1.5r_2^2-0.50r^2-r_1^3\,r^{-1})$,
$\rho=Q/[(4\pi/3)(r_2^3-r_1^3)]$; (c) $(\rho/2\varepsilon_0)(r_2^2-r_1^2)$, with ρ as in (b);
(d) yes   **97.** (a) 38 s; (b) 2.7×10^2 days   **101.** (a) 0.484 MeV;
(b) 0   **103.** -1.7

Chapter 25

CP   **1.** (a) same; (b) same   **2.** (a) decreases; (b) increases;
(c) decreases   **3.** (a) V, $q/2$; (b) $V/2$; q
Q   **1.** a, 2; b, 1; c, 3   **3.** (a) no; (b) yes; (c) all tie   **5.** (a) same;
(b) same; (c) more; (d) more   **7.** a, series; b, parallel; c, parallel
9. (a) increase; (b) same; (c) increase; (d) increase; (e) increase;
(f) increase   **11.** parallel, C_1 alone, C_2 alone, series
P   **1.** (a) 3.5 pF; (b) 3.5 pF; (c) 57 V   **3.** (a) 144 pF; (b)17.3 nC
5. 0.280 pF   **7.** 6.79×10^{-4} F/m^2 **9.** 315 mC   **11.** 3.16 μF
13. 43 pF   **15.** (a) 3.00 μF; (b) 60.0 μC; (c) 10.0 V; (d) 30.0 μC;
(e) 10.0 V; (f) 20.0 μC; (g) 5.00 V; (h) 20.0 μC   **17.** (a) 789 μC;
(b) 78.9 V   **19.** (a) 4.0 μF; (b) 2.0 μF   **21.** (a) 50 V; (b) 5.0×10^{-5} C;
(c) 1.5×10^{-4} C   **23.** (a) 4.5×10^{14}; (b) 1.5×10^{14}; (c) 3.0×10^{14};
(d) 4.5×10^{14}; (e) up; (f) up   **25.** 3.6 pC   **27.** (a) 9.00 μC;
(b) 16.0 μC; (c) 9.00 μC; (d) 16.0 μC; (e) 8.40 μC; (f) 16.8 μC;
(g) 10.8 μC; (h) 14.4 μC   **29.** 72 F   **31.** 0.27 J   **33.** 0.11 J/m^3
35. (a) 9.16×10^{-18} J/m^3; (b) 9.16×10^{-6} J/m^3; (c) 9.16×10^6 J/m^3;
(d) 9.16×10^{18} J/m^3; (e) ∞ **37.** (a) 16.0 V; (b) 45.1 pJ; (c) 120 pJ;
(d) 75.2 pJ   **39.** (a) 190 V; (b) 95 mJ   **41.** 81 pF/m   **43.** Pyrex
45. 66 μJ   **47.** 0.63 m^2 **49.** 17.3 pF   **51.** (a) 10 kV/m; (b) 5.0 nC;
(c) 4.1 nC   **53.** (a) 89 pF; (b) 0.12 nF; (c) 11 nC; (d) 11 nC;
(e) 10 kV/m; (f) 2.1 kV/m; (g) 88 V; (h) $-0.17\ \mu$J   **55.** (a) 0.107 nF;
(b) 7.79 nC; (c) 7.45 nC   **57.** 45 μC   **59.** 16 μC   **61.** (a) 7.20 μC;
(b) 18.0 μC; (c) Battery supplies charges only to plates to which it
is connected; charges on other plates are due to electron transfers
between plates, in accord with new distribution of voltages across
the capacitors. So the battery does not directly supply charge on
capacitor 4.   **63.** (a) 10 μC; (b) 20 μC   **65.** 1.06 nC
67. (a) 2.40 μF; (b) 0.480 mC; (c) 80 V; (d) 0.480 mC; (e) 120 V
69. 4.9%   **71.** (a) 0.708 pF; (b) 0.600 ; (c) 1.02×10^{-9} J;
(d) sucked in   **73.** 5.3 V   **75.** 40 μF   **77.** (a) 200 kV/m;
(b) 200 kV/m; (c) 1.77 μC/m^2; (d) 4.60 μC/m^2; (e) $-2.83\ \mu$C/m^2
79. (a) $q^2/2\varepsilon_0 A$

Chapter 26

CP   **1.** 8 A, rightward   **2.** (a)–(c) rightward   **3.** a and c tie, then b
4. device 2   **5.** (a) and (b) tie, then (d), then (c)
Q   **1.** tie of A, B, and C, then tie of $A+B$ and $B+C$, then
$A+B+C$ **3.** (a) top-bottom, front-back, left-right; (b) top-
bottom, front-back, left-right; (c) top-bottom, front-back, left-right;
(d) top-bottom, front-back, left-right   **5.** a, b, and c all tie, then d
7. (a) B, A, C; (b) B, A, C **9.** (a) C, B, A; (b) all tie; (c) A, B, C;

(d) all tie   **11.** (a) a and c tie, then b (zero); (b) a, b, c; (c) a and b
tie, then c
P   **1.** (a) 1.2 kC; (b) 7.5×10^{21} **3.** 6.7 μC/m^2 **5.** (a) 6.4 A/m^2;
(b) north; (c) cross-sectional area   **7.** 0.38 mm   **9.** 18.1 μA
11. (a) 1.33 A; (b) 0.666 A; (c) J_a **13.** 13 min   **15.** 2.4 Ω
17. $2.0\times10^6\ (\Omega\cdot\text{m})^{-1}$ **19.** $2.0\times10^{-8}\ \Omega\cdot\text{m}$ **21.** $(1.8\times10^3)°$C
23. $8.2\times10^{-8}\ \Omega\cdot\text{m}$ **25.** 54 Ω **27.** 3.0   **29.** 3.35×10^{-7} C
31. (a) 6.00 mA; (b) 1.59×10^{-8} V; (c) 21.2 nΩ **33.** (a) 38.3 mA;
(b) 109 A/m^2; (c) 1.28 cm/s; (d) 227 V/m   **35.** 981 kΩ **39.** 150 s
41. (a) 1.0 kW; (b) US$0.25   **43.** 0.135 W   **45.** (a) 10.9 A;
(b) 10.6 Ω; (c) 4.50 MJ   **47.** (a) 5.85 m; (b) 10.4 m   **49.** (a) US$4.46;
(b) 144 Ω; (c) 0.833 A   **51.** (a) 5.1 V; (b) 10 V; (c) 10 W; (d) 20 W
53. (a) 28.8 Ω; (b) 2.60×10^{19} s^{-1} **55.** 660 W   **57.** 28.8 kC
59. (a) silver; (b) 51.6 nΩ **61.** (a) 2.3×10^{12}; (b) 5.0×10^3; (c) 10 MV
63. 2.4 kW   **65.** (a) 1.37; (b) 0.730   **67.** (a) -8.6%; (b) smaller
69. 146 kJ   **71.** (a) 250°C; (b) yes   **73.** 3.0×10^6 J/kg   **75.** 560 W
77. 0.27 m/s   **79.** (a) 10 A/cm^2; (b) eastward   **81.** (a) 9.4×10^{13} s^{-1};
(b) 2.40×10^2 W   **83.** 113 min   **85.** (a) 225 μC; (b) 60.0 μA;
(c) 0.450 mW

Chapter 27

CP   **1.** (a) rightward; (b) all tie; (c) b, then a and c tie;
(d) b, then a and c tie   **2.** (a) all tie; (b) R_1, R_2, R_3 **3.** (a) less;
(b) greater; (c) equal   **4.** (a) $V/2$, i; (b) V, $i/2$ **5.** (a) 1, 2, 4, 3;
(b) 4, tie of 1 and 2, then 3
Q   **1.** (a) equal; (b) more   **3.** parallel, R_2, R_1, series   **5.** (a) series;
(b) parallel; (c) parallel   **7.** (a) less; (b) less; (c) more
9. (a) parallel; (b) series   **11.** (a) same; (b) same; (c) less; (d) more
13. (a) all tie; (b) 1, 3, 2
P   **1.** (a) 0.50 A; (b) 1.0 W; (c) 2.0 W; (d) 6.0 W; (e) 3.0 W; (f) supplied; (g) absorbed   **3.** (a) 14 V; (b) 1.0×10^2 W; (c) 6.0×10^2 W;
(d) 10 V; (e) 1.0×10^2 W   **5.** 11 kJ   **7.** (a) 80 J; (b) 67 J; (c) 13 J
9. (a) 12.0 eV; (b) 6.53 W   **11.** (a) 50 V; (b) 48 V; (c) negative
13. (a) 6.9 km; (b) 20 Ω **15.** 8.0 Ω **17.** (a) 0.004 Ω; (b) 1
19. (a) 4.00 Ω; (b) parallel   **21.** 5.56 A   **23.** (a) 50 mA; (b) 60 mA;
(c) 9.0 V   **25.** $3d$ **27.** 3.6×10^3 A   **29.** (a) 0.333 A; (b) right;
(c) 720 J   **31.** (a) -11 V; (b) -9.0 V   **33.** 48.3 V   **35.** (a) 5.25 V;
(b) 1.50 V; (c) 5.25 V; (d) 6.75 V   **37.** 1.43 Ω **39.** (a) 0.150 Ω;
(b) 240 W   **41.** (a) 0.709 W; (b) 0.050 W; (c) 0.346 W; (d) 1.26 W;
(e) -0.158 W   **43.** 9   **45.** (a) 0.67 A; (b) down; (c) 0.33 A; (d) up;
(e) 0.33 A; (f) up; (g) 3.3 V   **47.** (a) 1.11 A; (b) 0.893 A; (c) 126 m
49. (a) 0.45 A   **51.** (a) 55.2 mA; (b) 4.86 V; (c) 88.0 Ω; (d) decrease
53. -3.0% **57.** 0.208 ms   **59.** 4.61   **61.** (a) 2.41 μs; (b) 161 pF
63. (a) 1.1 mA; (b) 0.55 mA; (c) 0.55 mA; (d) 0.82 mA; (e) 0.82 mA;
(f) 0; (g) 4.0×10^2 V; (h) 6.0×10^2 V   **65.** 411 μA   **67.** 0.72 MΩ
69. (a) 0.955 μC/s; (b) 1.08 μW; (c) 2.74 μW; (d) 3.82 μW
71. (a) 3.00 A; (b) 3.75 A; (c) 3.94 A   **73.** (a) 1.32×10^7 A/m^2;
(b) 8.90 V; (c) copper; (d) 1.32×10^7 A/m^2; (e) 51.1 V; (f) iron
75. (a) 3.0 kV; (b) 10 s; (c) 11 GΩ **77.** (a) 85.0 Ω; (b) 915 Ω
81. 4.0 V   **83.** (a) 24.8 Ω; (b) 14.9 kΩ **85.** the cable   **87.** $-13\ \mu$C
89. 20 Ω **91.** (a) 3.00 A; (b) down; (c) 1.60 A; (d) down; (e) supply;
(f) 55.2 W; (g) supply; (h) 6.40 W   **93.** (a) 1.0 V; (b) 50 mΩ
95. 3   **99.** (a) 1.5 mA; (b) 0; (c) 1.0 mA   **101.** 7.50 V
103. (a) 60.0 mA; (b) down; (c) 180 mA; (d) left; (e) 240 mA;
(f) up   **105.** (a) 4.0 A; (b) up; (c) 0.50 A; (d) down; (e) 64 W;
(f) 16 W; (g) supplied; (h) absorbed

Chapter 28

CP   **1.** a, $+z$; b, $-x$; c, $\vec{F}_B=0$ **2.** (a) 2, then tie of 1 and 3 (zero);
(b) 4   **3.** (a) electron; (b) clockwise   **4.** $-y$ **5.** (a) all tie; (b) 1 and
4 tie, then 2 and 3 tie
Q   **1.** (a) no, because $\vec{v}$ and $\vec{F}_B$ must be perpendicular; (b) yes;

(c) no, because $\vec{B}$ and $\vec{F}_B$ must be perpendicular
3. (a) $+z$ and $-z$ tie, then $+y$ and $-y$ tie, then $+x$ and $-x$ tie (zero);
(b) $+y$ **5.** (a) $\vec{F}_E$; (b) $\vec{F}_B$ **7.** (a) $\vec{B}_1$; (b) $\vec{B}_1$ into page, $\vec{B}_2$ out of page;
(c) less **9.** (a) positive; (b) $2 \rightarrow 1$ and $2 \rightarrow 4$ tie, then $2 \rightarrow 3$ (which
is zero) **11.** (a) negative; (b) equal; (c) equal; (d) half-circle
P **1.** (a) 400 km/s; (b) 835 eV **3.** (a) $(6.2 \times 10^{-14} \text{N})\hat{k}$;
(b) $(-6.2 \times 10^{-14} \text{N})\hat{k}$ **5.** -2.0 T **7.** $(-11.4 \text{ V/m})\hat{i} - (6.00 \text{ V/m})\hat{j} +$
$(4.80 \text{ V/m})\hat{k}$ **9.** $-(0.267 \text{ mT})\hat{k}$ **11.** 0.68 MV/m **13.** 7.4 μV
15. (a) $(-600 \text{ mV/m})\hat{k}$; (b) 1.20 V **17.** (a) 2.60×10^6 m/s;
(b) 0.109 μs; (c) 0.140 MeV; (d) 70.0 kV **19.** 1.2×10^{-9} kg/C
21. (a) 2.05×10^7 m/s; (b) 467 μT; (c) 13.1 MHz; (d) 76.3 ns
23. 21.1 μT **25.** (a) 0.978 MHz; (b) 96.4 cm **27.** (a) 495 mT;
(b) 22.7 mA; (c) 8.17 MJ **29.** 65.3 km/s **31.** 5.07 ns
33. (a) 0.358 ns; (b) 0.166 mm; (c) 1.51 mm **35.** (a) 200 eV;
(b) 20.0 keV; (c) 0.499% **37.** 2.4×10^2 m **39.** (a) 28.2 N;
(b) horizontally west **41.** (a) 467 mA; (b) right **43.** (a) 0; (b) 0.138 N;
(c) 0.138 N; (d) 0 **45.** $(-2.50 \text{ mN})\hat{j} + (0.750 \text{ mN})\hat{k}$ **47.** (a) 0.10 T;
(b) 31° **49.** $(-4.3 \times 10^{-3} \text{ N·m})\hat{j}$ **51.** 2.45 A **55.** (a) 2.86 A·m²;
(b) 1.10 A·m² **57.** (a) 12.7 A; (b) 0.0805 N·m **59.** (a) 0.30 A·m²;
(b) 0.024 N·m **61.** (a) -72.0 μJ; (b) $(96.0\hat{i} + 48.0\hat{k})$ μN·m
63. (a) $-(9.7 \times 10^{-4} \text{N·m})\hat{i} - (7.2 \times 10^{-4} \text{ N·m})\hat{j} + (8.0 \times 10^{-4} \text{N·m})\hat{k}$;
(b) -6.0×10^{-4} J **65.** (a) 90°; (b) 1; (c) 1.28×10^{-7} N·m
67. (a) 20 min; (b) 5.9×10^{-2} N·m **69.** 8.2 mm **71.** 127 u
73. (a) 6.3×10^{14} m/s²; (b) 3.0 mm **75.** (a) 1.4; (b) 1.0
77. $(-500 \text{ V/m})\hat{j}$ **79.** (a) 0.50; (b) 0.50; (c) 14 cm; (d) 14 cm
81. $(0.80\hat{j} -1.1\hat{k})$ mN **83.** -40 mC **85.** (a) $(12.8\hat{i} + 6.41\hat{j}) \times$
10^{-22} N; (b) 90°; (c) 173° **87.** (a) up the conducting path; (b) rim;
(c) 47.1 V; (d) 47.1 V; (e) 2.36 kW **89.** $(mV/2ed^2)^{0.5}$ **91.** $n = JB/eE$

Chapter 29

CP **1.** b,c,a **2.** d, tie of a and c, then b **3.** d,a, tie of b and c (zero)
Q **1.** c,a,b **3.** c,d, then a and b tie (zero) **5.** a,c,b
7. c and d tie, then b,a **9.** b,a,d,c (zero) **11.** (a) 1, 3, 2; (b) less
P **1.** (a) 3.3 μT; (b) yes **3.** (a) 16 A; (b) east **5.** (a) 1.0 mT;
(b) out; (c) 0.80 mT; (d) out **7.** (a) 0.102 μT; (b) out
9. (a) opposite; (b) 30 A **11.** (a) 4.3 A; (b) out **13.** 50.3 nT
15. (a) 1.7 μT; (b) into; (c) 6.7 μT; (d) into **17.** 132 nT
19. 5.0 μT **21.** 256 nT **23.** $(-7.75 \times 10^{-23} \text{ N})\hat{i}$ **25.** 2.00 rad
27. 61.3 mA **29.** $(80 \mu\text{T})\hat{j}$ **31.** (a) 20 μT; (b) into **33.** $(22.3 \text{ pT})\hat{j}$
35. 88.4 pN/m **37.** $(-125 \mu\text{N/m})\hat{i} + (41.7 \mu\text{N/m})\hat{j}$ **39.** 800 nN/m
41. $(3.20 \text{ mN})\hat{j}$ **43.** (a) 0; (b) 0.850 mT; (c) 1.70 mT; (d) 0.850 mT
45. (a) -2.5 μT·m; (b) 0 **47.** (a) 0; (b) 0.10 μT; (c) 0.40 μT
49. (a) 533 μT; (b) 400 μT **51.** 0.30 mT **53.** 0.272 A
55. (a) 4.77 cm; (b) 35.5 μT **57.** (a) 2.4 A·m²; (b) 46 cm
59. 0.47 A·m² **61.** (a) 79 μT; (b) 1.1×10^{-6} N·m **63.** (a) $(0.060 \text{ A·m}^2)\hat{j}$;
(b) $(96 \text{ pT})\hat{j}$ **65.** 1.28 mm **69.** (a) 15 A; (b) $-z$ **71.** 7.7 mT
73. (a) 15.3 μT **75.** (a) $(0.24\hat{i})$ nT; (b) 0; (c) $(-43\hat{k})$ pT; (d) $(0.14\hat{k})$
nT **79.** (a) 4.8 mT; (b) 0.93 mT; (c) 0 **83.** $(-0.20 \text{ mT})\hat{k}$
87. (a) $\mu_0 ir/2\pi c^2$; (b) $\mu_0 i/2\pi r$; (c) $\mu_0 i(a^2 - r^2)/2\pi(a^2 - b^2)r$; (d) 0

Chapter 30

CP **1.** b, then d and e tie, and then a and c tie (zero) **2.** a and b
tie, then c (zero) **3.** c and d tie, then a and b tie **4.** b, out; c, out; d,
into; e, into **5.** d and e **6.** (a) 2, 3, 1 (zero); (b) 2, 3, 1
7. a and b tie, then c
Q **1.** out **3.** (a) all tie (zero); (b) 2, then 1 and 3 tie (zero) **5.** d
and c tie, then b,a **7.** (a) more; (b) same; (c) same; (d) same (zero)
9. (a) all tie (zero); (b) 1 and 2 tie, then 3; (c) all tie (zero) **11.** b
P **1.** 0 **3.** 30 mA **5.** 0 **7.** (a) 31 mV; (b) left **9.** 0.198 mV
11. (b) 0.796 m² **13.** 29.5 mC **15.** (a) 21.7 V; (b) counterclockwise **17.** (a) 1.26×10^{-4} T; (b) 0; (c) 1.26×10^{-4} T; (d) yes;
(e) 5.04×10^{-8} V **19.** 5.50 kV **21.** (a) 40 Hz; (b) 3.2 mV

23. (a) $\mu_0 iR^2\pi r^2/2x^3$; (b) $3\mu_0 i\pi R^2 r^2 v/2x^4$; (c) counterclockwise
25. (a) 13 μWb/m; (b) 17%; (c) 0 **27.** (a) 80 μV; (b) clockwise
29. (a) 48.1 mV; (b) 2.67 mA; (c) 0.129 mW **31.** 3.68 μW
33. (a) 240 μV; (b) 0.600 mA; (c) 0.144 μW; (d) 2.87×10^{-8} N;
(e) 0.144 μW **35.** (a) 0.60 V; (b) up; (c) 1.5 A; (d) clockwise;
(e) 0.90 W; (f) 0.18 N; (g) 0.90 W **37.** (a) 71.5 μV/m; (b) 143 μV/m
39. 0.15 V/m **41.** (a) 2.45 mWb; (b) 0.645 mH **43.** 1.81 μH/m
45. (a) decreasing; (b) 0.68 mH **47.** (b) $L_{eq} = \Sigma L_j$, sum from $j = 1$
to $j = N$ **49.** 59.3 mH **51.** 46 Ω **53.** (a) 8.45 ns; (b) 7.37 mA
55. 6.91 **57.** (a) 1.5 s **59.** (a) $i[1 - \exp(-Rt/L)]$; (b) $(L/R) \ln 2$
61. (a) 97.9 H; (b) 0.196 mJ **63.** 25.6 ms **65.** (a) 18.7 J; (b) 5.10 J;
(c) 13.6 J **67.** (a) 34.2 J/m³; (b) 49.4 mJ **69.** 1.5×10^8 V/m
71. (a) 1.0 J/m³; (b) 4.8×10^{-15} J/m³ **73.** (a) 1.67 mH; (b) 6.00 mWb
75. 13 μH **77.** (b) have the turns of the two solenoids wrapped in
opposite directions **79.** (a) 2.0 A; (b) 0; (c) 2.0 A; (d) 0; (e) 10 V;
(f) 2.0 A/s; (g) 2.0 A; (h) 1.0 A; (i) 3.0 A; (j) 10 V; (k) 0; (l) 0
81. (a) 10 μT; (b) out; (c) 3.3 μT; (d) out **83.** 0.520 ms
85. (a) $(4.4 \times 10^7 \text{ m/s}^2)\hat{i}$; (b) 0; (c) $(-4.4 \times 10^7 \text{ m/s}^2)\hat{i}$
87. (a) 0.40 V; (b) 20 A **89.** (a) 10 A; (b) 1.0×10^2 J **91.** (a) 0;
(b) 8.0×10^2 A/s; (c) 1.8 mA; (d) 4.4×10^2 A/s; (e) 4.0 mA; (f) 0
93. 1.15 W **95.** (a) 20 A/s; (b) 0.75 A **97.** 12 A/s **99.** 3×10^{36} J
101. 1.15 μWb

Chapter 31

CP **1.** (a) $T/2$; (b) T; (c) $T/2$; (d) $T/4$ **2.** (a) 5 V; (b) 150 μJ
3. (a) remains the same; (b) remains the same **4.** (a) C,B,A; (b) 1,
A; 2, B; 3, S; 4, C; (c) A **5.** (a) remains the same; (b) increases;
(c) remains the same; (d) decreases **6.** (a) 1, lags; 2, leads; 3, in
phase; (b) 3 ($\omega_d = \omega$ when $X_L = X_C$) **7.** (a) increase (circuit is
mainly capacitive; increase C to decrease X_C to be closer to resonance for maximum P_{avg}); (b) closer **8.** (a) greater; (b) step-up
Q **1.** b,a,c **3.** (a) $T/4$; (b) $T/4$; (c) $T/2$; (d) $T/2$ **5.** c,b,a **7.** a
inductor; b resistor; c capacitor **9.** (a) positive; (b) decreased (to
decrease X_L and get closer to resonance); (c) decreased (to increase
X_C and get closer to resonance) **11.** (a) rightward, increase (X_L
increases, closer to resonance); (b) rightward, increase (X_C decreases,
closer to resonance); (c) rightward, increase (ω_d/ω increases, closer to
resonance) **13.** (a) inductor; (b) decrease
P **1.** (a) 1.17 μJ; (b) 5.58 mA **3.** (a) 6.00 μs; (b) 167 kHz;
(c) 3.00 μs **5.** 45.2 mA **7.** (a) 1.25 kg; (b) 372 N/m;
(c) 1.75×10^{-4} m; (d) 3.02 mm/s **9.** 7.0×10^{-4} s **11.** (a) 6.0;
(b) 36 pF; (c) 0.22 mH **13.** (a) 0.180 mC; (b) 70.7 μs; (c) 66.7 W
15. (a) 3.0 nC; (b) 1.7 mA; (c) 4.5 nJ **17.** (a) 275 Hz; (b) 365 mA
21. (a) 356 μs; (b) 2.50 mH; (c) 3.20 mJ **23.** (a) 1.98 μJ;
(b) 5.56 μC; (c) 12.6 mA; (d) $-46.9°$; (e) $+46.9°$ **25.** 8.66 mΩ
29. (a) 95.5 mA; (b) 11.9 mA **31.** (a) 0.65 kHz; (b) 24 Ω
33. (a) 6.73 ms; (b) 11.2 ms; (c) inductor; (d) 138 mH **35.** 89 Ω
37. 7.61 A **39.** (a) 267 Ω; (b) $-41.5°$; (c) 135 mA **41.** (a) 206 Ω;
(b) 13.7°; (c) 175 mA **43.** (a) 218 Ω; (b) 23.4°; (c) 165 mA
45. (a) yes; (b) 1.0 kV **47.** (a) 224 rad/s; (b) 6.00 A; (c) 219 rad/s;
(d) 228 rad/s; (e) 0.040 **49.** (a) 796 Hz; (b) no change;
(c) decreased; (d) increased **53.** (a) 12.1 Ω; (b) 1.19 kW
55. 1.84 A **57.** (a) 117 μF; (b) 0; (c) 90.0 W; (d) 0°; (e) 1; (f) 0;
(g) $-90°$; (h) 0 **59.** (a) 2.59 A; (b) 38.8 V; (c) 159 V; (d) 224 V;
(e) 64.2 V; (f) 75.0 V; (g) 100 W; (h) 0; (i) 0 **61.** (a) 0.743; (b) lead;
(c) capacitive; (d) no; (e) yes; (f) no; (g) yes; (h) 33.4 W
63. (a) 2.4 V; (b) 3.2 mA; (c) 0.16 A **65.** (a) 1.9 V; (b) 5.9 W; (c) 19 V;
(d) 5.9×10^2 W; (e) 0.19 kV; (f) 59 kW **67.** (a) 6.73 ms; (b) 2.24 ms;
(c) capacitor; (d) 59.0 μF **69.** (a) -0.405 rad; (b) 2.76 A;
(c) capacitive **71.** (a) 64.0 Ω; (b) 50.9 Ω; (c) capacitive
73. (a) 2.41 μH; (b) 21.4 pJ; (c) 82.2 nC **75.** (a) 39.1 Ω; (b) 21.7 Ω;
(c) capacitive **79.** (a) $0.577Q$; (b) 0.152 **81.** (a) 45.0°; (b) 70.7 Ω

83. 1.84 kHz   **85.** (a) 0.689 μH; (b) 17.9 pJ; (c) 0.110 μC
87. (a) 165 Ω; (b) 313 mH; (c) 14.9 μF   **93.** (a) 36.0 V; (b) 29.9 V;
(c) 11.9 V; (d) -5.85 V

Chapter 32

CP   1. d, b, c, a (zero)   **2.** a, c, b, d (zero)   **3.** tie of $b, c,$ and d,
then a **4.** (a) 2; (b) 1   **5.** (a) away; (b) away; (c) less   **6.** (a) toward;
(b) toward; (c) less
Q   1. 1 a, 2 b, 3 c and d **3.** a, decreasing; b, decreasing
5. supplied   **7.** (a) a and b tie, then c, d; (b) none (because plate
lacks circular symmetry, $\vec{B}$ not tangent to any circular loop);
(c) none   **9.** (a) 1 up, 2 up, 3 down; (b) 1 down, 2 up, 3 zero
11. (a) 1, 3, 2; (b) 2
P   1. $+3$ Wb   **3.** (a) 47.4 μWb; (b) inward   **5.** 2.4×10^{13} V/m·s
7. (a) 1.18×10^{-19} T; (b) 1.06×10^{-19} T   **9.** (a) 5.01×10^{-22} T;
(b) 4.51×10^{-22} T   **11.** (a) 1.9 pT   **13.** 7.5×10^{5} V/s
17. (a) 0.324 V/m; (b) 2.87×10^{-16} A; (c) 2.87×10^{-18}
19. (a) 75.4 nT; (b) 67.9 nT   **21.** (a) 27.9 nT; (b) 15.1 nT
23. (a) 2.0 A; (b) 2.3×10^{11} V/m·s; (c) 0.50 A; (d) 0.63 μT·m
25. (a) 0.63 μT; (b) 2.3×10^{12} V/m·s   **27.** (a) 0.71 A; (b) 0; (c) 2.8 A
29. (a) 7.60 μA; (b) 859 kV·m/s; (c) 3.39 mm; (d) 5.16 pT   **31.** 55 μT
33. (a) 0; (b) 0; (c) 0; (d) $\pm 3.2 \times 10^{-25}$ J; (e) -3.2×10^{-34} J·s;
(f) 2.8×10^{-23} J/T; (g) -9.7×10^{-25} J; (h) $\pm 3.2 \times 10^{-25}$ J
35. (a) -9.3×10^{-24} J/T; (b) 1.9×10^{-23} J/T   **37.** (b) $+x$;
(c) clockwise; (d) $+x$ **39.** yes   **41.** 20.8 mJ/T   **43.** (b) K_i/B;
(c) $-z$; (d) 0.31 kA/m   **47.** (a) 1.8×10^{2} km; (b) 2.3×10^{-5}
49. (a) 3.0 μT; (b) 5.6×10^{-10} eV   **51.** 5.15×10^{-24} A·m^2
53. (a) 0.14 A; (b) 79 μC   **55.** (a) 6.3×10^{8} A; (b) yes; (c) no
57. 0.84 kJ/T   **59.** (a) $(1.2 \times 10^{-13}$ T$)$ exp$[-t/(0.012$ s$)]$;
(b) 5.9×10^{-15} T   **63.** (a) 27.5 mm; (b) 110 mm   **65.** 8.0 A
67. (a) -8.8×10^{15} V/m·s; (b) 5.9×10^{-7} T·m   **69.** (b) sign is
minus; (c) no, because there is compensating positive flux through
open end nearer to magnet   **71.** (b) $-x$; (c) counterclockwise;
(d) $-x$ **73.** (a) 7; (b) 7; (c) $3h/2\pi$; (d) $3eh/4\pi m$; (e) $3.5h/2\pi$;
(f) 8   **75.** (a) 9; (b) 3.71×10^{-23} J/T; (c) $+9.27 \times 10^{-24}$ J;
(d) -9.27×10^{-24} J

Chapter 33

CP   1. (a) (Use Fig. 33-5.) On right side of rectangle, $\vec{E}$ is in
negative y direction; on left side, $\vec{E} + d\vec{E}$ is greater and in same
direction; (b) $\vec{E}$ is downward. On right side, $\vec{B}$ is in negative z
direction; on left side, $\vec{B} + d\vec{B}$ is greater and in same direction.
2. positive direction of x **3.** (a) same; (b) decrease   **4.** a, d, b, c
(zero)   **5.** a
Q   1. (a) positive direction of z; (b) x **3.** (a) same; (b) increase;
(c) decrease   **5.** (a) and (b) $A = 1, n = 4, \theta = 30°$ **7.** a, b, c **9.** B
11. none
P   1. 7.49 GHz   **3.** (a) 515 nm; (b) 610 nm; (c) 555 nm;
(d) 5.41×10^{14} Hz; (e) 1.85×10^{-15} s   **5.** 5.0×10^{-21} H   **7.** 1.2 MW/m^2
9. 0.10 MJ   **11.** (a) 6.7 nT; (b) y; (c) negative direction of y
13. (a) 1.03 kV/m; (b) 3.43 μT   **15.** (a) 87 mV/m; (b) 0.29 nT;
(c) 6.3 kW   **17.** (a) 6.7 nT; (b) 5.3 mW/m^2; (c) 6.7 W   **19.** 1.0×10^{7} Pa
21. 5.9×10^{-8} Pa   **23.** (a) 4.68×10^{11} W; (b) any chance
disturbance could move sphere from directly above source—the
two force vectors no longer along the same axis   **27.** (a) 1.0×10^{8} Hz;
(b) 6.3×10^{8} rad/s; (c) 2.1 m^{-1}; (d) 1.0 μT; (e) z; (f) 1.2×10^{2} W/m^2;
(g) 8.0×10^{-7} N; (h) 4.0×10^{-7} Pa   **29.** 1.9 mm/s   **31.** (a) 0.17 μm;
(b) toward the Sun   **33.** 3.1%   **35.** 4.4 W/m^2 **37.** (a) 2 sheets;
(b) 5 sheets   **39.** (a) 1.9 V/m; (b) 1.7×10^{-11} Pa   **41.** 20° or 70°
43. 0.67   **45.** 1.26   **47.** 1.48   **49.** 180°   **51.** (a) 56.9°; (b) 35.3°
55. 1.07 m   **57.** 182 cm   **59.** (a) 48.9°; (b) 29.0°   **61.** (a) 26.8°;
(b) yes   **63.** (a) $(1 + \sin^2 \theta)^{0.5}$; (b) $2^{0.5}$; (c) yes; (d) no   **65.** 23.2°

67. (a) 1.39; (b) 28.1°; (c) no   **69.** 49.0°   **71.** (a) 0.50 ms; (b) 8.4 min;
(c) 2.4 h; (d) 5446 B.C.   **73.** (a) (16.7 nT) sin$[(1.00 \times 10^{6}$ m$^{-1})z +$
$(3.00 \times 10^{14}$ s$^{-1})t]$; (b) 6.28 μm; (c) 20.9 fs; (d) 33.2 mW/m^2; (e) x;
(f) infrared   **75.** 1.22   **77.** (c) 137.6°; (d) 139.4°; (e) 1.7°
81. (a) z axis; (b) 7.5×10^{14} Hz; (c) 1.9 kW/m^2 **83.** (a) white;
(b) white dominated by red end; (c) no refracted light
85. 1.5×10^{-9} m/s^2 **87.** (a) 3.5 μW/m^2; (b) 0.78 μW;
(c) 1.5×10^{-17} W/m^2; (d) 1.1×10^{-7} V/m; (e) 0.25 fT   **89.** (a) 55.8°;
(b) 55.5°   **91.** (a) 83 W/m^2; (b) 1.7 MW   **93.** 35°   **97.** cos$^{-1}(p/50)^{0.5}$
99. $8RI/3c$ **101.** 0.034   **103.** 9.43×10^{-10} T   **105.** (a) $-y$; (b) z;
(c) 1.91 kW/m^2; (d) $E_z = (1.20$ kV/m$)$ sin$[(6.67 \times 10^{6}$ m$^{-1})y +$
$(2.00 \times 10^{15}$ s$^{-1})t]$; (e) 942 nm; (f) infrared   **107.** (a) 1.60; (b) 58.0°

Chapter 34

CP   1. $0.2d, 1.8d, 2.2d$ **2.** (a) real; (b) inverted; (c) same
3. (a) e; (b) virtual, same   **4.** virtual, same as object, diverging
Q   1. (a) a; (b) c **3.** (a) a and c; (b) three times; (c) you
5. convex   **7.** (a) all but variation 2; (b) 1, 3, 4: right, inverted; 5, 6:
left, same   **9.** d (infinite), tie of a and b, then c **11.** (a) x; (b) no;
(c) no; (d) the direction you are facing
P   1. 9.10 m   **3.** 1.11   **5.** 351 cm   **7.** 10.5 cm   **9.** (a) $+24$ cm;
(b) $+36$ cm; (c) -2.0; (d) R; (e) I; (f) same   **11.** (a) -20 cm;
(b) -4.4 cm; (c) $+0.56$; (d) V; (e) NI; (f) opposite   **13.** (a) $+36$ cm;
(b) -36 cm; (c) $+3.0$; (d) V; (e) NI; (f) opposite   **15.** (a) -16 cm;
(b) -4.4 cm; (c) $+0.44$; (d) V; (e) NI; (f) opposite   **17.** (b) plus;
(c) $+40$ cm; (e) -20 cm; (f) $+2.0$; (g) V; (h) NI; (i) opposite
19. (a) convex; (b) -20 cm; (d) $+20$ cm; (f) $+0.50$; (g) V; (h) NI;
(i) opposite   **21.** (a) concave; (c) $+40$ cm; (e) $+60$ cm; (f) -2.0;
(g) R; (h) I; (i) same   **23.** (a) convex; (b) minus; (c) -60 cm;
(d) $+1.2$ m; (e) -24 cm; (g) V; (h) NI; (i) opposite   **25.** (a) concave;
(b) $+8.6$ cm; (c) $+17$ cm; (e) $+12$ cm; (f) minus; (g) R; (i) same
27. (a) convex; (c) -60 cm; (d) $+30$ cm; (f) $+0.50$; (g) V; (h) NI;
(i) opposite   **29.** (b) -20 cm; (c) minus; (d) $+5.0$ cm; (e) minus;
(f) $+0.80$; (g) V; (h) NI; (i) opposite   **31.** (b) 0.56 cm/s; (c) 11 m/s;
(d) 6.7 cm/s   **33.** (c) -33 cm; (e) V; (f) same   **35.** (d) -26 cm; (e) V;
(f) same   **37.** (c) $+30$ cm; (e) V; (f) same   **39.** (a) 2.00; (b) none
41. (a) $+40$ cm; (b) ∞ **43.** 5.0 mm   **45.** 1.86 mm   **47.** (a) 45 mm;
(b) 90 mm   **49.** 22 cm   **51.** (a) -48 cm; (b) $+4.0$; (c) V; (d) NI;
(e) same   **53.** (a) -4.8 cm; (b) $+0.60$; (c) V; (d) NI; (e) same
55. (a) -8.6 cm; (b) $+0.39$; (c) V; (d) NI; (e) same   **57.** (a) $+36$ cm;
(b) -0.80; (c) R; (d) I; (e) opposite   **59.** (a) $+55$ cm; (b) -0.74;
(c) R; (d) I; (e) opposite   **61.** (a) -18 cm; (b) $+0.76$; (c) V; (d) NI;
(e) same   **63.** (a) -30 cm; (b) $+0.86$; (c) V; (d) NI; (e) same
65. (a) -7.5 cm; (b) $+0.75$; (c) V; (d) NI; (e) same   **67.** (a) $+84$ cm;
(b) -1.4; (c) R; (d) I; (e) opposite   **69.** (a) C; (d) -10 cm; (e) $+2.0$;
(f) V; (g) NI; (h) same   **71.** (a) D; (b) -5.3 cm; (d) -4.0 cm; (f) V;
(g) NI; (h) same   **73.** (a) C; (b) $+3.3$ cm; (d) $+5.0$ cm; (f) R; (g) I;
(h) opposite   **75.** (a) D; (b) minus; (d) -3.3 cm; (e) $+0.67$; (f) V;
(g) NI   **77.** (a) C; (b) $+80$ cm; (d) -20 cm; (f) V; (g) NI; (h) same
79. (a) C; (b) plus; (d) -13 cm; (e) $+1.7$; (f) V; (g) NI; (h) same
81. (a) $+24$ cm; (b) $+6.0$; (c) R; (d) NI; (e) opposite
83. (a) $+3.1$ cm; (b) -0.31; (c) R; (d) I; (e) opposite   **85.** (a) -4.6 cm;
(b) $+0.69$; (c) V; (d) NI; (e) same   **87.** (a) -5.5 cm; (b) $+0.12$; (c) V;
(d) NI; (e) same   **89.** (a) 13.0 cm; (b) 5.23 cm; (c) -3.25; (d) 3.13;
(e) -10.2 **91.** (a) 2.35 cm; (b) decrease   **93.** (a) 3.5; (b) 2.5
95. (a) $+8.6$ cm; (b) $+2.6$; (c) R; (d) NI; (e) opposite
97. (a) $+7.5$ cm; (b) -0.75; (c) R; (d) I; (e) opposite   **99.** (a) $+24$ cm;
(b) -0.58; (c) R; (d) I; (e) opposite   **105.** (a) 3.00 cm; (b) 2.33 cm
107. (a) 40 cm; (b) 20 cm; (c) -40 cm; (d) 40 cm   **109.** (a) 20 cm;
(b) 15 cm   **111.** (a) 6.0 mm; (b) 1.6 kW/m^2; (c) 4.0 cm   **113.** 100 cm
115. 2.2 mm^2 **119.** (a) -30 cm; (b) not inverted; (c) virtual; (d) 1.0
121. (a) -12 cm   **123.** (a) 80 cm; (b) 0 to 12 cm   **127.** (a) 8.0 cm;

(b) 16 cm; (c) 48 cm   **129.** (a) $\alpha = 0.500$ rad: 7.799 cm; $\alpha = 0.100$ rad: 8.544 cm; $\alpha = 0.0100$ rad: 8.571 cm; mirror equation: 8.571 cm; (b) $\alpha = 0.500$ rad: -13.56 cm; $\alpha = 0.100$ rad: -12.05 cm; $\alpha = 0.0100$ rad: -12.00 cm; mirror equation: -12.00 cm   **131.** 42 mm
133. (b) P_n **135.** (a) $(0.5)(2 - n)r/(n - 1)$; (b) right   **137.** 2.67 cm
139. (a) 3.33 cm; (b) left; (c) virtual; (d) not inverted
141. (a) $1 + (25 \text{ cm})/f$; (b) $(25 \text{ cm})/f$; (c) 3.5; (d) 2.5

Chapter 35

CP **1.** b (least n), c, a **2.** (a) top; (b) bright intermediate illumination (phase difference is 2.1 wavelengths)   **3.** (a) $3\lambda, 3$; (b) $2.5\lambda, 2.5$
4. a and d tie (amplitude of resultant wave is $4E_0$), then b and c tie (amplitude of resultant wave is $2E_0$)   **5.** (a) 1 and 4; (b) 1 and 4
Q **1.** (a) decrease; (b) decrease; (c) decrease; (d) blue   **3.** (a) $2d$; (b) (odd number)$\lambda/2$; (c) $\lambda/4$ **5.** (a) intermediate closer to maximum, $m = 2$; (b) minimum, $m = 3$; (c) intermediate closer to maximum, $m = 2$; (d) maximum, $m = 1$ **7.** (a) maximum; (b) minimum; (c) alternates   **9.** (a) peak; (b) valley   **11.** c, d **13.** c
P **1.** (a) 155 nm; (b) 310 nm   **3.** (a) 3.60 μm; (b) intermediate closer to fully constructive   **5.** 4.55×10^7 m/s   **7.** 1.56
9. (a) 1.55 μm; (b) 4.65 μm   **11.** (a) 1.70; (b) 1.70; (c) 1.30; (d) all tie   **13.** (a) 0.833; (b) intermediate closer to fully constructive   **15.** 648 nm   **17.** 16   **19.** 2.25 mm   **21.** 72 μm
23. 0   **25.** 7.88 μm   **27.** 6.64 μm   **29.** 2.65   **31.** $27 \sin(\omega t + 8.5°)$
33. $(17.1 \ \mu\text{V/m}) \sin[(2.0 \times 10^{14} \text{ rad/s})t]$ **35.** 120 nm   **37.** 70.0 nm
39. (a) 0.117 μm; (b) 0.352 μm   **41.** 161 nm   **43.** 560 nm
45. 478 nm   **47.** 509 nm   **49.** 273 nm   **51.** 409 nm   **53.** 338 nm
55. (a) 552 nm; (b) 442 nm   **57.** 608 nm   **59.** 528 nm   **61.** 455 nm
63. 248 nm   **65.** 339 nm   **67.** 329 nm   **69.** 1.89 μm   **71.** 0.012°
73. 140   **75.** $[(m + \frac{1}{2})\lambda R]^{0.5}$, for $m = 0, 1, 2, \dots$ **77.** 1.00 m
79. 588 nm   **81.** 1.00030   **83.** (a) 50.0 nm; (b) 36.2 nm   **85.** 0.23°
87. (a) 1500 nm; (b) 2250 nm; (c) 0.80   **89.** $x = (D/2a)(m + 0.5)\lambda$, for $m = 0, 1, 2, \dots$ **91.** (a) 22°; (b) refraction reduces θ **93.** 600 nm   **95.** (a) 1.75 μm; (b) 4.8 mm   **97.** $I_m \cos^2(2\pi x/\lambda)$ **99.** (a) 42.0 ps; (b) 42.3 ps; (c) 43.2 ps; (d) 41.8 ps; (e) 4   **101.** 33 μm
103. (a) bright; (b) 594 nm; (c) Primary reason: the colored bands begin to overlap too much to be distinguished. Secondary reason: the two reflecting surfaces are too separated for the light reflecting from them to be coherent.

Chapter 36

CP **1.** (a) expand; (b) expand   **2.** (a) second side maximum; (b) 2.5   **3.** (a) red; (b) violet   **4.** diminish   **5.** (a) left; (b) less
Q **1.** (a) $m = 5$ minimum; (b) (approximately) maximum between the $m = 4$ and $m = 5$ minima   **3.** (a) A, B, C; (b) A, B, C
5. (a) 1 and 3 tie, then 2 and 4 tie; (b) 1 and 2 tie, then 3 and 4 tie
7. (a) larger; (b) red   **9.** (a) decrease; (b) same; (c) remain in place
11. (a) A; (b) left; (c) left; (d) right   **13.** (a) 1 and 2 tie, then 3; (b) yes; (c) no
P **1.** (a) 2.5 mm; (b) 2.2×10^{-4} rad   **3.** (a) 70 cm; (b) 1.0 mm
5. (a) 700 nm; (b) 4; (c) 6   **7.** 60.4 μm   **9.** 1.77 mm   **11.** 160°
13. (a) 0.18°; (b) 0.46 rad; (c) 0.93   **15.** (d) 52.5°; (e) 10.1°; (f) 5.06°
17. (b) 0; (c) -0.500; (d) 4.493 rad; (e) 0.930; (f) 7.725 rad; (g) 1.96
19. (a) 19 cm; (b) larger   **21.** (a) 1.1×10^4 km; (b) 11 km
23. (a) 1.3×10^{-4} rad; (b) 10 km   **25.** 50 m   **27.** 1.6×10^3 km
29. (a) 8.8×10^{-7} rad; (b) 8.4×10^7 km; (c) 0.025 mm   **31.** (a) 0.346°; (b) 0.97°   **33.** (a) 17.1 m; (b) 1.37×10^{-10} **35.** 5   **37.** 3
39. (a) 5.0 μm; (b) 20 μm   **41.** (a) 7.43×10^{-3}; (b) between the $m = 6$ minimum (the seventh one) and the $m = 7$ maximum (the seventh side maximum); (c) between the $m = 3$ minimum (the third one) and the $m = 4$ minimum (the fourth one)
43. (a) 9; (b) 0.255   **45.** (a) 62.1°; (b) 45.0°; (c) 32.0°   **47.** 3

49. (a) 6.0 μm; (b) 1.5 μm; (c) 9; (d) 7; (e) 6   **51.** (a) 2.1°; (b) 21°; (c) 11   **53.** (a) 470 nm; (b) 560 nm   **55.** 3.65×10^3
57. (a) 0.032°/nm; (b) 4.0×10^4; (c) 0.076°/nm; (d) 8.0×10^4; (e) 0.24°/nm; (f) 1.2×10^5 **59.** 0.15 nm   **61.** (a) 10 μm; (b) 3.3 mm
63. 1.09×10^3 rulings/mm   **65.** (a) 0.17 nm; (b) 0.13 nm
67. (a) 25 pm; (b) 38 pm   **69.** 0.26 nm   **71.** (a) 15.3°; (b) 30.6°; (c) 3.1°; (d) 37.8°   **73.** (a) $0.7071a_0$; (b) $0.4472a_0$; (c) $0.3162a_0$; (d) $0.2774a_0$; (e) $0.2425a_0$ **75.** (a) 625 nm; (b) 500 nm; (c) 416 nm
77. 3.0 mm   **83.** (a) 13; (b) 6   **85.** 59.5 pm   **87.** 4.9 km   **89.** 1.36×10^4
91. 2   **93.** 4.7 cm   **97.** 36 cm   **99.** (a) fourth; (b) seventh
103. (a) 2.4 μm; (b) 0.80 μm; (c) 2   **107.** 9

Chapter 37

CP **1.** (a) same (speed of light postulate); (b) no (the start and end of the flight are spatially separated); (c) no (because his measurement is not a proper time)   **2.** (a) Eq. 2; (b) $+0.90c$; (c) 25 ns; (d) -7.0 m   **3.** (a) right; (b) more   **4.** (a) equal; (b) less
Q **1.** c **3.** b **5.** (a) C_1'; (b) C_1' **7.** (a) 4 s; (b) 3 s; (c) 5 s; (d) 4 s; (e) 10 s   **9.** (a) a tie of 3, 4, and 6, then a tie of 1, 2, and 5; (b) 1, then a tie of 2 and 3, then 4, then a tie of 5 and 6; (c) 1, 2, 3, 4, 5, 6; (d) 2 and 4; (e) 1, 2, 5   **11.** (a) 3, tie of 1 and 2, then 4; (b) 4, tie of 1 and 2, then 3; (c) 1, 4, 2, 3
P **1.** 0.990 50   **3.** (a) 0.999 999 50   **5.** 0.446 ps   **7.** 2.68×10^3 y
9. (a) 87.4 m; (b) 394 ns   **11.** 1.32 m   **13.** (a) 26.26 y; (b) 52.26 y; (c) 3.705 y   **15.** (a) 0.999 999 15; (b) 30 ly
17. (a) 138 km; (b) -374 μs   **19.** (a) 25.8 μs; (b) small flash
21. (a) $\gamma[1.00 \ \mu\text{s} - \beta(400 \text{ m})/(2.998 \times 10^8 \text{ m/s})]$; (d) 0.750; (e) $0 < \beta < 0.750$; (f) $0.750 < \beta < 1$; (g) no   **23.** (a) 1.25; (b) 0.800 μs
25. (a) 0.480; (b) negative; (c) big flash; (d) 4.39 μs   **27.** $0.81c$
29. (a) 0.35; (b) 0.62   **31.** 1.2 μs   **33.** (a) 1.25 y; (b) 1.60 y; (c) 4.00 y
35. 22.9 MHz   **37.** $0.13c$ **39.** (a) 550 nm; (b) yellow
41. (a) 196.695; (b) 0.999 987   **43.** (a) 1.0 keV; (b) 1.1 MeV
45. 110 km   **47.** 1.01×10^7 km   **49.** (a) 0.222 cm; (b) 701 ps; (c) 7.40 ps   **51.** $2.83mc$ **53.** $\gamma(2\pi m/|q|B)$; (b) no; (c) 4.85 mm; (d) 15.9 mm; (e) 16.3 ps; (f) 0.334 ns   **55.** (a) 0.707; (b) 1.41; (c) 0.414   **57.** 18 smu/y   **59.** (a) 2.08 MeV; (b) -1.21 MeV
61. (d) 0.801   **63.** (a) $vt \sin \theta$; (b) $t[1 - (v/c) \cos \theta]$; (c) $3.24c$
67. (b) $+0.44c$ **69.** (a) 1.93 m; (b) 6.00 m; (c) 13.6 ns; (d) 13.6 ns; (e) 0.379 m; (f) 30.5 m; (g) -101 ns; (h) no; (i) 2; (k) no; (l) both
71. (a) 5.4×10^4 km/h; (b) 6.3×10^{-10} **73.** 189 MeV
75. 8.7×10^{-3} ly   **77.** 7   **79.** 2.46 MeV/c   **81.** $0.27c$
83. (a) 5.71 GeV; (b) 6.65 GeV; (c) 6.58 GeV/c; (d) 3.11 MeV; (e) 3.62 MeV; (f) 3.59 MeV/c   **85.** $0.95c$ **87.** (a) 256 kV; (b) $0.745c$
89. (a) $0.858c$; (b) $0.185c$ **91.** $0.500c$ **93.** (a) 119 MeV; (b) 64.0 MeV/c; (c) 81.3 MeV; (d) 64.0 MeV/c   **95.** 4.00 u, probably a helium nucleus   **97.** (a) 534; (b) 0.999 998 25; (c) 2.23 T
99. (a) 415 nm; (b) blue   **101.** (a) 88 kg; (b) no   **103.** (a) 3×10^{-18}; (b) 2×10^{-12}; (c) 8.2×10^{-8}; (d) 6.4×10^{-6}; (e) 1.1×10^{-6}; (f) 3.7×10^{-5}; (g) 9.9×10^{-5}; (h) 0.10

Chapter 38

CP **1.** b, a, d, c   **2.** (a) lithium, sodium, potassium, cesium; (b) all tie   **3.** (a) same; (b)$-$(d) x rays   **4.** (a) proton; (b) same; (c) proton   **5.** same
Q **1.** (a) greater; (b) less   **3.** potassium   **5.** only e   **7.** none
9. (a) decreases by a factor of $(1/2)^{0.5}$; (b) decreases by a factor of 1/2
11. amplitude of reflected wave is less than that of incident wave
13. electron, neutron, alpha particle   **15.** all tie
P **1.** (a) 2.1 μm; (b) infrared   **3.** 1.0×10^{45} photons/s   **5.** 2.047 eV
7. 1.1×10^{-10} W   **9.** (a) 2.96×10^{20} photons/s; (b) 4.86×10^7 m; (c) 5.89×10^{18} photons/m²·s   **11.** (a) infrared; (b) 1.4×10^{21} photons/s
13. 4.7×10^{26} photons   **15.** 170 nm   **17.** 676 km/s   **19.** 1.3 V;

(b) 6.8×10^2 km/s   **21.** (a) 3.1 keV; (b) 14 keV   **23.** (a) 2.00 eV;
(b) 0; (c) 2.00 V; (d) 295 nm   **25.** (a) 382 nm; (b) 1.82 eV
27. (a) 2.73 pm; (b) 6.05 pm   **29.** (a) 8.57×10^{18} Hz; (b) 3.55×10^4 eV;
(c) 35.4 keV/c **31.** 300%   **33.** (a) -8.1×10^{-9}%; (b) -4.9×10^{-4}%;
(c) -8.9%; (d) -66%   **35.** (a) 2.43 pm; (b) 1.32 fm; (c) 0.511 MeV
(d) 939 MeV   **37.** (a) 41.8 keV; (b) 8.2 keV   **39.** 44°   **41.** (a) 2.43 pm;
(b) 4.11×10^{-6}; (c) -8.67×10^{-6} eV; (d) 2.43 pm; (e) 9.78×10^{-2};
(f) -4.45 keV   **43.** (a) 2.9×10^{-10} m; (b) x ray; (c) 2.9×10^{-8} m;
(d) ultraviolet   **45.** (a) 9.35 μm; (b) 1.47×10^{-5} W; (c) 6.93×10^{14}
photons/s; (d) 2.33×10^{-37} W; (e) 5.87×10^{-19} photons/s
47. 7.75 pm   **49.** (a) 1.9×10^{-21} kg $\cdot$ m/s; (b) 346 fm   **51.** 4.3 μeV
53. (a) 1.24 μm; (b) 1.22 nm; (c) 1.24 fm; (d) 1.24 fm   **55.** (a) 15 keV;
(b) 120 keV   **57.** neutron   **59.** (a) 3.96×10^6 m/s; (b) 81.7 kV
67. 2.1×10^{-24} kg $\cdot$ m/s   **71.** (a) 1.45×10^{11} m^{-1}; (b) 7.25×10^{10} m^{-1};
(c) 0.111; (d) 5.56×10^4 **73.** 4.81 mA   **75.** (a) 9.02×10^{-6};
(b) 3.0 MeV; (c) 3.0 MeV; (d) 7.33×10^{-8}; (e) 3.0 MeV; (f) 3.0 MeV
77. (a) -20%; (b) -10%; (c) $+15$%   **79.** (a) no; (b) plane wavefronts of infinite extent, perpendicular to x axis   **83.** (a) 38.8 meV;
(b) 146 pm   **85.** (a) 4.14×10^{-15} eV $\cdot$ s; (b) 2.31 eV   **89.** (a) no;
(b) 544 nm; (c) green

Chapter 39
CP   **1.** b, a, c   **2.** (a) all tie; (b) a, b, c **3.** a, b, c, d   **4.** $E_{1,1}$ (neither n_x
nor n_y can be zero)   **5.** (a) 5; (b) 7
Q   **1.** a, c, b   **3.** (a) 18; (b) 17   **5.** equal   **7.** c   **9.** (a) decrease;
(b) increase   **11.** $n = 1, n = 2, n = 3$ **13.** (a) $n = 3$; (b) $n = 1$;
(c) $n = 5$ **15.** b, c, and d
P   **1.** 1.41   **3.** 0.65 eV   **5.** 0.85 nm   **7.** 1.9 GeV   **9.** (a) 72.2 eV;
(b) 13.7 nm; (c) 17.2 nm; (d) 68.7 nm; (e) 41.2 nm; (g) 68.7 nm;
(h) 25.8 nm   **11.** (a) 13; (b) 12   **13.** (a) 0.020; (b) 20   **15.** (a) 0.050;
(b) 0.10; (c) 0.0095   **17.** 56 eV   **19.** 109 eV   **23.** 3.21 eV
25. 1.4×10^{-3} **27.** (a) 8; (b) 0.75; (c) 1.00; (d) 1.25; (e) 3.75; (f) 3.00;
(g) 2.25   **29.** (a) 7; (b) 1.00; (c) 2.00; (d) 3.00; (e) 9.00; (f) 8.00;
(g) 6.00   **31.** 4.0   **33.** (a) 12.1 eV; (b) 6.45×10^{-27} kg $\cdot$ m/s; (c) 102
nm   **35.** (a) 291 nm^{-3}; (b) 10.2 nm^{-1} **41.** (a) 0.0037; (b) 0.0054
43. (a) 13.6 eV; (b) -27.2 eV   **45.** (a) $(r^4/8a^5)[\exp(-r/a)]\cos^2\theta$;
(b) $(r^4/16a^5)[\exp(-r/a)]\sin^2\theta$ **47.** 4.3×10^3 **49.** (a) 13.6 eV;
(b) 3.40 eV   **51.** 0.68   **59.** (b) $(2\pi/h)[2m(U_0 - E)]^{0.5}$
61. (b) meter$^{-2.5}$ **63.** (a) n; (b) $2\ell + 1$; (c) n^2 **65.** (a) $nh/\pi md^2$;
(b) $n^2h^2/4\pi^2md^2$ **67.** (a) 3.9×10^{-22} eV; (b) 10^{20}; (c) 3.0×10^{-18} K
71. (a) $e^2r/4\pi\varepsilon_0 a^3$; (b) $e/(4\pi\varepsilon_0 ma_0^3)^{0.5}$ **73.** 18.1, 36.2, 54.3, 66.3,
72.4 μeV

Chapter 40
CP   **1.** 7   **2.** (a) decrease; (b)$-$(c) remain the same   **3.** A, C, B
Q   **1.** (a) 2; (b) 8; (c) 5; (d) 50   **3.** all true   **5.** same number (10)
7. 2, -1, 0, and 1   **9.** (a) 2; (b) 3   **11.** (a) n; (b) n and ℓ **13.** In addition to the quantized energy, a helium atom has kinetic energy; its
total energy can equal 20.66 eV.
P   **1.** 24.1°   **3.** (a) 3.65×10^{-34} J $\cdot$ s; (b) 3.16×10^{-34} J $\cdot$ s   **5.** (a) 3;
(b) 3   **7.** (a) 4; (b) 5; (c) 2   **9.** (a) 3.46; (b) 3.46; (c) 3; (d) 3; (e) -3;
(f) 30.0°; (g) 54.7°; (h) 150°   **13.** 72 km/s^2 **15.** (a) 54.7°; (b) 125°
17. 19 mT   **19.** 5.35 cm   **21.** 44   **23.** 42   **25.** (a) 51; (b) 53; (c) 56
27. (a) $(2, 0, 0, +\frac{1}{2})$, $(2, 0, 0, -\frac{1}{2})$; (b) $(2, 1, 1, +\frac{1}{2})$, $(2, 1, 1, -\frac{1}{2})$,
$(2, 1, 0, +\frac{1}{2})$, $(2, 1, 0, -\frac{1}{2})$, $(2, 1, -1, +\frac{1}{2})$, $(2, 1, -1, -\frac{1}{2})$ **29.** g
31. (a) $4p$; (b) 4; (c) $4p$; (d) 5; (e) $4p$; (f) 6   **33.** 12.4 kV   **35.** (a) 35.4 pm;
(b) 56.5 pm; (c) 49.6 pm   **39.** 0.563   **41.** 80.3 pm   **43.** (a) 69.5 kV;
(b) 17.8 pm; (c) 21.3 pm; (d) 18.5 pm   **45.** (a) 49.6 pm; (b) 99.2 pm
47. 2.0×10^{16} s^{-1} **49.** 2×10^7 **51.** 9.0×10^{-7} **53.** 7.3×10^{15} s^{-1}
55. (a) 3.60 mm; (b) 5.24×10^{17} **57.** (a) 0; (b) 68 J   **59.** 3.0 eV
61. (a) 3.03×10^5; (b) 1.43 GHz; (d) 3.31×10^{-6} **63.** 186
65. (a) 2.13 meV; (b) 18 T   **69.** (a) no; (b) 140 nm   **71.** $n > 3$;

$\ell = 3$; $m_\ell = +3, +2, +1, 0, -1, -2, -3$; $m_s = \pm\frac{1}{2}$ **73.** (a) 6.0;
(b) 3.2×10^6 y   **75.** argon   **79.** $(Ze/4\pi\varepsilon_0)(r^{-2} - rR^{-3})$

Chapter 41
CP   **1.** larger   **2.** a, b, and c
Q   **1.** b, c, d (the latter due to thermal expansion)   **3.** 8
5. below   **7.** increase   **9.** much less than   **11.** b and d
P   **3.** 8.49×10^{28} m^{-3} **5.** (b) 6.81×10^{27} m^{-3} eV$^{-3/2}$;
(c) 1.52×10^{28} m^{-3} eV^{-1} **7.** (a) 0; (b) 0.0955   **9.** (a) 5.86×10^{28} m^{-3};
(b) 5.49 eV; (c) 1.39×10^3 km/s; (d) 0.522 nm   **11.** (a) $1.36 \times$
10^{28} m^{-3} eV^{-1}; (b) 1.68×10^{28} m^{-3} eV^{-1}; (c) 9.01×10^{27} m^{-3} eV^{-1};
(d) 9.56×10^{26} m^{-3} eV^{-1}; (e) 1.71×10^{18} m^{-3} eV^{-1} **13.** (a) 6.81 eV;
(b) 1.77×10^{28} m^{-3} eV^{-1}; (c) 1.59×10^{28} m^{-3} eV^{-1}
15. (a) 2.50×10^3 K; (b) 5.30×10^3 K   **17.** 3   **19.** (a) 1.0; (b) 0.99;
(c) 0.50; (d) 0.014; (e) 2.4×10^{-17}; (f) 7.0×10^2 K   **21.** (a) 0.0055;
(b) 0.018   **25.** (a) 19.7 kJ; (b) 197 s   **27.** (a) 1.31×10^{29} m^{-3};
(b) 9.43 eV; (c) 1.82×10^3 km/s; (d) 0.40 nm   **29.** 57.1 kJ
31. (a) 226 nm; (b) ultraviolet   **33.** (a) 1.5×10^{-6}; (b) 1.5×10^{-6}
35. 0.22 μg   **37.** (a) 4.79×10^{-10}; (b) 0.0140; (c) 0.824   **39.** 6.0×10^5
41. 4.20 eV   **43.** 13 μm   **47.** (a) 109.5°; (b) 238 pm
49. (b) 1.8×10^{28} m^{-3} eV^{-1} **53.** 3.49×10^3 atm

Chapter 42
CP   **1.** ^{90}As and ^{158}Nd   **2.** a little more than 75 Bq (elapsed time is
a little less than three half-lives)   **3.** ^{206}Pb
Q   **1.** (a) ^{196}Pt; (b) no   **3.** yes   **5.** (a) less; (b) greater   **7.** ^{240}U
9. no effect   **11.** yes   **13.** (a) all except ^{198}Au; (b) ^{132}Sn and ^{208}Pb
15. d
P   **1.** 1.3×10^{-13} m   **3.** 46.6 fm   **5.** (a) 0.390 MeV; (b) 4.61 MeV
7. (a) 2.3×10^{17} kg/m^3; (b) 2.3×10^{17} kg/m^3; (d) 1.0×10^{25} C/m^3;
(e) 8.8×10^{24} C/m^3 **9.** (a) 6; (b) 8   **11.** (a) 6.2 fm; (b) yes
13. 13 km   **17.** 1.0087 u   **19.** (a) 9.303%; (b) 11.71%
21. (b) 7.92 MeV/nucleon   **25.** 5.3×10^{22} **27.** (a) 0.250; (b) 0.125
29. (a) 64.2 h; (b) 0.125; (c) 0.0749   **31.** (a) 7.5×10^{16} s^{-1};
(b) 4.9×10^{16} s^{-1} **33.** 1×10^{13} atoms   **37.** 265 mg
39. (a) 8.88×10^{10} s^{-1}; (b) 1.19×10^{15}; (c) 0.111 μg   **41.** 1.12×10^{11} y
43. 9.0×10^8 Bq   **45.** (a) 3.2×10^{12} Bq; (b) 86 Ci   **47.** (a) 2.0×10^{20};
(b) 2.8×10^9 s^{-1} **49.** (a) 1.2×10^{-17}; (b) 0   **51.** 4.269 MeV
53. 1.21 MeV   **55.** 0.783 MeV   **57.** (b) 0.961 MeV   **59.** 78.3 eV
61. (a) 1.06×10^{19}; (b) 0.624×10^{19}; (c) 1.68×10^{19}; (d) 2.97×10^9 y
63. 1.7 mg   **65.** 1.02 mg   **67.** 2.50 mSv   **69.** (a) 6.3×10^{18};
(b) 2.5×10^{11}; (c) 0.20 J; (d) 2.3 mGy; (e) 30 mSv   **71.** (a) 6.6 MeV;
(b) no   **73.** (a) 25.4 MeV; (b) 12.8 MeV; (c) 25.0 MeV   **75.** ^{7}Li
77. 3.2×10^4 y   **79.** 730 cm^2 **81.** ^{225}Ac   **83.** 30 MeV   **89.** 27
91. (a) 11.906 83 u; (b) 236.2025 u   **93.** 600 keV   **95.** (a) 59.5 d;
(b) 1.18   **97.** (a) 4.8×10^{-18} s^{-1}; (b) 4.6×10^9 y

Chapter 43
CP   **1.** c and d   **2.** e
Q   **1.** (a) 101; (b) 42   **3.** ^{239}Np   **5.** ^{140}I, ^{105}Mo, ^{152}Nd, ^{123}In, ^{115}Pd
7. increased   **9.** less than   **11.** still equal to 1
P   **1.** (a) 16 day^{-1}; (b) 4.3×10^8 **3.** 4.8 MeV   **5.** 1.3×10^3 kg
7. 3.1×10^{10} s^{-1} **9.** (a) 2.6×10^{24}; (b) 8.2×10^{13} J; (c) 2.6×10^4 y
11. -23.0 MeV   **13.** (a) 253 MeV; (b) typical fission energy is
200 MeV   **15.** (a) 84 kg; (b) 1.7×10^{25}; (c) 1.3×10^{25} **17.** (a) ^{153}Nd;
(b) 110 MeV; (c) 60 MeV; (d) 1.6×10^7 m/s; (e) 8.7×10^6 m/s
21. 557 W   **23.** 0.99938   **25.** (b) 1.0; (c) 0.89; (d) 0.28; (e) 0.019;
(f) 8   **27.** (a) 75 kW; (b) 5.8×10^3 kg   **29.** 1.7×10^9 y
31. 170 keV   **33.** 1.41 MeV   **35.** 10^{-12} m   **37.** (a) 4.3×10^9 kg/s;
(b) 3.1×10^{-4} **41.** 1.6×10^8 y   **43.** (a) 24.9 MeV; (b) 8.65 megatons TNT   **45.** (a) 1.8×10^{38} s^{-1}; (b) 8.2×10^{28} s^{-1} **47.** (a) 4.1
eV/atom; (b) 9.0 MJ/kg; (c) 1.5×10^3 y   **49.** 14.4 kW

51. $^{238}\text{U} + \text{n} \rightarrow {}^{239}\text{U} \rightarrow {}^{239}\text{Np} + \text{e} + \nu,\ {}^{239}\text{Np} \rightarrow {}^{239}\text{Pu} + \text{e} + \nu$ **55.**
(a) 3.1×10^{31} protons/m³; (b) 1.2×10^6 **57.** (a) 227 J; (b) 49.3 mg;
(c) 22.7 kW

Chapter 44

CP   **1.** (a) the muon family; (b) a particle; (c) $L_\mu = +1$
2. b and e   **3.** c
Q   **1.** b, c, d   **3.** (a) 1; (b) positively charged   **5.** a, b, c, d   **7.** d
9. c   **11.** (a) lepton; (b) antiparticle; (c) fermion; (d) yes
P   **1.** $\pi^- \rightarrow \mu^- + \bar{\nu}$ **3.** 2.4 pm   **5.** 2.4×10^{-43} **7.** 769 MeV
9. 2.7 cm/s   **11.** (a) angular momentum, L_e; (b) charge, L_μ;

(c) energy, L_μ **15.** (a) energy; (b) strangeness; (c) charge
17. (a) yes; (b)−(d) no   **19.** (a) 0; (b) −1; (c) 0   **21.** (a) K^+; (b) $\bar{n}$;
(c) K^0 **23.** (a) 37.7 MeV; (b) 5.35 MeV; (c) 32.4 MeV   **25.** (a) $u\bar{u}\bar{d}$;
(b) $u\bar{d}\bar{d}$ **27.** $s\bar{d}$ **29.** (a) Ξ^0; (b) Σ^- **31.** 2.77×10^8 ly   **33.** 668 nm
35. 1.4×10^{10} ly   **37.** (a) 2.6 K; (b) 976 nm   **39.** (b) 5.7 H atoms/m³
41. 4.57×10^3 **43.** (a) 121 m/s; (b) 0.00406; (c) 248 y
47. 1.08×10^{42} J   **49.** (a) 0.785c; (b) 0.993c; (c) C2; (d) C1;
(e) 51 ns; (f) 40 ns   **51.** (c) $r\alpha/c + (r\alpha/c)^2 + (r\alpha/c)^3 + \cdots$;
(d) $r\alpha/c$; (e) $\alpha = H$; (f) 6.5×10^8 ly; (g) 6.9×10^8 y; (h) 6.5×10^8 y;
(i) 6.9×10^8 ly; (j) 1.0×10^9 ly; (k) 1.1×10^9 y; (l) 3.9×10^8 ly
53. (a) ssd; (b) $\bar{s}\bar{s}\bar{d}$

I N D E X

Figures are noted by page numbers in *italics*, tables are indicated by t following the page number.

A

a_g (gravitational acceleration), 360, 360t
absolute pressure, 390
absolute zero, 515
absorption:
 of heat, 522–527, *523*
 photon, *see* photon absorption
absorption lines, 1206, *1207*
ac (alternating current), 903, 913
acceleration, 20–30, 283t
 average, 20
 centripetal, 76
 constant, *23*, 23–27, 24t
 free-fall, *27*, 27–28
 graphical integration in motion
  analysis, *29*, 29–30
 instantaneous, 20–22, *21*, 67–69
 negative, 21–22
 and Newton's first law, 95–98
 Newton's laws applied to,
  108–113
 and Newton's second law,
  98–101
 principle of equivalence (with
  gravitation), 374–375
 projectile motion, 70–75
 reference particle, 429
 relating linear to angular, *269*,
  269–270
 relative motion in one
  dimension, 79
 relative motion in two
  dimensions, 79–80
 rockets, 241–243, *242*
 rolling down ramp, *299*,
  299–300
 sign of, 21–22
 simple harmonic motion, 418,
  418
 system of particles, 220–223
 two- and three-dimensional
  motion, 79–80
 uniform circular motion, *76*,
  76–78, *77*, 133
 as vector quantity, 41
 yo-yo, 302
acceleration amplitude, in simple
  harmonic motion, 418
acceleration vectors, 41
accelerators, 818–819, 1334–1336,
  1336
acceptor atoms, 1264
acre-foot, 9
action at a distance, 630
activity, of radioactive sample, 1287
addition:
 of vectors by components, *46*,
  46–47, 49

 of vectors geometrically, *41*,
  41–42, *42*, 44
adiabat, 571, *572*
adiabatic expansion, 531–532, *532*
 ideal gas, 571–575, *572*
adiabatic processes:
 first law of thermodynamics
  for, 531–533, 532t
 summarized, *575*, 575t
adiabatic wind, 580
air:
 bulk modulus, 480–481
 density, 387t
 dielectric properties at 1 atm,
  732, 732t
 and drag force, 130–132
 effect on projectile motion, *73*,
  73–74
 electric breakdown, 646, *646*
 index of refraction at STP, 992t
 speed of sound in, 480–481,
  481t
 terminal speeds in, 131t
 thermal conductivity, 535t
 thin-film interference of water
  film in, 1067
air conditioners, 596
airplanes:
 projectile dropped from, 74
 turns by top gun pilots, 77–78
 two-dimensional relative
  motion of, 80–81
 vector components for flight, 44
airspeed, 90
alkali metals, 1235
alpha decay, 1289–1291, *1290*
alpha particles, 621, 705, *1277*,
  1277–1279, 1289
 binding energy per nucleon,
  1283
 magic nucleon number, 1299
 radiation dosage, 1296–1297
 in thermonuclear fusion,
  1324–1325
alternating current (ac), 903, 913
alternating current circuits,
  903–934
 damped oscillations in *RLC*,
  910–912, *911*
 forced oscillations, 912–920, *914*
 inductive load, *919*
 LC oscillations, 903–910, *904*
 phase and amplitude relationships, 920t
 power in, 927–929
 resistive load, *915*
 series *RLC* circuits, 921–926,
  922
 in transformers, 930–933

alternating current generator,
  913–914
 with capacitive load,
  916–918, *917*
 with inductive load,
  918–919, *919*
 with resistive load, 914–916, *915*
ammeters, 788, *788*
ampere (unit), 614, 746, 843
Ampère, André-Marie, 844
Ampere–Maxwell law, 944–945,
  949t
Ampere's law, 844–850
Amperian loop, *844*, 844–848
amplitude:
 alternating current, 920t
 current, *922*, 922–923, 926
 of emf in ac, 914
 exponentially decaying in *RLC*
  circuits, 911
 LC oscillations, 905
 simple harmonic motion,
  416–418, *417*
 waves, 447, *447*, 448, *448*
amplitude ratio, traveling
  electromagnetic waves, 976
amusement park rides:
 roller coasters, 21
 Rotor, 267–268
analyzer, 988
Andromeda Galaxy, 354–355, *355*
anechoic chamber, 513
angles, 45
 angle between two vectors, 54
 degrees and radian measures, 45
 vector, 43, *43*, 45
angled force, applied to initially
  stationary block, 128
angle of incidence, 991, *991*
angle of minimum deviation, 1005,
  1007
angle of reflection, 991, *991*
angle of refraction, 991, *991*
angular acceleration, 261, 283t
 relating to linear, *269*, 269–270
 rolling wheel, *299*, 300
 rotation with constant, 266–268
angular amplitude (simple pendulum), 426
angular displacement, *259*, 260, 265
angular frequency:
 circulating charged particle, 815
 damped harmonic oscillator,
  430–432
 driving, 914
 LC oscillations, 908–909
 natural, 433, 914
 simple harmonic motion,
  414–418, *417*

simple pendulum, 426
sound waves, 483
waves, 448
angular magnification:
 compound microscope, 1032
 refracting telescope, 1033
 simple magnifying lens, 1031
angular momentum, 305–318, 312t
 atoms, 1221, *1221*
 conservation of, 312–316,
  313, *314*
 defined, *305*, 305–316
 at equilibrium, 328
 intrinsic, 953, 954
 Newton's second law in angular
  form, 307–308
 nuclear, 1284
 orbital, 954, *955*, 1222–1224,
  1223, 1223t
 rigid body rotating about fixed
  axis, *311*, 311–312
 sample problems involving, 306,
  308–309, 315–316
 spin, 953–954, 1223t, 1224, *1225*
 system of particles, 310–311
angular motion, 259
angular position, *259*, 259–260,
  283t
 relating, to linear, 269
angular simple harmonic motion,
  423, 423–424
angular simple harmonic oscillator, *423*, 423–424
angular speed, 261, 262
 relating, to linear, 268–270
 in rolling, 295–297, *296*
angular velocity, 260–264, 283t
 average, 260–261
 instantaneous, 260
 vector nature of, 264–265, *265*
angular wave number, 447, 1171
 sound waves, 483
annihilation:
 electron–positron, 622, *622*,
  1338
 particle–antiparticle, 1338
 proton–antiproton, 1339–1340,
  1340t
annular cylinder, rotational inertia
  for, 274t
antenna, 974, *974*
antiderivative, 26
antihydrogen, 1338, 1340, 1340t
antimatter, 1310t, 1338–1339
antineutrino, 1292n
antinodes, *465*, 466, 467–468
antiparticles, 1338–1341, 1359
antiprotons, 1338
antisolar point, 994, *994*

SOME PHYSICAL CONSTANTS*

Speed of light	c	2.998×10^8 m/s
Gravitational constant	G	6.673×10^{-11} N $\cdot$ m^2/kg^2
Avogadro constant	N_A	6.022×10^{23} mol^{-1}
Universal gas constant	R	8.314 J/mol $\cdot$ K
Mass–energy relation	c^2	8.988×10^{16} J/kg
		931.49 MeV/u
Permittivity constant	ε_0	8.854×10^{-12} F/m
Permeability constant	μ_0	1.257×10^{-6} H/m
Planck constant	h	6.626×10^{-34} J $\cdot$ s
		4.136×10^{-15} eV $\cdot$ s
Boltzmann constant	k	1.381×10^{-23} J/K
		8.617×10^{-5} eV/K
Elementary charge	e	1.602×10^{-19} C
Electron mass	m_e	9.109×10^{-31} kg
Proton mass	m_p	1.673×10^{-27} kg
Neutron mass	m_n	1.675×10^{-27} kg
Deuteron mass	m_d	3.344×10^{-27} kg
Bohr radius	a	5.292×10^{-11} m
Bohr magneton	μ_B	9.274×10^{-24} J/T
		5.788×10^{-5} eV/T
Rydberg constant	R	$1.097\,373 \times 10^7$ m^{-1}

*For a more complete list, showing also the best experimental values, see Appendix B.

THE GREEK ALPHABET

Alpha	A	α	Iota	I	ι	Rho	P	ρ
Beta	B	β	Kappa	K	κ	Sigma	Σ	σ
Gamma	Γ	γ	Lambda	Λ	λ	Tau	T	τ
Delta	Δ	δ	Mu	M	μ	Upsilon	Υ	υ
Epsilon	E	ϵ	Nu	N	ν	Phi	Φ	ϕ, φ
Zeta	Z	ζ	Xi	Ξ	ξ	Chi	X	χ
Eta	H	η	Omicron	O	o	Psi	Ψ	ψ
Theta	Θ	θ	Pi	Π	π	Omega	Ω	ω

SOME CONVERSION FACTORS*

Mass and Density

$1 \text{ kg} = 1000 \text{ g} = 6.02 \times 10^{26} \text{ u}$

$1 \text{ slug} = 14.59 \text{ kg}$

$1 \text{ u} = 1.661 \times 10^{-27} \text{ kg}$

$1 \text{ kg/m}^3 = 10^{-3} \text{ g/cm}^3$

Length and Volume

$1 \text{ m} = 100 \text{ cm} = 39.4 \text{ in.} = 3.28 \text{ ft}$

$1 \text{ mi} = 1.61 \text{ km} = 5280 \text{ ft}$

$1 \text{ in.} = 2.54 \text{ cm}$

$1 \text{ nm} = 10^{-9} \text{ m} = 10 \text{ Å}$

$1 \text{ pm} = 10^{-12} \text{ m} = 1000 \text{ fm}$

$1 \text{ light-year} = 9.461 \times 10^{15} \text{ m}$

$1 \text{ m}^3 = 1000 \text{ L} = 35.3 \text{ ft}^3 = 264 \text{ gal}$

Time

$1 \text{ d} = 86\,400 \text{ s}$

$1 \text{ y} = 365\frac{1}{4} \text{ d} = 3.16 \times 10^7 \text{ s}$

Angular Measure

$1 \text{ rad} = 57.3° = 0.159 \text{ rev}$

$\pi \text{ rad} = 180° = \frac{1}{2} \text{ rev}$

Speed

$1 \text{ m/s} = 3.28 \text{ ft/s} = 2.24 \text{ mi/h}$

$1 \text{ km/h} = 0.621 \text{ mi/h} = 0.278 \text{ m/s}$

Force and Pressure

$1 \text{ N} = 10^5 \text{ dyne} = 0.225 \text{ lb}$

$1 \text{ lb} = 4.45 \text{ N}$

$1 \text{ ton} = 2000 \text{ lb}$

$1 \text{ Pa} = 1 \text{ N/m}^2 = 10 \text{ dyne/cm}^2$
$\qquad = 1.45 \times 10^{-4} \text{ lb/in.}^2$

$1 \text{ atm} = 1.01 \times 10^5 \text{ Pa} = 14.7 \text{ lb/in.}^2$
$\qquad = 76.0 \text{ cm Hg}$

Energy and Power

$1 \text{ J} = 10^7 \text{ erg} = 0.2389 \text{ cal} = 0.738 \text{ ft} \cdot \text{lb}$

$1 \text{ kW} \cdot \text{h} = 3.6 \times 10^6 \text{ J}$

$1 \text{ cal} = 4.1868 \text{ J}$

$1 \text{ eV} = 1.602 \times 10^{-19} \text{ J}$

$1 \text{ horsepower} = 746 \text{ W} = 550 \text{ ft} \cdot \text{lb/s}$

Magnetism

$1 \text{ T} = 1 \text{ Wb/m}^2 = 10^4 \text{ gauss}$

---

*See Appendix D for a more complete list.